THIRD INTERNATIONAL CONFERENCE ON SPACE STRUCTURES

Proceedings of the Third International Conference on Space Structures, held at the University of Surrey, Guildford, UK, 11–14 September 1984.

THIRD INTERNATIONAL CONFERENCE ON SPACE STRUCTURES

Edited by

H. NOOSHIN

Space Structures Research Centre, University of Surrey, UK

ELSEVIER APPLIED SCIENCE PUBLISHERS
LONDON and NEW YORK

ELSEVIER APPLIED SCIENCE PUBLISHERS LTD
Ripple Road, Barking, Essex, England

Sole Distributor in the USA and Canada
ELSEVIER SCIENCE PUBLISHING CO., INC.
52 Vanderbilt Avenue, New York, NY 10017, USA

British Library Cataloguing in Publication Data

International Conference on Space Structures
(3rd: 1984: Guildford)
Third International Conference on Space Structures.
1. Space frame structures
I. Title II. Nooshin, H.
624.1 TH1635

ISBN 0-85334-309-8

WITH 146 TABLES AND 2023 ILLUSTRATIONS

Printed in Great Britain by Galliard (Printers) Ltd, Great Yarmouth

PREFACE

The present volume contains a collection of 180 papers on various aspects of analysis, design and construction of space structures such as grids, barrel vaults, domes, towers, folded plates and tension structures. These papers are written by 294 experts from 30 different countries for presentation at the Third International Conference on Space Structures, University of Surrey, UK, 11–14 September 1984.

The volume is divided into eight parts. The first part consists of a number of general surveys beginning with a review of the whole space structure scene by Professor Z. S. Makowski. The second part consists of papers concerned with the study, classification and formulation of space structure configurations. Parts three, four, five and six contain papers relating to recent advances in analytical techniques and part seven consists of papers discussing the design, construction and performance of many actual space structures built in steel, aluminium, plastics, timber, fabric, concrete or a combination of these materials. Finally, part eight contains a number of papers dealing with architectural studies and developments.

The volume provides a vast amount of information on space structures on all fronts and is of direct interest to structural engineers, architects, contractors and space structure manufacturers. With regard to design and construction, many new ideas and innovations are presented and, in addition to discussing the well-established space structure systems, some new jointing techniques are introduced. On the analysis front, a number of important advances are reported. In particular, research results leading to a better understanding of collapse behaviour of space structures are presented and more elaborate techniques for consideration of geometric and/or material non-linearities are discussed. The part on configuration processing is a main feature of the volume, presenting significant progress in this new and exciting field of knowledge.

In addition to many authors who helped by producing faultless manuscripts, the Editor was assisted by a number of colleagues and friends in producing the present volume. In this respect, the Editor wishes to express his gratitude to Professor Z. S. Makowski, Dr L. Hollaway, Dr P. Mullord, Mr G. A. R. Parke, Mr P. J. Wicks, Mrs P. Elven and Miss S. Nicholls.

H. Nooshin

THIRD INTERNATIONAL CONFERENCE
ON SPACE STRUCTURES

ORGANISING COMMITTEE

Sir Harold Montague Finniston FRS	Chairman
Dr H Nooshin	Secretary of the conference, Space Structures Research Centre of the University of Surrey
Prof Z S Makowski	Head of the Department of Civil Engineering of the University of Surrey
Dr E H Mansfield FRS	
Mr R G Taylor	Consulting Engineer
Prof Sir A J Harris	Representing the Institution of Civil Engineers
Dr J Bobrowski	Representing the Institution of Structural Engineers
Mr N Royce	Representing the Royal Institute of British Architects
Dr I Dunstan	Representing the Building Research Establishment
Mr B E Twyford	Representing Guildford Borough Council

Overseas Members

AUSTRALIA: Prof H J Cowan
Mr G A Day
AUSTRIA:
Prof R J O Krapfenbauer
BELGIUM: Prof M van Laethem
Prof Ch Massonnet
BRAZIL: Prof O A Trindade
CANADA: Mr H G Fentiman
Dr D T Wright
CHINA: Dr He Guanqian
Prof Tien T Lan
COLOMBIA: Prof J A Padilla
Prof A Ramirez Rivera
CZECHOSLOVAKIA:
Prof R A Bares
Mr J Zeman
DENMARK: Mr Kaj Thomsen
EIRE: Dr J Gaughan
Mr O McNulty
FRANCE: Dr S du Chateau
Dr R Motro
GERMANY: Dr H Eberlein
Prof J Schlaich
GREECE:
Mr D C Haramidopoulos
HONG KONG: Prof Y K Cheung
HUNGARY: Prof L Kollar
INDIA: Mr Sharat Das
Mr Kamal N Hadker
IRAN: Dr A Behravesh
Dr M Haristchian
Dr A Kaveh
ISRAEL: Prof M Reiss
Mr U Stock
ITALY: Prof M Pagano
Prof A Sollazzo
JAPAN: Prof Koichiro Heki
Dr Fujio Matsushita
JORDAN: Dr U R Madi
MEXICO: Prof J Mirafuentes
Mr F Castano
NETHERLANDS: Dr P Huybers
Prof J Oosterhoff
NEW ZEALAND:
Dr J W Butterworth
Mr G Fletcher
POLAND: Prof J Brodka
Prof W Gutkowski
PORTUGAL: Prof A R G Lamas
ROMANIA: Prof V Gioncu
Prof M V Soare
SINGAPORE: Dr Y S Lau
Dr H S Parmar
SOUTH AFRICA: Mr H Lawrence
SPAIN: Prof J Margarit
Prof C Buxade
SWITZERLAND: Prof R Ekchian
Prof J K Natterer
SYRIA: Dr N G Kazma
USA: Mr R E Linder
Mr D L Richter
USSR: Prof Ulo A Tarno

CONTENTS

PART TWO
CONFIGURATION PROCESSING

PART FIVE
OPTIMIZATION

PART SIX
ANALYSIS OF TENSION STRUCTURES

PART SEVEN
DESIGN, CONSTRUCTION AND PERFORMANCE

LIST OF CONTRIBUTORS

Numbers following contributors' names are the paper numbers as given in the Contents.

SPACE STRUCTURES OF TODAY AND TOMORROW

A Brief Review of their Present Development and a Glance into their Future Use

Z.S. Makowski, Ph.D., F.I.C.E., F.Eng.

Professor of Civil Engineering
Space Structures Research Centre
University of Surrey, Guildford, England.

A. INTRODUCTION

In 1966, the author of this paper was asked by the Architectural Design Journal to prepare a survey of the development of three-dimensional structures. He finished his survey with this statement: 'Whatever the material in structures of the future, one thing is certain; more and more use will be made of three-dimensional structures. They will stay with us. Space structures are not a passing fashion...' (Ref. 20).

Eighteen years later, in 1984, it is a pleasure to find that his prediction has been proved to be correct. Especially during the last decade, space structures have influenced the architectural scene all over the world. Architects turned their attention to space structures because these systems gave them greater freedom of design and in many instances led to lower costs through prefabrication and standardisation of component parts. In their search for new forms, architects and engineers have discovered that space structures not only offer them many structural advantages, but also produce a striking simplicity of form, often an unconventional, but very pleasing, appearance and frequently unusual beauty. In short, space structures have become a part of Modern Architecture and Progressive Engineering.

The reasons for the present interest in space structures is due to several factors; the most important are:

(i) Introduction and widespread use of high-speed electronic computers;

(ii) development of highly efficient standardised connections, and

(iii) a truly remarkable amount of recent scientific research into the elastic and non-elastic behaviour of space structures and the determination of modes of their failure under excessive loading.

The impact of industrialisation on prefabricated space structures has proved to be most significant. It permits economical standardisation of the component parts and in the hands of a creative architect it offers ample flexibility and leads to very imaginative designs. Many progressive architects talk of the integration of the structure with architecture - and this is precisely what many well-designed space structures achieve nowadays.

The acceptance of various types of space structure in many parts of the world is the result of research and development work done during several decades by progressive designers, architects and engineers, who many years ago already appreciated the huge potential of space structures.

The early and highly original development work of Alexander Graham Bell, Max Mengeringhausen, Attwood, Robert le Ricolais, Buckminster Fuller, Stéphane du Château, Yoshikatsu Tsuboi, Fujio Matsushita, Don Richter, Wright, Castano and Fentiman provides typical examples of designers interested in practical applications, who, through their work, left an impact on the architectural and engineering acceptance and rapid spread of space structures.

Several theoreticians have been fascinated by the behaviour of space structures from the very beginning, however, at that time, these people were the exceptions. It is only during the last decade when the introduction of the electronic computer truly revolutionised civil engineering and so many theoreticians, mathematicians and structural engineers have turned their attention to the problems of elastic and inelastic analysis of space structures.

Architects were the first to appreciate the potentials of space structures and to develop various types of bracing used nowadays in skeleton space systems. There are a number of people whose names have to be mentioned here - they come from many countries - the contributions made by R. le Ricolais, H. Lalvani, J.D. Clinton, P. Pearce, D.G. Emmerich, K. Critchow, V. Dragomir, A. Gheorghiu as well as R. Motro are of real significance.

Many of these researchers dealt extensively with the morphologic aspects of space structure configurations and focussed the attention on the systematic, exhaustive generation of regular and semi-regular structures, their duals and compounds (Refs. 7 & 19). In some cases it is easy to trace this interest to the highly original work of the 'comprehensive designer', Richard Buckminster Fuller, the inventor of geodesic domes, responsible for the 'synergetic-energetic' geometry applied to prefabricated space systems.

An ideal prefabricated system would be a structure consisting of identical elements joined together by simple connectors, the same for all the joints. In such a case, the elements and connectors could be mass-produced with considerable economy. The dream of Konrad Wachsmann called aptly by him 'the turning point in building' (Ref. 36).

Practice shows that this can be achieved for various types of double-layer grids and braced barrel vaults, in which the surface can be developed.

Domes, however, do not belong to this class and many types of braced dome, although prefabricated, consist of a number of elements of varying length. The search for a regular type of bracing is responsible for the really phenomenal number of studies of regular polyhedra. These investigations, as a rule, have been done either by architects or applied mathematicians. During the last decade several engineers made scientific attempts to find the influence of the type of bracing upon stress distribution and a number of valuable papers have been published on the optimisation of space structures.

B. RECENT ARCHITECTURAL TRENDS

The wide acceptance of space frames is especially visible in the construction of large-span buildings covered with various types of double-layer grids. Though, originally, these structures were used almost exclusively for flat roof structures, some interesting trends can be observed within the last decade. Now architects are using this form of construction not only for roofs, but also for floors in multi-storey buildings.

Architects now also realise the possibility of using such systems for whole buildings in which the walls, floors and roof are constructed using modular prefabricated components, in most cases forming either a two- or three-way double-layer grid. This trend started some time ago in Germany and one of the very early examples is the space frame building designed by a German architect, Eckhard Schulze Fielitz, for the Ellor Church in Essen, built in the MERO system. There is no doubt at all that this structure proved to be a fore-runner of a large number of similar architectural solutions, used during the last decade, in other countries. Extremely pleasing architectural effects can be obtained using such concepts. One of the recent and strikingly elegant uses of a double-layer grid is the library building for the University of Rochester, New York. Another one is provided by the shopping centre in the Franklin Park Mall, Toledo, Ohio, U.S.A. where a whole building is constructed as a huge space frame. Similar examples of the exquisite beauty obtained through the use of space frames for the vertical, inclined and horizontal planes are provided by several shopping centres built in France during the last few years by S. du Château. The commercial centre, Saint Denis Laval, is especially convincing from architectural and structural points of view. Ten years ago probably it would have been impossible to analyse a structure of this complexity. Nowadays the use of matrix methods and electronic computers enable the designer to investigate the stress distribution in such systems without undue difficulty.

The architectural freedom in shaping the structural form through the use of modular components in various types of space frames is especially illustrated by numerous and very recent MERO structures built in Europe over large-span sports halls, cultural centres, churches and supermarkets. All these buildings, composed of prefabricated elements, use space frames not only for the roof, but also for other parts of the structure, often arranged in several layers, interconnected in a continuous fashion. The beautiful gossamer-like appearance of the space structure gives architects the opportunity to leave the structural framework exposed to view.

Within the last five years several large exhibition halls have been constructed in which the space frames have been used not only for the roof but also for the inclined walls, built as integral parts of the whole structure. One should mention here the Warsaw Building Exhibition in Poland, or an equally interesting structure over the railway station at Gatwick Airport, England.

In all these cases the structure is exposed or viewed through glazing.

A vast space frame now nearing its completion has been used for the huge convention and exposition centre in Manhattan, New York - it is a most convincing example of the great impact space frames have made upon contemporary architecture. This vast building, stretching 360 m along Manhattan's 11th and 12th Avenues, and costing over $200 million, will become the biggest steel space frame in the U.S.A. Two-way double-layer steel grid has been chosen for the walls, floors and roofs of this important structure.

One must also mention several extremely interesting pyramidal structures which have been built recently in England, France, Austria and the U.S.S.R., consisting of inclined interconnected segments of prefabricated double-layer grids. The old architectural form is being used again, though the material, methods of connection and erection technique are completely new.

The survey of the developments in China is especially revealing. Over the last ten years, space structures have been accepted by Chinese designers, and nowadays are being used in numerous applications, even for very large clear spans. The interest in space frames in China is best illustrated by the fact that in 1981 the first national code of practice for the analysis, design and construction of space frames was produced in China.

One must also bring to the attention the publication in 1984 of a comprehensive report on the Analysis, Design and Realisation of Space Frames by the I.A.S.S. Working Group on Spatial Steel Structures produced under the chairmanship of Professor Yoshikatsu Tsuboi of Tokyo University. This report also contains an important appendix on the effective rigidities of latticed plates written by Professor K. Heki.

A survey of developments in Eastern Europe also shows a growing interest in space structures. Numerous steel space frames have been built over the last decade in Romania, Poland, Czechoslovakia, Hungary and East Germany, in spite of the fact that officially steel is considered a 'deficit' material, and as such is normally replaced by reinforced concrete. Even in these countries, the use of steel space frames has proved to be highly competitive in many instances.

In the U.S.A., several extremely successful solutions have been obtained within the last decade in the field of large-span conservatories and botanical gardens covered with double-layer geodesic aluminium domes. A very well known earlier example was provided by the Climatron covered by an aluminium braced dome over the Missouri Botanical Garden in St. Louis, U.S.A. The dome, which has a clear span of 53.3 m and a rise of 21.3 m, is glazed with translucent acrylic plastic sheets fixed to the tubular aluminium framework. The more recent and most beautiful Triodetic dome over the Bloedel Conservatory in Vancouver, British Columbia, Canada, is another proof that it is possible to create an enclosed environment with a minimal consciousness of a shelter. This building encloses more than 300 varieties of tropical plants. Its architect, Mr. H. Wilson, who decided to use the dome, stated that: 'The dome concept dissolves the side walls and there is nothing to relate the size to. In most cases, the conservatories are really gigantic greenhouses in which you get the atmosphere of a tunnel and are very conscious of the height. The dome is a shelter, but it has the feeling of being part of the outside and open to the sky...'

Visitors to this remarkable structure agree that inside the dome the lush vegetation creates its own sense of space and environment. The plant displays are arranged according to the three climatic zones: tropical, temperate and arid.

One should, perhaps, also mention the attempts, especially in the U.S.A. and Canada, to develop new low-cost systems based on various types of timber domical structures, though there is still some prejudice towards the idea of a round house.

In spite of this, there are now quite a number of small building firms specialising in the construction of timber dome structures used primarily for a low-cost storage agricultural building, or even single- or double-storey houses.

Some extremely interesting configurations have been developed in the U.S.A. Here one must especially refer to the Hexadomes of Gene Hopster, consisting of prefabricated stressed-skin timber panels. In his system, a typical small family dome consists of 24 identical triangles and 3 trapezoids.

Timber was often used in the past as a structural material. The new interest in its potential, especially for large-span structures, is due to the improved techniques of lamination. The economic advantages of laminated timber have been recognised by several American firms responsible for the recent construction of braced domes and barrel vaults in this material, several of them of very considerable spans. The first example refers to the timber dome over the sports hall for the Montana State College in Bozeman, U.S.A. Its diameter is 91.5 m and the structure consists of 36 meridional timber ribs interconnected by a series of horizontal rings. The second example is provided by a spherical timber dome over a sports hall in Salt Lake City, with a clear diameter of 105 m. The type of bracing is the parallel lamella system, often used by the late Dr. Kiewitt for his steel domes. The biggest laminated timber dome was completed in 1983 in Tacoma, Washington, U.S.A. This all-timber triangulated dome has a clear span of 162 m, and a rise of 48 m. The glue-laminated timber elements are connected with steel-plate hubs and form triangles, about 13.7 m on a side. Among the reasons that the designers used wood, at a cost of $30 million, was simplicity of design, good acoustics of the interior and lack of snow melting or air-blower requirements of other designs.

C. SUSPENDED ROOF STRUCTURES

The general demand for large spans unobstructed by any intermediate columns has produced within the last decade a remarkable development in another form of three-dimensional structures, the suspended roof system. It is usual to divide them into three main types -

(a) cable net systems
(b) membranal structures
(c) air-supported structures.

The practice shows that these structures, often of unusual and exciting shapes, can frequently be wholly utilitarian and provide economic solutions especially for temporary structures of limited life. It was Maciej Nowicki's suspended cable roof over the Raleigh Arena, built in 1952, which proved to be a real turning point, drawing the attention of progressive designers to the possibilities offered by the cable systems.

The main credit for the development of suspension structures must go to Frei Otto and his lightweight structures research institute in Stuttgart University renowned for extensive architectural studies of such structures. René Sarger, at one time, was instrumental in drawing the attention of French architects to the potential of these structures. Many designers followed these ideas in various countries. In the U.S.A. several large span structures have been built over sports centres. The work of Lev Zetlin and his associates has to be specially mentioned. In Europe the activities of David Jawerth in the 1960s led to the construction of a number of large-span suspended cable structures in Sweden, Germany, Switzerland, France and England.

The huge flat roofed circular building in the Madison Square, New York, is probably one of the best examples of such systems. Its diameter is 120 m. One should, perhaps, also mention the famous Scandinavium in Gothenburg, Sweden, 105 m dia, a beautiful example of a saddle shaped cable net structure, as well as the outstanding example of a double-layer prestressed cable roof over the Palazzo dello Sporto in Milan, Italy, with a dia of 125 m.

Recent developments in modern fabrics are responsible for the renewed growth of interest in lightweight membranal structures. The introduction of strong fibre reinforced plastics fabrics has been instrumental in the widespread use of membranal structures.

Fabrics made of nylon, polyester, Kevlar, strengthened with glass fibres produce material suitable for doubly curved surfaces that are able to span up to 20 m. For larger spans, the fabric membranal structures are normally reinforced by prestressed steel cables supporting the flexible membrane.

During the last decade some extremely interesting projects have been carried out using pneumatic structures in which the fabric is prestressed by air pressure.

In a way these structures are the outcome of the great enthusiasm shown in the 1960s in the structural use of plastics, though many of the 'revolutionary' ideas of 'all-plastics' structures in the meantime have been abandoned.

A survey of recent developments in this field shows that during the last five years significant achievements have been obtained in the application of fabrics made of G.R.P. or nylon coated with polyester, polyvinyl chloride or fluorocarbon resins.

The exceptionally favourable high strength and lightweight of these weather-resistant membranes provide new exciting architectural opportunities for the structural designs. They are described in some detail in Ref. 25.

It is claimed by their manufacturers that fabric structures permit the most efficient enclosure of space. Their cost not only compares well with that of conventional structures, in some cases it may be even lower. These claims however normally refer to the cost of the roof enclosure and may not take into account the considerable cost of foundations.

In membranal structures the material resists only tension - it cannot resist compression or bending. The membranal structures are therefore prestressed, or air-supported or air-inflated. The basic shapes, most commonly used in practice, are the sphere and the cylinder, though various other shapes have been investigated and occasionally used.

There are basically three main types of air-supported structures -

(a) single-skin systems, in which the structure is supported by a small differential of pressure between the outside and inside, maintained by a continuous flow of low-pressure air supplied by fans.

(b) Systems in which the plastics cover is supported by a framework of inflated rubber tubing forming the basic framework of the structure.

(c) Double-skin systems consisting of interconnected separate pneumatic panels forming barrel vaults or domes. The research work of Kawaguchi and Murata in Japan has to be specially mentioned in this connection.

In the U.S.A. most of the development work has been done within the last decade by Birdair Structures, Owens Corning Fiberglass Corporation and Irvin Industries.

In Japan, where several significant novel techniques have been developed, the practical achievements in the field of membranal structures of the Tokyo Kogno Co. Ltd. should be specially acknowledged. The general public still considers that membranal structures are temporary, though many manufacturing firms claim that the predicted life of their structures will be at least 20 years. One could consider the 1970 World Exposition in Osaka, Japan, as the commercial introduction of membranal structures. Since that time numerous air-supported structures have been built over exhibition pavilions, sports centres, swimming pools, warehouses, etc.

In the U.S.A. the glass fibre fabric is usually coated with 'Teflon' made of fluorocarbon resins. Extensive laboratory tests show that this fabric is highly resistant to corrosive atmosphere, sunlight and ageing. The solar transmission of the fabric can be up to 15%, providing a pleasing diffused natural illumination during daylight.

The use of membranal plastics structures, after Japan and the U.S.A., is now being extended to Europe and especially to the Middle East.

The best known is the huge complex of 210 identical tent-like structures erected in 1978-1981 at the new Jeddah International Airport, made of heavy woven fibreglass fabric coated with polytetrafluorethylene. Each 45 m square tent unit is cable supported by means of steel pylons 45 m high. Lower edges of each tent are attached to cables spanning between supporting pylons at a height of 20 m. The whole roof covers 425,000 square metres and accommodates 750,000 pilgrims during a six-week period of the annual Haj (or pilgrimage) to Mecca and Medina.

The design was carried out by Skidmore, Owings and Merrill and the roof fabric was manufactured by Owens Corning and Birdair Structures, Inc.

From a structural point of view, some even more interesting air-supported plastics membranal structures have been erected in the U.S.A. A huge air-supported dome of glass fibre fabric coated with 'Teflon' fluorocarbon resin over the Pontiac Silverdome covers an area of 255 m x 200 m.

Membranal structures can be built not only using plastics fabric; a roof over a sports centre at the Dalhousie University, Halifax, Nova Scotia in Canada, shows that thin steel membrane can be used with great success for large-span structures.

In fact, the Nova Scotia air-supported stainless steel roof is the world's largest example as the roof membrane covers an area of 91 m x 73 m with thickness of membrane being only 1.6 mm. The Sinoski stainless steel membrane roof built in 1979 combined great structural integrity and strength and enabled the architects to make the maximum use of space for athletics at the lowest possible cost. The use of nickel stainless steel membrane shows the superior characteristics of the material in relation to corrosion resistance, environmental attack and exceptionally good toughness to resist puncturing and tearing. In the case of the Dalhousie structure, Sinoski's design is based on membranal structure consisting of trapezoidal segments made in 1.6 mm thin steel membrane, a regular polygon for the central part of the roof and a peripheral compression ring. Special contraction joints have been developed by the designer to join the segments to the polygon and to each other and to bridge the gaps that would open up between them as the roof assumed under the internal air pressure a shallow domical profile.

The three basic components of compression ring, membrane segments and contraction joints have been shop-fabricated under controlled conditions. During field assembly all the pieces have been connected by means of welding. The designer claims that the all-welded steel membrane offers a permanence and durability not hitherto available in low cost air-supported roofing systems (Ref. 32). It should also be pointed out that unlike other air-supported systems, the Canadian structure shows no sign of loss of structural integrity in the event of air pressure being lost.

In such a case, the stainless steel membrane reverts to a suspended-type roof. The calculations show that even in this shape it is able to support its full design load.

One of the disadvantages of this form of construction over single-layer plastics membranal structures is that the stainless steel membranal system does not possess translucent properties. On the other hand, the fire resistance and strength of this form of construction are greater than those of more conventional roof structures.

D. ADVANCEMENTS IN STRUCTURAL ANALYSIS

The introduction of digital computers completely changed the approach used in the analysis of complex structures. Modern space structures consist, as a rule, of so many members that analysis by traditional techniques becomes extremely time-consuming and in most practical cases virtually impossible.

It has been known for some time that it is possible to describe the relationship between stresses and deformations even in very complex structures by using matrix algebra. However, the solution of the resulting simultaneous equations had to wait until the advent of high-speed digital computers. At the same time, it was realised that efficient computer utilisation required the introduction of matrix algebra to describe structural theory in a form suitable for computer use.

One of the main problems involved in the analysis of complex space structures is the sheer amount of data preparation for the computer analysis. Professor H.B. Harrison (Ref. 11) said recently that 'there is only one task worse than the preparation of data to describe a large three-dimensional structure and that is to verify the data'.

The work of Dr. H. Nooshin, at the Space Structures Research Centre of the University of Surrey, in developing the original concept of formex algebra, (Ref. 27) is considered by many specialists to be a real breakthrough in the algebraic representation and processing of highly sophisticated structural configurations. This technique is proving to be especially useful in describing the interconnection patterns and geometrical properties of double-layer grids, braced barrel vaults and domes. Use of formex algebra leads to the automated joint numbering scheme, so important in the data preparation necessary for the electronic computer analysis. Topological properties of the structure are related to the formex representation of its configuration through sets of rules and conventions, defined in detail in Ref. 28. Formices can be graphically represented through formex plots.

During the last ten years most important advancements have been achieved in the field of computer graphics. Their use helps the model generation and proves to be especially valuable in the verification of input data as well as the visualisation of response of the structure under the applied loading.

As a rule, the stresses and deflections are represented by means of plots.

The recent introduction of mini-computers produced a remarkable impact on small-size consulting offices, allowing easy access to computers for all designers. The mini-computers with the virtual memory has now removed many of the difficulties encountered previously in the analysis of complex engineering structures which at one time could only be carried out using large storage capacity computers.

It is interesting to note that in spite of all these developments plate analogies are still receiving continued interest from many theoreticians.

To the long list of theoreticians responsible for the development of plate and shell analogies in the 1960s, one now has to add the names of K. Heki, J.D. Renton, S. di Pasquale, A. Solazzo, M.V. Soare and L. Kollar, just to mention some who have published during recent years articles dealing with various aspects of plate analogies applicable to double-layer grids.

Refs. 21 & 24 contain a comprehensive comparison of stresses and deflections in eight different types of double-layer grids determined by the exact computer analysis and also obtained by the use of various plate analogies.

The German Constructional Steelwork Association produced in 1981 a publication on this topic, (Ref. 37). There are also numerous articles dealing with the same subject published by Japanese engineers.

Several publications have also been prepared on the discrete field analysis of double-layer grids. The work of D.L. Dean and W. Gutkowski must receive special mention in this context, (Refs. 8 & 9).

The interest in space structures in general and in double-layer grids in particular, and their acceptance for covering large spans, can be directly related to the publication of numerous papers dealing with various aspects of analysis, design and construction.

Many articles refer to experimental laboratory testing of small-scale models of double-layer grids, but only a very few give details of full-size tests. Field testing by its nature is quite difficult and also more expensive than the laboratory small-size tests. However, full-size tests are often essential as the acceptance tests required by the building authorities for new types of structures. The field measurements can be used to confirm the validity of theoretical analysis and to check the influence of rigidity of joints and cladding on the load-carrying capacity of the structure.

Double-layer grids are examples of prefabricated structures consisting of a large number of modular units and, as a rule, are highly statically indeterminate and therefore may be subject to additional stresses due to lack of fit introduced during the assembly.

In some systems, e.g. Space Deck, turnbuckles are used to introduce camber into the structure. This leads to prestressing and modifies the final stress distribution, but the determination of the magnitudes of the prestressing forces can be based only on an approximate assessment.

The modular units can vary slightly in their dimensions and the tolerance obtained during construction depends on the manufacuring technique, size of the units, types of the connector, etc.

The full-size tests enable the designer to assess the influence of some of the factors mentioned above.

Ref. 23 contains details of four full-size tests carried out on double-layer grid structures. The test on a Space Deck roof structure, having overall dimensions of 20.4 m x 18.0 m, carried out in 1972 under carefully controlled conditions at the National Tower Testing Station in Cheddar,Somerset, U.K., is of particular interest, especially as attempts were made to test the structure to failure. The structure proved to behave generally within the elastic range up to stiffness test loading of 122.5%. After reaching this level of loading several members started to yield. The maximum load applied to this structure was 206% of test load.

Complete collapse of the roof could not be achieved, though several members failed. The test demonstrated in no uncertain terms the 'fail-safe' characteristics of this type of double-layer grid construction.

One should also draw attention to load tests of several prototypes of the 'Mostostal' system of double-layer grids designed for industrial buildings. The tests were carried out in Warsaw, Poland, between 1975 and 1977, on structures covering areas 12 m x 18 m, 24 m x 24 m and 30 m x 30 m.

The measured deflections from the tests were in close agreement with those obtained by elastic linear analysis, though in general the experimental deflections were somewhat higher than the theoretical ones. In many bars the forces determined experimentally were in good agreement with the forces calculated by theory. However, there were quite a number of members where the correlation was less satisfactory; this applied especially to diagonal members connecting top and bottom layers, (Ref. 18).

Many designers believe that due to the inherent redundancy of double-layer grids and their assumed ability to survive the losses of several members or joints without losing the overall stability, structural failures of such systems in most cases can be linked to gross design errors or unauthorised changes introduced during the erection by the contractors.

In the case of the Hartford Coliseum the subsequent examination of the collapsed structure showed that -

'The detail of the bracing to several members rendered the bracing essentially ineffective, considerably reducing the capacity of the affected members', (Ref. 38).

E.A. Smith, in his report on the collapse behaviour of space trusses, refers to the Hartford Coliseum failure in this way -

'While several deficiencies existed in that ill-fated structure, the physical cause for the collapse appears to be due to the undercapacity of several compression members because of design and detailing mistakes, particularly in respect of the bracing connections...' (Ref. 30).

The full-size tests show clearly that the elastic linear analysis gives only an indication of the possible load-carrying capacity of the structures and that in cases of important structures non-linear analysis may become essential to find the more realistic ultimate load capacity.

Within the last decade several techniques have been developed which may be used in the non-linear analysis of double-layer grids, (Refs. 16 & 30). Also, in the last years an increasing number of researchers have turned their attention to the determination of the ultimate load carrying capacity of double-layer grids.

Ref. 34 contains an extremely useful review of the previous work on limit state analysis of double-layer grids. It is stated, however, that at present the limit state analysis is still in its infancy and in fact it is very difficult to find any theoretical method which gives acceptable and predictable results in relation to experimental data.

The research work carried out within the last decade at Melbourne University (Ref. 29) has resulted in the publication of several important studies on instability behaviour of various types of double-layer grids. These investigations concentrate on the inelastic response of 'brittle-type' members with their relatively sharp unloading characteristics.

The experimental tests, carried out by L.K. Stevens, L.C. Schmidt and P.R. Morgan, suggest that steel compressiveelements of slenderness ratio l/r in the range of 40 to 140 instead of the normally assumed elastic-plastic behaviour, may exhibit a severely reduced load-carrying capacity once they reach their peak load.

Theoretically, therefore, the use of the 'brittle-type' struts may lead to progressive collapse of the structure.

Several alternatives, to improve the space truss ductility, have been suggested, e.g. the use of eccentrically loaded struts or the application of force limiting devices (FLD), (Ref. 10), though at this stage these suggestions are of rather limited academic interest and no commercially available FLD systems have so far been developed.

The concept of 'brittle-type' struts has recently been investigated by E.A. Smith in an attempt to find the reasons for the collapse of the Hartford Space Truss Roof.

In a series of reports and papers produced during the last two years, he claims that according to his analysis, the fully stressed double-layer grids are vulnerable to progressive collapse; however, in his conclusions he states that designs utilising compression member overdesign do have adequate resistance to progressive collapse.

There is no doubt that a great deal of further research is needed towards a better understanding of the effects of random imperfections on the non-elastic buckling behaviour of double-layer grids. It is still very difficult to take into account accidental imperfections, e.g. initial curvature of component members, joint eccentricity, joint slip, variations in the cross-sectional areas, lack of fit, member-joint rotational behaviour, yielding of supports, etc. All these factors influence the stress distribution and make an accurate theoretical prediction of the load carrying capacity extremely difficult.

However, it is expected that, perhaps because of these difficulties, the interest in non-linear behaviour of structures will grow, and that the theoreticians will take up the challenge and resolve the problem of premature buckling of members and determine the influence of the failure of individual members on the final load capacity of the structure.

E. THE FUTURE POSSIBLE DEVELOPMENTS IN SPACE STRUCTURES

(i). The present systems of space structures show that for very large clear spans the deflections may become the controlling factor in the design of large exhibition halls, aircraft hangars, sports halls and shopping malls. The design of the diagonal grid concept used for the design of the two hangars at London Airport influenced several designers to use similar systems for even larger spans. The diagonal grid for the Singapore Airlines hangar is a typical example of this trend with a clear span exceeding 200 m, a remarkable structure designed by Dr. Y.S. Lau and built in 1983.

There is no doubt that even larger spans can be constructed, though probably triple-layer grids will be used. A detailed study of six different types of triple-layer grids have been carried out under the auspices of Constrado at the University of Surrey (Ref. 2). These studies show that the behaviour of triple-layer grids depends not only on the total amount of material used in their construction, but is also influenced by the type of bracing used for their structural framework. Their analytical and experimental studies highlight the function of the middle layer. It is shown that the middle layer is instrumental in stiffening the whole system, enabling it to take large suspended or unsymmetrical loads.

A number of triple-layer grids have already been built. The triple-layer grid covering the Denver Convention Centre, Colorado, U.S.A., is a good example of such applications. In this case the roof structure covers an area of 205 m x 72 m and contains more than 24,000 members. The roof tests at the corners on inverted pyramids, formed by extending the basic framework downward and forming multi-layered grid supports built from modular units of identical length.

Another very interesting triple-layer structure is the roof over the Franklin Park Pall, Toledo, Ohio, U.S.A. Several structures of this type have been built in the MERO System, and the Philips-Halle at the exhibition centre in Düsseldorf, West Germany is a good example, as well as the famous roof over the Atlantis hangar at the Frankfurt a.M. airport.

The large space frame used as a roof structure for a hangar at the Klöten Airport in Zürich, recently erected for the maintenance of Jumbo Jets, illustrates an even more sophisticated form of double-layer grid. It covers an area of 128 m x 129 m and is supported by only four supports.

The vertical reaction at each of the supports amounts to some 2,000 tonnes. Detailed studies of this structure have confirmed its remarkable rigidity as well as its ability to preserve its structural integrity in the event of damage due to internal explosion or fire.

(ii). One of the extremely interesting possibilities for the use of space frames is in the construction of highway bridges. Early examples of such an application have been given by the Space Deck System employed for the construction of demountable temporary bridges. During recent years, several permanent pedestrian bridges have been erected in the Triodetic and MERO systems.

A continuous five span structure of this type erected for the Siemens Co. in Stuttgart, West Germany, illustrates the remarkable versatility of the MERO system.

However, a most remarkable application in the field of highway bridges has been provided by the Tridilosa system developed in Mexico by Professor H. Castillo and further developed by his associate Mr. A. Calderon Ollivier.

The Tridilosa system is a typical example of low technology, derived from the realisation of the disadvantages of the conventional reinforced concrete construction. Being a practical designer, Professor Castillo noticed that the ratio of dead weight to useful live load in reinforced concrete strutures is unduly high. A great deal of the steel reinforcement and concrete is being used to support the dead weight of the structure. Castillo decided, therefore, to reduce the amount of concrete and to arrange his reinforcing bars in the form of a two-way double-layer skeleton grid reinforced with some 5 cm thick concrete slab in the top compressive zone.

This composite construction, consisting of three-dimensional grid with r.c. slab, has been used successfully for thousands of low-cost buildings in South America.

A modified version of this system has also been used for highway bridges. Over 50 large span highway bridges have been built in Mexico and various countries in South America within the last five years, some of them having a clear span of 50 m. These structures are examples of space frames consisting of steel prefabricated pyramidal units interconnected together into modified triple-layer grids.

The main advantages of this form of construction are the lower cost and the remarkable speed of erection, with no need for costly formwork (see Ref. 22). The prefabricated welded bridge units, some 20 m long, are manufactured in workshops and delivered by special articulated lorries to the building site. After interconnection of the units by welding, the three-dimensional framework of the bridge is rolled over the r.c. abutments into its final position.

Once the bridge grid is resting on supports, removable plywood panels are fixed to underneath the top layer, and concrete top deck slab is produced as an integral part of the structure.

The designers of these Tridilosa bridges claim that in some cases savings up to 60% have been achieved in comparison with conventional r.c. bridges. So far clear spans up to 50 m have been achieved.

(iii). A truly amazing field of application has appeared within the last few years as a result of the space exploration by the Americans and Russians. Various studies were produced by NASA on space colonies, gigantic solar power stations, as well as projected space platform applications. Several types of space erectable structures from earth fabricated members have been explored, resulting from considerable interest in the feasibility of constructing large three-dimensional structures in space.

The possibility of establishing a colony in space received a great deal of public attention. Various governments are investigating the concept of Satellite Solar Power Stations to provide exportable solar energy.

In 1976, NASA published its 'Outlook for Space' which forecasts large antennae and space structures by the year 2000 (Ref. 14). NASA carried out studies on techniques of construction in space. Its recommendations include the construction of 300 m platforms in space constructed as three-way double-layer grids. It is claimed that considerable progress has been made in studies of structural building blocks in space including deployable structures, erectable structures and assembly concepts of tetrahedral space trusses. Present studies of deployable structures (Refs. 12 & 13) suggest that transportation limitations will influence the size of deployable platforms or antennae and that platforms with spans of 250 m appear at present to be a practical maximum. Packaging densities and structural efficiencies are prime variables in such studies.

NASA is supporting research into the development of structural members which can be prefabricated on earth, efficiently packaged for orbital transportation and erected in space member-by-member to construct large antennae or space platforms (Ref. 17).

One of the developed concepts, as reported in Ref. 15, uses tapered columns, which are nested like paper cups in half-column sections for orbital transformation. On-orbit, the half-columns are joined and assembled into a large space structure.

According to NASA a manufacturing technique has already been developed for graphite-epoxy fibre reinforced columns capable of producing half-columns up to 10 m in length. Column building strength and vibration behaviour of nestable column members have been investigated as individual members (Ref. 3) and as sub-elements of larger truss assemblies. According to M.F. Card (Ref. 5) packaging studies indicate that mass critical payloads can be achieved with nested aluminium and graphite composite structures and that a shuttle-load of members would be sufficient to build square space trusses as platforms with a 300 m span.

Experiments have been carried out in the Skylab Space Suit by Langley Research Center personnel at the Marshall Space Flight Center Neutral Buoyancy Chamber in the U.S.A. on the construction of space platforms (Refs. 5 & 13). The tests show that the assembly of very large erectable space structures will require in space some form of automation. The Lockheed's automated assembly machine will probably provide the answer.

All these studies show that many of the so-called fantasies regarding space travel will be realised within the next decade with space structures playing a major role in the achievement of these dreams.

(iv). Experience shows that whereas remarkable progress has been achieved within the last decade in the construction of large span space structures, covering sports centres, exhibition pavilions and industrial buildings, the general public still demonstrates a visible reticence to accept the visionary designs of space towns. The trend towards the exploration of the potential of spaceframes in the macrostructure of towns, which was extremely strong in the 1960s among progressive town planners, did not lead to the same acceptance and realisation of structures in practice. In a way, this trend can be compared with the earlier visionary tendencies of some designers eager to develop all-plastics houses, often of very unorthodox shapes. The slow acceptance of the all-plastics houses has not only been because of high cost and low fire resistance of plastics materials. The general public seems to be very conservative in their outlook and architects will have great difficulty in persuading people to live in prefabricated all-plastics cells, even if, this could provide many advantages.

Progress in this area is, therefore, much slower, and so far we have no practical examples of 'space towns'. However, the interest of designers in such concepts still continues and several architectural studies have been recently carried out in the U.S.A. and France. The investigations of J.F. Gabriel and P. Pearce on the development of prefabricated space frames for multi-layer building systems are of special interest.

Gabriel rightly points out that there is nothing superior about the square as the generator of architectural space. He shows that regular hexagons could provide an even greater flexibility and an extremely interesting architectural potential of habitation based on prefabricated three-dimensional elements.
It is interesting to follow the development of similar ideas by Stéphane du Château incorporated in his famous pyramidal multi-layered Unibat structure for the Ville de Rennes ZUP-SUD-Triangle covering a large cultural social centre with a huge space frame. In this case the designer succeeded in creating open habitation spaces serviced with all the amenities, at really economic rates. In the Rennes structure, all internal partitions are located according to the needs and can be changed to suit the taste and inventiveness of the occupants, who use their own initiative to create a habitat to suit their individual desires (Ref. 6).

REFERENCES

1. M S ANDERSON, Instability of periodic lattice structures, Proceedings of the AIAA/ASME/ASCE/AMS 21st Structures, Structural Dynamics and Materials Conference, AIAA-80-0681-CP, Seattle, W.A., 12-14 May 1980.

2. U K BUNNI, P DISNEY and Z S MAKOWSKI, Multi-layer Space Frames, Constrado, London, 1981.

3. H G BUSH and M M MIKULAS, A nestable tapered column considerations for large space structures, NASA TM-X-73927, July 1976.

4. H G BUSH et al, Some design considerations for large space structures, AIAA Journal, Vol. 16, No. 4, pp. 352-359, April 1978.

5. M F CARD et al, Efficient Concepts for large erectable space structures, Large Space Systems Technology, NASA Conference Publications 2035, Vol. 11, pp. 627-656, 1978.

6. S du CHÂTEAU, The industrialisation of modular space structures, Chapter 13, pp. 381-407, Analysis, Design and Construction of Double-layer Grids, (Ed. Z S Makowski), Applied Science Publishers, London, 1981.

7. A GHEORGHIU and V DRAGOMIR, Geometry of structural forms, Applied Science Publishers, London, 1978.

8. W GUTKOWSKI, Regularne konstrukcje pretowe, Polska Akademia Nauk, Warsaw, 1973.

9. W GUTKOWSKI et al, Obliczenia statyczne przekryc strukturalnych, Arkady, Warsaw, 1980.

10. A HANAOR and L C SCHMIDT, Space truss studies with force limiting devices, Journal of the Structural Division, ASCE, Vol. 106, No. ST11, pp. 2313-2329, November 1980.

11. H B HARRISON, Minicomputer analysis of large space structures, Proceedings of the Australian Conference on Space Structures, AISC, Melbourne, pp. 11-15, 1982.

12. W L HEARD et al, Buckling tests of structural elements applicable to large erectable space trusses, NASA, TM-78628, October 1978.

13. W L HEARD et al, Structural sizing considerations for large space platforms, Proceedings of the AIAA/ASME/ASCE/AHS 21st Structures, Structural Dynamics and Materials Conference, Seattle, W.A., 12-14 May 1980.

14. A R HIBBS et al, Summary conclusions and recommendations outlook for space, NASA, SP-386, A forecast of space technology, Part 11, January 1976.

15. JACQUENUN et al, Development of assembly and joint concepts of erectable space structures, NASA CR-3131, August 1979.

16. D S JAGANNATHAN, H I EPSTEIN and P D CHRISTIANO, Non-linear analysis of reticulated space trusses, Journal of the Structural Division, ASCE, Vol. 101, No. ST12, pp. 2641-2658, December 1975.

17. E KATZ, Large space systems erectable structures assembly simulations, Final Report, NAS8-33420 Report SSD 790215, Rockwell International, December 1979.

18. Z KOWAL and W SEIDEL, An attempt of measurement of random forces in bars of a regular space structure, Proceedings of the 2nd International Conference on Space Structures, University of Surrey, Guildford, pp. 762-766, 1975.

19. H LALVANI, Transpolyhedra: Dual transformation by explosion-implosion, Lalvani, 1977.

20. Z S MAKOWSKI, Three-dimensional structures, Architectural Design, London, pp. 10/41, 1966.

21. Z S MAKOWSKI (Editor), Analysis, Design and Construction of Double-layer Grids, Applied Science Publishers Ltd., London, 1981.

22. Z S MAKOWSKI, Space Structures in Mexico, Building Specification, pp. 21-25, October 1981.

23. Z S MAKOWSKI, Full-size tests on space structures and development of their nodes, Proceedings of a Symposium on Shell and Space Structures, FTW - KU, Leuven, 1981.

24. Z S MAKOWSKI, Raumfachwerke, Beratungsstelle für Stahlverwendung, Düsseldorf, 1980.

25. Z S MAKOWSKI, Structural applications of plastics in Architectural and Civil Engineering, Proceedings of a Symposium on Plastics Technologies and Applications, United Nations Economic and Social Council, Amman, Jordan, 1982.

26. M MENGERINGHAUSEN, Komposition im Raum, Bertelsmann Fachzeitschriften Verlag, Gütersloh, 1983.

27. H NOOSHIN, Algebraic representation and processing of structural configurations, Computers and Structures, Vol. 5, pp. 119-130, 1975.

28. H NOOSHIN, Formex formulation of double-layer grids, Analysis, Design and Construction of Double-layer Grids, Applied Science Publishers Ltd., Chapter 4, pp. 119-183, 1981.

29. L C SCHMIDT et al, Ultimate load behaviour of a full-scale space truss, Proceedings of the Institution of Civil Engineers, Vol. 69, Part 2, pp. 97-109, March 1980.

30. E A SMITH, Collapse behaviour of space trusses, CE 82-144, University of Connecticut, July 1982.

31. M V SOARE and I H TOADER, Contributions to the analysis of triple-layer grids, Proceedings of the 3rd Conference on Steel Structures, Timisoara, pp. 232-239, October 1982.

32. J SPRINGFIELD and D SINOSKI, The air supported steel membrane roof at Dalhousie University, Halifax, Nova Scotia, Proceedings of the Canadian Structural Engineering Conference, 1980.

33. L K STEVENS et al, Ultimate load analysis and design of plate-like space trusses, Proceedings of the Australian Conference on Space Structures, AISC, Melbourne, pp. 1-10, 1982.

34. W J SUPPLE and I M COLLINS, Limit state analysis of double-layer grids, Analysis, Design and Construction of Double-layer Grids, Applied Science Publishers Ltd., Chapter 3, pp. 93-117, 1981.

35. C H THORNTON, Lessons learned from recent long span roof failures, Proceedings of a Symposium on Long Span Roof Structures, St. Louis, Missouri, ASCE, New York, pp. 89-95, 1981.

36. K WACHSMANN, The turning point in building, Reinhold Publishing Corporation, New York, 1961.

37. H WITTE, Einfache Regeln zur Vorbemesserung von Raumfachwerken, Merkblatt 110, Beratungsstelle für Stahlverwendung, Düsseldorf, 1981.

38. LEV ZETLIN, ASSOCIATES, INC., Report of the engineering investigation concerning the causes of the collapse of the Hartford Coliseum Space Truss Roof on 18 January, 1982, New York, 1982.

INTERNATIONAL EXAMPLES OF LATTICED TRIODETIC* STRUCTURES

H.G. FENTIMAN, President and Managing Director
Triodetic Structures Limited (International)
Ottawa, Canada

*Triodetic is a Registered Trademark of Triodetic Structures Limited Ottawa, Canada

This paper describes a number of latticed space frame structures constructed for different purposes, and to suit a variety of climatic conditions. Photographs, details and information are provided on structures built in Latin America, U.S.A., Canada, the U.K., Europe, Australia and the Middle East. The structures discussed include domes, single and multi-leaf hyperbolic parabaloids and free-formed structures, as well as 3-dimensional space grids. Information on structural components, connections, materials and finishes is given. The practical progress that has been made in the widespread use of such structures in different areas of the world is reviewed. It serves to illustrate the freedom of architectural design now being employed in utilizing these structures in unusual, yet economical enclosures.

GENERAL

It is not intended to discuss the details of analysis and design for space frames or shell structures, although some brief comment may be applicable. Special discreet methods of solution have been developed by Z.S. Makowski as in Ref 1, and Stephen du Chateau, see Ref 2. Further straightforward methods have been documented by D.T. Wright in Refs 3 and 4. Many other powerful computerized solutions for space structures have been developed, including those by C. Marsh and P. Kneen as in Ref 5, as well as by F. Castano, see Ref 6.

The purpose here is to discuss and illustrate the various types of latticed space frames. In this context, it is possible to generally classify space frame structures according to their basic physical form and structural behaviour. Two of the most common in general use are flat space frame plates as in Fig 1 and single or double curved latticed shells, as in Figs 2 and 9.

Fig 1 External Space Frame Structure - The Netherlands Pavilion (Expo '67) Montreal, Canada

STRUCTURAL MEMBERS

Many different types and shapes of structural members have been applied to space frames, including tee sections, angles and channels, in aluminum and steel. For the most part, however, tubular sections have been recognized as the most efficient on a strength to weight ratio; and since space frame structures are almost invariably exposed, the tubular element is most aesthetically pleasing and architecturally acceptable.

In Canada, some of the earliest structures were achieved at a time when construction was restricted to the use of aluminum tubular components, see Ref 7. There was, of course, an increasing demand for the use of tubular steel elements, and a variety of coatings and finishes on these products are now available. Both aluminum and steel components may be factory painted with a powder or wet paint finish in a wide range of colours. Some striking structures have been erected using anodized aluminum and stainless steel.

Fig 2 Eight Hypars - Ontario Place Forum, Toronto, Canada

JOINTS AND CONNECTORS

The lack of suitable connectors was, for many years, the principal difficulty in the realization of economical space frames and shells. While many architects and engineers felt it was a fairly easy problem to solve, the development of new and effective connections used in the joining of multiple elements in space, was relatively slow and much more difficult than anticipated.

Welded joints in various forms have been used with hub elements, as well as cases where weldments were used with no hub element at all. It was quite clear that mechanical joints could offer the advantage of pre-fabrication and simple site assembly, and would be ideal for this type of construction. Of the many types of mechanical joints that have been tried, only a few have been successful in terms of structural efficiency, aesthetics and economy.

A variety of nodes or connectors now exist which are suitable for different specialized purposes. In addition, techniques are also currently available to produce special nodes, which satisfy unusual requirements at an economical cost. Test specimens and actual elements (of aluminum and steel) have been subjected to tests to demonstrate various modes of behaviour under tensile, compressive and combined loadings. These tests have indicated an average efficiency of Triodetic joints in excess of 90% ultimate under direct tension loadings, with deviations of approximately 6%, see Ref 8. In addition, new and innovative details of connections between 3-dimensional structures or shell structures and the conventional forms of construction are constantly being developed, see Fig 3.

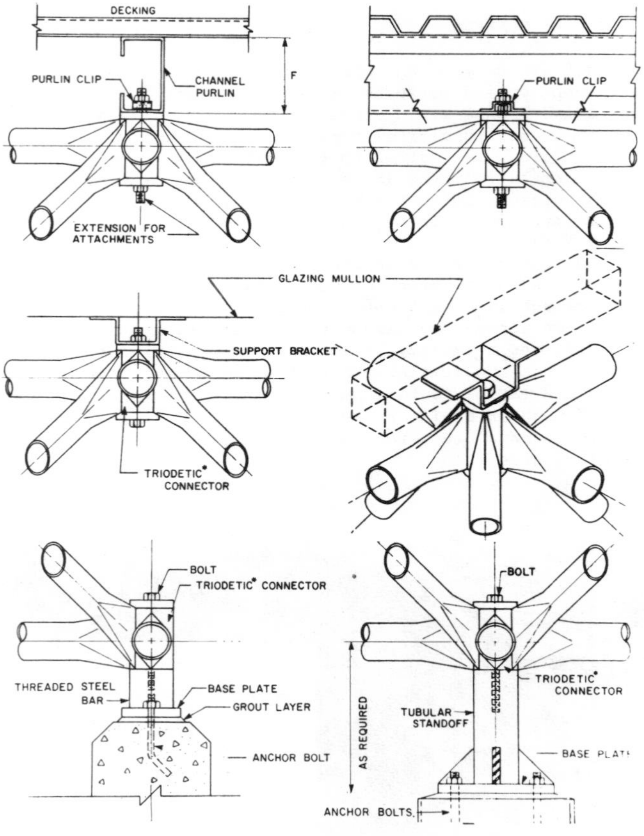

Fig 3 Typical Triodetic Connections - Decking, Glazing and Frame Support Details

It should be noted the connection design does not govern the design of space frame structures. Connections must be provided, however, that will resist the forces that arise in individual members and satisfy the overall needs of the structure. The overall structural behaviour dictates the requirements for connections.

SPACE FRAMES AND GRILLAGES

These structures behave similarly to 2-way concrete slabs or flat plates, depending upon support conditions and framing arrangements. A simple grillage (without face diagonals) has relatively little resistance to twisting as in Fig 4. Torsion or twist is most commonly resisted by the insertion of face diagonals in the upper and lower chords. Three-way grids, (i.e., upper and lower chords fully triangulated), add considerable rigidity against twists, see Fig 5. There are, of course, a wide variety of geometric forms that can be used in space frame structures, see Ref 9.

Fig 4 Two-way Grillage - 32 m x 32 m Sports Hall (Royal Saudi Air Force) Alkharj, Saudi Arabia

Fig 5 Three-way Space Frame - Sacred Heart Church, Ottawa, Canada

Up to the end of the 1960's, it was customary to base the design of a structure upon the behaviour of an analogous solid plate. In the 1970's, however, general purpose computer solutions became available, which continue to be used to the present day. As the lightness and elegance of these structures became recognized by architects, they have been frequently used on structures with a more modest span. Figures 6, 7 and 8 illustrate a fairly popular usage in the vertical and sloped positions. Enclosure is achieved by glazing with clear, solar or tinted glass.

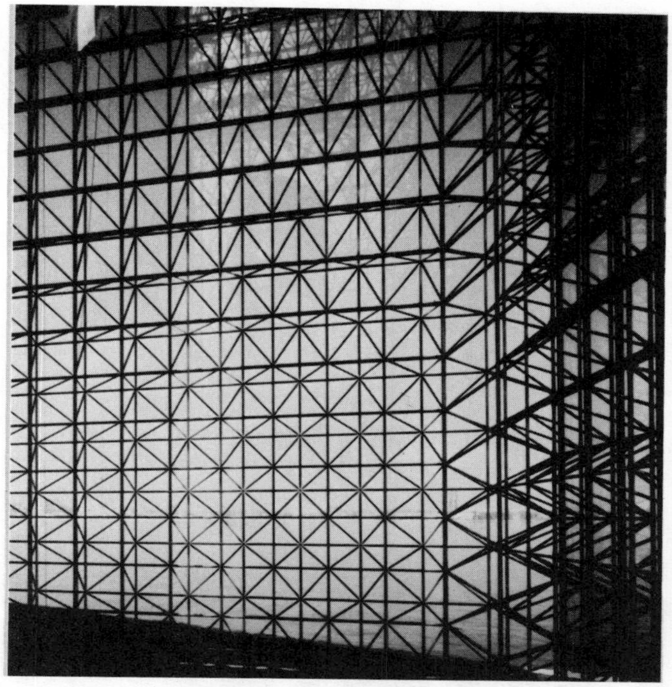

Fig 6 Vertical Glazed Space Frame - John F. Kennedy Library (Dorchester Bay) Boston, U.S.A.

Fig 7 Sloped Glazed Space Frame - Shopping Centre Entrance, Toronto, Canada

Fig 8 Glazed Space Frame Pyramid - Dolphin Leisure Centre Romford, England

LATTICED SHELLS

A great variety of shell forms are available. When using re-inforced concrete construction, the choice of form is greatly constrained by the expense and difficulty of erecting formwork. The introduction of light-weight metal latticed shells provides the advantage of greater structural efficiencies than flat plates or grillages, as they can usually carry external loads without bending in the primary structure.

The design of shell structures is relatively simple. Four conditions must be satisfied in order for a shell to act properly by developing only membrane stress, (shell action) first, it must be thin; second, it must be continuously curved; third, it must be properly supported so that reactions act tangentially to the shell at the boundary, and membrane displacements are allowed at the boundary of the shell; fourth, applied loading must be uniform or vary regularly without concentrations or abrupt discontinuities.

The most common varieties of such shells are still the dome, see Fig 9, the cylindrical shell, see Fig 10, and the hyperbolic parabaloid, see Fig 11. Recently, elegant concrete water tanks have been developed by Eng. Francisco Castano of Mexico using hyperboloids of revolution, formed with latticed metal frameworks, see Fig 12.

Fig 9 Dome - Civic Centre, Guayaquil, Ecuador. This Triodetic structure employs a geodesic pattern. It has a base diameter of 60 m and a height of 20.8 m. It has a spherical radius of 32 m.

Fig 10 Cylindrical Shell - Skating Rink, St. Mark's Academy, Southboro, U.S.A.

Fig 11 Hyperbolic Parabaloid, Niagara Falls, Canada

Fig 12 Hyperboloid of Revolution - Water Tank, Mexico

Space structures and reticulated shells have been associated with exotic and highly visible applications, but they are in no way confined to this type of use only. In fact, they have been successfully applied to many industrial projects. Space does not permit the presentation of complete statistical information on all the projects mentioned. However, it may be appropriate to provide some typical information on several structures, both industrial and architectural. See Figs 12, 13, 14 and 15.

Fig 13 Stores Building, New South Wales, Australia

Figure 13 illustrates a large storage structure consisting of 15 bays, each 30m x 30m, providing a total area of some 152.4m in length by 91.4m in width. Erected in situ from scaffolding, the roof incorporated some 34,000 tubes and 9,000 eight-slot connectors. The grid spacing on the top and bottom booms was 2.0m x 2.9m with a total grid depth of 1.8m. The structure was designed for a dead load of 39.06 kilograms per square meter, and a live load of 24.4 kilograms per square meter. Uplift, dead load plus wind load was 53.7 kilograms per square meter.

Fig 14 Thickner Tank, Grand Cache, Canada

The project shown in Fig 14 required a winter cover for an existing "coal wash" thickner tank. It was necessary that the project be completed before heavy snowfalls occurred, with the tank in operation during the construction.

The lamella-patterned dome was 56.39m in diameter with 64 base divisions and a height of 13.72m. The network was comprised of 2,312 galvanized steel tubes, each 217mm in diameter, with a wall thickness of 3.8mm. The structure was designed for a dead load of 48.8 kilograms per square meter, with a maximum of 219.7 kilograms per square meter for the snow load. In this case, seismic conditions had to be considered along with a maximum wind velocity of 128.8 k.p.h. Radial wood purlins and pre-cut sheets of plywood provided the enclosure for the network. A final weatherproofing was applied with asphalt shingles.

Fig 15 Maintenance Building (Abu Dhabi National Oil Co.) Abu Dhabi.
This all aluminium frame structure covers an area of 65m x 65m.

Fig 16 Metro Zoo - Toronto, Canada

Two of the main structures in this outstanding zoo are the Indo-Malayan and the African Pavilions. The two buildings consist of a series of inter-connected hyperbolic paraboloids tilted at various angles, and covering an area of approximately 8,361 meters. The hypar networks (including nine different types) total thirty-two in number, and are both square and diamond-shaped in plan, see Ref 10.

Fig 17 Arts Centre Spire, Victoria, Australia
Exposed gold anodized aluminium hyperbolic parabaloids form the base for the "landmark" spire of this distinctive structure.

Fig 18 University of Rochester, Rochester, U.S.A.

Fig 19 El Nilein Mosque, Khartoum

This unique and elegant structure consists mainly of a 34.9 m diameter reticulated single layer hemisphere of a lamella pattern. The cladding consists of a series of diamond-based pyramids, which reduce in size towards the apex. These pyramids were fabricated from stucco-embossed aluminum sheet and fastened to the main triangulated framework, see Ref 11.

Fig 20 First Federal Bank, Chicago, U.S.A.

Fig 21 Sports Palace, Mexico City, Mexico
Shown under construction, this structure combines a dome with hypar elements. The total span is 150m with the hypars used for infilling and to assist in stabilizing the main arch elements.

Fig 22 Melitta-Werke Tank Cover, Berlin, Germany

Fig 23 Shell of Arbitrary Shape, Toluca, Mexico

Fig 24 100,000 Tonne Grain Silo, Monterrey, Mexico

Fig 25 Hospital Atrium, Calgary, Canada 25 m x 40 m

Fig 26 Ontario Place Cinesphere, Toronto, Canada

These figures cover a very small sampling of projects constructed. A detailed description of the structural design, fabrication and erection of any structure of significant magnitude could be the subject of a full-length paper as in Refs 10 and 12.

ACKNOWLEDGEMENTS

This paper reflects the work of Triodetic Structures Limited and accredited licensed associates. The author would especially like to acknowledge the important contributions of Dr. D.T. Wright, Eng. Francisco Castano, William J. Vangool, P.Eng. and Baco Contracts Limited for their assistance in the development and use of the Triodetic system internationally.

REFERENCES

1. MAKOWSKI, Z.S.; "Steel Space Structures"; Michael Joseph, London; 1965

2. DU CHATEAU, S.; The SDC Structural System, International Association for Shell Structures; Colloquium on Hanging Roofs, Continuous Metallic Shell Roofs and Superficial Lattice Roofs; Paris, 9-11 July, 1962. Amsterdam, North-Holland Publishing Co., New York, Interscience Publishers (John Wiley & Sons, Inc.); 1963

3. WRIGHT, Dr. D.T.; "A Continuum Analysis for Double Layer Spaceframe Shells"; International Association for Bridges and Structural Engineering, Memoires, v. 26, 1966

4. WRIGHT, Dr. D.T.; "Membrane Forces and Buckling in Reticulated Shells"; ASCE, St. Div., V.91, n. ST1, Feb., 1965

5. MARSH, C., and KNEEN, P.W.; "Aluminum Space Frame for Parque Anhembi"; ASCE National Structural Engineering Meeting, Baltimore, Maryland, U.S.A.; April 1971

6. CASTANO, F., and WRIGHT, Dr. D.T., International Congress on the Application of Shell Structures in Architecture; Mexico, September, 1967

7. WRIGHT, Dr. D.T., and FENTIMAN, A.E.; "The Design and Construction of a 60 foot Paraboloidal Antenna"; IRE Canadian Convention, Toronto, Ontario, October, 1958

8. WRIGHT, Dr. D.T.; "Static Tests of TRIODETIC* Structural Connections"; Publication pending

9. KNEEN, P.W.; "Geometric Forms for Flat Double Layer Space Frame Structures"; 2nd International Conference on Space Structures; Guildford, England, September, 1975

10. BERGMANN, R., & FENTIMAN, H.G.; "Design and Construction of Multi-Leaf Shell Structures Using Interconnected Hyperbolic Paraboloids of Latticed Tubular Members"; 2nd International Conference on Space Structures, Guildford, England, 1975

11. ELLIOTT, W.A.; Section 25, "TRIODETIC* Domes"; Analysis, Design and Construction of Braced Domes, Granada Publishing, publication pending

12. FENTIMAN, H.G.; "TRIODETIC* Connections in Space Frame Structures"; I.A.S.S. Synposium; Part II, "Tension Structures and Space Frames"; Tokyo and Kyoto, Japan, 1971

THE SPACE DECK SYSTEM

S. C. BAIRD, C.ENG. M.I.Struct.E., F.F.B., M.A.S.C.E.

Technical Director, Space Decks Limited
Chard, Somerset, England

SPACE DECKS LIMITED manufacture one of the longest established and proven Space Frame Systems available today. The first structure being erected in Burma in 1954 and since then the Company has designed, fabricated and erected approximately 3000 space frame structures world wide, varying in complexity from the very simple rectangular canopy to multilayer and sloping structures.

INTRODUCTION

SPACE DECK is a factory made, pre-fabricated steel Space Frame System based on repetitive use of modular inverted pyramidal units connected by solid steel high tensile tie bars. Its most common application is in double layer format using square on offset square configuration.

Recent developments at SPACE DECKS have considerably increased the range of space frames manufactured so that now designs for most space frame projects are possible.

It is to be assumed that in general the use and application of space frames are understood and appreciated, hence the following paper will emphasise and explain the direct advantages and applications of the SPACE DECK system.

Photo 1 Al Dhiafa Exhibition Centre Riyadh

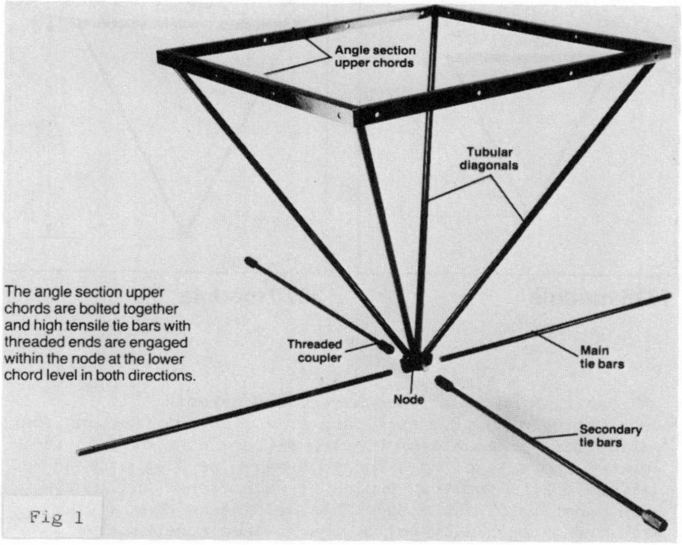

Fig 1

Some of the advantages are:-

a) Additional purlins are not required to support the roof or floor decking as the decking is fixed directly to the space frame.

b) Fully adjustable tie bars in both directions enable the exact design camber requirements to be achieved.

c) Standard units are pre-fabricated and held in stock ready for immediate delivery to site.

d) Site assembly time is considerably reduced over the conventional chord and joint space frame systems by the use of pre-fabricated pyramidal units.

DESCRIPTION OF SPACE DECK UNITS

Upto 1980 the most commonly used SPACE DECK unit has been the "1212" unit but with recent development the range of units now available includes:-

"1275"	-	1200 sq x 750 deep pyramidal unit
"1212"	-	1200 sq x 1200 deep pyramidal unit
"1515"	-	1500 sq x 1500 deep pyramidal unit
"2020"	-	2000 sq x 2000 deep pyramidal unit

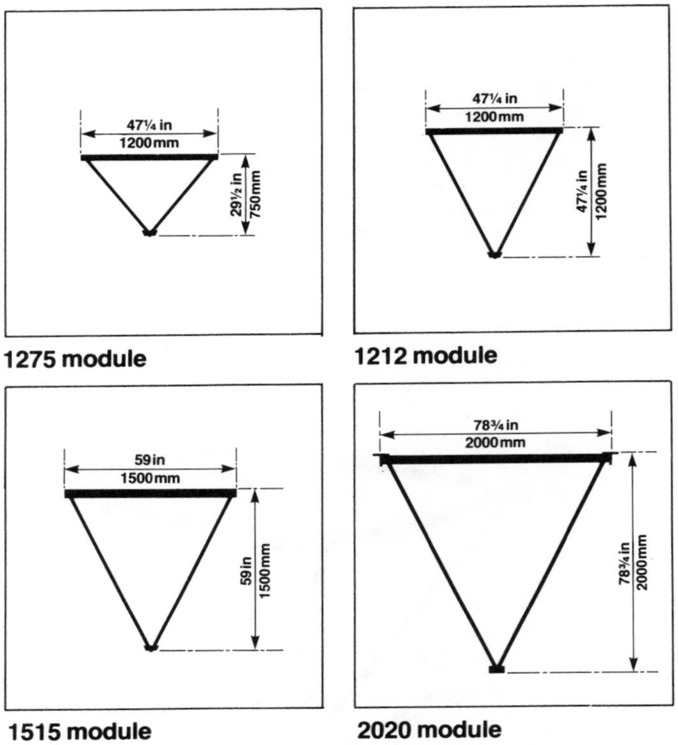

Fig 2

The unit is an inverted square-based pyramid (Fig 1) consisting of a mild steel angle section top tray and four mild steel diagonals or bracing members connected at the bottom chord level by a forged boss. The diagonals are welded to the top tray angles at each corner and to the flat section of the boss. This welding is done at the factory in specially built jigs to ensure repetitive accuracy.

The forged boss at bottom chord level is threaded in both directions to receive the tie bars. (Fig 3)

These tie bars are high tensile steel rods which are threaded at one end with a left hand thread and at the other end with a right hand thread. When in place these tie bars form the bottom chord member of the space frame.

For reasons of assembly two types of tie bars are used -

a) the Main Tie and

b) the Secondary Tie.

The main tie screws directly into the boss forging whereas the secondary tie is attached to a cross stud in the boss

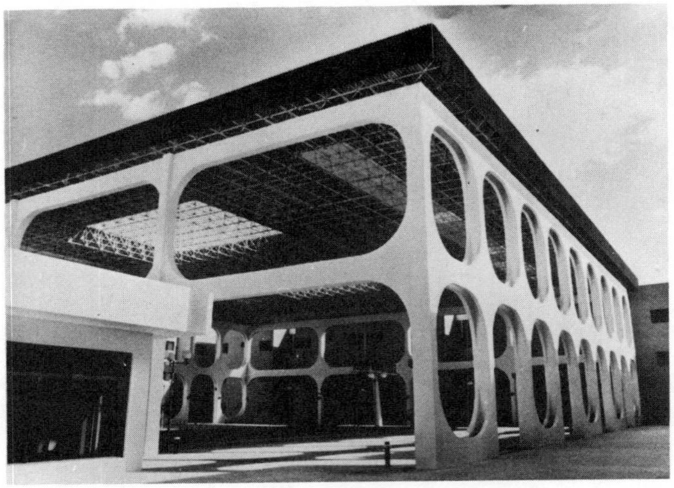

Photo 2 Swimming Pool Abu Dhabi

by means of a tapped hexagonal coupler. Both types of tie bar, which are usually of equal strength, have the opposite ends threaded differently. (Fig 3)

This feature, unique to the SPACE DECK system, of left and right hand threaded tie bars provides a turnbuckle facility which effectively reduces/increases the centre to centre dimension of adjacent bosses thus allowing the camber to be adjusted to suit the required configuration. In this way camber configurations ranging from shallow domes, with or without level perimeters, to simple barrel vaults can be provided with tie bar components of a single length and it is therefore unnecessary to identify individual chord members.

The advantages of this standardisation being:-

a) Design/Detailing time

b) Fabrication speed/stocking of standard components

c) On site it becomes unnecessary to find a specific tie bar for a specific location.

The same should be said of the standard pyramidal unit which again does not need to be individually identified due to its standard and therefore interchangeable format.

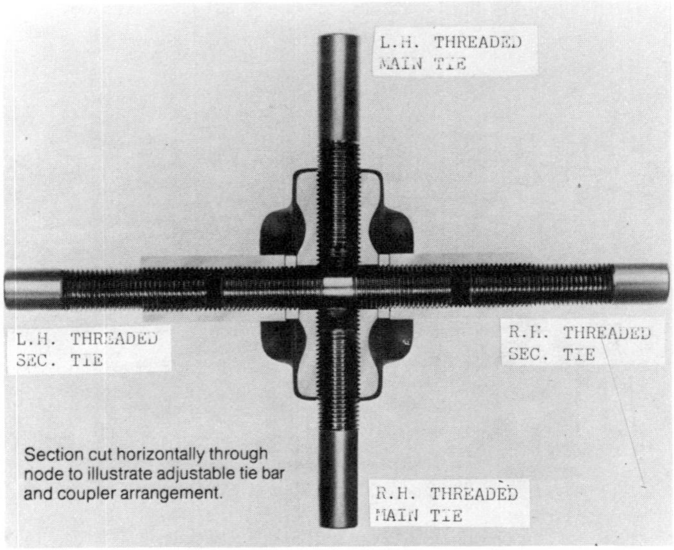

Fig 3

EXPLODED VIEW OF 2·400m SQUARE SECTION OF STANDARD 1212 SPACE DECK

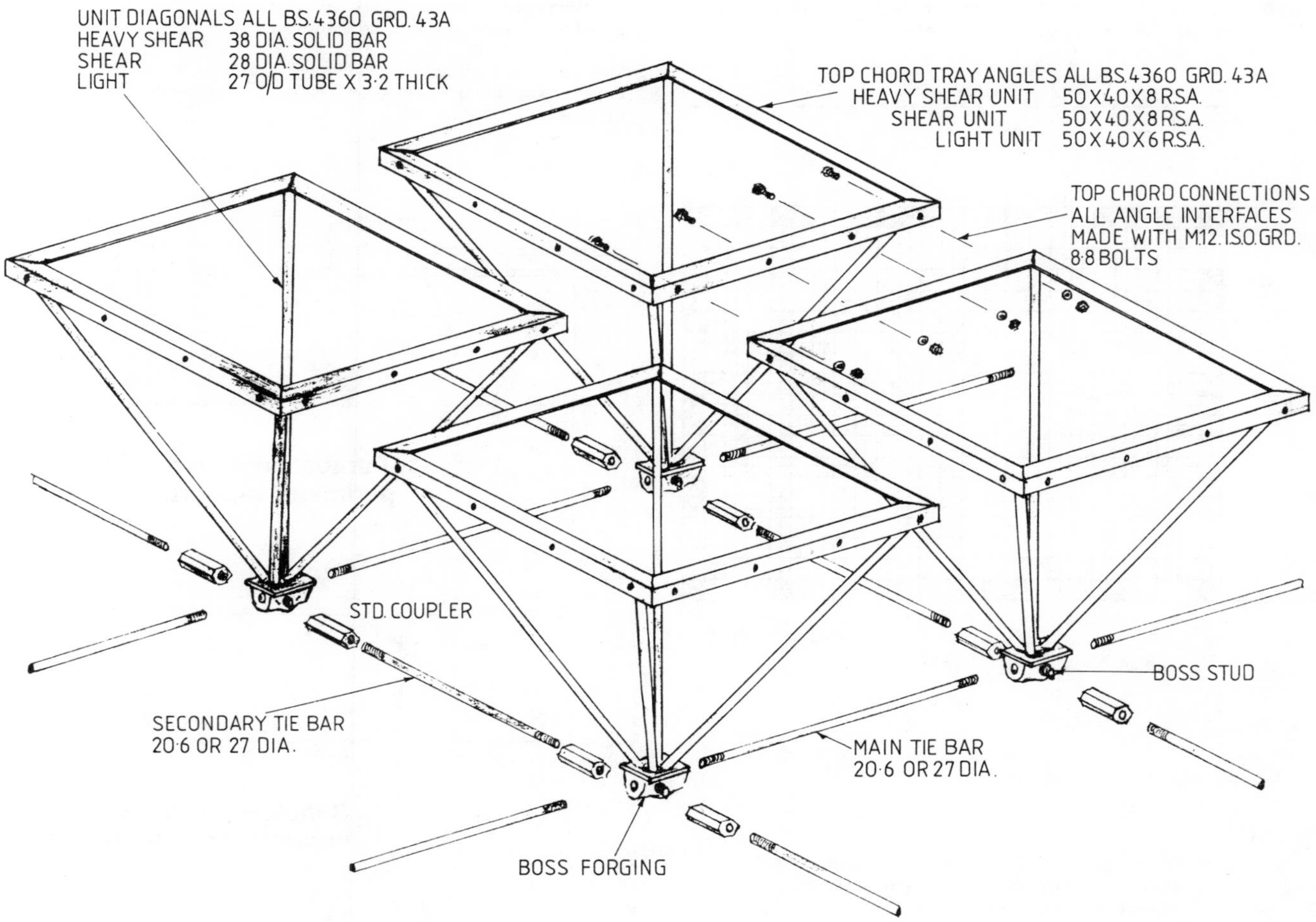

Fig 4

Each range of units (i.e. 1275, 1212, 1515, 2020) has three types of unit each type having the same bending capabilities but varying in shear capacity:-

1) Light Unit

2) Shear Unit

3) Heavy Shear Unit

In application then, the SPACE DECK components are constructed on the drawing as a series of square modules to build up areas of a flat roof or floor structures. (Fig 4) As the strength of the individual tie bars and pyramids are constant the required strength of the overall structure is adjusted by varying the tie bar and pyramid density. So where less than 100% of units is required then the spans are infilled with top chord trays. Selective positioning of various shear units is required, these coming in the main adjacent to support positions where the build up of load requires a stronger unit.

The additional strength in a shear unit is obtained by substituting the previously tubular section of the diagonals with solid mild steel bar.

The procedure for selecting and positioning the units is:-

1) select type of unit relative to span conditions referring to performance envelopes

2) design the structure to determine density of units required and also the positioning of the shear units.

The design is carried out by SPACE DECKS Engineering Department and is dependent upon the support conditions applicable. These being broken down into two alternatives generally.

1) Plates

Clear span areas in which support for the structure is provided at the perimeter edges only. The plate application is the most structurally economical and is the one often used in Sports Halls etc.

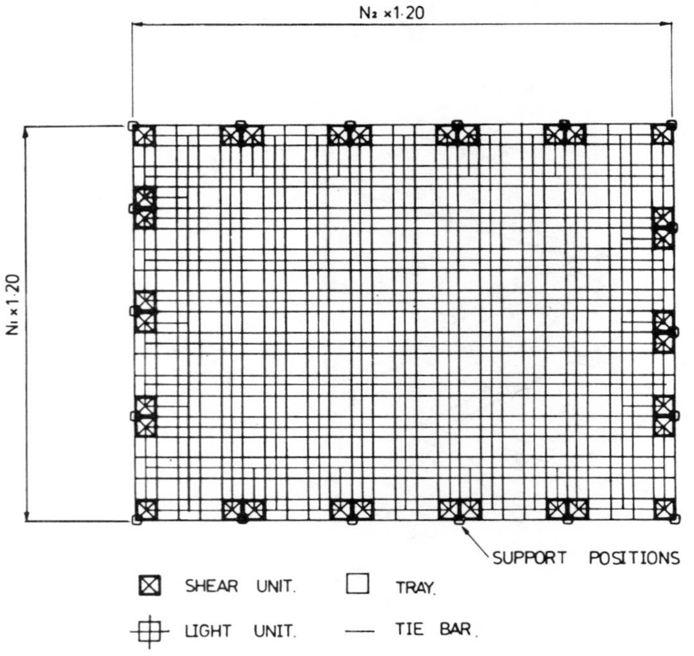

Fig 5 Typical Plate

2) Grids

That is where the SPACE DECK structure is supported at the four corners only. Useful for large areas where a minimum of internal columns is required.

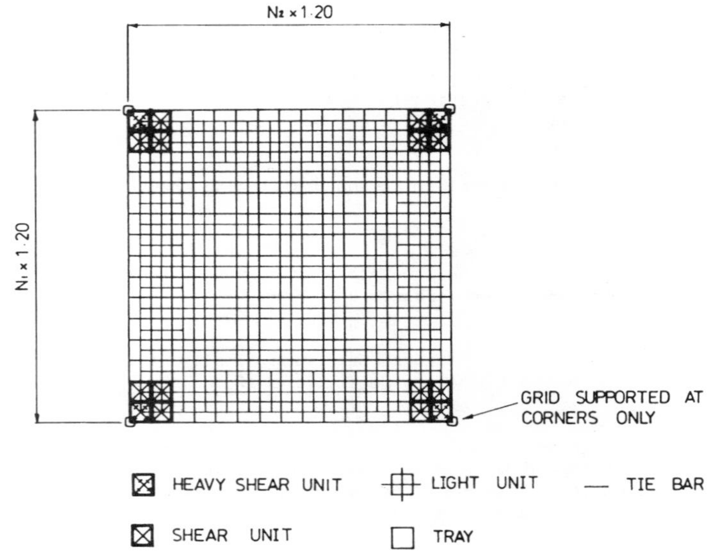

Fig 6 Typical Grid

The complete roof structure is then supported preferably at the top chord level by another SPACE DECK standard component called a cap head.

Support may also be taken at the bottom chord level when a mansard edge detail is required.

The cap head supports and connects the SPACE DECK to the supporting structure - be it an R.C. ring beam or column or a steel stanchion.

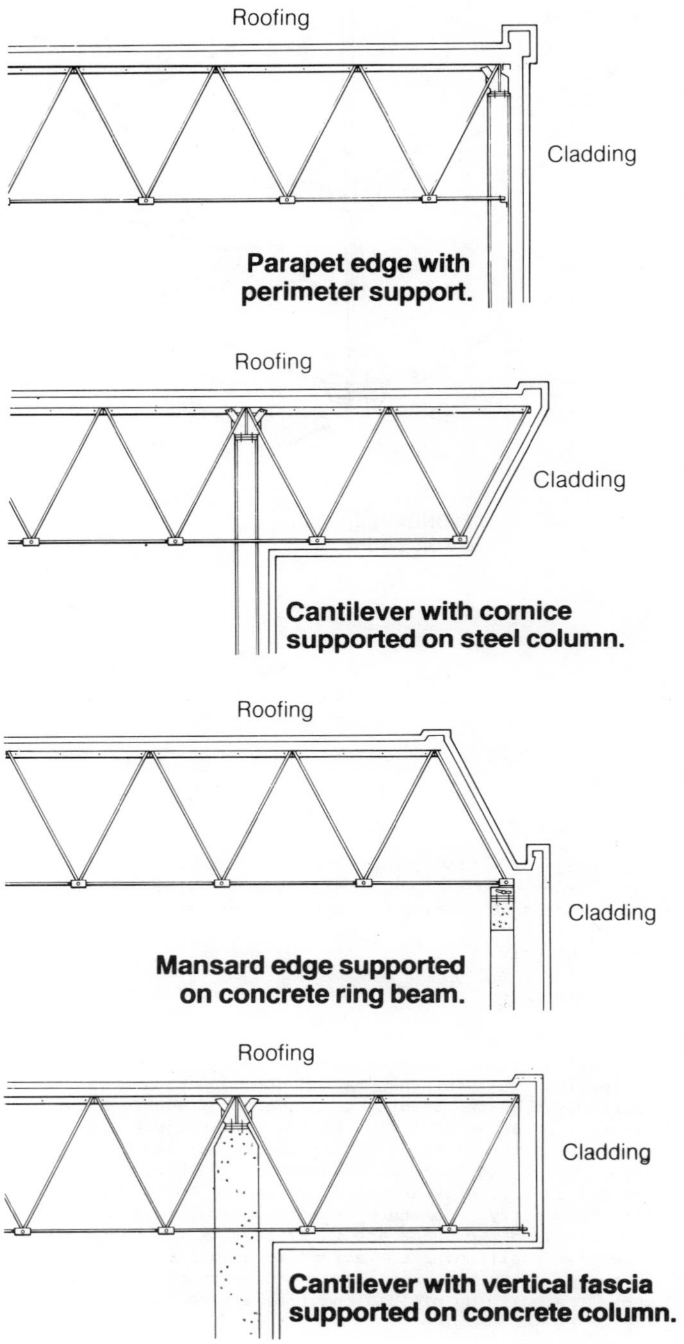

Fig 7 Typical Support Conditions

FABRICATION AND PRODUCTION

Manufacture of all components is carried out at SPACE DECKS main factory in Chard, Somerset, England. Components are fabricated to form the pyramidal units in addition to fabrication being carried out in the main factory at Chard there are local factories situated in Saudi Arabia, Nigeria, Venezuela and Italy.

With few variations in the types of units, it is possible to fabricate the standard SPACE DECK pyramid units in advance of receiving an order and to hold in stock in the company's large warehouses. When an order for a relatively straightforward building is received then the units can be delivered directly to site out of stock.

Even a job with non-standard units on it presents little problem as the percentage of non-standard units on a job is generally small and these special units can be quickly made to be delivered with the main despatch of standard units.

Photo 3 Rotary Transfer Trunnion Machine. By utilising six boring and two threading positions produces a SPACE DECK node in nineteen seconds.

Photo 4 Degreasing unit in the flow line production at the Chard factory.

Photo 5 University of Ilorin Nigeria

Fabrication is on a production line basis. The components are shot blasted and then treated with a clear lacquer prior to fabrication. The components are then positioned in specially built multi-axial jigs and held firmly and accurately in position by pneumatic rams.

While in the jigs they are finally welded using a semi-automatic metal-arc inert gas (M.I.G.) welding process. This uses a solid wire electrode which is shielded by a stream of carbon dioxide gas and has good penetration properties. Using variable position pneumatic jigs ensures the operator can work in a downhand position thus ensuring the quality of the welds.

This mass production technique produces a very accurate finished structure as the risk element of fabrication inaccuracies is almost eliminated.

After fabrication the pyramidal units are loaded onto a conveyor line and taken through the electrostatic spray booth before being oven dried.

Prior to painting the threaded parts of the units are protected with grease and plastic caps are fitted.

For local fabrication (i.e. in countries outside UK) the components are shot blasted then treated with a weldable zinc phosphate Epoxy Blast Primer prior to despatch.

Completed units and components are transferred to the adjacent warehouse and stored. Units stack on top of each other conveniently and economically for storage and transportation by road or sea.

Photo 6 For overseas contracts unit transportation is very economical. Over 750 sq metres of roof can be loaded onto one standard 40'0" flat rack.

ASSEMBLY AND ERECTION PROCEDURE

There are two methods of assembling the units:-

1) ground level assembly and then lifted into place by crane
2) insitu assembly by hand and temporarily supported by scaffolding.

1) Ground Level Assembly

Ideally there should be a flat area adjacent to the site where the units can be laid out to suit the pattern of SPACE DECK used. (Fig 8) The units are actually laid out upside-down - that is similar to a true pyramid - this way up the bottom chord tie bars can be easily positioned and it keeps the threaded parts of the system well away from the ground thus preventing any damage to the threads by the soil.

Each row of units is attached together in this upside-down position to form long Vee Beams and tie bar lengths are adjusted at this stage to ensure that the correct amount of camber is induced into the structure.

The rows of Vee Beams are then turned over by hand into their correct position - that is so they are now sat on their bosses and then the top chords are joined together by bolts and the bottom chords are joined together using the secondary tie bars. These secondary tie bars also being adjusted in length to ensure the correct amount of camber is induced into the structure.

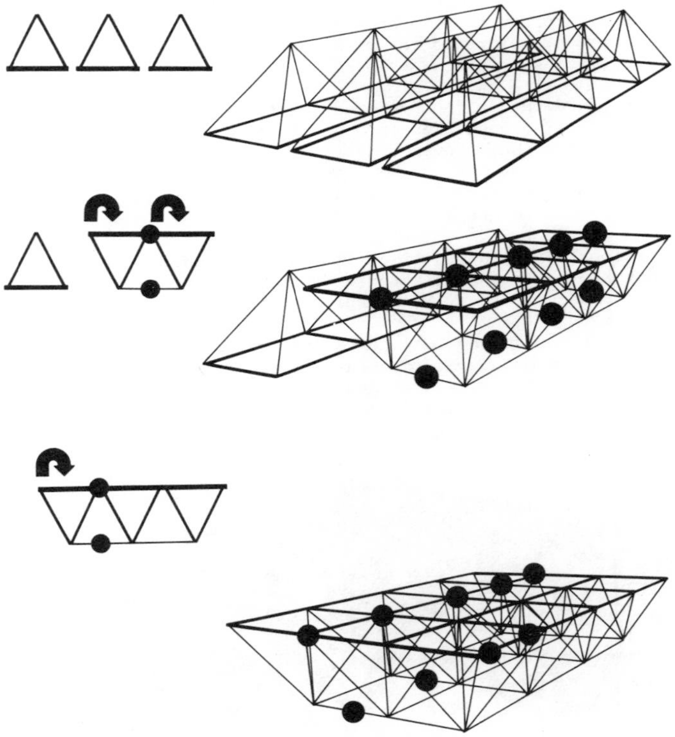

Fig 8 Ground Level Assembly Procedure

Photo 7 Sahari Centre Jeddah

Areas of SPACE DECK are now built up in this way - the sizes of the assembled areas being determined by the support centres, available cranage and location on site. When completed the areas of SPACE DECK are lifted into place by crane onto the supporting structure.

The advantage of assembling the roof on the ground in small sections and then lifting is that where site access is good the size of crane can be kept to a minimum and often there will be no requirement for a special large crane to erect the Space Frame

Another very important advantage of assembling the roof deck on an area adjacent to the site is that the main contractor is able to continue his work unimpeded whilst the remainder of the SPACE DECK is being assembled off site. The effect to his site program is limited to hoisting the assembled SPACE DECK into its final position.

2) Insitu Assembly by Hand

Where crane access is totally impossible then you can still use the SPACE DECK system because it can be erected insitu by hand without the use of any mechanical aids.

The relatively light weight of individual units means they can be man-handled into position and bolted at the top chord and the tie bar fixed at the bottom chord level. This is best done off mobile scaffolding. Each unit is individually positioned until the whole roof area is covered. Each time a tie bar is fixed in place its length is checked to ensure the correct camber will be obtained.

A possible alternative to the insitu assembly by hand is a cross between the crane erected roof assembly and the the hand erected roof. This method is one of assembling the Vee Beam sections on site directly beneath their final position and then using small hoists to lift them into position.

Each of the erection methods described is so simple that they present little problem for the main contractor to use his existing staff to do the work under the direction of a SPACE DECK erection supervisor.

On a site where crane erection is possible but access to the site for erecting the SPACE DECK is limited or a very tight site program is required then the SPACE DECK plates can be assembled off site and stacked on top of each other until required. This way the actual erection time of the building is minimal which has obvious advantages to the main contractor.

DESIGN AND DEVELOPMENT

One of the advantages of a system that uses standard units and components is that the strength of each individual member can be established by testing.

When collated this information is used in the quality control of the units ensuring each unit manufactured is of the required strength.

With this information we can provide design data in the form of performance envelopes. So for say a 1212 unit that is the 1200 deep unit - as the unit properties are constant - then for a given load condition we can determine the maximum possible span.

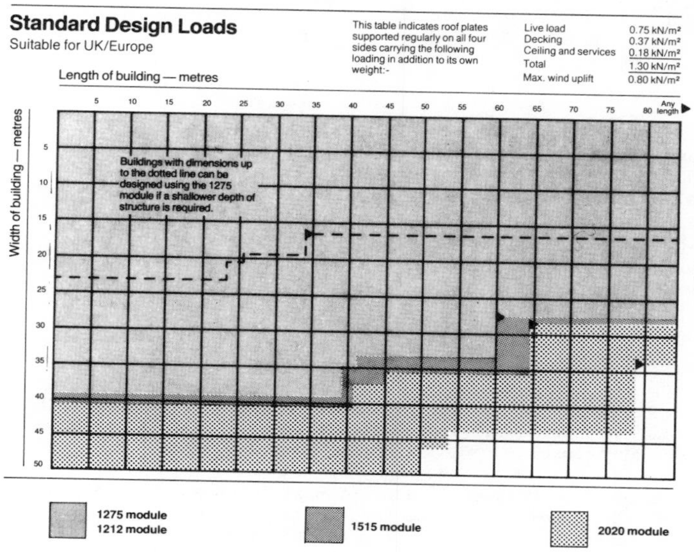

Fig 9 Typical Load/Span Table (UK Conditions)

SPACE DECKS LIMITED are one of the few Space Frame manufacturers to have carried out a full scale test through to failure. This full scale test was carried out on behalf of SPACE DECKS LIMITED by an independant testing authority in England.

The size of the tested structure was 20.4 m x 18.0 m and was assembled out of the shallower 1275 units. The test consisted of applying a simulated U.D.L. by means of hydraulic rams and linkages to the top chord members. In addition to taking readings of vertical deflections, strain gauges were also used in recording individual members.

It was intended that the test should investigate two main features of the structure:-

1) flexural behaviour in terms of Appendix 'A' to BS449 part 1 - that is stiffness test.

2) the ultimate load carrying capacity of the structure by establishing the load necessary to cause failure.

For the stiffness test the applied load was the design load + ½ the superimposed load. The results for this test showed that the structure satisfied the specific requirements for a recovery of 80% of the maximum deflection recorded during the 24 hour period.

Photo 8 Full Scale Test Structure (20.4 m x 18.0 m)

However the most interesting result was obtained from testing the structure to failure. At almost precisely twice the total design loading three main tie bars at midspan position failed and the structure sagged noticably at the centre.

The load was then removed and a visual inspection revealed that in addition to the broken ties, diagonal members had buckled.

The loading was then re-applied to the now damaged structure up to the stiffness test load value (that is total design load + ½ superimposed load). This load was sucessfully held by the now damaged structure.

The load was then increased to 133% of the design load before the next failure occurred - this being another main tie bar.

The loading was again removed and the damage to the structure inspected. After which the now defective structure was still able to carry upto 129% of its design load before another failure occurred.

After this third test testing was discontinued.

The test proved that even with individual members failing at nearly 2 x design load requirements the structure redistributed the forces and was able in a damaged condition to carry in excess of its design load requirement.

This full scale test has given us great confidence in backing the design theory we use.

Photo 9 Sloping Space Frame at LBJ Plaza U.S.A.

RECENT TRENDS AND DEVELOPMENTS OF SPACE TRUSSES IN CHINA

TIEN T. LAN

Senior Research Engineer, Chinese Academy of Building Research
Beijing, China

The state-of-the-art review of the developments of space trusses in China is presented in the paper. Since sixties space truss has found wide applications as roof system for buildings which is demonstrated by different types of space truss from medium to long spans, with emphasis on their special features. The erection and construction techniques of space truss are then described. Simple and effective methods for assembly work have been carried out either in the elevated position or on the ground. Recent trends of the development of space truss are also discussed.

INTRODUCTION

In the early sixties, a new type of space structure, plate-like steel space truss was developed in China. After more than ten years of continuous development and promoting, space truss roofs not only appear in large cities in China, but also in small towns, industrial and mining districts. The reason that space truss attracts so much popularity is its convenience in fabrication and erection. Innovation in the technology for erection has been introduced so that space truss can be built economically without using large and special equipments. Architects also appreciate the flexibility in shape and form provided by the space truss. Building plans of square, rectangular, circular or polygonal shapes can all be formed with the modules of space truss, which satisfies the architectural and functional requirements.

The first space truss in China was constructed in 1964 as the roof of a gymnasium for Shanghai Teacher's College. It consists of 252 square pyramids and covers an area of 1300 m^2. Since then space trusses spanning up to 110 m are widely used to cover sports stadia, assembly halls, factory buildings, warehouses, theaters, railway stations etc., of which the sports halls consist the majority. Maximum economical effect could be obtained when space truss is used in buildings with square or nearly square plans and supported at perimeters. Besides, for those buildings, such as airplane hangars, which need opening on one side, space truss supported along three edges and free on the other edge is a promising form.

Space truss is an appropriate form of construction in seismic area. During the catastrophic Tangshan earthquake of July 28, 1976, all space trusses of long and medium spans located in Tianjin (8° intensity) and Beijing(6° intensity) withstood the seismic shock very well and suffered no damage. While no space truss was built in Tangshan then, several steel roof trusses possessing proper bracing systems survived earthquake. Thus space truss which has better horizontal stiffness than ordinary truss system will provide an exceptional ability to resist earthquake shock. In the rehabilitation of Tangshan city many public and industrial buildings were designed with this new type of structure.

Since 1973, Chinese Academy of Building Research associated with a number of design, constructional and educational organizations was engaged in the compilation of Specifications for Design and Construction of Space Trusses. The Specifications sum up the practical experiences and research findings in type-choosing, analysis and design, fabrication and erection of space truss, which not only reflect comprehensively the technical achievements in our country but also assimilate advanced experiences from other countries. The document was approved by the State Bureau of Building Construction and effective from May 1, 1981 as a national code.

ENGINEERING PRACTICE

The specifications recommend 12 types of commonly used space trusses. They are classified into following groups and named accordingly. The names in the parenthesis indicate the terminology suggested in some English literature.

1. Composed of lattice trusses
   (1) Two-way orthogonal lattice grids, Fig 1 (a)
   (Square on square with no skew diagonals)
   (2) Two-way diagonal lattice grids, Fig 1 (b)
   (3) Two-way diagonal skew lattice grids, Fig 1 (c)
   (4) Three-way lattice grids, Fig 1(d)

2. Composed of square pyramids
   (1) Orthogonal square pyramid space grids, Fig 1 (e)
   (Square on square offset)
   (2) Orthogonal square pyramid space grids with openings, Fig 1(f)
   (Square on square offset with internal openings or Square on larqer square)
   (3) Square pyramid space grids of checkerboard pattern, Fig 1 (g)

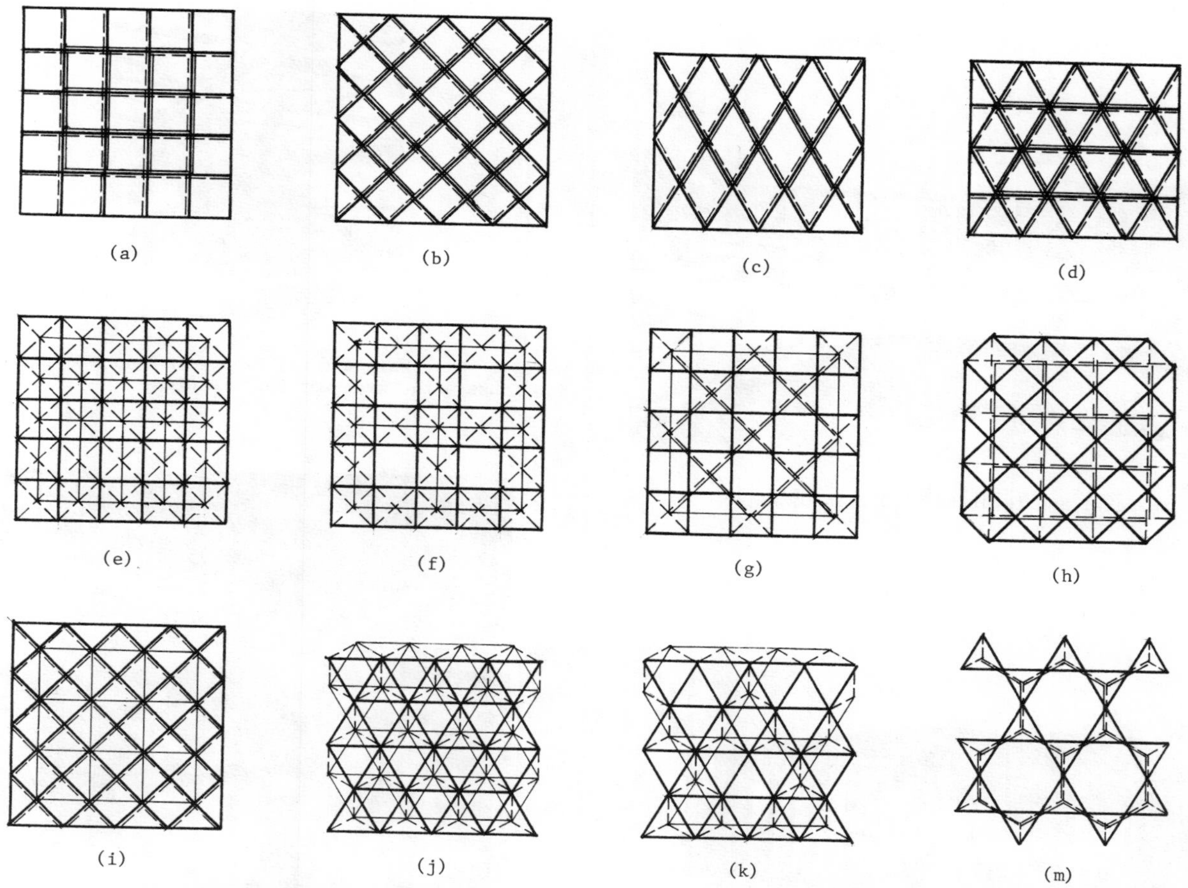

Fig 1 Types of Commonly Used Space Trusses

(Square on square offset diagonally with internal openings or square on diagonal)

(4) Diagonal square pyramid space grids, Fig 1 (h) (Diagonal square on square with internal openings or diagonal on square)

(5) Square pyramid space grids with star elements, Fig 1 (i)

3. Composed of triangular pyramids

(1) Triangular pyramid space grids, Fig 1 (j) (Triangle on triangle offset)

(2) Triangular pyramid space grids with openings, Fig 1 (k) (Triangle on triangle offset with internal openings)

(3) Triangular pyramid space grids of honeycomb pattern, Fig 1 (m) (Obayashi Truss)

Space trusses composed of lattice trusses are mostly used in long span roof structures. Two-way orthogonal lattice grids were evolved from lattice trusses that intersected at right angles to each other. It has been designed and constructed recently to cover Shanxi Sports Hall in Xian, the largest one in northwestern China. The roof structure is 90 m x 66 m in rectangular plan with clipped corners and consists of 4.9 m deep trusses at 6 m centers. All tubular members are connected by welded hollow spheres of 500 mm diameter which are reinforced with annular diaphragms. The total steel weight amounts to 47 kg/m^2. This type of space truss has also been used in medium span sports halls and training halls from 30 m to 60 m with good economical effects.

Two-way diagonal lattice grids were first used to roof the Capital Sports Hall, Beijing in 1968. The stadium covering an area 99 m x 112 m, is the largest unobstructed interior space in our country and seats more than 10,000. The space truss was constructed with a module of 4.7 m and depth of 6 m. Using 16Mn steel (yield strength 3500 kg/cm^2) angle sections, the total weight of the roof structure was more than 700 tons giving an average of 65 kg/m^2. The construction of the first long span space truss showed that this type of structure could meet all functional requirements and compete successfully with more conventional systems. Similar examples of employing diagonal lattice grids in sports halls were found in Fuzhou, Fujian Province (54 m x 67.5 m) Lanzhou, Gansu Province (50 m x 50 m), Jinan, Shandong Province (62.7 m x 74 m), Harbin, Heilongjiang Province (50.4 m x 61.6 m) and Shijiazhuang, Hebei Province (70.4 m x 83.2 m).

Three-way lattice grids are interconnected by lattice trusses in three directions. The triangular modules thus formed are adaptable to spanning odd shaped areas. For example, the space truss over the Shanghai Cultural Hall was fan-shaped, covering an area of 7800 m^2 (Fig 2) and the Jiangsu Sports Hall was octagonal, covering 6090 m^2. It was also adopted for the circular roof of Shanghai Sports Arena, the longest span in China as shown in Fig 3. The roof structure, with 110 m diameter, projected 7.5 m from 36 reinforced concrete supporting columns around the perimeter. The space truss is 6 m deep with a triangular module of 6.1 m and constructed with tubular sections of 16Mn grade steel. The weight of the lattice grids, including members and joints, was 47 kg/m^2. Adjacent to Shanghai Sports Arena, a new Swimming Hall was built

Fig 3 Shanghai Sports Arena, 110 m in Diameter

Fig 2 Shanghai Cultural Hall. Fan-shaped Space Truss with Three-way Lattice Grids

Fig 4 A Maintenance Hangar. Space Truss spans 40 m over the Hangar Door

Fig 5 Liuzhou Sports Stadium Covered with 45 m x 60 m Space Grids

recently employing the same type of space truss but with variable depth. The plan is an unsymmetrical hexagon with axes measuring 95 m and 90 m respectively which gives an area of 6000 m^2. Each triangular grid is 5.2 m long and the depth varies from 4 m at the perimeter to 7.6 m at the center to provide for drainage as well as the necessary stiffness for the roof.

The earliest space truss constructed in China was the type of orthogonal square pyramid space grids with top chord square grid offset over a bottom chord square grid, it was used later in several medium span roofs. Examples could be cited as the roof for Hangzhou Theater covering an area of 31.5 m x 36 m and Zhabei Sports Hall, 45 m square, in Shanghai. If certain elements in the internal grids are eliminated alternately then orthogonal square pyramid space grids with openings could be obtained. The hangar roof of Shanghai Air Sports Club is a typical example which covers an area of 27.3 m x 35.1 m with a top chord module of 1.95 m and depth of 2.2 m. Each erecting unit is a truncated pyramid with square base of 3.9 m and connected by top chord members. This type of grid is also suitable for column-supported multiple span buildings.

Probably the type of space truss most widely used in China is the diagonal square pyramid space grids, with top chords on diagonal grid lines and bottom chords on orthogonal, and elements of internal grids eliminated in a checkerboard pattern. This type of space grids was first adopted as a roof over Science Auditorium at Tianjin in 1966. Since then it has been used to cover a considerable number of medium and short span buildings, such as cinemas, workers' clubs, dining halls, workshops etc. Furthermore, a maint e nance hangar at Beijing, which covers an area of 54 m x 48 m and requires to provide hangar door along the 54 m span, also adopted this type of space grids with triple layer along the door opening. The roof structure was designed for a load of 250 kg/m^2 and also two suspended traveling cranes of capacities 5 and 3 tons, giving a steel consumption of 48 kg/m^2. Fig 4 shows another airplane hangar of similar type, the span being 40 m over the hangar door. The largest diagonal square pyramid space grids were completed in 1978 as the roof to cover a sports stadium in Liuzhou, Guangshi Zhuang Autonomous Region (Fig 5). The structure measures 45 m x 60 m with a grid depth of 3 m and a top chord module of 3.5 m. The members are made of steel angles of the Grade A3 (yielding strength 2400 kg/cm^2) and connected by cruciform gusset plates, the weight of the structural steel being 26.8 kg/m^2.

The square pyramid space grids of checkerboard pattern is similar to the above type except the directions of the chord system are reversed. They have been adopted effectively in two similar dining halls for miners in

Datung, Shanxi Province with a steel consumption of 9 kg/m^2 only. The roof measures 18 m x 24 m in plan with depth of 1.3 m and supports are at node points of top chord resting on reinforced concrete peripheral beams. Another example is a plywood workshop in Taiyuan, Shanxi Province. The whole roof, 144 m long, consists of 8 units, each 18 m x 18 m in size and 1.5 m deep. Each unit is designed individually as a space truss supported on four corners.

Space grids composed of triangular pyramids were used only to a limited extent. A remarkable example of triangular pyramid space grids with openings is given by the Tanggu Railway Station in Tianjin, which is circular in plan with 47 m diameter. The top chord module of space grids is 3.4 m and depth is 3 m. The roof structure is constructed with tubular members and hollow spherical joints, giving a steel consumption of 30.4 kg/m^2.

ERECTION AND CONSTRUCTION

The erection of space truss can be classified into two main types, depending on where the assembly work is carried out, either in the elevated position or at the ground level. A method belonging to the first type is to fabricate members or elements on the ground and then assemble them in the elevated position. During construction full or partial scaffolding is required so that the assembly of members or elements may proceed in the high place and the whole space truss completed in its final position. This method was mainly employed in the construction of long and medium span space trusses. Such as the lattice grids for Capital Sports Hall were assembled from prefabricated truss elements of 1-3 panels and connected by high-strength bolts in the air as shown by Fig 6. Two tower cranes were used to transport and hoist the structural elements. Several medium span space trusses were constructed from individual members by welding in those district where heavy lifting equipments were not available.

Another method is to erect the space truss in assembling units. Space truss devided into strips or blocks is fabricated at ground level, then hoisted by ordinary equipment into its position, so that the work at high place can be minimized. Temporary scaffolding is established at the junction of units when necessary. These units are then interconnected to form a whole space truss. During the construction of Zhabei Sports Hall in Shanghai, the space grids were devided into 3 stripped units, each measuring 15 m x 45 m and weighing about 18 tons, which were erected by mobile cranes to the required position. It was estimated that erecting cost was only 1/26 of the cost of space truss.

An innovative method of assembling space truss in the elevated position is the sliding technique introduced in recent years. Separate strips of the space truss are fabricated at ground level and then hoisted to one end of the roof. Stripped elements may also be assembled on a platform at one end. The supports of the assembling unit are laid on the sliding rails on the roof with or without rollers. Drawn by hoisting engines, the stripped elements slide into position according to a predetermined sequence and then form the final structure. This method was employed during the construction of space grids for Zhenjiang Sports Hall, Jiangsu Province. The space grids measured 45 m x 55 m were devided into 7 sliding units of 7.86 m x 45 m . Each sliding unit was fabricated on the ground in two halves and then transported and hoisted to a platform at the end of the hall for subassembly. The supports of the space grids rested directly on the sliding rails along two sides of the building. After a sliding unit was moved to a certain distance, another unit was attached to it and both units slided together. The units were thus composed successively, when they reached to the other end of the building the space truss was completed. Another possible sequence of assembling is to slide the stripped element separately to their final position and then connected together to form the complete structure.

Because of the low efficiency and safty problem involved with the assembly work in the elevated position, especially for long span structures, the other type of erection is evolved so as to assemble the space truss on the ground. A method is developed to lift the space truss bodily which is then displaced into position in the air. In the work site the whole space truss is fabricated and assembled offset from its final horizontal position to enable the supporting columns erected prior to lifting. Then the space truss is lifted to a level above the column top and then displaced horizontally so that the supports may rest directly on the column. For large scale space trusses, specially made derrick masts are used as the support and electric winches as the lifting force. This method was first used in 1970 to erect a fan-shaped space truss weighing 285 tons for Shanghai Cultural Hall. Later it was used in a larger scale to lift a circular space truss for Shanghai Sports Arena with a diameter of 125 m. The roof structure has a total weight of 702 tons. 6 derrick masts of 50 m high were established to lift the space truss at 24 lifting points, using 12 winches of capacity 10 tons. When the strucutre was lifted 0.5 m above the column top, by tightening and loosening the front and rear pulley blocks respectively on the derrick mast, a tangential displacement was possible, and the whole space truss rotated 2° 5' clockwise in the air. The supporting joints were aligned with the column top and the space truss was lowered and seated on its final position. The newly completed space truss for Shanghai Swimming Hall with total weight of 560 tons was also erected by such lifting technique using 4 derrick masts (Fig 7).

The other way of erecting the space truss bodily is through the jack-up or lift-up mehod. Their common feature is the horizonatl position of the structure remains the same during the process of lifting. The equipments used are mostly of the conventional type, such as screw or hydraulic jacks, jacks for slip form, lifting machines for lift-slabs, etc. It is economical to make full use of the load-bearing columns of the structure as the supporting points for erection, thus avoiding heavy hoisting equipments.

By using the jack-up mehod, supports of the space truss, which is assembled at ground level, can be established directly on the position of columns without offset. Jacks are used to raise the structure to the roof level. For the erection of circular space grids for Tanggu Railway Station, 6 screw jacks of 30 tons capacity were used and timber sleepers were piled up as supports. A total weight of 78 tons, including the 47 m diameter space truss, was jacked up to a height slightly above the roof, then surrounding columns were erected and the space truss descended to its final position of 11.2 m high. (Fig 8)

Recently with the development of the lift-slab technique in multistorey buildings, lifting machines may also be utilized in the erection of space truss by such technique. A large size space truss covering the Shanxi Sports Hall was erected by 26 eletric lifting machines for lift-slab construction as shown in Fig 9. When the space truss was assembled and ready for raising, suspension rods reaching down from the lifting machines on top of the roof were attached to the supports through upper and lower transverse beams. All lifting machines were synchronized, operating form a central control console. The whole roof structure weighing 380 tons was lifted to 21 m high at the speed of 4 m per work shift. The lift-up method of space truss can also utilize the center hole hydraulic jacks for slipform construction. In the construction of the space grids for a sports stadium in Liuzhou, all welding and assembling work as well as the laying of timber roofing boards were carried out at the ground level. Metal lifting collars were established at the supports and connected with columns through supporting beam and jack rods. Motivated by hydraulic jacks the structure climbed to a height of 11 m.

An innovative new technology has been introduced in our construction practice, hydraulic jacks were used to haul up the space truss while lifting the slipforms for columns or walls. The space truss assembled on the ground was attached to a lifting mechanism which was connected with the jack rods in the slipform. When the slipforms of reinforced concrete columns go up the space truss serving as a working platform is also lifted simultaneously. Concreting of the columns is finished as soon as the space truss reaches the roof level. The first space truss employing this new technique was executed in 1975 for the roof of a theater in Shangyao, Jiangxi Province. The total weight including the space truss and constructional load was over 100 tons, and manual screw jacks were used for lifting. It was employed in a much larger scale for the maintainance hangar shown in Fig 10. The total load for lifting was 340 tons. 192 center hole hydraulic jacks were used in the columns. The actual time for lifting was only five days and nights for the structure to reach a height of 19 m.

Fig 6 Capital Sports Hall. Assembling of Space Truss in High Place

Fig 7 Shanghai Swimming Hall. Space Truss Assembled on Ground and Lifted up by 4 Derrick Masts

Fig 8 Tanggu Railway Station. Erection of Space Truss by Jack-up Method

Fig 9 Shanxi Sports Hall. Space Truss Erected by 26 Lifting Machines for Lift-slab Construction

Fig 10 Space Grids of a Maintenance Hangar Lifted up with the Slipforming of Concrete Columns

RECENT TRENDS

In China, the popularity of space truss is still increasing. A noticeable trend of the application of space truss is the roofing of multi-span industrial buildings. In order to create flexibility in production, factory buildings with a large unobstructed area and a minimum interference from internal supports are required. Thus, space truss with large column spacings is appropriate for this purpose. Several single storey factory buildings have been built. For example, in the newly developed industrial district of Tangshan city, the assembly shop of Tangshan Rolling Stock Pant, 72 m x 222 m in plan, was covered by orthogonal square pyramid space grids with openings. The 18 m square space truss was cladded with reinforced concrete roof slabs and suspended with 2 tons travelling cranes at all spans. More recently, a huge project for the carpet production shop of Dongfeng Chemical Works in Beijing is under construction. The four main shops, with an area of 40,000 m^2, use 18 m x 18 m space grids as basic units to cover the roof. (Fig 11). More than 400 tons of circular hollow tubes are used for the members and connected by bolted spherical joints, the steel weight amounts to 12.8 kg/m^2.

Fig 11 18 m x 18 m Space Grids Used in Multi-span Factory Buildings

As concrete is a traditional building material in China, reinforced concrete slab is used extensively as the cladding of space truss. Consequently, it is reasonable to consider the concrete slab as a part of the top chord and act integrally with the steel members of space truss. An interesting example of this type of structure is the triangular pyramid space grids of honeycomb pattern built for the roof of a dining hall in Xuzhou, Jiangsu Province. The rib of the reinforced concrete roof slab is acting as the top chord of the space truss, while bottom chord and diagonal member are steel tubular sections. The roof measures 21 m x 54 m in plan as shown in Fig 12 . This composite system has been designed as roofs for several projects and is now being extended into floor construction of multi-storey buildings. Our research findings justified this composite action of the slab and space truss. A simplified method of analysis by converting the concrete section into top chord has been studied.

Fig 12 Composite Construction of R.C. Slab and Steel Space Truss for a Dining Hall

The advent of electronic computers has greatly reduced the work in the analysis and design of space truss. There are a number of special purpose computer programs for space truss in China, some of which can also execute automated design. Such program provides the function of generating of joint coordinates, numbering of joints and members as well as the selection of sectional areas. Thus for a given type of space truss, the imput data can be reduced to a minimum. Research has also been carried out on the optimization of space truss. Based on a predetermined geometrical configuration, optimun design of member sections has been worked to satisfy the fully-stress criteria. A study has been carried out on the optimum design of commonly used space trusses simply supported at perimeters with the module dimension and depth as design variables. In formulating the mathematical model for optimization, the weight and cost of members and joints, as well as claddings were all taken into consideration, and the total cost was taken as the objective function. It is expected that the optimization procedure will be applied not only in the search of most efficient member sections and geometrical configuration, but also in the selection of the best type of layout.

In order to promote the application of space truss, industrialization is a necessary step to assure economy and efficiency. Standard designs of space grids employing different types of connecting nodes have been worked out. The span varies from 18 m to 27 m for bolted spherical node and 18 m to 40 m for welded hollow spherical node with a maximum aspect ratio of 1:2. Designer can select the standard design for required span and loading, then the prefabricated modular elements can be obtained form specialized metal construction works. Besides, several types of spherical nodes are manufactured and supplied by order. For example, the Fenyang Metal Construction Industrial Coorperation in Shanxi Province can supply 36 kinds of hollow spherical node, the diameter varies from 160 mm to 500 mm with different thickness. There are also several factories which can produce bolted spherical nodes with diameters corresponding high strength bolts of M20-M30. High quality and economical effect will be achieved through the mass production of standardized nodes and members.

SPACE STRUCTURES IN THE NETHERLANDS SINCE 1975

IR J M GERRITS BI

Lecturer, Department of Architecture,
Delft University of Technology, the Netherlands.

The paper contains two reviews of space structures in the Netherlands during the period 1975-1983. The first review reflects development and research in the field of space structures, while in the second one deals with the principal space structures that were built in this period.

DEVELOPMENT AND RESEARCH 1975-1983

The development of space structures in the Netherlands during the period 1975-1983 was marked by the development of new systems for two-way double-layer grids and for domes. At the same time research and development of polyhedra for building structures take place.
The principal research consisted of setting up a framework for design and calculation rules of tubeconnections. Some research had been carried out to obtain a economical design of space frames using strips of enlarged stiffness ('hidden beams').

A. Two-way double-layer grids

In his graduate work at the department of Architecture at Delft University of Technology in 1973 A.C.J.M. Eekhout has tried to answer questions concerning the use of double-layer grids. In the process he analyzed a number of systems. His conclusions were:

- market research is largely neglected in the development of these systems - that is, no attempts are made to determine what architects would like to use these grids for.
- the systems depend on overly complex connectors and bar links. As a result the capital expenditure entailed by their production imposes such high costs on the product in the introduction stage that it is impossible to develop a market in the Netherlands.

With this in mind Eekhout developed a new space frame system, which he called Octatube. The Octatube system can be used for two-way double-layer grids and with a few modifications also for domes, using steel or aluminium bars (fig 1). In 1975 this system was used for the first time to build an industrial building with a flat double-layer grid. Since then the Octatube system has been applied in about 25 buildings with a total roof area of 55000 m^2.

1. The Octatube joint

The Dutch steel construction company Bailey was seeking to raise its sales potential, resulted in putting on sale of two space frames, namely Piramodul and Piramodul Large Span. The Piramodul system (fig 2) was developed by Bailey. The grid is, as far as the construction and the lay-out of the frame is corcerned, the same as the Unibat and Space Deck systems. The latter two systems are not sold on the Dutch market. The Piramodul system can be used for clear spans up to approx. 25 m. It was applied for the first time in 1975 as a flat double-layer grid in the construction of a high school. Since then it is used in about 16 projects with a total roof area of 8000 m^2.

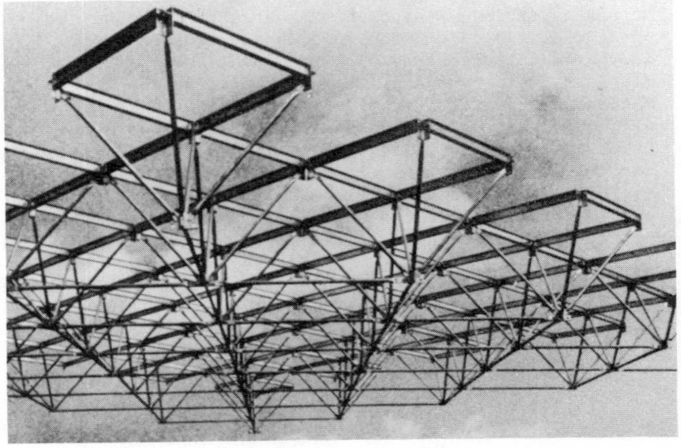

2. Piramodul half-octahedra

3. The RAI joint

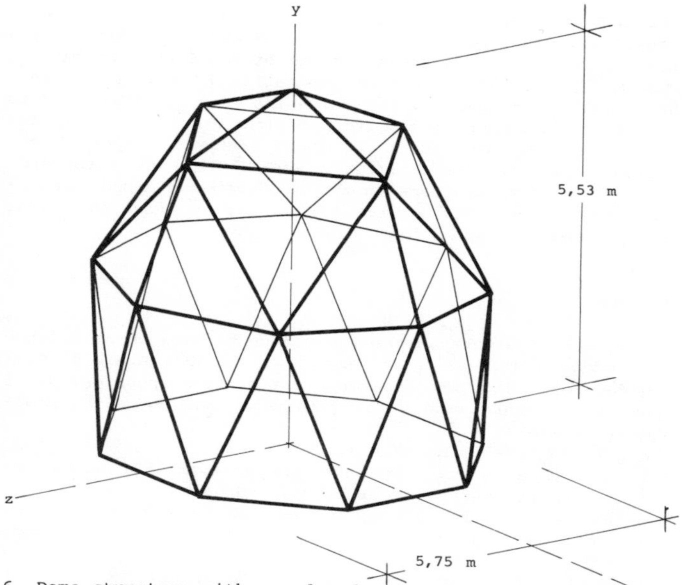

6. Dome structure with roundwood

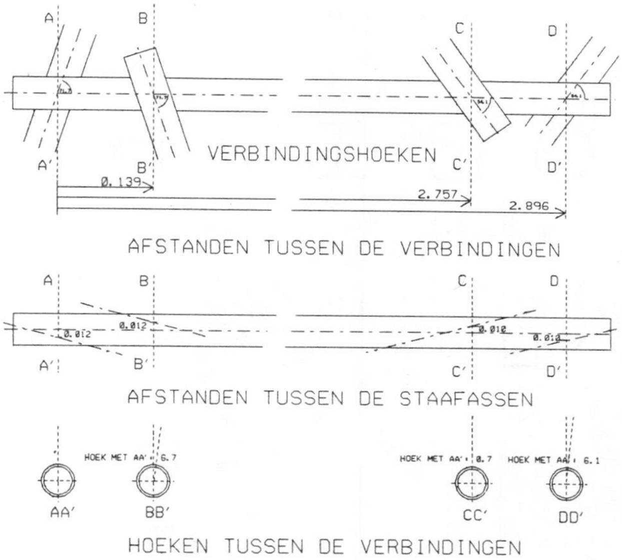

4. Working drawing for the Nomadome project in Amsterdam (plot)

7. A small office building in Mali, Africa

5. Nomadome on the festival of the future in Amsterdam

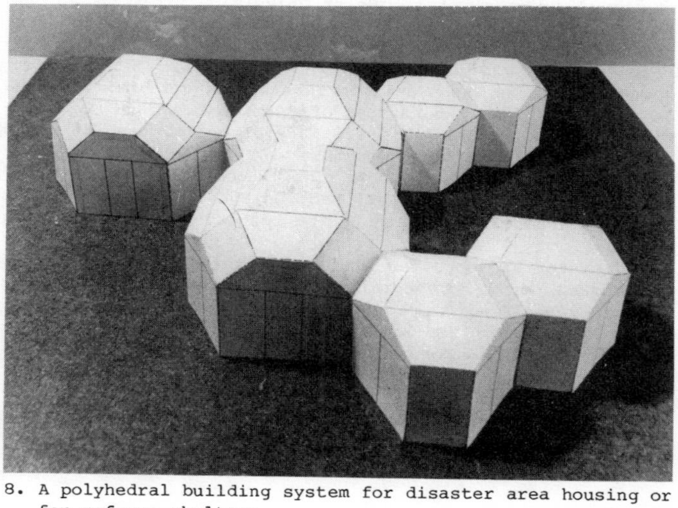

8. A polyhedral building system for disaster area housing or for refugee shelters

The DSBV bureau, consulting engineers and architects (J.W.B. Enserink), developed a two-way double-layer grid for the extension of the RAI Congress and Exhibition Centre in Amsterdam in 1982 (Delta, Randstad and Holland Hall, clear span 68, 68 and 98 m; ref 1). The RAI joint (fig 3) is related to the British Nodus system (connection bracing member and joint) and to the Dutch Octatube system (shape of the joint). The steel manufacturer was Bailey and afterwards they put this space frame on the Dutch market as the Piramodul Large Span system. This system can be used for clear spans between 50 and 100 m.

B. Domes

In his graduate work at the department of Management Science at Eindhoven University of Technology in 1974 A.H.J.M. Bijnen developed CAD/CAM computerprograms for a spherical steel spacial structure with a new jointtype, the so-called Nomadome (ref 2).
The aims in developing this system were:
- to design a sphere with the largest possible number of equal memberlengths (result: icosahedron).
- to avoid the joint altogether and to connect the members directly (result: eccentric member connections = Nomadome joint).

The Nomadome CAD system (of the firm Metaform) produces per strut a 4 x 4 matrix containing axial angles, connection angles, connection distances and axial distances for all four connecting struts (fig 4).
Two domes were built with this Nomadome system, namely one for an exhibition at Antwerp in 1976 (diameter approx. 5 m) and one on the occasion of 'The festival of the future' in Amsterdam in 1982 (diameter 10 m, height 7,4 m; 100 scaffolding pipes with a length of approx. 3 m and 200 universal joints; fig 5). A design for a house was made for the competition 'De Fantasie', Almere, in 1983 (diameter approx. 20 m).

By the Building Technology Group (Material Science) of the Delft University of Technology an experimental dome structure has been built, using roundwood members as the main structural components (fig 6; ref 3). The structure as a whole had the shape of an ellipsoid which is taller than it is wide. Actually this ellipsoid had been approximated by an octagonal antiprism for the bottom part and by a hemispherical part for the top part. It had a groundfloor plan with a diameter of 5,75 m and its height was 5,53 m. The surface was triangulated and it consisted of basically two different memberlengths. All dimensions and angles were calculated using two CAD-programmes. The structure was erected on the occasion of an exhibition at Amsterdam in 1983.

C. Polyhedra for building structures

A field of the interest of the Building Technology Group (Material Science) of the Delft University of Technology is the research and the development of the geometry of polyhedra orginates in the field of standardized building structures.
A result of this was a small office building in Mali, Africa in 1977 (fig 7). It consisted of 3 units based on the rhombicuboctahedron. The walls are made of hollow cement blocks and the roof structure consists of the termite-resistant wood of the Rhonier palm. The roof is covered with sheets of corrugated galvanized steel. Each unit has a raised central part, which acts as a chimney for better ventilation during hot days.
For disaster area housing or for refugee shelters the Group has worked out a polyhedral building system (fig 8), using sandwich panels with outer faces of oil-treated weather-resistant hardboard and with resin-impregnated honeycomb as the core material. Twelve different basic shapes and sizes have been adapted, facilitating the formation of a great number of building forms. A number of basic panels can be used for the construction of standard containers for air cargo transport. A prototype was built and tested in 1978 as a half truncated octahedron.

D. Research

In the period 1975-1983 several research programmes were completed regarding the static and fatigue behaviour of various types of welded plane girder joints of circular and rectangular hollow sections (fig 9). Further combinations of hollow sections with I- and U-shaped sections were investigated.
Although the joints tested are not commonly used for space structures, the results obtained can be used for the evaluation of the strength of welded joints in space structures.
The results derived from the above mentioned programmes were a basis for the design rules drafted by the International Institute of Welding (ref 4) and the new Dutch regulations (ref 5). The background information for these design rules is described in the book 'Hollow Section Joints' by J. Wardenier (ref 6).

Some research has been carried out to obtain a regular stress distribution in space frames and to make space frames more suitable to span rectangular areas. To obtain this goal 'hidden beams' of modules of enlarged stiffness are applied between the supports, so that the space frame will act as two systems of indepent elementary 'beams'. The stress distribution in all parallel 'beams' is the same. The calculations can easily be carried out by hand. Load transfer in x- and y-direction acts according to the wishes of the designer, just by changing the stiffnesses of the 'hidden beams'. The required stiffnesses of the beams are proportional to the required strength (ref 7).

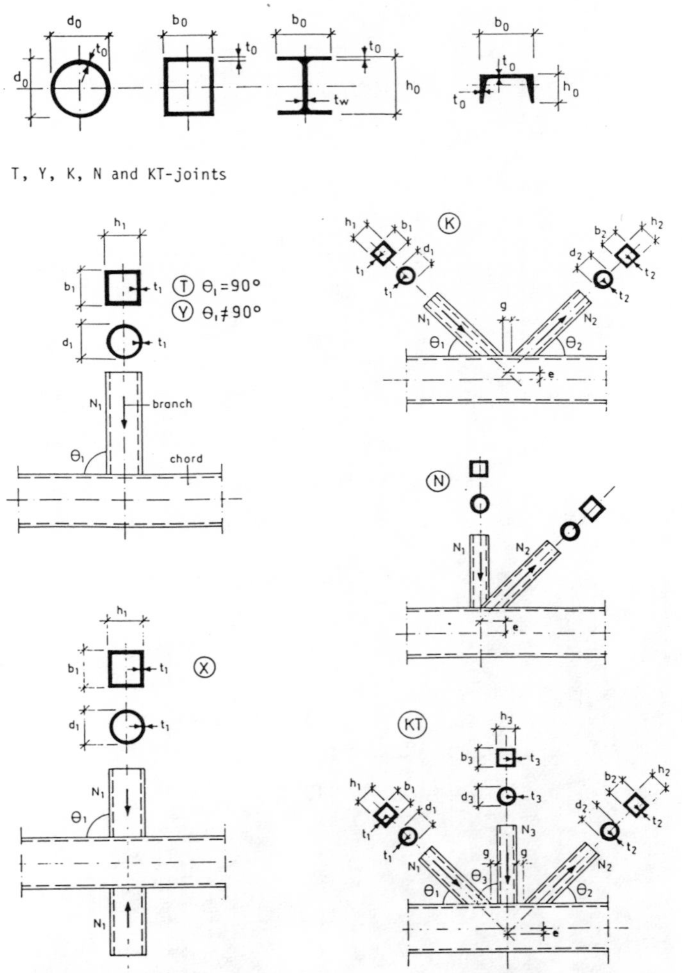

9. The joints covered have a CHS, RHS, I, H or channel section as cord and CHS or RHS section as branches

SPACE STRUCTURES BUILT IN 1975-1983

The aim of this review is to list the principal space structures built in the Netherlands during the period 1975-1983. What is the purpose of such a list? It will throw light upon the frequency of application of various structures, showing that certain types have come to the fore whereas others have receded into the background. It will permit conclusions as to the overall increase or reduction in the use of space structures, at least in the Netherlands.
Next, such conclusions may be followed by an inquiry into the causes. One anticipates that these will not only be associated with changes in the types of buildings or with shifts in cost patterns, but will also be found to depend on changing views in building design.
Lastly, the information obtained from this analysis should facilitate projections of the direction in which the building industry will move in developing new structures and extending present systems. It is clear that this will not result in a complete picture, if only because of the fundamental unpredictability of the creative process: new demands may inspire new techniques, and vice versa.

The list classifies the structures by their principal structural material.

A. Steel
Double-layer grids

Mero system
- Railway station in Breda, platformroof. Completed in 1975. Architectural and structural design: bureau Articon (I. Bak). Area 39 x 168 m. Mero double-layer grid (type: square on square) is supported by 16 regular polyhedra (half-octahedron) on concrete columns. The columns are at a distance of 24 m widthwise (cantilevers of 7,5 m) and 21 m lenghtwise (cantilevers of 10,5 m). Module 3 m. Construction height 2,12 m.
- Sports hall in Lelystad. Completed in 1982. Architectural design: Rijksdienst voor de IJsselmeerpolders (M. Voorwijk). Structural design: Mero Raumstruktur.
Area 48 x 48 m. Mero double-layer grid (type: square on square) is supported by 4 regular polyhedra (half-octahedron) on concrete columns. Module 3 m. Construction height 1,8 m.
- Railway station in Zaandam, platformroof (fig 10). Completed in 1983. Architectural design: bureau Articon (P.A.M. Kilsdonk). Structural design: Mero Raumstruktur. Area 91 x 46 m. Mero double-layer grid (type: square on square) is supported by 6 regular polyhedral structures. These structures are at a distance of 26 m widthwise (cantilevers of 10,5 m) and 31 m lengthwise (cantilevers of 12 m and 18 m). The cantilever of 18 m slopes under 45 degrees to a traverse. Module 2,56 m. Construction height 1,83 m.

Perfrisa system
- This Spanish system is also known as the Palc or Comelsa system. The Perfrisa system for which the worldwide concessionaire is Comelsa, was designed by the architect Francisco Javier Alcade Cilveti. Since 1976 this space frame is used in the Netherlands.
- Tennishall 'Het Steinerbos' in Stein. Completed in 1978. Architectural design: P.L.W. Peutz. Structural design: bureau Schrauwen. Area 36 x 72 m. Perfrisa double-layer grid, type: square on larger square. Top chord module 2,58 m. Construction height 1,8 m.
- Tennishall 'De Bloemen' in Castricum (fig 11). Completed in 1978. Architectural design: P.L.W. Peutz. Structural design: bureau Schrauwen. Two halls each 36 x 36 m. Perfrisa double-layer grid, type: square on square. Module 2,76 m. Construction height 1,95 m.

Nodus system
- Workshop of the Amsterdam underground railways in Diemen. Completed in 1978. Architectural design: P.H. van Rhijn (Public Works Amsterdam). Structural design: bureau Schrauwen. Area 32 x 90 m; 4 units each 20 x 32 m supported by lattice box-griders and facade-columns. Nodus dou-

10. Railway station in Zaamdam, platformroof

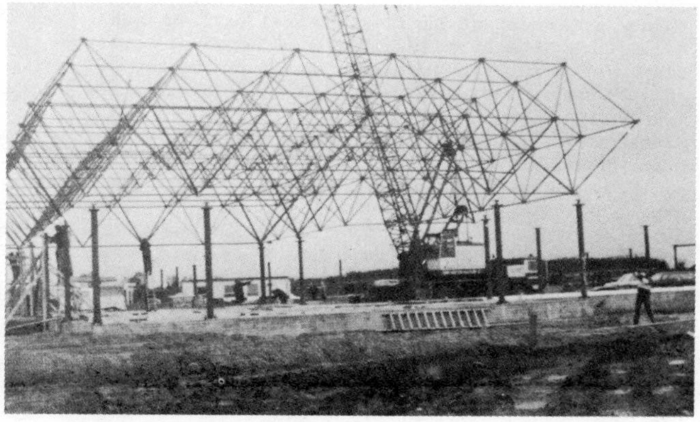

11. Tennis hall 'De Bloemen' in Castricum

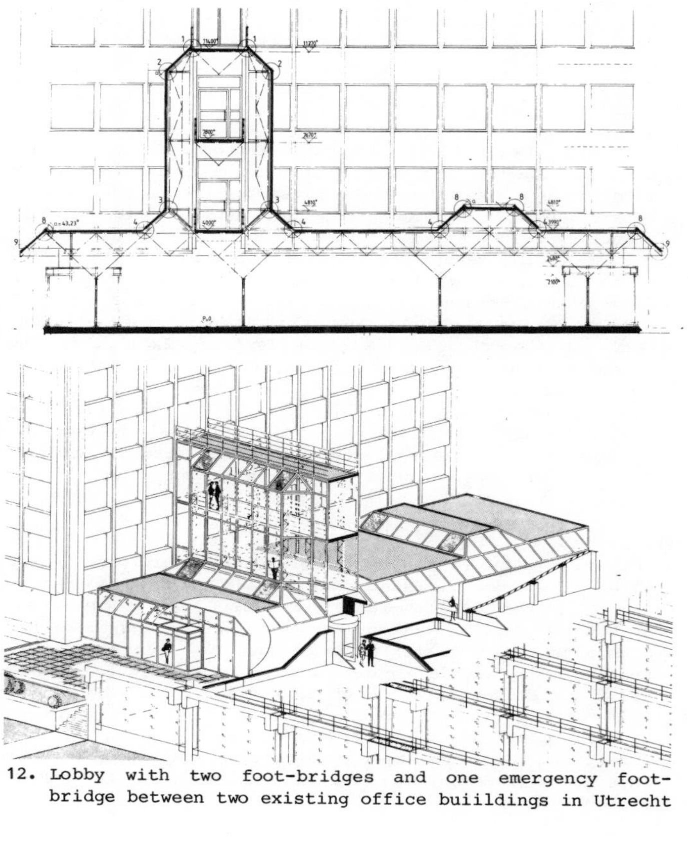

12. Lobby with two foot-bridges and one emergency footbridge between two existing office buiildings in Utrecht

ble-layer grid, type: square on larger square. Top chord module 2,9 m. Construction height 1,8 m.
- A hall in Zoetermeer. Completed in 1979. Architectural design: Municipal Works Zoetermeer (F. de Man). Structural design: bureau Schrauwen. Area 48 x 68 m. Nodus double-layer grid, type: square on larger square. Top chord module 2,4 m. Construction height 1,6 m.
  This hall has to function as a temporary shopping centre during the first five years. After that period the hall will be used as an events or sports hall; perhaps on an other location.
- Lobby with two foot-bridges and one emergency footbridge between two existing office buildings in Utrecht (fig 12). To be completed in 1984. Architectural and structural design: Rijksgebouwendienst (architect F. Groot, engineer S. Driessen). Area 26 x 10 m. Nodus double-layer grid (type: square on square) supported by 6 regular polyhedra (half-octahedron) on columns. Top chord module 2 x 1,9 m. Construction height 0,94 m.

Octatube system
- Tennishall 'Bakestein' in Zwijndrecht. Completed in 1978. Architectural design: P.W.L. Peutz. Structural design: Octatube engineering. Area 4940 m^2: 1 hall 35 x 51 m and 2 halls each 35 x 35 m in difference heights. Octatube double-layer grid, types: square on larger square and square on rectangle. Top chord module 2,7 m. Construction height 1,9 m.
- Wintergarden roof of the 'Holiday Inn' hotel near Leiden (fig 13). Completed in 1980. Architectural design: A.C.J.M. Eekhout. Structural design: Octatube engineering. Area 2200 m^2. Octatube double-layer grid (type: square on larger square) is supported by 10 regular polyhedra (half-octahedron) on concrete columns. Top chord module 3 m. Construction height 2,1 m.
- Three roofs at a customs station on the Dutch border in Hazeldonk. To be completed in 1984. Architectural design: bureau Benthem & Crouwel. Structural design: Octatube engineering. One roof 24 x 43 m. Octatube double-layer grid (type: square on larger square) is supported by 4 regular polyhedra (half-octahedron, height 4,8 m; spacing 14,4 x 33,6 m). Top chord module 2,4 m. Construction height 1,7 m. Two roofs each 24 x 51 m. Octatube double-layer grid (type: square on square) supported by 6 framework columns, spacing 18 x 15 m. Top chord module 3 m. Construction height 2,1 m.

Piramodul system
- The roof covering an inner court of a high school in Delft (fig 14). Completed in 1975. Architectural design: bureau Hendriks, Campman & Tennekes. Structural design: the Dutch firm Bailey. Area 22 x 22 m. Piramodul double-layer grid, type: square on diagonal. Top chord module 1,425 m. Construction height 1,1 m.
- A hall of the veterinary science block in Drachten. Completed in 1982. Architectural design: bureau Van Linge & Kleinjan. Structural design: bureau Van Wassenaar. Area 36 x 36 m. Piramodul double-layer grid (type: square on diagonal) is supported by 4 columns (spacing 22 x 22 m) and facade-columns. Top chord module 1,697 m. Construction height 1,2 m.
- The roof of an aula of a high school in Spijkenisse. Completed in 1982. Architectural and structural design: bureau Campman & Tennekes. Area 15 x 15 m. Piramodul double-layer grid, type: square on diagonal. Top chord module 1,185 m. Construction height 1 m.

RAI system
- The enlargement of the RAI Congress and Exhibition Centre with the Delta, Randstad and Holland Hall in Amsterdam (fig 15; ref 1). Completed in 1982. Architectural and structural design: bureau DSBV, consulting engineers and architects (architects A. Bodon and J.H. Ploeger; engineering and joint development J.W.B. Enserink). Total area 18750 m^2: Delta and Randstad Hall each 68 x 68 m, Holland Hall 98 x 98 m. RAI double-layer grid, type: square on square. Module 5,3 m. Construction height 3,75 m.
  Afterwards the Dutch steel construction company Bailey put this space frame on the market as the Piramodul Large Span system.

13. Wintergarden roof of the 'Holiday Inn' hotel near Leiden

14. Roof covering an inner court of a high school in Delft

15. The Delta, Randstad and Holland Hall of the RAI Congress and Exhibition Centre in Amsterdam

B. Aluminium

Double-layer grids

- Aluminium extrusions plant in De Lier. Completed in 1977. Architectural design: Mick Eekhout. Structural design: Octatube engineering. Area 4500 m^2. Octatube double-layer grid, type: square on diagonal. Clear span between columns 25 m. Top chord module 2,1 m. Construction height 2,5 m.
- Lobby of an office building in Nieuwegein. To be completed in 1984. Architectural design: Peter Gerssen. Structural design: Octatube engineering. Span 18 x 18 m triangular. Total surface 162 m^2. Tuball double-layer grid, type: square on square. Top chord module 1,8 m. Construction height 1,3 m.

Domes

- Sports centre in Zoetermeer. Completed in 1978. Architectural design: Municipal Works Zoetermeer (F. de Man). Structural design: Temcor USA and bureau Van Strien. Area 2700 m^2. Temcor system, hexagonal. Diameter 63 m. Rise 14 m.
- Miranda swimming pool in Amsterdam, dome (fig 16). Completed in 1979. Architectural design: bureau Baanders, Frencken, Wilgers & Verhey. Structural design: IBG International USA. Area 1600 m^2. Stiff-jointed frame dome. Diameter 45 m. Rise 12 m.
- A showroom with offices for the firm of Louwman & Parqui in Raamsdonkveer (fig 17). Completed in 1981. Architectural design: G.A.M. Roovers. Structural design: Temcor USA and bureau Interplan. Area 3900 m^2. Temcor system, hexagonal. Diameter 71 m. Rise 19 m.
- Sewage water-treatment plantcovering in Udenhout. Completed in 1982. Architectural and structural design: Octatube engineering. Clad with aluminium panels suspended from the structure. Diameter 15,5 m, height 3,9 m, module ca 3 m.
- Swimmingpool dome in Jeddah, Saudi Arabia. Completed in 1983. Architectural and structural design: Octatube engineering. Aluminium structure with suspended double membrane. Diameter 13,9 m, height 3,5 m.

Folding structures

- With the Canadian system Panelarch antiprismatic folding structures can be made using aluminium tetrahedra as structural components. This Canadian system is on sale in the Netherlands as the Aluboog system since 1978 and has been used in three small projects.
- A storage shed in Nunspeet (fig 18). Completed in 1978. Contractor: the firm Jan Kuipers. Area 6 x 25 m. Aluboog system, widthwise 5 standard tetrahedra (each 2,785 x 1,267 m). Height 5 m.

C. Wood

Suspended roof

- Sports hall of a high school in Amsterdam (fig 19). Completed in 1976. Architectural design: J.B. Ingwersen. Structural design: bureau Van Rossum and the firm Nemaho. Area 25 x 46 m. The double cylindrical barrel vault is formed by wooden suspended arches, centre to centre 5,1 m, which spans two times 22 m between concrete edge-beams and a concrete frame.

Domes

- A showroom with offices in Heerhugowaard. Completed in 1979. Architectural design: bureau Pino. Structural design: bureau Harder. Building area 1590 m^2. Three hinged arch-ribbed dome, diameter 33 m, rise 3 m.
- Dome on the library-roof of the Leiden University. Completed in 1982. Architectural design: Bart van Kasteel. Structural design: bureau D3BN (H.W. Bennenk). Area 145 m^2. Stiff-jointed frame dome. Diameter 14 m. Rise approx. 5 m.

Polyhedra for building structures

- Pile-houses in Helmond. Completed in 1975. Architectural design: Piet Blom. Structural design: bureau Beltman. A cube (= one house) is placed on one of its angular points. One house contains 250 m^3 with 90 m^2 floor area.
- Facet-houses in Gouderak (fig 20). Completed in 1981. Ar-

16. Miranda swimming pool in Amsterdam

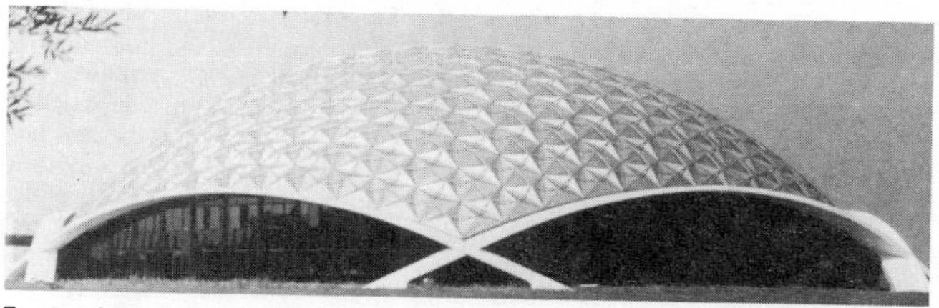

17. A showroom with offices for the firm of Louwman & Parqui in Raamsdonkveer

18. A storage shed in Nunspeet

19. Sports hall of a high school in Amsterdam

chitectural design: Peter Gerssen. Structural design: bureau D3BN (H.W. Bennenk). The shape of a house is based on a fourth part of a cube. One house contains 300 m^3 with 135 m^2 floor area.

- A structure for a roof over an inner court of a building of Leiden University in Leiden (fig 21). Completed in 1982. Architectural design: Tj. Dijkstra and bureau Loerakker, Rijnbout & Ruijssenaars. Structural design: Adviesbureau voor Bouwtechniek. The structure as a whole is a half-octahedron placed on its top; base approx. 13 x 13 m, height 10 m. The inner structure consists of regular polyhedra (tetrahedra and cubes).

D. Plastic

Membrane structures

- 'Parapluie Permanente' on a home for the elderly in Dongen. Completed in 1976. Architectural and structural design: Mick Eekhout and Egbert Tons. Area 18 x 10 m.
- Main entrance for the 'Floriade' exhibition in Amsterdam. Completed in 1982. Structural design: Mick Eekhout and Paul Verhey. The construction was afterwards sold to 'Ponypark' in Slagharen.
- A pavilion for the 'Floriade' exhibition in Amsterdam (fig 22). Completed in 1982. Architectural design: bureau Kruis & Partners. Structural design: bureau Kroon. Area 140 m^2. The membrane is made of a prestressed PVC-polyester fabric, supported by two laminated wooden parabolic three-hinged arches. Clear span of the arches 14 m, height 6,5 m.
- Tensigrity structure with membranes for the canteen of an electrical power plant in Amsterdam (fig 23). Completed in 1982. Architectural design: L. v.d. Horst and M. Eekhout. Structural design: Octatube engineering. The structure consists of 6 vertical masts, 4 horizontal yards and 3 horizontal poles connected with each other by means of prestressed cables. In summer the structure is completed by 3 prestressed PVC-polyester membranes. Maximum dimensions 30 x 15 x 8,5 m.
- Mobile shelter for popmusic programs of the Veronica broadcasting company, travelling since 1982 (fig 24). Structural design: Octatube engineering. Area 15 x 15 m on a triangular base, height 8,5 m. Can be erected within 4 hours.

Domes

- Two oxidation basins of the wastewatercleaning 'De Groote Lucht' in Vlaardingen are covered by glassfibre reinforced plastic (GRP) domes built in 1982 (fig 25; ref 8). The domes have a diameter of 48 m, and the height above the basin is 8 m. They are made of 36 identical segments, which are mounted on the edge of the basin. At the upperpart the segments are connected to a steel ring. This ring has a diameter of 6,5 m and is closed by 6 segments and a central part.
  An other GRP dome was built in 1983 in Amsterdam (planetarium) as a half sphere with a diameter of 25 m.
  These three GRP domes were designed and constructed by the Dutch firm Polymarin.

Inflatables

- Exhibition hall for travelling show 'Energie Expresse', touring around the Netherlands during 1983 and 1984. Architectural and structural design: Octatube engineering. The pneumatic structure is held up by a maximum overpressure of 47 mm watercolumn and takes the shape of a lightbulb (enlarged on scale 1:300). A socket is composed of a tensile membrane over 4 arches. The foundation is formed by water ballast cubes. Maximum length 35 m, height 15 m, width 20 m.
- A mobile roof of a swimming pool near Leiden (fig 26). To be completed in 1984. Structural design: Albers textielbouw and bureau Claessens. Area 23 x 60 m. The foldable infatable cushion roof is formed by a double flexible PVC-coated fabric. Inside the cushion there are web joists and an isolation foil. When the roof is closed, the overpressure in the cushion is a watercolumn of about 25 mm.

20. Facet-houses in Gouderak

21. A structure for a roof over an inner court of a building of Leiden University in Leiden

22. A pavilion for the 'Floriade" exhibition in Amsterdam

Conclusion

The list of the principal spacial structures built in the Netherlands in the period 1975-1983 must lead us to the conclusion that practically no interesting shells, folding structures, braced barrel vaults or suspended roofs were built. In our view the main causes are:

- the handicap of labour-intensive structures, for example concrete shells
- the architects are no longer interested in baroque structural shapes, for exemple hyperbolic paraboloidal shapes
- the present economical situation in the Netherlands.

The list also indicates that Dutch architects take a great interest in space frames and have a growing one for polyhedral and membrane structures.
They are genuinely interested in two-way double-layer grids and principally in the flat ones (types: square on square, square on larger square and square on diagonal). However here the stumbling-block is frequently the higher cost of construction in comparison with other, more traditional, structures. This higher cost is often the consequence of the fact that the architects make too little use of the specific properties (both structural and architectural) of the double-layer grids.
Some Dutch architects are getting more interested in polyhedra for building structures as a consequence of the smaller scale in architecture. They are looking for new shapes in combination with lower construction and heating costs in comparison to other more tradional building structures.
During the last years there is a growing interest of designers in membrane structures (shelters). This interest is based on the visual, temporary or mobile aspects of these light structures.

ACKNOWLEGDEMENTS

The author wishes to thank the firms and persons, mentioned in the article, for the information and the pictures they have supplied.

REFERENCES

1. J.W.B. Enserink. Space structures of the RAI Congress and Exhibition Building at Amsterdam. In: Space structures, proceedings of the third international conference. Applied Science Publishers Ltd., Barking, Essex. 1984.

2. A.H.J.M. Bijnen. Creation of information in the building process. In: Proceedings of the first international IFIP conference on computer applications in prodution and engineering CAPE '83. Amsterdam. North-Holland Publishing Company, Amsterdam. 1983.

3. P. Huybers. A timber pole dome structure. In: Space structures, proceedings of the third international conference. Applied Science Publishers Ltd., Barking, Essex. 1984.

4. IIW-XV-E. Design Recommendations for hollow section joints - Predominantly statically loaded - IIW Doc. XV-491-81 (revised). Welding in the World 20(1982)3/4.

5. SG-TC-18. Regulations for the design and calculation of tubular structures, RB'82. Staalbouwkundig Genootschap, Rotterdam. 1982.

6. J. Wardenier. Hollow section joints. ISBN 90 6275 084 2. Delft University Press, Delft. 1982.

7. W.J. Beranek and G.J. Hobbelman. Economic design of space frames using strips of enlarged stiffness. In: Space structures, proceedings of the third international conference. Applied Science Publishers Ltd., Barking, Essex. 1984.

8. J. Olthoff. The application of glassfibre reinforced plastic for large roofs. In: Space structures, proceedings of the third international conference. Applied Science Publishers Ltd., Barking, Essex. 1984.

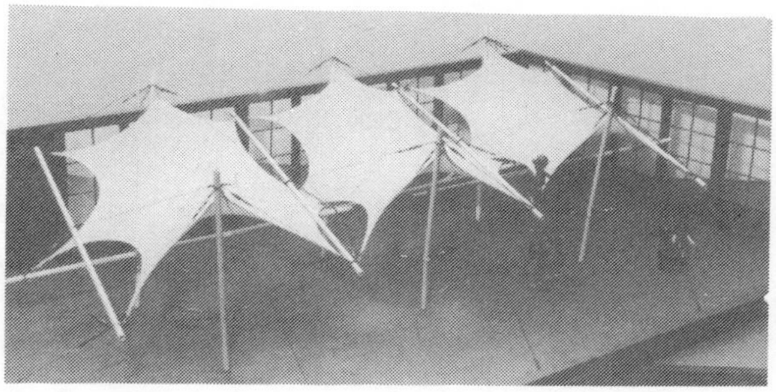

23. Tensigrity structure with membranes for the canteen of an eletric power plant in Amsterdam

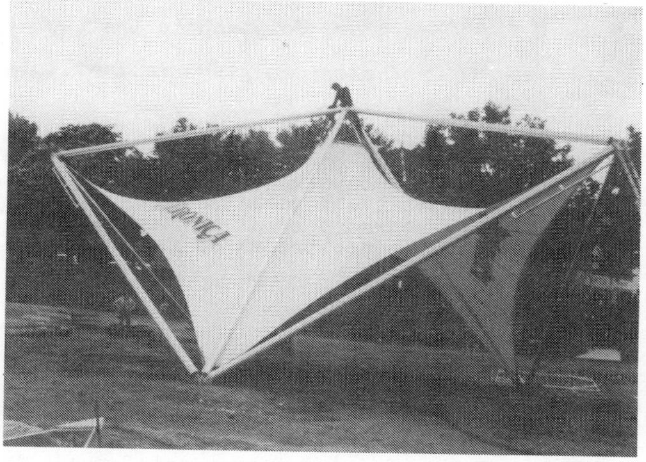

24. Mobil shelter of the Veronica broadcasting company

25. A dome of the wastewatercleaning 'De Groote Lucht' in Vlaardingen

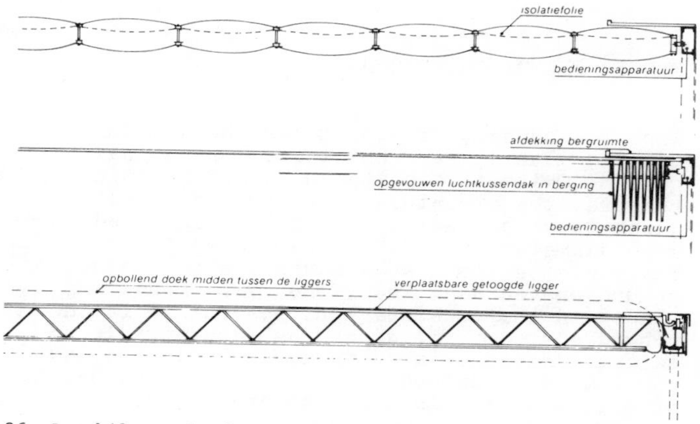

26. A mobile roof of a swimming pool near Leiden, sections

SOME RECENT EXAMPLES OF LATTICED SHELLS FROM ROMANIA

N BĂLUŢ[x], D PORUMB[x], V GIONCU[xx],
I TOADER[xxx], V ILLE[xxxx]

[x]Senior Research Engineers, Building Research Institute of Timişoara, Romania

[xx]Assoc.Prof.,"Traian Vuia" Polytechnic Institute of Timişoara, Romania

[xxx]Lecturer, Polytechnic Institute of Cluj-Napoca, Romania

[xxxx]Professor, Polytechnic Institute of Cluj-Napoca, Romania

The paper presents four types of latticed shells used for roof structures over greenhouses, sports halls, markets or stadium tribunes. They are respectively geodetic domes, barrel vaults and hypars (supported at the four corners or at a single point). The main component elements and joint details are presented for each type of structure.

INTRODUCTION

After a long period of designers' enthusiasm for reinforced concrete shells, their attention turned back to metal shell-like latticed structures. Their advantages over reinforced concrete shells are beyond any doubt: better possibilities of prefabrication and erection, lower dead weight and cost. In spite of this superiority, their application on a large scale is hindered by some aspects not solved properly so far: joint connections, the marked nonlinear behaviour and instability problems. This is why an effervescence can be observed all over the world in the efforts for solving these problems: practically each company, university or researcher has elaborated a joint connection system, a nonlinear computer program and a particular design philosophy (especially concerning the verification of stability).

In the framework of this general trend, theoretical studies and laboratory tests were carried out in Romania, see Refs 1, 2, 3, 4, 5, 6, by two research teams, one in Timişoara and one in Cluj-Napoca. The results of this research work were materialized in the design of some roof structures, which are presented herein.

EXOTIC PLANTS GREENHOUSES

Two greenhouses, one for palms and one for aquatic plants, were built for a botanical garden. Owing to architectural and functional reasons, the structures were chosen in both cases to be single-layer geodetic latticed domes with diameters of 24.6 m and respectively 13.2m. Their disposition takes advantage of the natural ground unevenness. Thus, the palm house was placed in the lower region, the change in the ground slope providing a greater available height. As can be seen from Fig 1, in the East zone, this structure is supported at the height of +1.5 m on a reinforced concrete girdle sustained by reinforced concrete frames. The rest of the palm house dome and the entire aquatic plants house dome are supported on cyclopean concrete continuous footings through reinforced concrete girdles. The geodetic domes were obtained by threefold and respectively twofold division of the regular icosahedron. There resulted 15 different types of members for the palm house dome and 5 types for the aquatic plants house dome. All members are steel tubes. The joints are spheres made by hot pressing from steel sheet, see Fig 2. Each sphere has two ring diaphragms inside and a lid. The provisional connection of the members is achieved by means of black bolts.

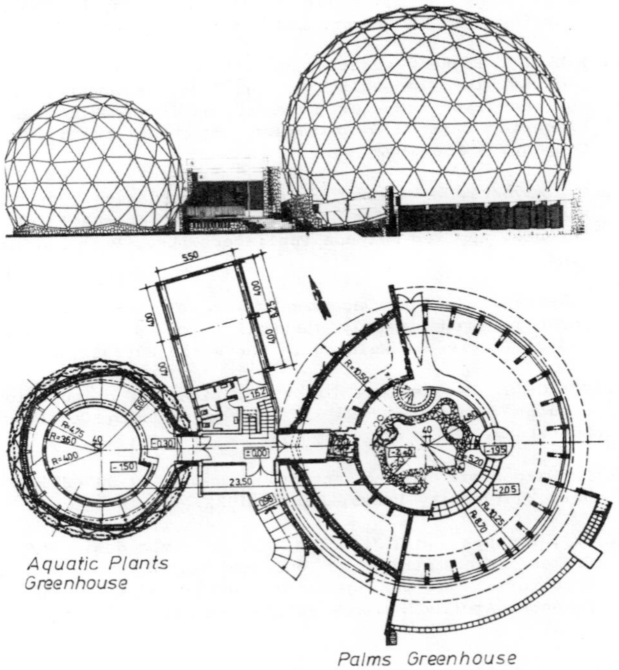

Fig 1

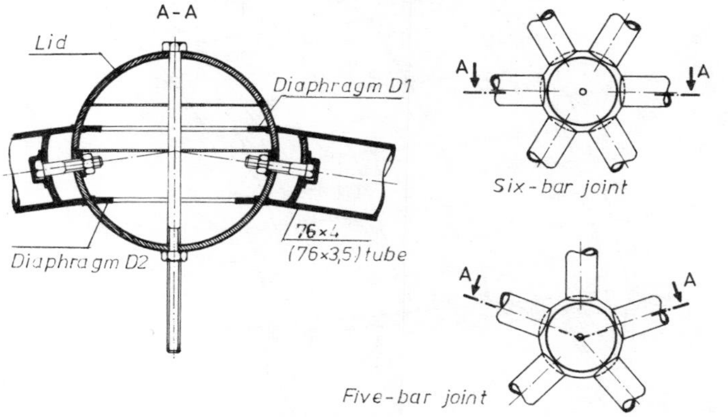

Fig 2

Fig 4

The erection of the structure was followed by the welding of the joint lids and then of the member ends. The conception of the joint connection aimed at the fulfilment of the following requirements:

a) simplicity of the member-by-member erection;
b) corrosion resistance of the structure subjected to weather action, ensured by water tight connections;
c) increased local and general stability, owing to rigid joint connections;
d) aesthetic aspect.

The long bolt, see Fig 2, is used on one hand for the temporary fixing of the lid before welding, and on the other hand for fastening the aluminium sash-bars disposed at the inner side of the dome structure (see Figs 5 and 6).

Fig 5

The analysis was performed for dead weight, snow and wind loads, using the computer program elaborated in Cluj-Napoca and presented in Ref 5. Aspects during erection are shown in Fig 4, the two greenhouses after completion in Fig 5 and joint details in Figs 3 and 6.

The steel consumption was 10 kg/m^2 for the bigger dome and 8.3 kg/m^2 for the smaller one.

Fig 3

Fig 6

SPORTS HALLS

The structure adopted for the roofs of some low-capacity sports halls (under 500 seats) was a latticed barrel vault. The plan view and the cross section are shown in Fig 7. The main parts of the structure are (see Fig 8):

a) the barrel vault composed of prefabricated plane trusses, using cold formed channels for the chords and tubes for the diagonals;
b) stiffening and skylight three-hinged latticed main arches, with triangular cross section and tubular chords and diagonals;
c) secondary stiffening arches, made of cold formed channels;
d) stiffening latticed edge beams, with triangular cross section and tubular chords and diagonals;
e) reinforced concrete isolated foundations of the arches;
f) reinforced concrete ties (placed under the floor) between the foundations, taking over the thrusts of the arches.

The design of this structure was based on the principles resulting from the theoretical studies presented in Ref 2. Thus, the longitudinal bars obtained by joining the channel chords of the adjacent trusses (see Fig 9) have a considerable bending rigidity in the radial direction. On one hand, this enables their use as purlins; on the other hand, the structural behaviour of the diagonals is improved. Even if one of these would fail or if some local overload would occur at a certain node, the longitudinal bars would act in the sense of a load redistribution and the damage would be limited. Another design principle presented in Ref 2 concerns the avoidance of coupling between the individual buckling of a diagonal with one of the other types of instability (line buckling or overall buckling), because the sensitiveness to geometric imperfections would be considerably increased in such cases. This can be achieved by a substantial differentiation of the two critical loads: either the diagonals are oversized and only general instability is relevant, or the diagonals are designed on basis of their individual buckling and the critical load corresponding to general instability is increased by means of additional measures. Since the latter alternative was found to be more economic, secondary stiffening arches were included into the structure, which resulted in a very considerable increase of its bearing capacity. A nonlinear analysis was carried out by using a computer program elaborated in Timişoara, see Ref 1. Beside dead weight, the main load considered to act upon the structure was unsymmetrical snow load, which makes instability to occur by snap-through buckling, not by equilibrium bifurcation.

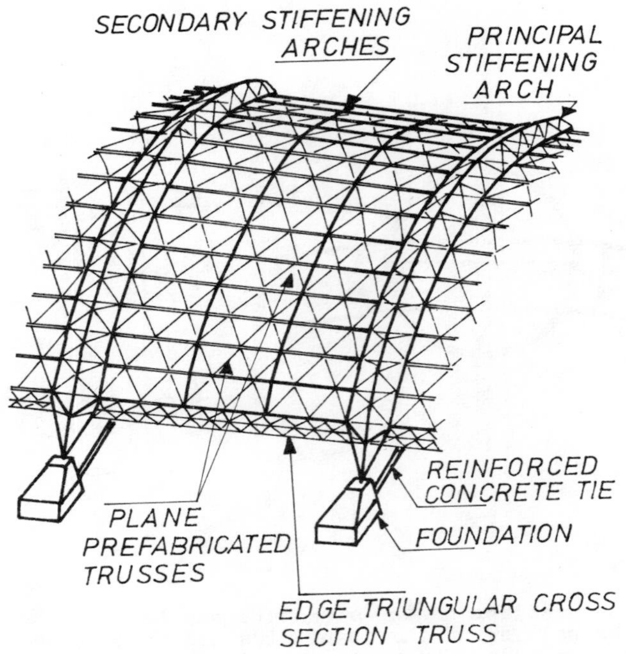

Fig 8

The steel consumption (without gable walls) is 22.5 kg/m^2.

Figures 10, 11, 12 show aspects during the erection.

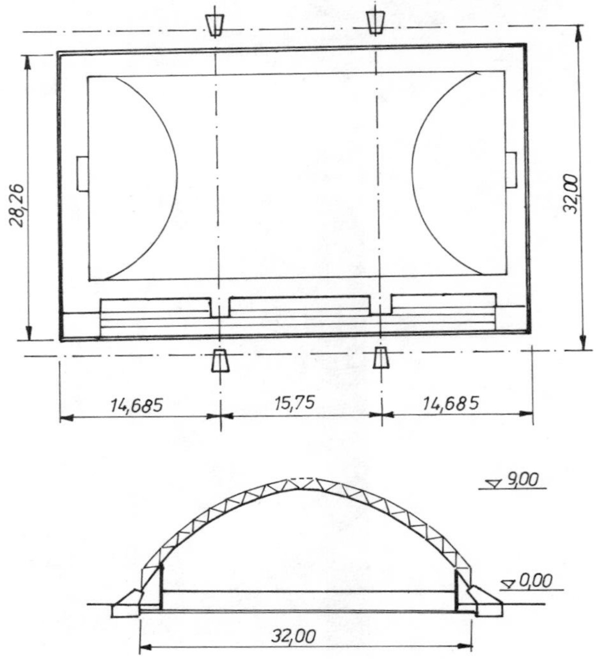

Fig 7

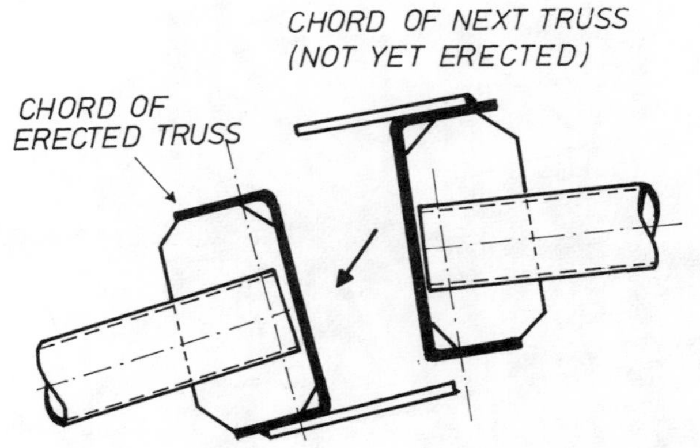

Fig 9

Fig 10

Fig 11

Fig 12

ROOF OVER A MARKETPLACE

A latticed shell consisting of multiple hypars, presented in Fig 13, was used for covering a 48 x 85 m marketplace. Figure 14 shows a component unit of the roof structure composed of:

a) the latticed shell made of straight tubes, disposed along the two families of generatrices

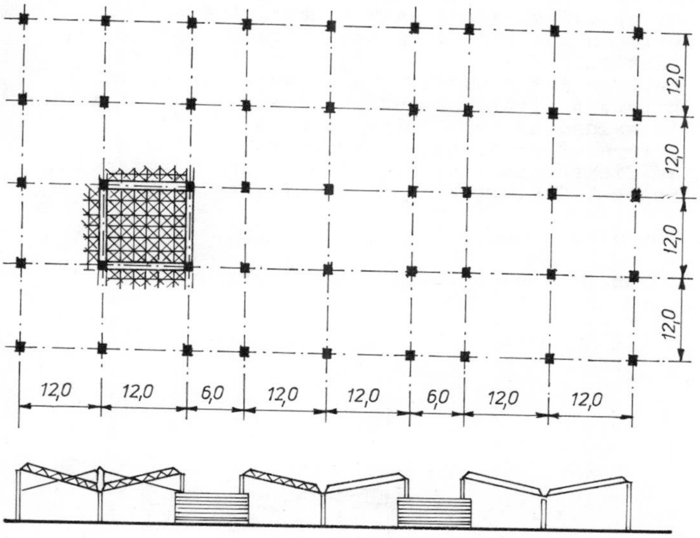

Fig 13

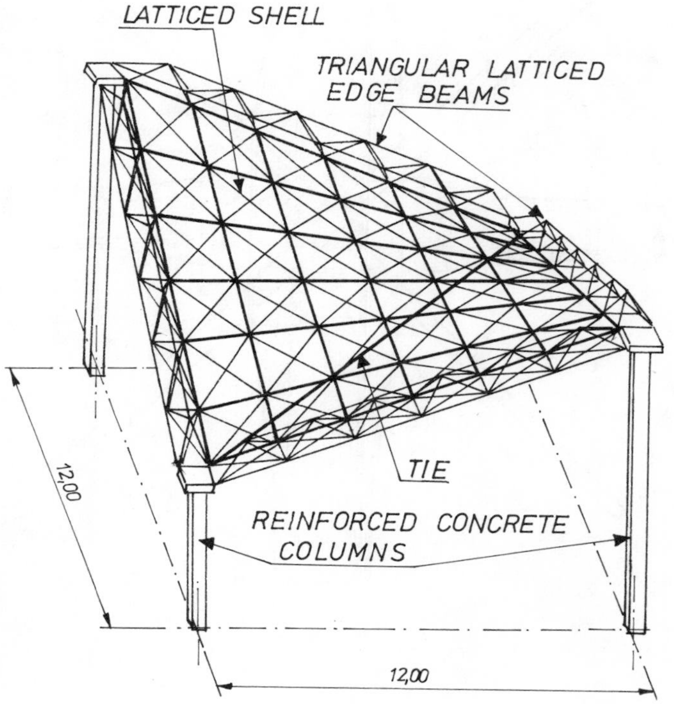

Fig 14

of the hypar and flexible rod diagonals along the two families of directrices, which ensure geometric indeformability for any direction of the applied loads;
b) triangular latticed edge beams, made of tubular members;
c) reinforced concrete columns, sustaining the shell unit at the four corners;
d) a steel tie joining the two lower corners of the shell and taking over the thrust,

A joint detail is presented in Fig 15. The tubes were flattened in order to avoid the necessity of splicing. Along each member, the angle between the planes of two consecutive flattenings is about 15°, which is the angle between two consecutive generatrices of the hypar. A cleat consisting of a quarter of a pipe is welded at each joint on each of the intersecting tubes, in order to provide a sufficient bending rigidity in the flattened region. The upper cleat is site welded.

The steel consumption is 17 kg/m^2.

Views of the roof are shown in Figs 16 and 17.

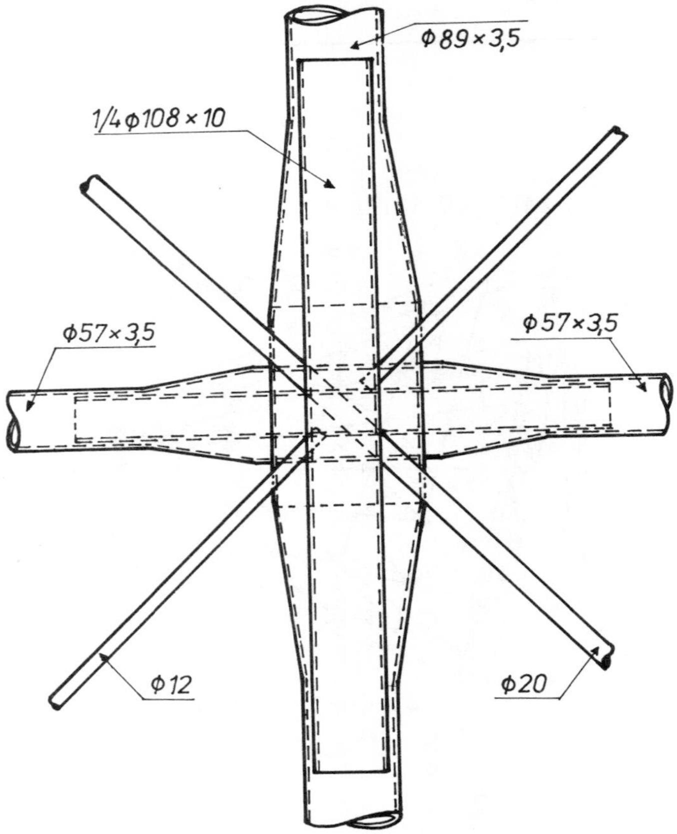

Fig 15

Fig 16

Fig 17

DESIGN OF THE CANOPY OVER A STADIUM TRIBUNE

The canopy over a football stadium tribune (see Fig 18) was designed in the form of several cantilever latticed hypar shells. The components of a roof structure unit are presented in Fig 19. Joint details are similar to those shown in Fig 15.

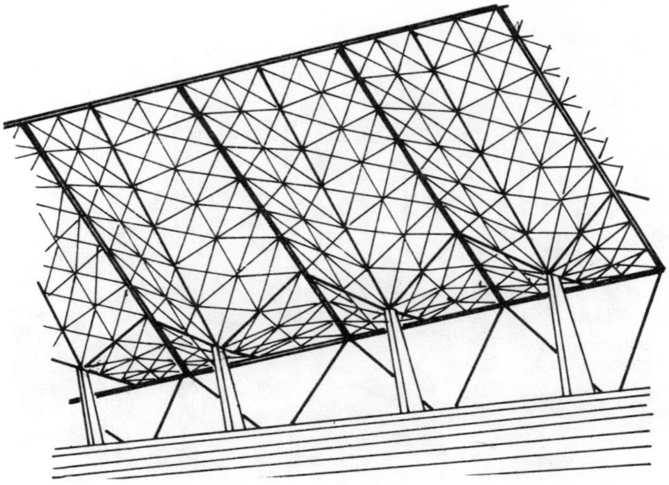

Fig 18

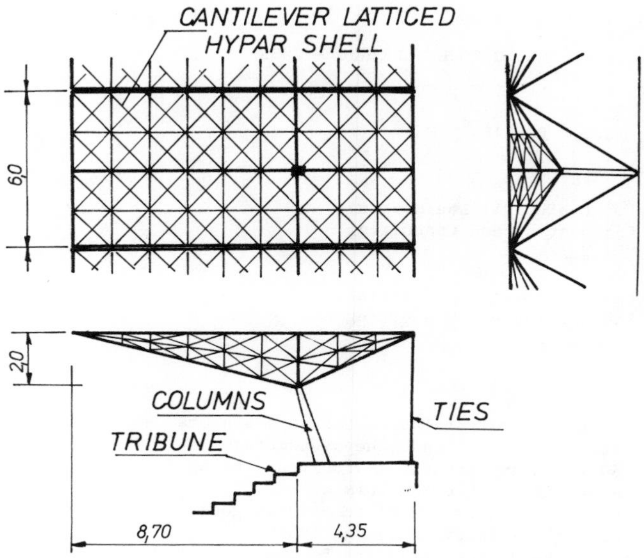

Fig 19

The steel consumption is 21.7 kg/m^2. In comparison to other possible structures of reinforced or prestressed concrete, the suggested solution proves to be very efficient and spectacular.

Views of the canopy model are shown in Figs 20 and 21.

Fig 20

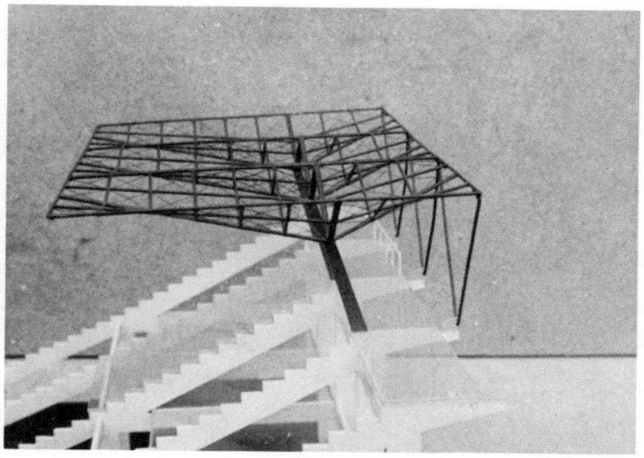

Fig 21

Unfortunately, the design was not materialized yet, owing to reasons not pertaining to the technical solution.

REFERENCES

1. N BĂLUŢ, D PORUMB and N RENNON, Nonlinear Analysis of Reticulated Shells (in Romanian), 2nd Conference on Metal Structures, Timişoara, October 1979, Vol 1, pp 88 - 102.

2. V GIONCU, N BĂLUŢ, D PORUMB and N RENNON, Instability Behaviour of Triangulated Barrel Vaults, Course on the Analysis, Design and Construction of Braced Barrel Vaults, Vol I, University of Surrey, September 1983.

3. V GIONCU, N BĂLUŢ, D PORUMB and N RENNON, Recent Examples of Steel Prefabricated Barrel Vaults from Romania - Their Analysis, Design and Construction, Course on the Analysis, Design and Construction of Braced Barrel Vaults, Vol II, University of Surrey, September 1983.

4. I TOADER, Application of the Stiffness Method for the Analysis of the Latticed Structures (in Romanian), Buletinul Stiinţific al Institutului de Construcţii Bucureşti, Anul XXIII, Nr. 1 - 4, 1980, pp 99 - 111.

5. I TOADER, Problems concerning the Analysis of Latticed Shell Roofs (in Romanian), Doctor's Thesis, 1981.

6. I TOADER, N JUNCAN and V ILLE, Aspects Concerning the Design and Achievement of Some Latticed Shells (in Romanian), Buletinul Stiinţific al Institutului de Construcţii Bucureşti, Anul XXI, 1978, pp 77 - 93.

PREFABRICATED STEEL SPACE STRUCTURES DESIGNED BY CHEMOPROJEKT PRAGUE - CSSR

R.RUSS[+], BSc, IASS, F.HODER[++], MSc, J.DOHNAL[++], MSc

[+] Consulting Engineer, Department of Civil Engineering, Chemoprojekt, The Design, Engineering and Consulting Corporation, Prague

[++] Structural Engineer, Department of Civil Engineering, Chemoprojekt, The Design, Engineering and Consulting Corporation, Prague

In recent years, the authors had an opportunity to design several prefabricated unit systems consisting of spatial and plane skeletal units made from steel sections. The systems are flexible in application and have been used in the CSSR for many structures widely differing in appearance and use. The systems enable large space structures rich in form and of high aestetic effect and economic value to be composed. These structures are of remarkably good properties such as economy, simple and rapid erection, high rigidity etc., and fully satisfactory in the point of functional requirements. The good properties can be gained partly by choosing a suitable form and module of units, partly by an inventive composition of units into structures. Information on prefabricated systems and brief description of several structures that have been constructed are given.

INTRODUCTION

Decisive stimulations for the design of structures and seeking for their new forms arise from the sphere of architecture. On the other hand, architecture is enriched by new technical solutions which can even result in contemporary architecture.

Contemporary spece structures follow classical space structures with a centenary tradition. The novelty is a purposeful design of the structure so that its spatial action is effectively realized in spite of the fact that all structures generally have a spatial effect.

The general trend in the industrialization of construction engineering is to introduce new progressive methods into both mass production and erection of structures.
In consequence of difficulties with energy supply the request for structures with low consumption of material had arised. From this point of view, the space structures represent an excellent means for an inventive engineering activity.

The prefabricated systems enable large space structures of small-scale units to be composed. An application of computers and new computing techniques sets the engineer practically free from a limitation to the field of the structural analysis of the system. In designing a structure, it is possible to concentrate oneself on the main aim that is to create a structure of modern style fully satisfying requirements as to function and economy.

Prefabricated unit systems which we have applied during recent years enable the constructions with above characteristics to be completed. At the design of prefabricated systems we considered also the erection of the structure in order to reduce the expenditure of work. Prefabricated unit systems which we have designed and used are listed in Table 1.

Figure 1. The erection of Pavilion S, Brno

The systems generally consist of space and plain skeletal units joined by an universal connectors. Connector of our system GYRO II enable up to nine members to be joined in one node. The connector contains only one erection bolt. The arrangement of the node is of basic importance for successful use of the system.

There are two main problems to be solved by the design of a node, i.e. the requirement of the rigidity of the structure and that of accuracy in units manufacturing. From this point of view there is a great difference between nodes "S" and GYRO. Both connectors fulfil the rigidity requirements. The older system "S" requires a high accuracy in the manufacture of units whilst the system GYRO allows the production of units by common way in fixtures. By using GYRO connectors, the labour necessary for erection dropped to 40 %.

The design of a prefabricated space structure is based on functional and technological requirements of the structure and on a catalogue of suitable unit system. Prefabricated system represent a certain restriction in this way because for fully economical solution of the given steel structure it is necessary to have available a system of a suitable module. That is why several various unit systems were designed /Tab. 1./.

Fig. 2.
"S" connector

Fig. 3.
GYRO connector

Type	System	Modules M1	M2	H
1	SP-VŽKG	1 400	1 212	750
	S-KSB	1 500	1 500	1 200
	GYRO I	1 500	1 500	1 200
	GYRO II	3 000	3 000	2 400
	SPPK-18	1 800	1 800	1 600
2	TETRA	2 800	2 800	2 500

Table 1. Prefabricated unit systems

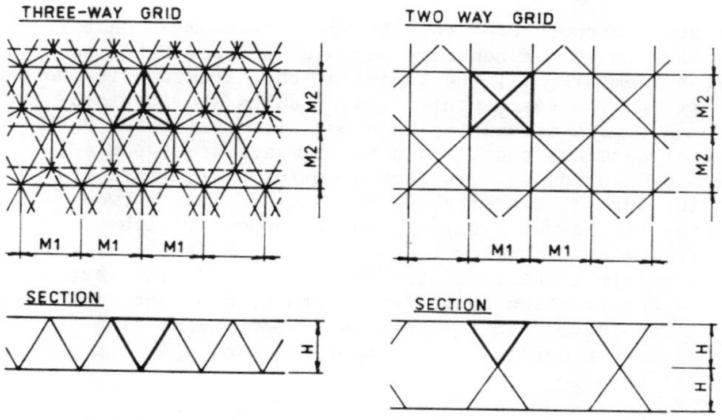

Figure 4. Geometrical system of the three-way and two-way grids

DESIGN

Designer's task at the space structure design is in choosing a suitable system of prafabricated units and in formulation of the geometrical arrangement of units in the structure. The next steps are the specification of loads and the formulation of individual loads combinations. After that, stress analysis of the structure on the level required, selection within the scope of prefabricated units and dimensioning of individual parts are provided by a specially developed programme for the electronic computer. Basic drawing documentation consists of erection drawings, the specification of prefabricated units and material data sheets, if necessary.

Construction	System	Covered Area m^2	Maximum Load $kN.m^{-2}$	Main Span m	Weight of Structure $kg.m^{-2}$
Storehouse Kutná Hora	GYRO II	2 808	2,15	29	33
Industrial Building Olomouc	GYRO II	7 328	3,19	24	27
Winter Stadium Uherské Hradiště	GYRO II	5 261	3,46	60	54
Winter Stadium Třinec	GYRO II	7 344	2,67	57	48
Meeting Hall České Budějovice	GYRO II	1 350	4,90	30	50
Meeting Hall Vyškov	GYRO II	945	6,50	30	57
Central Bus Station Brno	Welded Frame	12 064	14,00	20 x 12	89
Multipurpose Hall Brno	Cable	5 648	1,20	84	47
Winter Stadium Bratislava	Welded Grid	2 875	3,20	36	31
Sport Hall Litvínov	Welded Grid	2 955	2,40	45	38
Sport Hall Prešov	Welded Grid	6 003	3,50	54	42
Winter Stadium Spišská Nová Ves	TETRA	5 935	2,90	70	67
Sport Hall Most	GP	3 318	2,50	49	47
Shopping Centre	GYRO I	5 659	2,70	12	27

Table 2.

CONSTRUCTIONS

During recent years, we had an opportunity to design and supervise a number of spatial structures widely differing in appearance and use. In Table 2., some of these structures are listed and major parameters characterizing the realization results are given. The factors affecting the choice of the system for space structures are so differing that it was not possible to use prefabricated systems generally.

The roof structure for the Sport Hall at Prešov /Fig. 5/ represents the classical solution of the space structure. All members made of tubular sections are connected at the joints by welding to a spherical element. The element is a hollow steel sphere reinforced by diaphragm. The ratio of the structural module to the span is 1 : 18, the ratio of the construction depth to the main span is 1 : 15. Basic prefabricates are in this case members with ends ready made for butt-welding and spherical connectors. The members and spherical nodes have a high frequency. The structure is welded at the erection site which means a lot of welding work there.

The steel structure of the the Sport Stadium at Uherské Hradiště is on the contrary composed of prefabricated unit system GYRO II. The layout of the structure is shown in Fig. 6. The basic prefabricated units are latticed pyramides with triangular base and plain triangles both made of seamless tubes. Joining elements are two discs and one erection bolt in each node /Fig. 3/. There is no welding work at the site. Of the basic units, a two-layered, three-way spatial grid is composed, with stiffening ribs forming the skylights. The ratio of the basic module of units to the span is 1 : 20, the ratio of the construction depth /the depth of skylights not being considered/ to the span is 1 : 20. Supporting columns are connected to the roof frame by means of special heads.

The steel structure of the Winter Stadium at Spišská Nová Ves represents the largest space structure composed of mass produced units,in the CSSR. The layout of the structure is shown in Fig. 7. The roof frame is constructed from prefabricated latticed steel units forming a two-way, double-layered grid. The grid is assembled from pyramides with square base with their tips pointing to the central plane. The square network is completed in both surface planes with diagonal members in two perpendicular zones. The horizontal structure of the roof is supported by four shearing walls /wind bracings/ and four swinging columns. The structure is conncted to the columns in the central plane by means of distribution heads.

The space steel structure of the bus parking above the Bus Station in Brno is shown in Fig. 8.
The horizontal structure is designed as a framework grillage supported by a system of swinging columns and two lines of columns fixed in one direction in order to secure the stability of the structure in horizontal direction. The basic unit of the structure is a plane latticed lamella shown in Fig. 8. These units were joined by welding to a gusset tube at the site /Fig.12/.

The space structure of the circular Sport Hall in Brno is a prestressed cable structure composed of prefabricated elements, cables and elements of classical type /Fig. 9/. Thanks to the circular shape, all elements are repeated 42 times. A peculiarity of this structure is that cable elements have no rectification equipment for the erection. The required prestress of the structure was reached by bracing of central discs at the given distance and by inserting of vertical members of given size between two cable layers /Fig. 14/. The whole erection metod was simulated in a computer. The exact size of all elements were calculated by non-linear analysis.

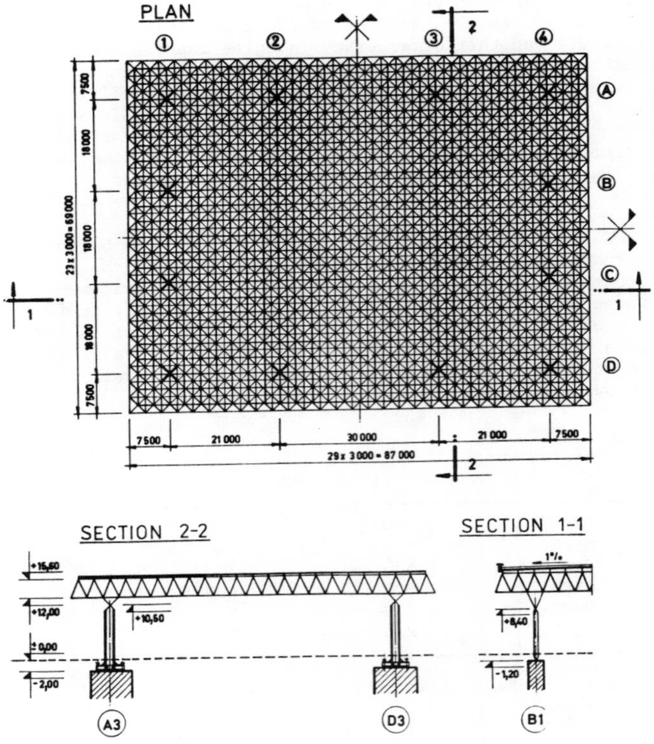

Figure 5. Sport Hall at Prešov, general arrangement

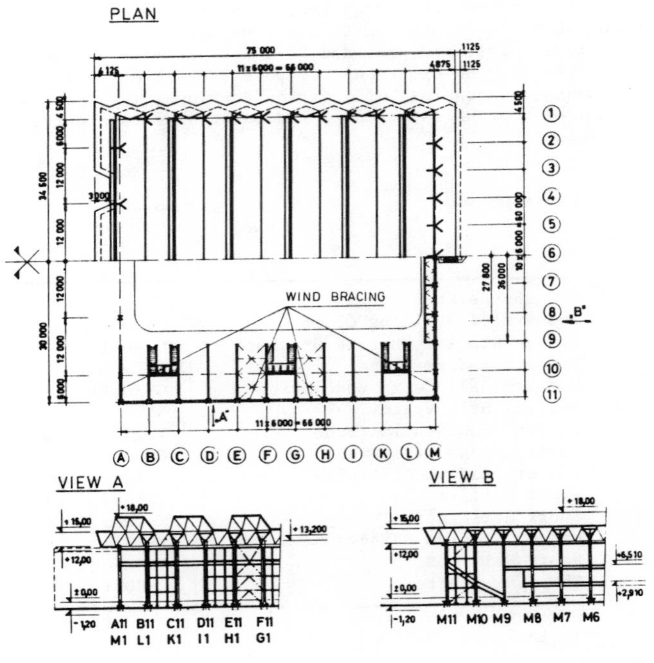

Figure 6. Winter Stadium at Uherské Hradiště

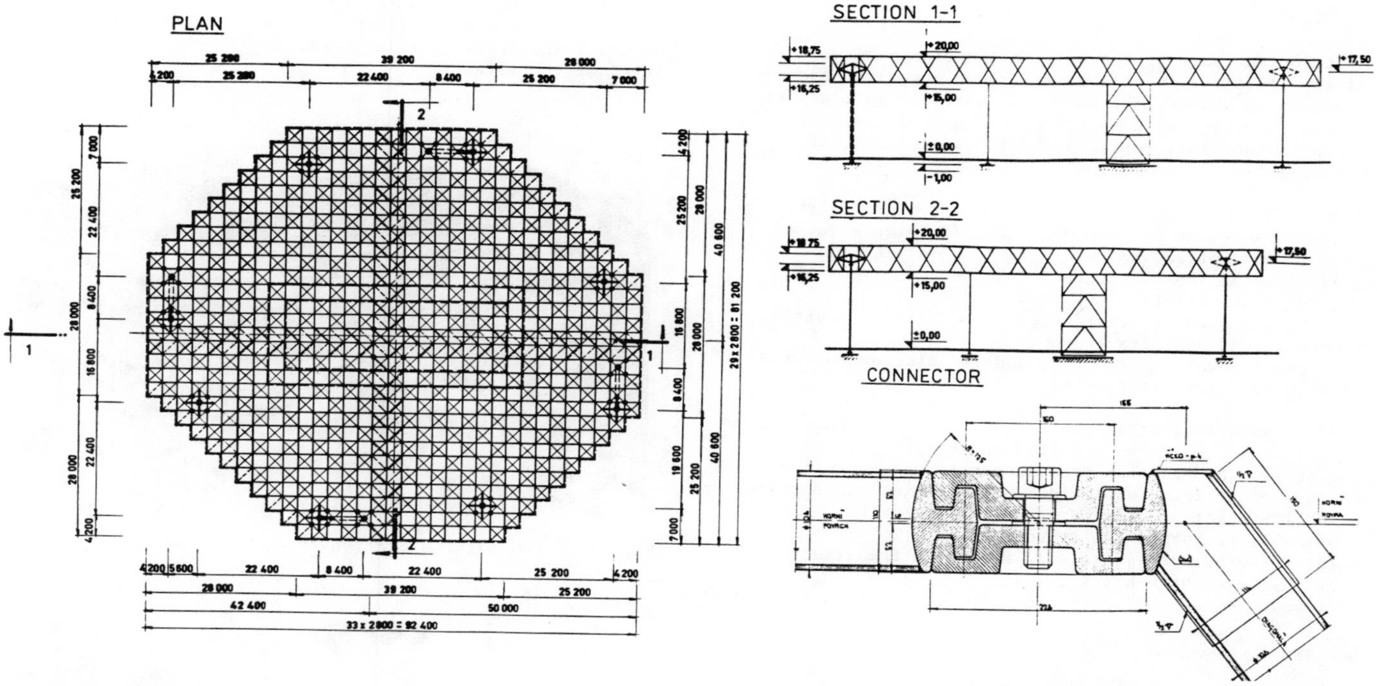

Figure 7. Winter Hall at Spišská Nová Ves - general arrangement

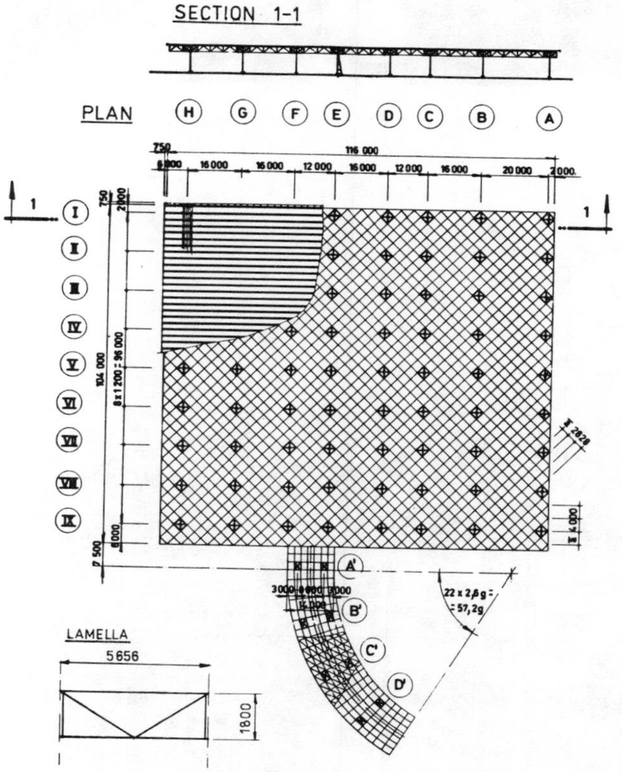

Fig. 8. The space frame for Central Bus Station in Brno

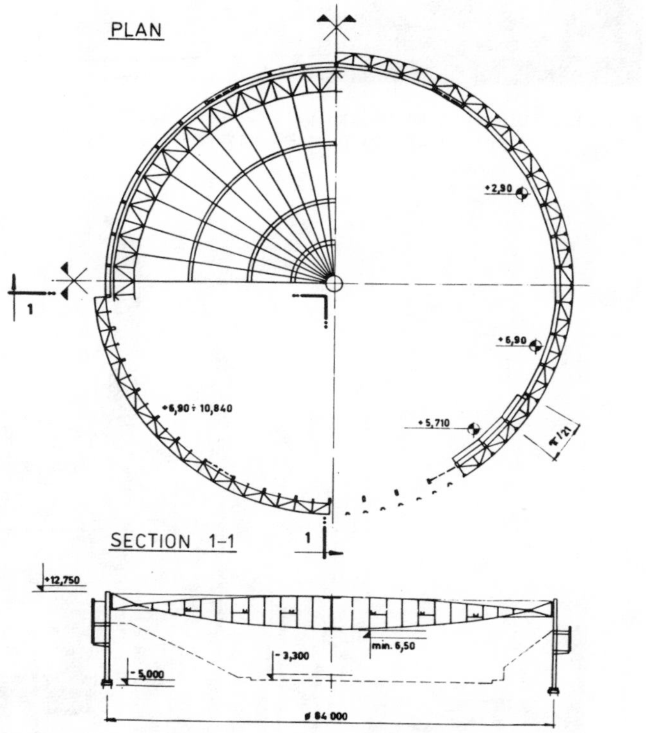

Fig. 9. Multipurpose Sport Hall in Brno

ERECTION

An important part of the realization of any structure is its erection. In principle we have suggested an erection in blocks. The aim is to assemble as big part of the structure as possible on the ground and then lift it up to the final position on its own columns, if possible.

Figure 13 shows the erection of the structure of the Winter Stadium at Uherské Hradiště. The simple method many times applied is:
- assembly of the structure on the ground,
- assembly of erection supports,
- lifting of the structure above its definite position,
- erection of definite supports,
- dropping down the structure and its placing onto the definite support bearings.

Another a very advantageous method of steel structure erection used for Brno Bus Station is evident from Figure 11. The erection procedure used was as follows:
- Erection of columns provided with special lengthenings for fastening of draw bars for lifting jacks,
- assembly of horizontal structures,
- fastening of twin-jacks and lifting of the structure to the level of bearings,
- dismounting of the jacks and lengthenings of the columns.

Fig. 10. Winter Stadium at Spišská Nová Ves, erection of space steel structure

At this assembly procedure, the roof structure of total size 104 x 116 m was divided into four blocks. These parts of the structure were interconnected by welding in definite level at the site.

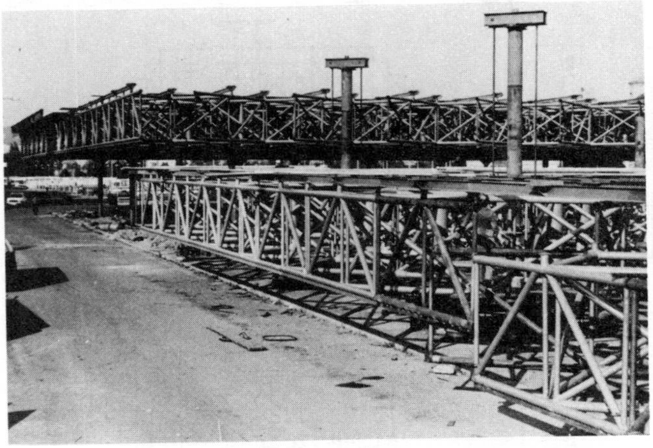

Fig. 11. Central Bus Station Brno, erection of space frame

Fig. 12. Central Bus Station Brno, detail of the node

Figure 13. Erection of the Winter Stadium at Uherské Hradiště

Figure 14. Multipurpose Sport Hall Náplavka in Brno, bracing of the discs, inserting of vertical members

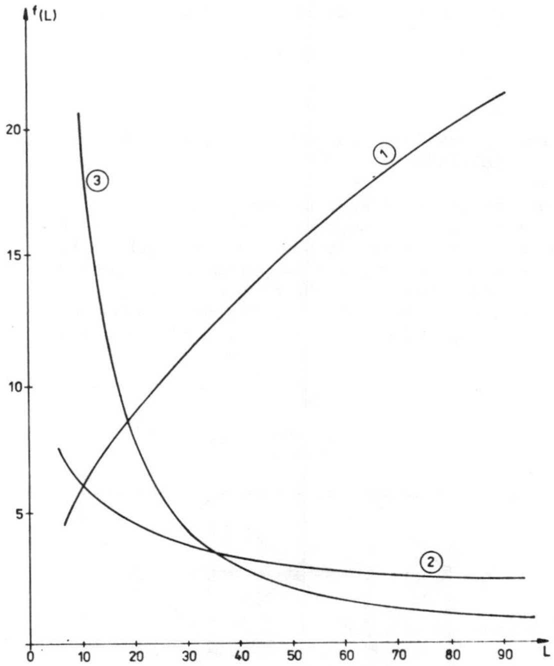

Fig. 15., Curves

1. $f/L/ = q \ /kg.m^{-2}/$

2. $10 \ . \ f/L/ = \dfrac{q}{L} \ /kg.m^{-3}/$

3. $š \ . \ 10^3 \ . \ f/L/ = \dfrac{q}{L^2} \ /kg.m^{-4}/$

q specific weight of the structure for a load of 1 $kN.m^{-2}$

L characteristic span

L^2 characteristic floor area defined by lines of columns

CONCLUSION

In our short contribution we have brought up the results of our design team engaged in designing of prefabricated space steel structures quite commonly. Our experience shows the advantage of using space structures. If designed well, they are very economical. In Fig. 15., relations obtained through evaluating the realized structures are grafically shown. The curve No. 3 shows that the specific consumption of steel decreases by enlarging of the span. The large span easily and economically available by using space structure increases economic value of the construction and makes possible more intensive utilization of the floor space.

Prefabricated unit systems have the advantage of mass production. It means that elements and details as well as shop drawings are repeated many times. These are assumptions for an automatic production of structures. The design of space structures requires inevitably the application of computors besides other aids used by the design team. Using of computers brings additional savings of material by an exact analysis and by designing of economical section of the members. According to our experience, there exists a contradiction between full use of standardized characteristics of the material and the real standard of production and erection of structures. This contradiction is the bigger, the more work is to be done at the erection site. Prefabricated space structures solve also partly this urgent problem.

ACKNOWLEDGEMENT

The authors are grateful to the Organizing Committee and especially to Professor Z.S.Makowski for the invitation to present some results of their work.
Thanks are also due to Mr. Z.Dvořáček and Mrs.B.Kaďourková that took part in the results submitted.

"MOSTOSTAL" SPACE GRIDS — DEVELOPMENT AND APPLICATIONS

J BRÓDKA[x], CEng, DSc, A CZECHOWSKI[xx], CEng, DSc, A GRUDKA[xxx], CEng
and J KORDJAK[xxx], ME

x Professor, xx Lecturer, xxx Assistant lecturer,

Metal Structures Research and Design Centre "Mostostal", Warsaw.

General principles of the MOSTOSTAL space frame system have already been mentioned at the Second International Conference on Space Structures in 1975, see Ref 1. Since that time the system has been developed and numerous double - layer roof structures have been designed and built. In this paper the MOSTOSTAL space frame system is briefly desecribed and a survey of most interesting projects is presented, giving an idea of the various possible applications of pyramidal space grids. Some conclusions are also included.

MOSTOSTAL SYSTEM

The MOSTOSTAL system is a prefabricated system of double - layer grids, designed as roof structures to be serial - produced in a specialized workshop. The system comprises two possible grid arrangements - so called "square on square" offset and "diagonal on square", which corrspond with two constructional subsystems being similar to PYRAMITEC and UNIBAT systems respectively, see Figs 1 and 2. Using some square pyramidal units as basic prefabricated elements, several double - layer grids have been designed for multi - bay industrial buildings and free - standing pavillions. Dimensional series are shown in Tables 1 and 2.

The roof grids indicated can be identified as standard ones in MOSTOSTAL system.

The essential difference between those two types of structure relays one the supporting system. In the case of multi - bay buildings seperate space grids, each stiffened by edge lattice girders, are supported at the corners on hollow section columns. In the case of pavillions a roof structure rests through circumferential beam on I - section posts spaced every 6,0 m along the boundary.

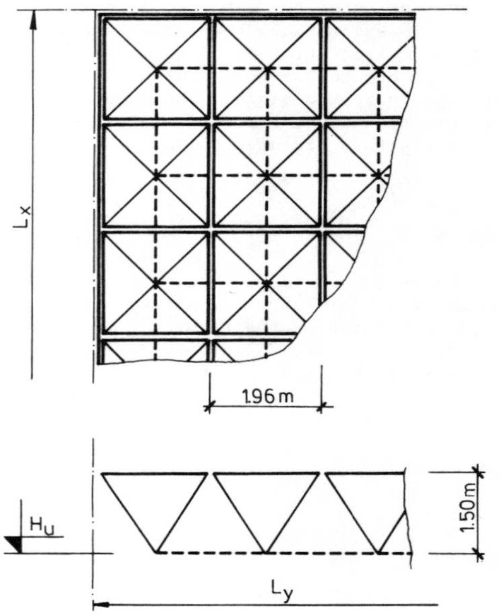

Fig 1

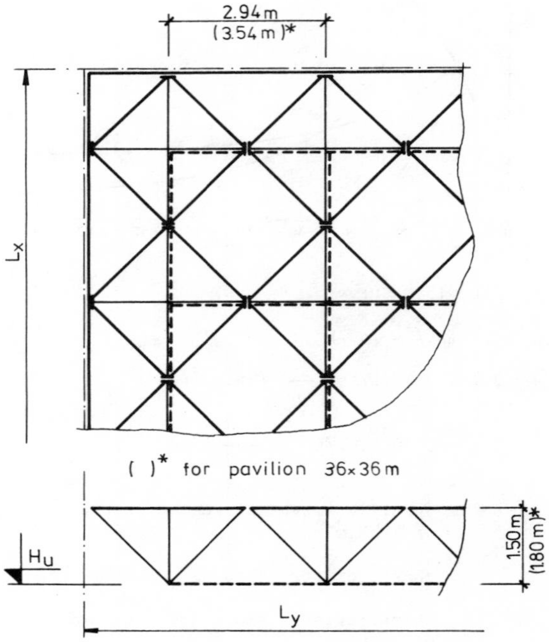

Fig 2

Before the MOSTOSTAL system was offered to the construction industry a full size prototype structures of both subsystems were built and tested, see Ref 2.

MOSTOSTAL system was developed and introduced progressively in accordance with investors' demands. Since 1974 quite

Table 1 Multi-grids

Spans $l_x \times l_y$ [m]	Heights H_u [m] 4.80	6.00	7.20
(24×24)	—	■ ¦ ◆	■ ¦ ◆
(24×18)	—	◆	◆
(18×18)	◆	◆	◆
(18×12)	◆	◆	◆

Table 2 Single grids-pavillions

36×36	◆	◆	◆
32×32	■	■	■
30×30	◆	◆	—

■ - 'square on square offset' arrangement

◆ - 'diagonal on square' arrangement

a number of small and large projects have been realised. Unfortunately some investments have been given up due to unfavourable economic situation. Standard double - layer grids were used to cover multi - bay hall buildings for electronic and textile industries, school gymnasia and assembly halls, shopping pavillions, see Fig 3. Furthermore besides this direct use of the MOSTOSTAL system grids, there are several examples of projects, which have been

Fig 3

prepared as one - off jobs, being beyond the standard range of application but using more or less prefabrication rules of the system. Such space grids seem to be of particular interest and this is the reason why the following part of this report is devoted to specially designed roof structures.

COFFEE - HOUSE IN WARSAW

The location of the modern coffee - house was chosen on the high bank of the Vistula River in the wood area. Author of the architectural concept wanted a structure to exibit itself an impressive appearance as well as to be harmonized with the natural environment. To fulfil those requirements a space frame form of construction was adopted, see Fig 4.

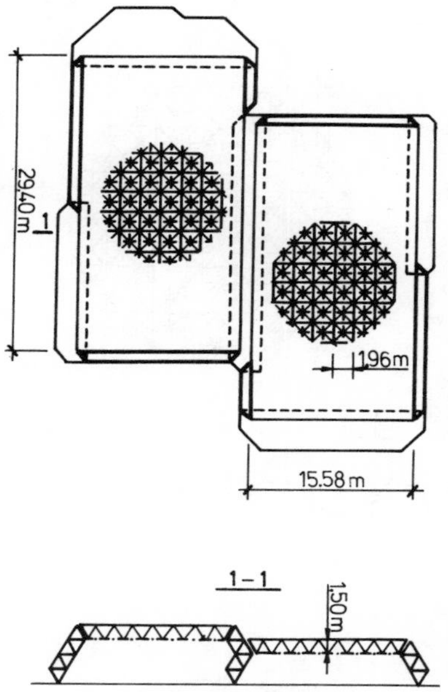

Fig 4

Finally designed structure consists of seven spatial blocks, two identical double-layer roof grids of rectangular pattern and five supporting space frames being arranged as a geometrical extension of the roof structure.

Roof grids have been designed using MOSTOSTAL system of prefabrication. Each of them is composed of pyramidal units, lower chords and mansard latticed girders specially added along the free edges. In the case of supporting space frame /not covered/ a different constructional solution was applied. The supporting blocks are welded tubular structures made from seperate members and spherical nodes.

The layout area of the coffee - house is 1150 m^2. The obtained steelwork mass, 53,5 kg/m^2, is rather considerable, but this is a tribute to the extraordinary shape of the structure.

SELF - SERVICE SHOPPING CENTRE "MEGASAM"

Another interesting example of a double - layer grid is a roof structure designed for a big shopping center in Warsaw, see Fig 5. A layout of the area to be covered was irregular. Taking this into account and also rather large spans in both directions, a diagonal offset grid was accepted as the most suitable form of construction. The roof structure has been designed in accordance with MOSTOSTAL system of prefabrication with some necessary changes in modular dimensions.

AVIARY IN POZNAN ZOO

The project of aviary for subartic birds in Poznań ZOO demonstrates a unique application of a single - layer

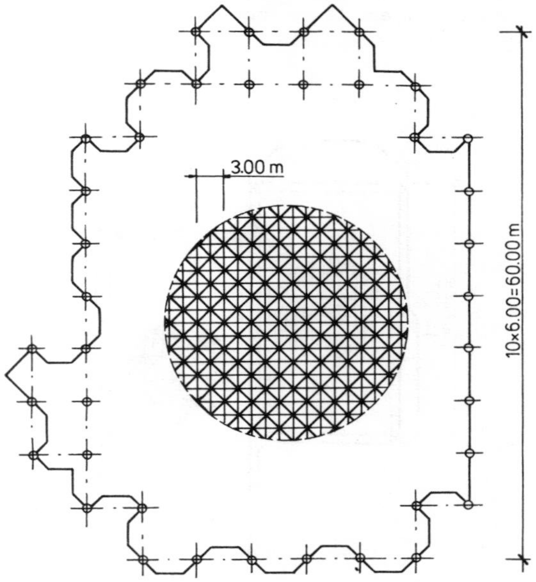

Fig 5

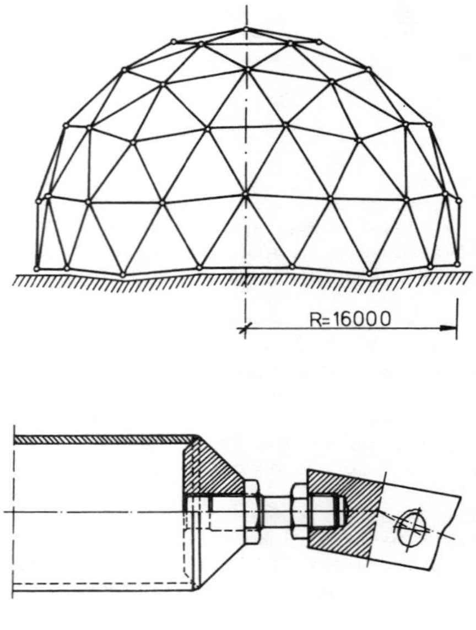

Fig 6

geodesic dome. This skeletal structure has been formed as a part of 180 - hedron grid, using tubular members of three lengths and nodal connections of three geometrical

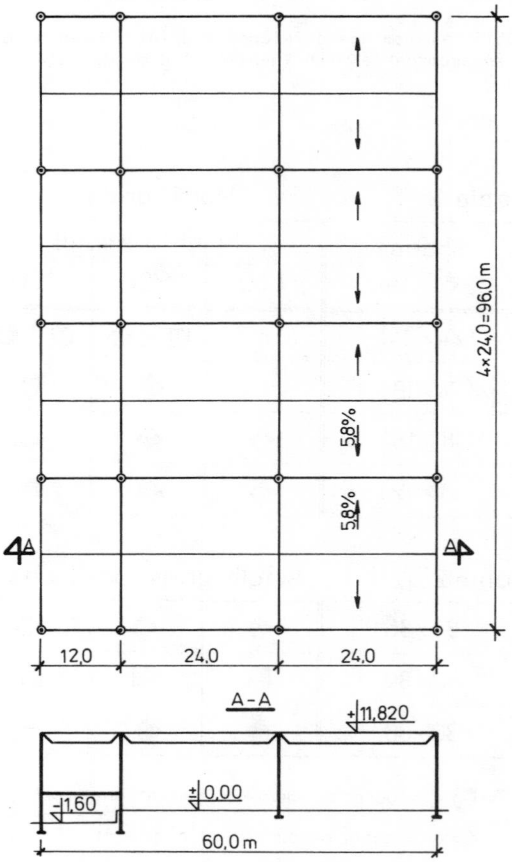

Fig 7

variants. An original jointing system has been proposed to make the construction easy erected and demountable, see Fig 6.

BRITISH PAVILLION AT INTERNATIONAL FAIR AREA IN POZNAN

British Pavillion is an example of multi - grid construction, see Figs 7 and 8. The regular layout of the exhibition hall comprises eight grids each 24 x 24 m and four half - reduced of 24 x 12 m. The diagonal on square arrangement has been adopted. All the grids are supported through edge lattice girders on columns. Grids 24 x 24 m were applied directly as standard MOSTOSTAL system space grids. The smaller ones were additionally designed.

Simplicity of manufacture due to the large amount of repeat fabrication work, no site welding, speedy assembly and erection using lift - slab technique all those factors proved the efficiency of the system used.
All phases of the building process proceeded consistently being finished in six months only.

RESEARCH AND EXHIBITION CENTRE OF MACHINE INDUSTRY "ITEKOMA"

Multi - funcional ITEKOMA complex comprises several hall buildings and 6 - storey office building on the area of over 23000 sq.m., see Fig 9. The roof structure for the hall buildings has been designed as double - layer MOSTOSTAL system space grid with diagonal on square arrangement. Roof space frame units of two sizes /30 x 30 m and 30 x

Fig 8

12 m/ have various loading and supporting conditionis depending on the location of the grid.

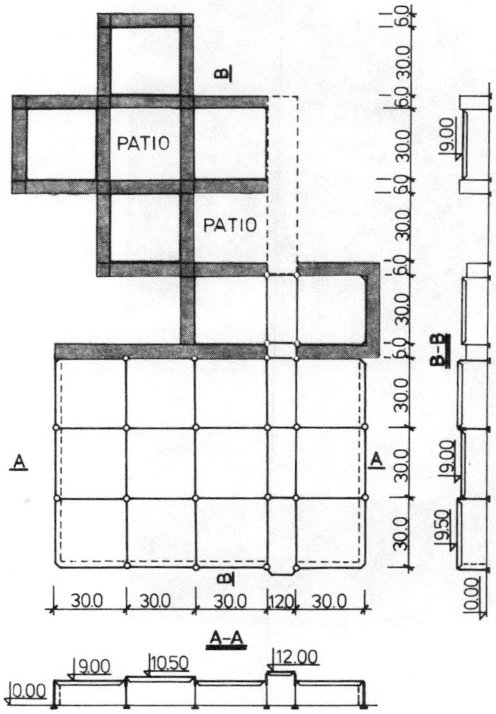

Fig 9

In the case of hall No 4A/B there are main columns spaced 30 m in both directions, secondary steel posts at the outer /cornice/ edge and also concrete walls on the remaining side of the building. The height of the hall is altered by the grid depth /1,5 m/, up and down from one grid to another. Thus it seemed adventegous to introduce coupled triple - chord lattice girders in lines of main columns to provide an additional elastic support for each grid and also to enlarge the rigidity of the whole structure. The lattice girder depth is then twice the grid depth /2 x 1,5 = 3,0 m/.

The ITECOMA complex is a very positive example of the solution, which combines an attractive architectural form and efficiency of the constructional system utilized. The rational choice of geometrical parameters of the structure and supporting system permitted a significant unification of components as well as a rather low steel consumption. The roof steelwork mass including lattice girders is a little less than 35 kg/m^2.

ROOF STRUCTURE OVER ANIMATION HALL IN BOROVEC HOTEL IN BULGARIA

So called Animation Hall in Borovec hotel had to be covered with a space frame supported at the boundry on posts spaced every 7,2 m, see Figs 10 and 11. Because of large snow loading /mountains region/ and a heavy concrete deck consequently presumed an extremely high loads had to be taken by the roof steelwork. The total design /enhaced/ loading was more than 700 daN/sq.m. Taking this into account as well as transportation requirements a triple - layer space grid has been finally accepted.

Two - way rectangular triple - layer grid with a module of 2,40 m and a depth of 2 x 1,80 = 3,60 m is composed of pyramidal units and chords in such way that upper and lower prefabricated pyramids, assembled apex downwards, are offset by half a module in both directions, thus making a very rigid space frame. The roof structure is simply and re-

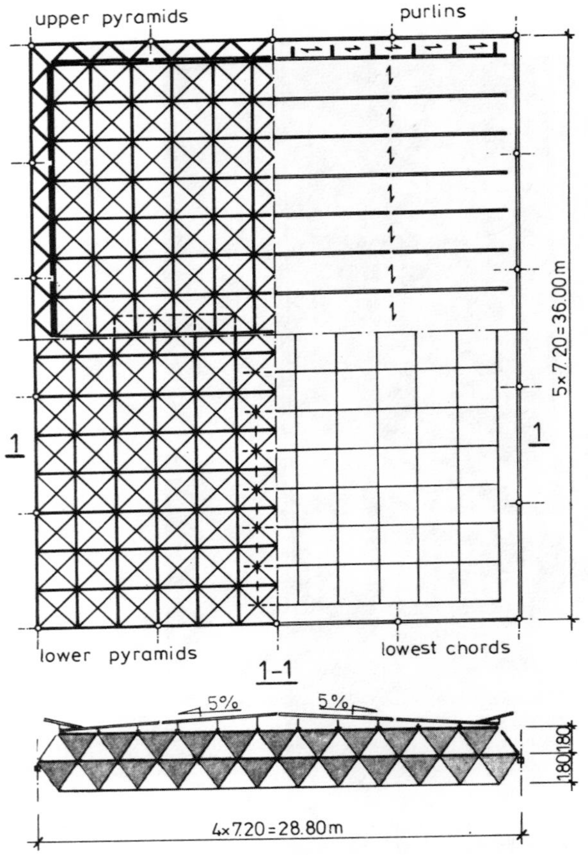

Fig 10

gularly supported at middle - layer joints, with mansard lattice girders added to provide the structural integrity of the frame.

Although there was some amount of welding on site to make upper pyramid apexes connected with middle - layer joints, the assembly and erection of the structure proved to be simple and fast. No tolerance problems were encountered. The roof steelwork was partitioned into five sub - assemblies, each of them was consistently assembled close to the side of the layout at the ground level. They were progressively lifted over the existing ring beams by cranes, then moved on special roller supports to their final positions and jointed together.

CONCLUDING REMARKS

The advantages of space frame grids have been well recognized. The architects are keen to make use of space frames, especially double - layer grids. As a rule they demand the structure to be exposed inside the building. The specific regular appearance and flexibility in choosing the geometrical form are not the only reasons of growing popularity of space grids.

Experience shows that space frame grids provide the economic answer to various design requirements such as large interrupted span, irregular layout and/or heavy loading.

Since computer technique is commonly available the engineer is now able to carry out a full structural analysis. The "spatial" philosophy in design has become a widely used practice.

The MOSTOSTAL system which is based on prefabricated pyramidal units has proved to be an attractive mean of construction as it combines an interesting appearance, versatility in use and structural efficiency.

Fig 11

REFERENCES

1. J BRÓDKA et al, Two - layered Grids for Mass Production in Poland, Proceedings, 2nd International Conference on Space Structures, University of Surrey, Guildford, 1975.

2. J BRÓDKA and A GRUDKA, Limit State Analysis and Full Scale Experiments of Double - layer Grids, Proceedings 3rd International Conference on Space Structures, University of Surrey, Guildford, 1984.

THE POWER-STRUT SPACE FRAME SYSTEM

D H GEBHARDT, PE

Power-Strut Division
Van Huffel Tube Corporation
Warren, Ohio U.S.A.

The Power-Strut space frame system was developed in the mid-1970's and has been marketed throughout the continental United States since 1977. Its patented construction consists of square tubular members joined by flat plate connectors. It is generally used to form double-layer grids of square-on offset-square configuration with typical module size being 122 and 152 centimeters (4 and 5-foot).

BASIC COMPONENTS

Design simplicity and ease of fabrication were primary objectives in the development of the Power-Strut space frame system. This was accomplished by bending the web member into an "S" shape with the distal portions being straight sections which are in parallel planes both to each other, as well as the chord members. The nodal connection can then be made through a flat connector plate. While this does present the disadvantage of eccentric loading at the nodal joint, it is not an insurmountable problem in "small module" (less than 160 cm) construction. When high shear loads must be transferred, a double diagonal can be inserted. Weld plate stiffeners can be added if required.

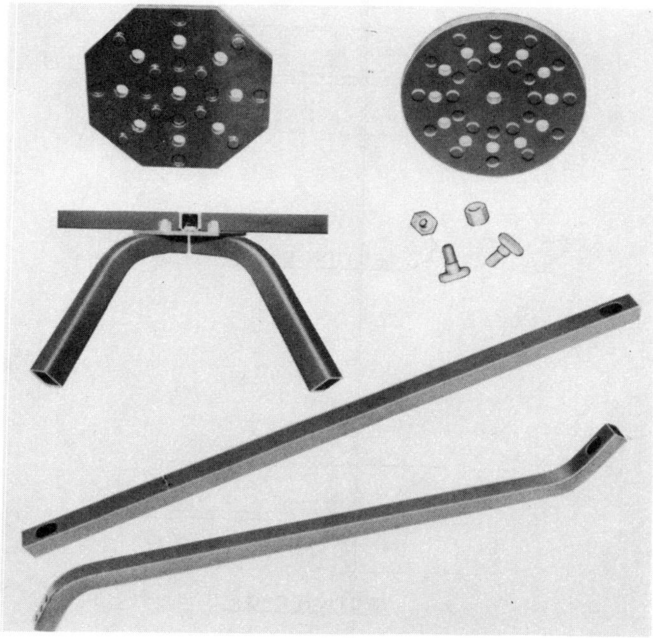

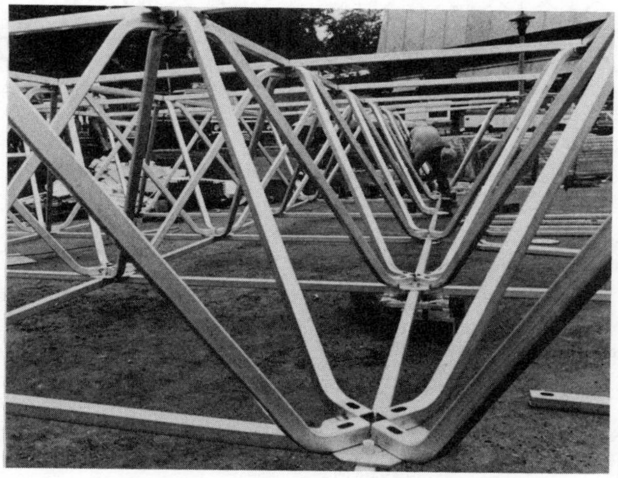

The overriding advantage lies in the uncomplicated methods of fabrication and assembly while maintaining an aesthetically pleasing appearance. Five basic components make up the standard system. The octagon connector plate is utilized in square module construction and the round plate in the triangular module.

Planning Modules

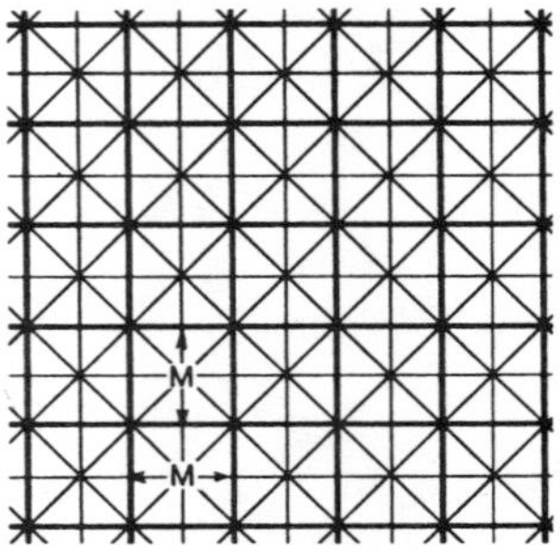

Square on Square Module M = 4′ or 5′

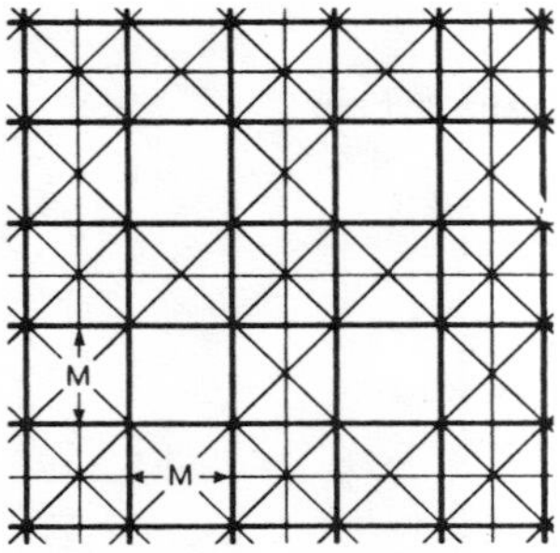

Square on Larger Square M = 4′ or 5′

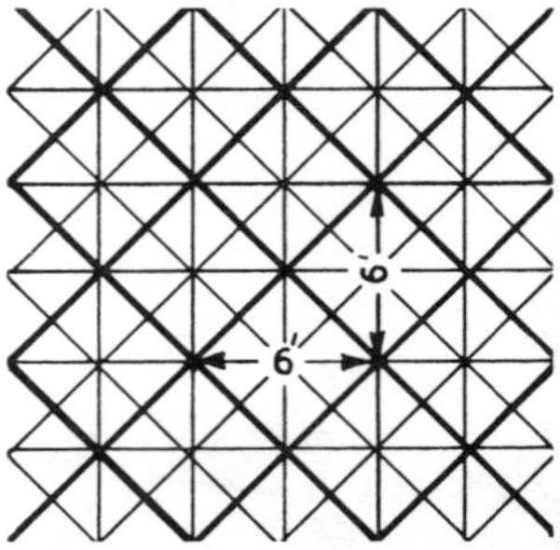

6′ Diagonal Module

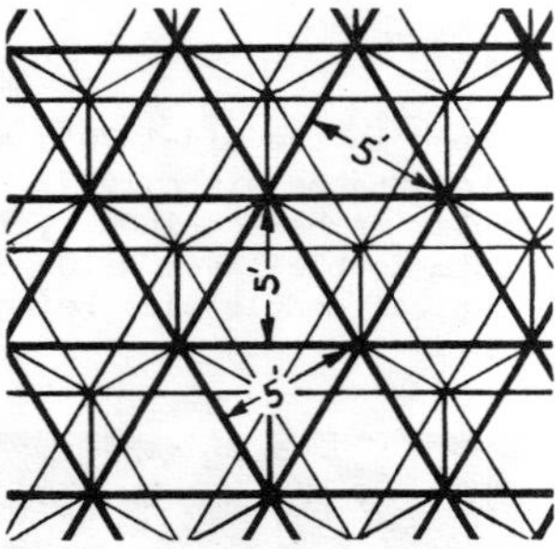

5′ Triangular Module

The system is offered in several planning modules; the most common embodiment being square-on-offset-square in 4-foot (122 cm) or 5-foot (152 cm) modules. These module sizes can also be employed in square-on-larger-square. Two other module configurations are available; a diagonal module measuring 6 feet (183 cm) and a triangular module of 5 feet (152 cm). Triple-layer grids and multiple-layer step-up conditions are easily accomplished with the flat plate connector. The intermediate chord is merely double punched and sandwiched between two plates with lower and upper grid web members attached to lower and upper plates respectively.

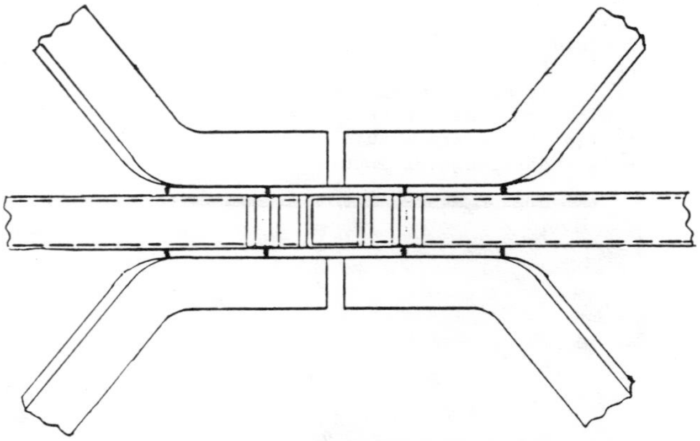

MULTI-LAYER GRID CONNECTION

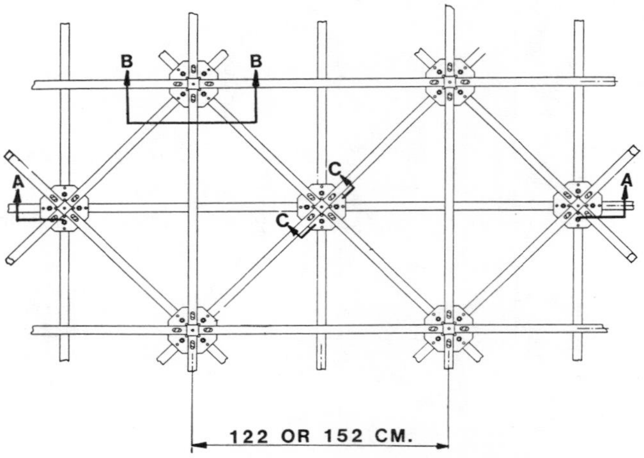

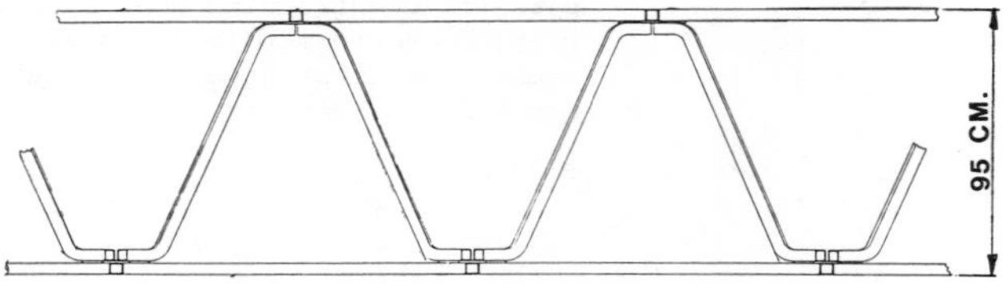

SECTION A-A (122 CM.)

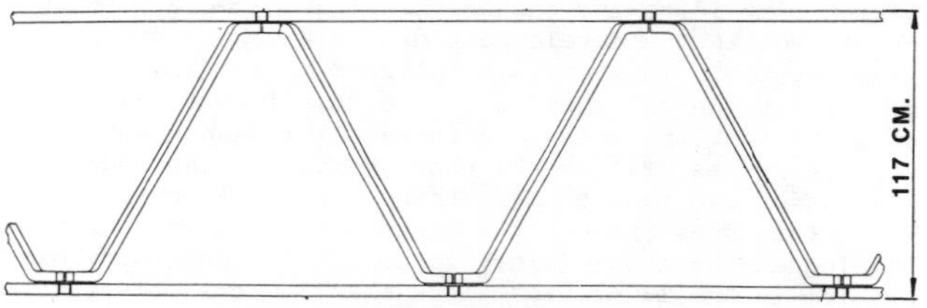

SECTION A-A (152 CM.)

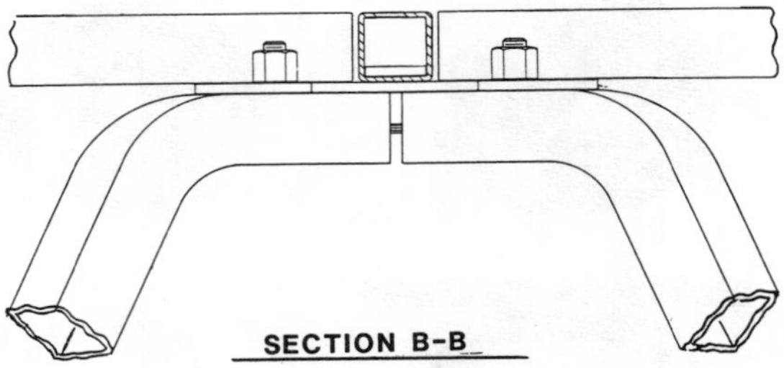

SECTION B-B

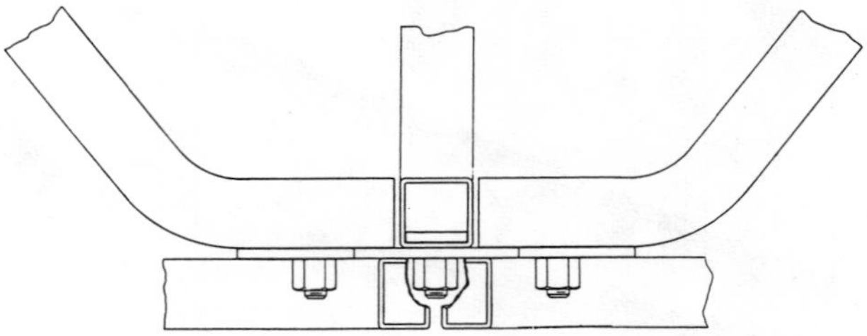

SECTION C-C

MANUFACTURING

The chord and web members are fabricated from electric welded tubing of square cross section. The tubing dimensions are 41 mm x 41 mm with a wall thickness of 2.7 mm. The top chord can be increased to a depth of 76 mm when higher combined axial and bending stresses are encountered. After welding, the tube is accurately cut to length by a double-shear process which leaves the ends burr free.

The web members are bent on a hydraulic bender with an internal mandrel. Both members are precisely punched utilizing a pinned fixturing devise to guarantee hole-to-hole accuracy and part-to-part consistency.

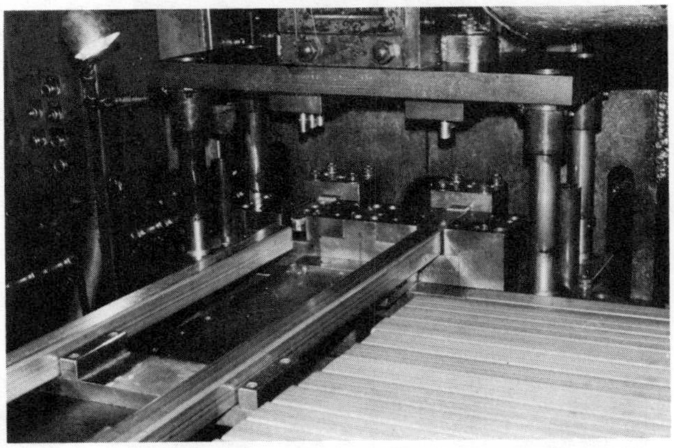

The octagon connector plates are blanked from mild steel 229 mm wide and 6.4 mm thick. A series of holes and formed shear connectors radiate at 45° increments. The shear lugs alternate to form 90° mating surfaces on the web and chord side respectively. The slab head shoulder bolt engages a counterbored nut. The bolt is inserted through an access hole on one side of the tube. The three-hole punching in the opposite wall of the tube engages the two shear lugs and shoulder bolt to form a close fitting metal-to-metal connection. All standard components are mass-produced and warehoused prior to project packaging and finishing.

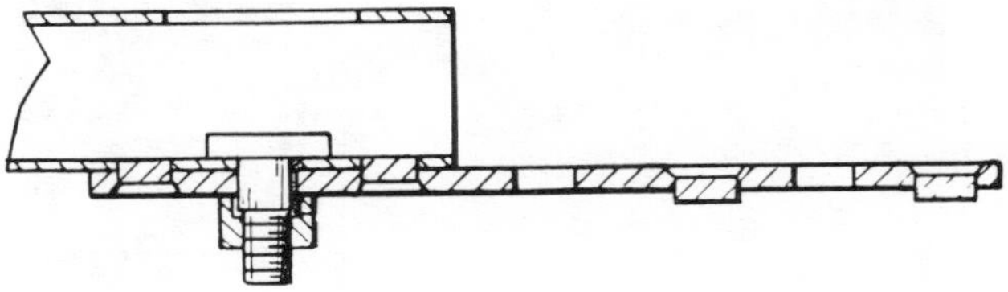

MATERIAL & FINISH

Durability of the resulting structure was given a high priority. All components have a galvanized substrate. The material utilized in the tubular members is zinc-iron alloyed steel. The connector plates are electrogalvanized after forming. Space frame components are then coated with a baked-on thermoset polyester powder applied by electrostatic spray. The end product has a smooth, hard, high gloss finish free of runs and sags. This tough, mar-resistant coating withstands impact without chipping or cracking. While almost any color is available on special contract, Power-Strut offers its space frame with a choice of twelve standard colors.

Hardware is also given special treatment. Nuts and bolts receive a heavy cadmium plating and are then bonded with a clear polymer coating.

DESIGN & ANALYSIS

Computer-aided methods are employed by Power-Strut to quote, plot, and analyze all space frame projects. Laboratory testing determines the allowable load carrying capacities of members and joints. Various types and stages of reinforcement can be used to satisfy different design parameters. Power-Strut space frame has general application where clear spans range from 6 m to 20 m.

ERECTION

The Power-Strut system provides ease of installation through uncomplicated construction. Unskilled labor can quickly become familiar with the standard repetitive connections. The combination of bolt and shear connectors enables the structure to be self-aligning.

Subassemblies are often mass-produced at the site, consisting of a connector plate and four (4) diagonal (web) members. These are then placed in the structure.

Three basic methods of erection:

1. In Place: Built-off scaffolding and with lift mechanisms.

2. Ground Assembly: Entire structure crane-lifted onto supports.

3. Partial Ground Assembly: Crane-lift segments with balance constructed in the air.

Depending on the module size and method of installation, the Power-Strut system is usually installed at the rate of 2 to 5 m^2 per man-hour.

Vertical lift and horizontal swing of 300 m^2 project built adjacent to site.

Construction around columns of a 600 m^2 project utilizing the lift-slab technique.

APPLICATIONS

Power-Strut space frames have been utilized throughout the United States in numerous applications. The primary uses are highlighted below:

a) Recreational Facilities:

Space frames have been provided for city parks, basketball pavilions, swimming pool covers, and recreational offices.

b) Hospital & Bank Construction:

Space frames have become a popular medium for covering hospital and emergency room entrances, as well as the drive-thru kiosks at branch banks.

c) Transit Facilities:

Park'N'Ride facilities have been constructed around metropolitan areas where commuters can leave their autos and take bus, subway, or train transportation into the city. Large transportation centers in downtown city settings have also utilized space frame structures.

d) Atriums:

Power-Strut space frame has been used on numerous construction projects involving atriums. Some of these have involved large hotel lobbies and intersections of shopping mall concourses. The following construction photo shows a multi-level space grid connecting high-rise office buildings.

e) Decorative Uses:

Power-Strut has furnished space frames for vertical and horizontal trellis structures, hanging chandliers, kiosks, and sunscreens. Some decorative frames have been furnished with nickel chrome finish.

f) Skylights & Glazing Support:

Clustered skylights are often supported on modular space frame structures. Vertical and sloped glazed wall support is also a common use.

g) Visitor Pavilions, Amphitheaters, Pedestrian Walkways:

Visitor centers are a popular application. A 1,000 m^2 canopy has been constructed at the Kennedy Space Center. Due to the corrosive salt air environment, the frame was hot dipped galvanized after fabrication, yellow dichromate rinsed, and polyester coated. This extraordinary finish enables the structure to be virtually maintenance free.

h) Renovation of Retail Centers:

Power-Strut space frame has been utilized on several interesting renovation projects. One involved the revitalization of a downtown shopping area of Elmira, New York, and consisted of 2,000 m^2 of pedestrian walkway interconnecting storefronts.

9,000 m^2 of frame was used to enclose an old retail center in Cheltenham, Pennsylvania, to form a renewed shopping mall. The 20 m x 260 m main concourse was ideally suited to space frame construction using a 12 m span with 4 m overhangs.

FUTURE CONSIDERATIONS

Architectural acceptance of space frame construction has grown rapidly in North America over the past two decades. Implementation of space grids to create visually pleasing structures will take on new and expanded applications. With over 100 projects installed, the Power-Strut Space Frame System has become a recognized factor in this marketplace.

PRESTRESSED MEMBRANE STRUCTURES

JOSE MIRAFUENTES GALVAN

Profesor, Laboratorio de Estructuras Laminares
Facultad de Arquitectura
Universidad Nacional Autonoma de Mexico.

The Laboratorio de Estructuras Laminares has built many prestressed membrane structures with tensile boundary members, as research work. The aim of this research is to obtain the optimum structural shapes in prestressed membrane structures with external loads. In the introduction, I present some definitions and the basic concepts of tensile structures for those who are not familiar with this speciality.

INTRODUCTION

Membrane structures may be classified in three types:
The inflated or pneumatic structures, the hydrostatically loaded, and the tent structures or prestressed membranes. "A membrane is a surface so thin that for all practical purposes it cannot resist compression, bending or shear but only tension", Ref 1.

A structure is prestressed if forces or stresses act on it when is not loaded.
A membrane can be analyzed as a structure on the basis of its state of stress or , formally, in accordance with its shape.
A membrane structure is stabilized by the counteracting tensions of opposing curvature of opposite sign, as a saddle surface or hyperbolic paraboloid, see Fig 1.

Membranes can be stretched between boundary cables. The prestress is developed as an internal stress reaction when external forces are applied to its boundary cables.

The saddle-shaped tents, were built with different curvatures, heights and spans, to compare stresses in the boundary cables and in the membranes and to prove the architectural efficiency, see Fig 1 and 2.

We study the normal force components generated by curvature and warpage and how the surface tension, that is originated by the prestress of the membrane, develops an internal stress reaction.

The prestressed membrane structures have almost the same characteristics as cable structures. A prestressed membrane or a cable net can be spanned within rigid edge beams and tensile boundary members in the same building, but the advantages of tensile boundary members are so, that the other types became obsolete.

The principal load-carrying members of membrane structures are the boundary cables, which can only be stressed in tension. In most of the cases such structures are supported by compressive masts, made of light structural tubes.

At the Laboratorio de Estructuras Laminares we have built and tested saddle membranes, varying the curvature. The first one was made with a curvature exactly like a soap bubble, but, from the point of an architectural view, it was not very efficient, because the masts were to high, and the area they covered was too small, see Fig 1.

The second one had a smaller curvature in order to cover more area, but the prestress at the boundary cables was so big, that it was necessary to reinforce the membrane at the ends, see 2.

Fig 1

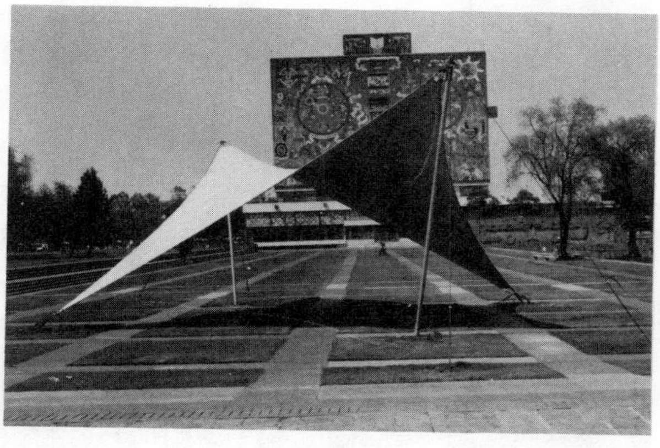

Fig 2

Fig 3 Music Pavillion in Aguascalientes, Mexico.

For this music pavillion, it was necessary to build a sound reflecting area, and conserve at the same time, the opposing curvature. This is a kind of membrane that slopes from horizontal to vertical. It was then necessary to design a central loop that made possible the curvature and the prestress.

The minimum surface is the shape of the most economical surface, soap bubbles are minimum surfaces, a soap bubble model can give an architect the basis to design a membrane structure. Architectural liquid films models are not made any more of soap solutions, but of water solutions with glycerin and sodium.

This minimal surface produces uniform surface tensions; "In reference to its prestress, a membrane is most efficient when its surface tension is uniform. Uniform surface tension occurs when, at any point on the surface, the principal lines of curvature are equal." See Ref 2.

A minimum surface is the surface of least area within a closed frame. For the design of membrane structures the knowledge of minimum surfaces is important, but minimum surfaces are not always the optimum structural shapes.

Theater at "Villa Universiada" for the university olimpic games. It is a parallel wave membrane with linear internal and perimeter supports, an additive series of saddle shapes between supporting and restraining ridges.
Length: 300 feet, width: 150 feet, mast height: 45 feet, see Fig 4.

Fig 4 Theater in Villa Universiada, Mexico.

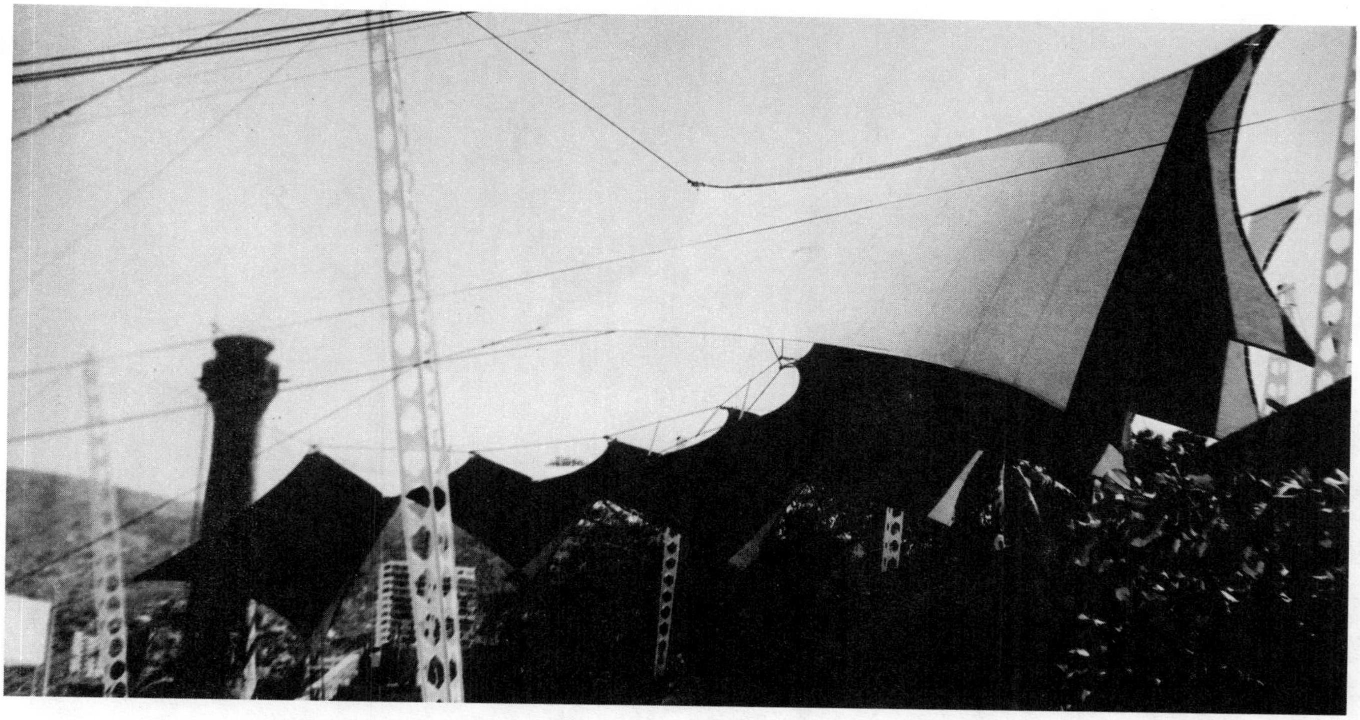

Fig 5 Dolphin-stadium in Acapulco, Mexico.

The radial-wave membranes are particulary strong, we used this shape for the cover of a dolphin-stadium in Acapulco, this structure has resisted wind of 250 kilometers/hour and Earthquakes of 7 grad in the scale of Richter.

The commercial center "Plaza Sedena" is a parallel wave membrane in polyester, length: 300 feet, width: 120 feet mast height: 18 feet. In Mexico City it never snows but we got two feet of hail in a hailstorm. This overweight deformed the structure because the supporting cables did not have the proper curvature due to the limitation of the mast height. The following morning the hail melted and the structure recovered its initial form, see Fig 6.

Fig 6 Commercial center "Plaza Sedena"

"When additional loads act, the minimum surface is not always the optimum shape. It is often necessary to change the form of the minimum surface in such a way that membrane zones requiring higher rigidity also have greater curvature" see Ref 3.

Lightweight membrane structures are adaptive structures, they have adaptable size, convertibility and portability. There is a fixed relation between height and span. For an exposition in Mexico City, we built small tents with saddle surfaces, but we had height limits in the masts, it was necessary to subdivide our hyperbolic-paraboloid-tents into modules which could be added to obtain a bigger covered area

In Villa de Guadalupe, Mexico, many hyperbolic-paraboloids were stretched in rigid edge beams for architectural reasons It was necessary to give a greater tension to the membrane so the result, was a more expensive structure.

Fig 7 National Palace in Mexico City

Membrane structures can span the greatest distances because their own weight is of little importance. When the proportion height-to-span is exceeded and the curvature cannot be achieved, it is possible to subdivide the membrane by the insertion of cables.

At the National Palace in Mexico City, we had to cover the 200 x 200 feet "patio". The masts could not be higher than 30 feet. To get the right curvature, it was necessary to subdivide the membrane in 12 sections, the best shape for this project was a wave-shaped prestressed membrane. The membrane was stretched between 6 parallel rows of high points, or ridge cables, alternating with 6 parallel rows of low points or valley cables.
The membrane has a saddle-shaped curvature and is the connection between the high and low cables.

The design process in our laboratory is a combination of architectural, engineering and mathematical systems. We begin with a textile architectural model. This architectural shape is corrected with a soap-film model. We obtain a similar configuration with other woven fabric model. This is a prestressed measuring model used to obtain the cutting patterns and the data for the computer analysis.

With this nearly exact data, the processor time of the computer is very short and cheap, and we can check shape patterns and stresses of the membrane, cables and masts.

There are many different systems to design membrane structures, the one who uses textile models is the most convenient for architects, but it is expensive and takes more time.
The geometrical design is very practical for membrane structures of definite shape and may be developed with a small programmable calculator or micro computer.

The computer designs have reached a great sophistication The works of the research institutes in Stuttgart, The Laboratory for Computer Graphics and Spatial Analysis in Harvard, The Formex Algebra of Dr. Nooshin -a mathematical system that may be used as a tool for representing and processing configurations- are remarkable, see Ref 4.

We use the programs MASL and PAM-Lisa from Eberhard Haug, see Ref 5, because in these programs there exists a great interaction between the architectural and engineering design.
These programs are specially valuable in the finding of geodesic membrane cutting patterns using 100% engineering (and not mathematical) tools.

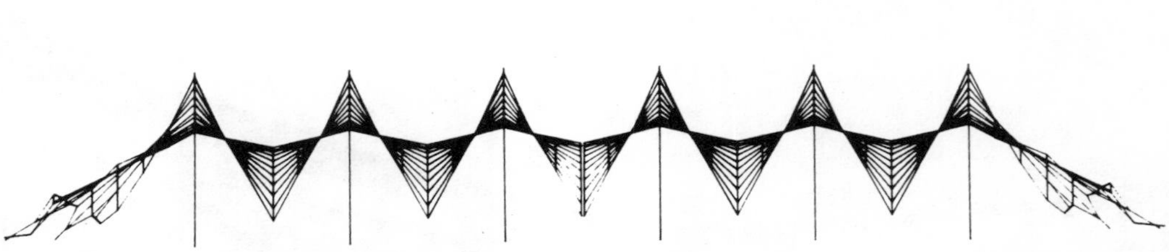

Fig 8   Computer design with the program MASL

Fig 9   Theatre at "Villa Universiada"

The manufacture of textile models is based on geometric measurements taken from minimal surfaces of soap film models, the data obtained from these measurements must be adapted to the textile model. Any mistake in the model measurement will be reflected in the construction 1:1, so it is convenient to check the cutting patterns in the computer analysis, see Figs 11 and 12.

Fig 11 Textile models

The measuring system for textile models must be very exact (about 1/10) milimeter). It is necessary to use a three dimention measuring table because of the anticlastic curvature of the membrane, see Fig 13.

Fig 13 3 D measuring table

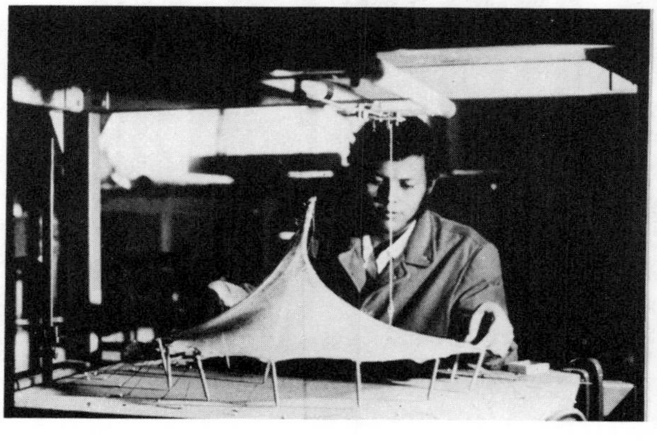

Fig 10   Stern-wave in Toluca, Mexico

Stern-wave shape for an exposition in Toluca, this radial wave-shape is a combination of 12 hyperbolic-paraboloids divided by ridges, with a central opening bordered by a steel cable. The shape is the most perfect structure in the equilibrium of counteracting tension. It is an almost exact minimum surface. Frei Otto built this shape for the first time in 1956 in Cologne, see Fig 10.

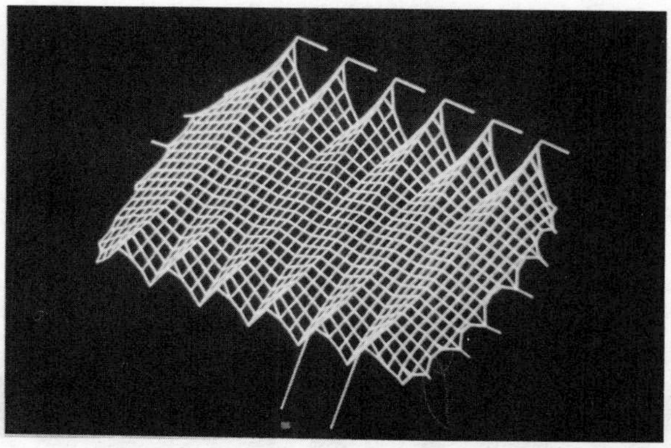

Fig 12 Computer design

Kindergarden in Iztacalco, Mexico.
This is a peaked-shape membrane with 1 central support point and 12 perimeter anchor points, length: 75 feet, width: 50 feet, mast height: 27 feet. Polyester membrane with steel cables at the edge, eye and ridge, see Fig 14.

Fig 14 Kindergarten in Iztacalco, Mexico.

Fig 15

For the study of the architectural shape of pneumatic structures, minimum surfaces can be approximated by rubber film, marked with a measuring grid, see Fig 16.

A systematic study of the pneumatic shape was carried on by the author for the First International Colloquium on Pneumatic Structures, on May 1967, under the direction of Frei Otto at the University of Stuttgart. These studies were continued in Mexico.

Models in plaster are very useful for the architectural design of pneumatic structures. The correct shape is the difference between a masterly piece of work and a poor ridiculous one. It is very easy to make models with a simple high pressure mechanism and rubber films.

REFERENCES

1. M SALVADORI, Structure in Architecture Prentice Hall, New York, 1966.
2. R L MEDLIN, Experimental Tension Structures, Washington University, St. Louis Missouri, 1970.
3. F OTTO, Zugbeanspruchte Konstruktionen Band 2 Ullstein Fachverlag, Berlin, BRD, 1966.
4. J SANCHEZ ALVAREZ, Formex Formulation of Structural Configurations, University of Surrey.
5. E HAUG, Analitical Shape Finding of Cable Nets, IASS Pacific Symposium, Tokio, October 1971.

Fig 16

Fig 17

THE DEVELOPMENT OF MEMBRANE STRUCTURES IN AUSTRALIA 1979 TO 1984

BY VINZENZ SEDLAK

Dipl.Ing.(Arch)(T.U.Graz), M.Phil.(Surrey)

Director, Lightweight Structures Research Unit,
University of New South Wales, Sydney, Australia

The historical background to the rapid development in Australia of membrane structures in recent years is given highlighting the role of University based research and development and its involvement with an emerging membrane structures industry as well as the role of the Membrane Structures Association of Australasia (founded 1981).
A selective review of recently constructed membrane structures forms the main part of the paper,with particular reference given to the Mobile Stage ,a large demountable canopy structure designed by the author which is now used for open-air concerts in Sydney.
An overview of the particular characteristics of the Australian market situation and their influences on the development of membrane structures and an outlook on major projects under construction concludes the paper.

Historical Development

The development of the field in Australia began as early as 1958 when Bert Bilsborough,then an employee of S.Walder Pty.Ltd.,-Sydney,then one of the largest and oldest tent and canvas goods fabricators in the country,constructed the first "Balloon House",a 12x6m,3m high air-supported structure (see Fig.1).
He utilised a newly developed single-face PVC coated Polyamide fabric by ICI Chemicals Australia.
Several larger standard air-structures followed between 1962 and 1977.
The decisive breakthrough from these early developments,however,did not occur until recently ,because of lack of acceptance of the,then,revolutionary building technique by the traditionally conservative Australian professions,the fabrication and material producing industry and subsequently the public at large.
Development in these early stages rested with a handful of enterprising small fabricators in the states of New South Wales,Queensland and Victoria and as their capacities for experimentation was limited by the market situation,membrane structures remained stagnant.

FIG 1 THE FIRST AUSTRALIAN AIR-HOUSE,1958

Since 1976 the situation began to change when a small group of professional educators joined by progressive fabricators developed a keen,renewed interest sparked by rapid progress in the field in Europe,USA and Japan during the late 60's and early 70's.Several smaller structures were constructed and the necessary technology for design,fabrication and construction was developed locally by experience and by overseas example and placed on a more professional footing.
During the 70's coated synthetic fabrics made substantial inroads into a traditionally canvas oriented market,supported by a strong group of canvas producers,and more fabricators commenced working with these materials.
Research and development work originated at Universities,in particular from the University of New South Wales,where Dr.Peter Kneen developed techniques for computer-aided shapefinding,-analysis and patterngeneration at the Department of Structural Engineering and the author conducted state-of-the-art reviews,-development of model-based formfinding techniques,testing of materials,project application studies and prototype construction at the Lightweight Structures Research Unit in the Faculty of Architecture.
The knowledge obtained was applied and checked in a range of practical applications providing continuing guidance and information to the developing local industry.
Against this background LSRU organised the first seminar on membrane structures in September 1981 which provided a first meeting place and forum for fabricators,material suppliers,-architects and engineers in government and professional practice as well as researchers and educators to present and to discuss developments and experiences.
After this event the "Membrane Structures Association of Australasia(MSAA)" was founded,thereby formalising and co-ordinating the interdisciplinary interests in the field.Since then the Association organises annual conventions and seminars,publicises the proper application of membrane structures and contributes to the establishment of design-guidelines through its working committees.

Recent projects

In the following several recent projects have been selected from the typical range of applications for membrane structures in Australia todate .A number of major projects presently under construction could not be presented but are listed.Main applications sofar can be found in Leisure-and Recreational Facilities,Sports-, Commercial-and Agricultural Facilities.

MELBOURNE UNIVERSITY MULTI PURPOSE SPORTS HALL

The single membrane, air supported structure on top of an existing sports administration building provides an enclosed indoor sports hall to supplement existing facilities (Figs.2&3). When compared with traditional building construction, the membrane structure provided substantial cost saving. The membrane covers an area of 730m2, is supported by a 0.25kPa (kN/m2) internal overpressure creating a prestress of 2.9kN/m and was designed for a wind velocity of 39m/sec.
The design called for penetration at different levels to provide access and links to adjacent building requiring careful sealing to maintain airpressure.
The mechanical airsupport system with back-up provides automatic control of fresh air, recirculated air and exhaust air and these controls are linked to the space heating system.
Anchorage is by high strength masonry anchors to the existing 280mm concrete roof/floor slab.
Completed in late 1980 the structure is an outstanding example of a membrane structure providing an economical, clear span vertical extension to an existing building.

Client:	University of Melbourne
Designer/Contractor:	Geodome Space Frames with Peter Kneen
Engineer:	B.J. O'Neil & Assoc.
Fabricator:	Geodome Space Frames Pty.Ltd.
Mechanical Engineers:	A.E. Smith & Son
Material Supplier:	Hoechst Australia Ltd.(Hammersteiner)

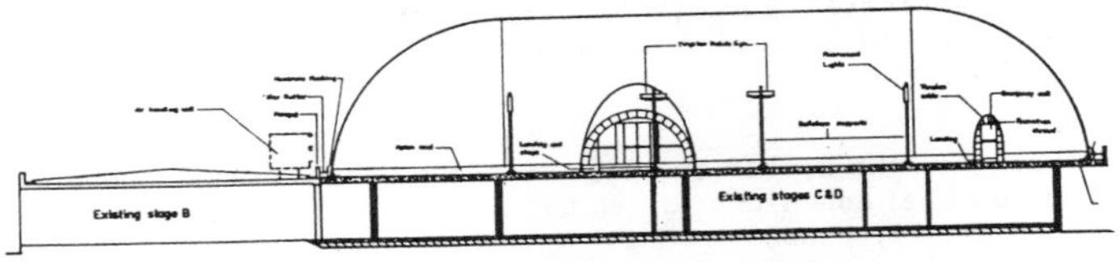

FIG 2 MELBOURNE UNIVERSITY SPORTS HALL,1980 (SECTION)

FIG 3 MELBOURNE UNIVERSITY SPORTS HALL,1980

TEMPORARY CANOPY FOR OUTDOOR EVENTS

The aim was to create a floating canopy for weather protection for open air theatre and concert performances for both,performers and audience while minimising any visual destruction and covering an area of approx. 500m2.
The structure was to be temporary and transportable to be used for a range of open air events at different locations. It was first erected for the Sydney Festival 1982 in December 1981 (Fig.4).
The membrane shape was first envisaged as an anticlastic, cable bordered ondulating surface with four perimeter high and low points at the cable inter sections.
The material used was white, plain weave PVC coated polyester 750g/m2 overall weight ("Airflex 750"by Nylex Corp.). Geometrically the surface is composed of eight identical individual hyperbolic paraboloidal panels over octagonal plan (four high and four low point corners) which allowed for efficient patterning.
The high points are supported by four 11m aluminium masts each of which is stabilised by a central spreader/cable stay system in order to reduce buckling length while decreasing material weight and cost. The low points are anchored directly to the ground using cables and dead man anchors. Clear spans approximately 27m.
A similar structure was built to shelter a resort home unit display on the Queensland Gold Coast.

Client:	Beatty and Beckett
Designer/Contractor:	Geodome Space Frames Pty. Ltd.
Engineer:	Peter Kneen Pty. Ltd.
Fabricator:	Covertex Pty. Ltd.
Material Supplier:	Nylex Corporation

FIG 4 TEMPORARY CANOPY FOR OUTDOOR EVENTS,1981

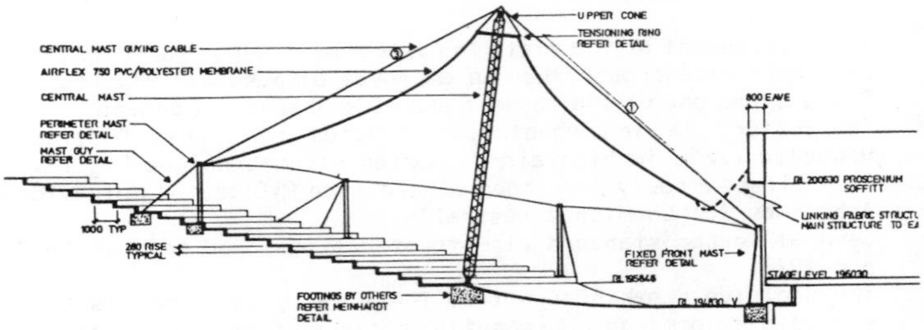

FIG 5 WOOLSHED AMPITHEATRE (SECTION),1983

WOOLSHED AMPHITHEATRE, EUROA, VICTORIA

The structure covers an outdoor seating area of 800m2 adjacent to a historic sheepshearing station in the Victorian countryside which is now used as a tourist attraction and where demonstrations and other public performances are conducted.
Due to the structurally suspect nature of the old building, the membrane roof, which is supported by a central 14.5m high mast and anchored via seven edge cables and perimeter masts to ground anchors, hand to be kept structurally independant (Fig.5).
The anticlastically shaped, conical membrane surface is subdivided into seven triangular panels radiating from the central high point. Each triangular panel is made up from parallel strips of material. The material used is a white, plain weave, PVC coated polyester fabric of 750g/m2 overall weight ("Airflex 750"by Nylex Corp.)and was joined by HF welding.
Maximum extensions are 40m and 25m respectively, with the longest edgecable spanning 30m at the interface to the old building which houses the stage. Along this edge a perimeter membrane gutter connects to the woolshed and provides water shedding. A clamped site joint facilitated membrane transport and assembly.
While a visual integration of the membrane roof with the existing building had not been attempted, the structure is both functional and economical and was designed to a small budget. It was completed in 1983 and received a 1983 MSAA Design Award.

Client:	Seven Creeks Trust
Designer/Contractor:	Geodome Space Frames Pty.Ltd.
Architect:	Roy Grounds
Fabricator:	Covertex Pty. Ltd.
Material Supplier:	Nylex Corporation

QUEEN STREET MALL SHADE UMBRELLAS, BRISBANE

Two shade umbrellas, each covering a plan area of 325m2 provide a visual attraction in the popular shopping mall in Brisbane's City Centre.
One of the parameters leading to the design specification was that the shade structures had to be selfsupporting and attachments to adjacent buildings were not permitted. The supports, which would intrude into the pedestrian traffic area, were to be kept at a minimum.
The material was an imported light blue PVC coated polyester fabric of 950g/m2 overall weight (Shelterite 8128 by Seaman Corp., USA).
The inverted conical canopies assembled from radial membrane strips extend some 18 x 18m in plan and are supported by a central diameter 800mm steel tube cantilevering 15m high out of the ground (Figs.6 & 7).
The membrane surface is stressed between a central low point, where it is attached to a ring, and its four perimeter edge cables which are attached to four outrigger struts cantilevering from the central tube.
These outrigger struts make ingenous use of loadsharing between tension cable and compression strut elements to transfer high forces at their ends resulting from windloads (overall uplift resultant 100kN based on design speed 50m/sec) while minimising on material and overall costs. The structure received a Honorable Mention at the 1983 Design Awards of MSAA.

Client:	Brisbane City Council
Design:	Robin Gibson & Partners (Architect) and McWilliam & Partners (Consulting Engineers to the Client)
Engineer:	Peter Kneen Pty.Ltd.
Fabricator/Contractor:	Geo Pickers (Brisbane) Pty. Ltd.
Material Supplier:	Vessel Engineering Pty.Ltd. (Seaman Corp.)

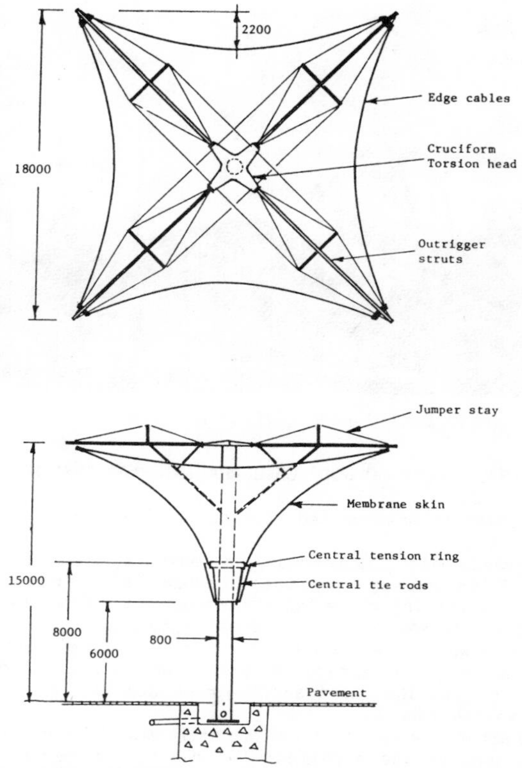

FIG 6 SHADE UMBRELLAS, BRISBANE, 1983
(PLAN AND ELEVATION LAYOUT)

PORTER BAY AQUATIC CENTRE, SOUTH AUSTRALIA

A permanent stressed PVC coated polyester fabric membrane roof of conical shape over square plan area 34 x 34m was erected in April 1984.
Supported from a centre mast and clamping ring the surface is bordered by 16 edge cables and tensioned to 16 cantilevered perimeter steel columns (Fig.8).
The 1.6 Mio leisure centre will, when completed in the latter part of 1984, house swimming pools, a gymnasium, spa and sauna and a coffee lounge and creche.
Being situated at the oceanfront the structure is exposed to onshore winds. Design wind speed was 45m/sec.

Client:	City of Port Lincoln
Architect:	Hannaford & Partners Pty. Ltd.
Engineer:	McWilliams & Partners
Fabricator/Contractor:	Geo Pickers (Brisbane) Pty. Ltd.
Material Supplier:	Hoechst Australia Ltd. (Hammersteiner Kunststoff)

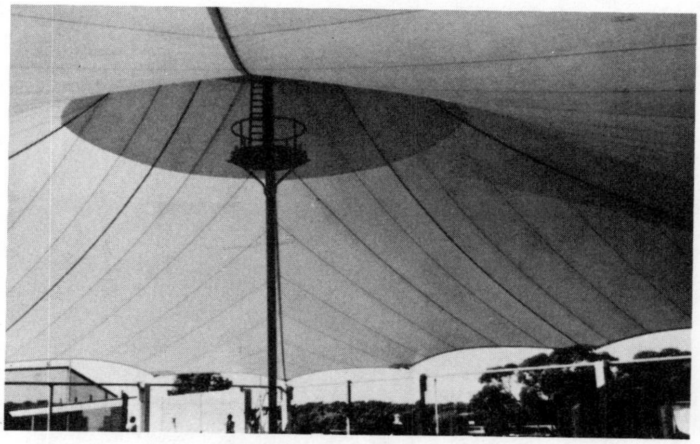

FIG 8 AQUATIC CENTRE, PORTER BAY, 1984

FIG 7 SHADE UMBRELLAS, BRISBANE, 1983

FIG 10 BANDSTAND, LANE COVE, SYDNEY, 1983

CENTRAL COURTYARD ROOF, CANBERRA INTERNATIONAL MOTOR INN, ACT.

A permanent prestressed roof canopy over an atrium courtyard (Fig.9).
The membrane is supported by a centrally located lattice mast which is stayed by 4 guy ropes. Secondary support is provided by a metal bale-ring at the top of the mast. The surface is prestressed to a saddle-shape. The beam boundary, formed by a galvanised steel tube to which the membrane is attached via a sleeve, is anchored at 1m centres to attachment points on the roof of the building surrounding the courtyard. The membrane is a white PVC coated polyester fabric.
The structure was the first one of a number of similar structures designed by the architect for the enclosure of atriums and constructed.

Client:	Canberra International Motor Inn
Architect/Designer:	Bryan Dowling
Fabricator:	KIB Konstruktion und Ingenieurbau GmbH

LLOYD REES BANDSTAND, LANE COVE PLAZA, SYDNEY

In 1982 the well known Australian artist and painter Lloyd Rees commissioned a new bandstand in consultation with Lane Cove Council which was to provide a focal point of the local shopping plaza and an feature for community entertainment and events. The landscape architect Harry Howard carried out the conceptual design. A stressed, domically shaped membrane over an existing elevated outdoor stage platform was designed, which was to be suspended from a domical framework of four intersecting stainless steel tubular arches, each one anchored to opposite corners of the 45m2 octagonal plan stage (see Fig.10).
The membrane surface is anticlastically shaped and suspended from the archframe at 17 points and tensioned via 8 catenary edge cables to the 8 arch springings. The cutting pattern is developed on the basis of a 16 fold radial symmetry and requires only three different shaped panels (Fig.11).
The membrane material is a white, PVC coated Polyester fabric, 1/1 plain weave, 750 g/m2 coated weight, UTS 53/44 KN/m, flameretardent to AS 1530 and stabilised for high UV exposure ("Airflex 750"). It was joined by 20mm seams, sewn and high frequency welded. Design wind loads were uplift of 1kPa resulting in a maximum membrane stress of 6.7 KN/m (Safety factor 6.6), design prestress was 4.3 KN/m.
Stressing is effected by adjustible galvanised chain at all supension points and s/s turnbuckles at anchorages. Point supports are provided by spherical galvanised steel caps and galvanised steel clamping plates fastened by a single ringbolt. The structure was erected in February 1983 and was awarded a 1983 Design Award of the Membrane Structures Association of Australasia.

Client:	Lane Cove Municipial Council
Architect:	Harry Howard & Associates
Designer/Engineer:	Surface & Spatial Structures (Vinzenz Sedlak Pty.Ltd.)
Engineer:	Miller, Milston & Ferris (archframe only)
Fabricator/Contractor:	B.W.Bilsborough & Sons Pty.Ltd.
Material Supplier:	Nylex Corp.

FIG 9 ATRIUM-COUTYARD ROOF, CANBERRA, 1981

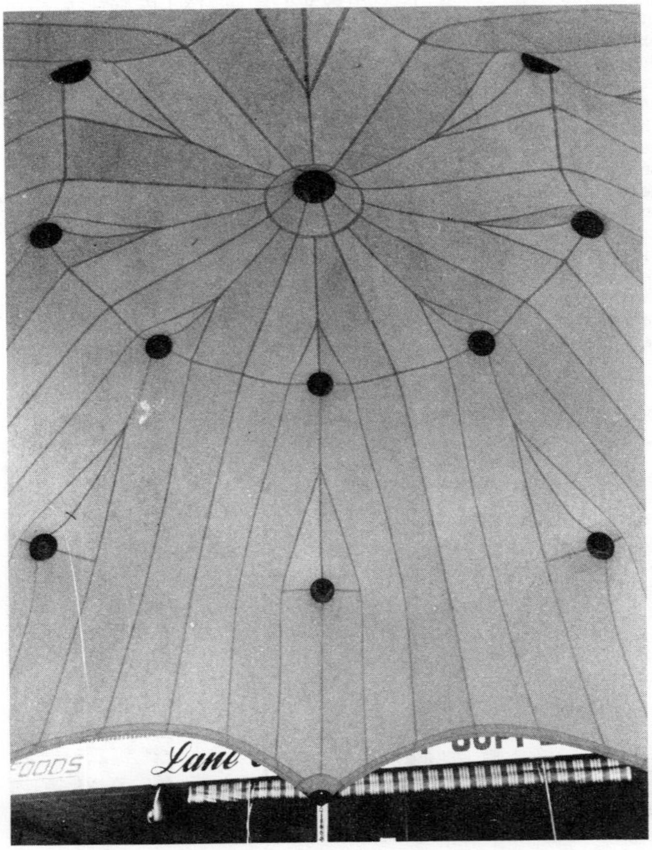

FIG 11 BANDSTAND CUTTING PATTERN

MOBILE STAGE FOR OPEN-AIR PERFORMANCES,SYDNEY,NSW

Since January 1984,the public of New South Wales has been provided with a new demountable stage covered by a large,sand coloured prestressed membrane.The building is a further addition to facilities provided for cultural activities and forms part of an ambitious cultural expansion programme currently pursued by the NSW Goverment under Premier Neville Wran.
The new "Mobile Stage" was first erected in the historic Domain Park,close to the City Centre,for the now well established and popular "Opera in the Park" performance featuring Dame Joan Sutherland,the Australian opera singer,and forming one of the main events of the Festival of Sydney (see Figs.12&13).
Held during the Australian summer(holiday)month the festival,-first introduced in 1978,encompasses a variety of cultural events and social entertainment for a range of age-groups and attracts each year large numbers of visitors and resident Sydneysiders alike.

Design

The design/construct tender called for a membrane roof canopy which would give protection to performers and equipment underneath against sun and driving rain while completely covering stage and overhead lighting structure.
The audience's view of the stage should not be obstructed and the building should have an attractive appearance.
Structurally,it was to be designed for high windloads ,while being readily demountable .Its components were to be readily stacked,transported and erected.
The Client plans to establish several sites in and around Sydney,some exposed to onshore winds,where the structure will be used for outdoor events such as concerts,theatre,ballet etc.
It was further specified,that the overall scheme was to be composed of 3 separate components :
a Stage (18.5x12m with 2.5m side-extensions),an Overhead Lighting Structure (18.7x12,7m and 11m high,above the stage),and the Membrane Roofcanopy.
The components were to be designed as individual units in order to allow for anticipated flexibility in the choice of components for different events.
The design of the canopy was based on the above brief in the following order of priority :

1.Cover the Overhead Structure as efficiently as possible integrating all user requirements with regard to protection and sightlines.
2.Design a structurally stable membrane surface in order to cater for high windloading involving adequate curvatures while minimising on the number of supports and anchorages in order to economise on installation time.
3.Design the membrane shape to be visually attractive and aesthetically pleasing resulting in a building which ,through its unique appearance would enhance its environmental setting and provide the organisers and sponsors of events with an attractive feature to maximise on spectator participation.
4.While observing all the above ,keep the structure as light as possible and its components simple in order to facilitate transport,storage and erection.

FIG 13 OPERA IN THE PARK PERFORMANCE,JAN 1984

FIG 12 MOBILE STAGE IN THE DOMAIN,SYDNEY,1984

Soon after design commenced it became clear that some of the tender requirements were contradictary and compromises had to be made in order to obtain a product that would satisfy the requirements as closely as possible.
The contradictions were :
Lightness against Size
Lightness against Structural Loading Requirements

Size resulted from the fact that the shape of the canopy had to be designed to accommodate the 11m high steel frame of the overhead lighting structure while allowing for adequate clearances in anticipation of large deflections of the membrane under high windloads (Integration of canopy and lighting structure by suspending a rigid steel lighting grid from four,internal masts was not permitted).
Structural loading requirements which specified a 156 km/h design windspeed with a 50 year return period (Australian Standard 1170,Part 2:Terrain Category 2,v=44m/s) were high given the anticipated temporary nature of the structure and resulted in weight penalties for masts,cables,anchorages and fittings thereby making speedy handling and erection more difficult to achieve.

The design developed from these parameters was an anticlastically shaped membrane surface,supported by an internal,diagonal cablessystem,composed of two intersecting maincables -which,in turn,are supported by four internal masts and a central "flying strut" at the intersection of these cables- connecting the four masttops and four maststaying cables,connecting the masttops to four perimeter groundanchorages.
The membrane is bordered by catenary edge cables and is tensioned to the four masttops utilising rosettes and,at the corner-intersections of edge cables,to nine groundanchorages at the perimeter (see Figs.14,15and16).Resulting overall dimensions were 35.9x26.1m,13.7m internal height at the centre,19.5m front mast height and 16.9m rear mast height.

FIG 14 FRONTAL VIEW OF MOBILE STAGE

FIG 18 INITIAL DESIGN UTILISING SUSPENDED POINTS FROM EXTERNAL MAST/CABLE SYSTEM

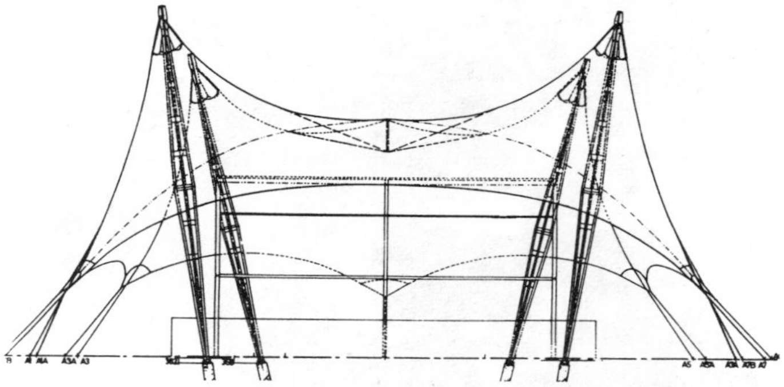

FIG 15 FRONT ELEVATION SHOWING OVERHEAD LIGHTING STRUCTURE

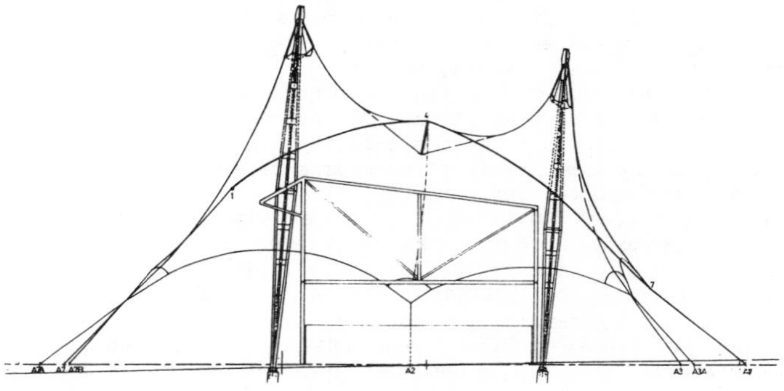

FIG 17 SECTION AND SIDE ELEVATION

FIG 16 INTERIOR,SHOWING CABLE SYSTEM AND CUTTING PATTERN

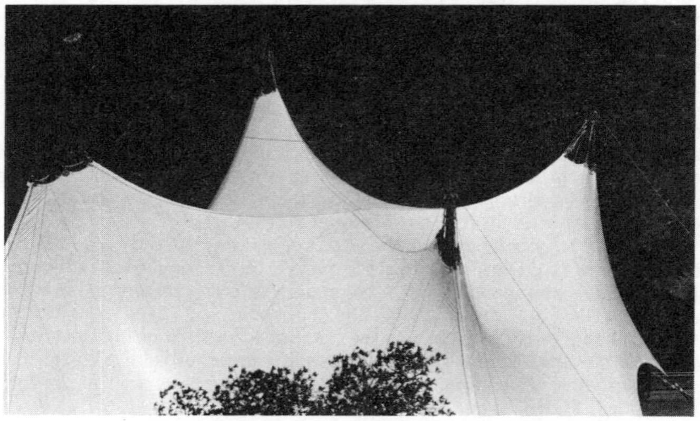

FIG 19 TOPVIEW OF MOBILE STAGE SHOWING CURVATURES AND ROSETTE SUPPORT

Initially,an alternative design utilising an external cable and mast support system to suspend a membrane surface at four high points (see Fig.18) was investigated,point support from internal masts was also considered.
The internal cable support system was chosen because it enabled tighter control of deflections within the surface and flatter entrance curvatures at the high points (see Fig.19). It was therefore possible to minimise the overall surface area required for stage coverage to 815 m2 by keeping the membrane surface closer to the overhead lighting frame and to reduce overall mast heights as well as overall ground coverage.Due to maststays being integrated in the cable support system for the membrane with the membrane surface itself providing lateral stability a reduction in the number of perimeter anchorages was achieved. Futhermore,by tilting the entire membrane surface back-wards,-the necessary central height for the large ,42m long egde cable over the front of the stage and unobstructed sightlines could be achieved,while utilising the backward slope of the roofplane of the lighting frame to reduce the overall height at the rear of the structure (see Fig.17).

During the various design stages the membrane roof was modelled from 1:200,1:100 up to 1:25 until a satisfactory balance between functional,aesthetic and structural requirements was achieved. Windtunneltests were carried out on a 1:100 model by John Howell at Vipac&Partners Laboratory ,Melbourne and windpressures were established for internal and external surfaces.Turbulence scale was matched to model scale and surface roughening was applied to achieve Reynolds number scaling.The average maximum free stream dynamic pressure (approx.3seconds gust) of 1.2KPa was used with the pressure co-efficients,corresponding to code requirements. Due to the steepness of the sides of the membrane surface some areas on the model were measured to have pressure co-efficients greater than 2.0,and large areas had nett pressure co-efficients in the order of 1.5.
The shape of the membrane surface was optimised on the 1:100 design model (Fig.20) after preliminary stress calculations based on experimental windloads were carried out and the geometry of the surface ,initially obtained by measurement from the 1:100 design model by co-ordinate mesurement of salient points

along cable catenaries and principal curvatures,was checked against a minimum energy soap-film shape by computer .This shape generation process by computer was carried out with LISA-a programme package developed by Dr.Haug-on one half of the surface (axial symmetry) by isolating the half surface into three individual anticlastic surfaces bordered by common (internal)support cables and edge cables which were measured from the model and assumed to be rigid boundaries for the purpose of obtaining the soapfilm shape (see Fig.21).

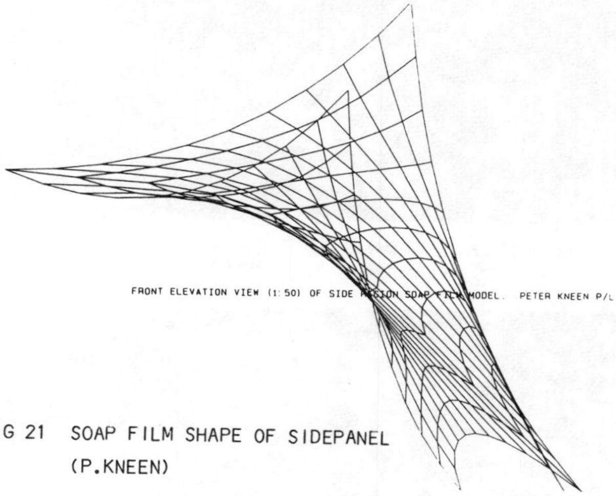

FIG 21 SOAP FILM SHAPE OF SIDEPANEL (P.KNEEN)

The resulting shapes where then used to obtain developed cutting patterns employing a program developed by Dr.Peter Kneen at the University of NSW.The patterns,which had been compensated for expected fabric elongation obtained from information on biaxial stress/strain behaviour of the chosen material,were then checked for fit by constructing a 1:25 scale model of one half of the structure (Fig.22).

Structural Design

The structure is designed to redistribute windforces by moving into a position of equilibrium between edge cable tensions,the main diagonal cable tensions and mast compressions.

Essentially,any high membrane stresses tend to be shared throughout the structure by large deformations resulting in load redistribution.Any increase in edge cable forces and hence anchorage loads is balanced by a combination of wind uplift and mast compression.

The main generators of the relatively high mast compressions is the lateral load on the steep corner areas of the membrane directly loading the main mast cables.The estimated maximum membrane stress of 12.5 KN/m under maximum windload leaves a factor of safety of 6 on ultimate fabric strenght (approx.80 KN/m).Initial average membrane prestress was calculated to be 1,5 KN/m.

Membrane

The material chosen was a PVC coated polyesterfabric,2/2 (panama)weave construction,900g/m2 overall weight,colour sand,flameretardent to AS1530 and stabilised for high UV exposure.Initially the fabric was to be supplied by a local coater,unfortunaltely the base cloth could not be woven in time and the material had to be imported from overseas.The final material was "Duraskin B127235"by Verseidag Coated Fabrics of W.Germany.

The material is joined by double row stitching with carbon black thread and overwelding (high frequency)with a protective,clear PVC strip (40mm seam).At stress points the membrane is reinforced.Tangential stressing reinforcement along edge cable pockets is provided by 75/5mm Polyester webbing sewn onto the fabric.

Membraneconnections and Fittings

In order to permit ease of handling during fabrication and erection/dismantling the membrane surface (total surface area 1340m2) was subdivided into four individual panels which were joined on site by aluminium plate clamping joints located alongside the ridges formed by maststay and main cables.All support cables are carried in sleeves to restrict movement.

At the rosette points the membrane edges are bordered by cables which are clamped to four galvanised steel clamping plates fastening the membrane corners and providing attachment for four

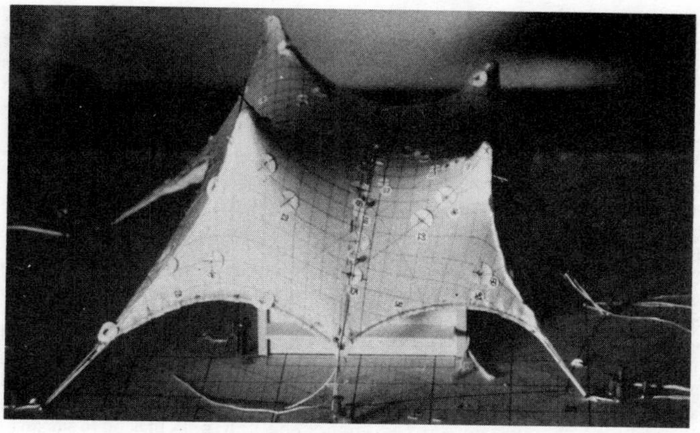

FIG 20 STRETCH FABRIC DESIGN MODEL 1:100

FIG 22 CUTTING PATTERN CHECK MODEL 1:25

At the rosette points the membrane edges are bordered by cables which are clamped to four galvanised steel clamping plates fastening the membrane corners and providing attachment for four support cables to the masttop.

At the intersections of edge cables at the membrane perimeter circular corner cut-outs teminate the membrane,while the edge cables are continued to common galvanised steel clamping plates which provide connection to the anchorage cables.Three types of corner plates occur.Tangential stressing cables connect the webbing reinforcement which terminates at the membrane edge to corner plates (Fig.23).

Cables

The cablesizes provided (maincables,supportcables for flying strut and smaller edgecables dia.26mm,maststay cables ,large front edgecable and all anchorage cables dia 36mm) allow for a factor of safety of 2.25 based on a UTS of 1770 MPa at maximum wind loading.In order to facilitate sleeving of edge cables with relative ease during erection and dismantling and to provide effective lenght adjustment with available fittings,a clamping joint was introduced in the vicinity of each corner plate.It utilises two galvanised steel "double base grips" fastened by eight high tensile bolts each and spaced apart to interconnect the edge cable to a short cable section attached to the corner fitting.This economical connection was developed after consideration of alternatives such as large turnbuckles and high strenght U bolts and was satisfactorily load tested to full design loads.

Masts

Design compression loads of 60 t for the front masts and 50 t for the rear masts were established including an allowance for possible future suspension of a lighting grid from the masts.The masts are galvanised mild steel built up struts from three circular hollow tube sections battened apart and spliced for transport and storage (Fig.24).Mastfootings are bored piers and the base plate has universal joint action to allow unrestricted mast movement during windloading allowing stress redistribution in cables and membrane.

Anchorage
Grouted high yield "Tempcore" reinforcing bars into bedrock were used for all ground anchors. The anchors are similar to large, 10m long splitpins with a 100 mm dia. loop above ground grouted into a 120mm dia. hole drilled at the angle of the anchorage cable.- Maximum design load was 230 KN.
Anchorage cables are connected to the anchors via twin galvanised steel anchorage bars which provide for incremental adjustment during stressing and allow for limited variations in groundlevels between different sites.

Erection
Once the anchor and mastpositions were confirmed and constructed, the masts were erected by crane and stabilised by temporary guying cables to temporary peg anchors. The membrane was unfolded, interconnected, cables were drawn through sleeves and all fittings were attached. Connected at the four high points to a pulley and lifting cable tirfor-hand winches were used to simultaneously lift the membrane into position. Stressing was introduced through perimeter anchorage cables with adjustments made at tangential stressing cables and rosette support cables.- The structure was dismantled after 2 months with geometry spot-checks made to be re-erected on the same site later in the year. At this occasion it is proposed to make final adjustments to the membrane shape to eliminate minor wrinkling and to obtain stress measurements on the anchorages.

Client:	New South Wales Premiers Department,- Division of Cultural Affairs (project administered by the Public Works Department, Special Projects Section)
Designer:	Surface & Spatial Structures (Vinzenz Sedlak Pty.Ltd.)
Engineer:	George Clark & Associates
Specialist Consultants:	Peter Kneen Pty.Ltd. (Cutting Pattern) Unisearch Ltd. (Model construction and testing) Vipac & Partners (Windtunnel tests) Stromeyer Ingenieurbau GmbH (Construction, erection and fabrication advisor)
Fabricator/Contractor:	B.W. Bilsborough & Sons Pty. Ltd.
Material Supplier:	Rheem Ltd., Woven Products Group
Cable and Fittings:	Bullivants Pty.Ltd. (NSW)
Steel work:	Simondsen & Edwards Pty. Ltd.
Main Contractor:	Brakell Products Pty.Ltd.

FIG 23 CORNERDETAIL SHOWING EDGE CABLE/MEMBRANE, TANGENTIAL STRESSING ARRANGEMENT AND CORNERPLATE

FIG 24 MASTS SHOWING MASTHEAD AND BASE DETAIL

Market Development

In comparison with established markets in Europe, Northamerica and Japan, the Australian market is by necessity small. However,- considering the relatively small population of 14 Mio with a density of only 5 inhabitants per square mile as against an average of 50 inhabitants per square mile in USA and 500 in Britain and Germany, and the present lack of export potential in the neighbouring region, market growth over the past five years has been significant.
Total estimated project values for 1984 are forecast in the order of A$ 2.8 Mio as against an estimated A$ 500,000 in 1979. Much of the upswing experienced is due to increased confidence by architects and engineers that membrane structures have finally emerged as true reputable alternatives to more conventional construction techniques because of their economy considering life cycle costs, appearance, the development of more durable and longer lasting fabric materials, such as PTFE coated fibreglass (the first structure utilising PTFE/Glass fabric was St. Anne's Church, Seaforth, Victoria, completed in 1982) and subsequent improvements to the established PVC coated polyester fabrics and last, but not least improved and professional design and construction techniques.
After the economic recession 1979-82 when major construction activity was severely curtailed, a number of major development projects in the sports-and leisure- and in commercial areas were commenced in 1983/84:
Yulara Tourist Resort, a completely new, self-contained tourist village in the Australian desert region employing membrane shade structures totalling approximately 14000 m2 covered plan area (Material PVC/PE)
NSW State Sports Centre, partial roof cover, 1050 m2 (Material PTFE/Glass)
Two Shopping Centres in Melbourne and Perth totalling 3590 m2 (PTFE/Glass)
A Casino at the Queensland Goldcoast, 1100 m2 (PTFE/Glass)
A number of medium size structures in PVC/PE have been completed:
Mobile Stage (815m2), NSW (see above)
Spraycover Roof at an Electric Powerstation, NSW (approx. 900m2)
Aquatic Centre, Port Lincoln, SA (approx. 800m2) (see above)
Agricultural Shade-Net Structures, QLD and VIC (approx. 9000m2)

At present PTFE/Glass structures are not being fabricated in Australia and must be imported from overseas. Local Fabricators are typically versatile, small businesses set up for welding of Vinyl coated materials, who depend on the full range of fabrication from synthetic to canvas material for domestic awnings,- truck tarpaulins etc. to membrane structures for their livelihood. This situation combined with present lack of investment enterprise after the recession makes setting-up of costly, specialised equipment and factory space required for large scale fabrication difficult at present.
As the market develops further this situation will no doubt change.

ACKNOWLEDGEMENTS
George Clark, Bernie Davies, Geodome Spaceframes Pty.Ltd. and Peter Kneen for making notes and illustrations available.

REFERENCES
E.Picker, V.Sedlak LS2 "Membrane Structures in Australia" Lightweight Structures in Architecture Series, University of New South Wales, 1982
V.Sedlak (Editor) LS3 "Membrane Structures" Lightweight Structures in Architecture Series, University of New South Wales, 1983
and Submissions for the 1983 Design Awards, Membrane Structures Association of Australasia, May 1983

Raumfachwerke als Mittel der individuellen Baugestaltung

Dr.-Ing., Dr.-Ing. E.h. Max MENGERINGHAUSEN, Würzburg

Dem Bau ebener Fachwerke der mittelalterlichen Bauten folgten die "unregelmäßigen Raumfachwerke" als erste Generation. Der Verfasser leitete mit den von ihm 1940 formulierten "Baugesetzen der regelmäßigen Raumfachwerke" eine zweite Generation von Raumfachwerken mit einheitlichen Stablängen ein, die den "Stil der Cubus-Segmente" begründeten. Mit der "Theorie der abgeleiteten Raumfachwerke" wurde eine dritte Generation von Raumfachwerken begründet, die durch Koordinaten-Transformation aus den regelmäßigen Raumfachwerken entstehen und beliebige Bauformen auch mit gekrümmten oder gebrochenen Umrißlinien ermöglichen. Mit einer Lehre für die "Komposition im Raum" wird die individuelle Baugestaltung mit Raumfachwerken aus industriell hergestellten Serienelementen als Baukunst der Zukunft vorgestellt.

EINLEITUNG

Der Verfasser hat im Anschluß an seine zahlreichen vorangegangenen Veröffentlichungen in seinem 1983 veröffentlichten Buch "Komposition im Raum" (Ref. 5) eine Theorie für eine "Baukunst der Zukunft" aufgestellt. Der nachfolgende Beitrag behandelt im Anschluß an einige Grundgedanken des Buches und die Veröffentlichung des Verfassers "Kompositionslehre räumlicher Stab-Fachwerke" anläßlich des ersten Kongresses über "Space structures" (Ref. 3) die Theorie der individuellen Baugestaltung mit Raumfachwerken aus industriell hergestellten Serienelementen.

WARUM INDIVIDUELLE BAUGESTALTUNG ?

Unter den drei Grundbedürfnissen des Menschen : "Nahrung - Kleidung - Wohnung" ist allein das Grundbedürfnis der Nahrung absolut unverzichtbar; hingegen sind die Bedürfnisse der Kleidung und Wohnung (oder Behausung) relativ verzichtbar. Jedoch ist für den Kulturmenschen der Gegenwart das Behausungs-Bedürfnis auch zu einer Lebens-Notwendigkeit geworden.

Die wachsende Zahl der Menschen hat die Behausung sogar zu einem Massenbedürfnis gemacht. Im Zeitalter der Industrie-Gesellschaften hat sich das Prinzip der Typisierung und Normung zu einem herrschenden Prinzip der Wirtschaft entwickelt. Dieses Prinzip hat daher auch im Hausbau Eingang gefunden. Hinzu kommt, daß jeder Baustoff von den Materialeigenschaften her eine "typische" Erscheinungsform für die Anwendung des Stoffes begünstigt und daß daher von den Baustoffen eine Tendenz zur Vereinheitlichung der Erscheinungsform ausgeht.

Es ist festzustellen, daß in der gesamten Natur jedes Lebewesen eine "typische" Erscheinungsform verwirklicht, zugleich aber eine individuelle, d.h. "persönliche" Note aufweist, die jedes einzelne Lebewesen von anderen Lebewesen sogar der gleichen Art deutlich unterscheidet. Alle Blätter einer Pflanze besitzen zwar ähnliche Gestalt, sind aber nicht deckungsgleich. So ähneln sich die Menschen jeder Bevölkerungsgruppe, sind aber nicht vollkommen identisch.

Jeder Mensch handelt daher auch in seiner Lebensführung einerseits "typisch" und zugleich "individuell". Und so ist er bestrebt, jedem einzelnen Bauwerk eine individuelle, d.h. auf die Umgebung abgestimmte Gestalt zu geben.

Dieses Bestreben nach Individualität steht im Zeitalter der Massen im Konflikt mit der Tendenz der Normung, die der Mensch in Fortsetzung der Entwicklungen der außermenschlichen Natur zu solcher Vollkommenheit geführt hat, daß genormte menschliche Erzeugnisse untereinander beliebig austauschbar sind.

DIE LÖSUNG DES KONFLIKTS ZWISCHEN INDIVIDUALITÄT UND NORMUNG

Seit die moderne Industrie begonnen hat, die Bedürfnisse der Massen durch genormte, d.h. austauschbare Produkte zu befriedigen, ist immer wieder versucht worden, auch ganze Häuser zu normen, ja sogar : ganze Städte aus genormten Häusern in einer genormten Ordnung zusammenzusetzen. Obwohl diese Idee in den Waben der Bienen und Wespen sogar ein natürliches Vorbild besitzt, wird diese Idee aber - mit Recht - ! von der überwältigenden Masse der Menschen abgelehnt, weil jeder einzelne das Bedürfnis hat, sich eine individuelle persönliche Umwelt zu schaffen und weil auch Lebensgemeinschaften bei Bauwerken, die einer Gruppe dienen, in diesem Bauwerk einen individuellen Ausdruck verwirklichen wollen.

Für den hier bestehenden Konflikt zwischen Massen-Bedarf und Individualitätsstreben ergibt sich als Folgerung nur eine Lösung : vollständig genormtes Fertighaus - "nein"; Bauen mit Fertig-Bauelementen in variabler Form - "ja". Mit anderen Worten : der Baustil der Zukunft für die gesamte Menschheit ist gekennzeichnet durch die Aufgabe, individuell gestaltete Gebäude mit industriell hergestellten, austauschbaren Serienelementen zu errichten.

DIE UNREGELMÄSSIGEN RAUMFACHWERKE ALS VORLÄUFER

In der Entwicklung der Lebewesen haben die Vögel die Kunst entwickelt, aus einzelnen Stäben (Halmen, Zweigen, Ästen) unregelmäßige Raum-Strukturen so zusammenzusetzen, daß außerhalb des eigenen Körpers eine tragende Raumstruktur entsteht. Dabei bestimmt "die Kon-Struktion" auch die äußere Gestalt, wie die Gerippe-Strukturen in den Körpern der Tiere und Menschen ihrerseits die äußere Erscheinungsform bestimmen (Bild 1). Die Menschen früherer Entwicklungsstufen haben in Fortsetzung der Nestbau-Technik der Vögel zuerst unregelmäßig geformte Bauwerke aus Zweigen und Ästen errichtet und dann im Mittelalter eine hochentwickelte Kunst des Baues von Fachwerkhäusern geschaffen. Die Fachwerke des mittelalterlichen Hausbaues zeichnen sich dadurch aus, daß einzelne (Holz-) Stäbe in wiederkehrenden Längen in ebenen Fachwerken zu Mustern so zusammengesetzt sind, daß die einzelnen Stäbe sich gegenseitig in einer Ebene abstützen und ein "steifes" Gerippe entsteht, dessen "Skelett-Lücken" durch Lehm und andere Baustoffe, später durch Steine ausgefüllt wurden. Aus heutiger Sicht gesehen, sind diese im Mittelalter entwickelten Fachwerke als "halbregelmäßig" zu bezeichnen, da innerhalb des gleichen Bauwerks zwar bestimmte Längen wiederkehren, jedoch der Längenbemessung keine festen und einheitlichen Regeln zugrunde liegen.

Auch als der Stahlbau im 19. Jahrhundert begann, für Dächer und Brücken Fachwerke aus Stahl (zuerst gegossen, dann gewalzt) zu errichten, wurden die Längen der Fachwerkstäbe individuell bemessen. Zwar wurden bei ebenen Raumfachwerken schon parallele Stäbe bevorzugt (Ref. 1) und daher größere Stückzahlen gleicher Stablängen ermöglicht; jedoch herrschte bei den Raumfachwerken die individuelle Bemessung der Stablängen vor. Dies galt auch noch 1892, als August FÖPPL sein grundlegendes Werk "Das Fachwerk im Raume" (Ref. 2) veröffentlichte, nachdem durch den Einsturz der aus ebenen Fachwerken zusammengesetzten Brücke über die Birs (Schweiz) eine Theorie über räumliche Fachwerke notwendig geworden war. Mit dem Werk von FÖPPL begann aber eine erste Generation wirklich räumlich gestalteter Fachwerke.

DIE REGELMÄßIGEN RAUMFACHWERKE MIT INDUSTRIELL HERGESTELLTEN SERIENELEMENTEN (Raumfachwerke der zweiten Generation)

Aus dem Bestreben, den Bau räumlicher Fachwerke durch industriell hergestellte Serienelemente zu rationalisieren, schuf FÖPPL-Schüler Max MENGERINGHAUSEN 1940 eine "Theorie der regelmäßigen Raumfachwerke". Diese Theorie ermöglichte es, Raumfachwerke mit Norm-Knoten und Norm-Stäben einer einzigen Größe auszuführen. Hierbei werden regelmäßige Polyeder aus Cubus-Segmenten mit gleichlangen Kanten zusammengesetzt (Ref. 3). Es ist möglich, für derartige Raumfachwerke aus genormten Bauelementen einen "Katalog der Raumstrukturen" aufzustellen. Solche regelmäßigen Raumfachwerke bilden die "zweite Generation". Im Kriege erfolgte erstmalig im großen Umfang der Einsatz mit der MERO-Bauweise. Da diese regelmäßigen Raumfachwerke erstmalig in der Geschichte des Bauwesens Elemente mit einer dreidimensionalen Normung verwendeten, erkannte der Deutsche Stahlbauverband 1972 diesen Fortschritt durch die erstmalige Verleihung der "Auszeichnung des Deutschen Stahlbaues" an.

Da die "regelmäßigen Raumfachwerke" aus Cubus-Segmenten zusammengesetzt sind (Fig. 2),wurde damit die Basis geschaffen für den "Baustil der Cubus-Segmente". Fig. 3 dokumentiert diesen Baustil in repräsentativer Form. Er gestattet sogar die Annäherung an die Kugel-Gestalt (Fig. 4). Dem Baugestalter ist dadurch die Möglichkeit gegeben, nicht nur Bauwerke individuell zu gestalten, die deutlich vom Cubus abgeleitet sind (Fig. 8 und 9), sondern auch Baukörper, die sehr individuellen Charakter besitzen (Fig. 6 und 7).

DIE "ABGELEITETEN RAUMFACHWERKE" ALS DRITTE GENERATION RÄUMLICHER FACHWERKE

Die Architekten haben seit der internationalen Bauausstellung Berlin 1957 ("INTERBAU Berlin 57") weltweit in großem Umfang von dem "Baustil der Cubus-Segmente" mit regelmäßigen Raumfachwerken Gebrauch gemacht. Gleichwohl haben die Architekten immer wieder den Wunsch wiederholt, auch beliebige Baugestaltungen mit gekrümmten Flächen auszuführen, die nicht von den Cubus-Segmenten abgeleitet sind. Der Verfasser hat diese Forderung als "Herausforderung" aufgegriffen und in dem Buch "Raumfachwerke aus Stäben und Knoten" (Ref. 4) 1975 die Theorie entwickelt, aus den regelmäßigen Raumfachwerken durch Koordinaten-Transformation beliebig gestaltete Körper und Flächen abzuleiten. Fig. 10 zeigt in diesem Sinne den Übergang von einem Quader mit der Kantenlänge 1 im Grundriß und der Höhe $\sqrt{2}$ in einen Cubus mit der Kantenlänge 1 im Grundriß und der Höhe 1 und umgekehrt. Fig. 11 zeigt die Umwandlung eines Vierflächners mit einer willkürlichen Höhe H_1 in einen regelmäßigen Tetraeder mit gleichlangen Kanten "a" und der Höhe $H_2 = a \cdot \sqrt{\frac{2}{3}}$. Umgekehrt kann aus den regelmäßigen Tetraedern durch Veränderung der Höhe und der Kanten sowie der Winkel ein beliebiges unregelmäßiges Vierflach gebildet werden. Die Verwirklichung dieser Theorie für die "Ableitung beliebiger Raumfachwerke" aus regelmäßigen Raumfachwerken mit der Koordinaten-Transformation in die Praxis ist durch die "komplementäre Technik" der elektronischen Datenverarbeitung (Computertechnik) möglich geworden.

Ein Beispiel für die Umwandlung eines regelmäßigen Halboktaeder (Fig. 9) in einen "gedrückten Halboktaeder" zeigen die Fig. 12 und 13, wobei man die Winkel der Stabanschlüsse an den Knoten verändert. Auch große beliebig gestaltete Strukturen können so aus "deformierten" Halboktaedern und Tetraedern zusammengesetzt werden, wie das Beispiel der Fig. 14 zeigt. Schließlich können auch Rundbauten mit Rotationssymetrie nach Fig. 15 und 16 aus deformierten Cubus-Segmenten gestaltet werden.

Darüber hinaus ist es möglich, auch einlagige räumliche Strukturen mit Hilfe der Theorie der abgeleiteten Raumfachwerke auszubilden, statisch zu erfassen sowie konstruktiv zu verwirklichen. Fig. 17 zeigt hierfür ein Beispiel in der Form eines "Hypar", d.h. in Form eines hyperbolischen Paraboloids. Wesentlich ist in diesem Falle, daß die Stäbe biegesteif an die Knoten angeschlossen werden, was bei dem MERO-System durch eine Spezial-Ausbildung der Verschraubung gelingt.

KOMPOSITIONSLEHRE FUER DIE INDIVIDUELLE BAUGESTALTUNG MIT RAUMFACHWERKEN AUS INDUSTRIELL HERGESTELLTEN SERIENELEMENTEN

Bereits Fig. 17 zeigt in idealer Weise die Ableitung eines räumlich verformten Fachwerks aus einem ebenen quadratischen Raster. Die Bauplanung auf der Grundlage einer Modulordnung ist ganz universell für das Bauwesen der Zukunft die Voraussetzung rationeller Bauausführung. So ergeben sich für das Bauen mit Raumfachwerken aus industriell hergestellten Serien-Elementen folgende Regeln :

1. Jede Bauplanung sollte von einem quadratischen Raster mit einem "Grund-Modul" ausgehen.

2. Als Grund-Modul gilt international allgemein nach der Deutschen Norm DIN 18000 "Maßordnung im Hochbau" sowie den entsprechenden ISO-Vereinbarungen und anderen nationalen Normen das Grundmaß von 100 mm sowie das Vorzugsmaß von 300 mm mit den daraus abgeleiteten Vielfach-Werten 600 mm, 1200 mm usw.

3. Bei Bauten mit Raumfachwerken müssen auch alle Ausbauelemente auf die Modul-Ordnung abgestimmt werden.

4. Durch die Bauorganisation und insbesondere durch die Bau-Verträge mit den einzelnen Unternehmern müssen Bau-Toleranzen für die Passung zwischen allen verschiedenartigen Bauteilen verbindlich festgelegt werden.

5. Bei Raumfachwerken, die aus industriell hergestellten Serienelementen auf der Baustelle montiert werden, sind die heute für den Stahlbau allgemein üblichen oder zugelassenen Bautoleranzen (etwa nach der Norm DIN 18203) nicht ausreichend, da bei großen Maßabweichungen "Montage-Zwängungen" entstehen, die die Standsicherheit gefährden können. Erfahrungsgemäß dürfen Fertigteile für Raumfachwerke höchstens Maßabweichungen im Bereich von ± 0,1 bis 0,2 mm besitzen.

6. Zur Förderung rationeller Bauausführung dient es, wenn der Bauplaner den Baukörper aus "Raumfachwerken der zweiten Generation", d.h. aus "regelmäßigen Raumfachwerken" im "Stil der Cubus-Segmente" entwirft.

7. "Abgeleitete Raumfachwerke", d.h. Raumfachwerke der "dritten Generation", ermöglichen beliebige Baugestaltung, bei denen auch gekrümmte Bau-Begrenzungsflächen (z.B. durch ebene Teilflächen angenähert) ausgeführt werden können; jedoch ist zu beachten, daß "abgeleitete Raumfachwerke" in der Regel höhere Baukosten verursachen, als "Raumfachwerke der zweiten Generation im Stil der Cubus-Segmente".

8. Das Hilfsmittel für alle Realisierungen von Raumfachwerken ist die EDV-Technik (Computertechnik), beginnend mit CAD (Computer Aided Design) und endigend mit CAP (Computer Aided Production).

9. Der Rationalisierung dient es, wenn der Bauplaner bereits bei seinem Bauwerks-Entwurf von Anfang an sich der Hilfe eines mit der Technologie der Raumfachwerke vertrauten Spezialberaters bedient und mit seiner Hilfe sowohl die Raumfachwerke selbst wie auch die technischen Einzelheiten des Anschlusses von Ausbauelementen an die Struktur der Raumfachwerke aus Knoten und Stäben plant.

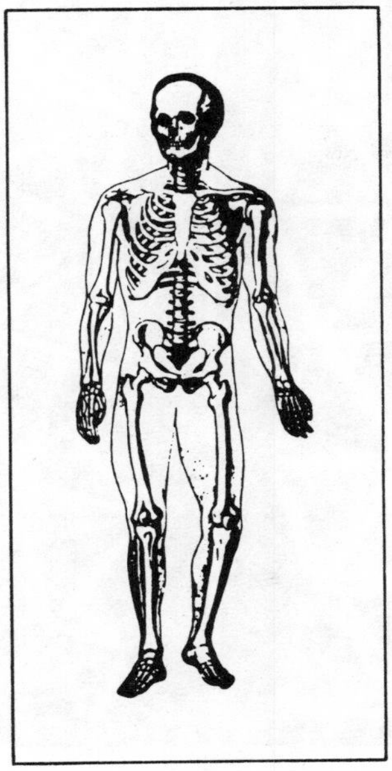

Fig. 1 Das Skelett bestimmt im wesentlichen die Gesamtgestalt des Menschen.

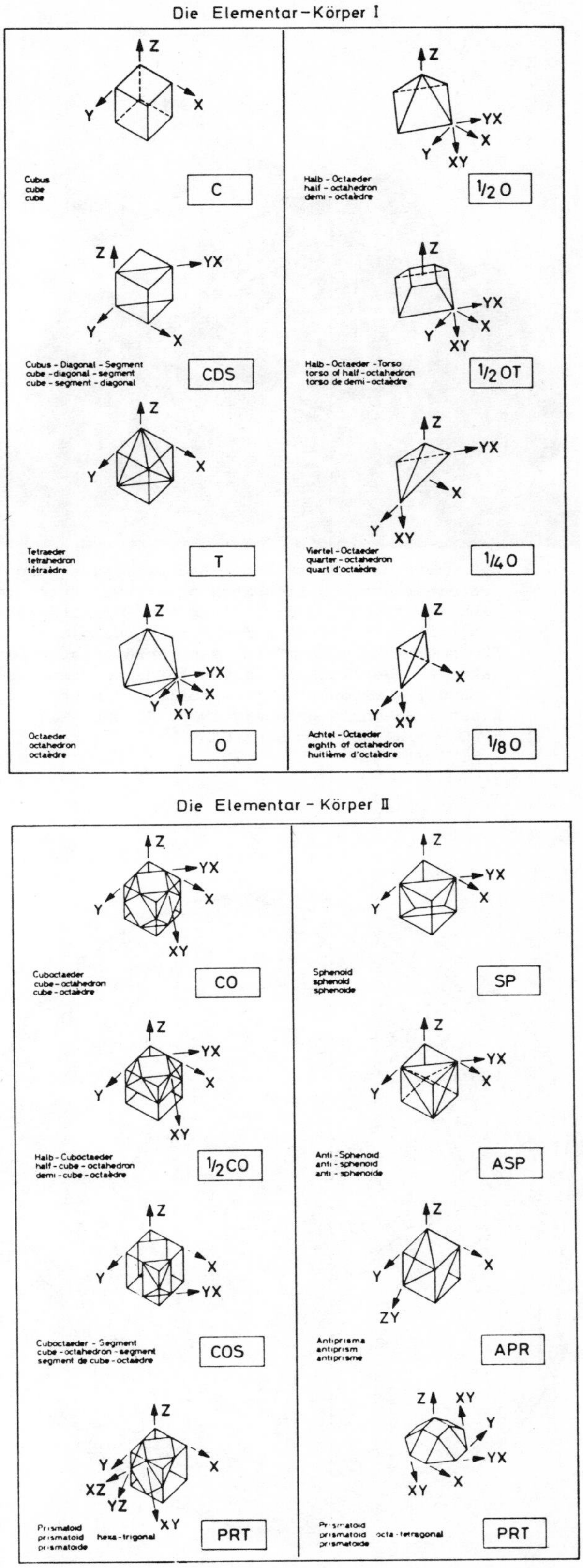

Fig. 2 Die Cubus-Segmente als Elementar-Körper für regelmäßige Raumfachwerke.

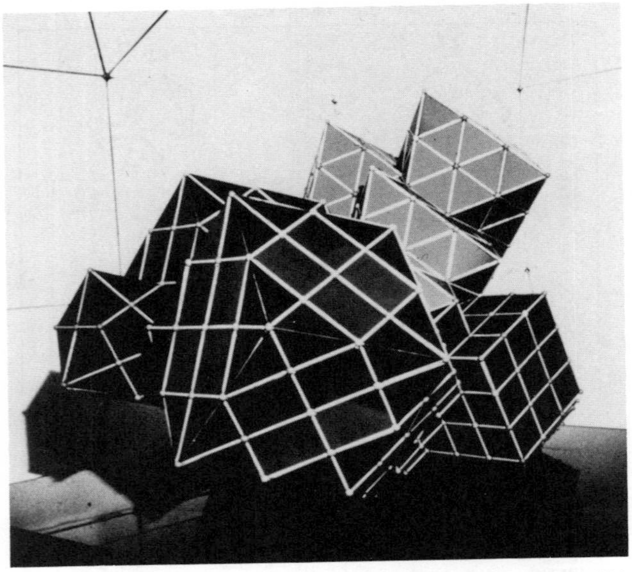

Fig. 3 Baukörper mit kristalliner Bauform, ausschließlich aus einer Art von Einheitsknoten, zwei Arten von Einheitsstäben und drei Arten von Einheitsplatten (oder deren Hälften) aufgebaut. - Beispiel einer "Komposition im Raum" für ein repräsentatives oder sakrales oder künstlerischen Aufgaben dienendes Gebäude. - Exponat der Ausstellung "Konstruktive Kunst", Biennale Nürnberg 1969. Entwurf:Max MENGERINGHAUSEN und Hans BAUER.

Fig. 5 Ausstellungsplastik zur Demonstration des "Plastischen Bauens mit Serien-Elementen" im Rahmen der Sonderausstellung der Gesamthochschule Kassel 1968. Entwurf : M. MENGERINGHAUSEN und H. BAUER. Beispiel einer individuellen Struktur aus Serienelementen mit Cubus-Segmenten.

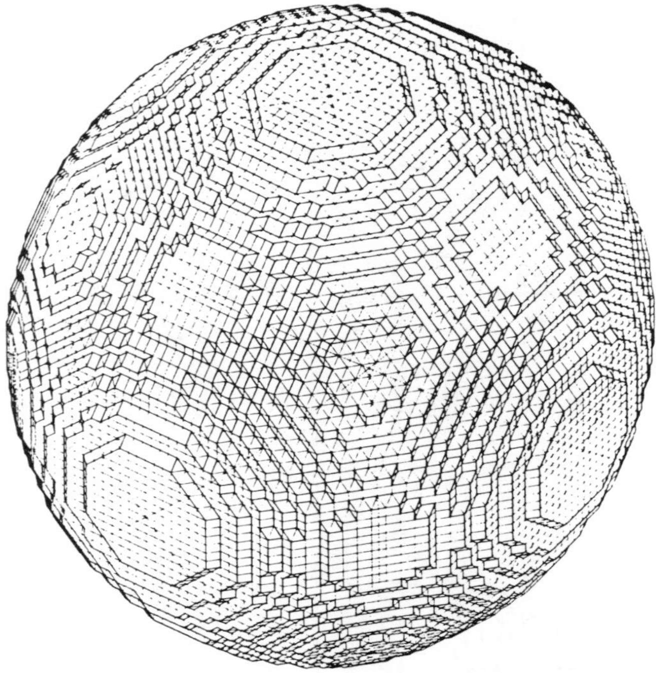

Fig. 4 Annäherungsform für eine Kugel, ausgeführt aus MERO-Normknoten und -Stäben sowie Norm-Platten und daraus geformten Cubus-Segmenten. - Plakat der Ausstellung der Gesamthochschule Kassel 1968. Entwurf : M. MENGERINGHAUSEN und H. BAUER.

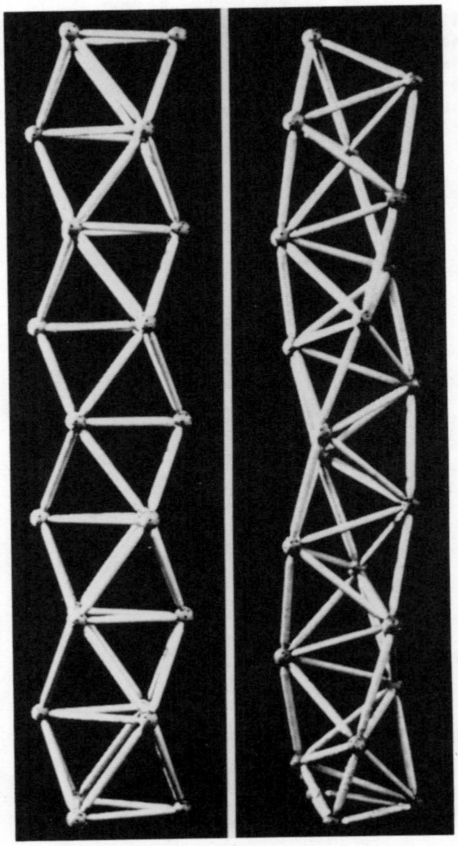

Fig. 6 Individuell gestaltete Türme aus einer Stablänge und einem Knotentyp. -
a) links : Körper aus Oktaedern
b) rechts : Körper aus Tetraedern.

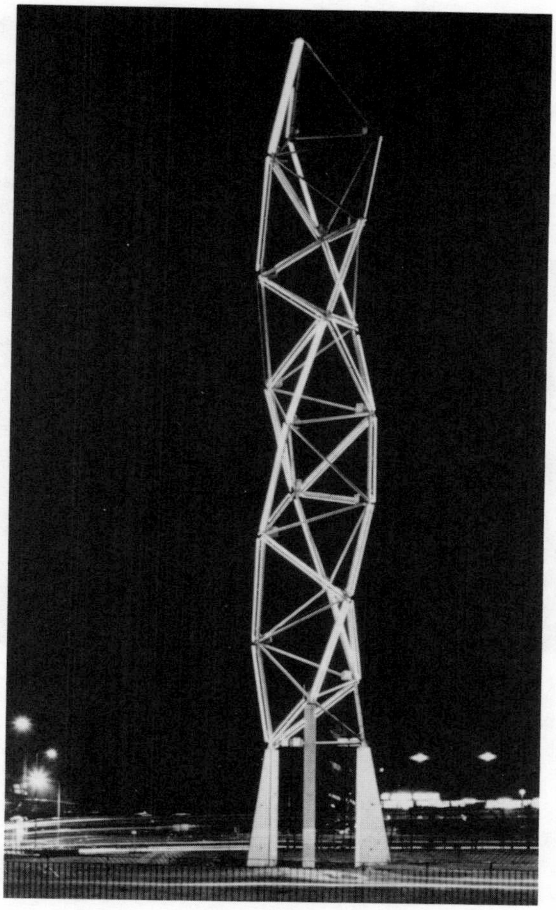

Fig. 7 Lichtturm in Edinburgh (GB) aus Normknoten und Normstäben, die Elementarkörper in Tetraederform bilden mit Leuchtstoffröhren. Architekt : Department of architecture of the city of Edinburgh district council.

Fig. 8 Außenansicht des Königlichen Luftfahrt-Museums in Hendon (GB) mit einem Dach in Form eines Cub-Oktaeder-Segmentes.

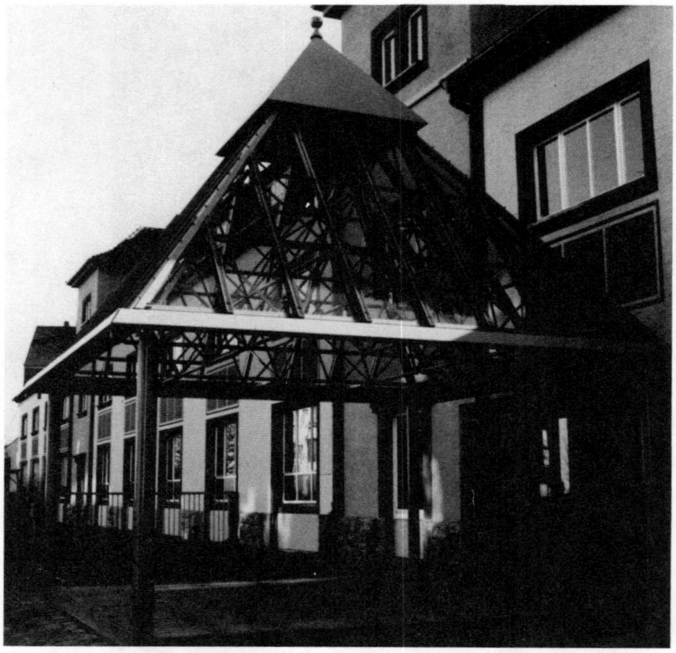

Fig. 9 Eingangsgestaltung der MERO-Hauptverwaltung Würzburg : Glasdach in Halb-Oktaeder-Form mit Normstäben und Normknoten. Architekt : P.RUDOLPH

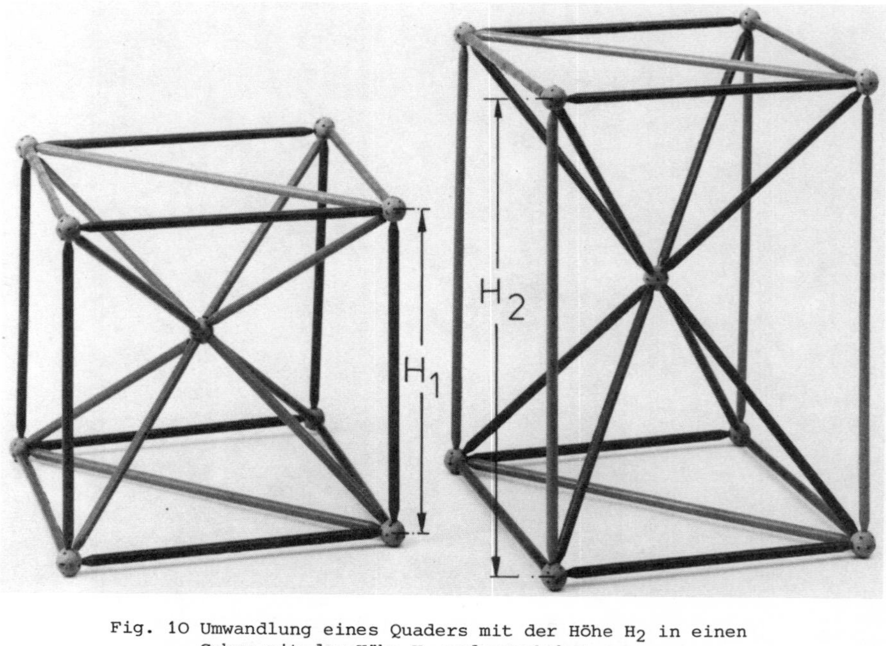

Fig. 10 Umwandlung eines Quaders mit der Höhe H_2 in einen Cubus mit der Höhe H_1 und umgekehrt.

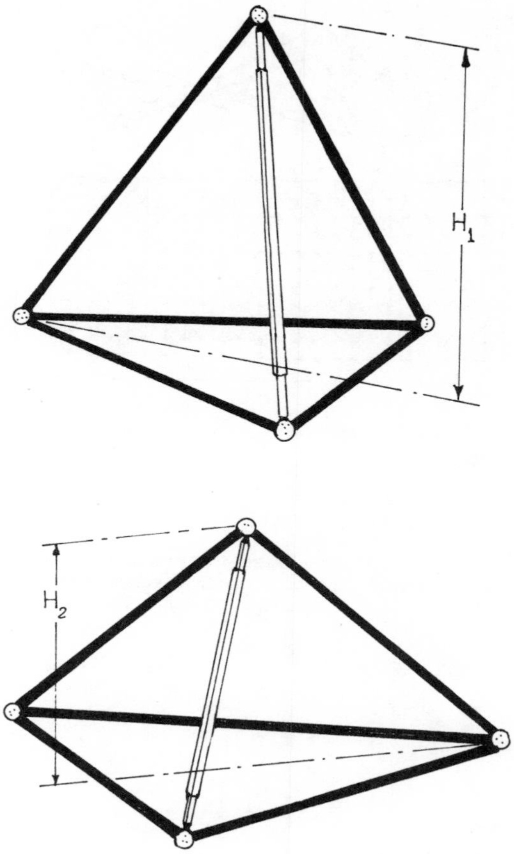

Fig. 11 Umwandlung von "Vierflächnern" durch "Transformation" der Höhen H_1 und H_2.

Fig. 12 Beispiel für ein Dach mit einem "gedrückten Oktaeder", ausgeführt mit "abgeleiteten Knoten" und "abgeleiteten Stäben" : Einfamilienhaus in Winterthur (Schweiz), Außenansicht.

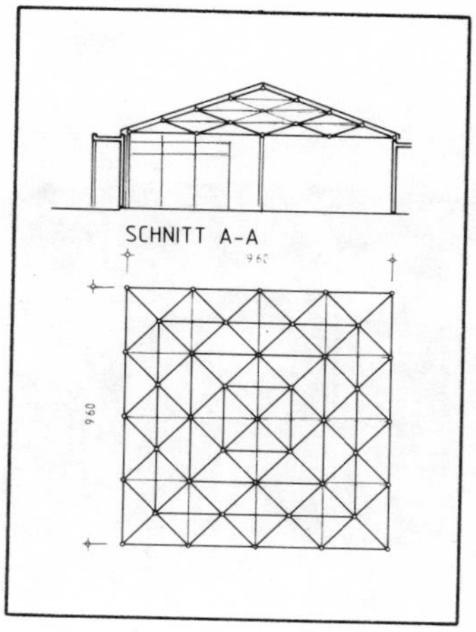

Fig. 13 Schematische Schnittzeichnung des Raumfachwerks für Fig. 12.

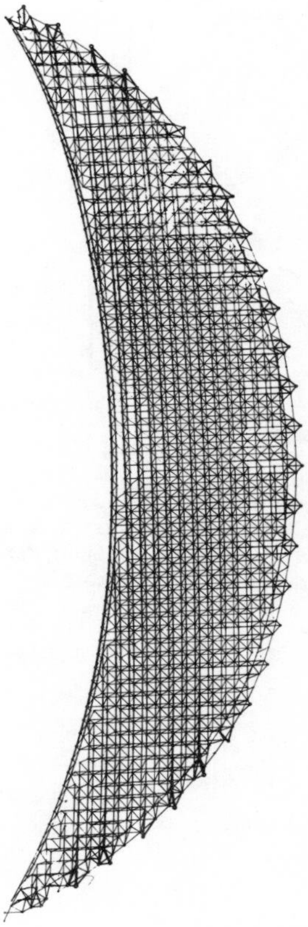

Fig. 14 Computer-Zeichnung des Raumfachwerks für das Stadion in Split (Jugoslawien) mit einem "abgeleiteten Raumfachwerk" aus Halb-Oktaedern und Tetraedern, durch Koordinaten-Transformation in drei Dimensionen gebildet. (Ref. 6).

Fig. 15 Individuell als Rundbau mit einem "abgeleiteten RFW" gestaltete Radar-Station in Ludlow (GB).

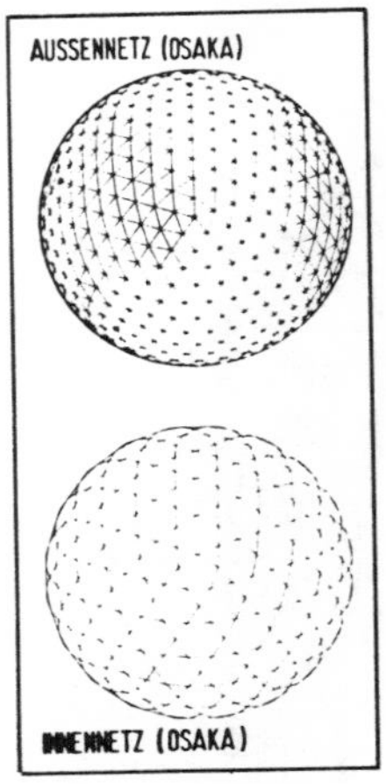

Fig. 16 Grundriß-Schema für Raumfachwerke einer Kuppel-Konstruktion mit Fünfeck im Zenit und abgeleitetem Sechseck-Netz (Computerzeichnung der MERO-Kuppel Osaka 1970). Geometrie : H. EMDE

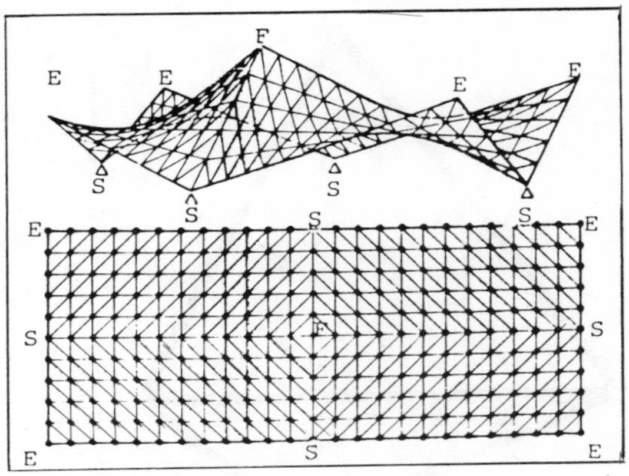

S = Stütz-Punkte

E = Eck-Punkte

F = Spitze in der Mitte

Fig. 17 Einlagiges Raumfachwerk in Form eines "Hyperbolischen Paraboloid" mit quadratischem Grundriß-Raster (Ausstellung "Constructa 78" Hannover).

REFERENCES

1. W. RITTER, Das Fachwerk, Zürich 1890

2. A. FÖPPL, Das Fachwerk im Raume, Leipzig 1892

3. M. MENGERINGHAUSEN, Kompositionslehre räumlicher Stabfachwerke in : Space Structures, Berichte der Londoner Konferenz von 1966, herausgegeben von R.M. DAVIES im Verlag BLACKWELL Scientific Publications, Oxford/Edinburg 1967

4. M. MENGERINGHAUSEN, Raumfachwerke aus Knoten und Stäben, 1. Auflage 1962, 7. Auflage Bauverlag Wiesbaden-Berlin 1975

5. M. MENGERINGHAUSEN, Komposition im Raum, die Kunst individueller Baugestaltung mit Serienelementen, Gütersloh 1983

6. Stadion Split (Jugoslawien), DBZ 8/81, S. 1145-1148, und acier-stahl-steel 44 (1979) Nr. 4, S. 121-126

SPACE STRUCTURES DESIGN WITH GRAPHIC WORKSTATION

S VALENTE

Department of Structural Engineering,
Politecnico di Torino, Torino, ITALY

Interactive computer graphics is a powerful technique which can be used to accelerate the space structure design. A graphic workstation, based on high speed processor (32-bit), is an efficient solution in the static model generation and in the stress analysis. Although this paper presents a technique for cylindrical vaults, several ideas can be extended to any kind of space structures.

INTRODUCTION

Most applications in the realm of computer-aided design (CAD), where computer graphics is used, necessitate a vast amount of computation, requiring either a powerful dedicated computer or an expensive high-speed link to a powerful 'host computer'. The solution of sharing the resources of the host computer with many users in order to make it cheaper to use, a concept that has proven so valuable in educational and commercial time-sharing systems, has not been applicable to CAD, see Ref 1. The major obstacle is the prohibitively long response time of the conventional time-sharing system. This response time is not a decisive factor in ordinary speed at which the user types in statements and data.

Conversely, the user of CAD system works with instruments such as, for example, lightpens, digitizers, etc. Very often, in space structure design, the user has to pick-up a graphic entity (point, segment or element), among some hundreds entities of the same type. This operation is accomplished moving a cursor on a graphic display, and has to be executed in a very short time, because the designer is waiting for the answer. Furthermore, it is useful to iterate the structural analysis several times in the same day, in case of large systems too (more than a thousand degrees of freedom). Consequently, a graphic workstation, with a powerful processor incorporated (32-bit), has been considered the most efficient solution.

In case of cylindrical vaults, the design sequence can go through the following steps:

1) Overall geometric model generation
2) Modification of geometrical and/or mechanical properties.
3) Stress analysis.

If some stresses are non allowable, the designer goes back to step 2).

OVERALL GEOMETRIC MODEL GENERATION

The section of a cylindrical vault is indicated in Fig 1.

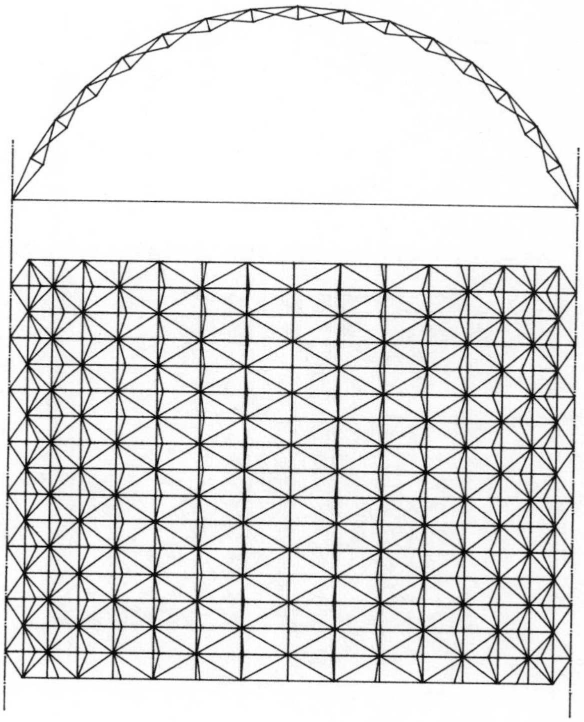

Fig 1

The overall geometric model generation goes through the fol_ lowing steps:

1) Creation and assembling of six-edges macro elements. The designer can change data reported in Fig 2, until the picture on graphic display is correct. For clarity, the 'shrinkage' technique is used in the assembled representation. When the image on graphic display is correct, the designer goes on to the next step.

2) Transformation of macro elements into rombic elements, see Fig. 3. This step can be articulated as follows :

   2.1) Rombic element generation across the top(label 1 in Fig 3).

   2.2) Insertion of a new strip of rombic elements between diagonals 3 and 4 (label 2 in Fig 3).

   2.3) The same as 2.2), until the rombic elements generation is completed.

   Near the diagonals, some rombic elements can be distorted. The designer can change data until the picture on graphic display is correct.

3) Transformation of rombic elements into:

   3.1) External trusses, which support climatic loads.

   3.2) Internal trusses.

4) Generation of standard support conditions(foundation, symmetry-planes, etc. ..)

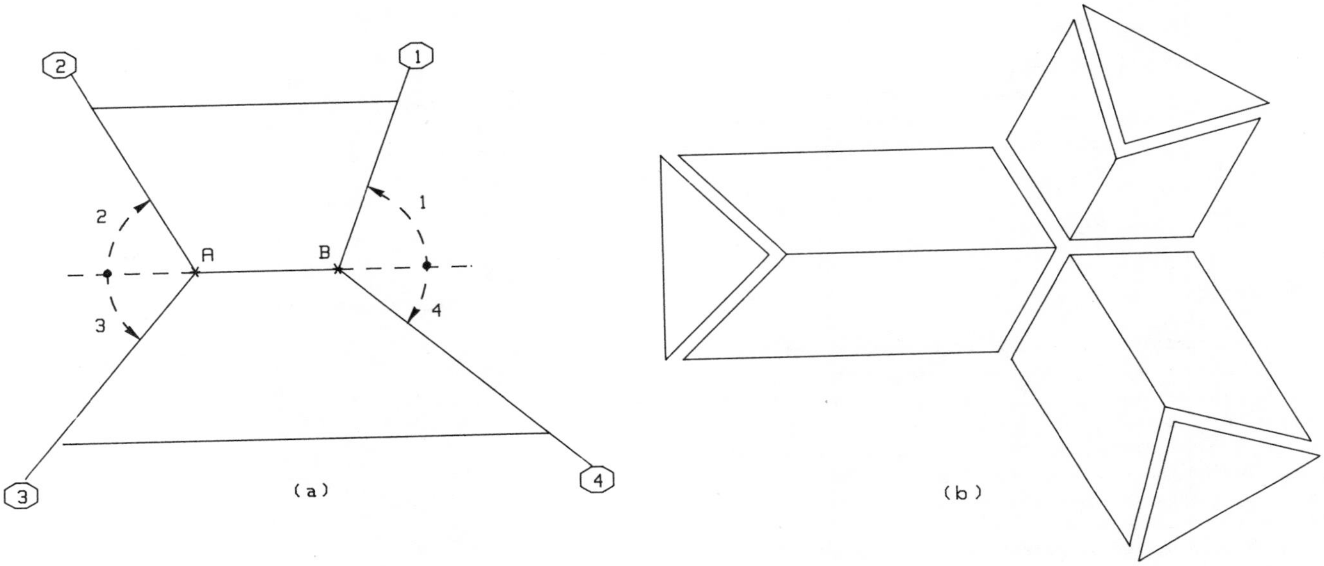

Fig 2

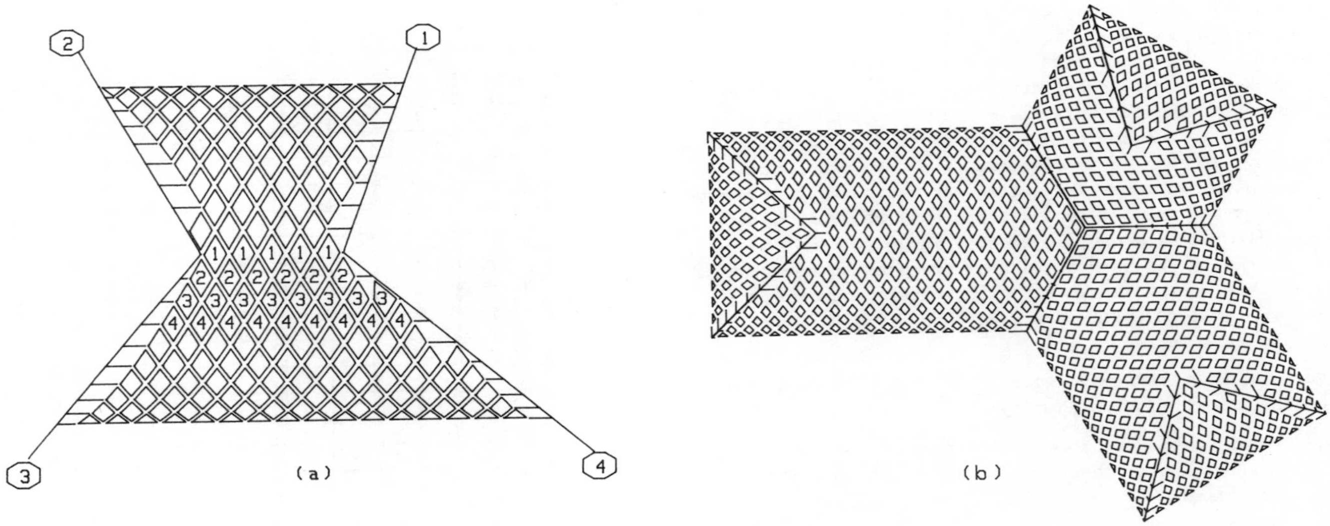

Fig 3

MODIFICATION OF GEOMETRICAL AND MECHANICAL PROPERTIES

The four steps for geometric model generation have to be executed sequentially, but the following steps can be executed in any sequence, pressing the appropriate key on a programmable function keyboard.

1) Creation of hollow in the vault by interactive deletions of trusses. In case of error, the user can restore the old situation. The overall vault is reported in Fig 4.

2) Vault reinforcing by interactive additions of trusses and section properties modifications.

3) Generation of snow, wind and seismic loading conditions. Loading conditions is checked on graphic display, see Fig 5.

4) Interactive creation/deletion of support conditions.

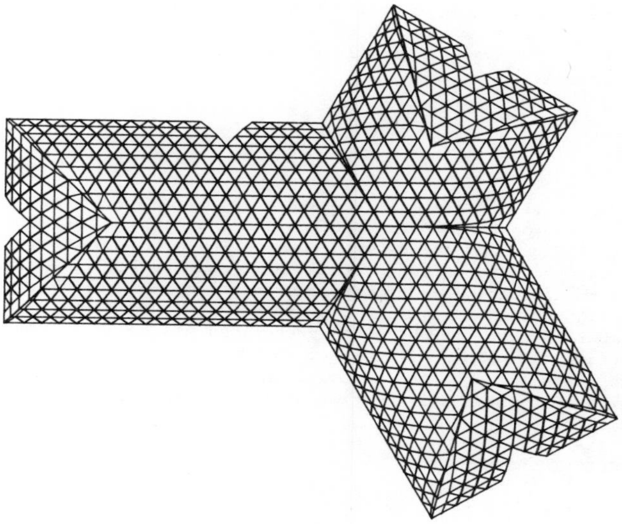

Fig 4

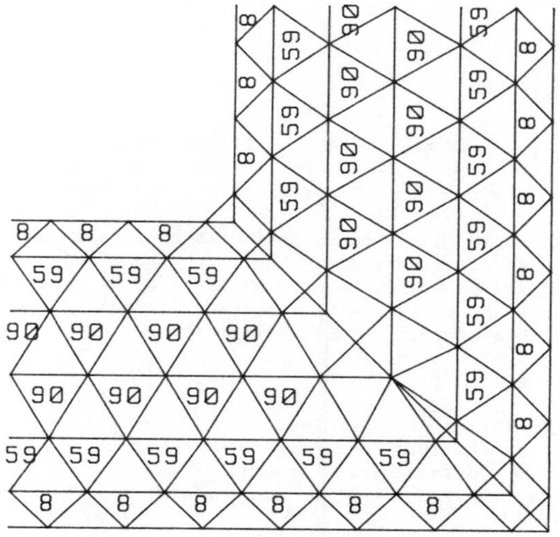

Fig 5

5) Evaluation of main and external memory necessary during the stress analysis.

6) Storing all information on external file.

7) Getting all information from external file.

Steps 1), 2), 3), 4), can be operated at any point of the structure. Consequently, in case of large vaults, zoom option is very useful.

STRESS ANALYSIS

The stress analysis program gets all information from the file created during step 6 above. In case of multi-tasking workstation, during the solution time, the user can work interactively on another problem, running in its own, 'partition'. Each partition is a virtual machine, which, with simple keyboard commands, can be attached to the system keyboard and display.

After stress examination, if the designer is not satisfied, he/she can go back to the previous paragraph.

COMPUTER PROGRAMMING

In recent years the subject of computer programming has been recognized as a discipline whose mastery is fundamental and crucial to the success of many engineering projects see Ref 2. The program described for static model generation of cylindrical vaults requires about 2500 statements to be coded in high-level language and, in the future, it can easily be extended to domes. Therefore it is thought a structured software design is useful, see Ref 3. The author feels that FORTRAN has outlived its use, in that it is far too restrictive in its flow-of-control constructs and data types. PASCAL, PL/I, APL, EXTENDED BASIC are considered to be more eligible languages to program a graphic system see Refs 1 and 4.

The program described has been coded using HP9000 BASIC language, which provides PASCAL flow-of-control constructs and a BASIC interactive environment. The key to provide such a feature is a run-time compiler that adds the functionality of an interpreter to the high-speed performance of a compiler. Consequently, the engineer gets programs written and running correctly in a shorter period of time than with traditional systems.

ACKNOWLEDGEMENTS

The author wishes to thank the Hewlett Packard Company for supporting the software development on graphic workstation model HP9020S

REFERENCES

1. W K GILOI, Interactive Computer Graphics, Prentice-Hall, 1978.

2. N WIRTH, Algorithms + Data Structures = Programs, Prentice-Hall series in automatic computation, 1976.

3. E YOURDON and L L CONSTANTINE, Structured Design, Prentice-Hall, 1979.

4. J D FOLEY and A VAN DAM, Fundamentals of Interactive Computer Graphics, Addison-Wesley, 1982.

STRUCTURAL CONFIGURATIONS FROM GEOMETRIC POTENTIAL

J W BUTTERWORTH

Senior Lecturer, Department of Civil Engineering,
University of Auckland, New Zealand

A potential function is defined for a set of nodes lying in a surface. Moving the nodes about on the surface until the potential is a minimum results in node positions which may be interconnected to create configurations of optimum regularity. The concept is illustrated using a spherical surface but may be applied to more complex surfaces.

INTRODUCTION

Aesthetic and practical requirements frequently dictate that structures should have a high degree of regularity in the disposition and sizes of their component parts. Although such regularity is readily achieved in structures of simple geometric form such as double layer grids and barrel vaults, structures whose nodes lie in a surface of non-zero Gaussian curvature, such as a sphere, present a slightly more difficult problem.

An elegant solution to the subdivision of a spherical surface is provided by the well known geodesic breakdowns, Ref 1, although there are drawbacks, such as the difficulty in truncating at the support level without disturbing the regularity (only even frequency alternate breakdowns provide a convenient equatorial truncation line, for example). There is also an inflexibility imposed on the designer by the geodesic patterns, restricting his choice to those particular symmetries that are rooted in the initial classical polyhedron on which the subdivision is based.

The spherical form also has limitations, such as an excessive volume to plan area ratio in the case of hemispherical truncations and low edge headroom in the case of cap truncations. To overcome these limitations designers have been turning to non-spherical surfaces such as ellipsoids and generalised ellipsoids of the form given by

$$(x/a)^u + (y/b)^v + (z/c)^w - 1 = 0 \qquad \ldots\ldots 1,$$

where a, b and c denote the radii of the generalised ellipsoid in the x, y and z Cartesian directions respectively and u, v and w define the form of the surface (for example, $u = v = w = 2$ defines an ellipsoid, and if in addition $a = b = c$ a sphere results).

Geodesic subdivisions of these generalised ellipsoids, if accomplished by projecting from a subdivided polyhedron onto the circumscribing ellipsoid will not generate the same regularity as in the case of a sphere and various modifications have been proposed, Ref 2, although no general and satisfactory solution appears to have been found.

A new approach to the problem of achieving a regular subdivision of a surface is to take a set of nodes, place them at random on the desired surface and assume that forces of mutual repulsion act between them. The nodes are then allowed to move on the surface until equilibrium is achieved and the resulting node positions interconnected to form the configuration. Any number of nodes may be used and any number of these may be given prescribed positions on the surface (at points of intended support, for example). The formulation of the method is presented from the view point of a potential function as set out in the following sections.

GEOMETRIC POTENTIAL

Let there be a set of n nodes numbered from 1 to n with associated weightings of m_1 to m_n and constrained to lie on a surface S.

Let d_{ij} denote the length of the chord joining nodes i and j.

The 'geometric potential' G is defined as

$$G = \sum_{i \neq j} m_i\, m_j\, d_{ij}^{-q} \qquad \ldots\ldots 2,$$

where q is a positive integer defining the power law of repulsion between nodes. For a given set of nodes and weightings G is thus a function of the node coordinates.

The value of G is now taken as a measure of the regularity of the node distribution and the set of node coordinates which minimises G will be optimum with respect to the given weightings and power law. The computational task then is to find this optimum set of coordinates.

MINIMUM GEOMETRIC POTENTIAL

The minimisation of G involves a nonlinear optimisation problem since G is a nonlinear function of the node coordinates. In addition it will generally be a constrained problem as the nodes will be constrained to lie on a surface or, more often, on a selected part of a surface.

In the special case of a spherical surface some simplifications are possible and this case is taken to demonstrate the process.

SPHERICAL SURFACE - UNCONSTRAINED PROBLEM

In terms of spherical coordinates, (θ, ϕ, r), the distance between two points

$$d_{ij} = [r_i^2+r_j^2-2r_ir_j(\cos\theta_i \cos\theta_j+\cos(\phi_i-\phi_j)\sin\theta_i \sin\theta_j)]^{\frac{1}{2}} \quad \ldots\ldots 3.$$

If the two points lie on a spherical surface then $r_i=r_j=r$, a constant, and Eqn 3 becomes

$$d_{ij} = \sqrt{2}\, r[1-\cos\theta_i \cos\theta_j-\cos(\phi_i-\phi_j)\sin\theta_i \sin\theta_j]^{\frac{1}{2}} \quad \ldots\ldots 4.$$

The geometric potential function may now be written as

$$G = (2r)^{-q/2} \sum_{i\neq j} m_i m_j [1-\cos\theta_i \cos\theta_j-\cos(\phi_i-\phi_j) \sin\theta_i \sin\theta_j]^{-q/2} \quad \ldots\ldots 5.$$

If the nodes are free to take up positions anywhere on the spherical surface the problem may be treated as an unconstrained optimisation in which the variables are the (θ,ϕ) coordinates of each node.

This unconstrained optimisation problem has been solved by the conjugate gradient method of Fletcher and Reeves, Ref 3. No claim is made that this is the best method for this problem, although it did prove to be more efficient than the variable metric method of Fletcher and Powell, Ref 4. Analytical derivatives of G are readily obtained (Appendix 1) and the method is straightforward from a programming point of view.

Trials were made by placing varying numbers of nodes at random on a spherical surface and then allowing the optimisation procedure to alter their coordinates until a minimum value of G was obtained. With equal weightings and a power law of q=2, low numbers of nodes generated the classical regular polyhedra and with higher numbers of nodes the geodesic patterns were obtained - for example:

Nodes	Resulting minimum G configuration
4	tetrahedron
6	octahedron
12	icosahedron
32	2-frequency triacon geodesic
42	2-frequency alternate geodesic

The geodesic patterns exist only for particular numbers of nodes - for example there is no geodesic pattern for node numbers lying between 12 and 32 or between 32 and 42. Minimisation of geometric potential is not, of course, restricted to these particular values and interesting patterns can be obtained for any number of nodes.

SPHERICAL SURFACE - CONSTRAINED PROBLEM

Although a configuration can be generated on a complete sphere and then only part of it used, a more useful approach is to generate the configuration directly on the portion of the sphere that is required for the structure. In this case a constrained optimisation technique will in general be required, however, it has been found that a number of useful cases can be handled by the same unconstrained technique used on the complete sphere.

The unconstrained optimisation procedure has been found to work when a ring of nodes is first positioned around the sphere with the remaining free nodes placed on the sphere above the fixed ring. The θ coordinate of the ring nodes is given a fixed value and the remaining node coordinates participate in the optimisation procedure. With an appropriate step-length parameter the free nodes remain 'fenced in' by the ring of fixed nodes and the desired solution is obtained.

EXAMPLE 1

Twelve nodes were fixed at 30° intervals around the equator (θ = 90°) of a sphere of unit radius with a further 18 nodes spaced at 20° intervals at θ= 30° and one node at the apex. The θ coordinate of the equatorial nodes was fixed and unit weights assigned to each node.

With a power law of q = 4 the configuration shown in Fig 1 was obtained with a value of G = 4881.329. The pattern of interconnection was subsequently added by hand. In Figs 1 to 5 the configurations are plotted on a polar coordinate system with θ varying linearly from the centre.

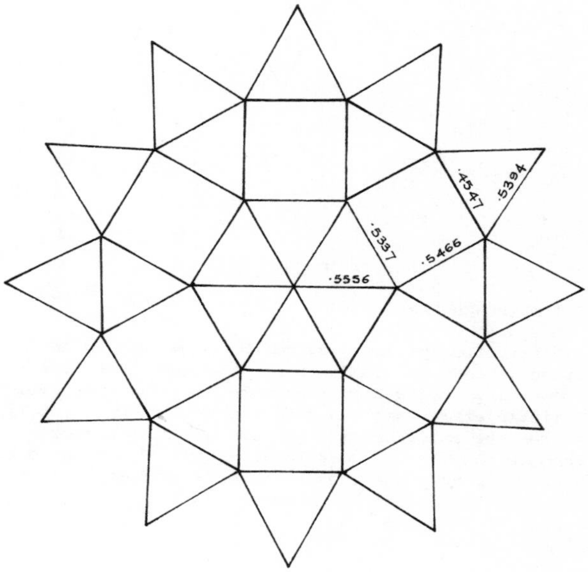

Fig 1 q=4, G (local minimum) = 4881

LOCAL MINIMA

As there is no guarantee that G is a unimodal function it is quite likely that local minima will be found. The use of different starting points for the optimisation procedure may help to overcome this, although a non-global minimum is quite likely to be a satisfactory solution anyway.

By way of illustration, a different starting point to the previous example results in convergence to the configuration shown in Fig 2, with a value of G = 4832.768. The configuration of Fig 1 is thus a local minimum only. However, with the connection patterns shown Fig 1 has five different member lengths compared with seven in Fig 2 and the ratio of longest to shortest member in Fig 1 is 1.222 compared with 1.279 in Fig 2. Which then is the better configuration?

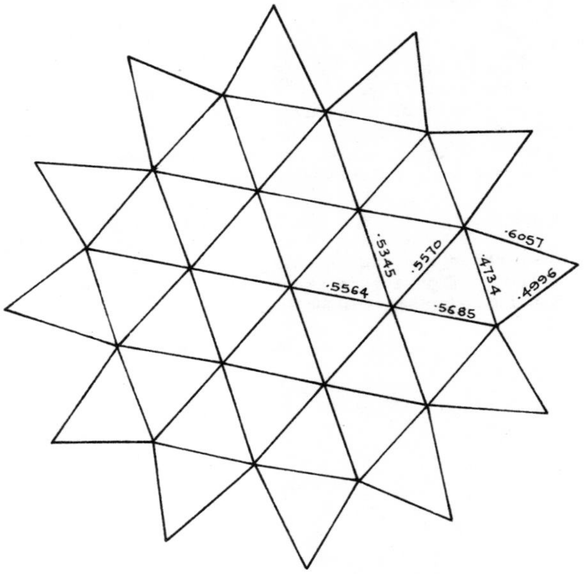

Fig 2 q=4, G (global minimum) = 4832

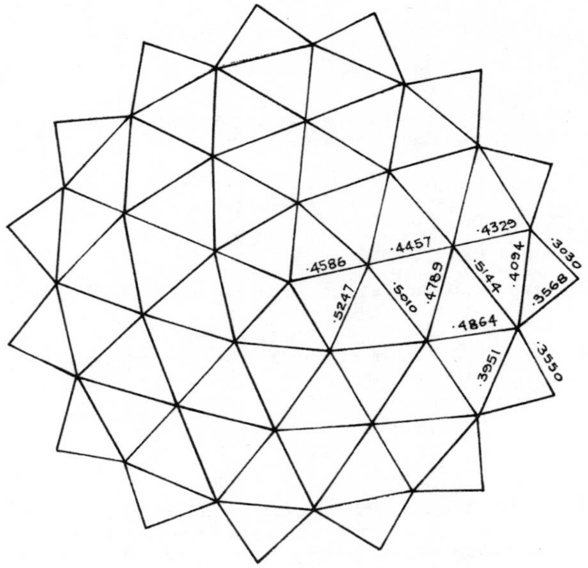

Fig 3 q=1, G (global minimum) = 1450

EFFECT OF POWER LAW

The contribution of one node to the geometric potential function depends on its distance from all other nodes. If the power law, q, is high then more distant nodes will have very little effect and the major contribution to G will come from the interaction of a node with its nearest neighbours only. The more local action of a high power is thus to force node spacings to be more uniform. This may be seen in the following examples.

EXAMPLE 2

A unit radius hemisphere with 15 nodes fixed at 24° intervals around the equator and one node fixed at the apex. A further 30 free nodes placed at equal intervals at $\theta = 45^{\circ}$, all weights = 1. With a power law q = 1 the configuration shown in Fig 3 was obtained (again, the interconnection pattern was added subsequently, by hand).

It is interesting to note that this example shows cyclic symmetry of order 5 compared with the order 6 shown in example 1. The number of different member lengths is 13 and the ratio of longest to shortest is 1.6977.

EXAMPLE 3

As for example 2 but with a power law of q = 2 resulting in the configuration shown in Fig 4.

The order 5 symmetry appears again with 13 different member lengths, but the ratio of longest to shortest member length decreases to 1.309.

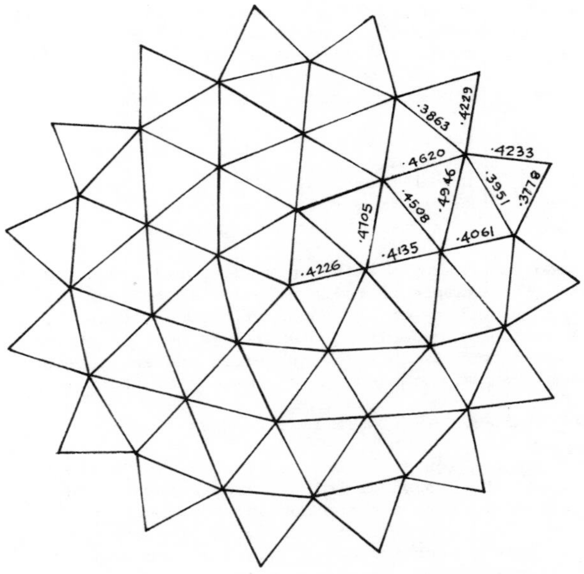

Fig 4 q=2, G (global minimum) = 2677

EXAMPLE 4

As for example 2 but with a power law of q = 4 resulting in the configuration shown in Fig 5. With the higher power law the ratio of longest to shortest member length drops further to 1.2018.

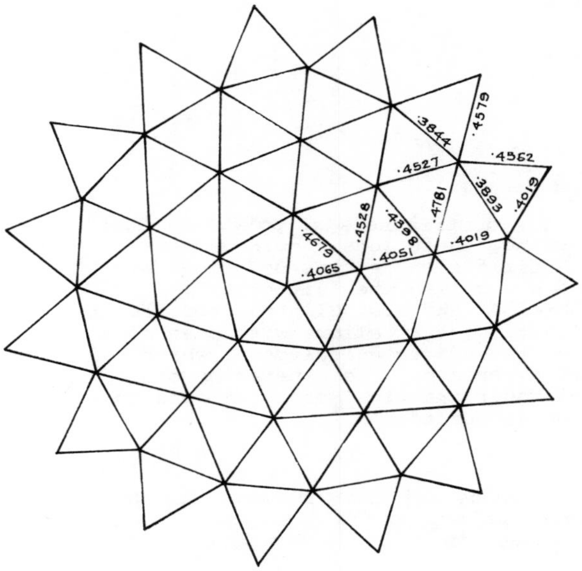

Fig 5 q=4, G (global minimum) = 17473

EFFECT OF NODE WEIGHTING

The purpose of including weightings for each node was to permit some control over the density of the node distribution. If, for example, greater node spacings are required in some region of the surface than node weightings in this area can be increased. Such variations could be required for structural or architectural reasons, of if the configuration was being generated for use as a finite element mesh then a sparser mesh could well be needed in areas of low strain gradient.

CONCLUSIONS

Results obtained to date suggest that minimisation of geometric potential provides a feasible way of creating configurations from a simple starting point. The ability to force these configurations to conform to prescribed support points without prejudice to their regularity is seen as an advantage.

Although the examples considered have been confined to spherical surfaces there is no conceptual difficulty in extending the application to other surfaces. Indeed, it is on the more general ellipsoid surfaces (Eqn 1) that the most useful applications are likely to be found.

REFERENCES

1. E POPKO, Geodesics, University of Detroit Press, 1968

2. H KENNER, Geodesic Math, University of California Press, 1976

3. R FLETCHER and C M REEVES, Function minimisation by conjugate gradients, The Computer Journal, 7, 1964

4. R FLETCHER and M J D POWELL, A rapidly convergent descent method for minimisation, Computer Journal, 6, 1963

APPENDIX 1 - DERIVATIVES OF G

When the geometric potential is defined by Eqn 4 the first order derivatives are given by

$$\frac{\partial G}{\partial \theta_k} = (2r)^{-q/2} \cdot q/2 \sum_{j=1,k-1} Q_{jk}[\cos(\phi_j-\phi_k) \sin\theta_j \cos\theta_k - \cos\theta_j \sin\theta_k]$$

$$+ (2r)^{-q/2} \cdot q/2 \sum_{j=k+1,n} Q_{jk}[\cos(\phi_k-\phi_j) \sin\theta_k \cos\theta_j - \cos\theta_k \sin\theta_j] \quad \ldots\ldots 6,$$

$$\frac{\partial G}{\partial \phi_k} = (2r)^{-q/2} \cdot q/2 \sum_{j\neq k} m_j m_k [\sin(\phi_j-\phi_k) \sin\theta_j \sin\theta_k] Q_j k \quad \ldots\ldots 7,$$

$$\text{where } Q_{jk} = m_j m_k [1-\cos\theta_j \cos\theta_k - \cos(\phi_j-\phi_k) \sin\theta_j \sin\theta_k]^{-\left(\frac{q+2}{2}\right)} \quad \ldots\ldots 8.$$

EXPANDABLE POLYHEDRAL STRUCTURES BASED ON DIPOLYGONIDS

Hugo F. VERHEYEN

Antwerpen, BELGIUM.

Various expandable systems can be constructed by symmetric sets of polygons, that are connected by their vertices. While transforming, these sets remain vertex-transitive in a group of symmetry operations. A first such set was described by R. Buckminster Fuller in many publications, and was called the "Jitterbug". Some more sets were discovered by R.D. Stuart (ref.4) and J.D. Clinton (ref.1). Introduced in this paper is a mathematical definition, that allows with precision to fully enumerate and classify these sets of vertex-connected polygons, which will be named "dipolygonids". Their theoretical importance, the generation of new classes of vertex-transitive polyhedra, and their practical use, the construction of expandable polyhedral structures will be discussed.

1. INTRODUCTORY DEFINITIONS

In the euclidian space E^3, A represents a rotation of finite order s, and P any point not being on the axis r_A. The image of the line segment $[P,A(P)]$ over the generated cyclic group gen{A} is a regular polygon, which is said to be produced in P by A. Clearly, the same polygon is also produced in P by A^{-1}.
If s=2, the polygon is called a digon, which is a degenerate polygon, composed of 2 collapsing edges. The plane ω, wherein the polygon lies, intersects r_A in M, the center of the polygon. In the triangle $\Delta\ P{\cdot}M{\cdot}A(P)$, the angle $\varphi = \measuredangle P{\cdot}M{\cdot}A(P)$ is called the central angle of the polygon. According to φ two types of regular polygons occur:

a) $\varphi = 2\pi/s \Leftrightarrow$ Convex polygon, denoted by {m} (m=s)
b) $\varphi = 2\pi.d/s$ ($d \in N_0^+$) $\Leftrightarrow$ Star polygon, denoted by {m} (m=s/d)

The rational number m is called the polygonal value of the rotation A.

2. DIPOLYGONIDS

Let A be a rotation of order s and PV m, and B of order t and PV n. The axes r_A and r_B intersect in a finite point O, and P is any point $\notin r_A \cup r_B$.
2.1. Definition:
the image of the line segments $[P,A(P)]$ and $[P,B(P)]$ over the generated group gen{A,B} is a dipolygonid, produced in P by the base {A,B}.

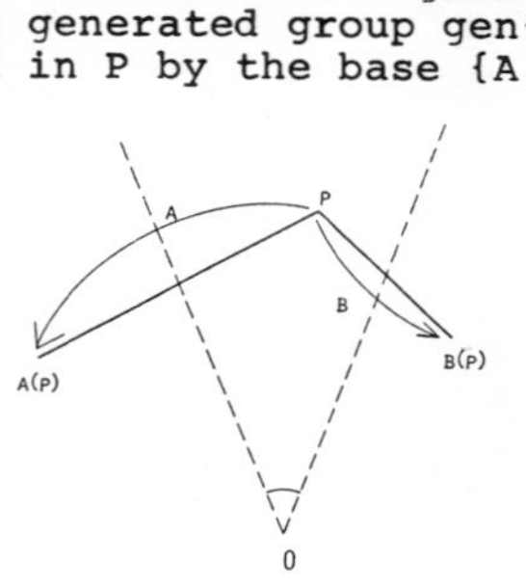

Fig.1

The polygons {m} and {n}, resp. produced by A and B in P, belong to the dipolygonid as a subset. The subset is called a dipolygon in P.

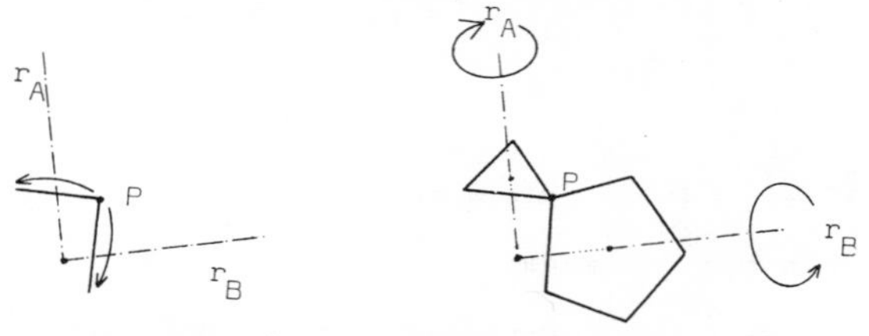

Fig.2: Produced dipolygon in P

It follows that a dipolygonid also is the image of a dipolygon over G=gen{A,B}. Hence, the dipolygonid is composed of regular polygons {m} and {n}, whereas each vertex is common to one {m} and one {n}. Figure 3 illustrates a stepwise process of the production of a dipolygonid from a dipolygon.

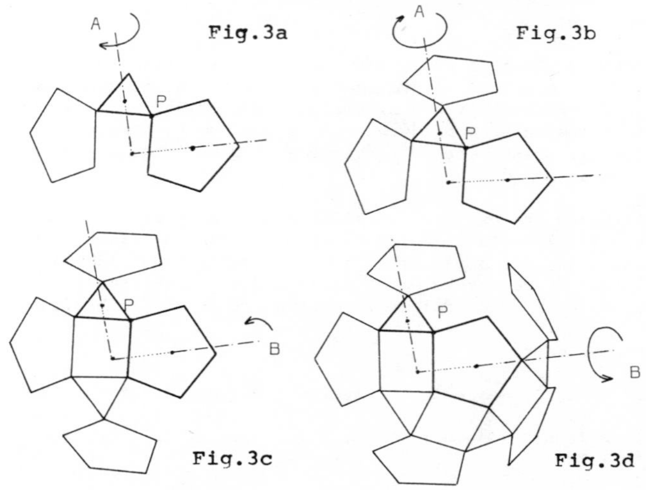

Fig.3a Fig.3b Fig.3c Fig.3d

2.2. Finite dipolygonids.
If G has finite order g, it is one of the following finite groups of rotations:

a) D_h: the dihedral group of order 2h;
b) T : the tetrahedral group of order 12;
c) O:: the octahedral group of order 24;
d) I : the icosahedral group of order 60.

The dipolygonid is then finite and contains:

- g vertices
- 2g edges
- g/s polygons {m}
- g/t polygons {n}

The types of bases {A,B}, such that G is finite, are listed in table 1. A and B are indicated by the order, while θ represents the sharp or right angle, formed by the axes r_A and r_B in O.

A	B	θ	gen{A,B}
2	2	$\frac{k}{h}$.180° (*)	D_h
2	3	20°54'19"	I
		35°15'52"	O
		54°44'08"	T
2	4	45°	O
2	5	31°43'03"	I
		58°16'57"	I
2	h	90°	D_h
3	3	41°48'37"	I
		70°31'44"	T
3	4	54°44'08"	O
3	5	37°22'39"	I
		79°11'16"	I
4	4	90°	O
5	5	63°26'06"	I

(*) $k \in N_o^*, \leq \frac{h}{2}$
k and h have no common factors > 1

The enumeration and classification of the finite types of dipolygonids follows from table 1.

The notation for a finite dipolygonid is

$$G[\{m\}^{g/s}\{n\}^{g/t}\ \theta]$$

a) Dihedral

$D_h[\{2\}^h\{2\}^h\ \frac{k}{h}.180°]$ ("Petrie" polygons)

$D_h[\{2\}^h\{\frac{h}{k}\}^2\ 90°]$

Examples:

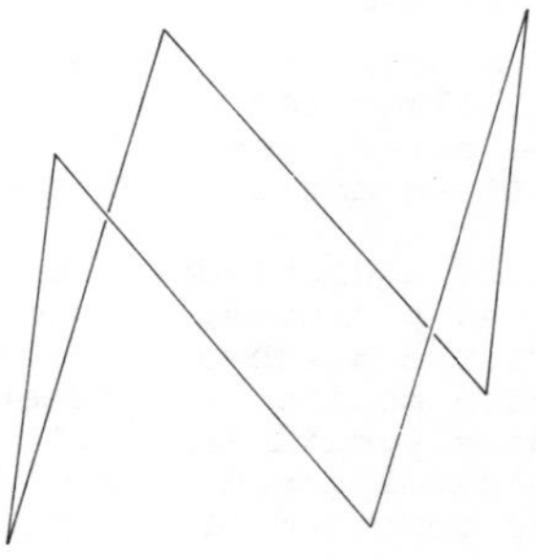

$D_3[\{2\}^3\{2\}^3\ 60°]$

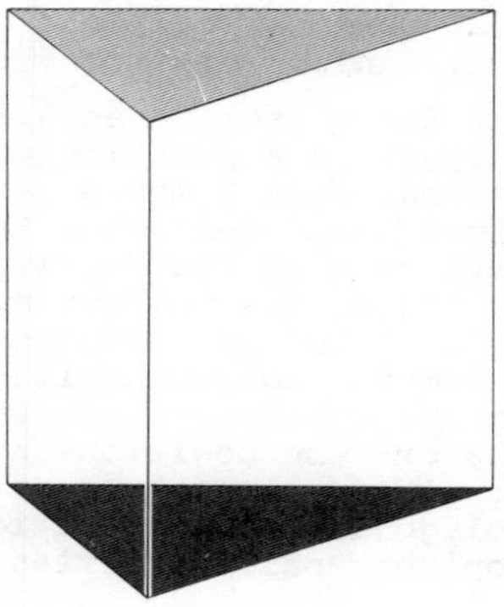

$D_3[\{2\}^3\{3\}^2\ 90°]$

b) Tetrahedral:

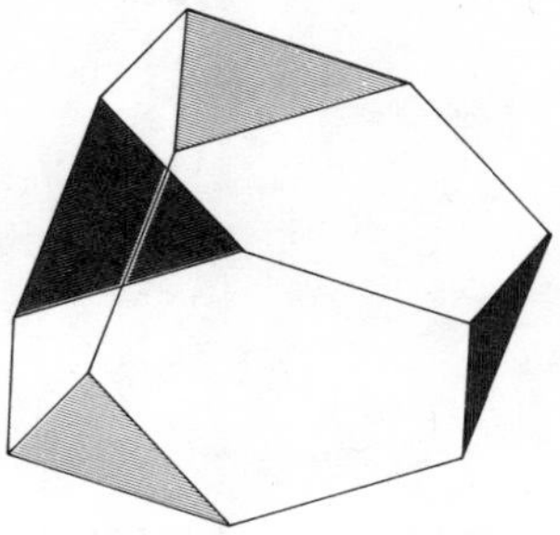

$T[\{2\}^6\{3\}^4\ 54°44'08"]$
(T_1)

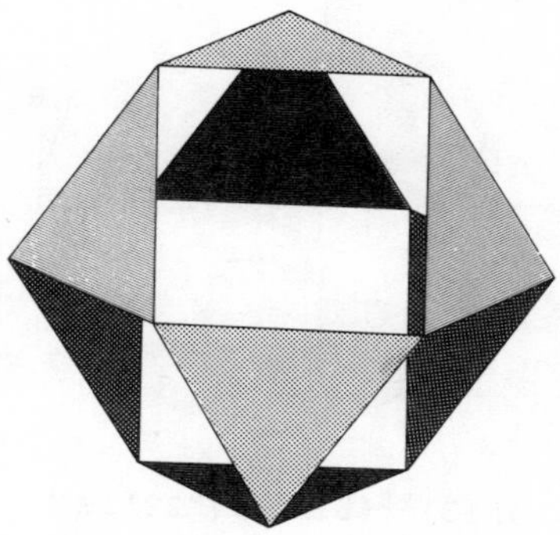

$T[\{3\}^4\{3\}^4\ 70°31'44"]$
(T_2) (*)

c) Octahedral:

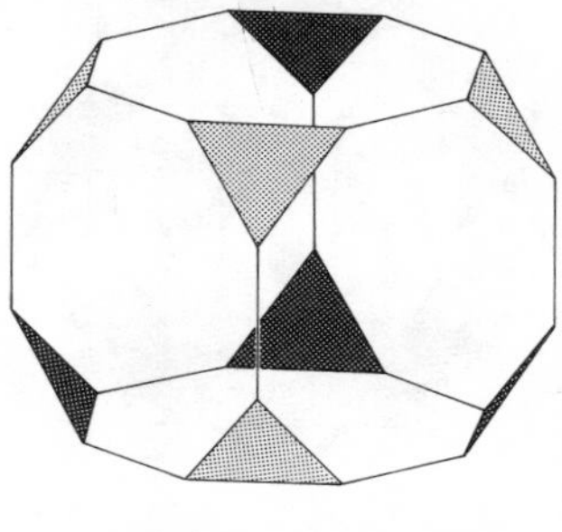

$O[\{2\}^{12}\{3\}^8\ 35°15'52"]$
(O_1)

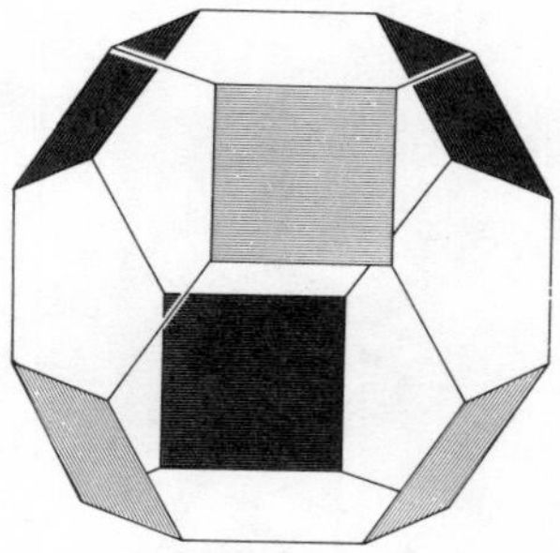

$O[\{2\}^{12}\{4\}^6\ 45°]$
(O_2)

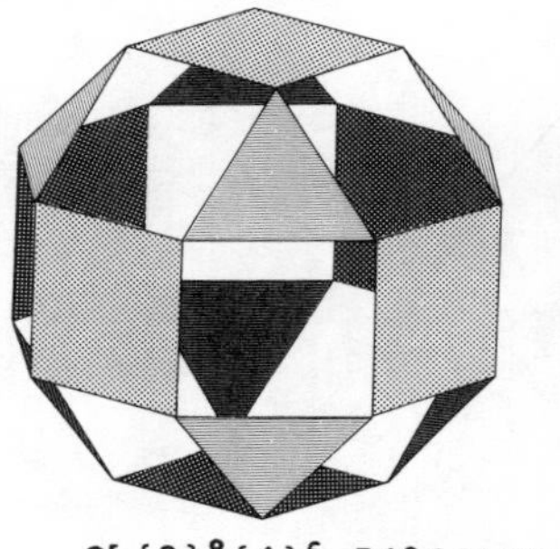

$O[\{3\}^8\{4\}^6\ 54°44'08"]$
(O_3)

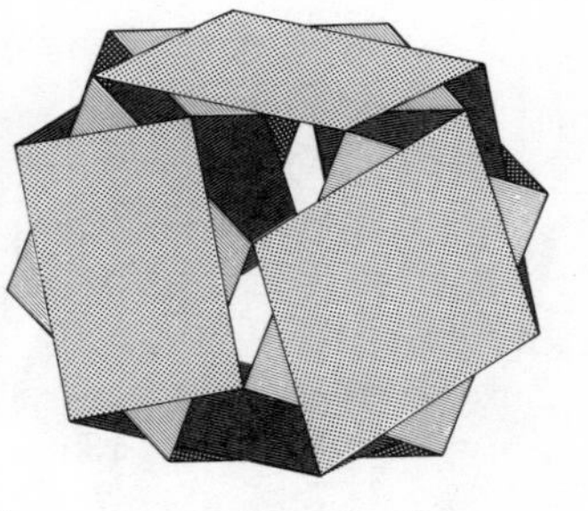

$O[\{4\}^6\{4\}^6\ 90°]$
(O_4)

d) Icosahedral:

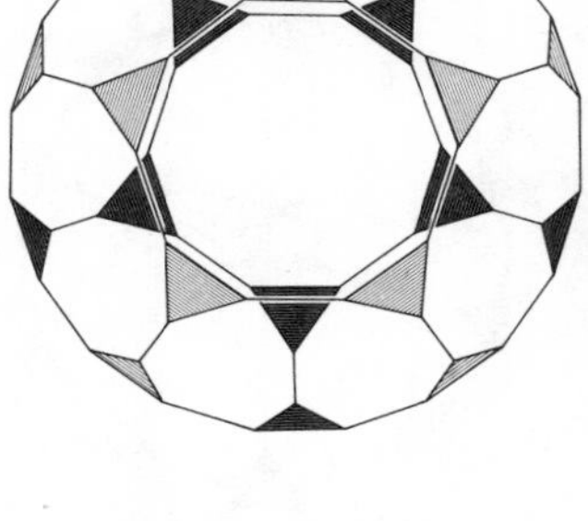

$I[\{2\}^{30}\{3\}^{20}\ 20°54'19"]$
(I_1)

$I[\{2\}^{30}\{5\}^{12}\ 31°43'03"]$
(I_2)

(*) R.B. Fuller's "Jitterbug"

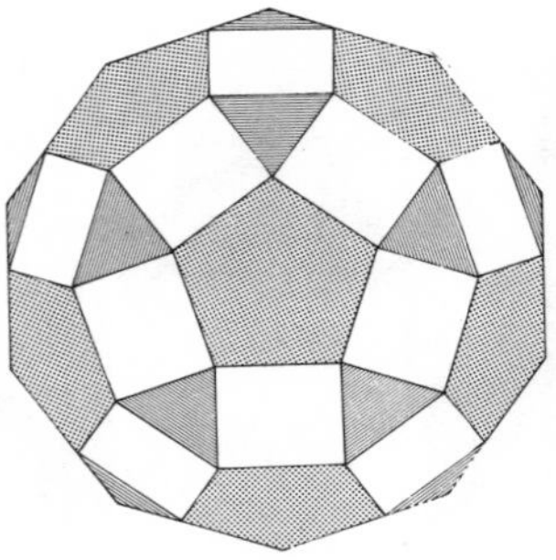

I[$\{3\}^{20}\{5\}^{12}$ 37°22'39"]
(I_3)

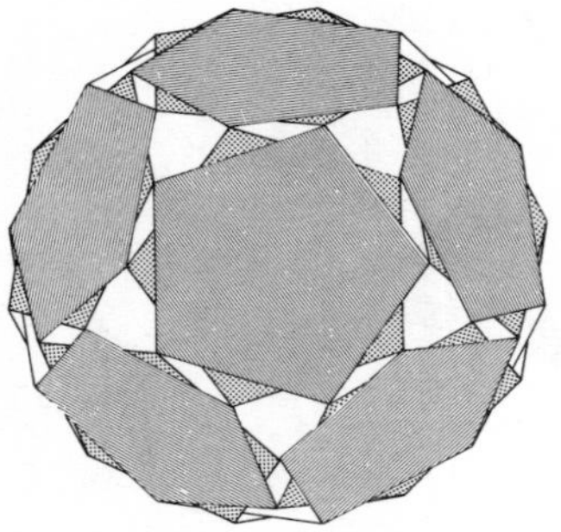

I[$\{5\}^{12}\{5\}^{12}$ 63°26'06"]
(I_4)

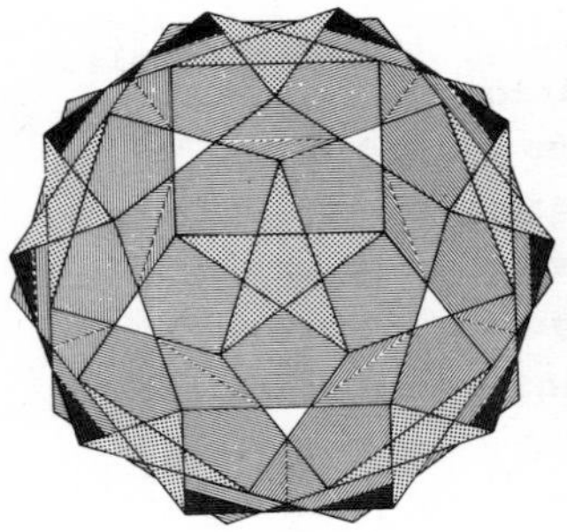

I[$\{5\}^{12}\{\frac{5}{2}\}^{12}$ 63°26'06"]
(I_{11})

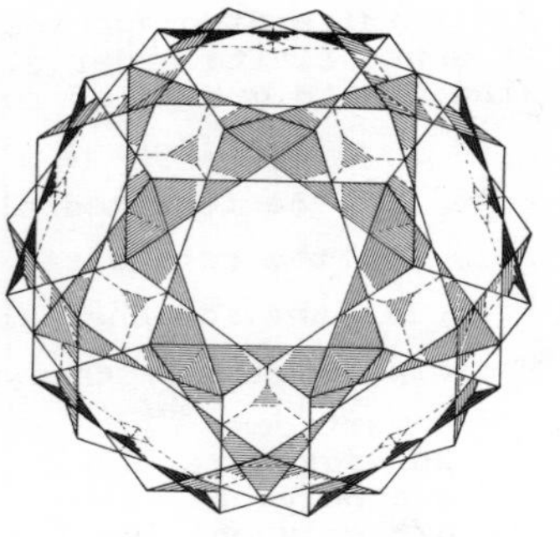

I[$\{3\}^{20}\{5\}^{12}$ 79°11'16"]
(I_{12})

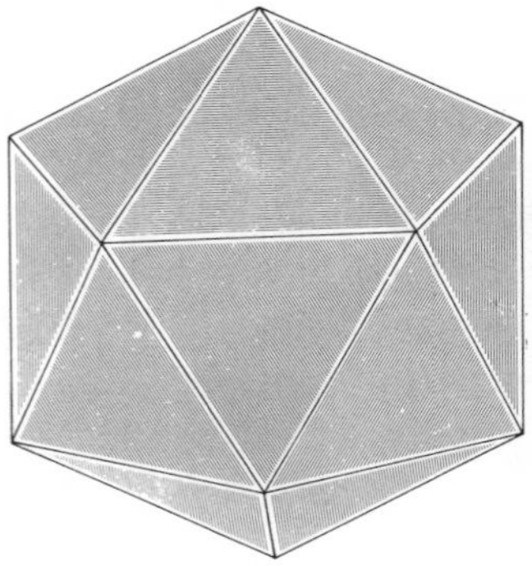

I[$\{2\}^{30}\{3\}^{20}$ 69°05'42"]
(I_5)

I[$\{2\}^{30}\{\frac{5}{2}\}^{12}$ 58°16'57"]
(I_6)

I[$\{3\}^{20}\{\frac{5}{2}\}^{12}$ 37°22'39"]
(I_{13})

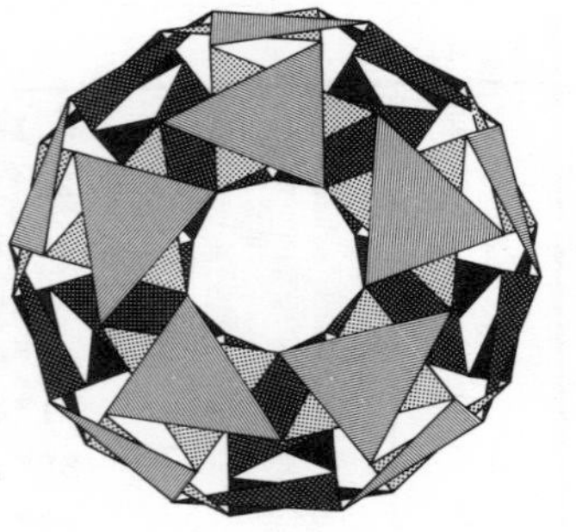

I[$\{3\}^{20}\{3\}^{20}$ 41°48'37"]
(I_{14})

I[$\{3\}^{20}\{\frac{5}{2}\}^{12}$ 79°11'16"]
(I_7)

I[$\{\frac{5}{2}\}^{12}\{\frac{5}{2}\}^{12}$ 63°26'06"]
(I_8)

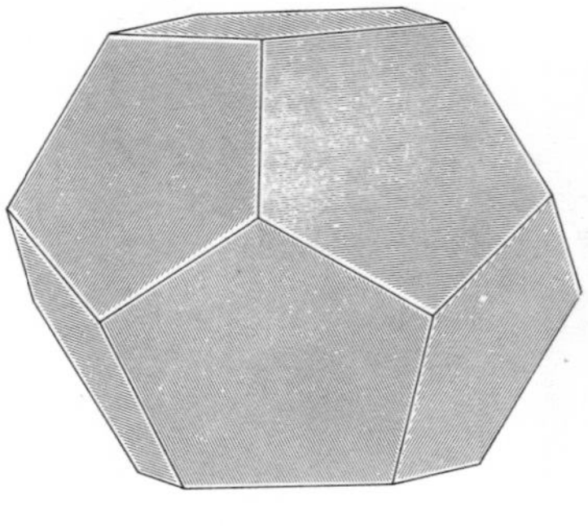

I[$\{2\}^{30}\{5\}^{12}$ 58°16'57"]
(I_9)

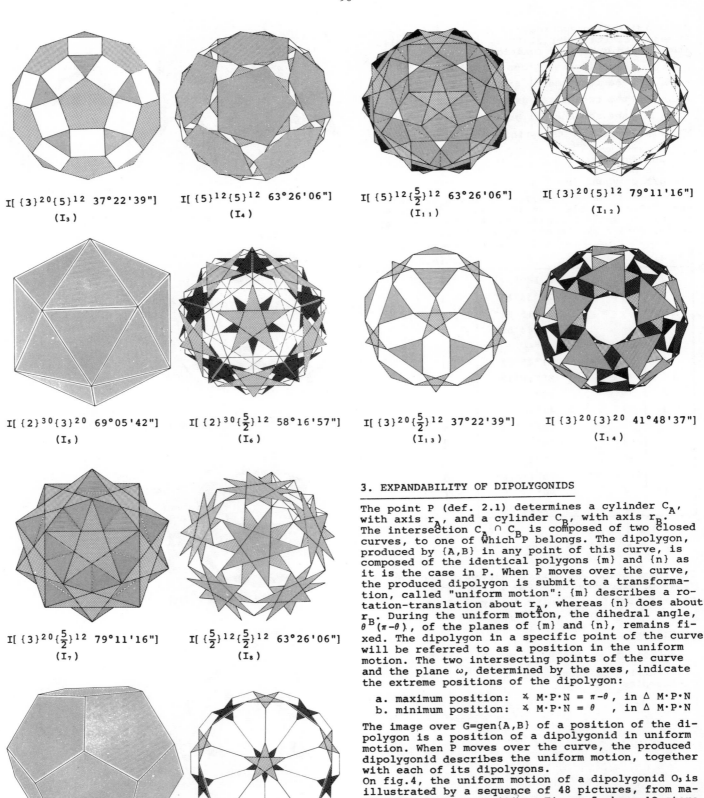

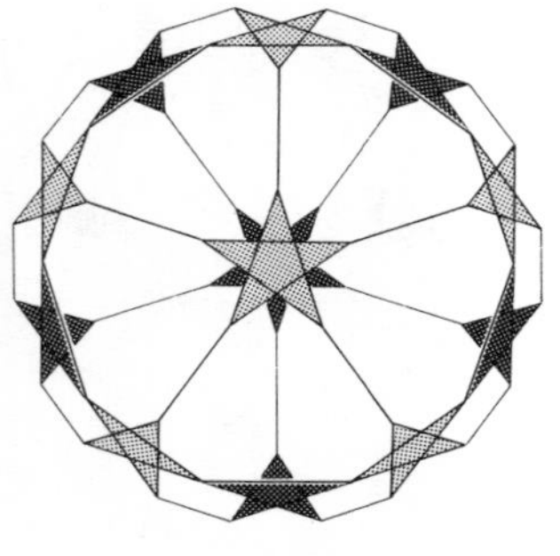

I[$\{2\}^{30}\{\frac{5}{2}\}^{12}$ 31°43'03"]
(I_{10})

3. EXPANDABILITY OF DIPOLYGONIDS

The point P (def. 2.1) determines a cylinder C_A, with axis r_A, and a cylinder C_B, with axis r_B. The intersection $C_A \cap C_B$ is composed of two closed curves, to one of which P belongs. The dipolygon, produced by {A,B} in any point of this curve, is composed of the identical polygons {m} and {n} as it is the case in P. When P moves over the curve, the produced dipolygon is submit to a transformation, called "uniform motion": {m} describes a rotation-translation about r_A, whereas {n} does about r_B. During the uniform motion, the dihedral angle, θ ($\pi-\theta$), of the planes of {m} and {n}, remains fixed. The dipolygon in a specific point of the curve will be referred to as a position in the uniform motion. The two intersecting points of the curve and the plane ω, determined by the axes, indicate the extreme positions of the dipolygon:

a. maximum position: $\measuredangle\ M\cdot P\cdot N = \pi-\theta$, in $\Delta\ M\cdot P\cdot N$
b. minimum position: $\measuredangle\ M\cdot P\cdot N = \theta$, in $\Delta\ M\cdot P\cdot N$

The image over G=gen{A,B} of a position of the dipolygon is a position of a dipolygonid in uniform motion. When P moves over the curve, the produced dipolygonid describes the uniform motion, together with each of its dipolygons.
On fig.4, the uniform motion of a dipolygonid O_3 is illustrated by a sequence of 48 pictures, from maximum to minimum position. Figure 5 shows 10 stereographic pictures of successive positions between the extreme positions of a dipolygonid I_{14}, with equal edge length (a uniform dipolygonid). This dipolygonid in all positions is composed of twenty coplanar pairs of triangles.

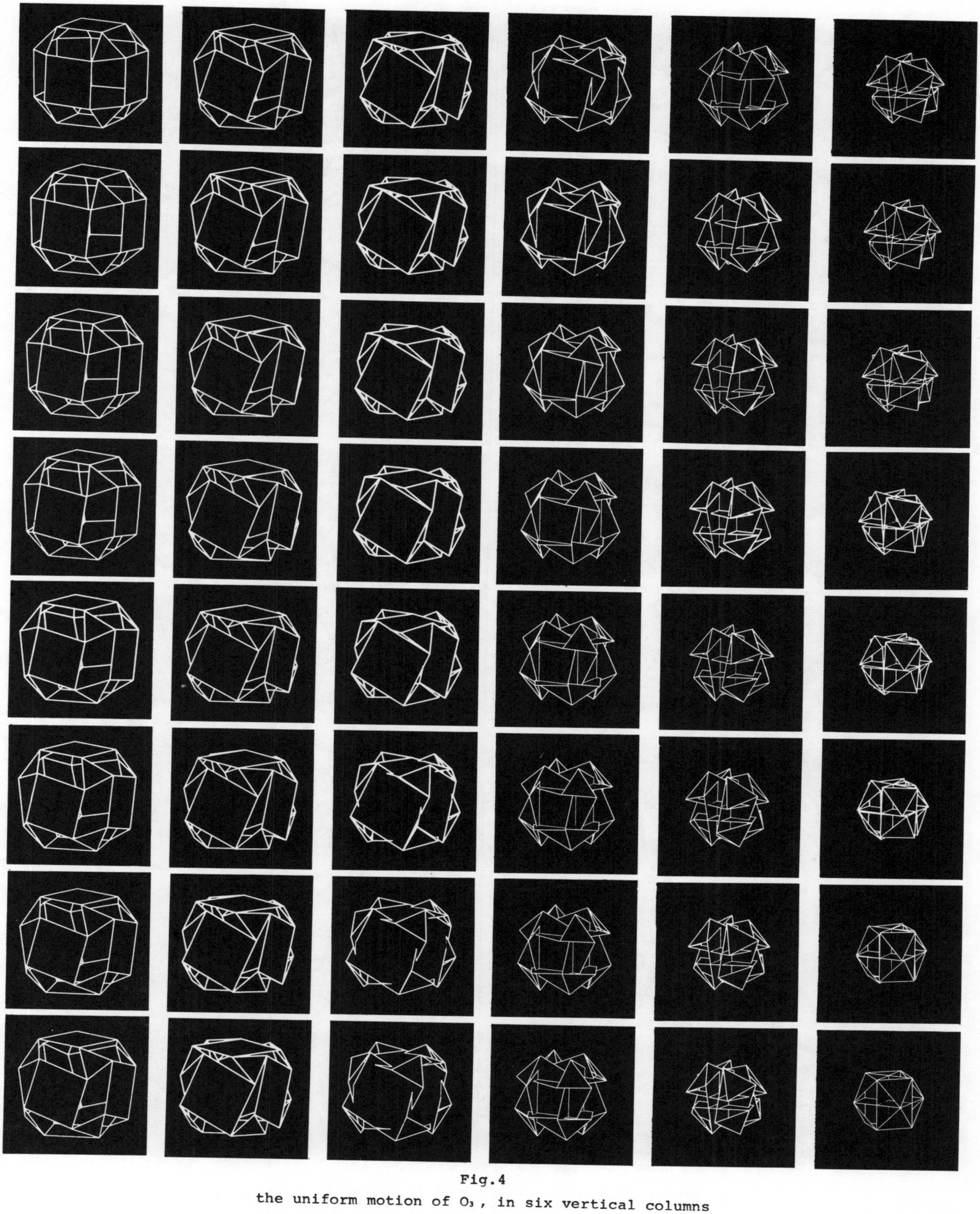

Fig.4

the uniform motion of O_3, in six vertical columns

Fig.5

ten stereographic pictures of positions of I_{14},
in two vertical columns

4. TRIADIC POLYHEDRA

A polyhedron is a finite set of polygons such that every side of each belongs to just one another (*) with the restriction that no subset has the same property (ref.2). The polyhedron is "uniform" if its faces are regular polygons, while each vertex can be transformed into any other by a symmetry-operation (vertex-transitivity in the symmetry group). In any position, a dipolygonid can be extended to a vertex-transitive polyhedron in G, having the same vertices as the very dipolygonid.
The line segment [A(P),B(P)] is the edge of a regular polygon {k}, produced by the rotation BA^{-1} in A(P), and which is called a structural polygon (ref.5). The triangle Δ A(P)·P·B(P) is a triadic triangle, and the set {{m},{n},{k}} is a triadic set of polygons. The triple of polygonal values G[m,n,k] is a basic triad, of which 14 exist, a-part of the dihedral basic triads (ref.5).
The image of a triadic set of polygons over G is called a tripolygonid, being a union of three dipolygonids, resp. composed of polygons {m} and {n}, {m} and {k}, {n} and {k}.
When adding the triadic triangles to the tripolygonid, a TRIADIC POLYHEDRON is obtained. If the order of BA^{-1} is u, the triadic polyhedron in that basic triad is composed of:

- g/s polygons {m}
- g/t polygons {n}
- g/u (structural) polygons {k}
- g triadic triangles

On fig.6, two positions out of the sequence on fig.4 have been extended to triadic polyhedra. (addition of 12 structural digons and 24 triadic triangles in basic triad O[2,3,4].)

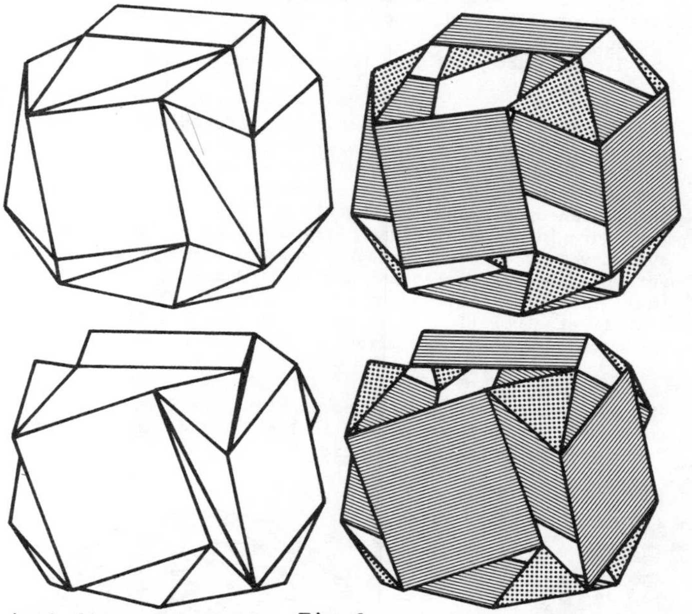

(triadic polyhedra) Fig.6 (positions of dip.)

Each type of triadic polyhedron is associated with a basic triad. Hence, there exist 14 types of triadic polyhedra in T,O and I. An interesting case of triadic polyhedron occurs when the three edge lengths become equal, i.e. when the triadic triangles become regular. Among the set of 75 uniform polyhedra (refs.2&3) the snub polyhedra are such. The triadic polyhedra in a basic triad are then considered as distorted versions of a snub polyhedron, representing that triad.(**)

(*) Skilling (Ref.3) allows an extension: an edge may belong to an even number of faces.
(**) Dihedral triadic pol. are distorted antiprisms.

5. EXPANDABLE POLYHEDRAL STRUCTURES AND TRANSFORMABLE SPACE FILLINGS

Models of dipolygonids can be built in cardboard, metal or plastic material, provided a well functioning system for connecting the vertices is used. After experiments with various materials, a vinyl plastic gave the most satisfying results, both in appearance, as in the assembling of the parts. These parts are double {3} , {4} , {5} or {5/2}. A special clip was designed to connect two or more vertices, while it can be removed easily and used again. The models prooved to be of educational value for the study of polyhedra, and transformations between these. Some of the models remain rigid in all positions (T_2, O_4, I_4, I_8), while others loose their rigidity in the maximum position, where they can be folded flat (I_{14}, $2{\times}T_2$, $2{\times}O_3$, $2{\times}I_3$, $2{\times}I_{13}$). The same parts can be used for multiples of dipolygonids (e.g. an expandable tetrahelix) or space fillings, that transform into other space fillings (cubes, tetrahedra + octahedra).
By adding pyramidal parts to the faces of a dipolygonid, various expandable polyhedral structures are obtained. Figure 8 shows a uniform I_4, with 12 pentagonal pyramids added to the outer faces, thus creating an expandable, small stellated dodecahedron (cardboard model). Figure 7 shows a cardboard model of a uniform I_8 (12 coplanar pairs of {5/2}).

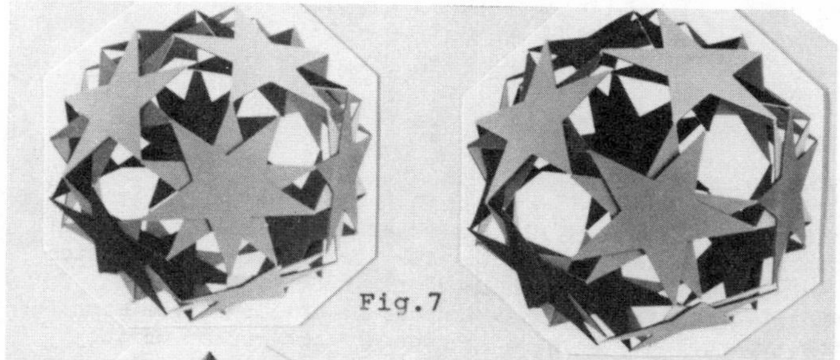

Fig.7

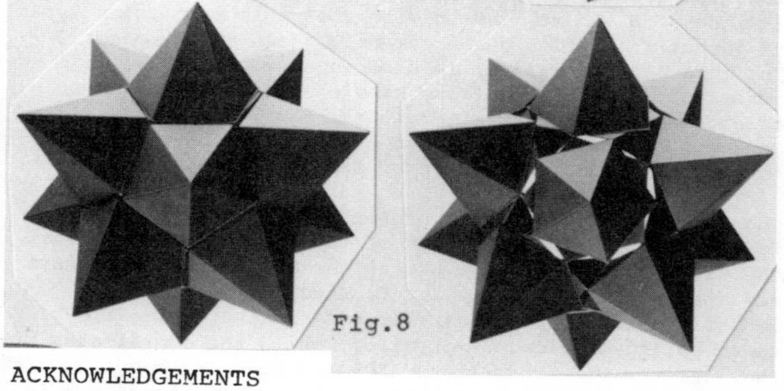

Fig.8

ACKNOWLEDGEMENTS

Special thanks must be extended to Dr.J.Skilling, DAMTP, University of Cambridge, England, who provided the computer graphics for figs.4&5.

REFERENCES

1. J.D. CLINTON, *A geometric transformation concept for expanding rigid structures.* NASA report NGR-14-008-002, 1970
2. H.S.M. COXETER, M.S. LONGUET-HIGGINS, J. MILLER *Uniform polyhedra*, Phil.Trans. 246A, 401-450 London, 1954
3. J. SKILLING, *The complete set of uniform polyhedra*, Phil. Trans. 278A, 111-135, London 1975
4. R.D. STUART, *Polyhedral and mosaic transformations*, Student publications of the School of Design, University of North Carolina, Raleigh, 1963
5. H.F. VERHEYEN, *Dipolygonids, mobile generators of uniform polyhedra*, University of Antwerp, preprint series UIA, 79-40, Antwerp 1979

FORMEX PROCESSING OF SYMMETRIC STRUCTURAL SYSTEMS

Jaime S. SANCHEZ ALVAREZ

Architect (University of Mexico), PhD (University of Surrey, England)

MERO-Raumstruktur GmbH & Co. Würzburg, West Germany

The present paper shows a way in which the mathematical system known as formex algebra can be used to represent and process structural systems, which can be symmetrically subdivided in a number of sub-structures. Various interconnection patterns are represented within a general scheme of formex formulation, which is also related to a set of geometric particulars. Formex formulation, due to its numerical nature, can be conveniently used to implement processes, such as data generation and graphics, in the computer aided analysis and design of space structures.

GENERAL SCHEME OF FORMULATION

The configuration of Fig 1 is used to introduce a general scheme of formex formulation for structural systems which can be subdivided in a number of symmetry parts.
The illustrated configuration can be obtained from mapping the triangulated faces of a regular icosahedron on the surface of a sphere, thus giving rise to a particular type of geometric configuration commonly known as 'geodesic'. The term 'configuration' is used in a general sense to refer to a collection of any kind of objects, such as an arrangement of structural elements. Also, while the concepts of formex algebra can be found in the references, the attention is centered in applications.

The type of object to be represented through the present formex formulation is the set of topological properties of the given configuration, whereas the geometric particulars are specified separately. Interconnection patterns and disposition of constituent elements in a structure are known as 'topological' properties, while the coordinates of nodal points with respect to a chosen coordinate system are referred to as 'geometric' particulars of the configuration.

The formex formulation for the above given configuration is organized in terms of the symmetries of the regular icosahedron. Thus, the whole system is represented through symmetric transformations of a basic sub-structure, for which topological and geometric particulars are specified in detail.

The 20 equal faces of the icosahedron are conventionally grouped in four regions, namely: North-Pole, North-Belt, South-Belt and South-Pole as it is indicated in Figs 1 and 2. Furthermore, Fig 2 shows the triangulated faces of the icosahedron mapped on the plane I1-I2. Also, the sub-structure labelled as Part-1 is identified as the basic symmetry part of the structure, whereas Parts 2 to 20 are interpreted as translations and reflections of the basic part. This organization proves to be of use later to identify the position of a particular sub-structure in the whole three-dimensional arrangement.

The purpose of graphically representing the given configuration on the plane as in Fig 2 is to simplify the basis for the formex formulation.

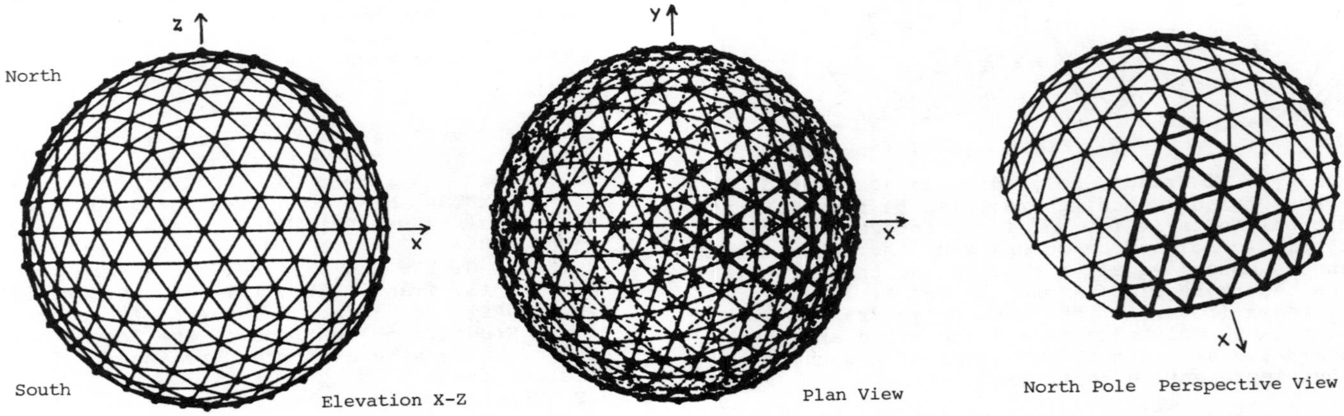

Fig 1 Geodesic configuration showing basic symmetry part in thick line

Here, the drawing is simultaneously interpreted as an interconnection pattern and as an 'intrinsic plot' of a formex that is capable of representing the topological entity. In the intrinsic plot, the nodal points at which the bar-like elements of the structure interconnect are identified by integer coordinates only. Little circles and dots represent nodes, while continuous line segments correspond to structural members. Also, the smallest modular translations of a triangular arrangement of bar-like elements within the basic sub-structure are given as 3 and 2 units, along the I1 and I2 Cartesian axes, respectively.

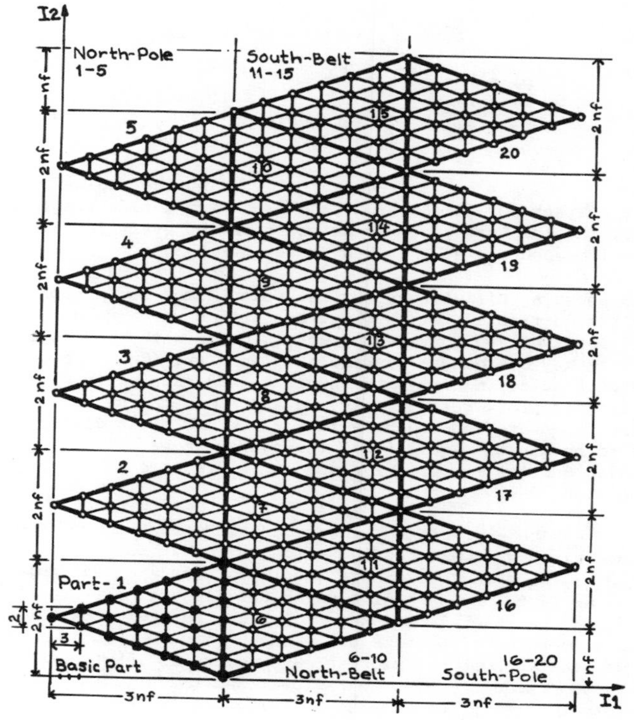

Fig 2

Some of the subsequently adopted forms of formex processing require that disposition of nodes and interconnection patterns are represented with different formices.
The general scheme of formulation covers this possibility, which also simplifies the eventual automated production of lists of node coordinates and structural members. Such lists constitute common forms of input data for further computer processing, for instance computer aided structural analysis.

If a formex F1 is taken to represent topological properties, either disposition of nodes or interconnection patterns, within the basic sub-structure, i.e. Part-1, then the corresponding topological properties of the remaining substructures, i.e. Parts 2 to 20, can be represented as functions of the formex F1. Thus,

the North-Pole, Parts 1 to 5, may be formulated as

$$NP = \sum_{k=0}^{4} \text{tran}(2,k*2nf):F1$$

where nf is the 'frequency' or number of segments in which the typical edge of the icosahedron is subdivided to triangulate the typical face. * indicates multiplication and 2nf is the measure of the basic sub-structure along direction I2 in the intrinsic plot of Fig 2. Σ— is used to indicate, in a short form, the concatenation or composition of a series of formices.

The North-Belt, Parts 6 to 10, may be represented as a reflection of the formex representing the North-Pole as:

$$NB = \text{ref}(1,3nf):NP$$

where 3nf is the measure of the basic part along direction I1 in the intrinsic plot.

The South-Belt, Parts 11 to 15, may be represented as a planar translation of the formex representing the North-Pole as:

$$SB = \text{tranid}(3nf,nf):NP$$

The South-Pole, Parts 16 to 20, may be also represented as a planar translation, this time of the formex representing the North-Belt as:

$$SP = \text{tranid}(3nf,nf):NB$$

Finally, the whole configuration may be given as the composition of the formices representing the North and South Regions as:

$$F = NP \# NB \# SB \# SP$$

THE BASIC SUB-STRUCTURE

The symbol F1 in the above formulation is clearly used in a general sense to represent the topological properties of the basic symmetry part. Thus, any kind of disposition of nodes or interconnection pattern can be assigned to the basic part via F1 or another symbol to obtain different topological arrangements for the (whole) system.

In this way, the set of 'basic' nodes represented with dots in Fig 2 may be given as

$$g1 = \sum_{i=0}^{nf} \sum_{j=0}^{i} \text{projid}(3i,nf-i+2j):[0,0,0]$$

where g1 replaces F1 in the general formulation and the signet [0,0,0] acts as a 'seed' or generant for the nodes.

The interconnection pattern of the basic part can be now conveniently obtained in terms of g1 as

$$F1 = \text{vin}\ (2,\sqrt{10}):g1$$

where, the vinculum function 'generates' all the two-plex cantles whose 'metrum' is between 2 and $\sqrt{10}$ and whose signets are all in g1. In Fig 2, a line segment represents a cantle and the metrum of this cantle is simply the length of the line segment.

GEOMETRIC PARTICULARS

The geometric definition of the configuration given in Fig 1 is obtained by numerically specifying the mapping of the previously treated topological properties on a sphere.

The three-dimensional Cartesian coordinate system in Fig 1 is the reference frame for the present coordinate specifications and its origin coincides with the centre of the chosen sphere. Thus, the coordinates x,y,z of a point whose signet is [I1,I2,I3] are given as

$$x = Lq1$$
$$y = Lq2$$
$$z = Lq3$$

where q1,q2,q3 are the coordinates of a point Q at the surface of the icosahedron and L is a positive factor, which 'projects' the point onto the sphere and is given as

$$L = \sqrt{(r + tI3)^2 / (q1^2 + q2^2 + q3^2)}$$

with $(q1^2 + q2^2 + q3^2) > 0$, because no point at the faces of the icosahedron coincides with the origin of the reference system. Here, r is the radius of the desired geodesic configuration, t is the structural height which adds to r and I3, the third index in a signet, is used as a 'layer' indicator.

The point Q(q1,q2,q3) is obtained from symmetry operations on a corresponding point P(p1,p2,p3) in the basic sub-structure. The number assigned to each sub-structure in Fig 2 determines the particular transformations for each part. The symmetry operations may be summarized as follows

$$Q = Rz(RyP)$$

where Ry and Rz denote non-commutative rotations of the point P around the y and z axes, respectively.
The corresponding transformation matrices are

$$Ry = \begin{bmatrix} \cos T & 0 & -\sin T \\ 0 & 1 & 0 \\ \sin T & 0 & \cos T \end{bmatrix}$$

with T = T1 for parts 1 to 5,
or = T1 + 2Tr for parts 6 to 10,
or = 180 - T1 - 2Tr for parts 11 to 15,
or = 180 - T1 for parts 16 to 20.
Also, Tr = atan $(nf\sqrt{3}/3S)$,
T1 = atan $(2nf\sqrt{3}/3S)$
and S = $nf\sqrt{(5 + 6\cos 72)/3}$, being geometric constants of the icosahedron expressed in terms of the frequency nf, as illustrated in Fig 3 below

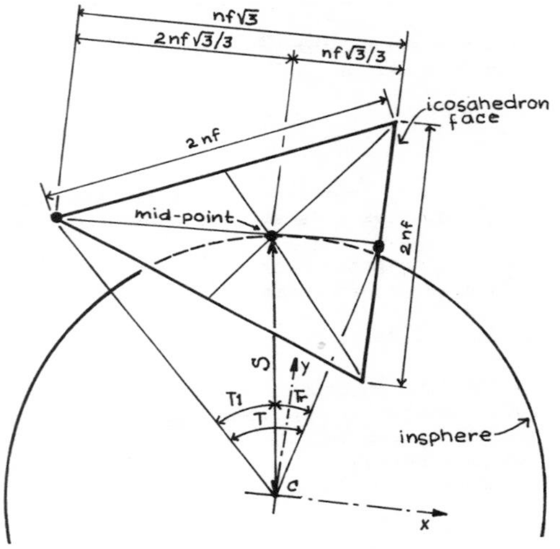

Fig 3

In a similar way

$$Rz = \begin{bmatrix} \cos G & -\sin G & 0 \\ \sin G & \cos G & 0 \\ 0 & 0 & 1 \end{bmatrix}$$

with G = (n-1)*72° for parts 1 to 10
or = (n-1)*72° + 36° , for parts 11 to 20.

The point P(p1,p2,p3) is in turn obtained as follows

$$p1 = \sqrt{3} * (I1 - nf*n1) / 3$$
$$p2 = I2 - (m1*nf + m2*2nf)$$
$$p3 = S$$

where nf = is the frequency of subdivision of the geodesic configuration

with

n1 = 2
m1 = 1
m2 = 0,1,2,3,4 for parts 1,2,3,4,5, respectively — for the North Pole

n1 = 4
m1 = 1
m2 = 0,1,2,3,4 for parts 6,7,8,9,10, respectively — for the North Belt

n1 = 5
m1 = 2
m2 = 0,1,2,3,4 for parts 11,12,13,14,15, respectively — for the South Belt

n1 = 7
m1 = 2
m2 = 0,1,2,3,4 for parts 16,17,18,19,20, respectively — for the South Pole

and S = $nf\sqrt{(5 + 6\cos 72)/3}$, being the radius of the insphere of a regular icosahedron with a side length of 2nf. The insphere touches the mid-point of every face in the icosahedron, as sketched above in Fig 3.

The above specifications for the point P may be interpreted in the intrinsic plot as the translations which make the mid-point of every symmetry part coincide with the origin of the x-y coordinate system. In plan view, the x and y axes coincide in turn with the I1 and I2 axes of the intrinsic plot, respectively. In addition, the sub-structures are reduced by a factor of $\sqrt{3}/3$ along direction I1 to obtain the equilateral triangular faces of the icosahedron and the former are placed at a distance p3 above the plane x-y.

As it may be appreciated, the geometric definition of the above given geodesic configuration is rather complex. However, once the coordinate specifications are available, they apply to any type of topological properties represented through the general formex formulation. Furthermore, the above coordinate specifications may be implemented in such a way that a computer process may require as input data only the frequency nf, the radius r and the structural heigth t, if any, of a geodesic structure to generate the necessary geometric information, such as node coordinates.

FURTHER INTERCONNECTION PATTERNS

The preceding formex formulation can be implemented in a similar manner, so that the frequency nf is the only input item required to generate the required topological information. Further features of the suggested processes may be the selection of sub-sets or cuts of the whole and choice among various combinations of topological arrangements.
As an illustration of the latter, the set of formices F2 to F11 are next formulated to represent a series of different interconnection patterns for the basic symmetry part and, as stated, these formices are interchangeable with F1 in the general formex formulation. Also, results of some geometric mappings are graphically recorded next to their corresponding formices.

Unless otherwise specified, formices in the sequel are obtained in the form

$$Fn = An \# Bn \# Dn,$$

where n is a subscript, A represents the 'ex-net' normally at the outer layer of the geodesic configuration, B represents the 'in-net' normally at the inner layer and D corresponds to the 'inter-net' or set of members interconnecting different layers. Also, when different line styles are used to plot the nets, continuous line is used for the ex-net, dashed line for the in-net and dotted line for the inter-net.

The formices g1 and F1 obtained to represent the basic disposition of nodes and the basic interconnection pattern of the configuration in Fig 1, are further operated on to obtain the following formices:

Formex F2. Fig 4

A2 = proj(3,1):F1

B2 = F1

D2 = vin(1,1):(g1 # proj(3,1):g1)

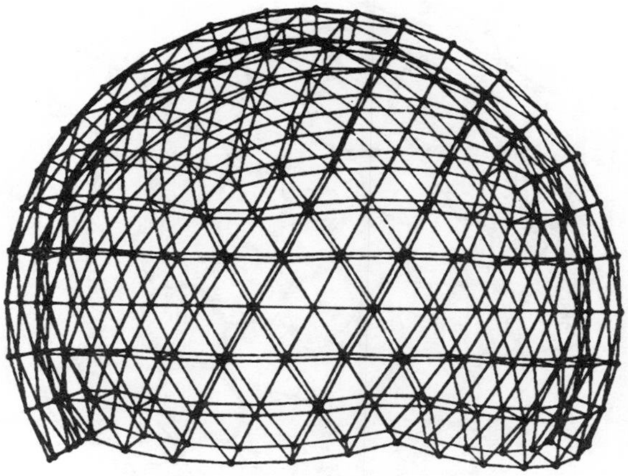

Fig 4 North Regions and South Belt. Elevation X-Z

Formex F3. Fig 5

A3 = proj(3,1):F1,

B3 = tran(1,2):lux(gx):F1

with $\quad gx = \sum_{j=0}^{nf} \text{projid}(3nf,2j):[0,0,0]$

and

D3 = vin($\sqrt{3}$,$\sqrt{3}$):g3 # vin ($\sqrt{5}$,$\sqrt{5}$):g3

with $\quad$ g3 = proj(3,1):g1 # tran(1,2):lux(gx):g1

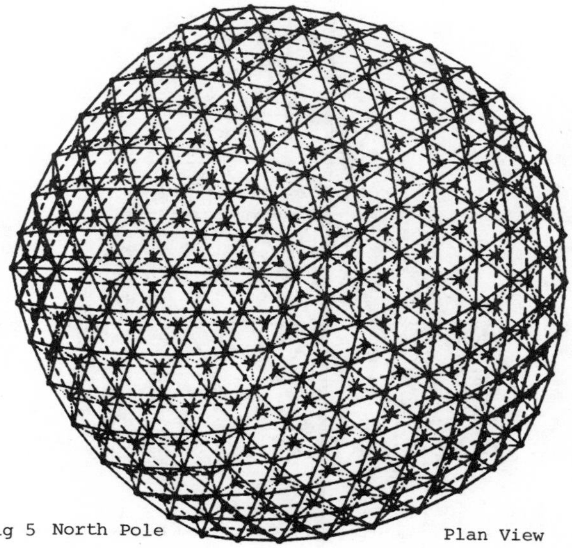

Fig 5 North Pole Plan View

Formex F4. Fig 6

A4 = A3

B4 = lux(gx4):B3

with $\quad gx4 = \sum_{i=0}^{2} \sum_{j=0}^{(nf/3)-i} \sum_{k=0}^{j} \text{projid}(9j+6i+2, nf-6k+3j):[0,0,0]$

where nf is a multiple of 3.

D4 = vin($\sqrt{3}$,$\sqrt{3}$):g4 # vin($\sqrt{5}$,$\sqrt{5}$):g4,

with $\quad$ g4 = lux(gx4):g3.

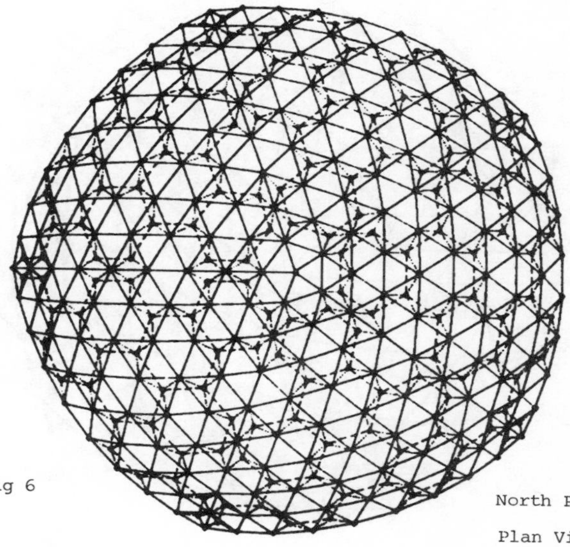

Fig 6 North Pole Plan View

Formex F5. Fig 7

F5 = lux(gx5):F1

with $\quad gx5 = \sum_{i=0}^{nf/2} \sum_{j=0}^{i} \text{projid}(6i,nf+4j-2i):[0,0,0]$

where nf is even.

The disposition of nodes may in turn be given as

g5 = lux(gx5):g1.

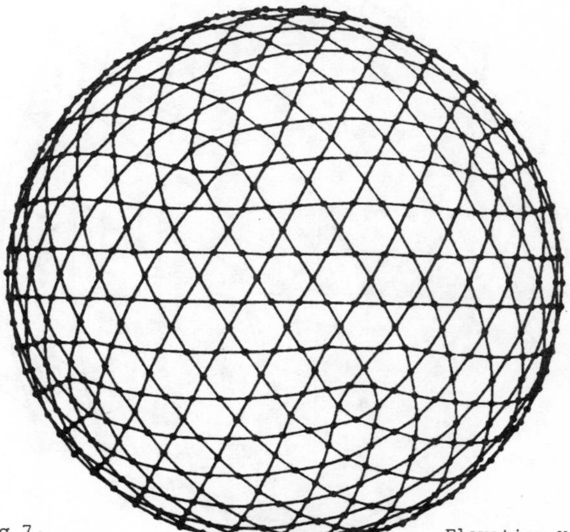

Fig 7 Elevation X-Z

Formex F6. Fig 8

A6 = proj(3,1):F5

B6 = dilid(2,2):F1

with an even frequency nf halved for F1.

D6 = dil(3,1/3):(vin$\sqrt{13}$,$\sqrt{13}$):g6 # vin ($\sqrt{19}$,$\sqrt{19}$):g6)

where g6 = proj(3,3):g5 # dilid(2,2):g1

with the even nf also halved for g1.

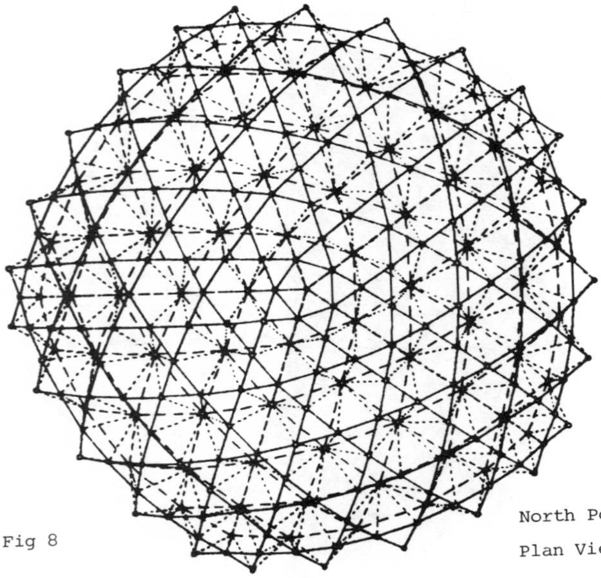

Fig 8

Formex F7. Fig 9

A7 = A6

B7 = F1

D7 = dil(3,1/3):(vin(3,3):g7 # vin($\sqrt{13}$,$\sqrt{13}$):g7 # vin($\sqrt{19}$,$\sqrt{19}$):g7)

where g7 = proj(3,3):g1 # g1.

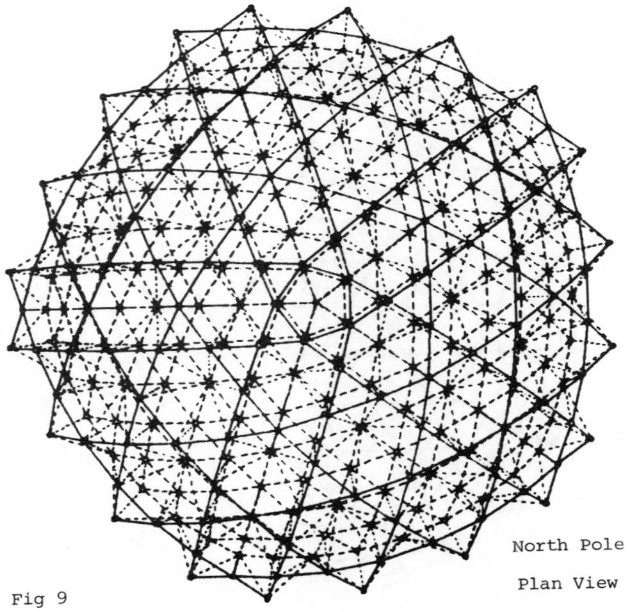

Fig 9

Formex F8. Fig 10

A8 = dil(2,1/2):vin(4,$\sqrt{20}$):dil(2,2):g8

where $g8 = \sum_{i=0}^{(nf/2)-1} \sum_{j=0}^{1} \sum_{k=0}^{i}$ projid(2j+6i+2, nf+2+4k-2j):[0,0,1]

with an even nf.

B8 = F5

D8 = vin($\sqrt{3}$,$\sqrt{5}$):(g5 # g8).

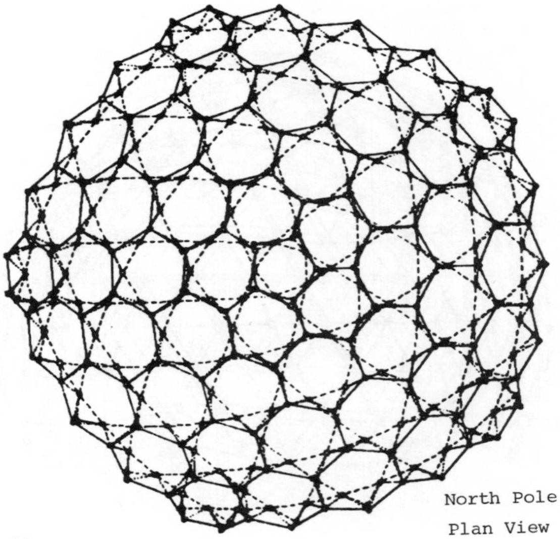

Fig 10

Formex F9. Figs 11 and 12

A9 = proj(3,1):F1

B9 = proj(3,0):dil(1/2,1/2):A8

D9 = vin($\sqrt{2}$,2):g9

where g9 = g1 # proj(3,0):dil(1/2,1/2):g8,

but with nf, any frequency, replacing nf/2 in g8.

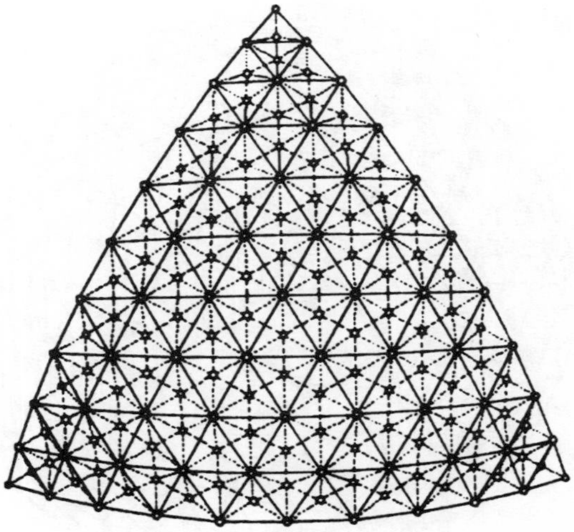

Fig 11 One Symmetry Part. View from above the mid-point

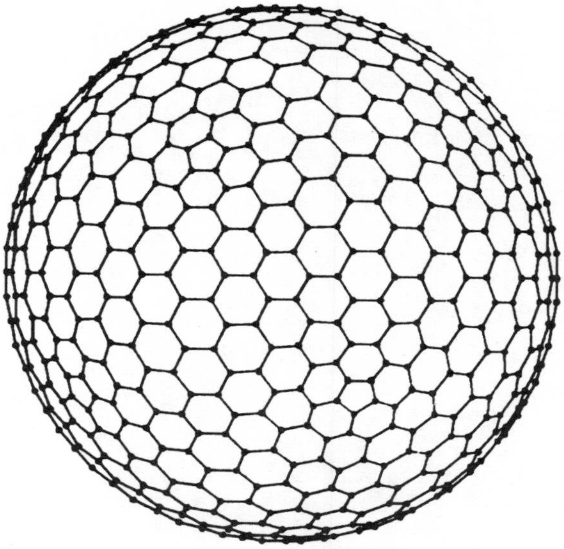

Fig 12 In-net B9 Elevation X-Z

Formex F10. Fig 13

F10 = D9

or $F10 = \sum_{i=0}^{1} \sum_{j=0}^{nf-1-i} \sum_{k=0}^{j} \text{tranid}(3j,2k-j):\text{ref}(1,3):b10$

with $b10 = \text{tran}(2,nf):\{[0,0,1;\ 2,0,0],\ [2,0,0;\ 3,1,1],[2,0,0;\ 3,-1,1]\}$

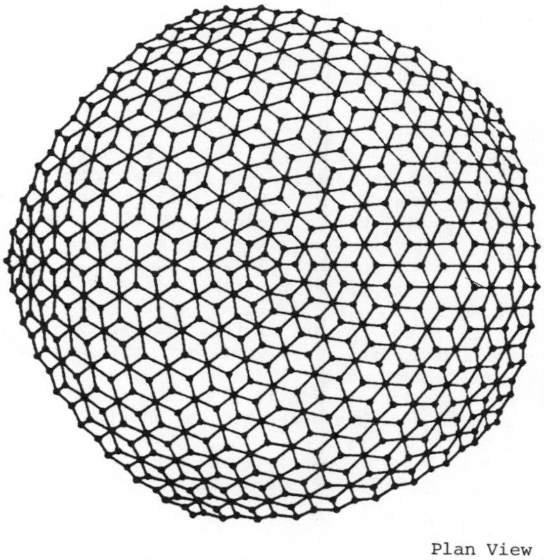

Plan View

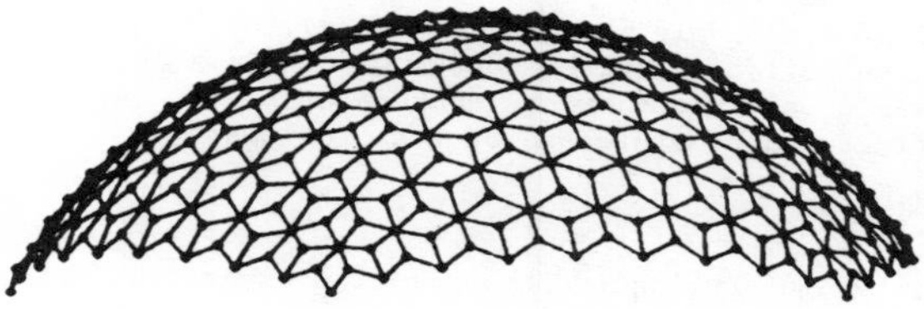

Elevation X-Z

Fig 13 North Pole

Formex F11. Fig 14

$F11 = \sum_{i=2}^{nf+1} \text{conex}(gi):F1$

with $gi = \sum_{j=1}^{i} \text{projid}(3i-3,nf-1+2j-i):[0,0,0]$

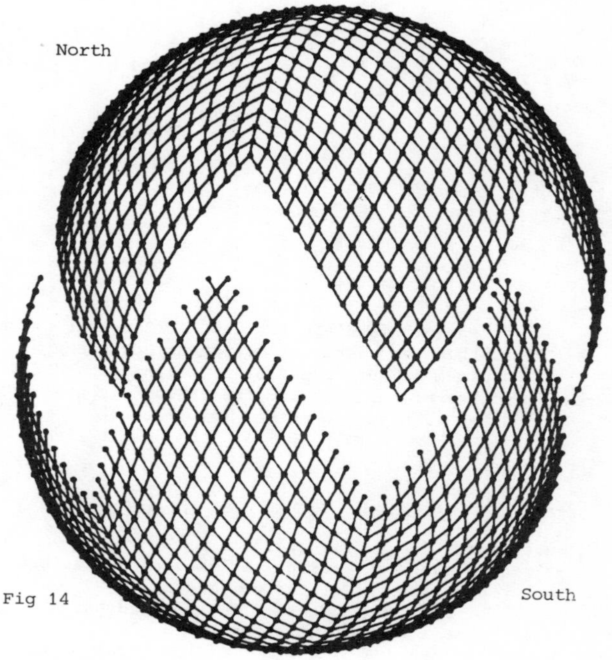

Fig 14

CONCLUDING REMARKS

The present paper aims to illustrate the application of formex algebra in the representation and processing of structural configurations and systems possessing certain symmetry properties are used to present the components of a processing strategy in some detail. For the sake of brevity, however, there are some special situations which are not explicitly solved here, such as the elimination of repeated elements at symmetry lines when joining sub-structures or the eventual interconnection of elements between symmetry parts. Finally, the included formulation does not yield fixed or unique forms of representing and processing structural configurations, but represents an attempt to show the convenience and advantages of formex algebra as an abstract tool.

REFERENCES

HARISTCHIAN, M. Formex and Plenix Structural Analysis. PhD Thesis, University of Surrey, England, 1980.

_______ Fortran Subroutines for Formex Functions. Space Structures Research Centre, U.of Surrey, July 1979.

NOOSHIN, H. Formex Formulation of Double Layer Grids. Analysis Design and Construction of Double Layer Grids, Chapt.4, pp.119-183, Ed.Z.S.Makowski, Applied Science Publishers Ltd., London, England, 1981.

_______ Formex Formulation of Braced Barrel Vaults. Notes to the course Analysis Design and Construction of Braced Barrel Vaults, pp.C1-C51, U.of Surrey,9.1983

SANCHEZ ALVAREZ, J. Formex Formulation of Structural Configurations. PhD Thesis, U.of Surrey, Engl.1980.

_______ Formex Formulation and Geometric Specification of Space Structures. Notes to the course Applications of Formex Algebra in Data Generation and Computer Graphics, U.of Surrey, England, Sept. 1982

_______ Formex Formulation of Braced Domes. Analysis, Design and Construction of Braced Domes. Ed.Z.S.Makowski, Applied Science Publishers, London, to appear: 1984.

OPTIMIZATION OF SPHERICAL NETWORKS FOR GEODESIC DOMES

T. TARNAI, Dr., C. Eng., Appl. Math.

Senior Research Fellow,

Hungarian Institute for Building Science

Budapest

This paper investigates how a triangular network of a given topology can be constructed on the sphere so that the number of different edge-lengths is a minimum. To find the networks containing even minimum number of different triangles and different nodes a combinatorial approach is used. If the number of triangles in the network is great, the combinatorial way becomes difficult. For these cases a new subdivision method is presented.

NOTATION

a_i	edge-length (chord factor),
b,c	pair of numbers generating the skew tessellation,
E	number of all edges of the polyhedron,
e_d	number of different edges of the polyhedron,
F	number of all faces of the polyhedron,
f	sum of the degrees of freedom of the nodal points on one third(sixth) part of a face of the icosahedron,
f_d	number of different faces of the polyhedron,
m	number of different edge-lengths determined by the degrees of freedom of the nodal points,
N	number of different networks with given number of different edge-lengths,
n	maximum number of different edge-lengths considering the icosahedral symmetries,
S(n,m)	Stirling number of the second kind,
T	triangulation number,
V	number of all vertices of the polyhedron,
v_d	number of different vertices of the polyhedron,
η	quotient of the longest and shortest edges of the polyhedron.

1. INTRODUCTION

Designers of geodesic domes endeavour to construct the dome with as little number of different kinds of construction elements as possible. McConnel (Ref. 7) has termed the network of a dome in a given topology "optimal", if the subdivision produces the minimum number of member lengths, panel types and node types. We use the term "optimal" in the same meaning.

The present paper examines how a spherical triangular network of a given topology can be constructed so that the number of different edge-lengths be a minimum. It will further be investigated, what kind of topology has to be chosen in which, in the case of a given number of different edge-lengths, the number of the triangles is a maximum. Since the solutions of these optimization problems are, in general, not unique, that network is considered the "best" one which has some extremal property, e.g. for which the quotient η is the smallest. (Importance of quotient η is treated in Refs 5 and 6). If the number of triangles is not too great, the method of investigation uses a combinatorial technique. However, with increasing the number of triangles the combinatorial way becomes more and more difficult. If the number of triangles is great, a new subdivision method is introduced instead.

The investigations are based on the skew, regular triangular tessellationson the regular icosahedron, which were discovered by Goldberg (Ref. 4) and independently by Caspar and Klug (Ref. 1).

2. TOPOLOGY OF THE NETWORK

Consider the regular tessellation of symbol $\{3,5+\}_{b,c}$ introduced by Coxeter (Ref. 3). This tessellation consists of equilateral triangles, five or more than five (i.e. six) at each vertex, some slightly folded, such that they cover and fill the polyhedral surface of the regular icosahedron $\{3,5\}$; the suffixes b,c indicate that a vertex of icosahedron can be arrived at from an adjacent one along the edges of the tessellation by b steps on the vertices in one direction then c steps after a change in direction by 60°. A part of this tessellation is shown by the continuous lines in Fig. 1 where the large equilateral triangle composed of dashed lines is a face of the icosahedron. A complete family of networks of this type may be seen in an ingenious figure of Clinton (Ref. 2).

The pair of integers b,c generating the tessellation determine the triangulation number T:

$$T = b^2 + bc + c^2$$

which gives the number of triangles lying on a face of the icosahedron (the area of a face of the icosahedron if the area of a triangle is equal to unity). By "blowing up" this tessellation onto a sphere we obtain a spherical triangle polyhedron, where numbers of edges, faces and vertices can be expressed by T as follows:

$$E = 30\,T, \quad F = 20\,T, \quad V = 10\,T + 2.$$

This type of triangular tessellation gives the topological base for our investigations.

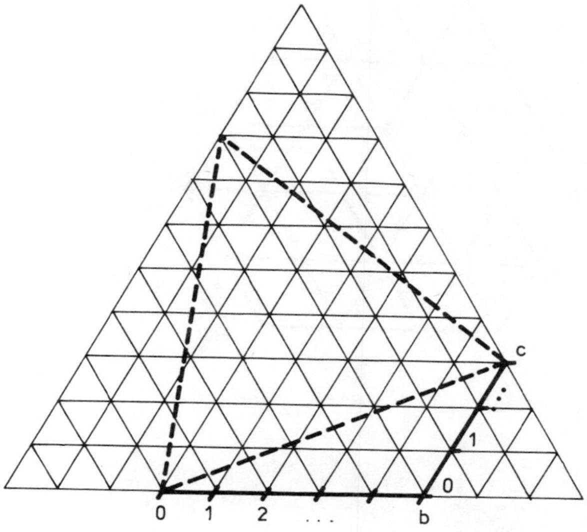

Fig. 1. The meaning of the Coxeter symbol $\{3,5+\}_{b,c}$

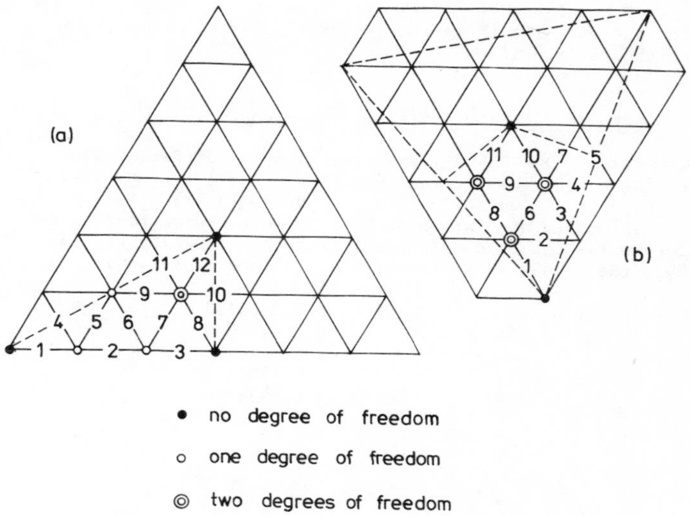

Fig. 2. Definition of numbers f, m, n for a network (a) with (b) without plane of symmetry

3. COMBINATORIAL WAY OF ANALYSIS

In order to decrease the number of different construction elements, the spherical triangular network is required to have rotational symmetries of the icosahedron; that is, 5-fold symmetry with respect to the vertices of the icosahedron, 3-fold symmetry with respect to the centres of the faces of the icosahedron and 2-fold symmetry with respect to the midpoints of the edges of the icosahedron. In this way, it is sufficient to execute the investigations only on one third part of a face of the icosahedron (Fig. 2(b)). In cases b = 0, or c = 0, or b = c, the network can have planes of symmetry. Presence of a plane of symmetry results in further simplification. Therefore in the forthcoming, in these cases it will be supposed that the network has a plane of symmetry, and thus it is sufficient to execute the investigations only on one sixth part of a face of the icosahedron (Fig. 2(a)). (This property is the reason why the networks with b = 0 or c = 0 or b = c can form separate classes, see Refs 2 and 12.)

Consider the network on one third (sixth) part of a face of the spherical icosahedron. Let n denote the maximum number of different edge-lengths of the network, considering the rotational (and mirror) symmetries of the icosahedron. We want to carry out the minimization by moving the nodal points. Since the nodal points can move on the spherical surface they, in general, have two degrees of freedom. However, if the network has a plane of symmetry and the nodal point lies on the plane of symmetry then it, in general, has one degree of freedom. A nodal point - which is identical to a vertex, a face centre, an edge midpoint of the icosahedron - has no degree of freedom. Let f denote the sum of the degrees of freedom of the nodal points on one third (sixth) part of a face of the icosahedron. The f degrees of freedom make possible to prescribe f conditions in order that f edges have lengths equal to those of certain other edges. So, the minimum number of different edge-lengths in a network of a given topology, in general, is

$$m = n - f.$$

However, it can occur that the length equality is realized for more than f edges. Thus, m is an upper bound of the minimum number of different edge-lengths: min $e_d \leqq m$. To illustrate these numbers, in Fig. 2(a) n = 12, f = 5, m = = 7; in Fig. 2(b) n = 11, f = 6, m = 5.

Now, we have a network of a given topology in which the edges are in n different positions and each edge has one of m different lengths. The question is: How many different networks with m different edge-lengths do exist? An approximation of this problem is to seek the number of ways such that one or more of the n positions correspond to one of the m different lengths. This generates a great number of different networks. According to combinatorial analysis this number is the Stirling number of the second kind, since the number of ways of putting n different things into m like cells, with no cells empty, is S(n,m), the Stirling number of the second kind (see Ref. 9). The Stirling numbers S(n,m), however, give only upper bound of the maximum number of geometrically possible networks, $N \leqq S(n,m)$, because actually not all of the networks defined by combinatorial way exist. But, knowing the geometrical conditions we can sharpen this bound. For instance, in case $b + c \geqq 3$ the networks have to be excluded at which the triangles joining in the vertices of the icosahedron are equilateral, and so $N \leqq S(n,m) - S(n-1,m)$. After determining the geometrical characteristics of all the possible networks we can chose the networks with minimum number of different faces and vertices.

The minimum number of different edge-lengths determined by the degrees of freedom of the nodal points and the number of different networks constructed by these edge-lengths are presented in Table 1 for some first values of b,c.

Table 1

b, c	T	n	f	m	N
1, 0	1	1	0	1	1
1, 1	3	2	0	2	1
2, 0	4	2	0	2	1
2, 1	7	4	2	2	4
2, 2	12	5	1	4	9
3, 0	9	4	1	3	5
3, 1	13	7	4	3	≦ 211
3, 2	19	10	6	4	≦ 26,335
3, 3	27	10	4	6	≦ 19,936

3.1 Network with one edge-length

Only one network of icosahedral symmetry can be constructed with one edge-length in the system $\{3,5+\}_{1,0}$ which is the icosahedron itself.

3.2 Network with two different edge-lengths

Networks of icosahedral symmetry with two different edge-lengths can be constructed in systems $\{3,5+\}_{1,1}$, $\{3,5+\}_{2,0}$ and $\{3,5+\}_{2,1}$. Due to Table 1 the system $\{3,5+\}_{2,1}$ gives the maximum triangulation number. In this system four kinds of networks can be constructed. These are shown by Fig. 3 where the edge-lengths (chord factors) are as follows:

(a) $a_1 = 0.4028362$ $a_2 = 0.4638569$ (b) $a_1 = 0.4124138$ $a_2 = 0.4744020$

(c) $a_1 = 0.4209745$ $a_2 = 0.4837981$ (d) $a_1 = 0.43014841$ $a_2 = 0.49383595$

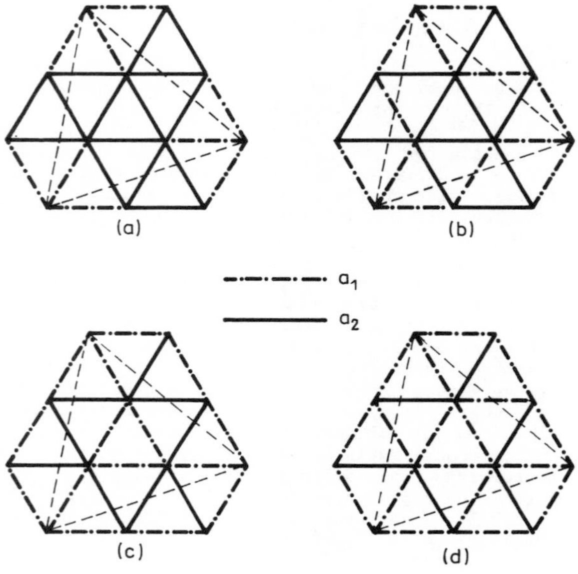

Fig. 3. Four solutions in $\{3,5+\}_{2,1}$ for $e_d = 2$

It is easy to see that the number of different faces in variants (a) and (d) is less by one than that in variants (b) and (c). Thus, variants (a) and (d) are optimal. If the further requirement is that the maximum edge-length be a minimum, then variant (a) is the best. But, if quotient η is required to be a minimum, then variant (d) is the best.

3.3. Network with three different edge-lengths

Due to Table 1, networks of isocahedral symmetry with three different edge-lengths can be constructed in systems $\{3,5+\}_{3,0}$ and $\{3,5+\}_{3,1}$. A recent investigation of spherical circle-packing (Ref. 11), however, has shown that there exists a network with three different edge-lengths in system $\{3,5+\}_{3,2}$ consisting of more triangles in spite of the fact that, with respect to the degrees of freedom of the nodal points, a network in this system can only be constructed without contradiction by not less than four different edge-lengths. A part of this network may be seen in Fig. 4 where the edge-lengths (chord factors) are as follows:

$$a_1 = 0.2357541 \quad a_2 = 0.2752134 \quad a_3 = 0.2913202$$

and $T = 19$, $f_d = 3$, $v_d = 4$. Fig. 5 shows a paperboard model of this polyhedron.

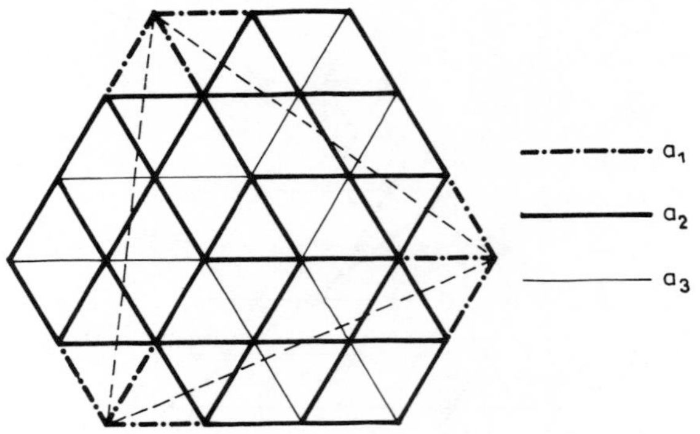

Fig. 4. Solution in $\{3,5+\}_{3,2}$ for $e_d = 3$

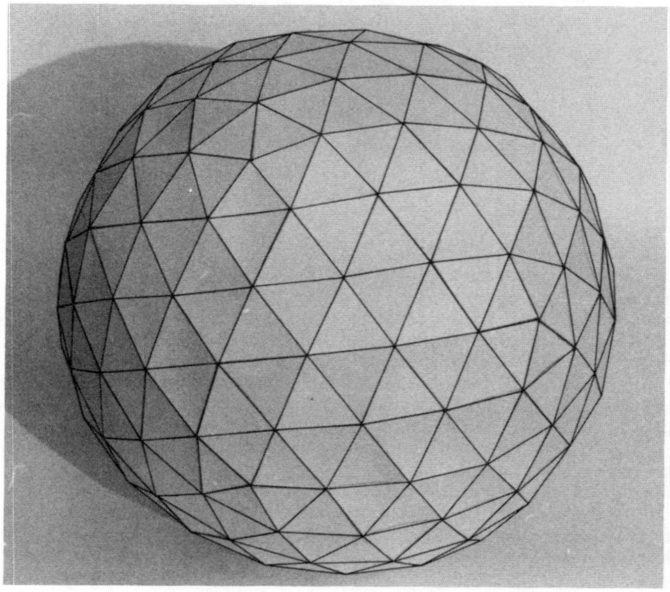

Fig. 5. Model of the polyhedron in $\{3,5+\}_{3,2}$ with e_d=3

4. A NEW METHOD OF SUBDIVISION

It is apparent from Table 1 that with an increase of b and c the number of geometrically possible networks N, in general, rapidly increases. The combinatorial technique becomes difficult because of the great number of the variants. Therefore we present a method of subdivision which exists for each $b \geqq 2$ and which, for any fixed e_d, in general, results in triangulation number T greater than that given by the other methods known so far. This method is developed in system $\{3,5+\}_{b,1}$ and uses the method suggested by the author (Ref. 10) and discovered independently by Pavlov (Ref. 8).

The new method is a generalization of the subdivision in system $\{3,5+\}_{2,1}$, shown by Fig. 3. Its principle is that, in the system $\{3,5+\}_{b,1}$, the edges of the icosahedron can always be covered by strips consisting of triangles (Fig. 6). If the nodal points on the boundary of a strip, apart from the vertices of the icosahedron, are fitted to two small circles whose planes are parallel, then the method suggested in Ref. 10 can be applied in the great "triangles" bordered by the strips. In order to minimize the number of different faces, only two kinds of edge-lengths are used for a strip in an analogy to those in Fig. 3. Thus, four variants of the strips can be defined (Fig. 7). It is easily seen that variant (d) of the strips can only be constructed

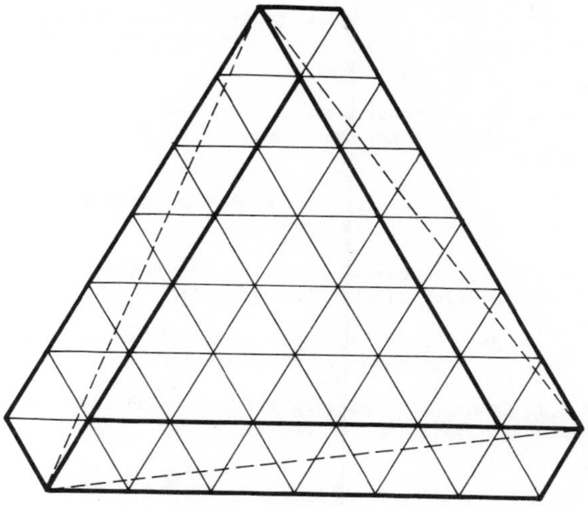

Fig. 6. The principle of the new method. The strips

of one kind of triangles and its replicas. (The other three contain two kinds of triangles.) For b = 4 a part of the network with strips due to Fig. 7(d) is shown in Fig. 8 where the edge-lengths (chord factors) are as follows:

$a_1 = 0.2485854 \qquad a_2 = 0.2899636$

$a_3 = 0.2779841 \qquad a_4 = 0.2657787$

and $e_d = 4$, $f_d = 4$, $v_d = 5$. This network with strips according to Fig. 7(a) has also been constructed. Its paperboard model may be seen in Fig. 9. Here $e_d = 4$, $f_d = 5$, $v_d = 5$.

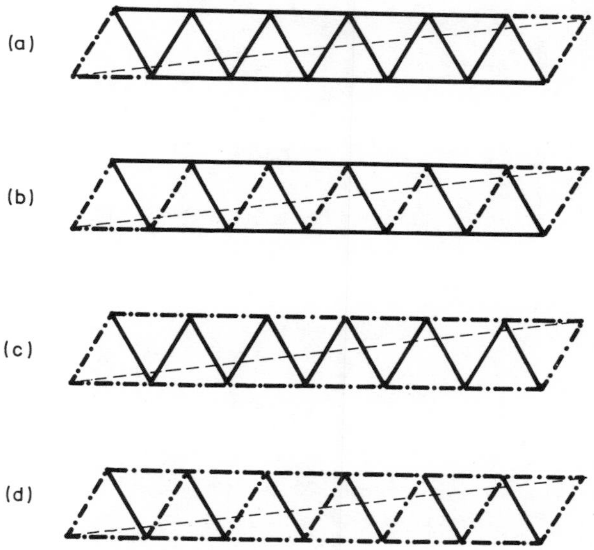

Fig. 7. Four ways of forming the strips

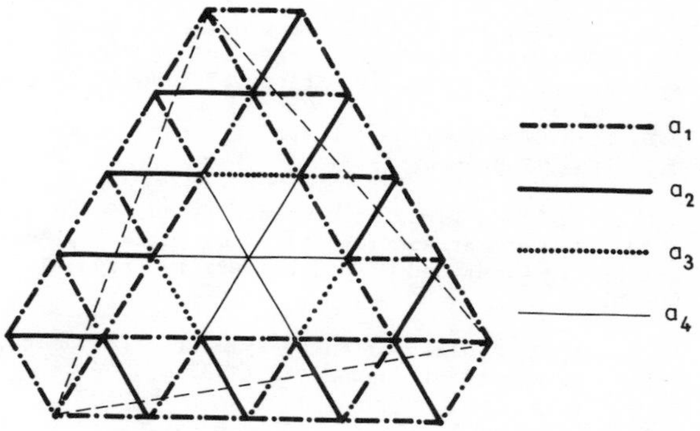

Fig. 8. Result of the new subdivision for b = 4 with strips (d)

Fig. 9. Model of the polyhedron for b = 4, with strips (a)

The new method of subdivision with strips according to Fig. 7(d) seems to result in optimal networks. For these networks, in case $b \geqq 2$, the following expressions are valid:

$$T = b^2 + b + 1,$$

$$e_d = b,$$

$$f_d = \begin{cases} \dfrac{4b - 4}{3}, & \text{if the remainder of b divided by 3 is 1,} \\ \dfrac{4b - 2}{3}, & \text{if the remainder of b divided by 3 is 2,} \\ \dfrac{4b - 3}{3}, & \text{if b is divisible by 3,} \end{cases}$$

$$v_d = \begin{cases} \dfrac{4b - 1}{3}, & \text{if the remainder of b divided by 3 is 1,} \\ \dfrac{4b - 2}{3}, & \text{if the remainder of b divided by 3 is 2,} \\ \dfrac{4b - 3}{3}, & \text{if b is divisible by 3.} \end{cases}$$

Here two elements of the same size, which can be carried into congruence only by mirroring, are considered different.

REFERENCES

1. D.L.D. CASPAR and A. KLUG, Physical Principles in the Construction of Regular Viruses, Cold Spring Harbor Symposia on Quantitative Biology 27 (1962), 1-24.

2. J.D. CLINTON, $(p,q+)^{\eta}_{b,c}$, Preprint, Kean College, Union, N.J., USA, September 1980.

3. H.S.M. COXETER, Virus Macromolecules and Geodesic Domes, A Spectrum of Mathematics (ed. J.C. Butcher), Auckland University Press and Oxford University Press, 1972, 98-107.

4. M. GOLDBERG, A Class of Multi-Symmetric Polyhedra, The Tôhoku Mathematical Journal 43 (1937), 104-108.

5. E.MAKAI Jr. and T. TARNAI, On Some Geometrical Problems of Single-Layered Spherical Grids with Triangular Network. Proc. of the 2nd Int. Conference on Space Structures, Guildford, 1975, 675-682.

6. E. MAKAI Jr. and T. TARNAI, Morphology of Spherical Grids, Acta Technica Hung. 83 (1976), 247-283.

7. R.E. McCONNEL, Notes on the Generation of Optimised Networks on Curved Surfaces, MERO Technische Berichte Nr. 3, Würzburg 1978.

8. G. PAVLOV. Compositional form-shaping of crystal domes and shells (in Russian), Arkhitektura SSSR, 1977 No 2, 30-41.

9. J. RIORDAN, An Introduction to Combinatorial Analysis, John Wiley, New York, 1958.

10. T. TARNAI, Spherical Grids of Triangular Network, Acta Technica Hung. 76 (1974), 307-336.

11. T. TARNAI, Geodesic Dome with Three Different Bar Lengths, Structural Topology No 8 (1983), 23-24.

12. M.J. WENNINGER, Spherical Models, Cambridge University Press, 1979.

FORMEX AND PLENIX STRUCTURAL ANALYSIS

M HARISTCHIAN, BSc, MSc, PhD

Lecturer, Department of Civil Engineering,

SHARIF University of Technology, Tehran, Iran.

This paper introduces a mathematical system which enables the structural analyst to carry out the processes of data generation and organization of large scale problems through algebraic expressions that are easily understood and manipulated. "Formices" and "plenices" are the two key concepts in the approach. In a structural system the topological aspects such as the interconnection pattern of the elements and general pattern of the joints are formulated through the concepts of formices; and all other aspects of the system such as the geometric particulars, intensity of the loads and material properties are expressed and classified in terms of plenices. The reader is assumed to be familiar with the concepts of formex algebra.

INTRODUCTION

This paper introduces a method for algebraic representation and processing of various pieces of information regarding a structural system. The method is based on the two fundamental concepts of formices and plenices.

The concept of formices, first introduced by H. Nooshin (see Ref 1), provides a powerful means for algebraic representation and processing of topological information regarding a structural system.

The concept of plenices finds its origins in the definition of a high-level programming language, PAVIC, by H. Nooshin, J.W. Butterworth and P.L. Disney of the University of Surrey. Through the concept of plenices the structural analyst can represent and organize all types of information relating a structural system in terms of a number of algebraic expressions.

The paper first gives a definition of the fundamental concepts regarding plenices, then an application of the concepts is exemplified in terms of a number of skeletal structures. The reader is assumed to be familiar with the concepts of formex algebra as in one of the references 2-4 together with Ref 5.

PLENICES

A "plenix" is an ordered collection of mathematical entities in the form

{P1,P2,..., Pn},

where $n \geq 0$ and where each one of the entities P1, P2,..., Pn is any mathematical entity such as a scalar, a matrix, ... or a well-defined algorithm. The entities P1, P2,...Pn are referred to as "principal panels" and a plenix that has n principal panels is said to be of "order" n.

Since a plenix is a mathematical entity in its own right, then a principal panel of a plenix may itself be a plenix having a sequence of principal panels each one of which in turn may be a plenix ... etc. For instance,

{1, true, [1; 2], 8.5},

{ [22; 10; 62] , [63,50,24], true},

{{'COMPRESSIVE', 'RED'},{'TENSILE', 'BLUE'}},

are plenices of order 4,3 and 2, respectively, where 'true' is Boolean entity and 'COMPRESSIVE',... and 'BLUE' represent four algorithms. Also,

{{1,{2,3},4}, {0,0,0}, 0, [0]}

is a plenix of the fourth order.

The special plenix with no principal panels, that is of the zeroth order, is referred to as the "empty" plenix, and it is denoted by {}.

Two plenices are said to be "equal" if and only if they are identical. The conventional equality symbol is used to indicate the relationship of equality between two plenices. Thus

{P1,{P2, P3}, P4} = {1,{2,2},3},

implies that P1=1, P2=P3=2 and P4=3. Also,

{P1, P1,{P2}, P3}={2,2,{true},{}},

implies that P1=2, P2=true, and P3={}.
The empty plenix is equal to itself. It is noted that there is a distinction between the empty plenix and the plenix whose only principal panel is the empty plenix.
Therefore {{}} is not equal to {}, that is
{{}} ≠ {} ,
{{}} is a plenix of the first order whereas the empty plenix is of the zeroth order. In general for any mathematical entity P, {P} ≠ P.
If P is a plenix, then a principal panel of a principal panel of P is referred to as a "subsidiary" panel of P. Furthermore, a principal panel of a subsidiary panel of P is referred to as a subsidiary panel of P. The word "panel" may be used to refer to both types of panels. In a plenix P, if α is the nth principal panel of the mth principal panel of the ... jth principal panel of the ith principal panel of P, then α is denoted as α = P{i,j,..., m,n},

where the sequence {i,j,...,m,n} is referred to as the "orderate" of α. In a plenix P, a panel α is said to be a "plecule" if and only if α itself is a plenix; and α is said to be a "pletom" if and only if α is a non-plenix entity such as a scalar or a matrix.
For example, all the principal panels of the plenix

P={{{7,8},9}, {true, false},{ 5 , 6},{} ,{{}} }

are plecules. Furthermore the subsidiary panels

P{1,1}={7,8} and

P{5,1} ={}

are plecules and all the other subsidiary panels of P are pletoms.

PLENIX COMPOSITION

Let P and Q be either non-plenix mathematical entities or be plenices {P1, P2,..., Pn} and {Q1, Q2,..., Qm}, respectively. Then, the composition of P and Q in the order of P,Q is defined as a plenix R such that

R={P,Q}, if both P and Q are non-plenix entities;

R={P,Q1,Q2,..., Qm}, if P is a non-plenix entity and Q is a plenix;

R={P1,P2,...,Pn, Q}, if P is a plenix and Q is a non-plenix entity; and

R={P1,P2,...,Pn, Q1,Q2,..., Qm} , if both P and Q are plenices.
The relationship between R,P and Q is written as
R=P#Q.
The symbol # is called and read "duplus" (Thus P#Q is read "P duplus Q"). The word "composition" refers to both the process of "composing" two plenices and the plenix that is the result of this process. For example, if

P=1,

Q= [true, false],

R={{3,4},1,2} and

S={1.3,[1.5,2],{}}; then

P#Q={1,[true, false]},

Q#P={ [true, false], 1},

P#R={1,{3,4},1,2},

S#Q={1.3,[1.5,2],{}, [true, false]} and

R#S={{3,4},1,2,1.3 ,[1.5,2],{}}.

Plenix composition is associative but, in general, is not commutative; thus for mathematical entities P,Q and R

P#(Q#R)=(P#Q)#R and

P#Q≠Q#P. Also,

P#{}={}#P=P.

Let Pi's (for i=m,m+1,...,n-1,n or i=m,m-1,...,n+1,n) be mathematical entities. Then, the notation

$\overset{n}{\underset{i=m}{\vdash}}$ Pi

referred to as the "libra notation" is used to represent the composition

$P_m \# P_{m+1} \# \ldots \# P_{n-1} \# P_n$, or

$P_m \# P_{m-1} \# \ldots \# P_{n+1} \# P_n$, respectively.

The symbol $\vdash$ is referred to as the "libra symbol"

and $\overset{n}{\underset{i=m}{\vdash}}$ reads "libra i, m to n".

For example, if

R= $\overset{0}{\underset{i=-2}{\vdash}}$ {i,i+1}, then

R={-2,-1} #{-1,0}#{0,1}, or

R={-2,-1,-1,0,0,1}.

If

S= $\overset{0}{\underset{i=-2}{\vdash}}$ {{i,i+1}}, then

S={{-2,-1}} #{{-1,0}}#{{0,1}}, or

S={{-2,-1},{-1,0},{0,1}}.

Note that
$\overset{0}{\underset{i=-2}{\vdash}}$ {i,i+1}≠ $\overset{0}{\underset{i=-2}{\vdash}}$ {{i,i+1}}.

If

T= $\overset{0}{\underset{j=-1}{\vdash}}${ $\overset{2}{\underset{i=1}{\vdash}}$ [1; 3i], 13j+1}, then

T= $\overset{0}{\underset{j=-1}{\vdash}}${{[1; 3],[1; 6]} , 13j+1}, or

T={{[1; 3], [1; 6]} ,-12,{[1; 3], [1; 6]} , 1}.

If

U= $\overset{2}{\underset{i=0}{\vdash}}$ $\overset{3}{\underset{j=1}{\vdash}}$($i^2$ +3j), then

U={3,6,9,4,7,10,7,10,13}.

PLENIX RELATIONS

The relationship of equality for plenices has already been defined. Further, a class of plenix relations is defined. Let P and Q be any two plenices, and let their panels be compared with one another irrespective of their types and values. If p and q denote any two panels of P and Q, respectively, then p and q are said to be "iso-image" of one another, if and only if p and q have identical orderates in P and Q.

Furthermore, if p is a plecule and p' denotes a panel of p, then p' is said to be a "sub-image" of q and q is said to be a "super-image" of p'. The word "image" may be used to refer to all types of images.

APPLICATIONS

Application of plenices to structural analysis is exemplified in terms of a number of structural analysis problems.

Example 1: Let it be required to prepare data for the linear static analysis of the rigidly-jointed flat grid of Fig 1. All the elements of the grid have a bending rigidity EI and a torsional rigidity GJ. The elements in the direction of X and Y axes have lengths L1 and L2, respectively. The grid is supported on four joints shown thus • in the figure. The supports are ball-joints which only restrain the vertical translations (normal to the plane of the grid). The grid is subjected to a vertical concentrated load of magnitude w at every joint lying within the dotted area.

To begin with, it will be necessary to describe the interconnection pattern of the elements, and this may be done in terms of formices. The formex formulation is carried out with the graphical arrangement of Fig 2. The convention is adopted that every joint of the grid is represented by a signet of grade two, and every element of the grid is represented by a two-plex cantle [t1; t2]. Assuming that the stiffness method of analysis will be used, then t1 corresponds to the "end 1" and t2 corresponds to the "end 2" of the structural element as used in obtaining the element stiffness matrices (the "direction" of an element is defined as going from t1 to t2).
Let the two formices F1 and F2 be obtained as

F1= $\overset{20}{\underset{j=0}{\vdash}}$ $\overset{19}{\underset{i=0}{\vdash}}$ tranid (i,j)¦ [1,1; 2,1] and

F2= $\overset{19}{\underset{j=0}{\vdash}}$ $\overset{20}{\underset{i=0}{\vdash}}$ tranid (i,j) ¦[1,1; 1,2].

Where, F1 and F2 represent the elements in the direction of U1 and U2 axes, respectively.

Then, a formex

F'=F1#F2

would represent the interconnection pattern of all the elements of the grid. However, the elements denoted by F1 and F2 are different in their geometric properties; and therefore, rather than composing them into a single formex F', they are put into a plenix

F={F1,F2}.

The plenix F, describes the interconnection pattern of the whole structure and at the same time it classifies the elements in two types. The panels of F may then be dealt with separately in conjunction with the appropriate properties for the corresponding type of the element. The next stage of the data preparation is concerned with the description of the geometric and material properties associated with the structural elements. The element lengths are represented by the plenix

L={L1,L2},

where L1 and L2 are the lengths of the elements represented by F1 and F2, respectively. The information regarding the orientation of the elements with respect to the X axis of the Cartesian coordinates system XYZ, is represented as

T={0,90}

here, the pletoms of T, in degrees, are the smallest angles between the positive direction of the elements with that of the X axis: 0 and 90 are the angles associated with the elements represented by F1 and F2, respectively. With appropriate assumptions, the data represented by T, may be used to obtain the element transformation matrices.

The material and geometric properties of the elements altogether, are represented by the plenix

Ω = {EI, GJ, L, T}.

Then, with the plenix

B= {F,Ω}

the description of the information regarding the structural elements may be assumed to be complete.

In relation to the joints of the grid, a formex representing the joints of the grid is generated first:

$$C = \lceil_{i=1}^{21} \lceil_{j=1}^{21} \text{ projid } (i,j) \mid [1,1].$$

C is an exclusive catena for the formices F1 and F2. In the way discussed in literature regarding formex algebra, C may be used to imply the conventional joint numbering scheme required by the stiffness method of the analysis.

For the external load, an ingot

$$g1 = \lceil_{i=0}^{9} \lceil_{j=0}^{15} \text{ projid } (i+8,j+2) \mid [1,\ 2]$$

represents the loaded joints. And a plenix

P= {g1, w}

may be used to denote both the loaded joints and the magnitude of the load applied at the joints. The support positions may be represented by

g1=lamid (11,11) ¦[4,4]; and a plenix

R= {g2,3},

would represent the constraint condition for the grid. In using this notation, it is assumed that every joint of the grid has three degrees of freedom, in the order of two rotations and a translation; and that the translation in the direction normal to the plane of the grid corresponds to the third degree of freedom. For the joints represented by g2, the third degree of freedom is fully constrained. A notation consistent with this could be used to indicate both the direction and the magnitude of the applied load:

P={g1,3,w}.

There are other possibilities: For instance, for the loads one could write

P={g1,[0,0,w]}; and for the constraints

R={g2, [false, false, true]}.

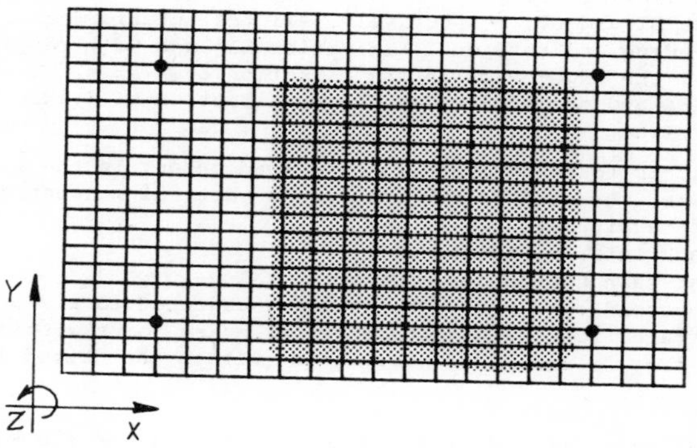

Fig 1

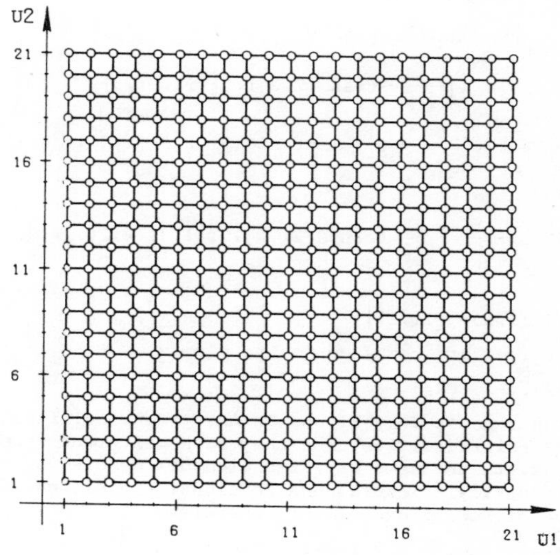

Fig 2

Here, the Boolean array denotes the type of the constraint ("true" for restrained, "false" for free).

Now that the pieces of data regarding the joints of the grid are avaialble, they can be combined together in a plenix

N= {C,P,R}.

Finally, a plenix

D={N,B}

may represent the required data for the analysis of the grid. D is referred to as a "data plenix" for the grid.

For a given system there are many data plenices. The choice of a constitution for the data plenix, is primarily determined by the manner in which the given information is classified; and the classification is determined by factors such as the ease with which one can relate various pieces of information to one another, the objectives of the analysis, the method of analysis and the type of the system.

The above mentioned example illustrates the manner in which plenices are envisaged to be used in data preparation. In the later examples of this paper, attention will be focused on particular points in obtaining a data plenix and occasionally, instead of a "complete" data plenix only certain panels of a data plenix will be constructed.

Example 2: Let it be required to prepare data for the linear static analysis of the rigidly jointed flat grid of Fig 3. All the edge-beams have bending and torsional rigidities EI and GJ, respectively. All the interior beams have bending and torsional rigidities EI' and GJ', respectively. The edge-beams in the direction of X axis, have a length 6a and the edge-beams in the direction of Y axis have a length 4a.

The grid is to be analysed for two load cases: In the first load case, the grid is subjected to a vertical concentrated load of magnitude:
w at each interior joint; and
0.5 w at each non-corner edge-joint.
In the second load case, the grid is subjected to a vertical concentrated load of magnitude u1 at every joint within the dotted area I; and u2 at every joint within the dotted area II.

To begin with, the interconnection pattern of the elements are described in terms of formices. The formex formulation is carried out with the graphical arrangement of Fig 4. The conventions of Example 1 are adopted. Let the formices F1, F2, F3 and F4 be obtained as

$F1= \text{lam}\,(2,13) \mid \mathop{⊑}_{i=0}^{15} \text{tran}\,(1,2i) \mid [1,1;\ 3,1]$,

$F2= \text{lam}\,(1,17) \mid \mathop{⊑}_{j=0}^{11} \text{tran}\,(2,2j) \mid [1,1;\ 1,3]$,

$F3= \text{lux}(g') \mid \mathop{⊑}_{j=0}^{11} \mathop{⊑}_{i=0}^{15} \text{tranid}(2i,2j) \mid \{[1,1;\ 2,2],[2,2;\ 3,3]\}$ and

$F4=\text{lux}(g') \mid \mathop{⊑}_{j=0}^{11} \mathop{⊑}_{i=0}^{15} \text{tranid}(2i,2j) \mid \{[1,3;\ 2,2],[2,2;\ 3,1]\}$,

where

$g'=\text{clovid}(11,15) \mid [11,11]$.

These formices are combined in the plenix

F={{F1,F2},{F3, F4}}.

The elements denoted by F1 and F2 have identical flexural and torsional rigidities; but they have different geometric properties. In the same way, the elements denoted by F3 and F4 have identical flexural and torsional rigidities, but they have different geometric properties.

The element lengths are represented by the plenix

$L=\{\{6a,4a\},\ 2\sqrt{13}a\}$.

The information regarding the angle of orientation of the elements with respect to the X axis, is denoted by the plenix

T={{0,90},{45,-45}}.

Similarly, the flexural and torsional rigidities of the structural elements are described by the planices

e={EI,EI'} and

s = {GJ,GJ'},

respectively. It may be noticed that the panels of the plenices s,e, T and L represent the properties associated with the structural elements denoted by their images in F. The properties of the elements, altogether are composed into the plenix

Ω = {e,s,L,T}. Then, with the plenix

B = {F,Ω},

the description of the information regarding the structural elements may be assumed to be complete.
The information relating the joints of the grid are treated as follows: The joints of the grid are represented by the ingot

$C = \text{nex}(F3\#F4) \mid \mathop{⊑}_{j=1}^{25} \mathop{⊑}_{i=1}^{33} \text{projid}(i,j) \mid [1,1]$.

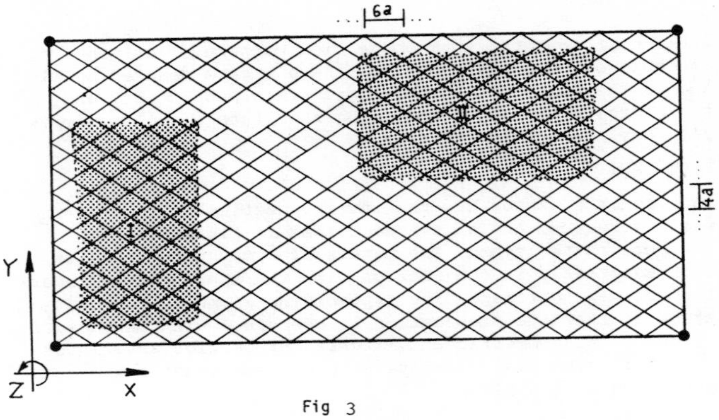

Fig 3

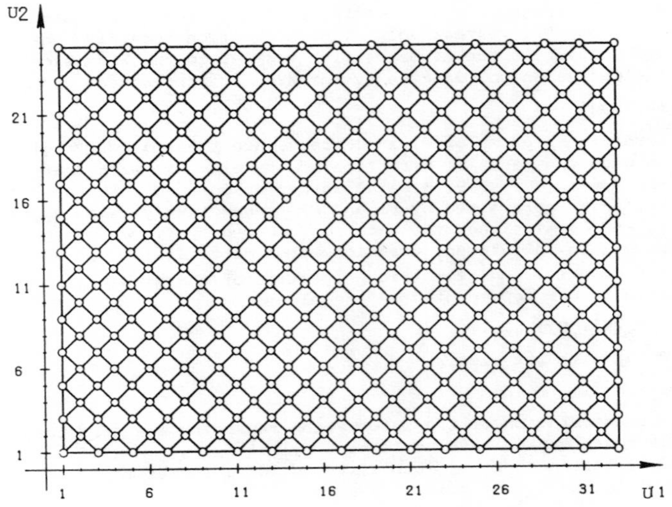

Fig 4

The load plenix for case 1, may be written as

P1 = {{g2,g3}, {[0,0,0.5w] ,[0,0,w]}}, where

g2 = lux(g1) ¦ nex(F1#F2) ¦ C and

g3 = lux(g1#g2) ¦C, where

g1 = lamid(17,13) ¦ [1,1].

Similarly, the load plenix for case 2, will be

P2 = {{g4,g5},{[0,0,u1],[0,0,u2]}} , where

$g4 = \mathop{⊑}_{i=3}^{5} \mathop{⊑}_{j=3}^{10} \text{tranid}(2i-3,2j-3) \mid \{[0,0],[1,1]\}$ and

$g5 = \mathop{⊑}_{i=1}^{6} \mathop{⊑}_{j=1}^{5} \text{tranid}(2i-2,2j-2) \mid \{[17,15],[18,14]\}$.

Example 3: Let it be required to produce a data plenix for the analysis of the flat grid of Fig 5. The elements of the grid have bending and torsional rigidities EI and GJ, respectively. The grid is subjected to a vertical concentrated load of magnitude $w(x,y)=\alpha x+\beta y$, at the joints lying within the dotted area. Here, α and β are constants and x and y are the coordinates of the joint with respect to the X and Y axes.
Let the three formices F,V and gl, represent the elements, the joints and the loaded joints of the grid respectively and be obtained with respect to the graphical arrangement of Fig 6 as

$$F = \text{nex}(V) \mid \mathop{\Big[}\limits_{i=0}^{10} \mathop{\Big[}\limits_{j=0}^{10} \text{tranid}(i,j) \mid \{[1,1;\ 2,1],\ [1,1;\ 1,2],\ [1,1;\ 2,2]\},$$

$$V = \mathop{\Big[}\limits_{i=1}^{11} \mathop{\Big[}\limits_{j=1}^{11} \text{projid}(i,j) \mid [1,1], \text{ and}$$

$$gl = \mathop{\Big[}\limits_{j=0}^{6} \mathop{\Big[}\limits_{i=0}^{7-j} \text{projid}(i+j+2,j+3) \mid [1,1].$$

Then the three plenices

$P = \{gl,\ [0,0,w(x,y)]\}$,

$C = \{V, [x(U1,U2),\ y(U1,U2)]\}$ and

$B = \{F,\ EI\ ,\ GJ\}$,

represent the data relating the loads, the coordinates of the joints, and the structural elements.

Where

$x(U1,U2)=13U1(13-U2)a/(169-U1U2)$ and
$y(U1,U2)=x(U2,U1)$.

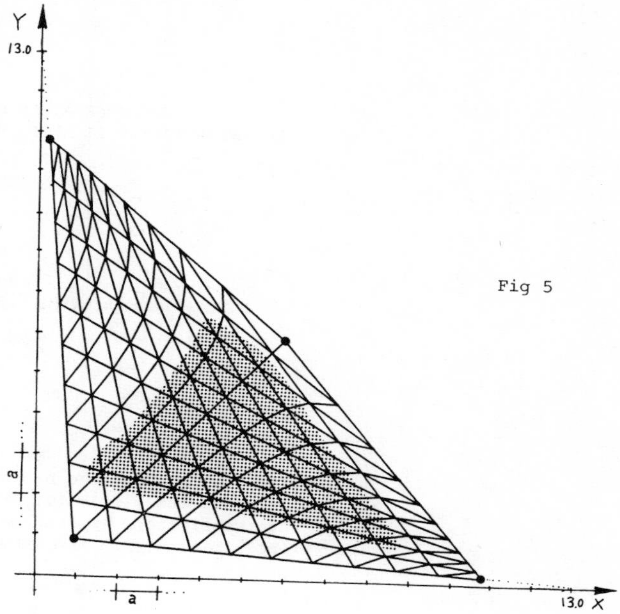

Fig 5

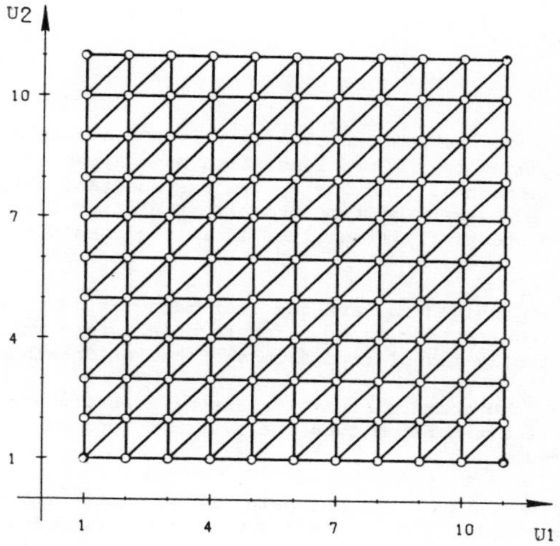

Fig 6

CONCLUDING REMARKS

It is shown that formices and plenices provide a powerful means for algebraic representation of various pieces of data regarding a structural system. The application of the concepts, however, extends far beyond the problem of data preparation: As an example, let L, EA and α be three plencies representing the lengths, the axial rigidities and the angles of orientation of the elements of a pin-jointed double-layer grid. Then by defining "link" between corresponding panels (say images) of these plenices, a plenix K which represents the stiffness matrices of various types of the elements may be obtained. Thus, in a functional rule
$K = f(L,EA,\alpha)$
plenices replace the conventional mathematical entities scalars and matrices. The concepts may also be particularly useful in obtaining automated graphical output, for instance, if F1, F2, and F3 represent the top-layer, the bottom-layer and the bracing elements of a double-layer grid, then a plenix

{{F1, "thick"},{F2, "dashed"},{F3, "thin"}}, or

{{F1, F2, F3},{"thick", "dashed", "thin" }}

may be used to obtain a layout of the grid with its elements drawn in "thick", "dashed" and "thin" lines.
The power of formices and plenices will become most evident for large and complex systems, be they architectural, electrical, hydraulic,...,or structural.

REFERENCES

1. H. Nooshin, Algebraic representation and processing of structural configurations, Computers and Structures, Vol 5, pp 119-130 1975.

2. Z.S. Makowski (Ed), Analysis, design and construction of double-layer grids, Applied Science Publishers 1981.

3. M. Haristchian, Formex and plenix structural analysis, Ph.D. Thesis, University of Surrey 1980.

4. J.S. Sanchez Alvarez, Formex formulation of structural configurations, Ph.D. Thesis, University of Surrey 1980.

5. H. Nooshin, Formex formulation of barrel vaults,(short course lecture notes), University of Surrey 1983.

A COMPUTER AIDED CONSTRUCTION SYSTEM FOR THE GENERATION OF GEOMETRY FOR RETICULATED SPACE STRUCTURES

Martin RUH

senior analyst
MERO - Raumstruktur GmbH & Co. Würzburg, West-Germany

The report shows some of the basic functions of the construction system GEOM and demonstrates its flexibility through two different complex examples. Due to the restrictions imposed on this paper, only a small part of the whole range of possibilities in the GEOM system are included.

1. INTRODUCTION

The development of the construction system GEOM has been strongly influenced by the problems which have appeared in practical situations [1] and the system has substantially shortened the investment of time and effort dedicated to the design and study of alternatives of reticulated structural systems.

The system is conceived to save the user mathematical formulation and requires the input parameters in a form which can be conveniently transferred from sketches or drawings.

Fundamentally, the geometric particulars of a reticulated space structure are defined once the system's topology has been formulated.

The GEOM system consists of four main parts, which are introduced in chapter 2 to 5 first and then two examples are presented to illustrate the application possibilities in chapter 6 and 7.

2. TOPOLOGY OF NODES

2.1 Nodes and Vectors

Experience has shown that almost all Geometries can be expressed in terms of integer type unit nets [2],[3]. Consequently, the topology of nodes is directly represented as characters, where blanks represent not given nodes. This enables a very fast listing of nodes topology in the form of printer plots and many possibilities for conditioning the subsequent generation of nets of bar-elements. The topology of nodes is produced in a two dimensional field, while the third dimension is related by means of a plane number. The two dimensional formulation has not restricted the system in any way and the decomposition of a problem in two dimensional units has proved to be easy to visualize and correponds with the human capabilities in this respect.

The generation of nodes in two dimensional unit nets through characters has definite advantages. It is not necesary to avoid the repeated generation of node or empty places, because it is physically impossible that more than one symbol occupies a location in the unit net. The elimination of unwanted nodes is archieved by generating empty places. Nodes can be arbitrarily added, deleted and newly generated at any time. Special symbols, such as slanted line, minus or plus, are predominantly used as auxiliary or dummy nodes to induce or prevent the generation of bars. Dummy nodes can be later disregarded in the actual geometry.

The fact that only discrete nodes should be generated in the unit net requires two linearly independent basic vectors (Ax,Ay) and (Bx,By) to span the basic grid net with an origin (Xo,Yo).

Commonly used basic nets (Fig 1) are automatically made available by the GEOM system, e.g.

Basic square grid net: (Ax,Ay)=(2,0) and (Bx,By)=(0,2)
Basic triangular grid net: (Ax,Ay)=(2,0) and (Bx,By)=(1,3)

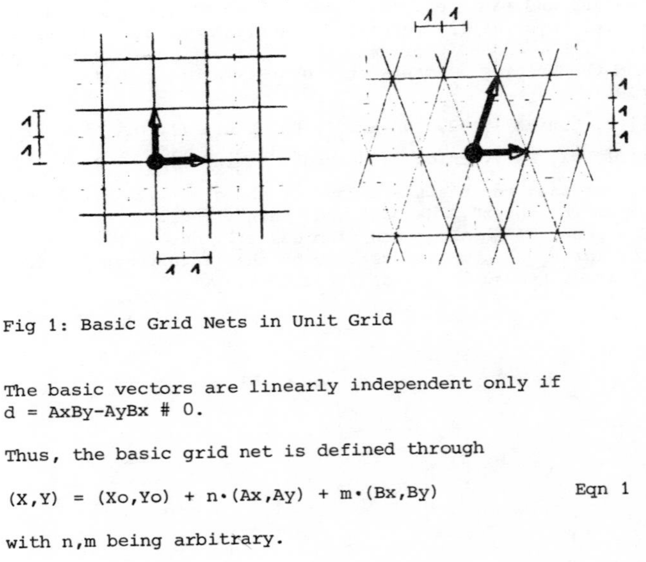

Fig 1: Basic Grid Nets in Unit Grid

The basic vectors are linearly independent only if d = AxBy-AyBx # 0.

Thus, the basic grid net is defined through

$$(X,Y) = (Xo,Yo) + n\cdot(Ax,Ay) + m\cdot(Bx,By) \qquad \text{Eqn 1}$$

with n,m being arbitrary.

Conversely, (X,Y) is determined in the basic grid net if No = Mo = 0

where No = mod(n,d) , Mo = mod(m,d)

and n = Ax(Y-Yo) - Ay(X-Xo)
m = Bx(Y-Yo) - By(X-Xo).

2.2 AREA Function

The most commonly applied formulation is AREA, which is an example of user convenient generation of nodes. Here, nodes are generated as set by the basic net but only within a closed polygonal field. A node series is built by giving the number of steps in the various directions, which are derived from the basic vectors (Fig 2). The user formulates the series with easy to learn code numbers, which represent the number N of steps, the step vector S and the quadrant direction Q:

Code number = N•100 + S•10 + Q with:

S :	vector definition :	Q :	signX, signY :
1 or 3	(Ax,Ay)	1	+ , +
2 or 4	(Bx,By)	2	- , +
5	2(Ax,Ay)	3	- , -
6	2(Bx,By)	4	+ , -
7	(Ax,Ay)+(Bx,By)		

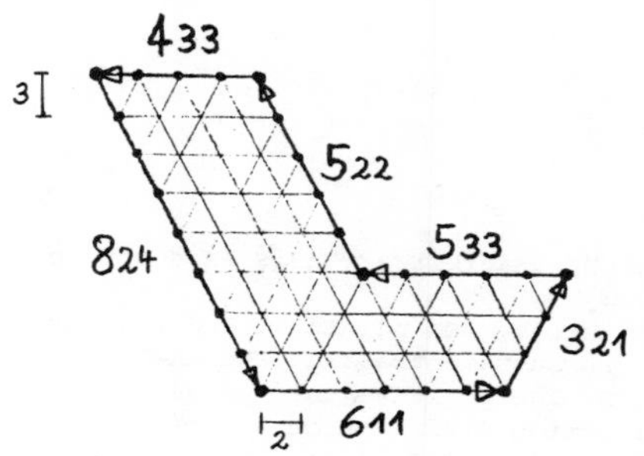

Fig 2: Polygonal Node Series in the Basic Triangular Net
(Ax,Ay) = (2,0) and (Bx,By) = (1,3)

2.3 CALL Function

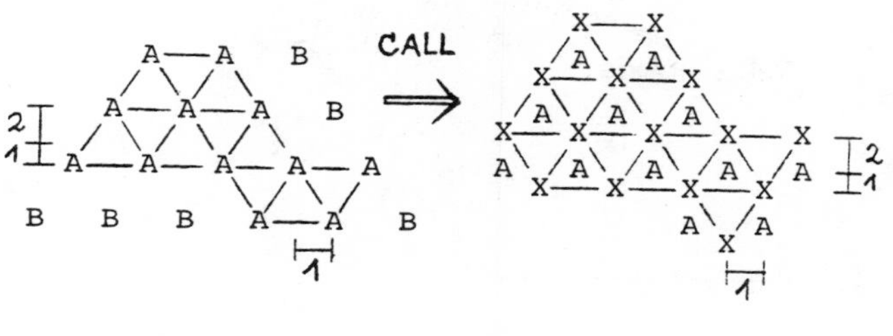

Fig 3: Generation with the CALL Function

A further interesting form of generation is the CALL formulation. It is based on previously generated nodes in the present or just closed working area to produce a modified form of topology through "and/or" two-dimensional vectors (Eqn 2). That is, nodes are generated only at places from which basic nodes can be reached through the given vectors. In Fig 3 left, for instance, only the nodes represented with an A are taken as a basis to generate the nodes represented with an X in Fig 3 rigth, through the associated vectors

XA = [(1,1) and (-1,1)] or (0,-2) Eqn 2

3. GEOMETRY OF NODES

3.1 Transformations

The generation of actual node coordinates can be performed through a set of available transformation algorithms, which include linear transformations, rotations, adaptations, bends as well as polar, cylindrical and other non linear transformations.

The transformations are not described as usual through the specification of mathematical functions, but by means of corresponding dummy points before and after the transformation.

This form of transformation description yields definite advantages, in particular for the linear transformations. The user can conveniently simplify the specification of transformations by appropriately selecting dummy points, while the computation of the transformation matrix and the linear translation is left to the program. The grouping of transformations refering to triangular and quadrangular planes, has proved to be very appropriate especially for the selection of node: For instance, transforming only the nodes which themselves or their projections are included in an arbitrarily chosen triangle are transformed. In case of the quadrangles, one of the dummy points has effect only for the selection.

Detailed Description of the Transformations:

3.1.1 Type 2/2 Transformation (Fig 4)

(straight to straight line trough P1 → Q1, P2 → Q2)
A node selection is possible, for example within the line.

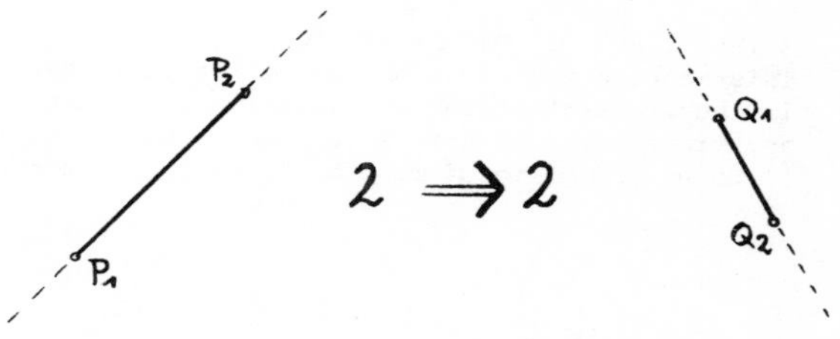

Fig 4: Transformation Type 2/2

3.1.2 Type 3/3 Transformation (P1→Q1, P2→Q2, P3→Q3)

It can represent a full space transformation (Fig 5).
A node selection is possible, for instance in the triangle, at the edge, etc.
The missing quantitied for the computation of the transformation in space are given by the distance between the basis plane P1P2P3 to the transformation plane Q1Q2Q3.

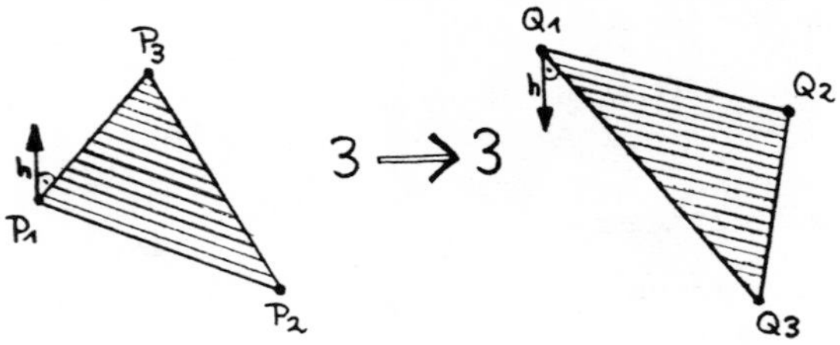

Fig 5: Transformation Type 3/3

3.1.3 Type 4/3 Transformation (Fig 6)

(P1 → Q1, P2 → Q2, P3 → Q3, P4)
P4 completes a quadrangle. The point has only effect in the selection parameter. The point must consequently lie on the plane P1P2P3 and has no significance for the transformation.

In all other respects it works exactly as for the transformation type 3/3.

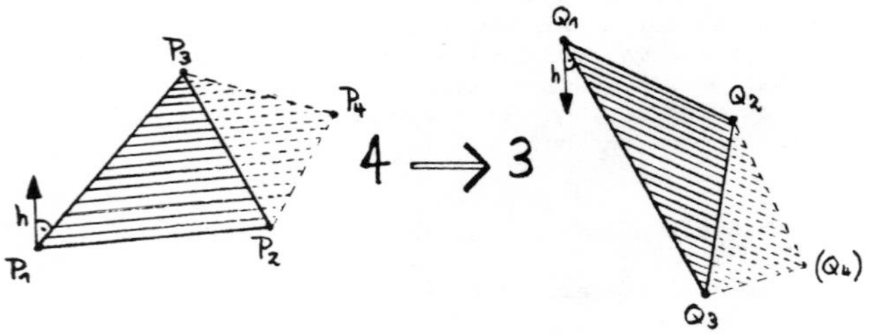

Fig 6: Transformation Type 4/3

3.1.4 Type 4/4 Transformation (Fig 7)

(P1 → Q1, P2 → Q2, P3 → Q3, P4 → Q4)
This mapping describes the general transformation in the space, that is, any rotation with stretching and translation. A node selection is not possible. P1 to P4 and Q1 to Q4 must be linearly independent.

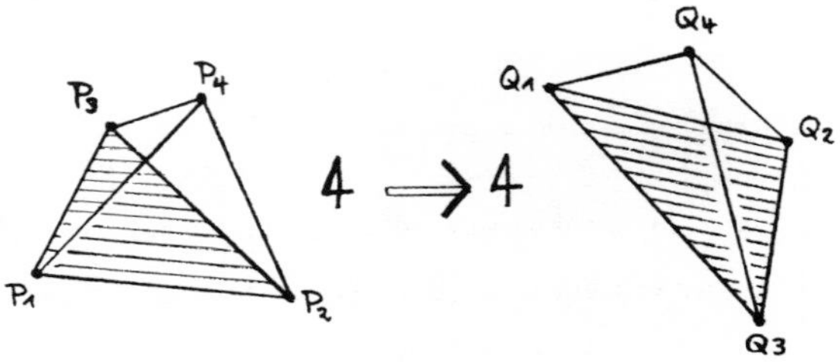

Fig 7: Transformation Type 4/4

3.1.5 Hypar-Transformation

The former transformations are completed with the Hypar-Transformation, whereby the given points P1 to P4 on a plane, normally a rectangle, are mapped to four corner points Q1 to Q4 a deformed quadrangle (Fig 8). Clearly a node selection corresponding to type 4/3 is possible.

If P is a point on the plane P1P2P3P4, then two straight lines g and f are determined which intersect at P. Line f divides the segments P1P2 and P3P4, while g divides P1P3 and P2P4 in equally proportional parts. The resulting points P12, P34, P13 and P24 are linearly transformed into Q12, Q34, Q13 and Q24. The resulting point Q is given as for the original point P, that is, as the intersection of the straight lines r = Q13Q24 and s = Q12Q34.

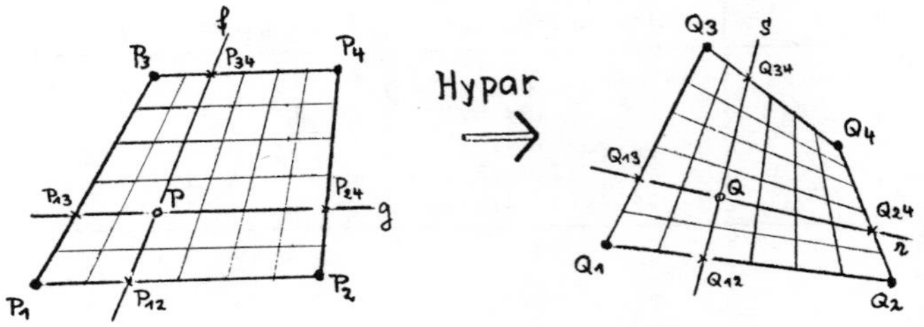

Fig 8: Hypar Transformation

3.2 Rotation

The rotation about an arbitrary axis is described as follows (Fig 9):
The axis is defined by two points P1 and P2, the O-angle trough a point P3 outside the axis and the rotation angle alpha in degrees. The positive turn is given with respect to the axis direction P1P2. A node selection is possible through an additional point P4 to complete a quadrangle. In the program, the rotation is interpreted in terms of the transformations type 3/3 or 3/4.

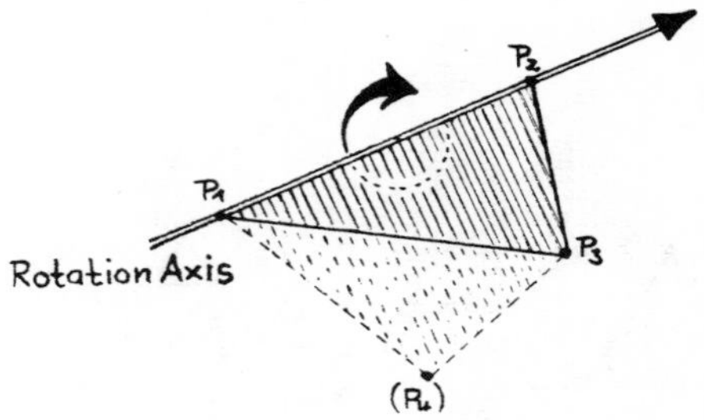

Fig 9: Rotation about any Axis

3.3 Corner Transformation

A somewhat unusual but important transformation is the Sharp Bend (Corner) Transformation. As in polar or cylindrical type transformations, Cartesian coordinates which are oriented in a linear net system are transformed into an open or closed polygon (Fig 10). The centre of the polygon is at the naught point, while the z-quantities remain unaltered. Problems such as interconnection of segments and overlapping are not discussed here.

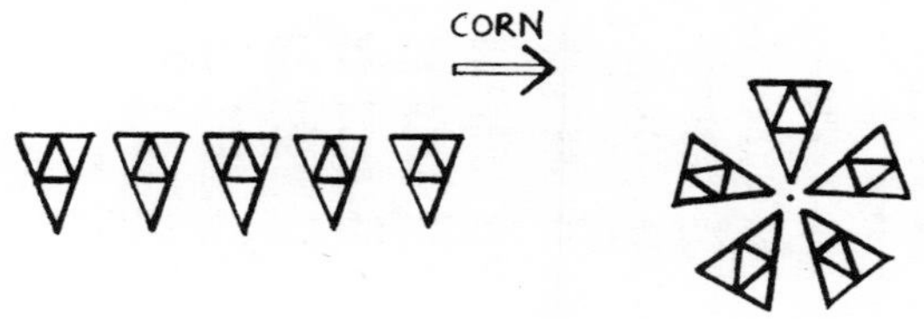

Fig 10: N-cornered Polygon Transformation

4. TOPOLOGY AND GEOMETRY OF BARS

4.1 Vectors and node search

The most common form of generation is given by specificating a set of vectors which are related to the previously generated node topology in the unit grids. In general, two to ten vector formulations are sufficient to cover even the most complicated practical nets.

Closed structures can be generated as unit nets by relating the net module to the rotation angle.

The generation of bar elements with the aid of the vector (DX,DY,DZ) at a random node (AX,AY,AZ) requires a very efficient search algorithm for a node (BX,BY,BZ) = (AX+DX,AY+DY,AZ+DZ).

In GEOM, the actual (unit grid or real) node coordinates of a structure are sorted in ascending order by

$$SA = 123 \cdot AX + AY + 4321 \cdot AZ \qquad \text{Eqn 3}$$

The value SB is computed after Eqn 3 for a node (BX,BY,BZ) under examination and a search by interval halving is carried out in the node table.
An average search in a system with 10000 joints requires some 13 to 14 interval steps.

4.2 Conditioning

In addition, the GEOM system offers two forms of conditioning the generation of bars. Namely, conditioning through vectors mid-points, which are defined as dummy nodes represented with special symbols, and a start-end vector conditioning, which relates to the end nodes of a member.

Examples for conditioning are:

- =/ generate only the bars having a dummy node which is given as the symbol / at the mid-point of the bar vector.
- Δ: do not generate the bars with dummy nodes given through the blocking symbol : at the mid-point of the bar vector.
- =AB generate only the bars whose node ends are represented through the characters A and B, in any order.
- ΔAB generate all bars, except those whose node ends are represented through the characters A and B, in any order.
- =AΔ generate only the bars, where one of the end nodes are represented with the character A (the remaining end node symbol is arbitrary).

A combination of conditions is also possible as it is shown next:

- =XXΔ* generate only the bars, whose both end nodes are represented with the symbol X without the symbol * at a bar's vector mid-point.

4.3 Equivalence

In some cases, it is meaningful to activate the actual node coordinates in the bar generator. The most important case is given by the EQUI function, which reduces nodes with identical coordinates to a single node and relates the generation of bars to the reduced number of nodes. This possibility makes the coupling of parts with different geometric particulars very convenient.

5. GENERATION OF LOADS

The generation of loads for the analysis of large structural normally requires vast amounts of time and work especially for wind loads. This situation is efficiently tackled in the GEOM system with the help of "influence areas". For the purpose, the surface to be loaded in the structure has to be defined. This can be done here through a convenient bar numbering, or the selection of geometric regions of the structure, or through the direct generation of the surface to be loaded. The PANE function creates all the triangular areas whose edges bars and assigns every node in a triangle with one third of the triangl's area. Rectangles are treated in a similar way, where only rectangles without diagonals are taken into account and whose four nodes lie, within a certain tolerance, on a plane.

There are three types of influence areas for choice (Fig 11).

5.1 Dead Weight Type

The first one is independent form direction and generates all triangles or quadrangles with their full area. This areas are appropriate to specify loads such as dead weight.

5.2 Live Load Type

The second type depends on the orientation of the loaded surface and generates all triangles or quadrangles as projected influence areas with respect to a given direction. For example, the projection of areas along the vertical direction is appropriate to specify loads such as snow or sand.

5.3 Wind Load Type

The third type is also direction dependent and creates influence areas for triangules or quadrangles, which are perpendicular to a defined direction. This type is used to specify wind loads.

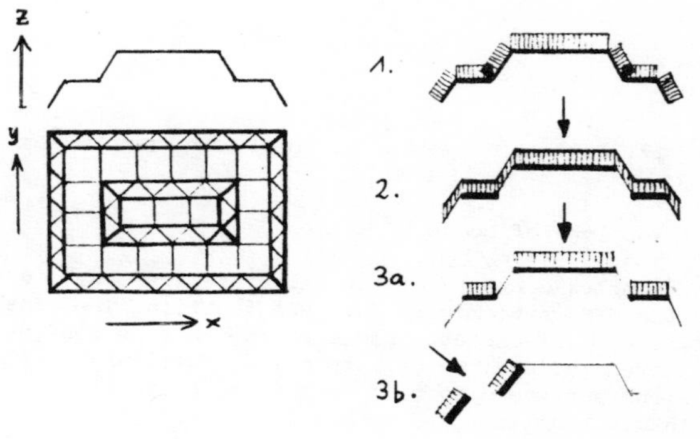

Fig 11: PANE Function

6. FIRST EXAMPLE

A reticulated space structure to roof a market hall is presented as the first example. Many conventional forms of "coding" this type of structure are time consuming and error prone although the internal structure is relatively simple. The problem arise mainly because of the complexity of the contours. This situation, however, is typical of today's applications of space structures. The surface geometry of the system is presented in Fig 12 to aid the visualization of the complex with its towers, recesses and protrusions.

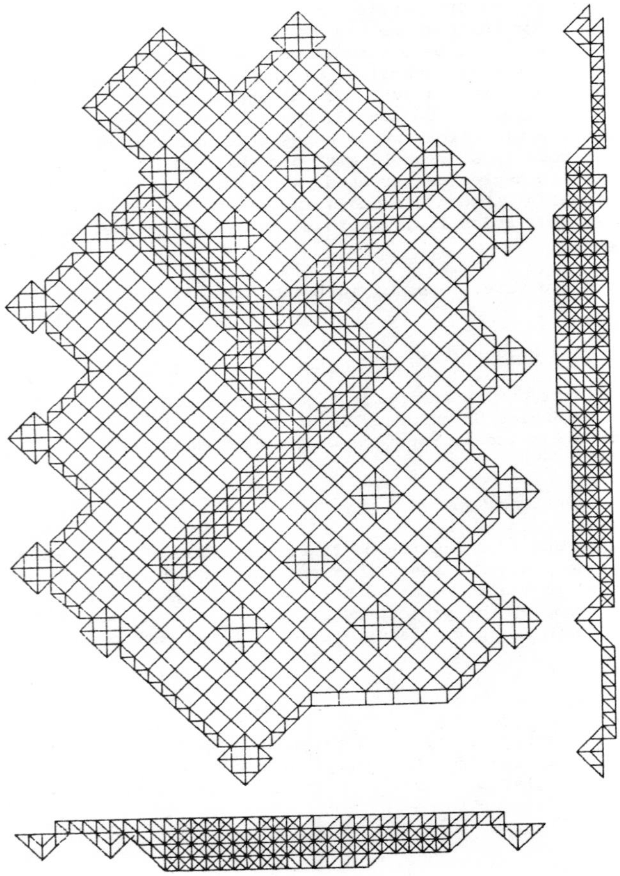

Fig 12: Surface Geometry for the First Example

6.1 Topology of Nodes

The complex contours of the structure are easily described in the GEOM system with the AREA function. The generation of towers and elevated parts of the roof is conveniently achieved through the CALL function. Particulars, such as various surface diagonals or omitted chord bars are marked with slanted lines and colons in the formulation of the node topology.

Printout fragments showing the coding for the first and second plane together with the corresponding printer plots complement the illustration of the method (Fig 13):

```
DISPLAY        0    70     0    80     4
 GRID         10     0     0    10
 AREATT  1     6    16   111   174   111   371   322   172   133   144   233   273   344
 LINE33  1     6    26   274   371   433   122   411   171   272   111  -273   133
 OKAY    1     3
 GRID          2     0     0     2
 AREAF/  1     5    15   511   574   611   522   511   422   171   411   422   171  2733  1044
 AREAF/  1     5    35  3111 -1622  533   522   633   544  1033   544   533   544   533   644
 AREA  R 1    27    47   411   422   433   444
 AREA  R 1    39    35   211   271   222   272   922   233   944   273   244   274
 LINE::  1    19    36   211  -311   211  -811  -522   211  -311   211
 LINE4/  1     5    15   511   574   611   522   511   422   171   411   422   171   411
 LINE4/  1    67    35  1622   533   522   633   544  1033   544   533   544   533  1644
 CALLFFT 4                                           1     1 0  -1     1 0  -1    -1 0    1   -1
 SINGXX  1    33     5  4491  6691  4491  6691  4491  6691  4491   492   69214096  4492
 SINGXX    18292  2293  4093  6093  4093  4493  6693  4493  6693  4493   494   69414098  4494
 OKAY    1     4
```

Plot record no. 1 and 2

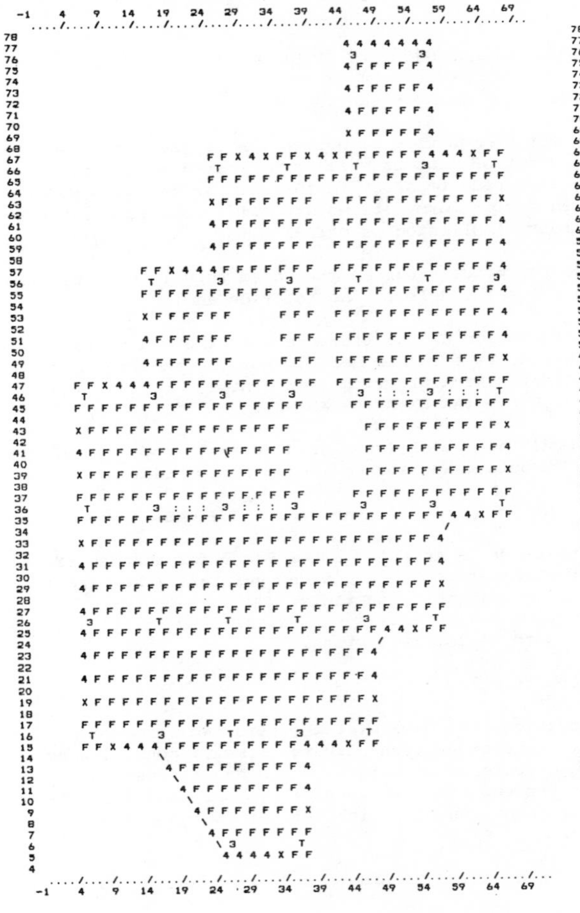

Fig 13: Fragments of Coding and Corresponding Characters

It should be noted, that the outstanding features of the above coding are the character representation and the possibility of conditioning the net generator.

6.2 Geometry of Nodes and Bars. Loads

The required node coordinates are obtained by factoring along x, y and z and rotating by -45 degrees about the z-axis.

The generation of bars turns out to be simple, because of the structure's internal regularity and the conditioned formulation.
The full geometry requires two different GRID formulations, which are complemented with four other lines corresponding to the surface geometry.

Type	Condition	Symm	Sect	Plane from	to	Vector		
GRID	:	48	1	4	8	2	0	0
GRID	XX	44	2	3	7	1	1	1
GRID	F =	48	1	4	8	2	0	0
GRID	=45	44	2	4	4	1	1	1
GRID	=X5	44	2	4	4	1	1	1
GRID	F	44	2	4	4	1	1	1

Symm indicates symmetry conditions for the bar generation, Sect are the cross section reference numbers.

The generation of loads follows with the help of the generated surface geometry and the PANE function. Figure 17 shows a perspective representation of the surface geometry.

7. SECOND EXAMPLE

The generation of a folded-plate dome is presented as a second example. In order to avoid the complicated calculation of the actual geometry, the basic parts of the structure are firstly generated in the unit grid and then brought to their true position with the aid of the TRAF function. The full system is completed taking advantage of the symmetry of the structure and the resulting symmetry parts are linked through the EQUI function.

7.1 Topology of Nodes

One sixth of the folded-plate structure is generated as a simple combination of triangular and square grid structures (Fig 14). In the unit grid, the important corner points at the top layer in each segment are specially identified as dummy nodes. The corresponding corner points in the actual geometry are similarly identified as auxiliary nodes. In this case, there are 35 corner points from which the only 14 different ones are obtained with a special program.

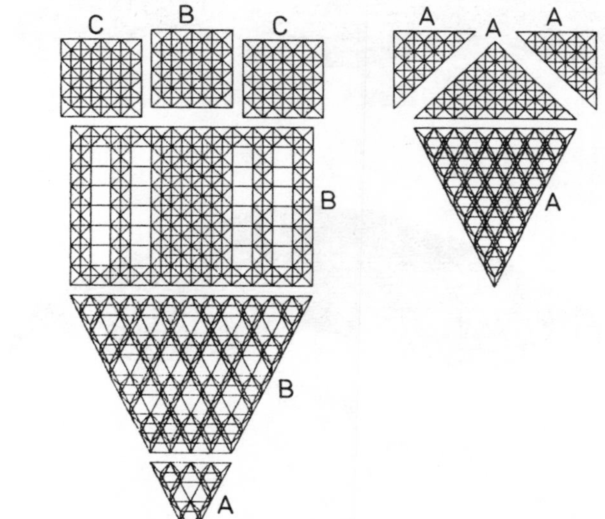

Fig 14: Basic Parts of the Folded-plate Dome

The extension of the basic parts to cover the whole dome (without the dummy nodes derived from the corners) was obtained by simple translations along the x-axis (Fig 15). The bar generator is later applied to the whole nodes system in its unit form.

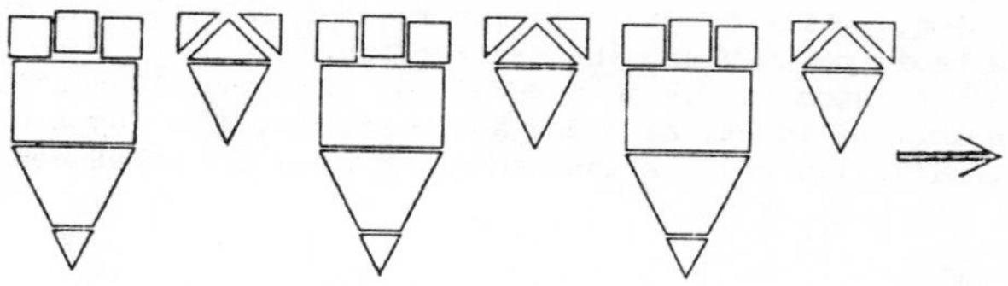

Fig 15: Development of the Basic Parts

7.2 Geometry of Nodes

The series of single parts along the x-direction is turned into a configuration of segments radially positioned, with respect to the naught point, via two CORN transformations (Fig 10). The unit grid corner dummy points are also included in this transformation and the structural height of the grid in z is adjusted to obtain the actual value. The stage can be seen in Fig 16.

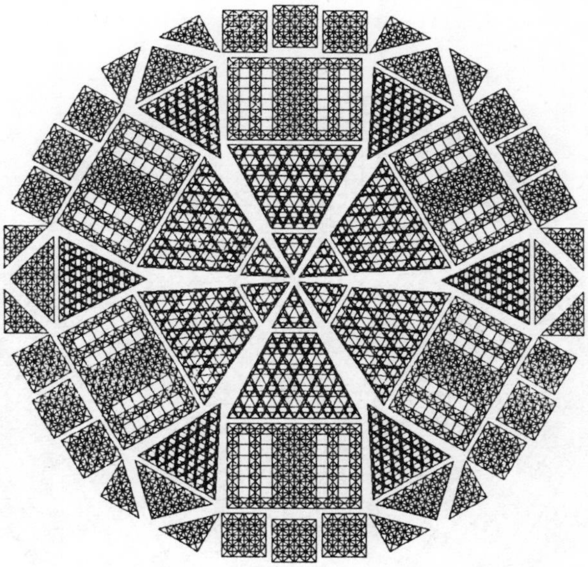

Fig 16: Radial Unit Form of the Basic Parts

The final form is obtained by transforming every single part with the aid of the transformations illustrated in Figs 5, 6 and 8. The corner points of a unit form in the radial configuration are taken as P-points, whereas the true corner points are taken as Q-points. Ten transformations suffice to complete the process, where the bottom layers are included in the selection areas (hence, special attention is required to prevent overlapping of parts during the transformations).
Figure 16 and 18 may be used to identify the necessary transformations: five 3/3 transformations, indicated with A (see Fig 14); three 4/3 transformations, marked with B and two hypar-transformations indicated with C. The results are summarized in Fig 18.

7.3 Topology and Geometry of Bars

The generation of all bars is achieved through six GRID functions only, since the generation works on the unit-topology, which is given as a series of translations of the basic parts along the x-direction.
The true geometry is thus obtained, but it still contains a number of identical nodes and bars. The EQUI function automatically reduces the redundant elements inbetween the coupled sub-structures to produce a single space structure.

7.4 Loads

The surface geometry is simply identified through the sub-division between top and bottom layers, in such a way that the necessary nodal loads for dead weight and snow can be readily obtained.

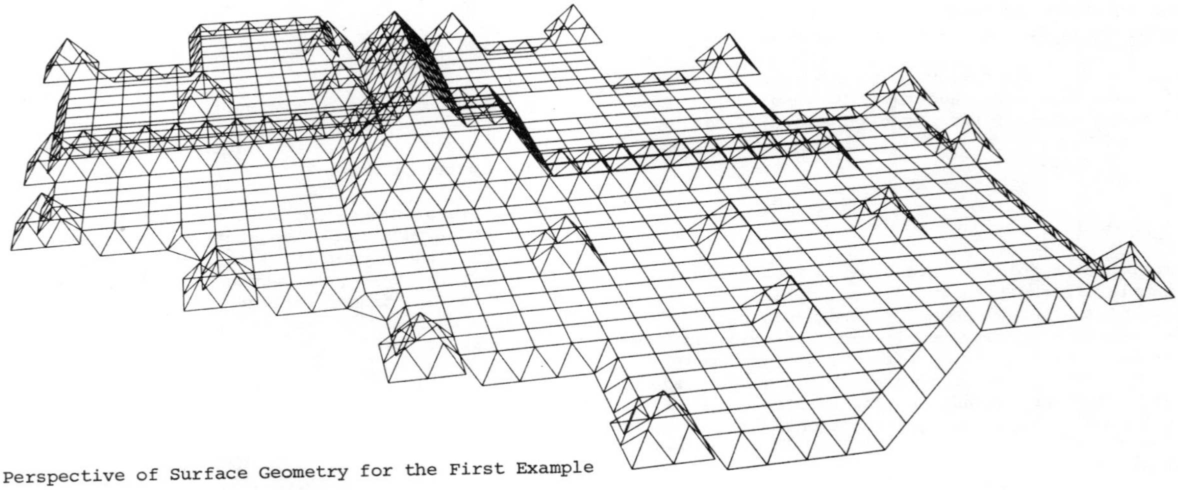

Fig 17: Perspective of Surface Geometry for the First Example

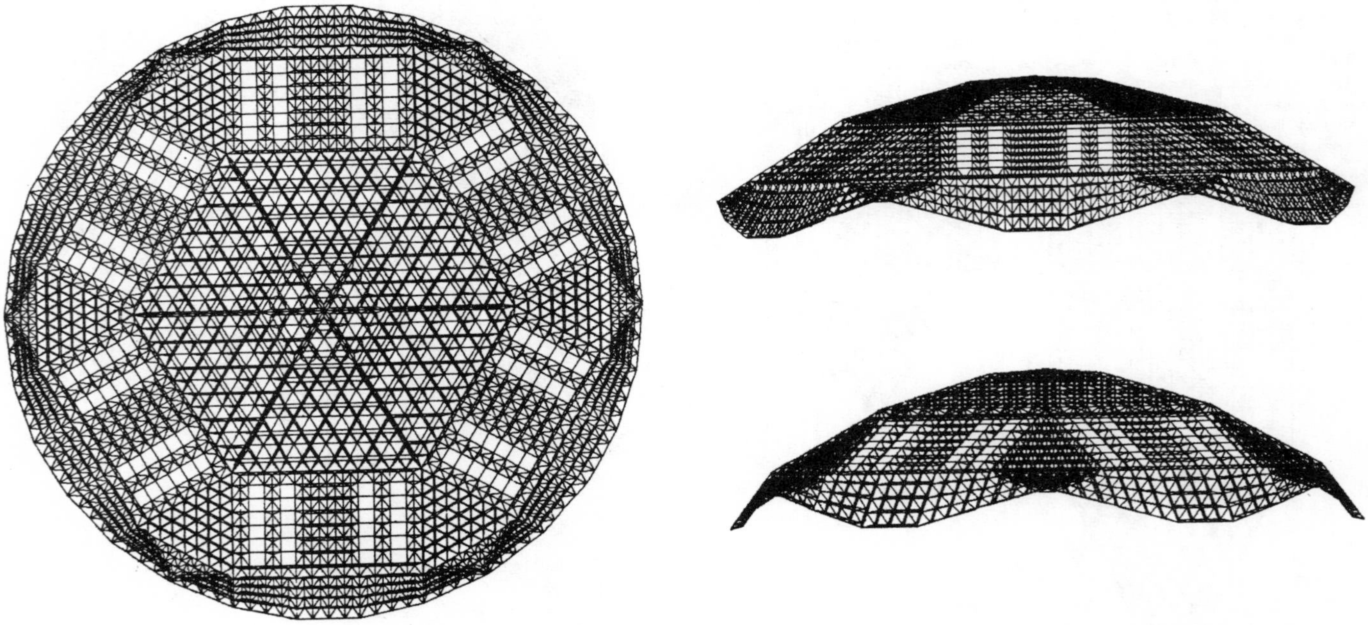

Fig 18: Actual Structure of the Folded-Plate Dome

8. REFERENCES

[1] M MENGERINGHAUSEN, Raumfachwerke aus Stäben und Knoten, Bauverlag GmbH, Wiesbaden und Berlin, West-Germany, 1975

[2] H NOOSHIN, Formex Formulation of Double Layer Grids, Publication Space Structures Research Centre, University of Surrey, England, September 1978

[3] H KLIMKE and M RUH, Darstellung eines Konstruktions-Systems zur Erzeugung der Geometrie von Knoten-Stab-Tragwerken, Internationaler FEM-Congress Baden-Baden, West-Germany, November 1981

FORMEX FORMULATION OF THE SPACE STRUCTURES OF S. DU CHATEAU

M.H. Yassaee

Research Student, Space Structures Research Centre, Department of Civil Engineering, University of Surrey, Guildford, England.

The objective of this paper is to illustrate the application of formex algebra in formulating the interconnection patterns of a number of structures. The concepts of formex algebra may be used to deal conveniently and efficiently with both the representation of the structural configurations and the automated generation of the related data. The structural configurations treated in this work are a selection of space structures designed by du Château. He has designed a large number of structures using single or double layer grids, domes and barrel vaults. His designs cover a wide range of structures such as swimming pools, supermarkets, railway stations, exhibition halls and multistorey buildings.

INTRODUCTION

In the past, the problem of data generation has been dealt with by various computer orientated techniques. However, these are designed to deal only with particular classes of problems. Formex algebra may be used in various ways to improve and considerably ease the generation of data for analytical purposes.

In this paper a selection of space structure configurations have been represented using formex algebra. The entire paper is focused on formex formulation of interconnection patterns, without considering the loading and the support conditions, etc.

The reader is assumed to have an understanding of the concepts of formex algebra as described in Ref 1.

The examples in this work were all processed at the Space Structures Research Centre, Surrey University, using the Formian interpreter - see Ref 2.

FORMULATION OF S. DU CHATEAU'S STRUCTURES

The structural configurations presented are a collection of elegant space structures designed by du Château who is well known for his valuable contributions to the development of space structures throughout the world.

Structure I - Tennis Court at Paris - Vaugirard

This three-directional grid of cylindrical shape was designed to cover the tennis court for the SNCF at Paris - Vaugirard. The hall is 36m long and 18m wide, comprising a simple SDC system grid.

For the purpose of formulation the following conventions are employed throughout the paper:

(1) Every joint of the structure is either represented by a signet of grade two or three.

(2) Every element of the structure is represented by a two-plex cantle of the second or the third grade.

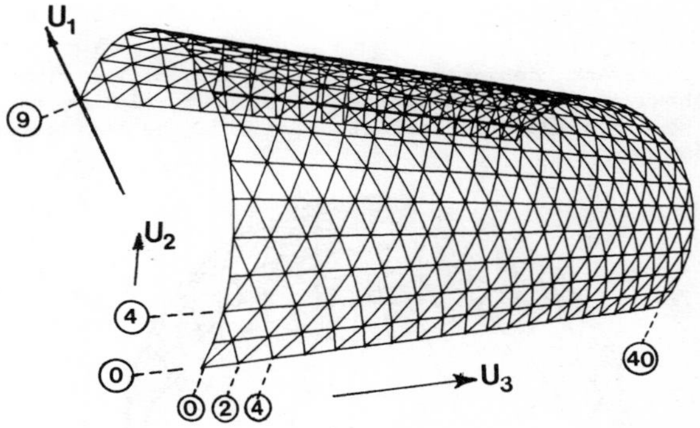

Fig 1

Figure 1 shows a view of the barrel vault under consideration whose joints are on a right circular cylinder with a radius equal to 9 units. A formex representing the interconnection pattern of the barrel vault relative to the indicated basicylindrical normat with basifactors of b_1=1m, $b_2=\pi/36$ and b_3=0.9m may be written as

$$c=\Gamma_{i=1}^{5}\, c_i$$

where

c_1=rinit(9,20,4,2)¦lamit(2,1)¦[9,0,0; 9,2,1]

representing the bracing elements of the barrel vault,

c_2=rinit (10,20,4,2)¦[9,0,0; 9,0,2],

c_3=rinit (9,19,4,2)¦[9,2,1; 9,2,3]

and

c_4=lam(3,20)¦rin(2,9,4)¦[9,2,0; 9,2,1]

represent the elements which are parallel to the generatrix of the cylinder and finally,

c_5=lam(3,20)¦rin(2,18,2)¦[9,0,0; 9,2,0]

representing elements along the normat lines U_3=0 and U_3=40.

Structure II - Ribbed Domes at Meaux Beauval

This structure consists of three adjacent domes, two of which are identical. Formex formulation of these may be done generically, that is, one may write the formulation in terms of one or more parameters and this would then provide a formulation for a family of configurations. A generic formulation for the interconnection pattern of a family of ribbed domes that have configurations similar to the dome of Fig 2 may be written as

RDIP(m,n)=pex¦rinit(m,n,1,1)¦E

where

$E=\text{rosit}(\tfrac{1}{2},1\tfrac{1}{2})¦[1,0,1;\ 1,1,1]$.

Here, RDIP (Ribbed Domes Interconnection Patterns) is the name given to the formulation and the parameter list specifies the order in which the parameters are to be given.

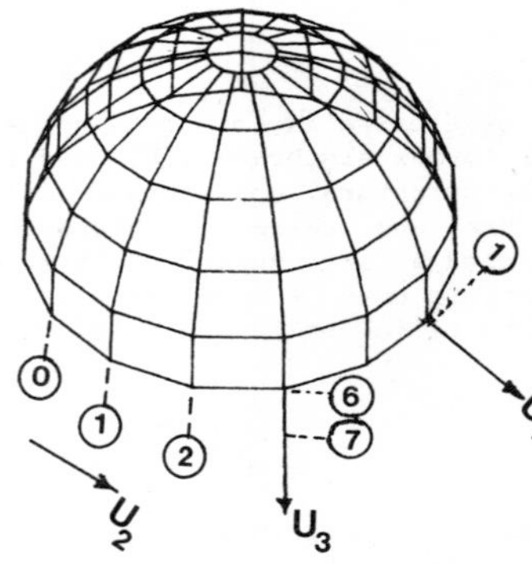

Fig 2

Thus

RDIP(17,5)

would represent the interconnection pattern of the ribbed dome shown in Fig 2. The formex is plotted relative to a 17-24 sect basispherical normat shown with the basifactor b_1=7.55m.

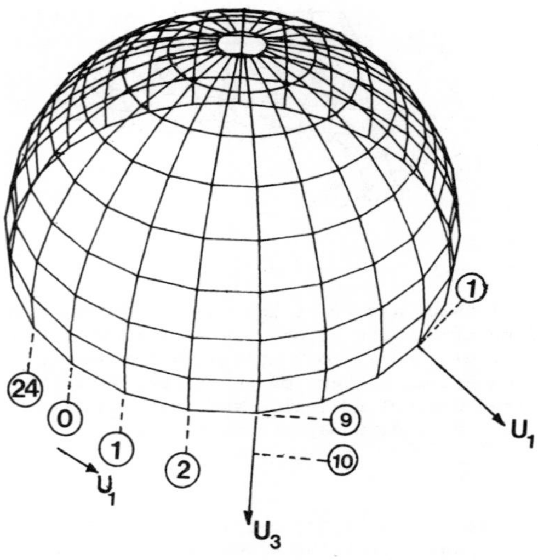

Fig 3

Similarly,

RDIP(25,8)

would be a formex representing the interconnection pattern of the dome shown in Fig 3. This formex is plotted relative to a 25-36 sect basispherical normat shown with the basifactor b_1=11.05m.

Structure III - University of Lyon Bron

This structure has been constructed using the square based pyramidal Unibat system and consists of four parts as shown in plan view in Fig 4.

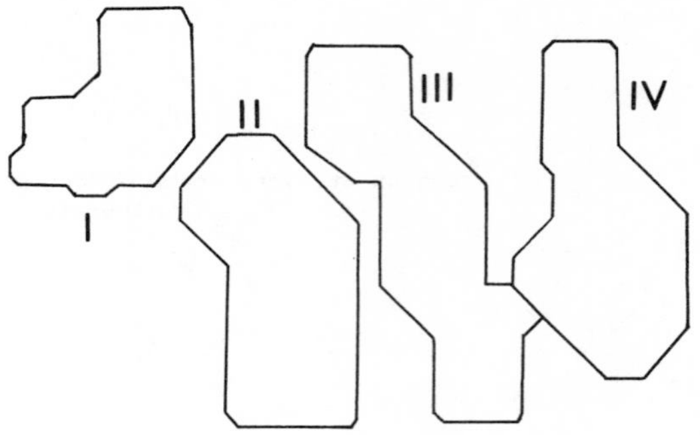

Fig 4

This is an interesting example which shows how a configuration with irregular boundary can be formulated through formices. Here, only the formex formulation of part I of the structure whose plan is shown in Fig 5 is discussed in detail. A convenient way of formulating the pattern is to adopt a superpansive approach by choosing

$E_1=\{[1,2,2;\ 2,2,1],\ [1,2,2;\ 2,1,2]\}$

and

$E_2=\{[2,2,1;\ 4,2,1]\}$

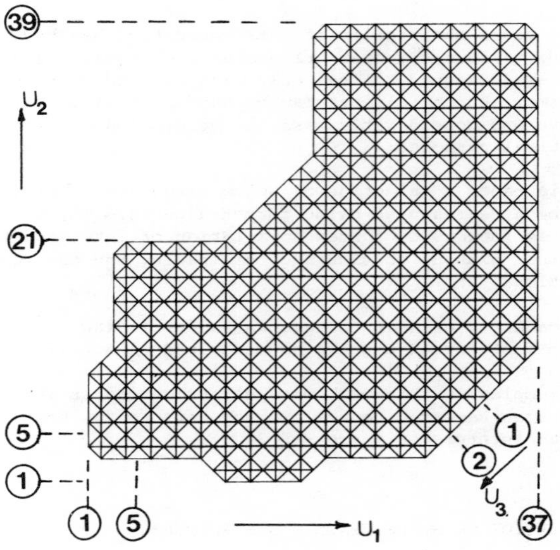

Fig 5

as generants, and writing

F_1=rinid(18,19,2,2)¦rosid(2,2)¦E_1

and

F_2=rinid(17,18,2,2)¦rosid(3,3)¦E_2.

Now, let,

$F_3=F_1$ # pex¦F_2.

A paritrifect N-plot of F_3 is shown in Fig 6.

Comparison of Figs 5 and 6 shows that the interconnection pattern represented by F_3 contains some superfluous elements. A formex representing the interconnection pattern of Fig 5 can be obtained from F_3 by removing these superfluous elements. To modify F_3, let J be an ingot that denotes all the joints of the structure; J may be obtained as:

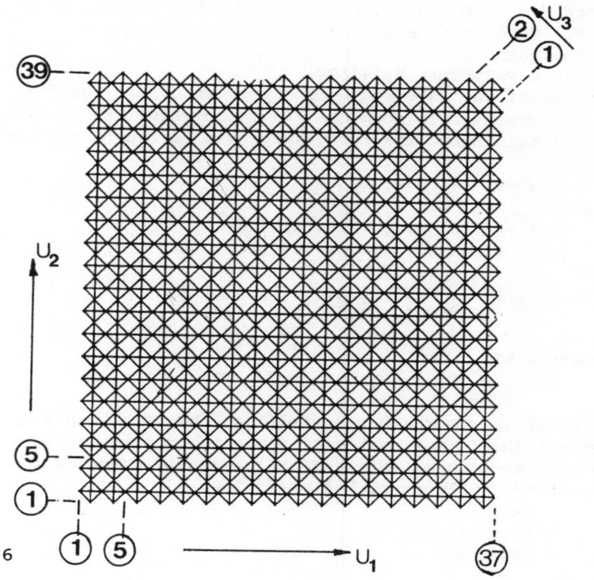

Fig 6

e ={[2,2,1], [2,1,2], [1,2,2]}
J_1=rinid(19,20,2,2)¦e

and

J =rel(P)¦J

where

rel(P)

denotes a function that selects only those signets of J_1 for which the perdicant P is true.

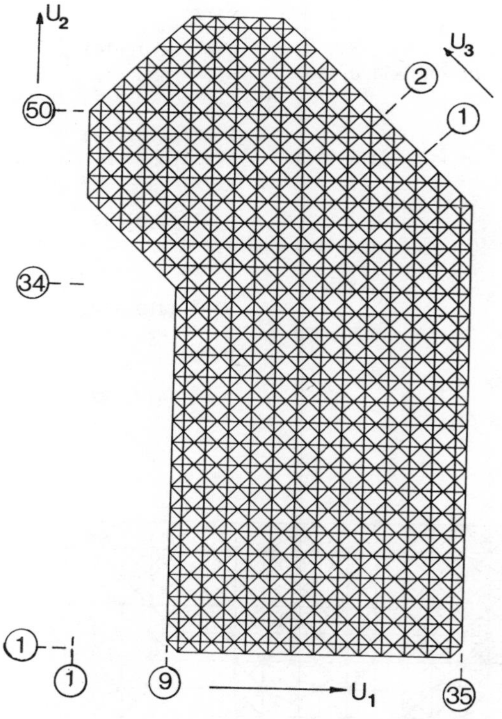

Fig 7

For example,

rel(U_2<20)

denotes a function that selects every signet of J_1 for which U_2<20.

The perdicant P for part I of the structure is defined as:

$P = P_1 \wedge P_2 \wedge P_3 \ldots\ldots \wedge P_5$

where

$P_1 = U_1 \geqslant 19 \wedge U_2 - U_1 \leqslant 9 \wedge U_2 \leqslant 21$,

$P_2 = U_1 \geqslant 3 \wedge U_2 - U_1 \leqslant 9$,

$P_3 = U_1 - U_2 \leqslant 25$,

$P_4 = U_2 \geqslant 3 \wedge U_2 + U_1 \geqslant 13$

and

$P_5 = U_2 \geqslant 3 \wedge U_1 - U_2 \leqslant 17$

in which the symbol $\wedge$ stand for the Boolean 'AND'. A formex representing the internal elements of the configureation of Fig 5 may be written as

F_4=nex(J)¦F_3.

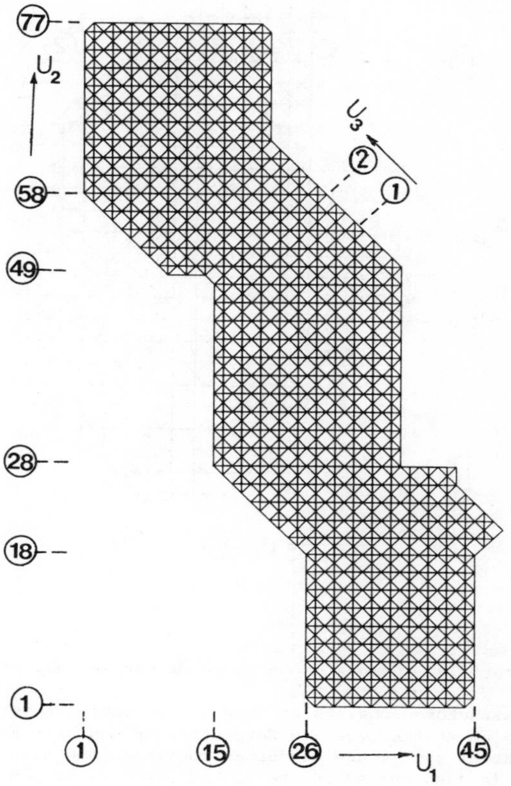

Fig 8

To complete the formulation, let

E_3=[1,4,2; 1,6,2],

then

B_1=rin(2,3,2)¦E_3,
B_2=rin(2,4,2)¦tranid(2,8)¦E_3,
B_3=rin(2,5,2)¦tranid(18,24)¦E_3

and

B_4=rin(2,13,2)¦tranid(36,8)¦E_3

would represent the elements at the boundary which are parallel to the U_2 - direction. Furthermore, let

E_4=[2,1,2; 4,1,2],

then

B_5=lam(1,15)¦rin(1,4,2)¦tran(2,2)¦E_4,
B_6=rin(1,3,2)¦tran(1,10)¦E_4,
B_7=rin(1,4,2)¦tranid(2,20)¦E_4

and

B_8=rin(1,8,2)¦tranid(18,38)¦E_4

would represent the element at the boundary which are parallel to the U_1 - direction. Thus

$$F = F_4 \# \left[\sum_{i=0}^{8} B_i \right]$$

is a formex representing the interconnection pattern of the configuration.

Other parts of this structure can be formulated in a similar manner. One may write a generic formulation and use an appropriate perdicant for each case. N-plots of these relative to paritrifect retronorms are shown in Figs 7-9.

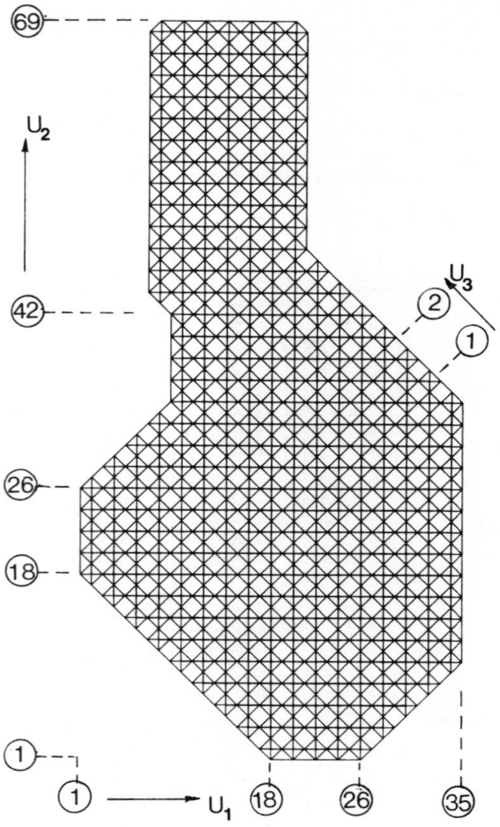

Fig 9

Structure IV - Gymnasium at Lycee de Carpentras, France

This structure consists of five barrel vaults covering the gymnasium at the Lycee de Carpentras in southern France. The barrel vaults are connected together, and each barrel vault is also curved in the longitudinal direction.

A formex E representing the interconnection pattern of the configuration relative to a paribifect retronorm, shown in Fig 10, can be formulated as follows:

GC_1=rinid(3,12,2,2)¦lam(1,3)¦[2,0; 4,2],
GC_2=lam(1,9)¦GC_1,
GC_3=rinid(7,12,2,2)¦rosid(3,1)¦[2,0; 4,0]

and

GC_4=GC_3 # GC_2

where GC_4 represents the interconnection pattern of one barrel vault with some elements doubly represented. Furthermore

GC_5=Pex¦rin(1,5,14)¦GC_4

represents the interconnection pattern of all the five barrel vaults. To complete the formulation, the elements that interconnect the five barrel vaults can be represented by

GC_9 =rin(1,6,14)¦GC_8

where

GC_8 =rin(2,13,2)¦[0,0; 4,0] #
rin(2,12,2)¦lam(1,2)¦[0,0; 4,2]

and

GC_{10}=rinid(2,12,74,2)¦[0,0; 0,2] #
rinid(2,13,74,2)¦[0,0; 2,0]

Subsequently

G=GC_9 # GC_5 # GC_{10}.

A formal retronorm that establishes the relationship between the formex that represents the configuration's interconnection pattern and the geometric shape of the structure can be described as follows:

$x = x_1$
$y = y_1 - RL(1 - \text{Cos } \alpha)$
$z = RL \text{ Sin } \alpha$

where

$\alpha = {}^{\pi}/_{24}(U_2 - 12)$

and where x_1 and y_1 are defined as follows:

If $(U_1 < 2)$	$\theta = {}^{\pi}/_{14}(U_1+12)$ $x_1 = x_2$ $y_1 = y_2$
If $(U_1 > 2$ and $U_1 \leqslant 16)$	$\theta = {}^{\pi}/_{14}(U_1-2)$ $x_1 = 12 + x_2$ $y_1 = y_2$
If $(U_1 > 16$ and $U_1 \leqslant 30)$	$\theta = {}^{\pi}/_{14}(U_1-16)$ $x_1 = 24 + x_2$ $y_1 = y_2$
If $(U_1 > 30$ and $U_1 \leqslant 44)$	$\theta = {}^{\pi}/_{14}(U_1-30)$ $x_1 = 36 + x_2$ $y_1 = y_2$
If $(U_1 > 44$ and $U_1 \leqslant 58)$	$\theta = {}^{\pi}/_{14}(U_1-44)$ $x_1 = 48 + x_2$ $y_1 = y_2$
If $(U_1 > 58$ and $U_1 \leqslant 72)$	$\theta = {}^{\pi}/_{14}(U_1-58)$ $x_1 = 60 + x_2$ $y_1 = y_2$
If $(U_1 > 72)$	$\theta = {}^{\pi}/_{14}(U_1-72)$ $x_1 = 72 + x_2$ $y_1 = y_2$

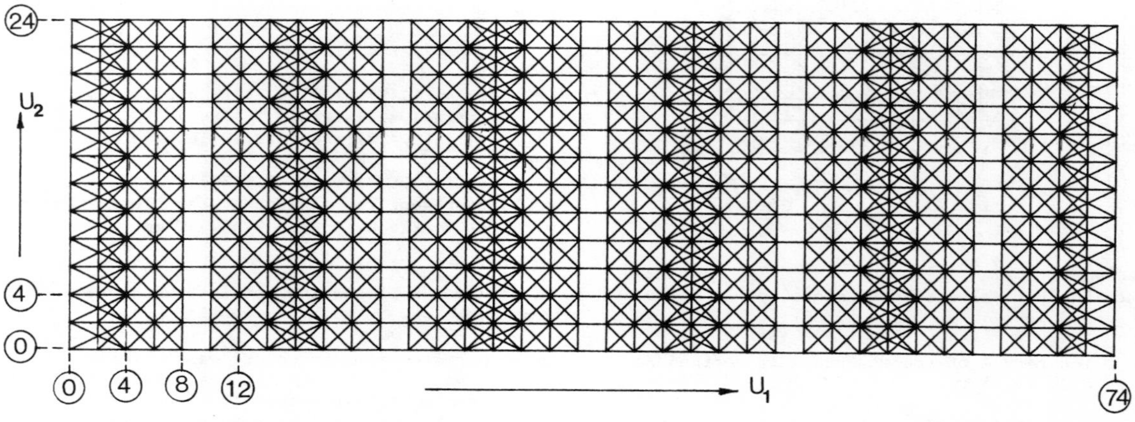

Fig 10

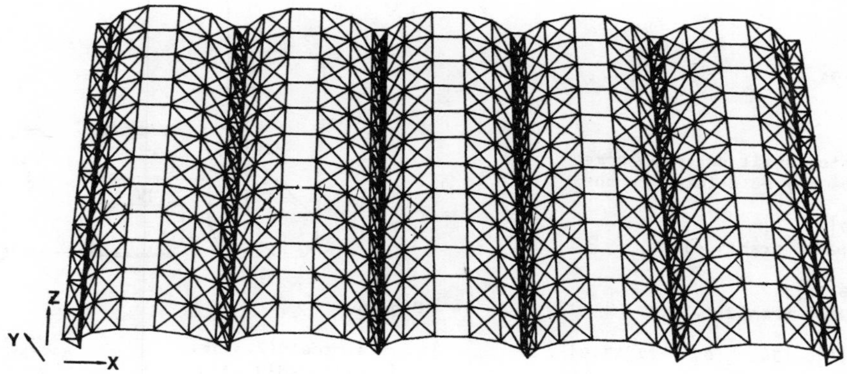

Fig 11

and where, in each case

x_2= -R Cos θ

y_2= R Sin θ

and R is the radius of each barrel vault. A view of the structure is shown in Fig 11.

Structure V - La Gare De Moubeuge

This structure has been constructed using the triangular based pyramidal Unibat system and has some 1500 elements. An N-plot of this structure is shown in Fig 12.

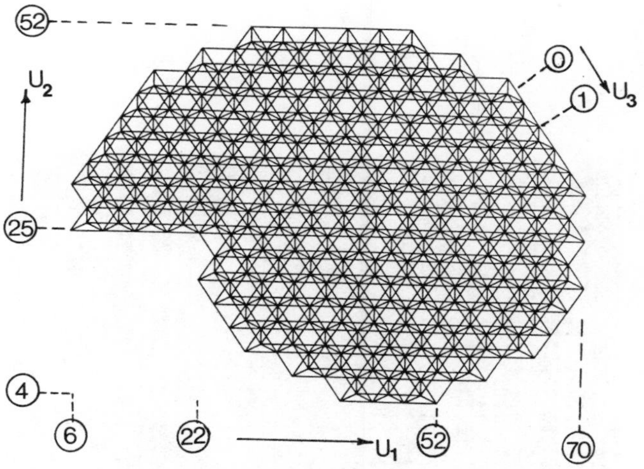

Fig 12

A convenient way of formulating this interconnection patter is again to adopt a superpansive approach by choosing

E_1={[4,4,1; 8,4,1], [6,5,0; 6,7,1]} #
lam(1,6)¦{[4,4,1; 6,7,1], [4,4,1; 6,5,0]} #
[6,5,0; 10,5,0] # lam(1,8)¦[6,5,0; 8,8,0]

as generant and

F_1=rinid(16,9,4,6)¦E_2

where

E_2=E_1 # tranid(2,3)¦E_1.

An N-plot of F_1 and a Z-plot of E_2 relative to a paritrifect retronorm are shown in Figs 13 and 14, respectively.

Comparison of Figs 12 and 13 shows that the interconnection pattern represented by F_1 contains some superfluous elements. A formex representing the interconnection pattern of Fig 12 can be obtained from F_1 by disposing of the cantles that relate to the superfluous elements.

The technique employed in formulating structure III may also be used to formulate the interconnection pattern of Fig 12. Thus an ingot J that denotes all the joints of the structure can be obtained as follows:

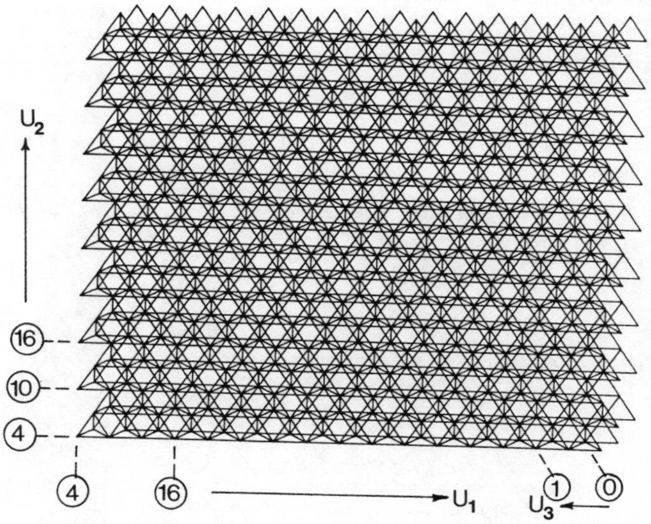

Fig 13

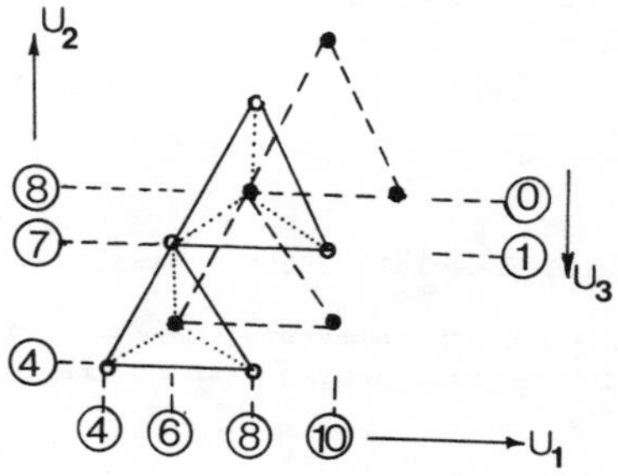

Fig 14

J_1=Med¦F_1

and

J =rel(P)¦J1

where perdicant P is given by

$P = P_1 \wedge P_2 \wedge P_3 \ldots\ldots \wedge P_6$

in which

$P_1=3U_1-2U_2 \leq 172$,

$P_2 = U_1 < 70 \wedge 3U_1 + 2U_2 < 272$,
$P_3 = 2U_2 - 3U_1 < 44 \wedge U_1 > 6$,
$P_4 = U_2 < 52$,
$P_5 = U_2 > 25 \wedge U_1 > 22$
and
$P_6 = U_2 > 18 \wedge 3U_1 + 2U_2 < 104$.

Subsuquently
F_2=nex(J)¦F_1
would represent a formex which still contains some superfluous elements. These may be deleted by the following operations:

GB ={[32,4,1], [34,5,0]},
GB_1=GB # tranid(-2,3)¦GB # tran(1,4)¦GB,
GB_2=lam(1,46)¦GB_1,
GT ={[18,47,0], [18,49,1]},
GT_1=GT # tranid(6,3)¦GT # tranid(2,3)¦GT,
GT_2=lam(1,38)¦GT_1
GG ={[6,29,0], [70,29,0], [70,23,0], [22,23,0]},
G =GG # GT_2 # GB_2
and
F =lux(G)¦F_2
is a formex representing the interconnection pattern of Fig 12.

Structure VI Immeuble de Bureau - Nimes

This structure is constructed using the triangular based Unibat system and has some 3300 elements. It is formulated in a similar manner to that of Structure V, by writing the formulation in a generic form and using a suitable perdicant. An N-plot of this structure is shown in Fig 15.

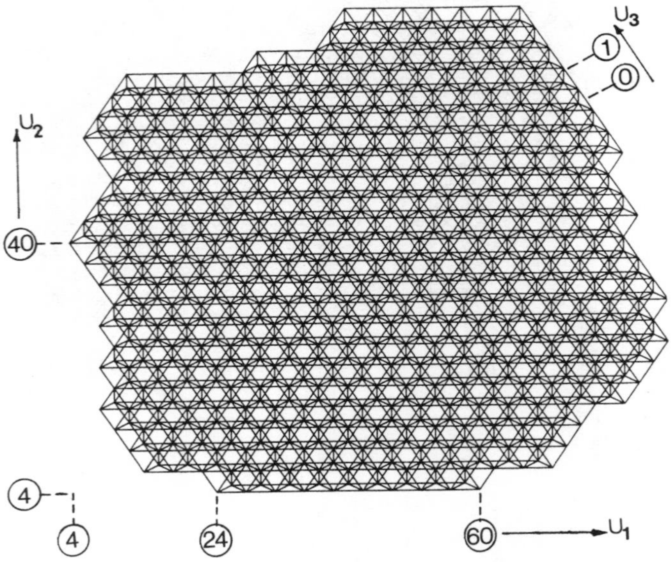

Fig 15

Structure VII - Commercial Centre - Laval

This structure covers a commercial centre in Laval and consists of four parts. A plan view of the complex is shown in Fig 16.

The generants for part IV are the same as those of Structure III. Therefore it can easily be formulated by writing an appropriate perdicant and using the formulation given for structure III. An N-plot of this part is shown in Fig. 22. Parts I-III have another type of generant. The technique employed for formulation is similar to that of Structure III. Here the formulation for part I is discussed in detail. Let

e ={[2,3,0; 3,3,1], [3,3,1; 2,4,1], [2,3,0; 3,4,0]},
g ={[2,2,1; 3,2,0]},

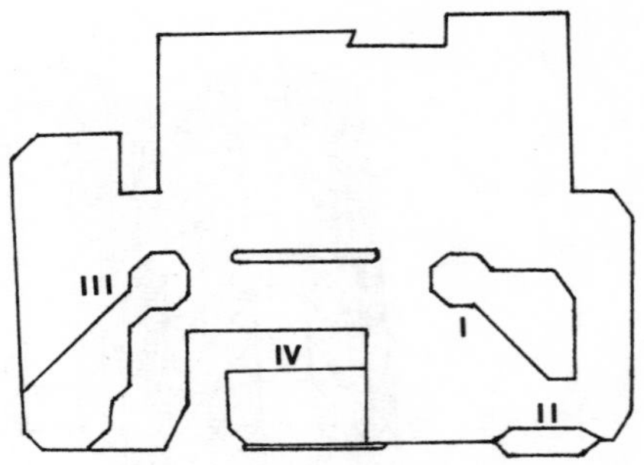

Fig 16

e_1=rosid(2,3)¦e,
g_1=rosid(3,2)¦g,
H =g_1 # e_1
and
F_1=rinid(26,24,2,2)¦H.

A Z-plot of H and an N-plot of F_1 are shown in Figs 17 and 18.

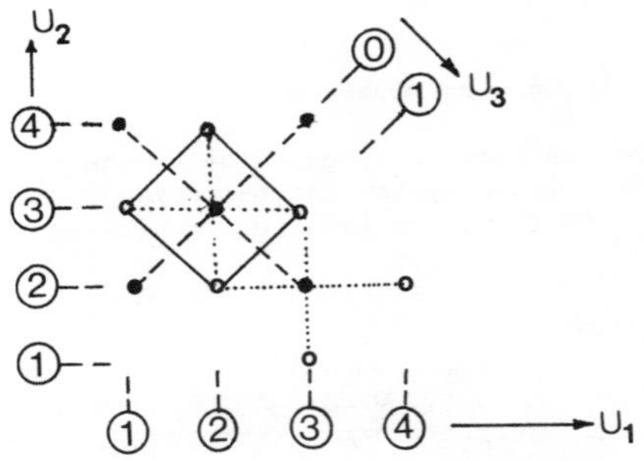

Fig 17

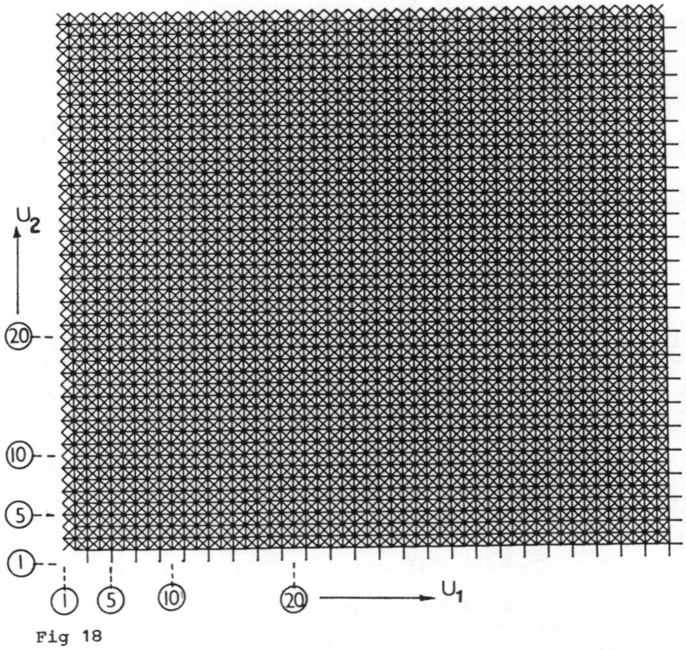

Fig 18

An N-plot of the interconnection pattern of part I is also shown in Fig 19. Comparison of Fig 19 and 18 shows that the interconnection pattern represented by F_1 contains some superfluous elements. These can be removed as follows:

J_1={[2,2,1], [3,3,1], [2,3,0], [3,2,0]},
J_2=rinid(26,24,2,2)¦J_1,

$J = \mathrm{rel}(P) \mid J_2$

and

$F = \mathrm{nex}(J) \mid F_1$

where

$P = (P_1 \wedge P_2 \wedge P_3 \; ..... \wedge P_5)$

in which

$P_1 = U_2 + U_1 > 46 \wedge U_2 > 31,$

$P_2 = U_2 + U_1 > 38 \wedge U_1 > 2 \wedge U_2 > 2$

$P_3 = U_2 - U_1 < 42 \wedge U_2 < 49,$

$P_4 = U_2 < 41 \wedge U_1 + U_2 < 64$

and

$P_5 = U_1 + U_2 < 88 \wedge U_1 < 52.$

Formex F represents the interconnection pattern of part I.

Formulation for parts II-III is carried in a similar manner and N-plots of these are shown in Figs 20 and 21.

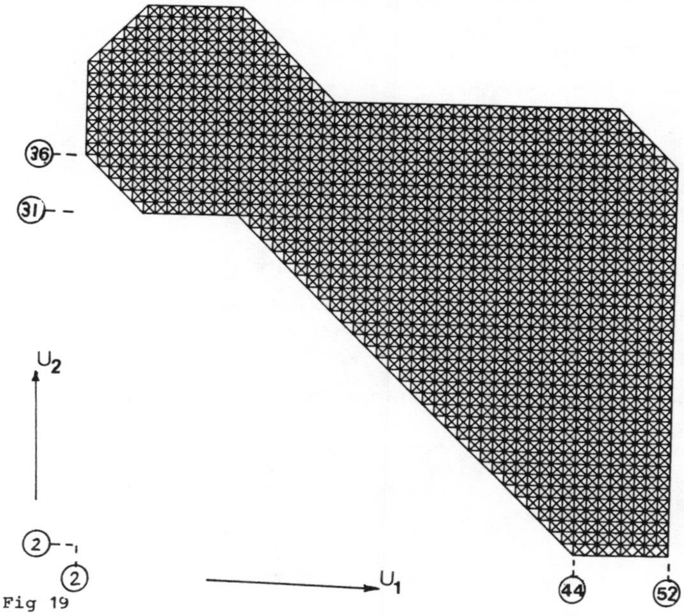

Fig 19

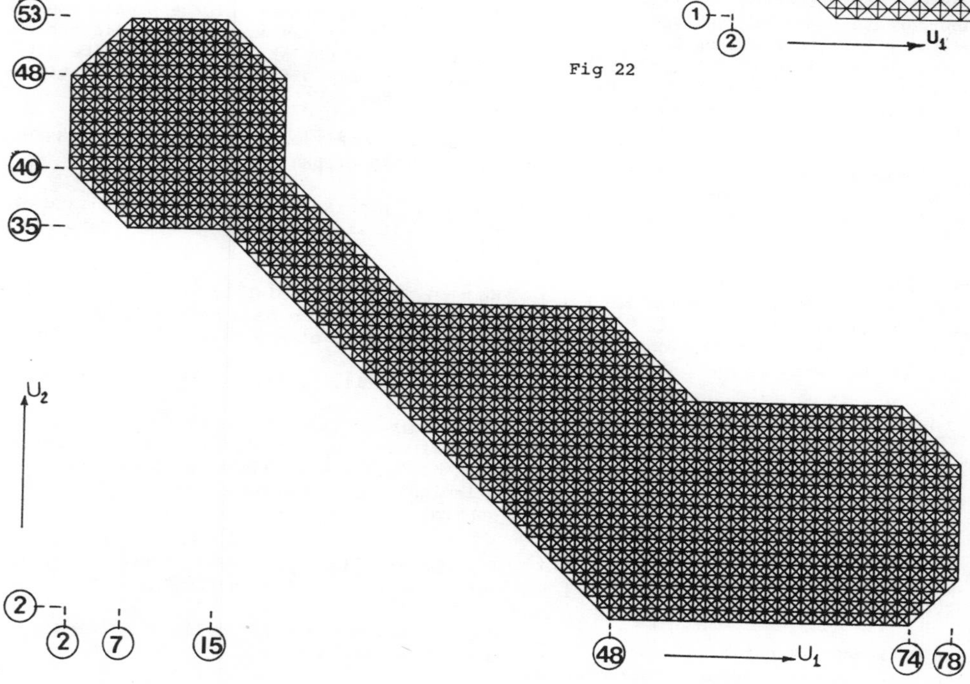

Fig 20

REFERENCES

1. Nooshin H. 'Formex Configuration Processing', Applied Science Publishers, 1984.

2. Disney P. & El-Labbar O. 'Introduction to Formian', Proceedings of the Third International Conference on Space Structures, 1984.

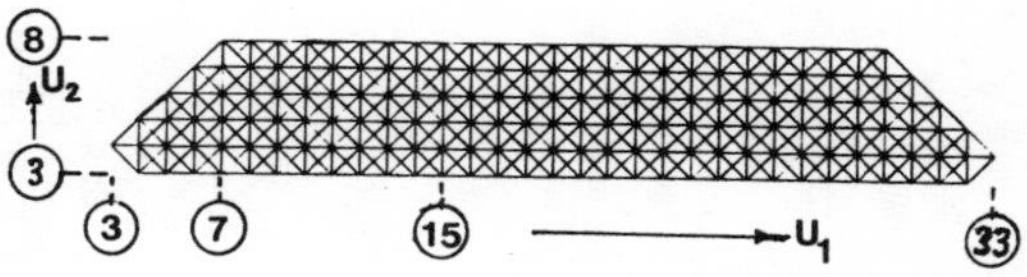

Fig 21

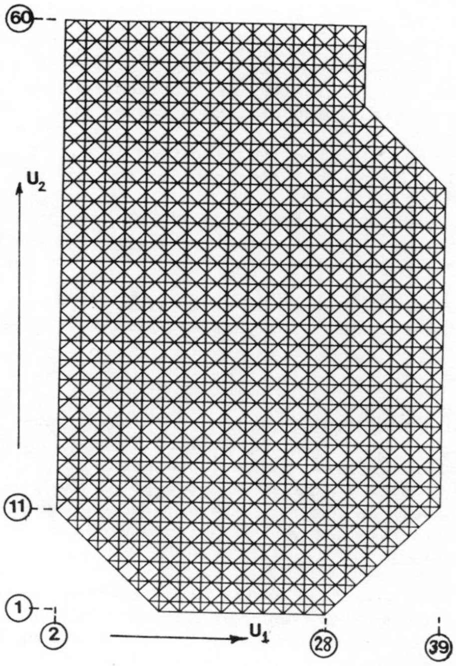

Fig 22

PLENICES AND STRUCTURES

ALFONSO RAMIREZ-RIVERA

Professor, Department of Civil Engineering,
National University of Colombia, Bogota

The enormous computing capacity available today, that make it possible to analyse structures of extraordinary size and complexity, is expected to be greatly increased in the near future with the advent of the fifth generation computers, much faster than today's machines.

To cope with the new situation it will be necessary to develop more powerful computer languages and in many cases to present the problems to the computer in quite different ways as those used until now.

In particular, the handling of vast amounts of data in the process of solving increasingly large problems will become a big problem by itself. Plenices are presented then as a tool for organizing the storage and data flow in the scale required by the new computing processes.

In what follows, the basic ideas of Plenices will be developed and their possible use in the study of problems of structural analysis and design will be shown.

INTRODUCTION

The fifth generation computers, a thousand times faster than today's machines, that will be in use in the course of the next two or three decades, are bound to produce dramatic changes in the size and type of the problems that can be solved in numerical terms, changes comparable to those produced by the advent of computers since the middle of the century.

To take full advantage of the new situation, it will be necessary to develop appropriate computer languages and more efficient ways of formulating the problems,specially in those aspects related to the handling of data. In fact, given the enormous size of the problems that can now be tackled,the handling of immense bodies of data can become a very big problem by itself.

Plenices are presented as a new mathematical tool for operating with large bodies of information and for the organization of its flow through a given process in the computer.

In the case of structures, the much more powerful computing facilities make feasible the treatment of many structural problems beyond the simple linear static analysis, and particularly the systematic incursion into the field of structural design by rational and automatic methods.

Not only the problems formulated by means of mathematical models similar to those used in structures could benefit from the new plenix technics, but also many others in which the handling of large bodies of data is an important consideration.

PLENICES

1- Basic Definitions.

Consider a set of various mathematical entities, for instance, magnitudes like real numbers, vectors, matrices, Booleans; algorithms such as the inversion of matrices, the solution of equations, etc., and arrange these elements in a sequence. The resulting arrangement give us the first idea of a PLENIX.

Plenices are written with curled brackets as,

$$\{P\} = \{P\{1\}, P\{2\}, \ldots, P\{i\}, \ldots, P\{n\}\}$$

where $P\{i\}$is the ith element of the plenix $\{P\}$.

Notice that any element of the plenix could be of quite different nature as any other. This represents an important property of plenices that permits the organization of different types of entities within the same arrangement.

The elements $P\{i\}$ are called PANELS and their number n is the ORDER of the plenix.

The EMPTY plenix is defined as a plenix containing no panels. It is considered of order zero and denoted as $\{\}$.

Another important point to notice in the idea of plenices is that a panel can be a plenix by itself. For instance, the panel $P\{j\}$ of a plenix $\{P\}$ could be the plenix

$$P\{j\} = \{Q\{1\}, Q\{2\}, \ldots, Q\{k\}, \ldots, Q\{m\}\}$$

The panels $P\{i\}$of a plenix $\{P\}$ are called the principal panels, and those of any other plenix contained in $\{P\}$, such as $Q\{k\}$, are called subsidiary panels of $\{P\}$. Both principal and subsidiary panels are referred to in general as panels of $\{P\}$.

A panel that is a plenix by itself is called a PLECULE, otherwise it is called a PLETOM.

It is important now to devise a method for locating a panel within a plenix.

The position of any panel in a plenix can be identified in a simple way be means of a numerical expression α, called the ORDERATE of the panel.

$$\alpha = \alpha(1), \alpha(2), \ldots, \alpha(i-1), \alpha(i), \ldots, \alpha(s)$$

where $\alpha\{i\}$ is a positive integer.

In fact, any panel is either a principal panel and in that case s = 1, or it belongs to a principal panel, which is then a plecule, for s = 1; in both cases $\alpha\{1\}$ specifies the position of that principal panel in the plenix.

If s > 1, then, the panel is again either a principal panel of the plecule identified by $\alpha(i)$, and in that case s = 2, or belongs to a principal panel of that plecule when s > 2, and in both cases $\alpha\{1\},\alpha\{2\}$ identifies that particular principal panel of the plecule.

In a similar way, $\alpha(1), \alpha(2), \ldots, \alpha(i-1)$, for $s \geq i$, specifies a plecule whose $\alpha(i)$ th principal panel contains the panel identified by α.

The panel of a plenix {P} identified by the orderate α is denoted as P $\{\alpha\}$.

The particular arrangement of elements in a plenix can be represented by means of a graph, called the DENDROGRAM, or in numerical way by means of a numerical expression called the DENDRONOM.

Consider for example the plenix

{P}= {A, {B, E, J, K}, {C}, {D, F {G, {H, I}}} }

A dendrogram of {P}is shown in Fig 1.

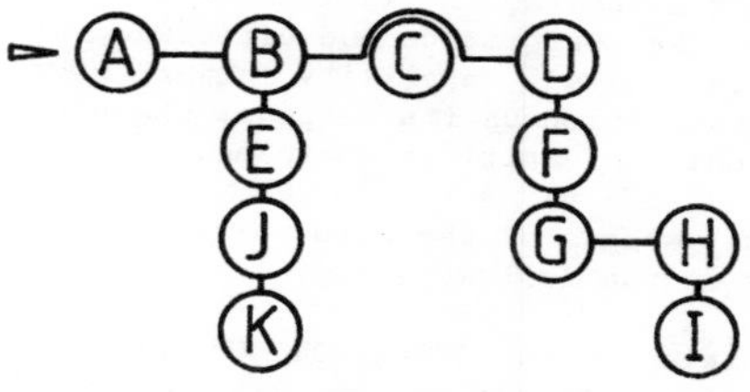

Dendrogram of {P}

Fig 1

For describing the structure of a plenix by means of the dendronom the following conventions are used,

a) If the jth principal panel of a plenix is a pletom, it is indicated by the number j alone. If it is a plecule the number j is enclosed in square brackets [j], or more explicity, it is shown as j [Ωj] , where Ωj is the dendronom of the plecule j.

b) The continuous sequence of integers

j, j + 1,..., k - 1,k

can be indicated as j : k

The dendronom of a plenix can then be written in many ways according to the degree in which the structure of the various panels is explicity indicated.

For example, if only the structure of the principal panels is indicated, the dendronom of P is then

ΩP = 1, [2], [3], [4]

If, on the other hand, we want to make explicit the structure of all its panels, the dendronom is

ΩP = 1, 2 [1:4], 3 [1] , 4[1, 2, 3[1, 2[1:2]]]

In the above example, the 4th principal panel is a plecule with dendronom 1, 2, 3 [1,2[1:2] and so it is shown as 4 [1, 2, 3 [1,2 [1:2]

In the same example, the panel P { 4, 3, 2 } which is the plenix {H, I }, that is, a plecule with dendronom 1, 2, is shown as 2 [1:2]

Note also that the first principal panel, A, is a pletom and so it is indicated by the number 1 alone, whereas the third principal panel, { C }, wich is a plecule, is indicated as 3 [1] , as a plenix of order 1.

The possibility of looking at particular portions of the information with various degrees of detail, according to the requirements of the current stage of the process under consideration, is another important property of plenices. It allows to deal with the pertinent information at a given stage, considering the rest of it only in a global way.

For example, the arrangement of joints and members in a skeletal structure can be referred to simply as the topological plenix {T} in a global way, or in a very detailed manner identifying each individual member and joint and their relations by considering in full detail the structure of { T }, or even in some intermediate degree of detail, all depending on the requirements of the design process.

The particular way in which the detail of the various panels is presented is referred to as an ASPECT of the plenix.

The idea of the aspects of a plenix is also important when establishing relations between two plenices or when defining plenix operations. In those cases, the plenices involved are presented under similar or congruent aspects.

Plenices that are expressed under the same aspect are said to be SIMILAR.

A continuous sequence of panels of a plenix P having also the structure of a plenix is called a SUBPLENIX of P . A subplenix can be identified by means of its dendronom with reference to the panels of the original plenix.

For example, the plenix,

{ {C}, {D, F, {G, { H }}} }

is a subplenix of the plenix { P} of Fig 1, and can also be written as

P{ 3 [1], 4 [1, 2, 3 [1, 2 [1]]] }

2- Panels.

Up to now we have examined the structure of a plenix and established the methods for making reference to one of its elements or to a group of them. In this section we will be concerned with the actual contents of the panels and with the assignement conventions to define them.

The ASSIGNEMENT is a non-symmetrical relation of identity by means of which new plenices can be created from previously defined ones on which specific mathematical operations are performed.

The following rules are used in relation to assignements in plenices,

a) The symbol = is used to indicate an assignment,

b) {A}= {B} defines a new plenix {A} as identical to a given plenix {B}.

c) If the panel P {α} of a plenix {P} is a plecule, the assignemet {A} = P{α} defines a new plenix {A} as identical to the plenix P{α}.

d) The assignement P{α} = A defines a new plenix {P} from a previous one, in which the panel P{α} becomes A. So, if A is not a plenix,P{α} becomes a pletom; otherwise, it becomes a plecule (the plenix A). In both cases, orderates different from α may be modified in the new plenix in relation to the corresponding ones in the former plenix.

For example, if

{Q} = {A, {X, Y, Z} }

the assignement P{3} = {Q}, where {P} is the plenix shown in Fig 1, will define a new plenix {P} as

{P} = {A, {B, E, J, K}, A, {X, Y, Z}, {D,F,{G,{H,I}}}}

The dendrogram of the new plenix {P} is shown in Fig 2.

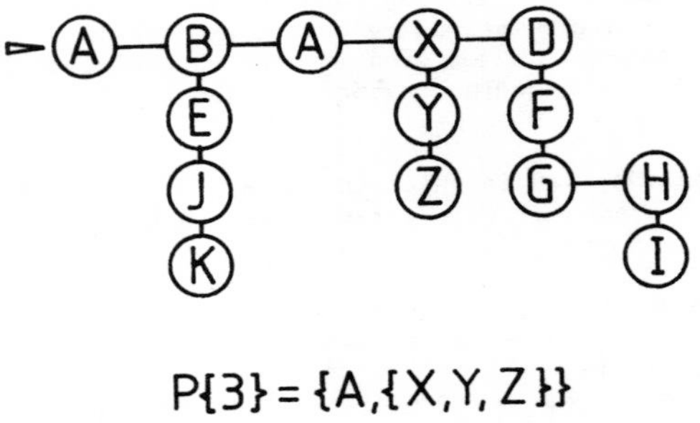

Fig 2

3- Plenix Operations.

Plenix operations can be defined by means of assignments. In fact, an algebra of plenices with a consistent system of operations can be established for a given type of problem in such a way that the results of the operations reflect the transformation of the data throughout the process of solution.

The general idea of a plenix operation can be stated as the definition of a new plenix {R} in such a way that each of its panels R{β} is a defined function of the panels of a set of plenices{P} (i), containing all the current information related to a particular stage in the process of solution of a problem.

For example, in the problem of the linear elastic analysis of skeletal structures by the stiffness method, a partial result at a certain stage of the process is the set of all the displacements of the joints. If the displacements are arranged in a plenix {F}, such plenix can be defined by a series of operations among the panels of the group of plenices containing all the information about the structure which is being analysed, such as its topological and geometrical information, the kinematical definition of the joints, the applied loads, etc. Such operations reflect the whole process of analysis.

The generic symbol ⋇, called NODUS is used to indicate plenix operations, with the following notation,

{P} = {Q} ⋇ {R}

where the plenix {P} is the result of a defined operation between the given plenices {Q} and {R}.

The above expression can also be interpreted as the action of PLENIX OPERATOR {Q} on a plenix {R}, producing as a result the plenix {P}.

The following are examples of plenix operators,

NULL OPERATOR. The null operator, denoted as {O} is defined as a plenix such that when operating on another plenix produces the empty plenix. So, if {P} is any plenix,

{O} ⋇ {P} = { }

IDENTITY OPERATOR. The identity operator, denoted as {I}, is defined as a plenix such that when operating on another plenix produces the same plenix on which it operates. So, if {P} is any plenix.

{I} ⋇ {P} = {P}

STRUCTURAL DESIGN

The structural design is a process of synthesis in the search for an optimum solution to a given problem, including a stage of analysis of defined structural configurations subjected to given distributions of static and dynamic loads, displacements and temperatures.

The rational process of design implies the existence of a body of information in the form of numerical data, equations, concepts, algorithms, etc., from which a synthesis can be made to achieve some prescribed design objectives. This body of information corresponds to the structural configuration and also to the analysis capability for assessing the influences of the various parameters on the performance and efficiency of the design product. This analysis capability is what gives to the structural design process its rational character,otherwise it would be only an empirical exercise.

The conceptual stages of the structural design process can be described in the following way.

a) The definition of the DESIGN OBJECTIVES. The elements of this definition can be arranged in two plenices, namely,

{F} The function of the structure, expressed usually by the organized information about the loading cases that the structure is expected to support.

{R} The functional requirements, normally expressed as strength, stiffness and economy requirements, when the structure is subjected to the loading cases that define its function.

In order to measure the strength it is necessary to define a mode of failure, whereas the stiffness is measured in relation to a set of prescribed maximum deformations.

The requirement of economy is applied to a set of structural configurarions that exhibit a satisfactory performance in relation to the strength and stiffnes requirements, by measuring the efficiency of those configurations in relation to a prescribed cost function

A possible form of the cost function could be the third order plenix.

{Q} = {Q1, Q2, Q3}

where Q1 is the panel that measures the influence of

the materials in the cost, Q2 that measures the influence of the fabrication and erection of the structure, and Q3 the panel measuring cost factors different from the materials and the fabrication. Q1 can be made proportional to the weight of the structure and Q2 given as a function of the number and type of connections.

The subject of the design objetives is then a proposed STRUCTURAL CONFIGURATION that could be analized.

Through the analysis of the structural configuration {S} it should be possible to predict its behaviour as determined by its performance {P} in relation to the strength and stiffness requirements, and by its efficiency {E} in relation to the requirement of economy. This can be expressed as,

$$\{P\} = \{f\} \divideontimes \{S\}$$

$$\{E\} = \{g\} \divideontimes \{S\}$$

If the plenix functions {f} and {g} can be estabished, the design process is reduced to a synthesis by means of an optimization algorithm that produces {S} as a function of {P}.

$$\{S\} = \{f^{-1}\} \divideontimes \{P\}$$

and the ideal structural solution corresponds then to a maximum of {E}.

The definition of the structural configuration and its analysis in terms of plenices will now be presented in more detail so that the use of plenices can be more fully apreciated.

PLENIX FORMULATION OF THE STRUCTURAL ANALYSIS.

In what follows, the plenix formulation of the structural analysis process will be shown in relation to the skeletal idealization of the structural model assuming a linear elastic behaviour.

The particular way in which the pertinent information is organized at each stage is shown only as an example, as the most convenient arrangement of plenices depends among other factors on the size and nature of the structural problem and on the computing facilities available.

Let us examine first the bodies of information that define a structural configurarion within the above limitations.

1) The TOPOLOGICAL plenix {T}, containing all the necessary information for identifying the joints and the way in which they are connected by the members. Very important advances have been made in the study of this aspect of the structural model with the development of the Formex algebra, that allows the generation of the topological information in a simple and convenient way.

2) The GEOMETRICAL plenix {G}, that quantifies the topological information in relation to a geometrical system of reference, for instance a three dimensional cartesian system of axes. {G} can then be obtained from {T} by a plenix operation that could be expressed as,

$$\{G\} = \{g\} \divideontimes \{T\}$$

where {g} is a plenix operator that produces {G} from {T} in a determined way. Formex algebra can be used again for an efficient formulation of this operation for computing purposes.

3) The MEMBER plenix {M}, organizes all the information referent to the members. The bodies of information containing material properties, cross sectional constants and geometrical specifications of member axes are organized in such a way that, for example, the stiffness matrix of each particular member identified in the topological plenix can be produced and arranged in a dynamic way at definite panels of {M}. The member plenix can be expressed as,

$$\{M\} = \{m\} \divideontimes \{ \{T\}, \{G\}, \{W, A, L\} \}$$

where {m} is a plenix operator that produces the member plenix from a plenix with panels {T}, {G} and {W,A,L},the last one containing the information referent to the elastic properties of the materials, W, the geometrical cross sectional properties, A, and the member axial information, L.

4) The JOINT plenix {J}, in which the internal kinematical information is contained, that is, the specification of the way in which the members meeting at each joint are connected. {J} can be obtained from {T} as,

$$\{J\} = \{j\} \divideontimes \{T\}$$

where {j} is a plenix operator that specifies the compatibility conditions at the joints for the particular family of structures. At this stage of the process the intrinsic degrees of freedom of the system are specified, resulting of course in an unstable arrangement which should be made stable in the following stage by defining the support conditions.

5) The SUPPORT plenix {H}, containig the external kinematical conditions of the system, that is, its relations with the earth or other structures in contact. {H} is obtained form {J} by means of a plenix operator {h} that specifies the degrees of freedom that are restricted at the supports.

$$\{H\} = \{h\} \divideontimes \{J\}$$

The above plenix operations include in general coordinate transformations between local and global systems of axes.

Once the plenix {S} is defined, a series of plenix operations, representing the process of analysis, can be carried out. The result is the expression of the performance and efficiency of the structural configuration in relation to the design requirements.

The performance of {S} can be expressed by a plenix operation of the type,

$$\{P\} = \{R1\} \divideontimes \{\{B\}, \{S\}, \{F\}\}$$

where {P} represents the standard of acceptable performance and {B} contains the booleans that determine the acceptability of {S} in relation to the functional requirements of strength and stiffness.

The efficiency of the structure is measured by a prescribed cost function, say

$$\{Q\} = R\{2\} \divideontimes \{S\}$$

from which the solution of the design problem can be established, corresponding to a minimum of {Q}.

As a final example, let us consider the organization of the data for the definition of the joints in the skeletal structural model.

For doing that the following definitions are introduced,

NODE. A node is defined as a set of four vectors at a given point, namely, a force, a moment, a linear displacement and a rotation.

A general expression of a node can be for instance the plenix,

$$\{N\} = [\ \{fi,ti\},\ \{mi,ri\}\ ] \qquad \text{for } i = 1:3$$

where fi, ti, mi, ri are the components of the vectors that define the node in a system of coordinates X1, X2,X3.

The idea of node is very useful for defining the concepts of member and joint in the structural model, and also for identifying the various types of structures used in practice.

In fact, particular families of structures, such as plane frames, space trusses, plane grids, etc., are characterized by a specific type of node defined by the number and type of its vector components that are intrinsically zero.

For example, the characteristic node of a space truss is

{N} = {{f1,t1}, {f2,t2}, {f3,t3}}

A simple member (bar) is then defined by a relation (stiffness matrix) between the force and moment components of two nodes located at the ends of the member and the corresponding linear and rotational displacements. The relation can be expressed with reference to a local or to a global system of coordinates.

A joint on the other hand is defined by a series of compatibility relations among the displacement components of a set of nodes located at the same point, wich also produce relations of addition among the force and moment components on the basis of which the conditions of static equilibrium are finally established with the externally applied loads.

Let m be the number of members forming a joint at a given point of a structure. The elements of the joint are then a cluster {C} of m nodes {Nj}.

{C} = { {N1}, {N2},...,{Nj},...,{Nm} }

If u is the number of displacement components of the characteristic node of the structure, the necessary condition for having a joint is that, if the nodes of the cluster are separated in two groups in any arbitrary manner, there are at least one node per group, and there exists at least one common displacement component between the displacements of both groups.

The maximum number of compatibility conditions is (m-1) u when the joint is a full joint. When the number of conditions is less than the maximum, the joint is a partial joint.

The number of degrees of freedom of the joint is then $v \geq u$.

The joint {J} can then be obtained from a plenix operator {ω} and the cluster {C} as follows,

$$\{J\} = \{\omega\} \divideontimes \{C\}$$

The operator {ω} produces the following effects,

1) Transform all the vectors of the cluster to the global system of coordinates.

2) Introduce the conditions of compatibility which reduce the total number of independent displacement components of the cluster to v, $v \geq u$.

3) Reduce by addition all the force and moment vector components of the cluster to v vectors in the directions of the degrees of freedom of the joint. These v force resultants are then going to equilibrate the externally applied forces and support reactions.

Finally the external kinematical conditions are applied, that is, the specification of whether the joint is a free joint or a support, which is done by introducing known values (zero or a small quantity) to the displacements in the directions of the supports.

ACKNOWLEDGEMENTS

The present work was developed at the S.S.R.C, University of Surrey, under the decisive guidance of Dr.H.Nooshin, to whom the author whishes to express his sincere gratitude.

GENERATIVE MORPHOLOGY OF TRANSFORMING SPACE STRUCTURES

Haresh LALVANI PhD

School of Architecture
Pratt Institute
Brooklyn New York
New York 11205
U.S.A.

A generalized "generative" classification of regular and semi-regular space structures is presented. All regular polygons, prisms, regular and semi-regular polyhedra, plane tesselations, prismatic and polyhedral space-fillings (packings), and their duals, can be systematically organized in analogous multi-dimensional periodic tables. These arrangements are based on various types of transformations of the 'fundamental regions' of the allowable symmetries of space structures. Known and new structures can be generated by these transformations within and between their order, dimension and symmetry. Space "grid" geometries can be systematically and exhaustively generated ; various methods for deriving the shapes of space grid elements are described.

INTRODUCTION

The past three decades have seen a multi-fold increase in the of 3-dimensional geometry for deriving structures that span and define architectural space (Refs 1-5). A variety of space structures and building systems have been developed some of which have been widely used. Alternative space structures can be developed from a systematic understanding of the geometry of space itself. This paper deals with the morphologic aspects of space structure configurations, and focusses on the systematic, exhaustive generation of regular and semi-regular structures, their duals and compounds. Some formal aspects are described first, followed by the generative aspects which extend to the geometric derivation of space "grids", and modular building systems. The broader concepts presented here have appeared elsewhere (Refs 6 and 7; also Refs 8 and 9); some of these are clarified and extended in this paper, and some others are presented for the first time.

'Space structures' are structures composed of topological elements of different dimensionalities. In architecture, the nature of built space restricts us to 3-dimensions where these elements are vertices (points), edges (lines), faces (planes) and cells (volumes or spaces). General principles of space structures can be studied from topology, symmetry, geometry and combinatorics. These aspects are of interest here.

1 SYMMETRY AND DIMENSION

The number of symmetrical structures around a point (0-dimensional), a line (1-dimensional), a plane (2-dimensional), and a cell (3-dimensional) are finite and are governed by allowable combinations of symmetry 'operations' on symmetry 'elements' (Refs 10 and 11). The symmetry operations are reflection, rotation, translation, inversion, and their combinations; the symmetry elements are rotation axes, mirror planes, translation axes, and centers of symmetry. The operations on elements generate 'classes' of symmetry. The derived structures are hierarchichal in dimension and can be respectively referred to as 0-, 1-, 2- and 3-dimensional structures.

The allowable rotation axes for space-filling symmetric structures are 1, 2, 3, 4 and 6. These numbers give the 'order' of space structures and correspond to 'primitive' polygons, namely, a paralellogram, rectangle, triangle, square and a hexagon, all of which have corresponding structures as analogues in each of the four dimensionalities we encounter in built space structures. For example, the square prism, a stacked square column, a square or box grid, and a cubic lattice are 0-, 1-, 2- and 3-dimensional analogs of order 4. The higher orders (7, 8, 9, 10, ... ∞) exist as 0- and 1-dimensional structures, and the 5-fold symmetry exhibits characteristic spatial arrangements.

2 FUNDAMENTAL REGION

Space structures are characterized by their unique 'fundamental regions'. For symmetric structures, this is the minimum spatial module bounded by symmetry elements, and generates the entire structure through repeated symmetry operations. The bounding planes intersect at rotation axes and centers of symmetry. The complete enumeration of all allowable combinations of rotation axes

TABLE 1

Order	Primitive Polygon	Polygons [P]	Prisms [P22]	Polyhedra [PQR]	Line-filling	Plane-filling	Space-filling
		0-dimensional			1-d	2-d	3-d
1	Parallogram	1	[122]	-	122(1)	122(2)	122(3)
2	Rectangle	2	222	<232>	222(1)	222(2)	222(3)
						[2222]	
3	Triangle	3	322	332	322(1)	322(2)	322(3)
						[333]	332(3)**
4	Square	4	422	432	422(1)	422(2)	422(3)
						[442]	432(3)**
5	Pentagon	5	522	532	522(1)	*	*
6	Hexagon	6	622	<632>	622(1)	622(2)	622(3)
						[632]	
	Circle	∞	∞22	<∞∞1>	∞22(1)	-	-
					∞∞1	∞∞11	

SYMMETRY IN RELATION TO ORDER & DIMENSION

*-non-crystallographic filling
**-also exist in combination

( )-indicates dimension
[ ]-conventional symmetry
< >-symmetries included elsewhere

surrounding the fundamental regions for all orders and dimensions provides a classification of all possible symmetries for space structures. These are known (Refs 10, 11 and 12) and are summarized in Table 1.

This paper focusses on derivation of space structures from all fundamental regions that are kaleidoscopic, i.e. bound by mirror planes, restricting us to the orders 2, 3, 4 and 6. The methods of structure-generation can be extended to other classes of symmetry. Kaleidoscopic symmetries are important as most building systems belong to this category and other classes can be derived from these.

3 KALEIDOSCOPES FOR SPACE STRUCTURES

The upper limit on space-filling kaleidoscopes is 6 mirrors (Ref 13) and the lowest is 1 producing bilateral symmetry. Between these limits the number of mirrors vary according to symmetry and dimension. From kaleidoscopes, space structures can be produced by placing topological elements within this region. Vertices, edges, faces, cells and their combinations could be placed in any fundamental region which "multiplies" in space to generate a space structure. By covering all unique placements within the fundamental regions, all space structures can be exhaustively generated. Different types of kaleidoscopes, their derivative space structures, and examples are summarized below.

3.1 0-Dimensional Kaleidoscopes (Figs 1-4)

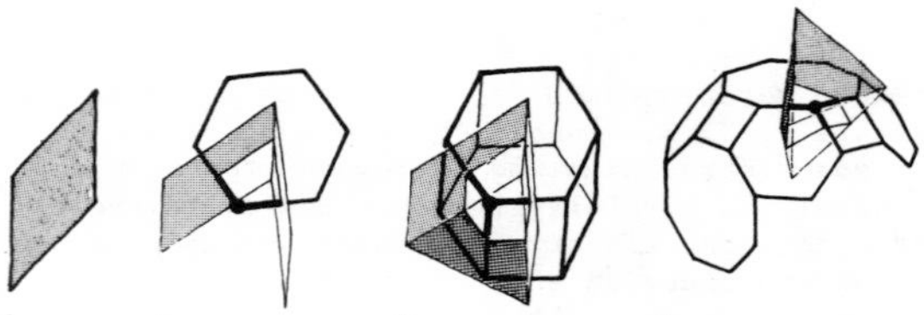

1 2 3 4

1 mirror (Fig 1) generates edges and digons of symmetry 1 and 122. Examples: struts, one-sided beams and columns, bilateral spaces.

2 mirrors (Fig 2) generate all polygons of symmetry P (P= 2, 3, 4, ... ∞), vertex stars and pyramids. Examples: one sided polygonal or pyramidal modules, domes and simple vaults.

3 mirrors (Fig 3 and 4) generate all prisms, bipyramids of symmetry P22, all polyhedra of symmetry 332 (tetrahedral) 432 (octahedral-cubic), and 532 (icosahedral-dodecahedral). Examples: prismatic polyhedral and star joints for space frames, 3-d live-in modules, geodesic and other spheres.

3.2 1-Dimensional Kaleidoscopes (Figs 5-7)

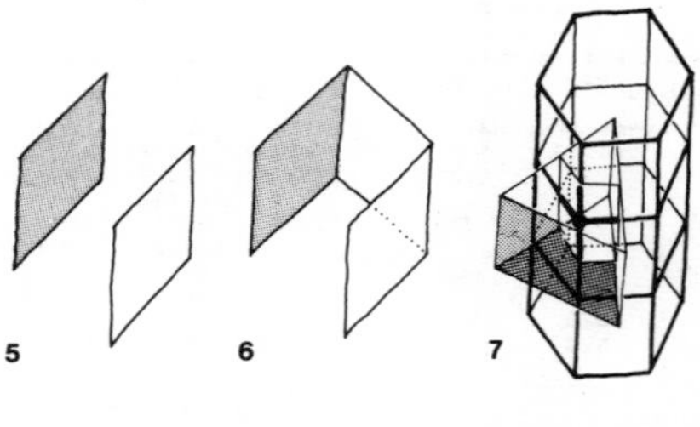

5 6 7

2 mirrors (Fig 5) generate one-sided bands of symmetry $1\infty\infty$. Examples: modular beams, columns and spaces with bilateral symmetry.

3 mirrors (Fig 6) generate one-sided bands of symmetry 22∞. Examples: as above.

4 mirrors (Fig 7) generate all prismatic, bipyramidal columns of symmetry P. Examples: as above, but polygonal sections.

3.3 2-Dimensional Kaleidoscopes (Figs 8-11)

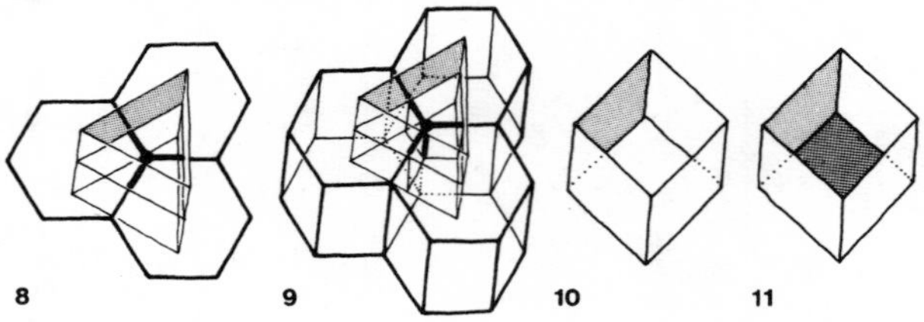

8 9 10 11

3 mirrors (Fig 8) generate all one-sided plane tesselations of symmetry 333 (triangular), 442 (square) and 632 (hexagonal). Examples: single-layered grids and tiles.

4 mirrors (Fig 9) generate one-sided plane tesselations of symmetry 2222. Examples: single-layered rectangular grids, spaces and tiles.

4 mirrors (Fig 9) generate all 2-sided plane tesselations of symmetries 333, 442 and 632. Examples: double- and triple-layered space grids (Refs 14-16), modular walls, floor tiles, spaces and bricks.

5 mirrors (Fig 11) generate two-sided plane tesselations of symmetry 2222. Examples: as above, but rectangular.

3.4 3-Dimensional Kaleidoscopes (Fig 12-16)

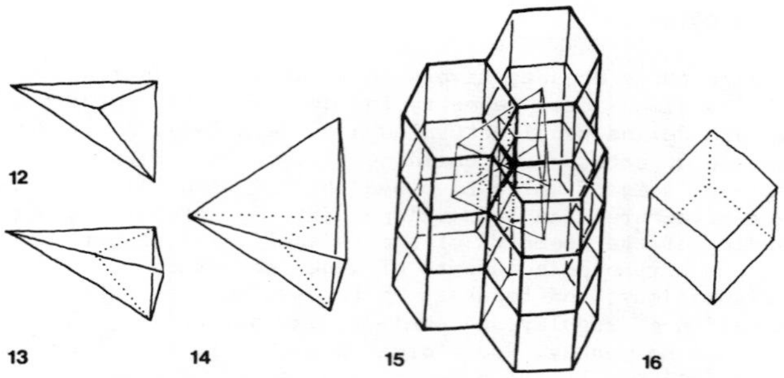

12 13 14 15 16

4 mirrors (Figs 12-14) generate all space-filling polyhedra, namely, the cubic (Fig 12), the cubic-tetrahedral (Fig 13) and the tetrahedral (Fig 14) packings. Examples: multi-directional spaces and frames like Mero, Unistrut, Universal Node, etc., 3-d habitats of Hecker, Tyng, Gabriel and others (Refs 17-19).

5 mirrors (Fig 15) generate all space-filling prisms of symmetries 333, 442 and 632. Examples: multi-layered prismatic grids and spaces.

6 mirrors (Fig 16) generate space-filling prisms of symmetry 2222. Examples: rectangular multi-layered spaces and frames.

4 TRANSFORMATIONS

Structures can transform from one to another, both between and within their dimensions and orders. Inter-dimensional and inter-order transformations are considered below. A special class of transformations that preserve symmetry by retaining both order and dimension are particularly interesting for space-structure-generation; these are considered seperately.

4.1 Inter-Dimensional Transformations (Figs 17-20)

For any symmetry, order-preserving transformations generate structures of different dimensions within the

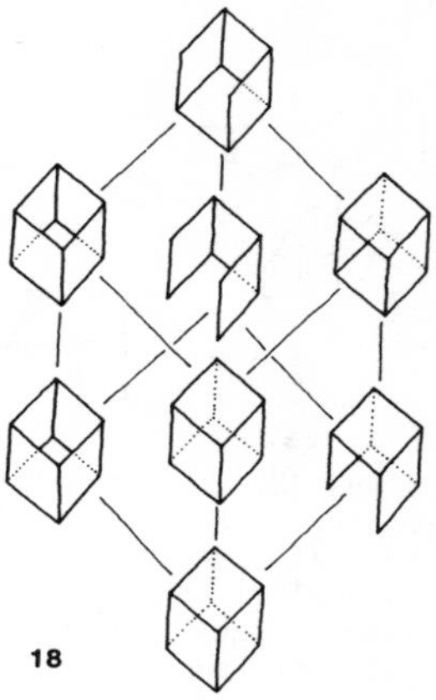

18

19

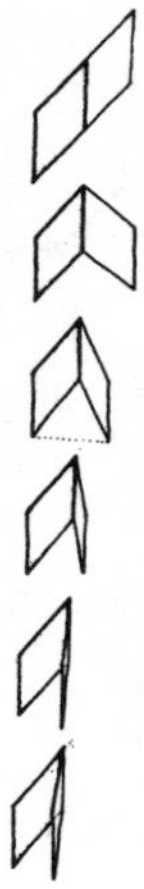

21

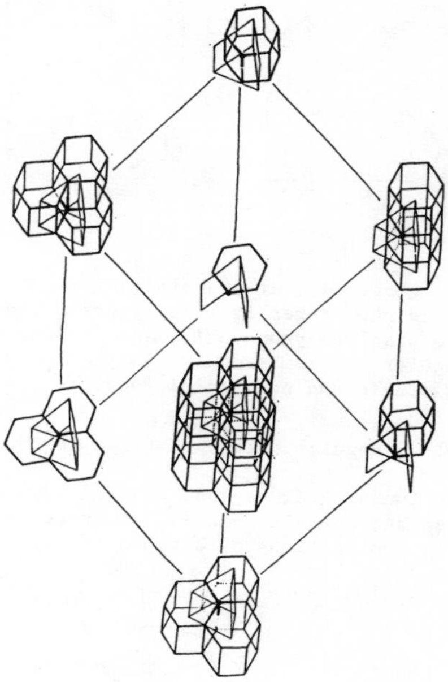

20

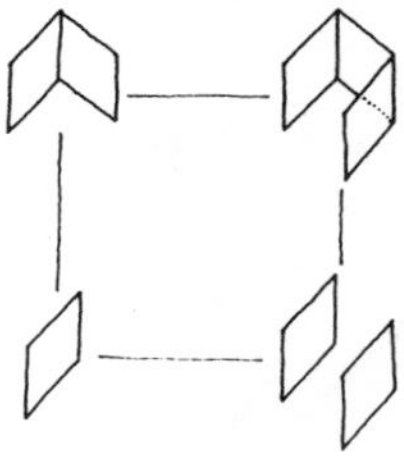

17

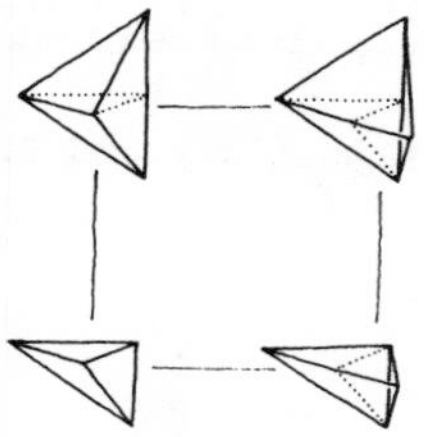

22

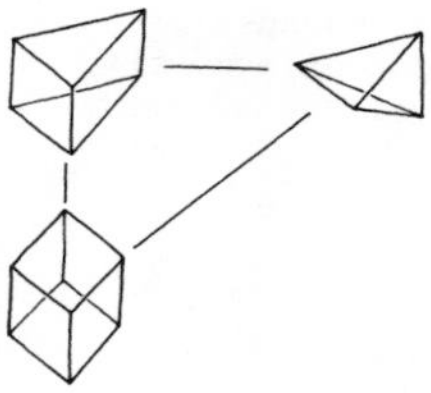

23

same order. From any space-filling kaleidoscopic region, lower dimensions can be systematically obtained by removing mirror planes (Figs 17-19), taking us from the highest dimension to the lowest. All combinations of such mirror-plane removal (in general, we remove faces of any fundamental region) produce 2^n structures, where 'n' is the number of mirrors in the space-filling kaleidoscope. All unique kaleidoscopes, including some topologically redundant ones, are generated. The kaleidoscopes can be arranged in n-space. Conversely, mirrors can be added combinatorially, taking us from the lowest to the highest dimension.

For order 2 kaleidoscopes, organization of mirror combinations between 3 and 1 in 2-space (ie. a square; Fig 17), and between 3 and 6 in 3-space (ie. a cube; Fig 18) are illustrated. Analogous transformations for prismatic kaleidoscopes (Fig 19) with mirror combinations between 2 and 6 follow. Derivative space structures for prismatic order 3 (Fig 20) show interdimensional transformations between a hexagon, a hexagonal prism, a hexagonal tesselation (a single-layered space grid), a "tower" of stacked hexagonal prisms, a single-layered packing of hex prisms (a double-layered space grid), and their multi-layered space-filling (-grid). The method extends to other orders by analogy.

4.2 Inter-Order Transformations (Figs 21-23)

Changes in order can be acheived by changing the geometry or topology of the fundamental region. Inter-order geometric transformations alter dihedral angles, and consequently the face-angles and edge-lengths of the fundamental regions. The natural order sequence of regular polygons (P=2, 3, 4, ...) are obtained by continious changes in the dihedral angle of a 2 mirror kaleidoscope (Fig 21, sequences 1-6), where this angle equals half of the central angle of polygons. The extension to prisms, prism stacks, plane tesselations and layered prism packings follow from Sec 4.1, ie. by adding mirror planes. For polyhedral packings, the 3 kaleidoscopes in Figs 12, 13, 14 can change by "doubling" the kaleidoscopes (Fig 22), ie. Fig 12 is half of Fig 13, which is half of Fig 14.

Inter-order topologic transformations result when topological elements of the fundamental region change to others, e.g. a face becomes an edge (f-e transformation), or an edge becomes a vertex (e-v transformation), or other transformations. Such transformations (Fig 23) take us from a rectangular kaleidoscope of order 2 to prismatic kaleidoscopes of orders 3,4 and 6, to the polyhedral space-filling kaleidoscopes of orders 3 and 4. Thus plane tesselations can become polyhedra, and space-filling prisms can become polyhedral packings; for space-fillings, the number of mirrors change from 6 to 5 to 4.

5 INTRA-SYMMETRY TRANSFORMATIONS

Within every symmetry, a variety of topological structures can be obtained by symmetry-preserving transformations. These are different types of 'explosions' and 'implosions', and are determined by combinatorial possibilities of movements of mirror planes "away from" (explosion) or "towards" (implosion) the centroid of the fundamental region (Figs 24 and 25). The fundamental regions change gnomonically, ie. the size changes leaving the shape invariant. As the regions change, the original centroid position occupies all and only available positions relative to the new fundamental regions. A vertex placement at the original centroid positions, and its repeated reflections over the new fundamental regions, produce all topological structures for that symmetry (see Figs 26 and 30). The number of structures produced equal 2^n, where 'n' is the number of mirror planes in the fundamental region. The resulting structures can be organised in n-space. The structures, and their transformations, are governed by algebraic set operations of full set, empty set, unions, intersections and complements (see note,Ref 20).

Conversely, a variant vertex position within a fixed fundamental region produces the same topological transformations; the latter method was discovered independently by Burt (Ref 21) and has been termed the "wandering vertex" method. We note that a vertex can be replaced by any topological element like an edge, face, cell, their combinations or groupings, leading to a generalised 'wandering polytope' as a determinant of space structure configurations. By analogy, the method extends to higher dimensions.

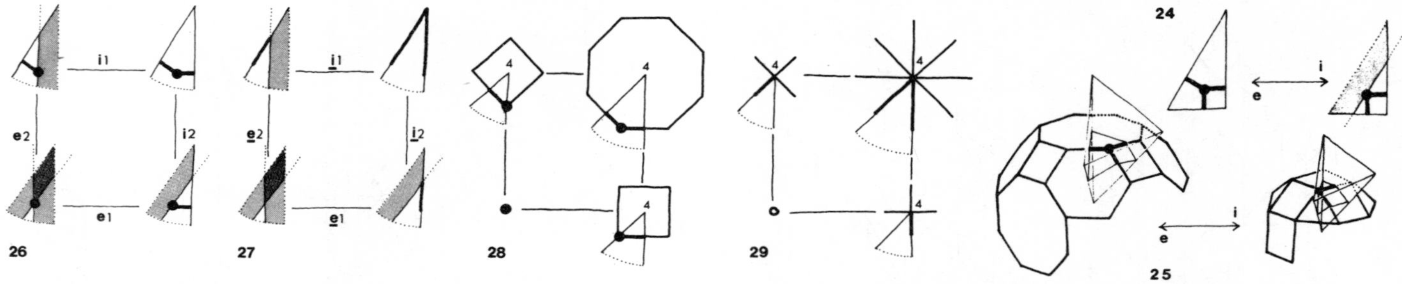

26 27 28 29 24 25

The generation of transforming regular and semi-regular space structures is briefly described. These are made up of 'regular' polygons only, and include regular prisms, regular and semi-regular polyhedra, plane tesselations, prismatic and polyhedral packings, and their duals.

5.1 **Regular Polygons** (Figs 26-29)

The fundamental region, bounded by 2 mirrors and an axis P (Fig 26, see also Fig 2), generates 4 polygons of symmetry P by combinations of 2 types of explosions, e1 and e2, or their complementary implosions, i1 and i2. The structures are arranged in 2-space. Corresponding duals (Fig 27) are vertex stars obtained by 2 pairs of dual transformations, 'inclusions' e̲1 and e̲2, and 'exclusions' i̲1 and i̲2; inclusions add edges, exclusions remove edges. Polygons of symmetry 4 and their duals are illustrated (Figs 28 and 29). Symmetries 2,3,4,5 and 6 generate all polygons and vertex stars for all regular and semi-regular polyhedra, plane tesselations and space-fillings.

5.2 **Regular and Semi-Regular Polyhedra** (Figs 30-38)

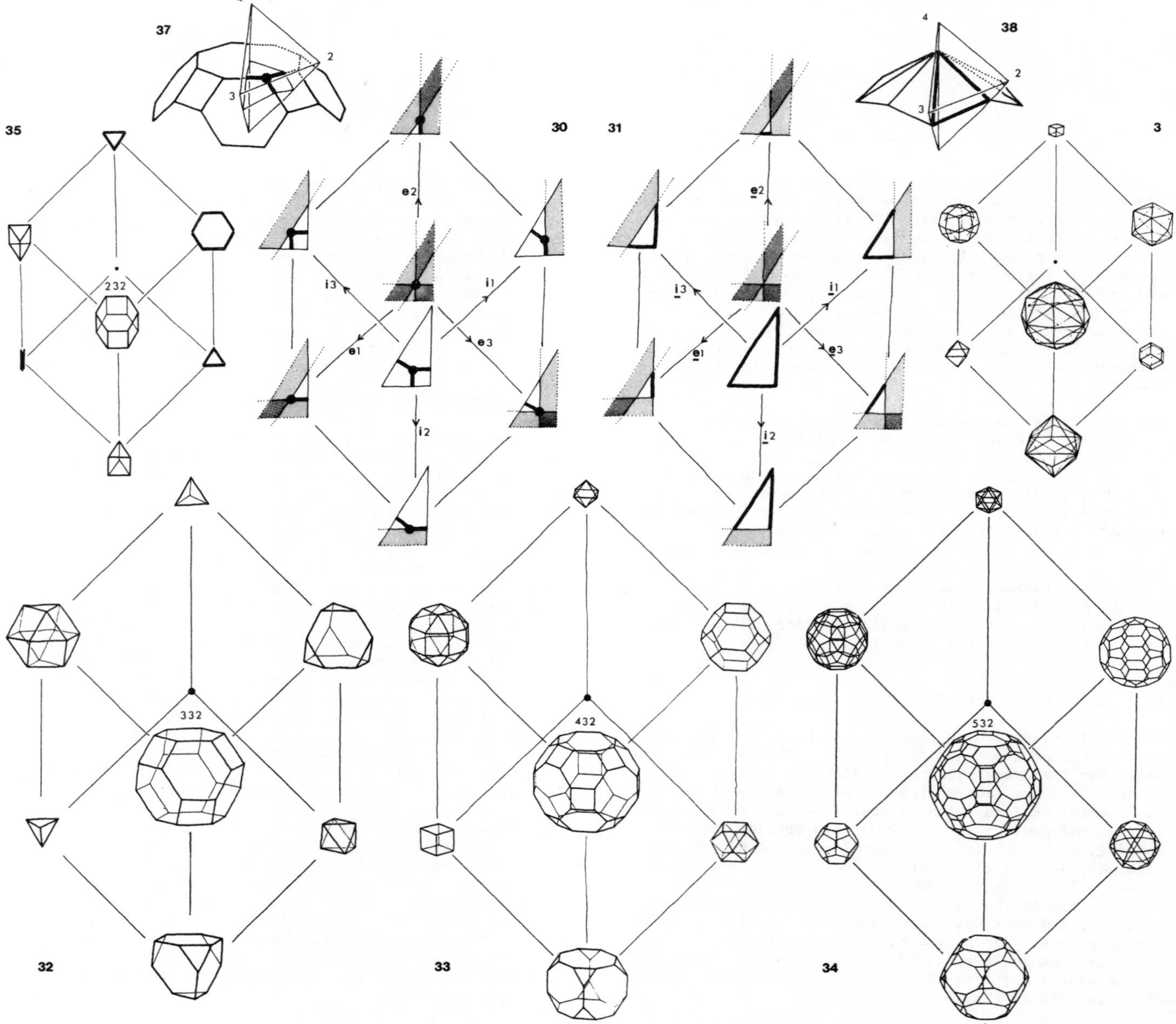

37 38 35 30 31 3 32 33 34

The fundamental region, bounded by 3 mirror planes and 3 axes (Fig 30), generates 8 polyhedra for every symmetry through combinations of 3 types of explosions, e1, e2 and e3, or their complementary implosions i1, i2 and i3. The 8 regions are arranged in 3-space. All regular-faced polyhedra of symmetries 332 (Fig 32), 432 (Fig 33), 532 (Fig 34) and prisms P22 (Fig 35; symmetry 322 shown) are generated. The corresponding 8 vertically regular duals (Fig 31; symmetry 432 illustrated in Fig 36) are derived from 3 types of dual transformations, inclusions e1, e2 and e3, or their complementary exclusions i1, i2 and i3. Polyhedron P1 of Figure 30 and its dual P2 of Figure 31 are shown partially in detail (Figs 37 and 38, respectively).

5.3 Plane Tesselations and Prism Packings (Figs 39-41)

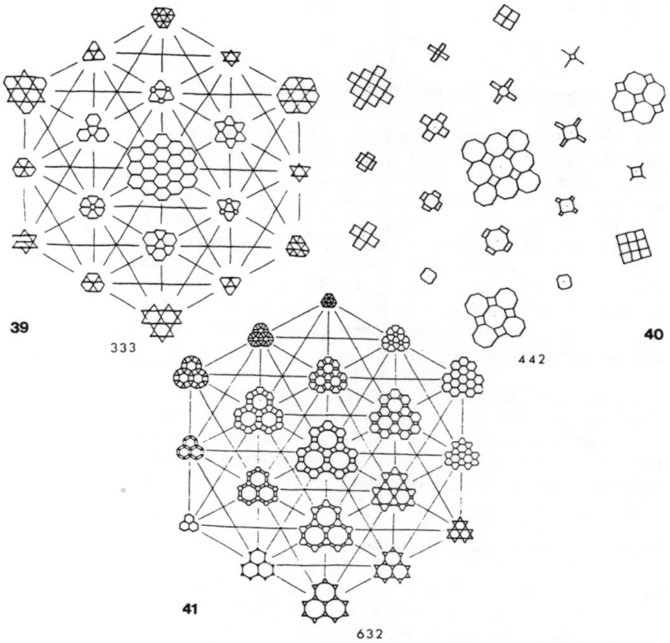

The fundamental region is analogous to polyhedra (Figs 30 and 31), and is bounded by 3 mirrors but 3 non-convergent axes. The transformations and the organisation of structures follow by analogy. Symmetries 333 (Fig 39), 442 (Fig 40) and 632 (Fig 41), corresponding to the orders 3, 4 and 6, are illustrated. The arrangement naturally extends to single-, double- and multi-layered packings of prisms and their duals.

5.4 Polyhedral Packings (Figs 42-48)

The fundamental region, bounded by 4 mirrors and 6 axes (see earlier Figs 12-14), generates 16 packings for each of the 3 polyhedral space-filling kaleidoscopes. These are based on combinations of 4 types of explosions, e1, e2, e3 and e4, or their complementary implosions, i1, i2, i3 and i4. The transformations are analogous to those of polyhedra and the fundamental region explodes and implodes gnomonically around the centroid to produce 16 unique vertex locations (Fig 42), each of which generate a polyhedral packing (Fig 43). The 16 structures can be arranged in 4-space (ie. a tesseract).

The derivation of the packing S1 of Figure 43 from the fundamental region S1 of Figure 42 (also, of earlier Fig 12) is shown (Figs 44 and 45). The fundamental region is treated as a space structure and duality is used to derive the packing; 'V' converts the region to a dual vertex , 'E' converts the 4 faces to 4 dual edges, 'F' converts the 6 edges to 6 dual faces and 'C' converts its 4 vertices to 4 dual cells. All fundamental regions of Figure 42 can thus be converted to the packings of Figure 43. The symmetry illustrated is for the cubic packings; tetrahedral-cubic

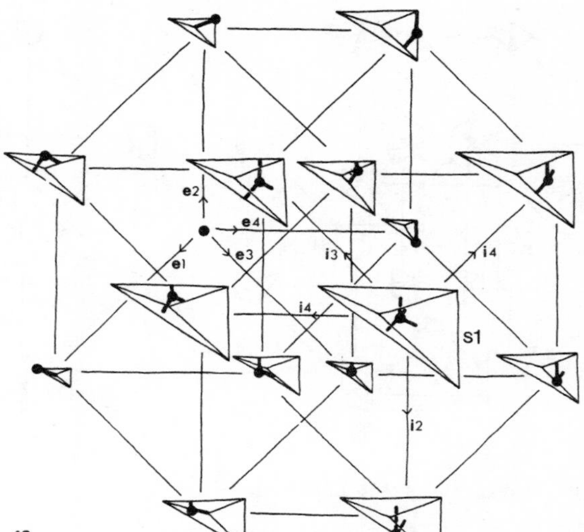

42

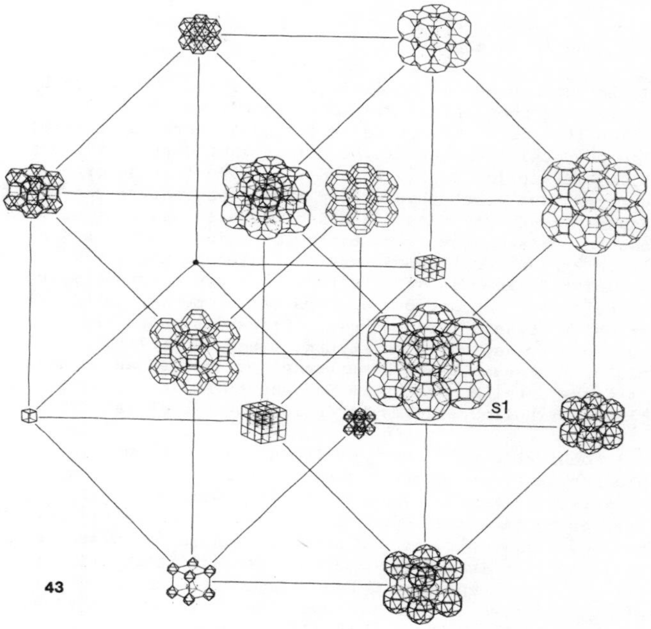

43

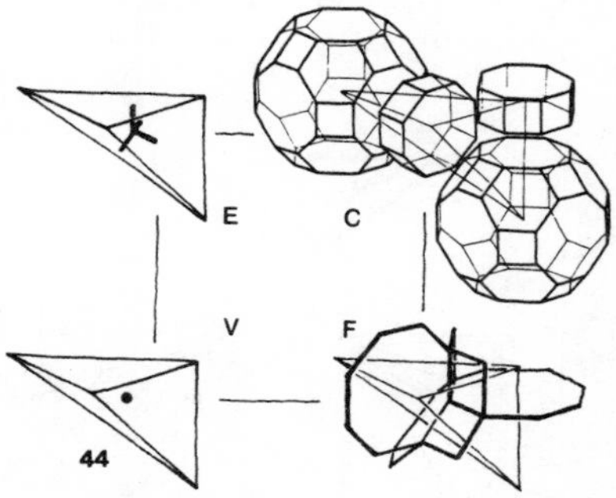

44

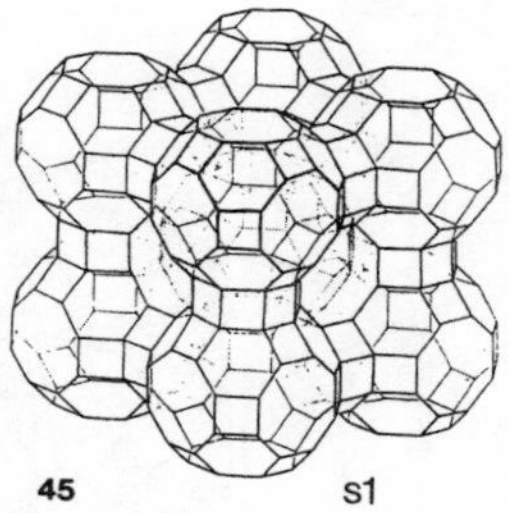

45 S1

and tetrahedral packings (earlier Figs 13 and 14) can be derived similarly.

The duals of packings are self-packing, using one "module" only to fill space. For the 3 kaleidoscopes of Figures 12, 13 and 14, the dual sets are illustrated in the three tesseracts of Figures 46, 47 and 48. (Figure 46 is the dual of Figure 43 where all cells are replaced by vertices in the other).

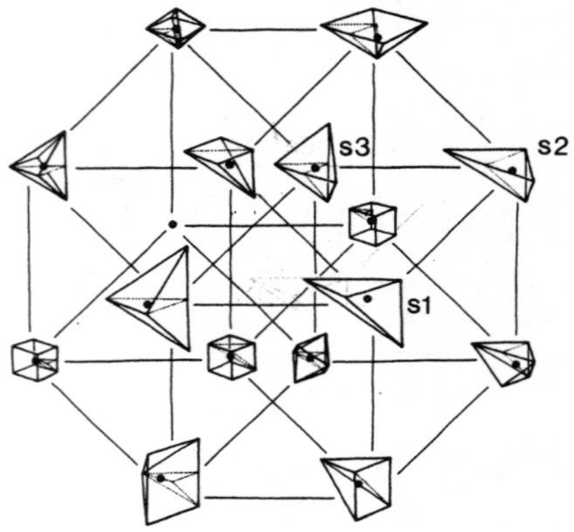

46

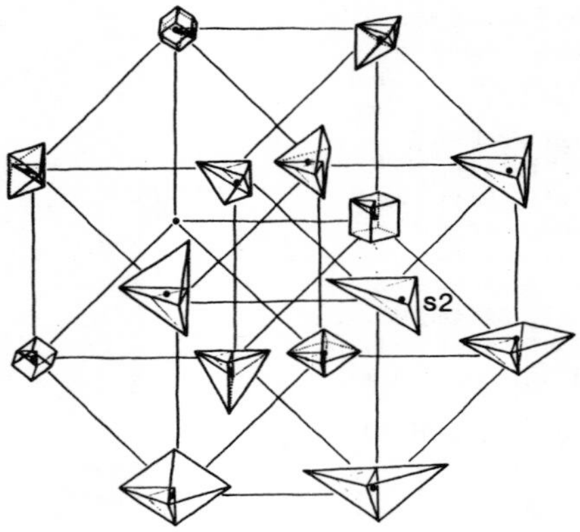

47

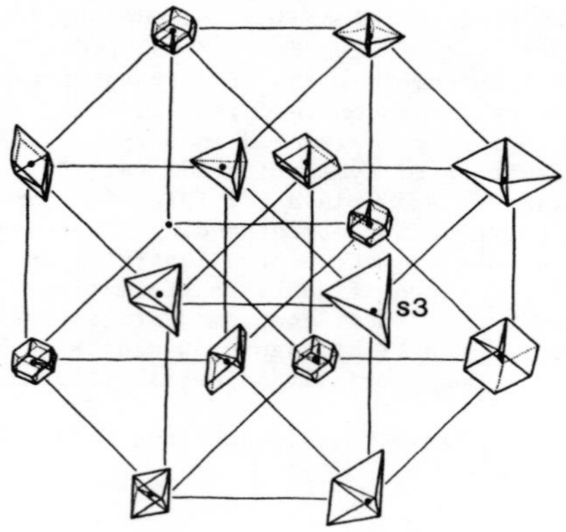

48

6 THE GENERATIVE SPACE GRIDS

Various methods for the derivation of space "grid" geometries and the shapes of their elements are described. These are a natural extension of earlier sections.

6.1 Built Structures

Built structures can be derived from space structures by converting topological elements, namely vertices (V), edges (E), faces (F) or cells (C), into physical elements. The physical elements can be straight or curved, rigid or flexible, solid or hollow, and can be in tension or compression; other physical or formal parameters can be added. A general classification of built space structures can be derived from these binary or trinary combinations. Space "grids" or "frames" are special-case modular building systems in architecture that are commnoly made of selected combinations of physical elements. "Skeletal" grids or "braced" frames have vertices as nodes or connectors, and edges as struts, beams or cables. Panels, slabs, "stressed skins", membranes or shells can be added as faces. Cells, when present, can be solid building blocks or hollow modules to enclose architectural space.

All combinations of these 4 elements,V, E, F and C, produce 16 types of structures which can be arranged in a tesseract. Examples include the following : Mero frame, made of vertices and edges,is a VE structure; a brick wall, made of cells only, is a C structure; a folded plate, made of faces only is a F structure; a stressed skin grid is a VEF structure, and so on. Each element in turn can be made of any of these 16 combinations. Their numbers are given by the well-known Euler-Schlafli equation,
$V + F = E + C + 1$, a relation particularly interesting for calculating the relative numbers of elements in building systems.

6.2 Generating Space Grid Geometries

All regular and semi-regular space structures described so far (Secs 5.1, 5.2, 5.3 and 5.4) can be converted into space grids. "Ball and stick" frames are the simplest derivations where the vertices become small spheres, and the edges are rods. Point grids, line grids, single- and double-layered grids, multi-layered and multi-directional space grids can be thus derived. Ball-and-stick grids for the 3 space-filling duals S1, S2 and S3 (Fig 46; S2 also in Fig 47, S3 in Fig 48) are shown in Figures 49, 50 and 51, respectively; their corresponding fundamental modules are shown alongside.

Various spatial transformations are used for the systematic derivation of space grids from the regular and semi-regular grids, and their duals. These methods are briefly described for plane- and space-filling symmetries, and are directly applicable to point and line grids.

6.21 Explosions and Implosions

Explosions of regular and semi-regular space structures generate new ways of subdividing space. These transformations are elaborated in Sections 6.31 and 6.32 since the method also provides a way of deriving shapes of elements for space grids.

6.22 Inclusion and Exclusion of Elements (Figs 52, 53)

Space structures can be derived from others by adding (inclusion) or removing (exclusion) topological elements. Vertices, edges, faces, cells or their combinations, can be systematically included or excluded; there are 8 such transformations. Of particular interest for the generation of space grids are edge-exclusions

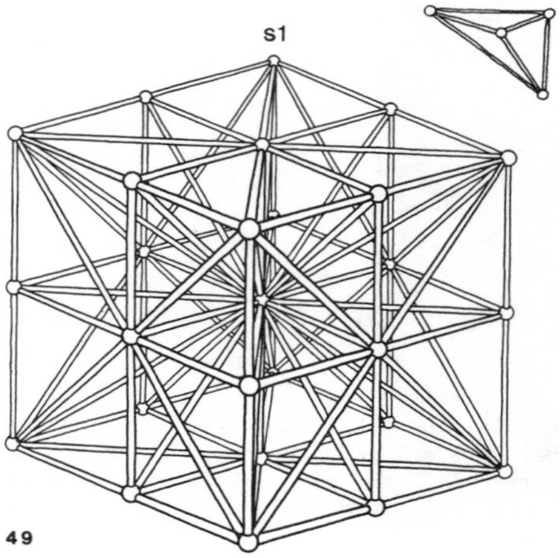

49

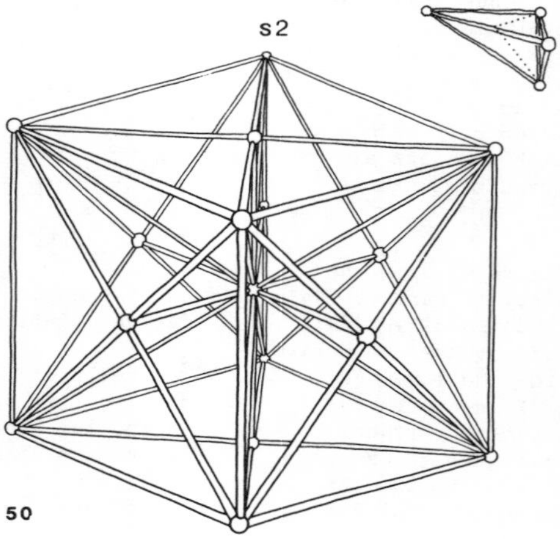

50

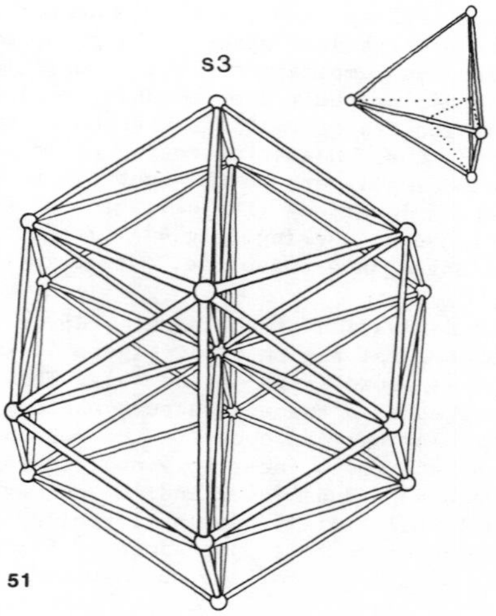

51

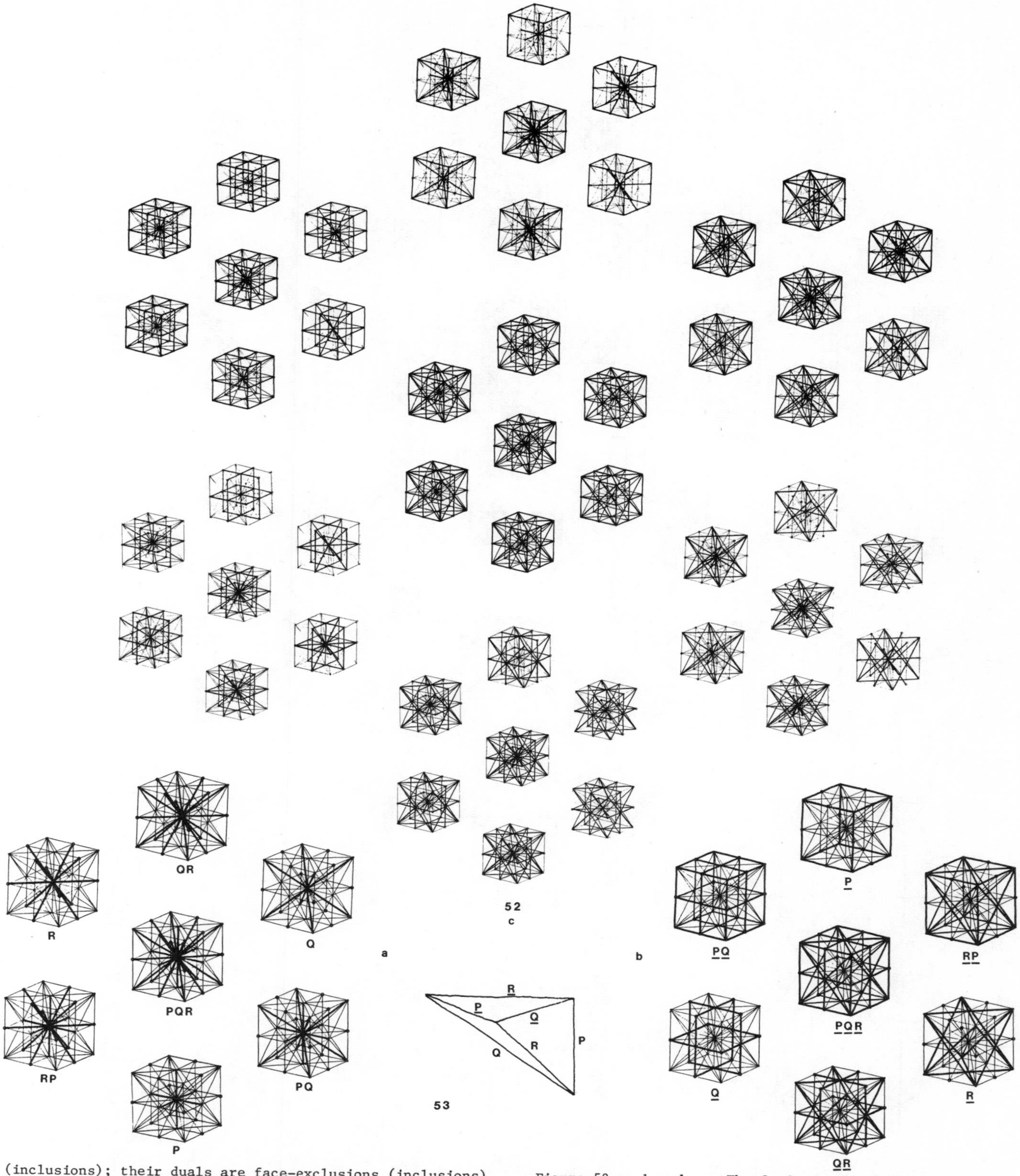

52
c

53

(inclusions); their duals are face-exclusions (inclusions) which, along with cell-exclusions (inclusions) generate some of the 'infinite polyhedra' (Ref 22).

Edge-exclusion is like strut-removal; inter-tranformations between space grids can be derived from combinations of strut removal or addition. All combinations for the multi-directional cubic grid of Figure 49 are illustrated in

Figure 52 a, b and c. The fundamental region has 6 types of struts, P, Q and R, and their complements, $\underline{P}$, $\underline{Q}$ and $\underline{R}$. (Fig 53). All combinations of these 6 edges generate 64 structures which can be arranged in 6-space. These structures include the 16 duals of Figure 46. The other two space-fillings (Figs 50 and 51), each having 6 edges in their fundamental regions, also generate 64 multi-directional space grids; the latter include the diamond

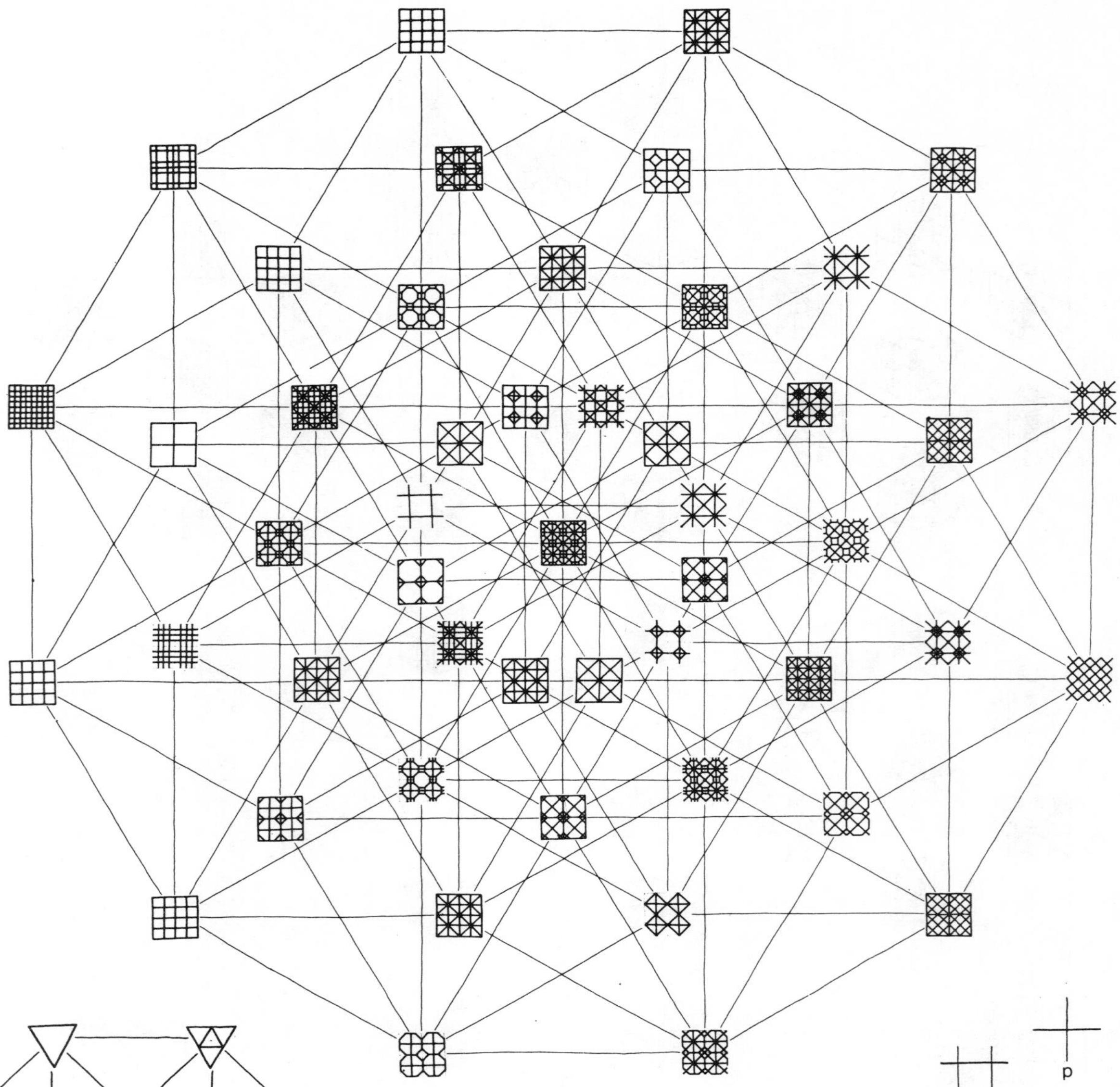

61

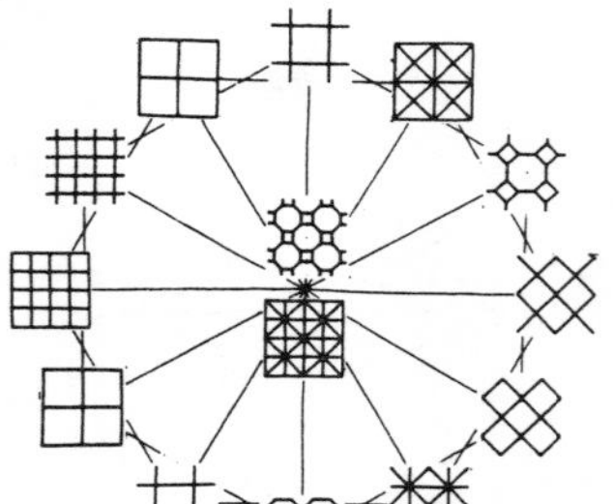

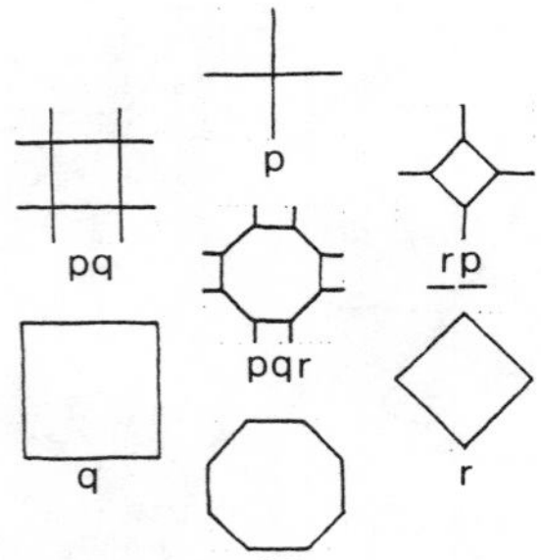

59

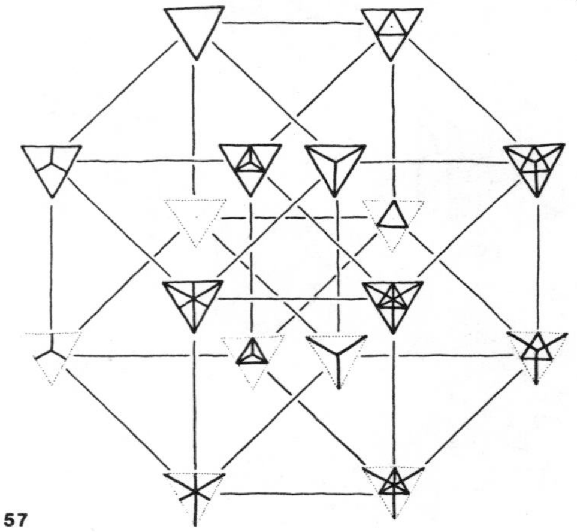

57

60

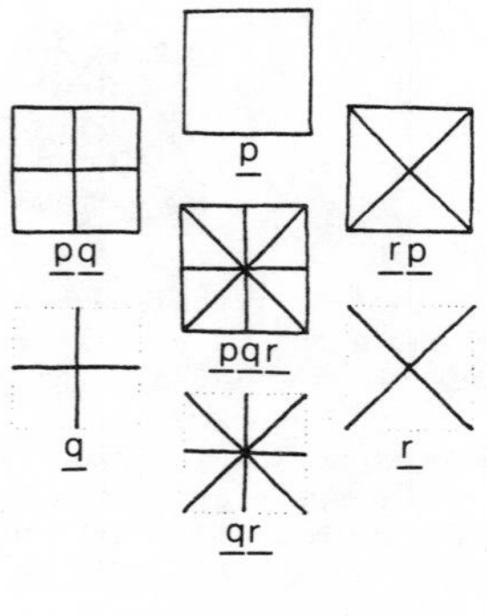

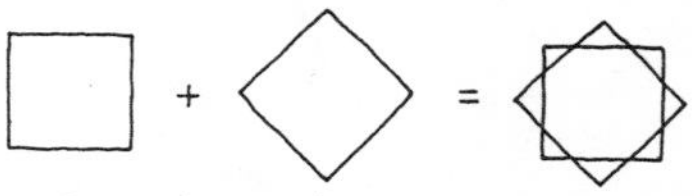

54

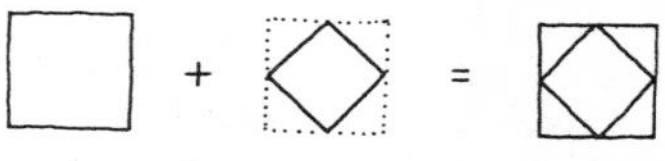

55

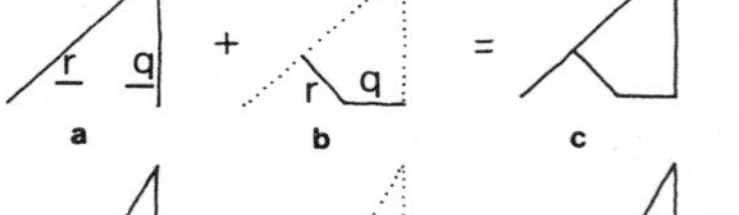

56

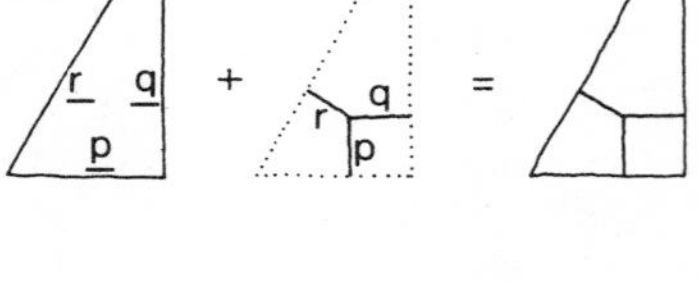

58

lattice combinations, though several structures are topologically redundant.

Of the 64 grids shown, the 7 structures of Figure 52a give the possible topology of multi-directional cubic nodes. Some examples are given; nodes from other packings can be similarly derived. The DK system and Emmerich's universal node are 6-connected P nodes, Catena is a 12-connected R node, Mero is an 18-connected PR node, Pearce's universal node is a 26-connected PQR node, and so on (see Refs 3 and 5 for nodes). Some of the double-layer grids use partial multi-directional nodes as they are derived by slicing space-fillings. Octaplatte and Triodetic are 9-connected R nodes, Unistrut and Varitec are 8-connected PQ nodes, and Nodus and Tridimatic are 8-connected PR nodes. The classification of space grids and node types is built-in in the generative method described.

6.23 Compound Structures

Compound structures are generated by superimposing structures over one another, or their duals, of the same or a matching symmetry. There are 2 geometric ways of making compound structures: by keeping the cell-size or the edge-length invariant. When the cell-size remains the same, the edge-length changes, and vice versa. Both methods are of interest in generating space grids. The generation of compound polygons, polyhedra, plane and layered grids is described, and extension to multi-directional space grids follows by analogy. We note that compounding is a type of inclusion, the dual of compound duals is an explosion.

6.231 Compound Polygons (Figs 54-57)

The two methods of generating compounds are shown for polygons of symmetry 4 (Figs 54 and 55). The edge-lengths of the two polygons (a) and (b) are equal in Figure 54; their cell-size, determined by (a) is the same in Figure 55 such that (b) fits in (a). Any two or more polygons, or their duals, can be compounded this way. The combinations of compounds can be derived from the fundamental region of polygons (Fig 56; a, b and c),, where (b) is the dual of (a), and (c) is their compound. (Compare with Figs 28 and 29.) All combinations of the 2 pairs of edges p, q, $\underline{p}$ and $\underline{q}$ generate 16 compound polygons which can by arranged in 4-space. Symmetry 3 is illustrated (Fig 57), and the method extends to all polygons of symmetry P. Their 16 dual structures follow.

6.232 Compound Polyhedra and Plane Grids (Figs 58-61)

Compound polyhedra and tesselations can be generated by analogy from the polygons. The fundamental region has 3 pairs of dual edges (Fig 58). Edges p, q and r generate sets of regular-faced polyhedra and plane nets described earlier (see Fig 30); edges $\underline{p}$, $\underline{q}$ and $\underline{r}$ produce the duals (Fig 31). These 2 sets for symmetry 4 (ie. 4-fold faces of 432 polyhedra & 442 plane net) are illustrated in Figures 59 and 60 keeping their cell-size constant. All combinations of these 6 edges generate 64 structures which can be arranged in 6-space. The 6-dimensional arrangement for intertransforming compound tesselations of symmetry 442 are illustrated (Fig 61 a,b); diagram 'a' is shown shrunken and can be overlaid on the 2nd inner ring of diagram 'b'; also compare diagram 'a' with Figures 59 and 60. The method extends to all symmetries of polyhedra, prisms and plane tesselations.

We note that the fundamental region can have 12 types of edges; all combinations of these generate 2^{12} (=4096) compound polyhedra or plane nets for any symmetry. Their arrangement in 12-space follows.

6.233 Layered Compound Grids (Figs 62-66)

Equivalent double-, triple- and multi-layered grids follow

62

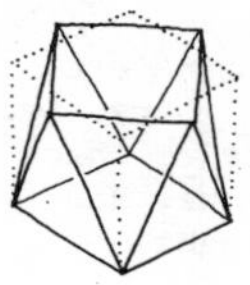

63

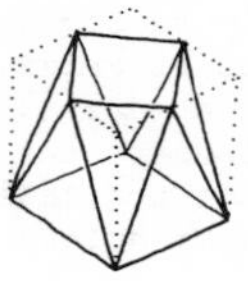

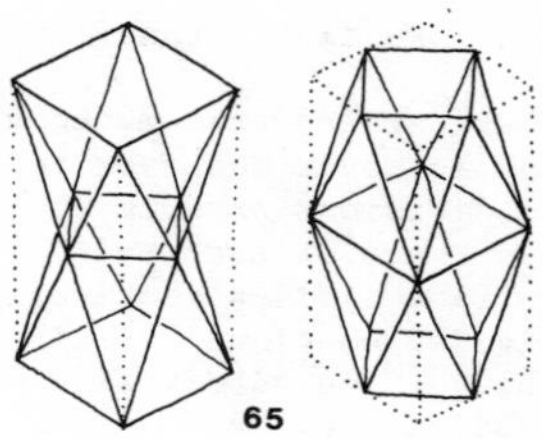

65

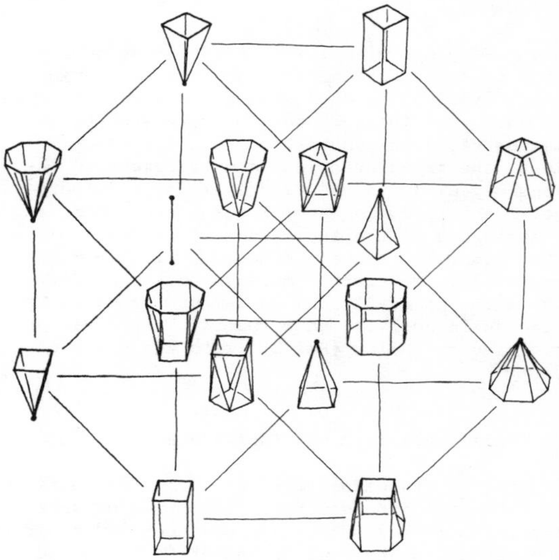

64

from compound plane grids by simply stacking layers over one another. Alternatively, 3-dimensional modules for layered grids can be generated by stacking any two polygons of the same symmetry spatially, but co-axially. The polygons (a) and (b) of Figures 54 and 55 generate the 3-dimensional modules of Figures 62 and 63, respectively. For the 4 polygons of Figure 28, 16 combinatorial stackings generate the 16 modules of Figure 64; their arrangement is in 4-space. Modules for symmetry 4 are shown, and these for symmetries 2, 3 and 6 can be similarly produced. All plane-filling symmetries can be produced by the allowable combinations (Secs 2 and 3.3) of these modules giving us double-layered grids. Reflection of these modules, above or below (Fig 65), produce modules for triple-layered grids. The dual vertex stars can be similarly combined to produce duals of the layered compounds.

The layered grids can be projected onto curved surfaces to produce different kinds of vaults, domes and other structures.

6.234 Multi-directional Compound Grids (Fig 66)

Multi-directional compound grids can be produced by superimposing the polyhedral packings over their duals. One example is shown (Fig 66); others can be systematically derived from the tesseracts of Figures 46-48, and their duals.

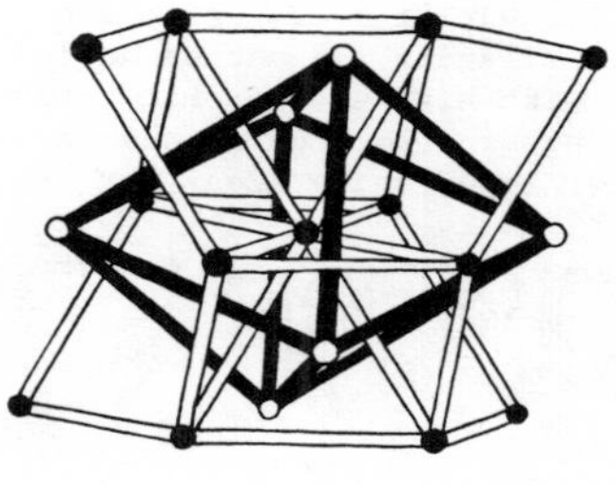

66

6.24 Multiple Transformations

There are several methods by which transformations can be applied over and over to continually subdivide space. All subdivisions produce different types of grids. There are two types of such multiple transformations : 1) where the surface divides continually, and 2) where the entire space divides continually. The two methods are analogous, and the surface subdivision can be seen as a local space subdivision.

6.241 Frequency

This is the simplest of multiple transformations, and is the method used for subdividing the face of a primary spherical polyhedron by subdividing its edge, as in a geodesic dome (Ref 16). There are 7 types of such subdivisions based on combinations of 3 types of inclusions; the transformations are analogous to the inclusions e1, e2 and e3 of Figure 31 for polyhedra and plane nets. Their duals are 7 inter-transforming geodesic spheres analogous to the explosions e1, e2 and e3 of Figure 30. Compound geodesic spheres follow from Sections 6.232 and 6.233. The method extends to the subdivision of space where the space-filling fundamental regions are subdivided analogously. Other spherical subdivisions have been developed by Fuller (Ref 23) with his great circle method.

6.242 Multiple Explosions and Inclusions (Fig 67)

Multiple surface explosions (Ref 24) transform existing vertices and edges into new faces by series of successive explosions. The method subdivides surfaces and generates various 'transpolyhedra' and 'transnets'. This and other recursive topological transformations on polyhedra and arbitrary nets are presently being explored to produce "smooth" surfaces (in collaboration with Hanrahan, Ref 25; see also Sec 6.33).

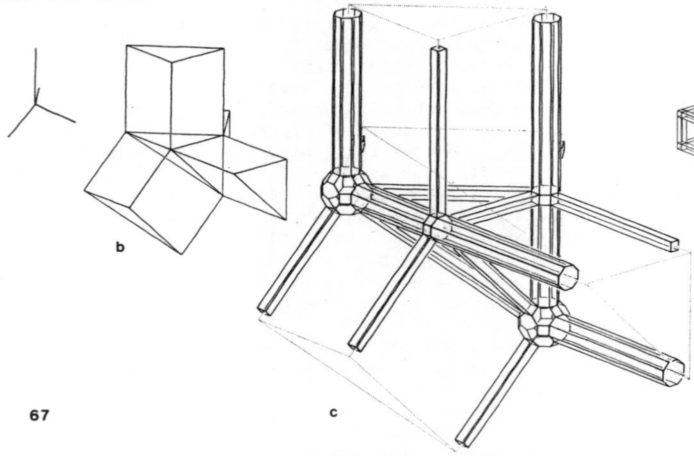
a b c

67

Multiple space explosions are an extension of regular and semi-regular explosions of space-fillings (see Sec 6.31 and 6.32). Every edge and vertex explode to new cells recursively (Fig 67, sequences a-c), making the structure increasingly "cellular" spatially. The extension to arbitrary space nets would follow from above. These transformations are somewhat similar to Mendelbrot's 'fractals' (Ref 26). During every explosion, the structure maintains a "self-similarity", a method anticipated by LeRicolais (Ref 4) with his 'automorphic' structures. Fractal space-fillings and polyhedra, and derivative fractal grids, offer interesting areas for futher study.

6.3 Shapes of Elements

Several methods for deriving the shapes of space structure elements are described. In contrast to the "ball and stick" space frames, shape-specific elements can be used. Such structures require shape-fitting nodes, struts, panels and cells which are "made for each other", a concept useful in generating architectural form and space. Various types of explosions-implosions are described to achieve this.

6.31 Regular Explosions-Implosions (Figs 68-71)

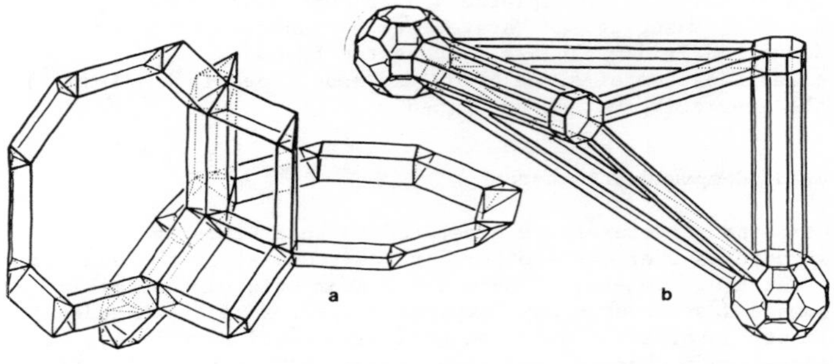
a b

69

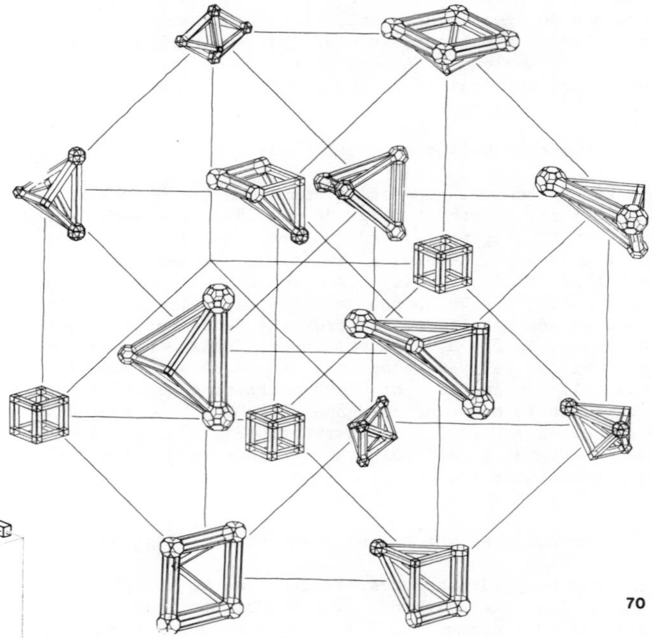

70

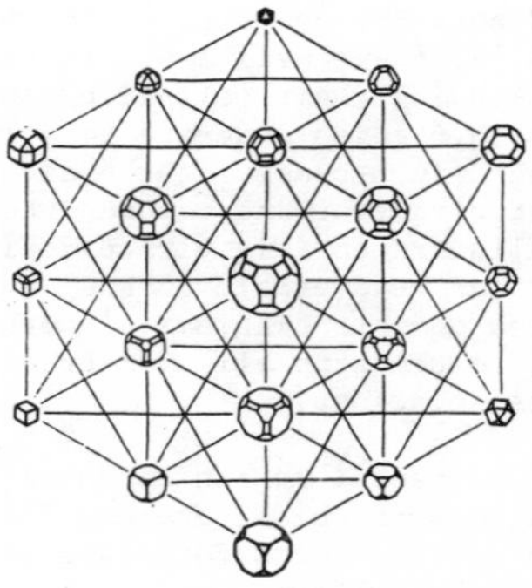

71

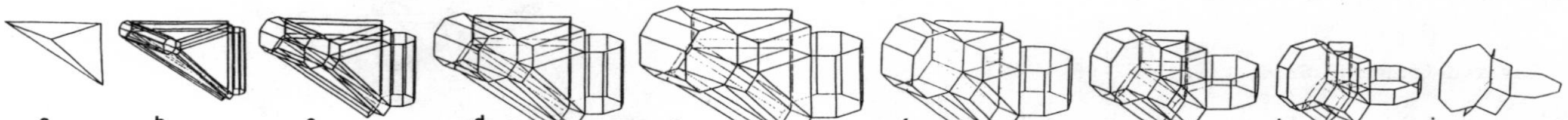

68 a b c d e f g h i

A fundamental module can be exploded keeping its faces invariant while its edges and vertices form new cells (Fig 68 , sequences a-e), and then imploded retaining the new vertex cells and diminishing both the edge cells and the original cell (sequences e-i). Animated sequences show this transformation, termed 'regular' explosion-implosion, which takes us from any structure to its dual (Ref 24). The method works in reverse, and all packings or their duals can be regularly exploded-imploded. Partial explosions (Figs 69, a & b) produce space frames with polyhedral nodes, and prismatic struts and panels. The partial explosion of Figure 46 (or of Figure 43) generates space frames of Figure 70 from the cubic packings. Their transforming polyhedral nodes (Fig 71, see also Fig 33) suggest transforming space frames with correspondingly transforming struts, panels and cells. The middle-stage explosion of the transformation (e, Fig 68) generates 'infinite transpolyhedra' as analogs of "infinite polyhedra" (Ref 22) and are particularly interesting as they produce two dual non-intersecting space labyrinths.

6.32 Semi-regular Explosions-Implosions (Fig 72)

If during the explosions, the faces of the original cell change shape, the transformations are termed 'semi-regular'. The regularly exploded node of Figure 69 could be semi-regularly exploded in various ways (Fig 72) to generate 8 interrelated nodes. We note their analogy to polyhedra of symmetry 332 of Figure 32. Applications to other nodes follow.

6.24 Dual Elements (Figs 75-77)

A structure could have radial or circumferential elements. For a polyhedron or a net, there are 4 possibilities (Fig 75, a-d): radial lines (a) or planes (b), and circum lines (c) or planes (d). These structures are inter-transformable, and with their 4 duals generate 16 compounds. The most interesting of these compounds are superimposed duals, e.g. a radial line structure with its circum plane dual. There are 4 such paired duals all of which have implications for stability especially when combined with tension and compression.

The concept extends to space-fillings where the number of compounds reduce to 4 (Fig 76, a-d) of which 2 (a and d) are paired duals. If a structure in tension is superimposed with its dual in compression, the number of paired space-filling duals is 4. One partially exploded compound, radial line with circum line, is illustrated (Fig 77). Its continued explosion produces a pair of non-intersecting space labyrinths. The dynamics of transformations between inter-transforming paired duals in tension and compression may provide insight into "dynamic equilibrium" processes.

These concepts are presently being explored further to develop a generalised, generative classification of built structures.

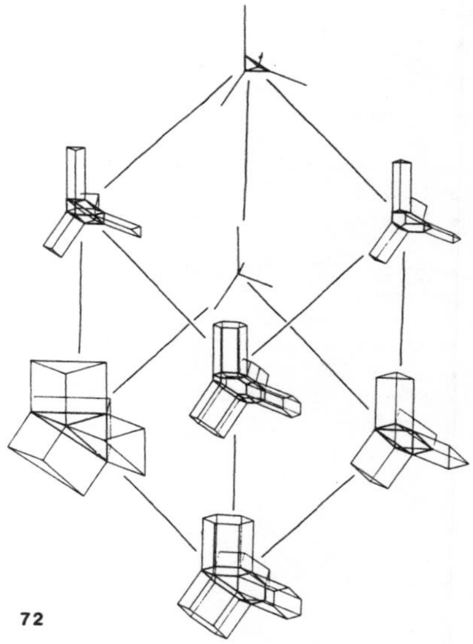

72

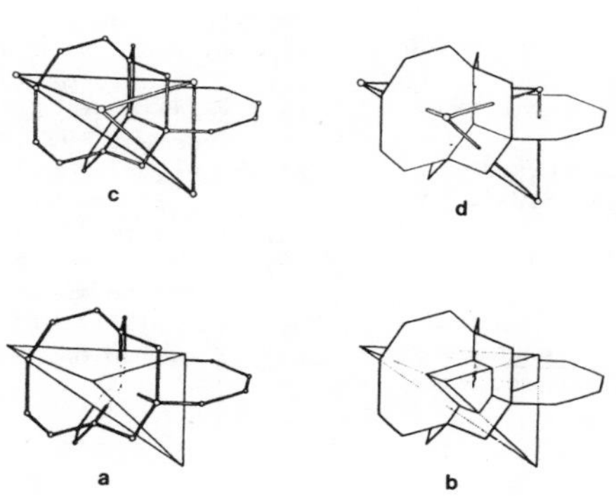

76

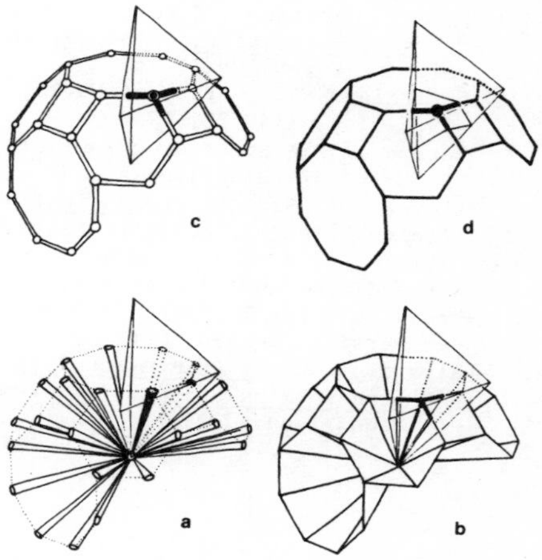

75

6.33 Multiple Transformations (Figs 73 and 74)

Multiple surface explosions of nodes and struts produce "smoother" joints and rods (Sec 6.242). The nodes become transpolyhedra (Fig 73). When the nodes and concurrent edges are treated as surfaces, explosions produce saddle surfaces (Fig 74, a-c) as analogs of 'saddle polyhedra' (Refs 27 and 19). Multiple explosions produce smoother saddles or saddle meshes for connectors.

74

73

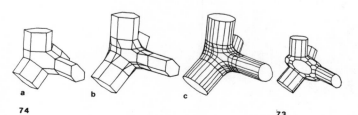

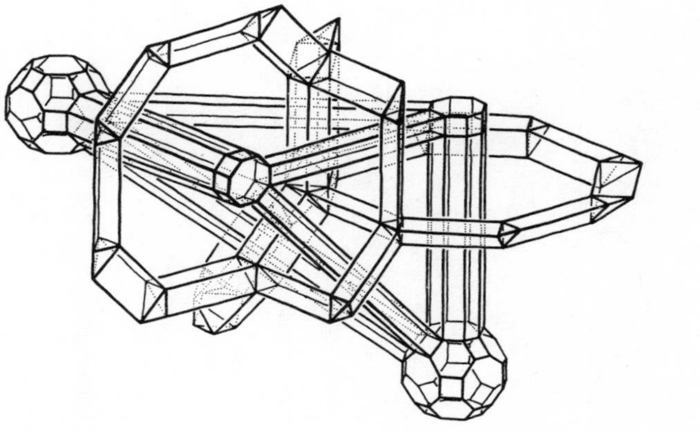

77

In Conclusion

This paper is aimed towards generalising some of the morphologic aspects of space structures. Space structures are continually transforming, and their transformations are governed by underlying patterns which are themselves transforming structures. Such patterns-underlying-patterns, termed 'meta-patterns' (Ref 28), are important in our unification of the laws of form and space. Such a unification leads to the concept of a "universal building system" as a inter-transforming system of systems. Meta-patterns are abstract structures that cross the boundries of disciplines, and for the space structures described, they are symmetrical. Present efforts include: their extension to other space structures, primarily, the 'infinite-','saddle-' and 'uniform-' polyhedra; extension to Laban's Space Harmony (Ref 29); and development of related computer animation films (Ref 30).

Acknowledgements

To the School of Architecture for their support; to R. Russ, G. Juergens, P. Plaza, J. Hersey, W. Nathans, T. Cramer and N. Katz for executing drawings at various stages of the project; to N. Katz and R. Russ for their assistance in the layout and production of this paper.

REFERENCES

1. Space Structures, Proceedings of the 1st International Conference on Space Structures, Blackwell, 1967.
   -Proceedings, 2nd International Conference on Space Structures, Univ. of Surrey, 1975.
2. Proceedings, World Congress on Space Enclosures (IASS), Montreal, Concordia Univ., 1976.
3. Structures Spatiales, Un Bilan, Techniques & Architecture, 309, 1976.
4. Le Ricolais, R., Survey of Works, Structural Research 1935-71, Zodiac 22, 1973.
   -Interviews in Structures Implicit and Explicit, Via 2, Univ. of Pennsylvania, 1973.
5. Makowski, Z. S., Structures Spatiales, Cahiers du Centre D'Etudes Architecturales, No. 14, Brussels, 1971.
   -Three-dimensional Structures, Zodiac 21, 1972.
   -Trends and Developments in Space Structures, University Lecture, Surrey, 1975.
6. Lalvani, H., Multi-dimensional Periodic Arrangements of Transforming Space Structures, PhD Thesis, Univ. of Pennsylvania, 1981.
7. -Structures on Hyper-Structures, Vol. 3 (series), Lalvani, 1982.
8. -Structures on Hyper-Structures (summary paper), Structural Topology No. 6, 1982.
9. -Patterns in Hyperspaces (exhibition catalog), Lalvani, 1982.
10. Buerger, M., Elementary Crystalography, M.I.T., 1968
11. Shubnikov, A. V. & Koptsik, V. A., Symmetry in Science and Art, Plenum, 1974.
12. Loeb, A., Color and Symmetry, Krieger, 1978.
13. Coxeter, H.S.M., Regular Polytopes, Dover (reprint), 1973.
14. Borrego, J., Space Grid Structures, M.I.T., 1968.
15. Du Chateau, S., Structures Spatiales, Cahiers du Centre D'Etudes Architecturales, No. 2, Brussels, 1967.
16. Fuller, R. B. and Marks, R., The Dymaxion World of Buckminster Fuller, Anchor, 1973.
17. Zodiac 19, 21, 22: selected works of Fuller, Neuman and Hecker, Tyng, Safdie, Kuhn, Minke, Semino, Burt and Critchlow, 1969, 72, 73.
18. Emmerich, D. G., Geometrie Constructive, Ecole Nationale Superieure des Beaux-Arts, Paris, 1970.
19. Pearce, P., Structure in Nature is a Strategy for Design, M.I.T., 1978.
20. The transformations are governed by DeMorgan's law; the latter was taught to the author by William Katavolos through his 'Correlations' model which uses primary colored cubes and rectangular blocks.
21. Burt,M., The Wandering Vertex Method, Structural Topology No.6, 1982.
22. Wachman, A., Burt, M., Kleinman, M., Infinite Polyhedra, Technion,Israel, 1974.
23. Fuller, R.B., Synergetics 1 and 2, McMillan, 1975,1979.
24. Lalvani, H. Transpolyhedra, Lalvani, 1977.
   -Transpolyhedra and Explosion-Implosion Principle for Dual Transformations, Proceedings, WCOSE-76 (IASS), Montreal, 1976.
25. Hanrahan,P. and Lalvani,H.,Recursive Topological Transformations, In-progress, 1983-.
26. Mendelbrot, B.B., The Fractal Geometry of Nature, Freeman, 1982.
27. Burt,M., Spatial Arrangements and Polyhedra with Curved Surfaces......,Technion, Israel, 1966.
28. The term is adopted from Gregory Bateson.
29. Through several workshops at Laban/Bartenieff Institute for Movement Studies; see LIMS News, Vol.6 No.2,1983.
30. In collaboration with Robert McDermott and Patrick Hanrahan at N.Y.I.T., 1983-.

FORMEX FORMULATION OF TRANSMISSION TOWERS

M HARISTCHIAN,* BSc, MSc, PhD and S MAALEK**, BSc, MSc

*Standards Division, Technical Bureau, Water Affairs,
Ministry of Energy, Islamic Republic of Iran.

**SANO Consulting Engineers, Tehran, Iran.

The paper illustrates application of formex algebra in formulating the interconnection patterns of transmission towers.The formulations may be used as data in conjunction with a suitable computer software for the purposes of structural analysis or may be employed in relation to automated graphics. In formulating an interconnection pattern, no consideration is given to such aspects as the jointing technique, support conditions and loads, furthermore, it is always assumed that a combination of reasons has given rise to the desirability of a particular configuration. The reader is assumed to be familiar with the concepts of formex algebra.

INTRODUCTION

Formex algebra, introduced by H Nooshin (See Ref 1) provides a powerful means for algebraic representation and processing of configurations of various kinds such as structural systems, electrical networks...and so on. Since its introduction, formex algebra has sucessfully been used in processing various types of configurations such as double - layer grids, domes, barrel vaults, finite element meshes and so on (Refs1-9). Through application to practical problems in various areas, formex algebra, has been enriched by newly evolved concepts, modifications and generalizations of the original concepts. The reader is assumed to be familiar with the concepts of formex algebra as described in one of the Refs 2 to 4 together with Ref5.

Transmission towers are an important type of structure in power industry. These towers usually consist of four faces, and such typical parts as the "body" of the tower, the "arms" and the horizontal "diaphragms". Because of the industrialized mass production of transmission towers,they normally consist of typical units. The paper starts with the formulation of the interconnection patterns of such typical units, then it is shown that how these simple formulations may be employed in representing the interconnection patterns of practical towers.

APPROACH

The typical tower units that are considered are mainly of two types: In the first type all four faces have identical interconnection patterns; in the second type, any adjacent faces are different, while every two opposite faces are identical.

To begin with, consider Fig 1 ,which shows a dimetric view of the "body" of a simple tower unit consisting of four identical faces. For clarity of the figure, only the elements on the two near - side faces are drawn and dotted lines are used to indicate the omitted repeating patterns (only the very top and bottom units are drawn). The horizontal thin lines which are not drawn in the U1U3 plane are not part of the configuration. Where U1U2U3 is the three-dimensional intrinsic coordinates system which is used for formex formulation of the configuration. For a tower unit of the frist type, projection regarding the U2U3 plane is identical to that of U1U3 plane, therefore, only projection on the U1U3 plane is drawn.

In obtaining formex formulations the following conventions are used:

(1) Every joint of the configuration is represented by a signet of the third grade [U1,U2,U3] with respect to the U1U2U3 intrinsic coordinates system.

(2) Every element of the configuration whose end - joints are represented by signets t1 and t2, is represented by the two - plex cantle [t1; t2] or [t2; t1].

Having established the general conventions, a formex T1 representing the interconnection pattern of the configuration of Fig 1, may be obtained as

$$T1 = \Big[_{i=1}^{n}\ \mathrm{tran}(3,2i-2) \mid \mathrm{rosid}(2,2) \mid E1$$

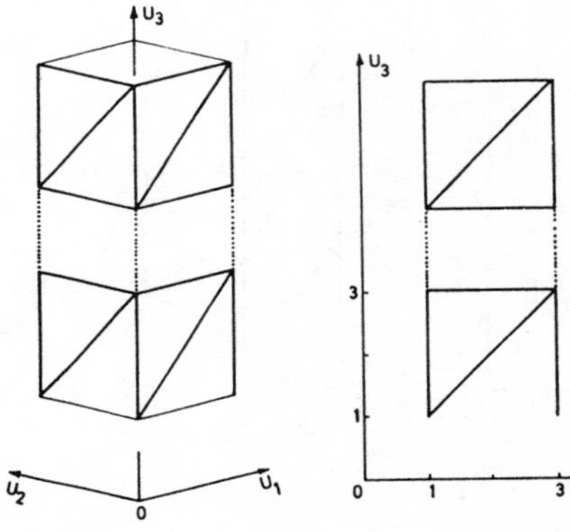

Fig 1

where n is the number of repeating units in the direction of the U3 axis, and

E 1 ={[1,1,1; 1,1,3],[1,1,3; 3,1,3],[1,1,1; 3,1,3]}.

A function Φ1(n , m , k) defined as

$\overset{n}{\underset{i=1}{\Big\lceil}}$ tran(3 , mi − m) ¦ rosid(k , k)

finds frequent application in the formex formulations of the tower units of the first type. With this notation

T 1 = Φ 1(n , 2 , 2) ¦ E 1.

A formex T 2 representing the interconnection pattern of the configuration of Fig 2 , may be obtained as

T 2 = Φ 1(n , 2 , 2) ¦ E 2

where Φ 1 is previously defined and

E 2 = e 2 # [1,1 1; 1,1,3]

where

e 2 =lamis(2 , 2) ¦ [1,1,1; 2,1,2].

A formex T 3 representing the interconnection pattern of the configuration of Fig 3 , may be obtained as follows

H 3 =lamit(2 , 2) ¦ {[1,1,1; 1,1,2],[1,1,2; 1,2,1]}

and

T 3 = $\overset{n}{\underset{i=1}{\Big\lceil}}$ tran(3,2i−2) ¦ (lam(2 , 2) ¦ e 2 # lam(1 , 2) ¦ H 3)

where n and e 2 are previously defined.
To formulate the tower units of the second type, it is appropriate to define a functional

T = $\overset{n}{\underset{i=1}{\Big\lceil}}$ tran(3,ki−k) ¦ (lam(2 , k) ¦ E # lam(1 , k) ¦ H)

or

T = Φ 2 (n , k) ¦ (E , H).

With this notation

T 3 = Φ 2(n , 2) ¦ (e 2 ,H 3).

A formex T 4 representing the interconnection pattern of the configuration of Fig 4 , may be obtained as follows

H 4 = H 3 # lam(2 , 2) ¦ [1,1,3; 1,2,3]

E 4 =[1,1,3; 3,1,3] # lam(3 , 2) ¦ [1,1,2; 3,1,1]

and

T 4 =Φ 2(n , 2) ¦ (E4 , H4).

A formex T 5 representing the interconnection pattern of the configuration of Fig 5 ,may be obtained as

T 5 =Φ 2(n , 2) ¦ ([1,1,3; 3,1,1],H 3)

where Φ 2 and H 3 are previously defined.

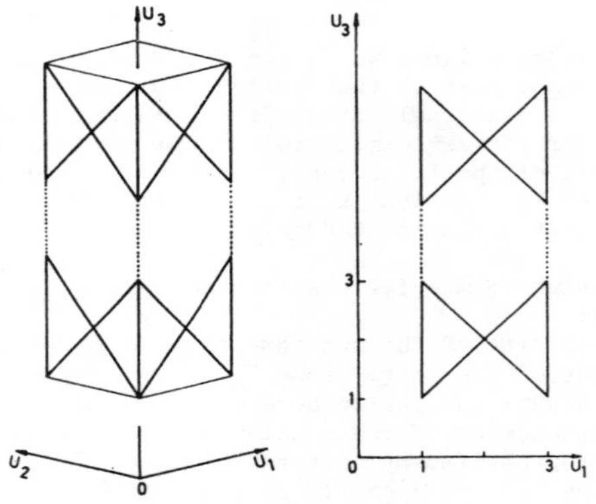

Fig 2

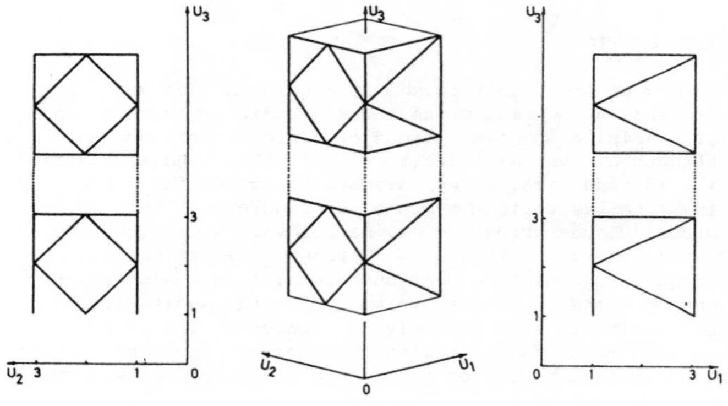

Fig 4

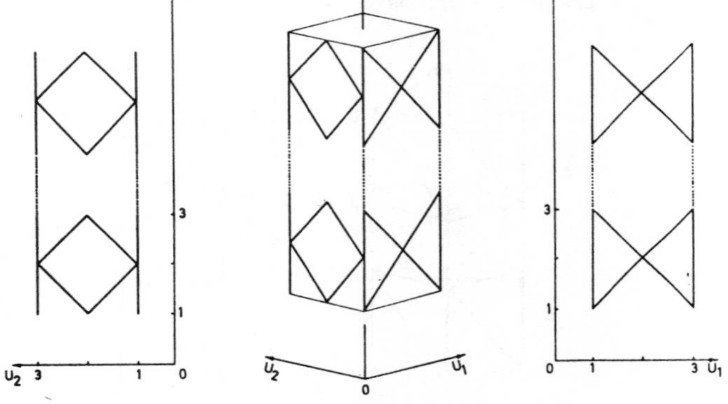

Fig 3

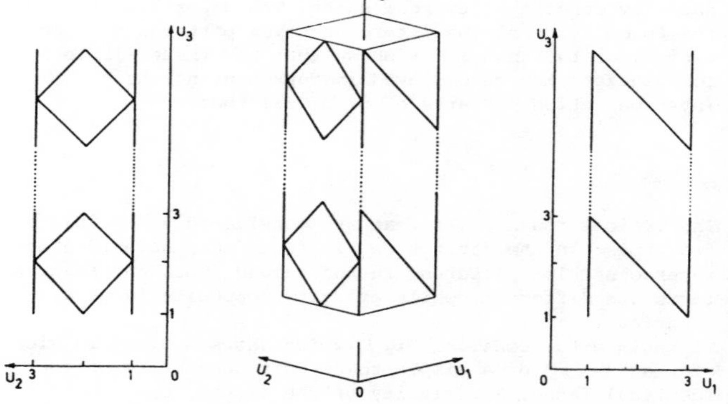

Fig 5

A formex T 6 representing the interconnection pattern of the configuration of Fig 6 , may be obtained as follows

e 6 = $\underset{i=0}{\overset{1}{\sqsubset}}$ clovis(1+i , 3-i) ¦[1+i,1,1+i; 1+2i,1,3]

E 6 =lam(1 , 3) ¦ (e 6 #[1,1,5; 3,1,3])

and

T 6 =pex ¦ Φ 1 (n , 4 , 3) ¦E 6 .

A formex T 7 representing the interconnection pattern of the configuration of Fig 7 , may be obtained as follows

e 7 =gamis(1 , 1) ¦{[1,1,1; 2,1,1],[2,1,1; 3,1,1]}

f 7 =clovis(2 ,2) ¦{[1,1,3; 2,1,2],[1,1,2; 2,1,2]}

E 7 =rosis(3 , 3) ¦ (e 7 #lux([3,1,2]) ¦f 7)

and

T 7 =pex ¦ Φ 1 (n , 4 , 3) ¦E 7 .

A formex T 8 representing the interconnection pattern of the configuration of Fig 8 , may be obtained as follows

e 8 = $\underset{i=0}{\overset{3}{\sqsubset}}$ tranis(i , i) ¦[1,1,1; 2,1,2]

f 8 =proj(1 , 1) ¦e 8

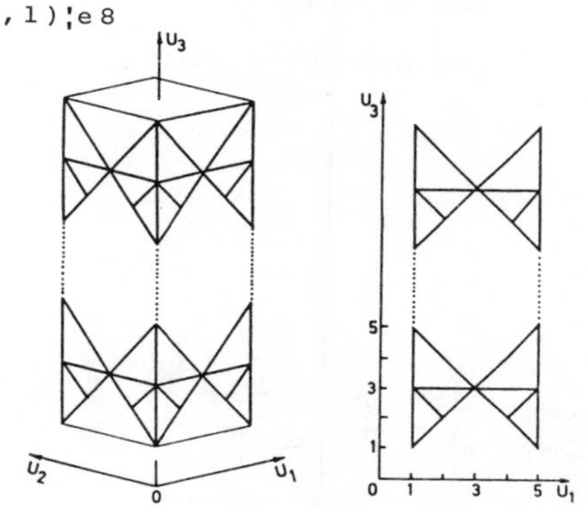

Fig 6

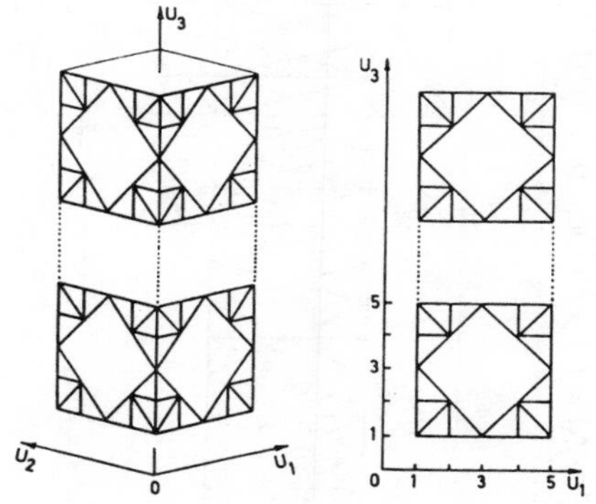

Fig 7

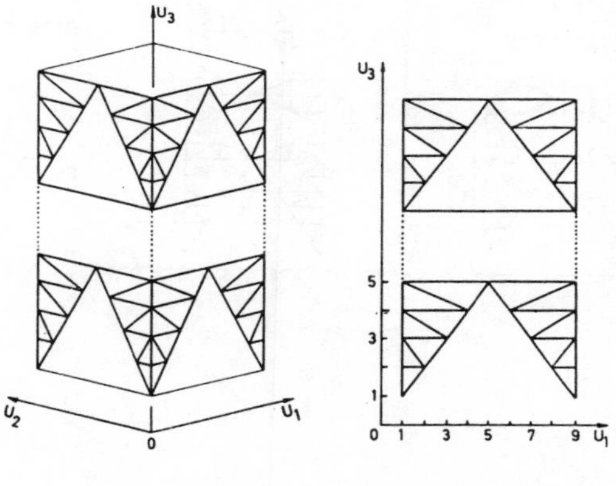

Fig 8

g 8 = $\underset{j=0}{\overset{1}{\sqsubset}}$ $\underset{i=0}{\overset{2+j}{\sqsubset}}$ [1,1,3+i-j; 2+i,1,2+i]

E 8 = e 8 # f 8 # g 8

and

T 8 =pex ¦ Φ 1(n , 4 , 5) ¦ lam(1 , 5) ¦E 8 .

A formex T 9 representing the interconnection pattern of the configuration of Fig 9 , may be obtained as follows

e 9 = gamis (2 , 2) ¦[2,1,2; 3,2,3]

f 9 =proj(1 , 3) ¦e 9 # e 9

g 9 =proj(2 , 1) ¦f 9 # f 9

E 9 = g 9 #{[1,1,1; 3,1,1],[3,1,3; 3,2,3]}

and

T 9 =T 5 # tran(3 , 2n) ¦lam(2 , 2) ¦E 9

where T 5 and n are previously defined.

A formex T 10 representing the interconnection pattern of the configuration of Fig 10, may be obtained as

T 10 = T 1 # tran(3 , 2n) ¦rosid(2 , 2) ¦[1,1,1; 2,2,3]

where T 1 is previously defined.

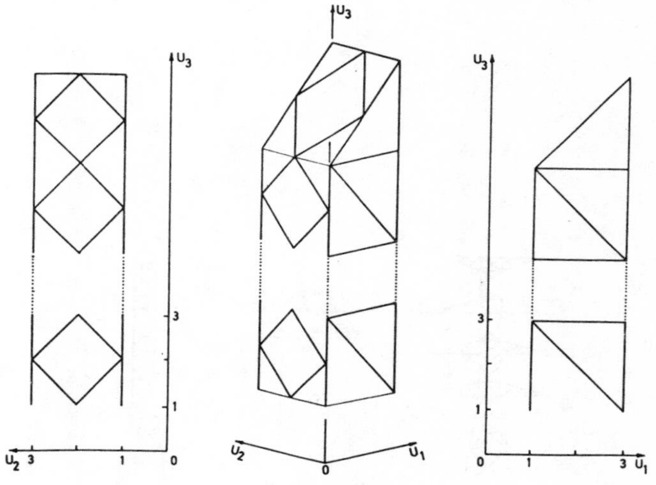

Fig 9

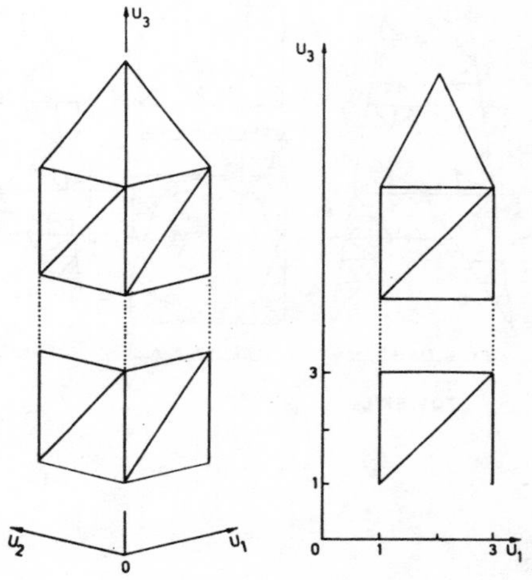

Fig 10

A formex T 11 representing the interconnection pattern of the configuration of Fig 11 ,may be obtained as follows

e 11 = clovis(1 , 6) ¦[1,1,5; 1,1,6]

f 11 = e 11 # $\left[\frac{1}{i=0}\right.$ lamis(3+3i ,6) ¦[1+4i,1,5; 3+3i,1,6]

g 11 = lux([5,1,7])¦f 11 #{[5,1,5; 7,1,5] , [6,1,6; 8,1,6]}

and

T 11 = pex ¦ rosid(7 ,7) ¦lam(1 , 7) ¦ (E 8 # g 11)

where E 8 is previously defined.

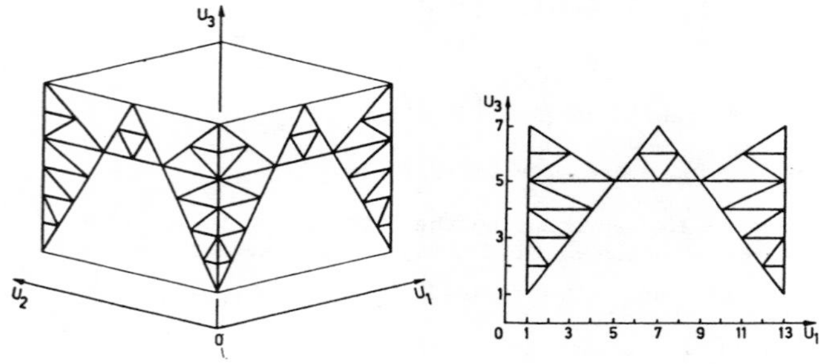

Fig 11

CASE STUDIES

Figure 12, shows a transmission tower. In order to produce a formex formulation for this tower,the graphical arrangements of Fig 13 , are used. Where, the U1U3 and U2U3 projections of the layout of the tower with respect to the three - dimensional intrinsic coordinates system U1U2U3,are drawn. The figure also shows projections of the horizontal bracings of the tower with respect to the U1U2 plane.

A formex TOWER 1 representing the interconnection pattern of the tower of Fig 12 ,may be obtained as

TOWER 1 = BT 1 # HT 1 # AT 1.

Where

BT 1 = $\left[\frac{5}{i=1}\right.$ Pi

represents the body(the four faces) of the tower, and

P 1 = tran(1 , 6) ¦ T 11

P 2 = tranad (4,-2 , 6) ¦ dilid(3 , 3)¦rel(EU 3 ≤17) ¦T 7(3)

P 3 = tranad (4,-2,16) ¦dilid(3 , 3) ¦T 6 (3)

P 4 = tranad(1,-5,28) ¦dilid(6 , 6) ¦T 2(2)

P 5 = tranad(1,-5,32) ¦dilid(6 , 6) ¦T 3(10).

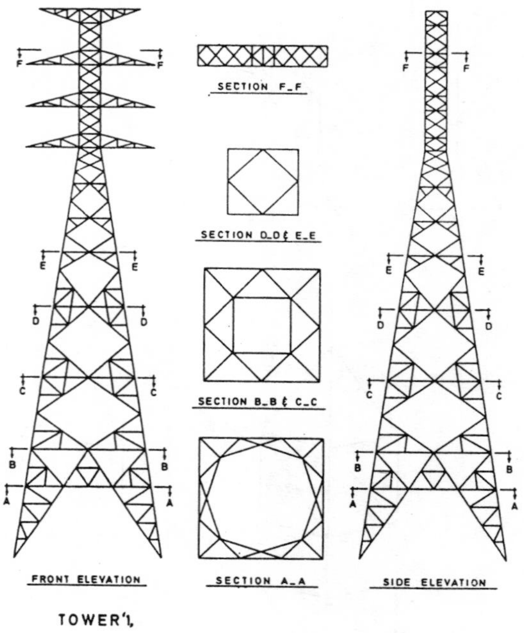

Fig 12

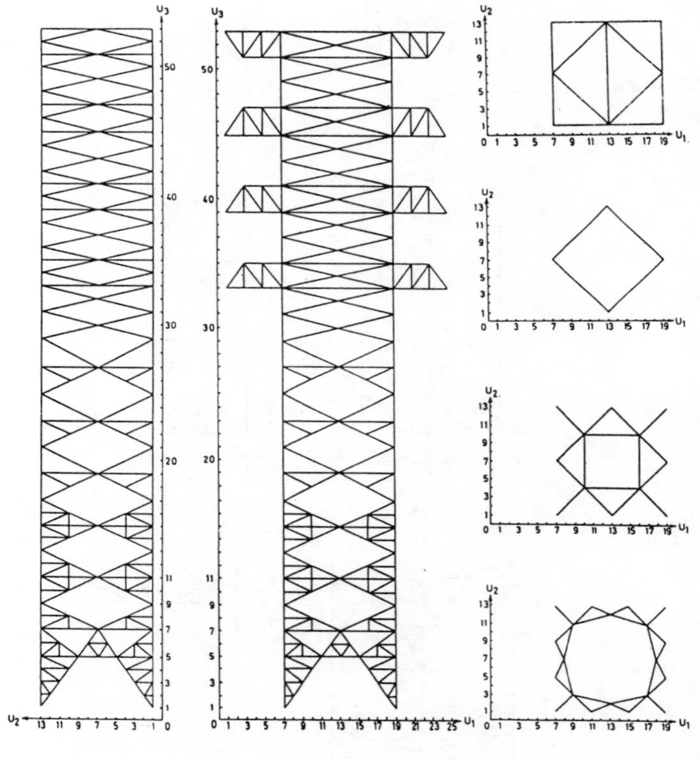

Fig 13

Here T 7(3) ,... and T 3(10) indicate the previously defined formices T 7 ,... and T 3 with the values of n=3,... and n=10, respectively.

HT 1 = $\underset{i=1}{\overset{4}{\sqsubset}}$ Si

represents the horizontal bracings with

S 1 = proj(3 , 5) ¦ b 1

S 2 = $\underset{i=0}{\overset{1}{\sqsubset}}$ proj(3,4i+7) ¦b 2

S 3 = $\underset{i=0}{\overset{1}{\sqsubset}}$ proj(3,4i+15) ¦b 3

S 4 = $\underset{i=0}{\overset{3}{\sqsubset}}$ tran(3,6i+33) ¦b 4

and

b 1 = rosid(13 , 7) ¦ ([7,1,0; 9,3,0] # lam(1 , 13) ¦w 1)

b 2 = rosid(13 , 7) ¦ (clovid(10 , 4) ¦[7,7,0; 10,4,0] # w 2)

b 3 = rosid(13 , 7) ¦ [7,7,0; 13,1,0]

b 4 = lam(3 , 1) ¦ (w 3 # rosid(13 , 7) ¦ lam(1 , 13) ¦[7,1,0; 13,1,0]).

Here

w 1 = {[9,3,0; 13,2,0],[13,2,0; 11,1,0],[11,1,0; 9,3,0]}

w 2 = [10,4,0; 10,10,0]

w 3 = b 3 # [13,1,0; 13,13,0].

Finally, the formex AT 1 , representing the arms of the tower is obtained as

AT 1 = $\underset{i=0}{\overset{2}{\sqsubset}}$ tran(3,6i+33) ¦ v1 # ref(3,26$\frac{1}{2}$) ¦vl

with

v 1 = lam(1 , 13) ¦ tranid (17 , -5) ¦ dil(2 , 6) ¦ver(3,1,1,2) ¦T9(2)

where T 9(2) is the previously defined formex T 9 with n=2.

Figure 14 ,shows the layout of a transmission tower. With the aid of the graphical arrangements of Fig 15,a formex TOWER 2 representing the interconnection pattern of the tower may be obtained as

TOWER 2 = pex ¦ (BT 2 # HT 2 # AT 2) .

Where

BT 2 = $\underset{i=1}{\overset{4}{\sqsubset}}$ Pi

represents the body of the tower, and

P 1 = tranis (8 , -1) ¦dil(3 , 2) ¦T 8(3)

P 2 = lam (1 , 13) ¦ rem(v) ¦w 1

P 3 = lam (1 , 13) ¦refis(17 , 33) ¦w 1

P 4 = refid (13 , 5) ¦w 2 # w 2

with

w 1 = tranad (14,-3,24) ¦dil(2 , 4) ¦T 9(3)

w 2 = tranad (6,-3,42) ¦dil(2 , 4) ¦T 10 (3)

v = lam(2 , 5) ¦lam(2 , 3) ¦[15,1,25; 13,1,25].

HT 2 = $\underset{i=1}{\overset{3}{\sqsubset}}$ Si

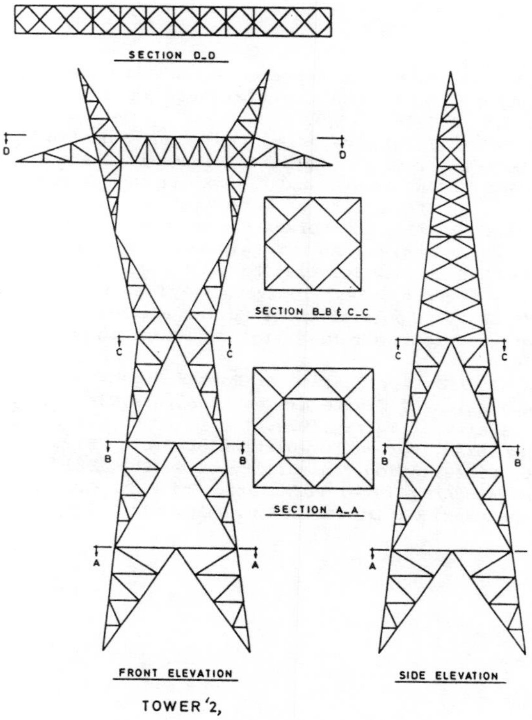

Fig 14

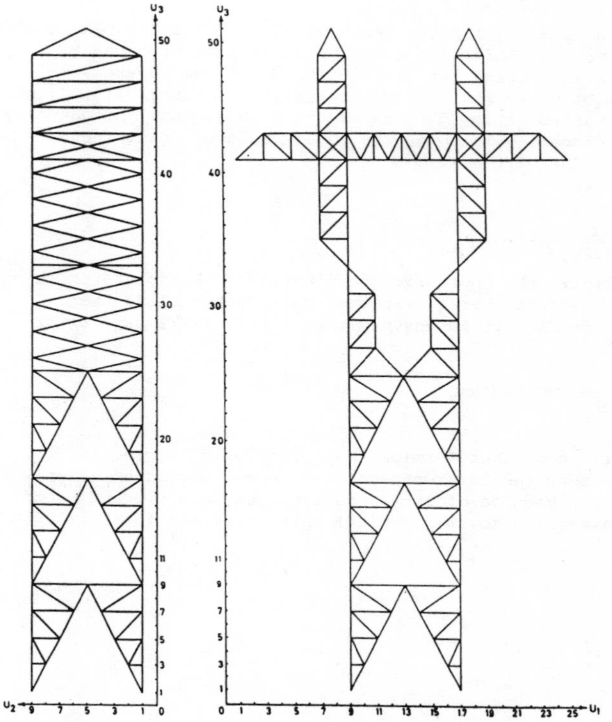

Fig 15

represents the horizontal bracings and the elements at the intersection of the "horizontal beam" and the body of the tower; and is obtained as follows

S 1 = proj (3 , 9) ¦b 1

S 2 = $\left[\frac{1}{i=0}\right.$ proj (3,8i+17) ¦b 2

S 3 = lam (1 , 13) ¦tranad (6,-3,40) ¦ dil(2 , 4) ¦b 3

and

b 1 = b 2 # rosid(13 , 5) ¦ [11,3,0; 11,7,0]

b 2 =rosid(13 , 5) ¦clovid(11 , 3) ¦[9,5,0; 11,3,0]

b 3 = T 2(1) # lamad(2,2,2)¦(gamid(1 , 1) ¦[1,1,1; 2,1,1] # [2,1,1; 1,2,1]).

Finally, the formex AT 2, representing the arms and the horizontal beam is obtained as

AT 2 = lam (1 , 13) ¦tranad(17,-3,41) ¦dil(2 , 4) ¦ ver (3,1,1,2) ¦T 9(2) # B

where

B = tranad(15,-3,41) ¦dil(2 , 4) ¦veris(2 , 1) ¦T 4(4).

In the above formulations T 8(3),... and T 4(4) indicate the previously defined formices T 8,... and T 4 ,with n=3,... and n=4, respectively.

GEOMETRY

In producing formex formulations of the transmission towers,the geometric particulars were not considered. However, once formulation of the interconnection pattern of a tower is in hand, the geometric coordinates of its joints may be obtained as a function of the intrinsic coordinates of the joints.
As an example, for the tower of Fig 14 , let XYZ be a three-dimensional Cartesian coordinates system, with the X , Y and Z axes being parallel to the U1,U2 and U3 axes, respectively. Furthermore , let[X1,Y1,Z1] and[X2,Y2,Z2] be the Cartesian coordinates of the line segment indicated by the signets[9,1,1] and [9,1,25] , respectively. Then, relations of the form

$$Z = \alpha U3$$

and

$$\frac{X - X1}{X2-X1} = \frac{Y - Y1}{Y2-Y1} = \frac{Z - Z1}{Z2-Z1}$$

will give the geometric coordinates of the joints on this line segment. Here, α is the scale factor along the Z - axis, and it is assumed that X1≠X2 , Y1≠Y2 , and Z1≠Z2.

CONCLUDING REMARKS

It is shown that formices may powerfully be used to represent the interconnection patterns of transmission towers. Because of the industrialized mass production of transmission towers, typical units are used in their construction. It is, therefore, appropriate to express the interconnection pattern of a transmission tower, in terms of the formulations regarding its constituent typical units.

With the aid of an appropriate interpreter, the formulations may be used to generate the required data regarding the structural system within the medium of an electronic computer. For instance, with the aid of the FORMIAN interpreter(now operational at the University of Surrey,SSRC), the formulation of the tower unit of Fig 1 , would be implemented by the following statements:

E 1 = {[1,1,1; 1,1,3],[1,1,3; 3,1,3], [1,1,1; 3,1,3]};

T 1 = LIB (I = 1,N) ¦TRAN(3,2*I-2) ¦ROSID(2 , 2) ¦E 1 ;

The paper is devoted to the formex formulation of the four - sided(transmission)towers; the approach, however,can easily be extended to deal with the triangular(three - sided), ...or circular(multi - sided) towers. For instance, in the expression for T 1, by replacing the "rosid" function by"clovid" and using the "remodition" function , formex formulation of a triangular tower is obtained. In the case of multi - sided towers(cooling towers...), a cylindrical coordinates system would prove suitable.

ACKNOWLEDGEMENTS

The authors wish to thank Dr H Nooshin, the innovator of the mathematics of formices and plenices for his encouragement and guidence, the Ministry of Energy of the Islamic Republic of Iran for the support, and SANO Consulting Engineers for producing the figures.

REFERENCES

1. H NOOSHIN, Algebraic Representation and Processing of Structural Configurations, Computers and Structures, Vol 5,pp 119-130 , 1975.
2. Z S MAKOWSKI(ED), Analysis , Design and Construction of Double - Layer Grids, Applied Science Publishers, 1981.
3. M HARISTCHIAN, Formex and Plenix Structural Analysis, PhD Thesis, University of Surrey ,1980.
4. J S SANCHEZ ALVAREZ, Formex Formulation of Structural Configurations,PhD Thesis, University of Surrey,1980.
5. H NOOSHIN, Formex Formulation of Barrel Vaults, (Short Course Lecture Notes), University of Surrey, Sept 1983.
6. S MAALEK, An Application of Formex Algebra to Dome Configurations, MSc Dissertation, University of Surrey , 1976.
7. K A KASHANI, Application of Formex Algebra in Data Preparation of Finite Element Analysis,MSc Dissertation, University of Surrey,1976.
8. O HOOSHYAR, Formex Formulation of Finite Element Meshes MSc, Dissertation, University of Surrey,1981.
9. M H YAASSAEE, Formex Formulation of S DU Chateau's Structures, MSc Dissertation, University of Surrey ,1981.

FORMEX FORMULATION FOR FRONTAL SOLUTION TECHNIQUE

C.M. Anekwe

Research Student, Space Structures Research Centre, Department of Civil Engineering, University of Surrey

Formex algebra and frontal solution techniques are essentially methods of data generation and solution of large number of simultaneous equations respectively. They can both be used for the stiffness method of structural analysis. The frontal solution technique requires that the equations are solved in a certain order to keep the frontwidth and thus the total number of arithmetic operations to a minimum. This means then that the efficiency of the technique is dependent on the size of the frontwidth of the stiffness matrix. The aim of this paper is to present the use of formex algebra to produce an efficient element numbering for the analysis. It is also extended to include an automatic renumbering of the elements which were produced by using formex algebra in the data generation, to minimise the frontwidth of the structure for a frontal solver.

INTRODUCTION

The solution of large structural systems by the computer has been posing both the problem of analysis with regard to storage requirement and computer time as well as the difficulty of giving to the computer a detailed description of the structure necessary for the analysis. The later forms the data needed for the analysis and includes the structural configuration, element properties and the node or element numbering depending on the solution technique being used. These problems led to two pronged directions of work to tackle them. Frontal solution technique represents the result of the work in the area of reducing storage requirement and computer time, while formex algebra is essentially in the realm of data generation. When these two aspects are linked together from the start of analysis an optimised and elegant solution of the problems is obtained.

FORMEX ALGEBRA

The task of giving the computer the data for small structures is simple and straightforward. It involves only inputting what is known the way the programmer intended it to be given. However, for large systems, the amount of information to be input to the computer makes the whole process time consuming and error prone. As a result of this and the realisation that the natural regularity of most structures and finite element meshes could be exploited, considerable interest in data generation has been shown by research workers in the last few years and is still continuing. Most of these methods are limited in scope and lack any form of generality. Some of them are in addition very complicated and cannot be easily used in the design office nor can they be easily adapted for use by designers.

A very powerful method which has revolutionised the whole process of data generation, is given a sound mathematical footing and simplicity in the formex algebra, a concept proposed firstly by H. Nooshin[1]. The process excels in its simplicity and universal applicability.

Formex algebra is a mathematical system used to process and organise a collection of entities or configurations. It includes abstract entities called formices and a set of rules used in their manipulation to organise the configurations. The word configuration is used in its widest sense and a structural system may be viewed as a collection of interrelated entities. The concepts of formex algebra may therefore conveniently and efficiently be used, to deal with both the representation of structural configurations and the automated generation of data.

Reference 3 gives the detailed description of the current definition of the concepts of formex algebra. It is therefore suggested that the reader familiarises himself with these concepts. Except for the constructs such as lib(i=0,3) used to represent

$$\sqsubseteq_{i=0}^{3}$$

all abbreviations, terminology and symbols are used as in Ref 3.

Also, the element numbers are written, as far as possible, at the middle of the elements while the node numbers are written near the nodal points in the figures.

FRONTAL SOLUTION TECHNIQUE

The frontal solution technique first presented by B.M. Irons[4], has proved to be an effective method for the solution of the large number of simultaneous equations, arising from the stiffness method of structural analysis.

The technique is basically Gaussian elimination, with a special housekeeping procedure for handling the structural stiffness matrices. The housekeeping procedure, enables element stiffness and load matrices to be transferred into the central memory sequentially from a file and the structure stiffness and load matrices are assembled and solved, a section at a time. This enables both small and large structures to be handled efficiently.

The reader is assumed to be familiar with the procedures of the frontal solution technique. It is therefore not repeated. The interest is on the necessity for frontwidth minimisation.

The frontal solution requires that the equations are solved in a certain order to keep the frontwidth and thus the total number of arithmetic operations to a minimum. For the technique, the frontwidth is controlled by the order in which the element matrices are introduced into the central memory. Apart from increased cost of solution, inefficiency in the order in which the elements are presented may cause a modest sized structure to become too large to be solved by the frontal solution technique.

In this solution technique, each element is given a unique element number, the location of each element is defined by its node numbers which must be specified in certain order. The element numbers for the structure may have omissions and need not start at one, however since frontal technique normally solves the structure equations according to ascending element numbers, it may be necessary to number the elements consecutively and across the narrow direction of the structure to minimise the frontwidth. Sometimes, however, it could be more convenient to number the elements to suit data generation. In this case, the frontwidth can be minimised by using a frontwidth minimisation technique.

The node numbers for a structure may also have omissions and do not have to start at one. The manner of node numbering is irrelevant to the frontal solution and is arranged to suit the data generation.

In general, the frontwidth of a structure will vary in size throughout the solution and the maximum frontwidth is important since it determines the size of the central memory required to solve the structure stiffness matrix.

It is also necessary to minimise the maximum frontwidth since the solution cost is proportional to the sum of the squares of the consecutive frontwidths throughout the solution. Most computer centres charge their customers according to an effective time t computed as

$$t = \alpha CP + \beta PP$$

where CP is the Central Processor time
PP is the Peripheral Processor time
α and β depend on charging system of the particular computer centre[5].

The frontwidth is determined by the order in which the elements are introduced into the central memory and this is controlled by the user. Considering for example, the finite element mesh for plain stress analysis shown in Fig 1 the elements are numbered across the narrow direction and using increasing element numbers as the order of solution, the frontwidth will be optimum.

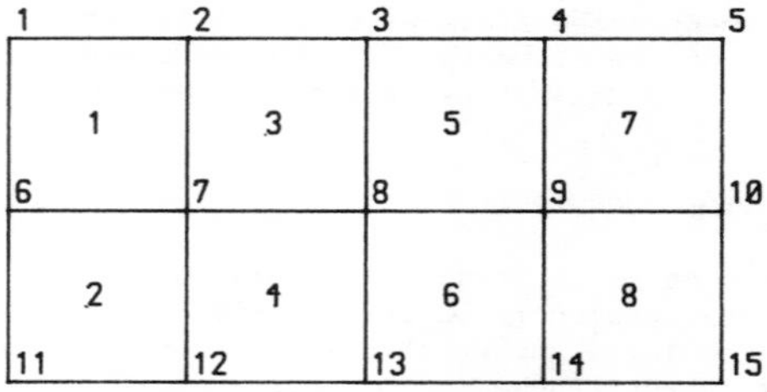

Fig 1

As each element is introduced into the central memory the solution process proceeds as in Table 1.

The front nodes in the central memory at each stage are shown in the second column of Table 1 and they constitute the active nodes which are partially assembled and those that are fully assembled, ready to be eliminated. The third column contains the nodes that are fully summed and eliminated. The frontwidth is shown in the last column.

The maximum frontwidth is 10 for this optimum solution order and it is not possible to reduce it any further.

Table 1

Element	Nodes in Central Memory (Front)	Nodes with Complete Equations that are Eliminated	Frontwidth (= No. of Nodes in CM x Degrees of freedom)
1	1,2,7,6	1	8
2	12,2,7,6,11	6,11	10
3	12,2,7,3,8	2	10
4	12,13,7,3,8	12,7	10
5	4,13,9,3,8	3	10
6	4,13,9,14,8	13,8	10
7	4,5,9,14,10	4,5	10
8	15,0,9,14,10	15,9,14,10	10

If an inefficient solution order was specified for the same structure as in Fig 2 by numbering the elements along the longer length, the frontwidth will be increased.

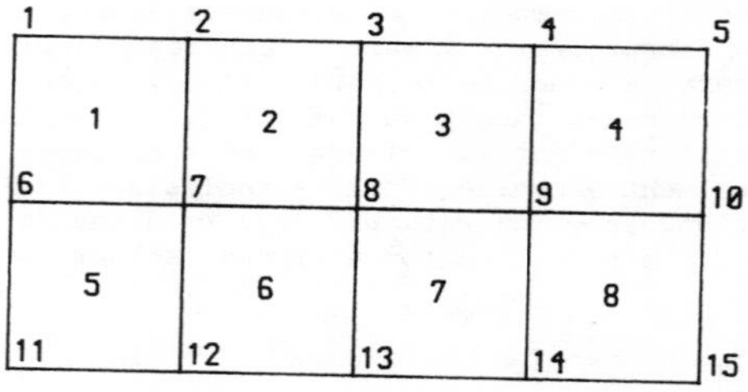

Fig 2

The frontal solution would have proceeded as in Table 2.

Table 2

Element	Nodes in Central Memory	Nodes with Complete Equations that are Eliminated	Frontwidth (= No. of Nodes in CM x Degrees of freedom)
1	1,2,7,6	1	8
2	3,2,7,6,8	2	10
3	3,4,7,6,8,9	3	12
4	5,4,7,6,8,9,10	4,5	14
5	12,11,7,6,8,9,10	11,6	14
6	12,13,7,0,8,9,10	12,7	14
7	14,13,0,0,8,9,10	13,8	14
8	14,15,0,0,0,9,10	14,15,9,10	14

The maximum frontwidth is now 14 and this requires more central memory and more computing time to produce a solution. The central memory requirements for the frontal solution are approximately

$$^{1}/_{2} \times (\text{maximum frontwidth})^{2}$$

and computing time is approximately

$$C \times \text{total degrees of freedom} \times \text{average frontwidth}$$

where C depends on the machine used[7]. The later solution would therefore be approximately twice as expensive as the former, hence the need to minimise the frontwidth.

FRONTWIDTH MINIMISATION

Since the efficiency of the frontal technique is dependent on the size of the frontwidth, it becomes very important to have techniques that could minimise it. There are many bandwidth minimisation techniques, yet it is more difficult to visualise the numbering of elements of a configuration to give a reasonably low frontwidth, than it is for numbering the nodes of the configuration to obtain a low bandwith. There is therefore, a greater need for an automated frontwidth minimisation technique.

Two methods are presented here. Both of these methods involve the rearrangement of the cantles of the formex which represent in an ordered form, the node numbers describing the interconnection pattern of the configuration. The first technique is directly dependant on being able to find an ingot which will produce a node numbering scheme that will result in a minimum bandwidth. The second method is not dependant on how the ingot is obtained.

THE FIRST METHOD

The minimisation of the frontwidth using the first method is done at the outset of the data preparation.

The procedure is as follows.

A formex, say F, describing the interconnection pattern of the configuration is obtained. It is unimportant how this formex is arrived at. The process most convenient for the layout of the configuration should be followed. The arrangement of its cantles is also not important. For example let the formex F representing an interconnection pattern be given by

F = {[1,1; 1,3; 3,1], [3,3; 5,3; 5,1], [3,1; 3,3; 5,1], [1,3; 3,3; 3,1]}

An ingot, say E, representing the nodal points of the configuration is then obtained. It is very important that the ingot is obtained across the narrowest width of the configuration. This will help to produce a minimum frontwidth required. If the narrowest width is not obvious and the dimensions of the configuration in all its directions are not equal, then this first method may not be adequate and the second method would be recommended. Also if there are midside nodes, then obtaining a low frontwidth with this method might be difficult. It is therefore recommended that the first method be used only when the configuration is simple and straightforward for obtaining the ingots of the nodal points, which will result to a minimum bandwith. For example, the ingot E of the nodal points of the configuration represented by formex F, may be given by

E = {[1,1], [1,3], [3,1], [3,3], [5,1], [5,3]}

The dictum of F with respect to E, say H is obtained. The formex H contains in an ordered form the node numbers describing the interconnection pattern of the configuration, that is, H = dic(E)'F.

Using examples of F and E above

H = dic(E)¦F = {[1;2;3], [4;6;5], [3;4;5], [2;4;3]}

Each cantle of the formex H represents an element in the configuration.

The rapported variant of each cantle is obtained based on the perdicant P1 given in brevic notation by

$U_i < U_{i+1}$ for i=1 to n
where n is the plexitude of the cantle

Therefore,

B = rav(P1)¦H = {[1;2;3], [4;5;6], [3;4;5], [2;3;4]}

The rapported sequation of the formex B is then obtained based on the perdicant P2 given also in brevic notation by

(U1<W1) OR (U1=W1 AND U2<W2) OR .....OR (U1=W1 AND U2=W2 AND ..... AND $U_{n-1} = W_{n-1}$ AND $U_n < W_n$)

Where n is the plexitude of the formex B. Thus

G = ras(P2)¦B

will produce the numbering scheme for the configuration. Continuing with the same example then

G = {[1;2;3], [2;3;4], [3;4;5], [4;5;6]}

In finite element method, the nodes are required to appear in certain order. It may therefore be necessary to rearrange the signets of each cantle of G in the form in which they are required for the particular analysis. This will invariably be as they appeared in the formex H. Therefore, for formices of plexitude greater than 2, the position of the cantles is determined by G but the arrangement of the signets in each cantle is determined by H.

The plexitude of the example is 3, which is greater than 2 therefore, the formex representing the interconnection pattern for analysis will be

G1 = {[1;2;3], [2;4;3], [3;4;5], [4;6;5]}

Incidentally, an examination of the effect of the transformation of B to G will appear to have been accomplished as follows: each cantle of B is read as a single number by ignoring the semi-colons separating the signets. The numbers which are so formed are then arranged in their order of ascending magnitudes. The result will be the formex G if the cantles are written again in their usual form.

As an illustration consider the configuration in Fig 3. The interconnection pattern may be formulated as

F11 = F1 # F2 # F3

where

F1 = lib(i=0,4)¦lib(j=0,3)¦tranid(2i,2j)¦rosid(2,2)¦[2,1; 3,2]

F2 = lam(2,5)¦lib(i=0,3)¦tran(1,2i)¦[2,1; 4,1]

and

F3 = lam(1,6)¦lib(j=0,2)¦tran(2,2j)¦[1,2; 1,4]

The ingot of the nodal points is given by

E11 = E1 # E2

where

E1 = lib(i=0,4)¦tran(1,2i)¦(E3 # E4)

E3 = lib(j=0,3)¦tran(2,2j)¦[1,2]

E4 = lib(j=0,4)¦tran(2,2j)¦[2,1]

and

E2 = ref(1,6)¦E3

H11 = dic(E11)¦F11 gives the node numbering of the structure. Based on this, the rapported variant B11 of H11 is obtained. Then the rapported sequation G11 is also obtained. The result will be that if the least numbered node of an element is 3 and that of another is 4 then the former is given an element number smaller than the latter. In other words the element with 4 as its least node number is assembled after that with 3. The final element number for the structure is as shown in Fig 3. This will produce an optimum frontwidth for this simple structure for the frontal solver.

Suppose the structure was inefficiently numbered by producing an ingot across the wider length of Fig 3 then an efficient element numbering is achieved as follows; the node numbers as inefficiently given are ignored and a formex representing pseudo-numbers for the nodes are produced as described in formex H. The node numbers produced in the formex H are called pseudo-numbers because they are only used to obtain element numbering scheme and are not actually used in the analysis. The rapported sequation of H is then obtained in the formex G. The formex H is then discarded and the structure analysed using the node numbers as inefficiently given but the orderate of the cantles of the formex G represents the element numbering scheme for the frontal solver. Therefore the element numbers will also be as in Fig 3 but the node numbers will be as inefficiently given.

Consider also the configuration shown in Fig 4. It could

be a triangular finite element mesh, with three nodes. The formex representing the configuration may be formulated as follows

```
   F  = lux(C1)¦lamid(9,5)¦lib(i=0,3)¦lib(j=0,1)¦tranid
                                           (2i,2j)¦F1
```

where

$$F1 = \{[1,1;\ 3,1;\ 1,3],\ [3,1;\ 3,3;\ 1,3]\}$$

$$C1 = lamid(9,5)\,\text{¦}\,\{[1,1],\ [9,1]\}$$

The ingot of nodal points is also given by

```
   E  = lux(C1)¦E1
```

where

```
   E1  = lib(i=0,8)¦lib(j=0,4)¦tranid(2i,2j)¦[1,1]
```

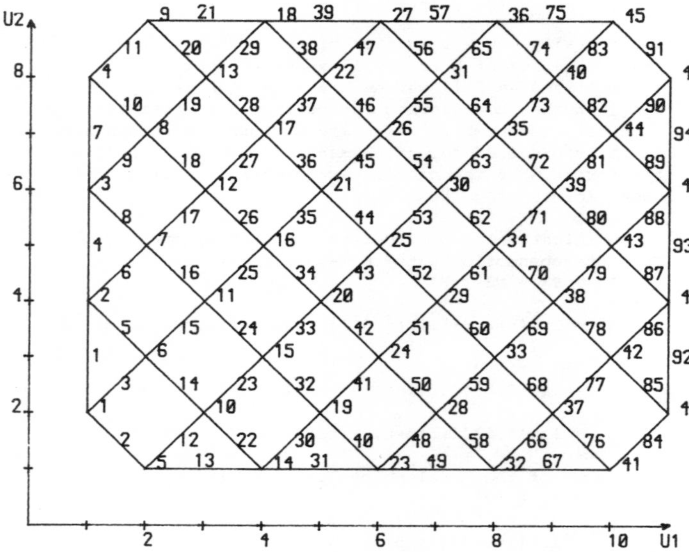

Fig 3

The element numbering as shown in Fig 4 will be used in the frontal routine for the analysis.

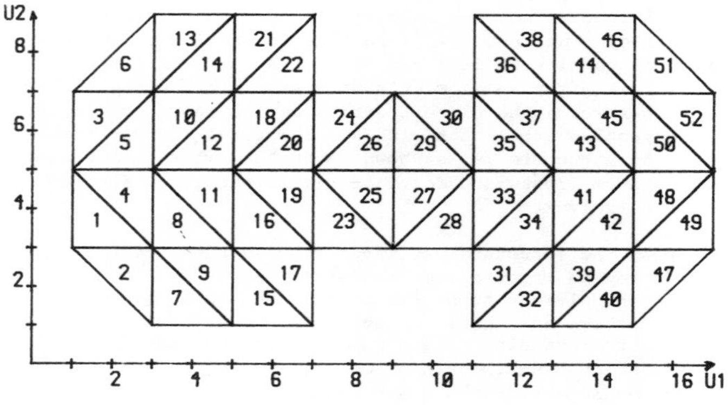

Fig 4

The steps for the first method may therefore be summarised as follows.

1) Obtain a formex F representing the interconnection pattern of the configuration
2) Obtain an ingot E representing the nodal points
3) Obtain H = dic(E)¦F
4) Obtain B = rav(P1)¦H
5) Obtain G = ras(P2)¦B
6) Rearrange signets of the cantles of G if necessary
7) The orderate of each cantle of G is the number of the element which it represents.

THE SECOND METHOD

The first method presented earlier is dependant on the ingot obtained for the node numbering for its efficiency. It therefore assumes the possibility of obtaining the ingot of the nodal points in such a way as to produce a low frontwidth based on it. For some configurations it might be difficult to obtain a good ingot and such configurations include complex structures and finite element meshes with midside nodes. Some renumbering techniques such as Ref 8 overcome this problem by using a bandwidth minimisation technique first. Based on the node numbering produced, the elements are then resequenced for the frontal solver.

However, the second method presented here does not involve any bandwidth minimisation. It uses the frontal solver's technique of wave-like flow through the configuration to find a minimum frontwidth growth path through the configuration.

The procedure is as follows:

Let a formex, F represent the interconnection pattern of a configuration and let E be an ingot of the nodal points. In this case, it is not important that the ingot of the nodal points be obtained across the narrowest width of the configuration. The user should employ the simplest means to obtain the ingot.

Furthermore let the formex H represent the dictum of F with respect to E. That is, H = dic(E)¦F. For example, let the formex H, which represents an interconnection pattern after the dictum has been obtained be given by

$$H = \{[1;4;5;2],\ [2;5;6;3],\ [4;7;8;5],\ [7;9;10;8]\}$$

When the dictum of F with respect to E is obtained, the formex H representing the interconnection pattern has cantles of the first grade, which represent the node numbers of the configuration in an ordered form. Each cantle of course represents an element in the configuration and the plexitude of each cantle represents the number of nodes on that corresponding element.

Each signet of H, say S, has attached to it a 'degree' given by the total number of distinct signets in the cantles which contain it. For example, the first signet of the first cantle is 1 and has a degree of 3, because there are 3 distinct signets (4,5,2) in the cantle which contain it. The second column of Table 3 gives the degrees of all the signets in H.

Table 3

Signet	Degree	Connected Signet	Associated Degree
1	3	4,5,2	17
2	5	1,4,5,6,3	21
3	3	2,5,6	15
4	5	1,5,2,7,8	25
5	7	1,4,2,6,3,7,8	29
6	3	2,5,3	15
7	5	4,8,5,9,10	23
8	5	4,7,5,9,10	23
9	3	7,10,8	13
10	3	7,9,8	13

Signets appearing in the same cantle are said to be 'connected'. This does not necessarily mean physical connection, but the nodes which they represent form the nodal points of one element which the cantle represent. The third column of Table 3 gives the 'connected' signets for each of the signets of the formex H.

To start the rearrangement of the cantles, the cantle with the lowest plexitude is considered first. If the formex representing the interconnection pattern is homogeneous, the signet of the lowest degree is then considered to belong to the first cantle. If there are more than one signet with the lowest degree and they belong to different cantles, then their 'associated degrees' are calculated.

The associated degree of a signet is the sum of the degrees of those signets connected to it. The associated degree of the signet 1 is the sum of the degrees of signets 4,5 and 2 which from column 2 of Table 3 is 5+7+5 = 17. The last column of Table 3 gives the associated degrees of the signets of the formex H. The signet with the lowest associated degree is then considered first signet and the cantle that contains it as the first cantle in the new formex G. If however, there are more than one signet with the same lowest associated degree, then the signet whose uniple is lowest is considered as the first signet. From Table 3, the signets with the lowest degree are 1,3,6,9 and 10. From these, the signets with the lowest associated degree are signets 9 and 10 and they belong to the same cantle. Therefore, the first signet will be any of them. The first signet would have been 9 if they do not belong to the same cantle since the uniple 9 is less than the uniple 10. The associated degree has the effect of probing the potential growth of the frontwidth beyond the present wavefront.

Once the first signet is established, then all cantles containing that signet are assembled consecutively in the new formex G according to their orderate in the formex H. Thus, if cantles of orderate 2 and 5 are the only cantles connected to this first signet then the cantle of orderate 2 takes the orderate 1 and that of orderate 5 takes the orderate 2. The first cantle in G for the example will be [7;9;10;8].

The signets in G which still appear in H and those that were just brought into G constitute what is known as 'current front'. For the example, the signets in the current front are 7,9,10 and 8.

All signets in the current front which do not appear in H and only in G are then removed. In the example when the cantle containing 9 is removed from H, the signets 9 and 10 do not appear in any other cantle and therefore, they will be removed from the current front.

The degree of the remaining signets without considering the signets in the current front and the signets which are in G is called the 'current degree' of the signets. The current degrees of the signets are then calculated for all the signets in the current front and those connected to them. The signet with the lowest current degree which is in the current front is then considered next. If more than one signet has this lowest current degree then, the associated degrees of those with lowest current degrees are calculated as outlined above. In calculating the associated degree, the current degree of the connected signets are also used. At this stage in the example, only 7 and 8 are the signets in the current front and their current degrees are calculated. Both have the same current degree of 2 and the same associated degree of 8.

The resulting signet with the lowest associated degree is then considered next. If there are still more than one signet with the same lowest associated degree and they belong to different cantles then, the signet among these which belongs to a cantle most recently brought into the formex G is considered next. All cantles in which it occurs are brought into G. The signets which do not appear in other cantles other than those in G are then removed from H. For example, the signets 7 (or 8) will now be considered and the cantle containing it will next be brought into G.

The cycle is then repeated until the current front becomes zero and the cantles would all have been brought into the formex G. The orderate of the cantles in this new formex G becomes the number of their corresponding elements and the order in which they will be introduced into the frontal solver routine. At the end of the process, the formex H would have been transformed into G.

G = {[7;9;10;8], [4;7;8;5], [1;4;5;2], [2;5;6;3]}

As an illustration consider the configuration shown in Fig 5. The configuration is taken from Ref 8 for comparison. Its formex formulation can be given by

F = F1 # F2 # F3

where

F1 = [1,1; 29,1] # lam(1,15)¦{[1,1; 1,9], [1,9; 1,13], [1,9; 3,9]}

F2 = lib(i=0,5)¦tran(1,4i)¦[3,9; 7,9]

F3 = lib(i=0,6)¦tran(1,4i)¦{[1,13; 5,13], [5,13; 3,9]},

and the ingot of the nodal points by

E = E1 # E2 # E3

where

E1 = lam(1,15)¦{[1,1], [1,9], [1,13]}

E2 = lib(i=0,5)¦tran(1,4i)¦[5,13]

E3 = lib(i=0,6)¦tran(1,4i)¦[3,9]

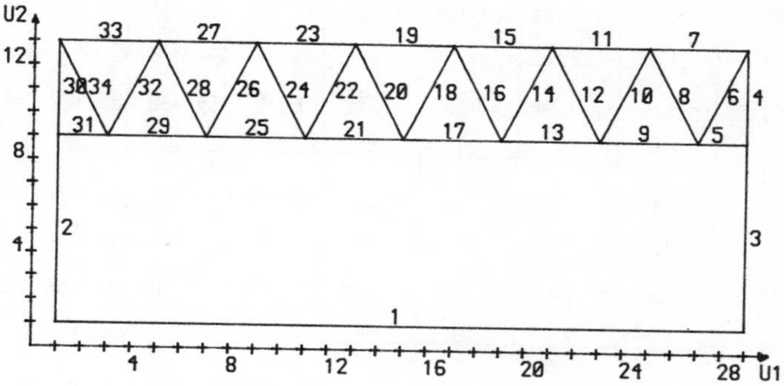

Fig 5

The element numbers as given will be used for the frontal solver routine. Fig 5 has a bandwidth of 6 after bandwith minimisation, but it has a frontwidth of 4 by this technique.

Consider also the configuration shown in Fig 6. It can be formulated as

F = lux(C)¦(F1 # F2)

where

F1 = lam(2,7)¦lib(i=0,2)¦lib(j=0,2)tranid(2i,2j)¦F11

F11 = [1,3; 3,1] # rosid(2,2)¦[1,1; 3,1]

F2 = lamid(11,7)¦lib(i=0,1)¦lib (j=0,2)¦tranid(2i,2j)¦F11

C = {[7,7], [11,1], [11,13]}

and ingot of nodal points E is given by

E = lux(C)¦E1

where

E1 = lib(i=0,7)¦lib(j=0,6)¦tranid(2i,2j)¦[1,1]

The configuration of Fig 6 has a frontwidth of 9 which is the same as the bandwith.

The configuration in Fig 7 is a finite element mesh, having 8 nodes per element. Its formex formulation can be obtained as

F = lux(C)¦F1

where

F1 = lib(i=0,4)¦lib(j=0,5)¦tranid(2i,2j)¦F11

F11 = [1,1; 2,1; 3,1; 3,2; 3,3; 2,3; 1,3; 1,2]

C = {[5,7], [8,9]}

and the ingot of the nodal points as

E = lux(C1)¦(E1 # E2)

where

E1 = lib(i=0,5)¦lib(j=0,12)¦tranid(2i,j)¦[1,1]

E2 = lib(i=0,4)¦lib(j=0,6)¦tranid(2i,2j)¦[2,1]

and

C1 = C11 # C12

C11 = {[5,6], [5,8], [7,8], [8,9]}

C12 = lib(i=0,2)¦tran(1,i)¦[4,7]

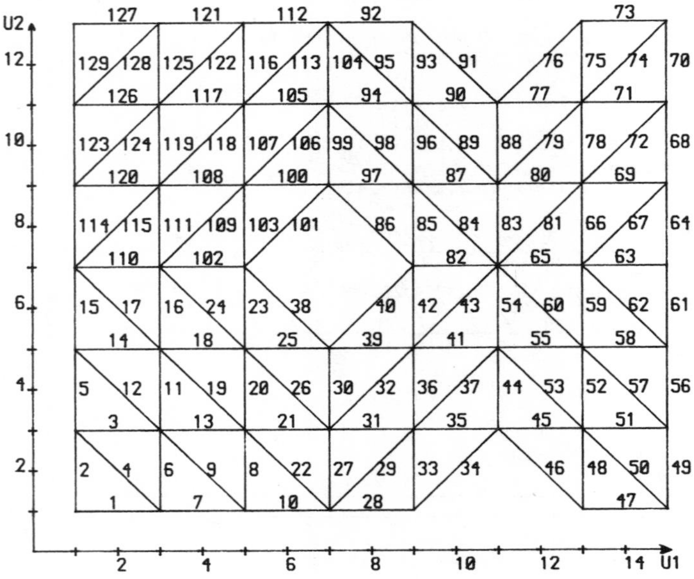

Fig 6

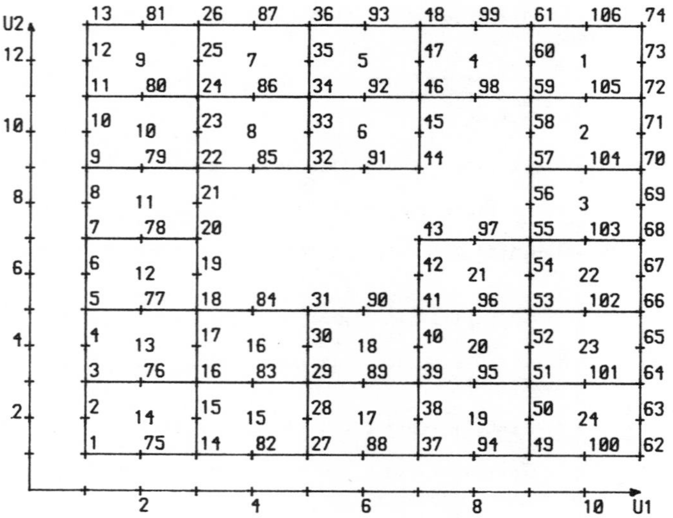

Fig 7

The node and element numbers for the frontal solver are as shown in Fig 7. The maximum frontwidth is 14 which is very much less than the bandwidth of 23. The bandwidth will however be less than 23 if it is reduced by a bandwidth minimisation technique.

The procedure for the second method may therefore be summarised as follows:

1) Repeat steps 1 to 3 of the first method.
2) Calculate the degree of all the distinct signets.
3) Obtain the signets with the lowest degree d_l. If there is only one signet with d_l move to step 6.
4) Calculate the associated degrees d_a of the signets with d_l and obtain signets with lowest associated degree d_{al}. If there is only one signet with d_{al} move to step 6.
5) Choose among signets with d_{al}, the signet with the lowest uniple.
6) Bring all the cantles containing the signet chosen or obtained into G and form the current front.
7) Calculate the current degree of the signets.
8) Obtain signets with the lowest current degree d_{cl} which is in the current frontwidth. If there is only one signet with d_{cl} move to step 11.
9) Obtain signets with lowest associated degree d_{al}. If there is only one signet with d_{al} move to step 11.
10) Choose among signets with d_{al} the signet most recently brought into G.
11) Bring all the cantles containing the obtained signet into G and modify the current front.
12) Repeat steps 7 to 11 until the current front becomes zero.

It has to be noted that the two techniques presented here do not guarantee the lowest frontwidth possible. However, the frontwidth obtained is always lower or equal to the lowest bandwidth and it derives its excellence in being related to the actual data generation using the concepts of formex algebra. Also the actual frontwidth for analysis will be the frontwidth obtained above multiplied by the number of degrees of freedom per node of the configuration.

CONCLUSION

The powerful technique of data generation using the concepts of formex algebra is linked to the frontal solution technique. This was done in two parts. Firstly, it was shown ways of generating the data to minimise the frontwidth. Secondly the concepts were used to minimise the frontwidth of an inefficiently numbered configuration. This was achieved by reordering the cantles to appear in the formex in such a way as to allow their associated elements to be introduced into the frontal solver routine with minimised increase of the frontwidth as the solution wave goes through the structure.

REFERENCES

1. Nooshin H. 'Algebraic Representation and Processing of Structural Configurations'. Computers and Structures, Vol 5, 119-130, 1975.
2. Nooshin H. 'Formex Formulation of Double Layer Grids in Analysis, Design and Construction of Double layer Grids' edited by Z.S. Makowski. Applied Science Publishers, 1981.
3. Nooshin H. 'Formex Configuration Processing'. Applied Science Publishers, 1984.
4. Irons B.M. 'A Frontal Solution Program for Finite Element Analysis'. Int. Journal for Numerical Methods in Engineering, Vol 2, 5-32, 1970.
5. Willoughby R.A. Ed. Proceedings of Symposium on Sparse Matrices and their Applications, IBM Watson Research Centre, Yorktown Heights, New York, 1969.
6. Melosh R.J. and Bamford R.M. 'Efficient Solution of Load Deflection Equations'. Journal of the Struct. Division Proceedings of the American Society of Civil Engineers Paper No 6510, 661-676, 1969.
7. LUSAS. 'Finite Element Stress Analysis System'. Users Manual 1980.
8. Razzaque A. 'Automatic Reduction of Frontwidth for Finite Element Analysis'. Int. Journal of Numerical Methods in Engineering, Vol 15, 1315-1324, 1980.

SPACEFRAME FORMS -- WHERE VISIONS TAKE SHAPE

WENDEL R. WENDEL*

*President, Space Structures International Corp.
Plainview, N.Y., U.S.A. 11803

Three factors have limited the growth and use of spaceframe systems for many architects and engineers. The first factor is their limited knowledge and experience in the use of spaceframes and the many forms that spaceframe systems can take. The second is the relatively limited computer design system available for verifying and experimenting with forms and shapes. The third factor has been the design inflexibility of many first generation spaceframe systems. In contrast, with the recent growth of more sophisticated Second Generation spaceframe systems and more advanced computer-aided design systems, the architect and engineer now have a new "technological palate" to be used for a new generation of design creativity and flexibility not previously possible.

INTRODUCTION

Space Structures International Corp. has been instrumental in assisting architects and engineers to develop an expanded "technological palate" in order that they may design and engineer more sophisticated climatic envelopes using spaceframe systems.

In our attempt to assist architects and engineers, we published a handbook entitled, "Spaceframe Basics," which provides the user with a step-by-step process to utilizing spaceframe systems such as the ORBA*HUB and OCTA*HUB in new and varied forms and shapes. The next step was the development of the SSCAD (Space Structures' Computer-Aided Design) system. This is a system that can be used to design, analyze, cost-out, detail, schedule and revise any complex spaceframe form in a matter of minutes rather than days. The third step was the development of two Second Generation systems, the OCTA*HUB and ORBA*HUB systems, see Ref 1. The OCTA*HUB and ORBA*HUB were designed around a production system rather than module size or component limitations to provide the hardware for the architect and engineer to create new forms with -- tools enabling their visions to take shape.

FORMS/GEOMETRIES/APPLICATIONS

HORIZONTAL - This is presently the most often used application of spaceframe systems and, unfortunately, the only application that architects and engineers have experience with or have seen. With horizontal forms, the geometry selected can be varied according to span, design load, cladding type, architectural requirements, and support conditions. A typical horizontal application is an atrium for an office or hotel where the beauty of the spaceframe is often complimented with the use of glass or acrylic glazing to bring in the natural light.

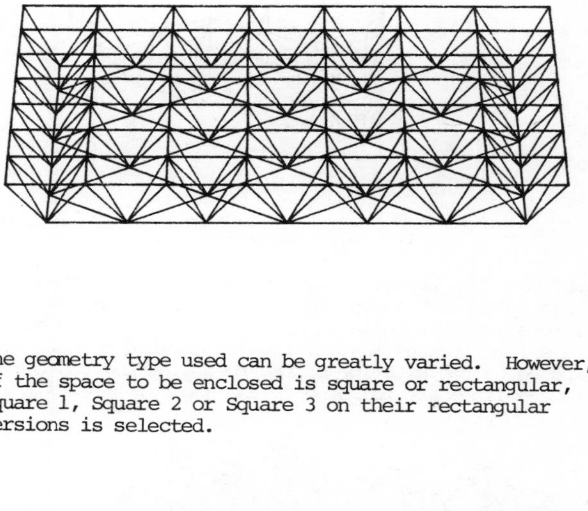

The geometry type used can be greatly varied. However, if the space to be enclosed is square or rectangular, Square 1, Square 2 or Square 3 on their rectangular versions is selected.

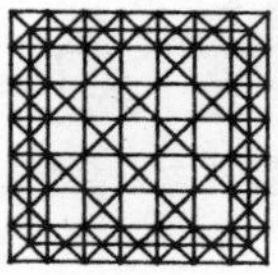

SQUARE 1

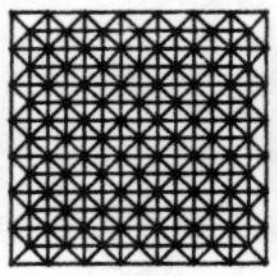

SQUARE 2

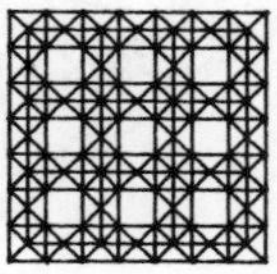

SQUARE 3

VERTICAL - With the growth of glass as one of the basic cladding materials in high-rise construction, there has been an increased demand for vertical spaceframes to support multi-story glazed areas.

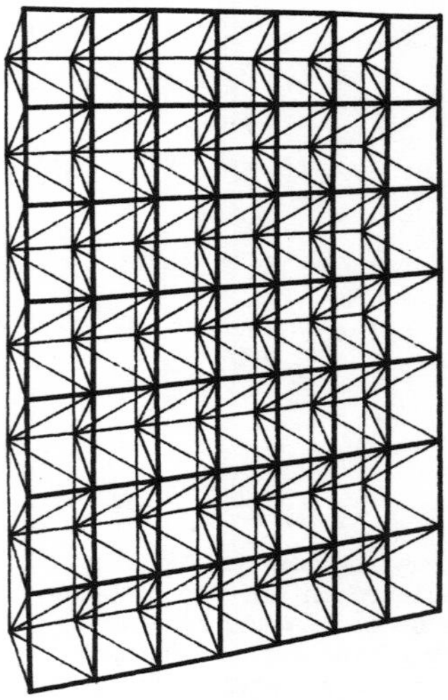

SQUARE 2 GEOMETRY

The module size chosen is often reflective of the glass size and the lines of the building, i.e., often one module per floor, approximately 10 feet/3.04 meters.

SLOPED - The use of sloped spaceframes has developed with the growth of multi-level offices and hotels. Often conceived as an exciting design feature, they can be located at an entryway, dining area or swimming pool area.

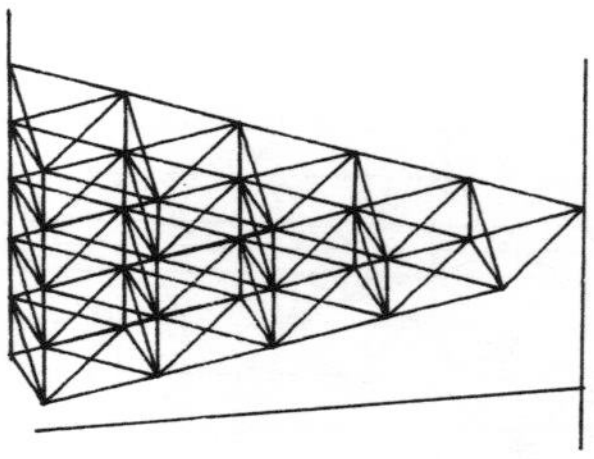

TRI 1 GEOMETRY

TRUSSED - The trussed form and shape is often used for large spans such as sports facilities, convention centers, exhibit halls or hangar applications. The truss form creates an exciting design when part is left exposed on the exterior of the roof cladding and the trussed lines are followed to the ground.

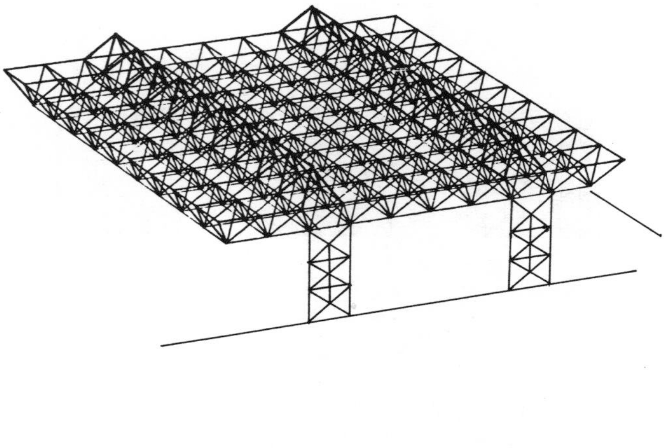

MULTI-LAYER - One of the versatilities of spaceframe design is the option of multi-layering a spaceframe for increased spanning capabilities. This form, like the trussed form, is used in large span applications such as stadiums, arenas, convention centers and hangars.

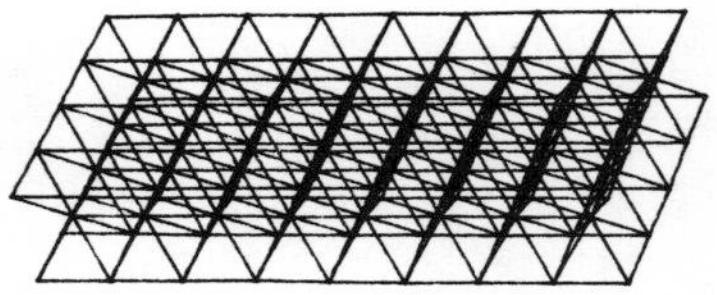

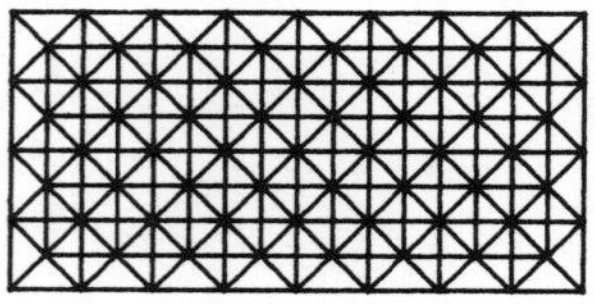

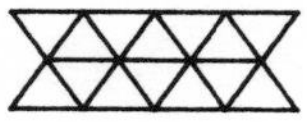

Four multi-layer approaches often used are:

Double or triple layering over the entire spaceframe area.

Trussing of the spaceframe in one direction either outward or inward of the main plane.

Perimeter trussing in both directions, which is an effective way to create a plate type of action to reduce both maximum bending moments and the amount of material required.

Localized triple or quadruple layers around support columns.

MULTI-PLATE - This form is ideally suited for large span applications where support is only possible on two opposite parallel walls. With this application the unit may be added to as additional enclosed space is required.

The multi-plate form has a "dual-depth" efficiency benefit -- the depth of the spaceframe itself, plus the depth of the sloped plates created.

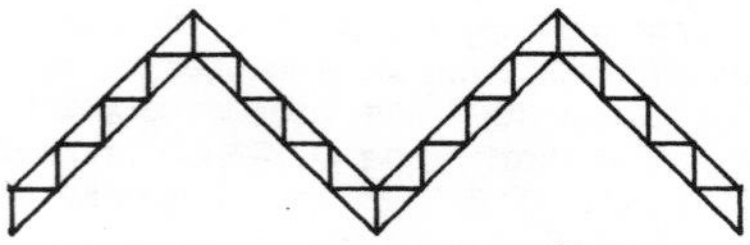

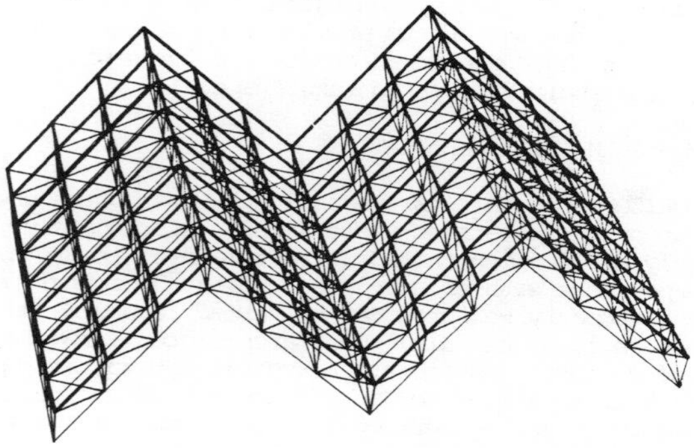

RIDGED - This form is a singular version of the multi-plate. The ridged form is suited for highly loaded areas such as northern areas with large snow loads. This form is often used in large atrium enclosures glazed in glass or acrylic or in sports facilities like ice skating rinks.

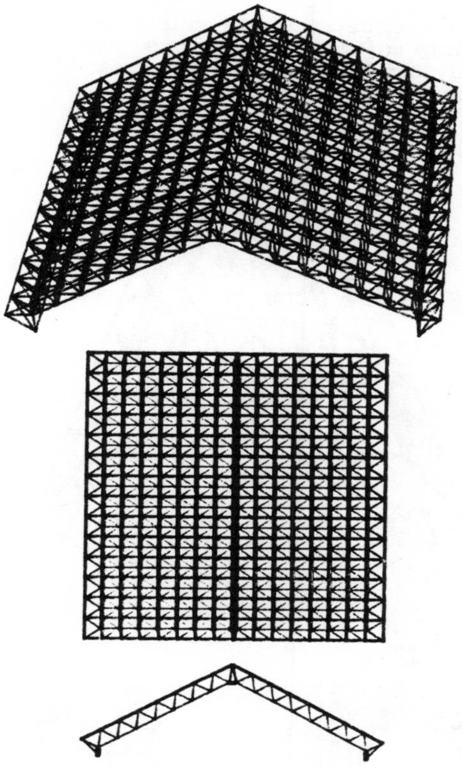

STEPPED - This form is typically used in glazed building applications where a large "cut out" in a building or an area between two buildings needs to be closed in. It is a visually exciting design solution whereby a dramatic transition occurs between the overhead horizontal roof and the vertical window wall.

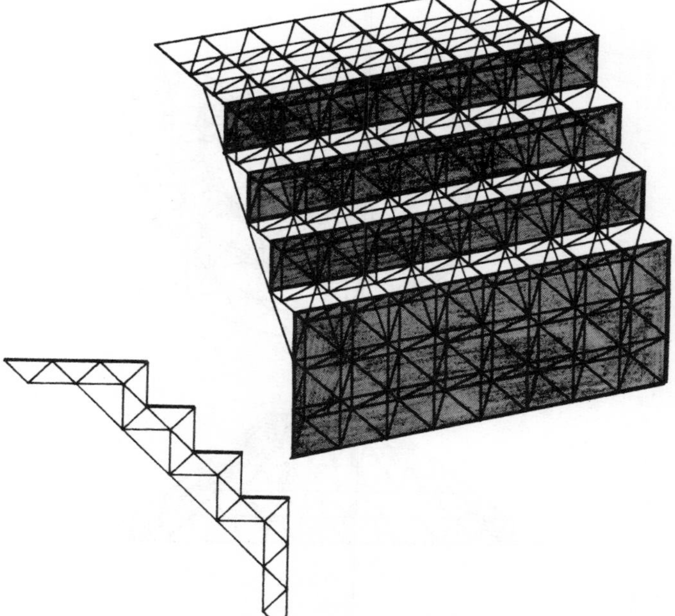

TOWERS - Towers of almost any height are possible with spaceframe forms. Towers can be either decorative such as a sculpture or functional such as a transmission tower. Towers can be of many different geometry types utilizing the basic triangular form.

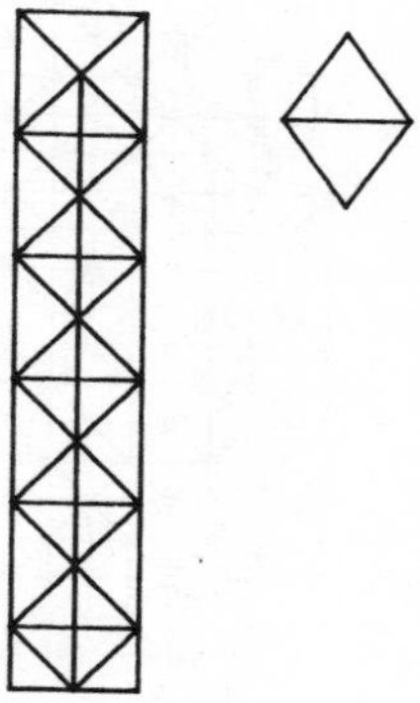

PYRAMID - This form, first used in large scale by Imhotep in 2650 B.C. in Egypt, has a social and historic background. The present application of pyramids include: specialty buildings such as exhibit centers, religious centers and atrium cuts of larger roofed areas.

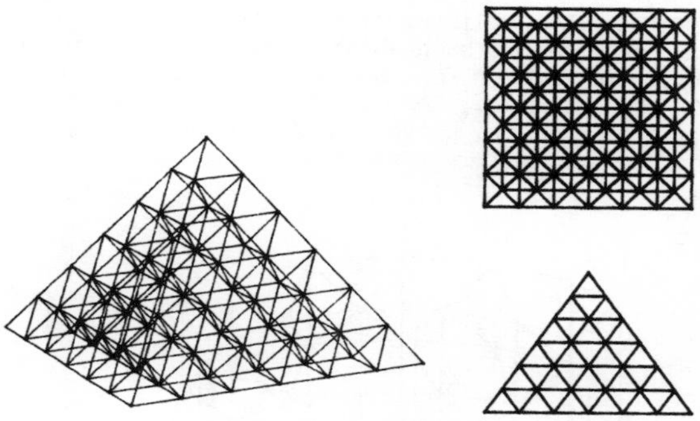

BARREL ARCH - This form is one of the most popular for large span applications due to its highly efficient use of material and, thus, relatively low cost. It has often been used in hangar applications, sports facilities, exhibit halls and convention centers.

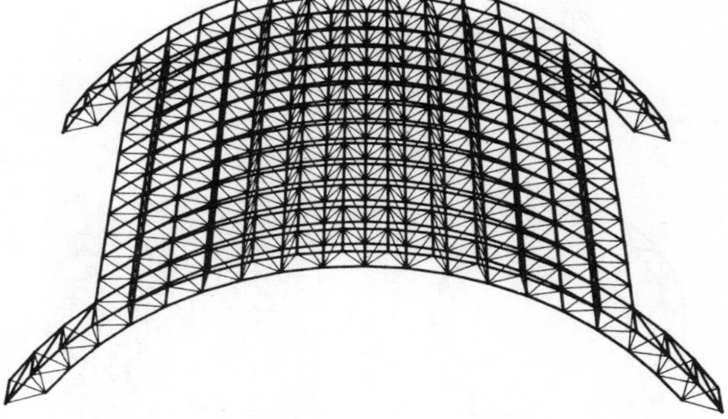

FOLDED PLATE - This form has unlimited variations and possibilities. Shown here is a horizontal form with the four corners folded down to act as columns.

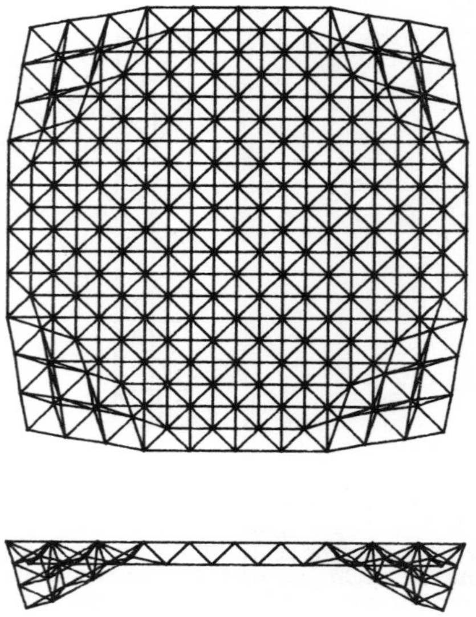

CYLINDRICAL - The cylindrical form has been used in special applications such as drilling derrick enclosures to provide protection for the equipment and personnel in hostile environments. The cylindrical shape may be used to support a dome shape over the center area or as a support for multi-story residential facilities.

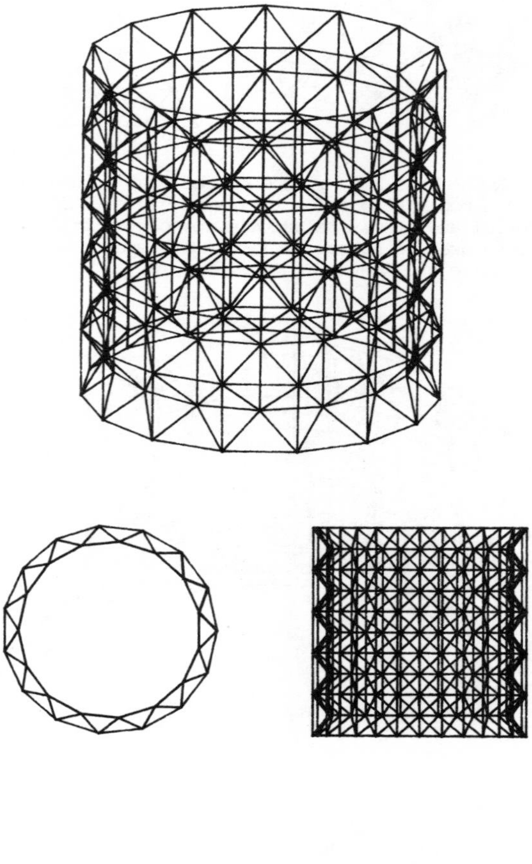

CONICAL - This unusual form's applications are just beginning to grow. The conical form can be used in a variety of ways which include: a partial shape to fit into a building, such as 90°, an entryway, in a whole shape as a theater or industrial storage facility or in a curved shape as a cool tower design.

Even with the complexity of the conical form's many different hubs and strut types, its use potential has been greatly increased with the development of computer-aided design and fabrication systems.

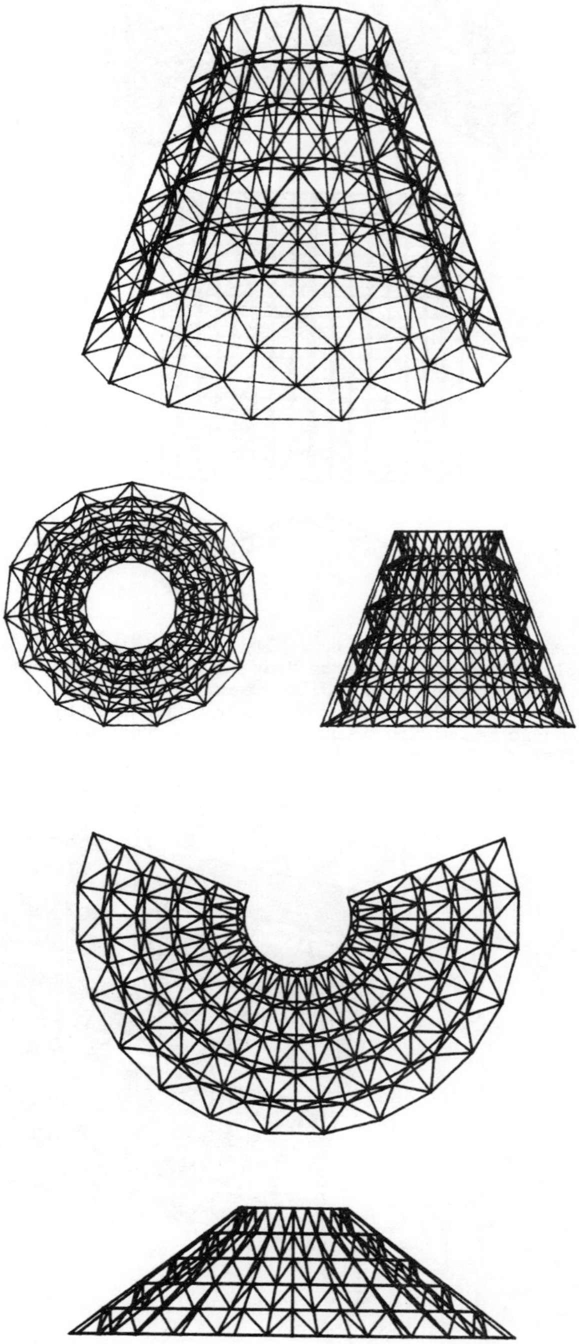

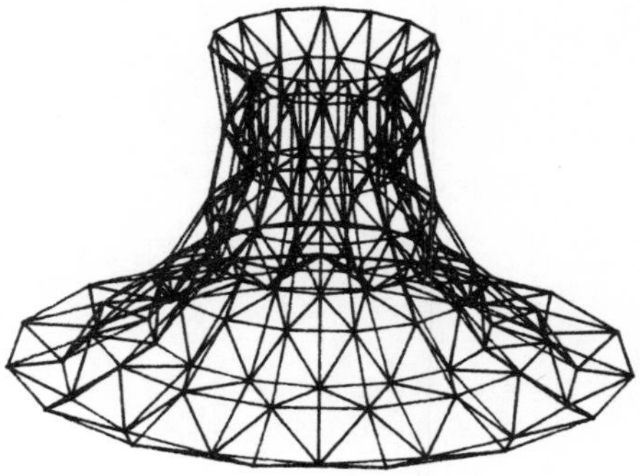

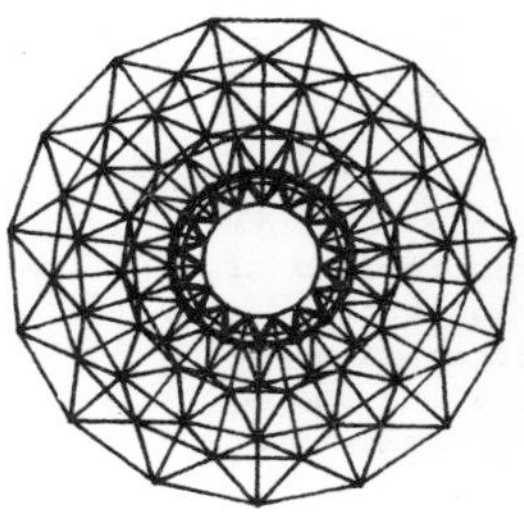

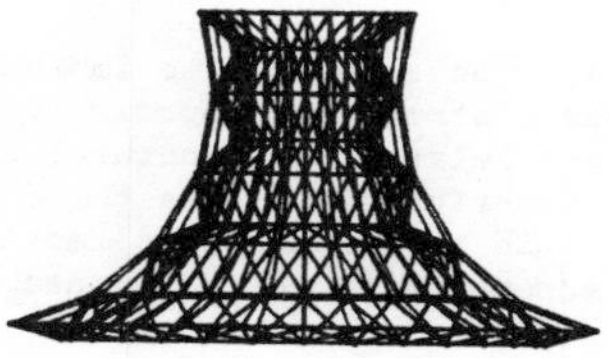

DOME - The dome form is the one used most often. Domes are used for large span applications such as stadiums or other sports facilities. The dome form is available in two basic types, single layer depth or double layer depth. The double layer is recommended for highly loaded and very large span structures.

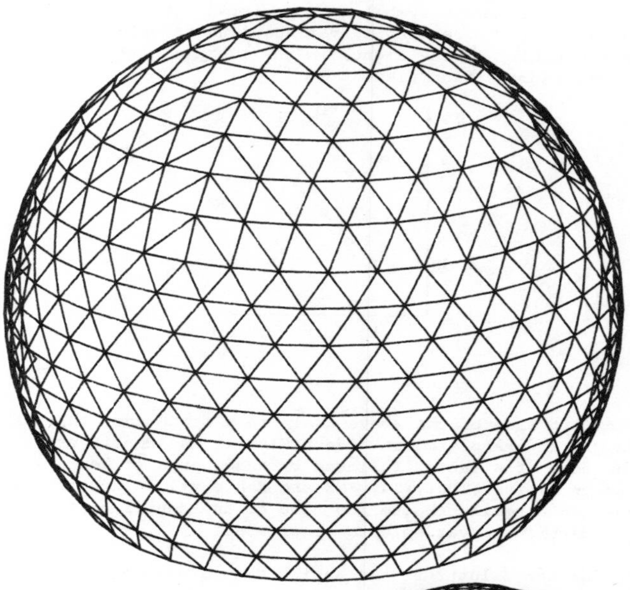

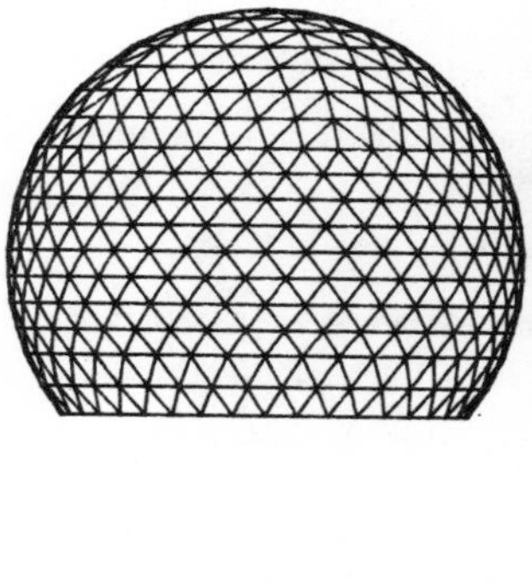

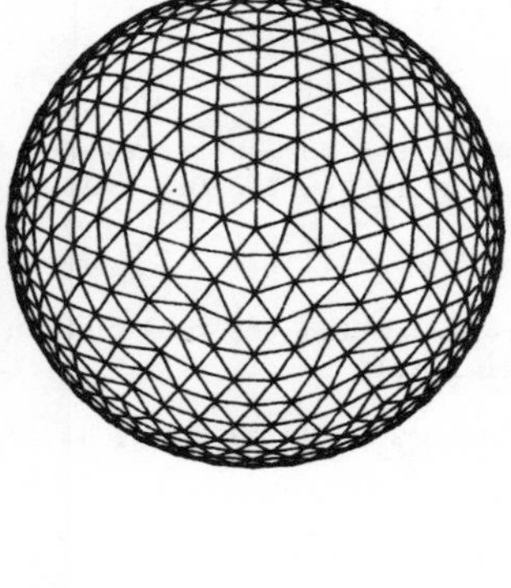

The STARDOME Stadium, the world's largest clear span enclosure, 765 feet/233 meters in diameter, is the proposed stadium for the City of Toronto. This dome features two layers with a 10 foot/ 3 meter deep truss to span the large distance. The unique feature of this form's application is the center section where the cladding is retractable to provide natural light and air.

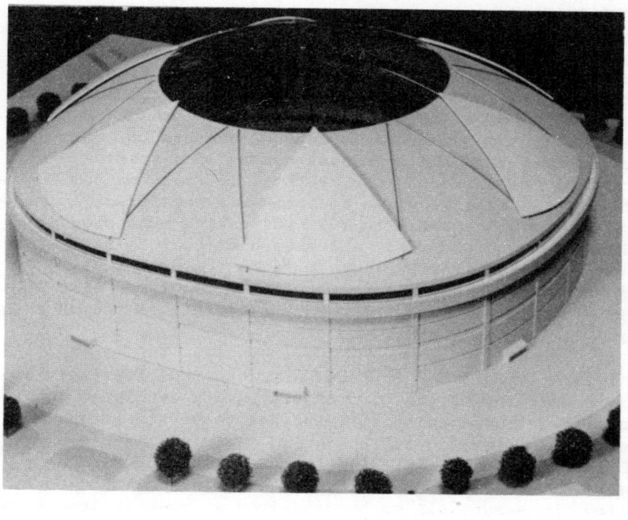

The above-mentioned forms are just a sample of the many variations and combinations that architects and engineers now have in their "technological palate" in order to make their "visions take shape." The new tools, software at the design end and the hardware available in physical systems, are just the beginning of a new Second Generation of spaceframe design and applications.

REFERENCES

1. "Spaceframe Basics, a Handbook for Spaceframe Design and Engineering," 3rd Edition by Wendel R. Wendel. Published by Publication Division, Space Structures International Corp., 155 Dupont Street, Plainview, New York 11803 (1984).

AN INTRODUCTION TO FORMIAN

P DISNEY and O EL-LABBAR

Space Structures Research Centre
University of Surrey

Formian is an interactive programming language which acts as a vehicle to implement the concepts of formex algebra. The language is designed to provide a structured approach to the problems of data generation, in particular to the generation of data related to structural configurations. Being modelled on formex algebra, the language allows powerful statements to be written in a concise yet readily understood manner. It also has simple to use graphics facilities and an editor built in, enabling problems of data generation to be accomplished in one programming environment.

INTRODUCTION

To date the use of formex algebra to solve practical data generation problems has been confined to writing programs utilizing these ideas to a high level language such as FORTRAN. Some examples of this approach can be found in Ref 1. By using a general purpose language of this type it is found that the simple structure of a formex formulation is hidden in a plethora of statements. These statements being required to translate the formex notation into a form acceptable to FORTRAN, making such programs time consuming to write, and more difficult to understand than the formex formulation on which they are based. Formian was devised to overcome these problems, allowing the use of statements which adhere closely to the notation used in formex algebra.

The paper is divided into two sections, the first giving a brief overview of the type of statements available in Formian. The second describing how these statements may be used in practice, illustrating their use with a number of examples. As this is an introduction to Formian, certain features of the language have been omitted to allow the essentials to be described at greater length; a complete description appearing in Ref 2.

It is expected that the reader has a knowledge of formex algebra, a full description of the algebra can be found in Ref 3.

THE ASSIGNEMENT STATEMENT

Consider the planar configuration composed of elements A, B, C and D as shown in Fig 1. Using the basic concepts of formex algebra, element A could be represented by the formex

$$[1,1;\ 2,2].$$

In Formian a construct such as

$$FA = [1,1;\ 2,2]$$

is referred to as an 'assignement statement' where the symbol = is called the 'assignment symbol'. The effect of the assignement statement is that the entity to the right of the assignement symbol is associated with the name on the left, thus creating a 'variable' whose value is that of the entity on the right, in the example given the formex representing element A.

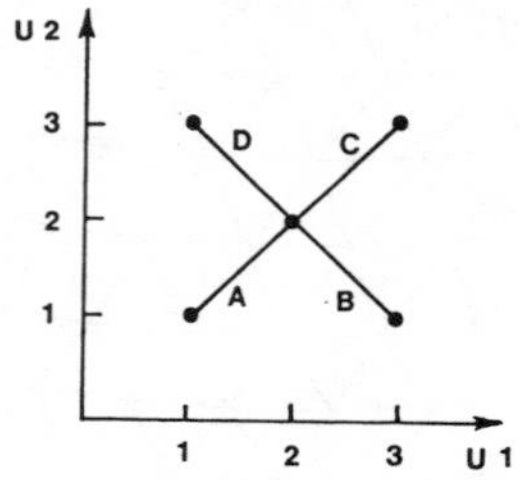

Fig 1

The left side of the above assignment statement, that is FA, is an example of an 'identifier'. An identifier takes the form of a letter followed by a sequence of letters and/or digits. Thus, T4, FMX and X are valid identifiers. One can normally choose the name of an identifier to suit the particular variable to be created.

To further illustrate the assignement statement, one may write

$$F1 = \{[1,1;\ 2,2],\ [3,1;\ 2,2]\}$$

where the formex represents the combination of elements A and B in Fig 1. Or, one may write

$$F2 = \{[1,1;\ 2,2],\ [3,1;\ 2,2],\ [3,3;\ 2,2]\}$$

representing the combination of elements A, B and C. Or, similarly

$$F3 = \{[1,1;\ 2,2],\ [3,1;\ 2,2],\ [3,3;\ 2,2],\ [1,3;\ 2,2]\}$$

where in this case the formex represents the combination of all four elements shown in Fig 1. The entities such as appear to the right of these statements are all examples of 'formex constants'. A variable which is associated with a formex constant is referred to as a 'formex variable'.

The simplest type of a formex is a single integer hence the following assignement statements

```
I   = 11
P77 = -32
KK  = 270
```

may be regarded as examples of creating formex variables. However, they can also be considered as the assigement of 'integer constants', thus creating 'integer variables'.

In Formian, an integer constant is defined as a sequence of one or more digits, optionally preceded by a plus or minus symbol. The value of an integer constant being the numerical value that results from the interpretation of the constant as a decimal integer number.

Real numbers are also used in Formian. These are referred to as floatal numbers and give rise to 'floatal constants' and 'floatal variables'. A floatal constant is one of the forms

I.D or IEJ or I.DEJ

where I and J are integer constants and D is a sequence of one or more digits. The value of a floatal constant is the numerical value that results from the interpretation of the constant as a decimal real number, where if I, D and J are -42, 55 and 2 respectively, then I.DEJ will become -42.55E2 which has a value $-42.55*10^2$.

FORMEX FORMATIONS

An assignement statement may also take the form

FN = [I,J; 2,2],

where I and J are integer variables. Now suppose that I=J=1. The above assignement statement would then be equivalent to

FN = [1,1; 2,2]

with FN representing element A in Fig 1. If, however, I=1, and J=3, FN would represent element D.

A construct of the form

[I,J; 2,2]

is referred to as a formex formation. A 'formex formation' may be defined as a formex in which one or more uniples are given as 'integer expressions'. For example, one may write

FAC = {[J-2,I; 2,2], [J,J; 2,2]},

where if I=1 and J=3 then the assignement statement would be equivalent to

FAC = {[1,1; 2,2], [3,3; 2,2]}.

OPERATIONS AND EXPRESSIONS

Formian provides six basic operations: addition, subtraction, multiplication, division, exponentiation and formex composition. The first five of these relate to operations on scalar quantities, ie, operations between single integer or floatal entities. The final operation is an operation on formices. These operations are represented by the following symbols

Addition	+	Subtraction	-
Multiplication	*	Division	/
Exponentiation	**	Formex Composition	#

An 'expression' is defined as a sequence of constants, variables or functions combined with operators and optionally parentheses which forms a meaningful mathematical expression.

In Formian two types of expression are used. They may be an arithmetic expression, which is a generic term and includes any expression which has as its operands scalar quantities. An arithmetic expression is either an integer expression or a floatal expression. The second type of expression is referred to as a formex expression.

ARITHMETIC EXPRESSIONS

An arithmetic expression is referred to as an integer expression, if it yields an integer value and is said to be a floatal expression if if yields a floatal value. In evaluating arithmetic expressions the following rules apply:

a) Arithmetic operators have the following precedence

Operator	Precedence
** (exponentiation)	Highest (evaluated first)
+ - (unary, ie, sign)	
/ *	
+ - (dyadic, ie, operator)	Lowest (evaluated last)

b) When two operators are of equal precedence, operations are evaluated from left to right.

c) Parenthesis may be used freely to alter the order of operator precedence. In which case expressions enclosed in parenthesis are evaluated independently of any preceding or succeeding operators.

FORMEX EXPRESSIONS

Consider the following assignement statement

F = FAB # FCD

where the symbol # is the formex composition operator 'duplus' and FAB and FCD are formex variables. Suppose that FAB had been assigned the formex representing elements A and B of Fig 1, and FCD the formex representing elements C and D. The effect of the statement is to perform the formex composition of the variables FAB and FCD such that F is assigned the formex

{[1,1; 2,2], [3,1; 2,2], [3,3; 2,2], [1,3; 2,2]}

which represents all four elements shown in Fig 1. The right-hand side of the assignement statement in the example above is a simple form of a construct referred to as a 'formex expression'. Further examples of formex expressions are shown below illustrating the use of the composition operator:

E = FAB # {[3,3; 2,2], [1,3; 2,2]}

F = [1,1; 2,2] # [3,1; 2,2] # FCD

G = [1,1; 2,2] # [3,1; 2,2] # [3,3; 2,2] # [1,3; 2,2]

It will be seen that in the examples above E, F and G are assigned the same value, ie, the formex representing all four elements shown in Fig 1.

FORMEX FUNCTIONS

The appearance and the use of formex functions in Formian conforms to the definition of such functions in formex algebra. To illustrate this, suppose a variable F had been assigned the formex representing P shown in Fig 2.

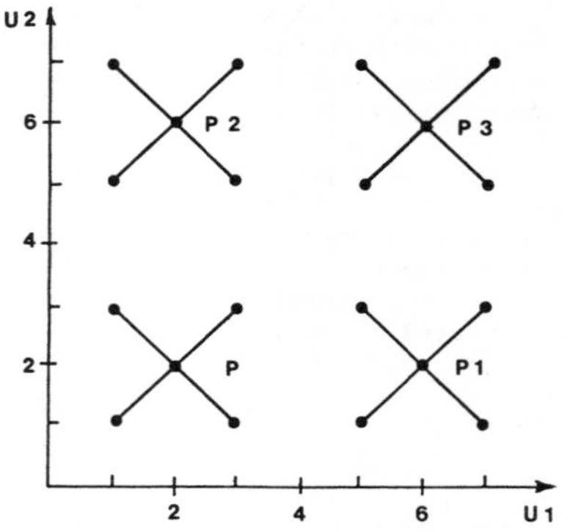

Fig 2

The formex variables representing P1 and P2 could then be created using the following assignement statements.

F1 = TRAN(1,4)¦F

and

F2 = TRAN(2,4)¦F

where the formex variable F1 represents P1 and F2 represents P2. The terms TRAN(1,4) and TRAN(2,4) are translation functions from formex algebra.

To create a formex variable to represent P3, one could use either of the following two assignement statements:

F3 = TRAN(1,4)¦ TRAN(2,4)¦F

or

F3 = TRAN(2,4)¦ TRAN(1,4)¦F.

The evaluation of such a sequence of nested functions proceeds from right to left. Thus, in the second example the translation in direction one preceds the translation along direction two.

If one wished to write the formex to represent the whole configuration shown in Fig 2, one could now simply use

FX = F # F1 # F2 # F3.

However, this may also be achieved by combining the ideas of formex functions and formex composition in a single statement. For example, by using the assignement statement

FX = F # TRAN(1,4)¦F # TRAN(2,4)¦F # TRAN(2,4)¦TRAN(1,4)¦F

Formex algebra is rich in functions and by choosing the most appropriate function often results in a simpler formulation. For example, the whole configuration shown in Fig 2 could also be formulated as follows

FX = ROSID(4,4)¦ROSID(2,2)¦[1,1; 2,2].

LIBRA COMPOSITION

Suppose one wished to generate the formex representing the configuration shown in Fig 3. One could proceed as before using successive translations and compositions to generate the formex. This would be a rather lengthy process and in practice the same effect may be achieved in a simpler fashion by using the concept of libra composition, using the libra operator of formex algebra.

This is one construct which in Formian differs in appearance to the notation used in formex algebra. In formex algebra the general form of a libra operator is written as

$$\Big[_{i=m}^{\,n}$$

whereas in Formian this construct has the appearance

LIB(i = m,n),

where the identifier i, is referred to as a 'libra variable' and where m and n are integer expressions.

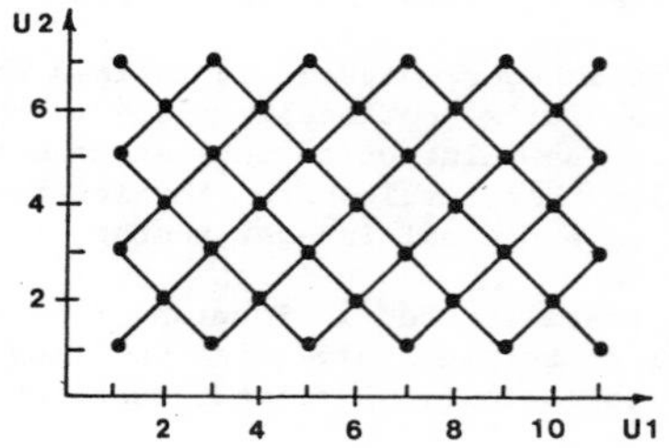

Fig 3

For example, suppose the formex variable F represents a single cross located at the bottom left-hand corner of Fig 3. To generate the bottom row of crosses, one could use the statement

BR = LIB(I=0,4)¦TRAN(1,I*2)¦F.

It will be noticed in this example that one of the canonical variables of the function is given as an integer expression.

To create the whole configuration, one could then use the statement

FC = LIB(J=0,2)¦TRAN(2,J*2)¦BR.

In this case the formex variable BR represents the bottom row of crosses and is translated along direction two. The same result may be achieved in one statement rather than two by using a nested libra composition, for example by using the following assignement statement,

FC = LIB(J=0,2)¦LIB(I=0,4)¦TRANID(I*2,J*2)¦F.

To further illustrate the use of formex functions and libra operators consider Fig 4. The configuration may be generated using the following assignement statements,

A = ROSID(2,2)¦[2,1; 1,2]
B = LIB(J=1,5)¦LIB(I=J,10)¦TRANID(I*2,J*2)¦A
C = B # REF(1,23)¦B
D = C # REF(2,13)¦C

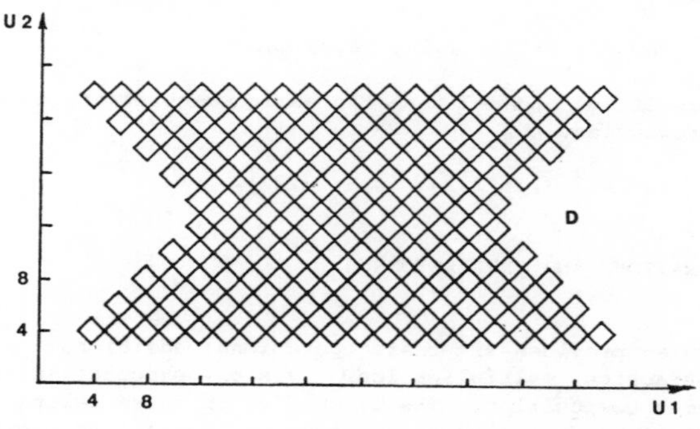

Fig 4

INFORMATION TRANSFER AND CONTROL STATEMENTS

Formian includes two other forms of statements. The first type are information transfer statements which initiate the transfer of information between the interpreter and the peripherals of the host computer. For example, to store or retrieve information from a disk unit or display data graphically on a plotter, etc. The second type are used to control the activities of the interpreter. Both these types of statements take the form of a keyword which is an identifier and represents the action to be performed, generally followed by a list of identifiers.

An information transfer statement is a KEEP, TAKE, PRINT, USE, DRAW or ERASE statement. A control statement is a RUN, EXIT or DELE statement. A brief description of each of these statements follows with the exception of the USE, DRAW and RUN statements which will be described in later sections.

KEEP STATEMENTS

It is often convenient to be able to store items such as formices or a sequence of formian statements permanently such that some days or maybe weeks later one can retrieve this information and use it again, either within Formian or perhaps to act as input data for a program written in another language. This can be accomplished by means of the KEEP Statement. A 'KEEP statement' is of the form

KEEP P1, P2, ..., Pn

where KEEP is a keyword and items P1, P2, ..., Pn represent a list of variables. The effect of the statement is to store separately the value of each of these variables on an area of permanent storage which is referred to as the 'repository'. This usually takes the form of a segment of disk space allocated to the user. The entities stored in the repository are referred to as 'covariables' and are stored under the name of their related variable. The variable itself remains unchanged after the process is completed. To illustrate this, consider the statement

KEEP F1, F2, FX

Where F1, F2 and FX are formex variables. The effect of the statement would be to store each of these items as a covariable in the repository.

TAKE STATEMENTS

The retrieval of items stored in the repository is accomplished by means of a 'TAKE statement'. This is of the form

TAKE P1, P2, ..., Pn

Where TAKE is a keyword and items P1, P2, ..., Pn are identifiers. The effect of the statement is to take each identifier in turn and search the repository to see if a covariable exists with the same name; if this is true the covariable is retrieved and a variable is created with the same name. Returning to the example given above,if one were to subsequently enter the statement,

TAKE F1, F2, FX

the effect of the statement would effectively be to create the formex variables F1, F2 and FX.

PRINT STATEMENTS

A 'PRINT statement' is a construct of the form

PRINT P1, P2, ..., Pn

where PRINT is a keyword and P1, P2, ..., Pn are variables. The effect of the statement is to print the value of these variables on the computer terminal.

ERASE STATEMENTS

An 'ERASE statement' is a construct of the form

ERASE C1, C2, ..., Cn

where ERASE is a keyword and C1, C2, ..., Cn are covariables. The effect of an ERASE statement is to erase each of the covariables C1, C2, ..., Cn together with their values from the repository.

EXIT STATEMENT

The 'EXIT statement' is of the form

EXIT

The effect of an EXIT statement is to terminate the current Formian session and return the user to the host computer's operating system.

DELE STATEMENTS

A 'DELE statement' is a construct of the form

DELE V1, V2, ..., Vn

where DELE is a keyword and V1, V2, ..., Vn are variables. The effect of a DELE statement is to delete each of the variables V1, V2, ..., Vn together with their values from the system.

FORMIAN GRAPHICS

When generating the data for a structure it is often useful to be able to display a geometric representation of the formices on a graphics device such as a plotter or graphics VDU. This provides a convenient way of displaying large quantities of data, allowing one to examine how the formulation is progressing and, incidentally, enabling any errors to be spotted quickly and at an early stage. Formian provides a number of facilities to achieve this.

DRAW STATEMENTS

Consider the following assignement statement

E = RINID(10,10,2,2)¦ROSID(1,1)¦{[0,0; 2,0], [0,0; 1,1]}.

The formex E may be regarded as an algebraic representation of the interconnection pattern of a configuration. Such an algebraic representation can be transformed into a geometric representation of an object by the use of a retronorm. The DRAW statement serves the dual purpose of transforming a formex using a retronorm and also drawing a picture of the object on a graphics device.

For example, subject to certain preliminary specifications described later in the section, the statement

DRAW E

would produce a basibifect N-plot of E, as shown in Fig 5, with basifactors b_1 and b_2, both being equal to 5 units in length.

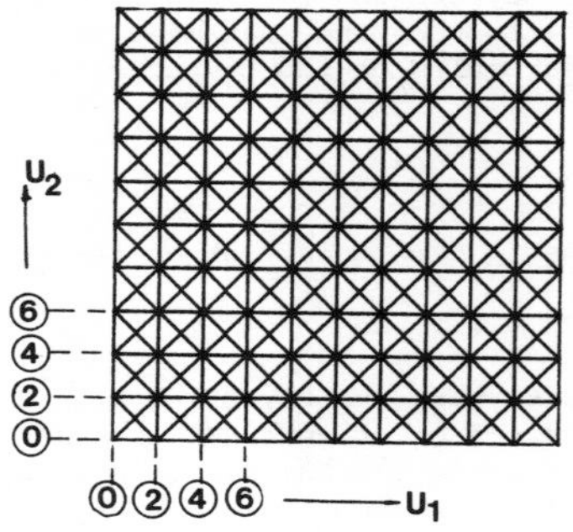

Fig 5

USE STATEMENTS

Formian implements each of the standard retronorms of formex algebra. One can select a particular retronorm by means of a 'USE statement'. For example, to change the basifactors used in the previous example one could enter

```
USE  BB(5,2)
```

where BB indicates the basibifect retronorm and 5 and 2 specify the values of the basifactors b_1 and b_2 respectively. If one were then to enter

```
DRAW E
```

then the plot shown in Fig 6 would be drawn

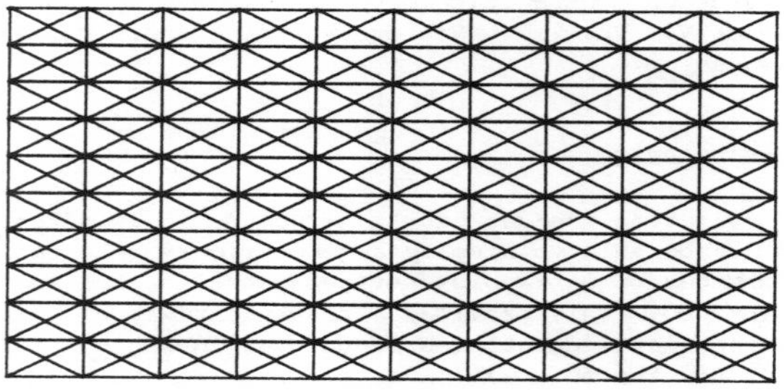

Fig 6

In addition to selecting the various retronorms USE statements are also used to select the other graphics facilities available in Formian. For example, suppose the drawing shown in Fig 6 were produced on a multi-pen plotter using pen 1 and subsequent drawings required the use of pen 2. One could enter

```
USE PEN(2)
```

and after the execution of this statement all drawings would be made using pen 2. In a similar fashion one can select a particular graphics output device with the statement

```
USE DEV(4)
```

where DEV indicates device and 4 might represent a colour graphics VDU.

At the start of a Formian session a default value for each graphics option is provided by the system. So, for example, drawings are output to device 1 which is referred to as the 'current' device. An option remains current until a new current option is selected by means of a USE statement. Thus, plots will be drawn using the same retronorm until another retronorm is selected to be current.

Suppose one wished to select a metribifect retronorm, one could enter

```
USE MB(2,2,1.3,1.3)
```

where MB indicates metribifect and the basifactors b_1 and b_2 are both equal to 2 and the metrifactors m_1 and m_2 are both equal to 1.3. If one subsequently entered

```
DRAW E
```

the drawing shown in Fig 7 would be output on the current graphical output device.

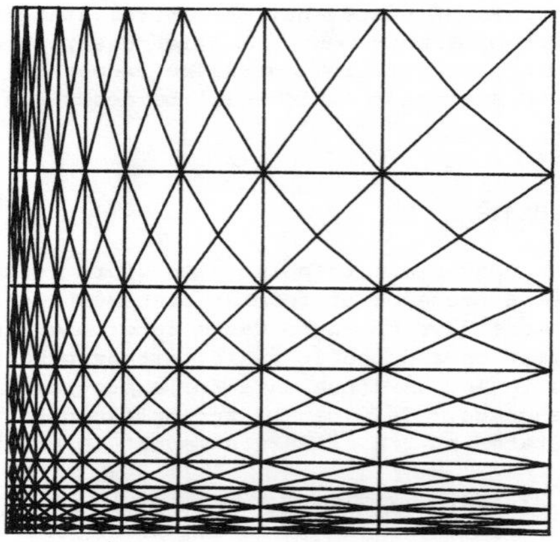

Fig 7

The figures shown so far in this section have been N-plots of the transformed formex. One can, however, view the object from any position; this is achieved by selecting an appropriate line of view relative to the coordinate system specified by the current retronorm. This may be achieved by a statement such as

```
USE  VL(10,10,20,0,0,0)
```

where VL indicates the viewline option and 10,10,20 represent a point from which the object is to be viewed referred to as the view point;and where 0,0,0 represents a point along the line of sight referred to as the view centre which is used to establish the direction of view. In addition, the plots shown so far have been axonometric images, one can also select a perspective view of the object with the statement

```
USE  PV
```

where PV indicates a perspective view. Thus, if one were to enter the following statements

```
F = ROSID(1,1)¦{[0,0,0; 2,0,0], [0,0,0; 1,1,0]}
SN = RINID(10,10,2,2)¦F
USE  PV, VL(5,-2, 2,5,0,0), BT(10,10,1)
DRAW SN
```

Then the drawing shown in Fig 8 would appear on the current graphics output device, BT indicates the basitrifect retronorm.

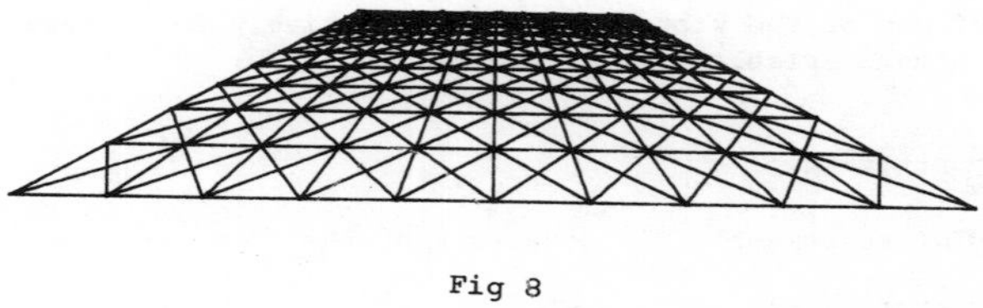

Fig 8

THE FORMIAN INTERPRETER

The program that processes Formian statements is referred to as the Formian interpreter. The next Sections illustrate how one may use the interpreter to create data for various structural configurations. Let us suppose that a user is seated at a computer terminal and has requested to use the Formian interpreter. The interpreter indicates that it is ready by printing at the terminal.

```
FORMIAN Mk 2

*
```

Showing the version of Formian to be used and displaying the symbol * which is referred to as the 'prompt'. The prompt appears at the start of a Formian session and subsequently each time the interpreter has completed a task and is ready to receive information from the user. Suppose one wished to generate the formex representation of a square on square double layer grid 10 bays long by 8 bays wide shown in Fig 10; this could be achieved by entering the following:

```
*BL = RINID(10,8,2,2)¦ROSID(2,2)¦[1,1,1; 3,1,1];
 TL = RINID(9,7,2,2)¦ROSID(3,3)¦[2,2,2; 4,2,2];
 BR = RINID(10,8,2,2)¦ROSID(2,2)¦[1,1,1; 2,2,2];
 F  = BL # TL # BR     Ⓐ
```

Where BL will become the formex variable representing the bottom layer, TL the top layer and BR the bracing members. It will be noticed that each statement is separated by the symbol ; which is referred to as the 'statement separator', and that the text is terminated by Ⓐ. The symbol Ⓐ is referred to as the 'accept directive' and is usually the escape character. Statements may be entered singly or as in this case in groups and until the interpreter receives an accept directive its sole task is to store the text as it is entered by the user. The interpreter at this stage is said to be in 'input mode'. On receipt of an accept directive the interpreter proceeds to process all the text entered since the appearance of the last prompt, the interpreter is then said to be in 'execution mode'.

In the example above, if all the statements have been properly constructed then on receipt of the accept directive the interpreter creates the formex variables BL, TL, BR and F. It then returns to input mode issuing a prompt to invite further information from the user. If, however, the interpreter detects an error in any of the statements, an error message is output to the user indicating the type of error and the statement at which it occurs and then returns to input mode.

Having created the formex representation of the grid which is now encapsulated in the formex variables BL, TL, BR and F one could now produce a drawing of the configuration by entering

```
* DRAW F     Ⓐ
```

which would result in a plot of the grid appearing on the current graphics device as shown below.

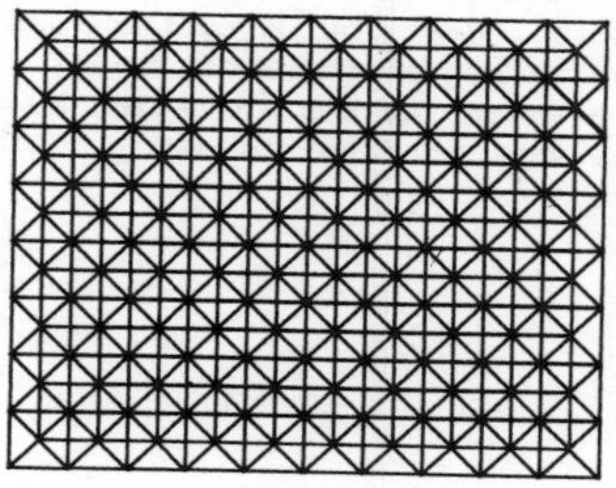

Fig 9

If, however, one wished to draw a perspective view of the configuration one would first need to enter

```
* USE PV,  VL(-100,-100,50,1,1,1)   Ⓐ
```

to change the default graphics option from an axonometric to a perspective view and define an appropriate viewline. Then by entering

```
* DRAW  F     Ⓐ
```

a perspective view of the configuration would be drawn as shown below:

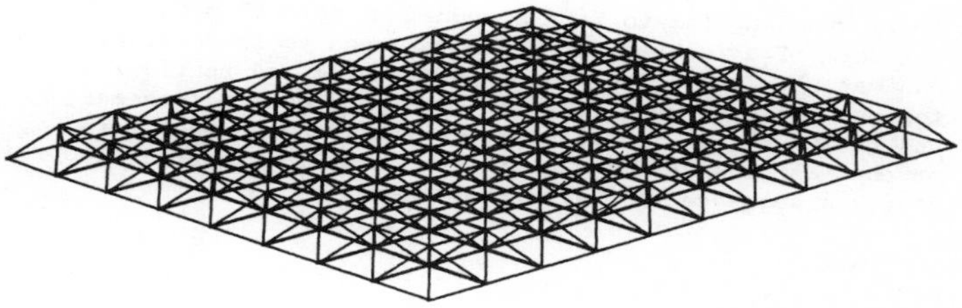

Fig 10

The formices could then be stored in the repository for later use in by entering

```
* KEEP BL, TL, BR, F     Ⓐ
```

One could then create a node numbering scheme and hence a member list for the configuration using the following assignement statments,

```
E1  = LIB(J=1,9)¦PROJ(2,2*J)¦[1,1,1];
E2  = LIB(J=1,8)¦PROJ(2,2*J)¦[2,2,2];
E3  = E1 # E2;
E4  = LIB(I=0,9)¦TRAN(1,2*I)¦E3;
G   = E4 # TRAN(1,20)¦E1;
NMS = DIC(G)¦F     Ⓐ.
```

Where NMS is a formex variable which represents the member list of the configuration and G provides a convenient means of obtaining the position of each node after the use of suitable geometric transformation. Thus, if one stores NMS and G using

```
KEEP NMS, G     Ⓐ
```

the formex NMS can then be used directly to provide the member list of the configuration, and after transformation G may be used to provide a list of node positions for an analysis program. In a similar fashion, one can use the dictum function to produce a description of the loading conditions for the configuration. A formex being generated for each group of nodes which are subject to the same loads.

For example, suppose the configuration was to be loaded with a unit load applied to each of the top layer nodes. A formex representing the node numbers of the top layer could be created as follows

```
L1 = RINID(10,8,2,2)¦[2,2,2];
P1 = DIC(G)¦L1;
KEEP P1     Ⓐ
```

The formex P1 representing the list of nodes could then be used in an analysis program relating the set of nodes either to a unit load as mentioned above or perhaps to a function to provide a more complex load pattern.

In a similar fashion, one could create a formex to represent the set of support nodes. For example, suppose the configuration were to be supported at each of the bottom layer edge nodes. A formex representing these node numbers could be created as follows

```
L2 = RIN(1,11,2)¦[1,1,1];
L3 = L2 # TRAN(2,16)¦L2;
L4 = RIN(2,7,2)¦[1,3,1];
L5 = L3# L4 # TRAN(1,20)¦L4;
P2 = DIC(G)¦L5      Ⓐ
```

where P2 is a formex variable representing the support nodes of the configuration.

STRING CONSTANTS

It is often useful to retain sequences of statements in the form of a 'scheme' as it is referred to in Formian. One may then store the scheme permanently and retrieve it for use whenever required. In order to describe schemes and their uses it is necessary to introduce another form of constant the 'string constant'. A string constant is simply a sequence of characters, therefore

```
ABCD
```
or
```
A = 3;  DRAW B;
```

are both examples of string constant. It will be noted that in the second example the string also represents two valid Formian statements. The assignement of a string constant is unlike previous forms of assignement statements and it is best illustrated by an example. Consider the following construct

```
ABCDEFG == EX1
```

This is an example of string assignement where ABCDEFG is a string constant, EX1 is an identifier which becomes a string variable and the string assignement symbol is ==. The effect of the statement being to assign the string to EX1, such that if one were to enter

```
* PRINT EX1          Ⓐ
```

then the string ABCDEFG would be output to the users terminal.

Strings may be used to provide captions for drawings and data, etc. But, in Formian they take an additional role of providing a convenient means of storing sequences of characters which represent statements. Suppose one were to enter the following

```
BL = RINID(10,8,2,2)¦ROSID(2,2)¦[1,1,1; 3,1,1];
TL = RINID(9,7,2,2)¦ROSID(3,3)¦[2,2,2; 4,2,2];
BR = RINID(10,8,2,2)¦ROSID(2,2)¦[1,1,1; 2,2,2];
F  = BL # TL # BR == PIGS          Ⓐ
```

The string variable PIGS would be created representing the statements used to generate the double layer grid shown in Fig 10.

RUN STATEMENT

To execute a sequence of statements which are held in the form of a string one uses the 'RUN statement'. This is of the form

```
RUN P1
```

Where RUN is a keyword and P1 a string variable. Suppose one were to enter

```
RUN PIGS     Ⓐ
```

Where PIGS is the string variable created in the previous example. The effect would be to interpret the string assigned to the variable PIGS as the text representing a sequence of valid Formian statements and proceed to execute them. Thus creating the formex variables representing the double layer grid shown previously in Fig 10. PIGS could then be stored in the repository and subsequently retrieved at any time and used in a RUN statement to achieve the same result.

The previous example illustrates the use of a simple scheme. In general a scheme consists of a sequence of Formian statements followed by a string assignement symbol and a scheme designator. Where a scheme designator is a string identifier optionally followed by a parameter list. Consider the following example

```
F1 = ROSID(2,2)¦{[1,1,1; 3,1,1], [1,1,1; 2,2,2]};
F2 = LIB(J=1,M)¦LIB(I=J,N)¦TRANID(I*2,J*2)¦F1;
DRAW F2 == GRID(M,N)      Ⓐ
```

where GRID(M,N) is a scheme designator and M and N are referred to as formal parameters. It will be noted that M and N are used in the second assignement statement as the upper limits for the libra variables I and J. If one were to enter

```
RUN  GRID(8,8)      Ⓐ
```

then the scheme would be executed with M=N=8 resulting in the following drawing appearing on the current graphical output device.

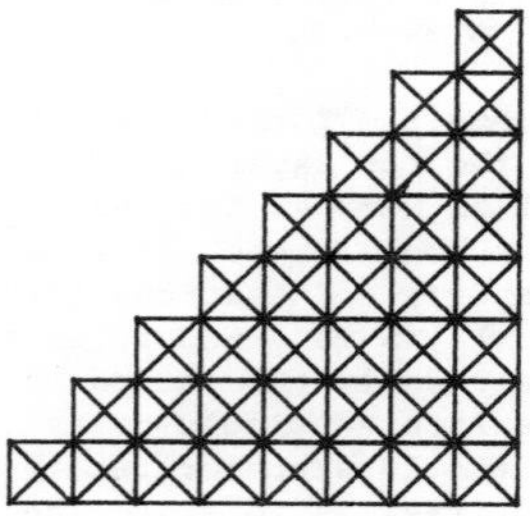

Fig 11

If, however, one were to enter

```
RUN  GRID(6,10)     Ⓐ
```

then M would receive the value 6 and N the value 10 resulting in the following plot appearing,

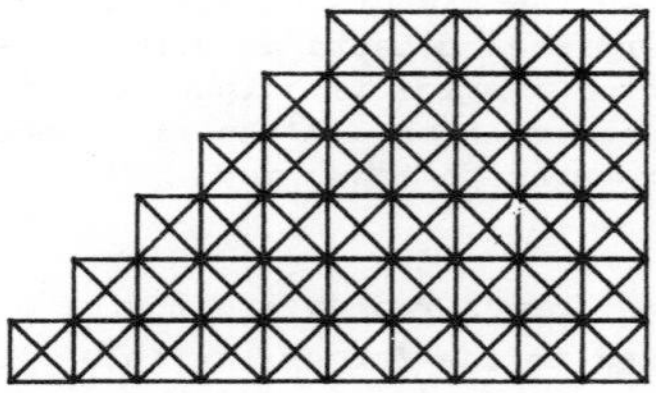

Fig 12

The examples shown so far have been of necessity simple to illustrate the facilities and use of Formian. Figs 13 and 14 have been included to show an example of the type of configuration which has been generated using this approach. A full description of the formex formulation for this problem may be found in Ref 1. Figure 13 shows a perspective view of the configuration and Fig 14 an N-plot of the same configuration.

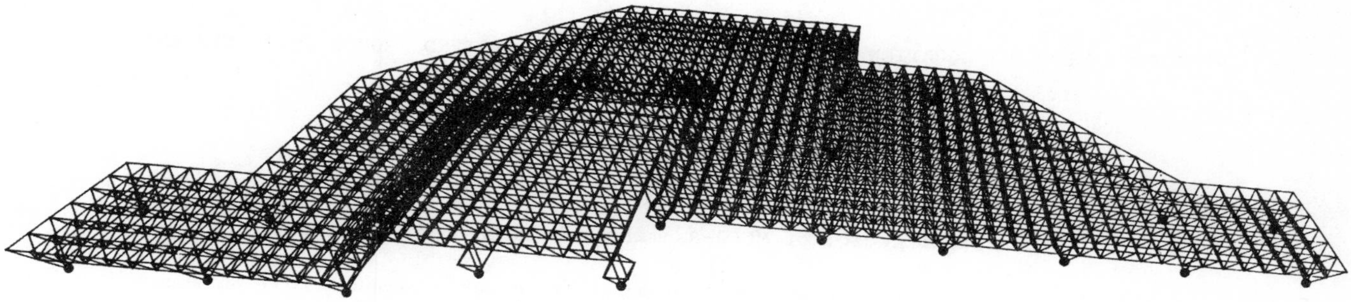

Fig 13

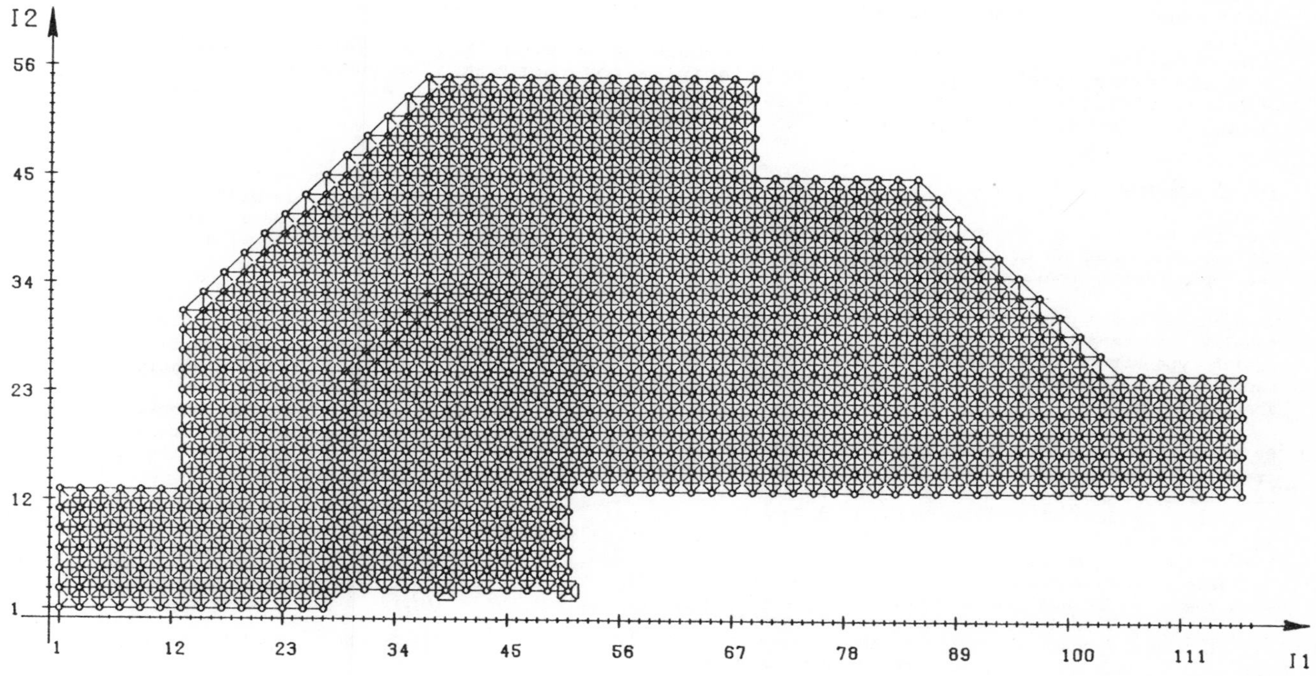

Fig 14

REFERENCES

1. M HARISTCHIAN, Formex and Plenix Structural Analysis, PhD Thesis, 1980, University of Surrey.

2. Dr H NOOSHIN, Formex Configuration Processing, Elsevier Applied Science Publishers, 1984

3. FORMIAN, Space Structures Research Centre, 1984.

STRUCTURE AND CONFIGURATION OF PLANAR SPACE-FRAMES - DERIVED FROM A CUBIC SYSTEM

Helmut EBERLEIN*, Dr. Ing. and Peter RUDOLPH**, Dr. Ing. habil.

* Technical Director of MERO-Raumstruktur GmbH & Co., Würzburg

** Head of Planning Department of MERO-Raumstruktur GmbH & Co., Würzburg
Lecturer, Department of Architecture, Technical University of Berlin

Space-frames have a wider range of application when used in combination with the same or different Space-frame types. This results in advantages of form, function and structure.

The derivation of the corresponding Space-packing modules from the various cubic lattices and the enunciation of the governing principles together with the techniques of repesentation constitute the basis for the practical applications which will be presented in the final section of this report.

1. INTRODUCTION

Over the last two decades space-frames have established themselves in structural engineering and developed into a separate field.

When an engineer speaks of a "space-frame" he is usually referring to a planar grid, i.e. externally a slab type structure but internally a double- or sometimes triple-layer two- or three-directional spatial truss.

By means of a terraced configuration of the space-frame a further dimension, i.e. the height is successfully exploited. This gives the opportunity of a high standard of creative structural design and is an economic solution where long spans are present. As well as satisfying the functional criteria such variations in shape are consistent with the specific analysis and design requirements. For example a folded space-frame roof enhances the architectural appeal of a building and results in improved utility and specific structural advantages.

Such structures are ideally derived from the <u>basic cubic grid</u> on account of its symmetrical properties. The objective of this report is to highlight the <u>rules and principles which govern the derivation</u> of the afore mentioned space-structures.
To this end, different practice - oriented designs will be considered. It is intended that the difficulties experienced by the uninitiated reader due to the confusing and nearly infinite variety of of structural permutations should be overcome by a few fundamental principles which will serve as a starting point for constructive design.

2. THE GEOMETRICAL DERIVATION OF SPACE-FRAMES

2.1 The Method of Derivation

A space-frame is made up of nodes and members or equivalently of fundamental "building modules" (e.g. semi-octahedra and tetrahedra) whose edges are defined by the members. They originate by means of cross sections through two non-coincident grid levels of any interlocking polyhedra.

We confine our attention to those cross sections with respect to grid levels made up of continuous identical polygons since these correspond to the structures which have practical application.

2.2 Three-Dimensional Packing

Here is denoted the continuous spatial sequence of the periodically recurring building blocks (interlocking space-packing modules).
The simplest forms of three-dimensional packings are derived from parallelohedra where complete interlocking is achieved by translating the space-packing modules. <u>No rotation</u> takes place.

Examples are:

 The hexahedron or cube,

 the hexagonal prism,

 the rhombic dodecahedron,

 the orthorhombic hexagonal dodecahedron,

 the trumcated octahedron.

2.3 The Cubic Grid, Properties of Symmetry of Cube

Of the depicted packing forms, the cubic form is the simplest. On account of the greatest number of symmetrical properties it is the most suitable form for practical building purpose.

2.3.1 A Code for Defining Surfaces in Space

To describe the precise orientation of grid areas or of the plane defined by a pair of particular chords of a space frame or of the faces of polyhedra the "Miller Indices Method" is adopted as used in crystallography.

In the orthogonal coordinate system X Y Z, the reference is the reciprocal value of the axial segments of the surface being considered.

2.3.2 THE POSSIBLE POSITIONS OF A CUBE IN SPACE

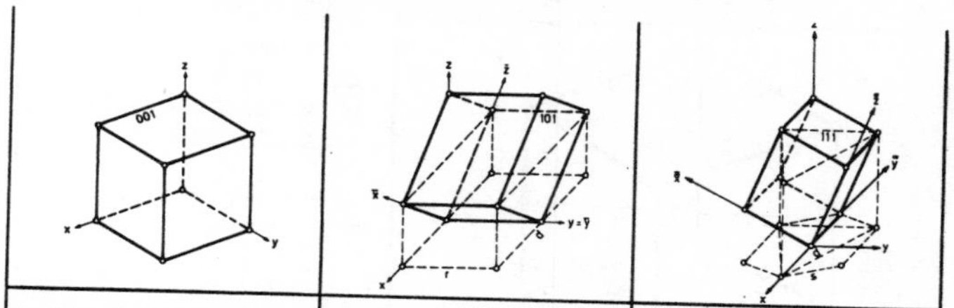

ON ITS LATERAL FACE	ON ITS EDGE	ON ITS CORNER
THE VERTICAL PROJECTION ON THE X-Y PLANE IS A SQUARE.	THE PROJECTION IS A RECTANGLE	THE PROJECTION IS A HEXAGON OF EQUAL SIDE LENGTH $S = \frac{R \times \sqrt{6}}{3}$

POSITION	LATERAL FACE	EDGE	CORNER
CORRESPONDING AXES OF SYMMETRY	3 x FOUR-NOTATED	6 x TWO-NOTATED	4 x THREE-NOTATED
SURFACE SYMBOLS	(001) (100)) 010)	(011) (0-11) (101) (-101) (110) (-110)	(111) (11-1) (1-11)(-111)
SELECTED CHARACTERISTIC SURFACE SYMBOL	001	101	111

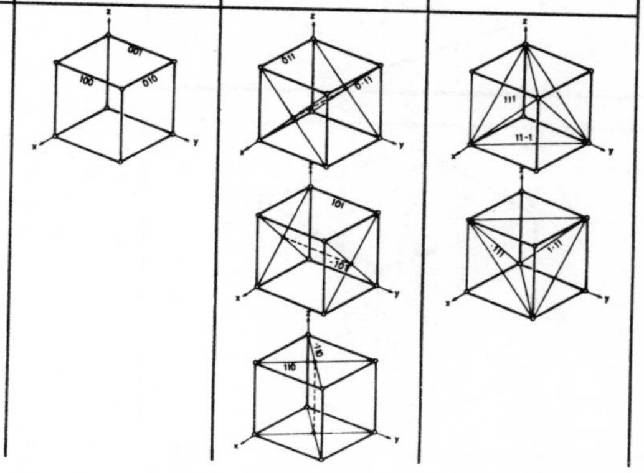

These three positions are of fundamental importance for deriving space-frames from the cubic grid. Each position with respect to its axes of symmetry is multi-notated because by means of a symmetra operation each position is revertable to the chosen plane for cunstruction.

POSITIONS OF AREAS IN THE CUBE

The accompanying grouping shows the different orientations derived from the principles outlined in 2.3.2 and 2.3.3. In each case the marked in grillage represents one chord of a space frame type derivated from the face centered cubic lattice.

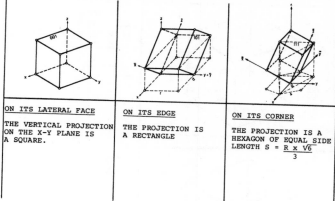

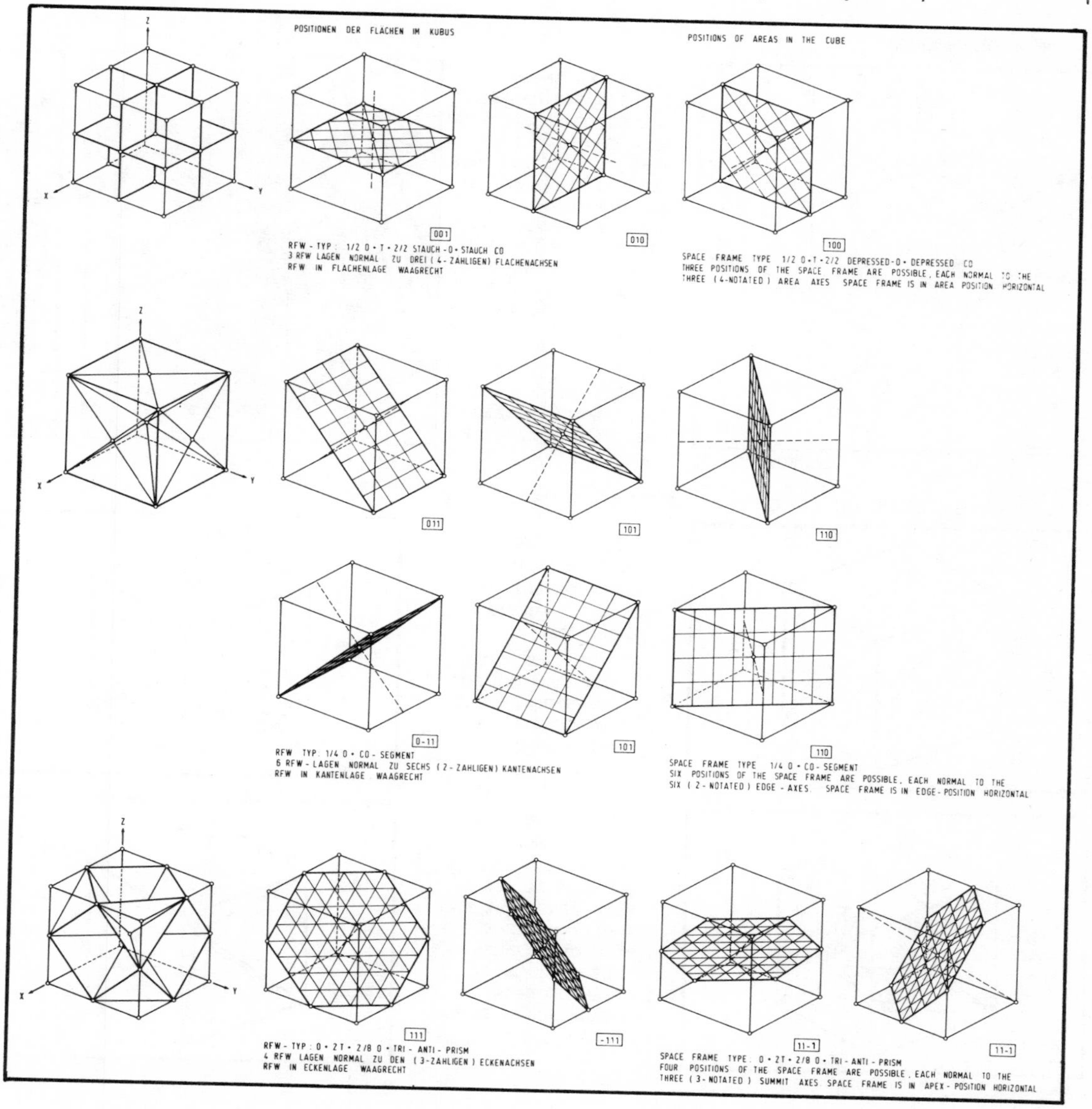

2.3.3. PROPERTIES OF SYMMETRY OF THE CUBE

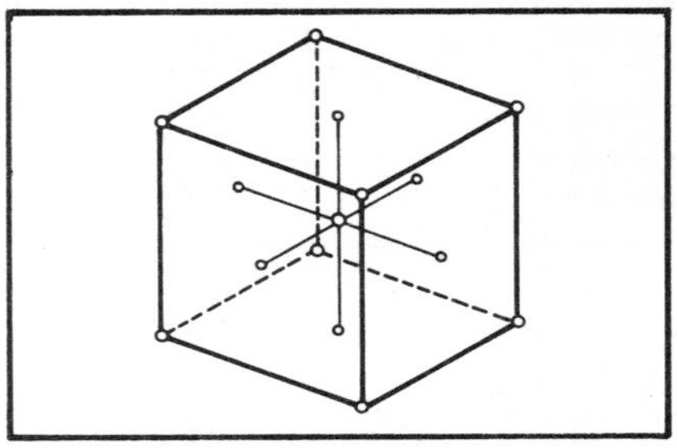

4-NOTATED PROPERTY OF THE 3 FACE AXES RUNNING THROUGH THE FACE CENTRES.

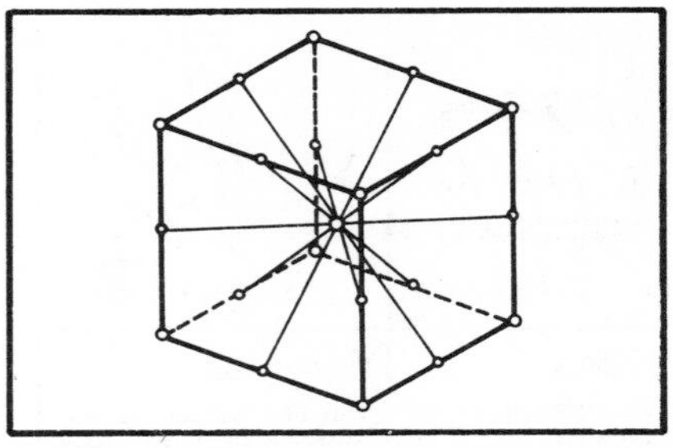

2-NOTATED PROPERTY OF 6 EDGE AXES RUNNING THROUGH THE EDGE CENTRES.

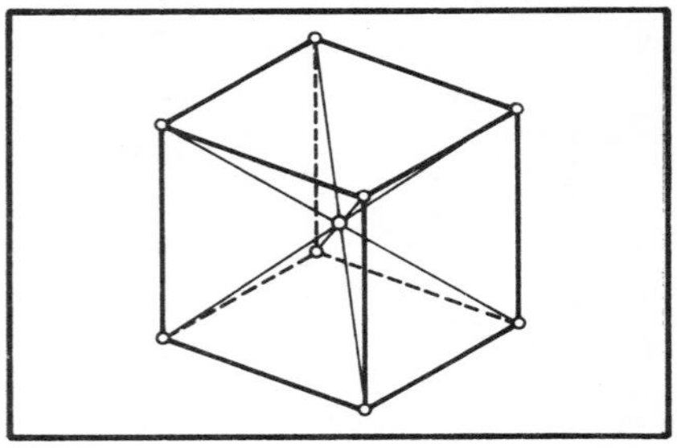

3-NOTATED PROPERTY OF THE 4 CORNER AXES RUNNING THROUGH THE CORNERS

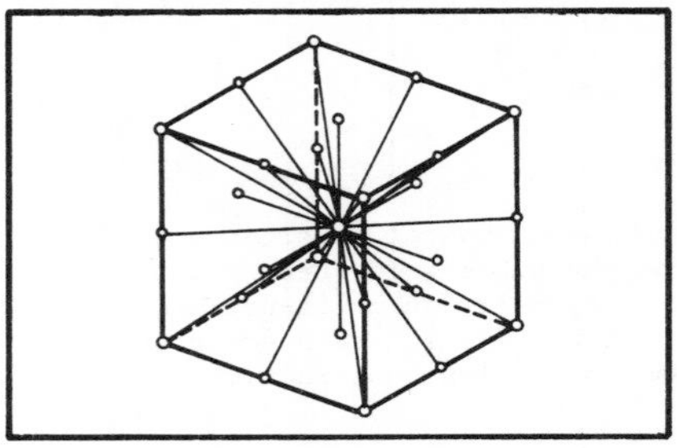

COMBINATION OF ALL AXES.

2.3.4 NODE GEOMETRY

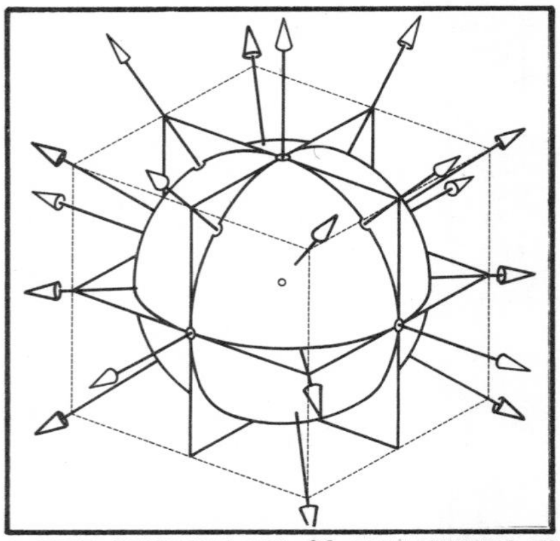

A NODE AS SHOWN IN 2.3.3 WHICH HAS BEEN DRILLED CORRESPONDING TO THE 3 FACE AXES AND THE SIX EDGE AXES (A TOTAL OF 18 CONNECTIONS) ALLOWS FOR A FACE-CENTRED CUBIC SPACE LATTICE. A NODE AS SHOWN IN 2.3.3 DRILLED CORRESPONDING TO THE MENTIONED 3 FACE AXES AND 4 CORNER AXES (TOTALLY 14 CONNECTIONS) ALLOWS FOR A BODY-CENTRED CUBIC SPACE-LATTICE.

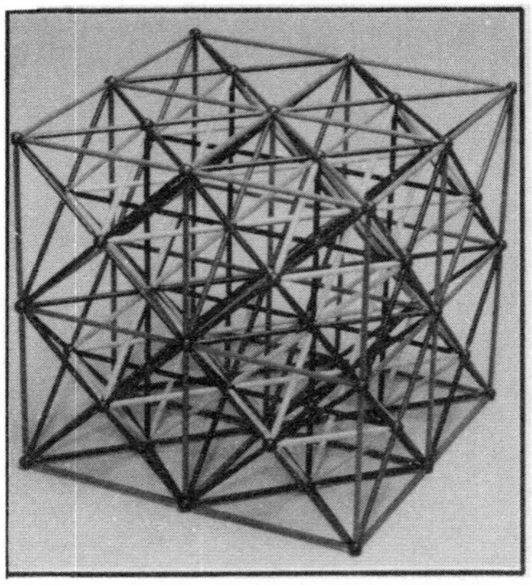

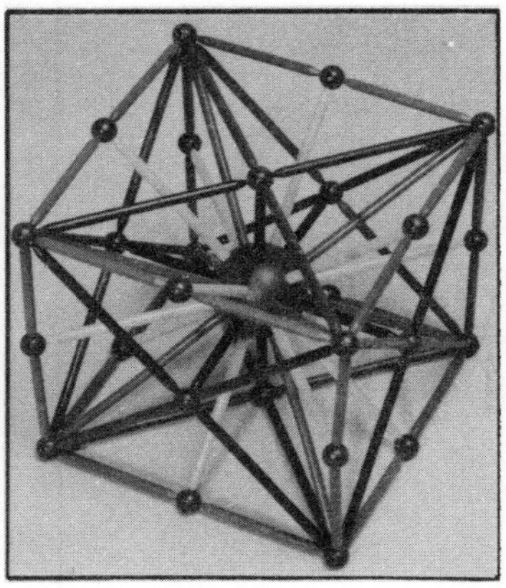

A NODE ENABLING ALL 26 CONNECTIONS IS SHOWN IN FIG.

2.3.5 CLASSIFICATION OF CUBIC LATTICE GRIDS

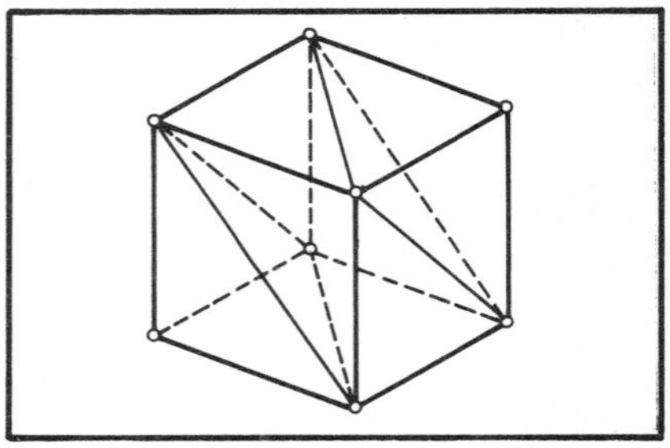

1. SIMPLE CUBIC LATTICE GRID
   INTERCONNECTIONS OF THE CORNERS BY EDGES AND
   A) ADDITIONAL CONNECTION OF THE CORNERS BY FACE DIAGONALS: TRI-ANTI-PRISM + 2 x 1/8 O

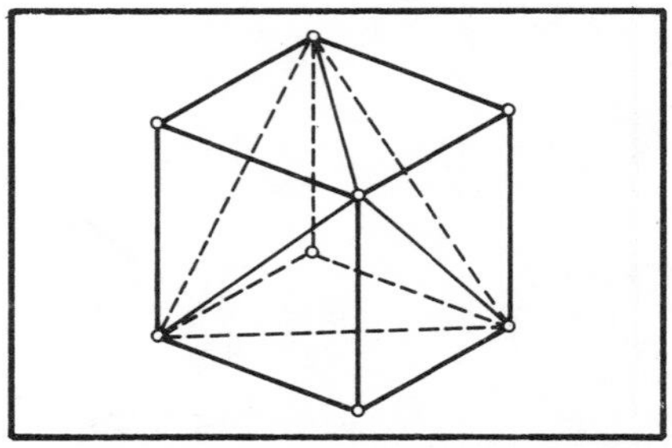

B) DIFFERENT CONNECTION OF THE CORNERS BY ALTERNATIVE FACE DIAGONALS:
TETRAHEDRON +4x1/8 OCTAHEDRON

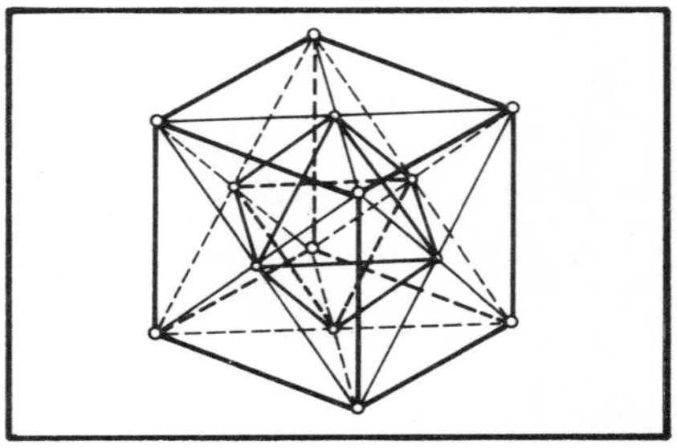

2. FACE-CENTRED CUBIC LATTICE GRID
   INTERCONNECTION OF THE CORNERS BY FACE DIAGONALS AND ADDITIONAL CONNECTION OF ADJACENT FACE CENTRES:
   OCTAHEDRON+4x1/8 OCTAHEDRON+4 TETRAHEDRA

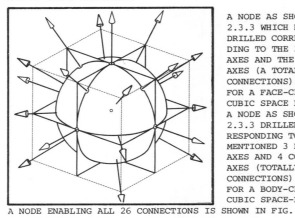

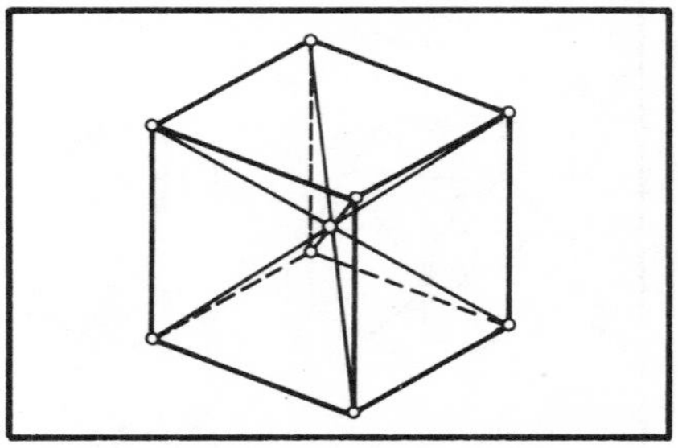

3. BODY-CENTRED CUBIC LATTICE GRID
   INTERCONNECTION OF THE CORNERS BY FACE DIAGONALS AND IN ADDITION CONNECTION OF ADJACENT FACE CENTRES:
   OCTAHEDRON+4x1/8 OCTAHEDRON+4 TETRAHEDRA

2.4 ABLEITUNG RÄUMLICHER FACHWERK-STRUKTUREN

2.4 DERIVATION OF SPACE STRUCTURES

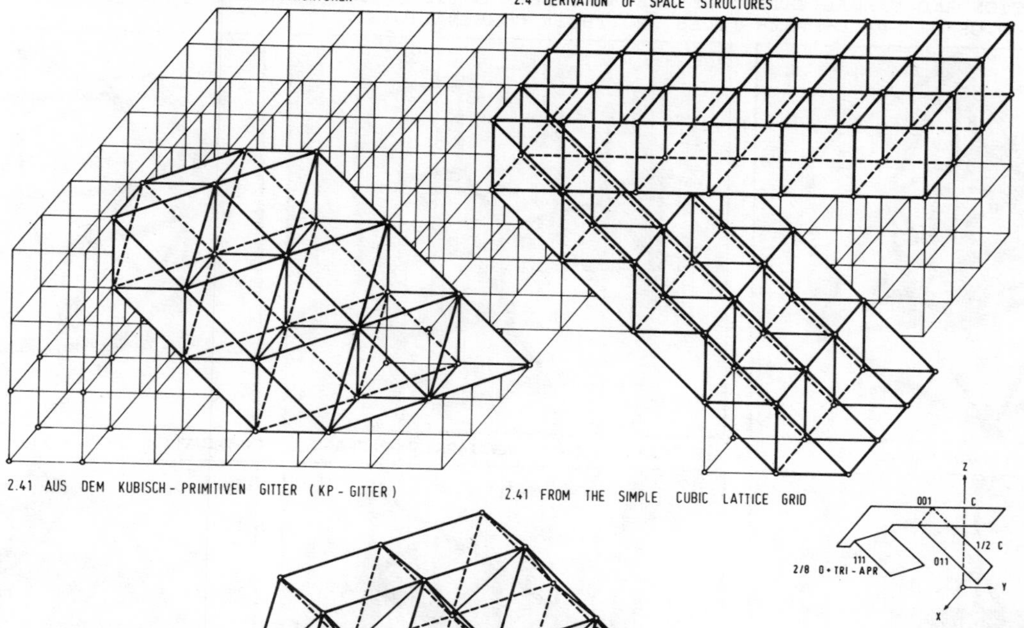

2.41 AUS DEM KUBISCH-PRIMITIVEN GITTER (KP-GITTER)

2.41 FROM THE SIMPLE CUBIC LATTICE GRID

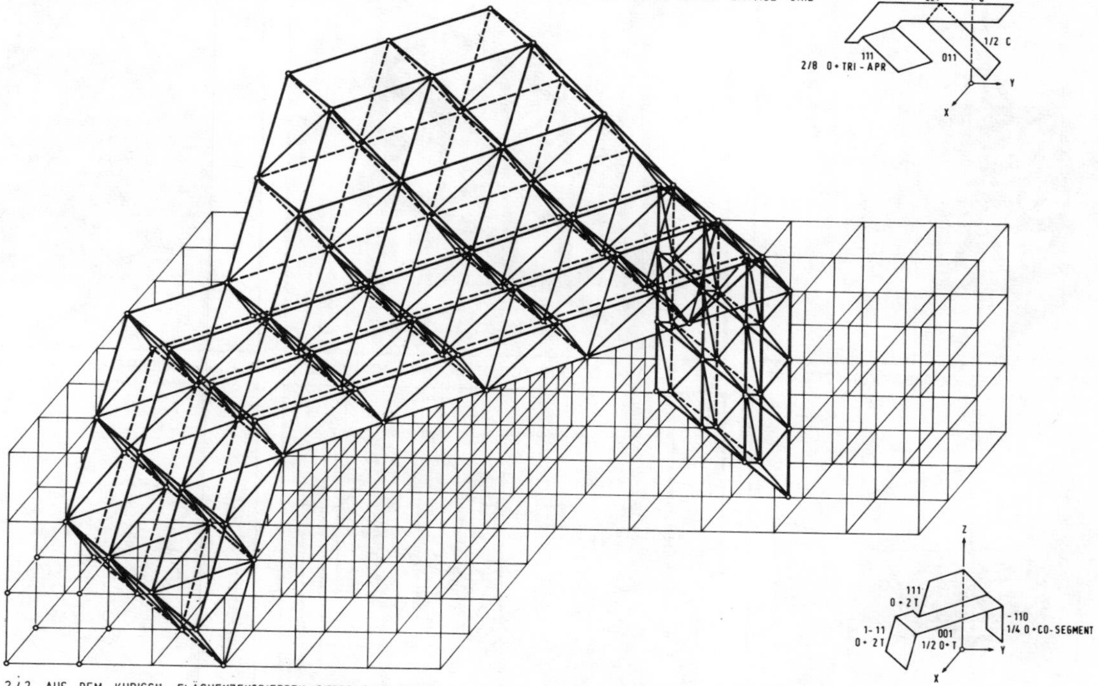

2.42 AUS DEM KUBISCH-FLÄCHENZENTRIERTEN GITTER (KF-GITTER)

2.42 FROM THE FACE-CENTRED CUBIC LATTICE GRID

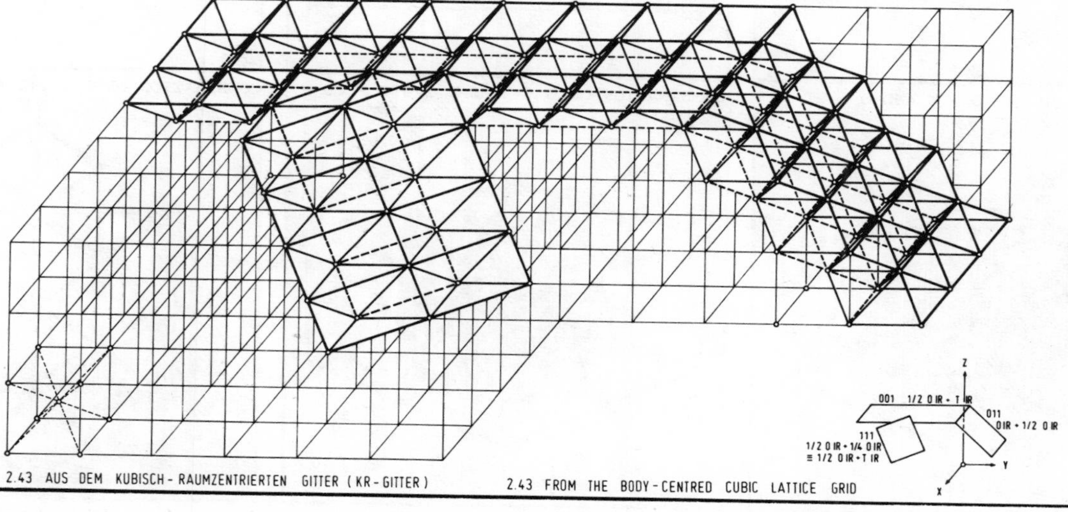

2.43 AUS DEM KUBISCH-RAUMZENTRIERTEN GITTER (KR-GITTER)

2.43 FROM THE BODY-CENTRED CUBIC LATTICE GRID

2.5 COMPOSITION AND VISUALIZATION OF THE SPACE-FRAME TYPES DERIVED FROM THE CUBIC LATTICE GRID

2.5.1 MODEL PHOTOS OF SPACE-FRAME TYPES AND THEIR COMBINATIONS

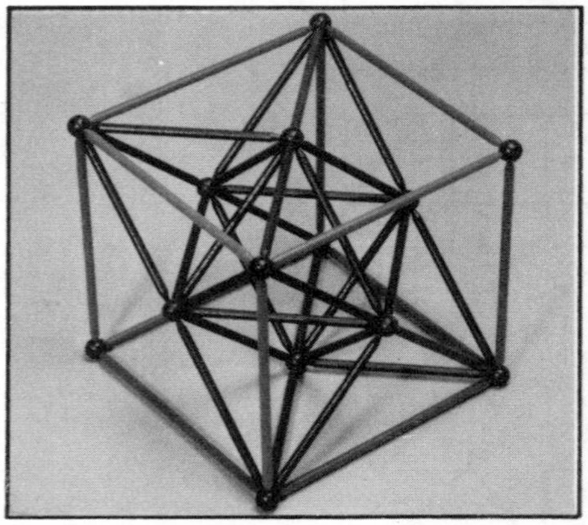

OCTAHEDRON/TETRAHEDRON-PACKAGE

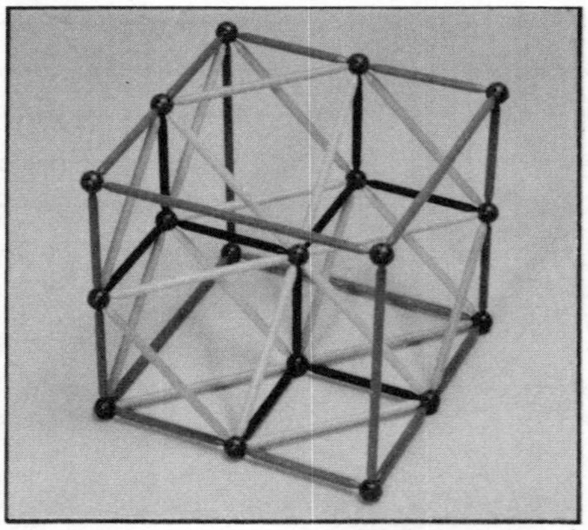

2/8 OCTAHEDRON/TRI-ANTI-PRISM-PACKAGE

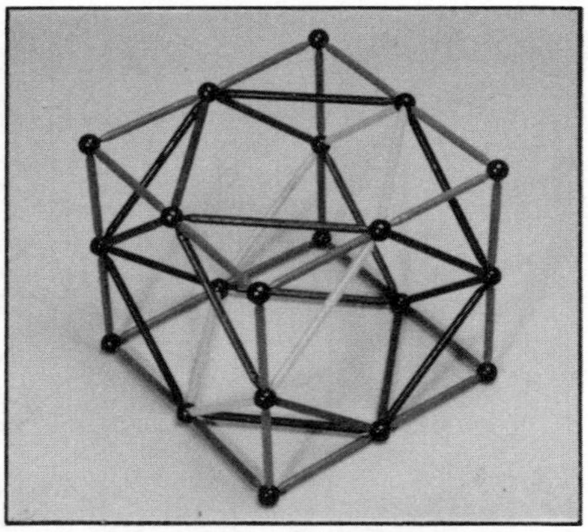

COMBINATION OF SIX 111-LATTICE GRIDS

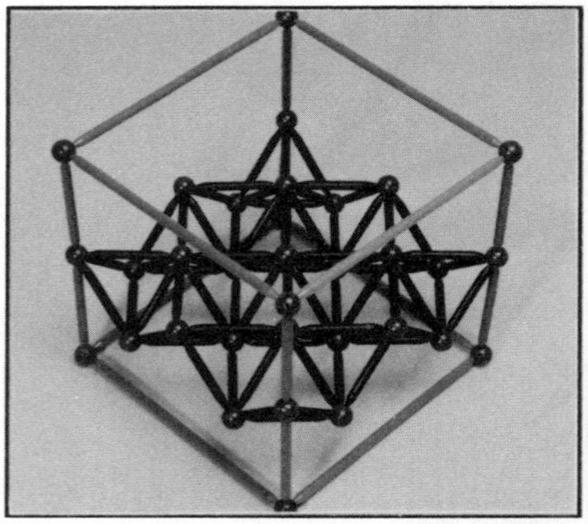

SPACE-FRAME TYPE 1/2 OCTAHEDRON+TETRAH.

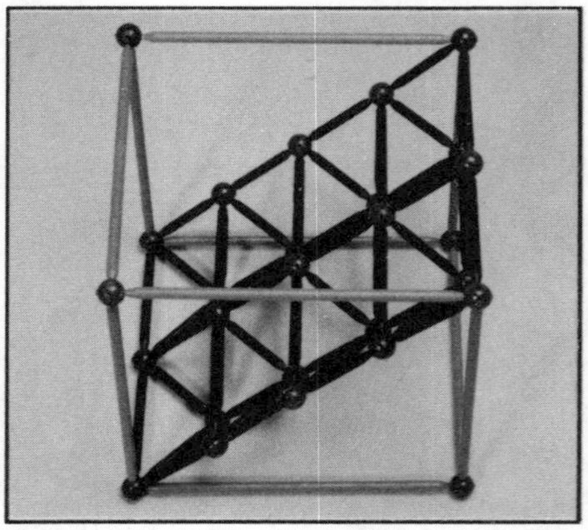

S.F.T. 1/4 OCTAH.+CUBE-OCTAH.-SEGMENT

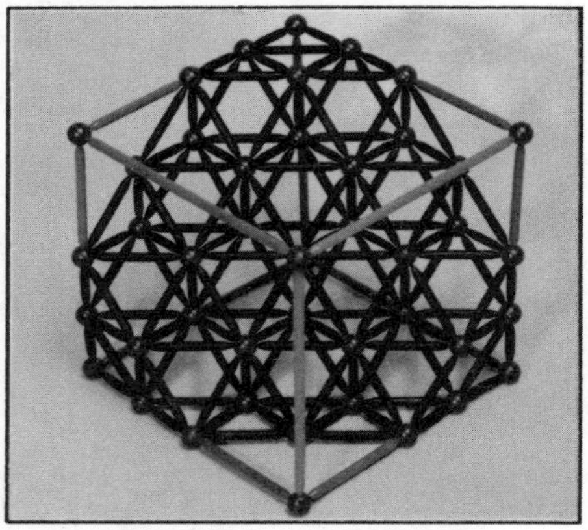

S.F.T. OCTAHEDRON + 2 TETRAHEDRON

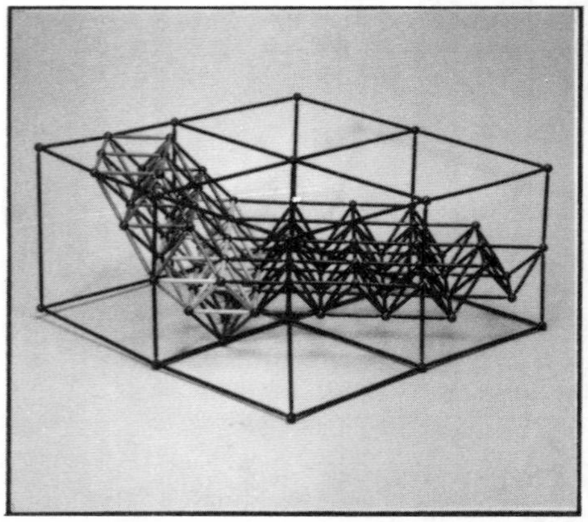

COMBINATION OF 1/2 O + T AND O + 2T

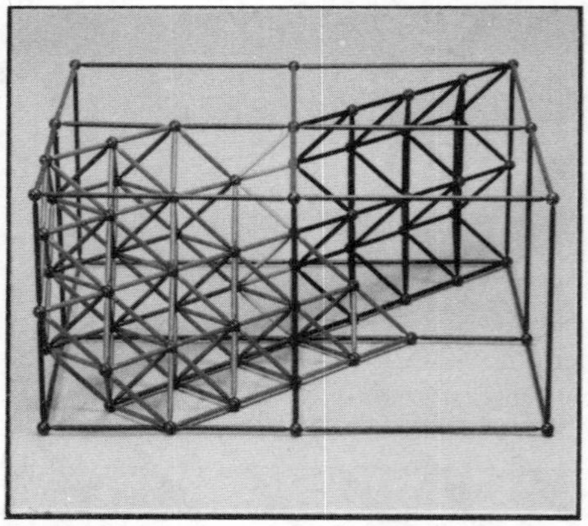

COMBINATION OF O + 2T AND 1/4 O + CO-S

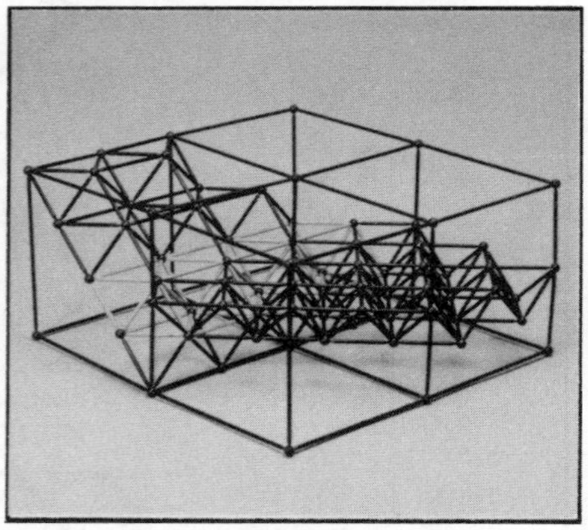

COMBINATION OF 1/2 O+T AND 1/4 O+CO-S

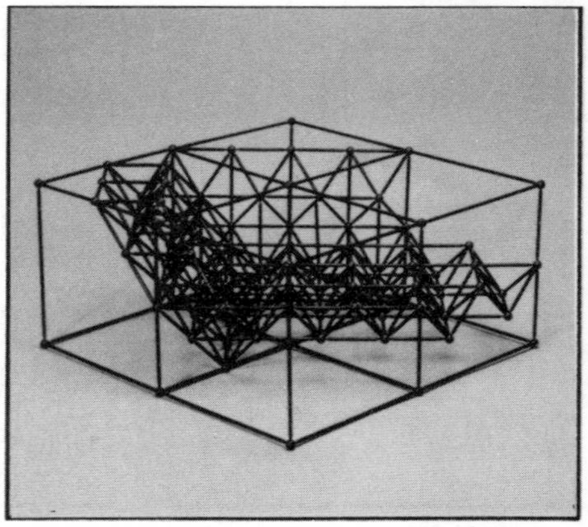

COMB. OF 1/2 O+T AND O+2T+1/4 O+CO-S

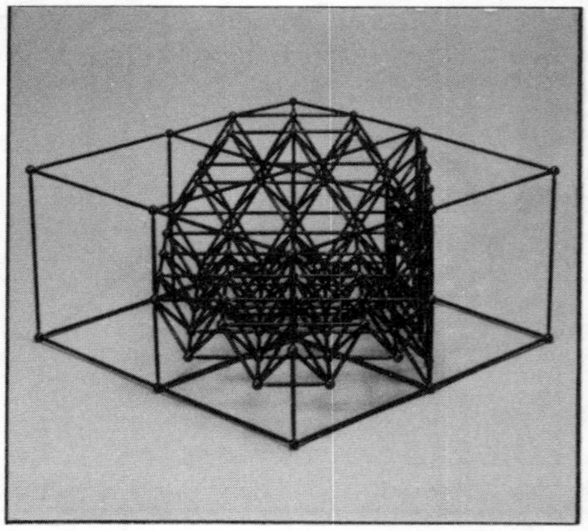

COMB. OF 1/2 O+T AND O+2T+1/4 O+CO-S

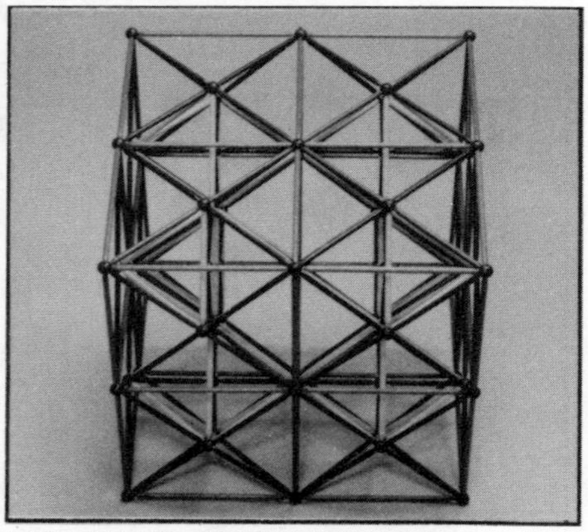

BODY-CENTRED CUBIC LATTICE GRID + O/T-LATTICE GRID IN EDGE-POSITION

2.5.2 PLANMETRIC AND AXONOMETRIC VIEW OF THE SPACE-FRAME LATTICE GRID SPACE-MODULES AND NODES

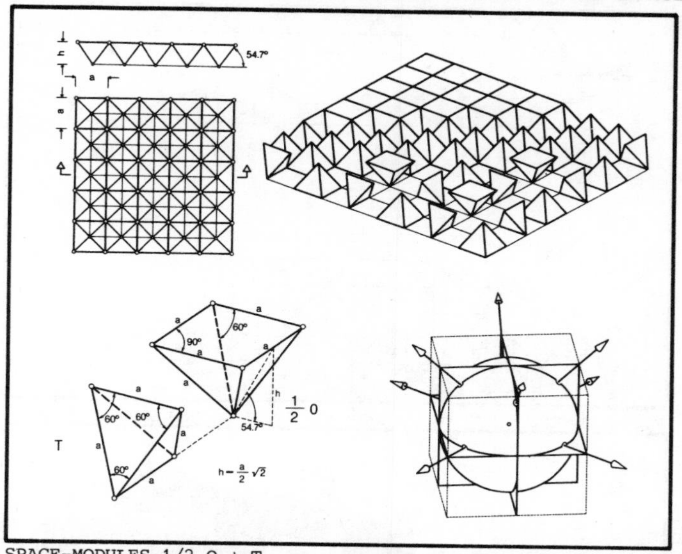

SPACE-MODULES 1/2 O + T

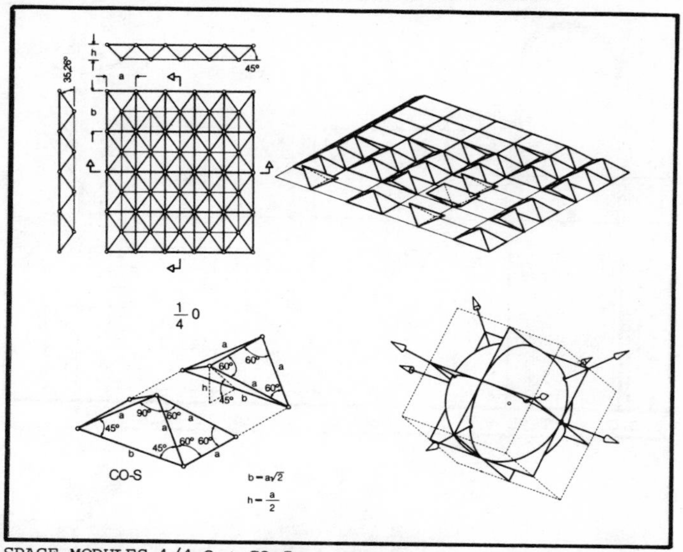

SPACE-MODULES 1/4 O + CO-S

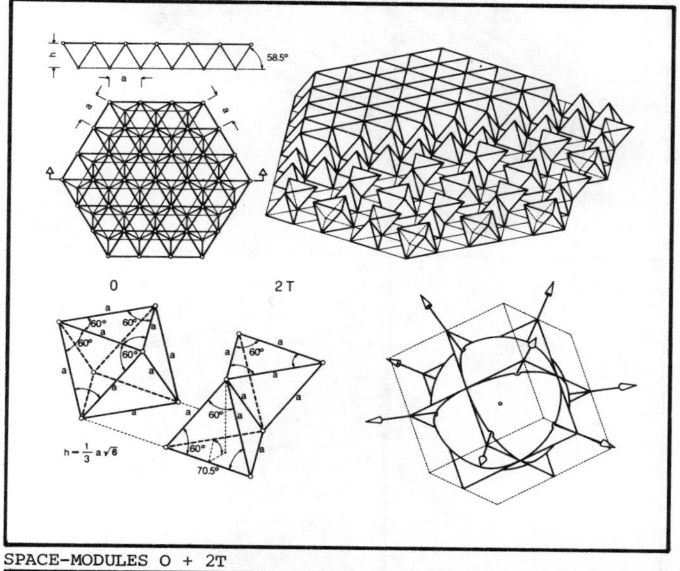

SPACE-MODULES O + 2T

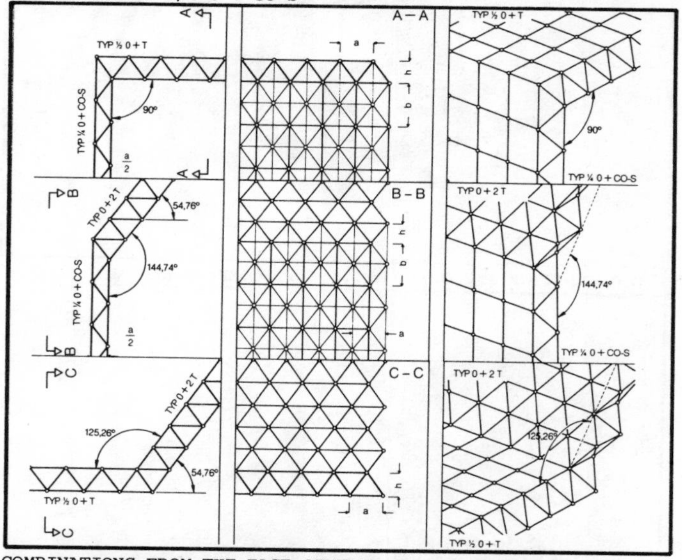

COMBINATIONS FROM THE FACE-CENTRED CUBIC LATTICE GRID

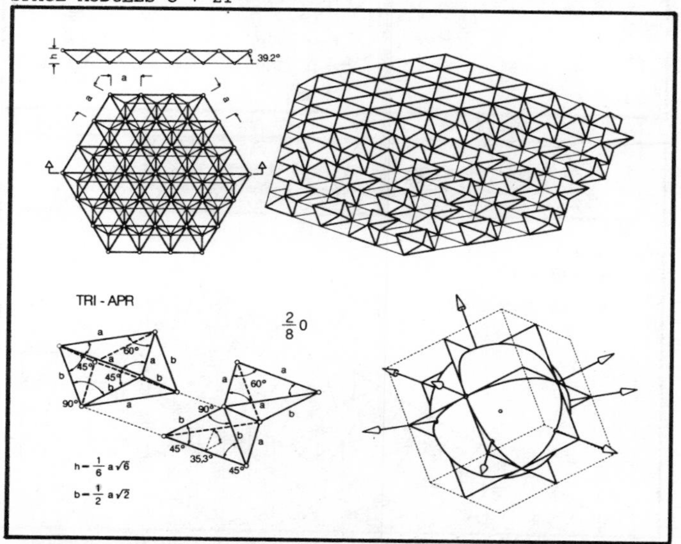

SPACE-MODULES 2/8 O + TRI-APR

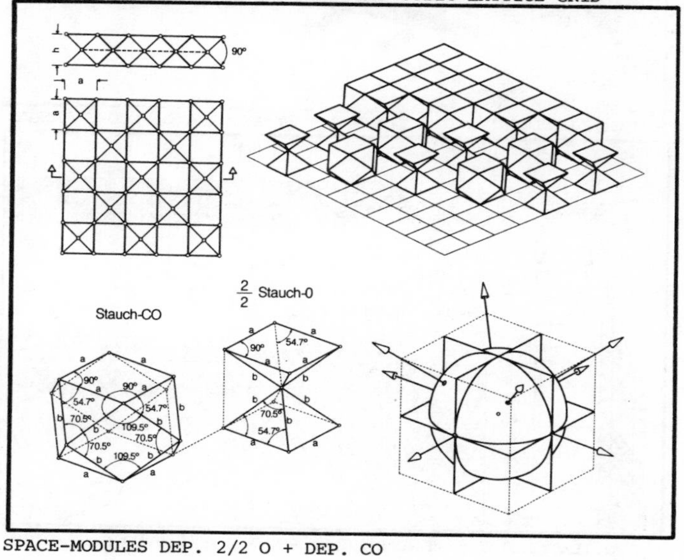

SPACE-MODULES DEP. 2/2 O + DEP. CO

2.6 REPRESENTATION OF SPACE-FRAME COMBINATIONS FROM DESIGN CONCEPTS

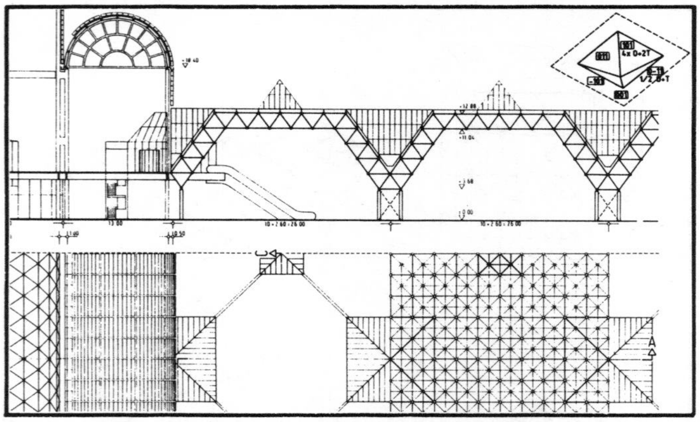

F.C.C. LATTICE 1/2 O + T AND O + 2T KUWAIT-FAIR

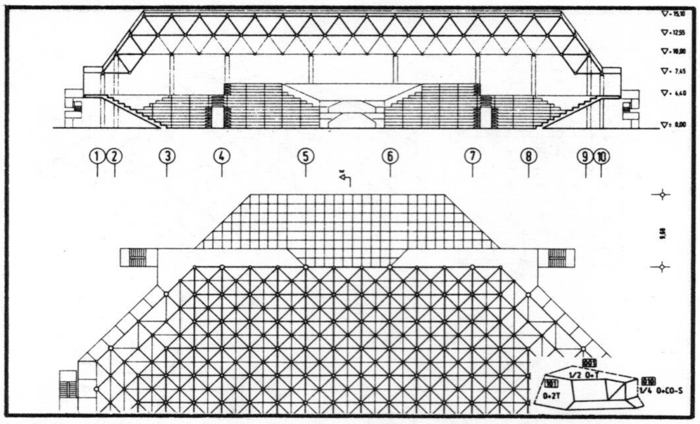

F.C.C. LATTICE 1/2 O + T AND = + 2T SPORTSHALL PROTOTYPE

F.C.C. LATTICE 1/2 O + T AND O + 2T COMPACT HANGAR

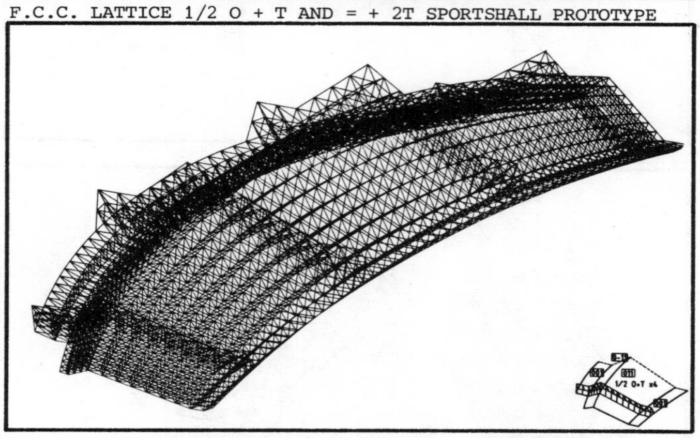

F.C.C. LATTICE TRANSFORM 1/2 O + T AND 1/2 O+T CDI STADIUM

F.C.C. LATTICE 1/2 O + T AND O + 2T AND 1/4 O + CO-S HANGAR

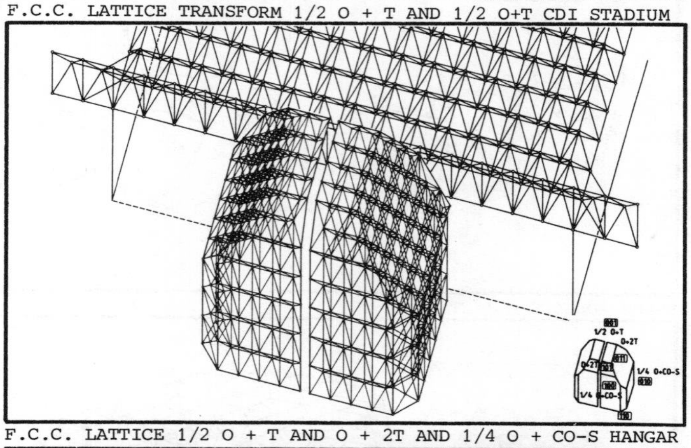

F.C.C. LATTICE 1/2 O + T AND O + 2T AND 1/4 O + CO-S HANGAR

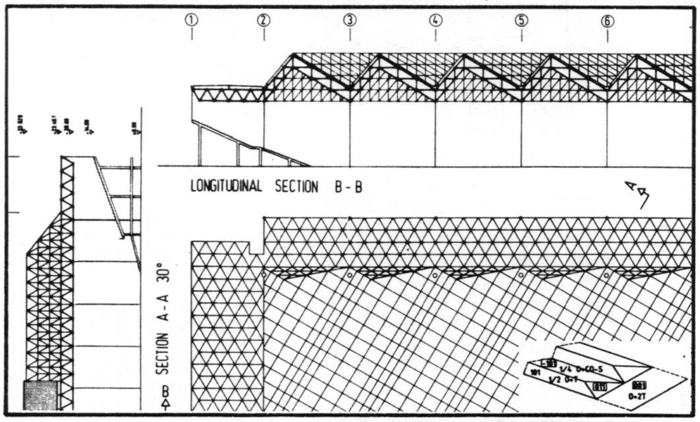

F.C.C. LATTICE O + 2T AND 1/2 O + T AND 1/4 O+CO-S STADIUM

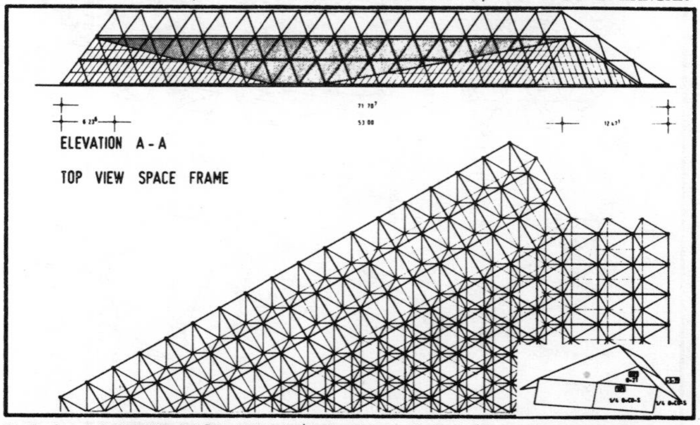

F.C.C. LATTICE O+2T AND 1/4 O+CO-S AND 1/4 O+CO-S TENNISC.

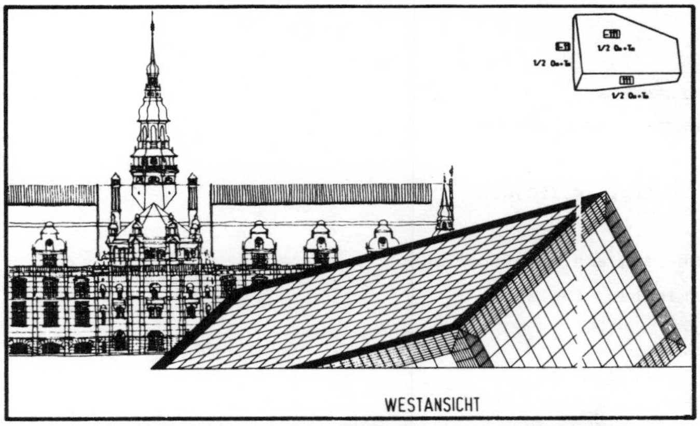

B.C.C. LATTICE 1/2 O_{IR} + T_{IR} x3 WASA MUSEUM STOCKHOLM

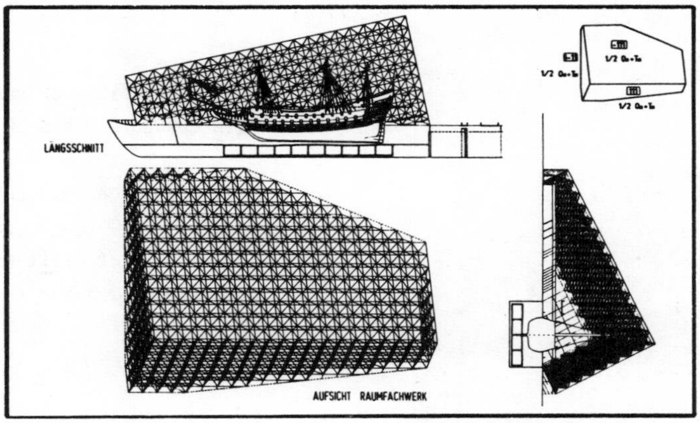

B.C.C. LATTICE 1/2 O_{IR} + T_{IR} x3 WASA MUSEUM STOCKHOLM

2.7 REPRESENTATION OF SPACE-FRAME COMBINATIONS FROM COMPLETED PROJECTS

F.C.C. LATTICE 1/2 O + T AND 1/4 O + CO-S CITY-BANK PARIS

F.C.C. LATTICE 1/2 O + T AND O + 2T SAVA HOTEL BEOGRAD

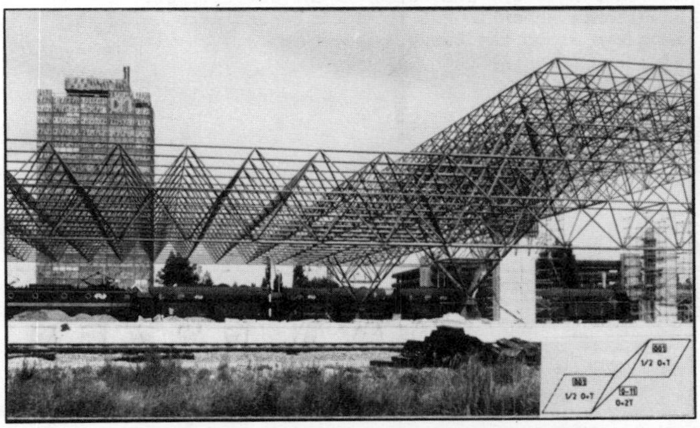

F.C.C. LATTICE 1/2 O + T AND O + 2T RAILWAY STATION

F.C.C. LATTICE 1/2 O + T AND O + 2T SWIMMING POOL

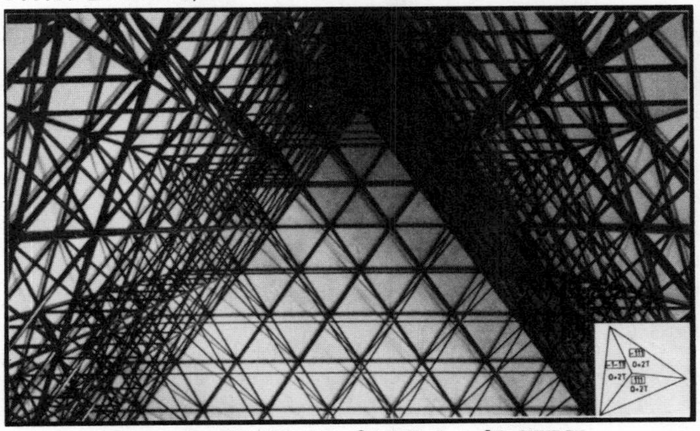

F.C.C. LATTICE O + 2T AND O + 2T AND O + 2T CHURCH

F.C.C. LATTICE O + 2T AND O + 2T AND O + 2T CHURCH

ANALYSIS OF SPACE FRAMEWORKS WITH CURVED MEMBERS OF GENERAL CROSS SECTION

D.J. JUST, B.Sc., Ph.D., C.Eng., M.I.C.E.

Lecturer, Department of Civil Engineering and Construction
University of Aston in Birmingham, England.

The precise 12 x 12 linear elastic stiffness matrix for a thin circularly curved prismatic beam with either symmetry or non-symmetry of cross-section, and subjected to both in-plane and out-of plane loading, is developed explicitly. The displacement transformation matrix and the method by which distributed loading is treated are then presented so that the element is able to be incorporated into a general space frameworks program. Finally the analysis is applied to the examination of a curved grillage and also to the solution of a curved space frame.

INTRODUCTION

Although the displacement method has been applied extensively to the solution of frames composed of straight members, its efficient application to the solution of structures containing curved members appears to be comparatively rare.

This sparse consideration of such elements may be due not only to the relative complexity of curved geometry but also to the fact that curvature in a beam may be treated by idealising the member into a number of straight segments. However, although such a procedure should indeed produce convergence to the correct result on increase of subdivision, the treatment is extravagant both in computation and in data preparation, besides always yielding an approximation.

In order to obtain an exact solution of the curved beam, thus eliminating the need for subdivision, the equations governing its behaviour must be solved in order to determine the precise modes of deformation. Once these have been obtained the exact stiffness matrix for the beam may be rapidly developed using a curtailed finite element process, and hence frameworks containing curved members rendered soluble with the same ease as those composed solely of straight elements.

FORMULATION OF STIFFNESS MATRIX

Construction of the stiffness matrix is dependent upon the determination of the precise manner in which the beam deforms under end-loading conditions. These displacement functions may be obtained by deriving the equilibrium equations for the beam, substituting the relevant stress resultant/displacement relationships, and then solving the resulting equations to obtain the modes of deformation. Although these functions could then be incorporated in a standard finite element procedure to obtain the stiffness matrix, such a process is unnecessary due to the fact that the use of exact displacement functions implies that equilibrium is satisfied not only at the nodes but also throughout the element, a situation which does not occur with the use of approximate functions where only nodal equilibrium can be obtained. Thus on invoking the derived displacement functions, the standard finite element derivation may be reduced and the coefficients of the stiffness matrix formulated in relatively simple terms.

a) Variation of stress resultants

Consider a thin circularly curved element of radius R and curved length δx, subjected to the positive stress resultants shown in Fig 1, the x co-ordinate being measured along the curve.

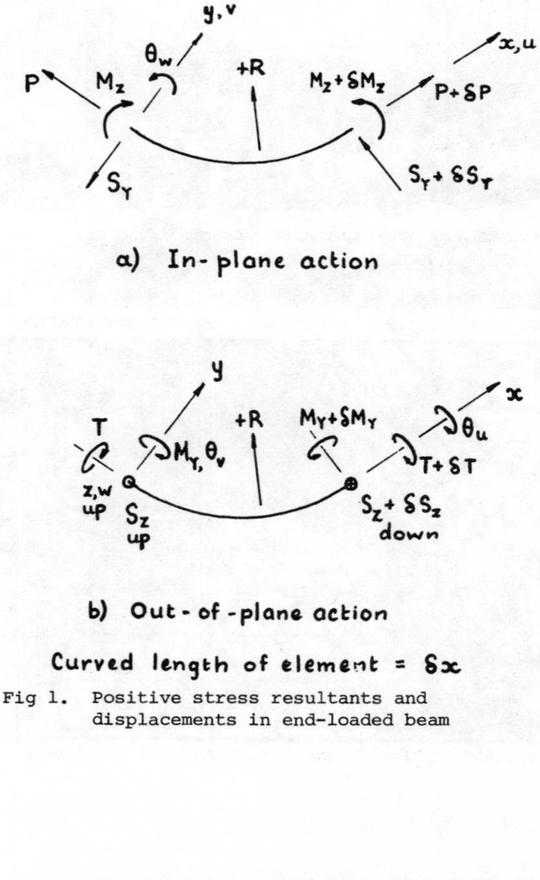

Fig 1. Positive stress resultants and displacements in end-loaded beam

From Fig 1a, consideration of radial, tangential, and rotational equilibrium gives

$$dS_Y/dx + P/R = 0 \qquad \text{......... 1a}$$

$$dP/dx - S_Y/R = 0 \qquad \text{......... 1b}$$

$$dM_Z/dx + S_Y = 0 \qquad \text{......... 1c}$$

Eliminating the axial force, P, from Eqns 1a and 1b, and substituting Eqn 1c yields

$$d^3M_Z/dx^3 + dM_Z/R^2dx = 0 \qquad \text{......... 2}$$

which on solution, and on letting $\psi = x/R$, gives

$$M_Z = A_1 + A_2 \sin\psi + A_3 \cos\psi \qquad \text{......... 3}$$

The variations of P and S_Y may then be obtained from Eqn 3 by using the relationships inherent in Eqns 1

$$P = Rd^2M_Z/dx^2 \qquad \text{......... 4}$$

$$S_Y = - dM_Z/dx \qquad \text{......... 5}$$

The relationships between the stress resultants S_Z, M_Y, and T may be similarly obtained by considering the equilibrium of the beam increment shown in Fig 1b.

Rotational equilibrium in the tangential and radial directions give respectively

$$T/R - dM_Y/dx + S_Z = 0 \qquad \text{......... 6a}$$

$$dT/dx + M_Y/R = 0 \qquad \text{......... 6b}$$

and since vertical equilibrium simply gives constancy of S_Z, differentiation of Eqn 6a and combination with Eqn 6b gives

$$d^3T/dx^3 + dT/R^2dx = 0 \qquad \text{......... 7}$$

which on solution yields

$$T = B_1 + B_2 \sin\psi + B_3 \cos\psi \qquad \text{......... 8}$$

The variations of M_Y and S_Z may then be obtained from this equation using the results obtained from Eqns 6a and 6b.

$$M_Y = -RdT/dx \qquad \text{......... 9}$$

$$S_Z = -Rd^2T/dx^2 - T/R \qquad \text{......... 10}$$

b) Stress resultant/displacement relationships

Referring to the conventions used in Fig 1, and considering a thin beam exhibiting sectional symmetry about either one or both of the axes y, z, the relationships between P and M_Z and the displacements u and v are given by (Ref 1)

$$P + M_Z/R = EA(du/dx - v/R) \qquad \text{......... 11}$$

$$M_Z = EI_Z(d^2v/dx^2 + v/R^2) \qquad \text{......... 12}$$

where A is the cross-sectional area, E is Young's modulus, and I_Z is the second moment of area about the z axis, while the relationships between T and M_Y and the displacements θ_u and w are (Ref 2)

$$T = GJ(d\theta_u/dx + dw/Rdx) \qquad \text{......... 13}$$

$$M_Y = EI_Y(d^2w/dx^2 - \theta_u/R) \qquad \text{......... 14}$$

I_Y being the second moment of area about the y axis, and GJ the torsional rigidity.

In considering a beam with non-symmetry of section, the product moment of area with respect to the y, z axes, I_{YZ}, becomes a further sectional property that must be considered, and in this case Eqns 12 and 14 must be replaced by (Ref 1)

$$M_ZI_Y - M_YI_{YZ} = EQ(d^2v/dx^2 + v/R^2) \qquad \text{......... 15}$$

$$M_YI_Z - M_ZI_{YZ} = EQ(d^2w/dx^2 - \theta_u/R) \qquad \text{......... 16}$$

where $Q = I_YI_Z - I_{YZ}^2$

Elimination of M_Y amd M_Z in turn from Eqns 15 and 16 then gives the results

$$M_Y = E\left[I_{YZ}(d^2v/dx^2 + v/R^2) + I_Y(d^2w/dx^2 - \theta_u/R)\right] \text{ ... 17}$$

$$M_Z = E\left[I_Z(d^2v/dx^2 + v/R^2) + I_{YZ}(d^2w/dx^2 - \theta_u/R)\right] \text{ ... 18}$$

c) Variation of displacements

The form of the displacement functions can now be obtained by substituting the expressions for the stress resultants into the stress resultant/displacement relationships. In this process Eqns 15 and 16 will be used so that results for both the symmetrical and non-symmetrical forms of cross-section are obtained, the symmetrical form being the special case when I_{YZ} vanishes.

Firstly the variation of v can be found by substituting the right hand sides of Eqns 3 and 9 into Eqn 15, giving

$$(A_1 + A_2\sin\psi + A_3\cos\psi)I_Y - (-B_2\cos\psi + B_3\sin\psi)I_{YZ}$$

$$= EQ(d^2v/dx^2 + v/R^2) \qquad \text{......... 19}$$

Since the constants A_1 ... B_3 are arbitrary, the solution of this equation may be written

$$v = a_1 + a_2\sin\psi + a_3\cos\psi + a_4x\sin\psi + a_5x\cos\psi \text{ ... 20}$$

where, in particular,

$$a_1 = A_1R^2I_Y/(EQ) \qquad \text{......... 21}$$

The function for u is now obtained using Eqns 11 and 20 and invoking Eqn 22, thus

$$P + M_Z/R = Rd^2M_Z/dx^2 + M_Z/R = A_1/R = EA(du/dx - v/R) \text{.. 22}$$

Hence,

$$du/dx = v/R + A_1/(EAR)$$

$$= Fa_1 + a_2(\sin\psi)/R + a_3(\cos\psi)/R + a_4\psi\sin\psi + a_5\psi\cos\psi \qquad \text{......... 23}$$

which upon integration yields

$$u = Fa_1x - a_2\cos\psi + a_3\sin\psi - a_4(x\cos\psi - R\sin\psi)$$
$$+ a_5(x\sin\psi + R\cos\psi) + a_6 \qquad \text{......... 24}$$

where $F = \left[1 + Q/(AR^2I_Y)\right]/R$

The functions for the vertical deflection, w, and the twist, θ_u, may now be found using Eqns 13 and 16 together with the stress resultant expressions.

Substitution of the right hand sides of Eqns 3 and 9 into Eqn 16 yields

$$(-B_2\cos\psi + B_3\sin\psi)I_Z - (A_1 + A_2\sin\psi + A_3\cos\psi)I_{YZ}$$

$$= EQ(d^2w/dx^2 - \theta_u/R) \qquad \text{......... 25}$$

while integration of Eqn 13 gives

$$\theta_u/R = (B_1x - B_2R\cos\psi + B_3R\sin\psi)/(GJR) - w/R^2 + \text{const} \qquad \text{......... 26}$$

Thus on substitution of Eqn 26 into Eqn 25 the equation for w is given by

$$d^2w/dx^2 + w/R^2 = xB_1/(GJR) + \left[(B_3I_Z - A_2I_{YZ})/(EQ) + B_3/(GJ)\right]\sin\psi$$
$$- \left[(B_2I_Z + A_3I_{YZ})/(EQ) + B_2/(GJ)\right]\cos\psi + \text{const} \text{ ..... 27}$$

which on solution produces

$$w = b_1 + b_2x + b_3\sin\psi + b_4\cos\psi + b_5x\sin\psi + b_6x\cos\psi \qquad \text{......... 28}$$

the constants b_1 ..... b_6 being independent of the constants a_1 ..... a_6 presented in Eqns 20 and 24.

The equation from which θ_u may be found can be obtained by differentiating and rewriting Eqn 13 as

$$d^2w/dx^2 = (B_2 \cos \psi - B_3 \sin \psi)/(GJ) - Rd^2\theta_u/dx^2 \quad .... \; 29$$

and substituting this result into Eqn 25 giving

$$d^2\theta_u/dx^2 + \theta_u/R^2 = A_1 I_{YZ}/(EQR) - \sin \psi \left[(B_3 I_Z - A_2 I_{YZ})/(EQ) + B_3/(GJ) \right] /R + \cos \psi \left[(B_2 I_Z + A_3 I_{YZ})/(EQ) + B_2/(GJ) \right] /R \quad .... \; 30$$

Noting that the coefficients within the square brackets are identical with those in Eqn 27, and that the first term on the right hand side may be written, by virtue of Eqn 21, as

$$a_1 I_{YZ}/(I_Y R^3)$$

the solution of Eqn 30 gives

$$\theta_u = c_1 \sin \psi + c_2 \cos \psi + a_1 I_{YZ}/(I_Y R) - b_5 \psi \sin \psi - b_6 \psi \cos \psi \quad .... \; 31$$

where c_1 and c_2 are dependent constants whose values may be deduced by substituting Eqns 13 and 17 into Eqn 9 and equating like coefficients. This process gives

$$c_1 = -b_3/R - 2E(I_{YZ}a_5 + I_Y b_6)/(EI_Y + GJ) \quad .... \; 32a$$

$$c_2 = -b_4/R + 2E(I_{YZ}a_4 + I_Y b_5)/(EI_Y + GJ) \quad .... \; 32b$$

thus enabling θ_u to be expressed as

$$\theta_u = a_1 U - b_3(\sin \psi)/R - b_4(\cos \psi)/R + b_5(H \cos\psi - \psi\sin\psi) - b_6(H \sin \psi + \psi \cos \psi) + a_4 V \cos \psi - a_5 V \sin \psi \quad .. \; 33$$

where

$$U = I_{YZ}/(I_Y R), \; H = 2EI_Y/(EI_Y + GJ), \; V = UHR$$

d) Stiffness matrix

Having derived the manner in which the deflections vary, the stiffness matrix relating the nodal forces to the corresponding nodal displacements may be assembled using a curtailed finite element process.

The sense of these nodal forces and displacements for a beam of curved length L is shown in Fig 2 where the values of the rotations are (Ref 3)

$$\theta_v = - dw/dx \quad .... \; 34a$$

$$\theta_w = dv/dx + u/R \quad .... \; 34b$$

Thus on substituting the functions for u, v, w, the rotations may be expressed as

$$\theta_v = - b_2 - b_3(\cos \psi)/R + b_4(\sin \psi)/R - b_5(\sin \psi + \psi \cos \psi) - b_6(\cos \psi - \psi \sin \psi) \quad .... \; 35$$

$$\theta_w = Fa_1 \psi + 2a_4 \sin \psi + 2a_5 \cos \psi + a_6/R \quad .... \; 36$$

The nodal displacements can now be related to the arbitrary constants a_1 ..... a_6 and b_1 ..... b_6 through a 12 x 12 matrix [C] which may be conveniently assembled in two stages.

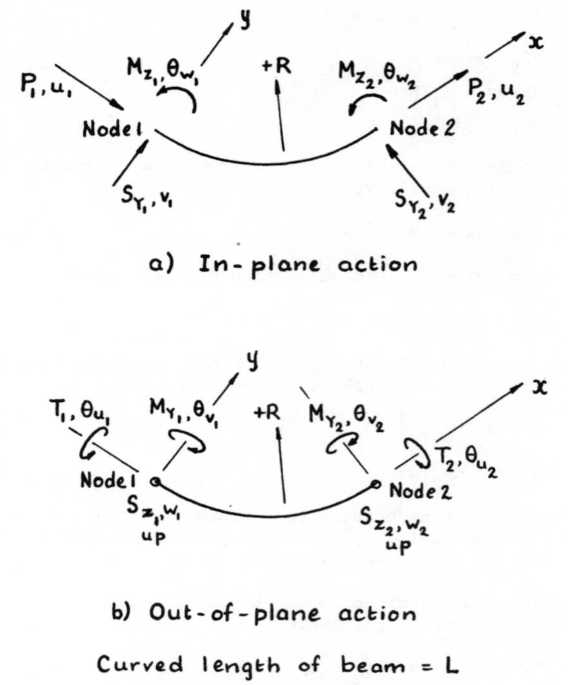

Fig 2. Nodal forces and displacements in curved beam.

Firstly, noting that the displacements u, v, θ_w depend only on a_1 .... a_6, and letting

$$s = \sin (L/R), \; c = \cos (L/R)$$

the in-plane nodal displacements may be expressed as

$$\begin{bmatrix} u_1 \\ v_1 \\ \theta_{w_1} \\ u_2 \\ v_2 \\ \theta_{w_2} \end{bmatrix} = \begin{bmatrix} & -1 & & & R & 1 \\ 1 & & 1 & & & \\ & & & & 2 & 1/R \\ FL & -c & s & Rs-Lc & Ls+Rc & 1 \\ 1 & s & c & Ls & Lc & \\ FL/R & & & 2s & 2c & 1/R \end{bmatrix} \begin{bmatrix} a_1 \\ a_2 \\ a_3 \\ a_4 \\ a_5 \\ a_6 \end{bmatrix}$$

or, more shortly,

$$\left[\delta_a\right] = \left[C_a\right]\left[a\right] \quad .... \; 37$$

Secondly, in considering the out-of-plane displacements it is observed that, in addition to depending upon b_3 ... b_6, the rotation θ_u is also dependent upon a_1, a_4 and a_5. Thus in order to assemble a 6 x 6 matrix similar to that shown by Eqns 37, θ_u must be replaced by λ such that, from Eqn 33,

$$\lambda = \theta_u - a_1 U - a_4 V \cos \psi + a_5 V \sin \psi \quad .... \; 38$$

Hence the nodal values of w, λ and θ_v may be written

$$\begin{bmatrix} w_1 \\ \lambda_1 \\ \theta_{v_1} \\ w_2 \\ \lambda_2 \\ \theta_{v_2} \end{bmatrix} = \begin{bmatrix} 1 & & & 1 & & \\ & & & -1/R & H & \\ & -1 & -1/R & & & -1 \\ 1 & L & s & c & Ls & Lc \\ & & -s/R & -c/R & Hc-Ls/R & -Hs-Lc/R \\ & -1 & -c/R & S/R & -s-Lc/R & -c+Ls/R \end{bmatrix} \begin{bmatrix} b_1 \\ b_2 \\ b_3 \\ b_4 \\ b_5 \\ b_6 \end{bmatrix}$$

or, more compactly,

$$\left[\lambda\right] = \left[C_b\right]\left[b\right] \qquad \text{......... 39}$$

The independent 6 x 6 matrices in Eqns 37 and 39 may now be inverted algebraically, producing the result

$$\begin{bmatrix} a \\ b \end{bmatrix} = \begin{bmatrix} C_a^{-1} & \\ & C_b^{-1} \end{bmatrix} \begin{bmatrix} \delta_a \\ \lambda \end{bmatrix} \qquad \text{......... 40}$$

in which the elements of $\left[C_a\right]^{-1}$ and $\left[C_b\right]^{-1}$ are α_{mn} and β_{mn}, m and n representing the row and column numbers respectively.

Finally, Eqn 40 must be rewritten so that $\left[\lambda\right]$, i.e. $\left[w_1\ \lambda_1,\ \theta_{v_1},\ w_2,\ \lambda_2,\ \theta_{v_2}\right]^T$, is replaced by $\left[\delta_b\right]$, i.e. by $\left[w_1,\ \theta_{u_1},\ \theta_{v_1},\ w_2,\ \theta_{u_2},\ \theta_{v_2}\right]^T$.

Observing from Eqn 38 that

$$\lambda_1 = \theta_{u_1} - a_1U - a_4V \qquad \text{......... 41a}$$

$$\lambda_2 = \theta_{u_2} - a_1U - a_4Vc + a_5Vs \qquad \text{......... 41b}$$

any particular constant b_m in Eqns 40 may be written

$$b_m = \beta_{m1}w_1 + \beta_{m2}\theta_{u_1} - \beta_{m2}(a_1U + a_4V) + \beta_{m3}\theta_{v_1} + \beta_{m4}w_2 + \beta_{m5}\theta_{u_2} - \beta_{m5}(a_1U + a_4Vc - a_5Vs) + \beta_{m6}\theta_{v_2} \qquad \text{......... 42}$$

Since

$$a_m = \alpha_{m1}u_1 + \alpha_{m2}v_1 + \alpha_{m3}\theta_{w_1} + \alpha_{m4}u_2 + \alpha_{m5}v_2 + \alpha_{m6}\theta_{w_2} \quad \text{.. 43}$$

Eqn 42 may be expressed as

$$b_m = \beta_{m1}w_1 + \beta_{m2}\theta_{u_1} + \beta_{m3}\theta_{v_1} + \beta_{m4}w_2 + \beta_{m5}\theta_{u_2} + \beta_{m6}\theta_{v_2} + \gamma_{m1}u_1 + \gamma_{m2}v_1 + \gamma_{m3}\theta_{w_1} + \gamma_{m4}u_2 + \gamma_{m5}v_2 + \gamma_{m6}\theta_{w_2} \quad \text{.... 44}$$

where

$$\gamma_{mn} = -\left[U\alpha_{1n}(\beta_{m2} + \beta_{m5}) + V\alpha_{4n}(\beta_{m2} + c\beta_{m5}) - Vs\alpha_{5n}\beta_{m5}\right] \qquad \text{......... 45}$$

Thus Eqns 40 may be rewritten as

$$\begin{bmatrix} a \\ b \end{bmatrix} = \begin{bmatrix} C_a^{-1} & \\ M & C_b^{-1} \end{bmatrix} \begin{bmatrix} \delta_a \\ \delta_b \end{bmatrix} = \left[C\right]^{-1}\left[\delta\right] \quad \text{...... 46}$$

in which any element of $\left[M\right]$ is γ_{mn}.

Now using Eqns 46 together with the stress resultant/displacement relationships enables the stress resultants along the beam and hence the nodal forces to be written in terms of the nodal displacements.

Denoting sin ψ and cos ψ by f and g respectively and letting

$$W = 2GJI_{YZ}/(EI_Y + GJ),\quad Z = 2(EQ + GJI_Z)/(EI_Y + GJ),$$
$$Q_R = Q/(I_YR)$$

the values of T and M_Z are given from Eqns 13 and 18 as

$$T = GJ/R\left[0\ 0\ 0\ -Vf\ -Vg\ 0\ 0\ 1\ 0\ 0\ -Hf\ -Hg\right]\left[C\right]^{-1}\left[\delta\right] \qquad \text{......... 47}$$

$$M_Z = E/R\left[Q_R\ 0\ 0\ Zg\ -Zf\ 0\ 0\ 0\ 0\ 0\ Wg\ -Wf\right]\left[C\right]^{-1}\left[\delta\right] \qquad \text{......... 48}$$

The value of M_Y can now be found from either Eqn 9 or Eqn 17, while P, S_Y and S_Z can be obtained from Eqns 4, 5 and 10 respectively, these stress resultants being then given by

$$M_Y = GJ/R\left[0\ 0\ 0\ Vg\ -Vf\ 0\ 0\ 0\ 0\ 0\ Hg\ -Hf\right]\left[C\right]^{-1}\left[\delta\right] \qquad \text{......... 49}$$

$$P = E/R^2\left[0\ 0\ 0\ -Zg\ Zf\ 0\ 0\ 0\ 0\ 0\ -Wg\ Wf\right]\left[C\right]^{-1}\left[\delta\right] \qquad \text{......... 50}$$

$$S_Y = E/R^2\left[0\ 0\ 0\ Zf\ Zg\ 0\ 0\ 0\ 0\ 0\ Wf\ Wg\right]\left[C\right]^{-1}\left[\delta\right] \quad \text{.... 51}$$

$$S_Z = GJ/R^2\left[0\ 0\ 0\ 0\ 0\ 0\ 0\ -1\ 0\ 0\ 0\ 0\right]\left[C\right]^{-1}\left[\delta\right] \quad \text{....... 52}$$

Since the nodal forces must be equal in magnitude to the nodal stress resultants and that their relative senses may be ascertained by inspection of Figs 1 and 2, the relationship between the nodal forces and displacements can be written directly as

$$\begin{bmatrix} P_1R^2/E \\ S_{Y_1}R^2/E \\ M_{Z_1}R/E \\ P_2R^2/E \\ S_{Y_2}R^2/E \\ M_{Z_2}R/E \\ S_{Z_1}R^2/(GJ) \\ T_1R/(GJ) \\ M_{Y_1}R/(GJ) \\ S_{Z_2}R^2/(GJ) \\ T_2R/(GJ) \\ M_{Y_2}R/(GJ) \end{bmatrix} = \begin{bmatrix} & & & Z & & & & & & & W & \\ & & & & -Z & & & & & & & -W \\ -Q_R & & & -Z & & & & & & & -W & \\ & & & -Zc & Zs & & & & & & -Wc & Ws \\ & & & Zs & Zc & & & & & & Ws & Wc \\ Q_R & & & Zc & -Zs & & & & & & Wc & -Ws \\ & & & & & & & -1 & & & & \\ & & & & V & & & -1 & & & & H \\ & & & V & & & & & & & H & \\ & & & & & & & 1 & & & & \\ & & & -Vs & -Vc & & & 1 & & & -Hs & -Hc \\ & & & -Vc & Vs & & & & & & -Hc & Hs \end{bmatrix} \left[C\right]^{-1}\left[\delta\right] \qquad \text{......... 53}$$

In the further development it is more convenient to rewrite Eqns 53 so that the left and right hand column vectors take the form

$$\left[P_1,\ S_{Y_1},\ S_{Z_1},\ T_1,\ M_{Y_1},\ M_{Z_1},\ P_2,\ S_{Y_2},\ S_{Z_2},\ T_2,\ M_{Y_2},\ M_{Z_2}\right]^T = \left[P\right]$$

$$\left[u_1,\ v_1,\ w_1,\ \theta_{u_1},\ \theta_{v_1},\ \theta_{w_1},\ u_2,\ v_2, w_2,\ \theta_{u_2},\ \theta_{v_2},\ \theta_{w_2}\right]^T = \left[\Delta\right]$$

respectively. These equations can then be written

$$\left[P\right] = \left[k\right]\left[\Delta\right] \qquad \text{......... 54}$$

where $\left[k\right]$ is the stiffness matrix.

It can be noted that $[k]$ is fully populated if I_{YZ} is non-zero, and composed of independent in-plane and out-of-plane portions when I_{YZ} vanishes. In both cases the matrix is symmetrical about the leading diagonal.

FORMULATION OF DISPLACEMENT TRANSFORMATION MATRIX

The local co-ordinate system in which the stiffness matrix $[k]$ has been derived is adequate only for the analysis of structural configurations in which there exist no discontinuities. Most structural forms, however, do exhibit discontinuities, the most common of these being at the joints of frameworks, and hence it is necessary to express the displacements occurring in the local co-ordinate system to those in a global system common to every element of the structure. The most common and usually the most easily interpreted system is one of rectangular Cartesian co-ordinates.

The relationships between the displacements in the two systems can be conveniently formulated through the product of two simple matrices.

Consider firstly a straight line element having nodes 1. and 2. contained in local axes p, q, r and lying in a global co-ordinate system X, Y, Z, as shown in Fig 3a. The relationships between the displacements $[Z]$ and $[X]$ in the local and global systems respectively are given by the well-known equations.

$$[Z] = \begin{bmatrix} V_o & & & \\ & V_o & & \\ & & V_o & \\ & & & V_o \end{bmatrix} [X] = [V][X] \qquad \ldots\ldots\ldots 55$$

in which

$$[Z] = [p_1, q_1, r_1, \theta_{p_1}, \theta_{q_1}, \theta_{r_1}, p_2, q_2, r_2, \theta_{p_2}, \theta_{q_2}, \theta_{r_2}]$$

$$[X] = [X_1, Y_1, Z_1, \theta_{X_1}, \theta_{Y_1}, \theta_{Z_1}, X_2, Y_2, Z_2, \theta_{X_2}, \theta_{Y_2}, \theta_{Z_2}]$$

and

$$[V_o] = \begin{bmatrix} l_p & m_p & n_p \\ l_q & m_q & n_q \\ l_r & m_r & n_r \end{bmatrix}$$

the elements of $[V_o]$ being the relevant direction cosines that the line element makes with the global axes.

Now consider a curved member whose nodes are those of the line element and which lies in the p - q plane as depicted in Fig 3b. The equations relating the displacements in the local co-ordinate system x, y, z of the curved member and the displacements in the p, q, r system are then given by (Ref 3)

$$[\Delta] = \begin{bmatrix} U_o & & & \\ & U_o & & \\ & & U_o^T & \\ & & & U_o^T \end{bmatrix} [Z] = [U][Z] \qquad \ldots\ldots\ldots 56$$

where

$$[U_o] = \begin{bmatrix} \cos(L/2R) & -\sin(L/2R) & \\ \sin(L/2R) & \cos(L/2R) & \\ & & 1 \end{bmatrix}$$

Thus the relationships between the local displacements $[\Delta]$ and the displacements in the global system $[X]$ are given by

$$[\Delta] = [U][V][X] = [A][X] \qquad \ldots\ldots\ldots 57$$

where $[A]$ is the displacement transformation matrix.

Use of $[A]$ enables the relationships between the nodal force vector $[L]$ referred to the global system and the corresponding displacement vector $[\Delta]$ to be expressed by the standard equation

$$[L] = [A]^T [k][A][\Delta] = [K][\Delta] \qquad \ldots\ldots\ldots 58$$

in which $[K]$ is the global stiffness matrix of the element.

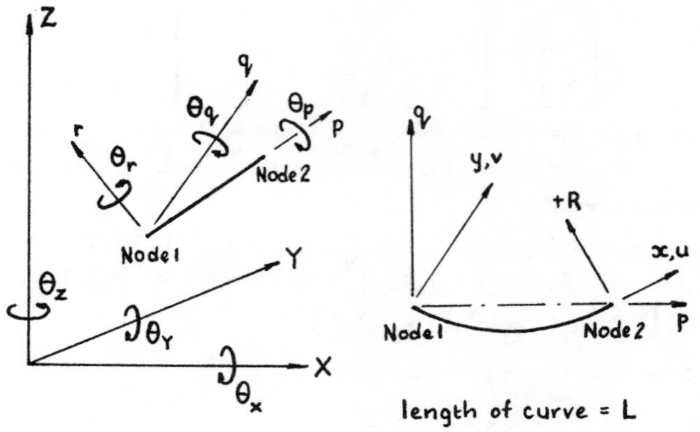

a) Line element b) Curved beam

Fig 3. Relationship of curved beam to local and global axes.

REPRESENTATION OF DISTRIBUTED LOADING BY EQUIVALENT NODAL FORCES

As long as the loading applied to the structure takes the form of nodal point loads, the derivations presented previously are completely adequate for the solution of curved structural elements.

In many practical situations, however, the elements are also subjected to distributed loading, such as self weight, and hence, in order to be amenable to the analysis presented, such loading must be replaced by equivalent nodal forces. The manner in which this is achieved has been well documented (Refs 3, 4) and consists of equating the work done by these nodal forces to the work done by the distributed loads when moving through the deflection profiles produced by nodal forces. Thus it is only deemed necessary to state the results applicable to curved elements.

The process may be conveniently initiated by considering a curved beam subjected to three unit point loads in each of the x, y, z, directions applied at a distance x along the member from node 1. The displacement functions for u, v, w, (Eqns 24, 20, 28) may be expressed as

$$u = \begin{bmatrix} Fx & -\cos\psi & \sin\psi & -x\cos\psi + R\sin\psi & x\sin\psi + R\cos\psi & 1 \end{bmatrix}\begin{bmatrix} a \end{bmatrix} = \begin{bmatrix} N_u \end{bmatrix}\begin{bmatrix} a \end{bmatrix}$$
$$v = \begin{bmatrix} 1 & \sin\psi & \cos\psi & x\sin\psi & x\cos\psi & 0 \end{bmatrix}\begin{bmatrix} a \end{bmatrix} = \begin{bmatrix} N_v \end{bmatrix}\begin{bmatrix} a \end{bmatrix}$$
$$w = \begin{bmatrix} 1 & x & \sin\psi & \cos\psi & x\sin\psi & x\cos\psi \end{bmatrix}\begin{bmatrix} b \end{bmatrix} = \begin{bmatrix} N_w \end{bmatrix}\begin{bmatrix} b \end{bmatrix} \quad \ldots\ldots\ldots 59$$

and the vectors of the equivalent nodal forces due to the three unit loads in the x, y, z directions then written respectively as

$$\left\{\begin{bmatrix} N_u \end{bmatrix}\begin{bmatrix} C \end{bmatrix}^{-1}\right\}^T, \left\{\begin{bmatrix} N_v \end{bmatrix}\begin{bmatrix} C \end{bmatrix}^{-1}\right\}^T, \left\{\begin{bmatrix} N_w \end{bmatrix}\begin{bmatrix} C \end{bmatrix}^{-1}\right\}^T$$

it being noted that the elements in these vectors take the same order as those included in the left hand vector of Eqns 53.

Having obtained the equivalent nodal forces for the unit point loads, the equivalent forces for any distributed loading condition can be obtained by suitable integration.

APPLICATION TO CURVED FRAMEWORKS

To demonstrate the scope of the analyses, two forms of curved structure are examined. The first of these is an annular grillage, as shown in Fig 4, supported at the nodes along its outer boundary and exhibiting inner curved beams of either rectangular or L-section form. Typical results are presented showing the effect of this variation of sectional geometry on deflections and bending moments.

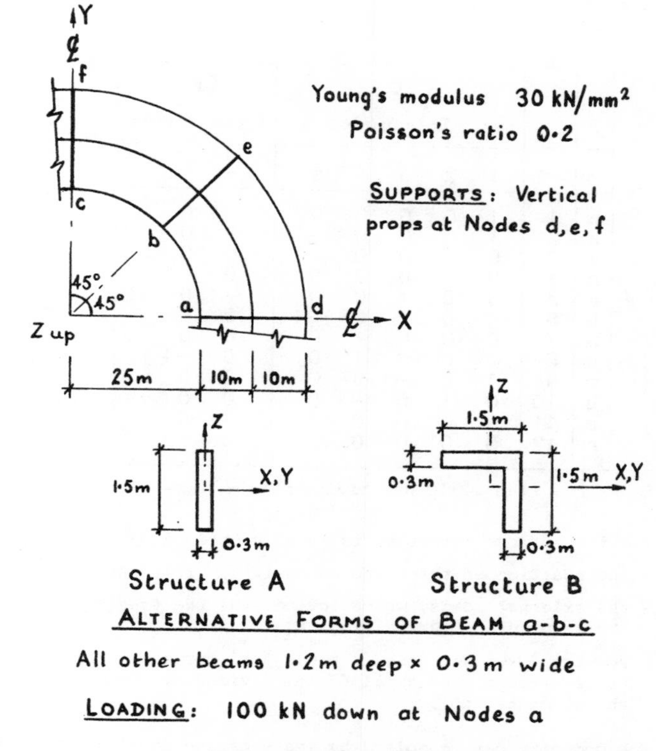

		Vert. defl^n, Z, (mm)	Horiz. defl^n, X, Y, (mm)	Vert. B.M. (kNm)	Horiz. B.M. (kNm)
Structure A	Nodes a	-155·1	0	694·6	0
	Nodes c	-103·6	0	84·3	0
Structure B	Nodes a	-119·6	-9·9 (x)	771·3	228·4
	Nodes c	-80·2	8·6 (Y)	139·3	196·2

Fig 4. Annular grillage

The second structure presented is a five-bay curved space frame composed of members of rectangular section, as shown in Fig 5. Under the loading considered, the predominant behaviour is that due to in-plane action in the cross-frames, as typified in the bending moment diagram shown.

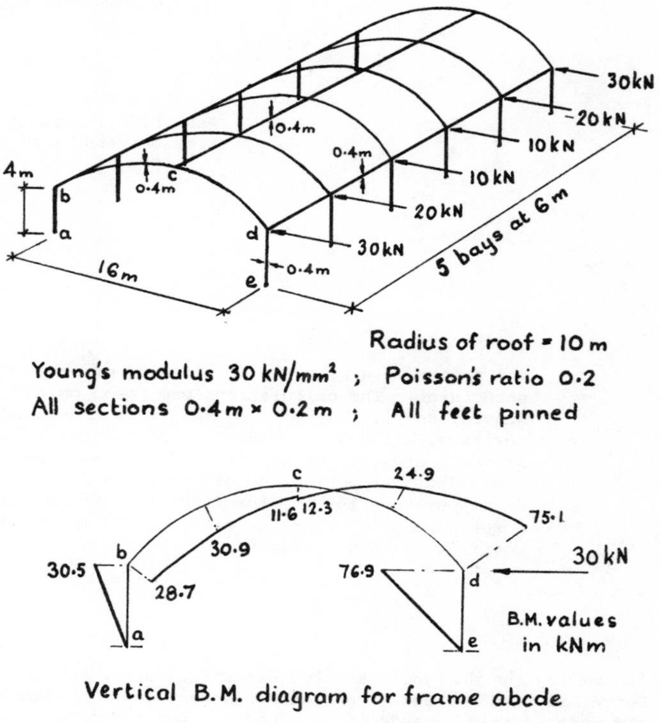

Fig 5. Curved space frame

CONCLUSIONS

The analysis and results presented show that, with the use of exact displacement functions, the solution of frameworks containing curved beams may be treated with the same efficiency as those composed only of straight elements.

It should be noted, however, that the effect of warping restraint is neglected. Such restraint may have a considerable effect on the behaviour of open sections (Ref 5), and hence the analysis developed is considered most applicable to those sections in which warping effects are of little importance.

REFERENCES

1. ODEN, J.T., Mechanics of Elastic Structures, McGraw-Hill Book Co. Inc., New York, N.Y., 1967.
2. SIDHU, J.S., The Application of Finite Element Theory to Curved Elements, Ph.D. thesis, University of Aston in Birmingham, England, 1973.
3. JUST, D.J., Circularly curved beams under plane loads, J. Struct. Div., ASCE, Vol.108, No.ST8, August, 1982.
4. JENKINS, W.M., Matrix and Digital Computer Methods in Structural Analysis, McGraw-Hill Book Co. Inc., New York, N.Y., 1969.
5. REDDY, M.N., Stiffnesses of circular arc-I-section girders, J. Struct. Div., ASCE, Vol.104, No.ST6, June, 1978.

THE FORMFINDING OF "MIXED STRUCTURES"

M. MOLLAERT

Ass. Prof., Department of Civil Engineering
Free University of Brussels (V.U.B.), Belgium

The paper deals with the formfinding of cable nets (stressed skin systems), grid shells and mixed structures. The calculations are based on the force density method, which has been developed by Scheck, Ref 1. The method has always been used with the restriction that all stresses should be positive, which is a natural condition for flexible materials. It will be demonstrated that compression arcs can be handled with the same ease.
The additional conditions of the force density method have been extended. A methodology has been developed to allow the formfinding of "form compatible" modules. Since part of them can be in compression and others in tension, the formfinding of grid shells, supporting a flexible roof, is possible by this approach.

Generally, the design of a structure starts after the geometry has been defined. For tensile structures for example the form largely depends on the load and the supports. The formfinding becomes part of the analysis.

THE FORCE DENSITY METHOD

The force density method is a FORMFINDING method, and does not require a full description of the geometry to start the equilibrium calculations of the structure.
The cable net, truss system or membrane structure is idealised as an assembly of linear pin joint elements. The nodes and branches are numbered, and the connectivity is specified as follows : the connectivity matrix contains for each branch a value 1 for its beginpoint and -1 for its endpoint (the beginpoints have the lowest nodenumbers), Figs 1, 2 and 3.

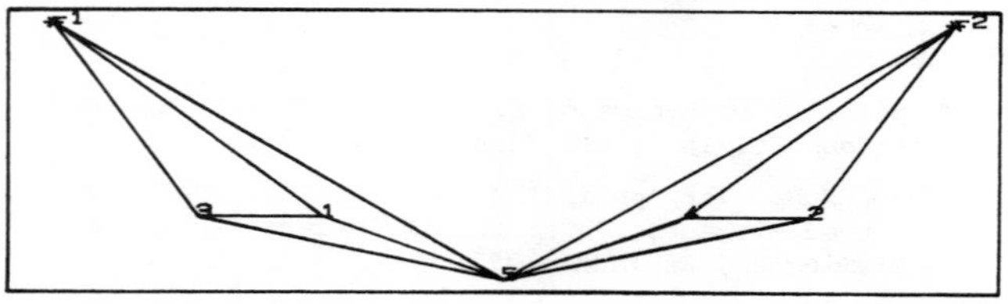

Fig 1 : Node numbers.

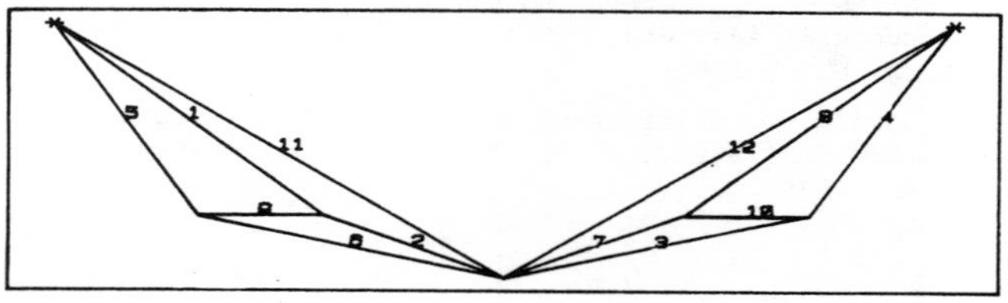

Fig 2 : Branch numbers.

		C					CF	
		free nodes					fixed nodes	
		1	2	3	4	5	−1	−2
b	1	1	0	0	0	0	−1	0
r	2	1	0	0	0	−1	0	0
a	3	0	1	0	0	−1	0	0
n	4	0	1	0	0	0	0	−1
c	5	0	0	1	0	0	−1	0
h	6	0	0	1	0	−1	0	0
n	7	0	0	0	1	−1	0	0
u	8	0	0	0	1	0	0	−1
m	9	1	0	−1	0	0	0	0
b	10	0	1	0	−1	0	0	0
e	11	0	0	0	0	1	−1	0
r	12	0	0	0	0	1	0	−1

Fig 3 : Connectivity matrix.

The form of the structure is totally defined by :

- the position of the fixed nodes {XF}, {YF}, {ZF}
- the external loads, which act only at the free nodes of the structure : {PX}, {PY}, {PZ}
- and the force densities : these are, for each cable or truss element, the axial forces divided by the length of the elements, {Q}.

The geometry can be solved by means of the equilibrium equations. These express that the internal forces have to balance the external loads in each free node :

$$[C^t.Q^*.C].\{X\} = \{PX\} - [C^t.Q^*.CF].\{XF\}$$
$$[C^t.Q^*.C].\{Y\} = \{PY\} - [C^t.Q^*.CF].\{YF\} \qquad (1)$$
$$[C^t.Q^*.C].\{Z\} = \{PZ\} - [C^t.Q^*.CF].\{ZF\}$$

C and CF are specified in Fig 4.

Or
$$[D].\{X\} = \{RX\}$$
$$[D].\{Y\} = \{RY\}$$
$$[D].\{Z\} = \{RZ\}$$

$*$: indicates that a vector is written as a diagonal matrix.

{X}, {Y}, {Z} are the coordinates of the free nodes, which can be solved from this linear system of equations.
These equations express that, if the fixed nodes, the loads and the connectivity stay unaltered, a different shape will be generated for each set of force densities. This makes the force densities suitable for the description of the form.

Figure 4 gives a general flow chart of the method.

The force density method.

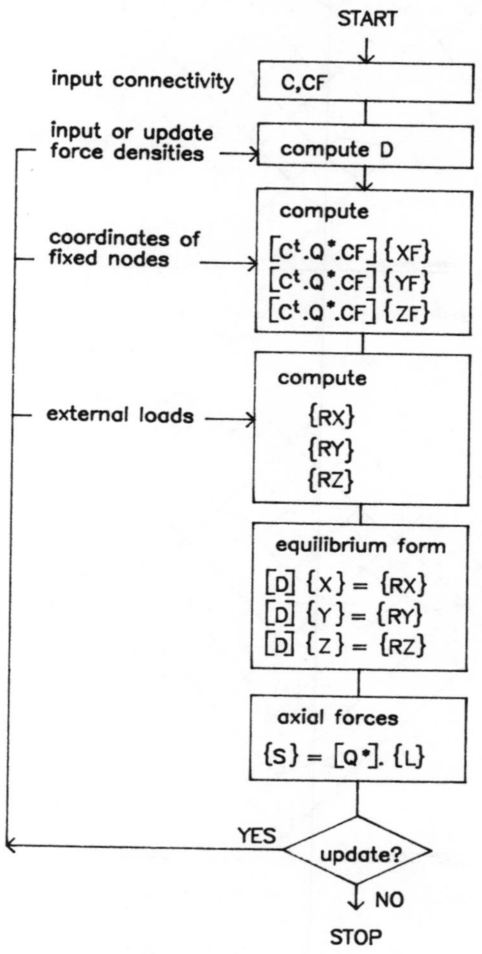

Fig 4 : General flow chart of the force density method.

The method can be used for the design of pending roofs. Figures 5 and 6 illustrate the effect of multiplying the force densities in the boundary elements by 5.

For a saddle shaped roof the force densities in the boundary elements have been inversed, which means that the roof is held up by four arcs, which act in compression. A totally different shape is generated. Figs 7 and 8.

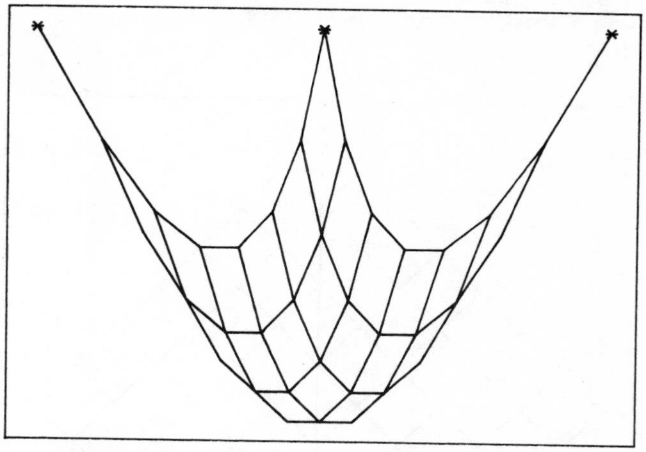

Fig 5 : Force density in the boundary elements : 4.

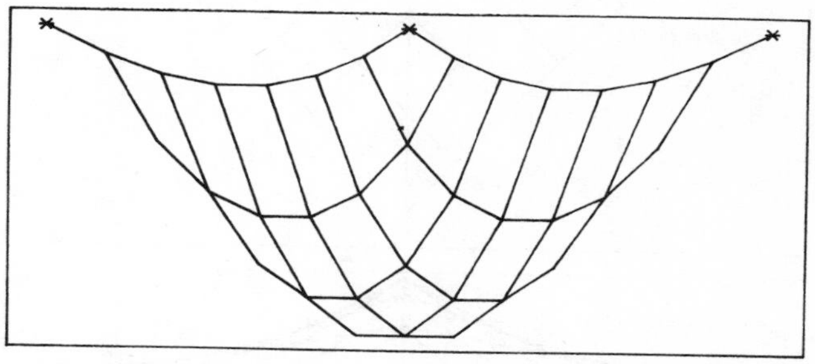

Fig 6 : Force density in the boundary elements : 20.

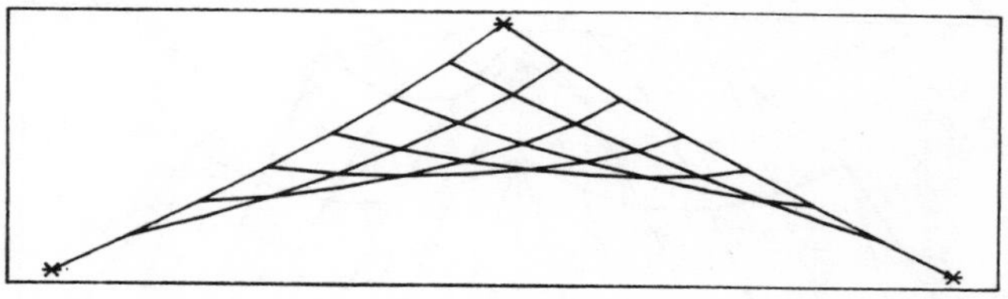

Fig 7 : Force density in the boundary elements : 20.

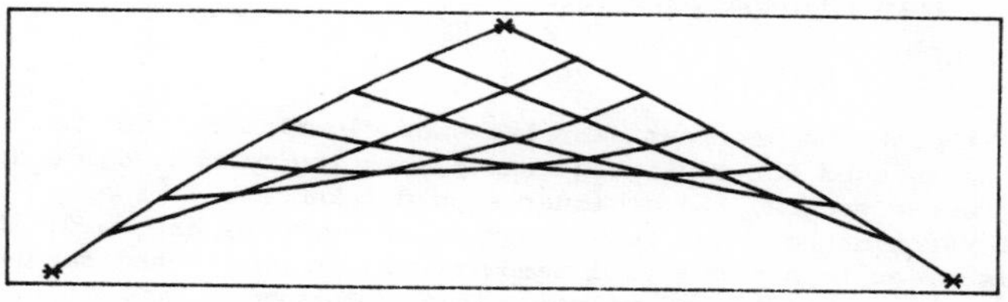

Fig 8 : Force density in the boundary elements : -20.

The pending roof of Figs 5 and 6, can be prestressed by one or two diagonal arcs. The additional linear elements were given a negative force density, Figs 9 and 10.

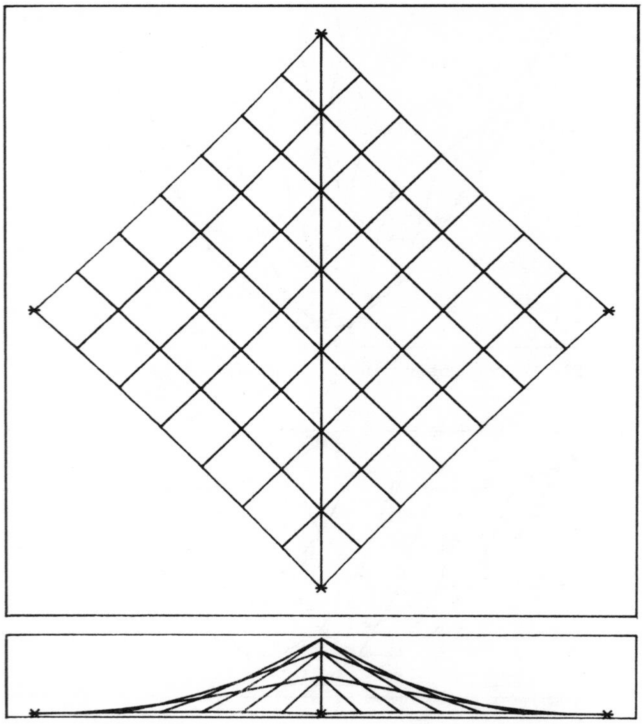

Fig 9 : One diagonal as supporting arc.

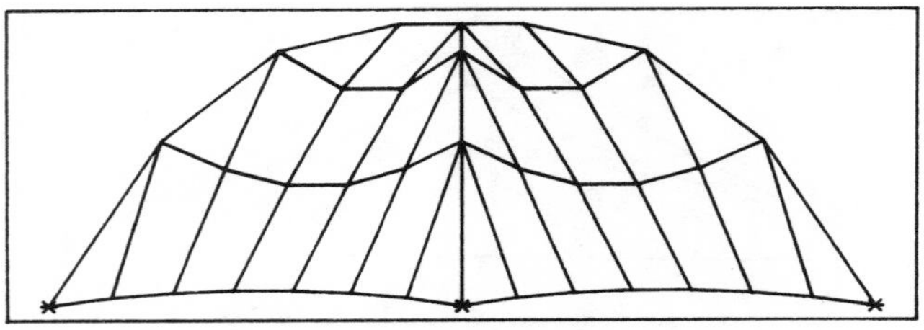

Fig 10 : Two supporting arcs.

Although the previous examples show clearly that the method can be used for configurations consisting of both cable and truss elements, the designer should take care : if only tension members (or only compression members) are used, it is clear to see that all members will be positioned smoothly between the fixed points. Once compression members are added very unconvenient forms can result from the equilibrium calculations.
Figures 11 and 12 illustrate an overlap of the structure as a result of decreasing the force density of only one element from -0.3 to -0.8.

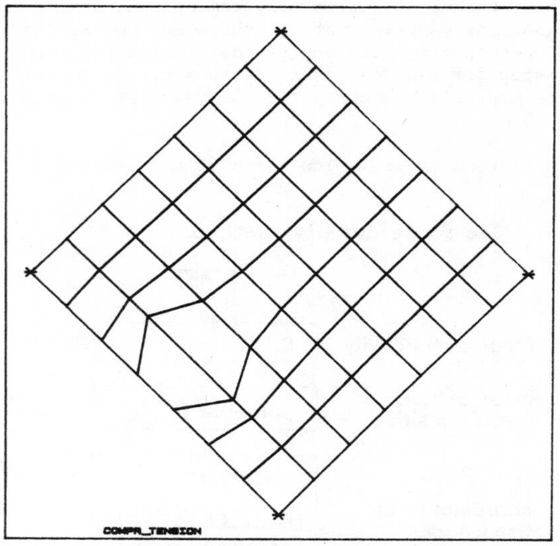

Fig 11 : Force density in compression element : -0.3.

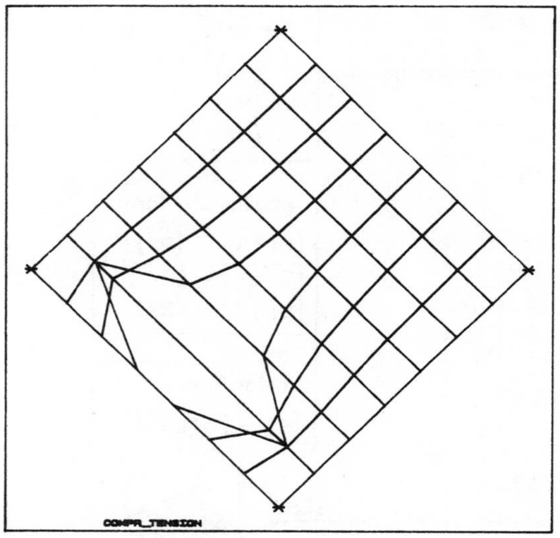

Fig 12 : Force density in compression element : -0.8.

ADDITIONAL CONSTRAINTS FOR THE FORCE DENSITY METHOD

1) Figure 13 illustrates the design of an anticlastic roof, calculated with uniform force density values. The roof pattern shows some irregularities near the boundary (slight buckling of the elements).

2) Figure 14 gives layout and front view of a design for a barrel vault.
The force densities of elements in similar positions were choosen equal. The unstrained lenghts are not uniform, which makes the construction very difficult.

3) Figure 15 shows the design of a hyperboloid, with its fixed nodes at level zero. Since the structure should stand above this level, the design must be modified.

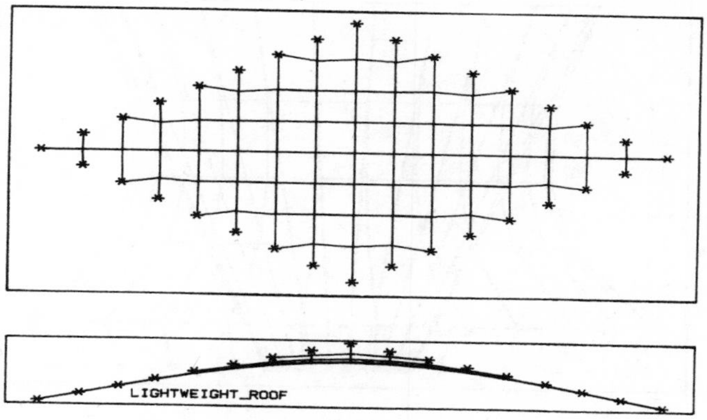

Fig 13 : Initial equilibrium form of an anticlastic roof.

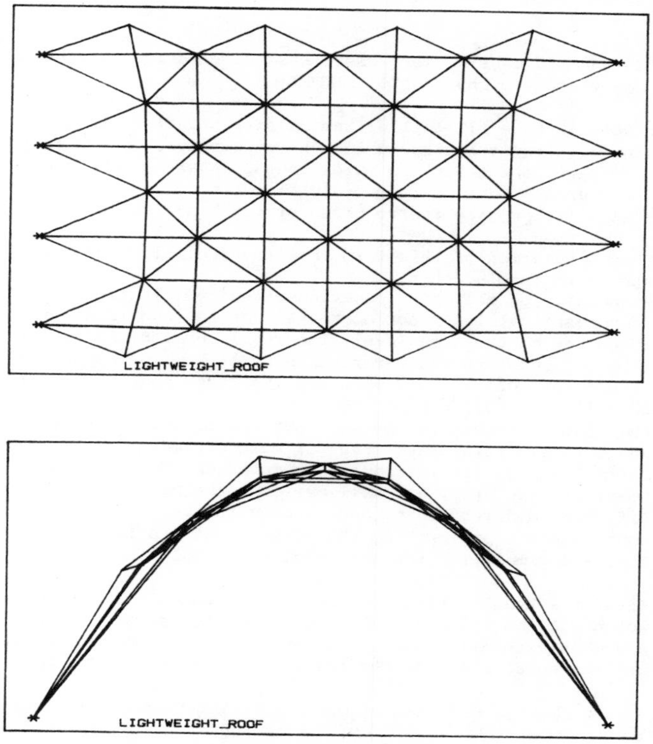

Fig 14 : Initial equilibrium form for a barrel vault.

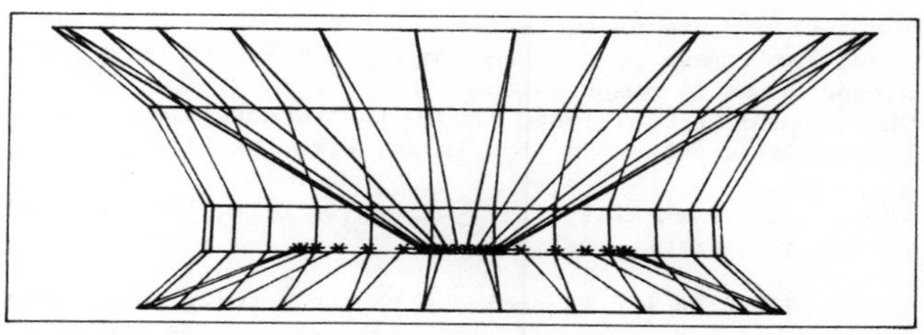

Fig 15 : Initial equilibrium form for a hyperboloid structure.

The force density method is a linear model, in which the force densities are choosen parameters. If specific design requirements have to be satisfied, the force densities must be calculated to fulfil these requirements. Mostly the additional constraints are non linear. The calculation of the force densities must resort to an iterative process.

A general flow chart is given in Fig 16.

Iterative process for the additional constraints

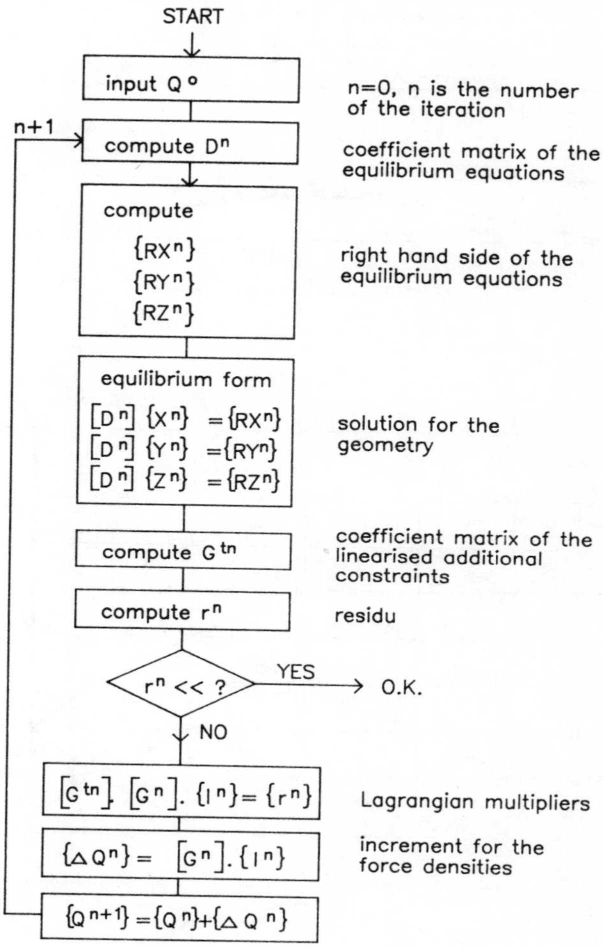

Fig 16 : Flow chart for the force density calculations with additional constraints.

The following additional constraints can be specified :

1) In the anticlastic roof of Fig 13, <u>all axial forces</u> were prescribed to a uniform value.

2) In the vault of Fig 14, <u>the unstrained length</u> of similar elements were forced to <u>have uniform values</u>.

3) The <u>height of the nodes</u> of the lowest and highest ring of the <u>hyperboloid has</u> been prescribed.

The method also allows to prescribe the strained length of some elements, the angle between elements, and the X- or Y-coordinate of some free nodes.

By the subsequent corrections of the force densities, (in the iterative process), the sign of some of them could be altered. An additional condition can be specified to prevent this.

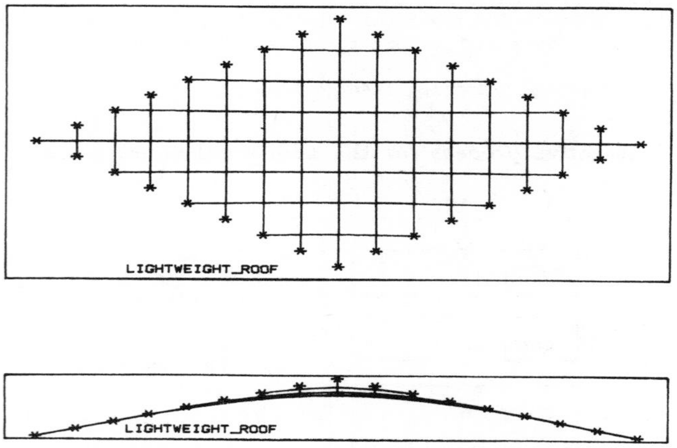

Fig 17 : Uniformly prestressed roof.

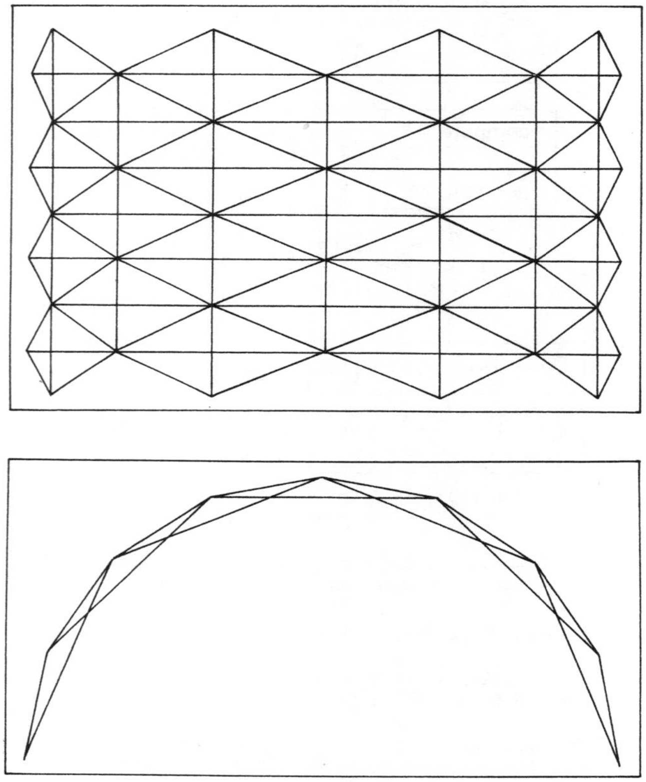

Fig 18 : Equilibrium form with prescribed initial lengths.

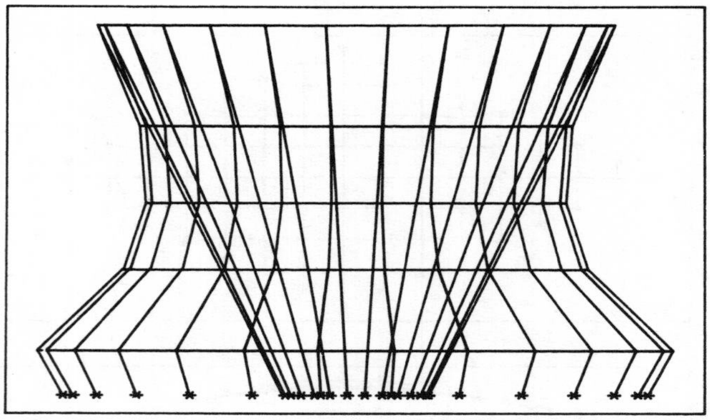

Fig 19 : Modified design for the hyperboloid.

THE DESIGN OF FORM COMPATIBLE MODULES

"In the old days we were restricted to classical shapes, now we can do anything". (BERGER)

Figure 15 is a first illustration of a structure, consisting of three sub-structures : the hyperboloid is the tensile part, the inner and outer cone both are compression structures. Each of them could be replaced by anticlastic shells, to provide the required strength against buckling.

The calculation of this form took the own weight of the structure into account.
Inspection of the third equation of system (1) reveals that if no vertical loads act on the structure, and if all fixed nodes are at the same level, the resulting form is flat :
This implies that the formfinding of lightweight space structures({PZ} = {O}), with all its fixed nodes at level zero ({ZF} = {O}) can not be handled.
The last statement is the most natural boundary condition, and both conditions are true for most projects.
The following will introduce a modular way for designing complex structures, and will overcome the mentioned problem. Different modules can be designed by the same additional constraints : prescribing the position of the boundary nodes, and simulating the action of the modules on each other.

First the tensile part of a roof is designed. The forces which will be introduced by the supporting structure are added as external loads. They simulate the two grid shells at the equal sides of the triangular structure, Fig 20.

It is clear that the size and the orientation of the boundary forces will largely influence the shape of the roof.

Secondly the supporting grid shells are calculated.
Figure 21 gives the initial form, which is the result of the equilibrium calculations for a "compression structure", with equal but opposite loads along the boundary which must be compatible with the edge of the tensile roof.

The author wants to emphasise, that the position of the fixed nodes must be choosen properly to ensure that all elements are in compression.
They should be positioned around the intersection point of the global resultant of the loads, with the floorplan.

Additional constraints are required to bring the boundary nodes at the right position, Fig 22.

Two grid shells are assembled with the tensile part, and can be analysed under different loadcases. The global structure is drawn in Fig 23.

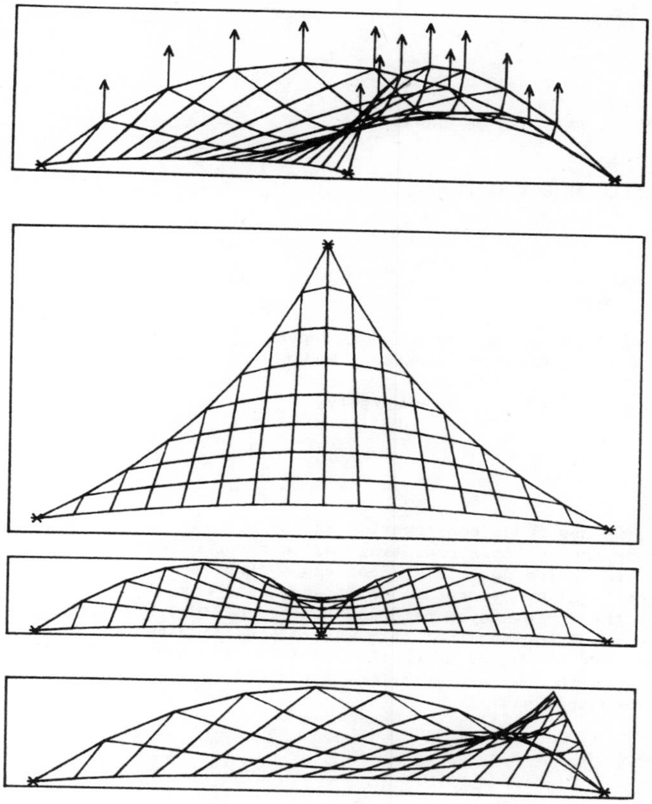

Fig 20 : Design of a tensile roof. All supporting nodes at level zero.

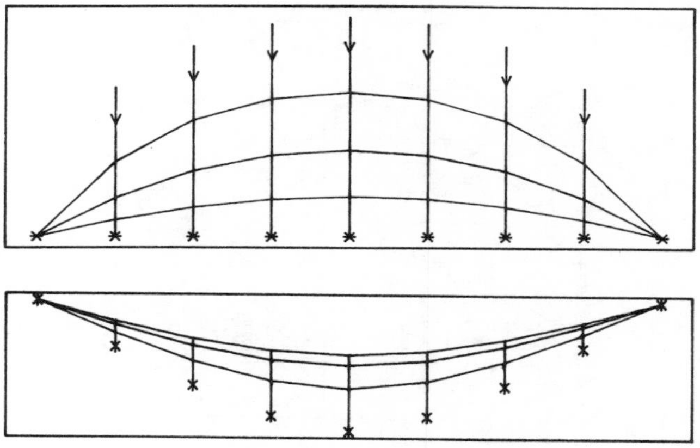

Fig 21 : Initial design for the grid shell.

The method can be automated and solves the problem of the formfinding of mixed structures in an efficient way. Since the modules are of smaller size, they require less memory and less computation time for the calculations. The method is well suited to be used in a computer-aided design environment.

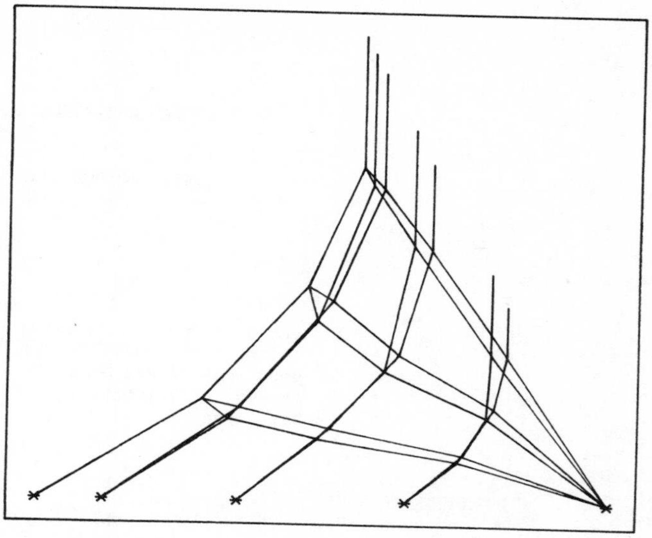

Fig 22 : Supporting grid shells which fit together with the tensile roof.

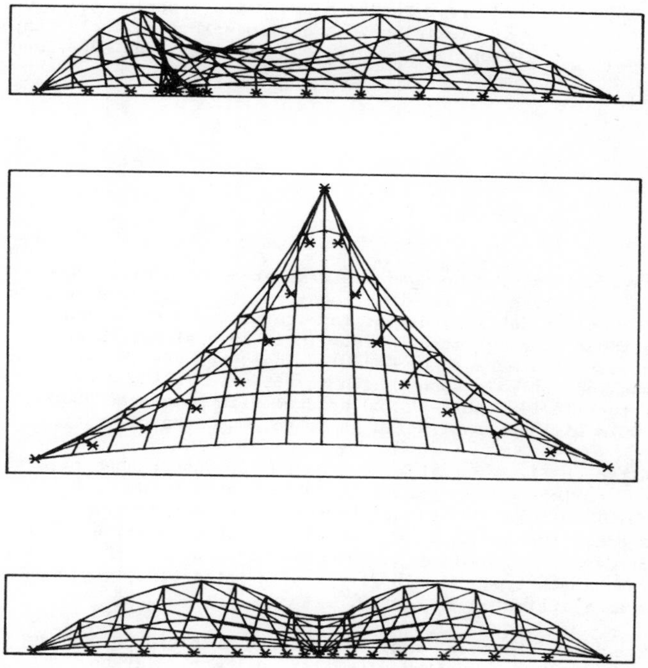

Fig 23 : Global structure : two grid shells, prestressing the tensile roof.

The author hopes that the possibility to incorporate compression members in the formfinding process will allow the designer to enlarge his "structural form" repertoire.

REFERENCES

1. SCHEK Y., The Force Density Method for formfinding and computation of general networks. Computer Methods in Applied Mechanics and Engineering 3, 1974.

THEORETICAL AND EXPERIMENTAL STRESS ANALYSIS OF CABLE-STAYED BRIDGE WITH MULTI-CELLULAR BOX GIRDER

J. NOGUCHI*, M. Eng., Y. TAIDO**, B. Eng., and H. NAKAI***, Dr. Eng.

* Consultant Engineer, Sogo Engineering Inc., Osaka Branch, Osaka, Japan
** Director, Third Osaka Construction Division, Hanshin Expressway Public Corporation, Osaka, Japan
*** Professor, Department of Civil Engineering, Osaka City University, Osaka, Japan

A long span (170m + 350m + 120m) cable-stayed bridge is now under construction on the highway along Osaka Bay, Japan. The main girder, suspended by the multiple radiating cables through the pylons, is a flat steel box girder with multicells. In designing this kind of space structure, there are many unclarified problems concerning the statical behaviors of a multi-cellular box girder which frequently adopts as the main girder of the cable-stayed bridges. This paper deals with the stress analysis of a multi-cellular box girder on the basis of thin-walled elastic beam theory. Furthermore, a few experimental studies are carried out by using the model girder made of plexiglass. A design method of cable-stayed bridge with multi-cellular box girder is proposed from these theoretical and experimental stress analysis.

1. Introduction

The cable-stayed bridges are aesthetic, economical and suitable type for the long span bridges in accordance with the recent developments of fabrications and erections methods of steel structures. There have already been constructed more than 30 cable-stayed bridges in Japan and the longest main span length is 355 m[1]. In the near future, a few cable-stayed bridges with main span 400 ~ 500 m will be constructed in Tokyo and Osaka districts.

Although it is possible to utilize various cross-sectional shapes to the main girder of cable-stayed bridges, a flat multi-cellular box girder is one of appropriate types against the aerodynamic stability. For example, Aji River Bridge in the highway along Osaka Bay, located in Fig. 1., has a such type of main girder with total span length 640 m as illustrated in Fig. 2.

There are, however, many unclarified problems concerning the statical behaviors of a flat multi-cellular box girder.

In order to examine these uncertain points, experimental and theoretical studies were undertaken on the model girders by using plexiglass. Based upon these date, a design code of cable-stayed bridge with the flat multi-cellular box girder is proposed herein. This paper presents these design methods.

2. Longitudinal Bending Stress

It is pointed out from the experimental study on the model girder that the shear lag phenomenon of a flat multi-cellular box girder is somewhat different from the ordinary box girder as shown in Fig. 3, where the following three significant properties can clearly be summerized;

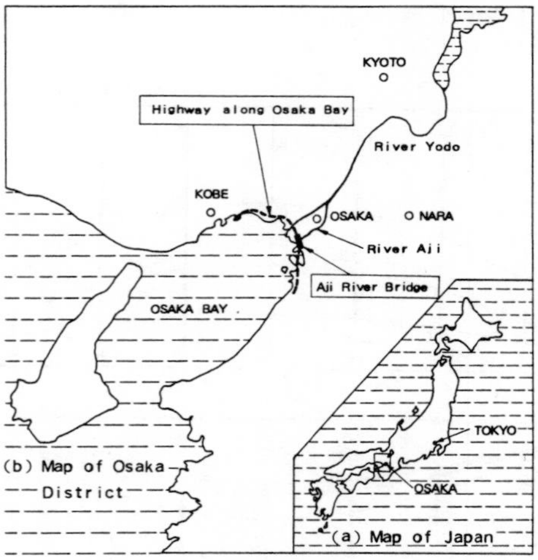

Fig.1 Location of Highway along Osaka Bay and Aji River Bridge

a) There occurs not only primary shear lag between web plates of each box cells, ① through ④, but also secondary shear lag throughout the multi-cellular box girder as is modified in fig. 3.

b) For primary shear lag, the longitudinal bending stress can be approximated by the parabolic curve with 4-th order as reported by P. J. Dowling et al[2].

c) The stress magnification at the exterior box cells ① and ④ due to secondary shear lag should, of course, be taken into accounts in design.

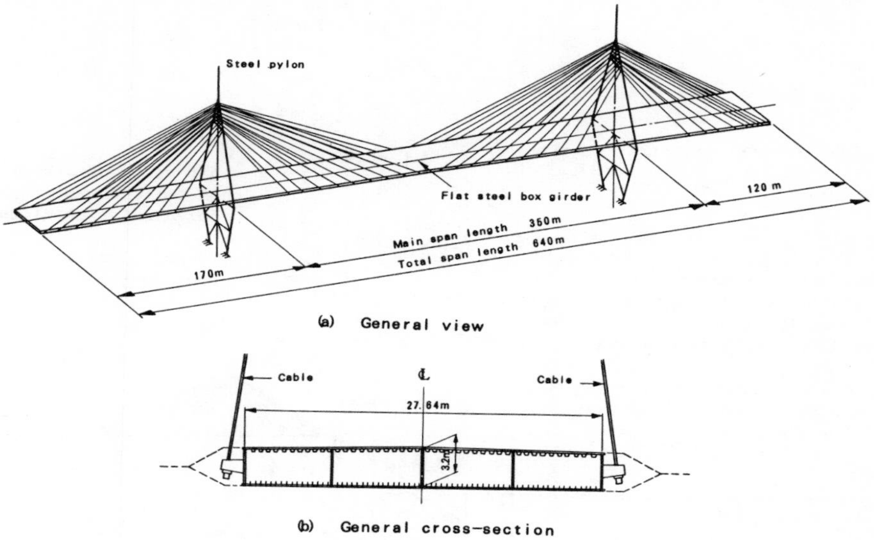

Fig.2 General view and cross-section of Aij River Bridge

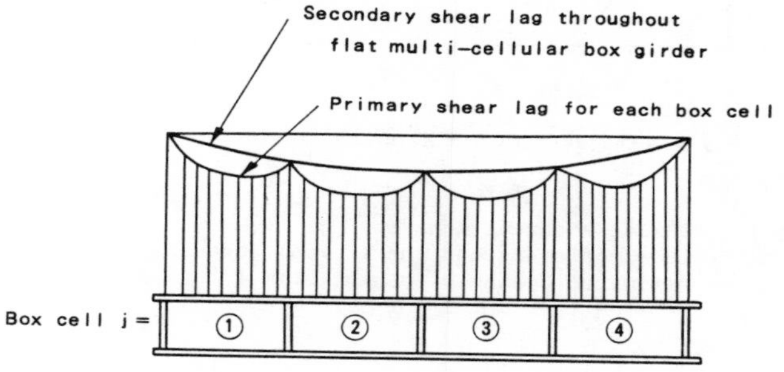

Fig.3 Primary and secondary shear lag phenomenon in flange plate of flat multi-cellular box girder

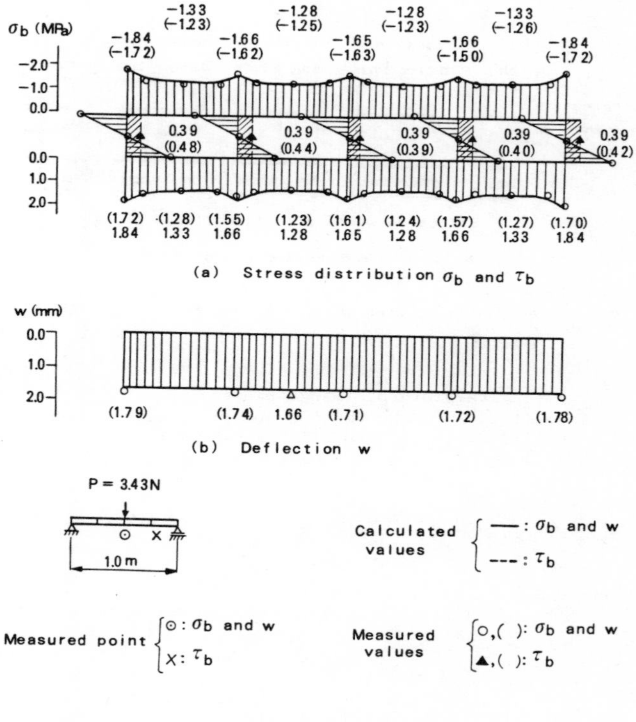

Fig.4 Longitudinal bending stresses and deflections

To clarify these phenomena analytically, a few numerical analyses were also carried out by the finite strip method[3]. Furthermore, an approximate method for evaluating shear lag and effective width of a box beam with single cell was developed by assuming the normal stress distribution in the flange plates as the parabolic curve with fourth order according to the method in reference 4). For the stress concentration due to secondary shear lag, the normal stress at the exterior box cell is modified by the magnification factor.

Fig. 4 illustrates the comparisons of test results with calculated ones for a model girder. Examination of the figure shows that the quite well agreements can be recognized between them.

By applying this method to a cable-stayed bridge, a design code for analyzing the longitudinal bending stress can be proposed as follows[5];

(1) Normal stress formula for longitudinal bending. Let us now idealize a flat multi-cellular box girder with box cells j = ①～④ into a multiple I-girder with I-girders i = 1 ～ 5 as shown in Fig. 5.

Then, the normal stress $\sigma_{b,i}$ for each girder i = 1, 2,···, 5 can be estimated as;

$$\sigma_{b,i} = \frac{M_i}{I_i} Z_i \quad \cdots\cdots\cdots\cdots\cdots (1)$$

where

I_i : geometrical moment of inertia under consideration of effective width of flange plates

z_i : fiber distance from considering point to centroid

and

$$M_i = \frac{I_i^*}{I_g^*} M_g \quad \dots\dots\dots\dots\dots\dots\dots\dots\dots\dots \quad (2)$$

: bending moment for each I-girder

M_g : bending moment acting upon overall the multi-cellular box girder

I_g^* : geometrical moment of inertia throughout the multi-cellular box girder without considering the effective width of flange plates.

I_i^* : geometrical moment of inertia for each I-girder without considering the effective width of flange plates.

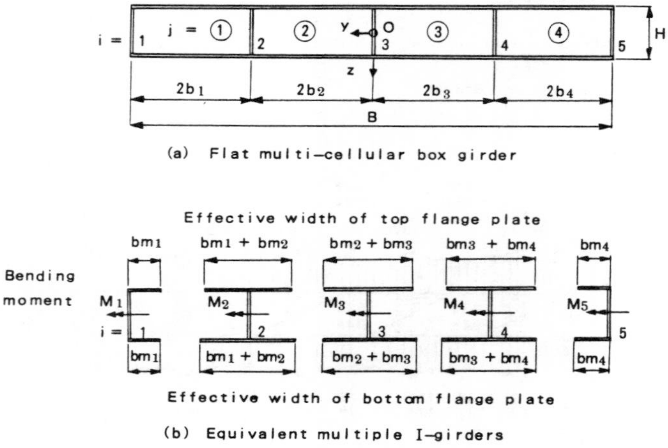

Fig.5 Flat multi-cellar box girder and equivalent multiple I-girders

(2) Effective width of flange plates

It is verified from the various numerical calculations that the effective width of flange plates $b_{m,j}$ for each box cells can practically be estimated as follows;

$$b_{m,j} = \left(1 - \frac{4\lambda}{5 + 3.8\lambda}\right) b_j \quad \dots\dots\dots\dots\dots\dots \quad (3)$$

where

b_j : half of web plate spacing for each box cells

$$\left.\begin{aligned} \lambda &= m_g/M_g \; : \text{for } m_g/M_g > 0 \\ &= 0 \qquad\;\; : \text{for } m_g/M_g \leq 0 \end{aligned}\right\} \quad \dots\dots\dots\dots\dots\dots \quad (4)_{a,b}$$

$$m_g = 1.7\, P\, b_j + 1.6\, p\, b_j^{\,2} \quad \dots\dots\dots\dots\dots\dots \quad (5)$$

: additional bending moment due to shear lag

P : intensity of concentrated load such as live load and reactions at interior support or junction points of cables.

p : intensity of distributed load

(3) Stress magnification factor

The stress magnification factor α due to secondary shear lag should be multiplied by Eq. (2) and α can be expressed as follows;

$$\left.\begin{aligned} \alpha &= 1.155 - 0.0031\,(\ell/b), \quad \ell/b < 50 \\ &= 1.000 \qquad\qquad\qquad\quad , \quad \ell/b \geq 50 \end{aligned}\right\} \quad \dots\dots \quad (6)_{a,b}$$

where

b : half width of web plates for the exterior box cells ① and ④

ℓ : equivalent span of cable-stayed bridge

and

ℓ_c : mid-span length of cable-stayed bridge shown in Fig. 6.

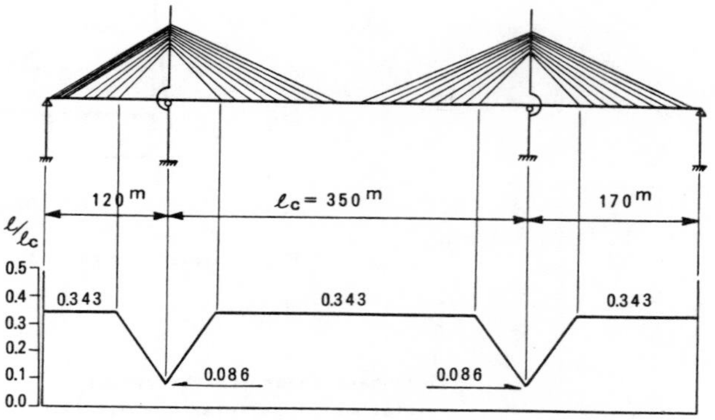

Fig.6 Equivalent span ℓ for cable-stayed bridge

(4) Flexual shearing stress

It is also confirmed that the flexual shearing stress τ_b can be estimated by the elemental beam theory as follows;

$$\tau_b = \frac{S}{A_w} \quad \dots\dots\dots\dots\dots\dots\dots\dots\dots\dots \quad (7)$$

where

S : shearing force due to flexure

A_w : gross cross-sectional area of web plate

3. Transverse Bending Stress

The transverse bending stress is also important in designing the flat multi-cellular box girder. In general, the diaphragms will obviously behave as the floor beam. For this case, the effective width of flange plates were also inquired through experimental researches on model girders and theoretical analysis based upon the fundamental theory of elasticity.

Fig. 7 shows these results in comparison with the values specified in a few design codes[6)~8)]. It is, then, discussed how to estimate an appropriate effective width of flange plate and the corresponding design method of the diaphragms is proposed by virtue of the following procedures[5)].

(1) Normal stress formula for transverse bending

The normal stress σ_B in the diaphragms due to transverse bending can be evaluated as follows;

$$\sigma_B = \frac{M_B}{I_D} Z \quad \dots\dots\dots\dots\dots\dots\dots\dots\dots\dots \quad (8)$$

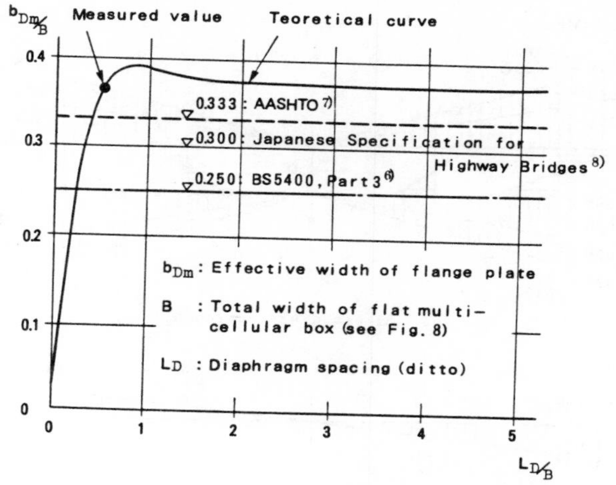

Fig.7   Effective width of flange plates in the transverse bending

where

I_D : geometrical moment of inertia of diaphragm taking into accounts of the effective width of flange plates

z : fiber distance from considering point to centroid

and

M_B : transverse bending moment acting upon the diaphragms regarding as the floor beam, which can be estimated by the grillage approach[9].

(2)   Effective width of flange plate

The effective width of flange plate b_{Dm} cooperating with the diaphragms can be determined by referring Fig. 8 as follows;

$$b_{Dm} = \frac{B}{3} \leq L_D \quad \dots\dots\dots\dots\dots\dots (9)$$

where

B : total width of flat multi-cellular box

L_D : spacing of intermediate diaphragms

(3)   Flexual shearing stress in diaphragm

The corresponding flexual sharing stress τ_b in the diaphragms can be found as follows;

$$\tau_B = \frac{S_B}{A_D} \quad \dots\dots\dots\dots\dots\dots\dots\dots (10)$$

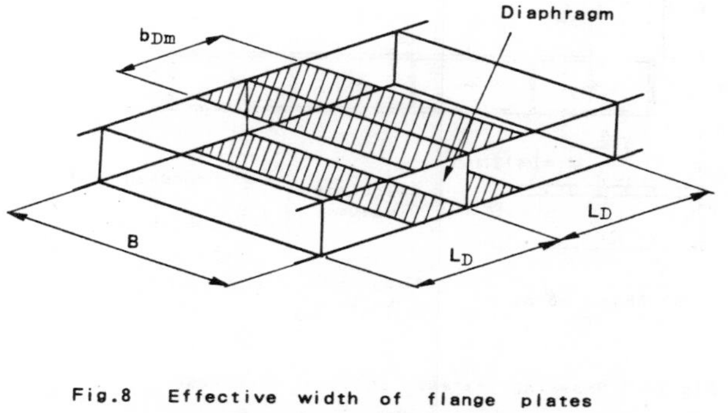

Fig.8   Effective width of flange plates cooperating with the diaphragm

where

S_B : Shearing force acting upon the diaphragm, which can also be estimated by the grillage approach[9].

A_D : Cross-sectional area of diaphragm

4.   Torsional and Warping Stress

The torsional bahaviour of a box girder can be treated by the fundamental equation of torsional angle θ as follows[10];

$$EI_w = \frac{d^4\theta}{dx^4} - GK\frac{d^2\theta}{dx^2} = m_T \quad \dots\dots\dots\dots (11)$$

where

EI_w : torsional warping rigidity

GK : pure torsional rigidity

It is, however, not yet inquired that the above equation can be applied to a flat multi-cellular box girder. Therefore, the experimental studies were now performed on the model girder as shown in Fig. 9. Observing this figure, it seems that the contributions of the interior web plates to the torsional warping are small enough to be ignored and almost all stresses are taken by the exterior web plates as well as top and bottom flange plates. Moreover, stress distributions and deflections vary linearly as are seen from calculated and measured values in Fig. 9.

Therefore, this type of flat multi-cellular box girder can be treated as a flat mono-cellular box girder with width B and depth H. Additional analyses were carried out for a mono-box girder with the ratio $B/H \geq 8$, the following design aids could be obtained[11].

(a)   Stress distribution σ_W and τ_s

(b)   Deflection $w = \theta \cdot y$

M_T 0.413N·m

2.0 m

Calculated values $\begin{cases} \text{—— : } \sigma_W \text{ and } w \\ \text{--- : } \tau_s \end{cases}$

Measured point $\begin{cases} \odot : \sigma_W \text{ and } w \\ \times : \tau_s \end{cases}$

Measured values $\begin{cases} \bigcirc : \sigma_W \text{ and } w \\ \blacktriangle : \tau_s \end{cases}$

—·— ; Deduced values by means of least square method
( )

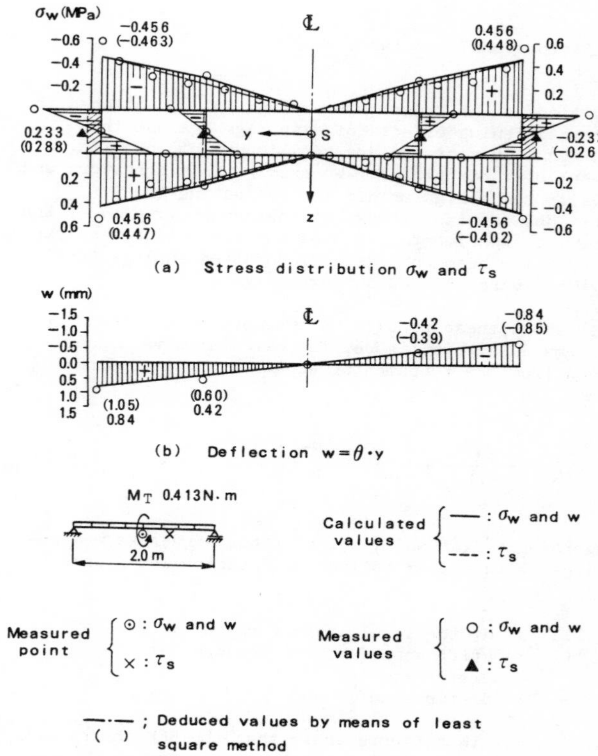

Fig.9   Torsional warping stresses and deflections

(1) Torsional warping stress

The torsional warping stress σ_w can be obtained as follows;

$$\sigma_w = -\frac{\eta H}{5K}\Delta T \quad \text{.............................} \quad (12)$$

where

K : pure torsional constant
H : girder depth (see Fig. 5)
ΔT : step of torsional moment, which can be estimated by the pure torsional theory[10]

$$\eta = -5 \quad \text{for } B/H \geq 8 \quad \text{..................} \quad (13)$$

: non-dimensional parameter depend upon the cross-section of box girder

Besides, Eq. (12) is completely coincided with the design code of BS5400, Part 3[6] by using Eq. (13).

(2) Shearing stress due to pure torsion

The shearing stress τ_s due to pure torsion can easily be found by Bredt's formula[10] as follows;

$$\tau_s = \frac{T}{2Ft} \quad \text{.............................} \quad (14)$$

where

T : pure torsional moment
F : cross-sectional area enclosed by the middle planes of top and bottom flange plates as well as both side of web plates
t : thickness of thin-walled member at considering point

5. Distorsional Warping Stress

In addition to the transverse bending as is mentioned in 3., the stiffening effects of diaphragm against the prevention of distorsion are very important in a flat multi-cellular box girder. To examine this behaviors and to develop the design method for evaluating the distorsional warping stress, strength and rigidity of the diaphragms, experimental studies and theoretical analyses based upon the beam on elastic foundation analogy (BEF) theory[11)~12)] were carried out extensively.

Fig. 10 shows these results on the section at mid-diaphragms. In this figure, the calculated values are obtained from the fundamental equation of distorsional angle θ, i.e.[12)~13)]

$$EI_{Dw}\frac{d^4\theta}{dx^4} + K_{Dw}\theta = \frac{m_T}{2} \quad \text{................} \quad (15)$$

by regarding a flat multi-cellular box girder as a mono-box girder with the exceptions of interior web plates, where

EI_{Dw}: distorsional warping rigidity
K_{Dw}: stiffness of box girder against distorsion
$m_T/2$: distorsional load

Examination of this figure shows that the BEF theory will able to estimate the distorsional warping stress σ_{Dw} correctly, but σ_{Dw} is much more smaller value than σ_b and σ_w.

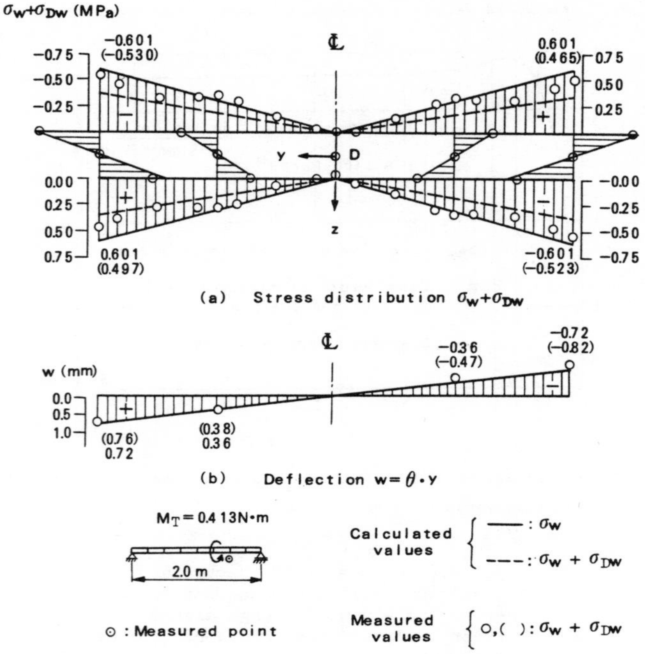

Fig.10 Distorsional warping stresses and deflections including torsional ones

On the other hand, the sharing stress τ_D in the diaphragm was also examined as shown in Fig. 11. From this figure, the shearing stress varies for each box cells and the value obtained by the BEF theory is more or less smaller than the measured values, so that much more accurate and practical method is proposed as detailed in the later.

Through these data and additional numerical calculations, a practical design method for determining distorsional stress in a flat multi-cellular box girder, strength and rigidity of the intermediate diaphragms can be proposed by the following manners[14)~15)].

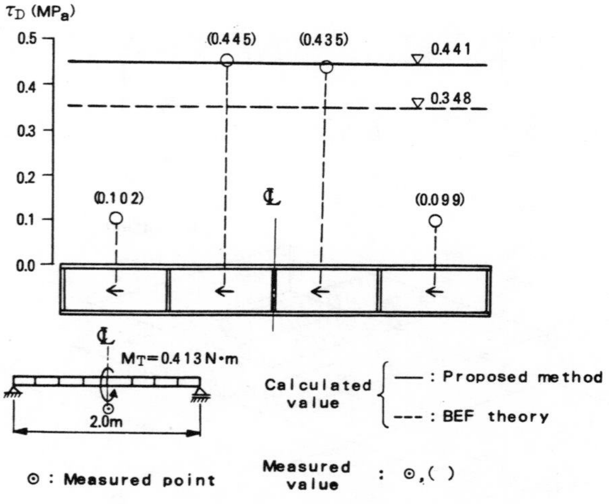

Fig.11 Shearing stress in the diaphragm due to distorsion

(1) Distorsional warping stress
The distorsional warping stress σ_{Dw} in a flat multi-cellular box girder can be neglected for the design in the case where the intermediate diaphragms with the solid web are arranged within the pitch $L_D/B \leqq 0.5$.

(2) Strength of intermediate diaphragms
The intermediate diaphragms should be designed to have enough strength against the shearing stress τ_D as follows;

$$\tau_D = \frac{T_D}{BHt_D} \qquad \text{(16)}$$

where

- H : depth of girder
- B : width of flat multi-cellular box
- t_D : thickness of diaphragm

and

$$T_D = \frac{1}{2}(M_T + m_T L_D) \qquad \text{(17)}$$

- m_T : distributed torque due to live load
- M_T : concentrated torque due to live load, reactions of supports and junction points of cables
- L_D : spacing of intermediate diaphragm (See fig. 8)

(3) Required stiffness for intermediate diaphragms
The intermediate diaphragms should have the following stiffness.

$$K_D \geqq 10\,\frac{EI_{Dw}}{L'^3_D} \qquad \text{(18)}$$

where

$$K_D = GBHt_D \qquad \text{(19)}$$

: stiffness of intermediate diaphragm

$$EI_{Dw} = \frac{E(BH)^2}{48(1+\beta)}\{A_b\beta + A_w(2\beta - 1)\} \qquad \text{(20)}$$

: distorsional warping rigidity

$$\beta = (A_d + 3A_w)/(A_b + 3A_w) \qquad \text{(21)}$$

: cross-sectional parameter

and

- G : shearing modulus of elasticity for steel
- A_d : cross-sectional area of top flange plate including stiffening ribs
- A_w : cross-sectional area of web plate
- A_b : cross-sectional area of bottom flange plate including stiffening ribs.
- L'_D : maximum spacing of intermediate diaphragm (L'_D=0.5B)

6. Design Criteria for Combined Stress

The design criteria for the combined stress in a flat multi-cellular box girder can be written on the basis of allowable stress method as follows[8];

$$\left(\frac{\sigma_x}{\sigma_a}\right)^2 - \left(\frac{\sigma_x}{\sigma_a}\right)\left(\frac{\sigma_y}{\sigma_a}\right) + \left(\frac{\sigma_y}{\sigma_a}\right)^2 + \left(\frac{\tau_{xy}}{\tau_a}\right)^2 \leqq 1.0 \qquad \text{(22)}$$

where

$$\sigma_a = \sigma_y/\nu \qquad \text{(23)}$$

$$\tau_a = \sigma_a/\sqrt{3} \qquad \text{(24)}$$

- σ_y : yield point of steel
- ν : factor of safety (= 1.5 - 1.7)
- σ_a : allowable normal stress
- τ_a : allowable shearing stress

and σ_x, σ_y and τ_{xy} can be given by;

$$\left.\begin{aligned} \sigma_x &= \sigma_b + \sigma_w \\ \sigma_y &= \sigma_B \\ \tau_{xy} &= \tau_b + \tau_s \end{aligned}\right\} \qquad \text{(25)}_{a\sim c}$$

Acknowledgments

In writing this paper, the authors are indebted to all the members of Technical Committee on Highway along Osaka Bay, Hanshin Expressway Public Corporation, for the valuable advices and suggestions. Our thanks are also due to; Mr. S. Fukuoka and Mr. H. Hayashi of Hanshin Expressway Public Corporation as well as Mr. S. Ohta of Osaka Municipal Office in conducting the experimental studies.

References

1) Nakai, H. and Kitada, T.: A Parametric Survey of Cable-stayed Bridge in Japan, Sino-American Symposium on Bridge and Structural Engineering, September 1982, Part I, Beijing, China
2) Moffatt, K. R. and Dowling, P. J.: Shear lag in steel box girder bridges, the Structural Engineer, No. 10, Vol. 53, October, 1975
3) Nakai, H., Taido, Y. and Ohta, S.: Analytical and Experimental Studies on Shear Lag in Multi-cellular Box Girder, Memoirs of Faculty of Engineering, Osaka City University, Vol. 24, Dec. 1983
4) Kondo, K., Komatsu, S. and Nakai, H.: Theoretical and Experimental Researches on Effective Width of Girder Bridges with Steel Deck Plates, Trans. of JSCE, No. 86, October, 1962
5) Nakai, H., Taido, Y. and Hayashi, H.: An Analysis of Shear Lag and Calculation of Effective Width of Flat Box Girder with Multicells, Proc. of JSCE, No. 340, December, 1983
6) British Standard Institution: BS5400, Part 3, Code of practice for design of steel bridges, April, 1982
7) Federal Highway Administration Office of Research and Development: Proposed Specification for Steel Box Girder Bridges, No. FHWA-80-205, January, 1980
8) Japanese Road Association; The Japanese Specification for Highway Bridges, February, 1980
9) Evans, H. R. and Shanmugam, N. E.: An Approximate Grillage Approach to the Analysis of Cellular Structures, Proc. of ICE, Part 2, Vol. 67, March 1979
10) Kollbrunner, C. F. und Basler, K.: torsion, Springer-Verlag, 1966
11) Nakai, H. and Tani, T.: An Approximate Method for the Evaluation of Torsional and Warping Stresses in Box Girder Bridges, Proc. of JSCE, No. 277, September, 1978
12) Wright, R. N., Abdel-Samed, S. R. and Robinson, A. R.: BEF Analogy for Analysis of Box Girder, Proc. of ASCE, Vol. 94, ST7, July, 1968
13) Dabrowski, R.: Gekrümmte dünnwandige Träger Springer - Velag, 1968
14) Sakai, F. and Nagai, M.: A Recommendation on the Design of Intermediate Diaphragms in Steel Box Girder Bridges, Proc. of JSCE No. 261, August 1977
15) Nakai, H. and Murayama, Y.: Distorsional Stress Analysis and Design Aid for Horizontally Curved Girder bridges, Proc. of JSCE, No. 277, September, 1978

ANALYSIS AND EXPERIMENT OF COMPOSITE SPACE SLAB-GRID STRUCTURE

HONG SHOU CHU *

*Engineer, Shanghai Industrial Building Design Institute

This paper presents a kind of new space structure - composite space slab-grid, carries out the structural analysis and puts forward the method of calculation. For the part of integral bending, the order of equations can be reduced with a simplified mathematical model using the method of equivalent substitution. Experimental results of a model test verifies the accuracy and reliability of the theory. The design and construction of a space slab-grid roof in a building is briefly reported.

INTRODUCTION

The composite space slab-grid studied in this paper is developed on the basis of conventional steel space grid with the steel top members substituted by precast reinforced concrete ribbed slabs which play the dual parts of the top members as well as the roof covering. It is no doubt more reasonable to use reinforced concrete members with an integrated slab for resisting compression forces and the construction cost may be reduced by bringing into full play the strength of material due to the dual capacities of the slabs. The general structure of the composite space slab-grid is shown in Fig 1.

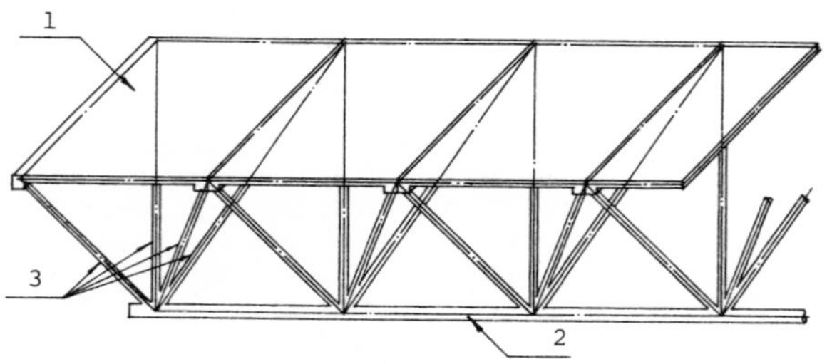

1. top slabs 2. bottom members
3. web members

Fig 1

BASIC ASSUMPTIONS AND FORMULATION

a) A dispersed method is adopted in calculating the member forces of the slab-grid, of which the total top member forces are assumed to be the sum of the moments due to integral bending of the whole grid and the local bending of the slabs. The top slabs are considered as simply supported on four sides for local bending in computation.

b) The ribbed slab is tee shaped in cross section with effective width of the flange taken as equal to the panel of the grid and even distribution of slab stress under integral bending is assumed.

c) The joints of the grid are all considered as pin-connected. In the top nodal joints, the axes of all members intersect at the centroid of the T-section of the slab.

Space composite slab-grid is a kind of three dimensional space structure composed of slabs and chord members. With the top slabs analogous to shell elements, the whole structure may be considered as a space truss stiffened by an elastic shell, and the computation may be carried out according to a model of elastic three dimensional composite shell and chord system. In order to avoid time consuming and expensive computation work in practice, a simplified computing model is recommended in this paper. In case of integral bending, a method of equivalent substitution is introduced. To replace elastic body with an equivalent space frame is the general form of equivalence, so is the equivalence for elastic shells. Thus the solution of displacements and internal forces can be carried out on computers of limited capacity with ordinary space truss programs, or by hand computation with available simplified charts in our country.

Suppose the respective strain energy of two static systems, the elastic body and the space frame, be expressed into $M = M(\{m_i\})$, $F = F(\{f_i\})$, in which $i = 1,2,......n$, m_i, f_i = the displacement vector of the ith node of the elastic body and the space frame, the static characteristics of the elastic body and the space frame are equivalent under equivalent loading, provided

$$\{f_i\} \equiv \{m_i\} , \quad F(\{f_i\}) \equiv M(\{m_i\}) \quad i = 1,2,......n \qquad ...... 1.$$

and the number of nodes is sufficiently large.

If the space frame and elastic body are subdivided into certain number of equal sub-frame elements and equal elastic elements, the strain energy of which be expressed by F' and M', the sum of F' or M' is equal to the strain energy of the space frame or the elastic body. If the number of sub-frames and elastic elements is sufficiently large and the following requirement fulfilled,

$$\{f_{i'}\} \equiv \{m_{i'}\} \ , \quad F'(\{f_i\}) \equiv M'(\{m_{i'}\}) \ , \qquad \ldots\ldots 2.$$

the problem of equivalence of whole structure will correspond to the equivalence of sub-frames and elastic elements.

Take any point P in the elastic body, u, v, w its displacement components, according to basic theory of elasticity, the relative displacement tensor will be

$$[E] = \frac{1}{2}\{[e] + [e]_d\} + [W] \qquad \ldots\ldots 3.$$

here

$$[E] = \begin{bmatrix} \frac{\partial u}{\partial x} & \frac{\partial v}{\partial x} + \frac{\partial u}{\partial y} & \frac{\partial u}{\partial z} + \frac{\partial w}{\partial x} \\ \frac{\partial v}{\partial x} + \frac{\partial u}{\partial y} & \frac{\partial v}{\partial y} & \frac{\partial v}{\partial z} + \frac{\partial w}{\partial y} \\ \frac{\partial u}{\partial Z} + \frac{\partial w}{\partial x} & \frac{\partial v}{\partial z} + \frac{\partial w}{\partial y} & \frac{\partial w}{\partial z} \end{bmatrix}$$

in which $[e]_d$ = the diagonal matrix same as $[e]$, $[W]$ = the matrix from rotation of the rigid body.

If the vector $\{x \ y \ z\}^T$ is applied for the line joining any two points in the elastic body before deformation, then the displacement vector will be

$$\frac{1}{2}\{[e] + [e]_d + 2[I]\}\{x \ y \ z\}^T + \{r_1 \ r_2 \ r_3\}^T$$

after deformation, in which $[I]$ = 3 x 3 unit matrix, $\{r_1 \ r_2 \ r_3\}$ = vector from rotation of the element as a rigid body. According to theory of elasticity, the energy density of any point in the elastic body is

$$U = \frac{E}{2(1+\nu)}\left\{\frac{1-\nu}{1-2\nu}(e_{11} + e_{22} + e_{33})^2 - 2(e_{11}e_{22} + e_{22}e_{33} + e_{33}e_{11}) + \frac{1}{2}(e_{12}^2 + e_{23}^2 + e_{31}^2)\right\} \qquad \ldots\ldots 4.$$

where E = Young's modulus, ν = Poisson's ratio and e_{ij} = the element in the ith line, jth row of the matrix $[e]$.

In case a differential parallelpiped element is taken from the elastic body, and the dimension parallel to ith axis ($i = 1,2,3$) is h_i, then the elastic strain energy of the element will be given as

$$M' = h_1 h_2 h_3 U \qquad \ldots\ldots 5.$$

Consequently, $\{M'\}$ may be expressed in terms of e_{ij} instead of $\{m_{i'}\}$. Similarly, $\{f_{i'}\}$ may also be expressed in terms of e_{ij}. Equation 2 is then equivalent to $\{F'(e_{ij})\} \equiv \{M'(e_{ij})\}$. When the sub-frames are small enough and all of the same size, the equation becomes the necessary and sufficient condition for mechanical equivalence.

On deriving the equivalent model of pin-connected sub-frame on the basis of the above theory, consider a parallelpiped sub-frame illustrated in Fig 2, where L_i, K_i ($i = 1,2,\ldots\ldots 24$) = the length before deformation and the stiffness of members, and ΔL_i = displacement increment of the ith member, then the elastic strain energy of the element will be

$$F_i' = \frac{1}{2}\sum_{i=1}^{24} \Delta L_i^2 K_i \qquad \ldots\ldots 6.$$

Substitute the different vectors of each member into Equ 6, and let K_1 = the stiffness of non-diagonal members, K_2 = the stiffness of diagonal members, the following equation can be obtained

$$F' = \frac{1}{2}\left\{4K_1(h_1^2e_{11}^2 + h_2^2e_{22}^2 + h_3^2e_{33}^2) + 4K_2\left[\frac{(h_1^2e_{11}+h_2^2e_{22})^2 + h_1^2h_2^2e_{12}^2}{h_1^2 + h_2^2} + \frac{(h_2^2e_{22}+h_3^2e_{33})^2 + h_2^2h_3^2e_{23}^2}{h_2^2 + h_3^2} + \frac{(h_1^2e_{11}+h_3^2e_{33})^2 + h_1^2h_3^2e_{13}^2}{h_1^2 + h_3^2}\right]\right\} \qquad \ldots\ldots 7.$$

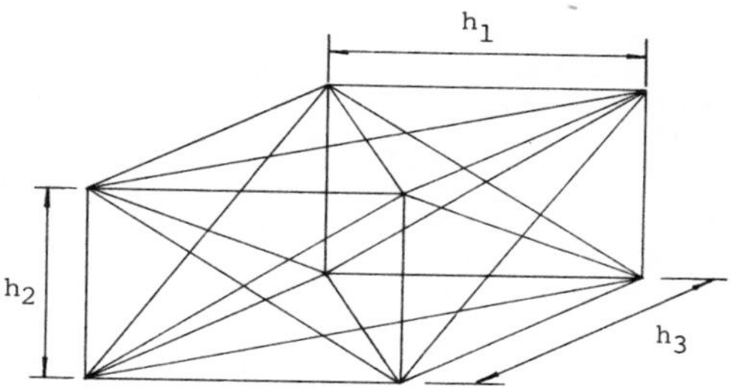

Fig 2

By the equivalent condition $F' = M'$, the following solutions will be obtained, $h_1 = h_2 = h_3 = h$, $\nu = 1/4$, $K_1 = Eh/10$, $K_2 = Eh/5$, which are the conditions that the equivalent sub-frames must satisfy. Previous results can also be directly applied to thin elastic shell with thickness H acting as membrane. Put $e_{13} = e_{23} = 0$, $e_{33} = -\frac{\nu}{1-\nu}(e_{11}+e_{22})$ and substitute into 4 and 5, $h_3 = 0$ and substitute into 7. Similarly, by $F' = M'$, the following results will be obtained $h_1 = h_2 = h$, $K_1 = 0$, $K_2 = 3EH/16$, $\nu = 1/3$.

Finally, the two adjacent members in Fig 2 will combine together and a pin-connected plane sub-frame will be formed as shown in Fig 3. The stiffness K of all the six members equals 3EH/8.

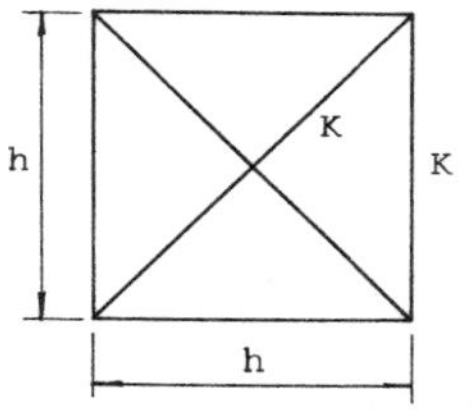

Fig 3

For material with Poisson's ratio $\nu = 1/3$, the results yielded demonstrate good accuracy by applying the above method of equivalent substitution. For reinforced concrete, the accuracy though not as good, however, the error is unremarkable. By converting the top slab including ribs of any of the space slab-grid into an equivalent pin-connected sub-frame as shown in Fig 3, calculating the internal forces due to the integral bending of the whole structure according to the model of space truss, and including the internal forces of top members due to local blending, the total forces in each member of the composite slab-grid will be obtained.

EXPERIMENT

The purposes of the model test are to check the correctness and accuracy of the basic assumptions and theory adopted, to observe the internal force distribution between the rib and the slab, to verify the even distribution of the stress across the slab width and the effective flange width of the T section and to observe the deformation of the structure, and its factor of safety.

Size of the model is 1 M x 6 M, height 0.5 ~ 0.59 M and panel width 0.5 M. The precast reinforced ribbed slabs are made of fine aggregate concrete and web and bottom members of seamless pipes. An analogous uniform load condition is applied on the slab surface. When the test load reaches 1.06 P (P is the theoretical value of the ultimate load of the structure - 8.47 T), the maximum stress in the bottom members reaches the yield point. When increase the load to 1.26 P, the structure is still capable of sustaining the excessively applied load, see Fig 4.

Fig 4

a) Deformation

The maximum deflection of the structure is 1/482 of the span under 1.06 P, and 1/357 under 1.18 P demonstrating the sound stiffness of composite space slab-grid. The theoretical deflection curves and corresponding experimental plots are shown in Fig 5 with good agreement. The maximum deflections of bottom members in different stages of loading are listed in Table 1.

b) Internal forces of the structure

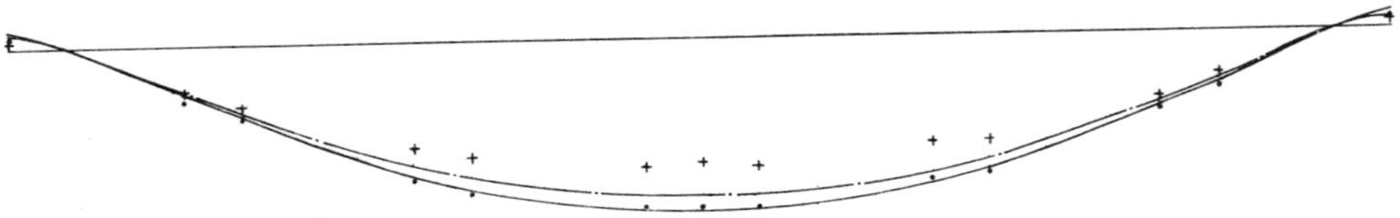

+ , · respective measured deflection during 1.06P, 1.11P
—,-·- respective theoretical deflection curve during 1.06P, 1.11P

Fig 5

The measured internal forces of the top and bottom members at the centre of span and those of maximum stressed web members in the inner side of the support are listed in tables 2 and 3. The internal forces in the slabs come to about 82% of the total internal forces of the top member, showing good joint action of the slab and the frame.

The strains across the slabs are shown in Fig 6, indicating an even distribution of slab stresses and thus verifying the assumed effective width of the flange of the top member as equal to the panel width of the grid.

The linear relation between internal forces of the top or bottom members at the centre of span with the loadings is shown in Fig 7. The maximum experimental errors in the balance of the longitudinal forces at the centre of the span and the vertical forces are respectively 0.5% and 8.6%. The result is accurate enough to check the equilibrium condition of all the internal and external forces.

loading/P*	theoretical value (cm) (1)	experimental value (cm) (2)	$\frac{(1) - (2)}{(1)}$ %
0.92	1.211	0.936	22.7
1.06	1.405	1.140	18.9
1.11	1.593	1.540	3.3

*P is the theoretical value of the ultimate load of the structure - 8.47 T.

Table 1

loading/P	experimental value of internal force in mid-span top member					internal force in mid-span bottom member		$\frac{(1)-(2)}{(1)}$ %
	in rib		in slab		total internal force (kg)	theoretical value (kg) (1)	experimental value (kg) (2)	
	internal force (kg)	percentage (%)	internal force (kg)	percentage (%)				
1.06	-6203	82	-1402	18	-7605	8422	7637	9.3
1.11	-6666	83	-1414	17	-8080	8792	8099	7.9
1.18	-7520	88	-1061	12	-8581	9481	8519	10.0

Table 2

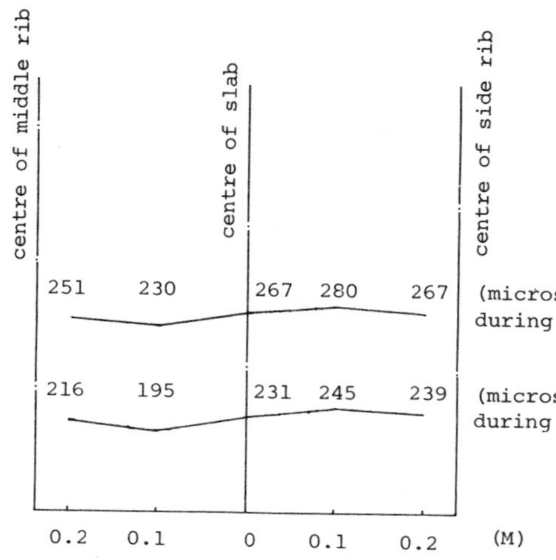

Fig 6

loading/P	theoretical value (kg) (1)	experimental value (kg) (2)	$\frac{(1) - (2)}{(1)}$ %
0.922	1008	833	17.3
1.060	1169	939	19.7
1.110	1233	1038	15.8

Table 3

c) Cracks

No cracks are found in the slab when the load on the test model reaches the theoretical ultimate value P. When the load is increased to 1.06 P, the stress in the bottom member at centre of span reaches the yield point, after then, visible cracks appear in the first time simultaneously on the surfaces of two of the slab panels. More cracks occur later on, however their widths are no more than 0.08~0.1 mm up to 1.26 P, and therefore unremarkable. The cracks may be due to the eccentricity existing in the intersection of the web members and the centroid of the ribbed slab during assembly.

d) Factor of safety of the structure

The experimental value (8.98 T) of the structural ultimate load exceeds the theoretical value by 6%. According to the Chinese Specification, the allowable design load will be 5.30 T, then there exists a factor of safety of 1.69. The fact that the test model still remains in working condition up to a load of 10.63 T, indicates the redistribution of the internal forces in an indeterminate structure of high redundancy.

ENGINEERING PRACTICE

The above mentioned type of composite space slab-grid and the design theory have been put into practice in a recently erected single story building, having a plan size of 36 M x 72 M, with 1 M over hanging roof around all supporting edges.

Orthogonal upright quadrangle pyramid elements are adopted to compose the slab-grid with a height of 1.975 M~2.45 M, and a panel width of 2 M. 2 M x 2 M precast ribbed slab panels are used for the top members and seamless pipes for the bottom and web ones. For the convenience of construction and erection, the roof is prefabricated into strips on the ground as shown in Fig 1. Every pair of strips are fabricated into one assembly unit to be hoisted up and seated on the support. They are finally connected into an integral structure by welding and careful filling of the in-panel

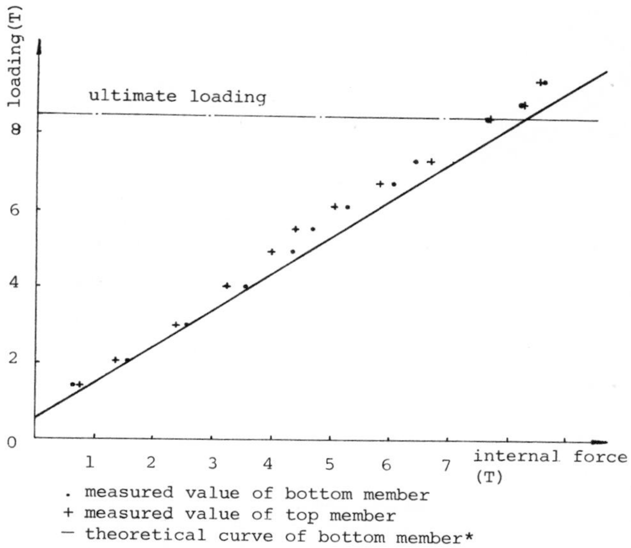

* The theoretical curves of the top member and the bottom member nearly coincide.

Fig 7

gaps and finally casting a thin layer of in-site fine aggregate concrete on the surface as illustrated in Fig 8. Some field test work such as deflection etc., has been carried out and the building completed with satisfaction not more than a year ago.

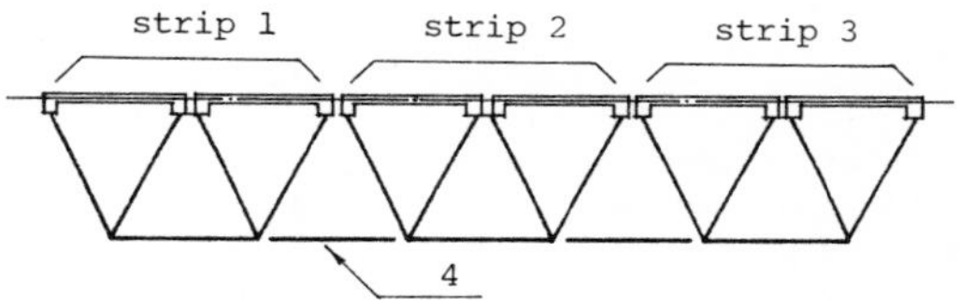

4 bottom member fabricated after the assembly units seated on support

Fig 8

By the above experience of engineering practice based on theoretical study, it may be seen that the application of slab-grid structure may be extended to large span floor structures in multistorey industrial buildings, with problems concerned different from that of roof system, such as greater floor loadings, vibration, continuous spans and particular requirements of construction details etc. Further studies are suggested to meet with such serviceability requirements.

ACKNOWLEDGEMENTS

The author wishes to thank her colleagues participating in the test, discussion and translation.

REFERENCES

1. Z S MAKOWSKI, Steel Space Structures, Michael Joseph, London, 1965.

2. C J RICHARDS, Mechanical Engineering in Radar and Communications, 1969.

3. Design and Construction Specification of Latticed Structures (JGJ 7-80), approved by Construction Bureau of PRC, Beijing, 1981.

A GRAPH THEORY FORMULATION OF STRUCTURAL ANALYSIS OF SPACE NETWORKS

J SUMEC DR, ING

Research worker, Institute of Construction and Architecture, Slovak Academy of Sciences, Bratislava, Czechoslovakia.

Mathematical theory of systems is used for analysis of a wide range of problems. Its special branch is theory of graphs used by the Author for analysis of framed structures. In addition to theoretical relationships and definitions the Author derives some topological matrices to be used for description of topological and algebraic properties of graphs. Highlighting the advantage of the use of graph theory for setting up algorithms of numerical analysis of framed structures as mechanical systems.

INTRODUCTION

The approach to many theoretical problems encountered in various scientific disciplines emphasises the need to develop a universal mathematical theory including the apparatus of solution. In the recent years, the systems' theory has been regarded as such a universal method of approach characterized by the input, the output and the state of the system.

An example of a simple system is a framed structure consisting of elements (bars and nodes) constituting a whole. In this case the action of the ambient environment on the structure e.g. loading by forces, temperature effects, corrosion etc. may be regarded as the input. The output are the state of stress and strain, the magnitudes to be found and which may depend on the time interval of the load action. The state (properly speaking the internal state) is the space of the functions defined by the given output.

This paper will show that the convenient tool of algorithmization of the numerical analysis of the frames is the use of the graph theory, a part of topology. Unlike many papers from the domain of statical and dynamic matrix analysis of frames, see Refs 1-4, the use of topological properties of the given algorithm allows to set up a compact algorithm of the treatment of the problem. The problem is in the closest analogy with those occurring in the theory of network problems, see Ref 5. Some applications of the theory of graphs to the analysis of trusses may be found in Refs 6-8.

For the numerical analysis of structures (e.g. for their statical and dynamic treatment) it is necessary, first of all, to define the mathematical model. Its setting up requires the departure from geometrical, mechanical and topological properties each of which being independent of the other two.

SOME CONCEPTS AND DEFINITIONS OF THE GRAPH THEORY

With regard to the fact that, but for some exceptions, we shall deal with oriented graphs, the attribute "oriented" will be omitted.

D e f i n i t i o n 1. By an oriented graph $\mathcal{H}=(U,\vec{H})$ we understand an ordered pair of non-empty infinite sets U and $\vec{H}$, where $\vec{H}\subset U\otimes U$. The elements of the set U (the nodal set) are the nodes of the graph $\mathcal{H}$ and the elements of the set $\vec{H}$ (the branch set) are the oriented branches of graph .

If the number of nodes of graph is small this graph may be represented by a diagram the nodes being assigned to points in a plane and the branches to arches or segment lines denoted by arrows. These branches depart from the node at which the branch "begins" and go to the nodes "at which the branch ends". A branch my begin and end at the same node. Then we speak of a loop of the graph $\mathcal{H}$. Fig 1 shows a graph $\mathcal{H}=(U,\vec{H})$. Circles denote the nodes and full lines indicated by arrows its oriented branches. According to Fig 1 $U=\{u_1,u_2,\dots,u_4\}$ and $\vec{H}=\{\vec{h}_1,\vec{h}_2,\dots,\vec{h}_7\}$.

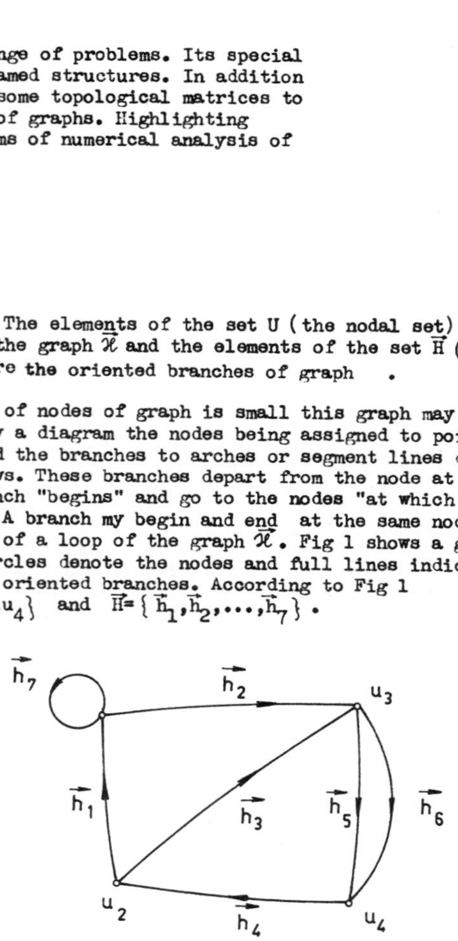

Fig 1

D e f i n i t i o n 2. Graph $\mathcal{H}=(U,\vec{H})$ is called finite if it has a finite number of branches and nodes.

Let $\mathcal{H}_1=(U_1,\vec{H}_1)$ and $\mathcal{H}_2=(U_2,\vec{H}_2)$ be two graphs and let $U_1\subset U_2$ and $\vec{H}_1\subset\vec{H}_2$. Then we speak of graph $\mathcal{H}_1$ being the subgraph of graph $\mathcal{H}_2$.

D e f i n i t i o n 3. Progression $\{u_1,\vec{h}_1,u_2,\vec{h}_2,\dots,\vec{h}_{k-1},u_k\}$ where $\vec{h}_i=(u_i,u_{i+1})$ is called a connection. A connection in which no node is repeated is referred to as a path. The number of branches in a connection or a path is called the length of the connection or the path.

Hence the path is the shortest connection.

D e f i n i t i o n 4. A circuit of length m in the graph $\vec{\mathcal{H}}$ is called the path which has m branches and in which the first and the last node coincide.

D e f i n i t i o n 5. Graph $\vec{\mathcal{H}}$ is called acyclic if it includes no circuit.

D e f i n i t i o n 6. Let $\vec{\mathcal{H}}=(U,\vec{H})$, then its oriented subgraph $\vec{\mathcal{S}}=(U_s,\vec{H}_s)$ will be called an oriented tree of graph $\vec{\mathcal{H}}$, if $|U_s|=|U|$ (i.e. if $\vec{\mathcal{H}}$ as well as $\vec{\mathcal{S}}$ have the same number of nodes) and if $\vec{\mathcal{S}}$ is strongly acyclic, i.e. if $\vec{\mathcal{S}}$ does not include any circuit even if the orientation of some of its branches is changed.

Hence there follows that

$$|\vec{H}_s|=|\vec{H}|-1 \qquad \text{...... 1.}$$

where $|\vec{H}_s|$ is the number of branches of tree $\vec{\mathcal{S}}$.

D e f i n i t i o n 7. The branches of tree $\vec{\mathcal{S}}$ will be called tree branches and the other branches chords.

From this definition there follows that the number of chords $|\vec{\mathcal{F}}|$ of graph $\vec{\mathcal{H}}$

$$|\vec{\mathcal{F}}|=|\vec{H}|-|\vec{H}_s| \qquad \text{...... 2.}$$

For a given tree $\vec{\mathcal{S}}$ of graph $\vec{\mathcal{H}}$ a basic circle $\vec{k}\in\vec{\mathcal{K}}$ may be assigned to each chords $\vec{t}\in\vec{\mathcal{F}}$, so that

$$|\vec{\mathcal{K}}|=|\vec{\mathcal{F}}| \qquad \text{...... 3.}$$

In general, the graph $\vec{\mathcal{H}}$ can contain several trees. Their number may be found in a purely algebraic way by means of a certain determinant, see Ref 11.

D e f i n i t i o n 8. We say that the node $u\in U$ is positively (negatively) incident with the branch $\vec{h}\in\vec{H}$ if the sense of the procedure from node u to branch $\vec{h}$ agrees (disagrees) with the orientation of branch $\vec{h}$.

Let $\mathcal{X}$ and $\mathcal{Y}$ be non-empty sets and let x and ω be representations for which the following is true

$$x : \mathcal{X} \longrightarrow \vec{H}, \qquad \omega : \mathcal{Y} \longrightarrow U$$

then the mapping x will be called the branch valuation $\vec{\mathcal{H}}$ and the mapping ω its node valuation.

D e f i n i t i o n 9. The output (input) degree of node $u\in U$ is given by a number of branches which depart from it (or enter it) in the sense of the orientation of the branches.

In setting up the algorithms of the numerical analysis of framed structures the nodes and the branches of graph $\vec{\mathcal{H}}$ will be denoted by Arabic numerals. The elements of the set U will be numerated successively by numerals $1,2,\ldots,|U|$ and the elements of the set $\vec{H}$ by symbols $\vec{1},\vec{2},\ldots,|\vec{H}|$. We shall always use only finite graphs $\vec{\mathcal{H}}$, where

$$U=\{1,2,\ldots,|U|\}, \qquad \vec{H}=\{\vec{1},\vec{2},\ldots,|\vec{H}|\}.$$

TOPOLOGICAL MATRICES OF ORIENTED GRAPHS

The statical and dynamic analysis of framed structures uses predominantly computers. For this reason, when setting up programs for practical use it is necessary to set up the algorithms in such a way as to make them suitable for digital computers. It appears that a very convenient tool for this purpose is the use of topological matrices allowing the description of the structures of the treated framed structures.

The first item of analysis is the description of mutual connections of bars with the nodes in matrix form. This description is very compact, comprehensible and advantageous for the setting up of the calculation model of the given system. It is well-known that graph theory is used today successfully e.g. for the treatment of some problems of electric and logical circuits for the solution of transportation problems etc.

Let us consider an arbitrary framed structure Ω (a mechanical system) given schematically in Fig 2a. The given structures will be associated with graph $\vec{\mathcal{H}}_\Omega$ (Fig 2b). The nodes of the associated graph $\vec{\mathcal{H}}_\Omega$ corresponding to the nodes of support of the mechanical system will be denoted with the same symbol and identified. The nodes of the associated graph $\vec{\mathcal{H}}_\Omega$ will be numerated. Similarly, the branches will be numerated, too, and oriented.

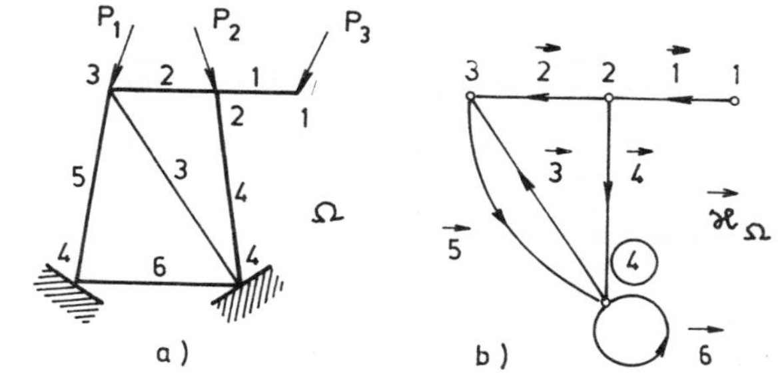

Figs 2a and 2b

The associated graph $\vec{\mathcal{H}}_\Omega$ shown in Fig 2b, gives a "topological information" on the given mechanical system. The supporting nodes of the structure correspond, as shown in Fig 2, to one node of graph $\vec{\mathcal{H}}_\Omega$ which will be called the "basic" node and its notation will be put into a cercle.

The toplogy of the mechanical system Ω will be described by incidence matrix $\vec{J}$ of graph $\vec{\mathcal{H}}_\Omega$. Its elements will be defined as follows

$$\vec{J}_{ij} = \begin{cases} 1 & \text{if the } \vec{i}\text{-th branch is positively incident with the j-th node,} \\ -1 & \text{if it is negatively incident,} \\ 0 & \text{otherwise.} \end{cases}$$

Hence for Fig 2b it is: $|U| = 4$, $|\vec{H}| = 6$ and

$$\vec{J} = \begin{bmatrix} {}^1\vec{J} \\ \\ {}^2\vec{J} \end{bmatrix} = \left[\begin{array}{cccc} 1 & -1 & 0 & 0 \\ 0 & 1 & -1 & 0 \\ 0 & 0 & -1 & 1 \\ \hline 0 & 1 & 0 & -1 \\ 0 & 0 & 1 & -1 \\ 0 & 0 & 0 & 0 \end{array}\right] \qquad \text{...... 4.}$$

We see that the rows of matrix $\vec{J}$ contain the elements 1,-1,0 and that the sum of elements in each row is zero. Hence it follows that any column of matrix $\vec{J}$ is a linear combination of the others. Hence the information on the incidence of arbitrary node with individual branches may be found immediately from the incidences of the other nodes. Hence, there follows that the rank of matrix $\vec{J}$ is

$$h(\vec{J}) = |U| - 1 \qquad \text{...... 5.}$$

Then, by omitting any column in matrix $\vec{J}$ the reduced incidence matrix $\vec{J}_r$ of graph $\vec{\mathcal{H}}_\Omega$ is obtained. In this case the last column was omitted, so that

$$\vec{J}_r=\left[{}^1\vec{J}_r, {}^2\vec{J}_r\right]^t = \left[\begin{array}{ccc|ccc} 1 & -1 & 0 & 0 & 1 & 0 \\ 0 & 1 & -1 & 0 & 0 & 1 \\ 0 & 0 & -1 & 0 & 0 & 0 \end{array}\right]^t \qquad \text{...... 6.}$$

According to Ref 9 each square regular submatrix of order $(|U| - 1)$ of incidence matrix $\vec{J}$ of the given graph $\vec{\mathcal{H}}_\Omega$ is a reduced matrix of some of its trees $\vec{\mathcal{S}}$ set up from $\vec{\mathcal{H}}_\Omega$ (Fig 3). Fig 3 and Eqn6 show that matrix ${}^1\vec{J}_r$ is a reduced incidence matrix of tree $\vec{\mathcal{S}}$. Using matrix $\vec{J}_r$ we can continue to calculate the total number of trees $|\vec{\mathcal{T}}|$ of graph $\vec{\mathcal{H}}_\Omega$ (Cf. Ref 11)

$$|\vec{\mathcal{T}}| = \det\left[\vec{J}_r^t\ \vec{J}_r\right]. \qquad \ldots\ldots\ 7.$$

Elements $(\vec{J}_r^t\ \vec{J}_r)_{ii}$ are sums of the input and output degrees of nodes $1,2,\ldots,|U| - 1$. In this case, as it is easy to find out, the total number of trees $|\vec{\mathcal{T}}| = 5$ and each tree $\vec{\mathcal{S}}_i$ is a strongly acyclic graph.

Let us now introduce the concept of the matrix of connections graph $\vec{S}$ of the nodes of tree $\vec{\mathcal{S}}$ with the basic node. The elements of matrix $\vec{S}$ will be defined as follows

$$\vec{S}_{ij} = \begin{cases} 1 & \text{if the } \vec{i}\text{-th branch of tree } \vec{\mathcal{S}} \text{ is positively incident with its } j\text{-th node with respect to path } (1,\vec{1},2,\ldots,U), \\ -1 & \text{if ... it is negatively incident ...}, \\ 0 & \text{otherwise.} \end{cases}$$

Then, in agreement with Fig 3 we have

$$\vec{S} = \begin{bmatrix} 1 & 0 & 0 \\ 1 & 1 & 0 \\ -1 & -1 & -1 \end{bmatrix} \qquad \ldots\ldots\ 8.$$

It can also be proved that there is an analogy between the product of matrix of paths and the incidence matrix, given in Ref 9

$${}^1\vec{J}_r\ \vec{S}^t = I \qquad \ldots\ldots\ 9.$$

where I is a unit matrix.

Another topological matrix is the basic circle matrix $\vec{\mathcal{K}}$. It will be set up by numerating the basic cercles (futher on only "circles") the number of which is $|\vec{\mathcal{F}}|$ in succession corresponding to the numeration order of the chords of graph $\vec{\mathcal{H}}_\Omega$ and by assigning them the same orientation as that of the chords. The elements of the matrix $\vec{\mathcal{K}}$ will be defined as follows

$$\vec{\mathcal{K}}_{ij} = \begin{cases} 1 & \text{if the orientation of the } \vec{i}\text{-th branch which is an element of the } \vec{j}\text{-th cercle is in agreement with the orientation of the } \vec{j}\text{-th circle}, \\ -1 & \text{if ... it is in disagreement ...}, \\ 0 & \text{otherwise.} \end{cases}$$

According to Fig 4 showing the circles of graph $\vec{\mathcal{H}}_\Omega$ there is

$$\vec{\mathcal{K}} = \begin{bmatrix} {}^1\vec{\mathcal{K}} \\ \\ {}^2\vec{\mathcal{K}} \end{bmatrix} = \begin{bmatrix} 0 & 0 & 0 \\ -1 & 0 & 0 \\ 1 & 1 & 0 \\ \hline 1 & 0 & 0 \\ 0 & 1 & 0 \\ 0 & 0 & 1 \end{bmatrix} \qquad \ldots\ldots\ 10.$$

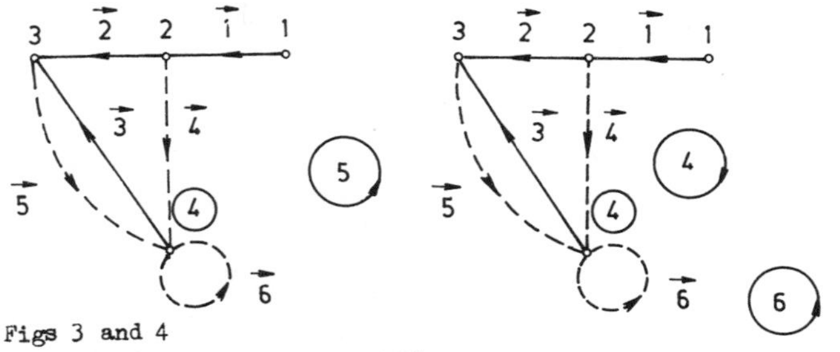

Figs 3 and 4

Relationship 10 shows that ${}^2\vec{\mathcal{K}} = I$. This equality will be true if the branches of tree $\vec{\mathcal{S}}$ "divide" the chords of graph $\vec{\mathcal{H}}_\Omega$. If it is not so, we have to renumber the branches, which correspond to the row-permutation in matrix $\vec{\mathcal{K}}$. For further derivations, this structure of matrix $\vec{\mathcal{K}}$ has considerable advantages. E.g. according to Ref 9 matrices $\vec{J}_r$ and $\vec{\mathcal{K}}$ are orthogonal i.e.

$$\vec{J}_r^t\ \vec{\mathcal{K}} = \emptyset \qquad \ldots\ldots\ 11.$$

Then

$$\begin{bmatrix} {}^1\vec{J}_r \\ {}^2\vec{J}_r \end{bmatrix}^t \begin{bmatrix} {}^1\vec{\mathcal{K}} \\ {}^2\vec{\mathcal{K}} \end{bmatrix} = {}^1\vec{J}_r^t\ {}^1\vec{\mathcal{K}} + {}^2\vec{J}_r^t = \emptyset$$

and from that

$${}^1\vec{\mathcal{K}} = -\left[{}^1\vec{J}_r^t\right]^{-1}\ {}^2\vec{J}_r^t \qquad \ldots\ldots\ 12.$$

or according to Eqn 9

$${}^1\vec{\mathcal{K}} = -\vec{S}\ {}^2\vec{J}_r^t \qquad \ldots\ldots\ 13.$$

It is obvious that the sub-matrix of circles ${}^1\vec{\mathcal{K}}$ may be obtained immediately in an algebraic way from a sub-matrix of the reduced incidence matrix ${}^1\vec{J}_r$, or from the matrix of connection $\vec{S}$, which will be multiplied from the right by submatrix ${}^2\vec{J}_r^t$.

There are other topological matrices which describe the topological properties of graphs such as e.g. the matrix of sections or the matrix of basic sections.

It means, however, that for the determination of the topological matrices and hence the whole topology of graph $\vec{\mathcal{H}}_\Omega$ it is sufficient to find its reduced incidence matrix $\vec{J}_r$. In addition to the description of the topological properties of the graph the topological matrices may be used to express some of its algebraic properties.

Let $\mathcal{X}$ and $\mathcal{Y}$ be sets of information on branches and nodes of graph $\vec{\mathcal{H}}_\Omega$ and let z be a $|U|$-dimensional vector where $z \in \mathcal{Y}$. The incidence matrix $\vec{J}$ of graph $\vec{\mathcal{H}}_\Omega$ is then its operator from $\mathcal{Y}$ to $\mathcal{X}$, i.e.

$$\vec{J}:\ z \in \mathcal{Y} \longrightarrow v \in \mathcal{X}\ , \text{ or } \quad v = \vec{J}z \qquad \ldots\ldots\ 14.$$

Let the $\vec{m}$-th branch of graph $\vec{\mathcal{H}}_\Omega$ be incident with nodes r and s, then there follows from the structure of matrix $\vec{J}$

$$v_m = \sum_{j=1}^{|U|} \vec{J}_{ij}\, z_j = z_r - z_s \qquad \ldots\ldots\ 15.$$

We see that the branch valuation is a function of the difference of the node valuations. A similar relationship holds for matrix $\vec{J}_r$

$$v_m = \sum_{j=1}^{|U|-1} \vec{J}_{r_{ij}}\ (z_j - z_{|U|}) \qquad \ldots\ldots\ 16.$$

It means that the branch valuation of graph may be obtained by means of the $(|U| - 1)$-dimensional vector z'

$$v = \vec{J}_r\, z' \qquad \ldots\ldots\ 17.$$

In general even the elements of vector z' may again be vectors. Departing from the orthogonality of matrices $\vec{J}_r$ and $\vec{\mathcal{K}}$ we may go on writing

$$\vec{\mathcal{K}}^t v = \vec{\mathcal{K}}^t\ \vec{J}_r\ z = \emptyset z' = 0 \quad \text{for } \forall z' \in \mathcal{Y}' \subset \mathcal{Y} \qquad \ldots\ldots\ 18.$$

Let $v^* \in \mathcal{X}$ be any element, then $\vec{\mathcal{K}}^t$ is an operator from $\mathcal{X}$ to $\mathcal{Z}$ (a set of information on cercles of graph $\vec{\mathcal{H}}_\Omega$).

$$\vec{\mathcal{K}}^t:\ v^* \in \mathcal{X} \longrightarrow v^o \in \mathcal{Z}\ , \text{ or } \quad v^o = \vec{\mathcal{K}}^t v^* \qquad \ldots\ldots\ 19.$$

where v^o is a $|\mathcal{K}|$-dimensional vector.
This operation may be used to get from the given branch valuation of the graph a cercle valuation.

From Eqns18 and 19 there is

$$\vec{\mathcal{K}}^t\,(v^* + v) = \vec{\mathcal{K}}^t(v^* + \vec{J}_r z') = \vec{\mathcal{K}}^t v^* + 0 = v^o \quad \ldots\ 20.$$

On the contrary, a graph valuation may be received from the cercle valuation of the graph. Let g be a $|\vec{\mathcal{K}}|$-dimensional vector where $g \in \mathcal{Z}$. Then

$$\vec{\mathcal{K}} : g \in \mathcal{Z} \longrightarrow f \in \mathcal{X} \quad , \text{ or } \quad f = \vec{\mathcal{K}}\, g \qquad \ldots\ldots \quad 21.$$

If the Eqn21 is multiplied from the left by operator $\vec{J}^t_r$ we have again an analogy with Eqn18

$$\vec{J}^t_r\, f = \vec{J}^t_r\, \vec{\mathcal{K}}\, g = \varnothing g = 0 \qquad \text{for } \forall g \in \mathcal{Z} \qquad \ldots\ldots \quad 22.$$

Let $f^{*} \in \mathcal{X}$ be an arbitrary element, then $\vec{J}^t_r$ is an operator from $\mathcal{X}$ to $\mathcal{Y}' \subset \mathcal{Y}$ i.e.

$$\vec{J}^t_r : f^{*} \in \mathcal{X} \longrightarrow f^{o} \in \mathcal{Y}' \subset \mathcal{Y} \quad , \text{ or } f^{o} = \vec{J}^t_r\, f^{*} \ldots\ldots \quad 23.$$

Similarly there follows from Eqns 22 and 23

$$\vec{J}^t_r\,(f^{*} + f) = \vec{J}^t_r\,(f^{*} + \vec{\mathcal{K}} g) = \vec{J}^t_r\, f^{*} + 0 = f^{o} \qquad \ldots\ldots \quad 24.$$

ALGORITHM OF CALCULATION OF FRAMED STRUCTURES USING TOPOLOGICAL MATRICES

In this part of the paper we wish to demonstrate the use of topological and algebraic properties of graph $\vec{\mathcal{H}}_{\Omega}$ associated with the given mechanical system Ω for the purpose of setting up an algorithm of statical calculation in the matrix form of the method of slopes and deflections. It is assumed

a) that neither the whole framed structure neither its part constitute a movable mechanism,
b) that Hooke's law is in vigour, that the material of the structure does not change in time, its mechanical properties and deform continuously,
c) that the influence of temperature or the subsidence of the supports is not considered,
d) that the Bernoulli-Navier's hypothesis is accepted and there is a perfect contact between the end-sections of the bars and the adjacent nodes.

Before starting the proper analysis of the problem let us have the following definition:

D e f i n i t i o n 10. Let $\vec{J}_r$ be the reduced incidence matrix of graph $\vec{\mathcal{H}}_{\Omega}$. Then matrix J^{*}_r is called the induced matrix to matrix $\vec{J}_r$ the elements of the matrix J^{*}_r being governed by the follwing relationships

$$\begin{aligned} \vec{J}_{r_{ij}} &= 1 \implies J^{*}_{r_{ij}} = [1,0]^t \\ \vec{J}_{r_{ij}} &= -1 \implies J^{*}_{r_{ij}} = [0,1]^t \\ \vec{J}_{r_{ij}} &= 0 \implies J^{*}_{r_{ij}} = [0,0]^t \end{aligned} \qquad \ldots\ldots \quad 25.$$

Cartesian co-ordinate space system denoted (X,Y,Z) and (x,y, z) will be chosen for the global co-ordinate system (GCS) and the local co-ordinate system (LCS), respectively. In the (LCS) the co-ordinate axes will be chosen in such a way as to make the x axis a tangent to the centroidal axis of the bar, the axis y and the axis z coincide with the main axes of inertia of the bar section.

To the given framed structure Ω we associate graph $\vec{\mathcal{H}}_{\Omega}$, set up its tree $\vec{\mathcal{S}}_{\Omega}$ and numerate the branches. All the support nodes of the structure both in the graph $\vec{\mathcal{H}}_{\Omega}$ and its tree $\vec{\mathcal{S}}_{\Omega}$ will correspond the only node - the basic node - denoted by 0. The branches of graph will be oriented as follows:

Let i and j be such nodes which are incident with branch $\vec{h}$. Hence, if $|i|<|j|$, branch $\vec{h}$ is oriented in the sense "from the node i to the node j". Each branch incident with the basic node will be oriented in the sense "from the basic node 0". Then the topological matrices $\vec{J}$, $\vec{J}_r$, and J^{*}_r of graph $\vec{\mathcal{H}}_{\Omega}$ are set up and the corresponding transformation matrix T_h for arbitrary bar $h \in \Omega$ derived. The bar has the property that

$$T_h : (x,y,z)_h \longleftarrow (X,Y,Z)_{\Omega} : T_h^{-1} = T_h^t \qquad \ldots\ldots \quad 26.$$

The condition equation of equilibrium in terms of unknown deformations of the nodes will be derived from the condition of minimum of potential energy of the system. E.g. according to Ref 12 the deformation of the x axis of any bar h due to the nodal load are expressed by

$$d_h(x) = [u(x), v(x), w(x), \vartheta(x), \xi(x), \eta(x)]^t = C_h(x)\, a_h \qquad \ldots\ldots \quad 27.$$

where $C_h(x)$ is the co-ordinate matrix of bar h,
a_h is the vector of coefficients of bar h.

The elements of vector a_h will be received by substituting the x co-ordinates of the end-points of the bar into Eqn 27 so that

$$a_h = C_h^{-1}(x_1, x_2)[{}^1d, {}^2d]^t_h \qquad \ldots\ldots \quad 28.$$

where $d_h(x_1,x_2) = [{}^1d, {}^2d]^t_h$ is the deformation vector of the end-point section 1 and 2 of bar h in (LCS). If curved bars occur in the framed structure that it is convenient to go over to parametrical expression of the co-ordinates.

The physical equations will be expressed by

$$\sigma_h(x) = E_h(x)\; \varepsilon_h(x) \qquad \ldots\ldots \quad 29.$$

where $E_h(x)$ is the matrix of the rigidity characteristics of bar h at point $x(x_1 \le x \le x_2)$,
$\varepsilon_h(x)$ - deformation vector of bar h at point x.

The deformation vector $\varepsilon_h(x)$ may be expressed by a convenient differentiation of Eqn 27

$$\varepsilon_h(x) = B_h(x)\, a_h \qquad \ldots\ldots \quad 30.$$

or as a deformation function of end-point sections 1,2 of bar h

$$\varepsilon_h(x) = B_h(x) C_h^{-1}(x_1, x_2)\, d_h \qquad \ldots\ldots \quad 31.$$

where $B_h(x)$ is the deformation matrix of bar h.

The total potential energy of bar $h \in \Omega$ in the (LCS) will be expressed in the form

$$\overline{\Pi}_h = \overline{\Pi}^i_h + \overline{\Pi}^e_h = \frac{1}{2}\int_{x_1}^{x_2} \sigma^t_h\, \varepsilon_h dx - \int_{x_1}^{x_2} q^t_h\, d_h\, dx - \sum_{l=1}^{\sigma} p^t_{hl}\, d_h \qquad \ldots\ldots \quad 32.$$

By substituting Eqns 29 and 31 into 32 and by transforming Eqn 32 form the (LCS) into the (GCS) we have

$$\overline{\Pi}_h = \overline{\Pi}^i_h + \overline{\Pi}^e_h = \frac{1}{2}\int_{x_1}^{x_2} D^t_h\, T_h\, [C^t_h]^{-1}\, B^t_h\, E_h\, B_h\, C^{-1}_h\, T^t_h\, D_h dx - \int_{x_1}^{x_2} Q^t_h\, T_h\, C_h\, C^{-1}_h\, T^t_h\, D_h\, dx - \sum_{l=1}^{\sigma} P^t_{hl}\, T_h C_h C^{-1}_h\, T^t_h\, D_h \qquad \ldots\ldots \quad 33.$$

where Q_h is continuous external load of bar h in the GCS,
P_{hl} - singular external load of bar h at point x_l,
D_h - vector of deformation of end-point sections of bar

The total potential energy $\overline{\Pi}_{\Omega}$ of a system Ω is obtained by introducing the corresponding compatibility conditions between the end-point sections of the bars and the corresponding nodes with the help of induced matrix J^{*}_r. Then we have due to Eqn 33

$$\overline{\Pi}_{\Omega} = \sum_{h=1}^{m} \overline{\Pi}_h - F^t\Delta = \frac{1}{2}\sum_{h=1}^{m}\int_{x_1}^{x_2} \Delta^t J^{*t}_{r_h}\, T_h\, [C^t_h]^{-1}\, B^t_h\, E_h\, B_h\, C^{-1}_h \times$$

$$\times T_h^t J_{r_h}^{*} \Delta \, dx - \sum_{h=1}^{m} \int_{x_1}^{x_2} Q_h^t T_h C_h C_h^{-1} T_h^t J_{r_h}^{*} \Delta \, dx -$$

$$- \sum_{h=1}^{m} \sum_{l=1}^{\sigma} P_{hl}^t T_h C_h C_h^{-1} T_h^t J_{r_h}^{*} \Delta - F^t \qquad \ldots\ldots \; 34.$$

Eqn 34 represents functional $\Pi_\Omega = \Pi_\Omega(\Delta, \Delta)$ which is a quadratic form with respect to Δ. The mechanical system Ω will be in equilibrium just if Π_Ω = min. The necessary conditions of stacionarity Π_Ω is

$$\partial \Pi_\Omega / \partial \Delta = 0 \qquad \ldots\ldots \; 35.$$

which conduces to a system of algebraic equations

$$K_G \Delta = R \qquad \ldots\ldots \; 36.$$

where K_G is the global rigidity matrix of the whole framed structure,
R - vector of the external load transformed to nodes.

For symbols used in Eqn 36 the following relations are true due to Eqns 34 and 35

$$K_G = \sum_{h=1}^{m} J_{r_h}^{*t} T_h [C_h^t]^{-1} \int_{x_1}^{x_2} B_h^t E_h B_h \, dx \, C_h^{-1} T_h^t J_{r_h}^{*}$$

$$\ldots\ldots \; 37.$$

$$R = \sum_{h=1}^{m} J_{r_h}^{*t} T_h [C_h^{-1}]^t \int_{x_1}^{x_2} C_h^t T_h^t Q_h \, dx - \sum_{h=1}^{m} \sum_{l=1}^{\sigma} J_{r_h}^{*t}$$

$$T_h [C_h^{-1}]^t C_h^t T_h^t P_{hl} - F$$

Matrix K_G is symmetric and positively definite which follows from the physical nature of the problem.

A very important property of matrix K_G is its banded form as it follows from the choise of tree $\vec{S}_\Omega$ from the associated graph $\vec{\mathcal{H}}_\Omega$. If an unsuitable tree is chosen (provided there exists a number of them and the nodes are numerated successively beginning from the basic node) the matrix K_G can have a dispersed form and in the case of large number of unknowns the higher capacities of the computer storage is importand.

The question is whether there exists a suitable numbering of the nodes of the associated graph $\vec{\mathcal{H}}_\Omega$ or a suitable tree $\vec{S}$ to result in a (p, p) - band matrix K_G

$$p = \max(p_o, 0) \qquad p_o = \max\{k-i,\ i,\ k,\ K_{Gii} \neq 0\}$$

where p is the smallest width of the band (in this case the optimum width). The trivial case of the graph $\vec{\mathcal{H}}_\Omega$ being strongly consistent is, of course, disregarded as in this case K_G is a full matrix.

In Ref 14 it is stated that for a general case of the form of framed structures no algorithm has been known hitherto (although such numbering of the nodes of graph $\vec{\mathcal{H}}_\Omega$ in the case of rectangular grid system is obvious). If the analysed structure had k internal nodes, it would be necessary to test all the k! possibilities which for large numbers is practically impossible. On the other hand, it is possible to use a convenient permutation of rows and columns or a simultaneous permutation of both rows and columns to reduce the matrix K_G to a (p,p)-band system where p is optimum. Ref 14 contains a theorem applicable to matrix K_G under our assumptions :

T h e o r e m 1. Let K_G be a square symmetric matrix of the n-th order which is of the (p,p)-band form. Let the internal nodes of graph $\vec{\mathcal{H}}_\Omega$ be numbered in sequence 1,2,...,n where n is the order of the matrix K_G. If a number $|i-j|$ is assigned to each branch incident with the internal nodes i and $j \in U$, then p is equal to the maximum of all the assigned numbers on the branches of graph $\vec{\mathcal{H}}_\Omega$.

Note : An internal node of graph $\vec{\mathcal{H}}_\Omega$ is such a node which is not the basic node.

In other words: to find the band form for a square symmetric sparse matrix of the n-th order is equivalent to the problem of the theory of finite graphs consisting in the notation of its internal nodes by mutually different numbers 1,2,...,i,j, ...,n in such a way as to make any of the branches $\vec{h} \in \vec{\mathcal{H}}_\Omega$ incident with the nodes i and j satisfy the condition

$$\max_{\forall \vec{h} \in \vec{\mathcal{H}}_\Omega} | i - j | = \min \qquad \ldots\ldots \; 38.$$

According to Ref 14 it has been proved that the problem of finding whether the given associated graph has the numbering of the internal nodes such that it gives the width of the halfband of the global rigidity matrix of the structure equal to p or less, belongs to the class of the so-called NP - full problems such as the problem of the travelling salesman.

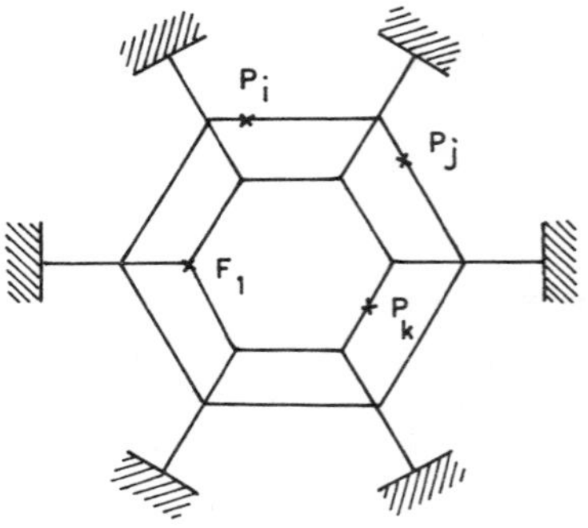

Fig 5

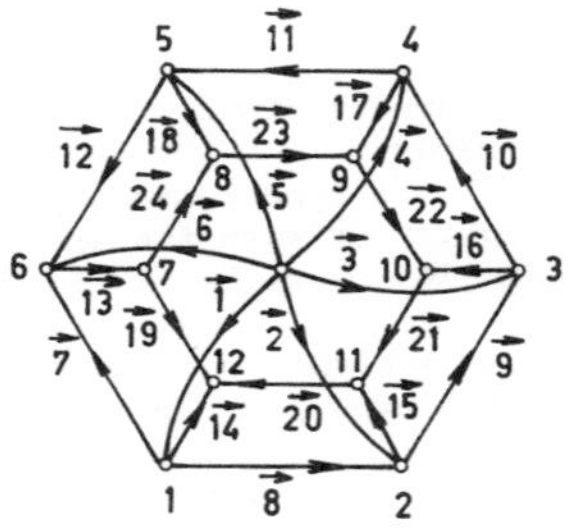

Fig 6

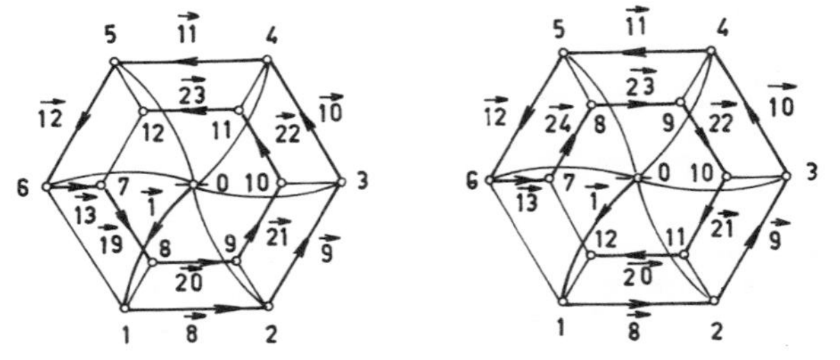

Figs 7 and 8

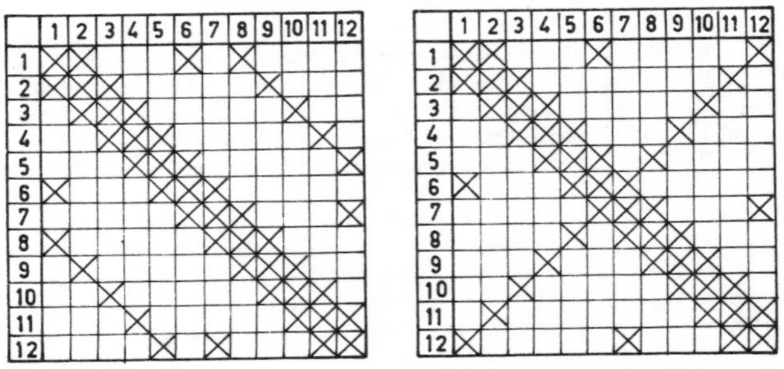

Figs 9 and 10

In literature several algorithms may be found for the renumbering of nodes when one numbering pattern is given and which yield a practically satisfactory solution such as the CM - algorithm (Cuthill - Mc Kee), the GPS - algorithm, see Ref 15 or the algorithm describes in Ref 16.

The example of the framed structure Ω (Fig 5) with associated graph $\vec{\mathcal{H}}_{\Omega}$ (Fig 6) consists in the setting up matrix $\vec{J}_r$ according to the suggested procedure and induced matrix J_r^*. Further, two trees $\vec{\mathcal{S}}_1$, $\vec{\mathcal{S}}_2$ (Figs 7 and 8) are drawn which conduce to a band or a sparse structure of the elements of the rigidity matrix K_G of the analysed structure as shown by Figs 9 and 10. The reduced incidence matrix $\vec{J}_r$ and the matrix J_r^* induces to the latter of the graph $\vec{\mathcal{H}}_{\Omega}{}^r$ have the form, see Ref 13.

Note: In Figs 9 and 10 the non - zero elements in matrix K_G are denoted with crosses.

CONCLUSION

The advantage of the presented method of analysis of framed structure using the linear theory of graphs, its topological and algebraic properties, the matrix algebra is that this method does not require the sign convention as the classical force and displacement methods. If error is ommitted during the calculation, its discovery is relatively quick requiring only the check of the reduced incidence matrix $\vec{J}_r$ of the given graph $\vec{\mathcal{H}}_{\Omega}$. All the other elements occurring in the algorithm of the analysis are generated automatically on the basis of the formulated procedures.

In comparison with the classical setting up of the global rigidity matrix K_G by means of congruent mappings (cf Ref 3) the advantage of the application of the graph theory relies in the fact that it does not require the setting up and the storage in the computer memory of the matrix expressing the conditions of compatibility of the deformation of nodes and the corresponding end-point sections of the bars.(Matrix a in Ref 3). In the demonstrated case it is enough to set up matrix $\vec{J}_r$ from which the non-zero ordered pairs of the corresponding induced matrix may be determined directly which according to Ref 3 is not possible. The mentioned non-zero ordered pairs suffice fully for the setting up of the global rigidity matrix.

Another procedure of the setting up of matrix K_G is based on the application of the global code numbers (Cf e.g. Ref 12). In this case it is necessary to set up the so-called localisation matrices for each element of the structure which results in very high - order matrices if the elements are numerous.

The Author assumed the perfect contact between the end-point section of the bars and the adjacent nodes. More general boundary conditions as well as the consideration of the temperature changes and the subsidence of the supports will be discussed in another paper by the Author ready for publication.

The studied algebraic properties of the graphs have a physical analogy. E.g. Eqn 14 is analogous to the conditions of compatibility of deformation in the nodes and Eqn 18 expresses the sum of the relative strains of a closed rigid frame to be zero.

A very important step of the used method is to find a suitable optimum tree $\vec{\mathcal{S}}_{opt} \in \mathcal{S}_{\Omega}$ conducing to the minimum band width in the global rigidity matrix. Such an optimum tree in the case of cyclically symmetric frames is a tree of a "helical" form.

The tree of graph $\vec{\mathcal{H}}_{\Omega}$ yields information on the number of the condition equations. In general if a structure comprises n nodes including r support nodes, n - r independent conditions of equilibrium may be written. The graph associated to such a structures has n + 1 - r nodes like its tree $\vec{\mathcal{S}}$. The number of the tree branches of tree $\vec{\mathcal{S}}$ of the given graph $\vec{\mathcal{H}}_{\Omega}$ is n - r , which corresponds to the independent equilibrium conditions.

REFERENCES

1. R K LIVESLEY, Matrix Methods of Structural Analysis, Oxford, 1964.

2. J S PRZEMIENIECKI, Theory of Matrix Structural Analysis, New York, 1968.

3. J H ARGYRIS, Recent Advances in Matrix Structural Analysis, Oxford, New York, 1964.

4. K CHOBOT, The Use of Matrix Calculus in Structural Mechanics, (In Czech), SNTL, Praque, 1967.

5. J P ROTH, An Application of Algebraic Topology: Kron's Method of Tearing, Quart of Appl. Math., 1959.

6. N C LIND, Analysis of Structures by System Theory, J. of the Struc. Div., ASCE, 1962.

7. W R SPILLERS, Network Analogy for Linear Structures, J. of the Engng. Mech. Div., ASCE, 1963.

8. T ODEN and A NEIGHBORS, Network-Topological Formulation of Analyses of Geometrically and Materially Non-Linear Space Framed, In: Space Structures, Oxford and Edinburg, 1967.

9. V BRÁT, Matrix Methods in Analysis and Synthesis of Space Bonded Mechanical Systems, (In Czech), Academia, Praque, 1981.

10. P BRUNOVSKÝ and J ČERNÝ, Foundations of Mathematical Theory of Systems, (In Slovak), Veda, Bratislava, 1980.

11. K ČULÍK, V DOLEŽAL and M FIEDLER, Combinatorial Analysis in Practice, (In Czech), SNTL, Praque, 1967.

12. V KOLÁŘ at al, Design of Planar and Space Structures using the Finite Element Method, (In Czech), SNTL, Praque, 1979.

13. J SUMEC, Use of Theory of Graphs for Analysis of Trusses, (In Slovak), Staveb. Čas., Bratislava, 1983.

14. M FIEDLER, Special Matrices and Their Use in Numerical Mathematics, (In Czech), SNTL , Praque, 1981.

15. N E GIBBS, W G POOLE and P K STOCKMEYER, An Algorithm for Reducing the Bandwidth and Profile of Sparse Matrix, SIAM, J. Numer. Anal., 1976.

16. M HORÁK, Minimization of the Matrix Bandwidth, (In Czech) In: A Seminar on the Finite Element Method and the Variational Methods, Plzeň, 1981.

WIND LOADING ON A MULTIPLE HYPERBOLIC PARABOLOID SHELL ROOF STRUCTURE

A J DUTT, PhD (Liverpool), BEng (Hons), CEng (London), MICE (London), MASCE (USA), FIAA (London), MIEM, MIES, PEng (S), PEng (M)

Senior Lecturer, Department of Building Science
National University of Singapore

The paper deals with the investigation of the wind loading on a multiple hyperbolic paraboloid (HP) shell roof by model tests in the wind tunnel. The roof of the model was a grouping of four similar HP shells in a "sawtooth" array and forming a square in plan. Wind tunnel experiments were carried out, wind pressure distribution and the contours of wind pressure on shell roof were determined for various wind directions. The average suctions on roof were computed and compared with that on a flat roof and single HP shell roof. The highest point suction encountered was -5q whilst the maximum average suction on the roof was -0.68q.

NOTATION

C_p	Pressure coefficient
L	a linear dimension fixing the scale
p	pressure on the surface of an object
p_s	static pressure
p_w	wind pressure
q	velocity pressure, $\frac{1}{2}\rho v^2$
v	wind velocity
μ	viscosity of air
ρ	air density
θ	angle of wind direction

INTRODUCTION

In the design of many space structures, the maximum loading that can be exerted by winds are of considerable importance but until now there are incomplete data for use in design. On account of the complicated nature of the conditions met with in practice, the determination of actual forces in any particular case would present a theoretical problem of extreme difficulty, and to obtain from full scale experiments on space structures, sufficient data which might be applicable to the general case would be a formidable task. The examination of models in the wind tunnel has thrown much light on the nature of the problem and by providing quantitative information will enable a more detailed study to be made in relation to many types of space structures.

Wind tunnel investigation on models of short and medium span structures have been carried out, see Refs 1, 2, 3, 4, 5 and 6. For a large majority of cases, wind tunnel tests are required for each individual structure shape and orientation. Thus there results a catalogue of pressure coefficients of each structure geometry. It is widely recognised that an extraordinary shape such as multiple hyperbolic paraboloid shell roof structure must be investigated in the scale model tests in the wind tunnel, because multiple HP shell roofs are being used extensively in recent years as they are one of the most economic structural means to cover a very large space with the minimum of supports so as to have an unobstructed floor space. Furthermore, they have the aesthetically pleasing shape. It was, therefore, considered essential to investigate multiple HP shell roof space structure in the wind tunnel.

THEORY

When an airstream blows against an object, the pressure, p, at any point on its surface may be regarded as consisting of two parts - the static pressure, p_s, which in a natural wind is the barometric pressure; and the excess $p-p_s$, caused by the presence of the object. This excess $p-p_s$, arises solely from motion of the air with reference to the model. It will be called simply the wind pressure and will be denoted by p_w. If there is no wind or no object present, then $p=p_s$ everywhere and the wind pressure everywhere is zero. The wind pressure may be either positive or negative or zero; that is p, which by definition of p_w is equal to $p_s + p_w$ may be either greater or less than p_s or equal to it. From dimensional reasoning it is shown that p_w is given by an expression of the form

$$\frac{p_w}{q} = f\left(\frac{VL\rho}{\mu}\right) = C_p \qquad \ldots 1$$

The expression applies only to geometrically similar bodies. The wind pressure, p_w could be measured in any convenient units, but there are advantages in using the wind velocity pressure, as the unit. For bodies with sharp corners C_p is practically independent of the wind speed and the size of the model that is $f\left(\frac{VL\rho}{\mu}\right)$ is a constant for any pressure point so that for a single value of it for any given shape of body of any size at any wind speed can be computed with the aid of a table of velocity pressures. The ratio C_p is a pure number independent of the units used so long as the pressures are all measured in the same units. In the presentation of data for the distribution of wind pressure over the exterior surface of the structure it is commonly assumed that the interior is at a constant pressure equal to static pressure, p_s; that is p_w is zero for the interior.

DESCRIPTION OF MODEL

The roof of the model as shown in Fig. 1 was a grouping of four similar hyperbolic paraboloid shells in a "sawtooth" array forming a square in plan. The length of the side of the square was 458 mm. Each individual unit of shell was also square in plan and the length of the side of this square was 229 mm. Each shell was tilted about the diagonal joining the low corners such that one of the high corners was at the same elevation as that of the low corners and the other high corner was at an elevation which was higher than the low corner elevation by two times the rise of the shell.

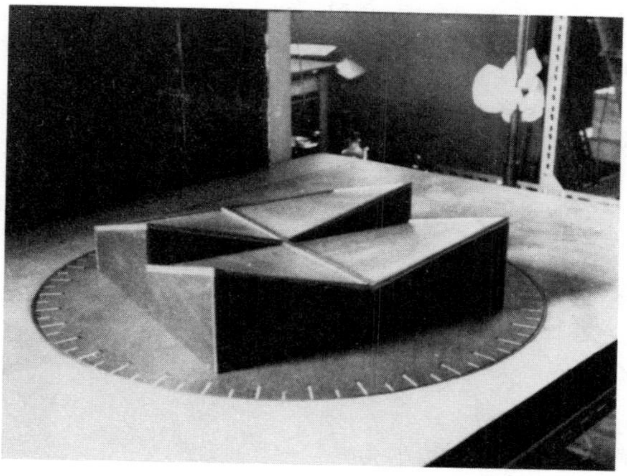

Fig. 1 Model of multiple HP shell roof structure in wind tunnel

The heights of the high corners of the shells were 140 mm and those of the low corners were 76 above the base of the model. The rise to span ratio of the shell was 1:10.2. The walls of the model was made of 6.35 mm plywood and the shell roof was made of obeche veneer strips since a certain amount of the twist was required in each strip. The outer edge beams consisted of 9.52 mm wide x 6.35 mm thick timber strips which projected over the roof and were glued and nailed along the outer edge of the shell group. The inner common edge beams between adjacent shells were 12.7 mm x 6.35 mm timber strips.

GENERAL PROCEDURE

One hundred and thirty pressure points were prepared on the surface of the model and were connected to a multitube manometer by means of pvc tubes passing through the interior of the model. The model rested on and was attached to the circular turntable as shown in Fig.1 which also showed the mounting in the open jet wind tunnel of the University of Liverpool. The working section of the tunnel was 1.54 m(horizontal) x 1.06 m (vertical). The surface of the turntable was 15 cm above the top of the tunnel floor at the outlet. The upstream edge of the table was feathered to a knife edge.

The general procedure in making the measurements of pressure distribution was as follows. The model was first set with the line joining the corners L and F parallel to the wind direction, an azimuth designated as 0° as shown in Fig. 2. A wind velocity of 15m/s was used. A complete set of observations were made for wind directions of 0°, 45°, 135°, 180° and 225°. Tests carried out at different wind speeds of the tunnel showed that the values of C_p were same for any particular pressure point.

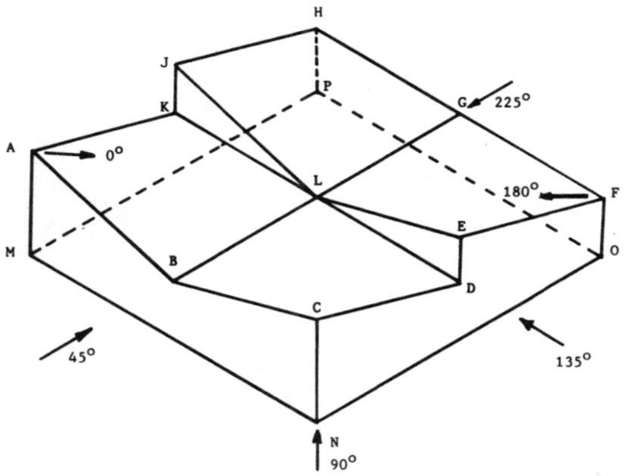

Fig. 2 Wind directions in relation to model

RESULTS

From the manometer readings, the values of p_w and C_p were calculated. By using the values of C_p for the pressure points, wind pressure contour diagrams were drawn for the multiple shell roof as shown in Figs. 3, 4, 5, 6 and 7. The average values of C_p were computed by a process of graphical integration. The results of this computation is shown in Table 1. The load on the roof may be obtained by multiplying the average value of C_p by the area of the roof and by the velocity pressure q. The maximum point suction encountered was -5q for a 180° wind and there were high point suctions adjacent to windward edge beams.

TABLE 1

The average values of C_p on roof for different wind directions

Wind Direction θ	C_p
0°	-0.44
45°	-0.68
135°	-0.63
180°	-0.19
225°	-0.37

COMPARISON WITH SINGLE SHELL ROOF AND FLAT ROOF

When comparison is made with a flat roof, see Ref 7, a highest point suction of -7q was encountered very close to the windward corner of a flat roof model, 75 mm high x 50 mm wide x 150 mm long with a corner-on-to wind flow. The maximum average suction on a 100 mm square flat roof was -0.5q, see Ref 8 and 9, as compared with -0.68q on a multiple shell roof. It was interesting to note that high suctions were found close to the parapets in all the roofs. They were, however, confined to very small areas of the roof and were likely to be affected by minor variations in the eaves details. In the case of HP shell wih edge beams such local suctions are not likely to be very significant as regards the design of whole roof structure because the corner is very stiff and could not be lifted without lifting the roof as a whole.

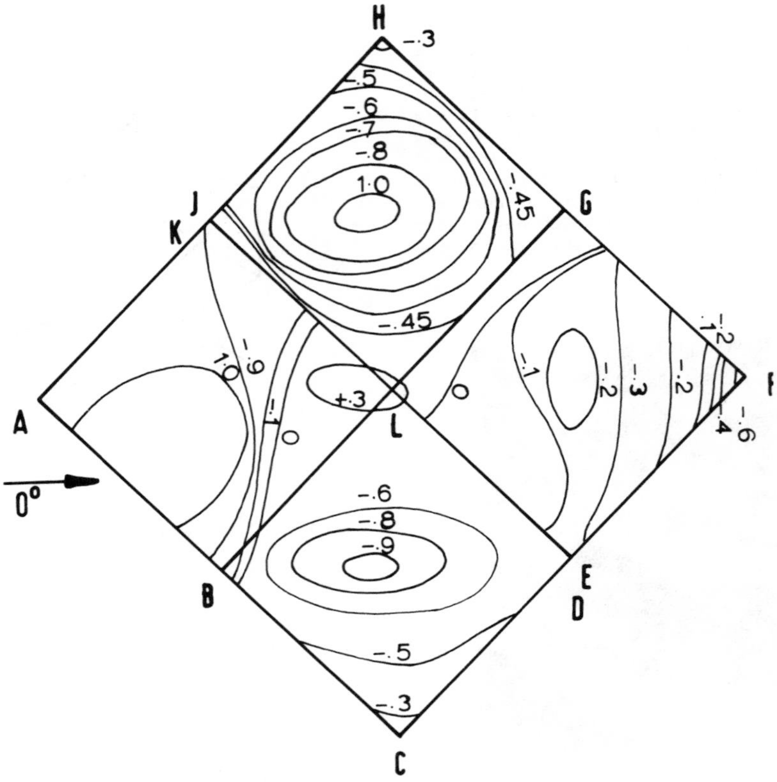

Fig. 3 Wind pressure contours: θ = 0°

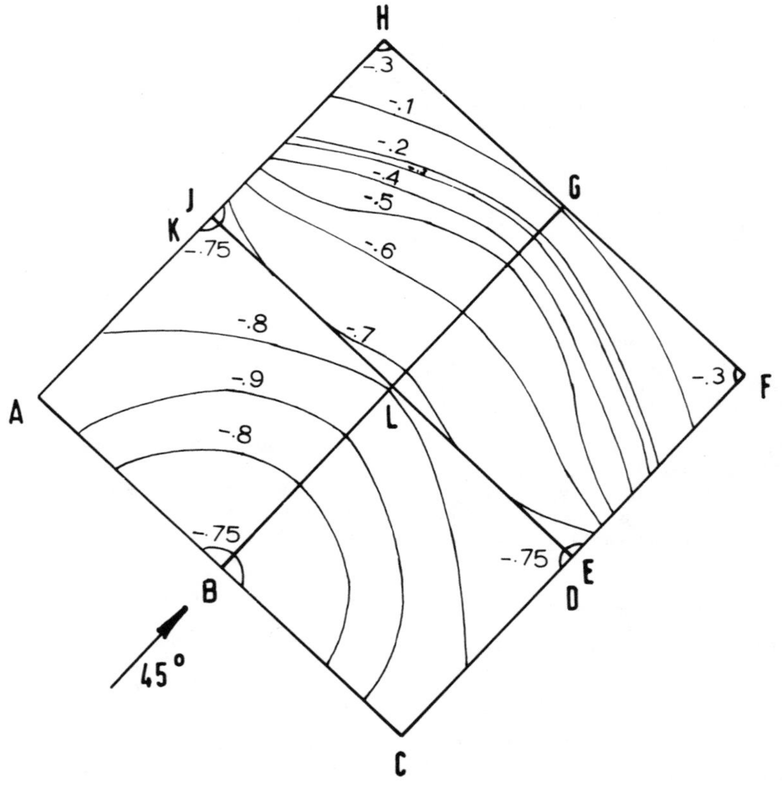

Fig. 4 Wind pressure contours: θ = 45°

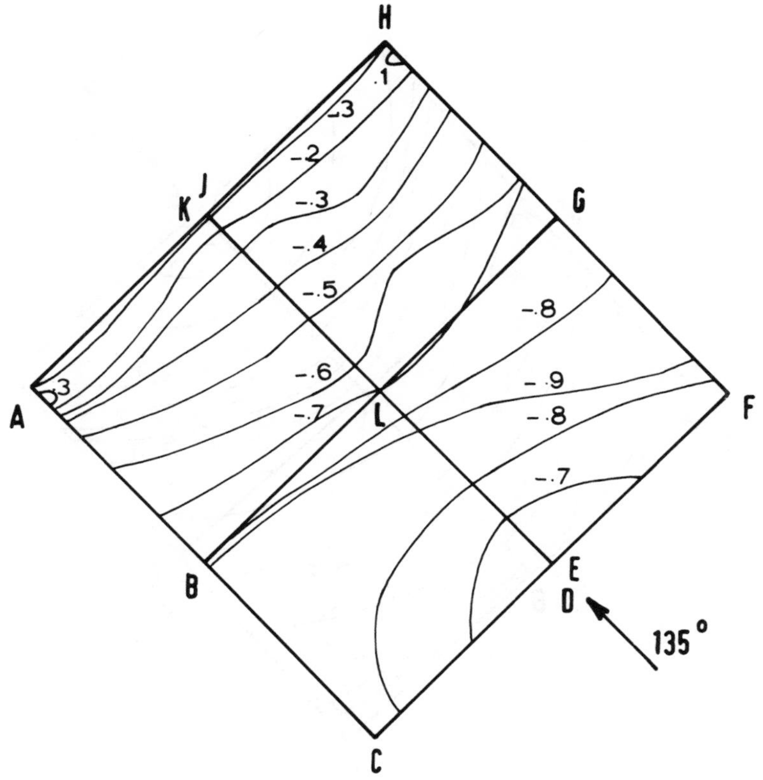

Fig. 5 Wind pressure contours: θ = 135°

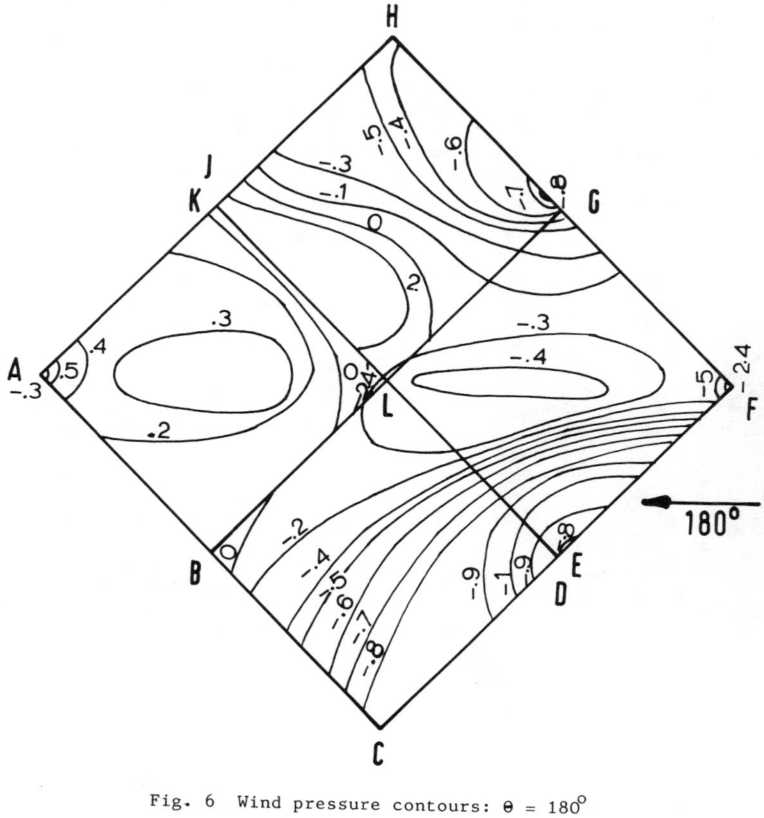

Fig. 6 Wind pressure contours: θ = 180°

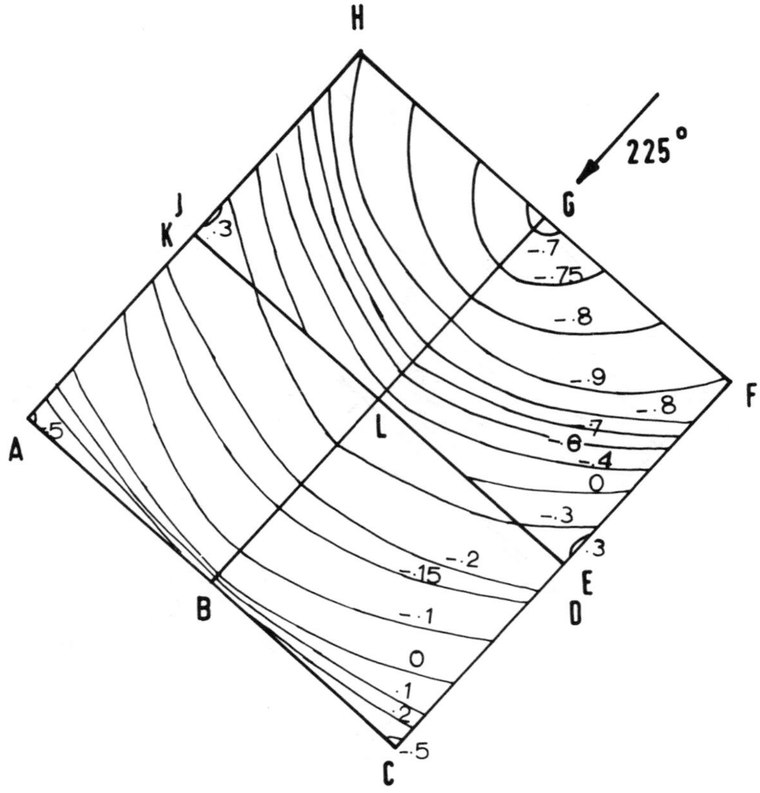

Fig. 7 Wind pressure contours: θ = 225°

A highest point suction of -5.5q was found on the windward high corner of a plain closed sided single HP shell roof, see Ref 10. The suction reduced to -0.6q in the case of a model with 25 mm high parapet. The maximum average suction found on a 300 mm square (in plan) single closed sided HP roof model with 162 mm high-corner height and 100 mm low corner height, was -0.88q, as compared with the maximum average suction of -0.68 q in the case of multiple shell. The maximum average suction found on a single open-sided shell roof of same dimension, as shown in Fig. 8, was -0.115q only. This showed that average suction on an open sided roof was much less that on a closed sided one.

Fig. 8 Model of open sided single HP shell roof structure in wind tunnel

CONCLUSIONS

The highest point suction encountered on a "sawtooth" array multiple HP shell roof was -5.0q and the maximum average suction on the roof was -0.68q. Effect of localised high suctions near the eaves and areas adjacent to the edge beams and parapets should be taken into account when detailing the roof coverings, e.g. felt or copper, since they may be ripped off by intense suction if not adequately secured. The pressure coefficients found in this investigation could be incorporated in the Code of Practice for wind loading and could be used for design purposes.

REFERENCES

1. R E AKINS, J A PETERKA and J E CERMAK, Averaged Pressure Coefficients for Rectangular Buildings, Proc. Fifth International Conference on Wind Engineering, USA, Pergamon Press, 1980.

2. A J DUTT, Oscillation Response of Structures in the Natural Wind, Proc. Fifth International Conference on Wind Engineering, USA, Pergamon Press, 1980.

3. A J DUTT, British Standard Code of Practice for Loading: Wind Loads - Discussion, The Structural Engineer, London, June 1970.

4. A J DUTT and W MOHAMAD, Wind Pressure Distribution on 'Minangkabau' Roof Building, Proc. Symposium on "Designing With the Wind", June 1981, CSTB, Nantes, France.

5. A G DAVENPORT, D SURRY and T STATHOPOULOS, Wind Loads on Low-rise Buildings, Final Report, The University of Western Ontario, Canada, November 1977.

6. W H MELBOURNE, The Relevance of Codification to Design, Proc. 4th International Conference on Wind Effects on Buildings and Structures, London, University of Cambridge Press, 1975.

7. C SALTER, Wind Loading on Flat Roofed Buildings, Engineering, October 1958.

8. J D HADDON, The Use of Wind Tunnel Models for Determining the Wind Pressure on Buildings, Civil Engineering and P W Review, April 1960.

9. A J DUTT, Wind Pressure Distribution on Hyperbolic Paraboloid Shell Roofs, Civil Engineering and P W Review, London, January 1971.

10. R E WHITEBREAD and M A PACKER, The Pressure Distribution on Hyperboloid Roof, National Physical Laboratory, U.K., April 1961.

FOURIER ANALYSIS OF TRUNCATED CONICAL SHELLS OF LINEARLY VARYING THICKNESS

MASASHI IURA, Dr. Eng.

Assistant Professor, Tokyo Denki University, Saitama, Japan

Fourier-series solutions of conical shells applicable to any prescribed boundary condition along circular and generator edges have been presented in the framework of linear theory. Utilizing the new parameter in place of the distance from the vertex, the fundamental equations with variable coefficients have been transformed into the partial differential equations with constant coefficients. The present general solutions of the resulting equations consist of the particular solutions due to normally distributed forces and the homogeneous solutions expanded into Fourier-sine and -cosine series in not only the circumferential but also the meridional directions.

INTRODUCTION

It is widely accepted that an analytical method has the significant advantage compared with the existing discrete numerical method in clarifying the characteristics of the fundamental equations and also the mechanical behavior of the shell. In this paper, an attempt is made to obtain the Fourier-series solutions of truncated conical shells of linearly varying thickness, which are applicable to any prescribed boundary conditions along circumferential and also meridional edges.

Truncated conical shells have been studied by a number of investigators. Flügge [1] has represented the rigorous fundamental equations of conical shells of linearly varying thickness and obtained the homogeneous solutions of the resulting equations in terms of three displacement components. Based on the development of power series, Wilson [2] has analyzed an asymmetrical bending problem of truncated cones with constant thickness. Chang [3,4] has obtained asymptotic solutions of conical shells of linearly varying thickness and constant thickness. Tsuboi and Ujiya [5] have worked on the bending problem of truncated conical shells utilizing the normal displacement component and the stress function. In case of a segment of cone, the existing Fourier-series solutions are applicable only to simple supports along the generator edges. This is owing to the fact that the existing solutions are expanded in the circumferential direction only. When the method of Fourier-series expansion is applied to a boundary-value problem for a two-dimensional problem, the solutions need to be expanded into the two directions in order to satisfy any prescribed boundary condition along edges.

Since, in the published papers, the distance from the vertex has been used as a meridional coordinate, the fundamental differential equations of conical shells have variable coefficients, see Ref 1. Consequently it has been thought difficult to expand the solutions in the Fourier-series both into circumferential and meridional directions. In this paper, with the help of the new parameter in place of the distance from the vertex, the rigorous differential equations with variable coefficients, derived by W. Flügge [1], are transformed into the partial differential equations with constant coefficients. Based on the resulting equations, the ordinary differential equations with constant coefficients are obtained by the method of separation of variables, in which the three displacement components are expanded into the Fourier-sine and -cosine series in the circumferential and also the meridional directions.

The solutions of the ordinary differential equations with constant coefficients are obtained by making use of the solutions of the characteristic equations which are, in general, reduced to the algebraic equations. In this paper, however, it is found useful to reduce the characteristic equations to the eigenvalue equations. The eigenvalues correspond to the characteristic values and furthermore the eigenvectors are used to obtain the relationships among the coefficients of the solutions. The constants of integration are determined so as to satisfy the prescribed boundary conditions along four edges.

BASIC EQUATIONS

The notation used in this paper follows that of Ref 1.

The basic equations of conical shells are essentially given by W.Flügge. When the thickness of shell t is independent of θ but proportional to s, the fundamental differential equations in terms of three displacement components, taking surface distributed loads into account, are represented by

$$\frac{(1-\nu)}{2} s^2 u^{\bullet\bullet} + u''\sec^2\alpha + (1-\nu)u + \frac{(1+\nu)}{2} s v'^{\bullet}\sec\alpha$$
$$+(2-\nu)v'\sec\alpha + w'\tan\alpha\sec\alpha + k[\frac{3}{2}(1-\nu)s^2 u^{\bullet\bullet}\tan\alpha$$
$$+3(1-\nu)su^{\bullet}\tan\alpha - 3(1-\nu)u\tan\alpha - \frac{(3-\nu)}{2} s^2 w'^{\bullet\bullet}\sec\alpha$$
$$-3(1-\nu)sw'^{\bullet}\sec\alpha + 3(1-\nu)w'\sec\alpha\,]\tan\alpha = -P_\theta s^2/D \quad \ldots\ 1\text{ a},$$

$$\frac{(1+\nu)}{2} s u'^{\bullet} sec\alpha - \frac{3}{2}(1-\nu)u' sec\alpha + s^2 v^{\bullet\bullet} + \frac{(1-\nu)}{2} v'' sec^2\alpha$$
$$+2sv^{\bullet} - (1-\nu)v + \nu s w^{\bullet} tan\alpha - (1-\nu)w tan\alpha + k[$$
$$\frac{(1-\nu)}{2} v'' tan\alpha sec^2\alpha - v tan\alpha - s^3 w^{\bullet\bullet\bullet} + \frac{(1-\nu)}{2} s w''^{\bullet} sec^2\alpha$$
$$-3s^2 w^{\bullet\bullet} - \frac{(3-\nu)}{2} w'' sec^2\alpha - s w^{\bullet} - w tan^2\alpha\ ]tan\alpha = - P_s s^2/D \qquad \ldots\ 1\ b,$$

$$[u' sec\alpha + \nu s v^{\bullet} + v + w tan\alpha]tan\alpha + k[- \frac{(3-\nu)}{2} s^2 u'^{\bullet\bullet} sec\alpha$$
$$-(3+\nu)s u'^{\bullet} sec\alpha + (3-5\nu)u' sec\alpha - s^3 v^{\bullet\bullet\bullet} + \frac{(1-\nu)}{2} s v''^{\bullet} sec^2\alpha$$
$$- 6s^2 v^{\bullet\bullet} + (2-\nu)v'' sec^2\alpha - 7sv^{\bullet} - v(1- tan^2\alpha)]tan\alpha$$
$$+k[s^4 w^{\bullet\bullet\bullet\bullet} + 2s^2 w''^{\bullet\bullet} sec^2\alpha + w'''' sec^4\alpha + 8s^3 w^{\bullet\bullet\bullet} + 4s w''^{\bullet} sec^2\alpha$$
$$+(11+3\nu)s^2 w^{\bullet\bullet} + 2w'' tan^2\alpha sec^2\alpha - (5-6\nu)w'' sec^2\alpha$$
$$-2(1-3\nu)s w^{\bullet} - w(1-tan^2\alpha)tan^2\alpha] = P_r s^2/D \qquad \ldots\ 1\ c,$$

where

$$(\ )^{\bullet} = \partial(\ )/\partial s \quad , \quad (\ )' = \partial(\ )/\partial\theta \qquad \ldots\ldots\ 2\ a,b,$$

$$D = Et/(1-\nu^2) \quad , \quad k = \delta^2/12 \qquad \ldots\ldots\ 2\ c,d,$$

$$t = \delta s \qquad \ldots\ldots\ 2\ e,$$

and u,v and w are the circumferential, meridional and normal displacements, α is the angle between a normal to the midsurface and the axis of the cone, θ and s are circumferential and meridional coordinates, E and ν are Young's modulus and Poisson's ratio, and P_θ, P_s and P_r, which have been neglected in Ref 1, are surface distributed loads per unit area in circumferential, meridional and normal directions respectively.

In order to derive the differential equations with constant coefficients, we introduce the new parameters x and y defined as

$$x = a\ lns + b \qquad \ldots\ldots\ 3\ a,$$

$$y = c\ \theta - 1 \qquad \ldots\ldots\ 3\ b,$$

where

$$a = \frac{2}{lns_1 - lns_o} \qquad \ldots\ldots\ 4\ a,$$

$$b = \frac{lns_o + lns_1}{lns_o - lns_1} \qquad \ldots\ldots\ 4\ b,$$

$$c = 2/\ \theta_o \qquad \ldots\ldots\ 4\ c,$$

in which s_o and s_1 denote the meridional coordinates of the circular edges, and θ_o denotes the central angle between the two generator edges as shown in Fig 1. Utilizing the parameters x and y, the boundaries of the shell are given by x= ±1 along the circular edges and y= ±1 along the generator edges.

With the aid of the parameters defined by Eqns 3, the fundamental differential equations can be rewritten by

$$\frac{(1-\nu)}{2}(u^{**} + u^{*}) - (1-\nu)u + u^{\acute{}\acute{}} + \frac{(1+\nu)}{2} v^{*\acute{}} + (2-\nu)v^{\acute{}}$$
$$+ w^{\acute{}} tan\alpha + k[\frac{3}{2}(1-\nu)(u^{**} + u^{*})tan\alpha - 3(1-\nu)\ u tan\alpha$$
$$+ \frac{(-3+5\nu)}{2} w^{*\acute{}} - \frac{(3-\nu)}{2} w^{**\acute{}} + 3(1-\nu)w^{\acute{}}\ ]tan\alpha =$$
$$- \frac{P_\theta}{D} exp[\frac{2(x-b)}{a}] \qquad \ldots\ldots\ 5\ a,$$

$$\frac{(1+\nu)}{2} u^{*\acute{}} - \frac{3}{2}(1-\nu)u^{\acute{}} + v^{**} + v^{*} + \frac{(1-\nu)}{2} v^{\acute{}\acute{}} - (1-\nu)v$$
$$+ \nu w^{*} tan\alpha - (1-\nu)w tan\alpha + k[\ \frac{(1-\nu)}{2} v^{\acute{}\acute{}} tan\ \alpha - v tan\alpha$$
$$- w^{**}_{**} - w tan^2\alpha + \frac{(1-\nu)}{2} w^{*\acute{}\acute{}} - \frac{(3-\nu)}{2} w^{\acute{}\acute{}}]tan\alpha =$$
$$- \frac{P_s}{D} exp[\ \frac{2(x-b)}{a}] \qquad \ldots\ldots\ 5\ b,$$

$$[\ u^{\acute{}} + \nu v^{*} + v + w tan\alpha\ ]tan\alpha + k[- \frac{(3-\nu)}{2} u^{**\acute{}} - \frac{3(1+\nu)}{2} u^{*\acute{}}$$
$$+ (3-5\nu)\ u^{\acute{}} - v^{***} - 3v^{**} - 3v^{*} + \frac{(1-\nu)}{2} v^{*\acute{}\acute{}} + (2-\nu)v^{\acute{}\acute{}}$$
$$- (1-tan^2\alpha)v\ ]\ tan\alpha + k[\ w^{**}_{**} + 2w^{***} + (-2+3\nu)w^{**}$$
$$- 3(1-\nu)w^{*} + 2w^{**\acute{}\acute{}} + 2w^{*\acute{}\acute{}} + w^{\acute{}\acute{}}_{\acute{}\acute{}} + 2w^{\acute{}\acute{}} tan^2\alpha - (5-6\nu)w^{\acute{}\acute{}}$$
$$- w(1- tan^2\alpha)tan^2\alpha\ ] = \frac{P_r}{D} exp[\ \frac{2(x-b)}{a}\ ] \qquad \ldots\ldots\ 5\ c,$$

where

$$(\ )^{*} = \partial(\ )/\partial\hat{x} \qquad \ldots\ldots\ 6\ a,$$

$$(\ )^{\acute{}} = sec\alpha\ \partial(\ )/\partial\hat{y} \qquad \ldots\ldots\ 6\ b,$$

$$\hat{x} = x/a \quad , \quad \hat{y} = y/c \qquad \ldots\ldots\ 6\ c,d.$$

It should be noted that Eqns 1 have variable coefficients, while Eqns 5 have constant coefficients.

It is assumed in this paper that the general solutions of Eqns 5 consist of particular solutions and homogeneous solutions expressed by

$$u = u^{\circ} + u^{\circ\circ} + u^{\circ\circ\circ} \qquad \ldots\ldots\ 7\ a,$$

$$v = v^{\circ} + v^{\circ\circ} + v^{\circ\circ\circ} \qquad \ldots\ldots\ 7\ b,$$

$$w = w^{\circ} + w^{\circ\circ} + w^{\circ\circ\circ} \qquad \ldots\ldots\ 7\ c,$$

where $(\ )^{\circ}$ denotes a particular solution, and $(\ )^{\circ\circ}$ and $(\ )^{\circ\circ\circ}$ denote homogeneous solutions expanded into the Fourier-series in the meridional and circumferential directions respectively.

PARTICULAR SOLUTIONS

Consider a conical shell subjected to a uniformly distributed normal load. The uniform load P_r can be expressed in the form of Fourier series in the circumferential direction as

$$P_r = \sum_{n=1,3,5..}^{N} P_n\ cos\lambda_n y \qquad \ldots\ldots\ 8,$$

where

$$P_n = \frac{2\ P_r}{\lambda_n}\ sin\lambda_n \qquad \ldots\ldots\ 9\ a,$$

$$\lambda_n = \frac{n\ \pi}{2} \qquad \ldots\ldots\ 9\ b.$$

Substituting Eqn 8 into Eqns 5 and noting that $P_\theta = P_s = 0$, the particular solutions due to uniformly distributed normal loads are obtained in the following form:

$$u^{\circ} = \sum_{n=1,3,5..}^{N} u_n\ exp[\ \frac{(x-b)}{a}\ ]\ sin\lambda_n y \qquad \ldots\ldots\ 10\ a,$$

$$v^{\circ}=\sum_{n=1,3,5..}^{N} v_n \exp\left[\frac{(x-b)}{a}\right] \cos\lambda_n y \qquad \text{...... 10 b,}$$

$$w^{\circ}=\sum_{n=1,3,5..}^{N} w_n \exp\left[\frac{(x-b)}{a}\right] \cos\lambda_n y \qquad \text{...... 10 c,}$$

where

$$u_n = P_n\,(r_{12}\,r_{23} - r_{13}\,r_{22})/(\hat{D}\,r_d) \qquad \text{...... 11 a,}$$

$$v_n = P_n\,(r_{13}\,r_{21} - r_{11}\,r_{23})/(\hat{D}\,r_d) \qquad \text{...... 11 b,}$$

$$w_n = P_n\,(r_{11}\,r_{22} - r_{12}\,r_{21})/(\hat{D}\,r_d) \qquad \text{...... 11 c,}$$

$$r_{11} = -c^2\lambda_n^2 \sec^2\alpha \qquad \text{...... 11 d,}$$

$$r_{12} = -\frac{(5-\nu)}{2}\,c\lambda_n \sec\alpha \qquad \text{...... 11 e,}$$

$$r_{13} = -c\lambda_n \tan\alpha \sec\alpha \qquad \text{...... 11 f,}$$

$$r_{21} = (-1+2\nu)\,c\lambda_n \sec\alpha \qquad \text{...... 11 g,}$$

$$r_{22} = 1+\nu - \frac{(1-\nu)}{2}\,c^2\lambda_n^2\,(1+k\tan^2\alpha)\sec^2\alpha - k\tan^2\alpha \qquad \text{... 11 h,}$$

$$r_{23} = (2\nu-1)\tan\alpha - k\tan\alpha - k\tan^3\alpha + k\,c^2\lambda_n^2\tan\alpha\sec^2\alpha \qquad \text{... 11 i,}$$

$$r_{31} = c\lambda_n\tan\alpha\sec\alpha - 6\,\nu\,kc\lambda_n\tan\alpha\sec\alpha \qquad \text{... 11 j,}$$

$$r_{32} = (1+\nu)\tan\alpha + k\left[-8 + \tan^2\alpha - \frac{(5-3\nu)}{2}\,c^2\lambda_n^2\sec^2\alpha\right]\tan\alpha \qquad \text{... 11 k,}$$

$$r_{33} = \tan^2\alpha + k\left[-2+6\nu - (1-\tan^2\alpha)\tan^2\alpha + (1-6\nu)c^2\lambda_n^2\sec^2\alpha - 2c^2\lambda_n^2\tan^2\alpha\sec^2\alpha + c^4\lambda_n^4\sec^4\alpha\right] \qquad \text{... 11 l,}$$

$$r_d = \det|\,r_{ij}\,|\;, \qquad \hat{D} = E\delta/(1-\nu^2) \qquad \text{... 11 m,n.}$$

HOMOGENEOUS SOLUTIONS

Consider the homogeneous solutions $u^{\circ\circ}$, $v^{\circ\circ}$ and $w^{\circ\circ}$, expanded into the Fourier-series in the meridional direction in the form

$$u^{\circ\circ}=\sum_{m=1,3,5..}^{M}\left[u^{\circ\circ}_{c,m}(y)\cos\lambda_m x + u^{\circ\circ}_{s,m}(y)\sin\lambda_m x\right] \qquad \text{...... 12 a,}$$

$$v^{\circ\circ}=\sum_{m=1,3,5..}^{M}\left[v^{\circ\circ}_{c,m}(y)\cos\lambda_m x + v^{\circ\circ}_{s,m}(y)\sin\lambda_m x\right] \qquad \text{...... 12 b,}$$

$$w^{\circ\circ}=\sum_{m=1,3,5..}^{M}\left[w^{\circ\circ}_{c,m}(y)\cos\lambda_m x + w^{\circ\circ}_{s,m}(y)\sin\lambda_m x\right] \qquad \text{...... 12 c,}$$

where

$$\lambda_m = \frac{m\,\pi}{2} \qquad \text{...... 13.}$$

Substituting Eqns 12 into the homogeneous parts of Eqns 5 leads to the following ordinary differential equations:

$$\frac{(1-\nu)}{2}\,a\lambda_m(u^{\circ\circ}_{s,m} - a\lambda_m u^{\circ\circ}_{c,m})(1+3k\tan^2\alpha) - (1-\nu)u^{\circ\circ}_{c,m} + u^{\circ\circ\,\prime\prime}_{c,m}$$
$$+\frac{(1+\nu)}{2}\,a\lambda_m v^{\circ\circ\,\prime}_{s,m} + (2-\nu)v^{\circ\circ\,\prime}_{c,m} + w^{\circ\circ\,\prime}_{c,m}\tan\alpha$$
$$-3(1-\nu)ku^{\circ\circ}_{c,m}\tan^2\alpha + \frac{(-3+5\nu)}{2}\,a\lambda_m k w^{\circ\circ\,\prime}_{s,m}\tan\alpha$$
$$+\frac{(3-\nu)}{2}\,a^2\lambda_m^2 k w^{\circ\circ\,\prime}_{c,m}\tan\alpha + 3(1-\nu)kw^{\circ\circ\,\prime}_{c,m}\tan\alpha = 0 \qquad \text{... 14 a,}$$

$$-\frac{(1-\nu)}{2}\,a\lambda_m(u^{\circ\circ}_{c,m} + a\lambda_m u^{\circ\circ}_{s,m})(1+3k\tan^2\alpha) - (1-\nu)u^{\circ\circ}_{s,m} + u^{\circ\circ\,\prime\prime}_{s,m}$$
$$-\frac{(1+\nu)}{2}\,a\lambda_m v^{\circ\circ\,\prime}_{c,m} + (2-\nu)v^{\circ\circ\,\prime}_{s,m} + w^{\circ\circ\,\prime}_{s,m}\tan\alpha$$
$$-3(1-\nu)ku^{\circ\circ}_{s,m}\tan^2\alpha - \frac{(-3+5\nu)}{2}\,a\lambda_m k w^{\circ\circ\,\prime}_{c,m}\tan\alpha$$
$$+\frac{(3-\nu)}{2}\,a^2\lambda_m^2 k w^{\circ\circ\,\prime}_{s,m}\tan\alpha + 3(1-\nu)kw^{\circ\circ\,\prime}_{s,m}\tan\alpha = 0 \qquad \text{... 14 b,}$$

$$\frac{(1+\nu)}{2}\,a\lambda_m u^{\circ\circ\,\prime}_{s,m} - \frac{3}{2}(1-\nu)u^{\circ\circ\,\prime}_{c,m} - a^2\lambda_m^2 v^{\circ\circ}_{c,m} + a\lambda_m v^{\circ\circ}_{s,m}$$
$$+\frac{(1-\nu)}{2}\,v^{\circ\circ\,\prime\prime}_{c,m} - (1-\nu)v^{\circ\circ}_{c,m} + \nu a\lambda_m w^{\circ\circ}_{s,m}\tan\alpha - (1-\nu)w^{\circ\circ}_{c,m}\tan\alpha$$
$$+k\Big[\frac{(1-\nu)}{2}\,v^{\circ\circ\,\prime\prime}_{c,m}\tan\alpha - v^{\circ\circ}_{c,m}\tan\alpha + a^3\lambda_m^3 w^{\circ\circ}_{s,m} + \frac{(1-\nu)}{2}\,a\lambda_m w^{\circ\circ\,\prime\prime}_{s,m}$$
$$-w^{\circ\circ}_{c,m}\tan^2\alpha - \frac{(3-\nu)}{2}\,w^{\circ\circ\,\prime\prime}_{c,m}\Big]\tan\alpha = 0 \qquad \text{... 14 c,}$$

$$-\frac{(1+\nu)}{2}\,a\lambda_m u^{\circ\circ\,\prime}_{c,m} - \frac{3}{2}(1-\nu)u^{\circ\circ\,\prime}_{s,m} - a^2\lambda_m^2 v^{\circ\circ}_{s,m} - a\lambda_m v^{\circ\circ}_{c,m}$$
$$+\frac{(1-\nu)}{2}\,v^{\circ\circ\,\prime\prime}_{s,m} - (1-\nu)v^{\circ\circ}_{s,m} - \nu a\lambda_m w^{\circ\circ}_{c,m}\tan\alpha - (1-\nu)w^{\circ\circ}_{s,m}\tan\alpha$$
$$+k\Big[\frac{(1-\nu)}{2}\,v^{\circ\circ\,\prime\prime}_{s,m}\tan\alpha - v^{\circ\circ}_{s,m}\tan\alpha - a^3\lambda_m^3 w^{\circ\circ}_{c,m} - \frac{(1-\nu)}{2}\,a\lambda_m w^{\circ\circ\,\prime\prime}_{c,m}$$
$$-w^{\circ\circ}_{s,m}\tan^2\alpha - \frac{(3-\nu)}{2}\,w^{\circ\circ\,\prime\prime}_{s,m}\Big]\tan\alpha = 0 \qquad \text{... 14 d,}$$

$$\big[u^{\circ\circ\,\prime}_{c,m} + \nu a\lambda_m v^{\circ\circ}_{s,m} + v^{\circ\circ}_{c,m} + w^{\circ\circ}_{c,m}\tan\alpha\big]\tan\alpha + k\Big[$$
$$\frac{(3-\nu)}{2}\,a^2\lambda_m^2 u^{\circ\circ\,\prime}_{c,m} - \frac{3}{2}(1+\nu)a\lambda_m u^{\circ\circ\,\prime}_{s,m} + (3-5\nu)u^{\circ\circ\,\prime}_{c,m} + a^3\lambda_m^3 v^{\circ\circ}_{s,m}$$
$$+3a^2\lambda_m^2 v^{\circ\circ}_{c,m} - 3a\lambda_m v^{\circ\circ}_{s,m} + \frac{(1-\nu)}{2}\,a\lambda_m v^{\circ\circ\,\prime\prime}_{s,m} + (2-\nu)v^{\circ\circ\,\prime\prime}_{c,m}$$
$$-(1-\tan^2\alpha)v^{\circ\circ}_{c,m}\Big]\tan\alpha + k\Big[a^4\lambda_m^4 w^{\circ\circ}_{c,m} - 2a^3\lambda_m^3 w^{\circ\circ}_{s,m}$$
$$+(2-3\nu)a^2\lambda_m^2 w^{\circ\circ}_{c,m} - 3(1-\nu)a\lambda_m w^{\circ\circ}_{s,m} - 2a^2\lambda_m^2 w^{\circ\circ\,\prime\prime}_{c,m} + 2a\lambda_m w^{\circ\circ\,\prime\prime}_{s,m}$$
$$+w^{\circ\circ\,\prime\prime\prime\prime}_{c,m} + (2\tan^2\alpha - 5 + 6\nu)w^{\circ\circ\,\prime\prime}_{c,m} - w^{\circ\circ}_{c,m}(1-\tan^2\alpha)\tan^2\alpha\Big]$$
$$= 0 \qquad \text{...... 14 e,}$$

$$\big[u^{\circ\circ\,\prime}_{s,m} - \nu a\lambda_m v^{\circ\circ}_{c,m} + v^{\circ\circ}_{s,m} + w^{\circ\circ}_{s,m}\tan\alpha\big]\tan\alpha + k\Big[$$
$$\frac{(3-\nu)}{2}\,a^2\lambda_m^2 u^{\circ\circ\,\prime}_{s,m} + \frac{3}{2}(1+\nu)a\lambda_m u^{\circ\circ\,\prime}_{c,m} + (3-5\nu)u^{\circ\circ\,\prime}_{s,m} - a^3\lambda_m^3 v^{\circ\circ}_{c,m}$$
$$+3a^2\lambda_m^2 v^{\circ\circ}_{s,m} + 3a\lambda_m v^{\circ\circ}_{c,m} - \frac{(1-\nu)}{2}\,a\lambda_m v^{\circ\circ\,\prime\prime}_{c,m} + (2-\nu)v^{\circ\circ\,\prime\prime}_{s,m}$$
$$-(1-\tan^2\alpha)v^{\circ\circ}_{s,m}\Big]\tan\alpha + k\Big[a^4\lambda_m^4 w^{\circ\circ}_{s,m} + 2a^3\lambda_m^3 w^{\circ\circ}_{c,m}$$
$$+(2-3\nu)a^2\lambda_m^2 w^{\circ\circ}_{s,m} + 3(1-\nu)a\lambda_m w^{\circ\circ}_{c,m} - 2a^2\lambda_m^2 w^{\circ\circ\,\prime\prime}_{s,m} - 2a\lambda_m w^{\circ\circ\,\prime\prime}_{c,m}$$
$$+w^{\circ\circ\,\prime\prime\prime\prime}_{s,m} + (2\tan^2\alpha - 5 + 6\nu)w^{\circ\circ\,\prime\prime}_{s,m} - w^{\circ\circ}_{s,m}(1-\tan^2\alpha)\tan^2\alpha\Big]$$
$$= 0 \qquad \text{...... 14 f.}$$

In order to solve the homogeneous differential equations 14 we may postulate the solutions in the form

$$u^{\circ\circ}_{c,m} = A_m \exp(\,\eta_m\,\hat{y}\,) \qquad \text{...... 15 a,}$$

$$u^{\circ\circ}_{s,m} = B_m \exp(\,\eta_m\,\hat{y}\,) \qquad \text{...... 15 b,}$$

$$v^{\circ\circ}_{c,m} = C_m \exp(\,\eta_m\,\hat{y}\,) \qquad \text{...... 15 c,}$$

$$v^{\circ\circ}_{s,m} = D_m \exp(\,\eta_m\,\hat{y}\,) \qquad \text{...... 15 d,}$$

$$w^{\circ\circ}_{c,m} = E_m \exp(\,\eta_m\,\hat{y}\,) \qquad \text{...... 15 e,}$$

$$w^{\circ\circ}_{s,m} = F_m \exp(\,\eta_m\,\hat{y}\,) \qquad \text{...... 15 f.}$$

Introducing Eqns 15 into Eqns 14 and using the condition that the exponential functions do not become zero yield the following equations:

$$[\, R_{i,j} \,]\{\, d \,\} = 0 \quad (i,j = 1,2,\ldots,6) \qquad \ldots\ldots 16,$$

where

$$\{\, d \,\}^T = \{\, A_m , B_m , \ldots , F_m \,\} \qquad \ldots\ldots 17,$$

and $R_{i,j}$ denotes a coefficient including η_m, ν, λ_m, k and α. Some of the coefficients are written as

$$R_{1,1} = R_{2,2} = -\frac{(1-\nu)}{2} a^2\lambda_m^2(1+3k\tan^2\alpha) + \eta_m^2 \sec^2\alpha - (1-\nu)(1+3k\tan^2\alpha) \qquad \ldots\ldots 18\text{ a},$$

$$R_{1,2} = -R_{2,1} = \frac{(1-\nu)}{2} a\lambda_m (1+3k\tan^2\alpha) \qquad \ldots\ldots 18\text{ b},$$

$$R_{1,3} = R_{2,4} = (2-\nu)\eta_m \sec\alpha \qquad \ldots\ldots 18\text{ c},$$

$$R_{1,4} = -R_{2,3} = \frac{(1+\nu)}{2} \lambda_m\eta_m \sec\alpha \qquad \ldots\ldots 18\text{ d}.$$

In general the characteristic values η_m are obtained from the characteristic equations which are represented, using the condition that the constants A_m to F_m have the non-trivial solutions, in the form

$$\det|\, R_{i,j} \,| = 0 \qquad \ldots\ldots 19.$$

In this case, it is a tedious task to derive the algebraic equations for η_m from Eqn 19 and also the relationships among coefficients of $u^{\circ\circ}_{c,m}, \ldots , w^{\circ\circ}_{s,m}$ from Eqn 16.

In this paper, based on the eigenvalue equations derived from Eqn 16, the characteristic values η_m are obtained as the eigenvalues. In order to transform Eqn 16 into the eigenvalue equations, we introduce the notation defined by

$$G_m = \eta_m A_m , \qquad H_m = \eta_m B_m \qquad \ldots\ldots 20\text{ a,b},$$

$$I_m = \eta_m C_m , \qquad J_m = \eta_m D_m \qquad \ldots\ldots 20\text{ c,d},$$

$$K_m = \eta_m E_m , \qquad L_m = \eta_m F_m \qquad \ldots\ldots 20\text{ e,f},$$

$$M_m = \eta_m K_m , \qquad N_m = \eta_m L_m \qquad \ldots\ldots 20\text{ g,h},$$

$$O_m = \eta_m M_m , \qquad P_m = \eta_m N_m \qquad \ldots\ldots 20\text{ i,j}.$$

The process of the derivation of the eigenvalue equations by utilizing Eqns 20 and Eqn 16 is not difficult. As a result, we obtain the following eigenvalue equations:

$$[\, \hat{S}_{i,j} \,]\{\, \delta \,\} = \eta_m \{\, \delta \,\} \quad (i,j = 1,2, \ldots ,16) \qquad \ldots 21.$$

where

$$\{\, \delta \,\}^T = \{\, A_m , B_m , \ldots , P_m \,\} \qquad \ldots\ldots 22,$$

and $\hat{S}_{i,j}$ denote constants including ν, λ_m, k and α. The detailed expressions for $\hat{S}_{i,j}$ are not given here, since their forms are too lengthy. From Eqn 21 the sixteen eigenvalues $\{\, \eta_{1,m} , \eta_{2,m} , \ldots , \eta_{16,m} \,\}$ are obtained for each m. In general, the resulting eigenvalues consist of real numbers and complex numbers. It is assumed here that all the eigenvalues are composed of real parts and imaginary parts, which do not become zero. The solutions in other cases are given in a similar manner described later without any difficulty. When the eigenvalues are complex numbers, they are represented for each m as

$$\eta_{\substack{2j-1,m \\ 2j}} = \kappa_{j,m} \pm i\, \mu_{j,m} \qquad (j=1,2, \ldots , 8) \qquad \ldots 23,$$

where $\kappa_{j,m}$ and $\mu_{j,m}$ are real numbers and $i=\sqrt{-1}$. The relationships among coefficients of $u^{\circ\circ}_{c,m}, \ldots , w^{\circ\circ}_{s,m}$ are determined from the eigenvectors $\{\, d \,\}$ expressed by

$$\{\, A_{j,m}, B_{j,m}, \ldots , F_{j,m} \,\} = \{\, \hat{\kappa}^1_{j,m} \pm i\, \hat{\mu}^1_{j,m} , \hat{\kappa}^2_{j,m} \pm i\, \hat{\mu}^2_{j,m} , \ldots , \hat{\kappa}^6_{j,m} \pm i\, \hat{\mu}^6_{j,m} \,\} \qquad \ldots 24,$$

and other eigenvectors such as $\{\, G_{j,m}, \ldots , P_{j,m} \}$ are apparently of no use. Using Eqns 12, 15, 23 and 24, the homogeneous solutions $(\;)^{\circ\circ}$ take the form

$$u^{\circ\circ} = \sum_{m=1,3,5..}^{M} \sum_{j=1}^{8} [\, (A_{2j-1,m} \cos\mu_{j,m} \hat{y} + A_{2j,m} \sin\mu_{j,m} \hat{y}) \exp(\kappa_{j,m} \hat{y}) \cos\lambda_m x + \{(R^1_{j,m} A_{2j-1,m} + S^1_{j,m} A_{2j,m}) \exp(\kappa_{j,m} \hat{y}) \cos\mu_{j,m} \hat{y} + (R^1_{j,m} A_{2j,m} - S^1_{j,m} A_{2j-1,m}) \exp(\kappa_{j,m} \hat{y}) \sin\mu_{j,m} \hat{y}\} \sin\lambda_m x] \quad \ldots 25\text{ a},$$

$$v^{\circ\circ} = \sum_{m=1,3,5..}^{M} \sum_{j=1}^{8} [\, \{\, (R^2_{j,m} A_{2j-1,m} + S^2_{j,m} A_{2j,m}) \exp(\kappa_{j,m} \hat{y}) \cos\mu_{j,m} \hat{y} + (R^2_{j,m} A_{2j,m} - S^2_{j,m} A_{2j-1,m}) \exp(\kappa_{j,m} \hat{y}) \sin\mu_{j,m} \hat{y} \,\} \cos\lambda_m x + \{\, (R^3_{j,m} A_{2j-1,m} + S^3_{j,m} A_{2j,m}) \exp(\kappa_{j,m} \hat{y}) \cos\mu_{j,m} \hat{y} + (R^3_{j,m} A_{2j,m} - S^3_{j,m} A_{2j-1,m}) \exp(\kappa_{j,m} \hat{y}) \sin\mu_{j,m} \hat{y} \,\} \sin\lambda_m x \,] \quad \ldots 25\text{ b},$$

$$w^{\circ\circ} = \sum_{m=1,3,5..}^{M} \sum_{j=1}^{8} [\, \{\, (R^4_{j,m} A_{2j-1,m} + S^4_{j,m} A_{2j,m}) \exp(\kappa_{j,m} \hat{y}) \cos\mu_{j,m} \hat{y} + (R^4_{j,m} A_{2j,m} - S^4_{j,m} A_{2j-1,m}) \exp(\kappa_{j,m} \hat{y}) \sin\mu_{j,m} \hat{y} \,\} \cos\lambda_m x + \{\, (R^5_{j,m} A_{2j-1,m} + S^5_{j,m} A_{2j,m}) \exp(\kappa_{j,m} \hat{y}) \cos\mu_{j,m} \hat{y} + (R^5_{j,m} A_{2j,m} - S^5_{j,m} A_{2j-1,m}) \exp(\kappa_{j,m} \hat{y}) \sin\mu_{j,m} \hat{y} \,\} \sin\lambda_m x \,] \quad \ldots 25\text{ c},$$

where

$$R^l_{j,m} = \frac{\hat{\kappa}^{l+1}_{j,m}\, \hat{\kappa}^1_{j,m} + \hat{\mu}^{l+1}_{j,m}\, \hat{\mu}^1_{j,m}}{(\hat{\kappa}^1_{j,m})^2 + (\hat{\mu}^1_{j,m})^2} \quad (l = 1,2,\ldots,5) \quad \ldots 26\text{ a},$$

$$S^l_{j,m} = \frac{\hat{\kappa}^1_{j,m}\, \hat{\mu}^{l+1}_{j,m} - \hat{\kappa}^{l+1}_{j,m}\, \hat{\mu}^1_{j,m}}{(\hat{\kappa}^1_{j,m})^2 + (\hat{\mu}^1_{j,m})^2} \quad (l = 1,2,\ldots,5) \quad \ldots 26\text{ b},$$

and $A_{j,m}$ (j=1,2,..,16) are constants of integration determined from the boundary conditions of the shell.

Turning now to the remaining homogeneous solutions $u^{\circ\circ\circ}$, $v^{\circ\circ\circ}$ and $w^{\circ\circ\circ}$ expanded into the Fourier-series in the circumferential direction, we consider the solutions in the form

$$u^{ooo} = \sum_{n=1,3,5..}^{N} [\, u_{c,n}^{ooo}(x)\, cos\lambda_n y + u_{s,n}^{ooo}(x)\, sin\lambda_n y\,] \qquad \ldots 27\ a,$$

$$v^{ooo} = \sum_{n=1,3,5..}^{N} [\, v_{c,n}^{ooo}(x)\, cos\lambda_n y + v_{s,n}^{ooo}(x)\, sin\lambda_n y\,] \qquad \ldots 27\ b,$$

$$w^{ooo} = \sum_{n=1,3,5..}^{N} [\, w_{c,n}^{ooo}(x)\, cos\lambda_n y + w_{s,n}^{ooo}(x)\, sin\lambda_n y\,] \qquad \ldots 27\ c.$$

Substituting Eqns 27 into the homogeneous parts of Eqns 5 and utilizing the method of separation of variables lead to the ordinary differential equations. In this case, the resulting equations consist of two sets of simultaneous differential equations, one for the variables $u_{c,n}^{ooo}$, $v_{s,n}^{ooo}$ and $w_{s,n}^{ooo}$, and the other for the variables $u_{s,n}^{ooo}$, $v_{c,n}^{ooo}$ and $w_{c,n}^{ooo}$. In order to solve these differential equations, we may postulate the solutions in the form

$$u_{c,n}^{ooo} = \hat{A}_n\, exp(\hat{\eta}_n \hat{x}) \qquad \ldots\ldots 28\ a,$$

$$v_{s,n}^{ooo} = \hat{B}_n\, exp(\hat{\eta}_n \hat{x}) \qquad \ldots\ldots 28\ b,$$

$$w_{s,n}^{ooo} = \hat{C}_n\, exp(\hat{\eta}_n \hat{x}) \qquad \ldots\ldots 28\ c,$$

$$u_{s,n}^{ooo} = A_n^o\, exp(\eta_n^o \hat{x}) \qquad \ldots\ldots 29\ a,$$

$$v_{c,n}^{ooo} = B_n^o\, exp(\eta_n^o \hat{x}) \qquad \ldots\ldots 29\ b,$$

$$w_{c,n}^{ooo} = C_n^o\, exp(\eta_n^o \hat{x}) \qquad \ldots\ldots 29\ c.$$

Introducing Eqns 28 into the resulting simultaneous differential equations for $u_{c,n}^{ooo}$, $v_{s,n}^{ooo}$ and $w_{s,n}^{ooo}$, and cancelling out the exponential functions, we have three homogeneous algebraic equations for the three unknown constants $\hat{A}_n$, $\hat{B}_n$ and $\hat{C}_n$. Letting the determinant of the equations vanish, in order to have nontrivial solutions, results in the following characteristic equation for $\hat{\eta}_n$:

$$\begin{vmatrix} \hat{T}_{1,1} & \hat{T}_{1,2} & \hat{T}_{1,3} \\ \hat{T}_{2,1} & \hat{T}_{2,2} & \hat{T}_{2,3} \\ \hat{T}_{3,1} & \hat{T}_{3,2} & \hat{T}_{3,3} \end{vmatrix} = 0 \qquad \ldots\ldots 30,$$

where

$$\hat{T}_{1,1} = \frac{(1-\nu)}{2}(\hat{\eta}_n^2 + \hat{\eta}_n)(1+3k\tan\alpha) - 1 + \nu - \lambda_n^2 sec^2\alpha$$
$$- 3(1-\nu)\, k\, tan^2\alpha \qquad \ldots 31\ a,$$

$$\hat{T}_{1,2} = \frac{(1+\nu)}{2}\hat{\eta}_n\lambda_n\, sec\alpha + (2-\nu)\,\lambda_n\, sec\alpha \qquad \ldots 31\ b,$$

$$\hat{T}_{1,3} = \lambda_n\, tan\alpha\, sec\alpha + \frac{(-3+5\nu)}{2}k\,\lambda_n\,\hat{\eta}_n\, tan\alpha\, sec\alpha$$
$$- \frac{(3-\nu)}{2}k\lambda_n\hat{\eta}_n^2 tan\alpha\, sec\alpha + 3(1-\nu)\, k\lambda_n tan\alpha\, sec\alpha \qquad \ldots 31\ c,$$

$$\hat{T}_{2,1} = -\frac{(1+\nu)}{2}\lambda_n\hat{\eta}_n sec\alpha + \frac{3}{2}(1-\nu)\,\lambda_n sec\alpha \qquad \ldots 31\ d,$$

$$\hat{T}_{2,2} = -\frac{(1-\nu)}{2}\lambda_n^2(1+k tan^2\alpha)\, sec^2\alpha + \hat{\eta}_n^2 + \hat{\eta}_n - 1 + \nu$$
$$- k tan^2\alpha \qquad \ldots 31\ e,$$

$$\hat{T}_{2,3} = -k tan^3\alpha + \nu\hat{\eta}_n tan\alpha - (1-\nu) tan\alpha - k\hat{\eta}_n^3 tan\alpha$$
$$- \frac{(1-\nu)}{2}k\lambda_n^2\hat{\eta}_n^2 tan\alpha sec^2\alpha + \frac{(3-\nu)}{2}k\lambda_n^2 tan\alpha sec^2\alpha \quad \ldots 31\ f,$$

$$\hat{T}_{3,1} = \lambda_n tan\alpha sec\alpha + \frac{(3-\nu)}{2}k\lambda_n\hat{\eta}_n^2 tan\alpha sec\alpha$$
$$+ \frac{3}{2}(1+\nu)\, k\lambda_n\hat{\eta}_n tan\alpha sec\alpha - (3-5\nu)\, k\lambda_n tan\alpha sec\alpha \quad \ldots 31\ g,$$

$$\hat{T}_{3,2} = [\, 1 - k(1-tan^2\alpha) - (2-\nu)\, k\lambda_n^2 sec^2\alpha + \nu\hat{\eta}_n - 3k\hat{\eta}_n$$
$$- \frac{(1-\nu)}{2}k\hat{\eta}_n\lambda_n^2 sec^2\alpha - 3k\hat{\eta}_n^2 - k\hat{\eta}_n^3\,]\, tan\alpha \qquad \ldots 31\ h,$$

$$\hat{T}_{3,3} = tan^2\alpha - k(1-tan^2\alpha)\, tan^2\alpha + (5-6\nu)\, k\lambda_n^2 sec^2\alpha$$
$$- 2k\lambda_n^2 tan^2\alpha sec^2\alpha + k\lambda_n^4 sec^4\alpha - 3(1-\nu)\, k\hat{\eta}_n$$
$$- 2k\hat{\eta}_n\lambda_n^2 sec^2\alpha - (2-3\nu)\, k\hat{\eta}_n^2 - 2k\hat{\eta}_n^2\lambda_n^2 sec^2\alpha$$
$$+ 2k\hat{\eta}_n^3 + k\hat{\eta}_n^4 \qquad \ldots 31\ i.$$

In a same way, from the condition that the constants A_n^o, B_n^o and C_n^o have non-trivial solutions, we have

$$\begin{vmatrix} \hat{T}_{1,1} & -\hat{T}_{1,2} & -\hat{T}_{1,3} \\ -\hat{T}_{2,1} & \hat{T}_{2,2} & \hat{T}_{2,3} \\ -\hat{T}_{3,1} & \hat{T}_{3,2} & \hat{T}_{3,3} \end{vmatrix} = 0 \qquad \ldots\ldots 32,$$

where it should be noted that η_n^o in place of $\hat{\eta}_n$ is introduced in Eqns 31. It is clear that the same characteristic values are obtained from Eqn 30 and 32. In this paper, however, the characteristic values $\hat{\eta}_n$ and η_n^o are also obtained from the eigenvalue equations, which can be derived from Eqn 30 and 32 in the same manner discussed in the derivation of the homogeneous solutions $(\ )^{oo}$. Consequently the homogeneous solutions $(\ )^{ooo}$ take the form

$$u^{ooo} = \sum_{n=1,3,5..}^{N}\sum_{j=1}^{4} [\, (\hat{A}_{2j-1,n} cos\tau_{j,n}\hat{x}$$
$$+ \hat{A}_{2j,n} sin\tau_{j,n}\hat{x}\,) exp(\sigma_{j,n}\hat{x})\, cos\lambda_n y$$
$$+ (\, A_{2j-1,n}^o cos\tau_{j,n}\hat{x} + A_{2j,n}^o sin\tau_{j,n}\hat{x}\,)\times$$
$$exp(\sigma_{j,n}\hat{x})\, sin\lambda_n y\,] \qquad \ldots\ldots 33\ a,$$

$$v^{ooo} = \sum_{n=1,3,5..}^{N}\sum_{j=1}^{4} [\{(\, R_{j,n}^{o1}\, A_{2j-1,n}^o + S_{j,n}^{o1}\, A_{2j,n}^o\,)\times$$
$$exp(\sigma_{j,n}\hat{x})cos\tau_{j,n}\hat{x} + (\, R_{j,n}^{o1}\, A_{2j,n}^o - S_{j,n}^{o1}\, A_{2j-1,n}^o\,)\times$$
$$exp(\sigma_{j,n}\hat{x})sin\tau_{j,n}\hat{x}\} cos\lambda_n y + \{(\, \hat{R}_{j,n}^1\, \hat{A}_{2j-1,n}$$
$$+ \hat{S}_{j,n}^1\, \hat{A}_{2j,n}\,) exp(\sigma_{j,n}\hat{x})cos\tau_{j,n}\hat{x} + (\, \hat{R}_{j,n}^1\, \hat{A}_{2j,n}$$
$$- \hat{S}_{j,n}^1\, \hat{A}_{2j-1,n}\,) exp(\sigma_{j,n}\hat{x})sin\tau_{j,n}\hat{x}\}\, sin\lambda_n y\,]$$
$$\ldots\ldots 33\ b,$$

$$w^{ooo} = \sum_{n=1,3,5..}^{N}\sum_{j=1}^{4} [\{(\, R_{j,n}^{o2}\, A_{2j-1,n}^o + S_{j,n}^{o2}\, A_{2j,n}^o\,)\times$$

$$\exp(\sigma_{j,n}\hat{x})\cos\tau_{j,n}\hat{x} + (R^{\circ 2}_{j,n} A^{\circ}_{2j,n} - S^{\circ 2}_{j,n} A^{\circ}_{2j-1,n})\times$$
$$\exp(\sigma_{j,n}\hat{x})\sin\tau_{j,n}\hat{x}\}\cos\lambda_n y + \{(\hat{R}^{2}_{j,n}\hat{A}_{2j-1,n}$$
$$+\hat{S}^{2}_{j,n}\hat{A}_{2j,n})\exp(\sigma_{j,n}\hat{x})\cos\tau_{j,n}\hat{x} + (\hat{R}^{1}_{j,n}\hat{A}_{2j,n}$$
$$-\hat{S}^{2}_{j,n}\hat{A}_{2j-1,n})\exp(\sigma_{j,n}\hat{x})\sin\tau_{j,n}\hat{x}\}\sin\lambda_n y\,] \qquad \ldots\ldots 33\text{ c},$$

where $\hat{R}^{l}_{j,n}$, $\hat{S}^{l}_{j,n}$, $R^{\circ l}_{j,n}$ and $S^{\circ l}_{j,n}$ are constants determined from the eigenvectors, and $\hat{A}_{j,n}$ and $A^{\circ}_{j,n}$ are constants of integration determined from the boundary conditions of the shell.

BOUNDARY CONDITIONS

In the expression of the displacements u ($u=u^{\circ}+u^{\circ\circ}+u^{\circ\circ\circ}$), v ($v=v^{\circ}+v^{\circ\circ}+v^{\circ\circ\circ}$) and w ($w=w^{\circ}+w^{\circ\circ}+w^{\circ\circ\circ}$), there exist the $16\times[(M+1)/2 + (N+1)/2]$ unknown constants which are determined so as to satisfy the prescribed boundary conditions along the four edges. For a matching with the boundary conditions along a boundary, the functions $\exp(\alpha_1 y)$, $\exp(\alpha_1 y)\cos(\alpha_2 y)$ and $\exp(\alpha_1 y)\sin(\alpha_2 y)$ must be expanded into the Fourier-series in the interval, $-1 \leq y \leq +1$ along the line x=const.. For the boundary conditions along the line y=const., the same functions of x must be expanded into the Fourier-series.

As an example, consider the boundary condition u=0 along the lines x=+1 and y=+1. Following the manner described above, the displacement u can be written in the form

$$u=\sum_{n=1,3,5..}^{N} [\,\hat{U}_{c,n}(x,m)\cos\lambda_n y + \hat{U}_{s,n}(x,m)\sin\lambda_n y\,] \qquad \ldots 34\text{ a},$$

$$u=\sum_{m=1,3,5..}^{M} [\,\mathcal{U}_{c,m}(y,n)\cos\lambda_m x + \mathcal{U}_{s,m}(y,n)\sin\lambda_m x\,] \qquad \ldots 34\text{ b}.$$

In order to satisfy the prescribed boundary condition u=0 along x=+1, we have the following equations:

$$\hat{U}_{c,n}(+1,m) = 0 \qquad (n=1,3,5,\ldots,N) \qquad \ldots\ldots 35\text{ a},$$

$$\hat{U}_{s,n}(+1,m) = 0 \qquad (n=1,3,5,\ldots,N) \qquad \ldots\ldots 35\text{ b},$$

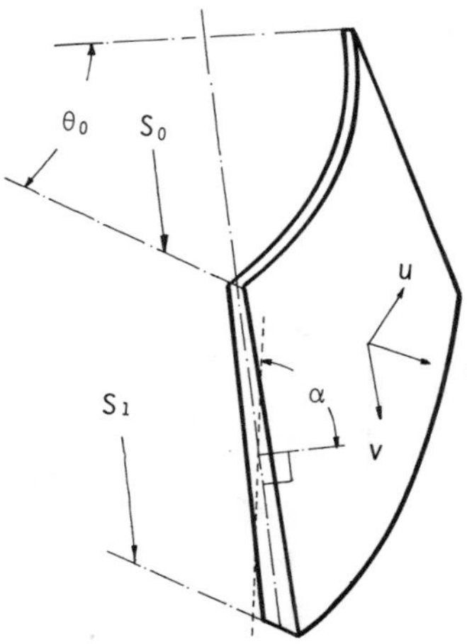

Fig. 1 Conical Shell

and as the boundary condition u=0 along y=+1, we have

$$\mathcal{U}_{c,m}(+1,n) = 0 \qquad (m=1,3,5,\ldots,M) \qquad \ldots\ldots 36\text{ a},$$

$$\mathcal{U}_{s,m}(+1,n) = 0 \qquad (m=1,3,5,\ldots,M) \qquad \ldots\ldots 36\text{ b}.$$

In this way the boundary conditions along the four edges are used to determine the constants of integration.

AN EXAMPLE

For purpose of illustration, take a truncated unclosed conical shell, the boundary conditions of which are given by

$$u= v= w= \partial w/\partial x= 0 \qquad \text{at } x=\pm 1 \qquad \ldots\ldots 37\text{ a},$$

$$u= v= w= \partial w/\partial y= 0 \qquad \text{at } y=\pm 1 \qquad \ldots\ldots 37\text{ b}.$$

The following material and geometrical constants are used:

$\alpha = \pi/3$ (rad.) , $\theta_0 = \pi/3$ (rad.)

$\delta = 1/180$, $E = 2.1\times10^{6}$ (t/m^2)

$\nu = 1/6$, $P_r = 1.0$ (t/m^2)

$s_0 = 24$ (m) , $s_1 = 30$ (m)

Since it is confirmed in this paper to be sufficient to take the fifteen terms of Fourier-series expansion, we take M = N = 15. The numerical results for the normal displacement w at y=0 are shown in Table 1.

Table 1. Normal Displacement w at y=0

x	w (m)	x	w (m)
-1.0	0.00000000 D-00	0.0	0.84534345 D-03
-0.9	0.43192624 D-04	0.1	0.83856663 D-03
-0.8	0.14584190 D-03	0.2	0.80750548 D-03
-0.7	0.27531310 D-03	0.3	0.75114605 D-03
-0.6	0.40928160 D-03	0.4	0.66893223 D-03
-0.5	0.53368506 D-03	0.5	0.56181336 D-03
-0.4	0.64057049 D-03	0.6	0.43368758 D-03
-0.3	0.72610504 D-03	0.7	0.29320965 D-03
-0.2	0.78889201 D-03	0.8	0.15587975 D-03
-0.1	0.82865942 D-03	0.9	0.46249945 D-04
		1.0	0.00000000 D-00

REFERENCES

1. W Flügge, " Stresses in Shells ", 2nd Ed., Springer, 1962.

2. B Wilson, Asymmetrical Bending of Conical Shells, J.Eng. Mech. Div., ASCE, No.86, pp.119-139, 1960.

3. C H Chang, On the Solution of Conical Shells of Linearly Varying Thickness Subjected to Lateral Loads, Int. J. Solids and Structures, No.3, pp.177-190 , 1967.

4. C H Chang, An Asymptotic Solution of Conical Shells of Constant Thickness, AIAA J., Vol.5 No.11, pp.2028-2033, 1967.

5. Y Tsuboi and K Ujiya, Approximated Solutions of Bending Stresses in Truncated Conical Shells, Proc. of A.I.J., No.117,1965.

SOME NEW RESULTS OF THE EQUIVALENT CONTINUUM METHOD FOR SPACE FRAMES

Dr. techn. L. KOLLÁR

Chief Structural Engineer,
Budapest City Council's Architectural and
Town Planning Office BUVÁTI
Budapest, Hungary

The paper summarizes the main results of research achieved by the author since the 2nd International Conference on Space Structures. The following items will be reported on:
- -Continuum equations of the "hexagonal on triangular" grid;
- -Approximate methods for double-layer grids;
- -Continuum equations of timber lattice shells.

1. INTRODUCTION

This paper aims to summarize the main results of research achieved since the Second International Conference on Space Structures /Ref 3/ in the field of the equivalent continuum theory. These results concern, on the one hand, double-layer grids, and, on the other hand, single-layer lattice shells, mostly made of timber.

The crucial point in using the continuum method is to develop the continuum equations for the grid in question. In the following we shall indicate the main steps of derivation and present the equations themselves.

2. THE "HEXAGONAL ON TRIANGULAR" GRID /Fig 1/

The main features of this type of grid are the following:
- -its hexagonal chord is unstable in its own plane, i.e. it can develop the deformations shown in Fig 2 without any resistance;
- -the grid is stiffened against torsion by reticulated tubes of trapezoidal cross section, running in three directions /Fig 3/.

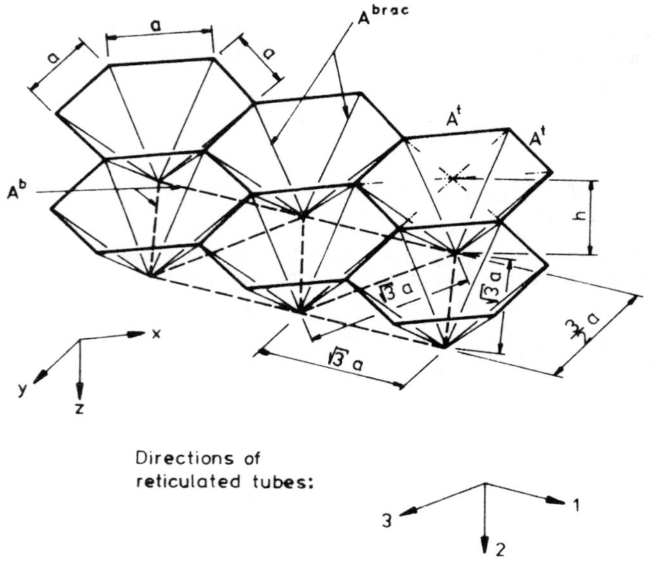

Fig 1

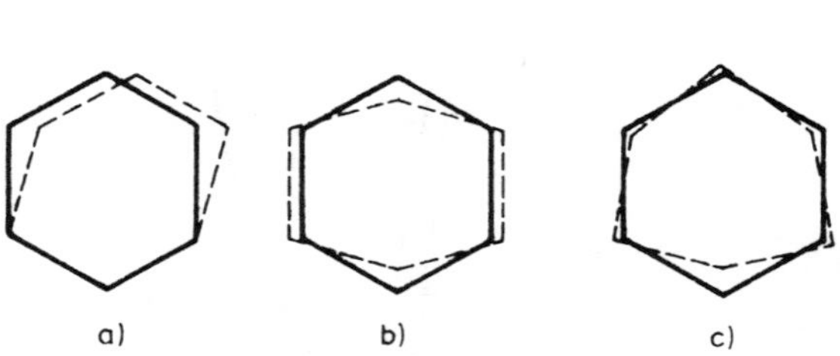

Fig 2

The stress and deformation states of the grid are described by a deflection function w and a stress function F. This latter defines a complementary stress state which develops in the triangular chord plane. These two functions have to satisfy the coupled differential equations /Ref 5/:

$$\frac{1+k^*}{k^*}\underline{\underline{D}}_w^*\left(\underline{\underline{B}}+\underline{\underline{B}}_{tubes}\right)\left(\underline{\underline{T}}^b+\underline{\underline{T}}^t+\frac{1}{k^*}\underline{\underline{T}}^b_{shear}\right)^{-1}\left(\underline{\underline{T}}^b\underline{\underline{D}}_w w-\frac{1}{h}\underline{\underline{D}}_F F\right)=p, \qquad /1/$$

$$\underline{\underline{D}}_F^*\left(\underline{\underline{T}}^b+\underline{\underline{T}}^t+\frac{1}{k^*}\underline{\underline{T}}^b_{shear}\right)^{-1}\left[h\left(\underline{\underline{T}}^t+\frac{1}{k^*}\underline{\underline{T}}^b_{shear}\right)\underline{\underline{D}}_w w+\underline{\underline{D}}_F F\right]=0. \qquad /2/$$

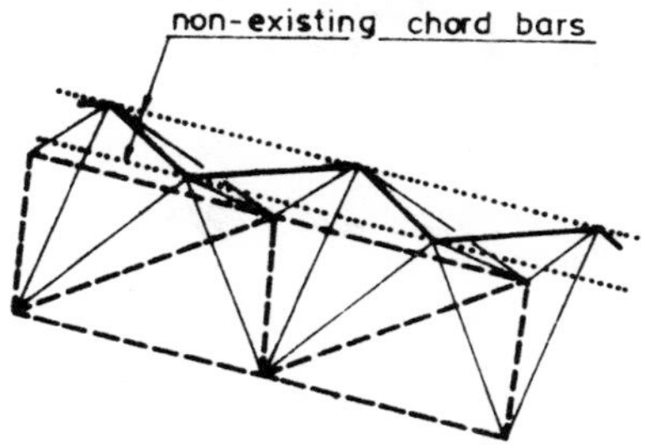

Fig 3

Here the following notation has been used:

h height of the grid /Fig 1/

$k^* = \frac{1}{2\sqrt{3}} \frac{A^b}{A^t} + \frac{\alpha_h^3}{\sqrt{3}} \frac{A^b}{A^{brac}} + \frac{1}{2}$ proportionality factor

$\alpha_h = \sqrt{\frac{h^2}{a^2} + 1}$ ratio of the lengths of bracing and top chord bars

a side length of a hexagon /Fig 1/

A^b area of bottom chord bars /Fig 1/

A^t area of top chord bars /Fig 1/

A^{brac} area of bracing bars /Fig 1/

E modulus of elasticity of the bars

$$\underline{\underline{T}}^t = \frac{EA^t}{2\sqrt{3}a} \begin{bmatrix} 1 & 1 & 0 \\ 1 & 1 & 0 \\ 0 & 0 & 0 \end{bmatrix}$$ tensile stiffness matrix of top chord /3/

$\underline{\underline{T}}$ is defined by the relation:

$$\underline{\underline{n}} = \underline{\underline{T}}\,\underline{\underline{\varepsilon}} \quad ,$$

with $\underline{\underline{n}} = \begin{bmatrix} n_x \\ n_y \\ n_{xy} \end{bmatrix}$ as the vector of the in-plane forces,

and $\underline{\underline{\varepsilon}} = \begin{bmatrix} \varepsilon_x \\ \varepsilon_y \\ \gamma_{xy} \end{bmatrix}$ as the vector of the deformations.

$$\underline{\underline{T}}^b = \frac{EA^b}{4a} \begin{bmatrix} 3 & 1 & 0 \\ 1 & 3 & 0 \\ 0 & 0 & 1 \end{bmatrix}$$ tensile stiffness matrix of bottom chord /4/

$$\underline{\underline{T}}^b_{shear} = \frac{EA^b}{4a} \begin{bmatrix} 0 & 0 & 0 \\ 0 & 0 & 0 \\ 0 & 0 & 1 \end{bmatrix}$$ auxiliary matrix

$$\underline{\underline{B}} = h^2 \frac{k^*}{1+k^*} \underline{\underline{T}}^t$$ bending rigidity matrix of the grid due to the tensile stiffnesses of the chords;

$\underline{\underline{B}}$ is defined by the relation:

$$\underline{\underline{m}} = -\underline{\underline{B}}\,\underline{\underline{\varkappa}} \;,$$

with $m = \begin{bmatrix} m_x \\ m_y \\ m_{xy} \end{bmatrix}$ as the vector of the bending and twisting moments, /5/

and $\underline{\underline{\varkappa}} = \begin{bmatrix} \varkappa_x \\ \varkappa_y \\ 2\varkappa_{xy} \end{bmatrix}$ as the vector of the curvatures and /double/ twist. /6/

$$\underline{\underline{B}}_{tubes} = \frac{GI_t}{6a} \begin{bmatrix} 1 & -1 & 0 \\ -1 & 1 & 0 \\ 0 & 0 & 1 \end{bmatrix}$$ bending rigidity matrix of the grid due to the reticulated tubes; /7/

$$GI_t = \frac{2\sqrt{3}Eh^2}{\frac{1}{A^t} + \frac{2\alpha_h^3}{A^{brac}} + \frac{3\sqrt{3}}{A^b}}$$ torsional stiffness of one reticulated tube; /8/

$$\underline{\underline{D}}_W = \begin{bmatrix} \frac{\partial^2}{\partial x^2} \\ \frac{\partial^2}{\partial y^2} \\ 2\frac{\partial^2}{\partial x \partial y} \end{bmatrix}$$ differential operator;

$$\underline{\underline{D}}_F = \begin{bmatrix} \frac{\partial^2}{\partial y^2} \\ \frac{\partial^2}{\partial x^2} \\ -\frac{\partial^2}{\partial x \partial y} \end{bmatrix}$$ differential operator;

p load acting on the grid.

The solution also has to comply with the boundary conditions, which consist of two conditions for w and two for F all along the boundary.

3. SIMPLIFIED CONTINUUM ANALYSIS FOR DOUBLE-LAYER GRIDS

If the tensile stiffness matrix of the top chord is not k times that of the bottom chord, or if the grid is stiffened against torsion by reticulated tubes /as in the cases of the "square with diagonals on square offset" grid /Ref 3/, or of the grid described in Sect 2/, the governing differential equations of the grid become rather complicated. In order to simplify the continuum analysis and make it more suitable for preliminary design, we have developed two kinds of approximate analysis, which will briefly described in the following. Both methods make the analysis simpler by eliminating the stress function altogether, thus arriving at a single fourth-order differential equation for the deflection w.

3.1. SIMPLIFIED EQLIBRIUM ANALYSIS /Ref 4/

If some rigidities of the top and bottom chord planes have ratios different from the others, then the neutral planes of the corresponding /bending or torsional/ moments do not coincide, so that compatibility is violated. This may also happen if the torsional rigidity is provided by lattice tubes.

From this viewpoint, the stress function is required to restore compatibility, i.e. to provide a unique neutral plane of all bending and twisting effects. Thus, if we want to eliminate the stress function, we have to tolerate the violation of compatibility. This means that we calculate every bending and torsional rigidity for itself, disregarding the different positions of their neutral planes /axes/. This procedure automatically means that those rigidities that have no pairs in the other chord are to be entirely neglected.

By so doing we generally calculate greater forces in the bars connected with rigidities taken into account than the actual ones, while in bars forming the neglected rigidities forces much smaller than the real ones /or possibly even zero forces/ are obtained.

As an example we show the procedure on the "hexagonal on triangular" grid, treated in Sect 2 /Fig 1/.

The top chord due to its unstable character /Fig 2/, can only resist normal forces equal in both direction. Consequently, the bending moments in both directions have also to be equal. This fact is expressed by the form of the tensile stiffness matrix of the top chord /3/, which ensures that due to any combination of ε_x and ε_y normal forces $n_x=n_y$, equal in both directions, arise.

In order to obtain equal bending moments in both directions, we have to consider the "hydrostatic" part of the tensile rigidity matrix of the bottom chord, i.e.

$$\underline{\underline{T}}^b_{hydr} = \frac{EA^b}{2a}\begin{bmatrix}1 & 1 & 0\\ 1 & 1 & 0\\ 0 & 0 & 0\end{bmatrix} , \qquad /9/$$

which gives for any combination of ε_x and ε_y normal forces $n_x=n_y$ of the same magnitude as those obtained with the real rigidity matrix /4/ from the deformation part $\varepsilon_x^{hydr}=\varepsilon_y^{hydr}=\frac{\varepsilon_x+\varepsilon_y}{2}$ ./The remaining part of the rigidity matrix $\underline{\underline{T}}^b$ has no counterpart in the top chord, so that this has not to be considered./

From $\underline{\underline{T}}^t$ and $\underline{\underline{T}}^b_{hydr}$ we compute the /hydrostatic/ bending rigidity matrix of the chords:

$$\underline{\underline{B}}_{chords} = \frac{Eh^2}{2a}\,\frac{\frac{A^t}{\sqrt{3}}A^b}{\frac{A^t}{\sqrt{3}}+A^b}\begin{bmatrix}1 & 1 & 0\\ 1 & 1 & 0\\ 0 & 0 & 0\end{bmatrix} . \qquad /10/$$

The sum of the two rigidity matrices /7/ and /10/ yields the complete rigidity matrix for the grid. Introducing the rigidity elements into the equation of the orthtropic plate

$$B_{11}w'''' + 2Hw''^{\cdot\cdot} + B_{22}w^{\cdot\cdot\cdot\cdot} = p \qquad /11/$$

/with the notation $'=\partial/\partial x$ and $^{\cdot}=\partial/\partial y$/, and observing that

$$H = B_{12}+2B_{33} \qquad /12/$$

/with B_{33} as the half torsional rigidity of the grid, due to the double twist appearing in the curvature vector $\underline{\underline{\kappa}}$, see Eqn /6//, we arrive at the differential equation

$$B_{11}\left(w''''+2w''^{\cdot\cdot}+w^{\cdot\cdot\cdot\cdot}\right)=p \qquad /13/$$

which corresponds to an isotropic plate.

We still have to deal with the problem: how to convert the bending and twisting moments of the equivalent plate into those of the grid, if they have different $B_{12}/B_{11}=\nu$ ratios /and also different B_{33}/B_{11} ratios/, while their H/B_{11} ratio is identical, see Eqn /12/.

Using the superscripts g and e for grid and equivalent plate respectively, we can express the curvatures and double twist, κ_x, κ_y and $2\kappa_{xy}$, of the equivalent plate by its moments:

$$\kappa_x = -\frac{m_x^e-\nu_e m_y^e}{B_{11}(1-\nu_e^2)} , \qquad /14a/$$

$$\kappa_y = -\frac{m_y^e-\nu_e m_x^e}{B_{11}(1-\nu_e^2)} , \qquad /14b/$$

$$2\kappa_{xy} = -\frac{m_{xy}^e}{B_{33}^e} . \qquad /14c/$$

Since the grid develops the same deformations as the equivalent plate, we can introduce Eqns /14a,b,c/ into the expressions of the moments of the grid:

$$m_x^g = -B_{11}(\kappa_x+\nu_g\kappa_y) = m_x^e\,\frac{1-\nu_g\nu_e}{1-\nu_e^2} + \nu_g m_y^e\,\frac{1-\nu_e/\nu_g}{1-\nu_e^2} , \qquad /15a/$$

$$m_y^g = -B_{11}(\kappa_y+\nu_g\kappa_x) = m_y^e\,\frac{1-\nu_g\nu_e}{1-\nu_e^2} + \nu_g m_x^e\,\frac{1-\nu_e/\nu_g}{1-\nu_e^2} , \qquad /15b/$$

$$m_{xy}^g = -B_{33}^g 2\kappa_{xy} = \frac{B_{33}^g}{B_{33}^e}m_{xy}^e . \qquad /15c/$$

In the case of the "hexagonal on triangular" grid the moment vector /5/ of the grid thus obtained, $\underline{\underline{m}}^g$, will partly be taken by the chords, according to the rigidity matrix /10/, and partly by the reticulated tubes, corresponding to the rigidity matrix /7/. Hence we have to write

$$\underline{\underline{m}}_{chords} = \underline{\underline{B}}_{chords}\left(\underline{\underline{B}}_{chords}+\underline{\underline{B}}_{tubes}\right)^{-1}\underline{\underline{m}}^g , \qquad /16a/$$

and

$$\underline{\underline{m}}_{tubes}=\underline{\underline{B}}_{tubes}\left(\underline{\underline{B}}_{chords}+\underline{\underline{B}}_{tubes}\right)^{-1}m^g . \qquad /16b/$$

Expanding Eqn /16a/ yields:

$$\underline{\underline{m}}_{chords}=\begin{bmatrix}1/2 & 1/2 & 0\\ 1/2 & 1/2 & 0\\ 0 & 0 & 0\end{bmatrix}m^g \qquad /17/$$

From the moment vector $\underline{\underline{m}}_{tubes}$ /16b/, with the components $m_{x,tubes}$, $m_{y,tubes}$, $m_{xy,tubes}$, we still have to compute the twisting moments m_{t1}, m_{t2}, m_{t3} of the reticulated tubes running in the directions 1, 2, 3 /Fig 1/. Uniting these in the vector

$$\underline{\underline{m}}_t=\begin{bmatrix}m_{t1}\\ m_{t2}\\ m_{t3}\end{bmatrix} , \qquad /18/$$

it can be shown, see Ref 7, that $\underline{\underline{m}}_t$ can be expressed by $\underline{\underline{m}}^g$ by the following relation:

$$\underline{\underline{m}}_t=\begin{bmatrix}0 & 0 & \frac{4}{3}\\ -\frac{1}{\sqrt{3}} & \frac{1}{\sqrt{3}} & -\frac{2}{3}\\ \frac{1}{\sqrt{3}} & -\frac{1}{\sqrt{3}} & -\frac{2}{3}\end{bmatrix}m^g . \qquad /19/$$

Knowing $\underline{\underline{m}}_{chords}$ and $\underline{\underline{m}}_t$, the bar forces can be computed in the usual way.

3.2. SIMPLIFIED COMPATIBILITY ANALYSIS

As a counterpart to the procedure outlined in Sect 3.1, we may choose another way to eliminate the stress function. By fixing a unique neutral plane, common for all bending and twisting effects, we ensure the compatibility of the deformations but violate the equilibrium: since the unique neutral plane does not lie in the centroid of some pairs of rigidities, the corresponding forces arising in both chords will not be of equal magnitude, they consequently give not only couples but also horizontal resultant forces.

We can appropriately fix the position of the common neutral plane either by choosing the position of the actual common neutral plane of several rigidities or by maintaining the torsional "neutral plane" of the lattice tubes. It is expedient, then, to decompose the rigidities of one of the chords into two parts, I and II, in such a way, that Part I be k times the rigidities of the other chord plane, taken as a basis, k being fixed by the requirement that the chosen neutral plane coincides with the centroids of the rigidities of the basis chord plane and of Part I of the other plane. Part II then contains the "excess" rigidities which furnish horizontal "excess" forces with no pairs in the other chord /Ref 7/.

With this method generally forces greater than the exact ones are obtained in the bars forming the excess rigidities.

Due to the limited length of this paper we cannot show the application of the compatibility analysis on an example here. We only mention that, according to comparative computations, the bar forces obtained by the simplified equilibrium and compatibility analiyses give, in most cases, an upper and a lower bound for the exact values of the bar forces.

4. CONTINUUM EQUATIONS OF LATTICE SHELLS

Lattice shells, made of continuous timber laths running in two directions /Fig 4/, present a special problem. They develop forces that are also present in continuous shells /Fig 4/ but,

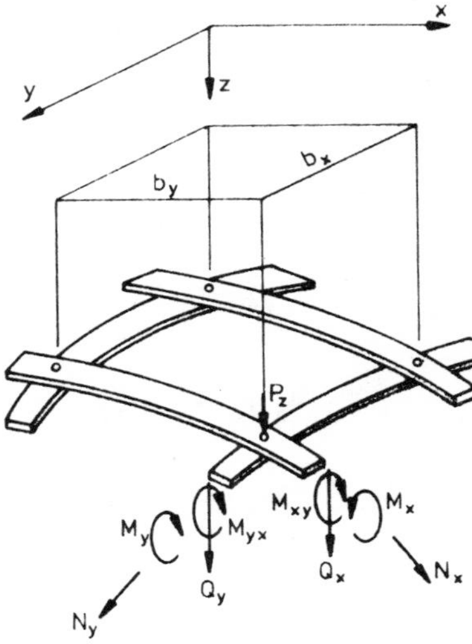

Fig 4

in addition, they also develop bending moments and shearing forces in the tangential plane of the shell surface /Fig 5/ which have no equivalents in the corresponding continuous shell.

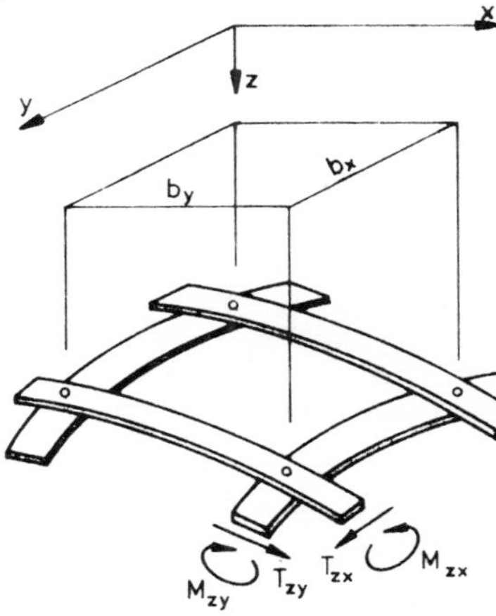

Fig 5

Although the shearing forces T_{zx}, T_{zy}, acting in the tangential plane /Fig 5/, seemingly correspond to the membrane shearing forces n_x, n_y of a continuous shell, there exists actually a fundamental difference between them. T_{zx} and T_{zy} are not connected to each other by a relation similar to $n_{xy}=n_{yx}$, but are independent of each other, since they are the shearing forces of two independent beams. They are, however, directly connected with the normal forces of the perpendicular laths, i.e. the normal force N_y /or N_x/ increases at a joint by the same amount as T_{zx} /or T_{zy}/ decreases.

We assume a shallow shell surface, so that we can consider the network of the laths as rectangular. Neglecting the elongations of the laths, and considering the joints of the continuous laths as hinged in the tangential plane, we can write the following equilibrium equations for the lattice shell regarded as a continuum /see Figs 4 and 5/:

$$\frac{T^{\bullet}_{zy}}{b_y} = n'_x \quad ; \qquad \frac{T'_{zx}}{b_x} = n^{\bullet}_y \quad ; \qquad /20a,b/$$

$$n_x\left(z''+w''\right) + n_y\left(z^{\bullet\bullet}+w^{\bullet\bullet}\right) + q'_x + q^{\bullet}_y + p_z = 0 \quad ; \qquad /20c/$$

$$m'_x + m^{\bullet}_{xy} = q_x \quad ; \qquad m'_{xy} + m^{\bullet}_y = q_y \; ; \qquad /20d,e/$$

with $z(x,y)$ as the ordinate of the shell surface and w as the displacement perpendicular to the surface.

All these internal forces, except for n_x and n_y, can be expressed by the x,y directed displacements u, v and by w in the usual way, see e.g. in Ref 1. We shall show this for the shearing forces T_{zx} and T_{zy} because they do not appear in the classical shell theory. They are to be defined as the shearing forces of the lath-beams in the usual way:

$$T_{zx} = -EI_{zx} v''' \quad ; \qquad T_{zy} = -EI_{zy} u^{\bullet\bullet\bullet} \quad ; /21a,b/$$

with EI_{zx} and EI_{zy} as the bending stiffnesses of the laths in the tangential plane of the shell. We also mention that we have to introduce half the sum of the torsional rigidities of the two lath rows referred to unit width as the torsional rigidity of the continuum.

Due to the assumed inextensionity of the laths the normal forces n_x and n_y are not directly connected to the longitudinal displacements of the laths. The two eqations thus missing will be replaced by the conditions of inextensionality itself. We only write here these conditions in a form valid for small displacements:

$$u' - w\left(z''+w''\right) = 0 \quad ; \qquad v^{\bullet} - w\left(z^{\bullet\bullet}+w^{\bullet\bullet}\right) = 0 \qquad /22a,b/$$

More general relations, valid for large displacements, are found in Ref 2.

We introduce Eqns /20d,e/ into /20c/, and we express the internal forces /except for n_x and n_y/ by the displacements u,v,w in Eqns /20a,b,c/. Taking also Eqns /22a,b/ into consideration we arrive at five equations with five unknowns u, v, w, n_x, n_y, which represent the continuum equation system of the lattice shell /Ref 6/.

REFERENCES

1. Flügge, W.: Stresses in Shells. Springer, Berlin/Heidelberg/New York, 1973.

2. Hegedüs, I.: Computation of the Streched Network Shape of Timber Lattice Shells. To be published in: Acta Techn. Acad. Sci. Hung.

3. Kollár, L.: Analysis of Double-Layer Space Trusses by the Equivalent Continuum Method. 2nd International Conference on Space Structures. Dept. of Civil Eng., University of Surrey, Guildford, England, 1975. pp.73-76

4. Kollár, L.: Simplified Continuum Analysis for Preliminary Design of Space Frames. IASS Conf. on Lightweight Shell and Space Structures for Normal and Seismic Zones. 1977, Alma-Ata, USSR. MIR Publishers, Sect.1, pp. 153-170.

5. Kollár, L.: Continuum Method of Analysis for Double Layer Space Trusses of "Hexagonal over Triangular" Mesh. Acta Techn. Acad. Sci. Hung. 86 /1978/, pp. 55-77.

6. Kollár, L.: Continuum Equations of Timber Lattice Shells. Acta Techn. Acad. Sci. Hung. /at press/.

7. Kollár, L.-Hegedüs, I.: Analysis and Design of Space Frames by the Continuum Method, Akadémiai Kiadó, Budapest-Elsevier, Amsterdam, to be published in 1984.

EXPERIMENTAL STUDY, THEORETICAL ANALYSIS AND DESIGN OF A SHEET-FRAMED SPACE STRUCTURE

H L ZHAO*, Z Z WU **, MS, CE, PhD and J L LIU ***

* Lecturer, Department of Civil Engineering
Nanjing Institute of Technology

** Professor, Department of Civil Engineering
Nanjing Institute of Technology

*** Senior Engineer Design Institute
Nanjing Chemical Industrial Company

Associated with a certain engineering project in northern China, a model sheet-framed parabolic cylindrical sheel structure is tested under four different load and displacement conditions. The model structure is primarily analyzed by the method of finite elements. Both results indicate that the cross ribs play an important role in carrying moments on the structure while the plate elements do reveal greater potential ability in resisting axial forces, which should not be ignored in design practice. Typical curves are presented for comparison, Measures for improving accuracy of results are discussed. Low consumption of steel and simplicity in construction reveals superiority of this type of structure.

INTRODUCTION

Sheet-framed structures are space structures composed of panel units with plate-elements and rib-elements. Much analytical work has been undertaken on long-span framed structures, but most of them deal primarily with the structural behaviour of the framed skeleton members themselves in resisting loads. However, the present experimental results show that the plate-elements do reveal greater potential ability in resisting axial forces, which should by no means be ignored in current design practice. The theoretical analysis herein has been made mainly by the method of finite element. The composite elements which consist of triangular plate and rib members are used in computation.

MODEL TESTS

In order to investigate the structural behaviour of this type of space structure, associated with a certain engineering project in northern China, a test model is made of synthetic glass with a scale of 1:25 in over-all dimensions and of 1:15 in sectional details. The model has a constant cross section of a second order parabola with horizontal projection of 2.11 m (chordwise) by 1.92m (longitudinal) and a rise of 0.91 m Fig.1. Its cylindrical surface is built up with a total of 224 pieces of 120x210mm panel units each of which has two heavy cross ribs in addition to four edge stiffening ribs glued on a 2 mm rectangular flat plate Fig. 2 . The whole model is hinge-supported on 18 bearings along two exterior longitudinal edges, and is also connected to a heavy arch of 20x120mm in section along each exterior circumferential edge Fig. 3 .

Due to geometric property of symmetry, measuring points are located only on one quarter the area of the model. Readings of all the 89 measuring points or 314 strain gages on the model are taken and printed by a Japanese SD-520A digital strain indicator and automatic scanning boxes. Joint displacements are measured with dial gauges. pneumatic pressure is applied to realize uniform point loading which is transmitted from triangular pressure distributing plates underneath the rubber bag to the nodal points on the model through transmitting rods of different lengths.

The complete testing program includes the following conditions:

(1) Uniform normal pressure --- formed by covering a pneumatic pressure bag in close contact with the shell top surface.

(2) Uniformly distributed vertical point loads --- formed by transmitting pneumatic pressure through vertical rods on whole shell top surface.

(3) Half surface horizontal point loads --- formed by transmitting pneumatic pressure through horizontal rods on half shell top surface.

(4) Simultaneous settlement of two diagonally extreme corner supports --- by jacking up the opposite end of each foundation grill; and relative chordwise horizontal movements of both rows of edge supports --- by jacking one foundation grill horizontally inward.

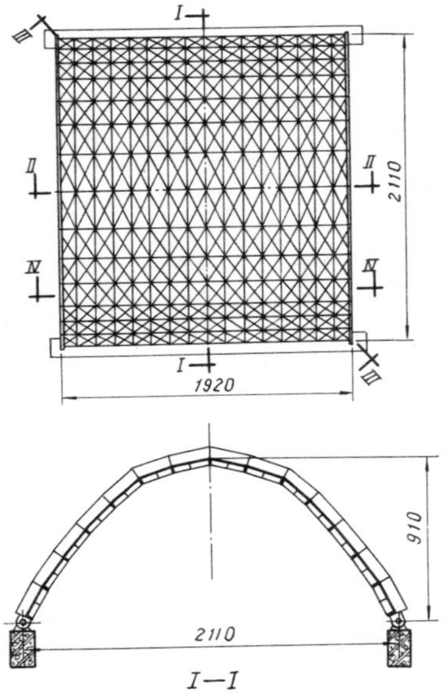

Fig. 1 Plan and sectional views of the model

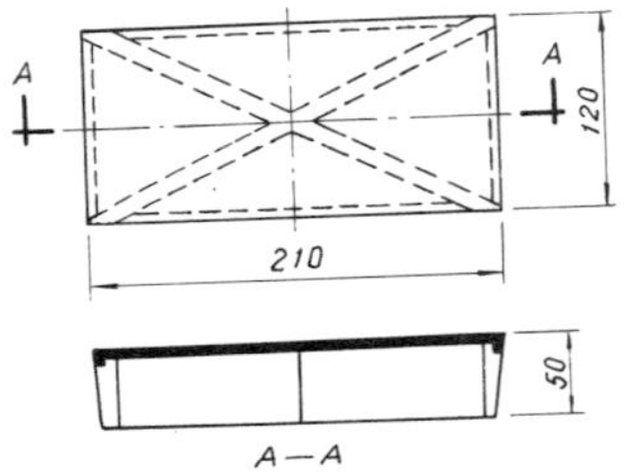

Fig. 2 Panel unit with cross-ribs and edge ribs

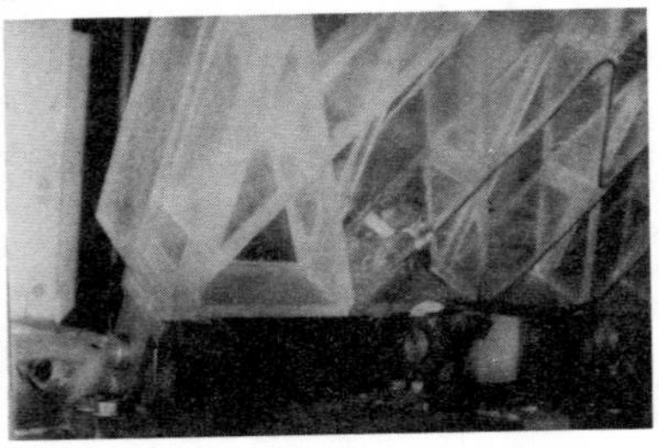

Fig. 3 The hinged support of the model

Under each loading condition, five identical test runs are made at specified time and loading intervals and hence five sets of readings are taken for comparison. Besides, for each fixed loading, readings at each specified time interval are also taken to find out creep effect of the model material upon the recorded readings, which with will be used for further corrections. Fig. 4 shows the test set-up for uniform normal pressure and uniformly distributed vertical point loads respectively.

Fig. 4a Test set-up under uniform normal pressure

Fig. 4b Test set-up under uniførmly distributed vertical point loods

THEORECTICAL ANALYSIS

Method of finite element is essentially adopted for analysis of the model structure. The composite element which consists of a triangular plate element and rib elements is chosen as a unit in analysis. The plate element has both plane stress and flexural actions, while the rib elements will resist moments, shears and axial forces.

(1) Plate element --- plane stress action
Fig. 5 shows a triangular element with nodal points i, j, m, denoted in counter-clockwise direction. The element is assumed to have linear plane displacement functions u,v, as follows:

$$\begin{Bmatrix} u \\ v \end{Bmatrix} = \left[IN_i^p, \quad IN_j^p, \quad IN_m^p \right] \left\{ \delta_p \right\}^e$$

where u,v, are displacement components of the element in the direction of the coordinate axes x and y respectively; the shape functions are

$$N_i^p = (a_i + b_i x + c_i y)/2A \qquad (i,j,m,)$$

in which, $a_i = x_j y_m - x_m y_j$, $b_i = y_j - y_m$, $c_i = -x_j + x_m$ (i,j,m) and the area of the triangle

$$A = \frac{1}{2} \begin{vmatrix} 1 & x_i & y_i \\ 1 & x_j & y_j \\ 1 & x_m & y_m \end{vmatrix}$$

with $x_i, y_i, \ldots x_m, y_m$ showing the rectangular cartesian coordinates of respective nodal points i,j,m, of the triangular element; the unit matrix of second order

$$I = \begin{bmatrix} 1 & 0 \\ 0 & 1 \end{bmatrix}$$

and the joint displacement vector of the element

$$\left\{ \delta_p \right\}^e = \left[u_i, \ v_i, \ u_j, \ v_j, \ u_m, \ v_m \right]^T$$

where u_i, v_i, ... u_m, v_m are displacement conponents of the respective joints of the triangular element Fig. 5.

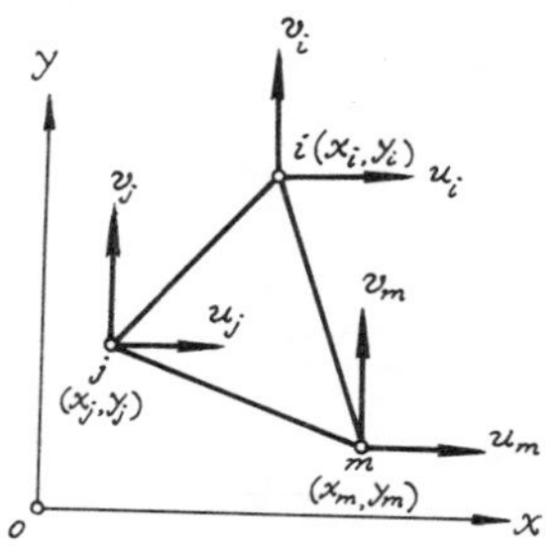

Fig. 5 Plane stress action of the plate element

(2) Plate element --- flexural action

A third degree polynomial function is assumed as plate deflection function w,

$$\{w\} = [N_i^b,\ N_{xi}^b,\ N_{yi}^b,\ N_j^b,\ N_{xj}^b,\ N_{yj}^b\ N_m^b\ ,\ N_{xm}^b\ \ N_{ym}^b]\{\delta_b\}^e$$

where the shape function for flexural case in area coordinates

$$\left.\begin{aligned} N_i^b &= L_i + L_i^2L_j + L_i^2L_m - L_iL_j^2 - L_iL_m^2 \\ N_{xi}^b &= b_jL_i^2L_m - b_mL_i^2L_j + \tfrac{1}{2}(b_j - b_m)L_iL_jL_m \\ N_{yi}^b &= c_jL_i^2L_m - c_mL_i^2L_j + \tfrac{1}{2}(c_j - c_m)L_iL_jL_m \end{aligned}\right\} \quad (i,j,m,)$$

here the area coordinates of each nodal point of the triangular element

$$\begin{Bmatrix} L_i \\ L_i \\ L_m \end{Bmatrix} = \frac{1}{2A} \begin{bmatrix} a_i & b_i & c_i \\ a_j & b_j & c_j \\ a_m & b_m & c_m \end{bmatrix} \begin{Bmatrix} 1 \\ x \\ y \end{Bmatrix}$$

where a_i, b_i c_i,... a_m, b_m, c_m and A have the same meaning as shown before.
The joint displacements vector of the element become

$$\{\delta_b\}^e = [w_i, \theta_{xi},\ \theta_{yi}, w_j, \theta_{xj}, \theta_{yj}, w_m, \theta_{xm}, \theta_{ym}]^T$$

where $w_i, \theta_{xi}, \theta_{yi}, \ldots w_m, \theta_{xm}$, θ_{ym} are deflections and angular rotations of the respective joints of the element as shown in Fig. 6.

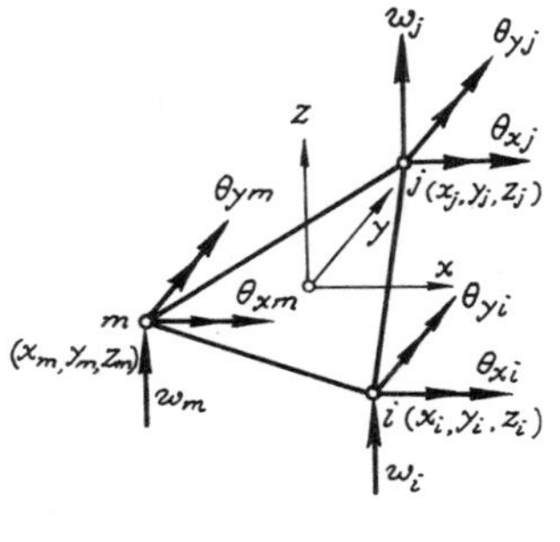

Fig. 6 Flexural action of the plate element

(3) Relationship between principal and subordinate degrees of freedom at a joint

At each end of a rib element, there are six joint displacements $\{\delta\}$ and six related joint forces $\{F\}$ as shown in Fig. 7. They are expressed in the matrix form

$$\{\delta\} = [u,\ v,\ w, \theta_x, \theta_y,\ \theta_z]^T$$

$$\{F\} = [X,\ Y,\ Z,\ M_x,\ M_y,\ M_z]^T$$

where u,v,w and θ_x, θ_y, θ_z are displacement components and angular rotational vectors in the direction of coordinate axes x,y,z, respectively; and X,Y,Z, and M_x, M_y, M_z, are their corresponding force components and monent vectors. The element stiffness matrix of the rib element can be found in Ref 1 (P79) and will not be given here. As this structure is composed of both plate elements and rib elements, the concept of principal and subordinate degrees of freedom is introduced in the computation, i. e. in a structure, the degrees of freedom of one or several joints are dependent upon the same of another joint. By this means, the total number of joint equilibrium equations in the whole system is not only greatly reduced but the possible error due to numerical ill-conditioning is also diminished

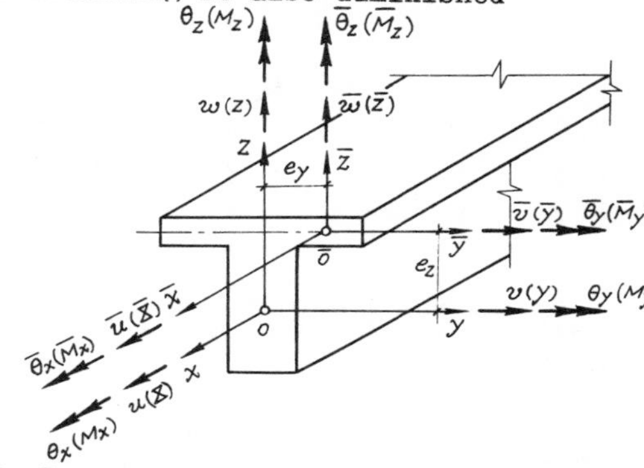

Fig. 7 Transformation of coordinates of the rib element

In this case, assuming the joints O and $\bar{O}$ lying in the same y-z plane, let $\bar{O}$ be a principal joint (e.g. a joint in the plate element) with its coordinate axes $\bar{O}x$, $\bar{O}y$ and $\bar{O}z$; and let O be a subordinate joint (e.g. a joint in the rib element) with its coordinate axes Ox, Oy and Oz.

At principal joint $\bar{O}$:

$$\{\bar{F}\} = [\bar{X},\ \bar{Y},\ \bar{Z},\ \bar{M}_x,\ \bar{M}_y,\ \bar{M}_z]^T$$

and $$\{\bar{\delta}\} = [\bar{u},\ \bar{v},\ \bar{w}, \bar{\theta}_x,\ \bar{\theta}_y,\ \bar{\theta}_z]^T$$

At subordinate joint O:

$$\{F\} = [X,\ Y,\ Z,\ M_x,\ M_y,\ M_z]^T$$

$$\{\delta\} = [u,\ v,\ w, \theta_x,\ \theta_y,\ \theta_z]^T$$

The relations between two system will be

$$\{\delta\} = [\lambda_e]\ \{\bar{\delta}\}$$

$$\{\bar{F}\} = [\lambda_e]^T\{F\}$$

It is apparent from Fig 7 that the transformation matrix becomes

$$[\lambda_e] = \begin{bmatrix} 1 & 0 & 0 & 0 & -e_z & e_y \\ 0 & 1 & 0 & e_z & 0 & 0 \\ 0 & 0 & 1 & -e_y & 0 & 0 \\ 0 & 0 & 0 & 1 & 0 & 0 \\ 0 & 0 & 0 & 0 & 1 & 0 \\ 0 & 0 & 0 & 0 & 0 & 1 \end{bmatrix}$$

Therefore, the principal joint forces become

$$\{\bar{F}\} = [\lambda_e]^T\{F\} = [\lambda_e]^T[K][\lambda_e]\{\bar{\delta}\} = [\bar{K}]\{\bar{\delta}\}$$

where e_y and e_z are projections of the distance between joints O and $\bar{O}$ on y- and z- axes respectively.
in which $[K]$ expresses the stiffness matrix of an element at the subordinate joint, and $[\bar{K}]$ expresses the modified stiffness matrix of the same element at the principal joint after translation.

$$[\bar{K}] = [\lambda_e]^T [K] [\lambda_e]$$

COMPARISON AND DISCUSSION

Curves plotted along some typical sections showing agreement on both test results and theoretical analysis are self-explainatory. Due to limited space herein, Figs 8-12 show some typical deflection and internal forces distribution curves of the structure under different loading conditions.

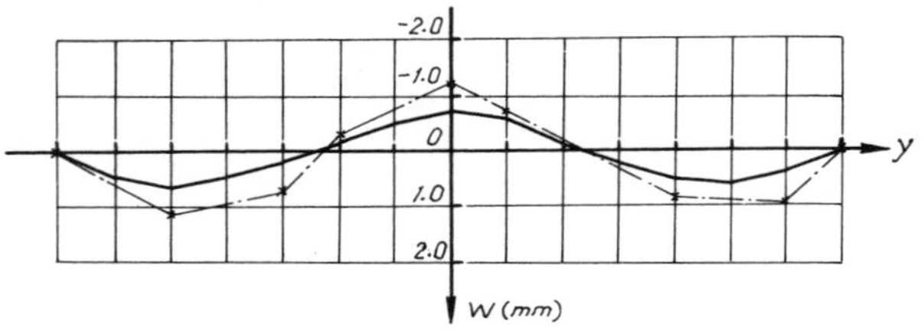

Fig. 8 Displacement component w along cross section I - I under 200 kg./m² uniform normal load

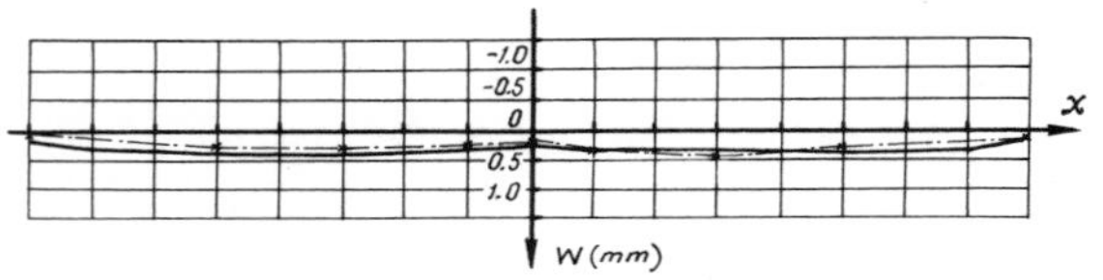

Fig. 9 Displacement component w along longitudinal section IV-IV under 600 kg/m² uniform vertical load

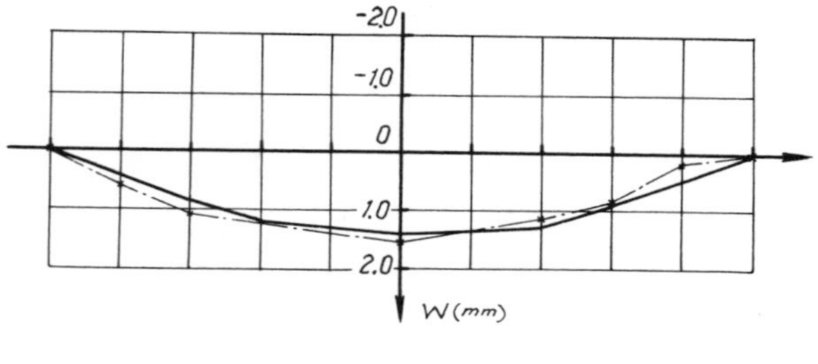

Fig. 10 Displacement component w along diagonal section III-III under supports horizontal movement

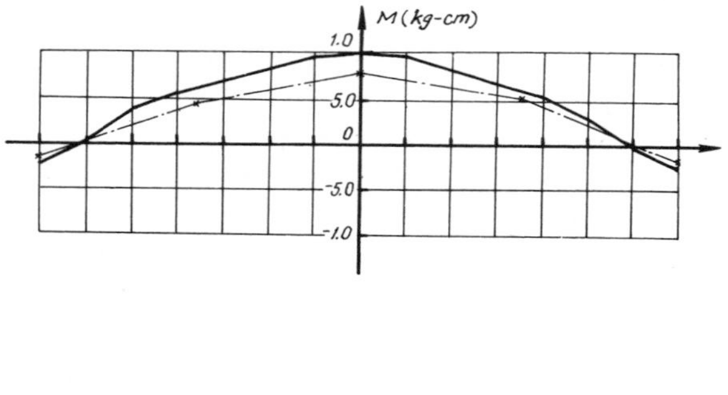

Fig. 11 Bending moment in main ribs along diagonal section III-III under supports horizontal movement

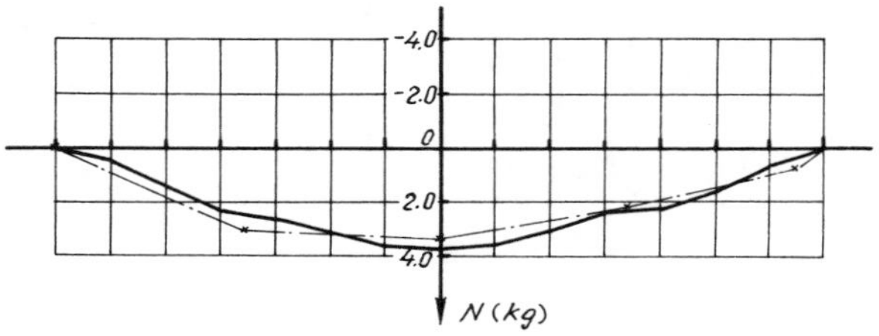

Fig. 12 Axial force in main ribs along diagonal section III-III under supports horizontal movement

——— theoretical results

—·— test results

In the most cases, both results look fairly agreeable each other. However, in certain cases, the discrapency may be nearly 30%. The causes which are responsible to such differences are essentially attributed to the following considerations.

(1) The shear deformation of rib elements has been disregarded in the finite element analysis, which is really an important factor for such ribs with a depth-joint spacing ratio of 1/2.4. An alternative calculation reveals that its effect is as large as 10% upon the deflection w of the structure particularly in the case of distributed normal load action as shown in Fig. 8.

(2) Due to limited capacity of the computer used (Siemens, BS1000 model), each panel unit is simply divided along ribs into four triangular elements, which makes totally 224 elements altogether. Furthermore, constant strain triangular elements are chosen for simplicity to compute plane stresses in plate, which certainly could otherwise be better by increasing adequately the total number of triangular elements. A close estimate suggests at least 10% discrapency in plane stress values due to these causes.

(3) By using of combined plate and rib finite elements, each straight rib is considered as a single element, in which the variation of axial force in between two ends is actually ignored. This is also a factor which causes some inaccuracy in rib forces.

(4) The synthetic glass selected as the model material has a modulus of elasticity $E=3.33 \times 10^4$ kg./cm² and Poisson's ratio $\mu=0.40$. It gives comparatively large deformation under fixed load within the elastic limit and has some creep effect varied with time,.Even though, the calibration work taken in due time intervals, however, does not eliminate all creep effect upon test readings any way.

(5) The dissimilitude in boundary conditions between the model structure and theoretical assumptions also causes some differences on distribution of deflections and forces especially in the region near boundaries of the model. In short, the composite triangular elements which are chosen for theoretical analysis have described basically the mechanical behaviour of the model structure, but their shortcomings can be well improved by the following resorts:

(1) Taking into account the shear deformation of the rib member to establish more accurate element stiffness matrix.

(2) Selecting more effective finite elements instead of the constant strain triangular elements.

(3) It may be more effective by using anisotropic finite strip method to solve such type of ribbed shell structure.

PRACTICAL APPLICATION

The rib-reinforced space structure investigated herein is chosen as roof structural system of a large automatic controlled warehouse which comprises three identical sheet-framed second order parabolic cylindrical shells, separated by expansion joints, each of which has a chord span of 50m, with a rise of 22,36m and 48m in length Fig. 13. By selecting the reasonable arch axis, the sectional moments are greatly reduced to nearly 4% of the maximum moment in a pre-stressed reinforced concrete three-hinged rigid frame with π-shape mambers of the same span length and loadings. Space action of this structure makes its internal forces more evenly distributed than that in plane structure. The interaction between plates and ribs boosts up load carrying capacity of the structure considerably.
Furthermore, the present structure has better economical index in material comsumption per unit area of the building than the others as compared in Tab. 1.
In order to guarantee the quality of fabrication and assembly of the panel units, the construction engineers precast 3 x 5m rectangular panel units with cross ribs in factory each of which weighs 5 ton (only 1/12 the hoisting weight of the prestressed π-section three hinged rigid frame) and assemble them on special scaffolds in the building site.
No large trucks and heavy hoisting equipment are necessary this method of construction have been well experienced by many construction companies in this country.
The details adopted at joints and connections between panel units are shown in Fig. 14.

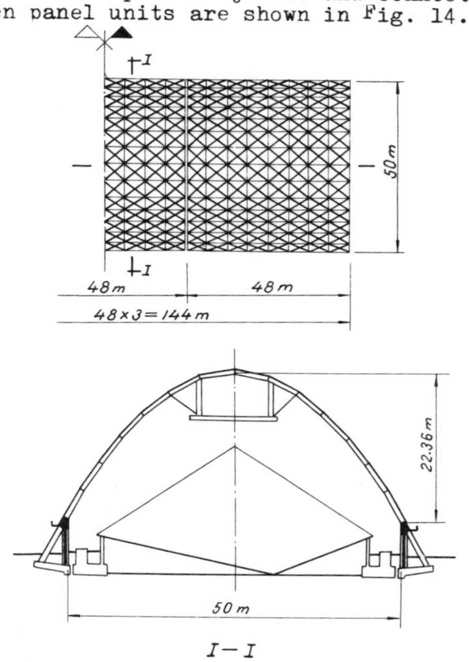

Fig. 13 A certain engineering project.

TABLE 1

Name & type of structure	Dimensions in plan	loading (kg/m²)	Steel Comsumption per unite area (kg/m²)
R.C. sheet-framed parabolic cylindrical shell structure	50×144m	600	27
Alton Gymnasium pyramid gridwork (England)	69×54m		33
Jiangsu provincial Gymnasium gridwork (China)	76.8×88.7m	240	53.5
Brno Gymnasium Suspanded Cable (Czechoslovakia)	84m in diameter	70	54
Froshilovgrad BUS Terminal station pyramid gridwork (USSR)	Equalateral triangle with 66m sides and 933m bottom	270	63
Wisconsin, Dn. County Memorial Hall Schwedier dome (U.S.A)	95m in diameter	70	83
Osaka International Exhibition Centre, Exhibition hall pyramid gridwork (Japan)	108×292m		109

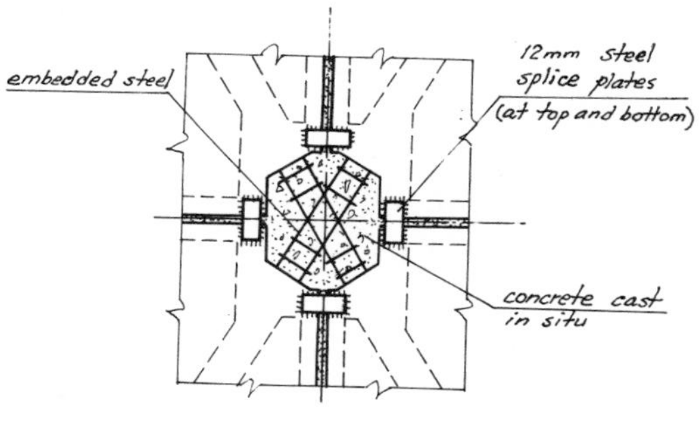

Fig. 14 Details of joint connetions

CONCLUSION

An experimental study and theorectical investigation on a model space structure composed of plate and rib elements has been successfully performed. In spite of those shortcomings discussed before, the analysis, however, shows that the model structure has a maximum deflection nearly 1/2000 of the span length under uniform normal loads and nearly 1/20000 of the same under uniform vertical loads, which fully indicates that this structure does have a greater stiffness in resisting loads. The ratio of internal forces in longitudinal direction to that in circumferencial direction is approximately 1:5. The average axial forces carried by the plate element even reach 60% of

that by main ribs, which provides a underlying potemtical of the structure and should never be overlooked in design.
Furthermore, the analysis alse shows that a maximun relative settlement of 1/700 of the span length and a horizontal support movement of 1/500 span length do not produce much internal stresses in the model structure. All these results obtained from both experimental and theoretical analyses have proved that such type of sheet-framed structure is, indeed, a stiffer, evenly strong and reliable space structure.

ACKNOWLEDGEMENT

The authors wish to express their indebtedness to the colleagues in Division of Structural Mechanics and Structural Laboratory of Nanjing Institute of Technology, for their kind help in performing the tests, especially to Messrs. G Y Xue and Z Z Chang, the graduate students in Structural Mechanics, who had been working in this project for their MS degree theses, to Nanjing Chemical Industrial Design Institute for its contributions of funds to this investigation, to Messrs. B X Xu, X H Bao and D R Xu, the Senior Design Engineers thereof who had given continued interest and encouragement in studying this type of space structure, and to the senior masters of Nanjing Folks Arts and Crafts Factory for their skillful techniques in fabricating such an excellent lucid model.

REFERENCES

1 J. S. Przemieniecki: Theory of Matrix Structural Analysis, 1968
2 O. C. Zienkiewicz: The Finite Element Method, third edition
3 I. Holand & K. Bell: Finite element Methods in Stress Analysis, 1972
4 Y. K. Cheung Finite Strip Method in Structural Analysis, 1976
5 S. A. Ambaretsumyam: Theory of Anisotropic Shells, Moscow, 1961
6 S. G. Lehenitski: Anisotropic Plates, Moscow, 1957

THE UNISTRUT CAD/CAM COMPUTER PROGRAM FOR THE "MERO" SPACE FRAME SYSTEM*

TODD SMITH

Space Frame Department
Unistrut Division, GTE Products Corporation

This article outlines the basic considerations taken when writing the MEROUSA CAD/CAM computer program and its impact on space frame design and fabrication. Highlights of the program, problems encountered & solved, experience gained, and benefits are presented. The basic node manufacturing system and computer generated fabrication data are illustrated.

In 1975 UNISTRUT Division obtained the exclusive rights from Mero-Raumstrukur GmbH & Co. to sell and manufacture the MERO system in North America. From 1975 to 1979, UNISTRUT only acted as a sales outlet and imported the entire frame material from West Germany. In 1979, UNISTRUT started manufacturing in the United States and has since then steadily expanded its capabilities.

The MERO space frame system is a proprietary structural system consisting of pre-engineered round steel tubes and spherical nodes.

The spherical node is to be faced, drilled and tapped at any specified location to accept a MERO tube. The MERO tube has a turnable-threaded-male-fastener on each end. These two features result in MERO system's flexibility of constructing a wide variety of geometric shapes and forms -- flat, curved, arched, or domed -- for unique application as each is custom designed and fabricated. The design of a universal joint of the MERO system is thus fully realized in its logical appearance, even though detailing and fabrication prior to the CAD/CAM system approach was very complex. This article highlights the progress made by UNISTRUT in MERO node design and fabrication.

This author developed the UNISTRUT MERO CAD/CAM Computer Program and named it MEROUSA**. The program was started in mid 1980 and by late 1981 it was ready for use as a design tool for MERO project engineering. Being a totally new program, it is designed for the state-of the- art hardware and file handling technique for optimal CAD/CAM functions.

When asked what the most important consideration is when writing a CAD/CAM program, I respond that one must design such a program so that it not only is a totally integrated system for the various operations of the CAD/CAM functions, but that it must also be a flexible system so that these functions can be changed later without a major alteration of the basic architecture. The key word in such a CAD/CAM system is change.

When one needs to update a hole size in a node the conventional method is to simply change the drawing. With the CAD/CAM system, not only the drawing, but the NC tape must also be updated. Therefore, the change must be made at a point prior to the node drawing function. If the program was not written with this in mind, adding this capacity later may be very difficult because the basic structure of the program may have to be changed also. One more example to illustrate this. In the past if the end module had to be 5 inches shorter to fit the site, the designer simply ignored this in the analysis and left the problem to the detailer. In a CAD/CAM system the 5 inches must be set at input stage. I believe it is clear that a CAD/CAM system must consider the entire structure from the initial design to final detailing.

To build a successful working CAD/CAM system, two areas must be looked at very closely and be resolved before start-up. One, how to allow the user to make changes in the structure and two, how the flow and protection of fabrication data moves from start to job site.

In the MEROUSA system the problem was solved by the formation of a formatted eighty column Fabrication Data File and by the writing of a simple subroutine to allow the user to make such changes as: nodes size, member type, and part number. After all engineering and basic changes are made, the program forms the Fabrication Data File. This file is then copied to another "User Account" and then all the changes are made there. The Fabrication Data File serves as input for all computer aided manufacturing functions and any changes to account for local details can be made by editing the Fabrication Data File.

*MERO is a registered trade name

**MEROUSA is a CAD/CAM program consisting of many subprograms. These are GEOM, FORCE, BEMS, KNOT (node selection), STUE (parts listing), BOHR, DRAW, PRICE and WEIGHT list. The program is housed in the GTE Computer Labs in Waltham, MA and is being remotely operated through portable terminals on a TSO command and TELENET communication. For the MERO node design and fabrication, the BOHR program checks on joint clearances, creates node data for the design and generates a set of NC (numerical control) line sheets or paper tapes for use by an NC machine.

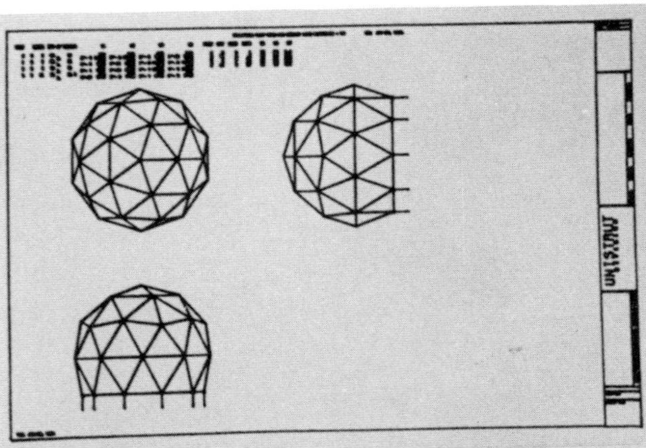

Illustrated is the computer generated data for fabrication of the MERO System.

Above: Computer plot of the frame including bill of material (member part nos. and nodes not shown for clarity).

Below: UNISTRUT operation sheet for manufacture of members.

Top Right: UNISTRUT BOHR sheet for checking the nodes.

Bottom Right: NC listing for manufacturing the node.

NOTE: After production of first node, the node is compared to BOHR sheet.

COMPUTER RUN DATE MON JAN 23, 1984

UNISTRUT — OPERATION SHEET — GTE

JOB NUMBERS	JOB TITLE	DATE
252 PA=345	SCULPTURE PLAY DOME	JAN 23, 1984

PART NO	QUANITY	TUBE
1	30	D1
L4= 478.	L3= 532.	210507

CODE	OPERATION	TRACKING
10 ( )	CUT TO LENGTH	( )
20 ( )	GRIND CONES	( )
30 ( )	ASSEMBLE CONES	( )
40 ( )	TACK AND WELD	( )
50 ( )	CHIP WELDS AND MARK	( )
60 ( )	PLACE IN SHIPPING RACKS & BAND	( )

FAB DATA

BOLT	M20* 5.6	210210
WASHER	W20* 8.8	210398
SLEEVE	SM 30/22	210315
PIN	5*29	210415
CONE	K60* 1.6	210115
TUBE	D1 TUBE	210010

TESTING REQ 3% MX PER PROCEDURE BULLETIN NO. SF-20-040

BOHR VERSION USA SCULPTURE PLAY DOME
MON JAN 23, 1984 PAGE 1

NODE PART NUMBER 1
NUMBER OF NODES .... 6 FINISH SPEC FS 352
INITIAL REF. THREADED HOLE .. 12 COSX= 0.00 COSY= 0.00 COSZ= 1.00

MERO NODE TYPE NUMBER	NODE NUMBER EXAMPLE	INSIDE AREA (MM)	OUTSIDE AREA (MM)	NUMBER OF BEAMS	NUMBER OF ADD.HOLES
3	1	76	85	5	0

ORIGINAL NODE — ONLY FOR REFERENCE NODE

NUMBER OF BORE HOLE	DIAMETER OF HOLE	HORIZONTAL ANGLE U	VERTICAL ANGLE V	BEAM NUMBER	CROSS SECTION
1	20	0.00	15.86	1	D1
2	20	72.00	15.86	2	D1
3	20	144.00	15.86	3	D1
4	20	216.00	15.86	4	D1
5	20	288.00	15.86	5	D1

COMPUTOR RUN DATE - MON JAN 23, 1984

```
N1(MSG,SIZE30TY6PART1LOADCODE20.000)
O2G0T10200M6
N3G70
O5G0X19149Y100000Z120000B0S1150F105M3
N7B0
N8G0X19149Y54645Z15354
N9G1X851S1150F210
N10G0Z70000
N12B180000
N13G0X19149Y54645Z15354
N14G1X851S1150F210
N15G0Z70000
O16G0T21102M6
O17G81X0Y54645Z-9000R15354S500F47B180000M13
N18G0Z70000
O20G81X0Y54645Z-9000R15354S500F47B0M13
N21G0Z70000
O22G0T4012M6
O23G84X0Y54645Z-6500R15354S203F137B0M13
N24G0Z80000
O26G84X0Y54645Z-6500R15354S203F137B180000M13
N27G0Z80000
N29(MSG,PLEASEROTATETABLEWHENREADY)
N29M2
N500(SHI,B91)
N501(SHO,B0)
O503G0T10200M6
N504G70
N506G0X125000Y87500Z125000B0S1150F105M3
N508M70
N509M70
N510M70
N511G0X97207Y87500Z37737B16000
N513G1X78909S1150F210
N514G0X107207
N515G0Z45354
N518M72
N519M78
N520M78
N521G0X97207Y87500Z37737B16000
N523G1X78909S1150F210
N524G0X107207
N525G0Z45354
N528M72
N529M71
N530M76
N531G0X97207Y87500Z37737B16000
N533G1X78909S1150F210
N534G0X107207
N535G0Z45354
N538M71
N539M74
N540M74
N541G0X97207Y87500Z37737B16000
N543G1X78909S1150F210
N544G0X107207
N545G0Z45354
N548M70
N549M77
N550M72
N551G0X97207Y87500Z37737B16000
N553G1X78909S1150F210
N554G0X107207
N555G0Z45354
O557G0T20175M6
O559G81X78058Y87500Z-15000R37737S500F57B16000M13
N560G0Z75354
N562M71
N563M74
N564M74
O565G81X78058Y87500Z-15000R37737S500F57B16000M13
N566G0Z75354
N568M72
N569M71
N570M76
```

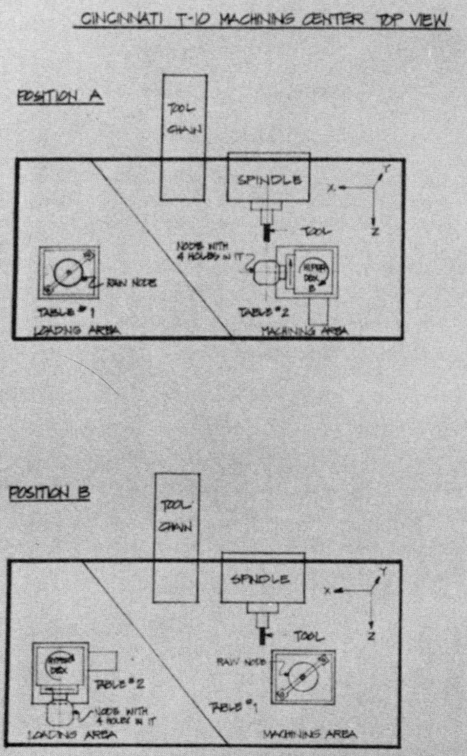

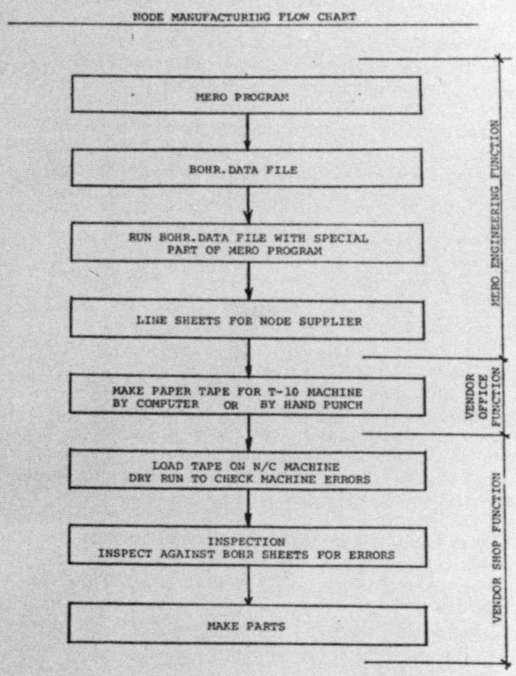

BASIC OPERATION

1. Initial set-up
2. As machine is tooling the node on Table #2, raw node is mounted on Table #1 (machine is in position A).
3. When machine is finished machining part on Table #2, machine rotates into position B and starts to drill the first 4 holes.
4. At this time the finished part on Table #2 is removed and inspected. Then a part that has been through Table #1, machine cycle is loaded on Table #2.
5. When machining is finished on Table #1, machine will rotate back into position A.
6. Machine now starts tooling remainder of the holes in the node on Table #2.

The cycle is now complete. The next step is number 2, repeat cycle until production of all parts is done.

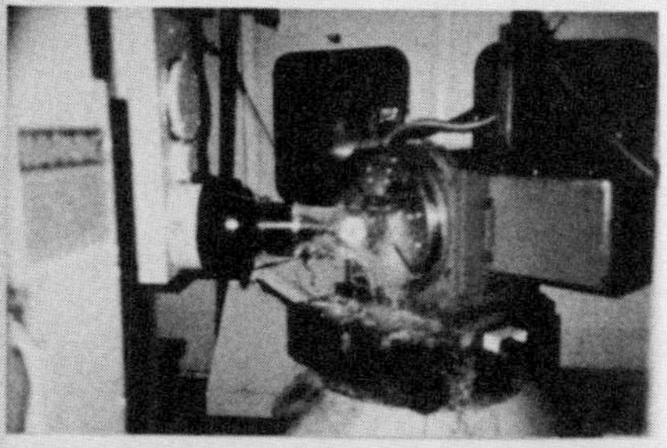

Illustrated is the MERO node fabrication on the Cincinnati Milacron, Model T-10 Machining Center.

Above: Diagram of the node machining process.

Top Right: Node manufacturing flow chart showing the flow of information from start to finish node.

Below Right: Pictures of actual machine in operation.

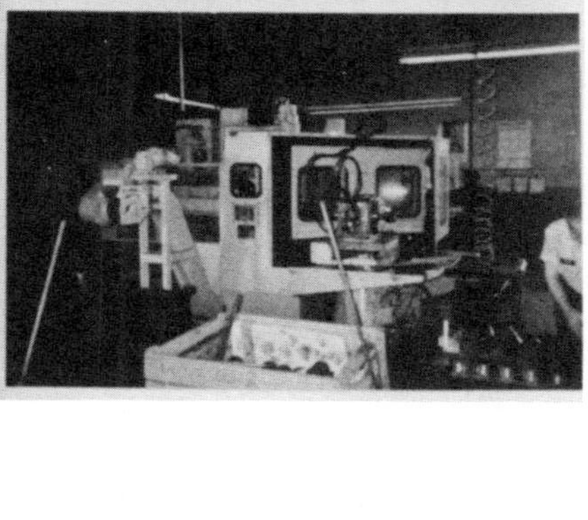

In the MERO space frame system, there are ten node sizes ranging from 1½" diameter to 10" diameter. Each node will require holes drilled, faced and tapped for tube member connection. There are eight different bolt sizes ranging from 12mm diameter to 64mm diameter. The number of tapped holes for the node varies from a few to as many as 18, which can receive 18 tube members converging into a single joint. This makes the manufacturing of the MERO node a natural use of CAD/CAM. The node is machined on a five axis NC machine center made by Cincinnati Milacron, model T-10. The machine has its own tool box and shuttle tables that will take a steel ball in the raw and completely fabricate with tapped holes and faced in just a few minutes.

For MERO space frame projects every node has a part number with its own NC tape, which can literally add up to thousands of NC tapes for many jobs a month. If these tapes were to be inputted by hand, it would take too long and not be practical to use an NC machine to fabricate them.

My personal experience in writing the subroutine to generate NC tapes took about six months. I wish to point out some interesting observations. The actual programming time to write the code was minimal, but the time spent learning the machine process and the NC language is where the six months was consumed. The first step was to decide upon an approach to the problem. Two approaches were available. One was to write the NC code in the MEROUSA program and the other was to generate the input file for ANVIL (ANVIL is a major CAD/CAM program with an NC generation function). Upon investigation, using ANVIL was not practical for two reasons. One, it was menu driven and therefore each node would have to be hand inputted, and two, there was no practical way to make ANVIL and the machining process that we wished to use compatible.

My first job was to learn the NC language. After I learned enough NC language and the basics of the machining process that the manufacturing operator wanted to use, I was able to write a very basic NC file for the MERO node production. Up to this point, I was mostly writing codes at my desk, away from the actual manufacturing site. The next step was to test the file on the NC machine. By this time, I was already two months into the project. I was thinking to myself that I was almost done, but in reality I had just touched the surface. Meanwhile, the manufacturing group was setting up the machine and buying tooling. Upon testing of the first node with the NC tape, generated by the MEROUSA program, it became clear that writing software away from the actual manufacturing site just would not get the job done. The node was made all right, but there were so many unnecessary moves that major revisions in computer logic were needed. For the next two months I wrote at the machine shop, spending hours learning the machine process and correcting my program so the final NC code would be efficient and would earn the respect and trust of the machine operator. One thing I learned was that to have a successful CAD/CAM system, the programmer and the machine operator must work side by side as a team. This team effort is a pre-requisite for a successful CAD/CAM system. It is important for the computer programmer to understand the machining process and equally important for the machine operator to understand the basic computer program logic so he can anticipate problems before they occur, such as a machine collision.

Since October 1982, more than 10,000 nodes have been made. comprising of over 1000 individual part designs using the MEROUSA program. This experience has taught us the following facts:

1) Quality control is now a simple inspection of the first piece per NC tape. If it is correctly made the rest will be correct.

2) The Designer retains total control over the part.

3) Design time has been greatly reduced.

4) Scrap factor for the node production is now below 0.5%.

5) Cost of production is comparable with the conventional methods but the quality is superior.

In conclusion, when starting a CAD/CAM system the basic program organization has to be thoroughly thought out before writing begins. Design and manufacturing needs must be considered instead of just programming execution efficiency. The MEROUSA program accomplishes the design and manufacturing needs by writing a subroutine to allow itemized part changes and the generation of a Fabrication Data 80 column file which can be edited. This system has been proven to work well in both serving the needs of manufacturing and Design/Detail personnel.

ACKNOWLEDGEMENTS

The author wishes to acknowledge the MERO-Raumstruktur Gmbh Co. computer program called MERO. The author has not seen the MERO source code but was a user of the MERO Program. There are many parts of the two programs that have similar functions. The author kept the same names for these parts to give credit to the MERO Program.

RESEARCH ON THE GEOMETRICAL UNCHANGEABILITY AND THE STRESS ANALYSIS OF HONEYCOMB TRIANGULAR PYRAMIDAL PLATE TYPE GRID TRUSSES

WU JIAN SHENG*, LIU XI LIANG**, LIU ZUE REN*** and YANG FENG DI****

* Professor, Architectural Design Institute,
Tianjin University, Tianjin.

** Associate Professor, Department of Civil Engineering,
Tianjin University, Tianjin.

*** Engineer, Architectural Design Institute,
Tianjin University, Tianjin.

**** Lecturer, Architectural Design Institute,
Tianjin University, Tianjin.

This article deals with the theoretical analysis of the geometrical unchangeability of a honeycomb triangular pyramidic plate type grid structure supported along the periphery and computations on 30 combinations of different loading conditions and different support restraints have been worked out with a result pointing out that the geometric unchangeability of this system only depends upon the restraint conditions of supports and no need to increase any member; and the kinematic analysis of the geometrical unchangeability in space may be simplified to the kinematic analysis of a plane truss formed by the upper chords only.

INTRODUCTION

Space grid trusses have many advantages and are widely used in roof structures in domestic structural design. Among all types of grid trusses the honeycomb triangular pyramidic plate type grid truss has the least number of members converging on each joint, only six members, and better economical results. We therefore intend to use this system for the roof structure of a cinema. But people think differently of the geometric unchangeability of this system; some hold that this system itself is geometrically changeable thus requiring extra members and joints to make it geometrically unchangeable, while others hold just the opposite views. For this reason, we, integrating specific project design, have made a number of theoretical analyses on the geometric unchangeability and stress analyses of this system and computations have been worked out for 30 combinations of different loading conditions and different support restraints.

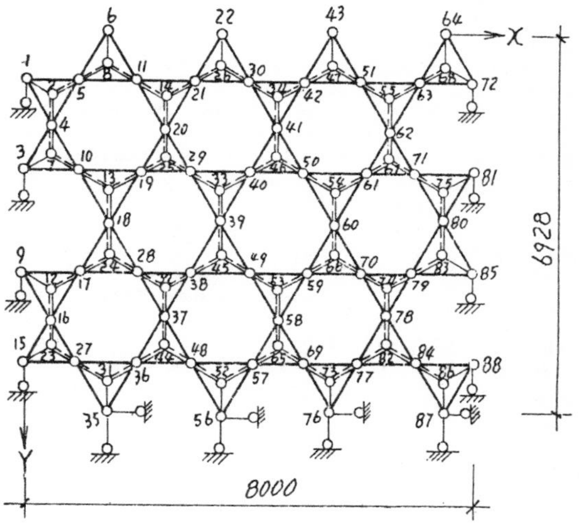

Fig 1. Honeycomb Triangular Pyramidic Plate Type Grid Truss -- Plan and Joint Number
Legends ——— Upper Chords
- - - - Lower Chords
——— Web Members

COMPOSITE FEATURES OF THIS SYSTEM

Fig 1 shows the plan of the arrangement of this system in a rectangular plane. It can be seen from Fig 1 that the basic unit that forms the grid structure is a triangular pyramid with four sides (Fig 2) as well as the lower chord which joins the apexes of triangular pyramids. (Fig 3) The apexes of the base triangles of triangular pyramids are mutually joined to form the upper chord system while the edges of the pyramids form the web member system.

If through point e (Fig 3) we draw a straight line mn perpendicular to the plane efg, the structural role that the lower chord fg plays is to place restrictions on the relative rotations about straight line mn of the two adjacent pyramids; this, in fact, restricts the joining point e of pyramids which is in the upper chord plane to displace along the plumb line.

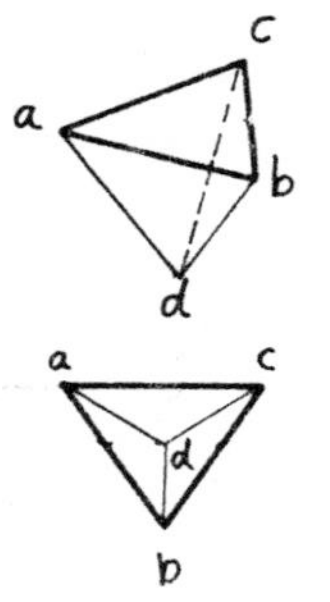

Fig 2 Triangular Pyramid Unit

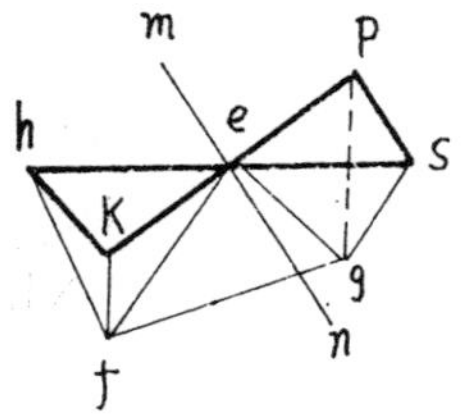

Fig 3

ANALYSIS OF THE GEOMETRIC UNCHANGEABILITY OF THIS SYSTEM

If we want a grid structure to be one of geometric unchangeable system, besides the necessary conditions W (freedom) $\leq$ 0 of the kinematic analysis must be satisfied, the adequate conditions of the same must also be satisfied. These two aspects are analyzed as follows:

1. The adequate Conditions of Geometric Unchangeability of this system

   To prove the geometric unchangeability of the structure, we may analyze in the following way.

1) The regular triangle of pyramid base in the upper chord plane is taken as a basic unit and treated as rigid plate (replacing the numbers of the pyramid base triangles with the numbers of pyramid apexes, i.e. the numbers of rigid plates). Owing to the three nonconcurrent horizontal link restraints at supports 87 and 88, rigid plate 86 forms a planar statically determinate system, (horizontal restraint at support 88 may be in X-direction but even so, the three links are not concurrent) point 84 becoming a fixed one; in this statically determinate system through the two horizontal links at support 76 and point 84 joining rigid plates 73 and 82 still forms a planar statically determinate system, point 69 becoming a fixed point; similarly, continuing the joining of rigid plates 52, 65, 31, and 44 still forms a planar statically determinate system, point 27 becoming a fixed point; later, through point 27 and the link in Y-direction of support 15 (not X-direction, for the produced line of the link should not pass through point 27, otherwise the structure is geometrically changeable) join rigid plate 23, this time point 16 becoming a fixed point; again from point 16 and the link of support 9 (may be in X-direction) join rigid plate 12 to form a statically determinate system, point 17 becoming a fixed point; similarly continue to join rigid plates 24, 32, 45, 53, 66 and 74 to form a statically determinate system, point 79 becoming a fixed point; point 79 and the horizontal link (may not be X-direction) of support 85 are joined to rigid plate 83 to form a statically determinate system, point 80 being a fixed point; point 80 and the horizontal link (may also be X-direction) of support 81 are joined with rigid plate 75, point 71 becoming a fixed point; similarly, join 67, 54, 46, 33, 25 and 13 to form a statically determinate system, point 10 becoming a fixed point; later, through point 10 and the Y-direction link of support 3 (may not be X-direction) join rigid plate 7, point 4 becoming a fixed point; from point 4 and the horizontal link (may also be X-direction) of support 1 join rigid plate 2 into a statically determinate system, and from fixed points 5 and 20 join rigid plates 8 and 14 into a statically determinate system, now support 6 being in the upper chord plane have no need of horizontal restraint links; similarly, continue to join rigid plates 26, 34, 47 and 55 to form a statically determinate system, point 63 becoming a fixed point; point 63 and the horizontal link (may not be X-direction) of support 72 are joined with rigid plate 68, making the whole grid structure a statically determinate system in the upper chord plane based on the restraint conditions of Fig 1.

2) Through the above analysis, it is ascertained that the whole grid structure forms a statically determinate system in the upper chord plane, and at this moment each pyramid base triangle uses 3 web members to form a pyramid, their point of intersection being the apex of the pyramid. Each apex has no relative displacement with respect to its pyramid base triangle. Up to now, each point of intersection of the pyramids in the upper chord plane of the structural system has no possibility of horizontal displacement but there is the possibility of vertical movement.

3) In order to restrict the vertical movement of the points of intersection of pyramids in upper chord plane, all apexes of these pyramids are joined together to form the lower chords of the structural system as shown in Fig 3.

4) Again at each support a vertical restraint link is used to restrict the vertical movement of a support joint.

   Through the analyses of (1), (2), (3) and (4), honeycomb triangular pyramid plate type grid structures formed with restraint conditions as Fig. 1 can be said as satisfying the adeguate conditions of geometric unchangeability in kinematic analysis and during analyzing there are no redundant members and restraints.

2. The Necessary Conditions of Geometric Unchangeability of this system

   Each joint of this system is supposed to be hinged, i.e. each joint has three freedoms. Suppose the whole structure is composed of I number of triangular pyramids and J number of supports. Owing to the characteristics of the structural composition that in the upper chord plane, pyramid base triangles are joined together through the apexes of the triangle, namely each apex of the triangle merging correspondingly with the apex of adjacent triangle, then the number of joints K (not including the side support joints) can be determined by the following formula:

$$K = \frac{3I - J}{2}$$

   Suppose the number of lower chords be L, from Fig 3 we can know that the number of lower chords is the same as the intermediate joints in the upper chord plane, namely L = K. Suppose the number of joints in the lower chord plane be M, from Fig 2 we can know that the number of lower chord joints is the same as the number of pyramids, namely M = I. Suppose the number of vertical links in the supports to be necessarily added be N, same as the number of support joints, namely N = J then the total number of joints of the whole structure is:

$$P = J + K + M = J + \frac{3I - J}{2} + I = \frac{5I + J}{2}$$

   The unknowns of the total displacement of the whole system are:

$$R = 3P = 3 \times \frac{5I + J}{2} = \frac{15I + 3J}{2}$$

The total number of members of the system is:

$$S = 6I + L = 6I + K = 6I + \frac{3I - J}{2} = \frac{15I - J}{2}$$

According to the kinematic analysis principle we know: when the total number of members is S; total joints P; total number of support restraint links B, the necessary condition of a geometric unchangeable system is $S \geqslant 3P - B$

$$B \geqslant 3P - S = \frac{15I + 3J}{2} - \frac{15I - J}{2} = 2J$$

Suppose the number of horizontal restraint links of supports in the upper chord plane be A, the total number of links at supports is

$$B = A + N = A + J \geqslant 2J$$

namely $A \geqslant J$

The above relation indicates that the system must rely on support restraints to keep itself a geometrically unchangeable restrained-by-others structural system. Besides the necessity of installing a vertical link at each support, it is necessary to properly arrange J number of horizontal restraint links in all the supports (of J number) according to the adequate conditions of kinematic analysis of geometrical unchangeability, to make the system a geometrically unchangeable, statical determinate system. As shown in Fig 1 the grid structure is composed of 32 pyramids and 16 supports,

Then number of pyramids $I = 32$
number of support joints $J = 16$
number of lower chord joints $M = 32$
number of vertical restraint links at supports $N = J = 16$
number of upper chord intermedinte joints $K = \frac{3 \times 32 - 16}{2} = 40$
number of lower chords $L = K = 40$
total number of joints $P = J + K + M = 16 + 40 + 32 = 88$
total displacement unknowns $R = 3P = 3 \times 88 = 264$
total number of members $S = 6I + L = 6 \times 32 + 40 = 232$

To ensure that the grid structure forms a statically determinate system, the minimum number of horizontal restraints at supports in the upper chord plane is

$$A = B - J = 3P - S - J = 264 - 232 - 16 = 16 = J$$

When $A = 16$, the system is statically determinate;
When $A > 16$, the system is statically indeterminate;
When $A < 16$, the system is geometrically changeable.

Through the analyses of adequate conditions and necessary conditions to this system we can know that the quantity and arrangement pattern of all the basic members of the system (upper chords, lower chords and web members) and vertical restraint links at supports follow a fixed law; only the quantity and arrangement pattern of the horizontal restraints at supports in the upper chord plane are changeable with the result that the whole grid structure may become geometrically unchangeable, statically determinate or indeterminate. Therefore the problem of the kinematic analysis of the geometrical unchangeability in space of the grid structure may be simplified as the planar kinematic analysis in the upper chord plane; only if the upper chords and the horizontal restraints at supports are geometrically unchangeable, the whole structure will be geometrically unchangeable; if not so, it is geometrically changeable.

In brief, whether the whole grid structure is geometrically changeable or unchangeable, statically determinate or indeterminate is entirely determined by the conditions of mutually connection of the members in the upper chord plane and the horizontal restraints at supports.

EXAMPLES OF COMPUTATION ANALYSIS

To prove the above kinematic analysis deduction using Fig.1 with the aid of a computer we have worked out 30 schemes of example computations with different combinations of loadings and horizontal restraints at supports and computation results of 15 major schemes are tabulated in Attached Table 1.

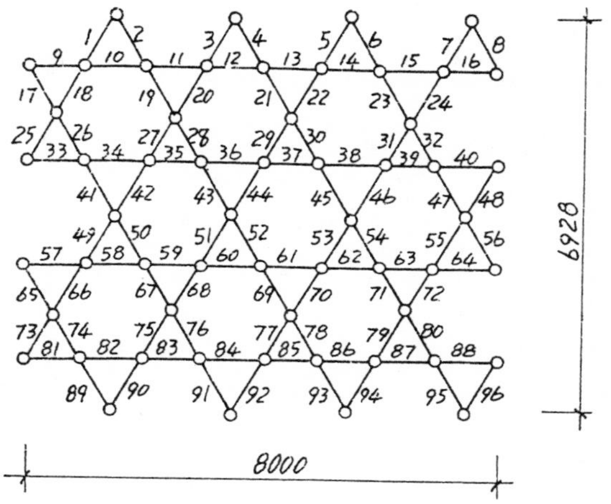

Fig 4 Upper Chord Numbers

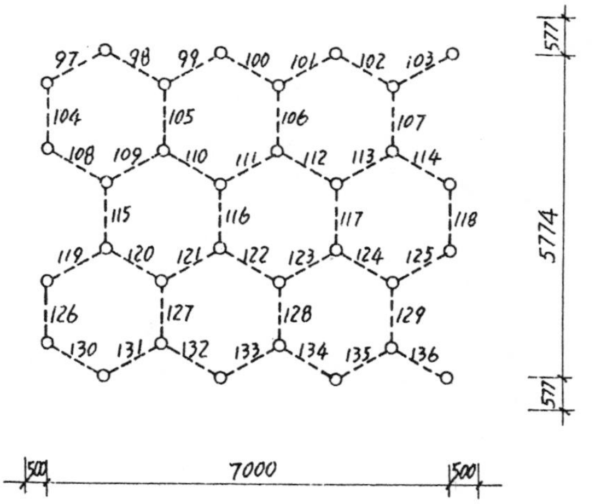

Fig 5 Lower Chord Numbers

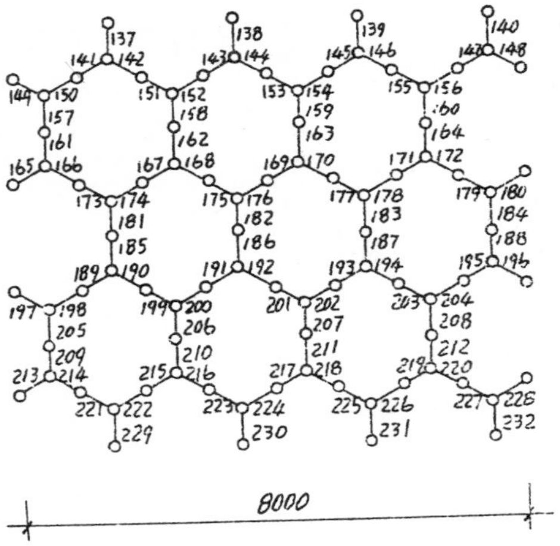

Fig 6 Web Member Numbers

From the Attached Table 1, with Figs 4, 5 and 6 we can know the following:

Schemes 1, 2, 3 and 4: this kind of restraint conditions judged by the above deduction is statically determinate system and the actual results of computation also proved so , because under vertical and horizontal loadings member internal forces all appeared to be normal values, and not affected by cross sections. If any member is removed, the deformations appear to be abnormal values (expressed in AAA, which in computer results represent a figure larger than 5 places, the unit being in centimeters)

Schemes 5, 6, 7 and 8 (Fig 7): this kind of restraint conditions judged by the above deduction is also statically determinate system and the actual results of computation also proved so, because under vertical and horizontal loadings, internal forces of members all emerged as normal values, not affected by cross sections. If any of the members is removed, deformation values become abnormal ones.

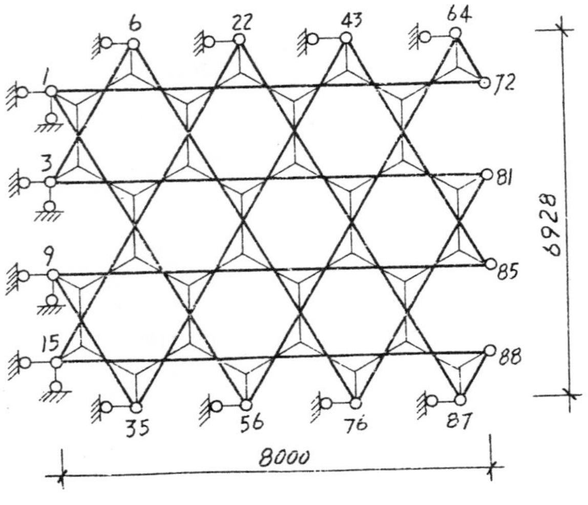

Fig 7

Schemes 9 and 10 (Fig 8): this kind of restraint conditions judged by the above deduction though does not lack horizontal restraints but they are improperly arranged. Using kinematic analysis principle for judgement, the upper chord plane is a geometrically changeable system, and the actual results of computation indicated that in this two-way supported diagram, under vertical loadings, (such as scheme 9) although the member internal forces and deformations merged as normal value, yet under horizontal loadings (such as scheme 10) deformation values became abnormal, explaining it is geometrically changeable.

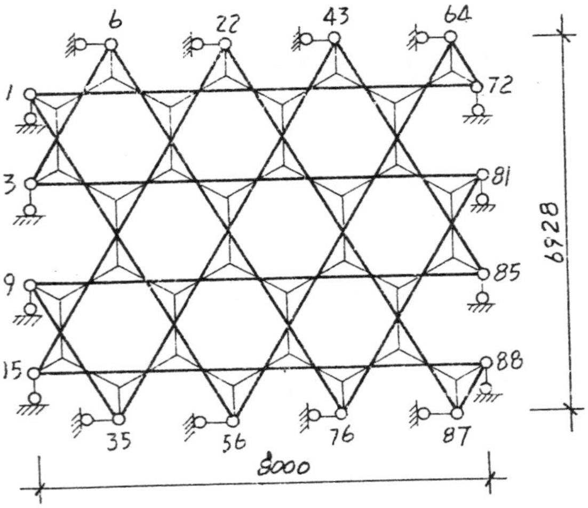

Fig 8

Schemes 11 and 12 (Fig 9): in the upper chord plane all the supports are two-way horizontally restrained, and by the above deduction and analysis $W < 0$ gives a statically indeterminate system; the actual results of computation give normal values of member internal forces and deformations.

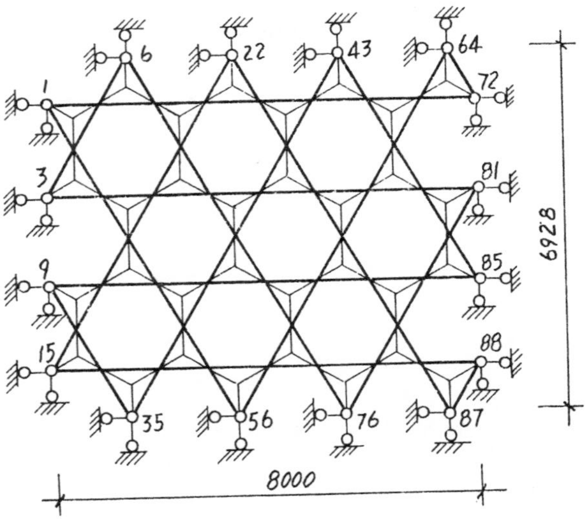

Fig 9

Schemes 13 and 14 (Fig 10): each support joint is of three-dimensioned elastically restrained condition; $W < 0$ gives a statically indeterminate system; in the

INTERNAL FORCE AND DEFORMATION

Table 1

Number of Scheme	Restraint Condition of Support	Loading Condition: Vertical U.C-0.35t L.C-0.15t	Loading Condition: Horizontal (1/10 Vertical) x	Loading Condition: Horizontal (1/10 Vertical) y	Section Area	Change of Section	Max. Internal Force (t): Upper Chord	Max. Internal Force (t): Lower Chord	Max. Internal Force (t): Web Member	Max. (cm) Deformation: X Direction	Max. (cm) Deformation: Y Direction	Max. (cm) Deformation: Z Direction
1	6.22.43.64-2 Way Restraint	↓			U.C.2#=3.49cm²; L.C.1#=5.78cm²; Webs3#=2.55cm²; 4#=0,5#=9.55cm²; 6#=8.29cm²		-2.510 (44)	4.491 (116)	2.294 (138)	0.237 (1)	0.302 (42)	0.784 (39)
2	35.56.76.87-3 Way Restraint	↓				37(2#→6#)	-2.510 (44)	4.491 (116)	2.294 (138)	0.237 (1)	0.302 (42)	0.779 (39)
3	Other Supports–2 Way Restraint (Fig. 1)	↓	→	→			-3.026 (67)	4.491 (116)	2.309 (138)	0.409 (1)	0.420 (42)	0.800 (39)
4		↓				53(2#→4#)	-2.147 (51)	3.595 (116)	2.060 (138)	AAA	AAA	AAA
5	72.81.85.88-2 Way Restraint	↓					-2.510 (43)	4.491 (116)	2.294 (138)	0.216 (80)	0.237 (71)	0.784 (39)
6	1.3.9.15-3 Way Restraint	↓				37(2#→6#) 116(1#→5#)	-2.510 (43)	4.491 (116)	2.294 (138)	0.216 (80)	0.237 (71)	0.758 (39)
7	Other Supports–2 Way Restraint (Fig. 7)	↓	→	→			-2.841 (51)	4.491 (116)	2.309 (138)	0.225 (80)	0.285 (85)	0.775 (39)
8		↓				37(2#→4#)	-2.304 (51)	3.954 (116)	2.114 (230)	AAA	AAA	AAA
9	All Supports–2 Way Restraint (Fig 8)	↓					-2.430 (67)	4.491 (116)	2.294 (138)	-0.142 (12)	0.153 (21)	0.733 (39)
10		↓	→	→			-2.582 (27)	4.461 (116)	2.310 (138)	AAA	AAA	0.730 (39)
11	All Supports–fixed Restraint (Fig. 9)	↓					1.232 (7)	4.491 (116)	2.294 (138)	-0.090 (12)	-0.112 (26)	0.548 (39)
12		↓	→				1.222 (33)	4.491 (116)	2.294 (138)	0.089 (75)	-0.112 (26)	0.548 (39)
13	All Supports–elastic Restraint (Fig. 10)	↓					2.455 (43)	4.491 (116)	2.294 (138)	-0.102 (7)	-0.109 (26)	0.780 (39)
14		↓	→				-2.478 (44)	4.491 (116)	2.294 (138)	0.375 (3)	0.210 (42)	0.780 (39)
15	Similar to 5-8 (Fig. 7)		→	→			1.707 (33)	0	0	0.186 (81)	-0.188 (43)	0.235 (39)

Internal Force (t)

Number of Scheme	Upper Chord 9	13	16	33	44	48	Lower Chord 97	100	103	108	116	118	Web Member 138	148	149	165	182	184
1	-0.421	-1.518	-0.277	-0.643	-2.510	-0.675	1.001	2.189	1.319	2.153	4.491	1.314	2.294	0.755	1.148	1.751	-0.226	-0.226
2	-0.421	-1.518	-0.277	-0.643	-2.510	-0.675	1.001	2.189	1.319	2.153	4.491	1.314	2.294	0.755	1.148	1.751	-0.226	-0.226
3	-0.077	-1.354	-0.262	-0.626	-2.584	-0.100	1.001	2.189	1.319	2.153	4.491	1.314	2.309	0.734	1.169	1.757	-0.242	-0.242
4	-0.700	-1.342	-0.258	-0.653	2.004	-0.499	0.821	1.954	1.229	1.884	3.595	1.213	2.060	0.706	1.018	1.559	0.201	-0.219
5	-0.421	-1.518	-0.277	-0.643	-2.510	-0.675	1.001	2.189	1.319	2.153	4.491	1.314	2.294	0.755	1.148	1.751	-0.226	-0.226
6	-0.421	-1.518	-0.277	-0.643	-2.510	-0.675	1.001	2.189	1.319	2.153	4.491	1.314	2.294	0.755	1.148	1.751	-0.226	-0.226
7	-0.084	-1.361	-0.269	-0.340	-2.844	-0.673	1.001	2.189	1.319	2.153	4.491	1.314	2.309	0.734	1.169	1.757	0.242	-0.242
8	-0.410	-1.261	-0.259	1.302	-2.234	-0.536	0.936	1.776	1.188	1.972	3.954	1.052	1.974	0.693	1.079	1.626	-0.281	-0.373
9	-0.240	-1.338	-0.096	-0.252	-1.728	-1.458	1.001	2.189	1.319	2.153	4.491	1.314	2.294	0.755	1.148	1.751	-0.226	-0.226
10	-0.214	-1.324	-0.071	-0.100	-1.780	-1.158	0.981	2.142	1.353	2.156	4.461	1.328	2.310	0.734	1.149	1.738	-0.214	-0.251
11	0.434	-0.663	0.578	1.064	-0.806	0.274	1.001	2.189	1.319	2.153	4.491	1.314	2.294	0.755	1.148	1.751	-0.226	-0.226
12	0.591	-0.685	0.420	1.222	-0.823	0.224	1.001	2.189	1.319	2.153	4.491	1.314	2.294	0.741	1.161	1.765	-0.226	-0.226
13	0.397	-1.494	-0.253	-0.588	-2.455	-0.685	1.001	2.189	1.319	2.153	4.491	1.314	2.294	0.755	1.148	1.751	-0.226	-0.226
14	-0.256	-1.534	-0.427	-0.425	-2.478	-0.670	1.001	2.189	1.319	2.153	4.491	1.314	2.294	0.741	1.161	1.765	-0.226	-0.226
15	0.855	0.855	0.855	1.707	1.704	0.949	0	0	0	0	0	0	0	0	0	0	0	0

actual results of computation each member internal force and deformation are of normal values.

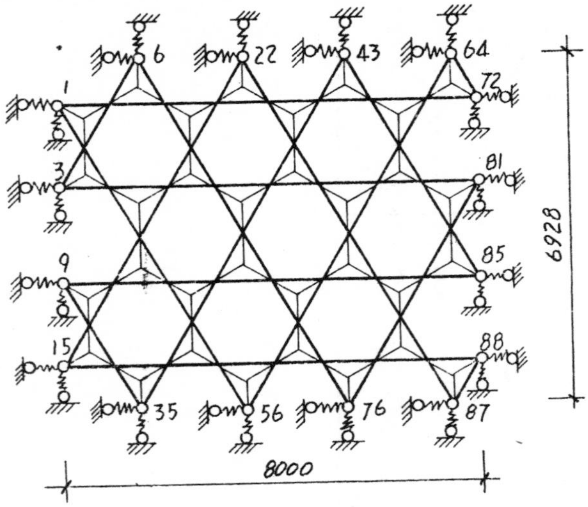

Fig 10

By analyzing and comparing with schemes 1, 5, 11 and 13, it can be seen that when loading remain constant, horizontal restraints of supports have no influence on the internal forces of lower chords and web members and only affect the internal forces of upper chords. To prove this, we removed a part of the horizontal restraints in scheme 11 to form the restraint conditions of Fig 7, using the horizontal reactions of supports as external loading to replace the removed support restraints and also removed the vertical loading for scheme 15 for computation; the results showed that only upper chords produce internal forces and the internal forces of lower chords and web members were all zero, which further proved that horizontal restraints have no effect on lower chords and web members.

CONCLUSION

Through the above theoretical analysis and actual computations we have reached the following conclusions:

1. We cannot say in general if a honeycomb triangular pyramid plate type grid structure is a geometrically changeable system or not, because it is closely connected with the restraint conditions of supports. With the restraint conditions of schemes 1 and 5, the whole structure is a statically determinate and geometrically unchangeable system; with the restraint conditions of scheme 10, the whole structure is a geometrically changeable system; while schemes 11 and 13 are of three-dimensioned fixed restraints and three-dimensioned elastic restraint, the whole structure is a statically indeterminate, and geometrically unchangeable system.

2. When each support has a vertical restraint, the judging if a system is geometrically changeable, statically determinate or indeterminate is decided by the conditions of mutual connection of members in the upper chord plane and the conditions of horizontal restraints. The kinematic analysis of geometrical unchangeability in space is simplified into planar kinematic analysis of the upper chord plane. So long as the arrangement patterns of upper chords and horizontal restraints are so planned as to make the upper chord plane a geometrically unchangeable system, no increase of any member at periphery is needed to ensure geometrical unchangeability.

3. In practice, the support restraints of a periphery - supported system are mostly three-dimensioned elastic restraints, presenting a geometrically unchangeable, statically indeterminate system; hence, no increase of any member at the periphery is necessary to ensure geometrical unchangeability.

REFERENCES

1. LIU XI LIANG and LIU YI XUAN, Design of Flat Space Frames, China Building Industry Publishing House, September 1979.

2. Guidelines for Structural Design and Construction of Space Frames, JGJ-7-80, China Building Industry Publishing House, 1980.

UNE VERSION AMELIOREE DES THEOREMES DES VARIATIONS STRUCTURALES

A.A. FILALI, P. LAC, P. MORLIER

Laboratoire de Génie Civil de
l'Université de Bordeaux I
I.U.T.A de Bordeaux
33405 TALENCE CEDEX

Les théorèmes des variations structurales, sous leurs différentes formulations plus ou moins élaborées, ont pour but de permettre le calcul d'une structure à partir de celui d'une structure voisine sans réanalyse complète. On présente ici une version améliorée de ces théorèmes qui, grâce à l'introduction des équations d'équilibre, conduit à un calcul dont les matrices ont la taille de la matrice de rigidité réduite de l'élément. L'application aux treillis tridimensionnels est explicitée.

I. INTRODUCTION

Les théorèmes des variations structurales donnent la manière de calculer les répercussions de la variation de raideur d'un élément sur les efforts et déplacements dans la structure toute entière ; ils fournissent la réponse d'une nouvelle structure sans calculer ni inverser la matrice de rigidité globale de celle-ci et ceci quelquesoit la variation de raideur de l'élément qu'il faut modifier.

MAJID et ELLIOT, voir Réf. 1, établirent ces théorèmes en 1973 pour des structures en treillis. Le raisonnement adopté par ces auteurs ne débouchant pas sur une formulation matricielle, leur application aux cadres rigides par CELIK voir Réfs 2, ne fut possible qu'en traitant séparément les variations de chacune des variables de base (section, moments d'inertie).

Finalement une formulation rigoureuse du problème fut proposée par ANTOUN et MORLIER, voir Réfs 3-4,. Les expressions qui en résultent sont matricielles : elles sont, d'une part, applicables à tout type de structure bi ou tridimensionnelles et, d'autre part, permettent le traitement simultané des variations de chacune des caractéristiques de l'élément ainsi que des variables topologiques.

Sous leur forme actuelle, l'application des théorèmes des variations structurales nécessitent de connaître la réponse de la structure sous les divers cas de chargement unitaire appliqués successivement suivant les degrés de liberté de l'élément à modifier et découle en un calcul matriciel de la taille de la matrice de rigidité élémentaire.

Le but de cette communication est de présenter une amélioration qui conduira à des calculs plus rapides : moins de cas de charge seront nécessaires et les calculs matriciels seront de la taille de la matrice de rigidité réduite. Ce résultat peut être obtenu en introduisant les équations d'équilibre des éléments dans la théorie proposée par ANTOUN et MORLIER.

II. EQUATIONS FONDAMENTALES

Notre étude portera sur une structure élastique chargée aux noeuds.

Sous le chargement réel, on note $M_i = (M_{il} ; M_{ir})^T$ le vecteur des efforts locaux exercés par le noeud origine i^l et par le noeud extrémité i^r sur la barre i et $\theta_i = (\theta_{il} ; \theta_{ir})^T$ les déplacements dans lesquels travaillent ces efforts.

II.1. Equations d'équilibre des noeuds

R_i étant la matrice de rigidité de l'élément i, nous avons

$$M_i = R_i \theta_i \qquad (1)$$

Appelant $- M_i^\circ = - (M_{il}^\circ ; M_{ir}^\circ)^T$ les forces de blocage des noeuds i^l et i^r dues au chargement extérieur, nous écrivons, suivant ANTOUN et MORLIER, l'équilibre des noeuds i^l et i^r sous la forme :

$$(R_i + R_i^s)\ \theta_i - M_i^\circ = 0 \qquad (2)$$

où R_i^s est une matrice de rigidité associée au reste de la structure.

Soit M_i^s les efforts définis par

$$M_i^s = R_i^s \theta_i \qquad (3)$$

On a alors $M_i + M_i^s = M_i^\circ$ (4)

II.2. Equilibre de l'élément

Les composantes de M_i doivent être en équilibre et dès lors ne sont pas linéairement indépendantes.

En ne considérant que le vecteur réduit des déplacements inconnus θ_i^r et le vecteur réduit des forces généralisées indépendantes associées M_i^r , Eqn 1 devient :

$$M_i^r = R_i^r \, \theta_i^r \qquad (5)$$

Le vecteur réduit des forces généralisées M_i^r est relié au vecteur complet par l'équation matricielle d'équilibre :

$$M_i = L_i \, M_i^r \qquad (6)$$

et selon le principe des forces virtuelles :

$$\theta_i^r = L_i^T \, \theta_i \qquad (7)$$

Il est alors facile de passer de la matrice de rigidité réduite à la matrice de rigidité complète par la relation :

$$R_i = L_i \, R_i^r \, L_i^T \qquad (8)$$

II.3. Matrice de répartition réduite

Pour le reste de la structure, la variation de raideur de l'élément i est équivalente à l'application de forces extérieures en équilibre aux noeuds i^L et i^r. Il est donc nécessaire pour la suite d'établir des relations liant les efforts M_i^s à de telles forces F_i° . Pour un tel chargement, Eqn 4 devient :

$$M_i + M_i^s = F_i^\circ \qquad (9)$$

Supposons que les forces appliquées soient en équilibre, c'est-à-dire de la forme :

$$F_i^\circ = L_i \, F_i^{\circ r} \qquad (10)$$

dès lors les composantes de M_i^s ne sont plus linéairement indépendantes et l'on peut définir $M_i^{s\,r}$ tel que

$$M_i^s = L_i \, M_i^{sr} \qquad (11)$$

et Eqn 9 devient, sous forme réduite,

$$M_i^r + M_i^{sr} = F_i^{\circ r} \qquad (12)$$

Sauf cas particulier d'hypostaticité d'une sous-structure associée à l'élément i, la matrice R_i^s est inversible, aussi peut-on écrire :

$$\theta_i = (R_i^s)^{-1} \, M_i^s \qquad (13)$$

En multipliant à gauche par L_i^T et se rappelant les relations (7) et (11) ,

$$M_i^{sr} = R_i^{sr} \, \theta_i^r \qquad (14)$$

$$\text{avec } R_i^{sr} = \{ L_i^T \, (R_i^s)^{-1} \, L_i \}^{-1} \qquad (15)$$

Substituons dans Eqn 12 M_i^r et M_i^{sr} par leurs valeurs données Eqns 5 et 14, on obtient :

$$\theta_i^r = (R_i^r + R_i^{sr})^{-1} \, F_i^{\circ r} \qquad (16)$$

soit en multipliant par R_i^{sr}

$$M_i^{sr} = \rho_i^{sr} \, F_i^{\circ r} \qquad (17)$$

avec la matrice de répartition réduite pour le reste de la structure :

$$\rho_i^{sr} = R_i^{sr} \, (R_i^r + R_i^{sr})^{-1} \qquad (18)$$

De même en multipliant par R_i^r :

$$M_i^r = \rho_i^r \, F_i^{\circ r} \qquad (19)$$

avec la matrice de répartition réduite pour l'élément i :

$$\rho_i^r = R_i^r \, (R_i^r + R_i^{sr})^{-1} \qquad (20)$$

Notons que

$$\rho_i^r + \rho_i^{sr} = I^r \qquad (21)$$

où I^r est la matrice identité réduite.

Selon Eqn 19, la construction numérique de la matrice ρ_i^r et de son complément à l'unité ρ_i^{sr} est aisée : en effet, il suffit d'imposer successivement à la structure, aux noeuds i^L et i^r extrémités de l'élément à modifier, les forces en équilibre :

$$F_i^{\circ r} = (1,0,\ldots,0)^T \; ; \; (0,1,0,\ldots,0)^T \; ; \; \ldots \; ; \; (0,0,\ldots,0,1)^T$$

Les efforts réduits M_i^r constituent alors les colonnes de ρ_i^r

III. THEOREMES DES VARIATIONS STRUCTURALES

Supposons que la barre i subisse une variation quelconque de raideur (R_i devient R_i^*), le reste de la structure et le chargement restant inchangés. Nous allons d'abord étudier les répercussions de cette variation sur les efforts M_i^r et les déplacements θ_i^r qui deviennent M_i^{r*} et θ_i^{r*}.

III.1. Variation des efforts et déplacements réduits dans la barre qui subit une variation de raideur : premier théorème

Supposons que la barre i subisse une variation de raideur $\Delta R_i^r = R_i^{r*} - R_i^r$. Le chargement restant inchangé, les composantes de ΔM_i^s sont en équilibre : en effet, d'après Eqn 4 :

$$\Delta M_i^s = - \; \Delta M_i \qquad (22)$$

Dès lors il est possible de définir un vecteur réduit ΔM_i^{sr} tel que

$$\Delta M_i^s = L_i \; \Delta M_i^{sr} = - \, L_i \; \Delta M_i^r \qquad (23)$$

Notons $\Delta M_i^r = M_i^{r*} - M_i^r$; compte tenu de Eqn 5, on a :

$$\Delta M_i^r = R_i^{r*} \, \theta_i^{r*} - R_i^r \, \theta_i^r \qquad (24)$$

soit, en ajoutant la quantité $(R_i^r \, \theta_i^{r*} - R_i^r \, \theta_i^{r*})$:

$$\Delta M_i^r = \Delta R_i^r \, \theta_i^{r*} + R_i^r \, \Delta\theta_i^r \qquad (25)$$

D'autre part, le reste de la structure restant inchangé, $\Delta R_i^s = 0$ et d'après les Eqns 13 et 23 et la définition Eqn 15 :

$$\Delta\theta_i^r = - \, (R_i^{sr})^{-1} \, \Delta M_i^r \qquad (26)$$

En y reportant cette valeur, Eqn 25 devient :

$$(R_i^{sr} + R_i^r) \; (R_i^{s\,r})^{-1} \; \Delta M_i^r = \Delta R_i^r \, \theta_i^{r*} \qquad (27)$$

soit, avec Eqn 24, et compte tenu de la définition (20) :

$$\theta_i^{r*} = (R_i^r + \rho_i^r \; \Delta R_i^r)^{-1} \; R_i^r \; \theta_i^r \qquad (28)$$

relation qui exprime le nouveau vecteur réduit des déplacements de l'élément i en fonction de la variation de la matrice de raideur réduite de ce même élément.

En multipliant à gauche l'expression (28) par R_i^{r*},on obtient la loi de transformation du vecteur réduit des efforts dans l'élément i :

$$M_i^{r*} = K_i^r \; M_i^r \qquad (29)$$

$$\text{avec } K_i^r = R_i^{r*} \; (R_i^r + \rho_i^r \; \Delta R_i^r)^{-1} \qquad (30)$$

La matrice K_i^r, qui est de la taille de la matrice de rigidité réduite, est appelée matrice de répercussion réduite de la variation de raideur de l'élément i sur lui-même. Pour calculer K_i^r , il suffit de connaître R_i^r , R_i^{r*} et ρ_i^r .

Existence de K_i^r : on peut affirmer que K_i^r existe si l'on ne modifie que les caractéristiques géométriques de l'élément ou si on lui ajoute des degrés de liberté : ces deux cas correspondent à une variation matricielle de raideur de l'élément de la forme :

$$R_i^{r*} = \beta_i \; R_i^r \qquad (31)$$

Sans perdre en généralité donc, la relation 30 se simplifie en introduisant la matrice β_i pour devenir :

$$K_i^r = \beta_i \{ I^r + \rho_i^r \; (\beta_i - I^r) \}^{-1} \qquad (32)$$

III.2. Variation des efforts et déplacements réduits dans le reste de la structure : deuxième théorème

Selon le premier théorème, la variation de raideur de l'élément i crée dans celui-ci une variation d'efforts $\Delta M_i^r = (K_i^r - I^r) \; M_i^r$ et d'après la relation d'équilibre 23, $\Delta M_i^{sr} = - \Delta M_i^r$.

Or selon la loi de répartition Eqn 17, un telle variation ΔM_i^{sr} est équivalente pour le reste de la structure à l'application d'efforts :

$$m_i^r = (\rho_i^{sr})^{-1} \; \Delta M_i^{sr} = - \; (\rho_i^{sr})^{-1} \; \Delta M_i^r \qquad (33)$$

aux extrémités de la barre i.

Dès lors, si on connaît la matrice $\rho_{j,i}^r$, dont les colonnes sont constituées des composantes des vecteurs réduits M_j^r sous les divers cas de charge $F_i^{\circ r}$, l'application du principe de superposition donne la loi de variation des efforts et déplacements réduits dans l'élément j :

$$M_j^{r*} = M_j^r + \rho_{j,i}^r \; m_i^r \qquad (34)$$

Et pour les déplacements réduits :

$$\theta_j^{r*} = \theta_j^r + (R_j^r)^{-1} \; \rho_{j,i}^r \; m_i^r \qquad (35)$$

Notons que l'on peut ainsi calculer la variation d'un effet quelconque E_k autre que ceux inventoriés dans les vecteurs M_i^r et θ_i^r ; il suffit de connaître la matrice $E_{k,i}$ associée à cet effet sous les divers cas de charge $F_i^{\circ r}$.

IV. EXEMPLES

Le premier exemple illustrera les divers cas de chargement nécessaires à l'application de notre théorie dans le cas d'une structure tridimensionnelle.

Le deuxième exemple traite du cas du treillis ; en effet, pour cette structure particulière les relations fondamentales sont scalaires et revêtent donc un certain attrait.

IV.1. Structure tridimensionnelle

Considérons un élément d'indice i d'une structure tridimensionnelle :

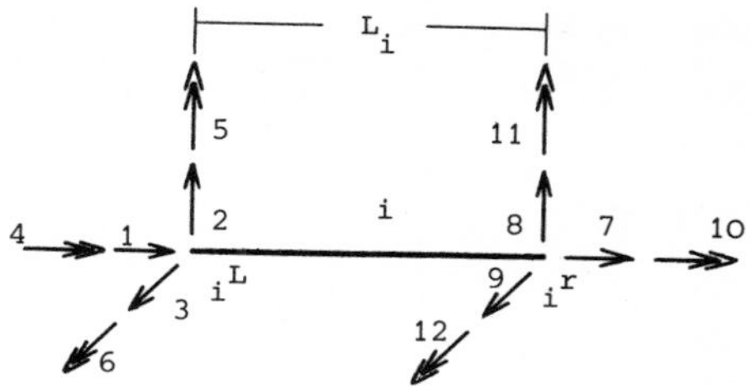

Compte tenu des conventions de signes adoptées sur la figure, les équations d'équilibre s'écrivent :

$$\left.\begin{aligned} F_7 &= - F_1 \\ F_8 &= - F_2 \\ F_9 &= - F_3 \\ F_{10} &= - F_4 \\ F_{11} &= - F_5 - l_i \; F_3 \\ F_{12} &= - F_6 + l_i \; F_2 \end{aligned}\right\} \qquad (36)$$

Supposons que l'on choisisse $M_i^r = (F_1 \; F_2 \; ... \; F_6)^T$, compte tenu de Eqn 36 la matrice d'équilibre L_i définie en Eqn 6 s'établit aisément et les divers cas de chargement nécessaires à la construction numérique de ρ_i^r (Eqn 19) sont les suivants :

1er cas de charge

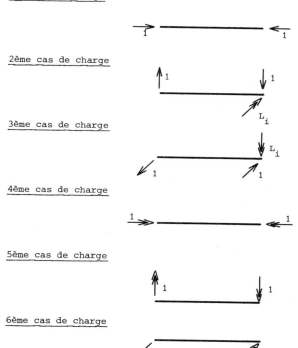

2ème cas de charge

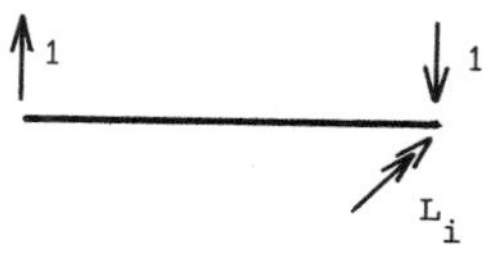

3ème cas de charge

4ème cas de charge

5ème cas de charge

6ème cas de charge

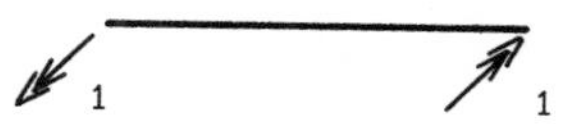

IV.2. Structures en treillis

Pour cette structure particulière, les relations fondamentales sont scalaires. Choisissons $M_i^r = N_{i1} = N_i$ de telle sorte qu'un effort soit positif lorsque la barre est comprimée.

Appelons f_i et f_{ji} les efforts dans les barres i et j lorsque l'élément i est comprimée par deux forces unitaires : les expressions (32) et (33) deviennent :

$$K_i^r = \frac{\beta_i}{1 + f_i\ (\beta_i - 1)} \qquad (37)$$

$$m_i^r = \frac{1 - \beta_i}{1 + f_i\ (\beta_i - 1)} \qquad (38)$$

avec $\beta_i = A_i^*/A_i$ où A_i^* est la nouvelle section de l'élément i.

Les nouveaux efforts dans la structure s'expriment simplement suivant :

$$N_i^* = \frac{\beta_i}{1 + f_i\ (\beta_i - 1)} N_i \qquad (39)$$

$$\forall_j \neq i \quad N_j^* = N_j + \frac{(1-\beta_i)\ f_{ji}}{1 + f_i\ (\beta_i - 1)} N_i \qquad (40)$$

V. CONCLUSION

En écrivant les grandeurs dans leur repère local lié aux éléments puis en introduisant les équations d'équilibre statique des barres, on a montré que pour passer de la solution d'une structure à une structure voisine, il suffit de connaître la réponse de la structure sous des cas de chargement unitaires combinés appliqués aux extrémités de l'élément à modifier. Il en résulte un calcul matriciel plus simple, de la taille de la matrice de rigidité réduite de l'élément.

BIBLIOGRAPHIE

1. K.I. MAJID and D.W.C. ELLIOT - Forces and deflections in changing structures, Struc. Engrs, vol. 51, n° 3,1973.

2. K.I. MAJID, M.P. SAKA and T. CELIK - The theorems of structural variation generalized for rigidly jointed frames, Proc.Inst.Civ.Engrs, part. 2, 1978.

3. N. ANTOUN - Théorèmes sur les structures voisines et applications, thèse n° 702, Université de Bordeaux I, 1981.

4. N. ANTOUN and P. MORLIER - Théorie des structures voisines, J.Mec.Theo.Appl., vol.2, n° 1, 1983.

SPACE FRAMES DESIGNED AS 'TWISTLESS CASE'

W J BERANEK * and G J HOBBELMAN **

* Professor, Department of Architecture,
Technological University, Delft.

** Senior Scientific Officer,
Department of Architecture,
Technological University, Delft.

A point supported space frame under uniformly distributed load shows an irregular stress distribution, more or less as in a flat plate. A very regular stress distribution can be obtained by using a known plate solution: the 'twistless case'. Strips of modules of greater stiffness are applied between the supports, which act as 'hidden beams'. In this way a stress distribution is enforced in which the space frame acts as two systems of independent elementary beams. Load transfer in x- and y-direction is controled by the designer and equations are given to determine the required 'beam stiffnesses'. Calculations can easily be carried out by hand and some examples are given.

INTRODUCTION

The stress distribution in space frames of known dimensions can be determined by numerical methods, using computers with a large memory capacity. As a starting point the cross-sectional area of all members may be chosen the same. On the basis of the calculated forces in the members, the actual cross-sectional areas of all members are selected from a limited number of commercially available tubes. As a result the stiffer parts of the space frame will attract greater forces. A second calculation, and possibly an adaption of the cross-sectional areas, will be necessary.

For design purposes, however, the space frame can also be regarded as an isotropic or orthotropic plate with bending stiffnesses in x- and y-direction and with or without a torsional stiffness, depending on the arrangement of the members. A calculation can be carried out by means of the finite element method, using elements of much larger dimensions than those of the modules of the corresponding space frame. If desired, parts of the plate can be given greater bending stiffnesses. Especially in the case of an isotropic plate, the solution of various problems is already known from literature.

In the case of point supports, the stress distribution in the space frame resembles more or less the stress distribution in a flat plate or a flat slab. The bending moments and the shear forces in the plate reach extreme values around the supports, see Fig 1.
To obtain a more regular stress distribution, another solution can be used which is known from plate theory: the so called 'twistless case', see Ref 1.
If beams of properly chosen bending stiffnesses are applied between the point supports, the deflection of the beams will be the same as those of parallel plate strips, see Fig 3. The plate will then act as two systems of elementary beams, without any twisting moments. All plate strips in one direction will have the same stress distribution, see Fig 2. The most important thing, however, lies in the fact that the designer can control the load transfer in the two directions by changing the beam stiffnesses.

In the case of the space frame, 'hidden beams' can be created by applying modules of greater stiffness between the supports. These 'hidden beams' also provide the required strength and will show a similar moment distribution as the parallel strips of normal stiffness. So a solution is obtained with an almost uniform stress distribution which can easily be calculated by hand.

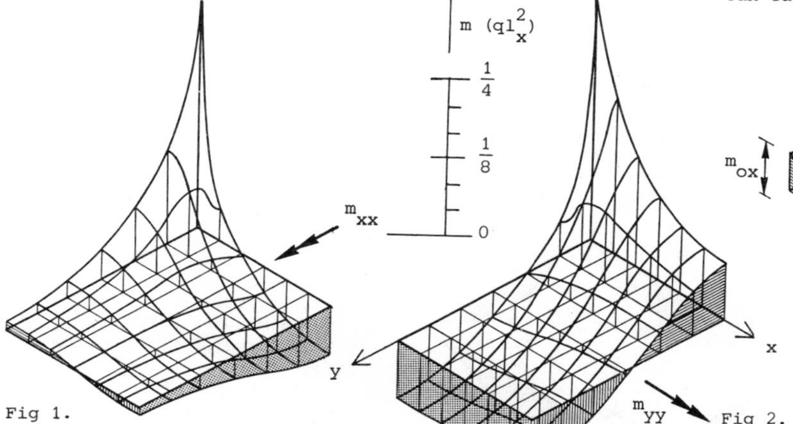

Fig 1.
Moment distribution in a flat plate.
$\lambda = l_y/l_x = 1.4$

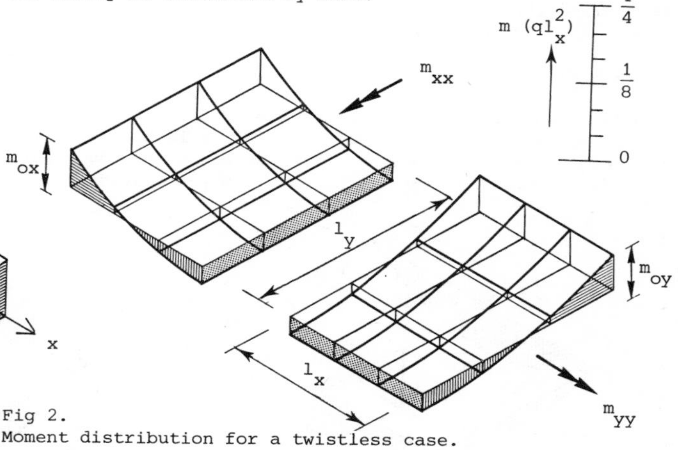

Fig 2.
Moment distribution for a twistless case.
$\lambda = l_y/l_x = 1.4, \quad \alpha = \lambda^2/(1+\lambda^2)$, so $m_{ox} = m_{oy}$.

TWISTLESS CASE FOR AN ISOTROPIC PLATE

The relationship between moments and curvatures in an isotropic plate is given by:

$$\begin{vmatrix} m_{xx} \\ m_{yy} \\ m_{xy} \end{vmatrix} = -D \begin{vmatrix} 1 & \nu & 0 \\ \nu & 1 & 0 \\ 0 & 0 & 1-\nu \end{vmatrix} \begin{vmatrix} \dfrac{\partial^2 w}{\partial x^2} \\ \dfrac{\partial^2 w}{\partial y^2} \\ \dfrac{\partial^2 w}{\partial x \partial y} \end{vmatrix} \qquad \ldots\ldots 1.$$

The stress distribution is governed by the differential equation:

$$\frac{\partial^4 w}{\partial x^4} + 2\frac{\partial^4 w}{\partial x^2 \partial y^2} + \frac{\partial^4 w}{\partial y^4} = \frac{f}{D} \qquad \ldots\ldots 2.$$

in which:

$D = Et^3/12(1-\nu^2)$ = flexural rigidity of the plate
E = modulus of elasticity
ν = Poisson's ratio
t = plate thickness
f = intensity of uniformly distributed load

To solve the differential equation, the boundary conditions have to be fulfilled. If edge beams are regarded, one has to do with elastically supported and elastically built in edges and the flexural rigidity as well as the torsional rigidity of the beams has to be taken into account.
For an edge parallel to the y-axis one finds for the bending of the edge beam, see Ref 2:

$$B\frac{\partial^4 w}{\partial x^4} = D\frac{\partial}{\partial x}\left\{\frac{\partial^2 w}{\partial x^2} + (2-\nu)\frac{\partial^2 w}{\partial y^2}\right\} \qquad \ldots\ldots 3.$$

For the torsion of the edge beam one finds:

$$-C\frac{\partial}{\partial y}\frac{\partial^2 w}{\partial x \partial y} = D\left\{\frac{\partial^2 w}{\partial x^2} + \nu\frac{\partial^2 w}{\partial y^2}\right\} \qquad \ldots\ldots 4.$$

in which:

B = flexural rigidity of the beam
C = torsional rigidity of the beam

To obtain the required solution for the twistless case, the edge beams should have no torsional rigidity, i.e. $C = 0$. The flexural rigidity should be chosen in such a way that the deflection surface of the plate will become a translation surface, i.e. a surface obeying the equation:

$$w = F(x) + G(y) \qquad \ldots\ldots 5.$$

In Fig 3 two examples of a translation surface are given. It is easily seen that all parallel strips in the plate in x-direction have the same deflected shape as the edge beams in x-direction. The same is true for the y-direction.

As all mixed derivatives of Eq 5 are zero, Eq 3 will become:

$$B\frac{\partial^4 w}{\partial x^4} = D\frac{\partial^3 w}{\partial x^3} \qquad \ldots\ldots 3'.$$

In Fig 3a the bending moments perpendicular to the edges are zero, so Eq 4 becomes:

$$\frac{\partial^2 w}{\partial x^2} + \nu\frac{\partial^2 w}{\partial y^2} = 0 \qquad \ldots\ldots 4a.$$

In Fig 3b an interior panel of a continuous floor is regarded. Now the boundary is a line of symmetry and Eq 4 is changed into:

$$\frac{\partial w}{\partial x} = 0 \qquad \ldots\ldots 4b.$$

The ideal situation is obtained when = 0. In that case the bending and twisting moments in the plate become:

$$m_{xx} = -D\frac{\partial^2 w}{\partial x^2} = -D\,F''(x)$$

$$m_{yy} = -D\frac{\partial^2 w}{\partial y^2} = -D\,G''(y) \qquad \ldots\ldots 6.$$

$$m_{xy} = -D\frac{\partial^2 w}{\partial x \partial y} = 0$$

The bending moments in x-direction are only dependent on the deflections in x-direction and the bending moments in y-direction only on the deflections in y-direction.
All twisting moments in the plate are equal to zero, which explains the expression 'twistless case'.
So the stress distribution is exactly the same as if the plate would consist of two independent systems of beams in x- and y-direction.

A similar behaviour is assumed in the strip method of Hillerborg. But in that case it is an ultimate load calculation, which gives a lower bound solution as only the equilibrium is regarded and the conditions of compatability are violated. In the twistless case, however, the conditions of equilibrium and compatability are both fulfilled.

When $\nu \neq 0$, the twisting moments remain zero, but there is an interaction between the bending moments in x- and y-direction. In the case of Fig 3a, the plate strips in x- and y-direction will no longer act as independent beams. This is due to the fact that in Eq 4a the value of ν affects the boundary condition. But in the case of Fig 3b the solution is independent of the value of ν, as ν does not appear in Eq 4b. In the differential equation 2 the value of ν is only explicitly present in the flexural rigidity D of the plate. So only the absolute value of the deflections is influenced by ν, but the ratio of the deflections between the various points of the plate remains unchanged. This means that interior panels of a continuous slab will always be able to act according to the twistless case, but that some disturbances may occur in edge panels.

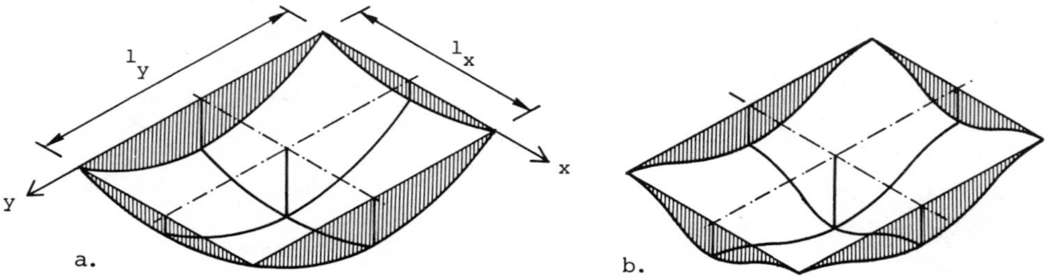

Fig 3.
Plate deflections in the shape of a translation surface.
a. single panel:
bending moments perpendicular to the edges of the plate are zero.
b. interior panel:
rotations perpendicular to the edges of the plate are zero.

Determination of the required bending stiffnesses of the beams for an interior panel of a continuous floor.

Intensity of the load	f	
load transfer in x-direction	$f_x = \alpha f$	...... 7.
load transfer in y-direction	$f_y = (1-\alpha) f$	

To ensure that the deflection of a strip of the plate corresponds with the deflection of a parallel beam, the ratio of load and stiffness should be the same for both. In Fig 4, a plate strip and a beam in x-direction are regarded. As the load intensity of the beam is equal to $f_y l_y$, the condition becomes:

$$\frac{f_x b_y}{\frac{1}{12} b_y t^3} = \frac{f_y l_y}{B_x}, \text{ or } B_x = \frac{f_y}{f_x} \frac{1}{12} l_y t^3 \qquad \text{...... 8.}$$

Instead of regarding a plate strip with width b_y, which transfers its load in x-direction, the width l_y of the whole panel can be regarded.
As 'panelstiffnesses are defined:

$$P_x = \frac{1}{12} l_y t^3$$
$$P_y = \frac{1}{12} l_x t^3 \qquad \text{...... 9.}$$

Now Eq 8 for the beam stiffness in x-direction becomes:

$$B_x = \frac{f_y}{f_x} P_x = \frac{1-\alpha}{\alpha} P_x \qquad \text{...... 10a.}$$

In a similar way the beam stiffness in y-direction becomes:

$$B_y = \frac{f_x}{f_y} P_y = \frac{\alpha}{1-\alpha} P_y \qquad \text{...... 10b.}$$

From Eqs 10a and 10b follows the general condition for the twistless case:

$$B_x . B_y = P_x . P_y \qquad \text{...... 11.}$$

The bending moment diagrams of plate and beams have a parabolic shape, as in the case of a continuous beam over many supports, see Fig 2 and Fig 4.

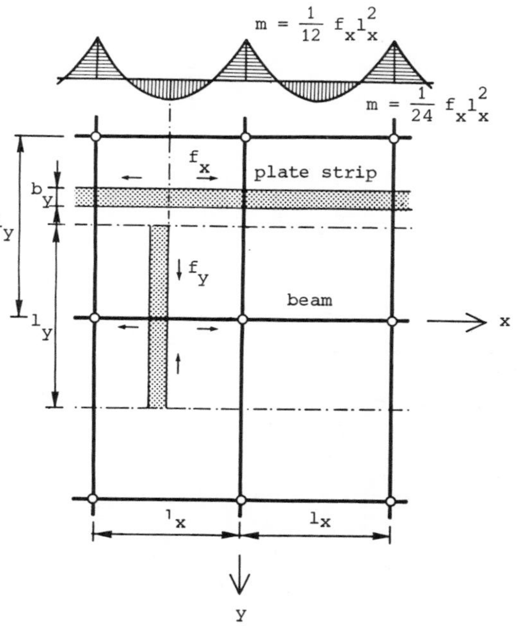

Fig 4.
Twistless case for an interior panel of a continuous floor.

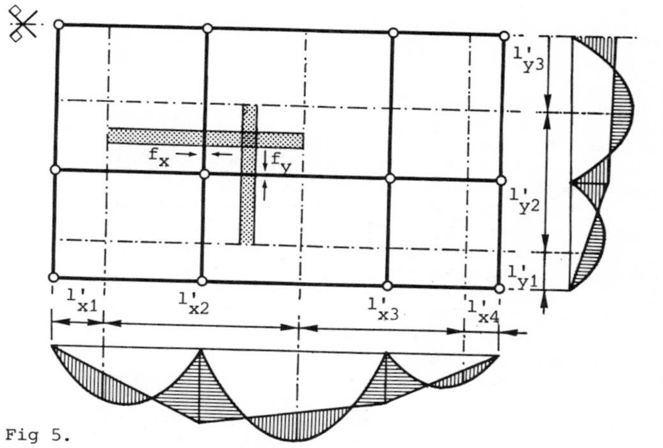

Fig 5.
Moment distribution in a floor with unequal panels, according to the twistless case. The moment distribution is shown for $\alpha = 0.5$.

Even in the case of a rectangular groundplan and floor panels of varying dimensions, the floor can be calculated as a twistless case, provided that the edge beams have no torsional stiffness and that one and the same coefficient α is chosen for all panels, see Ref 3.
This situation is shown in Fig 5. The slab acts in both directions as continuous beams over rigid supports and this action is independent of the choice of the coefficient α. The location of the maximum bending moments is then fixed as well as the load transfer to the various beams.
For the determination of the various 'panelstiffnesses', the areas are now regarded which transfer the load to one set of beams, see Fig 5.

$$P'_x = \frac{1}{12} l'_y t^3$$
$$P'_y = \frac{1}{12} l'_x t^3 \qquad \text{...... 9',}$$

and the general condition becomes:

$$B_x . B_y = P'_x . P'_y \qquad \text{...... 11'.}$$

A line load on the beam can also easily be taken into account. In all calculations it is assumed that the plate does not act as the flange of a beam, see Fig 6a. In the case of a normal beam, as in Fig 6b, the bending stiffness of the beam has to be determined including an effective flange width.
To obtain a more or less equal load transfer in x- and y-direction, the bending stiffnesses of the beam, and consequently the heights, have to be relatively low. So the flange effect has a minor influence on the bending behaviour of the plate as tests on elastic models and models of reinforced concrete have shown, see Ref 3.
'Hidden beams' as in Fig 6c have no substantial effect on the beam stiffness of reinforced concrete slabs, but are fully active in the case of space frames, see Fig 6d.

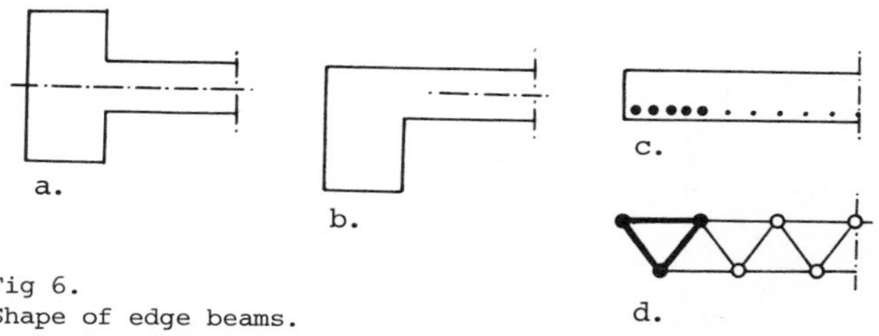

Fig 6.
Shape of edge beams.
a. beam symmetrical in respect to the neutral plane of the plate; no flange effect.
b. normal beam, considerable flange effect.
c. 'hidden beam' in a reinforced concrete slab, the bending stiffness is small compared with the strength.
d. 'hidden beam' in a space frame; the bending stiffness is proportional to the strength.

TWISTLESS CASE FOR AN ORTHOTROPIC PLATE

To ensure that a space frame can act as a twistless case, it is necessary that 'hidden beams', preferably without torsional stiffness, can be applied between the supports. This means that most of the commonly used member arrangements are suitable, see Fig 7a,b.
When square grids are applied, the space frame can not be regarded as an isotropic plate as the torsional stiffness is lacking. The bending stiffnesses in x- and y-direction, however, are the same and there is no interaction between the bending moments in x- and y-direction.
In the case of Fig 7c the bending stiffness is the same in all directions and there exists also a torsional stiffness, just as in an isotropic plate. There is also interaction between the bending moments in x- and y-direction.

In certain cases it might be economic to apply different bending stiffnesses in x- and y-direction, so this possibility will be taken into account. Membrane action is neglected, i.e. the 'neutral plane' is assumed to be the middle plane of the space frame.
In the customary theory of orthotropic plates the shear stiffness of the plate is neglected, leading to the following relationship between moments and curvatures, see Ref 2.

$$\begin{vmatrix} m_{xx} \\ m_{yy} \\ m_{xy} \end{vmatrix} = - \begin{vmatrix} D_{xx} & D' & 0 \\ D' & D_{yy} & 0 \\ 0 & 0 & D_{xy} \end{vmatrix} \begin{vmatrix} \frac{\partial^2 w}{\partial x^2} \\ \frac{\partial^2 w}{\partial y^2} \\ \frac{\partial^2 w}{\partial x \partial y} \end{vmatrix} \qquad \text{...... 12.}$$

The stress distribution is governed by the differential equation:

$$D_{xx} \frac{\partial^4 w}{\partial x^4} + 2(D'+D_{xy}) \frac{\partial^4 w}{\partial x^2 \partial y^2} + D_{yy} \frac{\partial^4 w}{\partial y^4} = f \qquad \text{.. 13.}$$

For the grid geometries of Fig 7a,b and similar geometries one finds:

$$D' = 0$$

$$D_{xy} = 0$$

So for these geometries the twistless case is fully applicable for interior panels as well as edge panels.
For geometries like the one in Fig 7c, however, one finds that $D' \neq 0$ and $D_{xy} \neq 0$. Now the coefficient D' is implicitely present in the governing differential equation and one may expect for interior panels and edge panels some interaction between the bending moments in x- and y-direction.

In trusses and space frames the shear stiffness is determined by the deformation of the bracings and may not be neglected. The influence of the shear stiffness on isotropic plates is treated in Ref 2 and on orthotropic plates in Ref 4.
A simplified solution for beams is given in Ref 5, but this solution can not easily be extended to orthotropic plates and consequently not to space frames. The assumptions of Ref 5, however, are exactly valid for trusses, see Fig 8:
- Plane sections remain plane (due to bending).
- Plane sections are tilted towards the neutral axis (due to shear forces).

In an attempt to obtain solutions according to Eq 5, these results can be used as plate strips and parallel beams both will act as elementary beams. So a description of the behaviour of beams or trusses under the combined action of bending moments and shear forces might be useful.

The relationship between internal forces and deformations is given by:

$$M = EI \frac{d\theta}{dx} \qquad \text{...... 14,}$$

$$Q = K \left(\frac{dw}{dx} + \theta\right) \qquad \text{...... 15.}$$

in which:
M = bending moment
Q = shear force
w = deflection
θ = rotation of a plane section towards the z-axis
EI = bending stiffness of the beam
K = shear stiffness of the beam
f = uniformly distributed load per unit length

The stress distribution is governed by the differential equation:

$$EI \frac{d^4 w}{dx^4} = f - \frac{EI}{K} \frac{d^2 f}{dx^2} \qquad \text{...... 16.}$$

The quantities M, Q and θ can all be expressed in w:

$$M = - EI \left\{ \frac{f}{K} + \frac{d^2 w}{dx^2} \right\} \qquad \text{...... 17,}$$

$$Q = - EI \left\{ \frac{1}{K} \frac{df}{dx} + \frac{d^3 w}{dx^3} \right\} \qquad \text{...... 18,}$$

$$\theta = - \frac{EI}{K} \left\{ \frac{1}{K} \frac{df}{dx} + \frac{d^3 w}{dx^3} \right\} - \frac{dw}{dx} \qquad \text{...... 19.}$$

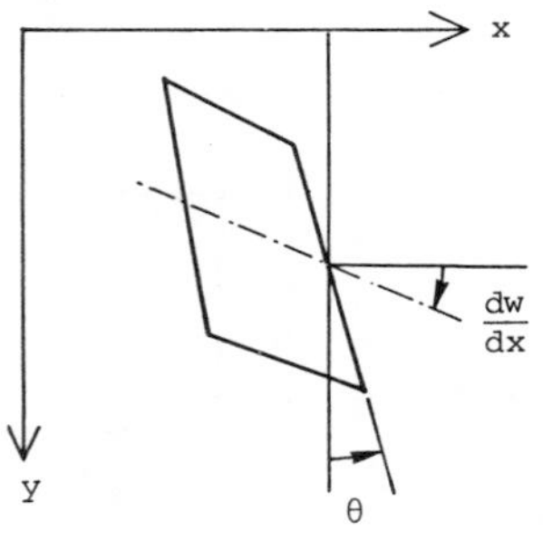

Fig 8.
Deformation of an element due to bending and shear.

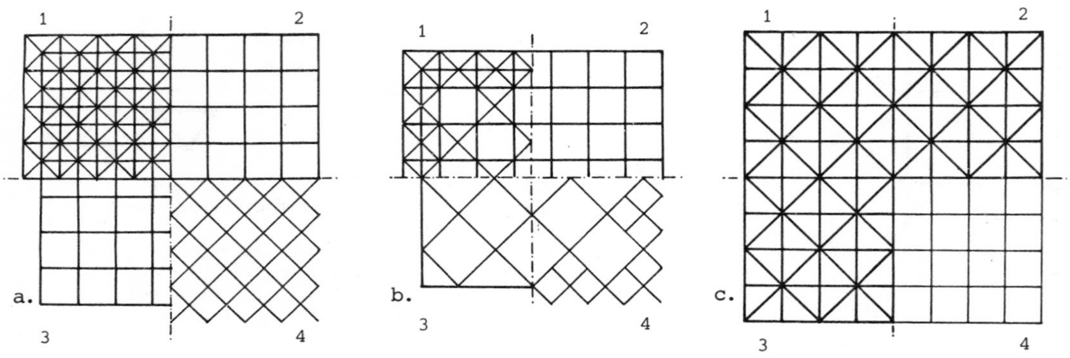

Fig 7.
Some grid arrangements.
a. square on square offset
b. square on diagonal
c. triangle

1 general arrangement
2 top chords
3 bottom chords
4 bracings

APPLICATION OF THE TWISTLESS CASE TO A SPACE FRAME

The calculation will be carried out for an offset grid, according to Fig 7a. To find equations for the twistless case, a strip of modules has been regarded as a 'beam', transferring the load in one direction only, see Fig 9a. A square module has been chosen with all members of equal length. For this statically determinate beam, the cross-sectional area of the members does not influence the stress distribution. In Fig 9b a cross-section through a space frame is shown, together with the cross-sectional areas which belong to an elementary 'plate strip'. In Fig 9c this plate strip has been isolated and the nodal forces due to its own weight have been resolved as shown. If all members are the same, in all transverse chords tensile forces occur with magnitude $0.25\ F\sqrt{2}$. To determine the other member forces, the plate strip can be regarded as two inclined plane trusses as shown in Fig 9d. The chords have a cross-sectional area of $0.5\ A_c$ and the bracings a cross- sectional area of A_b. The loads in the nodal points are equal to $0.25\ F\sqrt{6}$. In this way the calculation of the space frame has been reduced to the calculation of one single plane truss, which can easily be carried out by hand, see Fig 11.

To obtain equations for the required stiffnesses of the 'hidden beams', the regarded panel of the space frame is divided in m strips in x-direction and n strips in y-direction, each having the width of one module, see Fig 10. The 'hidden beams' consist of one or two strips with the same width as the parallel 'plate strips'. In Fig 10 the situation is shown where the 'hidden beams' consist of two 'beam strips'. For all strips each top chord belongs half to one strip and half to the adjacent strip, see Fig 9b. The bottom chord always belongs to one and the same strip. The four bracings per module belong for their whole cross-sectional area to the strips in x-direction and also to the strips in y-direction. Their deformations in x- and y-direction no not influence each other.

The bending stiffness of the strips is only determined by the cross-sectional area of the chords and the distance between them. The cross-sectional area of the bracings has some influence on the deformations, but has no influence on the moment distribution in interior panels and edge panels. The bending stiffness of a plate strip is denoted by B and the bending stiffness of a beam strip by $\bar{B}$:

$$B_x = \frac{1}{2} A_x h^2 \qquad B_y = \frac{1}{2} A_y h^2 \qquad \text{...... 20,}$$

$$\bar{B}_x = \frac{1}{2} \bar{A}_x h^2 \qquad \bar{B}_y = \frac{1}{2} \bar{A}_y h^2 \qquad \text{...... 21.}$$

The calculation is carried out for an uniformly distributed load f. As the beam strips have a much larger cross-sectional area than the plate strips, extra loads $\bar{f}_x$ and $\bar{f}_y$ on the beam strips in x- and y-direction respectively have been taken into account. So the loads are:

	Plate strips	Beam strips
x-direction:	$f_x = \alpha f$	$f_x + \bar{f}_x$
y-direction:	$f_y = (1-\alpha) f$	$f_y + \bar{f}_y$

To determine the bending stiffness of the 'hidden beam' in x-direction, the ratio of load and stiffness of a plate strip has to be equal to the same ratio of the two beam strips, together forming the 'hidden beam', see Fig 10:

$$\frac{f_x b_y}{B_x} = \frac{f_y l_y + (f_x + \bar{f}_x).2b_y}{2\bar{B}_x}$$, so one finds for $\bar{B}_x$:

$$\bar{B}_x = \frac{\frac{n}{2} f_y + f_x + \bar{f}_x}{f_x} B_x = \rho_{2x} B_x \qquad \text{...... 22,}$$

and in a similar way for $\bar{B}_y$:

$$\bar{B}_y = \frac{\frac{m}{2} f_x + f_y + \bar{f}_y}{f_y} B_y = \rho_{2y} B_y \qquad \text{...... 23,}$$

and consequently, using Eqs 20 and 21:

$$\bar{A}_x = \rho_{2x} A_x, \qquad \bar{A}_y = \rho_{2y} A_y \qquad \text{...... 24.}$$

If only one beam strip is used to form the 'hidden beam', the equations for the coefficients ρ_{1x} and ρ_{1y} become:

$$\rho_{1x} = \frac{nf_y + f_x + \bar{f}_x}{f_x}, \qquad \rho_{1y} = \frac{mf_x + f_y + \bar{f}_y}{f_y} \qquad \text{...... 25.}$$

If a single panel is regarded with just one beam strip along the edges, the coefficients ρ'_{1x} and ρ'_{1y} are obtained by replacing m and n in Eqs 25 by: $(m+1)/2$ and $(n+1)/2$ respectively.

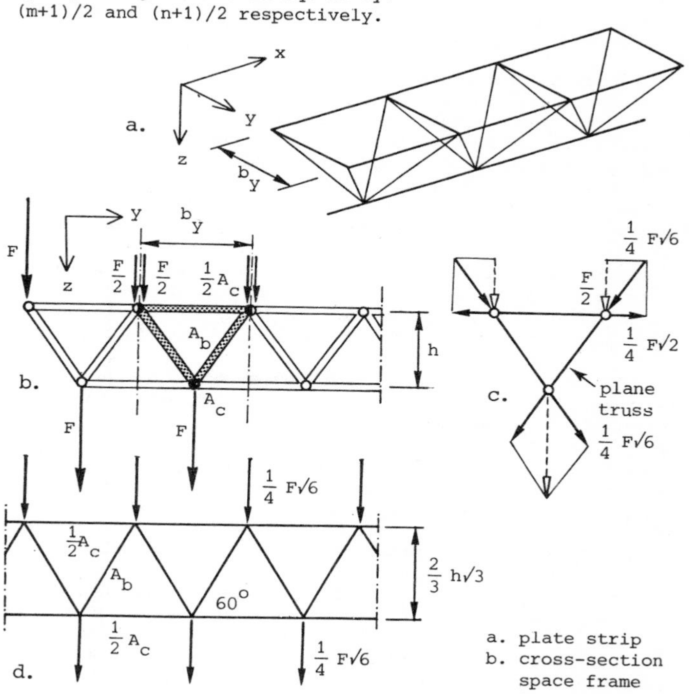

Fig 9.
Plate strip regarded as a 'beam'.

a. plate strip
b. cross-section space frame
c. isolated plate strip
d. plane truss.

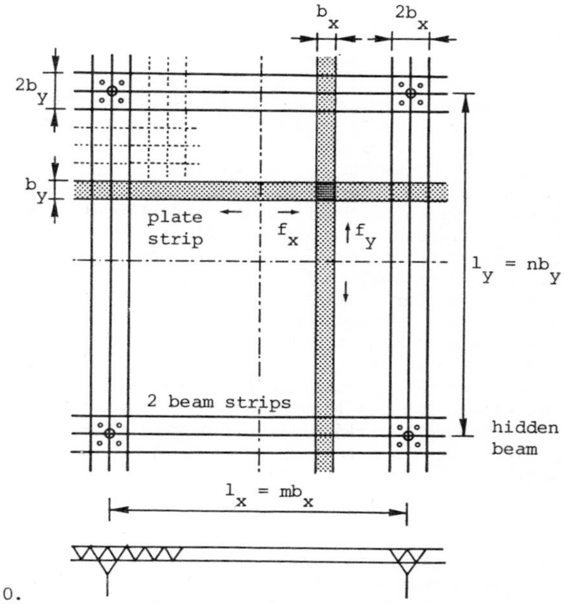

Fig 10.
Internal panel of a space frame acting according to the twistless case (two beam strips per 'hidden beam').

EXAMPLES

To check the given equations, some square and rectangular panels under uniformly distributed load have been analyzed with the help of the computer program ICES STRUDL (square panels with square modules and rectangular panels with square and rectangular modules). Some results are shown in the Figs 12 and 13. Note that the ratio of force and cross-sectional area is the same for all members, so all stresses are the same. The results can be compared with those of an elementary truss, which behaves as a simply supported or a continuous plate strip, see Fig 11. For this last case the situation is shown where all members just possess their required cross-sectional area, so that M/EI = constant and the ratio M_{neg}/M_{pos} = 3.

In Fig 12 the forces in top and bottom chords are given for a single panel. The computer results show a very slight deviation from the theoretical equal stress distribution. This is due to the fact that the unbalanced horizontal forces along the edge strip cause equal tensile forces in all top chords and no forces in the bottom chords, see Fig 9b,c. So the space frame tries to obtain constant curvatures in x- and y-direction, which is opposed by the stiffer beam strips. An adaption of the theoretically required cross-sectional areas to commercially available cross-sectional areas does not essentially alter the stress distribution.
It is also possible to use the same cross-sectional areas for the two top chords of one beam strip, see Fig 12b (*).

In Fig 13 the forces in top and bottom chords are given for an interior panel. Now all results coincide exactly with the theoretical values. For interior panels the total design moment of $0.125\ f\ l_y l_x^2$ has to be divided over the negative moment in section A-A and the positive moment in section B-B. The ratio of these moments depends of the stiffness ratio of the various chord members of one 'plate strip'. This ratio M_{neg}/M_{pos} will globally vary between the values 2 (all chords members with the same cross-sectional area) and 3 (cross-sectional area of the chords adapted to the finally occuring member forces), see Fig 11c. But whatever cross-sectional areas are chosen, the calculation of one plane truss will provide all the required information.
It will be quite clear that the use of varying cross-sectional areas in one 'plate strip' will not disturb the behaviour of the space frame as a twistless case, provided that the ratio ρ between the corresponding members of beam strip and plate strips remains the same.

CONCLUSIONS

A space frame will behave exactly as two independent systems of trusses in x- and y-direction as long as the proper stiffness ratio between the corresponding members of beam strip and plate strips is maintained.
An adaption of the theoretically required cross-sectional areas to commercially available cross-sectional areas does not essentially alter the stress distribution.
In rectangular panels equal design moments in x- and y-direction can be obtained for the 'plate strips' by applying the proper stiffnesses of the beam strips.
All means which are used for continuous beams to obtain a more uniform stress distribution, are also applicable to space frames, e.g. reduction of the end span (l' = 0.33 l√6) or application of a cantilever (l" = 0.67 l√6).
If chess board loading is applied, the four hidden beams per panel behave as rigid (simply supported) edges.
If alternate bays are anti symmetrically loaded, two opposite hidden beams per panel behave as rigid supports.

REFERENCES

1. R H WOOD, Studies in Composite Construction; Part II, The Interaction of Floors and Beams in Multi-Storey Buildings. Research Paper No 22, Building Research Stat.
2. S TIMOSHENKO and S WOINOWSKY-KRIEGER, Theory of Plates and Shells, Mc Graw-Hill Book Company.
3. W J BERANEK, Berekening van Platen; deel 2, Krachtswerking in rechthoekige platen, 1976. TH Delft, Bouwkunde.
4. K GIRKMANN and R BEER, Anwendung der verschärften Plattentheorie nach Erich Reissner auf orthotrope Platten. Österreichisches Ingenieur Archiv, 12 (1958) Nr 1/2.
5. C HARTSUIJKER, Buiging en afschuiving, Rapport 02-81-01. TH Delft, Civiele Techniek.

a. numbering of chords and bracings.
b. forces F_b in the bracings:
$$|F_{bi}| = \Big(2(n+1)+1\Big)P$$

c. forces F_c in the chords:
$$|F_{ci}| = \frac{1}{2}\ P\ \overset{i}{\underset{1}{\Sigma}}\ F_{bi}$$

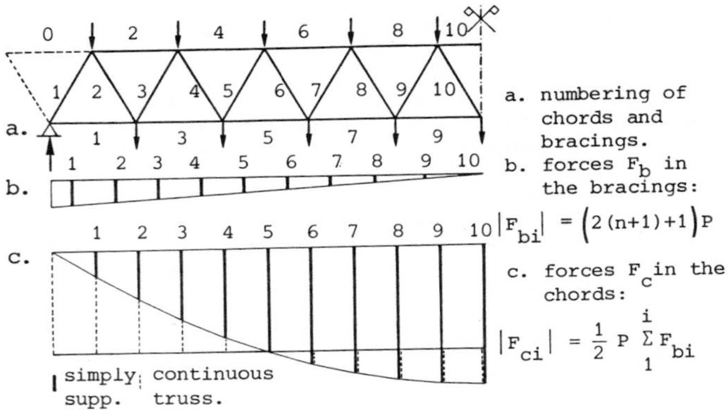

Fig 11.
Plane truss under own weight (force P per nodal point).

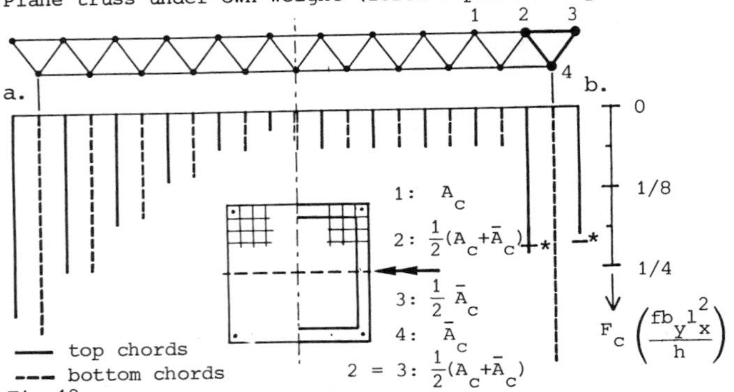

Fig 12.
Single panel; forces in top and bottom chords in section B-B. (A_c = constant)
a. panel without hidden beams.
b. twistless case, (m = n = 10; α = 0.5; ρ = 6.5).

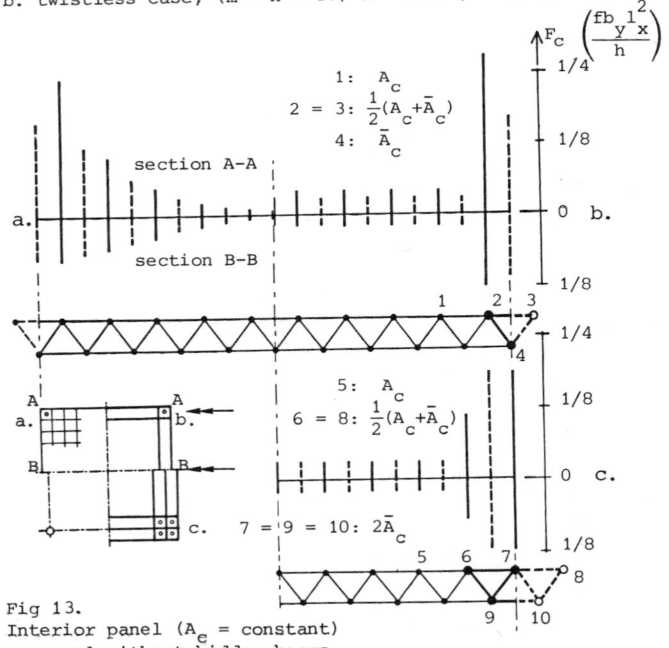

Fig 13.
Interior panel (A_c = constant)
a. panel without hidden beams.
b. twistless case, single beam strip (α = 0.5; ρ = 11).
c. twistless case, double beam strip (α = 0.5; ρ = 6).

GRAPH THEORETICAL APPROACH FOR BAND WIDTH AND FRONT WIDTH REDUCTIONS

A KAVEH*, MSc, PhD, AFIMA and K RAMACHANDRAN**, MSc, PhD, MICE, MIE

* Associate Professor, Iran University of Science and Technology, Tehran.

** Research Fellow, Department of Civil Engineering, Imperial College, London.

This paper discusses various methods of ordering nodes, (finite) elements and cycles of a mathematical model of a structure using graph theory concepts, and relates the algorithms published recently to the authors original work. Further aims of this paper are to identify the relationship between the ordering of a cycle basis and that of a cocycle basis of a graph model, and to introduce some new ideas. Finally, it has been shown that the use of a graph model of the corner node incidence matrix gives better band-widths at less cost.

INTRODUCTION

Both the stiffness method and the flexibility method of structural analysis require the solution of a set of linear system of equations of the form $\underset{\sim}{A}x = b$, where the matrix $\underset{\sim}{A}$ is symmetric, positive definite and usually very sparse. The time required to solve these equations by the banded matrix technique, a popular method among structural analysts, is directly proportional to the square of the bandwidth of the matrix concerned. Since the solution of these equations forms a large percentage of the total computational effort required for structural analysis, it is not surprising that much attention is being paid to the optimisation of the bandwidth of these sparse matrices. With the development of non-linear structural analyses the bandwidth reduction demands high priority because of the iterative nature of non-linear analysis techniques. Another powerful method for solving these equations as met in finite element applications, is the frontal technique of Irons, Ref 1, which is said to have advantages over the banded matrix technique when the finite elements have many side nodes. In this method finite elements are assembled and entered into the solution one at a time. The basis of the frontal technique is to keep a node in active storage until the contributions to the stiffness coefficients are computed and then to discard it. The efficiency of the frontal technique therefore depends on the numbering of the finite elements. Thus it can be seen that a proper automatic numbering scheme of the elements of a mathematical model of the structure which may be nodes, finite elements (in stiffness method) or cycles (in flexibility method) is necessary for both the banded solution technique and the frontal technique to reduce the computational cost substantially. This paper is concerned with the problems associated with the numbering in these two methods and it is not intended to discuss other solution techniques which may or may not require renumbering for efficiency. The problem of profile reduction is also not given much attention in this presentation.

Basically there are two approaches available for band width reduction, viz direct and indirect methods. In the direct method, rows and columns of $\underset{\sim}{A}$ are interchanged according to certain criteria until no further improvement on the bandwidth can be achieved. A complete set of references on both direct and indirect methods can be found in Ref 2. Though the direct methods are attractive due to this simplicity, the computational effort required is too high. The indirect method uses graph theoretical concepts in some form or other to re-number the elements of the graph corresponding to the matrix $\underset{\sim}{A}$ automatically. The authors interest in re-numbering was initially motivated by their experience with shortest route trees (SRT), which they used to generate a cycle basis of a graph model associated with a skeletal structure, see Refs 3, 4, 5 and 6. The observation that the maximum number of vertices in any level (contour) - see definition - of a SRT represents the possible maximum difference between any two adjacent vertex numbers and the challenge to develop an efficient method of reducing the band width of any flexibility matrix led to further independent research by the authors. But it has been subsequently observed that Cuthill and McKee (Ref 5) were the first to use the graph theory in re-numbering and to use a SRT without recognising it to be such. However the authors were the first to propose methods for (1) re-numbering the nodes based on the re-numbered finite elements, (2) re-numbering the finite elements based on re-numbered nodes for front width reduction and (3) re-numbering the cycle basis for flexibility analysis. Unfortunately these methods have escaped the attention of other researchers and credit is being now given to others who re-discovered these simple methods much later, See Refs 8, 9 and 10.

The purposes of this paper are to discuss several concepts of ordering of elements of a graph, to relate the algorithms published recently to the authors original work and to present some new ideas. Special attention is given to the role of SRT in the nodal decomposition and the problems in selecting suitable trees are discussed with examples.

DEFINITIONS

Though the definitions given here are standard (Ref 13), for completeness and clarity they are reproduced below.

A graph S consists of a set N(S) of elements called vertices (nodes) and a set M(S) of elements called edges (members) together with a relation of incidence, which associates with each edge two nodes (called its ends). A graph is connected if every pair of nodes are joined by a path. A subgraph of S is a graph having all its nodes and edges in S.

Two nodes of S are adjacent if these nodes are the ends of an edge. An edge is called incident to a node if the node is an end of the edge. The valency (degree) of a node n_i, denoted by val (n_i), is the number of edges incident to n_i.

A simple path, P_i is a set of one or more distinct edges which can be ordered thus ab, bc, gh where nodes in different positions are distinct, i.e. the simple path may not intersect itself. The length $L(P_i)$ of a simple path is the number of edges in the path. A path P_i between any two nodes is called a shortest path (shortest route) if for any other path P_j between the nodes $L(P_i) \leq L(P_j)$. There may be a number of shortest paths of length greater than one between two nodes. A shortest path is also called a geodesic. The diameter of a connected graph is the length of any longest geodesic. The distance between any two nodes is the length of the shortest path joining them.

A cut-set is a set of edges of S such that the removal of these edges from S results in a disconnected graph. A cocycle is a minimal cut-set. A simple cycle (circuit) is a simple path the two end nodes of which coincide. The vector space associated with the simple cycles (called cycle hereafter) is called the cycle space and that corresponding to cut-sets is called cocycle basis.

The dimension of a cycle space is $b_1(S) = E - N + 1$ and that of a cocycle space is $b_1^*(S) = N - 1$ where E and N are the number of vertices and edges of S. Cycle adjacency matrix is a $b_1(S) \times b_1(S)$ consisting of elements C_{ij} where $C_{ij} = 0$ if cycles C_i and C_j have no common edge

$C_{ij} = 1$ if cycles C_i and C_j have at least one common edge.

Cocycle adjacency matrix is a $\overset{*}{b}_1(S) \times \overset{*}{b}_1(S)$ matrix with elements C_{ij} where

$C_{ij} = 0$ if cocycles $\overset{*}{C}_i$ and $\overset{*}{C}_j$ have no common edge

$C_{ij} = 1$ if cocycles $\overset{*}{C}_i$ and $\overset{*}{C}_j$ have at least one common edge.

A connected subgraph of S which has no cycles is called a spanning tree or simply tree of S.

A shortest route tree (SRT) is a tree rooted at any node, called starting node or root with this special property:

the path from any node to the root through the tree is a shortest path.

A SRT decomposes the node set of S into subsets according to their distances from the root. This is called primary nodal decomposition. Each subset is called a level (contour) of the SRT. The number of nodes in each contour is called the width of the contour and the height of the tree is defined as the number of contours. The maximum width of the contours is called the width of the tree. The longest SRT is one with maximum height and the narrowest SRT is one with the minimum width. The longest SRT also gives the diameter of S.

Multi roots shortest route tree (MRSRT) is similar to SRT but with more than one root and possess the following property: the path from any node to one of the roots through the tree is a shortest path.

A node connectivity graph of a finite element model (FEM) is a graph whose vertices are in 1-1 correspondence with the nodes of the FEM and two vertices are connected by an edge if both nodes belong to the same finite element.

An associate natural graph of a FEM is a graph whose vertices are in 1-1 correspondence with the finite elements of FEM and two vertices are connected by an edge if both finite elements have a common interface (edge or surface).

A modified associate natural graph of a FEM is a graph whose vertices are in 1-1 correspondence with the finite elements of FEM and two vertices are connected by an edge if both finite elements have at least a common node.

Bandwidth of a matrix is defined as max $b_i + 1$ where b_i is the number of columns from the first non-zero in the row to the diagonal.

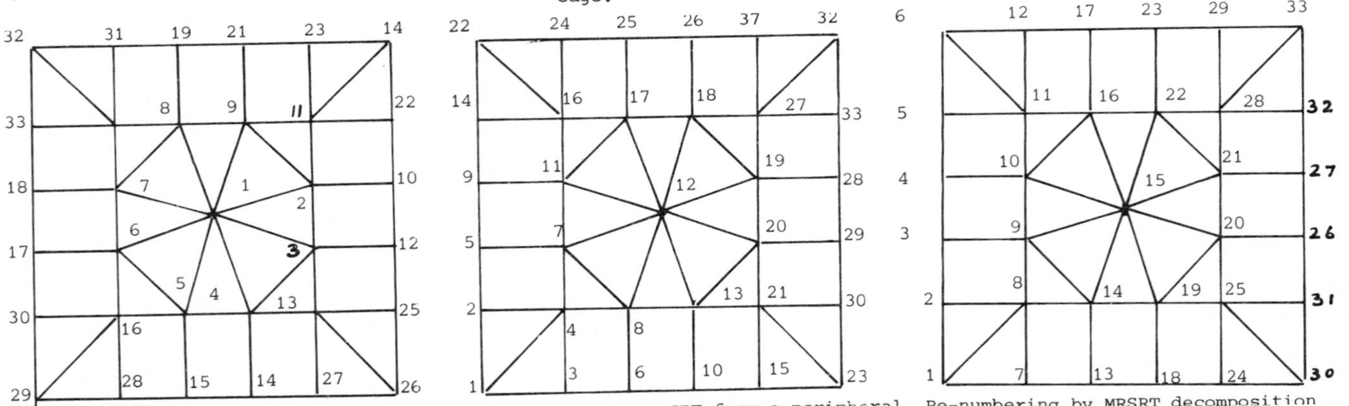

Re-numbering by a SRT from an interior vertex
Height of tree = 3
Width of tree = 12
Bandwidth = 14

Re-numbering by a SRT from a peripheral vertex
Height of tree = 6
Width of tree = 9
Bandwidth = 10

Re-numbering by MRSRT decomposition
Height of tree = 5
Width of tree = 6
Bandwidth = 8

Fig 1 Space Structure

BASIC PRINCIPLE IN NUMBERING

The underlying principle in re-numbering is to group the vertices of a graph model into sets so that all the vertices in each set are connected to one or more vertices in the preceding and the following sets. The authors called this process a nodal decomposition. The obvious way to start a preliminary nodal decomposition is to span an SRT over all the vertices. The concepts of the longest SRT (diametral SRT) and the narrowest SRT and first used by the authors and a full description can be found in Refs 3 and 4. It has been observed by the authors that what is required for optimal numbering is the narrowest tree and not the diametral tree, though usually both are the same. Methods are given in our earlier work for the determination of the diameter and an approximate diameter of a graph and suggestions have been made therein to improve the nodal decomposition by the use of MRSRT. The idea behind the above methods is to find a suitable root (roots for MRSRT) and in most of the cases analysed, the root is found to be one of the peripheral vertices, Fig 1.

So far our discussion had been general and the concepts discussed above are applicable to any connected graph. Therefore as long as there is one to one correspondence between the elements of the mathematical model of the structure (for example nodes, finite elements, cycles) and the vertices of a graph, the elements of the model can be re-numbered using the algorithms developed to re-number the vertices of a graph. However an understanding of the topology of the mathematical model helps to identify the relationship between the ordering of the nodes of a finite element model and that of the finite elements. For a moment assume that the nodes of a finite element model have been numbered to give the optimum bandwidth and that the finite elements are re-numbered in the ascending order of (a) the average nodal numbers of the nodes in the element or (b) the least node number in the element. It is easy to see such a numbering strategy would produce a finite element adjacency matrix (defined similar to cycle adjacency matrix) or an associate natural graph of the FEM with narrow bandwidth. This will in turn give optimal or sub-optimal frontwidth, Fig 2.

It is therefore now conceivable that an associate natural graph or a modified associate natural graph can be used advantageously to re-number the nodes of a finite element model. In this method the vertices of the corresponding associate graph are first re-numbered using the algorithm described in this paper, and the nodes of the finite elements are then re-numbered in the ascending order of the finite element numbers. This approach can be considered as the inverse process of the previous approach, Fig 3. The same principle applies to the numbering of cycles for optimising the bandwidth of a flexibility matrix. In the literature a number of claims have been made by different authors as to the efficiency of their approaches. It is interesting to note that all these methods fall into one of the above categories and the efficiency of the method is dependent on the problem concerned (dependent on the connectivity of the elements, number of nodes in an element etc) contrary to these claims.

It is considered prudent to include a brief summary of the relevant published work in chronological order before outlining our algorithm. After the appearance of Cuthill-McKee's paper in 1969, Collins (Ref 11) published an improved numerical algorithm without providing any basis. The difference between the GIPS method (Ref 8) and the authors is the use of two SRT rooted at the ends of a pseudo diameter instead of a single tree. The algorithm proposed by Razzaque (Ref 9) is not different from that in Ref 3. The two-step approach of Fenves and Law (Ref 10) does not differ from the authors except in minute details of ordering the nodal variables numbering after selecting the appropriate finite element. The authors doubt if the two trees of GIPS method is more efficient than the algorithm described below, considering the extra time required in their second and third algorithms. However, the authors do not have experience with the GIPS algorithm to draw definite conclusions and the work is progressing at Imperial College and Iran University to develop an efficient program for re-numbering.

ALGORITHM

Step 1 Span an SRT from any vertex of the graph and note the vertices on the last level (contour); observe the width of the tree.

Step 2 Choose one of the vertex of the lowest valency on the last level (root) and span an SRT; observe the width of the tree.

Step 3 Select a vertex on the last level in Step 1 adjacent to the root in Sept 2 and span an MRSRT from both these vertices; observe the width of the tree and if this width is less than or equal to the width obtained in Step 2, go to Step 4. If not try another vertex on the last level and repeat 3, otherwise go to Step 5.

Step 4 Repeat Step 3 with another vertex adjacent to the roots already selected and in the last level of Step 1; observe the width of the new tree and repeat Step 4 if the new width is not greater than the previous width till the number of roots is equal to the revised width.

Step 5 Start re-numbering the vertices level by level as follows: re-number the roots sequentially starting from one; select the vertices connected to the vertices in the previous level in the numerical sequence and re-number them; if more than one vertex is connected, re-number the vertices in the ascending order of their valencies.

AN IMPROVED METHOD USING CORNER NODE INCIDENCE MATRIX

The computational requirement of the algorithm described is a function of the number of vertices and edges in the graph model with the complexity of the connectivity having an indirect influence. It is therefore worthwhile to look for other families of graphs which have fewer vertices and edges that can be used directly or indirectly for re-numbering the original graph. The concepts of associate natural graph and the modified associate natural graph belong to this family of graphs but the main drawback with both these graphs is the extra effort required to construct these graphs, which may be quite large in certain cases. The use of corner node incidence matrix to re-number the finite elements or cycles is very useful particularly for finite elements with midside nodes and interior nodes. It is believed that this approach is novel to the re-numbering field, though it has been used in Ref 3 as early as 1973. In this method the corner nodes of the finite elements are first re-numbered and then either they are modified taking into account the number of midside nodes and interior nodes (if any) for banded matrix solutions or they are used to re-number the finite elements for frontal solution as discussed in earlier sections. This is best illustrated by an example, Fig 3. In the limited number of examples studied, this method gives smaller bandwidth and frontwidth than the other methods at less computer cost. Note that the bandwidth obtained by this method is 48 (incidentally 48 is the optimum bandwidth as well) compared to 72 obtained with the use of an associate natural graph, Ref 10.

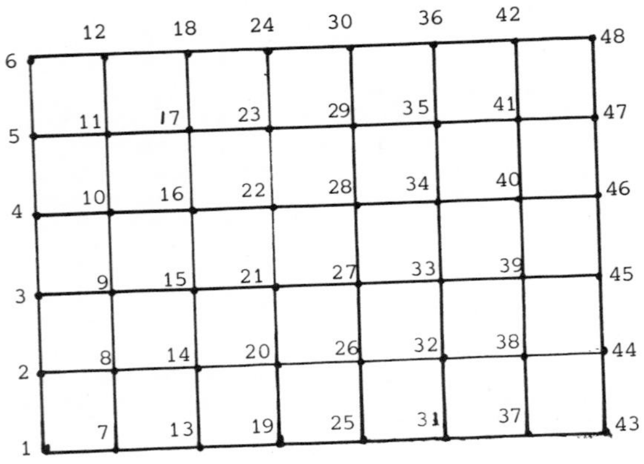

Optimal Nodal Numbering
Bandwidth = 7

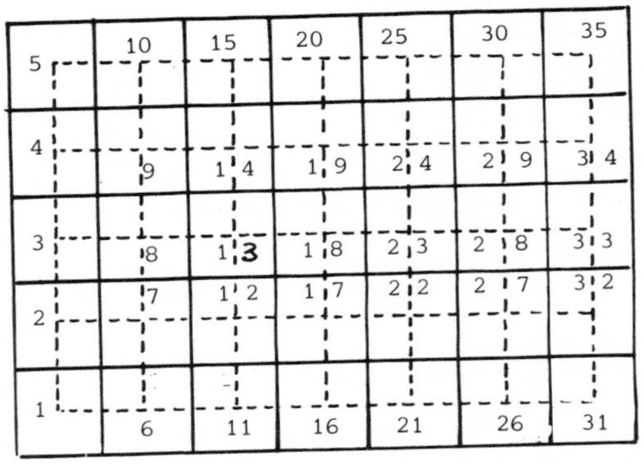

Optimal Finite Element Numbering and Optimal Nodal Numbering of Associate Natural Graph, Frontwidth = 8

Fig 2 Ordering of Finite Elements According to Nodal Numbering

EXAMPLE 1

A simple space structure whose optimum bandwidth is 7 is shown in Fig 1. A sensible numbering scheme without the use of any complicated programming gives a bandwidth of 11. A bad selection of root increases it to 15. Authors algorithm reduces it to 10 with an SRT and to 8 with a MRSRT.

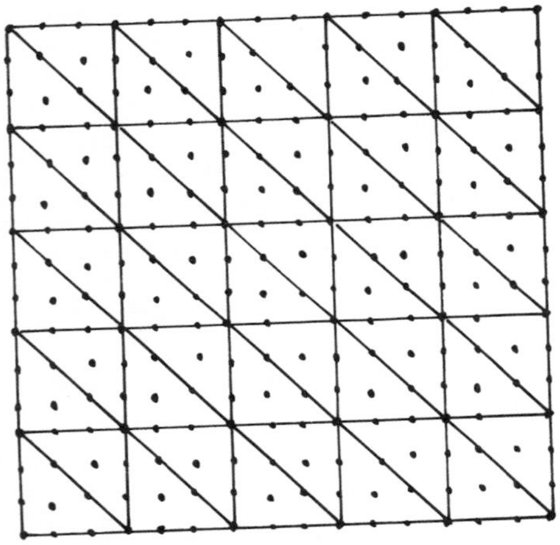

Finite Element Model

EXAMPLE 2

Fig 3 shows an FEM with 256 nodes and 50 finite elements. The associate natural graph, modified associate graph, and the corner node incidence matrix are used to re-number the finite elements. The results show that the use of corner node incidence matrix gives the lowest bandwidth and the frontwidth.

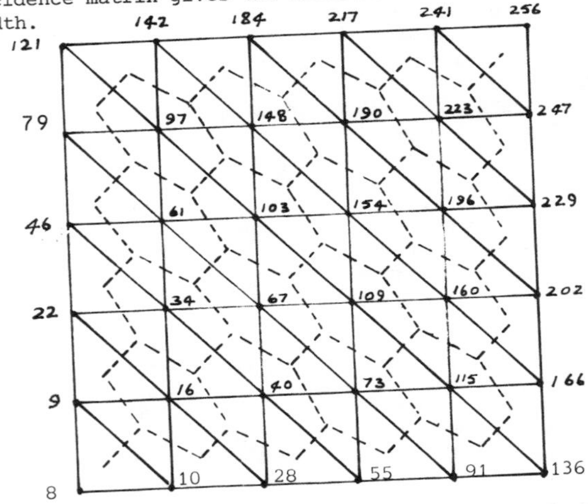

Re-numbering with Associate Natural Graph (ANG)
Bandwidth = 72

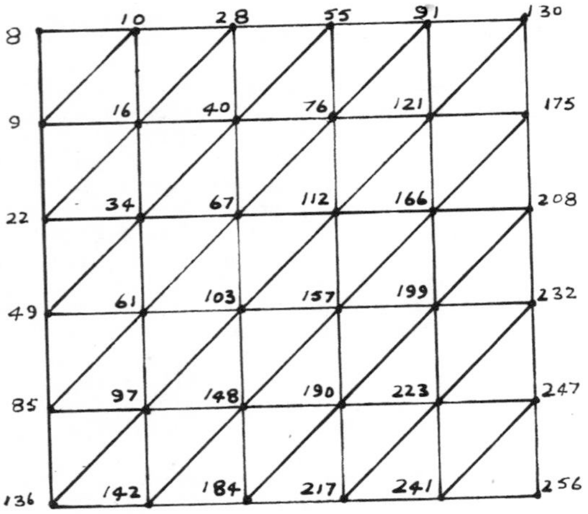

Re-numbering with Modified ANG (not drawn here)
Bandwidth = 54

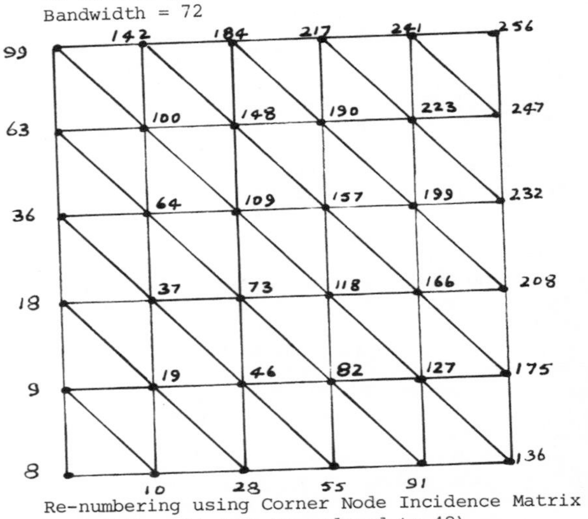

Re-numbering using Corner Node Incidence Matrix
Bandwidth = 51 (can be reduced to 48)

Fig 3 Use of ANG, Modified ANG and Corner Node Incidence Matrix to Re-number the Nodes
(only corner nodes are shown for clarity)

FRONTWIDTH REDUCTION REFINEMENTS

In the previous sections, general principles governing the numbering of the finite elements for frontwidth reduction are described. It is not difficult to give examples where such approaches fail to give the optimum frontwidth, Fig 4, and further research suggested that the width of the associate natural graph is not sufficient to ensure optimum frontwidth. What is required to optimise the frontwidth is to optimise the modified width which may be defined as the maximum number of edges incident with the vertices in any level as opposed to the number of vertices alone as in the width. Further improvement can be obtained by using a weighted associate natural graph which would give due consideration for the number of nodes in a finite element instead of a normal associate natural graph. More details about frontwidth reduction can be found in Ref 3. The concept of modified width gives a frontwidth of 21 in Example 2.

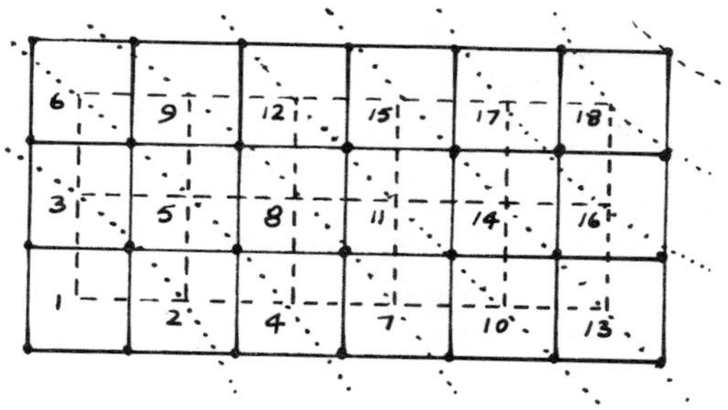

Width = 3, Modified Width = 5, Frontwidth = 9

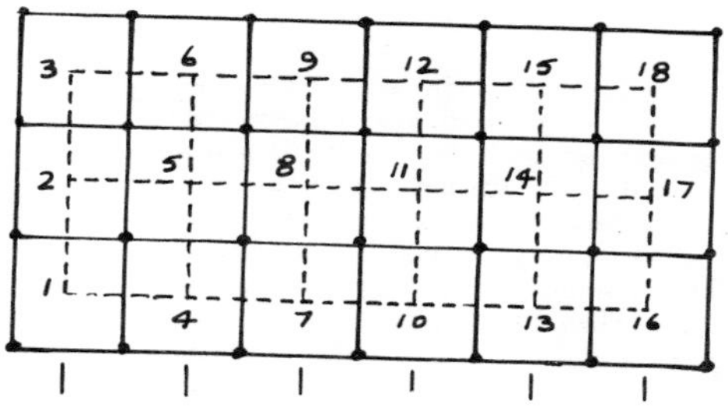

Width = 3, Modified Width = 3, Frontwidth = 6

Fig 4 Use of Modified Width

CONCLUSIONS

The application of graph theoretical concepts in bandwidth and frontwidth reduction has been described in this paper with examples. In this context, the algorithms recently published on re-numbering have been compared with the authors original work and it has been shown that the understanding of the topology of the mathematical model is helpful in deciding on the most efficient method for re-numbering the nodes and the finite elements (cycles). It has also been shown that the optimum nodal numbering leads to optimum finite elements (cycle) numbering if the methods proposed in this paper are used. The new idea of using the graph model of the corner node incidence matrix to reduce the bandwidth of a stiffness matrix of a finite element model has been illustrated with an example. Also the concept of a modified width of a shortest route tree has been introduced to optimise the frontwidth. The methods proposed in this paper besides their efficiency are highly adaptable to both the banded technique and frontal technique.

ACKNOWLEDGEMENTS

The authors would like to thank Dr A.C. Cassell and Mr J.C. de C. Henderson for their help in the early stages of the research reported here. The encouragement and facilities given by Mr M.J. Baker, Professors P.J. Dowling, and J. Munro for the preparation of this paper are also gratefully acknowledged.

REFERENCES

1. B M IRONS, A Frontal Solution Program for Finite Element Analysis, Int. J. Num. Meth. Engng. 2, 1970, pp. 5-32.

2. G C EVERSTINE, A Comparison of Three Resequencing Algorithms for the Reduction of Matrix Profile and Wavefront, Int. J. Num. Meth. Engng, 14, 1979, pp. 837-853.

3. K RAMACHANDRAN, Compatibility Linear Analysis of Skeletal Structures with an extension to the Finite Element Methods, PhD Thesis, London University, 1975.

4. A KAVEH, Application of Topology and Matroid Theory to the Analysis of Structures, PhD Thesis, London University, 1974.

5. A C CASSELL, J C de C HENDERSON and A KAVEH, Cycle Bases for the Flexibility Analysis of Structures. Int. J. Num. Meth. Engng, 8, 1974, pp. 521-528.

6. A C CASSELL, J C de C HENDERSON and K RAMACHANDRAN, Cycle Bases of Minimal Measure for the Structural Analysis of Skeletal Structures by the Flexibility Method, Proc. Roy. Soc., London, A350 (1976), pp. 61-70.

7. E CUTHILL and J McKEE, Reducing the Bandwidth of Sparse Symmetric Matrices, Proc. 24th Nat. Conf. ACM, 1969, pp. 157-172.

8. N E GIBBS, W G POOLE and P K STOCKMEYER, An Algorithm for Reducing the Bandwidth and Profile of a Sparse Matrix, SIAM J. Numer. Anal., Vol 12, No. 2, 1976 pp. 236-250.

9. A RAZZAQUE, Automatic Reduction of Frontwidth for Finite Element Analysis, Int. J. Num. Meth. Engng, 15, 1980, pp. 1315-1324.

10. S J FENVES and K H LAW, A Two-Step Approach to Finite Element Ordering, Int. J. Num. Meth. Engng, 19, 1983, pp. 891-911.

11. R J COLLINS, Bandwidth Reduction by Automatic Re-num Re-numbering, Int. J. Num. Meth. Engng, 6, (1973), pp. 345-356.

12. A KAVEH, Generalised Cycle Basis for the Flexibility Analysis of Structures, Proc. of the 2nd Int. Conf. on Space Strs, Edited by W J Supple, Surrey, England, 1975.

13. F HARARY, Graph Theory, Addison-Wesley Publishing Company, 1969.

PATTERN SOLVER FOR SPACE FRAME STRUCTURES

CHAN Hon-Chuen* and C. Falzon**

* Senior Lecturer, Department of Civil Engineering
University of Hong Kong

** Research Student, University of Hong Kong

This paper describes a new method for organizing the memory storage for the solution of the system of equations generated in the computer analysis of space frame structures.

The existing solution methods commonly used, the 'Skyline Method' and the 'Frontal Method', are very efficient already compared with the 'Band Method'. But with the very sparse nature of the stiffness matrix of a space frame, there are still quite a considerable amount of unnecessary storage space and operations for processing or for intricate house-keeping involved in these two methods.

The method proposed here will take full advantage of the extremely sparse property of the stiffness matrix for a space frame and will pre-determine the sequence of operation and all the essential storage requirement before the elimination procedure actually starts. Only those sub-matrices need to be filled up with numbers during the assembly stage and those coefficient locations which will become non-zero during the elimination process will be allocated in the memory. Storage addresses and house-keeping of the elimination procedure are organized in the same simple form as the member array which is standard input data. Hence, near optimum efficiency in storage requirements and execution time can be achieved and better economy in analying space frame structures is possible with this method.

The method has been implemented for out-of-core as well as in-core memory. Numerical examples are given to demonstrate the big saving in memory space and computer time of this method in comparison with the 'Skyline Method'.

INTRODUCTION

As the capability of computing facilities keeps on improving, it seems that the size of systems of equations generated from engineering problems, such as that produced in the analysis of space structures, keeps on growing at a comparable rate. Search for more and more efficient methods for solving large systems of equations has been a continuous effort. Some of the more popularly adopted methods are the 'Banded Solver', the 'Skyline Solver' and the 'Front Solver'.

The 'Band Solver' uses the fact that during the elimination and back substitution phases, the matrix of coefficients will not alter beyond the point where the last non-zero coefficient occurs during the assembly of the original equations. Hence only that portion of the matrix within this "band" will be stored and operated on. However, within the band there can be a large number of zero coefficients which are needlessly stored and unnecessarily operated on.

The 'Skyline Solver' is a refinement of the above method in which the 'band-width' is not taken as constant but is allowed to vary depending upon the position in the assembled equations of the last non-zero coefficient of each equation. Although it is an improvement on the 'Band Method', it is still possible that a large number of coefficients are needlessly stored or operated on.

The 'Front Solver' is mainly an out-of-core solver in which as many member stiffness matrices as the assigned in-core storage will permit to be input at a time in order to eliminate a few variables which can be eliminated from the sub-set. A small amount of in-core memory is sufficient to solve a large system of equations like those created in finite element methods. However, this method requires fairly complex "house-keeping" techniques as well as a great deal of data transfer between the backing memory and the in-core memory.

The proposed solver, though also based on Gauss elimination principle, employs a completely different data structure organization. The central idea is that only the required data are stored in a compact form and the addresses of these stores and the sequence of necessary operations are pre-determined and contained in a simple member array to ensure near optimum efficiency of memory space and execution time.

CONCEPT OF THE PATTERN METHOD

The concept of the pattern method is similar to the idea of the packed storage together with an indexing system. The packed storage idea is not a novel one and has been employed by many research workers in this field. The indexing system employed here utilizes the usual member array with some modifications to incorporate the addresses of all required memory stores during the elimination process. Moreover, the modified member array is also used to monitor the sequence of elimination so as to attain the optimum efficiency in the elimination process and in the storage requirements.

In the analysis of a space frame structure by the stiffness method, consider a member whose ends are denoted by node numbers i and j. The coordinates of the nodes will be input into a node array and the node numbers for i and j and the type of section of the member will be contained in a member array.

Usually the node numbers i and j are entered randomly in the member array. But in the present method they must be arranged in a particular order in such a way that the i's are always smaller than the j's; the i's are arranged in ascending order, and the j's within each i are also arranged in ascending order. This can be achieved either by using a sorting subroutine on a randomly arranged member array or by inspection when the member array is first entered. For example, the original member array of the structure with its nodes numbered as shown in Fig 1, after arranging in the proper order, is as shown in Fig 2(a).

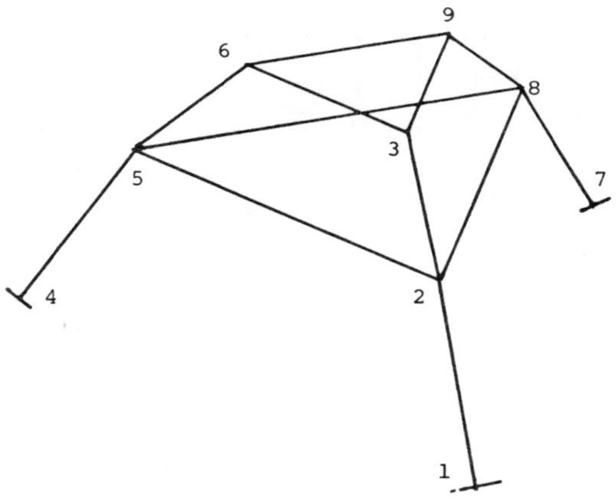

Fig 1 Space frame example for illustration

Node i	Node j	Type no.
1	2	1
2	3	1
2	5	2
2	8	2
3	6	3
3	9	3
4	5	1
5	6	1
5	8	2
6	9	3
7	8	1
8	9	1

(a) Original Member Array

Node i	Node j	Type no.
1	2	1
2	3	1
2	5	2
2	8	2
3	5	0
3	6	3
3	8	0
3	9	3
4	5	1
5	6	1
5	8	2
6	9	3
7	8	1
8	9	1

(b) Intermediate Member Array

Node i	Node j	Type no.
1	2	1
2	3	1
2	5	2
2	8	2
3	5	0
3	6	3
3	8	0
3	9	3
4	5	1
5	6	1
5	8	2
5	9	0
6	8	0
6	9	3
7	8	1
8	9	1

(c) Modified Member Array

Fig. 2 Member Array

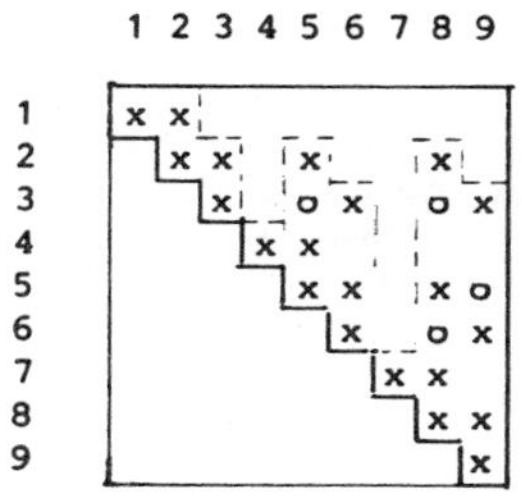

Fig 3 Full Stiffness Matrix

The member array also registers the addresses of the off-diagonal stiffness sub-matrices of the structure. During the assembly of the structural stiffness matrix, the member stiffness sub-matrices will go to the proper stores in the full stiffness matrix in accordance with the addresses shown in the original member array. The full entries for the illustating example is as shown in Fig 3 in which the x's indicate the original member stiffness sub-matrices. If JT denotes the total number of nodes in the structure and MT, the total number of members in the structure, it is noted that there are JT number of diagonal sub-matrices; and in the upper triangle, there are MT number of off-diagonal sub-matrices. Thus, altogether there are (JT + MT) entries after assembly. The coefficients in other locations in the full structural matrix are all equal to zero.

In reducing a set of linear simultaneous equations by Gauss elimination method, many of those initially zero coefficients will become non-zero during the elimination process. But the locations of these created non-zero coefficients can be determined beforehand.

Consider in a set of linear equations the unknown x_e with coefficient a_{ee} is being eliminated from the equation i. The coefficient a_{ij} in equation i will be in general changed to a new value given by:-

$$a'_{ij} = a_{ij} - \frac{a_{ei} * a_{ej}}{a_{ee}} \qquad (\text{since } a_{ie} = a_{ei})$$

It is obvious from this expression that when eliminating an unknown x_e the coefficient a_{ij} in any equation i and column j in the stiffness matrix will be affected only if both the coefficients a_{ei} and a_{ej} are not equal to zero. If either a_{ei} or a_{ej} is zero, then a_{ij} will not be affected by the elimination of x_e. That is to say, if a_{ij} is initially a zero coefficient before the elimination of x_e it will remain zero. This fact can be used to predict, before the elimination procedure starts, the locations in the full set of equations which will become non-zero during the elimination process.

As mentioned earlier, the member array has been used to indicate the addresses of the entries of non-zero sub-matrices during the assembly of the stiffness matrix. Now it can be conceived that with some modification it can also be used to indicate the addresses of those locations originally with zero sub-matrices but then changed into non-zero sub-matrices during the elimination process. Furthermore, since each new non-zero coefficient might cause other non-zero coefficients to be created, it must be incorporated in its proper place in the member array when considering the elimination of the next unknown.

The technique for modifying the original member array to incorporate the addresses of the created non-zero coefficients can be best explained with the illustrating example shown above.

By starting with node number 1, it is obvious that since only a_{12} is non-zero, therefore no new non-zero sub-matrix will be created due to the elimination of all the unknowns related with the degrees of freedom at node number 1 from all the following equations. Next, consider node number 2. There are three existing non-zero sub-matrices a_{23}, a_{25} and a_{28}. Therefore, in eliminating the unknowns related to node 2, besides some diagonal sub-matrices, the coefficients in a_{35}, a_{38} and a_{58} will be affected. Since a_{35} and a_{38} are initially zero, it is necessary to provide memory stores for the created non-zero coefficients at addresses 3-5 and 3-8. On the other hand, a_{58} is initially a non-zero sub-matrix and memory locations have already been assigned, so no new memory stores need be allocated. In order to establish the addresses of the newly created non-zero sub-matrices, two fictitious dummy members 3-5 and 3-8 with no real sectional properties are inserted in the member array as shown in Fig 2(b). As a result, node 3 now has four non-zero off-diagonal sub-matrices a_{35}, a_{36}, a_{38} and a_{39}. The elimination of the unknowns at node 3 will affect the coefficients in a_{56}, a_{58}, a_{59}, a_{68}, a_{69} and a_{89}. Two more new non-zero coefficients, a and a , are created. Hence, two more dummy members, 5-9 and 6-8, are inserted in the member array at this juncture. When proceeding on with the elimination of nodes 4 to 9, no more new memory stores are required other than those already provided for. Thus, the addresses of all the memory locations required for the whole elimination process have been entirely determined as given in the ultimate modified member array shown in Fig 2(c).

THE ALGORITHM FOR MODIFICATION OF THE MEMBER ARRAY

The above-described procedure for modifying the original member array to incorporate all the dummy members for registering the addresses of the created non-zero sub-matrices can be easily performed by a computer subroutine. The algorithm of the procedure is as follows:-

Let E1(m), E2(m) and TS(m) denote the three columns in the member array which contain the node number i, the node number j and the type number of the member respectively. Let R1 and R2 denote the row numbers of the first occurrence and last occurrence of the same node number in column E1(m) and let RT be the running total number of rows of the modified member array. The algorithm is as shown in Fig 6.

THE PACKED STORAGE

With the modified member array indicating the addresses of all sub-matrices needed for the entire elimination procedure, it is possible to store the sub-matrices in a compact manner to save storage space.

Three real arrays are specified for storing the data in the elimination process:

1. A real array DG(JT,D) to store the upper triangle of the diagonal sub-matrices; where JT = total number of nodes and D = DF(DF+1)/2 with DF denoting the degrees of freedom at each node.
2. A real array CT(RT,C) to store the off-diagonal sub-matrices in the upper triangle of the stiffness matrix; where RT = total number of rows in the modified member array, and C=DF*DF.
3. A real array BL(JT,B) to store the load vectors; where B = DF*LC with LC denoting the number of loading cases.

These three arrays can be illustrated diagramatically as shown in Fig 4.

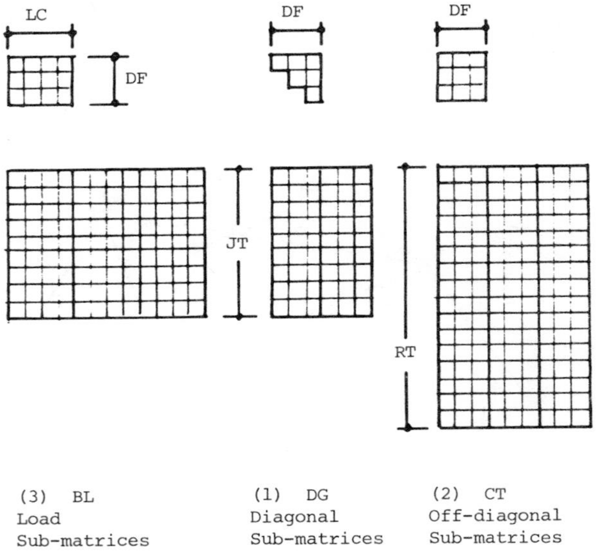

Fig 4 Storage arrays

EXECUTION OF THE SOLVER

The generation of the member stiffness matrices will be conducted in the order of the records of the member array. The diagonal sub-matrices will be stored in DG and the off-diagonal sub-matrices stored in CT at the locations corresponding to the row numbers in the member array. The load vectors will be stored in BL in accordance with the ascending node numbers.

When a '0' is encountered for the section type number of a member signifying that it is a dummy member, the generation and assembly sub-routine will be skipped. Boundary conditions or prescribed displacements can be inserted in the load array and the diagonal stiffness sub-matrices according to the node number and in the off-diagonal stiffness sub-matrices according to the addresses given in the member array.

Now that the memory stores for the generated data and the required intermediate computed data have been allocated and the data structure has been properly organized for the efficient execution of the elimination procedure, in accordance with the sequence set out in the member array the unknowns at each node are then eliminated one by one. Calculations need be carried out only on the existing and anticipated non-zero sub-matrices; hence, most of the checking and operations on zero coefficients generally performed in other methods have been largely avoided in the method. Exact amount of memory stores required for the elimination process has been allowed for, and the calculated data can be put in the memory locations at the appropriate addresses as pre-determined from the modified member array. Finally back-substitution is conducted in the reverse order of the sequence of the modified member array.

If not enough memory is available in the core to solve the system of equations by in-core memory alone, an off-core version of this method has been developed for solving large system of equations on micro-computers.

Given the number of available in-core memory spaces for storing the coefficients and load vectors, it is possible to determine, by scanning through the modified member array, the maximum number of members that can be assembled and the nodes that can be eliminated at each transfer, and the total number of transfers required for solving the whole problem before any processing has actually taken place. No additional 'house-keeping' operations are required except those for marking the locations of the modified member array where transfers should be made. The first subscripts in the three arrays DG, CT and BL will be changed accordingly during the forward elimination and back-substitution.

NODE NUMBERING TECHNIQUES

The way that the nodes are numbered in a problem is all important to the generation of an efficient modified member array and hence to all the procedures that follow. It is observed that in this method the greater the difference between the node numbers of the two ends of a member, the further down the modified member array is the effect of reducing a particular unknown carried. This has a bearing both on the number of memory stores and the execution time required. It appears that one way in keeping down the total number of stores is to make the difference between the node numbers at ends i and j of each member as small as possible. If a numbering scheme for a problem is optimal for the 'Skyline' method, it will always require less memory stores in the proposed method; but it is not necessary that it is also the optimal numbering scheme for the proposed method. A couple of different ways of numbering have been tried. While some of the results obtained are very enlightening, no definite rules can as yet be arrived at. The following numerical example may be able to illustrate this point.

NUMERICAL EXAMPLE

In order to demonstrate the saving in memory spaces of this method as compared with the 'Skyline' method as well as the effect of using different node numbering schemes on the actual memory requirements, a skeleton dome structure with 72 members and 48 joints each with 6 degrees of freedom is studied.

Numbering scheme A was designed to be an optimal numbering scheme for the 'Skyline' method and numbering scheme B was designed to be more favourable to the proposed method. The resulting structural stiffness matrices are fully plotted as shown in Fig 5A and Fig 5B respectively.

Note that in these figures:
- x - denotes a non-zero sub-matrix created during the assembly of the member stiffness sub-matrices;
- o - denotes a non-zero sub-matrix created at some stage of the elimination process; and
- --- the dotted line indicates the skyline of the extreme sub-matrices.

The comparison of the memory sizes required for storing the stiffness matrix by the two methods under the two numbering schemes is given in the following table:-

Method	Total Number of Stores Required			
	Numbering Scheme A		Numbering Scheme B	
	Sub-matrices	Real numbers	Sub-matrices	Real numbers
Proposed method	295	9900	224	7344
Skyline method	363	12384	417	14292

It can be seen that in scheme A the saving is 20% and in scheme B, nearly 50%. When comparing the optimal of the proposed method to the optimal of the skyline method the saving is 40%.

CONCLUSIONS

The proposed solver disregards all null sub-matrices and stores only those generated and created non-zero sub-matrices in packed arrays. Hence, maximum economy in memory space requirements is ensured

The elimination process is scheduled in such a way that computations are performed straightly on the non-zero sub-matrices. Thus, it requires minimum execution time. The schedule is pre-determined and the steps are recorded simply in the form of an member array.

When compared with the 'Skyline Solver', this solver always requires less memory spaces and computer time.

The out-of-core version of this solver is comparable with the 'Front Solver' in its capability in handling large systems of equations with a small in-core memory but without the intricacy of its 'house-keeping'. It is particularly suitable for use in micro-computers.

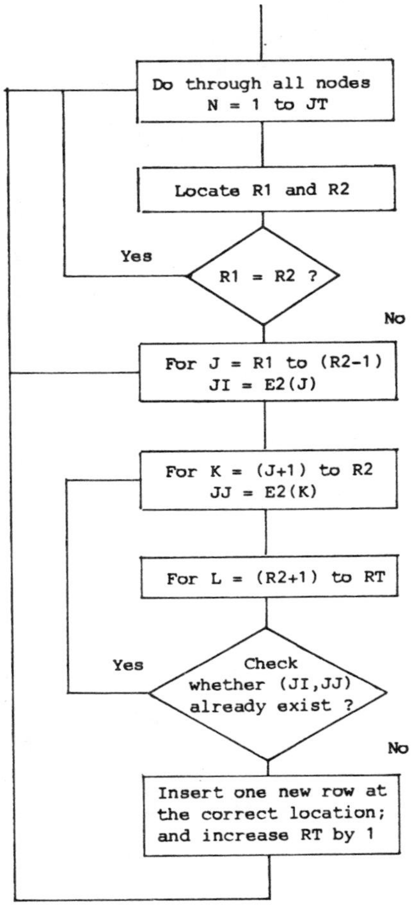

Fig 6 Flow chart for modification of member array

REFERENCES

1. Klans-Jurgen Bathe and Edward L. Wilson, 'Numerical Methods in Finite Element Analysis', Prentice-Hall, 1976.
2. L. Fox, 'An Introduction to Numerical Linear Algebra', Oxford University Press, 1964.
3. Y.K. Cheung, 'Finite Strip Method in Structural Analysis Pergamon Press, 1978.
4. Y.K. Cheung and M.F. Yeo, 'A Practical Introduction to Finite Element Analysis', Pitman, 1979.
5. E.L. Wilson, 'Role of Small Computer Systems in Structural Engineering', Seventh Conference on Electronic Computation ASCE, New York, 1979, pp.331-339.
6. B.M. Irons, 'A Frontal Solution Program for Finite Element Analysis', Int. Journal for Numerical Methods in Engineering, Vol. 2, No. 1, 1970, pp. 5-32.
7. K.E. Atkinson, 'An Introduction to Numerical Analysis', John Wiley & Sons, 1978.
8. Alan Jennings, 'Matrix Computation for Engineers and Scientists', John Wiley & Son, 1977
9. C. Falzon and H.C. Chan, 'Pattern Solver For Framework/ Network Numerical Models', Research Report, September 1983, Department of Civil Engineering, University of Hong Kong.

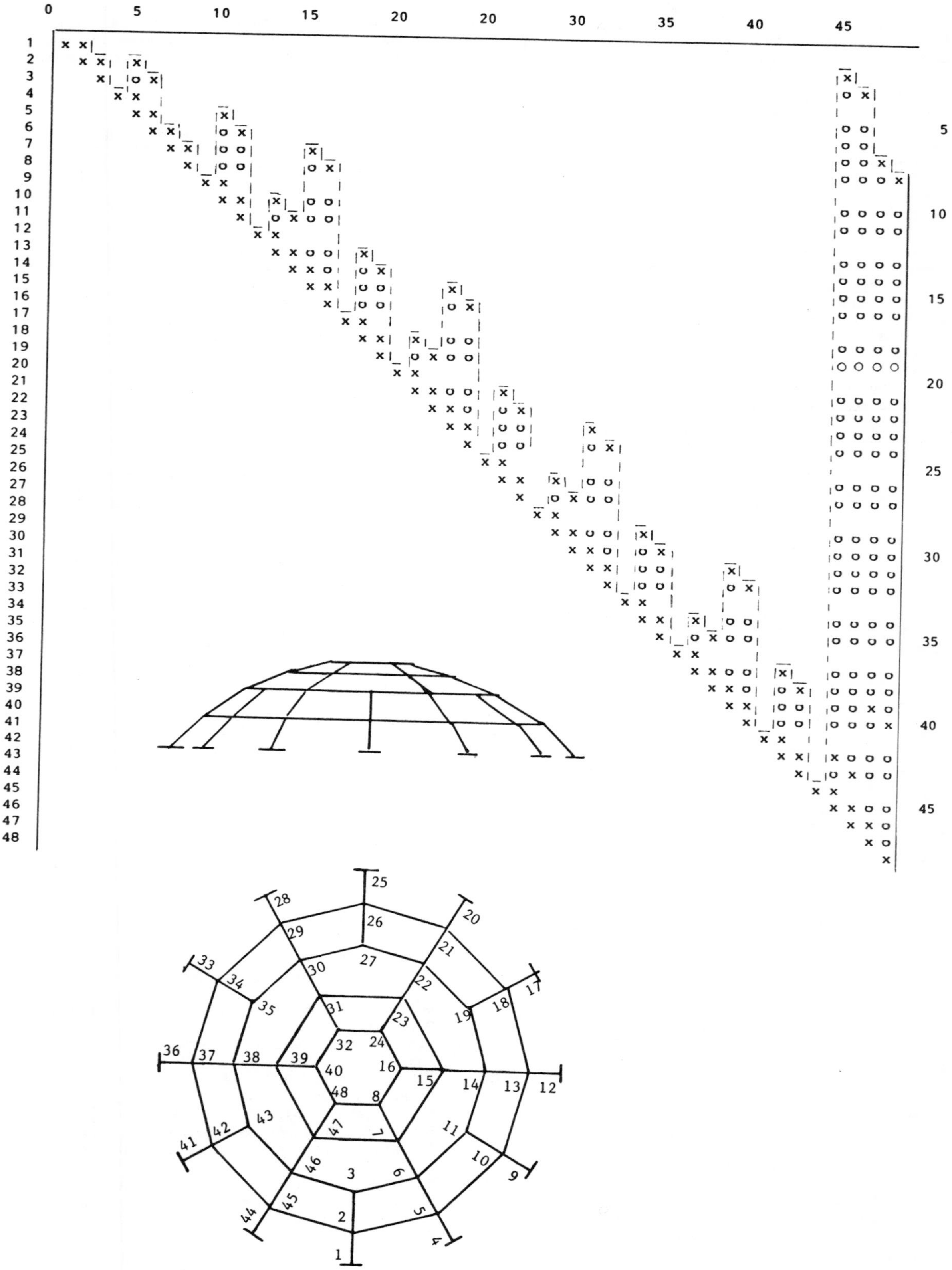

Fig. 5A Numbering Scheme A Non-zero Sub-matrices

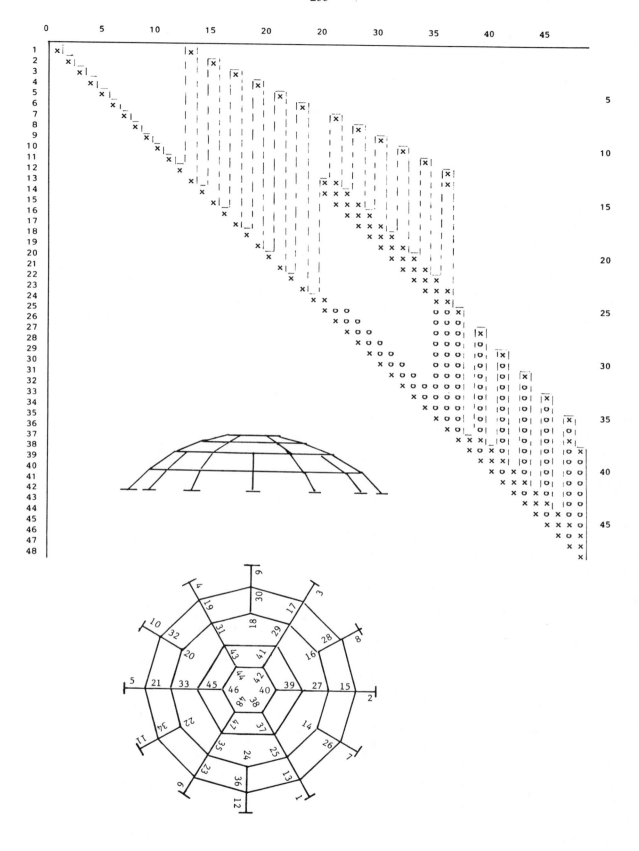

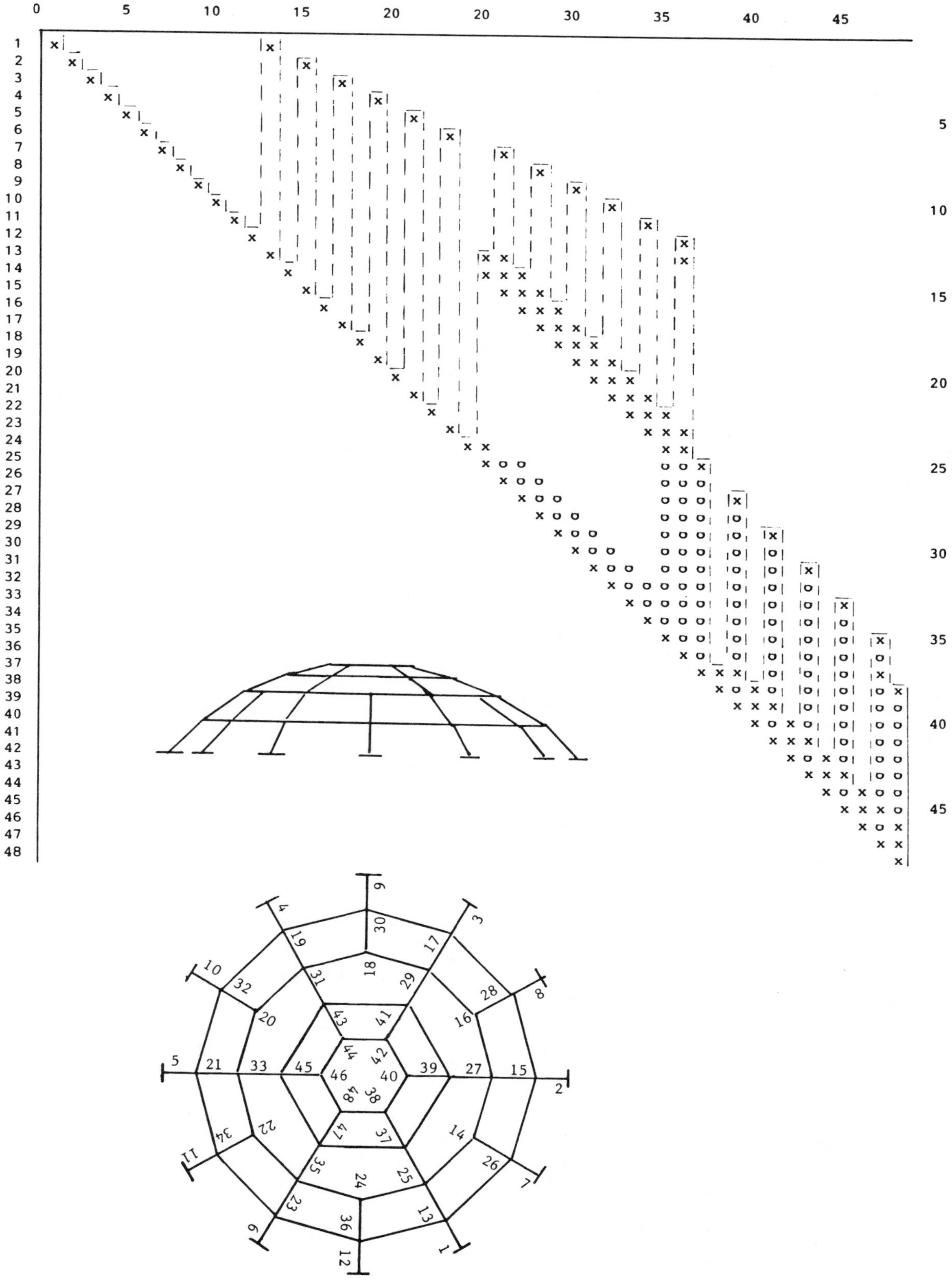

Fig. 5B Numbering Scheme B Non-zero Sub-matrices

NON-PRISMATIC FOLDED PLATES ANALYSIS USING SUB-PARAMETRIC HIGHER ORDER FLAT SHELL ELEMENTS

Y K CHEUNG* and H C CHAN**

* Professor, Department of Civil Engineering, University of Hong Kong.

** Senior Lecturer, Department of Civil Engineering, University of Hong Kong.

Classical elasticity methods for the analysis of prismatic folded plates are generally well recognised. But when it comes to the analysis of non-prismatic folded plates, the finite element method is undoubtedly most appropriate because of its versatility of modelling the irregular geometric shapes and boundary conditions.

This paper describes the derivation and formulation of a series of sub-parametric higher order flat shell elements which are particularly suitable for applying in the analysis of non-prismatic folded plate structures. Each piece of the triangular or trapezoidal plane in a non-prismatic folded plate can be modelled by just one or two of the proposed higher order elements and yet very accurate results can be obtained. Hence the amount of input data can be much reduced. The versatility and accuracy of these elements are demonstrated through the analysis of an example of non-prismatic folded plate structure.

INTRODUCTION

During the last couple of decades, folded plate roof system has been popularly adopted for large span structures as an alternative to domes, cylindrical shells or barrel vaults because it is an easier and more economical type of construction. Many interesting forms can be created by combining rectangular, triangular and trapezoidal shaped plates together in a variety of patterns producing sculpture-like non-prismatic folded plate structures which are especially aesthetical.

While classical elasticity methods are available for the analysis of prismatic folded plates, very little literature has been published in the English language on the analysis of non-prismatic folded plates, other than those by approximate numerical methods, probably due to their complicated geometry and structural behaviour. With the advent of the finite element method and due to its versatility in modelling complicated shapes and boundary conditions, it has been widely accepted as a common suitable method for the anlaysis of folded plate structures.

Most of the simple flat shell elements used are formed by combining an in-plane displacement function to another bending displacement function of a different order resulting in incompatibility of the in-plane and lateral displacements at the fold lines. Besides, usually a large number of elements are required to model a folded plate, calling for a great deal of data preparation and computer time. Efforts have been made in forming more sophisticated flat shell elements for analysing non-prismatic folded plates which are to some extent successful in reducing the discontinuities and the number of elements required, but those elements are difficult to form. The proposed subparametric higher order elements described herein employ same higher order displacement polynomials for u, v and w in their formulation; hence conformity is ensured at the fold lines. The higher degree of variability of the stress field in the element has made it possible to model a folded plate with a small number of elements and yet giving very accurate results.

FORMULATION OF HIGHER ORDER SUBPARAMETRIC FLAT SHELL ELEMENTS

In deriving a flat shell element for the analysis of folded plate structures, it is necessary to consider both the in-plane action as well as the bending action in the element. With the common methods of formulation of the in-plane action and the bending action for a plate element, it is difficult to maintain the compatibility of the three displacements u, v, and w at the fold lines between the elements. In order to achieve compatibility at the fold lines, the displacement functions for the displacements in the u, v and w directions should be similar. It is possible to have similar polynomial displacement functions for u, v and w by employing Mindlin's bending theory for moderately thick plates. The theory assumes that the normals to the mid-surface of the plate before deflection remain straight but not necessarily normal to the mid-surface after deflection. Hence, besides the normal displacement w, two other separate variables, namely, the rotations of the plate about the x-axis and about the y-axis, α and β respectively, are required to specify the bending and shear deformations of the plate.

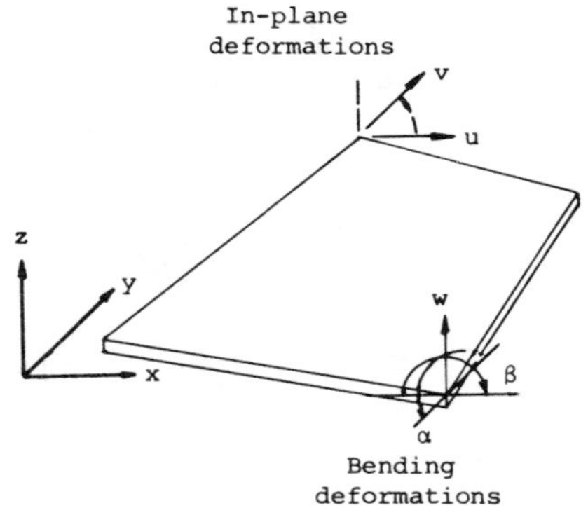

Fig 1 Flat shell element with nodal displacements in local coordinates

Consider a higher-order flat shell element in its element coordinate (x,y) as shown in Fig 1. The in-plane and out-of-plane displacements at any point in the element will be related to the displacements of all the nodes in the element by

$$u = \Sigma L_i u_i$$
$$v = \Sigma L_i v_i$$
$$w = \Sigma L_i w_i$$
$$\alpha = \Sigma L_i \alpha_i$$
$$\beta = \Sigma L_i \beta_i$$

with i = 1 to n, the number of nodes in the element where L_i's are displacement relationship functions in terms of the natural coordinate (ξ,η) as shown in Fig 2. They are given by

$$\{L_1\ L_2\ L_3\ \ldots\ L_i\ \ldots\ L_n\} = [P][C_N^{-1}]$$

where $[P] = [1\ \xi\ \eta\ \xi^2\ \xi\eta\ \eta^2\ \xi^3 \ldots]$ is an n-term polynomial in ξ and η and $[C_N^{-1}]$ is the inverse of the matrix of the natural coordinates of the nodes.

Figure 3 gives the terms of the polynomials for higher order elements with number of nodes of 13, 17, 21 and 25.

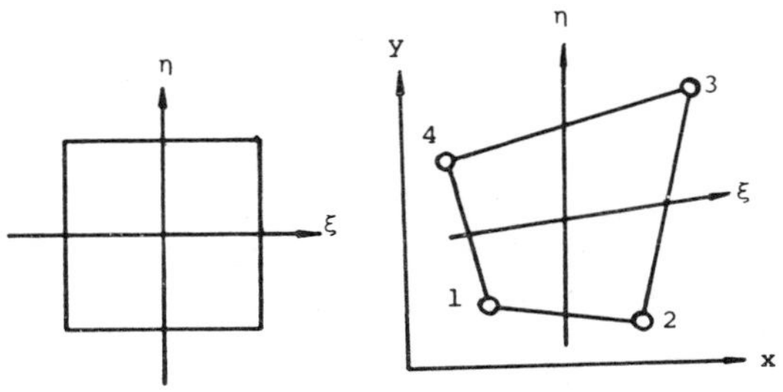

(a) Parent element in natural coordinates

(b) 4-node mapping element in local coordinates

Fig 2

For modelling non-prismatic folded plates, flat shell elements of non-rectangular shapes are preferred. The subparametric mapping technique is employed to convert the higher order rectangular shaped parent elements into "quadrilaterals". Here ony 4 corner nodes are needed for mapping and the natural coordinates $(\xi, \eta)_m$ of these 4 corner nodes are given as $(-1, -1)_1$, $(1, -1)_2$, $(1, 1)_3$ and $(-1, 1)_4$ for the nodes m = 1, 2, 3 and 4 respectively. Hence the corresponding mapping functions are

$$M_1 = \frac{1}{4}(1 - \xi)(1 - \eta)$$
$$M_2 = \frac{1}{4}(1 + \xi)(1 - \eta)$$
$$M_3 = \frac{1}{4}(1 + \xi)(1 + \eta)$$
$$M_4 = \frac{1}{4}(1 - \xi)(1 + \eta)$$

Any point (x,y) in the quadrilateral element is related to a corresponding point (ξ,η) in the parent square element by the following expression

$$x = \sum_{1}^{4} M_m x_m$$

$$y = \sum_{1}^{4} M_m y_m$$

where $(x,y)_m$ for m = 1 to 4 denote the coordinates of the 4 corner nodes of the quadrilateral in its local coordinates.

No. of nodes	Terms in polynomials	Element type
Side ξ 5 Side η 3 Total 13	1 $\xi \quad \eta$ $\xi^2 \quad \xi\eta \quad \eta^2$ $\xi^3 \quad \xi^2\eta \quad \xi\eta^2$ $\xi^4 \quad \xi^3\eta \quad \xi^2\eta^2$ $\xi^4\eta$	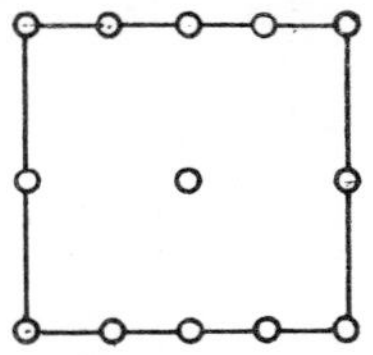
Side 5 Total 17	1 $\xi \quad \eta$ $\xi^2 \quad \xi\eta \quad \eta^2$ $\xi^3 \quad \xi^2\eta \quad \xi\eta^2 \quad \eta^3$ $\xi^4 \quad \xi^3\eta \quad \xi^2\eta^2 \quad \xi\eta^3 \quad \eta^4$ $\xi^4\eta \quad \xi\eta^4$	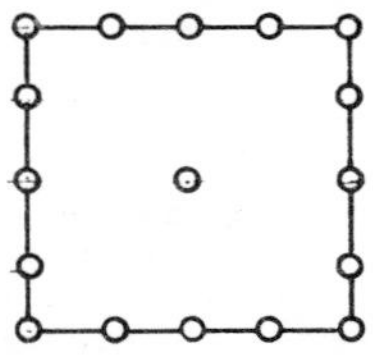
Side 5 Total 21	1 $\xi \quad \eta$ $\xi^2 \quad \xi\eta \quad \eta^2$ $\xi^3 \quad \xi^2\eta \quad \xi\eta^2 \quad \eta^3$ $\xi^4 \quad \xi^3\eta \quad \xi^3\eta^2 \quad \xi\eta^3 \quad \eta^4$ $\xi^4\eta \quad \xi\eta^4$ $\xi^4\eta^2 \quad \xi^3\eta^3 \quad \xi^2\eta^4$ $\xi^4\eta^4$	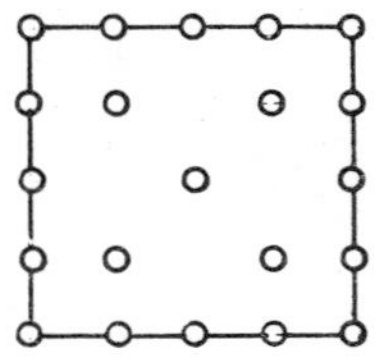
Side 5 Total 25	1 $\xi \quad \eta$ $\xi^2 \quad \xi\eta \quad \eta^2$ $\xi^3 \quad \xi^2\eta \quad \xi\eta^2 \quad \eta^3$ $\xi^4 \quad \xi^3\eta \quad \xi^2\eta^2 \quad \xi\eta^3 \quad \eta^4$ $\xi^4\eta \quad \xi^3\eta^2 \quad \xi^2\eta^3 \quad \xi\eta^4$ $\xi^4\eta^2 \quad \xi^3\eta^3 \quad \xi^2\eta^4$ $\xi^4\eta^3 \quad \xi^3\eta^4$ $\xi^4\eta^4$	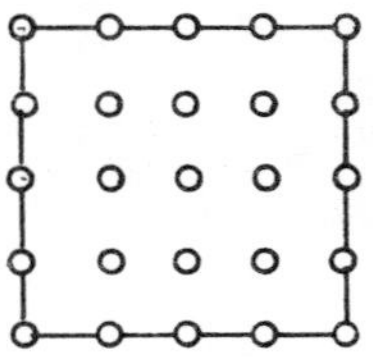

Fig 3 A family of higher order parent flat shell element

In a linear-elastic analysis, the in-plane action and the bending action in a flat shell element are independent of each other and hence they can be dealt with separately.

The in-plane strains are

$$\{^P\varepsilon\} = \begin{Bmatrix} \dfrac{\partial u}{\partial x} \\ \dfrac{\partial v}{\partial y} \\ \dfrac{\partial u}{\partial y} + \dfrac{\partial v}{\partial x} \end{Bmatrix} = \begin{bmatrix} \dfrac{\partial L_i}{\partial x} & 0 \\ 0 & \dfrac{\partial L_i}{\partial y} \\ \dfrac{\partial L_i}{\partial y} & + \dfrac{\partial L_i}{\partial x} \end{bmatrix} \begin{Bmatrix} u_i \\ v_i \end{Bmatrix}$$

$$= \sum_{i=1}^{n} [^PB_i]\ \{^P\Delta_i\}$$

where $[^PB]$ is the strain-displacement relationship matrix for plane stress and $\{^P\Delta\} = \{u\ v\}$ are the nodal in-plane displacements.

The bending strains are

$$\{^b\varepsilon\} = \begin{Bmatrix} \dfrac{\partial\beta}{\partial x} \\ -\dfrac{\partial\alpha}{\partial y} \\ -\dfrac{\partial\alpha}{\partial x} + \dfrac{\partial\beta}{\partial y} \\ \dfrac{\partial w}{\partial x} + \beta \\ \dfrac{\partial w}{\partial y} - \alpha \end{Bmatrix} = \begin{bmatrix} 0 & 0 & \dfrac{\partial L_i}{\partial x} \\ 0 & -\dfrac{\partial L_i}{\partial y} & 0 \\ 0 & -\dfrac{\partial L_i}{\partial x} & \dfrac{\partial L_i}{\partial y} \\ \dfrac{\partial L_i}{\partial x} & 0 & L_i \\ \dfrac{\partial L_i}{\partial y} & -L_i & 0 \end{bmatrix} \begin{Bmatrix} w_i \\ \alpha_i \\ \beta_i \end{Bmatrix}$$

$$= \sum_{i=1}^{n} [^bB_i]\ \{^b\Delta_i\}$$

where $[^bB]$ is the strain-displacement relationship matrix for plate bending
and $\{^b\Delta\} = \{w\ \alpha\ \beta\}$ are the nodal normal and rotational displacements.

The differentials $\frac{\partial L_i}{\partial x}$ and $\frac{\partial L_i}{\partial y}$ occurring in the above expressions involve differentiation of a displacement function L_i, which is given in the natural coordinates ξ and η with respect to a variable in the local coordinate x or y. Hence they cannot be carried out directly or explicitly. However, since x and y are functions of ξ and η, the differentials can be obtained as follows:-

$$\frac{\partial}{\partial\xi} = \frac{\partial x}{\partial\xi}\cdot\frac{\partial}{\partial x} + \frac{\partial y}{\partial\xi}\cdot\frac{\partial}{\partial y}$$

$$\frac{\partial}{\partial\eta} = \frac{\partial x}{\partial\eta}\cdot\frac{\partial}{\partial x} + \frac{\partial y}{\partial\eta}\cdot\frac{\partial}{\partial y}$$

or

$$\begin{vmatrix} \dfrac{\partial}{\partial\xi} \\ \dfrac{\partial}{\partial\eta} \end{vmatrix} = \begin{vmatrix} \dfrac{\partial x}{\partial\xi} & \dfrac{\partial y}{\partial\xi} \\ \dfrac{\partial x}{\partial\eta} & \dfrac{\partial y}{\partial\eta} \end{vmatrix} \begin{vmatrix} \dfrac{\partial}{\partial x} \\ \dfrac{\partial}{\partial y} \end{vmatrix} = |J| \begin{vmatrix} \dfrac{\partial}{\partial x} \\ \dfrac{\partial}{\partial y} \end{vmatrix}$$

where $|J|$ is the Jacobian matrix relating the natural coordinate derivatives to the local coordinate derivatives. From above

$$|J| = \begin{vmatrix} \sum_1^4 \dfrac{\partial M_m}{\partial\xi} x_m & \sum_1^4 \dfrac{\partial M_m}{\partial\xi} y_m \\ \sum_1^4 \dfrac{\partial M_m}{\partial\eta} x_m & \sum_1^4 \dfrac{\partial M_m}{\partial\eta} y_m \end{vmatrix}$$

By inverting,

$$\begin{vmatrix} \dfrac{\partial}{\partial x} \\ \dfrac{\partial}{\partial y} \end{vmatrix} = |J^{-1}| \begin{vmatrix} \dfrac{\partial}{\partial\xi} \\ \dfrac{\partial}{\partial\eta} \end{vmatrix}$$

Denoting the four coefficients in the inverse of the Jacobian by

$$[J^{-1}] = \begin{bmatrix} I_{11} & I_{12} \\ I_{21} & I_{22} \end{bmatrix}$$

the differentials of the shape functions with respect to the local coordinates can be expressed as

$$\frac{\partial L_i}{\partial x} = I_{11}\frac{\partial L_i}{\partial\xi} + I_{12}\frac{\partial L_i}{\partial\eta} = Q_i(\xi,\eta)$$

$$\frac{\partial L_i}{\partial y} = I_{21}\frac{\partial L_i}{\partial\xi} + I_{22}\frac{\partial L_i}{\partial\eta} = R_i(\xi,\eta)$$

Hence the strain-displacement relationship matrices $[^pB_i]$ and $[^bB_i]$ can be expressed as functions of ξ and η as follows:-

$$[^pB_i] = \begin{bmatrix} Q_i & 0 \\ 0 & R_i \\ R_i & Q_i \end{bmatrix} \quad \text{and} \quad [^bB_i] = \begin{bmatrix} 0 & 0 & Q_i \\ 0 & -R_i & 0 \\ 0 & -Q_i & R_i \\ Q_i & 0 & L_i \\ R_i & -L_i & 0 \end{bmatrix}$$

The stiffness coefficients of the quadrilateral element will be given by combining

$$^pk_{ij} = [\int_{-1}^{+1}\int_{-1}^{+1}\ {^pB_i^T}\ .\ {^pD}\ .\ {^pB_j}\cdot\det J]\,h\ d\xi\cdot d\eta \text{ and}$$

$$^bk_{ij} = [\int_{-1}^{-1}\int_{-1}^{+1}\ [^bB_i^T\ .\ {^bD}\ .\ {^bB_j}\cdot\ \det J]\ d\xi d\eta$$

where ^pD and ^bD are the plane stress elasticity matrix and the bending elasticity matrix respectively.

Expanding the above expressions in full, the stiffness coefficients will be the double integrals of the terms of the following matrix:-

$\{Q_i\}\ E_x\ [Q_j]\ h$ $+\ \{R_i\}\ G\ [R_j]\ h$	$\{Q_i\}\ E_v\ [R_j]\ h$ $+\ \{R_i\}\ G\ [Q_j]\ h$	0	0	0
	$\{R_i\}\ E_y\ [R_j]\ h$ $+\ \{Q_i\}\ G\ [Q_j]\ h$	0	0	0
		$\{Q_i\}\ S_1\ [Q_j]$ $+\ \{R_i\}\ S_2\ [R_j]$	$-\{R_i\}\ S_2\ [L_j]$	0
	Sym.		$\{R_i\}\ D_2\ [R_j]$ $+\ \{Q_i\}\ D_3\ [Q_j]$ $+\ \{L_i\}\ S_2\ [L_j]$	$-\{R_i\}\ D_v\ [Q_j]$ $-\{Q_i\}\ D_3\ [R_j]$
				$\{Q_i\}\ D_1\ [Q_j]$ $+\ \{R_i\}\ D_3\ [R_j]$ $+\ \{L_i\}\ S_1\ [L_j]$

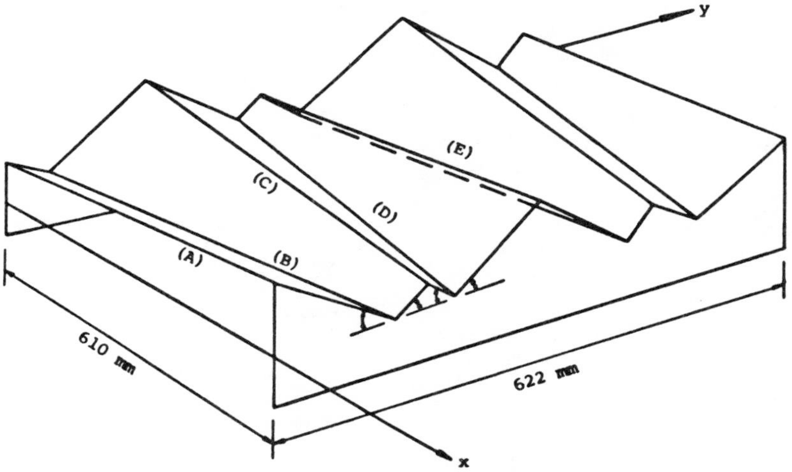

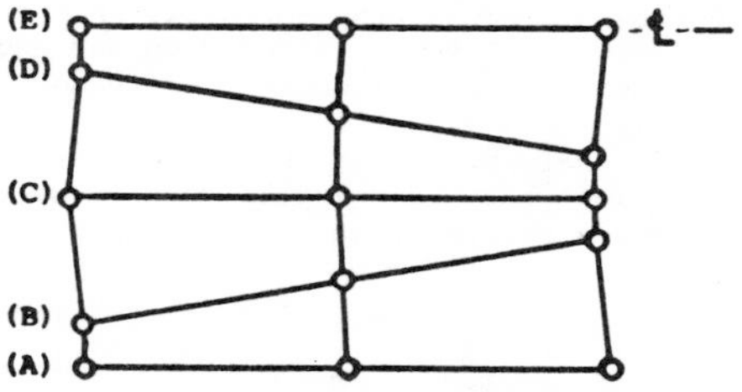

8-element idealization of half of the folded plate

All plates taper from 152.4 to 50.8 mm All angles = 40°; t = 1.6 mm
E = 71.8 x 10^6 kN/m^2; ν = 0.33 Vertical line load on ridges = 409 N/m

Geometry of a non-prismatic fold plate example

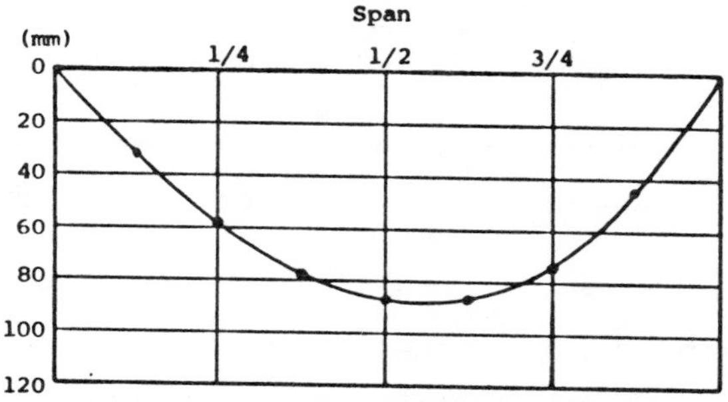

Vertical deflection, fold line (C)

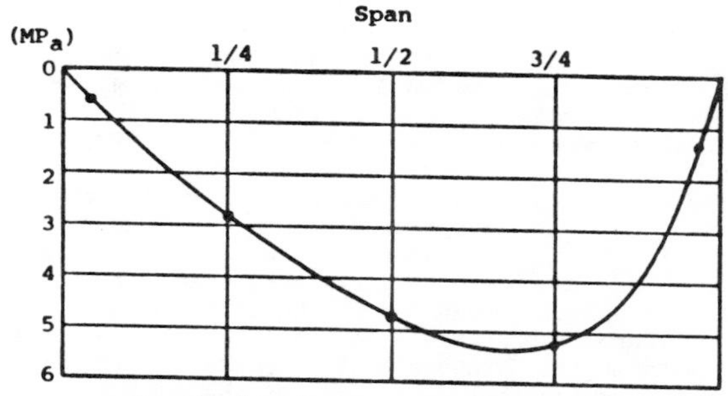

Longitudinal stress, fold line (C)

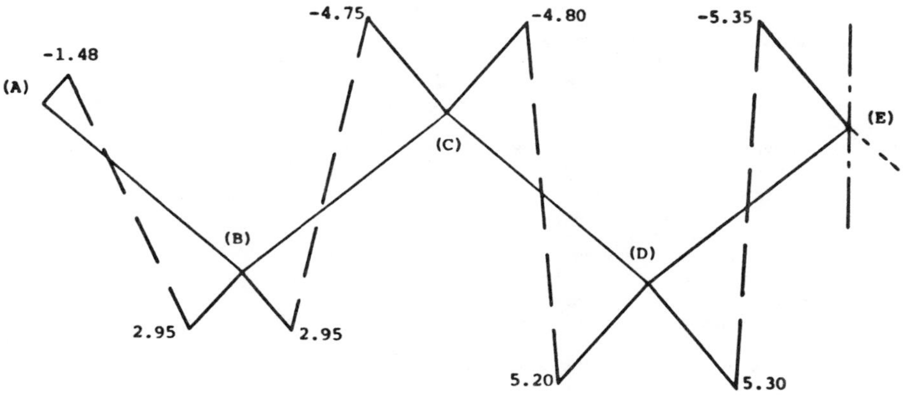

Longitudinal stress across mid-span section

Fig 4 Numerical example

where the various E, G, D, S are the usual elasticity constants and h is the thickness of the plate.

Evaluation of the integrals is done by using Gauss-Legendre numerical integration method.

These coefficients correspond of the five degrees of freedome of displacements u, v, w, α, β in the local coordinates of the flat shell element as defined previously. A fictitious rotation in the xy-plane with the corresponding stiffness coefficients being put to zero or infinitesimal value is added to the above matrix in order to complete the six degrees of freedom in space.

The transformation of the element stiffness matrix in local coordinates to the global coordinates is then carried out by means of axes rotational transformation as usual:-

$$[K_{global}] = [T^T][K_{local}][T]$$

The load vector for a uniformly distributed load q is given by

$$\{F_i\} = \int_{-1}^{+1} \int_{-1}^{+1} q[L_i^T] \det J \cdot d\xi \cdot d\eta$$

NUMERICAL EXAMPLE

The simplicity and accuracy of applying the herein derived higher-order subparametric flat shell elements in modelling folded plate structures as compared with other types of elements can be fully demonstrated in the analysis of a non-prismatic folded plate structure with a geometry and dimensions as shown in Fig 4. Half of the example structure was modelled by using eight 17-node elements, i.e. two elements for each piece of trapezpidal shaped fold plane. The results of vertical deflections, and longitudinal stresses along a ridge fold line and the longitudinal stresses across the mid-span transverse section as shown in Fig 4. It was noticed that the results so obtained by the present method are in excellent agreement with the 32-element mesh solution for the same problem presented in Ref 6; and yet less than a quarter of the number of unknowns required by the latter were employed. There is practicaly no significant difference or discontinuity in the stresses at the boundaries between the elements showing that conforming at the fold lines is very satisfactory.

CONCLUSION

The method presented here enables the stiffness matrix for higher order subparametric quadrilateral flat shell elements with any number of nodes to be formulated and generated automatically in a computer programme. This formulatin takes into account of the in-plane and bending strains as well as the transverse shear strains. Hence the effect of shear deformation, which is increasingly appreciable as the plate thickness increases, can be included in the analysis without any further effort. Folded-plates of irregular shapes, which are difficult to idealize and to analyse with the existing methods, can be readily and easily modelled and analysed by using the present quadrilateral flat shell elements. Furthermore, unlike the so far existing plate elements with the question of incompatibility along the fold lines, the proposed elements produce a conforming boundary because the same shape function is employed in all displacement vectors. It has been demonstrated that complicated non-prismatic folded-plate structures can be idealized by using just a few of these higher order flat shell elements and yet very accurate results are obtained. Therefore the data preparation can be significantly simplified and the computing effort considerably reduced in the analysis of folded-plates.

ACKNOWLEDGEMENT

The authors wish to extend their appreciation to Mr. P.M. Wong, a former research student, and Dr. G. Tham. a lecturer in the Department of Civil Engineering, University of Hong Kong, for their assistance in computer programming and analysis work.

REFERENCES

1. H.R. Evans and K.C. Rockey, A Critical Review of the Methods of Analysis for Folded Plate Structures, Proc. Instn. Civ. Engrs., 1971, Vol. 49, pp. 171-192.

2. Report of the Task Committee on Folded Plate Construction, Phase I Report on Folded Plate Construction, Journal of the Structural Division, ASCE, 1963, Vol. 89, ST6, pp. 365-405.

3. K.C. Rockey and H.R. Evans, A Finite Element Solution for Folded Plate Structures, Proceedings of Conference on Space Structures, Oxford, Blackwell, 1967, pp. 165-188.

4. Y.K. Cheung, Folded Plate Structures by Finite Strip Method, Journal of the Structural division, ASCE, 1969, Vol. 95, ST12, pp. 2963-79.

5. G.M. Lindberg, T.M. Hrudey and G.R. Cowper, Refined Finite Elements for Folded Plate Structures", Proceedings, Specialist Conference on Finite Elements in Civil Engineering, Montreal, Canada, June 1972, pp. 205-232.

6. J.E. Beavers and F.W. Beaufait, Higher Order Finite Element for Complex Plate Structures, Journal of the Structural Division, ASCE, 1977, Vol. 103, No. ST1, pp. 51-69.

7. Y.K. Cheung, P.M. Wong and H.C. Chan, Generation of Higher Order Subparametric Bending Elements, Engineering Structures, 1980, Vol. 2, No. 1, pp. 2-8.

8. H.C. Chan and Y.K. Cheung, On the Formulation of Higher Order Conforming Flat Shell Elements, Proceedings of the International Conference on Finite Element Methods, August 1982, Shanghai, China, pp.121-125.

A FOLDED PLATE SPACE STRUCTURE

P.K.K. LEE, D. HO, L.G. THAM and H.W. CHUNG

Department of Civil Engineering,
University of Hong Kong.

A folded plate reinforced concrete structure covering an area of 32m × 36m is proposed for a games hall. The roof of the structure consists of V-shaped beams supported on walls which are slightly inclined and folded in the same manner. The structure has a span of 28m at roof level and an unobstructed headroom of 9.4m. There are no stiffening diaphragms at the open ends. The structure was analysed for various load combinations by means of computer programs based on the finite strip and the finite element methods. Apart from analytical results, this paper also describes particular considerations involved in the design of the structure and special attentions to be observed during construction.

INTRODUCTION

The proposed games hall at a housing estate in Shatin, Hong Kong, is a reinforced concrete folded plate structure covering an area of 32m × 36m. The structure provides cover for a basketball court and a spectator stand. This games hall forms part of the public facilities provided by the government in the public housing estate.

The roof of the structure is a folded plate consisting of 9 V-shaped, thin-walled, reinforced concrete beams each 4m wide and 2m deep. The same folded form continues from the roof to the side walls which are slightly inclined, (Fig.1). The cross-sections of the roof beam and the side wall are shown in Fig.2. The structure has a span of 28m at roof level and an unobstructed headroom of 9.4m. There are no stiffening diaphrams at the open ends. The entire folded plate structure is simply supported on a set of 20 hinges. The foundation consists of strip footings supported on concrete piers which rest on bed rock some 20m below. The strip footings are connected by a series of ground beams.

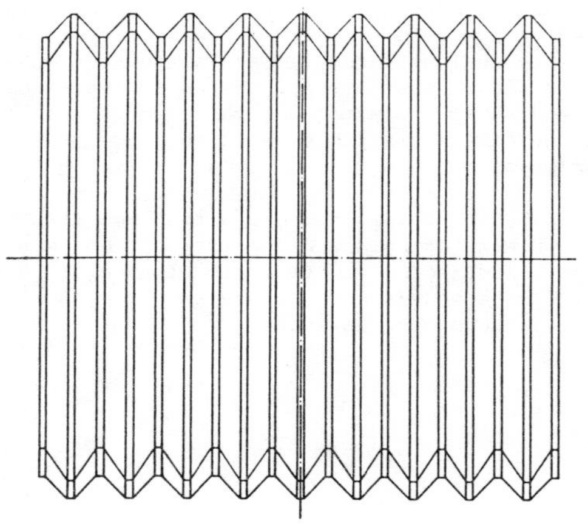

PLAN

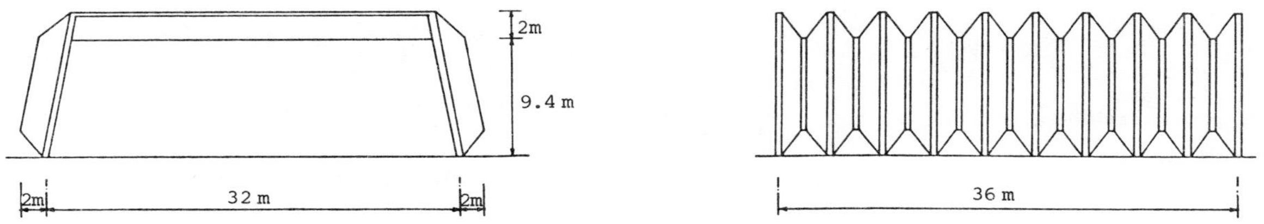

END ELEVATION SIDE ELEVATION

Fig 1 : General View of the Games Hall

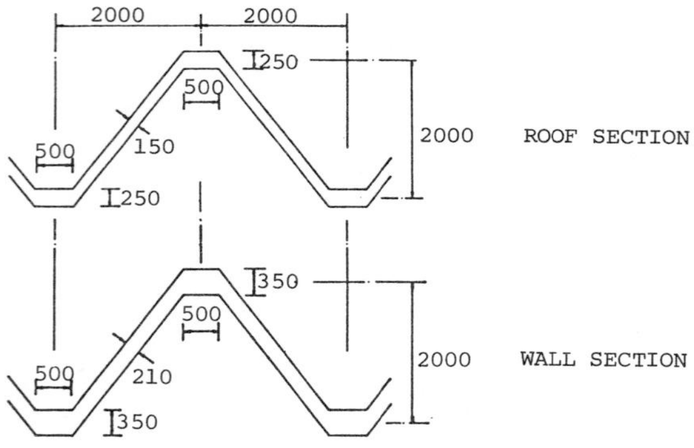

Fig.2 : Roof and Wall Sections (All dimensions in mm)

STRUCTURAL ANALYSIS

It can be seen from Fig.1 that the slightly inclined folded plate walls not only provide supports to but are also joined monolithically with the folded plate roof. This form of folded plate portal poses stress concentration problems at the roof-wall junction which is quite unique in itself and does not occur in ordinary folded plate roofs simply supported at the ends. Furthermore, the need for window and access openings from the sides reduces the supports for the entire structure to a set of 20 support "points". Hinge supports were assumed at these locations and special considerations would be required to select a suitable design for these highly stressed supports.

In order to study in more detail the behaviour of this special form of structure under load, the analysis was performed in two stages. At the preliminary stage, a finite strip analysis (Ref 1) was carried out on the folded plate roof only. The purposes of the analysis were to study the gradual diminution of the edge effects and to compare the stress distribution in the edge bays and the interior bays.

In this analysis, the folded plate roof was assumed to be simply-supported at both ends but free to deflect at the edge bays. A uniformily distributed load of 12.5 kN/m^2 was applied. Three models each consisting of 8, 4 and 2 bays respectively have been analysed.

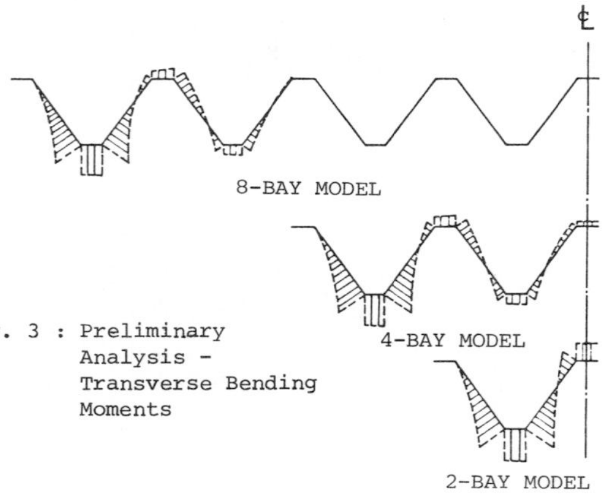

Fig. 3 : Preliminary Analysis - Transverse Bending Moments

The analysis showed that the edge effect diminishes rapidly when moving away from the edge. Take the transverse bending moment in the web of the V-shaped beam as an example (Fig.3). Comparison indicates that this moment in the penultimate bay is reduced to less than one third of the corresponding moment in the edge bay and becomes almost negligible in the third bay from the edge. It was also observed that longitudinal stresses and bending moments in the edge bay are very similar in the three models analysed. It was concluded that the 2-bay model is sufficiently accurate for use in the analysis of the edge bay.

Based on the findings of the preliminary analysis, a refined finite element analysis was carried out on two typical bays, that is, the edge bay and the interior bay.

For the edge bay, a two bay model was used. Due to symmetry, only one bay of the 2-bay model was considered. Along the two lines of symmetry, one at mid-span and another between the two adjacent bays, lateral rotation and movement were restrained. The support conditions were simulated by a roller-on-wall system, The remaining side of the structure was allowed to deflect without restraint.

The behaviour of the interior bay was assumed to resemble a typical bay within a chain of infinite number of bays connected together. As all the interior bays behaved in exactly the same manner, symmetry can be assumed along both sides of the typical bay where both lateral rotation and deformation will be prevented.

The element used in the finite element analysis was a two-dimensional isoparametric element with 8 nodal points. Each node has six degrees of freedom. (Ref.2) Due to symmetry in the geometry of the structure, it was only necessary to consider one half of each bay in the analysis. The half bay was divided into 105 elements with 56 elements in the roof and 49 elements in the wall-column.

The concrete properties adopted in the analysis were:

Modulus of elasticity	25 kN/mm
Poisson's ratio	0.15
Coefficient of expansion	$1 \times 10^{-6}/^{o}C$

Both the edge and the interior bay model were analysed for three different combinations of loading. The dead load (DL) included the weight of concrete, finishes, insulations and other miscellaneous fixtures. The imposed load (LL) on the plan area of the roof was taken as 0.75 kN/m . The wind load (WL) was similated by a horizontal pressure of 1.2 kN/m acting on the elevation of the side walls. Effects due to possible foundation settlement and thermal movement (FT) have also been included.

The three loading combinations considered were:

Case I: 1.4 DL + 1.6 LL

Case II: 1.2 DL + 1.2 LL + 1.2 WL

Case III: 1.4 DL + 1.6 LL + FT

It was assumed in the analysis that the hinges at the supports were not capable of taking tension though lateral movements were prevented. In the differential settlement analysis, it would therefore be possible for a support to be free of all reactions whilst the bay was supported entirely by the remaining non-sinking supports. In the edge bay, only the outer support was assumed to settle with respect to the inner support due to the layout of the foundation piers.

In the thermal movement analysis, the effect of an annual temperature variation of 36°C which produced an estimated contraction of 6mm of the roof member was considered.

RESULTS OF ANALYSIS

The analysis shows that stresses obtained in loading case I is more critical than those obtained in loading case II. This is obvious because the self-weight of the structure is much greater than the imposed load and wind load.

Under the action of uniformly distributed load the interior bay simply behaves as a portal frame with a V-shaped roof beam and a pair of V-shaped columns. Fig.4 shows the inplane stresses and the transverse bending moment acting in the inclined web of the roof beam. The stresses acting in the wall-column are shown in Fig.5.

In the edge bay, the roof member also behaves as a V-shaped beam in bending. However, as the outer edge of the beam is free to deflect, a substantial transverse cantilever moment is developed in the web of the roof beam (Fig.6). Fig.7 shows the inplane stresses acting in the wall-column. It is interesting to observe that the edge column is subjected to bending only and does not take any vertical load. The entire bay is therefore, supported by the inner column.

Under loading case III, it was found that the stress patterns in the walls of both the edge bay and the interior bay are almost the same as the stress pattern in the walls of the edge bay under the action of loading case I except

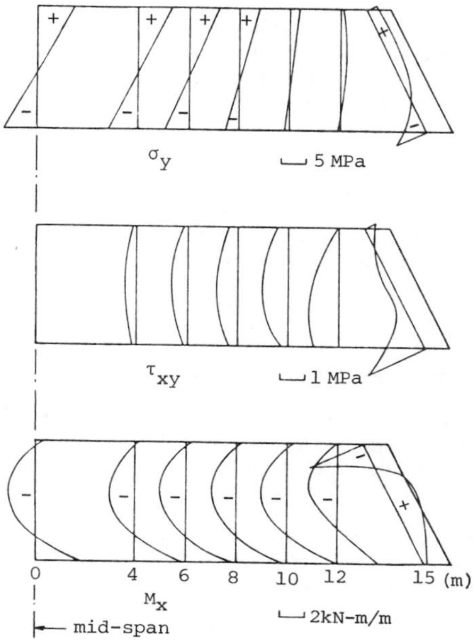

Fig.4 : Interior Bay - Inclined Web of Roof Beam

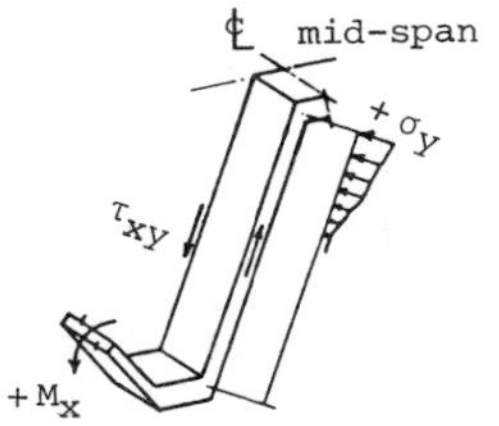

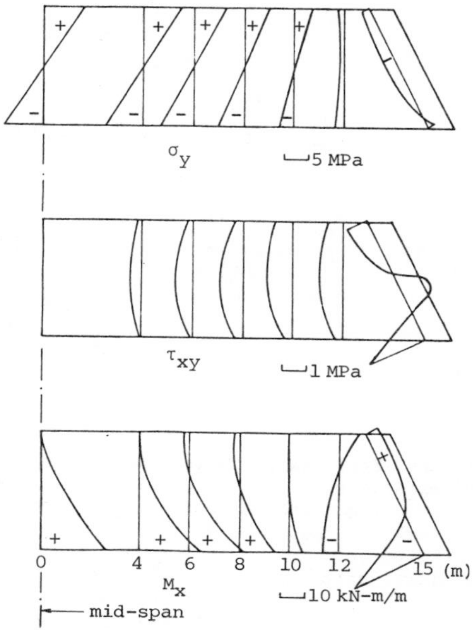

Fig.6 : Edge Bay - Inclined Web of Roof Beam

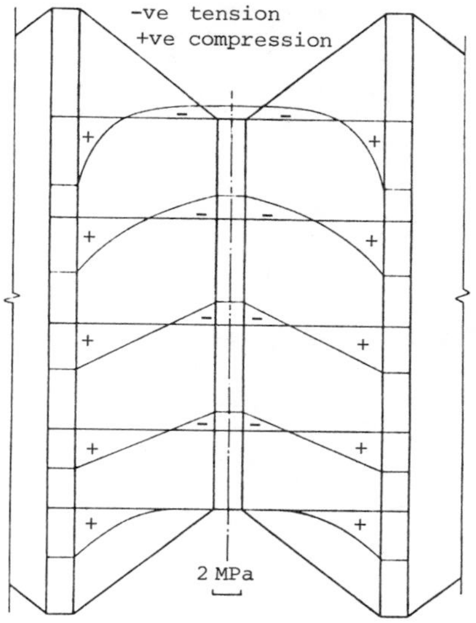

Fig.5 : Interior Bay - Inplane Stresses in Wall Column

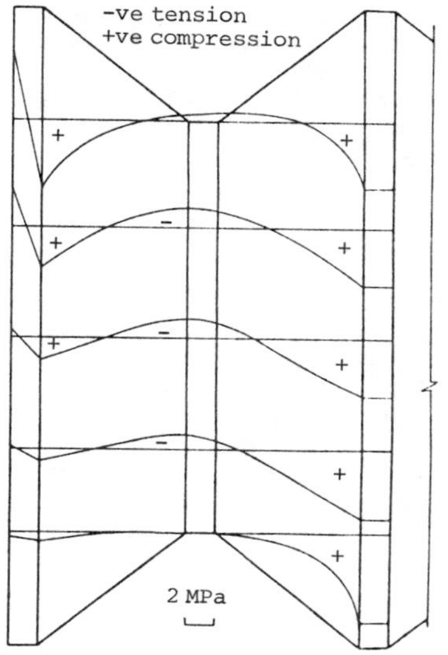

Fig.7 : Edge Bay - Inplane Stresses in Wall Column

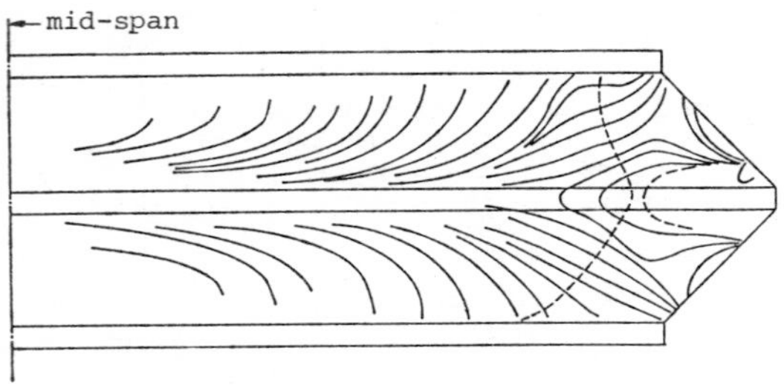

Fig. 8 : Principal Stress Trajectories - Roof

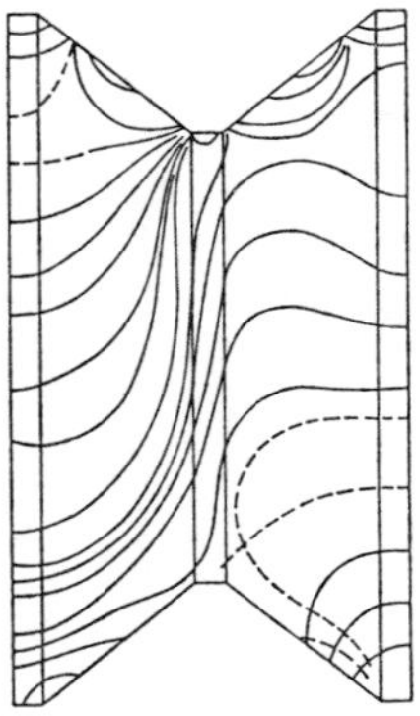

—— Tensile Stress

--- Compressive Stress

Fig. 9 : Principal Stress Trajectories - Wall

that in the latter case the stresses are always greater. Effects of differential settlement and temperature variation in the V-shaped roof beam are negligible in both typical bays.

Stress distribution is more complicated at the roof-wall junction. As expected, stress concentration will occur in this region. Fig.8 & Fig.9 show respectively the stress trajectories in the roof member and the wall of the edge bay subjected to loading case I. It can be seen that large stress concentration occurs at the obtuse corner of the web of the roof beam. Similar concentrations of bending and torsional moments are also observed at the roof-wall junction. However, the effects of such stress concentrations are very localized and disappear almost entirely at about 1m to 1.5m away from the junction.

Finally, the deflected shapes of the interior and edge bay subjected to loading case I are shown in Fig.10 and 11 respectively.

DESIGN

With the inplane stresses and moments obtained from the finite element analysis, the folded plate structure was designed in reinforced concrete. Grade 30 concrete and high yield deformed bars were used.

A comparison of stresses in the roof beams indicates that the beam in the interior bay under loading case I is most critical except for transverse bending moment in the edge bay. Identical reinforcement has been provided for all the roof beams with additional transverse reinforcement provided in the inclined web of the edge bay to cater for the large transverse cantilever moment which occurs near the free edge.

Reinforcement in the wall-columns have been designed to resist the stresses in the inner half of the wall-column of the edge bay under loading case I. The adequacy of the columns have also been checked against shear and torsion.

To cater for the stress concentrations at the roof-wall junction, reinforcement capable of resisting a tensile stress of 10 MPa has been provided in two directions within a band of 2m wide along both sides of the junction. Additional band of reinforcement has also been provided along the lower inclined edge of the walls.

As the entire weight of the structure rests upon the 20 hinge supports, the design of the hinge support is of great importance. Bearing pads with dowels have been adopted in the design and the support has been designed to take compression only. Links at close spacings have been provided for the lowest 2m of the columns immediately above the bearing pads in order that a triaxial state of stresses can be achieved.

The estimated maximum central deflection of the roof beams is about 50mm. Allowing for shrinkage and creep of concrete and 1:100 fall required architecturally, a precamber of 200 mm has been specified.

Since the folded plate is designed as a continuous structure, the sequence of removal of formwork deserves careful consideration. There should be no overstressing in any part of the structure as a result of formwork striking. It was recommended that removal of formwork should commence from the middle of the structure and proceed gradually towards the open ends in a symmetrical manner. The concrete in adjacent bays should have attained its designed strength before the formwork can be loosened in any bay. External finishes should not be put on until all the formwork has been removed and the structure has set in under its own weight.

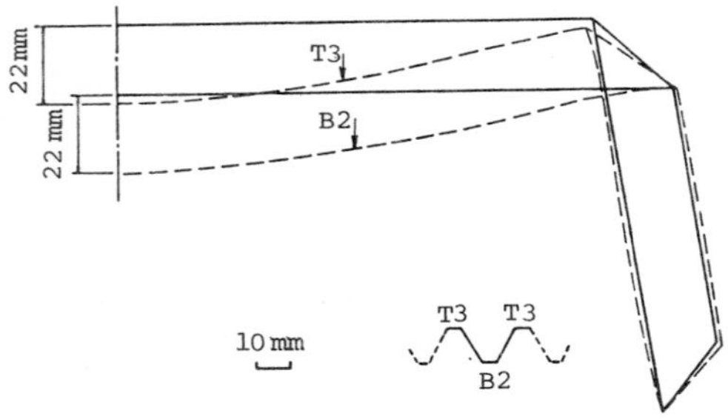

Fig. 10 : Interior Bay - Deflection

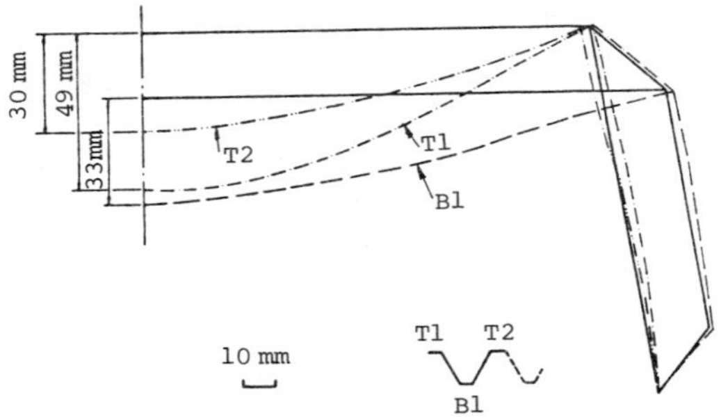

Fig. 11 : Edge Bay - Deflection

CONCLUSION

The folded plate space structure is simple in configuration and architecturally pleasing. It is an effective structure to cover an open area of large size. For the particular structure under consideration, member sizes are mainly determined by construction method rather than strength requirements.

The behaviour of the structure resembles quite closely to that of portal. Stress concentrations are localized in areas along the roof-wall junction and at the supports. Transverse moment due to free edge is substantial in the edge bays but diminishes rapidly towards the interior bays. Due to the unexceptionally high stiffness of the structure, a small differential settlement at a support may nullify the reaction there.

The erection and removal of formwork require special attention. Formwork striking schedule should be well planned so that no overstressing would occur in any part of the structure. It is interesting to note that in spite of the special form of the structure the unit cost for concreting inclusive of formwork is only slightly higher than ordinary reinforced concrete.

ACKNOWLEDGEMENTS

The investigation reported above was commissioned by the Hong Kong Housing Authority whose permission for publication of the paper is gratefully acknowledged.

REFERENCES

1. Y K CHEUNG, Finite Strip Methods in Structural Analysis, Pergamon Press, 1976.

2. E HINTON and D R J OWEN, An Introduction to Finite Element Computation, Pineridge Press, 1979.

LINEAR COMPLEMENTARITY PROBLEMS IN ELASTIC-PLASTIC-FRACTURING TRUSSES

E. BARONI*, A.BOVE**, B. LEGGERI***

* Adjoint Professor, Department of Constructions, University of Florence, Italy

** Associate Professor, Department of Constructions, University of Florence, Italy

*** Adjoint Professor, Department of Constructions, University of Florence, Italy

Several formulations of structural problems by linear complementarity methods are presented, including any type of linear constraint conditions on both stresses and displacements. The pre-solution with respect to unconstrained variables allows to minimize the dimensions of the final problem. If the tableau matrix is symmetrical, the Kuhn-Tucker conditions allow to express equivalent variational inequality problems, where some sort of energy forms may be recognized. Such procedure may no longer be applied when friction constraints are specified, since the tableau is unsymmetrical. In such cases, some "enlarged" linear complementarity problems or variational inequalities are discussed: if the given problem admits a solution, this solves also the so obtained enlarged problem, although vice versa does not hold. Such remarks are still applied to elastic-fracturing structures, in order to solve the resulting linear nearly-complementarity problems.

INTRODUCTION

A great deal of interest has been recently devoted to the enhancing of numerical problems concerning inelastic materials or/and unilateral constraint conditions. These can be included in a standard linear complementarity problem (LCP) which is stated as:

$$\begin{cases} M\,x + p = y \\ x\,,\ y \geqslant 0 \\ x^t\,y = 0 \end{cases}$$

Several applications of the LCP to structural analysis are outlined in Ref 1, where the main contributions are quoted within specific fields of investigations (plasticity, for istance). Furthermore, an extension of LCPs to the nonlinear analysis of contact problems was attempted in Ref 2, by means of a step-by-step LCP, where linear constraints on displacements only are specified. Some remarks about the possibility and the meaning of an unsymmetric tableau matrix M are given in Ref 3, together with a simple numerical example where rigid blocks are connected by elastic-plastic-fracturing truss elements. As a matter of fact, the several numerical procedures available for searching a solution of a LCP will not be investigated in detail in this paper. They mostly depend on the particular characteristics of the tableau, which finally results from the initially stated constraint conditions.
A fairly simple solution technique may be successfully used when only sign restrictions are obtained in the final formulation and other assumptions are satisfied. Thus, an application of fixed-point methods and the Hildreth-d'Esopo procedure is described in Ref 4, where truss structures are investigated.
In this paper we shall try to suggest a comprehensive LCP formulation of linear-constrained structural problems. Such approach is strictly confined to truss-like structural models, so that the constraint conditions in terms of stresses may be always expressed by linear inequalities. Otherwise, nonlinear conditions (e.g. relating to principal directions) can be specified by piecewise linear ones.
Furthermore, we shall give a variational formulation which is equivalent to a given LCP, via Kuhn-Tucker conditions. Some final remarks allow to find some sort of energy variational inequality even in the case of unsymmetrical tableau, when K.T. conditions may no longer be applied. In these cases, an available approach in terms of QP and LC problems may still be searched, as will be explained in the last section.
Finally, we agree that much attention ought to be payed to the computational problems outlined in this paper, and in particular to the discussion of both feasibility and optimality of the solutions, if any exists. We shall only quote about this matter the contributions in Ref 5.

FORMULATION

i) LCP with constraint conditions on internal and boundary stresses only.

We shall consider the case of discretized truss-like structures, where any type of linear constraint conditions on the stresses are introduced.
A direct LCP formulation will be achieved by stating equilibrium, kinematical relationships and other related inequalities. Afterwards, an energy formulation will be given.
Let us denote by X and Z the internal and boundary forces, respectively. A set of global equilibrium equations may be expressed by:

$$A\,X + B\,Z + F = 0 \qquad \ldots\ldots \quad 1.$$

where A and B are the configuration matrices and F is a nodal load vector (parametric LCPs will not be considered here). Any set of constraint conditions in terms of internal or boundary forces will be always stated by the following inequalities:

$$C\,Z + q = w \qquad \qquad \ldots\ldots\ 2.$$
$$w \geqslant 0$$

$$L\,X + b = y \qquad \qquad \ldots\ldots\ 3.$$
$$y \geqslant 0$$

Many different plasticity conditions or simply sign restrictions on either stress may be expressed by the above-stated formulae. Particular cases of sign restrictions or simple homogeneous restrictions of the stresses will be discussed later. We shall only remark that C and L ("constraint configuration matrices") may be boolean, sparse and they account for Coulomb-friction restraints, according to Ref 3.

On the other hand, we must add to Eqn 1 the corresponding kinematic relationships between nodal displacements x and the stress unknowns :

$$A^t\,x + K\,X + d_x = s_1 \qquad \qquad \ldots\ldots\ 4.$$

$$B^t\,x + H\,Z + d_z = s_2 \qquad \qquad \ldots\ldots\ 5.$$

where K and H are symmetrical strictly positive-definite matrices of the elastic coefficients. In truss-like structures, they are also diagonal matrices, but this is no necessary assumption here. Indeed, we might account for very stiff internal or - more likely - boundary elements, by requiring only semi-positive-definiteness of K and H ; however, what follows may be better expressed under the usual hypothesis of nonsingularity of such matrices, so that we shall denote the unique inverse matrices by:

$$E = K^{-1} \quad ; \quad F = H^{-1} \qquad \qquad \ldots\ldots\ 6.$$

Moreover, we must point out that d_x and d_z are the settlements or distortions in internal and boundary elements, respectively (these are known loads in terms of displacements or strains).

In unconstrained elastostatics, it is well-known that the right-hand side terms of Eqns 4 and 5 vanish, due to the compliance between kinematical and elastic deformations. This is no longer true in constrained problems, where yielding may occur: plastic strains, fractures or other yield effects (including any slipping due to the friction) may all be viewed as residual terms of those equations. Such unknown residua are here denoted by s_1 and s_2 and are nullified if the specified constraints have no effect on a given stress component. It should be remarked that s_1 and s_2 may express only the 'resulting' effects of all the possible constraints. They actually do not correspond to the single constraints (as may be seen checking the dimensions of A,B,L,C matrices), but they only account for the total amount of unbalanced residua in each of the afore stated Eqns 4 and 5, and hence in the corresponding structural elements. So we must introduce suitable matrices, in order to project the space of constraints into the space of the stress components. We shall denote the so obtained relationships by:

$$L'\,u = s_1 \qquad \qquad \ldots\ldots\ 7.$$

$$C'\,v = s_2 \qquad \qquad \ldots\ldots\ 8.$$

$$u \geqslant 0 \qquad \qquad \ldots\ldots\ 9.$$

$$v \geqslant 0 \qquad \qquad \ldots\ldots\ 10.$$

Here, the used symbols L' and C' suggest that in many cases these are the transpose matrices of L and C ; nevertheless, in particular types of plasticity conditions and also when friction is taken into account, such relationship is no longer true. We shall adopt such notation, with the agreement that in most cases $L'=L^t$, $C'=C^t$ (see also Ref 3). Furthermore, u and v must be nonnegative variables (as stated in Eqns 9 and 10), just in the same way as the corresponding w and y in Eqns 2 and 3 are sign restricted. In other words, if a given constraint becomes active, there is a $w_k=0$ or $y_i=0$. The evolution of the corresponding yielding mechanisms leads to: $u_k > 0$ or $w_i > 0$. In each single j-th element such yielding strains are properly summed up to give the residuum vector in Eqn 4 or 5. Actually, such yielding phenomena may be foreseen as acting in opposite directions: i.e. crushing and tensile yield; yet each constraint is stated separately and requires a couple of u,v or u,y variables.

For istance, a yielding truss model with n and k constraints on compressive and tensile stresses respectively,may be defined by the following non-zero elements in the constraint configuration matrix L:

$$L = L_{ij} \ ; \ i=1,\ldots,n+k+m \ ; \ j=1,\ldots,n+1$$

$$i \leqslant n \longrightarrow L_{ii}=1$$

$$n < i \leqslant n+k \longrightarrow L_{i,i+1}=1$$

There are n-2xl elemnts where both upper and lower bounds for stresses are defined. The lower m rows of L account for friction conditions.

Finally, an orthogonality condition must be stated, since for each constraint condition at least one variable vanishes : u or y, v or w. From a purely mechanical point of view, this means that only if the so-called "slack" variable y or w is zero, then there is the possibility of a corresponding yield strain (plastic strain and so on) denoted by u or v, respectively. Dually, only if yielding does not occur, the slack variables may be non-zero and so the constraints are inactive. Since all these are nonnegative variables, we obtain:

$$u^t y + v^t w = 0 \qquad \qquad \ldots\ldots\ 11.$$

ii) A variational inequality problem with constraint conditions on internal and boundary stresses only.

The preceding equations 1 to 11 may be collected here, assuming also that:

$$M' = M^t \ ; \quad L' = L^t \ ; \quad C' = C^t$$

We then get:

$$A\,X + B\,Z + F = 0$$
$$K\,X + A^t x - L^t u = 0$$
$$H\,Z + B^t x - C^t v = 0$$
$$L\,X + b = y$$
$$C\,Z + q = w$$
$$u\ ,\ v\ ,\ w\ ,\ y \ \geqslant 0$$
$$u^t y + v^t w = 0$$

Such are the Kuhn-Tucker conditions of the following variational inequality problem:

$$\frac{1}{2}\,X^t K X + \frac{1}{2}\,Z^t H Z = \min$$

subjected to:
$$A\,X + B\,Z + F = 0$$
$$L\,X + b \geqslant 0$$
$$C\,Z + q \geqslant 0 \qquad \qquad \ldots\ldots\ 12.$$

This expression may be viewed as the complementary energy

principle where stress components are linearly constrained.

iii) Formulation of a LCP with constraint conditions on both stresses and displacements.

Extending the preceding remarks and denoting by M the constraint configuration matrix relating to the nodal displacements, we obtain:

$$\left[\begin{array}{ccc|ccc} O & A & B & -M' & & \\ A^t & K & 0 & & -L' & \\ B^t & 0 & H & & & -C' \\ \hline M & & & & & \\ & L & & & 0 & \\ & & C & & & \end{array}\right] \left[\begin{array}{c} x \\ X \\ Z \\ \hline h \\ u \\ v \end{array}\right] + \left[\begin{array}{c} F \\ d_x \\ d_z \\ \hline g \\ b \\ q \end{array}\right] = \left[\begin{array}{c} \\ O \\ \\ \hline p \\ y \\ w \end{array}\right]$$

$$\left[p^t,\ y^t,\ w^t\right] \geqslant 0$$

$$\left[h^t,\ u^t,\ v^t\right] \geqslant 0$$

$$p^t\ h + y^t\ u + w^t\ v = 0 \qquad \text{...... 13.}$$

This can easily be expressed in compact form:

$$\left[\begin{array}{c|c} S & -Q' \\ \hline Q & 0 \end{array}\right] \left[\begin{array}{c} r_1 \\ \hline r_2 \end{array}\right] + \left[\begin{array}{c} a \\ \hline b \end{array}\right] = \left[\begin{array}{c} 0 \\ \hline r_3 \end{array}\right]$$

$$r_2, r_3 \geqslant 0$$

$$r_2^t\ r_3 = 0 \qquad \text{...... 14.}$$

Anyway, in such LCP only two out of the three unknown vectors are explicitly constrained (otherwise: there is a ready zero complementary solution above the r_3 subvector). So, it is fairly useful to solve with respect to the only unknown which is not sign restricted, and take advantage by the nonsingularuty of the S submatrix: such hypothesis states no more than the existence and uniqueness of the initial unconstrained elastic problem.
We obtain so:

$$r_1 = S^{-1} Q' r_2 - S^{-1} a$$

$$Q S^{-1} Q' r_2 - Q S^{-1} a + b = r_3$$

$$r_2, r_3 \geqslant 0$$

$$r_2^t\ r_3 = 0 \qquad \text{...... 15.}$$

Thus we have obtained a standard LCP where the tableaû is

$$T = Q\ S^{-1}\ Q'$$

and the known vector (external loads and bound values) is:

$$q = b - Q\ S^{-1}\ a$$

When no constraint is active, the trivial solution coincides with the unconstrained elastic one; that is:

$$r_3 = b - Q\ S^{-1}\ a \geqslant 0 \quad ; \quad r_2 = 0 \quad ; \quad r_1 = -S^{-1}\ a$$

However, the practical solution of Eqn 13 does not require to compute the inverse matrix of S (i.e. to find all the influence coefficients in terms of both stresses and displacements). In fact, a partitioned form of such inverse is known by linear elasticity, so that (recalling what has been stated in Eqn 6) we get the following LCP :

$$\left[\begin{array}{c|c|c} MR^{-1}M' & MR^{-1}AEL' & MR^{-1}BFC' \\ \hline LEA^tR^{-1}M' & LEL'-LEA^tR^{-1}AEL' & -LEA^tR^{-1}BFC' \\ \hline CFB^tR^{-1}M' & -CFB^tR^{-1}AEL' & CFC'-CFB^tR^{-1}BFC' \end{array}\right] \left[\begin{array}{c} h \\ \hline u \\ \hline v \end{array}\right] +$$

$$\left[\begin{array}{c} g + MR^{-1}F - MR^{-1}AEd_x - MR^{-1}BFd_z \\ \hline b - LEA^tR^{-1}F - LEd_x + LEA^tR^{-1}AEd_x + LEA^tR^{-1}BFd_z \\ \hline q - CFB^tR^{-1}F + CFB^tR^{-1}AEd_x - CFd_z + CFB^tR^{-1}BF \end{array}\right] =$$

$$= \left[\begin{array}{c} p \\ \hline y \\ \hline w \end{array}\right] \geqslant 0 \quad ; \quad h,u,v \geqslant 0$$

$$p^t\ h + y^t\ u + w^t\ v = 0 \qquad \text{...... 16.}$$

Where we have set: $R = AEA^t$, so that the previous form of LCP requires as a first step only the pre-solution of Eqn 13 by the standard displacement method.

iv) Variational inequality problem with constraint conditions on both stresses and displacements.

Once again, assuming that the tableau is a symmetric matrix (i.e.: $M'=M^t$, $C'=C^t$, $L'=L^t$), the Kuhn-Tucker conditions allow to state a correspondence between a given LCP and an equivalent quadratic programming problem (QPP) and vice versa.
Since we have dealt in Eqns 13 and 14 with a quite general case, let us consider a global energy form, where both stresses and displacements are included.
In long form we obtain:

$$\frac{1}{2}\left[h^t\ u^t\ v^t\right] \left[\begin{array}{ccc} MR^{-1}M^t & MR^{-1}AEL^t & MR^{-1}BFC^t \\ & LEL^t-LEA^tR^{-1}AEL^t & -LEA^tR^{-1}BFC^t \\ \text{sym} & & CFC^t-CFB^tR^{-1}BFC^t \end{array}\right] \cdot$$

$$\cdot \left[\begin{array}{c} h \\ u \\ v \end{array}\right] + \left[h^t\ u^t\ v^t\right] \left[\begin{array}{c} g+\dots \\ b-\dots \\ q-\dots \end{array}\right] = \min$$

$$\text{subjected to: } \left[h^t\ u^t\ v^t\right] \geqslant 0 \qquad \text{...... 17.}$$

From a comparison between such QPP and Eqn 12 some remarks arise: firstly, the number of unknowns is rather increased (they are as many as the initially-stated constraint conditions, which involved displacements too). Furthermore, we have reduced all the possible constraints to sign restrictions only.
Since such problems are well handled by relaxation-type algorithms (e.g. Hildreth-d'Esopo), it will be worthwhile to remark that the tableau T is now nothing more than the strictly-positive-definite matrix S^{-1} , although this is "scaled" by the configuration matrices Q.This may help to properly formulate such constraint conditions.

A suitable choice of the form by which the constraint conditions will be expressed is indeed a valuable step in the improving of the actual solution technique; here we will point out that the final LCP may always include as many unknowns as the formerly stated constraint conditions.
In many cases only the constraining of some stress variables - say X - is required, so that we shall set:

$$\begin{bmatrix} 0 & A_0 & A_c \\ A_0^t & K_0 & 0 \\ A_c^t & 0 & K_c \end{bmatrix}\begin{bmatrix} x \\ X_0 \\ X_c \end{bmatrix} + \begin{bmatrix} F \\ 0 \\ 0 \end{bmatrix} = \begin{bmatrix} 0 \\ 0 \\ L'u \end{bmatrix}$$

$$LX_c + b = y$$

$$u^t y = 0$$

$$u, y \geqslant 0$$

Once again a pre-solution method leads to a more compact LCP:

$$(L E_c L' - L E_c A_c^t R^{-1} A_c E_c L') u + (b - L E_c A_c^t R^{-1} F) = y$$

$$u , y \geqslant 0$$

$$u^t y = 0$$

where:

$$E_c = K_c^{-1}$$

$$R = A_o K_o^{-1} A_o^t + A_c E_c A_c^t$$

and also:

$$x = R^{-1} (F + A_c E_c L'u)$$

$$X_c = - E_c A_c^t x + E_c L'u$$

$$X_o = - E_o A_o^t x$$

v) Unsymmetrical tableau: how to solve these problems by variational inequalities.

The Kuhn-Tucker conditions do not allow to formulate an equivalent QP problem, corresponding to a given LCP with unsymmetrical tableau. This can be viewed as a drawback, since specific methods for the solutions of QP problems (see, for instance, Ref 6) may no longer successfully employed. On the other hand, a QP formulation may be considered as a statement involving some sort of energy and this may improve the understanding of such problems.
As a matter of fact, let us suppose to have obtained an LCP of the form:

$$M x + p = y$$

$$x , y \geqslant 0$$

$$x^t y = 0$$

$$M \neq M^t$$

What follows allows to still treat the LCP by QP methods let us consider:

$$x^t M x + p^t x = s = \min$$

subjected to:

$$x \geqslant 0$$

$$M x + p \geqslant 0$$

such constrained minimum is clearly nonnegative and gives the solution of the original LCP if and only if s = 0. Otherwise, a suitable algorithm can give the minimum amount of what could be denoted as "constrained work", which shows that no solution exists: this means that any feasible set of complementary variables does not fit the ortogonality condition.

vi)Elastic-fracturing structural models by QP.

A particular problem is given by the formulation of fracturing-constraint conditions; by such term we shall denote truss-like elements that behave elastically until fracturing occurs (or any type of sudden brittle failure); afterwards, they keep a quite neglectable stiffness. The constraint conditions, together with the other already stated equations, are:

$$A X + F = 0$$

$$A^t x + K X = u$$

$$u \geqslant 0$$

$$X \leqslant c$$

$$u^t X = 0$$

By siply setting: z=c-X we get what follows:

$$(EA^t R^{-1} AE - E) u + (c + EA^t R^{-1} F) = z$$

$$u , z \geqslant 0$$

$$u^t z - u^t c = 0 \qquad \ldots\ldots 18.$$

Once again we want to formulate a QP problem, as in the previous section, which is not equivalent to Eqn 18. However, the QP that will be obtained has the property to admit only positive solutions. The vanishing of the following scalar s implies that the original problem has been solved:

$$u^t (EA^t R^{-1} AE-E) u + u^t (q-c) =$$

$$= u^t Mu + u^t p = \min = s$$

subjected to: $u \geqslant 0$

$$Mu+p \geqslant 0$$

We could simply call such a problem " enlarged equivalent" QP (EEQP), since it may admit solutions not satisfying the original nearly-complementary problem (which in this case has no solution at all). However, if the original problem has any solution, this satisfies also the EEQP. The final matter is only in the checking of the nonnegative minimum s. It is possible now to obtain a LCP equivalent to the EEQP, via Kuhn-Tucker (just the same was possible in the preceding section, too).
The "enlarged equivalent" LCP (EELCP) may now be written as:

$$M u + p = y \geqslant 0$$

$$2Mu - Mt + p = v \geqslant 0$$

$$u,t \geqslant 0$$

$$u^t v + y^t t = 0$$

The final check must be performed on the complementary variables; actually, the EELCP solves the given nearly-complementary problem only if the following also holds:

$$y^t u = 0 \quad (\text{or, equivalently: } v^t t = 0)$$

which is indeed an orthogonality condition formerly required but not explicitly stated in the enlarged problems.

REFERENCES

1. R.W.COTTLE: Numerical Methods for Complementarity Problems in Engineering and Applied Sciences, in: Comp. Meth. in Appl. Sci. and Eng., 37, Berlin, 1979.

2. A.BOVE, S.BRICCOLI BATI, B.LEGGERI: Geometrically nonlinear Analysis of Contact Problems in Elastic Continua, ZAMM-58, T 197-198 (1978).

3. A.BOVE, M.PARADISO, G.TEMPESTA: Problemi di complementarietà lineare nell'analisi di strutture discretizzate a comportamento elasto-plasto-fragile (with English summary), 6th Nat. Congress AIMETA, II, 267-275, Oct. 1982.

4. A.BOVE, M.PARADISO, G.TEMPESTA: Relaxation Techniques for Variational Inequality Problems in Structural Analysis, 9th IKM Int. Kongr. über Anwend. der Math. in den Ingenieurwissenshaften, Weimar (DDR), Heft 5, 78-81, 1981.

5. M.L.BALINSKI, R.W.COTTLE Eds: Complementarity and Fixed Point Problems, Amsterdam, 1978.

6. J.L.KUESTER, J.H.MIZE: Optimization Techniques with Fortran, New York, 1973.

WIND INDUCED LOADING OF BARREL VAULT STRUCTURES

N TOY*, BSc, PhD and T A FOX, BSc**

* Lecturer, Department of Civil Engineering
University of Surrey, Guildford U.K.

** Research Student, Department of Civil Engineering
University of Surrey, Guildford U.K.

Structural loading due to wind effects still represents a considerable unknown in the barrel vault design procedure, and, since it is the airflow around these structures which gives rise to the loading, it is of fundamental importance that the engineer understands the aerodynamic properties of the building's geometry. This report, therefore, presents results from an initial investigation of the airflow in the near wake region of two and three dimensional barrel vault models.

The wind tunnel tests involved pulsed wire anemometer traverses behind four models, each having a different chord/length ratio, in order to assess what effect this dimensional change has on the airflow characteristics. The different flow patterns obtained indicate that many of the features identified within the near wake follow length related trends. Particular attention is then given to the significance of these variations with regard to both structural wind loading, and environmental design considerations.

INTRODUCTION

The use of the barrel vault in situations where space coverage is required with a clear internal span, has become increasingly popular in recent years. This development can be attributed to a significant reduction in the self-weight of these structures, due mainly to the utilization of both new materials and numerical analysis techniques. However, although many barrel vaults have been constructed in the past decade, both on a large and small scale, the influence of the wind still represents one of the greatest unknowns in their design procedure.

When considering wind loading problems the designer generally requires information on the relationship between the wind environment at the chosen site, and the forces it induces upon the structure. In order to predict such loads, most structural engineers will turn first to the relevant codes of practice for the particular country in which they work. Such documents as the Russian Code (U.S.S.R., BC & R II-A. 11-62), Indian Code (IS:875-1964) and the U.K. Wind Loading Handbook, Ref 1, provide mean surface pressure distributions for a variety of shapes under specific wind regimes. For the case of the barrel vault, it is possible to find some information in these codes about wind pressures on a curved roof, either upon an enclosure wall or sitting directly on the ground. This then gives the designer a limited indication of the forces acting upon the structure.

Wind tunnel tests carried out at Surrey, Ref 2, using both two and three dimensional semi-cylindrical models, suggested that much of the wind loading information provided by these codes for curved surfaces is an over-simplification. It was found that the pressure coefficients and pressure distributions predicted for specific airflows, were of an approximate character, and not as precise as might be desired for the understanding of wind forces acting upon a barrel vault structure.

Wong, therefore, proceeded to carry out further experimental work to determine the characteristics of this pressure distribution, and from her initial observations it is possible to determine values for the coefficient of drag (C_D) and, the coefficient of lift (C_L), for each model tested. Some of these are presented in Table A, and clearly indicate that the structural loads of uplift and drag resulting from wind effects vary depending on the models length to chord ratio.

TABLE A: Relationship of C_D and C_L values to model length for rough surface barrel vaults in the smooth boundary layer. (Data from Ref 1).

Model length/chord	C_D	C_L
2 Dimensional	0.39	0.93
2:1	0.41	0.47
1:1	0.44	0.42

Re $\cong 10^5$

e/D: $7.3 \times 10^{-3} \sim 1.3 \times 10^{-2}$

Uplift $= C_L \times \frac{1}{2}\rho \tilde{U}^2 \times D \times L$

Drag $= C_D \times \frac{1}{2}\rho \tilde{U}^2 \times \frac{D}{2} \times L$

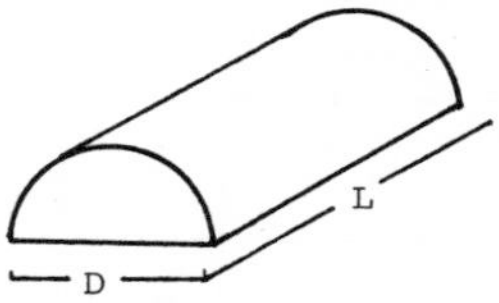

Since a change in one of the characteristic dimensions of a barrel vault, in this case its length, has been shown to alter the surface pressure distribution and therefore the degree of structural loading. It is of fundamental importance that the designer understands the aerodynamic effect of the building's geometry when considering wind loading problems.

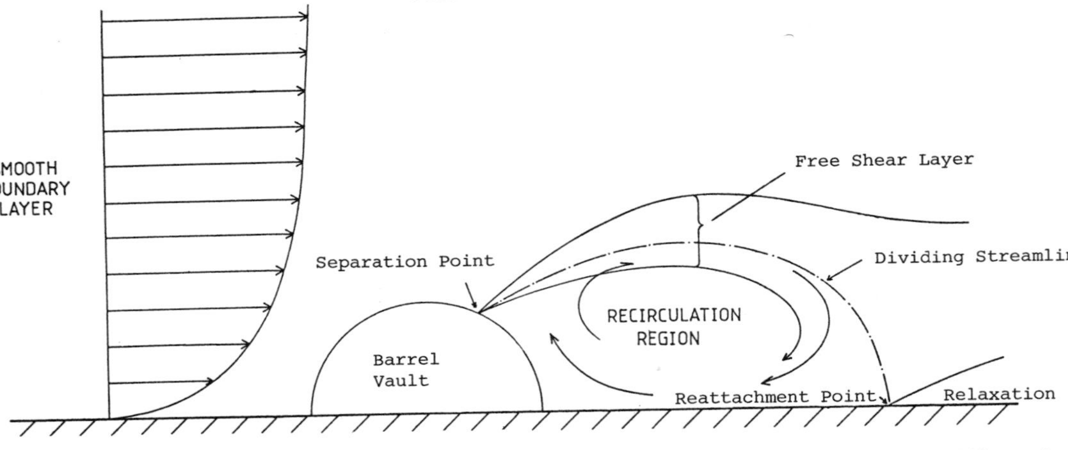

FIGURE 1. Two Dimensional Flow Phenomena.

Figure 1 illustrates the two dimensional flow phenomena associated with the basic barrel vault shape. This is seen to be extremely complex with three identifiable flow regions. The separation point marks the begining of recirculation, which is divided from the freestream by the growing shear layer. Beyond the reattachment point the recirculation region ceases to exist, being replaced by relaxation and a corresponding redevelopment of the undisturbed boundary layer. This whole pattern becomes even more complex when consideration is given to the three dimensional case. Yet, little significant research has been done on the parameters affecting these phenomena and their relationship to the resulting wind loads.

This paper is therefore concerned with the experimental determination of the air flow in the wake of barrel vault models immersed in a simulated atmosphere boundary layer. Particular attention is given to what effect the length of the model has on the nature and parameters of the recirculation region. Comments are also made on the relationship between the experimental data obtained and the requirements of the designer.

EXPERIMENTAL ARRANGEMENT

Four barrel vault models were the object of the laboratory work, all having a circular arc cross-section with subtended angle at the centre of 180°, and all constructed from plastic pipe of 0.15m diameter. One model was made with its length equal to the height of the tunnels working section so that the two dimensional flow condition could be studied. For the observation of three dimensional flow effects, each of the other models had a different length, giving length to chord ratio values of 1, 2 and 3, as shown in Table B.

TABLE B: Model Dimensions

Subtended angle, θ = 180°
Radius of curvature, R = 0.075m
Chord, Sx = 0.15m
Rise/Chord, R/Sx = 0.5
Blockage Ratio = 7.0%
Roughness Ratio = $7.3 \times 10^{-3} \sim 1.3 \times 10^{-2}$

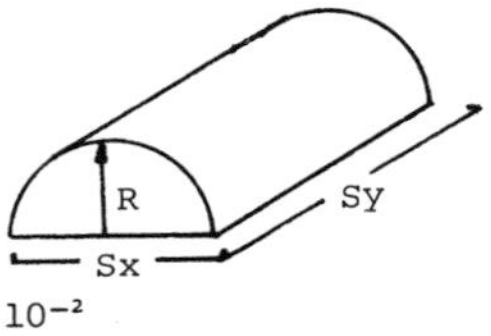

Model:	2 Dimensional	3:1	2:1	1:1
Length, Sy	1.372m	0.45m	0.3m	0.15m
Length/Chord Sy/Sx	9.147	3.0	2.0	1.0

Simulation of the Reynold's number associated with full size barrel vaults in the field was approximated by the use of a surface roughness technique. This involves artificially roughening the models surface by the use of sand and adhesive to promote the transition of subcritical to supercritical flow at lower Reynold's numbers. Thus allowing the flow pattern of the full scale condition to be adequately simulated in a low speed wind tunnel. This method has been successfully used in previous work described in Ref 2.

Each model was investigated in the Department of Civil Engineering's low speed, blow down, open circuit wind tunnel, the details of which can be found in Ref 3, previously presented at this conference. The 'smooth' boundary layer was generated for all the experimental work, with a reference freestream air velocity ($\hat{U}$) , of 6 m/s, giving a Reynold's number based on model diameter of $R_e = 0.6 \times 10^5$.

For measurement of the local velocity (U), a pulsed wire anemometer is used, since it is capable of detecting both the positive and negative velocities to be found in the turbulent wake of each of the models (a detailed account of the probe is given in Ref 4). This instrument is mounted on the three dimensional traversing mechanism located in the working section of the tunnel and capable of positioning at any point in the flow. Movement of this traversing mechanism is controlled via a Commodore Pet 3032 micro-computer, which is also used for pulse generation as well as data acquisition and analysis.

DISCUSSION

Velocity profiles and turbulence intensity plots were obtained at regular intervals within the wake region of each model and these are shown for the 2:1 ratio as Fig 2(a) and Fig 2(b) respectively.

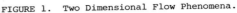

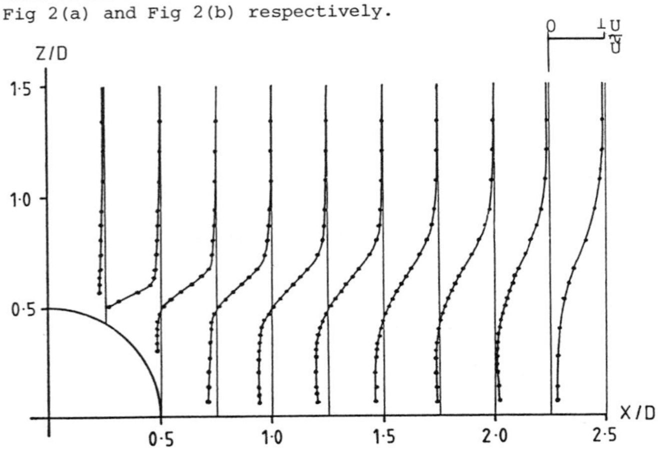

FIGURE 2(a) Velocity Profiles for Model Length/Chord Ratio 2:1.

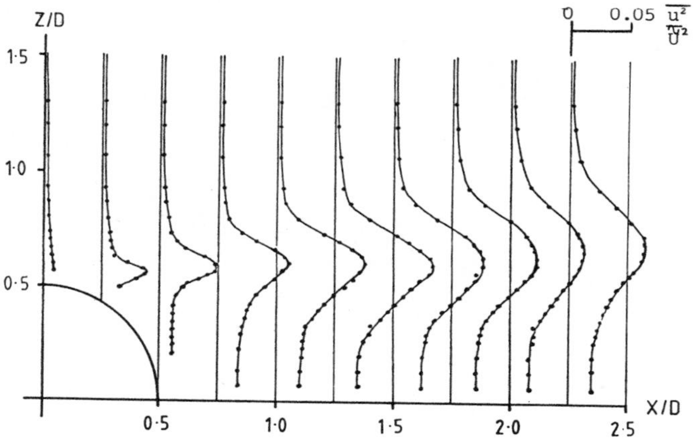

FIGURE 2(b). Turbulence Intensity Profiles for Length/Chord Ratio 2:1.

Streamline drawings, which illustrate the general flow pattern in two-dimensions, were produced for each model by operating on the velocity profiles using the streamline function;

$$\psi = \int U/\tilde{U}\, d(Z/D)$$

and these are shown for the 2:1, 3:1 and Two-Dimensional models as Figs 3(a), (b), and (c) respectively.

From these drawings it is possible to recognise clearly the general flow phenomena illustrated in Fig 1.

The recirculation regions are well developed, and although the separation and reattachment points were not directly measured, their positions can be estimated with reasonable accuracy.

The location of the zero streamline can be taken as representing the limit of the recirculation region. This line has been plotted for each of the four models on the same axis, Fig 4. In the case of the two dimensional model a long and relatively flat recirculation region is observed with a reattachment point at about four diameters downstream. As the model length is then decreased, with all other dimensions and properties kept constant, the separation point moves downstream on the models surface, and the reattachment point moves upstream on the tunnel wall.

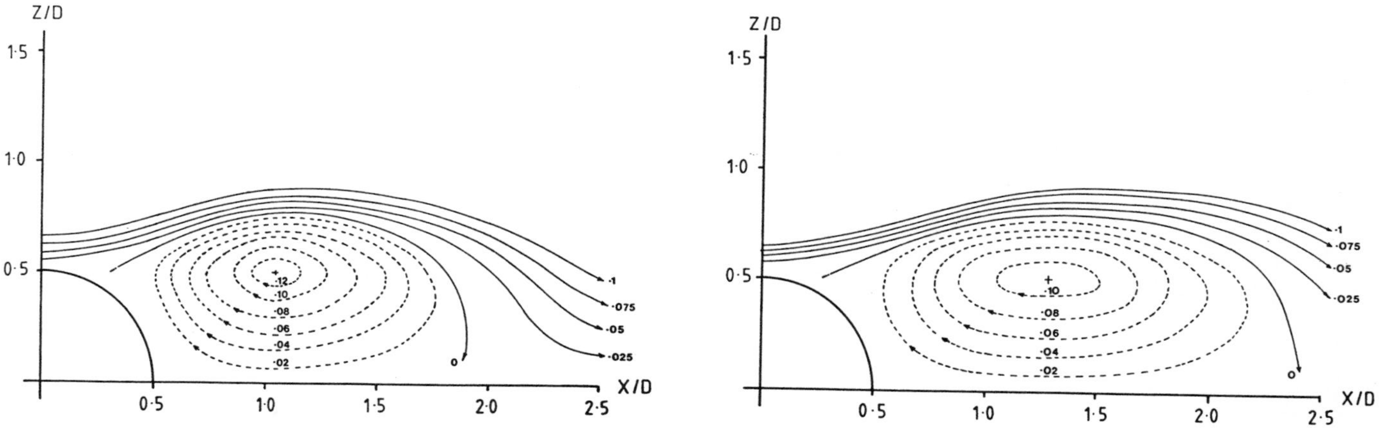

(a) Model Length/Chord Ratio 2:1

(b) Model Length/Chord Ratio 3:1

FIGURE 3 Streamline Drawings.

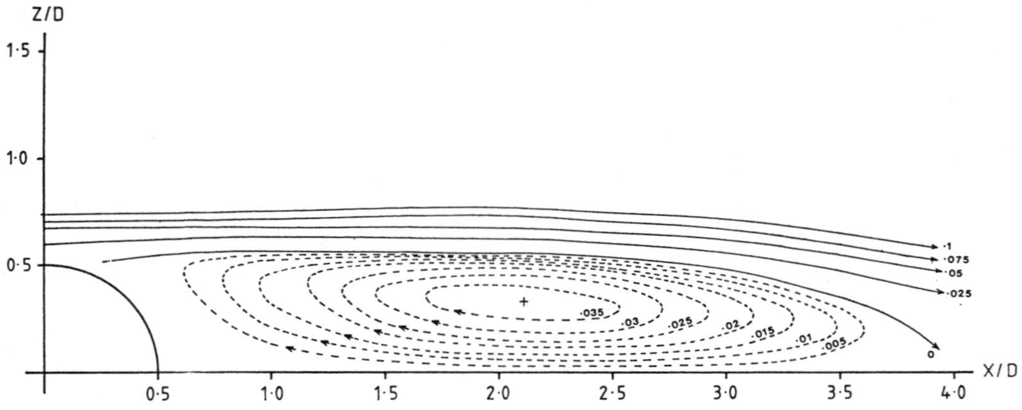

(c) Two Dimensional Model

FIGURE 3 Streamline Drawings

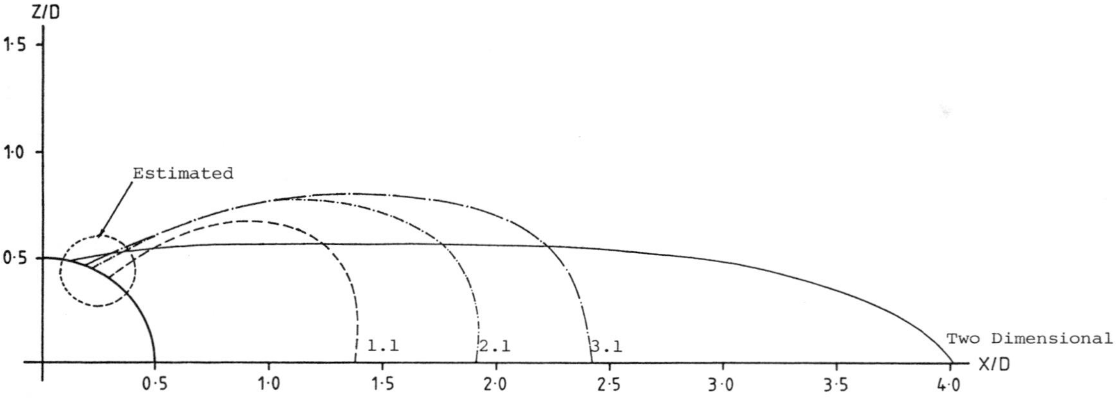

FIGURE 4 Positions of zero streamlines for Models of Different Length/Chord Ratio.

Fig 5 indicates that for the three dimensional models, this movement of the reattachment point appears to be linearly related to the models ratio, although further work is needed to corroborate this result.

It is also noted from the streamline drawings that with a decrease in model length, there is a decrease in the physical area of the recirculation region (Fig 4), and an increase in the maximum value of the streamlines within that area (Fig 3). In other words, the intensity of air flow in the region increases for decreasing model length. This distinct effect can be confirmed by reference to the turbulence intensity plots, which show that the peak values of turbulence obtained in the recirculation region also increase with decreasing model length. A possible explain-ation of this is that as the models become more three-dimensional there is increased air entrainment into the recirculation region from air flow around the ends of the model. This then leads to increased turbulence within the wake as intense mixing takes place, leading to higher peak values for smaller lengths.

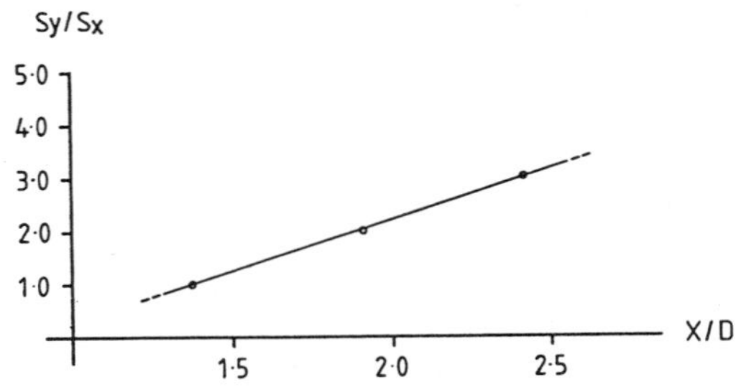

FIGURE 5 Length/Chord Ratio Against Reattachment Point

Reference to table A shows that an increase in the value of drag, and a decrease in the value of lift accompanies this decrease in model length. This relationship may be due to the combined effect of the movement of the separation point on the models surface, and to the change in the pressure distribution on the rear of the model, which is in turn related to the intensity of the recirculation region. Further experimental work is required to define more accurately this relationship between near wake intensity and surface pressure distribution. Other useful design information can be obtained by further inspection of the turbulence intensity plots. From these it can be seen that the level of turbulence generated in the recirculation region close to the wall may be as much as 5-times greater than that found in the boundary layer if the model is not present. This result could be important for the comfort of pedestrians passing through the near wake of the structure.

Peak intensities, however, of about 20-times greater than those recorded without the model are to be found in the separating shear layer. Fig 6 shows the position of these peak intensities for each set of profiles obtained by experiment. In the case of the three-dimensional models, as the length is decreased, the position of the peak values moves towards the tunnel wall. For comparison, the growth of the shear layers in which these peaks occur are plotted for each model on Fig 7. (The upper and lower limits are taken as described in Ref 5). The designer should be aware of the position of this region and its associated intensity of turbulence before considering the wind loads applied to structures sited close to the rear of a barrel vault.

CONCLUDING REMARKS

It is hoped that the results from this work may provide useful, initial information to the designer regarding the properties of the air flow around a simple barrel vault, which lead to wind induced loading. For a more detailed understanding of the nature of this relationship, future work should include wake traverses along lines other than that central to the model. This would yield information on the shear layer development at the ends of the models and provide an indication of the three-dimensional size of the recirculation region.

Consideration should also be given to the affect upon the air flow of openings and appendages to the structure. Such details would be useful to the designer when account is to be taken of the emission of pollutents, provision of ventilation and comfort of pedestrians using the building.

Finally, more data from experimental work is required to determine the precise effects that the turbulence in the near wake has on buildings placed downstream of the barrel vaults, structural loading that may not necessarily be taken into account by the designer.

ACKNOWLEDGEMENTS

The authors would like to express their gratitude to Professor Z S Makowski for the use of the laboratory facilities and to Mr E Savory for his advice and assistance throughout the project.

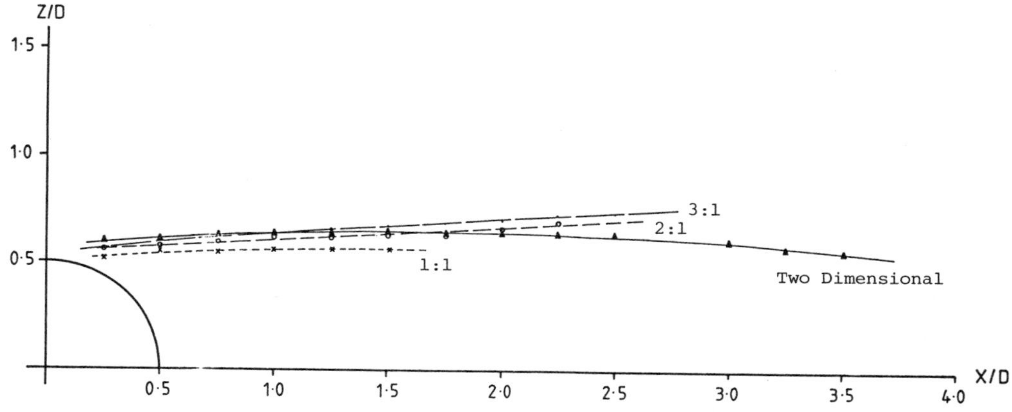

FIGURE 6. Positions of Peak Turbulence Intensities for Models of Different Length/Chord Ratio

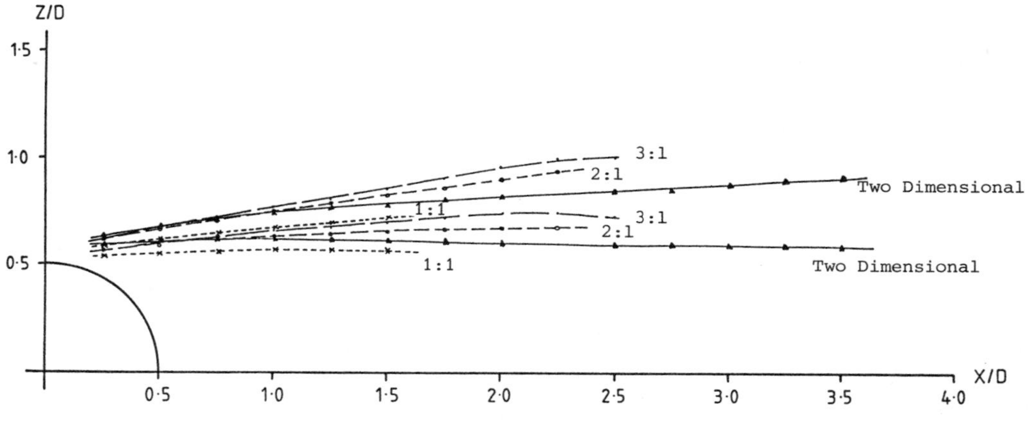

FIGURE 7. Developing Separated Shear Layers for Models of Different Length/Chord Ratio

NOTATION

C_D - Coefficient of drag
C_L - Coefficient of lift
D - Model diameter
H - Model Height
R - Model Radius
R_e - Reynolds number
Sx - Length of model in x-direction
Sy - Length of model in y-direction
$\underset{\sim}{U}$ - Mean local velocity
U - Mean freestream velocity
X - Cartesian co-ordinate
Y - Cartesian co-ordinate
Z - Cartesian co-ordinate
θ° - Subtended angle
Ψ - Streamline function

REFERENCES

1. C W NEWBURY and K J EATON, "Wind Loading Handbook." Building Research Establishment Report, Department of the Environment Building Research Establishment, London, 1974.

2. C W WONG, "The Structural Behaviour of Braced Barrel Vaults with Particular Reference to Wind Effects." Ph.D Dissertation, Department of Civil Engineering, University of Surrey, 1981.

3. E SAVORY and N TOY, "Investigation of the Wind Loading on Domes in Turbulent Boundary Layers." Third International Conference on Space Structures, University of Surrey, 1984.

4. L J S BRADBURY and I P CASTRO (1971), "A Pulsed-Wire Technique for Measurements in Highly Turbulent Flows." J F M 49, 657-91.

5. B S CHEUN, "Separated Shear Layers Behind Two-Dimensional Square-edged Bodies." Ph.D Dissertation, Department of Civil Engineering, University of Surrey, 1981.

INVESTIGATION OF THE WIND LOADING ON DOMES IN TURBULENT BOUNDARY LAYERS

E SAVORY* and N TOY**

* Research Student, Department of Civil Engineering
University of Surrey, U.K.

** Lecturer, Department of Civil Engineering
University of Surrey, U.K.

Although wind loading is a major part of the total load acting on dome structures, to date very little research has been undertaken in order to realistically quantify these forces. A greater understanding of the wind loads and of the nature of the airflow around these structures can only assist in increasing the accuracy of modern design techniques.

The work presented is a wind tunnel investigation of the flow around a hemispherical dome immersed in two thick turbulent boundary layers of different velocity and turbulence intensity profiles. Surface pressure distributions on the dome are illustrated, along with turbulence intensity profiles in the near wake of the structure. In addition, some of the main features of the flow regime near the hemisphere are described.

INTRODUCTION

Wind loading is a significant part of the total load acting on dome structures and so it is important for the structural engineer to have some knowledge of the magnitudes and directions of these loads. The cladding of a dome is normally made from lightweight material, such as aluminium, hence the size of local peak wind pressures must be taken into account during the design process. It is worth remembering that, even for a heavy, steel-frame hemispherical dome, the magnitude of the local peak lift force, near the top of the building, generated by a wind-speed of 50 m/s, is approximately twice the dead load of the structure. This ratio may be greater than ten for the case of an all-aluminium dome (see Ref 1).

The information found in Codes of Practice concerning the magnitudes of pressure coefficients on domes can be misleading. In some cases the coefficients for the domes are those for a sphere in uniform flow and do not relate to a hemisphere immersed in the Earth's natural boundary layer. The limited amount of experimental work that has been undertaken in order to obtain realistic pressure distributions on domes includes investigations reported in Refs 2, 3 and 4. Data from Refs 2 and 3 has been incorporated in the U.K. Wind Loading Handbook (Ref 5). However, in all the aforementioned research, the test conditions were such that in only a few of the measurements were the models fully immersed in a turbulent boundary layer, in most cases the model being subjected to nearly uniform flow. In the present work, the height of the model hemisphere was less than half of the thickness of the thinner of the two boundary layers, thus producing more realistic test conditions.

In addition to assessing the wind loading on the dome, the engineer should have some understanding of the nature of the flow in the wake of the structure. This is important for predicting:

(a) the likelihood of discomfort to personnel
(b) any ventilation problems
(c) the dispersion of pollutants
(d) the effect on any downwind buildings, particularly the wind loading on them.

Experimental investigations of hemisphere wakes have been presented in Refs 6, 7, 8 and 9. Most of the velocity and vorticity measurements undertaken in Ref 7 were made in the far wake and it was found that the system of trailing vortices persisted for more than fifty hemisphere diameters downstream of the model. In recent work at the University of Surrey (Refs 8 and 9) the near wake region, within the first two diameters downstream of the dome, has been studied.

In the present work, some of the features of the wind loading on a hemisphere and of the flow in the wake of the structure are examined.

EXPERIMENTAL DETAILS

The investigations were undertaken using the low-speed, blow-down, open circuit wind tunnel in the Department of Civil Engineering at the University of Surrey. The tunnel, shown diagramatically in Fig 1, has working section dimensions of 1.372m height x 1.067m width x 9.0m length.

Two thick turbulent boundary layers of different mean velocity and turbulence intensity profiles were generated on the wall of the tunnel (Fig 2) using a barrier fence and a set of vorticity generators similar to those described in Ref 10. The barrier fence "trips" the oncoming flow and thereby removes some momentum from the flow in the region near the wall. The elliptical vorticity generators produce in their wakes a graded distribution of turbulence as well as taking additional momentum from the flow. For the "smooth" boundary layer the wall of the tunnel was maintained in its smooth state, whereas for the "rough" boundary layer the wall was artificially roughened by a method similar to that described in Ref 11. The roughness elements provide continuous production of turbulence near the wall. Details of the boundary layers are given in the table below.

TABLE 1 Boundary Layer Details

	Thickness δ (mm)	Displacement Thickness δ_* (mm)	Momentum Thickness θ (mm)	δ_*/θ	δ_*/δ	θ/δ
SMOOTH	258	34.23	25.03	1.368	0.133	0.097
ROUGH	367	65.66	44.39	1.479	0.179	0.121

The usual definition for the boundary layer thickness has been employed here, that is the distance from the wall to where the mean velocity is 99% of the freestream velocity. The displacement thickness and momentum thickness have been calculated from the mean velocity profile using the conventional expressions given in the notation.

For a full size dome, the Reynolds number is of the order 10^8 to 10^9 which is in the supercritical Reynolds number range. Any wind tunnel tests should, therefore, be undertaken in this range, if possible. The highest Reynolds number that could be attained in these experiments was 1.6 x 10^5 which is in the subcritical regime. By surface roughening the model near its critical Reynolds number, flow separation was induced and a pressure distribution created similar to that for a model in super-critical flow. This technique was utilized successfully in previous work using barrel vault models (Ref 12).

The test model was a hemisphere of 225mm diameter, artificially roughened with sand giving a relative roughness of e/D = 0.001. The size of this model would represent a dome of 160-225m in diameter, giving a model ratio of 1/700-1/1000. All of the experiments were undertaken at a Reynolds number of 1.6 x 10^5 based on dome diameter and a freestream velocity of 10.3 m/s.

Pressure distributions were measured by tappings on the dome connected to a Setra transducer via a Scanivalve switching mechanism.

Turbulence intensities were measured by a single hot-wire anemometer, Disa 55M10 unit with a P11 probe. The positions of separation of the flow from the surface of the dome and of reattachment on the ground plane behind the model were located using the heated element technique described in Ref 9.

Operation of all the instrumentation, including data acquisition and analysis, was performed on-line using a PET microcomputer.

RESULTS AND DISCUSSION

The mean pressure distributions on the surface of the hemisphere in the smooth and rough boundary layer are shown in Fig 3(a) and Fig 3(b) respectively. The pressure coefficient contours have been plotted as if the dome is three-dimensional and so they are in the correct position with respect to the dome outline. With increased wall roughness and, hence, increased boundary layer turbulence intensity there is a decrease in the positive pressure on the front face of the hemisphere, a small pressure increase on the rear face, and a decrease in the magnitude of the peak suction at the top of the dome. The U.K. Wind Loading Handbook (Ref 5) gives a value of +0.6 for the maximum pressure coefficient on the leading face and -1.0 for the coefficient at the top of the dome.

The overall drag and lift coefficients for the two cases, obtained by integrating the pressure distributions, are given in the following table:

TABLE 2 Drag and Lift Coefficients

	Drag Coefficient C_D	Lift Coefficient C_L
SMOOTH	0.20	0.35
ROUGH	0.18	0.25

The drag and lift forces may be computed from the following equations:

Drag Force = Drag Coefficient x Freestream Dynamic Pressure x Projected Area of Dome

$$= \frac{\pi}{16} . C_D . \rho . \tilde{U}^2 . D^2$$

Lift Force = Lift Coefficient x Freestream Dynamic Pressure x Projected Area of Dome

$$= \frac{\pi}{8} . C_L . \rho . \tilde{U}^2 . D^2$$

The effect of increasing the boundary layer turbulence level is to reduce the drag and lift forces. At a short distance downstream of the lateral centre-line of the hemisphere the flow separates from the surface of the model to form a complex recirculation region in the near wake. Fig 4 shows a simplified sketch of some of the main features of the flow around a hemisphere. A horse-shoe vortex forms immediately upstream of the hemisphere and curves around the dome to form trailing vortices which interact with the flow shed from the near wake recirculation region. The location of the separation line on the dome has been approximately determined from the pressure distribution by using the construction suggested in Ref 13. The validity of this approximation has been examined, in the smooth boundary layer case, by using the heated element probe described in Ref 9. Fig 3(a) shows good agreement between the separation lines obtained by the two techniques. Separation from the surface of the dome occurs slightly further downstream in the rough boundary layer than in the smooth profile and this causes the small reduction in the drag coefficient.

Figure 5 shows the approximate line of reattachment, of the separated flow, on the ground plane, in the smooth boundary layer, located using the heated element probe. The length of the recirculation region on the wake centre-line, measured from the centre of the hemisphere, is 0.9D for the smooth boundary layer case and 1.0D for the rough boundary layer. In experiments using models mounted on a flat plate which generated a very thin boundary layer (refs 14 and 15) it was found that the reattachment length for a smooth sphere was 2.5D and for a hemisphere-cylinder combination (overall height equal to the hemisphere diameter) was 2.0D.

Figures 6(a), (b) and (c) show longitudinal turbulence intensity profiles obtained from single hot-wire anemometer measurements made at three heights above the wall in the near wake of the dome in the smooth boundary layer. Each figure gives profiles at radii of 0.6D, 0.8D and 1.0D, measured from the centre of the dome, plotted using polar coordinates. The sharp-peaked turbulence profiles coincide with the shear layer caused by the separation of the flow from the surface of the hemisphere. With increasing radial distance from the dome and increasing distance from the wall, the positions of the peak intensities move towards the longitudinal centre-line, indicating the curvature of the shear layer. In addition, with increasing radial distance, the peaks spread and become less pronounced due to mixing of the turbulent fluid. It may be seen from Fig 6(c), showing the profiles obtained very near the wall, that the magnitudes of the peak intensities are still considerably greater than the prevailing boundary layer turbulence level. This is significant since it is at this level that people will be traversing the wake of the dome.

The rough boundary layer profiles, not presented here, are similar to those in Fig 6 except that the peak intensities tend to smear out more rapidly due to the higher level of the prevailing boundary layer turbulence.

ACKNOWLEDGEMENTS

This work has been supported by a Science and Engineering Research Council Studentship awarded to E Savory.

NOTATION

A - Projected area of dome

C_D - Drag coefficient (Drag force/$\tilde{p}$A)

C_L - Lift coefficient (Lift force/$\tilde{p}$A)

Cp - Pressure coefficient ($(p-p_o)/\tilde{p}$)

δ - Boundary layer thickness

δ_* - Displacement thickness ($\int_0^\infty (1-\frac{U}{\tilde{U}})dz$)

θ - Momentum thickness ($\int_0^\infty \frac{U}{\tilde{U}}(1-\frac{U}{\tilde{U}})dz$)

D - Diameter of dome

e - Size of sand roughness on surface of dome

e/d - Roughness ratio

μ - Dynamic viscosity of air

ρ - Density of air

p - Pressure on surface of dome

p_o - Static pressure

$\tilde{p}$ - Freestream dynamic pressure ($\frac{1}{2}\rho\tilde{U}^2$)

R - Distance from centre of dome

Re - Reynolds number ($\rho D\tilde{U}/\mu$)

U - Local mean velocity

$\tilde{U}$ - Freestream mean velocity

u - Velocity fluctuation in x-direction

X,Y,Z Cartesian coordinates axes

Ψ - Yaw angle

REFERENCES

1. S BAKER, A comparison of the codes of practice used in different countries for the determination of wind loads on domes, Course on the Analysis, Design and Construction of Braced Domes, Volume 1, University of Surrey, 1981.

2. F J MAHER, Wind loads on basic dome shapes, Journal of the Structural Division, American Society of Civil Engineers, Volume 91, Number 3, pp 219-228, 1965.

3. J BLESSMANN, Pressures on domes with several wind profiles, Proceedings of the 3rd International Conference on Wind Effects on Buildings and Structures, Tokyo, pp 317-326. 1971.

4. S TANIGUCHI, H SAKAMOTO, M KIYA and M ARIE, Time-averaged aerodynamic forces acting on a hemisphere immersed in a turbulent boundary layer, Journal of Wind Engineering, Volume 9, pp 257-273, 1982.

5. C W NEWBERRY and K J EATON, Wind Loading Handbook, Building Research Establishment Report, H.M.S.O.,1974.

6. W JACOBS, Flow behind a single roughness element, Ing. Archiv., Volume 9, pp 343-355, 1938.

7. A C HANSEN, Vortex containing wakes of surface obstacles, Ph.D Thesis, Colorado State University, 1976.

8. N TOY, W D MOSS and E SAVORY, Wind tunnel studies on a dome in turbulent boundary layers, Journal of Wind Engineering, Volume 11, pp 201-212, 1983.

9. N TOY and E SAVORY, Turbulent shear flow in the near wake of a hemisphere, Proceedings of the 8th Biennial Symposium on Turbulence, University of Missouri-Rolla, pp 16.1-16.12, 1983.

10. J COUNIHAN, The structure and the wind tunnel simulation of rural and adiabatic boundary layers, Symposium on External Flows, University of Bristol, 1972.

11. I P CASTRO, Two rough wall boundary layers in the 0.9m x 0.75m wind-tunnel, C.E.G.B. R/M/N 1006, August, 1978.

12. C W WONG, The structural behaviour of braced barrel vaults with particular reference to wind effects, Ph.D Thesis, University of Surrey, 1981.

13. H J NIEMANN, Stationary wind load on hyperbolic cooling towers, Proceedings of the 3rd International Conference on Wind Effects on Buildings and Structures, Tokyo, pp 335-344, 1971.

14. S OKAMOTO, Turbulent shear flow behind sphere placed on plane boundary, Proceedings of the 2nd Symposium on Turbulent Shear flows, pp 16.1-16.6, 1979.

15. S OKAMOTO, Turbulent shear flow behind hemisphere-cylinder placed on ground plane, Proceedings of the 3rd Symposium on Turbulent Shear Flows, pp 16.11-16.16, 1981.

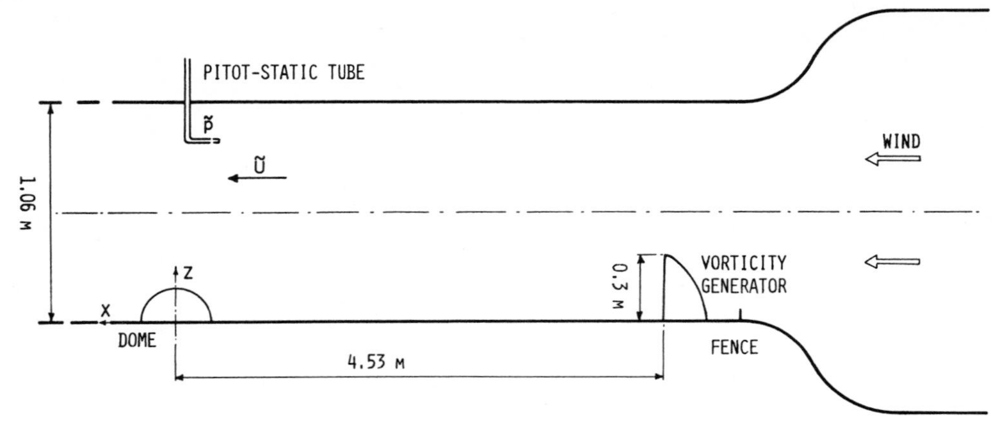

FIG 1 DIAGRAMMATIC LAYOUT OF 1.37m x 1.07m WIND TUNNEL

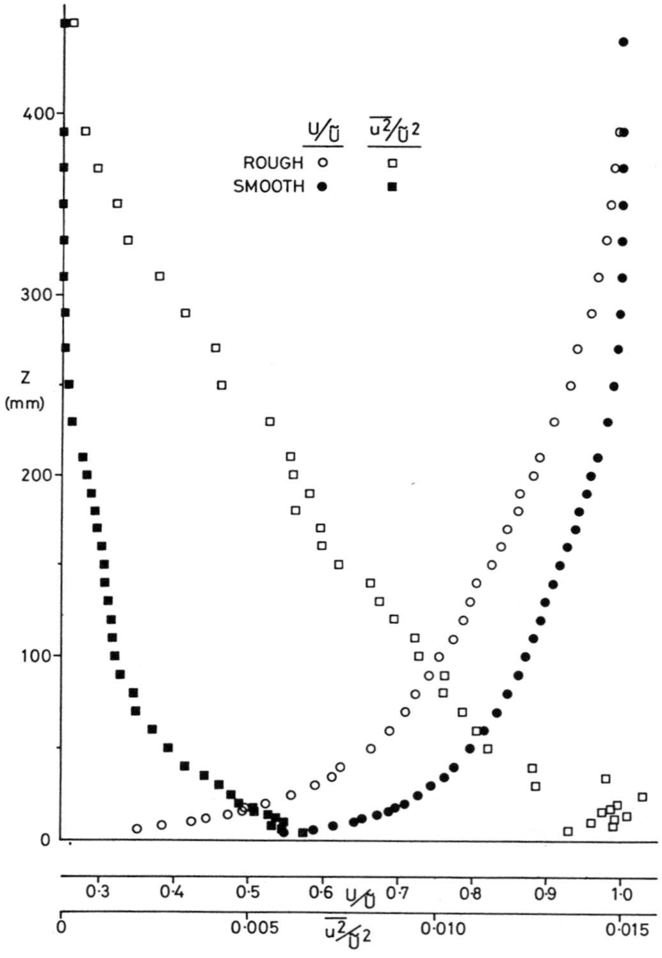

FIG 2 MEAN VELOCITY ($U/\tilde{U}$) AND TURBULENCE INTENSITY ($\overline{u^2}/\tilde{U}^2$) PROFILES FOR THE SMOOTH AND ROUGH BOUNDARY LAYERS

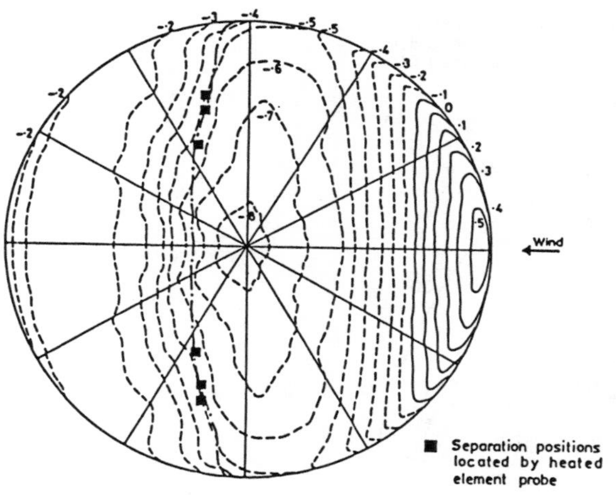

(a) SMOOTH BOUNDARY LAYER

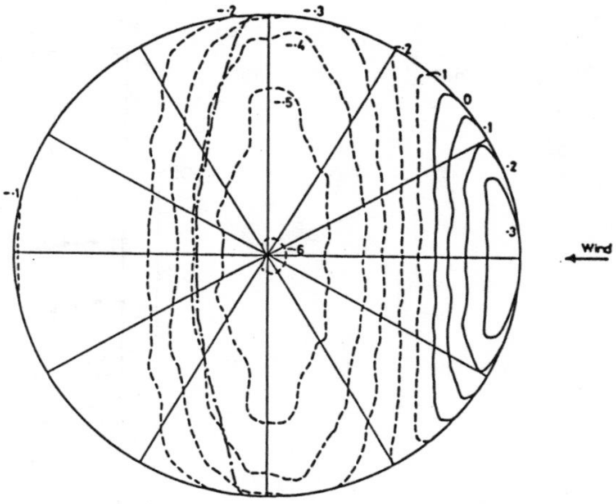

(b) ROUGH BOUNDARY LAYER

FIG 3 LINES OF EQUAL PRESSURE OVER THE DOME IN THE TWO BOUNDARY LAYERS

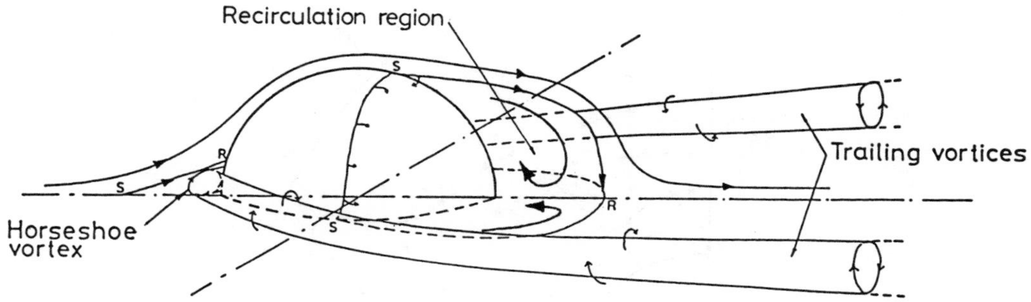

FIG 4 THE FLOW REGIME IN THE NEAR WAKE OF A HEMISPHERE

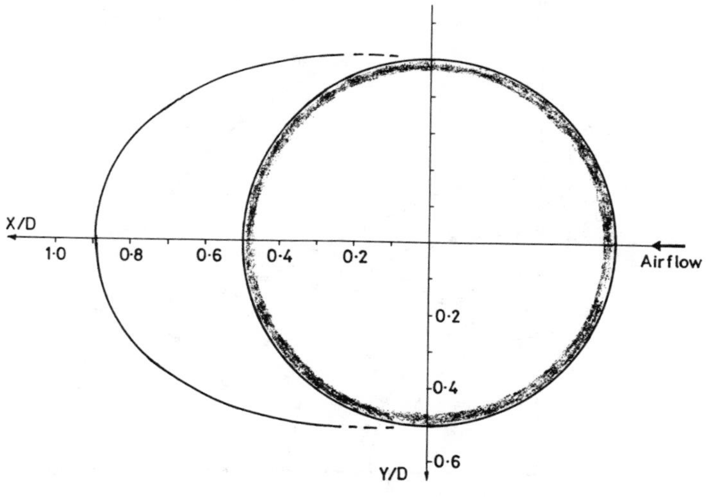

FIG 5 PROFILE OF THE NEAR WAKE RECIRCULATION REGION FOR THE SMOOTH BOUNDARY LAYER SHOWING THE LINE OF REATTACHMENT ON THE GROUND PLANE

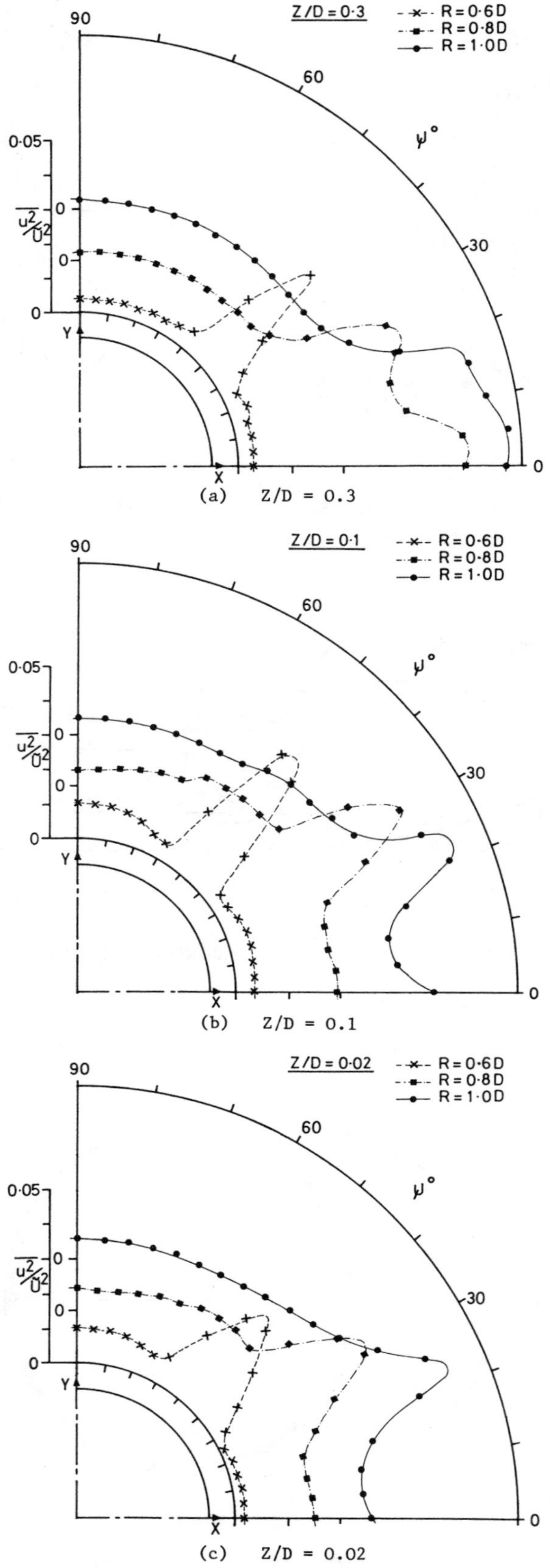

FIG 6 NEAR WAKE TURBULENCE INTENSITY PROFILES IN THE SMOOTH BOUNDARY LAYER

FORMS AND FORCES IN TENSEGRITY SYSTEMS

R. MOTRO, DE

Maître-Assistant, Civil Engineering laboratory
Université des Sciences et Techniques du Languedoc
Montpellier, France

The relationship between forms and forces in tensegrity systems are examined with a view to future optimization. The notions of relational structure and form are given in agreement with the vocabulary of systems theory. The paper describes a way of using graph theory to build up the relational structure of tensegrity systems. The "critical" self-stressed form is obtained by a dynamic relaxation method. A general theoretical model and an experimental prototype for the case of "simplex" show the mechanical behaviour of tensegrity systems.

INTRODUCTION

A small number of publications discuss tensegrity systems. The present paper is the result of an approach at formalisation which begins at the design stage and reaches that of the study of mechanical behaviour once the form liable to be self-stressed has been determined.

DEFINITIONS

The "relational structure" of a tensegrity system is defined qualitatively by the list of its members and their terminal nodes. Definition of the qualitative and quantitative characterics of the members added to that of the relational structure determines the "overall structure" of the system. Its "form" refers here to the projection of the overall structure in three-dimensional geometrical space: it takes all the geometrical characteristics of the members into account including their position in space.

OPTIMIZATION OF RETICULATED SYSTEMS

After the early work of Maxwell, Ref 1, and Michell, Ref 2 optimization studies on the form of plane and then spatial reticulated systems became more numerous.

It should be noted with Topping, Ref 3, that the taking into account of the relational structure of reticulated systems took place very late in all the optimisation methods put forward; although, for the moment, calculation algorithms make it possible to eliminate certain members or even certain nodes, the introduction of new nodes and/or members into an initial relational structure presents difficulties that have not yet been altogether overcome.

However, all authors agree that taking the relational structure into account as an optimization variable leads to considerable advantages as regards the objective function (cost or weight of the systems).

The design aspect of the relational structure of a reticulated system made up of rigid members in compression and tension poses no other problems than those of classic geometry. This is not the case for systems made up of bars and cables in which the relational structure must meet certain criteria. Tensegrity systems belong to the latter category and are the subject of this publication.

TENSEGRITY SYSTEMS

In reticulated systems, the members are connected in such a way that they are only subjected to simple compressive or tensile stress. The idea of separating the tension members from the compression members during the design of these systems gave birth to tensegrity systems.

It seems certain that the "father" of these systems was the sculptor K Snelson, Ref 4, who worked at Black Moutain College with the painter W de Kooning and R B Fuller at the beginning of the 1950's. Taking the inspiration from Snelson's work, R B Fuller gave a definition of tensegrity systems in 1951 :

"*A tensegrity system is established when a set of discontinuous compressive components interacts with a system of continuous tensile components to define a stable volume in space.*"

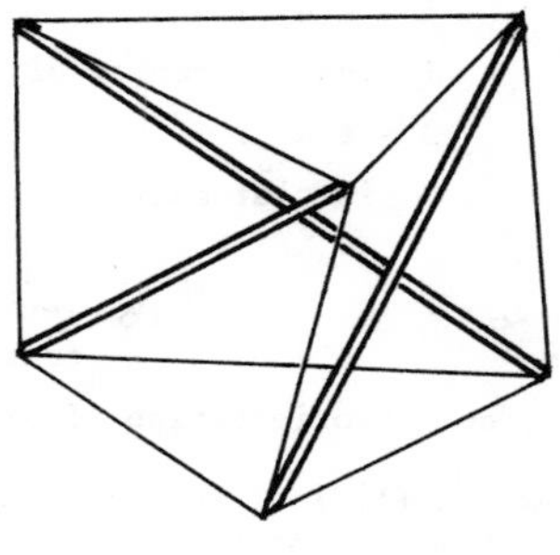

Fig 1. Simplex

The smallest reticulated space system which fits this definition is made up of 3 bars and 9 cables; this is the "Simplex", Fig. 1. An account of current knowledge about these systems is given elsewhere, Ref 5.

These systems form a stable volume in space in the absence of any external action and the equilibrium of the nodes is given in the first order by the matrix formula :

$$[K] \cdot |D| = |W| \qquad \ldots 1.$$

in which $[K]$ is the matrix of rigidity of the system
$|D|$ is the displacement vector of nodes
$|W|$ is the vector of external actions; in this case it is identically zero, giving :

$$[K] \cdot |D| = |0| \qquad \ldots 2.$$

For self-stressing to occur, the determinant of the matrix of rigidity must be zero in order to obtain a solution other than zero vector for $|D|$. In this case, the components of the system are either compressed or under tension with the overall stresses in equilibrium.

Tensegrity systems therefore belong to the "critical configuration" class of reticulated systems. Any calculation associated with them must include the effect of geometrical non linearity and that of the terms of self-stress (initial stresses). The form of such systems is not defined a priori: the spatial position of nodes must satisfy the zero condition of the determinant of $[K]$.

STUDY MODELS OF TENSEGRITY SYSTEMS

The study of tensegrity systems consists of :

- determination of the "relational structure", a qualitative phase carried out using graph theory.
- search for the form meeting the self-stressed condition; dynamic relaxation can be used to obtain this form.
- study of the mechanical behaviour under the application of external actions for various levels of self-stressed.

a) Determination of the relational structure

(i) Reticulated systems

Any reticulated system can be represented by a graph whose vertices x_i correspond to the nodes and arcs $(x_i x_j)$ to the bars u_{ij}.

As I is lower than J in the numbering chosen, the following can be written :

under compression $I = f(u_{ij})$ ... 3.
I initial extremity of u_{ij}
$J = s(u_{ij})$ ... 4.
J final extremity of u_{ij}

under tensile stress :
$I = s(u_{ij})$ ... 5.
I final extremity of u_{ij}
$J = f(u_{ij})$ ... 6.
J initial extremity of u_{ij}

I
u ij
J
Compression
I
u ij
J
tension

Fig. 2 Convention for orientation of arcs

The Boolean matrix (A_{ij}) of the graph associated with the reticulated system is as follows :

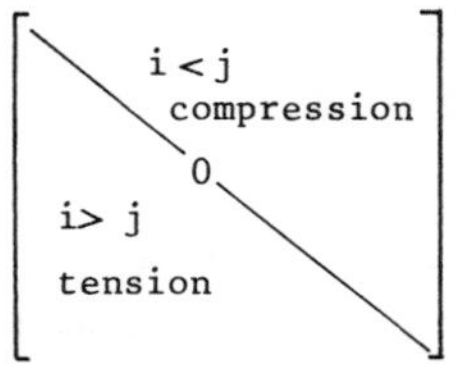

Fig. 3 Boolean matrix of the graph associated with the reticulated system

In this matrix :

$$a_{ij} = 1 \quad \text{if arc } u_{ij} \text{ exists} \qquad \ldots 7.$$

$$a_{ij} = 0 \quad \text{if arc } u_{ij} \text{ does not exist} \qquad \ldots 8.$$

If the internal half-degree of a vertex x_k (the number of arcs with their initial extremity at x_k) is noted $d^+(x_k)$, and if the external half-degree of the same apex is noted $d^-(x_k)$, it is possible to separate tension and compression components, the corresponding half-degrees being d_t and d_c, and to obtain :

$$d^+(x_k) = d_t^+(x_k) + d_c^+(x_k) \qquad \ldots 9.$$

and $$d^-(x_k) = d_t^-(x_k) + d_c^-(x_k) \qquad \ldots 10.$$

with:

$$d_c^+(x_k) = \sum_{j=k+1}^{n} a_{kj} \qquad \ldots 11.$$

$$d_t^+(x_k) = \sum_{j=1}^{k-1} a_{kj} \qquad \ldots 12.$$

$$d_c^-(x_k) = \sum_{i=1}^{k-1} a_{ik} \qquad \ldots 13.$$

$$d_t^-(x_k) = \sum_{i=k+1}^{n} a_{ik} \qquad \ldots 14.$$

(n is the number of vertices in the graph)

(ii) Tensegrity systems

Assuming a single compression component to be incident at each node, the three main ideas of R B Fullers's definition can be translated into terms of half-degrees.

+ Discontinuity of the compression system

$$d_c(x_k) = d_c^+(x_k) + d_c^-(x_k) = 1 \qquad \ldots 15.$$

which implies :

$$d_c^+(x_k) = 1 \text{ and } d_c^-(x_k) = 0 \qquad \ldots 16.$$

or

$$d_c^+(x_k) = 0 \text{ and } d_c^-(x_k) = 1 \qquad \ldots 17.$$

+ Continuity of the tension system

$$d_t(x_k) = d_t^+(x_k) + d_t^-(x_k) \geq 2 \qquad \ldots 18.$$

+ Necessary stability condition

$$d_t(x_k) \geq 3 \qquad \ldots 19.$$

This condition includes that of Eqn 18. in spatial systems. It is necessary but insufficient.

(iii) Spherical tensegrity systems

When the cables form the edges of a polyhedron homeomorphic to a sphere the relational structure has interesting characteristics.

In 1934 Rademacher and Steinitz showed that any convex polyhedron can be represented by a planar graph of degree of connexity 3. *The graph of the tension components of a spherical tensegrity system is therefore a planar graph.*

It is also possible to break down the vertices's group X into two sub-groups :

Y : initial extremities of compression components
Y' : terminal extremities of compression components

and to check by means of the expressions of half-degrees (Eqns 11 to 14) that :

$$Y \cup Y' = X \quad \text{and} \quad Y \cap Y' = \emptyset \qquad \ldots 20.$$

The arcs and vertices of the graph representing the compression components alone form a *bipartite graph making perfect coupling between the n vertices* of the overall graph associated with spherical tensegrity systems.

In order to draw up the relational structure of a spherical system with "n" vertices, it is necessary to determine a planar graph with "n" vertices and a perfect coupling. The choice of the order of these operations depends on the number of vertices. When "n" is too large ($n \geq 10$), it is preferable to :

1/ determine a triangular planar graph represented in a linear manner, Fig 4, giving then a representation with rectilinear arcs, Fig 5.

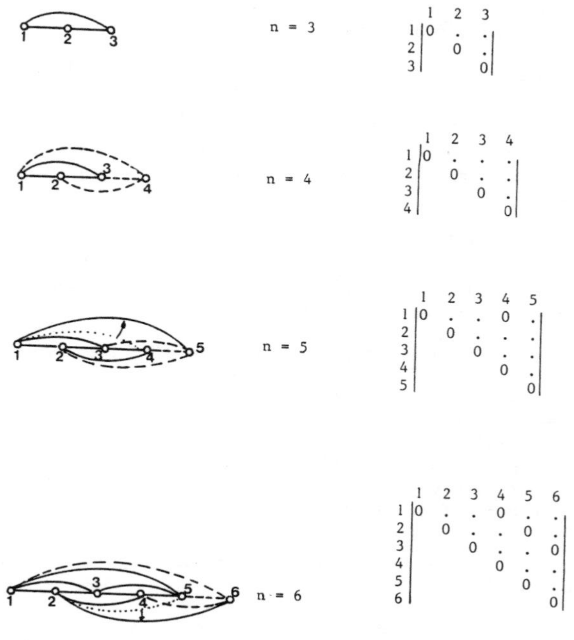

Fig 4. Construction of a linear representation of triangular planar graphs.

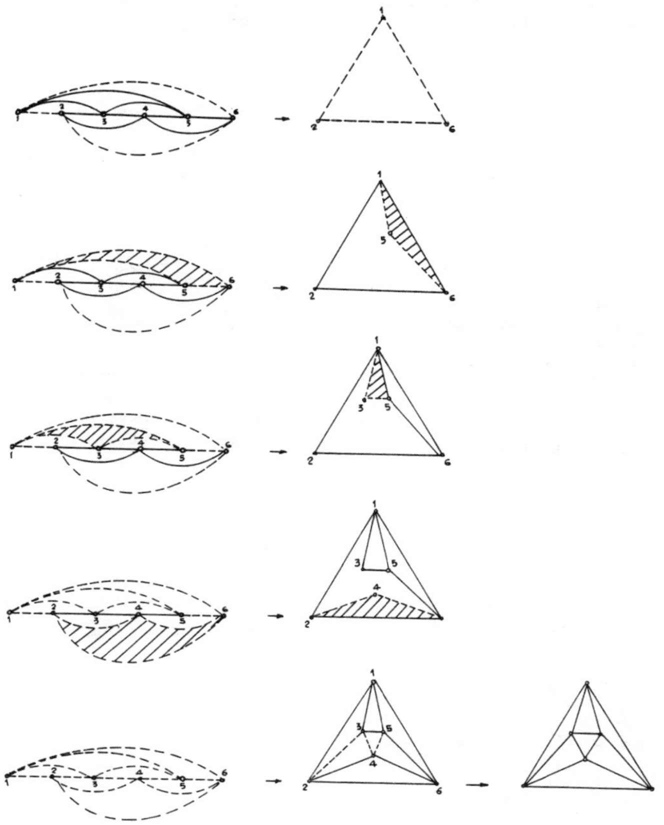

Fig 5 Representation's algorithm for rectilinear arcs

(The first triangle plotted corresponds to arcs which separate the linear representation of the planar graph from the rest of the plane. The vertices must then be positioned using an edge which has already been plotted: vertex 5 is connected to vertices 1 and 6; edges (1,5) and (1,6) are plotted in such a way that no vertex is located inside the cross-hatched triangle. Once they have been transcribed, the links of the plotting of the planar graph are plotted using dotted lines on the linear plotting).

2/choose a perfect coupling which obeys a preestablished pattern wich can lead to the elimination of certain edges of the planar graph. For example, it is possible to define a pattern imposing tension components connected to the same compression component forming a diamond, Fig 6.

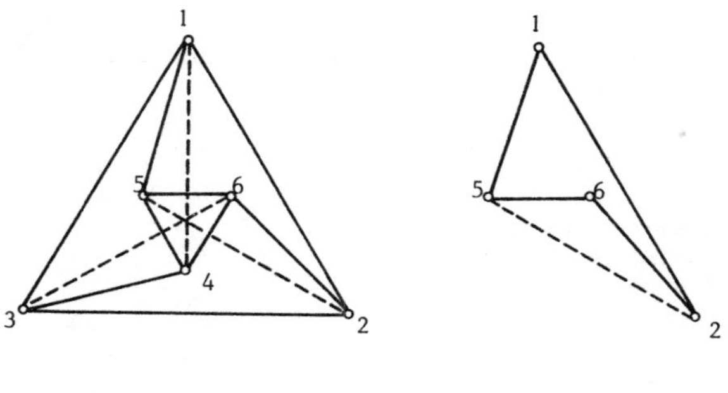

Fig 6 Obtention of the Simplex graph using diamond pattern

There is not a single solution for systems with twelve vertices. Application of the method proposed leads, for example, to the relational structures of three systems : expanded octahedron, Fig. 7, truncated tetrahedron, Fig. 8, hexagonal prism, Fig. 10.

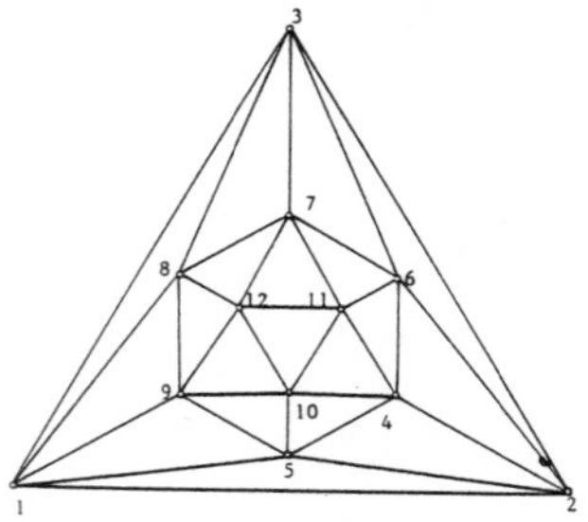

a) Graph with twelve 5-degree vertices

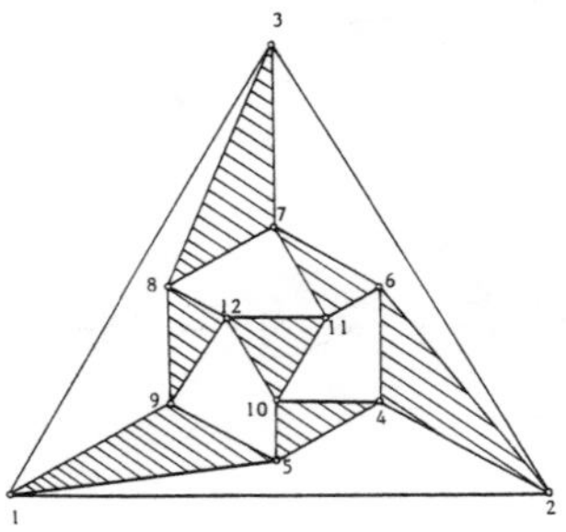

b) Graph with twelve 4-degree vertices

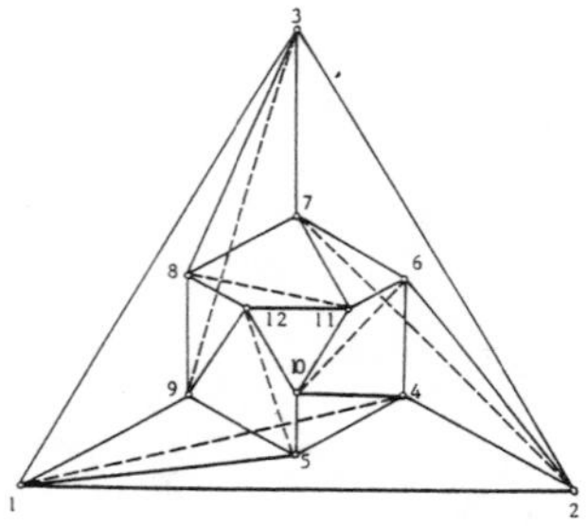

c) Relational structure of expanded octahedron

Fig 7 Expanded Octahedron

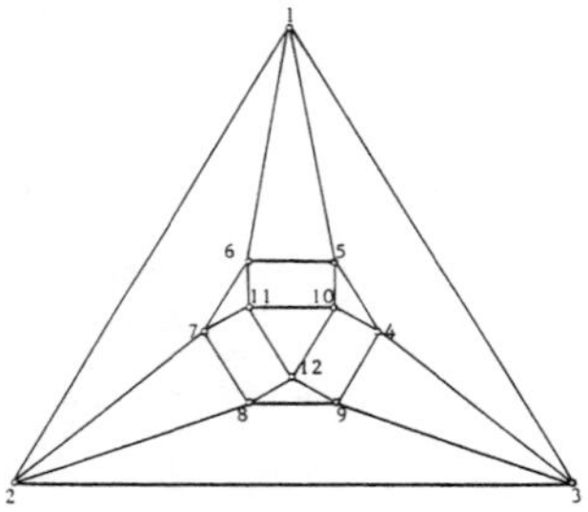

a) Graph with twelve 5-degree vertices

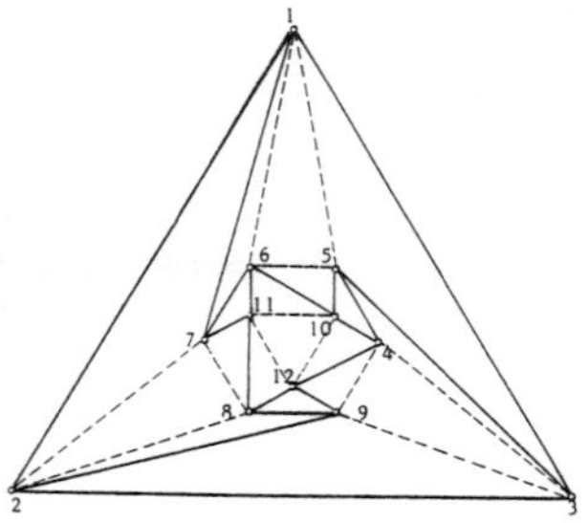

b) Reduction of the degree of the vertices

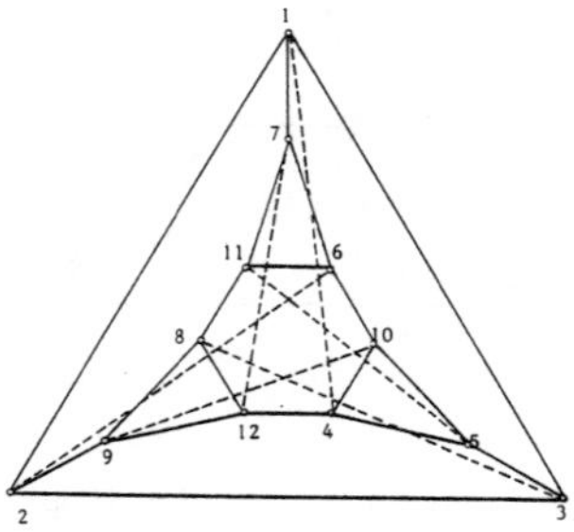

c) Relational structure of truncated tetrahedron

Fig. 8 Truncated Tetrahedron

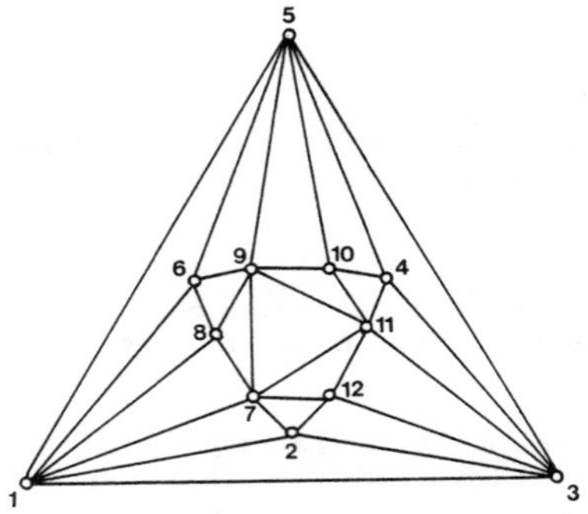

a) Planar graph with twelve vertices: six 6-degree vertices six 4-degree vertices

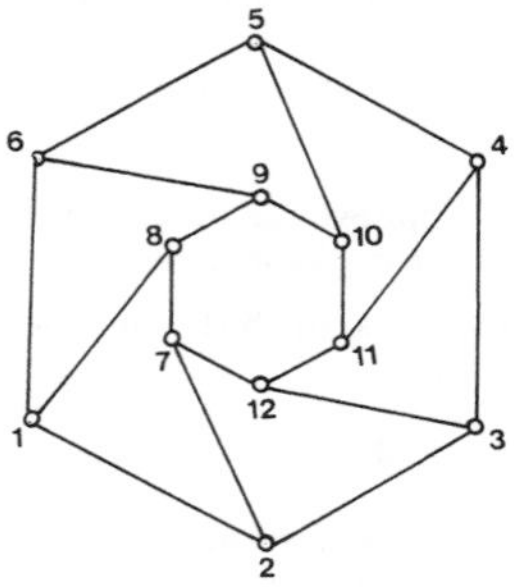

b) Planar graph with twelve 3-degree vertices

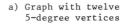

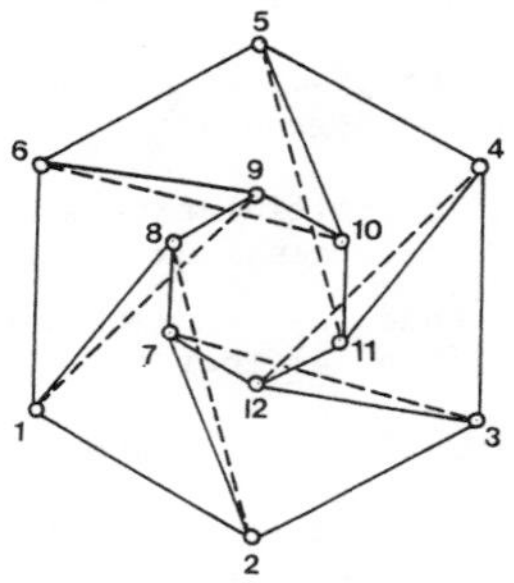

c) Relational structure of hexagonal prism

Fig 9 Hexagonal Prism

RESEARCH INTO FORM USING DYNAMIC RELAXATION

The transition from the relational structure of the self-stressed form is carried out by introducing the relative dimensions of the tension and compression members. "r" is taken as the ratio of dimensions :

$$r = \frac{\text{length of compression component}}{\text{length of tension component}} \qquad ...21.$$

As J C Maxwell, Ref 6, has suspected, the rigidification of a critical configuration can be obtained as soon as $r > r_{critical}$. Any excess results in strain of all the components and thus in the accumulation of internal self-stressing energy. The system behaves as a mechanism at any value of $r < r_{critical}$.

Seeking the possible form or forms involves calculating the value of $r_{critical}$. We have used the dynamic relaxation method. The dynamic equation in the case of a system with "n" degrees of freedom is as follows :

$$[M]\,|\ddot{v}(t)| + [C]\,|\dot{v}(t)| + [K(t)]\,|v(t)| = |0| \qquad ...22.$$

$[M]$ matrix of masses concentrated at the nodes
$[C]$ damping matrix
$[K(t)]$ rigidity matrix in function of time
$|v(t)|$ vector of the positions of the nodes at instant t
$|\dot{v}(t)|$ velocity vector
$|\ddot{v}(t)|$ vector of accelerations assumed constant during Δt

Equation 22 has no right-hand side value since there is no external action.

Solution is carried out by giving increment values to r, see eqn. 21. The static state resulting from the damped movement is sought for each interval of time Δt.

The result is independant of the values chosen for matrices $[M]$ & $[C]$ which are assumed to be diagonal. A test on the state of stress of the components then makes it possible to know wether the self-stressed form has been attained or not.

Search for form in the case of Simplex is illustrated in Fig. 8

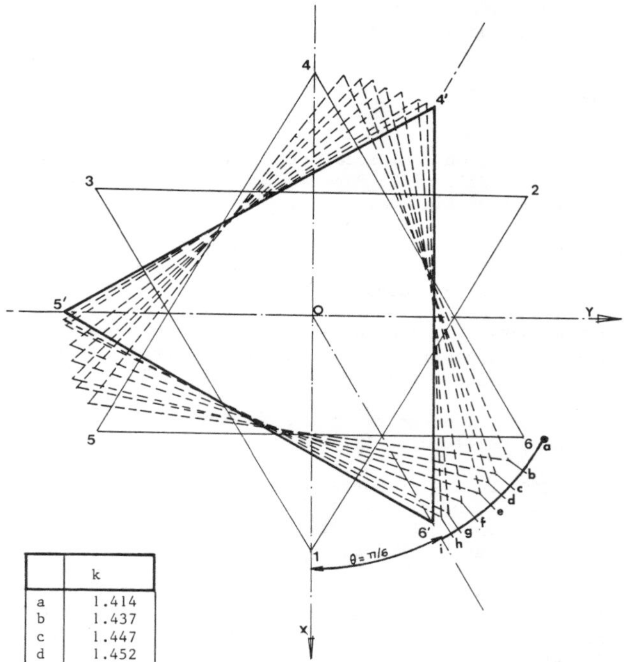

	k
a	1.414
b	1.437
c	1.447
d	1.452
e	1.457
f	1.464
g	1.466
h	1.467
i	1.468

Fig 8. Search for form: Simplex (each position corresponds to a value of "r")

$r_{critical} = 1.468$

Triangle 123 is assumed to be fixed and the self-stressed position occurs here for the seventh incrementing of r. The difference between this solution and the theoritical solution which can be calculated in this elementary case is less then 1%.

STUDY OF MECHANICAL BEHAVIOUR

(i) Mathematical model

In the presence of external actions, the equilibrium of the system is calculated in the stable deformed state under the load applied (second order calculation). The matrix equation to be solved is as follows :

$$[K_{sr}]|D_r| = |W_r| \qquad \ldots 23.$$

where $[K_{sr}]$ = Matrix of rigidity reduced to terms related to the degrees of liberty

$|D_r|$ = Vector of unknown displacements

$|W_r|$ = Vector of actions applied in the directions of the degrees of liberty

Taking the terms of non linearity into account leads to putting $[K_{sr}]$ into the following form :

$$[K_{sr}] = [K_{LD}] + [K_S] \qquad \ldots 24.$$

$[K_{LD}]$ is the matrix of large displacements. Its terms are functions of the rigidity of the bars, the actual coordinates of the nodes and the law of tangential behaviour.

$[K_S]$ is the matrix of initial stresses. Its terms depend on the actual stresses in the components and the actual coordinates of the nodes.

The system described by eqn 23 is non-linear and is the subject of incremental resolution.

(ii) Physical model

A prototype of the Simplex has been used, Fig. 9.

Single compression type loading was carried out using cast iron weights and levers. Mesearument of displacements was carried out simultaneously using comparators, strain gauges and a set of three theodolites which formed the triangular base required for sighting.

(iii) Results of the behaviour study

Comparison of the experimental and theoritical stresses confirmed the validity of the theoretical model (comparison carried out for self-stressing close to zero)

Figures 10 and 11 show the great flexibility of the system and its progressive stiffening under load.

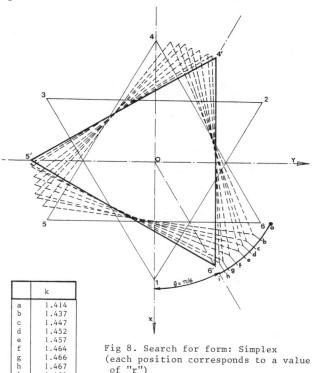

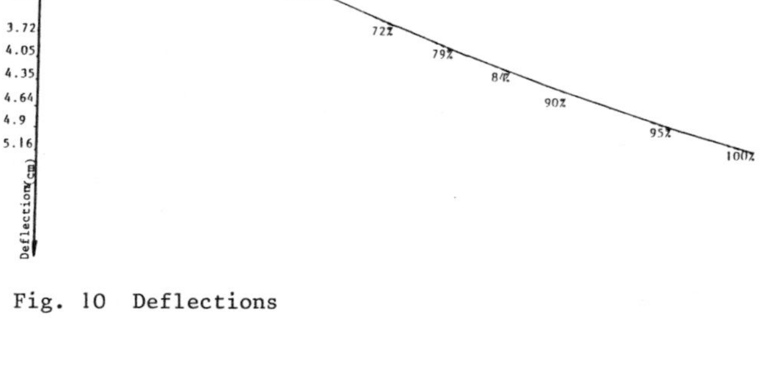

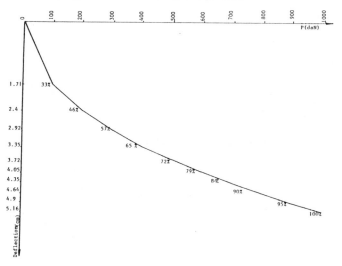

Fig. 10 Deflections

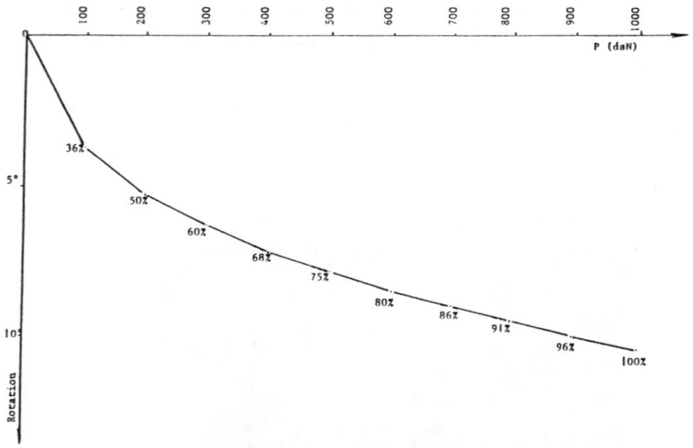

Fig 11 Rotations

The theoretical study reports the stiffening of the Simplex at different degrees of self-stressing; this stiffening is shown by the reduction in deflection, Fig 12, and rotation Fig 13. Homogeneisation of the stresses in the cables as a whole was also observed.

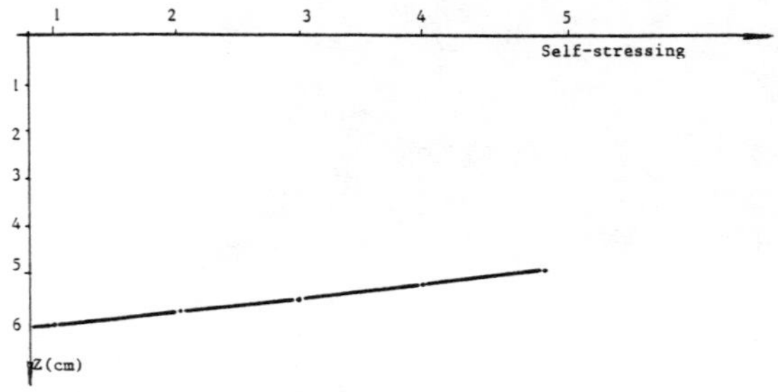

Fig 12 Deflections for different degrees of self-stressing

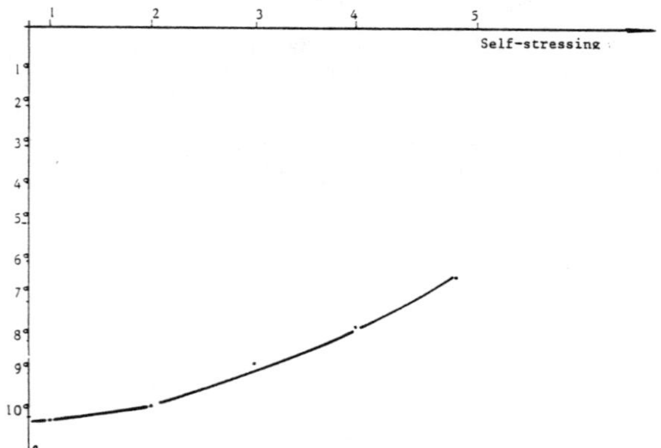

Fig 13 Rotations for different degrees of self-stressing

CONCLUSION

The present work is a preliminary approach to the various problems posed by tensegrity systems. It currently concentrates on the field of the determination of the relational structures and that of dynamic behaviour.

REFERENCES

1. J C MAXWELL, On reciprocal figures, frames and diagrams of forces, Scientific Papers, Cambridge, University Press, United Kingdom, Vol 2, 1890, pp. 175-177

2. A G M MICHELL, The limits of economy of materials in frame structures, Philosophical magazine, series 6, Vol 8 N° 47, 1904, pp. 589-597.

3. B H V TOPPING, Shape optimization of skeletal structures: a review; Journal of the Structural Division, ASCE, Vol 109, N°8, August 1983, Paper 18185, pp. 1933-1951.

4. K SNELSON, Tensegrity Masts, Shelter Publications, Bolinas, Californie, 1973.

5. R MOTRO, Formes et Forces dans les systèmes constructifs Application au cas des systèmes réticulés spatiaux autocontraints, Thèse d'Etat, Université des Sciences et Techniques du Languedoc, juin 1983.

6. C R CALLADINE, Buckminster Fuller's "Tensegrity" Structures and Clerk Maxwell's rules for the construction of stiff frames, International Journal of Solids and Structures, 1978.

7. A S DAY, A general computer technique for form finding for tension structures, Symposium I.A.S.S. : Development of form, Morgantown, 1978.

FINITE ELEMENT ANALYSIS OF BARREL VAULTS

A B SABIR*, BSc, PhD, and A R BOAG**, BSc, MSc.

* Senior Lecturer, Department of Civil and Structural Engineering, University College, Cardiff, Wales.

** Graduate Engineer, Scott Wilson, Kirkpatrick and Partners.

A finite element solution to the problem of barrel vaults is presented. The solution utilizes a rectangular curved element whose shape functions are based on satisfying exactly the strain free rigid body modes of displacements and on independent strain assumptions insofar as it is allowed by the compatibility equations. This cylindrical curved element is used in conjunction with exact three dimensional beam and arch elements to obtain solutions to practical barrel vault problems. Barrel vaults of several aspect ratios of length to width and having different types of boundary conditions along the straight edges are first considered. The influence of restraining the straight edges by beams of varying rectangular cross sections on the deflections and stresses is obtained. The case of replacing the rigid diaphragms by a more aesthetic assembly of circular arches and columns is also considered.

INTRODUCTION

Cylindrical shell roofs are normally designed using two distinct approaches. The first approach is based on avoiding the complicated mathematical considerations and an emphasis is made on the physical action of the shell. Usually an attempt is made to model the behaviour of the barrel vault to that of a slab, a beam or an arch. The second approach is based on a mathematical formulation for circular cylindrical shells and the solution to the resulting differential equations is achieved for practical barrel vault problems by making certain simplifying assumptions, see Refs 1 and 2. One of these assumptions is associated with expressing the applied load on the shell by Fourier expansions. The others are of a physical nature. No radial displacement at the curved ends of the shell is allowed and the edge beams are not allowed to offer any horizontal or twisting restraints.

In the present paper a finite element solution to the problem of practical barrel vault problems is given. None of the above mentioned simplifications is made but the accuracy of any finite element solution will depend on the performance of the different types of elements used. In earlier investigations on the suitability of the available finite elements for curved structures it was revealed that to obtain satisfactory converged results, the finite elements based on independent polynomial displacement functions require the curved structures to be divided into a large number of elements, see Ref 3. To gain understanding of the problem of selecting and devising suitable shape functions for curved structures, a detailed investigation was carried out on simple circular arches. An account of basic considerations in the selection of displacement fields was examined numerically for thin, thick, shallow and deep structures. Shortcomings highlighted in this way were found to be largely removed if assumed strains, rather than displacement functions were used, see Ref 4. This work lead to the development of a new class of simple and efficient finite elements suitable for arches of all proportions. The shape functions satisfy exactly the requirement of strain-free rigid body displacements and are based on independent strains rather than on independent displacements. The superiority of the strain elements were further demonstrated by investigating the eigen-value problem of in-plane free oscillations of circular rings, see Ref 5. The work was further extended to develop elements for arches deforming out of the plane of curvature, see Ref 6. The opportunity was taken in this work to develop high order strain elements as well as elements requiring only the essential external degrees of freedom. An improvement in the performance of the high order elements was also obtained by statically condensing the unnecessary internal degrees of freedom at the element level.

This approach was then applied to devise a new class of finite elements for cylindrical shells, see Ref 7. A rectangular element was first tested by applying it to the analysis of the familiar pinched cylinder and cylindrical shells under gravity loading. The results converged rapidly for displacements, Ref 7 and for stresses, Ref 8. Further tests were carried out to investigate the ability of these elements in predicting the high stresses in the neighbourhood of applied concentrated loads, Ref 9. The loads considered were either radial or axial forces as well as moments about tangents to the circular cross section. The results obtained were not only in agreement with those of Forsberg and Flugge, Ref 10 but when plotted for the complex parameters defining proportions of the shell and flexibility as suggested by Calladine Ref 11, their general forms corresponded closely with theoretical predictions.

The above mentioned cylindrical element was formulated using Timoshenko's shell equations, Ref 12. More recently considerable attention has been given to formulations using Sanders-Koiter shell equations. A strain based rectangular curved element was therefore developed using Sanders-Koiter equations, Ref 13. The results obtained by using this new element differed only slightly when compared with those obtained in the previous work.

Practical cylindrical shell roofs utilize other structural

components such as beams and arches. The exact three dimensional beam and column elements are in existence and their stiffness matrices are commonly used in structural analysis. An exact three dimensional stiffness matrix for circular arches deforming within and out of the plane containing the curvature has also been developed in Ref 14. In that work the three dimensional beam and circular arch elements were combined to model and analyse cylindrical grillage roofs.

ELEMENTS USED IN THE ANALYSIS OF BARREL VAULTS

The cylindrical shell element used in the analysis is shown in Fig 1. A curvilinear system of co-ordinate axes are used, the element is rectangular in plan and with four corner nodes only. The degrees of freedom are the longitudinal displacement u, the circumferential displacement v, the normal displacement w and the two rotational degrees of freedom

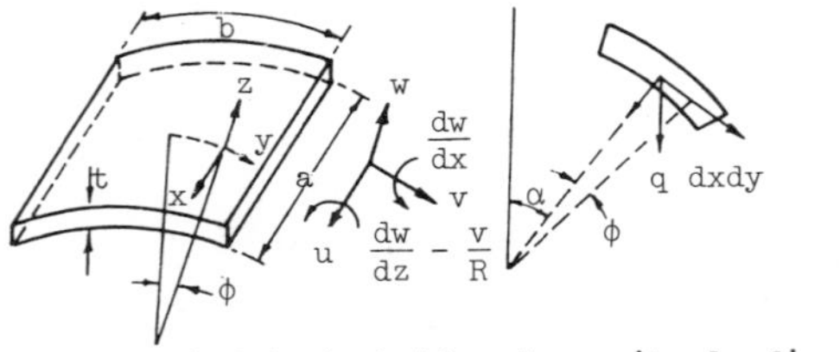

FIG. 1 Cylindrical shell and gravity loading

$\partial w/\partial x$ and $\partial w/\partial y - v/R$ where R is the radius of the shell. The stiffness matrix of such an element will be a matrix of (20 x 20) in size. The assumed displacement for such an element based on satisfying strain free-rigid body displacements and on simple independent strain assumptions insofar as it is allowed by the compatibility equations is given by

$$u = \alpha_2 R\cos\phi+\alpha_4 R\sin\phi+\alpha_5+\alpha_7 x+(\alpha_{11}R+\alpha_{19}R^3-\alpha_{20}R^2)\phi+\alpha_8 x\phi -\alpha_{17}R^3\phi^2/2-\alpha_{19}R^3\phi^3/6 \qquad (1)$$

$$v = (\alpha_1+\alpha_2 x)\sin\phi-(\alpha_3+\alpha_4 x)\cos\phi+\alpha_6-(\alpha_{19}R-\alpha_{20})Rx+\alpha_{16}R^2\phi +\alpha_{17}R^2 x\phi+\alpha_{18}R^2\phi^2/2+\alpha_{19}R^2 x\phi^2/2 \qquad (2)$$

$$w = -(\alpha_1+\alpha_2 x)\cos\phi-(\alpha_3+\alpha_4 x)\sin\phi+\alpha_9 R-\alpha_{16}R^2+(\alpha_{10}R-\alpha_{17}R^2)x -\alpha_{18}R^2\phi-\alpha_{12}x^2/2-\alpha_{19}R^2 x\phi-\alpha_{13}x^3/6-\alpha_{14}x^2\phi/2-\alpha_{15}x^3\phi/6 \qquad (3)$$

The stiffness matrix K_s is calculated in the usual way of the finite element method and for further details see Ref 7.

The straight beam or column element used in the analysis is shown in Fig 2. The element has six degrees of freedom at each end, three transulational u,v,w and three rotational Θ_x, Θ_y and Θ_z. This element allows bending about the two principal directions, axial change in length and twisting of the cross section. A Cartesian three directional system of co-ordinate axes are shown in Fig 2a and is used to derive the general stiffness matrix K_b for the beam. The displacement functions used for the calculations of the stiffness matrix for bending about the two principal planes are

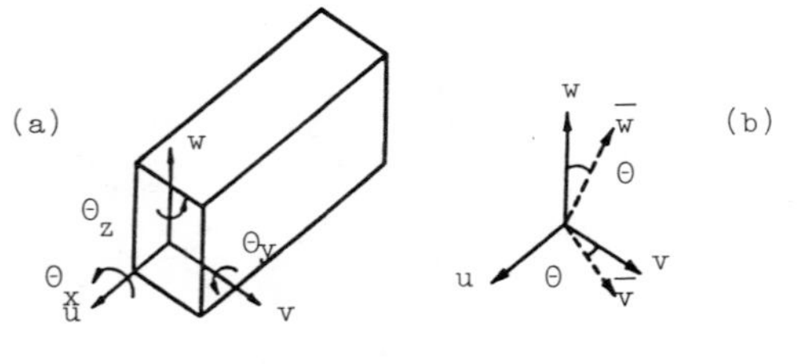

FIG. 2. Beam element and co-ordinate axes

$$v = a_1+a_2x+a_3x^2+a_4x^3 \text{ and } w = a_5+a_6x+a_7x^2+a_8x^3 \qquad (4)(5)$$

and for the axial and torsional stiffnesses of the beam

$$u = a_9+a_{10}x \quad \text{and} \quad \Theta_x = a_{11}+a_{12}x \qquad (6)(7)$$

The curved beam element used in the analysis is shown in Fig 3. The element has also six degrees of freedom at each node. The three degrees of freedom associated with deformation of the curved beam within the plane containing the curvature are v,w and Θ_x and the remaining three degrees

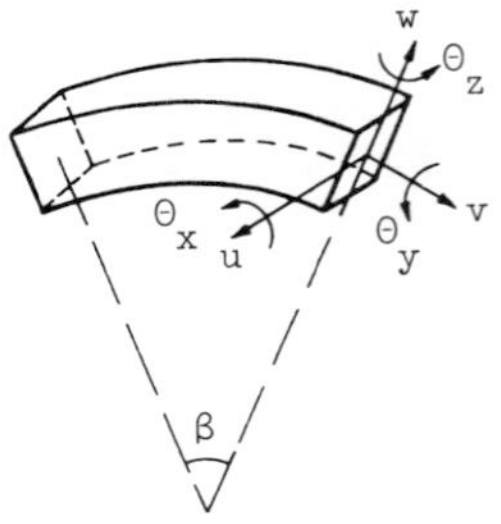

FIG. 3 Arch element

of freedom u, Θ_z and Θ_y are associated with deformation out of this plane. The derivation of the stiffness matrices for inplane deformation K_{ai} and out of plane deformation K_{ao} are based on the curvilinear coordinate axes shown. The displacement functions used for calculating K_{ai} are given by, see Ref 4.

$$w = b_1\cos\phi+b_2\sin\phi+b_4\{\cos\beta(1-\cos\phi)-(1+S^2)\phi\sin\phi/2\} +b_5\{-\sin\beta(1-\cos\phi)+(1+S^2)(\sin\phi-\phi\cos\phi)/2-b_6(1-\cos\phi) \qquad (8)$$

$$v = -b_1\sin\phi+b_2\cos\phi+b_3+b_4\{-\phi\cos\beta+\sin\phi(\cos\beta+\tfrac{1}{2}) -S^2\sin\phi/2-(1+S^2)\phi\cos\phi/2\}+b_5\{\sin\beta(\phi-\sin\phi) +\cos\phi+(1+S^2)\phi\sin\phi/2\}+a_6(\phi-\sin\phi) \qquad (9)$$

where $S=I_x/AR$, I_x is the relevant second moment of area, A is the cross sectional area and β is the angle subtended by the element.

The stiffness matrix for an arch element deforming out of its plane of curvature is given by, see Refs 14 and 15.

$$\begin{bmatrix} M^{-1} & NM^{-1} \\ NM^{-1} & M^{-1} \end{bmatrix}$$

where

$$|M| = \frac{R^3}{EI_y}\begin{bmatrix} (\rho f+a & (\rho a-a)/R & (-\rho e-b)/R \\ & (\rho c+a)/R^2 & (1-\rho)b/R^2 \\ \text{symmetric} & & (\rho a+c)/R^2 \end{bmatrix}$$

and $\rho = EI/GJ$, $a = (\beta-\sin\beta\cos\beta)/2$, $b = \sin^2\beta/2$

$c = (\beta+\sin\beta\cos\beta)/2$, $d = \sin\beta-c$, $e = 1-\cos\beta-b$

$f = \beta-2\sin\beta+c$

and

$$|N| = \begin{bmatrix} -1 & 0 & 0 \\ R(\cos\beta-1) & \sin\beta & \cos\beta \\ R\sin\beta & -\cos\beta & -\sin\beta \end{bmatrix}$$

ASSEMBLY OF THE STRUCTURAL STIFFNESS MATRIX

Having obtained the stiffness matrices for each of the three essential components namely the shell, beam or column and arch the overall structural stiffness matrix is obtained by satisfying compatibility and equilibrium at the nodes. This is carried out in such an order so that a minimum band width is obtained. In satisfying the compatibilities of the nodal degrees of freedom it is essential that

(i) All the elements have a mutual co-ordinate system.

(ii) All the elements have the same degrees of freedom.

In the present analysis both the shell and the arch elements have a mutual curvilinear system of co-ordinate axes while the beam or the column elements are in Cartesian co-ordinates with vertical and horizontal axes. The beam element is transformed into the shell and arch system before assembly. The transformation procedure is a well known one and it can be shown that for example if the beam stiffness matrix in the vertical and horizontal set of axes is K_b then the transformed stiffness matrix in a system inclined by an angle Θ, see Fig 2b will be given by $\overline{K}_b$ where

$$\overline{K}_b = |C|^T \; |K_b| \; |C| \qquad (10)$$

and the transformation matrix C at each end is

$$\begin{bmatrix} 1 & 0 & 0 & 0 & 0 & 0 \\ 0 & \cos\Theta & \sin\Theta & 0 & 0 & 0 \\ 0 & -\sin\Theta & \cos\Theta & 0 & 0 & 0 \\ 0 & 0 & 0 & \cos\Theta & 0 & \sin\Theta \\ 0 & 0 & 0 & 0 & 1 & 0 \\ 0 & 0 & 0 & -\sin\Theta & 0 & \cos\Theta \end{bmatrix}$$

As we have already seen that the shell element has only the five essential external degrees of freedom per node while both the arch and the beam elements have six nodal degrees of freedom. Combination of these three essential components is achieved by expanding the shell element to occupy a size of 24x24 matrix shown below.

K_{11}	K_{12}	K_{13}	K_{14}
K_{21}	K_{22}	K_{23}	K_{24}
K_{31}	K_{32}	K_{33}	K_{34}
K_{41}	K_{42}	K_{43}	K_{44}

original 20x20 shell stiffness matrix

K_{11}	K_{12}	K_{13}	K_{14}
K_{21}	K_{22}	K_{23}	K_{24}
K_{31}	K_{32}	K_{33}	K_{34}
K_{41}	K_{42}	K_{43}	K_{44}

Expanded 24x24 matrix

The elements of the rows and columns corresponding to the sixth degree of freedom are made equal to zero. In this way the addition of the arch and beam elements are made by a simple superposition procedure after they are also expanded into spaces of 24x24 in size e.g. at a corner of the barrel vault where a shell element is to be combined to beam and arch elements as shown in Fig 4. The assembly leads to the matrix shown below, the elements of the beam and the arch stiffness matrices are added to the spaces corresponding to nodes 2 and 4 and 3 and 4 respectively.

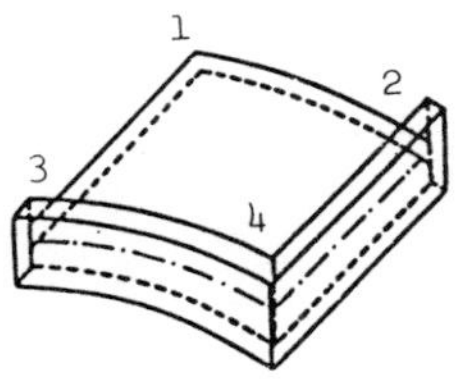

K_{s11}	K_{s12}	K_{s13}	K_{s14}
	K_{s22} $+K_{b22}$	K_{s23} $+K_{a11}$	K_{s24} $+K_{b21}$ $+K_{a12}$
symmetric		K_{s33}	K_{s34}
			K_{s44} $+K_{b11}$ $+K_{a22}$

FIG. 4 Assembly of shell, arch and beam elements

Unlike other methods used in the finite element analysis of structures where elements possessing different degrees of freedom are employed, the present method does not suppress any of the degrees of freedom. For example, the variation in the longitudinal displacement u of the circular end of the shell is allowed to contribute to the bending of the arch out of its plane of curvature and the variation in the circumferential displacement v of the straight edges is also allowed to bend the edge beams in a similar manner. These modes of bending are associated in part with the unsuppressed sixth degree of freedom of the arches and the edge beams.

LOAD VECTOR FOR SELF WEIGHT

The nodal applied loads are calculated either by a lumping process or more accurately by obtaining a consistent load vector. In the present analysis we used the latter method. For a shell under its own gravity self weight of q per unit surface area of the shell we first calculate the work done by the distributed load on the deformations of the element which can be shown to be given by

$$q \int_{-\frac{b}{2}}^{\frac{b}{2}} \int_{-\frac{a}{2}}^{\frac{a}{2}} \{w \sin(\alpha-\phi) + v \cos(\alpha-\phi)\} \, dxdy \qquad (11)$$

where α and ϕ are shown in Fig 1. This work is then equated to the work done by the nodal forces on the nodal displacement to obtain the necessary load vector.

COMPUTER PROGRAM

The program used to obtain the results given in the present paper is the NODAL solution routine, Ref 16. This routine is coded in FORTRAN, it assembles the overall structural matrix in a way that a minimum band width is obtained and the resulting linear algebraic equations are solved using the wellknown Gaussian elimination method. By making judicious use of 'backing-store', in the form of disc or magnetic tape units, it is possible to solve large sets of equations. The solution routine performs the triangularization of the coefficients corresponding to one node at a time. Only a triangle of such coefficients in the form of a vector needs to be stored in the core of the central processor unit at any time. In this way computers with small storage capacities can be used and the other main advantage is that since the arithmetical operations are performed on the coefficients corresponding to one node at a time, any additional nodal stiffness due to other structural components such as the arches, the beams and the columns can be added and dealt with easily. Furthermore the assembly process mentioned earlier in the paper leads to the existence of zero terms in the diagonals corresponding to the sixth degree of freedom at all the internal nodes of the shell. The NODAL solution routine has a facility to deal with this problem by not carrying out the triangularization on the coefficients of the rows and corresponding columns of any of the degrees of freedom. This is done in conjunction with an identification vector which is used also for satisfying boundary conditions. The output of the program was arranged to give the nodal deflections as well as the components of stresses at the middle of the elements.

PROBLEMS CONSIDERED

(a) Barrel Vaults Without Edge Beams

Since the method used in the present work can deal with the analysis of barrel vaults having any prescribed boundary conditions along the straight edges. Barrel vaults with free or completely fixed straight edges and supported by rigid diaphragms at the ends are analysed to show the level of deformations and stresses set up for these two limiting cases. Furthermore vaults of three aspect ratios of width to length are considered. They are referred to as short, intermediate and long shells. The dimensions considered are given in the following Table. Young's modulus for concrete was taken to be 20,688 x 10^3 kN/m^2. The applied load was taken to consist of the self weight of 1.8 kN/m^2 together with a snow and limited access uniform load of 3 kN/m^2.

	Short	Intermediate	Long
Radius	7.62m	7.62m	7.62m
Length	3.50m	15.24m	22.00m
Thickness	75.00mm	75.00mm	75.00mm
Angle subtended (ϕ)	40°	40°	40°

Convergence tests for deflections and stresses were first carried out to establish the meshes into which the cylindrical part of the vaults are to be divided so that acceptable levels of accuracy of results is obtained. These meshes are used throughout the subsequent analysis. It was generally found that a mesh made of square elements and consisting of no more than eight circumferential and twelve longitudinal elements in one quarter of the cylinder is acceptable for the most critical dimensions used.

For the shells with free straight edges the maximum radial and circumferential deflections occur at the middle of the free edge while the maximum longitudinal deflection occurs at the corner of the shell. The largest of these deflections is the radial one, it amounted to half, thirteen and twenty five times the thickness for the short, intermediate and long shells respectively. When the straight edges are restrained either by hinging or fixing them to rigid supports the maximum radial deflection occurred at the centre of the shell and its value was drastically reduced to about one fortieth, one sixth and one fifth of the thickness for the short, intermediate and long shells respectively when the straight edges were fixed. The above results show the importance of the introduction of restraints along the straight edges when practical barrel vault roofs are designed and constructed.

Of interest is the variation of the various stress components set up in the shell. Figure 5 shows the variation of the longitudinal force N_x per unit length along the central arc of the shell when the straight edges are free. It is seen that this force has large maximum tensile values at the free edge of about 200,1100 and 1700 kN/m for the short, intermediate and long shells respectively. These tensile forces are balanced by compressive forces having maximum values of about 40,250 and 390 kN/m. For the long shell the maximum compressive force is at the crown while for the intermediate and short shells are at about ϕ = 22° and 26° respectively. In this figure we also give the variations of the circumferential force N_y per unit length for the intermediate shell. The corresponding values of N_y for the short and long shells vary only slightly from those for the inter-

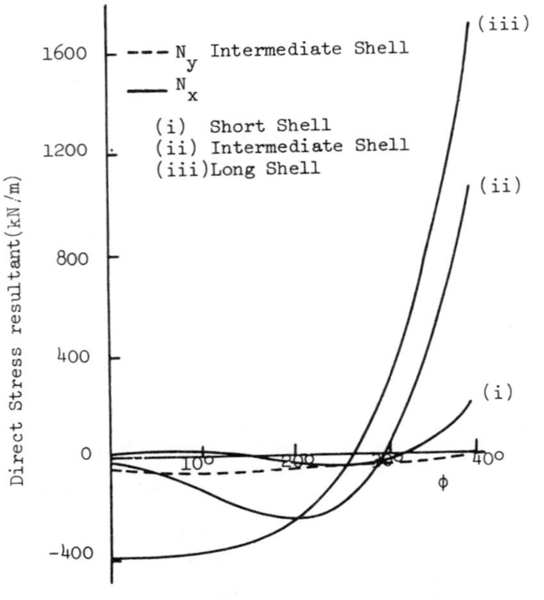

FIG. 5 Variation of N_x and N_y along central arc of free shell

mediate shell. It is seen that N_y is compressive throughout and having a maximum value of about 50 kN/m and hence only of importance when designing short shells. For the intermediate and long shells N_y is of a smaller order of magnitude when compared with N_x. The above results show the essential difference in the structural behaviour of long and short shells. Long shells behave like beams where the moment of resistance is mainly supplied by the variation of N_x along any circular section while short shells resist the applied loads to a greater degree by the membrane circumferential thrust N_y.

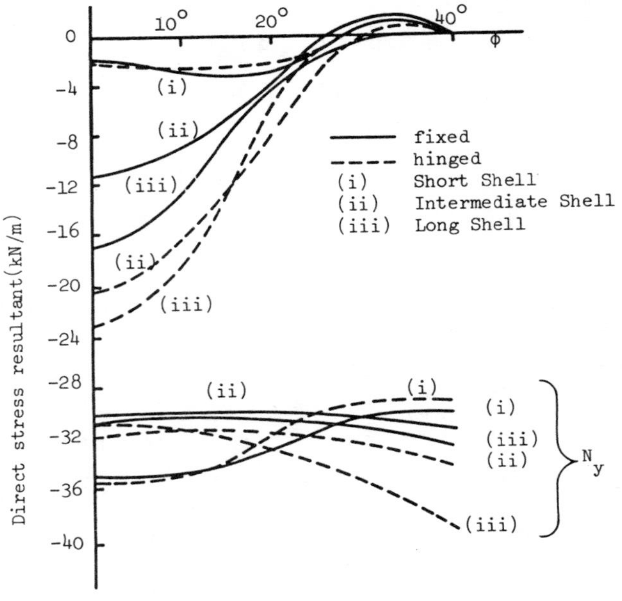

FIG. 6 Variation of direct stress resultant along central arc of shell

Figure 6 shows the variation of the direct forces N_x and N_y when the shells are restrained along the straight edges. It is seen that the tensile forces N_x near the edges have almost disappeared and that the compressive values of N_x away from the edges have reduced drastically for all the shells.

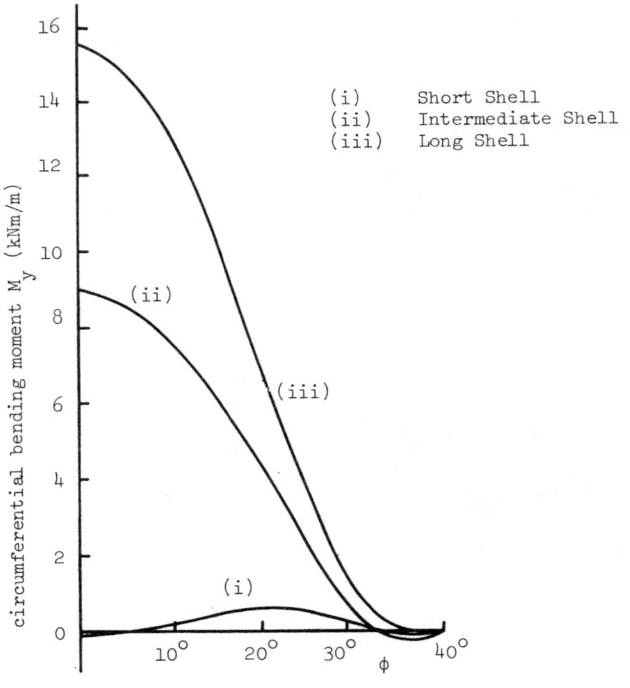

FIG. 7 M_y along central arc for free shell

For example for the long shell with free edges the value of N_x at the crown of 390 kN/m is reduced to about 17 kN/m for fixed edge and 23 kN/m for the hinged edge. Furthermore the circumferential compressive force per unit length N_y has larger values than N_x for all the shells considered. It has a more uniform distribution and it has become the critical direct force for design purpose. The other important stress component is due to the circumferential bending moment M_y. Its distribution along the central arc for the case of free straight edges is given in Fig 7.

The maximum values of about 17 and 19 kN/m for the long and intermediate shells occur at the crown while that for the short shell occur at about $\phi = 20^o$ with a maximum value of about 0.7 kN/m. Again if the shell is restrained along its straight edges this bending moment is reduced drastically. Figure 8 gives the variation for the cases when the straight edges are hinged and fixed. We note that for the long shell the maximum value of M_y is less than 1.0 and in the other cases is even smaller.

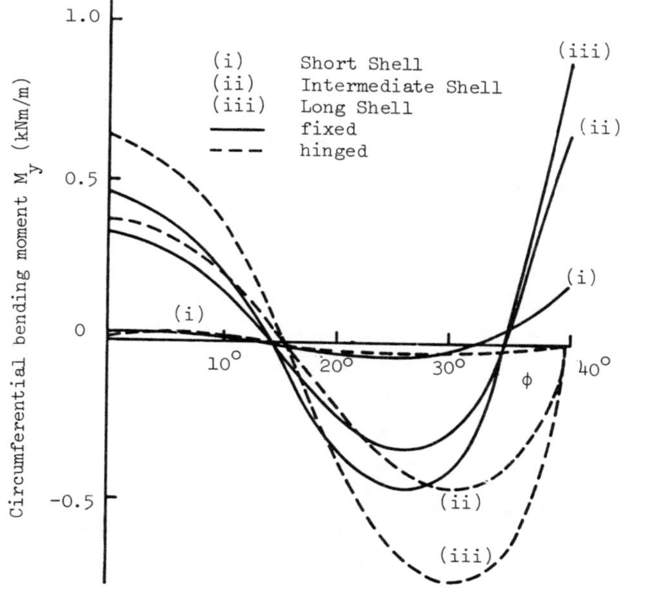

FIG. 8 Variation of M_y along central arc of shell

The accuracy of the above results were ascertained by a comprehensive comparison with other approaches including those given in Refs 1 and 2.

(b) BARREL VAULTS WITH EDGE BEAMS

The above results show the importance of the introduction of some form of restraints along the longitudinal edges of the barrel vaults. This is usually carried out in practice by employing edge beams of rectangular cross sections having some finite stiffness. The results of barrel vaults with edge beams will lie between the two limiting cases of free and fixed straight edges considered earlier. A finite element analysis was therefore carried out for the previously considered barrel vaults with edge beams having the following range of dimensions.

Case	Dimensions of edge beams (width x depth)
A	0.10m x 0.20m
B	0.10m x 0.30m
C	0.10m x 0.40m
D	0.15m x 0.50m
E	0.15m x 0.76m
F	0.20m x 1.00m
G	0.20m x 1.25m

The loading considered is the same as for the previous case together with the self weight of the respective concrete edge beams.

The intermediate shell having an edge beam of 0.15m x 0.76m was first analysed and a comprehensive comparison was made between the finite element results and those given by Gibson's (Ref 1) type of analysis. A computer program coded in FORTRAN has been written at Cardiff incorporating Gibson's equations for the design of concrete barrel vaults. A small sample of results from the two types of analyses are given in Table 1 for deflections as well as stress components along the central arc of the shell. The table shows that the two sets of results are in general agreement in the way deflections and stresses vary along the central arc but the numerical values in some places can differ considerably. N_x differs but generally by few percentages, at $\phi = 30^o$ the difference is as large as 25%. N_y appears to differ generally by almost 11% and the other important stress component due to M_y differs by up to 25% again at $\phi = 30^o$.

Owing to restriction of space we have decided to confine the results to the effect of the size of the edge beams on the maximum values of the stress components N_x, N_y and M_y along the central arc of the shell. For a more detailed report on the deflections and stresses over the entire shell see Ref 17.

ϕ	N_x(kN/m)		N_y(kN/m)		M_y(kNm/m)		W(mm)	
	FE	Gibson	FE	Gibson	FE	Gibson	FE	Gibson
0°	- 78.5	- 82.1	-43.1	-48.2	2.75	2.45	2.8	1.1
10°	- 97.5	- 95.3	-40.5	-45.1	1.87	1.51	- 0.2	- 0.9
20°	-122.3	-109.9	-32.3	-36.1	-0.14	-0.56	- 7.7	- 5.2
30°	- 55.6	- 74.4	-19.5	-21.1	-2.0	-1.58	-14.5	-15.2
40°	72.8	68.8	- 4.9	- 3.5	-2.27	0.61	-15.3	-17.2

Table 1 Comparison of the finite element results with those given by Gibson's type of solution

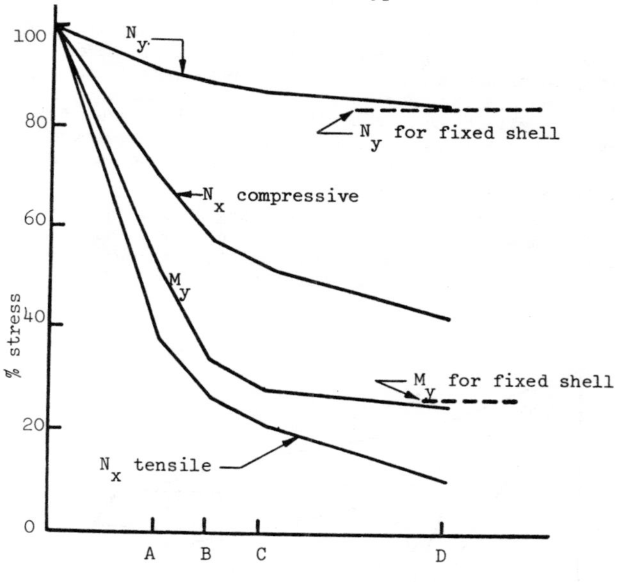

FIG. 9 Influence of edge beams on stresses for short shell

Figure 9 shows the effect of progressively increasing the size of the edge beam on these stresses for the short shell. The vertical axis is expressed as a ratio of the maximum stress occurring to that when the straight edges are free (i.e. without an edge beam). The horizontal axis is drawn to a scale proportional to the cross sectional area of the edge beams. It is seen that even a small edge beam of 0.1m x 0.2m will reduce the maximum value of the tensile and compressive stress resultant N_x to less than 40% and 70% respectively while the maximum compressive stress resultant N_y is reduced to 90%. The reduction in the level of all the stress component continues as the edge beam is increased in size. When the edge beam has a size of 0.15m x 0.50m (case D) N_y is reduced to about 84%. This is nearly equal to the maximum value of N_y when the straight edges of the shell are fixed. Furthermore with this size of edge beam the maximum value of M_y changes position from ϕ to 20° to ϕ = 40° and a further increase in size of the edge beam increases the torsional stiffness of the beam to such a state that M_y at the edge increases further. We may therefore conclude that for the short shell an edge beam of a size between cases C and D might be considered desirable. The maximum radial deflection will be as small as one ninth and one fifteenth of the thickness of the shell when edge beams cases C and D are employed.

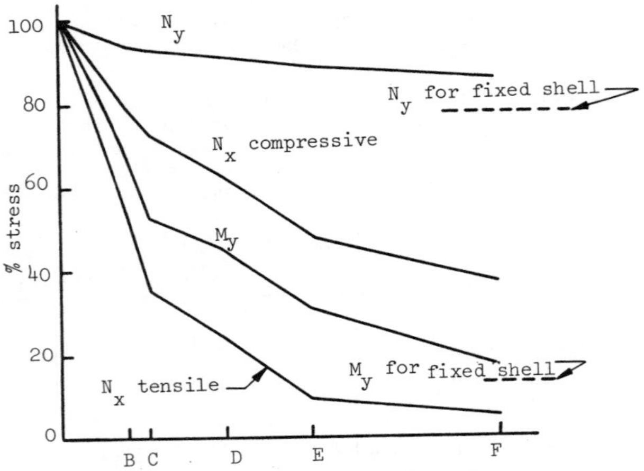

FIG. 10 Influence of edge beams on stresses for intermediate shell

Figure 10 gives the same results for the intermediate shell. It is seen that an edge beam of size 0.10m x 0.40m is required to reduce the tensile stress resultant N_x to less than 40% and similar consideration to those mentioned for the short shell helps to conclude that for the case of the intermediate shell a beam size of 0.20m x 1.00m (case F) will be desirable. N_y would have been reduced to only 10% more than that for the fixed edge.

Figure 11 gives the results for the long shell. An edge beam of 0.2m x 1.25m (case G) will reduce the stresses and deflections to an acceptable level and the maximum value of N_y will only be 3% more than that for the fixed edge.

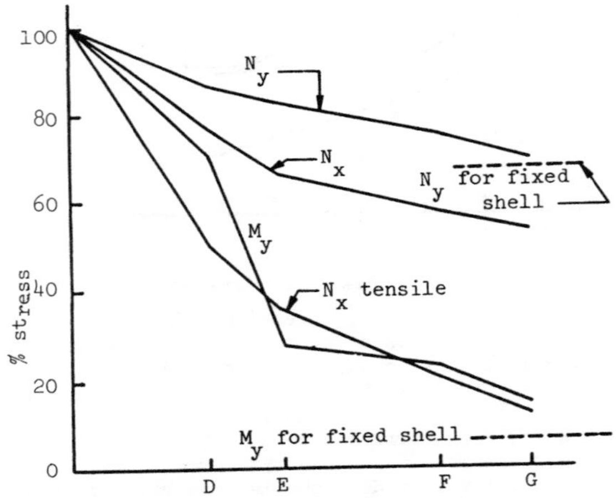

FIG. 11 Influence of edge beams on stresses for long shell

(c) BARREL VAULTS WITH CIRCULAR ARCHES

All the results given previously were for cases where the ends of the barrel vaults are supported by rigid diaphragms. These supports are normally assumed to restrain deformation within the plane of the diaphragms and offer no restraints out of this plane. A more aesthetic end support may be achieved if these diaphragms are replaced by an assembly of circular arches and corner columns as shown in Fig 12.

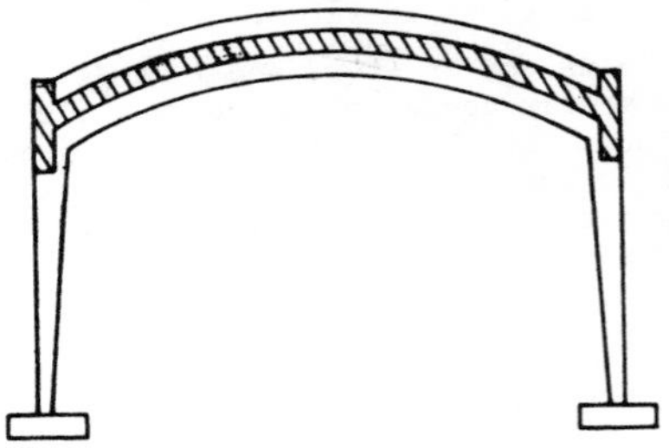

FIG. 12 Barrel vault with end circular arches and corner columns

A finite element analysis incorporating the arches and columns at the circular ends is carried out for all the shells considered previously. The dimensions of the arches were taken to be

Case	Dimension	R/depth
A	0.10m x 0.25m	30
B	0.10m x 0.30m	25
C	0.15m x 0.38m	20
D	0.15m x 0.50m	15
E	0.20m x 0.76m	10
F	0.20m x 1.52m	5

In this way the effect of progressively increasing the size of the arch on the level of deflections and stresses is studied. A small sample of results are given in Figs 13,14.

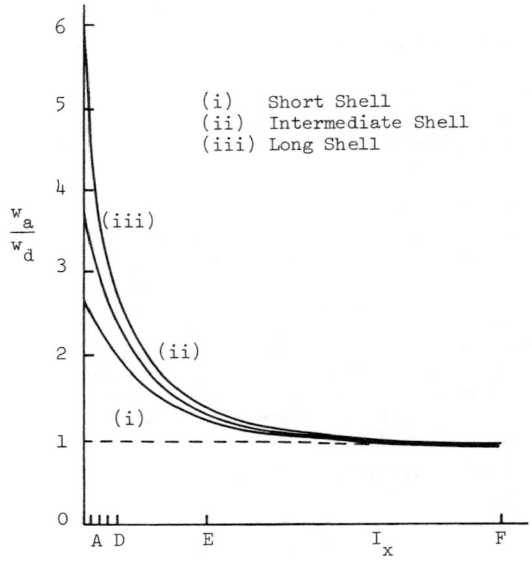

FIG. 13 Influence of size of arch on the radial deflection at crown

Figure 13 gives the results for the radial deflection at the crown of the barrel vault. The vertical axis represents the ratio of this deflection to that when end diaphragms are used and the horizontal axis is drawn to a scale proportional to the second moment of area, I_x, of the arches.

Each barrel vault was considered to have an edge beam of a depth of one twentieth of the length. It is seen that for all the cases of short intermediate and the long shell an arch (case E) of depth one tenth of the radius of the vault is capable of reducing the maximum radial deflection to within 5% of that when rigid wall diaphragms are used.

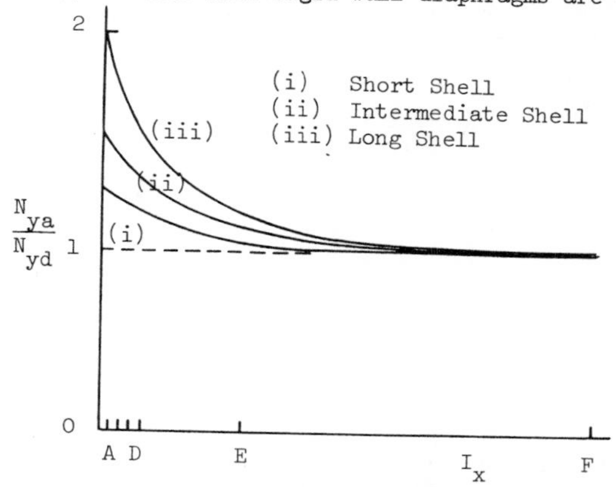

FIG. 14 Influence of size of arch on circumferential stress

Figure 14 gives similar results for the circumferential stress resultant N_y. Again we see that an arch of a depth of one tenth of the radius is sufficient to reduce N_y to within 5% of that when the rigid wall diaphragm is used. From the above sample of results together with the more detailed results given in Ref 17, we may conclude that an arch (case E) may be considered as a satisfactory replacement to the rigid wall support.

CONCLUSIONS

1. A reliable finite element solution is presented for practical barrel vault problems. A well tested cylindrical shell element together with exact beam and arch elements are utilized for this type of analysis.

2. A computer program coded in FORTRAN employing the NODAL solution routine is available at Cardiff for general use by industry for the analysis of such barrel vaults.

3. The analysis shows that the size of the edge beams required depend on the aspect ratio of the shell for the short shell of a beam of 0.15m x 0.50m can be considered adequate while for the intermediate and the long shells beams of 0.2m x 1.0m and 0.2m x 1.25m are required.

4. An effective replacement of the rigid diaphragm is achieved by the use of an arch having a depth of one tenth of its radius.

5. The method presented and the computer program have the facility of dealing with thickening of the shell near the edges and with barrel vaults having reinforced openings. The reinforcement can be in the form of beams and arches around the holes or by a grillage of beams and arches within the area of the openings. Due to restriction of space we decided not to report on these cases.

6. A computer program coded in FORTRAN employing Gibson's method of analysis is also available at Cardiff. This program is written to analyse barrel vaults with edge beams and to calculate the necessary reinforcements for the shell as well as the edge beams.

7. The method outlined in this paper can be utilized to analyse cylindrical shells with stiffeners in the form of beams and arches.

REFERENCES

1. J E GIBSON, Computer analysis of cylindrical shells, E&FN Spon Ltd. 1961.

2. C B WILBY, Design graphs for concrete shell roofs. Applied Science Publishers Ltd, London 1980.

3. D G ASHWELL and A B SABIR, Limitations of certain curved finite elements when applied to arches. Int.J. Mech. Sci., Vol. 13, p. 133, 1971.

4. D G ASHWELL A B SABIR and T M ROBERTS, Further studies in the application of finite elements to circular arches. In.J.Mech.Sci., Vol. 13, p. 507, 1971.

5. A B SABIR and D G ASHWELL, A comparison of curved finite elements when used in vibration problems. Journal of Sound and Vibration, Vol. 18, p.555, 1971.

6. A B SABIR, Stiffness matrices for the general deformation (out of plane and inplane) of curved beam members based on independent strain functions. The mathematics of finite elements and application II. Ed. J R Whiteman. Academic Press, 1975.

7. D G ASHWELL and A B SABIR. A new cylindrical shell finite element based on simple independent strain functions. Int.J.Mech.Sci., Vol. 14, p.171, 1972.

8. A B SABIR and F K RAMADHANI. A doubly curved rectangular shallow shell element for general shell analysis.In press.

9. A B SABIR and D G ASHWELL. Diffusion of concentrated loads into thin cylindrical shells. The Mathematics of finite element and application III. Edited by J R Whiteman, Academic Press, 1978.

10. K FORSBERG and W FLUGGE, Point load on shallow elliptical paraboloid. J.App.Mech. 33, p.575, 1966.

11. C R CALLADINE, Thin walled elastic shells analysed by Rayleigh Method. Int.J.Solids Structures 13, p.515,1977.

12. S TIMOSHENKO, Theory of plates and shells. McGraw-Hill 1940.

13. A B SABIR and T A CHARCHAFCHI, Curved rectangular and general quadrilateral shell elements for cylindrical shells. The mathematics of finite elements and application IV. Edited by J R Whiteman, Academic Press, 1982.

14. A B SABIR, Computer analysis of circular cylindrical grillage. Int.Conf. on the behaviour of slender structures. City University, Sept. 1977.

15. HWA PING LEE. Generalised stiffness matrix for a curved beam element. AIAA Journal, Vol. 7, No. 10, 1969.

16. A B SABIR, The NODAL solution routine for the large number of linear simultaneous equations in the finite element analysis of plates and shells. Finite elements for thin shells and curved members. Wiley 1976.

17. A R BOAG, Finite element analysis of barrel vaults with edge beams and central openings. M.Sc. Thesis, University of Wales, 1981.

THE DESIGN OF SPACE FRAMES SUBJECT TO DEFLECTION LIMITATIONS

K I MAJID, BSc, PhD, DSc, CEng, FICE, FIStructE

Professor, Department of Civil and Structural Engineering,
University College, Cardiff.

Excessive deformation can be the limiting design factor in the case of slender space frames covering large areas. In radio telescopes the main design criterion is its homological deflection. In this paper three methods are given for the economic or optimum design of such structures. These either use an iterative approach or the theorems of structural variations or non-linear programming. The design variables may be the cross sectional areas of the members and the shape of the actual structure. Examples given include a space grid, two radio telescopes, a dome and a transmission tower.

INTRODUCTION

The deflection of a space frame can often be the main design criterion. For instance when covering a large area. In the case of radio telescopes, the "back up" structure of the dish has to be light, easy to steer and its surface deformation remains parabolic.

Hitherto, design methods select the member sizes by intuition and then carry out an analysis to check their strength. This is a costly process because large structures require the repeated solution of a large number of simultaneous equations. Furthermore such an approach does not rectify excessive deflections. Increasing member sizes by intuition only aggravates the problem.

This paper reports on a number of methods for the design of space frames, subject to deflection limitation, developed in the last ten years by the author and his collaborators. These cover: (i) The design of structures in which the deflections at one or at a number of points are specified. The design method calculates the unknown member areas so that the stresses and the deflections are satisfactory. (ii) The homological design of radio telescopes using the theorems of structural variation, see Ref 1, or by an iterative method, see Ref 2. (iii) The design of space frames with variable shape, selected in such a manner as to produce the cheapest design.

DESIGN FOR SPECIFIED DEFLECTIONS

When solving the stiffness equations of a structure, the stiffness matrix $\underline{K}$ contains the products of unknown member areas A and joint deflections Δ. To proceed, the areas are assumed and the equations are solved for $\underline{\Delta}$. Often the deflections at certain points are specified by the design codes and it is possible to modify the stiffness equations, see Ref 2, so that the known deflections are substituted and the unknown areas are calculated. The areas are selected in such a way that ensures satisfactory deflections in the structure.

The contribution of a general pin ended member i, connecting joints j and k at its first and second ends, to the stiffness of joint j in the X direction is known to be:

$$W_{Xj} = a_i\, l_{Pi}\, (l_{Pi}\, x_j + m_{Pi}\, y_j + n_{Pi}\, z_j - l_{Pi}\, x_k - m_{Pi}\, y_k - n_{Pi}\, z_k) \qquad \ldots\ldots\ 1.$$

where a = EA/L, l_P, m_P and n_P are the direction cosines of the longitudinal P axis of the member and x, y, z are the joint deflections in X, Y and Z directions. The external load W_{Xj} is applied at j in the X direction. This equation can be rearranged either to calculate the unknown deflection x_j or the unknown area of the member. Thus, in the general case when a total of R members meet at j, the value of x_j is, see Ref 2:

$$x_j \quad [-W_{Xj} + \sum_{i=1}^{R} a_i\, l_{Pi}\, (m_{Pi}\, y_j + n_{Pi}\, z_j - l_{Pi}\, x_k - m_{Pi}\, y_k - n_{Pi}\, z_k)]/[-\sum_{i=1}^{R} (a_i\, l_{Pi}^2)] \quad \ldots\ldots\ 2.$$

Similar equations can be written for y_j and z_j.

On the other hand, if x_j is known or specified, Eqn 1 can be rearranged to calculate the area of the first group of members, thus

$$A_{g1} = W_{Xj}/[\sum_{i=1}^{R} \delta_{ir}\, b_i\, l_{Pi}\, (l_{Pi}\, x_j + m_{Pi}\, y_j + n_{Pi}\, z_j - l_{Pi}\, x_k - m_{Pi}\, y_k - n_{Pi}\, z_k)] \qquad \ldots\ldots\ 3.$$

where $b_i = E_i/L_i$. It is considered that the members are in n groups with areas A_{g1}, A_{g2}, ...A_{gr}...A_{gn} and that:

$$\delta_1 = A_{g1}/A_{g1} = 1,\ \delta_2 = A_{g2}/A_{g1},\ \ldots\ \delta_r = A_{gr}/A_{g1},\ \ldots$$
$$\ldots\ \delta_n = A_{gn}/A_{g1} \quad \ldots\ldots 4.$$

Equations similar to Eqn 3 can be also derived when y_j or z_j are specified. These equations can be solved by an iteration process to calculate either the unknown joint displacements or the unknown group areas. If required member grouping can be altered to suit the strength requirements.

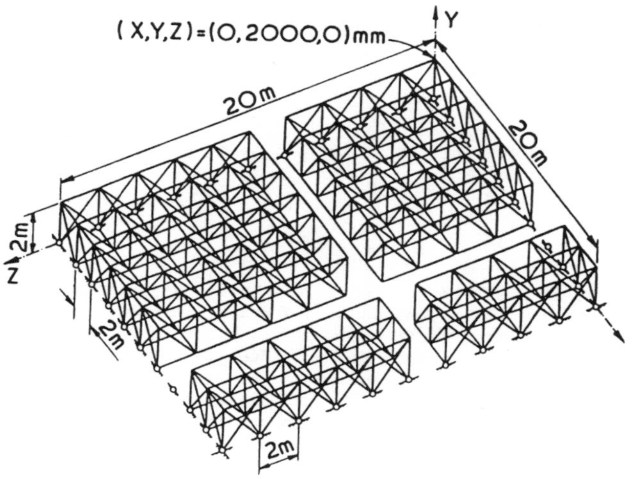

Fig 1 A pin jointed space structure with 781 members.

This method was used to design the space structure shown in Fig 1. This has 781 members divided into four area groups. Two alternative designs, one in high yield and the other in mild steel were produced. Each structure was carrying a vertical load of 10 kN at each joint together with a single concentrated load of 140 kN acting at the centre of the lower horizontal grid. The vertical deflection at the centre of the lower grid was limited to 55.55 mm, see Ref 3. The cheapest design was produced after only two cycles of iteration. This had a volume of 98.20×10^7 mm^3 with high yield steel and 119.26×10^7 mm^3 with mild steel. In the first case the vertical deflection governed the choice of the member areas while with the mild steel the strength requirements decided the sections.

It is possible to generalize the above equations so that the deflections at many joints are specified. For a total of t specified joint deflections, the members are divided into t independent area groups. However, it is more advantageous to have the members divided into more groups than the number of known deflections. For a total of n specified deflections, with n < t only n independent area groups can be obtained from the stiffness equations. The remaining t-n areas can be expressed in terms of the area of group 1, as explained earlier. It is now possible to calculate (see Ref 4) the area A_g of group g of members from :

$$A_g = \{W_{Xj} - \sum_{i=1}^{R_1} A_1\, b_i\, l_{Pi}\, c_{jk} - \sum_{i=1}^{R_c} A_m\, b_i\, l_{Pi}\, c_{jk}$$
$$- \sum_{i=1}^{R_o} \delta_{ri}\, A_1\, b_i\, l_{Pi}\, c_{jk}\}/\{\sum_{i=1}^{R_g} b_i\, l_{Pi}\, c_{jk}\} \quad \ldots\ldots 5.$$

where R_g is the total number of members in group g,

$$c_{jk} = l_{Pi}\, x_j + m_{Pi}\, y_j + n_{Pi}\, z_j - l_{Pi}\, x_k$$
$$- m_{Pi}\, y_k - n_{Pi}\, z_k\ , \quad \ldots\ldots 6.$$

$$R_d = \sum_{m=2}^{n} R_m = R_2 + R_3 + \ldots R_n\ , \quad \ldots\ldots 7.$$

$$R_o = R_{n+1} + R_{n+2} + \ldots + R_t \quad \ldots\ldots 8.$$

and

$$R_c = R_d - R_g \ \text{(with } g = 2,3,\ldots,n) \quad \ldots\ldots 9.$$

In this manner all the areas of the first n groups are calculated. Notice that these are independent equations and the area of a group is obtained from one, simply by substituting the current values of the other variables in it.

This iterative method was used to design radio telescopes subject to homological deformation. The surface deflections are first specified so that the deflected shape of the telescope is a predefined paraboloid. The aim is then to calculate the cross sectional areas of the member groups. For each known deflection a stiffness equation is rearranged, so that it calculates the area of one of these groups. This area is, however, dependent on other unknown joint deflections and group areas in a non-linear manner. To begin with, these unknowns are assumed and a group area is calculated. This is used in the next stiffness equation to calculate either an unknown deflection, or another group area and the process is repeated with all the stiffness equations. A more realistic set of variables thus become available to be used in the next cycle. The process is then repeated until the stiffness equations are satisfied with the final set of group areas.

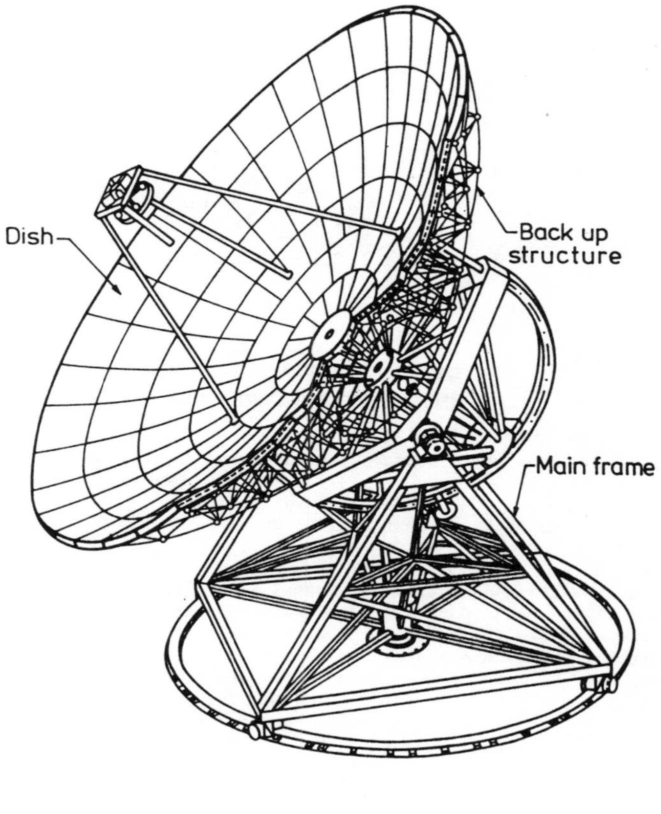

Fig 2 A diagram of the radio telescope.

The 15 m diameter radio telescope shown in Fig 2 was designed by this method. Details are shown in Fig 3.

The back up structure has 444 members, in 8 groups, arranged in a cyclically symmetrical manner. There are 132 joints and 12 supporting joints. The primary structure has 60 parabolic panels supported by 72 surface joints. A vertical load of 3 kN is thus acting at each inner surface joint. The members were divided into eight area groups. Mild steel with tensile and compressive design stresses of 0.25 kN/mm^3 and 0.2 kN/mm^2 was used in the design. The six surface joints in a typical grid were made to deflect, see Fig 3, by fixed amounts, thus :

$$\{ d_1 \quad d_2 \quad d_3 \quad d_4 \quad d_5 \quad d_6 \} =$$
$$\{-0.55 \quad -0.69 \quad -1.09 \quad -1.81 \quad -2.36 \quad -2.66\}\text{mm}$$

In seven cycles the volume of the structure was reduced to 206.7 × 10^6 mm^3. Throughout the process, the surface deflections controlled the sectional areas while the limit state for strength did not play a part.

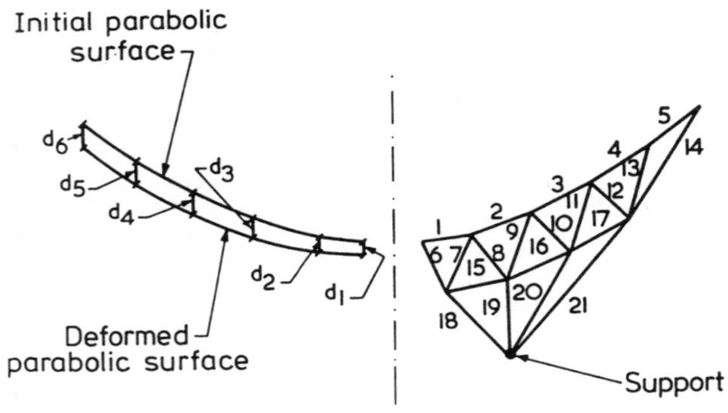

(a) Details of each cyclic grid

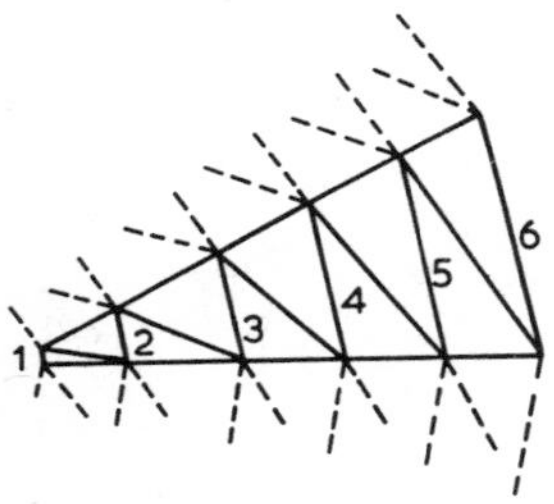

(b) Details of each sector showing the horizontal members.

Fig 3 Details of 15 m back-up structure.

THE THEOREMS OF STRUCTURAL VARIATIONS

Without a fresh analysis, see Refs 5 and 6, these theorems predict the exact forces and deflections throughout a structure when the material or cross sectional properties of a number or members are altered, or when these are totally removed. For pin jointed structures, let p_j be the force in member j due to the external loads and f_{ji} be the force in j due to unit axial loads acting on another member i. If the area of member i is changed from A_i to A_i', the new force π_j in j is $\pi_j = p_j + r_{\alpha i} f_{ji}$ where $r_{\alpha i}$ is the variation factor for the changing member i, given by :

$$r_{\alpha i} = - \alpha_i \, p_i / (1 + \alpha_i \, f_{ii}) \qquad \ldots\ldots \; 10.$$

Here $\alpha_i = \delta A_i / A_i = (A_i' - A_i)/A_i$, p_i is the axial force in i due to unit external loads and f_{ii} is the force in i due to unit external loads acting axially on it. The deflection ψ_t at any joint t is $\psi_t = x_t + r_{\alpha i} \chi_{ti}$, where x_t is the deflection at t before altering i and χ_{ti} is the deflection at t due to unit external forces acting at the ends of i. Al-Bakri, see Ref 7, extended these theorems to cover the simultaneous alteration of several members. He proved that when n members are altered differently, the force in the new member i is :

$$\pi_i = (1 + \alpha_i)(p_i + \sum_{j=1}^{n} r_{\alpha i} \, f_{ij}) \qquad \ldots\ldots \; 10.$$

and the new force π_k in member k which is kept unaltered is :

$$\pi_k = p_k + \sum_{j=1}^{n} r_{\alpha j} \, f_{kj} \qquad \ldots\ldots \; 11.$$

The new deflection ψ_t at a typical joint t is now :

$$\psi_t = x_t + \sum_{j=1}^{n} r_{\alpha j} \, \chi_{tj} \qquad \ldots\ldots \; 12.$$

Thus it is possible to predict the joint deflections throughout a structure when members are altered. This fact is used to design radio telescopes subject to homological deflection constraints. The procedure is to analyse the telescope back up structure, calculate the root mean square (RMS) deviation of the joints, at their new positions, from a best fit paraboloid and use the theorems of structural variations to alter the members to reduce the RMS errors to a minimum.

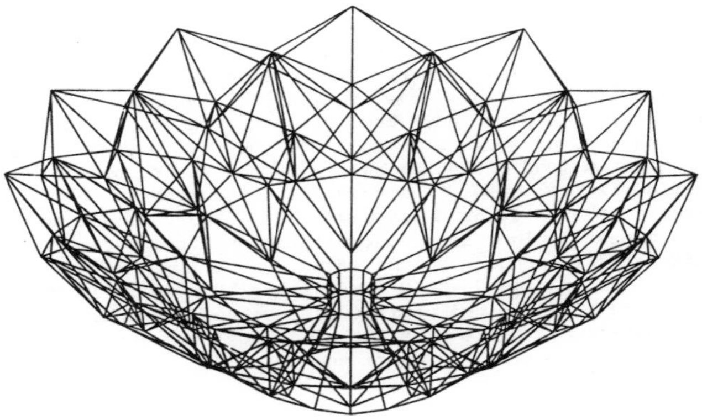

Fig 4 A 14.6 m diameter radio telescope.

The 14.6 m diameter structure shown in Fig 4 was designed in this manner. This has 156 joints and 612 members in 15 groups. A segment of it has six surface joints to be controlled. Figure 5 shows the distances between these nodes, from a best fit paraboloid of the initial trial structure, while Fig 6 shows the manner in which these nodes are influenced by unit loads applied externally at the ends of the members in various groups. The Figure is used to alter the member areas so that the RMS errors are reduced in the best possible manner.

OPTIMUM SHAPE DESIGN

The design of a structure invariably includes the choice of its shape. Often this is decided upon intuitively and then it may be modified in an arbitrary manner. In fact it is possible to make the shape of a structure as one of the

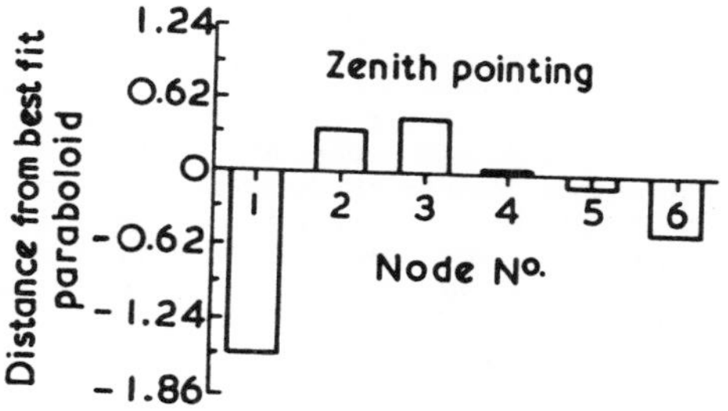

Fig 5 The objective vector.

design variables, to be decided upon during the design process, not using the engineers judgement, which may or may not be sound, but by using structural principals. The position of the joints, the length and the inclination of the members and thus the actual shape of the structure are all unknowns. Initially a trial structure is considered and the design process changes the number and the position of the supports, joints and the members until the final design has the least material cost.

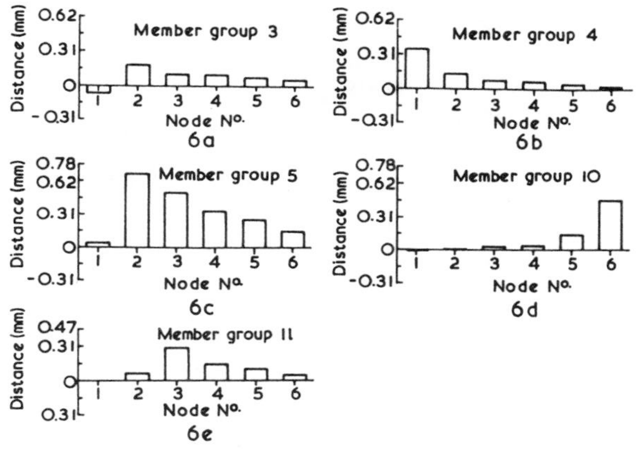

Fig 6 Influence of member unit loads on nodal deflection.

Because the member areas and the joint co-ordinates are variables, the stiffness matrix $\underline{K}$ is no longer square or symmetrical. It is also a function of these variables, thus

$$\underline{K} = f(\{\underline{A}_g\}\{\underline{C}\}) \qquad \text{...... 13.}$$

where the vector $\{\underline{A}_g\} = \{\underline{A}_{g2} \cdot\cdot \underline{A}_{g1} \cdot\cdot \underline{A}_{gG}\}$ lists the areas of the groups and for a total of Q joints, the variable co-ordinates of a given joint J are X_J, Y_J and Z_J and the vector $\{\underline{C}\} = \{X_1 \ Y_1 \ Z_1 \ldots X_J \ Y_J \ Z_J \ldots X_Q \ Y_Q \ Z_Q\}$ lists the co-ordinates of the joints. As the design operation progresses towards minimum material costs, several shapes become available. The engineer can therefore price these and select the one with the lowest fabrication or construction cost.

The stress $\sigma = p/A = E\delta/L$ in each member may be tensile or compressive. To cover both cases, two stress constraints are needed for each member, thus :

$$E([-\underline{D}^T \ \underline{D}^T]\{\underline{\Delta}\})/L \leqslant \sigma_{tw}$$
$$-E([-\underline{D}^T \ \underline{D}^T]\{\underline{\Delta}\})/L \leqslant \sigma_{cw} \qquad \text{...... 14.}$$

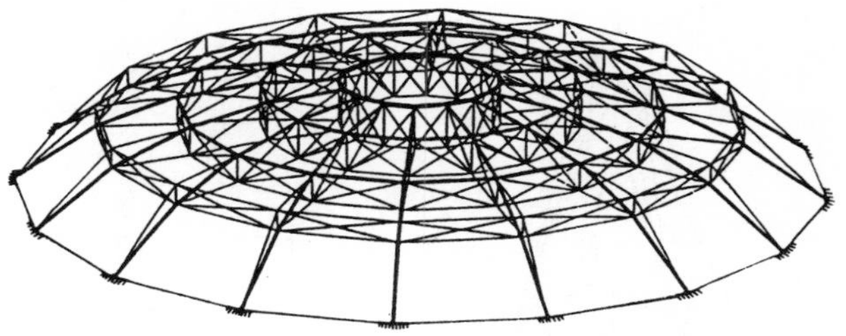

Fig 7 The final optimum dome.

where $\{\underline{D}\} = \{l_p \ \ m_p \ \ n_p\}$ and σ_{tw} and σ_{cw} are the limiting (characteristic) stresses for the member - taken from the new code, see Ref 3. The buckling strength σ_b is assumed to be inversely proportional with slenderness squared and for a tubular section, is taken to be $0.154\ \pi^2\ EA^{1.5}/L^2$.

The deflection at a point is decided upon by its distance from a fixed point, such as the origin. The upper and the lower limits on the deflection of a joint are thus variables of the form :

$$\phi_1\ (Y_J) \leqslant x_J \leqslant \psi_1\ (Y_J)\ ,$$
$$\phi_2\ (Z_J) \leqslant x_J \leqslant \psi_2\ (Z_J) \qquad \text{...... 15.}$$

where ϕ and ψ signify functions of the variables Y_J etc. The volume of the member groups is taken as the objective function to be minimised.

As a practical example a space frame to cover a circular hall 30 m in diameter was designed. It was required to leave a 6 m diameter skylight at the centre of the hall. The design was initiated with a flat circular space frame which was supported at 16 points along the circumference of a supporting rigid ring beam in the same plane as the bottom grid of the flat space frame. The self weight of the members were included as design variables and all the joints at top and bottom grid were also subject to imposed loads. All the unsupported joints were allowed to move vertically within a design region of ±2 m of their initial position. Stress and buckling constraints were imposed on all the members while the vertical deflection of the joints were not allowed to exceed diameter/360, i.e. 83.33 mm.

Making use of symmetry only (1/16)th of the structure was considered. The non-linear stiffness, stress and deflection constraints were linearized and "move limits" were imposed before using a simplex method to solve the linear programming problem. After each linear design, the shape altered and the process was repeated until eventually the dome shown in Fig 7 was obtained as the best shape with the least material cost. This has principal members consisting of radial arches connected by concentric rings. The vertical deflections in the final dome proved to be uncritical.

The design of this dome demonstrates that the design method can in fact guide the engineer to improve the design approach and produce a more efficient shape than that initially suggested by the designer.

Similarly the cranked transmission tower shown in Fig 8 was produced by the computer to be a cheaper alternative to an initial conventional trial structure with four straight posts.

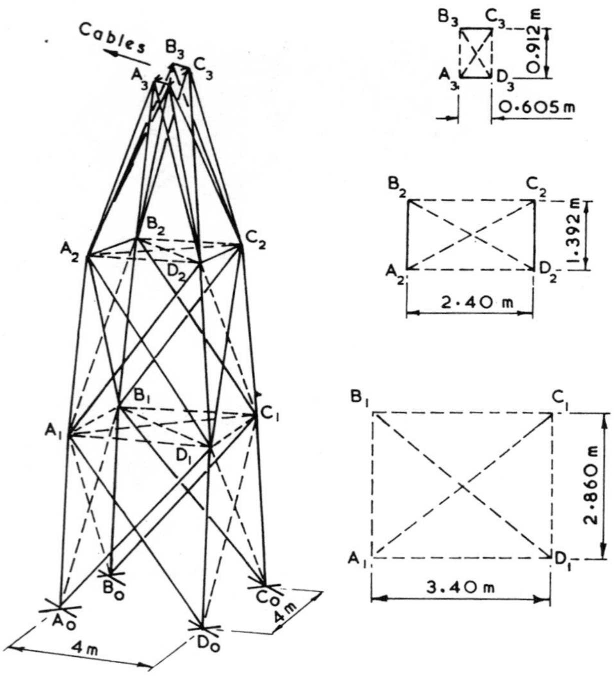

Fig 8 The optimum shape of the transmission tower.

CONCLUSIONS

Various methods were outlined for the design of space frames in which the joint deflections govern the design outcome. Details of these methods can be obtained from the writer. It is evident that to cope with deflections, an orthodox trial and error method is not sufficient and methods that can include these deflections as constraints are necessary.

REFERENCES

1. K I MAJID, Optimum Design of Structures, Newnes-Butterworths, Chapter 6, London, 1974.

2. K I MAJID and S OKDEH, Limit State Design of Pin-Jointed Space Structures, The Structural Engineer, Vol. 60B, No. 1, March 1982.

3. Draft Standard Specification for the Structural Use of Steelwork in Buildings. Part 1 : Simple Construction and Continuous Construction, London, British Standards Institution, 77/13808DC.

4. S P CHIEW, MSc Thesis, University College, Cardiff, 1983.

5. K I MAJID and D W C ELLIOTT, Forces and Deflections in Changing Structures, The Structural Engineer, 1973, Vol. 51, No. 3, March.

6. K I MAJID and D W C ELLIOTT, Topological Design of Pin-Jointed Structures by Non-Linear Programming, Proc.Instn.Civ.Engrs., Part 2, 1973, Vol. 55, March.

7. M A E AL-BAKRI, Optimum Design of Transmission Towers, PhD Thesis, University of Surrey, 1978.

A MICROCOMPUTER-AIDED DESIGN SYSTEM FOR LARGE SPACE STRUCTURES

JEAN-JACQUES JONATOWSKI*, BCE, MCE, PhD and DEAN KOUTSOUBIS**, BCE

* President, Multitech Computer Services, Inc.
Little Neck, N.Y., U.S.A. 11363

** Structural Engineer/System Manager, Space Structures International Corp.
Plainview, N.Y., U.S.A. 11803

Until recently, analyses of three-dimensional double-layer grid structures were performed using main-frame computers. Generation of analysis input data, structural design, and fabrication drawings were usually done manually. However, with the advent of super-fast microcomputers, it is now possible to perform, in a few hours, at one's desk, a complete analysis and design of very large space structures. This paper describes a microcomputer-aided engineering system specially developed for large three-dimensional structures.

INTRODUCTION

Until a year ago, Space Structures International Corp. was analyzing three-dimensional double-layer grid structures, of the type shown in Fig 1, using a well known general purpose finite element program through a computer service bureau. The program ran on the Cray-1 computer. Although its processing time was extremely fast, the actual computer turnaround time for a typical analysis was not. Turnaround time was affected by the following factors: computer execution priority, which in turn affected cost, system usage load on the service bureau, and delivery of output results. Because Space Structures did not have an on-site high speed printer and its location was somewhat distant from the service bureau, computer output results were delivered by messenger service and were normally available for review on the following day after submittal of the computer run. In addition to the slow turnaround, the cost of using a computer service bureau, which included batch processing, connect and storage charges became high.

In order to reduce costs, shorten the analysis turnaround time, and improve the quality of the design process, an extensive study was conducted to determine alternatives to the service bureau. Both present and future needs and requirements of Space Structures were defined in detail, and initial costs and schedules of return on investment for different systems were investigated. Finally, it was determined that a microcomputer with custom software would be more effective and economical.

Thus, a Hewlett-Packard Series 200 model 36 microcomputer with a 10 megabyte Winchester type disk drive, 180 cps line printer, and 8 pen plotter were acquired (Fig 2). The model 36 is a 16 bit microcomputer with a 32 bit internal architecture based on Motorola's MC68000 microprocessor.

Fig 1. Hyatt Space Truss being installed in New Jersey.

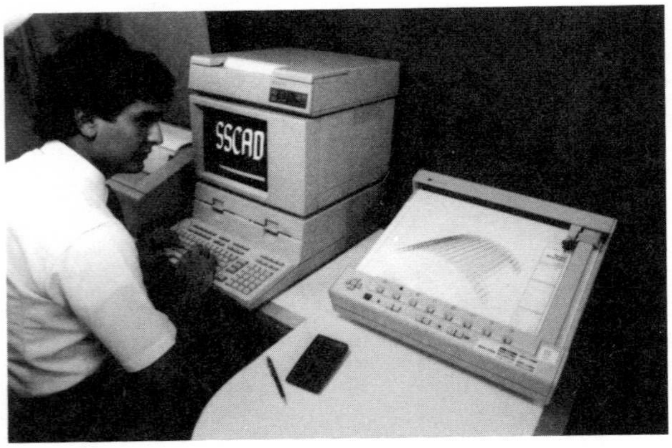

Fig 2. SSCAD Microcomputer System.

A three-dimensional truss analysis program, originally developed by Multitech Computer Services for Radio Shack's TRS-80 Model II, was converted to the HP Model 36. Custom pre- and post-processing modules were developed jointly by Space Structures International Corp. and Multitech Computer Services and added to the analysis program to form a totally integrated computer-aided engineering system. This system has been named "SSCAD" by the engineers of Space Structures.

The SSCAD System has not only eliminated the monthly service bureau cost, but has provided productivity gains of a magnitude exceeding our most hopeful projections. For a project consisting of geometry and load generation, an average of three analysis passes, member and connection design, and final drawing generation, what used to take approximately 245 hours was reduced to 21 hours (Ref 4).

SSCAD SYSTEM

The SSCAD System presently consists of the following major functions: modeling, analysis and design. The modeling function, as implied by its name, is used for modeling the structure. This includes not only geometry generation, but loading and support condition generation. The analysis function computes the node displacements, member forces and support reactions using finite element methods. The design function of SSCAD sizes the members, designs the connections and prepares final drawings. All three functions are integrated using a network of menus and specially programmed keys. The engineer has no commands to memorize because all the possible options available to him at any given time are displayed and, if necessary, explained on the screen. The advantage of SSCAD over other systems is its extensive utilization of graphics to communicate information to the engineer. By graphically generating and reviewing data, the engineer can input and interpret more information, more efficiently, with less error.

MODELING FUNCTION

Using a geometry generation subprogram, the engineer begins by inputting a minimum amount of information defining the type of space structure and its overall dimensions. This replaces the previous method of generating a geometry one component at a time. Mathematical algorithms have been developed which define the inherent characteristics of a variety of double-layer grid structural systems and geometry patterns. Structural systems currently available for automatic generation include flat planes, barrel arches, conical surfaces, and domes. Once a structural system has been chosen, the engineer has great flexibility to quickly and easily create any geometry pattern he may need to fulfill project requirements. Within seconds, the subprogram generates all the coordinates and connectivity for that geometry. Using a graphics subprogram, the engineer graphically displays the geometry. He can choose a plan view, an elevation or a perspective from any location in space. Figure 3 illustrates a perspective of a space truss after being generated on SSCAD. If the engineer needs to view the space truss in greater detail, he can easily zoom up on any section as shown in Fig 4. The engineer could then use a variety of subprograms to graphically modify the geometry for his particular project.

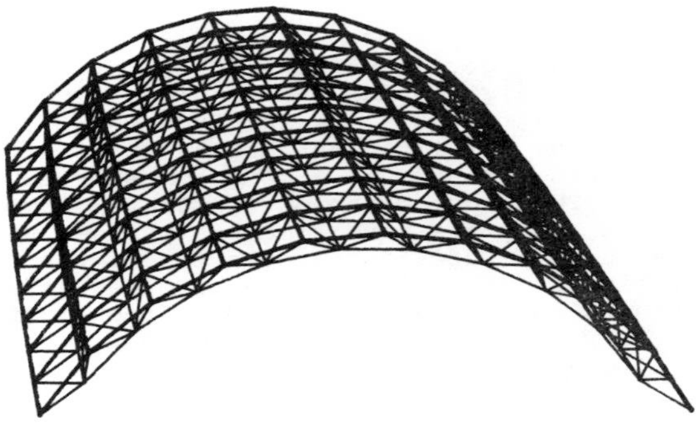

Fig 3. Perspective view of Barrel Arch Space Truss.

In order to quickly modify the geometry, each subprogram includes a graphical box which is superimposed over a chosen view. This box can be shifted, scaled, or rotated using the keyboard to isolate any part of the space truss within the box. Once the desired part of the space truss is within the box, the engineer can specify any of a number of different functions to be performed only on those space truss components within the box. These functions, among others, include deletion, creation, shifting, and rotation of space truss components. Figure 5 illustrates the use of the graphical box to delete the corners of a flat space truss. Three corners have already been deleted and the graphical box has been positioned over the fourth corner. The engineer would then press the proper key to execute the final deletion.

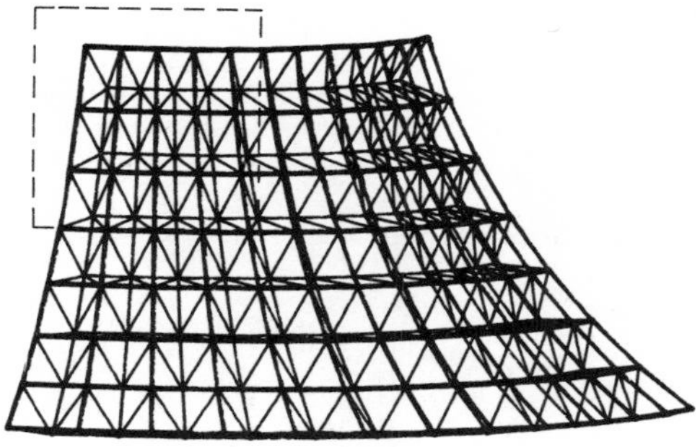

Fig 4a. Graphical "ZOOM" Box used to identify section of Conical Space Truss to be magnified.

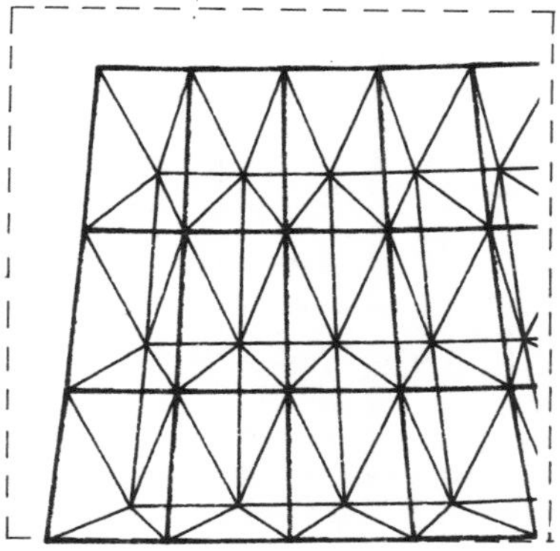

Fig 4b. Section of Conical Space Truss magnified through use of "ZOOM" Box to show greater detail.

Once the geometry has been completed, the engineer can graphically generate forces, supports and member properties in the same way he generated the geometry. For example, by specifying the type and magnitude of a particular force and isolating a section of the geometry with the graphical box, the engineer can generate the specified force on every component within the box. Once the loading information is complete, the engineer is ready to access the analysis function of SSCAD.

ANALYSIS FUNCTION

The analysis section of SSCAD consists of a three-dimensional truss analysis program utilizing the conventional finite element displacement method (Ref 1, 8). The efficiency of a finite element program is influenced by the method used for the solution of the equations. Some of the available procedures that could have been used range from iterative methods such as the Gauss-Seidel technique (Ref 5) to direct Gaussian elimination schemes. SSCAD utilizes the frontal equation solution technique originated by Irons (Ref 3). This procedure minimizes computer core size requirements and the number of arithmetic operations (Ref 2).

The size of the three-dimensional truss system that may be analyzed has been set to 3500 members, 1000 nodes, 6 load cases, 20 load case combinations, and a solution frontal width of 50 nodes. Using a 10 megabyte disk drive, there is ample space left to increase those limits if necessary. The analysis results consisting of node displacements, member forces and support reactions are automatically saved for post-processing.

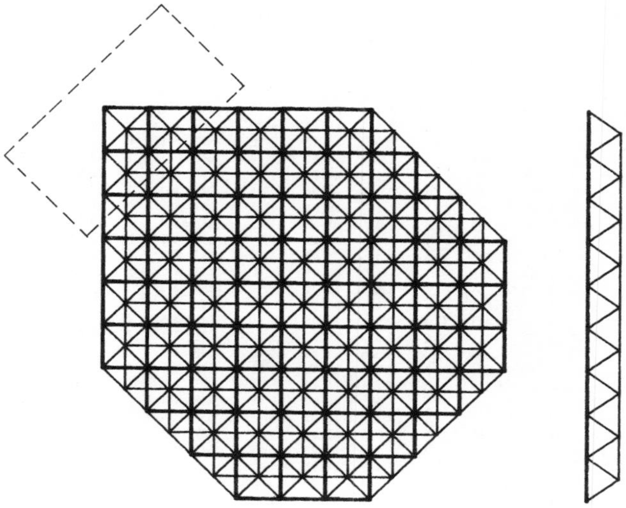

Fig 5. Use of Graphical Box to remove portion of structure.

After the analysis has been performed, the engineer may choose to review the analysis results before proceeding to design. The results may be presented numerically or graphically. Displacements can be presented by graphically displaying the deformed structure from any view using any degree of deflection magnification (Fig 6). Multi-colored plots of member forces, where each color represents a different magnitude range of member force, can be used to

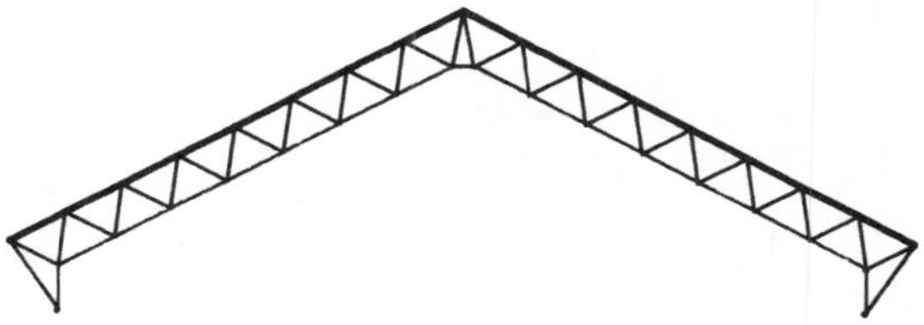

Fig 6a. Elevation of Ridged Space Truss before loading.

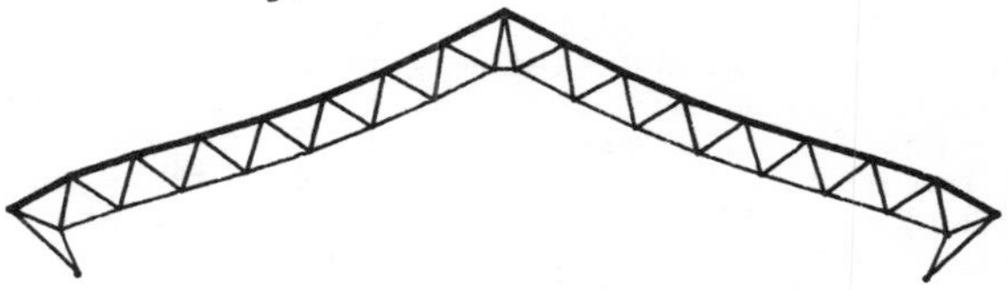

Fig 6b. Elevation of Ridged Space Truss after loading. 100X Exaggerated deflections assist in understanding behavior of structure.

Space Structures
INTERNATIONAL CORP.
155 Dupont Street
Plainview, NEW YORK 11803

STRUT NO/TYP	HUB CONNECT.	P	DESCRIPTION	MAT GRD	WEIGHT lbs/ft	(---LENGTH--- (inches) NOMINAL	EFF.	NET TENS. AREA	BOLT/PIN TYPE	(ALLOWABLE BOLT STRENGTH (kips)) SHEAR (1)	BEAR. IN EAR (2)	BEAR. STRUT (3)	(--------ALLOWABLE STRUT STRENGTH (axial=kips,BM=k-ft)--------) ALLOW. COMP. BEARING (4)	(------MAX. TENSION------) ALLOW. (5)	ACTUAL	MOMENT	LC	(----MAX. COMPRESSION----) ALLOW. (6)	ACTUAL	MOMENT	LC	(--STRESS RATIOS--) (%) BOLT	(--STRUT---) TENS.	COMP.
2641 D	647 766	1	2X2X0.083	1	2.135	60.00	54.00	.542	4-1/4 P	17.57	21.05	5.82		16.27	1.29	0.00	11	13.24	.36	0.00	4	22.1	7.9	2.7
2642 I	647 660	1	2X2X0.083	1	2.135	60.00	55.50	.542	4-1/4 P	17.57	21.05	5.82	28.26	16.27	.25	0.00	4	13.02	.63	0.00	11	4.4	1.6	4.8
2643 D	660 766	1	2X2X0.083	1	2.135	60.00	54.00	.542	4-1/4 P	17.57	21.05	5.82		16.27	.71	0.00	10	13.24	1.00	0.00	5	17.1	4.3	7.5
2644 I	766 767	1	2X2X0.083	1	2.135	60.00	55.50	.542	4-1/4 P	17.57	21.05	5.82	28.26	16.27	.07	0.00	10	13.02	.11	0.00	5	1.1	.4	.8
2645 D	660 767	1	2X2X0.083	1	2.135	60.00	54.00	.542	4-1/4 P	17.57	21.05	5.82		16.27	1.22	0.00	10	13.24	1.72	0.00	5	29.6	7.5	13.0
2646 D	706 768	1	2X2X0.083	1	2.135	60.00	54.00	.542	4-1/4 P	17.57	21.05	5.82		16.27	.83	0.00	10	13.24	1.33	0.00	5	22.8	5.1	10.0
2647 I	768 769	1	2X2X0.083	1	2.135	60.00	55.50	.542	4-1/4 P	17.57	21.05	5.82	28.26	16.27	.07	0.00	10	13.02	.11	0.00	5	1.1	.4	.8
2648 D	706 769	1	2X2X0.083	1	2.135	60.00	54.00	.542	4-1/4 P	17.57	21.05	5.82		16.27	1.01	0.00	10	13.24	.78	0.00	5	17.4	6.2	5.9
2649 I	706 719	1	2X2X0.083	1	2.135	60.00	55.50	.542	4-1/4 P	17.57	21.05	5.82	28.26	16.27	.26	0.00	4	13.02	.60	0.00	11	4.4	1.6	4.6
2650 D	719 769	1	2X2X0.083	1	2.135	60.00	54.00	.542	4-1/4 P	17.57	21.05	5.82		16.27	.66	0.00	11	13.24	.34	0.00	2	11.3	4.0	2.5
2651 I	769 770	1	2X2X0.083	1	2.135	60.00	55.50	.542	4-1/4 P	17.57	21.05	5.82	28.26	16.27	.25	0.00	11	13.02	.11	0.00	4	4.3	1.5	.9
2652 D	719 770	1	2X2X0.083	1	2.135	60.00	54.00	.542	4-1/4 P	17.57	21.05	5.82		16.27	2.03	0.00	11	13.24	.82	0.00	4	34.9	12.5	6.2
2653 I	719 732	1	2X2X0.083	1	2.135	60.00	55.50	.542	4-1/4 P	17.57	21.05	5.82	28.26	16.27	.11	0.00	11	13.02	.07	0.00	4	1.8	.7	.5
2654 D	732 770	1	2X2X0.083	1	2.135	60.00	54.00	.542	4-1/4 P	17.57	21.05	5.82		16.27	2.68	0.00	11	13.24	1.30	0.00	4	49.4	17.7	9.8
2655 I	634 771	2	2X2X0.109	1	2.733	60.00	55.50	.692	4-1/4 P	17.57	21.05	7.65	36.18	19.59	2.37	-.07	4	10.00	7.86	.29	11	31.0	12.1	78.7
2656 D	765 771	1	2X2X0.083	1	2.135	60.00	54.00	.542	4-1/4 P	17.57	21.05	5.82		16.27	1.28	0.00	4	13.24	3.50	0.00	11	60.0	7.9	26.4
2657 D	765 772	1	2X2X0.083	1	2.135	60.00	54.00	.542	4-1/4 P	17.57	21.05	5.82		16.27	.88	0.00	2	13.24	1.58	0.00	11	27.1	5.4	11.9
2658 I	647 772	5	2X2X0.165	1	3.836	60.00	55.50	.958	4-1/4 P	17.57	21.05	11.58	50.77	24.57	2.34	.24	2	8.73	7.16	.72	11	20.2	9.5	81.9
2659 D	766 772	1	2X2X0.083	1	2.135	60.00	54.00	.542	4-1/4 P	17.57	21.05	5.82		16.27	.53	0.00	4	13.24	1.56	0.00	11	26.8	3.3	11.8
2660 D	766 773	1	2X2X0.083	1	2.135	60.00	54.00	.542	4-1/4 P	17.57	21.05	5.82		16.27	1.16	0.00	5	13.24	.77	0.00	10	19.9	7.1	5.8
2661 I	660 773	3	2X2 1/8	1	3.050	60.00	55.50	.771	4-1/4 P	17.57	21.05	8.77	40.50	8.14	4.10	.91	5	6.03	5.58	.72	11	46.7	50.4	92.5
2662 D	767 773	1	2X2X0.083	1	2.135	60.00	54.00	.542	4-1/4 P	17.57	21.05	5.82		16.27	1.93	0.00	5	13.24	1.35	0.00	10	33.2	11.9	10.2
2663 D	768 774	1	2X2X0.083	1	2.135	60.00	54.00	.542	4-1/4 P	17.57	21.05	5.82		16.27	1.54	0.00	5	13.24	.96	0.00	10	26.4	9.5	7.3
2664 I	706 774	4	2X2X0.134	1	3.281	60.00	55.50	.827	4-1/4 P	17.57	21.05	9.40	43.44	13.91	3.24	.67	5	7.02	6.84	.72	11	34.5	23.3	97.4
2665 D	769 774	1	2X2X0.083	1	2.135	60.00	54.00	.542	4-1/4 P	17.57	21.05	5.82		16.27	.93	0.00	5	13.24	.98	0.00	10	16.9	5.7	7.4
2666 D	769 775	1	2X2X0.083	1	2.135	60.00	54.00	.542	4-1/4 P	17.57	21.05	5.82		16.27	.48	0.00	4	13.24	1.01	0.00	11	17.4	2.9	7.6
2667 I	719 775	4	2X2X0.134	1	3.281	60.00	55.50	.827	4-1/4 P	17.57	21.05	9.40	43.44	13.51	1.63	.69	5	7.02	7.00	.72	11	26.0	12.0	99.8
2668 D	770 775	1	2X2X0.083	1	2.135	60.00	54.00	.542	4-1/4 P	17.57	21.05	5.82		16.27	.94	0.00	4	13.24	2.28	0.00	11	39.1	5.8	17.2
2669 D	770 776	1	2X2X0.083	1	2.135	60.00	54.00	.542	4-1/4 P	17.57	21.05	5.82		16.27	1.19	0.00	4	13.24	2.63	0.00	11	45.2	7.3	19.9
2670 I	732 776	4	2X2X0.134	1	3.281	60.00	55.50	.827	4-1/4 P	17.57	21.05	9.40	43.44	19.22	2.13	-.34	4	6.77	6.50	.74	11	22.7	11.1	96.0
2671 I	771 772	1	2X2X0.083	1	2.135	60.00	55.50	.542	4-1/4 P	17.57	21.05	5.82	28.26	16.27	.43	0.00	11	13.02	.47	0.00	4	7.4	2.7	3.6
2672 I	772 773	1	2X2X0.083	1	2.135	60.00	55.50	.542	4-1/4 P	17.57	21.05	5.82	28.26	16.27	.42	0.00	11	13.02	.33	0.00	4	7.3	2.6	2.5
2673 I	774 775	1	2X2X0.083	1	2.135	60.00	55.50	.542	4-1/4 P	17.57	21.05	5.82	28.26	16.27	.68	0.00	11	13.02	.36	0.00	4	11.7	4.2	2.8
2674 I	775 776	1	2X2X0.083	1	2.135	60.00	55.50	.542	4-1/4 P	17.57	21.05	5.82	28.26	16.27	1.32	0.00	11	13.02	.59	0.00	4	22.6	8.1	4.6

MAT GRADE: 1=A500/GR C 2=A572/GR 65 3=A500.GR C

AVERAGE STRESS RATIOS FOR THE ENTIRE STRUCTURE 35.4 11.5 19.9

Fig 7. Sample Printout of Member and Connection Design.

identify different stress levels and their location in the structure. This information combined with nodal deflections are valuable for comprehending the behavior of the structure. They may identify subtle problems or prompt changes resulting in a more efficiently engineered structure which would not otherwise be noticed by reviewing numerical information.

DESIGN FUNCTION

In the design phase, the structure is designed to meet minimum strength requirements. Thus, every member is designed for the worst load cases according to AAI specifications (Ref 6) or AISC specifications (Ref 7) depending upon whether the material used is aluminum or steel. Member end strength, and number of bolts required are determined for each member. Similarly, every node is designed using Space Structures' proprietary data and fabrication procedures. All the design results are then saved for further manipulation.

After the design of the structure has been completed, the engineer may review the design results and modify any part of it to take into account such factors as availability and cost of material. Thus, an optimum design is obtained by taking into account not only engineerig strength requirements, but also availability, fabrication and installation costs, and architectural requirements.

Figure 7 shows a typical page of the printed design results. For each member, the worst tension and compression load cases with the actual forces are identified. The printout shows the size of members selected and its maximum allowable tensile and compressive carrying capacity. Also shown is the computer designed connection consisting of size and number of bolts required with its allowable capacity in shear and bearing. The stress ratios column on the right side of the table shows, for each member and its connection, the percentage of actual stress to allowable. This information gives an indication of reserve strength for the structure.

Similar to analysis, the design results may be presented graphically. Graphical presentation of the design results consist of plots of the structure using different colors for the selected member sizes. This information is very helpful for assembling and erecting the structure.

CONCLUSION

In conclusion, SSCAD has proven to be an effective tool in the engineering and design of three-dimensional double-layer grid systems. By taking advantage of the graphical and interactive capabilities of the microcomputer, and by tailoring functions specially for spaceframe applications, the engineer is freed from the number crunching and repetitive tasks and given the opportunity to do creative engineering on innovative structures.

REFERENCES

1. Gere, J.M., and Weaver, W., Jr., Analysis of Framed Structures, D. Van Nostrand Company, Inc., Princeton, N.J., 1965.

2. Hinton, E., and Owen, D.R.J., Finite Element Programming, Academic Press, London England, 1977.

3. Irons, B.M., "A Frontal Solution Program," Int. J. Num. Meth. Eng., Vol 2, 1970, pp 5-32.

4. Jonatowski, J.J. and Koutsoubis, D. "Microcomputer Engineering of Large Space Trusses," to be published in Proceedings of Third Conference in Computing in Civil Engineering, San Diego, CA, April 2-6, 1984.

5. Ralston, A., A First Course in Numerical Analysis, McGraw-Hill, New York, N.Y., 1965.

6. Specifications for Aluminum Structures, The Aluminum Association, Inc., Washington, D.C., 4th Ed., 1982.

7. Specifications for the Design, Fabrication and Erection of Structural Steel for Buildings, American Institute of Steel Construction, Chicago, IL, 8th Ed., 1978.

8. Zienkiewicz, O.C., The Finite Element Method in Engineering Science, McGraw-Hill, New York, NY, 1971.

SHAPE FINDING TECHNIQUES AND DESIGN OF THIN CONCRETE SHELLS

J L MEEK*, BE, BSc, MS, PhD and P HO**, BSc, CEng, MICE, MIStructE

* Associate Professor, Department of Civil Engineering, University of Queensland, Brisbane, Australia.

** Research Student, Department of Civil Engineering, University of Queensland, Brisbane, Australia.

The choice of ferro-cement as the construction medium will allow the use of very thin sections for shell structures. Because it is desirable to construct shells in which membrane stresses predominate it would appear advantageous to choose shapes which have been formed by membrane action only. Membrane shapes can be obtained by applying multiples of the dominating surface loads on a flat membrane acting as a tension structure. Inversion of the shell will provide for a structure in which under service loads compression stresses predominate. This paper describes the use of the computer analysis for such a design purpose, and the subsequent analysis of the shell for its service loads to estimate maximum stresses from both membrane and bending effects.

INTRODUCTION

The use of ferro-cement is as old as the discovery of portland cement itself, see Ref 1. It first seemed that the use of cement rich mortar with a matrix of finely divided steel wires produces a synergistic effect. It would appear however that the onset of cracking of the material is at stresses comparible with those of normal reinforced concrete, except that now the cracks are closely spaced and of very small width. Also because the volume of mortar is relatively low the effects of shrinkage should also be minimized. Originally ferro-cement was fabricated by tieing together a number of layers (up to 8 or 10) of wire mesh over a frame work of pipe frames (1 metre spacing) and a grid of 6 mm steel rods (100 mm spacing). The technique suffers tragically from an extremely high labour content which makes construction almost prohibitively expensive. Recently, however it has been realized that the wire mesh can be replaced by the inclusion of 5-7% by weight of steel fibre in the mortar. A basic frame work of high tensile (2-3 mm) steel wire provides the overall structural integrity in tension. The fibre enriched mortar may be applied to a framework of 2 or 3 layers of expanded metal lathe, or alternatively in mass production cast against form work. Then the labour component in the construction is virtually eliminated.

Practical applications of ferro-cement are shown in Figs 1 and 2. In Fig 1, the hyperbolic paraboloid shell is constructed from four precast panels, 3.5 m square by 15 mm thick using standard mesh reinforced ferro-cement. The shell has been in service 15 years. The water tank in Fig 2 has panels 1.8 m square, 12.5 mm thick. These panels have a single layer of 100 mm square wire mesh and the cement mortar contains 5% fibre steel.

In this paper the theory is presented for the simple generation of membrane shapes. Inversion of the shape will then produce a shell whose stress fields are essentially compressive in nature. The advantage of the compressive shell being that its durability does not depend to any large extent on the behaviour of the material in tension (cracking). The generation of shell shapes have been undertaken by the research works, Refs 2, 3 and 4. The application herein to the actual design of ferro-cement shells is believed to be novel.

Fig 1. Hyperbolic Paraboloid Shell

THEORY-SHAPE FINDING PROGRAM

The essential feature of the shape finding program is that the required shape is determined by successive applications of pressure to the membrane initially plane and of the plan form dimensions of the shell. The resulting 'w' deflections give a possible shell shape. The shapes will be progressively distorted and will be deemed suitable for use once the resulting membrane plus bending stresses in the inverted shell fall below a certain threshold level. For example for ferro-cement a value of maximum extreme fibre tension stress of 5 MPa may be used. Alternately the choice of shape may sometimes depend on the architectural requirements.

Fig 2. Water Retaining Shell.

The actual mechanics of the shape finding program are quite complicated because of the presence of large displacements, see Ref 11. If linearized finite element theory is used the stiffness equations at any load equilibrium level can be written.

$$(K_E + K_G)\Delta r = \Delta R \qquad ...1$$

In Eqn 1, K_E is the elastic stiffness of the current configuration, K_G the geometric stiffness resulting from the membrane stresses, Δr and ΔR are the increments of the node displacement and node load vectors respectively. In order to apply Eqn 1 successfully from the planar position where $K_E \equiv 0$, some temporary prestress must be used to ensure that $K_T > 0$ everywhere. Some experience is required with the level of prestress necessary to produce stable solutions near this zero deflected position.

In the present studies it was believed that the transverse membrane deflections are of primary importance, hence not only was the simple constant strain triangle chosen to model the membrane elasticity, but also convergence of the iterations was based on the norm of the increments of these deflections rather than on that of the out-of balance force vector. In order to avoid any complications involving large displacement analysis, it was decided that the formulation of the triangular element stiffness matrix based on the natural mode techniques, Ref 5, is the most appropriate. Only a brief resumé of the theory is included herein. The essential feature of the natural mode technique applied to the three node triangle in a state of plane stress is that its state of stress may be represented by three components, $< \sigma_{1c} \; \sigma_{2c} \; \sigma_{3c} >$, parallel to the sides 1, 2, 3 of the triangle, Fig 3. Stress component increments in the triangle will thus be directly additive, irrespective of its geometric disposition.

In Fig 3 the stress components $\tilde{\sigma}_T = < \sigma_{1c} \; \sigma_{2c} \; \sigma_{3c} >$ and their corresponding nodal force $\tilde{P}_T = < P_{32} \; P_{13} \; P_{21} >$ are shown together with the global force components $< P_{2x} \; P_{2y} >$ etc.

The relationship between $\tilde{\sigma}_T$ and $\tilde{P}_T$ is found to be,

$$\tilde{P}_T = At \left[\ell_D\right]^{-1} \tilde{\sigma}_T \qquad ...2$$

where, A = area of triangle, t the thickness and

$$\left[\ell_D\right] = \left\lceil \ell_{23} \; \ell_{31} \; \ell_{12} \right\rfloor, \; \ell_{ij} \text{ being length of side ij.}$$

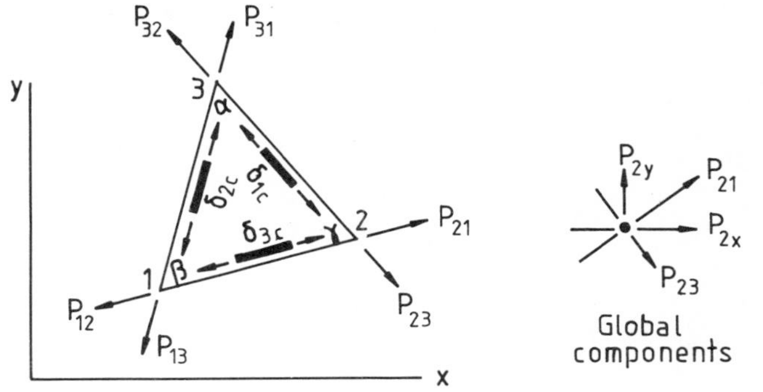

Fig 3. Constant Strain Triangle.

The relationship between natural strains and stresses for isotropic material is,

$$\tilde{\varepsilon}_T = \frac{1}{E} [T][f][T]^T \tilde{\sigma}_T \qquad ...3$$

where,

$$[f] = \begin{bmatrix} 1 & -\nu & 0 \\ -\nu & 1 & 0 \\ 0 & 0 & (1+\nu) \end{bmatrix} \qquad ...4$$

and

$$[T] = \begin{bmatrix} \cos^2\gamma & \sin^2\gamma & \sin\gamma \cos\gamma \\ \cos^2\beta & \sin^2\beta & \sin\beta \cos\beta \\ 1 & 0 & 0 \end{bmatrix} \qquad ...5$$

Writing Eqn 3 as,

$$\tilde{\varepsilon}_T = \left[f_T\right] \tilde{\sigma}_T \qquad ...6$$

Inversion gives the stiffness expression

$$\tilde{\sigma}_T = \left[f_T\right]^{-1} \tilde{\varepsilon}_T = \left[K_T\right] \tilde{\varepsilon}_T \qquad ...7$$

and

$$\tilde{\sigma}_T = \left[K_T\right] \left[\ell_D\right]^{-1} \tilde{\rho}_T \qquad ...8$$

where extension of each side $\tilde{\rho}_T = < \rho_{32} \; \rho_{13} \; \rho_{21} >$. Substitution of Eqn 8 into 2, gives the natural stiffness relationship,

$$\tilde{P}_T = \left[K_N\right] \tilde{\rho}_T \qquad ...9$$

where

$$\left[K_N\right] = (At)\left[\ell_D\right]^{-1}\left[K_T\right]\left[\ell_D\right]^{-1}$$

A further transformation is required to obtain the global stiffness matrix in terms of $< P_x \; P_y \; P_z >_i$ and $< U_x \; U_y \; U_z >_i$, i = 1, 2, 3. That is,

$$\tilde{P} = K_E \tilde{\rho} \qquad ..11$$

where,

$$\left[K_E\right] = [G]^T \left[K_N\right] [G] \qquad ..12$$

$$[G] = \begin{bmatrix} 0 & 0 & 0 & -\alpha_{23} & -\beta_{23} & -\gamma_{23} & \alpha_{23} & \beta_{23} & \gamma_{23} \\ \alpha_{31} & \beta_{31} & \gamma_{31} & 0 & 0 & 0 & -\alpha_{31} & -\beta_{31} & -\gamma_{31} \\ -\alpha_{12} & -\beta_{12} & -\gamma_{12} & \alpha_{12} & \beta_{12} & \gamma_{12} & 0 & 0 & 0 \end{bmatrix} \quad ..13$$

In Eqn 13,

$$\alpha_{ij} = \frac{x_j - x_i}{\ell_{ij}}$$

$$\beta_{ij} = \frac{y_j - y_i}{\ell_{ij}}$$

$$\gamma_{ij} = \frac{z_j - z_i}{\ell_{ij}} \quad ..14$$

It is seen that when the element is initially in the x y plane it has zero stiffness in the transverse, z direction. To provide this transverse stiffness in the non-zero load position, and to calculate $[K_T]$ the tangent stiffness in the deformed position, it is necessary to calculate the geometric stiffness which expresses change in global components for a rigid body displacement of the triangle. In the initial position a temporary prestress is applied to generate a non-zero $[K_G]$ matrix.

The geometric stiffness is calculated here in following Ref 5. That is, the geometric stiffness of a bar element can be shown to be,

$$\left[K_G\right] = \frac{P}{\ell} \begin{bmatrix} I_3 & -I_3 \\ -I_3 & I_3 \end{bmatrix} - \underset{\sim}{C}\,\underset{\sim}{C}^T \quad ..15$$

where, P = force in the bar, ℓ = length, I_3 = unit 3 x 3 matrix and $\underset{\sim}{C}$ is the vector of direction consines of the member w.r.t. the global axes. The geometric stiffness of the triangle is formed by simply coupling the three geometric stiffness of each of the pair of forces $< P_{jk}\ P_{kj} >$ where (j,k) are cyclic permutations of 1, 2, 3. Thus,

$$\left[K_G\right] = \sum_{i=1,2,3} \left[M_{ji}\right]^T \left[K_{Gji}\right] \left[M_{ji}\right] \quad ..16$$

where, j = cyclic permutation of i and

$$M_{21} = \begin{bmatrix} I_3 & 0 & 0 \\ 0 & I_3 & 0 \end{bmatrix}, \text{etc.} \quad ..17$$

SOLUTION TECHNIQUE FOR SHAPE FINDING PROGRAM

The load deflection path from the initially planar position is highly non-linear and some care is required for the successful application of Eqn 1 and the subsequent iterations on the residual force vector. It is found that for a given load increment the membrane takes up its transverse shape rather quickly. From that point on, the iterations appear to involve adjustment of in-plane stresses and distortions. Because in the present context interest is focused on the shape, convergence has been based on the magnitude of the norm of the transverse displacement increment.

Several iteration schemes were tested. In general it has been found that those techniques which iterate on both the load and displacement increments give the best results. A comparison of the various methods for both number of iterations and c.p.u. time (on the VAX-11/780) are shown in Table 1.

From Table 1 it is seen that the arc length method gives a considerable improvement over the schemes where the load step is kept constant in the iteration cycle. The above figures are for rather small structures and the c.p.u. times given would become more significant for large meshes. Because of the nature of the arc length method, in which a quadratic equation must be solved for the constraint on the load increment factor ($\delta\lambda$) the possiblity exists that in certain circumstances non-real roots will occur.

Such is the case for the present analyses in the initial load steps from the planar structure. In these cases, when they occur, the iteration was simply terminated and the next load increment applied. The number of steps where this occured is shown in brackets in the column 'n' of Table 1. Once the membrane has taken up a shape with a certain amount of curvature, the arc length method converges within a few iterations and is obviously a superior technique, see Refs 6 and 7 for details of the method. The initial 'arc length' evaluated at the beginning of the load increment is taken as constant throughout. This constrains the displacements at subsequent load increments and is the main reason for the success of the arc length method. The convergence criterion is set as .01% of the displacement norm.

RESULTS OF THE SHAPE FINDING ANALYSES

The shapes which will have the most practical application will probably be those which have either rectangular or triangular plan form. These shapes are useful in that they allow for a simple repetition of a pattern to cover a given plan area. The triangle, for example can be used for any fan-like structure. Herein, three examples are chosen to illustrate the technique.

(1) The rectangular panel under U.D.L., and

(2) under hydrostatic load,

(3) the triangular panel under hydrostatic load.

In cases where symmetry exists only portion of the structure has been analysed. The plots of the whole structure have been obtained by reflection about the axes of symmetry.

Examples of the use of the shape finding program are given in Figs 4, 5 and 6. In Fig 4, the results are given for a shell under U.D.L. This will be essentially the same as that for uniform pressure for the amount of deflection shown. The shell shape is for a 3 m square panel. A different situation exists in Fig 5, in which a square, 3 m panel is subjected to hydrostatic load. The panel shown here on its side has the top free. The last example is of a triangular panel under hydrostatic load. Such a panel could be used as the component part of a cone shaped reservoir.

ANALYSIS OF THE SHELL FOR BENDING STRESSES

In the present studies only a simple flat facit plate bending element has been incorporated in the analysis of the shell for service loads. The reasons for this are, (1) the primary shell shape was chosen for membrane stress action alone, so that the bending stresses should be secondary. (2) The shell has been generated with a relatively fine finite element mesh so that the coarse representation of the bending strains should be adequate. Our assumptions will be checked with a more refined shell element at a later Stage. The flat facit bending element used was derived in Ref 8. A brief resumé of theory is given herein. The element used in the shell analysis has for its degrees of freedom, displacements (u, v, w) at the triangle apices and w_n, the normal slope at the mid-points of the triangle sides. The bending element alone in its local x, y, z axes is shown in Fig 7.

Structure	Method								
	No Load Steps	Arc Length(6,7) CPU Sec	'n'	Newton Secant(9) CPU sec	'n'	Modified Secant(10) CPU sec	'n'	Newton Rhapson CPU sec	'n'
1. Square Plate (5 x 5 mesh)									
(a) Point load at centre	5	17.77	4(1)	52.66	49	44.10	40	70.12	22
(b) U.D.L.	3	6.76	1(2)	30.91	29	28.09	25	44.45	15
2. Rectangular Panel (3 x 6 mesh) Hydrostatic Load.	5	9.43	4(1)	35.38	45	29.96	42	27.65	18
3. Circular Dome (Non-Conservative Load) 21 Joints.	9	24.12	8(1)	60.66	64	60.26	62	84.70	35

'n' = total number of iteration cycles for all load steps

Table 1. Comparison of Non-Linear Solution Techniques for Shape Finding.

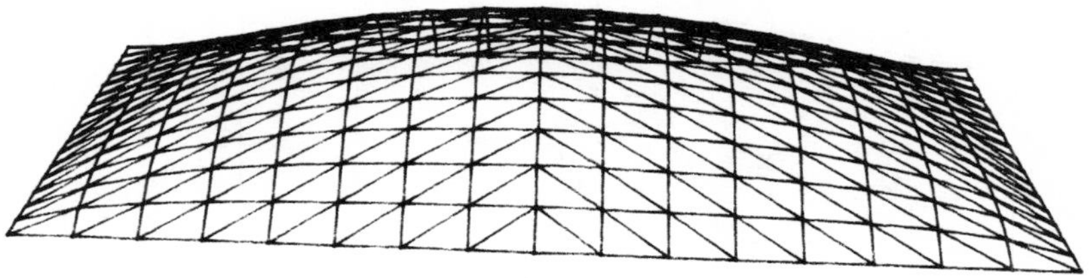

Fig 4. Square Panel Uniformly Distributed Load.

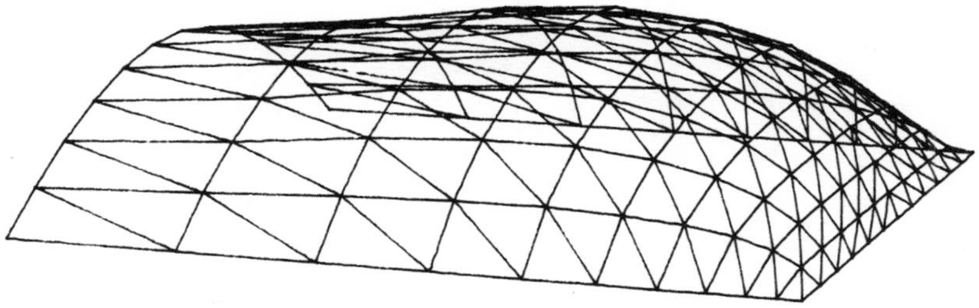

Fig 5. Panel Supported on Three Sides Hydrostatic Load.

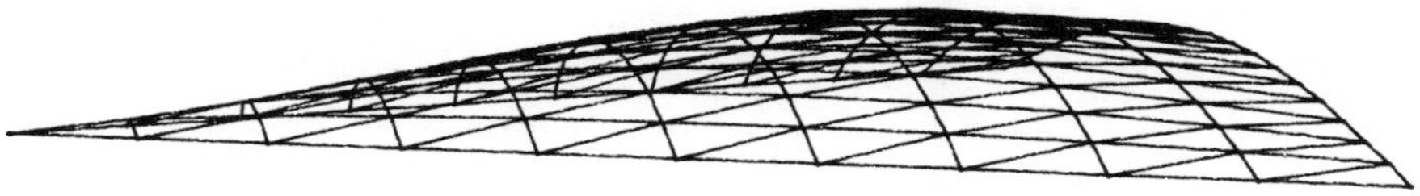

Fig 6. Triangular Panel Hydrostatic Load.

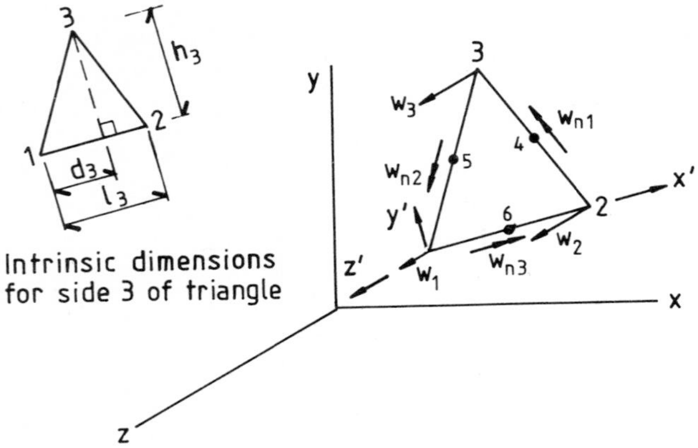

Fig 7. Nodal Arrangement for Bending Element.

For this element the transverse deflection is given

$$w = w^1 + w^R \qquad ..18$$

where w^R is the rigid body deflection and w^1 the deflection from this surface. Then in the present case, w^1 is a quadratic and w^R a linear surface. From Eqn 18,

$$w^1 = w - w^R \qquad ..19$$

using area coordinates ζ_i,

$$w^R = \begin{bmatrix} \zeta_1 & \zeta_2 & \zeta_3 \end{bmatrix} \begin{Bmatrix} w_i \\ w_j \\ w_k \end{Bmatrix} = \begin{bmatrix} \phi_1 \end{bmatrix}^T \tilde{w}_e \qquad ..20$$

Where $\tilde{w}_e$ is the nodal displacement vector, then

$$\frac{\partial w^R}{\partial n_i} = \frac{\partial [\phi_1]^T}{\partial n_i} \tilde{w}_e \qquad ..21$$

Substitution of Eqn 21 in Eqn 19 gives

$$\begin{Bmatrix} \dfrac{\partial w^1_4}{\partial n_1} \\ \dfrac{\partial w^1_5}{\partial n_2} \\ \dfrac{\partial w^1_6}{\partial n_3} \end{Bmatrix} = [T] \begin{Bmatrix} w_1 \\ \dfrac{\partial w^1_4}{\partial n_1} \\ w_2 \\ \dfrac{\partial w^1_5}{\partial n_2} \\ w_3 \\ \dfrac{\partial w^1_6}{\partial n_3} \end{Bmatrix} \qquad ..22$$

where,

$$T = \frac{1}{2A} \begin{bmatrix} -\ell_1 & 2A & -(d_1 - \ell_1) & 0 & d_1 & 0 \\ d_2 & 0 & -\ell_2 & 2A & -(d_2 - \ell_2) & 0 \\ -(d_3 - \ell_3) & 0 & d_3 & 0 & -\ell_3 & 2A \end{bmatrix} \qquad ..23$$

For the relative deflections, w^1, the orthogonal iterpolation polynominal for the mid-side rotations, is,

$$\{\phi\} = 2A\,[\ell]^{-1} \begin{Bmatrix} \zeta_1(1 - \zeta_1) \\ \zeta_2(1 - \zeta_2) \\ \zeta_3(1 - \zeta_3) \end{Bmatrix} \qquad ..24$$

In Eqn 24, A is the area of the triangle, and

$$[\ell] = \begin{bmatrix} \ell_1 & 0 & 0 \\ 0 & \ell_2 & 0 \\ 0 & 0 & \ell_3 \end{bmatrix}$$

To obtain the bending stiffness matrix for the element, first form $\left[\bar{K}_{elt}\right]$, where,

$$\left[\bar{K}_{elt}\right] = \int_{area} [a]^T\,[k][a]\,dV \qquad ..25$$

Where,

$$\{\tilde{\varepsilon}\} = [a]\,\{r\} = [a]\,\tilde{w}^1_e$$

where $\tilde{w}^1_e$ is the mid-side rotation vector, and since

$$\{\tilde{\varepsilon}\} = \begin{bmatrix} -\dfrac{\partial^2 w^1}{\partial x^2} \\ -\dfrac{\partial^2 w^1}{\partial y^2} \\ 2\,\dfrac{\partial^2 w}{\partial x \partial y} \end{bmatrix} \text{ thence,}$$

$$[a] = \frac{1}{A} \begin{bmatrix} b_1^2 & b_2^2 & b_3^2 \\ a_1^2 & a_2^2 & a_3^2 \\ -2a_1b_1 & -2a_2b_2 & -2a_3b_3 \end{bmatrix} [\ell]^{-1} \qquad ..26$$

where $a_i = x_k - x_j$, $b_i = y_j - y_k$; i, j, k in cyclic permatation of 1, 2 and 3. As in usual plate bending, the flexural rigidity is

$$[k] = \frac{Et^3}{12(1 - \nu^2)} \begin{bmatrix} 1 & \nu & 0 \\ \nu & 1 & 0 \\ 0 & 0 & \dfrac{1 - \nu}{2} \end{bmatrix} \qquad ..27$$

Because terms of [a] and thickness t are constant,

$$\left[\bar{K}_{elt}\right] = A[a]^T\,[k][a] \qquad ..28$$

For the nodal displacement vector given in Eqn 22, contragredience now gives the (6 x 6) stiffness matrix to be,

$$\left[K_{elt}\right] = [T]^T \left[\bar{K}_{elt}\right] [T] \qquad ..29$$

The transverse components must be transformed to the global coordinate system. This is best achieved by combining with the plane stress stiffness and transforming (u,v,w)' components simultaneously to (u,v,w) components.

RESULTS OF SHELL ANALYSES

For the various shells generated such as those shown in Figs 4, 5 and 6, analyses have been made for each shell shape for a suitable live load. Shell thicknesses of 15 and 20 mm have been investigated. The curves given in Fig 8 are the plots of these results for the principal membrane stresses which occur in the element together with the principal maximum bending stresses. Membrane stresses are compressive whereas bending stresses may be either tension or compression. In order to compare possible magnitudes of the extreme fibre stresses, the maximum and minimum stresses from the two effects have been calculated by,

$$\sigma_{max} = |\sigma_{membrane\ max}| + |\sigma_{bending\ max}|$$

$$\sigma_{min} = |\sigma_{membrane\ min}| - |\sigma_{bending\ max}|$$

These results should be conservative because the principal membrane and bending stresses will not always have the same orientation. These max/min stresses are shown in Fig 9. From Fig 9 it is seen that the aspect ratio (height/span) for this particular shell should be in the range .05 to .10 for best efficiency. Past .10, little benefit is gained from the extra curvature of the shell. Except for the very flat shell (where 20 mm thickness should be used), practical considerations will probably dictate the shell thickness.

In Fig 10, a similar graph of maximum and minimum stresses has been plotted for the panel for which, the live load is hydrostatic. It again clearly demonstrates the range of practical height/span ratios.

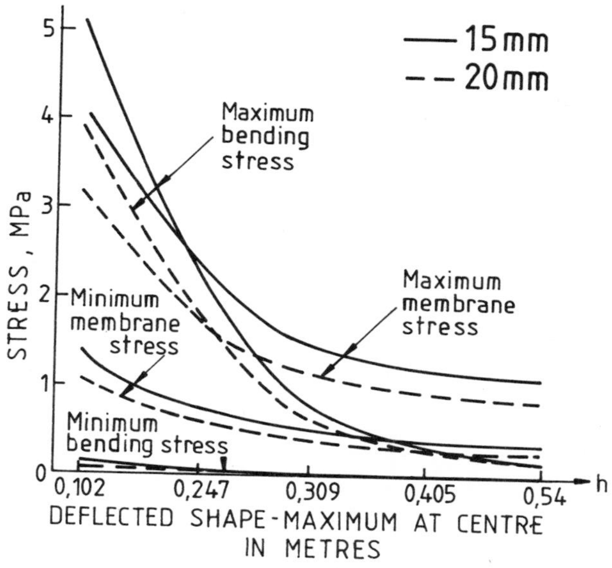

Fig 8. Membrane and Bending Stresses for 3 x 3 m Plate Under U.D.L. (density = 25 kN/m^3) and Live Load (5 kN/m^3)

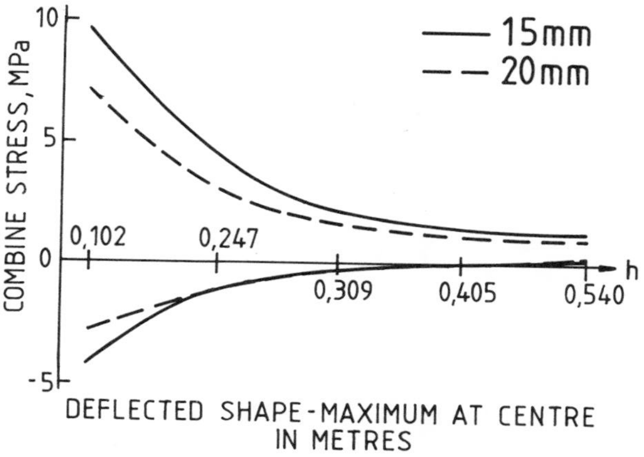

Fig 9. Combines Stresses Envelope for 3 x 3 m Plate Under U.D.L. and Live Load.

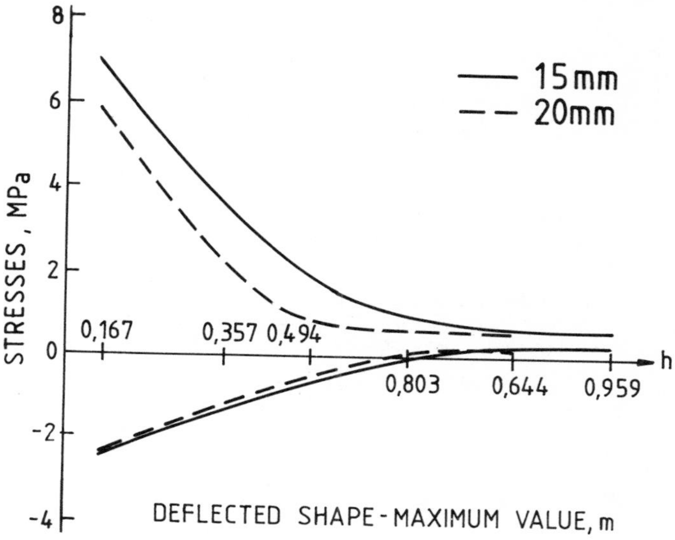

Fig 10. Combined Stress Envelope for Hydrostatic Panel. (Water density = 10 kN/m^3).

THE COMPUTER PROGRAMS

The programs are written in Fortran 77 running on the VAX-11/780 installed in the University of Queensland Prentice Computer Centre. They include Mesh Generator, Shape Finding, and Pre-shell and Shell Analysis programs plus Graphic plotting routines. The Mesh Generator is capable of generating two and three dimensional triangular mesh together with an optional bandwidth/profile reduction routine. The Shape Finding includes the iterative schemes already described with automatic/manual load increment capacity. Both in-core and out-of-core skyline equation solvers are available. Choice of Solution scheme is automatic depending on the initial core space. As three mid-side nodes are required in the shell analysis, the Pre-shell generates new mid-side nodes, rearranges nodal connectivity and ultilises the bandwith/profile reduction routine to obtain an optimal nodal arrangement. The Shell Analysis couples the membrane stiffness in the Shape Finding with the bending stiffness and outputs principle membrane and bending stresses, displacements and reactions. The mesh is checked and the perspective views of the shapes are plotted using the routines presented on the HP7221A Graphics Plotter. Although a hidden line algorithm is available, the plots are presented in full to show the boundaries. All inputs and outputs in each phase are compatible. The Shape Finding and the Shell Analysis programs check all input data before proceeding to the major computations.

DISCUSSION AND CONCLUSIONS

The paper has described a method whereby suitable shapes can be obtained for membrane shells of various plan form and subjected to uniform or hydrostatic pressure. Apart from any other consideration, these are lofted shapes. That is curves formed by any cutting plane will be smooth. It has been shown by the analysis of the inverted shell (membrane stresses are compressive), that the shapes choosen are suitable for application to thin, ferro-cement construction. The sizes choosen for the demonstration are relatively small (3 m x 3 m) and it would be of interest to determine the span limits for this class of Structure. The practical limit on the thickness of the shell is approximately 30 mm, and this will determine the maximum spans that may be obtained. The analyses of these shells is relatively elementary. It would be of significant interest to study their overload and collapse behaviour, either numerically or by experiment.

REFERENCES

1. P L NERVI, Structures, F W Dodge Corporation, New York, 1956.

2. P G SMITH AND E L WILSON, Automatic Design of Shell Structures, Journal of Structural Division, ASCE, ST1, pp 191-201, 1971.

3. G A MOHR, Design of Shell Shape Using Finite Elements, Computers and Structures, Vol 10, pp 745-749, 1979.

4. E HAUG AND G H POWELL, Finite Element Analysis of Non-linear Membrane Structures, Proceedings 1971 IASS Pacific Symposium Part II on Tension Structures and Space Frames, Tokyo and Kyoto, pp 165-175.

5. J H ARGYRIS, Recent Advances in Matrix Methods of Structural Analysis. Progress in Aeronautical Sciences, Vol 4, edited by D Küchemann and L H G Sterne, Pergamon Press, 1964.

6. M A CRISFIELD, A Fast Incremental/Iterative Solution Procedure That Handles "Snap Through". Computers and Structures, Vol 13, pp 55-62, 1981.

7. E RAMM, Strategies for Tracing the Non-linear Response Near Limit Points. Proceedings of the Eurpoean - U.S. Workshop in "Non-linear Finite Element Analysis in Structural Mechanics", edited by Wunderlich, E Stein and K J Bathe, Springer Verlag, pp 63-89, 1981.

8. L S D MORELY, The Constant-Moment Plate-Bending Element, Journal of Strain Analysis, Vol 6, No 1, pp 20-24, 1971.

9. M A CRISFIELD, A Faster Modified Newton-Raphson Iteration, Computer Methods in Applied Mechanics and Engineering, Vol 20, No 3, pp 267-278, 1979.

10. L ZHANG and D R J OWEN, A Modified Secant Newton Method for Non-linear Problems, Computers and Structures, Vol 15, No 5, pp 543-547, 1982.

11. R H KNAPP and R SZILARD, Non-linear Differential Equations for General Membrane Shells, Proceedings IASS Pacific Symposium Part II on Tension Structures and Space Frames, Tokyo and Kyoto, pp 155-156, 1971.

TREATMENT OF BRIDGES AS SPACE STRUCTURES

M. Shafik AGGOUR*, Ph.D. and M. Sherif AGGOUR**, Ph.D.

*Professor of Metallic Bridges, Department of Civil Engineering
Cairo University, Cairo, Egypt

**Associate Professor, Department of Civil Engineering
University of Maryland, College Park, Maryland, U.S.A.

Simply supported railway bridges of the same span but with various angles of skew and various numbers of intermediate transverse bracings were studied using the ordinary planar method of calculation and also by treating them as space structures. The effect of the settlement of one of the supports was also considered. From the comparison of the results of the two methods of solution, summarized in this paper, it is concluded that the design of a safe and economical bridge should consider it as being a space structure.

INTRODUCTION

In a truss or plate girder bridge, the traditional design method assumes that every structural element has only a certain definite carrying function to perform. In calculating the stresses in the different parts of a bridge, the practice is to split it up into a group of vertical and horizontal plane frames and trusses acting independently of each other in which the vertical loads are taken by the vertical main girders, and the horizontal loads by the horizontal wind bracings and then transmitted through the transverse bracings to the bearings of the bridge. In reality, the whole bridge acts as a space structure; the vertical loads produce stresses not only in the main girders, but also in the wind bracings and transverse bracings. With skewed bridges, the spatial interaction of main girders, wind bracings and transverse bracings has a more decisive influence on the load distribution than in the case of a square bridge, due to the much greater twisting moments produced at the end transverse bracings. This fact has been demonstrated by means of model and full scale testing of skewed bridges, Refs 1, 2, and 3. Thus it is important in structural design to consider the behavior of the structure as a whole, rather than as a collection of distinct elements, such that a true monolithic supporting structure is being considered. Also, by introducing intermediate transverse bracings, it is possible to relieve the two-end transverse bracings from the torsional moment they have to resist, and these intermediate transverse bracings begin to participate in the resistance. Thus the method of analysis should be able to take into account these intermediate transverse bracings.

Several simply supported skewed bridges of the same span, but with various angles of skew to the abutment line, and with various numbers of intermediate transverse bracings, have been treated in detail as space structures, Refs 4 and 5. In this paper the method of solution that treats a bridge as a space structure and provides a complete representation of the load distribution and stresses on various surfaces of a bridge due to vertical loading is presented. This is followed by a summary of a comparison of the results of the space frame calculations with those obtained by the ordinary planar method. The comparison deals with the effect of the various angles of skew to the abutment line presented in Ref 4 and the effect of intermediate transverse bracings presented in Ref. 5. In addition, by treating bridges as space structures it was possible to calculate the stresses due to settlement of the supports, whereas in traditional bridge design, stresses due to settlement of the support of simple span bridges cannot be calculated. The method of analysis for such calculations and an evaluation of the results are also presented.

THEORY

The carrying surfaces theory developd by Schwyzer, Ref 6 and subsequently demonstrated experimentally by Aggour, Ref 2, considers that the space structure is composed of several plane frames or trusses connected together at their lines of intersection. Along these lines the plane frames or trusses will act mutually on one another. The corresponding actions and reactions are called edge forces (or shears), since their lines of action in the case of plane frames with negligible lateral stiffness coincide with the edge. After the determination of these edge forces, the calculation of the space frame is carried back to the calculation of several plane frames or trusses loaded by external forces in their own planes.

The general shape of a bridge with two main girders, an upper and a lower wind bracing, and two-end transverse bracings is a closed polyhedron. If the main girders are trusses or plate girders with parallel chords, the polyhedron will possess six surfaces and 12 edges. For the equilibrium of each plane surface of the polyhedron, the external loads and edge forces acting upon this surface must satisfy the three conditions of equilibrium in the plane. If the bridge is supported by means of four vertical and three horizontal reactions, the total number of unknown reactions and edge forces is 7 + 12 = 19, the number of equations of equilibrium is 6 x 3 = 18, and therefore the space frame is once statically indeterminate.

By eliminating one of the four vertical reactions or by omitting either of the two-end transverse bracings or a diagonal one in either the upper or lower wind bracing, we obtain a statically determinate space structure. On the other hand, each additional transverse bracing increases the number of unknowns by one (four edge forces and three equations of equilibrium) so that a bridge with two main girders, two wind bracings, two-end transverse bracings and n intermediate transverse bracings supported by seven reactions is n + 1 times statically indeterminate. This means that the redundant edge forces can be obtained from n + 1 equations. This is true not only when all the component parts of the bridge are statically determinate plane plate girders or trusses, but also when they are statically indeterminate ones, because the additional indeterminacy of the plane trusses can be calculated separately.

Furthermore, skewed bridges are not symmetrical with regard to vertical planes, as are square bridges, but can in most cases be made symmetrical with regard to a vertical axis through the center of the bridge. By splitting up the acting vertical or horizontal loads into symmetrical and asymmetrical loadings the number of redundant forces can be reduced by one half.

BRIDGE DESCRIPTIONS

The investigations are carried out for one square and four single track railway plate girder skewed bridges, with two end transverse bracings and with different angles of skew. The analysis was then repeated for the skew bridges when an additional intermediate transverse bracing was introduced and repeated again with three intermediate transverse bracings introduced.

The general arrangement and cross sections of members of the main griders and wind bracings are the same in all cases. The elevation, plans, and cross sections of the skewed railway bridges treated are shown in Fig 1. They consist of two simply supported main girders of the plate girder type, 24.0 m span, 4.0 m apart. Two stiff horizontal bracings of the K-type are arranged between the upper and lower chords of the two main girders and are executed between the bearings of the two obtuse ends of the bridges. The chosen angles of skew, α, have slopes of 0.5 (Bridge I), 0.66 (Bridge II), 1.0 (Bridge III), and 2.0 (Bridge IV), respectively. Two transverse bracings are provided at the two obtuse ends of the bridges.

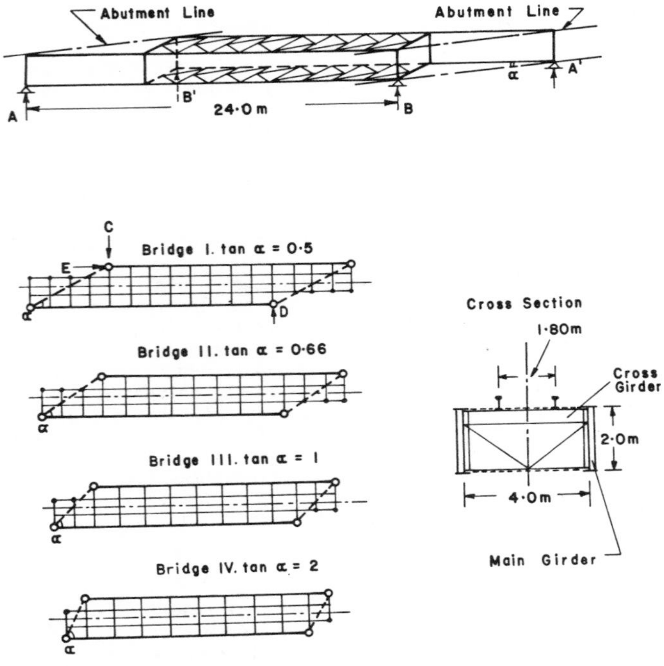

Fig. 1. Schematic of Bridge Studied

The section chosen for the main girder is a welded section consisting of a web plate 2,000 mm X 11 mm, two flange plates 500 mm X 30 mm, and two flange plates 400 mm X 12 mm. The cross-sectional areas of the members of the horizontal and transverse bracings are chosen sufficiently large in order that they will be adequate when treating the bridge as a space structure. The bridges have four bearings so chosen as to provide seven components of reactions.

In all the skewed bridges with various angles of skew to the abutment line, the influence lines for reactions, and bending moments are worked out for an axle load of 19.6 kN moving along the track. Since the bridges act as a single web girder beyond the bearings at the obtuse ends, the influence lines were modified to account for the direct transmittal of part of the load directly to the abutment. These influence lines are then used to obtain: (1) Reactions and bending moment diagrams due to a uniformly distributed dead load of intensity 22.54 kN/m of bridge; (2) reactions, and maximum positive and negative bending moment curves due to an equivalent uniformly distributed live load of intensity 96.82 kN/m of bridge, representing the effect of the train; and (3) the increase due to impact has been considered and added.

FORMULATION

The paper deals with three areas, effect of the angle of skew, effect of the number of intermediate transverse bracings and effect of settlement of one of the supports on the load distribution and stress on various surfaces of the bridge. The formulation for these three areas of concern are as follows:

1. Effect of angle skew: In this area of concern, the bridges have two end transverse bracings; the main system is obtained by separating the plane of the upper wind bracing from the plane of the left transverse bracing. The unknown edge force Y_1 is introduced as a redundant edge force along this cut edge as shown in Fig 2. The value of Y_1 is obtained from the condition that the total relative displacement along this edge must be zero, in other words:

$$\delta_{10} + Y_1\delta_{11} = 0 \qquad \text{...... 1.}$$

in which δ_{10} = the relative displacement of the main system along the cut edge, due to the actual external loads; and δ_{11} = the relative displacement of the main system along the cut edge due to two equal and opposite forces $Y_1 = 1$ at the separated edge.

$$\text{Therefore} \quad Y_1 = \frac{-\delta_{10}}{\delta_{11}} \qquad \text{...... 2.}$$

The displacements δ are calculated by the method of virtual work. For the plate girder, the effect of bending moment, shearing force, and thrust have been considered; for the members of the wind bracings and transverse bracings, only the axial forces are considered. For example, Fig 2 shows the straining actions of the different components for skewed bridge II (slope at end = 0.66) and Fig 3 shows the influence line for the redundant edge force Y_1 for the same bridge.

2. Effect of the Number of Intermediate Transverse Bracings: In this case, the solution was provided for bridges with two end and one intermediate transverse bracings and for bridges with two end and three intermediate transverse bracings. The four skewed bridges with two-end transverse bracings treated previously, are now provided with one intermediate transverse bracing as shown in Fig 4. The new system is supported again by four vertical and three horizontal reactions, represents a closed polyhedron

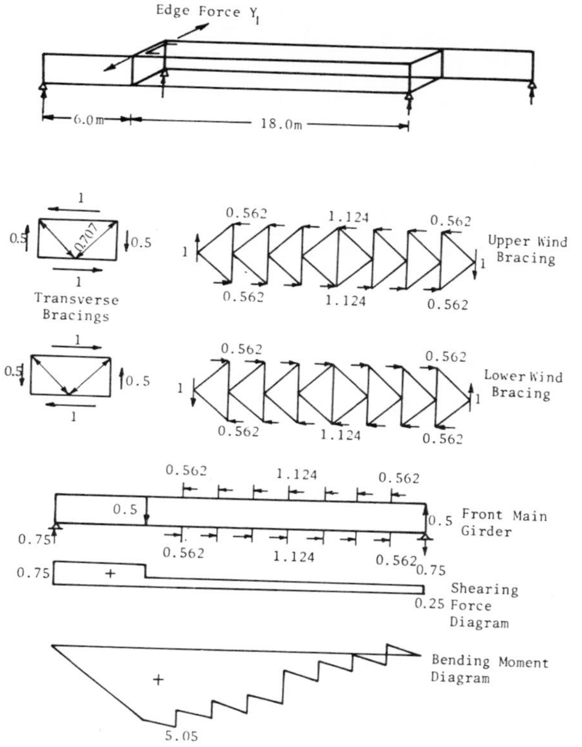

Fig. 2. Straining Actions due to Edge Force $Y_1 = 1$ (Bridge II)

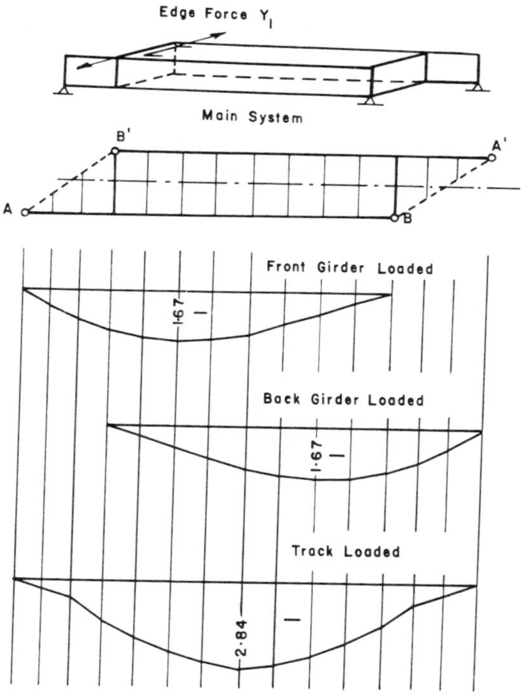

Fig. 3. Influence Line for the Edge Force Y_1 (Bridge II)

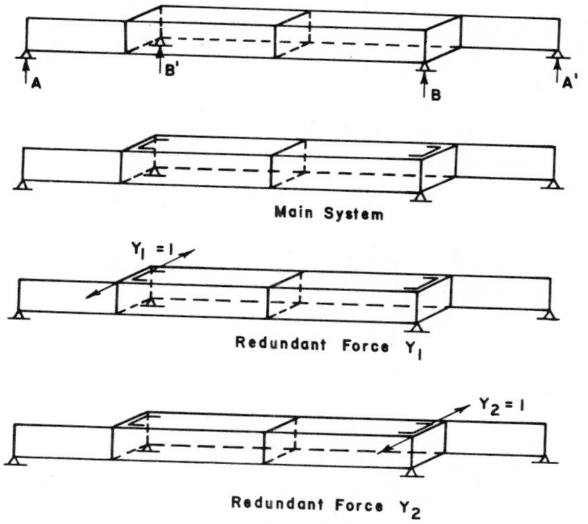

Fig. 4. Bridge with Three Transverse Bracings

with one intermediate transverse bracing and is thus a twice statically indeterminate constructional space structure. The main system is formed by separating the upper horizontal bracing from the two-end transverse bracings as shown in Fig 4. The redundant forces Y_1 and Y_2 along these two separated edges are found from the conditions that the total relative displacements along these two edges must be zero. In other words:

$$\delta_{10} + Y_1\delta_{11} + Y_2\delta_{12} = 0 \qquad \ldots\ldots 3.$$

$$\delta_{20} + Y_1\delta_{21} + Y_2\delta_{22} = 0 \qquad \ldots\ldots 4.$$

The four skewed bridges with two-end transverse bracings treated before are now provided with three intermediate transverse bracings at the middle and quarter points as shown in Fig 5. The new system is still supported by four vertical and three horizontal reactions, represents a closed polyhedron with three intermediate transverse bracings and is thus four times statically indeterminate. The main system is obtained by separating the upper wind bracing at four positions from the two-end transverse bracings one and two as well as from the central transverse bracing three as shown in Fig 5. These cuts are chosen symmetrically with respect to the vertical axis through the center of the bridge, thus any horizontal transverse shear acting on the upper wind bracing surface will work its way through the main system, to be transmitted to the lower wind bracing and then to the bearings. By splitting any vertical load acting at any panel point into two symmetrical and two asymmetrical half loads with respect to the vertical central axis of the bridge, and by solving these two separate cases of loading and adding the results, we get the effect of the unit load acting successively at the panel points of either the front or the back main girder. This method reduces the number of redundancy to one-half and consequently the mathematical calculations will be reduced.

3. Effect of Settlement of One of the Supports: For the calculation of the effect of the settlement of one of the supports, the main system is obtained by separating the plane of the upper wind bracing from the plane of the left transverse bracing. The unknown edge force Y_1, is introduced as a redundant edge force along this cut edge. The value of Y_1, is obtained from the condition that

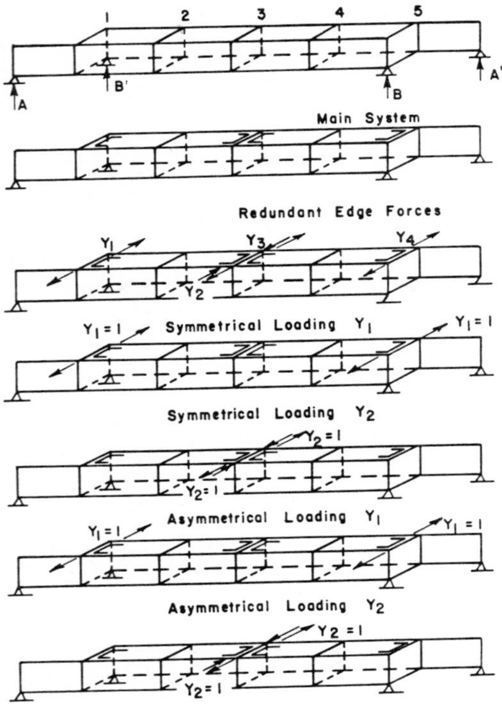

Fig. 5. Bridge with Five Transverse Bracings

$$F \cdot \Delta + Y_1 \delta_{11} = 0 \qquad \ldots\ldots 5.$$

in which F = vertical reaction at the support due to $Y_1 = 1$; Δ = assumed vertical settlement of the support; and δ_{11} = relative displacement of the main system along the cut edge due to two equal and opposite forces $Y_1 = 1$ at the separated edge.

Therefore $$Y_1 = \frac{-F \cdot \Delta}{\delta_{11}} \qquad \ldots\ldots 6.$$

Again, the displacement δ is calculated by the method of virtual work.

DISCUSSION AND EVALUATION

In order to compare the behavior of square and skewed bridges as space structures, a square and a skewed bridge with only two end transverse bracings and the same span have been calculated by Aggour, Ref 7. The influence line of the redundant edge force between the plane surface of the upper wind bracing and the end transverse bracings for the square and the skewed bridges due to vertical loads moving either in the plane of one main girder or along the center line of the bridge were calculated. In the square bridge, the influence line for a unit load moving in the plane of one main girder has negative ordinates on the left half and equal positive ordinates on the right half. All ordinates of the influence line for an axle load moving along the center line of the bridge are therefore zero. In skewed bridges, however, all the ordinates of the influence line for the edge force due to a load moving in the plane of one main girder have the same sign, and thus all the ordinates of the influence line for a load moving along the center line of the bridge are therefore nearly doubled.

It follows then, that in the calculation of square bridges with two main girders, the space frame action can usually be neglected. It is only of a certain importance in double track railway bridges and roadway bridges subject to eccentric loading. On the contrary, in all skewed bridges the actual forces in the main girders, wind bracings, and transverse bracings will differ from those obtained by ordinary planar analysis.

Effect of Angle of Skew

From the influence lines for the horizontal edge force Y_1 between the plane of the upper wind bracings and the plane of the end vertical transverse bracings, the influence lines for moments and reactions are then calculated. Figure 6 shows the curve of absolute bending moments for one main girder due to dead load, live load and impact. The influence lines for the vertical reactions at the acute and obtuse ends of all skewed bridges were calculated and the results for bridge I are shown in Fig 7. Detailed solutions are available in Ref 4 and 8. The following is a summary of the conclusions:

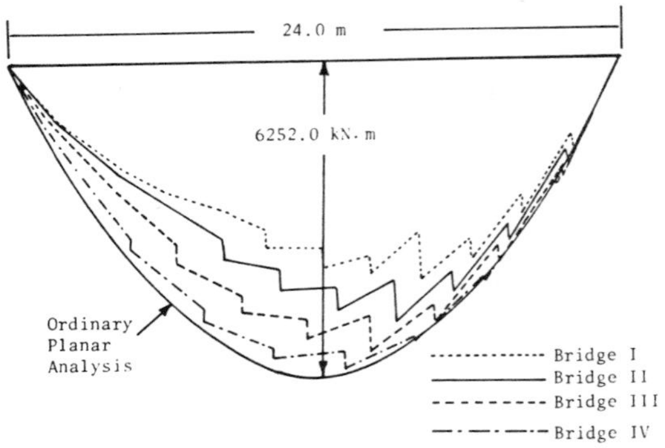

Fig. 6. Absolute Bending Moments

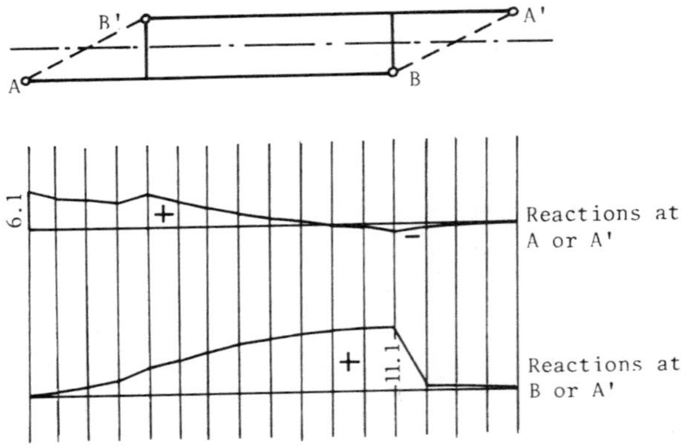

Fig. 7. Influence Lines for Vertical Reactions (Bridge I)

1. The bending moment diagrams for the main girders for all the skewed bridges have positive values all over the entire span of all the bridges, and the maximum ordinates in these diagrams do not exist at the center of the span, but are shifted towards the obtuse ends.

These maximum values of bending moment decrease with the decrease of the slope of angles of skew.

2. The absolute maximum bending moment in the main girders of a single track railway bridge, whose angle of skew is 45° or more, can be calculated with the ordinary planar analysis, neglecting the space frame action. The value of the bending moment obtained in this case is only 7.0% higher than that obtained by the space frame action.

3. The reactions at the supports of a skewed bridge are quite different from those calculated by ordinary planar analysis. At the acute ends of skewed bridges a relief in reactions is obtained, while at the obtuse ends, a considerable increase in reaction occurs. This indicates that the design of bearings and abutments must be according to exact space frame action, and that the ordinary planar analysis gives unsafe values at the obtuse ends.

4. End transverse bracings in skewed bridges are subject to sizeable additional horizontal forces, due to vertical loading of the bridge. These forces should be considered in their design. If stiffer wind bracings are used, the main girders forces are reduced and the wind bracings forces are increased. The wind bracings receive an additional force due to vertical loads. Calculation of the wind bracings by the ordinary planar analysis which neglects the space frame action, could lead to bracings which are unable to resist transverse shear in skewed bridges.

Effect of Number of Intermediate Transverse Bracings

The following is the summary of the conclusions obtained from solving the four bridges once with an intermediate transverse bracing and once with three transverse bracings. Detailed conclusions are presented in Refs 5 and 8.

1. By arranging one or three intermediate transverse bracings, the maximum values of the bending moment have a negligible change compared to those which have only twoend transverse bracings. Thus intermediate transverse bracings have a negligible effect on the load distribution, i.e., intermediate transverse bracings may be neglected in the space frame calculation when computing the stresses in the main girders. It is sufficient to consider the two-end transverse bracings only which will facilitate the space frame calculation.

2. By arranging any number of intermediate transverse bracings, the values of the reactions at either the acute or the obtuse ends do not differ from the corresponding reactions for the same bridge with only two-end transverse bracings.

3. By adding one intermediate transverse bracing the horizontal transverse shear acting on the end transverse bracings is nearly the same as that for the case of skewed bridges with two transverse bracings only. For a uniformly distributed dead load, no stresses are created in the intermediate transverse bracing. However, the intermediate bracing is fully stressed for a live load covering half the span of the bridge, and it may be designed to carry only one-quarter of the shear acting on the end transverse bracings. By introducing three intermediate transverse bracings, the horizontal shear acting on the end transverse bracing is slightly less than that obtained for bridges with one intermediate transverse bracing. For a uniform dead load the middle transverse bracing is not stressed at all while those situated at the quarter points of the bridge are slightly stressed. For a uniformly distributed live load, the three intermediate transverse bracings are nearly equally stressed with a horizontal shear equal to about one-sixth of the maximum shear on the end transverse bracing. The greater the number of intermediate transverse bracings the less they are stressed.

Effect of Settlement of One of the Supports

If a bridge is treated by the usual method of calculation, where the different plane surfaces are considered to act independently of each other, a settlement of one of the bearings in simple span bridges does not create any additional stresses. However, if the same bridge is treated as a space structure, where all carrier surfaces act together to resist all loads, a settlement of one of the bearings will give rise to additional stresses in all surfaces of the bridge. The following is the summary of the conclusions obtained from solving the four bridges (with two-end transverse bracings) for a settlement of 1.0 cm of one of the supports.

1. Square and skew simple span bridges consisting of closed polyhedrons are affected by differential settlement which may occur at the bearings of the bridge.

2. The maximum additional bending moment in the main girder due to a settlement of one of the supports of the simple span bridge increases with the decrease of slope of the angle of skew.

3. End transverse bracings in skew bridges are subject to additional horizontal forces created due to settlement of one of the supports and these forces decrease with the decrease of "tan α". Thus the wind bracings and transverse bracings are stressed more in square bridges than in skew bridges due to settlement.

4. In order to obtain a safe structure, stresses introduced due to differential settlement of supports of simple span bridges must be considered in their design. For the range of variables considered in this paper such stresses were of large magnitude. These stresses cannot be obtained by the usual method of calculation and must therefore be obtained by the space frame method.

CONCLUSION

From the discussion and evaluation presented we can conclude that a safe and economical design would be achieved if square and skew truss and plate girder bridges were treated in their design as space structures. In addition, the task of repairing old bridges or strengthening existing bridges to support ever increasing applied loads, would be accomplished in a more precise and economical way if they were analized as space structures, since such analysis would incorporate the existing conditions of the main girders, wind bracings and transverse bracings. An added advantage of treating bridges as space structures is the ability to calculate the stresses introduced due to settlement of the supports even for a single span bridge which cannot be accomplished in traditional methods of bridge design.

REFERENCES

1. Aggour, M. Shafik, "Treatement of Skew Truss Bridges as Space Frames," thesis presented to Cairo University, Cairo, Egypt, 1943, in fulfillment of the Degree of Master of Science.

2. Aggour, M. Shafik, "Theoretical and Experimental Study of Space Frame Action and Load Distribution in Square and Skew Bridges," thesis presented to Cairo University, Cairo, Egypt, 1947, in fulfillment of the requirements for the Degree of Doctor of Philosophy.

3. Hutter, G., "Skew and Curved Box-Girders, Theory and Research," Proceedings of the Seventh Congress, International Association for Bridges and Structural Engineering, Rio de Janeiro, Brazil, 615, 1964.

4. Aggour, M. Sherif and Aggour, M. Shafik, "Load Distribution in Skewed Bridges Treated in Space," Journal of Civil Engineering Design, Volume 1, No. 2, 1979.

5. Aggour, M. Sherif and Aggour, M. Shafik, "Skewed Bridges with Intermediate Transverse Bracings," Journal of the Structural Division, American Society of Civil Engineers, Vol. 105, No. ST8, August, 1979.

6. Schwyzer, H., "Staiche Untersuchung der aus ebenen tragflächen zusammengesetzten räumlichen Tragwerke," thesis presented to ETH, Zurich, 1920, in fulfillment of the requirements for the Degree of Doctor of Philosophy.

7. Aggour, M. Shafik, "Space Frame Action and Load Distribution in Skew Bridges," Proceedings of the Seventh Congress, International Association for Bridges and Structural Engineering, Rio de Janeiro, Brazil, 647, 1964.

8. Aggour, M. Sherif, "Effect of Partial End Restraint on the Design of Skew Railway Bridges as Space Structures," thesis presented to Cairo University, Cairo, Egypt, 1966, in fulfillment of the Degree of Master of Science.

Application of the Finite Element Method to the Analysis of a Continuum/Skeletal Double Layer Structure

V.G. ISHAKIAN[1] and L. HOLLAWAY[2]

[1] Dar Al-Handasah Consultants (Shair and Partners) London, U.K., formerly Department of Civil Engineering, University of Surrey, Guildford, Surrey, U.K.

[2] Department of Civil Engineering, University of Surrey, Guildford, Surrey, U.K.

The stiffness and strength of glass reinforced polyester folded plate structures can be increased by the use of skeletal members connected to the continuum components. Such skeletal members must have good unidirectional strength and stiffness and it is suggested that pultruded glass/polyester with 60-65 weight percent glass fibres would be a suitable material. A theoretical approach to the analysis of such structures by the finite element method is demonstrated. The analysis is used to design a full size skeletal/continuum roof structure manufactured from g.r.p. material.

1. INTRODUCTION

Glass reinforced polymer (g.r.p.) composites belong to the family of materials known as fibre reinforced matrix materials. The modulus of the glass fibre is relatively low compared with that of steel and when this component is combined with the low modulus polyester resin the composite has a modulus of elasticity value lying between the two component values and its actual value will be dependent upon the fibre-matrix volume fractions. To provide greater stiffness to the overall structure it is usual to fold the low modulus continuum g.r.p. system.

To provide even greater stiffness to the structural system it is possible to introduce skeletal members which could connect specific points in the continuum component. In the proposed composite structure the continuum would be manufactured either by land lay-up or a semi-mechanical process, using a chopped strand mat and polyester resin in which the fibre fraction would be between 30-35 weight percent; the elastic modulus would therefore be low. The members forming the skeletal part of the construction would require unidirectional strength and stiffness and would be manufactured by the pultrusion technique using continuous unidirectional rovings and polyester resin; the fibre volume fraction in this case would be about 65-70 weight percentage. The modulus of elasticity of the material, although higher than that of the continuum, would be less than one third that of steel and therefore within certain areas of the composite construction it would be possible to use a hybrid composite component consisting of glass and carbon fibre in the polyester matrix to increase the stiffness of particular members.

This paper examines, by using the finite element method, the behaviour of a glass reinforced polyester composite structure consisting of continuum plates which are manufactured from randomly orientated glass fibre in a polyester resin and skeletal members which are manufactured from the pultrusion technique. The double-layer grid is a three-way skeletal system situated at the top and bottom levels of the Vee continuum and the diagonal members lie in the plane of the continuum.

2. THEORETICAL ANALYSIS

Basic concepts and types of elements

For the analysis of continuum/skeletal systems it is desirable to use the Finite Element discretization in combination with the direct stiffness method. This procedure provides a systematic approach to the solution which is applicable to any shape, configuration and boundary condition. In addition it enables optimization to be made of the configuration and the proportion and the cross-sectional properties of both the components which form the composite.

To analyse the composite structure, it is necessary to combine the line and plate types of elements. The plate element used in this paper is a four noded rectangular one and the skeletal members are represented by a two node line element.

No discussion will be made of the necessary steps for the stiffness matrix derivation of the elements with an assumed displacement function across the element; these steps have been described elsewhere, Refs 1-6.

However, an important consideration arises regarding the degrees of freedom per node for the two types of elements when combined in one overall stiffness matrix.

For the system of equations to have a solution, compatibility should exist for each degree of freedom per node. The line element used in this paper is that of a two ended member in space, having six degrees of freedom per node (three translational and three rotational as shown in Fig 1a) it is derived from an assumed linear polynomial for the axial and torsional displacement and a cubic polynomial for the transverse bending displacements, consistent with the skeletal structural theory, Refs 2, 7 and 8. Consequently the plate element in space should have six degrees of freedom per node, and this is achieved by the uncoupled combination of inplane degrees of freedom (u, v, θ_z) and the out-of-plane bending action (w, θ_x, θ_y) shown in Fig 1b.

However, most thin plane finite element formulations include only two translational degrees of freedom for the inplane stress analysis (u and v) and neglect the inplane rotation θ_z (Fig 1b) since it is necessarily arbitrary, because there is no unique value of such rotation (apart from a rigid body movement) at a point in a two dimensional continuum. The moment which corresponds to it is not fully tractable to physical explanation and hence combining such formulations (u and v) with the bending action (w, θ_x, θ_y) results in an element with only five degrees of freedom per node.

A method is proposed by Zienkiewicz, Ref 1, which overcomes the singularity of the stiffness matrix because of the omission of the inplane rotation in plate element. This method inserts a fictitious set of inplane rotation stiffnesses, based on the fact that the real rotational stiffness is very high in comparison with the plate bending stiffness. This method is explained in Ref 9. However this approach lacks one important aspect, that is the inplane displacements field has two degrees of freedom whereas the bending displacement field has three. This makes it difficult to use the same shape functions for both the inplane stress and plate bending analysis.

The difference between the functional variations of the inplane displacement field and the transverse displacement field leads to gross violation of conformity between adjacent elements which do not lie in the same plane, Ref 10.

The utilization of the inplane rotation θ_z as an additional degree of freedom enables the same shape functions to be employed for both inplane stress and plate bending analysis. Such a formulation which achieves six degrees of freedom, is ideally suited to the analysis of three dimensional plate assemblies forming folded plate structures combined with skeletal systems.

Some formulations which include the inplane rotation θ_z as a degree of freedom have been developed by Mc Leod, Ref 11, Tocher and Hartz, Ref 12, Pole and Felippa, Ref 13 and Scordelis, Ref 14.

The formulation which is used in this investigation is that of Scordelis, Ref 14, which was initially developed for the analysis of box-girder bridges. The element used is a rectangular one of four nodes each having six degrees of freedom formed by the uncoupled combination of inplane degrees of freedom (u, v, θ_z) and out of plane bending degrees of freedom (w, θ_x, θ_y) it is important to mention here that the same shape functions are used in setting the displacement functions for both inplane and out of plane deformations. These functions combine a linear function, a beam rotation (i.e. a beam clamped at one end and a unit rotation applied at the other as shown in Fig 1b) and a beam displacement function (i.e. a beam clamped at both ends but with a relative displacement of unity between them). At each of the four nodes the inplane rotation θ_z is defined as the average of the rotations of the two adjacent sides of the element at any particular node as:

$$\theta_z = \frac{1}{2}\left[\frac{\partial v}{\partial x} - \frac{\partial u}{\partial y}\right]$$

2.1 Finite element formulations considering non-linear analysis

A general method for non-linear analysis is given by Zienkiewicz, Ref 1, in which he introduces a 'Tangent Stiffness Matrix' for the element which includes the sources of linearities as:

$$K_T = K_O + K_\sigma + K_L$$

where

K_O = linear elastic stiffness matrix;

K_σ = symmetric matrix dependent on the axial stress level and is called the 'Initial Stress Matrix' or 'Geometric Matrix'; this has been defined in Ref 15;

K_L = 'Initial Displacement Matrix' or 'Large Displacement Matrix' and contains only terms which are linear and quadratic in the displacement. This has been defined in Ref 15.

In situations such as perfectly straight struts, plates, shells, 'v' and 'box' sectioned members etc., under inplane stresses only, the 'Large deformation matrix' K_L is identically zero because of the absence of the coupling effect of bending stresses with the axial stresses. In these situations the non-linear behaviour initiates as a distinct bifurcation point followed by a non-linear post buckling path in the deflections and stresses. The distinct bifurcation load can be found by solving the typical eigen-value problem for the equation:

$$dF = (K_O + \lambda K_\sigma) \equiv 0 \qquad \text{.... (1)}$$

where λ is the load factor.

The buckling node is expressed by determining the corresponding eigenvector.

Theoretically folded plate structures interconnected to skeletal systems have very small stresses and these arise from the rigid connections under a loading system applied at the nodal joints. The start of a non-linear behaviour therefore can be reasonably predicted by the assumption of a linear load-deflection relation up to a distinct bifurcation point which is a function of the axial stresses only.

A computer program has been written in FORTRAN for a linear elastic analysis and the prediction of the first critical load by the reformulation of the global stiffness matrix at every load increment, according to Eqn 1, followed by the triangular decomposition; the critical load is reached as the number of negative pivots of the decomposed stiffness matrix changes from zero to one. A flow chart of the program is given in Fig 2.

3. THE COMPOSITE STRUCTURE

The global dimensions of the space structure are 20x25 m formed by the continuous connection of 18 Vee sectioned composite members. The latter members consist of Vee-shaped g.r.p. folded plates interconnected with g.r.p. pultruded tube members formed into a skeletal system. An experimental and analytical investigation has been undertaken by the authors in a geometrically similar system manufactured from g.r.p. but consisting of only one single Vee sectioned composite member; the results of these analyses have been reported in Refs 9 and 15.

The present Vee sectioned continuum member has a thickness of 6.0 mm, a width of 1388.88 mm and a depth of 1520.0 mm. The double layer grid system, which is the other component of the composite structure, is manufactured from 100 mm diameter pultruded tubes of wall thickness 4 mm. The two way grid at the top and bottom of the Vee continuum has diagonal members lying in the plane of the continuum. Fig 3a shows the plan and cross sections of the continuum/skeletal double layer grid and Fig 3b shows details of the interconnection between the skeletal members and the continuum folded plate members.

The structure was assumed to be position fixed in the vertical z direction but position free in the X and Y directions and direction free in X, Y and Z.

4. ANALYTICAL PROCEDURE AND RESULTS

The theoretical model is discretized into rectangular plate elements and line elements and because of the symmetrical loading and configuration only one quarter of the model was considered.

Figures 4a and 4b show the discretization and the joint numbering system respectively. The maximum number of joints is 1081. The theoretical analysis is divided into two parts:

(a) the analysis of the continuum and the skeletal component is two independent structural systems;

(b) the analysis of the two components above as one skeletal continuum structural system.

Figures 5 and 6 show diagrammatically the continuum and skeletal structural systems. The analyses were undertaken to study the comparative stiffness of two independent components and the relative contributions to the stiffness of the composite continuum/skeletal structure.

The values of the vertical deflection at the centre joint, node joint number 1081, for the two individual components are given in Table 1. It may be observed that the continuum system buckles with 796 possible buckling nodes of the plate whereas the skeletal system under the same loading condition is in a state of equilibrium and stability.

In the second part of the analysis in which the continuum/skeletal structure was analysed, three thicknesses of the continuum composite structure were considered; the dimensions of the skeletal system's members, as given in section 3, remain constant throughout. The results for the system, in which the continuum thickness are 3.0, 6.0 and 10.0 mm, are given in Table 2. The only system to have no buckling associated with it is that in which the continuum plate thickness is 10.0 mm.

The results in Table 3 show the variation of the axial forces against continuum thickness in the most highly stressed members at the centre of the grid. It may be seen that as the thickness of the continuum increases the axial forces in the skeletal structure component decrease which indicates that there is a redistribution of strain and hence stress between the two components of the composite structure.

5. DISCUSSION OF THE RESULTS

When the continuum and skeletal components are loaded separately the latter one is stiffer than the former when the continuum has a uniform thickness of 6.0 mm. The skeletal system is in a completely stable state as no negative diagonal element is present in the decomposed stiffness matrix but the continuum component has 796 negative diagonal elements indicating a state of post bifurcation deformation with 796 possible mixed local and global buckling modes of the plates. Consequently the maximum deflection of 118.5 mm (given in Table 1) and predicted by the linear analyses, is under-estimated.

It may be seen from Table 2 that the maximum deflection of the skeletal/continuum structure, in which the continuum plate thickness is 6.0 mm, deflects only one half of the value of that for the skeletal component alone (given in Table 1) and when the continuum plate thickness is 10.0 mm, the composite structure deflects only one third that of the skeletal system alone. The presence of negative diagonals in the decomposed stiffness matrix for two of the cases, shown in Table 2, indicates instability. However, because the skeletal system above is in a stable situation the continuum component apparently undergoes a series of local buckling deformations, but the overall composite structure is stable. The skeletal/continuum structure in which the plate thickness is 10 mm is a stable system and has a maximum deflection of 40 per cent of that of the skeletal system alone.

It is clear that as the bending stiffness of the skeletal/continuum structure is increased by increasing the plate thickness of the contintuum component, the resulting increase in the strain capacity of the continuum will cause a reduction of strain in the skeletal members. Table 3 shows the effect of this increase in thickness and gives a percentage reduction in axial force in particular members of the skeletal component of the composite. Although a 10 mm thick continuum plate is unrealistic from an economic viewpoint, considerable reduction is evident when a comparison is made with a practical value of 6 mm thick plate.

Observations

It has been shown that when a double layer skeletal system is combined structurally with a continuum system the stiffness of the composite structure is improved considerably over the skeletal system alone the reduction in maximum deflection is of the order of 50%. The continuum structure has two functions, first to act as a cover to the roof and secondly as a stressed skin system.

It has also been shown that for the design of a skeletal/continuum structure manufactured from a low modulus, elastic material, it is necessary to consider a buckling analysis to enable an accurate prediction of the thickness of the continuum and the cross sectional properties of the skeletal members.

REFERENCES

1. O C ZIENKIEWICZ, The Finite Element Method in Engineering Science, McGraw-Hill, 1971.

2. B NATH, Fundamentals of the Finite Element Method, Athlone, 1974.

3. K C ROCKY, H R EVANS, D V GRIFFITHS and D A NETHERCOT, The Finite Element Method, Crosby Lockwood, 1975.

4. C A BREBBIA and J J CONNER. Fundamentals of Finite Element Techniques for Structural Engineers, Butterworths, 1973.

5. S J FENVES et al, Numerical and Computer Methods in Structural Mechanics, Academic Press, 1973.

6. R H GALLAGHER, Finite Element Analysis Fundamentals, Prentice-Hall, 1975.

7. R K LIVESLEY, Matrix Methods of Structural Analysis, 2nd Edition, Pergamon Press, 1975.

8. J M GERE and W WEAVER Jnr., Analysis of Framed Structures, D. Van Nostrand, 1965.

9. V G ISHAKIAN and L HOLLAWAY, Application of the Finite Element Method to the Analysis of a Skeletal/Continuum G.R.P. Space Structure, Composites, April 1979, Vol. 10, No. 2, pp. 81-88.

10. K T WILLIAM, Finite Element Analysis of Cellular Structures, Ph.D. Thesis, University of California, Berkeley, 1969.

11. I A McLEOD, New Rectangular Finite Element for Shear Wall Analysis, Journal of the Structural Division, Proceedings of the American Society of Civil Engineers (A.S.C.E.), Vol. 95, March 1969.

12. H L TOCHER and B J HARTZ, Higher-order Finite Element for plane stress, Journal of the Engineering Mechanics, Proceedings of the American Society of Civil Engineers (A.S.C.E.), Vol. 93, August 1967.

13. G M POLE and C A FELIPPA, Discussions on New Rectangular Finite Element for shear wall Analysis, Journal of the Structural Division, Proceedings of the American Society of Civil Engineers (A.S.C.E.), Vol. 96, January 1970.

14. A C SCORDELIS, Analysis of Continuous Box Girder Bridges, Struct. Engng. and Struct. Mech. Report No. SESM 67-25 (University of California, Berkeley, U.S.A.), November 1967.

15. V G ISHAKIAN, Stability Analysis of Continuum Skeletal Fibre Matrix Systems, Ph.D. Thesis, University of Surrey, 1980.

Type	Vertical deflection at the centre Joint (1081) mm	Number of negative diagonals in the decomposed stiffness matrix
Skeletal System	71.65	0
Continuum System	118.56	796

TABLE 1 Results of deflections and stability state (Number of negative diagonals in the decomposed stiffness matrix)

Thickness of the continuum plate	Vertical deflection at the centre Joint (1081) mm	Number of negative diagonals in the decomposed stiffness matrix
3	43.53	480
6	35.46	34
10	28.9	0

TABLE 2 Results of deflections and stability state of the continuum/skeletal composite for the three different thicknesses of the continuum folded plate component and with constant cross sectional properties for the skeletal system component

Member		Skeletal system only (Parla, Table 2)	Axial forces (N) in the skeletal members of the composite structure for varying thicknesses of the continuum structural component			Percentage decrease in axial forces as thickness increase from 0-10 mm
		0	3.0 mm	6.0 mm	10.0 mm	
1004	1042	+ 8352.0	+ 8721.0	+ 8046.0	+ 7156.0	14.32
1042	1079	+ 8352.0	+ 8695.0	+ 8022.0	+ 4598.0	44.95
1077	1081	+ 31040.0	- 21280.0	- 17210.0	- 13650.0	56.02
928	932	- 30290.0	- 20640.0	- 16650.0	- 13180.0	56.48
928	965	- 47000.0	- 39260.0	- 32640.0	- 26900.0	42.76
965	1002	- 47000.0	- 39420.0	- 32820.0	- 27060.0	42.42
1002	1040	- 47000.0	- 39620.0	- 32000.0	- 27240.0	42.04
1040	1077	- 47000.0	- 39840.0	- 33240.0	- 26440.0	43.74
932	969	- 50920.0	- 39800.0	- 33120.0	- 27300.0	46.38
969	1006	- 50920.0	- 39980.0	- 33300.0	- 27460.0	46.07
1006	1044	- 50920.0	- 40200.0	- 33500.0	- 27660.0	45.67
1044	1081	- 50920.0	- 40420.0	- 33740.0	- 27880.0	45.25

TABLE 3 Variation of axial forces in the skeletal members as the continuum folded plate thickness increases

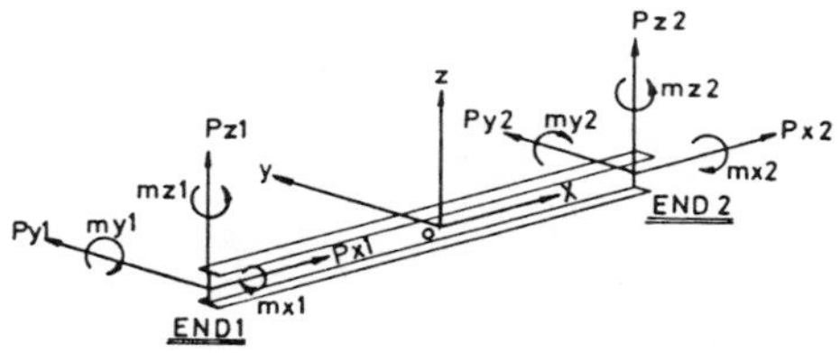

FIG 1-1A. Degrees of freedom for a skeletal member in space in local coordiates.

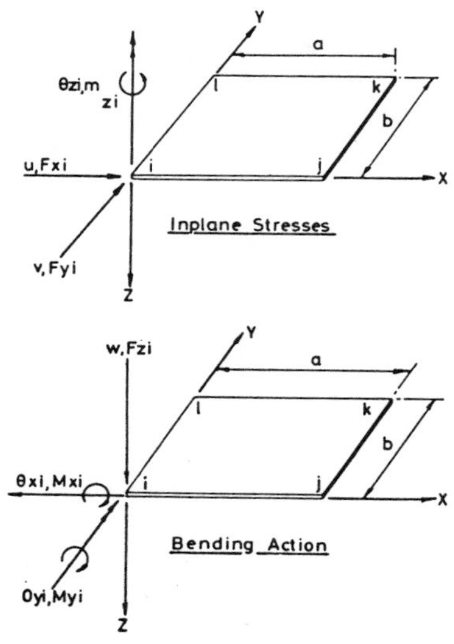

FIG 1 1B. Degrees of freedom for plate elements.

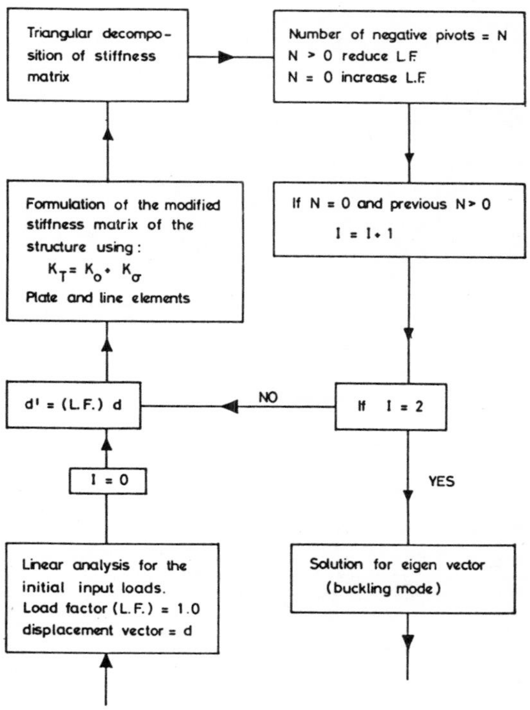

FIG. 2. Flow chart of the computer program for determining the buckling load and the corresponding buckling mode of the composite space structure.

Figure (3)

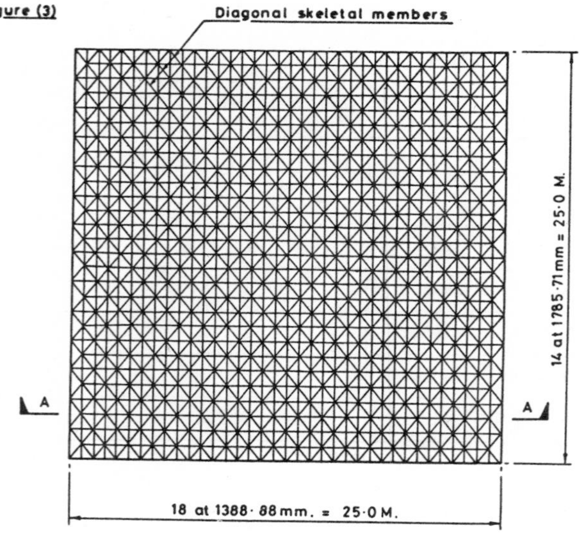

PLAN

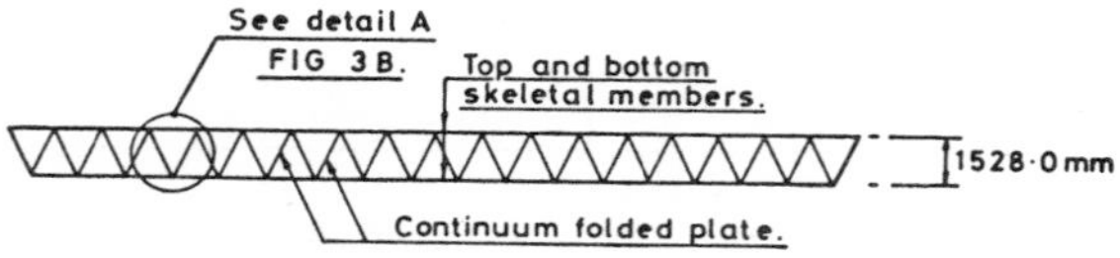

SECTION A—A

FIG.3A. Plan and cross-section of continuum/skeletal double layer grid.

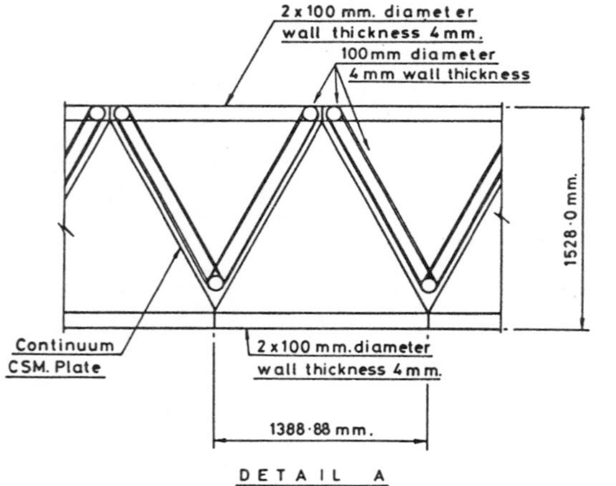

DETAIL A

FIG 3 B. Detailed dimensions of the interconnection between the skeletal members and the continuum folded plate

For plan and cross-section see FIG 3 A

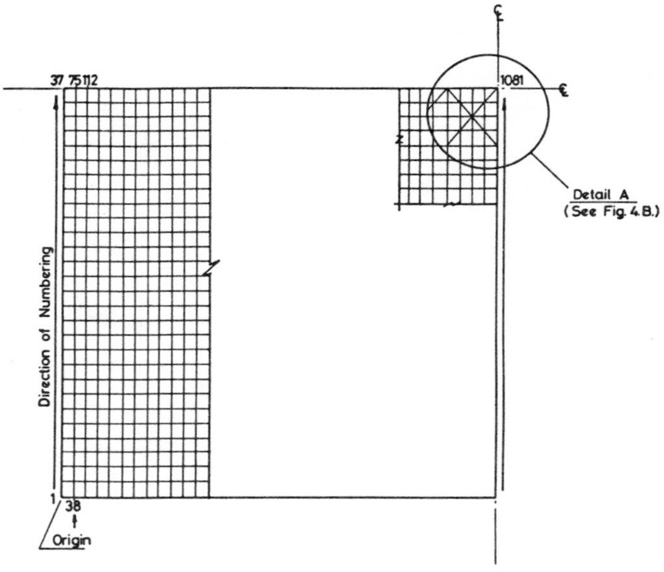

FIG. 4 A. Descretization and joint numbering direction for one quarter of the grid.

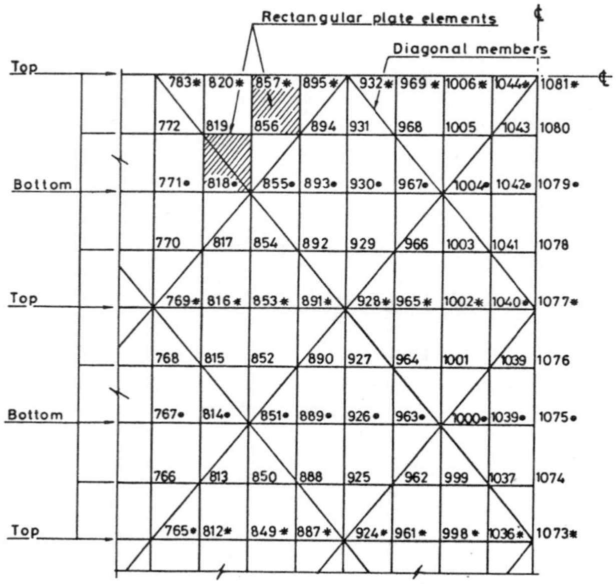

DETAIL A

* Numbers of the joints at the top of the cross-section.

• Numbers of the joints at the bottom of the cross-section.

FIG 4B Joint numbering system at the centre of grid resulting from the descretization of the continuum folded plate and the skeletal system.

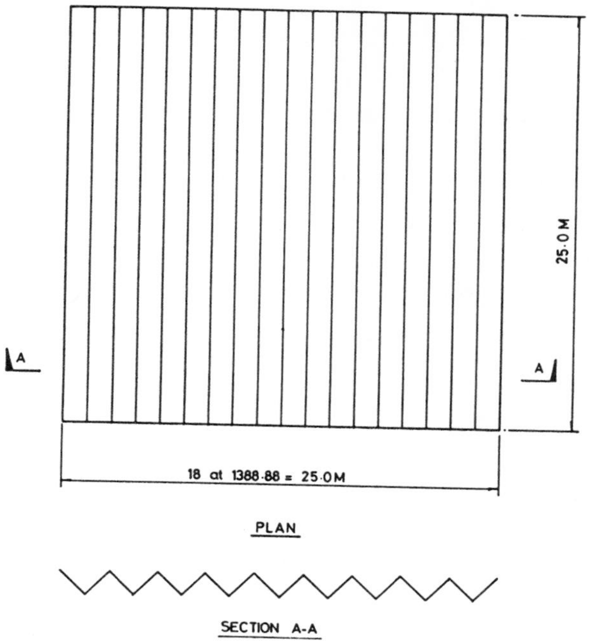

Fig. 5. Plan and crossection of the continuum folded plate considered to be the only component for covering the square span.

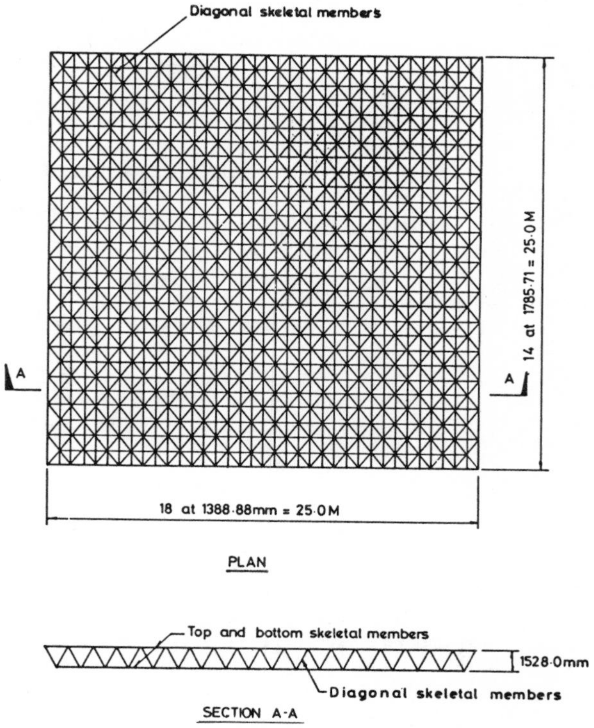

Fig. 6. Plan and crossection of the skeletal double layer grid.

ANALYSIS OF STRUCTURES INTERACTING WITH CORRUGATED PANELS AS A FRAME AND PLATE SYSTEM

J BRÓDKA[x], CEng, DSc, R GARNCAREK[xx], CEng, DSc, A GRUDKA[xxx], CEng

and B PIOTROWSKA - GRUDKA[xxxx], CEng

x Professor, xx Lecturer, xxx Assistant lecturer,

Metal Structures Research and Design Centre "Mostostal", Warsaw.

xxxx Assistant lecturer, Department of Architecture,

Warsaw Technical University, Warsaw.

Co-action of members of a supporting structure and cladding made from trapezoidally corrugated plates exists irrespective of the fact whether it has taken into account, or not, in the design of the building. The authors have devised a method of calculation if a supporting structure and corrugated sheeting considered as a space structure consisting of linear members and diaphragms. In such a space structure there are included all the existing parts of the skeleton, their mutual eccentric location, fixing of joints, as well as flexibility of connections. Analysis of three industrial buildings has been performed and a full-scale experiment has been carried out. It has been shown that there is a good agreement between the theoretical and experimental results.

INTRODUCTION

Engineers designing frame structures have intuitively felt long ago that frame structures co-act with cladding in carrying loads. However advantage has been taken of this co-action only rarely because of brittleness of cladding materials and lack of a reliable technique of connecting cladding to the frame. Use of corrugated plates has changed this situation and has necessitated an elaboration of respective methods of calculations.

Corrugated plates and steel skeleton are permanently connected to one another and cannot work separately.
This co-action exerts influence upon displacements and leads to a redistribution of internal forces.
The effects of this co-action are generally positive /but not always/, although they often lead to overloading of fasteners.

There is ample literature, see Refs 1 and 3, devoted to experimental and analytical approach to the solution of this problem. Design guidelines have been elaborated for members subjected to a simple static work, see Ref 8.
A good representation of roof surfaces slightly weakened by openings is a model of a diaphragm, see Ref 4, carrying loads applied in its plane. Action of a diaphragm is determined by flexibility being the sum of flexibilities of all linear and plate components and their connections. A simplified method of calculation of flexibility of a diaphragm has been given in Ref 8, and a more accurate one, in Ref 6.

Finite elements method makes possible an extremely accurate modelling of diaphragms taking into account an interaction of corrugated plates, fasteners and framing members. Analysed was a diaphragm with dimensions of 3.66 x 3.05 m, see Ref 7. Its properties have been modelled by means of 240 rectangular orthotropic plate element, 34 linear elements /struts/ and 79 fasteners. The whole model had about 400 nodal points each of them having two degrees of freedom.

Slightly simpler model has been presented in Ref 5.
In the model described in the above mentioned work all fasteners have been represented, but corrugated plates have been divided into a smaller number of elements. However, if one would like to use the described model for the really existing buildings of relatively small dimensions, then the number of nodes would amount to tens of thousands.

TECHNOLOGICAL AND PHYSICAL MODELS

In conventional designing of steel structures bracings /ties/ are used for carrying horizontal forces and for ensuring an overall stability of the structure. These bracings are situated in each plane of a roof and in gable and side walls as well. When corrugated plates are being used these braces can be replaced by roof and wall diaphragms. Corrugated plates will then carry shearing forces, and framing elements will carry normal forces. Wall diaphragms distinguished from the roof ones, usually have numerous openings with various framing rigidity. For this reason, rigidity of window and gate framings should not be taken into account in calculations.

A new method of calculation of frame structures and corrugated plates as a space system of linear members and diaphragms has been devised, see Ref 2.

In such a space structure there are represented all the existing components of skeleton and cladding, their mutual eccentric position, fixing of joints and flexibility of connections. Systems usually obtained for singlestorey industrial buildings of dimensions 12.0 x 38.0 m have about 450 frame elements, 240 panels with 400 nodes. Generally they are not symmetrical because of the arrangement of windows and gates in the longitudinal and gable walls.

Main columns, wall studs rafters, edge and wall beams, purlins, top chords of girders, branches of main laces columns have been regarded as frame element, the nodes of which have six degrees of freedom. The remaining linear members of the skeleton, such as diagonal bracings and bottom girder chords, main column web members, ties and bracing /roof and wall ones/ has been assumed as pin-ended bars with nodes having three degrees of freedom. Corrugated plates are represented by rectangular orthotropic panel with nodes having two degrees of freedom. Flexibility of joints calculated acc. to Ref 8 has been added to the flexibility of trapezoidal sheets. The elasticity matrix has been assumed in the following form

$$\begin{bmatrix} E_{xx} & 0 & 0 \\ 0 & E_{yy} & 0 \\ 0 & 0 & G \end{bmatrix}$$

Thus an independent work of corrugated plates under normal loads acting in two perpendicular directions has been assumed. This leads to the neglectance of interaction of displacements of these loads. Hawever, this simplification seems to be acceptable because of a small influence ot the above mentioned interaction.

BUILDING WITH COLUMN - BEAM SKELETON

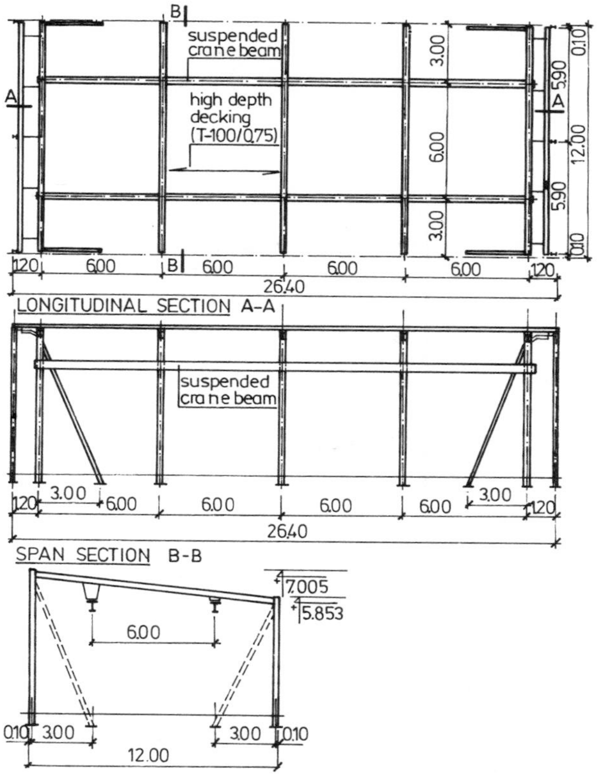

Fig 1

Skeleton of the building, see Fig 1, consists of frames with hinge joints. Stability of the building has been ensured by roof plate transmiting loads on end columns braced in two directions. Gable walls are spaced 1.2 m from end frames. Corrugated plates with 100 mm profile height and 0.75 mm wall tnickness are fastened to beams in each fold by "Hilti" fire pins. Self-supporting wall cladding consists of two layers. The inner layer has been made from cassettes with 1.0 mm wall thickness and the outer layer, from corrugated plates with 55 mm profile height and 0,75 mm wall thickness. Cassettes have been connected to columns by "Hilti" fired pins. Arrangement of windows, doors and gates is shown in Fig 2.

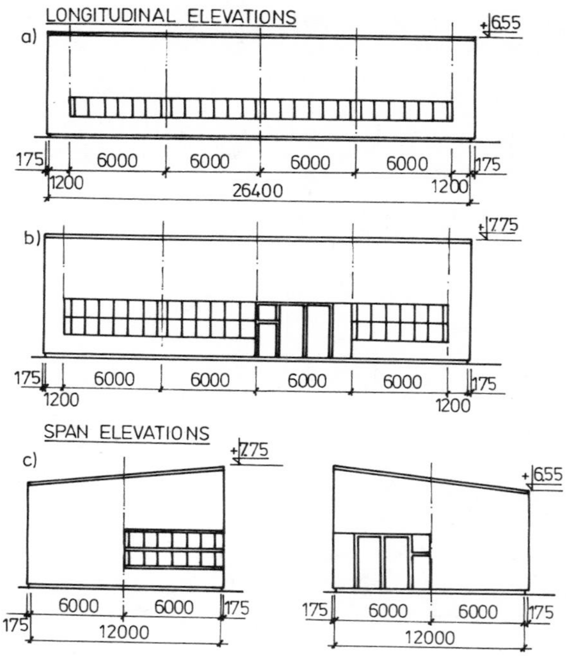

Fig 2.

The following characteristic loads act on the structure:

a/ Vertical load including dead weight of roof cladding, snow and installation, totaly 1,210 N/m^2, as well as dead weight of roof beams and crane beams suspended to the roof structure.

b/ Wind load equal to 450 N/m2 acting in the direction of longitudinal wall.

c/ Wind load as above acting in the direction of gable wall.

d/ Forces generated by overhead travelling crane with 23 kN lifting capacity.

Technological model of the building has been established using five statical schemes, which take into account to various extent the co-action of cladding members and skeleton in the calculations of the building. Scheme I, see Fig 3a, is a conventional one. Main columns are fixed in foundations. Roof and gable wall bracings have been added. In scheme II, see Fig 3b, co-action of corrugated plates with skeleton has been assumed but not confirmed by calculations. A qunatitative analysis of this co-action has been performed for schemes III, IV and V. Scheme III, see Fig 3c, takes into account the roof diaphragm. Scheme IV, see Fig 3d, takes into account both the roof and wall diaphragms. Between the roof and the walls there are gaps exerting negative influence on the roof diaphragm and its fasteners. These gaps have not been encountered in scheme V.

Calculations have been performed by finite element method using ICES programme and IBM 370 digital computer. Building properties have been modelled by 210 plate elements and 96 linear elements. Model had 332 nodes and no symmetry.

For elements of the main system, ie. for beams and main columns, the influence of the panels upon the magnitude of bending moments is neglectable. Differences do not exceed 1 per cent for beams and 3 per cent for columns. This influence is more distinguished in the distribution of longitudinal forces, which are increased in extreme beams. However, these forces are not great and are only of minor importance in design. More differentiated were forces in main columns.

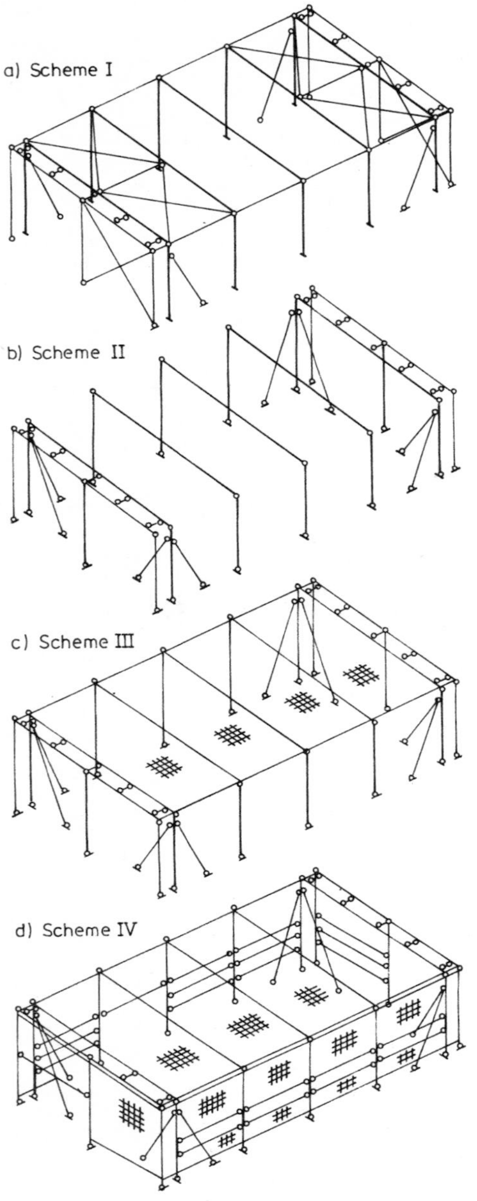

Fig 3

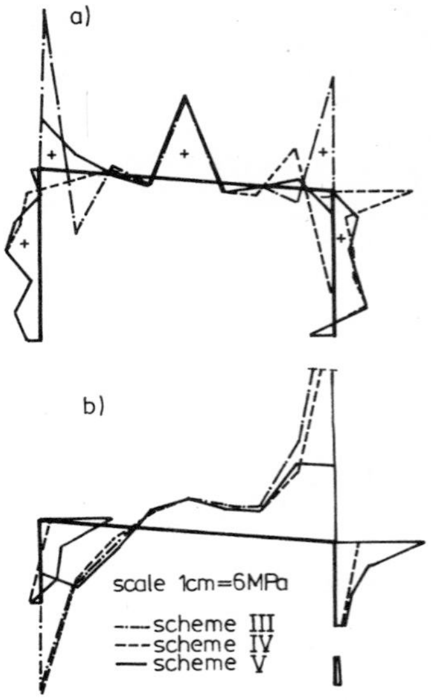

Fig 4

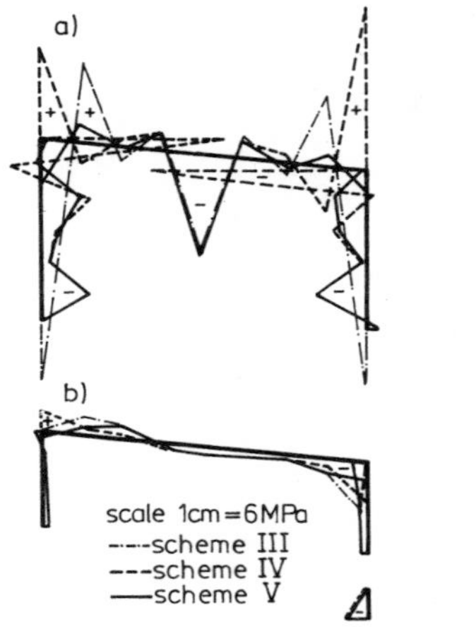

Fig 5

Total forces for columns differ by 7 per cent, and for extreme columns by 14 per cent. Inclined braces of corner columns are considerably overloaded. Total forces in braces of gable walls are greater by 22 per cent than design ones. Still more overloaded are braces of longitudinal walls. Total forces in these braces exceed by 55 per cent the design ones. This overloading has been mainly due to the vertical load. It is not always safe, particularly for fasteners.

A quantitative analysis of roof and wall panels has been performed for schemes III, IV and V. Exemplary results are presented in Figs 4, 5 and 6. Stress diagrams in trapezoidal sheets in the direction of generatrices relate to wind acting on longitudinal walls, see Fig 4 and to wind acting on gable walls, see Fig 5. Part a/ of the figures relates to the sections at the extreme frame on the side of the gable wall, whereas part b/ of the figures relates to the sections at the centre frame on the side of the gate.

Normal stresses in a direction perpendicular to generatrices of corrugated plate are negligible. Shear stresses due to vertical load are shown in Fig 6. Part a/ relates to the section between the end frame and gable wall, whereas part b/ relates to the section at the first intermediate frame on the side of the end frame. In accordance with the principles of finite element method stresses have been calculated in the nodes of the model. For normal stresses a conventional linear distribution has been assumed. For shear stresses a linear step-wise model has been assumed. The gable wall framing causes a strongly non-uniform distribution of normal stresses. On the other hand, disturbances and irregularities at the centre beam are caused by secondary deformations of the sheet existing in the case of a deflection of the beam. Connection of the roof plate to the wall plates results in a considerable reduction of maximum stresses in the trapezoidal sheets of the roof. Design capacity of trapezoidal sheet in both compression and bending taking into account the possibility of buckling is too small because of a strongly non-uniform normal stress distribution for schemes III and IV. On the contrary, it is completely sufficient for scheme V. Shear stresses in trapezoidal

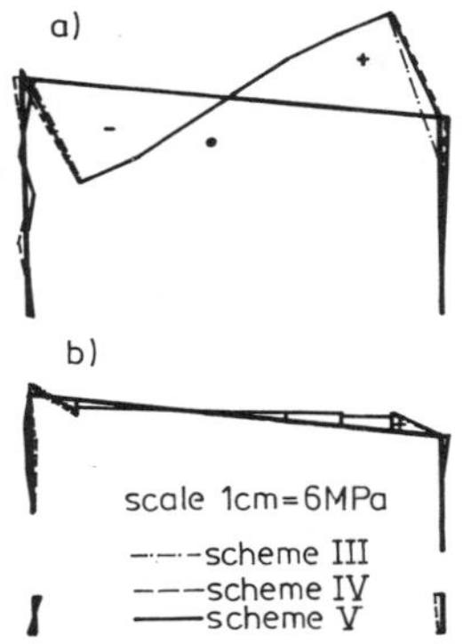

Fig 6

sheets do not cause in effect loss of stability of the plate.

The greatest loads act on the fasteners in the case of end columns and end beam. For schemes III and IV these loads exceed the safe ones. On the contrary, for scheme V they are considerably smaller than the safe ones.

Vertical deflections of main beams differ by 2 per cent for centre frame and up to 3 per cent for the end frame. Transversal horizontal displacements differ considerably for scheme I as compared with the results obtained for the remaining schemes. This difference is mainly due to the action of corrugated plates of the roof. Horizontal displacements of the tops of the higher columns of end frame have been diminished 34 times, and those of centre frame, 10 times /from 5.6 to 0.53 cm/. Longitudinal horizontal displacements of building are reduced mainly due to the action of bracings and, secondly due to the action of side wall cladding. Frame beam lateral deflection in mid-span /equal to the deflection of the centre column of gable wall/ taking into account corrugated plates of the roof is 12.8 cm and corresponds to the maximum horizontal displacement of the building. After consideration of co-action of side walls this deflection has been reduced 64 times to about 0.2 cm.

BUILDING OF COLUMN - GIRDER SYSTEM

Skeleton of the building, see Fig 7, consists of roof girders and laced columns fixed in foundations, as well as of purlins, ties, wall framework and roof and wall braces. Trapezoidal sheets with 55 mm height and 0.75 mm wall thickness have been used for roof and wall cladding. The arrangement of windows and gates is shown in Fig 6c /unshaded zones/. Corrugated plates are connected with supporting members by means of blind rivets of 5 mm diameter.

The following characterstic loads act on the structure:

a/ Vertical load including of dead weight of the roof cladding, purlins, bracings, snow and installation, totaly 1500 N/m^2, and also dead weight of a girder.

b/ and c/ wind load 550 N/m^2 acting in the direction of longitudinal or gable wall.

Technological model on the building has been compiled from three statical schemes which take into account to various extent three-dimensional behaviour of skeleton and corrugated plates, see Fig 8. In scheme I the skeleton is divided into plate, lattice, or beam members. In scheme II three dimensional behaviour of a framework systems has been taken into consideration. In both above mentioned schemes

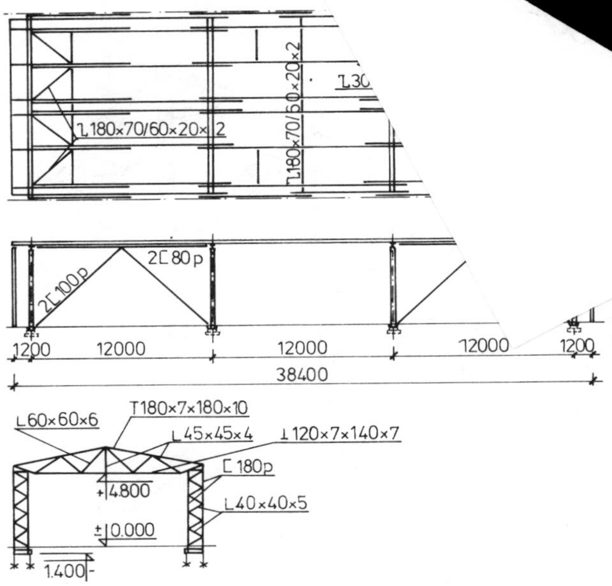

Fig 7

the existance of panels has been neglected. Scheme III has been obtained by superimposing panels formed by corrugated plates on scheme II. Between the roof and wall panels there is a gap.

For calculations by means of ICES program a previously described physical model has been used. Properties of the building have been modelled acc. to scheme III, by means of 94 plate members and 529 linear elements. The model thus obtained contained 265 nodes and not any axis of symmetry. Panels in Fig 8c are marked by shading the respective fields.

In the building of the type considered corrugated plates play, firts of all, the role of sheeting. However, a permanent conncection of this sheeting to skeleton causes a redistribution of forces in some bars or in some portions of the sheeting. Most significant reduction of forces occures in the top chords of the girder. For the wind load acting in the direction of gable walls forces in the top chords of the girder are reduced by 41 per cent. For this load the force in the outer branch of the column drops by 20 per cent. On the contrary, comparison of schemes I and II shows an overloading of the bottom flange up to 6 per cent and girder diagonal braces up to 15 per cent, as well as qualitative changes of effort of inner branches and diagonals of the main columns.

Moreover, comparison of schemes I and II also shows a qualitative change of character of work of middle purlins. Bending moments at the supports have been decreased by 7 per cent, and at mid-span by 19 per cent, but at the same time compressive longitudinal forces have appeared. Consideration of trapezoidal sheeting widens the range of this re-distribution and, simulteneously, bending moments caused by loads from the roof planes practically vanish. However, all the above mentioned phenomena have no influence upon the safety of purlins because the resultant stresses are considerably smaller than design strength.

In bracings of longitudinal walls for scheme I, where shortening of main columns have not been taken into account, the forces in bars under vertical load have zero values. Consideration of skeleton as a three-dimensional system reveals the presence of considerable forces. So in the case of braces beams hare been overloaded by 57 per cent, and diagonal bracings, by 65 per cent. After the corrugated cladding was considered this overloading has been increased up to 94 per

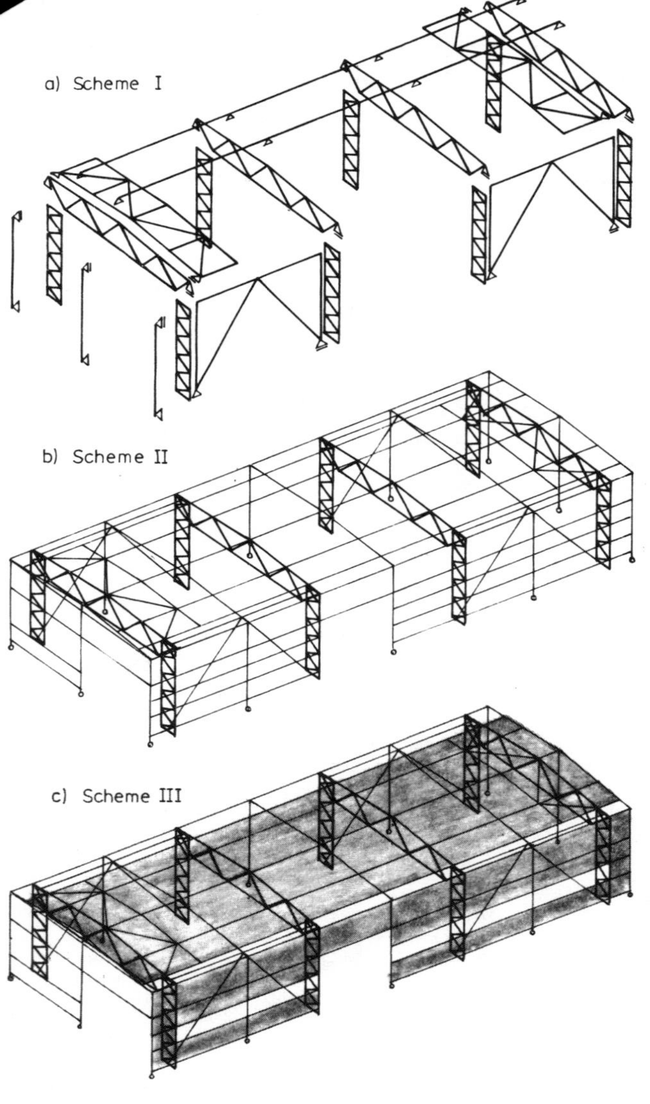

Fig 8

cent and 82 per cent. This relates to the action of wind in the direction of longitudinal walls. In the case of wind acting in the direction of gable walls this overloading is even greater. However, in the case of the building under consideration, this has no influence on the safety of building because the cross section areas of the bracings have been taken maximum slenderness permittee.

The vertical load has also a similar influence on the overloading of braces in the roof planes. The resultant forces due to this load and the wind load are several times greater than design forces calculated for the wind action only. However, this in not dangerous for the cross braces, but it can be dangerous for the other types of trusses. Consideration of corrugated sheeting leads in effect to increased forces in diagonal bracings and decreased forces in the chords. If sheeting has been used, then the overloaded braces can act in a post-critical state and thus are not nocessary.

Analysis of stresses in the panel members indicates that for the building under consideration normal stresses in corrugated sheeting due to bending and an additional deflection together with the top chords of the girder, or the deflection of columns are greater by 25 per cent than design strength. Additional stresses amount up to 41.5 MPa, for roof sheeting and up to 33.5 MPa for wall sheeting. These additional stresses cause an eccentric compression of sheeting apart from bending. Thus the top chords of the girders are underloaded and the sheeting with folds directed along the chords is overloaded. Shear stresses in the sheeting are very small.

The analysis performed has also shown that the main fasteners have been considerably /7 times/ overloaded this being due to additional stresses in trapezoidal sheeting and 25 per cent overloading of sealing fasteners.

Horizontal displacements of the building with consideration of corrugated sheeting are smaller by 30 per cent in the middle section and by 40 per cent for gable walls as compared with those obtained for the skeleton of the building only.

FULL SCALE EXPERIMENT

Full scale experiment has been performed on a building of an analogous construction as that shown in Fig 1. It had two bays 11.8 m wide and 38.4 m long. Wind load was simulated by rope in tension as shown in Fig 9. For the time of experiments the structure was divided into two parts: measuring /basic/ part and loaded part with roller bearings between them.

The building has been calculated as a three-dimensional system consisting of a skeleton and panels representing a model analogous to that previously described. The system had 240 plate elements and 348 linear elements with 334 nodes. Forces in this frame-panel system were of similar order as those in the structure shown in Fig 1, but for the columns of intermediate frames differences up to 25 per cent have been observed, particularly for the lower ones.

Before of investigations stated additional foundations had been made for anchoring the tension ropes. In the longitudinal wall special catches for inclined tension ropes were made. In the basic part of the building dial gauges were arranged. Forces in tension ropes were obtained by turnbueles. Force measurements were checked by a strain gauge dynamometer. Loads were applied stepwise with the horizontal component of force being changed from 2 to 28 kN. The forces in tension ropes differed from the nominal ones by - 5 to + 5 per cent and, in the case of large values, from - 5 to + 2 per cent.

The results of measurements of horizontal displacements are presented in Fig 10 where "o" means the results of the mea-

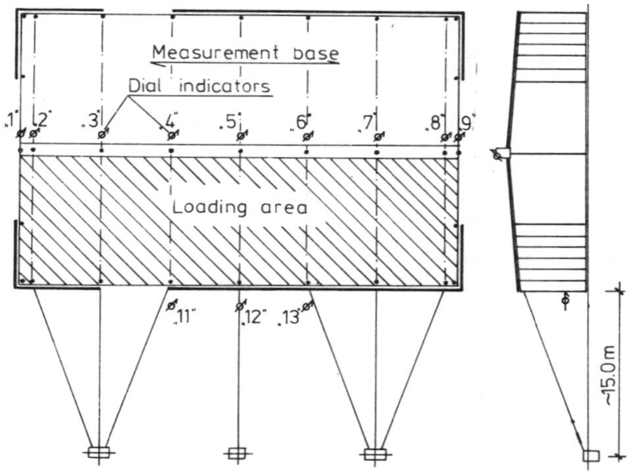

Fig 9

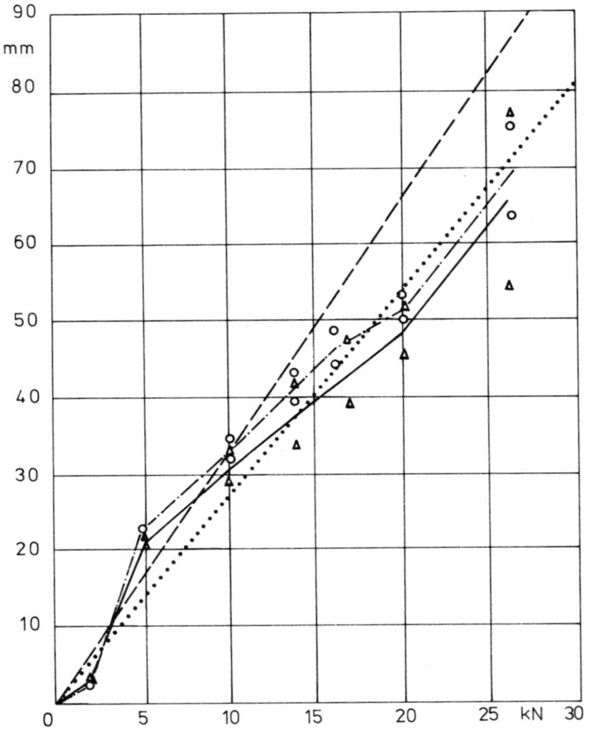

Fig 10

surements for midle frames and " Δ " the results of the measurements for gable walls. Horizontal displacements calculated theoretically for middle frames are shown with a dashed line, and for gable walls, with dotted line. Avarage displacements obtained from the investigations have shown with dot-and-dash line, and for gable walls, with a continuous line. For gable walls at small loads there are considerable differences between the displacements obtained from experimental investigations and theoretical calculations. This is the result of various structural imperfections. However, these values drop significantly when medium values are applied. In the case of considerable loads they are within the limits from - 18 up to + 12 per cent /the medium value being - 3 per cent/. On the other hand, for centre frames they are within the limits from - 26 up to - 14 per cent /the medium value being - 20 per cent/. Thus the structure investigated appeared to be more rigid than analytical three dimensional model.

Design wind load of the building investigated roughly corresponds to the load exerted by tension ropes equal to 14 kN. Then the theoretical horizontal displacements are as follows: for centre frames 45 mm /= H/133/, and for gable wall, 38 mm /= H/158/. On the other hand, medium displacements obtained from experimental investigations were 41 mm /= H/148/ and 38 mm /= H/158/ respectively. For the serviceability limit state displacements equal to H/150 = 40 mm can be assumed. In such a case the displacements obtained from experimental investigations would be higher than the theoretical ones by 2.5 to 12 per cent for centre frames.

During the investigations no damages of trapezoidal sheeting or fasteners have been noticed.

ACKNOWLEDGEMENTS

The authors wish to thank Mr. J Wojciechowski for the experimental evidence.

REFERENCES

1. J BRÓDKA, R GARNCAREK and K MIŁACZEWSKI, Corrugated plates in steel building /in Polish/, Arkady, Warsaw 1984.

2. J BRÓDKA, R GARNCAREK and A GRUDKA, Interaction of corrugated plates and steel frames /in Polish/, Archiwum Inżynierii Lądowej, 1981 no 2.

3. E R BRYAN, The stressed skin design of steel buildings, Constrado monograph, Crosby 1973.

4. J M DAVIES, Calculation of steel diaphragms behaviour, Journal of the Structural Division, July 1976.

5. H K HA, Corrugated shear diaphragms, Journal of the Structural Division, March 1979.

6. M I HUSSAIN and CH LIBOVE, Trapezoidally corrugated plates in shear, Journal of the Structural Division, May 1976.

7. A M NILSON and A R AMMAR, Finite element analysis of metal deck shear diaphragms, Journal of the Structural Division, April 1974.

8. European Recommendations for the Stressed Skin Design of Steel Structures, ECCS, September 1975.

NON-DESTRUCTIVE EVALUATION OF DAMAGE IN PERIODIC STRUCTURES

N. Stubbs *, B.A., B.S., M.S., Eng. Sc.D.

*Associate Professor, Departments of Construction Science and Civil Engineering

Texas A&M University, USA.

Two potential methods of damage evaluation are explored herein. The first method examines the hypothesis that the location of damage is only a function of the ratio of two characteristic frequencies. The hypothesis is utilized to locate the damage in a beam-like periodic structure. In the second method the location and the severity of the damage is evaluated on the basis of the magnitude of the error between the eigenfrequencies associated with the damaged structure and the predicted eigenfrequencies of the damaged structure. In this method, the changes in the eigenfrequencies are also predicted using an equivalent continuum of the discrete structure.

INTRODUCTION

After a structure has experienced a severe loading, it is desireable to assess the damage to the structure by using some form of non-destructive testing. Engineers may then use the results of such tests as the bases for future recommendations or repairs.

As damage accumulates in a structure, the stiffness of that structure decreases. Hence, qualitative damage growth can be monitored indirectly through stiffness changes (See Ref. 1). It can be shown that if more than one natural frequency of a structure is measured, before and after the damage has occured, the ratio of the changes in natural frequencies is only a function of the location of the damage (See Ref. 2). Therefore, if a ratio of frequency change for a damaged structure was obtained experimentally, locations where the theoretically predicted ratio is equal to the experimentally determined ratio are possible damage sites. This finding was demonstrated on several occasions by predicting the location of damage in specimens of advanced composite materials (See Ref. 1).

The overall objective of this research program is to investigate potential methods of non-destructively locating, and estimating the severity of, damage in large civil engineering structures. In this paper, two potential non-destructive methods of damage evaluation are explored. In the first method, the hypothesis that the location of damage is only a function of the ratio of two characteristic frequencies is applied to locate the damage in beam-like periodic structures. Damage is introduced at specific locations in the discrete system and dynamic analyses performed to determine eigenfrequencies and mode shapes. The discrete structure is modelled as an equivalent continuum and damage introduced into the latter at several locations. The resulting ratios of the eigenfrequency changes, of the damaged discrete structure and the equivalent continuum, are then compared. The validity of the hypotheses is established on the basis of the accuracy of the prediction of the location of damage. In the second method, the location and the severity of the damage is evaluated on the basis of the magnitude of the error between the eigenfrequencies associated with the damaged structure and the predicted eigenfrequencies of the damaged structure. In this method, the changes in the eigenfrequencies are also predicted using an equivalent continuum of the discrete structure.

THEORY OF DAMAGE EVALUATION

In this section, criteria for determining the location and severity of damage in lattice structures are developed. Only structures that have been damaged in one location will be considered here. The location will normally correspond to either a joint or a member of the structure. The approach used here is motivated by the approach proposed in Ref. 2 for the damage assessment of composite materials with the appropriate additions and modifications to account for the different class of structure under consideration here.

Damage Location Predictions

Consider a lattice structure defined by a mass matrix M and a stiffness matrix K. Assume that at some time t_o in the life of the structure M and K are known. Furthermore, suppose that at some time t_1 (greater than t_o) the structure experiences a hostile loading environment and is damaged at one location. Let position vector $\underline{r}$ denote the location of the damage, dm denote the change in mass at position $\underline{r}$, and dk denote the change in stiffness of the member. Then, if w_i is the eigenfrequency associated with mode i in the undamaged structure, the change in the frequency, dw_i, associated with the damage is hypothesized to be a function of $\underline{r}$, dk and dm. Mathematically,

$$dw_i = g_i(dk, dm, \underline{r}) \qquad ....1.$$

If the function is expanded about the undamaged point $(0,0,\underline{r})$, Eqn.(1) becomes:

$$dw_i = g_i(0,0,\underline{r}) + \partial g_i/\partial k(0,0,r)\partial dk + \partial g_i/\partial m(0,0r)dm \quad ....2.$$

Since there is no damage if dk = dm = 0, the first term in Eqn.(1) reduces to zero and Eqn.(2) becomes:

$$dw_i = g_{ik}(r)dk + g_{im}(r)dm \quad ....3.$$

where $g_{ik} = \partial g_i/\partial k$ and $g_{im} = \partial g_i/\partial m$

Using the same argument, the change in frequency of the jth mode dw_j resulting from the damage may be written:

$$dw_j = g_{jk}(r)dk + g_{jm}(r)dm \quad ....4.$$

If the damage results only in a change in stiffness while the mass of the structural member remains unchanged, then dm = 0 and Eqns. (3) and (4) may be combined to give:

$$dw_i/dw_j = g_{ik}(r)/g_{jk}(r) = h_{ij}(r) \quad ....5.$$

Thus if the foregoing conditions are satisfied, the ratio of frequency changes in any two modes is independent of the magnitude of the damage and depends only upon the location of the damage.

The location of damage in a structure may be determined by using Eqn.(5) as follows. Let $(dw_i/dw_j)_T$ be the measured ratio of the changes in the eigenfrequencies for any two modes i and j. Furthermore, let $(dw_i/dw_j)_P$ be the predicted ratio of frequencies for the structure assumed to be damaged at position $\underline{r}$. If the damage is assumed at successive locations and the measured and predicted ratios are compared, then the most likely position of damage in the structure is that value determined by the value of $\underline{r}$ which yields:

$$(dw_i/dw_j)_T - (dw_i/dw_j)_P = \text{minimum} \quad ....6.$$

Frequency Change Prediction

The predicted change in frequency of mode i may be found by performing successive dynamic analyses on the damaged structure and comparing the results with the undamaged structure according to Eqn.(6). However, although straight forward in concept, this method could be very costly especially for large structures with many degrees of freedom. A more efficient method of determining the change in frequencies, in terms of cost and time, is to estimate the frequency changes from the variations in the modal stiffnesses and masses for each mode.

The eigenfrequencies w_r (r = 1, 2, 3, ..., n) for an n-degree-of-freedom system are given by:

$$w_r^2 = K_r/M_r \quad ....7.$$

where K_r and M_r are the modal stiffness and modal mass, respectively, of mode r. If X_r is the rth mode shape vector of the system and K and M are the stiffness and mass matrices for the structure, respectively, then K_r and M_r are given by:

$$K_r = X_r^t K X_r \quad ....8.$$

and

$$M_r = X_r^t M X_r \quad ....9.$$

where superscript t denotes the transpose of the matrix. From Equations (7-9), the variation in the modal frequency, $d(w_r^2)$ is given by:

$$d(w_r^2) = dK_r/M_r - (K_r/M_r^2)dM_r \quad ....10.$$

where

$$dK_r = dX_r^t K X_r + X_r^t dK X_r + X_r^T K_r dX_r \quad ....11.$$

and

$$dM_r = dX_r^t M X_r + X_r^t dM X_r + X_r^t M dX_r \quad ....12.$$

If the conditions that no change in mass occurs and that second order contributions to the changes in modal masses and stiffnesses due to the incremental changes in mode shape vectors dX_r can be ignored, then Eqns. (10) and (11) reduce to

$$d[W_r^2] = dk_r/M_r \quad ....13.$$

and

$$dK_r = X_r^t dK X_r \quad ....14.$$

Estimation of Frequency Change For Damaged Continuous Systems

Equation (7) also holds for continuous systems. For example, let a beam be subjectd to bending and assume that the Euler-Bernouli assumptions prevail. Then, if A(x) is the cross-sectional area of the beam, m(x) is the mass density (per unit volume), EI is the bending rigidity, and X_r is the normalized mode shape vector for the rth mode; then

$$M_r = \int_0^L m(x)A(x)X_r^2(x)dx \quad ....15.$$

and

$$K_r = \int_0^L E(x)I(x)[X_r''(x)]^2 dx \quad ....16.$$

Considering only the effects of changes in stiffness on the variation in eigenfrequencies, and assuming that the structure is damaged in the neighborhood x such that $x_i < x < x_i + \Delta x_i$

$$dK_r = \int_{X_i}^{X_i+\Delta X_i} d[E(x)I(x)][X_r''(x)]^2 dx$$

$$= d(EI)[X_r''(x_i)]^2 \Delta x_i \quad ....17.$$

The functions $X_r(x)$ will depend upon the governing boundary conditions of the problem. Thus if a discrete structure is modeled as an equivalent continuum and existing solutions for X_r(r = 1, 2, ..., n) are known, changes in modal frequencies can be estimated using Eqns. (22) and 13).

Direct Estimation of Location and Severity of Damage

The location and severity of the damage may also be estimated directly by comparing the predicted eigenfrequencies with the measured values. The decision on the location can be made on the basis of some optimization scheme. One such scheme is presented herein. By definition the damaged frequency for any mode is given by

$$_Fw_i^2 = {_Iw_i}^2 + dw_i^2 \quad .. \quad ....18.$$

where $_Iw_i$ and $_Fw_i$ are the actual initial and final frequencies of modes i, respectively and dw_i is the actual change in frequency of mode i. Using Eqns. (13) and (17), the change in frequency of mode i may be expressed by an equation of the form:

$$d\bar{w}_i^2 = f(dk, r_o, X_r) \quad ....19.$$

where the bar denotes the predictive value, dk denotes the level of damage, r_o denotes the location of damage, and X_i is the mode shape vector for mode i. Using Eqns. (18) and (19), the estimate of the eigenfrequencies associated with the damaged state are then given by

$$ {}_{F}\bar{w}_i^{\,2} = {}_{I}w_i^{\,2} + d\bar{w}_i^{\,2} \qquad \ldots.20. $$

From Eqns. (18)and (20), the error in the prediction of mode i becomes

$$ x_i = {}_{F}\bar{w}_i - {}_{F}w_i \qquad \ldots.21. $$

and, if N modes are considered, the mean and the variance associated with the prediction are given by

$$ \mu = \frac{1}{N}\sum_{i=1}^{N} x_i \qquad \ldots.22. $$

and

$$ \sigma^2 = \frac{1}{N}\sum_{i=1}^{N} x_i^{\,2} \qquad \ldots.23. $$

Finally, let +c and −c represent a preselected tolerance of the error to the right and left respectively and assume that the error follows a Gaussian distribution. Then, the combination of dk and r_o that gives the greatest probability for the error being within the designated tolerances, defines the most likely damage location and damage severity. Note that the probability that the error is within the range of tolerance $p(\underline{c} < x < \underline{c})$ is given by

$$ p(-c \le x \le c) = \frac{1}{\sigma\sqrt{2\pi}} \int_{-c}^{c} e^{-[(x-\mu)^2/2\sigma^2]}\, dx \qquad \ldots.24. $$

The main advantage of this formulation stems from the fact that both the standard deviation and the mean of the error are combined into a single unambiguous measure. Statistically, this approach can be viewed as a two-tailed test. Hence, the greater the probability that the error lies within the stated tolerances, the smaller the error between the predicted frequencies and the measured values. Furthermore, if two competing sets of data posses the same standard deviation, the data set with the smaller mean would yield the larger probability.

APPLICATION OF THE METHOD

In this section, the method outlined in the previous section is used to locate the damage and to estimate the magnitude of the same damage in a hypothetical periodic structure. The structure under consideration here has been selected primarily to demonstrate the method and not to represent an actual lattice structure. Finally, any discussion relative to experimental techniques or instrumentation is precluded at this stage in the development.

Problem Description and Analysis

The selected structure, a plane truss with ten panels and forty degrees of freedom, is shown in Fig. 1. All of the panels have identical geometric and elastic properties. All panel members have the same elastic and geometric properties which are listed in the accompanying Figure.

Four damaged cases are considered here. The first case corresponds to that in which the upper chord member in the third panel sustains a level of damage which is equivalent to a loss in stiffness of one hundredth the value of the undamaged stiffness. The three remaining cases correspond to losses in stiffness of magnitudes 0.1k, 0.5k, and 0.7k (where k is the undamaged value) of the same upper chord member in the third panel, respectively.

Five dynamic analyses were performed (one analysis for the undamaged structure and four analyses for the damaged cases), assuming a lumped-mass model and the resulting model shapes and eigenfrequencies were recorded. The objective of the ensuing analysis was to predict the location of damage and the magnitude of the damage using only the eigenfrequencies of the structures in the undamaged and damaged states, and the frequency variation predicted by the equivalent continuum model.

A simply-supported Bernoulli-Euler beam was selected to predict the location and severity and the damage in the truss structure. The bending eigenfrequencies (w_r), mode shape vectors (X_r) and modal masses (M_r), r = 1, 2, 3, 4, ..., are given by (See Ref. 3):

$$ w_r = (\pi r/L)^4 (EI/mA)^{\frac{1}{2}}, \qquad \ldots.25. $$

$$ X_r(x) = \mathrm{Sin}\, r\pi x/L, \qquad \ldots.26. $$

and

$$ M_r = mAL/2 \qquad \ldots.27. $$

In the expressions above, the origin for the beam coincides with the support at the first panel of the truss. The bending stiffness (EI) of the beam was determined using standard techniques from the area of discrete-continuum modelling (See Ref. 4) and is given by

$$ EI = Ah^2/2 \qquad \ldots.28. $$

where A is the area of the chord member and h is the depth of the truss. Using Eqns. (27) and (17), the variation in the frequency change becomes:

$$ dw_r^{\,2} = 2d(EI)(r\pi/L)^4\, \mathrm{Sin}^2 r\pi x'/(mAL/2) \qquad \ldots.29. $$

where

$$ x' = x/L \qquad \ldots.30. $$

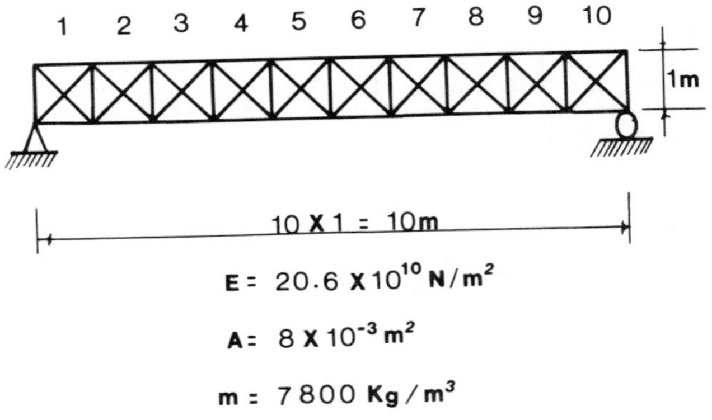

On observing that

$$dw_r^2 = 2w_r dw_r \qquad ....31.$$

the ratio of changes in frequencies for mode i and j, assuming the structure to be damaged at location x, is approximated by

$$dw_i/dw_j = i^2\int_x^{x+\Delta x}\sin^2 i\pi x'dx/j^2\int_x^{x+\Delta x}\sin^2 j\pi x'dx \qquad ....32.$$

The ratio of changes in eigenfrequencies as outlined in Eqn.(6) was used to locate the damage. For each damage case, only the first four bending modes were utilized. This restriction stems from the finding that the Euler-Bernoulli beam can predict only the first four modes with reasonable accuracy (See Ref. 5). The following ratios were computed for the actual structure and the beam model: dw_2/dw_1, dw_3/dw_1, dw_4/dw_1, dw_3/dw_2, dw_4/dw_2 and dw_4/dw_3. In addition, the proposed method presented in Eqns. (18) to (24) for estimating the goodness of fit of the predicted level of damage was also utilized. A computer program was written to facilitate the processing of the data.

RESULTS OF ANALYSIS

The results of the dynamic analyses are summarized in Table 1. Typical results comparing the theoretical ratios of frequency changes and those ratios obtained from Table 1 are presented in Tables 2-4. Typical results summarizing the most likely location and magnitude of the damage, in terms of a probability, are shown in Table 5. Finally in Table 6, the actual damage inflicted in each case is compared with the most probable prediction given in the Tables. Note that the loss in stiffness corresponds to the fractional loss in the second moment of area of the truss section which in turn is based on the assumption that the loss in stiffness is proportional to the loss in member area. Note also that because of symmetry about the midspan of the structure, only results for half the structure are listed.

TABLE 1 - EIGENFREQUENCIES OF UNDAMAGED AND DAMAGED STRUCTURE

STRUCTURE	DAMAGE LOCATION (PANEL)	DAMAGE MAGNITUDE (FRACTION OF STIFFNESS)	EIGENFREQUENCIES (RADIANS PER SECOND)			
			w_1	w_2	w_3	w_4
Truss	-	None	138.96	567.48	1,003.02	1,582.36
Beam	-	None	148.50	594.00	1,336.00	2,376.00
Truss	3	0.01	138.93	567.12	1,002.76	1,582.31
Truss	3	0.1	138.63	563.71	1,000.22	1,581.81
Truss	3	0.5	136.27	538.84	938.30	1,578.34
Truss	3	0.7	133.52	513.53	968.50	1,575.02

TABLE 2 - ACTUAL RATIOS OF CHANGES IN EIGENFREQUENCIES FOR THREE CASES

DAMAGE LOCATION (PANEL)	DAMAGE SEVERITY (FRACTION OF STIFFNESS)	VALUES OF RATIOS		
		dw_2/dw_1	dw_3/dw_1	dw_3/dw_2
3	0.01	12.00	8.67	0.72
3	0.1	10.16	8.42	0.83
3	0.5	10.66	7.34	0.69
3	0.7	9.92	6.35	0.64

TABLE 3 - RATIOS OF CHANGES IN EIGENFREQUENCIES FOR DAMAGED BEAM

DAMAGE LOCATION	COMPUTED RATIOS		
	dw_2/dw_1	dw_3/dw_1	dw_3/dw_2
1	30.16	207.36	6.88
2	24.45	116.26	4.76
3	15.48	27.00	1.74
4	6.53	3.14	0.48
5	1.01	20.99	20.88

TABLE 4 - ACTUAL VERSUS PREDICTED LOCATIONS BASED ON THE RATIOS OF CHANGES IN EIGENFREQUENCIES

ACTUAL DAMAGE LOCATION (PANEL)	DAMAGE SEVERITY (FRACTION OF STIFFNESS)	PREDICTED DAMAGE LOCATION FOR GIVEN RATIOS		
		dw_2/dw_1	dw_3/dw_1	dw_3/dw_2
3	0.005	3	4	4
3	0.05	4	4	4
3	0.25	4	4	4
3	0.35	4	4	4

TABLE 5 - PROBABLE LOCATION AND SEVERITY OF DAMAGE

PREDICTED SEVERITY OF DAMAGE (LOSS IN SECTION STIFFNESS)	LOCATION OF DAMAGE (PANEL)				
	1	2	3	4	5
Inflicted Damage = 0.005 EI					
0.001	0.73468	0.75865	0.77946	0.69762	0.79133
0.002	0.72776	0.52862	0.89381	0.52250	0.77761
0.003	0.64855	0.35568	0.95553	0.37535	0.64365
0.004	0.54885	0.26003	**0.95813**	0.28420	0.50482
0.005	0.46024	0.20308	0.90259	0.22628	0.40119
0.006	0.38960	0.16597	0.79566	0.18716	0.32806
0.007	0.33476	0.14012	0.67712	0.15921	0.27560
0.008	0.29197	0.12113	0.57385	0.13838	0.23676
0.009	0.25807	0.10660	0.49121	0.12228	0.20712
0.010	0.23081	0.09517	0.42636	0.10946	0.18387
Inflicted Damage = 0.05 EI					
0.010	0.69683	0.73310	0.73543	0.67065	0.75190
0.015	0.70862	0.64184	0.79554	0.60133	0.77351
0.020	0.69946	0.52725	0.85147	0.51529	0.75513
0.025	4.67237	0.43072	0.89735	0.43742	0.70498
0.030	0.63324	0.35756	0.92884	0.37397	0.63869
0.035	0.58829	0.30289	0.94491	0.32371	0.57003
0.040	0.54238	0.26142	**0.94634**	0.28388	0.50669
0.045	0.49848	0.22928	0.93318	0.25198	0.45140
0.050	0.45807	0.20382	0.90479	0.22605	0.40430
0.055	0.42167	4.18323	0.86235	0.20467	0.36449
Inflicted Damage = 0.25 EI					
4.175	0.71442	0.47621	0.90826	0.48216	0.74759
0.200	0.69007	0.41735	0.92958	0.43179	0.70545
0.225	0.66125	0.36881	0.94451	0.38837	0.65865
0.250	0.62982	0.32883	0.95330	0.35125	0.61101
0.275	0.59736	0.29571	**0.95638**	0.31954	0.56507
0.300	0.56507	0.26801	0.95387	0.29234	0.52225
0.325	0.53377	0.24436	0.94554	0.26889	0.48314
0.350	0.50396	0.22468	0.93097	0.24855	0.44785
0.375	0.47594	0.20751	0.91001	0.23077	0.41619
0.400	0.44982	0.19260	0.88307	0.21515	0.38786
Inflicted Damage = 0.35 EI					
0.400	0.65906	0.36831	0.93147	0.38802	0.65606
0.425	0.64218	0.34452	0.93783	0.36600	0.62993
0.450	0.62469	0.32314	0.94233	0.34582	0.60390
0.475	0.60684	0.30386	0.94500	0.32732	0.57836
0.500	0.58887	0.28645	**0.94586**	0.31035	0.55358
0.525	0.57096	0.27067	0.94491	0.29476	0.52976
0.550	0.55325	0.25632	0.94211	0.28042	0.50701
4.575	0.53586	0.24324	0.93743	0.26720	0.48540
0.600	0.51890	0.23128	0.93081	0.25498	0.46494
0.625	0.50241	0.22030	0.92222	0.24368	0.44562

TABLE 6 - PREDICTED VERSUS ACTUAL SEVERITY OF DAMAGE

LOCATION OF DAMAGE (PANEL)	PREDICTED SEVERITY	ACTUAL SEVERITY	ERROR
3	0.004	0.005	0.20
3	0.040	0.050	0.20
3	0.275	0.250	0.10
3	0.500	0.350	0.43

Discussion of Results

The results of the dynamic analysis (Table 1) show several expected trends. As shown, the damage is always accompanied by a decrease in modal frequency. The magnitude of the frequency decrease increases with the severity of the damage. Finally, the frequency variation in a given mode is clearly influenced by the location of the damage.

The first method of locating the damage, i.e. the method based on the ratio of changes in the eigenfrequencies (See Table 2), incorrectly predicts the location of the damage in all cases (See Table 3). Note that when the damage was in the third panel all ratios predicted the location to be in the fourth panel. Note also that this negative result, as far as the predictive capability of the ratios is concerned, does not contradict the findings in Ref. 2 or the theoretical development presented in the earlier portion of this paper. In both of the above instances, it has been assumed that the eigenfrequencies, mode shapes, modal masses and modal stiffness were derived from the same structure. Here the mode shapes, modal masses and modal stiffness have been developed on the basis of an equivalent continuum which approximated the behavior of the actual structure.

The second method used here to evaluate the damaged structure, i.e. the method based on the minimization of the error between the predicted eigenfrequencies of the damaged structure and those of the actual structure, correctly predicts the location of the damage in all cases. Furthermore, this method also accurately predicts the order of magnitude of the severity of the damage. For the cases considered in this work, the error in the prediction of the severity of the damage ranged from ten to approximately forty percent.

SUMMARY AND CONCLUSIONS

Two potential methods of evaluating damage in a discrete periodic structure have been presented herein. The first method, which is based on the hypothesis that the location of damage is only a function of the ratio of the changes in any two eigenfrequencies and in which the changes in eigenfrequencies are computed using an equivalent continuum to model the discrete structure, fails to predict unambiguously the location of the damage. The second method, which is based on the direct minimization of the error between the eigenfrequencies of the actual damaged structure and those eigenfrequencies predicted using the same continuum model discussed in the first model, not only successful predicts the location of damage in all cases but also accurately estimates the order of magnitude of the severity of the damage.

Although the results obtained here using the second method appear convincing, several additional avenues must be investigated before the method is endorsed. First the assumption that the error in the prediction follows a normal distribution must be further investigated. Obviously, the shape of the distribution affects the area under the curve for a given level of tolerance. Secondly, if as few as four modes are to be used to predict the damage, then the statistical analysis must be adjusted to account for small samples. This adjustment was not considered herein. Finally, the capability to predict the location and severity of the damage at other locations must be confirmed and other vibrational modes should be introduced into the analysis.

ACKNOWLEDGMENTS

The author acknowledges with gratitude the financial support of LeRoy Callender P.C., Consulting Engineers New York, Project 1685, which made this work possible. The assistance of Darrell Rials who developed the computer programs is also greatly appreciated. Finally, the author appreciably acknowledges Carolyn Thompson for her assiduous effort in typing the manuscript.

REFERENCES

1. T.K. O'BRIEN, Stiffness Change as a Non-destructive Damage Measurement, Mechanics of Non-Destructive Testing, ed. W.W. Stinchcomb, Plenum Press, New York, 1980, pp. 101-121.

2. P. CAWLEY and R.D. ADAMS, The Location of Defects in Structures from Measurements of Natural Frequencies, Journal of Strain Analysis, Vol. 14, No. 2, 1979 pp. 49-57.

3. R.R. CRAIG, Structural Dynamics: An Introduction to Computer Methods, John Wiley and Sons, New York, 1981.

4. N. STUBBS and H. FLUSS, Continuum Modelling of Discrete Structures, Recent Advances in Engineering Mechanics and Their Impact on Civil Engineering Practice, Vol. 1, eds. W.F. Chen and A.D.M. Lewis, ASCE, 1983, pp. 475-478.

5. S. ABRATE and C.T. SUN, Wave Propagation in Plane Trusses and Frames, Recent Advances in Engineering Mechanics and Their Impact on Civil Engineering Practice, Vol. 1, eds. W.F. Chen and A.D.M. Lewis, ASCE 1983, pp. 351-354.

FORMULAS FOR DOUBLY CANTILEVERED PLATES

Donald L. DEAN* and Mohammed E. TAWFIK**

* Owner, Pyramid Electronics, Sarasota, Florida USA

** Assistant Professor, Al-Azhur University, Cairo, Egypt

Using double sinusoidal series plus boundary functions, exact and rapidly converging formulas are derived and numerically illustrated for the elastic analysis of rectangular plates which are: a) simply supported along two adjacent edges, b) free along the other two adjacent edges, c) arbitrarily loaded on the interior and along the free edges and d) modeled according to classical fourth order plate theory. The practical design formulas can be routinely modified to formulas for more complex two-dimennsional structures with unnatural boundary conditions such as: a) plates modeled by higher order theories, b) discrete roof systems such as trussed plates and grids and c) mixed discrete-continuous structures such as ribbed or waffle plates.

INTRODUCTION

One of the most aesthetically dramatic space structures is the flat rectangular plate or grid which is loaded out-of-plane and simply supported along only two adjacent sides (see Fig. 1). These structures also present a challenge to the stress analyst who has only limited or classical training in multi-dimensional members or a straightforward Fourier analysis, which requires that two opposite sides be simply supported. A doubly cantilevered plate-like structure must be torsionally stable; thus, many types of trussed plates can not be used in this manner.

The object of this paper is two-fold. First, a relatively simple and rapidly converging double series solution will be presented for the exact analysis of an elastic homogeneous plate which is: a) simply supported along two adjacent sides, b) free on the other two adjacent sides, c) arbitrarily loaded on the interior and along the free edges and d) modeled according to classical fourth order plate theory (5). Second, using the classical plate model as an object problem, the authors hope to demonstrate how, with the addition of simple boundary functions, double sinusoidal series can be used to derive practical design formulas for more complex two-dimensiional structures with unnatural boundary conditions, such as: a) elastic plates modeled by higher order theories, including Reissner's sixth order theory (4), b) discrete systems, including grids and trussed plates and c) mixed discrete-continuous structures, including ribbed or waffle plates.

The techniques used here to extend use of Fourier series to cover unnatural boundary conditions, i.e., conditions not satisfied by each term of the series, are not fundamentally new. However, they are not covered in classical texts on the subject and, due in part to the current preoccupation with empirical and/or numerical procedures in our profession, seem in danger of being lost to practioners. Also, the authors were unable to find in the literature an analysis of the object problem which converged well enough to consitute a practical solution. For example, in 1953 Fletcher and Thorne (3) presented plate solutions which, in theory, covered all possible combinations of boundary conditions on the four sides of a rectangular panel with arbitrary interior load distributions. In 1963 Wilson (6) presented a solution to the doubly cantilevered and propped corner plate with a uniform interior load. While both references use solution series which theoretically converge to exact answers, their failure to extract the first degree or rigid body terms resulted in series which converge too slowly for practical design office use. That deficiency will be corrected here to, in effect, remove the classical limitations as to boundary conditions on the use of Fourier series for the analysis of structures. The resulting formulas are simple enough and the algorithm required to find the unknown coefficients of the boundary displacements (the only part of the solution not given in closed form) converge so well that programmable 'pocket' calculators or inexpensive personal micro computers serve as more than adequate computational aids for design and optimization studies.

SOLUTIONS IN TERMS OF BOUNDARY DISPLACEMENTS

The final solution must satisfy the mathematical model on the interior, Eq. 1, plus the following boundary conditions: 1) zero deflection and normal moments along $x=0$ and $y=0$, 2) zero normal moments and given shear resultants (zero if $P(x)$ and $\overline{P}(y)$ are zero) along $x=a$ and $y=b$, and 3) given corner reaction (zero if P^c is zero) at $x,y = a,b$.

$$(D_x^2 + D_y^2)^2 w(x,y) = \frac{1}{D} q(x,y) \tag{1}$$

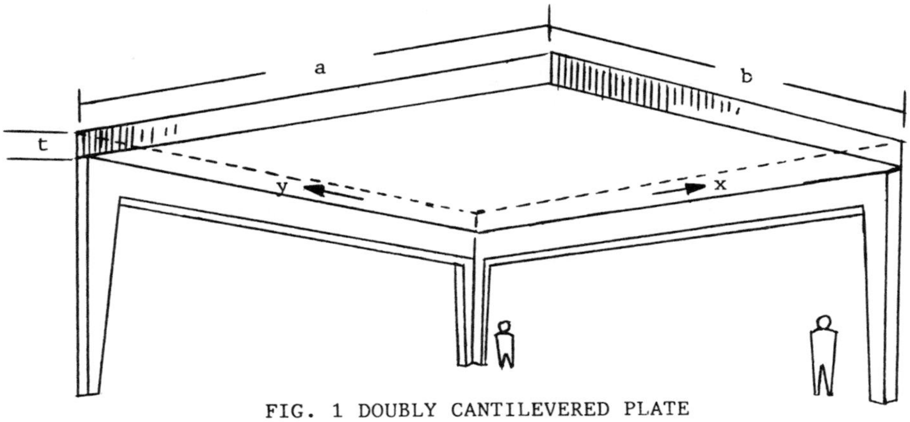

FIG. 1 DOUBLY CANTILEVERED PLATE

A double sine series satisfies condition 1) term by term, but not conditions 2) and 3). On the other hand, if we were given a solution which satisfied the above model exactly, one could routinely expand it into a double sine series which would fit the field exactly everywhere except along the boundaries x=a and y=b . Furthermore, if the 'rigid body' terms (actually two conoids and a hyperbolic paraboloid in this two-dimensional field) were extracted as separate algebraic or single series terms, the resulting series would converge rapidly. Thus, one only needs to find this rapidly converging double series plus boundary displacements directly for an exact solution which satisfies the equation and the natural as well as the unnatural boundary conditions.

To use a Fourier series solution for a mathematical model with one or more boundary conditions which cannot be satisfied in closed form by each term of the series, one adds to a particular solution, which has natural boundary conditions, a sufficient number of homogeneous solutions, which contain unknown boundary inhomogeneities. These inhomogeneities are then determined so as to force the solution to satisfy the unnatural boundary conditions. Here, these homogeneous functions will be written as simple double series plus boundary functions rather than the more complex and computationally sensitive single series 'Levy' solutions used in the cited references. That is:

$$w(x,y) = w^p(x,y) + w^h(x,y) + \bar{w}^h(x,y) + w^c(x,y) \qquad (2)$$

in which; 1) $w^p(x,y)$ is a particular solution to Eq. 1 for simple support conditions on all four edges;

2) $w^h(x,y)$ is a homogeneous solution to Eq. 1 for simple support conditions except for an arbitrary simply imposed

displacement along y=b (see Eq. 3a below); 3) $\bar{w}^h(x,y)$ is a homogeneous solution to Eq. 1 for simple supports

except along x=a (see Eq. 3b below); and 4) $w^c(x,y)$ is the well known pure torsion solution for the plate simply supported along x=0 and y=0 , free along x=a and y=b , but with a corner displacement imposed by a concentrated load at x,y = a,b.

$$w^h(x,b) = \sum_{i=1}^{\infty} W_i \sin \frac{i\pi x}{a} \qquad (3a)$$

$$\bar{w}^h(a,y) = \sum_{j=1}^{\infty} \bar{W}_j \sin \frac{j\pi y}{b} \qquad (3b)$$

$$W^c(x,y) = W^c \frac{xy}{ab} \qquad (3c)$$

The pure double series expression for Eq. 2 (not the best formula for computation due to slow convergence) is:

$$w(x,y) = \sum_{i=1}^{\infty} \sum_{j=1}^{\infty} (\overset{*}{q}_{ij}\overset{*}{B}_{ij} - \frac{2}{b}(-1)^i \bar{W}_j \overset{*}{\bar{C}}_{ij} + \frac{4W^c}{\alpha_i \beta_j}(-1)^{i+j})$$

$$(\sin \alpha_i x \sin \beta_j y) \qquad (4a)$$

in which $\overset{*}{q}_{ij}$ are the Fourier coefficients of the series for the normal interior load q(x,y) and:

$$\alpha_i = \frac{i\pi}{a}; \quad \beta_j = \frac{j\pi}{b}; \quad \overset{*}{B}_{ij} = \frac{1}{D} \frac{1}{(\alpha_i^2 + \beta_j^2)^2} \qquad (4b,c,d)$$

$$\overset{*}{C}_{ij} = \frac{\beta_j(\beta_j^2 + (2-\mu)\alpha_i^2)}{(\alpha_i^2 + \beta_j^2)^2} \qquad (4e)$$

$$\overset{*}{\bar{C}}_{ij} = \frac{\alpha_i(\alpha_i^2 + (2-\mu)\beta_j^2)}{(\alpha_i^2 + \beta_j^2)^2} \qquad (4f)$$

As alluded to above, a superior computational formula results by extracting the 'rigid body' or first degree terms out of Eq. 4a. That is, the fourth term or pure torsional component is replaced by the algebraic expression given by Eq. 3c and the second and third terms are written as double series plus a partial boundary function single series:

$$w^h(x,y) = \sum_{i=1}^{\infty} W_i (\frac{y}{b} - \frac{2}{b} \sum_{j=1}^{\infty} (\overset{*}{C}_{ij} - \frac{1}{\beta_j})(-1)^j \sin \beta_j y) \sin \alpha_i x \qquad (5a)$$

$$\bar{w}^h(x,y) = \sum_{j=1}^{\infty} \bar{W}_j (\frac{x}{a} - \frac{2}{a} \sum_{i=1}^{\infty} (\overset{*}{\bar{C}}_{ij} - \frac{1}{\alpha_i})(-1)^i \sin \alpha_i x) \sin \beta_j y \qquad (5b)$$

For those readers interested in duplicating the derivations in detail, it should be noted that while Eqs. 5a and 5b satisfy the respective boundary conditions at y=b and x=a as to deflection, they do not produce the desired zero normal moments. The single series portions contain only part of the boundary function used in the derivation. For example, the complete boundary function used to find $\overset{*}{\bar{C}}_{ij}$ (1,2) is as follows:

$$w^h(x,y) = \sum_{i=1}^{\infty} W_i\left(\frac{y}{b} - \mu\alpha_i^2\,\frac{y}{b}(b-y)\left(1+\frac{y}{b}\right)\sin\alpha_i x\right.$$

$$\left. - \frac{2}{b}\sum_{i=1}^{\infty}\sum_{j=1}^{\infty} W_i(\overset{*}{C}_{ij} - \overset{*}{E}_{ij})(-1)^j \sin\alpha_i x \sin\beta_j y\right) \qquad (6a)$$

$$\overset{*}{E}_{ij} = \frac{1}{\beta_j} - \frac{\mu\alpha_i^2}{\beta_j^3} \qquad (6b)$$

Equation 6a could be used in lieu of Eq. 5a to produce zero values for $M_y(x,b)$ but is otherwise inferior to the simpler formula for computations.

SOLUTION FOR BOUNDARY DISPLACEMENTS

The solutions given by Eqs. 3, 4 and 5 contained the unknown terms W_i, $\overline{W}_j$ and W^c which must be determined to complete the solution for the deflection and (through derivatives in accordance with standard formulas) stress resultant fields for a doubly cantilevered plate. As the torsional term, Eq. 3c, produces no edge shear resultants, we can first solve for W_i and $\overline{W}_j$ simultaneously and then determine W^c independently. Of course, for the case of a corner propped doubly cantilevered plate, $W^c = 0$ and thus finding W_i and $\overline{W}_j$ completes the solution. W_i and $\overline{W}_j$ are found by summing the shear resultants from w^p, w^h and $\overline{w}^h$ to zero, or in the case of edge line loads, to $p(x)$ and $\overline{p}(y)$, which are assumed to be given as the Euler coefficients of their respective series., i.e.,

$$p(x) = \sum_{i=1}^{\infty} P_i \sin\alpha_i x \; ; \qquad \overline{p}(y) = \sum_{j=1}^{\infty} \overline{P}_j \sin\beta_j y \qquad (7a,b)$$

For example, the boundary shear resultants along y=b due to w^p, w^h and $\overline{w}^h$, respectively, are:

$$V_y^p(x) = \sum_{i=1}^{\infty}\sum_{j=1}^{\infty} \overset{*}{q}_{ij}\,\overset{*}{C}_{ij}(-1)^j \sin\alpha_i x \qquad (8a)$$

$$V_y^h(x) = \sum_{i=1}^{\infty} S_i^h W_i \sin\alpha_i x \qquad (8b)$$

$$V_y^h(x) = -\frac{2}{a}\sum_{i=1}^{\infty}\sum_{j=1}^{\infty} \overset{*}{S}_{ij}\, W_j(-1)^{i+j} \sin\alpha_i x \qquad (8c)$$

in which $\overset{*}{S}_{ij}$ and its sum, S_i^h, see Ref. 1, are given by:

$$\overset{*}{S}_{ij} = (1-\mu)^2 D \frac{\alpha_i^3\beta_j^3}{(\alpha_i^2+\beta_j^2)^2} \qquad (9a)$$

$$S_i^h = \frac{1-\mu}{4} D\alpha_i^3\left(\frac{(3+\mu)\sinh\alpha_i b + 2(1-\mu)\alpha_i b}{\sinh^2\alpha_i b}\right) \qquad (9b)$$

The corresponding shear coefficients, $\overline{S}_j^h$, follow from Eq. 9b by induction, i.e., replace b by a and α_i by β_j.

Now, setting the shear resultants, $V_y(x,b)$ and $V_x(a,y)$, equal to the imposed edge loadings, gives the following equations to be solved for W_i and $\overline{W}_j$:

$$S_i^h W_i - \frac{2}{a}(-1)^i \sum_{j=1}^{\infty} \overset{*}{S}_{ij}\overline{W}_j(-1)^j = -P_i - \sum_{j=1}^{\infty} \overset{*}{q}_{ij}\overset{*}{C}_{ij}(-1)^j \qquad (10a)$$

$$-\frac{2}{b}(-1)^j \sum_{i=1}^{\infty} \overset{*}{S}_{ij} W_i(-1)^i + \overline{S}_j^h \overline{W}_j = -\overline{P}_j - \sum_{i=1}^{\infty} \overset{*}{q}_{ij}\overset{*}{\overline{C}}_{ij}(-1)^i \qquad (10b)$$

Having determined W_i and $\overline{W}_j$ from Eqs. 10a,b, W^c can then be found by setting $R(a,b) = P^c$, i.e.,

$$2(1-\mu)\, D\, D_x D_y\, w(x,y)\Big|_{(a,b)} = P^c \qquad (11a)$$

$$W^c = \frac{a\,b\,P^c}{2(1-\mu)D} - b\sum_{j=1}^{\infty} \overline{W}_j\beta_j(-1)^j - a\sum_{i=1}^{\infty} W_i\alpha_i(-1)^i$$

$$-\sum_{i=1}^{\infty}\sum_{j=1}^{\infty}\alpha_i\beta_j\left(ab\overset{*}{q}_{ij}\overset{*}{B}_{ij}(-1)^{i+j} - 2aW_i\left(\overset{*}{C}_{ij} - \frac{1}{\beta_j}\right)(-1)^i\right.$$

$$\left. - 2b\overline{W}_j\left(\overset{*}{\overline{C}}_{ij} - \frac{1}{\alpha_i}\right)(-1)^j\right) \qquad (11b)$$

Unfortunately, a closed formula solution to the simultaneous summation equations for W_i and $\overline{W}_j$, Eqs. 10a,b, is not available at this time. A rapidly converging algorithm is available for numerical results and the step of finding W_i and $\overline{W}_j$ is the one and only open form procedure required to analyse a doubly cantilevered plate. (A similar step is required in the Fourier analysis of other and more complex structures with multiple unnatural boundary conditions.) This single open form step in an otherwise closed form analysis presents no major difficulty providing the basic solution has been written in a form which converges so rapidly that only a small set of W_i and $\overline{W}_j$ are needed. For example, the pure double series expressions for the first degree terms in Eqs. 3c, 5a and 5b, see Eq. 4a, require the use of hundreds of terms to yield the accuracy given by i,j = 1,(1),5 in the preferred mixed series-algebraic forms.

Equations 10a,b are well conditioned for an iterative solution which converges rapidly. It seems that any reasonable algorithm produces satisfactory results. Once an initial set of values is determined, alternate back substitutions of $\overline{W}_j$ into Eq. 10a for improved values of W_i and substitutions of W_i into Eq. 10b for improved values of $\overline{W}_j$, etc. produce exceedingly accurate results after only 2 or 3 cycles. One simple approach to getting initial values, $W_i^{(0)}$ and $\overline{W}_j^{(0)}$, is to truncate both sums to a single term and solve Eqs. 10a and 10b as algebraic equations. This works satisfactorily. However, the authors used the following approach: Use Eq. 10b to eliminate $\overline{W}_j$ from Eq. 10a and truncate only the series on W_i to find a very good initial value for the dominate term in W_i, say $W_1^{(0)}$. This is then substituted into Eq. 10b to find all $\overline{W}_j^{(0)}$, which, in turn, are back substituted into Eq. 10a to find the remaining $W_i^{(0)}$ to complete the initial cycle. Having found W_i and $\overline{W}_j$ to the desired accuracy, all other plate deflection and stress quantities can then be found in closed form by use of Eqs. 3, 4, and 5 and the standard formulas for stress resultants.

NUMERICAL EXAMPLES

As a first numerical example to illustrate the above we will consider analysis of a square doubly cantilevered plate of reinforced concrete with a harmonic interior load equal to the first term in the expansion of a uniform load equal to 100 psf (4790 Pa.). In contrast to structures with natural boundary conditions, here a harmonic load will not produce a harmonic response. Thus, a single loading will normally result in a full series for the deflection field. The input data are as follows: $a = b = 240$ in. (6.1 m); $t = 8$ in. (203 mm); $\mu = 0.2$; $E = 2{,}000$ ksi (13.8×10^6 kPa); $P_i = \overline{P}_j = 0$; $\overset{*}{q}_{11} = 1.1 \times 10^{-3}$ ksi (7.6 kPa); and all other $\overset{*}{q}_{ij} = 0$. Some of the representatiave intermediate results are as follows:

$\overset{*}{C}_{11} = \overset{*}{\overline{C}}_{11} = 53.476$ in. (1.36 m) from Eqs. 4e,f;

$\overset{*}{S}_{11} = 2.4369$ k/in. (427 kN/m) from Eq. 9a;

$S_1^h = \overline{S}_1^h = 0.25765$ in. (1775 kPa) from Eq. 9b;

$W_1^{(0)} = 0.24825$ in. (6.3 mm) by truncating the sum on W_i in Eq. 10b (For comparison, the result by truncating both sums on W_i and $\overline{W}_j$ in Eqs. 10a,b is $W_1^{(0)} = 0.24784$ in.); and $\overline{W}_1^{(0)} = 0.24787$ in., $\overline{W}_2^{(0)} = -3.1608 \times 10^{-3}$ in., $\overline{W}_3^{(0)} = 7.9022 \times 10^{-4}$ in., $\overline{W}_4^{(0)} = -2.7343 \times 10^{-4}$ in. and $\overline{W}_5^{(0)} = 1.1690 \times 10^{-4}$ in. from Eq. 10b.

The results from the 3rd and final cycle are: $W_1 = \overline{W}_1 = 0.24832$ in., $W_2 = \overline{W}_2 = -3.3629 \times 10^{-3}$ in., $W_3 = \overline{W}_3 = 8.7957 \times 10^{-4}$ in., $W_4 = \overline{W}_4 = -3.1662 \times 10^{-4}$ in. $W_5 = \overline{W}_5 = 1.397 \times 10^{-4}$ in. from cyclic solutions to Eqs. 10a,b. These results then allow the 3rd unknown to be calculated as: $W^c = 2.1468$ in. (54.5 mm) from Eq. 11.

The above intermediate results for W_i, $\overline{W}_j$ and W^c are substituted into Eqs. 2, 3 and 5 to yield a closed formula for the deflection field, which, in turn may be operated on in accordance with standard formulae to yield closed formulas for the various stress resultants. Sample results for the deflection field are:

$w(\frac{a}{4},\frac{b}{4}) = 0.2338$ in. (5.94 mm); $w(\frac{a}{2},\frac{b}{2}) = 0.7980$ in. (20.27 mm); $w(\frac{a}{2},b) = w(a,\frac{b}{2}) = 1.3210$ in. (33.55 mm); $w(\frac{3a}{4},\frac{3b}{4}) = 1.4697$ in. (37.32 mm); $w(\frac{3a}{4},b) = w(a,\frac{3b}{4}) =$ $= 1.7896$ in. (45.46 mm); and $w(a,b) = W^c = 2.1469$ in. (54.5 mm).

For a second numerical example, consider a change in the data to that of a rectangular doubly cantilevered plate with a uniform interior load equal to 100 psf (4790 Pa). The change in data from the first example are: $b = 360$ in. (9.1 m); $t = 10$ in. (254 mm); $\overset{*}{q}_{11} = 1.1 \times 10^{-3}$ ksi (7.6 kPa), $\overset{*}{q}_{13} = 3.75 \times 10^{-4}$ ksi (2.6 kPa), plus nonzero values for all other $\overset{*}{q}_{ij}$ for both i and j odd. Some of the representative intermediate results are: $\overset{*}{C}_{11} = 54.787$ in. (1.39 m); $\overset{*}{\overline{C}}_{11} = 65.907$ in. (1.67 m); $\overset{*}{S}_{11} = 2.704$ k/in. (474 kN/m); $S_1^h = 0.4987$ ksi (3436 kPa); $\overline{S}_1^h = 0.1571$ ksi (1082 kPa); $\overline{W}_1^{(0)} = 0.5723$ in., $\overline{W}_2^{(0)} = -5.147 \times 10^{-3}$ in., $\overline{W}_3^{(0)} = 4.913 \times 10^{-3}$ in., $\overline{W}_4^{(0)} = -6.045 \times 10^{-4}$ in., and $\overline{W}_5^{(0)} = 5.035 \times 10^{-4}$ in.

Again, the 3rd cycle yields the final results:

$W_1 = 0.18749$ in., $\overline{W}_1 = 0.57265$ in.,
$W_2 = -3.12 \times 10^{-3}$ in., $\overline{W}_2 = -5.39 \times 10^{-3}$ in.,
$W_3 = 1.28 \times 10^{-3}$ in., $\overline{W}_3 = 5.05 \times 10^{-3}$ in.,
$W_4 = -2.61 \times 10^{-4}$ in., $\overline{W}_4 = -6.86 \times 10^{-4}$ in.,
$W_5 = 1.45 \times 10^{-4}$ in., $\overline{W}_5 = 5.53 \times 10^{-4}$ in. and
$W^c = 3.3850$ in. (86 mm).

Use of W_i, $\overline{W}_j$ and W^c in the deflection field equations yields the following results at the quarter points:

TABLE 1.--Plate Deflections for Doubly Cantilevered Plate

$(w(x,y)/t)$

y \ x	a/4	a/2	3a/4	a
b/4	.0353	.0678	.0971	.1249
b/2	.0632	.1219	.1753	.2261
3b/4	.0828	.1600	.2297	.2952
b	.0977	.1879	.2675	.3385

CONCLUSIONS

Design formulas are now available for optimization studies of doubly cantilevered rectangular plates with free or propped corner subjected to arbitrary loadings. They are exact, consistant with classical plate theory, but require only a low powered computational aid, such as a programmable calculator or a low cost personal micro-computer. The techniques used to circumvent the usual boundary condition limitations on the use of Fourier series, i.e., adding homogeneous solutions in the form of a series plus a partial or full boundary function, can also be routinely applied to more complex structures which have continuous, discrete or mixed discrete-continuous mathematical models.

ACKNOWLEDGEMENTS

The work leading to this paper was done while both authors were affiliated with the College of Engineering, Illinois Institute of Technology whose support is gratefully acknowledged. Figure 1 was drawn by G. Kaiser, instructor, Ringling Art School.

APPENDIX I.--REFERENCES

1. D L Dean, Discrete Field Analysis of Structural Systems, CISM #203, Springer-Verlag, Wein, 1976.

2. D L Dean, and W W Payne, Analysis of Reinforced or Cellular Panels and Decks, J. Mech. Sci. Pergamon Press, 1979

3. H J Fletcher and C J Thorne, Bending of Thin Rectangular Plates, 2nd US Natl. Cong. Appl. Mech., 1954.

4. E Reissner, On the Theory of Bending of Elastic Plates, Jr. Math. and Physics, 1944.

5. S Timoshenko and S. Woinowsky-Krieger, Theory of Plates and Shells, McGraw-Hill, New York, 1959.

6. B Wilson, Bending of Rectangular Plates with Adjacent Free Edges, Trend in Engrg., Univ. of Wash., 1963.

APPENDIX II.--NOTATION

The following symbols are used in this paper:

$\overset{*}{B}_{ij}$, $\overset{*}{C}_{ij}$, $\overset{\overline{*}}{C}_{ij}$ = double series coefficients of particular and homogeneous plate solutions;

a,b = dimensions of plate;

D = flexural stiffness of plate;

D_x , D_y = derivitatives with respect to index coordinates;

E = Young's modulus;

$\overset{*}{E}_{ij}$ = series coefficient for full boundary function;

i,j = series indices;

P^c = load at free corner of plate;

P_i , $\overline{P}_j$ = series coefficients of edge loads;

p(x), $\overline{p}$(y) = edge loads on plate;

q(x,y) = surface loading normal to plate;

$\overset{*}{q}_{ij}$ = series coefficients of surface load;

$\overset{*}{S}_{ij}$, S^h_i, $\overline{S}^h_j$ = shear series coefficients;

t = plate thickness;

x,y = plate coordinates;

$V^p_y(x)$, $V^h_y(x)$, $\overline{V}^h_y(x)$ = boundary shear resultants;

W^c = deflection at free corner of plate;

W_i, $\overline{W}_j$ = series coefficients of plate edge deflections;

w(x,y) = plate deflection;

$w^p(x,y)$, $w^h(x,y)$, $\overline{w}^h(x,y)$, $w^c(x,y)$ = components of plate deflections;

α_i , β_j = series parameters, $i\pi/a$ and $j\pi/b$;

μ = Poisson's ratio;

π = standard ratio.

USE OF LINEAR PARAMETRIC ELEMENT IN ANALYSING SPACE STRUCTURES

R. DELPAK*
V. PESHKAM**

* Principal Lecturer in Civil Engineering,
The Polytechnic of Wales, Wales, U.K.

** Research Assistant in Civil Engineering
The Polytechnic of Wales, Wales, U.K.

A new element has been developed which manifests some additional features compared with the standard beam element. This formulation retains the axial, flexural and torsional responses as well as possessing non-nodal internal degrees of freedom of optional number. The characteristics of every element could be changed by altering the input data rather than the element formulation. The same input data could specify the element to be used as a bar, a beam or a grillage member both in planar and space modes. This does not involve any loss of efficiency in either time or storage. The benefit becomes apparent in eigenvalue type analysis of space structures since the need for subdividing members into smaller elements is alleviated. Therefore the entire analysis is rendered cheaper owing to significant curtailment in structural degrees of freedom.

INTRODUCTION

The slender elastic bar members, were among the first to be used in practical finite element analysis. Owing to simple structural geometry, it was possible to determine the element stiffness matrix exactly, see Ref 1. Used as a space member, the element enjoys its total possible number of degrees of freedom (d.o.f.). However, when deployed in a specific role, e.g. a grillage or a truss element, one or more d.o.f. have to be suppressed. The above simplicity has proved advantageous in the following main areas:

(a) it facilitates the teaching of the finite element method (FEM) to students and laymen alike, since the concepts of discrete elements assembled together to form (sub)structures etc., can be visulised with relative ease, and

(b) it enables the codification of selected structural problems on most microcomputers.

However, the classical beam element, also suffers from a number of practical drawbacks, the most prominent of which are given as follows,

(i) in an actual structural design, the engineers often seek the salient values of the bending moment along the beam, so that the correct member size could be chosen. With the values of generalised forces estimated at the nodes, a further set of calculations are needed to determine the bending moment variation, to complete the design processes,

(ii) there are occasions in structural engineering, when the accurate determination of the deformed structural geometry is likely to be of some significance. For instance, the near correct deformed geometry is needed to determine the structural response to applied dynamic loads. Since the polynomial representation of any member cannot be improved beyond cubic, the analyst is compelled to increase the number of elements to have more nodes available. The consequents are blatantly obvious, since the engineer is penalised in both the cost and the choice of the machine used.

The present formulation is designed not to side-step the above difficulties but to solve them using a direct approach. The principles involved have been used by the authors in previous developments and are well proven in both fundamentals and scope, see Refs 2 and 3.

ELEMENT FORMULATION

A. Definitions and geometry - The present element is a part of the isoparametric family. It has a linear parent element, the nodes 1 and 2 of which are at $\xi = -1$ and $\xi = +1$ respectively, see Fig 1. The values of both ϕ_i and the corresponding derivative $\left(\frac{d\phi}{d\xi}\right)_i$ are defined such that,

$$\begin{cases} \text{at node } i=1,\ \xi=-1 \\ \phi_i = \phi_1 \text{ and } \left(\frac{d\phi}{d\xi}\right)_i = \left(\frac{d\phi}{d\xi}\right)_1 \end{cases} \quad \text{and} \quad \begin{cases} \text{at node } i=2,\ \xi=+1 \\ \phi_i = \phi_2 \text{ and } \left(\frac{d\phi}{d\xi}\right)_i = \left(\frac{d\phi}{d\xi}\right)_2 \end{cases}$$

Elsewhere at $-1 < \xi < +1$ on the parent element, the $\phi(\xi)$ functional is interpolated in terms of the above values, using the normalised function B_1 - B_4, Fig 2. The interpolation is given explicitly as,

$$\phi(\xi) = \Big\{\phi_1 \cdot \tfrac{1}{4}(\xi^3 - 3\xi + 2) + \left(\tfrac{d\phi}{d\xi}\right)_1 \cdot \tfrac{1}{4}(1-\xi)^2(1+\xi) + \phi_2 \cdot \tfrac{1}{4}(-\xi^3 + 3\xi + 2) + \left(\tfrac{d\phi}{d\xi}\right)_2 \cdot \tfrac{1}{4}(\xi - 1)(1+\xi)^2\Big\} \qquad \text{...... 1(a),}$$

or in terms of shape functions as,

$$\phi(\xi) = \sum_{i=1}^{2} \left\{\phi_i \cdot N_i(\xi) + \left(\tfrac{d\phi}{d\xi}\right)_i \cdot N_i'(\xi)\right\} \qquad \text{...... 1(b),}$$

where

$$N_1(\xi) = B_1,\ N_1'(\xi) = B_2,\ N_2(\xi) = B_3 \text{ and } N_2'(\xi) = B_4.$$

Fig 1

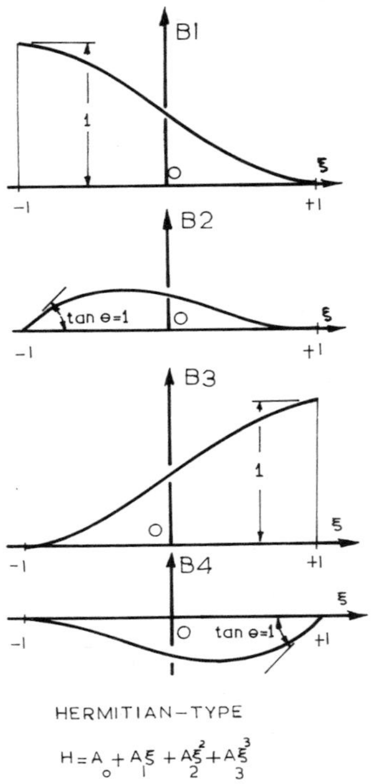

Fig 2

The slender prismatic element of Fig 3, has its origin at the mid-length. The points (ξ = -1, 0 and +1) can then be mapped linearly, into ($x = -\frac{L}{2}$, 0 and $+\frac{L}{2}$) where $dx = \frac{L}{2} \cdot d\xi$.

The geometry is hence defined to be,

$$x(\xi)=\sum_{i=1}^{2}\left\{x_i \cdot N_i(\xi) + \left(\frac{dx}{d\xi}\right)_i \cdot N_i'(\xi)\right\} \qquad \ldots\ldots 2.$$

The beam is assumed to possess all the usual following geometrical properties of area (A), the second moments of area (I_y, I_z) and the appropriate torsional parameter (J_x); y and z are the principal sectional axes.

B. Displacements - The element enjoys the availability of the following categories of displacements, the directions of which are indicated in Fig 4 and are all expressed in terms of ξ. The first category which has been used in previous formulations, see Ref 1, contains six d.o.f. per node and is redefined as follows.

(i) Linearly varying displacements along and about the x-axis, namely,

$$u(\xi)=\left\{u_1 \cdot \frac{1}{2}(1-\xi) + u_2 \cdot \frac{1}{2}(1+\xi)\right\} \qquad \ldots\ldots 3,$$

and

$$\theta(\xi)=\left\{\theta_1 \cdot \frac{1}{2}(1-\xi) + \theta_2 \cdot \frac{1}{2}(1+\xi)\right\} \qquad \ldots\ldots 4.$$

(ii) Cubically varying lateral deformations v and w, expressed in a unified form to incorporate the corresponding slope or rotation terms, so that,

$$v(\xi)=\sum_{i=1}^{2}\left\{v_i \cdot N_i(\xi) + (\theta_z)_i \cdot N_i'(\xi)\right\} \qquad \ldots\ldots 5,$$

and

$$w(\xi)=\sum_{i=1}^{2}\left\{w_i \cdot N_i(\xi) + (-\theta_y)_i \cdot N_i'(\xi)\right\} \qquad \ldots\ldots 6,$$

where $N_i(\xi)$ and $N_i'(\xi)$ are as defined in Eqn 1(a) and (b). Upon comparison of Eqns 1, 2, 5 and 6, it could be seen that the element is isoparametric and that all the desirable properties of such elements are, therefore, preserved intact, see Ref 4.

The second category of displacements are linked with internal or hierarchical d.o.f. These shape functions which are due to Irons could be generated to any practical arbitrary order and are called the surplus functions, see Ref 5. Typical plots indicating zeroth, first, ... and n-th order functions shown as S_o, S_1, ... and S_n are given in Fig 5. The surplus functions are used to represent the lateral displacements only, the specification of the order or the number of which, need not be the same for v and w. Indeed the choice of the types and the numbers used would depend on the manner and the direction of loading together with the general level of the accuracy required.

DETERMINATION OF ELEMENT MATRICES

A. Generation of $[K]^e$, $[M]^e$ and $\{P\}^e$ - The equations used to generate the various matrices, are taken from Zienkiewicz, see Ref 6. The relationships are quoted in their simplest form and the nomenclature of the above reference is retained where possible. The Gaussian integration procedure has been used throughout to form all the matrices which renders the displays of all the integral signs symbolic. The element stiffness and the mass matrices are estimated from,

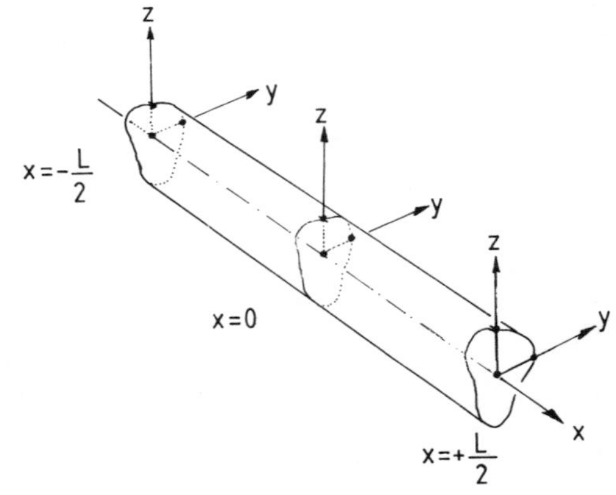

Fig 3

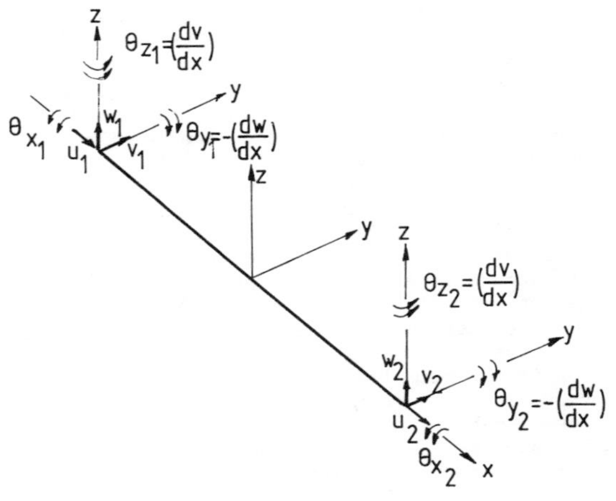

Fig 4

$$[K]^e = \int_v B^T D B \, dv \qquad \ldots\ldots 7,$$

and

$$[M]^e = \int_v N^T \rho N \, dv \qquad \ldots\ldots 8,$$

where symbols have their usual meaning. Neglecting the interal stresses and the self-weight, the load vector is given by,

$$\{P\}^e = \int_x N^T p(x) \, dx \qquad \ldots\ldots 9.$$

A maximum of linear variation is allowed for p(x) above, although theoretically any variation could be incorporated.

The pre-final size of the element matrices might require some clarification highlighted as follows. Assume that the engineer requires functions S_o, S_1, S_2 and S_3 for v-displacements and S_1, S_3 and S_5 for w-displacements.

The number of additional d.o.f. are 4 + 3 = 7, thus the total number of d.o.f. for the element is 12 + (7) = 19. This makes $[K]^e$ and $[M]^e$ to be 19 x 19 matrices. The above matrices and the load vector are transformed into global axes immediately after generation.

B. Static Condensation - In order to modify the size of the element matrices, to correspond to the total number of d.o.f. available, use could be made of the static condensation method. This concept was initially discussed by Irons and Guyan and was subsequently extended by others, see Refs 7, 8, 9 and 10. The process simply consists of a systematic means of reducing the size of the matrices involved. The method is outlined as follows. Let suffices B and S denote the corresponding basic and surplus d.o.f. The equations are then partitioned as,

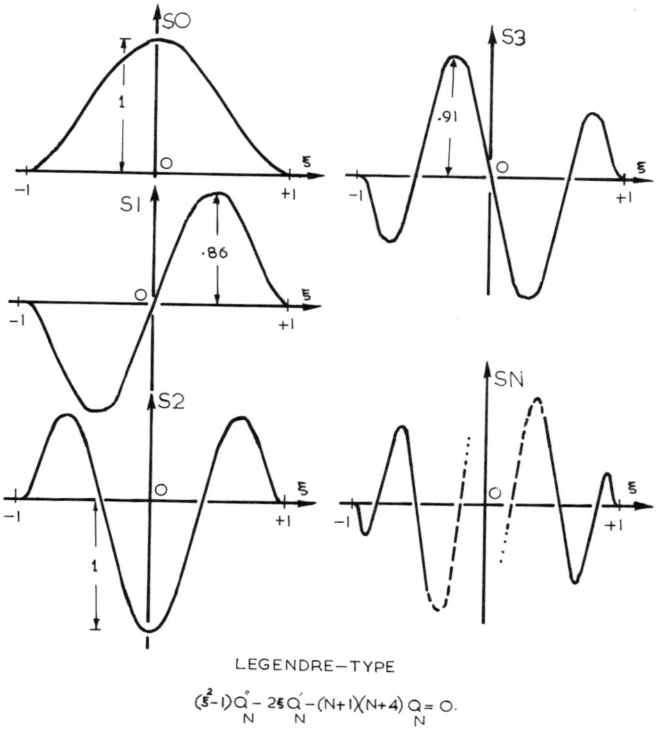

Fig 5

$$\begin{bmatrix} [K_{BB}] & [K_{BS}] \\ [K_{SB}] & [K_{SS}] \end{bmatrix} \cdot \begin{Bmatrix} \{a_B\} \\ \{a_S\} \end{Bmatrix} = \begin{Bmatrix} \{P_B\} \\ \{P_S\} \end{Bmatrix}$$

where d.o.f. corresponding to $\{a_S\}$ are to be eliminated. This gives,

$$\{a_S\} = [K_{SS}]^{-1} \cdot (\{P_S\} - [K_{SB}] \cdot \{a_B\}),$$

which upon substitution yields the condensed set of equations,

$$[K_{BB}]' \cdot \{a_B\}' = \{P_B\}',$$

where,

$$[K_{BB}]' = [K_{BB}] - [K_{BS}] \cdot [K_{SS}]^{-1} \cdot [K_{SB}],$$

and,

$$\{P_B\}' = \{P_B\} - [K_{BS}] \cdot [K_{SS}]^{-1} \cdot \{P_S\}.$$

An equivalent method is deployed for processing the stiffness and the mass matrices when dealing with the dynamic problems.

C. Assembly and housekeeping - The practising and the design engineers are often faced with the problems of neither having access to large machines nor affording overt CPU usage. The present programs have been developed with the above constraints in mind. In order to process the structural problems successfully on the small or the medium sized computers, special solution techniques have been devised. The structural analyst will have the following options.

(i) Apply the condensation process of the last section to

(a) static and dynamic problems, and
(b) pre and post assembled structural matrices.

Programs of standard capability could then be applied to the assembled equations, to solve either for static displacements or for eigenvalues and vectors.

(ii) Retain the surplus d.o.f. of each element for simplicity and accommodate them in the solution routine. An efficient program has been developed for this specific purpose, the modus operandi of which is depicted in Fig 6. As implied, only the nodal d.o.f. are appropriately arranged to assemble. The surplus d.o.f. are tolerated as 'free-standing'. Detailed algorithm of the latter technique is discussed in Ref 10.

RESULTS

The numerical examples presented are chosen in such a way as to display the elemental capabilities in the correct light and yet avoid encumbering the reader. Therefore, the inclusion of those structures with either a sizeable collection of members or a large number of d.o.f. have been avoided. The examples contain some features of both the dynamic and the static analyses. It is felt that there was little merit in citing any structure with truss-type action, since improvement has not been channelled into axial displacements. There has been deliberate self-discipline, to maintain the number of d.o.f. to a level capable of being managed by the smaller of microcomputers.

1. Natural frequencies of an encastré beam - The structure is modelled by one element only. The analysis with the conventional formulation ceases, when all four nodal d.o.f. are suppressed. Fig 7 is a convergence plot of the angular frequency ω against the total number of d.o.f. The graph is prepared on a semi-log scale, to accommodate the third and fourth natural frequencies. As can be seen, the convergence is almost instantaneous even at higher modes. The beam studied had the following geometrical and mechanical properties.

$L = 1.0$ m, $A = 0.12$ m^2 $\rho = 0.768 \times 10^4$ kg/m^3
$I_z = 0.4 \times 10^{-3}$ m^4 $E = 0.2 \times 10^{12}$ N/m^2

2. Natural frequencies of a cantilever - The structural parameters used in this case are identical to those of the previous example. The conventional beam element should be incapable of providing the second mode with one element representation. The plot is again carried out on a semi-log scale. The horizontal axis represents the number of S-functions specified in a single element. The rapid convergence of the frequencies to the correct values can be seen in Fig 8.

3. Static analysis of a loaded encastré beam - The intention here was to identify the structural response to (a) uniformly distributed load, and (b) linearly graded load. The S-functions were used sparingly as shown in Fig 9(a) and (b). The results for displacements and moments in both cases, are identical with the theoretical values. A single element representation, cannot yield any results owing to imposition of boundary conditions. The element properties used are as given in example 1.

4. Static analysis of a propped cantilever - This study is of particular interest, since the only available d.o.f. for the ordinary element, is a rotation term (in the form of either B2 or B4, Fig 2). The stationery point and the point of inflection are located invariably at $x=-L/6$ and $x=L/6$, respectively. Thus the deformed configuration of the beam is insensitive to any variation in load distribution. This would no doubt result in an incorrect variation of the bending moment and the shear force along the member. Thus the necessity will arise to subdivide the beam into many elements, in order to obtain an accuracy of a satisfactory order. Alternatively, the present formulation can respond to different variations of the distributed loads using a single element representation. The values of the displacement and the moment from exact and FE analyses are given in Fig 10(a) and (b). For purposes of comparison, the corresponding FE values from an ordinary beam element is also included. It can be seen that the present FE formulation yields exact solutions to both displacements and bending moments.

OBSERVATIONS AND CONCLUSIONS

From the above results, it can be seen that the following objectives have been fulfilled.

(a) Unlike the conventional beam element, the present formulation is capable of evaluating the displacements and the moments anywhere along the member (and naturally at the nodes as well).
(b) The added feature is thought to be cost beneficial, since the amount of additional formulation required is relatively small and quite simple to implement.
(c) The extra storage required for the initial generation of element matrices is negligible and at assembly level no excess storage is needed over the requirements of the usual element.
(d) The engineer is in control of the element characteristics, since the change in element properties is made at the data input level, making the need for retaining a library of various elements obsolete.
(e) Significant economies could be achieved by discarding the need for having to subdivide a member to create more nodes.

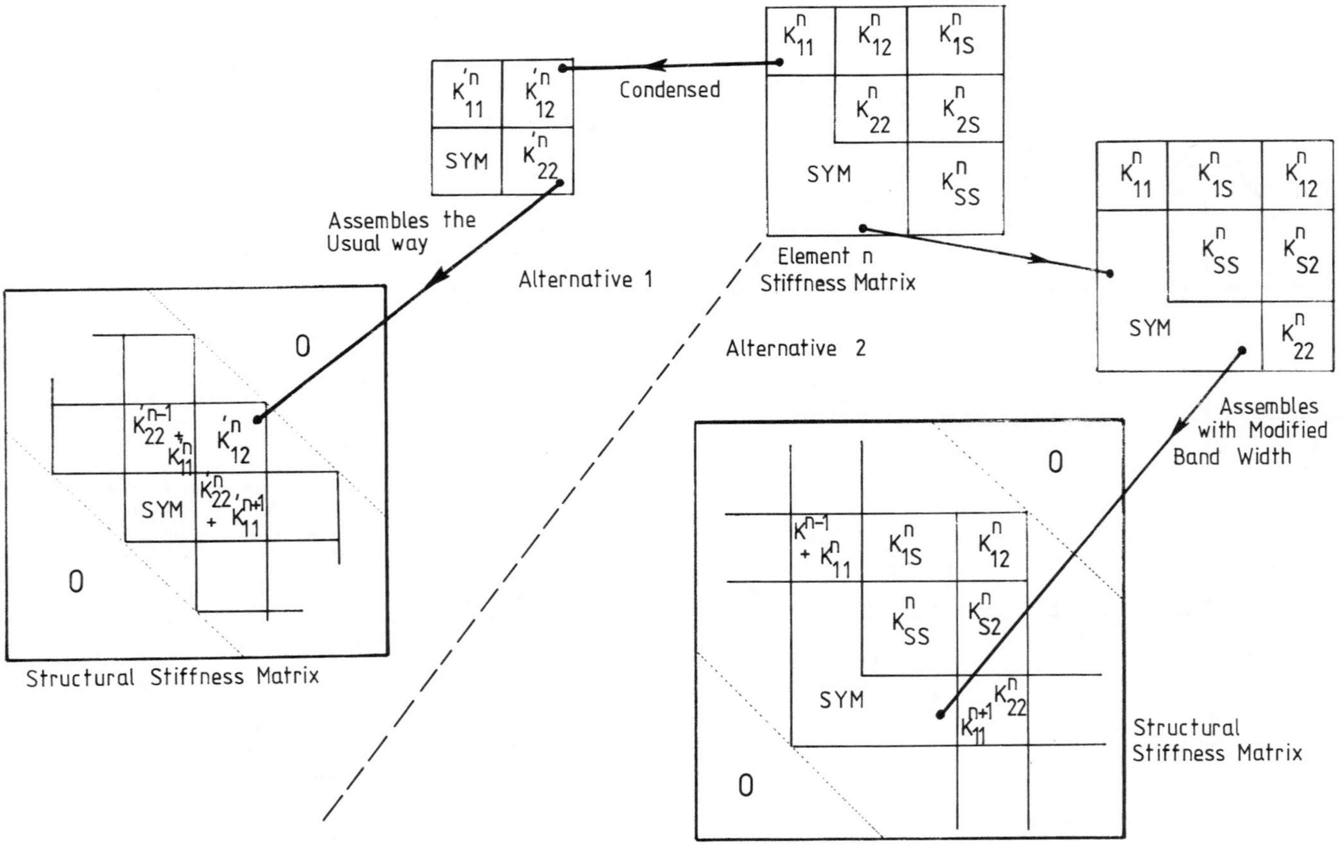

Fig 6

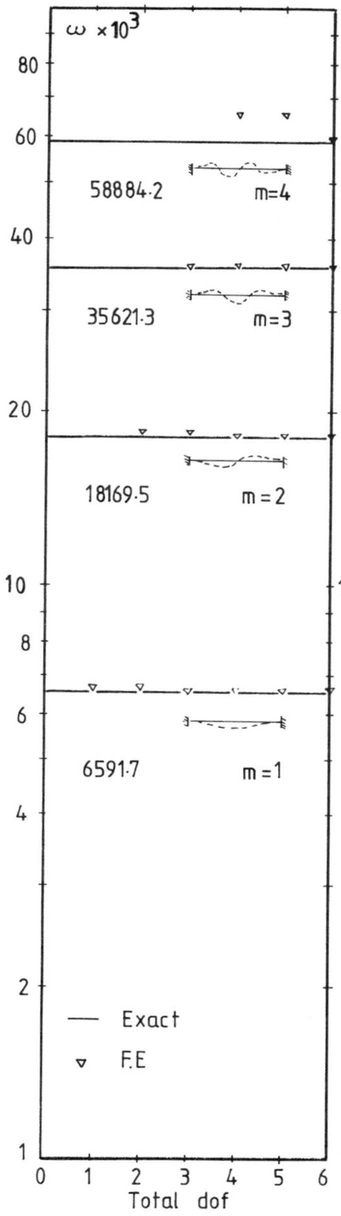

Fig 7

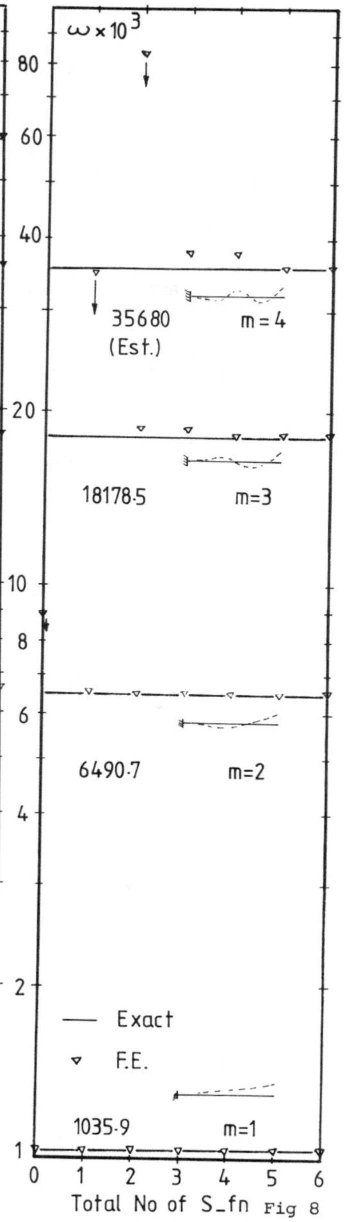

Fig 8

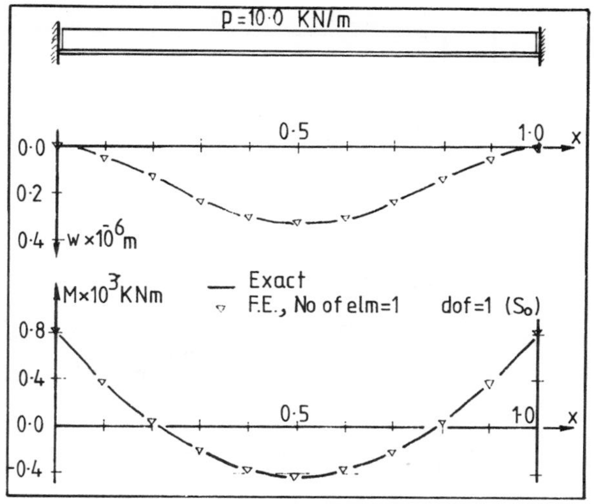

Fig 9a

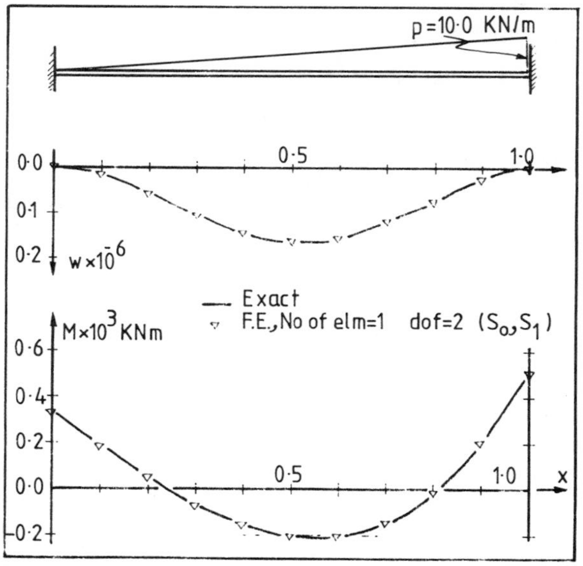

Fig 9b

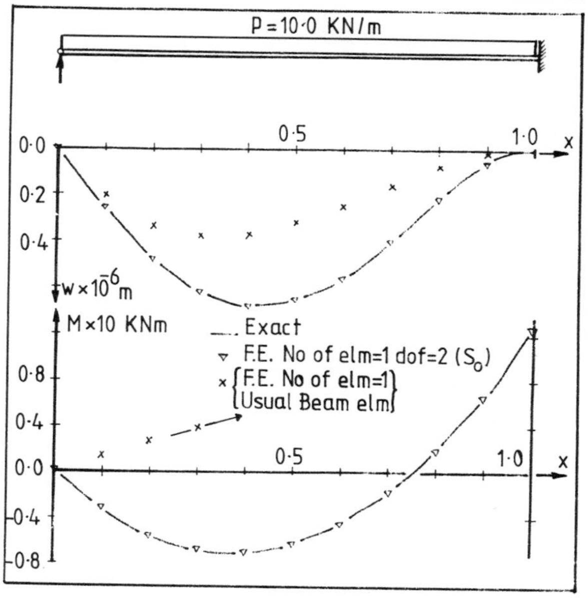

Fig 10a

(f) Selected and realistically sized problems could be run on most of the larger microcomputers owing to severe curtailment in the size of the storage required. This is due to
(i) element formulation, and
(ii) efficient solution programs.

ACKNOWLEDGEMENTS

The authors wish to thank Mr. R.D. McMurray, Head of Department of Civil Engineering and Building, in making his departmental facilities available for the present submission.
They are also grateful to Mrs. A. Davies for typing a difficult manuscript.

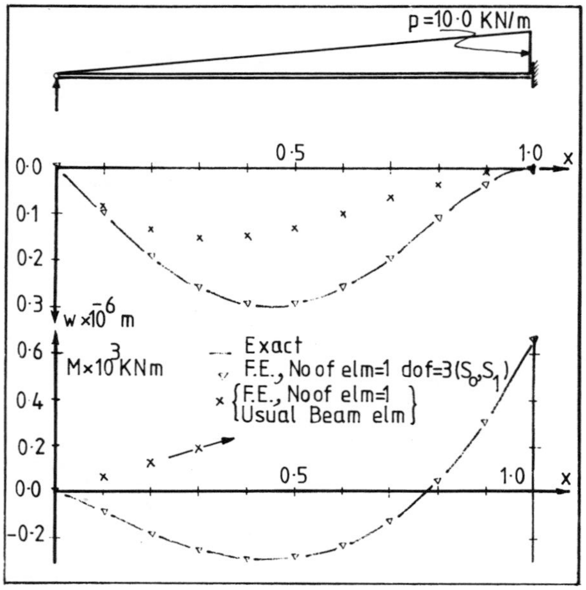

Fig 10b

REFERENCES

1. R K Livesley, Matrix Methods of Structural Analysis, Pergamon Press, Oxford, England 1964.

2. R Delpak, Axisymmetric Vibrations by the Curved Cone Element, Proc. Conf. Recent Advances in Stress Analysis, The Jnt. Brit. C'ttee for Struct. Anal., Roy. Aer. Soc. pp. 6-27, London, 1968.

3. R Delpak and V Peshkam, Vibration Analysis of Elastic Rotational Shells Using Microcomputers, Proc. of the 3rd Int. Conf. 'EngSoft III', Imp. Coll., pp. 481-494, London, April 1983.

4. B M Irons and A Razzaque, The Mathematical Foundations of the Finite Element Method with Applications to Partial Differential Equations, (Edited by A.K. Aziz), pp. 557-585, Academic Press, N.Y. and London, 1972.

5. B M Irons, Shape Functions for Elements with Point Conformity, A.S.M. 1204, Rolls-Royce Library.

6. O C Zienkiewicz, The Finite Element Method, McGraw-Hill, London and N.Y., 1977.

7. B M Irons, Eigenvalue Economisers in Vibration Problems, Journ. of the Roy. Aer. Soc., Vol. 67, pp. 526-528, London, 1963.

8. R J Guyan, Reduction of Stiffness and Mass Matrices, Jour. AIAA Vol. 3, No. 2, pp. 380, 1965.

9. V M Trbojevic and R R Kunnar and D C White, The Reduction of Degrees of Freedom in the Dynamic Analysis of Complex Three-Dimensional Structures, Proc. of Jnt. I.Struct.E.–B.R.E. Seminar on Dynamic Modelling of Structures, B.R.E. Herts, England, Nov. 1981

10. V Peshkam, Linear Static and Dynamic Analysis of Shells of Revolution, Dept. of Civ. Eng. Inter. Report, The Polytechnic of Wales, Wales, U.K., 1983.

EXACT FINITE ELEMENTS IN THE ANALYSIS OF SHELLS OF REVOLUTION

Zenon WASZCZYSZYN[x] and Janina PIECZARA[xx]

[x]Professor, Institute of Structural Mechanics,
Cracow Technical University, Poland.

[xx]Research-worker, Institute of Structural Mechanics,
Cracow Technical University, Poland.

The complete set of equations of the linear theory of thin shells is integrated in the Exact Finite Element (EFE). In this way no shape functions are needed and the number of degrees of freedom per FE is n=8 for Sanders' theory and n=10 for Reissner's equations. Derivation of the stiffness and nodal force matrices is briefly reported for the crown open and closed EFE. The algorithm of numerical integration, based on the midpoint finite difference method enables us to consider longer FE (up to the length of a whole shell) and use the above mentioned shell theories. A possibility of the application of these algorithms to the linear buckling analysis is also pointed out.

INTRODUCTION

In the Finite Element Method (FEM) the fields of displacements (and/or strains) are approximated by sets of admissible functions. The number of basic functions refers to the number of degrees of freedom of an element. The approximation is better if DOF number increases or the dimension of the finite element decreases. In the shell analysis it is rather difficult to derive such basic functions (shape functions) which fulfill fundamental criteria of FEM.

In Ref 1-5 so-called Exact Finite Element (EFE) was introduced for the static analysis of thin elastic shells of revolution. The main idea of EFE lies in the computation of the stiffness matrix K^e and the matrix of nodal forces F^e directly from the differential shell equations. In contrary to the Standard Finite Element (SFE) no shape functions are used in EFE and the number of DOF equals minimal number which has to be assumed in SFE. The accuracy of computation in EFE depends only on the accuracy of numerical integration of shell equations. Application of EFE assures continuity of appropriate quantities in FE nodes. Variable geometric and load parameters along the meridian of a shell of revolution can be easily taken into account.

The idea of EFE is not new. Exact analytical solutions were used to formulate the matrices of conical, cylindrical and circular plate finite elements - cf Ref 6-9 respectively. Such an approach is limited to a small class of plates and shells for which explicit analytical solutions exist.

In shells of revolution the set of ordinary differential equations is obtained after all unknow quantities are expanded in the Fourier series in the circumferential direction. In Refs 2,3 the Runge-Kutta method was applied to integrate numerically the canonic set of shell equations. Such a method implies the restraint of the FE length because of parasitic errors. For certain types of ordinary differential equations (sandwich shells, Reissner-type plates and shells of moderate thickness) Runge-Kutta-type methods cannot be used effectively. Because of that the midpoint finite difference method (MFDM) - cf Ref 10 - was applied. In Ref 3 the algorithm based on MFDM and adequate to the analysis of shells of revolution was worked out. In this algorithm the stiffness matrix and the nodal force matrix for many external loads are computed by onefold solving of the appropriate set of linear algebraic equations.

Such an approach makes it possible to get beyond the classical Love - Kirchhoff hypotheses (three-parameter theory). The algorithm was explored for integrating the Reissner-type equations (five-parameter theory of shells). In Ref 3 the analysis of moderately thick circular fundamental plate of high chimney was made. The algorithm was also used in Ref 4 to compute a circular shell with shear deformations taken into account.

Application of MFDM enables us to elongate EFE, even to the length of a whole shell - cf Ref 3. In Ref 5 the singular EFE was considered if one end of the meridian lay on the axis of symmetry (closed FE).

In the paper main ideas of EFE and algorithms supported on MFDM are presented. Both crown open and closed EFE are considered. Possible application of the EFE algorithm to the linear stability analysis of shells of revolution are briefly reported.

ASSUMPTIONS

The following assumptions are adopted :

a) The mid-surface is axially symmetric, bounded by parallel circles (lines of curvature).

b) Linear kinematic and equilibrium equations of thin shells are valid. Sanders' three-parametric theory or Reissner's five-parametric theory equations will be used.

c) Material is linearly elastic, homogeneous and isotropic.

d) Loads are statically applied and have one plane of symmetry.

MATRIX EQUATIONS OF EXACT FINITE ELEMENTS

On the basis of the assumptions the semi-analytical approach can be exploited. Sets of ordinary differential equations are obtained as a result of the application of the trigonometric series in the circumferential directions. For every harmonics j the canonic set of equation can be written down in the following form :

$$y' = A(x;j)y + p(x) \equiv f \quad \text{for } y,f \in R^n \qquad \ldots\ldots 1,$$

where $x \in [x_0, x_L]$ is the independent variable, measured along the meridian, y is the state vector of n components (n=8 for Sanders' theory, n=10 for Reissner's theory). In what follows the vector p(x) will be called the load vector.

In Appendix 1 the Reissner equations are shown in the canonic form. Appropriate set of Sanders' equations was shown in Ref 1.

The set of Eqns 1 formulates the two-point, symmetric BV problem. This set is completed by the boundary conditions.

$$B_b\, y_b = e_b \qquad \ldots\ldots 2,$$

where b = 0, L corresponds to the ends of the finite element. The matrix B_b is of dimension (n/2 x n) and the vector $e_b \in R^{n/2}$ corresponds to known components of the input and output vectors $y_b = y(x_b)$.

The derivative y' is approximated by the difference formula, symmetric with respect to the midpoint i + 1/2 and the right hand side of Eqn 1 is approximated by the arithmetic mean of functions f at points i and i + 1. Thus, the following algebraic relation is obtained :

$$y_{i+1} - y_i = (f_{i+1} + f_i)\Delta x_{i+1}/2 \qquad \ldots\ldots 3.$$

This approximation, of accuracy $O(\Delta x^2)$, enables us to apply a variable step Δx_{i+1} and to avoid additional, external nodes except the nodes i = 0,1,..., L for the L difference intervals.

If N vectors p are considered Eqn 1 is transformed into the following matrix equations

$$Y' = A(x;j)Y + P(x) \qquad \ldots\ldots 4,$$

where the matrices Y, P are of dimension (n x N). After taking advantage of Eqn 3 the set of linear algebraic equations is obtained :

$$C_i Y_i + D_{i+1}\, Y_{i+1} = G_i \quad \text{for } i = 0,1,\ldots,L-1 \qquad \ldots\ldots 5,$$

where the matrices C_i, D_i, G_i are computed at the points $x = x_i$ for fixed values of j = 0,1,...

The set of Eqns 5 is completed by the boundary conditions 2 in which the vectors y_b, e_b are changed by the matrices Y_b, E_b respectively :

$$\underset{(n/2 \times n)}{B_b} \quad \underset{(n \times N)}{Y_b} \quad = \quad \underset{(n/2 \times N)}{E_b} \qquad \ldots\ldots 6.$$

STIFFNESS AND NODAL FORCE MATRICES FOR OPEN EFE

The state vector y has components of the generalized strains q and generalized stresses r :

$$y = \{q, r\} \qquad \text{for } q,r \in R^{n/2} \qquad \ldots\ldots 7.$$

For the Sanders' theory the state vector has n = 8 components :

$$y = \{u, v, w, \beta \,\vdots\, \bar{n}_1, \bar{s}, \bar{t}, m_1\} \qquad \ldots\ldots 8a,$$

where the dash is added over symbols to underline that the equivalent edge forces are considered. In the case of the Reissner's theory (n = 10) the following state vector is used

$$y = \{u, v, w, \varphi_1, \varphi_2 \,\vdots\, n_1, s, t, m_1, m_s\} \qquad \ldots\ldots 8b,$$

where notation is explained in Appendix 1 and in Fig 1a.

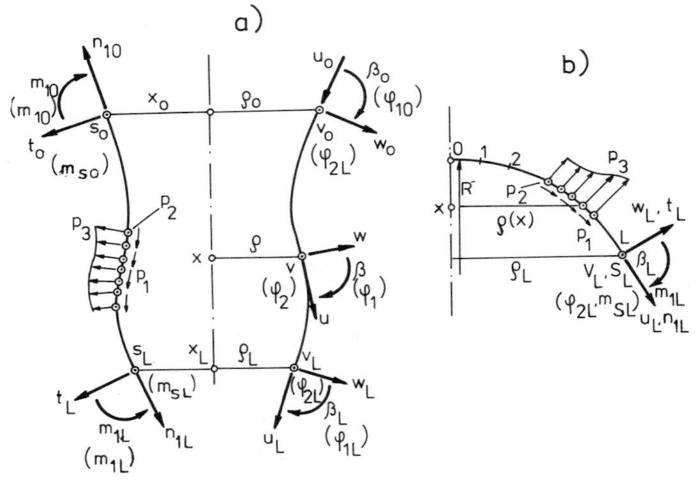

Fig 1

In the stiffness version of FEM the elements with clamped edges are considered. In such finite elements the matrices B_b are of the following form :

$$B_b = [\,I \,\vdots\, 0\,] \qquad \text{for } b = 0,L \qquad \ldots\ldots 9,$$

where the blocks I, 0 are of dimension (n/2 x n/2).

The stiffness matrix is computed for $p = 0$ and for unit values of edge displacements. The matrix of nodal forces is computed for m external loads under assumption of clamped edges of the finite element. Thus, for $N = n + m$ the matrices E_b are formulated as shown in Fig 2. The structure of the complete set of algebraic equations

$$HY = G \qquad \text{...... 10.}$$

is also shown in Fig 2.

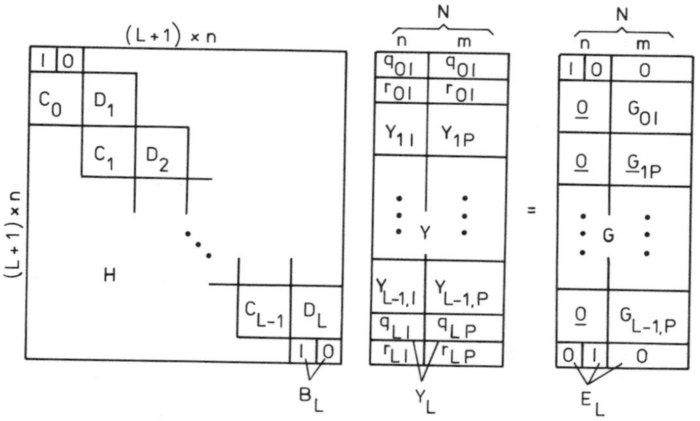

Fig 2

The matrix H is non-symmetric, so instead of Eqn 10 the computation is carried out for the following matrix equation

$$(H^tH)\ Y = H^tG \qquad \text{...... 11.}$$

The matrix H^tH is, of course, symmetric therefore standard procedures prepared for symmetric and banded matrices can be used in order to solve the linear Eqns 11.

From the solution Y the forces r_{bI} and r_{bP} are selected. After multiplication by the factors $c\pi\rho_b$ the symmetric stiffness matrix K^e and the nodal force matrix F^e are of the following form :

$$\underset{(n \times n)}{K^e} = c\pi \begin{bmatrix} -\rho_0 r_{0I} \\ \rho_L r_{LI} \end{bmatrix}, \qquad \underset{(n \times m)}{F^e} = c\pi \begin{bmatrix} -\rho_0 r_{0P} \\ \rho_L r_{LP} \end{bmatrix} \qquad \text{..... 12.}$$

In formula 12 the sign minus refers to transformation from the local to node coordinate systems and to the global forces (reactions)

$$R_b = \int_0^{2\pi} r_b \rho_b \cos^2 j\theta d\theta = c\pi\rho_b r_b \qquad \text{...... 13.}$$

The coefficient c is $c = 2$ for $j = 0$ and $c = 1$ for $j \geq 1$ and ρ_b are radii of parallel circles as shown in Fig 1a.

CLOSED CROWN EFE

In Fig 1b the closed EFE is shown, i.e the element whose crown point $x_0 = 0$ lies on the axis of symmetry. We assume that no concentrated forces and no displacement constraints are applied to the singular point $x_0 = 0$.

The vector f at the right - hand side of Eqn 1 is of the form

$$f_0 = A(0;j)\ y_0 + p_0 \qquad \text{...... 14.}$$

The components of this vector have to be computed as the limit $\lim f(x)$ for $x \to 0$. Since in the matrix A are factors $1/\rho^m$ therefore the vectors y_0 and p_0 are expanded into power series

$$\begin{aligned} y_0 &= y_{00} + y_{01}\rho + y_{02}\rho^2 + \ldots, \\ p_0 &= p_{00} + p_{01}\rho + p_{02}\rho^2 + \ldots, \end{aligned} \qquad \text{...... 15,}$$

where $\rho(x)$ is the radius of the parallel circle.

Uniqueness of solution requires $n/2$ independent components of all vectors y_{0k}. These independent components are combined as the initial value vector $z_0 \in R^{n/2}$.

It appears that for axial symmetry ($j=0$) the vector z_0 is composed only of components of the vector y_{00}. Subsequent harmonics $j > 0$ require components of y_{0k} for increasing k. In Table 1 of Appendix 2 the components y_{0k} are written for $j = 0,1,2$ and for simplified case $p_{0k} = 0$.

Compared with the open crown EFE Eqn 2 is omitted at the point $x_0 = 0$. Instead, the relation

$$y'_{1/2} \approx (y_{01} + 2y_{02}\rho + \ldots)\ \rho'|_{\Delta x_1/2} = (y_1 - y_{00})/\Delta x_1 \qquad \text{......16,}$$

leads to the equation

$$\underset{(n \times n/2)}{C_0}\ \underset{(n/2 \times N)}{Z_0} + \underset{(n \times n)}{D_1}\ \underset{(n \times N)}{Y_1} = \underset{(n \times N)}{G_0} \qquad \text{...... 17.}$$

The next equations for $i \geq 1$ are formulated according to Formula 5, and to Eqn 6 for $b = x_L$. In this way the set of $(L + 1/2) \times n$ linear equations is formulated for computation of Z_0 and Y_i for $i = 1,\ldots,L$.

The stiffness and nodal force matrices for the closed EFE are computed according to the following formula

$$\underset{(n/2 \times n/2)}{K^e} = c\pi\rho_L r_{LI}, \qquad \underset{(n/2 \times m)}{F^e} = c\pi\rho_L r_{LP} \qquad \text{.... 18.}$$

STATIC AND BUCKLING ANALYSIS OF A WHOLE SHELL

EFE can be used as superelements for whole parts of the structure. Such an example is shown in Fig 3a, where circumferential ribs and parts of the hyperboloidal shell are treated as EFE of numbers I,...,VI.

Assuming appropriate matrices B_b and E_b in Eqns 6 enables us to apply the presented algorithms to the static analysis of whole shells. For shells shown in Fig 3b,c the matrix B_0 for the Sanders theory is of the following form

$$\underset{(4\text{x}8)}{B_0} = [\ \underset{(4\text{x}4)}{0} \quad \underset{(4\text{x}4)}{I}\ ] \qquad \text{...... 19.}$$

The matrix B_L for a shell in Fig 3b is formulated by Formula 9. The case of a shell in Fig 3c corresponds to the matrix

$$B_L = \begin{bmatrix} 1 & 0 & 0 & 0 & 0 & 0 & 0 & 0 \\ 0 & 1 & 0 & 0 & 0 & 0 & 0 & 0 \\ 0 & 0 & 1 & 0 & 0 & 0 & 0 & 0 \\ 0 & 0 & 0 & 0 & 0 & 0 & 0 & 1 \end{bmatrix} \qquad \ldots\ldots \; 20.$$

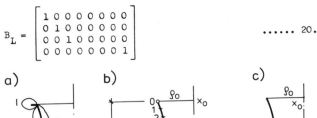

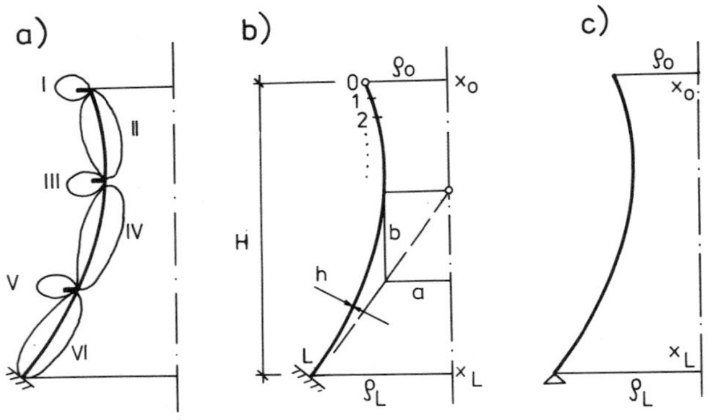

Fig 3

The static analysis of a whole shell treated as a super-element is now extended on linear buckling analysis of shells of revolution. For this purpose the 2nd order theory of shells has to be used. For instance non-linear shell equations from Ref 11 can be used to formulate the canonic set of incremental equations :

$$\Delta y' = A(x; j, \overset{o}{y})\, y \qquad \ldots\ldots \; 21.$$

If the solution of a prebuckling state is written in the form

$$\overset{o}{y} = \lambda\, \overset{o}{y} \qquad \ldots\ldots \; 22.$$

Eqn 10 takes the following form

$$(H_0 + \lambda H_1 + \lambda^2 H_2)\Delta y = 0 \qquad \ldots\ldots \; 23.$$

In the above equation the matrix H_1 depends only on forces $\overset{o}{n}_{\alpha\beta}$ and the matrix H_2 on rotations $\overset{o}{\varphi}_{\alpha}$. The matrices H_k for k = 0,1,2 are non-symmetric, therefore it is not possible to explore standard procedures of eigen-problem analysis. Transition to Eqn 11 raises the range of the eigen-value analysis.

That is the reason why now an algorithm is worked out for computation of the critical load parameter

$$\lambda_{cr} = \inf_{j} \lambda^{j}_{cr} \qquad \ldots\ldots \; 24,$$

where λ^j_{cr} is the lowest root of the nonlinear equation

$$\det \left| H^t H(\lambda; j, \overset{o}{y}) \right| = 0 \qquad \ldots\ldots \; 25.$$

TESTING OF COMPUTER CODES

The algorithms described above are the basis of subroutines and codes implemented on the computer CYBER 72. From among many tests only one is given below to illustrate the effectiveness of the proposed approach.

The shell of a hyperboloidal cooling tower and static scheme as has been shown in Fig 3 was analyzed for the following data : a = 27.5 m, b = 67.7m, ρ_0 = 28.47 m, ρ_L = 50.o m, H =120.0 m, h = 0.14m, wind load. The computation was carried out by means of 60 standard FE of 8 DOF per node and the length $(x_L - x_0)$ = 2.0 m. The computation was repeated for one EFE assuming 60 difference nodes and constand step Δx = 2.0 m. It turned out that for the most important harmonic j = 1 the coincidence of results was stated up to the three significant digits.

Now the computer code for the linear buckling analysis of cooling tower shell is being tested.

CONCLUSIONS

a) Advantages of EFE allowed them to be used in big computer codes - cf Ref 12 - 14.

b) Presented version enables us to extend significantly EFE, even to the length of a whole shell.

c) Various shell theories can be easily explored if the midpoint finite difference method is used.

d) Works developed now at the Institute of Structural Mechanics of the Cracow Technical University should give an answer about effectiveness of EFE application to the buckling analysis of shells of revolution.

REFERENCES

1. Z WASZCZYSZYN, A MŁODZIANOWSKI, Exact Finite Element for Thin Elastic Shells of Revolution (in Polish), Rozpr Inż, No 1, Vol XXV, 1977, 97-114.

2. E PIECZARA and Z WASZCZYSZYN, Generalized Stress Analysis in a Shell of Lenslike Reservoir (in Polish), Arch Inż Lad, No 4, Vol XXVI, 1980, 671-676.

3. J PIECZARA, K RZEGOCINSKA-PEŁECH and Z WASZCZYSZYN, Exact Finite Elements for Plates and Shells of Revolution (in Polish), Proc of XXVI Polish Civil Engng Conference, Krynica 1980, Vol 1, 223-229.

4. J PIECZARA, Exact Finite Elements for Elastic Shells of Revolution (in Polish), Proc of V Conference Comp Methods in Struct Mech, Karpacz 1981, Vol 2, 145-151.

5. J PIECZARA, Z WASZCZYSZYN, A Singular Exact Finite Element for Elastic Plates and Shells of Revolution (in Polish), VI Conf Comp Methods in Struct Mech Bialystok 1983, Vol 2, 81-88.

6. R R MEYER and M B HARMON, Conical Segment Method for Analysing Open Crown Shells of Revolution for Edge Loading, AIAA J, No 1, Vol 4, 1963, 886-891.

7. E P POPOV, J PENZIEN and ZUNG-AU-LU, Finite Element Solution for Axisymmetrical Shells, Proc ASCE, EM5, Vol 90, 1964, 119-145.

8. A A LAKIS, Cylindrical Finite Element for Analysis of Cylindrical Structures, Dept Mech Engng, Mc Gill Univ, Rep No 70-2, Montreal Canada, Nov. 1969.

9. G C PARDOEN, Static Vibration and Buckling Analysis of Axisymmetric Circular Plates Using Finite Elements, Int J Computers & Struct, No 3, Vol 2, 1973, 355-373.

10. W BEAFAIT and G W REDDIEN, Midpoint Difference Method for Analysing Beam Structures, Int J Comp & Struct,

No 6, Vol 8, 1978, 745-751.

11. J L SANDERS Jr., Nonlinear Theories for Thin Shells, Quart Appl Math, No 1, Vol 21, 1963, 21-36.

12. A GUMINSKI, J KRUPINSKI, A MŁODZIANOWSKI and Z WASZCZYSZYN, Computer Analysis of High Chimney Basement (in Polish), Inż i Bud, No 4, Vol XXXIV, 1977, 133-136.

13. A GUMINSKI and J PIECZARA, Static Analysis of Hyperbolic Cooling Towers on Non-Homogeneous Foundation, (in Polish), VI Conf Comp Methods in Struct Mech, Białystok 1983, Vol 1, 197-215.

14. O MATEJA, R KALUŻA, J PELA, Numerical Static Analysis of Shell Cooling Towers (in Polish), VI Conf Comp Methods in Struct Mech, Białystok 1983, Vol 2, 37-44.

15. J L SANDERS Jr., An Improved First-Approximation Theory for Thin Shells, NASA, Techn Rep R-24, 1959, 1-11.

16. S TIMOSHENKO and S WOINOWSKY-KRIEGER, Theory of Plates and Shells, Mc Graw-Hill, 1959, 166-169.

APPENDIX 1 : Equations of the Reissner theory for shells of revolution.

In order to obtain the symmetric tensors of generalized strains the following relations are assumed, according to Ref 15 :

$$e_{12} = e_{21}, \quad \bar{k}_{12} = (k_{12} + k_{21})/2 \quad \ldots\ldots \text{A1.}$$

Kinematic equations, taken from Ref 15, are transformed to the following form :

$$e_{11} = u,_1/A + w/R_1,$$
$$e_{22} = v,_2/\rho + Cu + w/R_2,$$
$$2e_{12} = v,_1/A - Cv + u,_2/\rho,$$
$$k_{11} = \varphi_1,_1/A, \quad k_{22} = \varphi_2,_2/\rho + C\varphi_1, \quad \ldots\ldots \text{A2.}$$
$$2\bar{k}_{12} = \varphi_2,_1/A + \varphi_1,_2/\rho - C\varphi_2 - k_s\varphi_n,$$
$$e_{13} = w,_1/A - u/R_1 + \varphi_1, \quad e_{23} = w,_2/\rho - v/R_2 + \varphi_2,$$

where : A - Lamé coefficient of the meridian, ρ - radius of the parallel circle, $C=\rho,_1/(A\rho)$ - function related to the covariant differientiation, $k_s=1/R_1 - 1/R_2$ - difference of curvatures, $2\varphi_n = [(\rho v),_1 - (Au),_2]/(A\rho)$ - angle of rotation around the normal to the shell midsurface. In these and following equations partial derivatives are pointed out by commas and subscripts, i.e. $(\ ),_1 = \partial(\ )/\partial x$, $(\ ),_2 = \partial(\ )/\partial$ where :

x - independent variable related to the meridian,
θ - angle on the parallel circle measured from the plane of symmetry of loads (assumption d).

On the basis of the principle of virtual work the following equations of equilibrium are obtained :

$$(\rho n_{11}),_1 + (An_{12}),_2 - \rho,_1 n_{22} + A(k_s m_{12}),_2/2 + $$
$$+ A\rho(t_1/R_1 + p_1) = 0$$

$$(\rho n_{12}),_1 + (An_{22}),_2 + \rho,_1 n_{12} - \rho(k_s m_{12}),_1/2 +$$
$$+A\rho(t_2/R_2 + p_2) = 0,$$
$$(\rho t_1),_1 + (At_2),_2 - A\rho(n_{11}/R_1 + n_{22}/R_2) + \quad \ldots\ldots \text{A3.}$$
$$+ A\rho p_3 = 0,$$
$$(\rho m_{11}),_1 + (Am_{12}),_2 - \rho,_1 m_{22} - A\rho t_1 = 0,$$
$$(\rho m_{12}),_1 + (Am_{22}),_2 + \rho,_1 m_{12} - A\rho t_2 = 0.$$

The boundary conditions at the edge x = const. are to be written down for the following quantities :

$n_{11} = n_1$		u		
$n_{12} - k_s m_{12}/2 = s$		v		
$t_1 = t$	or	w		.... A4.
$m_{11} = m_1$		φ_1		
$m_{12} = m_s$		φ_2		

Physical relations for the symmetric tensors $n_{\alpha\beta} = n_{\beta\alpha}$, $m_{\alpha\beta} = m_{\beta\alpha}$ and for the transverse shear forces t_α are assumed in the form :

$$Ehe_{11} = n_{11} - \nu n_{22},$$
$$Ehe_{22} = n_{22} - \nu n_{11},$$
$$Ehe_{12} = 2(1+\nu)n_{12},$$
$$Eh^3k_{11}/12 = m_{11} - \nu m_{22}, \quad \ldots\ldots \text{A5.}$$
$$Eh^3k_{22}/12 = m_{22} - \nu m_{11},$$
$$Eh^3k_{12}/12 = (1+\nu)m_{12},$$
$$Ghe_{13} = t_1,$$
$$Ghe_{23} = t_2.$$

Equations based on similar assumptions used in the theory of plates refer to E. Reissner - cf Ref 16, p. 166. That is why we call them the Reissner equations.

Because of the plane of load symmetry (assumption d) unknown functions are expanded into the Fourier series of cos or sin type :

$$\cos j\theta : p_1, p_3, u, w, \varphi_1, e_{11}, e_{22}, k_{11}, k_{22},$$
$$n_{11}, n_{22}, m_{11}, m_{22}, t_1, \quad \ldots\ldots \text{A6.}$$
$$\sin j\theta : p_2, v, \varphi_2, e_{12}, k_{12}, n_{12}, m_{12}, t_2.$$

In such way the sets of ordinary differential equations can be obtained for subsequent harmonic j = 0,1.... After simple transformations of Eqns A2, A5, A6 the following set of canonic equations is obtained with respect to the edge quantities A4 :

$$u' = (n_1 - \nu n_{22})A/Eh - Aw/R_1,$$
$$v' = (s + k_s m_s/2)2A(1 + \nu)/Eh + jAu/\rho + \rho' v/\rho,$$
$$w' = At/Gh + Au/R_1 - A\varphi_1,$$
$$\varphi_1' = (m_1 - \nu m_{22})12A/Eh^3,$$

$$\varphi_2' = \left[(24/h^2 + k_s^2/2)m_s + k_s s\right] A(1 + \nu)/Eh +$$
$$+ jA\varphi_1/\rho + \rho'\varphi_2/\rho + (\rho'v + jAu)k_s/\rho,$$
$$n_1' = (n_{22} - n_1)\rho'/\rho - (s + k_s m_s)jA/\rho -$$
$$- (t/R_1 + p_1)A, \qquad \ldots\ldots \text{A7.}$$
$$s' = -2s\rho'/\rho - k_s m_s \rho'/\rho + jAn_{22}/\rho - (t_2/R_2 + p_2)A,$$
$$t' = -t\rho'/\rho - jAt_2/\rho + (n_1/R_1 + n_{22}/R_2)A - Ap_3,$$
$$m_1' = (m_{22} - m_1)\rho'/\rho - jAm_s/\rho + At,$$
$$m_s' = -2m_s\rho'/\rho + jAm_{22}/\rho + At_2.$$

In the above equations the circumferential forces are used :

$$n_{22} = Eh(jv/\rho + Cu + w/R_2) + \nu n_1,$$
$$m_{22} = (j\varphi_2/\rho + C\varphi_1)Eh^3/12 + \nu m_1, \qquad \ldots\ldots \text{A8}$$
$$t_2 = Gh(-jw/\rho - v/R_2 + \varphi_2).$$

In Eqns A7 - A8 the index j is ommited for the coefficients of Fourier series.

APPENDIX 2 : Coefficients of the expansion into power series of components of the state vector in the singular point for the Reissner theory

j	0	1	2
u_{00}	0	X_1	0
v_{00}	0	$-X_1$	0
w_{00}	X_1	0	0
φ_{100}	0	X_2	0
φ_{200}	0	$-X_2$	0
n_{100}	X_2	0	X_1
s_{00}	0	0	$-X_1$
t_{00}	0	X_3	0
m_{100}	X_3	0	X_2
m_{s00}	0	0	$-X_2$
u_{01}	$-X_1/R + X_2(1-\nu)/Eh$	0	$X_1(1+\nu)/Eh$
v_{01}	X_4	0	$-X_1(1+\nu)/Eh$
w_{01}	0	$X_3/Gh + X_1/R - X_2$	0
φ_{101}	$12X_3(1-\nu)/Eh^3$	0	$12X_2(1+\nu)/Eh^3$
φ_{201}	X_5	0	$-12X_2(1+\nu)/Eh^3$
n_{101}	0	$X_4/3 - 2X_3/3R$	0
s_{01}	0	$X_4/3 + X_3/3R$	0
t_{01}	0	0	$-X_3$
m_{101}	0	$X_5/3 + 2X_3/3$	0
m_{s01}	0	$X_5/3 - X_3/3$	0
u_{02}			0
v_{02}			0
w_{02}			$-X_3/2Gh + X_1(1+\nu)/2REh - 6X_2(1+\nu)/Eh^3$
φ_{102}			0
φ_{202}			0
n_{102}			$X_3/2R$
s_{02}			$X_4/2 - X_3/4R$
t_{02}			0
m_{102}			$-X_3/2$
m_{s02}			$X_5/2 + X_3/4$

where : R - radius of curvature at the point $x = 0$.

Table 1.

EXPERIMENTAL INVESTIGATION OF STRUCTURAL PLATES

Ülo A. Tärno, DSc

Professor, Department of Civil Engineering,
Tallinn Technical University, Tallinn.

The paper deals with the results of experimental investigations of structural plates. The studies were carried out on the 1 : 30 scale steel model. The model of a structural plate is made up of two directional grids. The basic pattern of a unit is the square. The paper consists of investigations of single and multi-span plates with various sizes of rectangular spans. Some variants of the plate with consoles are studied there. For the determination of stresses the method of electrotensometrics was used. Diagrams of internal forces and vertical displacements of structures in the form of the structural plate are given. The paper also presents some variants of structural plates with various lower layers.

INTRODUCTION

Some problems of the behaviour of the structural plate consisting of pyramidal units are presented.

The height of the unit pyramid is equal to the base and equal to 1/16 of the longer span of the structural plate. The plate has different constructional elements, supporting conditions and loading schemes. It is necessary to determine the internal forces experimentally in the space structural plates under complicated supporting and loading conditions.

The separate steel sheet elements of the 1:30 scale model (Fig. 1) were made with special stamps and had different stiffness depending on the place and direction of the elements.

The electrotensometric method was used for determining the internal forces of the cords of the structural plate. The suspended loads were used only at the lower joints of the structural plate.Aside the uniformly distributed load load for single spans and consoles of the plate were also used. The maximum load was determined by the conditions of elasticity and buckling of cords. The whole load of the plate was transferred to supporting elements with the help of four supporting rods fixed to the four lower joints of the structural plate. The longitudinal span is directed to the axes A, B and C, the cross-span - to the direction of the axes 1 and 2. There are presented the internal forces for the loads 10 N to the joint in the upper cords as well as the diagonals.

STRUCTURAL PLATE WITHOUT CONSOLES

A structural plate with a rectangular plan (maximum sizes 1200 x 934 mm), supported by four (model 1) or six (model 2) supports has been investigated. In the latter case a two span plate with a longitudinal and cross span relation of 2.67 was formed (Fig. 1, model 2). In the case of a rectangular one

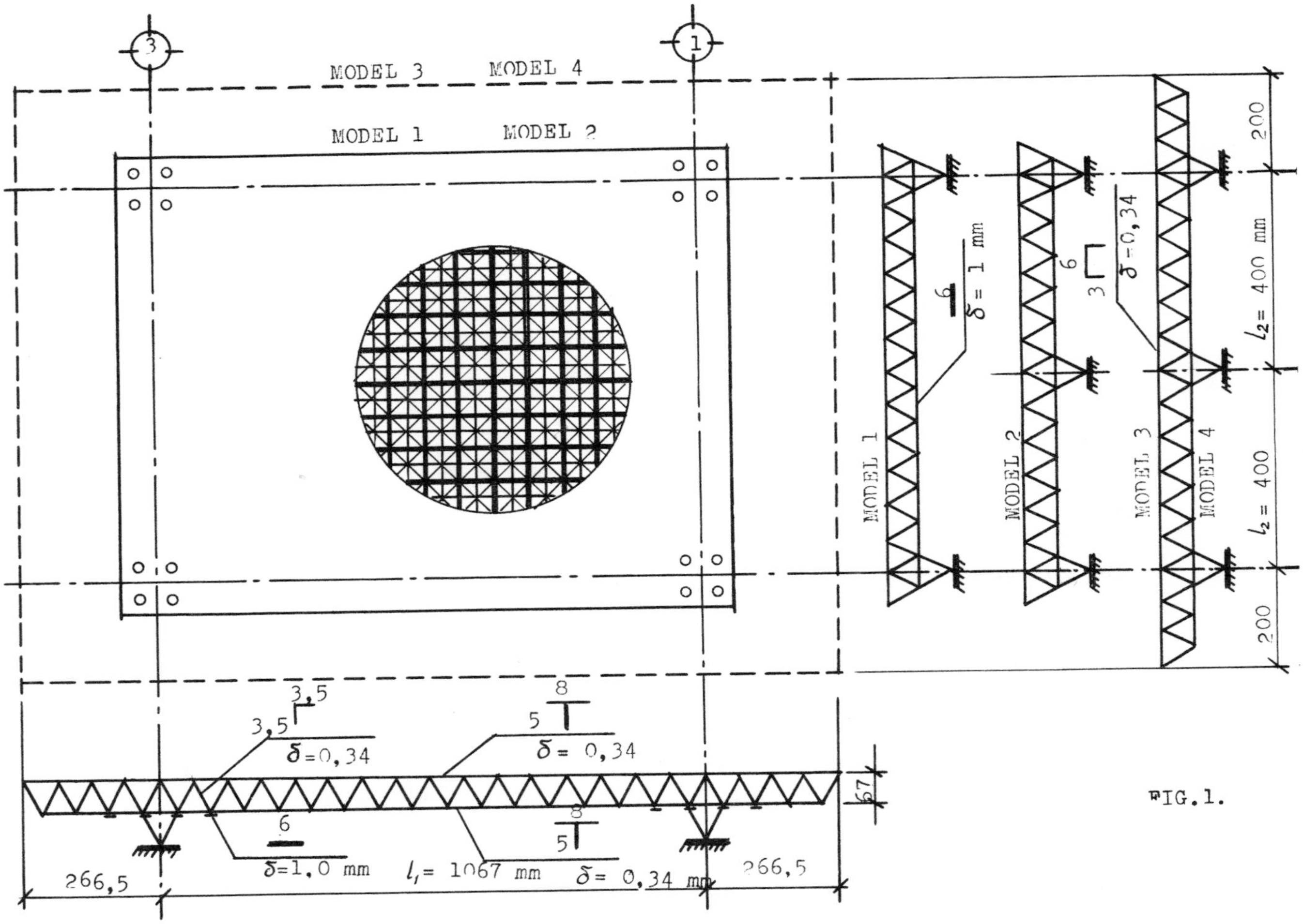
MODEL 3
MODEL 4
MODEL 1
MODEL 2
3,5
3,5
δ=0,34
8
5
δ= 0,34
6
δ=1,0 mm
8
5
l_1= 1067 mm
δ= 0,34 mm
266,5
266,5
67
6
δ= 1 mm
6
3
δ=0,34
200
l_2= 400 mm
l_2= 400
200

FIG.1.

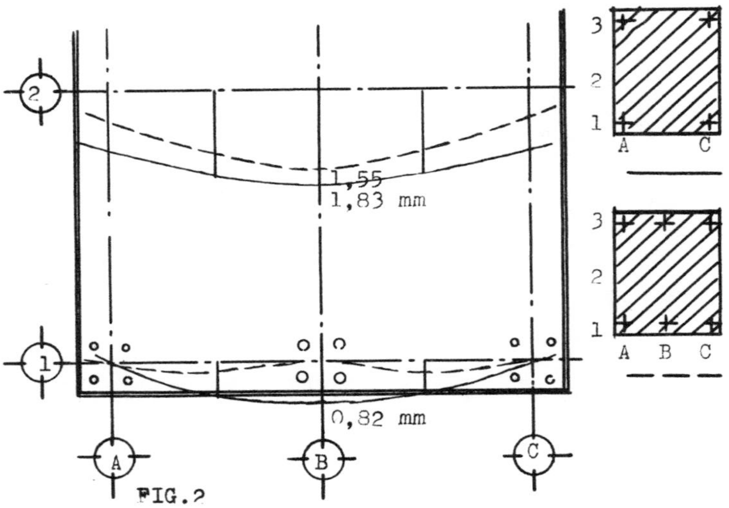
1,55
1,83 mm
0,82 mm
3
2
1
A
C
3
2
1
A
B
C

FIG.2

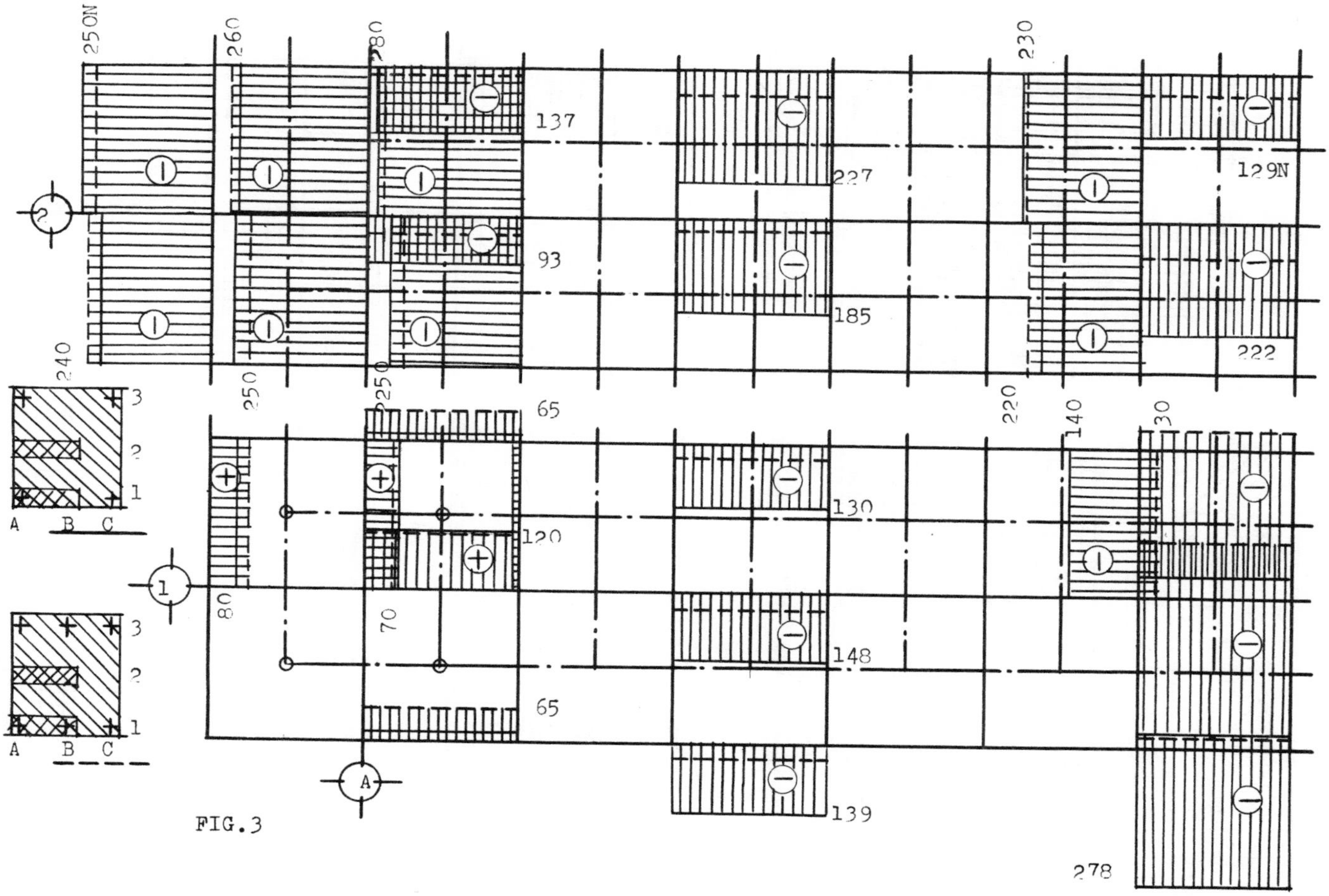

FIG.3

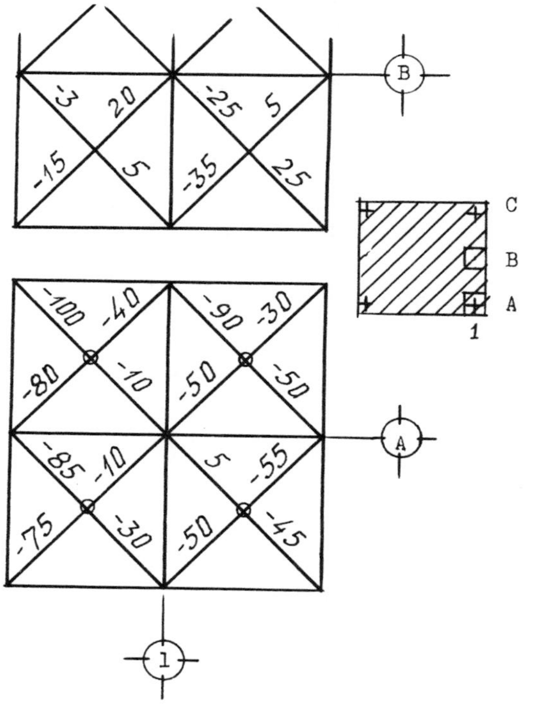

FIG.4

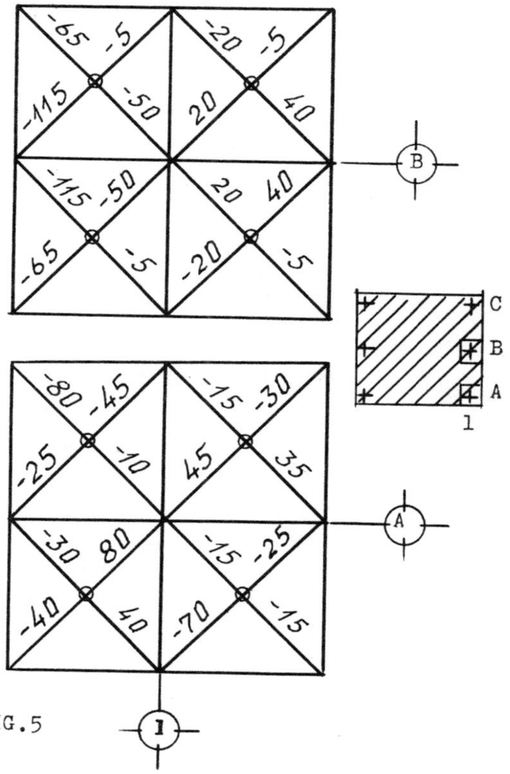

FIG.5

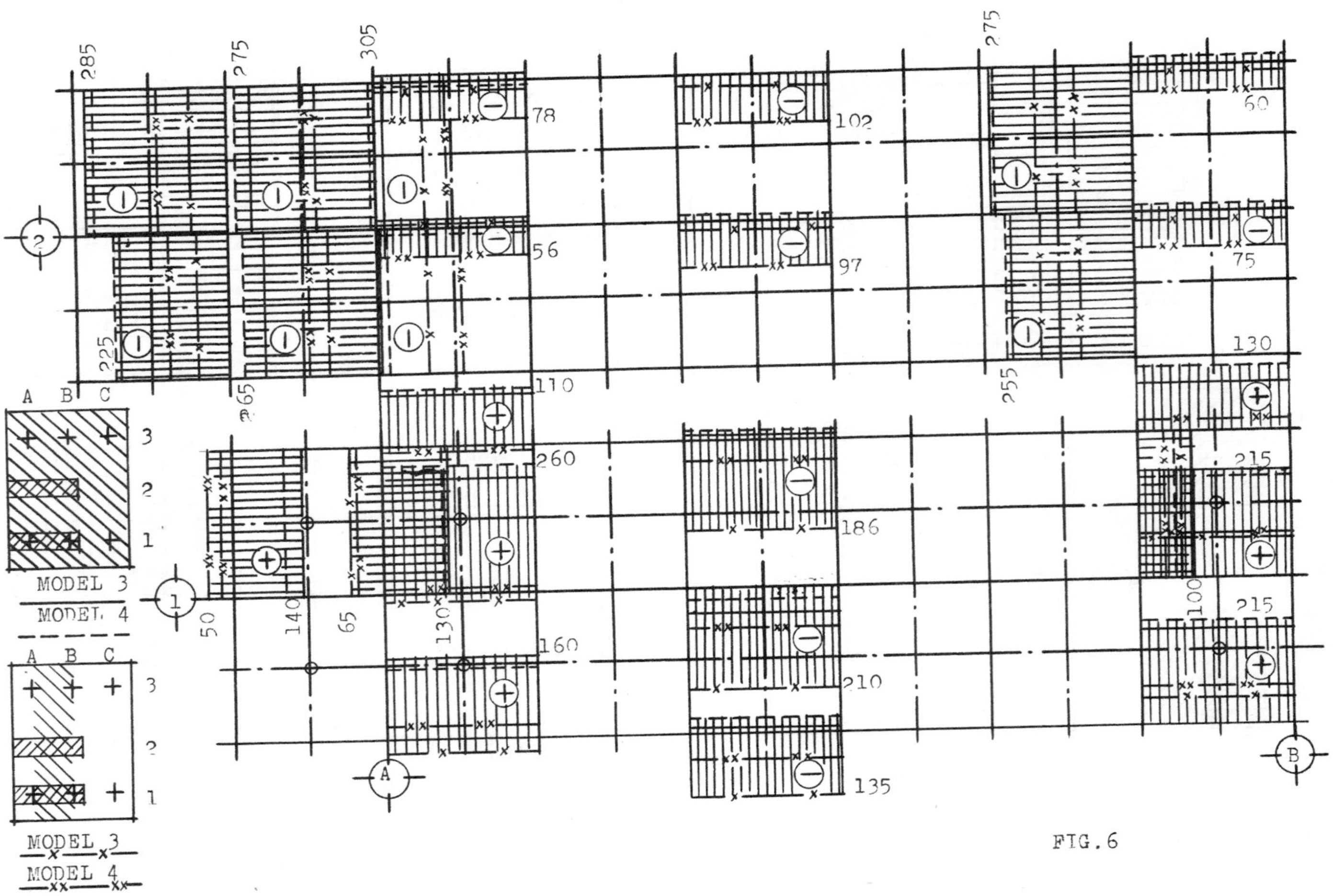

FIG.6

span plate supported by four supports the vertical displacements of the centre are about 1.2 times bigger than those found in a plate supported by six supports (Fig. 2). The use of additional supporting elements will also decrease the displacements in the spans between longitudinal supports, being $\frac{1}{890} l_1$ (model 1) and $\frac{1}{2000} l_1$ (model 2) respectively. The use of additional supporting elements will decrease the space behaviour of the structural plate (Fig. 3). The internal forces in the cross cords will decrease about 2.5 times at the centre of the span, while the importance of the longitudinal cords is increasing.

The supporting rods are nonuniformly loaded. The nearest to the centre of the plate rods of model 1 are more loaded (Fig. 4). In model 2 (Fig. 5) the internal supporting rods 1 and 4 on the central support B and the rods 1 and 3 on the external support A prove to be more loaded.

STRUCTURAL PLATE WITH CONSOLES

A structural plate (maximum size 1600x1200 mm) supported by six supports has been investigated. In the longitudinal direction we had a single span plate (Fig. 1) with consoles (0.25 l_1), in the cross direction a two span plate with consoles (o.5 l_2). A plate with consoles enables to utilize better the quality of all cords and to uniform the internal forces (also in the supporting rods).

On the basis of the preceding tests we can presume that the cross directed cords work intensively only in the zones of the supports. According to this circumtance the lower cross directed cords were inserted only into the four joints nearer to the supports in model 3.

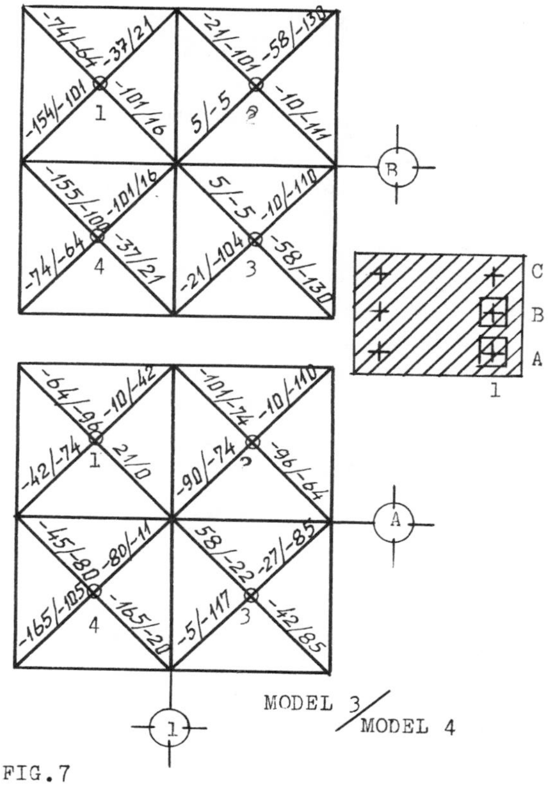

FIG.7

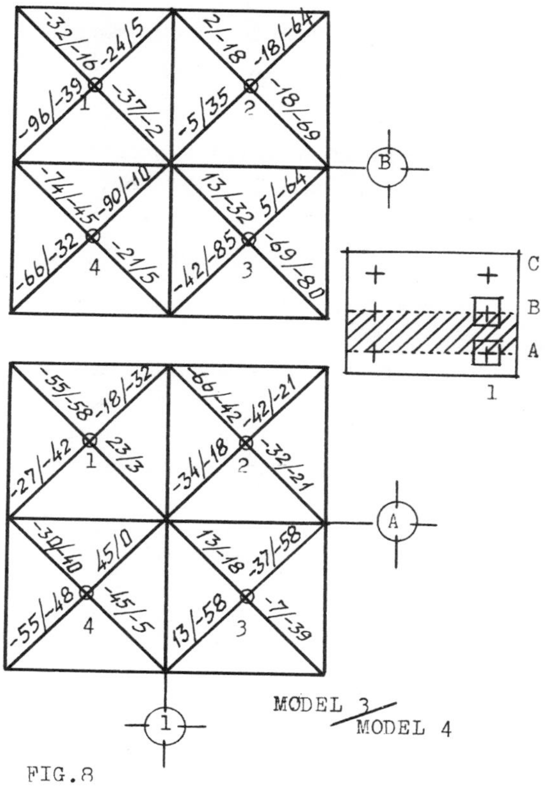

FIG.8

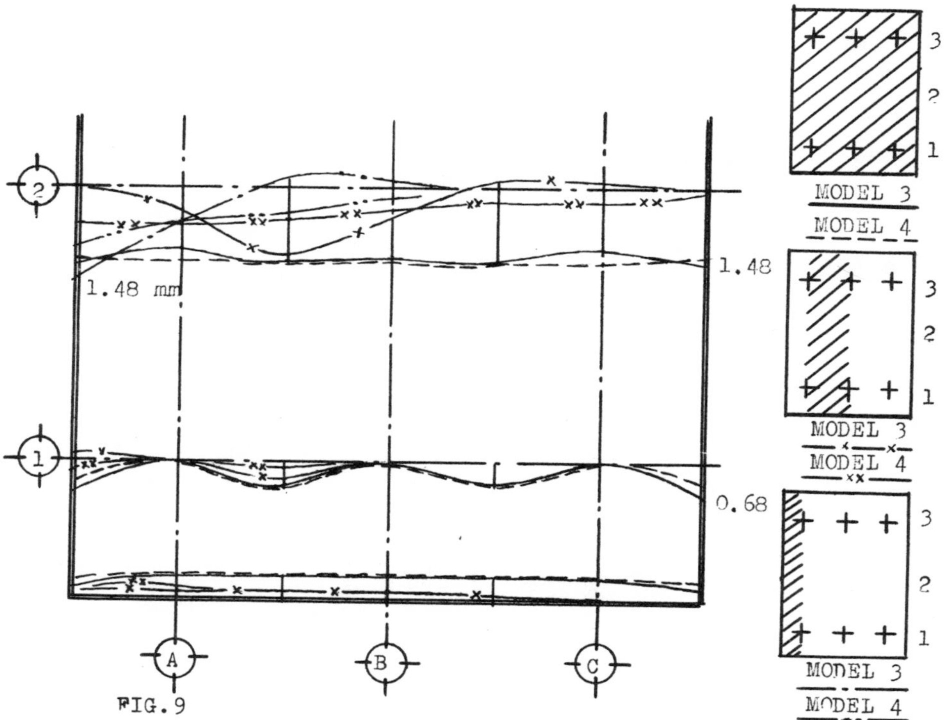

FIG.9

The control test (model 4) with all cross directed cords confirmed our presumptions. There is a good correspondence between the results of the tests of model 3 and model 4 under uniform loads. Under nonuniform loads there are some differences.

The diagonal cords of supporting joints (Fig. 7, 8) are nonuniformly loaded. As a rule, the diagonal cords inside the contour of the supports are less loaded.

The tension was found to maximum under uniform loads in the upper rods of the cross-trusses on the support and longitudinal trusses in the middle of the span (Fig. 6). In the support diagonals (Fig. 7, 8) the maximum compressive forces are produced under the influence of uniformly distributed loads.

The displacements for different loading schemes for models 3 and 4 are presented in fig. 9. There are no essential differences between models 3 and 4 under uniformly distributed loads. The use of additional cross cords will make the vertical displacements more uniform under the nonuniformly distributed loads.

With a wiew to using the materials economically the cross directed lower cords should be placed only in the zones of the supports. The longitudinally directed lower cords will work under tension and so the cross cords against buckling are not needed. The upper longitudinally directed cords, working under compression should have cross directed ties.

CALCULUS PROBLEM FOR DISCRETE SPACE STRUCTURES WITH ELEMENTS HAVING A RANDOM BEHAVIOUR

A LA TEGOLA*

*Professor, Department of Structures,
University of Calabria, Italy.

In this paper some procedures are supplied for the calculation of space reticular structures with random elements. In particular, the most adequate probability density functions are specified,which interpret the real characteristics of the random variables. As for the use of numerical simulation methods procedures for the generation of pseudo-random structures and formulas for the use of the small samples theory are shown, the validity of which is proved according to some theorems of statistics. Moreover, some computation programs useful for real applications are reported in the appendix.

INTRODUCTION

Modern techniques of construction allow to obtain very big space structures at relatively low costs, through the use of standard components which can be accurately controlled during their planning and production. At the same time advanced computation methods are required which take into account the statistic data relative to the behaviour of each element,in order to assure,within prefixed limits, the reliability of the global response of the structures.
Therefore very important theoretical problems have to be faced,concerning both the computation procedures and the probability analysis for the evaluation of the structure response resulting from all the random variables, under the effect of the distribution of given stresses.
In these papers we want to stress the most important questions relative to each problem giving information about the most adequate procedures to solve them. Anyway, because of their general character, it is advisable to consult specific works as far as the analytic development of each aspect is concerned, 'Refs 3, 4 and 5'.
A great importance is also given to questions concerning the statistical approach of random variables which influence the behaviour of each element. The structures' global response, which can be generally obtained only through numerical simulation methods, is justified by the help of statistical theorems which also assure the use of the small samples theory.

COMPUTATION PROBLEMS OF SPACE RETICULAR STRUCTURES

The computation of space reticular structures is generally done by making resort both to the dealing of discrete structures and of equivalent continuous structures. The second method presents bigger advantages as it allows to decrease the volume of the unknown quantities peculiar to the problem as much as one likes. This procedure cannot be utilized when the structure characteristics change locally. On the contrary, if on one hand the first method does not present limits as to the introduction of the structure specific characteristics, on the other hand it requires many initial parameters as well as the solution of equation systems with a very high number of unknown quantities. Unfortunately,when the structure is composed of elements the characteristics of which are not deterministic but random, the use of the equivalent continuous method cannot take into account the probability aspects. Therefore in those cases it is advisable to make resort to discrete structures. Problems to be faced concern the description of the structure itself, as it is necessary to make the calculation by computer through particular methods based on formex-algebra functions which present the advantage of supplying the inputs of the form in a very compact way.
In this case the methods and formulas proposed by Makowski, Z.S. and his collegues are very efficient, 'Refs 1 and 2'. They allow both the graphic description and the memorization of all the geometrical data useful for the automatic calculation of the stresses.
Very simple algorithms are enough, for example, to describe and memorize the structures shown in Fig 1 and 2, which are relative to barrel vaults and to reticular translation vaults with simple layer.
These simple programs, (see appendix 1), can be modified even to describe these structures in a different geometrical configuration when the effects of the geometrical variation on the stress state have to be taken into account.

Making resort to solution methods of incremental kind, instead of the coordinates X_{oi}, Y_{oi}, Z_{oi} of the generic joint in the unchanged configuration, new coordinates $X_i^{(j)} = X_{oi}+U_i^{(j)}$; $Y_i^{(j)}=Y_{oi}+V_i^{(j)}$; $Z_i^{(j)}=Z_{oi}+W_i^{(j)}$ must be considered which take into account the displacement $U_i^{(j)}, V_i^{(j)}, W_i^{(j)}$ undergone by the joint at j step.

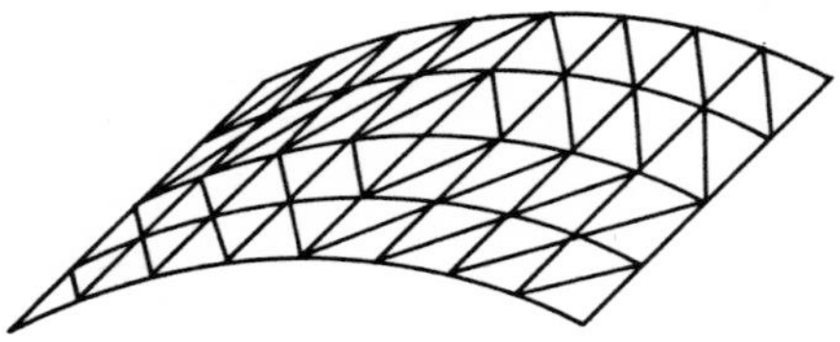

Fig 1

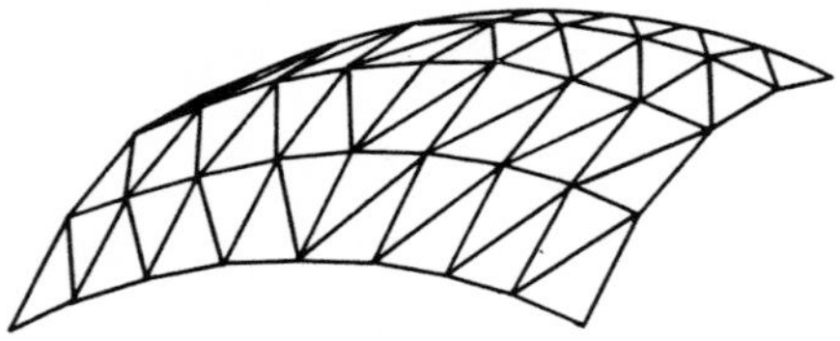

Fig 2

NATURE OF ALEATORY VARIABLES

Aleatory characteristics which are frequently present in each element and influence the structure behaviour are relative to strains and to imperfections of geometrical nature. Quantities intimately linked to the resistence of each element concern both the yeld stress of the material and the geometrical characteristics which influence the strain. Information about such variables can be attained through statistical analyses. Anyway it is evident that if X is one of these variables, the probability density function is of normal or gaussian kind:

$$F(X) = \frac{1}{\sigma\sqrt{2\pi}}\, e^{-(X-\mu)^2/2\sigma^2} \qquad 1.$$

Where μ and σ are the mean and the standard deviation of X This function is symmetrical compared with the mean value and is defined for $-\infty < X < +\infty$, Fig 3.
As we think that for mass-production, enquiries on samples of high number can be carried out, it is enough to assume as values of the mean and of the standard deviation those of the sample mean and of the mean square deviation:

$$\mu = \sum_{1}^{n}{}_i \frac{X_i}{n} \qquad \sigma = \sqrt{\frac{\Sigma\,(Xi-\mu)^2}{n-1}} \qquad 2.$$

Moreover, for application purpose, it is necessary to know both the value of the maximum probability p,max and the values of X,min and X,max. X,min and X,max, taking into account their scarce probability of occurring, are not considered as important in the probable structure behaviour. As function F(X) is symmetrical compared with the mean value we have:

$$p{,}max = F(X{=}\mu) = \frac{1}{\sigma\sqrt{2\pi}} \qquad 3.$$

and:

$$\bar{F}(X) = \frac{F(X)}{p{,}max} = e^{-(X-\mu)^2/2\sigma^2} \qquad 4.$$

If we fix the probability $1-\Theta$ of the variables which are not essential for the structure strain, the values of X,min and X,max can be evaluated through:

$$\Theta = \int_{Xmin}^{Xmax} F(X)dx \qquad 5.$$

Considering the symmetry of F(X) we have also:

$$\frac{\Theta}{2} = \int_{\mu-\Delta x}^{\mu} F(X)dx \qquad 6.$$

and we can write:

$$Xmin = \mu - t(\Theta)\sigma \qquad Xmax = \mu + t(\Theta)\sigma \qquad 7.$$

where $t(\Theta)$ is Student's coefficient.
If Θ=95% we have t=1,964. Fig 4.

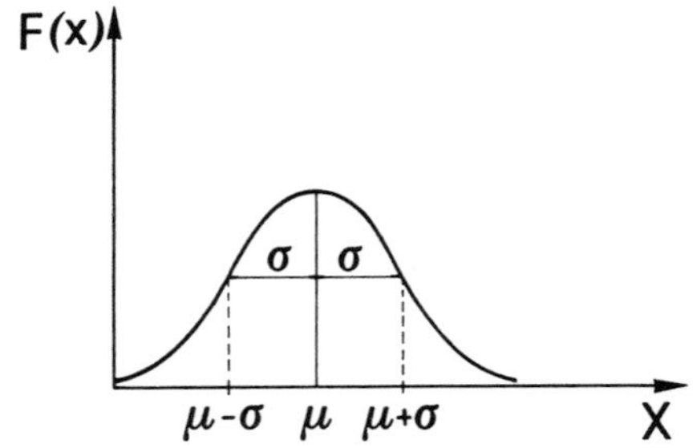

Fig 3

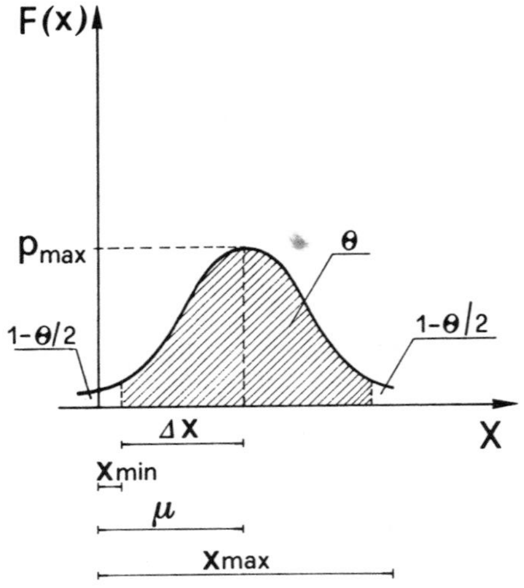

Fig 4

As for the description of imperfections concerning other characteristics of the elements, as for example the imperfect rectilinear character of rods, Fig 5., the Gamma distribution function has proved to be the one which better explains the real phenomenon. In this case we have:

$$F(\delta) = \frac{1}{\Gamma(a+1)b^{a+1}}\, \delta^a\, a^{-\frac{\delta}{b}} \qquad 8.$$

see Fig 6

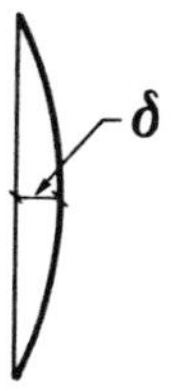

where the mean value δ_m and the variation σ_δ^2 show to be linked to the coefficients a and b by the relations:

$$b = \frac{\sigma_\delta^2}{\delta_m} \qquad a = \frac{\delta_m^2}{\sigma_\delta^2} - 1 \qquad 9.$$

Therefore even this kind of distribution has to be considered as known when we know two statistical parameters δ_m, σ_δ which can be evaluated through experiences on sufficiently numerous samples.

Fig 5

The value of p,max is obtained from the condition:

$$\frac{\partial y}{\partial \delta} = 0 \qquad 10.$$

being: $y = \delta^a e^{-\frac{\delta}{b}}$ 11.

which gives: $\delta = a\,b$ 12.

and then:

$$p{,}max = \frac{1}{\Gamma(a+1)b^{a+1}} (ab)^a e^{-a} \qquad 13.$$

and also:

$$\bar{F}(\delta) = \frac{F(\delta)}{p{,}max} = \frac{\delta^a e^{-\frac{\delta}{b}}}{(ab)^a e^{-a}} = \left(\frac{e\delta}{ab}\right)^a e^{-\frac{\delta}{b}} \qquad 14.$$

Given:

$$\alpha = \frac{\delta_m}{\sigma_s} \qquad \beta = \frac{\delta}{\delta_m} \qquad 15.$$

we can also write:

$$\bar{F}(\delta) = \left(\frac{\alpha^2\beta}{\alpha^2-1}\right)^{\alpha^2-1} e^{\alpha^2-\beta\alpha^2-1} \qquad 16.$$

In order to determine the value of $\delta{,}max$, having probability $(1-\Theta)$ irrelevant for the structure strain, Fig 7 , it is necessary to evaluate the integral:

$$\Theta = \int_0^{\delta_{max}} F(\delta)d\delta \qquad 17.$$

In this case we have: $\delta{,}min=0$ 18.

while $\delta{,}max$ can also be put in the form:

$$\delta_{max} = \delta_m + K(\Theta)\sigma_\delta \qquad 19.$$

$k(\Theta)$ being a coefficient reported in function of Θ in statistics handbooks.

The strain of a structure in which the elements are defined by random quantities is of course of random nature as well. If, for example, we want to know the probable collapse load of such a structure, this load is represented by a random variable for the description of which it is necessary to know the probability distribution function. For this purpose, the "Central limit theorem " of statistics can be useful.

As the structure strain depends on the value of the aleatory variables which define each element, as these variables are independent among them, given the high number of elements and consequently of variables which are necessary to define the structures, independently from the probability distribution functions of each variable, the validity conditions of the "Central limit theorem " can be considered as satisfied and so the limit of the structure strain tends to the normal distribution. Such a condition is very important as in this case, in order to know all the information useful for the applications, it is enough to obtain information on the mean value μ_r and on the standard deviation σ_r of the strain. That is to say, it is possible to know, for example, the probability of an expected strain under given actions (for example the collapse probability under given loads) or even the loads to give in order to provake a certain event with given probability (for example, given the collapse probability limit, what are the loads which determine it).

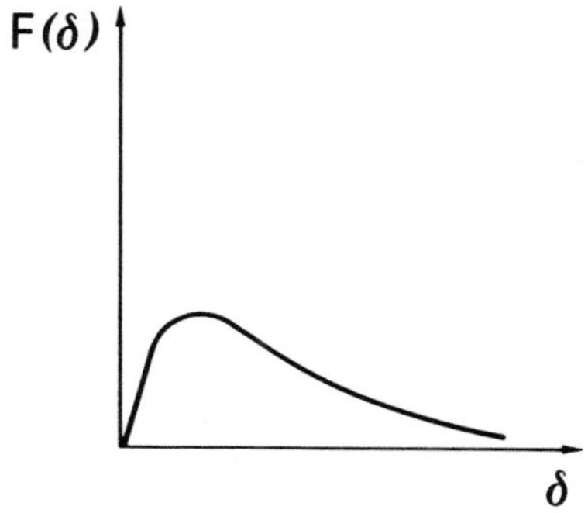

Fig 6

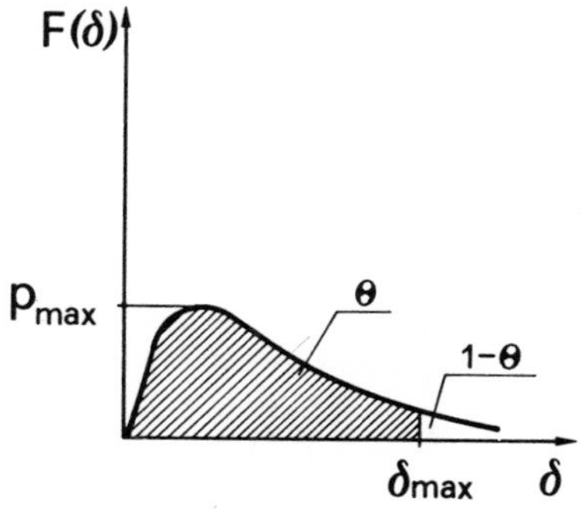

Fig 7

DETERMINATION OF THE RESPONSE OF STRUCTURES WITH RANDOM ELEMENTS

The computation of the response of a structure with random elements can be made in direct way only in some very particular cases. In general and when the structures are statically indeterminate, the direct approach is practically impossible. Methods which allow to supply useful indications are based on numerical simulation procedures. The validity of these methods, in reference to structures, has already been shown in 'Refs 3, 4, 5 and 6'. Such methods require to generate, through Montecarlo-like procedures, pseudo-random structures belonging to the family of random structures having elements of which the p.d.f.of the characteristics are known.

The authomatic generation program is very simple as it requires the generation of couples of pseudo-random numbers with uniform distribution in the 0-1 interval. If N_1, N_2 is the generated couple and we have:

$$F(N_2) \leq N_1 \qquad 20.$$

F(X) being the distribution function of the variable X

under examination, we have: $N_2=X$ 21.

linking the value X to the quantity which defines the aleatory characteristic of the generic element. If we obtain instead:

$$F(N_2) > N_1 \qquad 22.$$

the experience is continued until (20) is satisfied. The same procedure is then applied to each variable of all the elements. The expression of distribution density functions in reduced form compared to p,max and with truncation of tails is necessary, as in this way the function itself is defined in a square domain having unitary side Fig 8. Generating in this way the pseudo-random structure, its response is evaluated by considering it as deterministic. Programs for the generation of random variables with gaussian disribution and Gamma distribution are shown in appendix 2. The generation of a n number of structures allow to calculate the value of the mean response Rm and the value of the square deviation s_r. As in the "Central limit theorem " the response tends to have the normal distribution as limit, the small samples theory with $n < 30$ may be utilized in order to estimate the values required by the response.

As for very high n R response having p probability of occurring is given by:

$$R = \mu_r + t(p)\sigma_r \qquad 23.$$

where t(p) is Student's coefficient, the evaluation of R for $n < 30$ can be made by evaluating both μ_r and σ_r with a desired confidence interval Θ .

The small samples theory affirms that the best evaluation of μ_r is given by the sample mean:

$$R_m = \frac{\Sigma Ri}{n} \qquad 24.$$

and the best evaluation of the standard deviation is given by the average square deviation:

$$s_r = \sqrt{\frac{\Sigma(R_i-R_m)^2}{n-1}} \qquad 25.$$

Through the application of the significance test to these values, Student's test to the mean and χ^2 test to the standard deviation respectively, we obtain:

$$R_m - t_{n,\Theta}\frac{s_r}{\sqrt{n}} \leq \mu_r \leq R_m + t_{n,\Theta}\frac{s_r}{\sqrt{n}} \qquad 26.$$

and:

$$s_r\sqrt{\frac{n-1}{\chi^2_{n,p2}}} \leq \sigma_r \leq s_r\sqrt{\frac{n-1}{\chi^2_{n,p1}}} \qquad 27.$$

where p_1 and p_2 are linked to the confidence interval Θ by the relation:

$$\Theta = p_2 - p_1 \qquad 28.$$

as Fig 9 shows.

Therefore the evaluation of R is given by:

$$R = R_m - K(n,\Theta,p)s_r \qquad 29.$$

with:

$$K(n,\Theta,p) = \frac{t_{n,\Theta}}{\sqrt{n}} - t_\infty(p)\sqrt{\frac{n-1}{\chi^2_{n,\Theta}}} \qquad 30.$$

The $K(n,\Theta,p)$ values are given in table 1(see appendix 3).

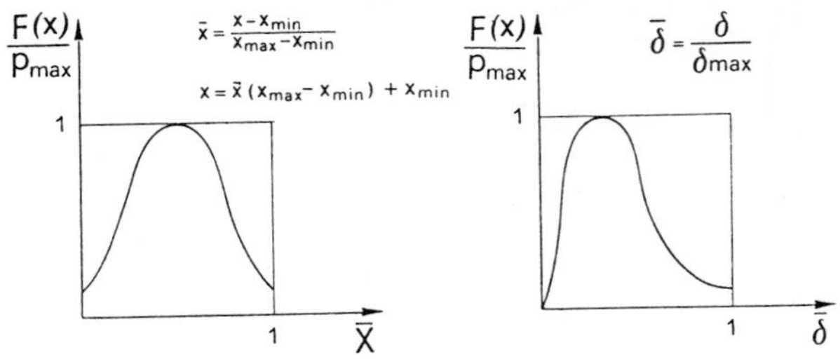

Fig 8

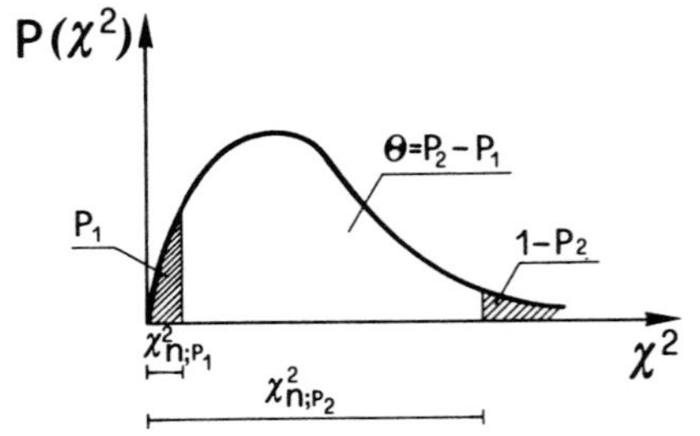

Fig 9

FINAL CONSIDERATIONS

The methods we have shown allow to analyse any kind of reticular space structures in a complete way. They require the availability of an appropriate computation program for deterministic structures and the use of an adequate computer for the dimensions of the problem to deal with. One of the program's qualifications must be a sufficient speed, as for each structure it is necessary to analyse a sample of 30 elements to be used in the determination of the probable distribution function of the response of the structures themselves.

REFERENCES

1. Z S MAKOWSKI, (ed.), Analysis, Design and Construction of Double-Layer Grids, Applied Science Publishers Ltd., 1981.

2. H NOOSHIN, Formex Formulation of Barrel Vaults, Course on the Analysis, Design and Construction of Braced Barrel Vaults, University of Surrey, Department of Civil Engineering, Vol I , 1983.

3. A LA TEGOLA, Il carico di collasso delle strutture con imperfezioni random, Giornale del Genio Civile, Roma 1979.

4. A LA TEGOLA, Simulation Methods for the Analysis of Pin-jointed Structures with randomly imperfect Members, Innovative Numerical Analysis for the Applied Engineering Sciences, Montrèal, June 1980.

5. A BADALA' and A LA TEGOLA, Sul carico di collasso delle

strutture costituite con materiale a resistenza random, Rep. n.40, Department of Structures, University of Calabria, 1981.

6. A LA TEGOLA, Probability analysis of Braced Barrel Vaults composed of random strenght elements, Course on the Analysis, Design and Construction of Braced Barrel Vaults, University of Surrey, Department of Civil Engineering, Vol II, 1983.

7. J R BENJAMIN and C A CORNELL, Probability, Statistics, and Decision for Civil Engineers, Mc Graw-Hill, Inc 1970.

APPENDIX 1

Program for the description and inputs of reticular barrel or translation vaults.

```
      DIMENSION NI(7600),NF(7600),MCN(2,2601),XMCN(3,2601)
      DIMENSION GX(27),GY(27),TX(27),TY(27),X(51),Y(51)
      DIMENSION Z(51,51),ALUNG(7600)
      READ(5,100)N,M
100   FORMAT (2I2)
      IF(N.GT.50.OR.M.GT.50) GO TO 99
      CALL PARI(N,ID)
      IF(ID.EQ.0) GO TO 99
      CALL PARI(M,ID)
      IF(ID.EQ.0) GO TO 99
      IF(N.GE.2.AND.M.GE.2) GO TO 1
99    STOP
1     EPS=0.0001
      READ (5,101) A,B,H,RA,RB
101   FORMAT (F9.4)
      AD=A/2
      BD=B/2
      NE=N+1
      ND=N/2
      NL=ND+1
      ME=M+1
      MD=M/2
      ML=MD+1
      IF(RA.EQ.0.0.AND.RB.EQ.0.0) GO TO 2
      IF(RA.EQ.0.0.AND.RB.GT.0.0) GO TO 3
      IF(RA.GT.0.0.AND.RB.EQ.0.0) GO TO 4
      CALL VELA(AD,ND,RA,EPS,GX,TX,ALX)
      CALL VELA(BD,MD,RB,EPS,GY,TY,ALY)
      CALL COORD(GY,ML,Y,ME)
      CALL COORD(GX,NL,X,NE)
      DO 62 J=1,ML
      DO 62 I=1,NL
      HZ=H-TX(I)-TY(J)
      Z(ND+I,MD+J)=HZ
      Z(ND+I,ML+1-J)=HZ
      Z(NL+1-I,ML+1-J)=HZ
      Z(NL+1-I,MD+J)=HZ
62    CONTINUE
      GO TO 12
2     ALX=A/N
      ALY=B/M


      DO 50 I=1,NE
50    X(I)=ALX*(I-1)
      DO 51 J=1,ME
51    Y(J)=ALY*(J-1)
      DO 52 J=1,ME
      DO 52 I=1,NE
52    Z(I,J)=H
      GO TO 12
3     ALX=A/N
      DO 53 I=1,NE
53    X(I)=ALX*(I-1)
      CALL VELA(BD,MD,RB,EPS,GY,TY,ALY)
      CALL COORD(GY,ML,Y,ME)
      DO 55 I=1,NE
      DO 55 J=1,ME
      IF(J.GT.ML) GO TO 6
      Z(I,J)=H-TY(ML+1-J)
      GO TO 55
6     Z(I,J)=H-TY(J-MD)
55    CONTINUE
      GO TO 12
4     ALY=B/M
      DO 57 J=1,ME
57    Y(J)=ALY*(J-1)
      CALL VELA(AD,ND,RA,EPS,GX,TX,ALX)
      CALL COORD(GX,NL,X,NE)
      DO 59 J=1,ME
      DO 59 I=1,NE
      IF(I.GT.NL) GO TO 8
      Z(I,J)=H-TX(NL+1-I)
      GO TO 59
8     Z(I,J)=H-TX(I-ND)
59    CONTINUE
12    CONTINUE
      CALL TEL8N,M,NTE,NN,NI,NF,MCN)
      DO 61 L=1,NN
      K1=MCN(1,L)
      K2=MCN(2,L)
      XMCN(1,L)=X(K1)
      XMCN(2,L)=Y(K2)
      XMCN(3,L)=Z(K1,K2)
61    CONTINUE
      CALL DIST(XMCN,NTE,ALUNG,NI,NF)
      CALL GRAF(NI,NF,NTE,ALUNG,XMCN)
      STOP
      END
      SUBROUTINE PARI(IND,K)
      I=1
3     K=0
      IF(I.EQ.IND) GO TO 2
      I=I+1
      K=1
      IF(I.EQ.IND) GO TO 2
      I=I+1
      GO TO 3
2     CONTINUE
      RETURN
      END
      SUBROUTINE VELA(CD,KD,R,EPS,EP,ET,CD1)
      DIMENSION EP(1),ET(1)
      KL=KD+1
      K=KL+1
      IC=5
      IF(R.LE.CD) STOP
      EP(1)=0.0
      ET(1)=0.0
      EP(K)=CD
      ET(K)=R-(SQRT(R**2-CD**2))
      DG=SQRT((EP(K))**2+(ET(K))**2)
      DG1=DG/KD
3     CONTINUE
      DO 1 I=2,KL
      EP(I)=EP(I-1)
      ET(I)=0.0
      DX=10.
      DO 2 L=1,IC
      DX=DX*0.1
14    EP(I)=EP(I)+DX
16    CONTINUE
      ET(I)=R-(SQRT(R**2-(EP(I))**2))
      IF(L.EQ.IC) GO TO 1
      DG=SQRT((EP(I)-EP(I-1))**2+(ET(I)-ET(I-1))**2)
      IF((DG-DG1).GT.EPS) GO TO 12
      IF((DG-DG1).LT.(-EPS)) GO TO 14
      GO TO 1
12    EP(I)=EP(I)-DX
      IF(L.EQ.IC) GO TO 16
2     CONTINUE
1     CONTINUE
      DF=SQRT((EP(K)-EP(KL))**2+(ET(K)-ET(KL))**2)
      IF(DF.LE.EPS) GO TO 18
      DF1=DF/KD
      IF(EP(K).GT.EP(KL)) GO TO 6
      DG1=DG1-DF1
      GO TO 3
6     DG1=DG1+DF1
      GO TO 3
18    CD1=DG1
      RETURN
      END
      SUBROUTINE COORD(A,K,B,I)
      DIMENSION A(1),B(1)
      DO 5 L=1,I
      IF(L.GT.K) GO TO 10
      B(L)=A(K)-A(K+1-L)
      GO TO 5
10    B(L)=A(K)+A(L-K+1)
5     CONTINUE
      RETURN
      END
      SUBROUTINE DIST(XMCN,NTE,ALUNG,NI,NF)
      DIMENSION XMCN(3,1),ALUNG(1),NI(1),NF(1)
      DO 1 I=1,NTE
      K=NI(I)
      J=NF(I)
      ALUNG(I)=SQRT((XMCN(1,K)-XMCN(1,J))**2+
     1               (XMCN(2,K)-XMCN(2,J))**2+
     1               (XMCN(3,K)-XMCN(3,J))**2)
1     CONTINUE
      RETURN
      END
      SUBROUTINE GRAF(NI,NF,NTE,ALUNG,XMCN)
      DIMENSION NI(1),NF(1),XMCN(3,1),ALUNG(1)
      WRITE(6,55)
      DO 12 I=1,NTE
      K=NI(I)
      J=NF(I)
      WRITE (6,50) I
50    FORMAT(' ELEMENTO = ',I3)
      WRITE (6,51) K,J
51    FORMAT(' NODO INIZIALE =',I3,' : NODO FINALE =',I3,';')
      WRITE (6,52)
52    FORMAT(' NODO          X             Y            Z')
      WRITE (6,53)K,XMCN(1,K),XMCN(2,K),XMCN(3,K)
      WRITE (6,53)J,XMCN(1,J),XMCN(2,J),XMCN(3,J)
53    FORMAT(' ',I3,'   ',3F12.4)
      WRITE (6,54) ALUNG(I)
54    FORMAT(' LUNGHEZZA =',F10.4)
      WRITE(6,55)
```

```
55    FORMAT(' *******************************************')
12    CONTINUE
      RFTURN
      END
      SUBROUTINF TEL(NC.NP,NTE,NN,NI,NF,MCN)
      DIMENSION NI(1),NF(1),MCN(2,1)
      NNC=NC+1
      NNP=NP+1
      NPD=NP/2
      NCD=NC/2
      NEO=NC*NNP
      NEV=NNC*NP
      NEI=NC*NP
      NTE=NEO+NEV+NEI
      NN=NNC*NNP
      K1=NCD+1
      DO 10 L=1,2
      DO 20 IP=1,NPD
      DO 30 K=1,K1,NCD
      IF(K.EQ.1) GO TO 32
      J=NC
      GO TO 34
32    J=NCD
34    CONTINUF
      DO 40 IC=K,J
      IE=IC+(IP-1)*NC+(L-1)*NC*NPD
      IF(K.EQ.1) GO TO 42
      IF(L.EQ.2) GO TO 44
36    NI(IE)=IC+(IP-1)*NNC+(L-1)*NNC*NPD

      NF(IE)=IC+1+IP*NNC+(L-1)*NNC*NPD
      GO TO 40
42    IF(L.EQ.2) GO TO 36
44    NI(IE)=IC+1+(IP-1)*NNC+(L-1)*NNC*NPD
      NF(IE)=IC+IP*NNC+(L-1)*NNC*NPD
40    CONTINUE
30    CONTINUF
20    CONTINUE
10    CONTINUF
      DO 56 IC=1,NNC
      DO 60 IP=1,NNP
      IF(IC.EQ.NNC) GO TO 52
      IEO=(IP-1)*NC+IC+NEI
      NI(IEO)=(IP-1)*NNC+IC
      NF(IEO)=NI(IEO)+1
      K=NI(IEO)
      MCN(1,K)=IC
      MCN(2,K)=IP
      I=NF(IEO)
      MCN(1,I)=IC+1
      MCN(2,I)=IP
52    IF(IP.EQ.NNP) GO TO 54
      IEV=(IP-1)*NNC+IC+NEI+NEO
      NI(IEV)=(IP-1)*NNC+IC
      NF(IEV)=IP*NNC+IC
      K=NI(IEV)
      MCN(1,K)=IC
      MCN(2,K)=IP
      I=NF(IEV)
      MCN(1,I)=IC
      MCN(2,I)=IP+1
60    CONTINUE
54    CONTINUF
56    CONTINUE
      RETURN
      END
```

APPENDIX 2

Subroutine for the generation of random variables with gaussian p.d.f.

```
      SUBROUTINE GEN2(X1,X2,SIGMA,ANHU,XMAX,XMIN)
      COMMON T1,SIGMA,ANHU
      EXTERNAL F
      T1=SECNDS(0.0)
      DEX=XMAX-XMIN
10    X1=RANDOM(0)
      X2=RANDOM(1)
      X=X2*DEX+XMIN
      IF(F(X).LE.X1) RETURN
      R=X2
      GOTO 10
      END
      FUNCTION F(X)
      COMMON T1,SIGMA,ANHU
      A=(X-ANHU)**2/(2*SIGMA**2)
      F=EXP(-A)
      RETURN
      END
      FUNCTION RANDOM(I)
      COMMON T1
      IF(I.FQ.0) GOTO 2
      DO 1 I=1,1000
1     CONTINUF
2     DELTA=SECNDS(T1)
      T11=ALOG(DELTA)
      T12=ABS(T11-AINT(T11))
      RANDOM=T12
      RETURN
```

Subroutine for the generation of random variables with Gamma p.d.f.

```
      SUBROUTINE GEN1(X1,X2,A,B,XMAX)
      COMMON T1,A,B
      EXTERNAL GAMMA
      T1=SECNDS(0.0)
10    X1=RANDOM(0)
      X2=RANDOM(1)
      XS=X2/XMAX
      IF(GAMMA(XS).LE.X1) RETURN
      R=X2
      GOTO 10
      END
      FUNCTION GAMMA(XS)
      COMMON T1,A,B
      E=EXP(-X^/B)
      F=EXP(1)*XS/(A*B)
      F=F**A
      GAMMA=E*F
      RETURN
      END
      FUNCTION RANDOM(I)
      COMMON T1
      IF(I.FQ.0) GOTO 2
      DO 1 I=1,1000
1     CONTINUF
2     DELTA=SECNDS(T1)
      T11=ALOG(DELI TA)
      T12=ABS(T11-AINT(T11))
      RANDOM=T12
      RETURN
```

APPENDIX 3

Table 1

n	ϑ=95%					ϑ=90%				
	1%	2%	3%	4%	5%	1%	2%	3%	4%	5%
1	82.54	73.94	68.47	64.36	61.01	41.71	37.35	34.58	32.50	30.80
2	17.11	15.40	14.31	13.49	12.82	11.94	10.74	9.97	9.40	8.93
3	10.26	9.24	8.60	8.12	7.72	7.97	7.17	6.67	6.29	5.98
4	7.93	7.15	6.65	6.29	5.97	6.47	5.82	5.41	5.11	4.85
5	6.75	6.09	5.66	5.34	5.08	5.67	5.11	4.75	4.48	4.25
6	6.04	5.44	5.06	4.78	4.54	5.18	4.66	4.33	4.08	3.88
7	5.57	5.01	4.66	4.40	4.18	4.85	4.36	4.05	3.82	3.63
8	5.23	4.70	4.37	4.12	3.92	4.60	4.14	3.84	3.62	3.44
9	4.96	4.46	4.15	3.91	3.72	4.40	3.96	3.67	3.46	3.28
10	4.75	4.28	3.97	3.74	3.56	4.25	3.82	3.54	3.34	3.17
11	4.52	4.12	3.83	3.61	3.43	4.13	3.71	3.44	3.24	3.07
12	4.45	4.00	3.71	3.50	3.32	4.02	3.60	3.34	3.15	2.99
13	4.32	3.89	3.61	3.40	3.23	3.93	3.52	3.27	3.07	2.92
14	4.22	3.79	3.52	3.31	3.15	3.85	3.45	3.20	3.01	2.86
15	4.13	3.71	3.44	3.24	3.08	3.78	3.39	3.14	2.95	2.80
16	4.05	3.64	3.38	3.18	3.02	3.72	3.34	3.09	2.91	2.76
17	3.99	3.58	3.32	3.12	2.96	3.67	3.29	3.04	2.86	2.71
18	3.92	3.52	3.26	3.07	2.91	3.62	3.24	3.00	2.82	2.67
19	3.86	3.47	3.21	3.02	2.86	3.58	3.20	2.97	2.79	2.64
20	3.82	3.42	3.17	2.98	2.83	3.53	3.16	2.92	2.75	2.60
21	3.76	3.38	3.13	2.94	2.79	3.50	3.13	2.90	2.72	2.58
22	3.72	3.34	3.09	2.91	2.76	3.47	3.11	2.87	2.70	2.56
23	3.68	3.30	3.06	2.88	2.73	3.43	3.07	2.84	2.67	2.53
24	3.65	3.27	3.03	2.85	2.70	3.41	3.05	2.82	2.65	2.51
25	3.62	3.24	3.00	2.82	2.68	3.38	3.02	2.80	2.63	2.49
26	3.54	3.22	2.98	2.80	2.65	3.35	3.00	2.77	2.60	2.47
27	3.55	3.18	2.95	2.77	2.62	3.32	2.97	2.75	2.58	2.44
28	3.53	3.16	2.93	2.75	2.61	3.31	2.96	2.79	2.57	2.43
29	3.50	3.14	2.90	2.73	2.59	3.29	2.94	2.72	2.55	2.42
30	3.47	3.11	2.88	2.71	2.56	3.27	2.92	2.70	2.54	2.40

STRUCTURING: A PROCESS OF MATERIAL DILUTION

J. G. PARKHOUSE, MA, CEng, MICE

Senior Engineer, Atkins Research and Development

The concept that structuring is a process of material dilution is introduced by first considering a simple structural cell and showing that its stiffness can be represented by a piece of solid material. Comparing the representative material properties with the actual material properties reveals a process of dilution. The process is shown to be recursive. Structural performance is related to representative material properties, and the Strut Problem is used to illustrate how the dilution concept can be utilised in design.

INTRODUCTION

Advances in structural analysis over recent years have been associated with the formulation of matrix methods, Ref 1, the development of finite element theory, Ref 2, and the use of computers to execute the programs embodying these techniques. Elastic stresses anywhere in any structure can now be calculated as accurately as required using commercially available computer software. In the early 1940s, Hrennikoff, Ref 3 and McHenry, Ref 4, were not in this position, but they were able to analyse frameworks and perceived that there was an analogy between solid material and a regular lattice having the same envelope, and they utilised it to analyse 2 and 3d stress problems.

Modern finite elements can be viewed as a development of Hrennikoff's framework method. No analogy is evident, as stiffness matrices derived from continuum theory of elasticity have long been substituted for frameworks, but structural analogies are still being found useful. Bridge decks are often modelled as beams or grillages of beams, Ref 5, and space frames as plates, Ref. 6. When complex structures like these can be modelled in a simpler way, their analysis is simpler.

This paper considers the converse of Hrennikoff's framework analogy, namely that every regular structure is analogous to a solid piece of material having nearly the same envelope. Analysis of the solid model is simpler though possibly not so accurate, and the derived material properties of the "material" of the solid analogue may be considered as the properties of the structure. Using photographic terminology Hrennikoff was zooming in to his structure, looking for accuracy in analysis. The converse is to zoom out to take a distant view, to consider the continuum properties of the structure as a whole. These are relevant at the design stage when considering structural form.

Comparison of these structural properties with the continuum properties of the material of which the structure is made suggests that structuring is a process of material dilution. The second section introduces this concept with a development of simple beam theory, focusing on the performance of structural members. Then structuring is looked at from a material viewpoint in the third section. Finally, both views are united in considering whole structures and their performance.

SIMPLE BEAM THEORY

A repetitive structure lies along the line AB of Fig 1a parallel to the x axis. A single cell of length L lies between Sections 1 and 2. This cell is supported at Section 1 and loaded at Section 2 by an axial force P_2, a shear Q_2 and a moment M_2. When the load distribution within the whole cell is known, the axial extension u_2, the transverse displacement v_2 and the rotation of the section θ_2 are derivable by Castigliano's theorem as:

$$u_2 = \partial U/\partial P_2 \qquad v_2 = \partial U/\partial Q_2 \qquad \theta_2 = \partial U/\partial M_2 \qquad \ldots\ldots 1.$$

U, the internal energy, is a function of P_2, Q_2 and M_2 and the geometry of the structural cell. These three equations may be written as:

$$\begin{bmatrix} u_2 \\ v_2 \\ \theta_2 \end{bmatrix} = \begin{bmatrix} f_{11} & f_{12} & f_{13} \\ f_{21} & f_{22} & f_{23} \\ f_{31} & f_{32} & f_{33} \end{bmatrix} \begin{bmatrix} P_2 \\ Q_2 \\ M_2 \end{bmatrix} \qquad \ldots\ldots 2.$$

or simply as $d_2 = FR_2$ ...... 3.

where d_2 is the displacement vector, R_2 is the load vector and F is a flexibility matrix.

By the reciprocal theorem F is symmetric, i.e. $f_{21} = f_{12}$ etc. Therefore the unit cell has six flexibilities: f_{11}, f_{22}, f_{33}, f_{12}, f_{13} and f_{23}.

If the cell were symmetrical about AB, reflection of the cell about this axis, as shown in Fig 1b, would not affect the flexibility at end 2. Related to the cell's local axis system:

$$\begin{bmatrix} u_2 \\ -v_2 \\ -\theta_2 \end{bmatrix} = \begin{bmatrix} f_{11} & f_{12} & f_{13} \\ f_{12} & f_{22} & f_{23} \\ f_{13} & f_{23} & f_{33} \end{bmatrix} \begin{bmatrix} P_2 \\ -Q_2 \\ -M_2 \end{bmatrix} \qquad \ldots\ldots 4.$$

from which it follows that:

$$f_{12} = f_{13} = 0 \qquad \ldots\ldots 5.$$

If the cell were symmetrical about a perpendicular axis CD, reflection of the cell about this axis, as shown in Fig 1c, would not affect the flexibility at end 2 either. Then:

$$\begin{bmatrix} u_2 \\ L\theta_2 - v_2 \\ \theta_2 \end{bmatrix} = \begin{bmatrix} f_{11} & f_{12} & f_{13} \\ f_{12} & f_{22} & f_{23} \\ f_{13} & f_{23} & f_{33} \end{bmatrix} \begin{bmatrix} P_2 \\ -Q_2 \\ M_2 + Q_2 L \end{bmatrix}$$

from which it would follow that:

$$f_{12} = \tfrac{1}{2}Lf_{13} \qquad \text{...... 6.}$$

and $$f_{23} = \tfrac{1}{2}Lf_{33} \qquad \text{...... 7.}$$

Hence, with both these axes of symmetry, there are only three flexibilities, f_{11}, f_{22} and f_{33} with the other three defined by Eqns 5 and 7. Cells with these properties will be referred to as orthotropic.

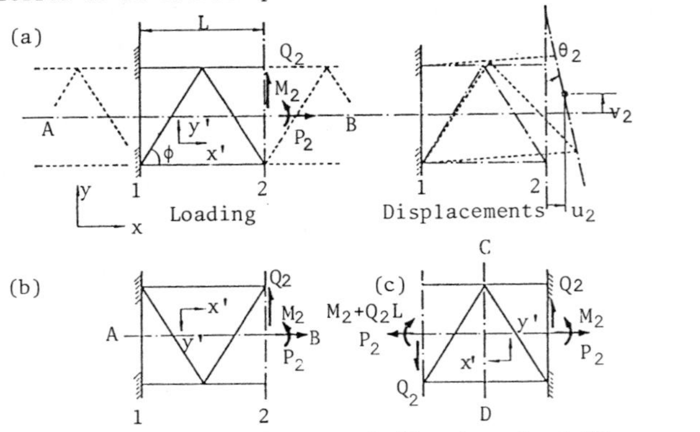

Reflection about AB — Reflection about CD

Fig 1

The frame shown in the unit cell of Fig 1 is not symmetric about AB, yet does behave orthotropically when the framework is pin-jointed. Because it is statically determinate, member loading is simple to calculate and the displacement vector relates to the points where the loading is applied. For such cases Eqn 3 is exact for small displacements. For redundant cell structures assumptions need to be made about the internal distribution of load in order to be able to calculate F. Related to this difficulty is an uncertainty about the meaning of the displacement vector; for example, an initially plane section will not generally remain plane when loaded.

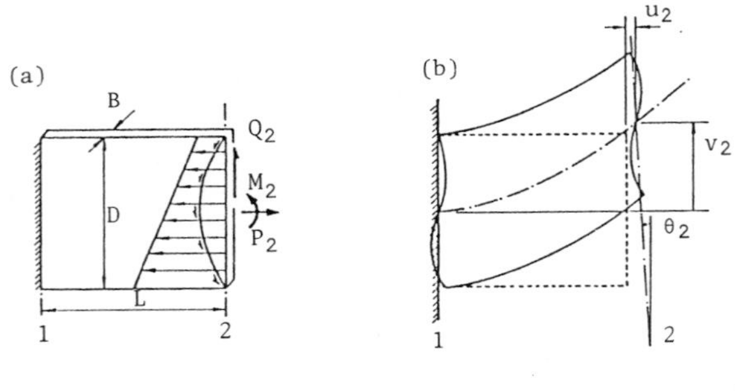

Fig 2

Consider a solid rectangular section as an example of a redundant cell structure. As shown in Fig 2a, the cell has dimensions L, B and D. The loading at end 2 is assumed to be distributed through the cell according to simple beam theory as illustrated. Using this distribution to determine U, the displacement vector at end 2 may be calculated from Eqn 1. The displacement vector describes a displaced position for section 2, but cannot describe how the section is curved as illustrated in Fig 2b, due to the parabolic variation of shear stress. It is consistent with simple beam theory to ignore this incompatibility and to assume sections always remain plane, even under shear.

The three flexibility coefficients f_{11}, f_{22} and f_{33} can be calculated from Eqn 1 as:

$$\begin{aligned} f_{11} &= L/EA \\ f_{22} &= 1.2L/GA + L^3/3EI \\ f_{33} &= L/EI \end{aligned} \qquad \text{where:} \qquad \text{...... 8.}$$

E = Young's modulus of the material
G = Shear modulus of the material
A = Sectional area, BD
I = "Inertia" of the section, $BD^3/12$

The flexibility matrix of any orthotropic cell of length L may be written in the following form:

$$\begin{bmatrix} L/k_1 & 0 & 0 \\ 0 & L/k_2+L^3/3k_3 & L^2/2k_3 \\ 0 & L^2/2k_3 & L/k_3 \end{bmatrix} \qquad \text{...... 9.}$$

where k_1, k_2 and k_3 are principal stiffnesses. For the solid rectangular section:

$$\begin{aligned} k_1 &= EA \\ k_2 &= GA/1.2 \\ k_3 &= EI \end{aligned} \qquad \text{......10.}$$

From the flexibility matrix F, which relates d_2 to R_2 by Eqn 3, a transfer matrix T_L can be derived which relates d_2 and R_2 to d_1 and R_1. d_2 and R_2 are concatenated to form a state vector S_2 of six elements which fully defines the displacements and internal reactions at section 2. S_1 is defined similarly. The sign convention used is consistent with Fig 1, showing displacements and forces in a positive sense. The forces are those which the greater x side applies to the lesser x side. When the cell is orthotropic, the state vectors are related as follows:

$$\begin{bmatrix} u_2 \\ v_2 \\ \theta_2 \\ P_2 \\ Q_2 \\ M_2 \end{bmatrix} = \begin{bmatrix} 1 & 0 & 0 & L/k_1 & 0 & 0 \\ 0 & 1 & L & 0 & L/k_2-L^3/6k_3 & \tfrac{1}{2}L^2/k_3 \\ 0 & 0 & 1 & 0 & -\tfrac{1}{2}L^2/k_3 & L/k_3 \\ 0 & 0 & 0 & 1 & 0 & 0 \\ 0 & 0 & 0 & 0 & 1 & 0 \\ 0 & 0 & 0 & 0 & -L & 1 \end{bmatrix} \begin{bmatrix} u_1 \\ v_1 \\ \theta_1 \\ P_1 \\ Q_1 \\ M_1 \end{bmatrix} \qquad \text{......11.}$$

$$\text{or: } S_2 = T_LS \qquad \text{......12.}$$

By multiplying the two matrices term by term, it can be shown that, taking L as the only variable in T_L:

$$T_XT_Y = T_{X+Y} \qquad \text{......13.}$$

for any values of X and Y. Therefore the transfer matrix for n cells strung together is given by:

$$(T_L)^n = T_{nL} \qquad \text{......14.}$$

Hence, Eqn 11 defines the transfer matrix of a member made up of any number of cells when L is redefined as the member's length. It follows that the deflections of such a member at the ends of each cell can be described by simple beam theory, and for orthotropic celled members, the flexibility of the member is defined by the three stiffnesses k_1, k_2 and k_3 which are properties of the constituent material and the cell geometry.

The same set of stiffnesses defines the structural behaviour of a family of different cross-sections and cell structures. It would seem appropriate to represent all these different sections by the simplest, such as a circular, square or rectangular solid section. Denoting the representative properties by a prime, the axial and flexural properties are represented correctly when:

$$\begin{aligned} E'A' &= k_1 = EA \\ E'I' &= k_3 = EI \end{aligned} \qquad \text{......15.}$$

$$\therefore I'/A' = k_3/k_1 = I/A$$

$$\therefore r'^2 = k_3/k_1 = r^2 \qquad \text{......16.}$$

Where r denotes the radius of gyration. Hence, the radius of gyration of the representative solid section can be determined, and after choosing a shape for the section, its size can also be determined. Then A' is known, and from Eqn 15, E' can be calculated.

The area A in the axial stiffness EA usually has physical significance as the net sectional area of material which is axially oriented; in a lattice member for example, the bracing does not contribute significantly to A. It will be convenient to introduce:

$$i = E/E' = A'/A \qquad \text{......17.}$$

While material on its own makes up the area A, both material and space make up A', which is why it is appropriate to call i a dilution factor. Fig 3 shows various sections with their representative solid sections shown shaded. The

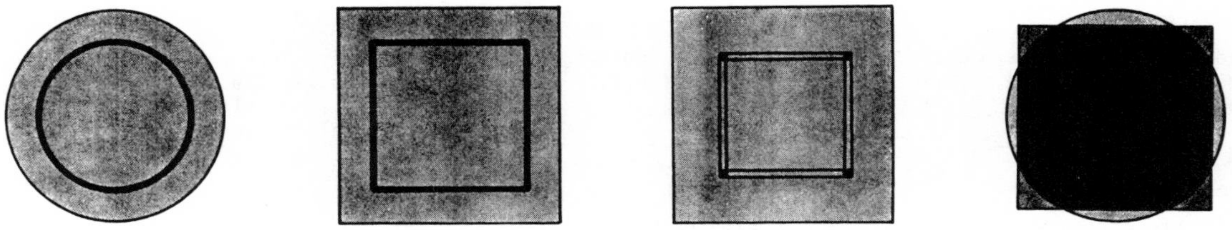

Section:	Circular tube dia. d. thk. t	Square tube side d, thk. t	Square lattice side d, txt	Solid square side d
Representative Section:	Solid circle dia. D	Solid square side D	Solid square side D	Solid circle dia. D
D/d	$\sqrt{2}$	$\sqrt{2}$	$\sqrt{3}$	$2/\sqrt{3}$
Dilution factor	$\frac{1}{2}d/t$	$\frac{1}{2}d/t$	$0.75(d/t)^2$	$\pi/3$

Fig 3

principal geometric relationships and dilution factors are written beneath each.

The shear stiffness k_2 can be linked to the representative section by calculating a shear modulus G', or else it can be linked to k_1 or EA in the following way:

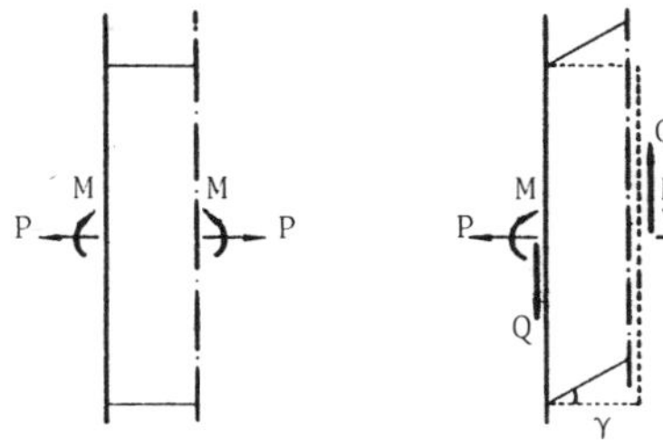

Fig 4

Fig 4 shows a short length δL of representative beam, whose behaviour is governed by Eqn 11. Since δL is small, any terms associated with L^2 or L^3 can be ignored, and therefore moments do not induce significant sway. An axial load P, tensile being positive, is applied to the section, together with a moment M, when a shear Q is subsequently applied which produces a sway rotation of γ. If the internal load distribution produced by the sway is unaffected by the loads P and M, then the increase in internal energy due to the sway is given by:

$$\delta U = \tfrac{1}{2}k_2\gamma^2\delta L \qquad \text{...... 18.}$$

Work done by the force Q is given by:

$$\delta E_Q = \tfrac{1}{2}Q\gamma\delta L \qquad \text{...... 19.}$$

Work done by the applied loads P and M is given by:

$$\delta E_P = P\delta L(1-\cos\gamma) = -\tfrac{1}{2}P\gamma^2\delta L \qquad \text{...... 20.}$$

If the strain produced by P is ε, then:

$$P = k_1\varepsilon \qquad \text{...... 21.}$$

Equating δU to $\delta E_Q + \delta E_P$, the apparent sway stiffness k_s is given by:

$$k_s = Q/\gamma = k_2 + k_1\varepsilon \qquad \text{...... 22.}$$

Hence, when ε is zero, $k_s = k_2$, but as P increases, so does the apparent sway stiffness. This tension stiffening effect has been shown to account for torsional stiffening of bars under tension, Ref 7. Eqn 22 is of particular relevance to structural members when k_2 may be much smaller than k_1. Then a sway instability is predicted when:

$$\varepsilon = -\varepsilon_s \text{ say } = -k_2/k_1 \qquad \text{...... 23.}$$

because k_s vanishes at this axial strain.

It is therefore suggested that the shear stiffness k_2 should be regarded as $k_1\varepsilon_s$. For solid members made of isotropic material, ε_s is always greater than 0.4. At such strains, simple first and second order theories are not applicable, so Eqn 23 does not predict instablity in such materials. Lightly braced lattice members may have quite low values of ε_s, at which strain Eqn 23 does predict instability, however stocky the member might be.

Flexural instability of members is described by Euler's equation as occurring at the Euler load P_e given by:

$$P_e = \pi^2 EI/L^2 \qquad \text{...... 24.}$$

ignoring the effects of shear flexibility. For a solid circular section of diameter D, Eqn 24 may be written as:

$$\varepsilon_e = \frac{\pi^2}{16}\frac{D^2}{L^2} \qquad \text{...... 25.}$$

revealing that the Euler strain is a function only of slenderness L/D, and is not influenced by the properties of the material within the member.

The effect of shear flexibility on member stability is considered in Ref 8. Manipulation of the formula given in this reference yields a critical buckling strain, ε_c, given by:

$$\frac{1}{\varepsilon_c} = \frac{1}{\varepsilon_e} + \frac{1}{\varepsilon_s} \qquad \text{...... 26.}$$

If ε_c is the critical strain a member is designed to, then attention must be given to the critical strains of the component members of that member, to ensure that they are also sufficiently stable at ε_c.

In this section, it has been shown how orthotropic structuring of an orthotropic material into a line member can be represented by an equivalent solid line member made of a derived hypothetical orthotropic material, having more dilute properties than the original material. Compressive performance has been shown to be limited by sway and flexural instabilities.

MATERIAL DILUTION

The structuring of material implies the formation of surfaces separating the space into those parts occupied by material and those parts that are not. When material and space are mixed together in this way, the mixture may be regarded either as voided or as dispersed material; either way, the material has to be shaped, and this process can be described as structuring. The words "nodal", "lineal" and "surface" have been chosen to describe the formation of a void or a piece of material about a point, a line or a surface respectively.

Suppose material is dispersed uniformly in space, lineally, with all lines of material oriented parallel to the x_1-axis. If the boundaries of this space are displaced and distorted by a uniform strain field, defined by (ε_{11}, ε_{22}, ε_{33}, γ_{12}, γ_{13}, γ_{23}) then the stress in the material is given by:

$$\sigma_{11} = E\varepsilon_{11} \qquad \text{...... 27.}$$

where E is its Young's modulus. The whole volume can be represented by a solid material stressed only in the x_1 direction by a stress:

$$\sigma_{11}' = E'\varepsilon_{11} = mE\varepsilon_{11} \qquad \text{...... 28.}$$

where:

m is the proportion of the total volume occupied by material. m will be referred to as a material concentration factor. The other five components of stress are zero.

Suppose material is dispersed uniformly in planes parallel to the x_1 and x_2 axes, then the representative properties of the solid filling the whole space are given by:

$$E_{11}' = mE_{11}$$
$$E_{22}' = mE_{22}$$
$$G_{12}' = mG_{12}$$
$$\nu_{12}' = \nu_{12} \qquad \text{...... 29.}$$

The density of the representative material ρ' is also related to that of the actual material ρ by:

$$\rho' = m\rho \qquad \text{...... 30.}$$

and the dilution factor i, as defined in Eqn 17, can have six values:

$$i_{11} = E_{11}/E_{11}' = 1/m$$
$$i_{22} = E_{22}/E_{22}' = 1/m$$
$$i_{12} = G_{12}/G_{12}' = 1/m$$
$$i_{33} = i_{13} = i_{23} = \infty \qquad \text{...... 31.}$$

When the space is strained in the x_3 direction, there is no resistance; no material is stressed. This may be considered a structuring inefficiency and η may be considered a structural efficiency factor defined as:

$$\eta = 1/mi \qquad \text{...... 32.}$$

At best η can be one. When m is one, there is no structuring, just solid material, and all six ηs are one, indicating no inefficiency. Any structuring of material will introduce a degree of inefficiency. In the example above, three ηs equal unity, but three are zero. This example is only efficient for transmitting loading in the x_1x_2 plane.

By combining material dispersed in several different orientations, the structuring becomes more practical, possessing stiffness in all directions. Its representative properties may be simply calculated by transforming the local stiffness matrix of each set of material to a global axis system. Summing them and inverting the resulting stiffness matrix yields the representative flexibility matrix, and hence the required properties. Material property transformations are described in Refs 9 and 10.

Fig 5a shows a random distribution of straight lines on a plane. This configuration usefully illustrates some general principles of structuring. Consider it first as being a section through a set of planes parallel to the x_1 axis. When material is dispersed about these planes, it forms a prismatic structure with axisymmetric properties. Using the method described above, it can be shown that:

$$E_{11}' = mE_{11}$$
$$E_{22}' = mE_{22}/(3-2\alpha\nu_{12}^2)$$
$$G_{12}' = mG_{12}/2$$
$$\nu_{12}' = \nu_{12}$$
$$\nu_{23}' = (1-2\alpha\nu_{12}^2)/(3-2\alpha\nu_{12}^2)$$
$$G_{23}' = mE_{22}/8(1-\alpha\nu_{12}^2) \qquad \text{...... 33.}$$

where $\alpha = E_{22}/E_{11}$

from which the following material efficiencies may be deduced:

$$\eta_{11} = 1$$
$$\eta_{12} = 1/2$$
$$\eta_{22} \simeq 1/3 \qquad \text{...... 34.}$$

Along the principal axis, the only inefficiency is a factor of 2 on shear performance, which will result in ε_s' being half of ε_s. This form of structuring is the most efficient means of diluting material in one direction. Timber and all prismatic members share these performance characteristics to some extent, but cross grain properties are sensitive to curvatures and discontinuities in the sectional geometry.

The compressive strength of such a structure may be governed by instability in its members. A long panel of width L and thickness t, with its long sides simply supported, has a critical compressive strain, when loaded lengthways of (Ref 8):

$$\varepsilon_c = \frac{\pi^2}{3(1-\nu^2)}\frac{t^2}{L^2} \qquad \text{...... 35.}$$

which is dependent on the slenderness λ defined as:

$$\lambda = L/t \qquad \text{...... 36.}$$

Slenderness itself is dependent on m. Fig 5 shows several regular tessellations. In order to achieve the maximum possible ε_c for a given m, no member should be more slender than any other. Assuming each has equal slenderness, λ can be shown to be inversely proportional to m. The value of λm for each tessellation is given in the figure. Honeycomb offers the lowest λm, and so offers the highest ε_c for a given material concentration.

(a) Random (b) Square (c) 3-Way (d) Honeycomb

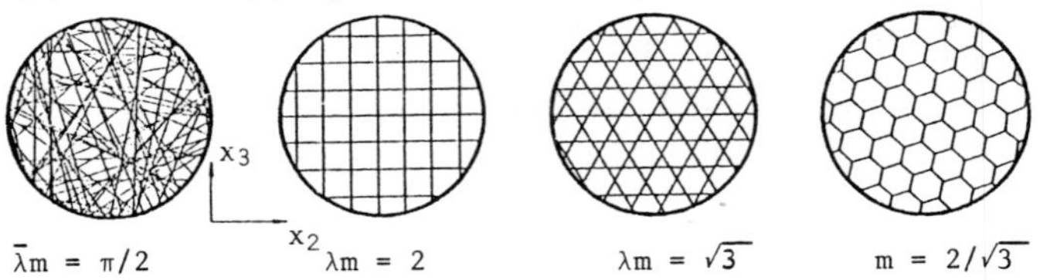

$\bar{\lambda}m = \pi/2$ $\lambda m = 2$ $\lambda m = \sqrt{3}$ $m = 2/\sqrt{3}$

Fig 5

For the random tessellation the $\bar{\lambda}$ in $\bar{\lambda}m$ is a mean slenderness and inevitably there must be scatter about the mean. However, $\bar{\lambda}$ is not much smaller than the λ for the three way lay-up which indicates that it is likely to be a lower bound of λ for this type of continuous tessellation.

A slice of this structuring, w deep in the x_1 direction, represents a panel of lineal members. Apart from the random orientation of its members, this is typical of a panel of bracing. When w is much smaller than $\bar{\lambda}$, E_{22} is the only material property that affects the performance of the panel and it can be seen from Eqn 33 that:

$$\eta_{22} = 1/3$$
$$\nu_{23}' = 1/3 \qquad \text{...... 37.}$$

Instability of this panel may be governed by λ. Eqn 24 may be rewritten as:

$$\varepsilon_c = \pi^2/12\lambda^2 \qquad \text{...... 38.}$$

for a rectangular sectioned member, ignoring shear flexibility.

The structuring just described may be considered as a material perforation; either as a deep perforation of a volume or as a shallow perforation of a surface. In both cases, the effects are approximately described by the following equations:

Efficiency	η	$= 1$ or $1/3$
Slenderness	λ	$\simeq \frac{1}{2}\pi/m$
Buckling strain	ε_c	$\simeq 2m^2$ or $\frac{1}{2}m^2$ ...... 39.

A dilute, low density structure is achieved when m is small. However, it is likely to have instability problems according to the last equation. Suppose a material concentration factor of 1/100 were sought, but the corresponding buckling strain of approximately 0.02% (Eqn 39) were unacceptable. A possible solution would be to dilute in two stages, each time by a factor of 1/10; the two stages are on different scales, one stage structuring the material out of which the other stage is made. A detail of both stages together is shown in Fig 6. The buckling strain would then be calculated by Eqn 39 as 2%.

This example introduces the recursive nature of the structuring process. Since structural material has the same type of properties as material, structured material can itself be structured. Indeed, structures can be structured, and the divide between material and structure is not sharply defined. This process is not confined to two stages of

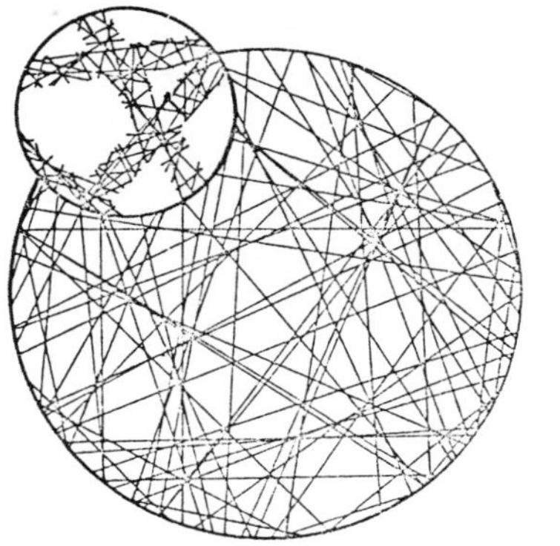

Fig 6

structuring. If Eqn 39 defines the process for one stage, for n stages:

$$\text{Material concentration factor, } m_n = m^n$$
$$\text{Efficiency, } \eta_m = 1 \text{ or } 1/3^n \qquad \ldots\ldots\ 40.$$

λ and ε_c are as defined by Eqn 39 for each stage of structuring. Notice that the efficiency can stay as unity for prismatic structuring, but the price to pay is in the geometrical intricacy and thinness of material.

The structured material just considered is not isotropic. When the planes of material are oriented completely randomly, the resulting structure is isotropic and may be shown to have the following properties:

$$E' \simeq \tfrac{1}{2}mE$$
$$\nu' \simeq 1/5 \qquad \ldots\ldots\ 41.$$

Each plane is covered with random lines of intersection, so would look like Fig 5a. It can be shown that the mean slenderness $\overline{\lambda}$, based on the mean side length of all panels divided by the common thickness t, is again given by:

$$\overline{\lambda} = \tfrac{1}{2}\pi/m \qquad \ldots\ldots\ 42.$$

which is equal to $\overline{\lambda}$ for prismatic perforation, as given in Fig 5a. All planes of material can therefore be perforated, as a further stage of structuring, each plane being left with m times its original material. The structure then becomes a random lattice of members with thickness t and mean slenderness given by Eqn 42.

The material concentration factor for this second stage is $\tfrac{1}{2}$m, not m, due to the fact that every lattice member lies on the intersection of two planes. Hence, another stage of dilution has been achieved, without a change in slenderness or a drop in buckling stress, and without increased geometric intricacy or reduction in thickness. The other properties of this isotropic lattice are:

$$E' = mE/6$$
$$\nu' = 1/4 \qquad \ldots\ldots\ 43.$$

The effects of isotropic latticing are approximately described by the following equations.

$$\text{Efficiency} = 1/6$$
$$\text{Slenderness} \simeq \pi/2\sqrt{2m}$$
$$\text{Buckling strain } \varepsilon_c \simeq 2m/3 \qquad \ldots\ldots\ 44.$$

Finally, the effect of structural shear on member flexure will be considered. Fig 7 shows a node in a structure with members spanning in directions OX, OY and OA. If all nodes are identical and the structure is under uniform strain, the nodes assume displacements which are compatible with the strain of the structure regarded as a material. Figure 7 shows the extreme case of maximum shear strain γ coinciding with the orientation of members OX and OY. When members OX are rigid, members OY suffer the maximum end rotations possible, which is γ.

It can be shown that the flexural strain in a prismatic member subjected to equal end rotations γ is always much less than γ, for all slenderness ratios and for all axial strains.

The peak shear strain is never greater than twice the peak

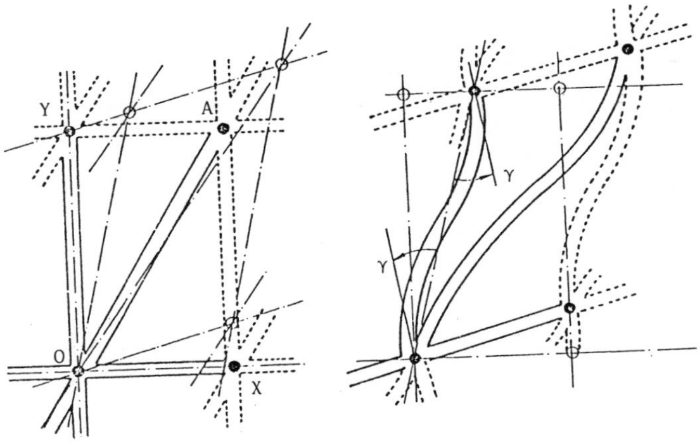

Fig 7

direct strain. Hence, in triangulated structures with prismatic members, member flexure is never a dominant effect when it is caused solely by structural shear.

This section has demonstrated how spatial and surface structuring can be described in terms of continuum mechanics without reference to any boundaries which would identify structures. Structuring is simply related to patterns and to their proportions. It is recursive; structuring can itself be structured, and the utilisation of high strength material in diluted form is likely to require more than one stage of structuring. One stage of structuring results in lineal dispersion of material about a surface, as might be achieved by perforating a solid surface. It also results in surface dispersion of material in space. If these surfaces are themselves perforated in a second stage of structuring, then lineal dispersion in space, ie latticing, is achieved.

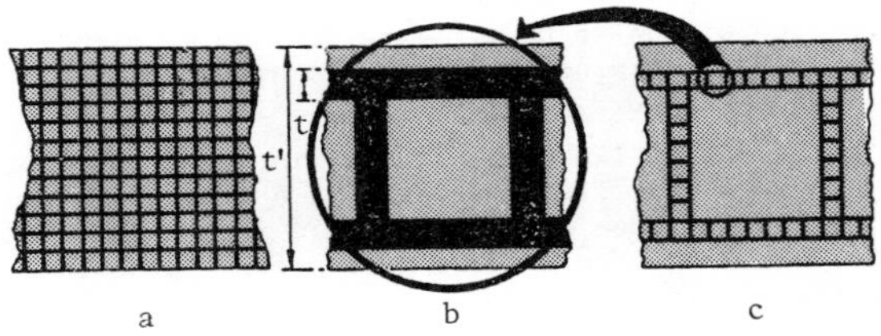

Fig 8

STRUCTURAL PERFORMANCE

The last section dealt with material dilution without reference to structure boundaries, and consequently only considered multiple cell, not single cell, structuring. An example of multiple cell structuring is shown in Fig 8a. The shaded envelope is its representative solid plate. Fig 8b shows the single cell structure which shares the same representative material and solid plate thickness; A and I are the same for both, and the slendernesses of the members and ε_c are almost identical. The difference is one of scale; the multiple cell section has thinner members and many more of them. The greater intricacy of multiple cell structuring is the reason why single cells are usually chosen to cover the shortest dimension of a structure.

Fig 8c, which shows single cell second order structuring, has been drawn to illustrate the difference between multiple cell and higher order structuring. The second order structure is intricate, but has a far smaller A and I than the other two. ε_c is unchanged, but dilution has been achieved. The multiple cells of Fig 8a do not contribute to dilution any more than the single cell of Fig 8b.

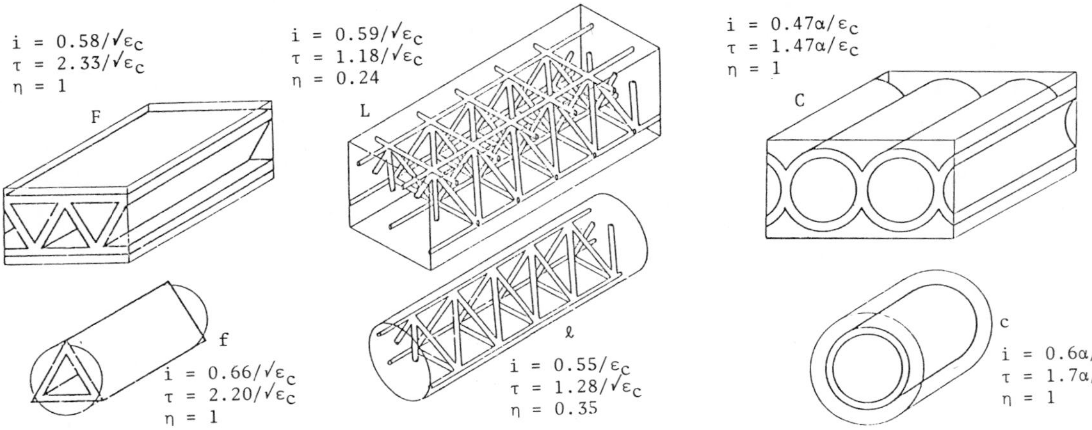

Fig 9

Two material thicknesses are relevant to each stage of structuring; that of the representative material and that of what will be referred to as the "parent" material. In the first stage of structuring, the parent material is actual material, whilst in subsequent stages it is the representative material of the previous stage. The thickness t' of the representative material is always much bigger than the thickness t of the parent material, and a thickness ratio τ will be defined as:

$$\tau = t'/t \qquad \ldots\ldots 45.$$

The effect of curvature on structural performance is generally adverse, reducing stiffness and strength by introducing bending and torsion into members. Curved shells however, can offer exceptional performance. Cylindrical shells are locally more stable than prismatic members made out of flat plate. Ref 8 gives the local buckling strain ε_c of a cylinder of radius R and thickness t as:

$$\varepsilon_c = t/R\sqrt{3(1-\nu^2)} \qquad \ldots\ldots 46.$$

In practice, the buckling strain is considerably less than indicated when t/R is very small, Ref 11. This loss of buckling performance may be described by multiplying the right hand side of Eqn 46 by α. α may not be much smaller than unity at high strains, and falls to around 0.1 at buckling strains of 10^{-5}, Ref 11.

All unit cells can be linked to form a line or a surface, depending on whether cells are linked in one or two directions. Linking in three directions would form a solid. The laterial linking of cylinders does present practical and theoretical difficulties, but so as not to make an exception of this structural form, it will be assumed that a cylinder may be joined across a diameter to two adjacent cylinders to produce surface structuring.

Three types of cell are illustrated in Fig 9, which represent typical ways of structuring flat plate, beams and curved plate. Each type of structuring can generate surfaces or lines of hierarchical structuring. It is convenient to assign a letter to a particular cell geometry and to use upper and lower case to differentiate between surface and line generation. Accordingly, the letters F, f, L, ℓ, C and c have been designated to the three cell types of Fig 9.

The F cell structure is prismatic, made up of flat equilateral plates. The lattice cell, L, is made up of members of two different lengths all having the same diameter. Members parallel to the surface in L have the same length. Diagonal members are slightly longer. The C cell is a cylindrical tube. Beneath the illustrations the properties of structuring, i, τ and η, are given in terms of ε_c. These have been derived by the techniques already described.

We are now in a position to calculate i, τ and η in terms of ε_c for any structuring using F, L or C and any higher order combinations of them. The effect of combining them hierarchically is found by multiplying together the contributions of each of the stages. For example, if material is formed into a double-skinned deck structure (F), which is then wrapped into a cylinder (c) and a number of these cylinders are formed into a lattice beam (ℓ), and the slenderness of members in each stage is such that ε_c is the same for each, then:

$$\text{Dilution factor, } i = i_\ell i_c i_F$$
$$\text{Thickness ratio, } \tau = \tau_\ell \tau_c \tau_F$$
$$\text{Efficiency, } \eta = \eta_\ell \eta_c \eta_F \qquad \ldots\ldots 47.$$

The resulting structural properties are given by:

$$E' = E/i$$
$$t' = t/\tau$$
$$\rho' = \rho/\eta i \qquad \ldots\ldots 48.$$

and ε_c is the local buckling strain in the principal direction throughout the structure. Conceptually, we have at our disposal a continuous range of materials of known pro-

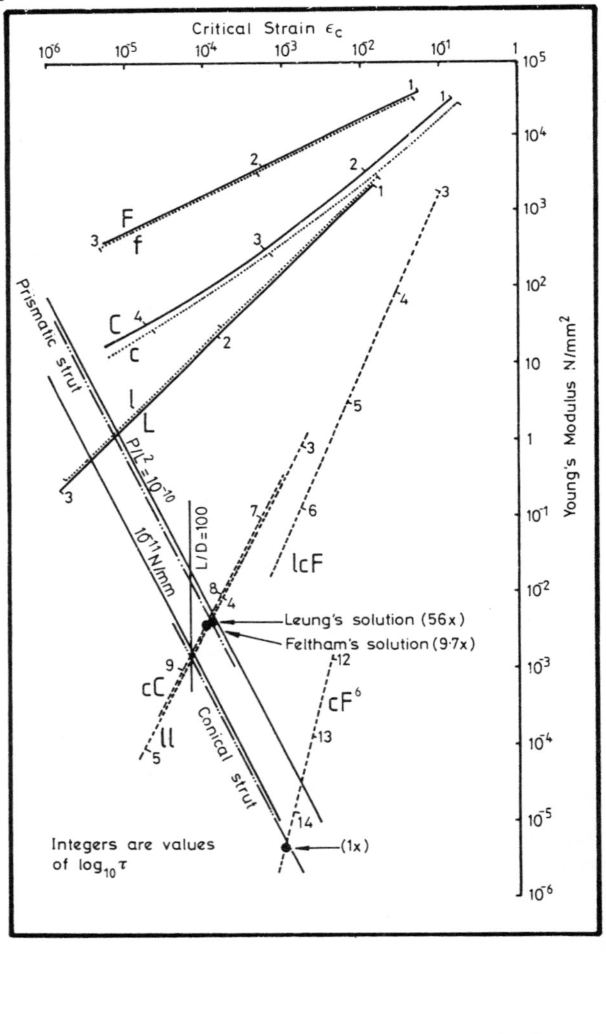

Fig 10

perties. Choosing an appropriate structural form is akin to material selection.

Fig 10 illustrates the structuring process. Critical strain ε_c is measured horizontally, and Young's modulus E' of the representative solid material is measured vertically both to log scales. For convenience, an actual material will be considered which has an E of 10^5 N/mm^2, making the top of the graph correspond to i = 1, and E' = 10^4 corresponding to i = 10 etc. Using the expressions for i in Fig 9, six lines can be drawn. The integers are values of $\log_{10}\tau$ and represent the degree of intricacy of the structuring. Notice that there is very little to distinguish the upper and lower case lines from each other. In the absence of all imperfection, the lines C and c practically coincide with L and ℓ. The curving of C and c is the result of choosing realistic values of α.

The line corresponding to the example of structuring in the order ℓcF (Eqn 47) is also shown. Because of the log scale it can be constructed by suspending the ℓ, c and F lines consecutively down from the strain axis.

Establishing that conceptually there is a full spectrum of materials has been an analytical process. The design process is different as a result; the designer has a wider choice of materials for the design of a solid structure. For example, consider designing a strut of length L to carry a force P. If the structure loading coefficient, P/L^2 (Ref 12) is small, then the strut is likely to be highly structured and overall buckling will be a principal parameter. If the strut is a solid cylinder of diameter D:

$$P = \pi D^2 E' \varepsilon_c / 4 \qquad \ldots\ldots\ 49.$$

If ε_c is given by Eqn 25, ignoring the effect of shear flexibility, then eliminating D from Eqn 49:

$$P/L^2 = 4E' \varepsilon_c / \pi \qquad \ldots\ldots\ 50.$$

This equation defines a family of diagonal straight line contours on Fig 10, two of which are illustrated. Where any such line cuts proposed structured material lines, there is a "just stable" solution. There will always be a choice between simple structures, working at lower strains using more material and more intricate structures using less material working at higher strains. There may of course be a stiffness criterion in the design that overrides that of strength.

Reproduced on this page is the Strut Problem. It was first published in "New Civil Engineer" (Ref 13) beside an article which introduced the concept of structuring as a process of material dilution. 20 entries were received from a readership of nearly 50,000.

THE STRUT PROBLEM

Deep in space where there is no extraneous influence, two equal and equally charged 10m diameter spheres are prevented from moving apart by a 1mm diameter wire, which in this situation is stressed to its working load. The wire is made of 'materium', a hypothetical uniform material. Its original length of 1km is extended under load by 1m.

Now assume the charge of one sphere is reversed. How many times more materium would be the minimum needed to form a strut to keep the spheres apart and what form of structure would it take if:

(a) the minimum manufacturable thickness of materium were 1mm?

(b) there is no lower limit of thickness?

Assume materium has no manufacturing constraints. Any shape and size can be fabricated free of imperfections. Its elastic properties and limiting stress are the same in compression and tension.

The strut must be designed to be stable.

The design criterion for the Strut Problem is effectively given by a structure loading coefficient of 7.85×10^{-11} N/mm^2 on Fig 10, labelled "prismatic strut", and a target strain of 10^{-3}. D. Leung won the first part of the problem, by designing 1mm wire into a latticed lattice ℓℓ structure. I. Feltham produced the winning solution to the second part, with the lightest valid entry. He considered the wall of his cylindrical tube to be made up of tubular straws. At 9.7 times as heavy as the tie, his was the lightest valid solution. He was the only entrant to utilise material dilution in his calculations.

Both solutions are marked on Fig 10; Leung's as ℓℓ and Feltham's as cC. As permitted by the conditions of the problem, α has been taken as one, which is why cC is straight and much further down than would be anticipated from the positions of the curves c and C above. Both struts have diameters close to 10m, the sphere diameter. The vertical line "L/D = 100" indicates that to the left of this line prismatic struts are satisfactory. To the right of it some tapering will begin to be necessary, and eventually a fully tapered strut will be required, which will need extra dilution at midspan. The design line for a conical strut with solid 1mm diameter ends is illustrated.

Had Feltham gone to two further orders of structuring, cC^3, he would have achieved a fully stressed strut which used no more material than the tie, a factor of one solution. The author's factor of one solution was cF^6, and for the first part he proposed a stayed strut, being an unstressed 1mm wire framework staying the fully stressed 1mm tie wire in compression. Freed from compressive strain, the stays do not themselves need staying, and with the greater freedom of structural form admitted, a strut using only 30 times more material was possible, compared with Leung's factor of 56. All solutions are illustrated in Ref 14.

CONCLUSION

This paper has demonstrated that structuring is a process of material dilution; that there is no clear divide between structure and material, and that the handling of structured material is an analytical process, not a design process. In introducing these ideas, the aim has been to lay a firm theoretical foundation and not to draw any practical conclusions.

ACKNOWLEDGEMENTS

The author wishes to thank his wife Nina, his colleagues at Atkins Research and Development, Dr. John Parkinson and the editorial staff of New Civil Engineer, and Sir Alan Harris for his advice and encouragement.

REFERENCES

1. R K LIVESLEY, Matrix Methods of Structural Analysis, 2nd ed., Pergamon Press, 1975

2. O C ZIENKIEWICZ, The Finite Element Method, 3rd ed., McGraw-Hill, 1977.

3. A HRENNIKOFF, Solution of Problems of Elasticity by the Framework Method, J. of Appl. Mech., Dec. 1941.

4. D McHENRY, A Lattice Analogy for the Solution of Stress Problems, J. of Inst. Civ. Eng., Dec. 1943.

5. E C HAMBLY, Bridge Deck Behaviour, Chapman and Hall, 1976.

6. Z S MAKOWSKI, Ed. Analysis, Design and Construction of Double-Layer Grids, Applied Science Publishers, 1981.

7. M A BIOT, Increase of Torsional Stiffness of a Prismatic Bar due to Axial Tension, J. of Appl. Phys., Vol. 10, 1939.

8. S P TIMOSHENKO and J M GERE, Theory of Elastic Stability, 2nd Ed., McGraw-Hill, 1961.

9. R L BISPLINGHOFF, J W MAR and T H H PIAN, Statics of Deformable Solids, Addison-Wesley, 1965.

10. A H NAYFEH and M S HEFZY, Effective Constitutive Relations for Large Repetitive Frame-Like Structures, Int. J. Solids & Structures, Vol. 18, No. 11, 1982

11. C G FOSTER, On the Buckling of Thin-Walled Cylinders Loaded in Axial Compression, J. of Strain Analysis, Vol. 16, No. 3, 1981.

12. H L COX, The Design of Structures of Least Weight, Pergamon Press, 1965.

13. J PARKINSON, Structures Analysed Anew, New Civil Engineer, No. 513, 21st Oct. 1982.

14. J PARKINSON, Ultimate Strut Solution, New Civil Engineer, No. 521, 23/30 Dec. 1982.

A SMALL STRAIN LARGE ROTATION THEORY AND FINITE ELEMENT FORMULATION OF THIN CURVED LATTICE MEMBERS

DR. L.F. BOSWELL

Department of Civil Engineering,
The City University, London, England

Accounting for the effects of curvature is one of the difficulties in the large deformation formulation of the behaviour of thin beam or lattice members. This is particularly so when the finite element method is used with elements having rotational degrees of freedom since moments referred to a fixed axis are non-conservative. An exact two-dimensional large rotation theory is proposed and alternative Lagrangian formulations of the theory presented. The theory has been incorporated into a family of curved beam elements which have been specially developed.

INTRODUCTION

Space and lattice frames which are constructed from inter-connected curved and straight thin beam members provide considerable flexibility and architectural effect when used to span large areas.

The linear analysis of such frames may be carried out conveniently using the stiffness method, for instance. If the individual members are flexible, however, and displacements not insignificant, then a geometrically non-linear analysis is appropriate. Such members whether curved or straight may be considered to undergo small strains, but large rotations.

In this paper an exact two-dimensional large rotation theory, which is based on an intrinsic co-ordinate system, has been developed. Four alternative Lagrangian formulations of the theory have been presented for comparison.

The incremental equilibrium equations which have been formulated are incorporated into a family of two-dimensional thin curved beam finite elements. The constraint technique has been used to develop the elements with a convective co-ordinate system. The elements may be used to analyse curved lattice members.

The approach may be extended to include the large deformation analysis of three-dimensional lattice structures.

TWO DIMENSIONAL LARGE DEFORMATION CURVED BEAM THEORY

Internal virtual work in terms of the Green strain

Using usual notation and referring to Fig. 1 the Green Strain tensor in two dimensions is

$$\varepsilon = \frac{1}{2}\left(\begin{bmatrix} \frac{\partial \bar{R}}{\partial x} \cdot \frac{\partial \bar{R}}{\partial x} & \frac{\partial \bar{R}}{\partial x} \cdot \frac{\partial \bar{R}}{\partial y} \\ \frac{\partial \bar{R}}{\partial y} \cdot \frac{\partial \bar{R}}{\partial x} & \frac{\partial \bar{R}}{\partial y} \cdot \frac{\partial \bar{R}}{\partial y} \end{bmatrix} - \begin{bmatrix} 1 & 0 \\ 0 & 1 \end{bmatrix} \right) \qquad 1$$

in which

$$\frac{\partial \bar{R}}{\partial x} \cdot \frac{\partial \bar{R}}{\partial x} = (1 - K\,y)^2(1 + 2e)$$

$$\frac{\partial \bar{R}}{\partial x} \cdot \frac{\partial \bar{R}}{\partial y} = (1 - K\,y)\hat{g}.\hat{n} = 0 \qquad 2$$

$$\frac{\partial \bar{R}}{\partial y} \cdot \frac{\partial \bar{R}}{\partial y} = \hat{n}.\hat{n} = 1$$

K is a measure of the physical curvature and y is the distance of a point from the reference line. The Green Strain measure of the reference is given by

$$e = \frac{du}{dx} + \frac{1}{2}\left(\frac{du}{dx}\right)^2 + \frac{1}{2}\left(\frac{dv}{dx}\right)^2 \qquad 3$$

Also, using convective differentiation, the natural base vector after deformation is

$$\hat{g} = \frac{d\bar{r}}{dx} = \left(1 + \frac{du}{dx}\right)\hat{x} + \frac{dv}{dx}\hat{y} \qquad 4$$

and

$$n = \frac{\vec{N}}{|N|} = \frac{-dv/dx\ \hat{x} + (1 + du/dx)\hat{y}}{(1 + 2e)^{\frac{1}{2}}} \qquad 5$$

Substituting Eqn. 2 into Eqn. 3 provides only one non-zero strain component ε_{xx} which is defined as

$$\varepsilon'_t = \varepsilon_{xx} = (1 - Ky)^2\left(\frac{1}{2} + e\right) - \frac{1}{2} \qquad 6$$

In a Total Lagrangian frame of reference the internal virtual work expression is given by, see Ref. 1

$$\delta W_{int} = \int_{L_o} \int_A \delta\varepsilon_o'^T S_o (1 + 2e)^{1/2} dA\, dL_o \quad 7$$

where S_o is the 2nd Piola-Kirchoff stress. The integration is over the initial area A and the initial length L_o. The expression given by Eqn. 7 is exact. Defining a new curvature term χ which is an explicit function of the displacement gradient as

$$\chi = -(1 + 2e)^{3/2} K \quad 8$$

by taking the variation of the strain in Eqn. 6, Eqn. 7 becomes

$$\delta W_{int} = \int_{L_o} \left[M\delta\chi + \tilde{N}\delta e \right] dL_o \quad 9$$

where the stress resultants P_o and M are given by

$$P_o = \int_A (1 - K\, y) S_o dA, \quad M = \int_A (1 - K\, y) y\, S_o dA \quad 10$$

and

$$\tilde{N} = (1 + 2e)^{1/2} P_o + 2K(1 + 2e)^{1/2} M \quad 11$$

The axial force measure $\tilde{N}$ is a function of both the strain e and the curvature K. Whilst e can be assumed small compared to unity, the curvature K cannot always be neglected without introducing an error. This is particularly so for problems where both curvatures and rotations are large and wherein the effect of axial force is of importance. Retaining this term in the virtual work expression, however, results in an unsymmetrical tangent stiffness matrix and hence it is neglected. This amounts to assuming direct proportionality between the 2nd Piola-Kirchoff stress and the Green-Lagrange strain in the virtual work expression and neglecting the coupling between the axial and bending stress resultants. Thus, the strain at any point is defined in terms of the generalised strain resultants as

$$\varepsilon_o' = \begin{bmatrix} 1 & y \end{bmatrix} \begin{Bmatrix} e \\ \chi \end{Bmatrix} = H\varepsilon_o \quad 12$$

and the stress is

$$S_o = E\, \varepsilon_o' = E\, H\varepsilon_o \quad 13$$

where E is Young's modulus.

The internal virtual work expression now takes the following form

$$\delta W_{int} = \int_A \delta\varepsilon_o^T \int_A H^T E\, H\, dA\, \varepsilon_o\, dL_o = \int_{L_o} \varepsilon_o^T\, \bar{D}\, \varepsilon_o dL_o \quad 14$$

in which $\bar{D}$ is the rigidity matrix.

Internal Virtual Work in terms of the Conventional Strain

The geometrical measures of strain as unit stretch and angle change are

$$\varepsilon_{ii} = \left(\frac{\partial\bar{R}}{\partial x_i} \cdot \frac{\partial\bar{R}}{\partial x_i}\right)^{1/2} - 1 \quad i=1,\ 3; \quad j=1,\ 3$$

$$\gamma_{ij} = \frac{\partial\bar{R}}{\partial x_i} \cdot \frac{\partial\bar{R}}{\partial x_j} \quad i \neq j \quad 15$$

where the shear strains γ_{ij} are assumed to be small. This assumption is valid for the thin beams considered here for which the shear strain is assumed zero. From Eqns. 2 and 15 it can be seen that the only non-zero strain term is ε_{xx}, as before, and is given by

$$\varepsilon_t' = \varepsilon_{xx} = (1 - Ky)(1 + 2e)^{1/2} - 1 \quad 16$$

and its variation is

$$\delta\varepsilon_t' = -y(1 + 2e)^{1/2}\delta K + (1 - Ky)(1 + 2e)^{-1/2}\delta e \quad 17$$

The strain displacement relations can be approximated from Eqn. 6 by using the binomial series expansion and neglecting third and higher order terms in displacement gradients, so that

$$\varepsilon_t' \simeq \begin{bmatrix} 1 & y \end{bmatrix} \begin{Bmatrix} \frac{du}{dx} + \frac{1}{2}\left(\frac{dv}{dx}\right)^2 \\ -\frac{d^2v}{dx^2}\left(1 - \frac{du}{dx}\right) + \frac{dv}{dx}\frac{d^2u}{dx^2} \end{Bmatrix} \quad 16a$$

Eqn. 16a can also be obtained by writing the Green strains in terms of deformation and rotation tensors and neglecting second and higher order terms in strains. In what follows, however, Eqn. 16 may be used in its exact form.

The strain components are redefined as follows:

$$\varepsilon_t^* = \begin{Bmatrix} e^* \\ \chi^* \end{Bmatrix} = \begin{Bmatrix} (1 + 2e)^{1/2} - 1 \\ -(1 + 2e)^{1/2} K \end{Bmatrix} \quad 18$$

Taking variations and substituting gives

$$\delta\varepsilon_t' = \delta e^* + y\delta\varepsilon^* = \begin{bmatrix} 1 & y \end{bmatrix} \begin{Bmatrix} \delta e^* \\ \delta\chi^* \end{Bmatrix} = H\delta\varepsilon_t^* \quad 19$$

The stress S_t is assumed to be proportional to the strain ε_t'. Therefore, from Eqns. 16 and 18

$$S_t = E\, \varepsilon_t' = E((1 + 2e)^{1/2} - 1 - K(1 + 2e)^{1/2} y) = EH\varepsilon_t^* \quad 20$$

The curvature term χ^* is then defined in terms of the explicit function of displacement gradients χ as

$$\chi^* = \frac{\chi}{1 + 2e} \quad 21$$

Therefore

$$\delta\varepsilon^* = \begin{Bmatrix} \delta e^* \\ \delta\chi^* \end{Bmatrix} = \begin{bmatrix} \frac{1}{(1 + 2e)^{1/2}} & 0 \\ \frac{-2\chi}{(1 + 2e)^2} & \frac{1}{(1 + 2e)} \end{bmatrix} \begin{Bmatrix} \delta e \\ \delta\chi \end{Bmatrix} = H^*\delta\varepsilon \quad 22$$

From Eqns. 19 and 20 the internal virtual work expression now takes the form

$$\delta W_{int} = \int_{V_t} \delta\varepsilon_t'^T\, S_t\, dV_t = \int_{L_t} \delta\varepsilon_t^T H^{*T} \bar{D} \varepsilon_t^* dL_t \quad 23$$

where $\bar{D}$ is the rigidity matrix.

This virtual work expression is equally applicable to both the Updated and Total Lagrangian formulations.

TOTAL LAGRANGIAN FORMULATION BASED ON GREEN STRAIN MEASURE (TLG)

The strains at a general point are given in terms of the displacement gradients as

$$\varepsilon'_o = H\left(\begin{Bmatrix} \frac{du_o}{dx} \\ \frac{d^2v_o}{dx^2} \end{Bmatrix} + \begin{Bmatrix} \frac{1}{2}\left(\frac{du_o}{dx}\right)^2 + \frac{1}{2}\left(\frac{dv_o}{dx}\right)^2 \\ \frac{dv_o}{dx}\frac{d^2u_o}{dx^2} - \frac{du_o}{dx}\frac{d^2v_o}{dx^2} \end{Bmatrix}\right) = H\{\varepsilon^o_o + \varepsilon^L_o\} \quad 24$$

in which ε^o_o is the infintesimal strain and is written, in a finite element representation, in terms of the nodal variables a_o as

$$\varepsilon^o_o = \begin{Bmatrix} \frac{du_o}{dx} \\ -\frac{d^2v_o}{dx^2} \end{Bmatrix} = \bar{B}_o\, a_o \quad 25$$

and ε^L_o is the nonlinear strain which can be written as

$$\varepsilon^L_o = \frac{1}{2}\begin{bmatrix} \frac{du_o}{dx} & \frac{dv_o}{dx} & 0 & 0 \\ -\frac{d^2v_o}{dx^2} & \frac{d^2u_o}{dx^2} & \frac{du_o}{dx} & \frac{dv_o}{dx} \end{bmatrix}\begin{Bmatrix} \frac{du_o}{dx} \\ \frac{dv_o}{dx} \\ -\frac{d^2v_o}{dx^2} \\ \frac{d^2u_o}{dx^2} \end{Bmatrix}$$

$$= \frac{1}{2}\bar{B}_L(a_o)a_o = \frac{1}{2}A_\theta\theta_o \quad 26$$

In a finite element representation defining the vector θ_o in terms of the nodal degrees of freedom a_o we have

$$\theta_o = \left\{\frac{du_o}{dx}, \frac{dv_o}{dx}, -\frac{d^2v_o}{dx^2}, \frac{d^2u_o}{dx^2}\right\}^T = G_o a_o \quad 27$$

Taking variations with respect to the nodal variables, the strain-displacement matrix B is given by

$$B = B_o + B_L(u_o) = H\left[\bar{B}_o + \bar{B}_L(a_o)\right] = H\bar{B} \quad 28$$

The tangent stiffness matrix becomes

$$K_T = \int_{V_o} B^T D\, B\, dV_o + \int_{V_o} \frac{\partial B}{\partial u} S_o\, dV_o = (K_o + K_L(a_o)) + K_\sigma \quad 29$$

where

$$\bar{S}_o = \int_{A_o} H^T S_o dA_o = \{P_o, M_o\}^T \quad 30$$

is the vector of initial stress resultants composed of an axial force P_o and a bending moment M_o.

The explicit form of the initial stress stiffness matrix K_σ is obtained from Eqns. 25, 26 and 27 as follows

$$K_\sigma = \int_{L_o} \frac{\partial \bar{B}^T}{\partial a}\bar{S}_o\, dL_o = \int_{L_o} \frac{\partial \bar{B}_L}{\partial a}\bar{S}_o\, dL_o = \int_{L_o} G_o^T \bar{P}_{oi}\, G_o\, dL_o \quad 31$$

where $\bar{P}_{oi}$ is the initial stress resultant matrix.

Using the tangent stiffness matrix and the residual nodal forces, the displacement increments Δa^i_o are evaluated. The total displacements are then obtained as

$$a^{i+1}_o = a^i_o + \Delta a^i_o \quad 32$$

The incremental strain resultants are defined by

$$\Delta\varepsilon^i_o = \left[\bar{B}_o + \bar{B}_L(a^i_o) + \frac{1}{2}\bar{B}_L(\Delta a^i_o)\right]\{\Delta a^i_o\} \quad 33$$

The increments of the stress resultants are given by

$$\Delta\bar{S}^i_o = \bar{D}\,\Delta\varepsilon^i_o \quad 34$$

and the total stress resultants are

$$\bar{S}^{i+1}_o = \bar{S}^i_o + \Delta\bar{S}^i_o \quad 35$$

From which the nodal residual forces are evaluated as follows

$$-\psi^{i+1} = R - \int_{L_o} \bar{B}^T \bar{S}^{i+1}_o\, dL_o \quad 36$$

where R is the vector of applied equivalent nodal forces and

$$\bar{B} = \bar{B}_o + \bar{B}_L(a^{i+1}_o)$$

Using the new total displacements Eqn. 32 and the new total stress resultants Eqn. 35 the tangent stiffness matrix can be reformed and hence used with the residuals given by Eqn. 36 to obtain a new set of displacement increments.

UPDATED LAGRANGIAN FORMULATION BASED ON GREEN STRAIN MEASURE (ULG)

The procedure follows the previous TLG formulation but the reference is now with respect to the configuration at time t. By replacing the subscripts with t the relevant equations are obtained.

The nodal residuals to be used to evaluate a new set of displacement increments thus become

$$-\psi^{i+1} = R - \int_{L_t} \bar{B}^T_t\, \bar{S}^{i+1}_t\, dL_t \quad 37$$

COMBINED UPDATED AND TOTAL LAGRANGIAN FORMULATION BASED ON GREEN STRAIN MEASURE (UTLG)

In this formulation the coordinates are updated at the beginning of each load increment only. The iterations within the increment are carried out using the Total

Lagrangian formulation. As a consequence of this, the displacements from the beginning of the increment until converence (a_c) must be stored.

Thus, the strain displacement matrix is defined as

$$B = B_t + B_L(u_c) = H\left[\bar{B}_t + \bar{B}_L(a_c)\right] = H\bar{B}_c \qquad 38$$

where t refers to the beginning of a new increment. The tangent stiffness matrix now takes the following form

$$K_T = \int_{L_t} \bar{B}_c^T \bar{D} \bar{B}_c \, dL_t + \int_{L_t} G_t^T \bar{P}_{ti} G_t \, dL_t$$

$$= (K_t + K_L(a_c)) + K_\sigma \qquad 39$$

The matrix G_t is given by Eqn. 27 but is evaluated at the beginning of a new increment only and $\bar{P}_{ti}$ is the initial stress matrix.

The displacements a_c are evaluated as follows

$$a_c^{i+1} = a_c^i + \Delta a_t^i \qquad 40$$

with their value being zero at the beginning of each load increment.

The incremental strain resultants are defined as

$$\Delta\varepsilon_t^i = \left[\bar{B}_t + \bar{B}_L(a_c^i) + \frac{1}{2}\bar{B}_L(\Delta a_t^i)\right]\{\Delta a_t^i\} \qquad 41$$

and the nodal stress residuals are given by

$$-\psi^{i+1} = R - \int_{L_t} \bar{B}_c^T \bar{S}_t^{i+1} \, dL_t \qquad 42$$

TOTAL LAGRANGIAN FORMULATION BASED ON CONVENTIONAL STRAIN MEASURE (TLC)

The variation in stress is defined from Eqns. 20 and 22 as

$$\delta S_o = E\,H\,H^* \,\delta\varepsilon_o = D\,H\,H^* \,\delta\varepsilon_o \qquad 43$$

The strain resultant measures ε_o and their variations $\delta\varepsilon_o$ are given in terms of the displacement gradients and the nodal variables a_o by

$$\varepsilon_o = \begin{Bmatrix} \dfrac{du_o}{dx} \\ -\dfrac{d^2v_o}{dx^2} \end{Bmatrix} + \begin{Bmatrix} \dfrac{1}{2}\left(\dfrac{du_o}{dx}\right)^2 + \dfrac{1}{2}\left(\dfrac{dv_o}{dx}\right)^2 \\ \dfrac{dv_o}{dx}\dfrac{d^2u_o}{dx^2} - \dfrac{du_o}{dx}\dfrac{d^2v_o}{dx^2} \end{Bmatrix}$$

$$= \varepsilon_o^O + \varepsilon_o^L = \bar{B}_o a_o + \frac{1}{2}\bar{B}_L(a_o)a_o = \bar{B}_o a_o + \frac{1}{2}A_\theta \theta_o$$

$$\delta\varepsilon_o = \bar{B}_o \,\delta a_o + \bar{B}_L(a_o)\,\delta a_o = \bar{B}\,\delta a_o \qquad 44$$

Thus from these relations and Eqn. 23 the non-linear equilibrium equations are

$$\psi = \int_{L_o} \bar{B}^T H^{*T} \bar{S}_o \, dL_o - R = 0 \qquad 45$$

and the tangent stiffness matrix, $K^T = \partial\psi/\partial a$ is given by

$$K_T = (K_o + K_L(a)) + K_\sigma + K_\sigma^* \qquad 46$$

in which K_σ^* is an additional initial stress stiffness matrix.

The components of K_T can be found , see Ref. 2, and the incremental equilibrium solved for displacement increments Δa_o^i and total displacements a_o^{i+1}, Eqn. 32.

ASSOCIATED FINITE ELEMENT FAMILY

The formulations which have been presented are based on the use of convected coordinates and associated element formulations must include these coordinates. Whilst the theory is based on material coordinates at every point along element reference line, the instrinsic coordinates need to be defined only at the integration points.

The correct definition of such a coordinate system is achieved by using the constraint technique. In the constraint technique geometric relationships are used to define discretely the shape of the element before deformation and the variation of variables referred to the instrinsic coordinate system. The elements presented are based on the constraint technique applied to displacement finite element models, in which independent interpolation of the displacements and rotations (or displacement derivatives) is used.

Three curved beam elements have been developed using the constraint technique, see Ref. 2. These are:

1. ISOBEM1 element which is a non-conforming curved beam element with C^0 continuity of in-plane displacement and C^1 continuity of the out-of-plane displacement. An additional internal degree of freedom is introduced to ensure continuity between the reference line and the cross-section centroid for eccentric elements.

2. ISOBEM2 element which is a non-conforming beam element with C^1 continuity of both displacement components.

3. SUBBEAM element which is a beam element with C^1 continuity for both displacement components and complete conformity of the displacement field.

EXAMPLES

A number of examples have been solved to assess the performance of the elements and the numerical effectiveness of the alternative formulations for geometric non-linearity.

Symmetrical Buckling of Two-Hinged Deep Arch

The buckling response of a two-hinged deep arch, Fig. 2 has been obtained previously, see Ref. 3, using the Total Lagrangian formulation and paralinear elements. Half the arch was idealised with eight elements. Sixty equal displacement increments with Newton-Raphson iterations were applied up to a total crown displacement of 0.36 R. Fig. 3 shows a plot of the central load versus deflection for the zero beam element. It can be seen from this figure that the results are in very close agreement with the results of Ref. 3. Table 1 provides values of the buckling loads obtained from all solutions.

Cantilever under Pure Moment

This example was chosen to demonstrate the capability of the theory and of the elements to deal with very large rotations.

The cantilever, see Fig. 4, was modelled by six equal elements. The moments were applied in thirty increments to bend the cantilever into a complete circle. The results shown in Fig. 4 compare favourably with the results of the exact solution, Ref. 1. Fig. 5 shows the displacements at the free end obtained from six subbeam elements.

The Post-bifurcation of a simply supported beam

This problem which has been presented in Ref. 4 is one of the classical elastica problems. The simply supported beam, see Fig. 6, has a slenderness ratio of 2.81×10^3 and was modelled by ten equal elements. Ten equal displacement increments of L/S were applied at the pinned end. An initial imperfection was assumed in order to initiate buckling. The undeformed geometry was defined by a sine curve. Fig. 6 shows the load-deflection curve obtained by the subbeam elements.

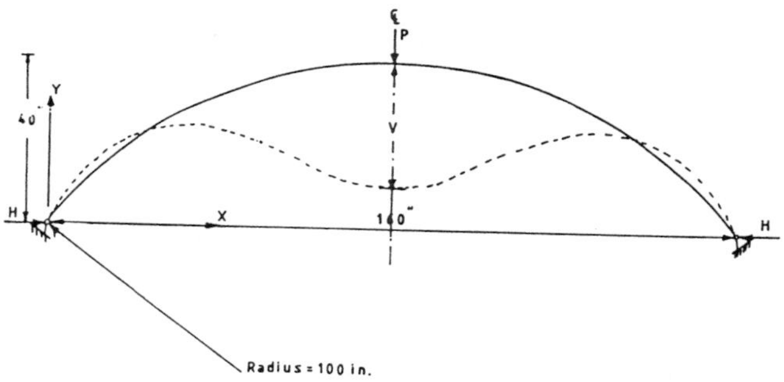

Fig. 2 Two-hinged deep arch geometry

CONCLUSION

An exact two-dimensional large rotation theory for thin beam or lattice members has been proposed. Four alternative Lagrangian formulations of the theory have been presented. The theory has been incorporated into a family of finite elements and may be used to solve the large displacement problems which may occur in lattice or space structures.

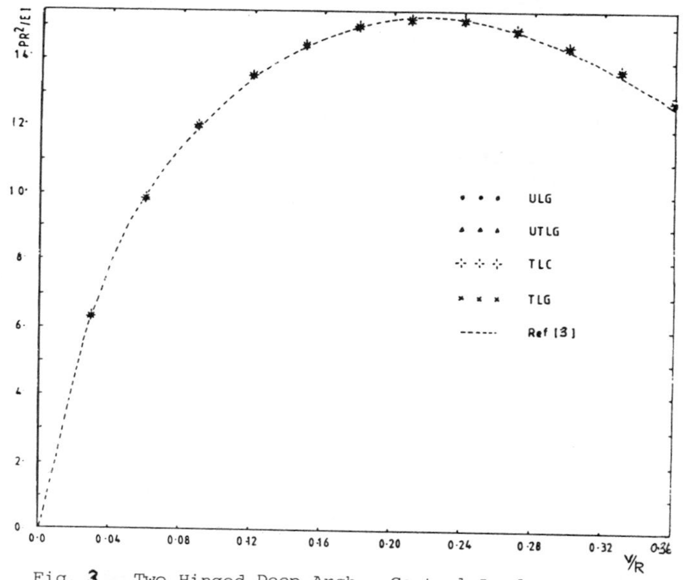

Fig. 3. Two Hinged Deep Arch - Central Load versus Deflection

REFERENCES

1. Marcelo Epstein and David W. Murray, 'Large Deformation In-Plane Analysis of Elastic Beams', Computers and Structures, Vol. 6, pp. 1-9, (1976).

2. Mohamed, A.E., 'A small strain large rotation theory and finite element formulation of thin curved beams', Ph.D thesis, The City University, London, 1983.

3. R.D. Wood and O.C. Zienkiewicz, 'Geometrically Non-linear Finite Element Analysis of Beams, Frames, Arches and Axisymmetric Shells', Computers and Structures, Vol. 7, pp. 725-735, (1977).

4. H.D. Hibbit, E.B. Becker and L.M. Tayler, 'Non-linear Analysis of some Slender Pipelines', Computer Methods Appl. Mech. Engrg., 17/18, pp. 203-225, (1979).

Element Type	Buckling Load ($\times \frac{EI}{R^2}$)			
	ULG	UTLG	TLC	TLG
ISOBEM 1	15.23	15.31	15.06	15.04
ISOBEM 2	15.23	15.23	15.31	15.29
SUBBEAM	15.25	15.25	15.26	15.25

These closely agree with the value of 15.3 $\frac{EI}{R^2}$ (and 15.2 $\frac{EI}{R^2}$) given in ref. [3].

Table 1. Values of buckling loads

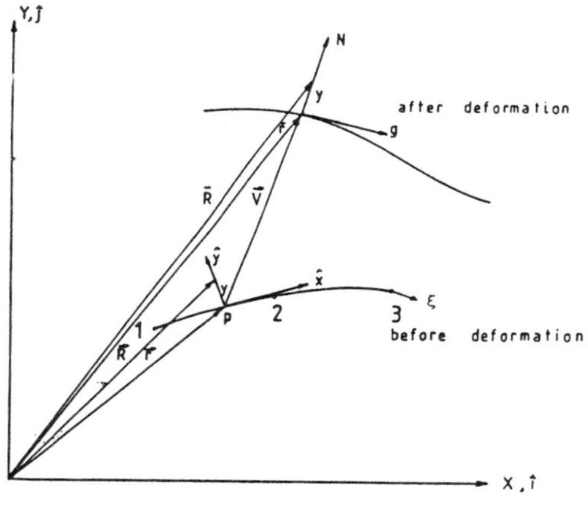

Fig. 1 Geometry of Deformation

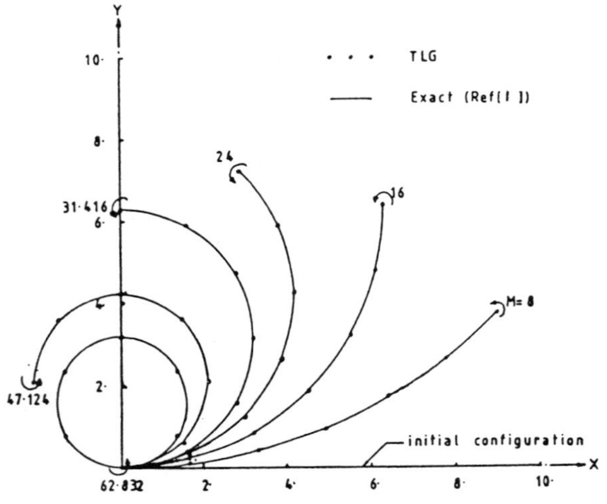

Fig.4 Cantilever subjected to a pure moment

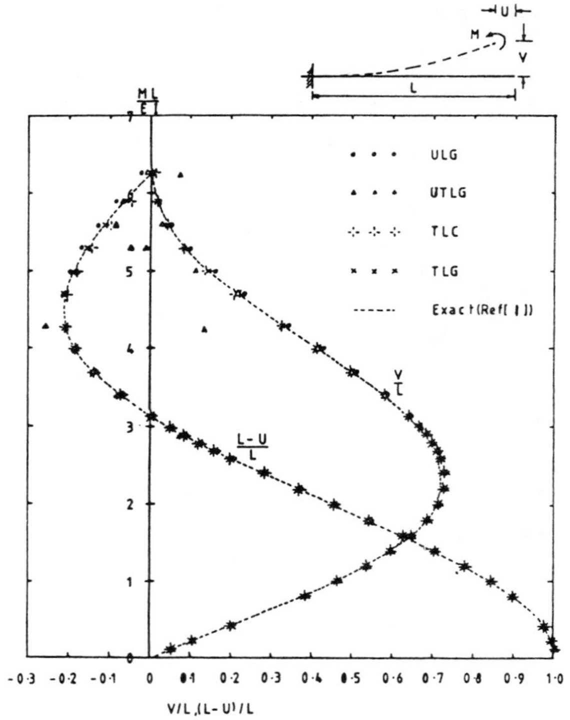

Fig.5 Free end displacement of cantilever

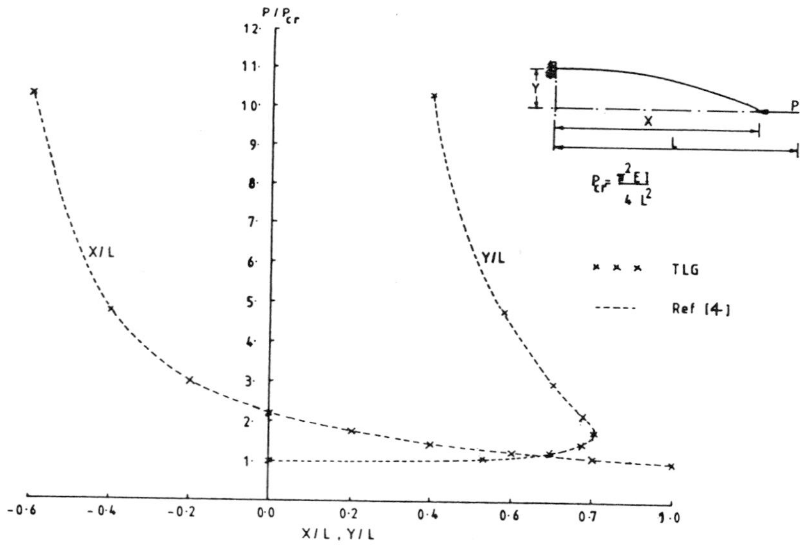

Fig. 6 Load deflection curve of simple supported beam

THE EFFECTS OF DAMAGE ON CIRCULAR TUBULAR COMPRESSION MEMBERS

D R GRIFFITHS*, BSc and H G WICKENS**, BSc

* Lecturer, Department of Civil Engineering, University of Surrey.

** Experimental Officer, Department of Civil Engineering, University of Surrey.

A number of the tubes have been subjected to a controlled amount of damage applied by a drop hammer with the tubes either supported as beams or directly under the impact point. Their performance in axial compression was compared with control samples and further comparison was made with measurements of the effect of the damage and analytical methods from other research methods.

INTRODUCTION

Structural engineers are being asked with increasing frequency to examine the integrity and load carrying capacity of damaged structural members and to predict the damage capability of members at the design stage. Information of direct relevance to the designer of space structures is limited as most of that published is either extremely complex or applicable only to very thin walled tubes.

The Heavy Structures Group at the University of Surrey has undertaken, as an introduction to this field of research, a program of compression tests on a series of commonly available circular steel tubes with geometries suitable for use in space structures. Specimens were damaged by impact with a knife edge, likely to cause the worst type of damage. Two support conditions were considered for the tubes under impact damage; first with direct bearing under the hammer and second with the tube simply supported at its ends. All the tests were to be on tubes of identical length with the damage inflicted at the midpoint. In the initial program only the effect of damage on axial compression loading of the tubes was investigated. Later work will examine the effect of damage on flexural and torsional members.

In parallel with the test work a study of theoretical analyses was undertaken to examine ways of relating impact force, damage and loss of performance. Many of the methods were found to be unsuitable for use in conjunction with the test program as they required very accurate modelling of the damaged zone that was too complex to relate to the tests. However one method [1,2] for relating the damage to the loss of performance of tubular members was found to be suitable. The analysis could be simply applied and required little information about the damaged area by making some basic assumptions about the dent profile.

RESEARCH INTO UNSTIFFENED TUBULAR COLUMNS

Research into unstiffened columns has been led by the Aerospace and Offshore Industries but unfortunately much of their work has been of little relevance to this project. In the Aerospace field research concentrated on very thin walled members with diameter to thickness (D/t) ratios greater than 90, where structures are sensitive to small levels of imperfection and local buckling dominates. The Offshore Industry has been examining structures with D/t ratios between 30 and 90, still greater than those commonly used in space structures. In these structures, failure may be caused by local or column buckling depending on the length of the member and the level of imperfection.

One experimental program, undertaken by Smith, Kirkwood and Swan [3], compared quarter scale model tests with elastoplastic beam column theory. The analysis was computer based and was reported as giving satisfactory correlation with test figures. Sixteen columns were tested; four control specimens, four eccentrically loaded and the remainder after damage had been inflicted; of these one had bend damage only, three were bent and dented and four were dented only. Their D/t ratios ranged between 30 and 90 and their slenderness ratios between 60 and 100. All of the tubes were stress relieved before being damaged with a pressure applied semi-circular indentation head whilst supported on rubber lined cradles. The paper concluded that tests on quarter scale models could give satisfactory results provided sufficient care was taken scaling the geometry of the members.

Smith, Somerville and Swan [4] extended the work by comparing full scale tests on initially undamaged bracing members, with D/t ratios between 30 and 40, from an offshore structure with one fifth scale members. Four full scale tests were performed, two on undamaged members and two after controlled amounts of damage had been inflicted; one with bending only and the other with denting and bending. Four parallel tests were carried out on the small scale stress relieved tubes. It was noted that tubes with a bending deformation approximately 0.003 times their length (L) could lose 20% of their strength while others with dent deformations of 0.12 times their diameter (D) exhibited a 30% reduction. A tube with a bend of 0.005L and a dent of 0.12D lost over 45% of its initial strength.

A theoretical and experimental program was reported by Taby, Moan and Rashed [5] presenting a stress model based on a yield line collapse mechanism. The analysis calculated the uniform compression stress in the tube to achieve a full plastic hinge at the middle of the dent. A computer program was developed to predict the performance of the dented tubes, which were allowed no bending damage. Twenty one tests were conducted on damaged tubes with dent depths of: 0.02D, 0.05D and 0.1D and the dent positioned 0.125L from the centre of the tube. Tubes with D/t ratios between 40 and 64 were found to lose 30% of their ultimate strength with a 0.05D dent and 40% if the dent was increased to 0.1D. Thus the results indicated a greater strength reduction than that presented by Smith et al[4] which may have been due to the greater diameter/thickness ratio of the tubes tested.

Ellinas[1] in his theoretical paper, has linked the stress model presented by Taby et al[5] to beam column theory and the requirements of the Norwegian Offshore Design Code (DNV) and has produced a lower bound solution based on simple calculations only. The method is suitable for denting and bending or denting or bending only and has been checked with experimental work from other papers.[3,4,5] This work, was followed by a joint paper with Walker[2] which simplified some of the equations for use with Diameter/thickness ratios between 30 and 90, and slenderness ratios between 40 and 100. Good correlation was achieved with test data showing the method to be suitable for use by the designer in estimating the reserve strength of damaged members since only knowledge of the dent depth and the level of initial bend in the members is required in addition to the normal parameters.

TEST EQUIPMENT (see Fig 1)

A damage rig was designed and constructed to inflict blows of varying intensity on tubes up to 100 mm in diameter, allowing variations in support conditions and shape of impact head. Its base was constructed from four I sections capable of withstanding impact shock whilst providing fixing points for the tube supports and the central tower. The two end blocks allowed variations in span up to 1.3 m and were designed to act as either a fixed or simple support. They consisted of two machined blocks which could be clamped around the tube and, for the fixed end condition, bolted through a spacer block to the base of the rig. For the simple support the spacer was replaced by a rocker unit designed to locate the specimen longitudinally and transversely. Separate supports were designed for each diameter of tube so that the top was at a fixed height. Additional blocks were manufactured so that the tubes could be supported under the impact point; either on a flat plate or in a semi-circular cradle. The central tower was fabricated from two sections of channel, to guide the drop hammer and support its associated equipment. Four tie rods were used to triangulate the tower to the base to ensure verticality and improve rigidity. Rods mounted on the channels acted as guides for the hammer and fixing points for the release mechanism which could be positioned to give drop heights of up to 2.5 m. The hammer had a mass of 20 kg and was fitted initially with a 90° toughened knife edge head mounted at right angles to the axis of the supported tubes. Facilities were available to change the shape and orientation of the hammer and to increase its mass to 30 kg.

A separate rig was constructed to facilitate measurement of the initial straightness of the tubes and the damage done to them. This allowed the tubes to be supported horizontally between a pair of ball seatings while a displacement transducer mounted on a trolley was guided along the top of the tube. The trolley allowed lateral movement of the transducer so that bent tubes could be measured along their deflected centreline.

Axial compression tests on the tubes were carried out on the Heavy Structures Laboratory's 500 kN screw driven 'Satec' universal testing machine. They were conducted at a constant crosshead velocity set to give a strain rate of approximately 3 microstrain/second. Accurate displacements were measured by a dial guage mounted between the crossheads of the machine. Where specimens were expected to fail in excess of 500 kN tests were performed in a 3,000 kN 'Denison' compression testing machine (4A stub columns) or a 100 Ton 'Avery' universal testing machine (4A undamaged full length specimens).

TUBES

A range of small diameter tubes with suitable geometries for use in space frames were selected. All were coldfinished seamless mild steel supplied to BS 980, 1950, CDS2 (now BS 6323, part 4, CFS 3BK), and thus were uniform along their length, had not been annealed and had high inbuilt stresses. The specification required a minimum yield stress (σy) of 360 N/mm², an ultimate stress (σult) of 450 N/mm², and the elongation at fracture to exceed 6%. Four diameters of tube were selected 25 mm, 50 mm, 75 mm and 100 mm with a wall thickness of 3.25 mm. The range was extended with different thickness tubes at 25 mm and 50 mm diameter matching the Diameter/thickness ratio obtained on the larger diameter tubes. The yield stress has been taken as the 0.2% proof stress which was checked by stub column tests where three samples of each section size were cut to a length three times the tube diameter with their ends machined square and loaded to failure between a pair of hardened steel platens. Little variation (approximately 1%) was observed between samples of the same section however the mean values for the different sections ranged between 468 N/mm² and 607 N/mm². It was noted that the ultimate stress of the stub columns was approximately 20% above the yield stress.

TEST PROGRAM

The test program required three pairs of tubes 1.2 m long to be cut from each section size. The tubes were matched such that each pair had similar lack of straightness properties. The first pair acted as control samples, the second were damaged in the drop rig whilst supported under the point of impact and the third were subjected to an identical damage force but were simply supported. Detailed measurements were carried out on all the tubes before they were loaded to failure in axial compression.

The preparation of the tubes involved squaring their ends, degreasing and marking a measuring grid on their surface. The grid comprised four longitudinal lines at 90° intervals and five circumferential lines at the ends, middle or quarter points. Additional points were marked on the tubes to be damaged on their top and bottom longitudinal lines relative to the impact point.

Measurements for the diameter and surface straightness were recorded for each of the grid intersection points but the wall thickness was measured only at the four grid points at each end of the tube. The lack of straightness of each tube was taken as the largest displacement along each datum line fixed by the end points.

The tubes to be damaged were set up in the rig with the supports adjusted for the tube diameter and the impact head release mechanism positioned to give the required drop height. No attempt was made to stop the head bouncing on the tube but an estimate of the rebound height was noted. The tubes were then returned to the measuring rig and the damage assessed. The measuring lines which had been set in the top and bottom positions for impact were monitored to check length of dent and bending respectively. Depth and width of dent (cut length) and side bulging of the tubes were also measured and all are recorded in table 2. The dent depth was defined as the mean diameter less the minimum diameter and the side bulge as the maximum diameter less the mean diameter of the tube. Two damage functions were also defined: the dent depth divided by the tube diameter and the maximum bend deflection divided by the tube length.

The specimens were tested in axial compression between a pair of ball seatings with load and end displacement recorded throughout the test. Loading was continued for a short while after the initial failure to give an indication of the failure mode. The tubes were visually inspected for failure damage on removal from the test rig.

NOTES ON THEORETICAL ANALYSIS

The method used for predicting the ultimate stresses was based on that presented by Ellinas[1,2] but included more accurate equations for the section properties of the tubes. The basic method was not intended for use with tubes with diameter to thickness ratios as low as those tested and therefore in places different equations had to be substituted for the damaged sections. The stress model[5] on which the calculations were based is outlined in Fig 2 and the ultimate stress (σud) was evaluated by solving the quadratic equation:

$$\left(\frac{1}{\sigma_e}\right)\sigma_{ud}^2 - \left(1 + \alpha_o\lambda_d + \frac{A_d e_d}{Z_d} + \frac{f_y}{\sigma_e}\right)\sigma_{ud} + \left(f_y + \sigma_{pd}\frac{A_d e_d}{Z_d}\right) = 0 \quad \ldots\ldots (1)$$

This general equation was constructed from a series of equations which required knowledge only of dent depth and out of straightness in addition to standard design parameters where:-

i) the imperfection parameter, α_o:

where the out of straightness d_o is less than 0.0015L

$$\alpha_o = 0.001167 + 0.875\,\frac{\sigma_y}{E}$$

where d_o is greater than 0.0015L

$$\alpha_o = \frac{\sqrt{2}\,d_o}{L} - 0.001 + 0.875\,\frac{\sigma_y}{E}$$

ii) the slenderness parameter, λ_d:

$$\lambda_d = \frac{L}{r_d} - 0.2\pi\sqrt{\frac{E}{\sigma_y}}$$

iii) the effective yield stress of the reduced section, f_y:

$$f_y = (\sigma_y - \sigma_{pd})\,\frac{A_d}{A} + \sigma_{pd}$$

iv) the plastification stress, σ_{pd} (taken from Ref 5):

$$\sigma_{pd} = \sigma_y\left[\sqrt{\left(4\left(\frac{c}{t}\right)^2 + 1\right)} - 2\left(\frac{c}{t}\right)\right]$$

where $c = \frac{2}{3}\,d_d$ (from Ref 1).

The section properties of the undented part of the damaged section were calculated from equations given in ref 6 where, referring to Fig 2:

i) the position of the neutral axis with reference to the bottom of the section NA:

$$NA = R\left[1 - \frac{2\,\mathrm{Sin}\,\theta}{3\,\theta}\left(1 - \frac{t}{R} + \frac{1}{2 - t/R}\right)\right]$$

ii) the second moment of inertia, I_d:

$$I_d = R^3 t\left[\left(1 - \frac{3}{2}\left(\frac{t}{R}\right) + \left(\frac{t}{R}\right)^2 - \frac{1}{4}\left(\frac{t}{R}\right)^3\right)\left(\theta + \mathrm{Sin}\,\theta\,\mathrm{Cos}\,\theta - \frac{2\,\mathrm{Sin}^2\,\theta}{\theta}\right) + \left(\left(\frac{t}{R}\right)^2 \frac{\mathrm{Sin}^2\,\theta}{3\theta(2-t/R)}\left(1 - \left(\frac{t}{R}\right) + \frac{1}{6}\left(\frac{t}{R}\right)^2\right)\right)\right]$$

The damage parameters, the theoretical performance data and the test results are recorded in tables 2 and 3. The theoretical and test results are quoted as a factor calculated to be the ultimate stress divided by the yield stress where the ultimate stress is the maximum load divided by the undamaged cross-section area and the yield stress is that found in the stub column tests.

INITIAL INSPECTION AND DAMAGE RESULTS

Averaged results for each pair of tubes are given in table 1; with the diameter and thickness quoted in terms of their mean values and coefficients of variation and the out of straightness as a proportion of the tube length. The variation in thickness around the tube circumference was approximately 5%, but was consistent throughout the tube length. The out of straightness for all of the tubes was better than 0.0006 of the tubes length with the exception of two of the 1C sections. Mean values of better than 0.002 were recorded for the initial ovality $((D_{max} - D_{min})/D_{mean})$ of the tubes.

There was considerable bounce of the hammer following the initial impact: on average 10% of the drop height with a maximum of 25% for one of the simply supported tubes. These tubes typically produced more bounce than those directly supported and bounce height also increased with decreasing tube diameter. It was possible for the tube to be subjected to six separate impacts before the hammer came to rest but the affect of the multiple impact was considered outside the scope of this work with the damage taken to be due to total transfer of energy.

The form of damage had four principle components (see Fig 3), cut length, dent area, tube flattening and tube bending. In most sections the cut length was little more than a surface mark indicating the length of hammer contact with the tube however on the small diameter thick walled tubes with direct support (section 1A, 1B and 2A) the cut had the shape of a machined groove across the tube. The appearance of the dented area was similar in all cases, only the size of the area changed. The dented zone approximated to a square with one diagonal along the cut and the other along the axis and the four edges defined by changes in the curvature of the tubes surface. Flattening of the tubes, the deformation of the cross section into an eclipse, could not be defined becuase of the dent, however the widening or side bulge of the tube was measured. It was noted that the bulge was in the order of one third of the dent depth, although the ratio varied with the wall thickness; with greater bulging of the thin walled tubes. The base flattening of the supported tubes was noted to be approximately one half the side bulge. Longitudinal bending occurred in all the damaged specimens due to hinging action at the dent however this movement was small in contrast to the bending produced in the members damaged whilst simply supported. The relative size of the dent and the bend in the simply supported tubes was governed by the diameter and wall thickness of each specimen, as the 1 m span was constant. Thin walled tubes exhibited greater dents and small diameter tubes were more bent.

Typical damage profiles are given in Fig 4 showing the top and bottom surface profiles of both supported and simply supported tubes.

RESULTS OF THE AXIAL COMPRESSION TESTS

A typical load/displacement plot for an undamaged member is shown in Fig 5. The plot was linear to approximately 70% of the ultimate load, followed by a smooth reduction in rate of change of load up to failure. The load then dropped suddenly to a secondary plateau as the specimen sprang into a bent configuration without any changes in its end displacement or evidence of local buckling. The 4A tubes were slightly different in that the load reduced steadily with increasing end displacement, due either to the low slenderness ratio of the tubes or to the lower stiffness of the machine testing them. Variations in the ultimate load averaged 13% between pairs increasing for the smaller diameter tubes to a maximum value of 34%. The secondary or buckled failure loads, however, were much more consistent. The variations in failure load were due to the differences in initial straightness of the tubes within the allowable tolerance of 0.0015 with moment amplification having greater effect on the thinner tubes.

The mode of failure of the damaged tubes with relatively small amounts of damage, ie dents below 0.1 of the diameter and out of straightness less than 0.0015 of the length, were similar to undamaged members, but with the ultimate value reducing in proportion to the level of damage. As damage increased the snap through buckling failure mode was replaced by an accelerated loss in load with no secondary failure plateau. The tubes then exhibited local buckling failures initiated at the dents. The stiffness of the damaged tubes reduced as the out of straightness exceeded 0.0015 due to the more pronounced effect of moment amplification and, as the tube was already in a buckled condition, the load/displacement plot tended towards the secondary plateau of the undamaged tube test. Where bending damage on the directly supported tubes was small and the tubes were thick walled so that local buckling was unlikely, as in section 1A and 2A, there was a significant decrease in the failure load of the simply supported section; approximately 20%. However in the majority of the cases, where either the bending exceeded 0.0015, or the dent was able to initiate local buckling, failure loads were similar for identical impact energies regardless of the support condition. Variations between identically damaged members were significantly lower than for the undamaged specimens; averaging 5% for the directly supported case and 3% for the simple support. The maximum variation for the directly supported tubes, 20% for 1A was due to a large relative difference in out of straightness but at a very low level where the value was influenced by initial imperfections.

COMPARISON BETWEEN THEORY AND EXPERIMENT

The experimental results are compared with the theoretical predictions in Figs 6 to 8 considering the three damage conditions separately. The undamaged theoretical values agree with the values obtained using the Norwegian Design Code on which the damage theory is based. The practical results show improvements on the calculated values due to the initial imperfections of the tubes being within tolerance levels. As diameter increases the effect of moment amplification reduces and the safety margin decreased. Indeed the 4A samples did not behave as well as predicted but this was possibly due to the surprisingly high yield stress of the tubes which was measured on a different test apparatus.

The experimental results for both sets of damaged tubes were approximately 40% higher than the predicted values with the exception of section 1A and 1B when directly supported and 1A, 1B and 2A when simply supported. The exceptional results all related to tubes with large diameter thickness ratios, outside the range recommended by Ellinas.[1,2] However as the diameter thickness ratio decreased the 40% safety margin was only significantly reduced if the tubes were, in addition, badly bent; 1B, both cases and 2A simply supported. Again the 4A results may have been affected by the yield stress.

CONCLUSIONS

i) Insufficient tests were conducted to relate impact damage to impact energy and a further test programme would be needed. The importance of bending means that such tests cannot be conducted on shorter specimens.

ii) Variations in undamaged tubes were high due to the allowable tolerance in bending deformation and its effect on moment amplification which also resulted in practical safety factors on tubes with low section moduli. Further analysis will be done using the measured imperfections.

iii) Tube damage was reasonably consistent and rendered initial imperfections insignificant resulting in more uniform test data.

iv) Small variations in straightness, within allowable tolerances, particularly in thick walled tubes showed significant effect on results such that knowledge of even small bend damage is critical to prediction of loss of strength. At damage levels where the tube is outside allowable bending tolerance or is badly dented the significance of the type of damage reduces and loss of axial compressive strength was related to impact load.

v) Ellinas's theory seems to allow an additional safety margin of 40% within its stipulated range of tubes. As diameter/thickness ratio reduces below the applicable value the safety margin decreases particularly for high levels of bending.

Figure 1. Layout of damage rig.

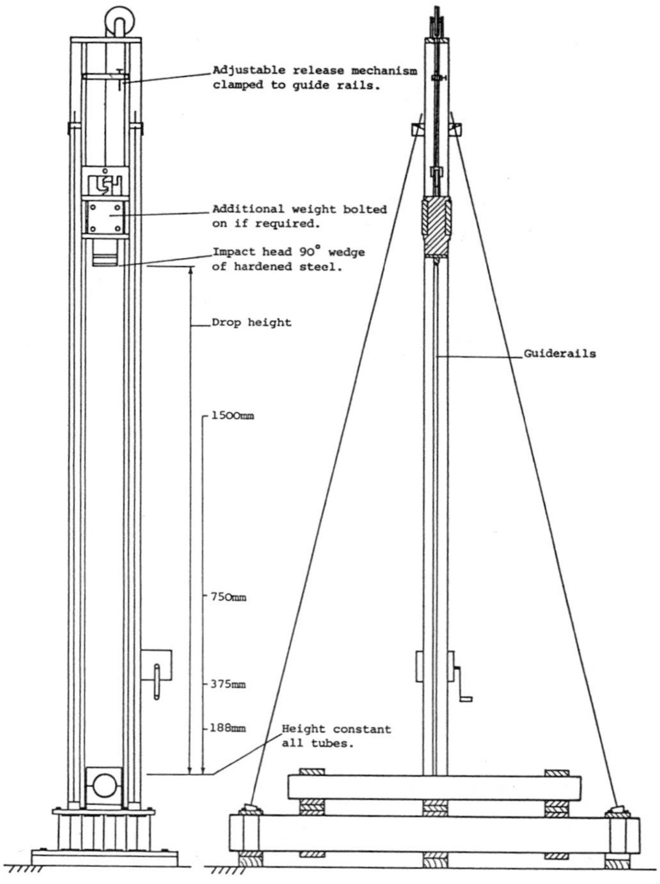

REFERENCES

1) ELLINAS C.P., Ultimate Strength of Damaged Tubular Bracing Members. Submitted to J. Structural Division, ASCE, 1983.

2) ELLINAS C.P. and WALKER A.C., Effects of Damage on Offshore Tubular Bracing Members. Presented at the IABSE Colloquium on Ship Collisions with Bridges and Offshore Structures, Copenhagen, 1983.

3) SMITH C.S.,KIRKWOOD W. and SWAN J.W., Buckling Strength and Post-Collapse behaviour of Tubular Bracing members including Damage effects. BOSS '79, Imperial College, London, England, pp 303-326, 1979.

4) SMITH C.S., SOMERVILLE W.L. and SWAN J.W., Residual Strength and Stiffness of Damaged Steel Bracing Members. 13th Annual Offshore Technology Conference, Houston, Texas, 4-7 May 1981, Paper No. OTC 3981, pp 273-282, 1981.

5) TABY J., MOAN T. and RASHED S.M.H., Theoretical and Experimental study of the behaviour of Damaged Tubular Members in Offshore Structures, Norwegian Maritime Research, No. 2/1981, pp 26-33.

6) ROARK J.R. and YOUNG W.C., Formulas for Stress and Strain, 5th Edition, pp 69, McGraw-Hill, Inc. 1975.

Figure 2. Calculation details.

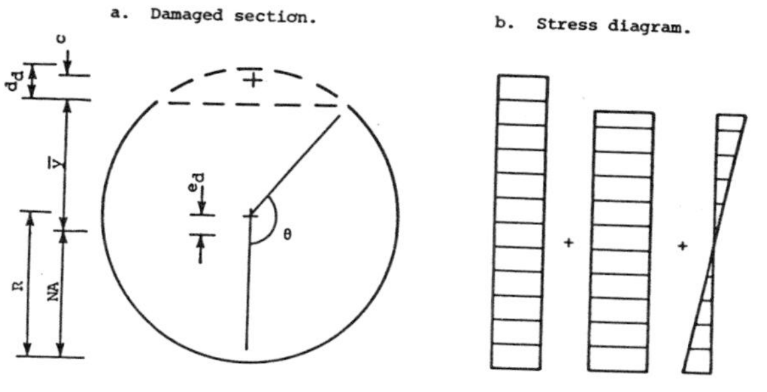

σ_{pd} + σ_{cd} + σ_{bd} = σ_y

σ_{pd} Plastification stress
σ_{cd} Compression stress
σ_{bd} Bending stress
σ_e Euler stress
σ_y Yield stress
c Centroid of damaged section
E Young's modulus (200kN/mm² used)

R Outside radius of tube
NA Neutral axis of reduced section
e_d Eccentricity of centroid
A_d, I_d, r_d and Z_d are properties of the reduced section (the undamaged part of the damaged section)

Table 1

Averaged initial imperfections of the tube pairs

Test Reference [1]	Diameter mean (mm)	Diameter variation†	Thickness mean (mm)	Thickness variation†	Out of Straightness Length d_0/L*
1AU	25.353	0.00023	3.21	0.01310	0.00049
1AS	25.353	0.00035	3.20	0.01561	0.00036
1AB	25.357	0.00040	3.21	0.01673	0.00054
1BU	25.395	0.00025	1.63	0.00816	0.00041
1BS	25.390	0.00020	1.63	0.01260	0.00047
1BB	25.396	0.00035	1.63	0.01352	0.00041
1CU	25.323	0.00053	0.89	0.01682	0.00079
1CS	25.325	0.00108	0.89	0.01453	0.00048
1CB	25.324	0.00072	0.89	0.01548	0.00076
2AU	50.867	0.00033	3.28	0.00547	0.00038
2AS	50.884	0.00047	3.28	0.01628	0.00015
2AB	50.891	0.00039	3.28	0.02193	0.00028
2BU	50.742	0.00049	1.66	0.01423	0.00031
2BS	50.753	0.00081	1.65	0.00856	0.00029
2BB	50.737	0.00050	1.66	0.01243	0.00025
3AU	76.123	0.00041	3.24	0.00828	0.00013
3AS	76.120	0.00030	3.25	0.02144	0.00010
3AB	76.115	0.00042	3.24	0.01172	0.00010
4AU	101.572	0.00046	3.30	0.02440	0.00028
4AS	101.617	0.00025	3.30	0.02737	0.00021
4AB	101.568	0.00057	3.27	0.01949	0.00019

† coefficient of variation
* tube length L = 1200 mm
1 The last letter indicates the damage condition of the tubes:
U undamaged
S directly supported
B supported as a beam

Table 2

Averaged damage condition of the tube pairs

Test Reference	Hammer Drop (mm)	Hammer Bounce (mm)	Dent depth d_d (mm)	Dent depth d_d/D	Out of straightness deformation d_0 (mm)	Out of straightness d_0/L*	Cut length (mm)	Dent Length (mm)	Side Bulge (mm)
1AU	0	0	0.00	0.000	0.59	0.00049	0.0	0	0.00
1AS	375	10	2.10	0.095	1.16	0.00097	14.6	28	0.15
1AB	375	100	0.16	0.007	4.80	0.00400	4.6	0	0.00
1BU	0	0	0.00	0.000	0.50	0.00041	0.0	0	0.00
1BS	375	10	4.74	0.199	7.05	0.00588	21.7	108	1.51
1BB	375	100	1.72	0.072	17.16	0.01430	10.2	52	0.73
1CU	0	0	0.00	0.000	0.95	0.00079	0.0	0	0.00
1CS	188	25	5.43	0.222	5.96	0.00496	23.3	175	2.35
1CB	188	50	3.62	0.148	13.36	0.01113	17.9	100	1.40
2AU	0	0	0.00	0.000	0.46	0.00038	0.0	0	0.00
2AS	1500	50	5.68	0.119	1.82	0.00152	32.8	155	1.54
2AB	1500	200	1.79	0.038	7.51	0.00626	16.9	30	0.61
2BU	0	0	0.00	0.000	0.37	0.00031	0.0	0	0.00
2BS	750	50	8.05	0.164	2.76	0.00230	38.8	248	2.97
2BB	750	50	5.59	0.114	7.77	0.00647	32.4	150	1.71
3AU	0	0	0.00	0.000	0.15	0.00013	0.0	0	0.00
3AS	1500	125	5.89	0.077	0.99	0.00083	40.7	200	1.83
3AB	1500	150	4.39	0.060	2.74	0.00228	33.5	170	0.96
4AU	0	0	0.00	0.000	0.34	0.00026	0.0	0	0.00
4AS	1500	200	4.77	0.048	0.68	0.00056	44.5	210	0.66
4AB	1500	300	4.88	0.050	1.23	0.00103	43.8	170	0.74

* tube length L = 1200 mm

Table 3

Averaged measured design parameters and compression test results for the tube pairs

Test Reference	Diameter / Thickness	Slenderness Ratio l/r †	Reduced slenderness ratio Initial λ †	Reduced slenderness ratio Damaged λ_d †	Yield stress σ_y N/mm²	Ultimate stress / Yield Stress Theory	Ultimate stress / Yield Stress Test
1AU	7.9	158.84	2.537	2.537	504	0.143	0.231
1AS	7.9	158.74	2.536	3.024	504	0.143	0.177
1AB	7.9	158.81	2.537	2.619	504	0.143	0.141
1BU	15.6	149.01	2.416	2.416	519	0.156	0.271
1BS	15.6	148.86	2.414	3.337	519	0.116	0.132
1BB	15.6	148.91	2.415	2.789	519	0.113	0.122
1CU	28.5	145.27	2.481	2.481	576	0.148	0.232
1CS	28.4	145.14	2.479	3.542	576	0.101	0.126
1CB	28.4	145.13	2.479	3.188	576	0.094	0.128
2AU	15.5	76.30	1.193	1.193	483	0.534	0.773
2AS	15.5	76.31	1.194	1.474	483	0.357	0.471
2AB	15.5	76.40	1.195	1.309	483	0.395	0.380
2BU	30.6	73.81	1.166	1.166	492	0.552	0.714
2BS	30.7	73.72	1.164	1.531	492	0.265	0.395
2BB	30.6	73.85	1.166	1.429	492	0.268	0.387
3AU	23.5	48.30	0.744	0.744	468	0.827	0.902
3AS	23.4	48.32	0.744	0.871	468	0.497	0.748
3AB	23.5	48.25	0.743	0.843	468	0.541	0.763
4AU	30.8	35.66	0.626	0.626	607	0.876	0.862
4AS	30.8	35.64	0.625	0.697	607	0.604	0.793
4AB	31.0	35.70	0.626	0.699	607	0.598	0.782

† length taken between centres of ball seatings: approximately 1260 mm

Figure 3. Measured features of damaged tubes.

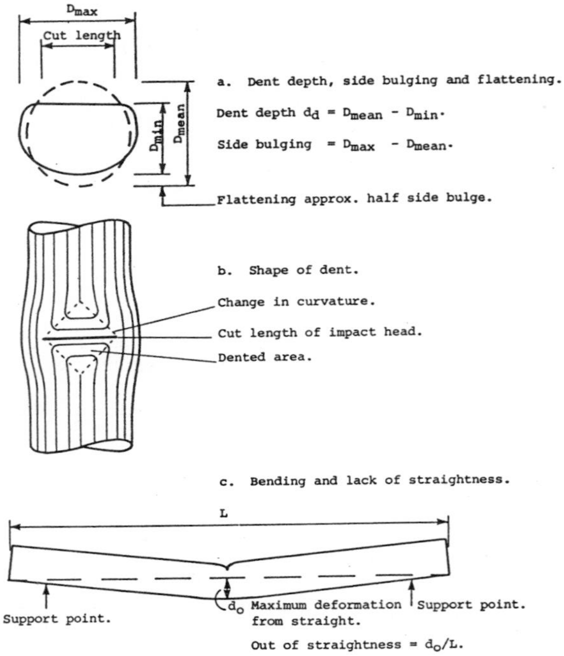

Figure 5. Typical load/deflection plots for axial compression on damaged and undamaged tubes of identical size. Tube section 2A.

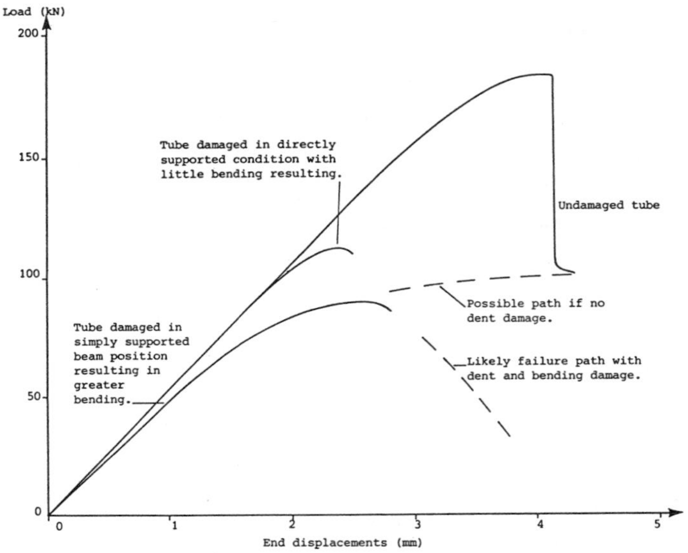

Figure 4. Top and bottom surface profiles of the damaged tubes. Different support conditions, for blows of equal energy.

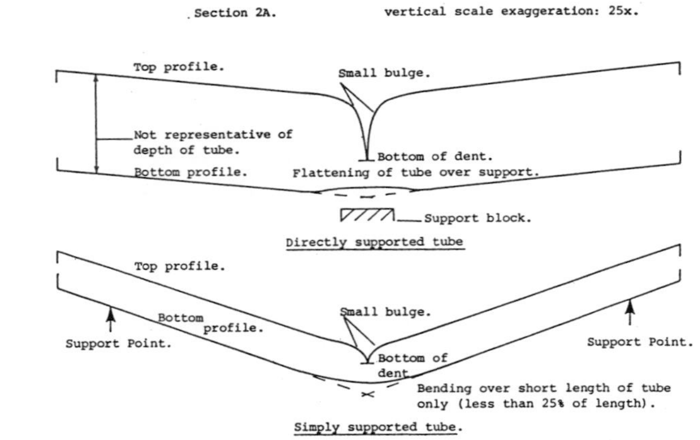

Figure 6. Comparison between experimental and theoretical results for the ultimate compressive stress of the undamaged tubes.

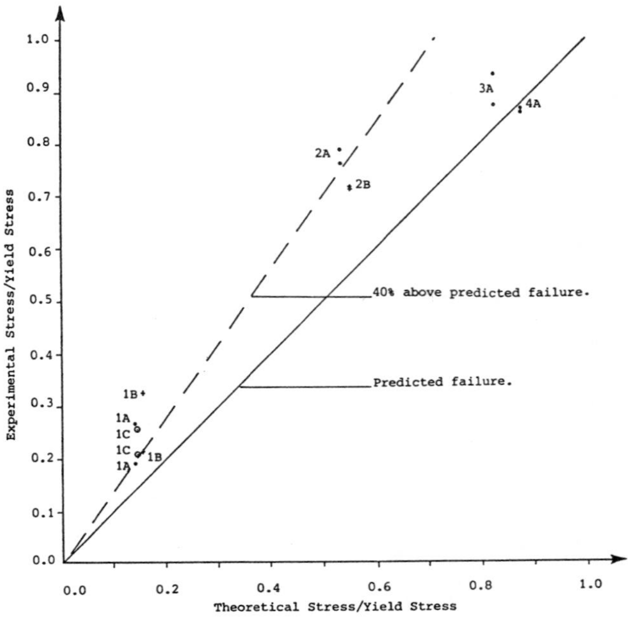

Figure 7. Comparison between experimental and theoretical results for the ultimate axial compressive stress of the tube damaged while directly supported.

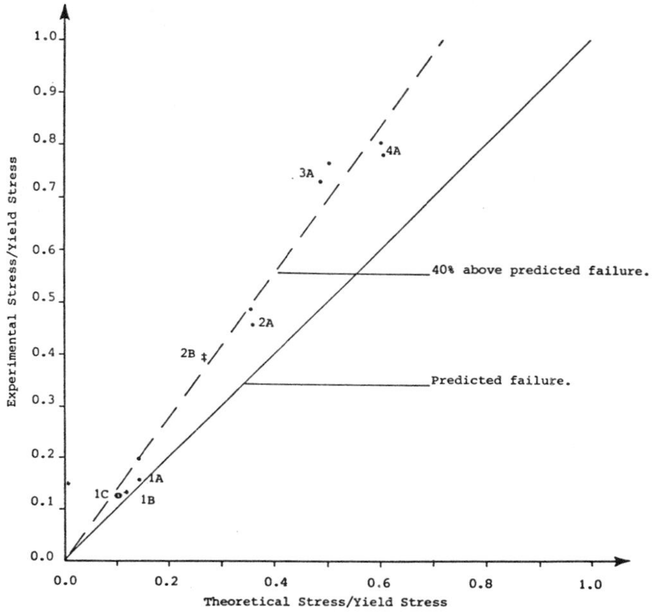

Figure 8. Comparison between experimental and theoretical results for the ultimate axial compressive stress of the tube damaged while supported as a beam.

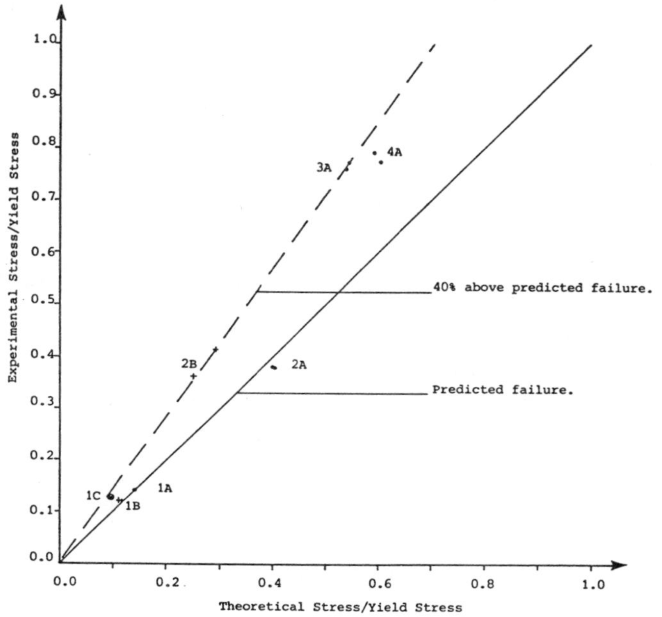

MATERIAL EFFECTS ON MILD STEEL STRUT STABILITY

P R MORGAN*, L C SCHMIDT**, and W A RHODES***

* Senior Lecturer, Department of Civil Engineering, University of Melbourne, Australia.

** Professor, Department of Civil & Mining Engineering, University of Wollongong, Australia.

*** Design Department, Australian Iron & Steel Pty.Ltd., Port Kembla, Australia.

The strength enhancement attainable through cold working is often used in the production processes of the structural sections used in space trusses. Unfortunately the improved capacities due to strain hardening and strain aging can be severely reduced by the so-called Bauschinger effect upon strain reversal. A programme of mild steel strut tests has been carried out for solid rectangular and hollow circular sections with slenderness ratios varying from 20 to 180 in order to investigate these effects in a controlled way. This work has highlighted significant losses of strut capacity when struts are made from material prestrained in tension by different amounts up to 10% strain.

INTRODUCTION

Strut stability has attracted the attention of researchers for many reasons, and these have ranged from fundamental issues like stress conditions at the microscopic level across critical cross sections, to the more applied variety like the definition of easily derived empirical moduli accurate enough for use in the prediction of collapse loads. The problems are many and varied because of the range of variables; initial imperfections, slenderness, yield strength, material type and conditions, and loading rate being amongst the most important. There are, however, other factors that need investigation and that have received little attention so far.

Mild steel tubular struts are a case in point. Although their overall behaviour is well documented from many tests, the practical influence on their strength of strain hardening, strain aging and the Bauschinger effect is not clear. This theme will be developed herein after a brief review of the individual components of the problem as they relate to mild steel.

Strain hardening (or work hardening) is the phenomenon of increasing strength with strain as plastic deformation occurs past the yield point; it has been explained in standard texts using the terminology of dislocation theory (Ref 1). Figure 1 (after Ref 3a) presents the increasing strength schematically for a bar in tension. If, while strain hardening is occurring in a test, the test is interrupted and unloading occurs followed by reloading as shown, the reloading path taken by the test specimen depends on the time since unloading and any temperature changes which have been applied to the material. Any strength increase, ΔY, reflects strain aging which occurs when carbon and nitrogen atoms migrate and pin dislocations (Ref 2). The migration occurs slowly with time at room temperature, but may be accelerated by aging at higher temperatures. It is possible to fully age or even overage a steel. Figure 1 illustrates the strength increases and reduced ductility that can occur. The influence of strain aging has been exhaustively reviewed in Refs 3a and 3b.

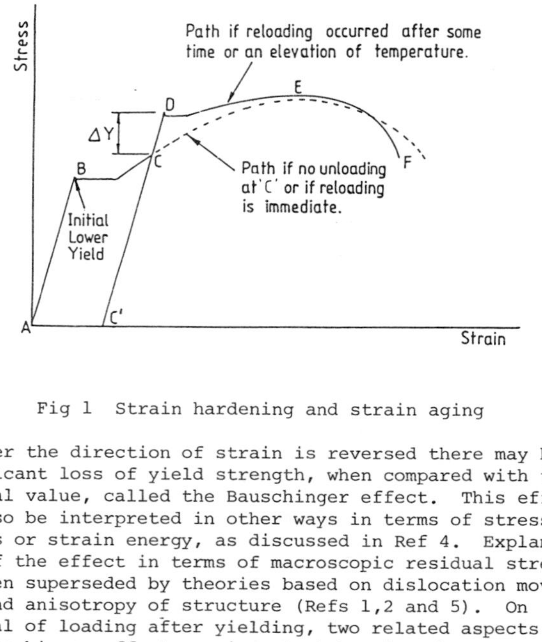

Fig 1 Strain hardening and strain aging

Whenever the direction of strain is reversed there may be significant loss of yield strength, when compared with the original value, called the Bauschinger effect. This effect can also be interpreted in other ways in terms of stresses, strains or strain energy, as discussed in Ref 4. Explanation of the effect in terms of macroscopic residual stresses has been superseded by theories based on dislocation movement and anisotropy of structure (Refs 1,2 and 5). On reversal of loading after yielding, two related aspects of the Bauschinger effect are important herein, these are the reduced yield stress levels achieved and the reduction of stiffness. The full lines in Fig 2 present this situation schematically.

In Fig 2 path ABCP includes the Bauschinger effect; path ABCDQ includes strain hardening and the Bauschinger effect; path ABCEFGHR includes strain hardening, strain aging and the Bauschinger effect. Curves labelled C',D', and H' are shifted images of the corresponding parts of the virgin curve and are to be used as reference curves.

It is clear that the base stress against which a Bauschinger

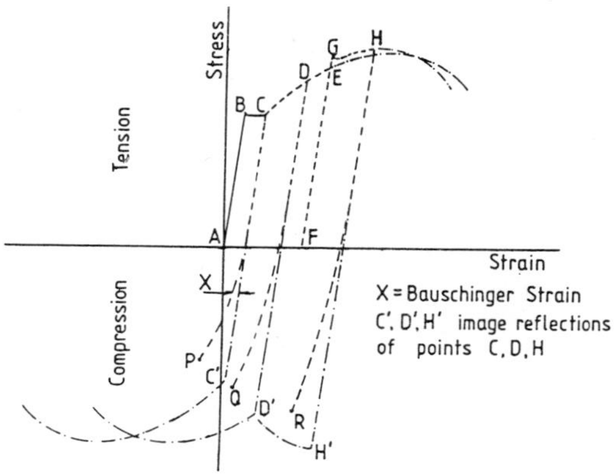

Fig 2 Bauschinger effect, strain hardening and strain aging

effect may be measured depends on the previous stress and temperature history of the material, which may also include the two effects already mentioned, strain hardening and strain aging. Figure 2 has been extended using dotted lines to make this point clear.

In Ref 6, a series of tests was reported of the sensitivity of the Bauschinger effect to the state-of-stress and the definition of the yield criterion. As different amounts of prestraining were applied in tension or in compression before stress reversal, the tests included the influence of strain hardening. Bauschinger loss was observed whether the prestraining was in tension or compression and it depended on the stress magnitudes obtained from the non-linear reverse loading curve, so it depended critically on the definition used for yield. It was further observed that moduli measured from the reverse loading curve were sensitive to prestrain conditions. As these moduli are the basis for Tangent or Reduced Modulus strut capacity predictions the stress history during the manufacture of steel struts is seen to be important. This observation was also noted in Ref 7. The conclusions on the relative importance of the Bauschinger effect and strain hardening are of interest but no information is given as to the influence (if any) of strain aging on the results.

In Ref 8, tensile and compressive tests were conducted on material prestrained in tension to different extents. The tests included material orientation as a variable, relative to prestrain direction, and the tests enabled the interaction of Bauschinger effect and strain hardening to be observed. The Bauschinger loss (at all reverse load yield definitions less than about 0.60% offset) heavily outweighed the gain in strength owing to strain hardening and strain aging. Unfortunately no mention was made of strain aging or test timing, and yet from the stress-strain plots presented strain aging must have made a significant contribution to increased strength. Reference 7, in a series of related tests, again did not mention the influence of strain aging, comments being restricted to the strong influence of the Bauschinger effect on yield strength and modulus of elasticity. The small amounts of prestrain (less than 1%) caused difficulty in some of the reversed loading tests because of the discontinuity of the strain in the Luder strain region but did eliminate strain hardening as a variable.

Turning attention to the strut as a structural element, limited work has been reported which includes these effects. In Ref 9 an extensive series of pin-ended mild steel strut tests (12.7mm square section) was reported on specimens which had been prestrained in tension by amounts up to 1.16%. It was concluded that tensile prestrain considerably reduced (sometimes by more than 50%) the buckling capacity of struts; the reduction was attributed to the Bauschinger effect. This reduction was confirmed by applying both Tangent and Double Modulus theories (see Ref 10) to the reverse loading material property plots which displayed the Bauschinger effect. The range of prestrain selected was lower than the value (1.4%) at which strain hardening commenced, which eliminated one variable, but at the expense of uncertainty within each test specimen about the uniformity of the prestrain, which was within the lower yield straining (Luder) plateau. No mention was made of the existence or influence of strain aging.

The influence of prior flexural prestrain on the stability of structural steel struts of 19.1mm x 12.7mm cross section was reported in Ref 11. Prestrain was induced by repeated flexural deformation, and the estimate of the extent of prestrain ranged from about ½ to 2 per cent for the differently prepared specimens. Although reverse bending was applied in the less strained specimens the extent of the strain ensured that not all the cross-section would have yielded but a proportion of the extreme fibres would have been strained past the yield strain. The specimens subjected to greater prestrains were straightened using a procedure which could not guarantee that large (about 2%) prestrains had occurred in both tension and compression, and the depth of penetration of yielded material was such that the Bauschinger effect may have been applicable to less than one half of the cross section. Significant loss of load axial load capacity was reported, increasing with the amount of flexural prestrain. The influence of strain hardening was almost excluded by restricting prestrains to about 2%, but the effect of strain aging can only be conjectured as the time elapsing between prestraining and final testing was not stated.

The foregoing work and the importance of struts in space trusses have led to the pilot tests reported herein. Space trusses often have mild steel tubular members, and truss capacity is usually limited by the behaviour of the compressive elements. In the manufacture of tubes flexural and tensile prestrains are usually applied. As a consequence during structural use strain aging occurs at ambient temperatures. The extent and consequent structural influence of strain hardening and aging in this process are uncertain. As both effects are interactive with a Bauschinger loss during later strut loading there is uncertainty about the overall effect on strut load capacity, and the margin of safety against collapse. This paper reports tests on rectangular and tubular mild steel struts where these related material effects have been deliberately included in a controlled way.

EXPERIMENTAL PROGRAMME

The programme of tests was devised to demonstrate variations in the influence of strain hardening, strain aging, and stress direction (the Bauschinger effect). To achieve this, and at the same time provide data for limited theoretical analyses, required both material and strut tests on virgin and prestrained material with specified strain aging treatments. Two cross sections were selected for study; rectangular solid 25mm by 12mm of 250 grade mild steel and circular tube 60.3mm outside diameter by 2.9mm wall thickness of 200 grade mild steel. The rectangular section provided potential correlation with earlier work (Refs 9,11), while the tubular section extended the work to include members more appropriate to space trusses or any full scale structures.

Material Tests

Material tests in tension and compression were performed to furnish initial yield properties and behaviour in the inelastic range for both rectangular and tubular sections. The material used for both types of section was a semi-killed mild steel that has an intermediate propensity for strain aging. Details of the tests are summarised in Table 3.

Of the material tests (refer to Table 3) typical results will be given. Figure 3 presents examples of each type of test performed, and permits some comments to be made at the material level. Additional comment about the influence on strut behaviour will be made later.

TABLE 3

Material Tests Performed

Aging* (Hours @ 100°C)	Test Type (c=Comp.) (t=Tension)	Tensile Prestrain (%) and Section (R = rectangle, T = tube)							
		0		3		7		As defined (%)	
		R	T	R	T	R	T	R	T
0 No aging	c	√	√	√		√			√10
	t	√	√	√		√			
2 Full aging	c			√		√			√14
	t			√		√			

*Some isolated tests were carried out after different amounts of aging at ambient temperatures. These are reported in the text.

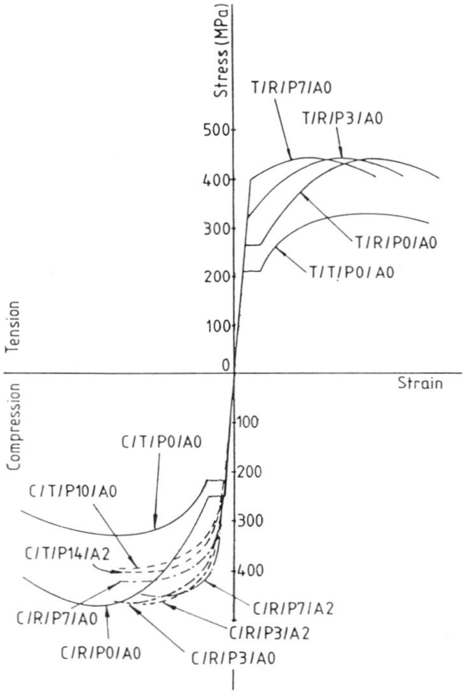

Fig 3 Typical material stress-strain curves

In Fig 3 the tests are identified by four codes: in the first, C=compression, T=tension; in the second, R=rectangle, T=tube; in the third, Px=tensile prestrain of value x per-cent; in the fourth, Ay=aging time of y hours at 100°C. The material tests in compression were 69mm long for the rectangular sections and 150mm long for the tubular sections. The influence of strain aging is seen clearly from the plots for the rectangular section tests carried out for that purpose (for example plots C/R/P7/A0 and C/R/P7/A2). The influence of strain aging on the tubular material has not been presented graphically but was clearly discernible in a form like that represented by path FG in Fig 2; whether the aging occurred at room temperature over long periods or at 100°C for two hours, typical strength increases were approximately 8-10 per cent.

Strain hardening was evident in both tensile and compressive tests as shown in Fig 3. The percentage increase of strength achieved was greater for the rectangular sections than the tubular sections for similar elongations and was greater under compressive testing owing to the changes of geometry during squashing of the short specimens. Peak stresses achieved were between 70 and 100 per cent higher than initial yield stresses, and most of this increase was achieved at relatively low (3 to 5 per cent) elongations. As the stress level attained before unloading gives an indication of the yield stress likely on future reloading, rather significant increases in yield strengths could be expected on reloading, although the Bauschinger loss would counteract the increases achieved by strain hardening and any strain aging if the reloading reversed the strain direction.

The Bauschinger effect can be seen in Fig 3 in several of the plots (eg. C/T/P10/A0 and C/R/P7/A0). It is evidenced by the absence of a definite yield point and the rounding of the stress strain curve from stress levels lower than the original virgin yield stress. Strain hardening is also evident on these same plots as, despite the rounding just mentioned, the stress achieved at reasonably low restrains is higher than that for the same strain on the virgin curves although no strain aging has occurred. Finally strain hardening, strain aging and the Bauschinger effect are clearly discernible in combination in Fig 3 in plots C/R/P7/A0 and C/R/P7/A2 for example. Here a strength increase is achieved first due to strain hardening (a 7% prestrain) and then due to strain aging (2 hours at 100°C) and yet the curve displays the rounding typical of a Bauschinger loss at any defined percentage offset strain.

Strut tests

a) Tubular strut.

Two series of tests (Series 1T and 2T) were performed on pin-ended tubular struts of slenderness ratios varying from 20 to 150. The details of these series are presented in Tables 4 and 5. The prestrain value of 3% was chosen to ensure that the steel was uniformly strained beyond the value at which strain hardening occurred over the entire length (approximately 1.4%). The value of 7% was chosen so as to generate the near-peak load capacity of each member without causing excessive geometric change of cross section. The effects due to the degree of strain hardening should be illustrated by choice of these significantly different values.

The two largest prestrains shown in Table 5 for series 2T are approximate as elongations have been deduced from ram movement of the testing machine and the nominal value of Young's Modulus for steel; nevertheless the less slender struts received more prestrain than the longer struts. The peak load achieved in any test was taken as the basis for comparison with other results.

Figure 4 is the simplest presentation of these results and for comparison purposes the theoretical Euler and Tangent Modulus load capacities have also been plotted. The Tangent Modulus capacities were computed for series 2T using the detailed properties of curve C/T/P14/A2 from Fig 3. Figure 4 masks the rather significant effect of cross sectional area changes due to prestraining for strut series 2T, but does include the small increases in slenderness ratio caused. It is also clear that at low slenderness ratios (<60) the capacity of the prestrained and aged struts (Series 2T) is greater than that of the virgin struts (Series 1T), indicating that the influence of strain hardening (due to prestrain) and strain aging (2 hours at 100°C) has exceeded the Bauschinger reduction. To test the wider applicability of this phenomenon the test program was extended to include rectangular sections, and a wider range of the important variables of strain hardening and strain aging. These results will now be reported before indicating the normalisation procedures used to account for the area changes mentioned above and the different grades of steel used.

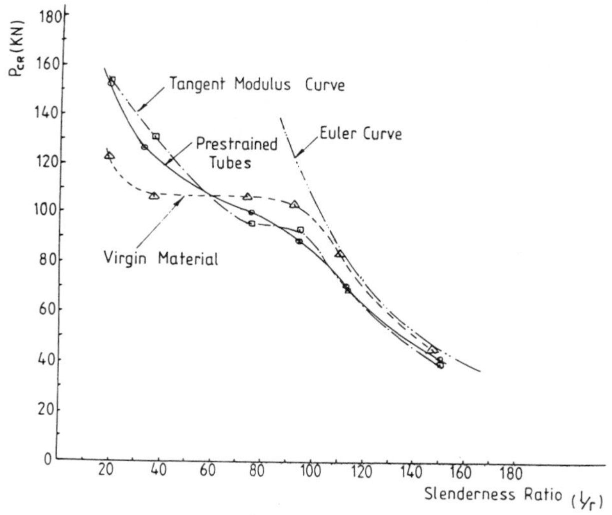

Fig 4 Tubular strut tests - load versus slenderness ratio

b) Rectangular solid strut

Five series of strut tests (series 1R to 5R) were performed on solid rectangular sections each including slenderness ratios ranging from 20 to 180. Details of these series are given in Table 4. It can be seen from Table 4 that test Series 1R and 5R were planned to provide comparisons with Series 1T and 2T respectively. Series 3R provided a prestrain intermediate between that given by Series 1T and 2T, while Series 2R and 4R included this intermediate prestrain but allowed the influence of aging time at 100°C to be identified.

TABLE 4

Strut Series R (Rectangular) and T (Tubular)

Strut Series	Prestrain (%)				Strain Aging	
	0	3	7	Specified Percent	None	2 hours @ 100°C
1R	✓				✓	
2R		✓			✓	
3R		✓				✓
4R			✓		✓	
5R			✓			✓
1T	✓				✓	
2T				7-12 See Table 5		✓

TABLE 5

Series 2T Prestrain (%) Versus ℓ/r

ℓ/r	19	37	74	92	111	148
% prestrain	12	11	9	8½	7	7

Figure 5 presents the raw results for the rectangular struts in a manner which parallels Fig 4 and the tubular struts. The prestrained struts have a comparable ultimate load capacity to that of the virgin struts for low slenderness ratios. In the intermediate slenderness range the rectangular prestrained struts achieve a lower percentage of the virgin strut capacities than the tubular struts, as can be seen on Figs 4 and 5 for slenderness ratios of about 80.

The rectangular struts provide some additional information about the amount and influence of strain aging as no prestrained tubular struts were tested without strain aging.

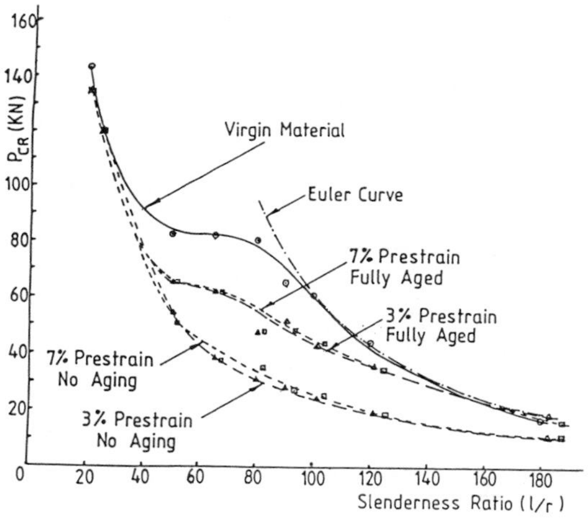

Fig 5 Rectangular strut tests - load versus slenderness ratio

As would be expected the load capacities were higher with strain aging; what was not expected was the size of the increase in capacity due to aging. The increase of strength for the rectangles (of the order of 30% to 60% in the intermediate slenderness range) was significantly greater than that which could have been expected on the basis of the strain aging strength increments (8-10 percent) seen in both the tensile and compressive tests of the tubes.

c) All struts - normalised data.

Comparisons of strut behaviour are complicated by the different steel grades, 200 for the tubes and 250 for the rectangles. This difference has been overcome in Fig 6 where the stress axis has been normalised by dividing the strut critical stress by the relevant yield stress. Two comments must be made about this procedure. Firstly as the prestraining was large there were significant dimension and area reductions from the virgin material. For all struts made of prestrained material the dimensions after prestraining were used to compute slenderness ratios and stresses at strut collapse (σ_{CR}). In addition, the yield stress (σ_Y) to be used in the normalisation procedure requires definition. For the virgin strut tests the lower yield stress (curve C/T/PO/AO or C/R/PO/AO from Fig 3 as appropriate) from the virgin compression tests was used, but for prestrained struts the compressive stress-strain curve did not display a definite yield point because of the Bauschinger effect. This

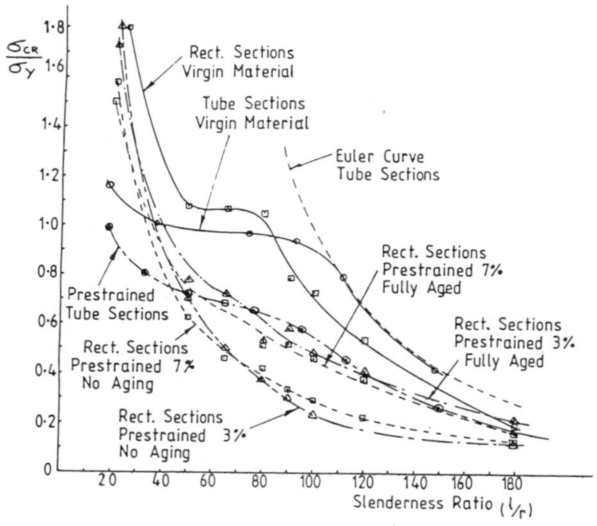

Fig 6 All strut tests - normalised on σ_Y

effect can be seen in all compression curves in Fig 3 except the two just cited. For any strut series with prestrain, a 0.2% offset strain was used to define the yield stress σ_Y. Thus the yield stresses used to normalise each strut series were different and have been tabulated in Table 6. It

TABLE 6

Yield Stresses (σ_Y) and Transition Length Slenderness Ratio $(\ell/r)_T$ used for Normalisation

Strut Series	σ_Y (MPa)	$(\ell/r)_T$
1T	207	99
2T	311	80
1R	248	89
2R	249	89
3R	278	84
4R	261	87
5R	295	82

should be noted how sensitive the strut plots are to the percentage offset strain used to define σ_Y. In Fig 7, Series 2T has been plotted to the same scales as Fig 6 but for 0.05, 0.10, 0.20 and 0.50% offset strain definitions of σ_Y.

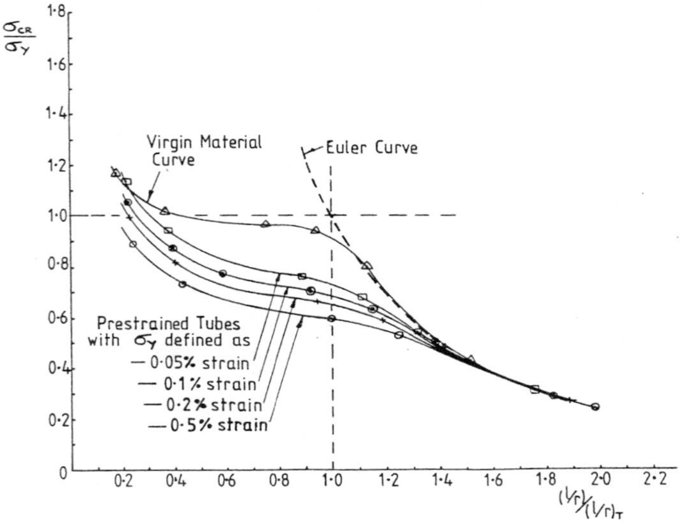

Fig 7 Influence of yield definition on strut curves

Three distinct bands of test results are evident in Fig 6; an upper band of virgin tests (Series 1R, 1T and the Euler curve for 1T), a lower band of prestrained struts without strain aging (Series 2R and 4R), and a middle band of prestrained struts with strain aging (Series 3R, 5R and 2T). These bands appear as a common factor in the tests, indepently of strut cross section.

The question of comparison between struts over all series is potentially blurred by differences on the slenderness ratio axis caused by the factors of cross-sectional shape and transition length differences due to different yield values. In Fig 6 the curves have been fitted to the experimental results by inspection. In Fig 8 the normalization has been carried out by dividing the actual slenderness ratio by the transition length slenderness ratio, where the Euler critical stress is equal to the yield stress σ_Y; the ordinate is plotted as critical stress, σ_{CR}. As the transition length slenderness ratio is different for each strut series the values used in the normalisation are tabulated in Table 6. This plot highlights the physical changes in strut behaviour expected at the transition length but unfortunately blurs the three clear bands of behaviour (virgin, prestrain with aging, and prestrain with no aging) seen in Fig. 6.

As a final comparison, therefore, the same data have been presented in Fig 9 where both axes are normalised, the critical stress axis as in Fig 6 and the slenderness ratio axis as in Fig 8. The values given in Table 6 are used for the normalisation. The plots in Fig 9 preserve the three distinctive bands of behaviour commented upon in Fig 6 and also the transition length characteristics observed in Fig 8.

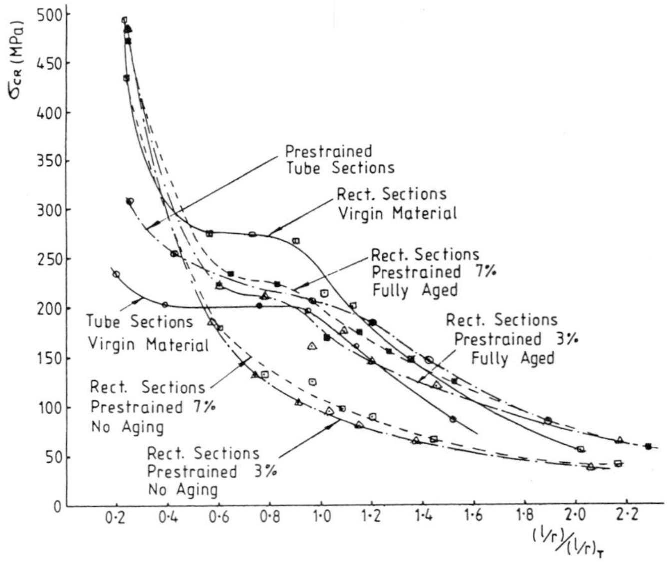

Fig 8 All strut tests - normalised on $(\ell/r)_T$

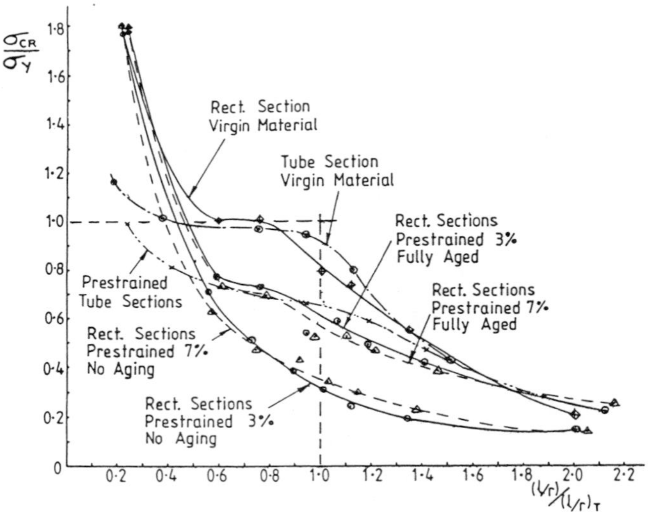

Fig 9 All strut tests - normalised on σ_Y and $(\ell/r)_T$

DISCUSSIONS AND CONCLUSIONS

The primary objective of the work reported herein was to obtain the ultimate load capacities of pin-ended columns subject to prior prestraining in tension, and to compare these results with those obtained from as-received members.

The results presented in Figs 4 and 5 show that the prestrained struts, whether strain aged or not, have a lower load capacity than the as-received struts in the structurally significant range of slenderness ratios, 40-180.

The Tangent Modulus calculation for the tubular struts for Series 2T as shown in Fig 4, was in reasonable agreement with the observed results for Series 2T, highlighting the significance of the loss of modulus associated with strain reversal. The reduction of the tangent modulus with increasing compressive strain is an aspect of the Bauschinger effect that is of prime significance in these tests.

As indicated in Fig 2, strain hardening and strain aging lead to an enhancement of strength when strains are of the same sign as the prestrain. On reversal of strain, however, the curvature of the stress-strain plot means that the tangent modulus is reduced which, in turn, leads to a reduction in the ultimate carrying capacity of the struts. This reduction occurs even though the yield point of the material may have been increased on reversal relative to that of the virgin material. The extent of the yield level increase, however, depends on the accepted definition of yield for the continuous stress-strain behaviour on reversal.

The arbitrary definitions of yield stress, as used in Fig 7, lead to a wide variation in the graphs of results on normalised axes. However, the 0.2 offset strain has been used throughout and forms the basis of comparison in Fig 9, where the prestrained and fully-aged rectangular and tubular struts have a comparable normalised critical stress. The results for these struts are below the normalised critical stress of the as-received struts for slenderness ratios greater than 40.

The results of the rectangular section struts in Fig 5 indicate that the level of prestraining (3% to 7%) has a small effect on strut strength. The influence of strain aging, however, seems to be significant, especially in the region of the transition length for the struts. At an ℓ/r of 80 the load capacity, relative to the as-received strut, of the 3% and 7% prestrained struts with no aging was about 40%, whereas the load capacity of the prestrained struts with full aging was about 60%.

The Bauschinger effect is inherently related to strain hardening. For the rectangular section struts the reduction in buckling load for non-aged specimens is in general slightly higher for 3% prestrain, at which 55-60% reductions occur, than for 7% prestrain, at which reductions of 50-55% occur. It appears that the influence of the Bauschinger effect has been reduced slightly by the increased tensile strain, even though strain hardening causes the effect.

The combined influences of strain hardening, strain aging, and the Bauschinger effect, lead to slight changes (5%) in load capacity for slenderness ratios less than 20. In the significant range of 40-180 the Bauschinger effect plays a much more dominant role owing to the associated reduction in tangent modulus that leads to a greater propensity for buckling to occur.

The mild steel used throughout was semi-killed. In view of the significance of strain aging in the tests reported herein it would be of value to repeat the series with a rimming steel, which is more susceptible to strain aging than the semi-killed steel used.

In conclusion it is seen that the combined effects of the three variables have a substantial influence on strut carrying capacity. Careful consideration needs to be given to these variables especially when utilizing structural sections formed by newer or modified manufacturing processes. This caution is apt when the strength of such sections is likely to be judged on yield stress as the only relevant material property.

ACKNOWLEDGMENTS

The authors wish to thank their respective Departments for providing resources to carry out the work. Thanks are due to Mr. Kingsley Davis, Tubemakers of Australia Ltd., Unanderra, and Dr. Roger Hazell, Department of Mechanical Engineering, University of Melbourne, for their constructive comments during the course of the work.

REFERENCES

1. N K POLAKOWSKI and E J RIPLING, Strength and Structure of Engineering Materials, Prentice-Hall Inc., 1966.

2. F A McCLINTOCK and A S ARGON, Mechanical Behaviour of Materials, Addison-Wesley Publishing Company Inc., 1966.

3(a) J D BAIRD, Strain Aging of Steel - A Critical Review, Part I; Iron and Steel; May 1963, pp.186-192. Part II; Iron and Steel; July 1963, pp.368-374. Part III(Cont'd); Iron and Steel; August 1963, pp.400-405.

3(b) J D BAIRD, The Effects of Strain-Aging Due to Interstitial Solutes on the Mechanical Properties of Metals, Metallurgical Reviews, Review 149; 1971, pp.1-18 (An Iron and Steel Institute Joint Activity).

4. A ABEL and H MUIR, A New Look at the Bauschinger Effect, Metals Australia, September 1972, pp.267-271.

5. L M BROWN, Orowan's Explanation of the Bauschinger Effect. Scripta Metallurgica, 11, (2), February 1977, pp.127-131.

6. S T ROLFE, R P HAAK and H H GROSS, Effect of State-of-Stress and Yield Criterion on the Bauschinger Effect. Journal of Basic Engineering; Transactions of the ASME; September 1968; pp.403-408.

7. K J PASCOE, Strength of Cold-Formed Cylindrical Steel Plates, Journal of Strain Analysis; Vol.6, No.3, 1971; pp.167-176.

8. K J PASCOE, Directional Effects of Prestrain in Steel. Journal of Strain Analysis; Vol.6,No.3,1971,pp.181-184.

9. P C PARIS, The Bauschinger Effect on Columns. Journal of Applied Mechanics; September 1956, pp.479-480.

10. F BLEICH, Buckling Strength of Metal Structures. McGraw-Hill, N.Y., 1952.

11. M N PAVLOVIC and L K STEVENS, The Effect of Prior Flexural Prestrain on the Stability of Structural Steel Columns, Engineering Structures; Vol.3, No.2, April 1981; pp.66-70.

DEVELOPMENT OF NUMERICAL PROCEDURES FOR ANALYSIS OF COMPLEX STRUCTURES

K. K. GUPTA
D. Sc. (Eng.), Ph.D., B.E.

NASA Ames Research Center
Dryden Flight Research Facility
Edwards, California, U.S.A.

The paper is concerned with the development of novel numerical procedures for the solution of static, stability, free vibration and dynamic response analysis of large, complex practical structures. Thus, details of numerical algorithms evolved for dynamic analysis of usual non-rotating and also rotating structures as well as finite dynamic elements are presented in the paper. Furthermore, the article provides some description of a general-purpose computer program STARS specifically developed for efficient analysis of complex practical structures.

INTRODUCTION

Current numerical structural analysis techniques essentially involve two distinct yet unified solution procedures. First the continuum is discretized by a suitable method such as the finite or the dynamic element procedure yielding sets of sparse matrix equations. The second step in the analysis process is concerned with the efficient solution of these matrix equations by appropriate solution schemes that fully exploit such matrix sparsity.

A large majority of structures, usually encountered in any relevant branch of engineering, is required to withstand dynamic, time dependent loading functions as the primary design consideration. Therefore, a suitable dynamic response analysis capability, involving solution of the associated free vibration problem as a vital preliminary, that ensures economical and accurate structural synthesis is of utmost importance for the safe design of various practical structural systems. A general matrix equation of equilibrium pertaining to spinning structures may be given as below,

$$\underline{M}\ddot{\underline{u}} + \underline{C}\dot{\underline{u}} + \underline{K}\underline{u} = \underline{P}(t) \qquad \ldots 1,$$

in which $\underline{M}$, $\underline{K}$ and $\underline{C}$ are the inertia, stiffness, damping and Coriolis matrices, respectively, $\underline{P}(t)$ being the external time dependent loading function. The associated free vibration equation of motion of the structure spinning at a uniform rate Ω is formulated as (Ref 1),

$$[\underline{K}(1+i^*g) + \omega(\underline{C}_c+\underline{C}_d) + \omega^2\underline{M}]\underline{q} = \underline{\emptyset} \qquad \ldots 2,$$

where $\underline{K} = \underline{K}_E + \underline{K}_G + \underline{K}'$ and in which,

$\underline{K}_E$ = elastic stiffness matrix
$\underline{K}_G$, $\underline{K}'$ = geometrical, centrifugal force matrices, respectively, each being a function of Ω^2
$\underline{C}_c$ = Coriolis matrix, a function of Ω
$\underline{C}_d$ = viscous damping matrix
$\underline{q}$ = amplitude of u
i^* = imaginary number, $\sqrt{-1}$

and in which the natural frequencies ω and the associated vectors occur in complex conjugate pairs. The various individual free vibration problems may all be derived from Eqn 2 and may be identified next:

Undamped free vibration of non-spinning structures

$$[\underline{K}_E - \omega^2\underline{M}]\underline{q} = \emptyset \qquad \ldots 3,$$

and,

$$[(\underline{K}_E + \underline{K}_G) - \omega^2\underline{M}]\underline{q} = \emptyset \qquad \ldots 4,$$

for pre-stressed structures.

Buckling

$$[\underline{K}_E - \gamma\underline{K}_G]\underline{q} = \emptyset \qquad \ldots 5,$$

γ being the buckling load.

Undamped free vibration of spinning structures

$$[\underline{K} + \omega\underline{C}_c + \omega^2\underline{M}]\underline{q} = \emptyset \qquad \ldots 6.$$

Structural discretization, on the other hand, is usually achieved by the finite element method in which the displacement field $\underline{u}$ within an element is expressed in terms of its undetermined nodal deformation values $\underline{U}$,

$$\underline{u} = \underline{a}\underline{U} \qquad \ldots 7,$$

$\underline{a}$ being the shape function matrix. Such a relationship is strictly valid for static cases only since for general dynamic problems the matrix $\underline{a}$ is not unique being a function of the entire time history of nodal displacements (Ref 2). However, in the special case of harmonic motion such as free vibration, $\underline{a}$ is a functon of instantaneous nodal displacements as well as the frequencies of harmonic motion. The matrix a is next expressed in series form in ascending powers of the natural frequency ω, which in turn produces the relevant matrices in similar form. The resulting equation for free vibration is obtained as,

$$[\underline{K}_{\emptyset} - \omega^2 \underline{M}_{\emptyset} - \omega^4 (\underline{M}_2 - \underline{K}_4) - \ldots] \underline{q} = \underline{\emptyset} \quad \ldots 8,$$

in which $\underline{K}_{\emptyset}$ and $\underline{M}_{\emptyset}$ are static stiffness and inertia matrices whereas the matrices $\underline{M}_2$ and $\underline{K}_4$ constitute dynamic correction terms. While in the finite element method (FEM) only the first two terms in Eqn 8 are retained for analysis, the dynamic element method (DEM) utilizes the higher order matrices for the solution process; the associated quadratic matrix equation may be written as,

$$(\underline{A} - \lambda \underline{B} - \lambda^2 \hat{\underline{C}}) \underline{q} = \underline{\emptyset} \quad \ldots 9,$$

with $\lambda = \omega^2$ and when it may be noted that the matrices $\underline{A}$, $\underline{B}$ and $\hat{\underline{C}}$ are positive definite in nature for most practical problems, being also of highly banded configurations. Employment of the DEM procedure (Ref 3,4) utilizing Eqn 9 is known to considerably accelerate the convergence characteristics of structural free vibration analysis when compared to the corresponding FEM solution.

This paper presents the details of an efficient, generalized eigenproblem solution procedure based on a newly developed Lanczos algorithm, which is followed by some descriptions of the DEM technique. Several numerical examples are also presented that depict the relative efficacy of the solution techniques described in the paper.

DEVELOPMENT OF NUMERICAL PROCEDURES FOR FREE VIBRATION ANALYSIS OF STRUCTURES

Eigenproblem solution

A unified numerical procedure, based on a combined Sturm sequence and inverse iteration method, was developed earlier for efficient eigenproblem solution of Eqn 2, details of which have been presented earlier in Ref 1. For the special case of undamped free vibration pertaining to non-spinning structures as defined by Eqn 3, Ref 5 presents a review of associated solution techniques including a simultaneous iteration method; also a recenty developed solution algorithm based on the Lanczos method (Ref 6,7) proves to be rather efficient.

In this paper description of a unified Lanczos algorithm is presented, in some detail, for the solution of the generalized eigenvalue problem expressed by Eqn 2. Thus, firstly the associated undamped equation of free vibration represented by Eqn 6 is rewritten as,

$$\left[\begin{array}{c|c} \underline{M} & \underline{\emptyset} \\ \hline \underline{\emptyset} & \underline{K} \end{array}\right] \left[\begin{array}{c} \dot{\underline{q}} \\ \hline \underline{q} \end{array}\right] + \left[\begin{array}{c|c} \underline{\emptyset} & -\underline{M} \\ \hline \underline{M} & \underline{C} \end{array}\right] \left[\begin{array}{c} \ddot{\underline{q}} \\ \hline \dot{\underline{q}} \end{array}\right] = \underline{\emptyset} \quad \ldots 1\emptyset,$$

which may also be described as,

$$\underline{B}\underline{y} + \underline{A}\dot{\underline{y}} = \underline{\emptyset} \quad \ldots 11,$$

where,

$$\underline{y} = \left[\begin{array}{c} \dot{\underline{q}} \\ \hline \underline{q} \end{array}\right] \quad \ldots 11a.$$

Assuming that solution of Eqn 11 is expressed as $\underline{y} = e^{\omega t}$, it takes the form,

$$(\underline{B} + \omega \underline{A}) \underline{y} = \underline{\emptyset} \quad \ldots 12,$$

which may be rearranged as,

$$(\underline{B} - \lambda \underline{A}^*) \underline{y} = \underline{\emptyset} \quad \ldots 13,$$

where $\underline{A}^* = i^* \underline{A}$ is a Hermitian matrix, the roots $\lambda = i^* \omega$ being real and $i^* = \sqrt{-1}$. The roots of the original Eqn 12 are simply obtained as λ / i^*, which along with the eigenvectors occur in complex conjugate pairs. A similar procedure may be developed in connection with the dynamic element formulation represented by Eqn 9 which is rewritten as,

$$\left[\begin{array}{c|c} \hat{\underline{C}} & \underline{\emptyset} \\ \hline \underline{\emptyset} & \underline{K} \end{array}\right] - \lambda \left[\begin{array}{c|c} \underline{\emptyset} & \hat{\underline{C}} \\ \hline \hat{\underline{C}} & \underline{M} \end{array}\right] \left[\begin{array}{c} \dot{\underline{q}} \\ \hline \underline{q} \end{array}\right] = \underline{\emptyset} \quad \ldots 14,$$

having the following form,

$$(\underline{E} - \lambda \underline{F}) \underline{y} = \underline{\emptyset} \quad \ldots 15.$$

The solution of Eqn 15 is effected in the same manner as that of Eqn 13, the computations in this case being performed in real arithmatic.

Associated Lanczos method employs the following numerical procedure:

Step 1 To initialize basic variables

$\underline{y}_{\emptyset} = \underline{\emptyset}$

$\underline{v}_{\emptyset}$ = a unit vector

$\beta_{\emptyset} = (\underline{v}_{\emptyset}^T \underline{A}^* \underline{v}_{\emptyset})^{1/2}$

$\tau_{\emptyset} = \tau_1 = \varepsilon$

$S_{\emptyset}, S_1, K_1 = (n+6)\,\varepsilon$

n being the order of matrices and ε is a small number denoting tolerence limit for root convergence accuracy.

Step 2 Basic Lanczos procedure

(a) Form $\underline{y}_j = v_{j-1} / \beta_j$

(b) Solve $(\underline{B} - \lambda \underline{A}^*) \underline{v}_j = N_{i+1} \underline{A}^* \underline{y}_j$ to yield $\underline{v}_j$; N_{i+1} is a normalizing factor

(c) Recalculate $\underline{v}_j = \underline{v}_j - \underline{y}_{j-1} \beta_j$

(d) Calculate $\alpha_j = \underline{v}_j^T (\underline{A}^* \underline{y}_j)$

(e) Form $\underline{v}_j = \underline{v}_j - \underline{y}_j \alpha_j$

(f) Calculate $\beta_{j+1} = (\underline{v}_j^T \underline{A}^* \underline{v}_j)^{1/2}$

Step 3 Update τ, monitoring components of Lanczos vector in direction of Ritz vector, appropriately. Update K, monitoring loss of orthogonality, appropriately. If $K_{j+1} < \sqrt{\varepsilon}$, repeat Step 2.

Step 4 Solve eigensystem of $\underline{L}_j$, a tridiagonal matrix with components $\alpha_1, \ldots, \alpha_j$ and $\beta_2, \ldots, \beta_{j+1}$. Quit if enough eigenvalues have converged.

An inverse iteration procedure (Ref 1) may next be adopted to solve the damped free vibration problem given by Eqn 2. The above formulation, developed for the solution of Eqn 13 is also applicable for solving Eqn 14 associated with the dynamic element technique by replacing $\underline{A}^*$ and $\underline{B}$ by $\underline{F}$ and $\underline{E}$ matrices, respectively.

This newly developed algorithm fully exploits matrix sparsity and employs only the upper symmetric halves of the relevant $\underline{K}$, $\underline{M}$, $\underline{C}$ and $\hat{\underline{C}}$ matrices, enabling economical and efficient solution of complex practical problems.

Centrifugal forces in elements of spinning structures

The derivation of $\underline{K}_G$ matrix in Eqn 2 due to spin is dependent on the evaluation of nodal centrifugal forces in the finite elements utilized for the idealization of the spinning structure. Figure 1 shows a typical triangular shell element rotating along an arbitrary axis with a spin rate Ω_R, having components Ω_X, Ω_Y and Ω_Z in global X, Y and Z coordinates, respectively. Assuming that the shape function matrix of Eqn 7 is expressed as,

$$\underline{a} = \underline{R}\underline{Q}^{-1} \quad \ldots 16,$$

in which,

$\underline{R}$ = portion of shape function matrix, which is a function of coordinate x, y

$\underline{Q}$ = portion of shape function matrix, a function of local coordinate of nodes

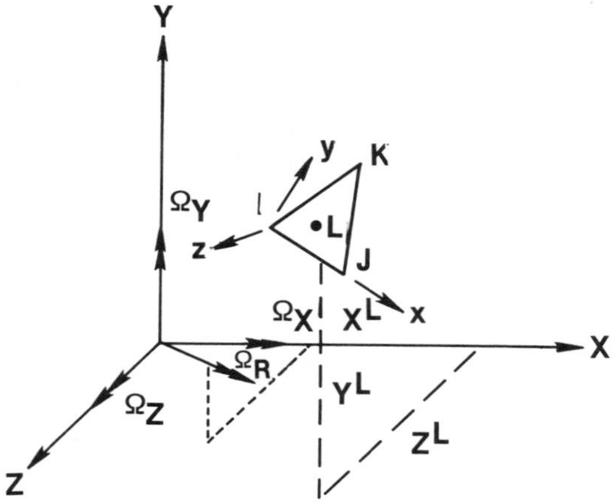

Fig 1. A Triangular Plane Finite Element Rotating Around an Arbitrary Axis

Furthermore, defining the element nodal forces vector in the local coordinate system (x,y,z) as,

$$\underline{f} = (f_x^1, f_y^1, f_z^1, \ldots, f_x^3, f_y^3, f_z^3)^T \qquad \ldots 17,$$

expressions for such nodal centrifugal forces in the planar local x- and y- directions due to spin rates along the global X-, Y- and Z- directions are derived as follows:

X-axis (Ω_X)

$$\underline{f} = \rho\Omega_X^2 t[Q^{-1}]^T \int_{\emptyset}^{YK}\int_{LX}^{HX} R^T [XM]\,dxdy \qquad \ldots 18,$$

Y-axis (Ω_Y)

$$\underline{f} = \rho\Omega_Y^2 t[Q^{-1}]^T \int_{\emptyset}^{YK}\int_{LX}^{HX} R^T [YM]\,dxdy \qquad \ldots 19,$$

Z-axis (Ω_Z)

$$\underline{f} = \rho\Omega_Z^2 t[Q^{-1}]^T \int_{\emptyset}^{YK}\int_{LX}^{HX} R^T [ZM]\,dxdy \qquad \ldots 2\emptyset,$$

The total element nodal force is obtained as,

$$\underline{f} = \underline{f}(X) + \underline{f}(Y) + \underline{f}(Z) \qquad \ldots 21,$$

in which,

$$[XM] = \begin{bmatrix} M_x(M_x x + M_y y + YI) + N_x(N_x x + N_y y + ZI) \\ M_y(M_x x + M_y y + YI) + N_y(N_x x + N_y y + ZI) \end{bmatrix} \qquad \ldots 22,$$

$$[YM] = \begin{bmatrix} L_x(L_x x + L_y y + XI) + N_x(N_x x + N_y y + ZI) \\ L_y(L_x x + L_y y + XI) + N_y(N_x x + N_y y + ZI) \end{bmatrix} \qquad \ldots 23,$$

$$[ZM] = \begin{bmatrix} L_x(L_x x + L_y y + XI) + M_x(M_x x + M_y y + YI) \\ L_y(L_x x + L_y y + XI) + M_y(M_x x + M_y y + YI) \end{bmatrix} \qquad \ldots 24,$$

[DIR] = direction cosine matrix

$$= \begin{bmatrix} L_x & M_x & N_x \\ L_y & M_y & N_y \\ L_z & M_z & N_z \end{bmatrix} \qquad \ldots 25,$$

where,

XI,YI,ZI = coordinates of node I in the global coordinate system (GCS)

t = element thickness

The element nodal forces may then be transformed into the GCS as below,

$$\underline{F} = \underline{D}^T \underline{f} \qquad \ldots 26,$$

in which $\underline{D}$ is the element direction cosine matrix. In plane stresses in various elements are next obtained by solving,

$$\underline{K}\underline{U} = \underline{P} \qquad \ldots 27,$$

where the vector $\underline{P}$ is obtained by combining the forces $\underline{F}$ for all structural elements; the geometrical stiffness matrix $\underline{K}_G$ is then derived by standard procedures.

The above formulation is general in nature and similar expressions have also been derived for quadrilateral elements. Such a formulation for line elements is obviously quite straightforward in nature.

Finite dynamic elements

A rectangular hexahedron finite dynamic element has recently been developed that involves derivation of higher order dynamic correction terms. Utilizing such an element natural frequency analysis was performed for a cube with a clamped surface; Fig 2 presents the convergence characteristics of two typical roots obtained by employing both the dynamic as well as the usual finite element method. It may be observed from the diagram that employment of dynamic elements effects improved solution convergence characteristics when compared with the relevant finite element analysis. In this connection it may also be noted that solution of the associated quadratic matrix formulation given by Eqn 9 has been achieved by expending about the same computational effort as that required for Eqn 3 in a finite element analysis.

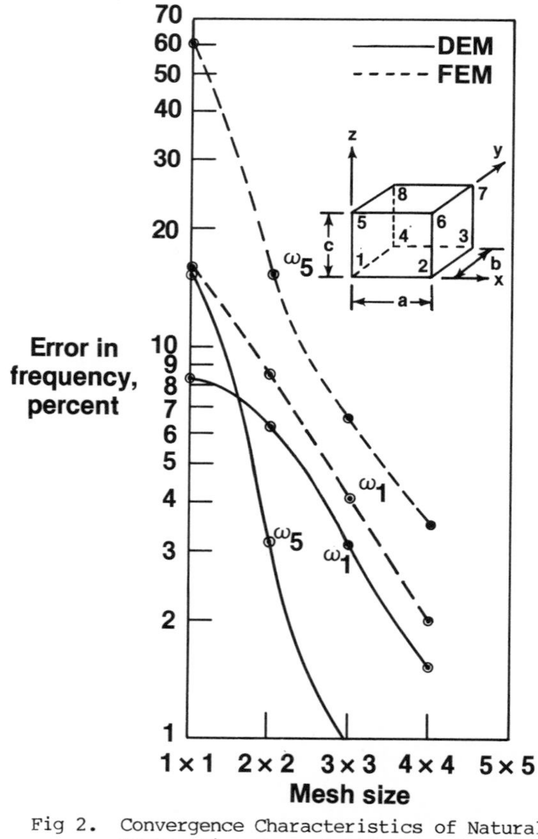

Fig 2. Convergence Characteristics of Natural Frequencies for a Solid Element

STARS - a general purpose computer program

The recently developed computer program STARS (STructural Analysis RoutineS) (Ref 8) is primarily an inter-active, graphics-oriented, discrete elememt routine for static, stability, free vibration and dynamic response analysis of damped and undamped structures including rotating systems. The novel procedures described in this paper have been incorporated in the program and Fig 3 provides an overview of the same, the size of which is tightly controlled and the current VAX 11-780 version written in FORTRAN language consists of less than 10,000 programmed instructions.

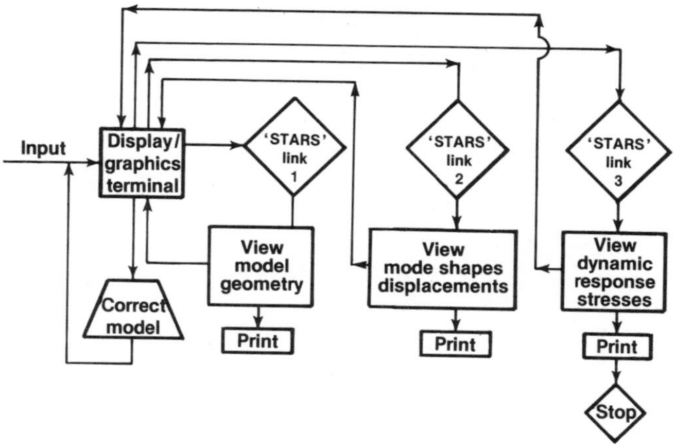

Fig 3. STARS Overview

Essential features of the program are highlighted below,

- A general-purpose digital computer discrete element program
- Elements: various bars, beams, plates, shells and solid elements
- Geometry: truss, frame, plane stress/strain, plate, shell, solid and any suitable combination thereof
- Analysis: static, buckling, free vibration and dynamic response analysis
  dynamic analysis of spinning structures
  analysis of damped and pre-stressed structures

followed by some special features,

- Random data input
- Matrix bandwidth minimizer
- Automatic node and element generation
- General nodal deflection boundary conditions
- General pre-, post-processors

The program has been successfully utilized for the analysis and design of a variety of engineering structures including the forward swept wing (FSW) X-29 aircraft components.

NUMERICAL EXAMPLES

General shell

Figure 4 depicts the finite element model of a cylindrical shell with aspect ratio equal to 1. A free vibration analysis was performed by the STARS program and the first few natural frequencies in parametric form is depicted in Table I, in which the definitions of the various terms are as follows:

ω = natural frequency

a,b = side lengths (=L for a square plate)
D = plate rigidity
t = plate thickness
ρ = mass density

Mode Number	Non-dimensional Parameter $\gamma = \omega a^2\sqrt{\rho t/D}$
1	10.60
2	16.99
3	30.65
4	42.23
5	47.68
6	65.45

Table I. Natural Frequencies of a Cylindrical Cantilever Shell

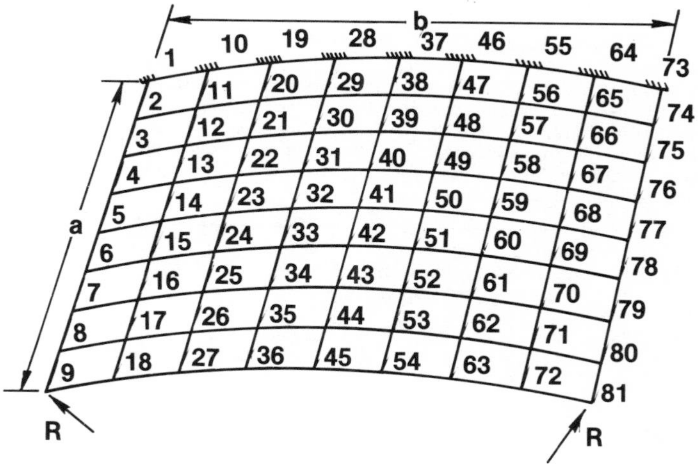

Fig 4. Finite Element Model of a Cylindrical Shell

Spinning cantilever plate

Figure 5 shows the finite element model of a plate rotating along an arbitrary axis with a uniform spin rate Ω_R.

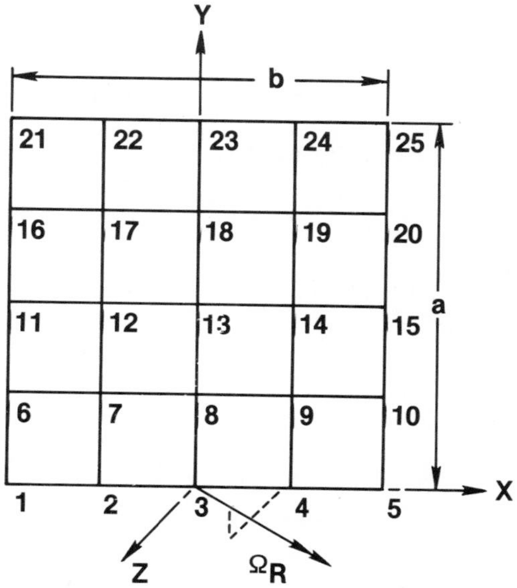

Fig 5. Square Cantilever Plate

A free vibration analysis of the non-spinning structure yielded a first natural frequency (ω_1) value of 3.59 expressed in non-dimensional parametric form. Subsequent analysis of the plate was performed for the case of $\Omega_Z=0.8\omega_1$ as well for $\Omega_R=0.8\omega_1$ having components $\Omega_X=\Omega_Y=\Omega_Z=0.8\omega_1/\sqrt{3}$ and such results are presented in Table II.

Mode Number	Natural Frequency Parameter $\gamma = \omega L^2\sqrt{\rho t/D}$ $\Omega_Z = 0.8\omega_1$	$\Omega_R = 0.8\omega_1$ ($\Omega_X=\Omega_Y=\Omega_Z=0.8\omega_1/\sqrt{3}$)
1	10.6103	7.4377
2	16.4093	13.4362
3	29.5585	26.4286
4	32.9242	30.3492
5	39.2103	36.1341
6	58.3640	56.2620

Table II. Natural Frequency Parameters of a Spinning Square Cantilever Plate

Helicopter structure

A coupled helicopter rotor-fuselage system, Ref 8, is shown in Fig 6 along with structural mass and stiffness distributions. Appropriate free vibration analysis of the structure with a spinning rotor and a non-rotating fuselage was performed utilizing the STARS program and such results are presented in Table III.

Mode Number	Natural Frequencies spin rate (rad/sec) $\Omega_Y=0$	$\Omega_Y=10$	Mode Shape
1,2,3	0.0	0.0	rigid body
4	4.645	11.83	rotor 1st antisym. bending
5	5.093	11.90	rotor 1st sym. bending
6	23.088	23.19	fuselage 1st bending
7	27.93	36.98	rotor 2nd antisym. bending
8	28.22	38.16	rotor 2nd sym. bending
9	38.40	39.29	rotor 3rd antisym. bending

Table III. Natural Frequencies of a Helicopter Structure

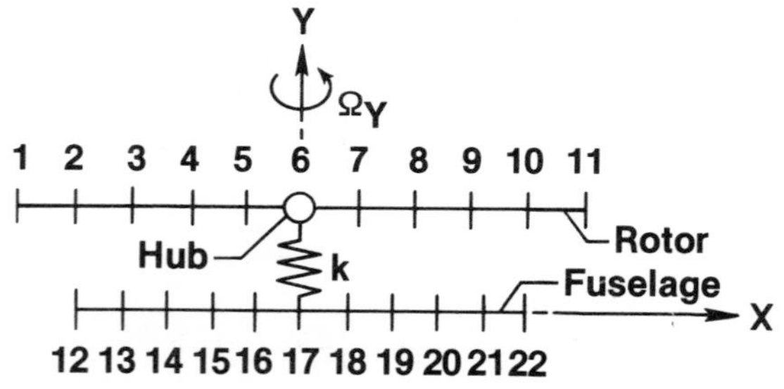

a) Discrete Element Model

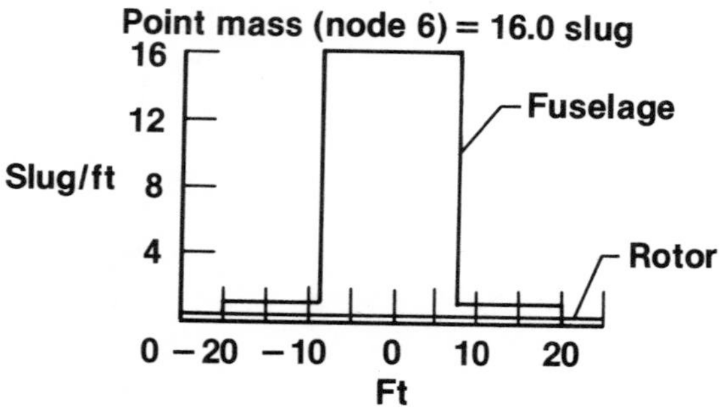

b) Structural Mass Distribution

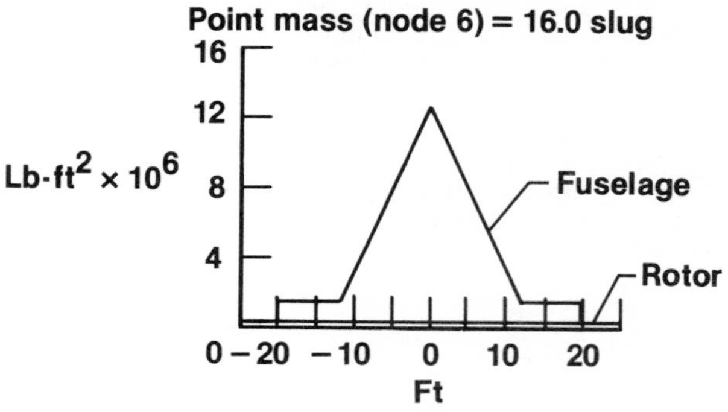

c) Structural Stiffness Distribution

Fig 6. Coupled Helicopter Rotor-Fuselage System

The LDR concept

A feasibility study is currently being pursued at NASA for developing a Large Deployable Reflector (LDR) for infrared and submillimeter astrophysical observations in the spectral range of 30μm to 1 mm wavelength. The major objectives of the proposed project may be summarized as follows (Ref 9),

1) To detect radiation from cooled or highly red-shifted phenomena such as newly forming stars and planetary systems located in the cold depths of optically opaque galactic clouds.

2) To receive ancient signals from galaxies or such clusters located at the very edge of the universe.

3) To observe fluctuations in the cosmic background radiation enabling a study on the time and nature of galaxy birth; the signals from the first matter such formed will help in understanding the evolution of the universe to its current state.

4) A study of the cosmic background radiation will also assist in determining the evolutionary fate of the universe.

Figure 7 shows a LDR system concept having a 20m diameter on-axis cassegrain telescope to be lifted by the shuttle to place it in Earth orbit.

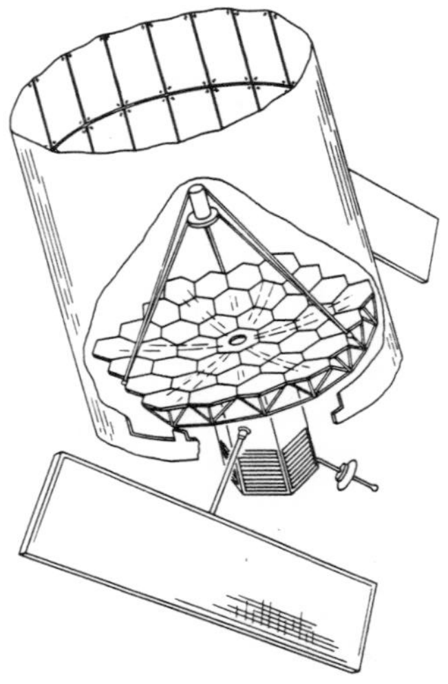

Fig 7. LDR System Concept

A large number of hexagonal mirror segments (Ref 10) will be utilized to form the primary reflectors. Main structural elements are expected to be made of graphite/epoxy tubes built in tetrahedral truss arrangement. A detailed system analysis involving structres, dynamics, optics, thermal effects and control is being undertaken involving the STARS program (Ref 8), among others. A deployment sequence (Ref 10) for the LDR is shown in Fig 8.

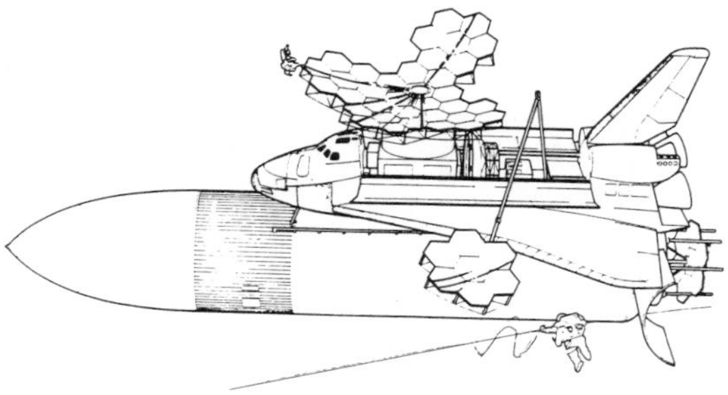

Fig 8. LDR Construction

CONCLUDING REMARKS

A number of novel numerical procedures has been presented in this paper for efficient dynamic analysis of complex, practical structures. Thus the Lanczos formulation pertaining to the general eigenvalue problem proves to be effective for large, sparse matrix systems. Also the formulation for finite element centrifugal forces is general in nature. A brief description of the STARS program is presented along with relevant numerical examples; Further applications of the STARS program in the area of space structures such as the LDR concept is also discussed in some detail.

ACKNOWLEDGEMENTS

The author wishes to thank Mr. R.J.Knight, Deputy Chief, Aerostructure Branch, for his encouragement and help in this effort. Thanks are also due to Ms. R.Pellicciotti, Research Assistant, for her exellent effort in the preparation of this manuscript.

REFERENCES

1. K K GUPTA, Development of a Unified Numerical Procedure for Free Vibration Analysis of Structures, International Journal for Numerical Methods in Engineering, Volume 17, Pp 187-198, 1981.

2. J S PRZEMIENIECKI, Theory of Matrix Structural Analysis, McGraw-Hill, New York, 1968.

3. K K GUPTA, On a Finite Dynamic Element Method for Free Vibration Analysis of Structures, Computer Methods in Applied Mechanics and Engineering, Volume 9, Pp 105-120, 1976.

4. K K GUPTA, Development of a Finite Dynamic Element for Free Vibration Analysis of Two-Dimensional Structures, International Journal for Numerical Methods in Engineering, Volume 12, Pp 1311-1327, 1978.

5. A JENNINGS, Matrix Computation for Engineers and Scientist, John Wiley and Sons, New York, 1978.

6. B N PARLETT and D S SCOTT, The Lanczos Algorithm with Selective Orthogonalization, Mathematics of Computation, Volume 33, Number 145, Pp 217-238, 1979.

7. B NOUR-OMID et.al., Lanczos Versus Subspace Iteration for Solution of Eigenvalue Problems, International Journal for Numerical Methods in Engineering, Volume 19, Pp 859-871, 1983.

8. K K GUPTA, STARS - A General-Purpose Finite Element Computer Program for Analysis of Engineering Structures, NASA Reference Publication, 1984.

9. M K KIYA et.al., A Technology Program for the Development of the Large Deployable Reflector for Space Based Astronomy, Technology for Space Astrophysics Conference: the Next 30 Years, Danbury, Connecticut, October 4-6, 1982, (AIAA SPIE conference).

10. R PITTMAN and C LEIDICH et.al., A Modular Approach to Developing a Large Deployable Reflector, SPIE conference, San Diego, August 21-25, 1983.

AN INVESTIGATION INTO THE COLLAPSE BEHAVIOUR OF DOUBLE-LAYER GRIDS

I M COLLINS, BSc, PhD

Power-Strut Division, Van Huffel
Tube Corporation, Warren, Ohio U.S.A.

The interconnection patterns and large number of members utilised in practical double-layer grids usually result in their being highly statically indeterminate. Although such structures are redundant and are usually made of ductile material, they do not necessarily possess reserves of strength beyond their elastic limit because their inelastic behaviour may involve the plastic buckling of struts. This paper presents a technique for analysing such structures in their inelastic range of behaviour. Strut post buckling behaviour is modelled on theoretically derived curves. A series of thorough tests on model double-layer grids is reported.

INTRODUCTION

This paper is a summary of a study performed at the Space Structures Research Centre, University of Surrey, see Ref 1.

The inelastic behaviour of double-layer grids is not usually considered in the design process because it is argued that due to their high degree of indeterminacy these structures, if made of ductile material, will possess load carrying capacities in excess of their elastic limit loads and therefore appropriately factored linear analyses will provide conservative designs. However, this premise relies upon the ability of members that reach their elastic limit load to maintain that load under increasing deformation. In the case of most practical double-layer grids, this is not the case.

If the elastic limit load of such a structure is exceeded, because external loads are transmitted primarily as axial forces, and because practical members are neither very stocky nor very slender, then the plastic buckling of struts may be involved in the post-elastic structural behaviour. Member instability due to plastic-buckling causes a sudden loss of member load carrying capacity which results in a redistribution of internal forces within the structure. If the structure is capable of withstanding this redistribution, then it may be able to support increased external load. However, if the redistribution causes other members to fail, then the "collapse" may become progressive and the structure may possess no reserve of strength beyond its elastic limit load.

Therefore, in order to establish design procedures that will give rise to economic double-layer grid structures which possess inelastic reserves of strength, it is necessary to develop techniques of analysis which accurately predict inelastic structural behaviour. Analyses are best verified by making comparisons with experimental data.

A TECHNIQUE OF ANALYSIS

The inelastic behaviour of double-layer grids involves the sequential failure of individual members which may result in overall collapse if the structure becomes a mechanism. However, "collapse" may be considered to have occurred if deflections have become excessive. Conventional ultimate load methods do not usually provide information on deflections and therefore analysis techniques are required which trace the sequential failure of members and hence the structure's load-displacement behaviour from elasticity to collapse. Such techniques should account for the post-buckling behaviour of struts.

A technique adopted by several authors, Refs 2, 3, 4, is the "member removal" method in which members upon attainment of their elastic limit loads are removed from the structure and replaced by equivalent nodal forces. These methods adopt a strut model which allows a failed compression member to shorten at a constant residual load, Fig 1. The residual load is taken as same part of the strut critical load. Schmidt, Ref 4, shows that the theoretical inelastic behaviour of a structure in which members buckle is highly dependent on their post-buckling residual loads. Supple, Ref 5, 6, shows how the structural behaviour is also dependent on the slope of the post-buckling curve. Therefore, in order to accurately model variations in the form of strut post-buckling curves (both residual loads and slope), a "modulus modification" technique of analysis was developed. An early version of this approach is reported elsewhere, Ref 6.

The pin-jointed member behaviour model used in this technique is shown in Fig 2. The load-displacement behaviour is modelled by a series of linear approximations. These linear idealisations allow a step-by-step type of analysis to be utilised in which the overall structural behaviour is considered as a series of appropriately factored linear analyses. In each successive analysis, the member moduli change in accordance with the idealised member model. This approach was chosen in preference to an iterative analysis technique because it was felt that the theoretical results from a less complex structural model/analysis should be compared with experimental observations before engaging in a more rigorous computing exercise. Should the proposed analytical technique be found to provide sufficiently accurate predictions of structural behaviour, then it gains a potential advantage over more complex techniques for use in engineering practice.

Figure 2 shows the typical member model to consist of six linear "phases" representing the entire tension-compression loading path. The model provides for elastic unloading from any position on this path. Each linear analysis in the "collapse" sequence is performed using the direct stiffness method. In this way, the stiffness contributions of failed members to the structure's overall stiffness are used in the prediction of the structural load-deflection behaviour. Numerical problems can occur in the solution under two conditions. Perfectly elastic-plastic member behaviour cannot be modelled as this requires zero member stiffness and if adopted in the solution would constitute a member removal. Similarly, perfectly rigid strut post-buckling behaviour cannot be modelled as this requires infinite negative member stiffness.

The exclusion of iterative techniques means that any internal stress redistributions caused by geometric changes are not considered. This omission is acceptable provided the total joint deflections are very small in comparison to the overall structural dimensions. This limits the theory to structures of high stiffness such as double-layer grids in which only small deflections are necessary to induce high stresses. Therefore, results obtained from this approach which indicate large deformations should be used with extreme caution.

Figure 3 shows the flow diagram for the analysis. A failed member is one which has either buckled or yielded. The governing factor in obtaining the structural behaviour is the treatment of the failed members' incremental stresses which are calculated at each step of the solution. The incremental stresses must conform to the member behaviour model. Strut plastic buckling behaviour is such that in all cases, it exhibits a negative modulus; i.e., as a strut shortens it carries less load, hence incremental stresses must be positive. Post-yield tension behaviour is such that the member shows a positive modulus; i.e., as the member lengthens it carries slightly more load, again incremental stresses must be positive. Failed members are considered in two categories. Of "M" failed members, there are N (M) which have just attained their failure load and M-N which are on some part of their post-yield or post-buckling path. The structural behaviour is governed by the group of N members. In the first analysis step after reaching their failure load, these N members must demonstrate a positive stress increment as they cannot unload elastically. This may necessitate a reversal in the direction of the applied external load (which would imply a reduction in the structure's load carrying capacity). Now, the remaining M-N failed members are checked to ensure that their incremental stresses conform to the member's position on the member model. If any non-conformities are found, then elastic unloading of some, or all, of the M-N members is indicated. This necessitates a change in member moduli and a new set of displacements for this step in the solution must be obtained. Successive linear analyses are performed until a mechanism is formed or a predetermined displacement is exceeded. The maximum load carrying capacity of the structure is obtained from the overall load-displacement behaviour curve.

The inelastic structural behaviour of varied reticulated structures in which members buckle plastically has been shown to be highly dependent on the form of the member post-buckling curves, Refs 4, 5, and 6. Further study of these curves was required.

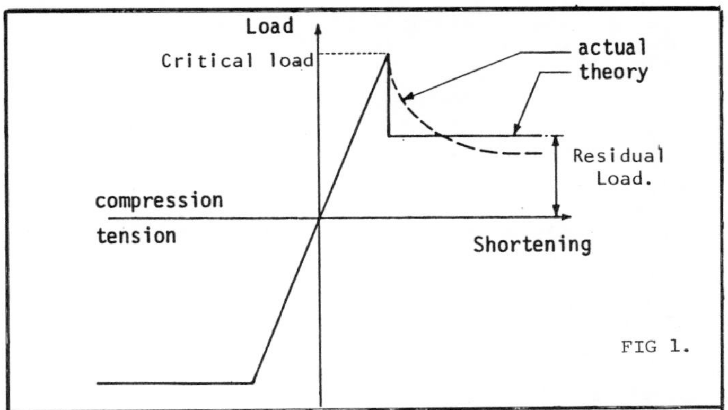

FIG 1.

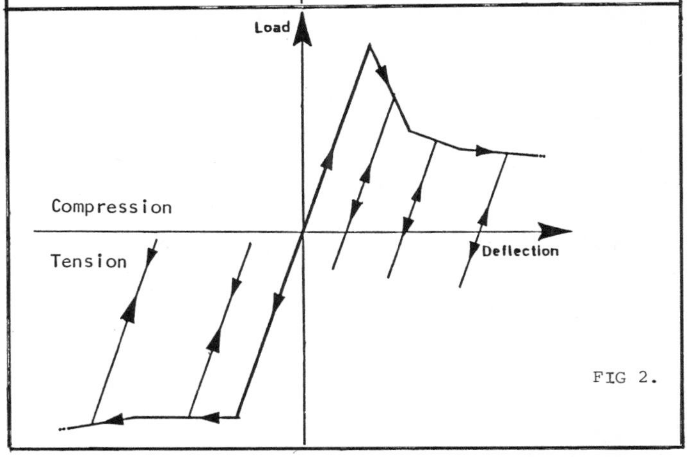

FIG 2.

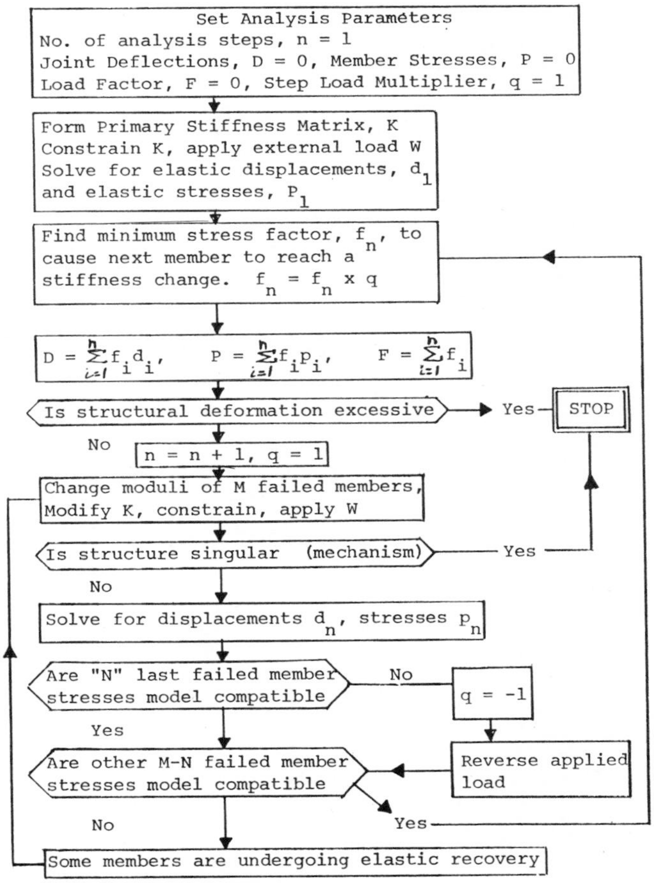

FIG 3. ANALYSIS FLOW CHART.

PLASTIC BUCKLING OF STRUTS

Various types of section are used in the design of double-layer grids. This paper limits itself to the treatment of the circular hollow section which is used in many patented double-layer grid systems.

Consider a pin-ended strut of length L and area A acted upon by an axial compression P which produces an end-shortening δ, Fig 4. Assume the material to have an elastic modulus of E and a yield stress of σ_y. The squash load of the strut is $P^* = A\sigma_y$. It is assumed that the bowing of the strut is sinusoidal with an amplitude at the mid-length of a.

An approach that can be adopted to find the post-critical behaviour of the member (the P-δ relationship) is that due to Paris, Ref 7. The axial shortening of the strut after buckling is taken to be the sum of the shortening due to direct stress and that due to flexure, Eqn 1

$$\delta = \frac{PL}{AE} + \int_0^L (ds - dx) = \frac{PL}{AE} + \frac{1}{2}\int_0^L \left(\frac{dy}{dx}\right)^2 dx \qquad \ldots\ldots 1.$$

$$y = a \sin\left(\frac{\pi x}{L}\right) \qquad \ldots\ldots 2.$$

The deflected shape of the strut, Eqn 2 is used to evaluate the flexural shortening and hence Eqn 3 is obtained.

$$\delta = \frac{PL}{AE} + \frac{(a\pi)^2}{4L} \qquad \ldots\ldots 3.$$

If the moment of resistance of the central cross section is M, then conditions of equilibrium at the mid-length of the member provide that

$$M = Pa \qquad \ldots\ldots 4.$$

Hence the post-buckling behaviour, Eqn 3 can be written in general form as

$$\delta = \frac{PL}{AE} + \frac{\pi^2}{4L}\left(\frac{M}{P}\right)^2 \qquad \ldots\ldots 5.$$

The moment of resistance, M, is governed by the geometry of the section being considered and by the yield stress of the material. Consider the case of a thin walled tube of radius R, wall thickness t, and cross sectional area A (= 2 π R t), Fig 5. If the stress distribution at the tube mid-length is as indicated, then it can be shown that

$$P = 2Rt\sigma_y(\pi - \theta) = P^*\left(1 - \frac{\theta}{\pi}\right) \qquad \ldots\ldots 6.$$

$$M = 4\sigma_y R^2 t \sin\frac{\theta}{2} \qquad \ldots\ldots 7.$$

The geometry parameter θ can be eliminated and Eqns 5, 6, 7, combine to provide the required P-δ relationship

$$\delta = \frac{PL}{AE} + \frac{1}{L}\left[\frac{P^*}{P} R \cdot \cos\left(\frac{\pi}{2} \cdot \frac{P}{P^*}\right)\right]^2 \qquad \ldots\ldots 8.$$

Figure 6 shows the post-buckling behaviour curves for struts of different slenderness ratios. References 1, 8, give a more detailed derivation of the presented theory and extends the work to thick walled tubes, fixed-ended struts, and also investigates the effects of yield stress magnitude on the curves. The theories developed indicate that end fixity increases the strut residual load factor (which is taken as the residual load divided by the squash load) and that the slope of the post-buckling curve is less severe than for a pin-ended strut of identical length. The effect of increasing the yield stress for a given length strut is to reduce the residual load factor and to increase the slope of the post-buckling curves.

An extensive range of accurate tests on tubular steel struts was performed to investigate experimentally actual strut post-buckling behaviour, Refs 1 and 9. Eight strut slenderness ratios were tested with values ranging from 50 to 120 in intervals of 10. For each slenderness, three tests were conducted at three different rates of strain (10^{-5}, 5 x 10^{-5}, 10^{-4} strain/second) giving a total of 72 compression tests. The tests were conducted using computerised measuring equipment.

The results obtained indicated that for any given slenderness ratio, the strut post-buckling behaviour was very repeatable despite variances in buckling loads. As each strut must possess slightly different geometric and material imperfections, the repeatability of the post-buckling curves indicates that such behaviour is less imperfection sensitive than the buckling load. The strain rates tested appeared to have minimal effect on the strut post-buckling behaviour. The strut end condition used in the tests was obtained by flattening the tube ends and then clamping the flattened sections in the testing machine compression jaws. (This end condition was adopted to correspond to the end conditions being used in a prototype model double-layer grid.) The degree of end restraint thus provided lies between the limiting conditions of the pinned and fixed ends. Comparisons of theoretical and experimental results showed that in virtually all cases (the exceptions being at slenderness ratio 50), the experimental curve was bounded by the pinned end and fixed end theoretical curves. A reason that could account for the shortest columns not demonstrating similar behaviour is that shorter struts squash and do not buckle. Therefore, there is a lower limit to the value of the slenderness ratio which can be used in the theory. A value of kL/r = 50 would appear to be a not unreasonable lower limit from the data observed.

A typical result is shown in Fig 7. It is concluded that sufficiently accurate predictions of the post-buckling behaviour of tubular struts can be obtained from this theory.

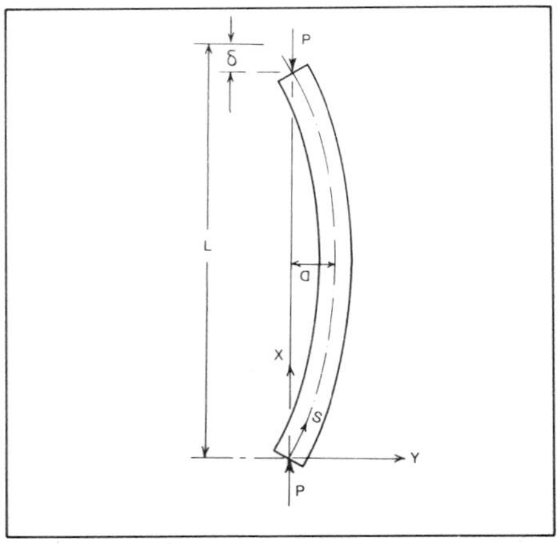

FIG 4

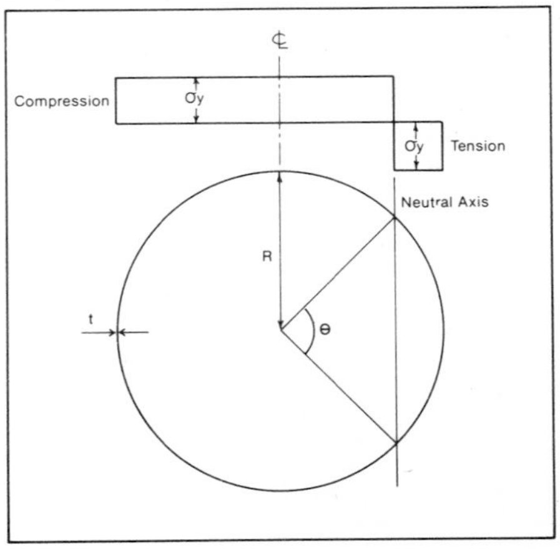

FIG 5

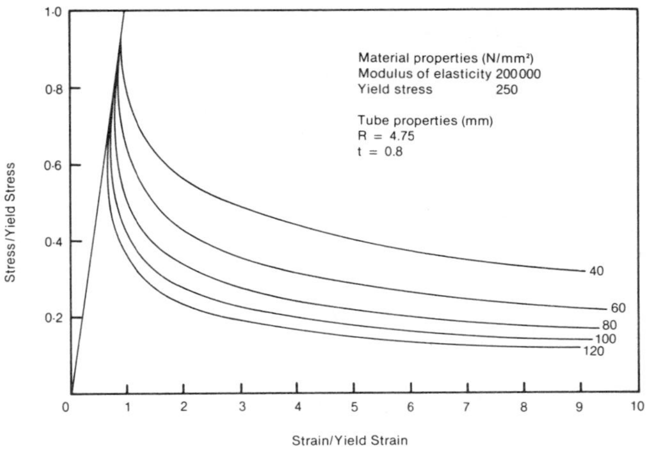

FIG 6. POST-BUCKLING CURVES FOR PIN-ENDED STRUTS WITH L/r OF 40 TO 120

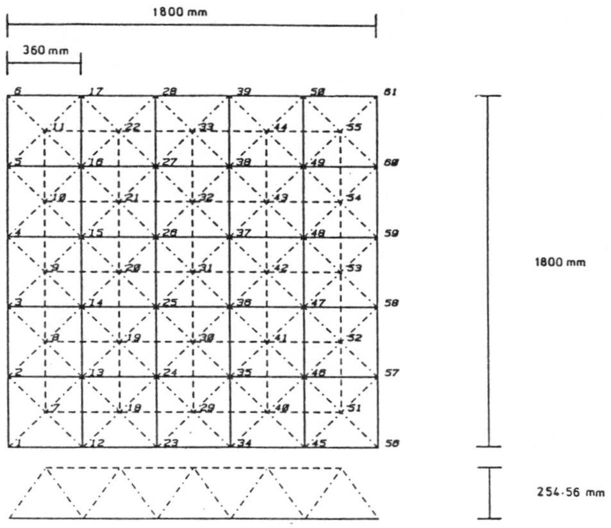

FIG 8. MODEL GEOMETRY AND NODE NUMBERS.
Typical Member A = 22mm^2, r = 3.09 mm, E = 203000 N/mm^2.
Yield Stresses 285 N/mm^2 (Tests 1,2,3) 375 N/mm^2 (Test 4).

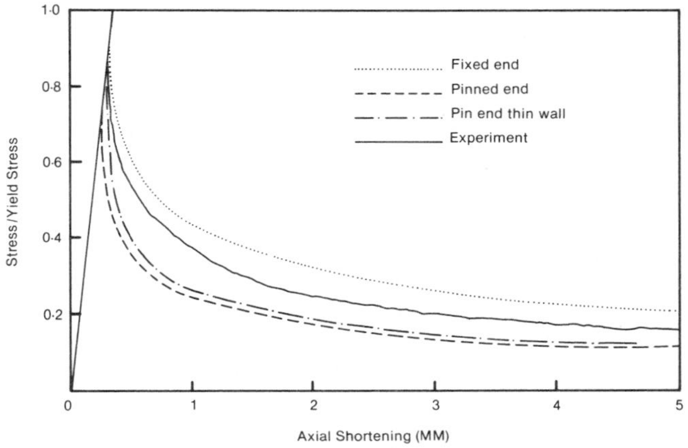

FIG 7. POST-BUCKLING CURVES FOR STRUT WITH KL/r = 100, YIELD STRESS = 245 N/mm^2, R = 4.75 mm AND t = 0.8 mm

FIG 9. TEST SET-UP.

AN EXPERIMENTAL STUDY

In contrast to the number of analytical studies on the collapse behaviour of double-layer grids, there are relatively few reported experimental studies. Many of these experimental studies have been conducted on commercially available double-layer grid systems which possess inherent mechanical imperfections such as lack of fit, eccentricity of connection and joint slippage under load. Although these tests are of great importance, as they do represent actual structural systems, in order to investigate the validity of a proposed analytical technique more precise experimentation and test structures are required.

Four tubular steel double-layer grids were constructed to the geometry and size shown in Fig 8. Extreme care was exercised in their manufacture to eliminate as far as practicable such factors as joint eccentricity, lack of fit stresses, residual stresses, non-homogeneity of material, and geometry imperfection. A special welded joint system was developed and final nodal geometries were better than ± 0.5 mm in elevation and ± 0.25 mm in plan. The precision taken in the manufacture of the grids and in the preparation of the test rig used in the experiments are given in Ref 1.

Each test was performed with the structure being simply supported on roller bearings at the corner joints (Nodes 1, 6, 56, 61, Fig 8). Three symmetric loading conditions were investigated with the fourth test being a repeat of the third test but using a grid with different material properties. The experiments were conducted under computer control using a closed loop loading system capable of applying unidirectional displacement increments of less than 0.1 mm to the models via two hydraulic actuators. A data logging system collected information on joint displacements, member forces, applied loads, and also monitored room temperature. Extensive calibration of load cells, strain gauges, and displacement transducers was performed before testing. A grid under test is shown in Fig 9.

EXPERIMENTAL OBSERVATIONS

The observed elastic behaviour of all the tests was linear and symmetric, reflecting the precision of the manufacturing and testing procedures. The pin-jointed analysis was found to predict with good accuracy the elastic behaviour of the rigid-jointed model. This is because the structure, although rigidly jointed, transmits loads primarily as axial forces.

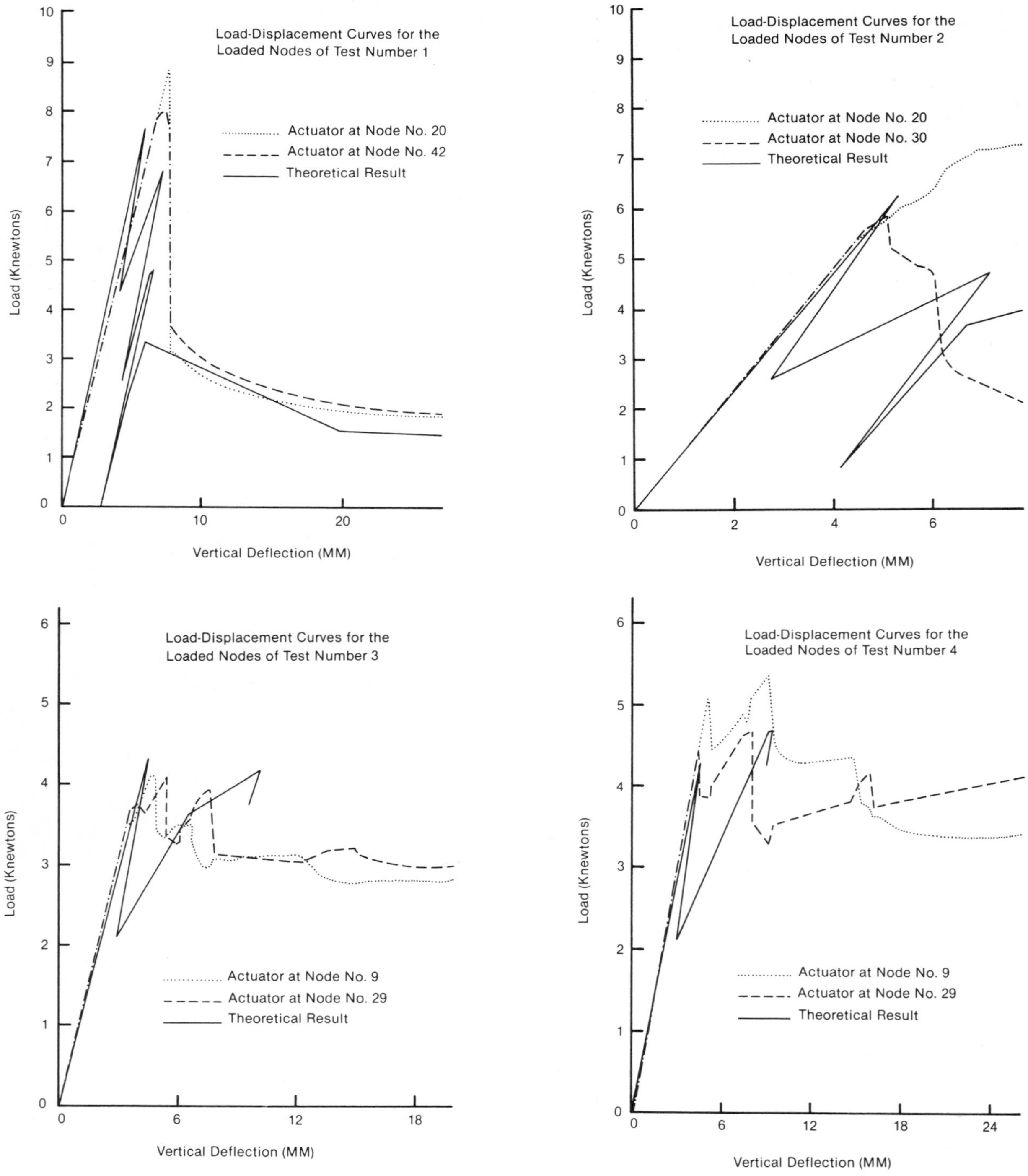

FIG 10. TEST STRUCTURE LOAD-DEFLECTION PLOTS

TABLE 1. ULTIMATE LOAD VALUES (NEWTONS)

(Experimental Value of Ultimate Load is the mean of the two actuators.)

TEST NO.	1	2	3	4
Loaded Nodes	20, 42	20, 30	9, 29	9, 29
Experimental Ultimate Load	8,410	5,868	4,059	4,861
Theoretical Ultimate Load	7,653	6,300	4,343	4,821
Percentage Error of Theory	-9.0	+7.4	+7.0	-0.8

FIG 11. PLASTIC HINGE -- TEST NO. 1

Table 1 shows the loading conditions considered and the ultimate loads obtained both experimentally and theoretically. It should be noted that the theory correctly predicted whether or not a reserve of strength existed beyond the elastic limit load of the structure.

The inelastic behaviour of the structures was found to be non-symmetric. The good summetry achieved in the elastic range of behaviour indicates that loss of symmetry in the inelastic range of behaviour is due to structural imperfections and not experimental inaccuracy. The effect of these imperfections is to cause variations in strut buckling loads and hence failures do not occur in a perfectly symmetrical manner (assuming a symmetric situation). The theoretical and experimental load-displacement curves of the loaded joints are given in Fig 10. The broken lines are the load-displacement curves of the two loaded joints and the solid line is the theoretical behaviour. The variance of the broken lines indicate loss of symmetry in the member failure patterns. However, in Test 1 symmetry was lost and regained; and in Test 3 the early inelastic behaviour remained symmetric. Tests 2 and 4 did not exhibit such behaviour.

Test 1 is of particular interest as it demonstrated the phenomenon of sudden progressive collapse, which is sometimes referred to as "the domino effect." The initial member to buckle was member 9-10, Fig 8, then without any further application of displacement, a structural mechanism formed by the successive buckling failures of members 20-21, 31-32, 42-43, and 53-54. Figure 11 shows this hinge. The entire failure sequence lasted less than 30 seconds and was both sudden and catastrophic. The formation of this hinge meant that symmetry about one axis was maintained and load-displacement behaviour of the two actuators and the theoretical result show good agreement. In this instance, a single compression member failure was shown to be catastrophic.

Detailed discussion of the tests and corresponding theoretical results are presented elsewhere, Ref 1. Within the limitations of comparing tests performed under unidirectional displacement control with a theory in which displacements are dictated by member failure, the analysis technique was considered to provide a sound basis for the investigation of double-layer grid collapse behaviour.

CONCLUSIONS

Accurate, reliable experiments were conducted with which analyses may be compared. More experimental investigation in this field of study is required to provide further data on actual structural behaviour. The effect of inevitable variations in strut buckling loads has a significant effect on the structural inelastic behaviour and should be investigated thoroughly. The presented technique of analysis and post-buckling theories provide a suitable approach to such investigations.

ACKNOWLEDGMENTS

The study was made possible by a Research Studentship Award from the Science Research Council. The author is indebted to the staff of the Space Structures Research Centre at the University of Surrey for their technical expertise, advice, and friendship. He would also like to thank Ms. Shirley Kirnec for her care in the preparation of this paper.

REFERENCES

1. I M COLLINS, Collapse Analysis of Double-Layer Grids, PhD Thesis, University of Surrey, 1981.

2. L C SCHMIDT, P R MORGAN and J A CLARKSON, Space Trusses with Brittle-type Strut Buckling, J. Structural Division, ASCE, July 1976.

3. G G ROY, A R TOAKLEY and L K STEVENS, Elastic-Plastic Analysis of Triangulated Frameworks, Building Science, Vol. 6, 1971.

4. L C SCHMIDT, Member Buckling Characteristics and Space Truss Behaviour, IASS Congress on Space Enclosures, Concordia University, Montreal, July 1976.

5. W J SUPPLE, A Plastic Collapse Model for Space Trusses, Conf. on Stability Problems in Engineering Structures and Components, University College Cardiff, Sept. 1978.

6. W J SUPPLE and I COLLINS, Limit State Analysis of Double-Layer Grids, Analysis Design and Construction of Double-Layer Grids (Ed. Z S MAKOWSKI), Applied Science Publishers, 1978.

7. P C PARIS, Limit Design of Columns, J. Aeronaut. Sci., Jan. 1954.

8. W J SUPPLE and I COLLINS, Post-critical Behaviour of Tubular Struts, Engineering Structures, Oct. 1980.

9. I COLLINS and W J SUPPLE, Experimental Post-Buckling Curves for Tubular Struts, Space Structures Research Centre Report, University of Surrey, 1979.

10. L C SCHMIDT and B M GREGG, A Method for Space Truss Analysis in the Post-Buckling Range, Int. J. for Numerical Methods in Engineering, Vol. 15, 1980.

AN INVESTIGATION OF HYPAR SHELLS

J.E. GIBSON, MSc, PhD, DSc, FICE, FIStructE, MIASS

Head of the Department of Civil Engineering
The City University, London

Hyperbolic paraboloids form a favourite method for roofing sports halls, churches and the like and may be constructed from reinforced concrete, timber or single layer steel grids. The analysis of such structures in the linear and non linear elastic domains may be relatively easily carried out, but little is known of their failure mode. It is the purpose of this investigation to study theoretically and with experimental models this ultimate failure mode for concrete, timber and steel hyperbolic paraboloid structures. However, in this present paper only the micro concrete hypar models will be examined in any detail.

INTRODUCTION

The main part of this paper deals with the structural behaviour of a micro concrete model of a typical type of hyperbolic paraboloid roof which is supported on two buttresses and cantilevers out to high points on either side. The analysis of such types of shell roof has occupied the attention of many authors and it is not intended in this work to deal with the linear to non linear elastic analyses as such, although some typical results from a large finite element package will be given. The main study is that of the failure of the roof under various loading conditions and the eventual aim of the complete investigation is to compare this with the failure of a geometrically identical hypar manufactured from timber and finally with that of a steel space frame consisting of a single layer grid of the same hypar profile and dimensions.

CONSTRUCTION OF THE MICRO CONCRETE HYPAR MODEL

In order to construct the micro concrete model it was found necessary to build a timber box with two high and two low corners giving the correct rise and plan required and then to laminate the surface with thin timber laths or strips using the box edges as guides. The surface was thus built up of straight thin sections which formed the well known straight line generators for this type of hyperbolic paraboloid. To give strength to the surface a second layer of strips was glued to the first set in the orthogonal direction. Having prepared this timber form, in order to allow ease of stripping the surface was covered with a sheet of Neoprene which was stretched to fit the hypar surface and which was then held in position by side laths.

Standard 76 x 76m cold worked steel mesh of 3mm diameter wire was then cut and laid on the surface and formed to the correct shape.

The shell was cast using a standard mix for micro concrete and the required thickness of 12mm was achieved by tamping with a steel bar against side forms of the same thickness located along the edge of the timber form. The thickness achieved was fairly uniform and of reasonable surface finish. The shell as cast was allowed to cure under dampened light sacking for some days and was then removed from the mould with comparative ease.

Fig. 1 Micro Concrete Hypar and Loading Frame

Because of the comparative weakness of the tapered high points it was decided at this point to strengthen up the shell by incorporating edge beams along all edges. This was achieved by attaching suitable timber side forms to give an edge beam of dimensions 50mm depth by 37mm width reinforced with two 10mm diameter bars. This ring beam considerably stiffened up the shell and allowed it to be transported without damage.

At this stage the shell had to be mounted on abutments in the test frame and it was decided that the type of abutment that should be used should be a combination of the types used in practice, namely rigidly fixed abutments together with a tie bar located across the low points. It was considered to be of some interest to allow a choice of either of these conditions to be used, and to this end the abutments were manufactured of RHS welded to flat plates at one end and so shaped at the other end so as to support the low corners of the shell. Holes were drilled across diagonal corners of the RHS so that a tie bar could be located through the two sections and locked off by special nuts bearing against the sides (shown in Fig. 1).

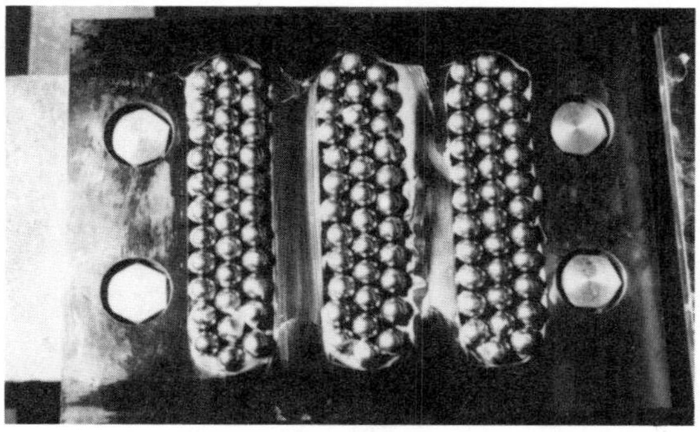

Fig. 2 Roller Bearing Support for Abutment

One of the abutments was firmly bolted to the test frame whilst the other rested on ball bearings (shown in Fig.2) thus allowing it to move longitudinally in the direction of the axis of the bar. Foil strain gauges mounted on the tie bar determined any stress communicated to it. The stresses in the tie bar were measured by the data logging system shown in Fig. 3.

The hypar was then located on the abutments and firmly grouted in position using the same micro concrete mix, the tie bars having been suitably covered with rubber tubing. Once the concrete had set the tubing was withdrawn leaving the bar free to move.

When tests were to be conducted in which the abutments were to be rigidly fixed, the roller bearings could be removed under the movable abutment which was then rigidly bolted down to the testing frame.

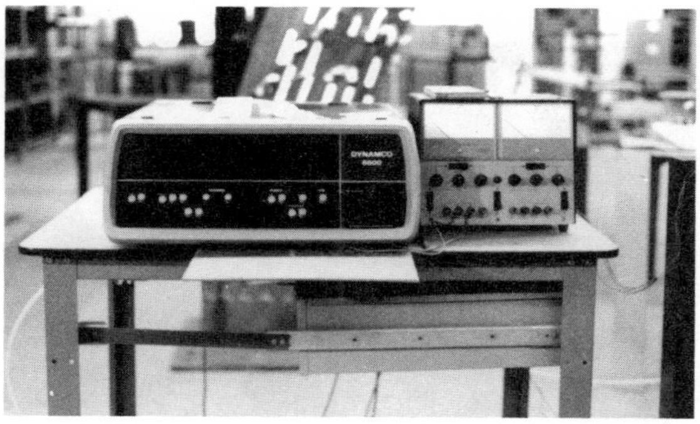

Fig. 3 Data logger

METHOD OF TESTING

In the series of tests of the concrete and timber hypars two types of loading were applied, the first being uniform pressure over the whole shell surface as this more closely resembled the type of loading found in practice, i.e. dead loading which is dominant plus an overload uniformly distributed to account for snow and the like. The second type of loading was that of point loading which allowed an investigation of the effects of local loading, this being of some theoretical interest.

The uniform load was applied by means of a pressure bag to the surface of the model. This bag was made from a special plastic that could be welded using hot air and was formed to the necessary hyperbolic shape. It was placed on the upper surface of the shell and held in position by a thick layer of polystyrene. The shuttering mould, having been cleaned and varnished, was located at the top of the polystyrene and locked in position by four tie rods bolted down to the main bed of the testing frame. Each tie rod incorporated a proving ring which checked the total load applied to the hypar surface once the plastic bag had been pressurized from a compressed air source. A water manometer coupled to a needle pressure valve served as a control system for applying pressure to the bag.

The complete system is shown in Fig.1 which shows the shell with the timber restraining frame and the four tie rods and proving rings. The deflections of the shell surface under loading were measured by dial gauges graduated in .01mm. These are also shown in Fig. 1. For simplicity these were located over only one quarter of the model surface as may be seen and consisted in all of nine gauges evenly distributed over the quarter surface (as shown in Fig. 5) with a tenth gauge located at the opposite high point which served as a check.

A series of tests were carried out in which the deflections of the ten points were measured for increasing load in the linear elastic range. The results indicated that the model behaved in a fairly symmetrical manner and these will be presented elsewhere. It was found, however, that it is impossible to produce collapse of the shell by this type of loading due to increasing damage to the restraining box. As this box was required again in testing the timber shell it was decided to discontinue this form of testing and use point loading as a means of producing failure.

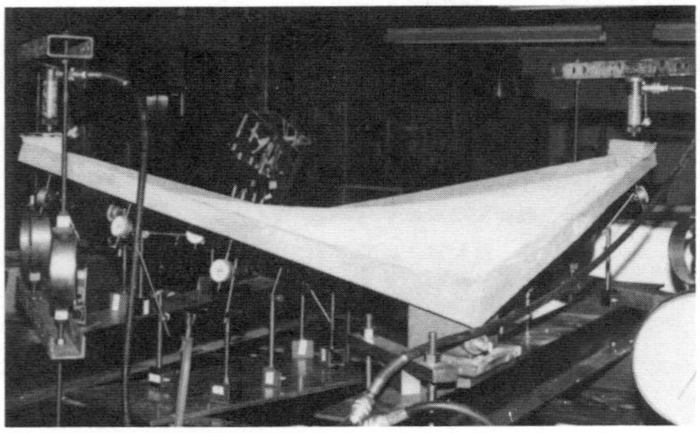

Fig. 4 Point Loading Mechanism on Model

To this end a new method of loading was used as shown in Fig. 4. It was decided to apply the point loads at the two high points of the structure as this allowed a relatively easy use of the proving ring tie rod set up. As is shown a hydraulic ram was mounted on a bar which in turn was supported by two proving ring tie rods on either side both being rigidly attached to the testing frame.

Load was applied simultaneously to the two cylinders from a hand pump using a Budenborg gauge as a control monitor and the proving rings to accurately determine the load. Again the dial gauges were located at the previous nine points and the tenth point at the other high point was used as a check on the symmetry of loading. Results of this test are given in Fig.6.

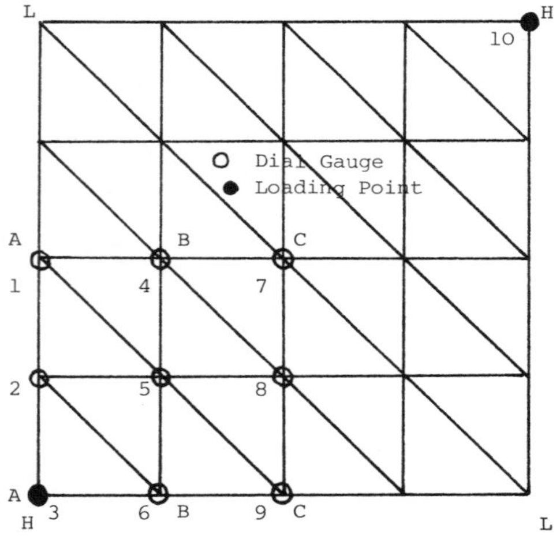

Fig. 5 Finite Element Mesh and Dial Gauge Positions

ANALYSIS OF HYPAR UNDER VERTICAL POINT LOADS AT THE HIGH POINTS

The preliminary analysis of the hypar shell consisted of using a triangular bending and membrane finite element with six degrees of freedom at each node. In this analysis the effect of the edge beam was ignored and two equal point loads were applied at the high points 1 and 10, shown in Fig. 5. The abutments (marked L,L in Fig. 5) were for the initial analysis assumed to be fixed and in all, thirty two elements were used to cover the whole shell; the output presented both bending and membrane stress resultants as well as deflections normal to the shell surface.

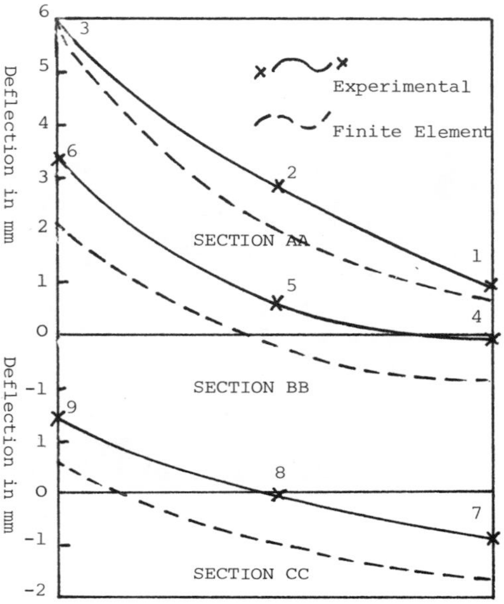

Fig. 6 Deflections of Shell Surface under End Point Loading

The results of this analysis are shown in Fig. 6 as dashed lines for the three sections AA, BB and CC as marked in Fig. 5. These results are comparative as will be explained later.

The actual experimental values of the deflection for two point loads of 200 lb each are shown in Fig. 6 as the full lines joining the experimental values indicated by crosses at the dial gauge points. As the theoretical values of the deflection were calculated on an assumed modulus of elasticity, in order to get any relevant comparison it was dediced to adjust the theoretical value of deflection at point 3 in Figs. 5 and 6 by multiplying the actual theoretical value by an appropriate factor which would yield a value close to the experimental value of 600 x .01 mm. Using this factor the remaining eight theoretical values were multiplied by this and these yielded the dashed theoretical lines shown in Fig. 6.

It may be seen that in so doing an element of comparison between theory and experiment may be made as follows. The resulting theoretical values at points 2 and 1 in Fig. 5 yielded the dashed line for theory which gave a not unreasonable comparison with the experimental values.

Repeating this operation for section BB it is seen that the theoretical line predicts a reversal of deflection half way along the section which of course is now shown by the experimental values.

Again in the case of section CC, theory predicts a rapid reversal of deflection not exhibited in the experimental curve which does however indicate a deflection reversal half way along.

Nevertheless the shape of all pairs of theoretical and experimental curves is remarkably similar.

It may well prove that when a more detailed finite element analysis involving more elements and including the effect of the edge beam is conducted at a later date, better comparison may be shown to exist.

As indicated, the main interest in this part of the investigation lay in the prediction of the actual collapse of the shell. This was adjudged to take place by the formation of yield lines running diagonally across the shell, i.e. from point 2 to 6 and from points 1 to 5 to 9 and so on, indicating that the tension steel had yielded, the points referring to Fig. 5.

In point of fact examining the 2,6 and 1,5,6 sections etc. assuming yield of the steel, it was found that the section 1,5,6 was the weakest and predominant cracking in this region did occur on the model, the cracks extending into the edge beams. The theoretical point loads to produce this crack were calculated to be 740 lb. The corresponding load from the experimental values was 800 lb, so that the type of failure postulated was in this case reasonably correct.

TIMBER AND STEEL HYPARS

Shortage of space precludes any lengthy statement on the timber and steel hypars save for an explanation of their construction.

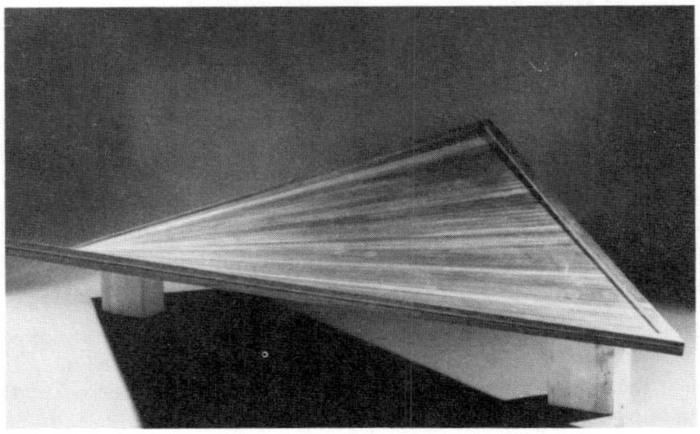

Fig. 7 Timber Hypar Shell

The timber hyperbolic paraboloid is shown in Fig. 7 and it was constructed from laths 25mm by 3mm in two layers which were firmly fixed together using a standard urea formaldehyde adhesive. In order to stiffen the shell edge again two extra layers of laths were fixed on the top and bottom edge surfaces. As the geometry of this timber shell is identical to that of the micro concrete model the restraining box used in the previous testing was again used to produce uniform pressure on the surface of the timber model. Testing is continuing.

Fig. 8 Single Layer Hypar Grid

The steel space frame shown in Fig. 8 was formed by welding the outer frame in relatively heavy sections, namely 25 x 25mm square bright mild steel. This formed a base which was sufficiently rigid to allow the single grid members of 6mm x 6mm bright mild steel to be welded into the required single grid hyperbolic paraboloid surface.

The model will be analysed as a whole, namely the long columns supporting the high points as well as the square base will be included, the loading being nine vertical point loads at the inner nodes of the single layer grid. The initial experiment will involve testing the frame as it appears in Fig. 8 using nine loading cylinders at the inner nodes. Deflections at the nodes will be measured as well as bending moments and axial forces.

Deflections at the nodes will be measured by dial gauges or L.V.D.T.'s and pairs of strain gauges will be located at selected points to measure bending and axial strains.

ACKNOWLEDGEMENTS

The author wishes to thank the technical staff of The City University and Mr. S. Ziaie for their considerable help in this project.

AN ELASTIC-PLASTIC BIFURCATION MODEL

P J WICKS

Space Structures Research Centre
Department of Civil Engineering
University of Surrey

In the field of elastic stability use is frequently made of rigid link-spring models to illustrate both limit-point and bifurcation-point behaviour. Equilibrium configurations and stability criteria are then investigated through the medium of total potential energy. Unfortunately, use of the latter precludes the study of non-conservative effects such as material plasticity. The present paper introduces plasticity into a rigid link-spring model by taking the spring material property as a bi-linear force-displacement relationship. The lack of conservation of energy being achieved through the use of a friction block. Force-displacement configurations for the model is established using simple ideas of equilibrium and compatibility of displacement. The effects on the postbuckling of the model of two factors are investigated: a geometric imperfection, and variation of the yield point of the material.

INTRODUCTION

Investigators in the field of elastic stability have frequently used models consisting of rigid links and springs to elucidate the phenomena associated with highly complex real structures. An expression for the total potential energy of such models may be written as the sum of elastic strain energy stored and the change in potential of any loading. The well-known principles of stationary total potential energy may then be used to establish equilibrium and hence the load-deflection response of the system (see Ref 2).

Such an approach depends, for its applicability, on the loading being conservative and the material behaviour remaining elastic throughout some load-deflection history. This is acceptable for some thin plate and shell structures and for shallow arches and domes which tend to buckle elastically. When considering columns however, a variety of behaviours may be expected depending on the slenderness ratio. A very slender steel column may undergo quite large out-of-straight bowing before any onset of material plasticity. In such circumstances the column exhibits stable-symmetric bifurcational behaviour, albeit of a highly "neutral" kind, before the material exceeds its elastic limit. As the slenderness ratio is reduced to values nearer the intersection of the Euler-Yield curve, (Ref 1), mild steel columns tend to exhibit a brittle, load-shedding, type of behaviour associated with material plasticity and a highly unstable load-deflection response. The failure load, corresponding to the initiation of this latter behaviour, for a column of given slenderness ratio, is very difficult to predict (witness the profusion of proposed failure stress curves in the last 80 years). When such compression members are assembled into Space Structures their individual behaviour has serious implications for the ultimate load behaviour of such structures. Such effects as sequential collapse, as members fail catastrophically, shedding load to other members causing further members to fail and so on, may occur.

The validity of the aforementioned models in elucidating some structural behaviour would, therefore, seem to depend on introducing some inelastic material behaviour. In the present paper this is achieved by introducing a friction block into a spring-rigid-link model, thus obtaining a bi-linear material force-displacement relationship. Since the non-conservative nature of the system precludes the stationary total potential energy type of approach to the model, a more traditional equilibrium and compatibility method is adopted. The effects of two factors on the equilibrium behaviour of the model are then investigated. The first is a geometric type imperfection, of the type favoured by elastic stability theorists. The second effect investigated is the variation of the yield point of the constituent spring material.

MATERIAL BEHAVIOUR

We consider the mechanical system shown in Fig 1. This consists of a rigid block B, of unit weight, resting on a friction plane. We assume that the coefficients of static and kinetic friction between the block and the plane are both the same value of μ. The block is joined to a rigid support A by means of a spring of stiffness αk, where α is a constant. The point C moves under controlled displacement conditions at a constant velocity. The spring BC, of stiffness k, then acts as a force transducer. We seek the relationship between the force in this spring, F and a full range of displacements of C, Δ (which include movement of the block B) for steady-state conditions.

We note that the block B will remain stationary up until the force in the spring BC is sufficient to overcome the frictional force between the block and the plane. Hence, there is a limiting value of displacement Δ^*, of the point C up to which the force-displacement relationship for the model is represented solely by that for the spring BC. This limiting value of displacement is given by

$$\Delta^* = \frac{\mu}{k} \qquad .....1.$$

Hence, for values of displacement between the limits $0 < \Delta < \Delta^*$ the force-displacement relationship for the model is

$$F = k\Delta, \qquad 0 < \Delta < \Delta^* \qquad .....2.$$

For values of displacement $\Delta^* < \Delta$ the block B will have moved and the spring AB will make a contribution to the behaviour of the model. Under steady-state conditions the plane will exert a constant friction force μ on the block B. At a given time let the displacement of the block be Δ' and the force in spring AB be F', for equilibrium of the block

$$F - F' - \mu = 0 \qquad .....3.$$

where F is the force in spring BC. If the displacement of point C at this time is Δ, we have from the force-displacement relationships for the individual springs

$$F = k(\Delta - \Delta') \qquad .....4.$$

$$F' = \alpha k\Delta' \qquad .....5.$$

From Eqn 4

$$\Delta' = \Delta - \frac{F}{k} \qquad .....6.$$

this into Eqn 5 gives

$$F' = \alpha k(\Delta - \frac{F}{k}) \qquad .....7.$$

substituting for F' from Eqn 7 into Eqn 3 yields

$$F = \alpha k(\Delta - \frac{F}{k}) + \mu \qquad .....8.$$

Rearranging Eqn 8 we arrive at the force-displacement relationship for the model for values of Δ in excess of Δ^* as

$$F = \frac{\mu}{(1 + \alpha)} + \frac{\alpha}{(1 + \alpha)} k\Delta, \qquad \Delta^* < \Delta \qquad .....9.$$

We note that the two lines given by Eqns 2 and 9 intersect at the point

$$F^* = \mu, \quad \Delta^* = \frac{\mu}{k} \qquad .....10.$$

For convenience we introduce the non-dimensionalised quantities,

$$\sigma = \frac{F}{kh} \qquad \varepsilon = \frac{\Delta}{h} \qquad .....11.$$

(where h is a length to be introduced in the sequel) whereupon Eqns 10, 2 and 9 become

$$\sigma^* = \varepsilon^* = \frac{\mu}{kh} \qquad .....12.$$

$$\sigma = \varepsilon, \qquad 0 < \varepsilon < \varepsilon^* \qquad .....13.$$

$$\sigma = \frac{\sigma^*}{(1 + \alpha)} + \frac{\alpha}{(1 + \alpha)} \varepsilon, \qquad \varepsilon^* < \varepsilon \qquad .....14.$$

The full range of force-displacement response for the system is shown schematically in Fig 2 for various values of α ranging from 0 to ∞. A bi-linear relationship is achieved with material "yielding" occuring at the point (σ^*, ε^*): the position of this point may be altered by varying the coefficient of friction or the spring stiffness. Taking a value of $\alpha = 0$ results in an elastic-perfectly plastic behaviour, the block B then moving under the restraining action of the frictional force alone without the benefit of the spring AB. For a value of $\alpha \to \infty$ the effect is to render the block immobile and the force-displacement behaviour of the system reverts to perfectly elastic, involving the stiffness of BC alone.

EQUILIBRIUM AND COMPATIBILITY EQUATIONS FOR THE MODEL

Figure 3a shows the single degree-of-freedom model for which the material characteristic, derived previously, is to be used. It consists of a rigid link AC, smoothly hinged to a rigid support at A and carrying a vertical point load of P and a small horizontal point load of f at C. It is restrained against rotation by means of the material represented schematically by DB. This is pinned to a rigid support at D and pinned to a collar B which may slide smoothly along the link AC and DB is constrained such that it remains horizontal.

The model is shown in a displaced configuration in Fig 3b and it is presumed that the loading is applied in such a way that steady-state conditions prevail and AC rotates at a constant angular velocity.

If at a given time the rotation of the link is θ, as shown, we require to find, for initially fixed values of f, the relationship between the displacement θ and the applied load P.

If the force in the material at this value of displacement is F, then taking moments about A for ABC gives

$$Fh \sec^2 \theta - PL \sin \theta - fL \cos \theta = 0 \qquad .....15.$$

The compatibility condition for the horizontal movement of the point B, Δ is given by

$$\varepsilon = \frac{\Delta}{h} = \tan \theta \qquad .....16.$$

and the value of θ, θ^* at which yielding in the material occurs is, from Eqn 12

$$\theta^* = \tan^{-1} \frac{\mu}{kh} = \tan^{-1} (\sigma^*) \qquad .....17.$$

Thus, for $0 < \theta < \theta^*$ the material behaviour is given by Eqn 2 and substituting for Δ from Eqn 16 into Eqn 2 gives

$$\frac{F}{kh} = \tan \theta, \qquad 0 < \theta < \theta^* \qquad .....18.$$

and for θ^* θ we may substitute for Δ from Eqn 16 into Eqn 9

$$\frac{F}{kh} = \frac{\sigma^*}{(1 + \alpha)} + \frac{\alpha}{(1 + \alpha)} \tan \theta, \qquad \theta^* < \theta \qquad .....19.$$

Rewriting the equilibrium equation, Eqn 15 as

$$\frac{F}{kh} \sec^2 \theta - \frac{PL}{kh^2} \sin \theta - \frac{fL}{kh^2} \cos \theta = 0, \qquad .....20.$$

introducing the non-dimensionalised quantities

$$\lambda = \frac{PL}{kh^2}, \qquad \emptyset = \frac{fL}{kh^2} \qquad .....21.$$

and use of the compatibility Eqns 18 and 19 yields the force-displacement relationships for the model as

$$\lambda = \sec^3 \theta - \frac{\emptyset}{\tan \theta}, \qquad 0 < \theta < \theta^* \qquad .....22.$$

$$\lambda = \frac{\sigma^*}{(1 + \alpha)} \cdot \frac{\sec^2 \theta}{\sin \theta} + \frac{\alpha}{(1 + \alpha)} \sec^3 \theta - \frac{\emptyset}{\tan \theta}$$

$$\theta^* < \theta \qquad .....23.$$

IDEALISED BEHAVIOUR

The idealised behaviour for the model is obtained from the previous Eqns 22 and 23 by setting the imperfection parameter $\emptyset$ equal to zero. Upon doing this we obtain

$$\lambda = \sec^3 \theta, \qquad 0 < \theta < \theta^* \qquad .....24.$$

$$\lambda = \frac{\sigma^*}{(1 + \alpha)} \cdot \frac{\sec^2 \theta}{\sin \theta} + \frac{\alpha}{(1 + \alpha)} \sec^3 \theta, \qquad \theta^* < \theta \qquad .....25.$$

As can be seen from Fig 4 the elastic response of the model, as represented by Eqn 24, is of the stable-symmetric bifurcation type, Ref 2. The behaviour continues in this form until the displacement reaches a value of θ^*, whereupon the load-deflection response is governed by Eqn 25, and the model branches into an unstable plastic mode.

The effect of varying the value of θ^* on this plastic birfurcation point is also shown in Fig 4. Curves are drawn for values of $\alpha = 0$ and $\theta^* = 2, 4, 6, 8$ and 10 degrees. These correspond to values of σ^* of 0.052, 0.070, 0.105, 0.141 and 0.176 respectively. Denoting the plastic bifurcation load by λ^*, this may easily be calculated by substituting for θ^* into Eqn 24 and Table 1, shows the values of λ^* corresponding to the values of θ^* mentioned above.

σ^*	θ^*	λ^*
0.035	2	1.002
0.070	4	1.007
0.105	6	1.017
0.141	8	1.030
0.176	10	1.047

Table 1

Figure 5 shows the effect of varying the spring stiffness ratio α; taking values of α equal to 0, 0.1, 0.5, 1.0 and 5.0 for a fixed value of θ^* equal to 4 degrees.

IMPERFECT BEHAVIOUR

For the ideal behaviour, variation of the yield displacement θ^* marginally increases the value of load λ^*, at which branching into an unstable plastic mode from a stable elastic mode occurs. The effect of increasing α is to produce a less unstable plastic branching path. The most unstable occurs for the elastic-perfectly plastic material behaviour arising from $\alpha = 0$.

The plastic bifurcation point, therefore, represents a limit on the load carrying capacity of the model and must be affected by both the presence of an imperfection parameter and the yield behaviour of the material. This necessitates the study of the force-displacement Eqns 22 and 23.

Figure 6 shows plots of the load-displacement relationships from these equations for values of $\alpha = 0$ and $\emptyset = 0.01$. The elastic portion arising from Eqn 22 assumes the expected form by taking the ideal equilibrium path as its asymptote. A number of plastic branching paths are shown for values of $\theta^* = 1, 2, 3, 4$ and 5 degrees, corresponding to σ^* values of 0.017, 0.035, 0.052, 0.070 and 0.087 respectively. From the diagram it can be seen that there is a marked reduction in value of the plastic bifurcation load of the imperfect system as the yield displacement is reduced.

We consider now the effect on the plastic bifurcation load of maintaining a fixed yield displacement and varying the imperfection parameter. The equilibrium paths arising from Eqns 22 and 23 shown in Fig 7 are for values of $\alpha = 0$ and $\theta^* = 3$ degrees ($\sigma^* = 0.052$). Five values of $\emptyset$ are taken as 0, 0.005, 0.01, 0.02 and 0.03 and as can be seen the plastic branching load is dramatically affected by the imperfection parameter.

The interrelationship between plastic bifurcation load λ^*, material yield force, σ^* and imperfection parameter $\emptyset$ may be obtained by considering Eqn 22 in conjunction with Eqn 17. Subsituting θ^* into Eqn 22, bearing in mind Eqn 17 and hence that

$$\sec \theta^* = (1 + \sigma^{*2})^{1/2} \qquad .....26.$$

we have

$$\lambda^* = (1 + \sigma^{*2})^{3/2} - \frac{\emptyset}{\sigma^*} \qquad .....27.$$

Figure 8 shows the variation of λ^* with σ^* for values of $\emptyset$ equal to 0.005, 0.010, 0.020, 0.030, 0.040 and 0.050 as can be seen the plastic bifurcation load of the ideal system can be significantly reduced by the presence of an initial imperfection. If the yield force of the constituent material is reduced then this is also likely to reduce the value of load at which plastic branching occurs.

CONCLUSIONS

The conclusions may be summarised as follows:

1. By introduction of a friction block between the two springs of different stiffness we are able to simulate, under displacement controlled conditions, a bilinear material behaviour. By varying the ratio of the spring stiffness, behaviours analogous to and varying between elastic-perfectly plastic and perfectly elastic material characteristics may be achieved.

2. Upon using this material characteristic in a particular model we find that, in the ideal case, the load-deflection behaviour is initially of the elastic stable-symmetric point-of-bifurcation type. The amount of post-elastic buckling strength of the model is limited by the presence of a further branching point on the elastic equilibrium path. At this stage the material yields and branching into a possibly highly unstable equilibrium path occurs. This would indicate, for a system comprised of a number of elements which exhibited this type of behaviour, that the first element to reach this state, under displacement controlled conditions, would tend to shed load onto the other elements. One may then find oneself in a sequential collapse situation.

3. The value of load at which this plastic branching occurs represents the load carrying capacity of the model. The interaction between the yield "stress" of the material and an initial imperfection is such that the load carrying capacity of the model can be reduced to a small fraction of its ideal elastic buckling load.

4. The form of the post-plastic buckling equilibrium is unaffected by either the yield stress or the imperfection but is governed by the spring stiffness ratio.

REFERENCES

1. I M Collins, PhD Thesis, University of Surrey, 1981.

2. J M T Thompson and G W Hunt, A General Theory of Elastic Stability, John Wiley, 1973.

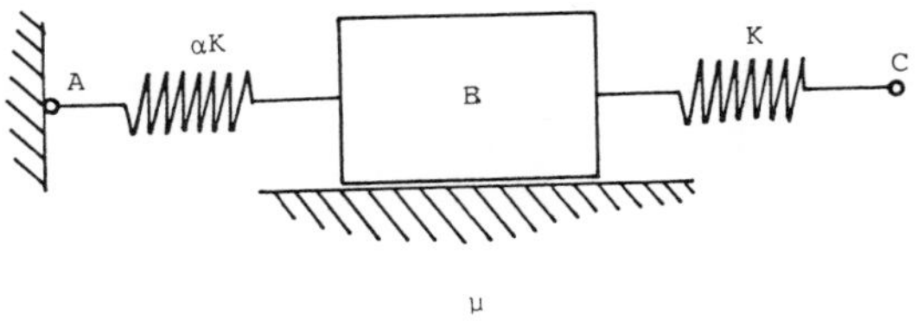

Fig 1

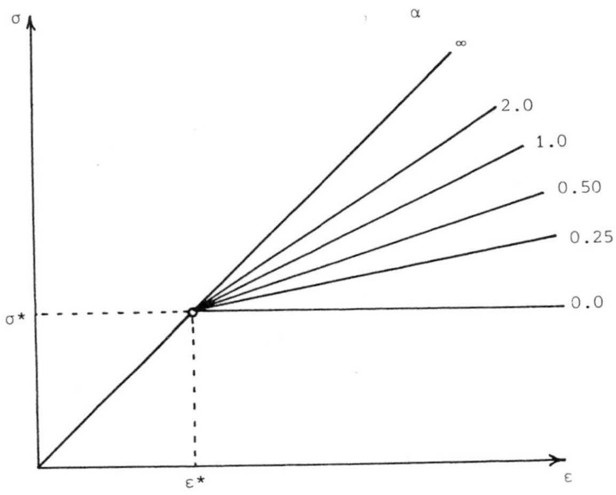

Fig 2

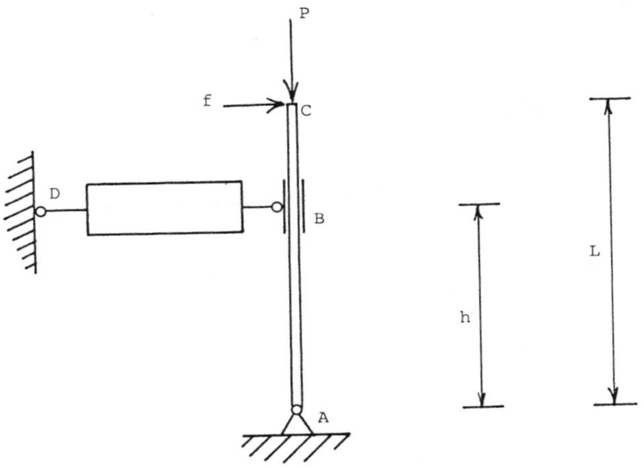

Fig 3a

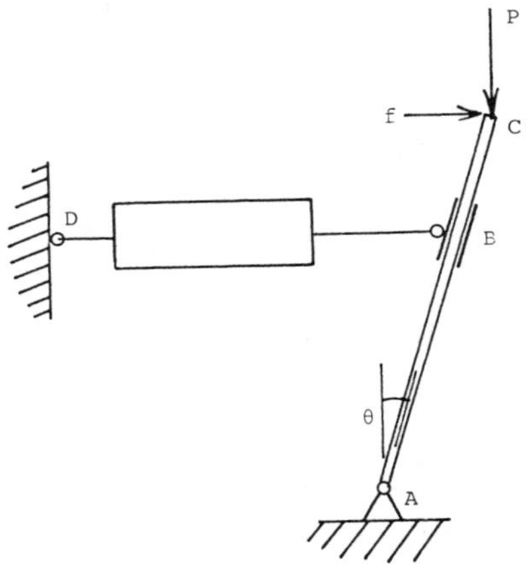

Fig 3b

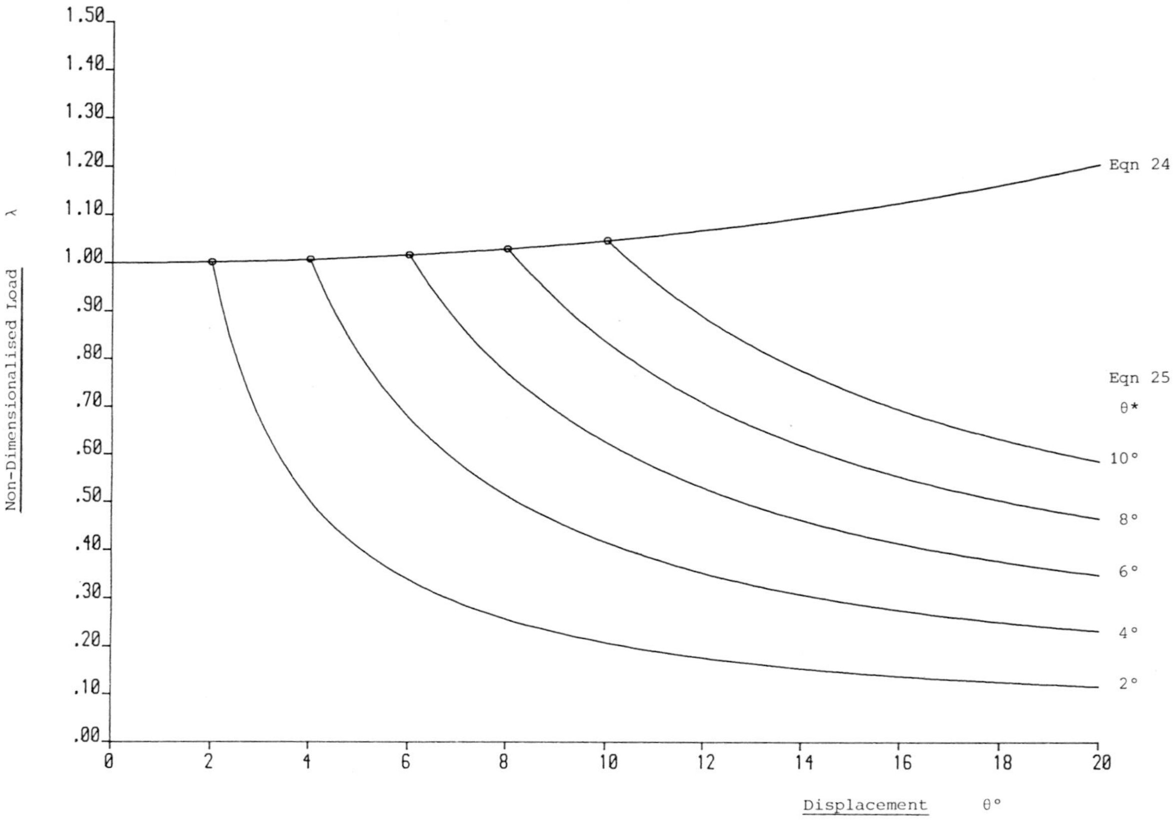
1.50
1.40
1.30
1.20
1.10
1.00
.90
.80
.70
.60
.50
.40
.30
.20
.10
.00
Non-Dimensionalised Load λ
0
2
4
6
8
10
12
14
16
18
20
Displacement θ°
Eqn 24
Eqn 25
θ*
10°
8°
6°
4°
2°

Fig 4

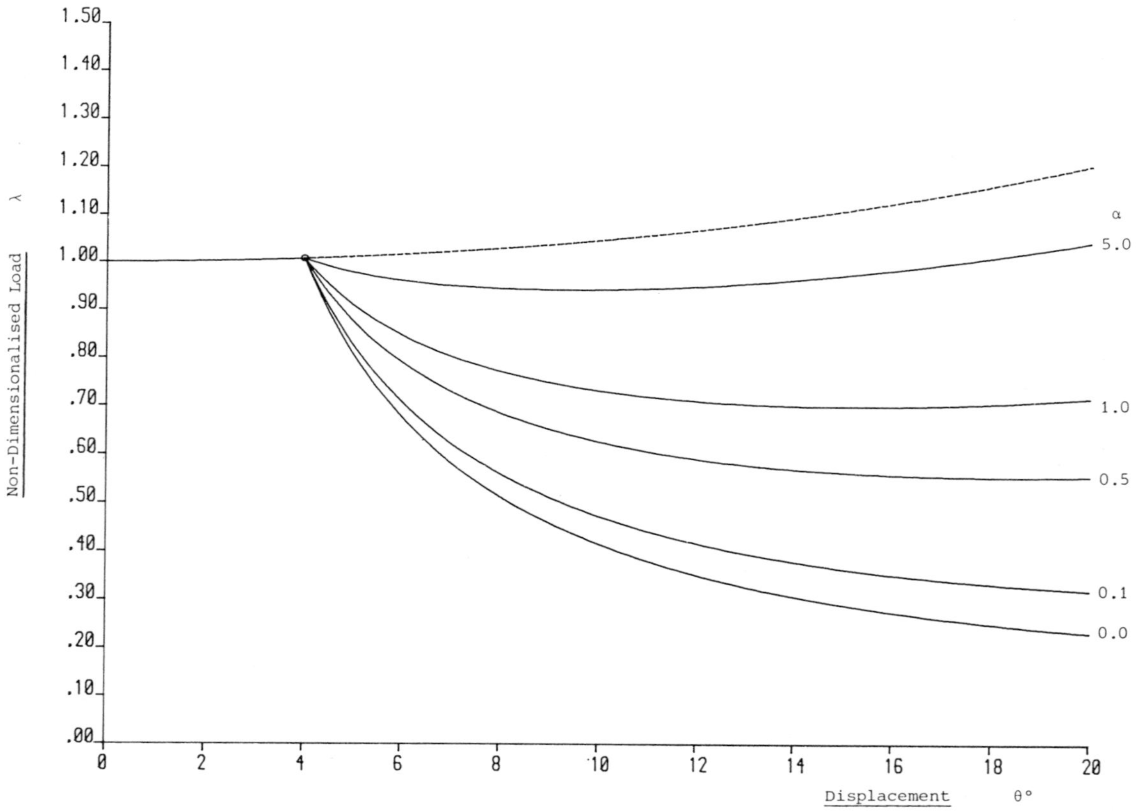
1.50
1.40
1.30
1.20
1.10
1.00
.90
.80
.70
.60
.50
.40
.30
.20
.10
.00
Non-Dimensionalised Load λ
0
2
4
6
8
10
12
14
16
18
20
Displacement θ°
α
5.0
1.0
0.5
0.1
0.0

Fig 5

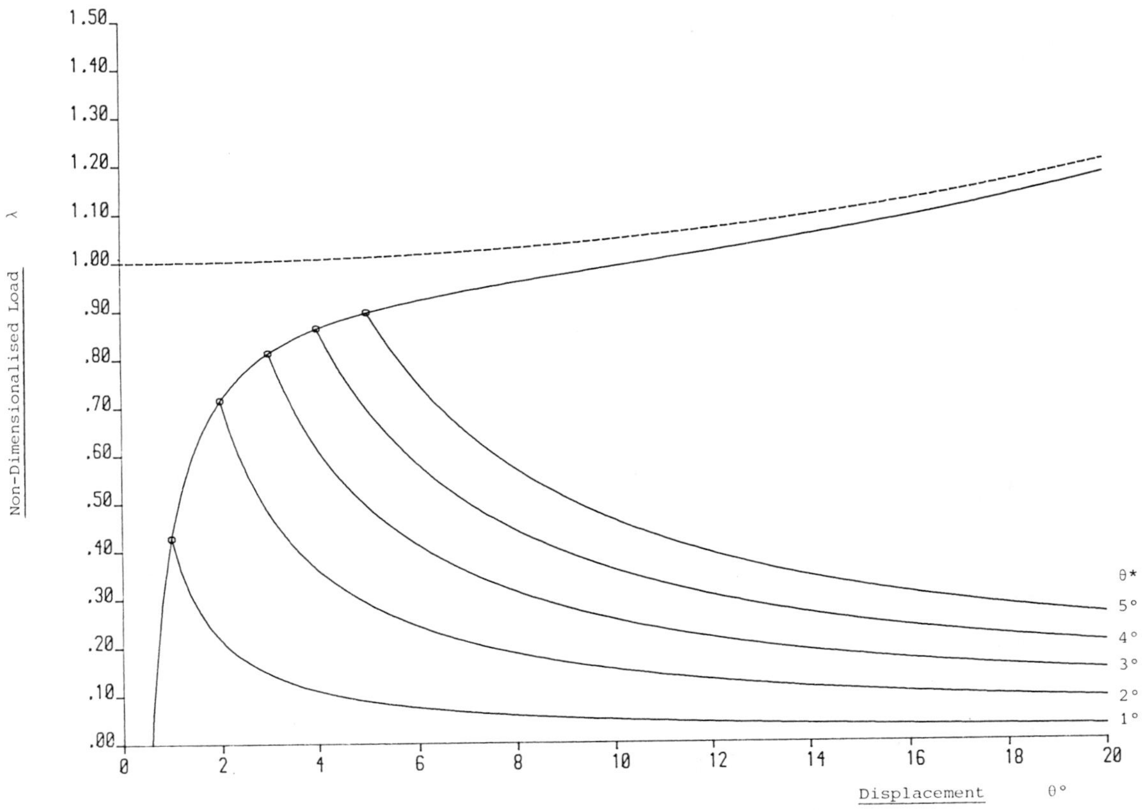
1.50
1.40
1.30
1.20
1.10
1.00
.90
.80
.70
.60
.50
.40
.30
.20
.10
.00
Non-Dimensionalised Load λ
0
2
4
6
8
10
12
14
16
18
20
Displacement θ°
θ*
5°
4°
3°
2°
1°

Fig 6

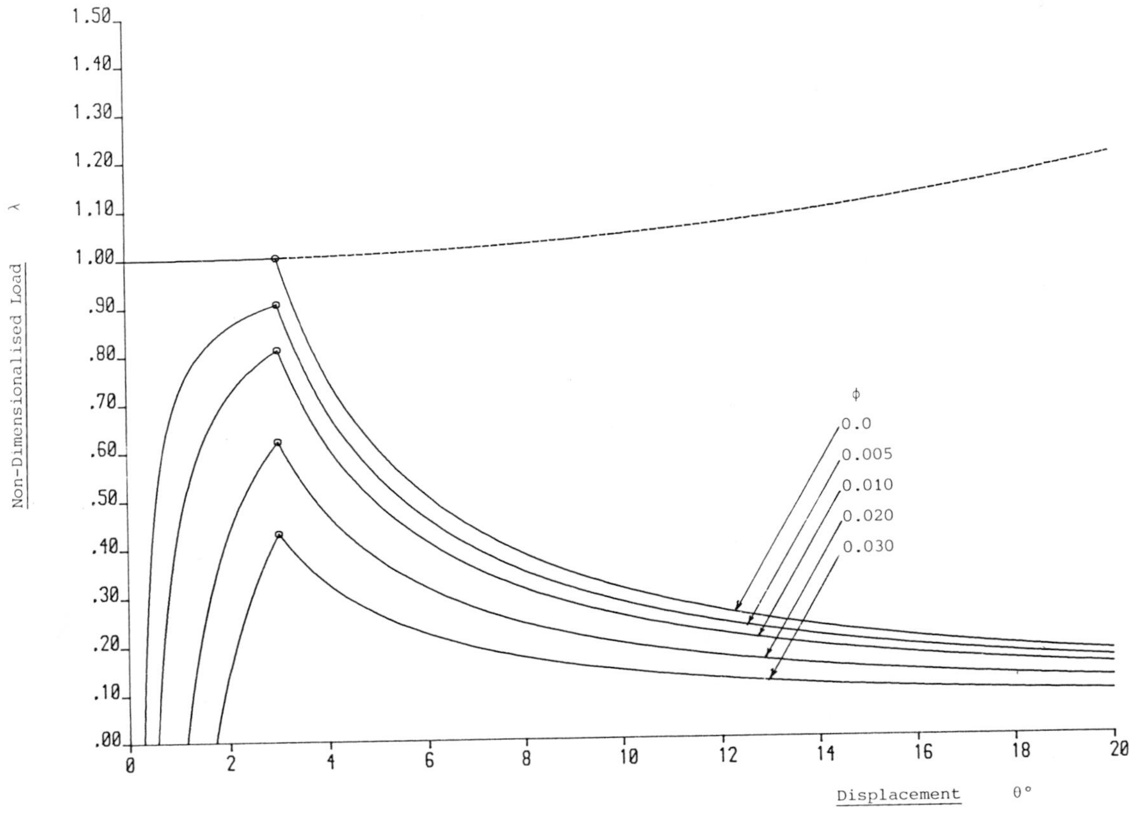
1.50
1.40
1.30
1.20
1.10
1.00
.90
.80
.70
.60
.50
.40
.30
.20
.10
.00
Non-Dimensionalised Load λ
0
2
4
6
8
10
12
14
16
18
20
Displacement θ°
φ
0.0
0.005
0.010
0.020
0.030

Fig 7

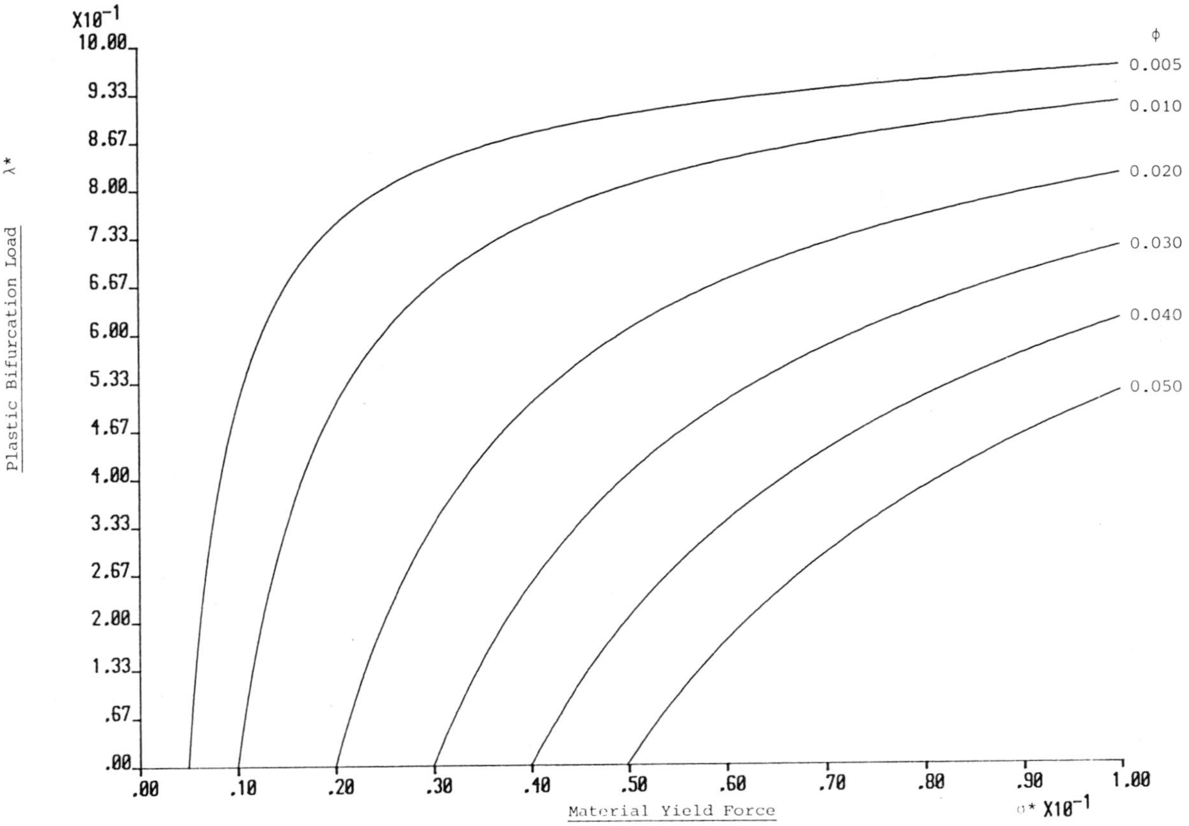
X10^-1
10.00
9.33
8.67
8.00
7.33
6.67
6.00
5.33
4.67
4.00
3.33
2.67
2.00
1.33
.67
.00
Plastic Bifurcation Load λ*
.00
.10
.20
.30
.40
.50
.60
.70
.80
.90
1.00
Material Yield Force
σ* X10^-1
ϕ
0.005
0.010
0.020
0.030
0.040
0.050

Fig 8

THE EFFECT OF JOINTS ON THE STRENGTH OF SPACE TRUSSES

T SAKA* and K HEKI**, Dr. Eng

* Lecturer, Department of Architecture and Building Engineering, Osaka City University, Japan.

** Professor, Department of Architecture and Building Engineering, Osaka City University, Japan.

The elastic buckling loads of space trusses are derived from using the slope-deflection equation for members under axial forces, including both the effects of dimensions and rigidities of the joints. Further the load-carrying capacity for double-layer grids is estimated by the analysis as the corresponding continuum plates, considering the effective strength determined by the elastic buckling of constituent members. These theoretical results are in good agreement with the experimental results for octahedral trusses and a type of double layer space trusses.

INTRODUCTION

Space trusses are widely used as long-span structures. Considering the jointing system of the space trusses, the simplicity and preciseness of assembly are required for the erection of structural frameworks, and bolted jointing systems are used more than welded jointing systems. The bolted jointing systems are developed by many engineers and most of them are prefabricated systems consisting of modular units, where special connectors are used, see Ref 1.

In the stress analysis for space trusses of various jointing systems, the joints are usually treated as pin-nodes. The axial forces of the members obtained as a pin-node truss are good approximate values for practical use, even if there are rotational restraints at the joints. But the strength of space trusses is not always equal to that of the corresponding pin-jointed trusses. Not only the strength of the joint itself but also the rigidity of jointing give the effects on the strength of a truss.

As for the buckling strength, the elastic stability analysis for the rigidly- and gusset-connected space trusses has been presented, see Ref 2. The behaviour of the space trusses composed of a great number of members is often calculated by the limit state analysis method, where the strut load-displacement characteristics are assumed as discrete linear portions, and idealized models of the compression member characteristics have varieties according to the authors, see Refs 3 and 4. But, for the trusses with a higher degree of indeterminacy, it was reported that the ultimate strength obtained by the above limit state analysis made a considerable difference from experimental load carrying capacities, see Ref 5. From the macroscopical point of view, The collapse loads for double-layer grids are derived by the method of the limit analysis as continuum plates proposed by the present authors, see Ref 6, but it was pointed out that its validity was not clearly reported yet, see Refs 4 and 7.

In this paper, for the purpose of establishing the method to estimate the strength of practical space trusses, an analytical method is proposed for the elastic buckling strength, which determines often the load carrying capacity of a truss, in consideration of the effects of both rigidity and dimension of joints according to the jointing system. Where the influence of the instability caused by axial forces is considered and each constituent member is assumed to consist of a uniform middle part, rotational springs, and rigid end parts.

A simple regular octahedral space frame is treated to investigate the effect of joints on the buckling strength theoretically, and to prove the validity of the proposed method by model tests.

Next, parallel square mesh grids are considered as an example of the space trusses of the practical type. The results of theoretical analyses and that of experiments are compared. As for theoretical methods, the elastic buckling analysis using the proposed model is considered for the exact values, and the continuum analogy method using the effective rigidity and effective strength is done for the approximate values. By the continuum analogy method, the elastic limit loads and the collapse loads from the yield line theory are estimated.

METHOD OF THEORETICAL ANALYSIS

1. Assumptions

In the present theoretical analysis, the following assumptions have been made.

1) The theoretical model of each constituent member in space trusses consists of a uniform middle part(a base member), rotational springs and rigid end parts as shown in Fig 1.

2) Deformations of structures are small, at least until critical conditions are reached.

3) Coupling effects of bending and twisting and effect of warping deformations due to twisting of members are considered negligibly.

4) Deformations due to shear of members are considered negligibly.

5) Members are straight and remain elastic, and their sections are uniform and biaxial symmetry.

6) Directions of principal axes of the rotational springs, which connect the base member and rigid end parts, coincide with that of the member.

7) There are no loads directly acting on the member.

2. Fundamental equations

The slope-deflection formula for a member is derived considering axial compression or axial tension.

The end displacements and forces of the member are shown in Fig 1.

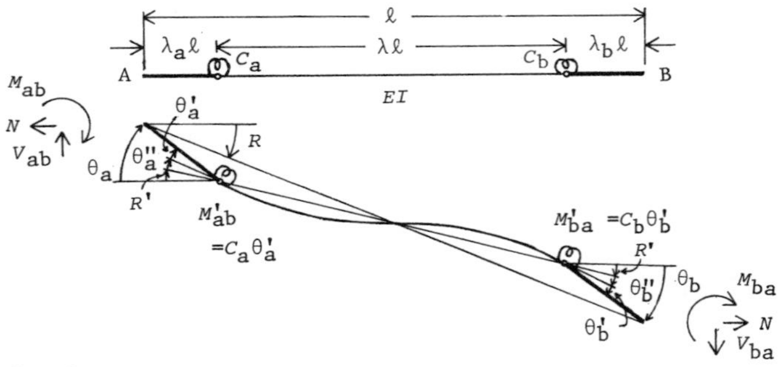

Fig 1

The relations of the end moments and the relative end rotations for the base member are given as follow;

$$\left.\begin{aligned}\theta''_a &= \frac{\lambda\ell}{EI}(a'M'_{ab}+b'M'_{ba})\\ \theta''_b &= \frac{\lambda\ell}{EI}(b'M'_{ab}+a'M'_{ba})\end{aligned}\right\} \quad \cdots\cdots 1,$$

where the coefficients a' and b' are shown in Table 1.

Table 1

N	$N>0$	$N=0$	$N<0$
Z	$\lambda\ell\sqrt{N/EI}$	0	$\lambda\ell\sqrt{-N/EI}$
a'	$\dfrac{\cosh Z}{Z\sinh Z}-\dfrac{1}{Z^2}$	$\dfrac{1}{3}$	$\dfrac{1}{Z^2}-\dfrac{\cos Z}{Z\sin Z}$
b'	$\dfrac{1}{Z\sinh Z}-\dfrac{1}{Z^2}$	$-\dfrac{1}{6}$	$\dfrac{1}{Z^2}-\dfrac{1}{Z\sin Z}$

Denoting the rigidities of the rotational springs as C_a and C_b , Eqn 2 is obtained.

$$\left.\begin{aligned}\theta'_a &= \frac{1}{C_a}M'_{ab} = \frac{\lambda\ell}{EI}S'_aM'_{ab}\\ \theta'_b &= \frac{1}{C_b}M'_{ba} = \frac{\lambda\ell}{EI}S'_bM'_{ba}\end{aligned}\right\} \quad \cdots\cdots 2,$$

where $\quad S'_a = EI/\lambda\ell C_a$ and $S'_b = EI/\lambda\ell C_b \quad \cdots\cdots 3.$

Next, the geometrical relations and equilibrium conditions for the member and rigid end parts are expressed as Eqns 4 and 5, respectively.

$$\left.\begin{aligned}\theta_a &= R'+\theta'_a+\theta''_a\\ \theta_b &= R'+\theta'_b+\theta''_b\\ R &= \lambda R'+\lambda_a\theta_a+\lambda_b\theta_b\end{aligned}\right\} \quad \cdots\cdots 4,$$

$$\left.\begin{aligned}&V_{ab} = V_{ba} = -(M_{ab}+M_{ba})/\ell+NR\\ &M_{ab}-M'_{ab}-\lambda_a\ell\theta_aN+V_{ab}\lambda_a\ell = 0\\ &M_{ba}-M'_{ba}-\lambda_b\ell\theta_bN+V_{ba}\lambda_b\ell = 0\end{aligned}\right\} \quad \cdots\cdots 5.$$

Using Eqns 1, 2, 4 and 5, two simultaneous equations for M_{ab} and M_{ba} are got. The solution of these is

$$M_{ab} = K^r\{\alpha^{rs}\theta_a+\beta^{rs}\theta_b+(\alpha^{rs}+\beta^{rs})R\} \quad \cdots\cdots 6,$$

where

$$K^r = EI/\lambda^2\ell \quad \cdots\cdots 7,$$

$$\alpha^{rs} = \frac{1}{\lambda}\left[\frac{(a'+S'_a)\lambda_a^2+(a'+S'_b)(1-\lambda_b)^2-2b'\lambda_a(1-\lambda_b)}{(a'+S'_a)(a'+S'_b)-(b')^2}+\lambda_a(1-\lambda_b)Z^r\right] \quad \cdots\cdots 8,$$

$$\beta^{rs} = \frac{1}{\lambda}\left[\frac{(a'+S'_a)\lambda_a(1-\lambda_a)+(a'+S'_b)\lambda_b(1-\lambda_b)-b'(\lambda+2\lambda_a\lambda_b)}{(a'+S'_a)(a'+S'_b)-(b')^2}+\lambda_a\lambda_bZ^r\right] \quad \cdots\cdots 9,$$

$$Z^r = \lambda^2\ell^2N/EI \quad \cdots\cdots 10,$$

and a similar expression for M_{ba} .

3. Analytical procedure

The effects of joints on the axial deformation and twisting deformation of the member can be considered by the equivalent rigidities.

The stiffness matrix of the constituent member with uniform and bi-symmetrical section is derived from Eqn 6 and the conditions of equilibrium, as the matrix of (12,12)-type. The incremental stiffness matrix for a space frame as a whole is obtained by the transformation matrix to general coordinate system and by the compositions of the stiffness matrices of members. The governing equations of the space frame are made by cosidering boundary conditions. Its stiffness matrix is a square matrix and the function of the axial forces in members. The buckling load is found as the first load corresponding to zero values of this determinant when loads are increased. And eigenvectors corresponding to unstable state are calculated by the method of Givens-Householder.

OCTAHEDRAL SPACE FRAMES

1. Theoretical analysis

As a simple example of space trusses, a regular octahedral space frame shown in Fig 2 is treated. This space frame is composed of all the same members.

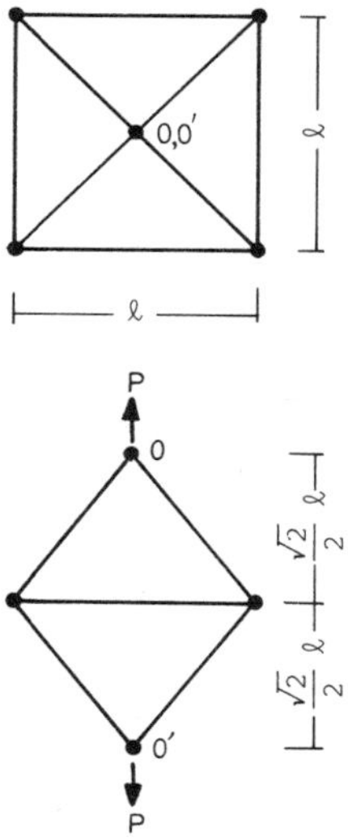

Fig 2

Under loading as shown in Fig 2, the axial forces in the members are statically determined.

Varying the dimensions and rotational rigidities of the joints, the elastic buckling loads for the space frames of octahedral type are calculated by using the present theoretical method, where it is assumed that the connecting springs are no restraints against twisting moments, considering the jointing system used to the model tests. The analytical results are shown in Fig 3, where the Euler's buckling load of pin-jointed members is used as the standard.

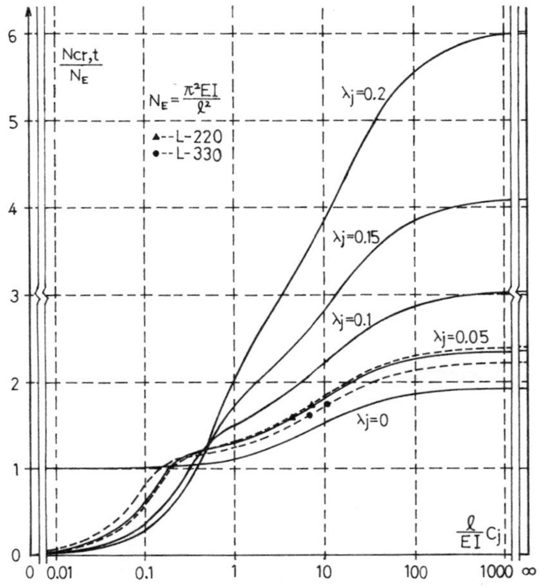

Fig 3

In many cases, the buckling loads of the space frames become greater than that of the corresponding pin-jointed trusses by the effects of joints. But if the rotational rigidities of joints with finite dimensions are too small, the buckling loads become smaller than that of pin-jointed case.

2. Model experiments

Two kinds of specimens denoted as L-220 and L-330 are used, of which distance between adjacent nodal points are 220mm and 330mm, respectively, see Fig 2. The bolted jointing system is used as shown in Fig 4.

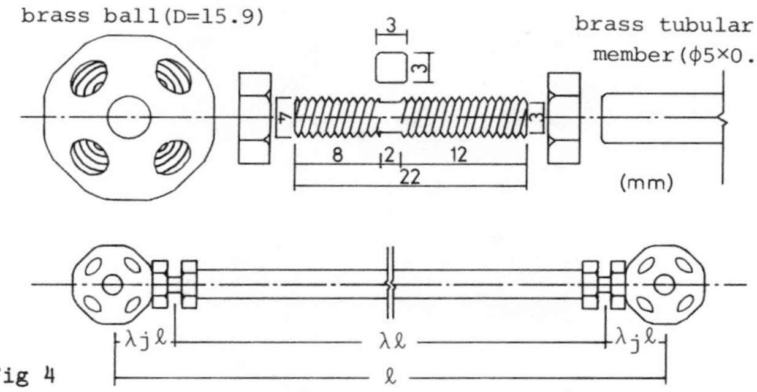

Fig 4

The mechanical properties of the brass tubular members and ball joints were investigated. Table 2 summarizes their properties derived from the test results, where $C_{j,1}$ and $C_{j,2}$ are rotational elastic rigidities of joints at the original loading and the secondly loading or after, respectively, the following value to $\pm$ in brackets denotes the ratio of variation, and the dimension of the joint $\lambda_j \ell$ is 11.9mm.

Table 2

EA (kgf)	$EI_y = EI_z$ (kgf mm^2)	$C_{j,y} = C_{j,z}$ (kgf mm)	
		$C_{j,1}$	$C_{j,2}$
1.13×10^5 (1 ± 0.011)	2.93×10^5 (1 ± 0.08)	5.7×10^3 (1 ± 0.16)	9.4×10^3 (1 ± 0.08)

Tests of the models were carried out under quasi-static deformation control. For some test models, the load-displacement relationships are shown in Fig 5, and a typical mode of buckling is in Photo 1.

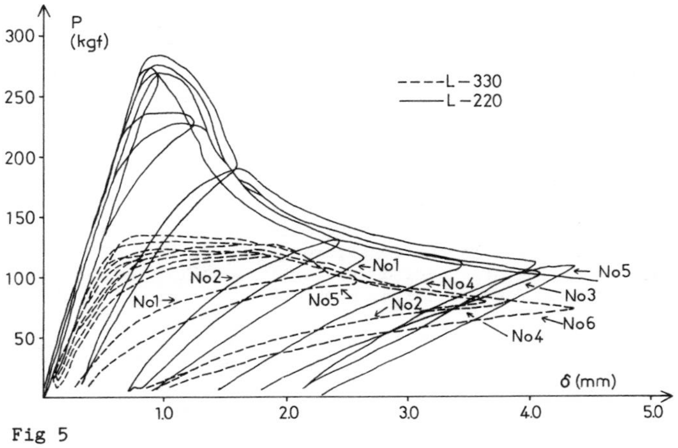

Fig 5

Table 3 shows the experimental ultimate loads $P_{cr,e}$ and the ratios of the axial forces of compressive members derived from $P_{cr,e}$ to Euler's buckling load of pin-jointed members.

Table 3

		L-330									L-220							
	Test number	No.1	No.2	No.3	No.4	No.5	No.6	No.7	No.8	mean	No.1	No.2	No.3	No.4	No.5	No.6	No.7	mean
Experimental values	$P_{cr,e}$ (kgf)	128.	122.	119.	122.	132.	120.	123.	128.	124. (1±0.034)	274.	275.	236.	269.	279.	244.	283.	266. (1±0.064)
	$N_{er,e}/N_E$	1.71 ±0.03	1.63 ±0.03	1.59 ±0.03	1.63 ±0.04	1.77 ±0.04	1.61 ±0.03	1.65 ±0.03	1.71 ±0.04	1.66 (1±0.034)	1.63 ±0.03	1.64 ±0.04	1.40 ±0.03	1.59 ±0.04	1.65 ±0.03	1.45 ±0.04	1.68 ±0.04	1.58 (1±0.064)
Theoretical values	$N_{cr,t}/N_E$	1.62 at $C_{j,1}$ 1.74 at $C_{j,2}$									1.60 at $C_{j,1}$ 1.74 at $C_{j,2}$							

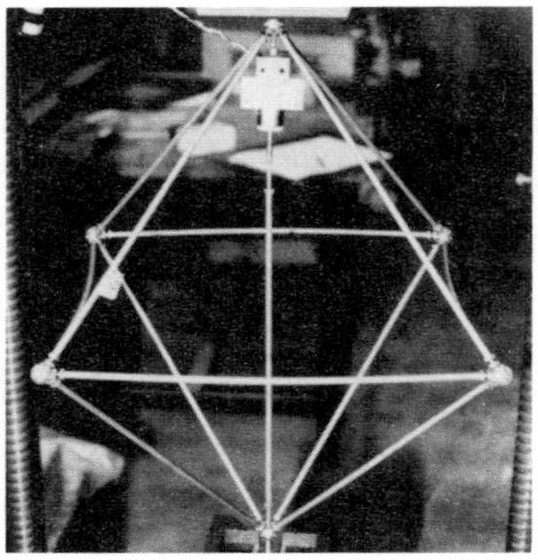

Photo 1

3. Comparison between theoretical and experimental results

The theoretical analyses are carried out using the member properties in Table 2, where the dimensions of joints of L-220 and L-330 are 0.054 ℓ and 0.036 ℓ , respectively. The values are written in Table 3 together with experimental values and also plotted in Fig 3.

The theoretical values furnished a close correlation with experimental values. Accordingly, it may be concluded that the elastic buckling loads for space trusses can be estimated by the present method.

PARALLEL SQUARE MESH GRIDS

1. Effective strength determined by elastic buckling of the constituent members

One of the present authors has already proposed the method to derive the effective strength for the rigidly connected lattice structures which subjected to the elastic buckling of constituent members, see Ref 8. The strength of a PSM grid as shown in Fig 8 is derived from the above mentioned method. This grid is composed of all the same members.

Varying the rigidities of rotational springs under the constant dimensions of joints λ_j =0.036, the effective strength of the PSM grid is shown as Fig 6, where M^{11} and M^{22} are equivalent bending moments. The equivalent twisting moment is not considered, because the corresponding pin-jointed grid is internally unstable under the action of the twisting moment, see Ref 6.

The buckling modes corresponding to the effective strength are drawn in Fig 6. Further varying the dimensions and rotational rigidities of the joints on the stress condition $M^{11} = M^{22}$, the effective strength is shown as Fig 7.

2. Elastic buckling loads of the structure as a whole

Two types of the simply supported square PSM grids under uniform loading shown in Fig 8 are considered and denoted as P-33 and P-22, respectively. These grids are composed of members and joints with the properties of Table 2.

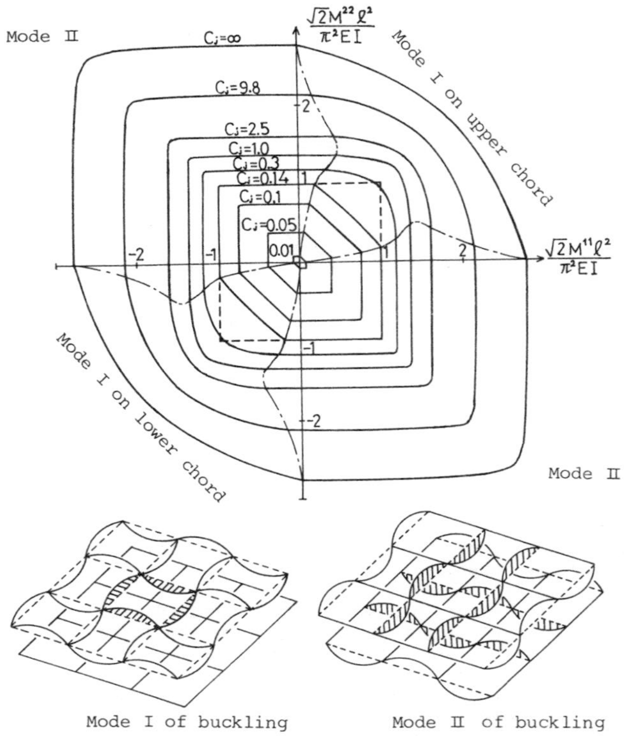

Fig 6

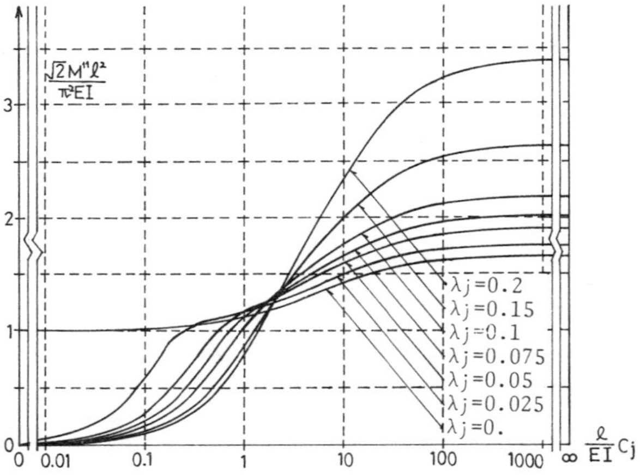

Fig 7

The elastic buckling analyses are carried out using the properties in Table 2. The calculated elastic buckling loads for P-33 and P-22 are shown as $P_{e,1}$ and $P_{e,2}$, respectively, corresponding to $C_{j,1}$ and $C_{j,2}$, in Figs 11 and 12. The mode of buckling for the space grid of P-33 is in Fig 9.

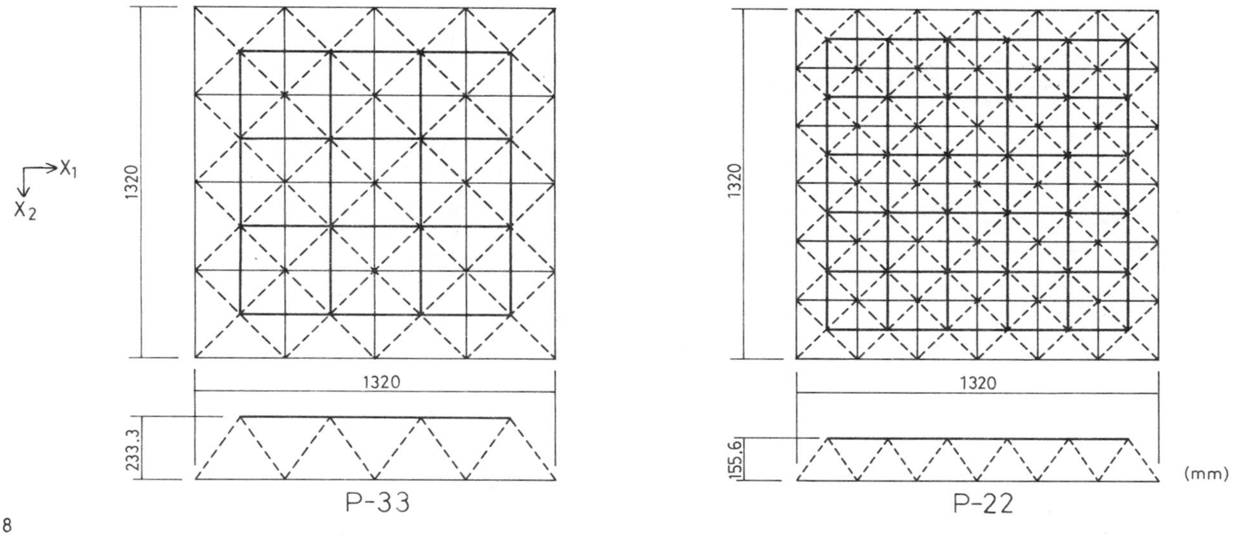

Fig 8

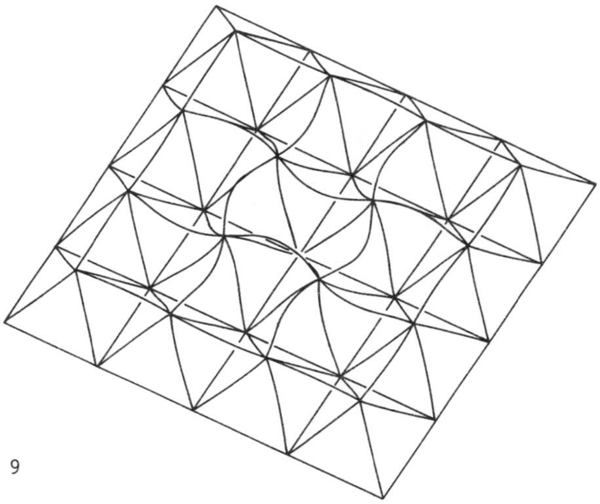

Fig 9

3. Elastic limit loads and collapse loads derived from the effective strength

a) Effective strength of P-33 and P-22

The effective strength of P-33 and P-22 are shown in Fig 10.

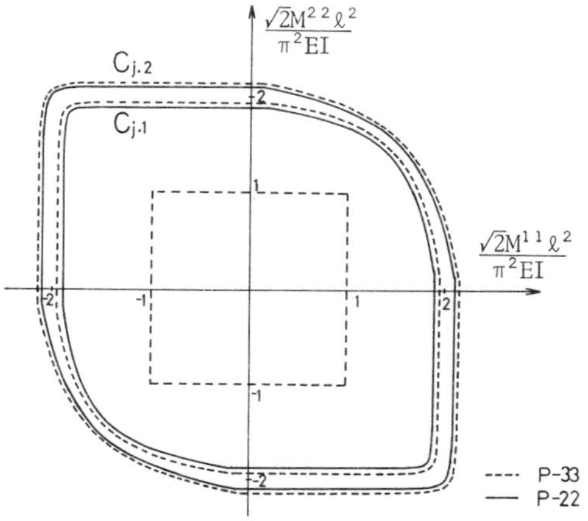

Fig 10

b) Elastic limit loads

Elastic limit loads of P-33 and P-22 are approximately estimated by the results of the elastic stress analysis as an orthotropic continuum plate and the effective strength. These results are shown as $P'_{e,1}$ and $P'_{e,2}$ in Fig 11 and 12.

c) Collapse loads

If the effective strength is assumed as the yield strength, collapse loads of P-33 and P-22 are approximately determined by the limit analysis of the equivalent continuum plates, see Ref 3. When the effective strength at $M^{11}=0$ or $M^{22}=0$ is used as the plastic moment, which is the upper bound of strength, the analytical results are shown as $P_{u,1}$ and $P_{u,2}$ in Figs 11 and 12. If the effective strength at $M^{11}=M^{22}$ is used as the plastic moment, The results are shown as $P'_{u,1}$ and $P'_{u,2}$ in Figs 11 and 12.

4. Model experiments

The truss models of P-33 and P-22 are constructed by using the same structural parts with the foregoing truss models of octahedral type. Tests were carried out under loading of tournament systems, applied loads being measured with load cells, and deflections being done with differential electrometers.

The test results of the load-deflection relationships at the central node are shown in Figs 11 and 12, and also a typical mode of buckling of P-33 is shown in Photo 2.

Photo 2

5. Comparison between theoretical and experimental results

The experimentally obtained P-δ relationships for P-33 and P-22 are shown In comparison with the theoretical results as Figs 11 and 12, respectively.

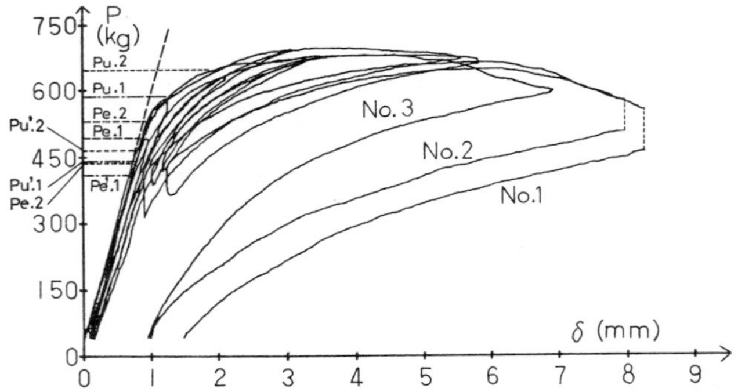

Fig 11

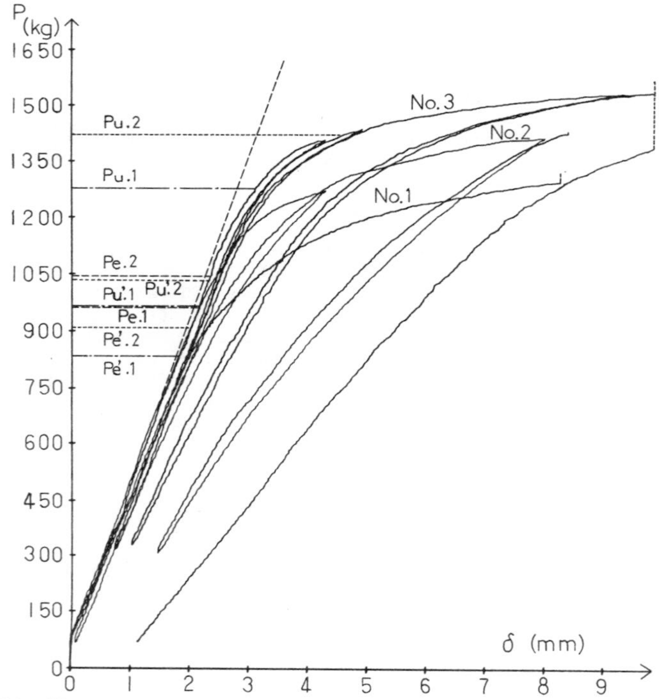

Fig 12

The derived elastic buckling loads and approximate collapse loads furnished closely correlative results with the experimental results. Approximate elastic buckling loads were lower than the experimental results or the theoretical elastic buckling loads. But the ratio of the elastic limit load to the elastic buckling load of P-22 is close to 1 than the ratio of P-33. Accordingly if the mesh is fine, a good correlation will be expected. The elastic buckling modes derived from the present analysis and the yield lines modes of cross-type assumed in the limit analysis could be confirmed by the experiments.

CONCLUSIONS

The effects of dimensions and rigidities of joints on the strength of the space trusses are studied by proposing a new theoretical model. Further the buckling behaviour of space trusses is investigated by model tests. Main conclusions are as follows;

1) The elastic buckling loads for the space trusses can be estimated by the proposed analytical method. Because these theoretical values showed good agreement with the experimental values.

2) In many cases, the buckling load of the elastically-connected space truss becomes greater than that of the corresponding pin-jointed truss by their effects of the joints. But if the rotational rigidities of the joints with finite dimensions are too small, the buckling loads become smaller than the latter.

3) The effective strength of the PSM grids was determined by the elastic buckling of the constituent member, including the effects of joints.

4) The elastic limit and ultimate load-carrying capacities of the PSM grids can be estimated approximately by utilizing the effective strength. Also, the ultimate load-carrying capacities derived from the limit analysis as equivalent continuum plates furnished good correlative results with the experimental results.

ACKNOWLEGEMENTS

The authors wish to thank Mr. H Tsuji for his valuable assistance during the experimental work. Thanks are also due to Mr. M Murakami for his valuable advice on the computer programming and Mr. S Abe for assisting the numerical computation. And this work is supported in part by the Ministry of Education of Japan under the Grant in Aid for Scientific Research.

REFERENCES

1. Z S MAKOWSKI(Editor), Analysis, Design and Construction of Double-Layer Grids, Applied Science Publisher, London, 1981.

2. J D RENTON, Stability of Space Frames by Computer Analysis, Proc. ASCE, Journal of the Structural Division, August 1962.

3. L C SCHMIDT, P R MORGAN and J A CLARKSON, Space Trusses with Brittle-Type Strut Buckling, Proc. ASCE, Journal of the Structural Division, July 1976.

4. W J SUPPLE and I COLLINS, Limit State Analysis of Double-Layer Grid, ANALYSIS, DESIGN and CONSTRUCTION of DOUBLE-LAYER GRIDS(Ed. Z S Makowski), Applied Science Publisher, London, 1981.

5. L C SCHMIDT, P R MORGAN and A HANAOR, Ultimate Load Testing of Space Trusses, Proc. ASCE, Journal of the Structural Division, June 1982.

6. T SAKA and K HEKI, Limit Analysis of Lattice Plates, Proc. 1971, IASS Pacific Symposium Part II on TENSION STRUCTURES and SPACE FRAMES, Architectural Institute of Japan, 1972.

7. C E MASSONNET, The Collapse of Struts, Trusses and Frames, COLLAPSE the buckling of structures in theory and practice(Ed. J M T Thompson and G W Hunt), Cambridge University Press, 1983.

8. K HEKI, The Effective Strength of Rigidly Connected Lattice Structures Determined from Elastic Instability of the Constituent Members, Transactions of the Architectural Institute of Japan, March 1983. (in Japanese)

ELASTOPLASTIC BEHAVIOUR OF FLAT GRIDS

N DIANAT BSc,MSc,PhD

Standards Division, Technical Bureau , Water Affairs,
Ministry of Energy, Islamic Republic of Iran.

The paper reports the elastoplastic behaviour of four grid structures recorded throughout an experimental study. The recorded data can be reliably used to evaluate the validity of the analytical techniques. The grid structures were made to high standard of precision from stress - free bars. Grids were gradually loaded under computer control far beyond the elastic limit. In addition to the electrical measuring systems close range photogrammetry was used to record the deformed shape of the grids. The grid structures were analysed numerically and results compared with experimental values , every experimental curve being plotted along with the theoretically predicted curve. This comparison provides an example of how these experiments may be used to assess the validity of any relevent theory.

INTRODUCTION

A considerable amount of work has been done on the study of elastic behaviour of grids but little has appeared on the elastoplastic behaviour and collapse behaviour of this type of structures . A survey of the literature dealing with these aspects of behaviour of grids has been presented by the author in Ref 1 .

In the present work the experimental study of elastoplastic behaviour of four grid structures is reported,the aim being to establish facts which could be used to evaluate the validity of analyical techniques.

Analytical techniques deal with mathematical models.Simulation of a structure with a mathematical model is based on some idealizations and simplifying assumptions.Behaviour predicted by these techniques therefore deviates from the actual behaviour and we need to assess these deviations by the aid of observations on real structures.

Experimental data on elastoplastic behaviour of a structure could be obtained as it is loaded gradually beyond the elastic range up to the point of collapse.This is a costly process.Care should be taken therefore to obtain sufficient and accurate data on the initial structure as well as on its behaviour under load.In the present study the test grids were constructed with care; and recent developements in the experimental analysis techniques were employed for accurate measurement and automatic data acquisition.

To design and carry out the experimental study successfully, some knowledge of elastoplastic behaviour of grids proved necessary.A piecewise linear analysis technique was therefore employed Refs 2 and 3 .

During the test for each grid,a large number of measurements (17000 - 21000) were recorded and 6 - 12 pairs of stereo - photographs taken.The recorded values were processed by a computer and a series of "Load - Displacement" and "Load - Moment" curves were plotted.Also from the stereo photographs, "Level Contours" were obtained.

GENERAL REMARKS AND LAYOUT

In the design of the experimental set-up the main considera -tions were:

a) To obtain information over a wide range of grid struc - tures regarding the difficulties and the effort needed in experimental studies.
b) To have accurate information about the grid structures and the boundary conditions.
c) To obtain accurate measurements during the test.
d) To record sufficient information during the test.

The experimental set-up of which Fig 1 is a picture, consisted of grid structures;test frame ; loading system;and measurement systems.

Four grid structures of the same layout but of two different member cross - sections were tested under two different loading arrangements.The layout was determined by the number of the supports and by analytical studies.In order to eliminate any residual stresses which might be caused by imperfections in the supports it was decided to have three supports and by analytical studies on some triangulated three - way grids,the layout of Fig 2 was chosen. This layout had three axes of symmetry, and the reliabitity of measurements could be assured by comparison of measurements on symmetrical joints and members.

The overall dimensions of the grids and the cross - sectional dimension of the members were determined thus:

a) Structures to be small enough to be made and handled under laboratory conditions and large enough to get reliable measurements.
b) The load - carrying capacity of the grid to be well within the range of the avaliable loading system.
c) To eliminate the possibility of lateral buckling.
d) The difference between the torsional rigidities of the two sections to be large.

The joints had to have very good rigidity (fixity) and their dimensions had to be relatively small.Studies on different types of joints proved that,regarding the dimensions of the grid and members cross - section, welded joints were the only answer.

Necessary information about the grid consisted of data on its geometry, its material and residual stresses throughout the grid. To obtain this information accurately was an important part of the experimental study, because these data provided a sound base for the evaluation of the experimental results. There was no easy way of measuring residual stress. Stress - free bars were therefore welded, to construct the grids, and care taken to minimize the stresses which might have been induced in the process of welding.

Fig 1 The experimental set - up

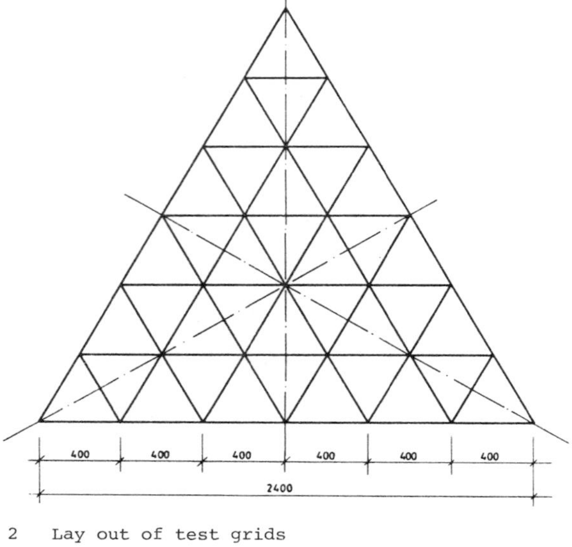

Fig 2 Lay out of test grids

MATERIAL AND MEMBERS' PROPERTIES

Bright mild steel bars obtained from the same batch for the same cross - section were used as members of the grid structures. Bright finish which means accuracy on the cross - sectional dimensions, is produced by cold rolling. This process increases the yield stress and hardness of the steel, and induces some residual stresses, which in this study were undesirable and therefore the material had to be annealed. Stress relief annealing was carried out in an air circulating electric furnace. Members which had been cut and machined to size, were positioned vertically in the furnace to avoid deformation at high temperatures. Bars were kept at 650°C for one hour then left in the furnace to cool down to room temperature. The Stress - Strain curves derived from tension tests on one of the materials before and after annealing are given in Fig 3.

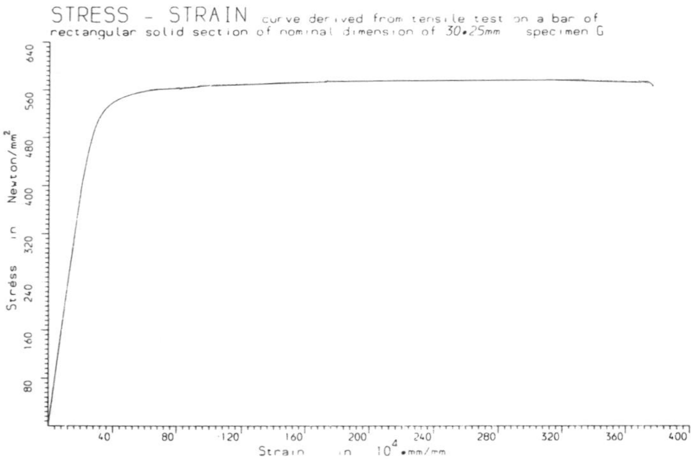

(a) **This is a test on the material before being stress relieved**

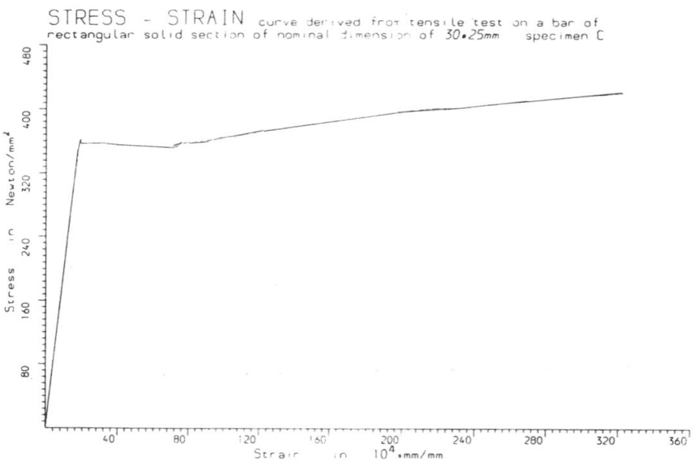

(b) **This is a test on the material after stress relieving**

Fig 3

To obtain the necessary information about the mechanical properties of the materials: tension, torsion and bending tests, were carried out. Considering the effect of the strainning rate on the value of the yield stress, this rate was controlled in order to have compatible results from different tsets. In a test on a complex structure in which, obviously, not all parts will deform with the same strain rate, the strain rates should be kept within a range , in which the variation of yield stress is minimum. The detailed discussion of tests are given in Ref 2. Here the main considerations and results are presented.

Tension Test - Tension tests were carried out to obtain the following information about the materials:

a) The value of modulus of elasticity, E.
b) The value of poission's ratio, ν.
c) The upper and lower values of the yield stress $\sigma y u$ and $\sigma y \ell$ respectively.
d) The stress - strain curve.

Fig 4 shows the experimental set-up. It consist of: test m chine, test speciemen, strain transducers, recording and control system. Specimens were full size and taken at random from the annealed bars for the grid. A typical result of tension tests are shown by Fig 3(b).

Torsion Test - Torsion tests were carried out to obtain the following information about the materials:

a) The value of shear modulus of rigidity, G.
b) The value of yield stress at shear, τ_y.
c) The torque - twist curve.

Fig 5 shows the experimental set up, consisting of: testing machine, test specimen; twist gauges; recording and control system.

The mechanical properties of the material obtained from tension and torsion tests on six specimens of each material are presented in Table 1. As it may be seen, the difference between the two values of G, for each material one obtained from tension tests and the other from torsion tests, is less than one percent.

Bending Tests - Bending tests were carried out in order:

a) To obtain the moment - curvature curves for both member types.
b) To calibrate the strain guages which were to be used for the measurement of bending moment in grid members.

The test set-up consisted of: test frame; test specimen; loading system; transducers; reccording and control system.
Fig 6 is a picture of the set-up. A typical Moment - Curvature curve is presented in Fig 7. These curves as well as the stress - strain and the torque - twist curves were plotted with the Calcomp Microfilm plotter.

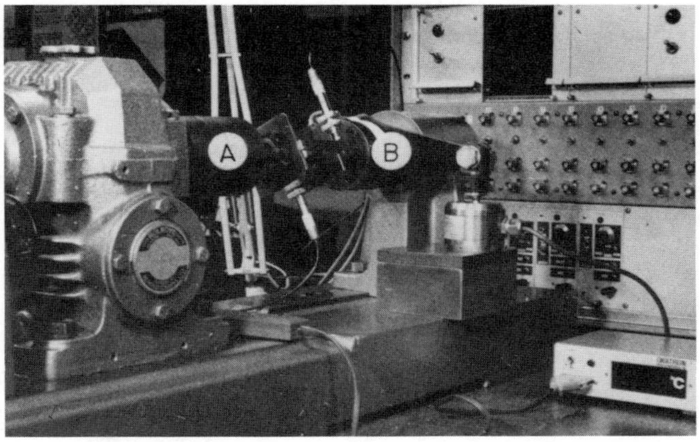

Fig 5 Torsion test set - up .

Fig 6 Bendung test set-up .

Fig 4 Tension test set-up .

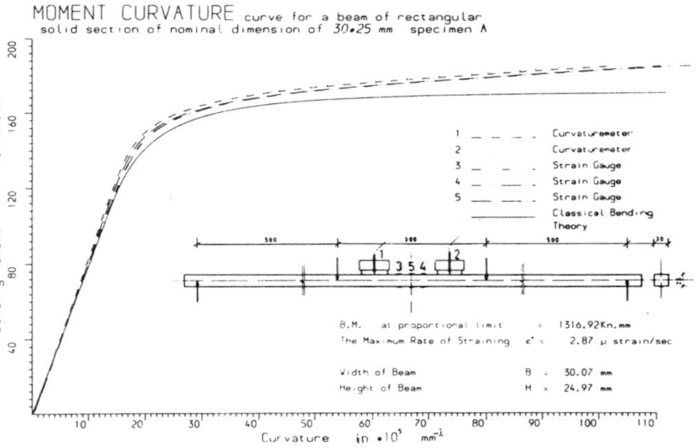

Fig 7 .

Table 1 Mechanical Properties of the Materials

X-Section		TENSION TESTS					TORSION TESTS	
		$\sigma_{y\ell}$ N/mm^2	σ_{yu} N/mm^2	E N/mm^2	ν	G N/mm^2	τ_y N/mm^2	G N/mm^2
16x35	Mean	196.6	206.5	199500	0.274	78330	113.4	77670
	S.D.	3.85(2.0%)	3.37(1.8%)	164.00(0.8%)	0.007(2.5%)	516.00(0.7%)	8.9(7.8%)	1370.00(1.8%)
25x30	Mean	365.9	337.5	201000	0.282	78670	200.7	78830
	S.D.	7.0 (1.9%)	8.6 (2.3%)	632.00(0.3%)	0.001(0.4%)	516.00(0.7%)	4.95(2.5%)	752.00(0.9%)

CONSTRUCTION OF THE GRIDS

Stress - free bars were to be welded to each other to from the grids such that :

a) The resulting geometry to match the design geometry within the close tolerances.
b) The residual stresses throughout the members and joints to be negligible.

Considerations involved in satisfying the above conditions are as follows:

a) The volume of the weld metal to be minimized and balanced symmetrically about the neutral plan of the grids.
b) Weld passes on both sides of the neutral plan to be balanced.
c) Work to be positioned for downhand and horizontal welding.
d) Grids to be divided into smaller units and proper jigs (fixtures) to do used.
e) A welding sequence to be followed.

All these conditions were satisfied, grids were divided into ten units of four types. A jig was designed for assembling all four types. To assemble these units to form the grids another jig was necessary. The joints were the smallest possible with the minmum weld metal balanced symmetircally. Details about the construction of the grids is given in Ref 2 . Here is to mention that the accuracy of ± 0.3 mm on the member length, ± 0.6 mm on the x,y coordinates of joints, and ±1 mm on the overall dimentions had been achieved, while the flatness check with a 2m straight edged after machining the excess weld showed a tolerance of - 0.4 mm.

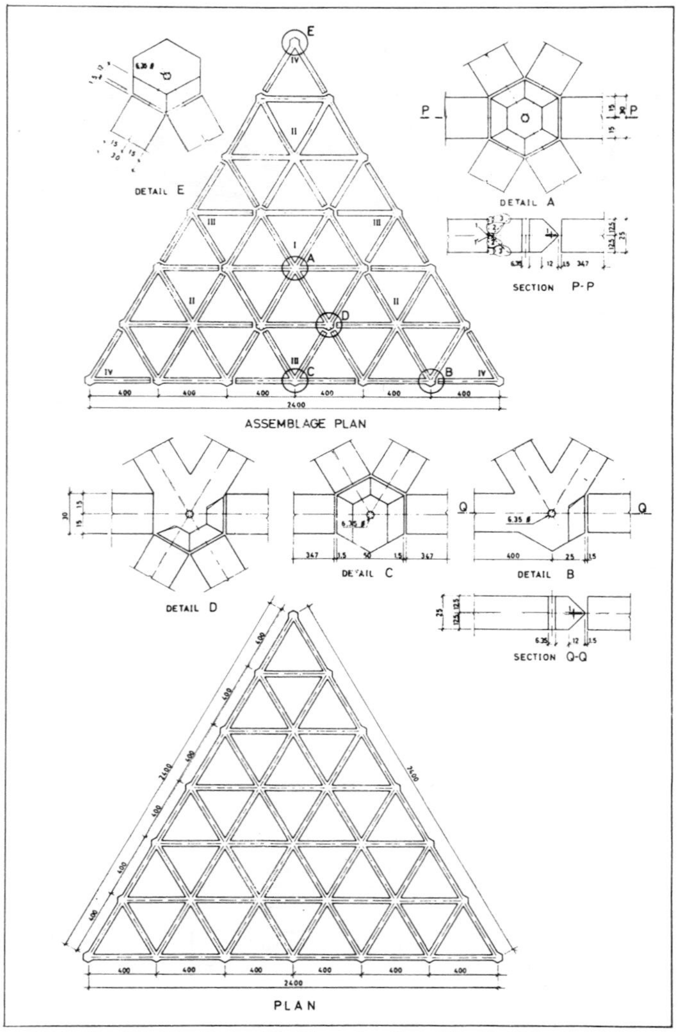

Fig8 **Drawing of a test grid with member X-section 30x25 mm**

TEST FRAME AND LOADING SYSTEM

The test frame consisted of a base, three columns, three supports and fixtures for the displacement transducers, Fig 9. The base was designed to be very stiff, as it was used as the origin for displacement measurements. In the supports the z translation of the supported joint was fully constrained while its x,y translations and all rotations were almost frictionless, and the reaction of the support during the test passed through the joint's center.

Fig 9 Test Frame .

The load consisted of three forces with the following characteriatics:

a) Having constant proportion to each other .
b) Controllable to produce strain rates in the desired range.
c) To be controlled by computer.

The equipment available comprised three hydraulic actuators together with hydraulic power pack and a console housing separate controlling and monitoring facilities of each actuator. The specification of the hydraulic actuators was as follows:

Maximun capacity of static testing	25 KN
Total working stroke	250 mm
Accuracy of load control and measurement	±0.5%
Accuracy of displacement control and measurement	±1 mm

To have loads of constant proportion, two of the actuators were set to load control and both made to follow the output load signal of the third, the proportional relationship of the three loads being adjustable. To meet the other two conditions the third actuator was set to displacement control and was fed by a signal controlled by the computer. The load was applied by pulling the grid downwards.

MEASUREMENT SYSTEMS

Quantities to be measured during the test were as follows:

a) The load values at loaded joints.
b) The displacements of a number of points on the grid (sufficient to give the grid's deformed shape).
c) The internal moments at some sections of certain members.

Because of the number of quantities to be measured and the speed of the test, electronic measuring systems were mainly employed; in addition close - range photogrammetry was used to record the deformed shape of grid in some stages during the test.

ELECTRONIC SYSTEMS

The electronic measuring systems comprise three basic parts; the transducer, the signal - conditioning equipment and the recording equipment. In all the measuring systems used for the different quantities in this work, the same recording equipment was employed, while each system had different transducers and signal conditioning equipment. Recording equipment as may be seen in Fig 10 cansisted of 100 input channels with the necessary switching, a precision d.c amplifier (all housed in cabinet A), a digital voltmeter B past paper tape punch C and page printer D. The analogue signal input from a channel was scaled by the amplifier, measured by the digital voltmeter and recorded on punched paper tape for use in subsequent analysis, and/or printed. The channel switching, scaling, measurement and recording were controlled by a minicomputer E. About 15-17measurements per second could be acquired by this arrangment.

Fig 10

A precision load cell was used to convert the value of the load applied by each actuator to an electrical signal. The value of the vertical displacement of each loaded joint was converted to a signal by the LVDT type displacement transducer of the actuator, these signals were amplified by the signal conditioning facitities of the acutators system and fed to the recording system via a system of buffer amplifier, filter and divider.

LVDT (Linear Variable Differential Transformer) type displacement transducers were used for the measurement of vertical displacement of grid joints; 22 transducers having three different stroke ranges (±1", ±2" and ±4") and accuracy grades A1, A1 and A(BSI), were used. For each group of transducers of the same stroke range an amplifier was employed to enable their ouput signals to be independently scaled. Within a group of transducers therefore a scanner was necessary. When a transducer was not selected it was fed, with a d.c warming current to eliminate self - heating errors. The final output for each transducer was calibrated at 20°C using a set of slip gauges ground to an accuracy of ±0.02 mm and the slope of the line fitted to the points of measurement was used as the calibration factor to convert the output data into the displacement in mm.

For measurement of bending moments precision electric resistance strain gauges were used. One pair of gauges was installed on the top and bottom surfaces of the member at the section where the bending moment was to be measured. A half-bridge was used having two active gauges, the bridge was completed by a highly stable half-bridge within the logging system. The bridge was energized by a constant current, while the measurement was taking place. The main reason for carrying out the bending tests described earlier was to calibrate the output of this system, to bending moment. The results of these tests were used in the analysis to convert curvature (or strain) into bending moment.

PHOTOGRAMMETRY

A pair of Zeiss Stereometric cameras (SMK - 40) was employed. The cameras were fixed to a gantry and centred on the grid structure at a height of 2.15 m above it, giving stereo coverage of the entire area of the grid in one overlap, with a base - to - depth ratio of 1/5.3. The fiducial marks and a number of the photo pair, and a label on the test frame indetifying the grid and the stage of test appeared on the potographs. In order to form a digital model of the grid structure at any stage. The coordinates of a sufficient number of points must be known. To dothis the following points were marked on the grids:

a) Center of joints.
b) Four points on top of each loading head.
c) Eighteen points on each member at six sections.

Control was obtained from 11 control points attached to the test frame. Coordinates of these points were determined from a trilatration survey and precise levelling. The ground coordinates of about 420 points were calculated having measured their photo coordinates by a Zeiss Stecometer C.

The accuracy of the ground coordinates could be judged in the following ways:

a) Comparing the ground coordinates of control points obtained by trilatration surveying and levelling with the values obtained from the photogrammetry. The root mean square error on all points was about 1mm in x,y and z coordinates.
b) Comparing the vertical displacement of each joint obtained by photogrammetry with the value obtained by electrical transducers. Table 2 is a comparison of this type.

Table2 **Vertical displacements at joints of grid 4. at two stages of loading obtained by photogrammetry and electrical transducers.**

Joint No.	First stage		Second stage		Joint No.	First stage		Second stage	
	photo-grammetry	electrical transducer	photo-grammetry	electrical transducer		photo-grammetry	electrical transducer	photo-grammetry	electrical transducer
2	21.7	22.95	46.6	47.49	15	40.4	38.67	84.2	82.17
3	40.4	39.8	86.6	86.4	17	39.8	39.55	77.8	76.91
4	48.9	47.41	109.2	106.18	18	30.6	28.56	54.7	53.34
5	40.7	40.16	87.7	86.57	19	32.8	31.89	69.4	67.82
6	25.2	23.13	49.	47.77	20	37.2	36.50	73.0	72.01
8	17.9	17.03	34.9	34.09	21	37.0	36.43	70.0	68.97
9	35.5	34.79	73.4	72.9	22	31.6	32.24	58.5	59.35
12	36.5	34.4	70.4	68.84	23	26.8	26.43	54.0	52.818
13	17.9	17.42	31.1	31.29	24	28.3	27.67	53.2	51.90
14	29.1	28.47	59.9	59.91	27	15.1	15.276	26.8	26.21

THE TESTS AND RESULTS

The grid structures were tested under computer control. An Alpha 16 mini- computer under program control served this puropose. The program was written in BASIC and called machine code routines to control the data logger and the loading system. Test was under displacement control, and to obtain sufficient information it was decided that the loading period be 60 seconds while 15-17 seconds was needed for each set of measurement. The displacement applied at each period of loading was 1.5 mm. This gave a range of straining rate compatible with the strain rates during the tests on the materials.

During tests for each of the grids 1,2 and 3 some 17 000 measurements and for grid 4 some 21 000 measurements were obtained and values recorded on punched paper tape. Also the deformed shapes of each grid at certain stages were recorded by stereo pairs of photographs (6-12 pairs of photographs were taken from each grid). The data recorded on punched paper tape were processed, the values of loads, displacements and internal moments calculated and a series of " Load - Displacement " and "Load - Moment" curves plotted. From each pair of stereo photographs the coordinates of a number of points on the grid were obtained and the " Level- Contours" for the deformed grids at the corresponding stages of loading were plotted by microfilm plotting facilities.

Load - Displacement curves. A graph of this type is shown on Fig 11. Such graphs consist of one or more (up to six) experimental curves and one theoretical curve plotted as a full line. Each of the experimental curves shows the variation of vertical displacement of a joint of the grid structure with load during the test. All joints for which load - displacement curves are given on this graph are marked in the sketch at the bottom right corner of the graph. Also the type of line identifying each curve is specified. As may be seen. These joints are symmetric and the theoretical curve is common to all of them. The data for this curve were obtained from theoretical analysis and, to make it comparable with the experimental curves, the value of displacement due to self - weight, was subtracted. In some cases the vertical displacement of certain joints happend to be larger than the working range of the displacement transducers used. So the corresponding load - Displacement curve of each was plotted up to the range limit. For grids 1,2,3, and 4 the numbers of graphs of this kind are 6,6,14 and 14 presented in Ref 2 .

Load - Moment curves. A graph of this type is shown on Fig 12. Such graphs consist of one or more (up to six) experimental curves and one theoretical curve plotted as a full line. each of the experimental curves shows the variation of bending moment at a member section of the grid structure with load during the test. All members for which the load moment curve at a section is given on this graph, are marked on the sketch at bottom right corner of the graph. Also the type of line identifying each curve is specified. As may be seen, these member sections are symmetric and the theoretical curve is common to all of them. The data of this curve were obtained from theoretical analysis, and to make it comparable with the experimental curves,the value of moment due to self - weight, was subtracted. For grids 1,2,3 and 4 the number of graphs of this kind are 4,5,7 and 9 presented in Ref 2 .

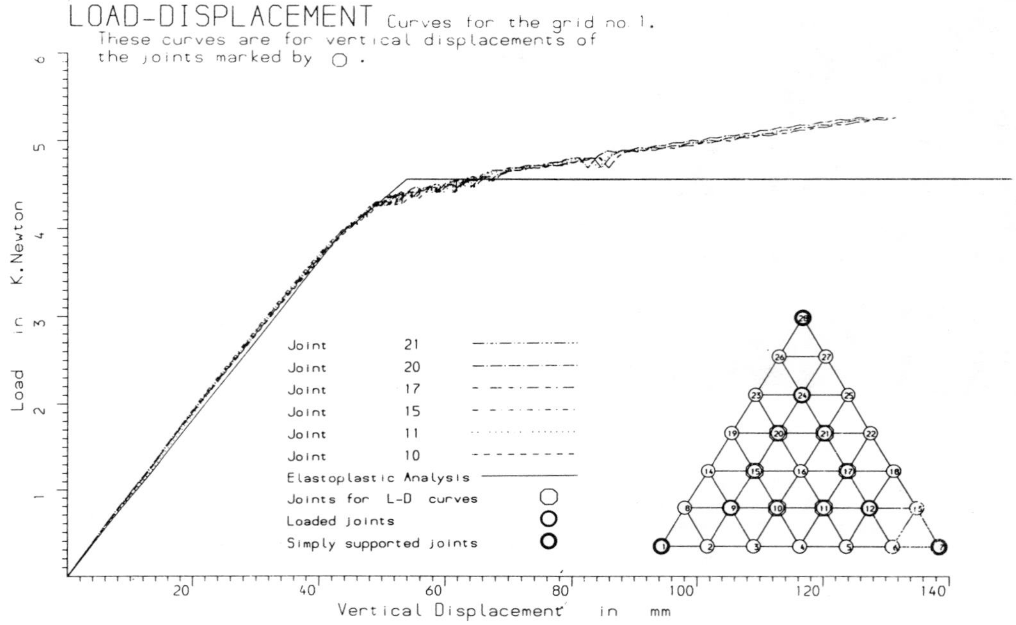

Fig11

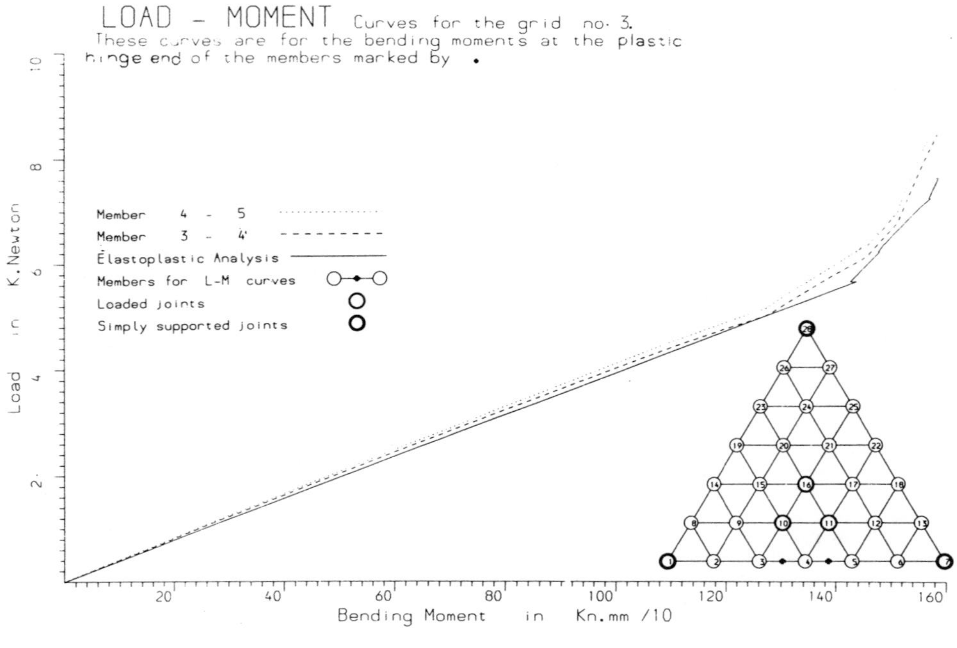

Fig12

Level - Contours A typical graph of this type is given on Fig13. For grids 1,2,3 and 4 the numbers of graphs of this kind are 2,1,2 and 3 presented in Ref2 .

History of Plastic Hinge Formation. The position of plastic hinges and the sequence of their formation in one of the grids are given on the sketch of Fig14. This is obtained from theoretical analysis, where it was assumed that plastic hinges were formed in the vicinity of joints. During the test,visual inspection was carried out to check the position of plastic hinges and the sequence of their formation, this was helped by the oxide scale produced on bars as a result of annealing. The sequence was the same as predicted by theory, but it was difficult to determine the exact position of the plastic hinges. However, at each section which was predicted as a plastic hinge, large plastic deformation was detected when the strain gauges were present.

ACKNOWLEDGEMENTS

The author wishes to express his gratitude and sincere thanks to Dr H Nooshin for his continuous help, advice and encouragement throughout the work which was carried out under his supervision in the Space Structures Research Centre.Also thanks are due to many people which without their expert help the experimental work presentad here would have not been possible. Also thanks are due to the Ministry of Energy of the Islamic Repulic of IRAN for the support.

REFERENCES

1. N DIANAT, Plastic Analysis of Flat Grids, Msc dissertation, University of Surrey, 1974.
2. N DIANAT, Elastoplastic Behaviour of Flat Grids, PhD thesis, University of Surrey, 1979.
3. G A MORRIS and S J FENVES, Elastic - Plastic Analysis of Fremeworks, Journal of struct. Div, ASCE, Vol.96, ST.5, 1970.

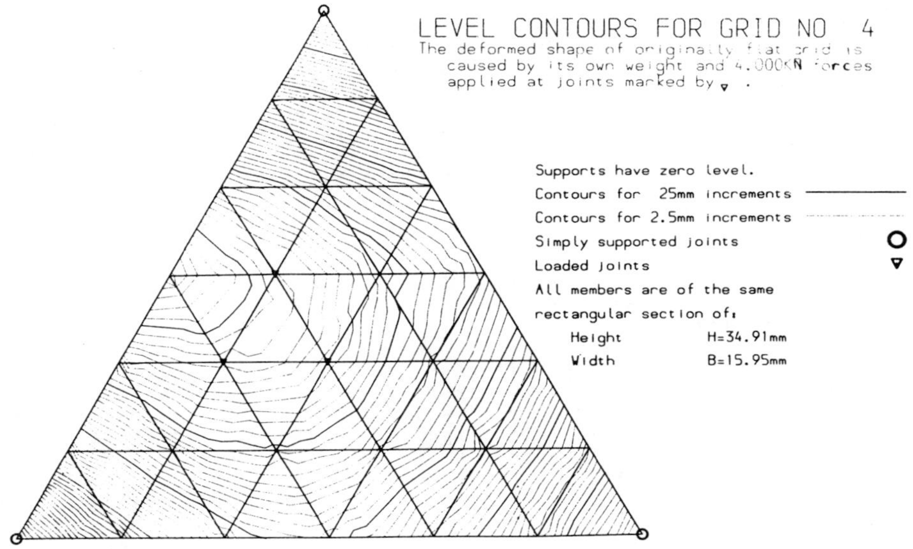
LEVEL CONTOURS FOR GRID NO 4
The deformed shape of originally flat grid is caused by its own weight and 4.000KN forces applied at joints marked by ▽ .
Supports have zero level.
Contours for 25mm increments
Contours for 2.5mm increments
Simply supported joints
Loaded joints
All members are of the same rectangular section of:
Height H=34.91mm
Width B=15.95mm

Fig13

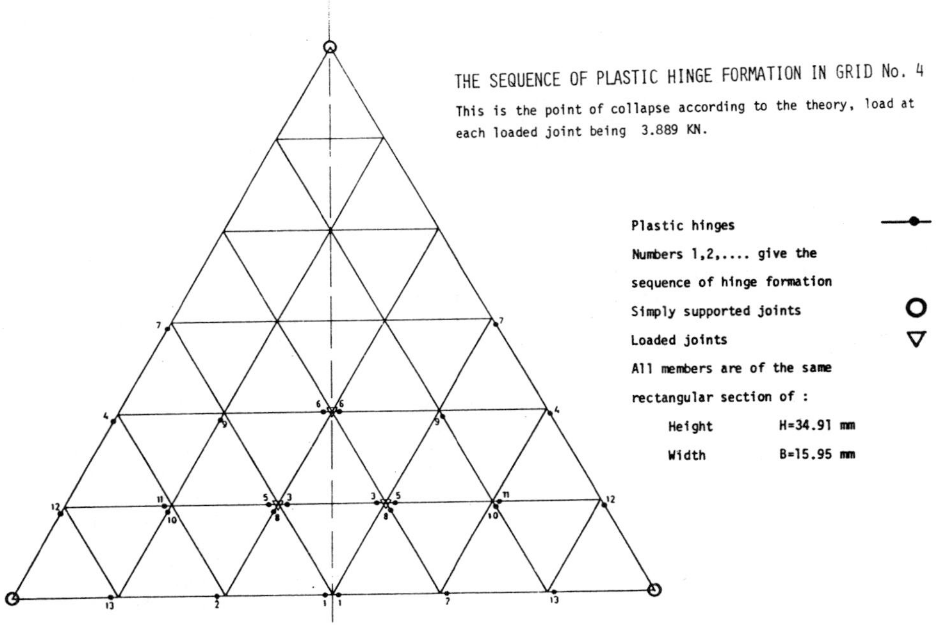
THE SEQUENCE OF PLASTIC HINGE FORMATION IN GRID No. 4
This is the point of collapse according to the theory, load at each loaded joint being 3.889 KN.
Plastic hinges
Numbers 1,2,.... give the sequence of hinge formation
Simply supported joints
Loaded joints
All members are of the same rectangular section of :
Height H=34.91 mm
Width B=15.95 mm

Fig14

SEISMIC ANALYSIS OF A SPACEFRAME

S. R. ROBERTSON, PhD

Department of Mechanical and Energy Engineering

University of Lowell

Lowell, MA 01852 USA

A simple planar spaceframe is analyzed for the purpose of exploring the use of Levy and Wilkinson's method for representing a seismic design spectrum by an equivalent time history excitation. The resulting time history is then used with the finite element code ADINA to model the response of the spaceframe. A free vibration analysis is performed first in order to study the effects of roof load and support conditions on the natural frequencies of the structure. This is followed by a dynamic analysis that uses the artificial ground motion time history to excite the structure.

INTRODUCTION

The seismic events that a structure will encounter during its lifetime cannot be stated with certainty. The frequency of occurrence, the duration and character of seismic events are statistical and the statistics are different for different geographical regions. Raw data from seismographs are usually converted to a seismic spectrum. A spectrum is the maximum response envelope of a single degree of freedom oscillator to the seismic event as the natural frequency of the oscillator is swept over a range of frequency. Plots are normally made for various values of oscillator damping. Even for a particular geographical region the spectra will vary from event to event so it is necessary to collect and process data over many years in order to specify a design spectrum with any confidence[1].

Traditionally the major concern regarding seismic excitation has been with respect to horizontal ground motion and buildings more than two stories high. Such buildings act in a similar manner to a cantilever beam where the foundation corresponds to the fixed end of the beam and quite often have bending-like modes of vibration in the frequency range of from 1 to 10 Hz which is the range where most damage due to seismicity is likely to occur. In the case of vertical excitation, traditional structures are very rigid and therefore have high natural frequencies which are not likely to be excited by a seismic event.

Long span planar spaceframe structures present a different situation. The roof is usually quite rigid horizontally, although the total structure of which it is a part may not be. Vertically, however, the roof can be quite flexible as is the case for the structure studied here. Thus, design procedures for traditional structures are probably not appropriate for long span structures when seismicity is important.

DEVELOPING A FORCING FUNCTION FROM A DESIGN SPECTRUM

A design spectrum is a statistically determined envelope for 85% or 90% of the expected seismic events for a particular geographic location. Figure 1 shows a seismic spectrum based on earthquakes that have occured in California[1]. The lower curve is the ground motion envelope normalized for a peak ground acceleration of 1g. The other curve is the response spectrum for damping 0.10 that of critical damping. This graph is called a tripartite plot because three different peak responses can be determined for one frequency. The plot has four logarithmic coordinates. The horizontal one is frequency, the vertical one velocity, the skew one with positive slope acceleration and the skew one with negative slope displacement.

There is no unique acceleration time history for a given spectrum but an infinite number of possibilities. The problem is to select a forcing function that will produce the same spectrum as the given spectrum. In addition to spanning the frequency range of the given spectrum, it must have enough frequency content to be able to excite any modes that lie within that range. Levy and Wilkinson[2] have developed a method for generating time histories from spectra which are rich in all frequencies in the range of interest. The time history is of the form

$$H(t) = F(t)\sum_i (-1)^i G_i \sin(w_i t)$$

The frequencies w_i are chosen such that the response curves of oscillators having adjacent values of frequency overlap at the half power points. These values are automatically chosen by the computer code developed for generating the time history. F(t) is an envelope function that varies slowly in time and places limits on the magnitude of the time history. An initial guess for the coefficients G_i is made using values from the design spectrum for the corresponding values of frequencies w_i. The equation

$$\ddot{X} + 2w(C/C_c)\dot{X} + w^2 X = -A(t) \quad ,$$

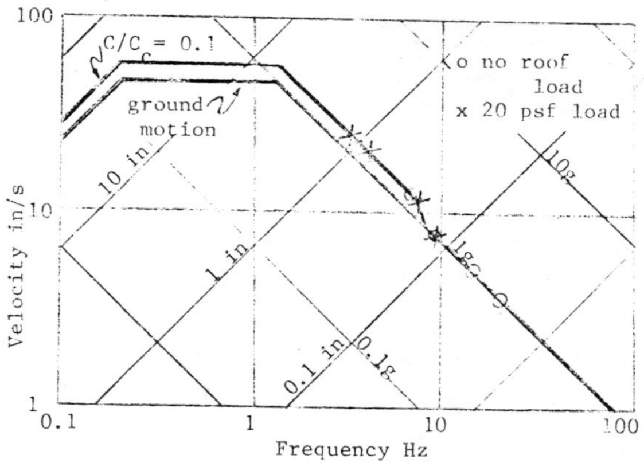

Figure 1. Tripartite plot of a seismic design spectrum.

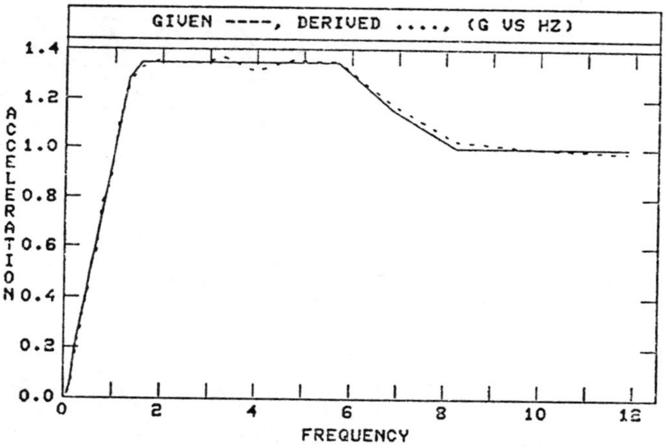

Figure 2. The given spectrum compared with the spectrum of the derived time history.

for a single degree of freedom oscillator, is then solved for each frequency w_i using the previous guess for H(t), where A = H(t) is the ground acceleration. C/C_c is the ratio of damping to critical damping and w the natural frequency of the oscillator. The peak value $S_i = |\ddot{X}|max$ is saved for each w_i. This new spectrum is compared to the given spectrum and the coefficients G_i in the time history function are suitably updated. The process is repeated until the time history is able to reproduce the given spectrum within a specified tolerance.

A code was developed to perform the above computations. Given the design spectrum, the envelope function and the frequency range, the code automatically selects the frequencies for the time history with the correct spacing and then generates the acceleration time history. For the response spectrum shown in Figure 1 the acceleration time history function has 31 frequencies from 0.05 to 11.87 Hz. The spectrum is reproduced in Figure 2 using linear scales and displays acceleration vs frequency. Note the close agreement between the given spectrum and the spectrum for the derived forcing function. The acceleration time history is shown in Figure 4, the frequencies and coefficients are given in Table I and the envelope is given in Table II.

TABLE I

FOURIER COEFFICIENTS

i	w Hz	G	i	w Hz	G
1	0.05	0.0	17	0.92442	0.63975
2	.06	.00005	18	1.10931	.28324
3	.072	.00067	19	1.33117	.51439
4	.0864	.00991	20	1.59740	.40470
5	.10368	.05340	21	1.91688	.40103
6	.12442	.07114	22	2.30026	.64551
7	.14930	.15264	23	2.76031	.74189
8	.17916	.05348	24	3.31237	.36840
9	.21499	.11610	25	3.97484	.43530
10	.25799	.21675	26	4.76981	.34613
11	.30959	.45264	27	5.72377	.53979
12	.37150	.31023	28	6.86853	.17262
13	.44581	.68540	29	8.24223	.08895
14	.53497	.36253	30	9.89068	.09873
15	.64196	.54263	31	11.86881	.12281
16	.77035	.17219			

TABLE II

ENVELOPE FUNCTION

t (s)	F(t)	t (s)	F(t)
0.0	0.0	10.0	0.13
2.0	.3	14.0	.10
4.0	.1	15.0	.11
5.0	.2	16.0	.06
6.0	.05	27.0	.10
9.0	.05	30.0	.00

SEISMIC ANALYSIS OF A SPACE FRAME

A planar spaceframe is used to demonstrate the results that can be obtained using the above procedure. The frame is shown in oblique view in Figure 3. It is 84.3 ft long, 24 ft wide and 5 ft deep. It is supported along the top edge at the far ends. This is an actual roof structure and the trusses vary in cross-sectional area. There are six different areal values for the trusses which are given in Table III.

TABLE III

MATERIAL PROPERTY SETS FOR TRUSSES

set	area sq-in	density $lb\text{-}s^2/in^4$	E lb/sq-in
1	0.850	0.000734	30.x106
2	1.417	↓	↓
3	1.729		
4	3.334		
5	4.480		
6	6.280		

The finite element code ADINA[3] was used for the analysis of the structure. A top view of the model is shown in Figure 4 with the node numbers included. In the actual structure the trusses are held together by connectors that are very rigid and have significant mass. The mass effect of these connectors is accounted for by applying concentrated masses to the structure at the node points. There are 74 nodes in the model which correspond to the 74 connectors in the structure. Table IV gives the nodal masses for the connectors.

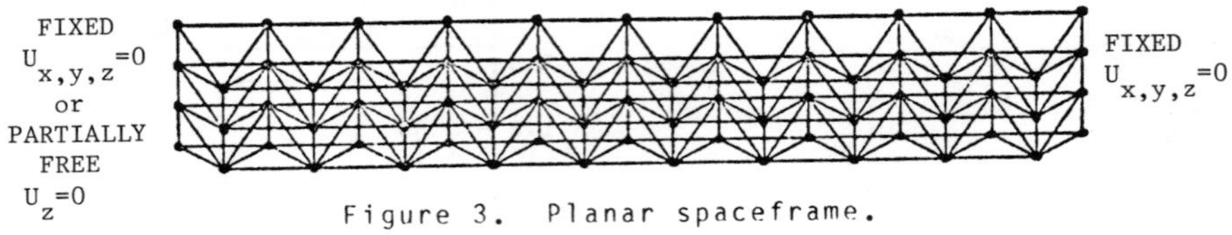

Figure 3. Planar spaceframe.

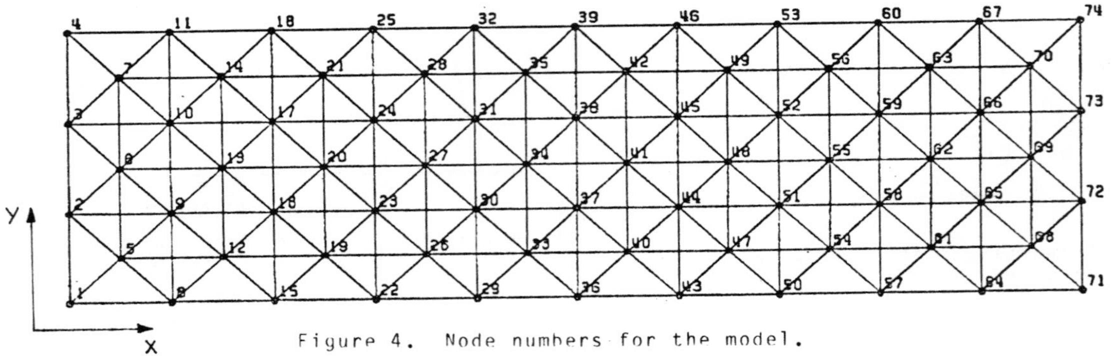

Figure 4. Node numbers for the model.

TABLE IV CONCENTRATED NODAL MASSES DUE TO CONNECTORS			
NODES	MASS $lb\text{-}s^2/in$	NODES	MASS $lb\text{-}s^2/in$
1 to 18	0.31	19 to 21	0.54
22 to 25	↓	27	↓
29 to 32		34	
36 to 39		41	
43 to 46		48	
50 to 53		54 to 56	
57 to 74		26,28,33,35, 40,42,47,49	0.87

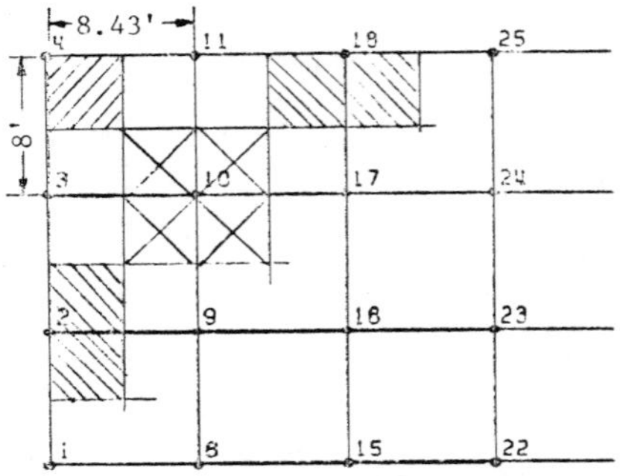

Figure 5. Roof load tributary areas.

In this structure the roof is attached to the frame using standoffs at each connector (node) on the top. The roof therefore causes no bending of the trusses. The roof is assumed to add no stiffness to the structure and therefore has only a mass effect. Any additional mass due to snow or water is added to the that of the roof. Figure 5 shows the tributary areas for typical nodes used to compute the added mass effect due to the roof and any additional load. This added mass is treated as a concentrated mass and is added to the concentrated mass due to the connectors. Table V gives the tributary areas, the loads and equivalent masses for the case of a 5 psf total roof load.

TABLE V Concentrated Mass for 5 psf Roof Load			
Node type	Area sq-ft	Load lbs	Mass $lb\text{-}s^2/in$
inner	67.44	337.	.873
side	33.72	169.	.436
corner	16.86	84.	.218

Because none of the trusses in the structure are subject to bending the truss element in ADINA was used to model the behavior of the structure. There are 240 truss elements in the model. Because of space limitations the data for all the elements cannot be reproduced here. A partial listing is given in Table VI. The material set number in Table VI refers to the corresponding set in Table III. Note that the data is not in the same format as input for ADINA. A complete listing is available from the author on request.

TABLE VI

Elem	Mat'l Set	Node 1	Node 2	Elem	Mat'l Set	Node 1	Node 2
1	1	1	2	2	2	1	5
3	1	1	8	4	1	2	3
5	1	2	5	6	2	2	6
7	1	2	9	8	1	3	4
9	2	3	6	10	1	3	7
.	.	.	.	.	.	.	.

The analysis is divided into two parts. The first part examines the free vibration of the frame for a range of roof loads from 5 to 20 lbs/sq-ft and for two sets of support conditions. For the first support condition, the top edges at the far ends are both constrained against movement in both the vertical and horizontal directions. For the second support condition one end is fully constrained and the other end is constrained in the vertical direction only. The second part examines the transient response of the frame using the derived seismic excitation function.

FREE VIBRATION ANALYSIS

Figure 6 shows the variation in frequency of the first 5 modes as the roof load increases from 0 to 20 lb/sq-ft. The first four modes are indicated on the seismic spectrum in Figure 1 for the cases of 0 and 20 lb/sq-ft loads. Note how the increased roof load causes the peak responses to shift to the left into a region where seismic damage is likely to occur. This would indicate that a full dynamic analysis should be performed to insure the safety of the structure. Table VII shows the effect on the natural frequencies of relaxing the support condition on one end for the case when the roof load is 20 lb/sq-ft. As should be expected the natural frequencies move down. Figure 7 shows the fourth mode when the top edges on both ends are fully constrained.

TABLE VII

Natural Frequencies (Hz)
Roof Load = 20 psf

Case 1: Both ends fully constrained on top edges
Case 2: Left end constrained in the vertical direction only right end fully constrained

Mode	Case 1	Case 2	Mode	Case 1	Case 2
1	3.08	1.58	2	3.99	3.22
3	7.16	3.59	4	7.93	7.41
5	9.60	9.29	6	12.72	9.50
7	15.34	12.72	8	17.09	12.98
9	18.22	16.20	10	19.20	18.20

SEISMIC RESPONSE

The direct time integration option using the Newmark beta method in ADINA was used to solve the problem. The excitation is applied to nodes 1,2,3,4,71,72,73 and 74. The application of an acceleration time history to the boundaries is very awkward with ADINA. It is far easier to apply a displacement time history. For this reason, the code that was written for generating the acceleration time history from the design spectrum also integrates that history to obtain the displacement time history shown in Figure 8.

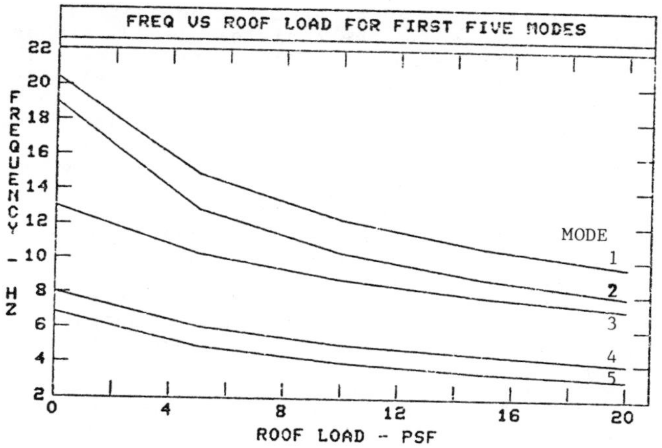

Figure 6. Frequency vs roof load when both ends are fixed.

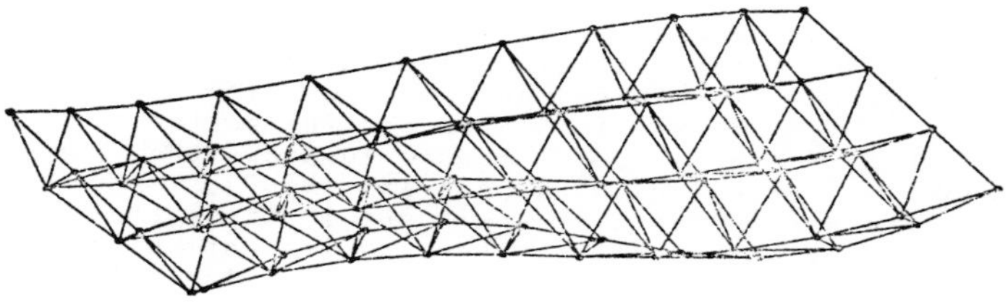

Figure 7. The fourth mode shape when both ends are fixed.

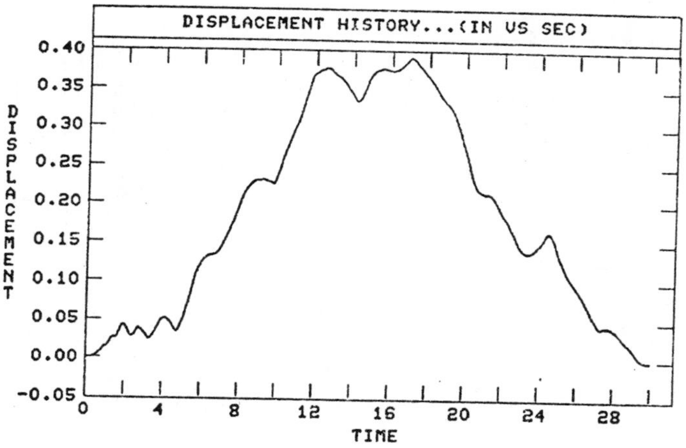

Figure 8. Applied displacement time history.

In the first case considered the boundaries are fixed at both ends, the roof load is 20 psf and the excitation is horizontal. Because there is such a wealth of computed data only the response of node 37, near the center of the roof on the top side, is considered. The excitation is applied by releasing the constraint in the y-direction on the supporting edge nodes and forcing them to follow the excitation. The structure is so rigid in the plane of the excitation that it follows the excitation very closely. In the vertical direction, however, there is deformation. Figure 9 shows the transient part of the relative vertical displacement at node 37 due to the horizontal excitation. (The relative displacement is the difference between the displacement at node 37 and the applied boundary motion and represents the deformation only without rigid body motion.) Clearly, some motion is induced but its magnitude is small, only 0.003 in. When superimposed on the static deflection due to the 20 psf roof load, which is about 0.77 in, it is not significant. The same excitation is applied in the vertical direction. Strictly speaking, the vertical excitation in a seismic event will be different from the horizontal excitation. However, the character will be similar, so for purposes of characterization this approach is acceptable. In this case the constraint in the z-direction is released at the supporting edge nodes which are then forced to follow the vertical excitation. Figure 10 shows the relative vertical displacement at node 37. Note that the displacement is now about 10 times that of the previous case. Further, it is continuing to grow at 10 secs into the event which indicates that a resonance is being excited. Still, the transient response is small compared to the static deflection due to the roof load.

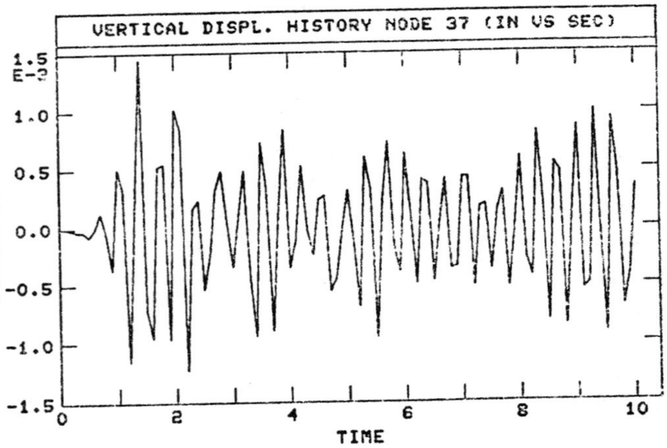

Figure 9. Relative vertical response of node 37 due to a horizontal excitation with a 20 psf roof load.

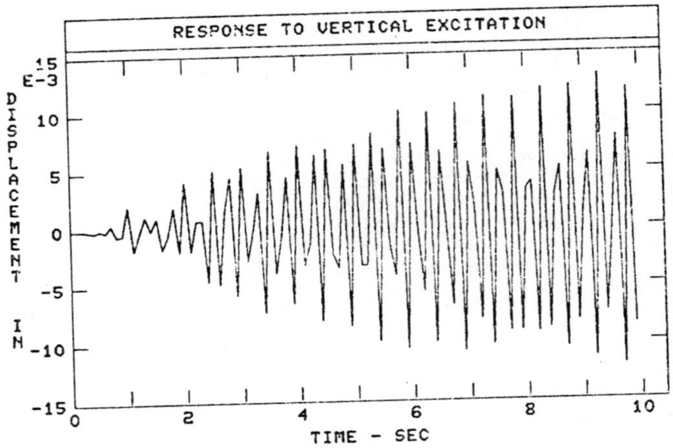

Figure 10. Relative vertical response of node 37 due to a vertical excitation with a 20 psf roof load.

Figure 11 shows the relative vertical response when the roof load is reduced to 10 psf. The character of the response is quite different from the previous case. The magnitude of the deformation is about 1/4 that of the previous case. This demonstrates that increasing roof load can shift the structure's natural frequencies into the frequency range where the seismic excitation has more energy.

The last case considered is the same as the one just discussed except that the constraints on the left supporting edge are all released. Because the natural frequencies for this case are lower than for the case when both ends are constrained it is to be expected that the response in this last case will be larger. This is born out by Figure 12 where the response is 3 times that of Figure 11.

DISCUSSION

The procedure demonstrated here started with a seismic spectrum from which a ground acceleration time history was constructed that would reproduce the given spectrum. A free vibration analysis of the spaceframe was then made for different roof loads and end conditions. It was demonstrated that such modifications could change the natural frequencies in such a way as to cause concern for its reliability during a seismic event. A full dynamic analysis of the structure was then performed for several of the cases examined in the free vibration study. The results supported the conclusions from the free vibration study that modifications that reduced natural frequencies would increase deformations. In general then, a free vibration study of a structure should be performed first. If there is any indication that the structure stands the risk of seismic damage a dynamic analysis should be performed. A typical dynamic analysis run took 2.5 cpu minutes on an IBM 3033 computer.

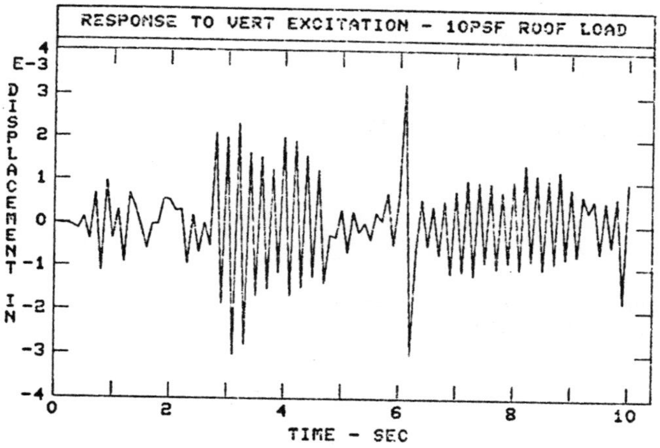

Figure 11. Relative vertical response of node 37 due to a vertical excitation with a 10 psf roof load.

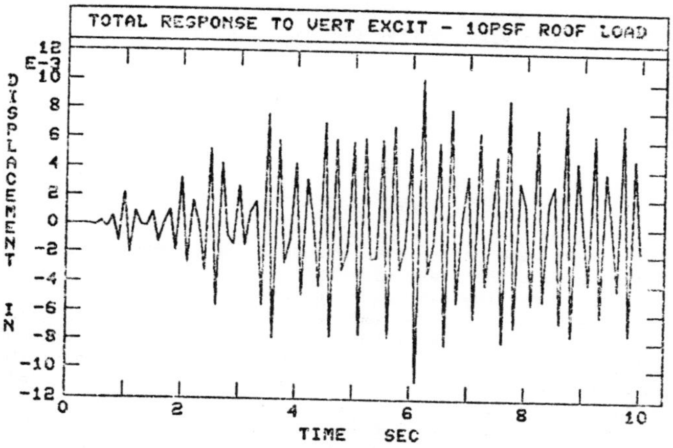

Figure 12. Relative vertical response of node 37 due to a vertical excitation with a 10 psf roof load and with no constraints on the left end.

ACKNOWLEDGEMENT

This work was performed at GTE Laboratories at the request of Mr. Sun Chien Hsiao of the UNISTRUT Building Systems Division of GTE. His technical support and encouragement during the course of the study are gratefully acknowledged.

REFERENCES

1. N.M. Newmark and E. Rosenbleuth, FUNDAMENTALS OF EARTHQUAKE ENGINEERING, Prentic-Hall, Englewood Cliffs, N.J., 1971.

2. S. Levy and J.P.D. Wilkinson, "Generation of Artificial Time-Histories, Rich in all Frequencies, from Given Response Spectra", Nuclear Engineering and Design 38(1976)241-251.

3. K.J. Bathe, "ADINA - A Finite Element Program for Automatic Dynamic Incremental Nonlinear Analysis", Report AVL 82448-1, MIT, 1975 (Rev 1978).

THE EFFECTIVE STRENGTH OF RIGIDLY CONNECTED LATTICE STRUCTURES

Koichiro HEKI, Dr. Eng.

Professor, Department of Architecture and Building Engineering,
Faculty of Engineering, Osaka City Uinversity, Japan.

For the prediction of strength of regular lattice structures of repetitive pattern constructed with rigidly connected congruent truss segments, the theoretical method to derive the effective strength in continuum expression determined by the elastic member buckling is shown, assuming that the buckling occurs as bifurcation from uniform state into repetitive modes. For examples, the effective strength of a latticed beam comparing with the exact solution for the whole beam, and those of two types of double-layer lattice plates are shown.

INTRODUCTION

For the analysis of lattice structures composed of a large number of structural units, the continuum treatment is one of the useful methods especially for the preliminary design. Using the effective rigidity, often called the equivaent rigidity, the stress state in macro point of view and the stability and vibration characteristics of the whole structure can be estimated. And, many papers on the effective rigidities of lattice structures have been presented, see Refs 1 through 6. The effective rigidities of some space structures will be summarised on the appendix of the IASS Recomendation for Spatial Steel Structures, see Ref 7.

To estimate the safety of lattice structures, excepting the stability of the whole structure or latticed members which may be examined by using the effective rigidity, usually the next two points are studied. First, the stress in every constituent member is referred to the allowable stress including the buckling effect assuming the joints are pin of no dimensions. Secondly, the strength of connections are checked.

But, for these procedure it is necessary to derive the stress state on every constituent member from the macro stress on latticed members, and to check every joint. Also usually the connections of latticed structures are not pin-jointed but rigidly or semi-rigidly jointed. Accordingly the actual strength is higher in many cases and lower in some cases than that calculated as ideal pin-jointed trusses.

So, the present author has proposed the concept of "the effective strength of latticed members" in terms of macro sectional forces, such as stress resultants and stress couples of continuum members. This effective strength may include all the effects of such items as the strength and ridigity of joints, the strength of members including the effects of imperfections, and the coupling effects between joints and members.

To know the effective strength of actual lattice structures, the experimental method would be best, especially for the latticed segments of mass production. As for the theoretical derivation of the effective strength, the method to derive it has been presented by the present author for simple cases which allow to assume that the strength of every constituent member is determined solely by the stress state of the member itself, see Ref 8. These assumptions can be applicable for the cases that the truss is pin-jointed or the strength is determined without the effects of joints under the condition that the members do not buckle and the joints are stronger than the members.

But, most of actual lattice structures are constructed by rigid or partially rigid connections of finite dimensions, for examples, by welding or bolting with balls or gusset plates. For such cases, the buckling strength of the constituent members depends not only on the stress state and the rigidity of the considering member but also on the rigidity, the connecting angles, and the stress states of the adjoining members which connect to the joints at both ends of the considering member.

So, this paper is focussed on the method to derive the effective strength determined by the elastic buckling which occurs as bifurcation from uniform state for regular lattice structures constructed with rigidly connected congruent latticed truss segments in repetitive pattern. And examples are shown for explanation of the method and for proof of the validity of this effective strength.

The theory for rigidly connected trusses may contain that of pin or elastic connection with connecting part of finite dimensions, because such trusses may be understand as the assembly of the composite struts which are one elastic prismatic bar with two rigid end parts connected by elastic springs. The pin-jointed truss is the case that every member has two springs of no rigidity without rigid regions at both the ends. For such cases, a separate paper will be presented at the 3rd International Conference on Space Structures under the title of "The Effects of Joints on the Strength of Space Trusses", see Ref 9.

ASSUMPTIONS

The assumptions used for the derivation of effective strength in this paper are as follows.

The object of research is the regular lattice structures constructed with congruent elastic lattice segments in the form of repetitive pattern of no imperfections. The additional assumption for the examples, not essential but for simplicity of formulae, is that the lattice structures are rigidly connected trusses and whose members are straight and of uniform section, neglecting the shearing and warping deformations.

The effective strength of lattice structures treated here is restricted to that determined by the elastic bifurcation buckling of constituent members in the repetitive modes from uniform prebuckling states.

FUNDAMENTAL DIFFERENCE EQUATIONS

The relation between the deformation of constituent members in one structural unit and the displacement of the joints is expressed as Eqn 1, where the displacements in the neighbouring structural units are expressed in terms of that of the considering unit using the shift operator.

$$\{d\} = [A]\{u\} \qquad \text{........1,}$$

where {u} and {d} are respectively the displacement vector and the deformation vector of one structural unit, and the matrix [A] is determined by the geometrical configuration of the structural unit as Eqn 2.

$$[A] = \sum_{\ell mn}\sum\sum[A_{\ell mn}]E_1{}^{\ell}E_2{}^{m}E_3{}^{n} \qquad \text{........2,}$$

where $[A_{\ell mn}]$'s are matrices of constant components and E_i denotes the shift operator in the i-direction. The number of summation operators corresponds to the number of directions wherein the adjoining units are connected, and is called as the degree of dimension of the lattice structure. For example, one dimensional structures are latticed beams and columns, two dimensional ones include lattice shells and plates, and the third corresponds to the three dimensionally expanded structures like junglegyms.

The equilibrium equations for the stresses and loads on one structural unit are represented in the form of Eqn 3 including the shift operator as similar as Eqn 1.

$$[B]\{N\} = \{P\} \qquad \text{........3,}$$

where {N} and {P} are the stress vector and the load vector correspoding to the deformation and displacement, respectively. The term "correspond" means that the scalar product of two corresponding quantities becomes the work.

The relation between two matrices [A] and [B] can be derived as follows. The principle of virtual work is applied for the virtual displacements satisfying the geometrical boundary condition as Eqn 4.

$$\sum_{stv}\sum\sum\{E_1{}^{s}E_2{}^{t}E_3{}^{v}\{u\}\}^T\{E_1{}^{s}E_2{}^{t}E_3{}^{v}\{P\}\}$$
$$-\sum_{pqr}\sum\sum\{E_1{}^{p}E_2{}^{q}E_3{}^{r}\{d\}\}^T\{E_1{}^{p}E_2{}^{q}E_3{}^{r}\{N\}\} = 0 \qquad \text{.........4,}$$

where, s, t, v and p, q, r are the discrete coordinates to denote the position of a structural unit. Consider the displacements which are nonzero on only one structural unit at s=t=v=0 of no contact with the boundary, the domain of p, q, and r is restricted as Eqn 5, because the components of deformation {d} defined in Eqn 1 are nonzero only for the term of {u} on the (0, 0, 0) unit.

$$p = -\ell, \qquad q = -m, \qquad r = -n \qquad \text{........5.}$$

Consequently, Eqn 4 becomes as Eqn 6.

$$\{u\}^T\{P\} = \sum_{\ell mn}\sum\sum\{E_1^{-\ell}E_2^{-m}E_3^{-n}[A_{\ell mn}]E_1{}^{\ell}E_2{}^{m}E_3{}^{n}\{u\}\}^T \cdot\{E_1^{-\ell}E_2^{-m}E_3^{-n}\{N\}\}$$
$$= \{u\}^T\{\sum_{\ell mn}\sum\sum[A_{\ell mn}]^T E_1^{-\ell}E_2^{-m}E_3^{-n}\{N\}\} \qquad \text{.....6.}$$

Dividing both sides of Eqn 6 by nonzero displacement {u}, and comparing with Eqn 3, we have the relation of Eqn 7.

$$[B] = \sum_{stv}\sum\sum[B_{stv}]E_1{}^{s}E_2{}^{t}E_3{}^{v} = \sum_{\ell mn}\sum\sum[A_{\ell mn}]^T E_1^{-\ell}E_2^{-m}E_3^{-n} \qquad \text{...7.}$$

This equation leads to the statement of a theorem: "the matrix of the equilibrium equation for lattice structures in difference expression is the transposed matrix of that of the relation between deformations and displacements by changing the sign of the power of the shift operators, and vise-versa". This theorem is applicable not only for the infinitesimal displacement state but also for the finite displacement state replacing the displacements and deformations by those increments.

The constitutive equation has the form of Eqn 8, where the effects of instability are considered.

$$\{N\} = [C]\{d\} \qquad \text{........8,}$$

The constitutive equations for a strut used in the following examples take the effect of axial force on the bending into consideration. But for simplicity some higher order terms, such as bowing effects and twisting buckling, are neglected. And the so called stability functions of Livesly and Chandler are used, Table 1.

Table 1 Stability functions

T	T > 0	0	T < 0
Z	$\sqrt{\frac{T\ell^2}{EI}}$	0	$\sqrt{-\frac{T\ell^2}{EI}}$
α	$\frac{Z\sinh Z - Z^2\cosh Z}{2\cosh Z - 2 - Z\sinh Z}$	4	$\frac{Z\sin Z - Z^2\cos Z}{2 - 2\cos Z - Z\sin Z}$
β	$\frac{Z^2 - Z\sinh Z}{2\cosh Z - 2 - Z\sinh Z}$	2	$\frac{Z^2 - Z\sin Z}{2 - 2\cos Z - Z\sin Z}$

The buckling difference equation for the bifurcation displacement {u} from uniform states becomes as Eqn 9, unless buckling occurs without {u}.

$$[B][C][A]\{u\} = \{0\} \qquad \text{.....9.}$$

This equation can be rewritten as Eqn 10.

$$\sum_{\ell mn}\sum\sum[K_{\ell mn}]E_1{}^{\ell}E_2{}^{m}E_3{}^{n}\{u\} = \{0\} \qquad \text{.......10,}$$

where, $[K_{\ell mn}] = [A_{pqr}]^T[C][A_{efg}]E_1^{e-p}E_2^{f-q}E_3^{g-r}$

$$\ell = e - p, \qquad m = f - q, \qquad n = g - r. \qquad \text{.......11,}$$

When joint rotations are taken as bifurcation displacement parameters for the truss, the Eqn 10 is valid under the condition that every axial force never reach the four times of the buckling load of the pin-jointed strut. Otherwise the strut buckles without end rotations.

BUCKLING EQUATIONS FOR REPETITIVE MODES

The repetitivity of buckling modes are represented generally in the form of Eqn 12, for the case of repetitive spans of p, q, and r in E_1, E_2, and E_3 directions, respectively, by Fourier expansion.

$$\{u(s,\ t,\ v)\} = \sum\sum\sum\{\{C_{jkh}\}\cos\eta - \{S_{jkh}\}\sin\eta\} \qquad \text{...12,}$$

where, $\eta = 2\pi(js/p + kt/q + hv/r)$ 13,

p, q, r: integer, indicating the repetitive span in E_1, E_2, and E_3 directions, respectively,
j, k, h: integer from 1 to p, q and r, respectively,

s, t, v: integer discrete coordinates,

$\{C_{jkh}\}$, $\{S_{jkh}\}$: arbitrary vectors of b components.

Introducing Eqn 12 into Eqn 9, we have the buckling equation 14, where the boundary conditions are implicitly assumed to be compatible with the repetitive mode.

$$\begin{bmatrix}[K_c] & -[K_s]\\ [K_s] & [K_c]\end{bmatrix}\begin{Bmatrix}\{C_{jkh}\}\\ \{S_{jkh}\}\end{Bmatrix} = \begin{Bmatrix}\{0\}\\ \{0\}\end{Bmatrix} \qquad ......14,$$

where,

$$\left.\begin{aligned}[K_c] &= \sum_{\ell}\sum_{m}\sum_{n}[K_{\ell mn}]\cos 2\pi(j\ell/p + km/q + hn/r)\\ [K_s] &= \sum_{\ell}\sum_{m}\sum_{n}[K_{\ell mn}]\sin 2\pi(j\ell/p + km/q + hn/r)\end{aligned}\right\} \qquad ......15.$$

EXAMPLE 1. PARALLEL CHORD LATTCE BEAMS

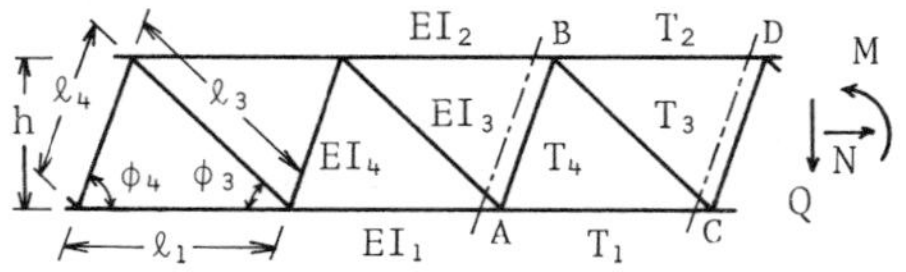

Fig 1 Parallel chord lattice beam and its stresses

A parallel chord lattice beam with V-type webs as shown in Fig 1 is treated as an example to derive its effective strength in terms of axial force N, bending moment M, and shear force Q. This effective strength is assumed to be determined by the elastic buckling in its own plane from uniform state, which means the stress states in all the structural units are equal. The bending rigidity EI_i's and axial force T_i's are denoted as shown in Fig 1. The macro stress, i.e. sectional forces, N, M, and Q, are represented in terms of the axial forces of constituent members T_i as Eqn 16.

$$\left.\begin{aligned}N &= T_1 + T_2\\ M &= (T_1 - T_2)h/2\\ Q &= T_3\sin\phi_3 = -T_4\sin\phi_4\end{aligned}\right\} \qquad ....16.$$

As for the bifurcation displacements the angular displacements of joints θ_A and θ_B are used. Then the equilibrium equations are established for joint rotations as Eqn 17. The relations between the deformation {d} and displacement $[\theta_A\ \theta_B]^T$ can be derived using the above mentioned theorm as Eqn 18.

$$\begin{bmatrix}1 & E^{-1} & 0 & 0 & 0 & E^{-1} & 1 & 0\\ 0 & 0 & 1 & E^{-1} & 1 & 0 & 0 & 1\end{bmatrix}\{N\} = \{0\} \qquad ....17,$$

$$\begin{bmatrix}1 & E & 0 & 0 & 0 & E & 1 & 0\\ 0 & 0 & 1 & E & 1 & 0 & 0 & 1\end{bmatrix}^T\begin{Bmatrix}\theta_A\\ \theta_B\end{Bmatrix} = \{d\} \qquad ....18,$$

where,

$$\{N\} = [M_{AC}\ M_{CA}\ M_{BD}\ M_{DB}\ M_{BC}\ M_{CB}\ M_{AB}\ M_{BA}]^T \qquad ..19,$$

$$\{d\} = [d_{AC}\ d_{CA}\ d_{BD}\ d_{DB}\ d_{BC}\ d_{CB}\ d_{AB}\ d_{BA}]^T \qquad ..20.$$

The elastic matrix for buckling is Eqn 21.

$$[C] = \begin{bmatrix}[C_1] & [0] & [0] & [0]\\ [0] & [C_2] & [0] & [0]\\ [0] & [0] & [C_3] & [0]\\ [0] & [0] & [0] & [C_4]\end{bmatrix} \qquad ....21,$$

where, $$[C_i] = \frac{EI_i}{\ell_i}\begin{bmatrix}\alpha_i & \beta_i\\ \beta_i & \alpha_i\end{bmatrix} \qquad ....22.$$

Introducing Eqn 17 through 22 into Eqn 9, the buckling differnce equation is as Eqn 23.

$$\begin{bmatrix}k_1[2\alpha_1+(E+E^{-1})\beta_1]+k_3\alpha_3+k_4\alpha_4 & k_3E^{-1}\beta_3+k_4\beta_4\\ k_3E\beta_3+k_4\beta_4 & k_2[2\alpha_2+(E+E^{-1})\beta_2]+k_3\alpha_3+k_4\alpha_4\end{bmatrix}\{u\} = \{0\} \qquad ....23,$$

where, $$k_i = (EI_i/\ell_i)/(EI/\ell) \qquad ....24,$$

$$\{u\} = [\theta_A\ \ \theta_B]^T \qquad ....25.$$

For the repetitive mode of p=1, i.e. uniform buckling mode, the condition of repetitivity is Eqn 26. Introducing the solution of Eqn 26 into Eqn 23, the buckling equation becomes as Eqn 27.

$$(E - 1)\{u\} = \{0\} \qquad ....26,$$

$$\begin{bmatrix}2k_1(\alpha_1+\beta_1)+k_3\alpha_3+k_4\alpha_4 & k_3\beta_3+k_4\beta_4\\ k_3\beta_3+k_4\beta_4 & 2k_2(\alpha_2+\beta_2)+k_3\alpha_3+k_4\alpha_4\end{bmatrix}\begin{Bmatrix}\theta_A\\ \theta_B\end{Bmatrix} = \begin{Bmatrix}0\\ 0\end{Bmatrix} \qquad ...27.$$

For the repetitive modes of p=2, the condition is Eqn 28, and its solutions are Eqn 29.

$$(E^2 - 1)\{u\} = \{0\} \qquad ....28,$$

$$\left.\begin{aligned}\{u\} &= \{u_0\}1^s\\ \{u\} &= \{u_0\}(-1)^s\end{aligned}\right\} \qquad ....29.$$

The first one is the same to the case of p=1, so the buckling equation for the second case is shown as Eqn 30.

$$\begin{bmatrix}2k_1(\alpha_1-\beta_1)+k_3\alpha_3+k_4\alpha_4 & k_4\beta_4-k_3\beta_3\\ k_4\beta_4-k_3\beta_3 & 2k_2(\alpha_2-\beta_2)+k_3\alpha_3+k_4\alpha_4\end{bmatrix}\begin{Bmatrix}\theta_A\\ \theta_B\end{Bmatrix} = \begin{Bmatrix}0\\ 0\end{Bmatrix} \qquad ..30.$$

The eigenvalue of Eqn 30 is smaller than that of Eqn 27. The numerical result for the truss of $\phi_3=\phi_4=\pi/3$ and k_i=constant is shown in Fig 2 as the strength surface, which is non-dimensionalized with the value of pin-jointed ones. This figure shows a quarter, because the strength surface is symmetric with respect to both the planes of the N-M plane and the N-Q plane.

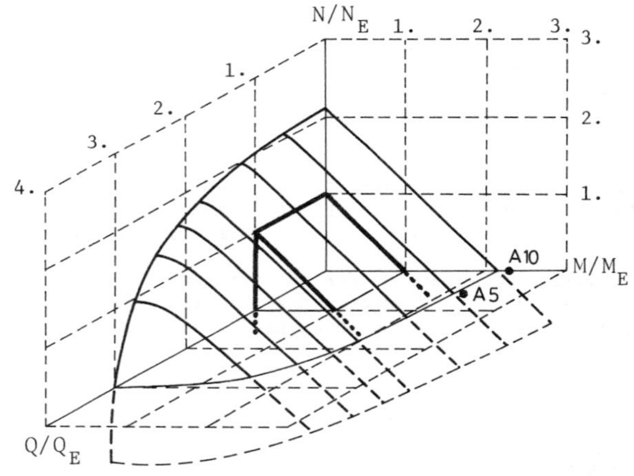

Fig 2 Strength surface of a parallel chord lattice beam

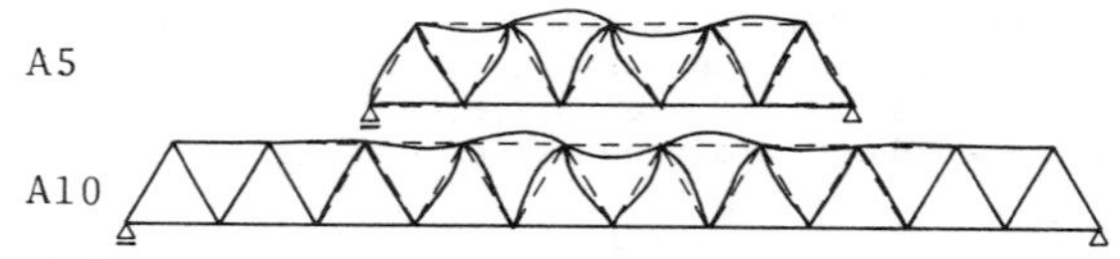

Fig 3 Parallel chord lattice beam of V-type webs

The inner thick lines show the strength surface of pin-jointed truss of the same members.

The comparison is made for the effective strength with the strength at the critical load of simply supported trusses of the same structural units loaded uniformly at the upper joints as shown in Fig 3. The maximum bending moments at the critical loads are plotted on Fig 2, showing slightly greater than the effective strength. This reason is clear, because the effective strength corresponds to the buckling moment for uniform state is smaller than that for the non-uniform state.

EXAMPLE 2. THREE WAY DOUBLE-LAYER LATTICE PLATES

The three way double-layer lattice plate as shown in Fig 4 is internally stable and its effective rigidity for stretching, bending, shearing, and their coupling effects has been published, see Ref 6. This truss is isotropic for bending and stretching rigidities. The effective strength for bending and twisting is derived here under the condition that in-plane forces and out-of-plane shear forces are not acting.

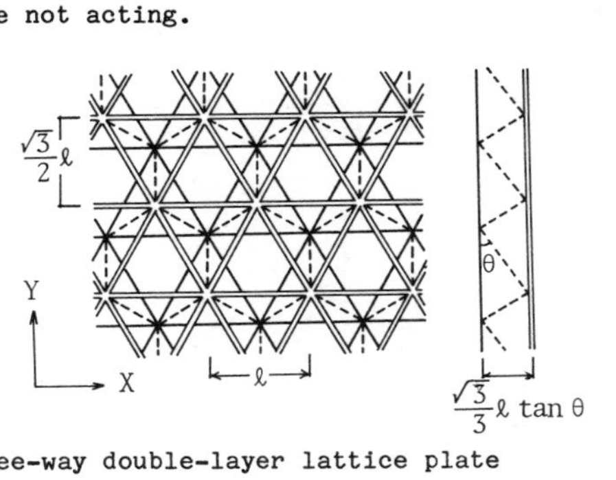

Fig 4 Three-way double-layer lattice plate

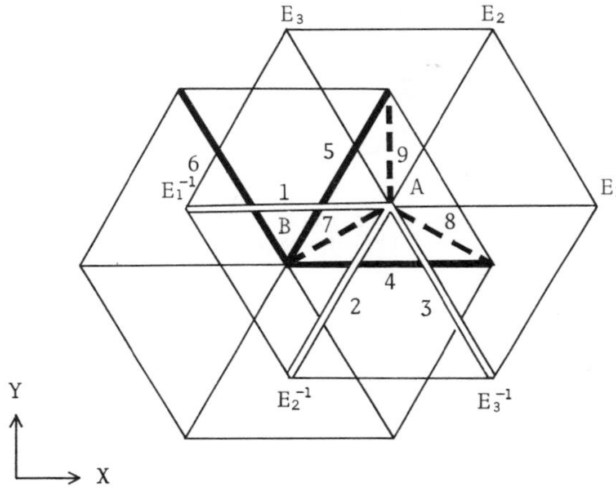

Fig 5 One structural unit of the lattice plate in Fig 4

The structural unit is defined in the form as shown in Fig 5. In this case the buckling modes represented by the joint rotations are important, so the angular displacement vector $\{\theta\}$ of six components as Eqn 31 on two joits in upper and lower surfaces are considered.

$$\{\theta\} = [\theta_{Ax} \ \ \theta_{Ay} \ \ \theta_{Az} \ \ \theta_{Bx} \ \ \theta_{By} \ \ \theta_{Bz}]^T \qquad ....31.$$

As for stresses or deformations, five components of bendings and twisting on every constituent strut are considered. For one structural unit of nine strut, the number of stress or deformation components is 45.

The constituent members are assumed to be uniform section of rotational symmetry, so the elastic matrix for the stress $\{N_i\}$ and deformation $\{d_i\}$ of every member becomes as Eqn 33, where all the surface members are equal, and the web members are equal too, respectively.

$$\left.\begin{aligned}\{d_i\} &= [d_t \ \ d_{Ax} \ \ d_{Bx} \ \ d_{Ay} \ \ d_{By}\ ]^T,\\ \{N_i\} &= [M_t \ \ M_{ABx} \ M_{BAx} \ M_{ABy} \ M_{BAy}]^T\end{aligned}\right\} \qquad ....32,$$

$$[C_i] = \frac{EI_i}{\ell_i}\begin{bmatrix} 1/(1+\nu) & 0 & 0 & 0 & 0\\ 0 & \alpha_i & \beta_i & 0 & 0\\ 0 & \beta_i & \alpha_i & 0 & 0\\ 0 & 0 & 0 & \alpha_i & \beta_i\\ 0 & 0 & 0 & \beta_i & \alpha_i\end{bmatrix} \qquad ....33.$$

By similar procedure as Example 1, we have the buckling difference equation as Eqn 34. Where, $E_3 = E_1^{-1}E_2$.

$$[K]\{\theta\} = \{0\} \qquad ....34,$$

where, the components of [K] are as follows, considering the equality of surface or web members, and the stress state of no axial forces on the webs.

$$\frac{EI_i}{\ell_i} = \begin{cases} K_c & \text{(for } i = 1, 2, \ldots, 6)\\ K_w & \text{(for } i = 7, 8, 9)\end{cases} \qquad ....35,$$

$$1/(1+\nu_i) = \mu \quad \text{(for } i = 1, 2, \ldots, 9)$$

$$\alpha_w = 4, \quad \beta_w = 2 \qquad ....36,$$

$$K_{11} = (K_c/4)[3\{2\alpha_2+\beta_2(E_2+E_2^{-1})+2\alpha_3+\beta_3(E_3+E_3^{-1})\} +\mu(12-4E_1-4E_1^{-1}-E_2-E_2^{-1}-E_3-E_3^{-1})] +(3K_w/2)(\mu\cos^2\theta+4+4\sin^2\theta),$$

$$K_{12} = K_{21} = (\sqrt{3}K_c/4)[-2\alpha_2-\beta_2(E_2+E_2^{-1})+2\alpha_3+\beta_3(E_3+E_3^{-1}) +\mu(E_3+E_3^{-1}-E_2-E_2^{-1})],$$

$$K_{13} = K_{31} = K_{46} = K_{64} = K_{23} = K_{32} = K_{56} = K_{65} = 0,$$

$$K_{22} = (K_c/4)[8\alpha_1+4\beta_1(E_1+E_1^{-1})+2\alpha_2+\beta_2(E_2+E_2^{-1})+2\alpha_3 +\beta_3(E_3+E_3^{-1})+3\mu(4-E_2-E_2^{-1}-E_3-E_3^{-1})] +(3K_w/2)(\mu\cos^2\theta+4\sin^2\theta+4),$$

$$K_{33} = K_c[2\alpha_1+\beta_1(E_1+E_1^{-1})+2\alpha_2+\beta_2(E_2+E_2^{-1})+2\alpha_3+\beta_3(E_3+E_3^{-1})] +3K_w(\mu\sin^2\theta+4\cos^2\theta),$$

$$K_{14} = (K_w/4)[(2+6\sin^2\theta-3\mu\cos^2\theta)(1+E_1)+6E_2],$$

$$K_{15} = K_{24} = -\ 3(K_w/4)\cos^2\theta(2+\mu)(1-E_1),$$

$$K_{16} = K_{34} = -\ 3(K_w/2)\sin\theta\cos\theta(2+\mu)(1-E_1),$$

$$K_{25} = (K_w/4)[(2\sin^2\theta-\mu\cos^2\theta)(1+E_1+4E_2)+6(1+E_1)],$$

$$K_{26} = K_{35} = -(K_w/2)\sin\theta\cos\theta(2+\mu)(1+E_1-2E_2),$$

$$K_{36} = K_w(2\cos^2\theta-\mu\sin^2\theta)(1+E_1+E_2) \qquad ....37.$$

The other K_{ij}'s are calculated by following way. Kpq's (p, q = 4, 5, 6) are those replaced the suffices i and j on α, β and K in the formulae for K_{ij}'s (i, j = 1, 2, 3) by i+3 and j+3, respectively. K_{ij}'s (i+j = 4, 5, 6) are those changed the sign of the power of the shift operator E_i in K_{ji}'s (i+j = 4, 5, 6).

For some modes of small buckilng mode units, the buckling equations for the case of $\ell_w=\ell$, and the numerical result for the truss of all equal members are shown as follows.

i) The modes whose pattern unit appears on two structural units are the next three cases, excepting the uniform mode, as Eqns 38-40.

$$(E_1-1)\{u\} = \{0\}, \qquad (E_2+1)\{u\} = \{0\} \quad \rightarrow \eta = \pi t \qquad ...38,$$

$$(E_1-1)\{u\} = \{0\}, \qquad (E_2-1)\{u\} = \{0\} \quad \rightarrow \eta = \pi s \qquad ...39,$$

$$(E_1+1)\{u\} = \{0\}, \qquad (E_2+1)\{u\} = \{0\} \quad \rightarrow \eta = \pi(s+t) \ ...40,$$

where, s and t are coordinates in E_1- and E_2-direction, respectively. These are cyclic in 120° rotation, so the first case is shown in Eqn 41.

$$\begin{bmatrix} A_1 & B_1 & 0 & F_1 & 0 & 0\\ & A_2 & 0 & 0 & F_2 & G\\ & & A_3 & 0 & G & F_3\\ & & & A_4 & B_4 & 0\\ & & & & A_5 & 0\\ \text{Symmetry} & & & & & A_6\end{bmatrix}\begin{Bmatrix}\theta_{Ax}\\ \theta_{Ay}\\ \theta_{Az}\\ \theta_{Bx}\\ \theta_{By}\\ \theta_{Bz}\end{Bmatrix} = \begin{Bmatrix}0\\0\\0\\0\\0\\0\end{Bmatrix} \qquad ......41,$$

where,

$$A_1 = K_c[2\mu+(3/2)(\alpha_2-\beta_2+\alpha_3-\beta_3)]+K_w(10+\mu/2),$$

$$A_2 = K_c[6\mu+2\alpha_1+2\beta_1+(\alpha_2-\beta_2+\alpha_3-\beta_3)/2]+K_w(10+\mu/2),$$

$$A_3 = K_c(\alpha_1+\beta_1+\alpha_2-\beta_2+\alpha_3-\beta_3)+K_w(4+2\mu),$$

$$B_1 = (\sqrt{3}/2)K_c(-\alpha_2+\beta_2+\alpha_3-\beta_3),$$

$$F_1 = (K_w/2)(2-\mu),$$

$$F_2 = (K_w/6)(14+\mu),$$

$$F_3 = (K_w/3)(2-2\mu),$$

$$G = -(2\sqrt{2}/3)K_w(2+\mu) \qquad \ldots\ldots\ldots 42.$$

A_4, A_5, A_6, and B_4 are got by adding 3 to the subscripts in A_1, A_2, A_3, and B_1, respectively.

ii) The modes whose pattern unit appears on three structural units are next four couples, excepting the uniform mode, as Eqns 43-46.

$$\eta = (2\pi/3)s, \qquad \eta = (2\pi/3)(2s); \qquad \ldots\ldots.43,$$

$$\eta = (2\pi/3)t, \qquad \eta = (2\pi/3)(2t); \qquad \ldots\ldots.44,$$

$$\eta = (2\pi/3)(s+t), \qquad \eta = (2\pi/3)(2s+2t); \qquad \ldots\ldots.45,$$

$$\eta = (2\pi/3)(s+2t), \qquad \eta = (2\pi/3)(2s+t) \qquad \ldots\ldots.46.$$

The three coupling modes of Eqns 43-45 are cyclic by 120° rotation and give higher buckling loads, so the buckling equation for the last coupling modes are shown as Eqn 47.

$$\begin{bmatrix} A_1 & B_1 & 0 & F & C & D & 0 & 0 & 0 & -C & F & G \\ & A_2 & 0 & C & -F & G & 0 & 0 & 0 & F & C & -D \\ & & A_3 & D & G & 0 & 0 & 0 & 0 & G & -D & 0 \\ & & & A_4 & B_4 & 0 & C & -F & -G & 0 & 0 & 0 \\ & & & & A_5 & 0 & -F & -C & D & 0 & 0 & 0 \\ & & & & & A_6 & -G & D & 0 & 0 & 0 & 0 \\ & & & & & & A_1 & B_1 & 0 & F & C & D \\ & & & & & & & A_2 & 0 & C & -F & G \\ & & & & & & & & A_3 & D & G & 0 \\ & & & & & & & & & A_4 & B_4 & 0 \\ & \text{Symmetry} & & & & & & & & & A_5 & 0 \\ & & & & & & & & & & & A_6 \end{bmatrix} \begin{Bmatrix} C_{Ax} \\ C_{Ay} \\ C_{Az} \\ C_{Bx} \\ C_{By} \\ C_{Bz} \\ S_{Ax} \\ S_{Ay} \\ S_{Az} \\ S_{Bx} \\ S_{By} \\ S_{Bz} \end{Bmatrix} = \begin{Bmatrix} 0 \\ 0 \\ 0 \\ 0 \\ 0 \\ 0 \\ 0 \\ 0 \\ 0 \\ 0 \\ 0 \\ 0 \end{Bmatrix} \qquad \ldots\ldots.47.$$

where,

$$A_1 = (K_c/4)(18\mu+6\alpha_2-3\beta_2+6\alpha_3-3\beta_3)+K_w(10+\mu/2),$$

$$A_2 = (K_c/4)(18\mu+8\alpha_1-4\beta_1+2\alpha_2-\beta_2+2\alpha_3-\beta_3)+K_w(10+\mu/2),$$

$$A_3 = K_c(2\alpha_1-\beta_1+2\alpha_2-\beta_2+2\alpha_3-\beta_3)+K_w(4+2\mu),$$

$$B_1 = (\sqrt{3}/4)K_c(-2\alpha_2+\beta_2+2\alpha_3-\beta_3),$$

$$C = -(\sqrt{3}/8)K_w(2+\mu),$$

$$D = -(\sqrt{6}/4)K_w(2+\mu),$$

$$F = -(1/8)K_w(2+\mu),$$

$$G = -(\sqrt{2}/4)K_w(2+\mu) \qquad \ldots\ldots.48.$$

A_4, A_5, A_6, and B_4 are got by adding 3 to the subscripts in A_1, A_2, A_3, and B_1, respectively.

The effective strength of the three way double-layer lattice plate as shown in Fig 4 constructed with the same rigid-jointed constituent members of rotationally symmetric section is reprsented as a surface in M_x-M_y-M_{xy} space as shown in Fig 6, where M's are shown in terms of the bending moments of equivalent continuum plates. This strength surface is made with 8 surfaces determined from the modes given by Eqns 38, 39, 40, and 46.

In Fig 6-8, three cases are compared in the same scale. Fig 6 shows the strength of rigid-jointed truss of the same members. Fig 7 shows that of the rigid-jointed truss with very small web rigidity or with pin-jointed web members. Fig 8 shows that of the pin-jointed truss plate of the same members. Fig 9 shows the buckling mode of Eqn 46 for M_x=M_y and M_{xy}=0.

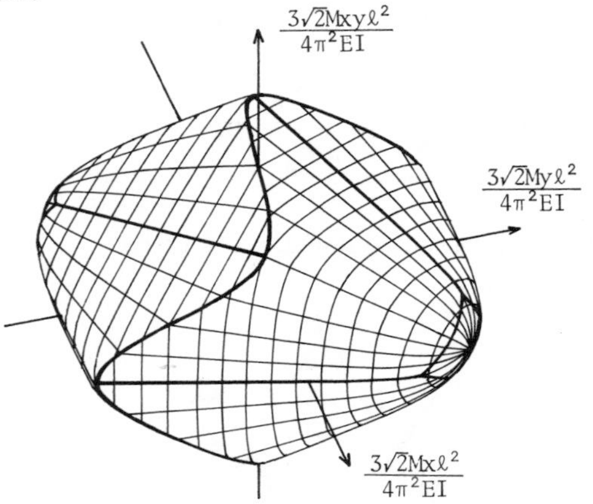

Fig 6 Strength surface of rigid-jointed three-way double-layer lattice plate of uniform members

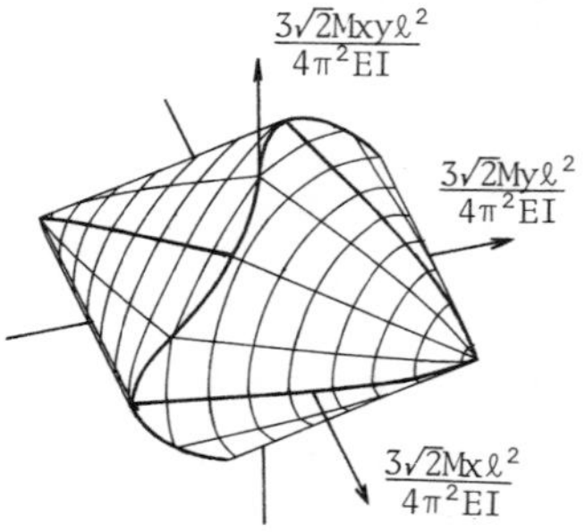

Fig 7 Strength surface for the case of rigid-jointed chords and pin-jointed webs

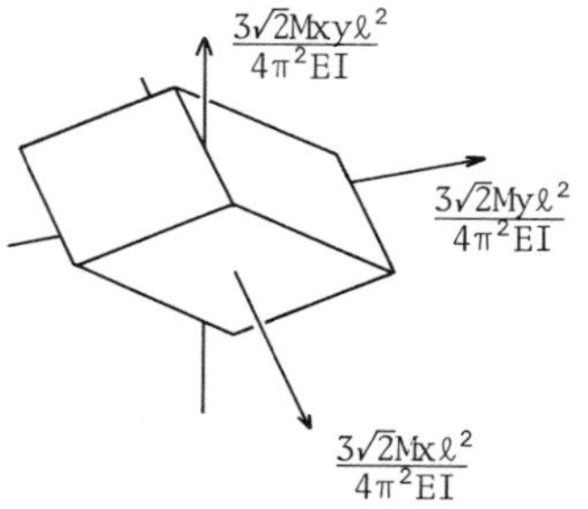

Fig 8 Strength surface of pin-jointed case

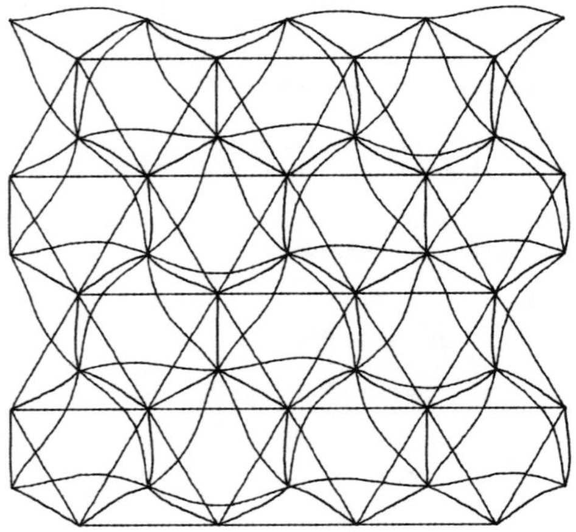

Fig 9 Buckling mode for M_x=M_y and M_{xy}=0.

EXAMPLE 3. DOUBLE-LAYER PARALLEL SQUARE LATTICE PLATES

Pin-jointed truss plates of the type shown on Fig 10 have no twisting rigidity as presented already, see Ref 7. Even for the rigid-jointed case the twisting rigidity is very small, so the effective strength for the bending moments, M_x and M_y, is derived under the condition of no other sectional forces, such as M_{xy}, N_x, Q_x, etc.

For this truss the modes repeated by two structural units show the lower buckling moments, whose modes are represented by Eqn 49.

$$\eta = \pi s, \quad \eta = \pi t, \quad \eta = \pi(s+t) \qquad ....49.$$

The equation for the lowest eigen value of the buckling equation for the mode of $\eta=\pi s$ is got by separation as Eqn 50, using the values α_y and β_y for the maximum compression acting on the chord members. Its mode is as Fig 12.

$$[3EI(\alpha_y-\beta_y)+GJ_w+12EI_w]\theta_1 = 0 \qquad ....50.$$

The equation for the lowest eigen value of the buckling equations for the modes of $\eta=\pi(s+t)$ is got by separation as Eqn 51. Its mode is as shown in Fig 11.

$$[EI(\alpha_x-\beta_x+\alpha_y-\beta_y)+GJ_w+4EI_w]\theta_2 = 0 \qquad ....51.$$

The strength curves on the M_x-M_y plane for the case of the same member length are as shown on Fig 13, where r denotes the rate, r=EI(web)/EI(chord).

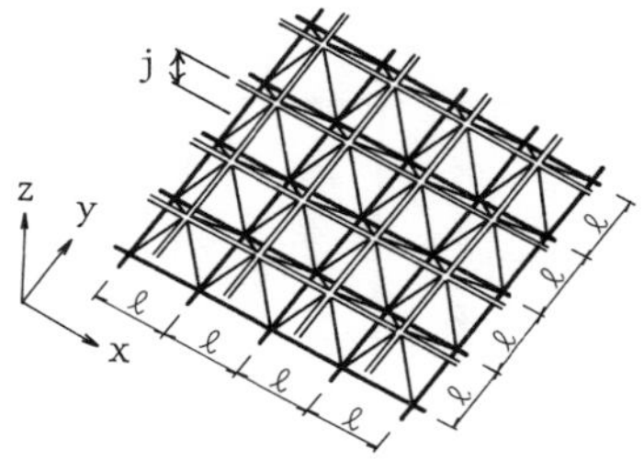

Fig 10 Parallel square double-layer lattice plate

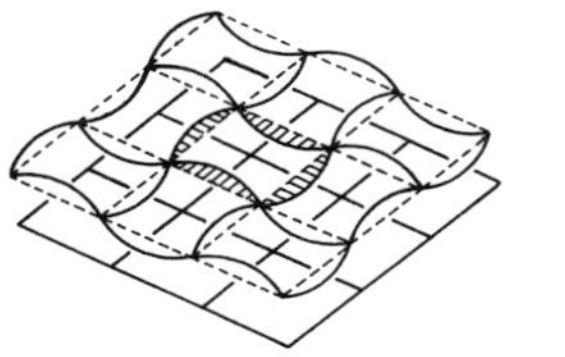

Fig 11 Mode I

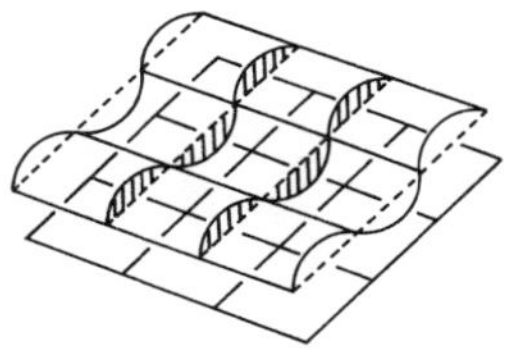

Fig 12 Mode II

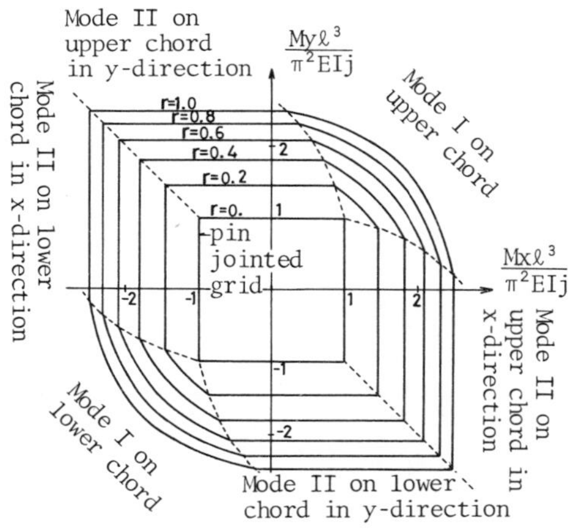

Fig 13 Strength curves on M_x-M_y plane

CONCLUSIONS

The effective strength of lattice structures, which is defined as the strength of the equivalent continuum, for the rigidly connected cases is treated theoretically, under the assumption that the grids are constructed with congruent latticed units in repetitive pattern and the strength is determined by the elastic bifurcation buckling of constituent members in the form of repetitive modes from uniform prebuckling state. And the main conclusions are as follows.

1) The matrix of equilibrium equation for lattice structures in difference expression is the transposed matrix of that of the relation between deformations and displacements by changing the sign of the power of the shift operators, and vice-versa.

2) The buckling equations for repetitive modes can be separated by the finite Fourier expansion, but the modes of the sine and cosine type of the same wave length are not separated.

3) The effective strength appeared usualy in the repetitive modes of small pattern unit.

4) The effective strength showed good result compared with that of the exact theory in the example of rigidly connected parallel chord truss whose load carrying capacity is determined by the elastic instability.

5) It is on the safe side that the maximum sectional force is reffered with the effective strength.

6) The effective strength of two types of double-layer lattice plates is shown.

AKNOWLEGMENTS

The author appreciates Mr. Murakami and Mr. Abe for their careful programming and computer operation, and that this work is supported in part by the Ministry of Education of Japan under the Grant in Aid for Scientific Research.

REFERENCES

1. F MATSUSHITA, Study on Skeleton Shells, Ann. Meeting Architectural Institute of Japan (in Japanese), 1954.

2. W FLÜGGE, Stresses in Shells, Springer, 1960, p.299-307.

3. K KLÖPPEL and R SCHART, Systematische Ableitung der Differentialgleichungen für ebene anisotrope Flächen-tragwerke, Stahlbau, 29, Feb. 1960, pp.33-43.

4. D T WRIGHT, A Continuum Analysis for Double-Layer Space Frame Shells, Publ. IABSE, 26, 1966, pp.593-610.

5. K HEKI, On the Effective Rigidities of Lattice Plates, Recent Researches of Structural Mechanics, 1968.

6. K HEKI and T SAKA, Stress Analysis of Lattice Plates as an Anisotropic Continuum Plates, Proc. 1971 IASS Symp. Tension Structures and Space Frames, pp.663-674.

7. K HEKI, Examples of the Effective Rigidities of Lattice Plates, Analysis, Design & Construction of Space Frames (Draft), IASS W. G. Spatial Steel Structures. 1979.

8. K HEKI and T SAKA, The Effective Strength of Double-Layer Grids in Continuum Treatment, IASS World Cong. on Shell and Spatial Structures, 1979, pp.4.123-4.137.

9. T SAKA and K HEKI, The Effects of Joints on the Strength of Space Trusses, 3rd Int. Conf. Space Struct., 1984, 9.

ANALYSIS OF COMPRESSION MEMBERS IN SPATIAL ROOF STRUCTURES

Benjamin KOO

Professor, Department of Civil Engineering
University of Toledo, Toledo, Ohio, U.S.A.

A spatial roof structure usually covers a large area and is composed of an assemblage of slender members. In the case of struts, the buckling effect is the main concern. This paper focuses on a generalized buckling behavior of compression members with particular attention to torsional-flexural buckling. For the purpose of engineering analysis, a generalized interaction equation is developed. Critical stress formulas for commonly used member shapes in the elastic and inelastic range are included.

INTRODUCTION

Space roof frames offer the possibility of providing a large unobstructed covered area, enhance the efficient use of structural materials, and provide great rigidity, inherent redundancy and a good safety factor. Consequently a growing number of investigators have shown an interest in these structures and have made engineering contributions such as an extensive bibliography (18), books (19), tests (10, 14, 15) and other publications (9, 10, 11, 20). This paper focuses on the behavior of slender compression members subjected to flexural and torsional-flexural buckling.

CRITICAL LOADS

With the usual general assumptions of an elastic column subjected to bending in the buckling mode, its cross-section undergoes translations, u and v, in the x and y directions of the principal coordinates and a rotation, ϕ, about the shear center, SC, as shown in Fig 1.

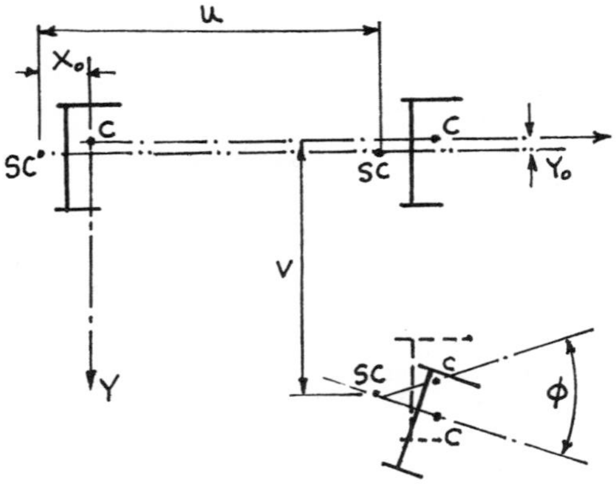

FIG 1

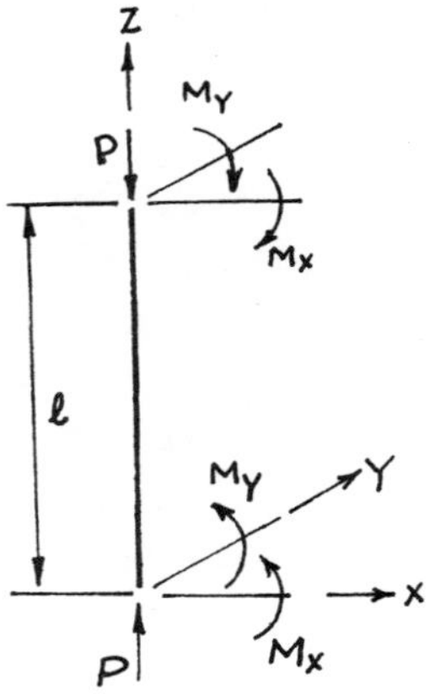

FIG 2

In reference to Fig 2, the general equilibrium equations (5, 6, 21), with all derivatives with respect to z along the member axis (letting $M_x = Pe_y$ and $M_y = Pe_x$), are:

$$E\,I_y\,u'''' + P(u'' + y_o\,\phi'') - Pe_y\,\phi'' = 0 \tag{1a}$$

$$E\,I_x\,v'''' + P(v'' - x_o\,\phi'') + Pe_x\,\phi'' = 0 \tag{1b}$$

$$E\,C_w\,\phi'''' - GJ\phi'' - P(e_y\beta_1 + e_x\beta_2 + r_o^{\,2})\phi'' + P(y_o - e_y)u'' - P(x_o - e_x)v'' = 0 \tag{1c}$$

$$\text{in which } \beta_1 = \frac{1}{I_x}(\textstyle\int_A y^3 dA + \int_A x^2 y dA - Zy_o \tag{2a}$$

$$\beta_2 = \frac{1}{I_y}(\textstyle\int_A x^3 dA + \int_A xy^2 dA - Zx_o \tag{2b}$$

$E\,I_x$ and $E\,I_y$ are the principal flexural rigidities, $E\,C_w$ and GJ are warping and St. Venant rigidity respectively x_o and y_o are the x and y coordinates of the shear center (SC) and r_o is the polar radius of gyration of the cross-section about its shear center.

By taking x, y and ϕ as the first harmonics of the buckling eigenfunctions of a member satisfying the hinged end boundary condtions of displacement of u, v and ϕ respectively, the Eqns (1a), (1b) and (1c) lead to the following characteristic equation.

$$(P_x - P)(P_y - P)[r_o^2 P_\phi - Pe_y\beta_1 - Pe_x\beta_2 - Pr_o^2] - P^2(y_o - e_y)^2$$
$$(P_x - P) - P^2(x_o - e_x)^2(P_y - P) = 0 \qquad (3)$$

where $P_x = \frac{\pi^2 EI_x}{(k\ell)^2}$; $P_y = \frac{\pi^2 EI_y}{(k\ell)^2}$ and $P_\phi = (GJ + EC_w\frac{\pi^2}{(k\ell)^2})/r_o^2$

k is the effective length factor, 1 for hinged-ends and 0.5 for fixed ends.

Equation (3) is the generalized expression for a member with combined bending and torsion. If the cross-section of the member is symmetrical about the y axis and the axial load acts in the plane, such that $x_o = e_x = 0$, then Eqn (3) is reduced to:

$$(P_y - P)[r_o^2 P_\phi - P(e_y\beta_1 + r_o^2)[-P^2(y_o - e_y)^2 = 0 \qquad (4)$$

This equation checks with Timoshenko and Gere (5).

Again, if no moments act on the member, i.e. $Pe_x = Pe_y = 0$ then the following equation results.

$$r_o^2(P_x-P)(P_y-P)(P_\phi-P)-Py_o^2(P_x-P)-Px_o^2(P_y-P) = 0 \qquad (5)$$

This equation checks with Chajes, Fang and Winter (2,3).

In order to demonstrate the relation of flexural and twist-bend buckling on critical load and to facilitate engineering analysis, an interaction equation is derived from Eqn (3).

$$\frac{P}{P_x} + \frac{P}{P_y} + \frac{P}{P_\phi}(k) - \frac{P^2}{P_xP_\phi}[k - k_x] - \frac{P^2}{P_yP_\phi}[k - k_y]$$
$$+ \frac{P^3}{P_xP_yP_\phi}[k - k_x - k_y] = 1 \qquad (6)$$

in which $k = (e_y\beta_1/r_o^2 + e_x\beta_2/r_o^2 + 1)$

$$k_x = (x_o - e_x)^2/r_o^2 \; , \; k_y = (y_o - e_y)^2/r_o^2$$

If the same special condition which produces Eqn (4) is imposed on the column, then Eqn (6) can be reduced accordingly to the following:

$$\frac{P}{P_y} + \frac{P}{P_\phi}[\bar{k}] - \frac{P^2}{P_yP_\phi}[\bar{k} - k_y] = 1, \text{ where } (\bar{k} = (\frac{e_y\beta_1}{r_o^2} + 1)) \qquad (7)$$

Likewise, under the same situation that results in buckling Eqn (5), the following equation can be derived from Eqn (6) as:

$$\frac{P}{P_x} + \frac{P}{P_\phi} - \frac{P^2}{P_xP_\phi}[1 - \frac{x_o^2}{r_o^2}] = 1 \qquad (8)$$

This equation again checks with references 2 and 3. Since the constant k, k_x, k_y and $\bar{k}$ are all related to the geometry of member cross-sections, the interaction Eqns (6) (7) and (8) present a fresh and clear approach showing the fundamental inter-relationship among all essential parameters used to evaluate the torsional-flexural buckling.

CRITICAL STRESS

In engineering design, the most important criterion is stress either in the elastic or inelastic range. The investigation of inelastic behavior centers on establishing the reserve strength between working and critical loads (12, 13). The normal stress, σ, for a member subjected to actions of an axial load, P, bending moments M_x and M_y, is expressed as:

$$\sigma = \frac{P}{A} + \frac{M_x y}{I_x} + \frac{M_y x}{I_y} \qquad (9)$$

The parameters, A, I_x, and I_y in Eqn (9) are constants depending upon the shape of the cross-section of a member. The most common and efficient shape for spatial roof structures is the circular tube.

1) Circular Tube

It has a large radius of gyration compared to the area and great torsional rigidity. According to Gerard and Beeker (8), the critical stress, F depends on the parameter, Z.

$$Z = \frac{\ell^2}{Rt}[1 - \mu^2]^{\frac{1}{2}} \qquad (10)$$

where ℓ = length, R = radius, t = thickness of the wall and μ = poisson's ratio = 0.3 for steel.

For a long column, Z is greater than 50 and

$$F = \frac{\pi^2}{2} E (\frac{R^2}{k\ell}) \qquad (11)$$

For a moderate length column, Z ranges between 2.85 and 50, and

$$F = C E \frac{t}{R} \qquad (12)$$

in which $C = 1/[3(1 - \mu^2)]^{\frac{1}{2}}$

In the inelastic range, the critical stress (22) becomes as:

$$F = a C E(\frac{t}{R}) \qquad (13)$$

where $a = 1.102 \; (\frac{E_s}{E})(\frac{E_t}{E})^{\frac{1}{2}}$

and E_s is the secant modulus and E_t the tangent modulus.

2) Doubly Symmetric Shapes (I-shape, Cruciform, Cold-formed channels back to back, etc.)

For a slender, axialy loaded member, the critical stress is:

$$F = \pi^2 E/(k\ell/r)^2 \qquad (14)$$

in which r = the smallest radius of gyration of the cross-section.

The torsional buckling stress can be determined and derived from Eqn. (3) as:

$$(P_x - P)(P_y - P)(P_\phi - P) = 0 \qquad (15)$$

and

$$F = \frac{P_\phi}{A} = \frac{1}{Ar_o^2}[GJ + EC_w(\frac{\pi}{k\ell})^2] \qquad (16)$$

In the inelastic range, the torsional-flexural buckling stress, σ_{tf} is shown as follows:

$$\sigma_{tf} = \frac{1}{2B}\left[\frac{P_x}{A} + \frac{P_\phi}{A} - \left[\left(\frac{P_x}{A} + \frac{P_\phi}{A}\right)^2 - 4B\frac{P_x P_\phi}{A^2}\right]^{\frac{1}{2}}\right] \qquad (17)$$

in which $B = 1 - \left(\frac{x_o}{r_o}\right)^2$

The tangent modulus of elasticity plays an important role in inelastic buckling and according to Bleich (4) and Chajes, Fang and Winter (3) is:

$$E_t = \frac{EF_y}{\sigma_{tf}}\left(1 - \frac{F_y}{4\sigma_{tf}}\right) \qquad (18)$$

$$F_x = \left(\frac{E_t}{E}\right)\left(\frac{P_x}{A}\right)$$

$$F_\phi = \left(\frac{E_t}{E}\right)\left(\frac{P_\phi}{A}\right)$$

where F_y is the yield stress.

When the residual stress is taken into consideration and the effective proportional limit is assumed to be one half of the yield point, then according to Johnston (23) and Winter (1), the critical stress can be conservatively approximated as:

$$F = F_y - \left(\frac{F_y^2}{4\pi^2 E}\right)\left(\frac{k\ell}{r}\right)^2 \qquad (19)$$

3) Singly Symmetric Shapes (Angles, Channels, T-Sections, etc.)

If the axial load is applied at the shear center, (i.e. $e_y = 0$), then Eqn (4) is reduced to the following:

$$P_{min} = \frac{1}{D}\left\{P_y + P_\phi - \left[(P_y + P_\phi)^2 - 2\,D\,P_y\,P_\phi\right]^{\frac{1}{2}}\right\} \qquad (20)$$

in which $D = 2\left(1 - \frac{y_o^2}{r_o^2}\right)$

Thus the critical stress is:

$$F = \frac{1}{D}\,(P/A) \qquad (21)$$

Then Eqn (21) checks with reference (7).

In the inelastic range, the critical stress can be shown to be:

$$F = F_y\left[1 - F_y/F_{elastic}\right] \qquad (22)$$

CONCLUSION

A generlized equation for evaluating torsional-flexural buckling load for slender compression member has been presented. An interaction equation is developed to facilitate engineering design and analysis using this method. The critical stress formulas in the elastic and inelastic range are included for the benefit of engineering professionals interested in design procedures that satisfactorily predict the strength of struts in spatial roof structures.

REFERENCES

1. Winter, G., "Commentary on the 1968 Edition of the Specification for the Design of Cold-formed Steel Structural Manual", Am. Iron and Steel Inst., 1970 Ed.

2. Chajes, A., and Winter, G., "Torsional-Flexural Buckling of Thin-Walled Members", Jl. of Structural Div. A.S.C.E. proceedings, Vol. 91, No. ST.4, Aug. 1965.

3. Chajes, A., Fang, P. J., and Winter, G., "Torsional-Flexural Buckling, Elastic and Inelastic of Cold-Formed Thin-Wall Columns", Cornell Engineering Research Bulletin 66-1, Aug., 1966.

4. Bleich, F., "Buckling Strength of Metal Structures", McGraw Hill Book Co., 1952.

5. Timoshenko, S. and Gere, J., "Theory of Elastic Stability", McGraw Hill Book Co., 1961.

6. Goodier, J. N., "Buckling of Compressed Bars by Torsion and Flexure", Cornell University Experiment Station Bulletin 27, December, 1941.

7. Yu, W. W., "Cold-Formed Steel Structures", McGraw Hill Book Co., 1973.

8. Gerard, G. and Becker, H., "Handbook of Structural Stability, Part IV-Buckling of Curved Plates and Shells", NACA, TN, August, 1957.

9. Chu, K. H., and Rawpetsreiter, R. H., "Large Deflection Buckling of Space Frame", Jl. of Structural Div. A.S.C.E., Vol. 98, No. ST12, Proc. paper 9955, Dec., 1972.

10. Hanaor, A. and Schmidt, L. C., "Space Truss Studies With Force Limiting Devices", Jl. of Structural Div., A.S.C.E., Vol. 106, No. ST11, Proc. paper 15808, November, 1980.

11. Holzer, S. M., Plaute, R. H., Somers, A. E., Jr., and White, W. S., III, "Stability of Lattice Structure Under Combined Loads", Jl. of Engineering Mechanics Div., A.S.C.E., Vol. 106, No. EM, Proc. paper 15322, April, 1980.

12. Jagannathan, D. S., Epstein, H. E. and Christiano, P. P., "Non-linear Analysis of Reticule Space Trusses" Jl. of Structural Div., A.S.C.E., Vol. 101, No. ST12, Proc paper 11799, December, 1975.

13. Robert, H., Dickel, T. and Renner, D., "Snap-through Buckling of Space Trusses", Jl. of Structural Div., A.S.C.E., Vol. 107, No. ST1, Proc. paper 15973, January, 1981.

14. Schmidt, L. C., Morgan, P. R. and Clarkson, J. A., "Space Trusses With Brittle Type Structural", Jl. of Structural Div., A.S.C.E., Vol. 12, No. ST7, July, 1976.

15. Schmidt, L. C., Morgan, P. R. and Clarkson, J. A., "Space Truss Design in the Inelastic Range", Jl. of Structural Div., A.S.C.E., Vol. 104, No. ST12, December, 1978.

16. Wolf, J. P., "Post Buckling Strength of Large Space Trusses", Jl. of Structural Div., A.S.C.E., Vol. 99, No. ST7, July 1973.

17. Smith, E. A. and Smith G. D. "Collapse Analysis of Space Trusses", Proceedings of a Symposium on Long Span Roof Structures, Oct., 1981, published by ASCE.

18. The Sub-Committee on Latticed Structures of the Task Committee on Special Structures of the Committee on Metals of the Structural Div. A.S.C.E., "Bibliography on Latticed Structures", Jl. of Structural Div., A.S.C.E., Vol. 98, No. ST7, Proc. paper 9055, July, 1972.

19. Davies, R. M., "Space Structures", The International Conference on Space Structures, University of Surrey, Surrey, England, Sept., 1966, The John Wiley & Sons, 1967.

20. Cuoco, D. A., "State-of-Art of Space Frame Roof Structures", Long Span Roof Structures, proceedings of A Symposium, the Committee on Special Structures of the Committee on Metals of the Structural Division, A.S.C.E., 1981.

21. Pekoz, T. B., "Torsional Flexural Buckling of Thin-Walled Section Under Eccentric Load", Cornell University Research Bulletin, No. 69-1, Sept., 1969.

22. Schilling, C. G., "Buckling Strength of Circular Tubes", Jl. of Structural Div., A.S.C.E., Vol. 91, No. ST5, Oct., 1965, Proc. paper 4520.

23. Johnston, B. G., "Guide to Design Criteria for Metal Compression Members", John Wiley & Sons, Inc., 1976.

DYNAMIC STABILITY OF SPACE STRUCTURES

Prof. Ing. J. SCHUN*, CSc and Dr. Ing. A. Tesár**, DrSc

* Lecturer, Chair of Steel and Timber Structures, Slovak Technical University, Bratislava CZECHOSLOVAKIA

** Scientific Worker, Slovak Academy of Sciences Bratislava, CZECHOSLOVAKIA

The present paper is concerned with nonlinear finite solution of dynamic stability behaviour of space structures in resonance regions of vibration. The development of reliable and efficient techniques for handling the dynamic stability behaviour is emphasized. An illustrative numerical solution associated with stability dynamic response of space structural systems is presented.

INTRODUCTION

In recent years space structures have acquired increased importance due to new developments in industry and technology. Economy and functional requirements have resulted in a drive for more daring design with better utilization of all structural and material properties. Urgent demands for thin space structures have made the engineers to face crucial problems of stability and buckling in static and especially in dynamic resonance regions. The buckling and postbuckling safety of such structures cannot be estimated unless the complete nonlinear spatial dynamic behaviour is taken into account. For the solution of such problems it is inevitable to consider all nonlinear interactions of effects of large deformation analysis as well as of inertia and damping parameters of space structures.
The purpose of this investigation is to study the nonlinear behaviour of space structures including the dynamic instability phenomena. The finite variant of transfer matrix method is adopted for numerical solution of the problem. Different numerical techniques for solving nonlinear problems are discussed. Special algorithms are used for automatic control of incremental operations. Particular attention is given to the development of reliable algorithms for instability behaviour. Numerical verification on real space structure is presented. Example illustrates the applicability of developed theoretical concepts. Studied is special case of dynamic instability behaviour of space structures.

BASIC CONCEPTS

The deformation path of the space structure is a continuous family of configurations H_t which takes the structure from an initial configuration H_o at $t=t_o$ to a final configuration H_F at $t=t_F$. The kinematics of deformation is described by two additional configurations, the current configuration H_1, corresponding to time $t=t_1$ and the neighbouring configuration H_2 $(t=t_2)$. It is assumed that H_2 is close to H_1, i.e. $t_2=t_1+\Delta t$, where Δt is a small quantity. In the updated Lagrangian formulation of motion the reference state is taken as the current configuration H_1 which is continuously updated throughout the entire deformation process. A new material reference frame is established at each stage along the deformation path. A major advantage of present formulation is its simplicity which provides an easy physical interpretation of generalized nonlinear stability behaviour of slender space structures.
The actual geometry of space surface is replaced by an assemblage of flat elements. Plate and stiffener elements may be combined to simulate stiffened space structure. The coupling between membrane and bending actions is obtained in the assemblage of the elements to form the complete space structure. For problems involving large displacements the stretching and flexural behaviour becomes coupled and this increases the complexity of formulation. However, in the updated Lagrangian description of motion, adopted herein, it is possible to arrive the simple formulation of present nonlinear case.

NONLINEAR ANALYSIS OF MOTION

The governing equations of motion for nonlinear space structure are given in the form

$$([M].\{\ddot{r}\})_t + ([C].\{\dot{r}\})_t + ([P].\{r\})_t = \{R\}_t \quad .... \ 1.$$

where the first term is the vector of inertia parameters, the second term is the vector of stress dependent nonlinear damping forces and the third term is the vector of internal, deformation dependent nonlinear forces. $\{R\}_t$ is the vector of external dynamic loads. For modified Newmark implicit time-integration scheme (see Ref 1), used herein, the pseudo-force linearization technique (see Ref 2), expressed as

$$([P].\{r\})_t = [K].\{r\}_t + ([N].\{r\})_t - \{\Delta F_{t+\Delta t}\} \quad .... \ 2.$$

is utilized. In Eqn 2 the term $([N].\{r\})_t$ is the vector of nonlinear pseudo-forces, $[K]$ is the incremental stiffness matrix, $\{r\}$ is the vector of displacements and $\{\Delta F_{t+\Delta t}\}$ is the local approximation error. In the application of this technique $([P].\{r\})_t$ is placed on the right side of Eqn 1 and the vector of nonlinear terms is treated as a pseudo-force vector. In the each time-step an estimate of $([N].\{r\})_t$ is computed and iterations are performed until $\{\Delta F_{t+\Delta t}\}$ converges to a prescribed tolerance norm. The major advantage of this technique is the fact that the effective stiffness matrix of the flowchart of modified implicit Newmark scheme need to be decomposed only once and only the right hand sides are modified to account for the nonlinear terms.

IMPLEMENTATION OF TRANSFER MATRIX METHOD

In transfer matrix method the information about analyzed mechanical simulation is modelled directly by algorithm of solution. The chaining of matrix multiplications together with relatively small requirements on core and storage of computers submit the possibilities for complex analyses of structural simulations characterized by general variability of physical parameters in various regimes of exploitation. With these realities connected improvement of economy of computer calculation compared with other finite concepts has cardinal importance for solution of nonlinear dynamic instability behaviour of analyzed slender space structures. The principles of the finite variant of transfer matrix method are known, see References 3 and 4, and will be no more discussed here.

DETERMINATION OF TRANSFER AND NODAL MATRICES

Generalized transfer matrix BK of analyzed space structures is constructed by diagonal assembly of transfer submatrices BKS of individual elements. Transfer submatrices are derived from corresponding linear and nonlinear stiffness matrices of rib and flat elements of applied theoretical simulation of space structure. Consider the longitudinal stiffener element shown in Fig 1, with force-deformation relations formulated in known stiffness matrices of linear and nonlinear beha-

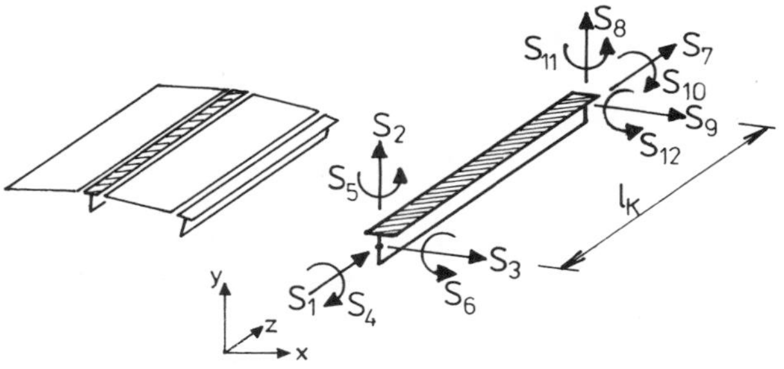

Fig 1

viour. Let us assume the subvectors of force and displacement components expressed as

$$\begin{aligned}\{p_{i-1}\} &= \{S_1, S_2, S_3, S_4, S_5, S_6\}^T, \\ \{p_i\} &= \{S_7, S_8, S_9, S_{10}, S_{11}, S_{12}\}^T, \\ \{d_{i-1}\} &= \{u_1, u_2, u_3, u_4, u_5, u_6\}^T, \\ \{d_i\} &= \{u_7, u_8, u_9, u_{10}, u_{11}, u_{12}\}^T.\end{aligned} \quad .... \ 3.$$

Taking account of these substitutions, the force-displacement couplings can be expressed as

$$\begin{Bmatrix} p_{i-1} \\ p_i \end{Bmatrix} = \begin{bmatrix} A & B \\ C & D \end{bmatrix} . \begin{Bmatrix} d_{i-1} \\ d_i \end{Bmatrix} \quad .... \ 4.$$

with adopted stiffness matrix

$$[k_L + k_G] = \begin{bmatrix} A & B \\ C & D \end{bmatrix} \quad .... \ 5.$$

The solution of system 4 there yields the following couplings of state components in initial and end point of analyzed stiffener element

$$\begin{Bmatrix} d_i \\ p_i \end{Bmatrix} = \begin{bmatrix} -B^{-1}.A & B^{-1} \\ C-D.B^{-1}.A & -D.B^{-1} \end{bmatrix} . \begin{Bmatrix} d_{i-1} \\ p_{i-1} \end{Bmatrix} . \quad .... \ 6.$$

This relation can be expressed as

$$\begin{Bmatrix} d_i \\ p_i \end{Bmatrix} = BKS. \begin{Bmatrix} d_{i-1} \\ p_{i-1} \end{Bmatrix} , \quad .... \ 7.$$

where BKS is the sought transfer submatrix of analyzed element. Submatrix BKS realizes the transfer of state vector from the initial into the end points of longitudinal discretization of structure. Submatrix BKS, for this case, has dimension (12,12).
Analogous analysis can be performed for determination of corresponding planar rectangular element shown in Fig 2.

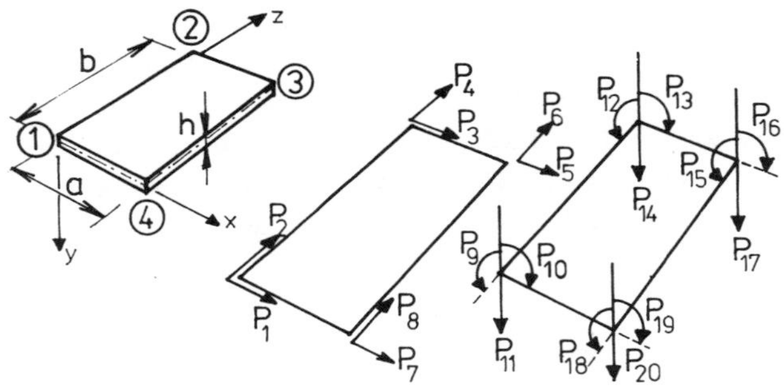

Fig 2

Generalized transfer matrix of spatial dynamic behaviour of analyzed space structures is illustrated as diagonal assembly of corresponding one-dimensional and flat elements. Scheme of matrix BK, with additional row-column 1) for loading vector, is shown in Fig 3.

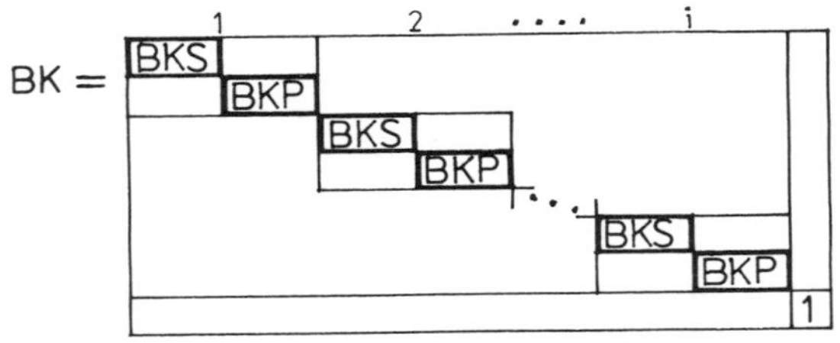

Fig 3

Generalized nodal matrix CK includes into calculation:

1. interactive couplings of individual elements,
2. coordinate transformations taking account of updated Lagrangian formulation,
3. nodal inertial parameters,
4. nodal loading or forcing parameters.

All of these influences were systematically dealt with in publication 3 and will be no more analyzed here.

INSTABILITY ANALYSIS OF SPACE STRUCTURES

A thorough understanding of the types of nonlinear behaviour likely to be sustained by space structures is essential for establishing reliable solution algorithms for instability analysis. From the various forms of structural instability behaviour four major approaches to the solution of instability phenomena have emerged:

- general nonlinear analysis,
- limit load calculation,
- analysis of bifurcation,
- postbuckling analysis.

A wide variety of numerical solution techniques have been suggested for solving nonlinear structural problems. The classification adopted herein divides the solution methods for general nonlinear analysis into the following four categories:

- direct minimization techniques,
- iterative methods,
- pure incremental methods,
- self-correcting and combined methods.

This mode of classification is closely related to the alternative mathematical formulations of the physical problem. The drifting from the true equilibrium path encountered in pure incremental methods may be crucial for the solution of present nonlinear problems. This is particularly true for structural instability analysis which may be highly sensitive to violation of static equilibrium. In order to obtain reliable solutions to such problems the incremental methods must be supplemented by a correction which restores equilibrium at each load level.

ALGORITHMS DEALING WITH LIMIT POINTS

The tracing of the load-deflection curve of a stability problem of space structures represents a formidable task. The major difficulty is associated with the singular incremental stiffness matrix that exists at the extremum points, indicating that an infinite increase in displacements takes place without any change of loading. The descending or unstable postbuckling path is characterized by indefinite stiffness matrix.

Several schemes for tracing the entire deformation path of snap-through problems have been suggested. The common feature of these methods is to introduce modifications or constraints that render the stiffness matrix positive definite in the postbuckling range. Another popular method for handling snap-through behaviour consists in incrementing displacements rather than loads. Rather than introducing modifications to keep the incremental stiffness matrix positive definite, one should adopt an equation solver that is capable of handling indefinite matrices. This can be accomplished by the general symmetric decomposition

$$K_I = L.D.L^T \qquad .... \quad 8.$$

where L is a lower triangular matrix with unit diagonal elements and D is a diagonal matrix containing the pivots of the elimination process. With the aid of Eqn 8 is possible to extend the technique of incrementing loads to the entire nonlinear zone. Occurence of instability points may be detected by checking the sign of determinant of the incremental stiffness matrix. Thus, from Eqn 8 there yields

$$\det K_I = \det D = d_1.d_2.d_3 \; ... \; d_M \qquad .... \quad 9.$$

where M is the dimension of the stiffness matrix and d_i (i=1,2,3, ... M) are the elements of diagonal matrix D. A procedure must be applied by which the automatic load incrementation techniques are adopted to deal with extremum points of the load-displacement curve. The alternative method is illustrated in Fig 4 and the basic steps involved are summarized below.

1. Apply a load increment Δp_{i+1} and compute the corresponding displacement vector Δr_{i+1}.

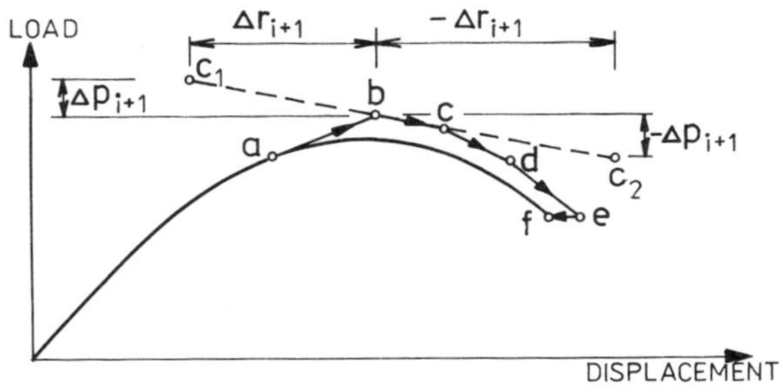

Fig 4

2. Compute the sign of the determinant of the incremental stiffness matrix. If the checking yields negative determinant, the sign of Δp_{i+1} and Δr_{i+1} must be reversed (see Fig 4).

3. In order to avoid excessive drifting of the solution in the vicinity of the extremum points, the load and displacement increments should be scaled according to the condition

$$\left|\left\{\Delta r_{i+1}\right\}\right| \leq \varrho \qquad \ldots . \quad 10.$$

where ϱ is prescribed value of the displacement norm.

4. One additional load step is carried out without equilibrium iteration to make sure that the peak has been passed.

The above procedure for finding the extremum points only applies to the incremental part of the solution technique. The nonlinear static and dynamic stability response of multi-degrees of freedom space system can be illustrated by the "current stiffness parameter", defined by

$$S_p = \frac{\Delta \dot{r}_o^T . R_{ref}}{\Delta \dot{r}^T . R_{ref}} \qquad \ldots . \quad 11.$$

where a dot denotes differentiation with respect to the loading parameter p. Replacing first derivatives by finite incremental quotients yields for load increment of number i

$$S_{p,i} = \frac{\frac{\Delta r_1^T}{\Delta p_1} R_{ref}}{\frac{\Delta r_i^T}{\Delta p_i} R_{ref}} = \frac{p_i . \Delta r_1^T . R_{ref}}{p_1 . \Delta r_i^T . R_{ref}} \qquad \ldots . \quad 12.$$

It was observed that the current stiffness parameter has the initial value of 1 for any nonlinear problem. A "softening" system is characterized by a value of S_p less than 1 while a system with increasing stiffness will have S_p greater than 1.

The selection of a proper time-integration method is another critical factor in the solution of stability response of space structures. The choice of the time discretization is problem dependent and hence a variety of methods are available for the purposes of the analysis. One area of research in time-integration methods is the development of schemes which permit different time integrators to be used simultaneously in different parts of structure. Operator splitting methods (see Ref 3) are efficient for solution of present problem. The Newmark implicit scheme adopted herein is combined with a splitting of transfer matrix such that each stage is integrated with the same size of the time-step. Hence, splitting in space (element-wise) and splitting in time are allowed. For nonlinear problems it is difficult to prove stability or convergence of solution. Common practice is to apply stability criteria derived for special cases and to accept the results if they appear stable. The use of operator splitting can sometimes simplify the development of stable methods.

NONLINEAR INTERACTIONS

The nonlinear interactions between stiffness, inertia and damping parameters come distinctly to the fore in resonance and stability analyses of space structures. Due to large deformations and high stress levels in such regions all of these parameters become variable and are interactively coupled. Especially, the damping will thus be a function of the stress state over the displacement dependence of incremental stiffness. The interactions of these parameters will be dealt with in this section. The equivalent hysteresis damping covering the total energy dissipation is utilized in this study. The specific work B of the total damping is expressed as a function of stress σ

$$B = J . \sigma^n \qquad \ldots . \quad 13.$$

where J and n are experimentally found in the linear and nonlinear regions (see Ref 3). The total work of damping is given as an integration of specific damping works of individual elements, for the used load and time steps. The variability of stresses causes that each particle of the structure has its own hysteresis curve contributing to the total damping of space structure. In the element volume V_g the work of damping is given as

$$B_g = \int_0^{V_g} B \, dV_g = \int_0^{\sigma_{max}} B \frac{dV}{d\sigma} \, d\sigma \qquad \ldots . \quad 14.$$

The maximal stress σ_{max} corresponds to maximal damping B_{max} and B_g can be expressed as

$$B_g = B_{max} . V_g . \varphi_1 \qquad \ldots . \quad 15.$$

with parameter φ_1 expressed as

$$\varphi_1 = \int_0^1 \left(\frac{B}{B_{max}}\right) \frac{d\left(\frac{V}{V_g}\right)}{d\left(\frac{\sigma}{\sigma_{max}}\right)} d\left(\frac{\sigma}{\sigma_{max}}\right) \quad \dots\dots 16.$$

The energy cumulated in the analyzed volume can be expressed as

$$U_g = \int_0^{V_g} \left(\frac{\sigma}{E}\right)^2 dV_g = \frac{1}{2} V_g \frac{\sigma_{max}^2}{E} \varphi_2 \quad \dots\dots 17.$$

with a nondimensional parameter φ_2

$$\varphi_2 = \int_0^1 \left(\frac{\sigma}{\sigma_{max}}\right)^2 \frac{d\left(\frac{V}{V_g}\right)}{d\left(\frac{\sigma}{\sigma_{max}}\right)} d\left(\frac{\sigma}{\sigma_{max}}\right) \quad \dots\dots 18.$$

The factor of damping for the analyzed element then can be expressed by parameter

$$\eta_s = \frac{B_g}{2.\pi.U_g} = \frac{E.D_{max}.\varphi_1}{\pi.\sigma_{max}^2.\varphi_2} \quad \dots\dots 19.$$

which is incorporated into the complex elasticity moduli

$$E = E_o.(1+i.\eta_s) \quad \dots\dots 20.$$

$$G = G_o.(1+i.\eta_s) \quad \dots\dots 21.$$

of the corresponding flat or one-dimensional members of space structure.

In the case of linear damping the ratio $\varphi_1/\varphi_2 = 1$. The resonance analyses of the nonlinear damping are based on the evaluation of parameters φ_1 and φ_2 for the given geometrical and stress relations due to Eqns 16 and 18. Iteration scheme is to be applied for the determination of the damping factors for each element of the utilized discrete simulation. In the first iteration step the stresses are determined, which yield the parameters $\varphi_1^{(1)}$, $\varphi_2^{(1)}$ as well as a preliminary damping factor $\eta_s^{(1)}$. The factor $\eta_s^{(1)}$ is the basis for the succeeding iteration steps with an evaluation of the modified stress states and the parameters $\varphi_1^{(i)}$, $\varphi_2^{(i)}$ and $\eta_s^{(i)}$. The iteration continues until the validity of applied convergence criterion expires.

In the presence of severe nonlinearities such as those encountered in the stability resonance behaviour of space structures, the convergence can be measured by studying the size of either the unbalanced forces or of the displacement increments. As a rational measure has been found to take the size of the incremental displacement correction as a means of determining when the iterative process should be terminated. The following error vector is introduced

$$\left\{\Delta r^{j+1}\right\} = \left\{\frac{\Delta r_1^{j+1}}{\Delta r_{1,ref}^j}, \frac{\Delta r_2^{j+1}}{\Delta r_{2,ref}^j}, \dots, \frac{\Delta r_m^{j+1}}{\Delta r_{m,ref}^j}\right\}^T \quad \dots 22.$$

where Δr_1^{j+1}, Δr_2^{j+1}, etc. are the changes of the displacement components during iteration number j+1 and m is the total number of non-zero components in the displacement vector Δr. Each component of the error vector is scaled by a reference value $\Delta r_{k,ref}^j$ to obtain a nondimensional measure. To measure the length or the size of the vector a modified Euclidean norm has been found adequate, expressed as

$$\left\|\left\{\Delta r^{j+1}\right\}\right\| = \sqrt{\frac{1}{s} \sum_{k=1}^{s} \left[\frac{\left\{\Delta r_k^{j+1}\right\}}{\left\{\Delta r_{k,ref}^j\right\}}\right]^2} \quad \dots\dots 23.$$

where the division by "s" has been introduced in order to obtain a measure that is not influenced by the total number of elements in the displacement vector. The aforementioned convergence criterion has a direct physical significance and provides a simultaneous indication of both displacements and stresses.

INITIAL IMPERFECTIONS

The initial imperfections were implemented into algorithm of transfer matrix method on the basis of the classical Marguerre's theory, see Ref. 3.

NUMERICAL RESULTS AND DISCUSSION

Studied is the special case of dynamic stability of stiffened shell members of space structures. Studied member is subjected to the simultaneous action of constant tension axial force combined with aperiodic inducing impulse in transverse direction. Present problem is interesting in space shell applications exposed to transverse and axial dynamic parametric loads. Until now, the dynamic stability was dealt with, first of all, in the members subjected to compression loads. Comparatively little attention was focused on the problem of dynamic stability of the members of space structures subjected to the action of parametric tension influences. However, this phenomenon in parametric vibrations can have the equivalent importance as in the aforementioned case. In compressed and tensioned zones, especially in resonance, buckling and postbuckling vibrations of space structures, there can arise the interactive couplings of both types of dynamic instabilities.

For solution of this problem it is inevitable to apply the present nonlinear analyses taking account of alternative initial imperfections which can distinctly influence the resulting response of space structure.

The physical and loading scheme of studied member is shown in Fig 5. The set of constant axial forces H acting in longitudinal stiffener of studied shell member is summed up in Table 1. The initial imperfections assumed in individual nodal points of applied discrete simulation of longitudinal stiffener are put together in Table 2. The moment of inertia of single stiffener is 0,98.10^{-4} m, the Young modulus is 2,1.10^{11} N/m^2, the length of one discrete element of studied stiffener is l_K=1,5 m, further loading parameters due to Fig 5 are: P = = 0,010 MN and a = 10. The boundary conditions: the longitudinal edges free, transverse edges jointed. The parameters of nonlinear material and structural damping were variable depending on the degree of nonlinearity. The problems of elastic-plastic dynamic behaviour were not considered in present study. Analyzed is the influence of variable axial forces on the resulting stability response of studied space member.

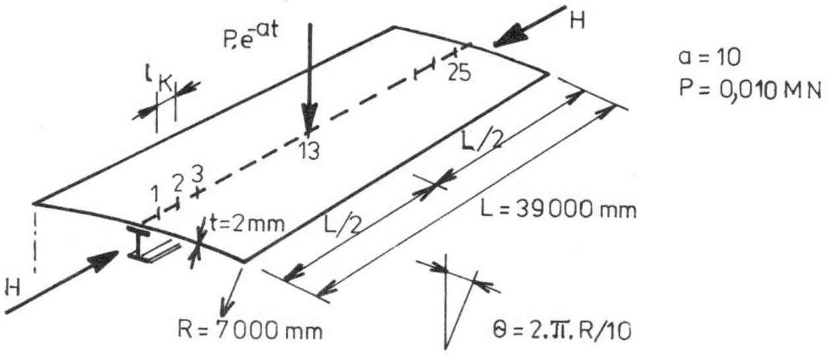

Fig 5

The results of realized numerical study are summed up in Figs 6, 7, 8, 9, 10, 11 and 12. The dynamic response

Table 1

H=1.10^4 N
H=1.10^5 N
H=6.10^5 N
H=1.10^6 N
H=1,5.10^6 N
H=1.10^7 N

Table 2

c=0,001 m /1/	c=0,008 m /8/
c=0,002 m /2/	c=0,009 m /9/
c=0,003 m /3/	c=0,010 m /10/
c=0,004 m /4/	c=0,011 m /11/
c=0,005 m /5/	c=0,012 m /12/
c=0,006 m /6/	c=0,013 m /13/
c=0,007 m /7/	symmetry until /25/

midspan of studied space member in the time-interval 0-4,9 sec is illustrated for various axial constant forces in Figs 6-11. Evident is the influence of axial stiffener-forces on the resulting time-history of bending deflection.

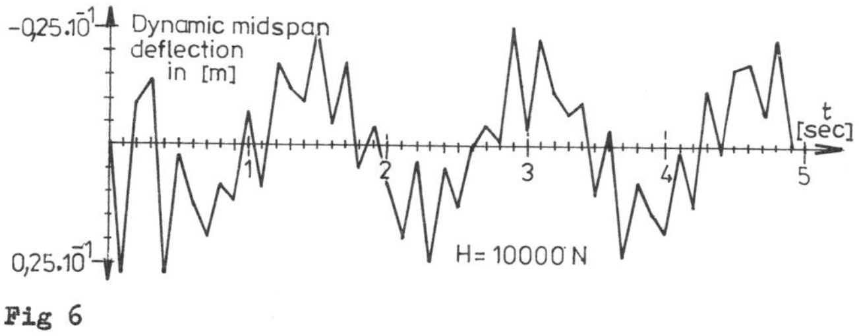

Fig 6

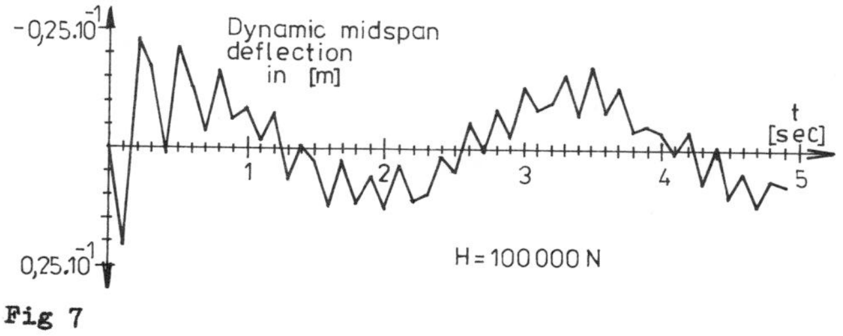

Fig 7

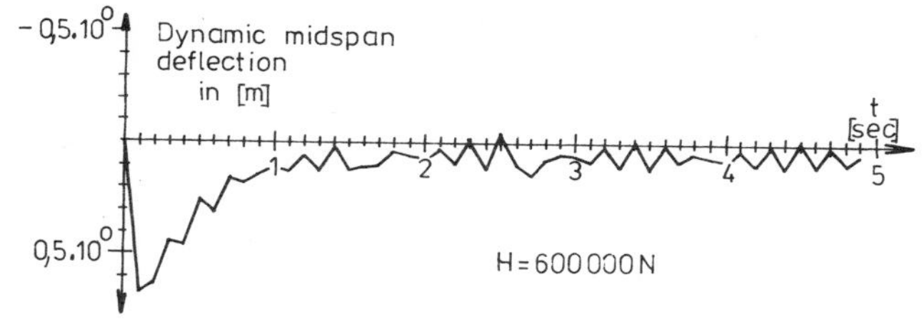

Fig 8

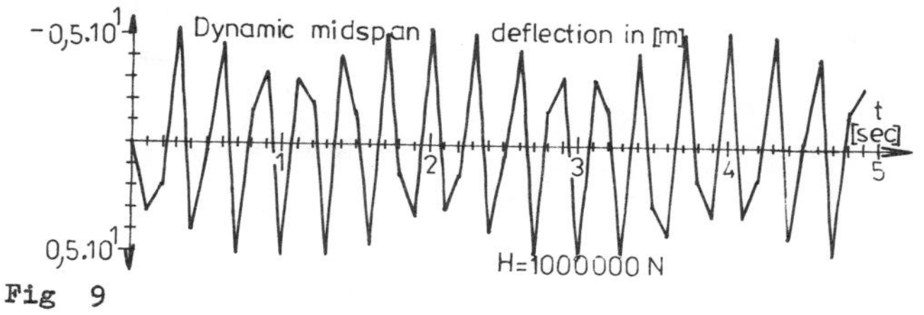

Fig 9

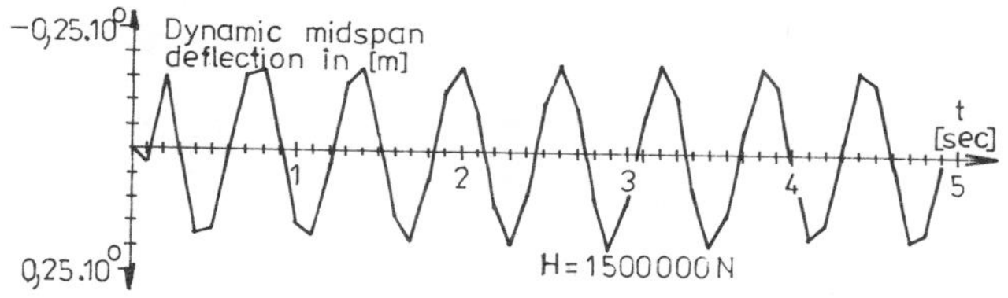

Fig 10

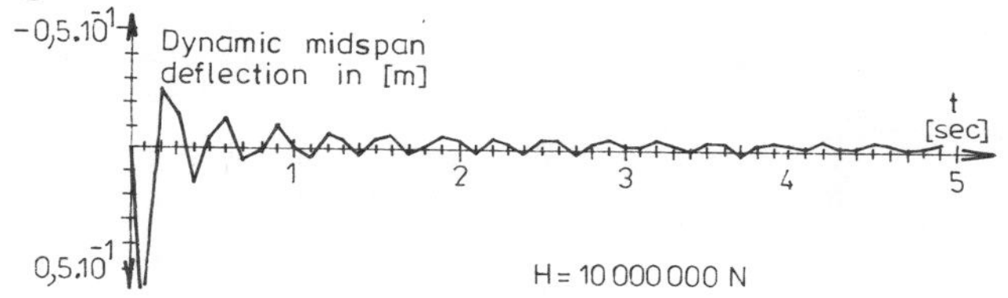

Fig 11

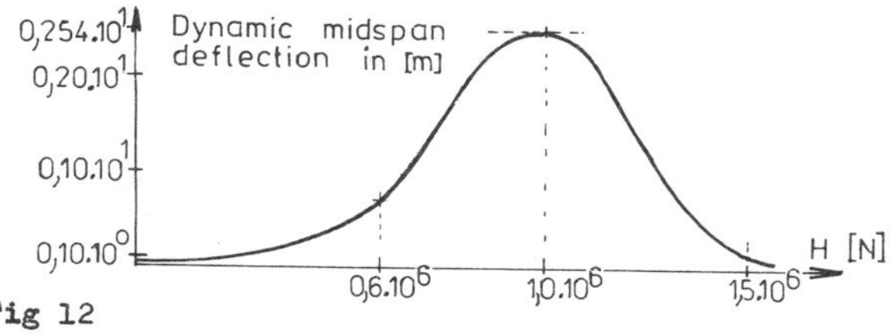

Fig 12

Registered was further the distinct increase of deflection amplitudes at some levels of axial loads H. In such zones there arises the dynamic instability of studied member due to the combined action of assumed loads. The continuous local zone of dynamic instabilities of present type is illustrated in Fig 12.

REFERENCES

1. N M Newmark, A Method of Computation for Struct. Dynamics. Proc. Am. Soc. Civ. Engrs., Vol. 85, 1959.
2. B O Almroth, Future Trends in Nonlinear Struct. Analysis. Computers and Structures, Vol. 10, 1979.
3. A Tesár, Vibration of Thin Shell Structures. ACADEMIA, Prague, 1984, Proc. of ČSAV.
4. A Tesár, Nonlinear Interactions in Resonance Response of Thin Shells. Computers and Structures /in print/.

GEOMETRICALLY-NONLINEAR BEHAVIOUR OF GRIDSHELLS

C. BORRI°, Dr.-Ing. and H.-W. HUFENDIEK°°, Dipl.-Ing.

° Res. Eng., Dipartimento di Ingegneria Civile,
Università degli Studi di Firenze
Florence, ITALY

°° Res. Eng., Inst. für konst. Ingenieurbau, III
Ruhr-Universität
Bochum, WEST GERMANY

On the basis of the Total-Lagrange-Formulation stiffness matrices are derived for a geometrically-nonlinear spatial beam element. Further, an octagonal gridshell (Fig 1) in several load cases is analysed. A linear stability analysis for this shell is implemented, thus allowing to obtain the first three critical load parameters and eigenvectors for each of the investigated load cases.

1. INTRODUCTION

In the last years, the diffusion of spatial gridshells has promoted a new interest for the nonlinear analysis of highly deformable structures.

As already known, in some instances nonlinear analysis of structures is necessary; this is the case when deformations are large enough to require the equilibrium equations to be written for the actual stressed configuration of the structure.

Gridshells are spatial structures made of straight beam elements connected by fixed-end joints. Their form is chosen by turning upside down a suspended net (Refs 1, 2 and 3) under a known load distribution (usually dead weight but no limits are set). Consequently, when the grid is subject to exactly the same load under which its form was determined, a membrane stress state prevails, while bending moments play a secondary role only, and the system, in spite of its lightness, is not highly deformable. However, under different load distributions, for which the grid is relatively unprepared, the bending moments prevail and deformations are large.

The linear analysis, carried out on an octagonal gridshell made of steel tubes (Ref 4), proved the influence of nonlinear effects on the final equilibrium configuration.

In general, nonlinear analysis is useful to investigate three types of problems:

I) analysis of the load-displacement diagram for small loads;
II) determination of critical load parameters;
III) analysis of the post-buckling behaviour.

The analysis of the load-displacement curve gives the main description of the evolution of the system with the increasing of loads.

2. BASIC NONLINEAR THEORY OF THE SPATIAL BEAM ELEMENT

A discretization into spatial straight beam elements is the best representation of the real model of a gridshell (Fig 2). The loads are assumed to be concentrated on the gridshells' joints; the material is supposed to be distributed along a grid of straight beams.

The derivation of this straight spatial element is mainly based on the following assumptions:

a) the thickness of the beam is small compared with its length;
b) stresses and strains, normal to the axes, are neglected;
c) the cross-section remane plane during deformation;
d) the material is assumed to be homogeneous, isotropic and hyperelastic;
e) shear deformations are neglected;
f) St. Venant torsion only is considered (i. e. warping effects are neglected).

Under these assumptions, we can derive a geometrically-nonlinear theory for a beam element (large displacements

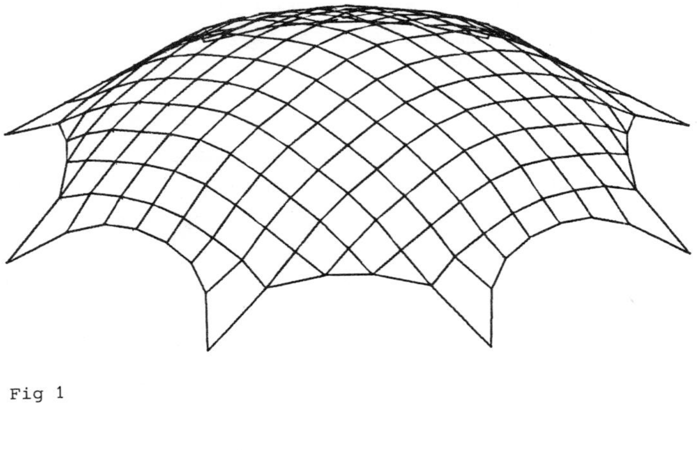

Fig 1

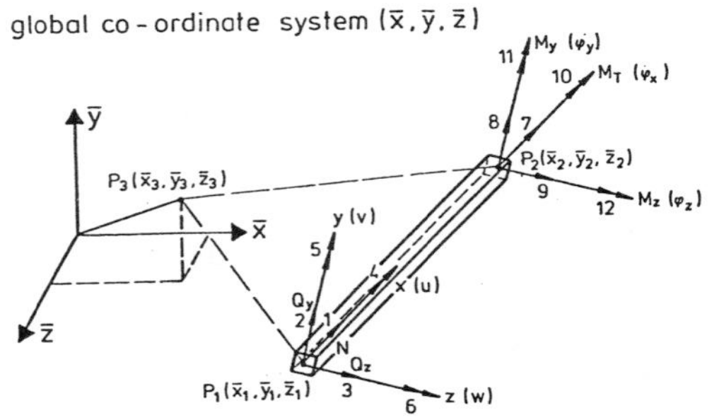

Fig 2

and small rotations) using the so called Total-Lagrange-Formulation (T. L. F.) (Refs 5, 6 and 7).

As represented in Fig 2, any beam element is described by the system of coordinates x, y and z of which the x-axis is parallel to the centre-line in the unstressed state, and the y-, z-axes are parallel to the principal axes of the cross-section. Using the T. L. F., in which all the variables are referred to the initial state of the system, the undeformed (initial) configuration, the current (fundamental) state and the adjacent state are schematically represented in Fig 3.

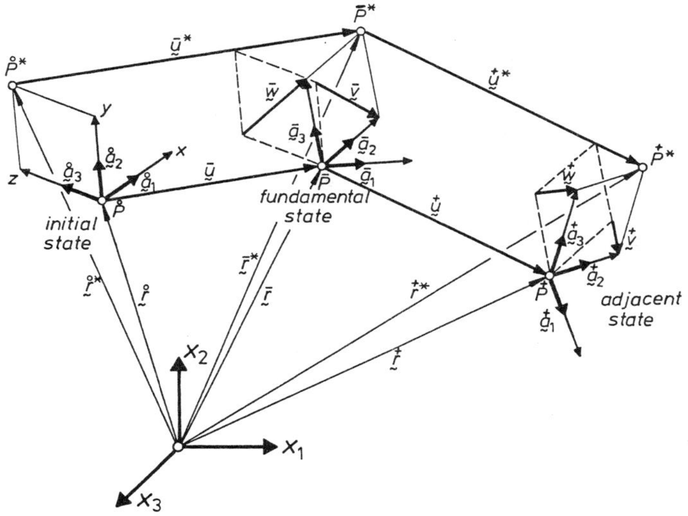

Fig 3

Following the T. L. F., the increment equation of the p-th element (after linearization) can be finally obtained:

$$\underline{K}_T^P \, \underline{v}^P = (\underline{K}_e + \underline{K}_g + \underline{K}_u) \, \underline{v}^P = \underline{P}_e - \underline{P}_i$$

where :

$\underline{K}_t^p$ = Tangent stiffness matrix

$\underline{K}_e$ = Linear elastic matrix

$\underline{K}_g$ = Geometric matrix

$\underline{K}_u$ = Initial displacement matrix $\underline{v}^p$ = nodal displacement vector

$\underline{P}_e$ = External load vector

$\underline{P}_i$ = Internal nodal forces

The integration of these matrices was carried out in a direct way, in relation to the nodal displacements. Several details on the analytical formulation are reported in Refs 8 and 9.

3. COMPUTATIONAL ALGORITHMS

Based on the derivation of No. 2, a calculation procedure has been built to be coupled with FEMAS (Finite Elemente Modulen Allgemeiner Strukturen, Ref 10) general programme. Performing a nonlinear analysis problem, the tangent stiffness matrix and the vector of initial nodal forces have to be built. The problem can be solved by reducing the global vector of out-of-balance forces to zero. This is done by an incremental and iterative procedure, fully described in the literature (Refs 4 and 5).

Describing solution algorithms for static nonlinear problems, the iterative method should be carefully selected; a good convergence of the procedure depends on this choice.

Some of the most common iterative procedures are described in Refs 3 and 4. Using the classical methods, such as Newton-Raphson, difficulties arise in critical zones (i. e. if $\underline{K}_t$ is singular) as easily deduced from Fig. 4, reported in Ref 11.

Following the Riks-Wempner procedure (Refs 12 and 13), iterations are made, starting from configuration P_1'; this is reached by the load increment from initial P_0, along the normal " n " to the tangent in P_1', until configuration P_1 (on the load-displacement curve). This is attained through intermediate positions P_1'' and P_1'''. In this way, the overpassing of zones with horizontal tangent is possible (Fig 5). Combination between different iterative algorithms during the same analysis is allowed because restart option is possible.

4. LINEAR STABILITY ANALYSIS

In the field of stability problems, different behaviours can be observed : "snap-through" behaviour, with an unequivocal local maxumum on the load-displacement curve, and "bifurcation" behaviour, when the main equilibrium branch is crossed by one or more secondary branches.

During the present research, the usual nonlinear analysis has been completed with the stability analysis. In particular, this as been done in order to confirm results of the load-displacement diagram obtained from nonlinear calculation.

Following the statical method (Euler method) buckling load level is attained when, beside the original equilibrium configuration, at least another one exists, infinitely close to it, at the same load level.

Therefore, considering an equilibrium configuration (fundamental state) of the system and assuming that a configuration 2 (adjacent state) is reached with a infinitely small perturbation, the equilibrium condition for state 2 can be written as :

$$\underline{K}_T(\bar{\underline{v}}) \cdot \dot{\underline{v}} = {}^2\underline{P}_e - {}^1\underline{P}_i \qquad \ldots. \ 1$$

where, as known

$\underline{K}_t(\bar{\underline{v}})$ = Tangent stiffness matrix in state 1

$\bar{\underline{v}}$ = Displacement vector

${}^2\underline{P}_e$ = Nodal load vector in state 2

${}^1\underline{P}_i$ = Internal forces in state 1

Now, if we write the conditions expressed in the definition

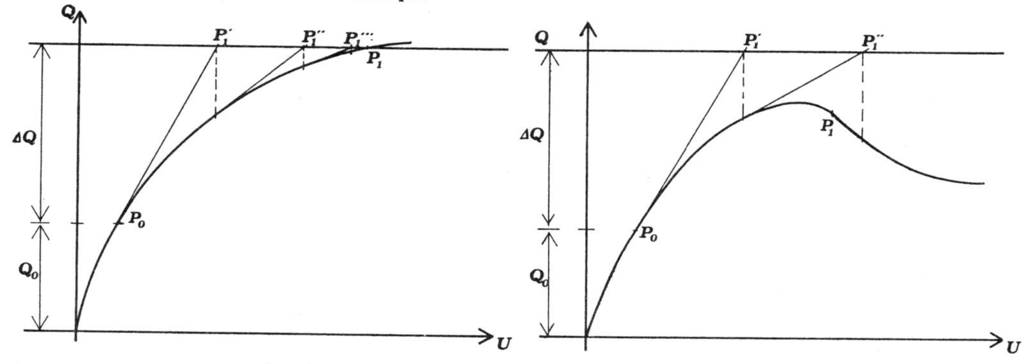

Fig 4 (Ref 11) (a) (b)

of the critical level according to the statical method, we obtain :

${}^{2}\underline{P}_e = {}^{1}\underline{P}_e$ (Equivalence of the external loads in the two infinitely close states)

${}^{1}\underline{P}_e = {}^{1}\underline{P}_i$ (State 1 equilibrium condition)

and together with Eqn 1 we get :

$$\underline{K}_T(\bar{\underline{v}}) \cdot \dot{\underline{v}} = 0 \qquad \text{.... 2}$$

Eqn 2 represents the condition of indifferent equilibrium for state 1. Vector $\overset{+}{\underline{v}}$ represents incremental displacements from state 1 to state 2, and gives the deformation values of the system when instability appears (modal shape). Eqn 2 can be solved either by studying the determinant or by facing an eigenvalue problem on the load parameter. This is expressed as :

$$\det\ (\underline{K}_e + \lambda\ \underline{K}_g) = 0 \qquad \text{.... 3}$$

where, as known :

$\underline{K}_e$ = Elastic component of the stiffness matrix

$\underline{K}_g$ = Geometric component of the stiffness matrix

λ = Scalar load parameter

Following the technique described in Ref 14, the linear eigenvalue problem has been solved.

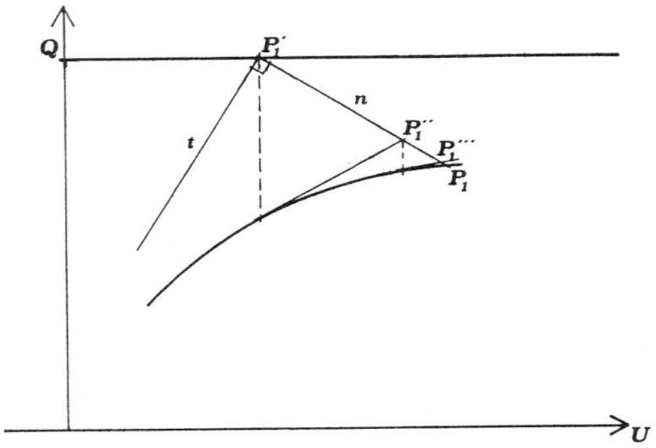

Fig 5

5. ANALYSIS OF RESULTS

The examined test structure is shown in Fig 1 (Ref 2). It is an octagonal based spatial grid (octagon side: 13 m, max height: 9 m); the tubular elements have the same cross-section (76 x 4.5 mm) and length; only boundary elements (arches) were chosen with greater cross-section (219 x 5 mm) and varying length. Cross-section dimensions were obtained from a linear analysis, in which six load cases were considered (Ref 4). The nonlinear analysis has been performed only for three of the mentioned load cases, namely :

a) snow load, uniformly distributed on the whole structure

b) concentrated load on the top joints

c) snow load uniformly distributed on half the structure.

In each of the three cases, the load-displacement has been plotted for the most relevant displacement parameters. Load parameter 'λ' multiplies the load condition (λ=1 for design load) taken to set structural dimensions in the linear analysis. For load conditions a) and b), taking into account the system' s symmetry with respect to two coordinate axes, only one fourth of the structure has to be analysed. For load condition c), analysis has been extended to one half of the structure, as by this load distribution a symmetry with respect to one axis only exists.

a) The curve pattern of vertical displacement at the top is plotted in Fig 6. It should be stressed that, due to numerical difficulties of convergence, calculation could not overpass the critical zone shown by the almost horizontal part of the curve, in spite of repeated efforts. Neither combination of different solution methods, nor sensible reduction of load increments could have a positive effect on the convergence of the procedure. Besides a certain threshold (load level), the presence of more relevant rotations added to "numerical" instability of the stiffness matrix could justify the behaviour of the calculation module. When λ=4.0 is approached, the critical zone appears and a relevant reduction of of load increments is then needed to allow convergence. The lack of an equilibrium downward path (at least within the field that could be investigated) suggests that no "snapping" phenomena are involved and that the curve is more likely to increase also in the so-called "post-buckling" field; this behaviour may be justified by the type of structural bonds and the resulting kind of deformation (Fig 7).

Stability analysis produced the values of the first three critical load parameters and the corresponding modal shapes (Fig 8 (a), (b) and (c)) :

$$\lambda^{(1)} \doteq 4.09 \qquad \lambda^{(2)} \doteq 5.16 \qquad \lambda^{(3)} \doteq 5.37$$

The first critical load value obtained from the analysis of the load-displacement curve is therefore confirmed.

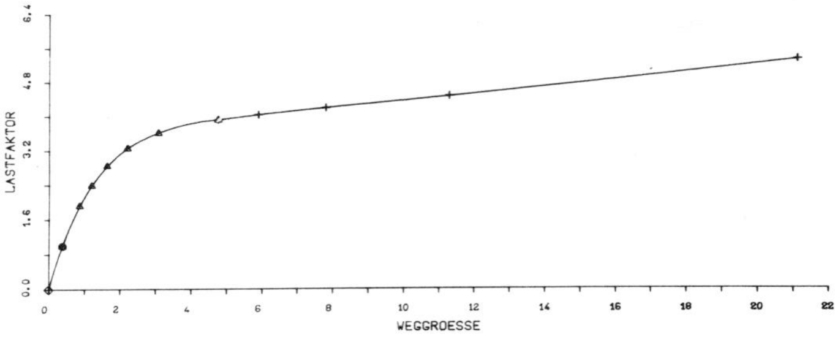

Fig 6

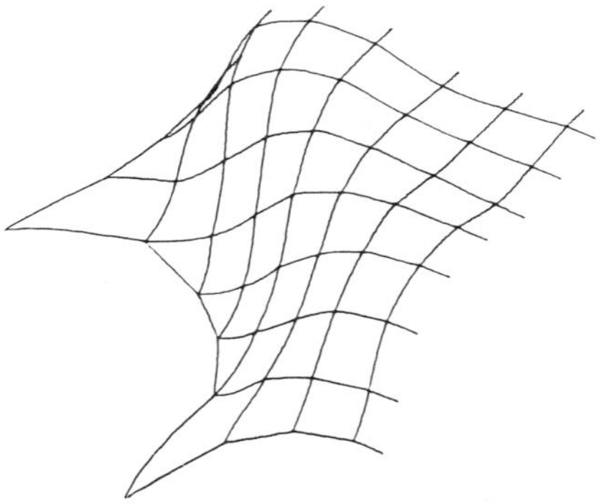

Fig 7

b) In this case, the analysis of the load-displacement curve has not produced the expected informations. As far as investigations could go, the load-displacement curve plotted for four nodal points (still referred to the vertical component) do not diverge from a straight path, as shown in Fig 9. Beyond a given load value, convergence could not be reached; however, nonlinear behaviour did not appear. To exclude that such a nonlinearity might appear only for some particular displacements, analysis has been extended to more nodal points further away from the loaded zone. Investigations show that the shell as a whole is not really influenced by load effects, which concentrate in the top joints. Moreover, the amount of displacements questions the exactness of the chosen calculation model (unsuited, as already said, when relevant rotations take place).

As to stability analysis, the obtained set of critical loads is :

$$\lambda^{(1)} \doteq 25.53 \qquad \lambda^{(2)} \doteq 33.77 \qquad \lambda^{(3)} \doteq 35.62$$

The corresponding modal shapes are represented in Fig 10.

c) For this analysis, the test has been redoubled, obtaining the structure reported in Fig 11 (a). Deformation due to snowload uniformly distributed on only one half of the structure (Fig 11 (b)) clearly shows the zone where upward vertical displacements occured. The load-displace-

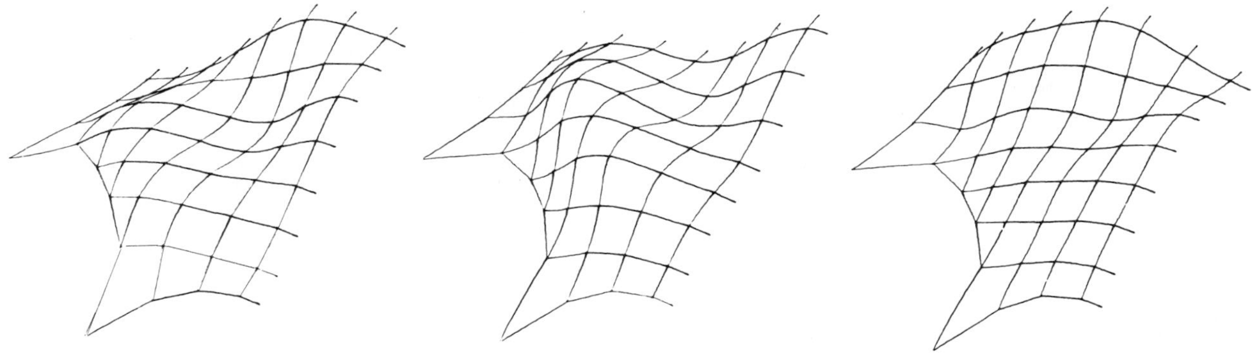

Fig 8 (a) (b) (c)

ment curve was studied in more nodal points (Fig 12) also for this load case. Contrary to load case a), collapse appears more suddenly when the load parameter approximates the value 4.0.

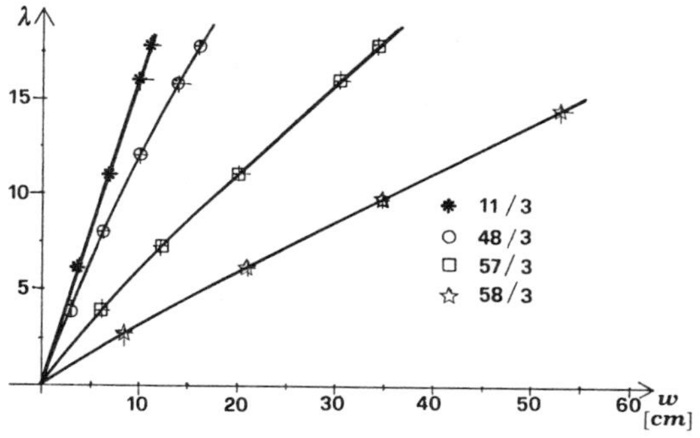

Fig 9

Considering the top joint, just before collapse a rigidity increase is observed. As in case a), stability analysis has fully confirmed the level of the first critical load parameter; the complete set of values is :

$$\lambda^{(1)} = 3.94 \qquad \lambda^{(2)} = 4.59 \qquad \lambda^{(3)} = 5.57$$

The corresponding modal shapes are plotted in Fig 13.

(At the PRIME 750 of the Institut für konst. Ingenieurbau, Ruhr-Universität Bochum, West Germany).

6. CONCLUSIONS

The results of this research offer useful material to discussion at different levels.

First, considering the load cases under investigation there could not be noticed any snap-through behaviour, although the linear zone has been very clearly overpassed. Moreover the investigated octagonal gridshell is likely to have characteristic reaching great deformations without sudden collapse. In addition it must be mentioned that the exactness of the computation module worsens as soon as large rotations appear; to improve the exactness, element subdivision could be applied.

It must also be pointed out that in both cases a) and c) the first critical threshold appears at the same load level. Although an unsymmetrical load influences the dimensioning of the gridshell enormously (because of the appearance of important bending and torsion moments), it obviously has not a great influence on the degree of stability.

Still, for evaluating selected dimensioning for linear analysis, it was anyway encouraging to see safety factor confirmed for the first critical load parameter, considering no secondary equilibrium branches exist. Their presence could be detected, for instance, by a so-called "nonlinear stability" analysis (Refs 15 and 16).

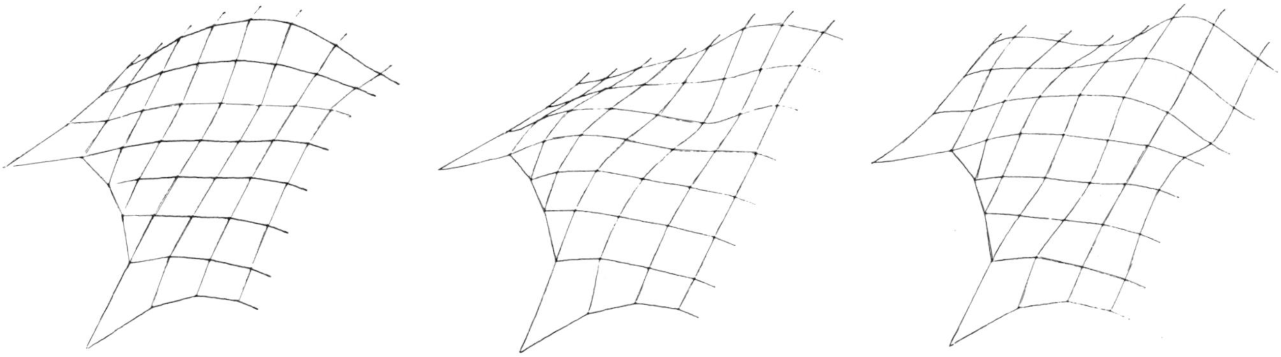

Fig 10 (a) (b) (c)

Finally, calculation times needed to perform nonlinear and stability analysis is reported for the three load conditions described above :

Load case	Nonlinear analysis	Stability analysis
a)	19' 18"	2' 21"
b)	49' 58"	2' 11"
c)	54' 14"	4' 32"

CPU - Times

AKNOWLEDGEMENTS

The research work was developed at the 3rd Chair of the Institut für konst. Ingenieurbau, Ruhr-Universität Bochum (West Germany). We would like to express thanks to Prof. W. B. Krätzig for his support and help, and for putting all the Institute's computation facilities at our disposal. Particular thanks to colleagues Dr.-Ing. H. Beem and Dr.-Ing. U. Eckstein concerning programmation and calculation.

We are also grateful to Prof. Augusti (Florence) for his

support and very important encouragement.

The finantial support of D.A.A.D (Deutscher Akademischer Austausch-Dienst, Bonn) and of C.N.R. (Consiglio Nazionale delle Ricerche, Roma) is also acknowledged.

REFERENCES

1. Gitterschalen, IL-10, Inst. für leichte Flächentragwerke, Universität Stuttgart, 1974.

2. C. BORRI, P. SPINELLI - Gusci grigliati: un metodo per la ricerca della forma, IV A.I.Me.T.A., Firenze, 1978.

3. H. J. SCHOCK - Zur Berechnung von Gitterschalen als biegesteife Stabwerke nach der Methode der finiten Elemente, in Ref 1.

4. C. BORRI - Gusci grigliati: la ricerca della forma e l'analisi statica, Bollettino degli Ingegneri, Firenze, 7/1978.

5. E. RAMM - Geom. nicht-lineare Elastostatik und finite Elemente, Habilitationsschrift, Universität Stuttgart, 1976.

6. K. J. BATHE, S. BOLOURCHI - Large displacement analysis of three dimensional beam structures, Int. Journ. for Numer. Methods in Eng., 14/1979, 961-986.

7. K. BRINK, W. B. KRÄTZIG - Geometrically correct formulation for curved finite bar elements under large deformations,in: WUNDERLICH, STEIN, BATHE - Nonlinear Finite Element analysis in Structural Mechanics, Ruhr-Universität Bochum, Springer Verl., 1981.

8. C. BORRI, H.-W. HUFENDIEK - Geometrical nonlinearity in Space Beam Structures, in preparation.

9. W. B. KRÄTZIG, Y. BASAR, U. WITTEK - Nonlinear behaviour and elastic stability of shells, in: Buckling of shells, Springer Verl., 1981.

10. H. BEEM, K. BRINK, B. LECHTLEITNER - FEMAS (Finite Elemente Moduln allgemeiner Strukturen), Inst. für konst. Ingenieurbau, III, Int. Arbeitsbericht, April 1980.

11. M. WESSELS - Das statische und dynamische Durchschlagsproblem der imperfekten flachen Kugelschale bei elastischer, rotationsym. Verformung, Inst. für Statik, TU Hannover, 23/1977.

12. E. RIKS - An incremental approach to the solution of snapping and buckling problems, Int. Journ. Sol. Structure, 15/1979, 529-551.

13. G. A. WEMPNER - Discrete approximations related to nonlinear theories of solids, Int. Journ. Sol. Structure, 7/1971, 1581-1599.

14. R. B. CORR, A. JENNINGS - A simutaneous iteration algorithm for symmetric eigenvalue problems, Int. Journ. for Num. Methods in Eng., 10/1976, 647-663.

15. B. BRENDEL - Zur geometrisch nicht-linearen Elastostabilität, Dissertation, Universität Stuttgart, 1979.

16. U. ECKSTEIN, R. HARTE, W. B. KRÄTZIG, U. WITTEK - Solution Strategies for Linear and Nonlinear Instability Phenomena for Arbitrarily Curved Thin Shell Structures, 7th Conf. on Struct. Mech. in Reactor Technology, L 6/2, Chicago, 1983, 163-170.

- - - - - - - - - - - - - - -

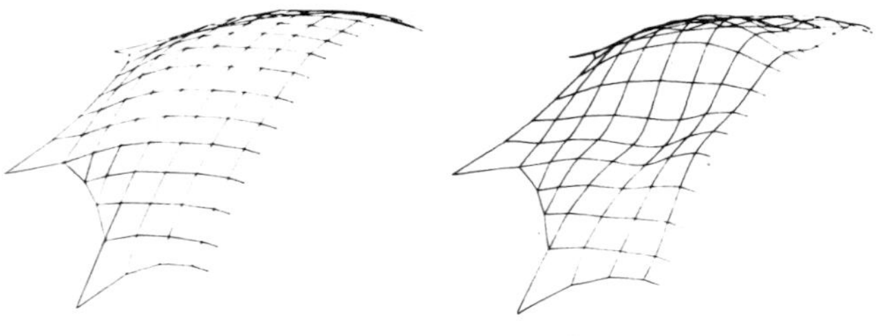

(a) (b)

Fig 11

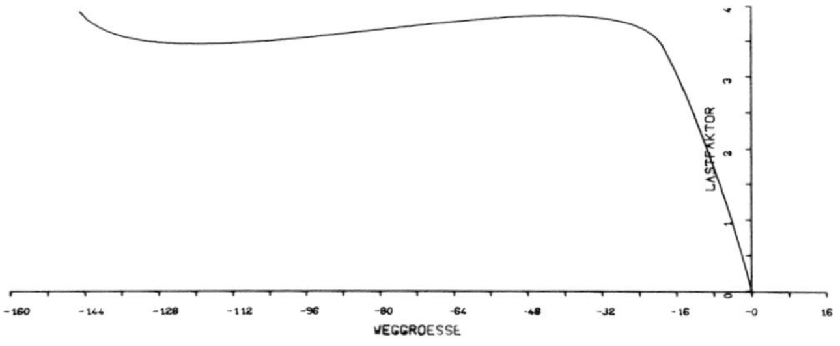

Fig 12

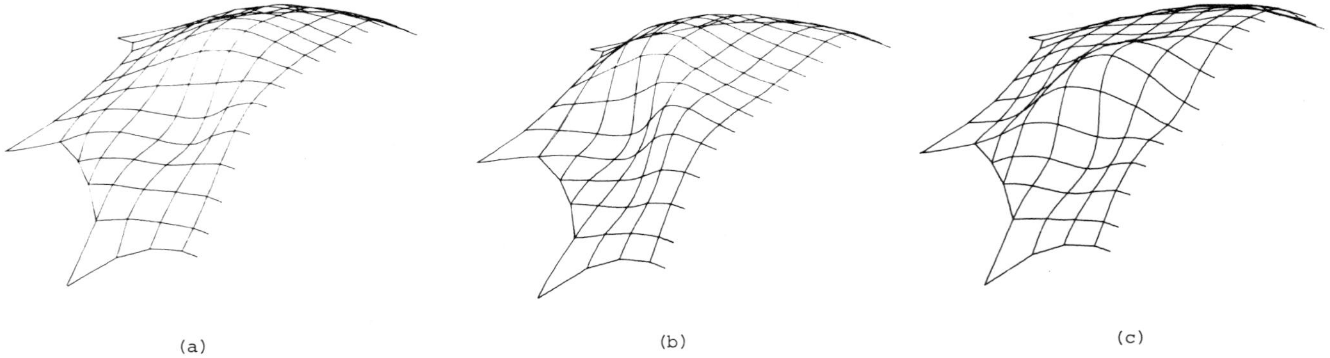

(a) (b) (c)

Fig 13

THE INFLUENCE OF SECOND ORDER VALUES IN THE NUMERIC ANALYSIS OF BAR-STRUCTURES APT TO SNAP-THROUGH

W GLABISZ, PhD

Assistant, Institute of Civil Engineering
Technical University of Wroclaw

In the present paper the author attempts to estimate the influence of second-order values in the numeric analysis of bar-structures apt to snap-through. For two applications of models of the construction analysed, the stiffness of elements is formulated, making allowances for second order values. Numerical testing of the formulated algorithms enabled us to determine the influence and significance of the analysed second order values on the results of the analysis of pre-buckling, buckling, and post-buckling phases in the work of the construction.

INTRODUCTION

Some constructions are peculiar in that under certain conditions they have two or more stable configurations of equilibrium. The transition from one configuration to the other is usually abrupt and is called a "snap-through". To this type of construction belong systems of bars which are analysed in this paper. Such structures are usually used as cover or elements of covers in large usable areas, see Refs 1, 3 and 6. Among the possible methods of analysis of the discussed bar-structure used in this paper is - apart from model analysis and shell analogy - the linear and non-linear numeric analysis based on discrete models.

We developed two models which, depending on the assumptions made, simulated the work of the type of construction considered. The first of the model is a truss made of rectilinear prismatic bars interconnecting at articulated joints. The other model is the system of bars with rigid nodes made of rectilinear prismatic thin-walled elements of arbitrary open cross-sections. These models, treated as ideal (free from imperfections) are subject to conservative transmitted static load described with only one parameter. The way these models behave under load represents the work of the construction for pre-buckling, buckling, and post-buckling ranges which, as regards the determination of safe load, optimal design, and consequences of possible catastrophe, are equally important.

Numeric analysis of the three distinct phases in the work of the construction with special allowances made for the influence and significance of the so called second order values is the main objective of this paper. One of the non-linear components of Green's strain tensor and not as yet considered terms of the functions which determine cross-sectional displacements of a thin-walled element are, in this paper, refered to as second order values.

SPACE TRUSSES

The subject of discussion is a superficial truss structure of arbitrary boundary conditions. The following, commonly used assumptions, were made:

- nodes of the structure are ideal spatial articulations,
- elements of the structure are rectilinear prismatic axisymmetric isotropic bars,
- bars - over the whole loading cycle - do not buckle nor reach the limit load-capacity,
- material, the bars are made of, has linear elasticity and its strains are small,
- large displacements of nodes are possible.

Axial strain of a bar - acc. to the definition of the Green's strain tensor - can be written as

$$\epsilon_{xx} = \frac{\partial U}{\partial x} + \frac{1}{2}\left[\left(\frac{\partial U}{\partial x}\right)^2 + \left(\frac{\partial V}{\partial x}\right)^2 + \left(\frac{\partial W}{\partial x}\right)^2\right] \quad \ldots\ldots 1,$$

where U,V,W are displacement functions of a bar along axes x,y,z, respectively, of a local orthogonal co-ordinate system of a bar. X axis of the reference system is in line with a longitudinal axis of a bar.

Generally, such expression for ϵ_{xx} is assumed which results from Eqn 1 neglecting the term

$\frac{1}{2}\left(\frac{\partial U}{\partial x}\right)^2$. This reduction leads to stresses in an element occuring during its rotation as a rigid body. Due to this, in the case of strongly geometrically non-linear problems, the above simplification may yield wide differences in solutions. In further considerations this term is taken into account and its effect as second order value on the solutions in pre-buckling, buckling, and post-buckling phases is studied.

By virtue of Castigliano theorem we get the relation between force and displacement for the nodes of elements. The matrix operator which transforms displacements into forces is the directional stiffness of an element [k]

$$[k] = [k]_E + [k]_{G1} + [k]_{G2} \qquad \ldots\ldots \; 2.$$

Matrix $[k]_E$ is an elastic stiffness of a bar while matrices $[k]_{G1}$ (linearly-dependent on displacements) and $[k]_{G2}$ (quadratic-dependent on displacements) are geometric stiffness of a bar and further are called stiffnesses of first and second order. Note that the first order stiffness does not form a symmetric matrix.

Incremental formulation of the problem leads to the relation between force and displacement increments. In this case the tangent stiffness of an element $[k]^S$ is the matrix operator

$$[k]^S = [k]_E + [k]^S_{G1} + [k]^S_{G2} \qquad \ldots\ldots \; 3.$$

As before, matrix $[k]_E$ is an elastic stiffness of a bar while matrices $[k]^S_{G1}$ and $[k]^S_{G2}$ form a geometric tangent stiffness of the first and second order.

Directional and tangent stiffnesses were described in the Lagrange reference system. In the modified Lagrange system the tangent stiffness of an element has the form

$$[k]^S_o = [k]_E + [k]^S_{G1} + [k]^S_{G2} + [k]_o \qquad \ldots\ldots \; 4.$$

Tangent stiffness term $[k]_o$ results from taking into account initial stresses in an element.

The equation obtained from the finite elements method and using directional stiffness has the form

$$[K]\,Q = F \qquad \ldots\ldots \; 5,$$

and the incremental equation is written as

$$[K]^S \;\Delta Q = \Delta F + R \qquad \ldots\ldots \; 6.$$

Matrices $[K]$ and $[K]^S$ are directional and tangent stiffnesses, respectively; Q and ΔQ are vectors of displacements and their increments while F, ΔF, and R are vectors of loads, load increments, and unbalanced nodal forces, respectively.

By taking advantage of the derived stiffnesses three algorithms of the solution were formulated and used to observe the paths of construction equlibrium within the displacement-load range. The first of the three algorithms consists in using Eqns 5 and 6 and is applied for pre-buckling range of the construction work, see Ref 4. This algorithm which, in this paper, is used for small load increments can be employed when determining the state of strain for the ultimate but pre-buckling vector of load. The remaining two algorithms based on Newton-Raphson procedure make use of tangent stiffnesses of an element described in Lagrange system and in the modified Lagrange system, see Ref 2. These procedures based on incremental Eqn 6 are employed for pre-buckling, buckling, and post-buckling phases of the construction work. The parameter of load was assumed to be the only variable controlling the calculations. In the instability ranges of such controlled calculations (ranges of limit points) - determined by the extrapolation of the determinant of a tangent matrix of stiffness - the modification of the Newton-Raphson method as proposed by Riks was applied, see Ref 7.

Test calculations were carried out for the Mises truss and for the Schwedler dome.

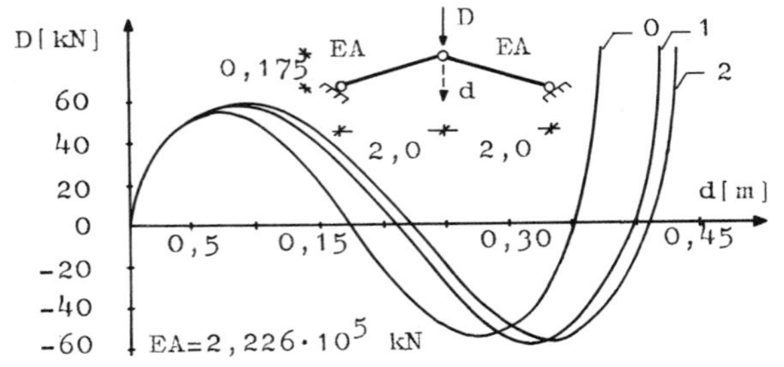

Fig 1

Fig 1 represents the solution for the Mises truss. Curve Ø was obtained with incremental methods realized by the second and the third algorithm. Curves 1 and 2 were plotted using the first algorithm in pre-buckling range and neglecting the analysed second order value to obtain curve 2.

From the numeric analysis of a wide range of truss systems apt to snap-through the following observations can be made:

- the analysis of pre-buckling phases can be performed assuming certain simplifications (neglecting the analysed second order value and second order stiffness) which does not affect the solution in any significant way,

- by assuming certain simplifications in pre-buckling phases, we overestimate the value of buckling load,

- studying into post-buckling phases requires precise solutions in pre-buckling and buckling phases; if the solutions in these two last phases deviate even slightly from the correct ones this may cause considerable disturbances (or divergence) of the solutions in post-buckling phases of the construction work,

- in consideration of the correctness of results and quick convergence, the incremental algorithm as formulated in the modified Lagrange system should be regarded as the best one.

The solution is based on the following assumptions:

- nodes of the structure are rigid, the rigidity of joint concerns rotations and shifts, at the same time the bar-node mount can be rigid or free due to warping,
- nodes, due to their structure, do not transfer warping,
- elements of the system are rectilinear prismatic thin-walled isotropic bars of arbitrary open cross-section,
- in the whole loading cycle the bars do not reach their limit load-capacity,
- material, the bars are made of, has linear elasticity and strains occuring in it are small,
- large displacements of nodes are possible.

The analysis of a bar element (a bar is divided along its longitudinal axis into two elements) is carried out adopting commonly used assumptions formulated by Vlasov, see Ref 8. Considering the assumptions made, the displacement functions for any point of the cross-section of an element can be written as

$$U(x,y,z) = \bar{U}(z) - (y-y_o)\,\Theta(z) - $$

$$-[\tfrac{1}{2}\,\Theta^2(z)(x-x_o)].$$

$$V(x,y,z) = \bar{V}(z) + (x-x_o)\,\Theta(z) - $$

$$-[\tfrac{1}{2}\,\Theta^2(z)(y-y_o)].$$

$$W(x,y,z) = W(z) - \left[\bar{U}'(z) + [\bar{V}'(z)\,\Theta(z)]\,.\right]x -$$

$$-\left[\bar{V}'(z) - [\bar{U}'(z)\,\Theta(z)]\,.\right]y - \Theta'(z)\,\bar{\omega} \qquad \ldots\ldots 7.$$

$\bar{U}$ and $\bar{V}$ are functions describe the displacement in the cross-sectional plane of an element while function W is derived for the displacement along the element's longitudinal axis z. The quantities related to the cross-section shear centre of (x_o, y_o) co-ordinates are denoted with a dash at the top. Constant $\bar{\omega}$ is a principal sectorial area co-ordinate for a pole in shear centre; primes are used to denote derivatives with respect to z.

The form of these functions can be obtained from simple geometric considerations or by Mac-Laurin series expansion of the functions obtained by Klöppel and Winkelmann at the same neglecting the terms of the third, fourth, and higher order, see Ref 5. In the description of the displacment functions, square brackets with a dot are used for those terms which have not as yet been considered in the formulation of thin-walled element stiffness. In our further considerations a parabolic form of functions $\bar{U}(z)$, $\bar{V}(z)$, $\Theta(z)$ and a linear form of functions W(z) are assumed.

Thin-walled element tangent stiffness making allowances for initial stresses can be written as

$$[k]^S = [k_o] + P^o\big[[k_1] + [k_1]^{\cdot}\big] +$$

$$+ M_x^o\big[[k_2] + [k_2]^{\cdot} + [k_2]_{.}\big] +$$

$$+ M_y^o\big[[k_3] + [k_3]^{\cdot} + [k_3]_{.}\big] +$$

$$+ \bar{B}^o\big[[k_4] + [k_4]^{\cdot}\big] \qquad \ldots\ldots 8,$$

where $[k_o]$ is an elastic stiffness matrix with the remaining matrices denoted with [k] and with indices forming the geometric tangent stiffness of an element. With dots those stiffness elements are denoted which formed after taking into account the analysed second order values - a dot at the top means those elements which formed after making allowances for $1/2\left(\frac{\partial W}{\partial z}\right)^2$ in the tensor of strain; a dot at the bottom denotes those elements resulting from a more complete description of displacements. P^o, M_x^o, M_y^o, B^o are initial axial force in an element, bending moments, and initial bimoment, respectively. The effect of a non-dilatational strain has been neglected in the analysis.

The algorithm of the solution is based on the incremental method of Newton-Raphson formulated in a modified Lagrange system. In the determination of internal forces in an element the method of Runge-Kuthego was used. This procedure finds its application in pre-buckling, buckling, and post-buckling phases in the work of the construction. In the range of limit points the algorithm is modified acc. to the procedure which, while maintaining the control of load, makes possible to pass instability regions of the method.

The developed program was tested on thin-walled Williams frame. Fig 2 represents curves for vertical displacements of the loaded node of the frame provided with constraints hindering its buckling from plane.

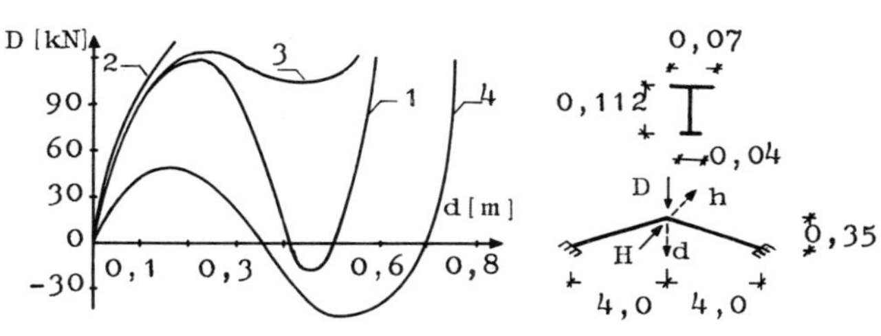

Fig 2

Curve 1 represents the basic solution. Curve 2 was plotted for the bar slenderness ratio four times smaller than that shown by the dimensions in the Fig 2. Curve 3 is for experimentally obtained path of frame stability, see Ref 9.

Curve 4 represents the solution for the Mises truss of the same form and axial stiffness as the frame considered. After having introduced the lateral load of the node (H), thus treating the problem as spatial, the dependence of the vertical, d, and horizontal, h, displacements of the node on its loading took the form as shown in Figs 3 and 4.

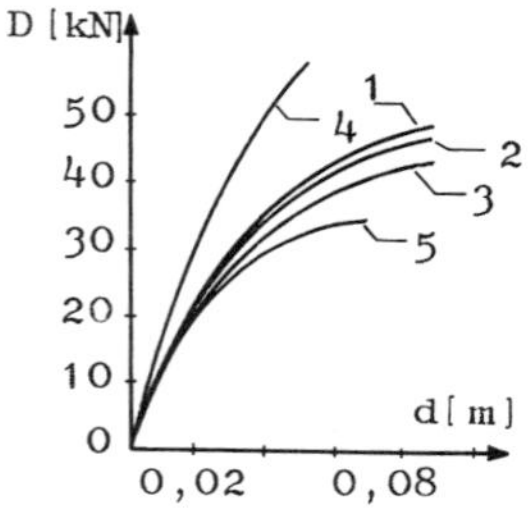

Fig 3

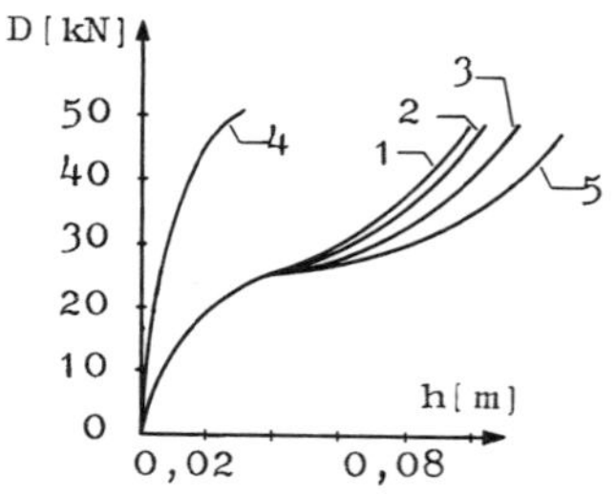

Fig 4

Curves obtained making no allowances for additional elements in the stiffness matrix are marked with 1; curves denoted with 2 are those plotted when taking into account all constituents in the Green's tensor component; curves 3 are those resulting from the analysis completed with additional terms in the functions describing displacements in an element. Curves 1, 2 and 3 were plotted for the lateral load equal to 1/10 of the vertical load. Curves 4 are equivalent to curves 1 with the stipulation that in this case the lateral load was 1/200 of the vertical load. The described diagrams were plotted for a rigid bar-node mount owing to the warping of bar ends; curves 5 represent the solution making allowances for second order values for free bar-node mount due to warping where the lateral load was 1/10 of the vertical load.

A number of various numeric tests showed that:

- the effect of second order values is insignificant in pre-buckling phase, while in post-buckling and, in particular, in buckling ranges (characterized by a considerable gradient of displacement increments) their quantitative significance was observed,

- the way a bar is mounted in a node is of importance due to the possibility of warping in case of considerable torsional deflexion of elements,

- treating the problem as plane one by excluding from the analysis certain degrees of freedom leads to wide differences compared to spatial solutions.

CONCLUSION

In the paper the author attempted to estimate the influence of second order values in the numeric analysis of bar-structures apt to snap-through. Bar-structures subject solely to static conservative transmitted load were considered.

In the analysis, the term - second order values - was given to the usually neglected component of Green's strain tensor and to not as yet considered terms of functions which determine the displacement of an element. When investigating into the effect of these values the stiffnesses of elements were formulated and then adopted to the analysis of two models of bar-structures described in this paper.

The tests, carried out for the developed algorithms enabled us to draw the following conclusions:

- the effect of the analysed second order values is negligible in pre-buckling ranges of the work of the construction,

- buckling ranges are clearly affected by additional terms in equations describing displacements; the effect of the component of strain tensor is negligible,

- in post-buckling ranges the effect of the component of strain tensor gets more significant although it may still be neglected,

- making allowances for second order values in the analysis results in decreasing the buckling load of the construction,

- the incremental method based on the tangent matrix of stiffness which was derived with making allowances for initial stresses in an element provided us with a particularly convenient instrument in the analysis of pre-buckling, buckling, and post-buckling phases in the work of the construction.

REFERENCES

1. Bibliography on Latticed Structures, Journal of the Structural Division, vol. 98, No ST7 , 1972.

2. R D Cook, Concepts and Aplications of Finite Element Analysis, John Wiley and Sons Inc., 1974.

3. W E Haisler and I A Stricklin, Development and Evaluation of Solution Procedures for Geometrically Nonlinear Structural Analysis, AIAA J., vol. 10, No 3, 1971.

4. D Jagannathan and H I Epstein and P Christiano, Nonlinear Analysis of Reticulated Space Trusses, Journal of the Structural Division, vol. 101, No ST12, 1975.

5. K Klöppel and E Winkelmann, Experimentelle und Theoretische Unter Suchungun Uberdie Traglast von Zweiachsig Aussermittig Gedrükten Stahlstäben, Der Stahlbau, vol. 3, No 2, 1962.

6. Latticed Structures: State-of-the-Art Report, Journal of the Structural Division, vol. 102, No ST11, 1976.

7. E Riks, An Incremental Approach to the Solution of Snapping and Buckling Problems, Int. J. Solids Structures, vol. 15, No 7, 1979.

8. V Z Vlasov, Thin-Walled Elastic Bars, Moscow, 1940.

9. F W Williams, An Approach to the Nonlinear Behaviour of the Members of a Rigid Jointed Plane Framework with Finite Deflections, Quart. I1. Mech. Appl. Maths., vol. XVII, 1964.

IDEALISING THE MEMBERS BEHAVIOUR IN THE ANALYSIS OF PIN-JOINTED SPATIAL STRUCTURES

Usama Rustom Madi*, BEng,MSc, PhD, DIC

* Assistant Prof., Civil Engineering Department, University of Jordan, Amman ,Jordan.

The constitutive relationship of the members, assuming that it has been obtained adequately either by experiment or by a theoretical model, needs to be idealised before use in an elastoplastic analysis of the structure to trace the structural response from the onset of loading until complete collapse. Various such idealisations have been suggested by researchers in the field. These idealisations are presented, studied and contrasted in an attempt to define rational basis for idealising the members' constitutive relations.

The effect of using increasingly more accurate piecewise linearised idealisations of the constitutive relations of the members is also investigated by means of example structures. A simple compression fan, a double layer grid and a braced mast are selected as example structures and their behaviour traced up to complete collapse using increasingly more accurate piecewise linearised idealisations of the members' constitutive relations .

INTRODUCTION

It is necessary to idealise the constitutive relations of the constituent members for conducting an elastoplastic analysis of the structure that traces its response up to complete collapse. This involves the derivation of such a relationship as well as idealising it in an easy to apply form (most conveniently a piecewise linearised form).

Various idealisations have been suggested by several researchers in this field. Figure 1 (Supple and Collins, 1978) shows some of the idealisations suggested. In this figure p is the axial compression force and X is the axial shortening.

Figure 1a shows elastic-perfectly plastic conditions assumed for both tension and compression. This idealisation is unrealistic for the compression case. Figure 1b shows the idealisation suggested by Cogan (1975) which assumes no load-shedding but includes 'failure bands' such that, if a member becomes critical, any other member whose load value lies within one of the bands is also assumed to have failed. Figure 1c suggests a certain 'plateau' after buckling followed by an instantaneous decrease in load to a residual load level. It is also possible to have a plateau of zero length, producing what Schmidt (1976) calls a brittle type of strut buckling. Dickie and Dunn (1975) combined a nonlinear elastic compression phase with the sudden post-critical load reduction to produce figure 1d. Supple (1978) in figure 1e and Lin (1977) in figure 1f used a simple structural model to examine the effect of variation of the post-critical slope of the member characteristic on structural collapse, the effects of member imperfections were also investigated. La Tegola (1974) in figure 1g and Karczewski (1977) in figure 1h proposed and used member characteristics with non-linear post-critical regions including non-linear unloading paths.

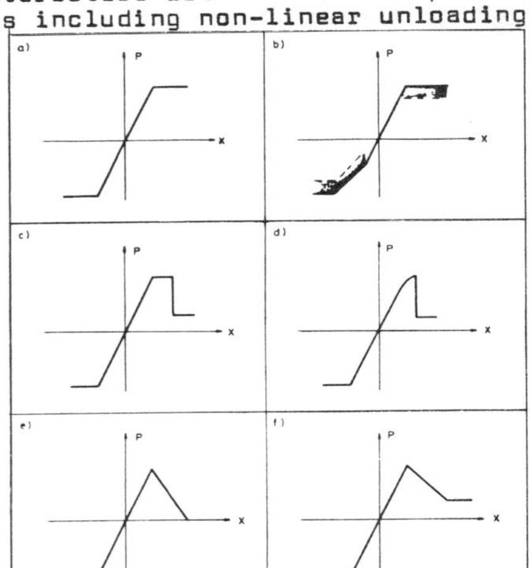

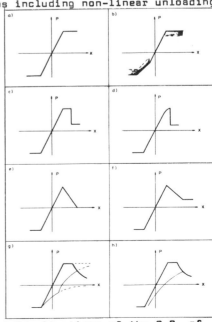

Fig. 1 Idealisations of the C.R. of Pin-Ended Members .

From this survey of idealisations, it becomes apparent that the problem of selecting an idealised constitutive relationship is far from being settled. This selection is of an extreme importance to any subsequent analysis as the validity of such an analysis is highly dependent on the validity of the idealisation. It appears that a large number of model tests and theoretical modelling are required to establish basis for selecting an idealised constitutive relation which gives predictive analysis results and is tolerable with regards to the complexity and cost of the analysis.

CONSTITUTIVE RELATION OF THE MEMBERS

By examining the results of some experimental and theoretical studies on the members' behaviour (Madi, 1981), a closer look at the idea of idealising the constitutive relationship of members in pin-jointed spatial structures can be developed as follows :

1. For the member under a tensile force, the behaviour is identical to that of a specimen under a direct tension test with the resulting well known linear elastic-perfectly plastic stress-strain diagram of mild steel. Therefore it can be assumed that a reasonable idealisation of the behaviour of the member in tension can be given by a linear elastic line followed by a perfectly plastic line at the yield load of the member.

2. For the member under a compressive force, the behaviour can be idealised in three phases:

   a. The stable phase. This phase starts with the loading of the member and ends when the load reaches its critical buckling load. Apart from very slender members with large initial imperfections, it seems that the whole of this phase could be idealised by a single straight line from zero load to the buckling load . The slope of this line can be taken to be the same as that of the same member in tension and is equal to $(\frac{EA}{L})$. It seems that, apart from very slender members with large initial imperfections, which are very seldom if ever encountered in real structures,the slope of this line is not affected by the member end condition or the initial imperfection. The other parameter required to fully describe this phase, namely its limit point (the buckling point) is contrastingly very sensitive to the member end condition and initial imperfection. The selection of this point in the idealised constitutive relationship is therefore subject to many factors that can be accounted for only by extensive testing and the use of statistical and probabilistic techniques.

   b. The softening phase. This phase starts with the buckling of the member and ends with the member retaining a residual load of relatively fixed value with further deformation. This phase is characterised by instability or loss of the member force with further deformation and it can be idealised by one or more straight segments of negative slope. In other words, it can be piece-wise linearised. The slope of the line or lines describing this phase is affected by the slenderness ratio of the member, its initial imperfection and end conditions. Furthermore, as observed by Smith et al (1979), it is possible in this phase for the axial deformation to tend to decrease with increasing central deflection. This is, of course, prevented from happening either by the testing machine in experiments on members, or by the stiffness of neighbouring members in real structures. To overcome this difficulty Smith et al (1979) suggested that, in the application of their predictive finite element model to the deriving of constitutive relations of members, the application of the external axial load to the member be carried out by means of springs. As such the member is always to have enough stiffness at its ends to prevent it from reducing its axial deformation. They did not , however, give any indication of a suitable value to be taken for that elastic stiffness. In experiments on members this stiffness must depend on the testing machine, its settings and the rate of deformation it imposes upon the member. In real structures, this stiffness depends on the method of joining the member to other members in the structure, on the stiffness and geometry of members sharing the end joints of the member under consideration and on their state of loading. In other words, the slope or slopes of this softening phase is not solely a function of the member properties but of the complete structure in which the member is embedded.

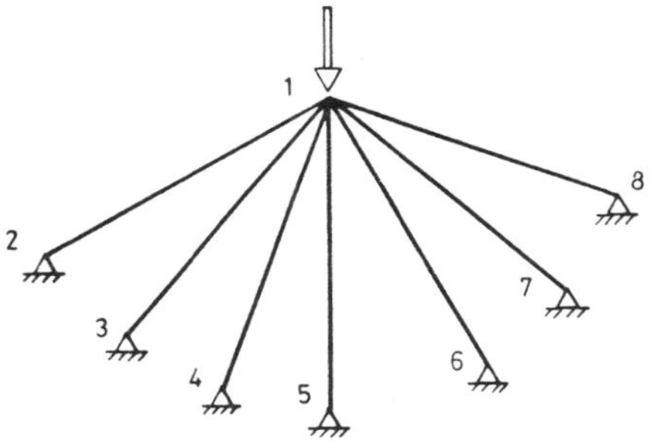

Fig. 2 Compression Fan

   c. The plastic phase. This phase begins at the end of the softening phase and is characterised by an almost constant axial residual force in the member , creating an end plateau in the idealised constitutive relationship. The value of this residual force is affected by the member slenderness ratio as well as its end conditions. The initial imperfection seems to have negligible effect, if any, on the value of this residual force.

EFFECT OF DIFFERENT IDEALISATIONS ON STRUCTURAL RESPONSE

To study the effect of idealising the constitutive relationship, the behaviour of example structures, was traced using different idealisations of the same constitutive relationship. The first structure chosen is the compression fan shown in figure 2. It was chosen becuase all of its members are under the effects of compressive axial forces, and, as such, is likely to show the most marked effect of the different idealisations of the member behaviour under compression on the overall behaviour of the structure. Figure 3 shows the constitutive relationship of the members of the compression fan as well as the different idealisations used in the subsequent analyses of the fan.

Idealisation 1 leads to a plastic limit analysis of the fan with the limit load of the member equal to its buckling load. The result of this analysis is a collapse load and an associated mechanism. If this mechanism is allowed to develop, the softening effect of the members results in the mechanism line shown in figure 4. Idealisation 2 leads to a plastic limit analysis of the structure with a limit load of the member equal to its residual load. The result of this analysis is a collapse load which is shown to be the load at collapse by all the other models and a collapse mechanism which is shown to be the collapse mechanism of the structure by all the other models. From figure 4, it becomes clear that a plastic limit analysis of the structure with the limit load being either the buckling load or the residual load is of little value in predicting the structural response to loading as it completely misses the actual structural response for almost all of

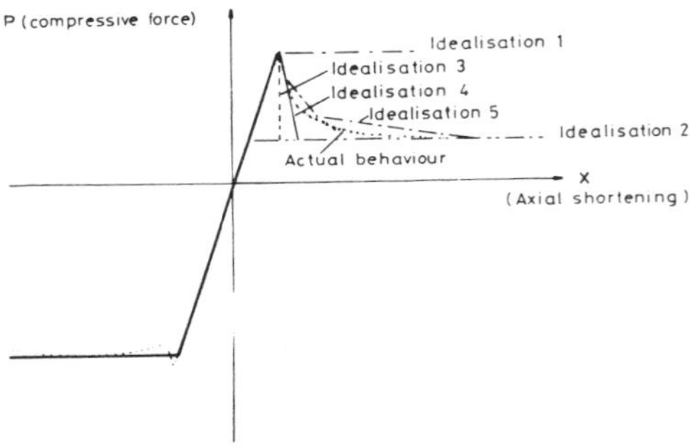

Fig. 3 Different Idealisation of the Member Constitutive Relations

its load history. The limit analysis involving the residual load is of some value in that it gives some useful information - the load at collapse and the collapse mechanism. However, this information is of little practical value since what is really required from any analysis process is to calculate the maximum load the structure can carry, to predict with some accuracy the structural response up to that load level and then to obtain the slope or sharpness of the initial post-buckling equilibrium path. Idealisation 3, which assumes an immediate decrease of the member load, without any corresponding deformation , to the residual load level, is of some historical importance as it is often associated with the member replacement technique for the elastoplastic analysis of spatial trusses suggested by Schmidt (1976) . However, it can be shown that Schmidt's technique has little relation with this idealisation (Madi, 1981).

This idealisation results in a very abrupt structural response after the buckling of the first member and ends with the same collapse load and mechanism as idealisation 2. Clearly, the assumption of this immediate decrease in the member force causes a similar decrease in the load the structure is carrying after the buckling of any of its members, since no deformation of the member is allowed while this apparent load shedding takes place, and accordingly the adjacent members are not given the opportunity of taking the load the buckled member is trying to shed. Idealisation 4 assumes a linear softening phase with a slope equal to the softening slope immediately after buckling. It predicts with reasonable accuracy the structural response up to the maximum load level and gives the slope of the resulting post-buckling curve, thus giving an idea about its abruptness and severity. However, with the increase of the deformation of the structure, the assumed softening, which is steeper than the actual one for the whole softening phase except its beginning, causes the kinematic response of the structure to be steeper than the actual one. This causes the structure to seem to fail with a smaller deflection than the actual one and with a steeper post-buckling path than in reality. Idealisation 5 assumes a piecewise linearised softening phase consisting of three straight lines with successively reducing slopes. It gives a much smoother structural response curve after initial buckling, a response which is closer to the actual behaviour of the structure. This shows that a better prediction of the structural response can be obtained by increasing the number of lines used in idealising the constitutive relationship, particularly in the softening phase. However, this adds to the complexity of and computation effort required by the structural analysis method employing the idealisation. A balance must, therefore, be achieved between the simplicity of the idealisation and the accuracy of the analysis. To examine further the effect of using increasingly more accurate piecewise linearised idealisations of the constitutive relations of the members, a double layer grid and a braced mast have been analysed using the two idealisations shown in figure 5. Idealisation(a) utilises a single line of fixed slope for the softening phase while idealisation(b) utilises three straight lines with successively reducing slopes.

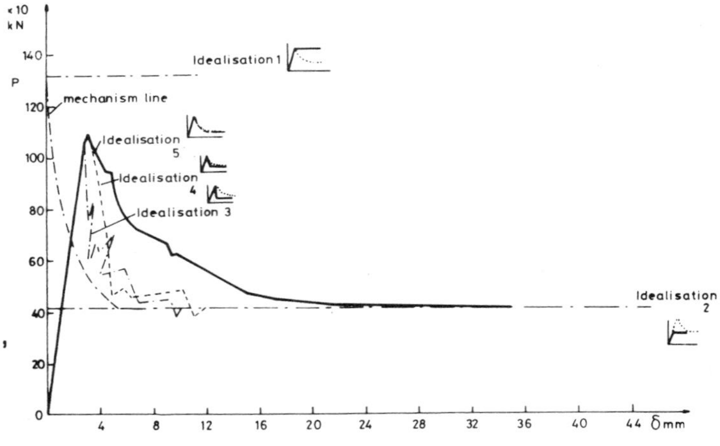

Fig. 4 Effect of Different Idealisations on Structure's Response.

The grid chosen consists of a (6x6) module double layer grid of the type 'square on square', supported on four points in its top grid. The plan and profile of the grid as well as the results of the analyses are shown in figure 6,7 and 8. As shown in these fugures, the behaviour of the double layer grid can be traced through its load history by a series of linear responses. The mast chosen is supported on its bottom four joints as shown in figure 9.

The behaviour of this braced mast is investigated for two loading cases. The results of this investigation are shown in figures 10 and 11. As shown by the four figures, the predicted behaviour using the two idealisations exhibits marked resemblance. This shows that idealisation (a) is adequate in predicting the pattern of structural behaviour under load. Idealisation (b) gives a smoother structural response which is more in conformity with the real behaviour. A choice between these two idealisations can therefore be based only on considerations of the cost and accuracy of the required analysis.

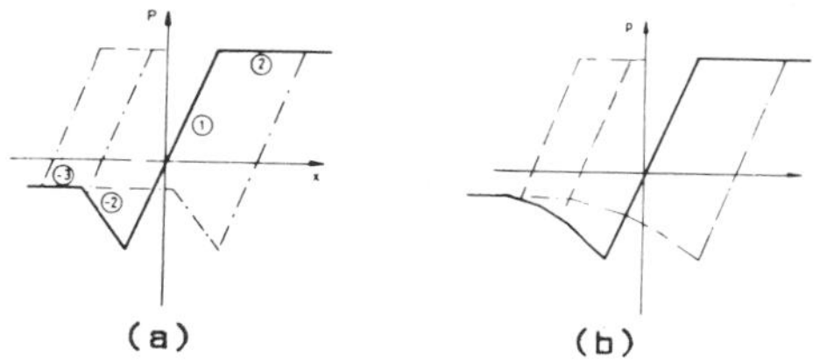

Fig. 5 Idealised Constitutive Relationship .

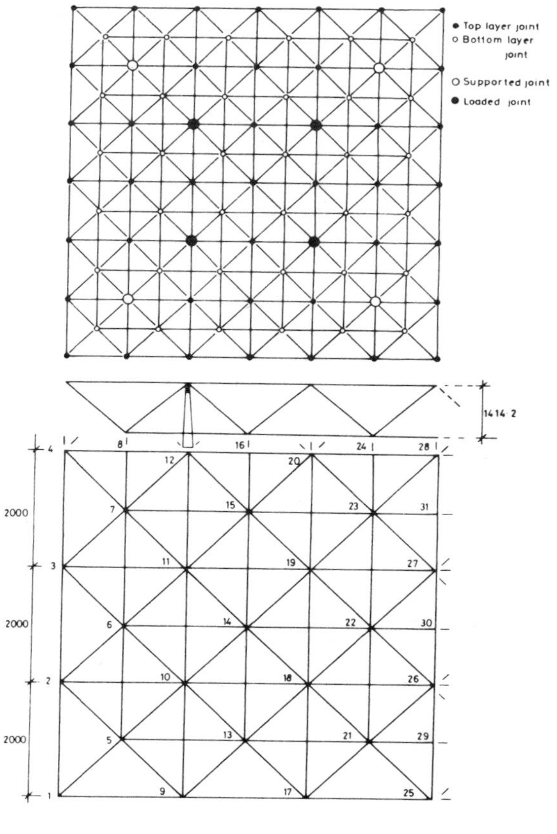

Fig. 6 Geometry of Double-Layer Grid

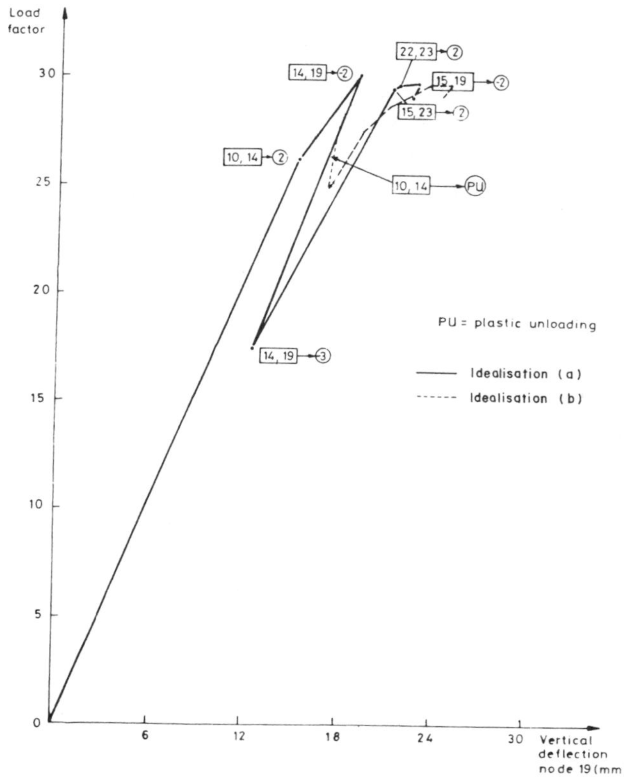

Fig. 7 Load-Deflection Plot for Four Symmetrical Point Loads .

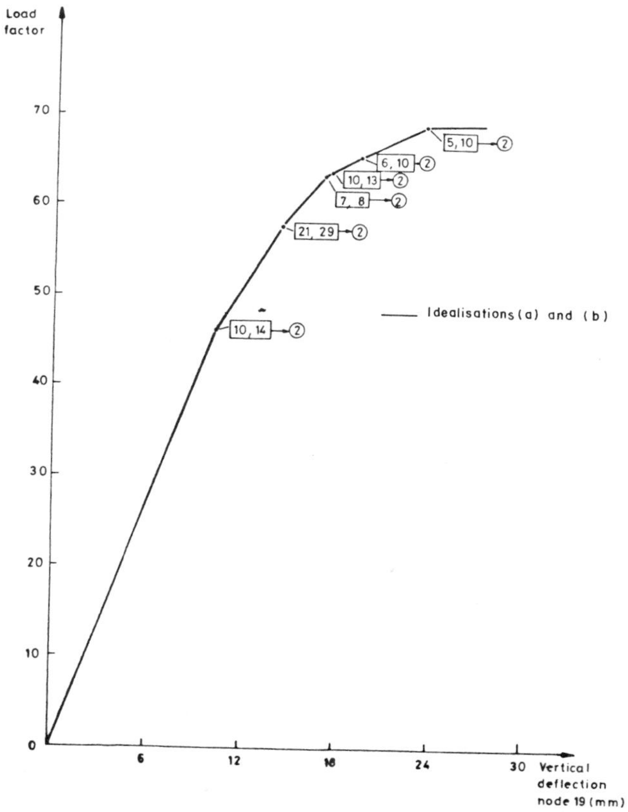

Fig. 8 Load-Deflection Plot for a Uniformly Distributed Load .

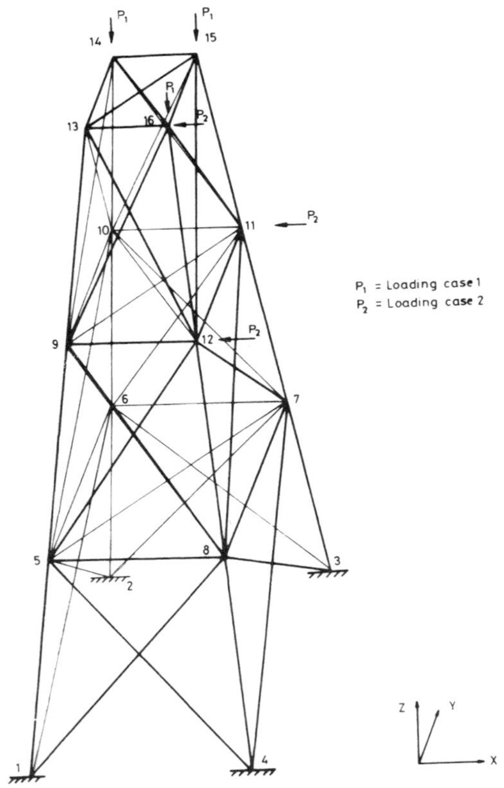

Fig. 9 Geometry of the Braced Mast

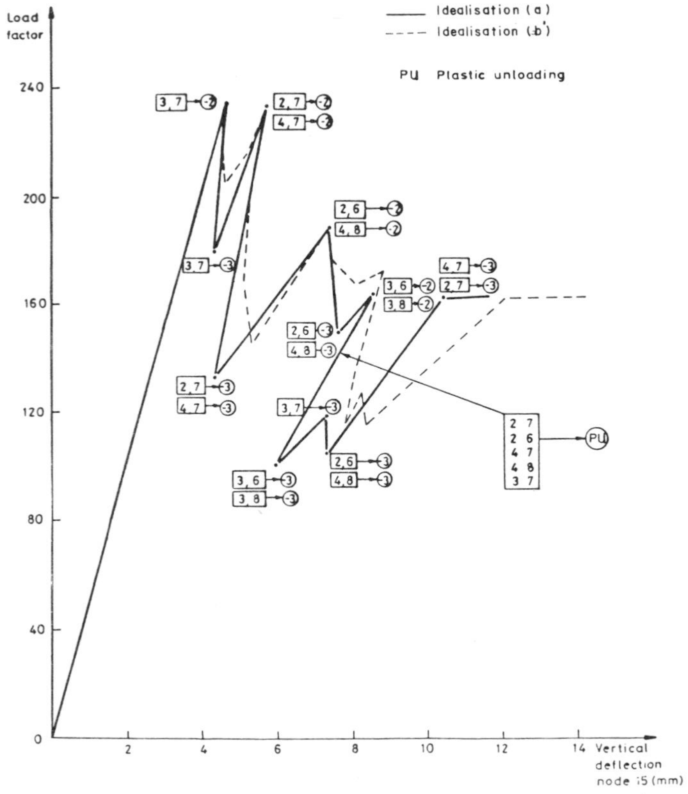

Fig. 10 Load-Deflection Plot for Load Case 1

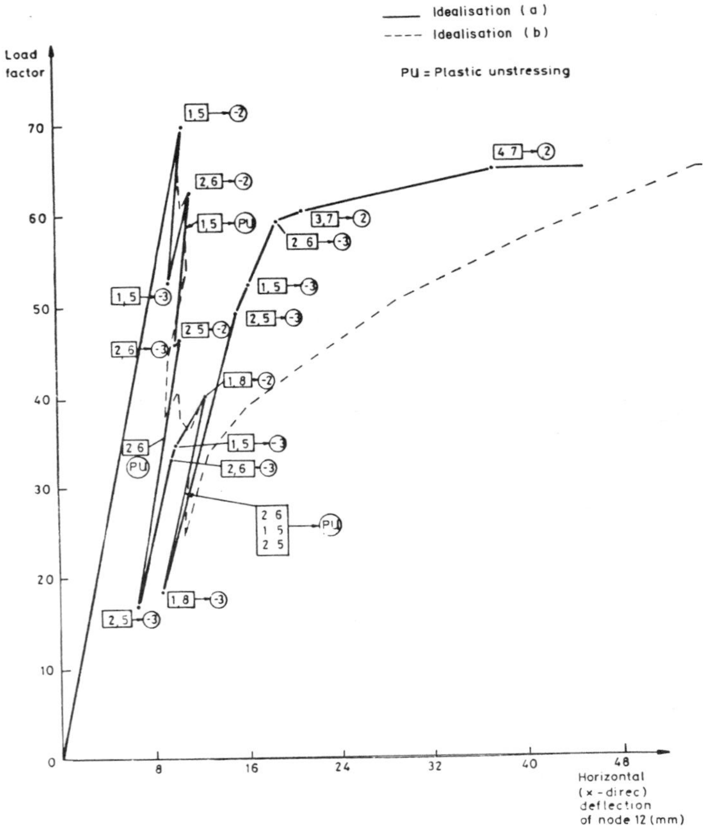

Fig. 11 Load-Deflection Plot for Load Case 2

CONCLUSIONS

It seems that still much research work is needed to formulate acceptable basis for idealising the constitutive relations of members for use in the elastoplastic analysis of pin-jointed spatial structures. The selection of a constitutive relation affects the accuracy and cost of this analysis. In general the behaviour of the member in tension can be idealised by a linear elastic-perfectly plastic response and in compression this behaviour falls into three phases; elastic, softening and plastic. The elastic and plastic phases may be idealised adequately by straight lines. Idealising the softening phase by a line of fixed slope is adequate in predicting the pattern of structural behaviour under load. Increasing the number of piecewise linearisations of the softening phase increases the accuracy and cost of the analysis.

REFERENCES

1. D. BENEDETTI, and V. IONITA, Non Linear Analysis of Space Trusses, Tech. Rept. 27, ISTC, Politecnico di Milano, 1974.

2. K. COGAN, A Limit State Analysis Program for Multi-Layer Grids, Proc. of the Second Int. Conf. on Space Structures, Univ. of Surrey , 1975.

3. A. LA TEGOLA, Analisi Incrementale di Strutture Reticolari Metaliche in Campo Plastico con Aste in Regime Post-Critico, Giornale del Genio Civile, Anno 110, Fasc 7-8-9 , 1972.

4. P.K. LI, An Analytical Investigation of Failure Patterns and Ultimate Loads of Pin-Jointed Space Structures, MSc thesis, Univ. of Surrey, 1977.

5. T.H. LIN, Theory of Inelastic Structures, John Wiley and Sons, 1968.

6. U.R. MADI, The Elastoplastic Analysis of Pin-Jointed Spatial Structures, PhD thesis, Univ. of London, 1981.

7. G. MAIER, Behaviour of Elastic-Plastic Trusses with Unstable Bars, Proc. ASCE, Eng. Mech. Division, Vol. 92, 1966.

8. Z.S. MAKOWSKI, and H.G. LEE., Study of Factors Affecting Stress Distribution in Double-Layer Grids of the Square and Diagonal Type, Arch. Sci. Review, Vol. 20, No. 4, 1977.

9. G. A. OLIVITO, A MARINETTI and A. BADALA, The Collapse Load of Pin-Jointed Networks of Randomly Imperfect Rods, Eng. Struc., Vol. 2, 1980.

10. G. PRETE, M. MEZZINA and A TOSTO, Automatic and Experimental Analysis for a Model of Space Grid in Elasto-Plastic Behaviour ,Proc. Second Int. Conf. on Space Structures, Univ. of Surrey,1975.

11. L.C. SCHMIDT, and P.R. MORGAN, Space Trusses with Brittle-Type Strut Buckling, Proc. ASCE, Struc. Division, Vol. 102, 1976.

12. L.C. SCHMIDT, P.R. MORGAN, A.J. O'MEGAHER and K. COGAN, Ultimate Load Behaviour of a Full-Scale Truss, Proc. ICE,69, 1980.

13. C.S. SMITH, W L KIRKWOOD and J.W. SWAN, Buckling Strength and Post-Collapse Behaviour of Tubular Bracing Members Including Damage Effects, Second Int. Conf. on Behaviour of Off-Shore Structures, Imperial College, 1979.

14. D.L. SMITH and J. MUNRO, Primal-Dual Programs of Plastic Analysis , Research Report SAM 72/2, Imperial College, 1972.

15. D.L. SMITH, and J. MUNRO, Plastic Analysis and Synthesis of Frames Subjected to Multiple Loadings, Proc. Int. Symp. Optimization in Civil Engineering, Univ. of Liverpool, 1973.

16. W.J. SUPPLE, A Plastic Collapse Model for Space Trusses, Proc. Inst. Phys. Conf. Stability Problems in Engineering Structures and Components, Univ. College, Cardiff, (1978a).

17. W. J. SUPPLE, and I COLLINS, Limit State Analysis of Double-Layered Grids, Proc. Short Course on the Analysis, Design and Construction of Double-Layer Grids,Univ. of Surrey, (1978b).

EVALUATION OF THE MOST UNSTABLE DYNAMIC-BUCKLING MODE OF PIN-JOINTED SPACE LATTICES

M. D. DAVISTER*, Ph.D. and S. J. BRITVEC**, Ph.D. (Cantab.), D.Sc. (habil.)

* Structural Engineer and Technical Specialist, U.S. Bureau of Reclamation, Denver, Colorado.

** Professor of Engineering Mechanics and Civil Engineering, University of Maine at Orono, Maine. (Formerly, Visiting Professor, University of Colorado, at Boulder.)

The usefulness of space structures made of slender pin-jointed members is strongly dependent on their statical stability in the near-critical range of loading, which is usually followed by a dynamic collapse caused by buckling. The dynamic collapse is characterized by a particular buckling mode involving the critically stressed members. The number of possible modes is huge, but the structure is most likely to "select" that mode which is the most unstable (the mode in which the largest amount of kinetic energy is released). In this paper two methods are presented for evaluating the most degrading (unstable) buckling mode of a hyperstatic pin-jointed lattice under conservative loading occuring during a dynamic collapse.

INTRODUCTION

Structural lattices of various shapes and sizes are finding many new applications in modern engineering. They combine a remarkable strength with a relatively low weight. On the ground they may be used as large reticulated shells, enveloped by a cover shell, as radio-antennas, etc. For the outer space uses they are being developed as radiometers and as large space stations and "permanent" satellites. In the off-shore and sub-sea technologies, the application of enveloped lattices is as deep-sea high pressure shells and habitats. In this paper, complex hyperstatic lattices, composed of three-dimensional polyhedral elements formed from quasi pin-jointed light, slender members are considered. By giving these members appropriate lengths, the lattice may be shaped into any flat or curved geometrical configuration as shown in Figs 1a and 1b.

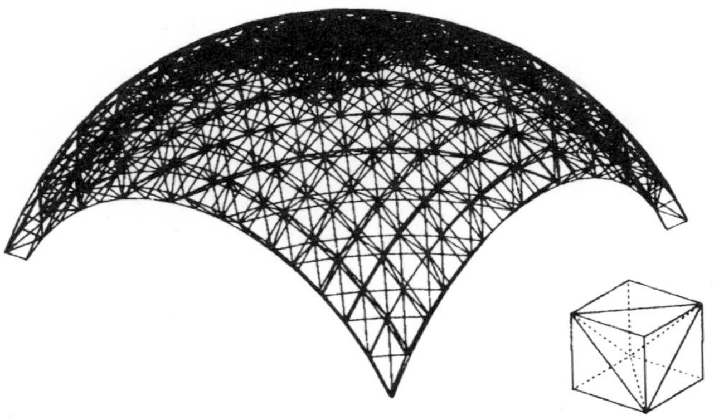

Fig 1a

Spherical Lattice Made of Cube-Terahedron Elements

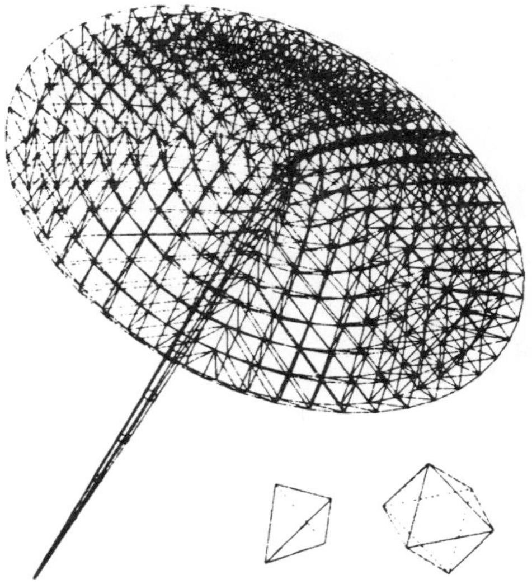

Fig 1b

Radiometer Large Space Structure Made of Tetrahedron-Octahedron Elements

On the ground, external loads are supposed to be transmitted to the lattice by the cover shell (not shown in Fig 1a), which is attached to it at the peripheral nodes and the entire structure is supported on a foundation along the shell boundaries.

In outer space, such lattices are unsupported but, in an interorbital maneuver, considerable inertial loads may be transmitted by various masses to the lattice of a large space structure, such as a radiometer, Fig 1b, at the nodal points in a quasi-static situation.

In either case, over prolonged or short periods, these loads may be proportionate to a load-parameter. In a suspended state, this parameter may be directly proportional to the thrust or the instantaneous acceleration in space. It is, therefore, relevant to study such large lattices under equilibrated proportionate loadings.

The response of slender pin-jointed structures, in general, under equilibrium conditions has been extensively studied by Britvec (Refs 1, 2, 5, etc.) and by Britvec and Davister (Refs 3 and 4) before, and it was shown that the statical response of such lattices is conditioned by both the most degrading post-buckling equilibrium modes of the perfect lattice and by the inherent geometrical imperfections which can modify the perfect response and result in nonlinear ascending-descending equilibrium paths corresponding to different post buckling modes. The equilibrium paths culminate in a peak load which is always situated on the stability boundary of the equivalent perfect system, the gradient of which, in pin-jointed systems, is always equal to *three times the gradient of the corresponding post-buckling equilibrium path* for the same lattice (Refs 1, 4, and 5).

This relationship is typified in Fig 2, where (W) represents a parameter of the external or the inertial forces under quasi-static conditions and ($\bar{s}$) represents a weighted average joint-displacement parameter defined in Refs 4 and 7.

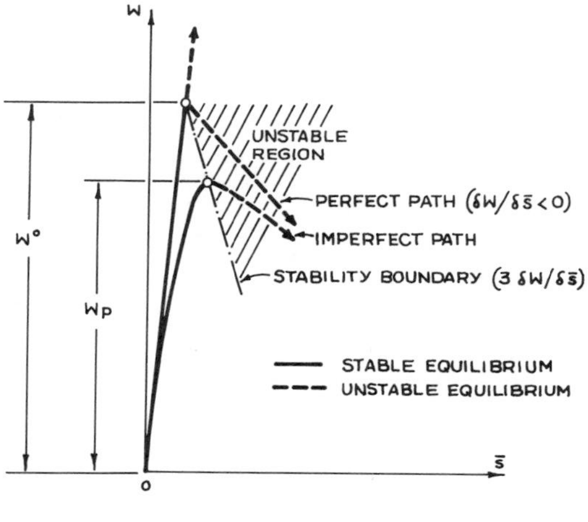

Fig 2

Typical Primary Equilibrium Paths of an Imperfection-Sensitive Pin-Jointed Lattice

During the initial stages of loading, the deformations are negligibly small and vary nearly linearly as shown. As (W) is increased, some members become critically stressed and begin to buckle. In a highly hyperstatic lattice, the critically stressed members are initially constrained from buckling freely so that these can support critical loads, as (W) continues to increase. This situation prevails until, in the required sequence, a sufficient number of members become critically stressed so that a partial or global kinematic buckling mechanism may develop. In this region of loading, the path is usually highly non-linear due to imperfection-sensitivity and the external loads are near their peak values, characterized by (W_p) in Fig 2. The system is now in its *ultimate* critical state.

Just as the pre-buckling portion of the stable equilibrium path is highly dependent on the initial linear elastic response of the structure characterized by Hooke's law, the *unstable* portion of the equilibrium path is highly dependent on the initial tangent of the induced buckling mode branching from the ultimate critical state. In the case of pin-jointed systems, numerous primary paths may branch from the ultimate critical state and each path is associated with a different kinematically admissible buckling mode. The path, associated with the steepest negative total potential energy gradient branching from the ultimate critical state, is of most interest as it represents the *most degrading mode* the structure may elicit under equilibrium conditions and under its sensitivity to imperfections.

In practice, *equilibrium conditions cannot be maintained in the structure beyond this ultimate critical state*, as the external loads are usually *independent* of the deformation of the structure beyond this point, so that under the limiting loads, exerted by the masses (dead loads) attached to the structure, the lattice fails by buckling in a *dynamic collapse*.

This collapse may be governed by one of several of its post-buckling modes. Under conservative loads, exerted by the masses attached to the nodes of the lattice in the ultimate critical state, this mode may be evaluated numerically even in very complex systems by two independent methods described in this article. These methods are based on the assumption that, essentially, the total potential energy change hypersurface of the equivalent perfect system, characterized by the independent coordinates of deformation (flexural contractions) of the buckling members, largely determines the choice of the most degrading collapse mode, this mode resulting in the largest release of kinetic energy under conservative conditions. So, the collapse is now determined by the most negative total potential energy gradient, on this hypersurface, while the system is in motion under constant external loads. This was observed to be the case in several tests on model pin-jointed systems (Refs 1, 5, 6, etc.). The need for a direct determination of the most degrading mode arises also from a numerical standpoint. The number of physically possible primary post-buckling modes either in motion or in equilibrium under modified external loads in a complex lattice may be in millions or billions so that it becomes exceedingly difficult to evaluate numerically all the modes in a given case. *A direct evaluation of the most likely collapse mode is, therefore, of considerable practical interest.*

The nature of this collapse, sudden and explosive or gradual, is dependent on the amount of total potential energy available for conversion. Also, the most likely mode to occur in this dynamic collapse in motion is *generally different* from the most degrading post-buckling mode under prescribed, variable, proportionate, external loads in post-buckling equilibrium, discussed in Ref 4.

STATEMENT OF THE PROBLEM

To evaluate the most degrading collapse mode in motion, we examine the total potential energy change hypersurface of the perfect system in the neighborhood of the ultimate critical state, see Ref 1, which was derived in matrix form in Ref 7 as

$$v = \frac{1}{2}\, \underset{\sim}{e}_c^T \underset{\sim}{E}_c \underset{\sim}{e}_c \qquad \ldots\ldots\ 1,$$

where the (1 x c) column vector $\{\underset{\sim}{e}_c\}$ represents the flexural shortenings in the critically stressed isostatic members and it contains (c) independent variables in post-buckling, while $[\underset{\sim}{E}_c]$ is a symmetric non-definite (c x c) matrix. We bound $(\underset{\sim}{v})$ in Eq 1 by a contour

$$\underset{\sim}{e}_c^T \underset{\sim}{B} \underset{\sim}{e}_c = C \qquad \ldots\ldots\ 2,$$

around the origin defined by an abritrary quadratic form where (C) is a constant. Then, it can be shown using the Lagrangian formulation, that the stationary values of (v) on the contour (C = 1) are equal to

$$v_m = \frac{1}{2}\,\frac{e_{cm}^T\, E_c\, e_{cm}}{e_{cm}^T\, B\, e_{cm}} = \frac{\lambda_m}{2} \qquad \text{...... 3,}$$

where (λ_m) are the eigenvalues of

$$[E_c - \lambda B]\, e_c = 0 \qquad \text{...... 4,}$$

and {e_{cm}}, m = 1, 2, ..., c, are the corresponding eigenvectors. The global minimum of (v) on this contour then corresponds to the least eigenvalue (λ_{min}).

However, the mode corresponding to (λ_{min}) is not necessarily the most degrading buckling mode, since the *kinematic constraints* may prevent the system from reaching the lowest possible energy level on this contour. These kinematic constraints are linear and, in this case, they are imposed by the inextensional theory of slender members which requires that the shortenings {e_c} of the statically determinate members be greater than or equal to zero, i.e.

$$e_c \geq 0 \qquad \text{...... 5a,}$$

and also by the flexural contractions {e_2} of statically indeterminate members*

$$e_2 = C_2\, C_{1c}\, e_c \geq 0 \qquad \text{...... 5b,}$$

Here, [C_{1c}] is a submatrix of the kinematic matrix [C_1^{-1}] while [C_1] and [C_2] are the submatrices of the kinematic matrix [C] of the lattice relating to statically determinate and indeterminate members, respectively, Ref 6. In the presence of kinematic restraints, the eigenvalue problem of Eqns 2, 3, and 4 is modified as follows**

Minimize

$$\frac{e_c^T\, E_c\, e_c}{e_c^T\, B\, e_c}$$

Subject to the contraints: 6,

$$e_c \geq 0$$

$$e_2 = C_2\, C_{1c}\, e_c \geq 0$$

The contour to be used to investigate the v-hypersurface is not unique, and several contours may be used so as not to bias the solution. Generally, the term ($e_c^T\, B\, e_c$) represents some measure of the shell distortions, and, if C = 1, the quadratic ratio in Eqn 6 may be considered as the change in (v) per unit value of this measure of lattice distortion, as well as, twice the total potential energy change of the system on the contour C = 1. For example, the overall shell distortions represented by this quadratic form may be characterized by the sum of the squares of the coordinate joint displacements (u_i), the sum of the squares of the flexural shortenings (e_α), or by the net change in the strain energy.

* For details see Refs 6 and 7.

** Incidentally, this problem is similar to the problem of determining the most degrading buckling mode in equilibrium (Ref 4) except that the matrix [B] is different in that case.

MINIMIZATION METHODS FOR EVALUATING THE MOST DEGRADING BUCKLING MODE

Two constrained minimization methods were developed and programmed for computer use to solve the problem of Eqn 6. Each method begins with an initial feasible (admissible) point on the contour. Subsequent points in the feasible region are computed, such that in the steps (k) and (k + 1),

$$\bar{v}^{\,k+1} < \bar{v}^{\,k} \qquad \text{...... 7,}$$

where

$$\bar{v}^{\,k} = \frac{\{e_c^k\}^T\, E_c\, e_c^k}{\{e_c^k\}^T\, B\, e_c^k} \qquad \text{...... 7a,}$$

and

$$\bar{v}^{\,k+1} = \frac{\{e_c^{k+1}\}^T\, E_c\, e_c^{k+1}}{\{e_c^{k+1}\}^T\, B\, e_c^{k+1}} \qquad \text{...... 7b,}$$

Also,

$$e_c^{k+1} = e_c^k + \alpha^k\, p^k \qquad \text{...... 8,}$$

where (α^k) is the optimal step-length in the search direction {p^k}. The change in ($\bar{v}$) in moving from the point {e_c^k} on the contour to the point {e_c^{k+1}} is denoted as ($\Delta\bar{v}$) and may be computed at the kth step as

$$\Delta\bar{v}^{\,k} = \bar{v}^{\,k} - \bar{v}^{\,k+1} \qquad \text{...... 9.}$$

The optimal step-length (α^k) minimizes ($\Delta\bar{v}$) in the direction (p^k).

The two methods developed differ primarily in the selection of the search direction (p^k) in Eqn 8. In the first method (p^k) is an *eigenvector* associated with the minimum eigenvalue of an abridged quadratic ratio in which the abridged form is obtained from the original quadratic ratio (or a transformation of the original form) by deleting the rows and columns associated with the active constraints of the matrices [E_c] and [B] in Eqn 6. In the second method, (p^k) is computed using the *method of conjugate gradients* (see Refs 8, 9, or 10).

It can be verified (Ref 7) that, if {p^k} is the eigenvector direction associated with the minimum eigenvalue, ($\Delta\bar{v}^{\,k}$) varies with (α) as shown in Fig 3a where (α^-) and (α^+) are the maximum step-lengths possible in the (+) and (-) directions, respectively, along the direction (p^k) through (x^k). It can also be verified that in this case the stationary point is always associated with a maximum value of ($\Delta\bar{v}^{\,k}$) and therefore (α^k) is either equal to (α^+) or (α^-) as shown in Fig 3a.

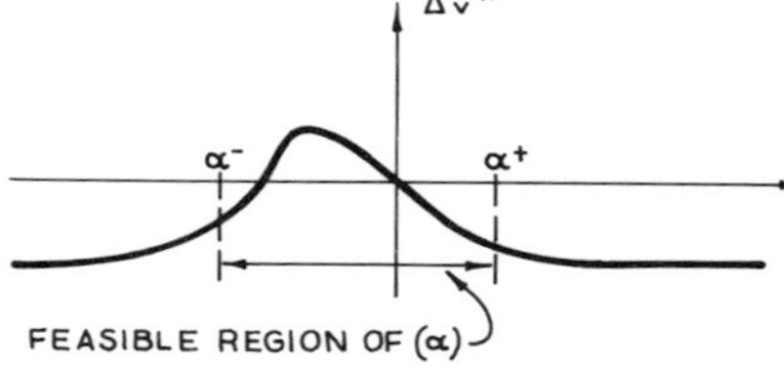

Fig 3a

Variation of ($\Delta\bar{v}^{\,k}$) with (α) in an eigenvector direction

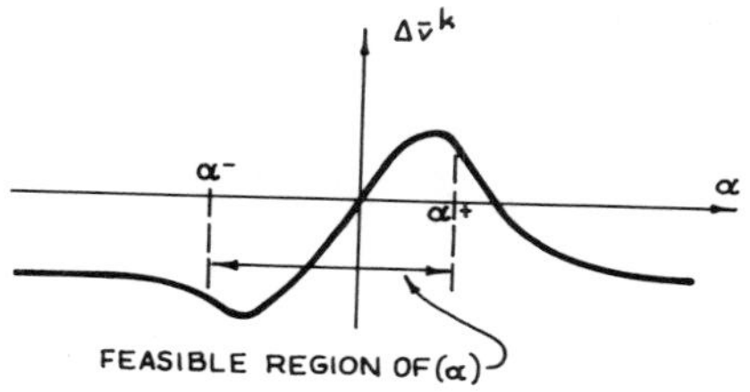

Fig 3b

Variation of $(\Delta\bar{v}^{\,k})$ with (α) in an arbitrary direction

When (p^k) is an arbitrary direction, (α) varies as shown in Fig 3b. The two stationary values may be evaluated exactly by solving a quadratic equation of the form

$$a\alpha^2 + b\alpha + c = 0 \qquad \text{...... 10,}$$

where the constants (a, b, and c) are defined in Ref 7. In this case (α^k) is either associated with a stationary point or with (α^+) or (α^-), if the minimum stationary value is in a non-feasible domain of (α). The procedures for computing (α^+) and (α^-) are also given in Ref 7.

Each time the optimal step-length is equal to (α^+) or (α^-), one of the constraints in Eqn 6 is said to be "activated" and the associated variable is called an *active constraint variable*. As a result of the two variations of $(\Delta\bar{v}^{\,k})$ with (α) depending on (p^k) (depicted in Figs 3a and 3b), it can be stated that each step in an eigenvector direction activates a constraint, while this is not necessarily true in the conjugate gradient minimization method. Also, under a set of active constraints, the point $\{e_c^k\}$ may be considered to be on one of the constraint-hyperplanes or at the intersection of (t) constraint hyperplanes, where (t) denotes the number of active constraints. In that case, it is necessary to move in directions "parallel" to the active constraints in order to remain in the kinematically admissible region of the v-hypersurface.

In the special case, where each of the (t) active constraint variables are contained in the vector of *independent* variables $\{e_c\}$, i.e. $e_{ci} = 0$ for i = 1, 2, ..., t, the problem of determining the search direction (p^k) parallel to the active constraint hyperplanes is greatly simplified. In this case, any vector expressible in the remaining (c-t) non-zero independent variables is parallel to the set of active constraint hyperplanes and that vector which minimizes (v) in the associated (c-t) dimensional subspace spanned by the non-zero independent variables is found by solving an eigenvalue problem. This eigenvalue problem is associated with the minimum stationary value of an *abridged* quadratic ratio obtained by deleting the rows and columns in the matrices $[E_c]$ and $[B]$ of Eqn 6 associated with the active constraint variables.

When one of the (t) active constraint variables are contained in the vector of *dependent* variables $\{e_2\}$, Eqn 6, one can interchange this variable with one of the non-zero independent variables. This is equivalent to changing isostatic systems by interchanging one of the isostatic members with one of the redundant ones. Mathematically this is accomplished by applying the transformation

$$e_c' = T\, e_c \qquad \text{...... 11,}$$

to the original problem in Eqn 6. Since $\{e_c'\}$ differs from $\{e_c\}$ by one variable, $[T]$ is of the form

$$T = \begin{bmatrix} c_1 & c_2 & c_3 & \cdot & \cdot & \cdot & c_c \\ 0 & 1 & 0 & \cdot & \cdot & \cdot & 0 \\ 0 & 0 & 1 & \cdot & \cdot & \cdot & 0 \\ \cdot & \cdot & \cdot & \cdot & \cdot & \cdot & \cdot \\ \cdot & \cdot & \cdot & \cdot & \cdot & \cdot & \cdot \\ 0 & 0 & 0 & \cdot & \cdot & \cdot & 1 \end{bmatrix} \qquad \text{...... 11a,}$$

where the coefficients $(c_1, c_2, ..., c_c)$ are obtained from *the relevant row of* the product matrix $[C_2\; C_{1c}]$ in Eqn 6. The matrix (T) is always selected to be non-singular and its inverse can be written by inspection. Then,

$$e_c = T^{-1}\, e_c' \qquad \text{...... 11b,}$$

and the problem in Eqn 6 can be written with respect to the new isostatic system characterized now by the vector $\{e_c'\}$. As a result of this transformation all of the active constraint variables are contained in the vector $\{e_c'\}$. Thus, (p^k) can be determined as discussed previously.

At each iteration (k), the point $\{e_c^k\}$ is checked for an optimal solution. This is accomplished by first computing the gradient vector $\{g^k\}$ on the v-hypersurface along the contour $(e_c^T\, B\, e_c = 1)$ which can be shown to be equal to

$$g^k = [E_c - \bar{v}^{\,k}\, B]\, e_c^k \qquad \text{...... 12,}$$

It was also shown in Ref 7 that an optimal solution is reached when the elements of $\{g^k\}$ associated with the active constraint variables are positive and the remaining elements of $\{g^k\}$ are zero. This is true, because the elements of $\{g^k\}$ associated with the active constraints are identical to the Lagrangian multipliers of the active constraint equations in the Lagrangian formulation, while the remaining elements, equal to zero, define a stationary point in the current subspace of the v-hypersurface. If in any iteration (k), the element of the gradient vector associated with any active constraint is negative, the associated variable is deleted from the set of active constraints and the current subspace of variables *increased* by one dimension.

NUMERICAL RESULTS

To verify the accuracy of the two minimization methods, six different model lattices were considered. Five of the models were a variation (obtained by stiffening selected members) of the lattice shown below, Fig 4, which has 506 members and 306 degrees of freedom and consequently a redundancy of 200 members. The remaining model (Model No. 4) included only the outer layer shown in Fig 4. Models No. 1 to 4 were subjected to hydrostatic loading, Model No. 5 was subjected to gravity loading, and Model No. 6 was subjected to wind loading. Each model was investigated in post-buckling to determine the most degrading collapse mode in motion by examining the total potential energy change (v) at finite deformations using four different contours surrounding the origin of the v-hypersurface. This resulted in 24 case studies. The numerical results obtained by applying both the eigenvector search-direction method and the conjugate gradient method to each of these case studies are presented in Table 1. As shown, there is very good agreement between the two methods and from these results it may be concluded that either method may be used to determine the most degrading buckling mode occuring during a dynamic collapse.

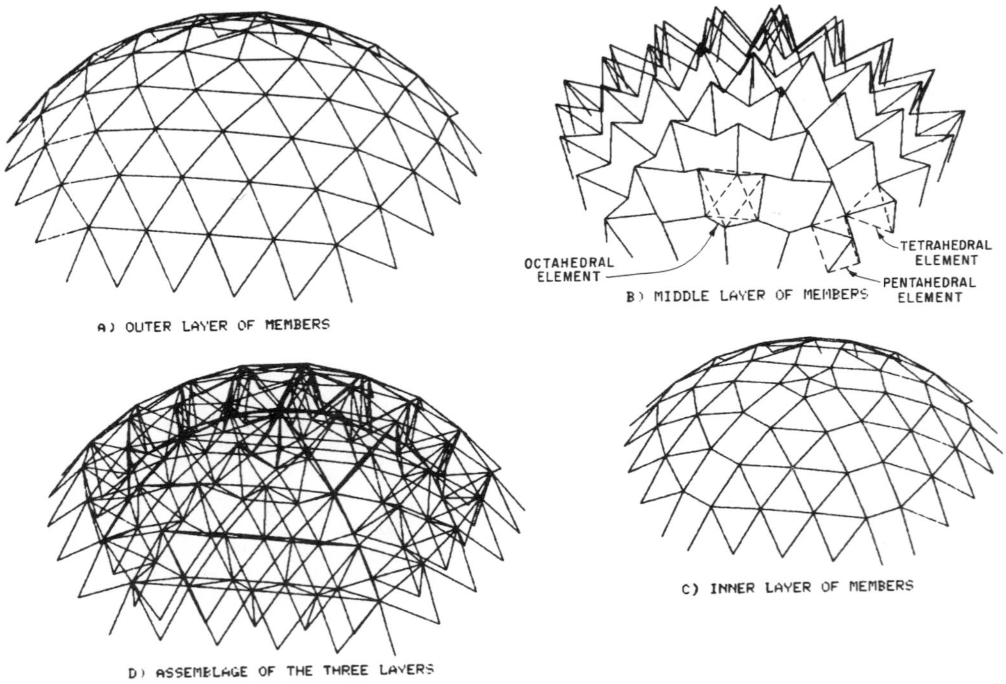

Fig 4

Model Lattice and the Three Divisions

Table 1

Comparison of results using two different methods to evaluate the most degrading mode which develops in motion

Contour (1)	Model No. (2)	Constrained minimum using eigenvalue descent method (3)	Constrained minimum using conjugate gradient method (4)
1	1	-2.395*	-2.395
	2	-2.823*	-2.823
	3	-4.048	-4.048
	4	-84.731*	-84.731
	5	-1.219	-1.219
	6	-4.536	-4.536
2	1	-63.286*	-63.286
	2	-74.349*	-74.349
	3	-99.733	-99.733
	4	-2367.624*	-2367.624
	5	-2.369	-2.369
	6	-5.397	-5.397
3	1	-16.519	-16.519
	2	-18.121	-18.121
	3	-52.226	-51.850
	4	-242.401	-242.401
	5	-1.615	-1.615
	6	-4.623	-4.623
4	1	-41.066	-41.071
	2	-48.313	-48.313
	3	-116.370	-116.368
	4	-242.437	-242.431
	5	-2.753	-2.753
	6	-9.169	-9.164

(*) indicates the constrained minimum is equal to the unconstrained minimum.

Also, it may be shown that the most degrading buckling mode may be determined by examining the v-hypersurface along any one of the four different contours, where the differences on these contours arise from different quantities used to characterize the overall distortions of the lattice. The contours listed in Table 1 and the associated distortion parameters are summarized below in Table 2.

Table 2

Contour	Associated Distortion Parameter
1	strain energy
2	sum of the squares of the flexural contractions
3	sum of the squares of all joint displacements
4	sum of the squares of the joint displacements at only the loaded joints

In order to compare the most degrading buckling modes evaluated on the four different contours, one can plot the flexural shortenings as demonstrated in Fig 5. Here, discrete points on the abscissa represent the 506 members making up the lattice, while the ordinate gives the relative magnitude of the flexural shortening in each of these members. From these plots one can determine that the modes are quite similar and consequently any one of the contours may be used.

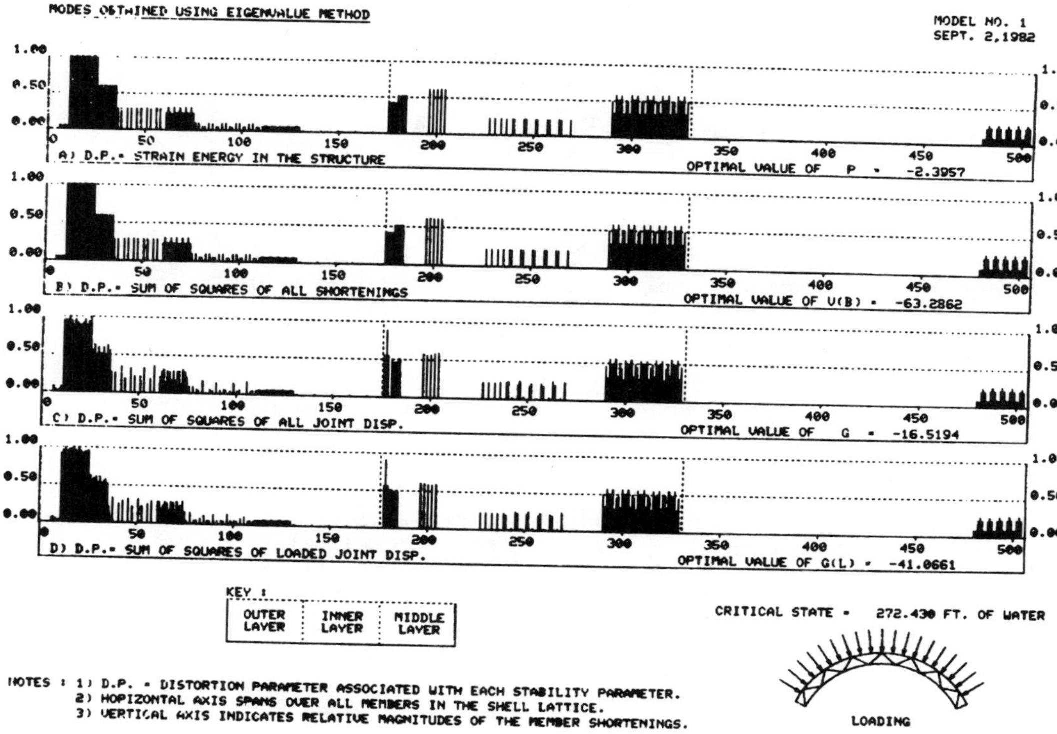

Fig 5

Comparison of the Most Degrading Buckling Modes on Different Contours

REFERENCES

1. Britvec, S.J., The Stability of Elastic Systems, Pergamon Press, Inc., New York, 460 pp., 1973.

2. Britvec, S.J., High-Pressure Shell in Off-Shore Engineering (I) - Existing Approaches, Their Limitations and Alternatives in the Design of High-Pressure Shells for Sub-sea Applications - The Post-Buckling Analysis of Reticulated Shells, Paper presented at the International Centre for Mechanical Sciences (CISM), Udine, Italy, September 1980.

3. Britvec, S.J. and M.D. Davister, High Pressure Shells in Off-Shore Engineering (II) - An Investigation of a Model Reticulated Shell Under Large Hydrostatic Pressures, Paper presented at the International Centre for Mechanical Sciences (CISM), Udine, Italy, September 1980.

4. Britvec, S.J. and M.D. Davister, Post-Buckling Equilibrium of Hyperstatic Lattices, Paper presented at the ASCE, EMD Specialty Conference, Purdue University, West Lafayette, Indiana, May 23-25, 1983.

5. Britvec, S.J., T. Manacorda, Z. Wesolowski, C. Wozniak, and B.R. Seth, Nonlinear Dynamics of Elastic Bodies, Edited by Z. Wesolowski, Springer-Verlag, New York, 1979.

6. Britvec, S.J. and D. Nardini, Some Aspects of the Nonlinear Elastic Behavior and instability of Reticulated Shell-Type Systems, Developments in Theoretical and Applied Mechanics, 8th SECTAM, Vol. 8, April 1976.

7. Davister, M.D., The Post-Buckling Equilibrium and the Dynamic Collapse of Complex Hyperstatic Pin-Jointed Lattices and Reticulated Shells, Ph.D. Dissertation, Univ. of Colorado, Boulder, 1983.

8. Fox, R.L., Optimization Methods for Engineering Design, Addison-Wesley Publishing Company, Menlo Park, California, 1971.

9. Fox, R.L. and M.P. Kapoor, A Minimization Method for the Solution of Eigenproblems Arising in Structural Dynamics, Proceedings, 2nd Conference of Matrix Methods in Structural Mechanics, Wright-Patterson AFB, Ohio, 1968.

10. Gill, P.E. and W. Murray, Editors, Numerical Methods for Constrained Optimization, Academic Press, New York, 1974.

BUCKLING ANALYSIS OF DOUBLE-LAYER GRIDS, ROOFS AND TRUSS-LIKE BEAMS

T. SUZUKI*, Dr Eng and T. OGAWA**, Dr Eng

* Professor, Department of Architecture and Building Engineering, Tokyo Institute of Technology, Japan

** Research Associate, Department of Architecture and Building Engineering, Tokyo Institute of Technology, Japan

The objective of this work is to obtain a new finite element scheme for the buckling analysis of double-layer grids, cylindrical roofs and truss-like beams. In the anlysis, a repeating unit composed of several members is considered to be one collective element. The essential feature which distinguishes this investigation from the conventional stiffness method is the separation of relative displacements and rigid-body motion of the element. Illustative examples are shown which confirm the accuracy and validity of the present method in the buckling analysis of multi-layer grids.

1. INTRODUCTION

There are various techniques of analysis that are currently applied to double-layer grid structures. Among them, the classical method of continuum mechanics or the method employing analogous continua has made remarkable achievement in the analysis of stress distribution in structures. However, it is difficult to express the instability phenomina such as plastic buckling of some compression members.

Recent advancement of the digital computer enables us to solve these problems in the direct methods. The finite element discretisation is one of the most popular approach which can take both the geometrical and the material nonlinear effect into account. However, in spite of these advantages, the finite element method requires very extensive amount of calculations to predict the nonlinear equilibrium path when the double-layer grids are composed of so many members. What makes the finite element analysis expensive in calculation efforts is the repetitive calculations of the stiffness matrix and inner forces for every point on the equilibrium path.

In this paper we developed a new finite element scheme that the repeating unit composed of several members is considered to be one collective element. The essential feature of the present work is the separation of the small relative displacement and the rigid-body motion of the element. Thus, the calculations of the tangential stiffness matrix and the inner forces can be generated by the small relative displacements. This causes the reduction of the computational time and the improvement of the accuracy as the total degrees of freedom can be reduced. In pursuing the plastic buckling of the compression members, such members that would buckle have some internal nodes at where the nonlinear stress – strain relations are evaluated. Several numerical examples are presented including truss-like beams, grids, and cylindrical roof structures undergoing large displacement.

2. FINITE ELEMENT FORMULATION

2.1 Main assumptions

(1) The double-layear grids are assumed to be composed of the repeating basic unit, as shown in Fig 1, which is a trigonal pyramid or a square-based pyramid consisting of rigidly jointed several members. This basic unit is considered to be one collective element.
(2) The nodal displacements of the element are separated into two parts, one is the relative displacements refered to the convected coordinate system and the other is the rigid-body displacements caused by the rigid-body motion of the element.
(3) The rigid-body displacements are evaluated taking into account the second order terms of the rigid-body motion.
(4) The external forces act on the nodes and no other lateral force exists.

2.2 Definition of the convected coordinate system

Figure 2 shows a trigonal pyramid element before and after deformation. The convected coordinate system is so defined that the x-axis coincides with the member I-J,

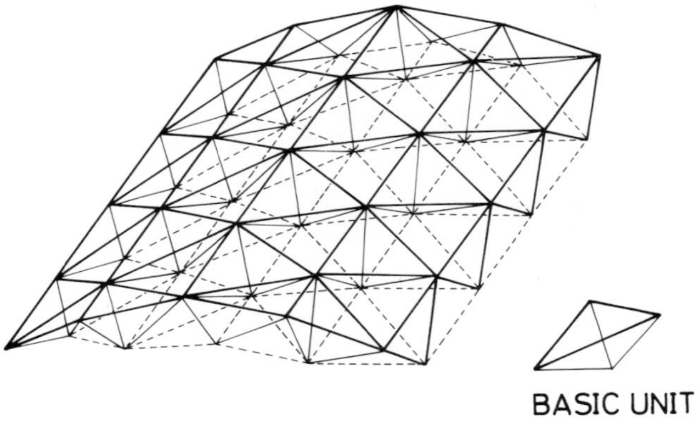

Fig 1 Three-way double-layer grids

the x-y plane contains a triangular plane I-J-K, and finally the z-axis is defined composing an orthogonal cartesian coordinate system.

Now suppose that such an element undergoes a rigid-body translation Λ and a rigid-body rotation Ω about its center D.

$$\Lambda^t = < U_1 \; U_2 \; U_3 > \qquad \ldots\ldots \; 1.$$

$$\Omega^t = < \theta_1 \; \theta_2 \; \theta_3 > \qquad \ldots\ldots \; 2.$$

where Ui and θi are the components of the vectors Λ and Ω referred to the coordinate system before deformation. (see Figs 3a and 3b.) The nodal displacement vectors $\mathbf{U}n^*$ and Θn^* induced by such a rigid-body motion should be represented by Eqns 3 and 4 as shown in Fig 4.

$$\mathbf{U}n^* = \Lambda + \Omega\times(\mathbf{P}n - \delta) + \frac{1}{2}\Omega\times(\Omega\times(\mathbf{P}n - \delta)) \qquad \ldots\ldots \; 3.$$

$$\Theta n^* = \Omega \qquad \ldots\ldots \; 4.$$

where $\mathbf{P}n$ and δ are the position vectors of the arbitrary node N and the center D referred to the origin O.

$$\mathbf{P}n^t = < Xn \; Yn \; Zn > \qquad \ldots\ldots \; 5.$$

$$\delta^t = < \delta_1 \; \delta_2 \; \delta_3 > \qquad \ldots\ldots \; 6.$$

Linearizing Eqns 3 and 4 in the iterative process of solution, these become

$$\mathbf{U}n^*(i) = \Lambda(i) + \mathbf{R}n\cdot\Omega(i) + \tilde{\mathbf{R}}n\cdot\mathbf{B}\cdot\Omega(i) \qquad \ldots\ldots \; 7.$$

$$\Theta n^*(i) = \Omega(i) \qquad \ldots\ldots \; 8.$$

In above equations, the suffix (i) denotes the i-th step of the iterative process. $\mathbf{R}n$, $\tilde{\mathbf{R}}n$ and $\mathbf{B}$ are the matrices, the components of defined by the coordinates of the position vectors $\mathbf{P}n$, δ and the values of rigid-body rotation. Applying Eqns 7 and 8 to all the nodes of the collective element, the following expression of the nodal displacement vector by the rigid-body motion is obtained.

$$\mathbf{U}^*(i) = \mathbf{H}\cdot\mathbf{D} \qquad \ldots\ldots \; 9.$$

where

$$\mathbf{U}^*(i)^t = < \mathbf{U}i^*(i)^t \; \Theta i^*(i)^t \; \mathbf{U}j^*(i)^t \; \Theta j^*(i)^t \; \mathbf{U}k^*(i)^t \; \Theta k^*(i)^t \; \mathbf{U}l^*(i)^t \; \Theta l^*(i)^t > \qquad \ldots\ldots \; 10.$$

$$\mathbf{D}^t = < \Lambda(i)^t \; \Omega(i)^t > \qquad \ldots\ldots \; 11.$$

$$\mathbf{H} = [\,\mathbf{I}\,,\; \mathbf{R} + \tilde{\mathbf{R}}\cdot\mathbf{B}\,] \qquad \ldots\ldots \; 12.$$

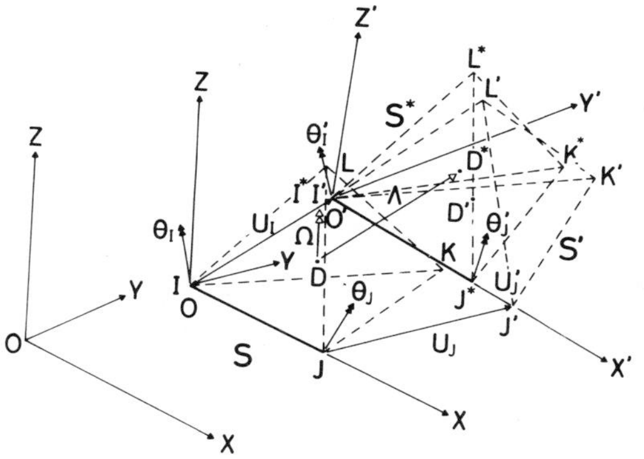

Fig 2 Coordinate system before and after deformation

In addition to these nodal displacement by a rigid-body motion, we defined the relative displacement vectors $\mathbf{U}'$ as shown in Fig 5. These relative displacement vectors are defined referred to the convected coordinate system after deformation.

$$\text{node i} \quad \mathbf{U}'_i{}^t = < \cdot \quad \cdot \quad \cdot \quad \theta'_{xi} \; \theta'_{yi} \; \theta'_{zi} > \qquad \ldots\ldots \; 13.$$

$$\text{node j} \quad \mathbf{U}'_j{}^t = < U'_{xj} \quad \cdot \quad \cdot \quad \theta'_{xj} \; \theta'_{yj} \; \theta'_{zj} > \qquad \ldots\ldots \; 14.$$

$$\text{node k} \quad \mathbf{U}'_k{}^t = < U'_{xk} \; U'_{yk} \quad \cdot \quad \theta'_{xk} \; \theta'_{yk} \; \theta'_{zk} > \qquad \ldots\ldots \; 15.$$

$$\text{node l} \quad \mathbf{U}'_l{}^t = < U'_{xl} \; U'_{yl} \; U'_{zl} \; \theta'_{xl} \; \theta'_{yl} \; \theta'_{zl} > \qquad \ldots\ldots \; 16.$$

2.3 Transformation of coordinate system

The nodal displacements by a rigid-body motion are described referred to the coordinate system before deformation. On the other hand, the relative displacements are defined referred to the convected coordinate system after deformation. To complete the description of the total displacement, it is needed to establish a coordinate transformation.

Coordinate cosine vectors $\mathbf{f}i$ (i=1,2,3) are expressed by following as shown in Fig 6.

$$\Omega\times\mathbf{e}i + \frac{1}{2}\,\Omega\times(\Omega\times\mathbf{e}i) = \mathbf{f}i - \mathbf{e}i \qquad \ldots\ldots \; 17.$$

where $\mathbf{e}i$ (i=1,2,3) are the unit base vectors referred to the coordinate system before deformation. Using Eqn 17, the following transformation matrix $\mathbf{T}$ is introduced.

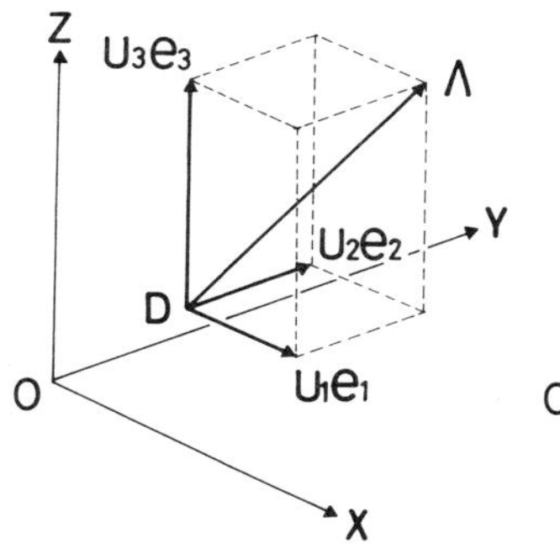

Fig 3a Components of rigid-body translation vector

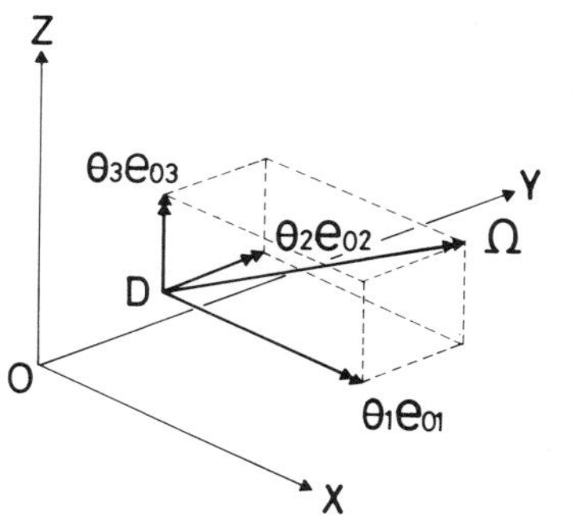

Fig 3b Components of rigid-body rotation vector

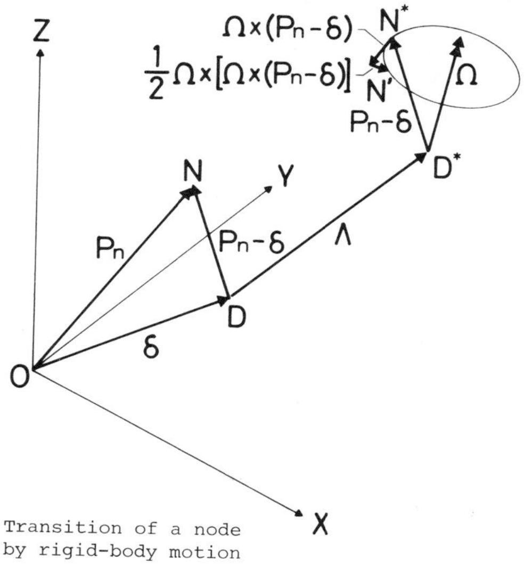

Fig 4 Transition of a node by rigid-body motion

$$T = \begin{bmatrix} 1-\frac{1}{2}(\theta_2^2+\theta_3^2) & -\theta_3+\frac{1}{2}\theta_1\theta_2 & \theta_2+\frac{1}{2}\theta_1\theta_3 \\ \theta_3+\frac{1}{2}\theta_1\theta_2 & 1-\frac{1}{2}(\theta_3^2+\theta_1^2) & -\theta_1+\frac{1}{2}\theta_3\theta_2 \\ -\theta_2+\frac{1}{2}\theta_1\theta_3 & \theta_1+\frac{1}{2}\theta_2\theta_3 & 1-\frac{1}{2}(\theta_2^2+\theta_1^2) \end{bmatrix} \qquad .....18.$$

The components of relative displacement vector $\mathbf{U}'$ referred to the convected coordinate system can be transformed to the components referred to the coordinate system before deformation.

$$\mathbf{U} = \mathbf{F}\cdot\mathbf{U}' \qquad .....\ 19.$$

where $\mathbf{F}$ is a transformation matrix composed of $\mathbf{T}$ in its diagonal. Replacing $\mathbf{U}'$ in Eqn 19 by $\tilde{\mathbf{U}}'$ which is defined by Eqns 13 to 16, we get

$$\mathbf{U} = \mathbf{F}^*\cdot\tilde{\mathbf{U}}' \qquad .....\ 20.$$

where

$$\tilde{\mathbf{U}}'^{t} = <\ \theta'_{xi}\ \theta'_{yi}\ \theta'_{zi}\ \theta'_{xj}\ \theta'_{yj}\ \theta'_{zj}\ U'_{xk}\ U'_{yk}\ U'_{xj}\ \theta'_{xk}\ \theta'_{yk}\ \theta'_{zk}\ U'_{xl}\ U'_{yl}\ U'_{zl}\ \theta'_{xl}\ \theta'_{yl}\ \theta'_{zl}\ > \qquad .....\ 21.$$

The matrix $\mathbf{F}^*$ is obtained by condensation of matrix $\mathbf{F}$. Linearizing Eqn 20 in the iterative procedure of increment, the following equation is obtained.

$$\mathbf{U}(i) = \mathbf{S}\cdot\Omega(i) + \mathbf{F}^*\cdot\tilde{\mathbf{U}}'(i) \qquad .....\ 22.$$

where the components of matrix $\mathbf{S}$ are the nodal displacements which are unknown in this state.

The total nodal displacement vector $\mathbf{Ue}$ referred to the coordinate system before deformation is obtained by adding Eqn 22 to Eqn 9.

$$\mathbf{Ue}(i) = \mathbf{U}^*(i) + \mathbf{U}(i)$$

$$= \mathbf{X}\cdot\tilde{\mathbf{U}}(i) \qquad .....\ 23.$$

$$\mathbf{X} = \left[\ \mathbf{I},\ \mathbf{R}+\tilde{\mathbf{R}}\cdot\mathbf{B}+\mathbf{S},\ \mathbf{F}^*\ \right] \qquad .....\ 24.$$

2.4 Tangential element stiffness matrix

In this subsection, the nonlinear stiffness matrix and the inner force vector of a collective element are derived on the basis of the principle of virtual work. The distribution of the relative displacements is assumed in the form of polynomials in the member coordinate system.

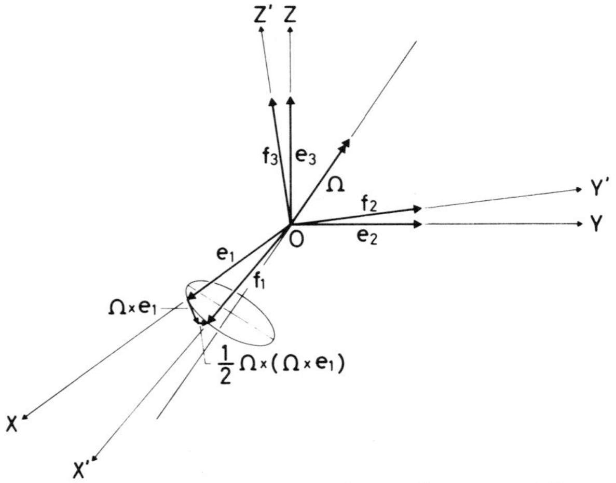

Fig 6 Translation of coordinate system

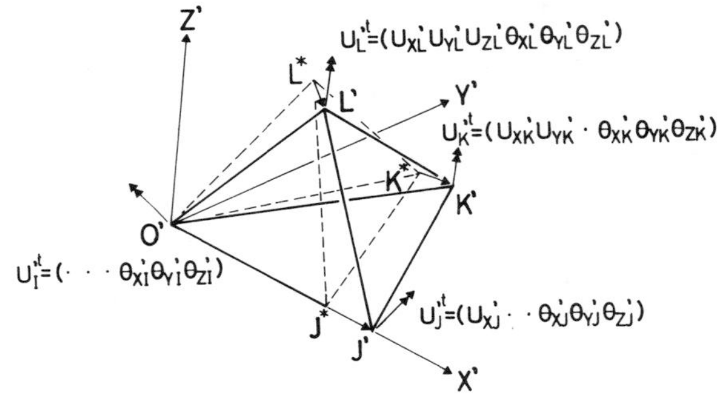

Fig 5 Relative nodal displacements.

$$\Delta\mathbf{u} = \begin{bmatrix}\Delta u\\ \Delta v\\ \Delta w\\ \Delta\phi\end{bmatrix} = \mathbf{Hmn}\cdot\Delta\alpha \qquad .....\ 25.$$

where $\Delta u, \Delta v$ and Δw are the translations in the member coordinate system and $\Delta\phi$ is the torsional angle of the section. $\mathbf{Hmn}$ is the shape function matrix composed of polynomials. The cubic displacement assumption is adopted for the lateral displacements Δv, Δw , and the linear one is adopted for the axial displacement Δu and $\Delta\phi$.

In each member of the collective element, the following equation of the virtual work should hold.

$$\int\{\Delta\sigma ij\cdot\delta eij^* + \sigma ij\cdot\frac{1}{2}\delta(\Delta u_{k,i}\cdot\Delta u_{k,j})\}\ dV$$

$$-\ \Delta Fi\cdot\delta\Delta u_i = \delta\Delta W \qquad .....\ 26.$$

where $\Delta\sigma$ and e^* are the incremental fiber stress and strain respectively, and σ is the initial fiber stress which forms the geometrical stiffness of the member. ΔF is the external force and ΔW is the unbalanced force produced in the iterative procedure of increment. Substituting Eqn 25 to Eqn 26, the following quadratic form of equation is obtained.

$$\delta\Delta\mathbf{u}^t\cdot\mathbf{Km}\cdot\Delta\mathbf{u} - \delta\Delta\mathbf{u}^t\cdot\Delta\mathbf{Fm} = \delta\Delta\mathbf{u}^t\cdot\mathbf{Rm} \qquad .....\ 27.$$

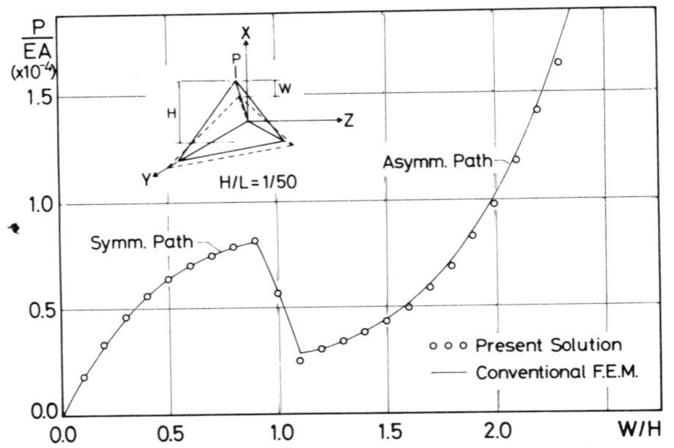

Fig 7 Load-deflection curve of one unit

where Km represents a tangential stiffness matrix of the member m-n. Summing up these stiffness matrices, a stiffness matrix Ku of the collective element can be expressed as

$$\delta\Delta\tilde{u}^t \cdot \tilde{K}u \cdot \Delta\tilde{u} = \delta\Delta\tilde{u}^t \cdot (\Delta F + Ru) \qquad ..... \ 28.$$

$$\tilde{K}u = \begin{bmatrix} \cdot & \cdot \\ \cdot & Ku \end{bmatrix} \qquad ..... \ 29.$$

$$Ku = \Sigma \ Tu^t \cdot Km \cdot Tu \qquad ..... \ 30.$$

$$\Delta\tilde{u}^t = < \Delta D^t \ , \ \Delta\tilde{U}'^t > \qquad ..... \ 31.$$

where Tu represents a transformation matrix from the member coordinate system to the convected coordinate system. Using Eqn 23 and a transformation matrix Ta which defines a fixed coordinate in space, we can get the final form of a stiffness equation as

$$K\tilde{u}^e \cdot \Delta qu = Ta^t \cdot X^{-1t} \cdot \ \Delta F + R\tilde{u}^e \qquad ..... \ 32.$$

$$K\tilde{u}^e = Ta^t \cdot X^{-1t} \cdot \tilde{K}u \cdot X^{-1} \cdot Ta \qquad ..... \ 33.$$

$$\Delta qu = Ta^t \cdot X^{-1t} \cdot \Delta\tilde{U}' \qquad ..... \ 34.$$

3. NUMERICAL EXAMPLES

3.1 Snap-through of one unit

Figure 7 shows a snap-through behavior of one unit in elastic range. The nodes in the base are simply supported though two of them can move smoothly on the base. As shown in Fig 7, the present solution gives good agreement with the result obtained by the conventional method . This indicates the validity of the finite element formulation by the relative displacements.

3.2 Cylindrical roofs

In this subsection, double-layer cylindrical roofs subjected to the uniformly distributed load are studied. Shell geometry and loading condition are shown in Fig 8. These shells are supported at the straight edges of the lower layer. As a geometrical parameter, we chose the stiffness of the diagonal bracing and the distance of the layers.

In Figs 9 to 11, the load-deflection curves and the deflection patterns at the peak are displayed. In case 1, the sectional area of the diagonal bracing is half of

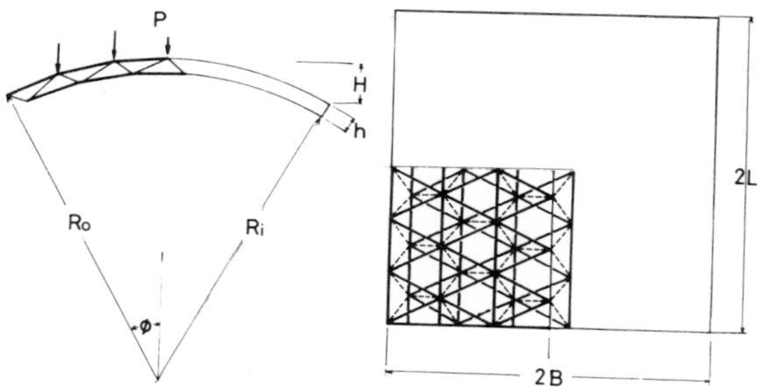

Fig 8 Shell geometry and loading

Table 1 Buckling strength and effective stiffness

	Case 1	Case 2	Case 3
$\bar{P}cr = (\frac{PR_0^2}{EI})$	729	933	1700
$t = \sqrt{3}\,h$ (cm)	0.3464	0.3464	0.8660
$E' = E\frac{2A}{\sqrt{3}bh}$ (t/cm^2)	72.7	72.7	29.1
$D' = \frac{E' t^3}{12(1-\nu^2)}$ (t·cm)	0.277	0.277	1.73

the chord member and the distance between the top layer and the bottom one is B/50. As shown in the deflection patterns, the local area including node 5 in top layer undergoes large deflection like a dimple buckling of single-layer shells.

In case 2, the diagonal bracing has the same sectional area as the chord member. As shown in Fig 10, the deflection patterns of the top and bottom layer resembles each other. This trend is recognized in Fig 11, where the distance of the layers is 2.5 times as large as the case 1. This fact indicates the importance of the diagonal bracing of double-layer structures in carrying load to each layer.

Table 1 shows the critical load of three cases with the effective stiffness as a plate. As the number of the repeating unit is not so large, the critical load is not exactly proportional to the effective stiffness. In more large-scale structures, this effective stiffness may give good indication of the critical load.

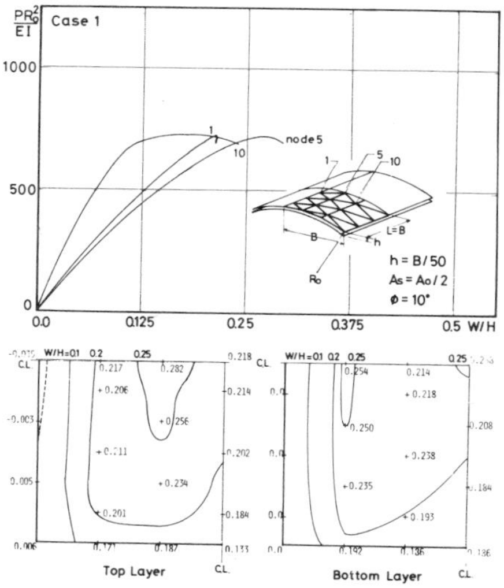

Fig 9 Load-deflection curves and deflection patterns (case 1)

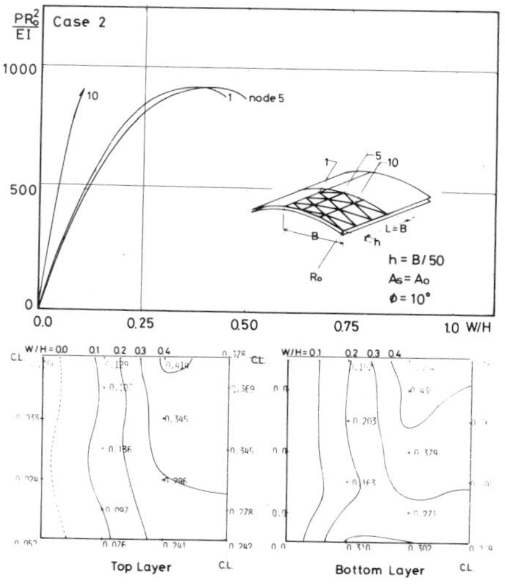

Fig 10 Load-deflection curves and deflection patterns (case 2)

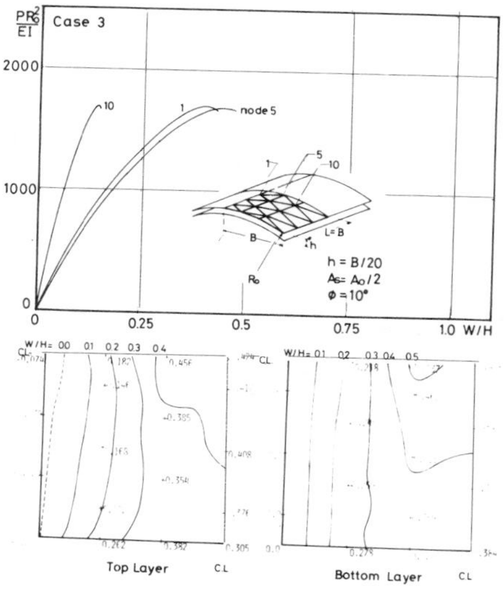

Fig 11 Load-deflection curves and deflection patterns (case 3)

3.3 Double-layer grid and truss-like beams

Figures 12a, 13a and 14a show the load- central deflection curves of a double-layer flat grid and truss-like beams. Their geometries and the supported nodes are shown in Figs 12b and 13b. Except the case in Fig 14, the stress-strain relations are assumed to be linear elastic. In Figs 12a and 13a, the applied load P is normalized by Pcl which represents a critical load that some members reach the Euler buckling load. As stated before, such members that would buckle have some internal nodes .

In these elastic analyses, the equilibrium path after member buckling is quite stable because of the re-distribution of the stresses. In both cases, the buckling load of these rigidly jointed grids is about three times as large as the buckling load of the pin-jointed ones. This is due to the restraint of the adjacent members. The deflection modes are displayed in Figs 12c and 13c. It is recognized here that the other units except the buckling members undergo almost rigid-body motion.

Figure 14a shows the load-central deflection curves of two truss-beams, the slenderness ratio of one is 60 (case 2-1) and the other is 100 (case 2-2). The geometry and loading condition are the same as shown in Fig 13b. In these cases, the stress-strain relations are assumed to be bi-linear and the applied load P in Fig 14a is normalized by the yield axial force Py.

As shown in Fig 14a, the critical load is higher in case 2-1 and the ratio of load-loosening is quite large in case 2-2. The critical load is about 1.0 to 1.3 times as large as the one defined by the empirical formula given by Johnson. This is considered to be caused by the reduction of the restraint accompanied by the yielding of the sections. Figs 14b and 14c show the deflection modes of the beams. It is noted here that the deflection modes become asymmetric compared with Fig 13c as the plastic buckling deformation advances in one member. This fact proves a sudden fall of the equilibrium path in Fig 14a.

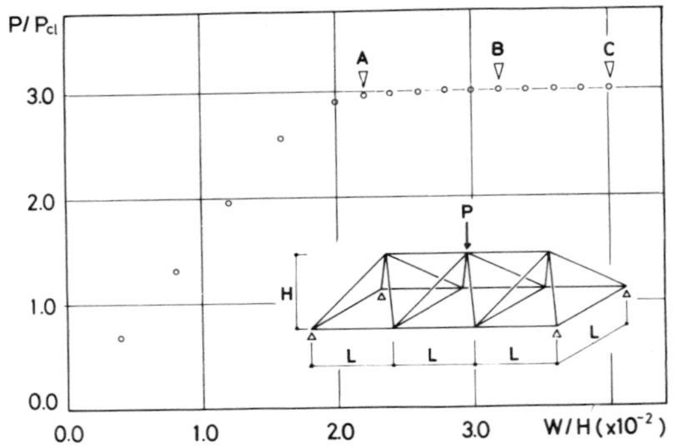

Fig 13a Load-deflection curve of a truss-like beam

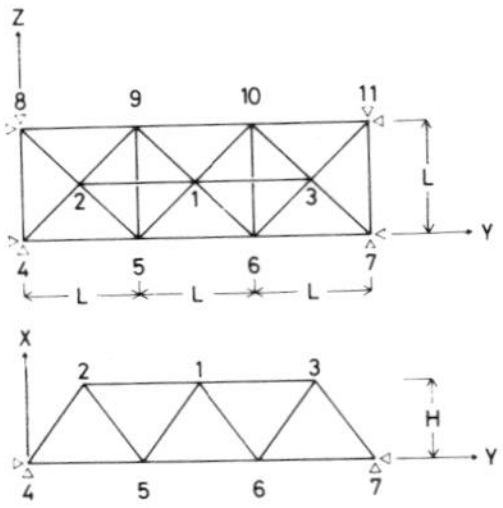

Fig 13b geometry

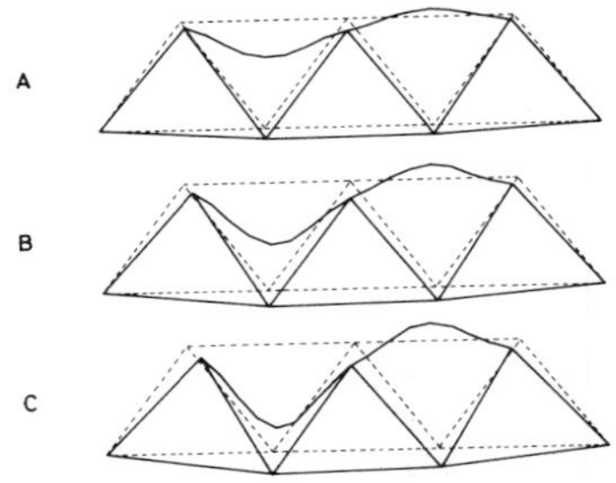

Fig 13c Deflection mode

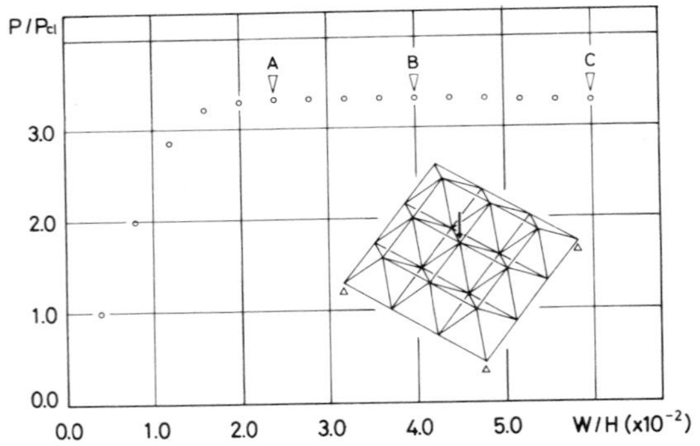

Fig 12a Load-deflection curve of a square grid

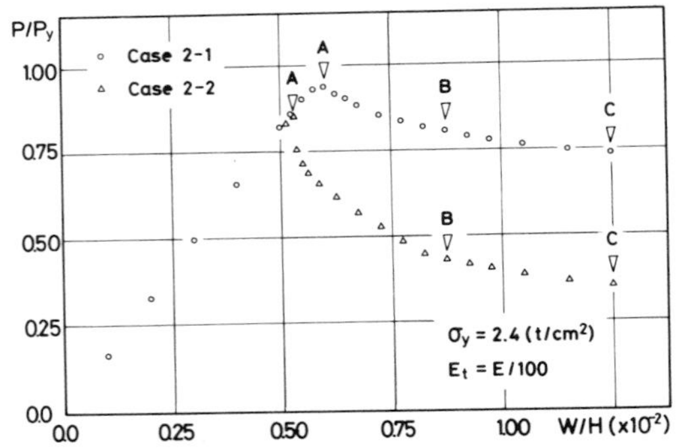

Fig 14a Load-deflection curve of a truss-like beam

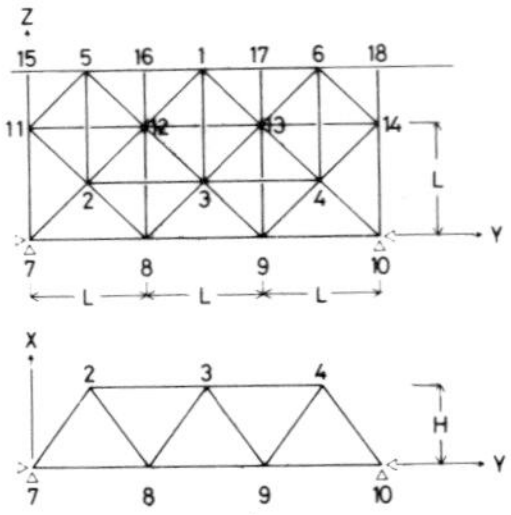

Fig 12b geometry

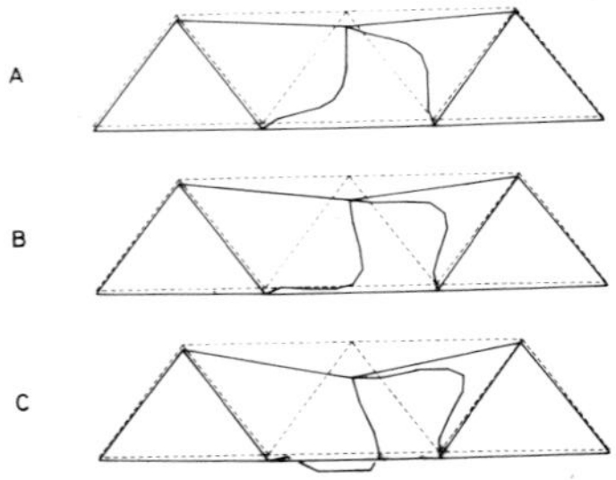

Fig 12c Deflection mode

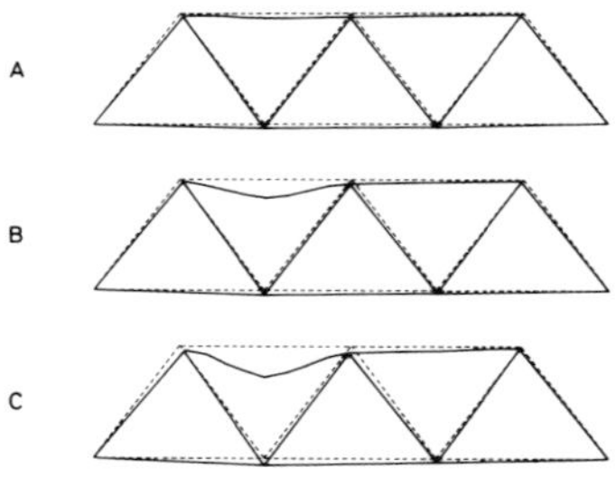

Fig 14b Deflection mode (case 2-1)

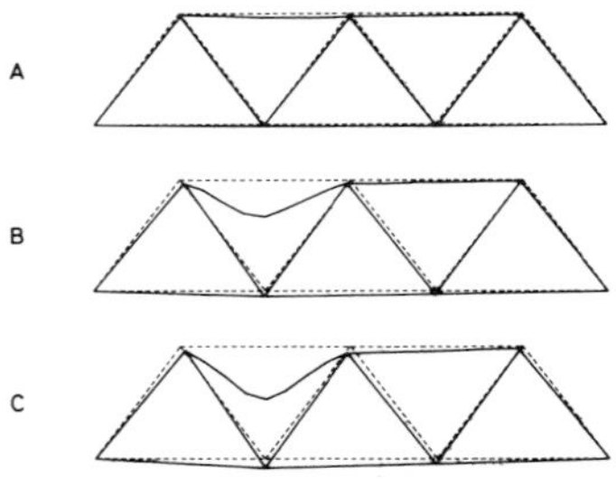

Fig 14c Deflection mode (case 2-2)

4. CONCLUDING REMARKS

In this paper, we developped a new finite element scheme for calculating the nonlinear tangential stiffness matrix and inner forces of double-layer grids. The analysis is performed introducing the concept of the collective element that is the repeating unit of double-layer grids, roofs and truss-like beams. The numerical examples presented in this papaer confirm the validity and the accuracy of the theoretical formulations.
As characteristic features of double-layer structures, following results are obtained.

1) The general buckling strength of double-layer roof structures mainly depends on both the distance between the layers and the stiffness of the bracing.
2) In the elastic analysis of a grid and truss-like beams, the post-buckling equilibrium path is stable because of re-distribution of the stresses.
3) The buckling strength of rigidly jointed members is about three times as large as the pin-jointed one due to the restraint of the adjacent members.
4) In the elasto-plastic analysis of truss-like beams, the post-buckling path falls quite suddenly and the asymmetric deflection increases like a bifurcation.

ACKNOWLEDGEMENTS

The computations in this paper were executed on Hitac M200 and M280 digital computer at Computer Center of Tokyo Institute of Technology.

REFERENCES

!. Z S Makowski Analysis, Design and Construction of Double-layer Grids, Applied Science Publishers 1981.

2. A H Nayfeh and M S Hefzy, Continuum Modeling of Three-Dimentional Truss-like Space Structures, AIAA Jour. Vol. 16, No.12, Dec. 1978.

3. T Matsui and O Matsuoka, A New Finite Element Scheme for Instability Analysis of Thin Shells, Int. J. for Num. Methods in Eng., Vol. 10, 1976.

4. M Fujimoto, M Iwata, F Nakatani and A Wada, Nonlinear Three-Dimentional Analysis of Steel Frame Structures, Proceedings of the Second International Conference of Space Structures, 1975.

ANALYSIS OF NONLINEAR AND COLLAPSE BEHAVIOR OF TM SPACE TRUSSES

S M H RASHED*, M KATAYAMA*, H ISHA*, I TODA**, K ODA** and T YOSHIZAKI**

* Century Research Center Corporation, Japan.

** Taivo Kogyo, Japan.

An analytical model of truss tubular members based on the idealized structural unit method is presented. Post buckling behavior is predicted by consideration of large deflection and the theory of plastic flow. An experimental study of members and member-joint interaction is summarized. Comparisons of theoretical predictions with experimental results are presented. A computer program, NOAMAS-T is described and an example of analysis is also presented.

INTRODUCTION

TM Trusses are composed of tubular members connected together through bolted spherical joints (globes), Fig 1. When subjected to extreme snow, wind or earthquake loads, highly compressed members may buckle. Tension members may yield and/or connection bolts may break. After the first of such failures, the structure may or may not carry further loads, depending on the nature of the structure and the applied loads. In truss design, in order to obtain a realistic evaluation of the safety of a truss, it is very important to accurately predict ultimate strength and post ultimate strength behavior until final collapse of the truss. Large deflections, plasticity and post buckling behavior of members should be taken into account.

A large deflection elastic-plastic finite element formulation (see Ref 1 for example) may be used to analyze such behavior to the required degree of accuracy. However, each member of a structure is to be divided into several finite elements and analyzing a large sized structure could be very expensive. To overcome this difficulty, several analytical and numerical models of tubular members subjected to compressive loads have been proposed such that one member is taken as one element. It is not intended to review these models in this paper, however some comments are made.

Several empirical models based on experimental or assumed load-shortening curves (see Refs 2 and 3 for example) are available. Those based on experiments may accurately represent the post buckling behavior of members. However, experiments are needed for each member geometry and material used in a structure. Such experiments are expensive and time consuming. Models based on assumed load shortening curves are approximate and sometimes risky to use.

Analytical models based on the moment-curvature relationship at a central plastic hinge inserted after buckling, Refs 4,5 and 6, give accurate results only for members with high slenderness ratios ($L_e/r>120$). A model based on the idealized structural unit method, Ref 7, takes account of plastic shortening as well as plastic rotation of the central plastic zone after buckling. It gives accurate results for members with lower slenderness ratios. The model, however, is developed for compression members subjected to end moments and lateral loads (as found in offshore tubular frames) and is too sophisticated to be used for simple tuss members subjected to axial compression only.

In this paper an analytical model of truss tubular members based on the idealized structural unit method is presented. In this model, large deflection and plasticity are taken into account. An experimental study is described and comparisons of theoretical predictions with experimental results are carried out. An example of analysis is also presented.

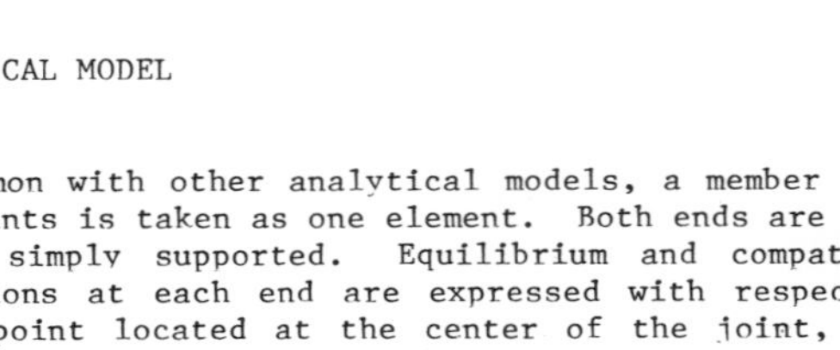

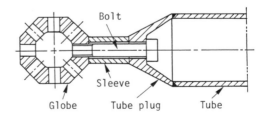

Fig 1 TM JOINTS

ANALYTICAL MODEL

In common with other analytical models, a member between two joints is taken as one element. Both ends are assumed to be simply supported. Equilibrium and compatibility conditions at each end are expressed with respect to a nodal point located at the center of the joint, Fig 2.

Three degrees of freedom are considered at each nodal point. Nodal displacement and nodal force vectors U and R may be expressed as follows:

$$U = [\, u_{xi} \;\; u_{yi} \;\; u_{zi} \;\; u_{xj} \;\; u_{yj} \;\; u_{zj} \,]^T$$
$$R = [\, R_{xi} \;\; R_{yi} \;\; R_{zi} \;\; R_{xj} \;\; R_{yj} \;\; R_{zj} \,]^T \qquad 1.$$

The incremental method with an updated large deflection elastic-plastic formulation is adopted. The nodal force-displacement relationship may be written in the element local coordinates as,

$$\Delta R = K \, \Delta U \qquad 2,$$

$$K = \begin{bmatrix} ku & 0 & 0 & -ku & 0 & 0 \\ & kg & 0 & 0 & -kg & 0 \\ & & kg & 0 & 0 & -kg \\ & & & ku & 0 & 0 \\ & \text{SYM} & & & kg & 0 \\ & & & & & kg \end{bmatrix} \qquad 3.$$

where, ku is the axial stiffness of the member,
$kg = -P/L$
L : Length of the member, and
P : the axial compressive force.

The term ku takes different forms depending on the condition of the member. The geometric term kg is considered in order to take account of overall large deflection and instability of the whole structure. Its value changes as P changes, however its form does not change with the condition of the member.

Idealization of Behavior

Tubular members used in TM truss systems have slenderness ratios (L/r) upto 150 and diameter thickness ratios (D/t) raging between 15 and 50. Local buckling of tube wall is not expected with D/t < 70, see Ref 8. Overall buckling and/or plasticity may however be expected. Ovalization of cross-section during buckling is also not expected, see Ref 9. These conclusions are also supported by observations during an experimental study summarized later in this paper.

Let the axial force-bending moment interaction relationship at a cross-section be considered. The first yield and full plastic relationships may be represented as in Fig 3. In this paper, an elastic perfectly plastic material is assumed (no strain hardening) and the full plastic interaction curve remains unchanged. When a simply supported member is subjected to an axial compression, it may reach its full plastic strength or may buckle depending on the slenderness ratio, material and initial imperfections of the member. Considering a slender member, because of initial imperfections (mainly residual stresses and out of straightness) a member starts to deflect slightly as the axial load increases. This creates bending moments along the member. At midlength, the axial load and bending moment may be represented by the curve ab in Fig 3. When the axial force approaches the buckling strength, deflection increases at a higher rate and at the buckling load, deflection increases at an almost constant axial force, curve bcd in Fig 3. At point d, the extreme fibers of the member at midlength yield and the axial force starts to decrease. The member continues to deflect while the bending moment at midlength continues to increase, and the plastic zone spreads approaching the full plastic condition as indicated by the curve de in Fig 3. The load path abcde may be idealized as ABCD.

When the member is subjected to axial tension, it may reach its full plastic strength or the connection bolts may get broken. If the member reaches its full plastic strength, it stretches while continues to carry a constant load. On the other hand, if one or both of the connection bolts gets broken, the axial force originally carried by the member is released.

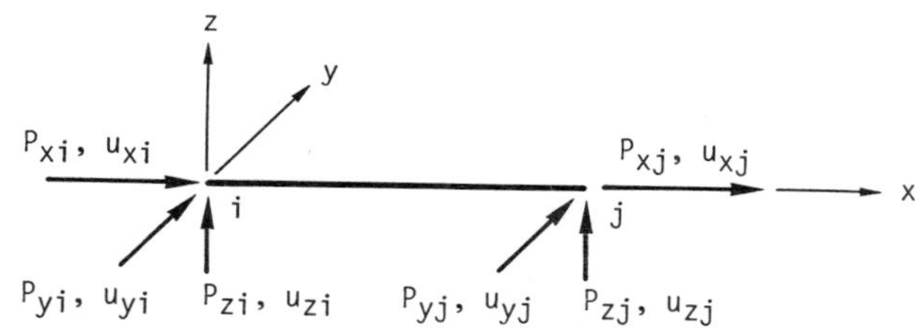

Fig 2 NODAL POINTS AND DEGREES OF FREEDOM

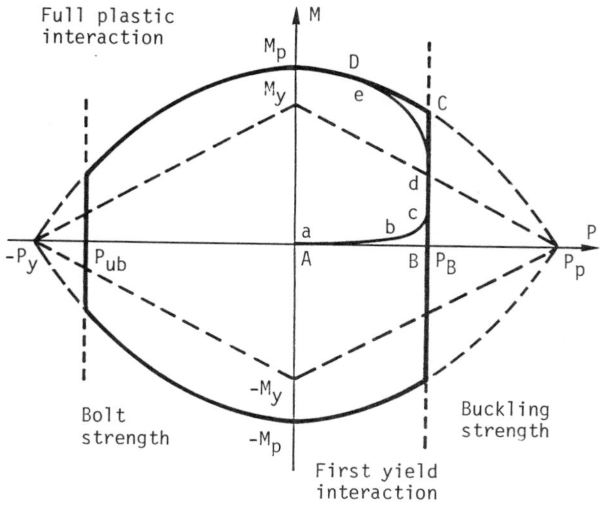

Fig 3 LIMITING STRENGTH AND BEHAVIOR IDEALIZATION

The combination of the buckling strength, full plastic strength and bolt strength represents a limiting strength interaction relationship of a member.

In the following, the term ku of the stiffness matrix K of Eqn 3, before, during and after buckling of a member, together with Equations for the limiting strength interaction relationship are derived.

Failure-Free Stiffness Matrix

Before buckling or yielding, the axial stiffness ku of Eqn 3 may be written as follows:

$$ku = EA/L \qquad 4.$$

where, E is the modulus of elasticity and A is the cross-sectional area.
This axial stiffness is assumed to be valid until the member reaches its limiting strength (buckling or yielding of member in compression / yielding of member or bolt fracture in tension).

Limiting Strength Condition

i) Buckling strength
Extensive studies of buckling strength have been performed in Lehigh University, Ref 10. Three column curves have been produced for different cross-sections, methods of manufacturing and steel grades. These curves are adopted in this study. The following two curves represent the buckling strength of welded tubes depending on the tube material.

Curve 1:

$$\frac{P_B}{P_p} = \begin{cases} 1 \text{ (yield level)} & (0 < \lambda < 0.15) \\ 0.990 + 0.122\lambda - 0.367\lambda^2 & (0.15 < \lambda < 1.2) \\ 0.051 + 0.801\lambda^{-2} & (1.2 < \lambda < 1.8) \\ 0.008 + 0.942\lambda^{-2} & (1.8 < \lambda < 2.8) \\ \lambda^{-2} \text{(Euler Buckling)} & (2.8 < \lambda) \end{cases}$$

Curve 2:

$$\frac{P_B}{P_p} = \begin{cases} 1 \text{ (yield level)} & (0 < \lambda < 0.15) \\ 1.035 - 0.202\lambda - 0.222\lambda^2 & (0.15 < \lambda < 1.0) \\ -0.111 + 0.636\lambda^{-1} + 0.087\lambda^{-2} & (1.0 < \lambda < 2.0) \\ 0.009 + 0.877\lambda^{-2} & (2.0 < \lambda < 3.6) \\ \lambda^{-2} \text{(Euler Buckling)} & (3.6 < \lambda) \end{cases} \qquad 5.$$

where $\lambda = \sqrt{\sigma_y/E}\, L/\pi r$ and $P_p = \sigma_y A$.

Curve 2 is used for A7, A36 or equivalent steels and curve 1 for steels with higher yield stress.

ii) Full plastic strength

In truss members, shearing stresses are very small compared to axial stresses and their effect on the plastic strength may be neglected. Integrating the full plastic axial stress over the cross-sectional area, the full plastic strength interaction function F^p of Fig 3 may be expresses in terms of the axial force P and the bending moment M as follows:

$$F^p = \frac{|M|}{M_p} - \cos\left(\frac{\pi}{2}\frac{P}{P_p}\right) = 0 \qquad 6.$$

where M_p and P_p are the full plastic bending moment and axial force respectively.

iii) Bolt strength

Ultimate tensile strength of bolts, P_{ub}, may be evaluated as follows:

$$P_{ub} = A_b\, \sigma_{ub} \qquad 7.$$

where, A_b is the effective cross-sectional area and σ_{ub} the ultimate tensile stress of the bolt material.

Buckling and Post Buckling Stiffness

When the increasing axial compressive load reaches the value P_B, it is assumed that the member buckles (deflects) at constant axial force until the bending moment at midlength satisfies the full plastic condition, point C in Fig 3. Due to deflection, an axial shortening takes place. During this stage, the axial stiffness of the member is equal to zero.

$$\overset{B}{ku} = 0 \qquad 8.$$

$\overset{B}{ku}$ may be substituted in Eqn 3 to obtain the buckling stiffness matrix.

After the internal forces at midlength of the member satisfy the full plastic condition, they cannot increase beyond it. However the member continues to deflect and plastic deformation takes place while the ratio of the internal forces changes. Noting the symmetry of a buckled member about its midlength, one half length, im, is considered. A coordinate system $\bar{x} - \bar{y}$ is defined as shown in Fig 4. At any load step the member is in equilibrium under an axial compressive force P along the x axis. An

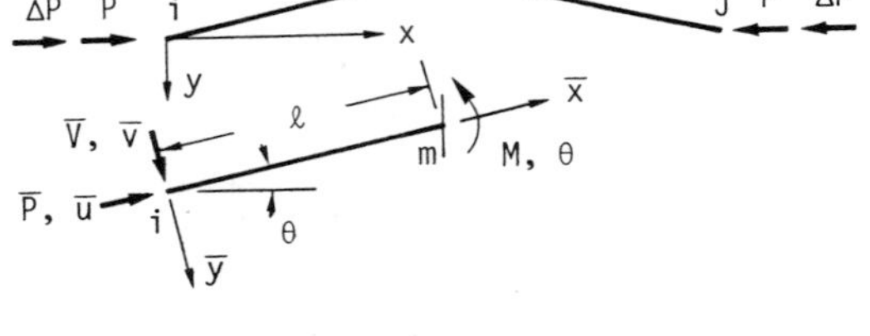

Fig 4 POST BUCKLING COORDINATES

increment ΔP is then applied. P and ΔP may be resolved in $\bar{x}$ and $\bar{y}$ directions as follows:

$$\bar{P} = P\cos\theta \qquad \Delta\bar{P} = \Delta P\cos\theta$$

$$\bar{V} = P\sin\theta \qquad \Delta\bar{V} = \Delta P\sin\theta$$

At node i, increments of displacements $\Delta\bar{u}$ in $\bar{x}$ direction and Δv in y direction takes place. They may be divided into elastic and plastic components as follows:

$$\Delta\bar{u} = \Delta\bar{u}_e + \Delta\bar{u}_p$$
$$\Delta\bar{v} = \Delta\bar{v}_e + \Delta\bar{v}_p \qquad 9.$$

$\Delta\bar{u}_e$ is caused by $\Delta\bar{P}$ and may be expressed as follows:

$$\Delta\bar{u}_e = \Delta\bar{P} \,/\, \overline{ku} \qquad 10.$$

where $\overline{ku} = EA/\ell$, $\ell = L/2$.

$\Delta\bar{v}_e$ may be expressed as follows:

$$\Delta\bar{v}_e = \Delta M \,/\, \overline{kv} \qquad 11.$$

where $\overline{kv} = (3EI/\ell^2) - \bar{P}/5$.

To evaluate $\Delta\bar{u}_p$ and $\Delta\bar{v}_p$, the theory of plastic flow is utilized. Assuming the plastic deformation to be concentrated at midlength of the member, the following relations may be written.

$$\Delta F^p = \frac{\partial F^p}{\partial M}\Delta M + \frac{\partial F^p}{\partial P}\Delta P = \phi_m\,\Delta M + \phi_p\,\Delta P = 0$$

and

$$\Delta\bar{u}_p = \lambda\,\phi_p, \quad \Delta\theta_p = \lambda\,\phi_m$$

From which

$$\Delta M = -(\phi_p/\phi_m)\,\Delta\bar{P} = -\frac{1}{n}\Delta\bar{P} \qquad 12.$$

and

$$\Delta\bar{u}_p = \Delta\theta_p \,/ n \qquad 13.$$

where $\Delta\theta_p$ is the plastic rotation at midlength. Equations 12 and 13 are represented in Fig 5.

The plastic component $\Delta\bar{v}_p$ of Eqn 9 is due to the plastic rotation $\Delta\theta_p$ and may be expressed as,

$$\Delta\bar{v}_p = \ell\,\Delta\theta_p \qquad 14.$$

The equilibrium condition may be written as follows:

$$\Delta M = \Delta\bar{V}\,\ell + \bar{P}\,(\Delta\bar{v}_e + \Delta\bar{v}_p) \qquad 15.$$

Substituting Eqns 11, 12 and 14 into Eqn 15, an expression for $\Delta\theta_p$ may be obtained. Equation 13 may then give $\Delta\bar{u}_p$ as follows:

$$\Delta\bar{u}_p = -\frac{\Delta\bar{P}}{n^2\,P\ell} - \frac{\Delta\bar{V}}{nP} - \frac{\Delta M}{\ell\,\overline{kv}\,n} \qquad 16.$$

Equations 10 and 16 may be added to obtain $\Delta\bar{u}$. Δu in x direction may now be obtained as,

$$\Delta u = 2\,(\Delta\bar{u}\cos\theta + \Delta\bar{v}\sin\theta) = \Delta P \,/\, \overset{P}{ku} \qquad 17.$$

where

$$\overset{P}{ku} = 1/\left[2\left(\frac{1}{\overline{ku}}\cos^2\theta - \frac{\overline{kv} - \bar{P}}{\ell n^2 P\overline{kv}}\cos\theta - \frac{2 + \ell n\tan\theta}{nP}\sin\theta\right)\right]$$

$\overset{P}{ku}$ is the post buckling axial stiffness and may be substituted in Eqn 3 to obtain the post buckling stiffness matrix.

Unloading is judged by the direction of plastic flow. When unloading occurs, $\Delta\bar{u}_p$, $\Delta\theta_p$ and $\Delta\bar{v}_p$ of Eqns 9 and 14 become equal to zero. Equations 10, 11 and 15 remain valid. Equations 12 and 13 are not valid any more. The post buckling unloading axial stiffness k_u^U may be derived as follows:

$$\overset{U}{ku} = 1/\left[2\left(\frac{\cos^2\theta}{\overline{ku}} + \frac{\ell\sin^2\theta}{\overline{kv} + P\cos\theta}\right)\right] \qquad 18.$$

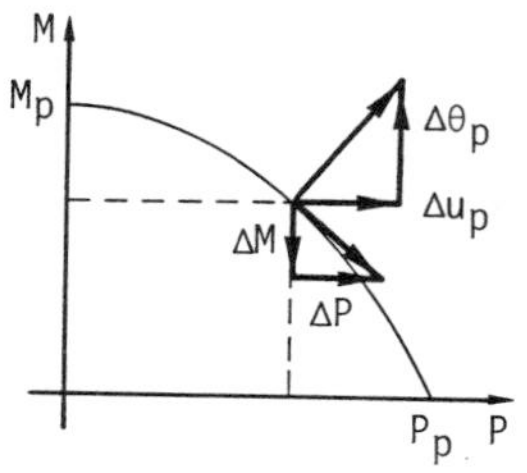

Fig 5 PLASTIC RELATIONS

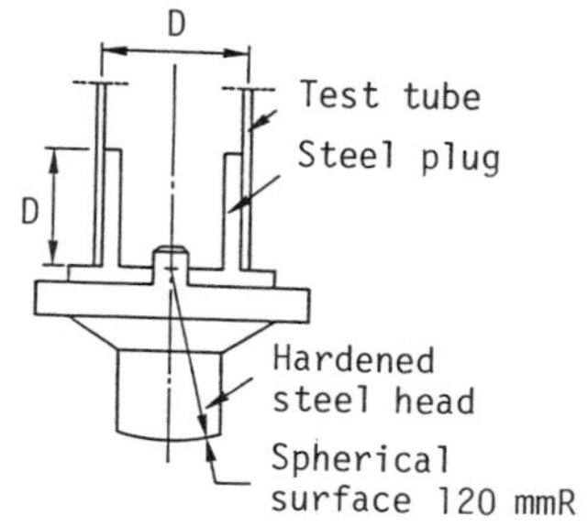

Fig 6 SPHERICAL SUPPORTS

EXPERIMENTAL INVESTIGATION

Eight simply supported tubular members and eight members bolted to ball joints prevented from rotation have been tested in compression, Ref 11. In the following, these tests, test results and comparisons with theoretical results are summarized.

Simply Supported Members (CCS Series)

Eight tubular members simply supported at both ends have been tested in axial compression. The aims of the tests were to confirm that the ultimate strength formula used in the theoretical analysis is appropriated for TM members as produced by Taivo Kogyo, and to check the theoretical post-buckling behavior against test results. Test models'dimensions, summarized in table 1, have been chosen to represent a wide range of actual members used in practice. Two models of each geometry have been tested to check the scatter of results. Models have been cut from long tubes taken from the existing stock (except models CCS 41 and CCS 42 which are not standard) in order to avoid any extra care in fabrication. These tubes have been rolled out of JIS G3444-STK 41 steel (mild steel) plates and automatically welded. One stub column (L = 3D) has been cut from each tube. Thickness, diameter and material properties have been surveyed. Wall thicknesses have been measured at 4 points at each end of models and stubs using a micrometer. Outer diameters have been measured at 3 stations along each model. 4 diameters spaced at 45 degrees have been measure at each station. Straightness of each tube has been measured in two perpendicular planes and the resultant maximum out of straightness has been evaluated. Stub columns have been tested to evaluate yield stresses and moduli of elasticity. These are reported in table 1 together with geometric properties.

As shown in Fig 6, each tube was fitted at its ends with sliding fit steel plugs to prevent premature local buckling at the ends, and spherical heads of hardened high strength steel to simulate simple supports. Axial compressive load was applied in a vertical universal testing machine under displacement control at an average axial strain rate of 3.5×10^{-7}/sec. Axial end shortening and lateral deflection at midlength in two perpendicular directions has been recorded by displacement transducers. Strains at stations at distance D from each end and at midlength have been measured. 4 electric strain gages at 90° were fitted at each station.

Experimental maximum loads are reported in table 1 and plotted in Fig 7 together with theoretical predictions. Experimental load-shortening curves are presented in Fig 8 together with those predicted by the present theoretical model. The scatter of experimental results has been found to be very small. Since first yield has been neglected in the theoretical model, extended shortening at the buckling load is observed specially with members of higher slenderness ratios where elastic deflection and the difference between first yield and full plastic conditions are large. The post buckling curve, however, quickly approaches the experimental curve. The experimental curves seem to indicate a little strength more than the theoretical ones. This could be because of small resisting moments at the supports and strain hardening. Satisfactory agreement of theoretical and experimental results may be, however, observed specially with members of lower slenderness ratios.

Table 1 CCS SERIES

Model	D	t	L	E	σ_y	D/t	λ	δ_0/L	σ_u^T/σ_y	σ_u^E/σ_y	σ_u^E/σ_u^T
CCS 11	42.77	2.193	1850	20830	38.90	19.50	1.771	.0006	.3063	.2888	.9428
CCS 12	42.83	2.125	1850	20830	38.90	20.16	1.766	.0002	.3078	.2971	.9652
CCS 21	60.55	2.984	2250	21083	38.70	20.29	1.372	.0002	.4766	.4971	1.043
CCS 22	60.67	2.989	2250	21083	38.70	20.30	1.369	.0002	.4783	.4862	1.017
CCS 31	89.14	4.045	2250	21384	37.75	22.04	0.999	.0003	.7456	.7380	.9922
CCS 32	89.04	4.045	2250	21384	37.75	22.01	1.000	.0002	.7448	.7701	1.033
CCS 41	89.17	2.636	2250	20863	34.18	33.83	0.947	.0004	.6445	.6703	1.040
CCS 42	89.01	2.634	2250	20863	34.18	33.79	0.949	.0004	.6435	.7119	1.106

D: Outer Diameter, T: Thickness, L: Length, E: Young's Modulus
σ_y: Yield Stress, λ: Reduced Slenderness Ratio, δ_0: Max. Out of Straightness
σ_u^T: Theoretical Max. Stress, σ_u^E: Experimental Max. Stress

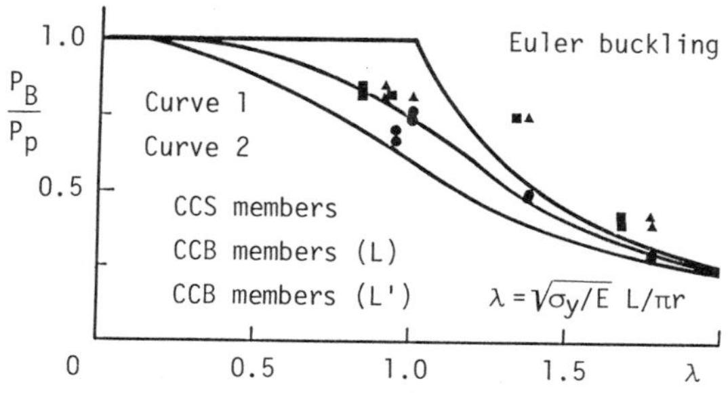

Fig 7 THEORETICAL AND EXPERIMENTAL BUCKLING STRENGTH

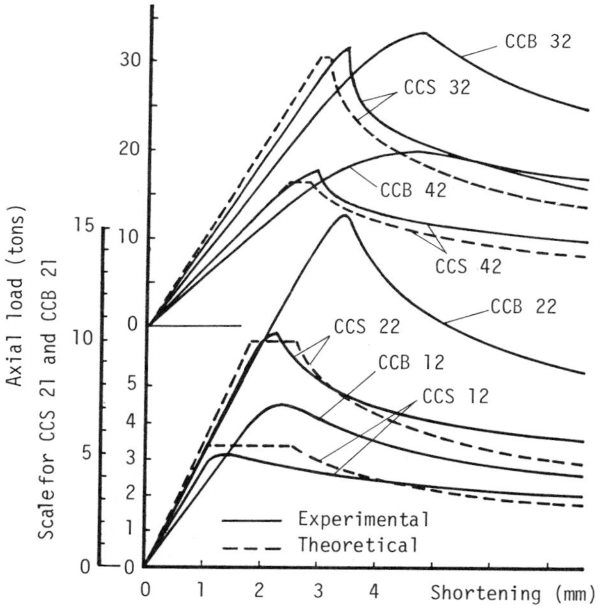

Fig 8 THEORETICAL AND EXPERIMENTAL LOAD-SHORTENING CURVES

Bolted Members (CCB Series)

Another eight tubular members bolted at both ends to ball joints prevented from rotation are tested to investigate the conditions of supports and the rotation ability of bolt connections after buckling of members. Dimensions has been chosen identical to the simply supported models for easy comparison. Models have been fabricated from tubes taken from the same lot as CCS series (except models CCB 41 and CCB 42) according to the tolerances used in actural practice. Geometry and material properties are investigated as in CCS series and summarized in table 2. Globes at both ends of each model have been machined flat as shown in Fig 9. Bolts of JIS G4105-SCM435 steel (σ_y = 80 kg/mm^2) have been used in the joints. The model is placed in the testing machine with the flat surfaces of globes against the machine heads thus preventing the globe from rotation. Shortening, lateral deflection and strains have been measured as in CCS series.

Table 2 CCB SERIES

Model	D	t	L	E	σ	D/t	λ	δ_0/L	σ_u^T/σ_y	σ_u^E/σ_y	σ_u^E/σ_u^{ES}
CCB 11	42.76	2.193	1850	20830	38.90	19.50	1.772	.0003	.3062	.3969	1.355
CCB 12	42.79	2.125	1850	20830	38.90	20.14	1.768	.0005	.3073	.4257	1.453
CCB 21	60.60	2.984	2050	21083	38.70	20.31	1.371	.0001	.4773	.7491	1.524
CCB 22	60.47	2.989	2050	21083	38.70	20.23	1.374	.0001	.4754	.7452	1.516
CCB 31	88.98	4.045	2250	21384	37.75	22.00	1.001	.0002	.7444	.8176	1.083
CCB 32	89.11	4.045	2250	21384	37.75	22.03	.9994	.0003	.7454	.8155	1.080
CCB 41	89.13	2.636	2250	22655	34.40	33.81	.9122	.0005	.7327	.8104	1.173
CCB 42	89.08	2.634	2250	22655	34.40	33.82	.9127	.0005	.7329	.8466	1.225

D: Outer Diameter, T: Thickness, L: Length, E: Young's Modulus
σ_y: Yield Stress, λ: Reduced Slenderness Ratio, δ_0: Max. Out of Straightness
σ_u^T: Theoretical Max. Stress, σ_u^E: Experimental Max. Stress
σ_u^{ES}: Mean Experimental Max. Stress of Corresponding CCS Models

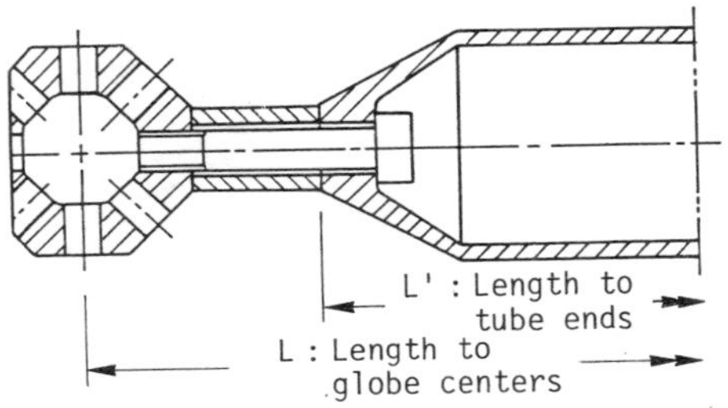

Fig 9 END FITTING OF CCB SERIES

Results of the tests are summarized in table 2 and Figs 7 and 8 together with results for simply supported members. An increase of 10% ∿ 50% over the buckling loads of simply supported members (CCS Series) and the theoretical simply supported buckling load calculated with L measured between the centers of the joints has been observed. If the length is taken between the ends of the tubes as shown in Fig 9, the buckling load is increased by about 0% ∿ 40% over the theoretical buckling load (see Fig 7). Similar increase of strength may be observed after buckling. Detailed investigation of behavior of joints and member-joint interaction in assembled truss structures is necessary before taking these increases into account in truss design.

After the tests, the bolt connections have been examined. No signs of failure or cracking have been observed in any bolt even with end rotations during the tests of more than 10 degrees. Actually many bolts have not even been bent. This may be attributed to the shortening of sleeves and deformation of the tube plugs under compression which almost free the bolts in their clearances. These observations suggest that bolt connections are able to undergo the rotations which take place after buckling.

PROCEDURE OF ANALYSIS AND COMPUTER PROGRAM

Using the model outlined above a computer program, NOAMAS-T, is completed. In this program, the usual procedure of incremental load analysis of nonlinear behavior is followed. First, the load-free structure is considered. The incremental stiffness matrix of each member is constructed and transformed into the global coordinates. The global incremental stiffness matrix of the whole structure is then assembled. After the boundary conditions are introduced, the first load increment is applied. The deformation of the structure is obtained and the internal forces in the members are evaluated. Newton-Rapson equilibrium iterations are performed until a convergence criterion is satisfied. Each member is then checked for buckling, plastification, or bolt fracture.

The coordinates are updated and a new stiffness matrix is constructed and transformed into global coordinates for each member after each load increment. The global stiffness matrix is reassembled and the next increment of loads is applied.

When buckling and/or plastification of a member or more, or fracture of a bolt or more are detected within a loading step, the load increment is scaled down to that just necessary to cause such failure. This prevents the internal force vectors from shooting out of the limiting strength interaction surfaces.

When bolt fracture is detected, the load which is carried by the concerned member is released and loaded in steps on the rest of the structure. The ultimate strength of the structure is detected by consideration of plastic deformation. The analysis may be continued in the post ultimate strength range to evaluate energy absorption.

EXAMPLE OF ANALYSIS

The roof truss shown in Fig 10.a is simply supported at 8 nodes as shown in the figure. A proportionally increasing snow load is applied to the structure. The load-central deflection curve is shown in Fig 10.b. The structure reaches its ultimate strength at a load several times as much as that at which first member buckling has occurred. The load-deflection curve suggests that the considered structure is not able to carry higher load than that of the first peak load.

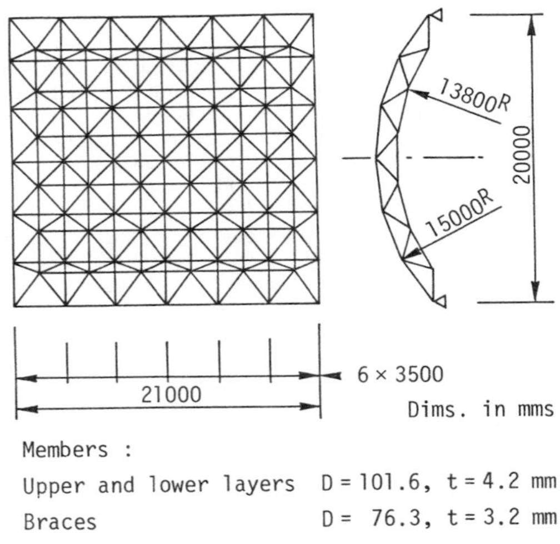

Fig 10.a EXAMPLE STRUCTURE

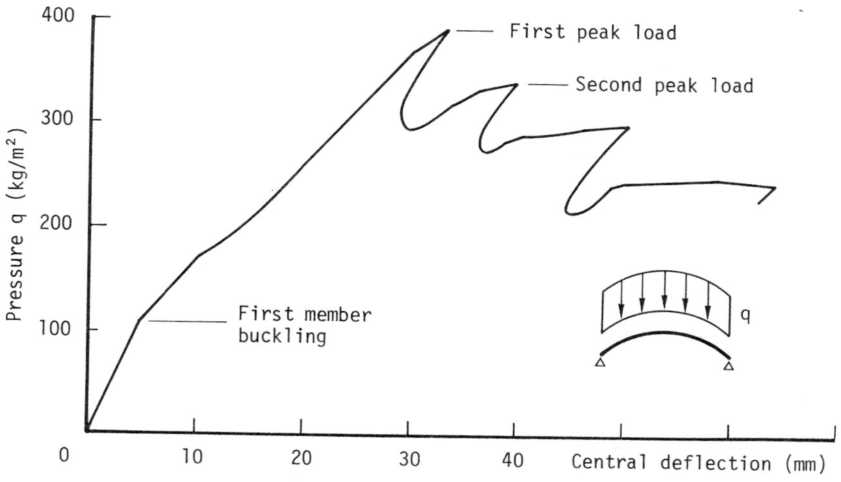

Fig 10.b LOAD-DEFLECTION CURVE OF EXAMPLE STRUCTURE

CONCLUSIONS

An analytical model of truss tubular members based on the idealized structural unit method is presented. Large deflection and plasticity are taken into account and the behavior of a member is represented by a set of strength conditions and a set of stiffeness equations. Comparisons with results of an experimental study have confirmed the accuracy and reliability of the model especially with respect to buckling and post buckling behavior of members of lower slenderness ratio ($\lambda < 1.5$). For members with higher slenderness ratios, consideration of first yield would improve the accuracy.

Experimental investigation of joint behavior and member-joint interaction have shown that bolt connections are able to undergo the large rotations which may take place after buckling. Some restraint may be expected from these bolt connections. However, further studies are necessary before considering this restraint in truss design.

Using the model presented in this paper, a computer program, NOAMAS-T, has been completed. An example analysis of a truss structure is presented.

Accuracy and reliability of the present model, specially with lower slenderness ratios, are in the same order as may be obtained by nonlinear finite element models, however, the required computer effort is very small compared to that required by these models.

REFERENCES

1. W F Chen and D A Ross, Tests of Fabricated Tubular Columns, Journal of the Structural Division, ASCE, March 1977.

2. B Mason and E P Popov, Cyclic Response Prediction for Braced Steel Frames, Journal of the Structural Division, ASCE, July 1980.

3. P W Marshall, Design Considerations for Offshore Structures Having Nonlinear Response to Earthquakes, Reprint, ASCE Annual Convention and Exposition, Chicago, October 1978.

4. A B Higginbotham, The Inelastic Cyclic Behavior of Axially-Loaded Steel Members, Dissertation, University of Michigan, Ann Arbor, January 1973.

5. R Nilforoushan, Seismic Behavior of Multi-story K-Braced Frame Structures, University of Michigan Research Report UMEE 73R9, Ann Arbor, November 1973.

6. P Singh, Seismic Behavior of Braces and Braced Steel Frames, Dissertation, University of Michigan, Ann Arbor, July 1977.

7. Y Ueda, S M H Rashed and K Nakacho, New Efficient and Accurate Method of Nonlinear Analysis of Offshore Tubular Frames (The Idealized Structural Unit Method), 3rd International Offshore Mechanics and Arctic Engineering Symposium, ASME, February 1984.

8. J G Bouwkamp, Buckling and Post-Buckling Strength of Circular Tubular Sections, Offshore Technology Conference, Paper number OTC 2204, 1975.

9. C S Smith, W Kirkwood and J W Swan, Buckling Strength and Post Collapse Behavior of Tubular Bracing Members Including Damage Effects, BOSS'79, Imperial College, London, 1979.

10. R Bjorhovde, Deterministic and Probabilistic Approaches to the Strength of Steel Columns, Ph. D. Dissertation in Civil Engineering, Lehigh University, Bethlehem, Pa., 1972.

11. S M H Rashed, Experimental Study of Buckling and Post Buckling Behavior of TM Truss Systems, Century Research Center Corp., Japan, March 1984.

BUCKLING OF STRUCTURES IN PROBABILISTIC CONTEXT

J P MUZEAU, M FOGLI and M LEMAIRE

Laboratoire de Génie Civil

Université de Clermont II, France.

A new approach to the phenomenon of buckling, in a probabilistic context, is developed in this paper. First, the random variables and the limit state associated with the collapse of the structure are defined, then the non-linear mechanical model is built. A Monte-Carlo method makes it possible to solve the associated probabilistic problem by simulation. It is applied to simple examples.

INTRODUCTION

The strength of compressive members in space structures is linked to the phenomenon of buckling which depends essentially on the imperfections. Their random nature suggests a probabilistic approach towards reliability in connection with a non-linear mechanical model.

The probabilistic analysis of the reliability is based upon a very precise strategy of which the principal steps are as follows :

- choice of random variables from the basic variables and the characterization of their laws,
- choice of a limit state of collapse for the structure,
- choice of a mechanical model which allows this state to be expressed,
- choice of a measure of reliability of the structure and the numerical model which is capable of evaluating it.

From the mechanical point of view, the real behaviour of the elements characterized by important displacements and the collapse of the compressive members subjected to combined bending and compression, must be taken into account.

The construction of a model which would take into account the initial imperfections from the simple notion of Euler's critical load would be therefore without significance. This is why it is necessary to construct a mechanical model best able to follow the non-linear geometrical behaviour of the elements. The notion of critical load is used solely in the discussion of examples as a parameter of reference.

This study is limited to the plane structures, the material of which obeys Hooke's law. Elastic-plastic behaviour does not imply, a priori, any particular difficulties. The cross-sections obey Navier-Bernouilli's principle.

After the development of these steps, several numerical applications are proposed.

CHOICE OF RANDOM VECTOR AND CHARACTERIZATION OF ITS LAW

Generally the random vectors are as follows :

$\mathbf{X}_I = (X_1, \ldots, X_p)^t$ the random vector of initial imperfections,
$\mathbf{X}_G = (X_{p+1}, \ldots, X_{p+q})^t$ the random vector of geometrical dimensions,
$\mathbf{X}_L = (X_{p+q+1}, \ldots, X_{p+q+r})^t$ the random vector of applied loads,
$\mathbf{X}_S = (X_{p+q+r+1}, \ldots, X_{p+q+r+s})^t$ the random vector of the strength of the materials.

The vector of the basic variables is :
$\mathbf{X} = (\mathbf{X}_I^t, \mathbf{X}_G^t, \mathbf{X}_L^t, \mathbf{X}_S^t)^t = (X_1, \ldots, X_n)^t$ with $n = p + q + r + s$.

It is a non-degenerate vector with values in $\mathbf{R}^n$, which satisfy the following hypotheses :

- its law is known and possesses the density f relatively to Lebesgue's measure on $\mathbf{R}^n$,
- it has second moments ; it belongs to the $\mathcal{L}^2((\Omega, \tau, P); \mathbf{R}^n)$-space of random square integrable vectors defined on the probability space (Ω, τ, P) with values in $\mathbf{R}^n$.

In fact, in the theory of probability, the random variables (r.v.) are defined on the probability space (Ω, τ, P) where Ω is an abstract set called the universal set, τ is a σ-field of subjects of Ω and P is a probability, i.e. an σ-additive application of τ on $[0,1]$. The characterization of Ω is of little interest to the engineer. Such a set can, moreover, accept several definitions for a given problem. It is essential, on the other hand, to know the laws of random variables. We can, for example, characterize Ω as the set of all envisageable observations of the random vector $\mathbf{X}$. Ω is therefore identified with $\mathbf{R}^n$, τ is the σ-field of Borel's set of $\mathbf{R}^n$ and P is the probability law with density f. An element ω of Ω is a realization of the vector $\mathbf{X}$, which is represented by the identical application : $\omega \rightsquigarrow \mathbf{X}(\omega) = \omega$. Such a probabilistic model is not unique.

CHOICE OF A LIMIT STATE OF COLLAPSE

Two limit states for the structure are considered ; one with relation to strength, the other with relation to the displacement.

.) Limit state of strength

Any section of a structure is, a priori, a critical section relative to chosen limit state of strength. However, it is possible to define a "limited" number of sections in which the limit state condition must be controlled. They are called control sections.

Let one structure (S), N control sections defined on (S) and R_j and S_j the actions which are resistant and act in the section A_j, we write :

$$R_j - S_j = G_j^R \circ \pi_j^n \circ \mathbf{X} \qquad \ldots\ldots 1.$$

The difference $R_j - S_j$ is a r.v. function of the set of the basic r.v.-s, G_j^R is a function with domain $\mathbf{R}^{q_j}$ (with $q_j \le n$) called the limit state function of section S_j and π_j^n is the projector from $\mathbf{R}^n$ to $\mathbf{R}^{q_j}$. The event of collapse associated with this section is written as :

$$\overline{E}_j^R = \{\omega \in \Omega \ : \ (G_j^R \circ \pi_j^n \circ \mathbf{X})(\omega) < 0\} \qquad \ldots\ldots 2.$$

The definition chosen from amongst several possibilities to characterize the event of the collapse of the structure is as follows :

$$\overline{E}_j^R = \bigcup_{j=1}^{n} \overline{E}_j^R \qquad \ldots\ldots 3.$$

This signifies that collapse of the structure (or, more precisely of its model) occurs when at least one control section has reached its limit state of collapse.

Equations 2 and 3 give :

$$\overline{E}^R = \{\omega \in \Omega \ : \ (G^R \circ \mathbf{X})(\omega) < 0\} \qquad \ldots\ldots 4.$$

where : $$G^R \circ \mathbf{X} = \min_{j=1,\ldots,N} G_j^R \circ \pi_j^n \circ \mathbf{X} \qquad \ldots\ldots 5.$$

Function G^R, defined on $\mathbf{R}^n$, is called the limit state function of the strength of the structure and equation $G^R(x_1,\ldots,x_n) = 0$ defines the boundary which separates the domain of collapse and the domain of reliability which are respectively :

$$\overline{D}^R = \{\mathbf{x} \in \mathbf{R}^n : G^R(\mathbf{x}) < 0\} \text{ and } D^R = \{\mathbf{x} \in \mathbf{R}^n : G^R(\mathbf{x}) \ge 0\} \qquad \ldots 6,7,$$

2) Limit state of displacement

In the same way, the control points of a structure (S) are defined as the points where the limit state of displacement condition must be controlled. Let $P_1,\ldots,P_M$, M controlled points defined on (S) and δ_K and $\overline{\delta}_K$ the calculated displacement and allowable displacement of point P_K in a given direction, the difference between δ_k and $\overline{\delta}_k$ is noted as before :

$$\delta_k - \overline{\delta}_k = G_k^\delta \circ \pi_k^n \circ \mathbf{X} \qquad \ldots\ldots 8.$$

G_k^δ defined on $\mathbf{R}^{q_k}$ is the limit state of displacement function relative to point P_k and, at this point, the event of collapse is written as :

$$\overline{E}_k^\delta = \{\omega \in \Omega \ : \ (G_k^\delta \circ \pi_k^n \circ \mathbf{X})(\omega) < 0\} \qquad \ldots\ldots 9.$$

Formally, Eqns 2 and 3 are analogous and the event of collapse is defined as :

$$\overline{E}^\delta = \bigcup_{k=1}^{M} \overline{E}_k^\delta \qquad \ldots\ldots 10,$$

This gives from Eqn 9 :

$$\overline{E}^\delta = \{\omega \in \Omega \ : \ (G^\delta \circ \mathbf{X})(\omega) < 0\} \qquad \ldots\ldots 11,$$

with $$G^\delta \circ \mathbf{X} = \min_{k=1,\ldots,M} G_k^\delta \circ \pi_k^n \circ \mathbf{X} \qquad \ldots\ldots 12,$$

G defined on $\mathbf{R}^n$ is the limit state of displacement function and the domains of collapse and reliability relative to this limit state are defined respectively as :

$$\overline{D}^\delta = \{\mathbf{x} \in \mathbf{R}^n : G^\delta(\mathbf{x}) < 0\} \text{ and } D^\delta = \{\mathbf{x} \in \mathbf{R}^n : G^\delta(\mathbf{x}) \ge 0\} \qquad \ldots 13,14.$$

MECHANICAL MODEL

We have studied the methods of non-linear structural analysis propounded by different authors as Jennings, Frey and Jouve (Refs 1, 2 and 3). The model of geometrically non-linear mechanical behaviour which we have constructed is based on the principal of Jennings' method. It allows us to obtain adequately precise results despite a discretization in a fairly small number of elements. Moreover it has the advantage that it does not necessitate numerical integration. The cost of calculation thus remains acceptable with a great number of simulations.

1) Expression of load and displacement vectors in different axes

Let XOY be the general fixed reference axes where the coordinates of the nodes, their displacement and the loads applied to each element of the structure are expressed as shown in Fig 1.

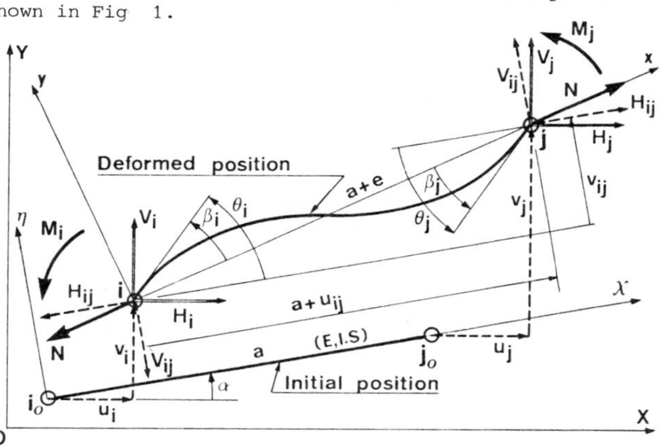

Fig 1

The initial length of the underformed element $i_o j_o$ is a and α is its orientation with reference to the axis OX. The vector **P** of loads applied to the nodes and the vector **U** of the displacements produced by them are :

$$\mathbf{P} = (H_i, V_i, M_i, H_j, V_j, M_j)^t \quad \text{and} \quad \mathbf{U} = (u_i, v_i, \theta_i, u_j, v_j, \theta_j)^t$$

The reference axes $\chi i_o \eta$ are attached to the initial position $i_o j_o$. Vectors $\mathbf{U}_r$ and $\mathbf{P}_r$, represent respectively the displacements and nodal loads, are expressed in these axes in the following form :

$$\mathbf{U}_r = (u_{ij}, v_{ij}, \theta_i, \theta_j)^t \quad \text{and} \quad \mathbf{P}_r = (H_{ij}, V_{ij}, M_i, M_j)^t$$

The first reduction in the number of unknowns corresponds to a rotation of angle α.

$$\text{Therefore } \mathbf{U}_r = \rho \mathbf{U} \text{ with } \rho = \begin{bmatrix} -\cos\alpha & -\sin\alpha & 0 & \cos\alpha & \sin\alpha & 0 \\ \sin\alpha & -\cos\alpha & 0 & -\sin\alpha & \cos\alpha & 0 \\ 0 & 0 & 1 & 0 & 0 & 0 \\ 0 & 0 & 0 & 0 & 0 & 1 \end{bmatrix} \qquad \ldots\ldots 15$$

Now, the local axes xiy are attached to the deformed position as is shown in Fig 1.

The vectors $\mathbf{U}_R$ and $\mathbf{P}_R$ of the displacements and loads are expressed as :

$$\mathbf{U}_R = (e, \beta_i, \beta_j)^t \quad \text{and} \quad \mathbf{P}_R = (N, M_i, M_j)^t$$

The accurate transformation of $\mathbf{U}_R$ into $\mathbf{U}_r$ is obtained by the following equations :

$$e = \sqrt{(a+u_{ij})^2 + v_{ij}^2} - a \qquad \text{..... 16.}$$

$$\beta_i = \theta_i - \text{arc tg}\,\frac{v_{ij}}{a+u_{ij}} \quad \text{and} \quad \beta_j = \theta_j - \text{arc tg}\,\frac{v_{ij}}{a+u_{ij}}$$

2) Calculation of the displacement variations

Equation 15 leads to : $d\mathbf{U}_r = \boldsymbol{\rho}\, d\mathbf{U} + d\boldsymbol{\rho}\, \mathbf{U}$

As the matrix $\boldsymbol{\rho}$ only contains terms in α, its variation is nil, giving :

$$d\mathbf{U}_r = \boldsymbol{\rho}\, d\mathbf{U} \qquad \text{..... 17.}$$

After calculation, the general expression for the variation of displacements is written in the form :

$$d\mathbf{U}_R = \mathbf{T}\, \boldsymbol{\rho}\, d\mathbf{U} \qquad \text{..... 18.}$$

$$\text{with : } \mathbf{T} = \begin{bmatrix} \frac{a+u_{ij}}{a+e} & \frac{v_{ij}}{a+e} & 0 & 0 \\ \frac{v_{ij}}{(a+e)^2} & -\frac{a+u_{ij}}{(a+e)^2} & 1 & 0 \\ \frac{v_{ij}}{(a+e)^2} & -\frac{a+u_{ij}}{(a+e)^2} & 0 & 1 \end{bmatrix}$$

3) Displacement field

The displacement field is linear in the χ-axis and cubic in the η-axis. For any point on the abscissa χ of the average of the element $i_o j_o$, the displacements are represented by vector $\mathbf{u}_G$ with components u_G and v_G :

$$\mathbf{u}_G = (\alpha_1 + \alpha_2\chi \;;\; \alpha_3 + \alpha_4\chi + \alpha_5\chi^2 + \alpha_6\chi^3)^t$$

Excepting a rigid movement, the displacements in the axes xiy after introducing boundary conditions are :

$$\mathbf{u}_G = \mathbf{N}\,\mathbf{U}_R \quad \text{with : } \mathbf{N} = \begin{bmatrix} \frac{x}{a} & 0 & 0 \\ 0 & x - \frac{2x^2}{a} + \frac{x^3}{a^2} & -\frac{x^2}{a} + \frac{x^3}{a^2} \end{bmatrix} \quad \text{... 19.}$$

4) Study of strains

Ignoring the second order term in u_G and assuming Navier-Bernouilli's general principal, Green's longitudinal strain ε_x, for any point M with coordinates x and y, gives :

$$\varepsilon_M = \frac{\partial u_G}{\partial x} + \frac{1}{2}\left(\frac{\partial v_G}{\partial x}\right)^2 - y\,\frac{\partial^2 v_G}{\partial x^2} \qquad \text{..... 20.}$$

Introducing $\mathbf{B}_1 = \left[\frac{\partial}{\partial x} \;;\; -y\frac{\partial^2}{\partial x^2}\right]$ and $\mathbf{B}_2 = \begin{bmatrix} 0 & 0 \\ 0 & \left(\frac{\partial}{\partial x}\right)^2 \end{bmatrix}$

leads to :

$$\varepsilon = (\mathbf{B}_1 + \frac{1}{2}\mathbf{u}_G^t\,\mathbf{B}_2)\,\mathbf{u}_G \quad \text{and} \quad d\varepsilon = (\mathbf{B}_1 + \mathbf{u}_G^t\,\mathbf{B}_2)\,d\mathbf{u}_G \quad \text{..21.}$$

Using Eqn 19 gives :

$$\varepsilon = (\mathbf{B}_o + \frac{1}{2}\mathbf{B}_L)\,\mathbf{U}_R \quad \text{and} \quad d\varepsilon = (\mathbf{B}_o + \mathbf{B}_L)\,d\mathbf{U}_R \quad \text{..... 22.}$$

with : $\mathbf{B}_o = \mathbf{B}_1\mathbf{N}$ and $\mathbf{B}_L = \mathbf{U}_R^t\,\mathbf{N}^t\,\mathbf{B}_2\,\mathbf{N}$

Calculations give :

$$\mathbf{B}_o = \left[\frac{1}{a} \;;\; \frac{4y}{a} - \frac{6xy}{a^2} \;;\; \frac{2y}{a} - \frac{6xy}{a^2}\right] \qquad \text{..... 23.}$$

$$\mathbf{B}_L = \left[0 \;;\; C^2\beta_i + CD\beta_j \;;\; CD\beta_i + D^2\beta_j\right] \quad \text{with: } C = 1 - \frac{4x}{a} + \frac{3x^2}{a^2}, \quad D = -\frac{2x}{a} + \frac{3x^2}{a^2}$$

5) Calculation of the secant stiffness matrix

For an element, the theorem of virtual works is written :

$$\delta\mathbf{U}_R^t\,\mathbf{P}_R = \int_V \delta\boldsymbol{\varepsilon}^t\,\sigma\,dv$$

or, introducing Eqn 22, the relationship being true for all cinematically admissible displacement becomes :

$$\mathbf{P}_R = \int_V (\mathbf{B}_o^t + \mathbf{B}_L^t)\,\sigma\,dv \qquad \text{..... 24.}$$

The material obeys Hooke's law, thus $\sigma = E\varepsilon$. Equation 22 gives :

$$\mathbf{P}_R = E\int_V \mathbf{B}_o^t\mathbf{B}_o\,dv\,\mathbf{U}_R + \frac{E}{2}\int_V \mathbf{B}_o^t\mathbf{B}_L\,dv\,\mathbf{U}_R + \int_V \mathbf{B}_L^t\,\sigma\,dv \qquad \text{..... 25.}$$

which can be written respectively : $\mathbf{P}_R = (\mathbf{K}_o + \mathbf{K}_{L1} + \mathbf{K}_{L2})\,\mathbf{U}_R$

If S is the section of the element and I its second moment of area, integration gives :

$$\mathbf{K}_o = E\begin{bmatrix} \frac{S}{a} & 0 & 0 \\ 0 & \frac{4I}{a} & \frac{2I}{a} \\ 0 & \frac{2I}{a} & \frac{4I}{a} \end{bmatrix} \quad \text{and} \quad \mathbf{K}_{L1} = \frac{ES}{60}\begin{bmatrix} 0 & 4\beta_i - \beta_j & 4\beta_j - \beta_i \\ 0 & 0 & 0 \\ 0 & 0 & 0 \end{bmatrix} \qquad \text{..... 26.}$$

Thus $\int_S \sigma ds = N$. If the elements are sufficiently small, the member axial tension is assumed constant along their length, giving :

$$\mathbf{K}_{L2}\mathbf{U}_R = N\int_0^a \mathbf{B}_L^t\,dx = \frac{Na}{30}\left[0 \;;\; 4\beta_i - \beta_j \;;\; 4\beta_j - \beta_i\right]^t \qquad \text{..... 27.}$$

The components N, M_i and M_j of vector $\mathbf{P}_r$ can be calculated by combining Eqns 25, 26 and 27 :

$$\mathbf{P}_R = \begin{bmatrix} N \\ M_i \\ M_j \end{bmatrix} = \begin{bmatrix} \frac{ES}{a}e + \frac{ES}{30}(2\beta_i^2 - \beta_i\beta_j + 2\beta_j^2) \\ \frac{2EI}{a}(2\beta_i + \beta_j) + \frac{Na}{30}(4\beta_i - \beta_j) \\ \frac{2EI}{a}(2\beta_j + \beta_i) + \frac{Na}{30}(4\beta_i - \beta_j) \end{bmatrix} \qquad \text{..... 28.}$$

By introducing N in the expressions of M_i and M_j, it is possible to calculate the secant stiffness matrix $\mathbf{K}_S$ from the relationship :

$$\mathbf{P}_R = \mathbf{K}_S\,\mathbf{U}_R \qquad \text{..... 29.}$$

6) Calculation of the tangential stiffness matrix

Differentiation of Eqn 24 with respect to the strains gives:

$$d\mathbf{P}_R = \int_V (\mathbf{B}_o^t + \mathbf{B}_L^t)\,d\sigma\,dv + \int_V d\mathbf{B}_L^t\,\sigma\,dv \qquad \text{..... 30.}$$

Using $\int_S \sigma\,ds = N$ with $dN = \int_S d\sigma\,ds$, Eqn 30 gives :

$$d\mathbf{P}_R = dN\int_0^a (\mathbf{B}_o^t + \mathbf{B}_L^t)\,dx + N\int_0^a d\mathbf{B}_L^t\,dx \qquad \text{..... 31.}$$

After integration, the tangential stiffness matrix $\mathbf{K}_T$ is obtained :

$$d\mathbf{P}_R = \mathbf{K}_T\,d\mathbf{U}_R \qquad \text{..... 32.}$$

in which $\mathbf{K}_T$ has the form :

$$\mathbf{K}_T = \begin{bmatrix} E\frac{S}{a} & E\frac{S}{30}(4\beta_i - \beta_j) & E\frac{S}{30}(4\beta_j - \beta_i) \\ & K_{T22} & K_{T23} \\ \text{Sym.} & & K_{T33} \end{bmatrix}$$

with : $K_{T22} = \frac{4EI}{a} + \frac{4ESe}{30} + \frac{ESa}{300}(8\beta_i^2 - 4\beta_i\beta_j + 3\beta_j^2)$,

$K_{T33} = \frac{4EI}{a} + \frac{4ESe}{30} + \frac{ESa}{300}(3\beta_i^2 - 4\beta_i\beta_j + 8\beta_j^2)$,

and : $K_{T23} = \frac{2EI}{a} - \frac{ESe}{30} + \frac{ESa}{300}(-2\beta_i^2 + 6\beta_i\beta_j - 2\beta_j^2)$

7) Tangential relationship expressed in the general axes

Equation 18 expresses $\mathbf{P}$ in relation to $\mathbf{P}_R$ from :

$$d\mathbf{U}_R^t \mathbf{P}_R = d\mathbf{U}^t \mathbf{P}$$

and gives : $\mathbf{P} = \rho^t \mathbf{T}^t \mathbf{P}_R$

Differentiating gives : $d\mathbf{P} = \rho^t \mathbf{T}^t d\mathbf{P}_R + \rho^t d\mathbf{T}^t \mathbf{P}_R$

or : $d\mathbf{P} = \rho^t (d\mathbf{T}^t \mathbf{P}_R + \mathbf{T}^t \mathbf{K}_T \mathbf{T} \rho d\mathbf{U})$

Expressing the product $d\mathbf{T}^t \mathbf{P}_R$ in the form $\mathbf{M} d\mathbf{U}_r$ gives the general relationship :

$$d\mathbf{P} = \rho^t (\mathbf{M} + \mathbf{T}^t \mathbf{K}_T \mathbf{T}) \rho d\mathbf{U} = \overline{\mathbf{K}}_T d\mathbf{U} \qquad \ldots\ldots 33.$$

with : $\mathbf{M} = \begin{bmatrix} M_{11} & M_{12} & 0 & 0 \\ M_{21} & M_{22} & 0 & 0 \\ 0 & 0 & 0 & 0 \\ 0 & 0 & 0 & 0 \end{bmatrix}$

$$M_{11} = (a+e)^{-4} \left[v_{ij} V_{ij} [v_{ij}^2 + 2(a+u_{ij})^2] - v_{ij}^2 (a+u_{ij}) H_{ij} \right]$$

$$M_{12} = M_{22} = (a+e)^{-4} [-v_{ij}^3 H_{ij} - (a+u_{ij})^3 V_{ij}]$$

$$M_{22} = \frac{a+u_{ij}}{(a+e)^4} \left[[(a+u_{ij})^2 + 2v_{ij}^2] H_{ij} - v_{ij}(a+u_{ij}) V_{ij} \right]$$

8) Algorithm for the calculation

The algorithm for the calculation is well-known and it is not necessary to repeat it in this paper. The resolution is effected by Newton-Raphson's method. The determinant sign for the tangential stiffness matrix is controlled to detect eventual snap-through. This model can be used for structures with large displacements like Lee's frame (Ref 4) for example, which has been used for an evaluation of its validity.

CHOICE OF A RELIABILITY MEASURE

The basic data are :

- the vector of the basic r.v.-s $\mathbf{X} \in \mathcal{L}^2((\Omega, \tau, P) ; \mathbf{R}^n)$ whose law is supposed to allow the density f relatively to Lebesgue's measure in $\mathbf{R}^n$
- the limit state function G, where $G = G^R$ or $G = G^\delta$ depending on whether one is interested in the limit state of strength or in the limit state of displacement.

If the pair $(\mathbf{X}, G)$ is known, this allows the definition of the events E and $\overline{E} \in \tau$ which characterize, respectively, the collapse and the reliability of the structure.

$$\overline{E} = \{\omega \in \Omega : (G \circ \mathbf{X})(\omega) < 0\} \qquad E = \{\omega \in \Omega : (G \circ \mathbf{X})(\omega) \geq 0\}$$

It can be deduced from this that the domains of collapse and reliability are defined as :

$$\overline{D} = \{\mathbf{x} \in \mathbf{R}^n : G(\mathbf{x}) < 0\} \quad \text{and} \quad D = \{\mathbf{x} \in \mathbf{R}^n : G(\mathbf{x}) \geq 0\}$$

To evaluate the reliability of the structure, the most natural step consists in calculating the probability of the occurrence of the event E, or, which amounts to the same thing (since $E \cup \overline{E} = \Omega$ et $E \cap \overline{E} = \Phi$), the probability of the event $\overline{E}$, which is called probability of collapse. These two probabilities are given by the formulae :

$$P(E) = \int_{\mathbf{R}^n} \mathbf{1}_D(\mathbf{x}) f(\mathbf{x}) d\mathbf{x} \quad \text{and} \quad P(\overline{E}) = \int_{\mathbf{R}^n} \mathbf{1}_{\overline{D}}(\mathbf{x}) f(\mathbf{x}) d\mathbf{x} = 1 - P(E)$$

where $\mathbf{1}_A(\mathbf{x})$ is the characteristic function, such that the function equals 1 if $\mathbf{x} \in A$ or equals 0 if $\mathbf{x} \notin A$.

In fact, this measure is difficult or impossible to evaluate with the necessary precision and it is generally preferable to substitute an index β, called the reliability index, of which there are several definitions. The best known are Hasofer-Lind's definition and Cornell's definition (Ref 5 and 6). For convenience of calculation, the second has been chosen for this study. Let us consider the r.v. $Z = G \circ \mathbf{X}$. This variable represents the margin of reliability of the structure relative to the criterion of collapse considered here. m_Z and σ_Z are, respectively, the mean and the standard deviation of Z. Cornell defines the index of reliability by:

$$\beta = m_Z \cdot \sigma_Z^{-1}$$

It is also shown that if $\mathbf{X}$ is gaussian and G is linear, then Cornell's and Hasofer-Lind's definitions coinside. In this case, index β can be linked biunivocally to the probabilities $P(\overline{E})$ and $P(E)$ by the relationship :

$$P(\overline{E}) = 1 - P(E) = \Phi(-\beta)$$

where $\Phi(\,.\,)$ is the repartition function of the reduced normal law in $\mathbf{R}$. Note, that for the envisaged study, it is impossible to determine m_Z and σ_Z exactly since G is not linear. To avoid this difficulty we could have tried, as certain authors (Refs 5, 6 and 7) to linearize G to obtain the explicit expressions m_Z and σ_Z and therefore β. This approximation runs the risk of being too imprecise, since G can be very non-linear and to avoid it, we chose a Monte-Carlo simulation technique to calculate β. It allows an acceptable compromise between precision and cost of calculation.

In fact, by simulation it is easy to construct a suit $\{\mathbf{x}_j ; 1 \leq j \leq m\}$ of (independant) realizations of the vector $\mathbf{X}$, of given density f (Refs 8 and 9), and the associated m-samples $\{z_j = G(\mathbf{x}_j) ; 1 \leq j \leq m\}$ of the realizations of the r.v. z. A statistical treatment of the latter makes it possible to obtain the estimations of m_Z and σ_Z and therefore of β. The estimations are even better if the number m of simulated realizations of Z is large. It is necessary next to define, for the estimated value of β a confidence interval at a level of significance fixed a priori.

NUMERICAL APPLICATIONS

Coupled with a Monte-Carlo simulation method like the previously described procedure, the non-linear mechanical model developed allows the realistic study of the incidence of initial faults with the reliability of structures. The following examples show the interest in such a step.

1) First application

Here, we are interested in a tubular steel column which in Young's modulus E is constant. It is clamped at one extremity and an axial load F is applied to the other. The column is discretized into four elements and the considered initial faults is the nodal displacement $\mathring{U}_4$ along the x-axis.

a) Choice of basic r.v.. The r.v. considered in this example are the fault $\mathring{U}_4$ and the elastic limit of the steel R_e. This gives $\mathbf{X} = (\mathbf{X}_I^t, \mathbf{X}_S^t)^t$ with $\mathbf{X}_I = (\mathring{U}_4)$ and $\mathbf{X}_S = (R_e)$. $\mathring{U}_4$ and R_e are assumed to be gaussian variables and are independant. It is also natural to assume $\mathring{U}_4$ has a zero mean value and its standard deviation, called σ_0, is taken as the parameter of the study. The numerical characteristics of variable R_e are : mean = 272.73 MPa ; standard deviation : 16.36 MPa. The other data are deterministic ; their values are given in the table in Fig 2.

b) Choice of limit state of collapse. In this application, two limit states are considered :

- a limit state of strength which corresponds to the effective stress passing the elastic limit in the control section 0 ;
- a limit state of displacement defined by the effective displacement of the control node 4 passing an admissible displacement $\overline{U}_4$ (with $\overline{U}_4 = L_1/250$) for this node.

The functions of the limit state G^R and G^δ are thus perfectly characterized.

c) *Results*. In order to give a view of the type of results which can be obtained by the proposed method, we have drawn the system of curves $\sigma_o \rightsquigarrow \beta^{\alpha}(\sigma_o)$ for each of the two limit states considered. α is an a-dimentional coefficient such that $F = \alpha P_{cr}$ and P_{cr} is Euler's critical load. Each point of construction of the system obtained has been determined from 500 simulations of vector $\mathbf{X}$. The curves obtained in Fig 3 show that the limit state of displacement is more rigid than the limit state of strength. Such results permit the resolution of problems of the type : for a given degree of dispersion of the fault, what is the load value which must not be passed (or load limit) so that the reliability level is at least equal to a value fixed before-hand ? Or : for a given degree of dispersion of the fault and given value F, what reliability can be expected for the structure ?

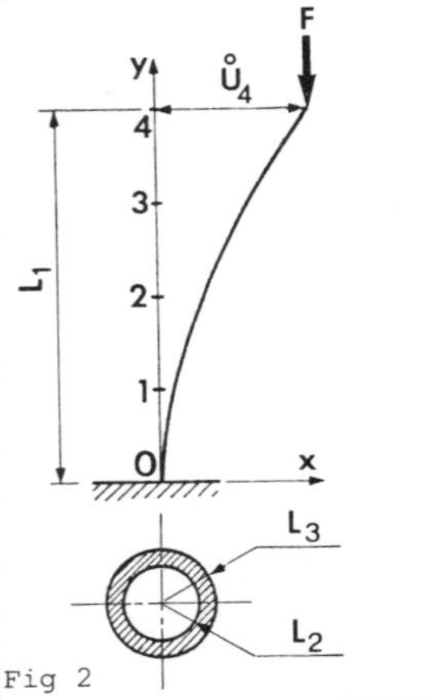

Data	Mean Value	Standard Deviation	
E	2.1 10^5 MPa	0	Deterministic data
P_{cr}	218 18 daN	0	
L_1	9.476 m	0	
L_2	0.1315 m	0	
L_3	0.1365 m	0	
$\mathring{U}_4$	0	σ_o	
R_e	272.73 MPa	16.36 MPa	

Fig 2

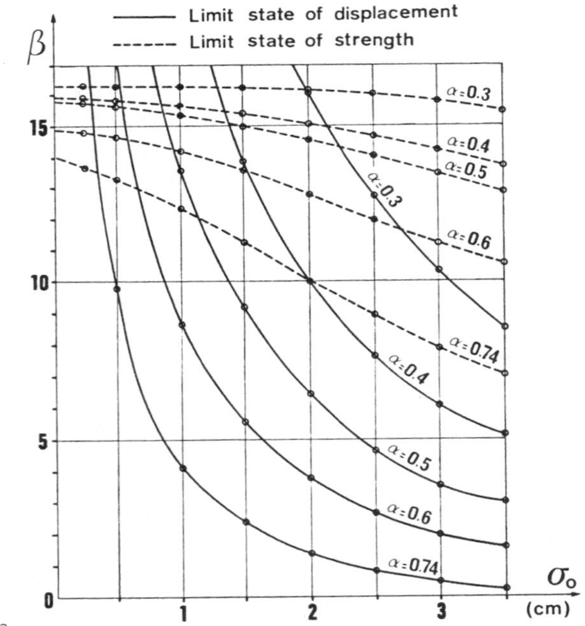

Fig 3

2) Second application

Taking the same example as before, this time considering a greater number of basic r.v.

a) *Choice of basic r.v.* . The r.v. considered in this example are : the elastic limit of the steel R_e, the dimensions L_1, L_2, L_3 and the initial faults $\mathring{U}_1$, $\mathring{U}_2$, $\mathring{U}_3$ and $\mathring{U}_4$ (see Fig 4). Thus $\mathbf{X}_I = (\mathring{U}_1, \mathring{U}_2, \mathring{U}_3, \mathring{U}_4)^t$, $\mathbf{X}_G = (L_1, L_2, L_3)^t$ and $\mathbf{X}_S = (R_e)$. Vector $\mathbf{X}$ is given by :

$$\mathbf{X} = (\mathbf{X}_I^t, \mathbf{X}_G^t, \mathbf{X}_S^t)^t = (\mathring{U}_1, \mathring{U}_2, \mathring{U}_3, \mathring{U}_4, L_1, L_2, L_3, R_e)^t$$

The numerical characteristics of these r.v. are given in the table in Fig 4. Also, the following hypotheses are made : $\mathbf{X}_I$, $\mathbf{X}_G$ and $\mathbf{X}_S$ are mutually independant, $\mathbf{X}_I$ is gaussian and has for its matrix of covariances $\Lambda_D = [\lambda_{Dij}]$ with $\lambda_{Dij} = \lambda_D(\mathring{U}_i, \mathring{U}_j) = 1 - 0{,}25|i-j|$ and $(i, j \in \langle 1,4 \rangle)$, the components of $\mathbf{X}_G$ are independant and are distribued according to a Log-normal law, and, finally, $\mathbf{X}_S$ is a gaussian variable.

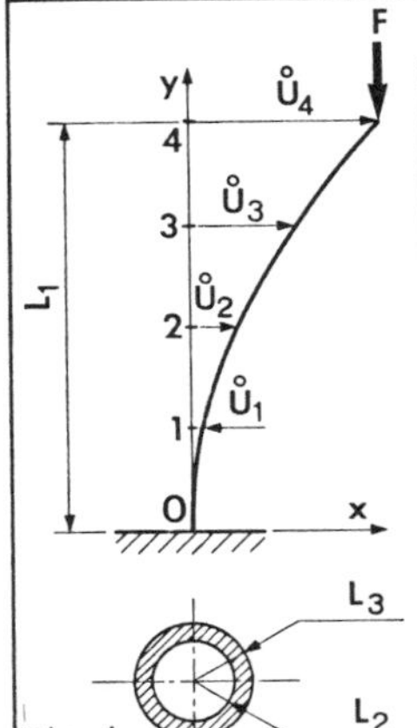

Data	Mean Value	Standard Deviation
E	2.1 10^5 MPa	0	Deterministic Data
P_{cr}	218 18 daN	0	
L_1	9.476 m	0.473 m
L_2	0.1315 m	0.0065 m
L_3	0.1365 m	0.0070 m
$\mathring{U}_i$(i=1,..., 4)	0	σ_o
R_e	272.73 MPa	16.36 m

Fig 4

b) *Choice of limit state of collapse*. In this application we will consider only the limit state of displacement defined in the previous application. The event of collapse is then :

$$\overline{E}^{\delta} = \{G^{\delta}(\mathbf{X}) = U_4 - \overline{U}_4 < 0\}$$

where U_4 is the displacement of node 4 in the x direction and $\overline{U}_4 = L_1/250$.

c) *Results*. The system of curves $\sigma_o \rightsquigarrow \beta^{\alpha}(\sigma_o)$ relative to the limit state of displacement considered is graphically represented in Fig 5. The comparison of curves obtained with those in Fig 3 show that the consideration of a greater number of basic r.v. has the effect of greatly diminishing the margin of reliability of the structure. This is a logical result and moreover can be formally shown in simpler examples. The importance of this diminution can also be explained by the fact that the variance of the additional variables has been chosen on a very large scale.

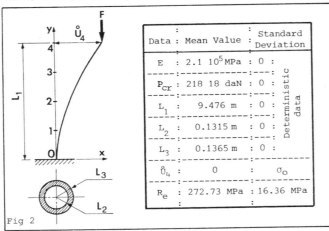

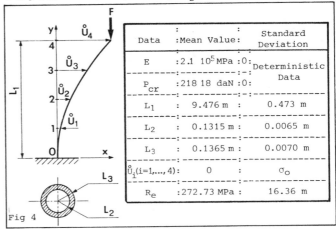

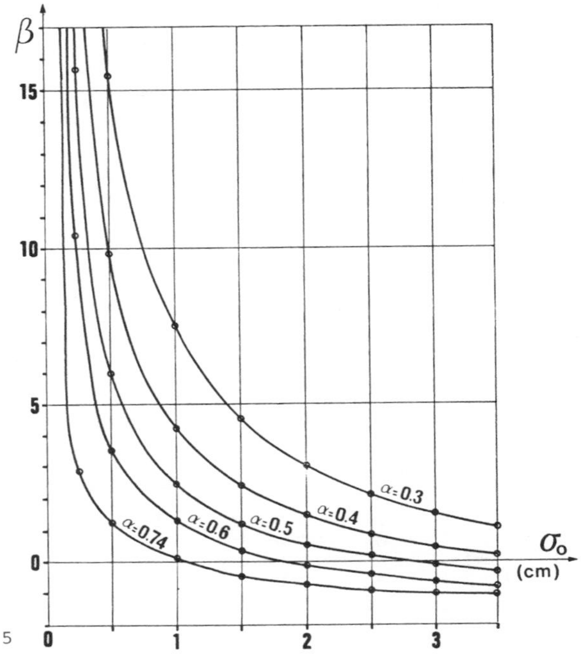

Fig 5

CONCLUSION

The introduction of random variables capable of representing the imperfections of the structures, in a non-linear design model, permits a new approach to the phenomenon of buckling.

The results obtained for the column show how typical curves can be drawn by linking the required level of reliability and the allowable rate of imperfection. The method has been applied equally to the study of a frame, for which the index of reliability (β) has been obtained for different data (Ref 9).

ACKNOWLEDGEMENTS

This study was sponsored by the French "Ministère de l'Urbanisme et du Logement".

REFERENCES

1. A. JENNINGS, Frame Analysis Including Change of Geometry. Proc. A.S.C.E., J. Struct. Div., vol 94, n° ST3, 1968, pp 627-644.

2. F. FREY, L'analyse statique non-linéaire des structures par la méthode des éléments finis et son application à la construction métallique. Thèse de Doctorat, Université de Liège, 1978, Belgique.

3. P. JOUVE, Contribution à l'étude du comportement non-linéaire des structures à barres. Thèse de Doctorat d'Etat ès Sciences Mathématiques, Université de Rennes, juin 1976, France.

4. S.L. LEE, F. MANUEL, E. ROSSOW, Large Deflections and Stability of Elastic Frames. Proc. A.S.C.E., J. Eng. Mech. Div., vol 94, n° EM2, 1968, pp 521-547.

5. A.M. HASOFER, N.C. LIND, An Exact and Invariant First Order Reliability Format. Proc. ASCE, J. Eng. Mech. Div., vol 100, n° EM1, 1974, pp 111-121.

6. C.A. CORNELL, A Probability-Based Structural Code. ACI Journal, vol 66, 1969, pp 974-985.

7. P. THOFT-CHRISTENSEN, J.D. SORENSEN, Structural Reliability Theory and its Applications. Springer-Verlag, Berlin Heidelberg New York, 1982.

8. P. BERNARD, M. FOGLI, Utilisation des méthodes de Monte-Carlo en sécurité structurale. Laboratoire de Génie Civil, Université de Clermont II (à paraître en octobre 1984).

9. M. FOGLI, J.P. MUZEAU, M. LEMAIRE, Recherche de la distribution probabiliste de la charge limite de flambement dans des systèmes de poutres, compte-tenu des défauts initiaux. Rapport final du contrat n° 80 71 546 00 223 75 01 entre le Ministère de l'Environnement et du Cadre de Vie et le Laboratoire de Génie Civil de l'Université de Clermont II, 1984.

AN INVESTIGATION ON THE ELASTOPLASTIC BEHAVIOUR AND COLLAPSE MECHANISM OF BRACED BARREL VAULTS

ZOU* HAO, HUANG* YOU MING

*Structural Engineers, Beijing Central Engineering & Research Incorporation of Iron & Steel Industry (CERIS)

This paper presents a study on the elastoplastic behaviour and collapse mechanism of single layer braced barrel vaults. In the analytic approach the rigidity of joints is taken into account and a computation model simulating an assembled structure from discrete space beam elements is assumed. In addition, on the basis of elastic analysis using rigidity matrix method, the principles of limit theory and finite element method with varying rigidities and incremental loading are applied. A method has been proposed, to find the form of collapse line pattern of the constituent elements in the vault. The paper also provides the results of a collapse test performed on a model of a braced barrel vault. Results of the experiment have shown a good approximation to those of the theoretical analysis.

INTRODUCTION

A single layer braced barrel vault is one of the various types of space frames and also an important system of shell structure.

A braced barrel vault is defined as the form resulting from approximating a solid shell surface by a framework of relatively short and closely spaced structural members regularly arranged in a geometric pattern. It has acquired the properties of a shell as well as that of a latticed structure. A braced barrel vault resists load in a manner similar to that of a thin shell. The main mode of resistance is by membrane action by which forces are carried from point to point by biaxial tension or compression and by shear in the plane of the shell. In addition to the membrane resistance, the structure has bending resistance to help resist loads.

Since a barrel vault is curved only in one direction, its Gaussian curvature K is equal to zero. Sometimes, for the purpose of easy fabrication, a braced barrel vault can be made up of linear members instead of curved ones, thus constituting a shape of polyhedral prism with stress behaviour quite similar to that of a vault with curved members. With the advantages of being reasonable in stress distribution, visually pleasing and economical in cost, steel braced barrel vaults have been extensively used for roof structures of long-span and modest-span buildings, such as indoor sports stadia, swimming pools, etc. and the popularity of braced barrel vaults is still on the increase.

There have been a lot of approaches to the analysis of braced barrel vaults in elastic range. But the analysis for the inelastic behaviour and the stability analysis are problems remaining unsolved and constantly receiving designers' attention. This paper presents a study on the elastoplastic analysis and the collapse mechanism of a braced barrel vault based on an engineering design and describes the performance of a model test. The method proposed in this paper is applicable, in principle, to other types of netty shells and space grid trusses.

ELASTOPLASTIC ANALYSIS OF SINGLE LAYER BRACED BARREL VAULTS

A. BASIC ASSUMPTIONS

a) An idealized elastic-plastic material is used having a simplified stress-strain curve, and no strain-hardening is considered. The stress-strain relation is linear until the yield stress is reached. After the yield stress has been reached, the stress keeps a constant value σ_y (yield point of the material).

b) The overall stability is ensured, and the elastoplastic analysis carried out according to the principles of the first-order analysis.

c) Simple loading is applied. That is, all the

components of forces are proportionally increased from zero to limit values.

d) The rigidity of joints is taken into account. Analysis of the structure is carried out using a computation model of an assembled structure from discrete space beam elements with rigid joints.

e) The exes of grid members intersect at joints and external loads are concentrated and acting on the joints.

f) The relative action of torsional shearing components is not considered.

B. ELASTIC ANALYSIS

1. BASIC EQUATION FOR ELASTIC ANALYSIS

Let $\{D\}$ be the column vector of generalized displacements (including linear and angular displacements), $\{F\}$ the column vector of generalized loads (including forces and moments) acting on the joints, and $[K]$ the overall stiffness matrix of the structure. The basic equation is

$$\{F\}=[K]\{D\} \qquad \ldots\ldots\ 1.$$

where $\{F\}=[X_1 Y_1 Z_1 M_{x1} M_{y1} M_{z1} X_2 Y_2 Z_2 M_{x2} M_{y2} M_{z2} \ldots\ldots X_n Y_n Z_n M_{xn} M_{yn} M_{zn}]^T$

$\{D\}=[U_1 V_1 W_1 \theta_{x1} \theta_{y1} \theta_{z1} U_2 V_2 W_2 \theta_{x2} \theta_{y2} \theta_{z2} \ldots\ldots U_n V_n W_n \theta_{xn} \theta_{yn} \theta_{zn}]^T$

The overall matrix of the structure in global coordinates is obtained by assembling and transforming the unit stiffness matrixes of all individual members. When joints loads are given, solving Eqn 1 will give the displacement of each joint.

2. DETERMINATION OF INTERNAL FORCES IN BRACING MEMBERS

The internal forces at the ends of any member can be worked out from the displacement components at the ends of that member, that is

$$\{\bar{F}\}_e=[\bar{K}]_e\{\bar{D}\}_e \qquad \ldots\ldots\ 2.$$

where $\{\bar{F}\}_e=[\bar{X}_i \bar{Y}_i \bar{Z}_i \bar{M}_{xi} \bar{M}_{yi} \bar{M}_{zi} \bar{X}_j \bar{Y}_j \bar{Z}_j \bar{M}_{xj} \bar{M}_{yj} \bar{M}_{zj}]^T$

$\{\bar{D}\}_e=[\bar{U}_i \bar{V}_i \bar{W}_i \bar{\theta}_{xi} \bar{\theta}_{yi} \bar{\theta}_{zi} \bar{U}_j \bar{V}_j \bar{W}_j \bar{\theta}_{xj} \bar{\theta}_{yj} \bar{\theta}_{zj}]^T$

$[\bar{K}]_e$=unit stiffness matrix of a space beam with rigid ends.

$$[\bar{K}]_e=\begin{bmatrix} \bar{K}_{11} & \bar{K}_{12} \\ \bar{K}_{21} & \bar{K}_{22} \end{bmatrix} \qquad \ldots\ldots\ 3.$$

and where,

$$\bar{K}_{11}=\begin{bmatrix} C_1 & 0 & 0 & 0 & 0 & 0 \\ 0 & C_6 & 0 & 0 & 0 & 0 \\ 0 & 0 & C_5 & 0 & 0 & 0 \\ 0 & 0 & 0 & C_2 & 0 & 0 \\ 0 & 0 & -C_4 & 0 & C_7 & 0 \\ 0 & C_3 & 0 & 0 & 0 & C_8 \end{bmatrix}$$

$$\bar{K}_{22}=\begin{bmatrix} C_1 & 0 & 0 & 0 & 0 & 0 \\ 0 & C_6 & 0 & 0 & 0 & 0 \\ 0 & 0 & C_5 & 0 & 0 & 0 \\ 0 & 0 & 0 & C_2 & 0 & 0 \\ 0 & 0 & C_4 & 0 & C_7 & 0 \\ 0 & -C_3 & 0 & 0 & 0 & C_8 \end{bmatrix}$$

$$\bar{K}_{12}^T=\bar{K}_{21}=\begin{bmatrix} -C_1 & 0 & 0 & 0 & 0 & 0 \\ 0 & -C_6 & 0 & 0 & 0 & -C_3 \\ 0 & 0 & -C_5 & 0 & C_4 & 0 \\ 0 & 0 & 0 & -C_2 & 0 & 0 \\ 0 & 0 & -C_4 & 0 & C_9 & 0 \\ 0 & C_3 & 0 & 0 & 0 & C_{10} \end{bmatrix}$$

$$C_1=\frac{EA}{L}, \quad C_2=\frac{GJ_x}{L}, \quad C_3=\frac{6EJ_z}{L^2}, \quad C_4=\frac{6EJ_y}{L^2},$$

$$C_5=\frac{12EJ_y}{L^3}, \quad C_6=\frac{12EJ_z}{L^3}, \quad C_7=\frac{4EJ_y}{L}, \quad C_8=\frac{4EJ_z}{L},$$

$$C_9=\frac{2EJ_y}{L}, \quad C_{10}=\frac{2EJ_z}{L}, \quad C_{11}=\frac{3EJ_y}{L^2}, \quad C_{12}=\frac{3EJ_y}{L^2},$$

$$C_{13}=\frac{3EJ_z}{L}, \quad C_{14}=\frac{3EJ_y}{L}, \quad C_{15}=\frac{3EJ_z}{L^3}, \quad C_{16}=\frac{3EJ_y}{L^3}.$$

E=modulus of elasticity in tension and compression,

G=modulus of elasticity in shear,

A=cross-sectional area of a member,

L=length of a bracing member,

J_x=torsional moment of intertia about X-axis,

J_y=moment of intertia about Y-axis,

J_z=moment of intertia about Z-axis.

If shearing deformations are considered, then

$$C_1=\frac{EA}{L}, \quad C_2=\frac{GJ_x}{L}, \quad C_3=\frac{6EJ_z}{L^2(1+V_y)},$$

$$C_4=\frac{6EJ_y}{L^2(1+V_z)}, \quad C_5=\frac{12EJ_y}{L^3(1+V_z)}, \quad C_6=\frac{12EJ_z}{L^3(1+V_y)},$$

$$C_7=\frac{(4+V_z)EJ_y}{L(1+V_z)}, C_8=\frac{(4+V_y)EJ_z}{L(1+V_y)}, \quad C_9=\frac{(2-V_z)EJ_y}{L(1+V_z)},$$

$$C_{10}=\frac{(2-V_y)EJ_z}{L(1+V_y)}, C_{11}=\frac{12EJ_z}{L^2(4+V_y)}, \quad C_{12}=\frac{12EJ_y}{L^2(4+V_z)},$$

$$C_{13}=\frac{12EJ_z}{L(4+V_y)}, C_{14}=\frac{12EJ_y}{L(4+V_z)}, \quad C_{15}=\frac{12EJ_z}{L^3(4+V_y)},$$

$$C_{16}=\frac{12EJ_y}{L^3(4+V_z)}, V_y=f_1\frac{12EJ_z}{GAL^2}, \quad V_z=f_2\frac{12EJ_y}{GAL^2},$$

where f_1 and f_2 are distribution coefficients of shearing stress over the member cross-section in directions Y and Z respectively.

In stability analysis, the geometrical stiffness should be taken into account.

C. BASIC METHOD OF ELASTOPLASTIC ANALYSIS-METHOD OF TRACKING DOWN COLLAPSE-LINE PATTERN

The main point of this method is that the procedure of loading on the vault is divided into several stages, each stage corresponding to an increment of loading for which the related internal stress and displacement are solved. The division of stages is identified by the failure or withdrawal from action of one or some bracing members in the assembled system, or by the appearance of certain limit states in which the bearing property of the system and its composition have been changed (for instance, plastic hinges may form at the ends of a member). By summing up the increments of stress and displacement of different stages, we can get the total stress and displacement of the structure in the inelastic range.

The first stage of the process is the elastic stage when all bracing members take part in action. In this stage, calculation is carried out using the elastic analysis as mentioned above and the maximum stressed elements are found. Under gradual increase of loading, the internal stresses increase in the same proportion, to the extent that the maximum stressed members will first reach a certain limit state and enter the plastic range. The other members remain working in the elastic range. According to the stress conditions of bracing members at various parts of the barrel vault, we may consider the following cases of limit state which the bracing members are liable to reach:

a) At one or both ends of a member the bending stress has reached the yield point and plastic hinge formed.

b) Under the action of tension or compression, the whole section of a member has reached the yield condition and the member has gone out of action.

c) The axial load in a compression member has exceeded the critical load, resulting in its buckling and withdrawal from action. The critical load is then taken as the stability load capacity of that member. But the geometrical stiffness must be taken into account in the stiffness matrix.

In the second stage of the process, the limit states of the maximum stressed members determined by elastic analysis have to be dealt with. Modify the end or ends of the member referred to in case (a) in accordance with the behaviour of the plastic hinge, or remove the members referred to in cases (b) and (c) so that they do not exist in the subsequent cycles of analysis. Thereby a new structural system is formed and the stiffness matrix of the original system is modified. The new system is taken as an elastic one with reduced degree of redundancy. This stepwise procedure is repeated until the following stituations occur: (1) the stiffness matrix of the structure becomes a strange matrix, or zero components appear along the diagonal of the matrix, indicating that a general or local mechanism is formed; (2) very large deformation of the structure is obtained; (3) the overall instability of the structure occurs. The method presented herein is suitable for analysis in all stages before the occurance of overall instability of the barrel vault.

D. THE GENERAL MODEL FOR COMBINATION OF INTERNAL STRESSES IN BRACING MEMBERS IN THE nth CYCLE OF ANALYSIS

$(n+1)$th cycle — (the nth cycle of compuation) the structural system modified $(n-1)$ times

$$\{\bar{F}_p\}_e - \{|\bar{F}^{(n-1)}|\}_e = \{\bar{F}_p^{(n)}\}_e$$

$$\{\bar{F}_u^{(n)}\}_e = [\bar{K}^{*(n-1)}]_e \{\bar{D}_u^{(n)}\}_e \quad P=1$$

$$a_n = \bar{F}_{pk}^{(n)} / \bar{F}_{uk}^{(n)}; \text{ member K reached } \bar{F}_{pk}$$

$$\{\bar{F}^{(n)}\}_e = \{\bar{F}^{(n-1)}\}_e + a_n \{\bar{F}_u^{(n)}\}_e$$

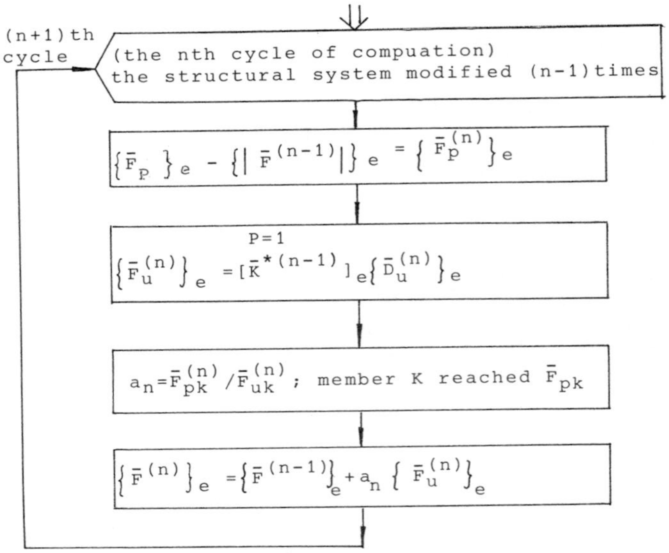

where $\{\bar{F}_p\}_e$ = column vector of ultimate load capacity of a bracing member;

$\{|\bar{F}^{(n-1)}|\}_e$ = column vector of internal force of a bracing member, obtained in the $(n-1)$th cycle of analysis;

$\{\bar{F}_p^{(n)}\}_e$ = column vector of remaining load capacity of a bracing member, in the nth cycle of anaysis.

$\{\bar{F}_u^{(n)}\}_e$ = column vector of internal force in a bracing member due to unit load P=1, obtained in the nth cycle of analysis;

$\{\bar{K}^{*(n-1)}\}_e$ = stiffness matrix modified $(n-1)$ times;

$\{\bar{D}_u^{(n)}\}_e$ = column vector of end displacement of a member due to unit load P=1, in the nth cycle of analysis;

$\bar{F}_{uk}^{(n)}$ = internal force in a member K due to unit load P=1, obtained in the nth cycle of analysis;

$\bar{F}_{pk}^{(n)}$ = the remaining load capacity of a member K, in the nth cycle of analysis;

$\{\bar{F}^{(n)}\}_e$ = column vector of total internal force in a member, obtained in the nth cycle of analysis;

a_n = parameter for loading increment in the nth cycle of analysis.

MODIFICATION OF STIFFNESS MATRIX AND DETERMINTION OF ULTIMATE LOAD CAPACITY OF BRACING MEMBER

A. MODIFICATION OF STIFFNESS MATRIX

a) CHANGING THE BEHAVIOUR OF PLASTIC HINGES

In case plastic hinges have appeared at one or both ends of a space beam element, it may be assumed that further incremental loading will not produce moment effect in any direction on the plastic hinges. Then, the unit stiffness matrix is modified accordingly.

For instance, a plastic hinge has appeared at end i of a member ij. We may let $\bar{M}_{xi}=\bar{M}_{yi}=\bar{M}_{zi}=0$, and substitute this into Eqn 2 to get

$$\left.\begin{aligned} \bar{\theta}_{xi}&=\bar{\theta}_{xj} \\ \bar{\theta}_{yi}&=\frac{3}{2L}\bar{w}_i - \frac{3}{2L}\bar{w}_j - \frac{1}{2}\bar{\theta}_{yj} \\ \bar{\theta}_{xi}&=\frac{3}{2L}\bar{v}_j - \frac{3}{2L}\bar{v}_i - \frac{1}{2}\bar{\theta}_{zj} \end{aligned}\right\} \quad \ldots\ldots 4$$

By substituting Eqn (4) back into Eqn (2), we can solve for the modified stiffness matrix of member ij with a plastic hinge at end i.

$$[\bar{K}]_{i^*j} = \begin{bmatrix} C_1 & 0 & 0 & -C_1 & 0 & 0 & 0 & 0 & 0 \\ 0 & C_{15} & 0 & 0 & -C_{15} & 0 & 0 & 0 & C_{11} \\ 0 & 0 & C_{16} & 0 & 0 & -C_{16} & 0 & -C_{12} & 0 \\ -C_1 & 0 & 0 & C_1 & 0 & 0 & 0 & 0 & 0 \\ 0 & -C_{15} & 0 & 0 & C_{15} & 0 & 0 & 0 & -C_{11} \\ 0 & 0 & -C_{16} & 0 & 0 & C_{16} & 0 & C_{12} & 0 \\ 0 & 0 & 0 & 0 & 0 & 0 & 0 & 0 & 0 \\ 0 & 0 & -C_{12} & 0 & 0 & C_{12} & 0 & C_{14} & 0 \\ 0 & C_{11} & 0 & 0 & -C_{11} & 0 & 0 & 0 & C_{13} \end{bmatrix} \quad \ldots\ldots 5$$

Then, $\{\bar{F}\}_{e,ij}=[\bar{X}_i\bar{Y}_i\bar{Z}_i\bar{X}_j\bar{Y}_j\bar{Z}_j\bar{M}_{xj}\bar{M}_{yj}\bar{M}_{zj}]^T$

With the same reasoning, let $\bar{M}_{xj}=\bar{M}_{yj}=\bar{M}_{zj}=0$ and $\bar{M}_{xi}=\bar{M}_{yi}=\bar{M}_{zj}=\bar{M}_{xj}=\bar{M}_{yj}=\bar{M}_{zj}=0$, we can also obtain the modified stiffness matrixes $[K]_{ij*}$ for the member with plastic hinge at end j, and $[K]_{i*j*}$ for the same member with plastic hinges at both ends i and j.

b) WITHDRAWAL FROM ACTION OF A BRACING MEMBER

There are two cases in which a bracing member may go out of action. Either the internal stress over the cross-section of a member has reached the yield point of the material, or the stability capacity of the member has been exceeded resulting in its buckling. In both cases, zero matrix should be used for that member in order to modify the stiffness matrix of the whole structure.

B. DETERMINATION OF LOAD CAPACITY OF A BRACING MEMBER

a) SIMPLIFICATION OF INTERACTIVE EFFECT

In the analysis of a braced barrel vault using the method proposed in this paper, we must first determine the load capacity of all individual members $\{\bar{F}_p\}_e$ and then work out, in different cycles, the parameters for loading increment a_n.

The general equation for the interaction between bending moment and axial force can be written as

$$t\left(\frac{|\bar{N}|}{|\bar{N}_y|}\right)^r + w\left(\frac{|\bar{M}|}{|\bar{N}_y|}\right)=1 \quad \ldots\ldots 6$$

where, t, w, r=shape factors determined by plastic analysis,

$\bar{N}$=axial load,

$\bar{N}_y$=ultimate load capacity of a member,

$\bar{M}$=bending moment about the major axis,

$\bar{M}_y$=ultimate resisting moment about the major axis.

In Eqn (6), if the dimensionless value $|\bar{N}/\bar{N}_y|$ is very small and approaches zero, the ultimate load capacity of member is determined by the bending moment; whereas, if $|\bar{M}/\bar{M}_y|$ is very small and approaches zero,the ultimate load capacity will be determined by the axial force. In case of possible buckling, the ultimate load capacity of a member may be the critical load $\bar{N}_{cr}$. Since $\bar{N}_{cr}<\bar{N}_y$, the member will withdraw from action in advance. In order to simplify the process of determining $\{\bar{F}_p\}_e$, we may use the results from elastic analysis in different stages, to predict the stress behaviour of the maximum stressed members, neglecting minor components and choosing $|\bar{N}/\bar{N}_y|=1$ (or $|\bar{N}/\bar{N}_{cr}|=1$) or $|\bar{M}/\bar{M}_y|=1$ for each cycle of computation.

b) LIMIT STATE OF A MEMBER SUBJECT TO OBLIQUE BENDING

If no torsional rigidity of bracing members is considered, the moment vector (torsional moment) along the axis of a member is equal to zero, and the moments at the ends of the member may be expressed by two vector components in the directions perpendicular to the member axis (one axis of local coordinates coincides with the member asis). In case the rigid end of a member is reduced to a plastic hinge, the condition of plasticity actually depends on the limit state of the member when it is subjected to oblique bending. In the calculation for the ultimate capacity to resist oblique bending of a member having a cross section with two axes of symmetry, the neutral axis of the cross section in elastic state is same as that in plastic state, and the plastic moment can easily be determined from the stress diagram in limit state. If the section has no axis of symmetry or only one axis of symmetry, there must be supplementary conditions for determining the position of neutral axis when the member is in limit state. One of such conditions is to form an equilibrium equation of forces along the member axis i.e. $\sum N=0$, from the stress diagram of the cross-section. Once the position of neutral axis has been determined, a relation of ultimate moment vectors in principal planes can be set up, or the ultimate resisting moment can be found, according to the ultimate stress diagram. Therefrom, the yield condition for the formation of plastic hinge is determined.

MODEL TEST

A. BRIEF DESCRIPTION OF MODEL TEST

Main dimensions of the test model were: span L=5.04m, width B=2m, total height f=0.53m, height of curved surface f_1=0.33m, height of longitudinal boundary beam h=0.2m, the angle subtended by the barrel arc $2\phi_k$=73°03'34". Trussed boundary beam and end cross arch were used. Fig 1 gives a general test layout.

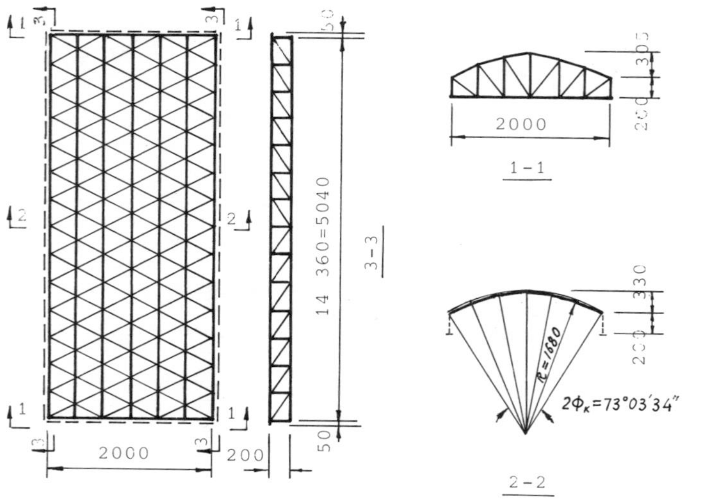

Test layout plan
Fig 1

The material used for the model of braced barrel vault was A3F steel. Angles L30×4, L25×4 and bar steel ϕ 20 were used for bracing members. Mechanical properties of the steel were taken from laboratory testing, the yield strength was 3000kg/cm^2 and the rate of elongation 24-27%.

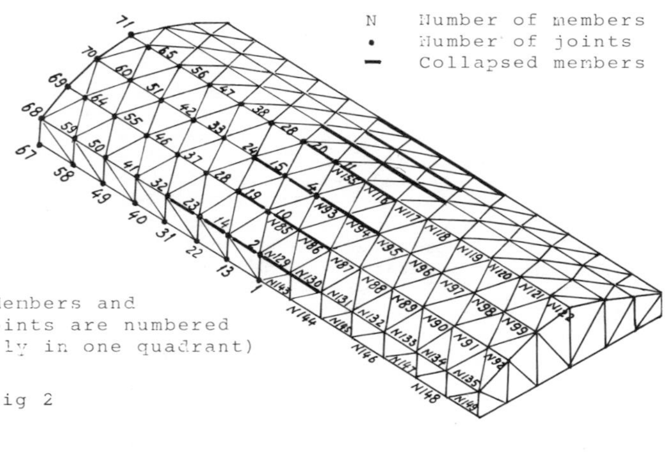

(Members and joints are numbered only in one quadrant)

Fig 2

Simple support bearings were provided at the four corners of the model. Loading was performed by 24 synchronons-acting hydraulic jacks and imposed onto the joints in the middle part of the model through distributing beams. Test was carried out by stages. Measurements were taken for joint displacements at 47 points and for stresses at another 152 points.

B. COLLAPSE PROCESS

According to theoretical analysis, the collapse line of the model was predicated to be N115-N93-N85-N86-143-N144 (Fig 2).

The first member that would fail was N115. The joint load under which the above mentioned members would go out of action was calculated to be 172kg. At that stage, the maximum calculated deflection at mid-span had already exceeded the allowable limit value and showed a tendency of significantly increasing although the load increment was small. Hence, the model was theoretically considered to have reached the state of collapse.

The following is a description of test results. When the joint load reached 140kg, slight curvature was observed on members N115,N93, N116.When the load was increased to 155kg, the measured internal force in member N115 was 4490kg, while its calculated force was 4865kg. Comparison between the two values showed a difference of 7.7%. Obvious twist was later observed in member N115 and its deformation was developing even though the loading on adjacent joints did not increase. The strain parameters of member N115 as displayed in the strain gauge began to drop quickly, and strains in adjacent members had increased to some extent. This phenomenon demonstrated that member N115 had gone out of action and internal stress redistribution followed. At that moment, the difference between the test result and the theoretical value of the joint load was 6.63%. When the load was increased up to 164kg, large flexural-torsional deformation successively appeared in members N93, N116, N85 and quickly developed. Afterward, some joints obviously

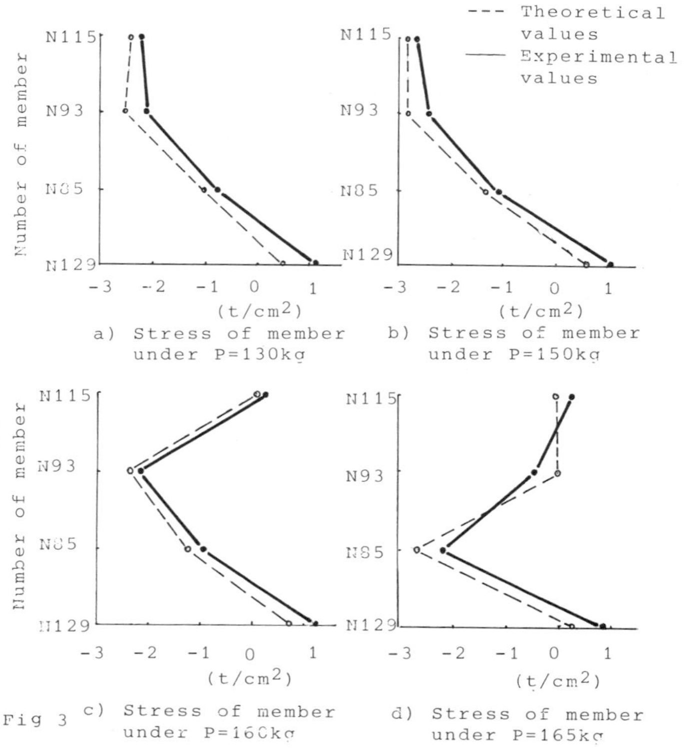

Fig 3

subsided with increasing deformation, the loading device was dislocated from the joints, the middle part of the curved surface of the model was deformed to have a wavy appearance and the structure finally collapsed. The results of experiment showed a good approximation to those of the theoretical analysis.

C. MEASUREMENT OF DEFORMATION AND STRESS

Comparison between the theoretical values and experimental results for member stresses is shown in Fig 3.

INVESTIGATION ON COLLAPSE MECHANISM

Collapse mechanism in a braced barrel vault is dependent on the shape features of the structure, its bracing arrangement and the behaviour of the constituent members. Generally, there are two modes of collapse.

The first mode of collapse is the loss of stability. This kind of collapse mostly occurs in the circumstances where a braced barrel vault has a relatively large span or width or where bracing members are too long and sparely arranged. There are three cases of instability, i.e. overall buckling, local buckling and buckling of an individual member. The three cases are closely interrelated. Buckling of individual members usually gives rise to local buckling and the overall buckling is an inevitable outcome of local buckling. The final collapse of a braced barrel vault is a phenomenon which usually happens when the external loading on it is increased up to the critical value. Due to the influence of lateral disturbing force and displacement, the structure will become unstable and sudden warping will take place.

However, this mode of collapse may be prevented by effective improvement on the detail of the structure. The classical solution of the stability analysis of equivalent continuous shells and other methods can also be used to check the critical load and critical stress.

The second mode of collapse is the collapse mechansism, i.e. the limit state collapse. If a braced barrel vault is ensured against overall instability, its constituent members may follow several different collapse-line pattern as external loading gradually increases and successively fail and withdraw from action, or lose their moment resisting capacity at member ends (plastic hinges are formed). Thereby, the degree of redundancy in the system of the structure decreases progressively. In an ideal design of a braced barrel vault which is statically indeterminate to nth degree, the degree of redundancy should be theoretically decreased by n+1 [the internal stress redistributed (n+1)times], in order that the structure may become unstable and collapse mechanisim may form. Then, the corresponding external load is the ultimate collapse load, and the corresponding displacement curve represents the form of ultimate collapse of the structure.

Nevertheless, the collapse of any barrel vault with a high degree of static indeterminacy usually occurs before the formation of a general collapse mechanism. Because the presence of initial imperfections can result in early appearance of local mechanism and make the value of collapse load decrease. When a local or general mechanism is created in a braced barrel vault, there will be very large deformation in the structure with a feature of steadily developing though external load does not increase. Under such condition, the structure is considered to be collapsed. In addition, the actual collapse line pattern and the stress redistribution are also related to the composition of bracing members and to the disposition of stiffness.

CONCLUSION

1. Braced barrel vault is a kind of space structure having a large safety reserve and good ductility. Even when some of its bracing members become inactive or fail, internal stress redistribution will take place and the structure can remain in action under further loading.

2. The collapse mechanism in a barrel vault is dependent on its shape features,bracing arrangment and the behaviour of its constituent members. Generally, there are two modes of collapse, namely, the loss of stability and the limit state collapse.

3. For the limit state collapse of a braced barrel vault, the elastoplastic analysis by stages, of tracking down collapse-line pattern of bracing members can be used to solve for the ultimate collapse load, the actual safety, the collapse line and collapse process, and the rule of redistribution of internal stress. Therefore, thorough understanding of the behaviour of barrel vaults can help to made better and more reasonable design of such structures.

ACKNOWLEDGEMENTS

The authors with to thank Mr Shen Nian Qiao, Senior Engineer of CERIS, for his reviewing the whole paper. Thanks are also due to Mr Liu Fu You, Senior Engineer of CERIS, for his cooperation in computer programming, and to their colleagues who have worked with them in the experiment.

REFERENCES

1. C K WANG, General Computer Program for Limit Analgsis, Proc ASCE, Vol.89 No.ST 6, 1963.

2. P G JR HODGE, Plastic Analysis of Structures, McGraw Hill, 1959.

3. A CHALL, A M NEVILLE, Y K CHEUNG, Structural Analysis, International Textbook Company 1972.

4. L C SCHMIDT, P R MORGAN, and J A CLARKSON, Space Trusses with Brittle-Type Strut Buckling, Journal of the Structural Division, ASCE, No.ST7, 1976.

EXISTENCE OF EQUILIBRIUM PATHS AT COMPOUND BRANCHING POINTS

P J WICKS

Space Structures Research Centre
Department of Civil Engineering
University of Surrey

It is well known that elastic systems possessing a set of coincident buckling loads are possible generators of a large number of post-buckling equilibrium paths. Although the contributing uncoupled modes of the system may themselves be stable, the coupled modes generated may be unstable, thus affecting the buckling load of the imperfect system. The present paper describes a general theory for completely symmetric idealised systems by which the existence of any coupled equilibrium paths may be predicted. The existence criteria are written in terms of polynomial coefficients in a Taylor expansion of the systems total potential energy. Quantitative algebraic forms for the equilibrium paths which do exist are established in terms of the energy coefficients, enabling the curvature with respect to load of these paths to be determined. The theory is applied to a number of examples.

INTRODUCTION

Broadly speaking, losses of stability through bifurcation may be classified into two groups. Distinct branching points, in which a single stability coefficient vanishes; and compound (or m-fold) branching points for which a number (m) of stability coefficients zero simultaneously.

Non-linear studies, in generalised co-ordinates, of the stability of elastic conservative structural systems was initiated by Thompson (Ref 9). He elucidated the basic principles, studying a distinct critical point. Providing an analysis of a snapping point, Thompson showed its relationship to a branching point: both require the vanishing of a stability coefficient, but the latter requires the zeroing of an additional energy coefficient.

Roorda (Ref 4) studied the effects of imperfections on systems which exhibit the two types of distinct bifurcational instability in the idealised case. These being the General (Asymmetric) and Special (Symmetric) types. Later Roorda (Ref 5) introduced two different types of imperfection into the symmetric type; what he termed first and second-order. The former have a very degrading effect on the critical load of the system and correspond to the imperfections used in Ref 4, whereas the latter have relatively unimportant consequences. In both Refs 4 and 5 Roorda provided experimental data, from tests on high tensile steel models, which verified the theoretical results.

Studies by these Authors had all been carried out using power series expansions of the total potential energy of the system. Thompson (Ref 10) used an alternative perturbation technique and rederived the equations for the two idealised distinct bifurcation points along with their corresponding imperfection sensitivities.

Investigators attention then turned from distinct critical points, to systems for which two bifurcational critical points approach and coalesce. Supple (Ref 7) considered an ideal two degree-of-freedom system whose total potential energy was symmetric in both generalised co-ordinates. He elucidated the forms of the coupling that may occur between the modes as their respective primary critical branching loads coincide. Supple pointed out that either two or four postbuckling equilibrium paths could branch from such a compound critical point.

Chilver (Ref 1) likewise studied an ideal two degree-of-freedom system. This confirmed that compound branching points of this type were possible generators of numbers of coupled equilibrium paths; and that to some extent, increasing the amount of symmetry in the system was likely to increase the number of such paths.

The next stage of development made by research workers was to attempt to generalise the results for distinct and two-fold branching points to m-fold points. Johns and Chilver (Ref 2) worked with an m-fold branching point subject to various amounts of symmetry. They derived formulae governing the minimum and maximum numbers of paths emanating from such a compound critical point. The formulae suggested that large numbers of paths were possible, particularly for the more symmetric systems; but the Authors derived no existence theorems for these paths.

Sewell (Ref 6) has reported his sophisticated general theory. This is used to determine the other paths passing through an m-fold critical point on a known segment of a given equilibrium path. Sewell demonstrated that his theory reproduces the results of Thompson (Ref 10) for the distinct critical point as a special case, and he verified the paths numbers formulae of Johns and Chilver (Ref 2). Johns (Ref 3) later studied completely symmetric compound branching systems in more detail, but no existence theorems were advanced.

It is obvious that compound branching situations have serious consequences for postbuckling behaviour and imperfection sensitivity.

In the sequel, a symmetric system is studied: one in which the total potential energy is an even function of all the displacement co-ordinates. The existence rules arise from consideration of EXISTENCE DETERMINANTS, the elements of which consist of energy coefficients. The formulae governing the minimum and maximum numbers of paths which can exist emerge naturally from the theory.

Quantitative expressions for the curvatures of the postbuckling equlibrium paths which do exist are derived. The theory is then applied to two systems: a thin rectangular plate and a strut on an elastic foundation.

TOTAL POTENTIAL ENERGY AND EQUILIBRIUM EQUATIONS

Following Thompson (Ref 9) we consider a conservative elastic structural system and suppose that a Total Potential Energy, W can be written for it. This is a function of a single loading parameter, Λ and a finite number, m, of generalised co-ordinates, H_i, which describe deflected shapes of the system. Thus:

$$W = W(\Lambda, H_i), \qquad i = 1,\ldots,m. \qquad \ldots\ldots 1.$$

Static equilibrium states of the system are defined by the m equations

$$\frac{\partial W}{\partial H_i} = W_i = 0, \qquad i = 1,\ldots,m \qquad \ldots\ldots 2.$$

where a subscript on W indicates partial differentiation of W with respect to the subscripted displacement variable.

We assume that there exists a solution to these equations, expressed in the parametric form.

$$H^E_i = H^E_i(\Lambda^E), \qquad i = 1,\ldots,m \qquad \ldots\ldots 3.$$

which represent points on a basic-state equilibrium path in the load-configuration space. We suppose that the H^E_i are single valued functions in Λ^E so that the basic-static path rises monotonically from the origin as the system is loaded from zero, thus experiencing no limit points.

Introducing the sliding, incremental co-ordinates λ and h_i ($i = 1,\ldots,m$) the energy expression now appears as:

$$W = W[\Lambda^E + \lambda, H^E_i(\Lambda^E) + h_i], \; i = 1,\ldots,m \qquad \ldots\ldots 4.$$

the incremental co-ordinates thus always measure from the basic state.

Finally, in order to diagonalise the quadratic form of the total potential energy in the incremental co-ordinates h_i, at all load levels, the following transformation of co-ordinates is adopted (after Roorda Ref 5),

$$h_i = \alpha_{ij}(\Lambda)u_j, \qquad i,j = 1,\ldots,m \qquad \ldots\ldots 5.$$

where $[\alpha_{ij}]$ is an orthogonal transformation matrix which is a function of load. This leaves the total potential energy as:

$$V(\lambda,u_j) = W[\Lambda^E + \lambda, H^E_i(\Lambda^E) + \alpha_{ij}(\Lambda^E)u_j] \qquad i,j = 1,\ldots,m \qquad \ldots\ldots 6.$$

The incremental co-ordinates, u_j, thus not only slide along the base-state equilibrium path but also rotate to allow for continuous diagonalization.

We confine our study to a system whose total potential energy is symmetric in the incremental displacements. Thus

$$V(\lambda,u_i) = V(\lambda,u_1,u_2,\ldots,-u_r,\ldots,u_m) \qquad i = 1,\ldots,m \qquad \ldots\ldots 7.$$

and r may assume any value between 1 and m.

We expand this function as a Taylor series about a point on the basic-state equilibrium path Λ^o, H^o_i ($i = 1,\ldots,m$) as follows

$$V = V_o + V_i u_i + V'\lambda + \frac{1}{2!}\{V_{ij}u_iu_j + 2V'_i u_i\lambda + V''\lambda^2\} + \frac{1}{3!}\{V_{ijk}u_iu_ju_k + 3V'_{ij}u_iu_j\lambda + 3V''_i u_i\lambda^2 + V'''\lambda^3\} + \frac{1}{4!}\{\ldots\} + \ldots \qquad \ldots\ldots 8.$$

Where as subscript on V represents partial differentiation of V with respect to the subscripted displacement variable and a prime partial differentiation with respect to load increment respectively. So, for example

$$V'_{ij} \equiv \frac{\partial^3 V}{\partial u_i \partial u_j \partial \lambda}$$

and derivatives are evaluated at the point Λ^o, H^o_i ($i = 1,\ldots,m$) ie, where $\lambda = u_i = 0$ ($i = 1,\ldots,m$) and a summation over all possible values of i,j and k from 1 to m in Eqn 8 is implied.

A number of the coefficients in this polynomial series will vanish for the following reasons:

Firstly, since the point Λ^o, H^o_i ($i = 1,\ldots,m$) is itself one of static equilibrium

$$V_i = V'_i = V''_i = \ldots = 0, \; i = 1,\ldots,m \qquad \ldots\ldots 9.$$

Secondly, because the total potential energy has been diagonalized in the incremental co-ordinates u_i

$$V_{ij} = \delta_{ij}V_{ij} \qquad \ldots\ldots 10.$$

where δ_{ij} is Kronecker's delta.

Finally, since symmetry of the total potential in all in displacement co-ordinates is assumed, all derivations involving odd powers of u_i must vanish. So, for example:

$$V_{ijk} = V'_{ijk} = V''_{ijk} = \ldots 0$$

$$V_{ijjj} = V'_{ijjj} = V''_{ijjj} = \ldots 0, \text{ etc.} \qquad \ldots\ldots 11.$$

In fact, choosing symmetric incremental displacement co-ordinates ensures diagonalisation of the quadratic form of the total potential energy.

We rearrange Eqn 8 with due regard to Eqns 9, 10 and 11, to obtain:

$$V = V_o + \frac{1}{2!} V_{ii} u_i^2 + \frac{1}{4!} \{V_{iiii} u_i^4 + 3V_{iijj} u_i^2 u_j^2\} + \ldots + \lambda\{V' + \frac{1}{2!} V'_{ii} u_i^2 + \ldots\} + \frac{\lambda^2}{2!}\{V'' + \frac{1}{2!} V''_{ii} u_i^2 + \ldots\} + \frac{\lambda^3}{3!}\{\ldots\} + \ldots \quad \ldots\ldots 12.$$

summation over all possible values of i and j from 1 to m. is implied but with the provision that $i \neq j$.

The V_{ii} in the above expression are commonly referred to as stability coefficents , Ref 1, and values of Λ^o at which any of these zero are critical (or buckling) loads.

We wish to study the special system in which the reference point Λ^o, $H^o{}_i$ ($i = 1,\ldots,m$) is one where all m stability coefficients vanish simultaneously, thus creating an m-fold branching point.

Hence, we allow:

$$V_{ii} = 0, \qquad i = 1,\ldots,m \quad \ldots\ldots 13.$$

this into Eqn 12 gives:

$$V = V_o + \frac{1}{4!} \{V_{iiii} u_i^4 + 3V_{iijj} u_i^2 u_j^2\} + \ldots + \lambda\{V' + \frac{1}{2!} V'_{ii} u_i^2 + \ldots\} + \frac{\lambda^2}{2!}\{V'' + \frac{1}{2!} V''_{ii} u_i^2 + \ldots\} + \frac{\lambda^3}{3!}\{\ldots\} + \ldots \quad \ldots\ldots 14.$$

We now ensure equilibrium of the system at positions adjacent to the basic-state equilibrium path by enforcing the total potential energy to be stationary with respect to the incremental displacement co-ordinates. That is:

$$\frac{\partial V}{\partial u_r} = 0, \qquad r = 1,\ldots,m \quad \ldots\ldots 15.$$

this gives the m equilibrium equations

$$V_r = \frac{1}{4!} \{4V_{rrrr} u_r^3 + 12V_{iirr} u_i^2 u_r\} + \ldots + \lambda\{V'_{rr} u_r + \ldots\} + \frac{\lambda^2}{2!}\{\ldots\} + \ldots = 0 \quad \ldots\ldots 16.$$

where r ranges from 1 to m, but adopts only one value per equation, and for convenience the rth co-ordinate has been abstracted from the summation over i. Thus i is summed from 1 to r-1 and from r+1 to m. Finally, we truncate Eqn 16 according to Johns and Chilver's criterion Ref 2 to give the m equilibrium equations governing initial post-buckling from the basic-state as:

$$u_r\{\frac{1}{3!} V_{rrrr} u_r^2 + \frac{1}{2} V_{iirr} u_i^2 + V'_{rr}\lambda\} = 0 \quad \ldots\ldots 17.$$

METHOD OF SOLUTION OF GOVERNING EQUILIBRIUM EQUATIONS

Upon inspection of Eqn 17, we conclude that solutions fall into three categories: firstly a solution involving all displacements remaining zero; secondly, solutions in which a single displacement varies whilst all others remain zero; thirdly, solutions where a number of displacements vary simultaneously whilst all others remain zero. Equations will be developed for equilibrium paths in category three, but their explicit solution will be dealt with in succeeding sections.

We have firstly the "trivial" solution

$$\lambda \neq 0, \; u_r = 0, \qquad r = 1,\ldots,m \quad \ldots\ldots 18.$$

this represents the basic-state equilibrium path or, in the case of the eigenvalue-type problem, an equilibrium path which is coincident with the load axis in incremental load-displacement space. Alternatively it can be regarded as a null vector in the displacement space.

In the second category are the uncoupled solutions (Ref 1) expressed as:

$$u_r \neq 0, \; u_i = 0, \qquad i = 1,\ldots,r-1, \; r+1,\ldots,m \quad \ldots\ldots 19.$$

where r may assume any value between 1 and m. Substituting Eqn 19 into Eqn 17 yields the incremental load-displacement relationship as,

$$\lambda = -\frac{V_{rrrr}}{6V'_{rr}} u_r^2 \quad \ldots\ldots 20.$$

This is the well-known symmetric point of bifurcation Ref 10. The coefficient V'_{rr} represents the rate of change of the stability coefficient with respect to load and is invariably negative (Ref 1). Hence whether a particular uncoupled equilibrium path has positive (negative) curvature with respect to load depends on whether the coefficient V_{rrrr} is positive (negative).

In load-displacement space these solutions represent parabolae which are wholly contained within the λ-u_r plane. In displacement space the uncoupled equilibrium path represented by Eqns 19 and 20 appears as a line coincident with the u_r axis. Such solutions may thus be represented by a vector of the form

$$\mathbf{u} \equiv \{u_1 \; u_2 \ldots u_r \ldots u_m\} = \xi\{0 \; 0 \ldots 1 \ldots 0\} \quad \ldots\ldots 21.$$

where ξ is a scalar. These ideas are summarised schematically in Fig 1.

We note that the curvature, but not the existence, of a particular uncoupled equilibrium path is influenced by the vanishing of the coefficient V_{rrrr}. We may conclude that, to the order of the analysis, the condition for an uncoupled path involving variation of the u_r displacement of this type to exist is that:

$$V'_{rr} \neq 0 \quad \ldots\ldots 22.$$

Broadly speaking the coefficients V'_{rr} are unlikely to vanish, so there will be m such uncoupled equilibrium paths emanating from this type of compound branching point.

The third category of solutions is obtained by allowing p of the displacements to vary simultaneously whilst the remaining m-p remain zero, where

$$2 \leq p \leq m \quad \ldots\ldots 23.$$

For convenience sake we relabel the non-zero displacements and their corresponding energy coefficients from 1 to p, and the zero displacements and their coefficients from p+1 to m. The p equilibrium Eqns 17 appear as:

$$\frac{1}{3!} V_{\alpha\alpha\alpha\alpha} u_\alpha^2 + \frac{1}{2} V_{\beta\beta\alpha\alpha} u_\beta^2 + V'_{\alpha\alpha}\lambda = 0 \qquad \text{.....24.}$$

where α adopts one value between 1 and p per equation and β is summed from 1 to p with the provision that $\beta \neq \alpha$.

Rearranging the first equation of this set gives:

$$\lambda = \frac{-1}{V'_{11}} \left\{\frac{1}{3!} V_{1111} u_1^2 + \frac{1}{2} V_{\beta\beta 11} u_\beta^2\right\} \qquad \text{.....25.}$$

and the δth gives:

$$\lambda = \frac{-1}{V'_{\delta\delta}} \left\{\frac{1}{3!} V_{\delta\delta\delta\delta} u_\delta^2 + \frac{1}{2} V_{\gamma\gamma\delta\delta} u_\gamma^2\right\} \qquad \text{.....26.}$$

equating Eqns 25 and 26 and cross-multiplying yields the p-1 equations:

$$V'_{\delta\delta}\left\{\frac{1}{3!} V_{1111} u_1^2 + \frac{1}{2} V_{\beta\beta 11} u_\beta^2\right\} =$$

$$V'_{11}\left\{\frac{1}{3!} V_{\delta\delta\delta\delta} u_\delta^2 + \frac{1}{2} V_{\gamma\gamma\delta\delta} u_\gamma^2\right\} \qquad \text{.....27.}$$

where δ adopts one value between 2 and p per equation, β is summed from 2 to p, and γ is summed from 1 to δ-1 and δ+1 to p per equation.

Equation 27 may be rearranged to:

$$V'_{\delta\delta}\left\{\frac{1}{3!} V_{1111} u_1^2 + \frac{1}{2} V_{\varepsilon\varepsilon 11} u_\varepsilon^2 + \right.$$

$$\left.\frac{1}{2} V_{\delta\delta 11} u_\delta^2\right\} = V'_{11}\left\{\frac{1}{3!} V_{\delta\delta\delta\delta} u_\delta^2 + \right.$$

$$\left.\frac{1}{2} V_{11\delta\delta} u_1^2 + \frac{1}{2} V_{\varepsilon\varepsilon\delta\delta} u_\varepsilon^2\right\} \qquad \text{.....28.}$$

collecting together the like displacements gives finally:

$$\{V_{1111}V'_{\delta\delta} - 3V_{11\delta\delta}V'_{11}\}\, u_1^2 + 3\{V_{\varepsilon\varepsilon 11}V'_{\delta\delta} -$$

$$V_{\varepsilon\varepsilon\delta\delta}V'_{11}\}\, u_\varepsilon^2 + \{3V_{\delta\delta 11}V'_{\delta\delta} - V_{\delta\delta\delta\delta}V'_{11}\}\, u_\delta^2$$

$$= 0 \qquad \text{.....29.}$$

where δ is as defined previously but now ε is summed from 2 to δ-1 and δ+1 to p.

We write possible solutions to Eqn 29 in the parametric form:

$$\mathbf{u} = \xi\, \mathbf{l} \qquad \text{.....30.}$$

where **u** has been defined in the first part of Eqn 21 and **l** is a vector of direction cosines with respect to the displacement axis,

$$\mathbf{l} \equiv \{l_1\ l_2\ l_3\ \dots\ l_p\} \qquad \text{.....31.}$$

ξ being a scalar. Taking the scalar product of the vector **l** with itself yields:

$$\mathbf{l.l} = 1 \qquad \text{.....32.}$$

additionally we may square each of the components of the vector **u** to give:

$$u_\alpha^2 = \xi^2 l_\alpha^2, \qquad \alpha = 1,\dots,p \qquad \text{.....33.}$$

substituting for u_α ($\alpha = 1,\dots,p$) into Eqn 29 yields the p-1 equations of the form:

$$\phi_\beta(l_\alpha^2) = 0 \qquad \beta = 1,\dots,(p-1),\ \alpha = 1,\dots,p \qquad \text{.....34.}$$

Since Eqn 32 is another equation in the squares of the direction cosines l_α^2, this equation in conjunction with Eqn 34 gives p equations in the p unknown l_α^2's.

NUMBER OF POST-BUCKLING EQUILIBRIUM PATHS AND EXISTENCE DETERMINANT

Explicitly Eqn 34 and 32 appear as:

$$\{V_{1111}V'_{\delta\delta} - 3V_{11\delta\delta}V'_{11}\}\, l_1^2 + 3\{V_{\varepsilon\varepsilon 11}V'_{\delta\delta} -$$

$$V_{\varepsilon\varepsilon\delta\delta}V'_{11}\}\, l_\varepsilon^2 + \{3V_{\delta\delta 11}V'_{\delta\delta} - V_{\delta\delta\delta\delta}V'_{11}\}\, l_\delta^2$$

$$= 0$$

$$l_\alpha^2 = 1 \qquad \text{..... 35.}$$

where δ and ε are as defined previously and α is summed from 1 to p, we note Eqn 35 represent p linear equations in the squares of the direction cosines l_α ($\alpha = 1,\dots,p$).

For brevity we rewrite Eqn 35 in the more concise form:

$$\mathrm{D}.\mathbf{w} = \mathbf{x} \qquad \text{..... 36.}$$

where the vectors **w** and **x** are given by:

$$\mathbf{w} = \{l_1^2\ l_2^2\ \dots\ l_r^2\ \dots\ l_p^2\} \qquad \text{..... 37.}$$

and

$$\mathbf{x} = \{0\ 0\ \dots\ 0\ \dots\ 1\} \qquad \text{..... 38.}$$

respectively, and the matrix **D** consists of the coefficients in Eqn 35. This matrix has the special property that its pth row consists entirely of 1's, by virtue of the relationship between the squares of direction cosines.

We refer to the determinant of the matrix **D** as the EXISTENCE DETERMINANT and for solutions to Eqn 36 to exist (in the form of squares of the direction cosines), this determinant must be non-zero. Some care must be taken at this stage because existence of solutions to Eqn 36 does not necessarily imply that the corresponding equilibrium paths exist. The latter can only exist if all the components of **w** are positive.

Let the determinant of D be denoted by $|\mathbf{D}|$ and the minor of the rth element of the pth row of $|\mathbf{D}|$ by D_{pr} then expanding $\mathbf{D}|$ by means of the pth row gives:

$$|\mathbf{D}| = \sum_{r=1}^{p} (-1)^{p+r} D_{pr} \qquad \text{..... 39.}$$

Suppose now we were to replace the rth column of $|\mathbf{D}|$ by the vector **x** and denote the resulting determinant by $|\mathbf{D}|'$. Expanding the latter by means of the pth row, (which is unaltered by the presence of **x**)because of the presence of the part column of zero's, we would find that all the minors corresponding to the pth row are zero, except for that associated with the p-r^{th} element. The only non-zero minor being equal to Dpr, as defined above.

This process of column replacement is merely the solution of simultaneous equations by Cramer's Rule, solutions existing providing $|\mathbf{D}|$ is non-zero.

We proceed to solve Eqn 36 assuming initially that $|\mathbf{D}|$ is non-zero. Solving using Cramer's Rule gives the rth element of the vector $\mathbf{w}$, w_r as:

$$w_r = \frac{(-1)^{p+r} D_{pr}}{|\mathbf{D}|} \qquad \text{..... 40.}$$

A necessary and sufficient condition for $w_r > 0$ for $r = 1,\ldots,p$ is, by virtue of Eqn 39, that all the cofactors of the pth row of the Existence Determinant are of the same sign, that is:

$$(-1)^{p+r} D_{pr} \quad 0 \text{ for all } r \qquad \text{..... 41.}$$

We may draw the conclusion that the non-vanishing of the Existence Determinant is a necessary but not sufficient condition for the existence of coupled post-buckling equilibrium paths involving simultaneous variation of p displacement co-ordinates. The necessary and sufficient condition is exemplified by inequalities Eqn 41.

We turn our attention now to, in the event of existence conditions being satisfied, and solutions of the form of Eqn 40 emerging, the number of post-buckling equilibrium paths which arise from such solutions.

For a particular choice of p displacement co-ordinates, varying simultaneously, a solution vector $\mathbf{w}$ exists in the form:

$$\mathbf{w} = \{l_1^2 \; l_2^2 \; \ldots \; l_r^2 \; \ldots \; l_p^2\} \qquad \text{..... 42.}$$

but the number of distinct, post-buckling equilibrium paths would be given by all possible sign combinations of:

$$l_1 \pm l_2 \ldots \pm l_r \ldots \pm l_p \qquad \text{..... 43.}$$

that is, there will be

$$2^{p-1} = \frac{1}{2} 2^p \qquad \text{..... 44.}$$

distinct displacement vectors after back-substituting for l into Eqn 30.

There are

$$\frac{m!}{(m-p)!p!}$$

different ways of choosing p from m, so the maximum number of post-buckling equilibrium paths possible for any choice of p is:

$$N_p = \frac{2^p m!}{2(m-p)!p!} \qquad \text{..... 45.}$$

But p may assume any value between 2 and m, so the total maximum number of paths over this range of p must be:

$$\sum_{p=2}^{m} N_p = \frac{1}{2} \sum_{p=2}^{m} \frac{2^p m!}{(m-p)!p!} \qquad \text{..... 46.}$$

Presuming all the uncoupled equilibrium paths established previously exist, then the total number of paths, N, must lie between the limits

$$m \leqslant N \leqslant \frac{1}{2} \sum_{p=2}^{m} \frac{2^p m!}{(m-p)!p!} + m \qquad \text{..... 47.}$$

This range may be rewritten as:

$$m \leqslant N \leqslant \frac{1}{2} \sum_{p=1}^{m} \frac{2^p m!}{(m-p)!p!} \qquad \text{..... 48.}$$

and further simplified to:

$$m \leqslant N \leqslant \frac{1}{2} (3^m - 1) \qquad \text{..... 49.}$$

noting that these numbers are in agreement with Sewell, Ref 6 and Johns and Chilver Ref 2.

SOLUTIONS AND POST-BUCKLING CURVATURES

We assume now, that for a particular choice of p displacement co-ordinates varying simultaneously, the existence conditions Eqn 41 are satisfied. We thus have solutions in the form:

$$w_r = l_r^2 = \frac{(-1)^{p+r} D_{pr}}{D} \qquad r = 1,\ldots,p \qquad \text{..... 50.}$$

substituting for l_r^2 into the elements of Eqn 30 yields the parametric relationship:

$$u_r^2 = \frac{(-1)^{p+r} D_{pr}}{D} \xi^2, \; r = 1,\ldots,p \qquad \text{..... 51.}$$

these equations enable us to write all the remaining s co-ordinates u_s $(s = 2,\ldots,p)$ in terms of the first co-ordinate u_1, say as:

$$u_s = \left[\frac{(-1)^{p+s} D_{ps}}{(-1)^{p+1} D_{p1}}\right]^{1/2} u_1, \qquad s = 2,\ldots,p \qquad \text{..... 52.}$$

with due regard to the fact that we have relabelled the co-ordinates. The coupled equilibrium paths all appear as pairs of straight lines, symmetrically disposed about the displacement axes when projected onto any u_s-u_r plane.

Taking the first equilibrium equation of Eqn 24 say, and substituting for $u_2, u_3, \ldots,$ etc, from Eqn 52 gives:

$$\frac{1}{3!} V_{1111} u_1^2 + \frac{1}{2} \sum_{s=2}^{p} V_{11ss} \frac{(-1)^{p+s} D_{ps}}{(-1)^{p+1} D_{p1}} u_1^2 + V'_{11} \lambda = 0 \qquad \text{..... 53.}$$

where the summation sign has been included for clarity. We rearrange this to:

$$\lambda = - \frac{1}{6V'_{11}} \left[V_{1111} + 3 \sum_{s=2}^{p} V_{11ss} \frac{(-1)^{p+s} D_{ps}}{(-1)^{p+1} D_{p1}} \right] u_1^2 \qquad \text{..... 54.}$$

Thus all coupled equilibrium paths in the load-displacement space when projected onto the λ - u_1 plane appear as parabolae. By virtue of Eqn 52 they appear as such when projected onto any load-displacement plane.

We note that the sign of the curvature of these parabolae is determined by the quantity within the bracket of Eqn 54 and that the cross-product derivatives V_{11ss} play a major role in determining the curvature sign. Fig 2 shows, schematically, some equilibrium paths arising from Eqns 52 and 54 both in displacement and load-displacement space.

EXAMPLES

Both a thin rectangular plate (Ref 8) and a strut on an elastic foundation (Ref 11) may be treated as symmetric two fold branching points.

In the former case coupled buckling is reported between two deflected waveforms involving n and m half-wave-lengths parallel to the loaded direction. This occurs at certain specified values of aspect ratio of the plate. From the equations of Ref 8 we may deduce the energy coefficients in Table 1, where

$$L(n,m) = \frac{(n^2 + m^2)}{nm} \quad \text{..... 55(a).}$$

$$K(n,m) = 6 + \frac{(n - m)^4}{\{(n+m)^2 + 4nm\}^2} \quad \text{..... 55(b).}$$

For the strut problem two-fold branching points occur for two deflected waveforms involving n and m half-wavelengths down the length. These occur at specified values of foundation stiffness. The energy coefficients for the strut from Ref 11 are given in Table 1.

In both cases the two generalised deflection co-ordinates are taken as non-dimensionalised amplitudes of the two contributing waveforms. In addition the wave number m is equal to n + 1.

Use of the foregoing theory with p = 2 results in the cofactors of the EXISTENCE DETERMINANT as follows

$$(-1)^3 D_{21} = V_{2222}V'_{11} - 3V_{1122}V'_{22} \quad \text{..... 56(a).}$$

$$(-1)^4 D_{22} = V_{1111}V'_{22} - 3V_{1122}V'_{11} \quad \text{..... 56(b).}$$

and the equation for the coupled equilibrium paths when projected onto the $\lambda - u_1$ plane is

$$\lambda = -\frac{1}{6}\left[\frac{V_{1111}V_{2222} - 9V_{1122}^2}{V_{2222}V'_{11} - 3V_{1122}V'_{22}}\right]u_1^2 \quad \text{..... 57.}$$

Writing Eqn 57 as

$$\lambda = C\,u_1^2 \quad \text{..... 58.}$$

the expressions pertaining to Eqn 56 and C in Eqn 58 are shown in Table 1 for both the plate and the strut.

From these we may draw the following conclusions regarding the two systems.

1. For the Plate

   i) Coupled equilibrium paths exist for all values of n and m.

   ii) Coupled equilibrium paths always have positive curvature with respect to load.

2. For the Strut

   i) No coupled equilibrium paths exist for n = 1.

   ii) Coupled equilibrium paths exist for all values of n > 1.

   iii) The coupled equilibrium paths which exist all have negative curvature with respect to load.

	PLATE	STRUT
V_{1111}	$\frac{6n}{m}L$	$6\pi^2 n^4(n^2 - 3m^2)$
V_{2222}	$\frac{6m}{n}L$	$6\pi^2 m^4(m^2 - 3n^2)$
V_{1122}	2K	$-4\pi^2 n^2 m^2(n^2 + m^2)$
V'_{11}	$-\frac{n}{m}$	$-2n^2$
V'_{22}	$-\frac{m}{n}$	$-2m^2$
$(-1)^3 D_{21}$	$6(\frac{m}{n}K - L)$	$12n^2m^2\pi^2(n^2m^2 - 3m^4)$
$(-1)^4 D_{22}$	$6(\frac{n}{m}K - L)$	$12n^2m^2\pi^2(n^2m^2 - 3n^4)$
C	$\frac{L^2 - K^2}{\frac{m}{n}K - L}$	$\frac{\pi^2 n^2}{2}\left[\frac{7n^4 - 2n^2m^2 + 7m^4}{n^2 - 3m^2}\right]$

Table 1

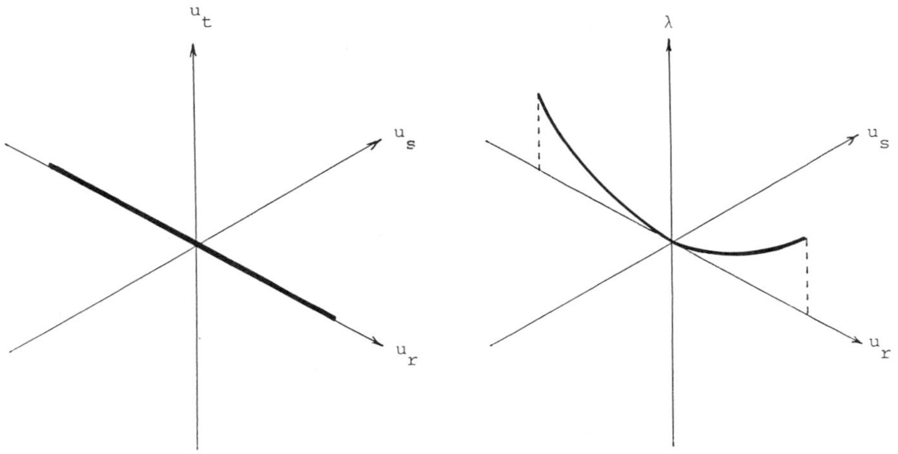

Fig 1

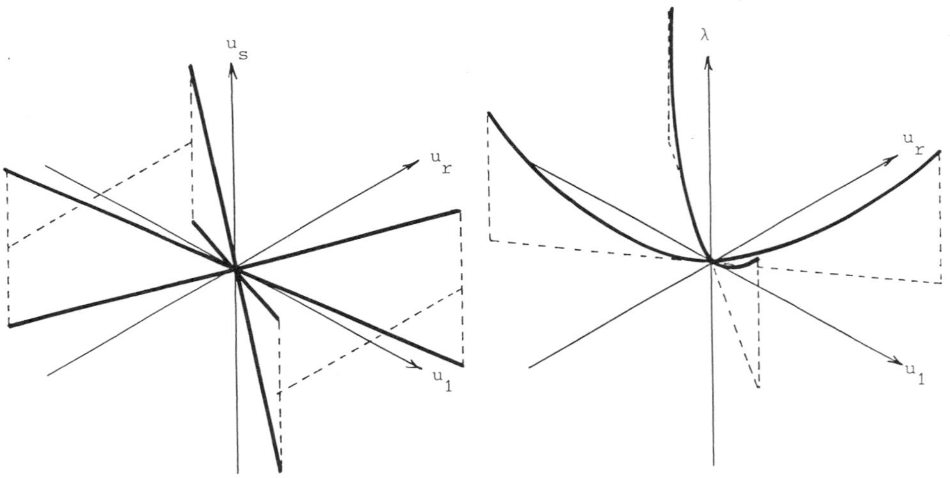

Fig 2

REFERENCES

1. A H Chilver, J Mech Phys Solids, **15**, 15, 1967.

2. K C Johns and A H Chilver, Int J Mech Sci, **13**, 899, 1971.

3. K C Johns, Eng Mech Div, Proc ASCE, **98**, EM4, 835, 1972.

4. J Roorda, Eng Mech Div, Proc ASCE, **91**, EM1, 87, 1965.

5. ibid , Int J Solids Structures, **4**, 1137, 1968.

6. M J Sewell, Proc Roy Soc London, A, **315**, 499, 1970.

7. W J Supple, Int J Mech Sci, **9**, 97, 1967.

8. ibid , Int J Solids Structures, **6**, 1243, 1970.

9. J M T Thompson, J Mech Phys Solids, **11**, 13, 1963.

10. ibid , J Mech Phys Solids, **13**, 295, 1965.

11. P J Wicks, PhD Thesis, University of Surrey, in preparation.

DYNAMIC RESPONSE ANALYSIS USING POLYNOMIALS REPRESENTING THE NONLINEARITIES

A. R. KUKRETI

Assistant Professor, School of Civil Engineering and Environmental Science,
University of Oklahoma, Norman, Oklahoma, U.S.A.

In this paper an algorithm is presented for nonlinear dynamic response analysis using the technique of mode superposition. The formulation is based on finite element theory for spatial dependence and Lagrangian concept is used to describe the motion. An "assumed linear" stiffness is used throughout the response solution and all the nonlinearities are grouped together with external dynamic loads. These are approximated by polynomials of time over small time increments. The discrete coordinates are transformed to modal coordinates based on the "assumed linear" stiffness of the whole system. The algorithm is compared to other implicit and explicit techniques for a beam problem.

INTRODUCTION

The nonlinear dynamic response problem requires much more computer time than the static problem because of having to advance the solution of each time increment over the entire response history desired. Thus, the development of more efficient, accurate and stable algorithms to solve nonlinear equations of motion is of continued interest.

Basically two approaches for response analysis of nonlinear dynamic systems have been suggested. In the first one, the nonlinearities are grouped together with known external dynamic loads and treated as generalized loads. These are approximated by a first order Taylor series about the previous time increment (Refs 1, 2 and 3). In the second approach, the equations of motion are written in incremental form and the behavior is assumed linear over small time increments. In this approach, for most part the work deals with the time integration operator of either implicit (Refs 4 and 5) or explicit type (Refs 6, 7 and 8).

During the last two decades the finite element method has been applied to a variety of structural dynamic problems. This can be inferred by the wide use of linear dynamic analysis of complex structures by the modal coupling methods (Refs 9, 10 and 11). But the mode superposition prinicipal can only be applied to a linear system, since the superposition principle is ipso facto inadmissable for nonlinear systems. However, Refs 12, 13 and 14 have reported its application to nonlinear systems.

In this paper a new approach to the problem is presented in which an "assumed linear" stiffness is used throughout the response analysis and all nonlinearities, which may be displacement and/or velocity dependent, are grouped with external dynamic loads and called pseudo-loads. The pseudo-loads are approximated by polynomials (of time variable) over small time increments used to advance the solution. Formulations are presented for linear and quadratic polynomial approximations. The approach can be easily automated and is suited for problems where discretization of the continuum into finite elements is appropriate.

FORMULATION

The equations of motion for a nonlinear dynamic system modeled by finite element theory can be written as:

$$[M]\{\ddot{x}\} + [C]\{\dot{x}\} + \{F(x)\} = \{P(t)\} \qquad \ldots 1.$$

in which $[M]$ and $[C]$ = mass and damping matrices, respectively; $\{\ddot{x}\}$, $\{\dot{x}\}$ and $\{x\}$ = time-dependent acceleration, velocity and displacement vectors, respectively; $\{F(x)\}$ = internal force vector considering material nonlinearities and nonlinear Green-Lagrangian strains; and $\{P(t)\}$ = given external dynamic load vector.

In this formulation, an "assumed linear" stiffness for each nonlinear element is considered, which may be taken as the linear part of the stiffness of that element, or as an average stiffness over the range of expected element displacements. Denoting, $[K]$ = assembled linear system stiffness matix, adding $[K]\{x\}$ on both sides of Eqn 1 and rearranging terms, yields

$$[M]\{\ddot{x}\} + [C]\{\dot{x}\} + [K]\{x\} = \{P(t)\} + \{G(x)\} \qquad \ldots 2.$$

in which $\{G(x)\} = [K]\{x\} - \{F(x)\}$ is called as the nonlinear force vector and $\{P(t)\} + \{G(x)\}$ is called as the pseudo-load vector. The following should be noted about Eqn 2: (i) The left-hand side is a linear equation and, hence, mode superposition technique can be applied; (ii) on the right-hand side, the vector $\{G(x)\}$ is dependent on the unknown displacement vector $\{x\}$ and so an iterative process is required to solve the problem; and (iii) since displacements are time dependent, so $\{G(x)\}$ also depends on time, i.e., $\{G(x)\} = \{G(x(t))\}$. In addition, the nonlinear force vector can also depend on velocity, i.e., $\{G(x(t), \dot{x}(t))\}$.

Let the total time, T, in which the response is desired, be divided into a sequence of time increments $[t_i, t_{i+1}]$ and let $h = t_{i+1} - t_i$ represent the time increment. One way to approach the solution of Eqn 2 is to approximate the pseudo-load vector by a polynomial of time, t, in each time increment, i.e.,

$$\{P(t)\} + \{G(x)\} = \{A_o\} + \{A_1\}t + \{A_2\}t^2 + \{A_3\}t^3 + \ldots \quad \text{for } o \le t \le h \qquad \ldots 3.$$

in which $\{A_o\}$, $\{A_1\}$, $\{A_2\}$, etc. are constant vectors. Thus, the algorithm presented is based on the assumption that the right-hand side of Eqn 3 represents the pseudo-loads on subintervals which can be made sufficiently small to assure convergence to the exact solution, in the limit. A more sophisticated polynomial may allow a larger time step to be used to advance the solution at the expense of a more complex iterative algorithm. As far as accuracy and computational efficiency are both concerned, the trade-off might be in favor of a slightly more sophisticated representation, for example, a quadratic over a linear one. To investigate these in the study presented, a linear and quadratic approximation were considered.

For a linear approximation in Eqn 3, vector $\{A_o\}$ represents the pseudo-load vector at $t = t_i$ and vector $\{A_1\}$ represents the gradient of the pseudo-loads during the time increment, h. For a quadratic approximation, Eqn 3 can be satisfied at three points, which may be selected as $t = t_i$, t_{i+1} and an arbitrary intermediate point $t = t_{in}$. It is convenient to change the origin at each time increment by defining a new variable $\tau = t - t_i$ for $o \le \tau \le h$. Thus, the interpolation points for linear and quadratic approximations of pseudo-loads are, $\tau = o,h$ and $\tau = o,nh,h$, respectively. Here "n" is a positive integer strictly between zero and one. Denoting, $\{P_o\} = \{P(o)\}$, $\{P_n\} = \{P(nh)\}$ and $\{P_h\} = \{P(h)\}$, and similarly defining $\{G_o\}$, $\{G_n\}$ and $\{G_h\}$. Then substituting $\tau = o,h$ for linear approximation and $\tau = 0,nh,h$ for quadratic approximation into Eqn 3, gives:

(i) For linear approximation

$$\{A_o\} = \{P_o\} \text{ and } \{A_1\} = (\{P_h\} + \{G_h\} - \{P_o\} - \{G_o\})/h \qquad \ldots 4.$$

(ii) For quadratic approximation

$$\{A_o\} = \{P_o\} + \{G_o\} \qquad \ldots 5a.$$

$$\{A_1\} = (\{P_n\} - \{P_o\} - n^2 (\{P_h\} - \{P_o\}) + \{G_n\} - \{G_o\} - n^2 (\{G_h\} - \{G_o\}))/nh(1-n) \qquad \ldots 5b.$$

$$\{A_2\} = (\{P_n\} - \{P_o\} - n(\{P_h\} - \{P_o\}) + \{G_n\} - \{G_o\} - n(\{G_h\} - \{G_o\}))/nh^2(1-n) \qquad \ldots 5c.$$

Modal superposition can now be applied to the homogeneous, undamped and linear form of the left-hand side of the equation of motion, given as Eqn 2. This equation can be used by using normal modes as the generalized coordinates by making the transformation

$$\{x(\tau)\} = [T]\{q(\tau)\} \qquad \ldots 6.$$

in which [T] = complete set of eigenvectors or mode shapes matrix corresponding to eigenvalues, $\lceil \omega^2 \rfloor$, satisfying the eigenvalue problem

$$[M][T]\lceil \omega^2 \rfloor = [K][T] \qquad \ldots 7.$$

The elements in [T] are displacement ratios. It is convenient to orthonormalize them with respect to the associated mass matrix, i.e., $[T]'[M][T] = [I]$, in which $[I]$ = unity matrix and $[T]'$ = transpose of [T]. The result of this mass orthonormalization is: $[T]'[K][T] = \lceil \omega^2 \rfloor$. Also considering the damping matrix proportional to the mass and the stiffness matrices, results in: $[T]'[C][T] = \lceil 2r\omega \rfloor$, in which for the jth mode r_j is the ratio of the damping coefficient of the mode and the critical damping coefficient of the system (i.e., c/c_r).

In view of the aforementioned identities, substituting Eqns 3 and 6 into Eqn 2; premultiplying the resultant equation by $[T]'$ throughout; and using $\{a_o\} = [T]'\{A_o\}$, $\{a_1\} = [T]'\{A_1\}$ and $\{a_2\} = [T]'\{A_2\}$, yields for the quadratic approximation

$$\{\ddot{q}(\tau)\} + \lceil 2r\omega \rfloor\{\dot{q}(\tau)\} + \lceil \omega^2 \rfloor\{q(\tau)\} = [T](\{a_o\} + \{a_1\}\tau + \{a_2\}\tau^2) \text{ for } o \le \tau \le h \qquad \ldots 8.$$

for the linear approximation, the term $\{a_2\}\tau^2$ on the right-hand side of Eqn 8 will not appear. Equation 8 represents a set of uncoupled linear second order differential equations with constant coefficients. Each of these uncoupled equations can be solved knowing the displacements and velocities of the generalized modal coordinates at the beginning of the time step, i.e., at $\tau = 0$. Each uncoupled equation of Eqn 8 is of the form

$$\ddot{q}_j(\tau) + 2r_j\omega_j\dot{q}_j(\tau) + \omega_j^2 q_j(\tau) = a_{oj} + a_{1j}\tau + a_{2j}\tau^2 \quad \text{for } 0 \le \tau \le h \qquad \ldots 9.$$

Consider that the displacement and velocity of the jth mode of vibration at $\tau = 0$ are denoted as $q_j(0)$ and $\dot{q}_j(0)$, respectively, and are known. Then the solution for displacement and velocity from Eqn 9 can be written as, respectively,

$$q_j(\tau) = \exp(-\alpha_j\tau)(\eta_j\cos\theta_j\tau + \gamma_j\sin\theta_j\tau) = \lambda_j + \pi_j\tau + a_{2j}\tau^2/\omega_j^2 \qquad \ldots 10a.$$

$$\dot{q}_j(\tau) = \exp(-\alpha_j\tau)((\gamma_j\theta_j - \alpha_j\eta_j)\cos\theta_j\tau - (\eta_j\theta_j + \alpha_j\gamma_j)\sin\theta_j\tau) + \pi_j + 2a_{2j}\tau/\omega_j^2 \qquad \ldots 10b.$$

in which

$$\alpha_j = r_j\omega_j;\ \theta_j^2 = \omega_j^2 - \alpha_j^2;\ k_j = \exp(-\alpha_j\tau)/\theta_j;$$
$$\lambda_j = (a_{oj}\omega_j^4 - 2(\alpha_j a_{1j} + a_{2j})\omega_j^2 + 8\alpha_j^2 a_{2j})/\omega_j^6;$$
$$\pi_j = (a_{1j}\omega_j^2 - 4\alpha_j a_{2j})/\omega_j^4;\ \eta_j = q_j(0) - \lambda_j; \text{ and}$$
$$\gamma_j = (\dot{q}_j(0) + \alpha_j q_j(0) - \alpha_j\lambda_j - \pi_j)/\theta_j \qquad \ldots 11.$$

Equations 10a and 10b can be written for each generalized modal coordinate and then all these equations can be grouped together into a single matrix equation of the form:

$$\begin{Bmatrix} \{q(\tau)\} \\ \hline \{\dot{q}(\tau)\} \end{Bmatrix} = \begin{bmatrix} [F'_{XX}] & [F'_{XV}] \\ \hline [F'_{VX}] & [F'_{VV}] \end{bmatrix} \begin{Bmatrix} \{q(o)\} \\ \hline \{\dot{q}(o)\} \end{Bmatrix} + \begin{bmatrix} [A'_X] \\ \hline [A'_V] \end{bmatrix}\{a_o\} + \begin{bmatrix} [B'_X] \\ \hline [B'_V] \end{bmatrix}\{a_1\} + \begin{bmatrix} [D'_X] \\ \hline [D'_V] \end{bmatrix}\{a_2\} \qquad \ldots 12.$$

in which $[F'_{XX}]$, $[F'_{XV}]$, $[F'_{VX}]$, $[F'_{VV}]$, $[A'_X]$, $[A'_V]$, $[B'_X]$, $[B'_V]$, $[D'_X]$ and $[D'_V]$ are diagonal matrices. For a linear approximationthe last term on the right-hand side of Eqn 12 will not appear.

The modal displacement and velocity vectors, $\{q(o)\}$ and $\{\dot{q}(o)\}$, respectively, required to start the solution at any time increment, can be computed by inverting Eqn 6. Noting that $[T]^{-1} = [T]'[M]$, these are given by: $\{q(o)\} = [T]'[M]\{x(o)\}$ and $\{\dot{q}(o)\} = [T]'[M]\{\dot{x}(o)\}$. Substituting these into Eqn 12, gives

$$\begin{Bmatrix} \{x(\tau)\} \\ \hline \{\dot{x}(\tau)\} \end{Bmatrix} = \begin{bmatrix} [F_{XX}] & [F_{XV}] \\ \hline [F_{VX}] & [F_{VV}] \end{bmatrix} \begin{Bmatrix} \{x(o)\} \\ \hline \{\dot{x}(o)\} \end{Bmatrix} + \begin{bmatrix} [A_X] \\ \hline [A_V] \end{bmatrix}\{A_o\} + \begin{bmatrix} [B_X] \\ \hline [B_V] \end{bmatrix}\{A_1\} + \begin{bmatrix} [D_X] \\ \hline [D_V] \end{bmatrix}\{A_2\} \qquad \ldots 13.$$

in which

$$[F_{XX}] = [T][F'_{XX}][T]'[M];\ [F_{XV}] = [T][F'_{XV}][T]'[M]$$

$$[F_{VX}] = [T][F_{VX}'][T]'[M]; \quad [F_{VV}] = [T][F_{VV}'][T]'[M]$$

$$[A_X] = [T][A_X'][T]'; \quad [A_V] = [T][A_V'][T]'$$

$$[B_X] = [T][B_X'][T]'; \quad [B_V] = [T][B_V'][T]'$$

$$[D_X] = [T][D_X'][T]'; \quad [D_V] = [T][D_V'][T]' \qquad \ldots 14.$$

For a linear approximation the last term on the right-hand side of Eqn 13 will not appear. Vector $\{A_o\}$ is constant both for linear and quadratic approximations, as can be seen from Eqns 4 and 5a. Thus, for a linear approximation there will be only one unknown vector, $\{A_1\}$, in Eqn 13, whereas for a quadratic approximation there are two unknown vectors, $\{A_1\}$ and $\{A_2\}$, in Eqn 13. These can be iteratively determined by the iterative procedure presented in the subsequent section.

ITERATIVE PROCEDURE FOR SOLUTION

The iterative procedure will be presented for a quadratic approximation first. The linear approximation is a special case of this. In this iterative procedure a relationship is first developed between the incremental value of nodal displacement vector, $\{\Delta x\}$, and the incremental nonlinear force vector, $\{\Delta G\}$, taken at two consecutive iterative cycles. As before, using the nomenclature: $\{x_n\} = \{x(nh)\}$, $\{x_h\} = \{x(h)\}$, $\{G_n\} = \{G(nh)\}$ and $\{G_h\} = \{G(h)\}$. Then, the incremental nonlinear force vector is defined as

$$\{\Delta G_n\}^k = \{G_n\}^{k+1} - \{G_n\}^k \qquad \ldots 15.$$

where superscripts k+1 and k denote the current iteration and the previous one, respectively. Similarly $\{\Delta G_h\}^k$, $\{\Delta x_n\}^k$ and $\{x_h\}^k$ are defined. In view of Eqns 5b, 5c, 13 and 14, the following relationship can be established:

$$\begin{Bmatrix} \{x_n\} \\ \text{----} \\ \{x_h\} \end{Bmatrix}^k = \left[\begin{array}{c|c} [L_{11}] & [L_{12}] \\ \hline [L_{21}] & [L_{22}] \end{array}\right] \begin{Bmatrix} \{\Delta G_n\} \\ \text{----} \\ \{\Delta G_h\} \end{Bmatrix}^k \qquad \ldots 16.$$

in which

$$[L_{11}] = ([B_x]_{\tau=nh} - n^2[B_x]_{\tau=h})/nh(1-n);$$

$$[L_{12}] = ([D_x]_{\tau=nh} - n^2[D_x]_{\tau=h})/nh(1-n);$$

$$[L_{21}] = (-[B_x]_{\tau=nh} + n[B_x]_{\tau=h})/nh^2(1-n); \text{ and}$$

$$[L_{22}] = (-[D_x]_{\tau=nh} + n[D_x]_{\tau=h})/nh^2(1-n) \qquad \ldots 17.$$

It should be noted that the elements of Eqn 17 have to be formulated only once for each time increment.

Similarly for a linear approximation the relationship between $\{\Delta x_h\}^k$ and $\{\Delta G_h\}^k$ can be established as

$$\{\Delta x_h\}^k = -n[B_x]_{\tau=h}\{\Delta G_h\}^k/h(1-n) \qquad \ldots 18.$$

The initial conditions $\{x_o\} = \{x(o)\}$ and $\{\dot{x}_o\} = \{\dot{x}(0)\}$ are given for the time increment as the values computed at the end of the previous time increment. For the first time increment these are given as the initial conditions of the problem. For each time increment the iterative procedure requires the steps given in Table 1.

NUMERICAL EXAMPLE

A beam of 20 in. span length and ends fixed is considered. The other properties of the beam are taken as: rectangular cross-section with 1 in. width and 0.2 in. depth, modulus of elasticity = 10^7 psi and mass density = 0.1 lbm/in.^{3}. The beam is idealized by using finite elements and the effect of large rotation of the beam upon the axial strain is considered in the finite element formulation. This results in a geometrically nonlinear problem. A concentrated load of 640 lb. is applied instantaneously at the midspan of the beam and is assumed to be constant over the complete response period. Due to symmetry in geometry and loading only half the beam needs to be considered.

First a parametric study was conducted to investigate the effect of h, [K] ("assumed linear" stiffness matrix), α, β, and n. When one of these parameters was varied the others were held constant. The alogrithm presented was then compared to the Newmark-Beta, Houbolt and Central Difference methods. For multi degrees-of-freedom (d.o.f.) cases, modal truncation was also applied to the algorithm presented and compared with Park Stiffly-Stable method (see Ref 15). In all the comparisons an exact solution is taken as the one which gives identical results by all procedures and a further decrease in the time increment, used to advance the solution, does not change the results within the significant digits considered. Also, a solution is considered unstable when the response results either become suddenly very large or the iteration fails to converge within 30 cycles.

For the parametric study the beam was idealized as two finite elements, resulting in a one d.o.f. problem. The total response time was taken as 0.08 seconds. All computer programs were written in FORTRAN IV language for IBM 3081 computer and were run under FORTVCLC compiler. An exact solution was obtained at h = $5\text{x}10^{-4}$ seconds both by the linear and quadratic approximations. The exact solution by Newmark-Beta, Houbolt and Central Difference methods was obtained at h equal to $1\text{x}10^{-4}$, $7.5\text{x}10^{-5}$ and $1\text{x}10^{-4}$ seconds, respectively. Thus, it can be seen that the algorithm presented gives an exact solution, even with quite a large time step (five times larger). For time increments greater than h = $5\text{x}10^{-5}$ seconds, the linear approximation results deviated more from the exact solution in comparison to the corresponding quadratic one. For example, the linear and quadratic approximations gave results for h = $7.5\text{x}10^{-4}$ seconds which had a maximum error in peak displacements of -15.5% and -5%, respectively, and a maximum phase shift of 4.7% and 5%, respectively. Negative sign implies a damped solution. At this time increment the Newmark-Beta and Houbolt methods gave results which had a maximum error in peak displacements of -6.4% and -28.6%, respectively, and a maximum phase shift of 43.8% for both. In general, it was observed that at time increments other than the one which produces an exact solution, the Houbolt method is too damped and so unsatisfactory. On the other hand, the Central Difference exhibits artificial attenuation and tends to become unstable as the total response time is increased. The Newmark-Beta method gives acceptable results for maximum response values, but there is a larger error in maximum deflection and phase shift in comparison to the algorithm presented and the exact solution. Unstable results were obtained by the Newmark-Beta, Houbolt, Central Difference, the linear approximation and quadratic approximation at h equal to $1\text{x}10^{-3}$, $1.525\text{x}10^{-3}$, $0.75\text{x}10^{-3}$, $1\text{x}10^{-3}$ and $1.75\text{x}10^{-3}$ seconds, respectively. Thus, it is seen that the algorithm presented with quadratic approximation yields stable approximate solutions at larger time increments.

The parameters K, α, β and n were varied for two typical time increments: h = $5\text{x}10^{-4}$ seconds, which produces an exact solution, and h = $7.5\text{x}10^{-4}$ seconds, which produces an approximate solution. The value of K was varied from 100 to 50,000 lbs/in. and α and β were varied from 0.1 to 10 for each time increment. For the quadratic approximation, it was found that the variations of these parameters neither effected the accuracy nor the CPU computer time for both the time increments. But for the linear approximation, it was found that the value of K effected the number of cycles required for convergence in each time increment and, hence, increased the CPU computer time. For large G(x) (=F(x) - Kx) the CPU computer time increased and for some cases even produced unstable solutions. For the quadratic approximation cases values of n equal to 0.2, 0.45, 0.5 and 0.85 were considered for the two time increments. The variation of n did not affect the response results obtained for the time increment h = $5\text{x}10^{-4}$ seconds (exact solution). For h = $7.5\text{x}10^{-4}$ seconds, the maximum error in peak displacement varied from +6.9% (for

Table 1. Iterative Procedure for Each Time Increment

Linear Approximation	Quadratic Approximation
1. Compute $\{P_o\}$, $\{P_h\}$ and $\{G_o\}$; $\{A_o\}$ from Eqn 4; and $\{a_o\} = [T]'\{A_o\}$.	1. Compute $\{P_o\}$, $\{P_n\}$, $\{P_h\}$ and $\{G_o\}$; $\{A_o\}$ from Eqn 5a; and $\{a_o\}^n = [T]^h \{A_o\}$.
2. Assume $\{G_h\} = \alpha\{G_o\}$, where α is a constant scalar to be chosen.	2. Assume $\{G_n\} = \alpha\{G_o\}$ and $\{G_h\} = \beta\{G_o\}$, where α and β are constants to be chosen.
3. Compute $\{A_1\}$ from Eqn 4.	3. Compute $\{A_1\}$ and $\{A_2\}$ from Eqns 5a and 5b, respectively.
4. Compute $\{a_1\} = [T]' \{A_1\}$.	4. Compute $\{a_1\} = [T]' \{A_1\}$ and $\{a_2\} = [T]'\{A_2\}$.
5. Compute $\{q(h)\}$ from Eqn 10a (note that $\{a_2\}$ does not appear in this case).	5. Compute $\{q(nh)\}$ and $\{q(h)\}$ from Eqn 10a.
6. Compute $\{x_h\}$ from Eqn 6.	6. Compute $\{x_{nh}\}$ and $\{x_h\}$ from Eqn 6.
7. Compute new $\{G_h\}$.	7. Compute new $\{G_{nh}\}$ and $\{G_h\}$.
8. Compute $\{\Delta x_h\}^k$ and $\{\Delta G_h\}^k$.	8. Compute $\{\Delta x_n\}^k$, $\{\Delta x_h\}^k$, $\{\Delta G_n\}^k$ and $\{\Delta G_h\}^k$.
9. Check if convergence is reached*, i.e., $\frac{\|\{\Delta x_h\}^k\|}{\|\{x_h\}^k\|} \leq$ tolerance. If not satisfied then set $\{G_h\}^k = \{G_h\}^{k+1}$ and compute new $\{x_h\}$ from: $\{x_h\}^{k+1} = \{x_h\}^k + \{\Delta x_h\}^k$. Then compute new $\{G_h\}$. Repeat steps 7 and 8. If the convergence is satisfied, then compute $\{\dot{x}_h\}$ and $\{\ddot{x}_h\}$ and set $\{x_o\} = \{x_h\}$ and $\{\dot{x}_o\} = \{\dot{x}_h\}$ as the initial conditions for the next time increment.	9. Check if convergence is reached*, i.e., $\frac{\|\{\Delta x_i\}^k\|}{\|\{x_i\}^k\|} \leq$ tolerance for i = n,h. If not satisfied then set $\{G_i\}^k = \{G_i\}^{k+1}$ for i = n, h and compute new $\{x_n\}$ and $\{x_h\}$ from: $\{x_i\}^k + \{\Delta x_i\}^k$ for i = n,h. Then compute $\{G_n\}$ and $\{G_h\}$. Repeat steps 7 and 8. If the convergence is satisfied, then compute $\{\dot{x}_h\}$ and $\{\ddot{x}_h\}$ and set $\{x_o\} = \{x_h\}$ and $\{\dot{x}_o\} = \{\dot{x}_h\}$ as the initial conditions for the next time increment.

* The norm can be any one of the following: $\|x\| = \max_{i=1,n} |x_i|$ or $\|x\| = (\sum_{i=1}^{n} x_i^2)^{1/2}$

n = 0.2) to -3.6% (for n = 0.85). This error was least (=+0.6%) at n = 0.5. Similar trend was observed at other time increments also. So it was concluded that n = 0.5 produces response solution closer to the exact one.

To investigate the algorithm presented for multi-d.o.f. problem and also the effect of modal truncation, the beam was modeled using three different meshes for half the span: 4-, 10- and 30-elements. For a mesh finer than 30-elements (for half span), results for displacement and velocity did not change. For the 30-element mesh the results were compared to the Park Stiffly-Stable method (Refs 15 and 16) and Adeli (see Ref 8), who have reported that this technique is competitive to Newmark-Beta, Houbolt and Central Difference methods.

For the 30-element mesh different time increments were considered to further define the exact solution. Such a solution was obtained by the linear and quadratic approximations at h equal to 3.5×10^{-5} and 1.0×10^{-5} seconds, respectively, and took a CPU computer time of 46.6 and 102.64 seconds, respectively, to obtain a response for a total time of 0.01 seconds. To further investigate the stability and degree of error in the solutions by both approximations, analyses were performed by increasing h in installments of 0.25×10^{-5} seconds starting from 1×10^{-5} seconds and going to 7.5×10^{-5} seconds. In all analyses the total response time was kept same, T = 0.01 seconds. The linear approximation was found to give unstable results for all h $\geq 2\times10^{-5}$ seconds, except at h = 3×10^{-5} seconds. The quadratic approximation gave stable results for all time increments except h = 5.5×10^{-5} seconds. The time station where iterations failed were noted and showed that they correspond to peak displacement locations. For solutions other than the exact one, at the same h, the linear approximation response results were more damped than the one which used a quadratic approximation. For h = 7.5×10^{-5} seconds the quadratic approximation response results exhibited a maximum error of -5% in peak displacement and a maximum phase shift of 1.8%. Thus, the algorithm with quadratic approximation is more stable, efficient and accurate for a larger size problem in comparison to the algorithm with linear approximation.

Modal truncation technique was then applied to the algorithm presented using the quadratic approximation. The total number of modes for the "assumed linear" 30-element mesh (for half span) is 59. The number of lowest modes which were included in a response analysis were successively taken equal to: 50-, 40-, 15-, 15, 10-, 8- and 5-modes. Modal truncation was applied taking h = 3.5×10^{-5} seconds and total response time, T = 0.01 seconds. It should be noted that h = 3.5×10^{-5} seconds and considering all 59 modes produces an exact solution. It was found that displacement and velocity solutions obtained by considering 8-modes or more were indentical to the exact solution. The 5-mode solution gave a maximum error of 3.1% in peak displacement and a maximum phase shift of 5.5%. The total CPU computer time for the 8- and 5-modes solutions was 36.78 and 33.97 seconds, respectively. In any linear response analysis, modal truncation is intended to reduce the CPU computer time for the response analysis. To study this for the algorithm presented, it was necessary to separate the time taken to formulate the input matrices, e.g., $[\phi]$, $[\omega^2]$, $[F'_{xx}]$, etc., from the total CPU computer time. For this problem this computational time was found to be 18 seconds. This is large because of the solution of a generalized eigenvalue problem with 59 d.o.f. This can be decreased if truncation based on frequency cut-off criterion of the "assumed linear" system is used and only the desired mode shapes and frequencies computed rather than the all. Thus, for the 8-mode solution (which is exact) the time taken for the response analysis only is equal to 18.78 (= 36.78 -

18.0) seconds. The similar response solution with all the 59-modes took 28.6 (=46.6 - 18.0) seconds. Thus, the modal truncation results in 34.3% CPU computer time savings for response solution without any loss in accuracy. For the 5-mode solution, which exhibits maximum 3.1% error in peak displacement, such savings in CPU computer time was found to be 44%.

By the Park Stiffly-Stable method, for the 30-element mesh, the exact solution was obtained at $h = 7.5 \times 10^{-6}$ seconds, which is 4.7 times larger than the h which produced an exact solution by the algorithm presented using a quadratic approximation. For other time increments, the response results by Park Stiffly-Stable method exhibited more error in peak displacements and phase shift. For example, the Park Stiffly-Stable method and the algorithm presented (with only 8-modes considered) gave results which had a maximum error in peak displacement of 6.2% and 2.8%, respectively, and a maximum phase shift of 16.4% and 1.8%, respectively. For smaller time increments and smaller total response time (e.g., $h = 3.5 \times 10^{-5}$ seconds, $T = 0.01$ seconds), the savings in CPU computer time for response analysis only by the algorithm presented over the Park Stiffly-Stable method was found to vary between 15% to 25%. For larger time increments (e.g., $h \geq 6 \times 10^{-5}$ seconds), both the methods took about the same CPU computer time. It was felt that for a larger total response time and larger time increment, the algorithm presented may require much less CPU computer time for response anaysis, which may compensate for the larger computational time required to perform the eigenvalue solutions. To investigate this, response analyses were performed with total time, $T = 0.04$ seconds, and h equal to 3.5×10^{-5}, 5.25×10^{-5} and 7.5×10^{-5} seconds. In all analyses 8-modes were considered. The savings in CPU computer time at these three time increments was found to be 19.2%, 17.9%, 17.2%. Thus, the algorithm presented with quadratic approximation and modal truncation is more efficient and accurate than the Park Stiffly-Stable method for this problem.

CONCLUSIONS

The results obtained support the conclusion that the algorithm presented using polynomials of time to replace the nonlinearities over small time increments and pseudo-mode superposition method is an efficient and practical method. It produces an exact solution at a time increment much larger than other competitive methods generally used. In the algorithm presented, the eigenvalue solution of the "assumed linear" system has to be performed once during the whole response analysis. It should be kept in mind that the frequencies obtained do not represent the true frequencies of the nonlinear system. Application of modal truncation can be attempted, which may lead to savings in computer time. As the total response time increases, the savings in computer time would increase and may compensate for the larger CPU computer time taken to perform the eigenvalue solution. The method presented is self starting.

ACKNOWLEDGEMENT

The author wishes to thank his graduate student Mr. Hadi I. Issa, who assisted in performing the computer runs.

REFERENCES

1. J R TILLERSON, Numerical Methods of Integration Applied in Nonlinear Dynamic Analysis of Shells of Revolution. NASA Report N71-10166, Texas A&M University, 1970.

2. J A STRICKLIN, J E MARTINEZ, J K TILLERSON, J H HONG and W E HAISLER, Nonlinear Dynamic Analysis of Shells of Revolution by Matrix Displacement Method. AIAA J., Vol. 9, No. 4, 629-636, 1971.

3. J A STRICKLIN, W E HAISLER and W A von RIESMANN, Large Deflection Elastic-Plastic Dynamic Response of Stiffened Shells of Revolution. J. of Pressure Vessel Tech., Vol. 85, 87-95, 1974.

4. J C HOUBOLT, A Recurrence Matrix Solution for the Dynamic Response of Elastic Aircraft. J. of Aero. Sc., Vol. 17, 540-550, 1950.

5. N M NEWMARK, A Method of Computation for Structural Dynamics. J. of Engr. Mech. Div., ASCE, Vol. 85, No. EM3, 67-94, 1959.

6. F B HILDERBRAND, Introduction to Numerical Analysis, McGraw-Hill, New York, 1956.

7. P T BOGGS, The Solution of Nonlinear Systems of Equations by A-Stable Integration Techniques. SIAM J. of Num. Analysis, Vol. 8, No. 4, 767-785, 1971.

8. H ADELI, Solution Techniques for Linear and Nonlinear Dynamics of Structures Modeled by Fininte Elements. Tech. Report No. 23, The John A. Blume Earthquake Engineering Center, Stanford University, Stanford, California, 1976.

9. W C HURTY, Dynamic Analysis of Structural Systems by Component-mode synthesis. Technical Report 32-350, Jet Propulsion Laboratory, Pasadena, California, 1964.

10. R L BAJAN and C C FENG, Free Vibration Analysis by the Modal Substitution Method. Presented at the Second AAS-AIAA Symposium, July 15-16, Denver, Colorado, 1968.

11. W A BENFIELD and R F HRUDA, Branch Modes Analysis of Vibrating Systems. AIAA J., Vol. 9, No. 7, 1255-1261, 1971.

12. H D RIEAD, Nonlinear Response Using Normal Modes. Presented at the AIAA 12th Aerospace Science Meeting, January 30 - February 1, Washington, D.C., 1974.

13. R E NICKELL, Nonlinear Dynamics by Mode Superposition. Computer Method in Appl. Mech. and Engr., (Netherlands), Vol. 7, No. 1, 107-129, 1976.

14. V N SHAH, G J BOHM and A N NAHAVANDI, Modal Superposition Method of Computationally Economical Nonlinear Structural Analysis. National Congress on Pressure Vessels and Piping, Paper No. 78-PVP, 1978.

15. K C PARK, An Improved Stiffly-Stable Method for Direct Integration of Nonlinear Structural Dynamic Equations. Presented at the Applied Mechanics Western Mechanics Conference, University of Hawaii, Honolulu, Hawaii, March 1975.

16. K C PARK, Evaluating Time Integration Methods for Nonlinear Dynamic Analysis. Applied Mechanics Symposia Series, ASME, 35-58, 1975.

DUCTILITY IN DOUBLE LAYER GRID SPACE TRUSSES

Erling A. Smith*, BSc, PhD

* Associate Professor, Department of Civil Engineering,

University of Connecticut, Storrs, CT, USA

Recent research indicates that double layer grid space trusses can collapse in an unstable apparently brittle manner. Several factors that contribute to this poor ductility are discussed and "Chordal Displacement Snap Through" is identified as a major contributor. Variations in the distribution in member sizes is numerically investigated for several hypothetical space trusses. It is shown that "Critical Member Underdesign" and "Compression Member Overdesign" strategies can improve space truss ductility, although safe, adequate and economic under/overdesign factors have yet to be established.

INTRODUCTION

The sudden collapse of the Hartford Space Truss roof (see Fig. 1) in 1978 indicated that Space Trusses can collapse in an unstable manner and might be vulnerable to Progressive Collapse. Previous research tends to confirm these observations. Using state-of-the-art methods, this study investigates Space Truss collapse behavior and a number of strategies to improve Space Truss ductility.

DUCTILITY OF SPACE TRUSSES

Experiments on double layer grid space trusses (5,6,11-15) show that these structures can develop large deformations but typically exhibit decreasing load capacity after the system proportional limit has been reached. A typical load-deflection response is shown in Fig. 2. Three situations are shown: curve OPQRSTU tracks the equilibrium path of the truss; curve OPQVSWU tracks the response of the truss tested in a monotonically increasing deflection environment in which the branches QV, SW are unstable "snap through"; curve OPQX plots the response of the truss tested in a monotonically increasing load environment in which the truss is unstable after the peak load is reached at Q. It can be seen then that in the usual load increasing environment the truss possesses little ductility but effectively has the "brittle" response OPQ.

PREVIOUS RESEARCH REVIEWED

The limited amount of experimental work published has shown that Space Trusses initially behave linearly but often fail to achieve predicted peak capacities and generally have unstable post linear behavior. Considerable variation between nominally identical members in space trusses possibly due to initial lack of fit forces has been reported (5,13).

Schmidt, Morgan and Clarkson (14) in 1978 presented results of trusses in which the tension chords were deliberately sized to yield long before buckling of the top chord occurs. Their results showed considerable and predictable ductility occurred after first yield. These results indicate the desirability of designing to prevent buckling occurring before considerable yielding has occurred.

In 1980, Schmidt et al. (15) reported test results of a large space truss also designed to favour tensile yield. The test failed to achieve the peak predicted load due to joint instability or "jack knifing" which occurred before sufficient tension members could yield to form a fold line. The jack knifing and the loss of stiffness in the attached members caused a load redistribution such that compression members reached their buckling load precipitating a "brittle" rather than ductile collapse.

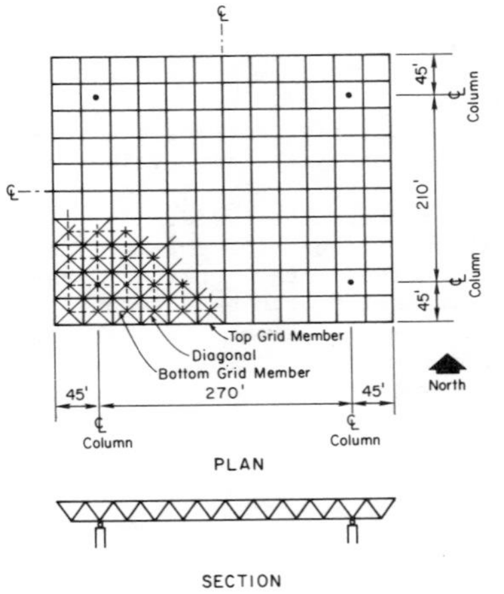

Fig. 1

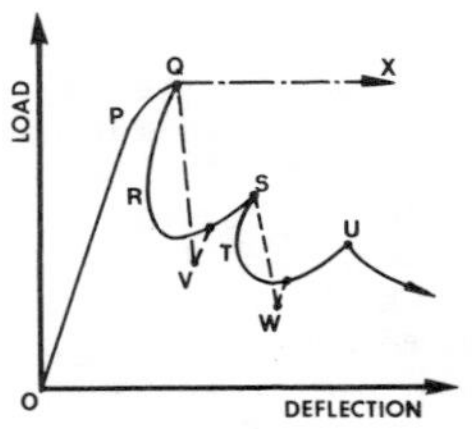

Fig. 2

Hanaor and Schmidt (6) in 1980 presented results of Space Trusses with and without Force Limiting Devices. FLD's are placed in series with critical members and are designed to have a predictable and stable compressive yield plateau well below the buckling load of the member.

Their tests of a determinate truss with FLD's gave good agreement with the theoretical predictions. However, for an indeterminate truss subassemblage, while the ultimate response was accurately predicted, the initial failure was approximately 75% of that predicted. Again, considerable variation occurred in nominally identical members of up to 25% deviation from the average value. Hanaor and Schmidt indicated that joint rotation and the corresponding change in member stiffnesses are probably the cause of the variation. The significance of the variation is that over-loaded members will fail before predicted and possible cause unfavourable force redistribution. FLD's improved the ductility although in a much less predictable way than with the determinate truss.

A full scale truss was tested with and without FLD's. Several of the joints were stiffened to prevent jack knifing. However, the continuity thus provided did not increase the actual buckling force of the members presumably because adjacent members were approximately critical simultaneously and thus unable to offer restraint.

THE MECHANISM OF FAILURE

The structural system response of a double layer grid is dependent on the behavior and interactions of its components. A typical member response is shown in Fig. 3(a) which shows that after initial linear behavior the member yields if in tension or buckles if in compression. After yielding in tension the member strain hardens whereas after buckling in compression the member "strain" softens. The consequences of these non linearities are that even for monotonically increasing proportional load the member forces change non linearly as shown in Fig. 2(b). When a member enters its strain softening regime in compression the rate of force redistribution can become severe and even unstable causing other members to buckle and enter their strain softening regimes. As described below, a "dynamic jump" can occur which can lead to total collapse of the system after the buckling of only one member. Thus collapse can rapidly propagate to develop "fold" lines somewhat analogous to yield lines in plates. Key factors in the collapse mechanism are the strain softening characteristics of the critical members and the ability of the structure to adequately redistribute the forces.

FACTORS AFFECTING DUCTILITY

The structural topology chosen for a double layer grid space truss has been demonstrated (8,10) to affect both the overall stiffness of the truss and the forces in the members. Several topologies exist although only a few are in common use, for example:

(i) parallel square grids, parallel spanning
(ii) parallel square grids, diagonally spanning
(iii) inclined square grids
(iv) parallel triangular grids

For the square grids, the top and bottom chords are square grids either parallel in case (i) and (ii) or inclined at 45° to each other in case (iii). The parallel square grids (i) and (ii) can be oriented such that the grids span parallel or span diagonally to the supporting boundaries. Renton (10) has shown that the parallel square grids have zero torsional stiffness and thus the load transfer mechanism is strictly two way spanning action in the two orthogonal directions of the grid. In contrast inclined square grids and parallel triangular grids have considerable torsional stiffness. The inclined square grid has been considered to be analogous to an isotropic plate (10) spanning in all directions and the triangular grids can be considered to span in at least three directions. The ability to span in more than two directions improves the force redistribution characteristics and the ductility of the space truss.

The member cross-section type affects the buckling load and the post buckling characteristics which affect the ductility of space truss.

In recent years, the maximum force capacity of compression members has been expressed by multiple column curves of the Structural Stability Research Council, SSRC, and of the European Convention of Construction Steelworks, ECCS (1,7). Each of the curves represents the maximum load for a particular type of cross section as a function of slenderness ratio expressed in the equation:

$$\gamma_{max} = \frac{P_{max}}{P_Y} = f(\lambda) \qquad .....1$$

in which P_{max} = maximum axial load,

P_y = squash load = A σ_y

$\lambda = \frac{1}{\pi}\sqrt{\frac{\sigma_y}{E}}\,\frac{\ell}{r}$; slenderness parameter

ℓ/r = slenderness ratio

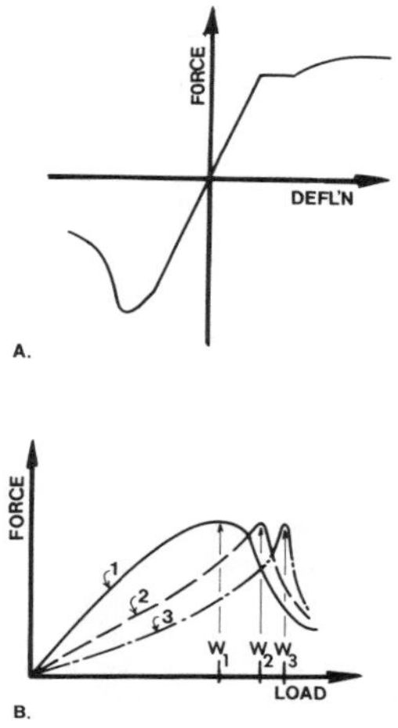

Fig. 3

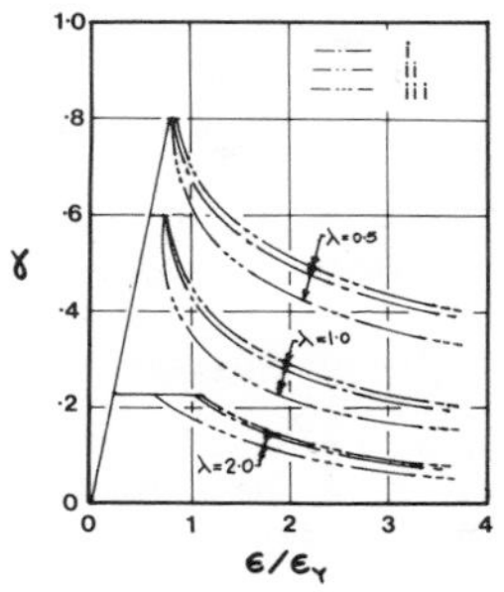

Fig. 4

The functions f(λ) are dependent on the section type and the range on λ.

The post buckling response is a function of the section type, the load level and of the slenderness parameter, λ. In the following it will be assumed that different section types are to be compared that have identical P_{max}, L, A and σ_y. In addition, it will be assumed that sections to be compared fall on the same column curve expressed in Eq. 1, and that the post buckled shape is sinusoidal with a plastic hinge at midlength. Hence the non dimensionalized shortening or apparent strain is given by

$$\varepsilon = \frac{P}{EA} + \frac{e^2\,\pi^2}{4\,L^2} \qquad .....2$$

in which $e = \dfrac{M_{P'}}{P}$ 3

maximum lateral deflection at midlength.

$M_p{}'$ = reduced plastic moment at midlength

Hence $\frac{e}{L} = \left(\frac{\mu}{\gamma}\right)\left(\frac{1}{L/r}\right)\left(\frac{1}{\alpha}\right)$ 4

in which μ = reduced plastic shape factor = $\left(M_p{}'\right)\Big/\left(\frac{\sigma_Y I}{C}\right)$

$\gamma = P/A\sigma_y \qquad \alpha = c/r$

c = distance between neutral axis and extreme fibre

r = radius of gyration

Substituting Eqs. 4 in 2 and simplifying gives

$$\frac{\varepsilon}{\varepsilon_Y} = \frac{\varepsilon}{\sigma_Y/E} = \gamma + \left(\frac{\mu}{2\gamma\lambda\alpha}\right)^2 \quad .....5$$

The apparent strain is thus a function of the load level, γ, slenderness parameter λ and the section properties μ and α. The reduced plastic shape factor, μ, is both section and load dependent while α is only section dependent. For example Table 1 shows α and μ for common cross section shapes. For the generation of the table it has been assumed that the enclosing dimensions of the cross sections are equal and for types ii and iii the plate thicknesses are small compared with the plate widths. For the I shape sections it was assumed that the total area of the flanges is equal to the area of the web.

Type	Cross Section Shape	α	μ	γ
i		1.732	$1.5 - 1.5\gamma^2$	<1
ii	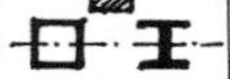	1.225	$1.125 - 1.5\gamma^2$	<0.5
			$1.5 - 1.5\gamma$	>0.5
iii	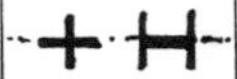	2.500	1.5	<0.5
			$6.0\gamma - 6.0\gamma^2$	>0.5

TABLE 1 Cross Section Parameters

Substitution of the equations in Table 1 into Eq. 5 gives apparent strain responses for the section types considered. Results are plotted in Fig. 4 which shows non dimensional load versus strain parameters for different section types and slenderness factors. The maximum load values were obtained using the equations for SSRC curve "2" which is similar to ECCS curve "b". In Fig. 4, it can be seen that the responses of section types i and ii are similar. In addition, it can also be seen that section type iii experiences a greater rate of softening than the other section types considered. The more rapid rate of softening of a member of type iii causes a more rapid rate of force redistribution within a space truss leading to earlier failures of adjacent members and thereby reducing the ductility of the space truss.

Examining Eq. 5, Table 1 and Fig. 4 indicates that sections in which the radius of gyration approaches the distance from the neutral axis to the extreme fibre for example square tubular sections, would contribute to improving space truss ductility.

Fig. 4 also shows the increase in capacity as member slenderness is decreased but the increase in the rate of softening in the post buckling regime. Examination of most column curves in the medium slenderness ranges shows that the maximum load is approximately inversely proportional to the slenderness whereas Eq. 5 shows that the rate of post buckling softening is inversely proportional to the square of the slenderness. Thus with decreasing slenderness, the member capacity increases but the member and possibly the system ductility decreases.

The analysis of double layer grid space trusses as trusses using initial geometry and neglecting continuity in the determination of the internal forces appears to be reasonably accurate. Continuity, however, can affect both the capacity and ductility of the members and hence the system. In the presence of continuity the compression grid, typically the upper grid, acts as a braced frame loaded at the joints. The bracing afforded by the diagonal web members and the tension grid prevents translation of the compression grid joints and depending on the joint configuration may provide rotational restraint to the compression grid joints. Continuity thus allows understressed members to restrain potentially overstressed members. The effective length of the critical member is reduced, thereby increasing the capacity but decreasing the ductility of the member and of the system. However in a truss designed to be "fully stressed" all the compression members buckle simultaneously and cannot restrain each other and thus continuity would have negligible effect for such trusses.

Eccentricities at the joints have two major effects, reducing both the stiffness and force capacity of the affected space truss members. The effective stiffness K' of a pin ended eccentrically loaded member is given by

$$K' = \left(\frac{AE}{L}\right) \Bigg/ \left[1 + \frac{1}{4}\left(\frac{e_o}{r}\right)^2 \left(\frac{1}{\cos^2\nu} + \frac{3\ \mathrm{tau}\ \nu}{\nu}\right)\right] \quad .....6$$

in which

$$\nu = \frac{\pi}{2}\sqrt{\rho}$$

$$\rho = \frac{P}{\pi^2 EI/L^2}$$

Typically, a value of $e/r < 0.1$ gives insignificant reduction in member stiffness and capacity.

For increasingly large values of eccentricity, the stiffness and the capacity of the member decrease at approximately the same rate. This behavior might be used to advantage in critical regions to soften the structure locally and thereby improve the overall ductility of the system.

The advantages of using this strategy are that a more favourable force distribution develops and that the far more ductile post maximum load softening response of the eccentrically loaded member can be utilized. Of course in the design of a real truss the effective axial stiffness must be assessed and included in the analysis and the capacity of the eccentrically loaded member demonstrated to be adequate.

On the other hand, eccentricities can give rise to unbalanced moments at the joints, causing their rotation and possible instability known as jack knifing. Therefore, a design employing deliberate eccentricities must ensure the joint configuration and attached members have adequate stiffness and strength to resist the moments developed.

Joint rigidity or compliance can affect both the stiffness, capacity and ductility of a space truss. If the axial stiffness of the joint is significantly different from that of the members and is different for chord and web members, these factors should be accounted for in the analysis since they will affect the force distributions. Differing joint compliances can be used to reduce the rate of loading on critical members to give more ductile response. Indeed, joints and member ends could be designed to reliably fail in a ductile manner before the critical members in a manner similar to Force Limiting Devices. However, as with FLD's, care must be taken to ensure the stability of this local mechanism.

As indicated above, a typical compression member in its post-buckling regime experiences strain softening, decreasing axial force resistance for increasing shortening. The adjoining structure, through the joints, maintains compatibility with the member. However because of the differing local stiffnesses of the member and the adjoining structure equilibrium might not be possible giving rise to Chordal Displacement Snap Through. This instability, described in detail elsewhere (2,3,16-18) causes the sudden reduction in force in the critical member and the immediate redistribution of the force into adjacent members. In severe circumstances, during the dynamic jump, members normally in compression can momentarily experience tension and vice-versa. CDST severely reduces space truss ductility and should be prevented or at least delayed. Critical Member Underdesign or Compression Member Overdesign - overdesign of all the compression members - can prevent CDST and improve system ductility. Critical Member Overdesign prevents CDST of the critical member and may improve system ductility.

Three common design types of truss are uniform trusses, fully stressed trusses and banded or strip design trusses.

To these three basic design types several modifications could be made that could affect the system ductility. Underdesign or overdesign modifications could be made to the critical member,to members adjacent to the critical member or to other members.

In the results described in a succeeding section, the effects of the above modifications are investigated. The most promising strategies as far as improving the structural ductility were "critical member underdesign" and "compression member overdesign" - overdesign of all the compression members by some factor. On the other hand, "critical member overdesign" reduced the system ductility although it did increase the system capacity.

More exotic modifications include Limited Slip Systems, Prestressing and Force Limiting Devices. For example, Limited Slip System would allow the critical member to take up a stable, designed and limited slack before receiving any of the load. After the slack has been taken up, the member then receives load at its normal rate.

A NUMERICAL INVESTIGATION

Several hypothetical space trusses were analysed to investigate the effect of some of the above parameters affecting space truss ductility. All trusses used the parallel square grid topology and cruciform cross sectioned members. To investigate the effect of slenderness, some of the trusses were assumed to have all their members braced at midlength. The primary variable investigated was the member size distribution and the effect of Critical Member Underdesign and Overdesign and Compression Member Overdesign strategies.

TRUSS DESIGN DETAILS

The trusses investigated were similar to that of the ill-fated Hartford Space Truss shown in Fig. 1 but had 12 x 12 bays each 30 feet (9.15 m) square and 21.21 feet (6.47 m) deep and were vertically supported at four lower joints each 1-1/2 bays in from each edge. Twenty different cruciform cross sections were developed using equal leg angles from 3 in. (76 mm) up to 8 in. (203 mm) leg size. For members using 8 and 6 in. (203 and 152 mm) angles, yield strengths of 50 ksi (344 N/mm^2) and for all other sections yield strengths of 36 ksi (248 N/mm^2) were used.

All members were assumed to be pin-ended at the truss joints and have torsional buckling suppressed. Some trusses had unbraced members and other had all members braced at midlength.

The three design types considered were:

I. Uniform - designs 1-5, Table 2
II. Strip - designs 6-9, Table 2
III. Fully Stressed - designs 10, 11, Table 2

Design type II was generated using an empirical strip design method similar to that used in Reinforced Concrete Flat Slab design (4). From the over twenty available cruciform sections, four were chosen for each of the four designs. Design type III was generated using repeated elastic analyses and design modification until no member was overstressed (19) and no member could have a smaller section without overstressing another member.

Table 2 shows that some of the designs were modified to investigate the effects of Critical Member Under/Overdesign and Compression Member Overdesign.

Designs 2 and 3 were obtained from Design 1 by replacing the first member to buckle with a member of 43% and 199% stiffness respectively. Design 5 was obtained from Design 4 with a member of 43% stiffness. Designs 8, 9 and 11 were obtained using Compression Member Overdesign. The compression members being overdesigned by an arbitrary factor of 4/3. Design 8 was designed for the same ultimate capacity as Design 7 which did not use Compression Member Overdesign.

The trusses were analysed using a Limit analysis "fold line" method and the Nonlinear analysis method (17). Only one load case was considered simulating increasing uniformly distributed gravity loading applied transversely to the plane of the truss. In order to investigate the vulnerability of Space Trusses to Progressive Collapse, two trusses designs 10 and 11 were reanalysed with the member of maximum strength removed.

RESULTS OF THE ANALYSES

Uniform Trusses

The responses shown in Figs. 5 and 7 of the two trusses 1 and 4 can be compared to investigate the effect of elastic and inelastic member buckling on Space Truss response. Comparing the member capacities with the corresponding Space Truss capacities in Table 1, the system capacity does increase with the member capacity but the ductility decreases.

Designs 1 and 4 were modified to produce designs 2, 3 and 5. Designs 2 and 5 employ Critical Member Underdesign strategy and design 3 uses Critical Member Overdesign strategy. It can be seen that the underdesigned trusses exhibit greater ductility than the overdesigned truss.

Strip Design Trusses

Designs 6 and 7 use identical members but Design 7 has the members braced at midlength. Design 8 is similar to and designed for the same ultimate capacity as Design 7 except that Compression Member Overdesign was used so that the factor of safety for the compression members was increased by a factor of 4/3. Design 9 was generated in a manner similar to that of Design 8 except that the design load was greater. Examination of Table 2 and curves A and B in Fig. 9 shows that Design 6 almost achieved its design capacity, whereas Design 7 grossly underachieved its design capacity. Further, Design 6 gave earlier warning of failure and exhibited greater post linear ductility than Design 7. These results confirmed the effects of member slenderness on space truss ductility and indicates that an unmodified Strip Design Truss might have unsatisfactory behaviour.

Examining the results for Design 8 in Table 2 it can be seen that it did achieve its designed ultimate capacity. Comparing curves B and C in Fig. 9 for Designs 7 and 8 respectively, it can be seen that Design 8 did give earlier warning of impending failure and exhibited slightly more post linear ductility than Design 7. In Fig. 10 (c) it can be seen that Design 8 did produce tension member failures by Compression Member Overdesign.

Design 9 also used Compression Member Overdesign and achieved its designed ultimate capacity, gave warning of impending collapse and as desired had a collapse mechanism controlled by the tension members.

It appears that the Strip Design method can be used to give Trusses with adequate factors of safety providing a strategy, such as Compression Member Overdesign is used to guarantee ductile tension member failures.

Fully Stresses Design Trusses

Examination of Table 2 shows that the trusses 10 and 11 did achieve their designed ultimate capacity since several of the members failed approximately simultaneously before any plastic hinges could form in the buckling members.

Design Identification Number	1	2	3	4	5	6	7	8	9	10	11
Number of Member Sizes	1	1	1	1	1	4	4	4	4	20	20
Design Method*	-	-	-	-	-	S	S	S	S	FS	FS
Bracing at Midlength	-	-	-	√	√	-	√	√	√	√	√
Critical Member Underdesign	-	√	-	-	√	-	-	-	-	-	-
Critical Member Overdesign	-	-	√	-	-	-	-	-	-	-	-
Compression Member Overdesign	-	-	-	-	-	-	-	√	√	-	√
Capacity** Design Service	-	-	-	-	-	26.5	50.0	50.0	75.0	85.0	63.8
Design Ultimate	-	-	-	-	-	45.0	85.0	85.0	128.	145.	108.
Nonlin. Analysis	31.5	27.5	33.4	63.1	56.2	40.5	63.9	84.7	124.	149.	117.

* S - strip, FS - fully stressed
** Uniformly Distributed Load in Pounds per Square Foot (1 psf = 47.88 N/m^2)

TABLE 2: Design and Analysis Data

However, when a plastic hinge did form in a buckling member, there were several member failures for no increase in load, leading to the formation of a collapse mechanism. The limited ductility and widespread member nonlinearity can be seen in Figs. 11(a) and (b) and 12(a) and (b).

PROGRESSIVE COLLAPSE OF FULLY STRESSED DESIGN TRUSSES

Figs. 11(c) and (d) and 12(c) and (d) show the responses and failure configurations of Designs 10 and 11 respectively analysed with the strongest (stiffest) member removed to simulate the effect of local damage on the capacity of the system.

It can be seen for both designs that there is a severe loss in capacity, damaged design 10 in Figs. 11(c) having a capacity less than the service load, whereas damaged design 11 in Figs. 11(d) having a capacity safely above its service load. All the responses show limited ductility; however, the tolerance of design 11 to safely accept local damage indicates the advantage of Compression Member Overdesign.

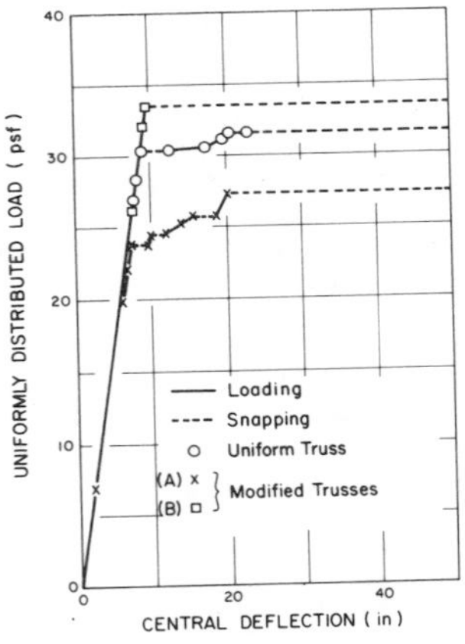

Fig. 5

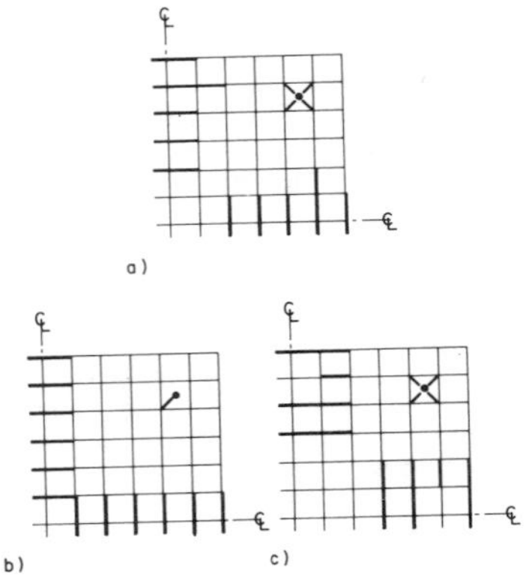

Fig. 6

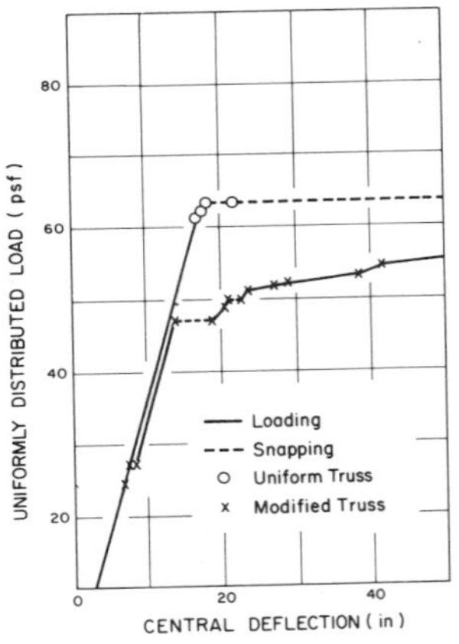

Fig. 7

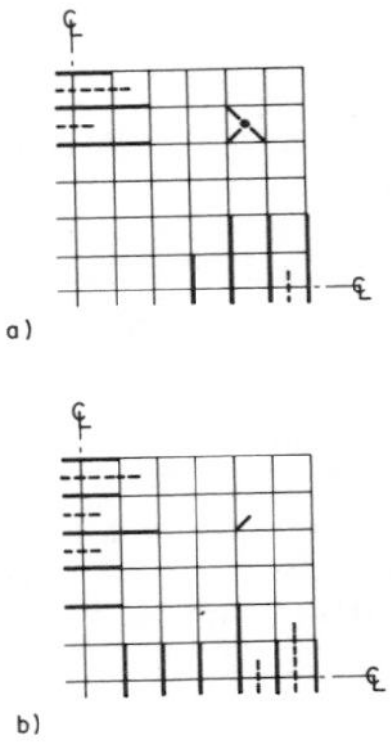

Fig. 8

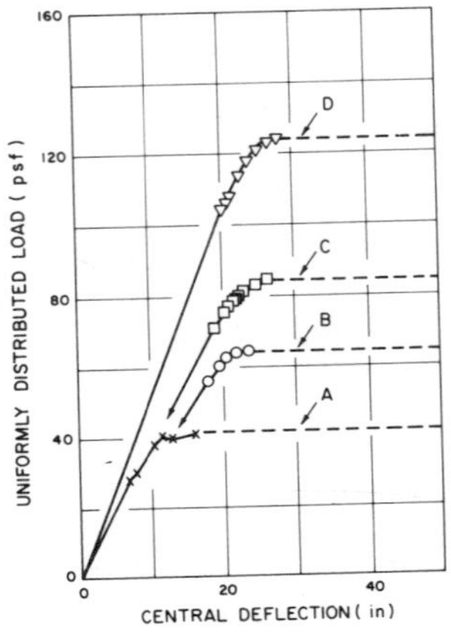

Fig. 9

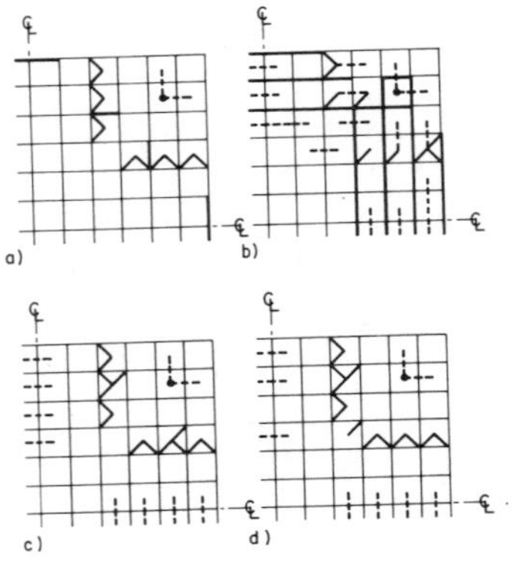

Fig. 10

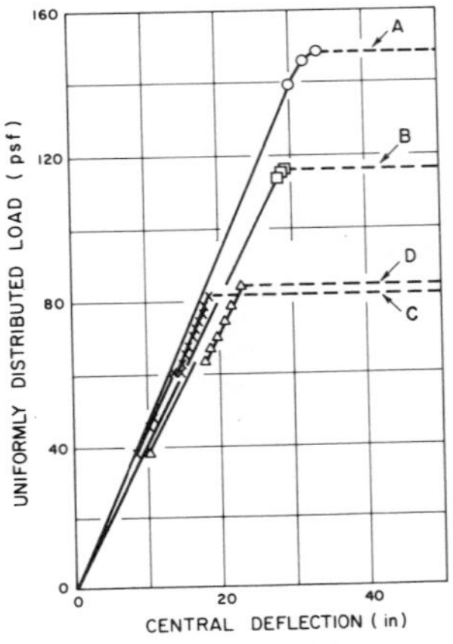

Fig. 11

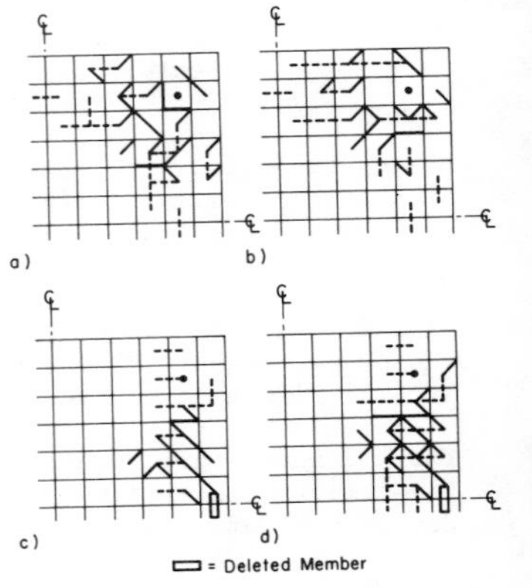

Fig. 12

CONCLUSIONS

Several promising strategies for improving space truss ductility have been investigated. Critical Member Under-design for Uniform trusses and Compression Member Over-design for Strip and Fully Stressed design trusses improve ductility. However, safe, effective but economic levels of over and under design factors have yet to be determined.

AKCNOWLEDGEMENTS

This research was supported by the National Science Foundation (USA) under Grant CME 7910303. Any opinions, findings, and conclusions expressed in this paper are those of the author and do not necessarily reflect views of the National Science Foundation.

REFERENCES

1. Beer, H. and Schultz, G., "Theoretical Basis of the European Column Curves," Construction Metallique, No.3, p. 58, 1970.

2. Davies, G. and Neal, B.G., "The Dynamical Behaviour of a Strut in a Truss Framework," Proceedings of the Royal Society, A253, pp. 542-562, 1959.

3. Davies, G. and Neal, B.G., "An Experimental Examination of the Dynamical Behaviour of a Strut on a Rigidly Jointed Truss Framework," Proceedings of the Royal Society, A274, pp. 225-238, 1963.

4. Ferguson, P.M., Reinforced Concrete Fundamentals, 4th Ed., Wiley, 1979.

5. Glogowczyka, K.Z. and Odrzansak, S.N., "An Attempt of Measurement of Random Internal Forces in Bars of a Regular Space Truss," 2nd International Conference of Space Structures, University of Surrey, England, September, 1975.

6. Hanaor, A., and Schmidt, L.C., "Space Truss Studies with Force Limiting Devices," Journal of the Structural Division, ASCE, November, 1980.

7. Johnston, B.G., ed., SSRC Guide to Stability Design Criteria for Metal Structures, 3rd Edition, John Wiley and Sons, New York, 1976.

8. Makowski, Z.S., Analysis, Design and Construction of Double Layer Grids, Halsted Press, John Wiley and Sons, New York, 1981.

9. Mezzina, M., Prete, G. and Tosto, A., "Automatic and Experimental Analysis for a Model of Space Grid in Elasto-Plastic Behaviour," 2nd International Conference of Space Structures, University of Surrey, England, September, 1975.

10. Renton, J.D., "General Properties of Space Grids," International Journal of Mechanical Sciences, Pergamon Press, Vol. 12, pp. 801-810, 1970.

11. Schmidt, L.C. and Hanoar, A., "Force Limiting Devices in Space Trusses," Journal of the Structural Division, ASCE, May, 1979.

12. Schmidt, L.C., Morgan, P.R. and Stevens, L.K., "The Influence of Imperfections on the Behaviour of a Space Truss," 2nd International Conference on Space Structures, University of Surrey, England, September, 1975.

13. Schmidt, L.C., Morgan, P.R. and Clarkson, J.A., "Space Truss with Brittle Type Strut Buckling," Journal of the Structural Division, ASCE, July, 1976.

14. Schmidt, L.C., Morgan, P.R. and Clarkson, J.A., "Space Truss Design in the Inelastic Range," Journal of the Structural Division, ASCE, December, 1978.

15. Schmidt, L.C., et al., "Ultimate Load Behaviour of a Full Scale Space Truss," Proceedings of the Institute of Civil Engineers, UK, March, 1980.

16. Smith, E.A. "Collapse Behavior of Space Trusses" Department of Civil Engineering Report CE 82-144, University of Connecticut, August, 1982.

17. Smith, E.A., "Space Truss Non Linear Analysis", Journal of the Structural Division, ASCE, April, 1984.

18. Smith, E.A. and Epstein, H.I., "Hartford Coliseum Roof Collapse: Structural Collapse Sequence and Lessons Learned," Civil Engineering - ASCE, April, 1980, pp. 59-62.

19. Specification for the Design, Fabrication, and Erection of Structural Steel for Buildings, American Institute of Steel Construction, Chicago, Ill., November, 1978.

THE COLLAPSE LOAD OF BARREL VAULTS IN THE ELASTO-PLASTIC RANGE

P LENZA and M PAGANO

Department of Civil Engineering
University of Naples, Italy

The interaction between plastic and critical elastic phenomena determines the type of collapse and the amount of load causing it. This work reviews these relationships and sets forth a theory suitable for their evaluation. Such a theory is then compared with the available experimental results. In reticulated structures the local phenomena's influence on the overall behaviour of the structure is also evaluated.

INTRODUCTION

For the uniformly distributed load condition the tensile stress of the vault centerline section corresponds to the "beam" model with sufficient approximation since the longitudinal shape of the deformation is almost linear. This depends on the fact that for such a particular load condition the edges do not exibit transverse displacements and hence the semicircular section remains basically the same.

Under different load conditions - and in particular for the dead weight condition - such a shape is far from being linear if the cross section is deformable and if the longitudinal edges are unconstrained. However it can become linear if internal constraints are introduced which can effectively oppose changes in the section shape.

A study [3] made by testing a steel barrel model by means of a number of internal constraints that limited the section deformation confirmed our theoretical prediction. Such an experimental approach was important in view of the subsequent developments since, in defining the overall instability models, the transverse deformability proved to be more relevant than the longitudinal one.

2. ELASTIC MODELS

2.1 Shape-degradation induced instability

The theory illustrated in a later study [4] permits us to draw the curve of the equilibrium conditions relating the existing load to the cross section deformation of semicircular vaults (Fig 1).

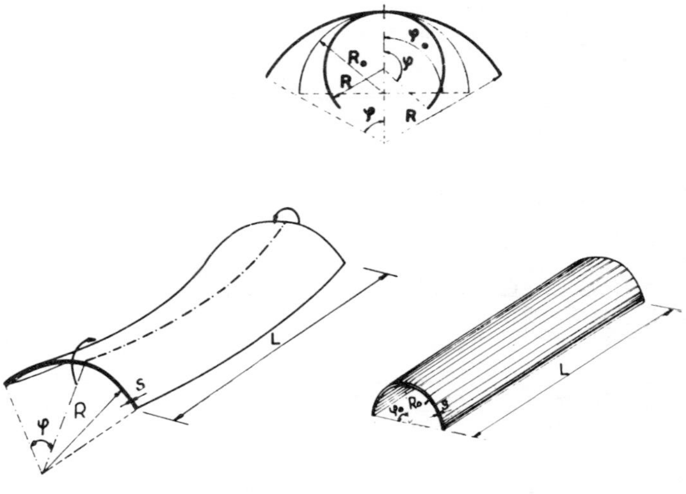

Fig 1 - Geometry and deformation of the structure: in the theoretical models the cross-section is assumed to deform by keeping its circular shape.

Assuming that the change in shape consists in a change in curvature, the following relationship is obtained:

$$Q_{E2} = E \cdot \left(\frac{1}{2} \frac{\psi}{\chi} \frac{R_o^4}{L^4} s \pm \sqrt{\left(\frac{1}{2} \frac{\psi}{\chi} \frac{R_o^4}{L^4} s \right)^2 + \frac{1}{1-\nu^2} \frac{\varrho}{\chi} \frac{s^4}{L^4} R_o^2} \right) \qquad \ldots\ldots 1$$

ρ, ψ and χ are function of the opening angle φ and their expression is given by the most common load distributions on the cross section.

For the dead load condition and for that of loads distributed along the edges, the section initially tends to close. Later when the deformation grows, the stresses due to the "Brazier's effect" prevails over the stress due to the load effect and thus the vault tends to open again reaching the original curvature radius Ro, for what is defined an "inversion" load value

$$q_i = \frac{\psi}{\chi} E\left(\frac{R_o}{L}\right)^4 \cdot s \qquad \ldots\ldots 2.$$

At a first stage expression (1) supplies two real and positive solutions bounded by the inversion load.
At a second stage angle φ grows as the load increases until the latter reaches its ultimate value (a critical value of second kind) and (1) supplies only one positive solution. Further increases in buckling can only occur when there are load reductions. Hence the ultimate load corresponds to the highest value taken by expression (1):

$$Q_{lim} = \text{MAX}\left[Q_{E2} = F(\varphi)\right] \qquad \ldots\ldots 3.$$

Such a theory assumes that the material has an indefinitely elasto-linear behaviour.

2.2 Eulerian instability

2.2.1 Longitudinal critical load

In this part as an alternative to the complex physical mathematical formulae available [6] [7] , an approximate but technically effective treatment is proposed.
In the barrel vault two "arches", that develop along the diagonals, as well as the corresponding tie-beams in the outbonds are identified, the latter being 2 φ/10 wide (Fig 2). The "arches" crown section is assumed to be 2/3φ wide. For a critical deflection curve having three waves, of which the central one accounts for 40% of the vault's total span, we have

$$Q_{E1} = C_{E1} \frac{E.I(\varphi).f(\varphi)}{L^4} \qquad \ldots\ldots 4,$$

expressing the ultimate load for the Eulerian instability, where f(φ) stands for the section rise, I'(φ) for the inertia of the ideal section of the arches and C_{E1} denotes a dimensionless numerical constant that in the present study has been assumed to be equal to 1263.

2.2.2 Transverse critical load

If we consider the cross section as an "arch", the critical condition for a three-wave deflection curve is reached with respect to a load supplied by

$$Q_{ET} = C_{ET} \frac{Es^3\, f(\varphi)}{l(\varphi)^3} \qquad \ldots\ldots 5.$$

The constant C_{ET} has been assumed to be equal to 5,14. For commonly applied geometries such a critical load results to be considerably greater than other loads taken into consideration in the same study. Therefore only rarely it effects the collapse of the structure.

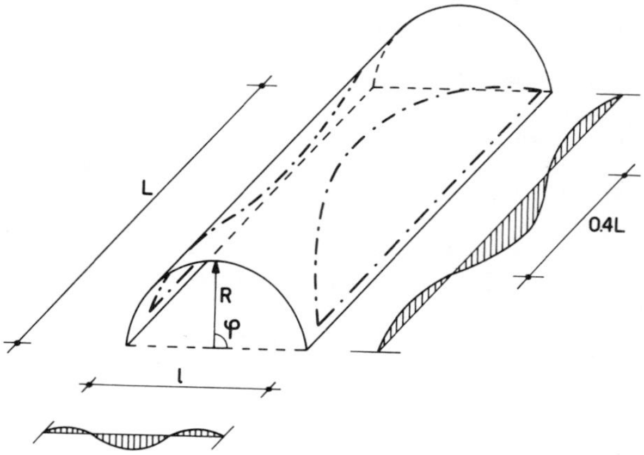

Fig 2 - The identification in the structure of longitudinal and transverse arches permits to define technically effective elastic and plastic theoretical models.

3 PLASTIC MODEL

When we consider the material having an ideal elasto-plastic behaviour, the ultimate plastic load is determined by using the same schematization employed in paragraph 2.2.1. The ultimate load is the one which corresponds to the total yield of the rise and is expressed by

$$Q_P = C_P \frac{\sigma_p .s.R(\varphi).\varphi.f(\varphi)}{L^2} \qquad \ldots\ldots 6,$$

where s stands for the vault thickness, σ_p for the plasticization ideal stress, and C_p for a numerical constant which is assumed to be equal to 3.2.
On the other hand we use the "beam" scheme for determining the load which triggers the plastic phenomenon along the edges of the section:

$$Q_{IP} = \frac{8\,\sigma_p .I(\varphi)}{y_{max}(\varphi).\, L^2} \qquad \ldots\ldots 7.$$

Curves 6) and 7) for each value of φ define an interval whose ends represent the onset and completion of the plasticization and thus enclose a band in which the entire plastic degradation of the structure develops.

4 DETERMINATION OF THE COLLAPSE LOAD

For each series of geometrical and mechanical parameters of the structure we can draw curves 1), 4), 5), 6) and 7), putting angle φ on the horizontal axis as the only parameter representing the section's deformation and the load on the vertical axis.

The main curve (1) represents the load-deformation link; the value of the collapse is given by its maximum point provided that none of the other phenomena described above occurs before such a value is reached. For example it might occur that the ascending portion of (1) intersects curve (4) yielding a lower collapse load and depriving of any physical significance the following portion of (1).

Figure 3 describes the theoretical behaviour of a polyvinyl chloride model tested in the laboratory under different load conditions. Curve Q_{E1} intersect the (1) before their points of maximum causing the collapse in the elastic field due to Eulerian instability with respect to load values supplied by the ordinates of the intersection points. In such a case the "plastic" curves 6) and 7) do not intervene because they are well above curves 1).

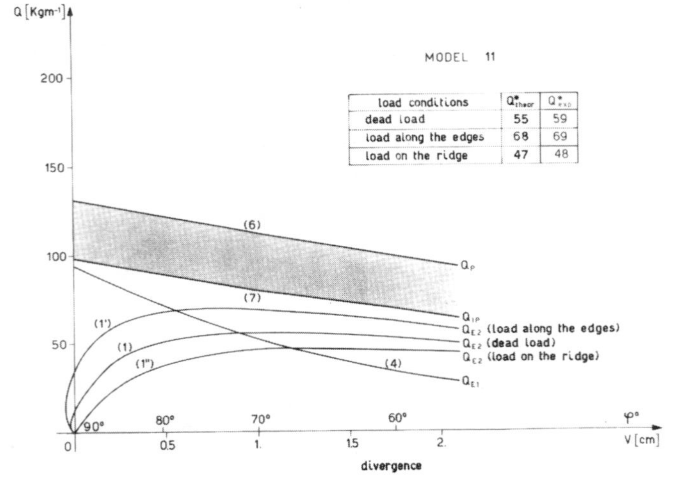

Fig 3 - Theoretical analysis of collapse of a p.v.c. model. Tests have confirmed the nature and relevance of the predicted elastic phenomenon.

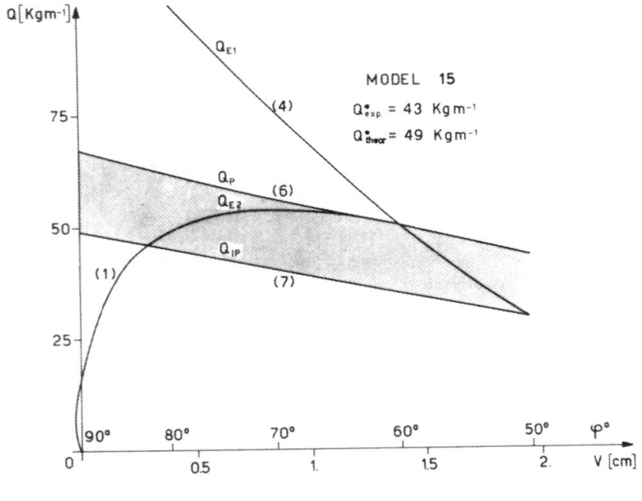

Fig 4 - Theoretical analysis of collapse of a steel model. Tests have confirmed the development of the theoretically predicted plastic phenomena.

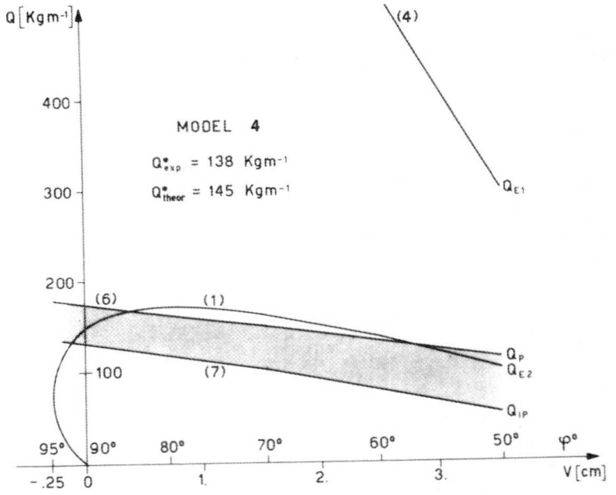

Fig 5 - Theoretical analysis of collapse of another steel model; in this case the plastic phenomenon is even more evident.

In fact the plastic phenomena are determined for the structure collapse when the plasticization band, bounded by curves 6) and 7), intersects the ascending portion of curve 1) before 4) and 5). The intersection level gives rise to different forms of collapse. Let us consider for example (Fig. 4 and 5) the theoretical behaviour of two experimental models in sheet steel; in the first case the plasticization is not completed because curve Q_p does not intersect the ascending portion of (1). Therefore the collapse occurs due to the instability of second kind in the elasto-plastic field. In the case of figure 5 plasticization becomes complete; and it is indeed such a phenomenon that causes the collapse.

In the cases characterized by the impact of plastic phenomena, it should be taken into account that the occurrence of the first plasticization phenomena change the subsequent shape of the elastic curve (1) due to the greater deformability of the section. In fact the collapse load will reach an intermediate value between the one corresponding to the actual attainment of the ultimate condition (determined without changing the (1)) and the value corresponding to the first plasticization phenomena. For purposes of practical applications the mean of the two values can be adopted as collapse load.

In the examples reviewed in this paper the transverse critical load expressed by the (5) does not appear because it is numerically very high. However for particular geometries (short vaults with large radius) it could also result to be determinant.

The comparative study of the curves that represent the ultimate phenomena permits to determine the collapse load. The validity of this theory will be evaluated in the following paragraph 6 on the basis of the experimental results.

5 APPROXIMATE DETERMINATION OF THE COLLAPSE LOAD

As concerns the so called long vaults ($\frac{R}{L} \ll 1$) the highest value taken up by 1) can be roughly expressed by

$$Q_{lim} = 0.660 \frac{E}{\sqrt{1 - \nu^2}} \frac{s^2}{L^2} R_o \qquad \ldots\ldots 3'$$

that supplies the ultimate load for the instability of second kind.

In the light of a rough evaluation we can neglect the effect of the section's transverse deformation when determining the ultimate loads for the Eulerian instability and the ultimate plastic load. In such a case it will suffice to use expressions 4, 5 and 6 considering φ as constant and equal to φ_0. So doing it is possible to immediately obtain four values of the ultimate elastic and plastic conditions calculated independently the one of the other, and consequently to take the lowest value as the collapse load. Such a rough evaluation may result particularly useful for guiding the design, keeping in mind that as far as collapse loads are concerned, high safety coefficients are usually adopted.

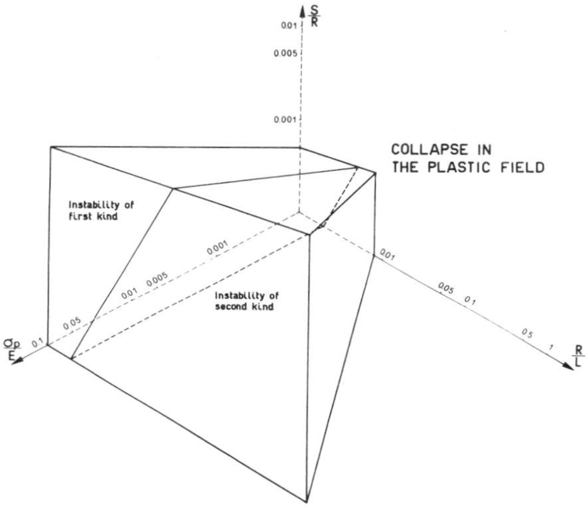

Fig 6 - Abacus for the prediction of the collapse type (elastic or plastic) for the most common geometries.

Figure 6 shows a three-dimensional abacus where the geometrical features of the structure and the mechanical characteristics of the material are represented on three cartesian axes in a dimensionless form and in a logarithmic scale. A domain is identified that encloses the points representing the structures for which the collapse is expected to occur in the elastic field due to instability of first or second kind. Outside such a domain the collapse is expected to take place in the plastic field.

Ribbings or other braces can make the transverse buckling negligible. In those cases (3') has no longer any meaning, while the approximations in expressions 4), 5) and 6) become rigorous since φ actually remains constant.

6 EXPERIMENTAL CONFIRMATIONS

The theory set forth in paragraph 4 was applied to the experimental results obtained by 24 steel, alluminium and polyvenyl chloride models with different geometries, shown in [4]. Table 1 summarizes the features of the models as well as the results of the tests and the theoretical analyses. The comparison between the theoretical and the experimental values of the collapse load is often fully satisfactory. Deeper differences are found in the aluminium models that are likely to be due to a scarcer knowledge of the material's plastic behaviour.

Finally, model 1 shows particular geometric characteristics (since it is quite a short vault) and the collapse occurred for transverse instability in the elasto-plastic range. Eventually the overall comparison between the set of theoretical and experimental data made by the methods set forth in [5] leads to the determination of an "Index of Concordance" equal to 70%.

The proposed theory is also supported by the observation that the experimentally determined deformation due to the collapse of the cross section is consistent with what stated by the theory.
Also the type of collapse envisaged by the theory is consistent with the experimental observations; in fact the yielding phenomenon occurred - often markedly - at the edges of the small vault in almost all tests except for polyvinyl chloride models in which the collapse took undoubtedly place in the elastic range. This enabled us to repeat the test on model 11 concerning, in addition to the dead weight condition, also those for the load distributed along the ridge and the loads distributed along the edges (tests n° 23 and 24). In both cases the collapse in the elastic range was obtained, as stated by the theory (see Fig.3).

7 RETICULATED VAULTS

The theoretical models illustrated here are applied to the reticulated vault by the definition of the "analogous" continuous surface, i.e. a surface exhibiting the same deformation under the same load also in the plastic range. It is also possible for a given lattice to determine the relationships that impose equality between the stretching and bending deformation of the reticulated modulus on the one hand and that of the continuous element.

The static analysis must be integrated by evaluating local phenomena due to the crisis of individual beams. With caution the collapse load can be obtained, following Rankine's method, by means of

$$\frac{1}{Q^*} = \frac{1}{Q'} + \frac{1}{Q''} \qquad \ldots\ldots 8$$

where Q' is the local ultimate load causing the collapse of a single beam and Q" stands for the collapse load of the analogous continuous vault.
The experiments made on three reticulated models (Fig.7) permitted to produce some first evaluations and comparisons. Models (denoted by A, B and C) have the same radius (R_o = 33 cm), the same span (L = 200 cm) and the same square mesh lattice (10 x 10 cm); the meshes are stiffened by a diagonal so arranged as to result generally stretched for vertical loads. The barrel beams on the contrary have a different diameter in the three models (0.5, 0.4 and 0.3 cm respectively). They are welded the one with the other and show unavoidable flows due to their small size. The beams eccentricities in the nodes are particularly evident since the models were constructed in superimposed layers in order

not to interrupt the continuity of the longitudinal, transverse and diagonal beams. Such a fact determined a transverse deformability of the experimental model much lower than the theoretical deformability as the test clearly demonstrated.

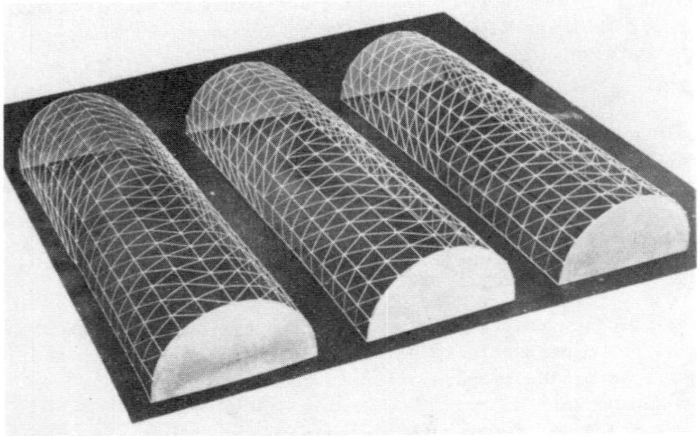

Fig 7 - Reticulated models of steel barrel vaults made of small cylindric bars, each having a different diameter.

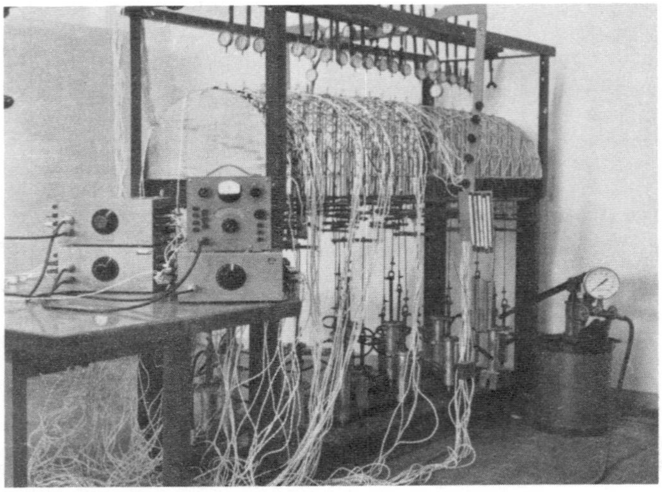

Fig 8 - A reticulated model during an experimental stage; electroresistant straingages permitted to detect the tensile stress in many beams.

Tests were performed (Fig.8) by estimating the tensile stress of the centerline beams at the end at the quarter of the span both by means of the tensostat and of electro-resistant straingages. The deformation detected was undoubted effected by the local bending phenomenon since the small size of the beams permitted the application of only one reading basis for section.

The comparison with the theoretical mean stress values of the beams obtained by finite elements models (Fig.9) actually reveals a qualitative consistence between theoretical and experimental stress behaviours.
The theoretical collapse values were determined by (8) using the theory of the analogous vault to search for the overall ultimate load (Fig.10).

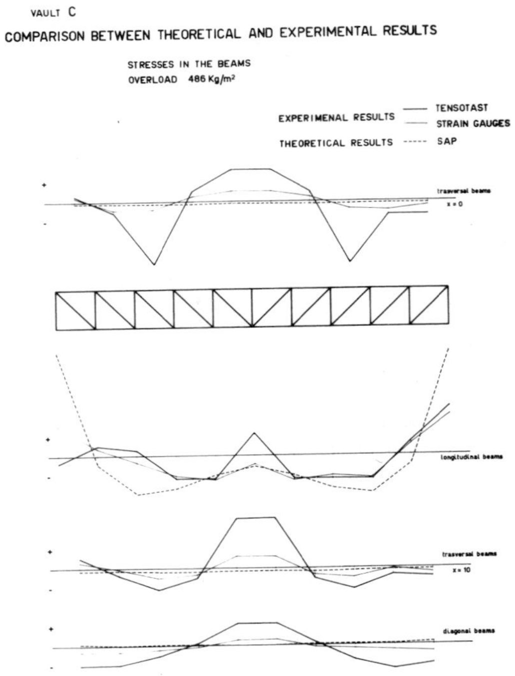

Fig 9 - Comparison between the experimentally measured stress in centerline beams of model C and that assessed by using the finite elements method.

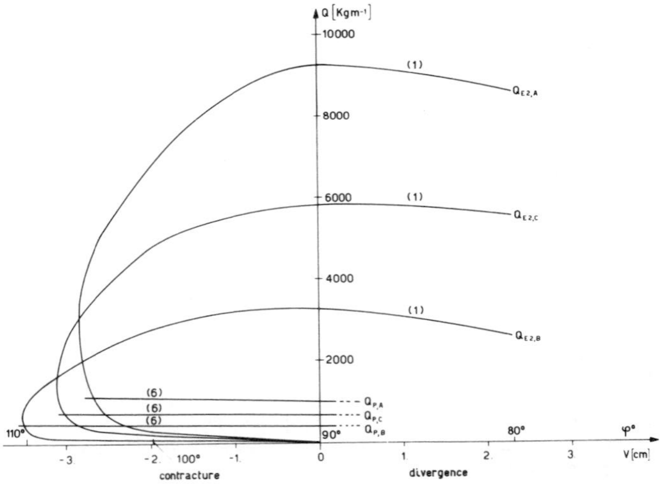

Fig 10- Theoretical analysis of the collapse of "analogous" continuous vaults in the experimental models A,B,C. For these geometries the cross section tends prevailingly "to strink". The crisis is expected to occur due to the plastic ultimate load.

Model	Q' Kgm^{-1}	Q'' Kgm^{-1}	Q^*_{theor} Kgm^{-1}	Q^*_{exp} Kgm^{-1}
A	1396	1160	633	1140
B	601	720	328	—
C	181	480	131	—

Models B and C were not loaded up to collapse in order to repeat the test ; therefore the only experimental value, though consistent with the theoretical one determined by (8), for the sake of caution, does not allow to evaluate the validity of the proposed theory.

8 CONCLUDING REMARKS

This paper confirms the relevance of non linear behaviour in the statics of barrel vaults. The cross section deformation is in fact a phenomenon that can already occur under operational conditions by affecting the tensile stress. At any rate it is always of great importance in the search for the collapse load not only when the latter takes place due to the section shape degradation (instability of 2nd kind), but also when plastic and/or Eulerian instability phenomena occur which accelerate the crisis.
For continuous vaults the comparison of the proposed theory with available experimental data has resulted to be promising.

Tests initiated on reticulated models do not yet allow a comparison with the theory. They applies to lattices by means of the definition of equivalent continuous structure and evaluate the local phenomena impacts on the overall behaviour separately. It is indeed in the evaluation of the influence exerted by such phenomena that the greatest operational difficulties arise.
Therefore research must aim at putting at the disposal of technicians, models that act directly upon the reticular structure taking into account their progressive and non linear deformation.

REFERENCES

1. O BELLUZZI , Scienza delle costruzioni, Vol.IV, Zanichelli Editore, Bologna 1955.

2. H LUNDGREN, Cylindrical shells, The Danish Technical Press, Copenhagen 1951.

3. M PAGANO, Contributo teorico-sperimentale al calcolo delle volte autoportanti, Atti Accademia Pontaniana, Vol.IV, 1953.

4. M PAGANO, R SANNINO, Contributo teorico-sperimentale al problema dell'instabilità delle volte travi, Giornale del Genio Civile, n°7-8-9, 1961.

5. P LENZA, Validità dei modelli per la progettazione di volte reticolari, Costruzioni Metalliche, n°1, 1977.

6. G KRALL, Moltiplicatore critico λ di una distribuzione di carico su una volta autoportante, Rend.Acc.Lincei, Vol.I, 1946.

7. G KRALL, D CALIGO, Moltiplicatore critico λ di una distribuzione di carico su una volta autoportante, Rend. Acc.Lincei, Vol.IV, 1948.

TABLE 1

MODEL N°	E $Kgcm^{-2}$	ν	σ_p $Kgcm^{-2}$	L cm	Ro cm	S cm	Q*teor Kgm^{-1}	Q* sper Kgm^{-1}	EXPERIMENTAL DEFORMATION OF CENTERLINE SECTIONS	$\varphi°$ OPENING ANGLE ENVISAGED AT COLLAPSE	TYPE OF COLLAPSE ENVISAGED BY THE THEORY
1	$2x10^6$	0,3	2525	130	25	0.07	3200	900	WITH LOBES	110°	plastic
2	"	"	2900	150	10	0.087	492	312	CONTRACTURE	94°	"
3	"	"	3300	200	15	0.07	598	452	"	100°	"
4	"	"	2770	200	10	0.05	145	138	DIVERGENCE	80°	"
5	"	"	2610	200	6.5	0.1	116	106	"	85°	"
6	"	"	2460	140	4.5	0.07	75	67	"	85°	"
7	"	"	2545	100	3.25	0.05	57	101	"	85°	"
8	"	"	2550	200	5.0	0.1	67	88	"	85°	"
9	"	"	2500	140	3.5	0.07	47	47	"	85°	"
10	$3x10^4$	0,35	462	100	3.5	0.2	19	29	"	85°	elastic
11	"	"	428	150	7	0.28	55	59	"	70°	"
12	$2x10^6$	0,3	3090	165	5.5	0.05	69	86	"	75°	plastic
13	"	"	3590	180	5.5	0.05	63	58	"	70°	elasto/plastic
14	"	"	3600	190	5.5	0.05	56	44	"	70°	"
15	"	"	3490	200	5.5	0.05	49	43	"	70°	"
16	"	"	2583	200	15	0.1	660	700	CONTRACTURE	98°	plastic
17	"	"	2185	200	9	0.1	192	448	DIVERGENCE	85°	"
18	$7x10^5$	0,36	1190	200	4.75	0.25	75	150	"	85°	"
19	"	"	1182	160	4.75	0.25	111	248	"	85°	"
20	"	"	1175	140	4.75	0.25	145	242	"	85°	"
21	"	"	1230	120	4.75	0.25	207	329	"	85°	"
22	"	"	1155	100	4.75	0.25	280	460	"	85°	"
23	$3x10^4$	0,35	428	150	7	0.28	46	48	"	70°	elastic
24	"	"	428	150	7	0.28	66	69	"	80°	"

DYNAMIC CHARACTERISTICS OF A SPACE REFLECTOR SUPPORT STRUCTURE

C. A. MAZZOLA*, M.S.; R. J. RECK**, Ph.D; and R. SHEPHERD#, D.Sc.

* Engineer/Scientist, McDonnell Douglas Astronautics Co., Huntington Beach, California, USA.

** Chief Technology Engineer, McDonnell Douglas Astronautics Co., Huntington Beach, California, USA.

\# Professor and Chairman, Department of Civil Engineering, University of California, Irvine, USA.

The need to provide large reflector structures suitable for use in a variety of space missions prompted the investigation of a representative three-dimensional space truss. The finite element model used idealizes the truss members as uniaxial rod elements with pinned end connections. In this paper comparisons of natural frequencies and mode shapes for various geometrical configurations and material properties are made. The accuracy of the model idealization is verified by comparing results for a uniformly loaded beam situation. It is concluded that the technique proposed is viable in the design of large reflector space structures.

INTRODUCTION

Missions beyond the earth's atmosphere require the design and construction of large space structures having low mass, large overall size, high dimensional stability and predictable vibration characteristics. Additionally, use of the space shuttle necessitates the development of deployable structural systems capable of being erected in space.

A basic representative conceptual solution to the problem is the use of a tetrahedral truss made of graphite-epoxy composite material. Whereas previous investigations (see Ref.1) simplified the analysis of such a structure by idealizing the truss framework as sandwich plates, in the study reported here a finite element representation was used. This confers the advantage of realistic modelling using appropriate elements and end boundary conditions.

The object of the study was to investigate the viability of the use of finite element analysis techniques to the design of large space structures of the type described.

STRUCTURAL CONFIGURATION

As an indication of a possible use for a large reflector structure, a conceptual layout of an advanced capability manned space platform is shown in Fig.1. The reflector is approximately 30 feet across with a dish radius of about 60 feet.

THE REFLECTOR STRUCTURE

The truss framework supporting the reflector is of the deployable type. Figure 2 shows the reflector and its support system in each of the deployed and stowed positions. The structural subsystem is made up of tetrahedral trusses. They consist of basic repeating elements connected together so that column members on one surface form a pattern of equilateral triangles. A similar pattern of inverted equilateral triangles is formed by the column members on the back surface of the framework. Diagonal columns, called "tripod" members, form the core of the structure and connect the two faces. Figure 3 shows the basic tetrahedral elements as well as the two patterns of triangles comprising the two surfaces of the structure. The complete truss consists of seven clusters of twelve tetrahedrons, namely one central cluster and six outer ones. In the stowed condition, the shaded clusters shown in Fig. 2 fold into the back of the cargo bay whereas the unshaded outer clusters fold forward.

The envisaged structural subsystem consists of 792 struts. The spherical nature of the truss structure necessitates the members on the back face (each 40.95 ± 1.23 inches) being longer than those or the face adjacent to the reflector dish. These, and the diagonal tripod members, are each 39.37 ± 0.92 inches long. The thickness of the framework, defined as the perpendicular distance between the surfaces, is 31.67 inches. The radii of the front and back spherical truss faces are 787.4 and 819.1 inches respectively.

MATERIAL ASPECTS

Major considerations in the choice of large space structures include transportation and fabrication costs. The first can be limited by reducing the total weight of the structure together with achieving a high packing density for the trip into orbit. Fabrication costs can be minimized by facilitating the ease of assembly of the structure by the use of simple elements and joints with a high degree of commonality. Nevertheless the most critical choice, affecting both performance and cost of a large space structure, is that of the material to be used.

The material properties critically affecting the structural performance of large space structures are the specific stiffness (elastic modulus divided by density), the coefficient of thermal expansion, the damping characteristics and the thermal and electrical conductivity. Other factors such as fatigue resistance,

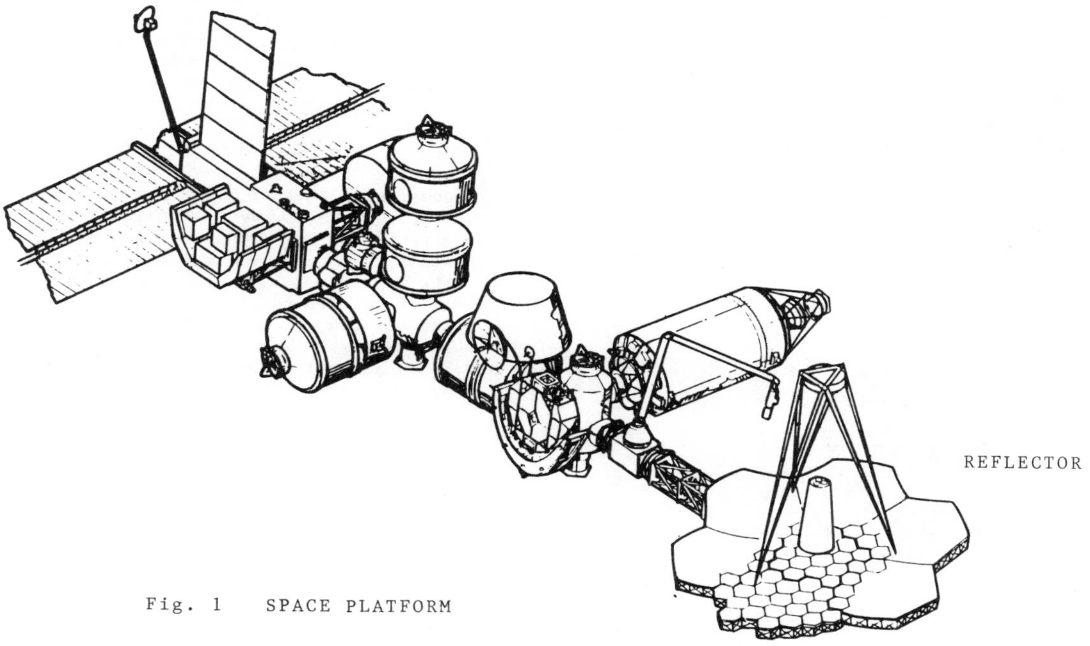

Fig. 1 SPACE PLATFORM

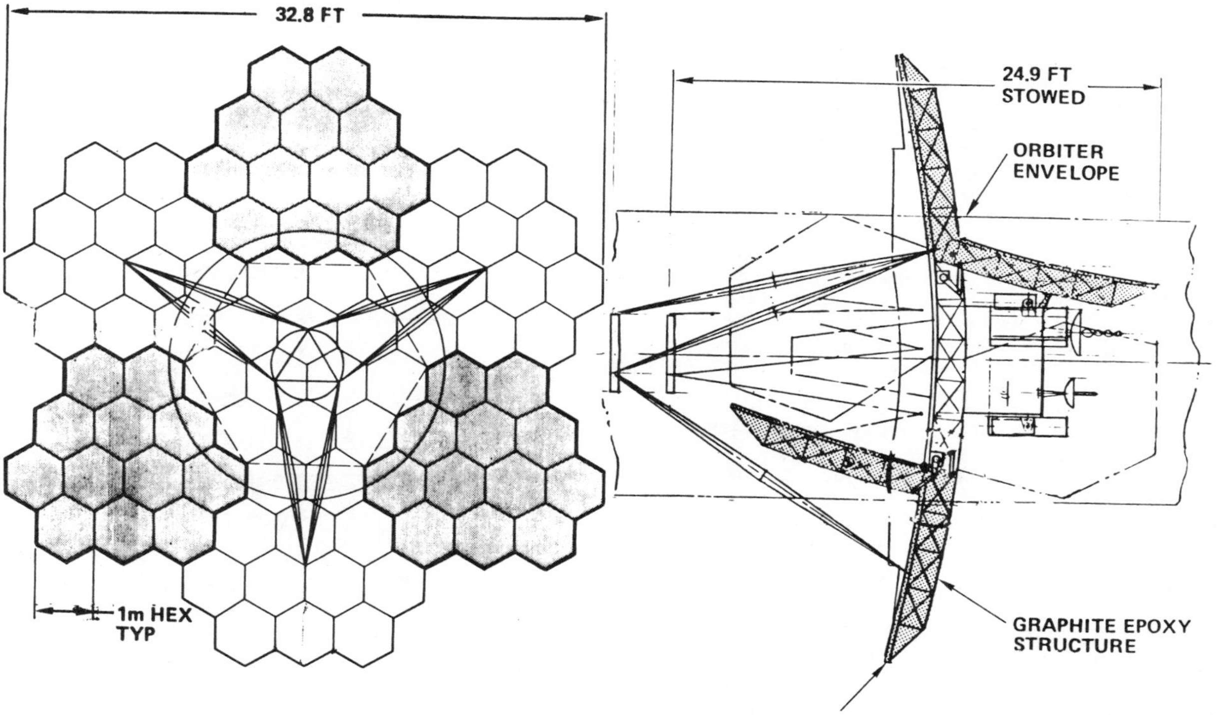

Fig. 2 LARGE DEPLOYABLE REFLECTOR STRUCTURE

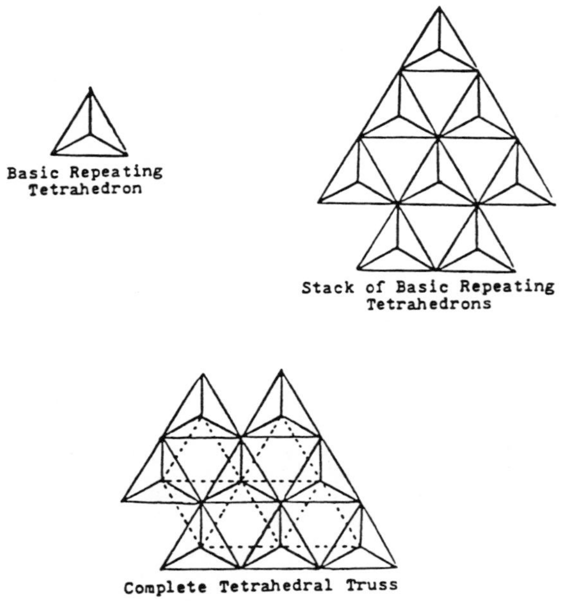

Fig. 3 BASIC REPEATING TETRAHEDRON ELEMENTS

thermal aging and response to ultraviolet radiation affect the longevity of a structure in space. Advanced composites compare favorably with metals in providing these required structural properties (see Refs. 2 and 3).

The use of advanced composites offers a further significant advantage. Other investigators (see Ref.4) have shown that the lower cost of aluminum is more than offset by the higher cost of transporting it to the extent that, assuming critical mass payloads, 70% more shuttle flights would be required to orbit a given area of aluminum structure than would be required for the same area of graphite-epoxy reflector (Ref.1). Moreover the costs of the newer materials is steadily decreasing as manufacturing methods improve.

Although not all aspects of the material choice have been resolved, for a general study of a typical large space structure it appears reasonable to anticipate the use of a graphite-epoxy composite as the primary structural material. For the investigation reported T3000 graphite/5208 Epoxy was selected. With a density of 0.058 $lb/in.^3$ and an elastic modulus of 10×10^6 $lb/in.^2$ a high specific stiffness is provided. The coefficient of thermal expansion of 0.1×10^{-6} in./in./°F is suitably low. Poisson's ratio is equal to 0.30. a thin walled tube section with an inside diameter of 2 inches and a wall thickness of 0.048 inch was chosen for each truss column member.

FINITE ELEMENT ANALYSES

In modelling the reflector support structure the truss was broken down into its column members components. Each member was idealized as a single rod element. As flexural deformation in an individual element is insignificant compared to the deformation of the entire framework, local column bending was ignored consistent with the choice of pinned end supports for all column members. The finite element code used did take into account transverse shear deformation between the two parallel truss faces.

The establishment of the model of the substructure, traditionally one of the biggest hazards of the finite element representation of a complex structure, was facilitated by the availability of an automated model building computer graphics system (PATRAN G) containing optimization features which compare the model so that it requires minimal computer time and memory at the analysis stage (see Ref.5)

For the purpose of the analysis only the reflector support framework was considered. All the columns comprising the two truss surfaces and the core (i.e. the tripod members) were modelled see, Fig.4. Since their effect on the stiffness of the structure is extremely small, the vertical columns connecting the reflector to the back truss face were ignored. Also the contribution to the stiffness of the framework of the reflector itself was considered insignificant. The aggravate weight of the reflector and vertical columns was distributed among the support framework members by increasing the effective density of all the column elements by a factor appropriate to account for this load redistribution.

The weights of the components considered are:

Component	Weight (lb.)
Support Structure Framework	565.3
Reflector	1017.6
Vertical Support Columns	56.5
Joint Allowance	56.5
TOTAL	1695.9

The analysis of the large space structure was undertaken using the EASE 2 (Elastic Analysis in Structural Engineering) code (see Ref.6). This allows for a maximum of six degrees of freedom at each node. In the interest of reducing computational costs the rotational degrees of freedom were deleted in the analyses undertaken. However, the validity of this simplification was subsequently verified as outlined below.

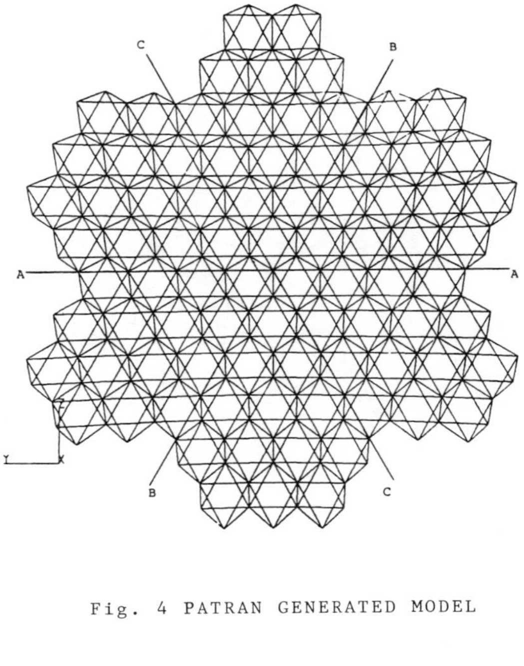

Fig. 4 PATRAN GENERATED MODEL

ANALYSIS RESULTS

Eigenvalues (Table 1) and eigenvectors (for mode shapes see Figs. 5 through 12) were determined for the first eight modes of free vibration under the self weight distribution outlined above.

Up to eleven iterations were found to be necessary to achieve specified convergence. Eigenvectors were computed, providing x, y and z translational components for each of the 205 node points and these results used to plot the mode shapes in the following manner. Deflections along three selected axes (shown on Fig. 4) were tabulated and normalized for each mode. The shapes were then plotted on these axes.

A series of parameter studies were undertaken to verify the accuracy of the model idealization. Material properties were varied, boundary conditions were changed and, for a single geometric configuration, the free vibration characteristics were computed when all six degrees of freedom at each mode were considered. Tables 2 through 4 summarize the results of these studies.

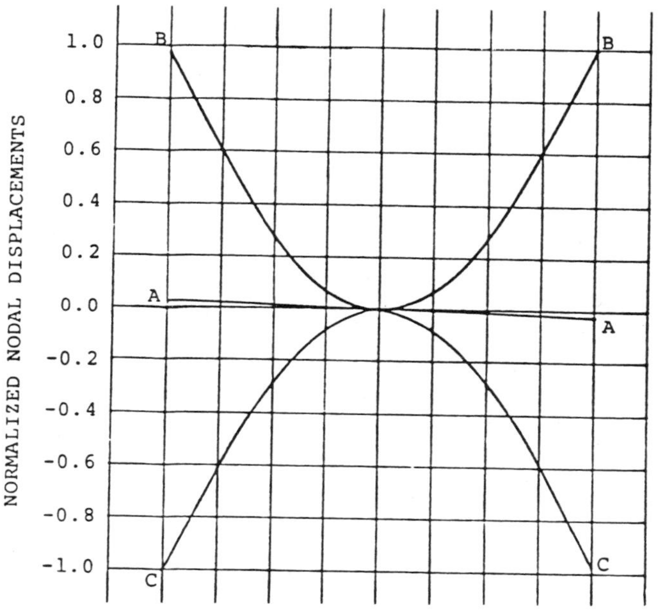

Fig. 5 FIRST MODE SHAPE

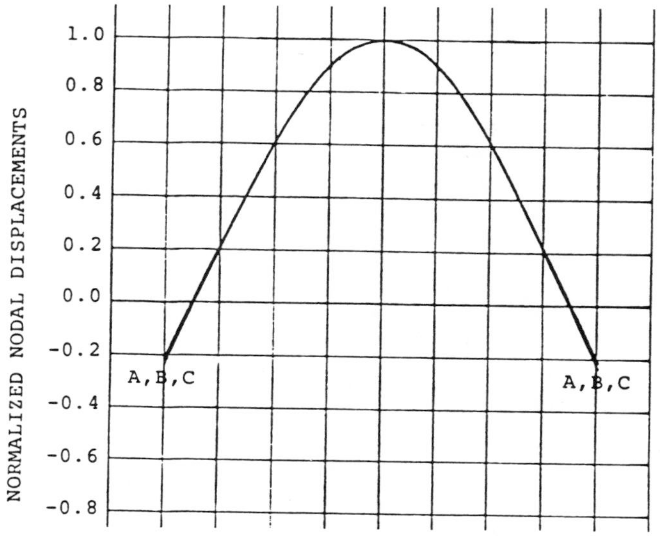

Fig. 7 THIRD MODE SHAPE

Table 1. Baseline Structure Modal Frequencies

Mode	Frequency (Cycles/Sec)
1	20.4595
2	20.4888
3	33.9519
4	36.3123
5	47.1268
6	58.6836
7	58.8495
8	63.3508

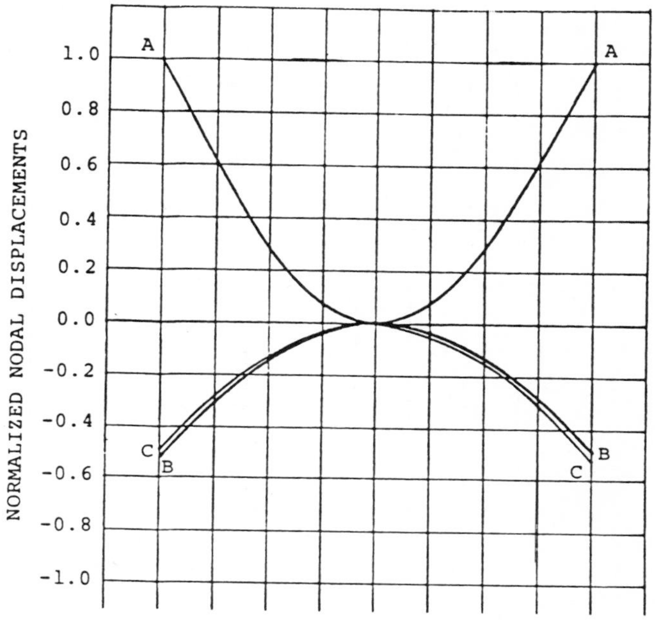

Fig. 6 SECOND MODE SHAPE

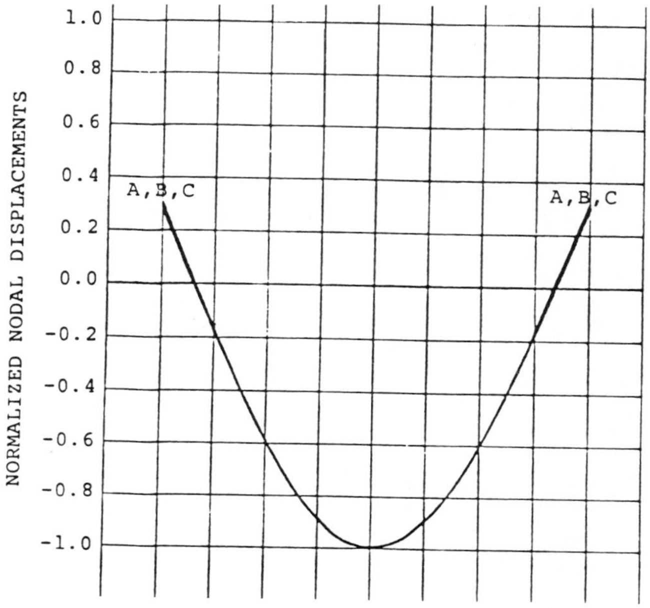

Fig. 8 FOURTH MODE SHAPE

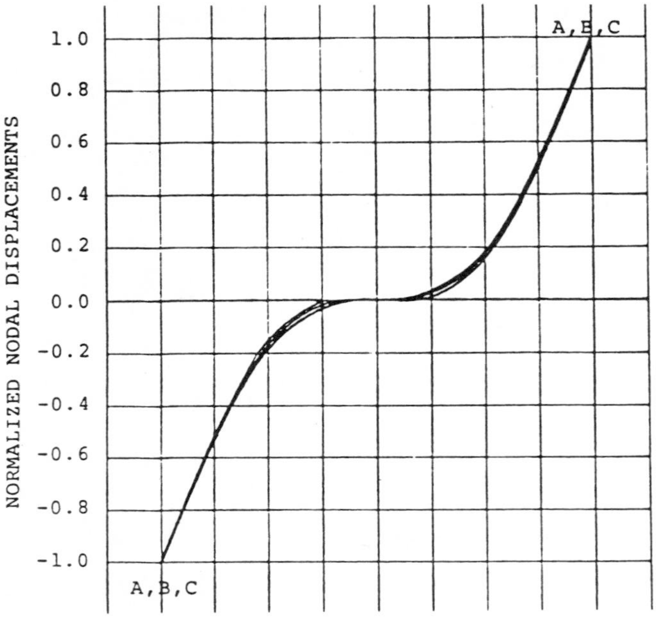

Fig. 9 FIFTH MODE

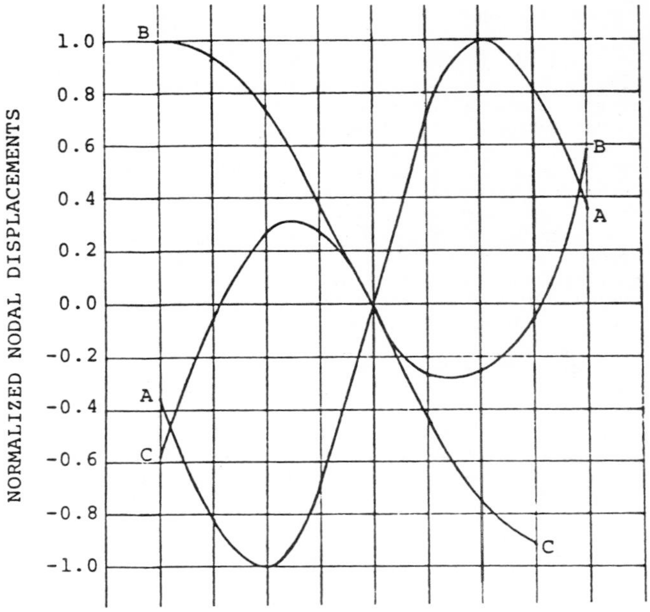

Fig. 11 SEVENTH MODE

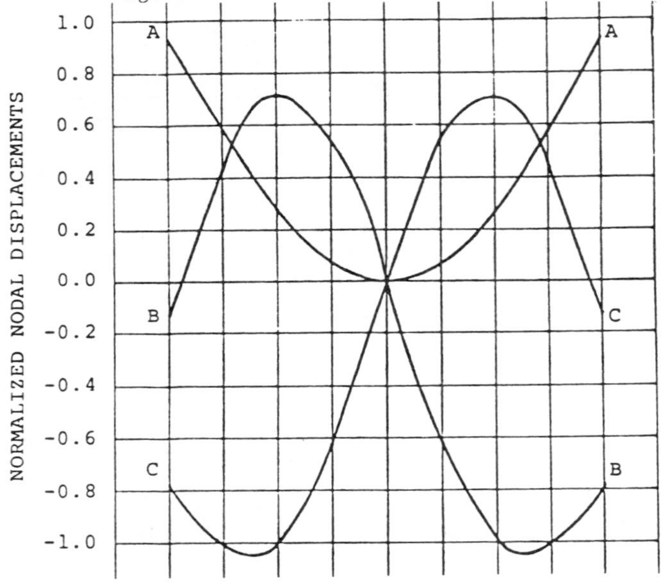

Fig. 10 SIXTH MODE

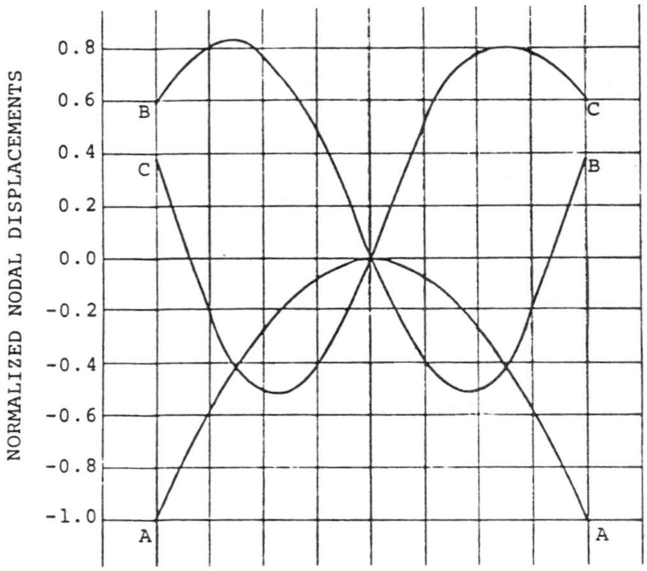

Fig. 12 EIGHTH MODE

Table 2. Stiffness Trade Study Assesssment

Mode	Baseline Frequency (Hz)	Trade Study Frequency* (Hz)	Frequency Ratio
1	20.4595	28.9342	1.4142
2	20.4888	28.9755	1.4142
3	33.9519	48.0152	1.4142
4	36.3123	51.3534	1.4142
5	47.1268	66.6474	1.4142
6	58.6836	82.9911	1.4142
7	58.8495	83.2257	1.4142
8	63.3508	89.5911	1.4142

*Trade study stiffness is double the baseline stiffness.

Table 3. Mass Trade Study Assessment

Mode	Baseline Frequency (Hz)	Trade Study Frequency* (Hz)	Frequency Ratio
1	20.4595	14.4671	0.7071
2	20.4858	14.4877	0.7071
3	33.9519	24.0076	0.7071
4	36.3123	25.6767	0.7071
5	47.1268	33.3237	0.7071
6	58.6836	41.4956	0.7071
7	58.8495	41.6128	0.7071
8	63.3508	44.7955	0.7071

*Trade study mass is double the baseline mass

Table 4. Nodal Degree of Freedom Assessment

Mode	Baseline Frequency (Hz)	Trade Study Frequency (Hz)	Frequency Ratio
1	20.4595	20.5513	1.0045
2	20.4888	20.6442	1.0076
3	33.9519	34.0206	1.0020
4	36.3123	36.4074	1.0026
5	47.1268	47.5130	1.0082
6	58.6836	59.0151	1.0056
7	58.8495	59.1378	1.0049
8	63.3508	63.7172	1.0058

NOTES:

1. Baseline configuration couples 3 degrees of freedom.
2. Trade study configuration couples 6 degrees of freedom.

CONCLUSIONS

The technique presented for the dynamic analysis of large space structures provides improved modelling capabilities compared with previous investigations. The finite element representation allows realistic simulation using appropriate elements and end boundary conditions.

The accuracy of the model was verified by testing the effect of altering the material properties and of varying the number of degrees of freedom allowed. Satisfactory agreement was obtained when the results were compared with those calculated for an equivalent uniformly loaded beam structure. An important dynamic characteristic, the appearance of multiple modes with similar frequencies, was observed. This can be attributed to the repetition of a single mode shape in two directions.

As a result of this study it is expected that the application of finite element techniques in the design of large space structures for use in space will prove both realistic enough to accurately predict important structural characteristics and will be simple enough to be manageable at an acceptable computer cost.

ACKNOWLEDGEMENTS

The work reported formed part of an investigation undertaken by the first author in the course of his M. S. studies supervised by the third author (see Ref.7).

Financial support and computing facilities were provided by the McDonnell Douglas Astronautics Company. The project was instigated by the second author, who also provided technical supervision.

REFERENCES

1. H. G. BUSH, M. M. MIKULAS and W. L. HEARD, Some Design Considerations for Large Space Structures, AIAA/ASME 18th Structures, Structural Dynamics and Materials Conference, San Diego, CA., May 1977.

2. HERCULES INCORPORATED, Advanced Composite Technical Data, Bulletin ACM7, Willmington, DE., January 1977.

3. K. L. REIFSNIDER, Fatigue in Composite Materials, AGARD Report No. 638, February 1976.

4. W. L. HEARD, H. G. BUSH and J. L. WALZ, Structural Sizing Considerations for Large Space Structures. Paper presented at the Large Space Systems Technology Conference, Hampton, VA., November 1980.

5. PDA ENGINEERING CORPORATION, Patran-G Users' Guide. Irvine, June 1980.

6. ENGINEERING/ANALYSIS CORPORATION, EASE2 Users' Manual, Lomita, CA., 1980.

7. C. A. MAZZOLA, Dynamic Response of a Large Space Reflector Support Structure, Unpublished M.S. Thesis, Structural Mechanics Report 83-2, University of California, Irvine, May 1983.

A LIMIT STATE DESIGN OF DOUBLE-LAYER GRIDS

G A R PARKE*, BSc, MSc, CEng, MICE, MIStructE

H B WALKER**, MUniv, CEng, MICE, MIMechE, MRAeS

* Lecturer, Space Structures Research Centre
University of Surrey

** Manager, Adivsory Service
Construction Steel Research and Development Organisation

The primary aim of this work is to determine if there are any likely economies associated with the limit state design of double layer grids. The redistribution of forces due to yield in lower chord tension members has been investigated and a pilot test on circular hollow sections undertaken.

A simplified analysis and design approach has been outlined and a test design presented. The study indicated that for square-on-square double-layer grids complying with the draft limit state British Standard 5950, and in which selective yielding of tension members is permitted, weight economies in the region of 20% are possible.

INTRODUCTION

To improve the economic viability of double-layer space grids it has become necessary to quantify and utilize in design any reserves of strength which are available in these generally highly indeterminant structures.

Two 'limit states' predominate in double-layer grid design when the structure is primarily subject to static loading. The first of these conditions is the serviceablity limit state which is associated with the functional suitability of the structure. The second limit condition, the ultimate limit state is concerned with the collapse of the structure which may be due to many factors including yielding, rupture or buckling of individual or groups of members.

For a large proportion of flexural structures the collapse load associated with the ultimate limit state can be determined by considering plastic collapse mechanisms. Provided the chosen collapse mechanism also yields a bending moment distribution which satisfies the lower bound theorem then a load factor against collapse can be obtained. Using this approach it is also possible to determine if the structure has attained a stable equilibrium condition and 'shaken down', or alternatively to assess if the structure is prone to incremental plastic collapse. This design procedure is only possible provided that member instability does not precede the formation of a mechanism and also that sufficient ductility is available in the individual members to permit the formation and rotation of plastic hinges.

Generally, it is not possible to apply the limit theorems of plastic analysis to determine the collapse load of space trusses. In these structures where axial forces predominate, compression member instability usually occurs before a limit moment of resistance is obtained in the structure.

LIMIT STATE DESIGN METHODS

To make use of the reserves of strength which exist in the majority of elastically-designed double-layer space grids, several design approaches are available for consideration. Structural member sizes can be arranged so that tensile chord yield or compressive chord collapse may occur in certain specified members, as the applied load approaches ultimate conditions, while the remaining memebers may behave elastically. Schmidt, (Ref 1), has shown from both analytical and experimental work that the ultimate collapse load for square-on-square and square-on-diagonal, double-layer space trusses is controlled by the post buckling characteristics of the compression members. For typical double-layer grids elastically designed this failure mode of compression member instability will generally finally occur irrespective of any tensile yielding which may have occurred in the tensile chord members. Consequently, to determine accurately the collapse load of either of these space truss types it is necessary to obtain the post-buckling characteristics of the actual compression members to be used in the structure and then to model the progressive buckling of these members along their respective post-buckling paths, until the complete structure has become unstable, (Ref 2 & 3). This is a difficult procedure to perfect because both strut stability and the associated post buckling characteristics are sensitive to a multitude of variables which include initial imperfections, material yield stress, loading rate, strain hardening, strain aging and also the previous stress history.

For structures of moderate size this analytical approach of following the post-buckling characteristics of failed compression members becomes a long procedure which, at present, tends to reduce its use as a design tool.

Additional experimental work by Schmidt and Hanaor, (Ref 4), on space trusses favouring both tensile yield and compression member instability, highlighted the importance of member-joint interaction on space truss behaviour. These investigations showed that there was a discrepancy between the behaviour of the members when they were tested individually and subsequently when they formed part of the structure. For tension members which are capable of yielding throughout their entire length, this discrepancy was not large enough to invalidate the use of tension chord yield as a means of improving the structural efficiency of double-layer space trusses. However, for compression member behaviour, Schmidt and Hanaor (Ref 4) found that for several members there was a large difference in the buckling load obtained from a member when tested individually and the buckling load measured in an identical member when it formed part of the structure. This discrepancy resulted from a reduction in joint stiffness, which occurred when one member at a joint buckled affecting the end conditions of the other compression members fixed to the joint. These differencies in compression member behaviour, controlled to a large extent by joint rigidities, further complicate the analytical modelling required to obtain an accurate assessment of the collapse load of the structure.

In order to reduce the computational requirements and to achieve a more favourable post yield deformation characteristic for the complete structure, the present work is concentrated on the design of space trusses so proportioned to allow tensile yield of chord members. Schmidt, Morgan and Clarkson, (Ref 5), have shown that as yield of the lower chord tension members proceed, compression chord forces increase at a much greater rate than the rate of increase in the imposed loading. As mentioned previously, this generally controls the ultimate collapse load of space trusses which is governed by both the initial buckling and the post buckling behaviour of the compression members. Consequently, it is not possible to consider the actual collapse of the structure as a 'limit state' if analytical modelling of compression member instability is to be avoided.

At present, compression member modelling does not provide a practical design method for, double-layer space frame grids. Consequently the approach of allowing yielding of selective tension members has been chosen with the aim of achieving a straight forward economical design method, which can be used in association with the current codes of practice.

PILOT TENSILE TESTS

To ascertain if the behaviour of structural hollow sections in tension is compatible with their individual material behaviour, a series of pilot tests was undertaken. Six circular hollow sections 76mm diameter and 1750mm long were tested in tension. Three of the members were grade 43C steel and the remaining three members grade 50C steel, both to BS4360, (Ref 6). Typical load V strain plots for the two tube types are shown in Fig 1. As expected, a good correlation exists between individual member and material behaviour. The grade 43C steel members have both a larger plastic plateau and strain hardening region than the grade 50C steel members.

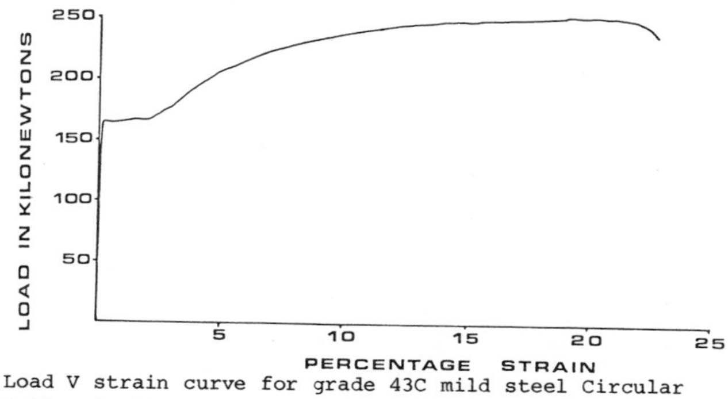

Load V strain curve for grade 43C mild steel Circular Hollow Sections

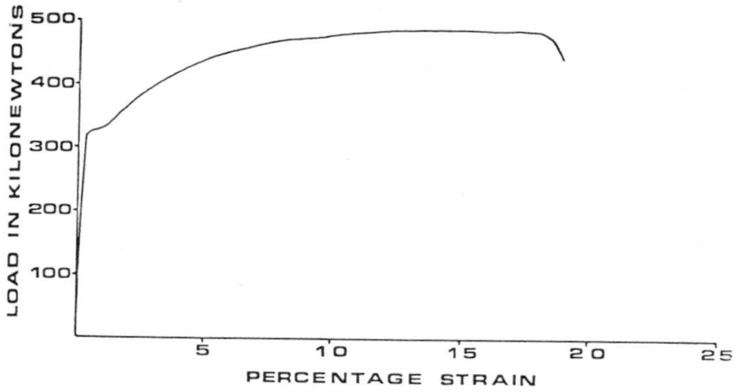

Fig 1: Load V strain curve for grade 50C high yield steel Circular Hollow Sections

Both tube types when tested yielded throughout their entire length, with the yield phenomenon for the mild steel tubes accompanied by the formation of Ludders' lines spiralling around the tube making an angle of about 45° with the axis of the tensile specimen. After a large amount of plastic flow had occured along these yield planes the member material strain hardened and the tubes finally necked and failed as shown in Fig 2.

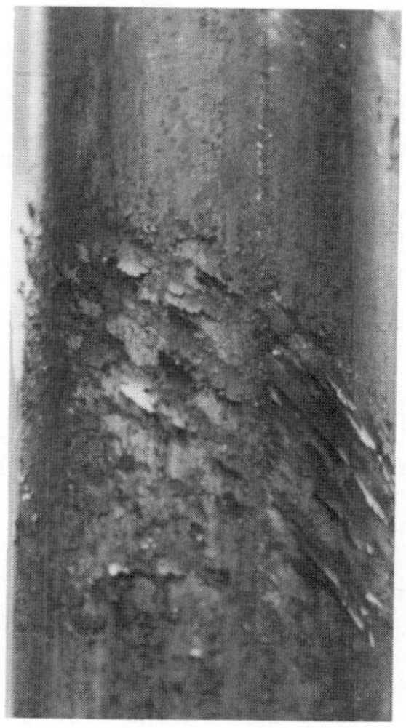

Fig 2: Yield of Circular Hollow Section

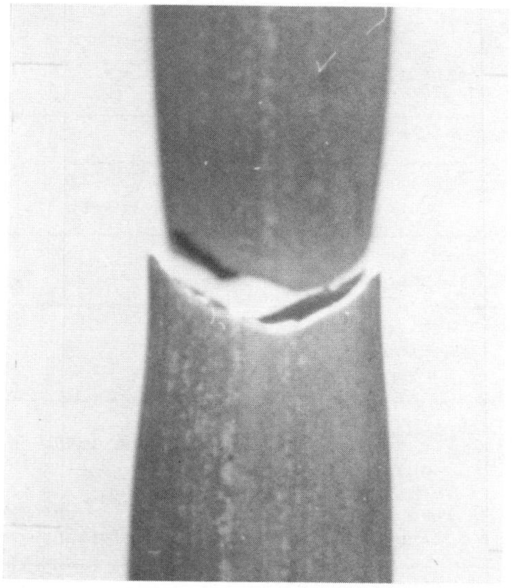

Necking and Failure of Circular Hollow Section

COMPARATIVE DOUBLE-LAYER GRID DESIGNS

In order to assess the amount of redistribution of forces which may occur in a square-on-square grid, due to tensile yield in a proportion of the lower chord members, a series of preliminary comparative designs have been undertaken. The square-on-square grid, shown in Fig 3, was first designed elastically to BS449, (Ref 7), to carry an imposed load of 1.5kN/m^2 with a load factor of 1.7 provided to prevent compression member instability and a load factor of 1.63 provided to prevent tension member yield. The grid members chosen for the preliminary designs were hot-rolled circular hollow sections in BS4360, grade 50C steel. The top chord, bottom chord and bracing members were treated as three independent groups of members with all members in one particular group having the same outside diameter but not necessarily the same wall thickness. The structure was considered to be pin-jointed with the effective length of compression members equal to their actual length. The joints were also assumed to be concentrically loaded.

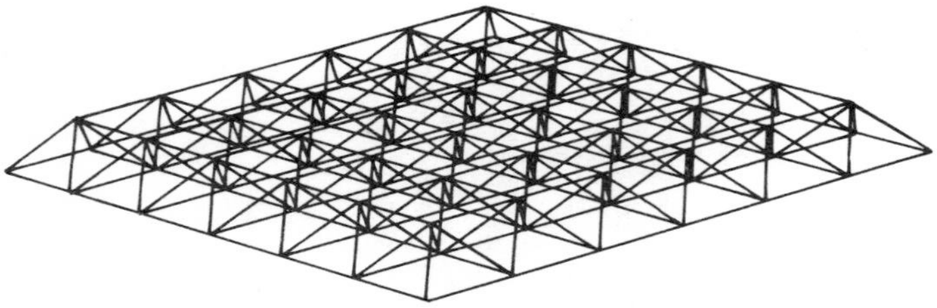

Fig 3: Square-on-Square Double-Layer Space Truss

The double-layer grid was then redesigned to be in accordance with the Draft British Standard BS5950, (Ref 8). This new limit state code for structural steelwork recommends a range of partial safety factors for different loading conditions shown in Table 1. For this second preliminary design, a dead load partial safety factor of 1.3 and an imposed load partial safety factor of 1.5 were used to assess the structural loading. This latter design, complying with the draft limit state code, resulted in a 10% reduction in weight when compared with the former load factor design.

Table 1: Partial Safety Factors

TYPE OF LOAD OR COMBINATION OF LOADS		γ_f
Dead Load:	Direct effect	1.3
	Countering overturning or uplift	0.9
Imposed Load:	In the absence of wind load	1.5
	Acting with wind load	1.2
Wind Load:	Acting with dead load only	1.3
	Acting with imposed load	1.2
	Acting with crane loads	1.2
Overhead travelling crane loads:		
	Vertical or horizontal crane load (considered separately)	1.5
	Vertical and horizontal crane load acting together	1.3
	Crane loads acting with wind load	1.2
Forces due to temperature effects:		1.2

In order to investigate further the possible potential of the grid due to plastic yield, the non-linear elastic plastic behaviour of the tension members in the structure was numerically modelled. This was achieved by incrementing the imposed load and as tensile members yielded they were removed from the system and replaced by nodal forces equivalent to the plastic load of the particular yielding member. The structure was then checked to ensure that stress reversals did not occur and that values of individual tension member strains were compatible with member characteristics. As tensile members yielded, the force in the most heavily stressed compression member increased at almost twice the rate of the increase in the imposed loading. When individual compression members approached instability the member area was increased but not the outside diameter of the particular tube. This increase in area was achieved by a change in wall thickness only, so that it would not be necessary to incorporate larger joints in the structure. This procedure was adopted and the load incremented until either the structure became a mechanism or the top chord compression member required an increase in external diameter.

For the particular grid considered, both of these prerequisites were reached simultaneously. Before instability was finally reached in the structure, almost 62% of the bottom chord tension members had yielded, see Fig 4, and the structure was capable of supporting a 70% increase in imposed loading with only a 14% increase in structural weight.

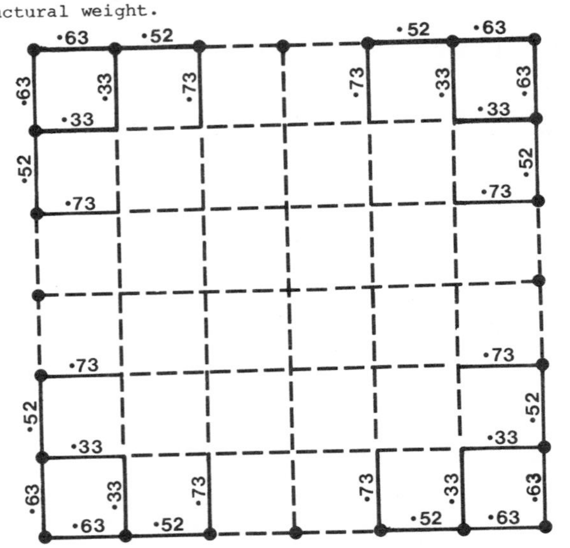

● Support Positions
– – – Yielded tension members.

Stress Ratio = $\frac{\text{Actual Stress}}{\text{Yield Stress}}$

Fig 4: Yield Pattern and Stress Ratio of Bottom Layer Members at Instability Limit State

In an attempt to both form and rationalise a simple design method, based on tensile chord yield, two design approaches were investigated. The first of these methods was to deliberately undersize the tension members after an elastic analysis and then proceed with the piece-wise method of incrementing imposed load along with the tension member replacement. For the second method the structure was elastically designed using the relevent partial safety factors and then a proposed pattern of yielded tension members was selected. These yielded members are removed from the structural system and replaced by the relevant nodal forces and then the modified structure is both re-analysed and re-designed. This subsequent design may yield different tension and compression member sizes which will then change the required nodal forces. The nodal forces are updated and the procedure repeated. For the particular grid investigated, the required equilibrium of the system was obtained after only three iterations. The second method proved to be the easiest of the two methods to implement and is outlined in Fig 5.

To ensure that deflections are not excessive at working load conditions it is also necessary to undertake a serviceability limit state design. Under working loads it may also be possible for a limited number of tension members to have yielded. This is acceptable, provided deflections are suitable, as there will be an adequate safety factor provided against compression member instability.

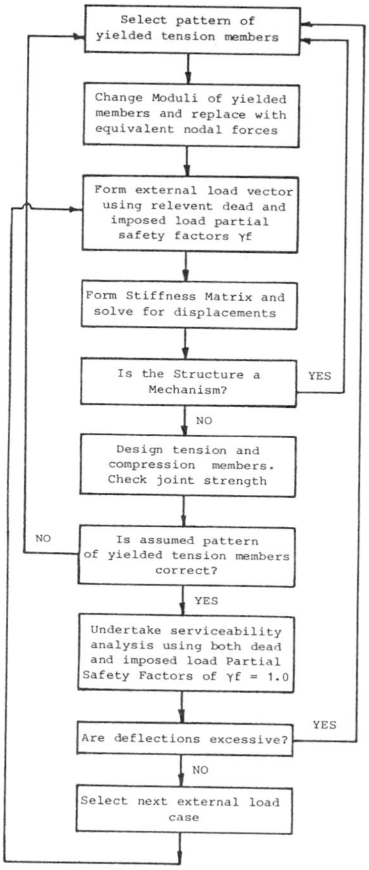

Fig 5: Analysis and Design Procedure

To illustrate the second design approach the double-layer grid, shown in Fig 3, was also preliminarilary designed with a partial safety factor of 1.3 and 1.5 for dead load and imposed load respectively. At the limit state of 'compression member instability', 60% of the lower chord tension members had yielded. By using this design method, summarized in Fig 5, a 10% reduction in weight was achieved in comparison with the structural weight obtained for the grid where redistribution of forces due to tensile chord yield was not permitted. At the serviceability limit state, where the partial safety factors for both dead and imposed load are reduced to 1.0, 14% of the lower chord tension members had yielded, as shown in Fig 6. For the particular grid investigated the final design was governed by the necessity to fulfil the serviceability requirements and prohibit excessive deflections under working load.

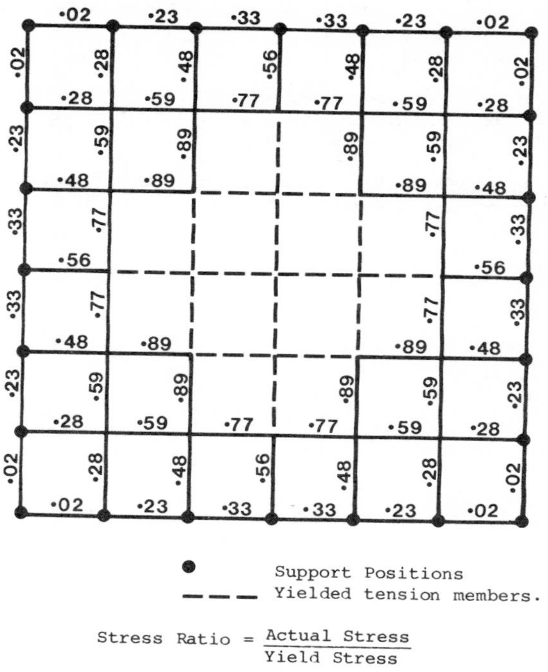

Fig 6: Yield Pattern and Stress Ratio of Bottom Layer Members at Serviceability Limit State

In order to determine if the effects of strain hardening should be included in the non linear analysis of the double layer grid, the maximum theoretical strain occuring in each yielded tension member was calculated. At the serviceability limit state, the maximum strain in the yielded tension members was in the region of 0.6%, while at the instability limit state the maximum strain had increased to 2.1%.

For mild steel circular hollow sections to BS4360, strain hardening occurs at approximately 2% strain while for higher strength steels this figure may reduce to 1%. Consequently, at working loads, strain hardening effects can usually be omitted but they should be considered at the instability limit state for the higher strength steels.

CONCLUSIONS

From the limited investigation undertaken at present, it does appear that certain economies are possible in the design of double-layer square-on-square grids favouring yield of lower chord tension members. In the example considered, a 10% reduction in weight was achieved by designing to the draft limit state British Standard 5950 (Ref 8) and an additional 10% saving in weight was obtained by allowing redistribution of forces due to tension member yield. Both the lower chord members and the lower chord joints could be reduced in size offering a worthwhile reduction in the overal cost of the structure.

REFERENCES

1. SCHMIDT, L.C. "Alternative Design Methods for Parallel-Chord Space Trusses", The Structural Engineer, August 1972, No 8, Volume 50.

2. SUPPLE, W.J. and COLLINS, I. "Limit State Analysis of Double-Layer Grids", Analysis, Design and Construction of Double-Layer Grids (Ed. Z.S. Makowski), Applied Science Publishers, 1978.

3. COLLINS, I.M. "Collapse Analysis of Double-Layer Grids", PhD Thesis, University of Surrey, 1981.

4. SCHMIDT, L.C. and HANOAR, A. "Force Limiting Devices in Space Trusses", Journal of the Structural Division, ASCE, May 1979.

5. SCHMIDT, L.C., MORGAN, P.R. and CLARKSON, J.A. "Space Truss with Brittle Type Strut Buckling", Journal of the Structural Division, ASCE, July 1976.

6. BRITISH STANDARD 4360. Weldable Structural Steels. London; British Standards Institution, 1979.

7. BRITISH STANDARD 449. The Use of Structural Steel in Building. London; British Standards Institution, 1969.

8. DRAFT STANDARD SPECIFICATION FOR THE STRUCTURAL USE OF STEELWORK IN BUILDINGS - BS5950 (B20 Draft). London; British Standards Institution, 1977.

THE LOAD-CARRYING CAPACITY OF SPACE TRUSS NODES COMPOSED OF FLAT STEEL PLATES

J A KARCZEWSKI[x], Dr.Sci., J CZERNECKI[xx], PhD, J A KÖNIG[xxx], Dr.Sci.

x Associate professor, Faculty of Civil Engineering, Warsaw Technical University, Poland.

xx Lecturer, Faculty of Civil Engineering, Warsaw Technical University, Poland.

xxx Professor in the Institute of Fundamental Technological Research, Warsaw, Poland.

Paper presents the results of experimental investigations of full-size space truss steel nodes aiming to determine their ultimate and/or shakedown loads. A numerical finite element method is, also, proposed to compute the ultimate loads. It is based on some simplifying assumptions confirmed by the experiments. The comparison of the theoretical and experimental ultimate loads shows a fair agreement. Some directions are given considering further investigations.

FORMULATION OF THE PROBLEM

The hitherto investigations on the load-carrying capacity of space truss nodes were mainly experimental ones, devoted to a particular structural system -see Ref.2, with no attempt for a generalization of the results. On the other hand, there exist some theoretical papers, -see Ref.1, trying to solve the title problem.

The present paper is a part of a long-term research programme devoted to space trusses investigations. After having worked out a set of digital computer programmes of the incremental elastic-plastic analysis of space trusses, -see Ref.3,4, it is natural to begin with the analysis of the truss nodes. In this case, also, the plastic approach may lead to considerable material savings as it is able to account for reserves neglected by the classical linear-elastic analysis.

The shapes of the space truss joints are,usually, rather complex ones. Therefore, any practically acceptable approach to the elastic-plastic analysis of them must be of approximate character and, thus, a certain experimental verification of the analysis is inevitable.

Keeping this in mind, a special experimental stand, allowing to test the full-size nodes, was assemblied in our laboratory a few years ago. Due to its versatility, nodes of various forms, loaded in various ways can be investigated.

We have begun with the nodes composed of flat sheets. Such nodes may be relatively easily analysed. Also, they are simpler to be investigated experimentally.

Moreover, it could be anticipated that the nodes of this type can be considered as a set of flat plates /without bending/, the interaction between them being possible to be simulated via appropriately defined boundary conditions of each one of the plates. Such an assumption enables to analyse each plate, separately, as an element in plane stress, by means e.g. of the finite element method or - if the ultimate state is considered - by means of the slip lines method.

As it will be seen further from the experimental data, the validity of such an approach has been confirmed in the tests for the whole range of the nodes response to various load combinations.

THE ANALYSIS OF A NODE

As mentioned above, every element of the node is considered separately and analysed in the plane stress state. The elastic-plastic incremental analysis, in our particular case, has been based on the following assumptions:

1. The node material is elastic, perfectly plastic obeying the Tresca yield condition with the associated flow rule. Thus the total strain is sum of the elastic strain and of the plastic strain. The former is to be calculated through the Hooke's law.
The non-zero increments of the plastic strain occur:

for $\sigma_x \sigma_y - \tau_{xy}^2 < 0$ if

$$(\sigma_x - \sigma_y)^2 + 4\tau_{xy}^2 = R_e^2 \quad \ldots\ldots 1$$

and they are given by the following formulae

$$\dot{\varepsilon}_x^P = 2\lambda(\sigma_x - \sigma_y), \quad \dot{\varepsilon}_y^P = 2\lambda(\sigma_y - \sigma_x),$$

$$\dot{\gamma}_{xy}^P = 16\lambda\tau_{xy}; \quad \ldots\ldots 2$$

for $\sigma_x \sigma_y - \tau_{xy}^2 > 0$ if

$$(\sigma_x - \sigma_y)^2 + 4\tau_{xy}^2 = [2R_e + |\sigma_x + \sigma_y|]^2 \quad \ldots\ldots 3$$

and they are given by

$$\dot{\varepsilon}_x^P = 4\lambda(\pm R_e + \sigma_x), \quad \dot{\varepsilon}_y^P = 4\lambda(\pm R_e + \sigma_y),$$

$$\dot{\gamma}_{xy}^P = 16\lambda\tau_{xy}. \quad \ldots\ldots 4$$

In both the cases $\lambda \geqslant 0$, R_e denoting the yield-point stress in pure tension.

2. The strains and displacements are small. The plate thickness change in the course of plastic deformations is neglected.

3. The loads are quasi-static and act within the plane of the plate.

4. No imperfections of the plate shape or of the loading are accounted for.

5. The boundary conditions of each plate analysed follow from some symmetry requirements. The influence of the stiffness of other plates is accounted for by increasing the thickness of the plate considered for 15 thicknesses of the respective, perpendicular plate, Fig.1 .

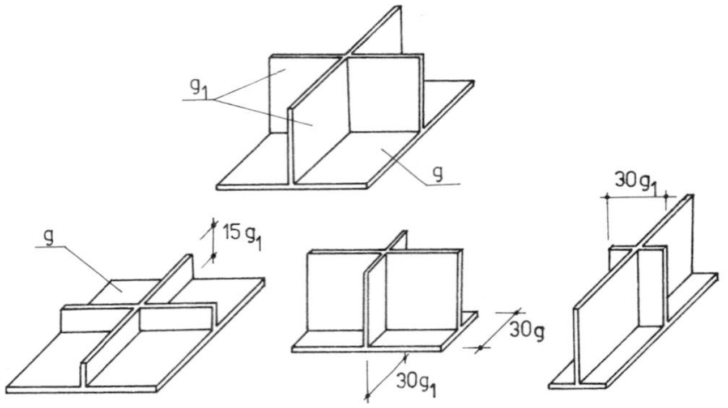

Fig 1

A finite element elastic-plastic algorithm using the constant stiffness matrix iterative procedure has been employed. Elements with internal nodes have been used and the element mesh has been condensed in the neighbourhood of expected stress concentrations. The ultimate load of a plate may be determined precisely due to the ill-conditioning of the equation system in the vicinity of that load. Therefore, the slip-lines method, based on the kinematical limit analysis theorem has been employed, additionally.

The magnitudes of the ultimate load of an individual plate, calculated in such a way, seem

Fig 2

to be in a fair agreement with the experimental data.

The load carrying capacity of the whole node, for given ratios between the loads of individual plates, is determined by the weakest one of them.

THE EXPERIMENTAL INVESTIGATIONS

The experimental arrangement we have at disposal, Fig 2, enables to load a node with up to 12 loads simulating the forces coming from the bars intersecting at the node. The cross-bars can be inclined under an arbitrary angle. The loads are executed by means of hydraulic jacks which can be regulated independently of each other or coupled into pairs. The maximum compressive load of a single jack is 320 kN, the maximum load in tension - 240 kN.

The main investigations have been carried out on specimens as shown in Fig 3, made of mild steel St 3 SX. There were four series of different load modes with 5 specimens in each one of them.

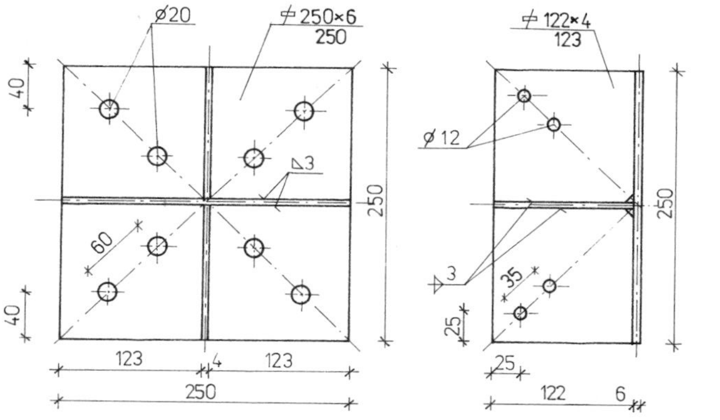

Fig 3

The strain state in each one of the flat plates has been recorded by means of the photoelastic coating made of epidian 2. To evaluate the principal directions, a special technique has been used, consisting of introducing a local self--equilibrated stress state in the vicinity of the point of interest. Additionally, strains have been recorded by means of electric resistance strain gauges. Some of them were placed so as to detect a possible bending of plates. The deformations in the neighbourhood of the screws, connecting the node with a strut, have been measured by means of dial gauges. Placement of all the measuring devices is shown in Fig 4.

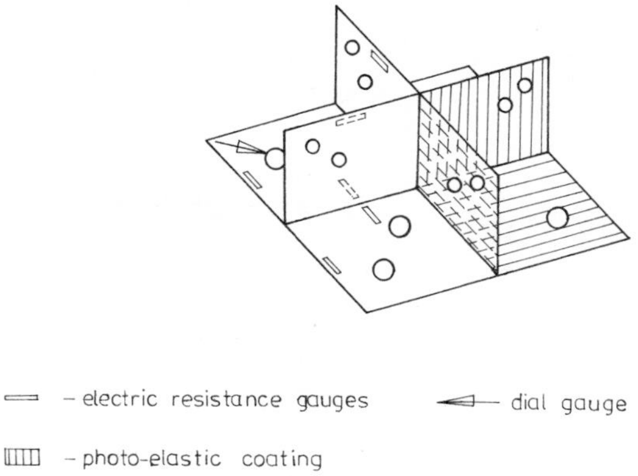

Fig 4

The load modes, Fig 5, have been selected so as to represent the most stringent situations of nodes of space truss roofings. The loads grew up slowly, in a quasi-static manner with decreasing

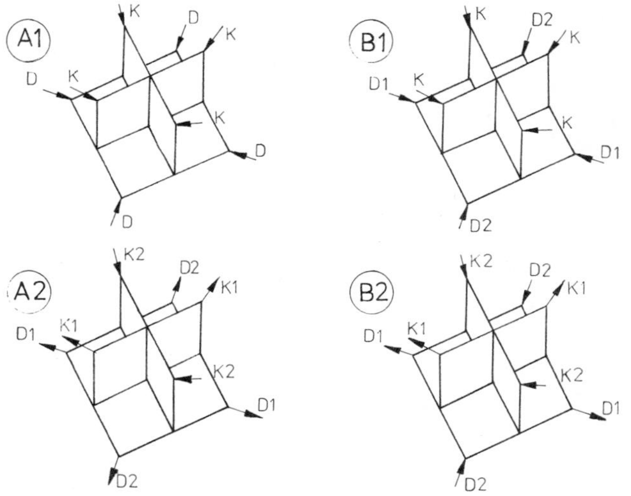

Fig 5

increments when approaching the expected ultimate load. On each level of the loading, the pattern of isochroms has been photographed and the data of electric strain gauges and dial gauges have been recorded. The development of yielded zones can be detected by observing the changes in the pattern of isochroms, Fig 6. An example of the deformations of the screw hole edge is given in Fig 7 whereas the difference of strains on the upper and lower faces of a horizontal plate is given in Fig 8. The latter clearly indicates that the bending effects have been negligible.

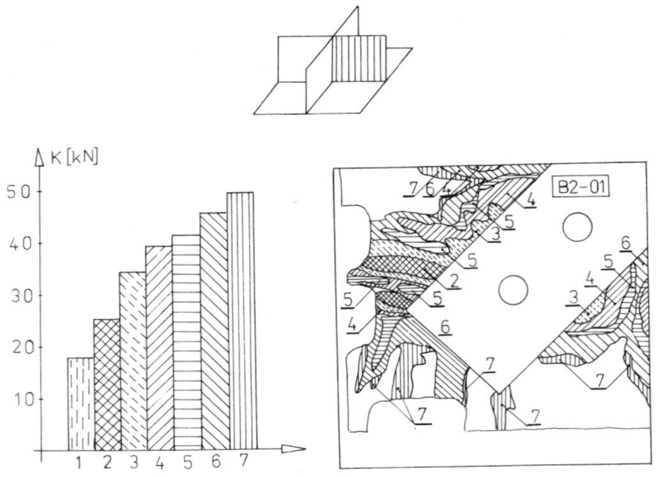

Fig 6

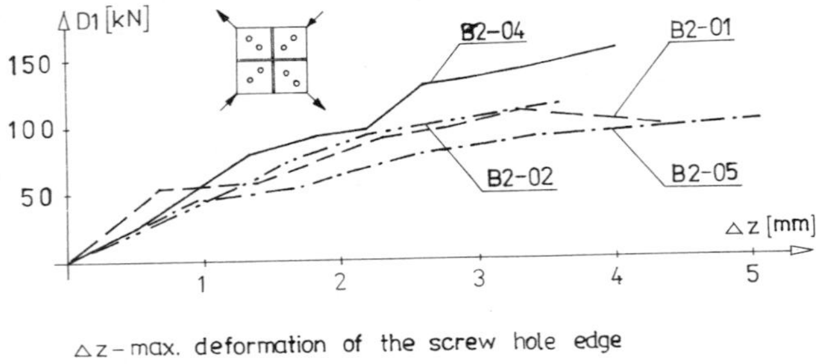

Fig 7

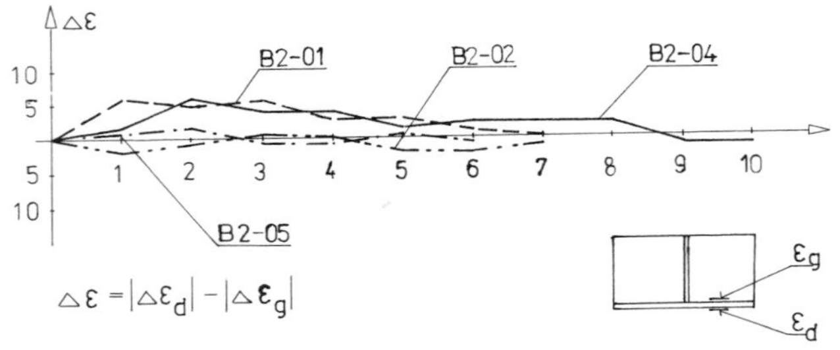

Fig 8

The ultimate load of a node has been attained at the instant when the strain instability could be seem. These loads, for the last two test series, are shown in Table 1. The theoretical magnitudes of these loads calculated as described in Section 2, are given there, also. Fig 9 shows an example of the finite element net employed.

Number of node	Limite load 1) analitical [kN]	experimental [kN]	Index of compatibility effects
A2-01	85,5	91,1	1,06
A2-02	85,5	96,2	1,12
A2-03	85,5	-•-	-•-
A2-04	85,5	102,8	1,20
A2-05	85,5	-•-	-•-
B2-01	85,5	94,3	1,10
B2-02	85,5	95,0	1,11
B2-03	85,5	-•-	-•-
B2-04	85,5	92,2	1,08
B2-05	85,5	103,1	1,21

1). – for force K2

-•- – damaged photo-elastic coating

Table 1

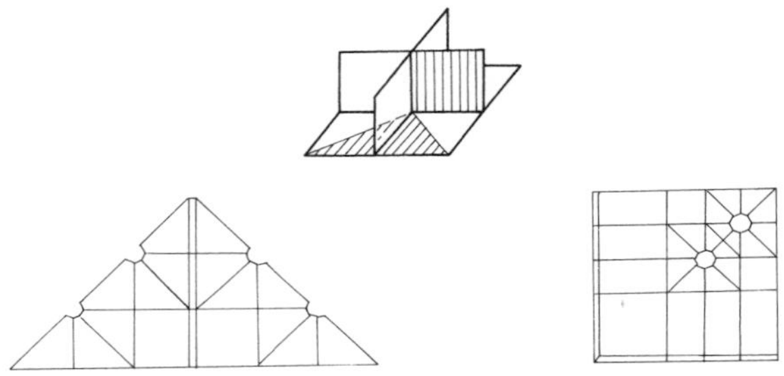

Fig 9

INVESTIGATIONS ON SHAKEDOWN OF NODES

The actual loads of real civil engineering structures are not constant nor grow up proportionally to a single parameter. Therefore, if a given structure has been designed taking the plastic strains into account, it must be safe against accumulation of permanent strains as well as against the alternating plastic strains what could result in low-cycle fatigue. There must be a guarantee that the plastic strain increments will cease, eventually, and the structure will respond in a purely elastic way. Such a state is called shakedown and codes of practice in several countries require to check if a given structure, designed by means of plastic methods, will shake down. All the said above applies to space trusses also.

Basing on the results of the monotonic loading tests, confirming the validity of the assumptions listed in Section 2, some experimental investigations have been performed on individual plates of the node considered in conditions of cyclic loading. These tests were aimed to detect the ability of the nodes to shake down to loads exceeding the elasticity limits. The load programmes are shown in Fig 10 and were applied to the horizontal plate of the node. The shakedown investigations consist of comparing two isochromatic patterns of the photoelastic coating after two subsequent loading cycles. Their identity indicates that the state of shakedown has been attained. Table 2 shows the results of the first series performed. The shakedown loads are compared with the ultimate load magnitudes, obtained in the proportional load tests. Fig 11 shows photos of

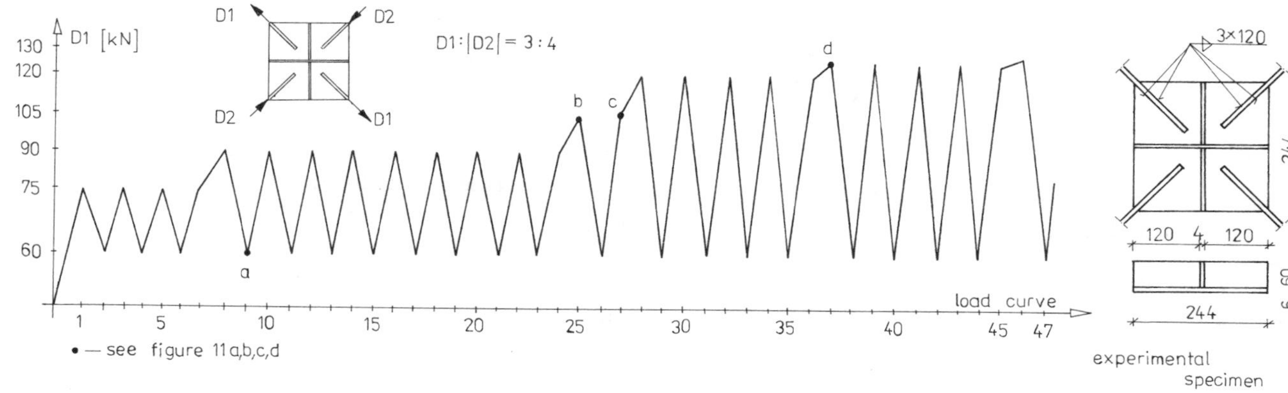

Fig 10

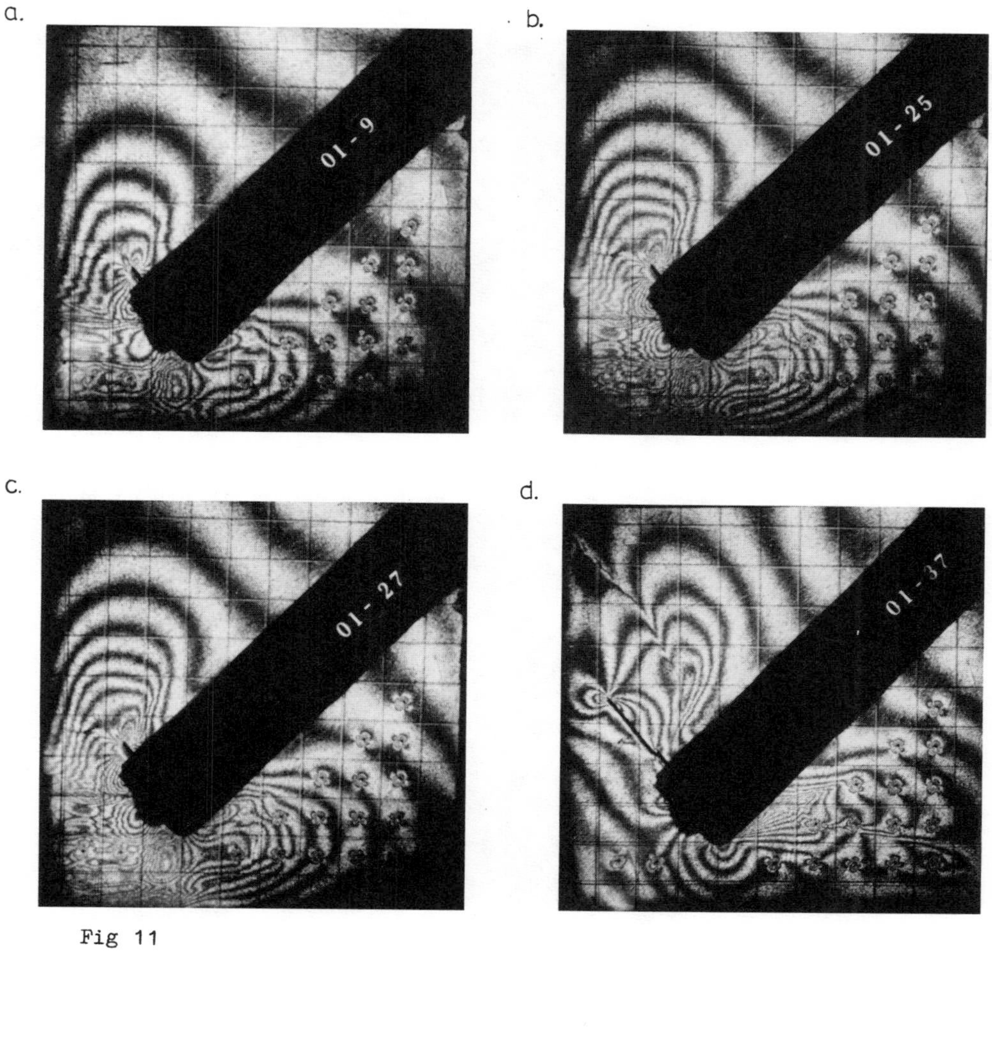

Fig 11

Number of node	Limite load		Index of compatibility effects
	analitical [kN]	experimental [kN]	
W – 01	115,7	127	1,10
W – 02	115,7	135	1,17
W – 03	115,7	120	1,04
W – 04	115 7	117	1,01
W – 05	115,7	120	1,04

Table 2

some isochromatic patterns. They correspond to the loads denoted in Fig 10.

Obviously, further shakedown investigations are needed, in particular - made on complete nodes.

CONCLUDING REMARKS

1. The experimental data confirm the validity of the simplifying assumptions made in the theoretical analysis. No important local bending has been detected and the ultimate load of an individual plate has not depended on the stress state of other plates of the node.

2. The theoretical magnitudes of the load-carrying capacity of the nodes seem to be in a good agreement with the experimental ones. The latters are for six up to twenty one percent higher what indicates that the analytical method proposed is safe. The additional safety margin is connected, probably, with the use of the Tresca yield condition and with the strainhardening of the steel.

3. The method of photoelastic coating seems to give reliable results. It is easy to be used. Moreover, without any essential change, it may be employed in the shakedown investigations of the space truss nodes.

4. Further investigations should tend to improve the measurment technique. In particular, to use the method of photoelastic coating on curvilinear surfaces as to investigate space truss nodes of other shapes. Another important problem - the interaction between the buckling of the strut and the node failure. Some investigations on this problem have begun.

REFERENCES

1. W SZCZEPAŃSKI, Statically admissible stress fields for the nodes of steel structures /in Polish/. IPPT Reports No 44/1977, Warszawa, 1977.

2. H B WALKER, The Nodus Space Frame System, Proc. 2nd Int. Conf. on Space Structures, 447-458.

3. J A KARCZEWSKI, Analysis of the elastic-plastic bar structures subjected to variable loads /in Polish/, Zesz.Nauk. Pol.Warsz.Bud.46 Warszawa 1976.

4. J A KARCZEWSKI, The Limit Load of Space Trusses. Arch. Inż. Ląd. 26, 247-258, 1980.

5. J A KARCZEWSKI, J A KÖNIG, On some basic problems of the theory of shakedown. Eng. Trans., 25 /1977/, 239-246.

INFLUENCE OF ELASTICALLY END-RESTRAINED BARS ON THE BEHAVIOUR OF CURVED LATTICED STRUCTURES

I H I TOADER[x], PhD, CEng and V GIONCU[xx], PhD, CEng

[x] Senior Lecturer, Polytechnic Institute of Cluj-Napoca, Romania

[xx] Associate Professor, "Traian Vuia" Polytechnic Institute of Timişoara, Romania

General stability functions, that include the effect of semi-rigid joints and an equivalent stiffness matrix, $[K_E]$, for member elastically restrained at the nodes is deduced. In a second order analysis the influence of the joint system on the behaviour of a single layered latticed structure - the triangulated barrel vault is studied.

INTRODUCTION

One of the most difficult problems in the design of latticed shells is the effect of partial restraint of the members, which depends on the solution adopted for joint connections. Thus, site welding provides perfect restraint, while some of the bolted joints behave like hinged connections. As a rule, the actual restraint degree for a given system has to be determined by laboratory tests. A rigorous analysis should take into account the influence of this restraint. Unfortunately, this implies serious difficulties (complex and time - wasting programs and consequently high costs). This is why the use of simplified methods can be preferable in many cases. The design of single-layer latticed shells requires the following verifications:

a) The design loads (i.e.the service loads multiplied by the load factors) should be less than the loads corresponding to either individual buckling of the members or general instability of the structure.

b) Deflections under service loads should not exceed the specified values.

The difficulty of considering the effect of partial restraint is the reason why most of the special-purpose programs presented in technical literature are based on the hinged connections assumption. In order to avoid coupling of the two above mentioned forms of instability, the critical load corresponding to general instability should be much greater than the one pertaining to individual buckling of the members, see Ref 4. Therefore, it is suggested that the critical load for general instability of a rationally designed structure be 4...5 times the service load, see Fig 1.

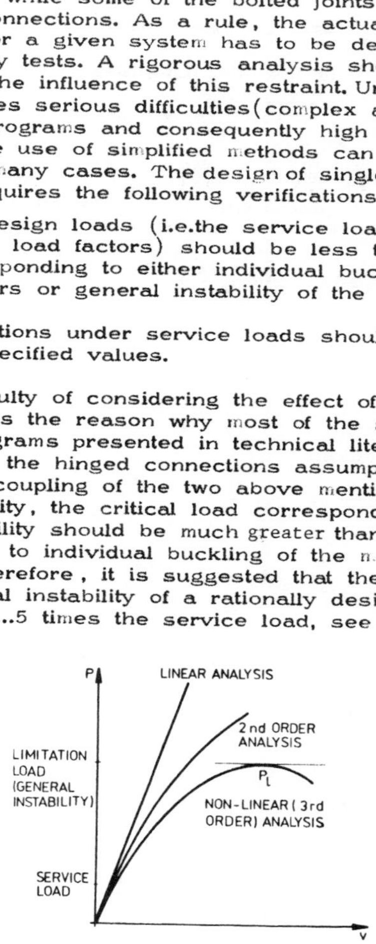

Fig 1

In such circumstances, the geometric nonlinearity of the structure under service loads is not so strong and a simplified variant of nonlinear analysis, namely the second order theory, can be used. This simplified approach allows the consideration of partial restraint effect.

EQUIVALENT STIFFNESS MATRIX

Figure 2 shows the bar of a space structure, partially restrained at the nodes, before and after deformation.

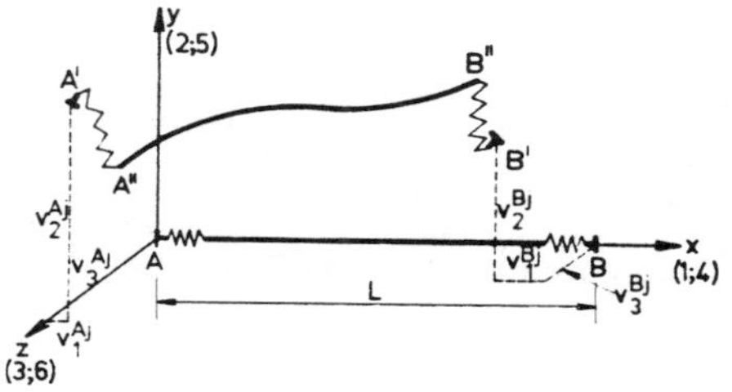

Fig 2

The following assumptions, consistent with the structural theory and with the geometry of latticed structures, will hold throughout the investigation of the straight member elastically restrained at the ends:

a) The structural material is linearly elastic.

b) The member has a double symmetric and constant cross section.

c) The structure and its component members are not affected by any imperfections.

d) Loads are acting at the joints.

e) The displacements of the member are small.

f) The effect of shear deflections is disregarded.

g) The torsional buckling and local instability do not occur.

The constants k_i of the spring stiffness $i(i=1,2...6)$ at joint which vary from 0 to ∞, may be replaced through the "fixity factors" η_i; which vary from 0 to 1, Fig 3. This replacing is advantageous for current available computer oriented methods, Refs 1,7 and 8, which lies in the basic relationship required for either stiffness or flexibility approach :

$$\eta_1 = \frac{k_1}{k_1 + \frac{AE}{L}} \qquad \eta_2 = \frac{k_2}{k_2 + \frac{12EI_2}{L^3}} \qquad \eta_3 = \frac{k_3}{k_3 + \frac{12EI_3}{L^3}}$$

$$\eta_4 = \frac{k_4}{k_4 + \frac{GJ}{L}} \qquad \eta_5 = \frac{k_5}{k_5 + \frac{4EI_2}{L}} \qquad \eta_6 = \frac{k_6}{k_6 + \frac{4EI_3}{L}} \qquad \text{......1,}$$

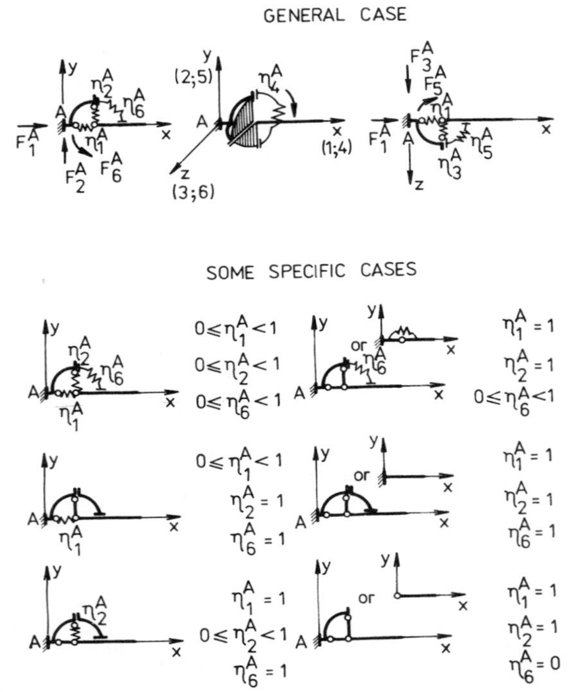

Fig 3

In Eqns1, AE is the axial stiffness, GJ is the torsional stiffness, EI_2 and EI_3 are flexural stiffness about y and z axis respectively and L is the bar length. The "fixity factor" η_i^A of the elastic connection i between member end Am and joint Aj is defined through the ratio between displacement V_i^{Am} of the bar end Am in the i direction and displacement V_i^{Aj} of the joint Aj in the same direction, provided that all the other displacements are zero, Refs 1,5,7 and 8.

The coordinate system Axyz of the bar AB, with x axis along the member length before its deformation, are shown in Figs 3 and 4.

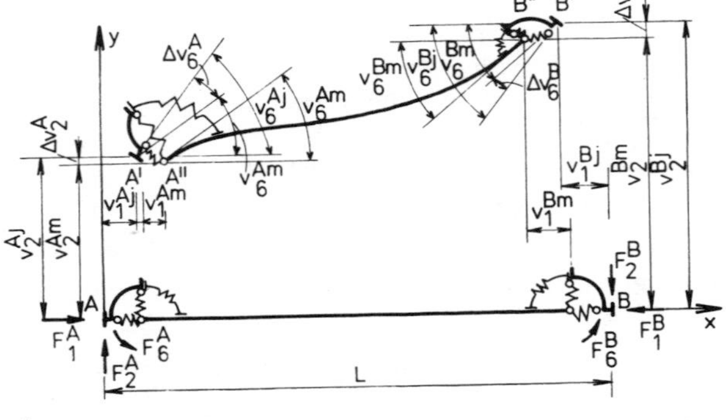

Fig 4.

The linear displacements in the plane Axy, are V_1^{Aj}, V_1^{Am}, V_2^{Aj}, V_2^{Am} and V_1^{Bj}, V_1^{Bm}, V_2^{Bj}, V_2^{Bm} and the angular ones are V_6^{Aj}, V_6^{Am} and V_6^{Bj}, V_6^{Bm}, Fig 4. The indices "j" and "m" are referred to joints and ends of the member, respectively.

The member end forces, are bending moments F_6^A and F_6^B, transverse shear forces $F_2^A = -F_2^B$ and the axial force $F_1^A = -F_1^B = N$.

In the second order theory the transverse shear forces and the rotations of the member ends are :

$$F_2^A = -F_2^B = \frac{1}{L}(F_6^A + F_6^B) + \frac{N}{L}(V_2^{Am} - V_2^{Bm}) \qquad \text{......2,}$$

$$V_6^{Am} = V_6^{AA}F_6^A + V_6^{AB}F_6^B + \frac{1}{L}(V_2^{Bm} - V_2^{Am}) \qquad \text{...... 3,}$$

$$V_6^{Bm} = V_6^{BA}F_6^A + V_6^{BB}F_6^B + \frac{1}{L}(V_2^{Bm} - V_2^{Am}) \qquad \text{......4,}$$

where V_6^{AA}, V_6^{BB}, V_6^{AB} and V_6^{BA} are flexibility coefficients of the member AB:

$$V_6^{AA} = V_6^{BB} = \frac{L}{3EI_3}\,\alpha(\nu) \qquad \text{......5,}$$

$$V_6^{AB} = V_6^{BA} = -\frac{L}{6EI_3}\,\beta(\nu) \qquad \text{......6,}$$

and $\alpha(\nu)$ and $\beta(\nu)$ are the well known functions depending on the axial force, Ref 8. For example, for axial compression force N, $\alpha(\nu)$ and $\beta(\nu)$ are :

$$\alpha(\nu) = \frac{3}{\nu}\left(\frac{1}{\nu} - \frac{1}{\tan\nu}\right) \qquad \text{......7,}$$

$$\beta(\nu) = \frac{6}{\nu}\left(\frac{1}{\sin\nu} - \frac{1}{\nu}\right) \qquad \text{......8.}$$

The variable ν is the dimensionless parameter :

$$\nu = L\sqrt{\frac{|N|}{EI_3}} \qquad \text{......9.}$$

The differences ΔV_i (for i=1,2...6), between joint and member end displacements are mathematically expressed by the relationships :

$$\Delta V_i^A = V_i^{Aj} - V_i^{Am} = \frac{F_i^A}{k_i^A} \qquad \text{......10,}$$

$$\Delta V_i^B = V_i^{Bj} - V_i^{Bm} = \frac{F_i^B}{k_i^B} \qquad \text{...... 11.}$$

with the notations :

$$d_2 = \frac{1}{\eta_2^A} + \frac{1}{\eta_2^B} - 1 \qquad \text{......12,}$$

$$e_2 = 1 - \frac{\nu^3}{12}(d_2 - 1) \qquad \text{......13,}$$

for $N \leqslant 0$, or

$$e_2 = 1 + \frac{\nu^3}{12}(d_2 - 1) \qquad \text{...... 14,}$$

for $N \geqslant 0$ and

$$a_2 = \alpha(\nu)\,e_2 \qquad \text{......15,}$$

$$b_2 = \beta(\nu)\,e_2 \qquad \text{......16,}$$

Equations 3 and 4 become :

$$\frac{6EI_3}{L}\,e_2V_6^{Aj} = \frac{1}{2}R_1F_6^A - \frac{1}{2}R_2F_6^B + \frac{6EI_3}{L^2}(V_2^{Bj} - V_2^{Aj}) \qquad \text{...... 17,}$$

$$\frac{6EI_3}{L}\,e_2V_6^{Bj} = -\frac{1}{2}R_2F_6^A + \frac{1}{2}R_1F_6^B + \frac{6EI_3}{L^2}(V_2^{Bj} - V_2^{Aj}) \qquad \text{......18,}$$

where R_1, R_2 and f_2 are

$$R_1 = d_2 + \frac{3e_2}{\eta_6^A} + f_2 \qquad \text{......19,}$$

$$R_2 = 2b_2 - d_2 + 1 \qquad \text{......20,}$$

$$f_2 = 4a_2 - 3e_2 - 1 \qquad \text{......21.}$$

It may be computed from Eqns 17 and 18 moments F_6^A and F_6^B :

$$F_6^A = \frac{4EI_3}{L} C_{1z}^A V_6^{Aj} + \frac{2EI_3}{L} C_{2z} V_6^{Bj} + \frac{6EI_3}{L^2} C_{3z}^A V_2^{Aj} - \frac{6EI_3}{L^2} C_{3z}^A V_2^{Bj} \qquad \text{......22,}$$

$$F_6^B = \frac{2EI_3}{L} C_{2z} V_6^{Aj} + \frac{4EI_3}{L} C_{1z}^B V_6^{Bj} + \frac{6EI_3}{L^2} C_{3z}^B V_2^{Aj} - \frac{6EI_3}{L^2} C_{3z}^B V_2^{Bj} \qquad \text{......23.}$$

In Eqns 22 and 23, C_{1z}^A, C_{1z}^B, C_{2z}, C_{3z}^A and C_{3z}^B are generalized stability functions :

$$C_{1z}^A = \frac{3e_2}{C_z} \left[(d_2+f_2)\, \eta_6^B + 3e_2 \right] \eta_6^A \qquad \text{......24,}$$

$$C_{1z}^B = \frac{3e_2}{C_z} \left[(d_2+f_2)\, \eta_6^A + 3e_2 \right] \eta_6^B \qquad \text{......25,}$$

$$C_{2z} = \frac{6e_2}{C_z} (2b_2 - d_2 + 1)\, \eta_6^A \eta_6^B \qquad \text{......26,}$$

$$C_{3z}^A = \frac{2}{C_z} \left[(2b_2 + f_2 + 1)\, \eta_6^B + 3e_2 \right] \eta_6^A \qquad \text{......27,}$$

$$C_{3z}^B = \frac{2}{C_z} \left[(2b_2 + f_2 + 1)\, \eta_6^A + 3e_2 \right] \eta_6^B \qquad \text{......28,}$$

where

$$C_z = \left\{ \left[f_2^2 - (2b_2+1)^2 \right] \eta_6^A \eta_6^B + d_2 \left[2(2b_2+f_2+1) \eta_6^A \eta_6^B + 3e_2 (\eta_6^A + \eta_6^B) \right] + 3e_2 f_2 (\eta_6^A + \eta_6^B) + 9e_2^2 \right\} \qquad \text{......29.}$$

Equation 2 becomes :

$$F_2^A = -F_2^B = \frac{6EI_3}{L^2} C_{3z}^A V_6^{Aj} + \frac{6EI_3}{L^2} C_{3z}^B V_6^{Bj} + \frac{12EI_3}{L^3} C_{4z} V_2^{Aj} - 12 \frac{EI_3}{L^3} C_{4z} V_2^{Bj} \qquad \text{......30,}$$

where

$$C_{4z} = \frac{1}{e_2} \left(\frac{C_{3z}^A + C_{3z}^B}{2} - \frac{\nu^2}{12} \right) \qquad \text{......31,}$$

is a generalized stability function too, Ref 2.

For bending moments F_5^A, F_5^B and transverse shear forces F_3^A and F_3^B, expressions similar to Eqns 22,23 and 30 may be obtained. The forces F_1^A, F_1^B, F_4^A and F_4^B only depend on correction factors α_{11}, and α_{44}

$$F_1^A = -F_1^B = \frac{EA}{L} \alpha_{11} (V_1^{Aj} - V_1^{Bj}) \qquad \text{......32,}$$

$$F_4^A = -F_4^B = \frac{GJ}{L} \alpha_{44} (V_4^{Aj} - V_4^{Bj}) \qquad \text{......33,}$$

where

$$\alpha_{11} = \frac{1}{d_1} \qquad \alpha_{44} = \frac{1}{d_2} \qquad \text{......34,}$$

and

$$d_1 = \frac{1}{\eta_1^A} + \frac{1}{\eta_1^B} - 1$$

$$d_2 = \frac{1}{\eta_4^A} + \frac{1}{\eta_4^B} - 1 \qquad \text{......35.}$$

The force - displacement relationships for the members elastically connected at the nodes, which express the twelve force components, $[F]$, acting on the member in terms of the twelve possible joint displacement components, $[V]$, can be written in the form of a matrix equation :

$$[K_E]\,[V] = [F] \qquad \text{......36,}$$

where the equivalent stiffness matrix, $[K_E]$, which takes into account the influence of the elastic connection on the behaviour of the bar, is displayed in Fig 5.

The variation of the general stability functions C_{1z}^A, C_{1z}^B, C_{2z}, C_{3z}^A, C_{3z}^B and C_{4z}, for some "fixity factors" of the elastic connection i=6 are shown in Figs 6, 7, 8 and 9.

For a rigid jointed structure the well known stability functions are obtained, Ref 2. For example for $N \leqslant 0$:

$$C_1^A = C_1^B = C_1 = \frac{1}{4} \frac{(\sin\nu - \nu\cos\nu)}{2(1-\cos\nu) - \nu\sin\nu} \qquad \text{......37,}$$

$$C_2 = \frac{1}{2} \frac{\nu(\nu - \sin\nu)}{2(1-\cos\nu) - \nu\sin\nu} \qquad \text{......38,}$$

$$C_3^A = C_3^B = C_3 = \frac{1}{6} \frac{\nu^2(\nu - \sin\nu)}{2(1-\cos\nu) - \nu\sin\nu} \qquad \text{......39,}$$

$$C_4 = C_3 - \frac{\nu^2}{12} = \frac{1}{12} \frac{\nu^3}{2(1-\cos\nu) - \nu\sin\nu} \qquad \text{......40.}$$

It should be noted that in the N = 0 case the general stability functions are reduced to correction factors for the linear analysis, Refs 1, 8 and 9:

$$C_{1z}^A = \frac{\eta_6^A}{C_z} (d_2 \eta_6^B + 3) \qquad \text{......41,}$$

$$C_{1z}^B = \frac{\eta_6^B}{C_z} (d_2 \eta_6^A + 3) \qquad \text{......42,}$$

$$C_{2z} = \frac{2\eta_6^A \eta_6^B}{C_z} (3 - d_2) \qquad \text{......43,}$$

$$C_{3z}^A = \frac{2\eta_6^A}{C_z} (1 + \eta_6^B) \qquad \text{......44,}$$

$$C_{3z}^B = \frac{2\eta_6^B}{C_z} (1 + \eta_6^A) \qquad \text{......45,}$$

$$C_{4z} = \frac{1}{C_z} (\eta_6^A + \eta_6^B + 2\eta_6^A \eta_6^B) \qquad \text{......46,}$$

where

$$C_z = d_2 (\eta_6^A + \eta_6^B + 2\eta_6^A \eta_6^B) - 3(\eta_6^A \eta_6^B - 1) \qquad \text{......47.}$$

$$
\begin{bmatrix}
\frac{EA}{L}\alpha_{11} & & & & & & -\frac{EA}{L}\alpha_{11} & & & & & \\
 & \frac{12EI_3}{L^3}C_{4z} & & & & \frac{6EI_3}{L^2}C^A_{3z} & & -\frac{12EI_3}{L^3}C_{4z} & & & & \frac{6EI_3}{L^2}C^B_{3z} \\
 & & \frac{12EI_2}{L^3}C_{4y} & & -\frac{6EI_2}{L^2}C^A_{3y} & & & & -\frac{12EI_2}{L^3}C_{4y} & & -\frac{6EI_2}{L^2}C^B_{3y} & \\
 & & & \frac{GJ}{L}\alpha_{44} & & & & & & -\frac{GJ}{L}\alpha_{44} & & \\
 & & -\frac{6EI_2}{L^2}C_{3y} & & \frac{4EI_2}{L}C_{1y} & & & & \frac{6EI_2}{L^2}C_{3y} & & \frac{2EI_2}{L}C_{2y} & \\
 & \frac{6EI_3}{L^2}C_{3z} & & & & \frac{4EI_3}{L}C_{1z} & & -\frac{6EI_3}{L^2}C_{3z} & & & & \frac{2EI_3}{L}C_{2z} \\
-\frac{EA}{L}\alpha_{11} & & & & & & \frac{EA}{L}\alpha_{11} & & & & & \\
 & -\frac{12EI_3}{L^3}C_{4z} & & & & -\frac{6EI_3}{L^2}C^A_{3z} & & \frac{12EI_3}{L^3}C_{4z} & & & & -\frac{6EI_3}{L^2}C^B_{3z} \\
 & & -\frac{12EI_2}{L^3}C_{4y} & & \frac{6EI_2}{L^2}C^A_{3y} & & & & \frac{12EI_2}{L^3}C_{4y} & & \frac{6EI_2}{L^2}C^B_{3y} & \\
 & & & -\frac{GJ}{L}\alpha_{44} & & & & & & \frac{GJ}{L}\alpha_{44} & & \\
 & & -\frac{6EI_2}{L^2}C^B_{3y} & & \frac{2EI_2}{L}C_{2y} & & & & \frac{6EI_2}{L^2}C^B_{3y} & & \frac{4EI_2}{L}C^B_{1y} & \\
 & \frac{6EI_3}{L^2}C^B_{3z} & & & & \frac{2EI_3}{L}C_{2z} & & -\frac{6EI_3}{L^2}C^B_{3z} & & & & \frac{4EI_3}{L}C^B_{1z}
\end{bmatrix}
\begin{bmatrix} v_1^{Aj} \\ v_2^{Aj} \\ v_3^{Aj} \\ v_4^{Aj} \\ v_5^{Aj} \\ v_6^{Aj} \\ v_1^{Bj} \\ v_2^{Bj} \\ v_3^{Bj} \\ v_4^{Bj} \\ v_5^{Bj} \\ v_6^{Bj} \end{bmatrix}
=
\begin{bmatrix} F_1^A \\ F_2^A \\ F_3^A \\ F_4^A \\ F_5^A \\ F_6^A \\ F_1^B \\ F_2^B \\ F_3^B \\ F_4^B \\ F_5^B \\ F_6^B \end{bmatrix}
$$

Fig 5

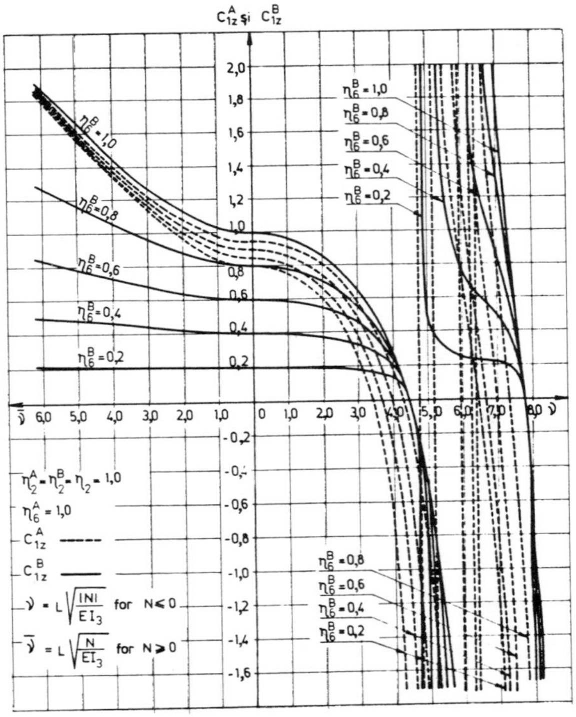

Fig 6

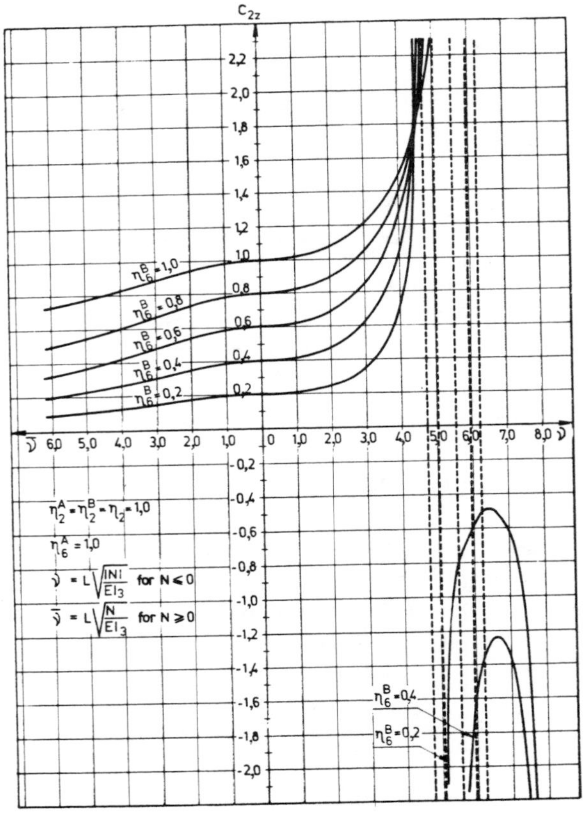

Fig 7

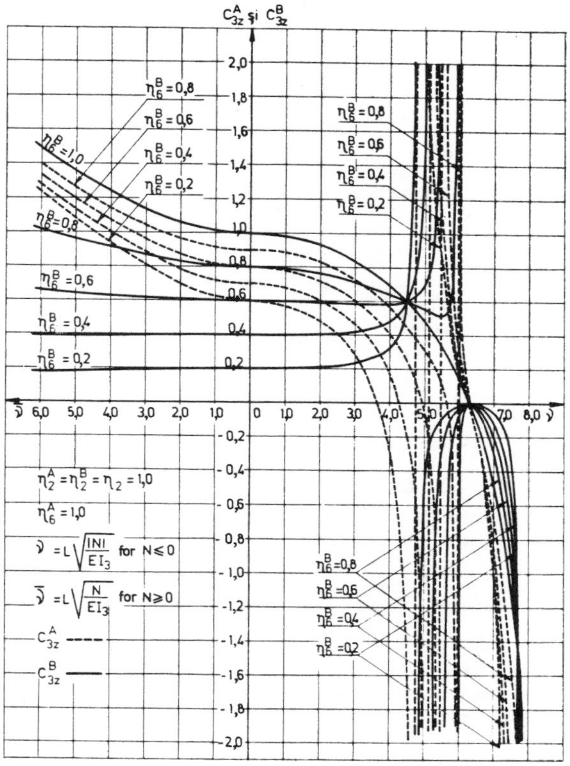

Fig 8

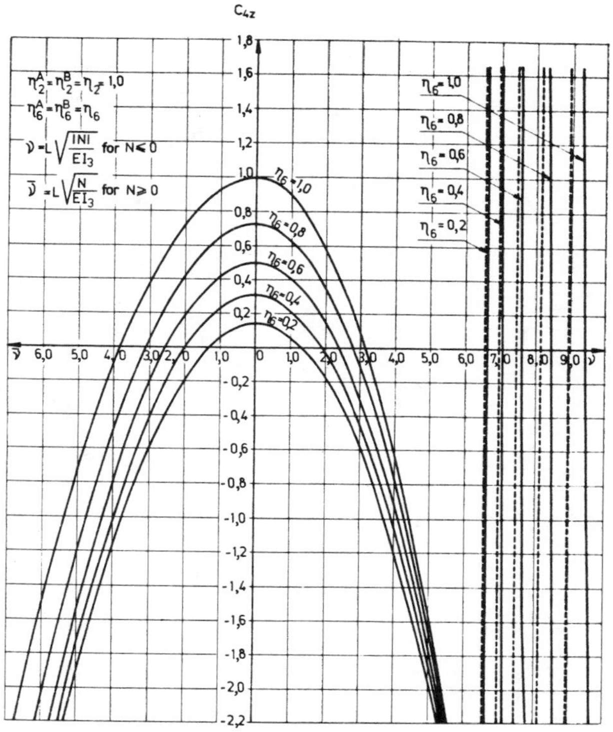

Fig 9

COMPUTER PROGRAM RET

The computer program RET, which uses in the calculations the equivalent stiffness matrix for members elastically restrained at the nodes, $[K_E]$, was developed by the first author for the second order analysis of space structures. The flow chart of the computer program RET is displayed in Fig 10.

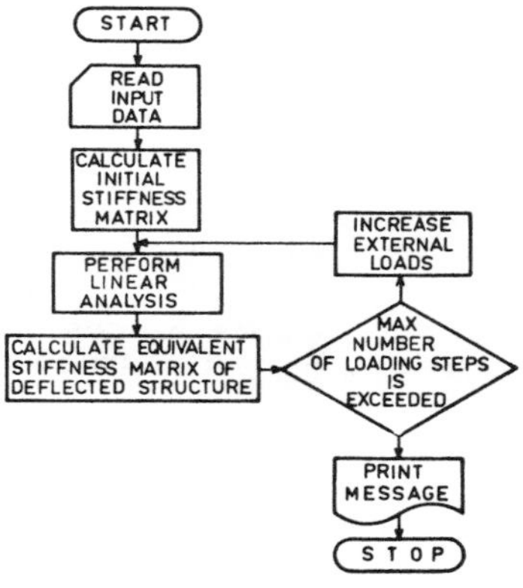

Fig 10

APPLICATIONS

In countries with a severe climate, as Romania and for light weight roofs, as the latticed ones, the loading conditions are mainly given by snow. The design codes specify for barrel vaults two load distributions not exceeding 60° : a uniformly distributed load or a highly unsymmetrical load, fig 11.

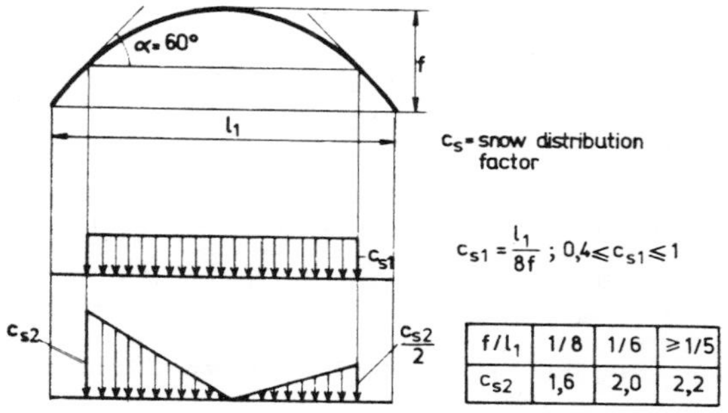

Fig 11

In the following we will analyse the influence of the joint system on the behaviour of the triangulated barrel vaults under the action of an unsymmetrical snow load, Fig 12 ,with the snow distribution factor C_{S2} = 2.2.

The steel members of the two analysed triangulated barrel vaults (the short and the long triangulated barrel vault) are all ϕ 76 x 4 mm tubes.

From Figs 12 and 13 it can be seen that the type of joint connection has a great influence especially on the behaviour of the long barrel vaults.

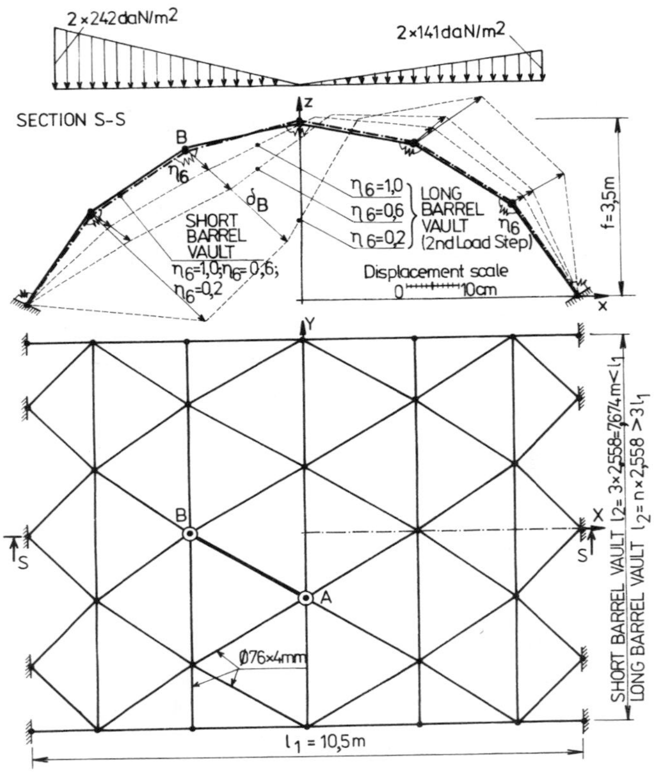

Fig 12

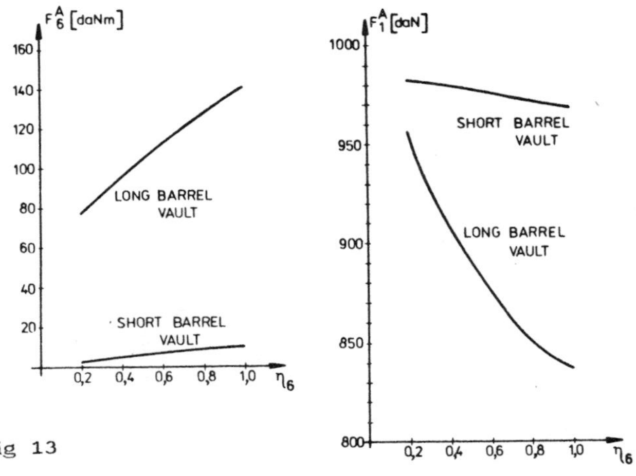

Fig 13

CONCLUSIONS

An equivalent stiffness matrix for members elastically restrained at the joints, $[K_E]$, in order to help to a better understanding of the latticed structures behaviour, was presented and used in a second order analysis for studying the influence of the joint system on the behaviour of latticed shells.

As the study has shown, the type of joint connection has a great influence, especially on the behaviour of long barrel vaults. For these latticed structures the joint system must ensure a perfect restraint of the member ends.

REFERENCES

1. P BOTIZAN Etablissement de la matrice de rigidité d'un element barre avec liaisons imparfaites aux neuds, Construction Metallique Nr 2, 1978.

2. S E FORMAN and J W HUCHINSON, Buckling of Reticulated Shell Structures, International Journal of Solids and Structures, volume 6, 1970.

3. V GIONCU and M IVAN, Buckling of Shell Structures (in Romanian), Editura Academiei, Bucureşti, 1978.

4. V GIONCU, N BALUT, D PORUMB and N RENNON, Instability Behaviour of Triangulated Barrel Vaults, Cours on the Analysis Design and Construction of Barrel Vaults, University of Surrey, September, 1983.

5. E LIGHTFOOT and A P Le MESSURIER, Instability of Space Frames Elastically Connected and Offset Member, 2nd International Conference on Space Structures, University of Surrey, England, 1975.

6. M V SOARE, Actual Problems of the Analysis and Design of Latticed Structures (in Romanian), Construcţii, Nr 8, 1980.

7. S S TEZCAN, Computer Analysis of Plane and Space Structures, Journal of the Structural Division, Nr 92 (ST2), 1966.

8. I H I TOADER, Generalized Stability Functions for Structural Frameworks, The 3rd International Colloqium on Stability, Timişoara, 16 October, 1982.

9. I H I TOADER, Problems of the Analysis of Latticed Structures, (in Romanian), Ph D Thesis, Cluj-Napoca, Romania, 1981.

10. D T WRIGHT, Membrane Forces and Buckling in Reticulated Shells, Journal of the Structural Division, ST 1, 1965.

BUCKLING ANALYSIS OF A SHALLOW DOME MANUFACTURED FROM PULTRUDED FIBRE/MATRIX COMPOSITES

L. HOLLAWAY AND A. RUSTUM*

* Department of Civil Engineering, University of Surrey

Composites manufactured from pultruded glass reinforced plastics (G.R.P.) can be made in both open and closed sections. The material can be produced with mainly orthotropic properties, the main strength and stiffness lying in the longitudinal direction of the members. Because of the high fibre content this material is linear up to failure but has relatively low axial and bending stiffness. The paper describes the linear and non-linear analytical and experimental testing of a shallow geodesic dome manufactured from G.R.P. where large deflection theory is applicable. Theoretical analysis of the dome under different boundary conditions is also undertaken.

The dome has a span of 3000 mm and a rise of 280 mm and is subjected to a central point load. The central deflection is controlled to enable the whole load displacement curve to be determined and to avoid the dynamic jump at snap through.

The non-linear theoretical analysis compares well with the experimental results and both methods of analysis show that there is a considerable difference in behaviour between linear and non-linear techniques for this class of structure.

The pultruded tube buckled severely under a high applied load but without any damage to the material until a crushing failure took place in two members connected to the central node. The size and stiffness of the nodes were a contributing cause of the failure.

1. INTRODUCTION

During the last two decades great technological advances have been made in the manufacturing processes and uses of polymer composites which have lead to their acceptance as high performance structural materials. The high strength to weight characteristics of the composites was one of the main factors that attracted the aerospace industry to them and thus in turn encouraged significant research and development to enable a wide range of highly efficient composites to be produced, Ref 1. The construction industry has benefited from the development work undertaken in the aerospace industry although acceptance of the material by the former industry has been much slower than by the latter. Consequently the field of composite materials in civil/structural applications is only in its infancy and only now has progress commenced.

This paper will describe the loading characteristics of a shallow dome made from glass reinforced polyester pultruded tubes, and will compare theoretical and experimental solutions within the linear and snap through deformation regions. The study forms part of a wider and more general investigation on the viability of the structural applications of this material. Post buckling analysis of double layer skeletal systems manufactured from pultruded fibre/matrix composites have been investigated from both theoretical and experimental viewpoint and have been reported in Ref 2.

2. THE THEORETICAL TECHNIQUE

2.1 INTRODUCTION

Before the introduction of digital computers in the early sixties the linear method of structural analysis was the only practical method for solving problems and most designers used the simplified and approximate methods to avoid laborious numerical computation. With the introduction of computers, during the sixties, considerable advances have been made in the field of structural mechanics. The direct stiffness method of linear analysis is now well established and widely used for linear elastic structural systems, Ref 3. The progress in materials technology opened a completely new field to engineers with the high strength lightweight composites being used in structural systems for which the linear analysis was no longer adequate and non-linear analysis had to be used.

For static problems, the non-linearities arise from two distinct sources, material non-linearities and geometric non-linearities, the latter being due to large displacements. This paper is limited to geometric non-linearity as it is the only non-linearity present in pultruded skeletal G.R.P. structures.

The majority of the finite element applications to geometric non-linear analysis of structures are based on either:

(a) the direct finite element methods

(b) the asymptotic methods.

The first investigator to apply the direct stiffness finite element method to geometrically non-linear structures was Turner et al in 1960, Ref 4. In this reference a step-by-step linearized incremental analysis procedure is described. However, since this earlier reference, intensive work has been carried out. References 5 to 11 include some of the contributions which are of considerable relevance.

The asymptotic method is an application and extension to Koiters perturbation method.

It is generally accepted that the direct finite element procedure is the more powerful of the above two as well as being a well-established method for a geometric non-linear and an instability analysis. In general most analytical techniques adopting this method may be undertaken as follows:

(1) A set of non-linear algebraic or differential equations is derived which represent a mathematical model of the structural behaviour.

(2) A solution to the system of non-linear equations is obtained. The different methods of solution used can be classified into an incremental method, an iterative-method or a combination of the two, Ref 12. If instability were to be involved extra efforts would be required to determine the critical point (either bifurcation or a limit point) and to overcome the problem of singularity associated with it. The post-buckling analysis would then be carried out.

Although several finite element programs are available and are being used for non-linear analysis, Refs 13, 14, the development of efficient programming techniques for complex geometries remains a difficult undertaking. The cost of operating a non-linear finite element analysis is considerable, consequently there is an intensive research effort, refs 15, 16, at the present time to provide efficient and economical techniques to undertake this task.

2.2 THE SNAP BUCKLING

The load deflection curve characteristic of shallow domes and arch type structures is shown in Fig 1. The loss of stability in this case is known as snap buckling in limit point. This curve shows the continuous line representing the stable section of the path whilst the dotted line indicates the unstable section. When an incremental load is applied to such a structure the load deflection curve will follow the section O to A then at the critical point A any increase in load will cause a dynamic jump from position A to position B. Similarly, when the structure is unloaded it will follow the load deflection-curve down to point C where it will snap to position D. If the displacements were controlled the structure, during loading, could be forced to follow the whole load deflection curve from the point A to C and thence to B.

In skeletal domes the snap buckling is usually local and occurs when one of the nodes snaps and the local curvature becomes negative. Local buckling often leads to general buckling as a result of the skeletal dome being unable to resist the shear loads. These loads are transmitted by the buckled unit to the neighbouring members thus causing the area of buckling to increase. General buckling and eventual failure could be caused by a series of local bucklings occurring at different locations affecting more than one node. The stability of such structures is affected by several factors such as the geometry, boundary conditions, type of loading, material properties. The mathematical treatment of the snap buckling is not appropriate to this paper as it falls outside the objectives, however Refs 17 and 18 contain the basic areas regarding the general theory of elastic stability where the snap buckling phenomenon is discussed.

2.3 THE COMPUTER PROGRAM

A linear and non-linear static analysis program for skeletal systems was developed to enable a research program to be undertaken into the response to loading of skeletal structures manufactured from pultruded tubular fibre reinforced polyester members. The non-linear behaviour is due to large deformations only. A perfectly linear elastic material behaviour up to failure was assumed and this was verified by experiment. As the whole load deflection curve was of interest, an incremental solution scheme with iterations at each load step was required, thus avoiding drifting from the real equilibrium path and, of greater importance, enabling the critical point to be dealt with (limit points only are dealt with). A method was used which is based on the Newton-Raphson technique, introduced by Riks, Ref 19 and subsequently modified by Crisfield, Ref 15. In this method an extra constraining equation is added to the usual n equilibrium equations fixing the incremental step.

A simplified non-linear element stiffness matrix was derived analytically which took into account axial strain only. The stiffness matrix consisted of three parts, these were:

(a) the elastic linear matrix K_E

(b) the initial stress matrix K_σ which is dependent upon the element internal stresses

(c) the initial displacement matrix K_d which is a function of element nodal displacements.

The program deals only with conservative loadings applied to the nodal points. A discretization of the members is required for structures having highly non-linear behaviour. The program makes use of the active column (skyline) storage technique which minimizes storage requirements, Refs 20, 21, and it is possible to use fixed or updated coordinate systems. When using updated coordinates, as is the case in the analyses of the present dome, K_d is neglected.

2.4 DERIVATION OF THE ELEMENT STIFFNESS MATRIX

The derivation of the element stiffness matrix is based on the simplified strain-displacement relationship:

$$\varepsilon_x = \frac{du}{dx} + \frac{1}{2}\left[\left(\frac{dv}{dx}\right)^2 + \left(\frac{dw}{dx}\right)^2\right] - y\frac{d^2v}{dx^2} - z\frac{d^2w}{dx^2}$$

This expression represents the longitudinal strain at a coordinate point (x, y, z) where u, v and w are displacements in x, y and z direction of a generic point on the member axis.

A linear displacement function in x was chosen for u and a cubic function in x for both v and w, this means that loads can only be applied to the nodal points, since this assumption restricts the member to carry only constant shear and linearly variable bending moment.

$$u = a_o + a_1 x$$

$$v = a_2 + a_3 x + a_4 x^2 + a_5 x^3$$

$$w = a_7 + a_8 x + a_9 x^2 + a_{10} x^3$$

The above ten coefficients are derived from boundary conditions and are expressed as functions of nodal displacements. Substituting the expression for displacement in the strain displacement relationship, we obtain:

$$\varepsilon_x = f\,(u_A,\ v_A,\ \ldots\ \theta_{BY},\ \theta_{BZ},\ x,\ y,\ z)$$

The strain energy of the element caused by a load increment is given by:

$$U = \iiint \int_{\varepsilon_o}^{\varepsilon+\varepsilon_x} \sigma\, d\varepsilon \quad dxdydz$$

$$= E \iiint \int_{\varepsilon_o}^{\varepsilon_o+\varepsilon_x} \varepsilon\, d\varepsilon \quad dxdydz$$

$$= E\varepsilon_o \iiint \varepsilon_x dxdydz + \frac{E}{2} \iiint \varepsilon_x^2 \ dxdydz$$

Where it is assumed that the member was in a deformed state, which is denoted by ε_o, before the load increment.

A substitution of ε_x in the strain energy expression, integrating and making use of Castigliano's theorem to derive the coefficients of the stiffness matrix, leads to:

$$KE = K_E + K_\sigma + K_d$$

A linear twist rate is assumed and the relation is added to the tangent stiffness matrix.

3. EXPERIMENTAL INVESTIGATION

3.1 THE MODEL

The three-way grid dome was fabricated from pultruded glass reinforced polyester composite members of 25 mm external diameter and a 2 mm wall thickness. The members chosen for the construction were as free from initial imperfection as possible. The glass fibre/polyester resin ratio by weight was 60/40. The dome was a portion of a sphere of diameter 9000 mm and had a span of 3000 mm and a rise of 280 mm to the central node. The dimensions of this test structure were consistent with the laboratory facilities and represented a large scale model.

The members of the dome were cut to size and a special technique was used to manufacture and assemble the nodal joints. Six members met at all internal joints and into each end of the members a milled cylindrical aluminium plug is placed from which flexible metal pins protrude to enter radially another milled cylindrically aluminium disc which forms the centre of the nodal joint. Fig 2a, b and c shows the operation to form this joint to this stage.
When every joint on the structure was partially made a theodolite and a level were used to adjust the dome to its correct geometry which was readily deformed by the application of a small pressure. Figure 3 shows the dome at this stage of manufacture. An adhesive was then used to fix the temporary nodes and to ensure a degree of rigidity of the structure; Fig 4 illustrates this stage of completion. A silicone rubber mould, shown in Fig 5 was manufactured from an aluminium disc and was used to cast a glass reinforced epoxy composite around the temporary nodes to complete the nodal joint. Figure 6 shows the completed node and Fig 7 shows the assembled model. The centre of curvature at the bottom of the six peripheral supper nodes was coincident with the centre of the node in order to allow rotations to take place about the same point and to avoid any eccentricity. A V-shaped steel channel, with adjustable supports, provided constraints against lateral displacements as shown in Fig 8.

3.2 TESTING ARRANGEMENT AND LOADING

The load was applied to the central node only, through a ball bearing and two load cells. The first load cell was connected to a monitoring device, to enable a constant check to be made on the applied load level, and the second load cell was connected to a data logging system to record the load readings. To control the displacements at the central node a manually operated mechanical jack, in series with a load cell and an aligning ball bearing, was placed under the loaded node. This latter load cell was also connected to the data logger. The two ball bearings on both sides of the loaded node were perfectly aligned thus ensuring that no appreciable rotational constraints were imposed. This arrangement was used as a displacement controlled actuator was not available. The displacement of the central node was controlled by the mechanical jack and the actual load applied to the structure was equal to the difference between the readings of the load cells above and below the node point. This arrangement, shown in Fig 9, allowed the whole load-deflection curve to be traced including the unstable equilibrium points; it prevented dynamic snapping from taking place.

3.3 INSTRUMENTATION

Electric resistance metal foil strain gauges were used with resistance of 120 ohms and gauge factor of 2.09. Selected members of the skeletal structure were strain gauged to enable axial and bending strains to be measured. At these points three gauges were used and were spaced equally around the member. Certain symmetric members were gauged to enable a study to be made of imperfections in the system which were unavoidably introduced during the assembly of the model.

Stroke potentiometric transducers were used to measure the vertical displacement of the central node and other selected nodes.

3.4 TESTING PROCEDURE

A series of preliminary tests were undertaken to adjust the supports and to check the instrumentation of the system. Subsequently the model was loaded twice; during the first test the magnitude of the load was such that snap of the central node took place and during the second test the structure was loaded to failure.

The loading sequence consisted of applying a high load through the hydraulic jack, this load was supported by the dome and by the mechanical jack; however, only a small part of it was supported by the dome initally. As the mechanical jack was lowered manually the share of the load taken by the dome was increased until the critical load value was reached. At this point, as the mechanical jack was further lowered, the load taken by the structure decreased until it reached a position of stable equilibrium and then commenced to increase its share again.

4. DISCUSSION OF THE RESULTS

The results are divided into two sections; section 4.1 deals with the analytical and experimental techniques where only one boundary condition is considered, section 4.2 investigates the analytical solutions only when the structure is under three different boundary conditions.

4.1 ANALYTICAL AND EXPERIMENTAL RESULTS FOR FULLY FIXED BOUNDARY CONDITIONS

All supports to the model were constrained against vertical and horizontal displacements and no rotational constraints were imposed. The results of the theoretical computer analysis are in graph form and the experimental results are represented by symbols on these graphs. Figure 10 shows the configuration and member discretization for the dome.

The sign convention is such that downwards displacements and compressive axial forces are assumed to be negative. Two sets of graphs have been presented, one of load against vertical displacement and the other of load against axial force. Graphs of load against bending moment for the analytical and experimental analysis agree closely as do the respective graphs of axial force against displacement and therefore only the latter relationship will be considered in this paper.

The theoretical and experimental load displacement curves for the central node are shown in Fig 11 and very good agreement is achieved with the percentage difference between the theoretical and experimental snap buckling less than 5%. In addition the experimental and theoretical values of the load-displacement relationship for the inner ring nodes, Fig 12, agree closely. The difference, however, was that in the pre-buckling state the experimental values are negative. This difference could have been caused by geometric imperfections or ineffective boundary conditions. Figure 18 shows the effect the different boundary conditions have on the behaviour of this node. From a consideration of the theoretical load deflection curves of the central and the inner ring nodes (viz. Figs 11 and 12) it may be concluded that the snap buckling affects only the central node.

The experimental strains in members A to K of Fig 10 have been converted to axial forces and plotted against the load factor. Figures 13 to 16 show the results. A symmetrical behaviour was assumed when undertaking the theoretical analysis and all the experimental results for symmetric members have been plotted on this theoretical curve. Figure 13 shows the load factor against the theoretical axial force for members connected to the central node and superimposed upon this curve are the experimental results for members A, B and C of Fig 10. Similarly Fig 14 shows the load factor against the theoretical axial force for the inner ring members and superimposed on this graph are the experimental results for members D E F. Figure 15 shows a similar theoretical curve and experimental results plotted on this graph for members G and H which are on the outer ring of the structure. The experimental relationship between the load factor and force in members I J and K is shown in Fig 16. In general the analytical analysis predicted with a good degree of accuracy the structure's performance but it must be stated that the assumption of symmetric behaviour was unrealistic.

The high values of axial force recorded in members C E H and K is an indication that the non-symmetric behaviour is the cause of the discrepancies between the analytical and experimental solutions.

The central unit of the model which consists of the central node and the six members connecting it did not sustain any damage during the first loading to snap and a comparison of the initial and final readings at no load condition confirmed this. During the loading cycle to failure and at a load in excess of that causing snap the two members C and L failed at their ends nearest to the central node. From previous investigations on the material and from an observation of the strain gauge readings on the members it was clear that failure initiated in the compression fibres of the composite. In addition it was clear that premature failure was caused by imperfections, first by eccentricities being caused in the central node due to members not meeting at the centre of the node, secondly by the excessive size and stiffness of this node and thirdly by the geometric imperfections in the members. It was noted that the value of force being taken by identical neighbouring members to those that failed and the compressive stress value in those members from the theoretical investigation were lower in value.

It was expected that the model would fail suddenly as the material contained a relatively high percentage of glass (60% by weight) and therefore the composite would follow more closely the stress-strain relationship and failure behaviour of the fibre which is a brittle material.

4.2 ANALYTICAL RESULTS FOR THREE DIFFERENT BOUNDARY CONDITIONS

The three boundary conditions considered were as follows:

(a) the six support nodes which are shown in Fig 10 were fixed in position only;

(b) the six support nodes were position fixed in the vertical direction and, in addition, node numbers 1, 127, 217 were position fixed in the horizontal direction;

(c) the six support nodes were position fixed in the vertical direction only.

In this section only the load ~ displacement curves will be considered. Figures 17 to 19 show these curves for the central node, the inner ring nodes and the outer ring unrestrained nodes respectively. It should be noted that in the second support boundary case, node point 50, for instance, will behave differently from node point 109 shown in Fig 10. Consequently only the node points in similar locations to that of node point 50 have been considered during the discussions of this support boundary case.

The perimeter numbers formed a tension ring beam when no horizontal support constraints were provided to the structure and consequently the value of the snap buckling load was sensibly the same for the three boundary conditions; this is shown in Fig 17. The general behaviour of the central node was similar in all cases considered here. However, there was a slight difference in the results for the structure with no horizontally constrained supports; these results showed a more flexible behaviour when the structure was under load and the snap buckling resulted after a more severe structural deformation occurred.

Although the deformation behaviour of the central node was essentially the same for all cases considered, the equivalent behaviour for the inner and outer ring nodes was influenced by the boundary conditions. The load ~ deflection curves for these nodes are similar in form but vary in magnitude and direction, particularly for the two extreme cases considered, viz. for the total horizontal constraint and for the total horizontal constraint and for the total horizontal freedom.

5. CONCLUSIONS

Shallow domes manufactured from pultruded glass reinforced polyester are sensitive to the changes in geometry resulting from deformations. Consequently a non-linear analysis is required to obtain details of their complete deformational behaviour. A comparison of the non-linear analytical and the experimental analyses shows that the vertical loads and general behaviour of the structure can be predicted with considerable accuracy.

It has been shown that pultruded glass reinforced polyester tubes recover completely after they have undergone large deformations and without sustaining any apparent damage. Because of the linear elastic nature of the material and its brittleness at failure the structure tends to collapse without warning.

Imperfections in structures can be caused by several factors and any combination of these can be incorporated into the analytical analyses, however to be able to understand the effects of particular imperfections it is necessary to be able to study them independently of other imperfections. In addition, the analysis of the structure, under difficult support conditions, related the effects of imperfections to the lack of horizontal constraints of the structure and thus enabled a greater understanding of the problem.

REFERENCES

1. C ZWEBEN, Advanced Composites for Aerospace Applications, A Review of Current Status and Future Prospects, Composites, Oct. 1981, pp. 235-240.

2. A RUSTUM and L HOLLAWAY, The Structural Behaviour of a Double-layer Skeletal System manufactured from Pultruded Fibre/Matrix Composites, Paper 39 of the Proceedings of the Reinforced Plastics Congress, Brighton, U.K., 1982.

3. O C ZIENKIEWICZ and Y K CHEUNG, Finite Element Method in Structural and Continuum Mechanics, McGraw-Hill, 1967.

4. M J TURNER, E H DILL, H C MARTIN and R J MELOSH, Large Deflections of Structures subjected to Heating and External Load, J. Aerospace Sci., Vol. 27, No. 2, 1960, pp. 97-106.

5. J A STRICKLIN, W E HEISLER and W A VON REISEMANN, Geometrically Non-linear Structural Analysis by Direct Stiffness Method, Journal of the Structural Division, ASCE, Vol. 97, 1971, pp. 2299-2314.

6. H D HIBBITT, P V MARCAL and J R RICE, A Finite Element Formulation for Problems of Large Strain and Large Displacement, International Journal of Solids and Structures, Vol. 6, 1970, pp. 1069-1086.

7. S A SAAFAN, Non-linear Behaviour of Structural Plane Frames, Journal of the Structural Division, ASCE, Vol. 89, 1963, pp. 557-579.

8. A JENNINGS, Frame Analysis including Change of Geometry, Journal of the Structural Division, ASCE, Vol. 94, 1968, pp. 627-644.

9. R H MALLETT and L A SCHMIT Jr., Non-linear Structural Analysis by Energy Search, Journal of the Structural Division, ASCE, Vol. 93, pp. 221-234, 1967.

10. J T ODEN, Calculation of Geometric Stiffness Matrices for Complex Structures, AIAA Journal, Vol. 4, No. 8, 1966, pp. 1480-1482.

11. O K KICIMAN and E P POPOV, Post-buckling Analysis of Cylindrical Shells, Journal of the Engineering Mechanics Division, ASME, Vol. 104, 1978, pp. 751-762.

12. W E HAISLER, J A STRICKLIN and F J STEBBENS, Development and Evaluation of Solution Procedures for Geometrically Non-linear Structural Analysis, AIAA Journal, Vol. 10, No. 3, 1972, pp. 264-272.

13. A K NOOR, Survey of Computer Programs for Solution of Non-linear Structural and Solid Mechanics Problems, Comput. Structures, Vol. 13, 1981, pp. 425-465.

14. C A BREBBIA, Finite Element Systems - A Handbook, A Computational Mechanics Centre Publication, Springer-Verlag, 1983.

15. M A CRISFIELD, A Fast Incremental/Iterative Solution Procedure that handles Snap Through, Comput. Structures, Vol. 13, 1981, pp. 55-62.

16. R A BROCKMAN, Economical Stiffness Formulations for Non-linear Finite Elements, Vol. 18, 1983, pp. 15-22.

17. J G A CROLL and A C WALKER, Elements of Structural Stability, The Macmillan Press Ltd., 1972.

18. J M T THOMPSON and G W HUNT, A General Theory of Elastic Stability, John Wiley & Sons, 1973.

19. E RIKS, An Incremental Approach to the Solution of Snapping and Buckling Problems, Int. J. Solids Structures, 15, 1979, pp. 524-551.

20. C A FELIPPA, Solution of Linear Equations with Skyline-stored Symmetric Matrix, Comput. Structures, Vol. 15, 1975, pp. 13-29.

21. E L WILSON and H H DOVEY, Solution of Reduction of Equilibrium Equations for Large Complex Structural Systems, Advances in Engineering Software, Vol. 1, No. 1, 1978, pp. 19-25.

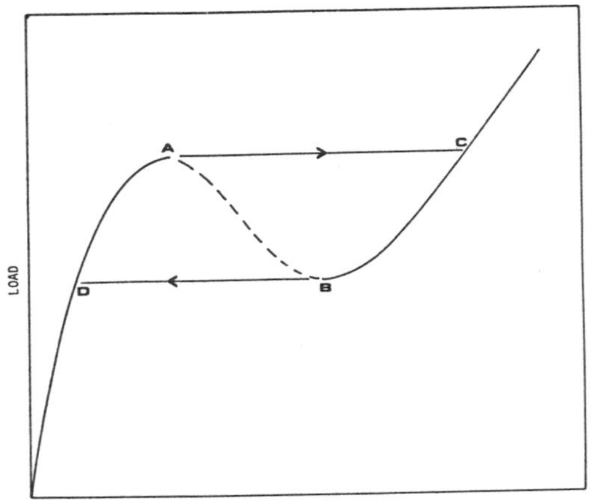

Fig. (1) Load-displacement relationship for the snap buckling

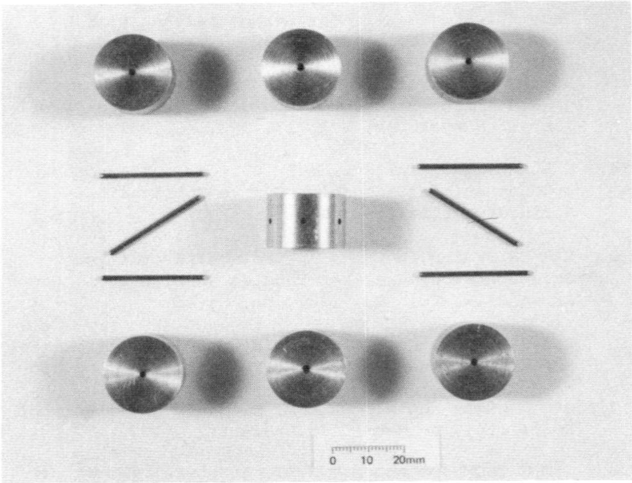

Fig.(2a) The components of the temporary node.

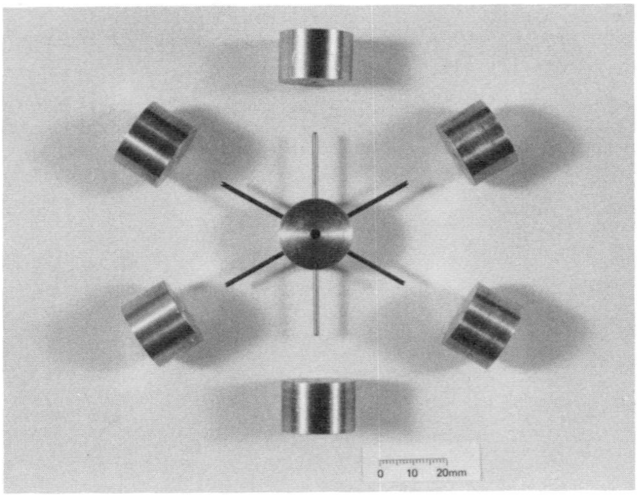

Fig.(2b) Position for connection of components.

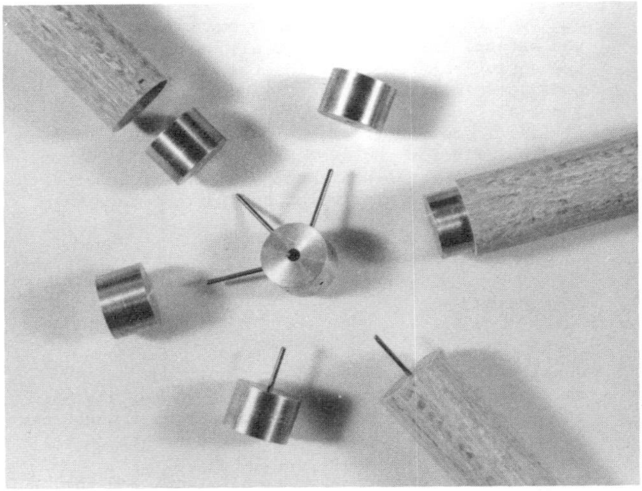

Fig.(2c) First members connected.

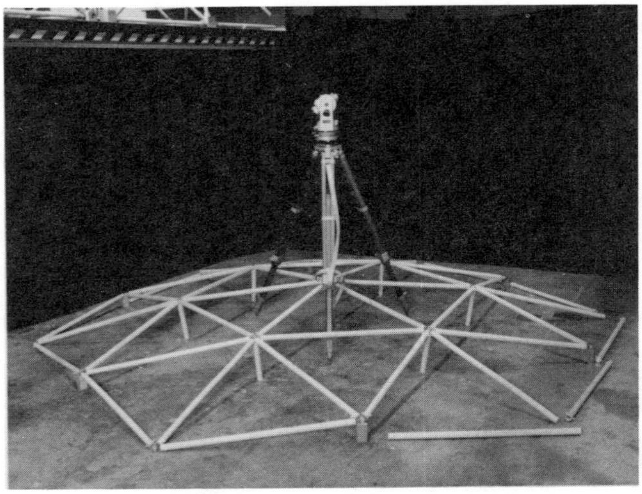

Fig.(3) Assembly of the dome.

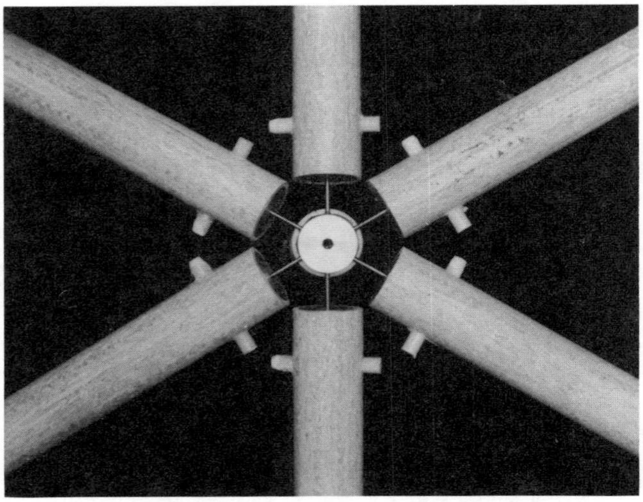

Fig.(4) Completed temporary node.

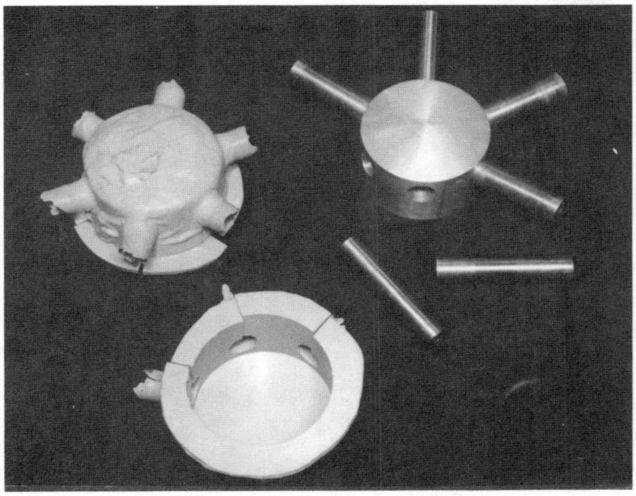

Fig.(5) The aluminium die and the silicon rubber mould.

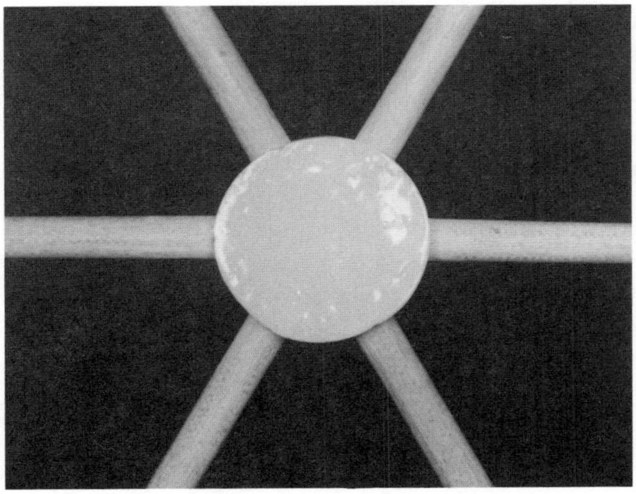

Fig.(6) Completed glass reinforced epoxy node.

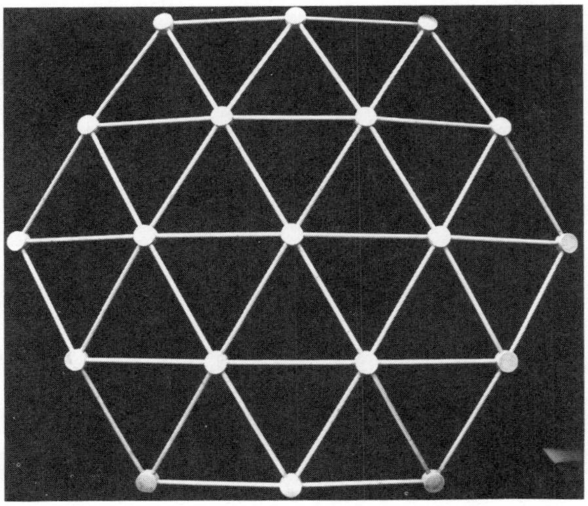

Fig.(7) The completed dome.

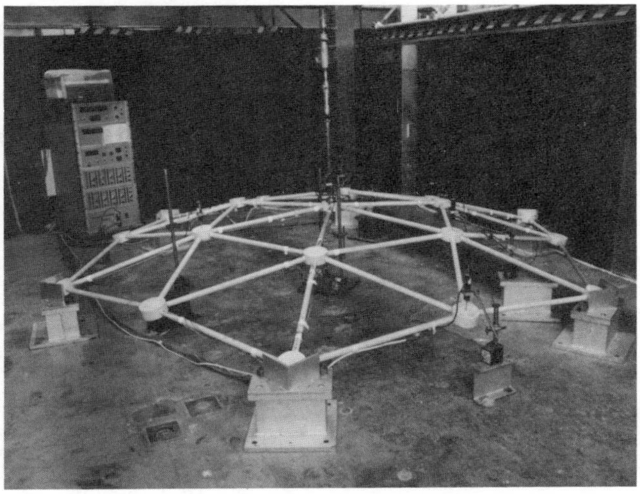

Fig.(8) The dome test arrangement.

Fig.(9) The displacement controlled testing arrangement.

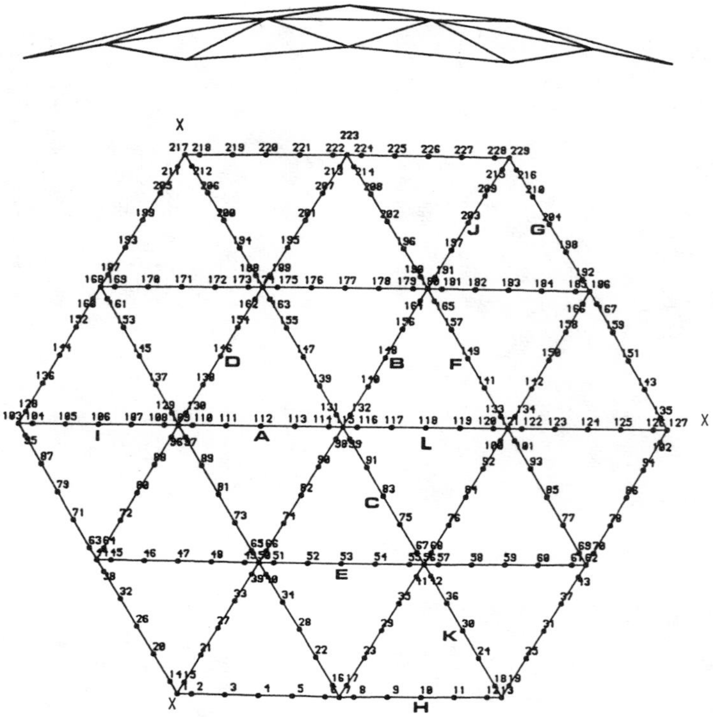

Fig. (10) Configuration and Member Discretization for the Dome

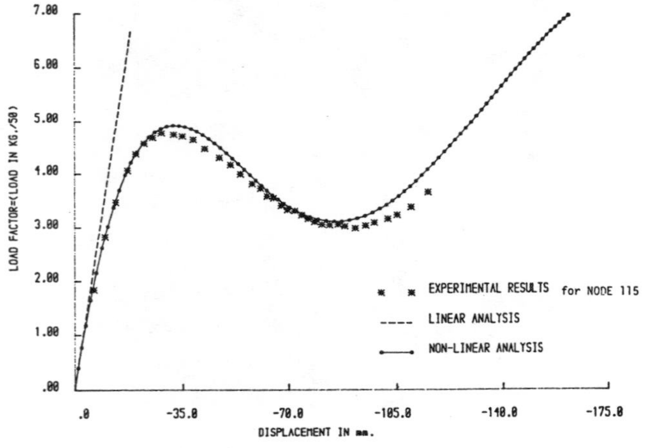

Fig. (11) Load Displacement Curve for Central Node

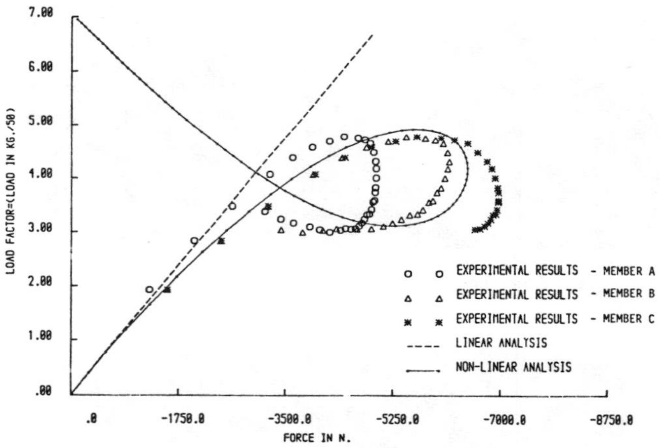

Fig. (13) Load Factor - Theoretical Axial Force for Members connected to Central Node with superimposed Experimental Results

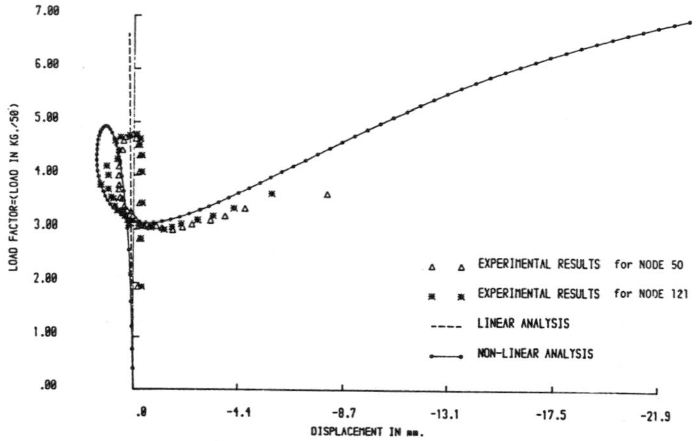

Fig. (12) Load-Displacement Curve for Inner Ring Nodes

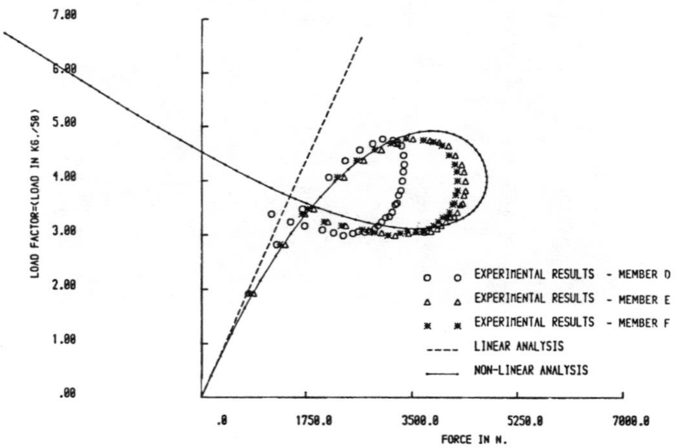

Fig. (14) Load Factor - Theoretical Axial Force for Inner Ring Beam with superimposed Experimental Results

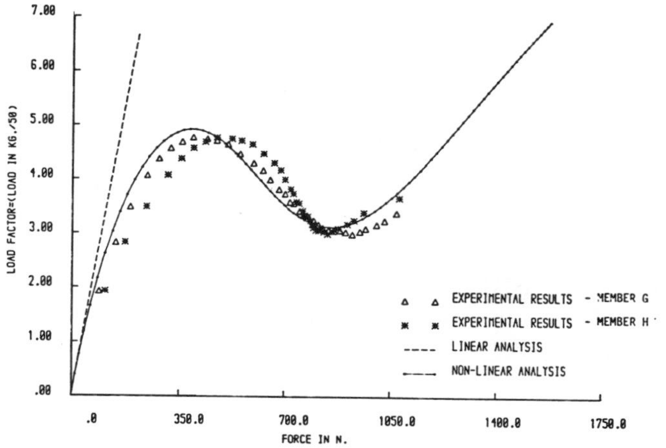

Fig. (15) Load Factor - Axial Force for Members G and H with superimposed Experimental Results

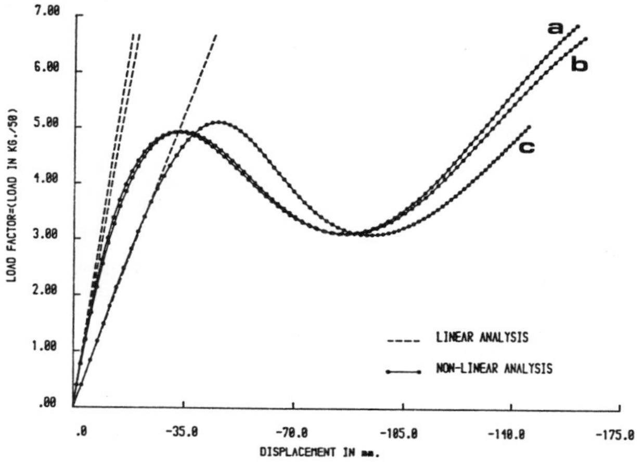

Fig. (17) Load Factor - Displacement curves for Central Node Point

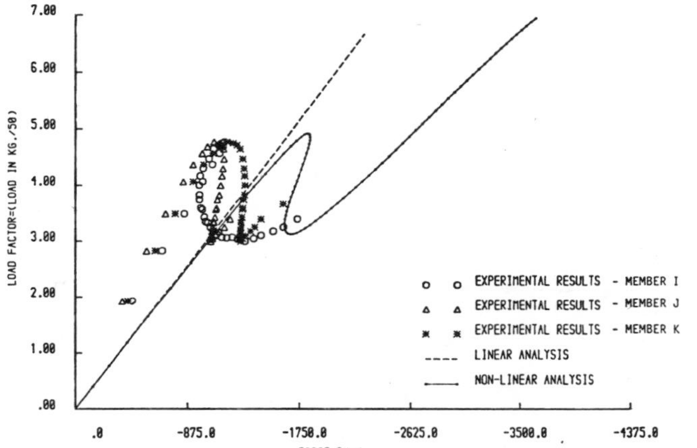

Fig. (16) Load Factor - Axial Force for members I, J and K with superimposed Experimental Results

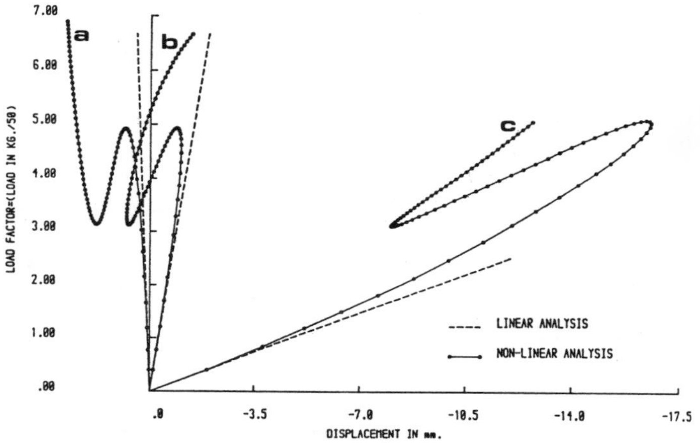

Fig. (18) Load Factor - Displacement for the Inner Ring Nodes

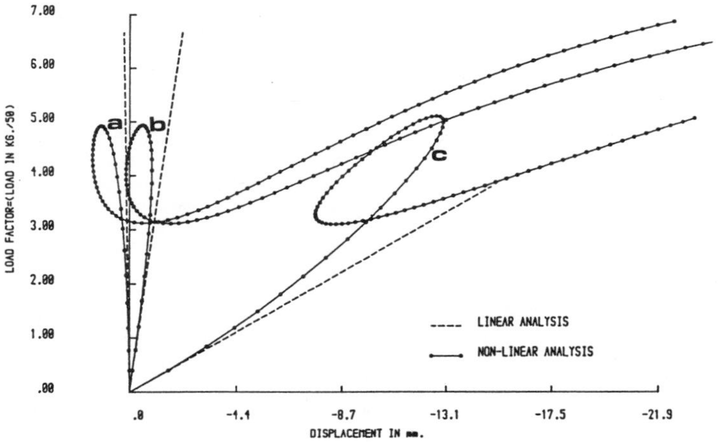

Fig. (19) Load Factor - Displacement Curves of Analytical Results for Outer Ring Nodes

LIMIT STATE ANALYSIS AND FULL SCALE EXPERIMENTS OF DOUBLE - LAYER GRIDS

J BRÓDKA[x], CEng, DSc and A GRUDKA[xx], CEng

x Professor, xx Assistant lecturer,

Metal Structures Research and Design Centre "Mostostal", Warsaw.

The paper deals with limit state analysis of a double-layer grid with a diagonal arrangement of top chords and also with test results of full-scale experiments. A space frame-panel system has been assumed as a calculation model taking into account the co-action of corrugated roof sheeting with the supporting frame structure. The P - Δ relationship in the elastic--plastic range of behaviour has been derived taking into account residual stresses and geometrical imperfection. Full-scale experiments were carried out on two double-layer grids. The bigger one /18 x 18 m/ was tested to failure and the smaller one /18 x 12 m/ was tested in the elastic range only with some controlled defects being imposed.

INTRODUCTION

The "MOSTOSTAL" system of double-layer grids is briefly presented in Ref 1. Before the system was introduced to a serial production two prototype grids, see Figs 1 and 2,

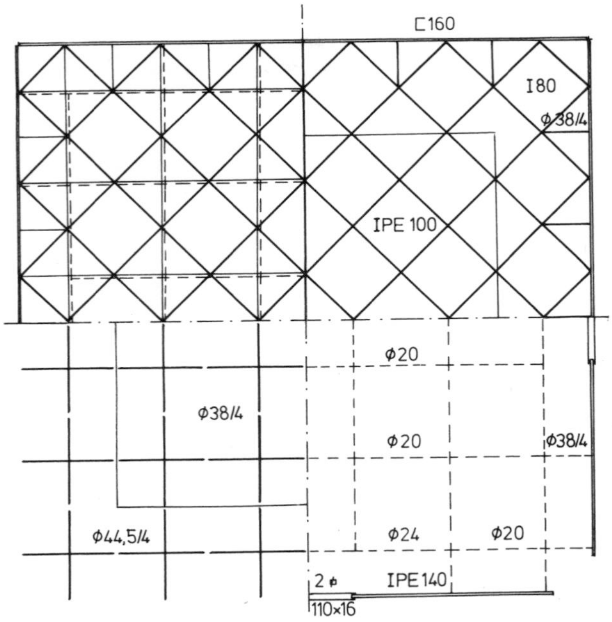

Fig 1

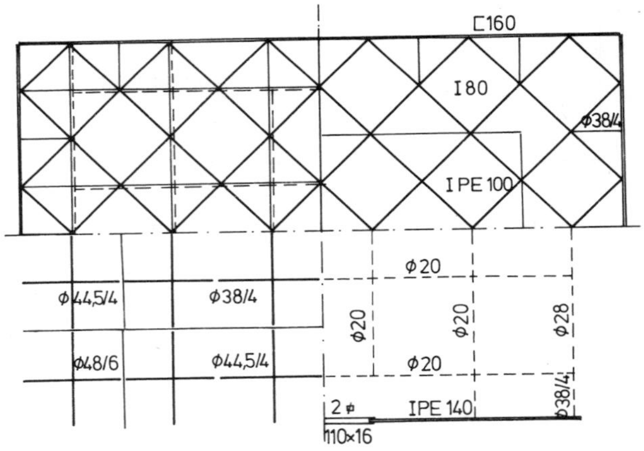

Fig 2

representing diagonal on square arrangement had been built and field tested. Both of them, 18 x 18 m and 18 x 12 m, were standard grids to be used as roof structure units for multi-bay industrial buildings.

The tests were aimed first of all at an experimental verification of strength properties of structural members designed on the basis of calculations. Another aim of the test was gaining some necessary experience of fabrication of individual components and their behaviour during transport and assembly work.

TEST

During experiments the following data were recorded:

a/ Vertical displacements of all joints of bottom layer.

b/ Vertical and horinzotal displacements of corner columns supporting the grid being tested.

c/ Vertical and horizontal displacements of middle joints of edge girders.

d/ Longitudinal strains of the middle cross section of the selected representative members.

For the measurements of vertical and horizontal displacements of columns and edge girders dial gauges were used. Deflections of the bottom layer have been recorded by means of rules with 1 mm accuracy of reading. Longitudinal strains were measured by means of resistance strain gauges. Three gauges equally-spaced over the perimeter were cemented on the chosen bottom and web members. Four gauges symmetrically spaced on the inner sides of the flange were cemented in the sections of top members made of IPE. Arrangement of measuring instruments and the members chosen for the measurements are shown in Figs 3 and 4. The numbers of the members shown in these figures correspond with the sequence of bars taken in the computer structural analysis.

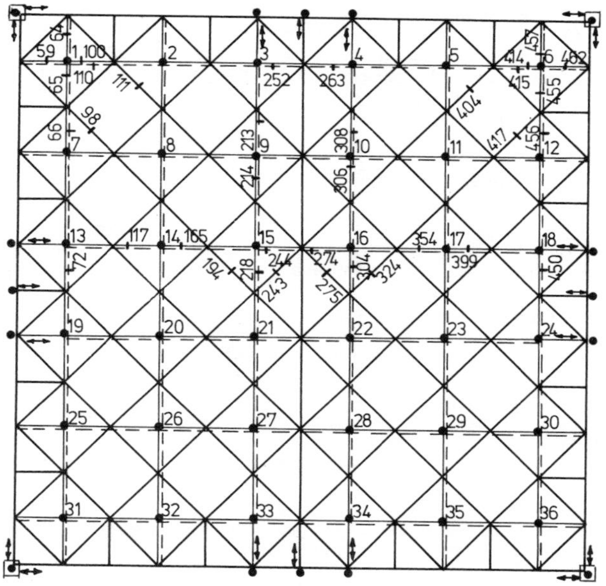

• measurement of vertical displacement
⟷ measurement of horizontal displacement
+ strain gauge measurements

Fig 3

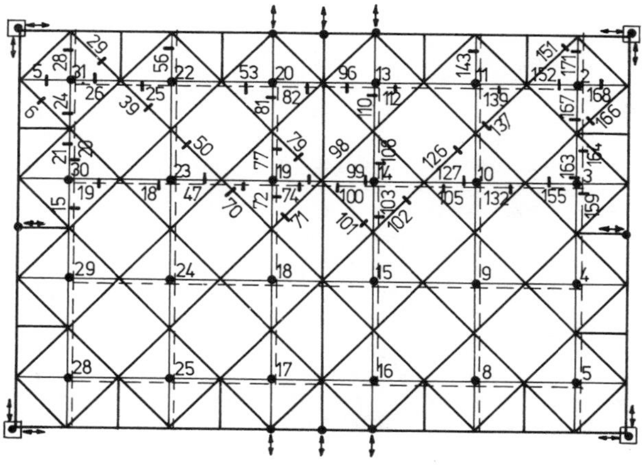

Fig 4

In the case of the grid 18 x 18 m the design component loads were simulated by putting down steel sheets of various thickness. It was intended to perform the measurements in the seven stages; as shown in Fig 5.

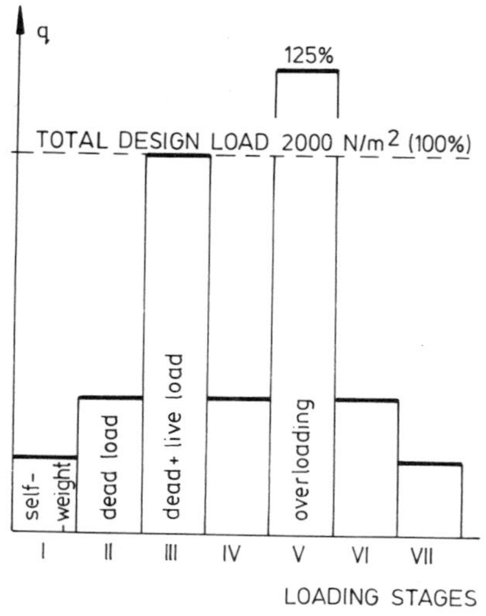

Fig 5

Under the load of stage V being applied considerable violently increasing deflections were observed on the rules and after a short time the structure had rested on protective supports. The electric resistance strain gauges showed that some members of the bottom layer had attained the yield point. Visual examination revealed considerable elongation of these members. Moreover, some rotation in the middle joints of the top layer, as well as buckling of slender members supporting corrugated plates in the roof ridge were also detected. On the contrary, no damage of assembly bolted joints and no damage of workshop welded joints were observed. Behaviour of the structure during overloading /stage V/ and the results of the visual examination have led to the conclusion that the decreased load-carrying capacity of the roof structure as compared with the design load has been due to the fact that wrong round bars of lower grade have been used in the bottom layer. This has been confirmed precisely by the results of material tests of the samples cut out from non-deformed members. The yield point of the samples tested was 210 MPa. The load of the structure while it rested on the protective supports was 2,302 N/m^2. This magnitude has been assumed as the ultimate load capacity of the grid tested.

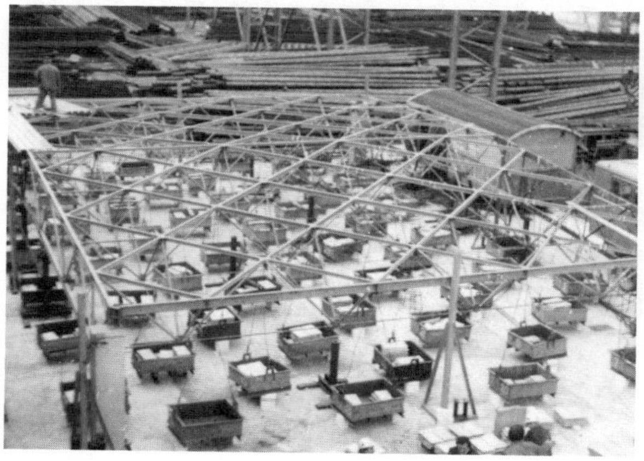

Fig 6

In the case of the grid 12 x 18 m loads corresponding with

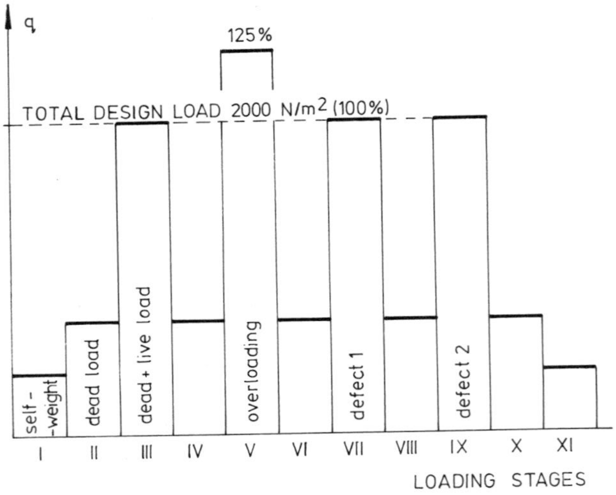

Fig 7

design ones were applied by suspending containers with a variable ballast to the joints of the top layer of the framework, see Fig 6. Measurements were performed in eleven stages, as shown in Fig 7. Stages from I to VI corresponded with the stages described for the grid 18 x 18 m. The additional measurements loading stages, VII-IX, were performed for the structure with defects.

Defect 1 was arranged as an elimination of the members of bottom layer and two vertical suspension members, which no longer carried the loads, see Fig 8. It included members No 82 and 96 shown in Fig 4 and the corresponding /symmetric/ members in another half of the structure. It was obtained by removal of bolts in the assembly joint connecting the members of the bottom layer. Defect 2 made eliminated further eight members of bottom layer. Additionally defected were members No 53, 77, 108 and 112 shown in Fig 4 and the corresponding ones in another half of the structure. It was obtained by removal of bolts in the bottom pyramid joint, to which the defected members had been attached.

Fig 8

Observations of the structural behaviour during experiments confirmed theoretical anticipations. No resistance strain gauge showed an attainment of the yied point. Defects of bolted joints and workshop welded joints were not observed either.

THEORETICAL ANALYSIS

The attainment of yield point by the most stressed member of the system, which is the basis for dimensioning of the structure within linearly elastic range, does not cause the collapse of a whole structure. It has still greater load-carrying capacity due to some additional reserve pertained to the remaining component members. Only yielding and/or instability in a definite number of members causes the transformation of a system into a mechanism and the attainment of ultimate load capacity. Thus the analysis becomes a "step by step" procedure, in which yielding members are consistently eliminated. This problem can not be solved without knowledge of P - Δ relationship for each member of the structure. Adequate determination of load carrying capacity of the structure depends to a considerable extent upon the assumed model of this relationship. Also considerable influence upon the accuracy of results in the case of analysis of the structures under consideration has taking into account corrugated deck being a bearing layer of the roof cladding. Permanent connection of corrugated plates with top members of the grid leads to a co-action of panel and bar members. The proper consideration of this co-axion in the theoretical model makes the results of this analysis closer to the real behaviour of the existing structure.

INFLUENCE OF CORRUGATED DECK

Co-action of the space grid and corrugated deck may be considered either locally, or globally.

The local co-action consists in stiffening of the top members of the roof, which results in an increased stability of these members. The advantage of this co-action may be particularly taken in the case of a double-layer grid with a diagonal arrangement of top members, thus members of both ways are efficiently restrained in the plane of the layer. For this reason calculation model of a top member in the plane of a diaphragm has been assumed in the form of a continuous beam on elastic supports, see Fig 9. The number of supports depends upon the spacing of fasteners being used to connect corrugated plates with top members. Solving such a system one will obtain an expression for the critical force:

$$P_{cr}=0{,}5\left[\frac{\pi^2 EJ_y}{L^2}+\frac{\pi^2 EAJ_\omega}{L^2 J_p}+\frac{AJ_s G}{J_p}+\frac{16AlK_\phi}{\pi^2 J_p}+\frac{16(2-\sqrt{2})}{\pi^2}\left(\frac{h^2 A}{4J_p}+1\right)K_x - \sqrt{\left[-\frac{\pi^2 EJ_y}{L^2}+\frac{\pi^2 EAJ_\omega}{L^2 J_p}+\frac{AJ_s G}{J_p}+\frac{16AlK_\phi}{\pi^2 J_p}+\frac{16(2-\sqrt{2})}{\pi^2}\left(\frac{h^2 A}{4J_p}-1\right)K_x\right]^2+\frac{256(2-\sqrt{2})^2 h^2 AK_x^2}{\pi^4 J_p}}\;\right] \qquad (1)$$

In formula /1/, besides geometrical properties of the member, there are two other magnitudes K_x and K_ϕ, characterizing flexibility of supports. K_x characterizes diaphragm rigidity of corrugated plates taking into account flexibility of main and intermediate fasteners. On the other hand, K_ϕ characterizes bending flexibility, being a function of bending rigidity of corrugated plates and local bending rigidity at the point of connection with the bar. It has to be noticed, that the second factor is of decisive influence and can be determined only from experimental investigations. Quantitative influence of corrugated plates on rigidity of the member depends obviously on the relationship between K_x and K_ϕ and geometrical parameters of the member itself. An exemplary increase of critical force versus corrugation height for top members made of IPE section, connected in every second fold, is shown in Fig 10. From the curve presented

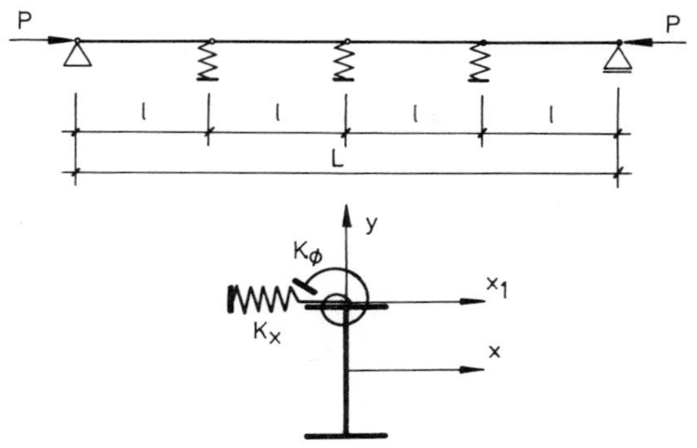

Fig 9

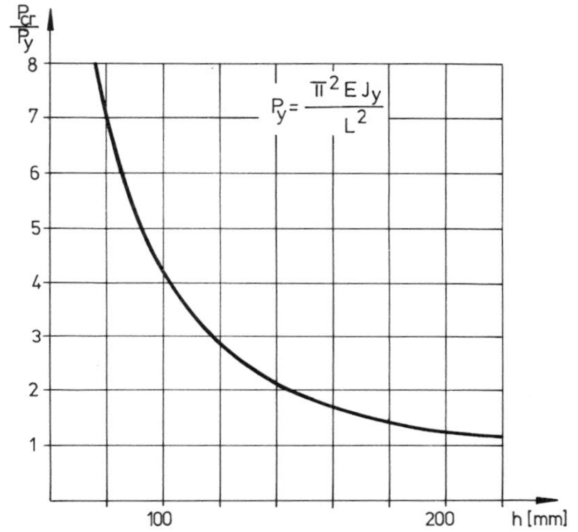

Fig 10

in this figure it can be seen that in the case of IPE 80 and IPE 100 sections being taken in the design top members rigidity is considerably higher. It proved to be higher than rigidity of these members in the plane perpendicular to the plane of top members.

Global co-action of the space grid and corrugated deck consists in a redistribution of internal forces in the structure. In order to determine quantitatively influence of this phenomenon a panel-frame model was assumed for calculations. Corrugated plates have been modelled on rectangular orthotropic panel members, for which the following matrix of elasticity is assumend

$$\begin{bmatrix} E_{xx} & 0 & 0 \\ 0 & E_{yy} & 0 \\ 0 & 0 & G \end{bmatrix}$$

The rigidity components have been determined in an approximate way with joint flexibility being added to panel flexibility. Calculations have been performed using finite element method, ICES-STRUDL II software and IBM 370 digital computer. The forces thus obtained and related to those previously determined for a space frame itself differ by 6 per cent for web and bottom members and by 10 per cent for top members with maximum forces in this group being lower by 14 per cent.

LIMIT STATE CALCULATION MODEL

In the limit state analysis the elastic-plastic behaviour of a single compressed member is of fundamental importance, particularly when post-buckling behaviour is considered. Because of complexity of this proboem various simplified models are used, differing from each another mainly in the way of assuming the force in a buckled member. Two extreme approaches may here be distinguished. In the first one it is assumed that the member after having reached ultimate force cannot carry any load any longer. Such a member is simply eliminated from the structure. Load-carrying capacity thus obtained is safe, but often it is considerably lower than the really existing one. In the second approach the member under compression is modelled in a similar way at the member in tension which means that after having reached critical force it is still able to carry that force. This model gives an unsafe evaluation of the ultimate load capacity. Therefore it has been decided to introduce more adequate model of member taking into account the initial deflection and residual rolling stresses as shown in Fig 11. Elastic-plastic analysis of an isolated member has been performed by

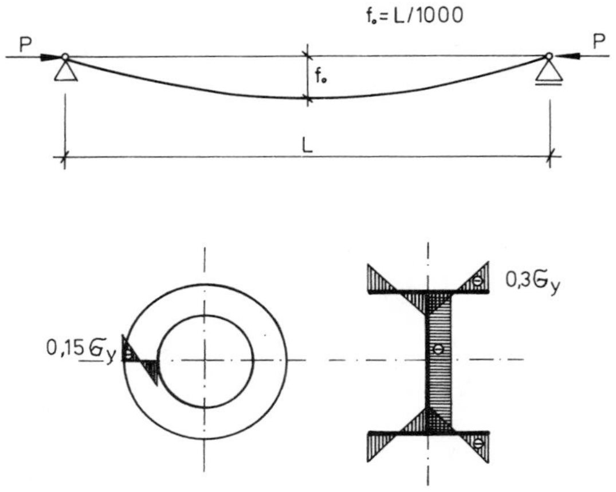

Fig 11

curvature method. In the behaviour of such a member three cases can be distinguished: elastic, one-side plastic and two-side plastic one. For the second and third cases transcedental equations are obtained. Solution of these equations for the assumed force P gives in effect the respective points of the curve P - Δ, see Fig 12, for the compression portion of the curve. Such an approach cannot be directly used in the structural analysis. For this reason, the determined P - Δ curve have been replaced by polynomials giving an explicit form of the function $\sigma / \varepsilon /$. Calculation of derivative of this function enables rigidity of the member to be determined in the individual stages of elastic-plastic behaviour. Depending on the accuracy required an approximation of curve $\sigma / \varepsilon /$ by means of tangentials involves division of this curve into a respective number of segments having constant derivative, that is corresponding with a constant rigidity of a member. Force corresponding with the end of each segment of the curve is a limit load of the member within respective rigidity range. Negative rigidity of the member after having reached ultimate force is represented by appropriately applied substitute force. Calculations are performed using step-by-step procedure, by applying unit loads and then determining minimum multipier of the load for which the given member or group of members will attain limit load corresponding with the given stage of analysis.

At that moment rigidity of the member is being changed, or in the case of tension members, the member is being elimina-

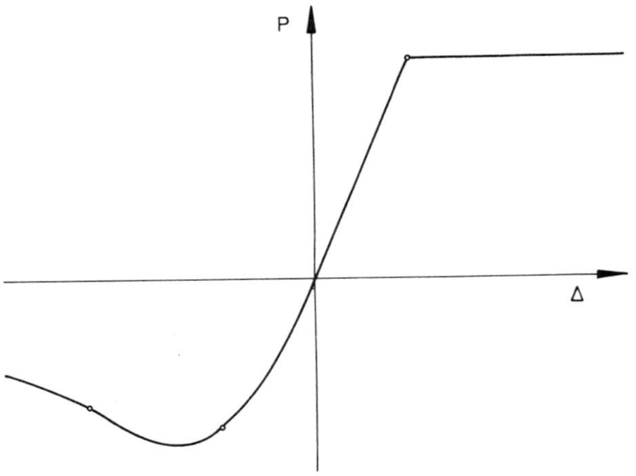

Fig 12

ted from further analysis. This procedure is repeated until a mechanism has been formed in the structure. In consequence, elastic-plastic analysis is reduced to consecutive static analyses which enables standard programmes to be used. The described method has been used for calculation of load-carrying capacity of the investigated roof grid 18 x 18 m. The load capacity thus obtained was 2,282 N/m^2, which corresponds with that obtained from full scale experiment with 1 per cent accuracy. It should be stressed, that in calculations an actually existing yield point obtained from material tests was assumed for bottom layer members, whereas for the remaining members a standard value was taken.

COMPARISON OF EXPERIMENTAL AND THEORETICAL RESULTS

Full presentation of a complete set of results obtained from full-scale experiments for both the grids investigated has not been possible because of a limited volume of this report. Therefore, the presentation is resrticted to comparison of deflections for the grid 12 x 18 m, and forces for the grid 18 x 18 m under design load /elastic range/ as well as for the grid 12 x 18 m with defects.

DISPLACEMENTS

Settlements of foundations and shortening of the columns supporting the roof structure were very small /practically negligible/ within the accuracy of measurements. Also the horizontal displacements of the columns were very small and did not have any influence on the work of the roof structure. On the other band, considerable "permanent" deflections of the space structure were obtained in spite of its elastic work under succeeding loads. This is clearly visible on the diagram of the load and deflection of the bottom layer joint, see Fig 13. Ranges of a considerable divergence of the loading and unloading curves and ranges of close proximity of them can here be distinguished. Permanent deflections do not increase during the next loading, if it is not greater than the previously applied one. Permanent deflections caused by overloading may be accepted as maximum for the structure under design load. Permanent deflections for grid 12 x 18 m amount to approx. 30 per cent of maximum deflections and 50 per cent of calculated ones. Vanishing of permanent deflections at the subsequent loading stages leads to a conclusion, that they are due to disappearance of clearance in bolted joints, straightening of members, etc. In consequence, they are not directly relevant to the loading function and, for this reason, they cannot be taken into account in comparative analysis of calculated and experimental deflections.

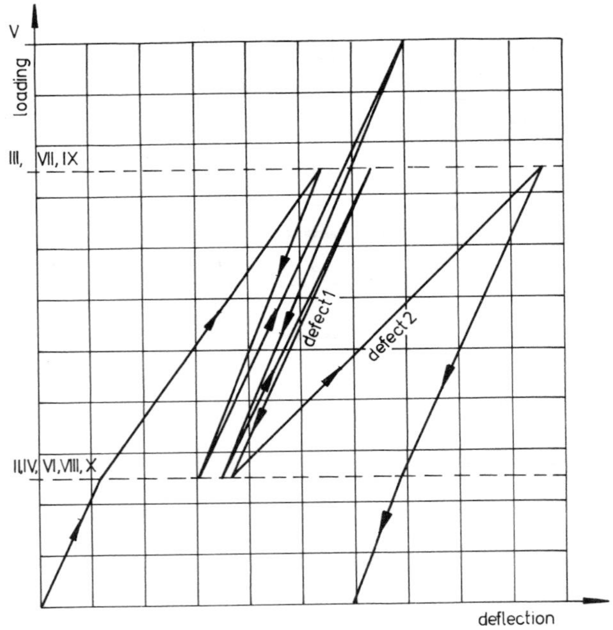

Fig 13

So-called recurring deflectione presented in table 1 should be recognized as representative ones in this respect. Data given in that table show a good agreement of results.

FORCES IN MEMBERS

Comparison of theoretical and experimental forces is performed in tables 2, 3 and 4. Percentage deviations of forces related to a theoretical value are also presented. In table 2 forces in the grid 18 x 18 m under the third stage loading are tabulated. In tables 3 and 4 forces for the same load in the case of the grid 12 x 18 m with defects 1 and 2 are given respectively. Taking into account the conditions of full-scale experoments on the building site, it can be find out that a good agreement between the theoretical and experimental results has been obtained.

CONCLUSIONS

On the basis of observations of full-scale experiments and theoretical analysis for both the elastic and elastic-plastic range of behaviour the following conclusions can be drawn:

a/ The experiments fully confirmed applicability of constructional solutions being accepted in design.

b/ Considerable permanent deflections of the roof structure consisting of prefabricated elements conncected by normal bolts must be taken into account in design and then during the exploitation of the structure.

c/ Taking into consideration in the design process the co-action of corrugated deck with space framework provides more accurate calculation model and enables top members weight to be.

d/ The method of evaluation of ultimate load capacity described herein has confirmed its applicability to the space structures under consideration.

Table 1

Joint number				Deflection /mm/ theoretical	Deflection /mm/ experimental				Deviation for a set of joints /%/
				Loading III					
2	5	28	31	11.2	10	11	10	15	3
3	4	29	30	16.3	16	17	16	16	0
8	11	22	25	17.7	17	16	16	16	- 8
9	10	23	24	21.2	22	21	21	21	0
13	16	17	20	21.9	20	21	26	19	- 2
14	15	18	19	24.6	23	24	23	23	- 5
				Leading V					
2	5	28	31	15.6	15	20	15	15	4
3	4	29	30	22.8	23	22	24	24	2
8	11	22	25	24.7	24	25	32	26	8
9	10	23	24	29.6	31	32	30	31	4
13	16	17	20	30.5	30	31	31	30	0
14	15	18	19	34.3	36	36	36	36	5
				Loading VII					
2	5	28	31	10.8	11	11	12	10	2
3	4	29	30	17.1	18	17	18	18	4
8	11	22	25	18.5	19	22	17	18	3
9	10	23	24	22.9	23	24	23	24	3
13	16	17	20	24.1	23	24	25	24	0
14	15	18	19	27.0	29	29	29	28	6
				Loading IX					
2	5	28	31	9.8	10	10	10	10	2
3	4	29	30	15.9	16	16	16	17	2
8	11	22	25	21.5	20	21	20	20	- 6
9	10	23	24	27.2	27	27	26	26	- 2
13	16	17	20	29.3	23	27	26	27	-12
14	15	18	19	36.7	33	34	33	31	-11

Table 2

Member kind	Member number	Member force /kN/ theoretical	Member force /kN/ experimental	Deviation /%/ for member	Deviation /%/ for a set of members
Top layer members	111	- 37.55	- 26.26	- 30	- 28
	404	- 37.55	- 27.46	- 26	
	194	- 52.41	- 45.42	- 13	- 3
	324	- 52.41	- 55.70	6	
	243	- 52.30	- 45.96	- 12	- 1
	275	- 52.30	- 58.40	11	
Web members	59	17.58	21.73	23	16
	462	17.58	19.24	9	
	100	- 17.97	- 21.38	19	10
	414	- 17.97	- 18.17	1	
	64	18.37	16.75	- 9	- 10
	457	18.37	16.39	- 11	
	165	- 8.07	- 14.35	77	47
	354	- 8.07	- 9.56	18	
	213	11.39	12.85	13	7
	308	11.39	11.66	2	
	214	- 10.74	- 11.06	3	- 6
	306	- 10.74	- 8.97	- 16	
Bottom layer members	66	25.79	25.72	0	1
	456	25.79	26.38	2	
	72	47.65	39.87	- 16	- 11
	450	47.65	44.14	- 7	
	218	47.36	39.78	- 16	- 16
	244	36.13	37.81	5	5
	252	48.85	51.57	5	6
	263	48.85	52.84	8	

Table 3

Member kind	Member number	Member force /kN/ theoretical	Member force /kN/ experimental	Deviation /%/ for member	Deviation /%/ for a set of members
Top layer members	29	- 32.10	- 36.61	14	- 2
	151	- 32.10	- 26.23	- 18	
	39	- 34.40	- 31.46	- 8	6
	137	- 34.40	- 41.76	21	
	50	- 41.07	- 49.23	19	15
	126	- 41.07	- 45.94	12	
	70	- 45.18	- 42.71	- 5	7
	102	- 45.18	- 54.07	19	
	71	- 44.96	- 52.43	16	2
	101	- 44.96	- 39.45	- 12	
	79	- 44.96	- 42.49	- 5	- 4
	98	- 44.96	- 42.95	- 4	
Web members	5	- 4.51	- 3.33	- 26	- 38
	168	- 4.51	- 2.22	- 50	
	28	43.39	- 30.41	- 30	- 19
	171	43.39	- 30.66	- 8	
	24	- 32.55	- 39.66	22	- 12
	167	- 32.55	- 16.63	- 49	
	21	24.91	27.79	11	- 3
	164	24.91	20.67	- 17	
	19	- 16.55	- 12.47	- 25	- 9
	155	- 16.55	- 17.82	7	
	56	14.90	18.81	26	19
	143	14.90	16.75	12	
	81	- 11.32	- 5.68	- 50	- 42
	110	- 11.32	- 7.47	- 34	
	26	- 6.32	- 5.03	- 20	10
	152	- 6.32	- 8.87	40	
Bottom layer members	20	45.64	47.43	4	- 4
	163	45.64	39,67	- 13	
	15	64.55	67.70	5	1
	159	64.55	62.42	- 3	
	18	7.05	9.23	30	16
	132	7.05	7.25	3	
	46	23.19	21.98	- 5	1
	105	23.19	24.84	7	
	74	25.44	28.79	13	12
	100	25.44	28.57	12	
	72	25.37	25.06	- 1	2
	103	25.37	26.43	4	
	77	20.81	22.20	6	0
	108	20.81	19.44	- 6	

REFERENCE

1. J BRÓDKA, A CZECHOWSKI, A GRUDKA and J KORDJAK, "Mostostal" Space Grids - A Development and Applications, Third International Conference on Space Structures, University of Surrey, Guildford, 11 - 14 Sept. 1984.

Tab. 4

Member kind	Member number	Member force /kN/ theoretical	Member force /kN/ experimental	Deviation /%/ for member	Deviation /%/ for a set of members
Top layer members	29	- 14.65	- 20.69	41	41
	39	- 13.12	- 15.92	21	29
	137	- 13.12	- 18.21	38	
	50	- 35.07	- 21.63	- 38	- 27
	126	- 35.07	- 29.74	- 15	
	70	- 51.41	- 54.07	5	- 1
	102	- 51.41	- 47.58	- 7	
	71	- 51.18	- 56.23	10	6
	101	- 51.18	- 52.45	2	
	79	- 51.18	- 50.83	- 1	- 13
	98	- 51.18	- 37.85	- 26	
Web mem bers	28	32.36	35.48	9	12
	171	32.36	39.92	23	
	24	- 27.64	- 33.82	22	- 1
	167	- 27.64	- 20.79	- 25	
	21	20.00	27.44	37	58
	164	20.00	35.99	80	
	56	52.29	56.65	8	- 2
	143	52.29	45.96	- 12	
	47	- 27.81	- 27.50	- 1	- 2
	127	- 27.81	- 26.60	- 4	
	81	7.80	11.06	41	39
	110	7.80	10.76	38	
Bottom layer members	20	42.57	43.12	1	- 7
	163	42.57	36.22	- 15	
	15	59.49	66.66	4	5
	159	59.49	63.39	6	
	25	- 3.06	- 1.31	44	8
	139	- 3.06	- 2.19	- 28	
	46	27.64	36.05	30	17
	105	27.64	28.79	4	
	74	42.62	37.37	- 12	- 5
	100	42.62	43.52	2	
	72	- 20.65	- 14.07	- 32	- 33
	103	- 20.65	- 13.62	- 34	

RANDOM ULTIMATE BEARING CAPACITY OF ELASTIC-PLASTIC SPACE TRUSSES

J W RZĄDKOWSKI PhD Eng

Adiunkt, Institute of Civil Engineering,

Technical University of Wrocław, Poland.

Until now, estimation of the reliability of elastic-plastic trusses has been usually conducted on the basis of semiprobabilistic estimation of distribution quantiles of the truss random ultimate capacity. For a full probabilistic estimation of the reliability, estimation of the cumulative distribution function of the truss random ultimate capacity is required. So far for redundant trusses, full probabilistic estimations of capacity have been conducted by using the Monte Carlo method. This paper presents a new method of full probabilistic estimation of the capacity of trusses made from elastic-plastic members.

INTRODUCTION

Knowledge of the cumulative distribution function $F[N_T]$ of the random ultimate bearing capacity N_T of a space truss is essential for accurate estimation of its reliability R, see Refs 1,3,6,10. N_T can be analysed by linear and stochastic programming methods. Algorithms of linear and stochastic programming are used for the estimation of the ultimate capacites associated with the particular kinematically sufficient mechanisms /KSM/ of collapse of bar structures, see Refs 1,2,4,7,8,10,11. To the most frequently used algorithms belong: the Simplex method and the Monte Carlo method . In many works when estimating the cumulative distribution function $F[N_T]$ plastic collapse mechanisms of the truss are authoritatively isolated or sets of weakened truss members are assumed a priori and the truss ultimate capacity is analysed by the step-by-step method, see Refs 9,12.

In the present paper, a method of estimating the ultimate capacity of non-linear redundant space trusses made from elastic-plastic members is presented. The method, called a Limit Equilibrium in Nodes /LEN/ method, is based on the kinematic theory of ultimate capacity.

DESCRIPTION OF THE LIMIT EQUILIBRIUM IN NODES /LEN/ METHOD

The fundamental assumptions of the Limit Equilibrium in Nodes /LEN/ method were given by Kowal, see Refs 5,6. A general form of the LEN method- for single-layered and multi-layered space trusses made from different types of elastic-plastic members is given in Ref 10.
The LEN method assumes that collapse of a space truss is caused by a formation of a plastic collapse mechanism /PCM/ which changes the static structure of the truss into a kinematic chain, see Ref 6,10. A plastic collapse mechanism of a space truss is a result of exhaustion of the capacity of the last member from such a minimum set A of members, the collapse of which changes the truss into a kinematic chain . Thus, each PCM is characterized by an ultimate capacity N_{PCM} being the function of capacities N which describe the members belonging to set A at the moment of a PCM formation. The value of capacity N of a given member is determined by the phase of plastic work in which this member is and by

the type of elastic-plastic model this member represents, i.e. by the relation between the force $S \rightarrow N$ in the member and the displacement δ of its ends. The most suitable for numerical calculations is the model of elastic-plastic member in the form of a segmental-linear approximation of the continuous function $\delta = \delta(S) : S \rightarrow N$.

When a PCM is formed, the members which belong to set A may be in different phases of plastic work, hence, the capacity N_{PCM} characteristic for the given PCM can have different values, depending on the actual combination of capacities N which describe the plastic work phases of the particular members from set A.

In the LEN method, the term limit state /LS/ of a truss denotes the moment of formation of a PCM characterized by:

1/ a specified set A of members which have reached ultimate capacity,

2/ a specified combination of capacities N in the members which form set A,

3/ a specified geometry of the truss structure,

4/ a specified ultimate capacity being the function of the capacity and geometry of members.

As the measure of ultimate capacity in the LEN method, load P - loading of the joints of a truss at the moment when a given PCM has been formed - is taken,

$$P_r = m_r P \qquad \ldots\ldots \; 1.$$

where:

P_r - loading of the r-th joint of the truss,
m_r - load factor of the r-th joint,
r - numeration of truss joints.

The load $P = N_{PCM}$ can be determined by solving a set of equations of the equilibrium of forces in the truss joints, specific for the given limit state /LS/ of the truss. Determination of all possible N_{PCM}s under specified restrictions allows to estimate the destribution function $F[N_T]$ of the truss ultimate capacity.

DETERMINATION OF THE GEOMETRY OF THE TRUSS IN THE LIMIT STATE

In order to solve the set of equations of the equilibrium of forces in the limit state of the truss, the geometry of the truss G_{PCM} at the moment of formation of a plastic collapse mechanism must be known. Since the geometry G_{PCM}, specific for a given PCM, is the unknown, it is determined in two steps.

The algorithm by means of which the coordinates of the truss in a given limit state, i.e. geometry G_{PCM}, are determined, is the following , see Ref 10, and Fig 1 a b c d,:

a/ it is assumed that the function $\delta = \delta S : S \rightarrow N$, making up a description of the elastic-plastic model of each of the members in set A, has an identical, as far as its shape and absolute values are concerned, image $\delta^* = \delta^*(S^*) : S^* \rightarrow N^*$ for load S^* equal in value to load S but having an opposite sign, see Fig 2 a b;

b/ for a known geometry G_o of the analysed truss, the capacity associated with the PCM formed by a given set of members A being in specified phases of plastic work is determined by the LEN method, assuming for the calculations that the truss is loaded with load P^* and the members are described by capacities N^*, see Fig 1 b;

c/ on the basis of the determined values of load P^* and forces S^* in the truss members and the known characteristics $\delta^* = \delta^*(S^*)$ of the members belonging to set A, the geometry G_o^* of the truss, i.e. the coordinates of the truss joints for the load $P^* = 0$, is determined by any of the available methods. The geometry G_o is the same as the geometry G_{PCM} of the truss at the moment of the PCM formation;

d/ for the analysed truss, for which the geometry $G_o^* = G_{PCM}$ has been determined, the capacity associated with a given PCM is found by the LEN method for a given combination of capa - cities N of the members which form set A.

The determined capacity N_{PCM} which characterizes a given plastic collapse mechanism takes into account the effect of the change in geometry in the limit state on the ultimate capacity of the truss.

A schematic diagram of the determination of the ultimate capacity N_{PCM}, which includes the change in the truss geometry is shown in Fig 1 a-d. Fig 2 shows the method of forming the characteristic $\delta^* = \delta^*(S^*)$ on the basis of function $\delta = \delta(S)$ which constitutes the elastic-plastic model of the truss members.

Numerical analysis of the geometry G_{PCM} specyfic for a given PCM may be done by means of the programs "GEOMEGR - GRW" presented in Ref 10.

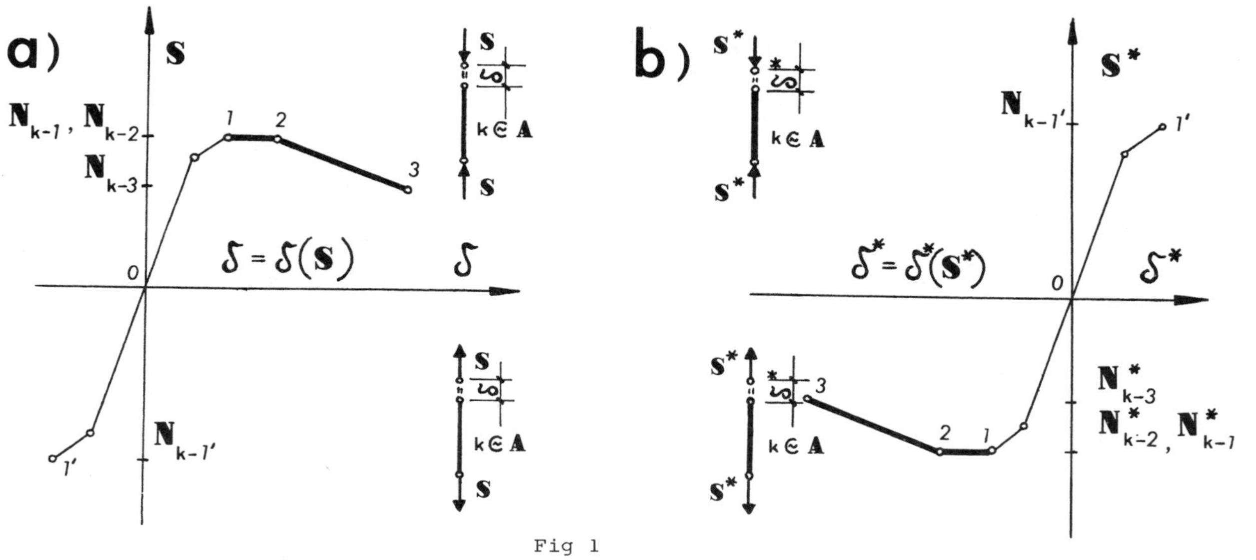

Fig 1

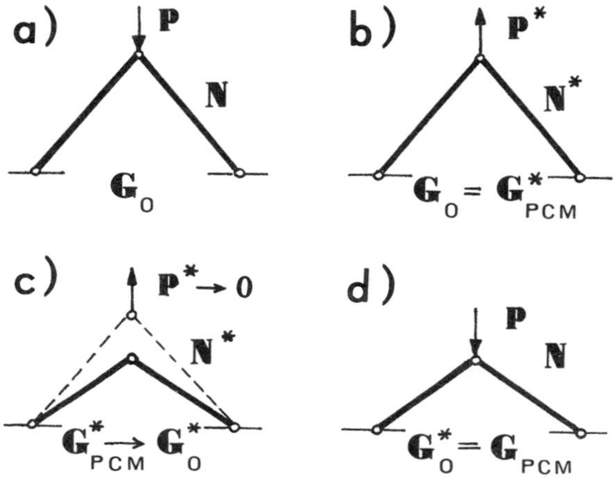

Fig 2

DETERMINATION OF PLASTIC COLLAPSE MECHANISMS IN ELASTIC-PLASTIC TRUSSES

In order to determine by the LEN method the cumulative distribution function $F[N_T]$ of the random ultimate capacity of an elastic-plastic truss, the analysis should be conducted according to the following algorithm given in Ref 10:

a/ determine all possible plastic collapse mechanisms /PCM/ of the truss;

b/ determine the capacities N_{PCM}, which characterize the particular PCMs, according to the algorithm presented above,

c/ determine the partial probabilities associated with the particular N_{PCM}s.

Until now for the determination of PCMs of elastic-plastic trusses, the Monte Carlo method combined with a repeated analysis of the truss by the step-by-step method has been used. The Monte Carlo method was applied to the random distribution of the weakened members in the truss and the ultimate capacity associated with the PCM characteristic for a given configuration of weak truss members used to be determined by the step-by-step method, see Refs 1,2.

In Ref 10, a new method of determining the PCM of trusses based on the Simplex algorithm combined with an analysis of the result by means of the theory of graphs is presented. The commonly used Simplex algorithm yields only PCMs described by the sets of capacities N_{k-1} determined by points "1" which separate the phases of plastic work of the members a, b, ..., k, ... $n \in A$. The remaining PCMs, formed by combinations of the capacities delimited by the points which separate the different phases of plastic work of the members a, b, ..., k, ..., $n \in A$ /except member "n", which being the last member of set A is characterized by capacity N_{n-1}/, are determined by analysing the graphs of the set of the limit equilibrium equations.

A schematic diagram of the graph analysis is shown in Fig 3, where broken line was used for the rejected branches of the graph.

Numerical analysis of plastic collapse mechanisms /PCMs/ and the capacities N_{PCM}s, which characterize the particular plastic collapse mechanisms space truss may be done by means of the programs "SIMPLEX - GRW" and "DENDRYT - GRW" presented in Ref 10.

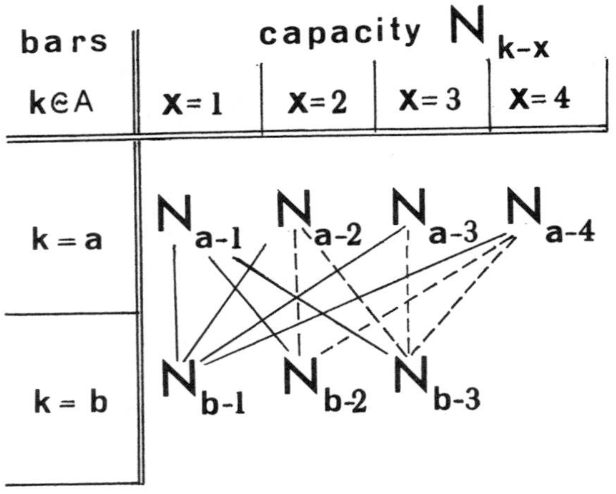

Fig 3

ESTIMATION OF THE CUMULATIVE DISTRIBUTION FUNCTION OF THE TRUSS RANDOM ULTIMATE CAPACITY

In the present paper, the ultimate capacities of the members which form set A are assumed to be random variables N of normal distribution. Thus, the ultimate capacities characterizing the particular PCMs are also random variables N_{PCM} since they are a function the the random capacities N of the members which belong to set A.

The cumulative distribution function $F[N_T]$ of the truss random ultimate capacity is determined numerically by tabulating the partial probabilities u_j associated with the capacities $N_{PCM\ j}$ which characterize the particular PCMs. Since the normal distribution is stable relative to the transformations, the variables N_{PCM} are also characterized by normal distribution.

In order to determine the partial probabilities, it is assumed that the expected value of the truss capacity $E[N_T]$ can be presented in the form:

$$E[N_T] = E[N_{PCM\ j=1}] - u_j\ D[N_{PCM\ j=1}] \quad \ldots\ldots\ 2.$$

where:

$E[N_T]$ - the expected value of capacity for the PCM characterized by the smallest value of N_{PCM} from among the set of capacities associated with the particular PCMs,

$j=1,2,..,m,..,p$ numeration of the capacities characteristic for the particular PCMs, ordered in a sequence in - creasing with the values of the particular N_{PCM}s,

$D[N_{PCM\ j}]$ - standard deviation of the $N_{PCM\ j}$ variable,

u_j - tabulated parameter of normal distribution.

The parameters $u_{j=m}$ of the remaining p-1 capacities N_{PCM} are determined from the equation:

$$E[N_{PCM\ j=1}] - u_{j=1}\ D[N_{PCM\ j=1}] = E[N_{PCM\ j=m}] - u_{j=m}\ D[N_{PCM\ j=m}] \quad \ldots\ldots\ 3.$$

The expected value of capacity $E[N_{PCM\ j=m}]$ is determined from an appropriate equation of the limit equilibrium of the truss loaded in the joints with forces P:

$$E[N_{PCM\ j=m}] = E[P_{PCM\ j=1}] = \Big(\sum_r m_r\ c_r^{(t)}\Big)^{-1} \times \Big(\sum_{i\in B} S_i\ c_i^{(t)} + \sum_{k\in A} E[N_k]\ c_k^{(t)}\Big) \quad \ldots\ldots\ 4.$$

and the standard deviation from the formula:

$$D[N_{PCM\ j=m}] = D[P_{PCM\ j=m}] = \Bigg(\Big(\sum_r m_r\ c_r^{(t)}\Big)^{-2} \times \Big(\sum_{k\in A} (c_k^{(t)})^2\ D^2[N_k]\Big)\Bigg)^{1/2} \quad \ldots\ldots\ 5.$$

where:

$c^{(t)}$ - directional cosines of the members and of the external load P re - lative to the axis t = X,Y,Z,

S_i - force in the "i" member belonging to the set B of members which do n not belong to set A,

N_k - random capacity of the "k" member from set A in a given plastic phase of work,

i - numeration of the members belong - ing to set B,

k - numeration of the members belong - ing to set A.

Partial probabilities for the assumed distribu - tion of the variable N_T are determined from the distribution tables:

$$\text{Prob}\ \{N_{PCM\ j=m} \geqslant P\} = \Phi(u_{j=m}) \quad \ldots\ldots\ 6.$$

The cumulative distribution function $F[N_T]$ of the random ultimate capacity is estimated numerically by tabulating successively according to numeration "j" all the N together with the cor - responding partial probabilities determined from formulas 4 and 6 :

$$F[N_T] \overset{\text{table}}{=} \text{Prob}\ \{N_{PCM\ j=m} < P\} \quad \ldots\ldots\ 7.$$

A NUMERICAL EXAMPLE

For the space truss shown in Fig 4 the cumulative distribution function of its ultimate capacity was estimated under the following assumptions :

a/ truss members are characterized by three phases of plastic work, as shown in Fig 5 and by parameters given in Tab 1;

b/ random ultimate capacities N which describe the models of member work are characterized by Gaussian distributions;

c/ distribution of the variable random ultimate capacity N of members is assumed to be Gaussian and is bilaterally cut off because of the quality control of the members.

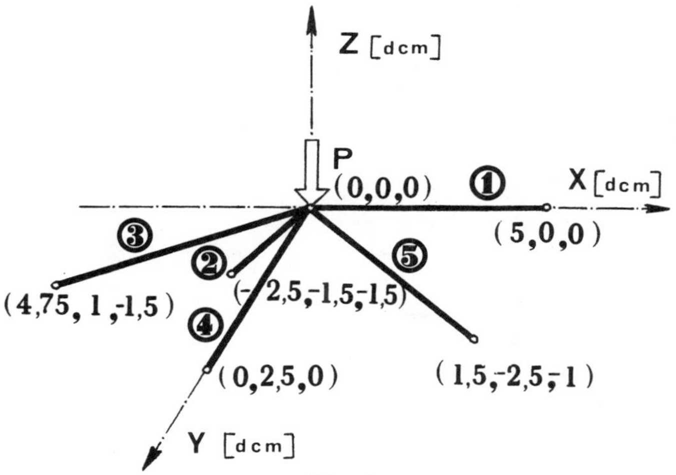

Fig 4

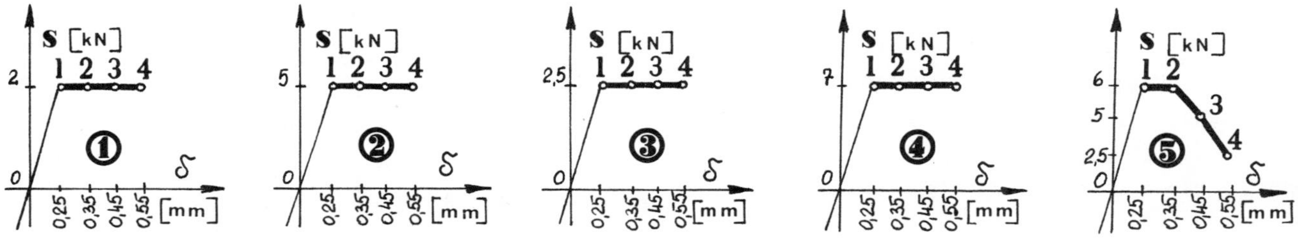

Fig 5

Tab 1

Bar No	E[N] [kN]	D[N] [kN]	$N_L = E[N] \mp 3\,D[N]$ [kN]	
1	2	3	4	
1	2,0	0,280	1,160	2,840
2	5,0	0,150	4,550	5,450
3	2,5	0,175	1,975	3,025
4	7,0	0,210	6,370	7,630
5	6,0 5,0 2,5	0,030 0,054 0,120	5,910 5,162 2,860	6,090 4,838 2,140

The results of the calculations are compiled in table 2. Column 1 of the table contains designations of consecutive PCMs, where: a, b, c - numbers of members which belong to set A and x - number of the plastic phase of a member; column 2 contains the expected values of capacity N_{PCM}; in column 3, the standard deviation of capacity N_{PCM} is given; column 4 contains the parameter of the normal distribution associated with $N_{PCM\ j=m}$; in column 5, the probability of occurrence of a given PCM under the imposed restrictions is given.

A graphic interpretation of the estimation of the truss cumulative distribution function $F[N_T]$ is shown in Fig 6.

Tab 2

N_{PCM} a−x, b−x, c−x	$E[N_{PCM}]$ [N]	$D[N_{PCM}]$ [N]	u	Prob $\{N_{PCM} \geqslant P\}$
1	2	3	4	5
2-**1**, 3-**4**, 5-**2**	4291	74,48	2	0,9772
2-**2**, 3-**1**, 5-**1**	4293	74,92	2,016	0,9781
2-**3**, 3-**1**, 5-**2**	4301	74,69	2,133	0,9835
2-**1**, 3-**3**, 5-**2**	4307	74,61	2,216	0,9867
2-**2**, 3-**1**, 5-**2**	4310	74,75	2,254	0,9879
2-**1**, 3-**2**, 5-**2**	4324	74,74	2,431	0,9925
2-**1**, 3-**2**, 5-**1**	4327	74,84	2,466	0,9932
2-**1**, 3-**3**, 5-**1**	4328	74,76	2,491	0,9936
2-**1**, 3-**1**, 5-**2**	4340	74,87	2,647	0,9959
2-**1**, 3-**1**, 5-**1**	4343	74,97	2,679	0,9963

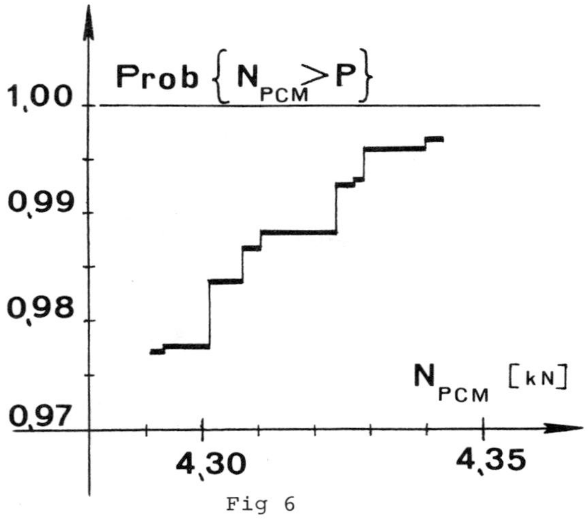

Fig 6

FINAL REMARKS AND CONCLUSIONS

The conducted analysis allows to formulate the following conclusions and observations:

1/ If an analysis of the random ultimate capacity of an elastic-plastic contains restrictions as to the range of plastic work of truss members, the occurrence of the particular PCMs is conditioned by the occurrence of an appropriate configuration of preliminary internal forces in the truss members.

2/ The population of capacity N_{PCM} obtained for the assumed restrictions of variable N and for random coordinates of loaded joint of truss shown in Fig 6 is divided into fractions delimited by the values of the coefficient of variation of capacity N_{PCM} and the values of the parameter of capacity distribution.

3/ Estimation of reliability of an elastic-plastic truss can be conducted by means of the Bonferroni-Boole inequality, using the formulas given in Refs 6, 10.

ACKNOWLEDGEMENTS

The author would like to thank Prof. Z. Kowal for his constructive suggestions and M.Sc. Z. Grycak for the help in numerical calculations. The author also would like to thank the Institute of Civil Engineering Technical University of Wrocław for sponsoring the present work.

REFERENCES

1. Augusti G., Casciati F., Reliability of Redundant Structures., Losowe obciążenia i nośność konstrukcji., KILIW PAN, Ossolineum 1979, pp. 9-65.

2. Baratta A., Non-linear Truss Reliability by Monte-Carlo Sampling., Applications of Statistics & Probability in Soil & Structural Engineering Proc. of 3-rd Int.Conf., Sydney 1979, pp.136-148.

3. Casciati F., Sacchi G., On the reliability theory of the structures., Meccanica Vol. 9, No. 4, Dec. 1974.

4. Čiras. A.A., Metody linejnogo programirowanija pri rasčete uprugo-plastičeskich sistem., Izd. Lit. po Stroilel'stvu., Leningrad 1969.

5. Dziubdziela W., Kopociński B., Kowal Z., Ultimate bearing capacity of structural systems with minimal critical sets having joint elements in pairs, Arch. of Mechanics 25,5/1975

6. Kowal Z., Random ultimate bearing capacity of domes with random-joint coordinates constructed from bars with random-ultimate bearing capacity., Applications of Statistics & Probability in Soil & Structural Engineering., Proc. 3-rd Internatl. Conf., Sydney. Australia. Jan.29- Feb.2, 1979., pp. 566-573.

7. Maier G., Metody programowania matematycznego w analizie konstrukcji sprężysto-plastycznych., Arch. Inż. Ląd. 21, 3, 1975.

8. Mezzina M., Prete G., Tosto A., Automatic and experimental analysis for a model of space grid in elasto-plastic behaviour., II-nd International Conf. on Space Structures. University of Surrey., Guildford 1975.

9. Murzewski J., Bezpieczeństwo konstrukcji budowlanych., Arkady, Warszawa, 1976.

10. Rządkowski J., Nośność graniczna jednowarstwowych kopuł prętowych obciążonych w węzłach ., PhD Dissertation, Wrocław, 1983.

11. Telega J.J., Zastosowanie programowania liniowego do wyznaczania nośności granicznej konstrukcji.,Mech.Teoret. i Stos., 1, 1971.

12. Tolman F.P., de Witte F.C., The Influence of Member Quality on the Safety of Space Trusses., International Association for Bridge and Structural Engineering., IX Congress., Amsterdam 1972.

STABILITY ANALYSIS OF ELASTIC SPACE TRUSSES

CZESLAW CICHON

Technical University of Cracow
31 - 155 Krakow, Poland

On the basis of FEM a method of the stability analysis of elastic space trusses has been worked out. Global instability as well as local buckling of truss members have been considered. An initial postbuckling behaviour has been analyzed applying an approximation according to Koiter's approach. The paper deals with the computation of the fundamental and postcritical paths of equilibrium related to single critical points. The numerical examples indicate that efficiency of the method seems to be satisfactory.

1. INTRODUCTION

In engineering practice the analysis of space trusses in the presence of large displacements is still under consideration. These structures, e.g. reticulated trusses are still used succesfully to design of large span building construction, see Ref 3. It is a well known fact that in case at such structures there is a possibility for instabilities to appear even in the elastic range, associated with relative large displacements, see Ref 5. Besides a global instability of the structure (of snap - through or bifurcation type) can also appear a local instability due to a buckling of the individual truss members. In certain situations it can be the reason for the rapid decrease of carrying capacity of the structure, see Ref 1.
In the paper the method of analysis of both types of instabilities of the structures is presented.

2. ASSUMPTIONS

The following assumptions are used :

1. The structure is a system of initially straight members joined in ideal - hinges.
2. Static and conservative loads are applied in the nodes of the structure.
3. The material obeys Hooke's law.
4. Strains are small but displacements can be large.
5. Buckling of the member can occur if $P = P_E$ where P_E is the Euler buckling load. The buckling plane (x,y) is established by the axes of the local cartesian coordinates (x,y,z).
6. After buckling of the member the Euler - Bernoulli hypothesis is valid.

3. KINEMATICS OF INDIVIDUAL MEMBER

3.1. Strain - displacement equation

Making use of the assumption 6 the strain - displacement equation for any point of the cross - section of the member is as follows,

$$E = E_o - ky \qquad \text{...... 1.}$$

On the basis of the assumption 4 and Fig 1 we can write formulae for the axial strain E_o and the curvature k. After some manipulations we find the forms,

$$E_o = u' + 0.5v'^2 + 0.125v'^4 \qquad \text{...... 2.}$$

$$k = d\alpha/ds = (1 + 0.5v'^2)v'' \qquad \text{...... 3.}$$

Here, u and v are the components of the displacement vector u(u,v,w) and the symbol ´ is used to denote x - derivative. The values of u^t and u^b, marked on Fig 1, are the shortening of the member due to the compression and the bending deflection, respectively.

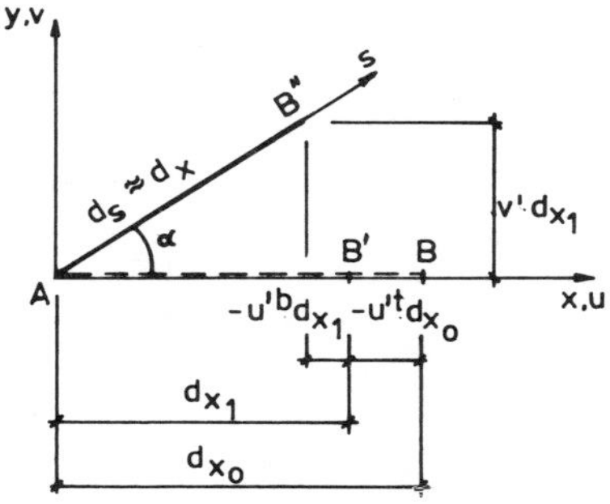

Fig 1

3.2. Incremental relationships

An incremental form of the local kinematic variables is derived in the framework of the Total Lagrange Approach (TLA) with the local moving coordinates related to an actual configuration of the member. It means that at time t the displacement vector u is calculated in relation to the initial configuration transformed to the actual configuration of the member, Fig 2b. The rigid motion of the structure is taken into account by an iterative procedure. The increments of the displacements are defined as a difference between the variables in the actual configuration at the time t and neighbouring configuration at the time t + dt.

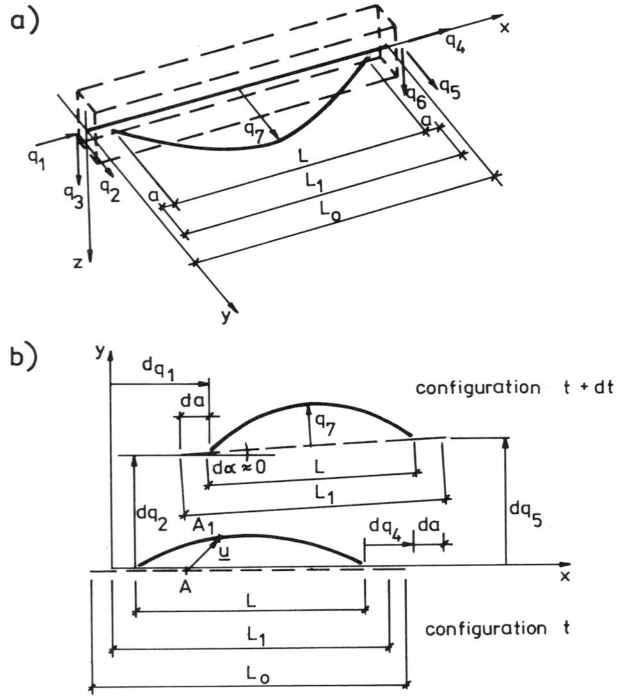

Fig 2

The increment of dE can be decomposed into two parts and written as,

$$dE = dE_L + dE_{NL} \qquad \ldots\ldots 4.$$

in which dE_L is linearly dependent on the increments of the displacements u and v and dE_{NL} is quadratically dependent on the increments of the displacement v,

$$dE_L = du' - v'(1 + 0.5v'^2)dv' - ((1 + 0.5v'^2)dv'' + v'v''dv')y \qquad \ldots\ldots 5.$$

$$dE_{NL} = -0.5(1 + 1.5v'^2)dv'^2 - (0.5v''dv' + v'dv'')dv'y \qquad \ldots\ldots 6.$$

4. PHYSICAL EQUATION

Taking into account the assumption 3 the physical equation has the form,

$$S = CE = S_o - Cky \qquad \ldots\ldots 7.$$

where C is the Young's modulus and $S_o = CE_o$ is the axial stress of the member. The incremental S - E equation is assumed in the linearized form,

$$dS = CdE_L \qquad \ldots\ldots 8.$$

5. INCREMENTAL VARIATIONAL EQUILIBRIUM EQUATION

The incremental wirtual work for an individual member has the form,

$$INT = \int_V (S + dS)\delta dE\, dv \qquad \ldots\ldots 9.$$

where the integral is extended over the volume of the member after deformation. Making use of Eqns 4 to 8 we can write the linear part of Eqn 9 as follows,

$$INT = I_1 + I_2 + I_3 + I_4 \qquad \ldots\ldots 10.$$

where the following notation has been introduced,

$$I_1 = CA\int_{L_1} du'\delta du'dx + CI\int_{L_1} dv''\delta dv''dx \qquad \ldots\ldots 11.$$

$$I_2 = AS_o\int_{L_1} (1 + 1.5v'^2)dv'\delta dv'dx \qquad \ldots\ldots 12.$$

$$I_3 = CA\int_{L_1} v'((1 + 0.5v'^2)^2 dv'\delta dv' + (1 + 0.5v'^2)(du'\delta dv' + dv'\delta du'))dx$$
$$+ CI\int_{L_1} (v''^2(1 + 0.25v'^2)dv''\delta dv'' + v''^2(1 + 1.5v'^2)dv'\delta dv' + v'v''(2 + v'^2)(dv'\delta dv'' + dv''\delta dv'))dx \qquad \ldots\ldots 13.$$

$$I_4 = AS_o\int_{L_1} (\delta du' + v'(1 + 0.5v'^2)\delta dv')dx + CI\int_{L_1} v''(1 + 0.5v'^2)(v'v''\delta dv' + (1 + 0.5v'^2)\delta dv'')dx \qquad \ldots\ldots 14.$$

in which L_1 is the length of the member under compression only and A is the cross - sectional area and I is the principal moment of inertia of the section relative to z axis.

The external wirtual work is given by equation,

$$EXT = P\,\delta dL \qquad \ldots\ldots 15.$$

where P is the axial load of the member in the configuration t + dt and dL is the increment of the length of the member.

Equating the right hand of Eqns 10 and 15 the following variational equilibrium equation is obtained,

$$I_1 + I_2 + I_3 + I_4 = P\,\delta dL \qquad \ldots\ldots 16.$$

6. DISCRETIZATION

It is assumed that the member has two nodes and seven degrees of freedom. The seventh DOF is the deflection of the middle section of the member in the buckling plane (x,y). The deflection can be approximated by the expression,

$$v = q_7 \sin(\pi/L_1)x \qquad \ldots\ldots 17.$$

Making use of Eqn 17 the total displacements u and v and their increments can be expressed as,

$$| u , v |^T = Nq \qquad | du , dv |^T = Ndq \qquad \ldots\ldots 18.$$

where N is the matrix of basic functions,

$$N = \begin{vmatrix} 1 - x/L_1, & 0 & , 0 , & x/L_1 , & 0 & , & 0 , & 0 \\ 0 & , 1 - x/L_1 & , 0 , & 0 & , x/L_1 & , & 0 , & \sin(\pi/L_1)x \end{vmatrix} \qquad \ldots\ldots 19.$$

and q and dq are the total and incremental displacements of the degrees of freedom of the member, respectively. The stress S_o can be found as,

$$S_o = CE_o = C(L_1 - L_o)L_o = C(dL/L_o + 2a/L_o) \qquad \ldots\ldots 20.$$

where a is the half of the symmetrical displacement of the member nodes due to the bending deflection,

$$a = (L_o/4L_1) \int_{L_1} v'^2(1 + 0.25v'^2)dx = \pi^2 L_o q_7^2/8L_1^2 + 3\pi^4 L_o q_7^4/128L_1^4 \qquad \ldots\ldots 21.$$

Finally, introducing Eqn 17 into Eqns 11 to 14 the following results are obtained,

$$I_1 = \delta dq^T(k_L)dq \quad I_2 = \delta dq^T(k_S)dq$$

$$I_3 = \delta dq^T(k_u)dq \quad I_4 = \delta dq^T(f) \qquad \ldots\ldots 22.$$

where k_L , k_S and k_u are the local matrices of the member: linear, initial stresses and initial displacements, respectively and f is the local vector of the nodalforces, see Ref 1.

The variational equilibrium equation, Eqn 15, now takes the form,

$$(k_L + k_S + k_u)dq = PB - f \qquad \ldots\ldots 23.$$

where the vector $B = | 1,0,0,-1,0,0,0 |$. This equation relating to the local moving coordinate system may be used for the analysis of the finite buckling deflections of the member.

7. EQUILIBRIUM EQUATIONS FOR STRUCTURE

The incremental equilibrium equations for the structure can be established by the standard assembly procedure. The external load is assumed to be one - parametric load,

$$R = pR_o \qquad \ldots\ldots 24.$$

where R_o is a reference load and p is the loading parameter.

Finally, the incremental equilibrium equations takes the form,

$$K_T\, dQ = dp\, R_o + U \qquad \ldots\ldots 25.$$

In this equation $K_T = K_L + K_S + K_u$ is the tangent stiffness matrix and U is the vector of unbalanced forces,

$$U = pR_o - F \qquad \ldots\ldots 26.$$

where F is the vector of the nodal internal forces of the structure. If the state t of the structure is in equilibrium then $U \equiv 0$, in the other case Eqn 25 is used to form a suitable procedure of iteration.

8. SOLUTION PROCEDURE

The solution of the problem can be obtained iteratively in the load - displacements space R^{N+1}, where N is the number of degrees of freedom of the structure. It means that instead of Eqn 25 the enlarged system of equations will be solved,

$$(K_T - R_o)dQ = U$$
$$T^T dQ_1 = dSP \qquad \ldots\ldots 27.$$

where vectors $dQ_1 = | dQ , dp |^T$ and T and parameter SP control the proces of computations. The parametr SP is related to the loading parameter, to any of the global displacement or the measure of the length of the tangent vector of the equilibrium path. Such an approach enables us to eliminate difficulties in computations in the vicinity of limit points. The solution procedure is presented in detail in Ref 4. The postcritical path is traced by an eigenvector of the tangent stiffness matrix (global buckling) or by an increment of the internal degree of freedom of a select member (local buckling).

As a convergence criterion for the iteration process the modified Euclidean norm for displacements has been adopted. A relative error is assumed to be $ER = 10^{-4}$.

The computer programme has been developed for the numerical studies. The programme is written in FORTRAN IV and implemented on a CYBER 70 computer.

9. NUMERICAL EXAMPLES

9.1. Postbuckling behaviour of the pinned column

The initial postbuckling behaviour of the column is determined by $dQ = 0$ and $Q_i = 0$ for $i = 1,2,\ldots,6$ and $Q_7 = 0$ which leads to the condition that $f_7 = 0$. If we now use Eqn 14 to solve this equation, we will obtain,

$$P/P_E = (1 + \pi^2 Q_7^2/2L_1^2 + 3\pi^4 Q_7^4/32L_1^4)/(1 + 3\pi^2 Q_7^2/8L_1^2) \qquad \ldots\ldots 28.$$

where $P_E = \pi^2 CI/L_1^2$ is the Euler buckling load.

Finally, after expansion of the denominator in the power series we can get the well known Koiter's approximation formulae, see Ref 2,

$$P = P_E(1 + \pi^2 Q_7^2/8L_1^2) \qquad \ldots\ldots\ 29.$$

This result indicates that the present method can be used for calculations of the postcritical behaviour of the structure if the deflections of the buckling member do not exceed ca $0.25L_o$.

9.2. Shallow truss dome

The shallow truss dome shown in Fig 3a was analyzed. The load versus the deflection of the load point for the entire loading path is shown in Fig 3b. The results obtained with the present method coincide with those obtained by the other authors, see Ref 3. In Fig 3c and in Tab 1 the results of the postcritical behaviour of the structure due to the buckling of the upper compressive members, N_1 to N_6, are shown. The decrease of the load after bifurcation of the structure is associated with the unloading in the lower compressive members, N_7 to N_{18}. It is interesting to note that the same situation takes place if the lower members buckled at first (it can be done by changing the geometric characteristics of the sections of the members). Then the upper members are unloaded. It means that in this structure it is impossible to buckle all the compressive members.

REFERENCES

1. CZ CICHON, Nonlinear Analysis of Elastic Trusses, Arch. Inz. Lad., 4, 1983 (in Polish).

2. C L DYM, Stability Theory and its Applications to Structural Mechanics, Noordhoff International Publishing - Leyden, 1974.

3. H ROTHERT, T DICKEL and D RENNER, Snap - Through Buckling of Reticulated Space Trusses, J. of Struct. Div., ASCE, Vol. 107, No ST1, Proc. Paper 15973, Jaunary, 1981, pp. 129 - 143.

4. Z WASZCZYSZYN, Numerical Problems of Nonlinear Stability Analysis of Elastic Structures, Comp. and Struct., Vol. 17, No 1, 1983, pp. 13 - 24.

5. CZ WOZNIAK and M KLEIBER, Nonlinear Structural Mechanics, PWN, Warszawa, 1982 (in Polish).

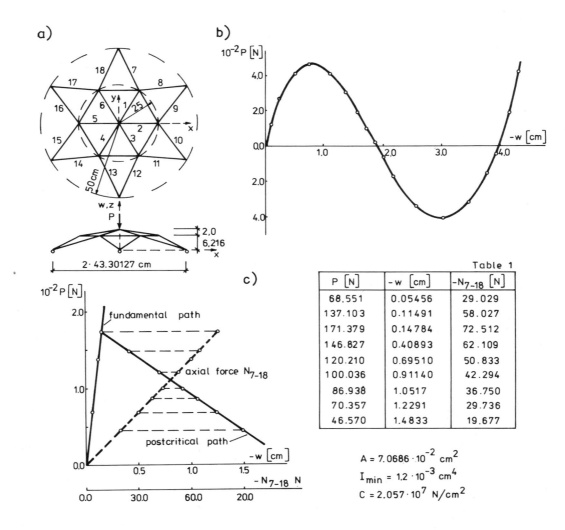

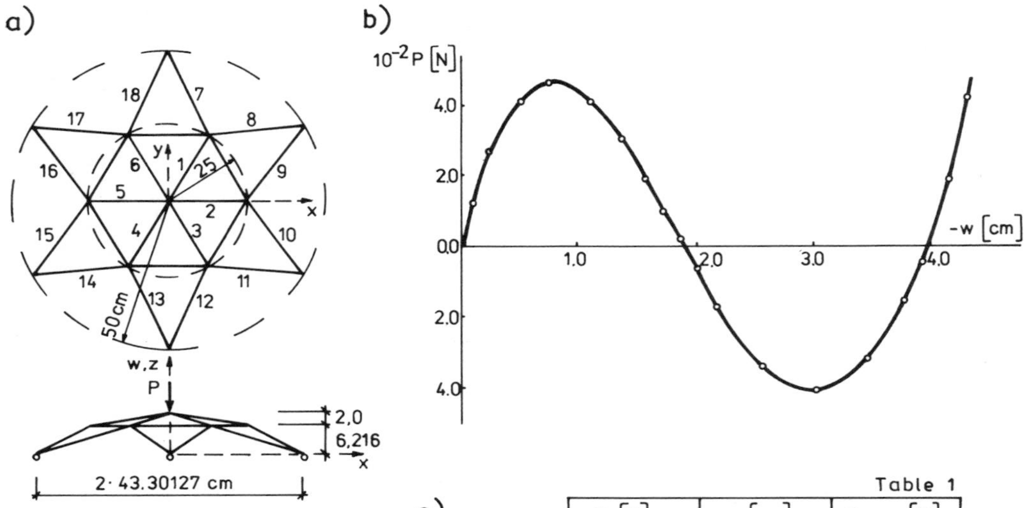

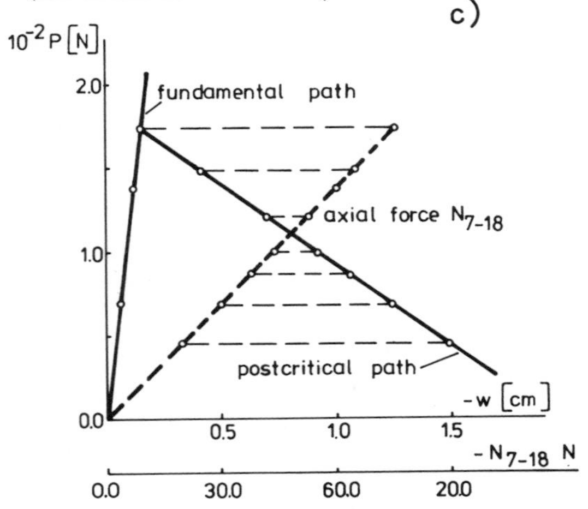

Table 1

P [N]	-w [cm]	$-N_{7-18}$ [N]
68.551	0.05456	29.029
137.103	0.11491	58.027
171.379	0.14784	72.512
146.827	0.40893	62.109
120.210	0.69510	50.833
100.036	0.91140	42.294
86.938	1.0517	36.750
70.357	1.2291	29.736
46.570	1.4833	19.677

$A = 7.0686 \cdot 10^{-2}\ cm^2$

$I_{min} = 1.2 \cdot 10^{-3}\ cm^4$

$C = 2.057 \cdot 10^7\ N/cm^2$

Fig 3

AN EXPERIMENTAL STUDY ON LOAD-BEARING CAPACITY OF STEEL TRUSS MEMBERS OF SPACE FRAMES

T. SUZUKI*, I. KUBODERA** and T. OGAWA***

* Professor of Tokyo Institute of Technology, Dr. of Eng.

** Chief Engineer of Structural Designing Dept., Tomoegumi Iron Works, Ltd.

*** Research Associate of Tokyo Institute of Technology, Dr. of Eng.

The numerical analysis of failure process for three-dimensional structures will require a great deal of expenses and labor. It is the purpose of this paper to outline a procedure simplifying the failure process analysis for the whole space structure. In this procedure, repeated load tests are performed on truss beams composing the space structure, the load-deformation diagrams are obtained and they are replaced with a simplified model. In the elastic-plastic range, it is assumed that the truss beam or basic unit behaves in accordance with the idealized model curve.

1. INTRODUCTION

The members of a large span structure are generally selected to resist against the most critical loading such as dead load and snow load. In many districts in Japan, not a few structures have collapsed owing to snow load beyond the design load. These structures including three-dimensional are generally designed in conformity to allowable stress design methods, and the behavior of these structures in the state of snow loading beyond the design condition and their ultimate strength are not fully realized. It is a matter of course that a knowledge of their behavior and ultimate strength is very useful to manage the safety of structures. For example, at the stage when the snow load on a structure approaches the ultimate load, it is possible to prevent the structure from collapsing by means of partial removing the snow or setting up supports.

In ordinary statically indeterminate structures including plane frames, even though some members are stressed exceeding their proportional limit, the redistribution of stresses is made and further load can be carried by other less heavily stressed parts of the structures. Especially, such tendency is remarkable in space structures with higher degree of redundancy. It is impractical to trace the behavior of space structures beyond their proportional limit to their collapse, through the overall numerical analysis requiring a great deal of expenses and labor. Therefore, the elastic-plastic behavior tests are performed of truss beams or unit trusses composing the space structure and the load-deformation diagram is expressed by means of a simple formula. Then, with this idealized curve, a simpler analytical method grasping the behavior of the whole structure is groped for.

This paper, as the first step, describes the idealized modeling of load-deformation relationship for truss beams. It is also applicable to plane frames. Thus, the ultimate analysis for space structures can be made by using the same method as plane frames.

2. OUTLINE OF THE EXPERIMENT FOR TRUSS BEAMS

In the experiment, a parallel-chord truss beam is regarded as a single-unit member forming the space structure. By applying concentrated loads to the beams repeatedly; the load-deformation curves were obtained. The experiment apparatus and the whole view are shown in Fig 1 and Photo 1, respectively. The test beams and the loading beam were rigidly connected with high strength bolts. Concentrated loads were applied to the rigid connections by using oil jacks. As to the end condition of the members, roller bearings were used so that both the ends can move and rotate freely in the direction of the member axis. The test beam was also restricted laterally at intermediate points. Concerning the lateral restraints, bearings were used at loading points, and Teflon resins and stainless steel plates were used at the adjacent panel points. They functioned smoothly. These lateral restraints induced the carrying capacity of truss beams to depend on the vertical buckling load of the chord members.

Loading tests were made, first, beyond the maximum load in the positive direction (the direction in which compressive stresses were set up in the upper chord); then, beyond the maximum load or to about 1.5 times the deformation for the proportional limit in the negative direction; and finally, causing the failure in the positive direction. It is a 1.5 times repeated loading. Experimental results show that the deformation in the range beyond the proportional limit concentrates on a certain panel with the maximum stress in the truss and the load-deformation relationship for the whole truss member can be explained by the characteristics of that panel.

Experimental parameters include the sectional shape of truss members, lacing pattern, slenderness ratio of compression members and shear-span ratio. As the materials of the test beam, JIS Standard SS41 structural carbon steels were used. Their tensile strength ranges from 41 to 52 kg/mm^2 and their mechanical properties are given in Table

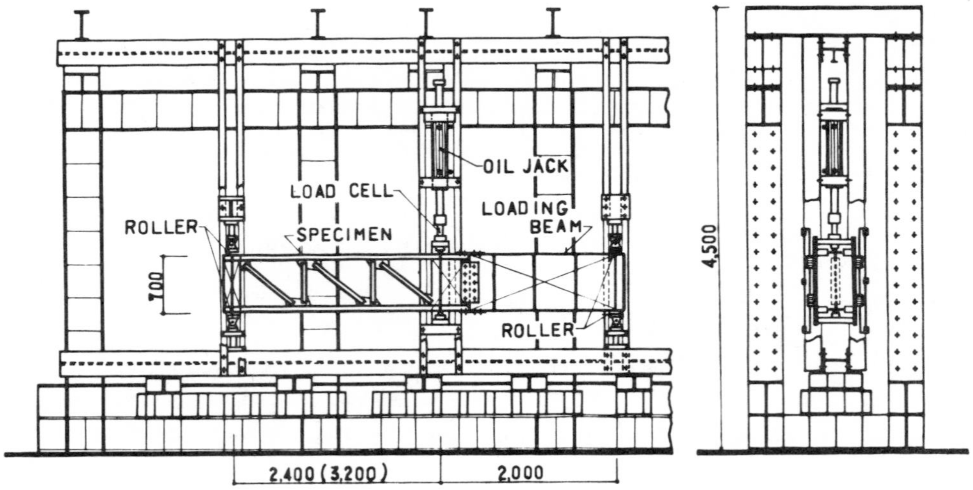

Fig 1

Photo 1

Table 1

Members	Yield Stress σ_y (t/cm^2)	Ultimate Strength σ_u (t/cm^2)	Elonga-tion (%)
L-65×65×6	3.357	4.736	24.10
L-75×75×6	3.059	4.619	27.63
CT-97×150×6×9	3.243	4.917	26.45

1. To investigate the fundamental behavior of an individual member forming the truss beam, stub column tests were performed. Stub columns whose slenderness ratio ranges from 15 to 50 were compressed without eccentricity on condition that both ends are hinged by means of oil-hydraulic rounded supports and the load-shrinkage diagrams were obtained. Photo 2 is a view of the test.

3. THE RESULTS OF STUB COLUMN TESTS

The load-shrinkage diagram for stub columns is roughly given in Fig 2. As the load increases, the Euler buckling occurs at point A, causing the buckling deformation and reducing the carrying capacity sharply to point B; but below point B, forming a comparatively stabilized curve; and at point C, bringing about a marked local buckling and reducing the carrying capacity further. The load ratio of point A to point B depends on the slenderness ratio of stub columns, and the deformation at point C depends mainly on the width-thickness ratio of members. However, in the case of JIS Standard steel shapes with their smaller slenderness ratio, the deformation at point C is 5 or 6 to 10 times as much of that at point A. Therefore, as far as the characteristics of truss beams is concerned, it will not be a serious deformation. From the above, replacing the experimental curve with an idealized model, an experimental equation (the broken line) can be obtained in Fig 2.

$$N_{mx} = \{1 - 0.4(\frac{\lambda_e}{\Lambda})^2\}/\{1 + \frac{4}{9}(\frac{\lambda_e}{\Lambda})^2\} \quad \text{....... (1)}$$

$$5\lambda_e^2 = (N_{mn})^{-2}\cos^2(\frac{\pi N_{mn}}{2}) \quad \text{............ (2)}$$

$$K_d = -0.15(\lambda_e - 0.5) \quad \text{.................. (3)}$$

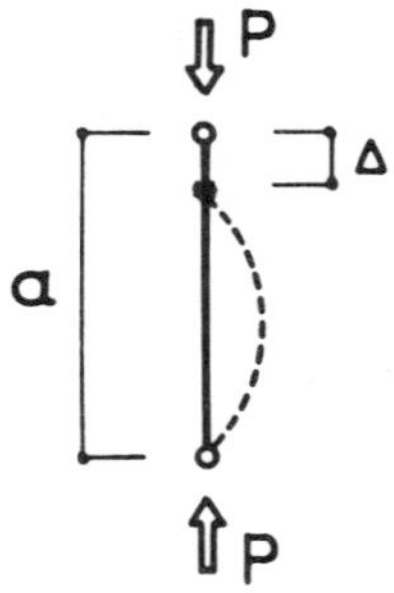

Photo 2

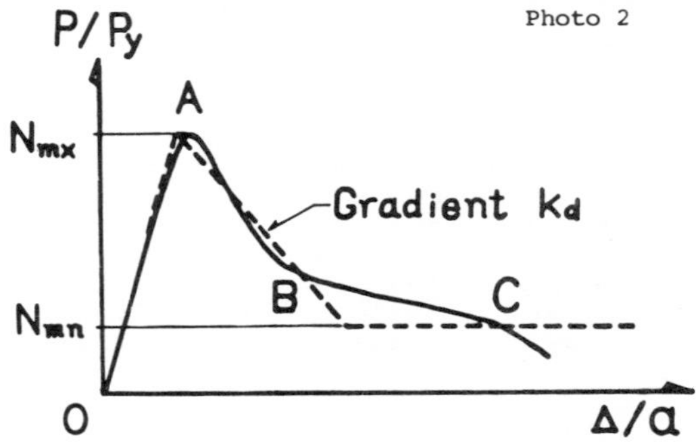

Fig 2

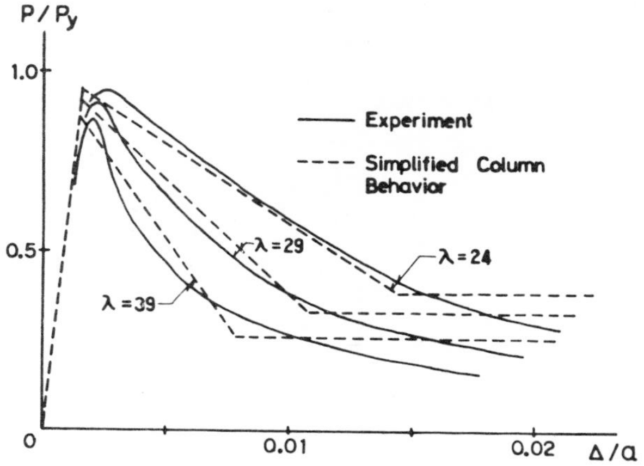

Fig 3

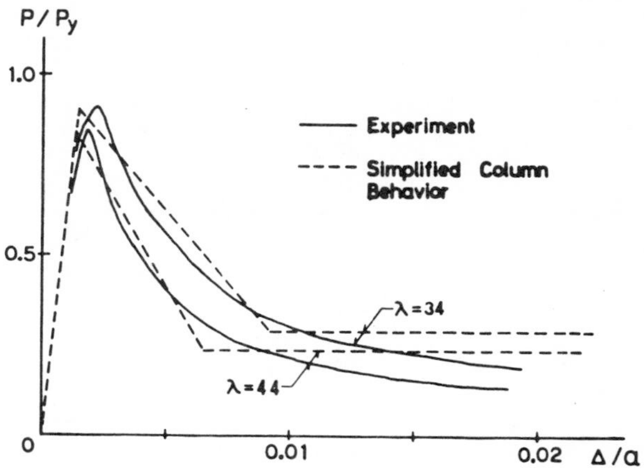

Fig 4

Where λ = slenderness ratio
ε_y = yield strain

$$\lambda_e = \sqrt{\varepsilon_y}\cdot\lambda \ , \ \Lambda = \frac{\pi^2}{0.6}$$

The results of stub column tests with every length are given in Figs 3 and 4. They show that Eqs (1) to (3) agree very well with the experimental results.
For above reason, the load-shringage diagrams of single-unit members composing truss structure are simplified as shown in Fig 5 by using Eqs (1) to (3).

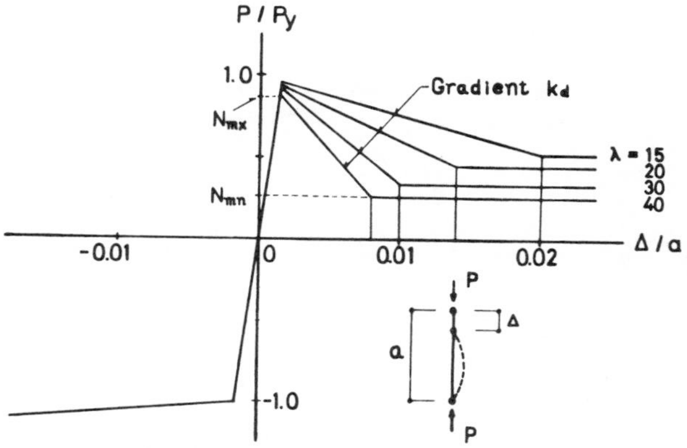

Fig 5

4. EXPERIMENTAL RESULTS OF TRUSS BEAMS

The load-deformation diagram for truss members by experiments is roughly given in Fig 6. In truss beams, due to the existence of the shearing force, the value of compressive normal forces is different from that of tensile normal ones. In Fig 6, if the load is applied to the truss beam in the positive direction, the maximum compressive normal force is produced in the upper chord member (member I) at the loading part and the buckling occurs at point A. After buckling, the carrying capacity reduces steeply to point B. If the load is applied in the negative direction, the maximum tensile normal force is produced in the same member, but the reduction of the carrying capacity is not recognized at point C. Then, due to the strain hardening, the carrying capacity increases to point C' where member II buckles. The load is unloaded at point C and loaded again to point D in the positive direction. Below point D, a stabilized curve is traced and a marked local buckling of member I occurs at point E, worsening the mechanical properties of the member.

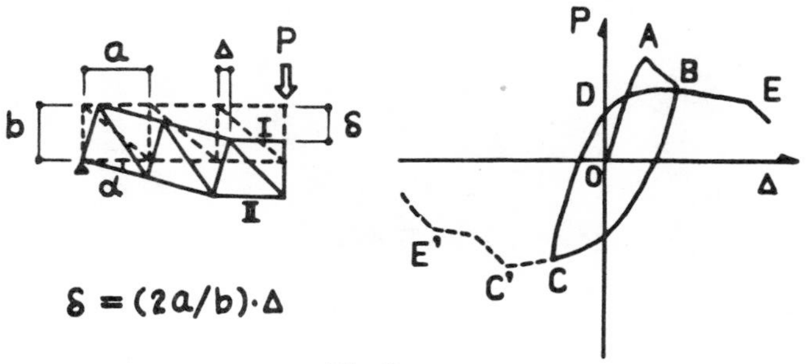

Fig 6

Suppose that the deformation characteristics of truss beams in the range above the proportional limit depend on only the deformation of a certain panel in a plastic zone, and the load-shrinkage diagram for the compression chord member in buckling failure is replaced with an idealized model described in the preceding paragraph. The deformation of the truss beam is related to the shrinkage of the compression chord member by the following equation, based on a few assumptions that the shrinkage of other members is smaller than that of the compression chord member and the joint translation angle (α) is very small.

$$\delta = (2a/b)\ \Delta \ \ldots\ldots\ldots\ldots\ldots\ldots\ldots\ldots \quad (4)$$

where δ = deformation of truss beam

Δ = shrinkage of the compression chord member

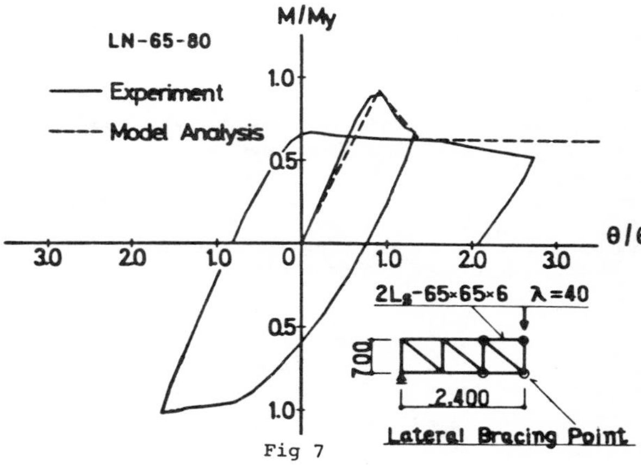

Fig 7

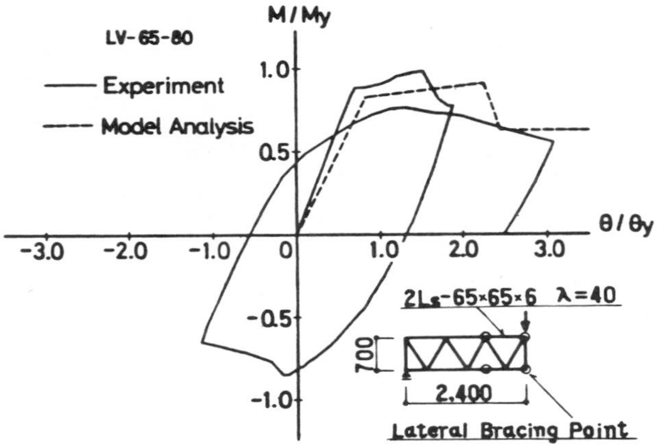

Fig 8

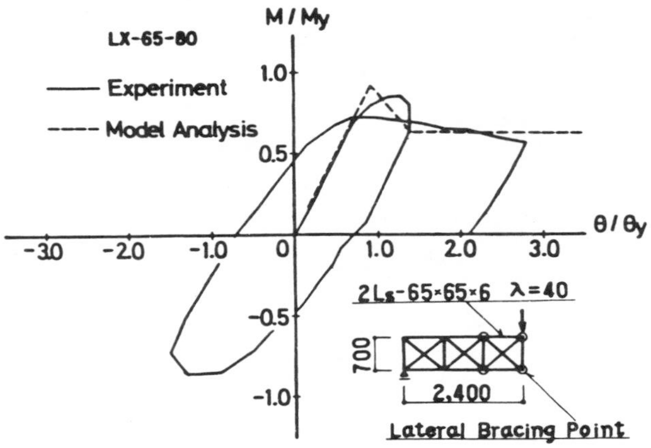

Fig 9

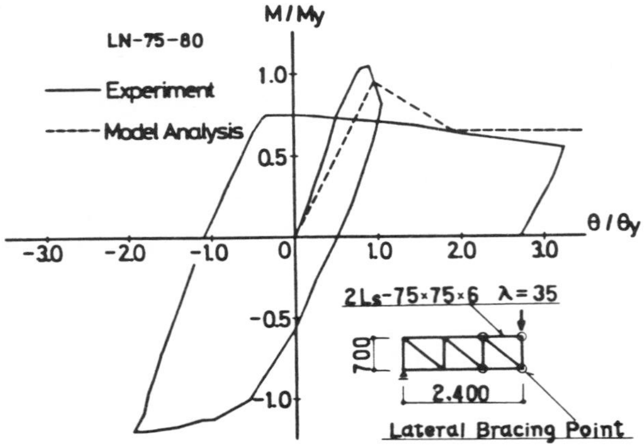

Fig 10

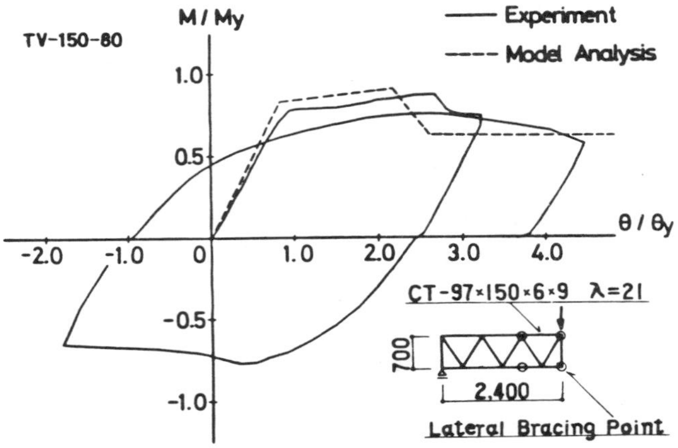

Fig 11

The load-deformation diagram by Eqn (4) is laid on the experimental results, as the broken lines in Figs 7 to 11. The broken lines indicate the condition in which the compression chord member with the loading part in buckling is assumed to have one hinged end and the other fixed end. The beams in which the tensile yielding goes first are assumed to increase the carrying capacity with rigidity of 1/100 of elastic rididity until the buckling of the compression chord member occurs. In the test beams with T-shaped sections, the effect of the torsional buckling on the chord members is so great that corrections are made for it by means of the effective slenderness ratio.

These results are given in Figs 7 to 11. In these figures, they are expressed as dimensionless by means of My and θy respectively, in which My is the moment at the loading point at the time when the chord member with the loading point of the truss beam begins to yield, and θy is the rotation angle at the member end.

Individual members composing the truss beam are reinforced with the gusset plates of lacings and have a certain degree of restraint effect. To reflect this effect on the above theory, the end condition of an individual member is considered to be one end hinged and the other fixed. This is proved by the fact that the idealized curves for the broken lines almost agree with the experimental results in Figs 7 to 11.

Based on the results presented herein, it is clarified that the characteristics of one truss panel in a plastic zone can explain the load-deformation relationship for the whole truss beam.

5. EXPERIMENTAL RESULTS OF PLANE FRAMES

A portal frame was tested. It consists of one truss beam and two solid-wall columns, on which a constant vertical load and a variable horizontal load were given. Fig 12 shows the experiment apparatus and Photo 3 shows the whole view. The materials of the tested frame are of JIS Standard SS41 and the beam was fabricated of 2L-65x65x6 (in mm) rolled shapes. The carrying capacity of the column is about two times that of the beam, this causing the frame to fail on the beam. The vertical loads applying to the three

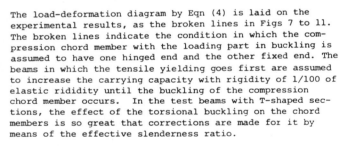

Photo 3

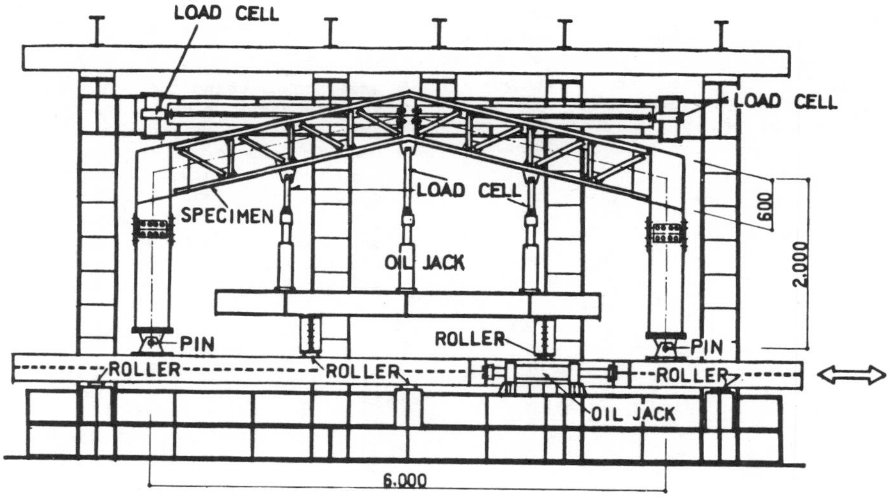

Fig 12

intermediate points of the beam were loaded increasingly until the beam-end moment reached a half of the yield moment and after then, were kept constant by using constant-loading oil jacks until the experiment was finished. The horizontal displacement of the frame was given by applying the horizontal force repeatedly to the middle of the beam as shown in Figs 12. The horizontal force H and horizontal deformation δ are expressed as dimension ton and cm, respectively.

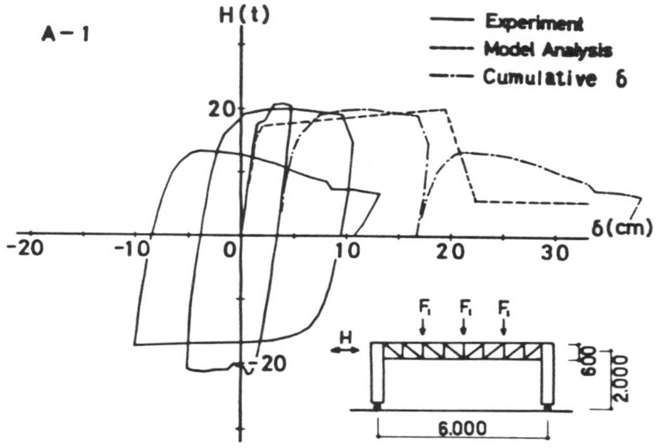

Fig 13

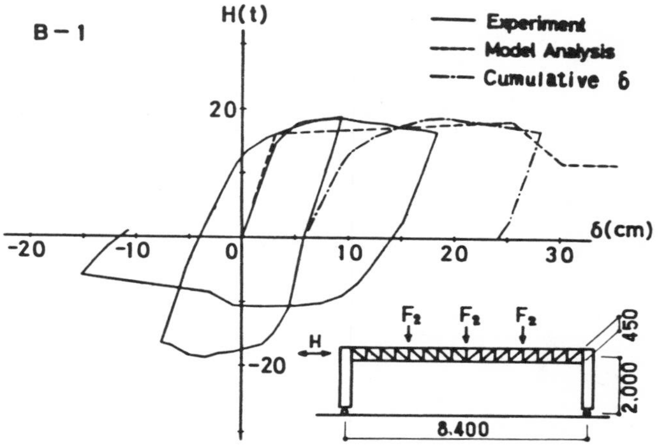

Fig 14

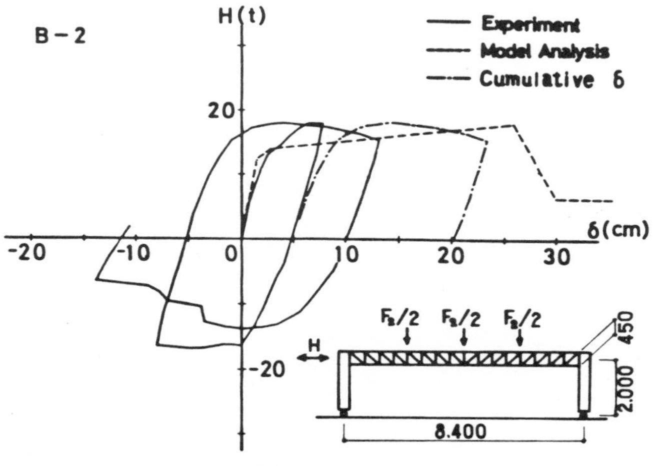

Fig 15

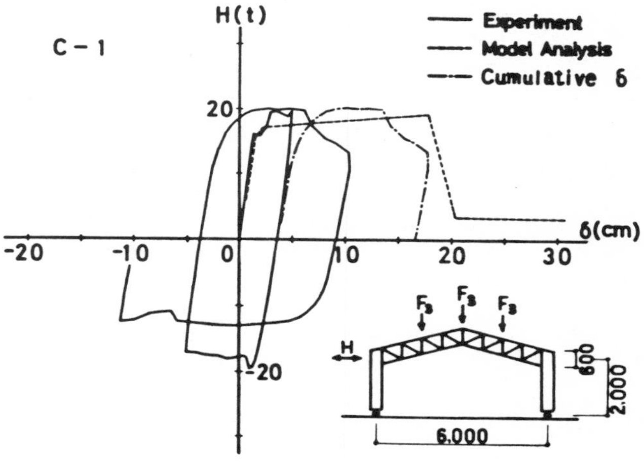

Fig 16

In this experiment too, the load-deformation relationship in the plastic range of the frame depends on the behavior of the beam-end panel in a plastic zone. On the basis that the behavior of the panel in a plastic zone has the quality of the idealized model as shown in Figs 7 to 11, the load-deformation diagram for the whole frame obtained by a simple analysis, in dashed lines is laid on the experimental results in Figs 13 to 16.

It is clear that the experimental results closely resemble the analytical results obtained by the idealized model and the behavior in the plastic range can almost be grasped by this analytical method in the plane frame too.

6. DEVELOPMENT INTO SPACE STRUCTURES

The failure process for space structures can easily be analyzed by using the analytical method of failure process for plane frames described above, but in the case of space structures, it is necessary to make the idealized model for plastic behavior of the three-dimensional basic unit. Suppose that an individual member composing the basic unit can show the idealized behavior given in Figs 5, defining the behavior of the basic unit beforehand by means of geometric analysis, and that the unit follows this rule in the plastic range of space structures. Then, the whole analysis can be made by combining it with the remaining elastic part.

EQUIVALENT STRUT THEORIES FOR POSTBUCKLED TRIANGULAR PLATES

R C GILKIE, PhD, MCSCE, FRSA

Associate Professor, Department of Civil Engineering
Technical University of Nova Scotia, Canada

A wide variety of space structures use stressed skin pyramidal units as basic elements. The triangular surfaces of these units operate largely in the postbuckled range. This paper shows that a number of effective width theories developed for buckled rectangular plates are applicable to the development of an equivalent strut concept for postbuckled triangular and trapezoidal plates. Results are compared to experimental analyses carried out on individual plate junctions as well as full pyramids.

INTRODUCTION

While many researchers have devoted attention to the post-buckling behaviour of rectangular plates, very little attention has been paid to the postbuckling behaviour of triangular plates. A wide variety of space structures use thin sheet pyramidal elements as a basic repeated form in their geometry.

Whether the pyramids have three, four or six sided bases, they are all made up of triangular or truncated triangular elements. The materials from which such structures are fabricated are usually thin sheet FRP, aluminum or steel. The very nature of the thin sheet triangular elements dictates that they operate in the postbuckled range through most of the load history of a structure.

Early attempts by the author to analyse such structures, see Ref 1, were based on the development of an equivalent strut concept where hexagonal base pyramidal units to be used in a grid structure were studied experimentally. Comparing the load-deformation characteristics of the thin sheet FRP pyramids with those of an idealized discrete member pyramid, with the thin sheet triangular wall elements transformed into equivalent struts; a graph of effective strut area versus load was developed for four different load-support conditions.

The concept was tested on a ten pyramid space grid, analyzed as a discrete element system, with encouraging results. Maximum stresses were predicted to within 15% and deflections were predicted to within 20% of the experimental values.

Early work by Von Karman, Sechler and Donnell, see Ref 2, indicated that rectangular plates, when in a postbuckled state, carried most of the load along two narrow strips at the simply supported edges parallel to the load direction. The ultimate load was independent of the width and length of the plate and approximately proportional to the square of the thickness.

A triangular plate can be considered as a plate with a variable width. Hence, it seems reasonable that a comparison can be made with the action of the buckled rectangular plate if the ultimate load is independent of width. An excellent review of attempts to describe the effective width of buckled rectangular plates is given by Jombock and Clark, see Ref 3. Most of the formulae developed were either empirical or semi-empirical in nature.

To test the equivalent strut theories developed for rectangular plates on triangular plates, a set of experiments was devised. A hexagonal based pyramid was loaded at its apex in one set of experiments; and in another, two adjacent triangular plates were loaded in compression and in tension along the junction between the plates. The results are compared and conclusions drawn.

EQUIVALENT STRUT THEORIES

Von Karman and Sechler showed that the ultimate load formula for a buckled rectangular plate was equal to

$$P_u = C\sqrt{E\sigma_y}\;t^2 \qquad \text{...... 1.}$$

where σ_y = the yield strength of the material and C lay between $\dfrac{\pi}{\sqrt{3(1-\mu^2)}}$ and $\sqrt{\dfrac{2}{1+\mu}}$

for two limiting cases of deflected shape. See Ref 2, and 'Figs 1(a) and 1(b)'.

In the process of calculating the values of C, it was shown that the expression for the effective width had the form:

$$b_e = Ct\sqrt{\frac{E}{\sigma_e}} \qquad \text{...... 2.}$$

where $b_e = 2W_e$, and σ_e is the stress on the effective width, see Fig 1(c).

Winter extended Von Karman's conclusions to show that they held for plates which were component parts of structural members as well as for plates tested individually. See Ref 4. He also found that the effective width formula worked for stresses below as well as at ultimate load. Hence Eqn 2 can be used between the limits of the critical buckling load and the ultimate load of the plate.

Combining the results of his tests with those given in Ref 2, Winter plotted values of C versus $\sqrt{\frac{E}{\sigma_e}}\,\frac{t}{b}$ and drew an averaging line through them to get an expression for C. This led to an expression for the effective width of the plate

$$b_e = 1.9t\sqrt{\frac{E}{\sigma_e}}\left[1-0.574\,\frac{t}{b}\sqrt{\frac{E}{\sigma_e}}\right] \qquad \ldots\ldots 3.$$

A different form of the expression for effective strut width was arrived at from the results of experiments carried out by Ramberg, see Ref 5. In his approach, the effective width was given in terms of the ratio of the critical stress to the stress at the simply supported edge.

$$b_e = b\sqrt{\frac{\sigma_{cr}}{\sigma_e}} \qquad \ldots\ldots 4.$$

The weakness with 'Eqns 2, 3 and 4' is that they require the previous knowledge of the edge stress before the effective width can be calculated.

A formula which avoids this drawback was developed by Jambock and Clark, see Ref 6. They arrived at the simple equation

$$b_e = b\,\frac{\sigma_{cr}}{\bar{\sigma}} \qquad \ldots\ldots 5.$$

where $\bar{\sigma}$ is the average stress on the gross area, σ_{cr} is the critical buckling load of the plate and b is the width of the plate. The value of σ_{cr} remains constant for any plate, and the value of $\bar{\sigma}$ can be derived from the equilibrium conditions of the plate. Regardless of the distribution of the stresses, the average stress over the gross area times the area must equal the total applied load.

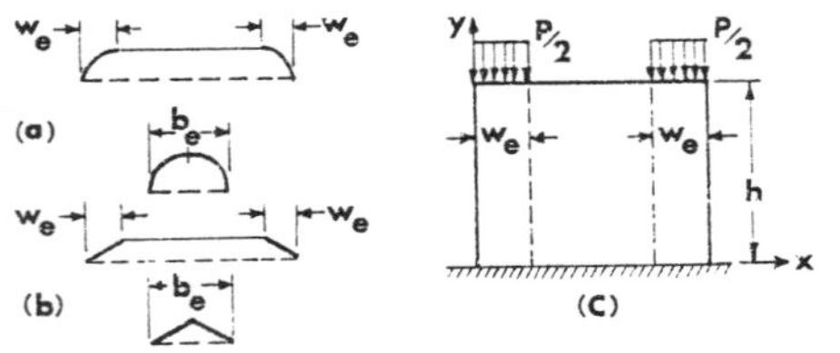

Fig 1

The Collocation Method, described by Klein in Ref 7, can be used to determine σ_{cr} for the triangular plates. The values of σ_e are obtained experimentally or taken as σ_y, the yield stress of the material.

Since the thin triangular plates may be deformed by loads perpendicular to the plate surface, it is necessary to consider the effect of initial out-of-plane deformation on the development of the equivalent strut. Hu found that for square plates, initial lateral imperfections led to insignificant errors in the effective width values when the average plate stress, $\bar{\sigma}$, was equal to 2 σ_{cr}. See Ref 7. It is expected that a similar result would apply to triangular plates when they are loaded to several times the critical buckling load.

In the author's opinion, the most useful form for the effective width equation is that given by Eqn 5 because all components are capable of being determined analytically. Hence it is proposed that an equation which takes into account the geometry of the trapezoidal or triangular plate will take the form:

$$b_e = b\cos\theta\left[\frac{\sigma_{cr}}{\bar{\sigma}}\right] \qquad \ldots\ldots 6.$$

where θ = the angle between the junction and the centerline of the plate. In summary, the equivalent strut theories which will be checked against the experimental results are:

Eqn 2(a) - Von Karman: $b_e = \frac{\pi}{\sqrt{3(1-\mu^2)}}\sqrt{\frac{E}{\sigma_e}}\;t$

Eqn 2(b) - Sechler: $b_e = \sqrt{\frac{2}{1+\mu}}\sqrt{\frac{E}{\sigma_e}}\;t$

Eqn 3 - Winter: $b_e = 1.9t\sqrt{\frac{E}{\sigma_e}}\left[1-0.574\,\frac{t}{b}\sqrt{\frac{E}{\sigma_e}}\right]$

Rqn 4 - Ramberg: $be = b\sqrt{\frac{\sigma_{cr}}{\sigma_e}}$

Eqn 5 - Jombock = $b_e = b\,\frac{\sigma_{cr}}{\bar{\sigma}}$

Eqn 6 - Gilkie: $b_e = b\cos\theta\left[\frac{\sigma_{cr}}{\bar{\sigma}}\right]$

EXPERIMENTAL STUDIES

The first set of experimental studies was performed on a hexagonal base perspex pyramid shown in Fig 2. The wall thickness was 3 mm. Tests were performed using two limiting support conditions; base continuously supported and clamped, and base clamped at the six plate junctions.

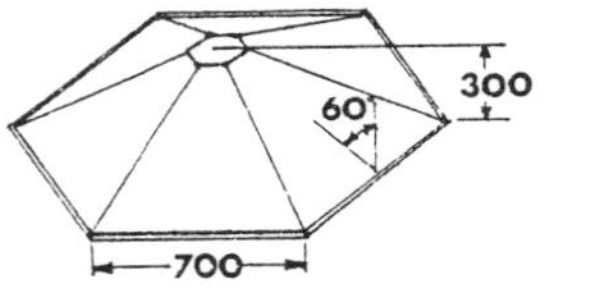

Fig 2

Strains were measured using 45° rosettes on both faces of one trapezoidal plate. The layout is shown in Fig 3. Load was applied vertically through a hydraulic jack and proving ring system. Details of these studies are given in Ref 9.

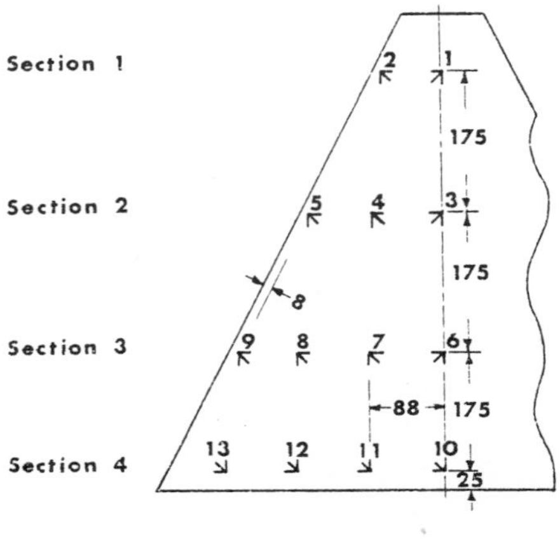

Fig 3

A second set of tests was performed on a model representing two adjacent triangular plates folded 110° to one another. Load was applied in compression along the axis of the junction by an Instron Universal Testing Machine. This technique was judged to closely resemble the action of load applied along the equivalent struts of a discrete member analogy to thin sheet pyramidal structures. In this way it was possible to compare the postbuckling behaviour of the triangular plates under two very different load application systems.

Utility grade aluminum, 1.62 mm thick, was used to fabricate the test model which is shown in Fig 4.

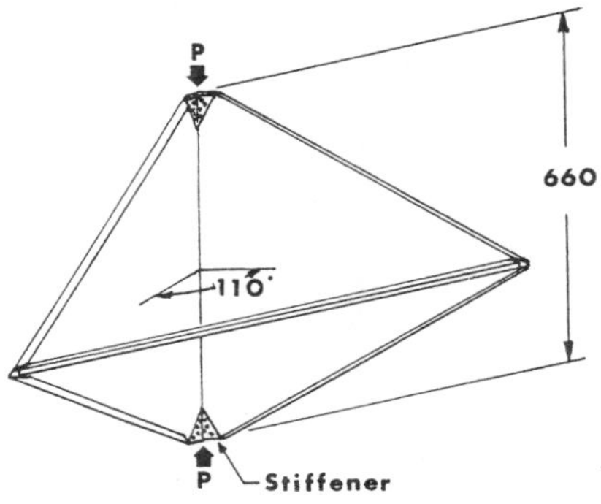

Fig 4

The free edges of the triangular plates were stiffened by folding strips 45 mm wide at an angle of 110° to the adjacent plate surface. This duplicated the edge stiffening effect of the adjacent triangular plate when such plates form a portion of a complete pyramidal unit.

An aluminum bar was connected between the two free corners of the triangular plates to maintain the fold angle at the plate extremities throughout the load sequence. The apexes of the triangular plates were truncated and stiffened to allow a 76 mm wide by 3.2 mm thick load application surface.

The arrangement of the strain gauge rosettes is shown in Fig 5. Details of the experimental setup are given in Ref 10.

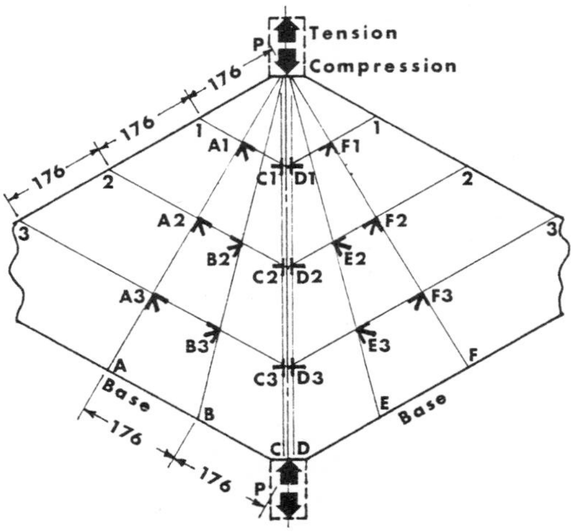

Fig 5

A third set of tests was performed on the model described above to determine if there was a reduction in the effective area of material contributing to an equivalent strut when the thin triangular plates were acting under tension load. Tabs, shown as broken line additions to the diagram of Fig 5 were used to clamp the two plate models into the Instron test machine.

EXPERIMENTAL RESULTS

The results of the tests on the hexagonal base perspex pyramid, base continuously supported and clamped, are discussed below.

The distribution of the axial stress perpendicular to the base across half the trapezoidal plate for four increments of load is shown in Fig 6. Prior to buckling (0.66 kN and 1.33 kN) the axial stress distribution was well predicted using the theory of the infinite wedge at Sections 1, 2 and 3 of the gauged plate. This theory and the coefficients used in the calculation of the stresses are given by Robak in Ref 11. The uneven distribution of axial stresses along the base was due to local effects of the clamping screws of the support system.

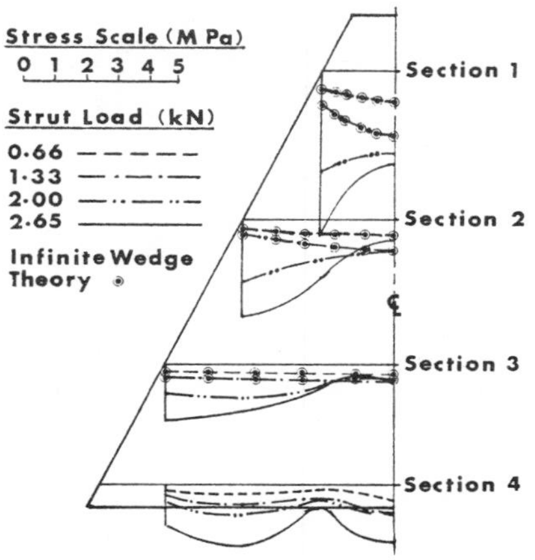

Fig 6

For vertical apex loads up to 1.33 kN, the horizontal axial stresses were so small as to be considered negligible.

After buckling had occurred, (2.00 kN and 2.67 kN) the axial stress along the junction increased rapidly and became several times greater than the stress along the centerline of the plate which remained almost constant. This was consistent with the behaviour of postbuckled rectangular plates. Horizontal tension stresses were developed in the middle of the plate in this load range. It is this tension which aids in stabilizing the junctions as the ultimate load of the pyramid is approached.

Horizontal compression stresses of a significant magnitude were developed near the loaded apex as an arch-like action developed in the top portion of the triangular plate.

A study of the load versus stress relationships indicated that buckling of the triangular plate occurred between 1.33 kN and 1.46 kN. This compared well with a value of 1.39 kN determined by the collocation method based on the assumption that the triangular plates were simply supported along the junctions with adjacent plates.

In contrast to the stress distribution found in buckled rectangular plates, where the vertical axial stress distribution remained the same along the length of the plate giving a constant effective width, the stress distribution varied along the length of the triangular plate. Hence the effective width of equivalent strut will vary along the length.

The axial stress distribution perpendicular to the base for the pyramid with corner supports only is shown in Fig 7.

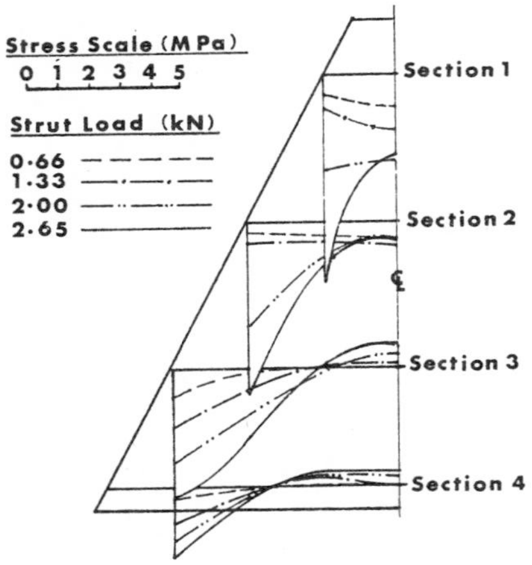

Fig 7

Near the apex, the stress distribution was very similar in magnitude and form to the model with continuous fixed supports. In the center region of the junction, more compression was transferred to the edge of the plate as the central region of the triangular plate first went into compression and then developed tension stresses perpendicular to the base. This accentuated the equivalent strut action.

A study of the load-stress diagrams for the various gauge locations produced no obvious critical buckling load range. As a result, only 'Eqns 2(a), 2(b), and 3' are used for comparisons between theoretical and experimental results for this support condition.

The results of the compression tests on the two adjacent triangular plates are discussed below.

The distribution of axial stress perpendicular to the bottom edges of the triangular plates, which would form the base of a thin sheet pyramid, is shown in Fig 8.

Although the stress at the junction is greater than along the centerlines of the triangular plates for all load increments, it is obvious that the stress along the junction increases more rapidly than at other gauge locations as the applied load increases. Also, near the centroids of the plates, the stress reaches a maximum value and remains approximately constant throughout the load range.

A study of the load versus stress relationships at gauge locations at the half height position on the junction between the two plates indicated that buckling occurred between 4.45 kN and 5.61 kN compared to a theoretical value of 6.34 kN obtained by the Collocation Method. This is consistent with expectations had there been any initial out-of-plane deformation in the triangular plates. A more complete description of the experimental results is given in Ref 10.

The details of the tension tests and the full analysis of the results are also given in Ref 10. The axial stresses perpendicular to the bottom edges of the triangular plates are shown in Fig 9. Here again, even though there is not a buckling phenomenon associated with tension load along the junction of the plates, there is a distinct concentration of stress along the junction between the plates. This suggests that an equivalent strut action can be developed for tension members as well as for compression members.

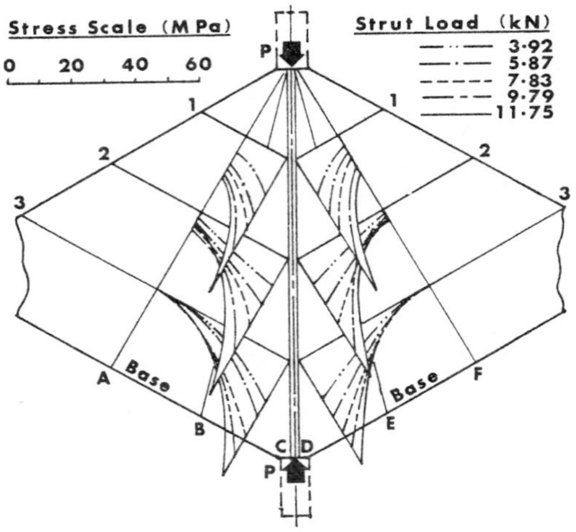

Fig 8

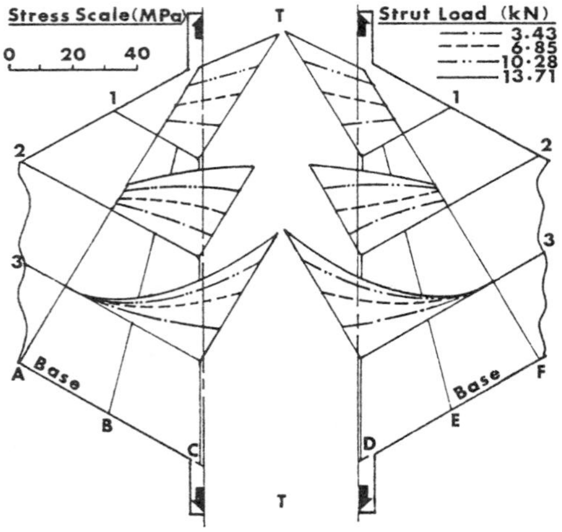

Fig 9

EQUIVALENT STRUT THEORIES-TRIANGULAR PLATES

Unlike the constant effective widths obtained for buckled rectangular plates, the effective widths for triangular and trapezoidal plates vary along the junctions. The average effective width was calculated by plotting the effective widths at gauged sections to a large scale, drawing a smooth curve through the data points, and determining the total area under the curve. This area was then divided by the length of the junction to determine the average effective width. A typical curve is shown in Fig 10.

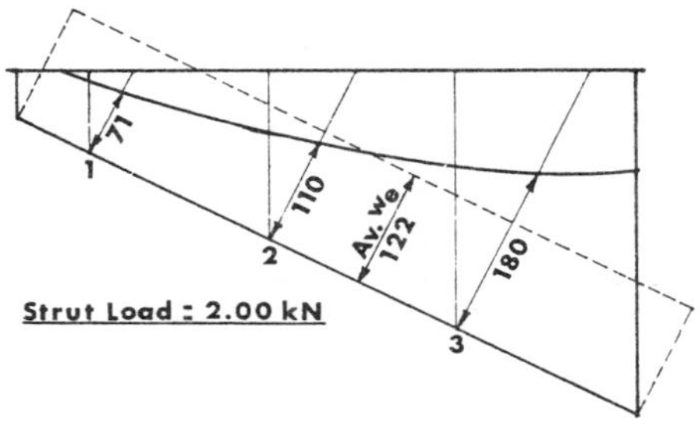

Fig 10

The average stress on the gross area, $\bar{\sigma}$ was obtained by graphical integration of the stress distribution curves of Fig 6. Equation 6 was also applied using theoretical values of σ_{cr} and $\bar{\sigma}$. The value of σ_{cr} was determined by the infinite wedge theory at the theoretical critical buckling load.

A convenient way to express the concept of the equivalent strut is to convert the effective width to effective area, A_e, by multiplying b_e by t, the thickness of the plate.

A comparison between the experimental, experimentally derived and theoretically derived values of equivalent strut effective areas is shown in Fig 11.

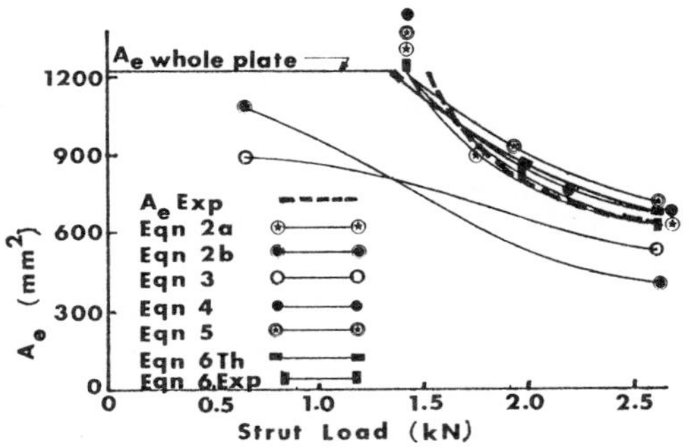

Fig 11

While several of the effective width formulas give reasonable predictions of the effective area of the equivalent strut for the pyramid with a clamped base, Von Karman's equation using experimentally derived values of σ_e and Gilkie's equation using theoretically derived values of σ_{cr} and $\bar{\sigma}$ are the most accurate.

Because of the difficulty in establishing the critical buckling load experimentally for trapezoidal plates of the pyramid with fixed corner supports, only 'Eqns 2(a), 2(b), and 3' were compared to the test results. This is shown in Fig 12.

All three equations gave exaggerated values of effective area at low loads but showed better correlation at higher loads. Winter's Eqn 3 and Sechler's Eqn 2(b) showed the best correlation.

The concept described by Von Karman, Ref 2, where a stress distribution such as the one shown in Fig 13(a) is replaced by a uniform stress over an effective width, see Fig 13(b), was also used to produce an equivalent strut of constant effective width for triangular plates loaded along the junction. The typical effective width diagram differed in form from one developed by the pyramid with a fixed base. See Fig 14. The total area under the curve was again divided by the length of the junction to produce the average effective width w_e. Twice the value of w_e times the thickness of the plate produced a value for effective cross-sectional area, A_e, of the equivalent strut.

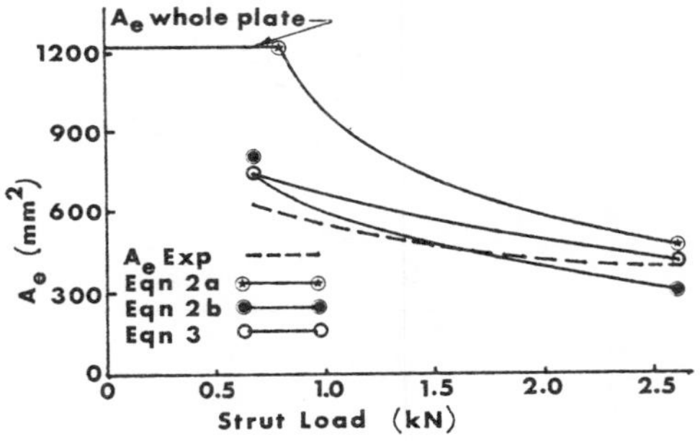

Fig 12

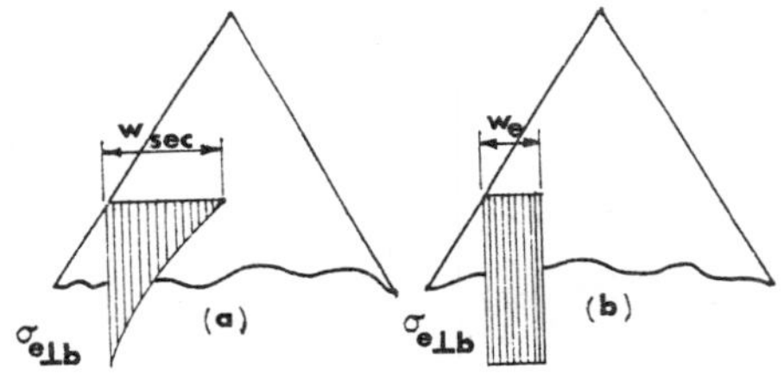

Fig 13

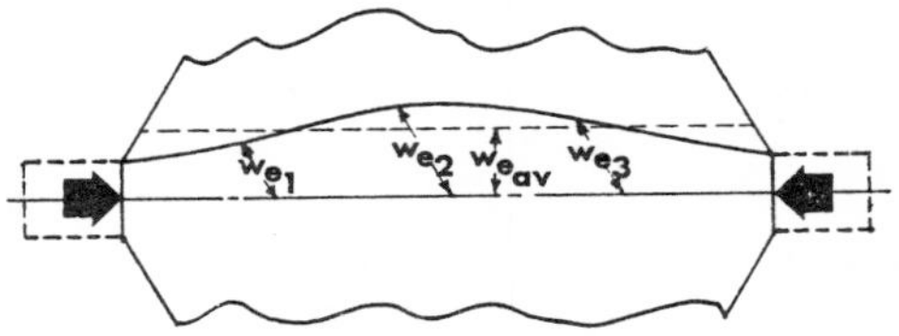

Fig 14

The effective width formulas, using stress data obtained from tests on the two plate model loaded in compression along the junction, are compared to the experimental results in Fig 15.

It was not possible to show the author's equation using only theoretically derived values of σ_{cr} because the infinite wedge theory does not apply to this type of loading.

As before, all equivalent strut theories were reasonably accurate at predicting the effective strut area at the higher loads. For this type of loading, the equations of Ramberg and Von Karman were the most accurate; although the equations of Winter, and the author using experimentally derived values of $\bar{\sigma}$ and σ_{cr} showed promise.

In contrast to the compression tests, the tensile edge stresses developed parallel to the junction of the triangular plates when tension was applied along the junction were proportional to load throughout the load range.

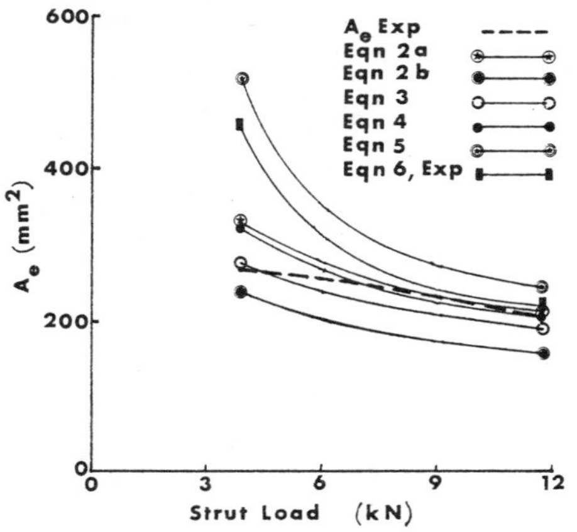

Fig 15

A method similar to that used for compression was used to determine the effective areas of equivalent struts in tension. The effective area was independent of load and so remained constant throughout the load range. A plot of the effective areas for the range of tension loads is shown in Fig. 16.

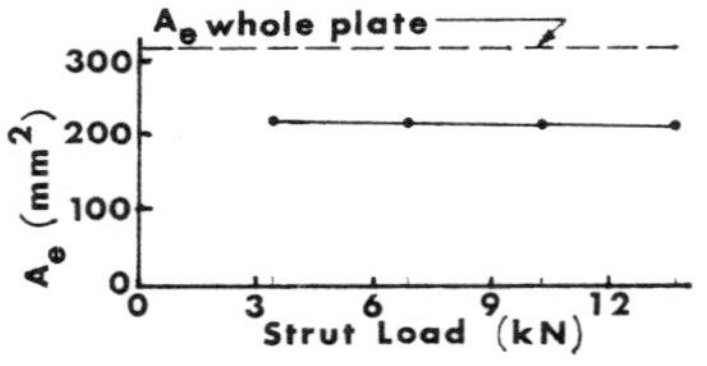

Fig 16

Previous studies, see Ref 9, assumed that the whole plate was active under tension load. The effective area for an equivalent strut would then be the area of one-third of each of the two adjacent triangular plates meeting at a junction divided by the length of the junction and multiplied by the plate thickness. Such a calculation for the test model gave an effective width of 203 mm. This compared to an experimentally derived effective width of 142 mm, or a thirty per cent reduction in effective strut area. Further studies are necessary to determine whether this reduction is affected by material properties or plate geometry.

CONCLUSIONS

The equivalent strut theories developed for postbuckled rectangular plates are useful in predicting equivalent strut behaviour in structural elements composed of thin triangular plates loaded in the postbuckled range.

Equivalent strut behaviour is also evident in triangular plates in tension but the effective width of strut appears independent of the applied load.

ACKNOWLEDGEMENTS

The author wishes to thank the Natural Sciences and Engineering Research Council of Canada for financial assistance in carrying out this work.

REFERENCES

1. R C GILKIE, A Comparison Between the Theoretical and Experimental Analysis of a Stressed Skin System of Construction in Plastics and Aluminum, Proc. International Conference on Space Structures, London, 1966, Blackwell, 1967.

2. T VonKARMAN, E E SECHLER and L H DONNELL, The Strength of Thin Plates in Compression, Trans. ASME, Vol. 54, No. 2, 1932.

3. J R JOMBOCK and J W CLARK, Postbuckling Behaviour of Flat Plates, Proc. ASCE, J. Struct. Div., Paper No. 2844, 1961.

4. G WINTER, Strength of Thin Steel Compression Flanges, Trans. ASCE, Vol. 112, 1957.

5. W RAMBERG, A E McPHERSON and S LEVY, Experimental Study of Deformation and of Effective Width in Axially Loaded Sheet Stringer Panels, Tech. Note 684, NACA, 1939.

6. J R JOMBOCK and J W CLARK, Postbuckling Strength and Effective Width of Flat Plates Subjected to Edge Compression, Report for the Column Research Council, October, 1957, USA.

7. B KLEIN, The Buckling of Simply Supported Plates Tapered in Planform, Journal of Applied Mechanics, (Series E), June 1956.

8. P C HU, E E LINDQUIST and S S BATDORF, Effect of Small Deviations From Flatness on Effective Width and Buckling of Plates in Compression, Tech. Note 1124, NACA 1946.

9. R C GILKIE, Pyramids in Light Weight Roof Systems, PhD Thesis, University of London, London, August 1967.

10. C V KUAN, Equivalent Strut Action in Thin Sheet Triangular Elements, M Sc Thesis, Technical University of Nova Scotia, Halifax, Canada, April 1983.

11. D ROBAK, The Structural Use of Plastics Pyramids in Double Layer Space Grids, Proc. International Conference on Space Structures, London, 1966, Blackwell, 1967.

RESPONSE OF SHALLOW HYPERBOLIC PARABOLOIDAL STRUCTURE ON FOUR SUPPORTS TO TRAVELING WAVE

Y GYOTEN*, Dr.Eng, T FUKUSUMI**, MS.Eng, H NOZOE***, MS.Eng

* President, Kobe University, Kobe

** Research Associate, Faculty of Engineering Kobe University, Kobe

*** Research Associate, Graduate School of Science and Technology, Kobe University, Kobe

Dynamic response of a large span structure which sustains spatially different ground motion has not yet been clarified enough. In this paper, hyperbolic paraboloidal structure which is framed with edge beams and supported at four points is considered, and the theoretical solution of the dynamic response to the three dimensional input excitation is presented. The influence of traveling wave on the response of the hyperbolic paraboloidal structure is investigated.

INTRODUCTION

In case of such a space structure with large span is under earthquake excitation, its supports sustain spatially different input motion because the structural length is comparable to the wave length. In this paper, hyperbolic paraboloidal structure (H.P. structure) as shown in Fig 1 is considered to sustain three dimensional different excitation at its supports. By employing the finite Fourier transformation method which was by authors developed the applicability to general boundary value problems, theoretical solution can be obtained in the form of double Fourier series. The static analysis for such hyperbolic paraboloidal and elliptic paraboloidal shells have been presented in the previous papers (Ref 1, 2 and 3).

In the first part of the numerical analysis, a concrete H.P. structure under static load is considered. The accuracy of results obtained by the presented method is examined in comparison with the results obtained by the finite difference method in Ref 4. In the second part of the numerical analysis, large span H.P. structure is considered. As for the static analysis, the variation of results with terms number taken into the calculation is examined. As for the dynamic analysis, H.P. structure is considered to sustain harmonic waves which propagate parallel to the diagonal axis of the structure. The influence of exciting frequency and the phase difference between each support on response of the H.P. structure is investigated.

ANALYSIS

An isotropic, visco-elastic hyperbolic paraboloidal shell flamed by edge beams on four supports which is forced to sustain the harmonic excitation in the three directions X, Y and Z is considered as shown in Fig 1. The equations of motion of a thin hyperbolic paraboloidal shells based on the Marguerre-Vlasov type shell thory are given as follows.

$$N_x' + \dot{N}_{xy} - \rho_s h\ddot{U} = 0 \quad \ldots\ldots 1,$$

$$\dot{N}_y + N_{xy}' - \rho_s h\ddot{V} = 0 \quad \ldots\ldots 2,$$

$$M_x'' + 2\dot{M}_{xy}' + \ddot{M}_y + 2cN_{xy} - \rho_s h\ddot{W} = 0 \quad \ldots\ldots 3,$$

where ρ_s and h are the mass density and thickness of the shell. A prime and dot used as a superscript indicate differential with respect to the x and y. U, V and W are the total displacements in the tangential directions, x and y, and normal direction, z, respectively, and are the functions of the x and y coordinates and time t. According to the method of separation of variables, as an example, U(x,y,t) is expressed as

$$U(x,y,t) = [\,u + u_g\,]e^{i\omega t} \quad \ldots\ldots 4,$$

where u(x,y) is written as u for the simplicity. $u \cdot e^{i\omega t}$ is considered as the relative displacement to the standard displacement $u_g \cdot e^{i\omega t}$ which does not generate any stress in the shell.

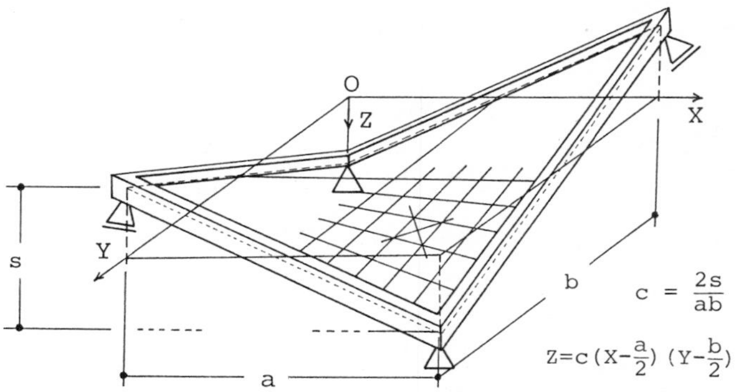

Fig 1 Analized hyperbolic paraboloidal structure

The shell stresses and moments are expressed as follows.

$$Nx=B(u'+\nu v^{\cdot}),\qquad Ny=B(v^{\cdot}+\nu u') \qquad \ldots\ldots 5,$$

$$Nxy=Gh(u^{\cdot}+v'-2cw) \qquad \ldots\ldots 6,$$

$$Mx=-D(w''+\nu w^{\cdot\cdot}),\qquad My=-D(w^{\cdot\cdot}+\nu w'') \qquad \ldots\ldots 7,$$

$$Mxy=-D(1-\nu)w'^{\cdot} \qquad \ldots\ldots 8,$$

$$Qx=-D(w'''+w'^{\cdot\cdot}),\qquad Qy=-D(w^{\cdot\cdot\cdot}+w''^{\cdot}) \qquad \ldots\ldots 9.$$

The complex stiffness B, Gh and D are expressed as

$$B=Eh/(1-\nu^2),\qquad Gh=(1-\nu)B/2,\qquad D=h^2B/12 \qquad \ldots\ldots 10,$$

in which complex rigidity E is assumed as $E=(1+2ih_d)E_0$, which is evaluated by using the damping constant h_d and Young's modulus E_0 of the shell.

Equation 1,2 and 3 are doubly transformed by the finite Fourier sin-cos, cos-sin and sin-sin transformations in the domain of $0\leq x\leq a$ and $0\leq y\leq b$, respectively, then the following simultaneous equations are obtained.

$$\begin{bmatrix} \alpha^2+\frac{1-\nu}{2}\beta^2-\frac{\rho_s h\omega^2}{B}, & \frac{1+\nu}{2}\alpha\beta, & (1-\nu)c\beta \\ & \beta^2+\frac{1-\nu}{2}\alpha^2-\frac{\rho_s h\omega^2}{B}, & (1-\nu)c\alpha \\ SYM & & \frac{h^2}{12}(\alpha^2+\beta^2)^2+2(1-\nu)c^2-\frac{\rho_s h\omega^2}{B} \end{bmatrix} \begin{Bmatrix} u_{sc}(\alpha,\beta) \\ v_{cs}(\alpha,\beta) \\ w_{ss}(\alpha,\beta) \end{Bmatrix}$$

$$= \begin{bmatrix} \alpha, & 0, & 0 & ,0 \\ \nu\beta, & -\frac{1}{B}, & 0 & ,0 \\ 0, & 0 & ,\frac{h^2}{12}[\alpha^3+(2-\nu)\alpha\beta^2], & \frac{\alpha}{B} \end{bmatrix} \left[\begin{Bmatrix} u_c(0,\beta) \\ N_s^{xy}(0,\beta) \\ w_s(0,\beta) \\ M_s^x(0,\beta) \end{Bmatrix} - (-1)^m \begin{Bmatrix} u_c(a,\beta) \\ N_s^{xy}(a,\beta) \\ w_s(a,\beta) \\ M_s^x(a,\beta) \end{Bmatrix} \right]$$

$$+ \begin{bmatrix} \nu\alpha, & -\frac{1}{B}, & 0 & ,0 \\ \beta, & 0, & 0 & ,0 \\ 0, & 0, & \frac{h^2}{12}[\beta^3+(2-\nu)\alpha^2\beta], & \frac{\beta}{B} \end{bmatrix} \left[\begin{Bmatrix} v_c(\alpha,0) \\ N_s^{xy}(\alpha,0) \\ w_s(\alpha,0) \\ M_s^y(\alpha,0) \end{Bmatrix} - (-1)^n \begin{Bmatrix} v_c(\alpha,b) \\ N_s^{xy}(\alpha,b) \\ w_s(\alpha,b) \\ M_s^y(\alpha,b) \end{Bmatrix} \right]$$

$$+ \frac{h^2}{6}(1-\nu)\alpha\beta \begin{bmatrix} 0, & 0, & 0, & 0 \\ 0, & 0, & 0, & 0 \\ 1, & -(-1)^m, & -(-1)^n, & (-1)^{m+n} \end{bmatrix} \begin{Bmatrix} w(0,0) \\ w(a,0) \\ w(0,b) \\ w(a,b) \end{Bmatrix} - \frac{\rho_s h\omega^2}{B} \begin{Bmatrix} u_{sc}^g(\alpha,\beta) \\ v_{cs}^g(\alpha,\beta) \\ w_{ss}^g(\alpha,\beta) \end{Bmatrix} \qquad \ldots .11.$$

The imaginary function $u_{sc}(\alpha,\beta)$, as an example, is defined as

$$u_{sc}(\alpha,\beta)=\int_0^a\int_0^b u(x,y)\sin\alpha x\cos\beta y\,dxdy \qquad \ldots\ldots 12.$$

where the parameter α and β are $\alpha=m\pi/a$ and $\beta=n\pi/b$.

The right hand terms of Eqns 11 are those of the imaginary functions along the four edges and the displacements w at four corners of the shell which are dealt as the unknown constants in this analysis and the imaginary functions of inertia forces expressed in terms of ug, vg and wg.

By the inverse transformation of $u_{sc}(\alpha,\beta)$, $v_{cs}(\alpha,\beta)$ and $w_{ss}(\alpha,\beta)$, the original functions $u(x,y)$, $v(x,y)$ and $w(x,y)$ are given in the form of double Fourier series as follows.

$$u(x,y)=\frac{2}{a}\sum_m u_{sc}(\alpha,\bar{o})\sin\alpha x+\frac{4}{ab}\sum_m\sum_n u_{sc}(\alpha,\beta)\sin\alpha x\cos\beta y \quad \ldots 13,$$

$$v(x,y)=\frac{2}{b}\sum_n v_{cs}(\bar{o},\beta)\sin\beta y+\frac{4}{ab}\sum_m\sum_n v_{cs}(\alpha,\beta)\cos\alpha x\sin\beta y \quad \ldots 14,$$

$$w(x,y)=\frac{4}{ab}\sum_m\sum_n w_{ss}(\alpha,\beta)\sin\alpha x\sin\beta y \quad \ldots\ldots 15.$$

Besides, derivatives of displacements are similarly expressed as the above, as an example, $w'(x,y)$ is given as

$$w'(x,y)=\frac{2}{ab}\{\sum_n w_{cs}'(\bar{o},\beta)+2\sum_m w_{cs}'(\alpha,\beta)\cos\alpha x\}\sin\beta y$$

$$=\frac{2}{ab}\sum_n\{w_s(a,\beta)+w_s(0,\beta)+2\sum_m[\alpha w_{ss}(\alpha,\beta)+(-1)^m w_s(a,\beta)-w_s(0,\beta)]\cos\alpha x\}\sin\beta y \quad \ldots\ldots 16.$$

Further, at the shell edge along x=0, u(0,y)) is given as

$$u(0,y)=\frac{1}{b}u_c(0,\bar{o})+\frac{2}{b}\sum_n u_c(0,\beta)\cos\beta y \quad \ldots\ldots 17.$$

and f(0,y) (Nxy(0,y), w(0,y) and Mx(0,y) is given as

$$f(0,y)=\frac{2}{b}\sum_n f_s(0,\beta)\sin\beta y \quad \ldots\ldots 18.$$

For the simplicity, the beam is considered to be bending type and sustain shell forces along the axis without eccentricity, and the inertia moment is not considered. In this case, the equations of motion of the edge beam parallel to the y axis are written as follows.

$$\dot{Mh}^{\cdot}+Nx(0,y)-\rho_b\ddot{U}=0 \quad \ldots\ldots 19,$$

$$N^{\cdot}+Nxy(0,y)-\rho_b\ddot{V}=0 \quad \ldots\ldots 20,$$

$$\dot{Mv}^{\cdot}+Qx(0,y)-\rho_b\ddot{W}=0 \quad \ldots\ldots 21$$

$$\dot{Mt}+Mx(0,y)=0 \quad \ldots\ldots 22.$$

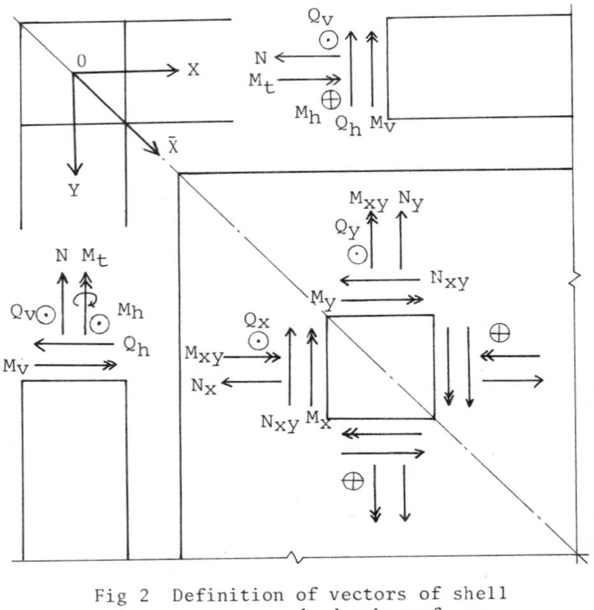

Fig 2 Definition of vectors of shell stresses and edge beam forces

in which ρ_b is the mass density of edge beam per unit length. N, Mh, Mv and Mt are the axial force, bending moment and the torsional moment of the edge beam. The directions of these vectors are shown in Fig 2. Further, from the equibriums of rotation, the shear forces Qh and Qv are given as follows.

$$Q_v=\dot{M}_v+M_{xy}(0,y), \qquad Q_h=\dot{M}_h \qquad \text{...... 23.}$$

Mh, N, Mv and Mt of the edge beam involved in Eqns 19, 20, 21 and 22 are governed by

$$M_h=-EI_h\bar{u}^{\cdot\cdot}; \quad N=EA\bar{v}^{\cdot}, \quad M_v=-EI_v\bar{w}^{\cdot\cdot}; \quad M_t=-J_t\bar{w}'^{\cdot} \qquad \text{...... 24,}$$

where $\bar{u}$, $\bar{v}$ and $\bar{w}$ are the displacements in the x, y and z directions of the edge beam, respectively. EIh and EIv, EA and Jt are the complex stiffness and are the bending stiffness, axial stiffness and tortional stiffness of the edge beam.

The equations of motion of the edge beam, Eqns 19 to 22 are transformed by such a combination of the finite Fourier cosine, sine, sine and sine transformations in the y direction, then the following equations are obtained.

$$(EI_h\beta^4-\rho_b\omega^2)\bar{u}_c(\beta)=EI_h\{\beta^2[(-1)^n\dot{\bar{u}}(b)-\bar{u}(0)]-(-1)^n\dddot{\bar{u}}(b)+\dddot{\bar{u}}(0)\}+N_c^x(0,\beta)+\rho_b\omega^2\bar{u}_c^g(\beta) \qquad \text{...... 25,}$$

$$(EA\beta^2-\rho_b\omega^2)\bar{v}_s(\beta)=-EA\beta[(-1)^n\bar{v}(b)-\bar{v}(0)]+N_s^{xy}(0,\beta)+\rho_b\omega^2\bar{v}_s^g(\beta) \qquad \text{...... 26,}$$

$$(EI_v\beta^4-\rho_b\omega^2)\bar{w}_s(\beta)=-EI_v\{\beta^3[(-1)^n\bar{w}(b)-\bar{w}(0)]-\beta[(-1)^n\ddot{\bar{w}}(b)-\ddot{\bar{w}}(0)]\}+V_s^x(0,\beta)+\rho_b\omega^2\bar{w}_s^g(\beta) \qquad \text{...... 27,}$$

$$J_t\beta^2\bar{w}'_s(\beta)=-J_t\beta[(-1)^n\bar{w}'(b)-\bar{w}'(0)]-M_s^x(0,\beta) \qquad \text{...... 28.}$$

The above Eqns 25, 26, 27 and 28 involve the displacements and its derivatives at the both ends of the edge beam, imaginary functions of the shell forces along the shell edge at x=0 and the inertia forces of the edge beam. By the inverse transformation of $\bar{u}_c(\beta)$ in Eqn 25, $\bar{u}(y)$ is given similar to Eqn 17, and by the inverse transformation of $\bar{f}_s(\beta)$ ($\bar{v}_s(\beta)$, $\bar{w}_s(\beta)$ and $\bar{w}'_s(\beta)$ in Eqns 26, 27 and 28) $\bar{f}(y)$ is given similar to Eqn 18.

The unknown constants dealt in this analysis are those included in Eqn 11 of the shell and those included in Eqns 25, 26, 27 and 28 of the edge beams, which amount to 4(2N+2M)+24 in total if the finite terms of expansion are taken as m=1,2,..M and n=1,2,..N. According to the continuity conditions of the shell edges and the edge beams and the conditions for nodal points at corners, the above unknown constants can be determined by solving the linear complex simultaneous equations.

NUMERICAL RESULTS AND DISCUSSIONS

In the numerical analysis, such a square (a=b) hyperbolic paraboloidal structure which is pin supported at the lower corners as in Fig 1 is considered.

For the first part, a specific concrete H.P. structure under the static load is numerically analized. Data used in this analysis are those in Ref 4, which are as follows: Young's modulus $E_0=3\times10^6$psi, span a=40 ft., rise s=8 ft., shell thickness h=3 in., edge beam section; 12 in.x12 in., edge beam load; 180 lb. per feet and uniform shell load; 67.5 lb per sq ft..

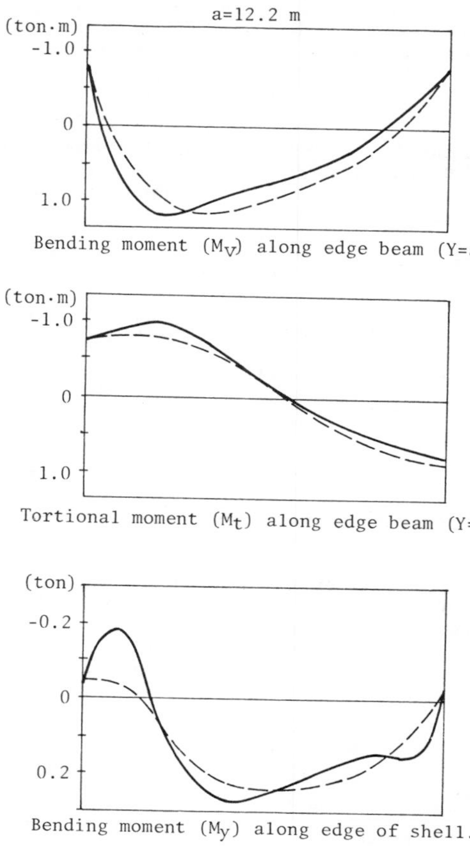
a=12.2 m
(ton·m)
-1.0
0
1.0
Bending moment (M_V) along edge beam (Y=a).
(ton·m)
-1.0
0
1.0
Tortional moment (M_t) along edge beam (Y=a).
(ton)
-0.2
0
0.2
Bending moment (M_y) along edge of shell.

Fig 3 Bending and tortional moments in edge beam and bending moment in shell edge.
(—— By this paper, – – – From Ref 4)

The results of bending moments Mv and My and tortional moment Mt obtained by the finite difference method (8x8 mesh) and those obtained by the presented method (N=20) are shown in Fig 3 for comparison. It is observed that the former shows rather smooth variation shape compared to the latter. It may be said that both results are sufficient for practical purpose.

In the second part of the numerical analysis, large spanned hyperbolic paraboloidal structure is considered in which shell structure is considered to be composed of considerably a number of unit grids which is able to be approximately taken as a continuum structure. Further, in case of isotropic lattice element in Fig 4 is supposed to be used as unit grid, equivalent properties are given as follows (Ref 1).

$$E=\frac{4}{3}\cdot\frac{E_0A_0}{\Delta x h}, \qquad \nu=1/3, \qquad h=\sqrt{3}D_s \qquad \text{....... 29.}$$

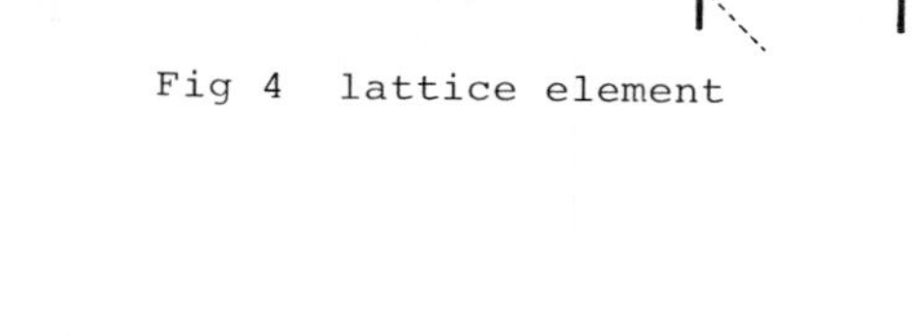

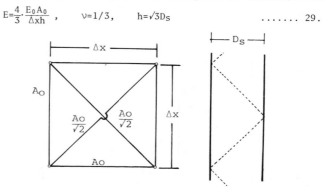

Fig 4 lattice element

Shear modulus G , in this analysis, is taken as $G=E/2(1+\nu)$ for convenience, which is different from that in Ref 1. As for the edge beam, square pipe ($D_b x D_b x t$) is considered,

$$I_h=I_v=(D_b^4-d_b^4)/12, \quad A=4(D_b-t)t, \quad J_t=GD_b^3t \qquad ..... 30.$$

Data used in the following analysis are taken as follows. a=40(m), s=0.01(1/m), $E_0=2.1\times10^7(t/m^2)$, $A_0/\Delta x=0.0025$(m), t=0.01(m) and hd=0.05. The depths of shell and edge beam are taken as same depth, D.

Four cases of H.P. structure are dealt in the numerical analysis, which are listed in the following Tab 1.

Table 1 Cases in the analysis

	Depth D (m)	Phase (deg)	$\rho s(\frac{kgs^2}{m^3})$	$\rho b(\frac{kgs^2}{m^2})$
Case 1	1.0	0.0	5.82	46.5
Case 2	1.0	90.0	5.82	46.5
Case 3	1.0	180.0	5.82	46.5
Case 4	2.0	0.0	11.6	93.0

Some numerical results of edge beam of Case 4 under the static load of the shell (ρsg) and edge beam (ρbg) are shown in the following Tab 2. Where terms number taken into account are changed as N=20, 15, 10, 5 and 2. In such a static analysis, accuracy for practical purpose may be obtained by taking about 10 terms in this solution.

Tab 2 Variation of results of edge beam with terms N

	Terms N	x/a 0.0	0.25	0.50	0.75	1.0
u $\times10^{-5}$(m)	20	-109	-76.2	-2.51	45.3	0
	15	-109	-75.9	-2.29	45.4	0
	10	-108	-75.1	-1.98	44.9	0
	5	-106	-73.6	-2.61	46.8	0
	2	-98	-55.2	-7.86	-14.1	0
v $\times10^{-5}$(m)	20	109	105	89.0	54.8	0
	15	109	105	88.9	54.8	0
	10	108	105	88.5	54.5	0
	5	106	103	87.4	53.6	0
	2	98	93.4	78.4	46.1	0
w $\times10^{-4}$(m)	20	0	76.4	106	74.0	0
	15	0	76.4	106	74.0	0
	10	0	76.6	107	74.2	0
	5	0	76.7	107	74.5	0
	2	0	78.2	109	76.3	0
M_h $\times10^{-1}$(t·m)	20	-69.1	-46.4	22.2	147	-503
	15	-64.4	-43.9	29.6	152	-417
	10	-82.3	-43.1	33.4	121	-397
	5	-56.0	-53.9	-10.3	214	-244
	2	-112	-23.9	78.6	23.9	-44.7
M_v $\times10^{-1}$(t·m)	20	-257	571	721	505	-249
	15	-261	570	729	504	-253
	10	-272	571	735	506	-265
	5	-290	576	746	514	-286
	2	-366	527	846	474	-334
M_t $\times10^{-1}$(t·m)	20	-257	-217	-6.37	203	249
	15	-261	-218	-6.49	204	249
	10	-272	-218	-6.60	204	265
	5	-290	-231	-7.31	218	286
	2	-336	-242	-8.54	232	334
N $\times10^{-1}$(t)	20	-6.38	-144	-413	-744	-1070
	15	-4.80	-144	-412	-743	-1070
	10	-2.93	-144	-409	-743	-1050
	5	10.2	-146	-408	-732	-1010
	2	-46.4	-138	-389	-685	-821
Q_h $\times10^{-2}$(t)	20	-29.3	62.8	87.8	128	-2890
	15	-48.0	68.8	86.6	127	-2890
	10	-29.3	62.8	87.8	128	-2890
	5	102	2.26	114	229	-3170
	2	-464	-362	683	-79.3	-4340

The following analysis are carried out to get insight into the excitation frequency and variation of phase angle at the supports of H.P. structure. Harmonic wave with displacement amplitude Zg and propagating velocity of Vg is considered to propagate parallel to the $\bar{X}$ axis which is shown in Fig 2. In this case, the excitation at support is given as $Z(\bar{X},t)=Zg\cdot e^{i\omega(t-\bar{X}/Vg)}$. In the analysis, the acceleration amplitude of the harmonic wave is taken as constant value of $1m/sec^2$ independent on the excitation frequency, where $Zg=1/\omega^2$.

In this analysis, since the structural shape and the excitation are symmetric, response of the structure becomes symmetric with respect to the $\bar{X}$ axis. The analysis was carried out for three different phase difference ($\sqrt{2}a/Vg$ of 0, $\pi/2$ and π) between two lower supports, which are listed in Tab 2 as Case 1 and Case 4, Case 2 and Case 3.

The frequency response curves of amplitude of the bending moment Mh of the edge beam at the lower support and the displacement amplitude normarized by Zg at the center of the shell are shown in Figs 5 and 6, respectively. Where terms number is taken as N=10.

It is observed that, in almost all of the frequency range, responses for Case 1, 2 and 3 are clearly varied whether phase difference of input excitation is existent (Case 2 and Case 3) or not existent (Case 1). The responses of Case 2 and Case 3 are greater than those of Case 1. At the low frequency range, the responses of Case 3 are most significant.

Resonant peaks appear at f=2.75Hz, 4.08Hz and 6.83Hz, which are called f_1, f_2 and f_3 for convenience. At the resonant frequency f_1 the response is unsymmetric with respect to the diagonal axis through the upper supports of the H.P. structure, whereas the responses are symmetric at the resonant frequencies of f_2 and f_3.

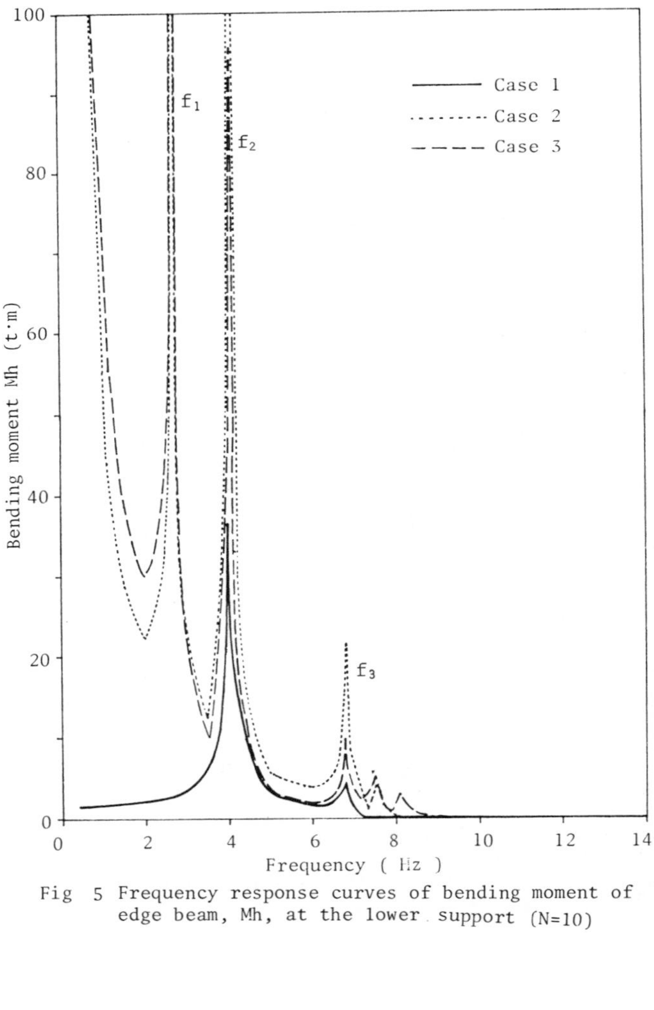

Fig 5 Frequency response curves of bending moment of edge beam, Mh, at the lower support (N=10)

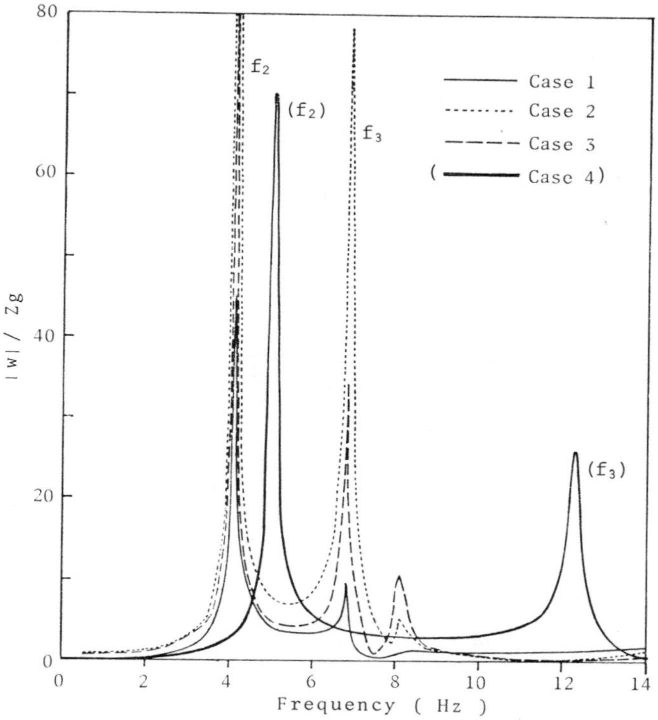

Fig 6 Frequency response curves of the shell displacement $|w|/Zg$ at the center of the shell (N=10)

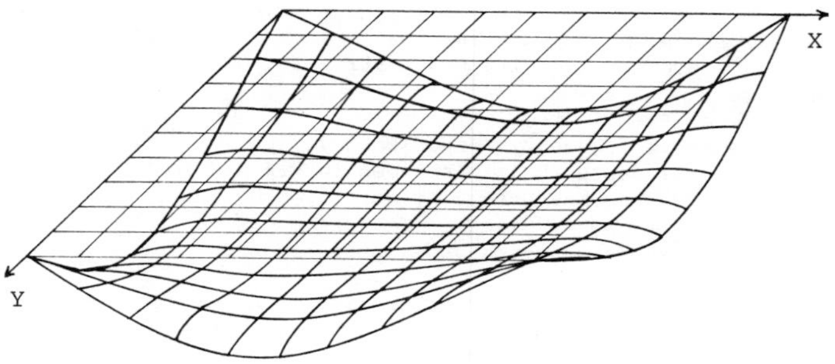

Fig 9 Displacement response $|w|/w_{max}$ at resonsnt frequency of f_2=4.085 Hz (case 1, N=20, W_{max} =0.097m)

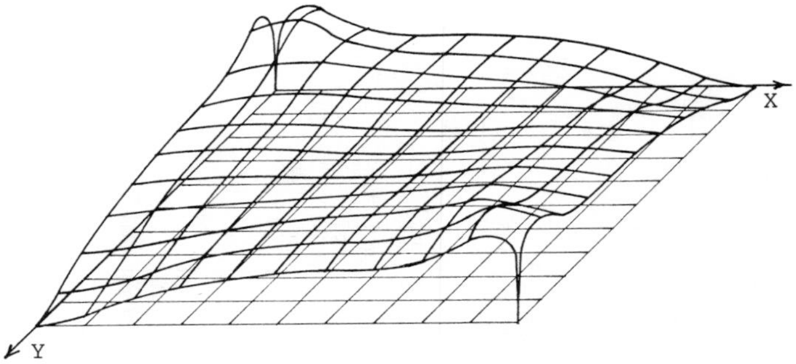

Fig 10 In-plane shear stress response $|Nxy|/Nxy_{max}$ at resonant frequency of f_2=4.085 Hz (case 1, N=20, Nxy_{max}=31.7t/m)

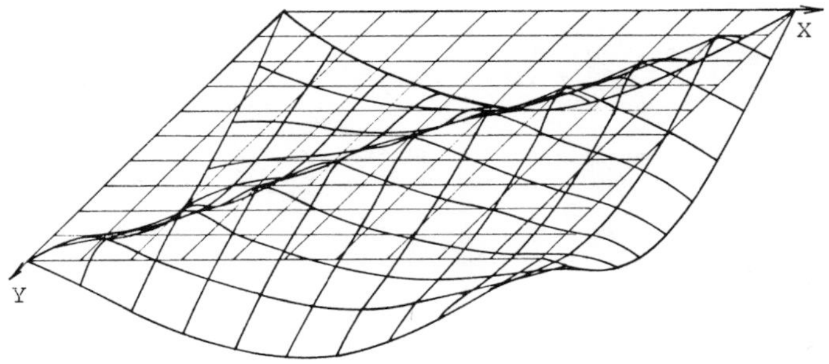

Fig 7 Displacement response w/w_{max} at resonant frequency of f =2.745 Hz (Case 3, N=20, w_{max}=1.07m)

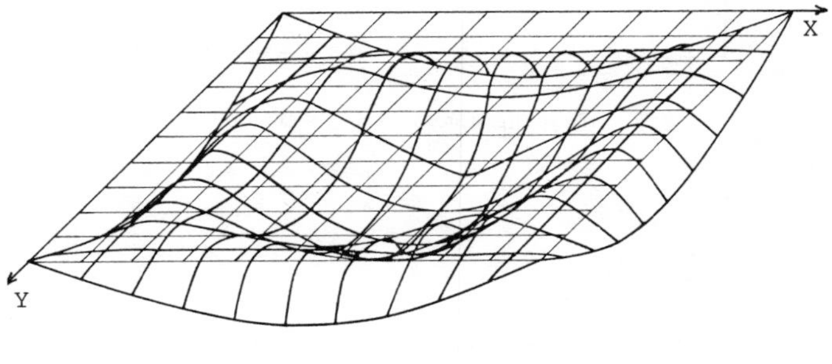

Fig 11 Displacement response $|w|/w_{max}$ at resonant frequency of f_3=6.825 Hz (case 1, N=20, w_{max}=0.0056m)

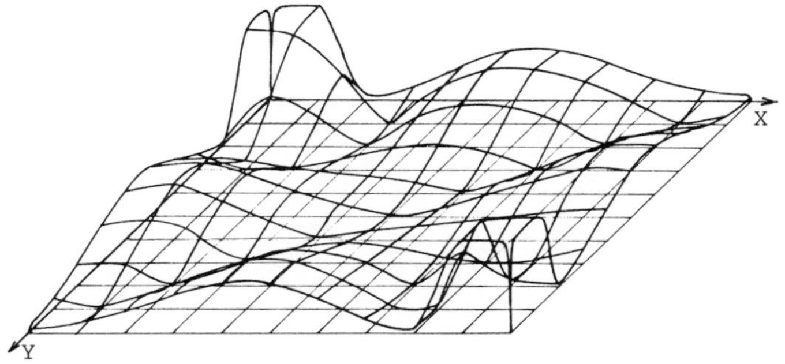

Fig 8 In-plane shear stress response $|Nxy|/Nxy_{max}$ at resonant frequency of f_1=2.745 Hz (Case 3, N=20, Nxy_{max}=104t/m)

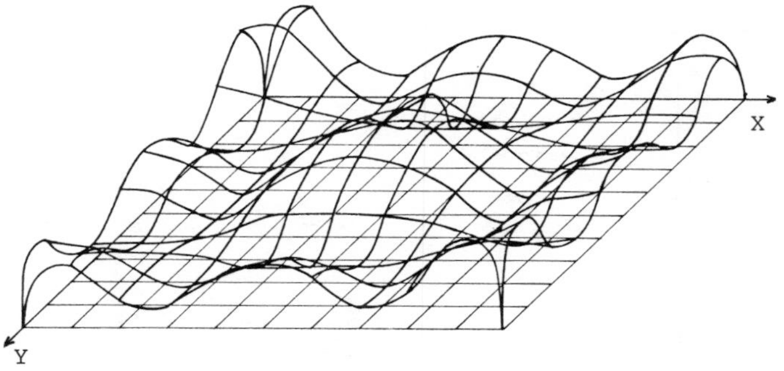

Fig 12 In-plane shear stress response $|Nxy|/Nxy_{max}$ at resonant frequency of f_3=6.825 Hz (case 1, N=20, Nxy_{max}=0.81t/m)

As can be seen in Fig 5, the resonant peaks appear at f_1 in Case 2 and 3, whereas , in Case 1, such a resonant peak which is accompanied with unsymmetric response does not appear.

Additionally, frequency response curve of $|w|/Zg$ for Case 4 with thicker depth (D=2m) is shown in Fig 6. In this case, resonant frequencies become higher (f_2=5.00Hz and f_3=12.3Hz) compared to those in Case 1. The effect of depth of structure can be observed as general tendency.

At f_1, for Case 3, the normarized amplitude of the displacement $|w|/w_{max}$ and the in-plane shear stress $|Nxy|/Nxy_{max}$ responses are shown in Figs 7 and 8, which are almost same for Case 2. At f_2 and f_3, for Case 1, the above results of $|w|/w_{max}$ and $|Nxy|/Nxy_{max}$ are shown in Figs 9 and 10 and in Figs 11 and 12 , which are almost same in Case 2 and 3.

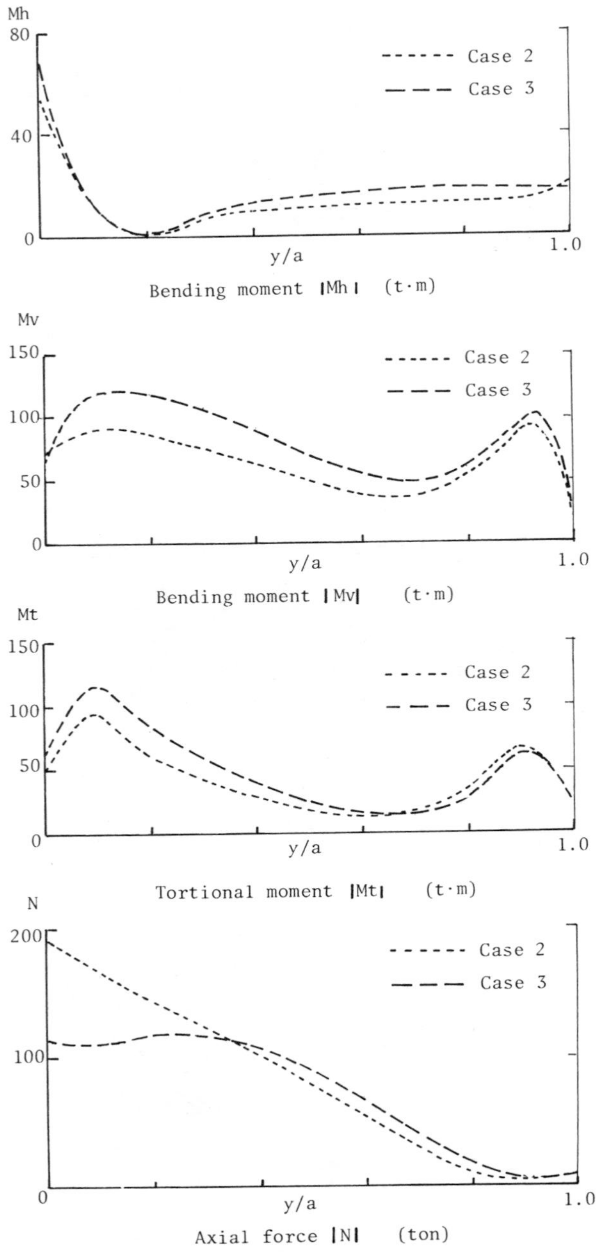

Fig 13 Responses in edge beam, |Mh| , |Mv| , |Mt| and |N| along the y axis (for Case 1 and 2, N=20)

From these Figures, it is observed that the maximum displacement w occurs at the midspan of the edge beam (at f_1) or at the center of the shell (at f_2 and f_3). Maximum in-plane shear stress occurs near the lower corners of the shell at every frequencies of f_1, f_2 and f_3.

As is evident from Figs 5 and 6, significant responses are generated in Case 2 and Case 3. In Fig 13, as an example, at the low frequency of f=2Hz, amplitude of responses of Mh, Mv, Mt and N along the edge beam are shown. It is observed that these responses become maximum near or at the lower supports.

CONCLUSION

In order to get insight into the influence of the spatially different ground motion on large span structure, hyperbolic paraboloidal structure which is elastically supported by edge beams is dealt in this paper. The presented thoretical solution is able to take into account the three dimensional responses to traveling waves.
The numerical analysis was carried out limiting to such a points supported hyperbolic paraboloidal structure in which shell is taken as isotropic and limiting to such a condition of harmonic wave excitation which propagate in horizontal plane parallel to the diagonal direction of structure.
From the results in this numerical analysis, it is concluded that the structural responses is greater under the excitation of phase difference than that under the uniform excitation. It becomes clear that it is important to consider the effects of spatially different excitation of traveling wave, when designing the earthquake resistant structures.

REFERENCES

1. Y.Gyoten et al, Analysis of latticed sallow hyperbolic paraboloidal shells by the finite Fourier transformation method, IASS Pacific Symp., 1971.

2. Y.Gyoten et al, Analysis of shallow hyperbolic paraboloidal shells by the finite Fourier transformation method, Memoirs of Faculty of Engrg., Kobe Univ., No.18, 1972.

3. Y.Gyoten et al, Analysis of shallow shells by the finite Fourier trans formation method, Proc.,22nd, JNCAM, 1974.

4. R.R.Russel and K.H.Gerstle, Hyperbolic paraboloid structures on four supports, ASCE, ST4, April, 1968.

5. S.D.Werner, L.C.Lee, H.L.Wong and M.D.Trifunac, Structural response to traveling seismic waves, ASCE,ST12,Dec.,1979.

6. R.V.Churchill, Operational mathematics, McGRAW HILL, 1958.

A MODIFIED FICTITIOUS FORCE METHOD FOR THE ULTIMATE LOAD ANALYSIS OF SPACE TRUSSES

H. KLIMKE, Dr.-Ing. and J. POSCH, Dipl.-Math.

MERO - Raumstruktur GmbH & Co. Würzburg, West-Germany

SYNOPSIS

A simple method for the ultimate load analysis of space trusses under proportional loading including temperature effects is presented. The practicability of the proposed method is shown through an application to fire resistance analysis of space trusses. The authors do not intend to give the solution of the combined problem of ultimate load and fire resistance analysis, but to encourage future work.

1. INTRODUCTION

The ultimate load analysis of trusses was first discussed by B.G.Neal [1]. He indicated the post critical behaviour of the compression members as the key problem of the ultimate load analysis.
Meissnest [2] was the first to consider the post critical behaviour of compression members of truss-girders. He determined stress-strain-curves for compression members with rectangular cross section. The extension to random sections and the application to space trusses was prevented by the limitations of computer resources at that time (1968).
Hensley and Azar [3] tackled the problem of the analysis of nonlinear trusses by a 'fictitious force' finite element technique, using the Ramberg and Osgood stress-strain relation for tension and compression.
Wolf [4] applied the 'fictitious force' technique to large space trusses. For the consideration of the post buckling behaviour of compression members he used a simple 'plastic hinge' model. This model is indeed a reasonable description of great deflections in the post buckling range (see Fig 3 and Tab 3).
Witte [5] described a modification (of the stiffness) algorithm, that extended the 'fictitious force technique'. The method provides a one step incremental technique for bilinear stress strain relations. The mechanical interpretation of the method is the search for a set of (prestressing) forces representing the unbalanced forces in failure (or plastic) members in the unchanged elastic system.
The method was extended by the author [6] to general nonlinear post buckling behaviour of compression members. The following section will introduce this method.

2. THE METHOD

2.1 The nonlinear behaviour of tubular compression members

Tubular members are most frequently used in space trusses. Their behaviour can be described by equilibrium equations, which relate the bending moment M and normal forces N to the stresses σ (Fig 1), and by a compatibility equation, which relates the bending moment M to the effective bending stiffness

$$B = \rho \cdot M \qquad (1)$$

where ρ is the radius of curvature. Eqn (1) is valid in the elastic and plastic range.

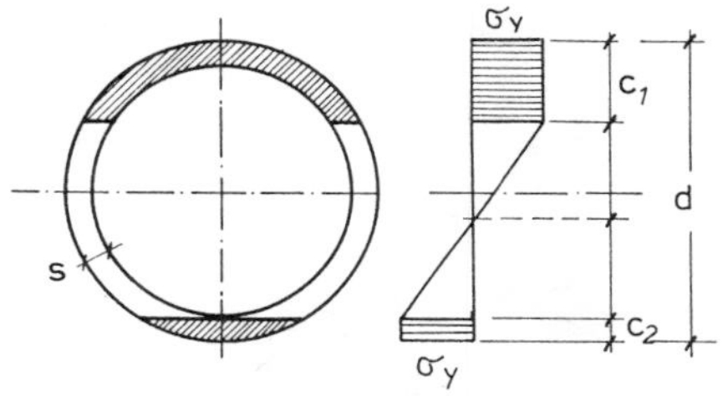

Fig 1: tubular cross section - stress distribution

The bending moment is defined by

$$M = N \cdot v = \sigma_a \cdot A \cdot v$$

where $N = \sigma_a \cdot A$ is the axial force, A the cross sectional area and v the lateral deflection of the member.

The M-B-relationship for a tubular section (60.3x2.9 mm) is shown in Tab 1. A $\bar{M}$-$\bar{B}$-Relation is adopted to make this table independent from material properties, where

$$\bar{M} = \frac{M}{\sigma_y \cdot A \cdot d}, \quad \bar{B} = \frac{B}{EI}, \quad \bar{N} = \frac{\sigma_a}{\sigma_y} \qquad (2)$$

Tab 1
curves of section resistance
profil 60,3 × 2,9

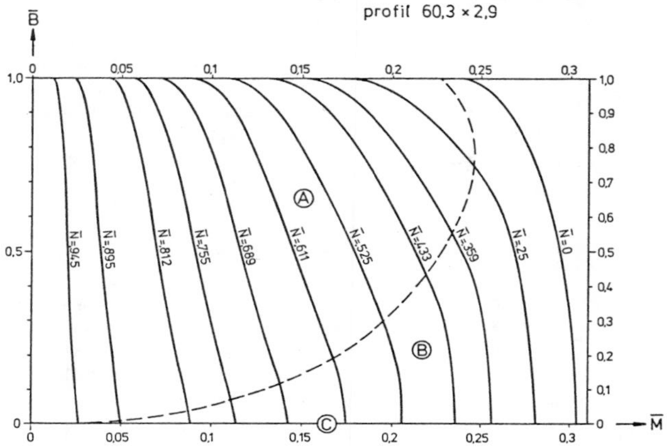

Region A indicates yielding only in the section of maximal compression, region B indicates yielding in compression and tension, C indicates the state of 'full plastic moment'.

The dotted line gives the values of maximum tensile stress equal to σ_y.

The M-B-relations of tubular sections are used to determine the stress-strain relations of compression members (Fig 2).

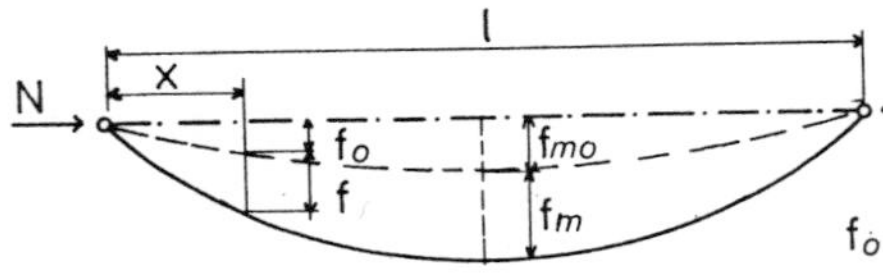

$$f_o = f_{mo} \cdot \sin\left(\frac{\pi x}{l}\right)$$

Fig 2: compression member with initial imperfection f_{mo}

The integration of the differential equation for member bending:

$$\frac{1}{\rho} = \frac{M}{B} = -\frac{d^2 f}{dx^2} \qquad (3)$$

leads to the curves of axial resistance shown in Tab 2.

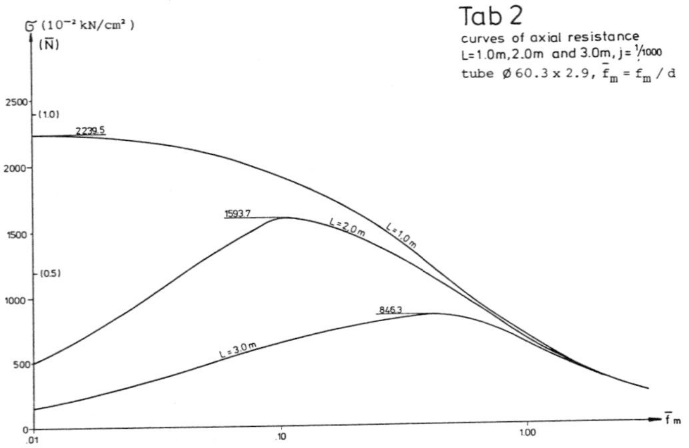

The axial deflection and the stress-strain relation can be obtained by d'Alembert's principle:

$$\varepsilon = \frac{N}{EA} + \frac{1}{2\ell} \int \frac{N \cdot v(x)}{B(x)} \, [v(x) + f_o(x)] \, dx \qquad (4)$$

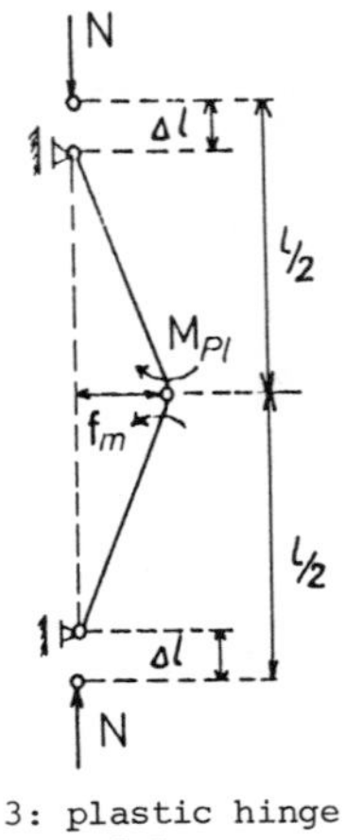

Fig 3: plastic hinge model

The integration process becomes extremely sensitve as plasticity increases in the postcritical range. These difficulties can be overcome by a relatively simple plastic hinge model after Fig 3. For the second equilibrium path d'Alembert's principle yields

$$\varepsilon = 1 - \sqrt{1 - \left(\frac{2f_m}{\ell}\right)^2} \left(1 - \frac{N}{EA} \sqrt{1 - \left(\frac{2f_m}{\ell}\right)^2}\right)$$

where $f_m = M_{pl}(N) / N$.

The typical stress-strain relations are shown in Tab 3, together with results obtained from a 'plastic hinge' model for the post critical behaviour.

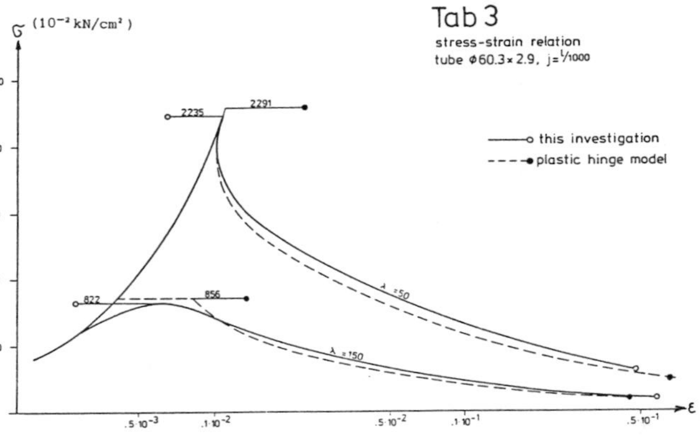

The curves of compact members ($\lambda = 50$) indicate a significant loss of load bearing capacity, accompanied by a sudden sag after reaching the ultimate load. For slender members ($\lambda = 150$), the loss after buckling is less significant. Early investigations assumed slender compression members keeping the buckling load in the post buckling range. The reliability of a relatively simple 'plastic hinge' model is demonstrated in Tab 3.

2.2 The nonlinear system behaviour, considering the post buckling reserves of compression members

2.2.1 The basic idea

The stress-strain relations of tension and compression members can be used to determine the axial load for a certain strain state ε.
The difference with respect to the related elastic load is given by the unbalanced force ΔS, which has to be redistributed in the remaining elastic structure. This can be done without changing the stiffness matrix of the system by prestressing the actual member i through the force V.
Using equilibrium equations for each member i in the plastic range, V_i can be obtained from Fig 4:

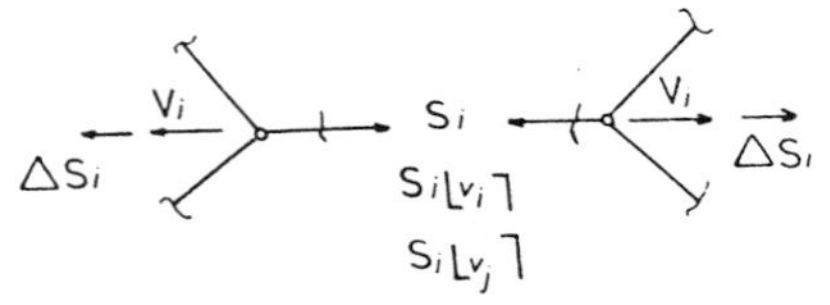

⌊⌉= due to

Fig 4: truss member in the post critical (buckling or plastic) range

For m members in the post critical range, the equilibrium equation for member i yields:

$$V_i - \sum_{j=1}^{m} S_i \lfloor V_j \rceil = \Delta S_i \qquad (6a)$$

$$S_i \lfloor V_j \rceil = V_j \cdot S_i \lfloor V_j = 1 \rceil \qquad (6b)$$

(6a) and (6b) give the final form of equation i:

$$(1 - S_i \lfloor V_i = 1 \rceil) \cdot V_i - \sum_{j \neq i} V_j \cdot S_i \lfloor V_j = 1 \rceil = \Delta S_i \qquad (6c)$$

All m equations can be rewritten in matrix form:

$$[I - S] \cdot \{V\} = \{\Delta S_i\} \qquad (7)$$

ΔS_i are the 'fictitious forces' as Hensley and Azar used them in [3].

The method represented by Eqn (7) offers the advantage that it converges in one single step for bilinear stress-strain relations, i.e. for tension members, and considerably accelerates convergence for radom nonlinear stress-strain relations, i.e. for compression members after Tab 3.

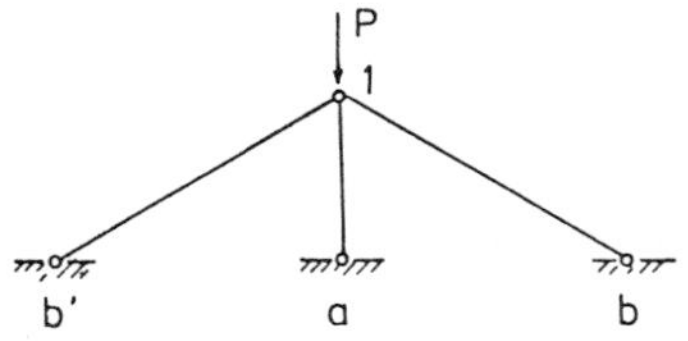

Fig 5: sample problem

The sample problem of Fig 5 will be used to demonstrate the different convergence behaviour of the 'fictitious force' and the proposed method: All members are assigned with the same slenderness ratio $\lambda = 50$ and cross sectional area A:

The theoretical solution can be obtained easily: In Tab 3, the critical load of member a1 is: $S_{a1} = 2235 \cdot A$, with the corresponding forces $S_{b1} = S_{b'1} = 559 \cdot A$, yielding to $P = 2794 \cdot A$. The system becomes instable when $S_{b1} = S_{b'1} = 2235 \cdot A$ with the correponding force $S_{a1} = 904 \cdot A$ yielding to an ultimate load of $P_u = 3139 \cdot A$.

The solution is verified by applying the fictitious force and the proposed method to the sample structure, loaded by $P = P_u$. The elastic solution is: $S_{a1} = 0{,}8 \cdot 3139 \cdot A = 2511 \cdot A$.

The iteration process is plotted in Tab 4 for the fictitious force method, while Tab 5 corresponds to the proposed method. Both methods give the correct answer for $S_{a1} = 904 \cdot A$, but the much better convergence of the proposed method (Tab 5) is obvious. In that sense, the proposed method can be interpreted as an improvement of the fictitious force approach.

Tab 4
sample problem
fictitious force method

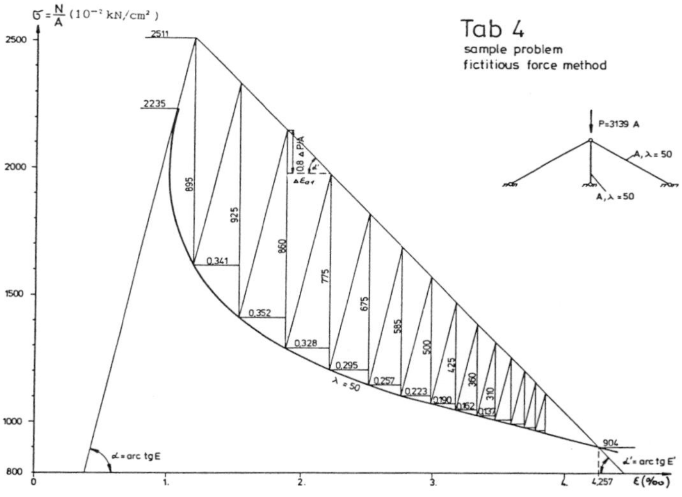

It should be noted, that the method is not limited to small deflection problems. The member forces from the loading and the member forces S of Eqn (7) can be obtained from a geometrically nonlinear analysis.
However the deflections related to the ultimate load of double (or more) layer space trusses can be considered as small and consequently a linear analysis is sufficient, as was stated in [2] and [3].

2.2 The problems related to the determination of the ultimate load of space trusses

2.2.1 A basic assumption

The ultimate load analysis of space trusses is normally performed for proportional loading. This assumption yields to a mechanism for the ultimate load in any case. To follow the history of structural failure and find the 'true' mechanism, the load must be decreased, as done by Collins [7].

It may still be questionable if this procedure makes sense from a practical point of view, but it is the only way to calculate a mechanism related to a plastic limit load which is less than the ultimate load. This problem is well known in the ultimate load theory of bending resistant frame structures (Ref [8] and Fig 6).

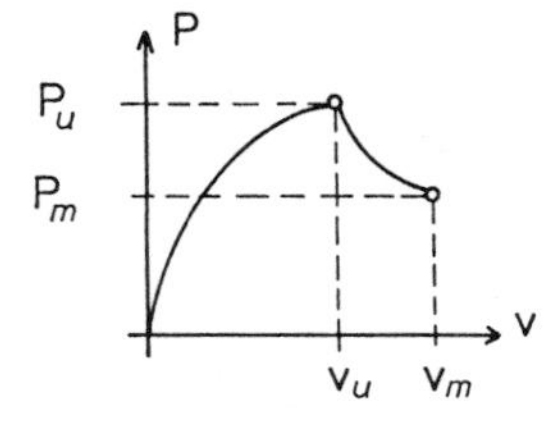

Fig 6: ultimate load and mechanism

The proposed method allows a very simple check for the mechanism in terms of Eqn (7):

$$\text{Det } [I - S] = 0 \qquad (8)$$

Tab 5
sample problem
proposed method

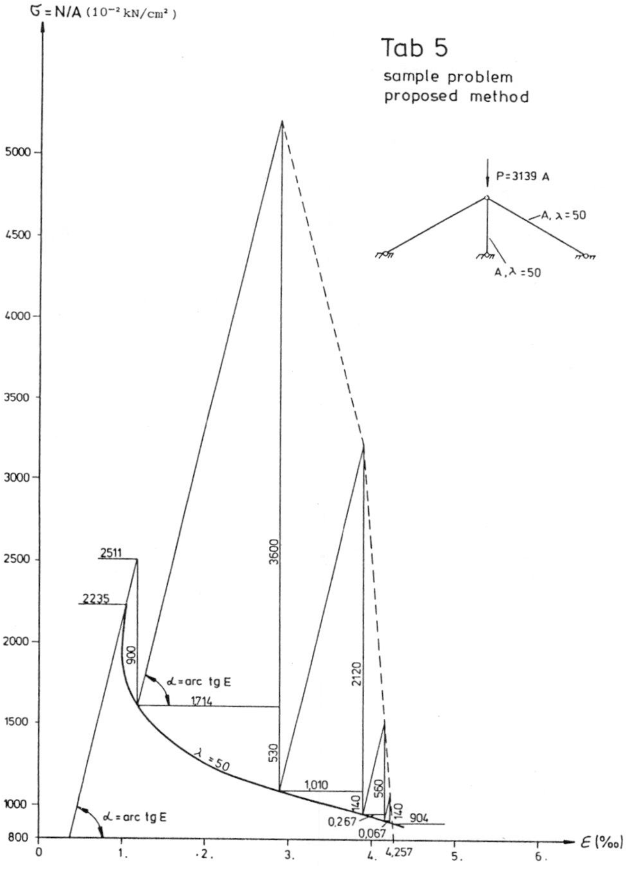

2.2.2 The effect of strain reversal

The consideration of the effect of strain reversal largely influences the loading history of members and consequently the ultimate load of the structure. Strain reversal can be understood as the unloading of a member in the post critical range. It follows form Eqn (6), that for one single member i with an unbalanced force ΔS_i

$$V_i - S_i \lfloor V_i \rceil = \Delta S_i + \sum_{j \neq i} S_i \lfloor V_j \rceil \qquad (9a)$$

Unloading occurs for

$$\sum_{j \neq i} S_i \lfloor V_j \rceil = - \Delta S_i \qquad (9b)$$

yielding to $V_i - S_i \lfloor V_i \rceil = 0$ (9c)

which is only possible for

$$v_i = 0 \tag{10}$$

Eqn (10) is used to identify the strain reversal.
For consecutive load increments, the unloading members are assumed to be elastic. This is an approximate assumption for compression members (Fig 7),

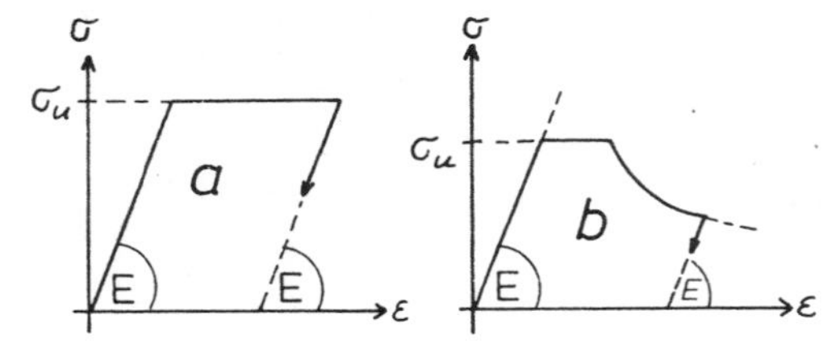

Fig 7: unloading of tension (a) and compression (b) members

3. APPLICATION OF THE PROPOSED METHOD TO THE FIRE RESISTANCE ANALYSIS OF SPACE TRUSSES

3.1 The basic task

The application of the presented ultimate load theory to the problem of fire resistance is an attempt to make use of the high redundancy of space trusses for fire protection. A research study was sponsored by the West German Government (BMFT) [9] to investigate the problems related to an analytical approach to the fire resistance of space trusses and to propose a detailed investigation program.

Consequently, an approximate method for the solution of the problem was used to study the influence of different parameters.

3.2 The approximation

To separate the ultimate load analysis from the (time dependent) heat distribution and to simplify the ultimate load analysis, the assumptions of uniform heat distribution and unrestrained expansion of the structure have been used.

3.3 The time dependent temperature

The time dependent gas temperature curve is given by the building codes (e.g. the ETK-curve after DIN 4102):

$$T_g(t) = 150\ell n(480t+1) \tag{11}$$

The steel temperature of the truss depends on the thickness of the material (e.g. the wall thickness of tubes commonly used for space trusses) and on the type of protection. Protection provided by means of water filled tubular members is of special interest in connection with space trusses.

A detailed analysis of the heat distribution in steel structures is commonly performed by the finite element method. For a uniformly heated structure, the average member temperature for tubular cross sections is determined by the differential equations for unprotected and water filled members, Eqn (12a) and (12b) respectively,

$$\gamma cs\,\frac{dT(t)}{dt} = \alpha(T_g(t) - T(t)) \tag{12a}$$

$$\gamma cs\,\frac{dT(t)}{dt} = \alpha(T_g(t) + T_w(t) - 2\,T(t)) \tag{12b}$$

where

- γ is the density [7800 kg/m^3 for steel]
- c is the specific heat [0.114 kcal/kg°C]
- α is the heat transmission coefficient [23 kcal/m^{2}h°C]
- s is the tube wall thickness [m]

Tab 6 shows the results of a time dependent heat analysis for a tubular truss member 76.1x2.9. For a steel temperature of 550°C, a fire resistance of about 15 min for the unprotected truss can be expected. Water filled members can improve the fire resistance, when the water is kept under 100°C.

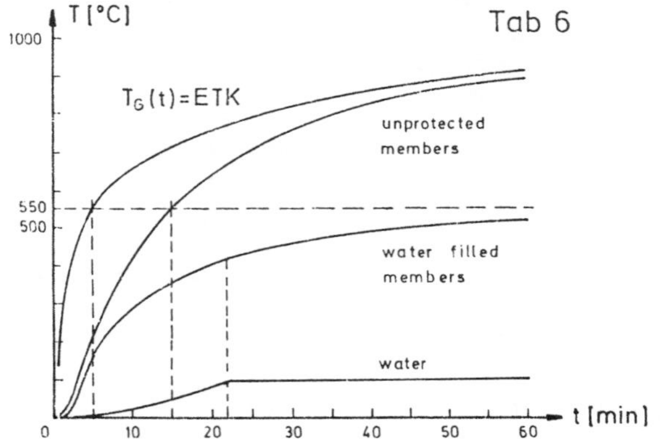

3.4 The analysis of the critical temperature

The temperature (°C) dependent material properties for Young's modulus E, the coefficient of thermal expansion α_T and the yielding stress σ_y are taken from [9]:

$$\frac{E(T)}{E(0)} = \begin{cases} 1-0.435\cdot10^{-4}(T-20)-0.103\cdot10^{-5}(T-20)^2 & [T\leq800°C] \\ (5-1.276\cdot10^{-2}T+0.128\cdot10^{-4}T^2)^{-1} & [T>800°C] \end{cases} \tag{13a}$$

$$\frac{\alpha_T(T)}{\alpha_T(0)} = 1 + T\,/\,3600 \tag{13b}$$

$$\frac{\sigma_y(T)}{\sigma_y(0)} = \begin{cases} 1-0.91\cdot10^{-5}(T-20)-1.988\cdot10^{-6}(T-20)^2 & [T\leq550°C] \\ (24.1-0.0881\cdot T+0.881\cdot10^{-4}\cdot T^2)^{-1} & [T>550°C] \end{cases} \tag{13c}$$

3.5 A sample problem

A three way double layer space truss after Fig 8 was designed for a working load of 1.0 kN/m^2 as a tubular structure. The vertical mid-point deflection is plotted in Tab 7.

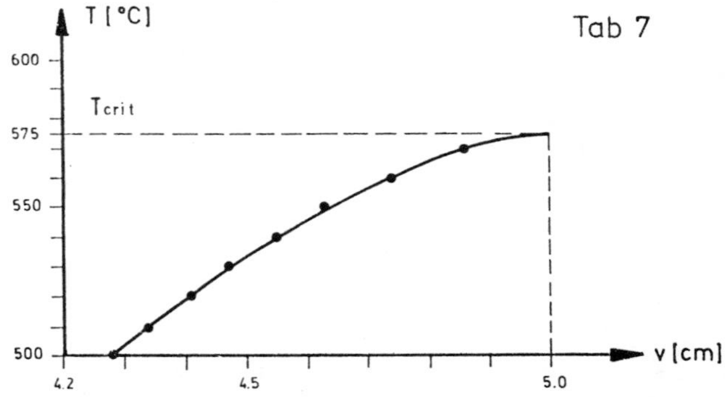

The failure of members began near the edges and proceeded into the truss. The ultimate load mechanism can be seen in Fig 8.

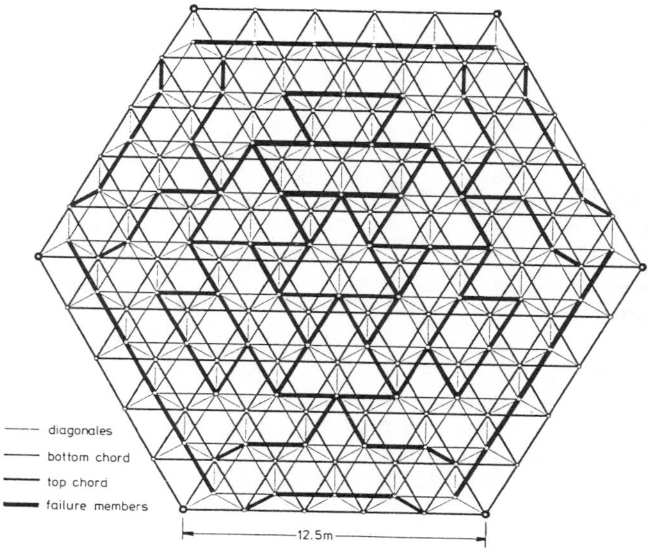

Fig 8: space truss mechanism pattern for $T_{crit} = 580°C$
(bilinear behaviour of compression members assumed)

To relate the different fire conditions in buildings to the standard fire (e.g. of the ETK) an 'equivalent time of fire resistance' can be determined by the fire regulations (e.g. by DIN 18 230 for industrial buildings).

The above calculated time of fire resistance can be incorporated in the design procedure (Fig 9) first discussed by Witteveen [10].

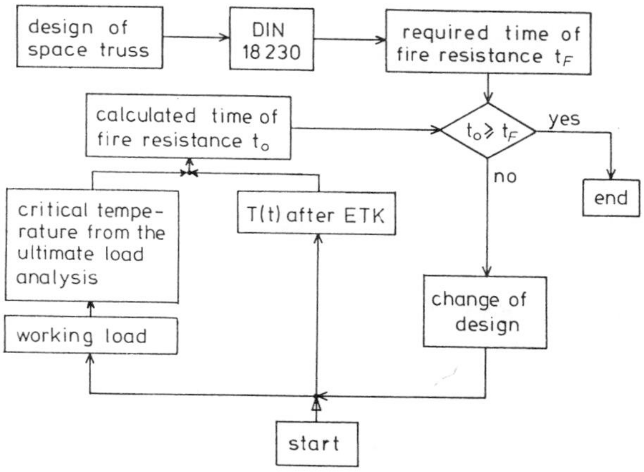

Fig 9: design procedure for a required fire resistance

4. CONCLUDING REMARKS

A relatively simple method for the ultimate load analysis of space trusses has been presented together with its application to the fire resistance analysis of space trusses.

Realizing that even unprotected space trusses may resist fire under certain conditions (Fig 10), the development of a safe fire design procedure will improve the application range of space trusses. The ultimate load analysis is the key operation in such a procedure.

Fig 10: The space truss of the Effelsberg (West-Germany) radiotelescope pavilion after fire

REFERENCES

[1] B G NEAL,
Die Verfahren der plastischen Berechnung biegesteifer Stahlstabwerke, Springer, Berlin, 1963

[2] H MEISSNEST,
Beitrag zur Berechnung der Traglast von Fachwerkträgern, Diss. TH Stuttgart, 1968

[3] R C HENSLEY and I J AZAR,
Computer Analysis of nonlinear truss-structures, ASCE-Proceedings, June 1968

[4] J P WOLF,
Elastisch-plastische Berechnungen großer Fachwerke im überkritischen Bereich, Schweizerische Bauzeitung 1973

[5] F C DE WITTE,
De invloed van ondeugdelijke staven op de veiligheid van vakwerkplaten, TNO Rapport No. BI-71-23, 1971

[6] H KLIMKE,
Berechnung der Traglast statisch unbestimmter räumlicher Gelenkfachwerke unter Berücksichtigung der überkritischen Reserve der Druckstäbe, Diss., TH Karlsruhe 1976

[7] I M COLLINS,
Collapse Analysis of Double-Layer Grids, PhD Thesis University of Surrey, GB, 1981

[8] U VOGEL,
"Methoden und Kriterien für eine wirtschaftliche Bemessung von Stahlrahmen bei Anwendung des Traglastverfahrens", Beitrag zu 'Theorie und Berechnung von Tragwerken' Springer Verlag, 1974

[9] TEICHEN, HAFNER, KLIMKE,
Brandverhalten kompletter Bauwerksysteme, Teil IV, 'Raumfachwerke', Projektdefinitionsstudie, Studiengesellschaft für Anwendungstechnik von Eisen und Stahl e.V., Düsseldorf, Projekt P86, Forschungsbericht 1982

[10] Fire Safety in Constructional Steelwork
ECCS Report III-74-2, Rotterdam, 1974

STATIC AND DYNAMIC ANALYSIS OF MULTILEVEL BUILDINGS: A COMPUTER AIDED BUILDING DESIGN METHODOLOGY

M MAJOWIECKI and F ZUCCARELLO***

*Istituto di Tecnica delle costruzioni, Facoltà di Ingegneria
Università di Bologna, Italy

**Computer Aided Design Laboratory
Bologna, Italy

This article illustrates a graphic interactive automatic package (SISMI) directed to the design and structural verification of multilevel space-framed buildings subject to seismic action. The programme is provided with both pre- and post-processors which make the adoption of a highly-efficient CAD methodology possible, especially in the input-output phases. Checking of the structural analysis results is easily done by graphic visualization of the displacements and stresses, thus permitting the structural engineer to make a rapid iterative and interactive design synthesis, understood as the reasoned sequence of successive steps of analytical checks.

1. INTRODUCTION

Antiseismic design of multilevel buildings calls for both structural and load-analysis schemes which are necessarily more sophisticated and complex than those required for static design.

In order to get a structural behaviour that conveniently reflect the constructional reality, the buildings must, as a rule, be modelled according to a spatial structural system which takes due account of the fact that the resultants of the overall actions do not pass through the stiffness gravity centres. Some buildings, in line with regulations in force, call for modal rather than simplified analysis through equivalent static loads.

The complexity and laboriousness of the structural calculations required by the above-mentioned schematizations, suggest the desirability of automatic structural calculation codes aimed at static and dynamic analysis of the building, understood as a spatial structural system made up of various substructures like (Fig 1):

- structure and space frame;
- bracing systems;
- foundation structure.

Many general static and dynamic analysis programmes of structures are available today (SAP, ADINA, NASTRAN, MARC, etc.) but their employment creates several difficulties which are discussed in detail in Ref 1.

In order to overcome this difficulty, a series of programmes has been developed in the last 15 years at the University of California (Berkeley) and aimed at analysing seismic-zone buildings.

These programmes, called ASCE, TABS, ETABS (Ref 2), etc., though "addressed", are of batch origin and, consequently, not suitable for design intended as programmed repetition of calculations permitting interactive modifications to be made whether reasoned or automatic (optimization).

The automatic calculation programme for seismic-zone buildings presented here aims at removing the limitations of the above-listed programmes. The main features of the code are:

- three-dimensional static and dynamic analysis of multilevel space-structured buildings;

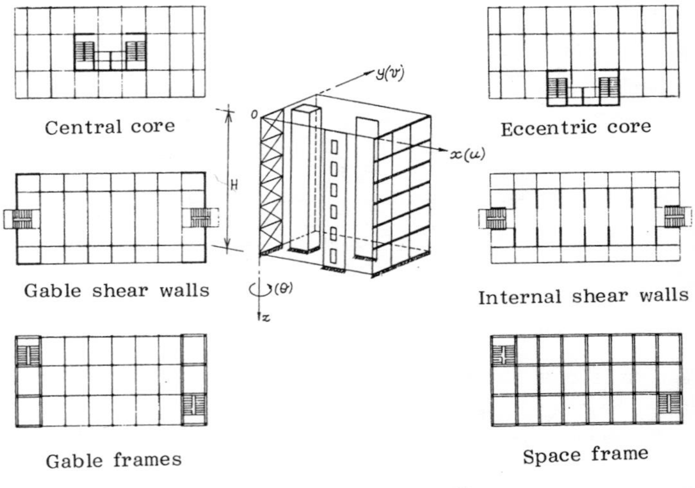

Fig 1 Integration of the bracing system

- integration of bracing systems (stair spaces, shear walls, party walls, etc.) by means of sub-structuring. Panel sub-structuring is schematized according to the equivalence theory, Ref 3;

- static condensation of the degrees of freedom considered unnecessary (slaves) when defining the structural theoretic scheme;

- completely interactive input-output organization in accordance with graphic techniques pertaining to CAD methodology;

- automatic and/or interactive design and verification of the sections and structural elements;

- automatic working drawings of reinforced concrete and steel in reinforced structures after interactive checking of the operator-designer.

2. PHYSICO-MATHEMATICAL MODEL

In the manner most generally foreseen by the programme the structure of a multilevel building is physically schematized by adopting the following elementary substructures:

- the 12 d.o.f. beam finite element;
- the panel, bracing element;
- the floor element;
- the foundation (footing or beam) on elastic soil.

The general outline of the SISMI programme permits the solution of space-framed structures without limiting the static scheme geometry. Generally this outline requires considerable computational effort (mainly bound to the unknown number = d.o.f.), which may be significantly reduced for most multilevel civil buildings, where the physico-geometric features of the structure permit adoption of the following, usual simplicying hypotheses:

- the floor plan is considered non-deformable in its own plane;

- the horizontal forces due to earthquake are considered as applied to the plan, a hypothesis equivalent to considering the masses concentrated and associated to several degrees of freedom (lumped mass parameters).

The first hypothesis involves having, at the floor node, a dependence bond between the six degrees of freedom per node of the rod elements and the displacement parameters relative to rigid floor movement. This bond, in displacement terms, is synthetized by Eqn 1:

$$\begin{Bmatrix} u \\ v \\ w \\ \phi_x \\ \phi_y \\ \phi_z \end{Bmatrix}_K = \begin{bmatrix} 1 & & & & & -DY \\ & 1 & & & & DX \\ & & 1 & & & \\ & & & 1 & & \\ & & & & 1 & \\ & & & & & 1 \end{bmatrix}_K \begin{Bmatrix} u_p \\ v_p \\ w \\ \phi \\ \phi \\ \phi_{zp} \end{Bmatrix}_K \qquad \ldots\ldots \; 1,$$

where:

$u;v;w$ = displacement of generic node K in the global reference system;

$\phi_x;\phi_y;\phi_z$ = rotations in K referred to the global axes;

$u_p;v_p;\phi_{zp}$ = degrees of freedom bound to the frame's rigid body movement;

$DX = X_k - X_p$;

$DY = Y_k - Y_p$;

$X_k;\ Y_k$ = coordinates of generic node K;

$X_p;\ Y_p$ = floor barycenter;

P = plan index (association index of K nodes pertaining to plan "p").

In terms of forces, the bond is expressed by Eqn 2

$$\begin{Bmatrix} F_x \\ F_y \\ F_z \\ M_x \\ M_y \\ M_z \end{Bmatrix}_{K;J} = \begin{bmatrix} 1 & & & & & \\ & 1 & & & & \\ & & 1 & & & \\ & & & 1 & & \\ & & & & 1 & \\ DY & -DX & & & & 1 \end{bmatrix} \begin{Bmatrix} F_{xp} \\ F_{yp} \\ F_z \\ M_x \\ M_y \\ M_{zp} \end{Bmatrix}_{K;J} \qquad \ldots\ldots \; 2,$$

where:

$F_x;F_y;F_z$ = forces associated to node K belonging to element K;J, connected to it;

$M_x;M_y;M_z$ = moments associated to node K belonging to element K;J connected to it;

$F_{xp};F_{yp};M_{zp}$ = forces and moments associated to node K belonging to element K;J, bound to the plane's rigid body movement;

K, J = identification indices of the beam elements.

From Eqns 1 and 2 it follows that 6 degrees of freedom per node are reduced to 3, independent of the rigid floor movement (two rotations and one displacement). The degrees of freedom activated by the suggested schematization are, therefore: 3 per node and 3 per floor.

In terms of the hardware technical characteristics and the solver available, it is or is not possible to carry out convenient static condensation of the degrees of freedom slaves w; ϕ_x; ϕ_y.

The "masters" degrees of freedom u_p; v_p; ϕ_{zp}, determined in the reduced solution, permit immediate "numeric control" with other very simple calculation programmes (CONVEN, Ref 8), where only 3 d.o.f.'s per floor are considered in determining the mathematical model.

The dynamic equilibrium of the structure is described by a system of well known second-order differential equations:

$$[M]\{\ddot{D}\} + [C]\{\dot{D}\} + [K]\{D\} = \{P(t)\} \qquad \ldots\ldots \; 3,$$

where:

$[M]$ = diagonal mass matrix (concentrated masses = 1st hypothesis);

$[C]$ = diagonal damping matrix (simplifying hypothesis);

$[K]$ = stiffness matrix;
$\{P(t)\}$ = applied loads (also as a function of time);
$\{D\}$ = displacement vector;
$\{\dot{D}\}$ = velocity vector;
$\{\ddot{D}\}$ = acceleration vector.

For the dynamic analysis in the programme presented here, modal analysis of Eqn 3 is provided for.

Since the calculations for dynamic analysis of the structures are more expensive than those for static analysis, a condensation of the masses is provided for in the general outline of the SISMI programme by means of the Guyan reduction method.

After determining the slaves d.o.f. for which the associated inertia forces are less important than the elastic ones transmitted by the master d.o.f.'s (e.g. determining the master d.o.f.'s for the highest mass/stiffness ratio) it is possible to process the "m" and "s" type submatrices, thus proceeding to separate the degrees of freedom desired (masters) from those to be eliminated (massless-slaves):

$$\left(\begin{bmatrix} K_{mm} & K_{ms} \\ K_{ms}^T & K_{ss} \end{bmatrix} - \lambda \begin{bmatrix} M_{mm} & M_{ms} \\ M_{ms}^T & M_{ss} \end{bmatrix} \right) \begin{Bmatrix} \bar{D}_m \\ \bar{D}_s \end{Bmatrix} = 0 \quad \ldots\ldots 4,$$

where:

m = degrees of master freedom to keep;

s = degrees of slave freedom to eliminate;

$\bar{D}_m$; $\bar{D}_s$ = eigenvectors of natural vibration modes (master type; slave type).

By eliminating $\{D_s\}$ from Eqn 4 we have:

$$\{\bar{D}_s\} = - [K_{ss}]^{-1} [K_{ms}]^T \{\bar{D}_m\} \quad \ldots\ldots 5;$$

now, by setting

$$\begin{Bmatrix} \bar{D}_m \\ \bar{D}_s \end{Bmatrix} = [T] \{D_m\} \quad \ldots\ldots 6,$$

where: $$[T] = \begin{bmatrix} 1 \\ -K_{ss}^{-1} \quad K_{ms}^T \end{bmatrix}$$

substituting in Eqn 4 and premultiplying by $[T]^T$ the condensed form is obtained:

$$([K_r] - \lambda [M_r]) \{\bar{D}_m\} = 0 \quad \ldots\ldots 7,$$

where the symmetrical condensed matrices are expressed by Eqns 8:

$$\underset{m \times m}{[K_r]} = \underset{mxn}{[T]^T} \underset{nxn}{[K]} \underset{mxm}{[T]}; \quad [M_r] = [T]^T [M] [T] \quad \ldots 8.$$

The condensation of the masses obtained in Eqn 7 makes it possible to economize in the number and, consequently, in the cost of solving the problem of the eigenvalue economizers.

In the case of civil buildings, the second simplifying hypothesis, relative to the approximation introduced in the distribution of the masses, restricts the building's seismic response. The masses of the building are assumed concentrated at the floor levels and associated to the movement parameters of this.

Keeping the above approximation in mind, the d.o.f. masters to preserve are those associated to floor movements (u_p; v_p; ϕ_{zp}). In this case, the following alternative procedure is foreseen in the programme to solve the generalized problem at the eigenvalues: having effected partition of the matrix in terms of the degrees of freedom, one writes:

$$\left(\begin{bmatrix} K_{mm} & K_{ms} \\ K_{ms}^T & K_{ss} \end{bmatrix} - \lambda \begin{bmatrix} M_{mm} & 0 \\ 0 & 0 \end{bmatrix} \right) \begin{Bmatrix} \bar{D}_m \\ D_s \end{Bmatrix} = 0 \quad \ldots\ldots 9,$$

where M_{mm} = matrix diagonal with masses concentrated only in the master nodes.

By eliminating $\{\bar{D}_s\}$ from Eqns 9,

$$\{\bar{D}_s\} = - [K_{ss}]^{-1} [K_{ss}^T] \{\bar{D}_m\} \quad \ldots\ldots 10$$

and, by sustitution in the remainder one gets:

$$([K_r] - \lambda [M_{mm}]) \{\bar{D}_m\} = 0 \quad \ldots\ldots 11$$

with

$$[K_r] = [K_{mm}] - [K_{ms}] [K_{ss}]^{-1} [K_{ms}]^T \quad \ldots\ldots 12.$$

With only three degrees of freedom per floor and not needing to condense $[M]$, calculation of $[K]$ may be obtained directly in the following manner without doing the onerous calculation of $[K_{ss}]^{-1}$:

$$\begin{bmatrix} K_{mm} & K_{ms} \\ K_{ms}^T & K_{ss} \end{bmatrix} \begin{bmatrix} I \\ D_s \end{bmatrix} = \begin{bmatrix} X_a \\ 0 \end{bmatrix} \quad \ldots\ldots 13,$$

where: I = matrix identity.

By eliminating $\{D_s\}$ from Eqn 13 one has:

$$[K_{mm}] - [K_{ms}] [K_{ss}]^{-1} [K_{ms}]^T = [X_a] \equiv [K_r] \ldots 14,$$

which is the very matrix 12 sought, obtained and accumulated in the X_a matrix by imposing unit movements $[I]$, associated at every degree of freedom to be preserved. Equation 13 is equivalent to solving 3 x p load conditions.

Having obtained the translating matrix $[K_r]$ (associated to the floor movements) it is possible to have the configuration of the vibration modes and the natural frequencies from the non-damped system expressed by

$$[M_r] \{\ddot{D}_r\} + [K_r] \{D_r\} = 0 \quad \ldots\ldots 15.$$

Where the problem at the eigenvalues to solve are written in the following manner:

$$[K_r] \{\Phi_r\} = \lambda [M_r] \{\Phi_r\} \quad \ldots\ldots 16,$$

Φ = eigenvectors (vibration modes).

The problem, the unknowns reduced to 3xp, is solved by the Jacobi method.

Once the solutions of the frequencies equation (eigenvalues) have been obtained,

$$|| [K] - \lambda [M] || = 0 \qquad \ldots\ldots 17$$

one gets the vector of the frequencies of the m modes of vibration

$$\{\omega\} = \{\omega_1; \omega_2; \omega_3; \ldots ; \omega_m\}^T ; \quad \lambda = \omega^2 \qquad \ldots\ldots 18.$$

The eigenvectors $\{\Phi_r\}$, amplitude or configuration of the modal vibration, are now obtainable with Eqn 16. The m vibration modes are orthonormalized with respect to the matrix of the masses so that:

$$[\Phi]^T [M] [\Phi] = [I] \qquad \ldots\ldots 19$$

thus,

$$[\Phi]^T [K] [\Phi] = [\omega^2] \qquad \ldots\ldots 20.$$

In accordance with Italian standards, the programme provides for dynamic analysis by means of a project response spectrum. To this end, as is well known, the decoupled equations of movement are expressed by:

$$\{\ddot{Z}\} + [\omega^2] \{Z\} = [\Phi]^T \{P\} a(t) \qquad \ldots\ldots 21,$$

with: $\{Z\}$ = response amplitude of each mode,

$$\{P\} = \begin{bmatrix} m_1 & \cos\alpha \\ m_1 & \text{sen}\,\alpha \\ J_1 & 0 \\ \vdots & \end{bmatrix} ; \quad \alpha = \text{seism input angle};$$

a(T) = g.C.R.(t); maximum soil acceleration in terms of the corresponding period of the structure for each vibration mode (response spectrum). From Eqn 21 one sees that the maximum response is given for each single vibration mode 'i' by:

$$Z_i^{max} = \frac{\{\Phi_i\}^T \{P\} a(T)}{\omega^2} \qquad \ldots\ldots 22.$$

We can now obtain the displacements in global coordinates by means of the following expression:

$$\{D_i\} = Z_{max} \{\Phi_i\} \qquad \ldots\ldots 23.$$

Having calculated the displacements for each vibration mode (in accordance with the regulations, the programme holds back only the first three nodes), we go back to the stresses, and these are made up through the following statistic combination law:

$$S_p = \sqrt{\sum_{i=1}^{3} S_i^2} \qquad \ldots\ldots 24,$$

S_p being the project stress and S_i the stress corresponding to mode i .

3. ELEMENTARY SUBSTRUCTURING

The building, as already mentioned, is schematized using the following elementary substructures:

- beam elements with 6 degrees of freedom per node;
- panel-bracing elements loaded in their own plane;
- floor elements loaded normally their own plane;
- foundation elements on elastic soil.

3.1. Beam elements

The stiffness matrix of a beam with 12 degrees of freedom is well known and permits consideration of axial and shear deformation.

The stiffness matrix may be suitably transformed in order to following operative possibilities:

- Static condensation of independent d.o.f.'s inside the rod element. If, for example, the extremity J of rod JK has to be hinged, rotation Φ_{JK} becomes independent of Φ_J and, in order to avoid its contribution, it is possible to condense Φ_{JK} before assembly. The elementary stiffness matrix remains 12x12 but the column and row corresponding to Φ_J is made up of zeros, thus eliminating its own contribution to bending stiffness of node J.

- Calculation of the elastically deformable length between internal alignment of the relative bays. The length between column axis and internal alignment are considered infinitely stiff, see Fig 2.

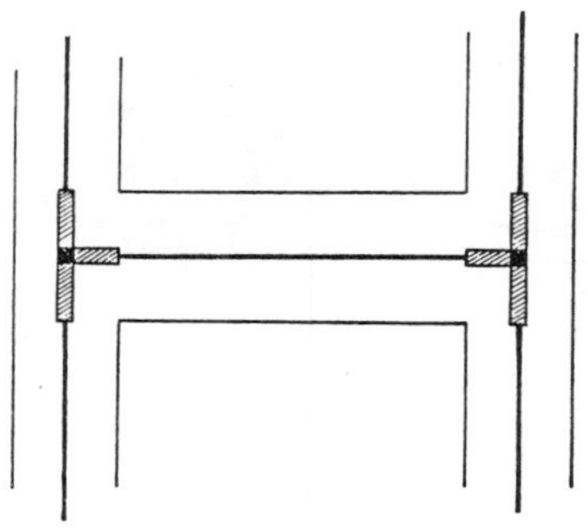

Fig. 2

- The possibility to consider states of initial strain/stress (pre-stresses, ΔT, yielding constraints, etc.).

The general elementary matrices (12x12), when subjected to the first simplifying hypothesis (see Section 2), must be transformed in the following manner: having called $[\Lambda]$ the transformation matrix in Eqn 1, the vector of the forces and the transformed stiffness matrices will be obtained from

$$\{P'\} = [T]^T \{P\} \qquad \text{with} \quad \underset{12\times12}{[T]} = \begin{bmatrix} \Lambda_1 & 0 \\ 0 & \Lambda_2 \end{bmatrix} \qquad \ldots 25,$$
$$\{K'\} = [T]^T [K] [T]$$

where: 1;2 = indices of the two nodes of the considered element.

3. 2. *Shear wall*

The shear wall element is schematized, passing from continuous to discrete, by means of the equivalence theory, Refs 3-4, like a grid of elementary rods of the type considered in Section 3.1. Through this schematization one can achieve:

- Essential, formal homogeneity in the resolution algorithm which operates on only one type of elementary matrices.

- A degree of precision in the results comparable to finite elements on the 3rd order, Ref 5.

- Versatility of elaboration of the standard wall substructures obtained separately through framed structure calculation methodology.

From the practical viewpoint the problem of the equivalence between a continuous and a discrete body consists of determining the geometric-mechanical characteristics of the model. The latter are obtained by equalizing the expressions of the deformation energy potentials of the two systems.

Considering the continuous model shown in Fig 3 for a plane state of tension, the deformation energy per unit of volume, expressed in terms of the strain tensors e_{11}; e_{22}; e_{12}, is:

$$U_o = \frac{E}{2(1-\nu^2)}\left[(e_{11})^2+(e_{22})^2+2\nu e_{11}e_{22} + + 2(1-\nu)(e_{12})^2\right] \quad \ldots\ldots\ 26.$$

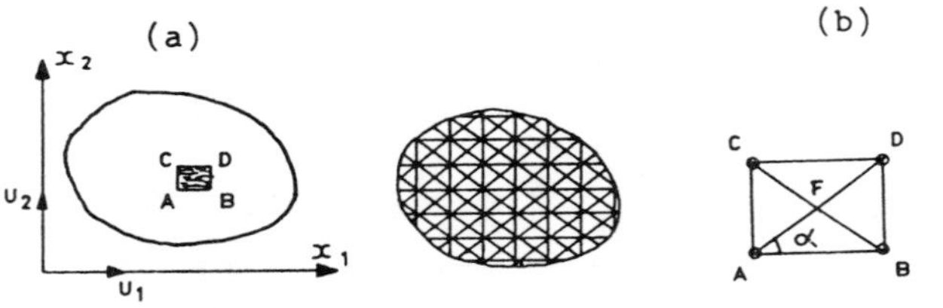

Fig 3

Considering wall thickness as h, one obtains, per surface unit:

$$(U_o)_h = U_o \cdot h \quad \ldots\ldots\ 27,$$

and if A_s = area of a floor element, the volume is

$$V = A_s \cdot h \quad \ldots\ldots\ 28.$$

The equivalence is obtained by making

$$A \cdot U_o = U' \quad \ldots\ldots\ 29,$$

U' being the deformation energy of the equivalent system which, in the case adopted by us, of a rectangular element with solid diagonal rods, Fig 3b, becomes:

$$U' = 2w'_{AB} + 2w'_{AC} + 2w'_{FB} + 2w'_{FD} + 4w''_{FD} \quad \ldots\ldots\ 30,$$

where w' indicates the deformation energy for normal and w" for bending stress.

After developing Eqns 26 and 30, equalizing the homologous coefficients, the equivalent characteristics of the model are obtained, expressed by the following equations:

$$\rho_{AF} = \frac{G \cdot V}{4\,\text{sen}^2\alpha\cos^2\alpha}$$

$$\rho_{AC} = \frac{G \cdot V}{2}\left(-\,\text{tg}^2\alpha + \frac{1+3\nu}{1-\nu}\right)$$

$$\rho_{AB} = \frac{G \cdot V}{2}\left(-\,\text{cotg}^2\alpha + \frac{1+3\nu}{1-\nu}\right) \quad \ldots\ldots\ 31,$$

$$\eta_{AF} = \frac{G \cdot V}{4\,\text{sen}^2\alpha\cos^2\alpha}\;\frac{1-3}{1-\nu}$$

where: ρ = E A L;
E = model elasticity modules,
A = area of model,
L = length of model,
η = EJ/L,
J = moment of inertia of model section.

The shear wall, with or without internal holes, can now be analysed within the sphere of the programme itself, and its stiffness matrix is condensed in the floor plane nodes (boundary), eliminating, in terms of a more compact overall solution, all the degrees of freedom inside the equivalente panel.

Condensation of the degrees of internal freedom corresponds to elimination by means of the Gauss method, which stops when only the internal d.o.f.'s have been eliminated (condensed). The process may be synthetized as follows:

Let $\{d_r\}$ be the "contact" or "frontier" d.o.f.'s with the main structure, to be preserved, and $\{d_e\}$ those to be eliminated. The element stiffness matrix, once partition has been done, is:

$$\begin{bmatrix} K_{rr} & K_{re} \\ K_{er} & K_{ee} \end{bmatrix} \begin{Bmatrix} d_r \\ d_e \end{Bmatrix} = \begin{Bmatrix} r_r \\ r_e \end{Bmatrix} \quad \ldots\ldots\ 32.$$

By solving the lower part for the unknown $\{d_e\}$, one gets:

$$\{d_e\} = -\left[K_{ee}\right]^{-1}\left(\left[K_{er}\right]\{d_r\} - \{r_e\}\right) \quad \ldots\ldots\ 33,$$

and, by substituting in the upper part of Eqn 32, one has:

$$\left[K\right]\{d_r\} = \{r\} \quad \text{where}$$

$$\left[K\right] = \left[K_{rr}\right] - \left[K_{ee}\right]^{-1}\left[K_{er}\right] \quad \ldots\ldots\ 34.$$

$$\{r\} = \{r_r\} = \left[K_{re}\right]^{-1}\{r_e\}$$

Matrix $[K]$ will be assembled conventionally in building the overall stiffness matrix.

3. 3. *Floor element*

As requested, it is possible to activate, as adopted for the wall element, a floor substructure, set within the 3-dimensional frame.

The equivalence method can also be employed, with great

versatility, for the floor element by adopting a model with equivalent girder lattice for which, after equalizing the respective deformation energies and saving all the transformations, the following geometric-mechanical equivalent characteristics are obtained.

We consider Fig 4 and think to neglect ν in the extensional terms.

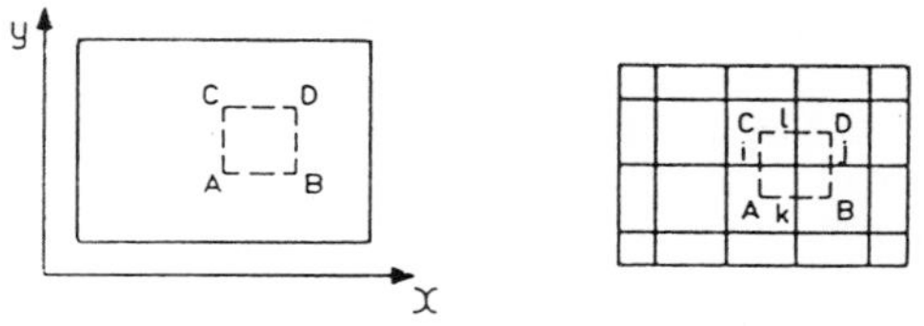

Fig 4

For the membrane effect:

$A_x = s(L_y/2)$, area of a rod having length x ;

$A_y = s(L_x/2)$, area of a rod having length y ;

$J_{px} = s(L_x^2 L_y)/24(1+\nu)$, flexional inertia in the plane of an x-length rod;

$J_{py} = s(L_x L_y^2)/24(1+\nu)$, flexional inertia in the plane of a y-length rod.

For the flexional effect:

$J_x = s^3(L_y/24)$, flexional inertia outside the plane of an x-length rod;

$J_y = s^3(L_x/24)$, flexional inertia outside the plane of a y-length rod;

$J_{tx} = s^3(L_y/12)$, flexional inertia of an x-length rod;

$J_{ty} = s^3(L_x/12\ (1+\nu))$, flexional inertia of a y-length rod,

where:

s = floor thickness;

L_x = length of equivalent element according to x ;

L_y = length of equivalent element according to y .

The floor substructure is calculated separately and is condensed at the level of the degrees of freedom relative to the plane nodes.

In the case of civil buildings, this schematization is very useful to determine distribution of the vertical loads in the frame beam elements.

3.4. Substructure foundation element on elastic soil

This element makes it possible, with sufficient engineering approximation with respect to the context of the hypotheses introduced in the reduction (constructive reality → mathematical model), to keep account of soil-structure interaction.

3.4.1. Footing element

The footing on elastic soil behaviour is represented by the following 6 x 6 matrix, directly associated with the relative node d.o.f.:

$$
\begin{array}{cccccc}
u & v & w & \phi_x & \phi_y & \phi_z
\end{array}
$$

$$
\left|\begin{array}{cccccc}
0 & 0 & 0 & 0 & 0 & 0 \\
0 & 0 & 0 & 0 & 0 & 0 \\
0 & 0 & a\,b\,k & 0 & 0 & 0 \\
0 & 0 & 0 & a\,b^3 k & 0 & 0 \\
0 & 0 & 0 & 0 & a^3 b\,k/12 & 0 \\
0 & 0 & 0 & 0 & 0 & 0
\end{array}\right|_K
$$

where:

$a;b$ = footing dimensions along x and y local reference axis;

k = soil elastic coefficient.

3.4.2. Beam element on elastic soil

According to Winkler's linear theory, the stiffness matrix for a beam element on elastic soil foundation can be written as follows:

$$
\left|\begin{array}{ccccc|cccc}
0 & 0 & 0 & 0 & 0 & 0 & 0 & 0 & 0 \\
0 & 0 & 0 & 0 & 0 & 0 & 0 & 0 & 0 \\
0 & 0 & EJ_y f'''(j) & 0 & EJ_y f'''(j) & EJ_y f'''(j) & 0 & EJ_y f'''(j) & 0 \\
0 & 0 & 0 & GJ_x^* g'(j) & 0 & 0 & GJ_x^* g'(j) & 0 & 0 \\
0 & 0 & -EJ_y f''(j) & 0 & -EJ_y f''(j) & -EJ_y f''(j) & 0 & -EJ_y f''(j) & 0 \\
0 & 0 & 0 & 0 & 0 & 0 & 0 & 0 & 0 \\
0 & 0 & 0 & 0 & 0 & 0 & 0 & 0 & 0 \\
0 & 0 & 0 & 0 & 0 & 0 & 0 & 0 & 0 \\
0 & 0 & -EJ_y f'''(k) & 0 & -EJ_y f'''(k) & -EJ_y f'''(k) & 0 & -EJ_y f'''(k) & 0 \\
0 & 0 & 0 & -GJ_x^* g'(k) & 0 & 0 & -GJ_x^* g'(k) & 0 & 0 \\
0 & 0 & EJ_y f''(k) & 0 & EJ_y f''(k) & EJ_y f''(k) & 0 & EJ_y f''(k) & 0 \\
0 & 0 & 0 & 0 & 0 & 0 & 0 & 0 & 0
\end{array}\right|_{K;J}
$$

12 x 12

where:

- elastic shape function

$$f(x) = C_1 \sin(\alpha x)\,\mathrm{sh}(\alpha x) + C_2 \left[\sin(\alpha x)\,\mathrm{ch}(\alpha x) - \cos(\alpha x)\,\mathrm{sh}(\alpha x)\right] ;$$

- torsional shape function

$$g(x) = \frac{h_t}{\mathrm{sh}(\alpha_t l)}\,\mathrm{sh}(\alpha_t x) + h_t x)$$

and

$h(x);\ h_t(x)$ = loading functions;

J_x^* = torsional inertia.

4. INTERACTIVE GRAPHIC PRE-CENTRAL-POST PROCESSOR

In the actual situation of development of sophisticated automatic machines and general programmes for structural calculation, the optimum relationship between man and machine seems to be achieved through the modern techniques of "interaction" between designer and electronic elaborator, which, in our field of application, are defined by various, evere increasingly known codes such as: CAD (Computer aided design); CAAD (Computer aided architectural design); CASD (Computer aided structural design); CAM (Computer aided manufacturing).

The interaction technique between man and machine emphasises the salient contributions of both parties, permitting the simultaneous achievement of the following objectives:

- Excellent relationship for the ANALYSIS phase (operation entrusted to the machine, using its potency, capacity and speed in automatic calculation of structures) and the SYNTHESIS phase (operation entrusted to the person in charge of control of validity of data, criticism of results and order of size, the latter obtainable through non-substitutive but integrative validity of approximated methods of calculation),

- Design optimisation through interactive means as a logical consequence of the design interactive cycle (Fig 5) which permits extremely simple and rapid modification of data and testing of the consequence of same through successive, repetitive testing according to the classic procedure of trial and error based on the experience of the designer who has the capacity to synthesize a considerable mass of data difficult to express as a mathematical problem.

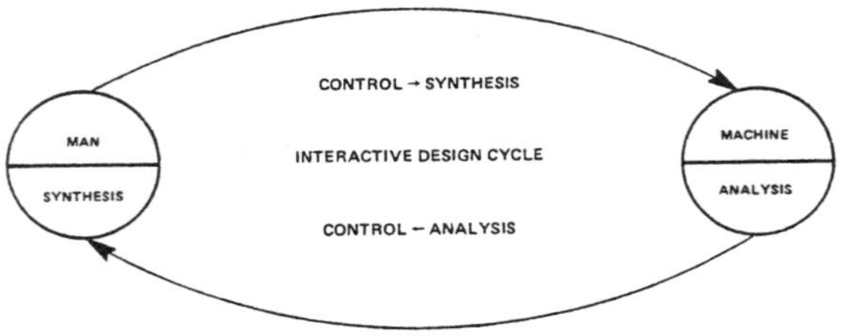

Fig 5

The interactive techniques are rendered possible through interactive machines (hardware) and interactive structural programme (software), completely innovative as regards the traditional batch technique in which the user may not open a computer aided procedure but is subjected to a passive relationship of data input and verification of results (output).

The importance of modern research and development of structural software lies here, in that they reduce the input-output procedure with convenient pre-elaboration of data (pre-processor) and post-elaboration of results (post-processor), Fig 6.

The two phases of pre- and post-elaboration of data and results, respectively, are achieved with suitable interfacing in hardware and software.

The pre-elaboration of data, through interactive techniques,

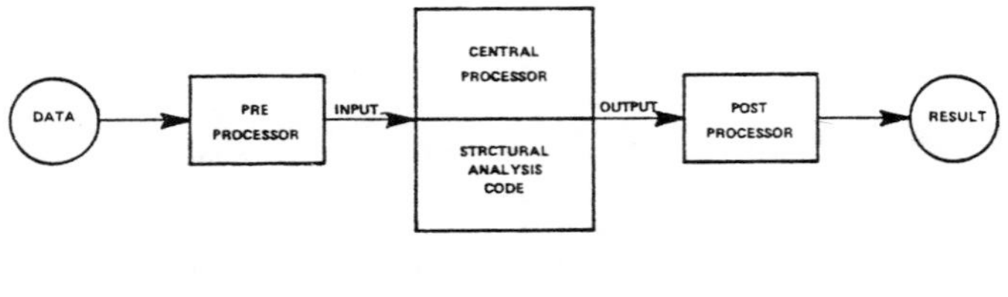

Fig 6

may be carried out mainly with:

- alphanumeric keyboard: hardware;
- luminous pen and joysticks: hardware;
- analog table or digitalizer: hardware;
- video-graph: hardware;
- menu and functions keyboard: software and hardware;
- programmes for automatic data generation: software.

The post-elaboration of results may be obtained through:

- video-graph: hardware;
- graphic table or plotter: hardware;
- programmes for graphic representation of results: software.

The modern availability of scientific languages for communication with machines and the new generations of extremely powerful mini-computers at ever-increasingly accessible cost permit the constitution of a man-machine combination according to an interactive configuration as outlined in Fig 7.

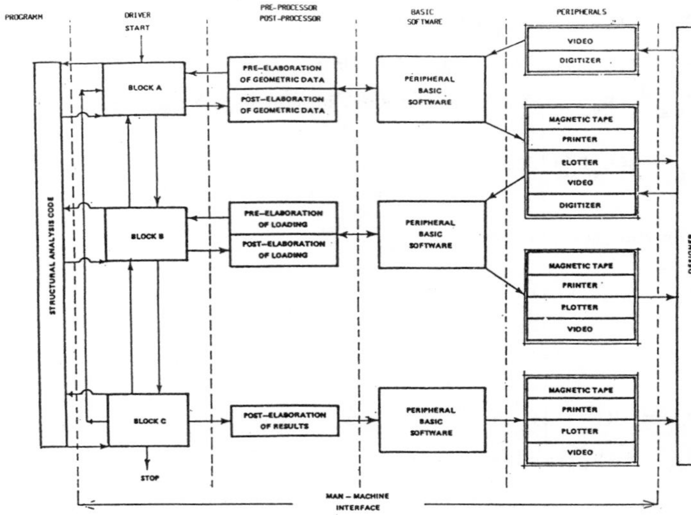

Fig 7

Through this interactive configuration (software-hardware interface), communication between man and computer takes place in human language, intending by this the collection of numerical data, symbols and graphics commonly used by man in relation to his particular professional technique.

The graphic type communication allowed by interactive techniques (IG) is substantially the most important in the field of structural engineering.

Data input takes place according to the natural operations mode of the designer by means of geometrical diagrams which may be read by the digitalizer (see further on). Data so reduced may be video controlled. Other data, such as the mechanical characteristics of the structure, may be introduced through the alphanumeric keyboard.

Once the calculation has been carried out, immediate visual information on the state of deformation and stress may be obtained graphically on the video or on the plotter.

The interactive graphic system may be used for designing, planning and verifying constructive details, allowing execution of changes and modifications visually controllable by making the figure under examination rotate and translate and, on obtaining the solution, allowing representation on paper on request.

To interact graphically one must communicate with the programme being executed with suitable hardware devices.

4.1. Definition of the data (PRE-PROCESSOR)

The data may be introduced graphically and/or alphanumerically and are classifiable as:

4.1.1. Geometric data

- Topological input and editing of the nodes and rods;
- input and editing of the bracing walls;
- input and editing of the foundation beams and footings;
- mechanical definition of the floors, the standard beam types, standard walls, and foundation elements.

4.1.2. Load data

- Loads on the nodes;
- loads on the rods;
- loads on the floors;
- loads produced by the wind;
- loads produced by the seism;
- definition of the load conditions;
- definition of combinations.

Software organization of the data is shown in the diagram of Fig 8

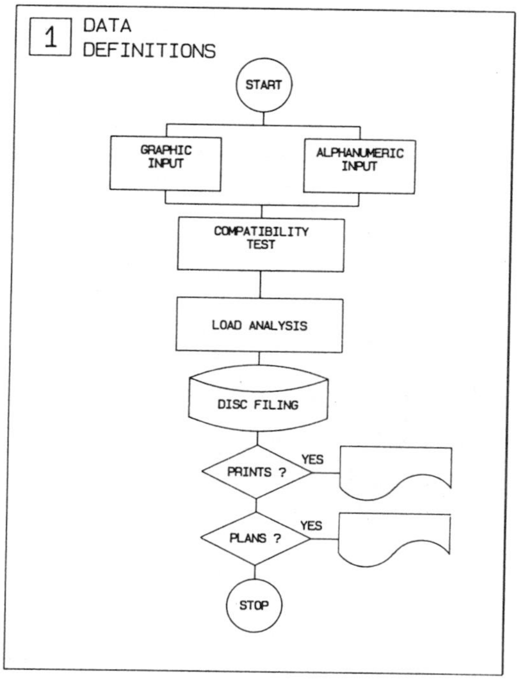

Fig 8

From the pre-processor module the entire execution of the work can be checked by an extremely simple graphic language. The controls may be digital on the terminal keyboard or selected by means of a menu on a graphic tablet. One can start by introducing the nodes at the base of the building, raise the columns, connect them by girders, copy several similar floors and at any time intervene, modifying rods, nodes, constraints etc. This may be done when working on schemes of plans, sections, perspectives of all or part of the structure, and simply indicating the elements to re-define or cancel.

The model includes girders, columns and rods however placed in space, floors, walls (also with holes), box walls and foundations. The sections for the beams to be set automatically are: rectangular, T-shaped, L-shaped and circular, but it is possible to introduce geometric characteristics of arbitrary section. It is also possible to have a directory of often-used sections: e.g. steel sections. The constraints may be defined at pleasure and re-defined at any time.

Load analysis is automatically carried out by the programme as regards dead loads, permanent and accidental distributed loads, loads due to wind and to seism (in the case of equivalent static calculation). There is, however, complete freedom to define or re-define the loads, choosing between concentrated loads, variable distributed loads, pre-strain and thermal variations.

All the data are, in any case, memorized on disk in a printable file constituting the input of the mathematical model.

4.2. Structural Analysis (Central Processor)

The solver used is organized with the following calculation phases:

- calculation of the masses, of the mass barycenters and moments of polar inertia of each floor (terms of the mass matrix);
- calculation of the local stiffness matrices, whether rigid at at the extremities or hinge to one or both nodes;
- calculation on the hypothesis of elastic soil behaviour (Winkler's theory);
- calculation of the local stiffness matrices for the continuous flat elements by static condensation of a discrete scheme of equivalent rods (Equivalence method);
- in the static case: calculation of the perfect fixed and joint reactions due to the loads;
- in the dynamic case: imposition of the unitary movement of the floors;
- assembly of the overall stiffness matrix of the structure with imposition of constraints (external constraints, infinitely rigid floors), anelastic settlement internal constraints, etc.;
- solution of the system of equations by the Gauss method to compute displacements. The programme uses a parking file on disk should the system of equations not be containable in the central memory (modified frontal method);
- in the dynamic case, calculation of the constraining reactions on the floors (terms of the translating stiffness matrix of the structure);
- solution of the generalized problem at the eigenvalues (Jacobi method) to determine periods and modes of structure vibration;
- calculation of the stresses on the extremity nodes of rods and panels. In the dynamic case the calculation is carried out for every vibration mode;
- in the dynamic case: calculation of the spectral accelerations according to the response spectrum of the Italian Regulations) for each vibration mode and superposing of the stresses.

Having done the calculation, the first control may be effected on the suitably amplified deformation and stress diagrams. Another useful instrument is given by the diagram that shows the variation in the intensity of participation of each vibration mode at each variation of the seism input angle. It is possible to define several combinations of static and dynamic load conditions.

The diagram of Fig 9 shows the analytical options of the SISMI programme, the logical flow-chart of which is indicated in Fig 10.

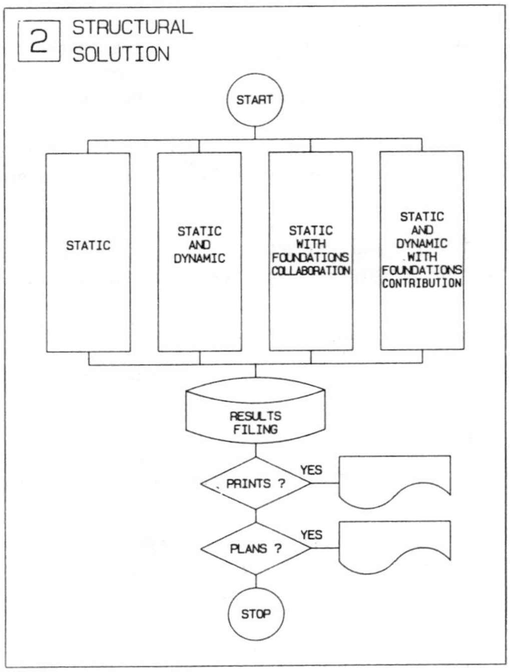

Fig 9

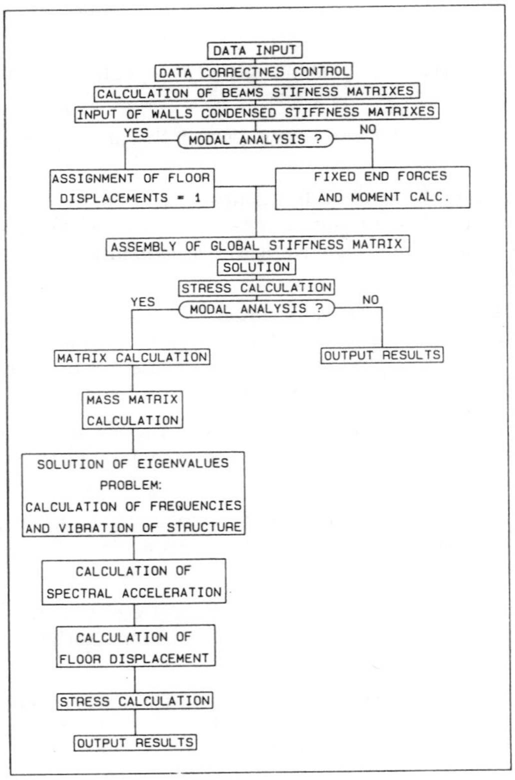

Fig 10

4. 3. Project and verification of structural elements

The project-control of the structural elements is done according to the allowable stress method and/or the semi-probabilistic limit design method.

Executive organization of the calculation and graphic procedure is illustrated in Fig 11.

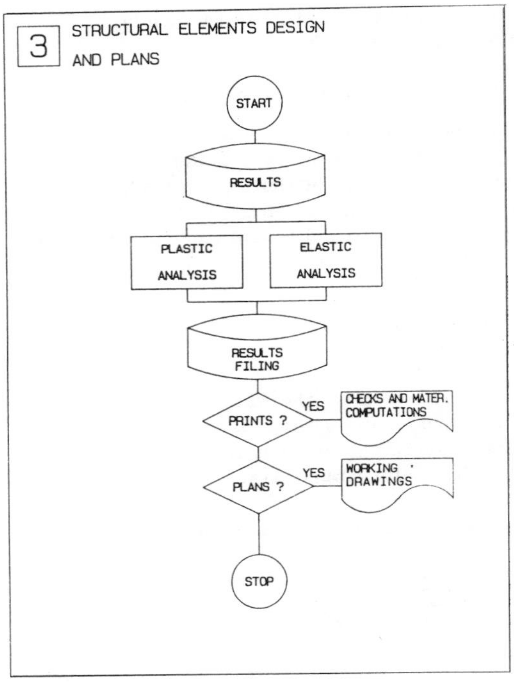

Fig 11

All the phases of pre-central and post-processing are interactively controlled by keyboard or video by means of a graphic menu illustrated in Fig 12.

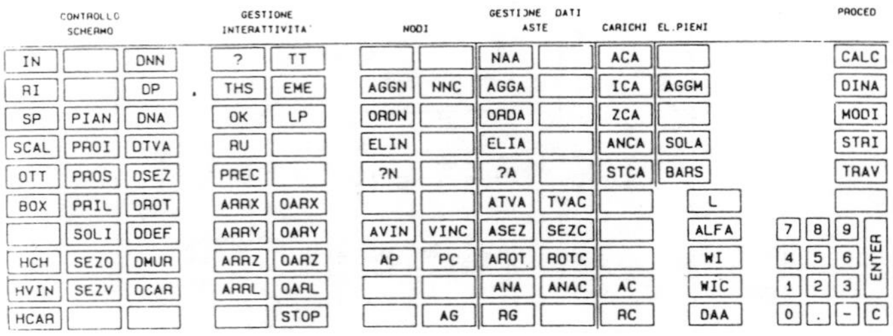

Fig 12

The control commands and executions are listed in Table 1.

The hardware system adopted consists of the following peripheral devices:

- Graphic video HP 1351;
- Graphic video HP 2648 with alphanumeric keyboard;
- Digitizer device HP 9111;
- Plotter device HP 7580 .

Table 1

PROGRAMME CONTROL

STOP	STOP	HE	information menu
?	Help	TT	check at keyboard
TD	graphic tablet check	TR	procedure check
THS	hard/software tracking	RV	run programme
LP	printing logical unit	DEUM	unit of measurement
AG	archive geometry	RG	recall geometry

SCREEN CONTROL

RI	redraw	OTT	optimization
SCAL	scale according to XYZ	SP	previous scale
IN	enlargement (zoom)		
PREC	precision trap	ARRY	rounding X
ARRY	rounding Y	ARRZ	rounding Z
ARRL	local rounding	OARX	rounding X origin
OARY	Y rounding origin	OARZ	rounding Z origin
OARL	local rounding origin		
HCH	letter size	HVIN	node constraint symbol dimension
HCA	load scale	HDEF	displacement scale
BOX	boxed portion	POQ	horizontal level Z
POS	horizontal floor level	PV	vertical plane
VPL	work plan view	PROS	perspective
PDV	point of view		
DNN	on-off node numbers	DS	on-off floor numbers
DNA	on-off bar numbers	DTVA	on-off bar constraints
DTSA	on-off bar section	DROT	on-off bar rotation
DFF	on-off fixed alignment	DA3D	on-off 3-D bar
DNPL	on-off footing numbers	DRPL	on-off footing rotation
DTPL	on-off footing type		
DMUR	draw walls	DCA	draw loading
DDEF	draw displacement	DMF	draw bending moment
DMT	draw torque moment	DSN	draw normal forces
DST	draw shear diagram	DINV	draw envelope
DBAR	draw barycenter		

DATA EDITING

NODES

AGGN	add node	COPN	copy nodes
VINC	current constraint	INC	fixed end constraint
LIB	free	AVIN	assign constraint
ELIN	remove node	NNC	current node number
SC	current floor	AS	assign floor
ASX	assign X to nodes	ASY	assign Y to nodes
ASZ	assign Z to nodes	?N	information on node

BARS

AGGA	add bar	COPA	copy bar
ORDA	bar	ELIA	remove bar
DESE	define section	SEZC	current section
ASEZ	assign section	TVAC	current bar constraint
ATVA	assign bar constraints	ROTV	current rotation
AROT	assign rotation	FFC	current fixed alignment
AFF	assign fixed alignment	NAC	current bar number
ANA	assign bar number	NAA	automatic bar numbering
?A	information on bar		

./.

WALLS-FLOORS

DESO	floor definition	AGGM	add wall
COPM	copy walls	ELIM	remove wall
NMC	current wall number	MC	current wall
?M	information on wall		

FOOTING (FOUNDATION)

AGPL	add footing	ORPL	place footing
ELPL	remove footing	DEPL	create footing
PLIC	current footing	ATPL	assign footing type
PLRC	current footing rotation	ARPL	assign footing rotation
PLNC	current footing number	ANPL	assign footing number
NAPL	automatic footing numbering	?P	information on footing

LØADS

AC	archive loads	RC	recall load
ZCA	zeroing loads	ICN	input joint-type load
ACN	assign node load	ICA	input joint-bar load
ACA	assign bar load	ANCA	load analysis
VENT	wind analysis	PESO	dead/load frame/wall

QUALIFIERS

/	assign XY coordinates	DX	increment along X
DY	increment along Y	L	previous node distance
ALFA	X axis angle	N	node
ALL	all	SOL	floor
WI	window	WIC	complementary window
DAA	window dimension		

CALCULATION

BAND	order nodes into *E	BARS	barycentric calculation
NMOD	number of vibration modes	MODI	seismic intensity and direction
CALC	static/dynamic calculation		

PRINTING

STGE	print geometry	STCA	print loads
STR	print partial results	STRI	print results
STFO	printi horizontal forces	STRE	print report

DESIGN/VERIFICATIONS

CRIT	project criteria	TRAV	beam project
STTR	print beam report	PILA	column project
STPI	print column report	PLIN	footing project
STPL	print footing report	STMU	print wall report

PLOTTERS

CARP	floor working drawing	PLPI	column working drawing
PLTR	beam working drawing	PLPI	footing working drawing

REFERENCES

1. E L WILSON and H H DOVEY, Three-dimensional analysis of building systems-TABS, Report N EERC 72-8, Dec.1972.
2. E L WILSON, J P HOLLINGS and H H DOVEY, ETABS, EERC, N 75-13, March 1979.
3. E ABSI, La théorie des equivalences et son application à l'etude des ouvrages d'art, Annales ITTTP N 218, Oct. 1972.
4. L C ZALESKI-ZAMENHOF, Methode du maillage orthogonal, Annales ITTP, N 331, Sept. 1975.
5. E ABSI and W PRAGER, A comparison of equivalence and finite element methods, Computer Methods in Applied Mechanics and Engineering, N 6, 1975.
6. M MAJOWIECKI, Tecniche interattive nella moderna metodologia della progettazione strutturale, Acciaio, n. 1, 1980.
7. M MAJOWIECKI, CAD per la progettazione strutturale, Pixel, settembre 1982.

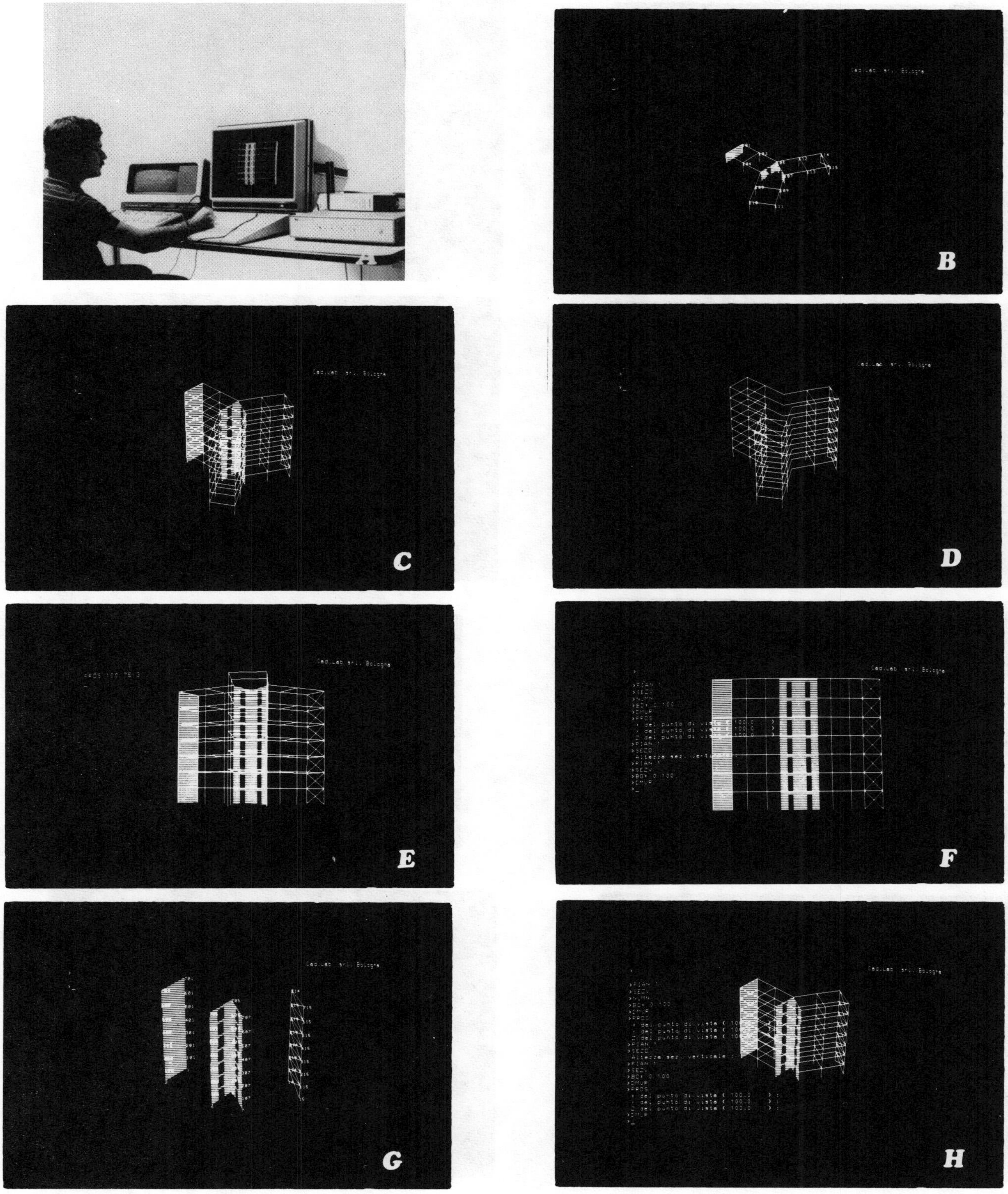

APPENDIX 1

A - Interactive hardware
B - Data elaboration of floor unit
C - Floor automatic generation
D - Visualization of framed elements only
E - Perspective view
F - Front view
G - Visualization of bracing system
H - Partial view (box command)

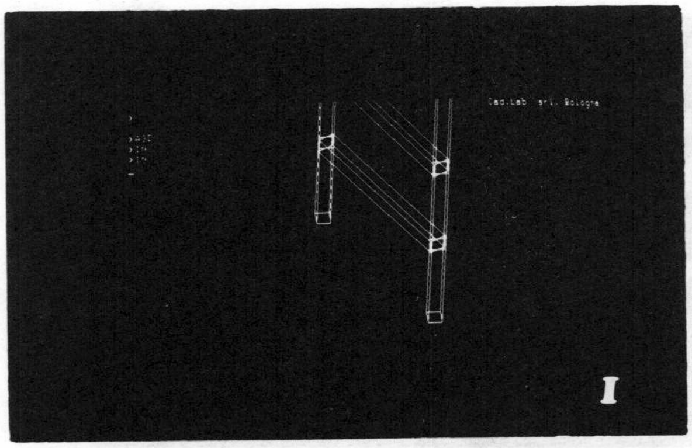

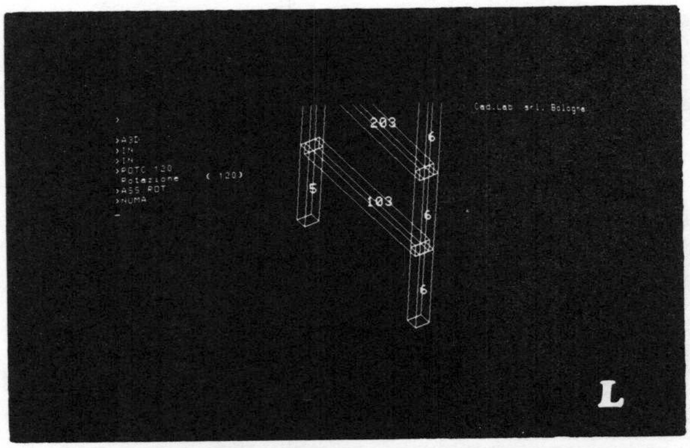

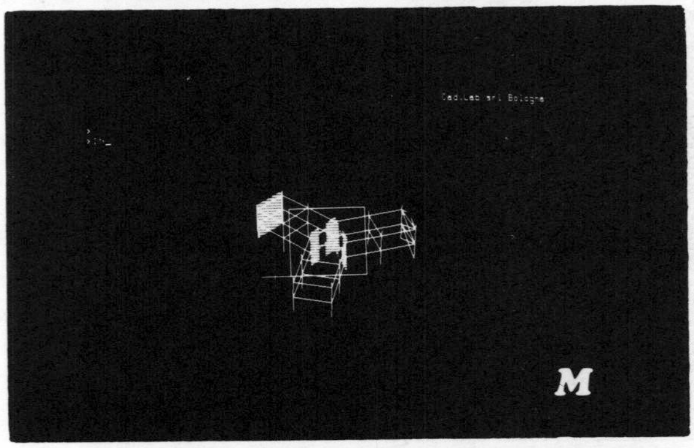

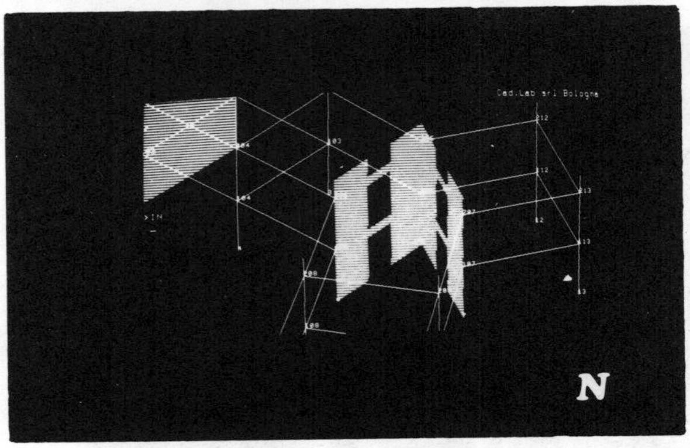

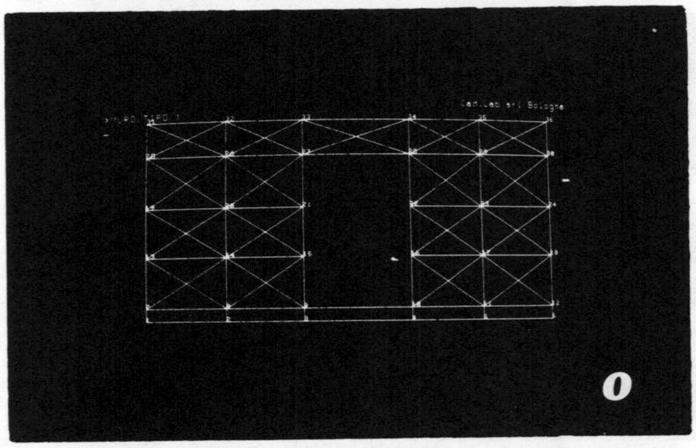

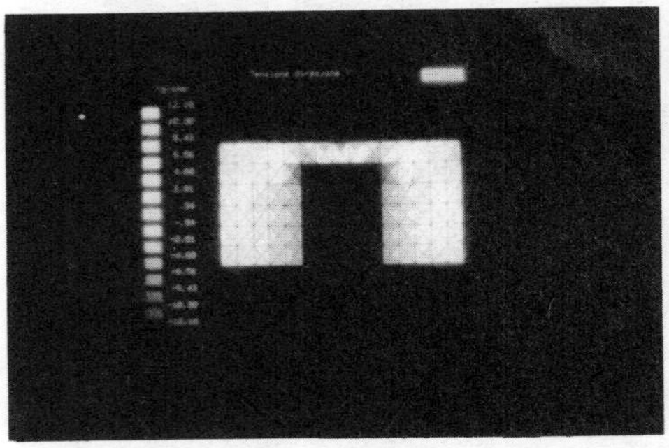

P

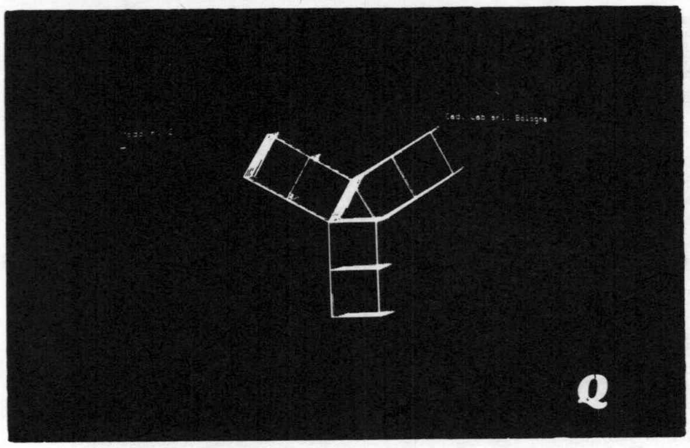

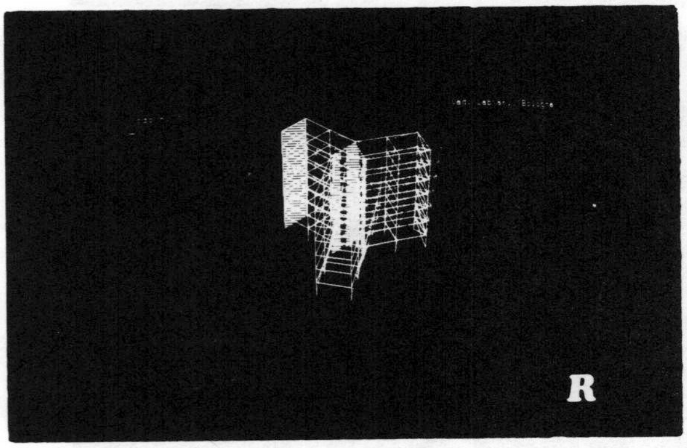

APPENDIX 1

I - Solid geometric control
L - Correction of relative rotation between beam and columns
M - Window
N - Zoom
O - Finite element or equivalent mesh for shear wall element (with holes)
P - Stress distribution on wall on color vieo terminal
Q - Modal displacement
R - Perspective view of modal displacement

INTERACTION OF RANDOM CRITICAL LOAD OF SPATIAL TRUSS JOINT LOADED AT THE JOINT AND ON THE BARS

Z KOWAL*, PhD, DSc, Eng and K RADUCKI**, MSc

* Professor, Dean of Department of Civil Engineering, Technical University of Kielce.

** Assistant, Department of Civil Engineering, Technical University of Kielce.

From the condition of minimum random energy of rods connected at the joint, formulae are derived to determine the mean value and the variance of the joint critical load P,q_{cr}. The structure is loaded with concentrated force P at the joint and with uniform force q on the rods. Interaction curves P,q_{cr} are plotted for mean critical loads of joint P,q_{cr} from the condition of the joint snap-through. The paper also presents the formation of the variability coefficient of the joint critical load in the function of the angle of rise of the joint under the assumtion of the constant variability coefficient of rod lengths.

INTRODUCTION

Random stability loss of bars has been investigated by Bolotin, Kowal, Murzewski, Refs 1,6,9 and others. The significance of this problem for engineering practice is commonly recognized, which is reflected by publications on this subject.

Random stability loss of joints in spatial structures has not been reported so far. Only have a few authors considered non-random stability loss of bar domes loaded at joints from the condition of joint snap-through, Refs 2,3,11. Such a critical load is characterized by a strong dependence on the angle of rise θ_o of the joint. This results directly from the solution of the classical, two-dimensional Mizes problem, Refs 4, 10, in which a non-random critical load of a two-armed, planar joint is determined from the condition of snap-through in the case of concentrated load at the joint.

Many bar dome failures in the USA and Europe (including Poland) were subject of analyses based, among other things, on the measurement of the structure geometry. It follows from geodesic measurements, Ref 8, that the real joint surface area deviations from the designed ones are considerable-particularly-in comparison with the angle of joint rise θ_o. The geodesic measurements of the joint coordinates in bar domes are difficult to make and are insufficiently accurate in comparison with the joint rise h.

A theoretical analysis of dome safety on the basis of the expected dome surface leads to a considerable overestimation of the safety of single-layer bar domes. The conception of estimating the random load of a dome with regard to the effect of the random load capacity of bars and random angle θ_o of joint rise is presented in Ref 5.

In the present paper, we shall be concerned with random stability loss from the condition of joint snap-through in a space truss constructed from n bars in the way shown in Fig 1. Apart from concentrated force load $P = nP_1$ at the joint, vertical, uniformly distributed arm load q was also taken into account because roofing in real domes is very often supported not only at joints but on bars as well.

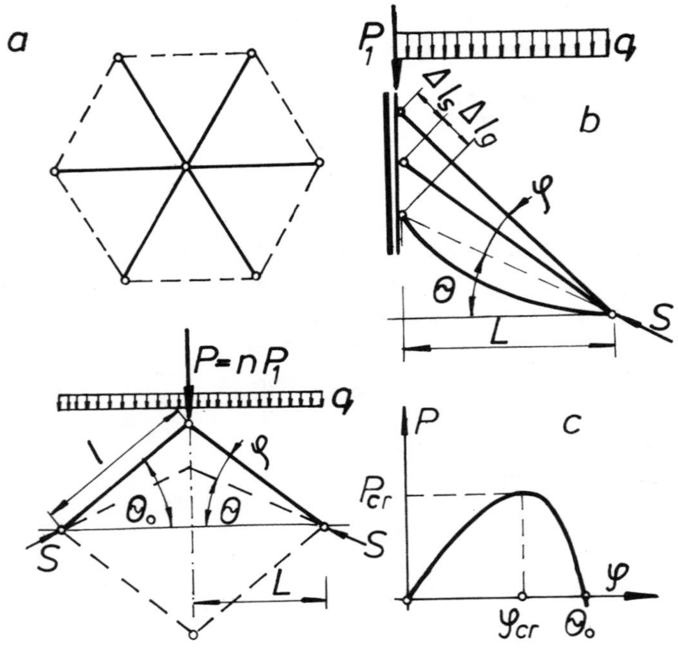

Fig 1

RANDOM STATIC EQUILIBRIUM PATH OF JOINT

Let's consider a structure made of n bars with a common vertex joint 1, satisfying the conditions of the structure symmetry and loading ,Fig 1a . Let the process of the joint stability loss be caused by the shortening of Δl of a bar ,Fig 1b .

$$\Delta l = \Delta l_s + \Delta l_q \qquad ....1,$$

where: Δl_s - shortening caused by a random axial force $S=S(\omega)$,
Δl_q - shortening caused by bending a bar with a random load $q=q(\omega)$,

Assume that random internal force $S=S(\omega)$ in the bar is less than both random critical load $S_{cr}=S_{cr}(\omega)$ of the bar and its random ultimate bearing capacity $S_u=S_u(\omega)$

$$S < S_{cr} < S_u \qquad ....2.$$

Random energy of the system will be used to determine the random static equilibrium path of the joint.

Displacement y of a bar caused by bending is determined on the basis of Ref 4 from potential energy V_1 of single bar bending.

$$V_1 = \int_{-0.5l}^{+0.5l} \left\{ K(y'')^2 - \frac{1}{2} S(y')^2 - qy \right\} dx \qquad ....3,$$

where: $K=EJ:l$ - flexural rigidity of a bar,
E - longitudinal modulus of elasticity of bar material,
J - moment of inertia of bar intersection,
y, y', y'' - transverse displacement of bar axis, and the 1st and 2nd derivatives,
F - area of bar intersection.

Mean axial load $\bar{S}_1$ of a bar is:

$$\bar{S}_1 = \frac{\bar{P}_1 + 0.5\bar{q}l}{\sin \bar{\theta}} \qquad ....4,$$

where: $\bar{\theta} = \bar{\theta}_o - \bar{\varphi}$,
$\bar{\varphi}$ - mean angular displacement,
$\bar{\theta}_o$ - mean angle of rise of an unloaded joint.

Taking into account boundary conditions $y(\pm 0.5l) = 0.$ and $y''(\pm 0.5l) = 0.$ and from minimum of energy $(\min V = 0)$, one obtains

$$y = \frac{B}{k^4} \frac{\cos kx}{\cos \frac{kl}{2}} + \frac{B}{2k^2} x^2 - \frac{B}{k^2}\left(\frac{1}{k^2} + \frac{l^2}{8}\right) \qquad ....5,$$

where: $B=q:EJ$,
$k^2=S:EJ$.

Shortening Δl_q of a bar caused by bending will be necessary for further considerations. It is determined from the formula:

$$\Delta l_q = \int_{-0.5l}^{+0.5l} (w')^2 \, dx = \frac{B^2 l^3}{12k^4} + 2\frac{B^2 l}{k^6} - 5\frac{B^2}{k^7} \operatorname{tg}(kl/2) +$$

$$+ \frac{B^2}{2k^6} \frac{1}{\cos^2 \frac{kl}{2}} \qquad ....6.$$

The expression $tg(kl/2)$ and $1/\cos^2(kl/2)$ can be conveniently arranged into a series. Taking into consideration 5 first terms, shortening Δl_q is given by the formula

$$\Delta l_q = \frac{q^2 l}{(EJ)^2}\left(\frac{1}{4480} + \frac{P_1 l^2}{3EJ\ \theta 8!}\right) \qquad \ldots .7.$$

Critical load $(P,q)_{cr} = (nP_1,q)_{cr}$ of the system from the condition of joint snap-through will be estimated from the full energy of a single bar

$$V_1 = 0.5K_F\left[L(\sec\theta_o - \sec\theta) - \Delta l_q\right]^2 -$$

$$- P_1 L(tg\,\theta_o - tg\,\theta) \qquad \ldots .8,$$

where: $K_F = EF:l$ - longitudinal rigidity of a bar, F-area of bar intersection.

The load equilibrium path $ql/2 + P_1$ of a bar, determined from the condition of minimum of energy, has the form

$$\frac{ql}{2} + P_1 = \theta EFJ\cos^2\theta_o \frac{E^2J^2\cos^7\theta_o(\theta_o^2 - \theta^2) - \frac{q^2L^6}{4480}}{E^2J^3\cos^8\theta_o + \frac{q^2L^6}{241920}\left[1 + \left(\frac{\theta_o}{\theta}\right)^2\right]F} \qquad \ldots 9.$$

From the investigation into the behaviour of the mean realization of random equilibrium path $\bar{P}_1(\theta)$ with fixed load $\bar{q}$, it follows that there occurs maximum $\bar{P}_1$, which results in unstable equilibrium. A slight /close to zero/ increase in load $\bar{P} = n\bar{P}_1$ of the joint over $\bar{P}_{cr} = n\bar{P}_{1cr}$ causes the joint to snap to the position shown in Fig 1. Load $\bar{P} = n\bar{P}_1$, at which unstable equilibrium will be reached, is the sought mean critical load $\bar{P}_{cr} = n\bar{P}_{1cr}$ of the joint /max $\bar{P} = \bar{P}_{cr}$/.

The operation of seeking for critical load may be carried out analytically by seeking for the first maximum of function $\bar{P}_1(\theta)$, or numerically by examining the course of function $\bar{P}_1(\theta)$. It follows from numerical investigation that the range in which the first maximum of load $\bar{P}_1(\theta)$ occurs is rather fuzzy, which requires that calculations steps should be refined for θ in the proximity of θ_{cr}, Fig 1c .

MEAN CRITICAL LOAD CONCENTRATED AT JOINT

The mean realization of concentrated critical load will be determined on the basis of the equilibrium path ,Eqn 9, assuming that q=0. Then one obtains as in Ref 10.

$$\bar{P}_1 = \frac{\bar{\theta}}{2} EF\left(\bar{\theta}_o^2 - \bar{\theta}^2\right) \qquad \ldots .10.$$

Maximum of function $\bar{P}_1(\theta)$ occurs at $\bar{\theta} = \bar{\theta}_{cr}$, which will be determined analogously as in Ref 10 from the relation , Eqn 11.

$$\frac{\partial P_1}{\partial\theta} = \frac{EF}{2}\left(\theta_o^2 - 3\theta_{cr}^2\right) = 0 \qquad \ldots .11,$$

we have: $$\theta_{cr} = \frac{\theta_o}{\sqrt{3}} \qquad \ldots .12.$$

Finally, mean critical load $\bar{P}_{1cr}$ of the joint will be determined from the formula

$$\bar{P}_{1cr} = \frac{EF\bar{\theta}_o^3}{3\sqrt{3}} \qquad \ldots .13.$$

The ultimate bar effort in the process of the joint stability loss cannot exceed the limit of strength or the yield point R_e of material and critical stress R_{cr}, determined from the condition of the bar stability loss

$$\sigma_u = \frac{S}{F} = \frac{\bar{P}_{1cr}}{F\sin\bar{\theta}_{cr}} = \frac{E\,\theta_o^3}{3\sqrt{3}\,\sin\frac{\theta_o}{\sqrt{3}}} \leqslant R_e, R_{cr} \qquad \ldots .14.$$

Ultimate rise $\bar{\theta}_o$ of the joint under load will be determined from the equation

$$\theta_o^3 - \frac{R_e\,3\sqrt{3}\,\sin\frac{\theta_o}{\sqrt{3}}}{E} = 0 \qquad \ldots .15.$$

Table 1. Limit angles of joint rise $\bar{\theta}_{ou}$ according to $\min|R_e, R_{cr}|$

min $[R_e, R_{cr}]$ [MPa]	7	28	63	86	112	142	175
θ_{ou} [Rad]	0.01	0.02	0.03	0.035	0.04	0.045	0.05
min $[R_e, R_{cr}]$	212	252	296	343	394	448	506
θ_{ou}	0.055	0.06	0.065	0.07	0.075	0.08	0.085

Table 1 shows exemplary limit angles $\bar{\theta}_{ou}$ of the joint rise for the yield point R_e from 175 MPa to 500 MPa and E=210000 MPa. The limit angles of rise of the joint from the condition of bar stability loss are different for various slenderness ratios of a bar.

In determining critical stresses R_{cr} from the

condition of the bar stability loss, i.e. from the known formula

$$R_{cr} = \frac{\pi^2 E}{\lambda^2} \quad \text{....16,}$$

one may use Table 1, inserting critical stress R_{cr} in place of the yield point R_e.

MEAN REALIZATION OF UNIFORMLY DISTRIBUTED CRITICAL LOAD FROM THE CONDITION OF JOINT SNAP-THROUGH

The mean realization of uniformly distributed critical load was also determined by the energy method on the basis of the equilibrium path $q(\Theta)$, Eqn 9.

The equilibrium path , Eqn 9, determined from the transformation of the equilibrium path $P(\Theta_o)$, 9 , in which q was assumed to be a variable, and P=0, has a confounded form

$$A_1 q^3 + A_2 q^2 + A_3 q = A_4 \quad \text{....17,}$$

where:

$$A_1 = \frac{1}{4x120960} \frac{Fl^9}{E^2J^3}\left(1 + \frac{1}{\alpha^2}\right),$$

$$A_2 = \frac{1}{4480} \frac{Fl^6}{EJ^2} \frac{\Theta_o}{\cos\Theta_o}\alpha \quad , \quad A_3 = \frac{1}{2}\cos\Theta_o \; ,$$

$$A_4 = \frac{EF}{2}\Theta_o^3\left(1 - \alpha^2\right)\alpha, \qquad \alpha = \frac{\Theta_{cr}}{\Theta_o}$$

Uniformly distributed critical load q_{cr} is determined analogously as concentrated critical load, which in this case can be most easily determined numerically.

It follows from a numerical analysis that whereas $J_1 \rightarrow \infty$, $Lq_{cr}/2 \rightarrow P_{1cr}$.

INTERACTION CURVES OF CRITICAL LOAD $(P_1, q)_{cr}$.

Interaction curves $(P_1, ql/2)_{cr}$ may be determined from relations 9 , 13 , and 17 .

Of particular importance are interaction curves for monomially directed loads P_1 and q. Each curve included in the family is determined by the angle Θ_o of joint rise.

Fig 2 shows a family of interaction curves $F(P_1/P_{1cr}, q/q_{cr}) = 1$ of which components P_1 and q of the critical set of loads $(P_1, q)_{cr}$ can be determined with the aid of formulae

$$\bar{P}_1 = k_p \bar{P}_{1cr} \quad \text{....18,}$$

$$\bar{q} = k_q \bar{q}_{cr} \quad \text{....19,}$$

where: $\bar{P}_{1cr}$ - mean concentrated critical load of the joint at q=0,

$\bar{q}_{cr}$ - mean uniformly distributed critical load at P=0.

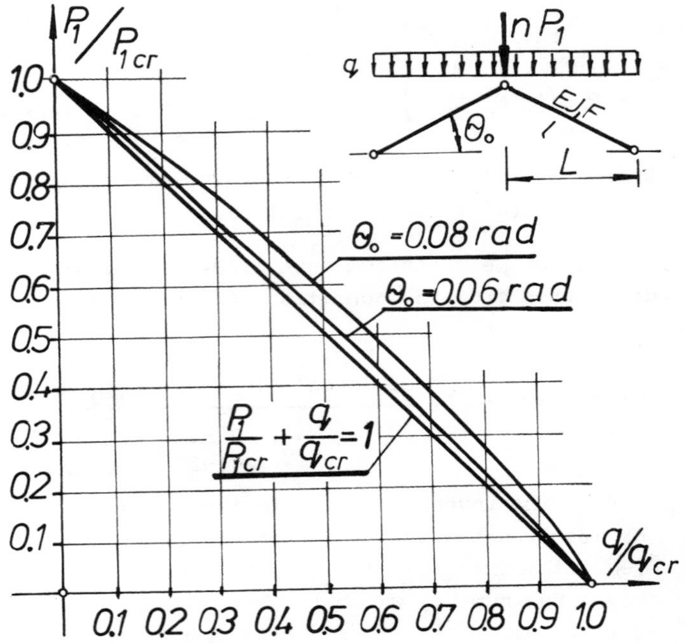

Fig 2

Interaction curves shown in Fig 2 are bounded from below by a straight line in the form

$$\frac{\bar{P}_1}{\bar{P}_{1cr}} + \frac{\bar{q}}{\bar{q}_{cr}} = 1 \quad \text{....20.}$$

The remaining interaction curves may be approximated by Eqn 21

$$\left(\frac{\bar{P}_1}{\bar{P}_{1cr}}\right)^{\beta_p} + \left(\frac{\bar{q}}{\bar{q}_{cr}}\right)^{\beta_q} = 1 \quad \text{....21.}$$

A slightly worse estimation from below, resulting from the equation

$$\left(\frac{P_1}{P_{1cr}}\right)^{\beta} + \left(\frac{q}{q_{cr}}\right)^{\beta} = 1 \qquad ....22,$$

seems to be more convenient for engineering practice.

Coefficients β_p and β_q depend on the rise angle θ_o.

Table 2. Coefficients β_p, β_q and β of interaction curves.

θ_o	0.03	0.04	0.05	0.06	0.07	0.08	0.09	0.10
β_p	1.00	1.0075	1.012	1.026	1.052	1.137	1.245	1.485
β_q	1.00	1.0078	1.014	1.036	1.074	1.201	1.375	1.624
β	1.00	1.0076	1.013	1.031	1.063	1.169	1.310	1.555

Table 2 shows coefficients β_p, β_q and β for different θ_o. It should be noted that coefficients β are approximately equal to mean coefficients from β_p and β_q

$$\beta = 0.5\left(\beta_p + \beta_q\right) \qquad ....23.$$

VARIANCE AND COEFFICIENT OF VARIATION OF CRITICAL JOINT LOAD

Deviations of joint positions from the expected surface have a very significant influence on the critical bearing capacity of bar coating, therefore we shall now consider the influence of variance $D^2(\theta_o)$ of the angle of joint rise on variance $D^2(P_1)$ of the joint critical bearing capacity.

By reducing the number of independent random variables in the relation $P_{1cr}=EF\theta_o/3\sqrt{3}$ to angle θ_o, we shall estimate variance $D^2(P_1)$ and standard deviation $D(P_1)$ of concentrated critical load from the known formula

$$D^2(P_1)=\left(\frac{\partial P_{1cr}}{\partial \theta_o}\right)^2 D^2(\theta_o) = \left(\frac{3P_{1cr}}{\theta_o}\right)^2 D^2(\theta_o) \qquad ....24,$$

where: $\dfrac{\partial P_{1cr}}{\partial \theta_o} = \dfrac{3EF\theta_o^2}{3\sqrt{3}} = \dfrac{3\bar{P}_{1cr}}{\theta_o}$,

$D^2(\theta_o)$ - variance of the angle of the joint rise.

The coefficient of variation m_p of critical load P_{1cr} is:

$$m_p = \frac{D(P_1)}{\bar{P}_{1cr}} = 3\,\frac{D(\theta_o)}{\bar{\theta}_o} = 3\,m_\theta \qquad ....25,$$

where: m_θ- coefficient of angle θ_o variation.

Variance $D^2(\theta_o)$ and standard deviation $D(\theta_o)$ of angle θ_o will be determined on the basis of bar length variance $D^2(l)$ using the relation:

$$\theta_o = h:l = \frac{\sqrt{l^2 - L^2}}{L} \qquad ....26,$$

then:

$$D^2\,\theta_o = \left(\frac{\partial \theta_o}{\partial l}\right)^2 D^2(l) = \frac{D^2(l)}{L^2}\,\frac{\bar{l}^2}{\bar{l}^2-L^2} = \frac{\bar{l}^2 D^2(l)}{\bar{h}^2\,L^2} \qquad ....27.$$

The variance coefficient m_θ of angle θ_o is

$$m_\theta = \frac{D\,\theta_o}{\theta_o} = m_l\,\frac{\bar{l}^2}{\bar{h}^2} = \frac{m_l}{\sin^2\bar{\theta}} = \frac{m_l}{\bar{\theta}_o^2} \qquad ....28,$$

where: $m_l = \dfrac{D(l)}{\bar{l}}$

Treating the rise angle θ_o as the only independent random variable, we eventually find the relation between the coefficient m_p of critical load and the coefficient of bar variation

$$m_p = 3\left(\frac{\bar{l}}{\bar{h}}\right)^2 m_l \qquad ....29.$$

Fig 3 shows a diagram of the dependence of the variation coefficient m_p of critical joint load from the angle of joint rise,assuming m_l= 0.0005 and 0.001. This is in agreement with the cutting mean error of cut of ca 0.5 and 1 mm for 1 m of the bar length.

CONCLUSIONS

In real domes, in which roofing is supported on bars and not only at joints, it is necessary to take into consideration the influence of bar bending on the critical bearing capacity of a joint. Bar bending caused by transverse load q decreases the critical bearing capacity of a joint.

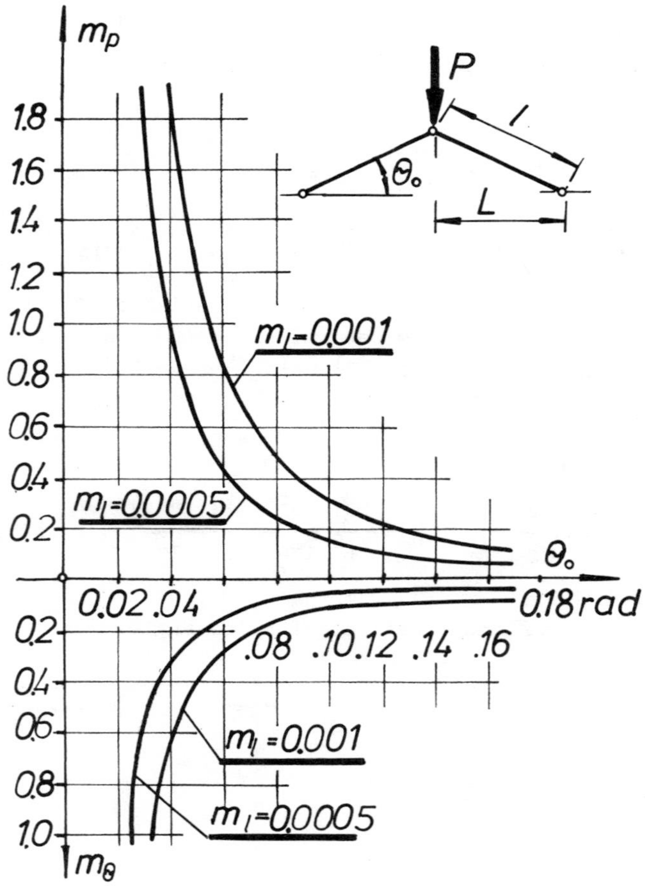

Fig 3

The corfficient of variation m_p of the critical joint load is strongly non-linearly increased together with the decreasing angle of joint rise θ_o, under the assumption of a constant coefficient of variation m_l of the bar length in a dome.

On the other hand, the dome safety decreases non-linearly together with the decreasing dome rise.

It is advisable to carry out systematic measurements of the coefficients of joints in bar domes in order to determine geometrical imperfections of the network of dome joints.

REFERENCE

1. W W Bołotin, Metody statystyczne w mechanice budowli, Arkady, Warszawa 1968.
2. J Głombik, Obciążenie krytyczne jednowarstwowych kopuł siatkowych o węzłach przegubowych, PhD Dissertation, Politechnika Śląska, Gliwice 1970.
3. K Klöppel, R Schardt, Zur Berechnung von Netzkuppeln, Der Stahlbau 5/1962, 12/1962.
4. W T Koiter, General theorems for elasto-plastic solids,Progress in Solids Mechanics, wol.1, North,Holland Publishing Company, Amsterdam 1961
5. Z Kowal, Random Ultimate Bearing Capacity of Doms with Random Joint Coordinates constructed from Bars with a Random Ultimate Bearing Capacity, JCASP 3, University of New South Wales Kensington, Australia 1978, Reports p.566-573.
6. Z Kowal, Parametry losowej wytrzymałości krytycznej prętów ściskanych i współczynniki wyboczenia, Arch.Inż.Lądowej 1/1981.
7. Z Kowal,A Biegus,J Cabaj, J Rządkowski, Hale o dużych rozpiętościach, W Bogucki,Poradnik Projektanta konstrukcji met., Arkady 1982.
8. S Mercik, J Mrozowski, K Tarnowski, Geodezyjne badania parametrów geometrycznych pewnych kopuł siatkowych, Konstrukcje Metalowe 2/74.
9. J Murzewski, Teoria nośności losowej konstrukcji prętowej, PWN, Warszawa 1976.
10. J Roorda, Problemy stateczności konstrukcji sprężystych, Współczesne metody analizy stateczności konstrukcji, Ossolineum 1981 .
11. V A Savyelev, Ustojčivost setčatych kupołov, Metalličeskije Konstrukcji. Roboty Škoły N S Streleckiego,Moskva 1966.

THE ANALYSIS AND TESTING OF A SINGLE LAYER, SHALLOW BRACED DOME

I.M. KANI *, M.Sc.   R. E. McCONNEL**, D.Phil., MICE   T. SEE ***, Ph.D.

* Research student, University of Cambridge U.K.

** Lecturer, University of Cambridge, U.K.

*** Lecturer, National University of Singapore

A nonlinear computer program is described which allows for the behaviour associated with changes in structural geometry and material properties, as well as that associated with bifurcation and limit point buckling. Brief details of the theory used in the program are given. The program is used to predict both the pre- and post-buckling response of a 24-member shallow braced dome model and the numerical results are compared with those of an experimental study. The influence of imperfections of node position and member shape on the numerical analysis will be discussed, and some consideration given to the implications of this work with respect to the analysis of larger domes.

INTRODUCTION

Single layer space structures must be curved to provide resistance to out-of-plane loads. However, structures with small rise-to-span ratios will show considerable nonlinear response before they reach their ultimate loads, and non-linear analyses are required for both the structural calculations and the stability analysis.

The nonlinear calculations are only possible with a suitable computer program, and this must be capable of considering three distinct sources of nonlinear behaviour. They are:

(i) Changes in structural geometry caused by finite nodal displacements.

(ii) Changes in geometry within the members affecting their stiffness characteristics (stability and bowing functions).

(iii) Changes in member stiffness due to nonlinear material properties.

Behaviour types (i) and (ii) are related, as (ii) can be ignored if (i) is implemented and a single member is cut up into a sufficient number of shorter members. In addition to these specific sources of nonlinear behaviour, the program must be capable of testing the stability of the equilibrium paths produced in the analysis of any structure.

Many computer programs capable of considering some or all of these points exist, but very few experimental results have been published against which the numerical results can be checked. Test results for a 36-member dome have been presented by Freeman (Ref 1), and Butterworth (Ref 2) has analysed and tested a 27-member dome. Recent work at Cambridge by See (Ref 3) has provided results for various domes, with a 90-member structure being the largest tested. However, all these tests have been confined to the elastic range, and post-buckling behaviour has not been investigated.

There is a need for an understanding of behaviour beyond the ultimate load, and this paper reports some early results in this range. A nonlinear inelastic program developed at Cambridge (Refs 3 and 7) will be discussed briefly, and its ability to handle the three types of non-linear behaviour detailed above will be demonstrated by the analysis of three simple example structures which demonstrate the various nonlinear forms of response. The program will then be used to predict the behaviour of a 24 member dome, and the numerical results compared with the results of an experimental study. The significance of including initial imperfections in the numerical analysis will be discussed.

NONLINEAR ANALYSIS

The object of any analysis is to obtain a displacement $\{\Delta\}$ such that the structure is in equilibrium with an applied load $\{P\}$. A sequence of solutions for increasing load will generate an equilibrium path in load displacement space. There are a large number of nonlinear iterative methods suitable for achieving this end, most of which proceed by repeating the following three steps:

(i) Equilibrium Check

For a current state of deformation $\{\Delta\}$, calculate the sum of the internal actions $\{Q\}$ at each node according to the following sequence;

$$\{\Delta\} \xrightarrow[\text{tibility}]{\text{(compa}} \{\varepsilon\} \xrightarrow[\text{strain)}]{\text{(stress-}} \{\sigma\} \xrightarrow[\text{librium)}]{\text{(equi-}} \{f\} \xrightarrow[\text{librium}]{\Sigma f,\ \text{equi-}} \{Q\} \quad \ldots 1$$

where $\{\varepsilon\}$, $\{\sigma\}$, and $\{f\}$ are the member strains, stresses and nodal forces respectively. Compare $\{Q\}$ with the current level of applied load $\{P\}$ to give the out-of-balance forces $\{P^{OB}\}$, where

$$\{P^{OB}\} = \{P\} - \{Q\} \quad \ldots 2$$

(ii) Convergence Test.

Check $\{P^{OB}\}$ against preset convergence limits.

- If the limits are satisfied, either terminate the calculation if the full load level has been achieved, or increase the load {P} by an amount equal to the next load increment, update the displacement, and return to Eqn 2.

- If the limits are not satisfied, update the displacements (as in (iii) below).

(iii) Displacement Update

If convergence is not satisfied an improvement $\{d\Delta\}$ on the displacement can be found with methods ranging from an estimate to a scaled value of $\{d\Delta\}$ calculated from

$$\{d\Delta\} = [K_T^{-1}] \{P^{OB}\} \quad \ldots 3$$

where $[K_T]$ is the *tangent stiffness matrix* (see below).

This is followed by updating the displacement

$$\{\Delta\} = \{\Delta\} + \{d\Delta\} \quad \ldots 4$$

and then returning to the equilibrium check (i).

The relationships in Eqn 1 can be nonlinear, but as no explicit statement of the relationship between $\{\Delta\}$ and $\{Q\}$ is required, the sequence of calculations represented is always possible (provided due account is taken of the non-commutative nature of large rotations in 3-dimensional space).

The tangent stiffness matrix $[K_T]$ at displaced state $\{\Delta\}$ relates the changes in load $\{dP\}$ which bear a linear relationship to changes in displacement $\{d\Delta\}$,

$$\{dP\} = [K_T] \{d\Delta\} \quad \ldots 5$$

The formulation of $[K_T]$ will be discussed below with the details of the computer program.

The vital step in the above sequence is the equilibrium check (i). Steps (i) and (iii) are quite different, although the equations relating bending to rotation in $[K_T]$ are very similar to those in Eqn 1 as the elements of $[K_T]$ are the first order terms (in $\{d\Delta\}$) of an incremental form of Eqn 1. Steps (i), (ii) and (iii) can be combined in many ways but only the incremental Newton-Raphson method will be described here, as it is the method employed in the computer program used to obtain the results presented in this paper.

It is of interest to note that if $[K_T]$ is independent of $\{\Delta\}$, then the terms in $[K_T]$ are constant, and are identical with the usual small displacement elastic stiffness matrix $[K_o]$. Eqn 5 is then valid for large displacements and reduces to give the normal non-iterative linear elastic stiffness method $\{\Delta\} = [K_o^{-1}] \{P\}$.

Incremental Newton-Raphson Method

If the external load is applied in increments, and Eqn 3 is used to reduce $\{P^{OB}\}$ to within the preset convergence limits at each load increment, then the *Incremental Newton-Raphson* method results, Fig 1.

When analysing structures with nonlinear material properties a modification of this method is necessary because of the variation in strain modulus with stress reversal. The equilibrium check (step (i)) is based, not on the total displacement from the zero stress condition, but on the increment in displacement from the previous converged position on the equilibrium path.

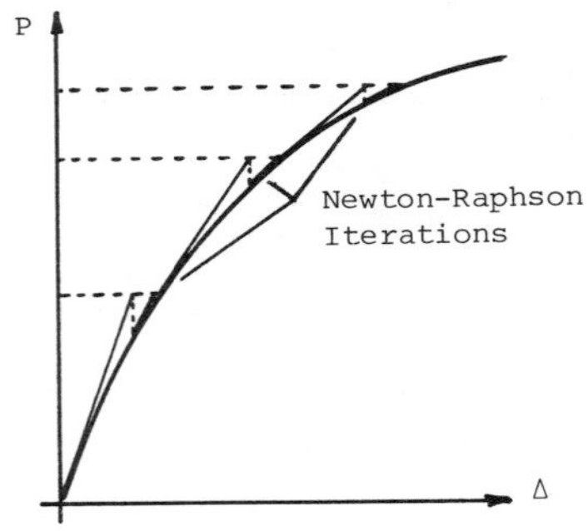

Fig 1 Incremental Newton-Raphson Method

Solution Control near Limit Points

In regions of flat (limit-point) or descending equilibrium paths, standard Newton-Raphson iteration schemes will not converge. Either displacement control (of a single degree-of-freedom) or the Riks-Wempner method (Ref 4) of iteration using a hyper-plane can be employed to overcome this problem. The displacement control approach can be derived either independently, or as a special case of a general technique described briefly here, and in more detail in Ref 5. This technique covers the Riks-Wempner method, as well as all other methods which impose a constraint on some or all of the displacement increments with a single constraint equation. Consider the following modified form of Eqn 5,

$$[K_T] \{d\Delta\} = \beta \{p\} + \{P^{OB}\} \quad \ldots 6$$

where $\{p\}$ is the normalised applied load vector, and β is a scalar that is the defined variable in load incrementation, but is an undefined variable in other displacement constraint methods. In the latter cases, the Gaussian elimination and back substitution are performed simultaneously but independently on both of the right hand side vectors, and the constraint equation on $\{d\Delta\}$ then yields β. The values of $\{d\Delta\}$ follow, and the external loads are updated so that a new value of $\{P^{OB}\}$ can be calculated based on the latest displacements. The iterative procedure is continued until the out-of-balance force vector vanishes, and both β and $\{d\Delta\}$ are small according to the chosen convergence criterion.

STABILITY AND BUCKLING

The methods described so far may be used to determine the equilibrium paths of space structures. However, there is no indication as to whether the path so found is stable. One way of determining this is to examine the appropriate derivatives of the total potential energy of the structure (Ref 6.).

As an alternative to the energy approach, the stability of a structure can be examined by consideration of the tangent stiffness matrix $[K_T]$. To do this, the normal solution algorithm is supplemented by a routine for calculating some of the eigenvalues and eigenvectors of the tangent stiffness matrix, and use is made of the information available either in a Gauss elimination or a Choleski decomposition of $[K_T]$. This approach is computationally easier to apply than energy based methods, requiring only the formulation of an accurate tangent stiffness matrix $[K_T]$. Unlike the energy methods this approach can be applied to relatively large space structures. Some of the results of this work (without proof) will be given here, but the original thesis (Ref 3) should be consulted for details.

Buckling Mode, Critical Points

At a critical point on an equilibrium path changes in displacement $\{d\Delta\}$ are possible without a corresponding change in the applied load $\{dP\}$ necessary to maintain nodal equilibrium, i.e. $\{d\Delta\} \neq 0$ for $\{dP\} = 0$. Substituting this into Eqn 5, $\{0\} = [K_T]\{d\Delta\}$, from which it follows that non-trivial (i.e. non-zero) solution(s) for $\{d\Delta\}$ exist only if the determinant of $[K_T]$ is zero, i.e. if $|K_T| = 0$. From a consideration of the spectral resolution of $[K_T]$ at a critical point it is found that at least one eigen-value of $[K_T]$ will be zero and the buckling mode is described by the corresponding eigenvector. Although this is only true for infinitesimal displacements away from the current position because of the assumption used for formulating $[K_T]$, critical points on an equilibrium path are often followed by unstable behaviour, so that a critical point is often associated with the onset of buckling.

Detection of Instability

From further consideration of the spectral resolution of $[K_T]$, it is possible to argue that for a general point on an equilibrium path to be stable, all the eigenvalues of $[K_T]$ at that point must be positive. This is equivalent to stating that $[K_T]$ must be *positive-definite* (Ref 7), so that for elastic structures this requirement is consistent with the energy based argument for stability based on a minimum in the total potential energy, Ref 8.

Since there is a one-to-one correspondence between the number of negative eigenvalues, and the number of negative diagonal terms (Ref 9) in the solution process (Gauss elimination or Choleski decomposition) of Eqns 3 and 5 it is possible to detect instability without calculating the eigenvalues of $[K_T]$ directly. By observing the appearance of these negative diagonal terms it is possible, without extra numerical manipulation, to detect when an equilibrium solution is in an unstable state. When negative diagonal terms appear the structure is re-analysed to allow for the influence of bifurcating post-buckling paths, as explained below.

Limit Point, Bifurcation Point

Two possible forms of behaviour exist in the vicinity of a critical point and these are:

(i) Limit point

In which the stiffness of the structure in the relevant buckling mode decreases progressively until there is zero stiffness at the critical point, Fig 2(a). Displacements in the direction of the buckling mode grow progressively larger under the influence of the applied load, so that at the limit point the buckling mode is evident in the calculated displacement of the structure. The buckling mode corresponds to the eigenvector whose associated eigenvalue is zero at the limit point. This form of behaviour is correctly analysed by any program that allows for changes in structural geometry caused by finite nodal displacements.

(ii) Bifurcation point

In which the fundamental equilibrium path is crossed by a separate secondary equilibrium path, Fig 2(b). Unlike the behaviour at a limit point the buckling mode is not now associated with the fundamental path but with the secondary path. The buckling mode associated with the secondary equilibrium path, described by the eigenvector whose corresponding eigenvalue is zero at the bifurcation point, is not evident in the calculated displacements of the structure along the fundamental path. As a result of this, equilibrium solutions may be found along the fundamental equilibrium path in the unstable state beyond the critical point. When this happens the solution process identifies the existence of negative eigenvalues. To follow the secondary equilibrium path the structure has to be *perturbed* from its fundamental path. This can be done through the introduction of imperfections in the initial geometry of the structure and by repeating the analysis with the 'imperfect' structure.

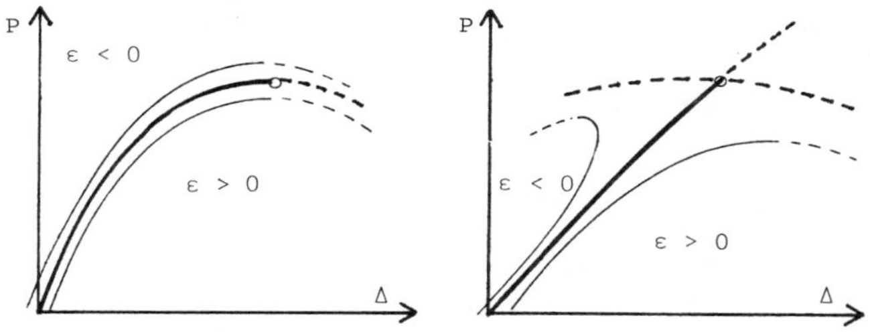

(a) Limit point (b) Unstable-symmetric bifurcation

Fig 2 ε represents an imperfection parameter.
o denotes critical point.
Dashed lines represent unstable equilibrium paths.

Initial Imperfections

It has been found (Ref 3) that alteration of the initial geometry of a structure exhibiting bifurcation behaviour by including small geometrical imperfections consistent with the buckling mode will precipitate a limit point failure described by the same mode. By this means bifurcation behaviour may be converted to limit point behaviour, so that real structural behaviour can be analysed numerically.

Since the magnitude and distribution of these imperfections in a real structure is unknown a suitable magnitude of initial imperfection has to be assumed for the analysis, and this might be based on known construction tolerances. The reduction in the collapse load from that of the structure with the original geometry to that of the structure with initial imperfections gives an indication of the degree of imperfection - sensitivity in the structure.

NONLINEAR COMPUTER PROGRAM

A general purpose frame analysis program has been developed which incorporates the features discussed above.

The form of the member stiffness matrix used to generate $[K_T]$ is described briefly here, and the ability of the resulting program to handle the three classes of nonlinear behaviour described in the introduction will be demonstrated by the analysis of a number of simple structures. For a rigorous derivation of $[K_T]$, Ref 5 should be consulted.

For space frames that are influenced by bifurcation behaviour before any limit points are reached, the sign of the determinant of the tangent stiffness matrix cannot be used to determine the direction of the load increment in the Riks-Wemper method (Ref 3). The present program uses the current stiffness parameter of Bergan (Ref 10) for this purpose.

Materially and Geometrically Nonlinear Member Tangent Stiffness Matrix

The principal of virtual work (Ref 11) can be used to formulate the stiffness matrix for the nonlinear beam element shown in adjacent displacement configurations in Fig 3. The following assumptions are made:

(1) Plane sections normal to the neutral axis remain plane and only the longitudinal stress and two shear stresses are non-zero.

(2) The element strains are small so the cross-sectional dimensions do not change.

The displacements and rotations of the element can be arbitrarily large, and isoparametric interpolation functions are used to describe the geometry and displacements. The material behaviour can be either elastic perfectly plastic or elastic linear hardening, although other post yield behaviours could be incorporated without major difficulty. These capabilities are described in detail in Ref 5, and both two noded and three noded versions of the nonlinear three dimensional beam column element are available.

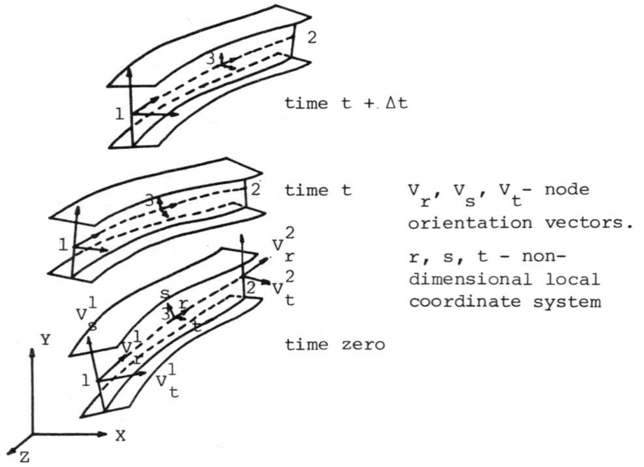

Fig 3 Displaced States of a Single Beam Element

An updated Lagrangian formulation is adopted by referring all variables at notional time t + Δt to those at notional time t, where Δt is an increment in time or loading. The materially and geometrically nonlinear elements of the tangent stiffness matrix are calculated by numerical integration using a technique similar to that employed to calculate the element equivalent nodal forces. For rectangular cross-sections, Gaussian Quadrature is used in three directions, but for other sections it is only used along the element length and integration over the cross-section is performed by establishing an adequate number of sampling points. Each sampling point is allocated a representative area of the cross-section and numerical integration carried out by simple summation over these points.

The three noded version can be used as a nonlinear curved beam element in the configuration at time 0 as shown in Fig 3. Both the von Mises and a simplified yield criterion are incorporated as described in detail in Ref 5. A method similar to that described by Owen in Ref 12 is used to maintain the yielded sampling points on the yield line or the yield surface. This is achieved by studying the sign and the absolute value of the total and current incremental iterative stresses at each sampling point, together with the appropriate yield criterion.

The theory described briefly here produces a general purpose nonlinear three-dimensional beam-column element. Geometric effects (stability and bowing functions) are intrinsically incorporated, as is demonstrated by the example analyses below.

Example Analyses

The three examples below can be verified by simple hand calculations, or by reference to the published data.

(a) Snap-Through of a Two Bar Arch

The simple plane arch structure shown in Fig 4 was analysed to demonstrate the capability of the program to handle large displacements and to pass limit points, with a frame made with either an elastic or an elasto-plastic material. The displacement control technique was used to pass limit points. In the first analysis the yield stress was assumed to be 20 kN/mm^2 so that the structure remained elastic throughout. To study the effect of plasticity, a lower yield stress of 5 kN/mm^2 was used in the second analysis. The structure yielded first in compression, and then yielded again in tension towards the end of the analysis. The program agrees well with the theoretical results shown.

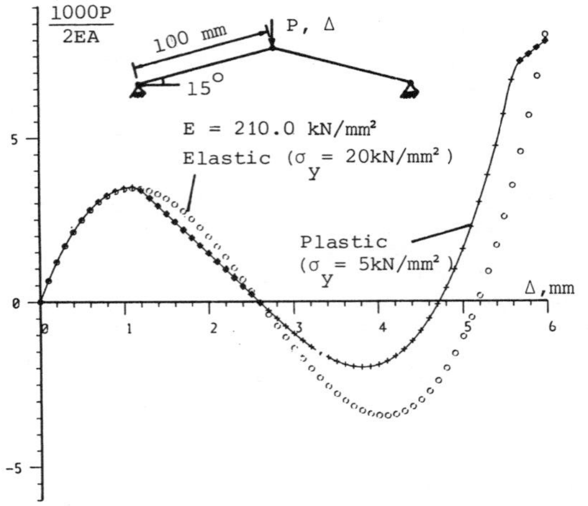

Fig 4 Snap-through of a 2 Bar Arch

(b) Plastic Collapse of a Portal Frame

The pitched portal frame shown in Fig 5 was analysed to demonstrate that the program can detect simple plastic collapse mechanisms. Material properties and cross-sectional dimensions are given in Fig 5 which also shows the vertical deflection of the ridge. The results of the numerical analysis agree well with those of a simple plastic collapse analysis when the influence of axial load on plastic moment capacity are considered.

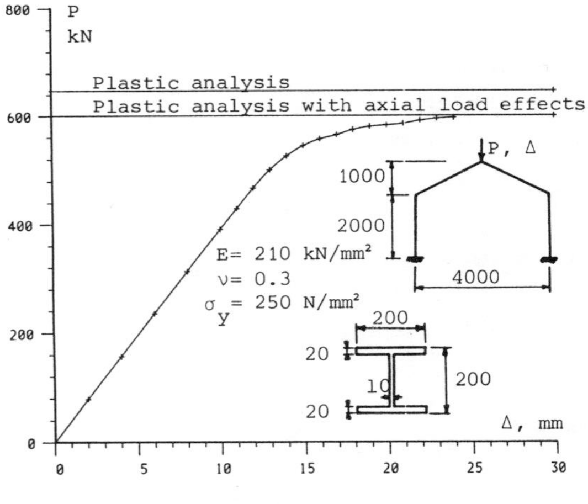

Fig 5 Plastic Collapse of a Portal Frame

(c) Euler Buckling of a Column

The ability of the program to detect member (Euler) buckling was tested by analysing the axially loaded column shown in Fig 6. One and two (3-noded) member models were used in turn to represent the column, and the numerical results are plotted in Fig 6.

Perfect (i.e. straight) members showed bifurcation (negative diagonal terms at 1.2 and 1.0 times the Euler load for the one and two member models respectively. Members with a 1 mm lateral imperfection at the central node produced the curves shown, these show the relationship between the bifurcation modes detected in the perfect analyses and the normal Euler buckling loads. The predictions of the Perry formulae are also plotted and it agrees well with the analysis using a two member model.

The results shown in Fig 6 indicate that a single member is inadequate for modelling member buckling. Therefore 2 members are needed in any region of a structure where significant axial load develops in the members. The calculated buckling load from the analysis with only one member is overestimated by 20%, but this is consistent with the results of analyses based on elements formulated with linearised stability functions (Ref 13). Thus a single 3-noded member would appear to have this level of ability to model individual member buckling.

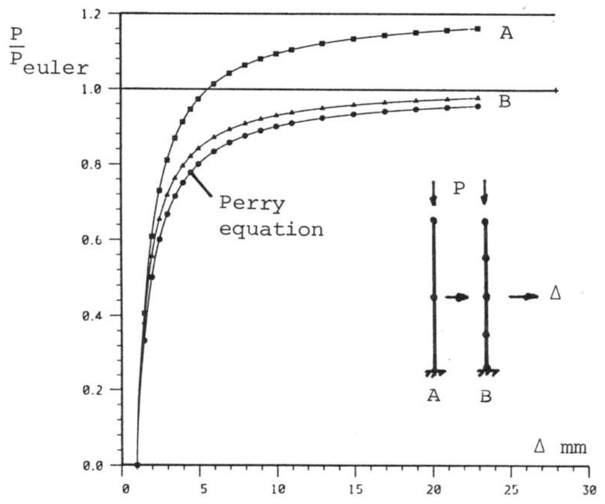

Fig 6 Euler buckling of a column

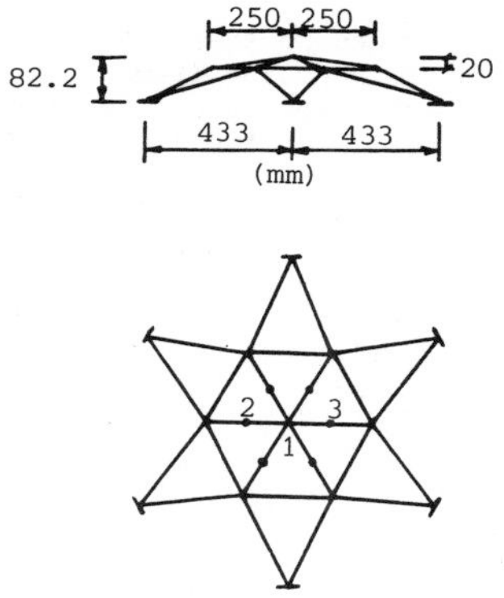

Fig 7 24-Member Dome

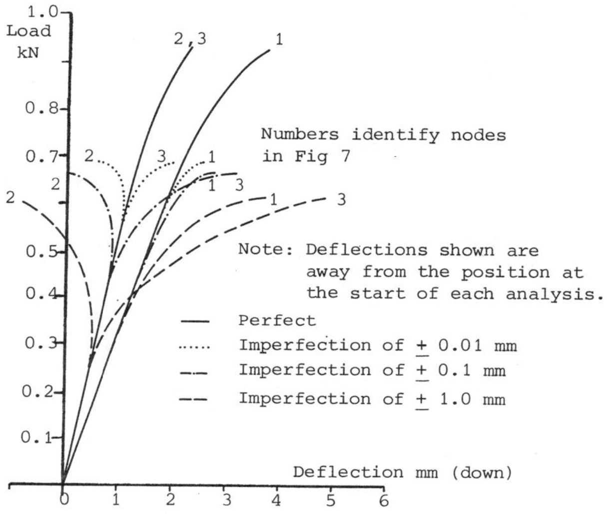

Fig 8 Equilibrium Paths of the 24-Member Dome

24-MEMBER DOME

A 24-member braced spherical dome (Fig 7) with rigid joints and fixed supports was tested and the results are compared with those of an analysis using the method discussed above.

Test Dome Details

The members of the dome were 4.76 mm $(\frac{3"}{16})$ diameter bright steel rods with a yield stress of 680 N/mm^2. The joints were 15 mm x 50 mm diameter steel blocks, with 16 mm deep holes drilled in the direction at which the members frame into the nodes. The members were secured with high strength glue (Araldite 2005). The dome was bolted to a plate which was supported by a 1.5 m high reaction frame which is not shown in the diagrams. The plate prevented the foundation nodes from spreading when the dome was under load. Vertical loading was provided by a screw jack connected to the bottom of the reaction frame through a load cell. This loading system enabled both the loading and the unloading of the structure to be observed. Dial gauges mounted on an independent frame were used to record deflections.

Numerical Analysis - Single Central Load

The dome was loaded with a single vertical point load at the crown, node 1 in Fig 7. An initial analysis of the structure, with perfect geometry (with one 3-noded element per member) and using only load incrementations, produced the fundamental equilibrium path shown in Fig 8. Negative diagonal terms appeared after a load of 0.7 kN had been applied. The response gradually became nonlinear, as significant nodal deflections developed, before convergence problems appeared at 0.85 kN. The yield stress was not exceeded.

Negative diagonal terms on the fundamental path indicated that bifurcation points had been passed. Since the members were slender and high axial loads were induced by the applied load, the ratios of applied axial force to Euler load for all members were inspected to determine which single 3-noded element members needed to be replaced by a pair of 3-noded elements to allow for a better representation of member buckling. The 6 members supporting node 1 were found to be critical, so these were modelled with a pair of members each and the analysis repeated. Two negative diagonal terms then appeared at 0.68 kN, and a third at 0.73 kN. On examining $[K_T]$ at 0.68 kN, two coincident eigenvectors were found which together represented rotation about any horizontal axis through node 1. This indicates that simple tilt about the central node with buckling of the 6 members supporting node 1 was the fundamental mode of the dome under a central point load.

To simulate the expected test behaviour, imperfections to excite the fundamental mode were added to the member geometry, and a number of analyses with load incrementation were made. Node 2 was raised by various magnitudes of imperfection, and node 3 lowered by the same amount, to give the dashed curves in Fig 8. Because displacement control was not used, all these program runs failed to converge as limit point behaviour developed at about 0.67 kN. The yield stress (680 N/mm^2) was not exceeded in any run.

Since the actual imperfections in the test model were small and could not be accurately measured it was necessary to determine a suitable imperfection magnitude. To do this it was noted that an imperfection of ± 0.1 mm was found to give results in general agreement with those of the experiment, so a run with this imperfection was made with the Riks-Wempner method employed to control the iterations to give the results shown in Fig 9. The post-buckling path is almost flat, and plasticity only occurred after a deflection of 10 mm had developed.

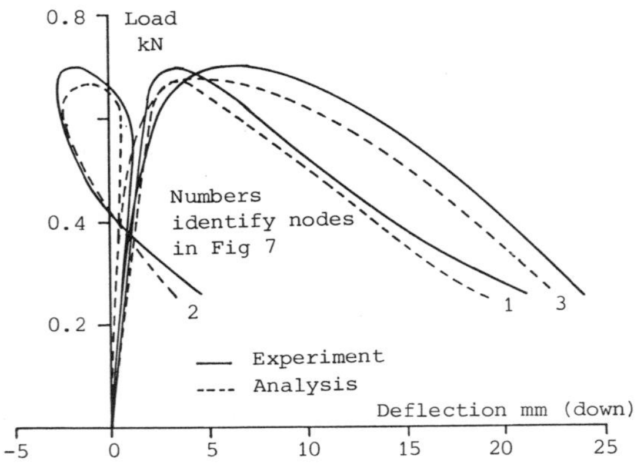

Fig 9 Experimental and analytical deflection of the 24-member dome

Experimental Test - Single Central Load

The numerical analysis predicted that buckling of some of the 6 members supporting the central node would dominate the collapse of the dome, and that vertical deflections would be the more dominant displacements. Dial gauges were therefore set to record the vertical movements of the joints, and of the mid-points of the 6 critical members.

The screw jack was used to increase the displacement in equal steps, leading to the experimental results shown in Fig 9. The dome tilted at 0.7 kN to give the deformed shape shown in Fig 10. After unloading, the dome retained very little permanent set, an observation consistent with the minimal amount of yielding indicated by the analysis.

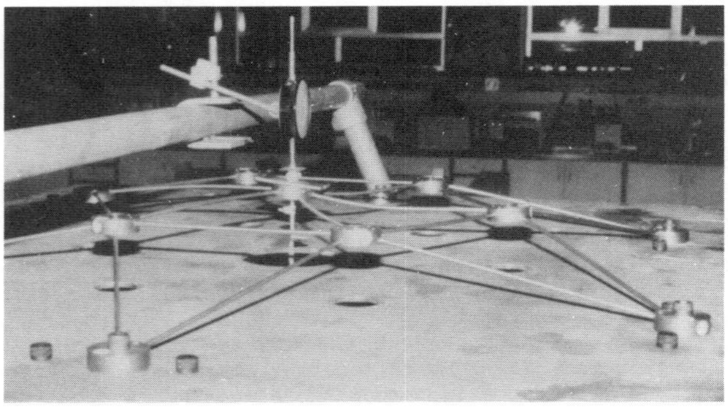

Fig 10 Dome after collapse with single central load

6.7 DISCUSSION

The test and analysis show the same general form, Fig 9, although the predicted failure load was 4% lower than the experimental value. Either the imperfections of ± 0.1 mm were too large to represent those in the dome, or the joint discs had a direct influence on the behaviour. The latter effect was probably critical, as analyses with 50 mm diameter totally rigid joints predicted a collapse load of 0.76 kN.

For a dome with very slender members under high axial compression the behaviour is dominated by member buckling, and this is precipitated by imperfections within the members, rather than by imperfections in joint position. This was clearly apparent in the analyses of the dome discussed above which were influenced much more by mid-member imperfections than by those in joint position. Where the compressive axial forces to Euler load ratios are much lower, as is the case with real domes, nodal snap-through or general buckling involving a consistent movement of a group of nodes is more likely. Therefore imperfections in nodal position would probably be critical in the analysis of a real structure, rather than member imperfections.

A real dome dominated by snap-through might be expected to have the rapidly dropping post-limit point behaviour characteristic of snapping problems, Fig 4. However, there are indications from a number of analyses of large pin-jointed domes that their behaviour is softer than that of the classic 2-bar arch. Introducing bending continuity (combined with plasticity) should further reduce the rate of load shedding, so that it is possible that real domes will have the favourable post-buckling behaviour exhibited by the experimental dome, Fig 9.

CONCLUSION

The results presented demonstrate the ability of the program described to track the pre- and post-buckling behaviour of an experimental dome. If due account is taken of the influence of finite joint size, and if the initial member imperfections could be measured, such a program should accurately predict the actual test deflections. The ability of the program to consider plastic behaviour of the members has not been fully tested in the test analysis because of the unexpectedly high yield stress of the bright steel rod.

REFERENCES

1. J R FREEMAN, The structural behaviour of cyclic, reticular domes, University of London, Ph.D. Thesis, 1975.

2. J W BUTTERWORTH, Non-linear analysis and stability of elastic skeletal systems, University of Surrey, Ph.D. Thesis, 1975.

3. T SEE, Large displacement elastic buckling of space structures, University of Cambridge, Ph.D. Thesis, 1983.

4. E RAMM, Strategies for tracing non-linear response near limit points, Europe-U.S. Workshop,Non-Linear Finite Element Analysis in Strucural Mechanics, Bochum, 1980.

5. I M KANI, Materially and geometrically non-linear static analysis of frames. Technical Report, Cambridge University Engineering Department, CUED/C-Struct/TR.104, 1983.

6. J M T THOMPSON and G W HUNT, A General Theory of Elastic Stability, John Wiley and Sons, 1973.

7. J H WILKINSON, The Algebraic Eigenvalue Problem, Clarendon Press, Oxford, 1965.

8 O C ZIENKIEWICZ, The Finite Element Method , McGraw-Hill, 1979.

9. W WITTRICK and F W WILLIAMS, A general algorithm for computing natural frequencies of elastic structures, Quarterly Journal of Mechanics and Applied Mathematics, Vo. XXIV, Part 3, pp. 263-284, 1971.

10. P G BERGAN, Solution algorithms for non-linear structural problems, Computers and Structures, Vol.12, pp. 497-509, 1980.

11. K J BATHE, Finite Element Procedures in Engineering Analysis, Prentice Hall, 1982.

12. D R J OWEN and E HINTON, Finite Elements in Plasticity, Theory and Practice, Pineridge Press, Swansea, U.K.,1980.

13. R K LIVESLEY, Matrix Methods of structural Analysis, Pergamon International Library, second edition, 1975.

MINIMUM ANALYTICAL REQUIREMENTS FOR THE DESIGN OF OFFSHORE WIND SENSITIVE STRUCTURES

F.E.S. WEST, M.Phil., M.I.C.E., M.I.Struct.E, M WeldI

Associate, PRC Engineering (UK) Ltd

This paper classifies tower structures mounted on offshore platforms in terms of their use and sensitivity to dynamic and fatigue effects from wind and seismic loading. From the various environmental and other data required to analyse such structures, alternative methods of analysis are commented upon and compared. From the comparison it is made evident that some dynamic design cases may be dismissed without undertaking more than a preliminary analysis and that most may be analysed satisfactorily using comparatively simple methods. Such simple design methods often have the advantage of indicating the overall condition of a structure more clearly than can more abstruse analyses. By such indications the need for further analysis or design modification may be assessed, or the adequacy of the design established.

INTRODUCTION

The accepted method used for the analysis of latticed tower structures is by means of a linear finite element stiffness analysis, initially for static loads but continued to provide eigenvalues, eigenvectors and modal frequencies and shapes. There is a growing and, in the view of the author, generally unnecessary, tendency to follow this with a dynamic response analysis using either a frequency or a time dependent function as the disturbing force. For structures of low and uneven stiffness the modal shape can influence resonant forces; provided the structure is of reasonable stiffness resonant forces will be negligible. Certainly in cantilevers of up to 250 metres when designed for stiffness the resonant component should be less than 5% of the total force. Re-examination of the form and of the principles adopted on the initial design is likely to be the method of improvement if wind induced resonance is shown to be a possibility, rather than to invoke analytical gymnastics to prove that something not very good may be just good enough. In any case the accuracy of knowledge on the environmental effects and the general theory is not high and has to be heavily supported by statistics developed from limited basic data.

The object of this paper is therefore to suggest simple methods which may be used to confirm the adequacy of a structure to fulfil its function against the dynamic forces and effects imposed upon it or to indicate that a re-think of the structure is necessary; rarely would further analysis be necessary.

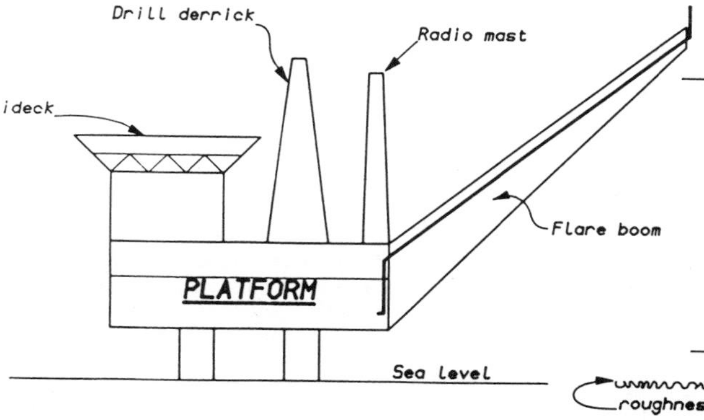

ELEVATION OF AN OFFSHORE PLATFORM

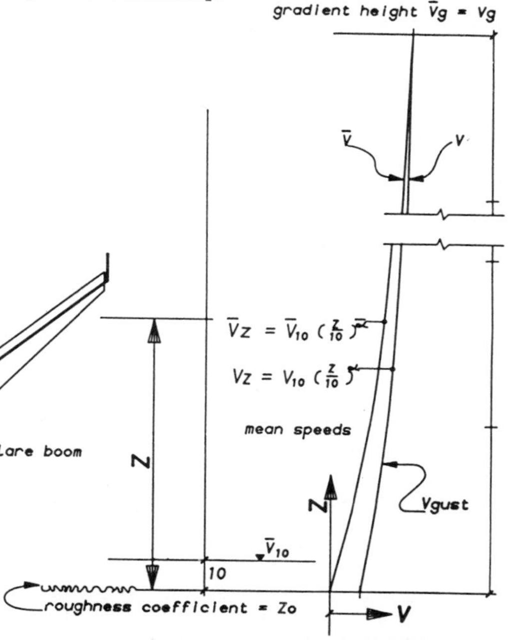

DEFINITIONS OF WIND FORCES

Fig. 1

The symbols used throughout the paper are those universally accepted in wind engineering but for convenience some of them are defined in Fig.1 which also indicates the types of lattice tower structure generally mounted on offshore drilling and production platforms.

LATTICED STRUCTURES FOUND ON OFFSHORE PLATFORMS

From the aspect of wind sensitivity these may be classified as follows:

1. Radio antenna towers. In these the principal drag forces are due to wind on the structural members with only a small proportion from the antennae. Invariably these towers have tightly limited allowable deflections specified and are thus very stiff overall. Tower height could be up to 70 metres, height of tip above sea level 100 metres. Towers have largely been superseded by dish antennae.

2. Flare booms and stacks. In these the flareline facility makes a significant contribution to the wind drag forces and no limits on deflection under load are specified. Boom lengths could be over 200 metres and height of tip above sea level well over 200 metres. They may be vertical, horizontal or at any intermediate angle.

3. Drill derricks. These structures are extremely robust and carry heavy equipment, piping and machinery loads such that they can hardly be described as wind sensitive structures. They are frequently covered with radiation shielding and in these cases the envelope of the structure rather than its individual members are exposed to wind forces.

In deep water platforms the last two types are invariably found - sometimes pairs of each are found on one platform. Radio towers are confined in general to the older platforms. It may be expected in general that the possibility of wind induced vortex shedding resonance on individual members must be investigated in design of the first two classes, the possibility of dynamic amplification from wind forces attaining significance in the second class and the possibility of resonance from vibrating machinery occurring on components of the third class. These are good but by no means infallible starting premises.

EFFECTS TO BE CATERED FOR IN DESIGN

1. Investigation of the design effects of the following forces must be made.

   a) Static gravitational due to structure, facility carried and ice*

   b) Quasi static due to wind forces

   c) Dynamic effects due to
      i) wind forces
      ii) seismic disturbance*
      iii) facility loads (it is unlikely that these will be significant)

   d) Time dependent effects of fatigue and corrosion

   (* where the location is liable to such effects)

   Treatment of the effects of (a) and (b) are well established and, providing the load input for the computer F.E. analysis is automated, is simple and not even time consuming. The items with which this paper is concerned are (c) and the fatigue part of (d).

2. The data required to analyse for the dynamic effects set out in 1(c) is as follows:

   a) Environmental

      (i) 1 year and 100 year return of 1 hour mean wind speed at 10 metres above sea level and its height power law index

      (i.e. $\bar{V}_{10}$ and α)

      (ii) Gust factors or 3 second and 15 second gust speeds as (i)

      (i.e. V_{10} and α for each)

      (iii) The average annual wind frequency rosette and cumulative probability diagram for wind, preferably to Weibull Scale with the equation to the curve.

      (iv) Expected seismic movements related to time or the seismic response of the main platform at points near the attachments of the boom or stack.

   b) Platform related

      (i) natural frequency ranges of the burner in operation

      (ii) possible frequencies from nearby mechanical operations (drilling is the only significant one)

      (iii) elastic characteristics of the main platform related to the points of support of the boom or tower preferably as stiffness matrices (they are rarely given in this form)

      (iv) damping characteristics of the tower

3. This is a formidable list of parametric requirements and let it be said immediately that it is rare indeed to receive all the data required, even for work in the North Sea which has very high standards of environmental data and platform analysis specification. What data is made available is frequently not in a convenient format. Other requirements such as stress levels, S/N curves, design life are given in the client's specification or derived in concert between the certifying authority, client and engineer.

   However, it is generally possible to obtain data which, even if not related to a specific small area of operations, can with some interpolation (or extrapolation, if need be) form a satisfactory basis for design of a particular structure. It should be noted that the uncertainty and inaccuracy of some of these parameters can have a significant effect on any calculations carried out.

METHODS OF ANALYSIS

The part played by research and the monitoring of the effects of wind forces on actual structures has very great importance, particularly when mathematical developments are subjected to the direct control of field and wind tunnel observations.

Of even greater importance is the conversion of the research into relatively simple analytical methods which allow the design engineer to practise his craft confidently but untrammelled by a burden of abstruse analysis. Furthermore, such methods should not obscure the logic of the design exercise.

The basis of such methods are at present being laid, the solid foundation being the classification of the sensitivity of structures under wind action proposed by Antony (see Ref.1) and amended by Wyatt and Harris (ibid) from which those appearing most apposite have been selected.

Class A Structures and/or elements which are stiff enough for wind effects to be determined by statics and small enough for the relevant wind information to be specified as a wind speed at a single point.

Class B Structures which are stiff enough for wind effects to be determined by statics, but large enough for the relevant wind information to be specified as multi-point data.

Class C As for B but with the complication that the shape of the individual structural load influence lines has to be considered in conjunction with multi-point wind data.

Class D Structures which are not stiff enough to be treated by static methods but require a full dynamic treatment.

Class E Structures for which the wind, aerodynamics and structural motion are inseparably combined to produce overall wind effects.

At this stage it is sufficient to say that these definitions are generally adequate but need some polishing to improve their clarity. They do not, however, take cogniscence of the fact that a structure of Class A or B can easily have components which need to be dealt with in Class D or E.

The papers from which these definitions are taken are somewhat full of generalities which, unproven, somewhat detract from their scientific content. For example,

"The usual variations of $\beta(z)$, $\bar{V}(z)$ and $\sigma(z)$ are such that the product of all three may often be assumed constant without any great loss of accuracy. Also in the particular case of the quasi-static structure, the use of the lowest dynamic mode shape $\mu(z)$ instead of the rigorously correct static influence function, $\gamma(z)$ often introduces no appreciable error."

Mathematically this may be a possibility but it implies that equivalent multi-point data may be used on any shape of structure, and states that the lowest dynamic mode shape (which generally is very close to the deformed shape under static load) is an approximation which OFTEN introduces no appreciable error, and both approximations can be used as "convenience" functions.

Nevertheless Harris's paper is a most logical and readable essay to provide a mathematical method of classification. The crux of the selection of a method of analysis for a particular structure is undoubtedly to establish its classification. In view of the fact that a structure and its component parts can occupy more than one classification, a purely mathematical classifying method, however logical and attractive to the academic mind, OFTEN would not provide a complete or an infallibly correct answer.

As the class of structure need not be decided before design commences, and certain design stages must be carried out regardless of the class of structure, it may well be better to embark on the preliminary stages, classifying the structure incidentally in the performance of the successive steps required.

This develops a logical train of thought from which any further analysis may be invoked depending on the final class of structure which emerges from the initial design stages. It is reiterated that these preliminary stages must be executed in any case and there is no wasted design effort. The object of the exercise, of course, is to ensure that the structure under study keeps below Class C if at all possible. The method suggested is as follows:

1. From inspection it may invariably be recognised that Radio antenna towers designed to specified low deflections together with drill derricks are in Class B. (In areas subject to seismic effects a dynamic response analysis will be required for the latter and possibly for the former).

2. By preliminary design approximations, particularly on base width to height ratios and approximate member forces and deflections, the probable scantlings are assessed and a member by member finite element model compiled. During the compilation of the model, members and components should be checked individually to ensure that they will not be subject to vortex shedding resonance or problems associated with interference effects from adjacent members. This can be carried out by the use of Natural frequency/ Strouhal frequency graphs (Fig.2) when these do not provide a clear decision, the damping parameters and interferences effects should be checked by ESDU Items 78006 and 79025 (Ref.2) or similar methods. This check will prevent the incorporation of elements which could otherwise be of Class E.

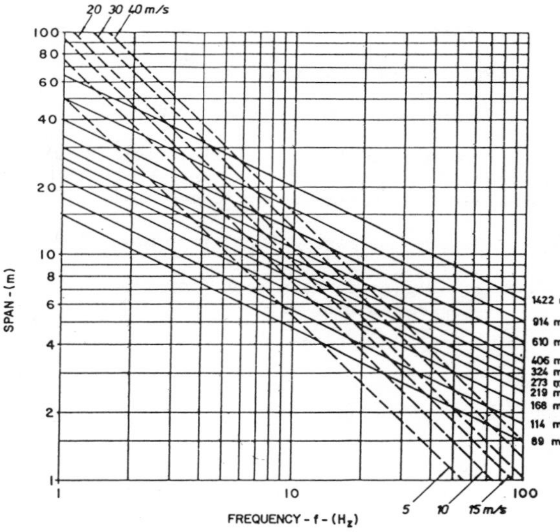

FIG. 2 NATURAL & STROUHAL FREQUENCIES AT S=0·2
PIN ENDED MEMBERS

3. The finite element analysis can now be carried out. Provided one uses an automatic pre-processor for the derivation of forces due to mean and fluctuating wind load the assessment of quasi-static wind forces is neither arduous, expensive nor time consuming. Having the finite element model to hand the next step of extracting eigenvalues, eigenvectors and, say, the first 10 or 15 modal shapes is also an automatic and economical process. If the lowest mode frequency is not less than 0.5HZ the structure almost certainly comes within Class B; if the lowest mode frequency is much less than 0.5HZ it may be worth having preliminary thoughts on how the structure may be stiffened. At this stage, if the structure has an Eiffel shape, the lowest mode shapes should be checked that their curvature is smooth throughout, with the radius of curvature increasing or decreasing gradually. It is possible (but unlikely) to have an Eiffel form

cantilever tower with good frequency characteristics, but still yielding high stresses under a fatigue analysis. The degree of curvature of the first mode shape can be used conveniently and with reasonable reliability to indicate if the transition of the internal forces through the structure will be even or not. A noticeable increase in curvature in one section of a structure indicates that this section has lower stiffness than optimum. By plotting mode shapes to a suitable scale a tendency becomes readily visible.

4. Having proceeded so far, the possibility of any part of the structure being of Class E is excluded by virtue of the tests done in preliminary design stages. The question of the general classification of the structure can, on a basis of lowest natural frequency, be decided as being either of up to Class B or as above Class B.

If the former case applies the general calculations on the stiffness and strength modes are complete.

If some doubt exists or the structure is clearly of Class C or above, the next stage is to assess possible dynamic magnification or resonant effects by an analytical method such as proposed by ESDU Item 76001 (Ref.2).

The method adopted by 76001 is to divide the wind forces into the following components:

a) Mean non-fluctuating ($\bar{E}$)

b) Fluctuating non-resonant ($\sigma_B(E)$)

c) Dynamic resonant ($\sigma_D(E)$)

and this method is that generally adopted.

For a) and b) the static analysis provides all the information required and this can be used to build up the required non-resonant forces, namely

$\bar{E}$ for the force/moment from mean wind

$\sigma_B(E)$ for the force/moment from fluctuating non-resonant wind

With regard to the resonant component $\sigma_D(E)$ many of the results obtained from the finite element analysis (e.g. natural frequencies and mode shapes) are directly usable. It should also be noted that in an automated formation of the external force vectors values of coefficients (such as the drag coefficients C_D) need not be presumed as constants but may be varied in accordance with any desired mathematic relationship.

Table 1 compares the maximum values for the mean, fluctuating and resonant components calculated for actual designs, normalised to the values of the fluctuating non-resonant component. Also contained in the table are values from an earlier analysis in which the mean and fluctuating non-resonant components had not been separated. Immediately below are the results of measurements taken by Dr Ing Aas-Jakobsen (Ref.3) of the same structure during a 20m/sec cross-wind. During these measurements a low periodic response occurred both in the wind direction and normal to it. The figure given in column (6) of the table represents the vector sum of these two responses consequently is not exactly equivalent to the other values quoted in column (6) but is reasonably representative.

The values given in rows 2 to 5 of the table formed part of an exercise to exhibit the influence of the lowest natural period on the resonant component; the actual lowest frequency on the particular structure was 0.88 HZ. A curve (Fig.3) of the resonant component plotted against frequency shows how steeply the value of resonance increases as the natural frequency reduces below 0.5 HZ. It must also be borne in mind that even at a frequency of 0.25 HZ the dynamic magnification factor due to resonance against other wind forces is only 1.08 and in the case of vector sum of the fluctuating forces the D.A.F. is only 1.01. It should be noted that the ESDU Item 76001 considers that the fluctuating and resonant act normally to each other (i.e. are added vectorially to find their combined effect) whereas the tests made by Dr Ing Aas-Jakobsen (Ref.3) show clearly that the resonant effects themselves can have two directional components. The combination of fluctuating and resonant components is therefore in actuality more complex than a straight summation of orthogonal components.

TABLE 1 COMPARISON OF MEAN, FLUCTUATING AND RESONANT COMPONENTS CALCULATED AND MEASURED ON ACTUAL PROJECTS

Structure Type	Length (m)	Lowest Mode Frequency (Hz)	Normalised Max.Calc.Moments			Vector Sum of (5) & (6)	% $\frac{(6)}{(4)+(5)}$	% $\frac{(4)+(7)}{(4)+(5)}$
			Mean	Fluctuating	Resonant			
(1)	(2)	(3)	(4)	(5)	(6)	(7)	(8)	(9)
45° Boom	90	0.74	1.89	1.0	0.015	1.0001	0.52	100.004
45° Boom (a)	106	2	1.94	1.0	0.064	1.002	2.18	100.07
45° Boom (b)	106	1	1.94	1.0	0.100	1.005	3.40	100.17
45° Boom (c)	106	.5	1.94	1.0	0.153	1.012	5.20	100.41
45° Boom (d)	106	.25	1.94	1.0	0.24	1.028	8.16	100.95
Stack	70	.91	2.25	1.0	0.052	1.0014	1.6	100.04
45° Boom (Calculated)	116	1.37	1.67	1.0	-	-	-	-
45° Boom (Measured)	116	1.29	1.67	1.0	0.10	1.005	3.4	100.17

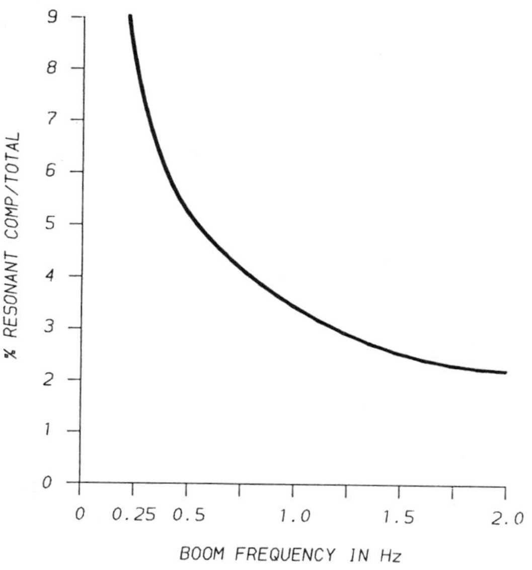

FIG.3 % RESONANT COMPONENT AGAINST FREQUENCY

Following the method suggested and bearing in mind the steepness of the rate of increase in resonance when the frequency drops below 0.5HZ it seems reasonable to adopt this value as a lowest value to contain structures in Class C.

In the author's opinion, if the site of the platform and structure is in a harsh environment, bearing in mind the possible cost in life and to the local ecology (apart from the cost of lost production and replacement of the structure in the event of failure) if one cannot be sure that the structure is contained within or better than Class C it is better to re-design, selecting a stiffer type of structure. The inaccuracy of the parameters available provides a less than desirable degree of certainty for the use of complex analytical methods, against this cantilever booms and towers of up to 250 metres in length can be easily and economically designed to have a lowest natural frequency of 0.5HZ or greater. However, in some cases it may be necessary to carry on and to assume that the structure will be either of Class D or E.

5. On the assumption that a Class D structure is involved, most of the finite element data produced so far remains valid.

Most commercially available structural analysis programs have a dynamic response capability; indeed the use of such has been necessary to carry out the analysis so far. The only problem facing the design engineer now is to develop time histories or a power spectral density tabulation compatible with the particular program to be used. To the author's knowledge no commercial program has evolved suitable pre-processing modules to their main analysis program to develop specifically dynamic wind loadings for direct input. On the other hand all main frame programs of which the author has experience can accept dynamic loadings both as time histories or power spectra.

The designer's attitude towards analysis must be to choose readily available, reliable and economic tools to carry out each particular task. Provided advantage may be taken of proven and well-documented recent research which can be applied to standard solution techniques, such analyses must be superior to the product of even generally accredited research by individuals or single centres of research.

The results of analysis so far has included the extraction of eigenvalues and eigenvectors which has provided the first step in the consideration of the general equation of motion. Provided a suitable model has been developed and the results saved they can be re-used in the full solution of the equation.

$$M\ddot{r} + C\dot{r} + Kr = R(t)$$

where M = the global mass matrix

C = the global damping matrix

K = the global stiffness matrix

R(t) = the time varying load vector

r = the displacement vector with time derivatives $\dot{r}$ and $\ddot{r}$

For the extraction of eigenvalues the values of C and R(t) will have been given zero values.

For the response analysis it is necessary to evaluate C and R(t). For boom or tower structures offshore particularly drilling derricks the solid area presented to the wind may vary appreciably with height. It may be assumed that only the lowest natural frequency modes need be considered, perhaps no more than two in each direction, but any number of frequencies may be considered using the method. If the same points are used in the calculation of aerodynamic damping as were used to define the eigenvectors, the calculation of the damping for each section can be carried out quite simply and added to the assumed value of structural damping. If the values for each section are found to be varying by less than ± 50% from the mean (as is most likely) this is within uncertainty adjudged for the parameter and the mean value can be taken throughout, making the damping matrix C(I). If the individual damping figures do vary greatly the damping matrix can be taken as a diagonal matrix (C).

The calculation to formulate power spectral densities and co-variance matrices is more complex and is probably best effected with a short pre-processor program based on ESDU Items 75001 and 76001. The analyses when completed will provide RMS values of translations, accelerations and member forces and stresses. These can be linked with forces due to mean wind speed and dead load putting the output into a single file and combining the forces to provide the maxima probable from the RMS values and their assumed deviations. Although the process is briefly described the formulation of the data for the dynamic analysis is a rather tedious procedure.

The only occasion of which the author has heard where this analysis was necessary was in the case of a shore based tower in which a defective design had slipped through the normally rigorous audit net. In this case the dynamic effects which included vortex shedding resonance had to be analysed very carefully to ascertain if the structure could last for sufficient time to allow a new one to be designed and built. This is the sole instance in which such an analysis was justifiable to the author's knowledge.

Structures of Class E should never be allowed to be built offshore, although suspension bridges and guyed masts (which are generally of Class E) can be most logical and economical forms when used ashore.

Because of the clean winds in the marine environment vortex shedding resonance can occur at much higher wind speeds and shedding frequencies than would be possible on land sites. This is the only Class E category which must be carefully guarded against for both single and groups of members.

6. All computer bureau finite element programs have adequate and generally well-documented seismic analysis capability. For the analysis of structures remote from the source of the seismic disturbance (i.e. the sea-bed) the required data for the analysis is not always so easy to come by.

If when the platform designers carry out the analysis for the main structure, they are briefed to provide an acceleration response spectrum for the location near to the supports of the platform mounted tower, this response spectrum can be applied to the tower directly and simply. Generally it is possible to obtain reasonable values even if they are not from exactly the required location.

7. Provided the tower is of Class A to C (i.e. has no significant resonant component) fatigue life can be assessed fairly directly from the Weibull Scale exceedance curve, the wind rosette and the agreed S/N curve and stress concentration factors. The presence of significant resonance immediately complicates the analysis as the actual stress is no longer proportional to the Weibull Curve. If the resonance is due to turbulence or buffeting to obtain an accurate spectrum of stress would involve many full dynamic analyses and it may be preferable to adopt a conservative dynamic magnification factor to the maximum recorded stresses or to reduce the permissible stress range by an equivalent factor. If resonance is due to vortex shedding at a finite wind speed, it is necessary to calculate the amplitude of stress on affected members; if this is greater than the fatigue stress limit, it is most probable that the critical design case will be of fatigue due to this resonance. It is simple to calculate from the Weibull Curve the probable duration of the relevant wind speed per annum and from this the number of stress reversals per annum; depending on the amplitude of stress calculated the ratio of actual to permissible cycles for this case alone can dominate the Miner's rule summation. This ratio can, of course, be added directly into the general summation for the stress spectrum derived from the Weibull Curve.

CONCLUSIONS

Perhaps the best summary of the author's opinion may be provided by the following flow chart, which indicates the limiting envelope of designs for offshore tower structures.

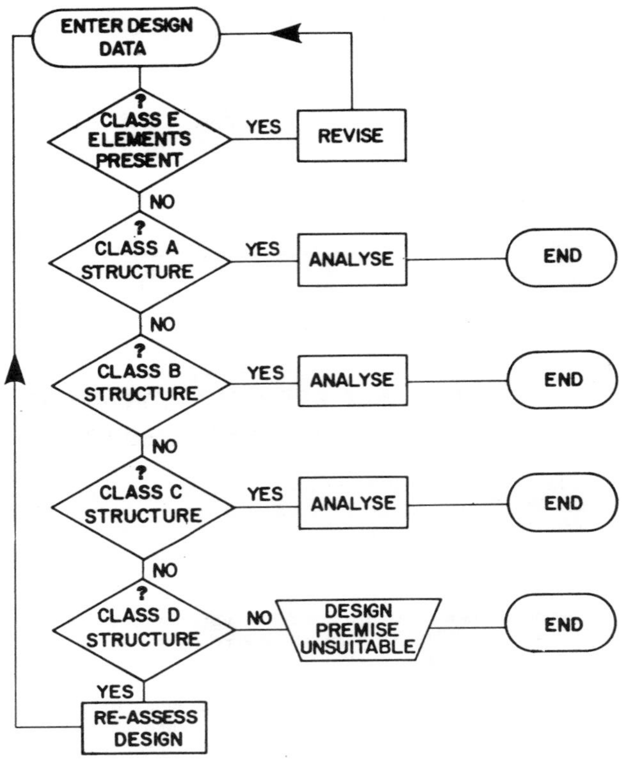

REFERENCES

1. Wind Engineering in the Eighties. Construction Industry Research and Information Association, London 1980.

2. Wind Engineering Vols. 1 to 4. Engineering Sciences Data Unit Ltd., London 1982.

3. DR ING A AAS-JAKOBSEN, "Dynamic Behaviour of the Statfjord "A" Flareboom", Dr Ing A Aas-Jakobsen A/S, Oslo January 1980.

A FINITE ELEMENT MODEL FOR DETERMINING THE CONSTITUTIVE RELATION OF A COMPRESSION MEMBER

USAMA R MADI*, DAVID LLOYD SMITH**

* Department of Civil Engineering, University of Jordan.

** Department of Civil Engineering, Imperial College of Science and Technology.

A simple finite element model for determining the constitutive relation of a compression member is presented. It is applied to a member of circular hollow section which is the most commonly used section in pin-jointed spatial structures. This model employs an incremental, first order, geometrically nonlinear analysis, presented in two formulations -- one suitable for force control, the other for displacement control. The model also accounts for various end constraints, the spread of plasticity and local unloading. A comparison is conducted between the constitutive relations obtained by this model and predictions obtained by other models as well as experimental evidence.

INTRODUCTION

Pin-jointed spatial structures are in general, highly statically indeterminate. Thus failure of a constituent member, or a group of members, of such a structure may not lead to overall failure: the structure may, indeed, remain integral and be capable of carrying increased load. Of course, an adequate prediction of this response, up to overall failure, will depend upon the accurate description of the post-failure behaviour of each member.

The obtaining of this constitutive relationship comprises two parts: one comparatively straightforward which deals with the behaviour of the member in tension, and one, relatively more complicated, which deals with the behaviour of the member in compression. Methods that have been used to overcome these difficulties and develop the constitutive relationships for members under axial compressive forces fall into two categories:

1. Numerical solutions of the beam-column equilibrium equations using elastoplastic moment-thrust-curvature relationships computed or assumed for the appropriate section geometry. The work of Jezek (1937), Paris (1954), and Supple and Collins (1980) are examples of techniques in this category. Such techniques, however, suffer from two basic weaknesses: they employ constitutive relations based on stress resultants and strain resultants and they concentrate the plastic effects in limited regions of the member. This tends to suppress the effects of the spread of plasticity inside the member from initial yield until collapse and possibly local unloading from playing their role in affecting the results of the analysis. They are simply constructed and easy to apply.

2. Incremental finite element methods. These techniques treat the member as a plane assembly of beam-column elements. They obtain the constitutive relation by carrying out incremental analysis upon the member subjected to external loads and for imposed displacements. The work of Smith et al (1979) is an example of the techniques in this category. The ability to account for various end constraint, spread of plasticity and unloading are the advantages of these techniques; the much larger computational effort required is clearly a drawback. The model to be presented belongs to the second category of techniques.

SELECTING A SIMPLE FINITE ELEMENT MODEL

A simple model for the constitutive relation of a member in axial compression can be formulated as follows:

(a) The member is divided along its length into a number of straight small elements.

(b) The analysis is carried out in small increments of load or displacement, updating all the parameters at the end of each increment.

(c) A first-order geometrically non-linear analysis is used.

(d) The local stress-local strain diagram shown in figure 1 is valid.

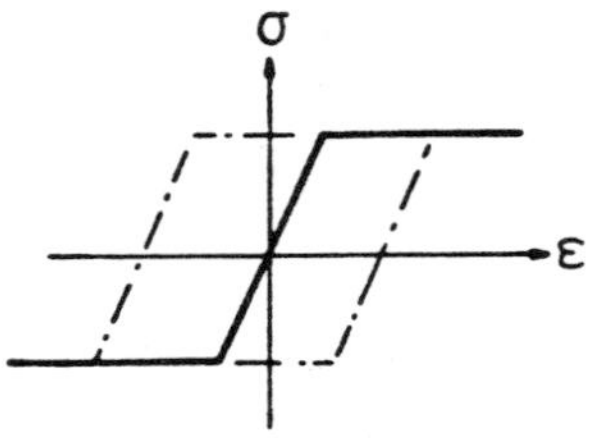

Fig. 1 Stress Strain Relationship

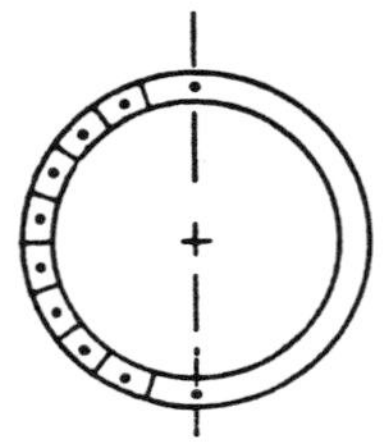

Fig. 2 Division of Section into Regions.

(e) The section of the member is divided into regions, a control point in every region being chosen as representative of the state of stress and strain in every part of that region.

(f) Each element is to be divided into two end elements, each having a length equal to half the length of the element. The current sectional properties of these end elements are to be based on the regions within these sections that are in an elastic state (either originally elastic or unloaded elastic) as the plastic regions do not offer any resistance to further plastic deformation.

Needless to say, a better approximation can be obtained by having more elements and smaller increments of load or displacement.

INCREMENTAL CONSTITUTIVE RELATIONSHIP OF THE ELEMENT

According to the assumptions of the previous section, the element takes the form shown in fig. 3,

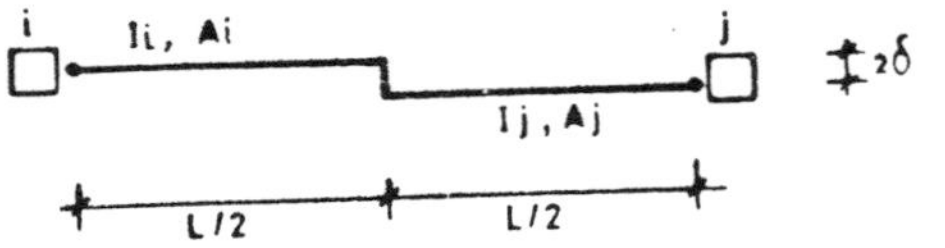

Fig. 3 A Typical Element.

Where I_i, A_i are the second moment of area and area respectively of the left section, taking into account only the elastic regions in this section, I_j, A_j are the second moment of area and area of the right section, taking into account only the elastic regions in this section, and 2δ is the distance between the two centroidal axes of the two sections taking into account only the elastic regions.

For this model, and ignoring the terms with higher power than one in δ , the following constitutive relation of the element, in stiffness form, can be derived.

$$\begin{bmatrix} \Delta M_i \\ -- \\ \Delta M_j \\ -- \\ \Delta P \end{bmatrix} = \frac{E}{L} \left[\begin{array}{c|c|c} \dfrac{8I_iI_j(7I_i+I_j)}{I_i^2+14I_iI_j+I_j^2} & -\dfrac{16I_iI_j(I_i+I_j)}{I_i^2+14I_iI_j+I_j^2} & \dfrac{2A_iA_j\delta(2I_i^2-10I_iI_j)}{(A_i+A_j)(I_i^2+14I_iI_j+I_j^2)} \\ \hline -\dfrac{16I_iJ_i(I_i+I_j)}{I_i^2+14I_iI_j+I_j} & \dfrac{8I_iI_j(I_i+7I_j)}{I_i^2+14I_iI_j+I_j^2} & \dfrac{2A_iA_j\delta(2I_j^2-10I_iI_j)}{(A_i+A_j)(I_i^2+14I_iI_j+I_j)} \\ \hline -\dfrac{2A_iA_j\delta(2I_i^2-10I_iI_j)}{(A_i+A_j)(I_i^2+14I_iI_j+I_j^2)} & \dfrac{2A_iA_j\delta(2I_j^2-10I_iI_j)}{(A_i+A_j)(I_i^2+14I_iI_j+I_j^2)} & \dfrac{2A_iA_j}{(A_i+A_j)} \end{array}\right] \begin{bmatrix} \Delta\theta_i \\ -- \\ \Delta\theta_j \\ -- \\ \Delta L \end{bmatrix} \quad (1)$$

That is:

$$\underline{\Delta X} = \underline{K}\ \underline{\Delta x}$$

FIRST ORDER GEOMETRICALLY NON-LINEAR ANALYSIS

A first order approximation to the large displacement elastoplastic analysis is one in which displacements appear linearly in the equilibrium equations. The validity of this approximation when the displacements are not excessive has been demonstrated by Smith(1977) who also gives the following static and kinematic descriptions of a plane member.

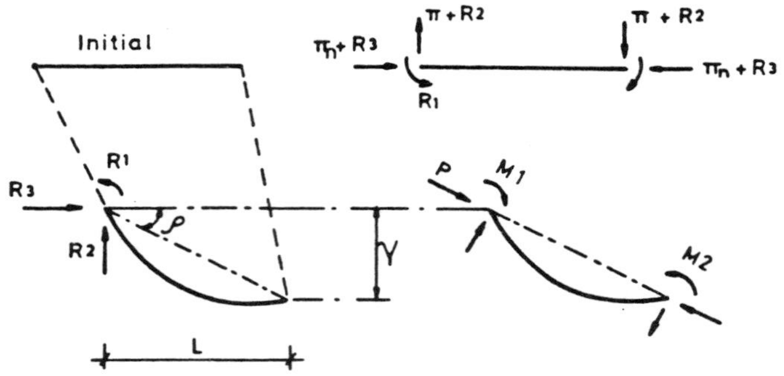

Fig. 4 A Plane Frame Member.

A member of length L is shown in both initial and displaced configurations in fig. 4. The member end loads $\underset{\sim}{R}$ can easily be expressed in terms of three independent stress resultants, M_1, M_2 and P. At the left hand end :

$$\begin{bmatrix} R_1 \\ R_2 \\ R_3 \end{bmatrix} = \begin{bmatrix} -1 & 0 & 0 \\ -\frac{1}{L} & \frac{1}{L} & -\rho \\ -\frac{\rho}{L} & \frac{\rho}{L} & 1 \end{bmatrix} \begin{bmatrix} M_1 \\ M_2 \\ P \end{bmatrix} \quad (2)$$

It is assumed that the rigid body rotation ρ is small and that the apparent shortening due to this rotation, to flexural bowing and to axial deformations are negligible, then it can be stated that :

$$\begin{bmatrix} R_1 \\ R_2 + \rho P \\ R_3 + \frac{\rho}{L}(M_1 - M_2) \end{bmatrix} = \begin{bmatrix} -1 & \cdot & \cdot \\ -\frac{1}{L} & \frac{1}{L} & \cdot \\ \cdot & \cdot & 1 \end{bmatrix} \begin{bmatrix} M_1 \\ M_2 \\ P \end{bmatrix} \quad (3)$$

Where the description now is based on the initial configuration. Thus the static description can be based on the undeformed geometry of the member provided that two Generalised force pairs $\pi = \rho P = \frac{P\gamma}{L} = S_\pi \gamma$ and $\pi_n = \frac{\rho}{L}(M_1 - M_2)$ are additionally imposed upon the member.

This gives the following descriptions for statics and kinematics of a structure, where superscript T denotes a transposed matrix :

Nodal:

$$\underline{Q} = \underline{O} = \begin{bmatrix} \underline{A}^T & \underline{A}_0^T & \underline{A}_\pi^T \end{bmatrix} \begin{bmatrix} \underline{X} \\ -\underline{\lambda} \\ -\underline{\pi} \end{bmatrix} \qquad \begin{bmatrix} \underline{x} \\ \underline{\delta} \\ \underline{\gamma} \end{bmatrix} = \begin{bmatrix} \underline{A} \\ \underline{A}_0 \\ \underline{A}_\pi \end{bmatrix} \underline{u} \tag{4}$$

STATICS KINEMATICS

Mesh:

$$\underline{X} = \begin{bmatrix} \underline{B} & \underline{B}_0 & \underline{B}_\pi \end{bmatrix} \begin{bmatrix} \underline{R} \\ \underline{\lambda} \\ \underline{\pi} \end{bmatrix} \qquad \begin{bmatrix} \underline{v} = \underline{O} \\ \underline{\delta} \\ \underline{\gamma} \end{bmatrix} = \begin{bmatrix} \underline{B}^T \\ \underline{B}_0^T \\ \underline{B}_\pi^T \end{bmatrix} \underline{x} \tag{5}$$

STATICS KINEMATICS

where $\underline{\pi} = \underline{S}_\pi \underline{\gamma}$, $\underline{f}_\pi \underline{\pi} = \underline{\gamma}$, $\underline{S}_\pi = \underline{f}_\pi^{-1} = \text{diag}\left[P/L\right]$

In order to be useful in the sort of analysis proposed, these relations must be put in incremental form. Therefore, taking one member in the nodal description, eq.(4) can be written

$$\underline{A}^T \underline{X} = \underline{A}_0^T \underline{\lambda} + \underline{K}_\pi \gamma \tag{6}$$

and

$$\begin{bmatrix} \tilde{x} \\ \tilde{\delta} \\ \gamma \end{bmatrix} = \begin{bmatrix} \tilde{A} \\ \tilde{A}_0 \\ \tilde{A}_\pi \end{bmatrix} \tilde{u} \tag{7}$$

where $\underline{A}$ is the nodal matrix of the undeformed element

$$\tilde{A} = \begin{bmatrix} \cdot & -\frac{1}{L} & -1 & \vdots & \cdot & \frac{1}{L} & \cdot \\ \cdot & \frac{1}{L} & \cdot & \vdots & \cdot & -\frac{1}{L} & 1 \\ 1 & \cdot & \cdot & \vdots & -1 & \cdot & \cdot \end{bmatrix} \tag{8a}$$

$\underline{X}^T$ is the vector of member stress resultants = $\begin{bmatrix} M_1 & M_2 & P \end{bmatrix}$

$\underline{x}^T$ is the vector of member strain resultants = $\begin{bmatrix} \theta_1 & \theta_2 & \Delta \end{bmatrix}$

and

$$\underline{K}_\pi = \begin{bmatrix} \cdot \\ \frac{P}{L} \\ \frac{M_1 - M_2}{L^2} \\ \cdot \\ -\frac{P}{L} \\ -\frac{(M_1 - M_2)}{L^2} \end{bmatrix} = \tilde{\underline{A}}_\pi \underline{S}_\pi \tag{8b}$$

and $\tilde{A}_\pi = \begin{bmatrix} \cdot & 1 & \cdot & \vdots & \cdot & -1 & \cdot \end{bmatrix}$

Expanding in incremental form:

$$\begin{bmatrix} \Delta \tilde{x} \\ \Delta \tilde{\delta} \\ \Delta \gamma \end{bmatrix} = \begin{bmatrix} \tilde{A} \\ \tilde{A}_0 \\ \tilde{A}_\pi \end{bmatrix} \Delta \tilde{u} \tag{8c}$$

and

$$\underline{A}^T \Delta \underline{X} = \underline{A}_0^T \Delta \underline{\lambda} + \underline{K}_\pi \Delta \gamma + \Delta \underline{K}_\pi \gamma \tag{9}$$

but $\Delta \underline{x}$ and $\Delta \underline{X}$ can be related through $\underline{K}$, the incremental constitutive stiffness matrix of the element

$$\Delta \underline{X} = \underline{K} \, \Delta \underline{x} \quad . \tag{10}$$

Matrix $\underline{K}$ is formed for the element of figure 3 by taking the mean of the current internal forces at its two ends.

So

$$\underline{A}^T \underline{K} \, \underline{A} \, \Delta \underline{u} = \underline{A}_0^T \Delta \underline{\lambda} + \underline{K}_\pi \underline{A}_\pi \Delta \underline{u} + \Delta \underline{K}_\pi \gamma \tag{11}$$

$\tilde{A}^T \tilde{K} \tilde{A}$	$\Delta \tilde{u}$	$= \tilde{A}_0^T \Delta \tilde{\lambda}$	$+ \tilde{K}_\pi \tilde{A}_\pi \Delta \tilde{u}$	$+ \tilde{\mathcal{A}} \tilde{K} \tilde{A} \gamma \; \Delta \tilde{u}$
The tangent stiffness matrix of the element	**displacement increment**	**load increment**	**First order large displacement correction matrix**	**Effect of existing load**

where

$$\tilde{\mathcal{A}} = \begin{bmatrix} \cdot & \cdot & \cdot \\ \cdot & \cdot & \frac{1}{L} \\ \frac{1}{L^2} & -\frac{1}{L^2} & \cdot \\ \cdot & \cdot & \cdot \\ \cdot & \cdot & -\frac{1}{L} \\ -\frac{1}{L^2} & \frac{1}{L^2} & \cdot \end{bmatrix} \tag{12, 12a}$$

Eqs.(12) may further be written

$$\{\tilde{A}^T \tilde{K} \tilde{A} - \tilde{K}_\pi \tilde{A}_\pi - \tilde{\mathcal{A}} \tilde{K} \tilde{A} \gamma\} \; \Delta \tilde{u} = \tilde{A}_0^T \Delta \tilde{\lambda} \tag{13}$$

$\{\tilde{A}^T \tilde{K} \tilde{A} - \tilde{K}_\pi \tilde{A}_\pi - \tilde{\mathcal{A}} \tilde{K} \tilde{A} \gamma\}$	$\Delta \tilde{u}$	$= \tilde{A}_0^T \Delta \tilde{\lambda}$
Incremental stiffness matrix of the element	**incremental end displacements**	**incremental end loads**

or, briefly

$$\underline{K}_m \, \Delta \underline{u}_m = \Delta \underline{w}_m \tag{14}$$

Then these element stiffness matrices are transformed and planted in the primary stiffness matrix of the assemblage.

This formulation can analyse the effect on the member of any loading system. In particular, for an axially loaded member, the axial displacement can be traced through the control of axial force,axial or transverse displacement.

ANALYSIS WITH FORCE CONTROL

The system equations assembled from (14) are

$$\underline{K} \, \Delta \underline{u} = \Delta \underline{w} \tag{15}$$

and the displacement increments $\Delta \underline{u}$ are determined from the force increments $\Delta \underline{w}$, the matrix $\underline{K}$ being updated in response to all previous increments. A scaling of the current increments induces the most critical region in the cross-section to become plastic, and the scaled increments are added to their current total values. Incremental drift is controlled by limiting the magnitude of the increments.

The region which was plastified in the immediately preceding increment must be conforming to the plateau in figure 1. If it is not, the force increment must be reversed since the member has become unstable. Further checks must be made that the incremental strains in all other plastified regions have the same sign as their current total values; contravention implies the local unstressing of such regions.

The process can be repeated as often as desired, or until a plastic collapse mechanism is formed.

ANALYSIS WITH DISPLACEMENT CONTROL

It may be required that the deformation process be fully controlled through a single displacement u_r. Then an increment Δu_r induces an increment of force Δw_r in the constraint controlling the displacement u_r. The system equations (15) can be partitioned to display the r th row and column of the matrix $\underline{K}$

$$\begin{bmatrix} \underline{K}_{uu} & \underline{K}_{ur} & \underline{K}_{ub} \\ \underline{K}_{ru} & K_{rr} & \underline{K}_{rb} \\ \underline{K}_{bu} & \underline{K}_{br} & \underline{K}_{bb} \end{bmatrix} \begin{bmatrix} \Delta \underline{u}_u \\ \Delta u_r \\ \Delta \underline{u}_b \end{bmatrix} = \begin{bmatrix} \Delta \underline{w}_u \\ \Delta w_r \\ \Delta \underline{w}_b \end{bmatrix}, \qquad (16)$$

and Δu_r can be exchanged with Δw_r

$$\begin{bmatrix} \underline{K}_{uu} & \underline{O} & \underline{K}_{ub} \\ \underline{K}_{ru} & -1 & \underline{K}_{rb} \\ \underline{K}_{bu} & \underline{O} & \underline{K}_{bb} \end{bmatrix} \begin{bmatrix} \Delta \underline{u}_u \\ \Delta w_r \\ \Delta \underline{u}_b \end{bmatrix} = \begin{bmatrix} \Delta \underline{w}_u \\ O \\ \Delta \underline{w}_b \end{bmatrix} + \begin{bmatrix} \underline{K}_{ur} \\ K_{rr} \\ \underline{K}_{br} \end{bmatrix} \Delta u_r . \qquad (17)$$

These are the system equations for displacement control. Of course, all the checks described previously for force control must still be applied. Greater accuracy, if so desired, may be had by performing an iterative analysis for the effects of a single increment, thereby correcting the lack of equilibrium after each iteration.

QUALITATIVE VERIFICATION OF THE MODEL

The finite element model is now used to represent the axial loading of a member of circular hollow section. It is important to compare the generally observed behaviour of such a member with the predictions of the model, as some of the most influential structural parameters are varied.

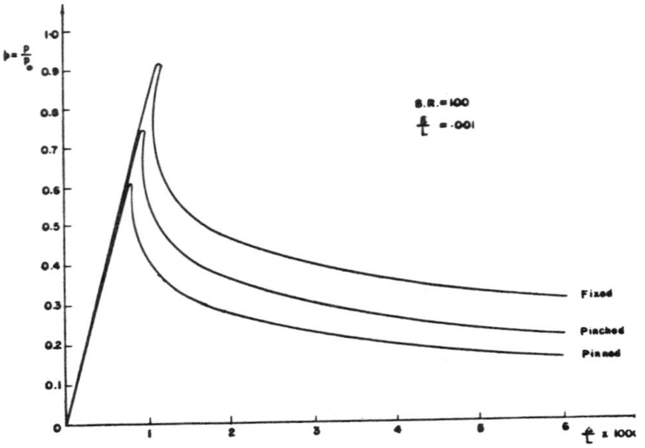

Fig. 5: Effect of Various End Conditions

An increase in the degree of end fixity is seen in figure 5 to produce the expected increases in both the failure load as well as the residual load (tail of the equilibrium path). In this figure, p is the ratio of the current axial load P to the squash load P_o, Δ is the total axial displacement and L is the length of the whole compression member. The imperfection in the member is not varied and is represented by an initial transverse displacement in the form of a sinusoid with central value ε.

The effect of varying the imperfection parameter ε is illustrated in figure 6. An increase in ε produces a marked decrease in the failure load, but only a minor decrease in the residual load is recorded. However, immediately after failure, the rate of load shedding is very sensitive to imperfection. Also, values of ε/L in excess of .001 seem to produce significant loss of axial stiffness in slender members just prior to failure.

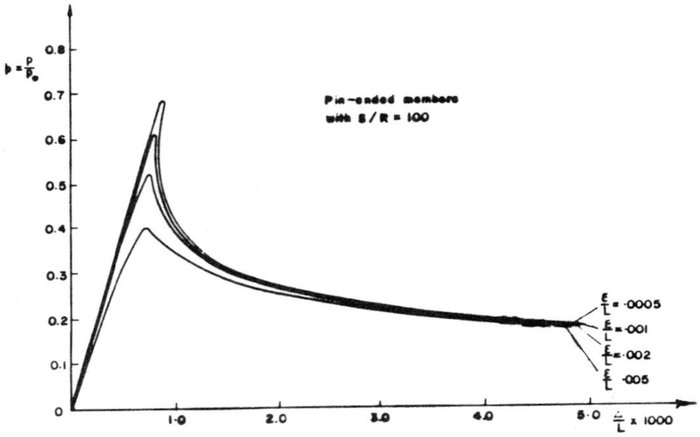

Figure 6: Effect of Increasing Imperfection.

An increase in member length, without variation in ε/L, thereby implies an increase in the imperfection ε as well as in the slenderness ratio (S.R.) of the member, and a reduction in its failure load is the cear inference, as figure 7 confirms. Concomitant reductions can also be observed in the immediate post-failure rate of load shedding and in the residual load.

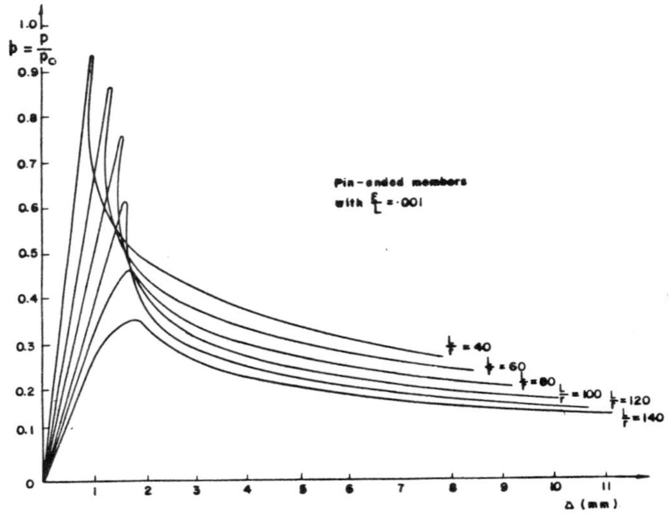

Figure 7: Effect of Increasing Member Length.

EXPERIMENTAL VERIFICATION OF THE MODEL.

It is essential to correlate this finite element model with experimentally obtained data. For this purpose, two sets of tests on compression members of circular hollow section have been chosen: these are the data obtained by De Donato (1968) on pin-ended members and the data obtained by Collins and Supple (1979) on members with connections made through pinched ends. It is necessary to mention that, since no direct measurements of the imperfection parameter were available, the value used in the model was chosen to give a critical load in close agreement with that observed experimentally.

DATA OF DE DONATO.

Data from De Donato's test on a member with a slenderness ratio (S.R.) of 90 are reproduced in fig. 8., together with the predictions of the finite element model.

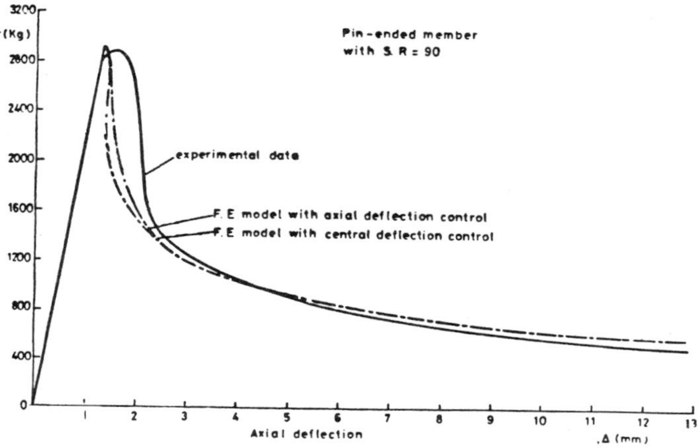

Fig. 8 Comparison with Experimental Data: Pinned Ends.

As shown in this figure, the model is generally in agreement with the experimentally observed behaviour. It gives good predictions of the initial axial stiffness of the member, initial post-buckling slope and the residual load. These member parameters with the failure load, are the important ones in any idealisation of its constitutive relation. It is also worth noting that the finite element model gives different predictions of the member behaviour in the immediate softening phase subsequent to buckling when the analysis is carried out with central and axial deflection controls. The apparent disparity between the experiment and the prediction of the model in the immmediate buckling range can possibly be attributed to experimental factors.

DATA OF COLLINS AND SUPPLE.

Collins and Supple carried out an extensive testing programme on pinched end tubular struts. The results of some members have been compared with the predictions of the model. Figure 9 shows a typical comparison for a member of slenderness ratic just in excess of 100. Again, the finite element model gives a generally acceptable representation, with a slight underestimate of the residual load.

CLOSING REMARKS

Establishing the response to loading of a spatial structure, until the occurrence of its ultimate collapse, requires an acccurate representation of the expected behaviour of each of its separate members. Such representation is acheived by the finite element modelling described herein.

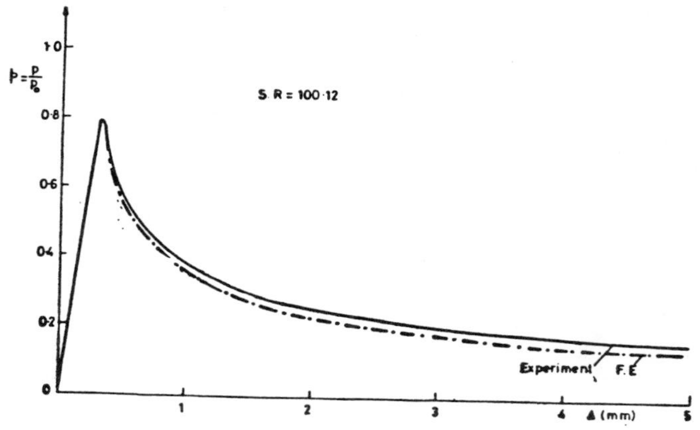

Fig. 9: Comparison with Experimental Data: Pinched Ends.

This model is relatively simple; yet it accounts for the nonlinear effects of moderately large displacements and the development of plasticity. In this latter respect, due allowance is made for the gradual spread of plasticity, both longitudinally and through the cross-section, as well as for the local unstressing (elastic recovery) of regions of the cross-section in which plastic flow had previously been occurring. Only uniaxial loading has been considered in this work, but the model is able to predict the effects of any general system of loading.

A verification of the finite element calculations by a comparison with published experimental evidence was undertaken. The results, although clearly not exhaustive, allow the tentative conclusion that the model is generally representative of the actual behaviour of an axially loaded compression member. It can therefore be used as a component in the analysis of spatial structures, and it can also be used to gauge the predictive capabilities of even simpler models for uniaxial compression members.

A more general constitutive model requires consideration of the uncertain nature of the member imperfections and of the material properties.

REFERENCES

COLLINS, I. and W.J. SUPPLE (1979), Experimental Post-Buckling Curves for Tubular Struts, Space Structures Research Centre, Univ. of Surrey.

DE DONATO, O. (1969), Idagine Sperimentale Sul Comportamento A Collasso Ed in Regime Instabile Di Elementi Tubolari Compressi, Instituto Lombardo di Scienze e Lettere (A) Vol. 103, 169.

JEZEK, K. (1937), Die Festigkeit von Druckstäben aus Stahl, Vienna, Springer.

MADI, U.R. (1981), The Elastoplastic Analysis of Pin-Jointed Spatial Structures, PhD thesis, University of London.

PARIS, P.C. (1954), Limit Design of Columns, J. Aero. Sci., Vol. 21, 1.

SMITH,C.S., WL. KIRKWOOD and J.W. SWAN (1979), Buckling Strength and Post-Collapse Behaviour of Tubular Bracing Members Including Damage Effects, Second Int. Conf. on Behaviour of Off-Shore Structures, Imperial College.

SMITH,D.L. (1979), First-Order Large Displacement Elastoplastic Analysis of Frames Using the Generalized Force Concept, Engineering Plasticity by Mathematical Programming, M.Z. COHN and G. MAIER (Eds), Pergamon Press.

SUPPLE, W.J. and I. COLLINS (1980), Post-Critical Behaviour of Tubular Struts, Eng. Struc., Vol. 2, 225.

PROGRESSIVE COLLAPSE OF WAREHOUSE RACKING

K. M. BAJORIA, B.E., M.Tech.*

R. E. McCONNEL, B.E., M.E., D.Phil. MICE**

* Research Student, University of Cambridge

** Lecturer, University of Cambridge.

Research to date on the structural behaviour of prefabricated proprietary jointed warehouse racking systems has almost exclusively considered only their static response to load, although it is known that local failures often progress (spread) away from the point of initiation. The results of work on this latter problem at Cambridge over the last 4 years, and based on experimental half-scale tests and two dimensional computer simulations, have been published recently. In this present paper, the earlier work is reviewed, and the need for a three dimensional computer simulation established. A suitable program is described, and calibrated against the published experimental results. This program should yield useful results if applied to the failure of conventional space structures, and such work is planned.

INTRODUCTION

Although prefabricated proprietary jointed warehouse racking systems (Figs 1 and 2) are not normally considered to be a form of space structure, they do have the main characteristics of space structures in that they are mass-produced light weight frames carrying their applied loads by axial action in relatively slender members. Racking is not usually designed in the conventional structural engineering manner, and in Europe at least does not have to be checked against the building codes, although racking consumes 90,000 tonnes of steel per annum in Britain alone. The Storage Equipment Manufacturers Association (SEMA) have produced a set of guide rules (Ref 1) based upon the cold formed addendum to BS.449, and a great deal of effort in Europe has gone into producing a European Racking Code which is presently in draft form. These design methods are purely static in approach, with no consideration given to the progression of a collapse given an isolated local failure. That there is ample scope for a collapse to progress is clearly apparent, given that racking structures can be up to 15 m (50 ft) high and an installation covering 5000 m^2 might carry 12000 tonnes or so of stored goods, with a live to dead load ratio of 20:1 or more. It is impossible to be certain of the number of annual collapses as both manufacturers and users are reluctant to report failures.

Fig. 1 High Bay Racking, Typical Joint (Dexion)

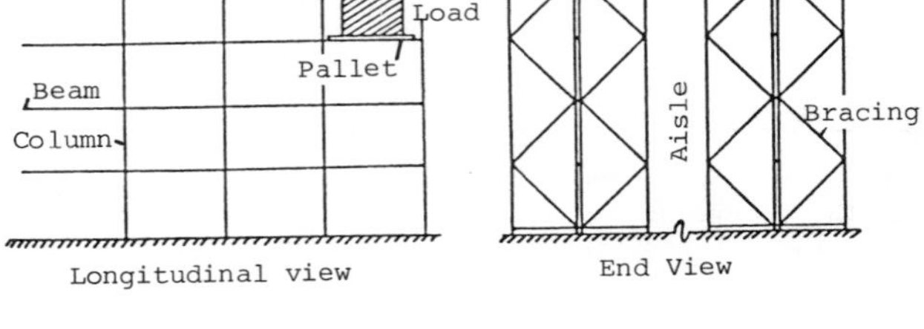

Fig 2 Typical Racking System

Although the Rack Manufacturers Association of America's design rules produced by Pekoz (Ref 2) make some reference to progressive collapse, the only research to date aimed specifically at this problem has been that carried out at Cambridge University since 1979. The main findings of the first 3 years of this work have been reported in detail in Refs 3 and 4 and will be reviewed briefly here. The need to extend the 2 dimensional analysis used in this earlier investigation will be established, a 3-dimensional program described, and its capabilities demonstrated by application to the experimental collapses presented in Ref 4. This paper will conclude with some consideration of the implications of this work for more conventional space structures.

PROGRESSIVE COLLAPSE STUDIES, 1980 - 82

Preliminary Considerations

The information available on racking collapses showed that some racking systems exhibited extensive collapses whereas other failures were confined to the region of initial damage. There was no evidence to indicate, or reason to believe, that the initiation of an extensive collapse was any different from that of a minor collapse. The obvious conclusion was that an extensive collapse progressed away from the point of initial damage. Attention was therefore directed towards the sequence of events following a single front bottom leg failure, most probably caused directly by fork-lift truck damage, but possibly resulting from simple overload. Given the low load factors employed and the dynamic nature of the problem, it is unrealistic to expect a fully loaded rack with a leg removed to be stable, so the study reported in Refs 3 and 4 was directed towards identifying those characteristics of racking and its stored loads which arrest, and those which promote, a potential progressive collapse.

Two-Dimensional Computer Simulation, Half Scale Collapse Tests

To achieve this end, a 2-dimensional dynamic analysis program capable of considering both geometric and material nonlinearity was developed at Cambridge. To calibrate this program and to obtain direct observations of collapses, half-scale racking installation (Fig 7) was set-up and tested. The more significant of these tests will be described in more detail later in this paper in relation to the results produced by the 3-dimensional analysis program reported below.

The above investigation highlighted the importance of the joint pull-out strength in determining whether a collapse would be confined, or spread. A special pull-out strength test was devised to evaluate this properly, and the test arrangement is shown diagramatically in Fig 3. Typical response curves for pull-out strength are shown in Fig 9, and the moment rotation curves of the same joints in the absence of axial tension are shown in Fig 10.

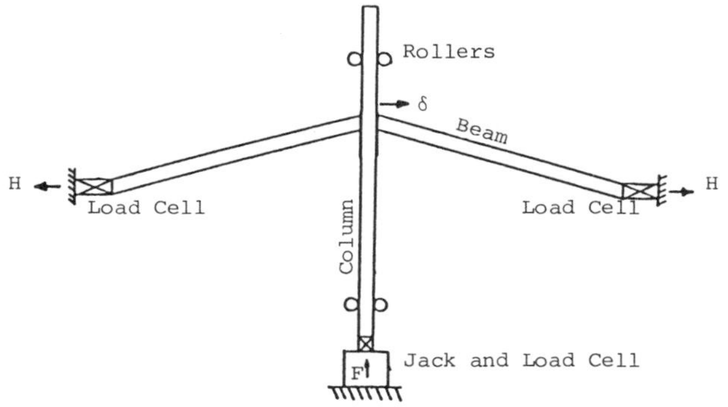

Fig 3 Joint Pull-Out Test

Collapse Sequences Following an Initial Single Bottom Leg Failure

Taken together, the results of the joint pull-out tests, the 2-dimensional computer program, and the half-scale tests, lead to the following collapse sequences being identified.

After bottom leg failure, a *joint rotation mechanism* forms in the two bays supported by that leg, Fig 4. The progress of this mechanism will be partially retarded by the cross-bracing to the leg behind that which has failed. It was assumed in the original study that this cross-bracing could not support the front leg, and that the joint rotation mechanisms would then proceed by drawing in the bays to either side. Tension would develop in the inclined beams, and one of the three sequences below would then develop.

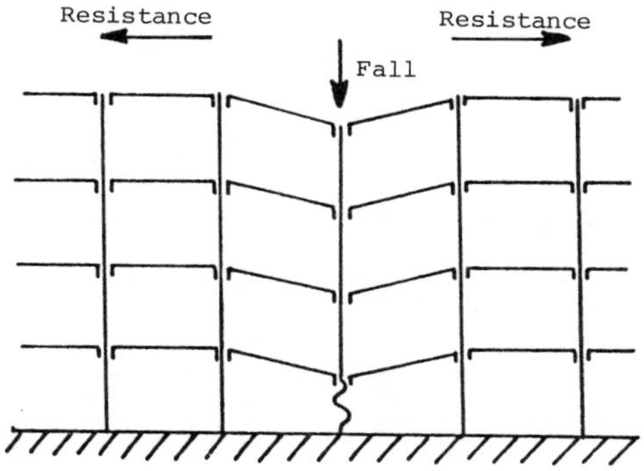

Fig 4 Joint Rotation Mechanism

(i) Confined Collapse.

If the tension developed in the inclined beams exceeded the joint pull-out capacity, these beams would separate from the rack and a confined collapse would result, Fig 5.

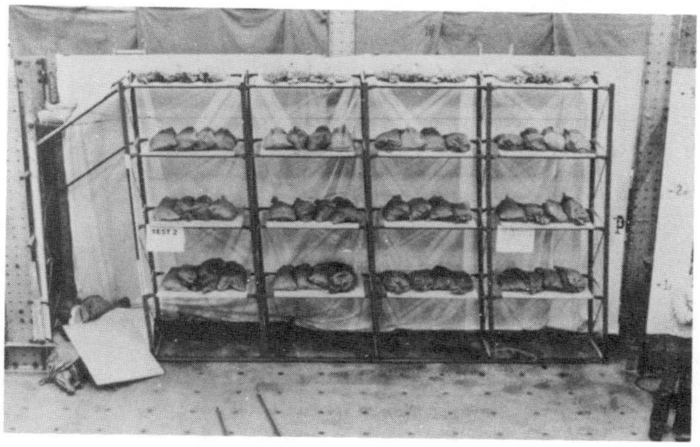

Fig 5 Confined Collapse

(ii) Successive Leg Failure

Fig 6 Successive Leg Failure

If the initial collapse is remote from a free end, the tension in the inclined beams will induce additional compression in the legs on either side of the failure. With high pull-out strength joints and dynamic magnification effects, this overload can easily exceed the column capacity, and one or both of the adjacent legs will fail. This process can then be repeated with the next adjacent leg, and successive leg failure will then occur to the end of the rack, Fig 6.

Fig 7 Bottom Leg Sway

(iii) Bottom Leg Sway
If the initial collapse is near the free end of a rack, the joint rotation mechanism and high pull-out strength joints can induce a bottom leg sway mechanism (to the free end), Fig 7.

Fig 8 Rigid Body Rotation of a Cross-Frame

Sequences (ii) and (iii) can lead to progressive collapse.

Participation of the Cross-Bracing - The Need for a Three - Dimensional Analysis

The above sequences provided an adequate initial understanding of the half-scale experimental collapses, but have one potentially important defect. The assumed non-participation of the cross-bracing necessary to justify the use of the 2-dimensional program, and predicting that the mechanisms of (ii) or (iii) above would proceed down the front legs only, was demonstrated to be incorrect even for extremely weak bracing. In particular, the non-failure of the bracing near the end of a rack in the bottom leg sway mechanism induced an out of the plane of the rack rigid body motion of the last 2 or 3 bays. This rigid body motion of an intact section of rack and its loads appears to be the most likely sequence of events that would allow a collapse to spread from one rack to another, Fig 8. To investigate this behaviour more thoroughly, a 3-dimensional dynamic computer simulation is obviously required, and a brief description of a suitable program, and its calibration against the experimental tests reported in Ref 4, is now presented.

THREE-DIMENSIONAL DYNAMIC ANALYSIS PROGRAM

The need for a general nonlinear, dynamic analysis program has been discussed above, and a suitable computer program has been developed. This analyses the nonlinear dynamic response of 3-dimensional space structures in which the geometry (Ref 5), the properties of the structure, and the arrangement of the loads are continually updated with time and displacement as the collapse proceeds.

The nonlinear (matrix) equation of dynamic equilibrium

$$\text{Inertial resistance} + \text{Damping resistance} + \text{Structural resistance} = \text{Applied loads} \quad \ldots \quad 1$$

is directly integrated using a step-by-step incremental solution technique.

A warehouse racking structure extensive enough to show progressive collapse tendencies will be extremely large, and the analysis of such a (3-dimensional) space structure will involve the formulation of large stiffness and mass matrices. These are generally too big for the high-speed storage of the computer, and peripheral storage must therefore be used to solve the dynamic equilibrium equations. Therefore the *Frontal Solution Technique* (Ref 6), which is an efficient out-of-core solution method, was adopted for the computer program. The assembly and the elimination of variables is performed simultaneously, resulting in a minimum core storage requirement, reduced arithmetic operations, and the most effective use of peripheral equipment.

A two-noded space frame element with three translational and three rotational degrees of freedom at each node, is used to model the rack components. Standard stability functions (Ref 7) are used to consider the effects of axial force on member stiffness. The most important feature of the racks, and also the detail where they differ most from other structural frames, is the joint between beam and upright, Fig 1. Though there are many different systems on the market, all consist of beams which hook into perforations in the upright. Rotationally flexible end connectors are incorporated in the beam element to model the nonlinear rotational flexibility, Fig 10, and the extension of the connectors due to 'hinging-out' of the joint hooks, Fig 9.

Lumped masses are used to represent the appropriate inertial properties of the system, and the damping matrix is formed as a linear combination of the mass and stiffness matrices,

$$[C] = \alpha \; [M] + \beta \; [K] \quad \ldots \quad 2$$

where α and β are the constants of proportionality obtained using the natural frequencies of vibration for a particular value of percentage critical damping γ. The relation used (Ref 8) is

$$\alpha + \beta \, \omega_i^2 = \omega_i \gamma_i \quad \ldots \quad 3$$

Though the shortest period of vibration of the structure (used for the selection of the time step (Δt) in the step-by-step integration scheme) changes as the collapse proceeds, the Newmark's (Ref 9) constant average acceleration method was found to be unconditionally stable. It also allows for different suitable time steps at various stages of the collapse analysis.

The subspace iteration method (Ref 8), which is most suitable for the solution of large eigen-problems at reasonable cost, has been used for the free vibration analysis. The least wanted variables are condensed out kinematically (Ref 10), before the free vibration analysis, to obtain reduced stiffness and mass matrices.

Test Analyses

The 3-dimensional program was tested by solving various simple structures reported in the literature. These tests include the analysis of various space frames (Ref 11), a nonlinear large displacement analysis of a three-pin arch, the buckling analysis of an euler column, and the plastic collapse analysis of a beam restrained at both ends. The joint pull-out strength test in Fig 3, and the free vibration response of a cantilever structure reported in Ref 8 were also successfully analysed using the program. The coding of Newmark's constant average acceleration method (Ref 9) for the direct step-by-step integration of the dynamic equations of motion, has been verified by the analysis of a problem documented in Ref 8.

3-DIMENSIONAL ANALYSIS OF THE HALF SCALE COLLAPSE TESTS

The ability of the computer program to predict the 3-dimensional collapse sequences reported in the earlier work (Refs 3 and 4) was investigated by matching the time-displacement histories of the program and the experiments (as recorded by two high speed cameras at three frames/sec., and one cine camera at 60 frames/sec.). The test rack was four bays high, five bays long and made of cold rolled shelving material, designed according to the SEMA code (Ref 1). The columns were 0.76 m long between beam levels, the beams spanning 1.15 m, and the front and back frames were a metre apart. Bags of sheet metal punchings on chipboard pallets were used to load the model. The test structure was restrained at one end, to represent a long rack with a collapse initiated close to a free end. The nonlinear characteristics of the hooked beam end connectors were determined experimentally by a cantilever test and a specially devised joint pull-out test, Fig 3. Figs 9 and 10

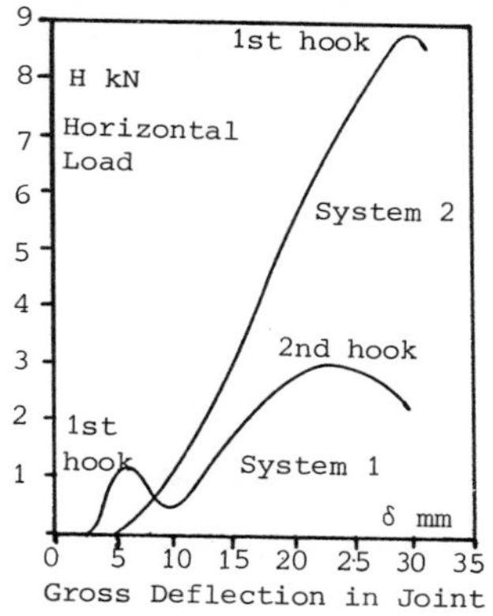

Fig 9 Joint Pull-Out Test

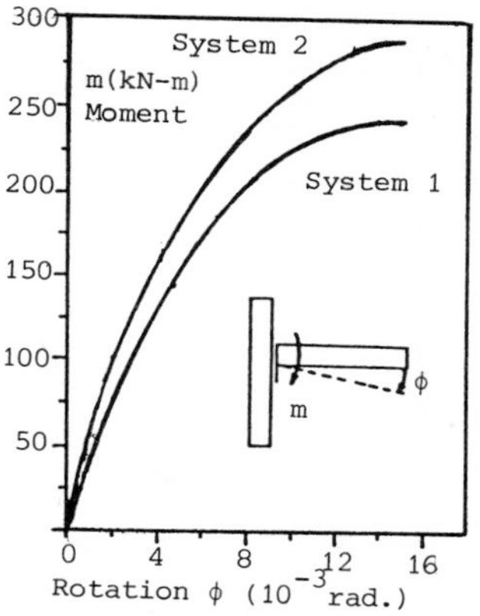

Fig 10 Joint Rotational Stiffness

show the behaviour of the two systems of joints used in the collapse tests described below. Although the bending characteristics of the two systems are similar, the pull-out strength of the system 1 connectors is very low compared to that of system 2.

Confined Collapse Test

The experimental collapse Test A, in which the system 1 joint connectors with low pull-out strength were used, is shown in the sequence of photographs in Fig 11. As the

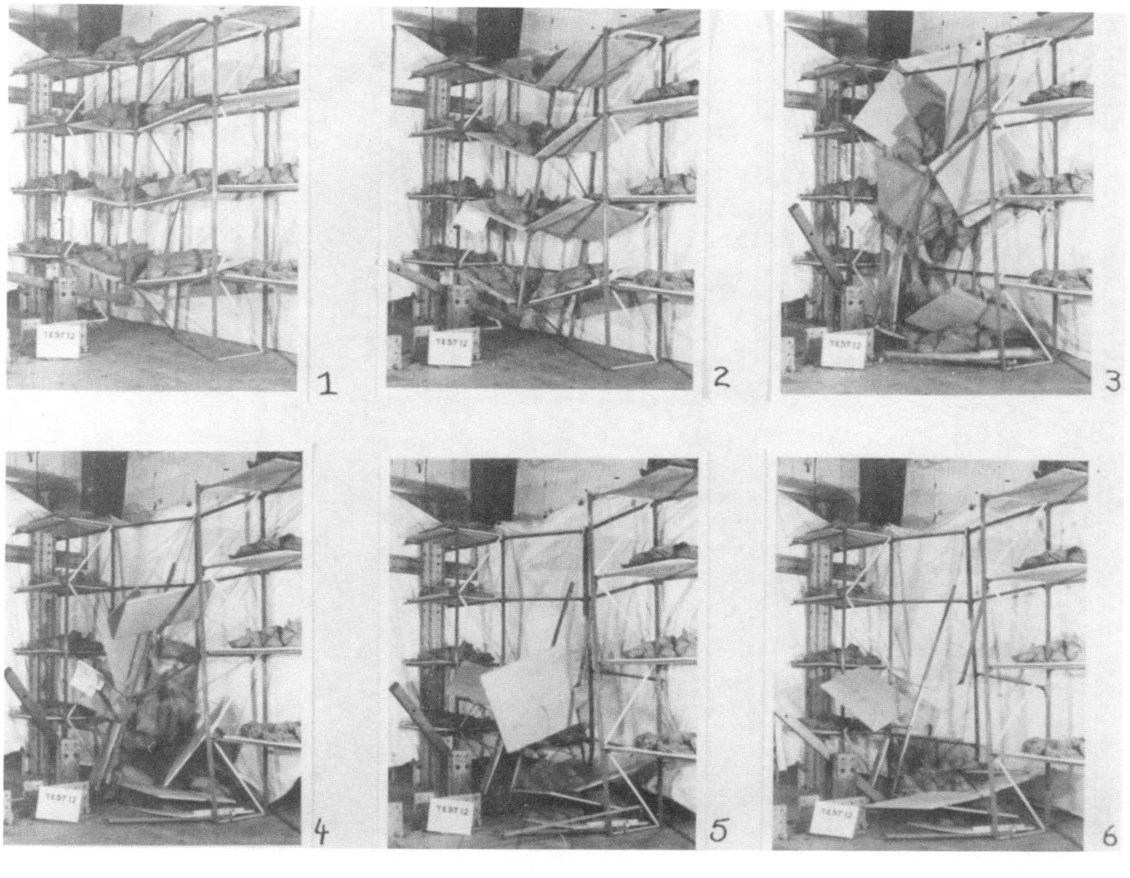

Fig 11 Confined Collapse

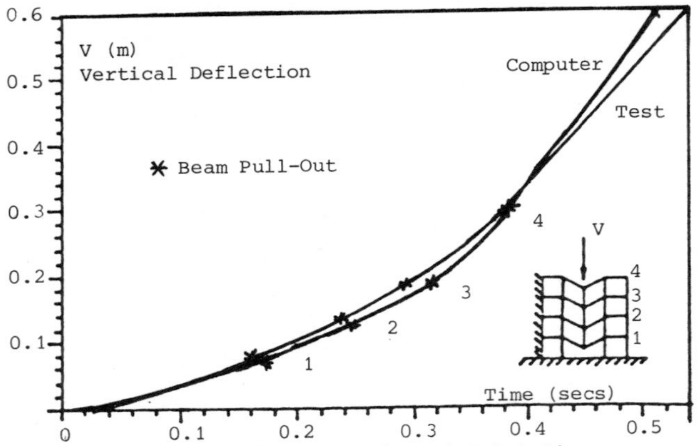

Fig 12 Test A - Deflection of the Joint Rotation Mechanism

joint rotation mechanism proceeded, the bracing in the central cross-frame failed, and the inclined pallets slipped forward. This was followed by the back column buckling over its full 3 m unsupported length, and the pallets beginning to rotate backwards as they slipped off the rear beams. At about this stage the front beams in the joint rotation mechanism pulled out of the columns, and the loads fell almost vertically to give the confined collapse shown.

The program predicts a similar sequence of events, and Fig 12 shows the close agreement between the computer simulation and the test collapse for the vertical displacement time-history of the joint rotation mechanism. Similar agreement between the experiment and the program exists for the other degrees of freedom, so that the ability of the program to model confined collapses was demonstrated.

Progressive Collapse Test

The system 2 joints with high pull-out strength were used in the second experimental collapse, Test B. It is a common practice in industrial warehouses to stand two racks back-to-back, therefore the structure in this test was passively restrained from behind. The results of the experimental collapse are shown in the sequence of photographs in Fig. 13. As the bracing in the central cross-frame failed, tension developed against the resistance of the surrounding rack in the front beams of the joint rotation mechanism. The back column was prevented from buckling backwards, and the motion was mainly to the front and down. The joint rotation mechanism did not separate from the adjacent bays because of the relatively high pull-out strength of the joints, so that a bottom leg sway mechanism developed in the front columns to the right of the initial failure. The severe lateral bending in these front columns meant that they could not sustain the axial loads imposed on them, and so they failed. As this failure is basically due to lateral sway rather than axial buckling, the front columns lost their axial capacity with minimal end shortening. The bracing to the right of the central cross-frame therefore did not fail, and so cross-sections of the type shown in Fig. 15 developed. The right hand cross-frames could now rotate as a unit out of the plane of the rack as shown in Fig 13. finally pulling the left hand bays (and the rest of the rack if it had existed) into the isle.

As with the confined tests, the 3-dimensional program predicts the sequence of events described below. Fig 14 shows the vertical displacement time curves of the joint rotation mechanism from both the experiment and the computer

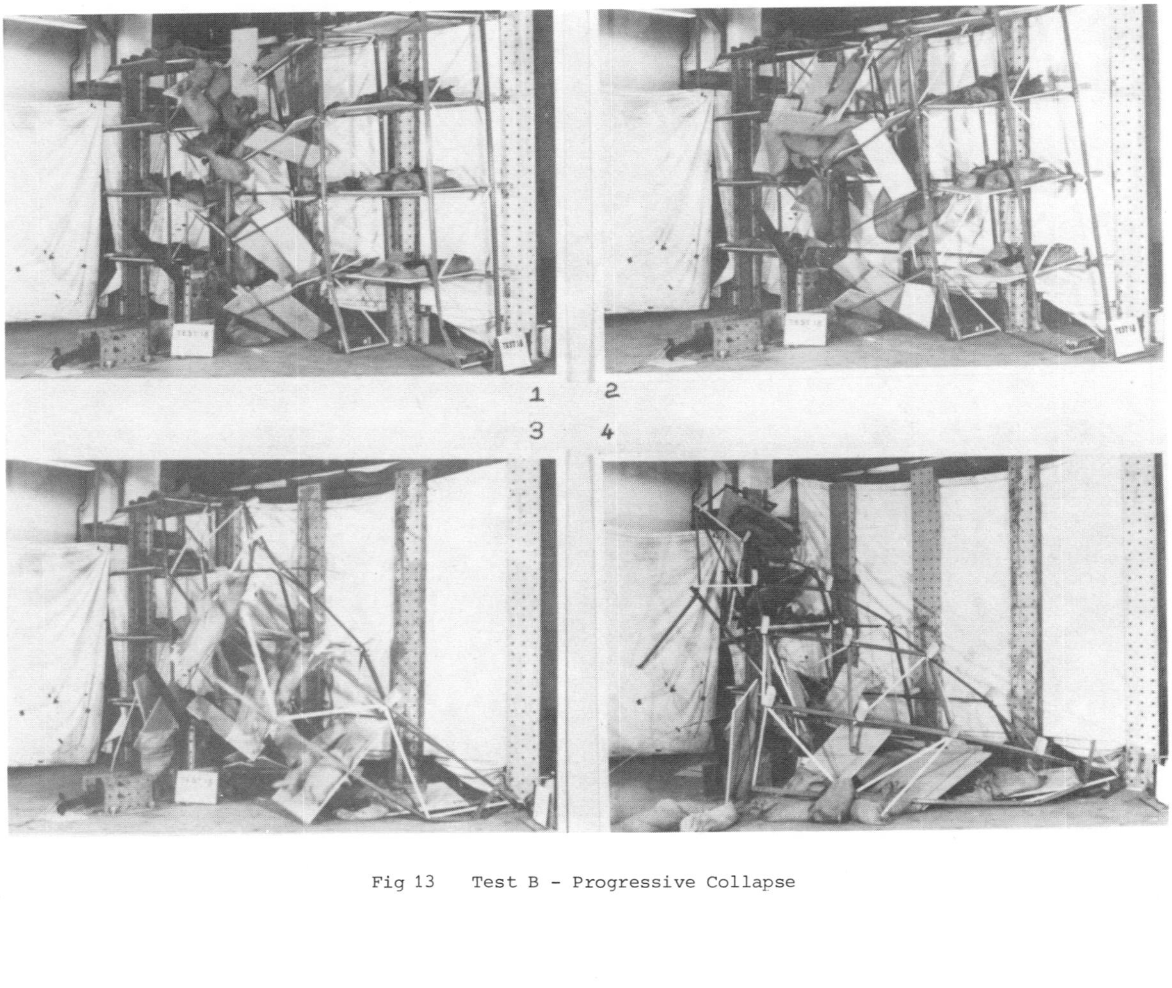

Fig 13 Test B - Progressive Collapse

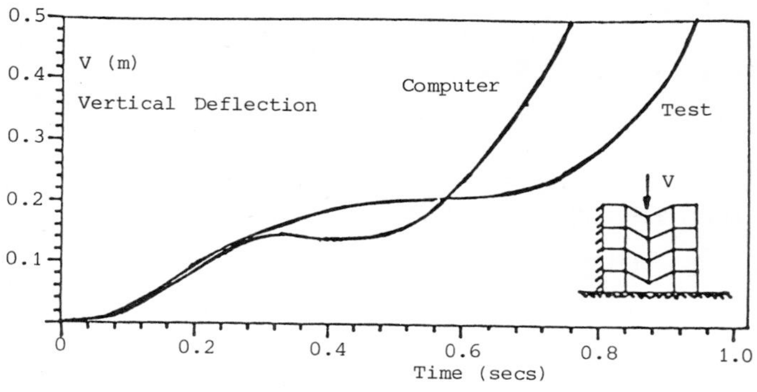

Fig 14 Test B - Deflection of the Joint Rotation Mechanism

analysis. Similar results for the out-of-plane rotation against time curves are presented in Fig 15. In the film of this collapse, a definite time-lag occurred between the formation of the joint rotation mechanism, and the rotation out-of-plane of the end bays. This time-lag appears as the flat portion of the response curve in Fig 14. The program showed a similar pause in the collapse sequence, and although this is shorter than that observed in the test, the basic sequence of events was adequately predicted by the program.

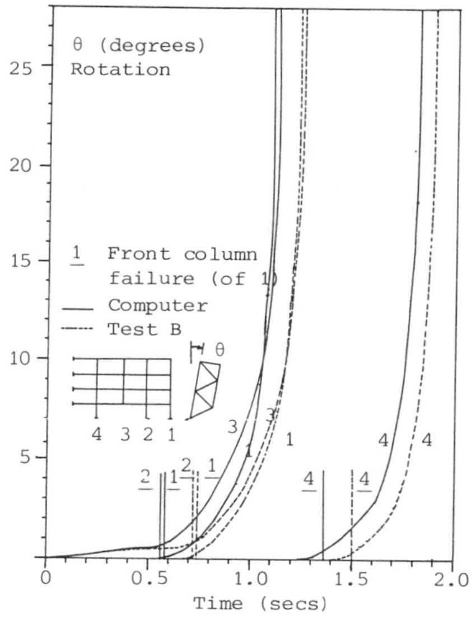

Fig 15 Test B - Out-of-plane Rotation

It should be noted from the final state of the structure shown in Fig 13, that if a second rack was positioned one aisle's width in front of the collapsing rack, the structural integrity of this second rack would have been seriously impaired. Given an isolated single member failure, a multiple repetition of the sequence of events in Test B could therefore lead to the total collapse of a racking installation.

CONCLUSIONS

The results presented in this paper have demonstrated the ability of sufficiently sophisticated nonlinear dynamic analysis program to follow the complex sequence of events involved in the progressive collapse of an extended framework. By considering the dynamic aspects of a collapse, the program was able to predict that increasing only the pull-out strength of the joints could transform a confined collapse into a progressive one. To date most collapse or limit-state analyses of space structures have been static calculations, although the similarity of a collapse spreading from an initial single member failure in a typical two-layer space truss to that of a progressive collapse in warehouse racking is obvious. While this analogy is far from complete, the significance of dynamic effects on the analyses reported in this paper would seem to have potentially important parallels in the collapse of space structures. For example, the simple static overload induced by a joint rotation mechanism in the rack legs adjacent to a failure would almost never cause a collapse to spread, and it is only dynamic effects which induce subsequent leg failures. Analogies are obviously possible in space structures (e.g. a situation where a single leg failure allowed a snow load to move), and applications of the program to the collapse of space structures is therefore planned. The well known problem of defining exactly how the axial and bending forces interact as a member fails, will still have to be considered. However the major advantage of using a dynamic program to study collapse, that of avoiding singular stiffness matrices, will obviously help off-set such problems.

ACKNOWLEDGEMENTS

Dexion Ltd. and IDC Consultants have supported and encouraged this work from the initial stages, and various racking manufacturers have helped with material and advice through SEMA. Thanks are due to these companies.

REFERENCES

1. SEMA, Code of Practice for the Design of Static Racking, Storage Equipment Manufacturer's Association, London, 1980.

2. Specification for the Design, Testing and Utilization of Industrial Steel Storage Racks, Rack Manufacturers Institute, Pittsburgh, U.S.A., 1979.

3. R. E. McCONNEL and S. J. KELLY, Structural Aspects of the Progressive Collapse of Warehouse Racking, The Structural Engineer, No. 11, Vol. 61A, November 1983.

4. S. J. KELLY, Ph.D. Thesis, University of Cambridge, 1982.

5. C. ORAN, Tangent Stiffness in Space Frames, Journal of Structural Division, ASCE, Vol. 99, No. ST6 (1973).

6. B. M. IRONS, A Frontal Solution Program for Finite Element Analysis, International Journal of Numerical Methods in Engineering, Vol. 6, 1970.

7. M. R. HORNE and W. MERCHANT, The Stability of Frames, Pergamon Press, London, 1965.

8. K. BATHE, Finite Element Procedures in Engineering Analysis, Prentice-Hall, New Jersey. 1982.

9. N. M. NEWMARK, A Method of Computation for Structural Dynamics, Journal of Engineering Mechanics Division, ASCE, Vol. 85, No. EM3, 1959.

10. R. J. GUYAN, Reduction of Stiffness and Mass Matrices, J. AIAA, Vol. 3, 1965.

11. W. Weaver and J.M. Gere, Matrix Analysis of Framed Structures, D. van Nostrand Co., London, 1980.

INVESTIGATION OF THE CAUSES OF HARTFORD COLISEUM COLLAPSE

CHARLES H. THORNTON, PhD, P.E.*

I. PAUL LEW, C.E.**

*President, Lev Zetlin Associates, Inc.

**Vice President, Lev Zetlin Associates, Inc.

The investigation of the collapse of the Hartford Coliseum was one of the most extensive investigations undertaken into the causes of the collapse of a building. The investigation required the development of state-of-the-art failure analysis methods. This paper presents the results of this investigation and reviews two proposed explanations of the cause of the collapse proposed by other investigators.

INTRODUCTION

At about four o'clock in the morning on January 18, 1978, six hours after 5500 basketball fans left the premises, the roof of the Hartford Coliseum in Hartford, Connecticut USA, completely collapsed during a freezing rain and snow storm. (See Fig. 1.) The massive long span space truss was over 21 feet deep and over 360 feet by 300 feet in plan and yet it collapsed under the weight of 18 pounds per square foot of snow and at about one-half the total load which would have caused first yielding in the weakest member.

The firm of Lev Zetlin Associates, Inc. was retained by the owner, the City of Hartford, to perform an independent investigation to determine the causes of the collapse. The investigation consisted of a survey of the wreckage, a review of design and construction data; a review of the meteorological data, metallurgical evaluations and testing, determination of actual weights, a design review and a collapse analysis of the roof space truss. The collapse analysis required the development of the post buckling behavior of the space truss individual members and the incorporation of their post buckling behavior into the overall space truss analysis. This post buckling collapse analysis represented one of the most in-depth analyses ever undertaken of the collapse of a space truss. As a result, new techniques were developed to facilitate this type of analysis.

The purpose of this paper is to present the results of the investigation and collapse analysis of the Hartford Coliseum roof structure. In addition, an overview of some of the alternate theories and opinions of the cause of the collapse is included. The complete report of the investigation entitled, "Report of the Engineering Investigation Concerning the Causes of the Collapse of the Hartford Coliseum Space Truss Roof on January 18, 1978" dated June 12, 1978, is available from the City of Hartford.(1)

DESCRIPTION OF THE COLISEUM AND ITS ROOF STRUCTURE

The Hartford Coliseum had overall roof plan dimensions of 366 by 306 feet. It was designed to seat approximately 12,500 spectators. The dominating feature of the Hartford Coliseum was the roof structural system. The roof structure consisted of a 21 foot deep offset orthogonal grid space truss. The space truss grid was based on 30 foot bay in both plan directions: The overall structural grid at the top chord was 360 feet by 300 feet. The bottom chord grid is offset one-half bay (15 feet) in both directions from the top chord. The space truss was supported on four pylons located 45 feet in from each exterior edge leaving a center clear span of 270 by 210 feet. This created a 45 foot overhang around the entire roof perimeter. (See Fig. 2.)

The space truss system was a "modified" version of the offset, double layer space truss system. It is called a modified system because the space truss has additional intermediate bracing members located on the four faces of the pyramid modules of the offset, double layer module system (See Fig. 3.) The intermediate members connect the midpoints of the main diagonals to the midpoints of the top chords (See Fig. 3.)

The actual roof framing consisted of 14 inch deep steel wide flange purlins which were posted down to the top chords of the space truss on a 15 ft. by 30 ft. post spacing (See Fig. 4.) Note that the post grid is 15 ft. by 30 ft. while the space truss is on a 30 ft. by 30 ft. grid.

The typical main members of the space truss had a cruciform configuration composed of four angles with equal legs spaced 3/4 inch to 7/8 inch apart back to back with intermittent spacers. The angle legs varied from 3-1/2 to 8 inches. The intermediate bracing members were 5 by 5 inch single angles.

The top chord members are classified into four types of members as follows:

Type A Exterior top chord members in the east-west direction with post located at midpoint.

Type B Exterior top chord member in the north-south direction without post at midpoint.

Type C Interior top chord member in the north-south direction without post located at the midpoint.

Type D Interior top chord member in the east-west direction with post at midpoint.

The main member connections at the panel point of the space truss consisted of horizontal and vertical gusset plates. The gusset plates were attached to the main cruciform members via the space provided between the four angles that comprised the main members (See Fig. 4.)

The centerlines of all the main members intersected at the same point in the main gusset plate connections.

The centerline axis of the intermediate bracing members did not intersect at the centerline axis of the main members. The main members were continuous, not interrupted, at the connection of the intermediate bracing. Instead, the intersection of intermediate braces and the main member were located about 12 inches below the main members. Thin bent plates were used to connect the intermediate braces to the main members.

STRUCTURAL BEHAVIOR

Two design features of the Hartford Coliseum space truss vary from those normally found in offset, double layer space truss roofs. These features are:

1. The separation of the roof purlin system from the space truss roof structure. This isolated the inherent diaphragm action of the roof skin from the top chord compression members in the space truss.

2. The introduction of intermediate bracing members connecting the midpoints of continuous main members (top chord and diagonals) within the plane of the faces of the pyramid module to reduce the unsupported length of the top chords and diagonals to one-half their panel point to panel point dimension.

Space trusses with such modifications are not inherently unstable provided successful execution of the details is accomplished so that the assumptions inherent in these modifications are satisfied. The following discussion examines the structural behavior apparently assumed in the structural design and compares the assumed behavior with the observed actual behavior with comments on the actual detailing, fabrication and construction of the space truss.

First, the use of an independent roof purlin system posted (or separated) from the top chords of the space truss was done to facilitate drainage, to eliminate the need for camber in the space truss and to eliminate flexural stresses in the top chord members due to uniform contact with the roof system.

Apparently, the choice of the typical member as a cruciform, a section that is weak in bending and torsion, was based on the design assumption that bending and torsion would be negligible in the space truss top chord members. The introduction of the post at the middle of Type A member and the absence of horizontal bracing at the midpoint inadvertently introduced significant flexural stresses in the cruciform members.

Second, the design assumed that all main top chord and diagonal members were fully braced by the intermediate bracing diagonals and bracing horizontals at the midpoint of the main top chord and diagonal members. This is confirmed by the close relationship between the actual design forces stated on the design drawings as compared to the allowable capacity of the main members assuming that the members are fully braced at their midpoint. Unfortunately, the assumption of fully effective midpoint bracing was never achieved in reality. The lack of effective bracing greatly reduced the capacity of the individual members and the entire space truss structure. Two cases of ineffective midpoint bracing existed: one on the interior and the other on the exterior. This weakness is further compounded when the member has a post located at the midpoint brace point.

The typical interior top chord member spanning in the north-south direction is called a Type C member. It does not have a post at its midpoint brace point. As discussed earlier, the actual detailing of the intermediate connection resulted in the intersection point of the intermediate bracing being below the centerline axis of the main top chord member (See Figs. 5 & 6). This configuration produced a partial brace or "soft spring" at the midpoint. A determination of the stiffness of the flexible plate and the intermediate bracing was performed. The resulting stiffness was then used as a "soft spring" to determine the capacity of the axial load of the Type C member.

The computer models used to determine the stiffness of the intermediate members were conservatively chosen to result in a stiffness somewhat greater than may have actually existed. This was done so as to not underestimate the actual strength of the top chord members.

The presence of the posts at the intermediate connection of the interior members spanning in the east-west direction, herein called the Type D members, was determined to unintentionally add sufficient stiffness to effectively brace the Type D members.

The lack of effective bracing is particularly apparent on the exterior top chord members. The exterior top chord Type A members span in the east-west direction. The exterior top chord members have intermediate diagonals bracing them in only one plane or direction; the inclined plane formed by the exterior face of the space truss. (See Fig. 7.) The intermediate bracing only provides a brace in the plane in which the bracing members lie. In a direction perpendicular to this plane, the intermediate diagonals offer no resistance to movement of the main top chord. Therefore, the member is unbraced for its full bay length. As a result the actual compressive load capacity of the exterior top chord members is only a fraction of the capacity required.

The capacity of the Type A member is further reduced by the addition of a post at the midpoint member. (See Fig. 8.) Perpendicular to the plane of the intermediate bracing there is negligible resistance to primary out-of-plane bending as well as buckling. The introduction of a vertical post at the intermediate bracing connection results in out-of-plane primary bending for which the cruciform member is particularly unsuited.

In summary, the lack of effective intermediate bracing was the primary cause for the collapse of the Hartford Coliseum. The next section describes the determination of the collapse load and a probable mode of collapse.

FAILURE ANALYSIS

A technically challenging aspect of the Hartford Coliseum investigation was the determination of the failure load and probable failure mode of the space truss. To perform the failure analysis it was first necessary to formulate the post buckling behavior of the members. Next, a method of incorporating the post buckling behavior of a member into an incremental elastic computer analysis was required. Finally, the actual computer analysis had to be incrementally carried up to the actual failure load.

The method used to link the post buckling behavior to the elastic computer analysis was through the development of a stress-effective strain curve. Within the elastic range the longitudinal shortening of the compressive member is primarily due to axial load. In the post buckling range the longitudinal strain is substantially effected by the flexural chord shortening of the member. The shortening of the chord length of the member can be then treated as an effective axial strain of the member for each increment of axial strain.

Once the effective axial stress strain curve for a member is developed it is possible to add corrective axial strain to the incremental stress-strain curve to simulate the stress-strain effective curve of the buckled member. This technique is a modification of a technique described by Wolf (2).

The first step in the failure analysis was to develop the post buckling behavior of the compressive top chord cruciform members. The deflected shape of an axially compressed bar with an initial imperfection (a beam-column) may be computed utilizing the analogy between the computations of statics and geometry. The angle changes per unit length are determined from the M-P-Ø relationships (See Fig. 9). Transverse displacements can be computed using the procedure popularized by Newmark.

The change of length of the member may be obtained by the angle change procedure (which gives the end displacements due to curvature), and adding the change in length due to elastic and inelastic axial strains. The latter component can be found by numerically integrating the values of E_a taken from the E_a-P-Ø curves.

The analysis is done for an imperfect column with initial bow and residual strains. The initial bow was assumed to be the span divided by 1000. Once the transverse displacements are converged for a given ratio of P/P yield, then the longitudinal chord displacement is computed including bow shortening. This procedure is repeated for various values of P divided by P yield until sufficient equilibrium positions are obtained. The resulting plot of axial stress versus effective axial strain curve is shown in Fig. 10. The downward branch is only valid when the member is part of a structure which can "relieve" the member of load, such as a stable space truss. The introduction of a mid-length spring brace increases the elastic buckling load of the member but as the movements in the post buckling range increase the effectiveness of the brace is diminished. This behavior is postulated in Fig. 10 wherein the upper curve assumes a fully effective brace and the lower curve is for a partially effective brace. In the lower case the buckling load is increased by the presence of the spring. However, as the post buckling distortion increases, the member behaves as if the spring is not present and the plot returns to the unbraced member curve. For each top chord member the post buckling plot was determined and the actual ultimate collapse load analysis of the space truss could then be performed.

The ultimate collapse load analysis of the space truss used an iterative procedure similar to Wolf (2). Incremental linear structural analyses were made for constant gravity loads with varying imposed strains in those members that have buckled, until an equilibrium position was obtained. The equilibrium position must satisfy the stress-effective strain relationship for each compressed member. Each increment produces a point on a plot of gravity load versus displacement. The gravity load was then increased and another equilibrium position was determined and the corresponding displacement plotted. This procedure was repeated until large displacement results from small increases in load. When the load-displacement curve becomes horizontal, the ultimate load is reached. Figure 11 shows the gravity load versus displacement curve for the midpoint top chord panel point of the space truss. The actual failure load occurred at approximately 16 to 18 psf of snow load. This correlates very closely with the measured and calculated snow load on the roof structure at the time of the failure (3). Figure 12 represents the buckled members in the quarter structure model at a snow load of 18 psf.

The mathematical model used in the analysis results in an initial fold forming in the north-south direction between column lines E and F. The actual initial fold can be seen in the aerial photograph to have formed in the north-south direction between lines D and E (Fig. 13). The reasons for this variation are unintentional and unpredictable variations in the actual forces in the structure at the time of the collapse due to fit up stresses, eccentrically located mechanical and equipment platforms and weights, and minor variations in the as-built structure causing it to be slightly unsymmetrical in its as-built configuration.

ALTERNATE EXPLANATIONS

Since the collapse of Hartford Coliseum, alternate theories of the collapse have been forwarded by various investigators. Two alternate theories for the collapse will be reviewed here in order to put the subject into perspective.

The first theory states that the failure of a weld securing the central scoreboard to the space truss initiated the collapse. The weld failure and the sudden impact of the falling scoreboard and framing created a large impact load which overloaded the structure.

This theory fails to recognize the inherent reductancy of a space truss. Local overstresses or failure will not fail a space truss. This was shown during World War II when direct bomb hits on space truss roofs of hangars and factories did not result in the collapse of the structure. Instead, a localized hole formed in the area where the bomb hit and the structure remained standing.

In the case of the Hartford Coliseum space truss, the localized overstress that could occur due to the impact forces caused by the falling scoreboard and framing could not propagate a failure of the overall structure. The total dead weight of the roof at the time of the collapse was 9,600,000 pounds. Comparing 9,600,000 pounds to the approximate weight of the scoreboard and framing at 25,000 pounds makes it highly improbable that this theory has any validity.

The second theory for the collapse is that torsional resistance of the cruciform members was over-estimated. This explanation has its basis in a misuse of a design simplification as a theoretical tool; the use of an equivalent radius of gyration for checking torsional buckling. The concept of an equivalent radius was developed by Gaylord (4) and Bleich (5) and it is valid as long as the boundary conditions for the bending mode and the torsional mode are the same (5). In other words, only if member is restrained both against torsion and bending at its boundary is the use of the equivalent radius of gyration valid as an approximation for torsional resistance.

For members such as utilized for the space truss of the Hartford Coliseum when the equivalent radius of gyration is calculated based on the ends being torsionally fixed and flexurally pinned, the calculated slenderness ratio (kl/r equivalent) will be double the actual slenderness ratio. Furthermore, the strength will be 1/4 of the actual strength. In actuality the torsional buckling strength is taken into account implicitly in the American Institute of Steel Construction (AISC) Code through its reductions in compressive strength for slender compression members.

Therefore, torsional buckling of the four angles was not a significant factor in the collapse of the Hartford Coliseum. However, the assumption of full composite moment of inertia for back to back angles with spacers needs further examination by the industry. These sections are really battened columns. Reductions in capacity of up to 15% are probable when the capacity is calculated using techniques recommended for batten columns (6).

In the opinion of the authors it is important for the designer to assure that the spacers are effective at each location about both axes when using cruciform members comprised of angles. A minimum number of bolts in each flange for a single spacer is required. When a double spacer is used it should be double at each location.

CONCLUSION

The primary cause of the collapse of The Hartford Coliseum space truss collapse was the inadequate bracing to the top chord members. The space truss failed at about one-half the total load expected to cause first yield of the critical member. A non-linear, incremental load collapse analysis confirmed the cause of the collapse. The collapse analysis was based on the post buckling behavior of the members. The analysis produced a collapse load within several pounds of the actual measured load on the structure at the time of the collapse.

REFERENCES

1. Lev Zetlin Associates, Inc. "Report of the Engineering Investigation Concerning the Causes of the Collapse of The Hartford Coliseum Space Truss Roof on January 18, 1978", June 12, 1978.

2. Wolf, J.P., "Post-Buckled Strength of Large Space Truss", Journal of the Structural Division, ASCE, Vol. 99, ST 7, July 1973.

3. Redfield, R.K., Tobiasson, W.N., and Colbeck, S.C., "Special Report 79-9 - Estimated Snow, Ice and Rain Load Prior to the Collapse of the Hartford Civic Center Arena Roof"; U.S. Army Corps of Engineers Cold Regions Research and Engineering Laboratory, Hanover, New Hampshire. USA, April 1979.

4. Gaylord, E. and Gaylord, C., "Design of Steel Structures", MCGraw Hill Co., 2nd Ed., 1972.

5. Bleich, F., "Buckling Strength of Metal Structures", McGraw Hill, New York 1952.

6. Lin, F.G., Glauser, E.C., and Johnson, B.G., "Behavior of Laced and Battened Structural Members", Journal of the Structural Division, ASCE, Vol. 96, ST 7, July 1970.

Fig. 1

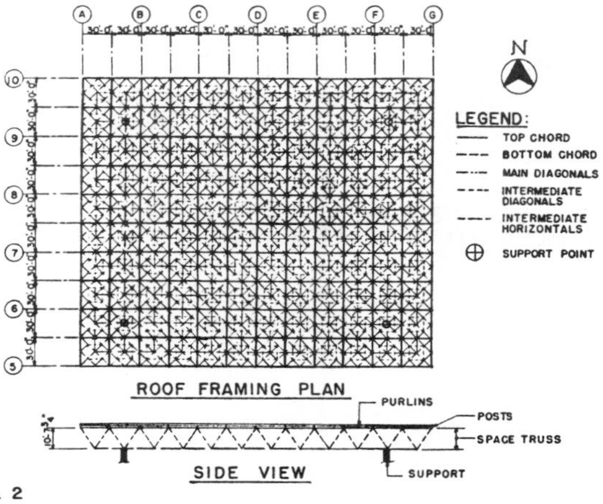

FIG. 2

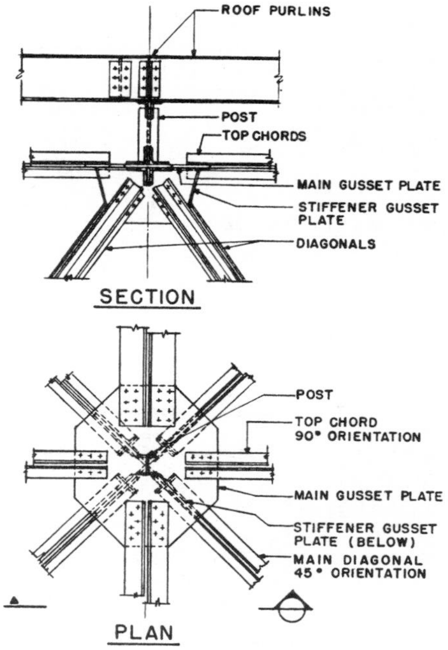

FIG. 4
INTERIOR TOP CHORD CONNECTIONS

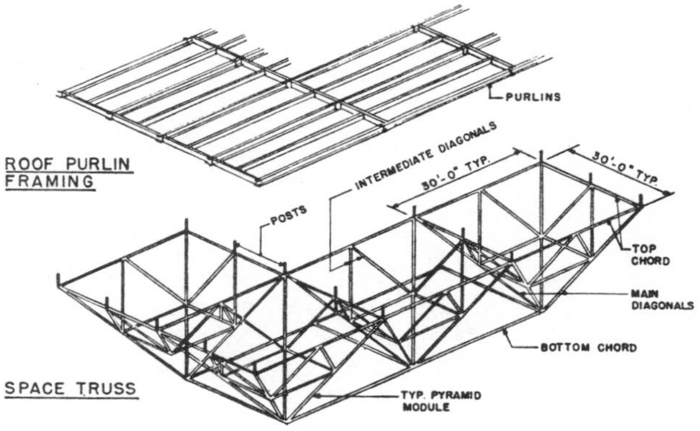

FIG. 3
SPACE TRUSS SYSTEM

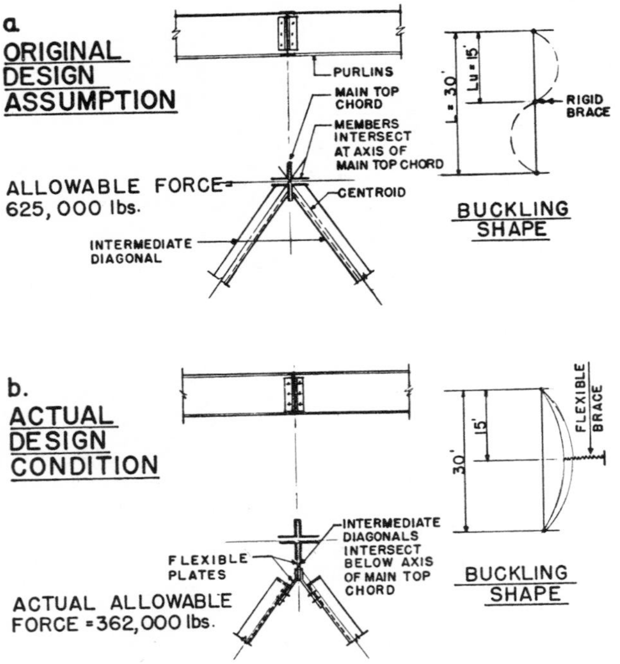

FIG. 5
TOP CHORD EAST-WEST DIRECTION
(EXAMPLE OF TYPE C MEMBER)

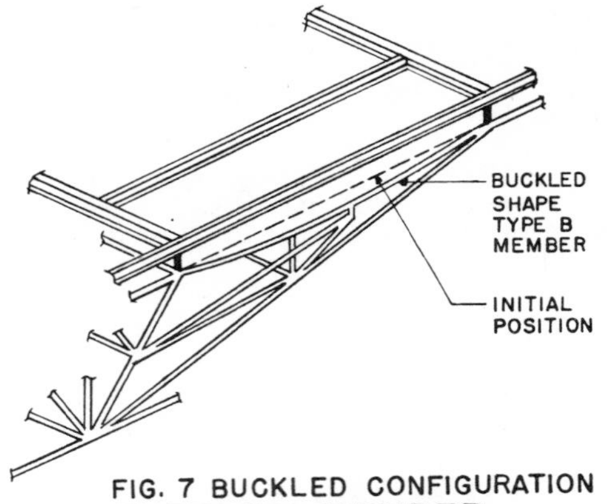

FIG. 7 BUCKLED CONFIGURATION
TYPE B - MEMBER

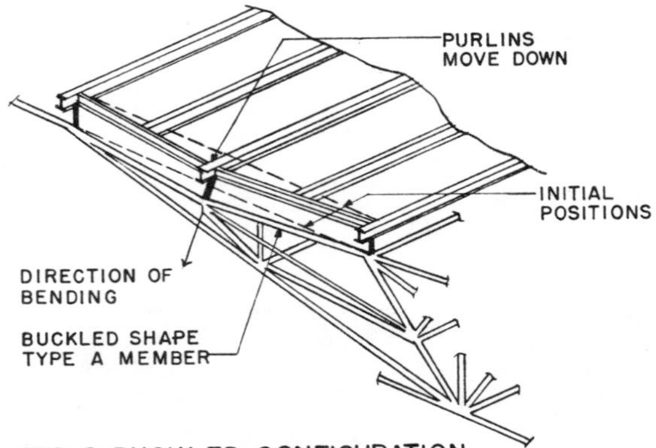

FIG. 8 BUCKLED CONFIGURATION
TYPE A - MEMBER

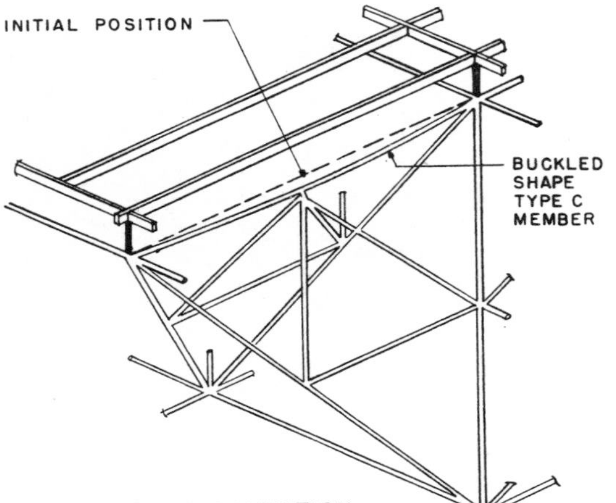

FIG. 6 BUCKLED CONFIGURATION
TYPE C - MEMBER

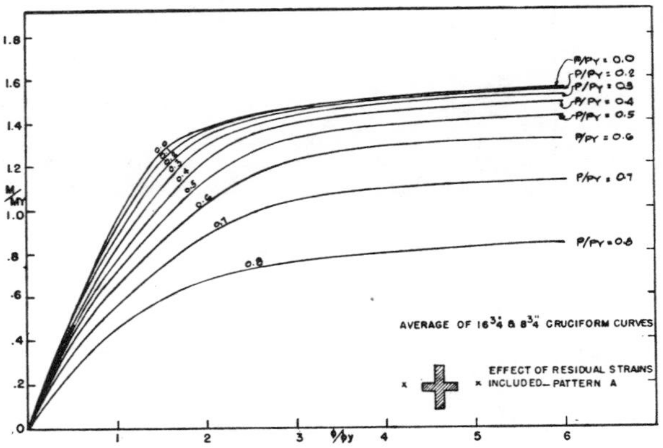

FIG. 9 AVERAGE MOMENT-AXIAL FORCE-CURVATURE RELATIONSHIP FOR CRUCIFORMS

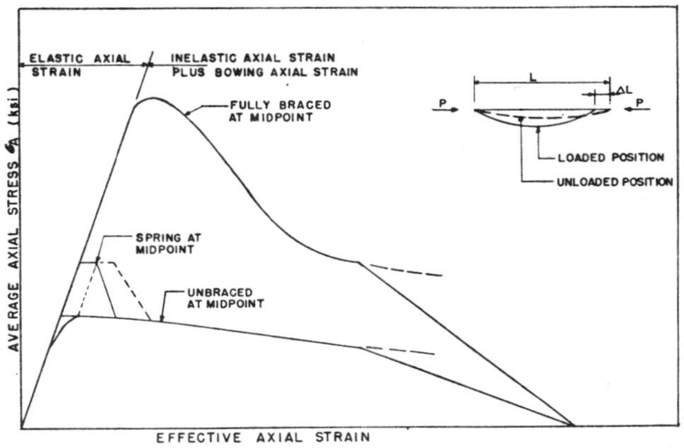

FIG. 10 AXIAL STRESS VS. EFFECTIVE STRAIN DIAGRAM

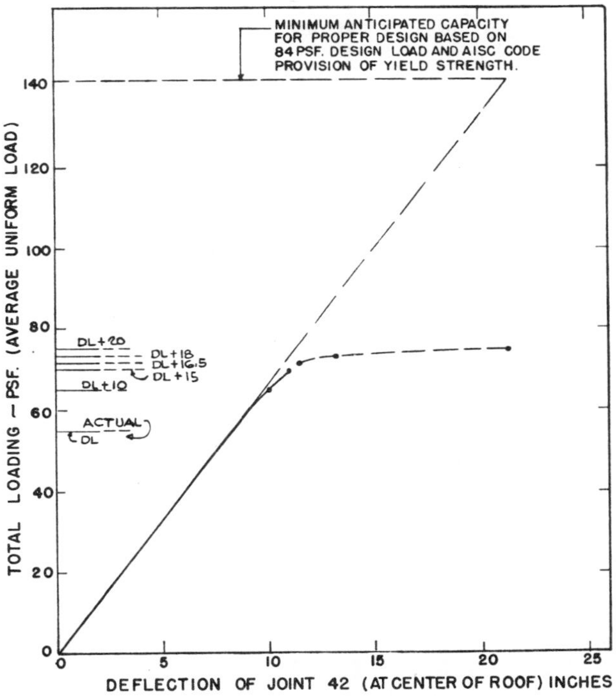

FIG. 11
TOTAL ROOF LOAD VS. DISPLACEMENT AT CENTER OF ROOF

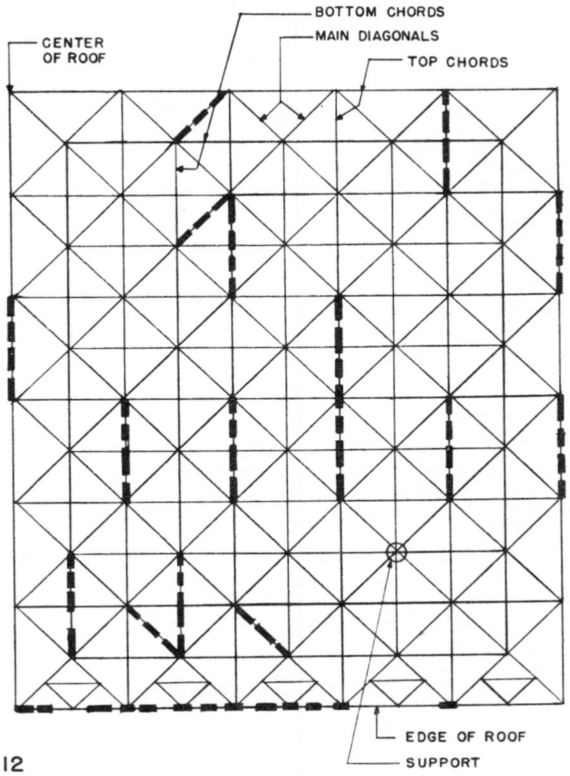

FIG. 12
BUCKLED MEMBERS AT LIVE LOAD = 18 P.S.F.

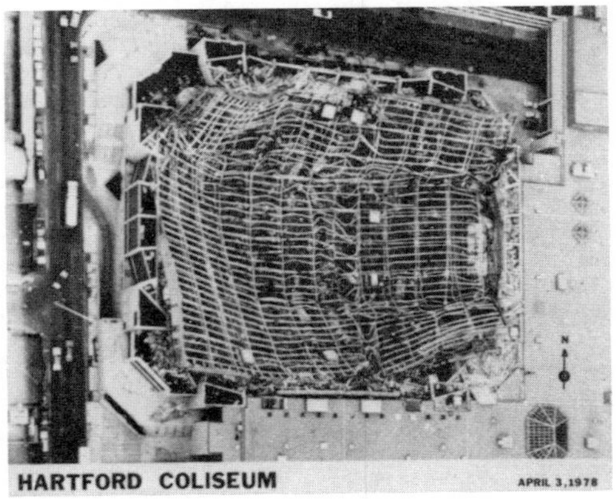

Fig. 13

DYNAMIC STABILITY OF THREE-DIMENSIONAL FRAME STRUCTURES

J. THOMAS* and B. A. H. ABBAS**

*Senior Lecturer, Department of Mechanical Engineering
University of Surrey, Guildford, Surrey. U.K.

**Associate Professor, College of Engineering

University of Basrah, Basrah, Iraq.

This paper presents a finite element solution for the dynamic stability problems of frame structures similar to the off-shore oil rig structures, subjected to periodic vertical and horizontal forces. Complete generality is maintained in the description of the structure co-ordinates. Natural frequencies of vibration, static buckling loads and regions of dynamic instability are determined.

Results of two different types of alternate loading on a structure, one vertical and the other horizontal, are presented. Comparison of the dynamic stability characteristics of the structure between these two types of loading condition are given, and it is shown that while one type of loading causes severe dynamic instability the other does have a fairly large stable region.

The dynamic buckling failure in the structure is not necessarily similar to that of the static buckling and may differ from the fundamental static buckling mode depending on the loading conditions.

The analysis and the results presented would make the designer conscious of the importance of the study of dynamic characteristics, especially the dynamic stability characteristics for the integrity of the structures under hostile environments.

INTRODUCTION

A large number of off-shore drilling units have been constructed since 1950. Some of these units have been lost or suffered damages. The high failure rate was not acceptable and continuous efforts have been made to improve the intergrity of the design. Although these failures were by no means exclusively of a structural nature, structural inadequacies have figured greatly in a number of failures or near failures.

These structures are designed for a variety of loading conditions. The loads have to be known in great detail for any extensive structural analysis to be carried out. The proper balance between the accuracy of load evaluation and structural analysis is fundamental in any structural design, and when the finite element method is used it is the load evaluation which sets the standard. The wave forces present the greatest problem in this connection.

In the Gulf of Mexico, the platforms were designed to withstand the biggest wave that can be expected in a 50-year period. The same reasoning was applied to the north sea, and some of the gas platforms in the southern sector were designed to withstand waves of maximum height of 49 feet and it is upto 100 feet in the northern sector of the north sea.

It is suggested here that the off-shore structures must be designed not only for the freak 50-year or 100-year wave, but for the continuous battering by the sea waves. The sea waves are assumed to be periodic in nature and their influence on the dynamic stability is investigated.

The problem of dynamic stability of plane frames has been investigated by Bolotin-see Ref.1. Stability coefficients and corresponding mass coefficients are determined for a number of predetermined system coordinates. The coordinate functions are taken as the static deflection forms of the frames under the influence of certain forces. These forces are assumed in such a way that the deflection forms sufficiently resemble the form of free or forced vibration of the frame. The construction of the basic equations for the chosen coordinates is preceded by many calculations. Furthermore, the choice of the coordinates may have a serious effect on the solution of the dynamic stability problems. For these reasons the range of application of the method reported by Ref.1 is almost limited to simple problems. A similar approach,with the same limitation was used by Roberts ,Ref.2, to study the dynamic stability of plane frames.

Gorzynski and Thornton, Ref.3, calculated a set of coefficients from which a stiffness matrix could be assembled for plane frames. The matrix was called the dynamic stability stiffness matrix. This terminology is inaccurate since the plane frames considered in Ref.3 was subjected to static loads only. Although the natural frequencies and the static buckling loads were extracted from the matrix, the problem of determining the regions of dynamic stability has not been considered.

This paper presents a finite element method for the dynamic stability analysis of three dimensional frame structures subjected to periodic axial loads or sea forces. Complete generality is maintained in the description of structure coordinates. Regions of dynamic instability, as well as, the natural frequencies and static buckling loads are determined.

The computer program developed is tested on a small scale model and the accuracy of the computed natural frequencies compared with the experimental results is illustrated.

FORMULATION OF ELEMENT MATRICES

For the beam element shown in Fig.1. the strain energy U is given by

$$U = \frac{1}{2}\int_o^L [AE(\frac{dU_z}{dz})^2 + EI_y(\frac{d^2Ux}{dz^2})^2 + EI_x(\frac{d^2Uy}{dz^2})^2 + GJ(\frac{d\theta_z}{dz})^2 \;]\; dz \quad ..1$$

where
E is the modulus of elasticity
A is the cross-sectional area
G is the modulus of rigidity
J is the St-Venant torsion constant
I_x and I_y are the moments of area of cross-section about the principal axes xx and yy respectively.

The potential energy V of the element is given by;

$$V = \frac{1}{2}P \int_o^L [\; (\frac{dU_y}{dz})^2 + (\frac{dU_x}{dz})^2 + \frac{I_p}{A}(\frac{d\theta_z}{dz})^2 \;]\; dz \quad 2$$

where
P is the axial load
I_p is the polar second moment of area of cross-section.

The kinetic energy T of the element is given by:

$$T = \frac{1}{2}\int_o^L [\rho A(\dot{U}_x^2 + \dot{U}_y^2 + \dot{U}_z^2) + \rho I_p \dot{\theta}_z^2]\; dz \quad 3$$

where ρ is the mass density of the material.
Assuming cubic polynomial expressions for U_y and U_x and linear polynomial expressions for U_z and θ_z of the forms

$$U_y = \sum_{r=o}^{3} a_r z^r \;;\quad U_x = \sum_{r=o}^{3} b_r z^r$$

$$U_z = \sum_{r=o}^{1} c_r z^r \;;\quad \theta_z = \sum_{r=o}^{1} d_r z^r \quad 4$$

and substituting into Eqn.1,2,and 3 and replacing the coeffifeients a, b, c, and d by the nodal displacement of the element, the elastic matrix $[k_e]$, the geometrix stiffness matrix $[k_g]$ and the inertia matrix [m] can be determined. The explicit form of the matrices are given in Ref.4. The nodal displacement vector for the element is:

$$[\; U_{z_i}\; U_{y_i}\; U_{x_i}\; \theta_{z_i}\; \theta_{y_i}\; \theta_{x_i}\; U_{z_{i+1}}\; U_{y_{i+1}}\; U_{x_{i+1}}\; \theta_{z_{i+1}}\; \theta_{y_{i+1}}\; \theta_{x_{i+1}}\;]^T \quad 5$$

MATRIX EQUATIONS FOR SPACE FRAMES

The three-dimensional or space frame is made up of interconnecting beam elements. The overall stiffness matrix $[K_e]$, geometrix matrix $[K_g]$ and inertia matrix [M] for the frame structure can be assembled from the element matrices $[k_e]$, $[k_g]$ and [m] and the knowledge of the geometric coordinates of the interconnections of the members.

The equations of motion of free vibration of the three-dimensional structure is given by

$$[[K_e] - p^2[M]]\{\zeta\} = 0 \quad 6$$

where p is the natural frequency
and $\{\zeta\}$ is the assembled displacement vector of the structure.

The equation of motion for the static stability of the structure is given by;

$$[\;[K_e] - P^*\;[K_g]\;]\{\zeta\} = 0 \quad 7$$

where P^* is the static buckling load.

The equation of motion for the dynamic stability of the structure is given by

$$[\;[K_e] - (\alpha + \tfrac{1}{2}\beta)\;P^*\;[K_g] - \frac{\omega^2}{4}\;[M]\;\;]\;\{\zeta\} = 0 \quad 8$$

where α and β are fractions representing the static and time dependent components of the load respectively and ω is the disturbing frequency.

APPLICATIONS TO AN OFF-SHORE OIL RIG

This section presents the application of the finite element solution to the dynamic stability problem of frame structures similar to the off-shore oil rig subjected to periodic vertical and horizontal forces. To verify the accuracy of the computed natural frequencies of vibration , a small scale-model has been built and free vibration tests were carried out. The model made of mild steel is shown in Fig.2. Welding is carried out carefully to minimise distortions and the residual stresses. Although the size and the complexity of the model were reduced to cut down the cost and the time necessary to build the model, the comparison of the experimental results with those obtained from the computer program will confirm the validity of the program.

RESULTS AND DISCUSSIONS

The coordinate axes and the displacement coordinates of the element employed in this analysis are shown in Fig.1. Six degrees of freedom are assigned to each node of the element, thus allowing complete freedom of displacements.

Table 1 shows the first ten frequencies of free vibration of the oil rig model. The agreement between the theoretical and experimental values is within 4%. The accuracy in the buckling results would also be of the same order since the formulation of the matrices are very similar.

Figure 3 shows the fundamental modeshape of free vibration of the oil rig model. The three views show the front elevation, the end elevation and the plan of the deflected model. To make the plan view clearer only the members of the lower horizontal frame are shown. It can be seen that the major deflections take place in the horizontal frames,giving an ovalised modeshape.

Table 2 shows the first ten static buckling loads of the oil rig model subjected to vertical loads as the first case and to horizontal loads at the midframe as the second case. The middle horizontal frame represents in this analysis the horizontal struts just beneath the water line in the splash zone which take the bulk of the sea wave forces. It can be seen from the table that buckling occurs at a much lower horizontal loads than in the case of vertical loading.

Figure 4 shows the fundamental static buckling modeshape of the model subjected to vertical loads. Major deformations take place in the x-direction while the deformations in the y- and z- directions are relatively small. The modeshape for the three horizontal frames are similar and only the lower frame is shown in the figure. The fundamental static buckling modeshape of the model subjected to horizontal loads at the middle horizontal frame is shown in Fig.5. As expected, the major deformation takes place at this frame while the other deformations are relatively small.

Figure 6 shows the regions of dynamic instability of the model subjected to periodic vertical loads. The intersection of the first region with ω/p_1 axis defines the fundamental mode of free vibration. In this case the value of ω/p_1 is 2 and that of β is zero. Inserting these values into Eqn.8 results $|[K_e] - p^2\;[M]| = 0$, which is the equation of free vibration. The fundamental static buckling mode is defined by the intersection of the first region with the β axis. This gives the value of ω/p_1 to be zero and that of β to be 2. Substituting these values into Eqn.8 gives $|[K_e] - P^* [K_g]| = 0$ which is the equation of static stability. From the figure it can be seen that as the disturbing frequency become equal to twice the fundamental natural frequency of the structure, the greatest part of the region becomes unstable even under the influence of periodic forces of small magnitude.

Figure 7 shows the regions of dynamic instability when the model is subjected to periodic horizontal loads acting at the middle horizontal frame. The loading conditions in this case excite very few modes of instability and it is seen from the figure that the fundamental modeshape of free vibration and static buckling mode are very similar. This can be verified from the obvious similarity between these modeshapes as shown in Figs.3 and 4.

CONCLUSIONS

This method of analysis provides a means of obtaining the regions of dynamic instability of three-dimensional frame structures subjected to periodic forces. The present approach is successfully applied to the analysis of an off-shore oil rig structure subjected to periodic sea wave forces. Agreement between the computed and measured frequencies of the experimental model is found to be very good.

Mode	Analytical frequency (Hz)	Experimental frequency (Hz)
1	78.4	81.0
2	113.4	101.0
3	115.3	108.0
4	118.1	112.0
5	118.8	117.0
6	130.2	122.0
7	132.5	124.0
8	132.8	133.0
9	138.5	141.0
10	146.5	147.0

TABLE 1 FREQUENCIES OF FREE VIBRATION OF THE OIL RIG MODEL

REFERENCES

1. BOLOTIN,V.V. The Dynamic Stability of Elastic Systems., Holden-Day Inc.1964

2. ROBERTS,T.A. "On the Dynamic Stability of Structural Systems using Conjugate Gradient and Search Techniques",Ph.D.Thesis, University of California,Los Angeles,1971

3. GORZYNSKI,J.W. and THORNTON,W.A. "Dynamic Stability Matrix for Planar Beams" Journal of tyhe Engineering Mechanics Division, ASCE 100,1974

4. THOMAS,J. and ABBAS,B.A.H. "Dynamic Stability of Space Frames by Finite Element Method Proc. 2nd International conference on space structure, University of Surrey, 1975

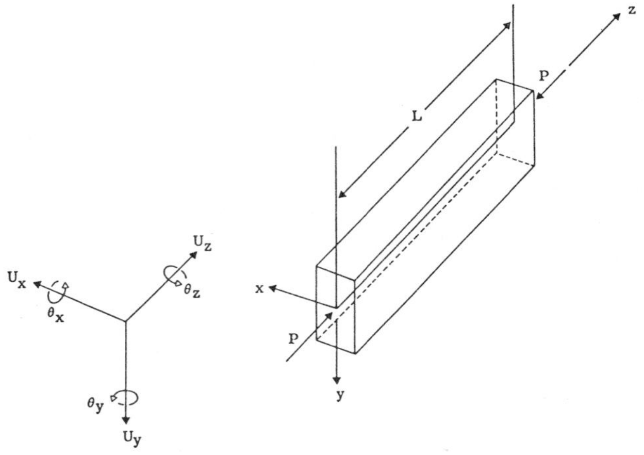

FIG. 1 A BEAM ELEMENT

Mode	Buckling loads case (1) lbf	Buckling loads case (2) lbf
1	103.2	26.0
2	105.7	89.6
3	106.3	89.7
4	106.8	100.9
5	108.7	106.2
6	110.8	109.6
7	112.1	113.3
8	116.4	142.0
9	131.0	149.5
10	205.5	151.8

TABLE 2 STATIC BUCKLING LOADS OF THE OIL RIG MODEL

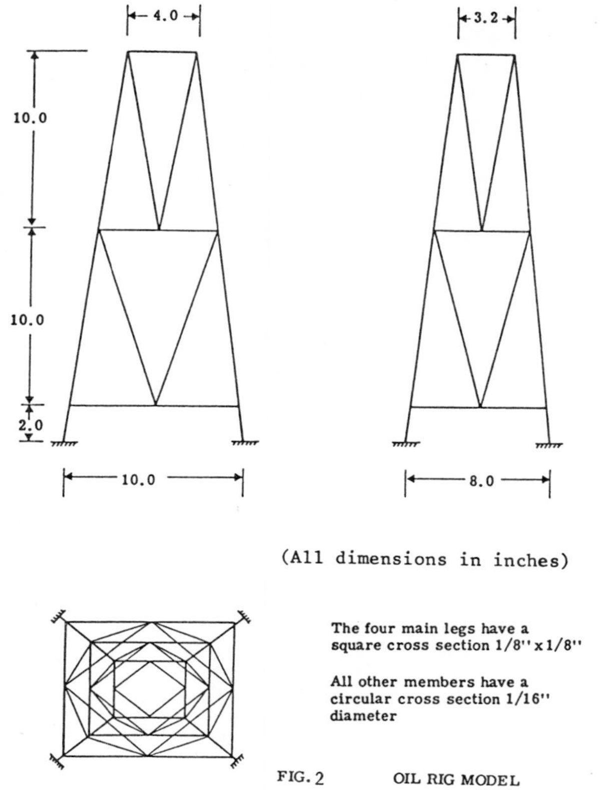

FIG. 2 OIL RIG MODEL

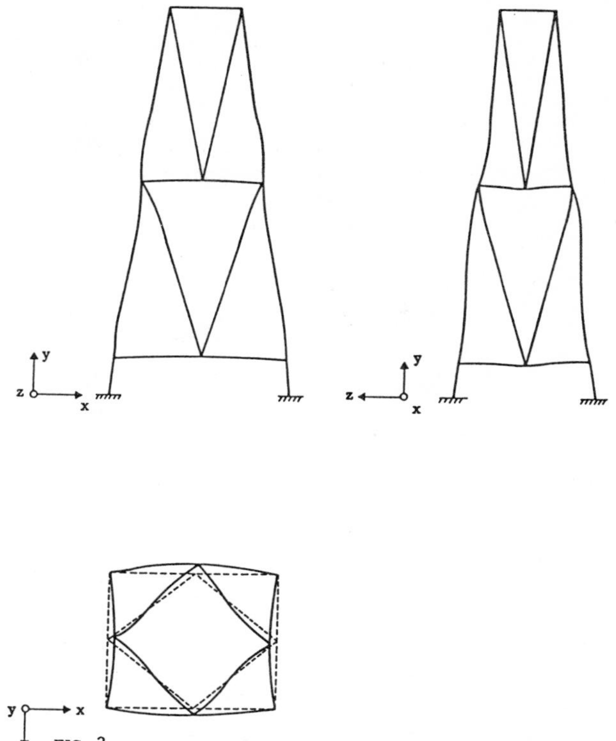

FIG. 3 FUNDAMENTAL MODE SHAPE OF FREE VIBRATION OF THE OIL RIG MODEL

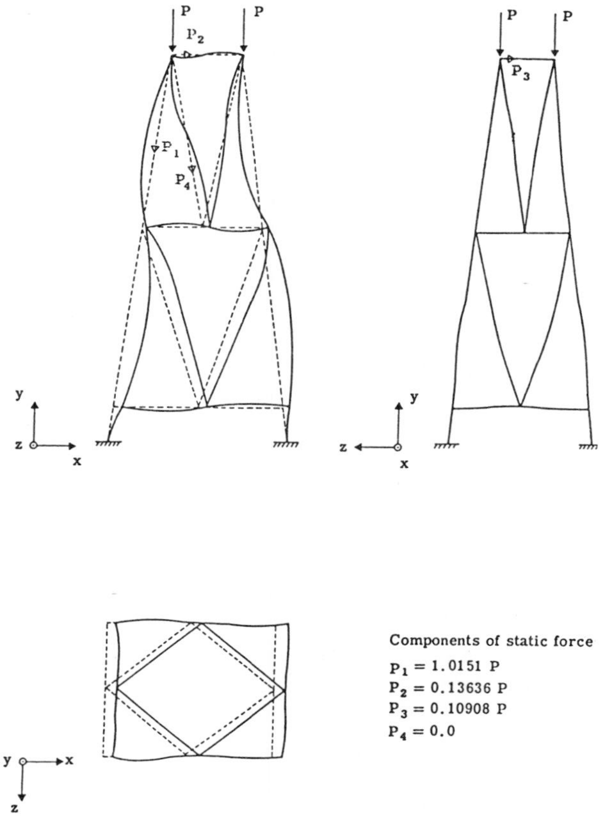

FIG. 4 FUNDAMENTAL BUCKLING MODE SHAPE OF THE OIL RIG MODEL SUBJECTED TO VERTICAL LOADS

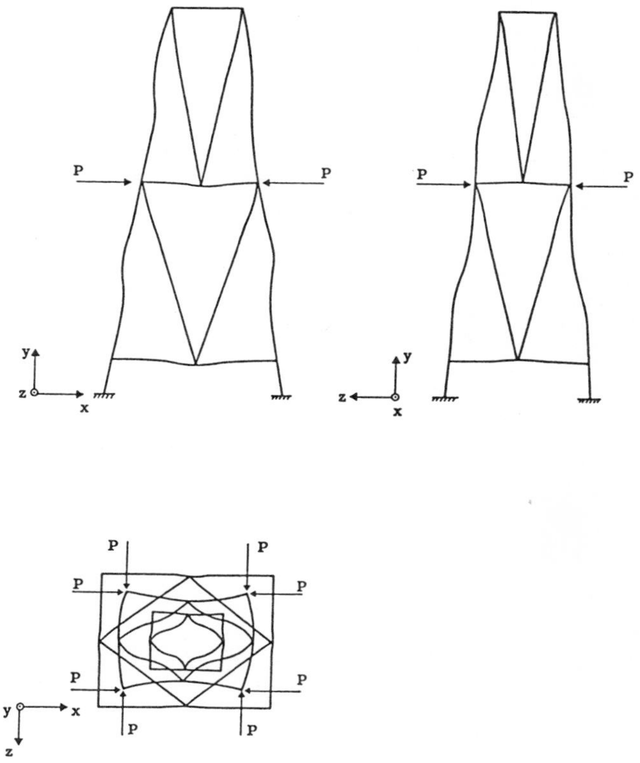

FIG. 5 FUNDAMENTAL BUCKLING MODE SHAPE OF THE OIL RIG MODEL SUBJECTED TO HORIZONTAL LOADS

$P_{(t)} = \alpha P^* + \beta P^* \cos \omega t$

$P^* = 103.2$ lbf Fundamental static buckling load

$p_1 = 78.4$ Hz Fundamental natural frequency

ω = Disturbing frequency

$\alpha = 0$

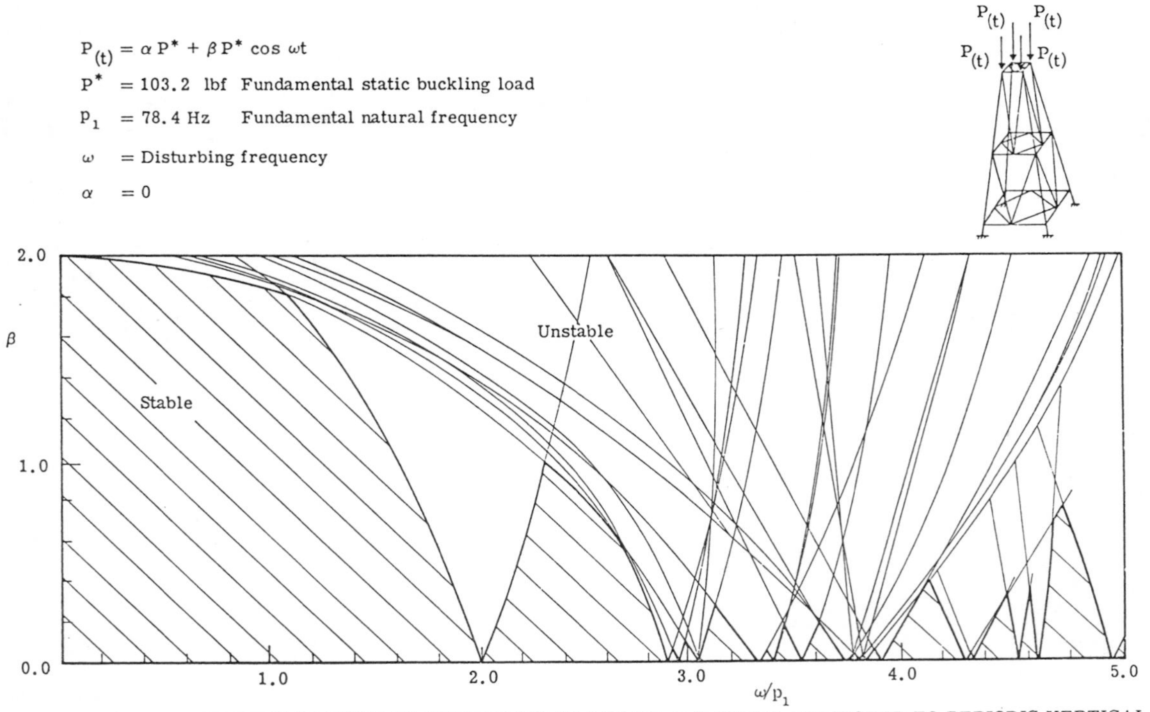

FIG. 6   REGIONS OF DYNAMIC INSTABILITY OF THE OIL RIG MODEL SUBJECTED TO PERIODIC VERTICAL LOADS

$P_{(t)} = \alpha P^* + \beta P^* \cos \omega t$

$P^* = 26.0$ lbf Fundamental static buckling load

$p_1 = 78.4$ Hz Fundamental natural frequency

ω = Disturbing frequency

$\alpha = 0$

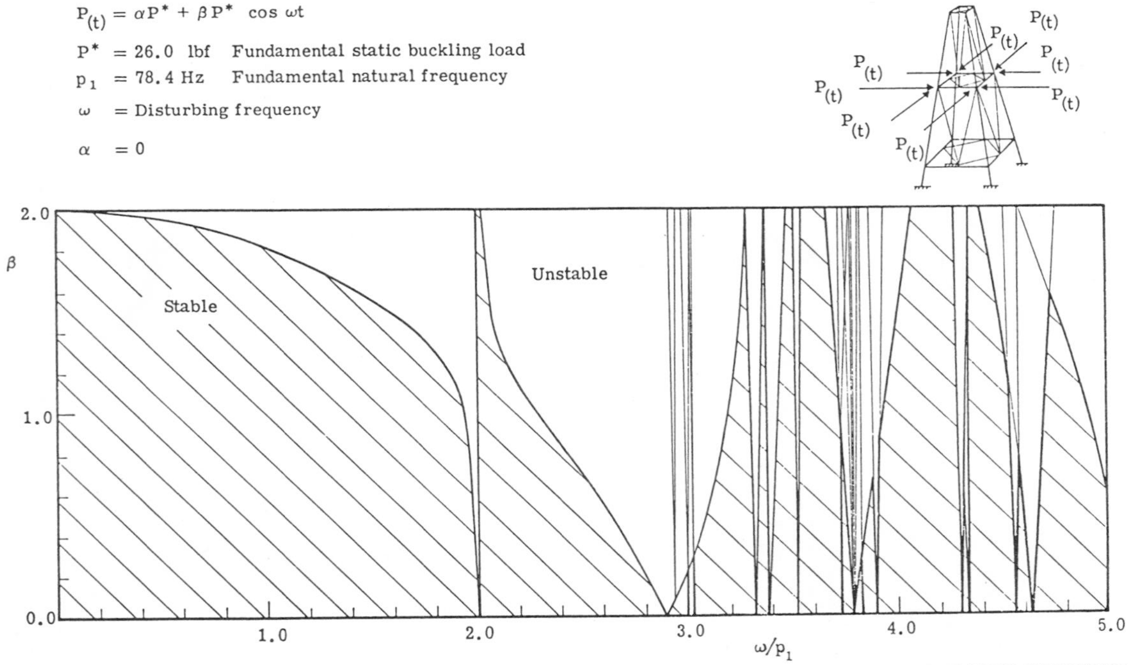

FIG. 7   REGIONS OF DYNAMIC INSTABILITY OF THE OIL RIG MODEL SUBJECTED TO PERIODIC HORIZONTAL LOADS

A REVIEW OF COLLAPSE ANALYSIS OF SPACE STRUCTURES

P Mullord

Space Structures Research Centre
Department of Civil Engineering
University of Surrey

This paper reviews the various approaches currently used in the collapse analysis of skeletal structures. Although the plastic mechanism and elastic stability approaches are mentioned, the paper is mainly concerned with the numerical approach to collapse analysis based on the stiffness method. The paper is intended for practising engineers and postgraduate students who are new to this subject.

INTRODUCTION

We can categorize the various techniques used for the non-linear analysis of skeletal structures into three main approaches. These are the Plastic Mechanism approach, the Elastic Stability approach and the Numerical approach. This paper reviews the application of these three approaches to the collapse analysis of skeletal structures with an emphasis on the use of the Numerical approach. The effect of initial stresses and deformation, the statistical approach to collapse analysis and the calculation of shakedown loads are also discussed.

PLASTIC MECHANISM APPROACH

Essentially this involves finding a plastic mechanism within a structure such that the structure is in equilibrium and no member carries a load greater than its squash load. A particular structure may have various mechanisms and the analyst must find the mechanism corresponding to the lowest applied load.

The plastic mechanism technique is best known when used as the yield line method for slabs or for the plastic design of portal frames. In these circumstances, the analyst normally chooses the plastic mechanism using past experience, the procedure can however be automated using the techniques of 'mathematical programming' or 'optimization'.

A plastic mechanism analysis can only accommodate stability effects in a very approximate manner. The buckling of a strut can be approximated rather crudely by using a lower yield stress in compression but this technique has considerable limitations. The buckling stress of a member varies with the effective length of the buckling mode which can be difficult to calculate if the mode involves several members, also the analysis assumes that the member continues to carry is squash load as the axial strain is increased. This assumption is difficult to justify when applied to space structures in general. The plastic mechanism approach should really be restricted to structures such as single layer structures where failure is dominated by member bending.

The solution of the mathematical programming problem is not trivial and can be very expensive. At present the plastic mechanism approach is not widely used for the analysis of skeletal structures.

ELASTIC STABILITY APPROACH

Stability analysis concerns the calculation of certain critical states at which the stiffness of a structure with respect to a small disturbing force becomes zero or negative. The stability problem is considerably simplified if we make the assumption that all displacements are insignificant when compared with the dimensions of the structure. This leads to the 'initial' or 'linearized' stability problem. If we assume that displacements are significant, which is essential for some problems, then the analysis is called a 'non-linear' stability problem.

Engineers are generally most familiar with the initial stability problem. This can be posed as a linear eigenvalue problem which has n solutions where n is the number of degrees of freedom considered in the analysis. The lowest few critical loads and their corresponding vectors or buckling modes can be computed at a reasonable cost.

An initial stability analysis will not necessarily produce the correct critical loads, in particular it will not detect snap through type critical loads at all. In general, the equilibrium and kinematic equations must refer to the deformed geometry of the structure immediately prior to buckling. A solution to this non-linear stability problem can be categorized as either a 'limit' type critical load, as with the snap through situation, or as a more conventional 'bifurcation' type critical load which involves switching from one equilibrium path to another in load/displacement space. As the solution of the non-linear problem is based on the deformed geometry, this problem can only be solved by using some sort of numerical approach.

All critical loads need to be interpreted with care. In practice some structures fail at loads well below their theoretical critical load while other structures carry loads which are well above the theoretical load. This leads to interest in post-buckling behaviour, imperfection sensitivity and the magnitude of initial imperfections. Useful information can only be obtained for particular structural configurations while in general critical loads of skeletal structures must be interpreted with care.

Plastic yielding can only be considered in a stability analysis very approximately by reducing the modulus of elasticity.

NUMERICAL APPROACH

The numerical approach to non-linear analysis uses incrementation and iteration to accommodate the changing stiffness of a structure as loading is applied. The iteration technique involves repeated improvements to displacement and stresses due to a given loading so as to reduce the error in the equilibrium equations. Unfortunately, there are circumstances when the iterations diverge and the analysis then fails.

With the incrementation technique the loading is applied in a number of small steps. As each step is applied the stiffness is modified according to the current stresses and displacements in the structure. With a pure incremental scheme the structure slowly drifts from equilibrium as the structure is loaded. This can be prevented by the use of iteration within each increment but this introduces the possibility of divergence of the iteration.

In a numerical non-linear analysis the kinematic and equilibrium equations can be established using the geometry of the deformed structure, the geometry of the undeformed structure or the geometry of the structure at the beginning of the current increment. The coordinate frame in which the equations are expressed can also refer to either one of these three geometries. If the kinematic and equilibrium equations refer to the deformed structure and the coordinate frame also refers to the deformed structure, then the analysis is called 'Eulerian'. An 'Updated Lagrangian' analysis is similar except the coordinate frame refers to the geometry of the structure at the beginning of the current increment, while a 'Total Lagrangian' analysis is also similar except the coordinate frame refers to the geometry of the undeformed structure.

Numerical non-linear analysis of general skeletal structures is invariably implemented on computers using the stiffness method of analysis.

ANALYSIS OF A SPACE TRUSS

The non-linear analysis of a space truss is a fairly straightforward procedure. The load/deflection curve for a pin ended strut can be obtained from either experiment or a sophisticated numerical calculation and then incorporated within a standard non-linear space truss program. A full non-linear analysis can be obtained at about five times the cost of a linear elastic analysis.

Engineers are accustomed to analysing rigid jointed, triangulated structures as pin jointed structures as the rigid joints have very little effect on the distribution of forces in the structure. However, the load carrying capacity of a strut varies considerably with the rotational end conditions which depend inturn on the geometry of and stress distribution in the adjacent members. Consequently a non-linear analysis of a rigid jointed triangulated structure must include the effect of the rigid joints.

ANALYSIS OF A SPACE FRAME

The non-linear analysis of a frame is a relatively complicated and expensive procedure. The procedure must consider the actual shape of a buckled bar and must also accommodate the progressive onset of plasticity through the cross-section and along the member.

The buckled shape of a rigidly jointed strut cannot be even approximated by the single cubic polynomial used in a linear analysis. A single member must be modelled by using either at least two cubic elements joined inline or alternatively a special element must be devised using at least one internal node.

To model the effect of material yield, the member must first be divided along its length into sub-elements. Within each sub-element the spread of plasticity through the cross-section is assumed to be constant. There are two distinct methods in use for handling the partially plastic cross-section. The first method subdivides the cross-section into sub-elements and assumes constant stress within in each subdivision. The second method uses an approximate yield criteria and flow rule expressed directly as a function of the stress resultants, that is the axial force, the bending moments, etc. This second method cannot accommodate torsional buckling and a different stress resultant yield criteria is required for different cross-section geometries. The use of this method can however lead to useful savings in computer time, storage requirements and data preparation.

The effects of cross-section distortion and warping are outside the scope of this review.

Non-linear analysis of a rigid jointed structure is clearly a relatively expensive procedure at present but programs are available and in use.

DYNAMIC STABILITY

Dynamic stability of space structures usually refers to the snap through buckling situation. The snap through phenomenon can be modelled as a transient analysis using small increments of time. At each time increment the equilibrium force residual is accommodated by accelerating the structure in space. Most snap through buckling problems can now be analysed using variable load loading strategies.

LOADING STRATEGY

If the loading is applied in increments of increasing applied load then the analysis breaks down as the maximum applied load is approached. The analyst cannot be sure that the maximum has been reached and the post maximum load response is not available. It is possible to increment past the maximum load if the structure is loading using prescribed displacements. This introduces however the severe restriction that the various prescribed displacements must be in a known ratio to each other. The latest loading strategies allow the magnitude of the applied load to vary as an additional degree of freedom and thereby permit the analysis to continue past the maximum load. This technique will also handle any snap through buckling mode which is significant enough to effect the overall response of the structure.

INITIAL IMPERFECTIONS

A numerical non-linear analysis will normally fail if the structure reaches a bifurcation point in the equilibrium path. To overcome this it is normal to give the structure an initial deflection in the direction of the anticipated collapse mode. If a fairly large initial displacement is used then this will make the analysis cheaper and also give a conservative estimate of the capacity of the structure.

If the computer power is available it is possible to run several analyses with various initial imperfections and thereby study the imperfection sensitivity of the structure concerned. It is also possible to input random imperfections and to study their effect on the collapse load.

SHAKEDOWN LOAD

Interest in collapse loads has been encouraged by the introduction of the limit state philosophy of structural design. There are however some reservations concerning the usefullness of collapse as a limit state because of the large and permanent deformations involved. The shakedown limit state is possibly more appropriate than collapse as plastic yielding is restricted to the first few applications of each load combination. Shakedown analysis is not at present generally available but in the meantime, a linear elastic analysis with suitable permissible stresses provides a conservative approximation.

NETWORK OPTIMISATION OF CURVED SPACE STRUCTURES

R. E. McConnel, B.E., M.E., D.Phil., MICE

Lecturer, Department of Engineering,
University of Cambridge,
U.K.

A general, rational approach for the generation of optimal networks for curved, pin-jointed space structures is presented. Optimal networks are here defined in terms of a minimum piece list of member lengths, node types, and cladding shapes. Optimisation is achieved by covering the given structural surface with a minimum number of different member lengths. Various possible grids are tried, and the optimal network chosen on a trial-and-error basis. The resulting nodal positions will be defined in various ways, the most common being to lie on the given surface and to be two known lengths from two known nodes. These various definitions have been incorporated in a computer program which establishes the nodal coordinates by an appropriate sequential calculation. This program is described, and the generation of various example geometries is presented.

INTRODUCTION

When consideration is given to the use of space structures for applications other than the well established double layer flat networks, a number of geometric and structural problems arise. The work reported in this paper was directed towards optimising the geometry of space structures acting primarily in compression, i.e. domes. However, there is no obvious reason why the basic approach outlined could not be applied to optimising tension structures.

Curved networks require freedom of joint angle, and with commercial space frame systems, this freedom is often associated with pin-jointed connections. Hence, the emphasis in this paper is on pin-jointed space structures, and unless otherwise noted, the terms network or structure in this paper refer to the pin-jointed form.

Any plane single layer pin-jointed network is obviously a mechanism, and (double) curvature must be introduced to provide out-of-plane stiffness. However, if the curvature is minimal (or is reduced by deflections), such structures are potentially unstable. Further a single layer pin-jointed compression structure must be determinate within its own surface to avoid in-surface mechanisms. This is usually ensured by triangulating the network. The conventional approach for calculating the stability of systems with bending continuity at the joints has been to use known thin shell solutions, Ref. 1. However, when pin joints are considered, this approach is invalid, and large deflection calculations are required for both the structural calculations and for the stability analysis. Fortunately, with the aid of modern computers large deflections calculations are now practical (e.g. Ref. 2), so that stable single layer alternatives to double layer structures are now a real possibility.

One major disadvantage of curved structures is that they will have generally increased piece lists compared with plane alternatives, i.e. more member lengths and cladding panel shapes in addition to the node types. A primary objective of any network genreation method should therefore be the production of a minimum piece list. In the past, the absence of a general method for the selection of an optimal piece list has made the rational consideration of curved space frames very difficult. To help resolve this problem, a new approach using member lengths to define node coordinates is presented in this paper. The method will be illustrated by application to some of the more common structural forms.

SELECTION OF THE OVERALL STRUCTURAL FORM

For most structures, the ground plan and general shape (elevation) are specified by the architect from functional and aesthetic considerations. Given the basic shape, there are two possible areas where the network may be optimised.

Single layer networks. Use of a single layer network is basically a question of stability, as many pin-jointed single layer structures are unstable. In such situations, and where an alternative shape is not possible, a double layer network must be employed. Generally a single layer network will be lighter and will have a smaller piece list than a double layer alternative. However, these advantages will be at the expense of larger deflections, more critical support conditions, and possible erection difficulties.

Simplified elevation curves. Generally, the simpler the elevation curve used, the smaller will be the piece list, e.g. a circle versus a parabola. Even if the elevation is specified, use can be made of this principle by replacing a higher order curve with a series of lower order ones, e.g. a parabola can be replaced by a series of circular arcs.

DIVISION OF THE CURVED SURFACE INTO FUNDAMENTAL REGIONS

With the structural form determined, the next step is to establish global symmetry lines on the curved surface, such that the surface is cut into a minimum number of *fundamental regions*, preferably of one type. For example on a sphere, these lines may be the arcs of any great circle. In many cases, triangular fundamental regions are sought as they readily subdivide into triangular (potentially stable) networks.

SUBDIVISION OF THE FUNDAMENTAL REGIONS BY MEMBER LENGTH

Once the structure has been divided into its fundamental regions, the further subdivision of these to produce a network of reasonable length members must be made. As noted in the introduction, for stability reasons this network is normally triangulated in the surface, and only networks of this type will be considered in this paper. It is of course impossible to cover a general surface with a triangulated network of equal length members. Further, there appears to be no direct, general method of achieving even an optimal network. Most published work concerns the spherical surface, and a survey of possible solutions for spherical domes is given by Makai and Ternai in Ref. 3 where the solutions discussed are those of the authors, and of Fuller. In Ref. 4, Emde has established an optimal network for 8 subdivisions of the icosahedron triangle, although this is not the best possible solution, as will be demonstrated later in this paper.

In contrast to the above methods, which generally use complex surface properties to achieve optimisation, the process of subdivision proposed in this paper is to define the nodal coordinates simply in terms of member length. By *trial and error*, a best (i.e. optimal) solution with a minimum number of different member lengths compatible with the surface being considered is found.

Member length is used as the structural element to be (directly) optimised, as experience has shown that reducing the number of member lengths also reduces the number of node types and panel shapes. Intuitive support for this approach can be obtained by considering a network on a spherical surface. In this case, members of equal length will sustain equal angles to the surface normals at each end. It follows that a network defined in terms of member length will generate as many node types as there are nodes surrounded by a distinct sequence of member lengths. Also, the number of panel shapes will equal the number of polygon types in the network. It clearly follows that if a regular network is established on a surface, minimisation of the number of member lengths will also control the number of node and panel types. A further, and perhaps the most important reason for optimising member lengths rather than node or panel types, is that member lengths are much easier to visualise in space than either node types or panel shapes. This is an important point when visualizing possible networks.

The number of networks to be considered by this trial and error approach increases with the number of side subdivisions of the boundary lines of the fundamental regions, and with the complexity of the panel shape. More possibilities are introduced if one of the boundary lines is not a symmetry line, but is either a free edge, or constrained in some manner, e.g. to lie on the ground.

Coordinate Specification

The location of a unique point in 3 dimensional space requires the specification of 3 independent coordinates.

For a general point, an individual coordinate can be fixed (e.g. on a boundary), or it can be determined by an appropriate independent equation. Any point can therefore be defined by a suitable combination of 3 fixed values and/or equations.

Generally, one of these equations will define the curved structural surface. Two further possible equations, fixed length from a given point, and constraint to lie a given distance from a defined plane, are now discussed.

Unknown point G (coordinates g_1, g_2, g_3,) is to be set a fixed distance AL from a given point A (coordinates a_1, a_2, a_3). By the standard equation for length between two points.

$$(g_1 - a_1)^2 + (g_2 - a_2)^2 + (g_3 - a_3)^2 - AL^2 = 0 \quad .....1$$

is the required relationship in G (g_1, g_2, g_3).

In 3 dimensional space, the general equation of a plane can be written,

$$Pa_1 + Qa_2 + Ta_3 - U = 0$$

where P, Q and T are arbitrary constants, U = 0 for planes through the origin and 1 for all other planes, and A (a_1, a_2, a_3) is a general point on the plane. For unknown point G (g_1, g_2, g_3) to lie distance PL from this plane, the required relationship is

$$Pg_1 + Qg_2 + Tg_3 = U \pm PL\ (P^2 + Q^2 + T^2) = 0 \quad .....2$$

Normally, the plane will be defined by 3 non-colinear points which establish P, Q, T and U.

Equations 1 and 2 have been used extensively to produce a wide range of practical structures - particularly Eqn. 1 which can be used directly to define spherical surfaces. Additional equations are obviously possible (e.g. parabolic surfaces), but have not yet been investigated.

Coordinate Calculation - Solution of Coordinate Definition Equation Sets

There are many possible combinations of fixed coordinates, surface equation, fixed length from a given point, and plane constraint sufficient to determine the coordinates of a general point G.

Only the following situations (which have arisen during actual problem solution) have been considered to date: node located by 3 lengths, node located by 1 known length and 3 unknown but equal lengths, node located by 2 lengths and 1 plane, node located by 1 length and 2 planes. For these 4 combinations, the 3 equations involved have been solved explicitly for g_1, g_2 and g_3 (the 3 coordinates of G). The final equation is generally quadratic (see Eqn. 1), so that alternative roots (solutions) must be considered.

Computer Program

Once a network has been selected, the equation sets described above can be used for the sequential calculation of the node coordinates. For any but the most trivial structures, these calculations are only possible with the computer, and a suitable program has been developed.

EXAMPLE CALCULATION OF THREE OPTIMAL NETWORKS ON SPHERICAL SURFACES

The generation of three different optimal triangular networks on a spherical surface is now considered. This surface was chosen as it is a standard situation which is well documented (Refs. 3 and 4), and because the spherical surface is easily established with Eqn. 1.

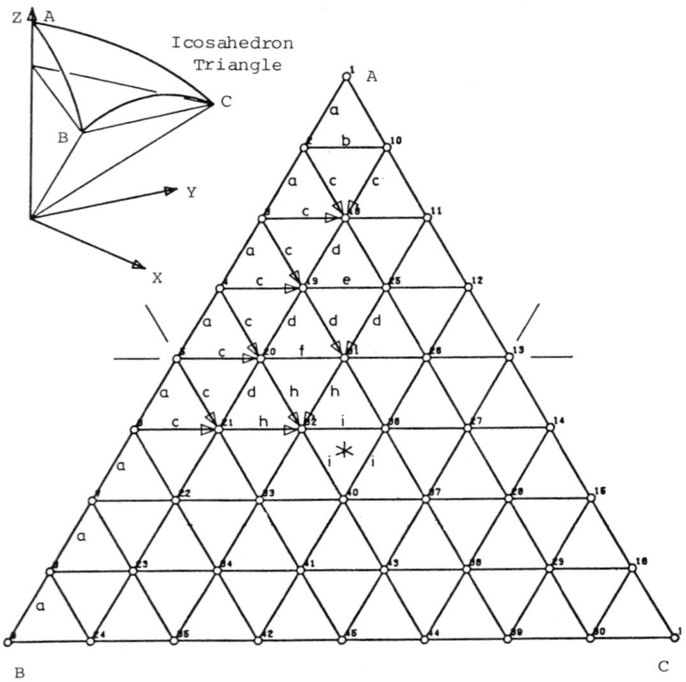

Lengths 9, Nodes 10, Panels 11.

Fig 1 C3/8 Subdivision of an Icosahedron Triangle

Number of lengths L, nodes N, and panels P for i subdivisions of icosahedron side:

$$L = i$$

$$N = i + \frac{(i-3)}{3}$$

$$P = i + \frac{(i-1)}{3}$$

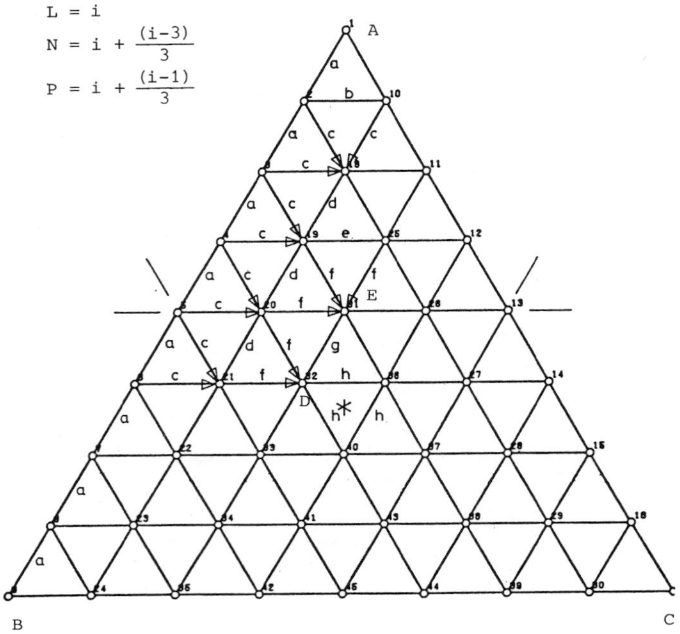

Lengths 8, Nodes 9, Panels 10.

Length range 1:1.23

Fig 2 C4/8 Subdivision of an Icosahedron Triangle

Subdivision of the Full Sphere

The classical icosahedron division of the sphere into 20 spherical triangles, each being identical, are the obvious fundamental regions to use if a large portion of a sphere is to be used as a dome.

To subdivide the fundamental regions, a triangular network is drawn on a typical icosahedron face, and trial member length arrays established, e.g. Figs. 1 and 2. Consideration here is restricted to 8 subdivisions of the icosahedron boundaries although the method is general. The network of small triangles in these drawings show the subdivision of the fundamental region, with letters a,b,c ... for member lengths. Arrows converging on a particular node indicate the members used to establish the coordinates of that node.

The requirements of geometric compatibility restrict the member length arrays, but there are still many possibilities. Optimising for a minimum number of different member lengths by trial and error, the layout shown in Fig. 2 (system C4) was the best solution found. This solution also had the least number of node and panel types of all those investigated, and it is applicable to any number of subdivisions of the icosahedron triangle - the piece list equations shown in Fig. 2 can be easily developed.

System C4 is in fact the best solution (in the terms defined in this paper) for subdivision of the icosahedron triangle known to the author. Evidence that this is the best optimal solution comes from Makai and Tarnai (Ref. 1) who developed the same system in an earlier investigation using a completely different approach. The existence of Tarnai's solution was unknown to the present author when the work reported here was performed. For comparison, the piece lists of three different solutions to the icosahedron problem are given in Table 1. The totals shown include entries for all theoretical length differences, and for both normal and mirror image node and panel types.

Subdivision	No. of Lengths	Length Range	No. of Node Types	No. of Panel Types
Fuller	20	1:1.37*	12	22
Emde	10	1:1.17	12	14
C4 (Tarnai)	8	1:1.23	9	10

* Estimated value

Table 1 COMPARISON OF TRIANGULAR ICOSAHEDRON NETWORKS FOR EIGHT SUBDIVISIONS.

An example dome using the C4 subdivision is shown in Fig. 4.

The Subdivision of a Spherical Cap

A spherical cap can be divided into spherical segments by any number of great circles through its crown.

Optimisation of a triangular network within a segment by trial and error produced the E1 network shown in Fig. 3 as the best solution. This is basically an open ended version of the C4 system, and the shape of the free edge BC is developed by the internal pattern of lengths.

System E1 is not as efficient as system C4 when the latter can be used, but the E1 system does have the advantage of being applicable to flat domes with orders of symmetry other than 5.

Domes with Defined Boundaries

Domes with defined boundary shapes are a class of structure for which network generation by member length can give relatively easily calculated optimal solutions compared with other methods.

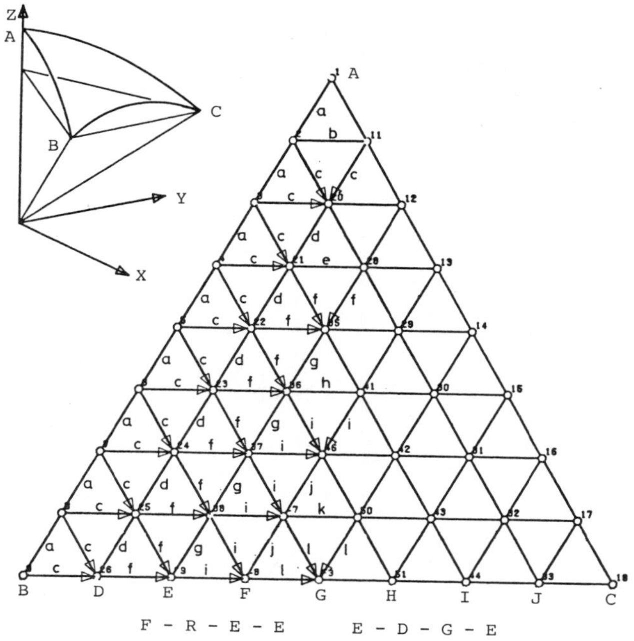

Lengths 12, Nodes 17, panels 15.

Length range 1:1.25

Fig 3 E1/8 Subdivision of an Icosahedron Triangle

For example, a spherical cap can be given a plane free edge (e.g. BC, Fig. 3) by requiring nodes E - G to be on the plane defined by nodes BDC. Such modifications introduce new member lengths, and this is a simple example of the unavoidable complications introduced by defined boundaries. A specification requiring a horizontal boundary BC introduces even more complication. However, optimal solutions are still possible, and the end (dome) section of the vault shown in Fig. 8 is a typical solution.

GEOMETRY GENERATION OF SOME STANDARD STRUCTURAL FORMS

Some results of the application of the method of the previous section to the optimisation of networks for some of the more common forms of curved space structures are now presented.

Barrel Vaults

a) Single Layer Barrel Vaults

As barrel vaults are developable surfaces (i.e. they can be formed from a plane) any plane networks can be curved into a vault. Such networks can have very small piece lists, but in the absence of bending stiffness they are mechanisms.

b) Double Layer Barrel Vaults

Any plane double layer network can be folded into a barrel vault of arbitrary cross-section by varying the length of either the top or bottom chord in the direction of curvature.

In particular, optimal barrel vaults can be generated with a circular cross-section by giving a uniform change in length to a chord in the direction of curvature. The vault so produced has only one new member length compared with the plane network, one more node type, and one more panel type.

The geometry generation method described in this paper has no great advantage when applied to this class of structure.

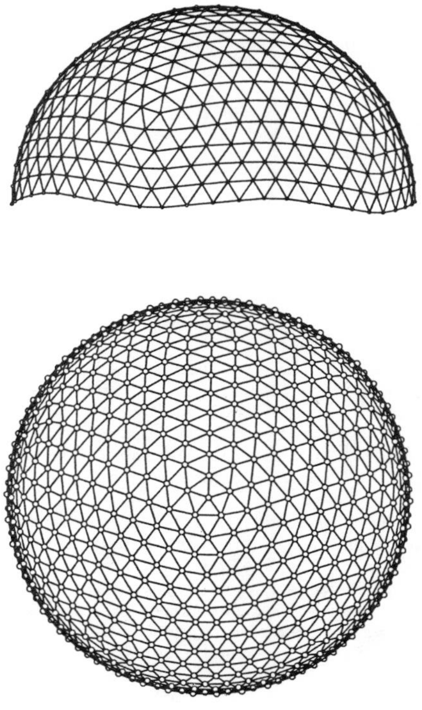

Fig 4 C4/10 Dome

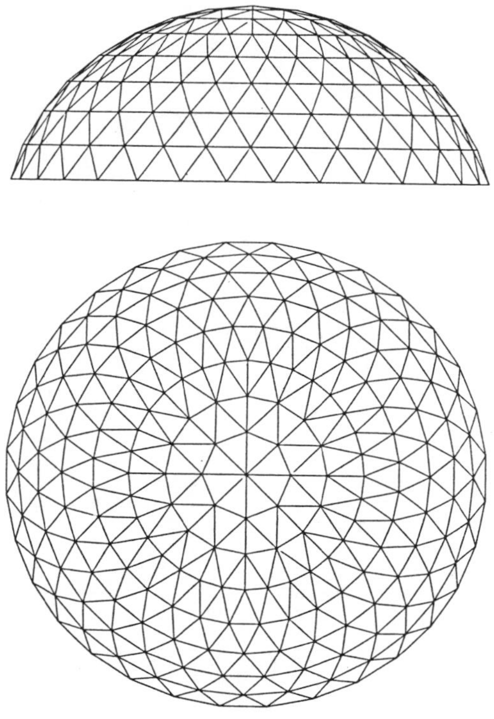

Fig 5 Ring-Net Dome

Barrel vaults of more complex section can often be optimised by replacing the true curve with a series of circular arcs.

Spherical Domes

a) Geodetic Single Layer Domes

The geometry generation of this class of structure has been discussed above in previous sections. A typical dome based on the C4 network of Fig. 2, but with 10 subdivisions to each boundary of the fundamental triangle, is shown in Fig. 4.

b) Single Layer Dome Caps

Shallow domes can be generated with the E1 network of Fig. 4.

c) Single Layer Ring-Net Domes

Ring-net domes can be readily optimised with the member length approach. For example, a geometry with constant length diagonals, and variable length ring members can be generated from the crown, Fig. 5.

It is possible that ring nets are inherently less stable than geodetic domes because of their greater member length range.

d) Double Layer Domes

The conversion of the single layer networks discussed above into double layer structures is not simple. Constructing dual networks (i.e. one layer from the other) is an obvious approach. The length method can be very useful in generating this form of network.

Non-Spherical Domes

a) Parabolic, Elliptic and Higher Order Surfaces

Unlike spherical surfaces, preliminary work has indicated that defining position by member lengths on more complex surfaces does not lead to explicit equations for the nodal coordinates. Although iterative solutions should be possible, this problem, and the fact that piece lists will be greater than those of spherical equivalents, indicate that the easiest, and possibly the best approach, is to use spherical approximations where possible.

b) Surface Shape Generated by Member Length

The concept of defining node coordinates in terms of member length can be used for the direct generation of non-spherical domes. Given two or more arbitrarily curved starting (symmetry) lines through the crown of the dome, a large number of length networks can be developed between these lines if the node points are not restrained to lie on some specified surface. The surface shape so generated will develop automatically from the length network.

This approach is perhaps best illustrated by a simple example. Consider the spherical solution, system C4 of Fig. 2, and define node D to be distance 'f' from the two nodes shown (as before), but to be distance 'd' from node E rather than on the spherical surface. Length type 'g' is now removed from the system, and the piece list has been improved at the expense of the spherical form of C4.

Three obvious problems arise with such an approach. Firstly, the number of possible solutions (already large for the sphere) becomes much greater with the additional freedom of a free surface. Secondly, and of fundamental importance, there is no guarantee that a solution will exist distance 'd' from E. Thirdly, although this approach reduces the number of member lengths and panel types, the number of node types will often increase.

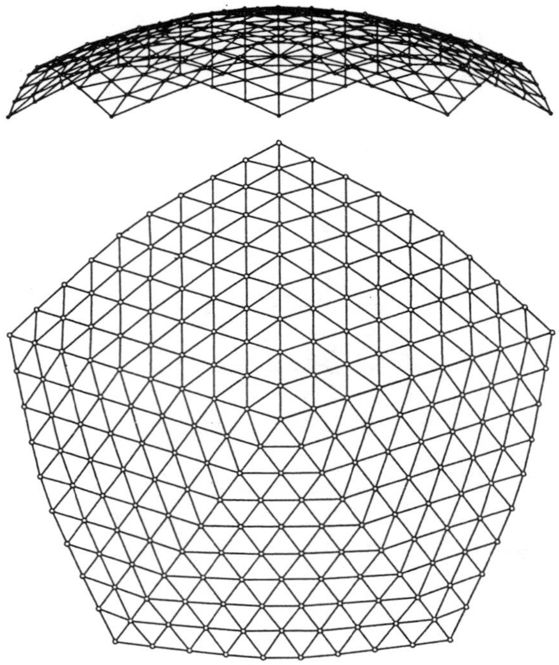

Fig 6 D12/9 Non-Spherical Dome

Applying this approach to networks developed from two 1/5th spherical symmetry lines of a standard shallow dome cap, produced many incomplete networks due to the defined lengths giving no solution for some nodes, and only two complete domes. The first of these was the single modification to the C4 system discussed above, and the second was the D12 system shown in Fig. 6. The D12 system has a larger piece list than the equivalent spherical C4 system (although it does require fewer panel types) and is a regular dome with a relatively high edge arch.

Domes on Rectangular Bases

Frameworks to cover rectangular areas will logically have rectangular panels. Pin-jointed single layer structures of this type are mechanisms, so the discussion of this section will be restricted to the generation of curved double layer networks based on the standard square-on-square space truss.

a) Domes Generated by Member Length

The member length coordinate generation approach can be used to generate double layer networks. For (statically) indeterminate networks with redundant members, this approach can be employed by generating a determinate sub-network (covering all nodes) and then completing the system with the remaining members of the necessary length.

As examples of this approach, two different determinate sub-networks of the standard square - on - square two layer space-frame are considered. If the generation is initiated from nodes lying on two orthogal symmetry lines, two (statically) determinate sub-networks are possible. Either all bottom chord and diagonal members, or all chord members and the diagonals in one direction, can be set equal in length. For the second case, the bottom chord lengths are set slightly shorter than the top chords to be compatible with the initial arcs. Both networks have been calculated with the initial nodes set on constant radius arcs. The dome-like structures produced have node angles variable throughout their 1/4 symmetry sections, so that every node in these regions is of a different type. The original square panels

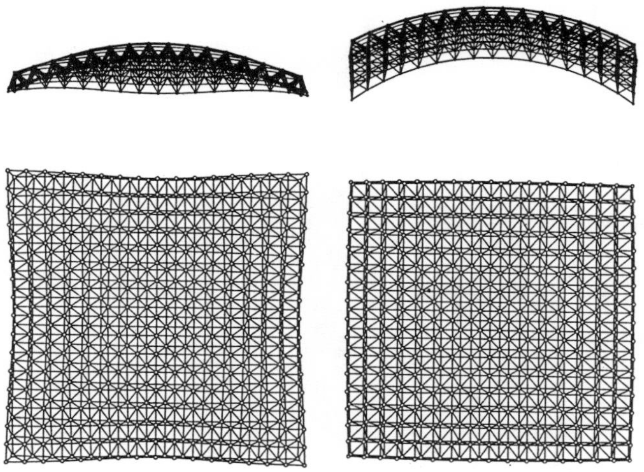

(a) Type A (b) Type B

Fig 7 Square Based Domes

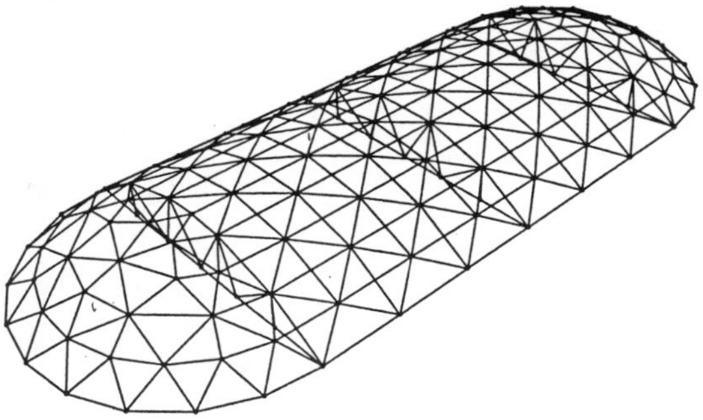

Fig 8 Dome-Barrel Roof

become progressively more distorted with distance from the initial rectangular symmetry lines, and with higher dome rise to span ratios. The first network generates a dome with "gull wing" edges which approximates a square dome with flat boundaries, Fig. 7(a). The second approach gives a regular dome with arched (almost circular) edges, Fig. 7(b).

b) Domes Generated by a Uniform Expansion (or Contraction) of One Chord

If the entire top chord is given a uniform expansion in both directions (or the bottom chord a contraction) it might seem that a dome with only two chord lengths can be generated. However, the dome developed is not stress free as the shape produced is not a simple surface, and the diagonals are of variable length. Although the approach may work for very shallow domes, any real elevation will produce prohibitive internal stresses. The true geometry for a stress free dome of this type can be obtained by sliding a vertical arc of given radius over an orthogonal arc of the same radius to form the chords. The diagonal lengths can then be calculated between the appropriate nodes.

That a simple expansion of the top chord does not produce a stress free dome is demonstrated by the fact that a 1% chamber rise developed by the correct approach produces a 6% range in the diagonal lengths. The original square chord panels become distorted parallelograms away from the central symmetry chord lines.

Dome Barrels

a) Single Layer Dome - Barrels

This is a very attractive form of structure, suitable for swimming pools, sports halls, etc. A very high level of network optimisation is possible, but unfortunately the barrel region is unstable. Various solutions to this problem are possible, such as providing stiff ribs and/or longer than normal members across the chords of the circular cross-sections, see Fig. 8. However, none of these solutions is entirely satisfactory.

b) Double Layer Dome - Barrels

The network optimisation for this class of structure is relatively simple, and they are stable.

Rotational and Hyperbolic Paraboloids

Both these networks contain straight line generators, so that optimisation by member length is of limited use. The rotational paraboloid can provide a smaller piece list than the hyperbolic type for the loss of some freedom in plan shape. Otherwise little optimisation appears possible, apart from using the basic symmetry whenever possible. The stability of this class of structure is an open question because of the straight line generators. The pin-jointed forms are mechanisms if the outer edges are not fully restrained, and it is possible that paraboloids act essentially as tension nets.

CONCLUSIONS

A new approach to the geometry generation of curved space frames based on member lengths has been presented. The application of this method to the well documented spherical surface reproduced the existing best network. In addition, this new approach permits the generation of optimised network for situations with additional geometry constraints, such as defined boundaries.

To date, only a small number of networks on specified non-spherical surfaces have been generated, and more investigations are planned in this field. Work on a parallel approach, that of defining length networks which generate their own surface shapes, is also only in the initial stages.

ACKNOWLEDGEMENTS

Much of the work reported in this paper was performed at MERO Raumstruktur, Würzburg, West Germany, and is reproduced with their kind permission.

REFERENCES

1. D. T. Wright. Membrane Forces and Buckling in Reticulated Shells. Journal of the Structural Division, ASCE, Vol. 91, No.ST1, Proc. Paper 4227, Feb. 1965, pp.173-201.

2. I. M. Kani, R. E. McConnel, T. See. The Analysis and Testing of Some Single Layer Shallow Reticulated Shells. Proceedings of the Third International Conference on Space Structures. University of Surrey, England, 1984.

3. E. Makai and T. Tarnai. On Some Geometric Problems of Single Layer Spherical Grids with Triangular Networks. Proceedings of the Second International Conference on Space Structures, University of Surrey, England, 1975.

4. H. Emde. Geometrie der KST:Struktur und Form. Course Notes, Geometry Seminar at MERO, Würzburg, April, 1977.

OPTIMUM DESIGN OF SPACE TRUSSES WITH BUCKLING CONSTRAINTS

M P SAKA* , MSc, PhD

* Associate Professor, Department of Civil Engineering

University of Karadeniz, Trabzon, Turkey

A practical optimum design algorithm of space structures subject to multiple load cases is presented. In the design problem displacement, stress, buckling and minimum size constraints are considered. The optimality criteria approach is used for the displacement constraints. The stress and buckling limitations are reduced to a lower bound on design variables and treated as similar to those of minimum size constraints.

INTRODUCTION

Structural optimization is now well established and widely applied to the solution of structural design problems following two decades of research. During this time, a large number of papers have been published dealing with a wide range of design problems, refs 1 and 2. The review of the algorithms presented so far reveals that there are mainly two different types of approaches to structural optimization problems from the practical engineering point of view.

The mathematical programming methods are extremly general and capable of handling any kind of structural design problem. Employing these algorithms such as linear, nonlinear, geometric or dynamic programming methods, it is possible to obtain the optimum cross sectional properties of members in a structure subject to stress, displacement, buckling and frequency constraints, see refs 3,4,5 and 6. Furthermore, it is also possible by virtue of these methods to take the configuration of the structure as a design variable which may be of special interest in design practice, see refs 7,8,9,10,11 and 12. However, the employment of mathematical programming methods in the design of large scale structures present some difficulties. Consequently, inspite of their generality, they become less effective in such design problems.

In contrast, the optimality criteria approaches, do not suffer from such drawback and their behaviours in obtaining the optimum solution do not change depending on the number of design variables. In these methods a prior condition is specified which is required to be satisfied in the optimum solution. This condition provides a basis in producing a simple recursion relationship for design variables, see refs 13,14,15 and 16. There are also other methods which make use of gathering respective advantages of both methods called mixed methods, ref 17.

In this paper, an optimality criteria approach for the optimum design of space structures under multiple loading cases is presented. The recurrence relationship derived from the optimality criteria for the displacement constraints is employed to obtain the new values of the design variables. The buckling constraints defined by various design codes are first expressed as nonlinear equations of design variables and these are then solved by Newton-Raphson algorithm to determine the lower bounds. As a result, it becomes possible to reduce the stress and buckling constraints to lower bounds on design variables.

OPTIMUM DESIGN PROBLEM

The optimum design problem of a pin jointed space structure can be written as follows :

$$\text{Min. } W = \sum_{k=1}^{ng} A_k \sum_{i=1}^{m_k} \rho_i L_i$$

subject to

$$\begin{aligned} &\Delta_{jl} \le \overline{\Delta}_{jl} && l=1,2,\ldots,nlc \\ &\sigma_{il} \le \overline{\sigma}^t_{il} \text{ or } \overline{\sigma}^b_{il} && j=1,2,\ldots,p \\ & && i=1,2,\ldots,nm \\ &A_k \ge \overline{A}_k && k=1,2,\ldots,ng \end{aligned} \qquad \ldots\ldots 1.$$

where A_k is the area of members belonging to grup k, m_k is the total number of members in group k, ρ_i, L_i is the density and length of member i, ng is the total number of groups in the structure. Δ_{jl} is the displacement of joint j and $\overline{\Delta}_{jl}$ is its upper bound under load case l, p is the number of restricted displacements, nlc is the total number of load cases. σ_{il} is the stress in member i, $\overline{\sigma}^t_{il}$ and $\overline{\sigma}^b_{il}$ is the allowable stress in tension and compression under the load case l, nm is the total number of members in the structure. $\overline{A}_k$ is the lower bound on the design variable k .

RECURRENCE RELATIONSHIP FOR DESIGN VARIABLES

If the optimum design problem given in eqn 1 is rewritten by considering the displacement constraints only the following is obtained.

$$\text{Min. } W = \sum_{k=1}^{ng} A_k \sum_{i=1}^{m_k} \rho_i L_i$$

subject to

$$g_{jl}(A_k) = \Delta_{jl} - \overline{\Delta}_{jl} \le 0 \qquad \begin{aligned} &l=1,2,\ldots,nlc \\ &j=1,2,\ldots,p \end{aligned} \qquad \ldots\ldots 2.$$

The Lagrange function $W(A_k, \lambda_{jl})$ for this problem can be written as :

$$W(A_k,\lambda_{jl}) = \sum_{k=1}^{ng} A_k \sum_{i=1}^{m_k} \rho_i L_i + \sum_{l=1}^{nlc} \sum_{j=1}^{p} \lambda_{jl}\, g(A_k) \qquad \text{......3.}$$

where λ_{jl} is the Lagrangian parameters. The necessary conditions for the local constrained optimum are obtained by differentiating eqn 3 with respect to the design variables A_k as :

$$\frac{\partial W(A_k,\lambda_{jl})}{\partial A_k} = \sum_{i=1}^{m_k} \rho_i L_i + \sum_{l=1}^{nlc} \sum_{j=1}^{p} \lambda_{jl} \frac{\partial g_{jl}(A_k)}{\partial A_k} = 0 \qquad \text{......4.}$$

With the use of unit load method, the j th displacement of the structure under the load case l can be expressed as :

$$\Delta_{jl} = \sum_{i=1}^{nm} \frac{F_{il}\overline{F}_{ij}L_i}{E_i A_i} \qquad \text{......5.}$$

where nm is the total number of members in the structure, F_{il} is the force in member i due to load case l, $\overline{F}_{ij}$ is the force in member i due to the unit load applied in the direction of the restricted displacement j; E_i, A_i and L_i are the modulus of elasticity, area and length of member i, respectively. Substituting eqn 5 into 2 and differentiating with respect to the design variable A_k leads to the following expression.

$$\frac{\partial g_{jl}(A_k)}{\partial A_k} = -\frac{1}{A_k^2} \sum_{i=1}^{m_k} \frac{F_{il}\overline{F}_{ij}L_i}{E_i} \qquad \text{......6.}$$

Using eqns 4 and 5, it follows that

$$\sum_{k=1}^{m_k} \rho_i L_i - \sum_{l=1}^{nlc} \sum_{j=1}^{p} \lambda_{jl} \frac{1}{A_k^2} \sum_{i=1}^{m_k} \frac{F_{il}\overline{F}_{ij}L_i}{E_i} = 0 \qquad \text{......7.}$$

Rearranging equation 7 by considering constants E and ρ throughout the structure, leads to

$$A_k = \left[\frac{1}{E\rho \sum_{i=1}^{m_k} L_i} \sum_{l=1}^{nlc} \sum_{j=1}^{p} \lambda_{jl} \left(\sum_{i=1}^{m_k} F_{il}\overline{F}_{ij}L_i\right)\right]^{\frac{1}{2}} \quad k=1,2,..,ng \qquad \text{....8.}$$

Equation 8 is the optimality criteria. It can be used to obtain the new values of design variables provided that the Lagrangian parameters are known. For a determinate space truss F_{il} and $\overline{F}_{ij}$ are constant which can be determined uniquely. However, for an indeterminate space truss F_{il} and $\overline{F}_{ij}$ are the functions of the design variables A_k. Numerical experiments have shown that these changes may be assumed to be small during the optimization process. Consequently, when the new values of design variables are computed, the structure is analysed with these values and, F_{il} and $\overline{F}_{ij}$ are updated.

RECURRENCE RELATIONSHIP FOR THE LAGRANGE MULTIPLIERS

The recurrence relationship for the Lagrange multipliers can be obtained from the constraint equation 2 as suggested in ref 16. Assuming that they are all equality constraints, they become

$$\Delta_{jl} = \overline{\Delta}_{jl} \qquad j=1,2,\ldots,p \qquad \text{......9.}$$

Multiplying both sides by λ_j^c and then taking the c th root and writing in recursive form gives the following relationship

$$\lambda_{jl}^{\nu+1} = \lambda_{jl}^{\nu} \left(\frac{\Delta_{jl}}{\overline{\Delta}_{jl}}\right)^{1/c} \qquad j=1,2,\ldots,p \qquad \text{.....10.}$$

where ν is the current and $\nu+1$ is the next iteration number. c is a preselected constant, known as a step size. In this work, it is found suitable to choose the value of 0.5 for c. In order to use eqn 10, it is required to select some initial values only for Lagrange multipliers. Hence, it becomes possible to carry out the optimum design of truss subjected to only displacement constraints by using the recursive relationship 8 and 10 iteratively.

STRESS CONSTRAINTS

If F_{il} is the force in member i under load case l and A_k is its area, the stress in that member must satisfy the following inequality

$$\sigma_i = \left|\frac{F_{il}}{A_k}\right| \le \overline{\sigma}_{il}^t \ \text{ or } \ \overline{\sigma}_{il}^b \qquad \text{.....11.}$$

in which $\overline{\sigma}_{il}^t$ and $\overline{\sigma}_{il}^b$ are the permissible stresses in tension and compression, respectively.
If member i is in tension, the lower bound for its area is obtained from the inequality 11 as

$$A_k \ge \frac{F_{il}}{\overline{\sigma}_{il}^t} \qquad \text{.....12.}$$

where $\overline{\sigma}_{il}^t$ is taken from the design codes depending on the steel used in the structure. For example AISC specifies that $\overline{\sigma}_{il}=0.6\,\sigma_y$ where σ_y is the yield stress.

BUCKLING CONSTRAINTS

If member i is in compression, the stress constaint turns out to be buckling constraint. The buckling of compression member i takes place either in elastic or plastic region depending on its slenderness ratio $\lambda_i=L_i/r_i$, where r_i is the radious of gyration. Since only the cross sectional areas are treated as design variables, it becomes necessary to express the radious of gyration in terms of areas. This can be achieved by applying the least square approximation to practically available sections such as pipes, angles, tees. This relationship has the form

$$r = aA^b \qquad \text{.....13.}$$

where a and b are the constants depending on the sections selected for the members. Lower bounds on areas are then obtained according to the design codes considered.

AISC CODE

According to AISC, ref 18, the buckling stress $\overline{\sigma}_{il}^b$ is computed as follows

If $\lambda_i > C_c$ elastic buckling $\quad \overline{\sigma}_{il}^b = \dfrac{12\pi^2 E}{23\lambda_i^2}$ .....14.

If $\lambda_i < C_c$ plastic buckling $\quad \overline{\sigma}_{il}^b = \sigma_y\left(1-\dfrac{\lambda_i^2}{2C_c^2}\right)/n \qquad n = \dfrac{5}{3}+\dfrac{3}{8}\dfrac{\lambda_i}{C_c}-\dfrac{\lambda_i^3}{8C_c^3}$ .....15.

where σ_y is the yield stress of the steel used and, $C_c = \sqrt{2\pi^2 E/\sigma_y}$ which has the value of 132 for the steel A7. For $\lambda = C_c$, both eqns 14 and 15 give the same $\overline{\sigma}_{il}^b$. Considering this fact, it is possible to determine a P_{cr} so that when the force in the member is greater than P_{cr}, plastic buckling takes place, otherwise the member buckles elastically. Thus

$$\lambda_i = C_c \rightarrow \sigma_i = \overline{\sigma}_{il}^b \rightarrow P_{cr} = \frac{12\pi^2 E A_{cr}}{23\, C_c^2} \qquad \text{.....16.}$$

where A_{cr} is the critical area value for the member obtained as

$$\lambda_i = \frac{L_i}{aA_{cr}^b} = C_c \rightarrow A_{cr} = \left(\frac{L_i}{aC_c}\right)^{1/b} \qquad \text{.....17.}$$

Finally, the lower bound on the area of compression member can be given as

$$\text{If } F_{il} < P_{cr} \quad \text{elastic buckling,} \quad A_k \ge \left(\frac{23F_{il}L_i}{12\pi^2 E a^2}\right)^{1/(2b+1)} \qquad .....18.$$

If $F_{il} > P_{cr}$ plastic buckling, A_k is obtained from the solution of the following nonlinear equation.

$$\alpha_1 A_k^{3b+1} - \alpha_2 A_k^{b+1} - \alpha_3 A_k^{3b} - \alpha_4 A_k^{2b} + \alpha_5 = 0 \qquad .....19.$$

where

$$\alpha_1 = 24\,C_c^3\,a^3\,\sigma_y\,,\quad \alpha_2 = 12C_c a L_i^2\,\sigma_y\,,\quad \alpha_3 = 40C_c^3 a^3 F_{il}$$

$$\alpha_4 = 9\,C_c^2 a^2 L_i F_{il}\,,\quad \alpha_5 = 3L_i^3 F_{il} \qquad .....20.$$

inwhich F_{il} is the compression force in member i under load case l, a and b are the constants. The eqn 19 is obtained by equating the member stress F_{il}/A_k to $\bar{\sigma}_{il}^b$ of eqn 15 and substituting $\lambda_i = L_i/(aA_k^b)$ and arranging it in terms of A_k. The nonlinear equation 19 can be solved by Newton Raphson method to obtain the lower bound of A_k.

DIN CODE

German code of practice DIN 4114, ref 19, is also similar to AISC code and the buckling stress is given as follows

$$\text{If } \lambda_i > \lambda_p \quad \text{elastic buckling} \quad \bar{\sigma}_{il}^b = \frac{\pi^2 E}{2.5\lambda_i^2} \qquad ......21.$$

If $\lambda_i < \lambda_p$ plastic buckling takes place and $\bar{\sigma}_{il}^b$ is obtained from the sulution of the following equation.

$$\lambda_i^2 = \frac{\pi^2 E}{\bar{\sigma}_{il}^b}\left[1 - \frac{m\bar{\sigma}_{il}^b}{\sigma_y - \bar{\sigma}_{il}^b} + 0.25\left(\frac{m\bar{\sigma}_{il}^b}{\sigma_y - \bar{\sigma}_{il}^b}\right)^2 - 0.005\left(\frac{m\bar{\sigma}_{il}^b}{\sigma_y - \bar{\sigma}_{il}^b}\right)^3\right] \qquad .....22.$$

where $m = 2.317(0.05 + \lambda_i/500)$, σ_y is the yield stress. $\lambda_p = 115$ for the steel St37. After the similar arrangements which is carried out in eqns 16 and 17 the lower bounds can be determined by the following expressions

$$\text{If } F_{il} < P_{cr} \quad \text{elastic buckling} \quad A_k \ge \left(\frac{2.5F_{il}L_i^2}{\pi^2 E a^2}\right)^{1/(2b+1)} \qquad ...23.$$

$$\text{If } F_{il} > P_{cr} \quad \text{plastic buckling} \quad A_k \ge \frac{1.5F_{il}}{\bar{\sigma}_{il}^b} \qquad ...24.$$

where $\bar{\sigma}_{il}^b$ is obtained by solving eqn 22.
It is shown that whichever design code is used, the stress or buckling contraints can be reduced to lower bounds on area variables as given in inequalities 12,18 and 19 or 12,23 and 24.

DESIGN PROCEDURE

The flow diagram of the optimum design algorithm is given in fig 1. The initial values of the area variables can be feasible or infeable. If desired, these can be choosen all equal to each other for simplicity. Load matrix is extended to cover the unit loadings in the direction of restricted displacements. Three different area values are computed for each group as a new value. The first one is obtained from the displacement constraints using eqn 8. The second one is the lower bound coming from the stress or buckling constraints. The third one is the minimum size constraint. The largest among them determines the new value of the design variable. The process is repeated until the convergenge is obtained in the values of area variables

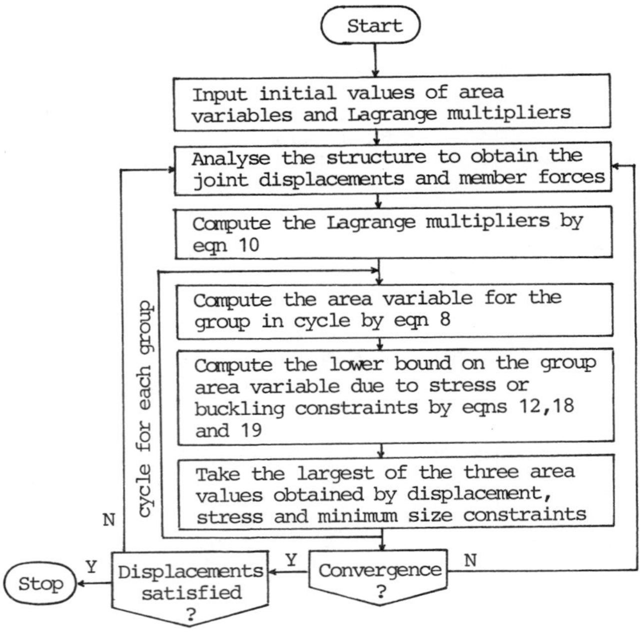

Fig 1

DESIGN EXAMPLES

The design algorithm developed is applied to the optimum design of three space trusses. The values of the constants a and b of eqn 13 are given in table 1, depending on the sections and design codes. It should be noted that in the computation of these constants the unit of centimeter is used.

Design codes		L	O	T	⅂Γ
DIN	a	0.4107	0.9195	0.6156	0.4356
	b	0.5350	0.5507	0.4009	0.5390
AISC	a	0.8338	0.4993	0.2905	0.5840
	b	0.5266	0.6777	0.8042	0.5240

Table 1

25 - BAR SPACE TRUSS

As a first example, the design of the space truss of fig 2, which is commonly used for comparision by various authors, is considered. The dimensions and the group numbers of the members are also shown in the figure. The load cases and the upper bounds imposed on displacements of joints are given in table 2.

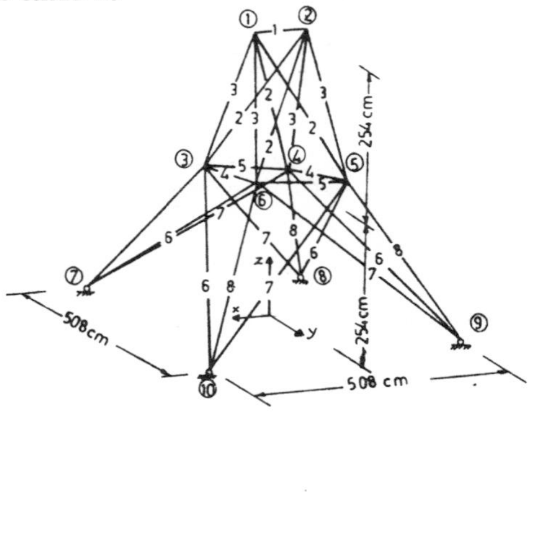

Fig 2

Load case	Joint Number	Loading (kN)			Upper bounds on displ. (mm)	
		x	y	z	x	y
1	1	4.54	45.4	-22.7	8.89	8.89
	2	0	45.4	-22.7	8.89	8.89
	3	2.27	0	0	-	-
	6	2.27	0	0	-	-
2	1	0	90.8	0	8.89	8.89
	2	0	-90.8	0	8.89	8.89

Table 2

The above truss is first designed by using the same material properties, displacement requirements and constant stress limitations as given in ref 17. The same optimum solution is obtained after 12 iterations.
The same truss considering the same limitations, is also designed by the proposed algorithm inwhich the AISC design code is employed for the buckling constraints. Two different types of section are used for members. The values of the initial and final designs are given in table 3 in relation to the sections. In both designs the minimum area requirement is taken as 6.45 mm^2. It is noticed that the buckling constraints are active in the design problem. Consequently in the final design, the stresses in members 7,8 and 25 are at their limit values, while the displacements of restricted joints are far from their bounds. In both designs, the optimum solution is obtained after 3 iterations.

Areas x10^{2}mm^2	Angle sections		Pipe sections	
	Initial	Final	Initial	Final
A_1	10.	0.064	10.	0.064
A_2	10.	9.30	10.	10.735
A_3	10.	10.058	10.	10.995
A_4	10.	0.064	10.	0.064
A_5	10.	1.553	10.	2.456
A_6	10.	6.678	10.	8.089
A_7	10.	11.586	10.	13.050
A_8	10.	9.537	10.	10.943
$Y_{1,2}$ (mm)	4.39	4.55	4.39	4.00
Volume x10^3 mm^3	84003	70417	84003	80884

Table 3

57 - BAR TRUSS

The cantilever space truss shown in fig 3 is subject to two load cases as given in table 4. The truss has 57 members belonging to 8 different groups. The modulus of elasticity is taken as 204 kN/mm^2. The upper bounds imposed on the displacements are also given in table 5. The allowable tensile stress is taken as 0.14 kN/mm^2. The pipe section is adopted for the members and AISC design code is employed for buckling constraints. The minimum size requirements are taken as 100 mm^2.

Load case	Joint Number	Loading (kN)			Upper boonds on displ. (mm)		
		x	y	z	x	y	z
1	19	0	0	-50	21	-	21
2	19	10	0	0	21	-	21

Table 4

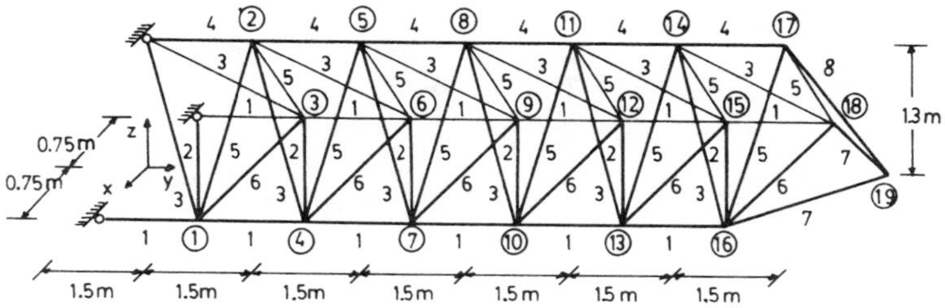

Fig 3

In the design problem the initial design point is selected to be infeasible and far from the optimum. Final design is obtained after 15 iterations, which is shown in table 5. The minimum volume of the truss is 8.775x10^8 mm^3. In the final design it is noticed that there is %0.5 excess in the vertical displacement of the joint 19. This is due to the selection of the infeasible initial design point. In this design problem, the displacement constraints are severe that govern the design process.

Areas x10^2 mm^2	A_1	A_2	A_3	A_4	A_5	A_6	A_7	A_8
Initial	10.	10.	10.	10.	10.	10.	10.	10.
Optimum	100.6	1.01	73.7	103.6	62.93	3.69	20.23	49.70

Table 5

244 - BAR TRANSMISSION TOWER

In the last example, the design of 244-bar transmission tower of fig 4 is considered to demonstrate the practical effectiveness of the design algorithm. Members of the truss are collected in 26 groups. The load cases and displacement limitations are given in table 6. The allowable tensile stress is 0.14 kN/mm^2. The angle section is used for the members. The truss is designed by considering both AISC and DIN codes. In each design the minimum areas are taken as 191.6 mm^2 and 174 mm^2, respectively. Optimum designs are obtained after 12 iterations and the resulting optimum values of areas are given in table 7. The iteration history is shown in fig 5. In the design problem, both buckling and displacement constraints are active. Consequently, both the vertical displacement of joint 17 and stresses in some of the members are found to be at their upper bounds.

Load case	Joint Number	Loading (kN)		Upper bounds on displ. (mm)	
		x	z	x	z
1	1	10	-30	45	15
	2	10	-30	45	15
	17	35	-90	30	15
	24	175	-45	30	15
	25	175	-45	30	15
2	1	0	-360	45	15
	2	0	-360	45	15
	17	0	-180	30	15
	24	0	-90	30	15
	25	0	-90	30	15

Table 6

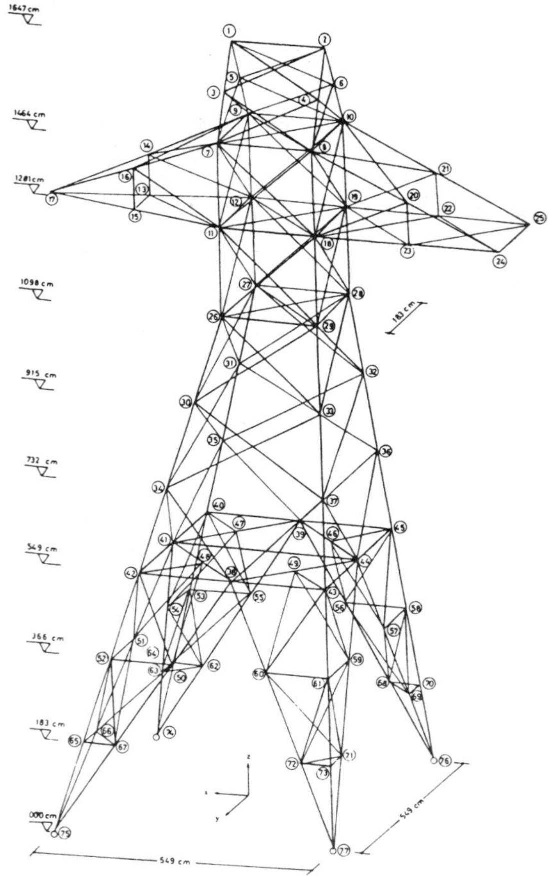

Fig 4

Areas mm^2	A_1	A_2	A_3	A_4	A_5	A_6	A_7	A_8	A_9
AISC	191.6	195	203	1515	309	2342	191.6	1896	191.6
DIN	174	398	486	1534	960	2590	174	2098	174
Areas mm^2	A_{10}	A_{11}	A_{12}	A_{13}	A_{14}	A_{15}	A_{16}	A_{17}	A_{18}
AISC	191.6	2743	881	191.6	191.6	5229	934	191.6	191.6
DIN	210	2815	2652	786	174	5709	1920	174	174
Areas mm^2	A_{19}	A_{20}	A_{21}	A_{22}	A_{23}	A_{24}	A_{25}	A_{26}	
AISC	191.6	4859	765	191.6	191.6	191.6	191.6	191.6	
DIN	174	5223	1386	174	174	174	174	174	

Table 7

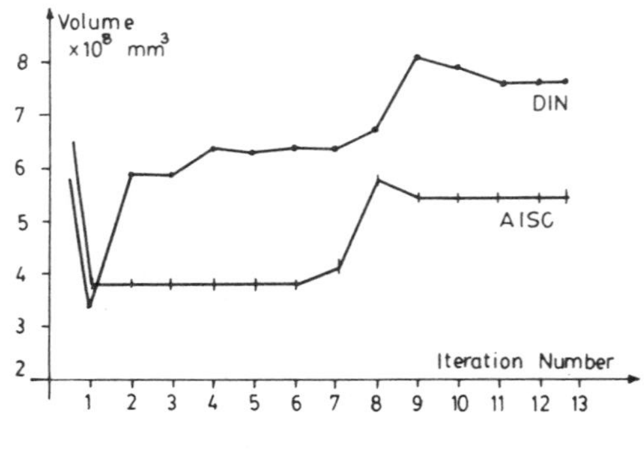

Fig 5

CONCLUSIONS

The optimality criteria approach developed can effectively be used in the design of large practical structures. The number of iterations to reach the optimum solution is relatively small. The algorithm is flexible and it works equally well with both fecasible or infeasible initial design points. In a design problem where the buckling constraints are dominant, the optimum solution may even be obtained in a small number of iterations. Although the space trusses are considered in this study, the algorithm can easily be extended to rigidly jointed space frames. The basic advantage of the technique is that it can easily be programmed as a subprogram and combined with one of the existing structural analysis programs.

REFERENCES

1. E J HAUG and J S ARORA, Applied optimal design, J.Wiley, 1981.
2. U KIRSCH, Optimum structural design, McGraw Hill, 1981.
3. M P SAKA, Optimum design of rigidly jointed frames, Comp. and Struc. Vol 11, 1980.
4. M P SAKA, Optimum design of grillages including warping, Proc. of Int. Symp. on optimum structural design, Univ. of Arizona, Tucson, USA, 1981.
5. A TEMPLEMAN, The use of geometric programming methods for structural optimization, AGARD lecture series, No.70, 1974.
6. J H LIN, Y W CHE and Y S YU, Structural optimization on geometrical configuration and element sizing with statical and dynamical constraints, Comp. and Struc. Vol. 15, 1982.
7. H V TOPPING, Shape optimization of skeletal structures J. Struc. Div.,Vol. 109, 1983.
8. K I MAJID and M P SAKA, Optimum shape design of rigidly jointed frames, Proc. of symp. on appl. of com. methods in Engr., Univ. of Southern California, USA, 1977.
9. K I MAJID, M P SAKA and T ÇELİK, The theorems of structural variations generalized for rigidly jointed frames, Proc. ICE, Vol 65, 1978.
10. K I MAJID, P STOJANOVSKI and M P SAKA, Minimum cost topological design of sway frames, The Structural Engineer, Vol 58 B, 1980.
11. M P SAKA, Shape optimization of trusses, J. Struc. Div. Vol 106, 1980.
12. M P SAKA, Minimum cost topological design of trusses, Proc. of Symp. on optimum structural design, Univ. of Arizona, Tucson, USA, 1981.
13. V B VENKAYYA, Design of optimum structures. Comp. and Struc., Vol 1, 1971.
14. J KIUSALAAS, Minimum weight design of structures via optimality criteria, NASA TN D-7115, 1972.
15. V B VENKAYYA, N S KHOT and L BERKE, Application of optimality criteria approaches to automated design of large practical structures, AGARD Second Symp. on structural optimization, Italy, 1973.
16. N S KHOT, L BERKE and V B VENKAYYA, Comparison of optimality criteria algorithms for minimum weight design of structures, AIAA Journal, Vol 17, 1978.
17. C FLEURY and M GERADIN, Optimality criteria and mathematical programming in structural weight optimization, Comp. and Struc., Vol 8, 1978.
18. AISC, Manual of Steel Construction,Seventh Edition,1970.
19. STAHL IM HOCHBAU, 13 Auflage, Verlag Stahleisen M B H, Düsseldorf, 1967.

OPTIMUM DESIGN OF DOUBLE-LAYER GRIDS

E. SALAJEGHEH

Department of Civil Engineering,
University of Kerman, Iran.

This work presents a method for optimization of the cost of double-layer grids using a nonlinear programming approach.The design problem is expressed as a constrained optimization problem in which the optimum values of the design variables are sought, while the design constraints are satisfied.The cross-sectional areas of the members of the double-layer grids are considered to be the design variables and so are the changes in the topology of the grids.The design constraints consist of bounds on member stresses, buckling of members, slenderness ratio of members and joint displacements.The constrained optimization problem is transformed into a sequence of unconstrained problems, each of which is, in turn, solved by a series of one-dimensional minimization problems.The unconstrained optimization method requires the partial derivatives of the cost function and the design constraints with respect to the design variables.The required derivatives are obtained by direct differentiation of the load-displacement relationships.

INTRODUCTION

To achieve an optimum design, a function which is the basis for choice between various acceptable designs (objective function) and the design requirements (design constraints) are expressed in terms of the parameters describing the structural system (design and preassigned variables). The design problem, can be posed as a constrained optimization problem, in which the minimum (or maximum) of the specified function is the ultimate objective, subject to the satisfaction of the design requirements.

Mathematically, an optimum design problem, may be stated as follows:Find the design vector X, such that the objective function f(X), is minimized or maximized, subject to the constraints

$$g_i(X) \geq o \;,\quad i=1,2,\ldots,I$$
$$h_i(X) = o \;,\quad i=1,2,\ldots,E,$$

representing inequality constraints (I) and equality constraints (E) on the variables.

There is a branch of mathematics that deals with the solution of these types of problems, known as mathematical programming.Among the techniques in this field of mathematics, nonlinear programming is of great importance as it is being used in conjunction with nonlinear functions, see Refs 1 and 2.The application of mathematical programming to some of the structural problems can be found in Refs 3 and 4.

STRUCTURAL OPTIMIZATION OF DOUBLE-LAYER GRIDS

In the present study, a general method for solving the nonlinear programming under consideration is used. The method is presented in Ref 5. First the constraint equations and the cost function are presented in terms of the design variables.Then computational aspects of the optimization method will be discussed in which a method is presented to evaluate the necessary derivatives of the functions under consideration. Also, a technique will be introduced by which the number of analyses of the structure can be reduced to one analysis only for each one-dimensional search.

PREASSIGNED VARIABLES

For the double-layer grids under consideration, the preassigned variables are as follows:

a) GRID CONFIGURATIONS- Considering that the number of cases which could be studied in this study is limited, it was thought more appropriate to concentrate on a relatively small number of basic cases (which, nevertheless, covers most practical cases). With this objective in mind, it was decided to consider four different configurations of two-way double-layer grids, from amongst all the possible double-layer grids, see Ref 6. These arrangements are shown in Fig 1. All the grids are chosen to have square shaped boundary and it is further assumed that the grids cover an area of 30m×30m.

b) SUPPORTS- It was assumed that the support conditions are the same for all the types of the grids consisting of columns that provide vertical constraints only.

c) LOADING- The external loading was considered to be a uniformly distributed load covering the whole of the top-layer. The intensity of loading is based on CP3 where allowance is made for imposed load, including snow, weight of joints and self weight of 150 kg/m^2.

d) STRUCTURAL CROSS SECTIONS AND JOINTS-Circular sections were assumed to be used for the members of the double-layer grids under consideration. In conjunction with the cross-sections being used, a suitable jointing system had to be considered. The Nodus jointing system was adopted which is a relatively new jointing system, introduced by the Tubes Division of the British Steel Corporation, see Ref 7.

DESIGN VARIABLES

The design variables are considered to be the cross-sectional areas of the members and some geometrical and topological parameters. In general, the cross-sectional area of each member can be taken as an independent design variable. In practical cases, however it is possible to reduce the number of area variables by declaring that certain members have identical cross-sectional properties. For the double-layer grids under consideration, it is assumed that all the members of the top-layer have a similar cross-sectional area A_t, members of the bottom layer have a different cross-sectional area A_b and the diagonal members have a third cross-sectional area A_d. Thus the design vector corresponding to cross-sectional areas X_A is defined as

$$X_A=[A_t, A_b, A_d]^T.$$

The design vector corresponding to geometric changes is considered as

$$X_B=[H_g, N_g]^T,$$

where H_g is the height of the grids and N_g is the number of pyramidal units along each side of the grids.

DESIGN CONSTRAINTS

Considering the assumptions made on the analysis of the double-layer grids, there are two possible loading conditions to which a double-layer grid member may be subjected. A member may be either subjected to combined stresses resulting from compressive forces and bending moments or it may be under the action of tensile forces and bending moments. For these two possibilities different design constraints will be considered.

For a member, subjected to compressive forces and bending moments the following constraints must be satisfied:

$$(g_c)_1=1-\frac{f_a}{F_a}-\frac{C_m f_b}{(1-\frac{f_a}{F'_e})F_b}\geq 0 \qquad \ldots.1,$$

$$(g_c)_2=\frac{C_m f_b}{(1-\frac{f_a}{F'_e})F_b}\geq 0 \qquad \ldots.2,$$

$$(g_c)_3=1-\frac{f_a}{F_y/\gamma}-\frac{f_b}{F_b}\geq 0 \qquad \ldots.3,$$

$$(g_c)_4=(\frac{kl}{r})/(\frac{kl}{r})_{c\ min}-1\geq 0 \qquad \ldots.4,$$

$$(g_c)_5=1-(\frac{kl}{r})/(\frac{kl}{r})_{c\ max}\geq 0 \qquad \ldots.5,$$

where g_c represents an inequality constraint for a compression member and

f_a= actual axial compressive stress,

f_b= actual maximum fibre compressive bending stress,

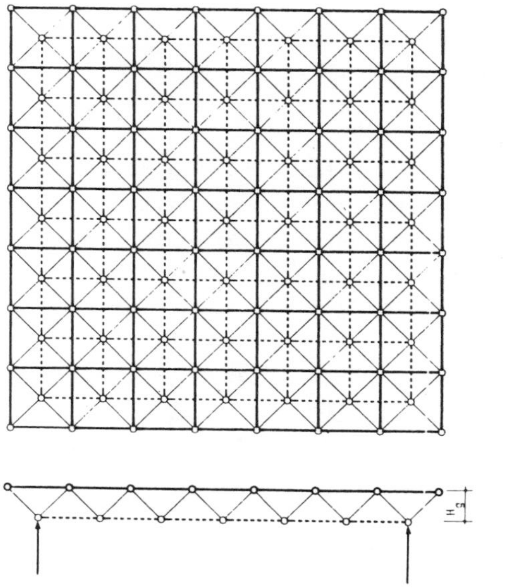

General layout of the *square-on-square* type of double-layer grids (N_g=7).

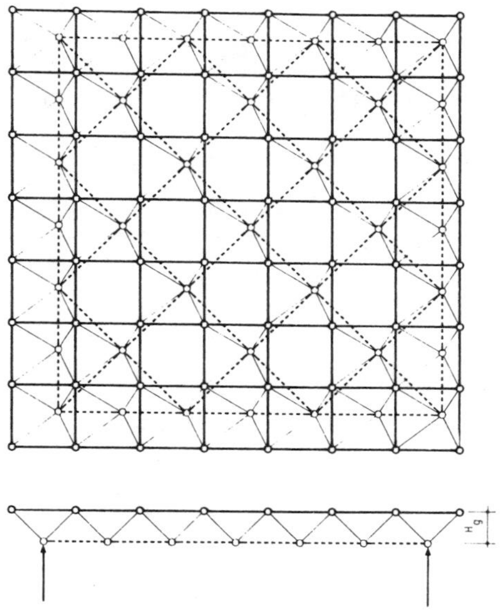

General layout of the *square-on-diagonal* type of double-layer grids (N_g=7).

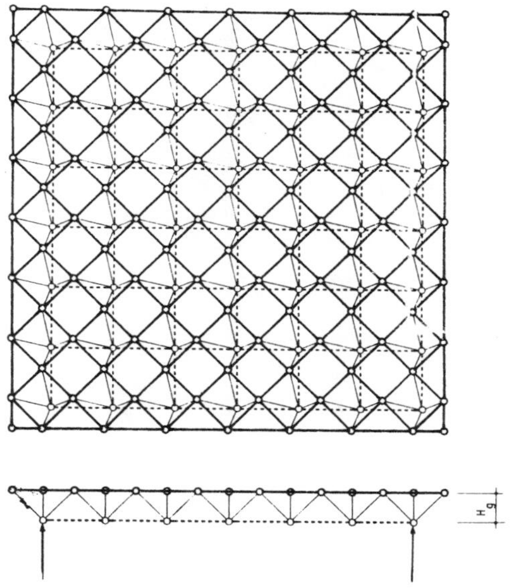

Fig 1

General layout of the *diagonal-on-square* type of double-layer grids (N_g=7).

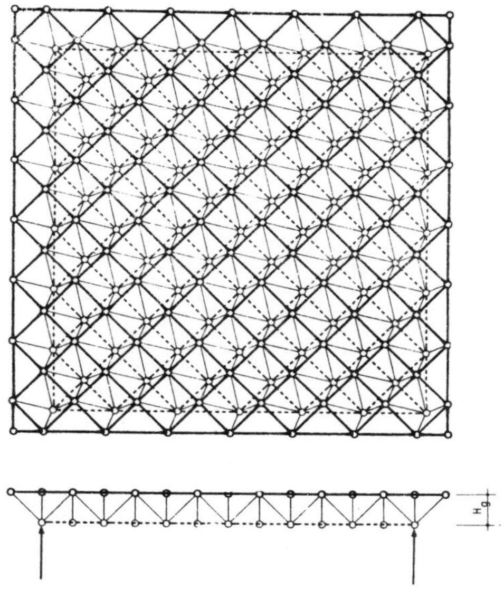

General layout of the *diagonal-on-diagonal* type of double-layer grids (N_g=7).

F_a= allowable axial compressive stress for axial force only,

F_b= allowable bending stress for bending moment alone,

F_y=yield stress of the material whose value is taken as 2400kg/cm,^{2}

F'_e= allowable Euler stress=F_e/γ where $F_e=\dfrac{\pi^2 E}{(kl/r)^2}$.

The value of γ in Eqn 3 is taken as 5/3, E is the modulus of elasticity (2.1×10^6 kg/Cm^2), kl/r is the slenderness ratio of the member, L is the member length, K is the effective member length and r is the radius of gyration in the plane of bending.

$(KL/r)_{c\ min}$ & $(KL/r)_{c\ max}$= minimum and maximum slenderness ratios considered for compression members,

C_m= a coefficient whose value shall be taken as 0.85, see Ref 8.

For a member subjected to tensile forces and bending moments, the following inequality constraints must be satisfied:

$$(g_t)_1 = 1 - \frac{f_a}{Fy/\gamma} - \frac{f_b}{F_b} \geq 0 \qquad ....6,$$

$$(g_t)_2 = \left(\frac{kl}{r}\right) / \left(\frac{kl}{r}\right)_{t\ min} - 1 \geq 0 \qquad .....7,$$

$$(g_t)_3 = 1 - \left(\frac{kl}{r}\right) / \left(\frac{kl}{r}\right)_{t\ max} \geq 0 \qquad .....8,$$

where g_t represents an inequality constraint for a tension member, $(kl/r)_{t\ min}$ and $(kl/r)_{t\ max}$ are the lower and upper bounds on the slenderness ratios of the tension members. The notation f_a is used for both members under tensile and compressive forces.

A constraint inequality which restricts the maximum deflection of the structure is also being considered. This constraint can be expressed as follows:

$$(g_5) = 1 - d_m/d_a \geq 0 \qquad .....9,$$

where d_m is the maximum calculated deflection and d_a is the maximum allowable deflection.

In addition to the above mentioned constraints, the local buckling of the member cross-sections and the lower/upper bounds on the height and the number of divisions of double-layer grids are also imposed. For the value of F_a, see Ref 8.

COST FUNCTION

The cost function is the cost of fabricated tubes, cost of painting, cost of delivery, cost of erection and cost of joints. The current price list was supplied by the British Steel Corporation (B.S.C.) and it can be found in Appendix A.

To present the cost function, suppose that the structure consists of a total of G groups of members. Let L_b be the total length of all the members in a group b, and let the cross-sectional area of members of that group be A_b. Then the weight of the structure W, is given by

$$W = \sum_{b=1}^{G} \gamma_b L_b A_b \qquad .....10,$$

where γ_b is the density of the material of group b. According to the data given in Appendix A, it follows that

cost of fabricated tubes $= w_1 W$
cost of painting $= w_2 W$
cost of delivery $= w_3 W$
cost of erection $= w_4 W$

where w_1, w_2, w_3 and w_4 may be deduced from Appendix A. Thus the overall cost of the structure (excluding the cost of joints) will be

$$C_m = wW = w \sum_{b=1}^{G} \gamma_b L_b A_b \qquad .....11,$$

where $w = w_1 + w_2 + w_3 + w_4$.

The cost of the joints does not depend on the weight of the joints but on the number and the size of the joints used. The joint cost is defined here as the total cost of the joints used for the top-chord and the bottom-chord, fork connectors at the bracing members and the stanchion assembly at supports. To evaluate the size of the joints, the following size range for circular hollow sections (CHS) has been produced by B.S.C., which shows the size of a joint for a specified range of sections, see Ref 7.

	joint size reference	24	30	35	45	55	66
CHS	outside diameter (mm)	60.3	76.1	88.9	114.3	139.7	168.3

Table 1 size range for CHS.

Since the joint size references and the CHS member sizes, are both discrete values, a continuous variation of one with respect to another must be found. To achieve this, using the discrete values of Table 1, a graph is plotted showing the variation of the joint size references with respect to the CHS member sizes (Fig 2).

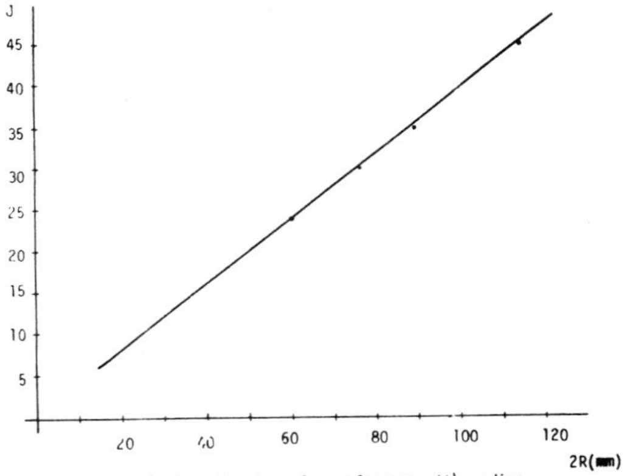

Variation of joint size reference with radius.

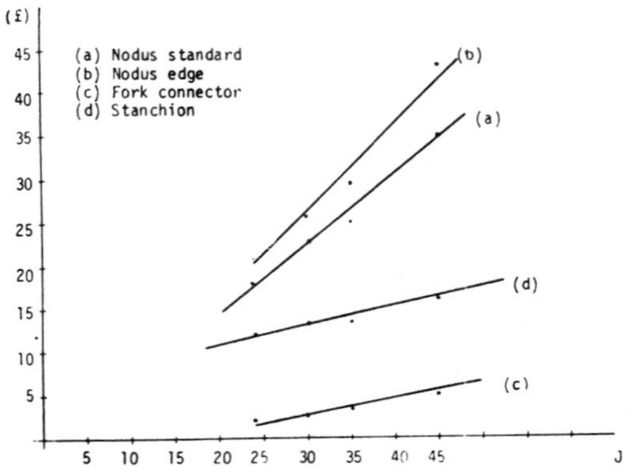

Fig 2 Variation of cost with joint size reference.

This graph indicates that the variation between these two parameters can be well approximated by a straight line as

$$J=u_1+u_2R \qquad \ldots.12,$$

where J indicates the joint size reference, R indicates the radius of the CHS chord member, u_1 and u_2 are two constants that can be evaluated by a least-squares first order polynomial approximation.

The same procedure can be adopted to express the relationship between the price of a joint C_j and the joint size reference J. Figure 2 shows the variation of C_j with respect to J for different joint types. Thus for a typical joint type, the following relation holds

$$c_j=u_3+u_4J \qquad \ldots.13,$$

where u_3 and u_4 are constants.

Substituting for J from Eqn 12, it follows that

$$c_j=u+vR \qquad \ldots.14,$$

where $u=u_3+u_4u_1$ and $v = u_4u_2$.

Now consider a double-layer grid consisting of a total of G groups of members with a typical group being referred to by b. Within a group b, it is assumed that there exists a total of T joint types each being referred to by t. Furthermore, let n_{bt} denote the number of joint types in the bth group and the tth joint type.
Then the total cost of the joints C_j will be

$$C_j=\sum_{b=1}^{G}\ \sum_{t=1}^{G}(u_{bt}+v_{bt}\ R_b)n_{bt} \qquad \ldots.15,$$

where u_{bt} and v_{bt} are constant coefficients, corresponding to a joint in the bth group and the tth joint type and R_b is the radius of the cross-section used for the bth group.

The total cost of the structure is the sum of the cost of the members and the cost of the joints as

$$C = C_m+C_j \qquad \ldots.16.$$

OPTIMIZATION PROCESS AND ITS COMPUTATION ASPECTS

In the last two Sections, the constraints and the objective function of the design problem were formulated in terms of the design variables.
Now the following design problem must be solved

minimize cost of the structure C

subject to

$(g_c)_{ib} \geq o, \quad i=1,2,3,4,5$
$\qquad b=1,2,\ldots,G$

$(g_t)_{ib} \geq o, \quad i=1,2,3$
$\qquad b=1,2,\ldots,G$

$g_\delta \geq o,$

where G is the number of groups of members. The cost function C and the design constraints g_c, g_t, g_δ can be expressed in terms of the cross-sectional areas

$$X_A=[A_1,A_2,\ldots\ldots.A_G].^T$$

Denoting the cost function C by $f(X_A)$ and the design constraints by

$$g_i(X_A) \geq o, \quad i=1,2,\ldots\ldots,I,$$

where I represents the total number of inequality constraints, then the optimum design problem of the double-layer grids can be stated as

minimize $f(X_A)$

subject to $g_i(X_A) \geq o, \quad i=1,2,\ldots\ldots,I \qquad \ldots17.$

To solve the above problem, a general solution procedure was presented in Ref 5 where the constrained problem (17) can be converted into a sequence of unconstrained problems as

$$P(X_A,R_K)=f(X_A)+R_K\sum_{i=1}^{I} 1/g_i(X_A), R_K > o, K=1,2,.. \qquad \ldots18.$$

Then the unconstrained functions (18) are minimized for a sequence of decreasing values of R_K. This minimization sequence converges to the solution of the constrained problem (17) as discussed in Ref 5. In this Section the method of solution is presented in the form of an algorithm.

Denoting X_A by X, the main steps of the algorithm are as follows:

1. Start with an initial feasible design point X_O satisfying all the constraints with strict inequality sign, that is
$$g_i(X_O) > o \text{ for } i=1,2,\ldots\ldots,I.$$
Set an initial value of $R_1 > 0$, also set $K=1$.

2. Minimize $P(X,R_K)$ by using the method of unconstrained optimization explained in Ref 5 and obtain the solution X_K^*.

3. Test whether X_k^*, is the optimum solution of the design problem, If so, terminate the process, otherwise, go to the next step.

4. Find the next value of parameter R_{k+1} as
$$R_{k+1}=\gamma R_k$$
where $\gamma < 1$.

5. Set the number of iteration as $K+1$, take the new starting point as

$X_o = X_k^*$ and go to step (2).

Although the algorithm is straightforward, there are a number of points to be considered in implementing the method. These are:

(a) Determination of a feasible starting point X_o.

(b) Choosing a suitable value of R_1.

(c) Evaluation of the necessary derivatives.

All these aspects are discussed in Ref 5.

REDUCTION IN NUMBER OF ANALYSES

The optimization process as explained is composed of finding a number of directions of search in the feasible region. Then along each of these directions the minimum point is obtained by comparing the values of Eqn 18. The constraint equations $g_i(X)$ are implicit functions of the design variables X and thus at every design point an analysis of the structure should be carried out to evaluate these constraints. For double-layer grids with a large number of joints and members, the demands on the computational time would have been a major problem if the grids had to be analysed for all the design points under investigation. An attempt is made to replace the implicit expressions by some approximate explicit functions along each direction of search. Then in the process of one-dimensional optimization all the functions are defined explicitly in terms of the design variables and no further analyses of the structure are carried out. It can be seen that the mean axial stress of each group of members and the maximum displacement $d_m(X)$ are the implicit functions.

To replace $d_m(X)$ by approximate functions, it is assumed that an initial vector of cross-sectional areas

$$X_o = [A_{o1}, A_{o2}, \ldots\ldots, A_{oG}]^T$$

is available. The following polynomial is considered for $d_m(X)$

$$d_m(X) \simeq d_m(X_o) + \sum_{b=1}^{G} (A_b - A_{ob}) \frac{\partial d_m(X_o)}{\partial A_b} + \frac{1}{2}\sum_{b=1}^{G}\sum_{c=1}^{G} (A_b - A_{ob})(A_c - A_{oc}) \frac{\partial^2 d_m(X_o)}{\partial A_b\, \partial A_c} \qquad \ldots.19.$$

To evaluate the derivatives in Eqn 19, let k, d and w be the stiffness matrix, the displacement vector and the external load vector of a linear structure, respectively. The relationship between the external loads and the joint displacements is given by

$$kd = w \qquad \ldots..20.$$

Differentiation of Eqn 20 with respect to the design variables A_b, $b=1,2,\ldots.,G$, yields

$$\frac{\partial k}{\partial A_b} d + k \frac{\partial d}{\partial A_b} = \frac{\partial w}{\partial A_b}, \quad b=1,2,\ldots.G \qquad \ldots..21.$$

The external load vector w is assumed to be independent of A_b, then

$$\frac{\partial w}{\partial A_b} = o \qquad \ldots..22.$$

Thus, from Eqn 21, it follows that

$$\frac{\partial d}{\partial A_b} = -k^{-1} \frac{\partial k}{\partial A_b} d, \quad b=1,2,\ldots.,G \qquad \ldots..23.$$

The right hand side of Eqn 23 consists of known quantities, therefore the derivatives of the displacement vector with respect to the design variables A_b, $b=1,2,\ldots,G$ can be evaluated. The details of the computational procedure for evaluating $\frac{\partial d}{\partial A_b}$, $b=1,2,\ldots,G$ are presented in Ref 5.

The second derivatives of the functions under consideration can be obtained by differentiation of Eqn 21 as

$$\frac{\partial^2 k}{\partial A_b\, \partial A_c} d + \frac{\partial k}{\partial A_b} \frac{\partial d}{\partial A_c} + \frac{\partial k}{\partial A_c} \frac{\partial d}{\partial A_b} + k \frac{\partial^2 d}{\partial A_b \partial A_c} = o \qquad \ldots..24,$$

for $b=1,2,\ldots.,G$

$c=1,2,\ldots.G$

where A_b and A_c represent two independent design variables. Since the design variables are the cross-sectional areas, then for a pin-jointed structure,

$$\frac{\partial^2 k}{\partial A_b \partial A_c} = o \qquad \ldots.25.$$

Thus, Eqn 24 will give rise to

$$\frac{\partial^2 d}{\partial A_b \partial A_c} = -k^{-1}\left(\frac{\partial k}{\partial A_b} \frac{\partial d}{\partial A_c} + \frac{\partial k}{\partial A_c} \frac{\partial d}{\partial A_b}\right) \qquad \ldots..26,$$

for $b=1,2,\ldots.,G$

$c=1,2,\ldots.,G$.

Similar expressions can be obtained for the stress functions, details of which is presented in Ref 5.

OPTIMUM GEOMETRY

The design variables considered to represent the geometry of the double-layer grids were the height of the grids H_g, and the number of divisions along *one side of the grids N_g. As far as the height of the grids is concerned, it can be incorporated in the optimization process explained in the previous Sections. The necessary derivatives of the functions under consideration with respect to H_g can be evaluated without great difficulty. Some approximate explicit relations can also be established to represent the variation of the design constraints as a function of H_g similar to those derived earlier. However, the design variable N_g is of a different nature in the sense that it is a discrete design variable and with a change in N_g, the number of members and joints in the structure are also changed. Thus it is not possible to establish an approximate relation by which the structural behaviour of the grids can be predicted. Furthermore, the derivatives of the objective function and the constraints with respect to N_g are not defined.*

Considering the above difficulties, a direct search method was developed to obtain values of H_g and N_g. The method is based on an extension of the technique explained in Ref 9, see also Ref 5.

RESULTS AND CONCLUSIONS

A computer program was developed based on the technique of optimization and it was successfully applied to the double-layer grids under investigation. The efficiency of the technique of optimization is studied in terms of the amount of computational work required for obtaining the optimum design point. Since the optimization method is a search method, then its efficiency is directly related to the number of search directions required to reach the optimum point. As previously explained, for each direction of search an analysis of the structure is needed, thus the total number of analyses that is necessary to obtain an optimum grid can be considered as a criterion for the efficiency and reliability of the method.

In the optimization process, the manner in which the cost of double-layer grids is reduced in terms of the number of analyses, is presented graphically. A coordinate system is chosen such that one of its axes shows the number of required analyses and the other axis shows the ratio of cost at each iteration to the optimum cost (C/C_{op}). The variation of the reduction in cost against the number of analyses is presented as a graph. For each double-layer grid with a given height H_g and a number of divisions along one side of the grid N_g such a graph can be presented. In this Section, however, only the graphs corresponding to the optimum geometry are given. Figure 3 illustrates the variation of the cost of the four types of the double-layer grids under consideration.

The values of H_g given in each figure, correspond to the grids that were found to have the least cost. The four graphs a,b,c, and d in each figure correspond to various initial design points. Let X_{op} be the optimum design vector and X_o be the initial design point and α be a scalar. The following relation may be written:

$$X_{op} = \alpha X_o.$$

Graphs a,b,c, and d are then for the cases where α takes the values 2,1.7,1.4 and 1.2, respectively.

In all cases, it can be observed that convergence to optimum solution occurs after 6 to 10 analyses of the grids. Obviously, the smaller the value of α,The less the number of analyses would be.

An interesting observation was that in most cases the optimum point coincides with the design point achieved from the fully stressed design, except for the cases where the deflection constraints are violated in the fully stressed design approach.It, therefore, can be concluded that for given values of H_g and N_g, it would be appropriate to accept the results of the fully stressed design,despite the fact that there is no proof which indicates the optimality of its results.Then the attention should be mostly focused on optimum geometry rather than finding the optimum cross-sectional areas.

The computational work was reduced substantially by replacing the implicit functions in the optimization problem by some approximate explicit functions.The effect of this replacement was that, the number of analyses required to complete each one-dimensional search,was reduced drastically. For example, each one-dimensional search would have required some 15 to 20 analyses.However,by introducing the technique explained earlier the number of analyses were reduced to one analysis only for each direction of search.

REFERENCES

1. Bazaraa, M.S., and Shetty, C.M., Nonlinear Programming, Theory and Algorithms, John Wiley & Sons, New York, 1979.

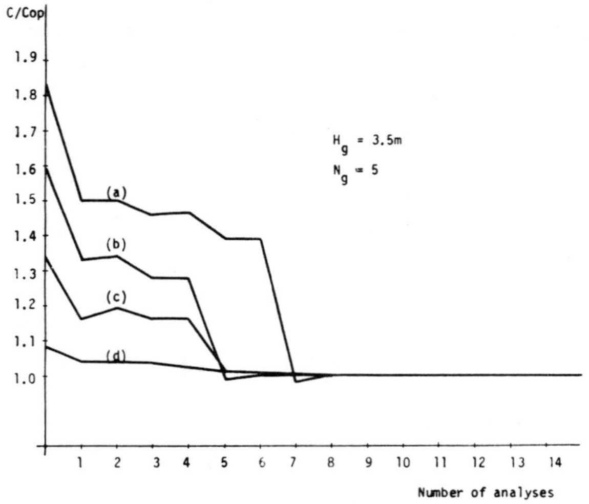

Variation of cost function with number of analyses for the *square-on-square* double-layer grid (with approximate functions).

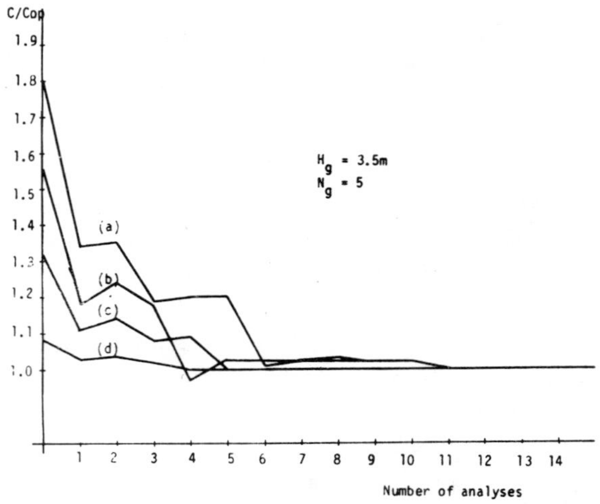

Variation of cost function with number of analyses for the *square-on-diagonal* double-layer grid (approximate functions).

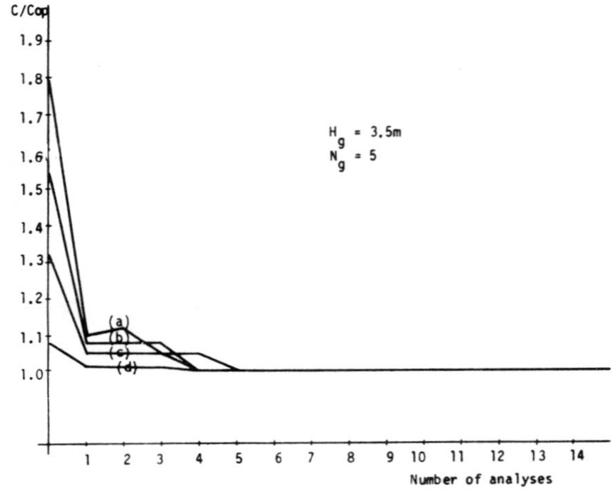

Fig 3

Variation of cost function with number of analyses for the *diagonal-on-square* double-layer grid (approximate functions).

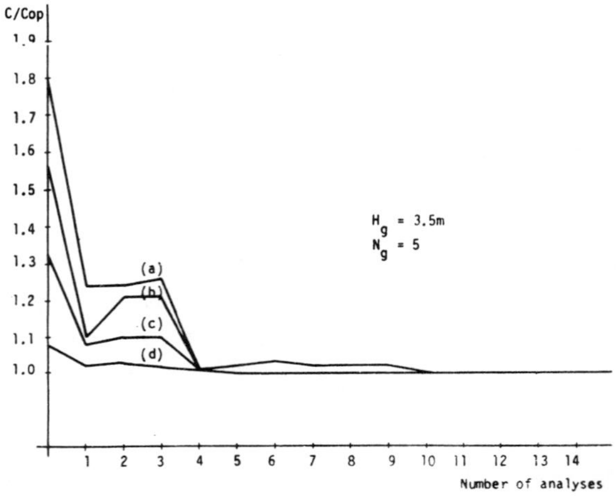

Variation of cost function with number of analyses for the *diagonal-on-diagonal* double-layer grid (approximate functions).

2. Fiacco, A.V., and McCormic, G.P., Nonlinear Programming: Sequential Unconstrained Minimization Techniques, John Wiley & Sons, New York, 1968.

3. Gallagher, R.H., and Zienkiewicz, O.C.(eds), Optimum Structural Design, Theory and Applications John Wiley & Sons, 1973.

4. Venkayya, V.B., Structural Optimization: A Review and Some Recommendations, Int.J.Num.Meth. Engng., Vol.13, pp.203-228, 1978.

5. Salajegheh, E., Optimum Design Of Double-layer Grids, ph.D. Thesis, University Of Surrey, 1981.

6. Makowski, Z.S., Steel Space Structures, Michael Joseph, London, 1965.

7. British Steel Corporation (Tubes Division), Nodus, Space Frame Grids, Parts 1 and 2, 2nd ed., 1978.

8. Manual Of Steel Construction, 8th ed., published by the American Institute Of Steel Construction, New York, 1979.

9. Kiefer, J., Optimum Sequential Search and Approximation Methods under Minimum Regularity Assumptions, J.Soce. Indust. Appl Maths., Vol. 5, pp. 105-136, 1957.

APPENDIX A

COST LIST (1980)

The following cost list has been supplied by the Constructional Steel Research and Development Organization (Constrado) and was used to evaluate the structural cost.

cost of fabricated tube	= £700 per tonne of tube.
Cost of painting average specification	= £65 per tonne of tube and joints.
Cost of delivery	= £60 per tonne of tube and joints.
Cost of erection	= £80 per tonne of tube and joints

OPTIMISATION OF SPACE TRUSSES USING NON-LINEAR BEHAVIOUR OF ECCENTRIC DIAGONALS

CEDRIC MARSH*
MOSTAFA RAISSI FARD**

* Professor, Centre for Building Studies
Concordia University, Montreal

**Research Associate, Concordia University, Montreal

By introducing eccentrically loaded compression diagonals, which possess a strong non-linear force-shortening relationship, it is possible to pre-establish the force distribution between a group of parallel chords to optimise the load capacity of the structure.

The method of analysis is described, the results of tests on a model space truss are given, and the optimum T-section used for the diagonals is developed.

INTRODUCTION

Today, methods of collapse analysis for highly redundant latticed structures are capable of treating any form of the relationship between force and extension. In practice, however, it is usually assumed that the relationship before the ultimate capacity is reached is linear, and it is the shape of the irreversible post-yielding or post-buckling relationship that controls the analysis. This paper shows that a much more valuable contribution can be made to the ultimate capacity by elastic geometric non-linearity prior to reaching the maximum capacity. This can only be developed effectively by using eccentrically loaded diagonal members, thus an efficient eccentric section is required to give the overall economy that collapse analysis promises. This paper deals with the development of such a section and the testing of a model structure to demonstrate the principles propounded.

FORCE/SHORTENING RELATIONSHIP OF STRUTS

A concentrically loaded straight strut shortens due to axial stress in a linear manner, until the critical force is reached. The strut then bows under constant force until yielding causes the resisting force to diminish with increased shortening.

Imperfections in a member affect the linearity, but, in general, a nominally straight bar in the economic range of slenderness ratios has an essentially brittle behaviour. The resisting force diminishes rapidly when the axial strain exceeds that at maximum capacity. The result is that there is little load carrying capability in excess of that which causes buckling in a main chord.

In the case of point supported space trusses, where heavier chords are used along the column lines, the buckling of lighter chords adjacent to the heavier chords can be tolerated and the loading increased until the primary chords fail. Only one primary chord need fail, however, to precipitate collapse. Usually there are a number of such chords in parallel and the objective is how to utilize as much as possible of the total capacity of a group of chords which, according to an elastic analysis, carry unequal forces.

An eccentrically loaded strut has a gross non-linear force/shortening relationship, and as the force approaches the ultimate capacity the axial stiffness diminishes. The result is that where forces are shared between diagonals the more highly loaded bars accept a decreasing portion of the total force to be carried, and the distribution of forces approaches uniformity.

It is not practical, or economical to design chords using eccentric connections, but it is relatively simple to arrange this for the diagonals.

Because the forces in the chords are dictated by the forces in the diagonals, by controlling the resisting forces in the diagonals the distribution of forces between the chords can be pre-determined.

ECCENTRIC STRUTS

Ideally, tubular shapes are used in space trusses, for both structural efficiency and aesthetics. Such structures will continue to be designed on the basis of simple elastic analysis, as no useful post-buckling capability is available.

When open sections are used, connected by bolting to reduce the cost of nodal details, the introduction of eccentric diagonals is possible and an efficient section for these members is required if the end product is to be of some economic advantage.

In unsymmetrical open sections, loaded concentrically, the thickness of the walls cannot be reduced below that at which torsional buckling governs. If the section is loaded through the shear centre the problem is overcome.

A T-section bolted through the table of the T is loaded through its shear centre. Failure may occur in one of three modes: yielding in tension, yielding or local buckling in compression, and flexural buckling about the axis of symmetry. The preferred mode is yielding in the extreme tension fibre, and the shapes are proportioned to achieve this.

The force-shortening relationship for such a pinned end T-section strut is given in Fig. 1.

It is to be observed that the non-linear behaviour is due to geometry changes only, the material remaining elastic up to the maximum capacity, considered to be the force that causes first yield. Beyond the maximum capacity, because the yielding is in tension there is no precipitous collapse, but a smooth reduction in resisting force which may be of value in extremis. In this study only the elastic range is utilized, thus removal of the load after the maximum load has been applied leads to full elastic recovery.

ANALYSIS

Collapse analysis by computer uses linear elastic analysis for each increment of load that causes the next strut to buckle. For chord buckling, the strut is removed if its capacity is assumed to remain constant, or its area is replaced by a negative area if the resistance of the strut is assumed to diminish linearly. The analysis is then repeated to determine the next increment, and is terminated when no further load can be carried.

For the eccentric diagonals the force-shortening relationship is approximated by two straight lines, Fig. 1. The choice of the lines is arbitrary, but it can be optimised by varying the choice until the load carried by the structure is a maximum.

The computer analysis uses the equivalent area of the member, i.e. that which gives the required slope of the force-shortening line using a constant elastic modulus. The equivalent area is changed as the forces exceeds the limit of the initial slope.

When any diagonal reaches its ultimate capacity the analysis is stopped. This is considered to be the maximum useful capacity of the structure. The structure is still elastic, and no reliance is placed on the somewhat unpredictable post-buckling behaviour.

No chord force exceeds the capacity of the member, unless it is a light chord adjacent to a heavy chord, in which case the post-buckling regime may be utilized.

A TYPICAL ANALYSIS

A square space truss supported on four columns is shown in Fig. 2.

For economy chords close to the boundary are of a heavier section. The change in force distribution between the chords effected by the introduction of eccentric diagonals, from that obtained from normal elastic analysis, is shown in Fig. 2. The increase in load capacity, in the particular example, was about 13%.

Eccentric diagonals in tension become more rigid as the load increases, but, as they do not control the design, this is neglected.

TEST MODEL

In Fig. 3 the arrangement of the test model is shown. The chords were steel angles and the diagonals were aluminum T-section. Two bolts were used at each joint.

Supported at four points, the model was loaded by hydraulic jacks at the three nodes on the centre line. Chord forces were measured by strain gauges.

Elastic analysis indicated that the central compression chord would carry approximately double the force in the side chords. Analysis for the truss with eccentric diagonals showed that at the limit the chords forces would approach to within 20%.

The theoretical and test values are illustrated in Fig. 4.

On removing the load the model returned to the original configuration.

By using eccentric diagonals the load carrying capacity of the model was increased by 30% over that predicted by normal elastic analysis for a specified maximum chord force.

As the load approached the limiting value, the bowing of the diagonals became conspicious, demonstrating that this type of member gives a visible sign of approaching distress well before the structure collapses. This is in contrast with trusses composed of axially loaded bars in which failure is unheralded and is precipitous.

OPTIMUM T-SECTION

As torsional buckling does not limit the breadth/thickness ratio of T-sections loaded through the shear centre, the maximum value of 50:1 permitted by the extrusion process is used.

Lips are added to the flanges to provide resistance to local buckling and to increase the rigidity for buckling about the axis of symmetry. The optimum proportions for the practical range of section are given in Fig. 5. This section, even though eccentrically loaded, is more efficient than any combination of angles axially loaded.

Use has been made of this section in a roof 130m x 130m with a clear span of 60m.

CONCLUSION

There have been many attempts to obtain a reliable means of utilizing the post-buckling capacity of chord members in space trusses but none has yet been presented that shows any significant improvement over a single cycle elastic analysis. Failure of the main chords precipitates collapse.

By introducing a form of "plasticity" in shear, rather than in bending, it is possible to create a "quasi-ductile" behaviour that avoids the risk of brittle failure, provides evidence of impending failure, while remaining fully elastic up to the limiting load, and results in an economically worthwhile improvement in the overall load carrying capability.

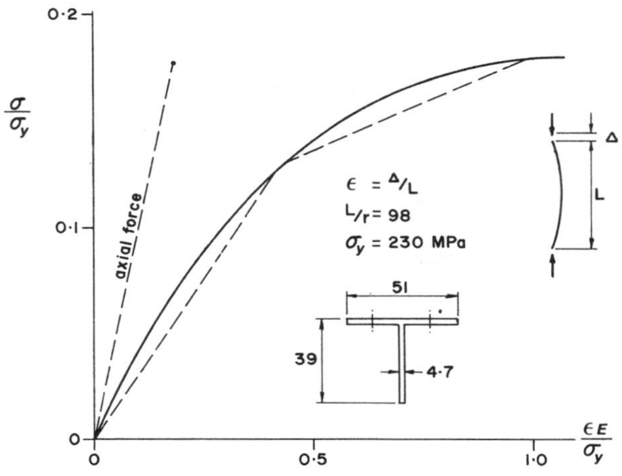

Figure 1. Typical Force-shortening of T-Strut

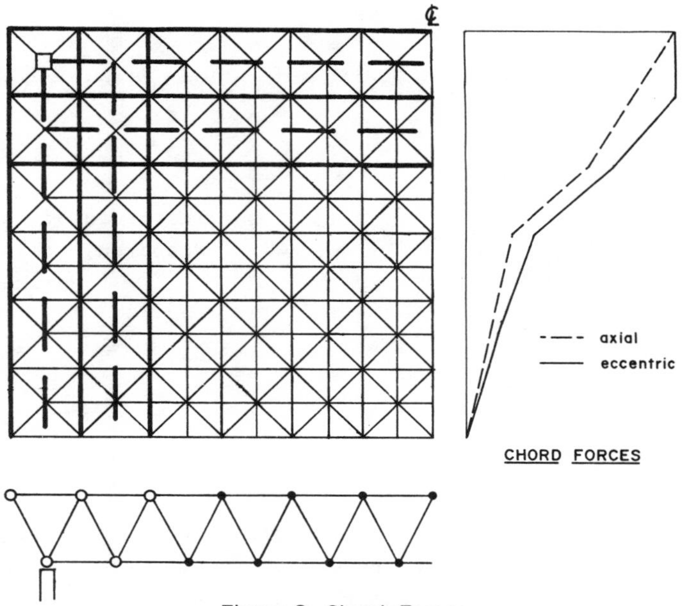

Figure 2. Chord Forces

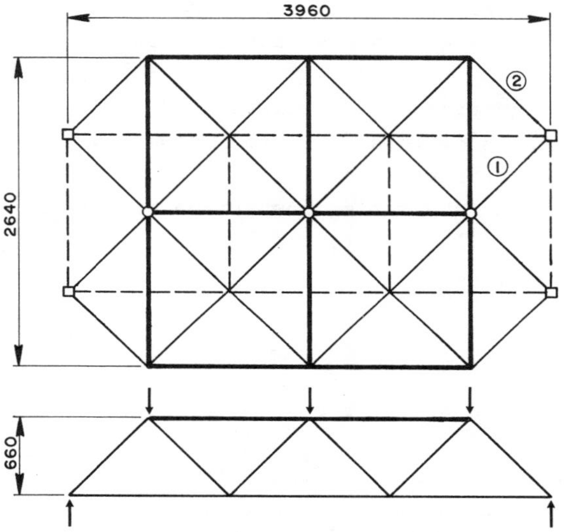

Figure 3a. Test Model

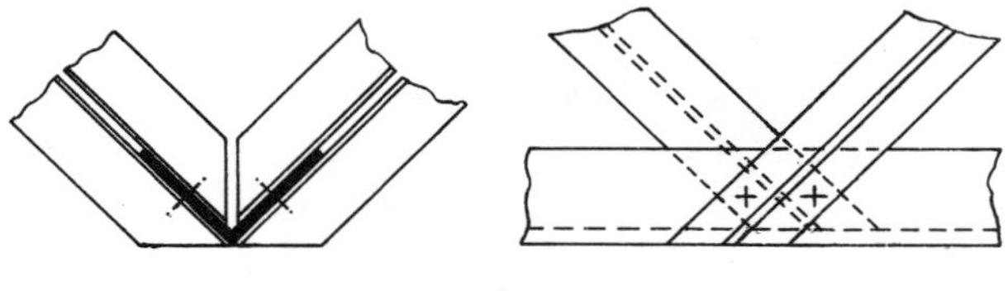

Figure 3b. Node Detail

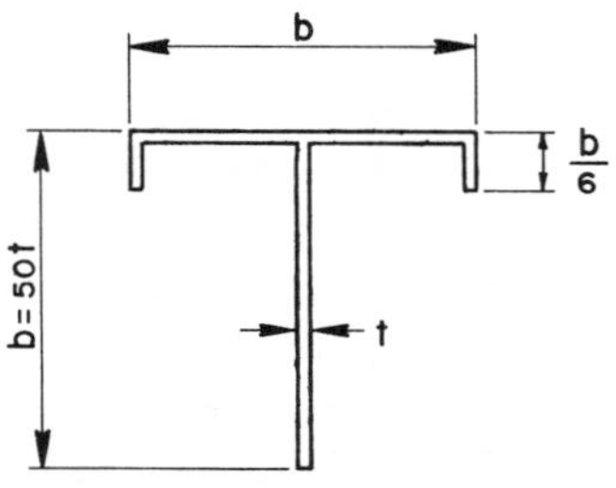

Figure 5. Optimum T-Section

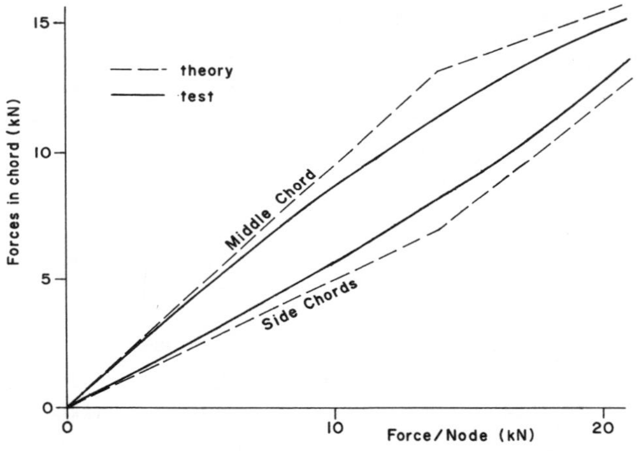

a) Chord Forces

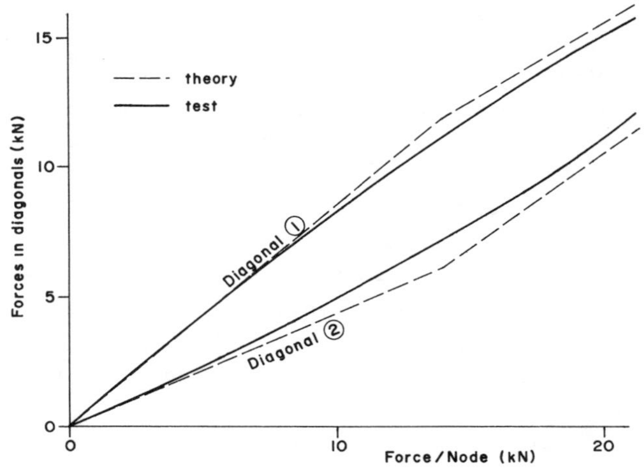

b) Diagonal Forces

Figure 4. Test Results

OPTIMUM DESIGN OF REGULAR SPACE STRUCTURES

J BAUER, W GUTKOWSKI and Z IWANOW

Institute of Fundamental Technological
Research, Polish Academy of Sciences,
Warszawa, Poland

The paper deals with the minimum weight design of regular space structures. In order to reduce the number of unknown joint displacements a semianalytical method is proposed to solve the problem.
It consists in applying the partial difference equations of equilibrium of a structural node and their solution by Galerkin method. This leads to a nonlinear programming problem with a reduced number of constraints.
The solution of practical problem together with an interesting comparison of two approaches based on finite and infinite sets of available struts section are presented.

INTRODUCTION

Space structures are in many cases composed of prefabricated elements i.e. struts and joints. Their different sizes constitute a finite set of values listed in a catalogue. Therefore the design of space structures consists in choosing appropriate elements from a given list in such a way as to obtain minimum cost of a structure under given constraints. For structural systems with only one type of joint the minimum cost problem leads to a minimum weight of total structure under constraints imposed on stresses, displacements, critical forces, natural frequencies etc. From the mathematical point of view this leads to discrete-continuous non-linear programming problems. This is because the listed elements constitute a finite set of values and stresses or displacements an infinite set.
The optimum design problems of space structures are numerically troublesome although there are already some, see Refs 4, 7 and 8, effective methods of their solution. In the case of regular space structures there is a possibility to overcome the major numerical difficulties by replacing a set of thousands of algebraic equations by a set of a small number of partial difference equations describing the equilibrium of typical node, see Refs 1, 2 and 6. Then solving these difference equations, for instance by applying the Galerkin method, we arrive at a nonlinear discrete-continuous programming problem with a small number of constraints conditions, see Ref 3.

PROBLEM FORMULATION

Let design a space structure which should achieve minimum weight subject to constraints in the form of equilibrium equations, maximum stresses and maximum displacements. Other words we have to find cross-sectional areas of bars for the considered structure in order to fulfill above conditions. However space structures are composed of hundreds or even thousands of elements and from practical point of view it is impossible to optimize the system assuming a different cross-sectional area for each member. For that reason designers usually divide the structure into several subsets or "equivalent classes" each of them being composed of elements having the same properties. Then the number of design variables is reduced to the number of assumed "equivalent classes" or subregions of the structure composed of the same elements. We start with objective function which in our case is

$$f=\sum_{s=1}^{so}\sum_{p=1}^{ps} c_p^s l_p^s \rightarrow \min \qquad \ldots\ \ldots\ 1,$$

where

c_p^s = weight of a unit lenght of the p-th bar in s-th equivalent class

l_p^s = lenght p-th bar in s-th equivalent class.

Next let define the constraint for our problem.

i/ We assume the space structure to be a regular system of bars and joints. If so we can write down three equilibrium difference equations for a node in upper plane of the structure and lower plane respectively. This gives us six equations of equilibrium in terms of six displacements. In many cases for which the number of segments of the structures are not very small we may simplify the problem by assuming some additional kinematic relations similar to the assumptions made in plate theory, see Ref 6. These additional assumptions allow to reduce mentioned six difference equations to one difference equation with respect to lateral displacement w.

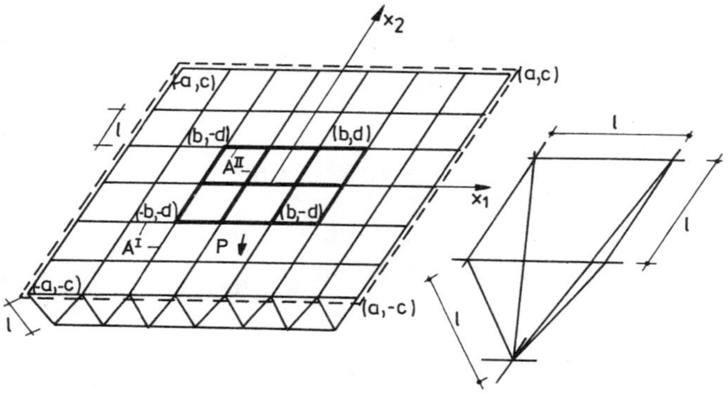

Fig 1

In the case of space structure of "Unistrut" system shown in Fig 1 this equation is

$$(\Delta^4_{2\,1} + \Delta^4_{2\,2})\, w(x_1,x_2) = \frac{8Pl}{EA^k}, \quad k = I, II \qquad \ldots\ \ldots\ 2,$$

where $\Delta^4_{2\,1}$ and $\Delta^4_{2\,2}$ denote fourth order difference operators with respect to two perpendicular axes x_1 and x_2. It can be shown in Ref 5 that such a simplifications give practically good results.

Solution of Eqn 2 for given boundary conditions gives then the constraint in terms of unknown cross-sectional areas of bars and unknown displacements.

ii/ Inequalities representing limitations on the displacements of choosen nodes

$$w(x_1^d, x_2^d) \leqslant w_o\ ,\ d=1,\ldots,d_o \qquad \ldots\ \ldots\ 3.$$

iii/ Inequalities representing limitations on the internal forces

$$N_p \leqslant N_o \qquad \ldots\ \ldots 4.$$

iv/ Constraints resulting from the fact that cross-sectional areas constitute a finite set of available values

$$A_p \in T \qquad \ldots\ \ldots\ 5,$$

where T is the list of available cross-section areas.

Objective function represented by Eqn 1 together with above constraints represented by Eqns 2,3, 4 and 5 constitute a discrete - continuous nonlinear programming problem. Below are presented solution of such a problem for simply supported space grid and a comparision with a solution obtained for a case when available cross section areas constitute a continuous set i.e. excluding constraint represented by Eqn 5.

OPTIMUM DESIGN OF SIMPLY SUPPORTED SPACE GRID

Let consider a rectangular space structure of Unistrut type simply supported on its boundaries shown in Fig 1, see Refs 9 and 10. We divide the structure into the two subregions composed of bars of the same cross-section area. Our task is to find cross-section areas AI and AII of bars in both subregions as a results of the solution of the problem formulated in previous section.

The objective function represented by Eqn 1 in our problem is the volume of all structural bars

/volume of structural joints is constant/

$$V = 8l\left[(ac-bd)A' + bdA''\right] \rightarrow \min \qquad \ldots\ \ldots\ 6.$$

The discussed above constraints are following

i/ The equilibrium equations in terms of lateral displacements of a node is

$$(\underset{2}{\Delta}{}_1^4 + \underset{2}{\Delta}{}_2^4)\,w(x_1,x_2) = \frac{8Pl}{EA^k} \quad , k = \mathrm{I}, \mathrm{II} \qquad \ldots\ \ldots\ 7.$$

Now we apply the Galerkin method to solve Eqn 7. First we express the function w in the form of a finite series of functions satisfying our boundary conditions

$$w(x_1,x_2) = \sum_j \sum_k C_{jk}\varphi_{jk}(x_1,x_2) \qquad \ldots\ \ldots\ 8,$$

with

$$\begin{aligned}\varphi_{jk} = &([x_1]_4 + 12[x_1]_3 + (28-6a^2)[x_1]_2 \\ &+ (8-12a^2)x_1 + 5a^4) \times ([x_2]_4 + 12[x_2]_3 \\ &+ (28-6c^2)[x_2]_2 + (8-12c^2)x_2 + 5c^4) \times [x_1]_j [x_2]_k\end{aligned} \qquad \ldots\ \ldots 9,$$

where

$$[x]_n = \underbrace{x(x-2)(x-4)\ldots\ldots(x-2n+2)}_{n}$$

It can be shown that φ_{jk} satisfy the boundary conditions. Let us confine to the first term of the series in Eqn 8. In this case this gives displacements which differ from the exact solution by no more than 2%.

Applying according to the procedure of Galerkin method we obtain

$$\begin{aligned}w(x_1,x_2) = &C_{oo}([x]_4 + 12[x_1]_3 \\ &+ (28-6a^2)[x_1]_2 + (8-12a^2)x_1 + 5a^4) \times \\ &([x_2]_4 + 12[x_2]_3 + (28-6c^2)[x_2]_2 \\ &+ (8-12c^2)x_2 + 5c^4)\end{aligned} \qquad \ldots\ \ldots 10.$$

and coefficent C_{oo} equal to

$$C_{oo} = \frac{\beta}{\alpha}\frac{1}{A^{\mathrm{I}}} + \frac{\gamma}{\alpha}\frac{1}{A^{\mathrm{II}}} \qquad \ldots\ \ldots\ 11,$$

where

$$\alpha = \sum_{x_1=-a}^{a}\sum_{x_2=-c}^{c}\left[\frac{E}{l}(\underset{2}{\Delta}{}_1^4 + \underset{2}{\Delta}{}_2^4)\varphi_{oo}\right]\varphi_{oo},$$

$$\beta = \sum_{x_1=-a}^{a}\sum_{x_2=-c}^{c} 8P\varphi_{oo} - \sum_{x_1=-b}^{b}\sum_{x_2=-d}^{d} 8P\varphi_{oo}, \qquad \ldots\ \ldots\ 12.$$

$$\gamma = \sum_{x_1=-b}^{b}\sum_{x_2=-d}^{d} 8P\varphi_{oo}$$

ii/ Let subject the lateral displacement w of the central joint of the structure to the constraint

$$w(0,0) \leq w_o \qquad \ldots\ \ldots\ 13.$$

Substituting expression represented by Eqn 11 into Eqn 10 we get for $X_1 = 0$; $X_2 = 0$ following constraint

$$25\left(\frac{\beta}{\alpha}\frac{1}{A'} + \frac{\gamma}{\alpha}\frac{1}{A''}\right)a^4c^4 \leq w_o \qquad \ldots\ \ldots\ 14.$$

iii/ Finally we have constraint as a result of the fact that available cross-sections constitute a finite set of four values, see Refs 9 and 10. Together i/, ii/ and iii/ constitute following nonlinear discrete programming problem:

$$f = 8l\left[(ac-bd)A' + bdA''\right] \rightarrow \min \qquad \ldots\ \ldots\ 15,$$

$$25\left(\frac{\beta}{\alpha}\frac{1}{A'} + \frac{\gamma}{\alpha}\frac{1}{A''}\right)a^4c^4 \leq w_o \qquad \ldots\ \ldots\ 16,$$

$$A', A'' \in \left[A_1, A_2, A_3, A_4\right] \qquad \ldots\ \ldots\ 17.$$

Let solve the our problem in two ways. Once assuming that our catalog is composed of four values only and secondly assuming that the list of available cross sections is infinite one. Introducing Boolean /zero-one/ variables the nonlinear discrete programming problem represented by Eqns 15, 16 and 17 is reduced to a linear one, see Ref 3 and now is possible to use one of numerical algorithms of logical programming.

For a constinuous programming problem represented by Eqns 15 and 16, applying Kuhn-Tucker optimality conditions, we get the set of three nonlinear equations with three unknowns: values of cross-sectional areas A^{I}, A^{II} and coefficient λ. In this particular case it is easy to solve the above set of equations and get in a explicit form expressions for sectional-areas of struts. For both cases, discrete and continuous programming problem, we consider two the same span square Unistrut space trusses, loaded by concentrated forces P = 529 kg for 4 ft and P = 924 kg for 5 ft module assemblies. These loading correspond to the same uniformly distributed load density of 400 kg/m^2 for both structures. As a constraint, assume the displacement of the central node of the structure equal to 3.0 cm, this being 1/400 of the structures span. Four different values of struts cross sections are given in the catalogue, taken from the Unistrut Standards as the most commonly used, see Refs 9 and 10. It is required to determine the value of A^{I} and A^{II} to give minimum total volume of the space truss with specified load, geometry, boundaries of subregions and with constraint on the maximum nodal displacement.

Numerical results for several alternative subregion configurations, i.e. for different values of "b" and "d", see Fig 1, are shown in Table 1.

Table 1

$a = c;\ b = d;\ W_o = 3.0$ cm

l = 1.22m, 4 ft, P = 592 kG					
b	A^{I}	A^{II}	A^{I}	A^{II}	f cm^3
	cm^2				
	C		D		D
2	6.99	14.16	7.62	7.62	743223.0
4	5.99	11.74	7.62	7.62	743223.0
6	4.57	10.0	3.81	13.24	702726.0
8	2.54	8.41	7.62	7.62	743223.0
10		7.57		7.62	743223.0

l = 1.52 m 5 ft, P = 924 kG					
b	A^{I}	A^{II}	A^{I}	A^{II}	f cm^3
	cm^2				
	C		D		D
4	4.12	8.76	3.81	10.69	431498.0
6	2.27	6.98	3.81	7.62	464514.0
8		6.03		7.62	594579.0

The catalogue A, see Refs 9 and 10

$$\begin{bmatrix} 0.591 & 1.182 & 1.657 & 2.052 \\ 3.81 & 7.62 & 10.69 & 13.24 \end{bmatrix} \begin{bmatrix} in^2 \\ cm^2 \end{bmatrix}$$

D - solution of discrete programming problem Eqns 15, 16 and 17

C - solution of continuous programming problem, Eqns 15 and 16

CONCLUSIONS

The presented semi-analytical method for discrete and continuous optimization of regular space structures shows the possibility of applying the method to a quite wide class of structural optimum design problems Using one of the well-developed analytical methods in solving the governing equations, we are able to transform a mathematical programming problem with a large number of constraints to a problem with a small number of constraints and design variables. The comparison between results for discrete and continuous programming problem, see Table 1, shows that in order to get a solution of discrete programming problem it can not apply results for continuous problem in a simple way i.e. using direct approximation but it needs to apply algorithms of integer or mixed programming.

REFERENCES

1. R R AVENT and D L DEAN, State of the Art of Discrete Field Analysis of Space Structures. Proc. 2nd International Conference on Space Structures, 7 - 16, Guildford, England, September /1975/.
2. J BAUER, J GIERLINSKI and W GUTKOWSKI, Semianalytical Solutions and Optimization of Space Truss. Proc. 2nd International Conference on Space Structures, 113 - 122,

Guilford, England, September /1973/.

3. J BAUER, W GUTKOWSKI and Z IWANOW, A Discrete Method for Lattice Structures Optimization, Engineering Optimization, vol.5, No 2, pp.121 - 128 /1981/

4. A CELLA and K SOOSAR, Discrete Variables in Sturctural Optimization, in Optimum Structural Design, Theory and Applications /Eds R H GALLAGHER and O C ZIENKIEWICZ/ 201 - 222, John Wiley and Sons, LONDON / 1973/.

5. L GAWKOWSKA, Optimization of Space Grids, Ph D Thesis, TECHNICAL UNIVERSITY OF SZCZECIN, 1982 /inPolish/.

6. W GUTKOWSKI, Mechanical Problems of Elastic Lattice Structures, in Progress of Aerospace Science /Ed D Kucheman/, 15, 181 - 216, Pergamon Press Ltd, Oxford and New York /1974/.

7. Z IWANOW, The Enumeration Method According to the Increasing Value of the Objective Function in the Optimization of Bar Structures, Bull.Acad.Polon.Sci.Serie Sci. Techn. 1982, 29, No 9 - 10, 482 - 486.

8. L A SCHMIT and C FLEURY, Discrete-Continuous Variable Structural Synthesis Using Dual Methods, AIAA Journal vol.18, No 12 /1980/, pp. 1515 - 1524.

9. UNISTRUT, Space Frame, Construction Procedures, Brochure SF-CP-3, Unistrut Corporation, Wayne, Michigan.

10. UNISTRUT, Space-Frame and Building Code Regulations, Brochure SF-BCR-3, Unistrut Corporation, Wayne, Michigan.

NON-LINEAR DISCRETE MODEL OF OPTIMIZATION OF SPACE TRUSSES

J A KARCZEWSKI, DSc, PhD, CEng, M ŁUBIŃSKI, DSc, PhD, CEng,
W M PACZKOWSKI, PhD, CEng

Associate professor, Warsaw Technical University, Poland

Professor, Warsaw Technical University, Poland

Lecturer, Technical University of Szczecin, Poland

This paper presents an effective method of guiding the process of optimization of two-layer space trusses of orthogonal networks.
The possibility of guiding practically the optimization of space trusses has been gained by introducing effective methods of searching through a set of permissible solutions, by the apropriate decomposition of a problem, and by the method of static analysis of the construction, which method reduces the time needed for calculating more than a hundred times as compared to rigorous methods.
The presented method allows for the optimization of regular space trusses of the number of struts LP $\in <250, 5000>$ and the number of nodes LW $\in <80, 1500>$ in 1/4 construction -if symmetry of structure and loading can be utilized - simultaneously with the analysis of seven decisive variables. The criterion of optimization may be the minimum capacity or weight of material as well as the comparative cost of the structure in which the truss is one of the constructional elements.

INTRODUCTION

Together with the development of automatization of calculations in engineering design emerged the possibility of shaping constructions in the best available way. Usually, during the process of optimization many variants of the designed system are analysed, and the most convenient of them for the assumed criterion is chosen. It is necessary for each of the variants to be analysed statically. The laboriousness of this depends on the size and complexity of the construction. The analysis of space trusses demands considerable efforts. The consideration of a large number of truss variants of proportions usually used in practical realizations is still impossible despite the continuous development of the calculation technique. The examples presented in most works display the difficulties connected with the time needed to analyse larger constructions - see Ref.6,9. Solved problems constitute a reason for further research - see Ref 1, 7, although only a few - see Ref 2, 4, 10, may be used in practice.
A review of optimal solutions of space trusses used to roof shows that investigations are carried out in three directions:
- geometrical shaping - see Ref 1, 2, 6,
- statical shaping - see Ref 4, 5, 8, 10, 11,
- constructional-technological shaping - see Ref 1, 3, 7, 10.

Some works include feature of geometrical shaping and of statical-resistant shaping - see Ref 1, 11 others present features of resistance-production shaping - see Ref 1, 10.
Presented method of optimization of space trusses which was worked out recently in WTU is a practical experiment of utilizing of our experiences with this problem.

FORMULATION OF THE PROBLEM

The minimum of the comparative cost of the hall

construction as a criterion of optimization was taken. The reference point may be any of the constructions belonging to the permissible area of the problem, in particular the initial construction. The value of the objective function for the reference point is

$$f(\bar{x}^{o},\bar{y}^{o}) = 1{,}0 \qquad \ldots\ldots 1.$$

In the problem of optimal forming halls roofing with space trusses the principle of decomposition is assumed. The construction is mentally divided into co-operating elements: the space truss, the supporting construction, roofing and the walls. There are distinguished local decisive variables which refer to respective object elements and global ones which are common to at least two elements. Global variables assume a proper choice of the direction of the search for the optimum. This makes it possible to divide the problem of optimization of hall into a couple of less complicated problems of element´s optimization. Each problem is solved independently, the coordinating factor being the homogeneous optimization criterion and global decisive variables-interlayer distance and internodes distance in the space truss /fig 1/.

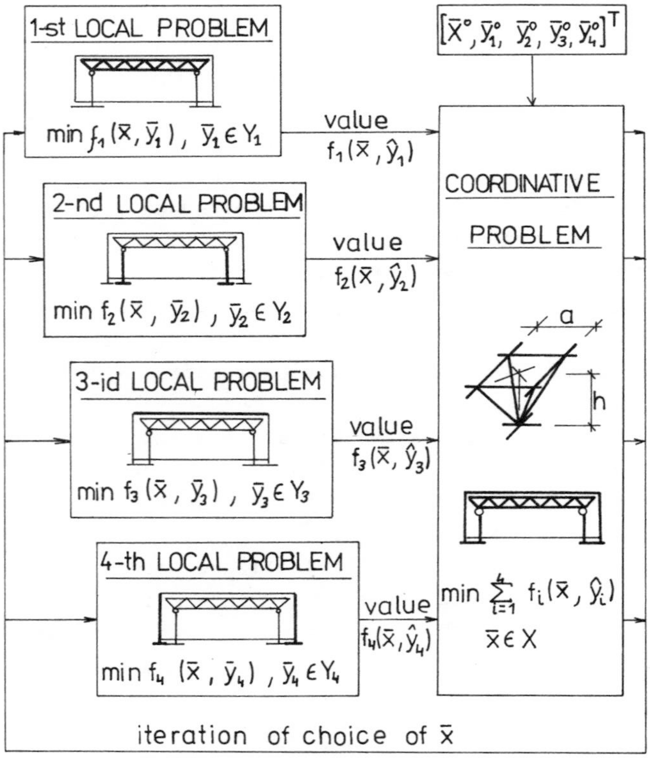

Fig 1

The general form of the optimization problem may be written in the following way

$$f(\hat{x},\hat{y}) = \min_{\substack{\bar{x}\in X\\ \bar{y}\in Y}} f(\bar{x},\bar{y}) =$$

$$= \min_{\bar{x}\in X} \sum_{i=1}^{6} \min_{\bar{y}_i\in Y_i} C_i \frac{K_i(\bar{x},\bar{y}_i)}{K_i(\bar{x}^{o},\bar{y}_i^{\,o})}; \qquad \ldots\ldots 2.$$

where

$$X = \{\bar{x}:\bar{g}_o(\bar{x}) \leqslant \bar{0},\quad \bar{h}_o(x) = \bar{0},\quad \bar{x}\in C\},$$

$$Y_i = \{\bar{y}_i:\bar{g}_i(\bar{x},\bar{y}_i) \leqslant \bar{0}, \qquad \ldots\ldots 3.$$

$$h_i(\bar{x},\bar{y}_i) = \bar{0},\quad \bar{x}\in C,\quad \bar{y}_i\in C\}$$

and

$$\bigcup_{i=1}^{6} Y_i = Y, \qquad \bigcup_{i=1}^{6} \bar{y}_i = \bar{y} \qquad \ldots\ldots 4.$$

$$f:(X\cup Y)\subset R^n,\quad \bar{g}_o:X\in R^n,\quad \bar{h}_o:X\subset R^n, \qquad \ldots\ldots 5.$$

$$\bar{g}_i:\ Y_i\subset R^n,\quad \bar{h}_i:\ Y_i\subset R^n$$

In eqn 2 ÷ 5 the following symbols are used:

- $f(\bar{x},\bar{y})$ - objective function of coordinating problem
- $K_i(\bar{x},\bar{y}_i)$ - objective function of local problem for fixed "i"
- K_1 - space truss´cost
- K_2 - supporting construction cost
- K_3 - roofing´s cost
- K_4 - cost of walls
- K_5 - maintenance cost
- K_6 - cost of designing of the hall
- $\bar{x}=[a,h]^T$ - vector of global decisive variables
- $\bar{y}$ - vector of local decisive variables
- $\bar{y}_i$ - vector of local decisive variables for fixed "i"
- $[\bar{x}^{o},\bar{y}^{o}]^T$ - vector of initial decisive variables
- X,Y,Y_i - sets of permissible vectors $\bar{x},\bar{y},\bar{y}_i$
- $\bar{g}_o,\bar{h}_o,\bar{g}_i,\bar{h}_i$ - limitation functions
- C_i - weight coefficients, where $\sum_{i=1}^{6} C_i = 1{,}0$.

The local objective function for the space truss has the form

$$F(x,y_1) = \sum_{ip=1}^{7} C_{ip} \frac{K_{ip}(\bar{x},\bar{y}_1)}{K_{ip}(\bar{x}^{o},\bar{y}_1^{\,o})}; \qquad \ldots\ldots 6.$$

where

- K_{1p},K_{2p},K_{3p} - cost of struts material of upper, mid and lower layers
- K_{4p} - cost of nodes material

K_{5p} - cost of connectors material
K_{6p} - cost of producing struts or prefabricated elements
K_{7p} - cost of producing modes or connector elements
C_{ip} - weight coefficients where $C_{ip} = C_1$

The space truss is described by vector of six values

$$\bar{x} = [\ I,J,K,L,M,N\ ]^T \qquad \ldots\ldots\ 7.$$

which can be assumed as decisive variables. The permissible area of the optimization problem is determined by the mechanical properties of used material, stability of the struts, usage conditions as well as constructional, technological and calculation conditions.
The parameters of optimization are: configuration of the struts, in a further passage called members, the way in which the construction is supported and loaded and the size of the truss projection. It is assumed that the truss´behaviour is linear-elastic, truss´structure does not colaborate with the roof cover, the load is acting on the construction only in nodes.

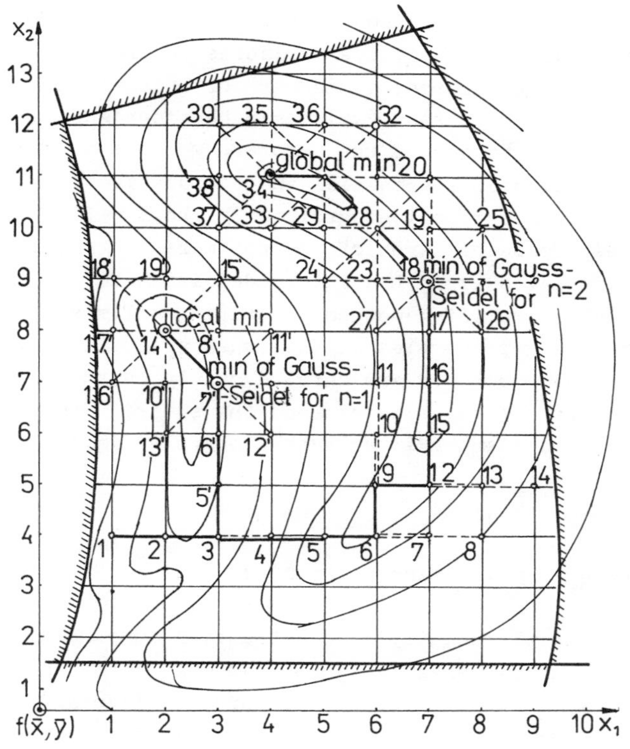

Fig 2

THE METHOD OF SEARCHING THROUGH THE PERMISSIBLE AREA

As a result of the analysis of the effeciency of some optimization methods - see Ref. 10, it was ascertained that in the case of decisive variables that give rise to a monotonic character of the objective function the most effective are the Gauss-Seidel and network methods.
For the rest of the decisive variables it is best to implement a full or decomposed search through variants.
In the modified Gauss-Seidel method the signal changing the direction of search is the appearance of not one, but n /n=2,3../ points, where the objective function does not improve its value - fig.2. Using $n > 3$ is pointless in the case of space trusses.
The deceptiveness of the Gauss-Seidel method in the points situated on the ridge and in the so called banana valleys has been eliminated by fusing it with the network method in one algorithm - fig.2. Due to the labouriousness of designing the construction, the algorithm is built in such a way that firstly one examines whether the decisive variables vector belongs to the permissible area and whether the construction has been analysed before. When these two conditions are fulfilled the designing and calculation of the objective function value takes place.

THE STATIC ANALYSIS OF THE SPACE TRUSS

The most important and one of the more laborious problems of the process of optimization of the space truss is the calculation of the inner forces acting in the members and of nodes displacements.
The computing difficulties are due to the fact that such constructions are highly hiperstatic. At the present level of the computing technology, the rigorous calculations of large constructions eg by the displacements method, considerably extend beyond the time available at computer centres.
The above mentioned calculation problems were overcome by introducing simplified static analysis, gained as the result of researches into the optimizing algorithm working according to more accurate known methods. This method shortens the time necessary for static calculations for the considered variant over 100 times, while the accuracy necessary for purposes of optimization remains sufficient.

The base for the determination of forces acting in the members of the construction with arbitrary values of decisive variables is the knowledge of the forces in members of initial and comparative trusses, which had been designed according to one of the accurate methods of static analysis. The initial and comparative trusses are constructions arbitrarly choosen from the permissible area of the optimization problem differing only in the internode distance.

In the simplified method of static analysis the following properties of the truss are utilized:

- bending moments are carried mainly by the upper and lower layers and the shearing forces are carried by interlayer members - fig.3.

$$N_{G2} = N_{G1} \cdot h_1/h_2 \; ;$$

$$N_{K2} = N_{K1} \cdot \sin\alpha_1/\sin\alpha_2 \; ; \qquad \ldots\ldots \; 8.$$

- flexural rigidity of construction supported and loaded exactly in the same way differing only in density of the network, is inversly proportional to the internode distance-supposing that the stresses in the members are taken in full advantage
- deflection of the truss does not depend on the internode distance - a

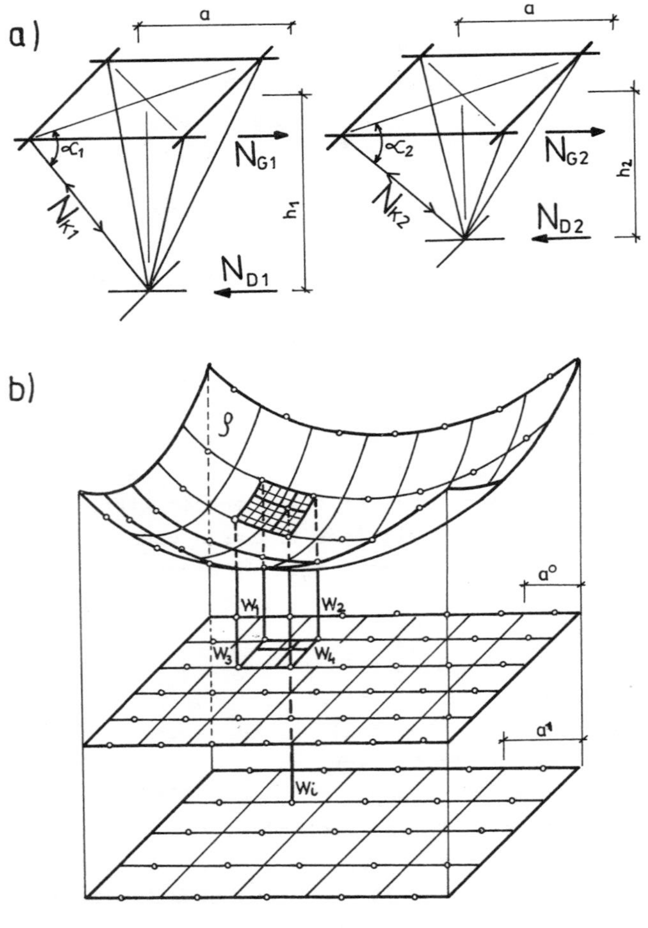

Fig 3

In numerical calculations, force surfaces which refer to the orthogonal location of members, are evaluated - fig.3a. Individual points of that surface are determined by attaching, in the middle of the strut value equal to that force which occurs in this member. The transformation of force surfaces coefficient is defined by comparing such surfaces of forces of the initial truss and of the comparative truss the forces of which are strictly determined. This coefficient is linear in character and depends on the internode distance - fig.3b.

Determination of forces for the truss with an arbitrary density of network is based on reading from the forces surface of the initial construction the value of the force for the coordinates of the member´s centre of gravity and multiplying it by the adequate value of the transformation coefficient. The values of the forces in the interlayer members are determined from the conditions for the node´s equilibrium.

It was ascertained from the analysis of numerical examples, that a change in the interlayer distance brings on average error of about 5%. A change in the internode distance brings on error of about 8%.

Due to discrete character of the members´ cross-section variations, relatively big differences in the value of inner forces neglect themselves in the process of determining the objective function value /2.2/. It is ascertained in the analysed cases that maximum differences in the constructions´weight do not exceed 3%. The fact that the method of analogy for the rigorous solution properly determines the direction of an objective function improvement and the optimum determined by using this method lies near of the optimum determined by the displacements method is of a great importance.

THE ALGORITHM OF THE OPTIMIZATION PROCESS

The general scheme of the optimization algorithm is shown in fig.4.

Analysis is done automatically by the program which is employing the approximate method of determining the forces in the truss´members with variable parameters.

There are three programs - see Ref 12 used for

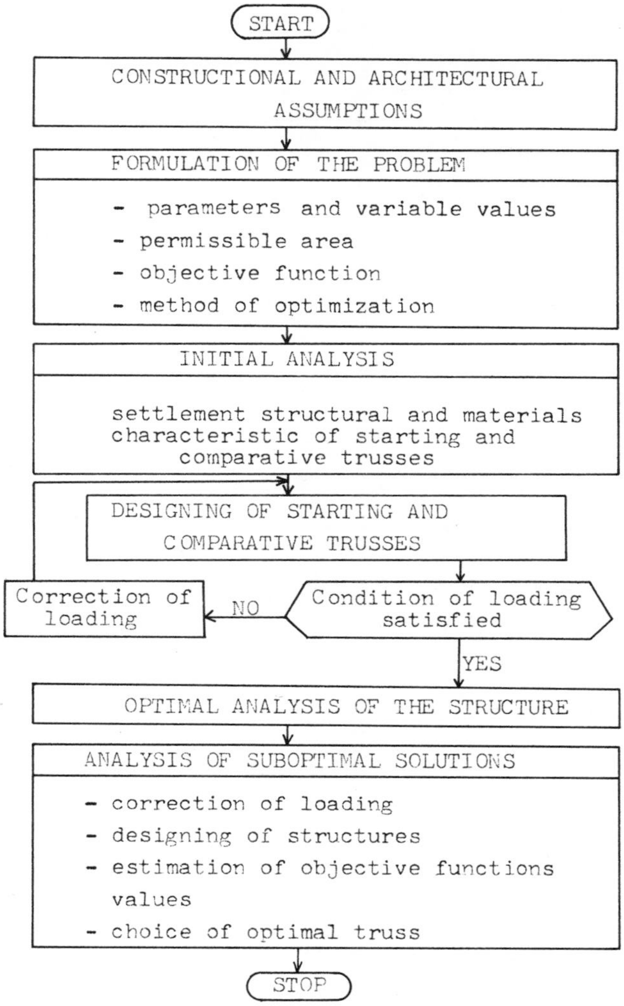

Fig 4

optimization computing. The program JC 12 based on the rigorous method of static analysis was choosen for the construction design. Due to the large number of data necessary for this program, the program DAG 1 adopted to the automatic generation of input data for the orthogonal truss was used - Both programmes DAG 1 and JC 12 operate one after the other in same computing cycle. Optimization is done by the program JC 6, which makes use of results of the initial and comparative constructions design made by the JC 12 program. The program JC 6 is written in FORTRAN IV and supposed to work on minicomputer PDP - 11/70. As a result of the analysis three semioptimal constructions are obtained. Due to the fact that the static analysis during optimization is carried out with approximate methods, the achived results should be verified. The semioptimal constructions should be analysed by a rigorous method. The optimal solution is choosen out of the semioptimal constructions after verification of them.

ANALYSIS OF THE OPTIMAL CONSTRUCTION

The aim of this analysis is the roofing for a commercial hall measuring 36 x 36 m with an inside diameter of 9,0 m /fig.5/.

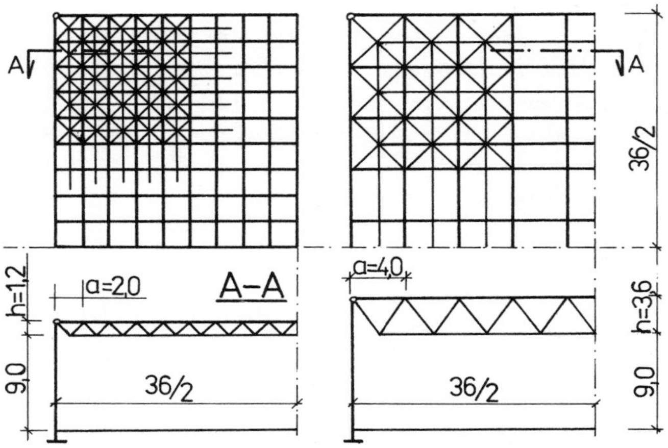

Fig 5

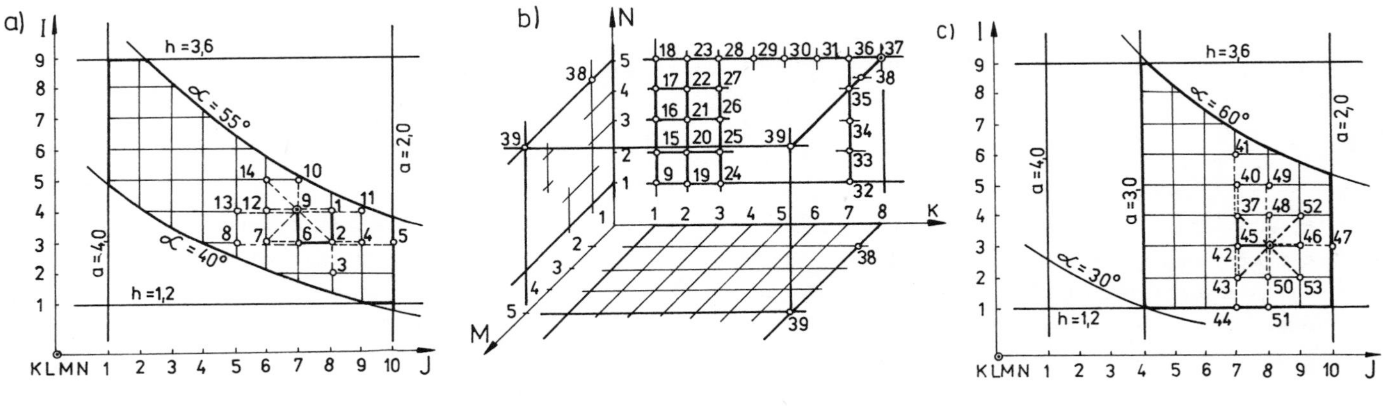

Fig 6

The truss is supported in the corners on four columns fixed in the foundation. The cover is made of corrugated zinc coated sheets protected from the cold by hard mineral wool. The total vertical load acting on the truss equals 1,7 kN/m^2.

The permissible area of the problem was taken as shown on Fig.6. If we suppose that the cross-sections of truss'members are determined optimally then there are 12.180 variants of constructions.

The criterion of optimization forms the comparative cost of the truss material, of the supporting construction and steel elements of the wall's construction.

The solution of the problem has been achieved after analysing 53 variants. The time CPU of the optimizing analysis using the minicomputer PDP-11/70 is equal to 10 min. 14 sek.

The three semioptimal constructions achieved as a result of JC 6 program are

1/ $[\bar{x}, \bar{y}]^T = [3,8,8,4,1,5]^T$

2/ $[\bar{x}, \bar{y}]^T = [2,8,8,4,1,5]^T$

3/ $[\bar{x}, \bar{y}]^T = [3,7,8,4,1,5]^T$

They are verified by the JC 12 program which implements displacement method. The parameters of the optimal truss remained identical, only the semioptimal solutions have been rearranged. Finally the dead weight of the truss is 313 N/m^2. The optimal height of the trusswork equals 1,8 m.

CONCLUSIONS

The method presented above enables the optimization of three dimentional orthogonal truss with members in layers with a span of up to 120 m. The implemented simplifications enable to make considerable extensions of the mathematical model of the construction. The possibility of a simultaneous analysis of seven project values as the decisive variables allows to design not only the three - dimensional truss, but also the whole construction of the hall. One of the most important elements of the optimization is the proper formulation of the problem.

The optimal formation of truss is a process which allows to achieve a considerable limitation of costs within applied calculations theories, mathematical models and technology assumptions.

Depending on the assumed criterion, function and the accuracy of the initial variant, the reduction of the costs achieved as the result of optimization varies from 2% to 30%. The designing of systems and trussworks of large spans is of particular importance.

On the other hand, the complex optimization of small construction designed individually is usually pointless, as it results in the rise of costs.

REFERENCES

1 Karczewski J.,Paczkowski W. Analiza optymalizacyjna kratownicy przestrzennej. Inż.i Bud. 3, 1979, s.96-98.

2 Motro R. Optimisation de structures spatiales et application a des grilles a double nappe. Construction Metallique. 2, 1976, p.24-36.

3 Biegus A., Kowal Z. Optymalizacja strukturalnych przekryć warstwowych metodami analizy wartości. V Konf.Nauk.-Techn. "Konstrukcje Metalowe", t.2, Warszawa 1974.

4 Karczewski J., Paczkowski W. Optymalizacja kratownicy przestrzennej metodą pełnego przeglądu. Inż.i Bud., 3, 1980, s.91-95.

5 Dzieniszewski W. Optimization of lattice rod structures. Arch.Mech.23, 2, 1971, p.223-248.

6 Lipson S.L., Gwin L.B. Discrete sizing of trusses for optimal geometry. J.Struct.Div. ASCE, vol 103, ST5, may 1977, p.1031-1046.

7 Eschenauer H. Arwendung der Vektoroptimierung bei räumlichen Tragstrukturen. Der Stahlbau, 4, 1981. s.110-115.

8 Bauer J., Gierliński J., Gutkowski W. Semi-analytical solutions and optimization of space trusses 2-nd Int.Conf. on Space Struct., Guildford 1975, p.113-122.

9 Gierliński J. Optymalizacja płyt siatkowych. Prace IPPT PAN, 5, 1975.

10 Paczkowski W. Optymalne kształtowanie regularnych kratownic przestrzennych. Praca doktorska, Politechnika Warszawska 1982.

11 Vanderplaats G.N., Moses F. Automated design of trusses for optimum geometry. J.Struct. Div.ASCE, vol 98, ST3, march 1972, p.671-690.

12 Karczewski J. and oth. Opracowanie metody sprężysto-plastycznej analizy przestrzennych układów prętowych wraz z uwzględnieniem przystosowania konstrukcji i elementów optymalizacji. Opracowanie w problemie węzłowym o7.1, zadanie 71.1. Warszawa 1980.

THE CONCEPTS OF NORM AND LIMIT IN STRUCTURAL OPTIMIZATION

A. BEHRAVESH, MSc, PhD

Assoc̸iated Professor, Department of Civil Engineering,

University of Tabriz, Tabriz, Iran

INTRODUCTION

Structural design is the process of determining the configuration and member dimensions of a structure, subject to specified performance requirements.

Traditional optimum design techniques require a reanalysis after each modification in the design variables. The techniques developed for modification of structural analysis have to be selected for each type of structure and modification.

The concept of norms has been used for the first time by Nooshin(1) for the study of digonal grids with members having torsional and shear rigidity, in the field of structural engineering. An extention of the concepts of norm and limit is employed in optimal structural design. The mathematical study consisting of several theorems is first presented and the process of optimization is described. The method is an iterative process with a high rate of convergence. Another application of the technique is presented for the analysis of large scale structures. For this purpose a more rapid and reliable iterative algorithm is given.

MATHEMATICAL STUDY

Let k_o,d_o and w be the stiffness matrix, the displacement vector and the external load vector for a linear structure, respectively. The relationship between the external loads and the joint displacements is given by:

$$k_o d_o = w \qquad \ldots\ldots 1.$$

Now, let the structure be modified in an arbitrary manner provided that:

1) The modified structure remains stable.

2) The modified structure has the same number of joints as the original structure and the number and the types of degrees of freedom at each joint remain unchanged.

3) The external load vector is not altered.

Examples of modifications consistent with the above restrictions (subject to the condition that the structure remains stable) are:

a) Inserting or removing members.

b) Changing the rigidities of the members.

c) Altering the positions of the joints.

The load-displacement relationship for the modified structure may be represented by:

$$(k_o + M)d = w \qquad \ldots\ldots 2.$$

Where $(k_o + M)$ and d are the stiffness matrix and displacement vector of the modified structure, respectively.

The precise values of the components of joint displacements of the modified structure can be found from the solution of Eqn 2 and the accurate values of the components of forces can then be obtained from the joint displacements and the force-displacement relations of the individual members. However, sometimes it is required to analyse a structure for a large number of cases in which some features of the structure are changing gradually. In such circumstances, one may think of analysing the structure for only a small number of basic cases and finding the approximate values of forces and displacements, for the other cases by extrapolation (or interpolation).

For the above type of problem the technique developed in the course of this paper will be found to be valuable.

The rest of this section is devoted to the explanation of a procedure for construction of a sequence of vectors, $d_1, d_2, \ldots, d_K, \ldots$ which is convergent to the displacement vector d of a modified structure, and deriving an upper bound for $\|d-d_K\|$. For this purpose 8 theorems are stated the proof of which can be found in Ref 2. In all the cases it is assumed that the original and modified structure are statically stable, and therefore the matrices k_O and $(k_O + M)$ are nonsingular.

THEOREM 1. For any sequence of non-zero scalars $\alpha_1, \alpha_2, \ldots, \alpha_K, \ldots$ the sequence of vectors $d_1, d_2, \ldots, d_K, \ldots$ obtained from the following relations

$$r_K = d_O - (I+K_O^{-1} M) d_{K-1} \qquad \ldots\ldots 3a.$$

$$d_K = d_{K-1} + \alpha_K r_K \qquad K=1,2,\ldots \qquad \ldots\ldots 3b.$$

converges to the displacement vector d of the modified structure provided that the $\lim_{K\to\infty} d_K$ exists (d_O is the displacement vector of the original structure.)

THEOREM 2. If $\|I-\alpha_K(I+K_O^{-1} M)\| < 1$, for $K=1,2,\ldots$, then the sequence of vectors d_K defined by relations 3 converge to displacement vector d. Where $\|I-\alpha_K(I+K_O^{-1} M)\|$ represents any norm of the matrix $I-\alpha_K(1+K_O^{-1}M)$.

THEOREM 3. All the eigenvalues of $(I+K_O^{-1} M)$ are positive.

THEOREM 4. The sequence of vectors d_K defined by the relations 3 converge to the displacement vector d if the following condition is satisfied.

$$0<\alpha_K < \frac{2}{\lambda_{max}} \qquad K=1,2,\ldots$$

Where λ_{max} is the largest eigenvalue of $(I+K_O^{-1} M)$, which has been proved to be non-zero and positive by the previous theorem.

THEOREM 5. For the sequence of scalars α_K which minimises the $\|I-\alpha_K(I+K_O^{-1} M)\|_{®}$ for $K=1,2,\ldots$, the sequence of vectors d_K defined by relations 3 converges to the displacement vector d.

THEOREM 6. For the sequence of scalars α_K which minimize $\|d-d_K\|_{®}$ for $K=1,2,\ldots$ and $R=(K_O + M)^{\frac{1}{2}}$, the sequence of vectors d_K defined by relations 3 converges to the displacement vector d.

THEOREM 7. An estimate of the rate of convergence of the sequence of vectors d_K defined by $d_K = d_{K-1}+\alpha_K r_K$ ....4a. is given by

$$\frac{\|d-d_K\|_{®}}{\|d\|_{®}} \leqslant \left(\frac{\lambda_{max}-\lambda_{min}}{\lambda_{max}-\lambda_{min}}\right)^K$$

$$\leqslant Max(|1-\lambda_{max}|, |1-\lambda_{min}|) \ldots\ldots 4b.$$

Where

$$r_K = d_O - (I+K_O^{-1}M) d_{K-1} \qquad \ldots\ldots 4c.$$

$$\alpha_K = \frac{r_K^T K_O r_K}{r_K^T (K_O+M) r_K} \left(\text{or } \alpha_K = \frac{2}{\lambda_{max} + \lambda_{min}}\right) \qquad \ldots\ldots 4d.$$

In THEOREM 4 and 5 the condition of convergence of the sequence of vectors d_K (defined by relations 3) is expressed in terms of the eigenvalues of the matrix $(I+K_O^{-1}M)$. Therefore, information about the eigenvalues of $(I+K_O^{-1}M)$ or even their upper bound and lower bound, as established below, provides valuable guideline for selecting the sequence of scalars α_K. However, if no information about the eigenvalues of $(I+K_O^{-1}M)$ can easily be obtained, the THEOREM 7, with an extra effort, provides a sequence of scalars α_K, independent of eigenvalues of $(I+K_O^{-1}M)$, such that the sequence of vectors d_K converges to the displacement vector d.

Now, considering that K_O and $(K_O + M)$ can be expressed in form VD_OV^T and VD_mV^T, respectively (for detailed information about diagonal Matrices D_O, D_m and rectangular matrix V see Ref 2), one can express the following theorem:

THEOREM 8. The eigenvalues of $(I+K_O^{-1} M)$, for a structure with varying cross-sectional properties, are bounded by the largest and smallest diagonal elements of $D_O^{-1}D_m$. It can be shown that if reducing any nonzero elements of diagonal matrix D_O to zero causes singularity in the matrix $K_O = VD_O V^T$, then the largest and smallest eigenvalue of $(I+K_O^{-1} M)$ are:

$$\lambda_{max} = \|D_O^{-1} D_m\|_3 \qquad \ldots\ldots 5a.$$

$$\lambda_{min} = 1/\|D_m^{-1} D_O\|_3 \qquad \ldots\ldots 5b.$$

The above theorems can be used for obtaining numerical values of displacements or forces when some features of a structure are changing gradually, and may be used in a qualitative manner for deriving different relationship or comparing the relative effects of different types of modifications. The discussion is so far quite general and applicable to any structure.

STRUCTURAL OPTIMIZATION AND THE CONCEPTS OF NORM AND LIMIT

In the above section, based on the concepts of norm and limit, a technique was developed by which the changes in the internal forces and displacements of a linear structure, produced by variations in some features of the structure, could be estimated. This technique finds immediate application in optimization processes. Any structural optimization process is evidently, concerned with the determination of the values of feature variables which result in the most favourable design. Such an end can be achieved by utilizing the technique developed in the paper.To wit,based on the results of a relatively small number of basic analyses,the internal forces and displacements of structure, corresponding to a large number of different values of the feature variables, could easily be estimated and the most favourable mode can then be found from direct comparison of the results.

The optimization procedures described in literature require that the merit function should have just one local extreme point in the region under consideration. However, in structural optimization there are examples that the merit function may have two or more local extreme points. As an example, consider a circular structural domain (see Fig 1a) which is simply supported. A point load P is to be transmitted to simple supports by means of three beams in flexure. All beams are to have the same cross-section and the merit function is the total weight of the structure. In this case, as illustrated in Fig 1, the merit function has two local minimums and a locus of minimum points. Therefore, it is often necessary to investigate crudely the whole region of the interest and then confine observation to the small region in which the only concern is the local behaviour of the merit function. The method suggested below is one way of achieving the purpose.

Let the members of a structure be divided into n groups, such that each group contains members having the same "variation factors', where a variation factor is a positive scalar which represents the rate of variation of a feature in members of a group. For example, if the cross-sectional area of a typical member in group i is changed from A_{oi} to A_{mi}, then $x_i = \frac{A_{mi}}{A_{oi}} > 0$ represents a variation factor for group i. A group with constant feature will have a variation factor equal to one. Subjected to different variations, a member may belong to more than one group, or a group may have more then one variation factor. Free grouping of members in a structure provides a means of easy application of any combination of variations on a structure.

Now suppose that displacement vector d_o is calculated for a linear structure with stiffness matrix $K_o = VD_oV^T$ (for detailed information about diagonal matrix D_o and rectangular matrix V see Ref 2) under external loading system w, i.e. $K_od_o = w$, and let the diagonal matrix D_o, which contains the cross-sectional properties of the members of the structure, be modified to D_m, then the stiffness matrix of the modified structure is given by VD_mV^T, and, the variation factors of members in group i can be expressed as: x_i=(diagonal elements of $D_mD_o^{-1}$ corresponding to the members of group i). Furthermore, denoting the maximum and minimum of x_i's for i=1,2,...,n by x_{max} and x_{min}, respectively, then

$$x_{min} = 1/\| D_m^{-1} D_o \| \qquad \text{...... 6a.}$$

and

$$x_{max} = \| D_m D_o^{-1} \| \qquad \text{...... 6b.}$$

If THEOREM 7, is used to estimate the displacements and the accuracy of the displacements are measured by $\| d-d_K \|_{®} / \| d \|_{®}$ and if this is less than a small value ε, then from the THEOREM 8 the feasible region for estimating the displacements is given by:

$$\frac{\| d-d_K \|_{®}}{\| d \|_{®}} \leqslant \left(\frac{\| D_o^{-1} D_m \| \; \| D_o D_m^{-1} \| -1}{\| D_o^{-1} D_m \| \; \| D_o D_m^{-1} \| +1} \right)^K \mathrm{Max}\left(|1-D_o^{-1} D_m|, \left|1-\frac{1}{D_o D_m^{-1}}\right| \right) \leqslant \varepsilon.$$

Substituting for $\| D_o^{-1} D_m \| = x_{max}$ and $1/\| D_o D_m^{-1} \| = x_{min}$ (see relations 6) into the above relations,

$$\left(\frac{x_{max} - x_{min}}{x_{max} + x_{min}} \right)^K \mathrm{Max}(|1-x_{max}|, |1-x_{min}|) \leqslant \varepsilon.$$

The region in which the values of x_{min} and x_{max} satisfying the above relation is shown in Fig 2. Confining the values of x_{max} and x_{min} to the square area, inside the feasible region, as shown in Fig 2, the following relations are obtained:

$$1 \leqslant x_{max} \leqslant 1 + \varepsilon^{\frac{1}{1+K}} \qquad \text{...... 7a.}$$

$$1-\varepsilon^{\frac{1}{1+K}} \leqslant x_{min} \leqslant 1 \qquad \text{...... 7b.}$$

These relations are of fundamental importance and, in the sequel, it will be demonstrated that they can be successfully applied to structural optimization.

Let the values of feature variables resulting in the most favourable design minimize function $F(x_1,x_2,...,x_n)$. The choice of a merit function $F(x_1,x_2,...x_n)$ out of a set of functions mainly depends on the question of relative importance of a criterion with respect to a set of criteria,the question of a criterion lending itself to easy definition, and the question of existing limitations,... etc. However, once a decision regarding the feature variables and merit function is made, then the optimization problem is defined. The next step will be direct comparison of designs and the technique introduced here will provide an efficient means of selection of a suitable deisgn. For example, if the set of variation factors $(x_1,x_2...,x_n)$ gradually change from $(a_1,a_2...,a_n)$ to $(b_1,b_2,...,b_n)$, then the minimization of the function $F(x_1,x_2,...,x_n)$ needs a tremendous amount of calculation to obtain the minimum value / or values of $F(x_1,x_2,...,x_n)$. Referring to relations 7, one can greatly reduce the amount of work as follows:

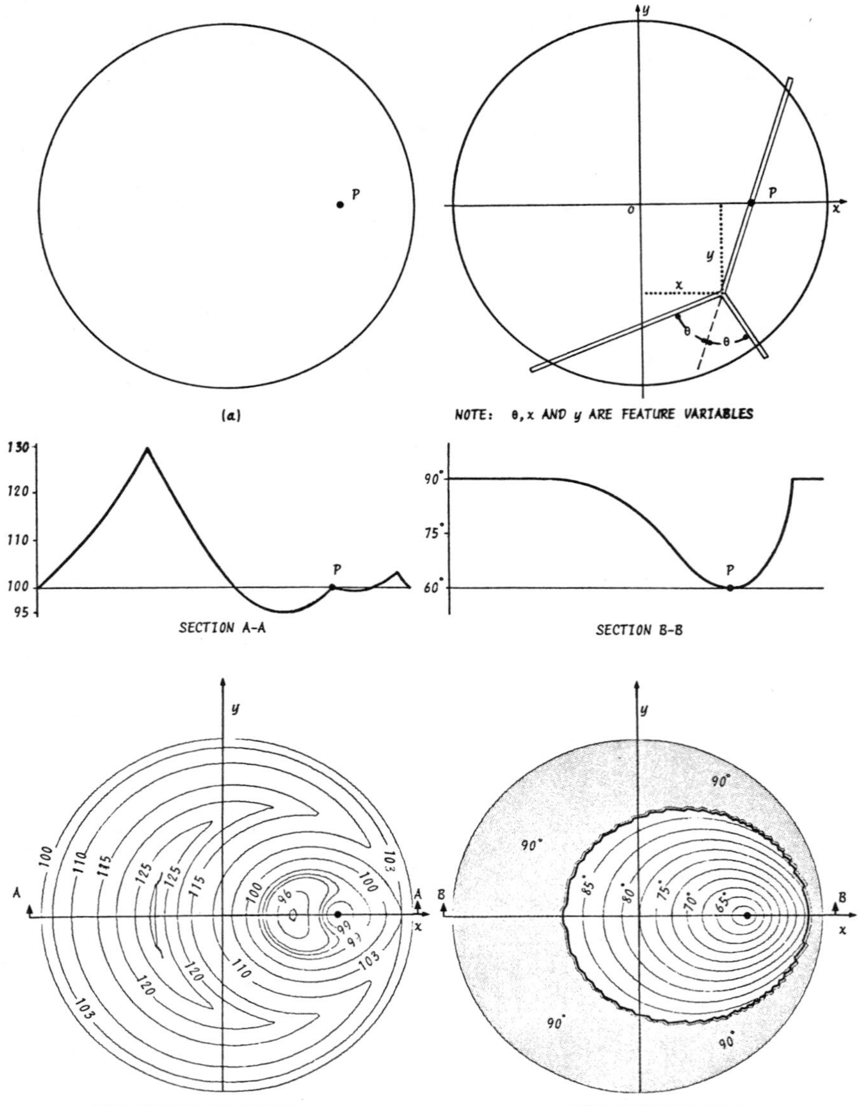

Fig 1

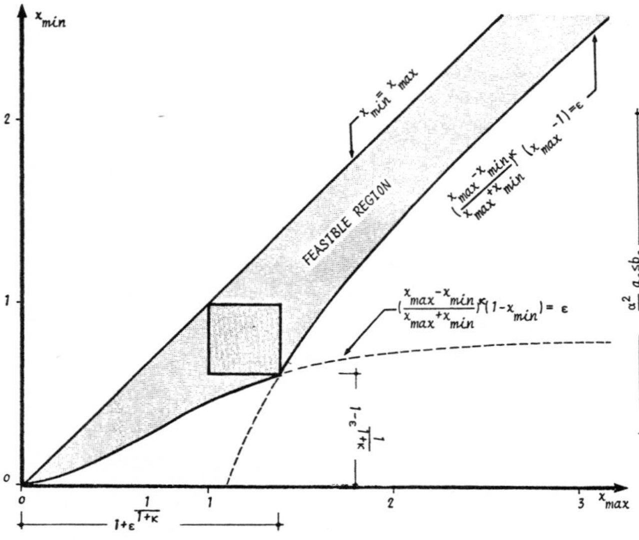

Fig 2

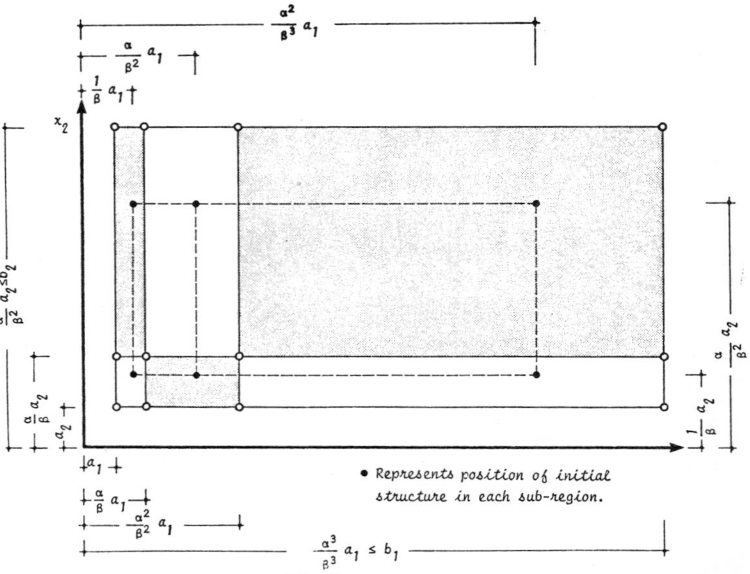

Fig 3

The whole region under consideration is divided into sub-regions provided that division in the direction of a typical variable x_i is

$$a_i \leqslant \frac{\alpha}{\beta} a_i \leqslant (\frac{\alpha}{\beta})^2 \quad a_i \leqslant \ldots\ldots \leqslant b_i \qquad \ldots\ldots 8.$$

where $\alpha = 1+\varepsilon^{\frac{1}{1+K}}$ and $\beta = 1-\varepsilon^{\frac{1}{1+K}}$. As an example, for the special case of n=2 the sub-regions are illustrated in Fig 3 and some of them are indicated by dotted areas.

From relations 7 and 8 it follows that each sub-region can be associated with an initial structure such that estimates for displacements are within the acceptable limit,

i.e. $\frac{\|d-d_K\|_{®}}{\|d\|_{®}} \leqslant \varepsilon$

This result enables the determination of the behaviour of function $F(x_1,x_2,\ldots,x_n)$ inside each sub-region with a desired accuracy. The repetition of this procedure for all sub-regions will give rise to the knowledge of the behaviour of the function $F(x_1,x_2,\ldots,x_n)$ throughout the region under consideration. It should be noted that in the above mentioned method the number of basic analysis are also optimal.

In general, for a linear structure with varying features, the local minimum points could be obtained through the following steps:

1) The stiffness matrix is written as the sum of a constant and a variable part and the force-displacement relation for modified structure is arranged in the form:

$$(K_o + \sum_{i=1}^{n} x_i D_i)d = w.$$

where D_i's are K_o-size constant matrices and a function of the initial cross-sectional properties and the interconnection pattern of the members in group i.
2) The initial structure is analysed under external loading system w.
3) A set of vectors is defined by relation $w_{io}=D_i d_o$, for $i=1,2,\ldots,n$ and the initial structure is analysed under loading systems $w_{1o},w_{2o},\ldots,w_{no}$.
4) Relations 7 are used to obtain the sub-regions in which the estimate for displacements are within the acceptable limit.
5) The displacements and forces in a structure are expressed as a function of d_o, $w_{1o},w_{2o},\ldots,w_{no}$ and then the function $F(x_1,x_2,\ldots,x_n)$ is evaluated.
6) Steps two to five are repeated to find the response surface of the function $F(x_1,x_2,\ldots,x_n)$ in the region under consideration.

OTHER APPLICATION

Considering THEOREM 7, the sequence of vectors

$$d_K = d_{K-1} + \alpha_K r_K \qquad \ldots\ldots 9a.$$

converges to displacement vector d, where

$$r_K = d_o - (I+K_o^{-1} M)d_{K-1} \qquad \ldots\ldots 9b.$$

and

$$\alpha_K = \frac{r_K^T K_o r_K}{r_K^T (K_o+M) r_K} \qquad \ldots\ldots 9c.$$

Let (K_o+M) represents the stiffness matrix of a given structure and, suppose (K_o+M) is decomposed into UU^T(U is an upper triangular matrix) such that:

$$U_{ii} \geqslant 10^{-n} \|K_o+M\| .$$

Where U_{ii} is the ith diagonal element of U,n is the number of decimal places to which matrix (K_o+M) is represented in the machine and $\|K_o+M\|$ represents any norm of (K_o+M). Now letting $K_o=UU^T$ the igenvalues of $I+K_o^{-1} M=K_o^{-1}(K_o+M)$ will be close enough to unity and, therefore from relation 4b

the ratio $(\frac{\lambda_{max} - \lambda_{max}}{\lambda_{max} + \lambda_{min}})^K$ rapidly tends to zero. Namely,the sequence of vectors d_K will converge to d rapidly.

Economical computation of d_K,r_K and α_K for successive K is in fact possible. From relation 9b

$$r_K = K_o^{-1} K_o d_o - K_o^{-1} K_o (I+K_o^{-1} M)d_{K-1},$$

or

$$r_K = K_o^{-1} [K_o d_o - (K_o+M)d_{K-1}] .$$

Substituting for $K_o d_o = w$ into the above relation,

$$r_K = K_o^{-1} [w-(K_o+M)d_{K-1}] .$$

Letting $w_{K-1} = w-(K_o+M)d_{K-1}$, it follows that

$$r_K = K_o^{-1} w_{K-1} \qquad \ldots\ldots 10a.$$

or

$$K_o r_K = w_{K-1} \qquad \ldots\ldots 10b.$$

Substituting for $K_o r_K$ from relation 10b into relation 9c,

$$\alpha_K = \frac{r_K^T w_{K-1}}{r^T (K_o+M) r_K} \qquad \ldots\ldots 11.$$

From relations 9,10 and 11, algorithm 1 can be achieved. Intermediate vectors w_K, r_K and d_K are included for computational convenience. Furthermore, the term $(K_o + M)d_{K-1}$ and $(K_o + M)r_K$ need not be evaluated by matrix multiplication, but would be constructed element by element. They in fact represent the component of external loading system caused by displacement vectors d_{K-1} and r_K in the structure.

REFERENCES

1 . Nooshin, H. "Diagonal Grids with Members having Torsional and Shear Rigidity", PH.D. Thesis, University of London, 1967.

2 . BEHRAVESH, A., " A Technique for Structural Optimization", PH.D. Thesis, University of Surrey, 1978.

THE ALGORITHM 1

start

Define (K_o+M), w, n, and $K=1$.

Decompose (K_o+M) into $UU^T = K_o$ under condition

$U_{ii} \geqslant 10^{-n} \|K_o+M\|$, $(K_o+M) \to UU^T$,

$d_{K-1} = (UU^T)^{-1} w$,

$w - (K_o + M)d_{K-1} \to w_{K-1}$

TEST $\|w_{K-1}\|$

$\|w_{K-1}\| \leqslant 10^{-n} \|w\|$ / \ $\|w_{K-1}\| > 10^{-n} \| w \|$

FINISH

$(UU^T)^{-1} w_{K-1} \to r_K$

$\beta = r_K^T w_{K-1}$

$(K_o+M) r_K \to w_{K-1}$

$\alpha_K = \beta/(r_K^T w_{K-1})$

$d_{K-1} + \alpha_K r_K \to d_K$

$K+1 \to K$

NOTE: The symbol " $\to$ " signifies "replaces".

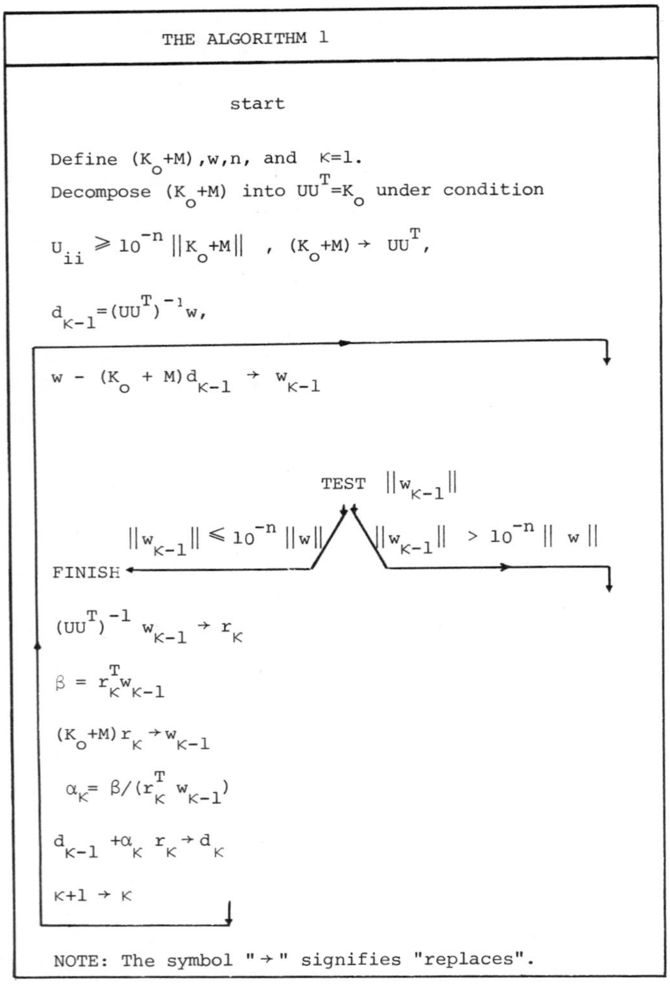

CONCLUSIONS

The method presented is simple efficient and applicable to structures of any configuration. In order to illustrate the reliability of the developed technique, some 65000 large scale single layer grids are studied. The results which are given in Ref 2 are prepared in the form of graphs to be used for optimal design of these structures.

ACKNOWLEDGEMENTS

The author is greatly indebted to Dr. H. Nooshin's guidance during his research in University of Surrey.

OPTIMAL DESIGN OF LONG-SPAN TRUSS GRIDS

G I N ROZVANY*, YEP K M**, and R SANDLER**

* Reader, Department of Civil Engineering, Monash University, Clayton, Victoria, Australia.

** Research Associates, Department of Civil Engineering, Monash University, Clayton, Victoria, Australia.

This paper is concerned with the optimal layout of long-span double-layered grids of constant depth consisting of a system of vertical plane trusses. A systematic method is developed for finding the minimum-weight layout for various types of boundary and loading conditions, taking the external load as well as the selfweight into consideration. In estimating the total weight of the trusses, it is assumed that the weight of the chords and web per unit length is proportional, respectively, to the absolute value of the bending moment and shear force acting on the truss.

The proposed method is based on theories of the first author and the late Professor Prager (Brown University). It is shown that for long-span truss grids the structural weight function is very sensitive to changes in the layout and, therefore, structural optimization achieves a significant saving in material consumption. Finally, a non-numerical computer algorithm for the selection of the optimal structural layout is described.

INTRODUCTION

During the late seventies, the late Professor William Prager and the first author developed a comprehensive theory for the optimization of the structural layout (Refs 1-9). Although it was originally formulated in the context of optimal plastic design, layout theory has been extended to optimal elastic design for given permissible stress, given compliance or given natural frequency (Refs 5, 10). More recently, the same theory was generalised to allow for selfweight (Ref 11). In this paper, optimal layout theory is applied to truss-grids of constant, prescribed depth. First, however, some basic principles of this theory are reviewed.

A BRIEF REVIEW OF OPTIMAL LAYOUT THEORY

Structural layout theory is based on two fundamental concepts, namely the Prager-Shield optimality criterion (Refs 12, 5, 13) and the notion of structural universe (Refs 4, 6, 9).

In optimal plastic design, the "specific cost" ψ or cost (or weight) per unit length, area, or volume can be expressed in terms of the generalised stresses $(\underline{Q})$ (i.e. local stresses or stress resultants) and then the total "cost" (Φ) is minimized subject to statical admissibility $(^{s})$:

$$\min_{\underline{Q}^{s}} \Phi = \int_{D} \psi(\underline{Q})\, d\underline{x} \qquad \text{... 1.}$$

where D is the structural domain referred to coordinates $\underline{x}$. As will be explained later, the same procedure may be used for the optimal elastic stress design of trusses and truss grids.

Considering for example a truss whose chords are parallel at a distance d from each other, the weight of the chords per unit length can be expressed as

$$\psi = \gamma|M|/\sigma_T d + \gamma|M|/\sigma_C d = a|M| \qquad \text{... 2.}$$

where γ is the specific weight of truss material, the generalised stress $Q_1 = M$ is the bending moment acting on the truss, σ_T and σ_C are permissible (or yield) stresses in tension and compression and $a = \gamma/\sigma_T d + \gamma/\sigma_C d$. Assuming that a suitable bracing ensures a reasonably constant value for σ_C, the specific cost function in Eqn 2 represents a good first approximation to the weight of chords per unit length. The total weight of the chords then becomes

$\Phi = \int_0^L a|M|\, dx$ where x is the distance along the truss axis and L is the truss length. In designing a truss plastically, the moments M(x) are required to satisfy the equilibrium condition

$$d^2M/dx^2 = -\,p(x) \qquad \text{... 3.}$$

and static boundary conditions only, where p(x) is the truss load per unit length. At a simple support (B), for example, the usual static boundary condition is $M(x)\big|_B = 0$.

In the elastic design of trusses and truss grids for a single load condition, the optimal moment value takes on a zero value at a sufficient number of points to render the structure statically determinate, unless a non-zero minimum cross-sectional area is prescribed for the chords. Vanishing chord members in optimal elastic design therefore ensure kinematic admissibility of the solution which becomes identical to the optimal plastic solution.

The extended version of the Prager-Shield condition (Ref 12) which is now also termed Prager-Shield-Rozvany theory in the literature (see Ref 14), can be expressed as an optimal strain-stress relationship

$$\text{on } D,\ \underline{q}^k = \underline{G}[\psi(\underline{Q}^s)] \quad \ldots 4.$$

where $\underline{q}$ is the generalised strain vector, $(^k)$ denotes kinematic admissibility (continuity of displacements) and $\underline{G}$ is the "generalised gradient" or "subgradient". The optimality criterion in Eqn 4 reduces a problem of optimal plastic design to a problem of (non-linearly) elastic analysis. This criterion is a necessary and sufficient condition for convex specific cost functions if at least one feasible solution exists.

Note: The "associated" strains $\underline{q}$ and displacements $\underline{u}$ given by the optimality criterion in Eqn 4 are fictitious quantities which do not necessarily represent actual elastic strains and displacements but facilitate greatly the optimization procedure. The above displacement field shall be termed the Pragerian displacement field.

Considering the truss example in Eqn 2 again, the relevant strain component is the vertical truss curvature

$$q_1 = \kappa = -\, d^2u/dx^2 \quad \ldots 5.$$

where u is the vertical truss deflection. The "subgradient" $\underline{G}$ denotes a collection of first derivatives of the specific cost function, $\underline{G}\ \psi\ (\underline{Q}) = (\partial\psi/\partial Q_1,\ \partial\psi/\partial Q_2,\ \ldots,\ \partial\psi/\partial Q_n)$ if $\psi(\underline{Q})$ is differentiable at a stress value $\underline{Q}$. At slope discontinuities of $\psi(\underline{Q})$, however, any convex combination of the slopes for the adjacent stress "regimes" may be taken (Refs 5, 13). Considering the truss example in Eqn 2, for example, the specific cost function $\psi = a|M|$ is shown graphically in Fig 1a and the corresponding subgradient $\kappa = G\psi$ indicated by Fig 1b in which

$$(\text{for } M > 0) \quad \kappa = a, \quad \ldots 6.$$

$$(\text{for } M < 0) \quad \kappa = -a, \quad \ldots 7.$$

$$(\text{for } M = 0) \quad -a \leqslant \kappa \leqslant a \quad \ldots 8.$$

It has been shown (Refs 15,11) that selfweight is automatically taken into consideration if the Prager-Shield condition in Eqn 4 is modified in the following form

$$\text{on } D,\ \underline{q}^k = \underline{G}[\psi(\underline{Q}^s)](1 + u) \quad \ldots 9.$$

where u is the vertical displacement.

The structural universe consists of all feasible (or potential or "candidate") members (Refs 4, 6, 9). Since the Prager-Shield condition gives a strain requirement (usually inequality) for vanishing generalised stresses (i.e. non-optimal members) also, see Figs 1a and 1b, its fulfilment for the entire structural universe constitutes a necessary and sufficient condition of layout optimality for convex specific cost functions.

OPTIMALITY CRITERIA FOR TRUSS-GRIDS

If the weight of the web is neglected, the specific cost of the trusses is moment-dependent (Eqn 2) and then Eqns 6-8 represent the relevant optimality criteria.

The combined specific cost (= weight per unit length) of the chords and web can be approximated by a moment and shear dependent specific cost function

$$\psi = a|M| + b|V| \quad \ldots 10.$$

where a and b are given constants, $Q_1 = M$ is the bending moment and $Q_2 = V = dM/dx$ is the shear force on the truss. The generalised strains corresponding to M and V are the curvature $q_1 = \kappa$ and shear strain $q_2 = \eta$ such that

$$u = u_M + u_V,\ \kappa = -\,d^2u_M/dx^2,\ \eta = -\,du_V/dx \quad \ldots 11.$$

where u is the total truss deflection, whilst u_M and u_V are its flexural and shear components. For the specific cost function in Eqn 10 the relevant optimality criteria furnished by Eqn 4 become

$$(\text{for } V > 0) \quad \eta = b \quad \ldots 12.$$

$$(\text{for } V < 0) \quad \eta = -\,b \quad \ldots 13.$$

$$(\text{for } V = 0) \quad -\,b \leqslant \eta \leqslant b \quad \ldots 14.$$

in addition to Eqns 6-8.

Finally, if selfweight of the trusses as well as external loads are taken into consideration, then in Eqns 6-8 and 12-14 "a" and "b" are replaced by a(1+u) and b(1+u), respectively. Then Eqns 5-9 and 11-14 furnish the following differential equation for the Pragerian deflection field:

$$-\,d^2u/dx^2 = \pm\, a(1+u) \pm b(du/dx) \quad \ldots 15.$$

where the double signs correspond to positive and negative values of M and V and for a zero value of M (or V) a (or b) is replaced by any value between a and -a (b and -b).

AN INTRODUCTORY EXAMPLE

Consider an elementary problem in which the structural universe consists of only two trusses, a cantilever truss having a length L and a simply supported truss having a length 4L (Fig 1c). A point load P at the intersection A of the trusses is to be supported by the two trusses such that P_1 is resisted by the cantilever and P_2 by the simply supported truss (Figs 1d and 1e) with

$$P_1 + P_2 = P \quad \ldots 16.$$

as an equilibrium condition. Assuming $P_1 > 0$, $P_2 > 0$, the moment and shear force diagrams for the two trusses are given by Figs 1f-1i. Assigning the cost coefficient values of a_1 and b_1 to the cantilever whilst a_2 and b_2 to the other truss, the components u_M and u_V of the Pragerian displacement field u for a system without selfweight are shown in Figs 1j-1m.

In considering specific values of the coefficients a_1, a_2, b_1 and b_2, the optimal solution will be determined by comparing the Pragerian deflection $u = u_M + u_V$ of the two trusses at their intersection A. Owing to the inequality conditions in Eqns 8 and 14 for M = 0 and V = 0, trusses with zero cross-sections may have a smaller deflection at A than the ones shown in Figs 1j - 1m.

Specific Examples without Selfweight

Moment-dependent cost. Considering the case $a_1 = 3$, $a_2 = 1$, $b_1 = 0$, $b_2 = 0$, the solution $P_1 > 0$, $P_2 > 0$ would give at the truss intersection (Figs 1j and 1k) $u^{(1)} = u_M^{(1)} = 3L^2/2$ and $u^{(2)} = u_M^{(2)} = 2L^2$. Since Eqn 4 requires kinematic admissibility (i.e. equal deflections at A), the solution with $P_1 > 0$, $P_2 > 0$ cannot be optimal. However, if we take $P_1 = P$, $P_2 = 0$ then by Eqn 8 (or Fig 1b) the value a_2 in Fig 1k is replaced by any quantity $\bar{a}_2$ whose absolute value is smaller than $a_2 = 1$. Adopting $\bar{a}_2 = 3/4$, we have by Figs 1j and 1k $u^{(1)} = u^{(2)} = 3L^2/2$ at the truss intersection. Since such a solution satisfies a sufficient condition of optimality (i.e. Eqn 4), $P_1 = P$, $P_2 = 0$ is in fact optimal, with vanishing truss members in the simply supported truss. Note that the considered solution is

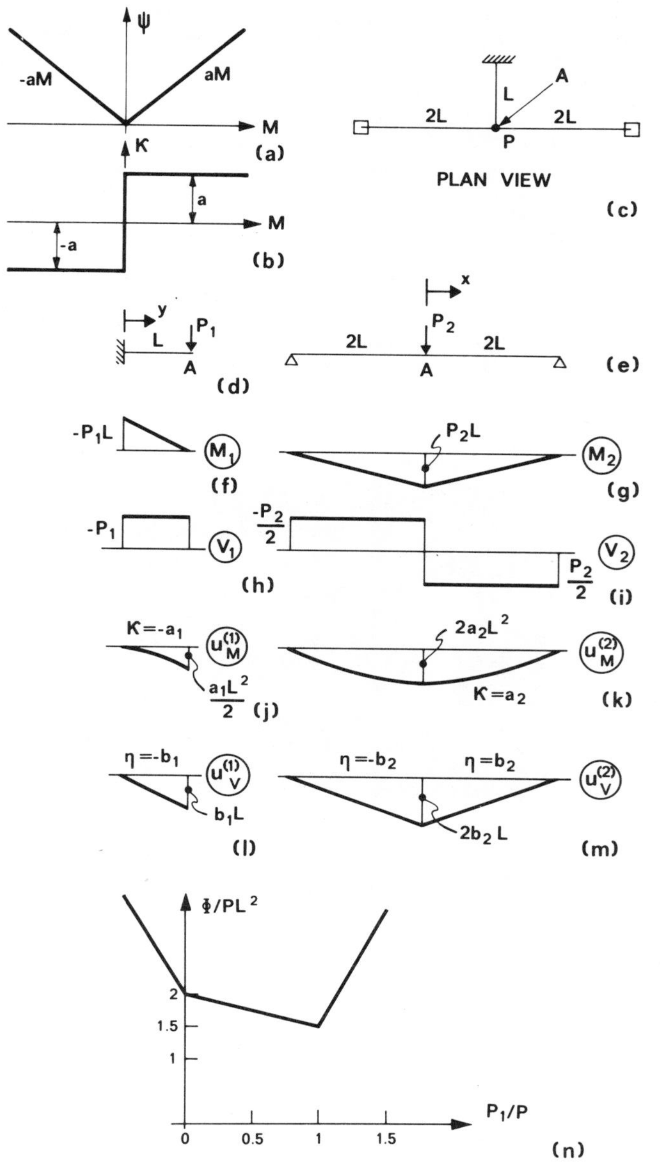

Fig 1 An elementary layout problem

statically determinate and hence it is valid for elastic as well as plastic design.

Note: In the particular case of $a_1 = 3$, $a_2 = 3/4$, Figs 1j and 1k furnish $u^{(1)} = a_1L^2/2 = u^{(2)} = 2a_2L^2$ for the truss intersection and hence by Eqn 4 any solution with $P_1 > 0$, $P_2 > 0$ is optimal in plastic design. However, in elastic design kinematic admissibility restricts the optimal solution to the limiting cases $P_1 = P$, $P_2 = 0$ and $P_1 = 0$, $P_2 = P$.

Moment and shear dependent cost. Let $a_1 = 3$, $a_2 = 1$, $b_1 = 3L$ and $b_2 = L$. Then Figs 1j - 1m furnish for the truss intersection $u^{(1)} = u_M^{(1)} + u_V^{(1)} = 4.5\ L^2$, $u^{(2)} = u_M^{(2)} + u_V^{(2)} = 4L^2$. Then the solution $P_2 = P$, $P_1 = 0$ is optimal, since for zero moment and shear in the shorter truss a_1 and b_1 in Figs 1j and 1ℓ can be replaced by $\bar{a}_1$ and $\bar{b}_1$ with $|\bar{a}_1| < a_1$ and $|\bar{b}_1| < b_1$ (see Eqn 14 and Fig 1b).

Check by independent calculations. Considering the foregoing problem with moment-dependent cost, it can be seen from the moment-areas in Figs 1f and 1g that the total cost is

$$\Phi = |P_1| a_1 L^2/2 + |P_2| 2a_2 L^2 = 3|P_1| L^2/2 + 2|P - P_1| L^2 \quad \text{... 17.}$$

The actual variation of the total cost Φ as a function P_1 is shown in Fig 1n in which the minimum value of Φ is indeed at $P = P_1$.

Similarly, by considering the areas of moment and shear force diagrams in Figs 1f - 1i, we have the total cost

$$\Phi = 4.5\ L^2 |P_1| + 4L^2 |P - P_1| \quad \text{... 18.}$$

confirming the minimum at $P_1 = 0$.

Pragerian displacement field as influence line for optimal cost. Once the Pragerian displacement field satisfying Eqn 4 or 9 is found, the minimum total cost Φ_{min} can also be determined from the dual statement

$$\Phi_{min} = \int_D [u\ p - (1+u)\ \hat{\psi}\ (\frac{q}{1+u})]\ dx \quad \text{... 19.}$$

where $\hat{\psi}$ is the complementary cost having the meaning

$\hat{\psi} = \int_0^{\bar{q}} Q(q)\ dq$ on the basis of the strain-stress relationship in Eqn 4. It can be seen from Fig 1b that $\hat{\psi} = 0$ for $|q_1| = |\kappa| < a$ (and for $|q_2| = |\eta| < b$). Hence the minimum total cost Φ_{min} can be calculated by taking the product of the point load P and the Pragerian displacement u_A at the intersection A. In the first example above, $u_A = a_1\ L^2/2 = 3L^2/2$ from Fig 1j and the total cost from Fig 1n (Eqn 17) is indeed $3PL^2/2$. Similarly, in the second example $u_A = 2a_2L^2 + 2b_2L = 4L^2$ and Φ_{min} by Eqn 18 with $P_1 = 0$ is $4PL^2$.

Allowance for Selfweight

Moment-dependent cost. For this case, Eqns 6-9 and end conditions furnish

$$-d^2u_1/dy^2 = -\ a_1(1+u_1),\ -d^2u_2/dx^2 = a_2\ (1+u_2),$$

$$u_1(0) = du_1(0)/dy = 0,\ du_2(0)/dx = u_2(\pm\ 2L) = 0,$$

$$u_1 = \cosh\ (\sqrt{a_1}y) - 1,\ u_2 = [\cos\ (\sqrt{a_1}x)/\cos\ (2\sqrt{a_1}L)] - 1,$$

$$u_1^A = \cosh\ (\sqrt{a_2}L) - 1,\ u_2^A = [1/\cos\ (2\sqrt{a_2}L)] - 1 \quad \text{... 20.}$$

This means that $P_1 = P$, $P_2 = 0$ is optimal if

$$\cosh\ (\sqrt{a_1}L) < \sec\ (2\sqrt{a_2}L) \quad \text{... 21.}$$

and $P_1 = 0$, $P_2 = P$ is optimal if the inequality in Eqn 21 is reversed.

Moment and shear dependent cost. From Eqn 15, after introducing $\beta_1 = \sqrt{4a_1^2 + b_1^2}/2$, $\beta_2 = \sqrt{4a_2^2 - b_2^2}/2$,

$$-\ d^2u_1/dy^2 = -\ a_1(1+u_1) - b_1(du_1/dy),$$

$$u_1(0) = 0,\ du_1(0)/dy = b_1,$$

$$(\text{for } x > 0)\quad -\ d^2u_2/dx^2 = a_2(1+u_2) + b_2(du_2/dy)$$

$$u_2(2L) = 0,\ du_2(0)/dx = -\ [1+u(0)]b_2$$

$$u_1 = [e^{b_1y/2}/\beta_1]\ [\beta_1 \cosh\ (\beta_1y) + \tfrac{1}{2}\ b_1 \sinh\ (\beta_1y)] - 1,$$

$$u_2 = \{e^{b_2(2L-x)/2}/[\beta_2\cos\ (2L\beta_2) - \tfrac{1}{2}\ b_2 \sin\ (2L\beta_2)]\}$$

$$[\beta_2 \cos\ (\beta_2x) - \tfrac{1}{2}\ b_2 \sin\ (\beta_2x)] - 1 \quad \text{... 22.}$$

Then $P_1 = P$, $P_2 = 0$ is optimal if $u_1(L) < u_2(0)$ or

$$(e^{b_1 L/2}/\beta_1)[\beta_1 \cosh(\beta_1 L) + 1/2\, b_1 \sinh(\beta_1 L)]$$
$$< \beta_2\, e^{b_2 L}/[\beta_2 \cos(2\beta_2 L) - 1/2\, b_2 \sin(2\beta_2 L)] \qquad \ldots 23.$$

and $P_1 = 0$, $P_2 = P$ is optimal if the inequality in Eqn 23 is reversed.

The above optimal solutions have been checked by rather lengthy independent calculations in which the actual total weight of the system was determined by integrating the specific cost over the two trusses.

Note: If the truss-grid contains a relatively small number of potential members, then a numerical method is suitable for optimizing its layout. However, if the number of potential members is extremely large, then an analytical approach based on a continuum model is much more appropriate. Whereas in the above example the structural universe consisted of two trusses only, most layout problems considered by the authors involves an infinite number of potential members.

THE OPTIMAL LAYOUT OF TRUSS-GRIDS

Problem Formulation

A vertical load system $p(x,y)$ acting over a horizontal plane domain D with coordinates (x,y) is to be transmitted to the boundary B of such domain by means of a system of intersecting parallel-chord trusses whose mid-surface is vertical. The distance between the two chords is prescribed and constant throughout. Considering any one point (x,y) of the domain D, the structural universe consists of potential trusses passing through such point in all horizonal directions.

Truss-grids Without Selfweight

In the case of a moment-dependent specific cost function $\psi = a|M|$ for all trusses, the weight of the web is neglected. In that case, Eqns 6 and 7 require a vertical curvature of $\kappa = a$ or $\kappa = -a$ along any truss of non-zero cross-section depending on the sign of the bending moment acting on the truss where $\kappa = -\partial^2 u/\partial w^2$, u is the vertical truss deflection and w is the distance measured along a particular truss. In addition, Eqn 8 requires that along any truss of zero cross-section [non-optimal or "vanishing" truss (Prager)], the curvature κ has an absolute value which is not greater than a. It follows that in the directions of optimal force transmission, the curvature takes on a directionally maximal or minimal value and hence optimal trusses may only occur in principal directions of the deflection field $u(x,y)$. Hence the optimal solution is associated with a deflection field such that (i) at any loaded point of D at least one principal curvature has an absolute value of "a" ($|\kappa_1| = a$ or $|\kappa_2| = a$); and (ii) a system of trusses along principal lines with an absolute curvature "a" can develop a statically admissible moment field whose sign coincides with those of the principal curvatures. This means that only the following five types of "optimal regions" may occur in a solution:

$$\begin{array}{lll}
S^+: & \kappa_1 = \kappa_2 = a, & M_1 > 0,\ M_2 > 0 \\
S^-: & \kappa_1 = \kappa_2 = -a, & M_1 < 0,\ M_2 < 0 \\
R^+: & \kappa_1 = a,\ |\kappa_2| < a, & M_1 > 0,\ M_2 = 0 \\
R^-: & \kappa_1 = -a,\ |\kappa_2| < a & M_1 < 0,\ M_2 = 0 \\
T\ : & \kappa_1 = a,\ \kappa_2 = -a, & M_1 > 0,\ M_2 < 0
\end{array} \qquad \ldots 24.$$

It will be seen that a rather complicated optimal layout problem has now been replaced with a relatively simple geometrical problem: the domain D must be covered with the five types of regions given in Eqn 24 such that (a) the deflection $u(x,y)$ is continuous and slope-continuous throughout, and (b) it satisfies the kinematic boundary conditions (along a built-in edge, for example, $u = \partial u/dx = \partial u/dy = 0$).

Using the above approach, a systematic method has been developed for determining directly the optimal truss-grid (originally: beam-) layout for most boundary and load conditions (Refs 1, 2, 4-6, 16). Moreover, a computer algorithm has been developed (Ref 17) for generating through purely analytical (non-numeric) operation the optimal truss-grid (or beam) layout for any boundary consisting of straight segments. Figure 2 indicates, for example, a computer-generated layout for a more complicated boundary condition. Naturally, such complex minimum weight layouts are not practical but they provide a basis of comparison for assessing the relative economy of practical designs.

Fig 2 A computer-generated optimal truss-grid layout.

In the case of moment and shear dependent specific cost functions, the weight of both the chords and the web is taken into consideration and then the optimality conditions in Eqns 5-8 and 11-14 apply along trusses. A comprehensive theory for this more refined formulation is also available and the minimum-weight layout for the specific cost function $\psi = a|M| + b|V|$ has been determined for a number of boundary shapes (Ref 18).

Truss-Grids with Selfweight

Long-span truss-grids subject to external load plus self-weight can be readily optimized using the modified Prager-Shield condition in Eqn 9. Considering first a moment-dependent specific cost function for the trusses, $\psi = a|M|$, the optimal regions will be the same as in Eqn 24 except that the quantity "a" is everywhere replaced by $a(1+u)$. Finally, if the specific cost function depends on both the moment and shear, $\psi = a|M| + b|V|$, then along all trusses the shear strains $\eta = -du_V/dw$ must also be taken into consideration. The latter are given by Eqns 12-14 in which "b" is replaced by $b(1+u)$. Because of the space limitation, only a simple example will be discussed in detail in this paper.

OPTIMAL LAYOUT OF A SIMPLY SUPPORTED CIRCULAR TRUSS-GRID WITH SELFWEIGHT

We consider a circular domain having a radius R with a simple line support along its boundary and subject to a uniformly distributed load p. It has been known for some time (see e.g. Ref 5) that if selfweight is neglected then the solution consists of a single S^+ region (Eqn 24). Adopting polar coordinates (r, θ) with the origin at the centre of domain, the Pragerian displacement field for the optimal solution is

$$u(r) = a(R^2 - r^2)/2 \qquad \ldots 25.$$

satisfying the boundary condition u(R) = 0 as well as the curvature condition in Eqn 24 : $\kappa = -\partial^2 u/\partial r^2 = a$ in the radial direction as well as in any other (w) direction $\kappa = \partial^2 u/\partial w^2 = a$. Then Eqn 24 implies that the total structural weight is equally optimal if trusses are placed in any arbitrary direction within the domain so long as the truss bending moments are everywhere positive. The minimum structural weight Φ_{min} can be readily determined from dual formulation (Eqn 19)

$$\Phi_{min}/2\pi = p \int_0^R ur\, dr = ap \int_0^R (R^2 - r^2)\, r dr/2 = apR^4/4 \qquad \ldots 26.$$

On the other hand, if selfweight is taken into consideration then the solution for a circular boundary becomes unique and always consists of radial and circumferential trusses. Such a layout is indicated in Fig 3 in which the radial (M_r) and circumferential (M_θ) moments are resisted by chords and the radial shear V is transmitted by web members in the radial direction. The optimization procedure is then reduced to determining the radial and circumferential moments within the equilibrium constraint

$$\partial^2(rM_r)/\partial r^2 - \partial M_\theta/\partial r = -r(p + \psi) \qquad \ldots 27.$$

For a purely moment dependent specific cost function, the solutions were given earlier (Refs 9,11).

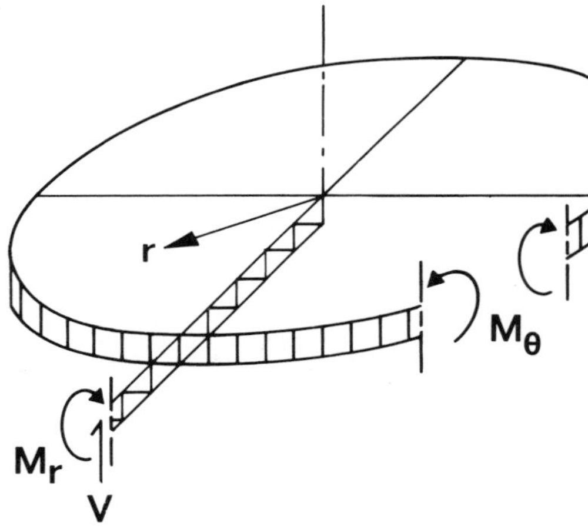

Fig 3 Axially symmetric truss grid.

Moment and Shear Dependent Specific Cost

In this section, symbols with overbars shall represent dimensional quantities and those without overbars non-dimensional ones. Using the specific cost function

$$\bar{\psi} = \bar{k}\,(|\bar{M}_\theta| + |\bar{M}_r|) + \bar{k}_1\,|\bar{V}| \qquad \ldots 28.$$

where $\bar{k}$ and $\bar{k}_1$ are given constants, the total "cost" (structural weight) $\bar{\Phi}$ can be determined by integration

$$\bar{\Phi} = 2\pi \int_0^R \bar{\psi}\, \bar{r}\, d\bar{r} \qquad \ldots 29.$$

Introducing the non-dimensional notation $M_i = \bar{M}_i\, \bar{k}/\bar{p}$, $k_1 = \bar{k}_1/\bar{k}\,\bar{R}$, $r = \sqrt{\bar{k}}\,\bar{r}$, $R = \sqrt{\bar{k}}\,\bar{R}$, $\Phi = \bar{\Phi}\,\bar{k}/\pi\bar{p}$, $\psi = \bar{\psi}/\bar{p}$ and $\kappa_i = \bar{\kappa}_i/\bar{k}$ $(i = r, \theta)$, Eqns 28 and 29 change into

$$\psi = |M_\theta| + |M_r| + k_1\, R\, [M_\theta - d(rM_r)/dr]/r,$$

$$\Phi = 2 \int_0^R \psi\, r\, dr \qquad \ldots 30.$$

Then the modified Prager-Shield condition in Eqn 9 furnishes the optimality criteria

$$(\text{for } M_\theta \gtrless 0) \qquad \kappa_\theta = -(du/dr)/r = (1+u)(\pm 1 + k_1\, R/r)$$

$$(\text{for } M_r \gtrless 0) \qquad \kappa_r = -d^2u/dr^2 = \pm (1+u) + k_1\, R(du/dr) \qquad \ldots 31.$$

For $M_\theta = 0$ the quantity ± 1 is replaced by any value between -1 and +1 and for $M_r = 0$ the term $\pm(1+u)$ can take on any value betwen -(1+u) and (1+u). Using the above optimality criteria, the layout of circular truss grids has been optimized and the corresponding optimal structural weights (as well as those of some non-optimal solutions) are indicated in Fig 4. All total weight values have been

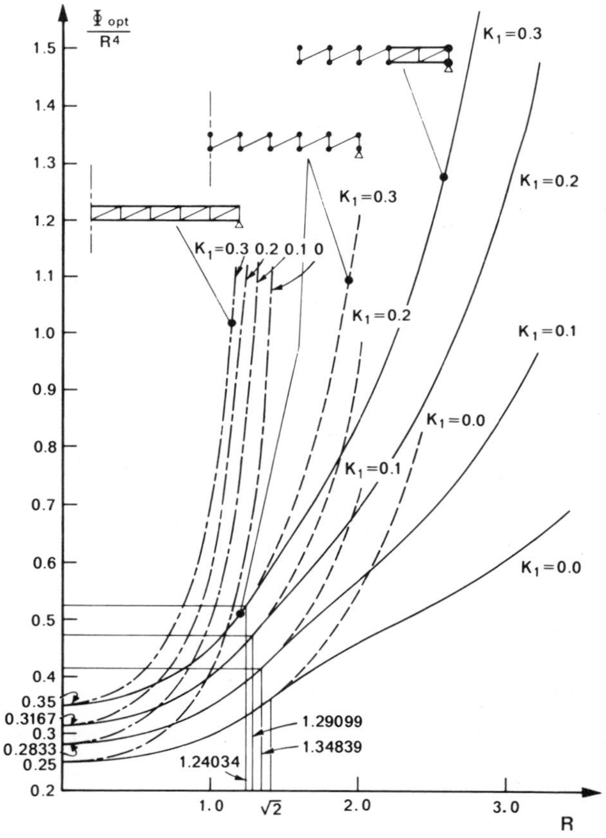

Fig 4 The total structural weight Φ of some optimal and non-optimal circular truss grids.

calculated using both primal formulation (Eqn 30) and dual formulation (Eqn 19); the two sets of results were in complete agreement.

It can be seen from Fig 4 that up to a limiting radius (e.g. $R = \sqrt{2}$ for $k_1 = 0$ and $R = 1.24034$ for $k_1 = 0.3$), the optimal layout (continuous lines in Fig 4) consists of purely circumferential chords. Beyond the limiting radius, the same solution (broken lines) becomes uneconomical, since the optimal solution consists of purely circumferential chords in the inner region and purely radial chords in the outer region. At the support, however, concentrated (heavy) circumferential chords balance the radial moments. It can also be seen from Fig 4 that purely radial trusses are highly uneconomical (dot-dash lines in Fig 4). Naturally, one-way truss systems would require some bracing against buckling of the compression chords but it has been established that the structural weight of bracing is relatively small compared to the primary structure.

All solutions for the above problems were obtained in a closed analytical form. For purely circumferential moments, for example,

$$M_\theta = e^{(r^2/2 + k_1 Rr)} - \sqrt{\pi/2}\, k_1 R\, e^{(r + k_1 R)^2/2} \left[\mathrm{erf}\left(\frac{r + k_1 R}{\sqrt{2}}\right) - \mathrm{erf}\left(\frac{k_1 R}{\sqrt{2}}\right)\right] - 1,$$

$$u = e^{-[r^2/2 + k_1 Rr - R^2(1/2 + k_1)]} - 1, \qquad \ldots 32.$$

where "erf" is the error function and then both primal and dual formulations furnish

$$\Phi_{min} = 2\int_0^R \psi\, r dr = 2\int_0^R ur\, dr$$

$$= 2\{e^{R^2(1/2 + k_1)} - 1 - k_1 R\sqrt{\pi/2}\; e^{R^2(1+k_1)/2} \left[\mathrm{erf}\left(\frac{R + k_1 R}{\sqrt{2}}\right) - \mathrm{erf}\left(\frac{k_1 R}{\sqrt{2}}\right)\right]\} - R^2 \qquad \ldots 33.$$

The above optimal solution becomes invalid when the deflection field u(r) in Eqn 32 violates the optimality condition for $M_r = 0$ [i.e. $-(1+u) + k_1 R(du/dr) \leq -d^2u/dr^2$] furnishing the following limitation for the purely circumferential solution

$$R \leq \sqrt{2/(1 + k_1)} \qquad \ldots 34$$

More complicated expressions represent other solutions in Fig 4.

Comparison of the Total Weight of Various Solutions. It can be seen from Fig 4 that for longer spans rather dramatic savings can be achieved by layout optimization. For $R = \sqrt{2}$, $k_1 = 0$, for example, the weight of the purely radial solution is 1317 per cent higher than that of the optimal solution. The savings become even greater for non-zero values of the shear cost coefficient k_1.

REFERENCES

1. W PRAGER and G I N ROZVANY, Optimization of Structural Geometry, in: A R BEDNAREK and L CESARI (Eds), Dynamical Systems, Academic Press, New York, 1977, pp 265-294.

2. W PRAGER and G I N ROZVANY, Optimal Layout of Grillages, Journal of Structural Mechanics, Vol 5, No 1, pp 1-18, 1977.

3. G I N ROZVANY and W PRAGER, A New Class of Optimization Problems: Optimal Archgrids, Computer Methods in Applied Mechanics and Engineering, Vol 19, No 1, pp 127-150, June 1979.

4. W PRAGER, Introduction to Structural Optimization, Springer-Verlag, Vienna 1974.

5. G I N ROZVANY, Optimal Design of Flexural Systems, Pergamon Press, Oxford, 1976. Russian edition: Stroiizdat, Moscow, 1980.

6. G I N ROZVANY, Optimality Criteria for Grids, Shells and Arches, in: E J HAUG and J CEA (Eds), Optimization of Distributed Parameter Structures (NATO ASI Series), Sijthoff and Noordhoff, Alphen aan der Rijn, 1981, pp 112-151.

7. G I N ROZVANY, A General Theory of Optimal Structural Layouts, Proceedings of the International Symposium on Optimal Structural Design, University of Arizona, Tucson, Arizona, 1981, pp 4.37-4.45.

8. G I N ROZVANY, Structural Layout Theory - The Present State of Knowledge, in: E ATREK et al., (Eds), New Directions in Optimum Structural Design, Chapter 7, pp 167-195, John Wiley & Sons Ltd, Chichester, 1984.

9. G I N ROZVANY, Extensions of Prager's Layout Theory, in: H ESCHENAUER and N OLHOFF (Eds) Optimization Methods in Structural Design, pp 103-110, Wissenschafts-Verlag, Mannheim, 1983.

10. N OLHOFF and G I N ROZVANY, Optimal Grillage Layout for Given Natural Frequency, Journal of the Structural Mechanics Division, ASCE, Vol 108, No EM5, pp 921-974, Jan 1982.

11. G I N ROZVANY and C M WANG, Optimal Layout Theory: Allowance for Selfweight, Journal of the Engineering Mechanics Division, ASCE, Vol 110, No 1, Jan 1984.

12. W PRAGER and R T SHIELD, A General Theory of Optimal Plastic Design, Journal of Applied Mechanics, Vol 34, No 1, pp 184-186, March 1967.

13. G I N ROZVANY, Variational Methods and Optimality Criteria, in: E J HAUG and J CEA (Eds) Optimization of Distributed Parameter Problems (NATO ASI Series), Sijthoff and Noordhoff, Alphen aan der Rijn, 1981, pp 82-111.

14. G STRANG and R V KOHN, Hencky-Prandtl Nets and Constrained Michell Trusses, Computer Methods in Applied Mechanics and Engineering, Vol 36, No 2, pp 207-222, Feb 1983.

15. G I N ROZVANY, Optimal Plastic Design: Allowance for Selfweight, Journal of the Engineering Mechanics Division, ASCE, Vol 103, No EM6, pp 1165-1170, Dec 1977.

16. G I N ROZVANY and R HILL, General Theory of Optimal Load Transmission by Flexure, in: Advances in Applied Mechanics, Vol 16, Academic Press, New York, 1976, pp 183-308.

17. G I N ROZVANY and R HILL, A Computer Algorithm for Deriving Analytically and Plotting Optimal Structural Layout, in: A K NOOR and H G McCOMB (Eds), Trends in Computerized Structural Analysis and Synthesis, Pergamon Press, Oxford, 1978, pp 295-300; also: Computers and Structures, Vol 10, No 1, pp 295-300, April 1979.

18. G I N ROZVANY, Optimal Beam Layouts: Allowance for Cost of Shear, Computer Methods in Applied Mechanics and Engineering, Vol 19, No 1, pp 49-58, June 1979.

DESIGN OF JOINTS IN STEEL SPACE STRUCTURES

T. Arciszewski, M.Sc., Ph.D

Department of Civil Engineering

Wayne State University, Detroit

An advanced approach to the design of joints in steel space structures is presented in this paper. A formal typology by coverings is proposed, and general models of joints are considered. A typologic table is also developed, one which can be used for review of standard types of joints and for the search for new types. For this purpose a nondeterministic optimization method is proposed, called "Stochastic Form Optimization," and nonhomogeneous Markov Chains are utilized for morphological analysis of joints.

INTRODUCTION

Two main stages can be distinguished in the design process of a joint: preliminary design (or analysis) and final design. The first stage has a qualitative character while the second has a quantitative one. Preliminary design is defined as a process of decision making, for the determination of the type of joint and variables included are qualitative. The subject of the final design is the analysis and optimization of quantitative variables.

The first stage is particularly important. The determination of a joint type has a global character, since it affects all subsequent local decisions in the final design and is decisive when the final appraisal of the developed joint is concerned. Usually only the designer's experience and, eventually, trial and error methods are applied here, often with unsatisfactory results. Methods of shape optimization should be useful in the preliminary design of joints, but shape optimization is still in an early stage of development; see Ref 1. Most applications so far can only generate the optimal shaping of a joint's cross-section, qualitative variables are usually fixed, and only sizing, quantitative variables are optimized. A fixed type of joint significantly reduces potential savings, which are expected to be much larger in the case where the type is also the subject of optimization; see Ref 2. Optimization of a joint of assumed type has a local character with respect to all possible solutions. Even in the case where several different types are analyzed and compared, this does not guarantee that the global optimum will be achieved.

These are reasons for proposing a new approach to the designing of joints. This approach utilizes formal descriptions of joints, their systems models, and a nondeterministic optimization technique.

TYPOLOGY OF JOINTS

A joint in a steel space structure can be described by a set of qualitative and quantitative design variables. Qualitative variables describe the general form of a joint and are discrete. Each variable has a certain number of feasible states. These variables are related to such incommeasurable properties as the kind of material used, but also include certain measurable properties of a discrete character which are decisive from the structural point of view. Quantitative variables are the dimensions of the joint, its specification, weight, cost, etc., and usually can be treated as continuous variables; see Ref. 3.

A type of joint is described by qualitative variables. It is assumed to be defined by a compatible combination of feasible states of qualitative variables, when for all variables one state is taken at a time. Such a morphological approach can be referred as "Typology by Coverings of Joints"; see Ref. 4,5.

Typology by coverings enables formal presentation of all considered qualitative variables and their feasible states in a tabular form called a typologic table. The general form of a typologic table is shown below.

	FEASIBLE STATES	
Qualitative Variables	1	2.....i.....k.....n.....p.....m
A	A1	A2....Ai....Ak....An
B	B1	B2....Bi....Bk
C	C1	C2....Ci....Ck....Cn....Cp....Cm
D	D1	D2....Di....Dk....Dn....Dp

In the rows of the typologic table successive qualitative variables and their feasible states are specified. Such a table allows a simple and formal type identification and has been developed for the purposes of optimization of joint types in steel space structures.

When a joint in its most complex is considered, also the end-pieces of bars should also be included. Three basic parts of a joint and their four connections can be distinquished.

Joints parts:

1. Joint elements

2. End-pieces of joint elements

3 End-pieces of bars

Connections:

1. Connections of joint elements

2. Connections of joint elements with joint end-pieces

3. Connections of end-pieces of joints with end-pieces of bars with bars

There are many types of joints which are not so complicated. Some of the specified parts do not occur in these and this is denoted by 0 in all rows of the typologic table of joints proposed in the paper. A general model of joint is given in Fig. 1A. Also, bars are shown although they are not parts of the joint.

From the structural point of view the most important function of a joint is the transmition and distribution of forces acting in all bars connected together at a given joint. Considering two bars, i and j, connected through the joint, their interaction can be interpreted as a subsequent transmission of forces by links of a chain of the interconnected parts of the joint, as shown in Fig. 1B. All links in such a basic chain, i.e., parts of the joint and connections, are functionally homogeneous, since their functions are simple and limited to interactions only. The joint will be a system of several such chains. Joint elements and their connections (shaded in Fig. 1B) unite these chains into a system, a set of functionally homogeneous elements which together perform a complex function: transmission and distribution of forces. A model of such a system with two basic chains is shown in Fig. 2. Both parts of the joint and connections are elements of the system. This model proves that the assumed structural division of a joint is correct and complete and can be used for practical purposes. A complete typologic table of joints is given on the next page. The joint is defined by 22 qualitative variables: joint parts are described by 13 variables and connections by 9 variables.

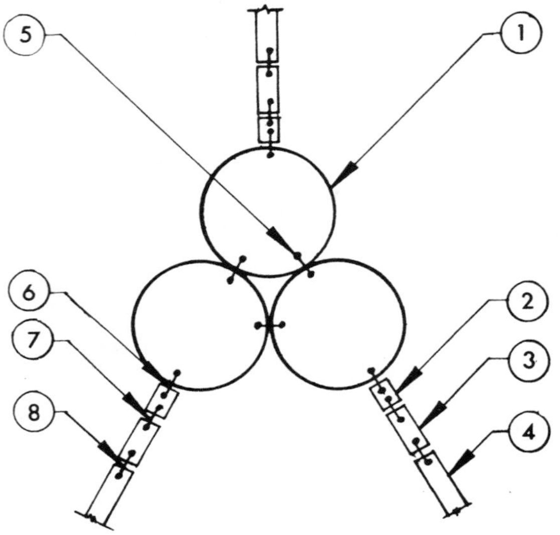

1. JOINT
2. END PIECE OF JOINT ELEMENT
3. END PIECE OF BAR
4. BAR
5. CONNECTION OF JOINT ELEMENTS
6. CONNECTION OF JOINT ELEMENT WITH JOINT END PIECE
7. CONNECTION OF END PIECE OF JOINT WITH END PIECE OF BAR
8. CONNECTION OF END PIECE OF BAR WITH BAR

FIG. 1A.

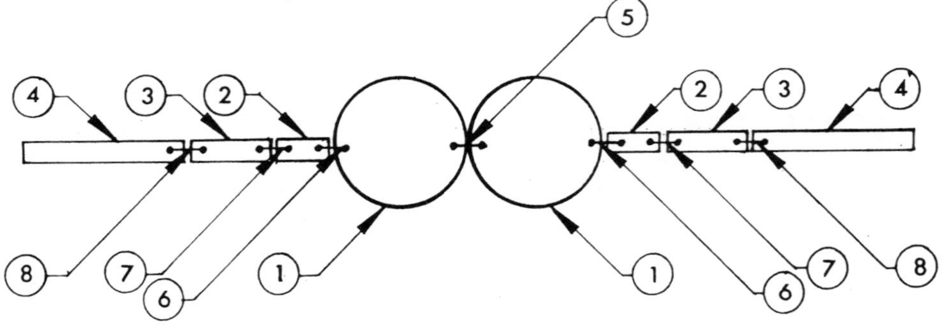

FIG. 1B.

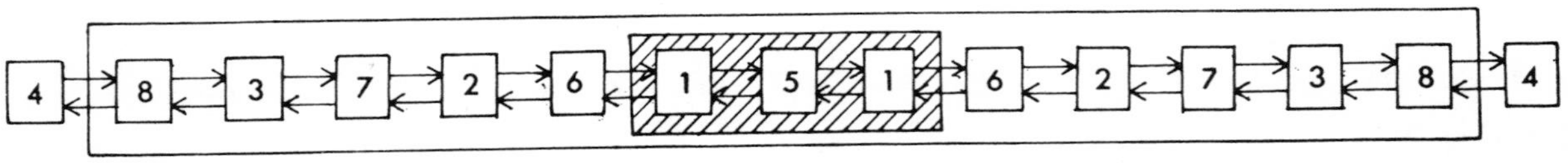

FIG. 1.

PARTS OF JOINT		FEASIBLE STAGES 1	2	3	4	5	6	7	8	9	10	11	12	13	14
ELEMENTS OF JOINT	A. NUMBER OF ELEMENTS	0	1	2	3	4	5	vary							
	B. FORM OF ELEMENTS	0	flat sheet	drawpiece	sphere	1/2 sphere	1/4 sphere	1/8 sphere	poly-hedron	1/2 poly-hedron	pipe	cylinder	cast	torus	crossbar
	C. No. OF PLANES IN POLYHEDRAL ELEMENTS	0	6	8	12	16	18	21							
	D. FILLING	0	hollow	solid	plastic										
	E. PRODUCTION	0	bending	forging	casting	extruding	squeez-ing	mechanic. working	cutting						
	F. MATERIAL	0	as bars	other than bars											
END PIECES OF JOINT	G. No. OF END PIECES FOR JOINT ELE.	0	1	2	3	4	vary								
	H. FORM OF END PIECES	0	flat sheet	**pipe**	tapped pipe	solid rod	tapped solid rod	angle bar	moulder	forged element	mount	moving mount	ear		
	I. PRODUCTION	0	bending	forging	casting	extruding	squeez-ing	mechanic. working	cutting						
	J. MATERIAL	0	as joint element	other than joint ele.											
END PIECES OF BARS	K. FORM	0	pipe	tapped pipe	flatten pipe	solid rod	tapped solid rod	flatten solid rod	flat sheet	forged element	tapped forged element	cast	tapped cast	**bolt**	
	L. PRODUCTION	0	bending	forging	casting	extruding	**squeez-ing**	lamina-tion	mechanic. working						
	M. MATERIAL	0	as bar	as joint element	other										
CONNECTION OF JOINT ELEMENTS	N. NUMBER	0	1	2	3	4	5	6	7	8	9	10	vary		
	O. TYPE	0	junction	weld	bolt	hi-resist. bolt	adhesive	cold compress	hot compress						
CONNECTION OF JOINT ELEMENTS WITH JOINT END PIECES	P. NUMBER	0	1	2	3	4	vary								
	Q. TYPE	0	junction	weld	bolt	hi-resist bolt	adhesive	cold compress	hot compress	native material	pin	groove connect.			
R. CONNECTION OF JOINT END PIECES WITH BAR END PIECES		0	junction	**weld**	bolt	hi-resist bolt	adhesive	cold compress	hot compress	native material	pin	groove connect.	pipe	tapped pipe	
S. CONNECTION OF BAR WITH ITS END PIECE		0	junction	weld	bolt	hi-resist bolt	adhesive	cold compress	hot compress	native material	pin	groove connect.			
CONN. OF JOINT ELE. WITH END PIECE OF BAR OR WITH BAR OR ONLY END PIECES OF BARS	T. TYPE	0	add. bolt	drawpiece	rod	tapped rod	pipe	**tapped pipe**	groove connect.						
	U. HOLD-UP OF CONNECTION	0	add. bolt	cold compress	hot compress	nut									
	V. POSITION OF HEAD	0	inside joint ele	inside bar	others										

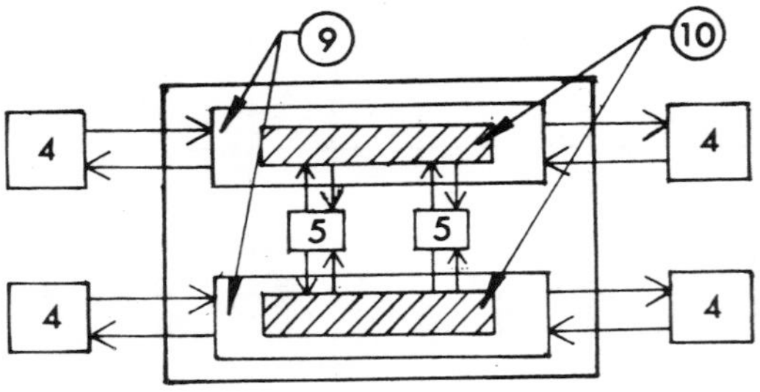

4. BARS
5. CONNECTION OF JOINT ELEMENTS
9. BASIC CHAINS
10. JOINT ELEMENTS AND THEIR CONNECTIONS IN BASIC CHAINS

FIG. 2

STOCHASTIC FORM OPTIMIZATION OF JOINTS

An optimization method called "Stochastic Form Optimization" has been developed for the generation of feasible types of a structural solution, their revision, and the choice of the optimal type. The method is applicable in preliminary design and can be used in the design of joints in steel space structures. For the complete description of the method see Ref. 5.

There are five main phases in the process of stochastic form optimization:

1. Construction of the typologic table
2. Random generation of a set of solution types to the structural problem
3. Feasibility analysis of generated solution types and elimination of infeasible solution types
4. Appraisal of feasible solution types
5. Choice of the optimal solution type

When the design of joints is considered, a typologic table, as described in the preceding Section, can be used to present qualitative variables describing the problem and to determine feasible states of these variables. Subsequent qualitative variables must be considered independently, and all intervariable relations are temporarily suspended. The typologic table of joints is a general model of the problem of structural shaping of joints, which enables the review of standard solutions to the problem and the generation of new solutions which might have been neglected in the traditional analysis because of their unconventional character.

The analysis of a completed typologic table starts with the generation of different combinations of feasible states, and for each qualitative variable only one feasible state is taken at a time. A stochastic model, a multistage nonhomogeneous Markov chain, is proposed for this generation, where stages represent subsequent qualitative variables. The initial probability vector can be assumed or generated by a computer; similarly, transition matrices are also proposed to be subsequently randomly generated by a computer. Using the above procedure a formal stochastic simulation of the morphological analysis can be performed completely by a computer, and a set of generated combinations of feasible states will be obtained. These combinations of feasible states must be analyzed and all incompatible combinations eliminated. Personal experience or formal restrictions regarding intervariable relations can be used to reduce the set of generated types to feasible types only.

In the case of joints the typologic table allows the generation of a large number of combinations of feasible states. The traditional approach, a systematic search of all combinations, becomes ineffective here, and even impossible. When these combinations are randomly generated by a computer their character does not depend on the assumed number of combinations. A researcher having the same task i.e., to determine a limited number of combinations, usually selects combinations reflecting his personal preferences, and this significantly reduces the probability of obtaining a new, unconventional type.

The generated feasible types are appraised and their relative values are determined. Formal appraisal models and the morphological box method are suitable for this purpose, as it was proposed by Ref. 6 for complex structural systems and in Ref. 7 developed in the appraisal method elaborated for steel space structures. Finally, when feasible types and results of appraisals are known, the optimal type can be determined.

APPLICATIONS

The developed typologic table can be used for a formal typology by coverings of existing types of joints and for purposes of stochastic form optimization in an innovative design, when the design objective is to develop a new type of joint.

In the first case all types of joints under consideration can be formally identified and their feasible states compared. In some cases such analysis and comparison are very useful and may even enable the development of new types of joints. When, for example, a well-known joint type MERO, described in Ref. 8,9, is formally identified (Fig. 3A) it can be easily observed that its development should be concentrated on end pieces of bars, which are here crutial but unfortunately relatively complicated.

Innovative design is usually regarded as the subject of heuristics, and optimization methods are considered inapplicable here. This is not the case when stochastic form optimization is concerned. The method has been developed for applications in preliminary design and is particularly effective in innovative design. The method was initially applied in the innovative design of wind bracings in tall building (see Ref. 5), but the results of its application in the design of joints are also promising. Fig. 3B shows a new type of joint, which resulted from the application of the method in the innovative design of joints in steel space structures.*

FINAL COMMENTS

The design of joints in steel space structures is still more an art than science. It requires structural experience, heuristic skills, and a very complex analysis, where traditional methods are usually ineffective. The proposed approach should improve this situaton: systematic analysis could be possible and computer generation of joint types should simplify the search for new types. However, there are still many problems, which require study. In particular, formal appraisal methods applicable in the preliminary design of joints should be developed further, in order to incorporate them into computer-aided stochastic form optimization.

* The joint was developed by M.M. Hanna, a graduate student at the Department of Civil Engineering, Wayne State University.

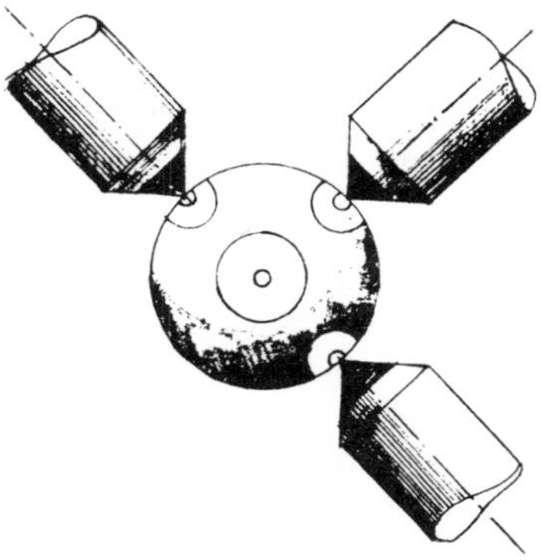

FIG. 3A

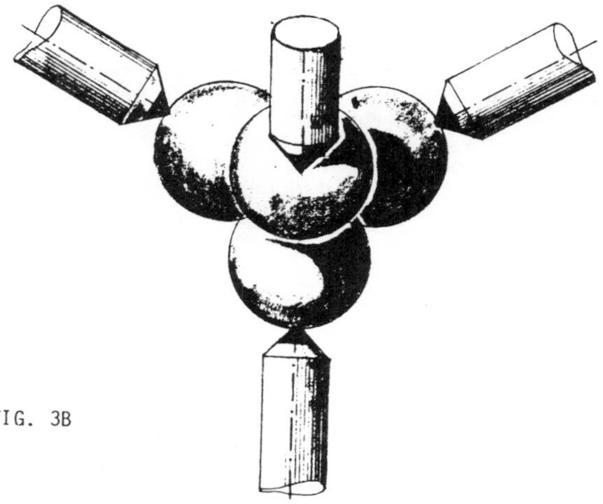

FIG. 3B

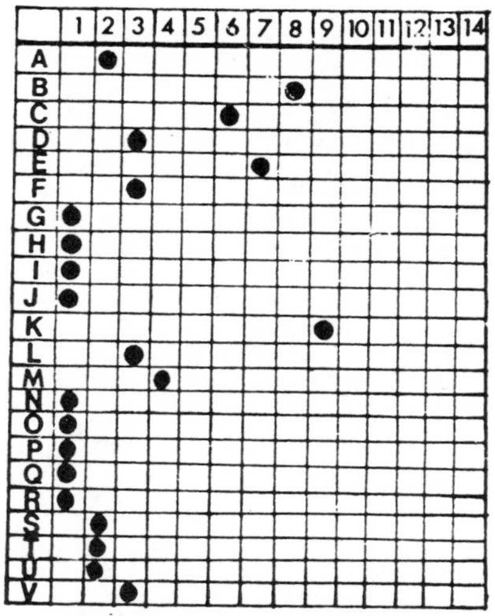

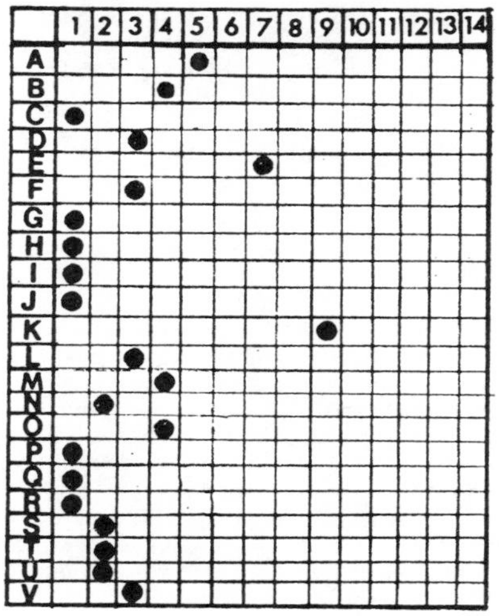

FIG. 3

ACKNOWLEDGEMENTS

The author wishes to thank Professor J. Brodka for all his inspiring and constructive remarks and suggestions regarding the subject of this paper.

REFERENCES

1. O.E. Lev, Structural Optimization: Recent Development and Applications, ASCE Publication, U.S.A., 1981.

2. O.E. Lev, Structural Optimization: A State of the Art Report, Proceedings of the Fourth Engineering Mechanics Division Specialty Conference, ASCE, Purdue, June 1983.

3. T. Arciszewski, Systems Approach to the Analysis of Steel Space Structures, in Polish, Proceedings of 3rd Conference on Heuristic Methods, Polish Cybernetic Society, September 1977.

4. T. Arciszewski, J. Kisielnicka, Morphological Analysis, in Polish, in the book, "Problem, Method, Solution. Technics of Creative Thinking," PWN-Pbl. House, Warsaw, 1977.

5. T. Arciszewski, Decision-Making Parameters and Their Computer-Aided Morphological Analysis by Means of Nonhomogeneous Markov Chains for Wind Bracings in Steel Skeleton Structures, in the book "Advances in Tall Buildings," Hutchinson Ross Publising Company, U.S.A. to be published in 1984.

6. T. Arciszewski, M. Lubinski, Methods of Analysis and Appraisal of Structural Systems, in Polish, Proceedings of Warsaw Technical University, March 1978.

7. T. Arciszewski, J. Brodka, Technical Quality of Structural Solution Appraisal Method, in Polish, Journal, "Invention and Rationalization," Warsaw, January 1978.

8. Z.S. Makowski, Steel Space Structures, Michael Joseph Ltd., England 1963.

9. J.S.B. Inffland, Preliminary Planning of Steel Roof Space Trusses, Journal of the Structural Division, Proceedings of the American Society of Civil Engineers, November 1982.

10. L.C. Schmidt, P.R. Morgan, Behaviour of Some Joint Design in Space Trusses Transactions of Institute of Engineering, August 1981, Australia.

MICHELL TO-DAY

J.M. LAGACHE° and P. MENET°°

° PSA Etudes & Recherches
La Garenne-Colombes, France.

°° Ecole Centrale des Arts & Manufactures
Châtenay-Malabry, France.

This paper presents hand made Michell frames, which have been born in Ecole Centrale, during the last two years, for the precise purpose of building windmills. The analytical Theory of Hencky-Prandtl nets played the role of an uncomparable preprocessor in the design scheme. It is known, however, that the theory suffers from a certain lack of generality. This is the reason why Michell Theory is re-analyzed, in the second part of the paper, by means of powerful Convex Analysis theorems. This results in a non-linear finite element approach to optimum geometries.

INTRODUCTION

Designers have to distribute materials over specified regions, in such a way that systems of loads are safely supported with minimum material consumptions. Numerical methods that proceed by successive gentle transformations of arbitrary starting contours, cannot be expected to work when the configuration of internal holes is unknown. Thus, write loads in the form $k\,L$, where L is a conventional unit-loading, and progressively take the load factor, k , to its critical value $\bar{k}$, beyond which no material system of any kind can support the loads. Optimum geometries at the beginning of the preceding path were first described, at the early date of 1904, in the celebrated paper by A.G.M. Michell, concerning optimal design of pin-jointed frames (Ref 1). Indeed, when the actual loads are small in comparison with $\bar{k}$, most of the feasible region is empty, and the material is concentrated on lines and surfaces ; at the considered low rate of inoccupation of the feasible region, transverse removals of material are licit, which would induce decreasements in the structural weight, if lines and surfaces were not placed in states of pure tension or compression. Considering, by example, that highly stressed natural systems, like human bones, are still truss-like systems, it is now suggested that further evolutions of optimum structures with increasing load factors, mainly consist in progressive rearrangements of internal holes, that should not drastically alter the overall shape and principal mechanical behaviour. A pragmatic method to track optimum geometries, in absenceof any mathematical model, involves successive phases of definition and perturbation of Michell fields ;construction of prototypes ; structural analysis and redesign (Fig 2). It was partially experimented by students of Ecole Centrale during the last two years, for the precise purpose of building windmill components, like the torque transmission aluminium and wood lever on Fig 1. Their experiments are related in Part I of the paper. Should Michell theory reduce to analytical computations of some Hencky-Prandtl nets, that the preceding scheme would be quite unefficient in complex real-world design problems, because of the complete dissymetry with powerful methods of structural analysis and design. It was judged of capital importance to show, in the second part of the paper, that a correct interpretation of Michell's ideas, results in a quite general non-linear finite element approach to optimum starting geometries.

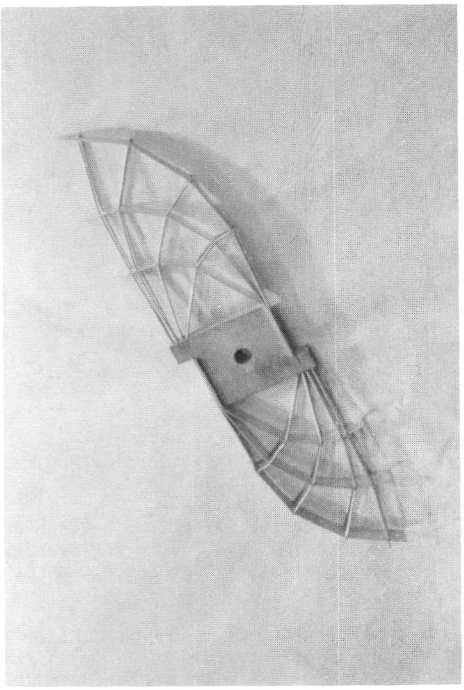

Fig 1. Aluminium and wood lever

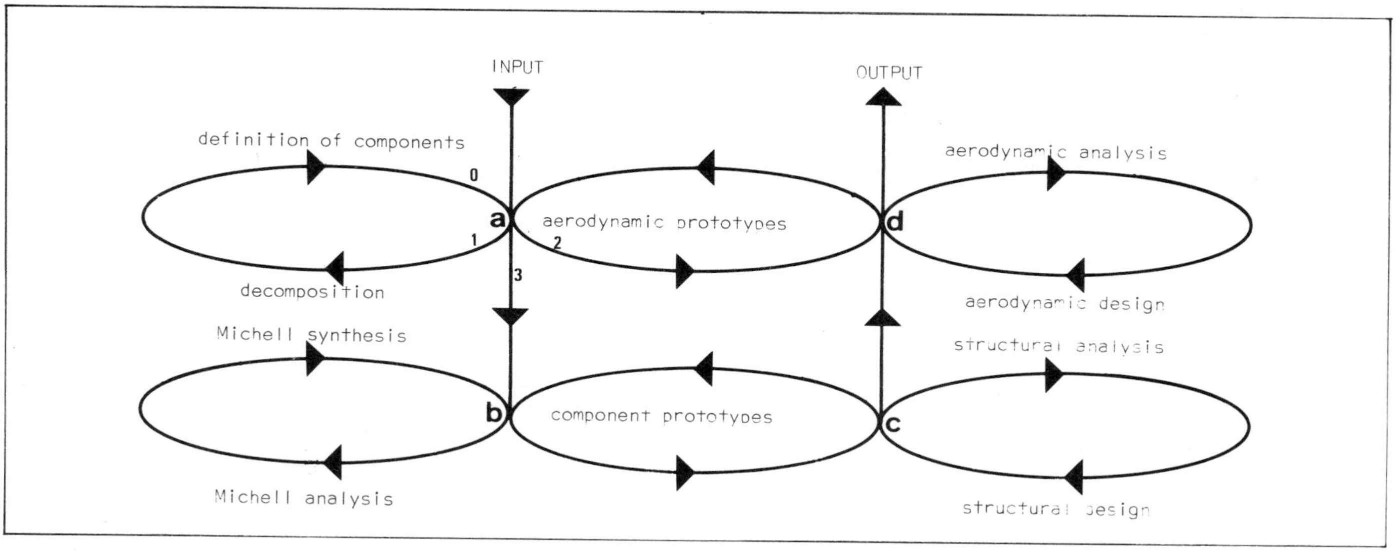

Fig 2. Design scheme

PART I : THE DESIGN OF WINDMILL COMPONENTS

In 1981-82 and 1982-83, students of Ecole Centrale have tried to design ultra-light windmills based on Michell's optimum geometries (Ref 1 to 13). Michell Theory was expected to yield good starting points for realistic designs. It was first decided that attention should be focused on the synthesis of mechanical components, in preference to complete windmills. Indeed, analytical Michell solutions concerning simple mechanical functions, were at once available, which was far from being the case for complex systems. It was also decided that practical experiments should be performed as soon as possible. Thank heaven, access to the school computer was pretty difficult, and there were but few risks that the project sunk into abyssae of pure computer-aided design. On the contrary, hand tools were immediately set at students' disposal, for the various purposes of cutting, sawing, piercing, joining, and welding. Rods of iron, wood, aluminium and plastics were bought. Thanks to the financial aid of the French Agency for Research Development (ANVAR), contacts were establishes for further material provisioning, and, also, to test forthcoming windmills in the wind tunnel of Ecole des Arts & Métiers. Students had not taken any course on design, and were quite unprepared to those various feedbacks that make the attraction and difficulty of engineering. Their first major discovery, by instance, was that not only structures, but structural environments themselses,- and especially precise values of loads,- were to be synthetized.

Figure 2 throws some light upon the labyrinth students stepped into, at the very moment when they decided to design windmills. The corresponding scheme was applied to the synthesis of Savonius windmills (Ref 4). Such low speed aerodynamic engines, whose principle is recalled on Fig 3, are generally used for waterexhaust. Benefic aerodynamic effects exert a natural control on rotation at high wind speeds, and small Savonius rotors are sometimes used to drive electric generators aboard sailing ships.

1. Decomposition Savonius rotors were decomposed in three mechanical parts : blades ; torque transmission components; and auxiliary structures, to fix blades, and prevent vertical bending of components. Triangularization of auxiliary structures (Fig 7) greatly facilitated practical assembly. It was later recognized that application of Prager's superposition principle (Ref 12) to the transmission of alternative loads to a rigid line, symbolizing the windmill spindle, leads to quite similar overall geometries.

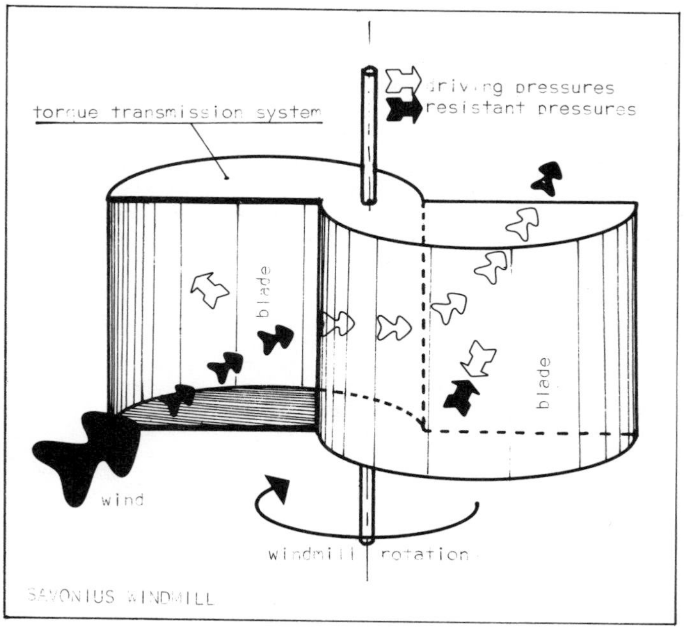

Fig 3. The Savonius windmill

Fig 4. Timber rotor with Rhodoid blades (2.40 m)

2. Michell analysis and design In plane torque transmission components that appear on Figs 4 , 7 , 8 , were derived from Michell's second example (Refs 1 to 13). The lever on Fig 1 was derived from the cycloidal Michell field within a narrow strip (Ref 3) , for the precise purpose of designing flat compact Savonius rotors. Various attempts were made, in conjunction with practical experiments, to design assembly solids at the junction of rods with the windmill spindle. Diagrams a , b , c , on Fig 5 , present the self-consistent strain-fields that are respectively associated with ideally rigid circular, polygonal, and rounded polygonal cores. Although not mentioned in Refs 10 , 11 , diagram b is clearly the right to be considered when one tries to define optimum frames with a finite number of supporting points. Very precise assembly details amazingly appear in the slightly perturbated strain-field of diagram c , as soon as dense networks in the vicinity of the rigid core, are converted to continuous plates, or, in other words, integrated to the rigid core itself. Notice that such a transformation is precisely the beginning of a possible transition from the Michell Truss to general optimum structures (Refs 15 , 16).

An example of realistic Michell design is given on diagram d . Dissymmetry from thick to thin lines once again results from practical assembly considerations. Because of the great number of nodes, Michell frames cannot be designed as ordinary pin-jointed frames, with well separated material segments, and sophisticated assembly nodes. Simplified systems of continuous fibers are highly preferable. To fix ideas, 1 cm^2 square fibers and 3 mm circular fibers were used in the construction of the timber component of Fig 3 . Holes were pierced in square fibers, providing outlets to circular fibers. The same method was used for the design of metallic components of Figs 1 , 4 , with the slight difference that holes were replaced by notches. To prevent excessive weakening of cross-sectional areas, it was necessary to choose as thin crossing members as possible. A convenient method to equilibrate the design, in such a way that thick and thin members played similar mechanical roles, precisely consisted in defining dissymetrical approximations of Michell fields. From Hencky's first theorem, Michell strain-fields can be decomposed in elementary line bundles of small constant angular magnitudes (right upper part on diagram b). The proposed method of drawing consists in assimilating bundles with their centre lines, each line being weighted by the corresponding angular magnitude. Once initial slopes and positions are known from the discussion of boundary conditions on the rigid core, an approximate network can be expanded, by iterative applications of the following discrete Hencky theorem : each time a line of the weight $\hat{a}$ meets with a line of the weight $\hat{b}$, the former is deviated by an amount of $\hat{b}$, and the later, by an amount of $\hat{a}$. Depending on the choice of bundles, the preceding procedure can generate either symmetrical or dissymetrical networks. On diagram d , origins of lines were taken along equi-angular spirals ab , ac , at respective angular intervals of 15° and 9° . Space restrictions are taken into account on Fig 1 , by successive reflections of the current line.

The preceding analysis finally results in the definition of a complete parametrized family of Michell designs. Parameters that characterize each element of the family, are, by example, the size of the rigid core, the number of wedges, the size of the small circles, the respective numbers of thick and thin lines, and the overall size of the structure. Synthesis of the corresponding geometries may be easily achieved by means of computer programs (Ref 17). Further reanalyses, in the dynamical scheme of Fig 2 , essentially consist in redefinitions of parameters, so as to find effective compromises between commercial member sizes and actual load intensity factors.

Notice that angular deviations keep constant along any fiber of diagram d , which is very helpful in practical realizations (Ref 18).

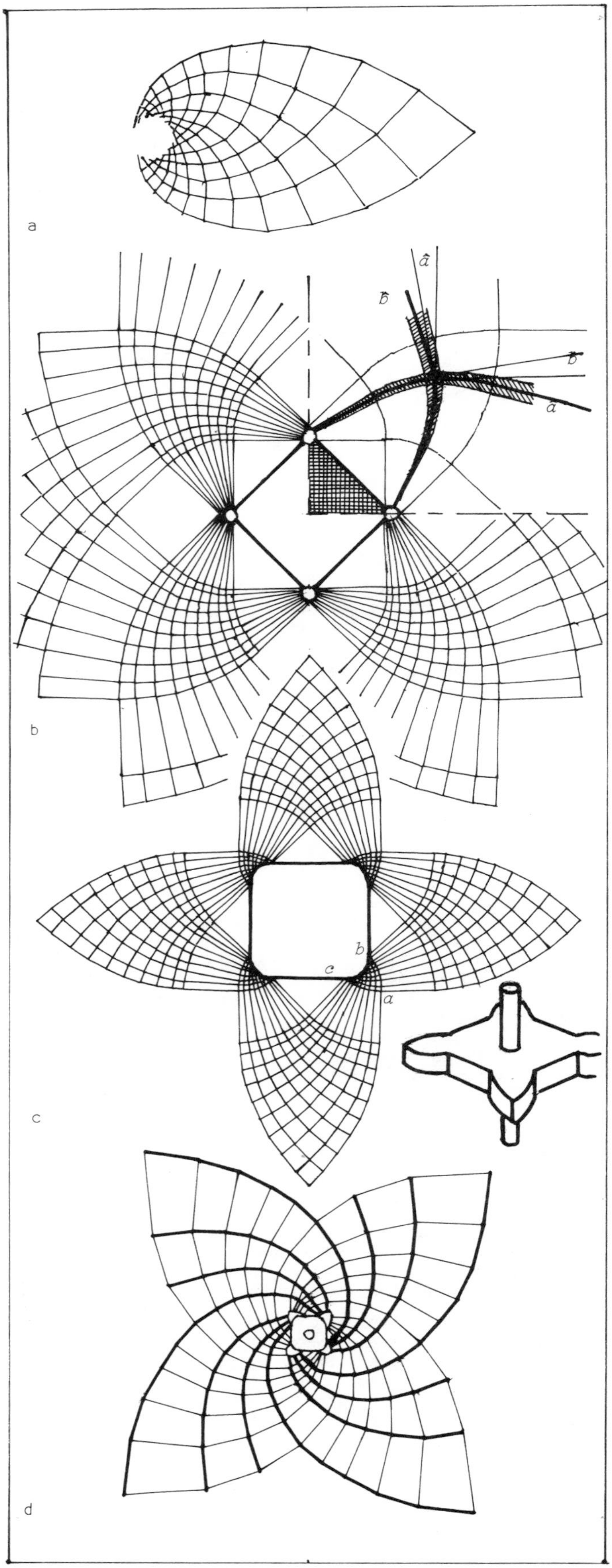

Fig 5 . Michell strain-fields

3. Structural analysis and design In addition to § 1, concerning preliminary decompositions of structures, it can now be remarked that a possible method to fit cross-sectional areas to precise values of loads, consists in sandwich superpositions of components. Decomposition of strain-fields in inequal bundles can have the same effect, by narrowing fibers at desired places of the structure. Numerical techniques of optimization would be perfectly suited to the discussion of such problems. The point, however, was not thoroughly investigated, because aerodynamic loads were unknown.
Nevertheless, experimental data about circular Savonius rotors, was taken from Ref 14. Simplified conventionnal aerodynamic loadings were defined. Efforts in structural members were evaluated, via Maxwell's diagrams, while stresses in the rigid cores, then considered as true elastic bodies, were computed by means of Brebbia's Boundary Element programs (Ref 19). An example of stress-field, relating to the aluminium lever of Fig 1, is shown on Fig 6.
To fix ideas, it was estimated that the critical wind speed for this later component, was about 100 mph ! Vibrations, fatigue, corrosion, being not taken into account in the considered simplified computations, no decision of remodeling was taken. Remark that numerical Shape Optimization methods (Ref 20) could be fruitfully applied to the redesign of assembly solids, not in view of saving infinitesimal quantities of material, but to decrease stress concentrations.

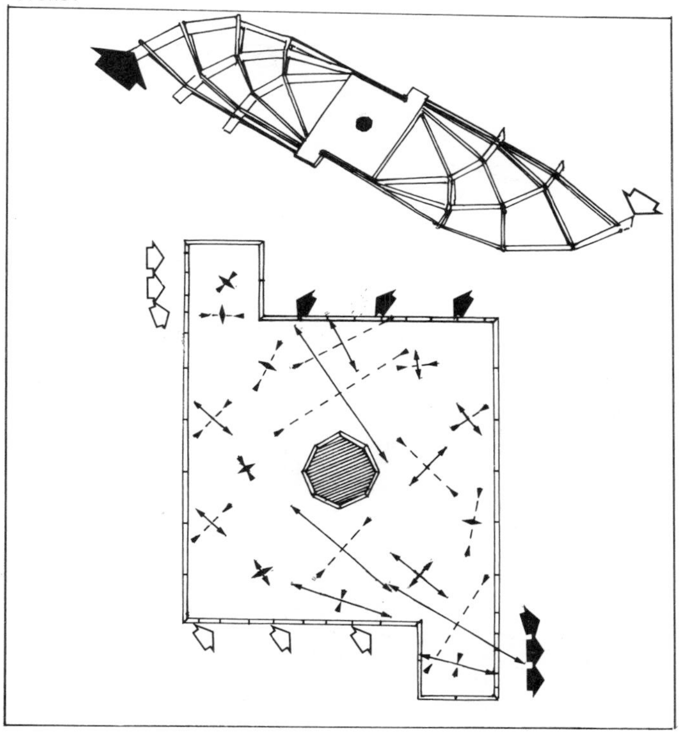

Fig 6. Stress analysis

4. Aerodynamic analysis and design Glass-fiber and resin coated rotors were first experimented in the wind-tunnel of Ecole des Arts & Mètiers, on 6/30/1983 (Figs 7, 8). Ball bearings and stay wires were placed at the tops of spindles, while bottoms were directly connected to the wind-tunnel torque-meter and tachometer. Structures well behaved under the traditional welcome blast. Further aerodynamic results were far from being satisfactory, and a new experiment is now on preparation. A special attention will now be paid to the drawing of blades in the vicinity of spindles, and also, to the practical obtention of smooth blade surfaces. Notice, as a provisory conclusion, that no special problem appeared, concerning the hand-made Michell components themselves.

Fig 7. 3 foot iron-framed and resin coated Savonius rotor

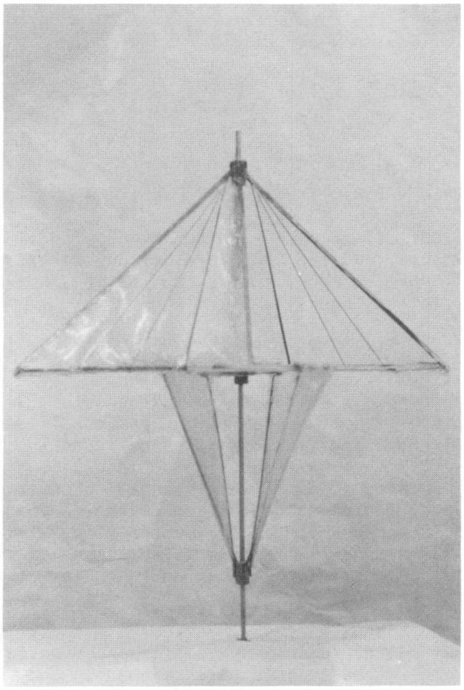

Fig 8. Triangular auxiliary structures in the 3 foot rotor

PART II : DEVELOPMENTS IN MICHELL THEORY

Should Michell Theory reduce to analytical computations of a few Hencky-Prandtl nets, that the scheme on Fig 2 would become unefficient, because of a complete dissymetry with powerful methods of structural analysis and design. Validity of Michell approach for complex problems is now considered.

Various problems of optimum structural topology, either in finite or infinite-dimensional backgrounds, are expressions of the unique abstract combinatorial problem that follows :

Given some positively homogeneous cost, f *, of a locally convex vector space,* X *, onto* $\overline{R}_+ = R_+ \cup \{+\infty\}$*, find*

$$(1) \qquad F(x) = \inf_{\substack{p \\ x = x_1 + \dots x_p}} f(x_1) + \dots f(x_p) \quad ,$$

or, in other words, find the decomposition of variable x *, in an arbitrary number of terms, that makes the summation of elementary costs a minimum.*

In Michell problem, by instance, X will denote the infinite-dimensional set of all possible systems of concentrated loads over the considered feasible region, and f will be defined as the cost of single bars :

$$(2) \quad f(x) = \begin{cases} |L|\,|a - a'| \,/\, s_o \;, & \text{when } x = (a,L;a',-L) \text{ is a self-equilibrated 2-force loading, at the ends of a feasible segment ;} \\ |L|\,|a - b| \,/\, s_o \;, & \text{when } x = (a,L) \text{ is a 1-force loading at end } a \text{ of a feasible segment resting at } b \text{ on the support ;} \\ +\infty \;, & \text{else.} \end{cases}$$

Owing to the fact that F , in Eqn 1 , is the indefinitely iterated infimal convolution (Refs 21 , 22) of functional f by itself, a straightforward proof of the following saddle-point theorem was given in Refs 23 , 24 :

Functional F *, in Eqn 1 , is proper, convex, and positively homogeneous ; its double Fenchel transform is the support functional of a closed convex set of the dual,* Y *, of* X :

$$(3) \quad F^{**}(x) = \underline{\lim}_{x' \to x} F(x') = \sup_{w \in C} x \cdot w \quad ,$$

with :

$$(4) \qquad C = \{ w \;;\; w \in Y \;;\; \forall z \in X \;:\; z \cdot w \leq f(z) \} \;.$$

As far as practical problems are concerned, functional F is finite-valued, and can therefore be considered as a norm. Equation 3 , then changes, from a continuity argument, to the strong mini-max equality :

$$(5) \qquad F(x) = \sup_{w \in C} x \cdot w \quad .$$

In Michell problem, Y is the set of all possible virtual displacements of the feasible region, and conditions (4) are the well-known restrictions on overall strains that define Maxwell-Michell virtual strain-fields. Equation (5) brings the proof that Maxwell-Michell concepts remain valid even when analytical solutions cannot be computed in a closed form. Further details, concerning existence and finite element approximations of solutions of the variational problem in Eqn 5 , can be found in Ref 25 .

Fig 9 illustrates a practical application of the preceding results to the approximate synthesis of a Michell truss within a rectangular region, symbolizing a human patella. The region was shared in triangular elements, and P^1-interpolations of the Michell fields led to finite-dimensional variational problems , with linear objectives and non-linear inequality constraints of the second degree in nodal displacements.

Michell constraints for three-dimensional problems are inequality constraints of the third degree (Ref 25). Efficient Fortran subroutines have been written to compute these constraints, and their gradients. They are presently incorporated to general non-linear programming algorithms. It is hoped that three-dimensional approximate Michell trusses will be very soon synthetized in that way.

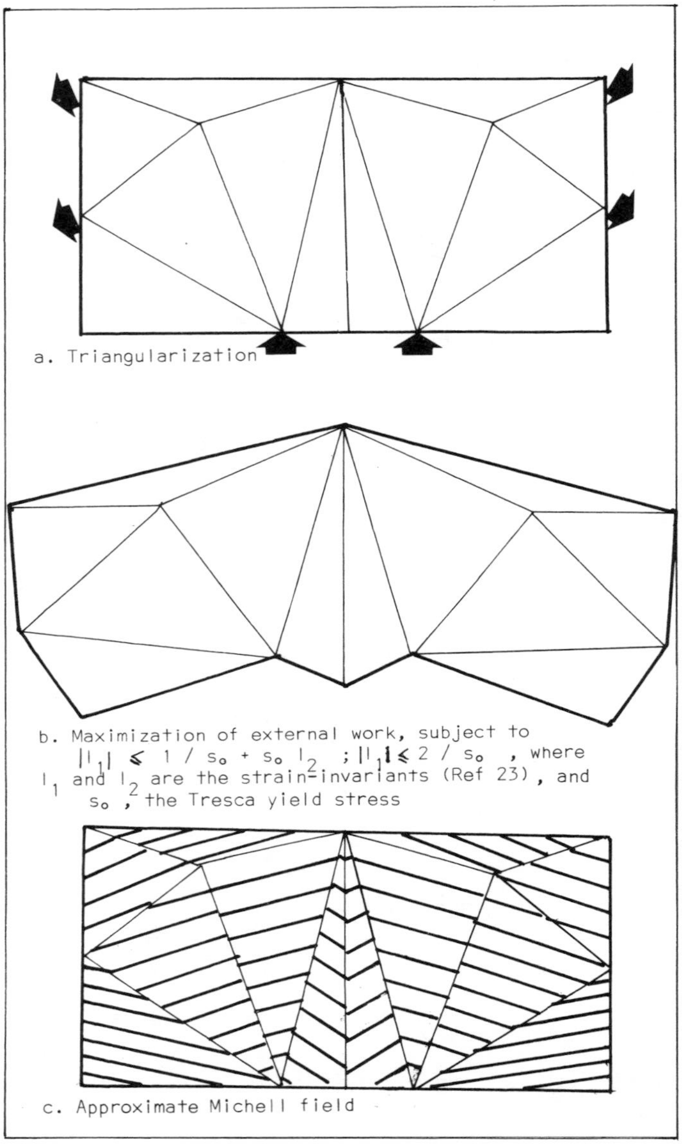

a. Triangularization

b. Maximization of external work, subject to $|I_1| \leq 1/s_o + s_o I_2$; $|I_1| \leq 2/s_o$, where I_1 and I_2 are the strain-invariants (Ref 23), and s_o , the Tresca yield stress

c. Approximate Michell field

Fig 9 . Numerical scheme.

REFERENCES

1. A.G.M. MICHELL , The Limits of Economy of Material in Frame-Structures. *Phil Mag* S6 , 8 , 589-597 (1904)
2. H.L. COX , *The Design of Structures of Least Weight*. Pergamon Press, Oxford (1965)
3. W.S. HEMP , *Optimum Structures*. Clarendon Press, Oxford (1973)
4. J.B.B. OWEN , *The Analysis and Design of Light Structures*. American Elsevier, New-York (1965)
5. W. PRAGER , *On a Problem of Optimal Design*. Brown University. Div of Appl Math. Tech Rep 38 (1958)
6. A.S.L. CHAN , *The Design of Michell Optimum Structures*. College of Aeronautics Report 115 .
7. G.A. HEGEMIER & W. PRAGER , On Michell Trusses. *Int J Mech Sci* 11 , 209-215 (1969)
8. H.S.Y. CHAN , Half-Plane Slip-Line Fields and Michell Structures. *Quart J Mech & Appl Math* XX , (4) , 453-469 (1967)
9. W.S. HEMP , Michell Framework for Uniform Load Between Fixed Supports. *Engineering Optimization* 1 , 61-69 (1974)
10. W. PRAGER , Nearly Optimal Design of Trusses. *Computers & Structures* 8 , 451-454 (1978)
11. W. PRAGER , Optimal Layout of Trusses With Finite Number of Joints. *J Mech Phys Solids* 26 , 241-250 (1978)
12. J.C. NAGTEGAAL & W. PRAGER , Optimum Layout of a Truss for Alternative Loads. *Int J Mech Sci* 15 , 583-592 (1973)
13. G.I.N. ROZVANY & W. PRAGER , Optimization of Structural Geometry. *Proc Int Symp on Dyn Syst, University of Florida*(1977)
14. M. BOTRINI , *Etude Aérodynamique d'une Eolienne Savonius*. Thesis, University of Aix-Marseille (1982)
15. R.V. KOHN & G.STRANG , Structural Design Optimization, Homogeneization & Relaxation of Variational Problems. *Proc of Conf on Disordered Media, NYU* , Springer Verlag (1981)
16. R.V. KOHN & G. STRANG , Hencky-Prandtl Nets & Constrained Michell Trusses. *Proc Int Symp on Opt Struct Design*, University of Arizona, Tucson (1981)
17. P. REISENTHEL, *Structures Ultra-Légères*. Internal Report, Ecole Centrale (1982)
18. E. BOUTROUX & B. LAFOUASSE, *Construction d'une Structure Ultra-Légère*. Internal Report, Ecole Centrale (1982)
19. C.A. BREBBIA , *The Boundary Element Method for Engineers*. Pentech Press, London (1972)
20. J. CEA , C. FLEURY , E. HAUG & B. ROUSSELET , *Shape Optimization*. University of Nice & INRIA (1983)
21. P.J. LAURENT, *Approximation & Optimisation*. Hermann, Paris (1972)
22. R.T. ROCKAFELLAR, *Convex Analysis*. Princeton University Press, Princeton (snd printing ; 1972)
23. J.-M. LAGACHE, Abstract Convolution and Optimum Layout. *Proc Euromech 164* , University of Siegen (1982)
24. J.-M. LAGACHE, Convolution & Géométrie Optimale. *Proc of GAMNI III Symp*, Pluralis, Paris (1983)
25. J.-M. LAGACHE, Treillis de Volume Minimal dans une Région Donnée. *J de Méc* 3 , 25-52 (1964)

STAYED COLUMN DESIGN FOR OPTIMUM INITIAL BUCKLING

W P HOWSON BEng PhD CEng MICE MIStructE*
F W WILLIAMS MA PhD CEng MICE MIStructE**

* Lecturer, Department of Civil Engineering and Building Technology, University of Wales Institute of Science and Technology, Colum Drive, Cardiff. CF1 3EU U.K.

** Professor, Department of Civil Engineering and Building Technology, University of Wales Institute of Science and Technology, Colum Drive, Cardiff. CF1 3EU U.K.

Stayed columns consist of a core with stay frames evenly spaced round it and are very efficient when P/L^2 is small, where P and L are the axial force and length of the column. A recent exact method for designing stayed columns to have a desired critical buckling load has been adapted, and applied to columns with the best stay frame pattern of ten previously studied. The results cover a significant range of columns in detail, and approximations enable a much wider range to be covered. Designs with optimum weight are included, but the range extends over many other designs to enable fabrication and other factors to be balanced against weight savings. The effects of residual prestress are considered, although no attempt is made to account for the sensitivity of the structures to initial imperfections.

PRINCIPAL NOTATION

A_N	representative stay area or leg cross-sectional area, for stayed column
A_p	as A_N, but for substitute plane frame
A_s	$2\pi t^2/3$, datum stay cross-sectional area
b	apex to core dimension of circumscribing parabola, see Fig 1(b)
B_i	intersection of core and axis of symmetry of a bipod, see Fig 1(b)
D_o	outer diameter of stayed column core
E	Young's modulus
F_b	2b/L, a design variable
J	number of bipods per stay frame
L	length of stayed column
M_i	attachment point between stays and core, see Fig 1(b)
P, P_c	axial force in stayed column, and its lowest critical value
R_b, R_o, R_s, R_w	respectively $W_b/W_u, W_o/W_u, W_s/W_u$ and W/W_u
t	core thickness
w	specific weight
W, W_b, W_o, W_s	weight of stayed column and, respectively, its bipods, core and stays
W_u	weight of unstayed column with same P_c and t as the stayed column
α_i	multiplier for initial prestress, with $\alpha_i=1$ giving unstressed stays when $P=P_c$
α_p	P_c/P_c^*
α_s	design variable which scales stay areas, see Eqn 1
α_t	t/t*
ε_o	core strain
*	superscript denoting values for a design covered by Fig 4

INTRODUCTION

References 1-7 demonstrate that stayed columns are competitively light when the dimensionless structural index P_c/EL^2 is low, where P_c is the lowest elastic critical value of the column's axial compressive load P, E is its Young's modulus and L is its length. Previous applications include their use as temporary supports and side booms from the mast of a derrick, described in Ref 1, while Refs 4, 7 and 8 propose their use for warship masts and as components for deep water offshore structures. Special forms, which lie outside the scope of this paper, are currently being explored for possible use in space and these are described in Refs 9 and 10.

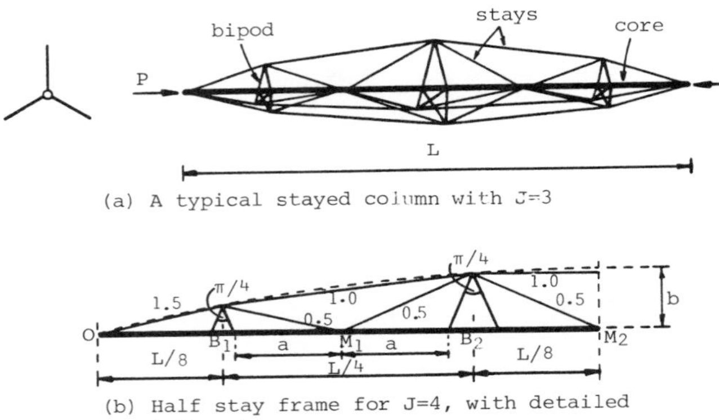

(a) A typical stayed column with J=3

(b) Half stay frame for J=4, with detailed information

Fig 1 Representations of stayed columns used.

In this paper we consider a method of designing stayed columns such that their initial buckling characteristics are optimised. The paper therefore presents many theoretical elastic critical buckling results for stayed columns with three identical stay frames equally spaced round a thin cylindrical core of thickness t, see Fig 1(a). J is the number of bipods per stay frame and symmetry about the column mid-length is assumed, so Fig 1(b) represents a stayed column with J=4. Thin lines denote pre-tensioned stays and thicker lines denote bipods. Stays are assumed to be wires or tie bars, because cables of equal extensional rigidity weigh more and so give a heavier column. For each stay frame the ends of the column and the apexes of the bipods lie on the dashed parabola of Fig 1(b). The stays are pin-ended and all other joints are rigid. The stay areas are given by stay area = $\alpha_s A_s$x(value by stay on Fig 1(b)), ... 1.

where α_s is a design variable and

$$A_s = 2\pi t^2/3. \qquad ...... 2.$$

It is shown later that the results can be adapted to cover stayed columns with more than three stay frames. Also the results cover J=1 to J=7 for stayed columns which relate to Fig 1 as described in the next paragraph.

The axes of symmetry of adjacent bipods are L/J apart and meet the core at $B_1, B_2, B_3, ...$. The stays support the core at points $M_1, M_2, M_3, ...$ which bisect the attachment points of adjacent bipod legs. Thus a on Fig 1(b) indicates equal lengths and $B_1M_1 \neq M_1B_2$. The stays can be visualised as a wire of area $\alpha_s A_s$ connecting the core ends via the bipod ends, plus wires of area 0.5 $\alpha_s A_s$ connecting O to M_1 via the intervening bipod end, M_1 to M_2 via the intervening bipod end, etc., so that every bipod end is connected directly to the core by two wires of area 0.5 $\alpha_s A_s$#. The bipod ends and core ends lie on a parabola with its apex at b from the core, see Fig 1(b).

Each of the main results presented consists of a minimum weight design for a particular value of P_c/EL^2 and t/L. Optimum designs use the smallest possible t, which will often be determined by practical considerations, e.g. corrosion. The datum results are for $P_c/EL^2=10^{-9}$, t/L = 10^{-4} and simply supported (i.e. ball-jointed) ends, but further results cover the separate effects of changing P_c/EL^2 and t/L, and of having built-in ends.

The main independent design variables were J, α_s, the core outer diameter D_o and F_b, the 'slenderness' of a stay frame, given by F_b = 2b/L.

The designs presented are based on critical buckling considerations alone, but core strains at buckling, ε_o, are given so that the onset of yield can be predicted for any chosen material.

The stay frame pattern of Fig 1 was the best of ten patterns compared previously in Ref 4, and use of only this one pattern yields much greater accuracy and far more extensive conclusions than before.

BASIC THEORY

The usual assumption is made (see Refs 2,3,6,7 and 11) that an optimum stayed column has all its stays taut for $P<P_c$ and just unstressed for $P=P_c$. Then Ref 11 (see particularly conclusion 14) gives theoretical proofs as follows.

1) Critical buckling cannot occur during the application of P_c so long as the column is stable under prestress alone and when P_c is fully applied.

2) Buckling under prestress alone can often be adequately and conservatively guarded against by checking that the axial forces in the bipod legs satisfy the inequality

axial force ≤ pin-ended Euler load. ...... 3.

3) P_c can be found exactly* by using a substitute plane frame which consists of the original P, the original core with its extensional rigidity made infinite, and a single stay frame which is identical to a stay frame of the original column except that

$$A_p = 1.5A_N \quad \text{and} \quad I_p = 1.5I_N, \qquad ...... 4.$$

where A is the area of any stay or leg, I is the second moment of area of any leg, the subscript p denotes a plane frame value and the subscript N denotes the corresponding stayed column value.

Hence, since the stays are unstressed when $P=P_c$ and so can be treated as beams of zero flexural rigidity, the plane frame program published in Ref 12 could have been used to find P_c, and the associated buckling mode, for a range of stayed columns subject to the safeguard of Eqn 3. However, a better solution lay in changing the program so that it could design a stayed column with the required P_c, using D_o, the external diameter of the core, as the only design variable. The following method results, and differs slightly from the one used in Ref 4.

The program of Ref 12 was iterative, with the design kept constant and P varied between iterations. Each iteration checked whether the frame was stable. P was increased if it was and decreased otherwise. Hence bisection enabled successive iterations to converge on P_c to specified accuracy.

The design program made the same check for stability at each iteration. D_o was decreased if the column was stable and increased otherwise, with P kept equal to the required value of P_c. This procedure was adopted because the column was inevitably stable for D_o very large and unstable for D_o very small. Hence initial bounds were established on the required value of D_o, and bisection was used to bound it to specified accuracy.

The stay areas were unchanged during any one design. Bipod legs of the substitute plane frame were also unchanged, being designed as hollow cylinders of thickness t and with their lengths equal to 180 times their radius of gyration, i.e. with a slenderness ratio equal to 180. Calculations for all the designs presented herein showed that such legs rarely violated Eqn 3 and that re-designing legs to avoid such violations would never have involved excessive weight penalties. Therefore legs derived from the results presented must be checked for adequacy under prestress alone and must be re-designed if Eqn 3 is violated.

PRIMARY RESULTS

Figure 2(a) was plotted for a substitute plane frame with three bipods, J=3, by changing the stay areas via α_s between successive designs, with all other design variables (except D_o) fixed as shown. Here $R_w=W/W_u$ and $R_o=W_o/W_u$, where W is the weight of a stayed column with the required values of P_c and t, W_o is the weight of its core, and W_u is the weight of an unstayed cylindrical column with the same P_c and t as the stayed column. The minimum weight design of Fig 2(a) is the one giving the lowest R_w, represented by the two points with the vertical dashed line between them. Figure 2(b) shows how this minimum weight design varied with $F_b(=2b/L)$, being obtained from Fig 2(a) and its equivalent for other F_b values. Thus the horizontal dashed lines from Fig 2(a)

The choice of the factor 0.5 was based on specimen results for two problems, which both showed that the weight obtained by using 0.5 was within 0.5% of the minimum weight which could be obtained by varying the factor. Similarly, altering the included angle of the bipods gave minimum weights within 0.5% of those given by the chosen value of $\pi/4$.

* Strictly speaking, the use of this substitute plane frame involves neglecting the stiffness of bipods against flexure out of their planes (see conclusion 18 of Ref 11), but a typical check result showed that the errors caused change the results presented imperceptibly, so that all results are exact to the accuracy of presentation.

give the two points shown at F_b=0.115 on Fig 2(b), etc. This procedure gave all the results presented and so the form of R_w versus α_s curves, e.g. the solid curve of Fig 2(a), is now discussed.

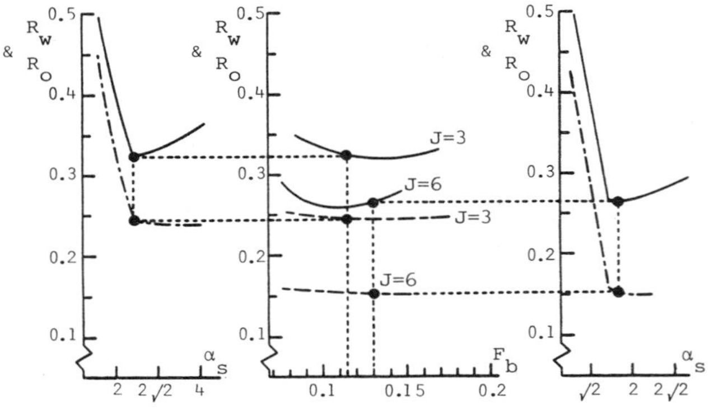

(a) J=3,F_b=0.115 (b) (c) J=6,F_b=0.13

Fig 2 Representative results for R_w (solid) and R_o (chain-dotted) for $P_c/EL^2=10^{-9}$ and $t/L=10^{-4}$. Note log scales for α_s.

All such curves were observed to have two smooth portions with discontinuous slope where they met. The left-hand portion corresponded to an overall buckling mode and the right-hand portion corresponded to a snaking mode, where the transverse core deflections for the former predominantly approximated a half sine wave between the column ends, whereas for the latter they usually occurred predominantly between, rather than at, the points B_1, M_1, B_2, M_2, etc., on Fig 1(b). Figure 3 shows typical calculated overall and snaking modes, to scale.

In Fig 2(a) R_w is least where the two portions of the R_w curve meet, and in this vicinity the left-hand portion is much steeper than the right-hand one. This typifies over half of the curves used to obtain the results presented and means that α_s can be increased above the optimum value to enable standard stay sizes to be used, but should not be decreased. Figure 2(c) typifies all the remaining curves, for which the minimum weight design was given by the right-hand portion and therefore corresponded to a snaking mode.

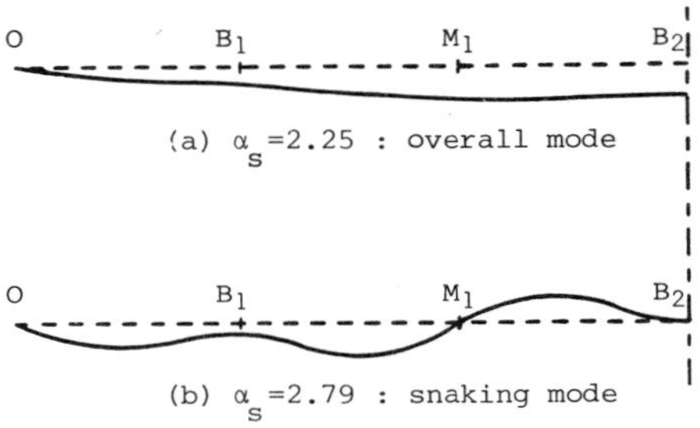

Fig 3 Modes for two designs of Fig 2(a), showing only the deflected and undeflected position of the half core length of Fig 1(b) because both modes are symmetric.

Enough points were obtained for all the curves used, e.g. Figs 2(a)-(c), to be drawn with confidence. This contrasts with Ref 4 in which only four points were available for plotting each Fig like Figs 2(a) and (b), with successive values of α_s for the former, and of F_b for the latter, doubled between successive points. Moreover Ref 4 needed so many such Figs to cover its ten stay patterns that plotting them proved prohibitive, and instead their minima were approximated by the lowest ordinate of each set of four points. Thus Ref 4 gave R_w values which were typically 8% too high and only gave crude indications of the corresponding values of α_s and F_b. Hence Fig 4 gives a much more comprehensive and accurate representation of optimum stayed column designs than Ref 4.

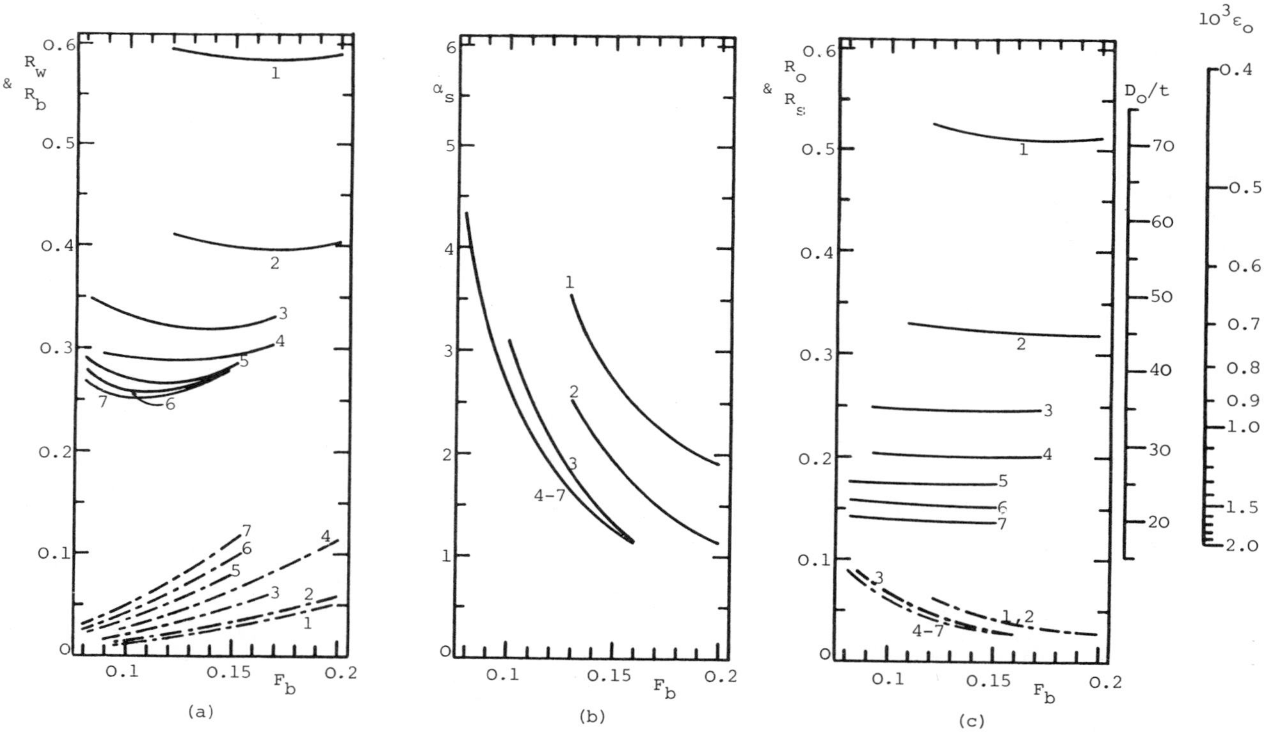

Fig 4 R_w, α_s, R_o, D_o/t and ε_o (shown solid) and R_b and R_s (shown chain-dotted) for nearly optimum stayed columns for $P_c/EL^2=10^{-9}$, $t/L=10^{-4}$ and the values of J shown by the curves. $W_u=4.31\times10^{-6}wL^3$ for all the curves.

Figure 4 was obtained as follows. The R_w and R_o curves of Figs 4(a) and (c), respectively, were obtained directly from Fig 2(b) and the equivalent curves for J=1,2,4,5 and 7. Figure 4(b) gives the corresponding α_s values. Thus the value α_s=2.34 for J=3 and F_b=0.115 on Fig 4(b) was obtained as the abscissa of the vertical dashed line of Fig 2(a), etc. W_s, the total weight of the stays, and W_b, the total weight of the bipods, were both readily calculated and gave the R_b and R_s values of Figs 4(a) and (c), where $R_b=W_b/W_u$ and $R_s=W_s/W_u$ so that $R_b+R_s+R_o=R_w$.

Since the results of Fig 4 all have $P_c/EL^2=10^{-9}$ and $t/L=10^{-4}$ they also all have $W_u=4.31\times10^{-6}wL^3$, from the definition of W_u, where w is the specific weight which the core, bipods and stays are assumed to share. Therefore the solid lines of Fig 4(c) give W_o/wL^3 to some scale, since $R_o=W_o/W_u$. But W_o can be found in terms of w,L,t and D_o from its definition, and since $t/L=10^{-4}$ this relates D_o/L, and hence D_o/t, to W_o/wL^3. Hence the D_o/t scale of Fig 4(c) was constructed for use with the solid lines, as was the core strain scale, $10^3\varepsilon_o$, which could be constructed because D_o/t and ε_o are related since P_c/EL^2 and t/L are fixed for Fig 4 and the stays are just slack when $P=P_c$.

When $P_c/EL^2=10^{-9}$ and $t/L=10^{-4}$, Fig 4 can be used to obtain near-optimum designs for any selected value of J between 1 and 7, e.g. when J=3 the curves give the lightest design as having $R_w\simeq0.318$ and $F_b\simeq0.14$ from Fig 4(a) and hence, from Figs 4(a)-(c) with F_b=0.14: $R_b\simeq0.043$, $\alpha_s\simeq1.55$, $R_o\simeq0.245$, $R_s\simeq0.035$, $D_o/t\simeq34.7$ and $\varepsilon_o\simeq0.95\times10^{-3}$. Note that the optimum value of F_b could not be estimated very accurately from Fig 4(a) and that all optimum F_b results presented share such approximations. Similarly the accuracy of all α_s values derived from Fig 4 is limited by the accuracy of estimation of the abscissae of the vertical dashed lines on Figs 2(a) and (c).

Figure 4 shows that D_o/t and R_w are relatively insensitive to F_b for near-optimum designs, but that the value of F_b chosen affects the required value of α_s considerably. Hence taking F_b as 0.11 instead of 0.14 in the previous example alters D_o/t negligibly and gives $R_w\simeq0.325$ instead of $R_w\simeq0.318$ (i.e. the column is only about 2% heavier), whereas α_s is increased from ≃1.55 to ≃2.60. This large increase in α_s is predictable because α_s and F_b respectively represent the area and 'lever arm' of the stays.

Therefore after fixing the number of bipods, J, the designer can choose F_b from a wide range of values and still get a nearly optimum column, but the value chosen determines α_s and D_o/t via Figs 4(b) and (c). Alternatively α_s can be chosen from a wide range of values and then F_b and D_o/t follow from Figs 4(b) and (c), e.g. a design with J=3 and α_s=1.3 has $F_b\simeq0.152$ and $D_o/t\simeq34.7$ while $R_w\simeq0.321$, so that it only weighs about 1% more than the optimum design (with $\alpha_s\simeq1.55$) given above.

The weight penalty involved in changing the core, bipods or stays of a chosen design can be calculated by hand from its values of R_o,R_b and R_s, e.g. increasing stay diameters by 20% increases R_w by $0.44R_s$, etc. Possible causes of such changes include using standard section sizes, using cables rather than wires or rods as the stays, using heavier legs when prudence or Eqn 3 requires them, and using empirical methods to allow for imperfections. Such changes can be made without reducing the column's capacity so long as they do not reduce any stiffnesses and the prestress is adjusted to ensure zero residual prestress at the P_c of the original column.

EFFECTS OF CHANGING DESIGN LOAD OR THICKNESS

The curves labelled 1 on Fig 5 duplicate the J=3 curves of Fig 4. The corresponding curves for $P_c/EL^2=3\times10^{-9}$ and $P_c/EL^2=\frac{1}{3}\times10^{-9}$ are labelled 3 and $\frac{1}{3}$ on Fig 5, respectively, and the corresponding curves for $t/L=2\times10^{-4}$ and $t/L=\frac{1}{2}\times10^{-4}$ are labelled (2) and ($\frac{1}{2}$). There is no R_o curve labelled 1 on Fig 5 because it was confusingly close to the curves shown.

Arguments given in the Appendix predict that Fig 4 can be used to design columns for a substantial range of values of P_c/EL^2 and t/L, as follows. Suppose that E and L are known and that P_c^* and t^* are values of P_c and t for which Fig 4 applies, so that $P_c^*=10^{-9}EL^2$ and $t^*=10^{-4}L$. Now suppose that a design is required for which

$$P_c=\alpha_p P_c^* \quad \text{and} \quad t=\alpha_t t^*. \qquad \ldots\ldots 5.$$

The prediction is that these values of P_c and t give a minimum weight design for which

$$R_w\simeq R_w^*,\quad D_o/t\simeq\alpha_p^{\frac{1}{3}}\alpha_t^{-\frac{4}{3}}(D_o/t)^* \quad \text{and} \quad \varepsilon_o\simeq\alpha_p^{\frac{2}{3}}\alpha_t^{-\frac{2}{3}}\varepsilon_o^*, \qquad \ldots\ldots 6.$$

while

$$\text{and} \quad \begin{aligned} F_b&=\{(\alpha_p R_s^*)/(\alpha_t R_b^*)\}^{\frac{1}{4}}F_b^* \\ \alpha_s&=\{(\alpha_p F_b^{*2})/(\alpha_t^2 F_b^2)\}\alpha_s^*, \end{aligned} \qquad \ldots\ldots 7.$$

where R_w^*,$(D_o/t)^*$,ε_o^*,R_s^*,R_b^*,F_b^* and α_s^* are the values given by Fig 4. Table 1 shows that the accurate results of Fig 5 confirm that Eqn 6 gives good predictions for wide variations of α_p and α_t. When obtaining Table 1, Fig 4 gave $R_w^*\simeq0.318$, $(D_o/t)^*\simeq34.5$, $\varepsilon_o^*\simeq951\times10^{-6}$, $F_b^*\simeq0.138$, $\alpha_s^*\simeq1.66$, $R_b^*\simeq0.041$ and $R_s^*\simeq0.037$.

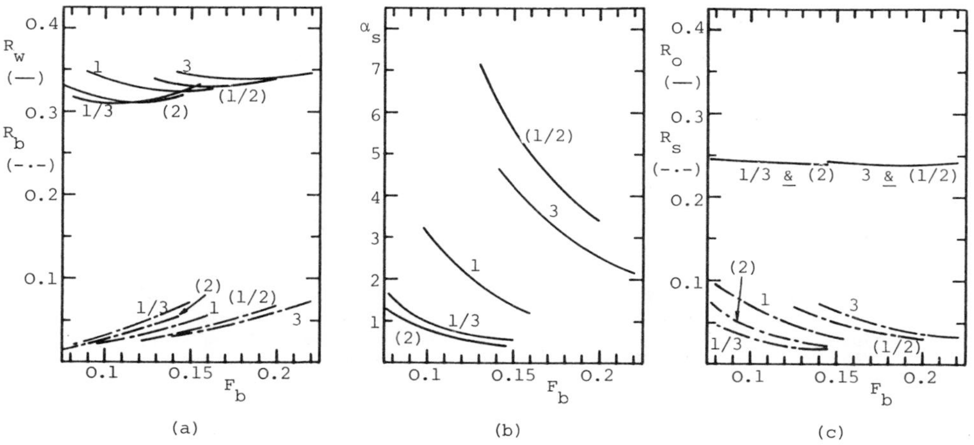

Fig 5 Effect on the J=3 curves of Fig 4 of multiplying P_c/EL^2 or t/L by factors, with the factors in brackets being applied to t/L.

α_p (see Eqn 5)	1/3	3	1	1
α_t (see Eqn 5)	1	1	1/2	2
R_w, correct	0.305	0.338	0.329	0.310
R_w, from Eqn 6	0.318	0.318	0.318	0.318
D_o/t, correct	23.8	48.1	84.0	14.1
D_o/t, from Eqn 6	23.9	49.8	86.9	13.7
$10^6\varepsilon_o$, correct	465	2028	1535	607
$10^6\varepsilon_o$, from Eqn 6	457	1978	1510	599
α_s, correct	0.91	3.21	5.00	0.60
α_s, from Eqn 7	1.01	3.03	4.94	0.62
F_b, correct	0.106	0.175	0.160	0.115
F_b, from Eqn 7	0.102	0.177	0.160	0.113

Table 1 Accuracy of Eqns 6 and 7 for J=3. The correct values are taken from Fig 5.

EFFECTS OF END CONDITIONS AND OF NUMBER OF STAY FRAMES

When one or both of the column ends were clamped instead of simply supported the ordinates of the R_w versus F_b curve for the problem specified by the caption of Fig 4 with J=3 never differed by more than about 1½% from the corresponding values for the J=3 curve of Fig 4(a). (Note that the W_u used to find R_w from $R_w=W/W_u$ was kept constant throughout, so that W_u was the weight of a simply supported/simply supported unstayed cylindrical column with the same P_c and t as the stayed column, even when the stayed column had one or both ends clamped). This surprising result is probably explained as follows. Figure 6 shows that clamping both ends altered the overall buckling portion of the curve greatly, but changed the snaking buckling portion imperceptibly, presumably because snaking modes usually predominantly involve the core portions OB_1, B_1M_1 and M_1B_2 of Fig 1(b) buckling as clamped/simply supported members, see Fig 3(b), and so clamping O affects B_1M_1 little and M_1B_2 even less. Also the slope of the snaking buckling portion is low and so the ordinate of its meeting point with the overall portion cannot be affected much by changes of the latter. Since the slope of the snaking portion was quite low where it met the overall portion for all the results presented, all the R_w values presented are probably still approximately correct when one or both column ends are clamped.

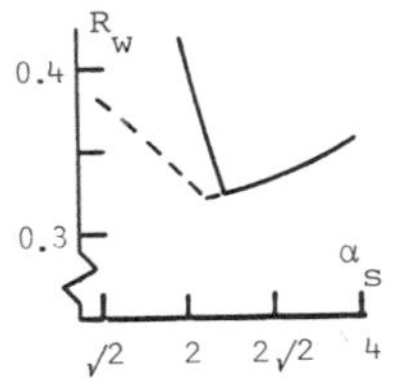

Fig 6 The R_w curve of Fig 2(a), shown solid, and the corresponding curve when the ends of the column are clamped, shown dashed.

Not unexpectedly, changing the end conditions of the simply supported/simply supported problem specified by the caption of Fig 4 with J=3 to clamped/free resulted in an optimum weight about 3.7 times heavier than that given by Fig 4.

The curves presented can also be used to design stayed columns with more than three stay frames equally spaced round the core. This is because Eqn 4 is still valid if the 1.5 is replaced by N/2 (see Ref 11) where N is the number of stay frames, so long as the denominator, 3, of Eqn 2 is replaced by N. Note that it was the thickness of a bipod leg of the substitute plane frame which had thickness t, i.e. A_p and I_p of Eqn 4 are for a hollow cylinder of thickness t. The cylinder was not very thin, but to a reasonable approximation thin cylinder theory and Eqn 4 give the thickness of the bipod leg of the stayed column as approximately 2t/N.

EFFECTS OF RESIDUAL PRESTRESS

So far, stays were assumed to become just unstressed when the axial force in the core attained its critical value, i.e. $P=P_c$. This requires the appropriate initial prestress pattern, which can be found by statics from P_c and the simple plane frame represented by Fig 1(b), with appropriate boundary conditions and member properties and with its core constrained to remain straight. Thus the prestress pattern needed for J=4 would be calculated from Fig 1(b) by making the small assumption of pin joints between the 4 legs and 6 core portions. This prestress pattern can be multiplied by α_i, so that all results presented above are for $\alpha_i=1$.

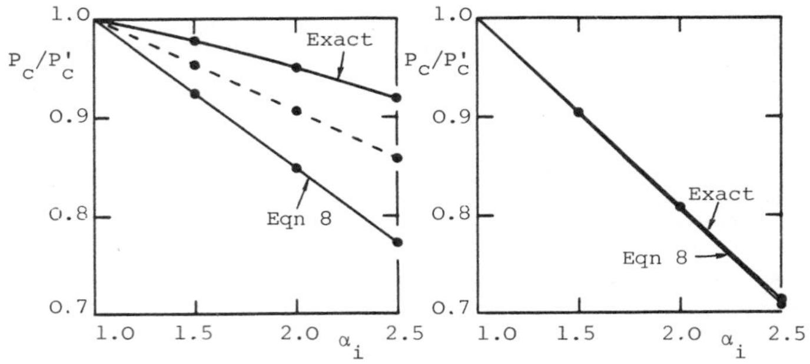

(a) α_s=2.25 (b) α_s=2.79

Fig 7 Effect of residual prestress on the designs of Fig 2(a) when α_s has the values shown.

When $\alpha_i>1$ the columns were analysed using a development of the plane frame program published in Ref 12. This program shared the exact nature of its predecessor and extended it to include tensioned stays, space frames and the checks required by Ref 13 to ensure convergence on P_c in the presence of dead and live loads, i.e. initial prestress forces and the forces caused by P. This program was applied to the type I substitute column of Ref 11, which gives exact results, to obtain the upper solid lines on Figs 7(a) and (b), where P_c' is the value of P_c when α_i=1 (i.e. it is the P_c of all the preceding results). The lower solid lines were obtained by hand from

$$P_c \simeq P_c' - (\alpha_i - 1)P_i, \qquad \ldots\ldots 8.$$

where P_i is the axial force in the core at O (see Fig 1(b)) under prestress alone when α_i=1, and Eqn 8 results from assuming that the axial force in the core at O when buckling occurs is independent of α_i and so always equals P_c'. The broken line on Fig 7(a) corresponds to the exact curve except that one of the stay frames of the type I substitute column described in Ref 11 was omitted, so that the results are those which would be obtained by using the plane frame used elsewhere in this paper and allowing for the effects of residual prestress. The corresponding line for Fig 7(b) lay between the two shown.

Figure 3 shows that Figs 7(a) and (b) respectively relate to designs governed by overall and snaking modes. If these results are typical, P_c is most sensitive to α_i when the governing mode is snaking rather than overall, and hand application of Eqn 8 gives excellent accuracy when the snaking mode governs and conservative results when the overall mode governs. Hence using $\alpha_i>1$, for columns designed to have coincident modes and α_i=1, should make the snaking mode govern and reduce P_c almost exactly in accordance with Eqn 8. Therefore it would be possible to prevent imperfections making stays slack before $P=P_c$ by substituting a suitably high value of α_i, an estimate of P_i and the required P_c in Eqn 8 to give P_c', and then using the various design procedures for $\alpha_i=1$ given earlier, but with P_c replaced by P_c'. If W' is the weight of the stayed column given by this modified design procedure, and since its weight would otherwise be W, application of $R_w \simeq R_w^*$ (see Eqn 6), $R_w=W/W_u$, the definition of W_u and thin cylinder theory give

$$W'/W \simeq (P_c'/P_c)^{1/3} \qquad \ldots\ldots 9.$$

Hence adopting the procedure only results in a small weight penalty, e.g. 7.7% when $P_c/P_c'=0.8$. However Eqn 9 will be slightly optimistic if increasing α_i results in heavier legs being needed to satisfy Eqn 3.

SUMMARY OF THE MAIN CONCLUSIONS

The stay frame pattern remained constant throughout, being the best of ten possible patterns considered previously in Ref 4, and the main conclusions drawn above are now summarised.

1) The primary results give complete details of good designs with $P_c/EL^2=10^{-9}$ and $t/L=10^{-4}$, see Fig 4.

2) They allow designers to choose convenient values of J and F_b.

3) Hence they enable designers to balance weight savings given by one choice of J and F_b against fabrication and other advantages given by an alternative choice, e.g. increasing J reduces the weight but probably increases the fabrication costs.

4) The designs could not be improved significantly by changes in the values of $\pi/4$ and 0.5 which were used for, respectively, the included angle of the bipod legs and the ratio of the cross-sectional areas of the 'diagonal' and 'outer' stays (see Fig 1(b)).

5) The weight penalties involved in altering the slenderness ratio of the bipod legs, adopting standard section sizes, using cables rather than wires or rods as the stays, making empirical allowances for imperfections, etc., can be approximated by hand (see the second paragraph above the one containing Eqn 5).

6) The primary results can be adapted to give good accuracy for widely varying values of P_c/EL^2 and t/L, by using Eqns 6 and 7.

7) All the columns considered buckled with overall or snaking modes, see Fig 3.

8) The designs presented are for simply supported/simply supported columns, except that the discussion of Fig 6 shows that clamping one or both ends has little effect on the designs.

9) The legs of a few of the designs presented may need stiffening to prevent buckling under prestress alone, e.g. see Eqn 3.

10) The designs covered can have any number of identical stay frames equally spaced round the core.

11) The effects of residual prestress are not too serious, particularly when the buckling mode is overall, see Fig 7.

12) Equation 8 gives manual estimates of these effects which are almost correct when the buckling mode is snaking and are conservative for overall modes.

ACKNOWLEDGEMENTS

This study was supported by the Marine Technology Directorate of the Science Research Council, and Dr. J.R. Banerjee made useful contributions.

REFERENCES

1. K -H CHU and S S BERGE. Analysis and design of struts with tension ties. *J. Struct. Div. Am. Soc. Civ. Engrs*, 1963,89,ST 1,Feb.,127-163.

2. H H HAFEZ *et al.* Pretensioning of single-crossarm stayed columns. *J. Struct. Div. Am. Soc. Civ. Engrs*, 1979,105,ST 2,Feb.,359-375.

3. I A HATHOUT *et al.* Buckling of space stayed columns. *J. Struct. Div. Am. Soc. Civ. Engrs*, 1979,105,ST 9,Sept.,1805-1822.

4. W P HOWSON and F W WILLIAMS. A parametric study of the initial buckling of stayed columns. *Proc. Instn Civ. Engrs*, Part 2,1980,69,June,261-279.

5. H R MAUCH and L P FELTON. Optimum design of columns supported by tension ties. *J. Struct. Div. Am. Soc. Civ. Engrs*, 1967,93,ST 3,June,201-220.

6. R J SMITH *et al.* Buckling of a single-crossarm stayed column. *J. Struct. Div. Am. Soc. Civ. Engrs*, 1975,101,ST 1,Jan.,249-268.

7. M C TEMPLE. Buckling of stayed columns. *J. Struct. Div. Am. Soc. Civ. Engrs*, 1977,103,ST 4,April,839-851.

8. F W WILLIAMS and W P HOWSON. Stayed columns cure undersea woes. *Offshore*, 1980,40,No.10,Sept., 144-147.

9. J R BANERJEE and F W WILLIAMS. Vibration characteristics of self-expanding stayed columns for use in space. *J. Sound Vib.*, 1983,90(2),245-261.

10. W K BELVIN. Analytical and experimental vibration and buckling characteristics of a pretensioned stayed column. *Proc. AIAA 23rd Structures, Structural Dynamics and Materials Conference*, 1982,Paper no. 82-0775CP.

11. F W WILLIAMS and W P HOWSON. Concise buckling analysis of stayed columns. *Int. J. Mech. Sci.*, 1978,20,No.5,299-313.

12. W P HOWSON. A compact method for computing the eigenvalues and eigenvectors of plane frames. *Adv. Engng Software*, 1979,1,No.4,181-190.

13. W H WITTRICK and F W WILLIAMS. An algorithm for computing critical buckling loads of elastic structures. *J. Struct. Mech.*, 1973,1,No.4,497-518.

APPENDIX - DERIVATION OF APPROXIMATE FORMULAE OF EQNS 6 AND 7

It is still assumed that P_c^* and t^* are values of P_c and t for which Fig 4 applies (so that $P_c^*=10^{-9}EL^2$ and $t^*=10^{-4}L$), that for $P=P_c^*$ and $t=t^*$ Fig 4 gives $D_o=D_o^*$, $R_b=R_b^*$, etc., and that the normal definitions can be used to give $W_b^*=R_b^*W_u^*$, etc. It is also assumed that P_c and t are altered to $\alpha_p P_c^*$ and $\alpha_t t^*$ without E or L being altered.

Equation 6 can be derived as follows. It was noted in the discussion of Figs 2(a) and (c) that all designs have a snaking buckling mode when $P=P_c$. Making the crude assumption that such modes are governed solely by the flexural rigidity of the core, and using thin cylinder theory, gives the approximation $D_o^3 t \propto P_c$. Hence if Fig 4 gives $D_o=D_o^*$ when $P_c=P_c^*$ and $t=t^*$, it follows that when $P_c=\alpha_p P_c^*$ and $t=\alpha_t t^*$ the diameter of the core is given by

$$D_o \simeq \alpha_p^{\frac{1}{3}} \alpha_t^{-\frac{1}{3}} D_o^*. \qquad \text{...... A1.}$$

The second of Eqns 6 follows from Eqn A1, by using $\alpha_t = t/t^*$.

Thin cylinder theory gives the approximation $W_o \propto D_o t$. Therefore when $P_c = \alpha_p P_c^*$ and $t = \alpha_t t^*$ it follows that $W_o \simeq (D_o/D_o^*)\alpha_t W_o^*$, so that substituting for D_o/D_o^* from Eqn A1 gives $W_o \simeq \alpha_p^{\frac{1}{3}} \alpha_t^{\frac{2}{3}} W_o^*$. Similarly it can be shown that $W_u \simeq \alpha_p^{\frac{1}{3}} \alpha_t^{\frac{2}{3}} W_u^*$, so that $W_o/W_u \simeq W_o^*/W_u^*$. Since W_o is the weight of the core and dominates the weight of the stayed column, the first of Eqns 6 follows.

The whole of P_c is carried by the core. Therefore thin cylinder theory gives the approximation $\varepsilon_o \propto P_c(D_o t)^{-1}$, so that when $P_c = \alpha_p P_c^*$ and $t = \alpha_t t^*$ it follows that $\varepsilon_o \simeq (\alpha_p/\alpha_t)(D_o^*/D_o)\varepsilon_o^*$. Then substituting for D_o^*/D_o from Eqn A1 gives the third of Eqns 6.

Equation 7 can be derived as follows. It has been noted that an overall mode often coincides with the snaking mode when $P = P_c$, see Fig 2(a), although Fig 2(c) shows that this is not always so. The conservative assumption of such coincidence of the overall mode is now made. Treating the column cross-section as solid, and ignoring everything except the outer stays, gives its second moment of area at mid-length as approximately $\alpha_s \pi t^2 b^2$, see Fig 1(b) and Eqns 1,2 and 4. Hence when $P_c = \alpha_p P_c^*$ and $t = \alpha_t t^*$ it follows that

$$\alpha_p = P_c/P_c^* = (\pi^2 E \alpha_s \pi t^2 b^2/L^2)/(\pi^2 E \alpha_s^* \pi t^{*2} b^{*2}/L^2)$$

$$= (\alpha_s \alpha_t^2 F_b^2)/(\alpha_s^* F_b^{*2}), \qquad \text{...... A2.}$$

where use has been made of the relationships $F_b = 2b/L$ and $F_b^* = 2b^*/L$. Since α_p, α_t, α_s^* and F_b^* are known, Eqn A2 relates α_s and F_b. Therefore any pair of values of α_s and F_b which satisfies Eqn A2 will give an adequate column. However since a minimum weight design is required, and as the core has already been determined by Eqn 6, α_s and F_b should be chosen to minimise the combined weight, $W_b + W_s$, of the bipods and stays. It is assumed that the slenderness of the bipod legs is kept constant as P_c and t are changed and that the relatively small effect which altering F_b has on the total length of all the stays can be ignored. These assumptions and Eqns 1, 2 and 4 give the approximations $W_b \propto F_b^2 t$ and $W_s \propto \alpha_s t^2$. Hence

$$W_b + W_s = (F_b/F_b^*)^2 \alpha_t W_b^* + (\alpha_s/\alpha_s^*)\alpha_t^2 W_s^*. \qquad \text{...... A3.}$$

Then substituting for α_s from Eqn A2, and differentiating with respect to F_b, shows that $W_b + W_s$ has its minimum value when

$$(F_b/F_b^*) = \{(W_s^*/W_b^*)(\alpha_p/\alpha_t)\}^{\frac{1}{4}}. \qquad \text{...... A4.}$$

The first of Eqns 7 follows because $R_s^* = W_s^*/W_u^*$ and $R_b^* = W_b^*/W_u^*$. Finally, Eqn A2 completes the derivation of Eqns 6 and 7, by giving the second of Eqns 7.

Despite the simple approximations used, Table 1 shows that Eqns 6 and 7 gave good results when tested. However the minima of the R_w versus F_b curves used to obtain Table 1 all happened to correspond to Figs like Fig 2(a) and so matched the coincident mode method assumed above (see between Eqns A1 and A2) when deriving Eqn 7. If the minima had corresponded to Figs like Fig 2(c) such good agreement would seem improbable, although the Eqns would probably still give a good choice of F_b and α_s because it was observed that all Figs like Fig 2(c) involved R_w changing little when F_b was varied about its optimum value.

MINIMUM-WEIGHT DESIGN OF FULLY STRESSED ARCHGRIDS

C M WANG*, W A M ALWIS* and G I N ROZVANY**

*Lecturers, Department of Civil Engineering, National University of Singapore, Kent Ridge, Singapore

**Reader, Department of Civil Engineering, Monash University, Clayton, Victoria, Australia.

The optimal shape of the middle surface of archgrids and cable networks was studied in the late seventies by the third author and the late Prof W Prager (Brown University). They considered the problem of transmitting a system of vertical loads over a plane horizontal domain to the boundary of the latter by means of a gridwork of interesting arches or cables. In their study, it was assumed that (i) each arch or cable is contained in a vertical plane; (ii) the orientation of such planes is restricted to two given directions at right angles, and (iii) the maximum permissible stress is prescribed. Rozvany and Prager also found that for distributed loads, the optimal solution consists of a "grid-like continuum" (Prager), having an infinitesimal arch (cable) spacing.

It was pointed out more recently by the first author that for certain boundary conditions some arches must take on a zero cross-section, if a discretized formulation is used. An optimality condition for such "vanishing arches" was suggested by the third author. The current paper discusses a numerical procedure which can handle vanishing arches.

INTRODUCTION

Although a number of publications have been devoted to the optimal design of isolated arches (e.g. Refs 1-10), the problem of two-way arch (and cable) systems was first studied extensively by German and Japanese research teams of Otto (Ref 11) and Argyris in Stuttgart, West Germany. Their investigation concerned (i) the establishment of fundamental classes of network topographies, (ii) the analysis of cable nets subject to large deformations, and (iii) the inversion of experimentally produced network shapes with a view to constructing fully stressed archgrids. However, the above research groups did not attempt optimization of archgrids and cable network until 1978 when Rozvany and Prager were commissioned by them to study this problem. These two researchers then derived (Ref 12) the most efficient (minimum-weight) solution for fully-stressed surface structures consisting of intersecting arches or cables of continuously varying cross-section. The considered structures are required to transmit a given single system of vertical forces over a horizontal domain to supporting lines along the boundaries of the latter. The vertical location of the forces is unspecified and all arches (cables) are restricted to vertical planes whose orientation is constrained to two directions at right angles. All arches (cables) are to develop the same constant and given longitudinal stress throughout the structure and the total weight of the system is to be a minimum.

An extension of the above problem to external loads plus selfweight was considered originally by Rozvany and Prager; this project was completed, after Prager's unexpected death, by Rozvany, Nakamura and Kuhnell (Ref 13). In further studies by Rozvany and Wang (Refs 9, 10 and 14), the optimization of the layout of archgrids and cable networks was considered. Such fully optimized surface structures have been termed "Prager-structures" and shown to constitute a special subset of Michell-frames (minimum-weight trusses, see Ref 15). It has been shown (Refs 9, 10, 14) that by restricting a minimum weight space truss to either purely tensile or purely compressive stresses, it reduces to a surface structure which is identical to a least-weight archgrid or cable network ("Prager-structure").

In order to facilitate their mathematical treatment, Prager and Rozvany (Ref 12) originally did not restrict their arches (or cables) to a single surface. However, the above papers (Refs 12 and 13) established that least-weight archgrids (cable networks) of prescribed maximum stress must satisfy three optimality criteria:

(i) equal elevation condition requiring arches (cables) to have the same elevation at their intersection thus forming a single middle surface for the system;

(ii) zero bending condition having the consequence that all arches (cables) are in pure compression (tension) and hence they take on the shape of a funicular for their (vertical) load;

(iii) the unit mean square condition requiring that the mean value of the square of the slope for each arch (cable) takes on a unit value.

A remarkable property of all least-weight fully-stressed arch-grids and cable networks (including "Prager-structures") is that their total weight is proportional to the sum of the products of external loads and their optimal elevations (Refs 9, 10, 14).

Rozvany and Prager (Ref 12) established that a for a uniformly distributed load, optimal archgrids have an infinitesimal arch-spacing and hence they form an "arch-grid-like continuum" (Prager). In deriving approximate solutions by discretization, however, the distributed load was replaced by a system of equidistant point loads. It was pointed out more recently by the first author that in such discretized solutions, some arches must take on a zero cross-section, because otherwise it would be impossible to satisfy all optimality conditions [conditions (i) to (iii) above] simultaneously. A relaxed optimality condition for such "vanishing arches" was suggested by the third author. The aim of this paper is to present a modified numerical procedure which can handle vanishing arches (cables).

It should be noted that the problems discussed do not take multiple load conditions (e.g. wind load), elastic instability of archgrids or "flutter" of cable networks into consideration. It was found, however (Ref 16) that the inherent instability of least-weight arch-grids can be eliminated by employing an efficient bracing system whose structural weight is relatively insignificant.

FULLY-STRESSED LEAST-WEIGHT ARCHES: PROBLEM FORMULATION AND OPTIMALITY CRITERIA

Before considering two-way arch-systems, the properties of optimal single arches of given maximum stress are discussed briefly.

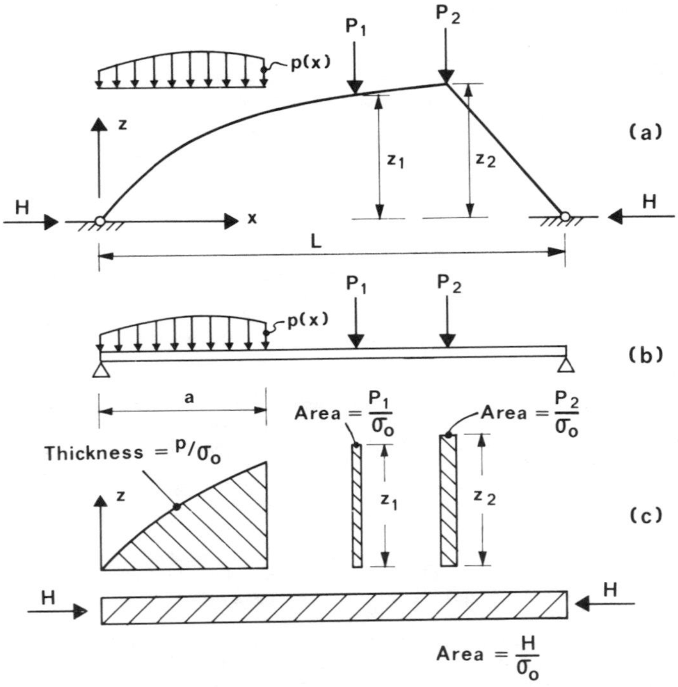

FIG 1 Optimal funicular arch.

Consider a fully stressed ("funicular") arch (Fig 1a) and its "equivalent beam" (Fig 1b, see Ref 12). Then zero moment and equilibrium conditions furnish

$$M' = S,\ p + S' = 0,\ p + M'' = 0 \qquad \ldots 1.$$

$$z = M/H,\ z' = S/H \qquad \ldots 2.$$

where S(x) and M(x) are the shear force and bending moment in the equivalent beam, H is the horizontal reaction of the arch, p(x) is the load, z(x) is the arch elevation and primes denote differentiation with respect to the horizontal co-ordinate x. It can be seen that the elevation (vertical distance from a horizontal reference plane) z of the arches is proportional to the bending moment M in the equivalent beam. The total weight of the arch can then be readily expressed as

$$W = \frac{\gamma H}{\sigma_o} \int_0^L [1 + (z')]\, dx \qquad \ldots 3.$$

subject to the equilibrium conditions in Eqns 1 and 2. In Eqn 3, γ is the specific weight of the arch (cable) material and σ_o is the permissible stress. The stationarity condition dW/dH = 0 and Eqn 3 readily furnish the unit mean square condition

$$\int_0^L (z')^2\, dx/L = 1.0 \qquad \ldots 4.$$

Restricting H to positive values, $H \geqslant 0$, an optimality condition by Rozvany (Ref 12, p 25) then yields for H = 0 ("vanishing arch"):

$$\int_0^L (z')^2\, dx/L \leqslant 1.0 \qquad \ldots 5.$$

Whilst Eqns 4 and 5 are quite sufficient for computational purposes, some of their consequences are mentioned herein. Equation 4 implies after transformations and integration by parts (Ref 12)

$$\int_0^L S^2\, dx = H^2 L = \int_0^L (M')^2\, dx \qquad \ldots 6a.$$

$$[M'M]_0^L - \int_0^L M''M\, dx = H^2 L \qquad \ldots 6b.$$

$$\int_0^L pz\, dx = HL. \qquad \ldots 6c.$$

This means that the unit mean square slope condition (Eqn 4) also implies that the product of the loads and their elevations must equal the product of the horizontal reaction (H) and the arch span (L), as derived by Rozvany and Prager (Ref 12). Moreover, if we replaced our original system with fully stressed vertical members transmitting the external loads to the foundation plane (Fig 1c) and a horizontal member transmitting only the horizontal reaction H then in the least weight solution the weight W_V of such vertical members would equal the weight W_H of the horizontal member:

$$W_V = W_H \qquad \ldots 7.$$

OPTIMALITY CONDITIONS FOR ARCHGRIDS

On the basis of proofs by Rozvany and Prager (Ref 12) together with the modified condition for vanishing arches in Eqn 5, optimality conditions for orthogonal archgrids can be derived readily. Let the co-ordinates along the structural domain D contained in a horizontal plane be x and y, the loads resisted by arches running in the x and y directions p_x (x,y) and p_y (x,y), the total external load p (x,y), the horizontal reactions in the x and y directions H_x and H_y and the elevation of the x-arches and y-arches z_x (x,y) and z_y (x,y). Then the optimal solution must satisfy the equal elevation condition

$$\text{on D},\quad z_x (x,y) = z_y (x,y) \qquad \ldots 8.$$

the zero bending and equilibriumg conditions

$$H_x z_{x,xx} = p_x,\ H_y z_{y,yy} = p_y,\ p_x + p_y = p \qquad \ldots 9.$$

and the unit mean square slope condition in Eqns 4 and 5.

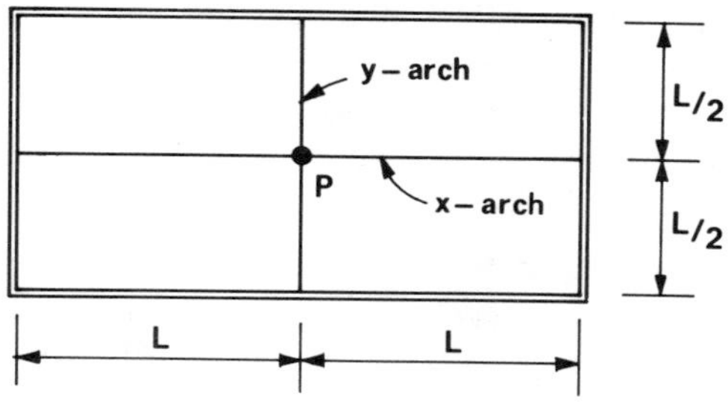

FIG 2 A Simple Example of a Vanishing Arch.

In Eqn 9, subscripts after comma indicate partial derivation with respect to the variables indicated.

It can be easily demonstrated on trivial examples that the above conditions could not always be satisfied for discretized systems if vanishing arches were not permitted. Consider for example a rectangular domain having a side ratio 1:2 (Fig 2) with a central point load P and a single x-arch as well as a single y-arch. Owing to the zero bending condition, both arches must consist of two straight segments of equal length. For a unit mean square slope (Eqn 4), the y-arch must have a constant slope of unity giving a maximum elevation L/2. Then the equal elevation condition in Eqn 8 restricts the x-arch to slopes of 0.5 which would violate Eqn 4 but not the modified condition (Eqn 5) for vanishing arches.

In summing up the above optimality conditions, we note that a rather complex variational problem has now been converted into the following geometrical one: it is necessary to find a middle surface whose mean square slope is unity along non-vanishing arches and does not exceed unity along vanishing arches.

COMPUTATIONAL PROCEDURE

The algorithm used by Rozvany and Prager (Ref 12) was based on an iterative procedure in which each cycle consists of two basic steps. First the horizontal forces are calculated on the basis of Eqn 4 for a given load distribution while relaxing the equal elevation condition (Eqn 8). Then for the new set of horizontal forces the equal elevation condition (Eqn 8) and equilibrium conditions (Eqns 9) are enforced (and thus a new set of p_x and p_y values obtained) while the horizontal forces are allowed to violate temporarily the unit mean square slope condition (Eqn 4). The procedure is terminated when the difference between the total weight values in two consecutive cycles is within a prescribed tolerance value.

In the above algorithm, the slope optimality condition was not expressed in the form of Eqn 4 but in the form of Eqn 6a. Since for "vanishing" arches H tended to zero, the program indicated automatically arches of zero cross-section, although it was not realised at the time (at least by the third author) that the solution would not satisfy the unit mean square condition (Eqn 4) which was only discovered later. In fact, Eqn 6a is valid for both non-vanishing and vanishing arches and thus, luckily, the algorithm by Rozvany and Prager did not break down for solutions with vanishing arches. In one example in Ref 12, the mean square slope of a vanishing arch is actually as low as 0.63 (Figs 13 and 14 of Ref 12). Since the enforcement of the unit mean square slope condition in one basic step of the Rozvany-Prager algorithm could lead to divergence of the solution, a modified improved algorithm is introduced in this paper, in which the simultaneous enforcement of the equal elevation constraint for all nodes is replaced by a relaxation procedure involving one node at a time.

The procedure to be used consists of the following steps.

(A) After replacing the distributed load by equidistant point loads, assume initially for all nodes (i, j)

$$P^x_{ij} = P^y_{ij} = P_{ij}/2 \qquad \ldots 10.$$

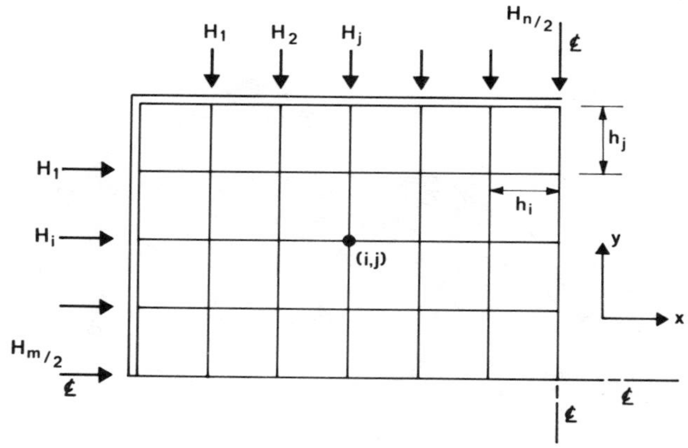

FIG 3 Discretized formulation.

where P^x_{ij} and P^y_{ij} are the loads carried by the x-arch and y-arch, respectively, at the node (i,j) (Fig 3) and P_{ij} is the external load at that node.

(B) Calculate the horizontal forces and elevations for a given load distribution (P^x_{ij}, P^y_{ij}) using discretized forms of Eqns 6a and 2 (see Appendix A).

(C) On completion of step (B), each arch has an optimal shape for the assumed loads resisted by it. Considering a given node, an increase (decrease) of the load at one particular node only would increase (decrease) the elevation at that node for a funicular arch that must satisfy the unit mean square slope condition. Considering an arch intersection (i, j), therefore, the load on the arch with a higher elevation must be decreased and vice versa. In the current step, the new values of P^x_{ij} and P^y_{ij} are calculated to achieve an equal elevation at the node (i, j), also satisfying temporarily the equilibrium and zero moment condition throughout the systems and supposing that the arch loads do not change at other nodal points. The new loads carried by the arches at node (i, j) are given by

$$P^x_{ij} = \alpha P_{ij}, \quad P^y_{ij} = (1-\alpha)\, P_{ij} \qquad \ldots 11.$$

Formulae for the factor α are derived in Appendix B.

Note that the above procedure is valid for both vanishing and non-vanishing arches.

(D) Repeat steps (B) nd (C) until

$$\left| P^{\alpha}_{ij}(k) - P^{\alpha}_{ij}(k-1) \right| < \delta \ (\alpha = x,y) \qquad \ldots 12.$$

where k is the serial number signifying the current cycle.

EXAMPLE. RECTANGULAR DOMAINS

The above algorithm has been tried out for various domain shapes. Owing to length limitation, however, only rectangular domains are considered in this paper.

The solution for a 20 x 30 rectangular archgrid is given in an isometric projection in Fig 4. The two vanishing arches (broken lines) have a mean square slope of 0.63394 (cf. Eqn 5). Note that the arches along the longer edges vanish. The number of such arches increases as we increase the side ratio of the domain.

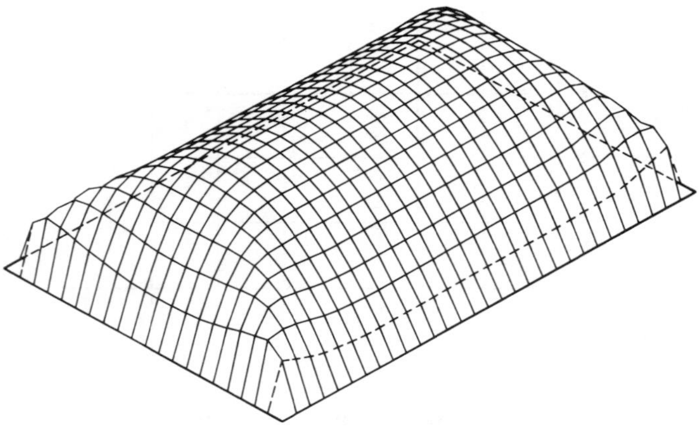

FIG 4 Rectangular Arch-Grid of Least Weight.

The exact number of vanishing arches for a given rectangular domain is difficult to determine directly because of the complicated non-linear relations involved. However, the third author (Ref 16) obtained the limiting value of the side ratio n/m for which all long arches vanish (i.e. the optimal solution is a one way system):

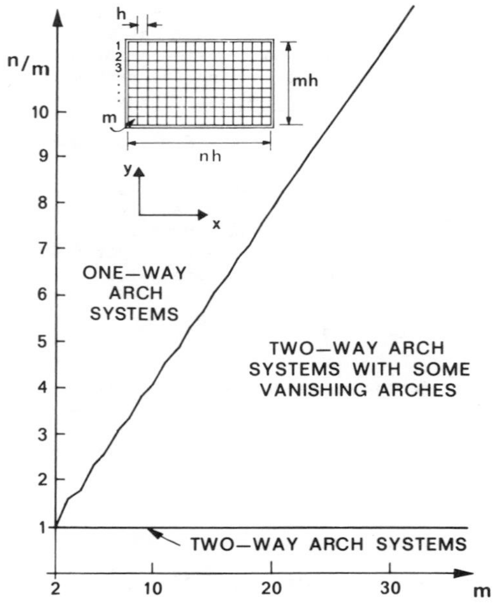

FIG 5 Side ratios for which all x-arches vanish.

$$n/m = \frac{4[(m-1)(r+1)/2 - \sum_{k=0}^{r} k]^2}{\sum_{k=0}^{r} (m-1-2k)^2} \quad \ldots 13.$$

where

$$r = [1 - (-1)^{m-1}](m-2)/4 + [1 + (-1)^{m-1}](m-1)/4 \quad \ldots 14.$$

and m is the number of gridlines across the shorter span. The above limiting ratios are represented graphically in Fig 5.

The above computer program can be easily modified to handle grid lines at 45° to the sides. The optimal solution for a 6 x 8 archgrid is given in Fig 6a and a comprison of the weight of optimal archgrids for various layouts (one way system as well as grid lines parallel and at 45° to the sides) in Fig 6b. Of the three layouts, arches parallel to the sides are the most efficient.

A comparison of the results using the Rozvany-Prager algorithm (bracketed numbers) and the algorithm proposed herein (numbers without brackets) is given in Table I for a

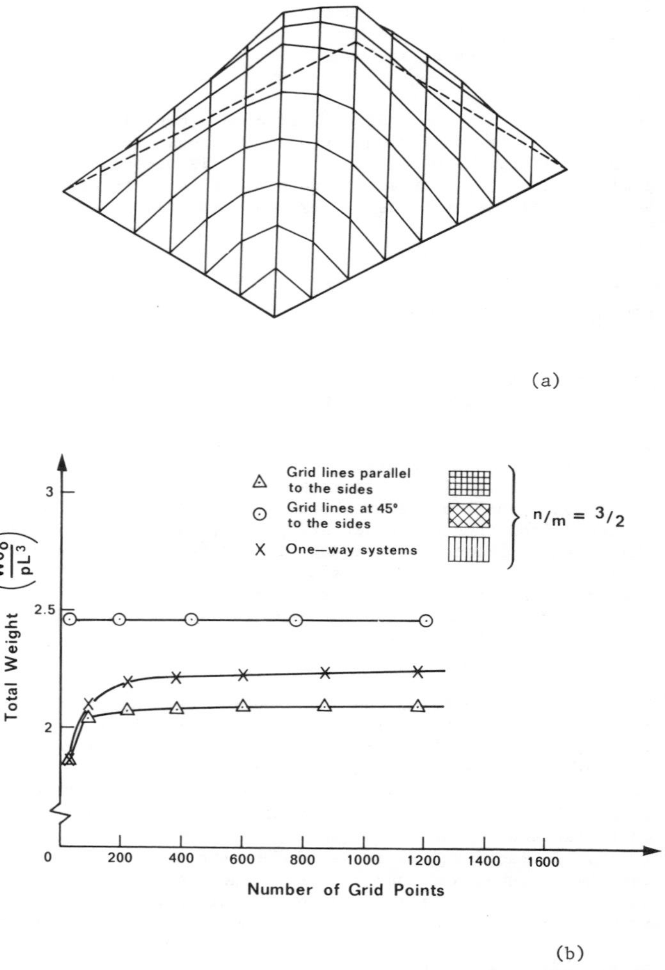

FIG 6 Optimal shapes of archgrid for grid lines at 45° to the sides (a) and weight comparison for various layouts (b).

$$\frac{W\sigma_o}{\gamma P L^3} = \begin{matrix} 225.133 \\ (225.216) \end{matrix}$$

HORIZONTAL FORCES H_x/P

i	1	2	3
	0 (0)	0 (0.03650)	0.31371 (0.31680)

HORIZONTAL FORCES H_y/P

j	1	2	3
	1.40651 (1.33475	1.68434 (1.68349)	1.70648 (1.70647)
	4	5	6
	1.70775 (1.70775)	1.70782 (1.70782)	1.70782 (1.70782)

ELEVATION OF NODAL POINTS z/L

i \ j	1	2	3	4	5	6	$\frac{\int(z')^2dx}{L_x}$
1	1.52523 (1.54243)	1.46835 (1.46856)	1.46411 (1.46411)	1.46387 (1.46386)	1.46385 (1.46385	1.46385 (1.46385)	0.397057 (0.397429)
2	2.33948 (2.33361)	2.34300 2.34308)	2.34221 (2.34222)	2.34216 (2.34216)	2.34216 (2.34216)	2.34216 (2.34216)	0.912197 (0.907638)
3	2.44276 (2.44041)	2.62395 (2.62381)	2.63431 (2.63430)	2.63490 (2.63489)	2.63493 (2.634493)	2.63493 (2.63493)	1.000000 (0.998224)
$\frac{\int(z')^2dy}{L_y}$	1.000000 (1.005398)	1.000000 (1.000088)	1.000000 (1.000002)	1.000000 (0.999996)	1.0000000 (1.000000)	1.000000 (1.000000)	

TABLE I A Comparison of results using the Rozvany-Prager algorithm (bracketed values) and the modified algorithm (unbracketed numbers)

6 x 12 arch grid. The tolerance value for the first method was $10^{-5} > [W(k) - W(k-1)]/W(k)$ where W(k) is the weight calculated in the last cycle. In the second method, the tolerance value for all nodes (see Eqn 12) was $\delta - 10^{-6}$. It can be seen that, as expected, the second method with the stricter tolerance conditions gives more accurate results.

REFERENCES

1. M M LEVY, La Statique Graphique et ses Applications aux Constructions, Gauthier Villars, Paris, 1874.

2. B BUDIANSKY, J C FRAUENTHAL and J W HUTCHINSON, On Optimal Arches, Journal of Applied Mechanics, Vol 36, pp 880-882, 1969.

3. I TADJBAKSH and M FARSHAD, On Conservatively Loaded Funicular Arches and their Optimal Design, Proceedings of IUTAM Symposium on Structural Optimization, Warsaw, 1973; Springer-Verlag. Berlin 1975, pp 215-228.

4. M FARSHAD, On Optimal Form of Arches, Journal of the Franklin Institute, Vol 302, pp 187-194, 1976.

5. W STADLER Uniform Shallow Arches of Minimum Weight and Minimum Maximum Deflection, Journal of Optimization Theory and Applications, Vol 23, pp 137-165, 1977.

6. R D HILL, G I N ROZVANY, C M WANG and K H LEONG, Optimization, Spanning Capacity and Cost Sensitivity of Fully Stressed Arches, Journal of Structural Mechanics, Vol 7, pp 375-410, 1979.

7. S L LIPSON and M I HAGUE, Optimal Design of Arches Using the Complex Method, Journal of the Structural Division, ASCE, Vol 106, pp 2509-2525, 1980.

8. G I N ROZVANY, C M WANG and M DOW, Arch Optimization via Prager-Shield Criterion, Journal of the Engineering Mechanics Division, ASCE, Vol 106, pp 1279-1286, 1980.

9. G I N ROZVANY and C M WANG, On Plane Prager Structures I, International Journal of Mechanical Sciences, Vol 25, pp 519-527, 1983.

10. C M WANG and G I N ROZVANY, On Plane Prager Structures II, International Journal of Mechanical Sciences, Vol 25, pp 529-541, 1983.

11. F OTTO, Grid Shells, Institute of Light Weight Structures, University of Stuttgart, 1974.

12. G I N ROZVANY and W PRAGER, A New Class of Optimization Problems: Optimal Archgrids, Computer Methods in Applied Mechanics and Engineering, Vol 19, pp 127-150, 1979.

13. G I N ROZVANY, H NAKAMURA and B KUHNELL, Optimal Archgrids: Allowance for Selfweight, Computer Methods in Applied Mechanics and Engineering, Vol 24, pp 284-304, 1980.

14. G I N ROZVANY, C M WANG and M DOW, Prager-Structures: Archgrids and Cable Networks of Optimal Layout, Computer Methods in Applied Mechanics and Engineering, Vol 31, pp 91-113, 1982.

15. A G M MICHELL, Limits of Economy of Material in Frame Structures, Philosophical Magazine, Ser 6, Vol 8, pp 589-597, 1904.

16. C M WANG, On Some New Classes of Optimal Structures, Ph D Thesis, Monash University, Clayton, Victoria, Australia, 1982.

17. G I N ROZVANY, Optimal Design of Flexural Systems; in English: Pergamon Press, Oxford, 1976; in Russian: Stroiizdat, Moscow, 1980.

APPENDIX I - PROPERTIES OF LEAST-WEIGHT ARCHES UNDER EQUIDISTANT POINT LOADS

Consider an arch having the span L = qh where h is the spacing of point loads $P_1, P_2, \ldots, P_{q-1}$. The condition in Eqn 6a (originally Ref 12) can then be expressed in a discretized form

$$H^2\, qh = h \sum_{i=1}^{q} S_i^2, \qquad \ldots \text{A1.}$$

where S_i is the shear force to the left of the point load P_i

(and S_q to the right of the last point load P_{q-1}).

Eqn A1 reduces to

$$H^2 q = \sum_{i=1}^{q} S_i^2 \qquad \ldots \text{A2}$$

where S_i can be readily calculated from elementary statics. The elevation z_t at any point load P_t can then be obtained from the simple geometric/static relationship

$$(z_i - z_{i-1})/S_i = h/H \qquad \ldots \text{A3}$$

furnishing

$$Z_t = \sum_{i=1}^{t} (z_i - z_{i-1}) = (h/H) \sum_{i=1}^{t} S_i = M_i/H \qquad \ldots \text{A4.}$$

where M_i is the moment in the "equivalent beam" at the point load P_i and $\sum_{i=1}^{t} S_i$ again follows from simple statics.

For symmetric load systems, for example,

$$S_i = \sum_{j=i}^{q^*} P_j, \quad \sum_{i=1}^{t} S_i = \sum_{j=1}^{t-1} j\, P_j + t \sum_{j=t}^{q^*} P_j \qquad \ldots \text{A5.}$$

can be substituted into Eqns A2 and A4, with $q^* = (q-1)/2$ if q is odd and $q^* = q/2$ if q is even.

APPENDIX II - DERIVATION OF THE DISTRIBUTION FACTOR α FOR TWO-WAY ARCHGRIDS

By equating the values of the elevations ($z_{ij}^x = z_{ij}^y$) of the x-arch and y-arch at the intersection (i,j) from Eqns A4 and A5, we obtain for symmetrically loaded arches the distribution factor (see Eqn 11):

$$\alpha = \frac{\bar{z}_{ij}^{x} - \bar{z}_{ij}^{y}}{P_{ij}\,(i\, h_x/H_x - j\, h_y/H_y)} \qquad \ldots \text{A6.}$$

where $\bar{z}_{ij}^{x}$ and $\bar{z}_{ij}^{y}$ are elevations for a modified loading in which $P_{ij}^x = P_{ij}$ and $P_{ij}^y = 0$, h_x and h_y are the nodal spacing in the x and y directions while H_x and H_y are the horizontal forces in the same directions.

ALGORITHM FOR OPTIMAL DESIGN OF REGULAR STRUCTURES

S G BYKOVSKI* AND L I KORSHUN**

* Lecturer, Minsk Polytechnical Institute.

** Assistant professor, Brest Civil Engineering Institute.

A multistage algorithm is suggested to solve the optimization problem of regular space bar structures. The division of this problem into subproblems is based both on ranging parameters and conditions limiting the parameter values and on taking into consideration the most essential of both of them at each stage. Firstly the optimal structure axes configuration is found. Then the problem of optimal material distribution is solved. At the end the violated displacement constraints are optimally satisfied. Mathematical programming, optimality criteria approaches and their combination are used to solve the problems of different stages. The results of each stage are used as a starting point for the next stage.

INTRODUCTION

This report deals with regular space bar structures produced by repeating elementary bar cells. Thus, given the overall structure sizes the sizes of elementary bar cell appear to be the parameters determing its axes configuration. The linear-elastic-deformed hinge-bar model of structure is adopted.

The task is to find the axes configuration of the structure and the material distribution between axes with due account of the minimum costs of the structure in question, its maintenance expenditures and the costs of related structures dependent on the parameters of the structure in question. The elementary cell sizes and cross-sectional areas of structure members are regarded to be independent variables. The variable values should meet the stress and buckling constraints of the members as well as joint displacement constraints. Besides, they must satisfy the equilibrium and compatibility conditions of the structure. In addition, the independent variables can range with certain limits. To top it all, conditions of structure technology such as conditions of unification and discontinuity of member cross-sectional area values can also be taken into consideration.

This problem is a nonlinear programming one. It is characterized by blurred dependences in objective and constraint functions on independent variables as well as discontinuity and multitude of variables. The direct solution of this problem turns out to be extremely difficult.

It is suggested to solve this problem in a stage-by-stage sequence based on both the preferential arrangement and subgrouping of variables and constraints with the most essential of both of them being taken into consideration at each stage.

OPTIMIZATION OF STRUCTURE AXES CONFIGURATION

The first stage aims at determining the

structure axes configuration to make it possible to calculate the minimum cost of the structure in question including its maintenance expenditures as well as the costs of related structures. The elementary cell sizes are regarded to be independent variables. Due to the structure regularity the number of independent variables is small and ranges from 2 to 5 depending upon the type of regularity. Independent variables can be changed either continuously or discontinuously within the limits imposed.

The number of structure members and material distribution between them both depend upon the structure axes configuration, i.e.the cross-sectional areas of the members are dependent on structure axes configuration in quality and quantity. Thus, the parameters describing the structure axes configuration are of higher level than cross-sectional areas of members. Therefore, the material expenditures for each axes configuration of structure is suggested to be determined only approximately proceeding from their theoretical expenditures multiplied by the coefficient taking into account the buckling, discontinuity and unification conditions of members and material expenditures for structure joints and supports. The theoretical material expenditures is calculated on the basis of the computation of the structure as a truss in which the cross-sectional areas of members are the same as well as the tensile and compressive permissible stresses of member materials.

At the first stage the optimization problem is formulated as following nonlinear programming one:

$$C = C(Z,F) \longrightarrow \text{minimum}; \quad (1)$$

$$Z_{min} \leqslant Z \leqslant Z_{max}; \quad (2)$$

$$Z \in Z^*; \quad (3)$$

$$K(Z,F^o)U = P; \quad (4)$$

$$F = F(U) = F(Z,F^o), \quad (5)$$

where C is the cost of the structure in question including maintenance expenditures and the costs of related structures which depend upon the axes configuration of the structure in question; Z is the M-dimensional vector of sizes of the elementary cell whose repetition constitutes the structure in question; Z_{min}, Z_{max}, Z^* are its respectively lower and upper limits and permissible values; F, F^o are the I-dimensional vectors respectively of cross-sectional areas and their initial values; K is the 3W x 3W stiffness matrix, U,P are the 3W x Q joint displacements and external joint load matrixies, respectively; I,W,M,Q are the number of structure members, joints (both of them variables depending upon vector Z), independent variables, loading cases.

Due to a small number of the independent variables the solution of the problem (eqs. 1-5) is suggested to be realized by means of one-dimensional minimizing search with its complexity rising due to an increase in the calculation volume. The solution is achieved in accordance with the following expression:

$$\min_{Z} C(Z,F) = \min_{Z_1} \ldots \min_{Z_m} C(Z_1,\ldots Z_m,F), \quad (6)$$

where $Z_1,\ldots,Z_m$ are the components of the vector Z.

Eq.6 indicates that the search of objective function minimum on parameter Z_1 demands finding its minimum on parameter Z_2 for each fixed value of parameter Z_1. This also demands to minimize the objective function on parameter Z_3 for each value totality of parameters Z_1 and Z_2. The last level is to minimize the objective function on parameter Z_m. It is a simple one-dimensional search.

The one-dimensional search on the parameters that can be changed discontinuously is an easy task as there is no problem in choosing the step length. This kind of search is sequentially to select the parameter values in direction that provides the objective function decrease. If the parameter is changed continuously the one-demensional search can be carried out with an effective method based on the quadratic approximation of the objective function.

The suggested method of optimizing the axes configuration of the structure is more efficient than the known method of coordinate optimization because the former provides finding the solution in one calculation cycle.

OPTIMIZATION OF MATERIAL DISTRIBUTION

The optimal axes configuration of the structure obtained at the first stage is fixed and the problem of optimal material distribution in the system with definite configuration of axes is solved at the second stage. The theoretical material expenditures for the structure is minimized. The equilibrium and compatibility conditions of the structure and the stress and buckling constraints of cross-sectional areas of its members are taken into consideration. The member cross-sectional areas are independent variables of this stage. Their values are also limited by lower bounds. The conditions of the discontinuity and unification of variable values can be taken into account too.

The second stage problem is suggested to be solved by combining the mathematical programming and optimality criteria approaches. The thing is that the mathematical programming approach to the optimal design of structures is the most common and the optimality criteria one is structure-oriented. Besides, the mathematical programming methods slow down their convergence at the optimum neighbourhood. On the contrary, the methods based on the optimality criteria approach have a better convergence at the optimum neighbourhood than away from it. That is why the former is used to attain the optimum neighbourhood and the latter is used for final determination of the member cross-sectional areas.

Firstly, the following nonlinear mathematical programming problem is formulated:

$$V = \sum_{i=1}^{I} F_i L_i \longrightarrow \text{minimum}; \tag{7}$$

$$\max\left(\frac{N_i^+}{\alpha_i R_i}, \frac{-N_i^-}{\varphi_i m_i R_i}, F_{\min_i}\right) - F_i \leq 0,$$

$$i = 1,\ldots,I, \tag{8}$$

where V is the theoretical material expenditures; L_i, F_i, $F_{\min_i}$ are respectively the length, cross-sectional area and its lower limit for the ith member; N_i^+, N_i^- are respectively the most possible (due to the loading cases) tensile and compressive forces in the ith member; α_i, φ_i, m_i are the coefficients taking into account respectively the possible weakness of member cross-section (due to holes, for example), buckling constraint and condition of importance for the i th member (according to the standards of metal structure design adopted in the USSR); R_i is the permissible stress in the i th member.

The following equilibrium and compatibility conditions of the structure in question should be added to eqs. 7,8 :

$$K(F)U = P. \tag{9}$$

To be solved, the constrained nonlinear programming problem (eqs. 7-9) is substituted by an equivalent sequence of unconstrained ones. Here the auxiliary objective function is made up by means of summing the initial one with penalty addings for violations of eqs.8. But at each step of the solution procedure eqs.9 are suggested to be satisfied as strict equations for every totality of independent variables. So the penalty addings for their violation are not necessary.

The minimization of the auxiliary objective function is carried out by the parallel tangent method that has been modified by the authors. Any other effective method of unconstrained minimization can be used here.

To limit reasonably the sequence of the unconstrained problems and to find final parameter values the optimal material distribution problem is formulated as the so-called reverse problem of structural mechanics using the optimality criteria approach. This is carried out after the optimum neighbourhood has been attained. Now the task is to find the member cross-sectional areas providing the satisfaction of eqs.8 as strict equations in addition to eqs.9. Thus the criteria of structure optimality is now one of the properties of its stressed-deformed state.

A part of the necessary optimality criteria for local optimum (Kuhn-Tucker conditions) is used to solve this problem. It results in the following recurrent expressions for iterative parameter recalculation:

$$F_j^{k+1} = \max\left\{F^k_{min_j},\ F_i^k\left[\frac{1}{F_i^k}\max\left(\frac{\left(N_i^+\right)^k}{\alpha_i R_i},\right.\right.\right.$$

$$\left.\left.\left.\frac{\left(-N_i^-\right)^k}{\varphi_i^k m_i^k R_i}\right)\right]^{\frac{1}{\eta}}\right\},\quad i = 1,\ldots I, \qquad (10)$$

where k is an iteration number; η is a parameter influencing the convergence rate.

Iterative use of eqs. 9,10 provides the final solution of the problem of optimal material distribution. The iterative process is characterized by fast convergence. The unification and discontinuity conditions of the member cross-sectional areas can be taken into account either at each iteration or after the termination of the iterative process. If the number of unified groups of members is predetermined the optimal unification problem should be solved in both cases.

The solution method of the optimal unification problem must provide the smallest increase in material expenditures compared with the ones for non-unified structure. It is a multistep process reducing the total number of different cross-sectional areas from initial value to predetermined one. At each step of the process this number is reduced by one.

The solution method is based on the problem feature and Bellman's optimality principle. The problem feature is that the task is to increase the cross-sectional areas of some structure members up to some values. But the number of these values is predetermined.

Each step of the solution process includes the following: member volumes are ranged according to their decreasing; differences between adjacent volumes are calculated; the step is terminated by putting together into one unified group those members which have the smallest difference between their volumes. The total step number is equal to the difference between the numbers of structure members and unified groups.

SATISFACTION OF DISPLACEMENT CONSTRAINTS

The displacement constraints are considered at the end of the solution of overall problem because they can be nonactive after the second stage. In this case their inclusion into the problem of the second stage in vain makes it complicated. But they have been violated after the second stage the optimal correction of member cross-sectional areas is carried out to deliver the structure joint displacement in accordance with their allowed values. The correction must provide the smallest increase of material expenditures.

The problem to be solved at the third stage is written as follows:

$$\Delta V = V - V_2^{opt} = \sum_{i=1}^{I} L_i\left(F_i - F_{2i}^{opt}\right) \longrightarrow \text{minimum}; \qquad (11)$$

$$\max_{q=1,\ldots,Q} \sum_{i=1}^{I} \frac{\bar{N}_{ij}N_{iq}}{E_i F_i} L_i - \left[Y_j\right] \leq 0,$$

$$j = 1,\ldots,J, \qquad (12)$$

where ΔV is the increase of material expenditures resulted in the third stage; V_2^{opt}, F_{2i}^{opt} are respectively the volume of material expenditures and cross-sectional area of the ith member both resulted in the second stage; N_{iq}, $\bar{N}_{ij}$ are forces in the ith member caused respectively by the qth loading case and a force equal to one applied in the jth direction; E_i is Young's modulus of the ith member material; $\left[Y_j\right]$ is the allowed joint displacement in the jth direction; J is the number of such directions. Eqs. 8,9, should be added to eqs. 11, 12.

Using Kuhn-Tucker necessary optimality conditions the following recurrent expressions for iterative recalculation of Lagrangian multipliers and member cross-sectional areas are received:

$$\tau_j^k = \tau_j^{k-1}\left(\frac{1}{[Y_j]}\max_{q=1,\ldots,Q}\sum_{i=1}^{I}\frac{\bar{N}_{ij}N_{iq}^k}{E_i F_i^k}L_i\right)^{\frac{1}{\beta}},$$

$$j = 1,\ldots,J; \qquad (13)$$

$$F_i^{k+1} = \max\left\{\frac{\left(N_i^+\right)^k}{\alpha_i R_i}, \frac{\left(-N_i^-\right)^k}{\varphi_{i}^{k} m_i^k R_i}, F_{min_i}^k,\right.$$

$$\left.F_i^k\left[\sum_{j=1}^{J}\tau_j\frac{\bar{N}_{ij}^k N_{iq^j}^k}{E_i\left(F_i^k\right)^2}\right]^{\frac{1}{\varkappa}}\right\}, \quad i = 1,\ldots,I \quad (14)$$

where τ_j is the Lagrangian multiplier conjugated with the jth displacement constraint; q^j is the number of loading case causing the largest displacement in the jth direction; β, $\varkappa$ are the parameters influencing the convergence rate.

Iterative use of eqs. 9,13,14 results in the optimal correction of the member cross-sectional areas simultaneously satisfying all displacement constraints violated. After the iterative process termination the discontinuity and unification conditions of the member cross-sectional areas should be taken into account again. But better results are provided if these conditions are satisfied at each iteration after calculation on eqs.14.

CONCLUSION

On a base of the suggested multistage algorithm for optimal design of regular space bar structures the computation procedures for all stages and common one have been developed. They have been realized in computer program The test calculation of several regular structures has shown the suggested algorithm advantage in comparison with traditional design approaches for these structures.

OPTIMIZATION OF TIMBER SPACE STRUCTURES

D.J. ROBINSON* and B.H.V. TOPPING**

* Industrial Associate, Shorts Brothers Ltd., Belfast and Ulster Polytechnic, Northern Ireland formerly Civil Engineering Research Student, University of Edinburgh.

** Lecturer in Civil Engineering, University of Edinburgh.

This paper illustrates how the sequential linear programming technique has been used to design timber framed structures in accordance with the British Code of Practice. The stress constraints were formulated using the Code and the objective was to obtain a minimum weight design by varying the member cross sectional properties and the coordinate positions of the joints of the structure. In this way, the geometry or shape of the structure as well as the member cross sectional properties were varied and the resulting design had not only a minimum weight with respect to member cross sectional properties but was also the most structurally efficient shape. The technique is first shown to be practical by application to a number of two dimensional examples. The algorithm is further generalized to account for the grouping of design variables, uniformly distributed loading, and multiple loading cases. Finally the use of the method in the design of space structures is illustrated by its application to a three dimensional frame and a rigidly jointed space dome structure.

INTRODUCTION

For the last fifteen years considerable research, as illustrated in Refs 1,2 and 3, has been undertaken in the use of nonlinear optimization techniques for large scale structural design. Until recently, progress in this area was limited by the capacity of the available computers. In the last five years, however, there have been many developments in computer hardware and many problems which earlier it was not feasible to solve are now within the reach of modern computer systems. Nonetheless, many of the methods for the optimization of skeletal structures are limited because they only allow for the variation of member cross sectional properties during the design process. As illustrated by the work in this paper, this rather restricted class of design algorithms may be generalized by also varying the geometry or shape of the structure. A review of these more general design algorithms may be found in Ref 4.

The type of structures that are examined in this paper is three dimensional rigidly jointed frames. The currently available nonlinear optimization algorithms are both efficient and numerically stable for the size of structures for which member forces and derivatives may be calculated using reasonable amounts of computer time. It has been found in previous work undertaken by the present authors (Refs 5 and 6) that the sequential linear programming method provides a suitable technique for the optimization of rigidly jointed frames.

Optimization techniques are often viewed as impractical by practising engineers because of the problems of relating them to design Codes of Practice. However the design algorithm utilized in this paper has been implemented for the elastic design of timber framed structures in accordance with the British Standard Code of Practice CP.112, Part 2, 1971, "The Structural Use of Timber", (Ref 7). The algorithm minimizes the weight of material used by varying the member cross sectional areas and the coordinate positions of the joints of the structure. The resulting design therefore has not only a minimum weight with respect to member cross sectional properties, but is also the most structurally efficient shape. The use of rectangular timber sections enables simplifications to be made which result in the use of the depth of section to specify the section size completely. The emphasis has been on a method which is practical, the results of which, complying with the Code of Practice, may be used with alteration. To this end, the optimization scheme has been written so that it may consider member cross sectional properties and joint coordinates that can be treated as grouped design variables and uniformly distributed loading. These features are important if the method is to be used realistically to solve practical structures. In structures such as multi-storey multi-bay frames, the shape of the structure is largely dictated by functional requirements. However, there are other structures, such as lattice shells, where there is more latitude in the shape.

PROBLEM FORMULATION

The structural idealization is assumed to be fully rigid and the predominant member forces are a combination of axial and bending forces. The member forces and joint displacements are evaluated by means of the usual linear elastic stiffness analysis.

Weight has been used as the objective function throughout this work. This may appear to be an oversimplification of the design problem, but it leads to many advantageous features in the formulation of the algorithm, particularly when deciding on the design variables. The weight is varied by changing the cross sectional dimensions of the members and also from changes of member lengths resulting from changes in the coordinates of the joints of the structure. The variables which alter the cross sectional dimensions of the members and the coordinates of the joints are called the design variables. The cross sectional properties which are important are the cross sectional area A and the second moment of area I. Considerable simplification of the problem is possible if both of these variables are expressed solely in terms of a single variable which defines both the cross sectional properties of each member. This was accomplished by assuming that the members have a rectangular cross section and that all members have the same breadth to depth ratio. The first assumption does not present any problems since most timber sections are rectangular. The second was selected to enable an investigation to be undertaken to assess the choice of section proportions used in timber framed structures. This assumption leads to advantages which become apparent when considering the examples discussed later. Notably, it allows greater use of the available standard sections and the member lateral stability requirements to be readily satisfied. Structures of the type where all the members are of constant breadth (for example roof trusses) are discussed in Ref 8. The design variables for the cross sections of the members are a vector of depths of the members. The cross sectional area of member A is given by the relationship: $A = b.d = \alpha.d^2$ where: b is the breadth of the section; d the depth of the section and α the breadth to depth ratio which is constant for the structure. The second moment of area of a member I, is given by the relationship: $I = bd^3/12 = \alpha d^4/12$.

The objective is the weight function:

$$\{W\} = \{L\}^T\{A\} = \alpha\{L\}^T\{d^2\} \qquad \text{..... 1.}$$

where: W is the weight of the structure, {L} the vector of member lengths; {A} the vector of member cross sectional areas and $\{d^2\}$ the vector of member depths squared. By partially differentiating this expression and utilizing the first term of a Taylor's series, the change in weight of the structure due to changes in member lengths and depths can be expressed thus:

$$\Delta W = \{\partial W/\partial d\}^T\{\Delta d\} + \{\partial W/\partial x\}^T\{\Delta x\} \qquad \text{..... 2.}$$

The objective function becomes ΔW, which must be minimized to decrease the volume of the structure by the greatest possible amount.

The deflection constraint for the jth degree of freedom is specified such that the absolute deflection must be less than a specified positive value. This is expressed as follows:

$$|\delta_{jn}| < |\delta_{jn}| \qquad \text{..... 3.}$$

where: δ_{jn} is the new actual displacement of the jth degree of freedom after the structure has been altered by $\{\Delta d\}$ and $\{\Delta x\}$; and δ^a_{jn} is the allowable or permissible displacement as specified in the Code of Practice. Using the first term in a Taylor's series the q displacement constraints for the complete structure may be written in matrix form thus:

$$[F]\{\{\delta\} + [\partial\delta/\partial d]\{\Delta d\} + [\partial\delta/\partial x]\{\Delta x\}\} \le [F]\{\delta^a\} \qquad \text{..... 4.}$$

where [F] is a q by q matrix with +1 or -1 terms on the leading diagonal and zeros elsewhere. The determination of the matrices $[\partial\delta/\partial d]$ and $[\partial\delta/\partial x]$ are given in Refs 8 and 9.

The structure is assumed to be loaded at the joints and the stress constraints are to be satisfied at both ends of each and every member. The stress constraint is of the same form as given in the Code but is expressed in terms of forces and moments rather than direct flexural stresses:

$$|P_{in}/P^a_{in}| + |M_{ikn}/M^a_{ikn}| \le R \qquad \text{..... 6.}$$

where: k signifies the end of the member to be considered (either 1 or 2); a the permissible or allowable values; i the ith member; n that these are new values after the structure has been modified; P the member axial force; M the member bending moment; and R is equal to 1.0 if the axial force is tensile and 0.9 if the axial force is compressive. The values of P^a_{in} and M^a_{ikn} are given the same signs as P_{in} and M_{ikn} respectively. The new member axial force P_{in} and the bending moment M_{ikn} may be expressed in terms of the current member axial force P_{ic}, and bending moment M_{ikc} by using the first term of a Taylor's series. These expressions for P_{in} and M_{ikn}, which are given in Refs 8 and 9, can be substituted in Eqn 6 giving a constraint in terms of current variables as follows:

$$\{A_{ik}\}^T\{\Delta d\} + \{B_{ik}\}^T\{\Delta x\} \le C_{ik} \qquad \text{..... 7.}$$

where:

$$\{A_{ik}\} = F\{P_i\{\partial M^a_{ik}/\partial d\} + M^a_{ik}\{\partial P_i/\partial_d\} + P^a_i\{\partial M_{ik}/\partial d\}$$
$$+ M_{ik}\{\partial P^a_i/\partial d\} - RP^a_i\{\partial M^a_{ik}/\partial d\} - RM^a_{ik}\{\partial P^a_i/\partial d\}\}$$

$$\{B_{ik}\} = F\{P_i\{\partial M^a_{ik}/\partial x\} + M^a_{ik}\{\partial P_i/\partial x\} + P^a_i\{\partial M_{ik}/\partial x\}$$
$$+ M_{ik}\{\partial P_{ik}/\partial x\} - RP^a_i\{\partial M^a_{ik}/\partial x\} - RM_{ik}\{\partial P_i/\partial x\}\}$$

and

$$C_{ik} = F\{RP^a_i M^a_{ik} - P_i M^a_{ik} - P^a_i M_{ik}\}$$

The stress constraints for both ends of all the members may be given in matrix form thus:

$$[A]\{\Delta d\} + [B]\{\Delta x\} \le \{C\} \qquad \text{..... 8.}$$

FORMULATING THE NONLINEAR PROBLEM IN LINEAR FORM

The design problem may now be expressed in the usual linear programming form thus:

$$\text{Minimise } \Delta W = \alpha(2\{Ld\}^T\{\Delta d\} + \{\partial L/\partial x.d^2\}^T\{\Delta x\})$$

subject to the following constraints:

$$[A]\{\Delta d\} + [B]\{\Delta x\} \le \{C\}$$

$$\text{and } [F]\{\delta\} + [\partial\delta/\partial d]\{\Delta d\} + [\partial\delta/\partial x]\{\Delta x\} \le F\{\delta^a\}$$

If the derivatives in these equations had been linear functions of the design variables $\{\Delta d\}$ and $\{\Delta \mathbf{x}\}$, then linear programming techniques would provide an exact solution to the above formulation. Owing to the nonlinear nature of these derivatives however, they must be solved iteratively by repeated application of the linear programming technique followed by an exact re-analysis. If this iteration is to be stable and converge to a solution, then limits on the amount by which the design variables may be moved or altered at each iteration must be imposed. In addition, the simplex linear programming algorithm must permit certain variables, such as joint coordinates, to be negative if required. The values of design variables at the end of each iteration should be such that the constraints are not normally violated and the objective function is reduced. In some problems the constraints will be violated, but provided this violation does not take the solution too far outside the design space the method should still converge to an optimum. The solution in the worst case will be a local optimum, and at best, a global optimum.

The restrictions on the changes in design variables required to ensure convergence may be expressed as constraints thus:

$$\{\Delta d^{min}\} \leq \{\Delta d\} \leq \{\Delta d^{max}\} \text{ and } \{\Delta x^{min}\} \leq \{\Delta x\} \leq \{\Delta x^{max}\}$$

The vector quantities $\{\Delta d^{min}\}$, $\{\Delta d^{max}\}$, $\{\Delta x^{min}\}$ and $\{\Delta x^{max}\}$ are referred to as move limits. From experience it has been found that it is not sufficient to include fixed move limits to ensure convergence. To avoid divergence the limits must be varied as the optimization progresses. Such move limits are termed adaptive. If adaptive move limits are not properly applied to sequential linear programming, then the design variables will frequently oscillate between two distinct values.

UNIFORMLY DISTRIBUTED LOADING AND OTHER MODIFICATIONS

The optimization scheme presented above only considers the effects of applied loading at the joints. If the design is to be practical, then uniformly distributed loading must also be considered. The modification of the optimization scheme required to include this loading must consider the following points:

(i) Inclusion of the loading in the analysis and in calculating the derivatives of the deflections and forces with respect to the coordinate variables.

(ii) Establishing the critical position, if any, of the maximum stressed position within the beam.

(iii) Calculating the derivatives of the moments of the maximum stresses position within the beam.

(iv) The inclusion of the extra constraints in the linear programming problem.

Details of these modifications may be found in Ref. 8. In addition, the design algorithm must be modified to enable the designer to specify that certain groups of members should have the same section size and that certain joints should have the same coordinates in particular directions. This modification has that added advantage that when specified it reduces the number of design variables and hence the size of the linear programming problem. The derivatives of the changes in joint deflections and member forces must be calculated with respect to the group member depths rather than each individual member depth. Details of these computational procedures together with those required for the design of structures subject to multiple loading cases are given in Ref 8. The algorithm must be further generalized for design of space structures as specified in Refs 9 and 10.

EXAMPLE STRUCTURES

In the following studies the members were of dry Douglas Fir and of cross sectional proportions given by $\alpha = 0.5$. The permissible stresses were as given in the Code of Practice CP112 where the adopted values depend on the particular conditions of service and the loading duration. The loading duration for all structures were considered to be long term and the minimum value of the modulus of elasticity of 6.6 kN/mm^2, as given in the Code, was used. The authors have considered (Ref 11) the difficulty in selecting a rational and efficient approach to the assessment of member effective lengths during an optimization scheme. In order to assist with the interpretation of the results in this paper, all member effective length ratios have been assumed to be equal to unity.

A detailed study of the use of the algorithm discussed in this paper when applied to portal frames subject to point loads is given in Ref 9. The study presented here is particularly concerned with the application of uniformly distributed loads (U.D.L.) to portal frames. The portal frame shown in Fig. 1 was subjected to a vertical U.D.L. of 500 N/m along the rafters.

The structure was first optimized, with the height h fixed,

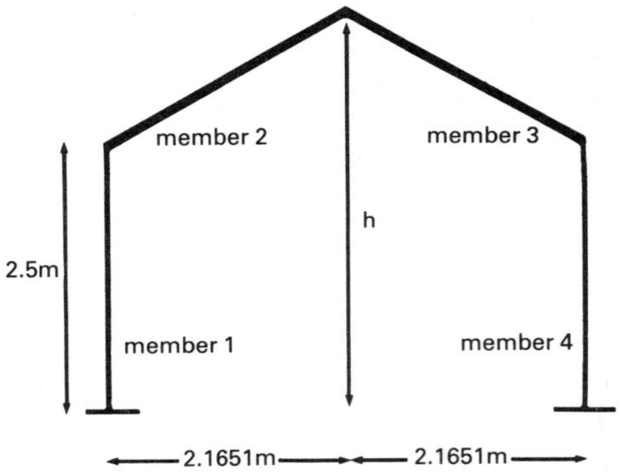

Fig. 1

first without a deflection constraint and then with a deflection constraint of 20 mm at the apex joint. Each of these cases was considered for several values of h enabling a graph shown in Fig. 2 of optimized volume against height to be plotted. For example, without a deflection constraint and fixing the height of the apex joint at 3.75 m the optimum member depths were d_1 = 91.40 mm, and d_2 = 90.89 mm,

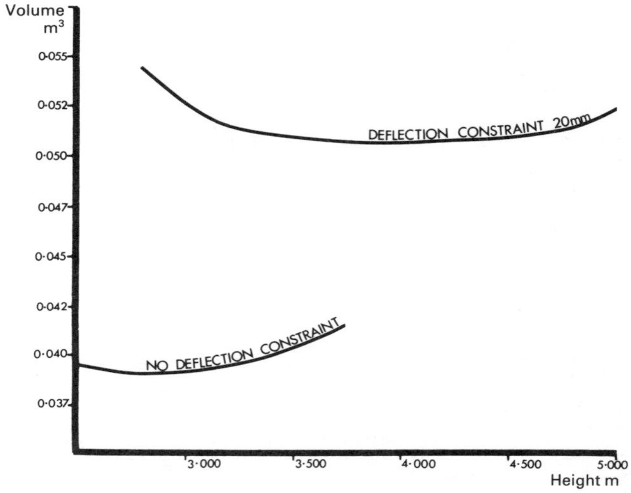

Fig. 2

the optimum structure volume being 0.0415m^3. The structure was subsequently designed considering the height of the apex joint as a design variable. The height of the apex joint in the optimum structures with no deflection constraint was 2.864 m (where d_1 = 91.82mm, d_2 = 90.58 mm and the volume = 0.0391m^3) whereas when a deflection constraint of 20mm was introduced it was 3.966m (where d_1 = 101.29 mm, d_2 = 97.71 mm and the volume = 0.0505 m^3). These designs agree with those that may be established from Fig 2. The inclusion of deflection constraints in the design of timber portal frames obviously represents an important limiting criterion since where included in these examples none of the stress constraints were critical.

The structure shown in Fig. 1 was reconsidered with a horizontal distributed load of 100 N/m on members 1 and 2 and a vertical distributed load of 100 N/m on members 2 and 3. If this structure was optimized without grouping members then the depth of members 1 and 4 would be different. This is hardly practical since it would not allow for the horizontal wind loading to act on the left hand side of the structure. The structure was first optimized with individual member size design variables and then with members 1 and 4 and members 2 and 3 grouped together as design variables. The height of the apex joint was fixed at 3.75m and the volume and member depths for the optimized structures were:

Members not grouped: $d_1 = 69.57$ mm, $d_4 = 87.42$ mm
(Individual member design variables) $d_2 = 48.81$ mm, $d_3 = 58.58$ mm
volume = 0.0226 m^3

Members grouped: $d_1 = d_4 = 78.57$ mm,
$d_2 = d_3 = 63.57$ mm
volume = 0.0255 m^3

The volume of the structure designed using grouped members is, as expected, greater since it is also designed for the reversal of the wind loading. The grouped member design may also be achieved by use of the multiple loading case facility with individual member size design variables by using the following load conditions:

Load Case 1: horizontal load 100 N/m members 1 & 2
vertical load 100 N/m members 2 & 3
Load Case 2: horizontal load 100 N/m members 3 & 4
vertical load 100 N/m members 2 & 3

A second more practical application of the use of the multiple loading case facility is illustrated by use of the following load conditions:

Load Case 1: vertical load 500 N/m members 2 & 3
Load Case 2: horizontal load λN/m members 1 & 2
vertical load λN/m members 2 & 3

The optimization was undertaken for a range of values of λ between 0 and 500N/m with the members grouped as before. Figure 3 shows how the member depths and volume of the optimized structure vary with the value of λ when the

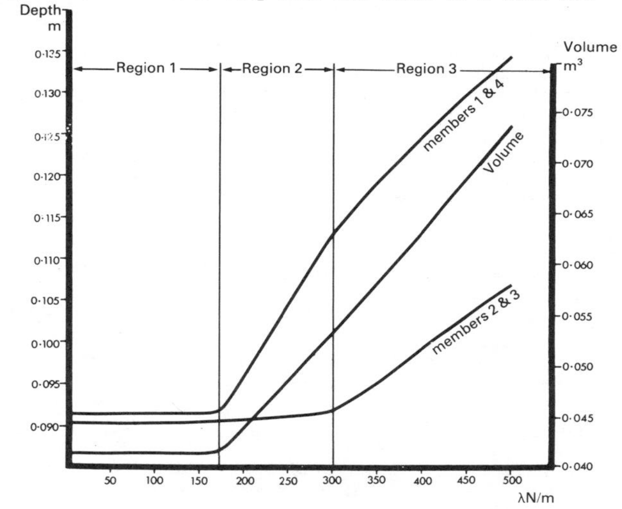

Fig. 3

height of the apex joint was fixed at 3.75 m. Region 1 corresponds to designs where only loading case 1 is a limiting factor while region 3 corresponds to designs where only loading case 2 plays a part. In Region 2, however, both loading cases contribute active constraints to the optimum design. When the height of the apex joint is included as a design variable and λ = 250 N/m (ie a value in Region 2) the optimized structure has a much reduced pitch. This has the effect of rendering loading case 2 inactive and the resulting design is the same as that obtained for the single loading case of 500N/m on the rafters. The designs shown in Fig 3 required on average 10 iterations of the design algorithm for convergence. The addition of the coordinate of the apex joint as a design variable increased the required number of iterations to 19. The ratio of the latter computer design time for the analysis, derivative calculations and linear programming was 4:3:1. This ratio demonstrates that the analysis and derivative calculations require a very large part of the computational effort compared with the linear programming solution.

The three dimensional portal frame shown in Fig 4 was designed assuming symmetry and only the numbered half shown in the figure was considered in the optimization. The

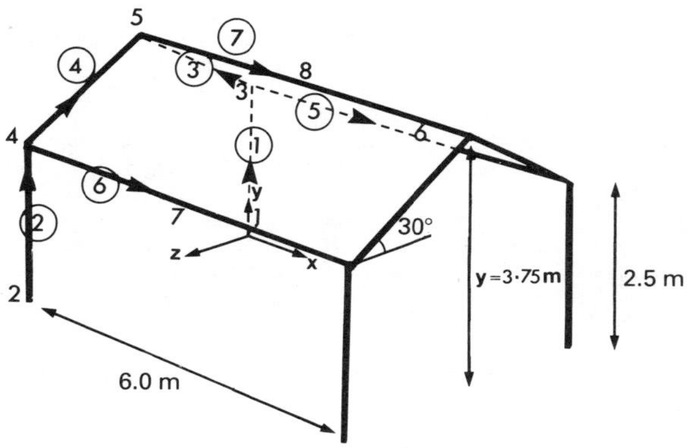

Fig. 4

members were grouped and the number of sectional design variables was 4. The initial member depths were all 0.30 m and the volume of the structure with a roof pitch of 30° was 1.71 m^3. The applied vertical loading was 1000 N/m for members 3,4,5,6 and 7. The structure was optimized assuming its shape to be fixed, with a number of different values for the y-coordinate of the apex joints. Figure 5 illustrates the variation in the volume of the optimized

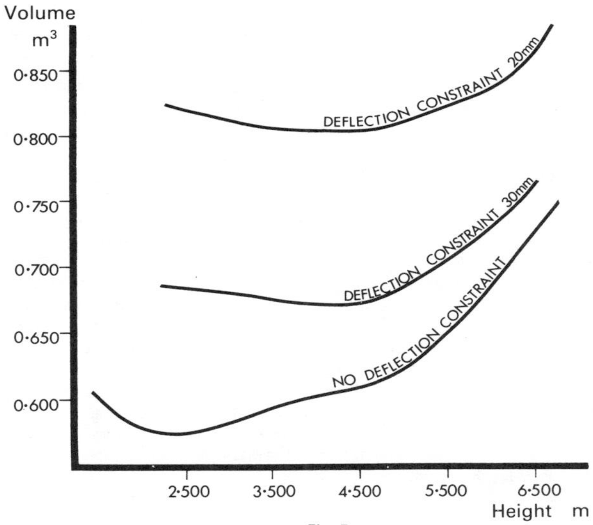

Fig. 5

structures with the increase in height of the apex joints. From this figure it may be concluded that without a deflection constraint, 2.5m is the optimum height for the apex joints giving a least weight volume of 0.5798 m^3. The structure was then re-optimized including the y coordinate of the apex joints as a design variable. The optimum solution coincided with that indicated by Fig. 4 the apex joints having a height of 2.5 m. This result is somewhat surprising since the roof has lost its pitch altogether. However, the flat roof gives a structure with least total member lengths and indicates that any saving achieved by increasing the roof pitch and hence reducing the bending moments in the structure is not as great as that obtained by reducing the length of the members. It was noted that the deflections under the design loadings for this structure were unacceptably large and as a consequence the structure was redesigned with deflection constraints of 20 mm and 30 mm in the y direction at joints 5,6,7 and 8. Figure 5 also shows the relationship between the volume of the optimized structure with deflection constraint and the height of the apex joints. The graph was prepared by optimizing a number of fixed shape structures with the apex joints at different heights. Both the volume of the optimum structure and the height of the apex joints increases with a reduction in the value of the deflection constraint. Although the volume of the optimized shape design is remarkably insensitive to changes in the height of the apex joints there was reasonable agreement between the optimum structures derived from Fig 4 and those designed using the height of the apex joints as a design variable. A change of height from 4.0 m to 4.5 m, with a deflection constraint of 20 mm, results in a change of optimum volume of 0.16% and a similar change from 4.5 m to 5.0 m results in a change in volume of only 0.063%. It is particularly encouraging that an optimum may be found by the variable geometry optimization approach when the problem is so insensitive. With a deflection constraint of 20 mm the optimum height of 4.9 m gives a reasonable pitch and would appear a sensible design.

The topology of the dome structure shown in Fig. 6 was first studies by Wakefield in Refs 12 and 13. In the present study, for efficiency, one quarter of the structure

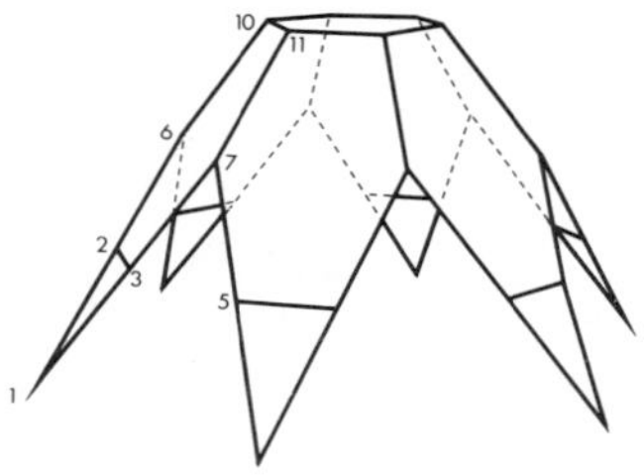

Fig. 6

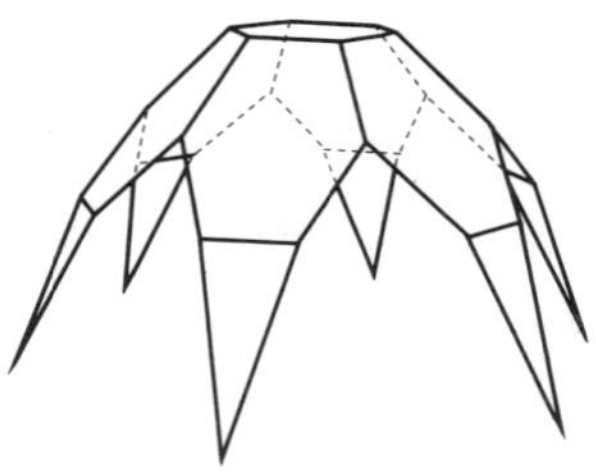
Fig. 7

was considered (i.e. twelve members) and hence only five design variables were required to specify the member depths. These depths were initially assumed to be 0.25 m and the volume 1.9375 m^3. The applied loading on all members was a vertical distributed loading of 5000 N/m. The structure was first optimized maintaining a fixed shape. The final volume of the structure was 0.9966 m^3.

The structure was then re-optimized allowing its shape to vary. This was accomplished by using two groups of coordinate design variables. The first consisted of the y coordinates of joints 2,3 and 5. The second consisted of the y coordinates of joints 6 and 7. The final volume of the structure was 0.6488 m^3 and the optimum shape is shown in Fig 7. This re-optimization resulted in a considerable saving in material owing to the changes in shape. This saving was not, as might be expected, the result of reduction in the lengths of the members since the total length of the members was increased by the optimization. The saving resulted primarily from the considerable reduction in the bending moments and consequent reduction in member depths.

DISCUSSION AND CONCLUSION

This paper has shown that the Sequential Linear programming method may be used to optimize two and three dimensional rigidly jointed structures subject to realistic design constraints based on the Code of Practice. The final designs for each of the examples considered appear to be optimal. However, it should be noted that the designs obtained by mathematical programming techniques are not always globally optimum solutions and may only be confidently identified as local minimums. In the example structures considered by the authors, there has been no evidence to suggest that the designs obtained by the method presented in this paper may be improved. Additional factors must be considered when judging the success of an automated design method to derive suitable designs for construction. Above all, the method must be able to cope with practical and realistic design situations. To this end, the inclusion of multiple loading cases, uniformly distributed loading and grouped design variables in the design method is particularly important. In addition, the formulation of the constraints from the Code of Practice is felt to be pertinent in showing that the Sequential Linear Programming Algorithm may be used to design practical structures.

It is clear that in the automated design of rigidly jointed frames it is the evaluation of the joint deflections, member forces and their derivatives that take most of the total computer time required to arrive at a solution. Therefore, regardless of how much computer capabilities increase, there will always be a limit on the size of the design which may be solved. The choice of mathematical programming method is not, however, independent of the limiting capabilities of the computer. It is particularly important to adopt a mathematical programming technique which requires the least number of evaluations of joint deflections, member forces and their derivatives.

Unfortunately, there is no way of predicting the convergence characteristics of a particular problem before the optimum solution is approached. The addition of extra deflection constraints may alter the characteristics of the solution totally. The convergence rate for difficult problems may, however, be improved by adjustment of the constraints which factor the move limits up and down. This is likely to be beneficial only if a number of similar designs are to be undertaken. However, designs which to the designer appear to be very similar may have very different solutions and convergence characteristics.

A factor that has a more predictable effect on the rate of convergence is whether or not joint coordinates are included among the design variables. The inclusion of variable coordinates in the problem generally decreases the rate of convergence and results in a more expensive shape optimization. This factor is accentuated since the derivatives with respect to the variable coordinates require considerable computational effort compared with the derivatives with respect to the member depths. It is, however, generally less expensive to perform a shape optimization than to undertake the optimization of the structure for a number of discrete joint coordinate values. When shape design is required, the Sequential Linear Programming Method appears to be a reliable and efficient method of undertaking the optimization.

Acknowledgements

The authors wish to thank Professor A.W. Hendry, Head of the Department of Civil Engineering and Building Science, University of Edinburgh for his support of this research. The authors would also like to thank the Department of Education, Northern Ireland for funding part of this work in the form of a research studentship for D.J. Robinson.

REFERENCES

1. Gallagher, R.H., Zienkiewicz, O.C., (eds.) "Optimum Structural Design - Theory and Applications", John Wiley & Sons, 1977.

2. Lev, O.E., (ed.), "Structural Optimization - Recent Developments and Applications", A.S.C.E., 1981

3. Vanderplaats, G.N., "Structural Optimization - Past, Present and Future", AIAA Journal, v.20, n.7, 992-1000, 1981.

4. Topping, B.H.V., "Shape Optimization of Skeletal Structures: A Review", Journal of Structural Engineering, A.S.C.E., v.109, n.8, 1933-1951, 1983.

5. Robinson, D.J., Topping, B.H.V., "The Use of Optimization Algorithms for Practical Microcomputer Aided Design", CIVIL-COMP83, Proc.Int.Conf. on Civil and Structural Engineering Software, Heathrow. London, November, 1983.

6. Topping, B.H.V., Robinson, D.J., "Selecting Nonlinear Optimization Techniques for Structural Design" to be published.

7. British Standards Institution, "The Structural Use of Timber", British Code of Practice, CP112: Part 2: 1971.

8. Robinson, D.J., "Optimization of Rigidly Jointed Frames", Ph.D. Thesis to be presented to the University of Edinburgh, March, 1984.

9. Topping, B.H.V., Robinson, D.J., "Optimization of Timber Framed Structures", to be published in the Journal of Computers and Structures, 1984.

10. Topping, B.H.V., Robinson, D.J., "Computer Aided Design of Space Frames", CAD84 - Proc. of the 6th Int. Conf. on Computers in Design Engineering, Butterworths, 1984.

11. Robinson, D.J., Topping, B.H.V., "Frame Optimization and Column Design", to be published.

12. Barnes, M.R., Topping, B.H.V., Wakefield, D.S., "Aspects of Form-Finding by Dynamic Relaxation", Proc. Int. Conf. on the Behaviour of Slender Structures, London, Sept. 1977.

13. Wakefield, D.S., "Dynamic Relaxation Analysis of Pretensioned Networks Supported by Compression Arches", Thesis submitted for the degree of Doctor of Philosophy, Department of Civil Engineering, The City University, London, 1980.

FORM-FINDING, ANALYSIS AND PATTERNING OF TENSION STRUCTURES

M.R. BARNES, B.Sc., M.Sc., M.A., Ph.D

Lecturer, Department of Civil Engineering,
The City University, and Consultant, Buro Happold, Bath

The paper briefly reviews methods appropriate for the non-linear analysis of tension structures indicating the controls necessary to achieve convergence. For interactive form-finding in which large but local residuals may be suddenly imposed, the method of Dynamic relaxation with kinetic damping is efficient since disturbances are not propogated throughout the structure. The application of the method to prestressed cable network, steel mesh and membrane structures is then discussed.

SOURCES OF NON-LINEAR BEHAVIOUR

Prestressed two-way cable networks forming anti-elastic single layer surfaces, and cable girders without shear bracing, are characterized as structural mechansims in that they contain insufficient members to form properly triangulated structural systems. Their stability of form is dependent on prestressing, and under live loading the variations in form are directly related to the variations in stress distribution induced by the loading. This mechanical flexibility is the principal source of non-linear behaviour. Another cause of non-linearity is due to geometric ill-conditioning: cable nets may be extremely shallow or even flat in certain areas, and thus for deformations normal to the surface the "geometric" stiffness associated with prestress is very significant. For prestretched cables, material non-linearity may be slight. However, if prestress levels are low (or curvatures high) some cables of the network may go slack under particular loading conditions; giving rise to on/off material non-linearity which must be accounted for in numerical analyses.

Prestressed woven fabric membranes, although containing a coating matrix between fibres, are subject to similar non-linearities since the stiffness of the coating matrix may be very low, giving rise to high bias deformations. In addition, the fabric material is highly non-linear due to the effects of crimp interchange between the warp and weft directions of the weave. For example, in using teflon glass fabrics at working prestress levels, an equal increment of stress in both warp and weft directions may result in a contraction of the warp direction.

MATRIX AND VECTOR METHODS OF ANALYSIS

Two principal groups of methods for non-linear iterative analysis of tension structures are Matrix and Vector methods (Ref. 3).

The majority of matrix methods can be characterized by the general recurrence equation:

$$\{\delta^{i+1}\} = \{\delta^i\} + [\alpha^i][K^i]^{-1}([\beta^i]\{R^i\} - \{C^i\}) \qquad (1)$$

where $\{\delta^i\}$ is the vector of nodal displacements at iteration i

$[K^i]$ is a matrix accounting for elastic and geometric stiffness

$\{R^i\}$ is the vector of residual nodal forces

$[\alpha^i]$ and $[\beta^i]$ are diagonal correction matrices and

$\{C^i\}$ is a correction vector which may depend on previous residuals and projected stresses.

The most widely used and stable analysis for geometrically non-linear problems is the Newton-Raphson method in which $[K^i]$ is the current tangent stiffness matrix, $\{C\}$ is null and, unless abnormal deflection increments or residuals are encountered, $[\alpha]$ and $[\beta]$ are identity matrices (Refs. 1,8,9). More rapid convergence may in some cases be obtained (Ref. 8) by using correction factors:

$$\alpha_j^i = \frac{k_{jj}^{i+1} + k_{jj}^i}{2k_{jj}^{i+1}} \quad \text{for the jth degree of freedom}$$

where the diagonal stiffness term k_{jj}^{i+1} is computed after determining the deflections $\{\delta^i\}$ using the tangent stiffness at iteration i. In other cases, when very low stiffnesses occur, it may be necessary to specify a maximum permissible deflection increment and scale down all increments if this maximum is exceeded (Ref. 9). The modified Newton-Raphson method, in which the stiffness is held constant, may be more efficient for softening structural systems or where non-linearity is small. For stiffening cable net systems, however, the initial out of balance forces may be too large under typical working loads and the analysis then diverges. This problem can be countered if estimates of tension changes can be made to artificially increase the overall stiffness (Ref. 10). In many cases, however, it may not be convenient to make these estimates.

Vector methods of numerical analysis which have been applied to tension structures fall into two categories: minimization and relaxation methods. Both can be characterized by the general recurrence equation:

$$\{\delta^{i+1}\} = \{\delta^{i}\} + \alpha^{i}\{\delta^{i}-\delta^{i-1}\} + \beta^{i}[S^{i}]\{R^{i}\} \qquad (2)$$

α and β are scalar iteration parameters which depend on the choice of solution method and [S] is a diagonal or block diagonal matrix. Thus at any stage the solution may proceed node by node in numerically uncoupled form. Consequently storage and coupled solution of an overall stiffness matrix is not required.

The simplest minimization process is the method of steepest descent in which α=O, β is the current step length and, if the process is unscaled, [S] is an identity matrix. In the classical point methods of iteration α is also zero, the elements of $[S^i]$ are $S_{jj} = (k_{jj}^{i})^{-1}$, the inverses of the current diagonal stiffness (reset at the end of each iteration in simultaneous methods), and β may be a relaxation parameter. A comparison of the above methods with those outlined below has shown them to be inefficient for geometrically non-linear problems (Ref.11).

The most widely used vector methods for analysis of tension structures are the Scaled Conjugate Gradient method (Ref. 4) and Dynamic Relaxation (Refs. 6,7,2) which for optimized convergence (Ref. 3) can both be expressed in the form of Eqn. (2).

In the scaled conjugate gradient method the elements of [S] are the inverses of the direct stiffness components ($S_{jj} = 1/k_{jj}$) (which for grossly non-linear problems may be up-dated at reinitialization stages), and

$$\alpha^{i} = \frac{\beta^{i} \, \|R^{i}\|^{2}}{\beta^{i-1} \, \|R^{i-1}\|^{2}}$$

where the current step length β^i may be determined either by linear search methods (Ref. 11) or by solving a cubic for the derivative of the total potential energy with respect to the step length (Ref. 4).

In one optimized form of dynamic relaxation (Ref. 3) the elements of the diagonal matrix [S] are

$$S_{jj} = \sum_{n=1,f} 1/k_{jn}$$

for the jth degree of freedom of a node having f displacement components, and

$$\alpha^{i} = \frac{\|\Delta\delta^{i}\|^{2}}{\|\Delta\delta^{i-1}\|^{2}}, \quad \beta^{i} = (\alpha^{i} + 1)$$

where $\{\Delta\delta^{i}\} = \{\delta^{i}\} - \{\delta^{i-1}\}$.

Alternatively α^i may be expressed physically as

$$\alpha_{i} = \frac{(1 - C/2)}{(1 + C/2)}$$

where C is the critical viscous damping per unit mass.

INTERACTIVE FORM-FINDING AND PATTERNING

By comparison with form-finding using initial data estimated only from a crude physical model or architectural sketch, the process of non-linear analysis of cable nets or membranes, starting from a defined geometry and prestressed state, presents few problems and can be solved using standard non-linear packages. Any of the methods briefly reviewed above can be used for the latter purpose; the matrix methods being probably more efficient for small bandwidth problems and the vector methods efficient for wide bandwidth problems. In form-finding, however, boundary shapes and internal support systems may be radically changed during the numerical investigation so that topology may be continuously altered, imposing very high residual forces. Frequently also zero stiffness situations associated with temporarily flat surface areas may occur. The matrix methods thus require a complete re-ordering of the stiffness matrix and all of the methods may require conditional controls on the magnitude of deflections and residuals at each iteration. In patterning, the form is more finely adjusted, but specified link lengths near all support areas will be reset which, although not radically affecting the overall shape, will entail suddenly imposed high residuals.

For the dynamic relaxation method an effective damping procedure suggested by Cundall (Ref. 6) for unstable rock mechanics problems, termed "kinetic damping", has been found to be entirely stable and rapidly convergent when dealing with large local disturbances. In this procedure no viscous damping terms are used; thus a recurrence equation for the velocity components of any node i in direction x at time t+Δt/2 may be set as:

$$V_{ix}^{t+\Delta t/2} = V_{ix}^{t-\Delta t/2} + \frac{\Delta t}{M_{ix}} \cdot R_{ix}^{t} \qquad (3)$$

Δt is a small time interval, M_{ix} is a "fictitious mass" component, and the ratio $\Delta t/M_{ix}$ is governed by the stability condition (Ref.3):

$$\Delta t < \sqrt{\frac{2M_{ix}}{S_{ix}}}$$

where S_{ix} is the row sum of the stiffness components at node i in the x direction.

The new geometry at time t+Δt is then:

$$x_{i}^{t+\Delta t} = x_{i}^{t} + \Delta t \cdot V_{ix}^{t+\Delta t/2} \qquad (4)$$

The new tension in any link m (cable or side of membrane elements) is:

$$T_{m}^{t+\Delta t} = T_{m}^{s} + \sum_{\substack{\text{elements e with} \\ \text{sides m containing m}}} \{K_{mn}^{s}\}_{e}\{\Delta L_{n}\}^{t+\Delta t} \qquad (5)$$

and the new residual forces are given by:

$$R_{ix}^{t+\Delta t} = P_{ix} \pm \sum_{\substack{\text{links m} \\ \text{connecting to i}}} \left(\frac{DX}{L}\right)_{m}^{t+\Delta t} (T_{m}^{t+\Delta t} \pm Q_{m}^{s}) \qquad (6)$$

whence updated velocity components are determined from Eqn. (3) and the iterative process continues.

In Eqns. (5) and (6) all factors with superscripts s can be specified controls during the process which may be applied to alter the form and stress distribution. For cable elements K_e^s is simply EA/L_o where L_o is the specified slack length, and for triangular membrane elements (Ref.2) $\{K_{mn}^{s}\}_e$ is the row natural stiffness corresponding to side m. T_m^s is the force in side m due to prestress in the element, $\Delta L_n^{t+\Delta t}$ is the current extension of link n from a specified initial state, $(DX/L)_m^{t+\Delta t}$ is the current direction cosine and Q_m^s may be a traction force in link m. The latter factor is of use when determining momentless boundary shear walls of fairly arbitrary shape. Along compression funiculars, the contribution of all link forces to residuals is reversed; the boundary being effectively determined as a tension funicular to the image of the adjacent surface (Refs.2,12).

During the process represented by Eqns. (3)-(6) the kinetic energy is traced. When a local peak is detected all velocity components are set to zero and the process is re-started from the current geometry and continued through further (decreasing) peaks until the energy of all modes has been dissipated and static equilibrium is achieved.

This vector process is ideally suited to interactive form-finding with radical boundary changes since initial energy dissipation and convergence in the altered zones is very rapid and thus disturbances are not significantly propogated throughout unaltered regions.

UNIFORM MESH, GEODESIC AND PRINCIPAL CURVATURE NETS

The most commonly used type of cable net is the uniform mesh net, in which typically the grid spacing is 0.5m. The form-finding of such nets preferably starts with a conceptual physical model which, after fibreglassing, may subsequently be used for wind tunnel testing. The physical model is also useful in providing an initial estimate of link topology for numerical form-finding. To simplify automatic data generation this may conveniently commence from a flat net in which the internal grid links have a high elastic stiffness (K_e in Eqn.(5)) in order to maintain the uniform grid spacing. The links adjacent to boundary or internal support systems, which initially will be grossly extended to meet with supports, may either be controlled by specifying a constant tension ($K_e = 0$ in Eqn. (5)) or by using a low value of elastic stiffness and adjusting their slack lengths to control the form. Usually a combination of both controls will simplify the process. Boundary scallops can be similarly adjusted using low EA values and specified lengths and tensions which accord with the required scallop radii and prestress in the surface network.

During the form-finding process(and load analysis checks) the use of colour graphics to display differing stress levels is convenient. By this means a complete picture of the stress system is displayed and necessary adjustments to the net topology, and boundary or internal support systems, can be made to achieve the required shape and prestress distribution. The use of automatic controls during the numerical procedure may be helpful in simple cases, but more commonly the overall shape, prestress distribution, topology and boundary cable connections must be interactively adjusted.

For large nets, or closely spaced steel meshes, such as that used for Munich Aviary (shown during construction in Fig. 1), the numerical idealization (Fig.2) must be comparatively coarse; each link representing a number of mesh lines. If spatial co-ordinates for setting out the boundary wall have been specified by the architect then the mesh must be adjusted during form-finding and patterning to mate with (cubic) splines fitted through these co-ordinates. Additionally the perpendicular offsets from all boundary nodes to adjacent grid links must be adjusted to comply with the grid spacing. A much simpler procedure may be followed if only the plan geometry of boundary walls is specified together with approximate vertical ordinates. In this case the mesh can be continued beyond the structure plan to a fictitious boundary. Subsequently the real boundary can be projected vertically to obtain the spatial co-ordinates of the boundary intersections.

For precise patterning, corrections to the grid lengths of the numerical idealization must then be made to allow for curvatures. This may be achieved by fitting splines through the grid traverses and shortening the slack grid lengths by the difference between surface arc and chord lengths. The final stage of patterning entails more precise boundary and mast support zone analyses, a typical example of which is shown in Fig. 3. The zone extremities in this case can be defined by cubic interpolation through the node points of the overall analysis. The vector of reactions at the mast top support, obtained from the local zone analysis, can then be projected to define the mast base co-ordinates. If these differ from those assumed for the overall analysis, the slack lengths or EA values assumed for the support ties can be adjusted until the zone analysis mates with the overall analysis.

Fig. 1

In geodesic nets, cables follow lines of minimum distance over the surface so that there is no shearing component between consecutive links in any cable. Such nets can be derived by holding the specified cable tensions constant in Eqn. (5) (with K_e set to zero). In practical terms they may be less useful than uniform mesh nets. However, geodesics are also those lines which a flat tape of material can follow over the surface without shearing, and this has relevence in the context of cladding systems.

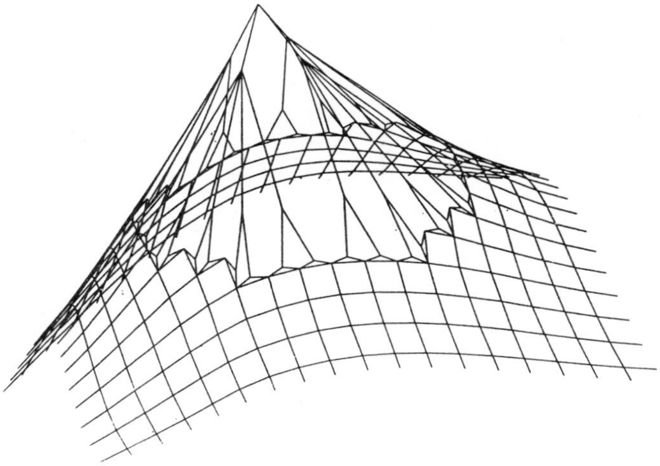

Fig.3

Principal curvature nets suffer the same drawback of non constant grid lengths as geodesic nets. However, by definition the twisting curvature in panels formed by the cable grid is zero, which may be helpful when stiff cladding panels are to be used. Fig. 4 shows the structural scheme for the enclosed "Cascades" botanic gardens at Ventnor. To achieve the level of transparency required polycarbonate cladding is proposed. The structure is sited in a land-slip area and thus differential settlements must be allowed for. This, together with the inherent flexibility of the network system, requires that the junctions between panels must accommodate shearing and separation; the panels themselves, however, will move as rigid body elements.

One approach to obtaining a principal curvature network is to derive, initially, a minimum surface structure (away from mast point supports) by idealizing the surface into triangular membrane elements in which stress is held constant. If the triangular elements are then assigned elastic stiffness and a uniform displacement is applied at each node of the structure in a locally normal direction, the resulting principal strains coincide with the directions of principal curvature. The principal trajectories can then be drawn on the surface in such a way that each panel tends to be square (with mesh lines crossing approximately at right angles as in a "flow-net"). The spacing of the grid must obviously vary but the tension in any link is approximately proportional to the cross spacing. Thus tension coefficients (T/L in Eqn. (5)) throughout the net should be constant and, using the topology so found, a cable net analysis can be carried out to obtain the equilibrium shape. In practice certain areas of the network will be required to depart from either the minimum surface shape, particularly in mast zones, or from principal curvature trajectories, for example where spacing becomes too wide - usually in flat or inflected surface areas. These alterations can be made by suitable adjustment of the specified tension coefficients.

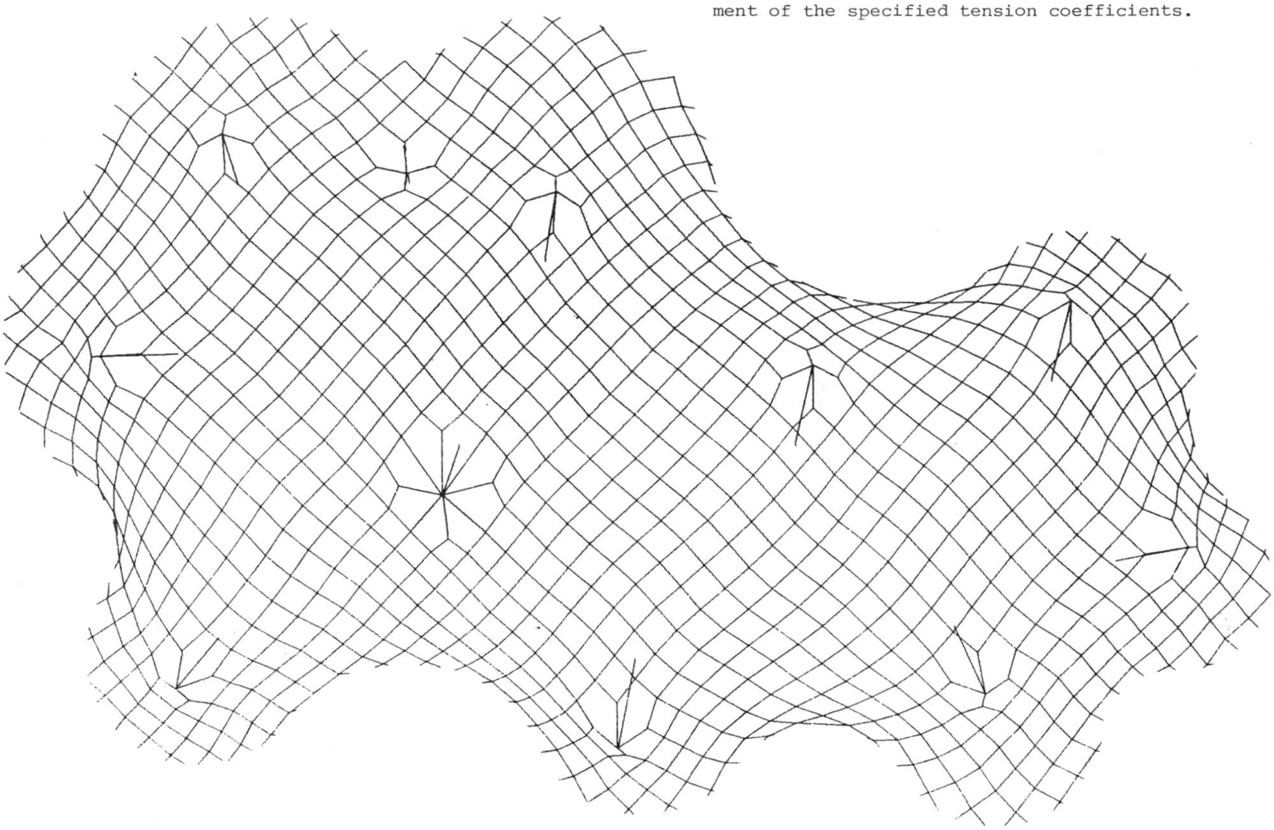

Fig. 2

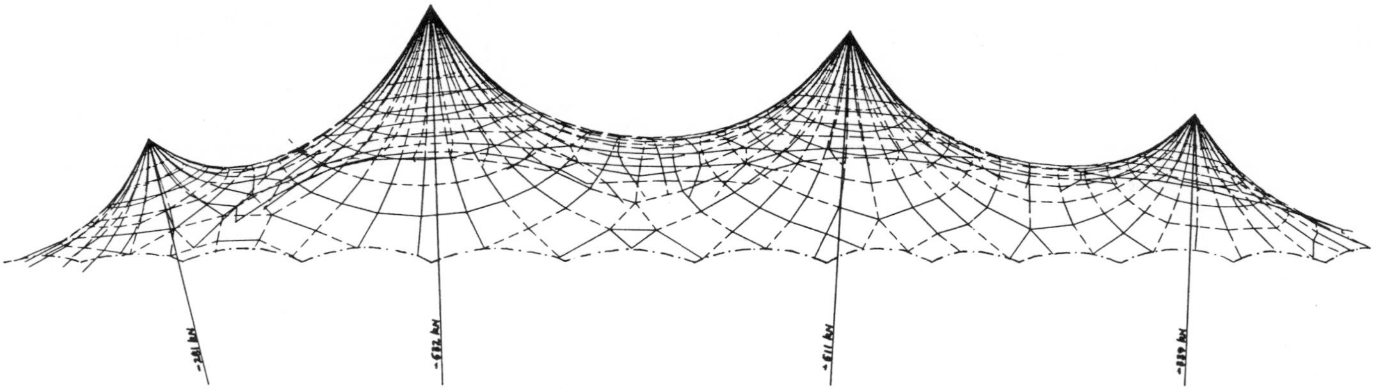

Fig. 4

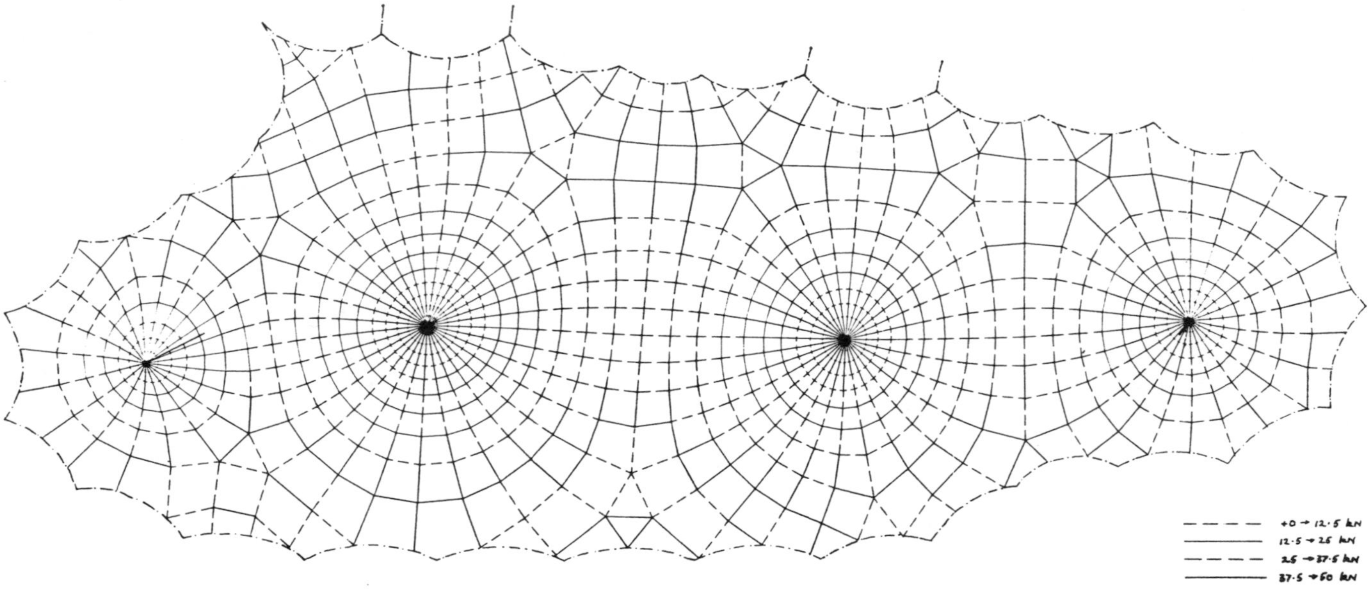

PRESTRESSED MEMBRANES

The shape of minimum surface membranes can be determined using a sufficiently fine idealization of warped quadrilateral elements (Ref.9) or triangular elements in which stresses are held constant. In the latter case the tensions along edge links are given by:

$$T_i = \frac{\sigma t \ell_i}{2\tan\alpha_i} \qquad (7)$$

where α_i is the apex angle opposite side i.

For variable stress elements, with stresses σ_x and σ_y corresponding to the intended fabric weave directions, the side tensions are:

$$\begin{Bmatrix} T_1 \\ T_2 \\ T_3 \end{Bmatrix} = V.[G]^T \begin{Bmatrix} \sigma_x \\ \sigma_y \\ \tau_{xy} \end{Bmatrix} \quad \text{(where } \tau_{xy} = 0 \text{ for form-finding)} \qquad (8)$$

V is the element volume and [G] is a transformation matrix relating orthogonal (x,y) strains to the element side strains (Ref. 2).

For analysis of the derived membrane under live loading the (3x3) element "natural" stiffnesses are given by:

$$[k_e] = V.[G]^T[D][G] \qquad (9)$$

where [D] must be a simple orthotropic stress matrix.

To account for crimp interchange effects, however, the stresses may be expressed as polynomial functions of strains:

$$\sigma_x = C_1 + C_2\varepsilon_x + C_3\varepsilon_y + C_4\varepsilon_x\varepsilon_y + C_5\varepsilon_x^2 + \ldots..$$

In this case, using a vector method (in which equilibrium and compatibility conditions are separated), Eqn. (8) can be used to update side tensions.

If one weave direction corresponds with a side in each element, and assuming shear stiffness is negligible, computing time and storage can be considerably reduced. In order to account for the orthotropic behaviour of the weave during load analyses, the idealization of the surface during form-finding should be adjusted to achieve this; the main aim being to simultaneously derive the overall form and patterning arrangement.

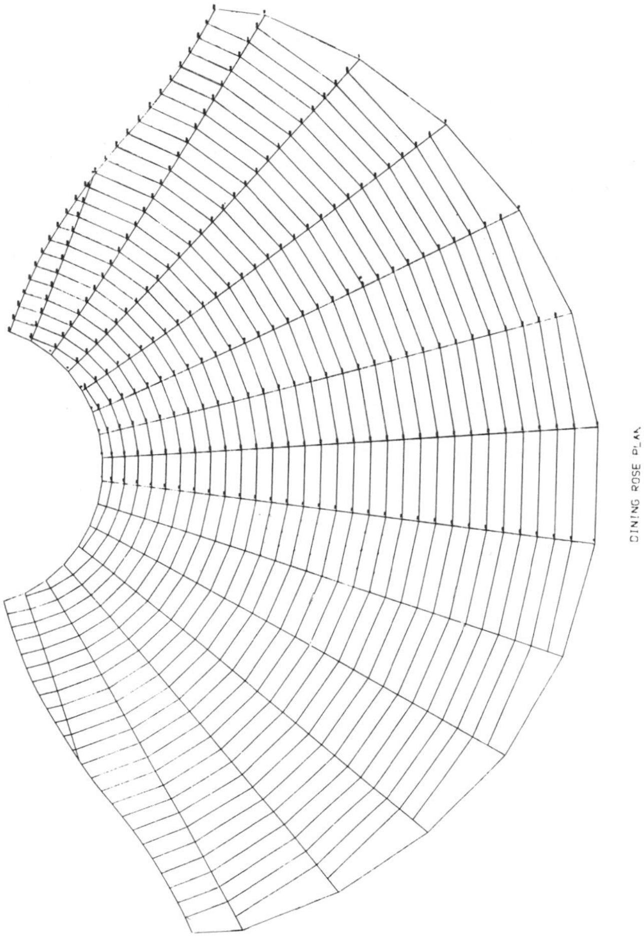

Fig. 6

Fig. 5 shows an elevation of the element idealization for one of the PTFE glass prestressed membranes used in the Diplomatic Club in Riyadh. The arrangement of reinforcing radial cables is shown in Fig. 6. This figure also shows the weft direction of the weave in each panel which was adjusted during form-finding to lie along hoop directions. Adjustments were made by fitting splines through the ridge line and wall intercept nodes and amending their positions (and the ridge link lengths) so that the bisecting angle between a particular control hoop line and ridge lines was equal on each side of each panel. The other hoop lines were then adjusted along the ridge splines so that the perpendicular distance, within the surface, between adjacent hoops was constant. This process was carried out automatically so that there was no need for interactive control. The program system, however, had to allow for changes in topology during the process. The above procedure achieves cutting patterns which for each radial panel (away from the wall intercepts) are as nearly as possible symmetric. Having obtained the final form corresponding with the specified membrane and ridge cable prestress, the panel and edge scallop shapes must then be compensated to allow for curvatures and the stretch that must occur from the initially flat unstressed state (Fig. 7). Biaxial tests for each batch of material are necessary in order to obtain these stretch compensations.

For more arbitrary shapes than essentially radial systems, patterning in which panel edges follow geodesic lines over the surface are ideal from the point of view of structural function and patterning economy. A procedure for obtaining membrane forms and patterns in such cases has been given by Williams (Ref. 13). In this process equilibrium equations are set only in directions normal to the surface with the in plane arrangement of nodes being adjusted geometrically to achieve geodesic control lines over the surface. An alternative procedure suggested by Haug is to incorporate elastic or constant tension strings within the membrane surface. These strings will then slip into geodesic positions provided that they have adequate tension in comparison with the membrane stress resultants. For the case of uniformly stressed membranes, however, the string tensions must be very low in order not to affect the overall form. In this case convergence will be slow or not possible. Wakefield has suggested a procedure to counter this problem which consists of using elastic strips to control the in plane movement of nodes, but subtracting their components of force normal to the surface when calculating nodal residuals.

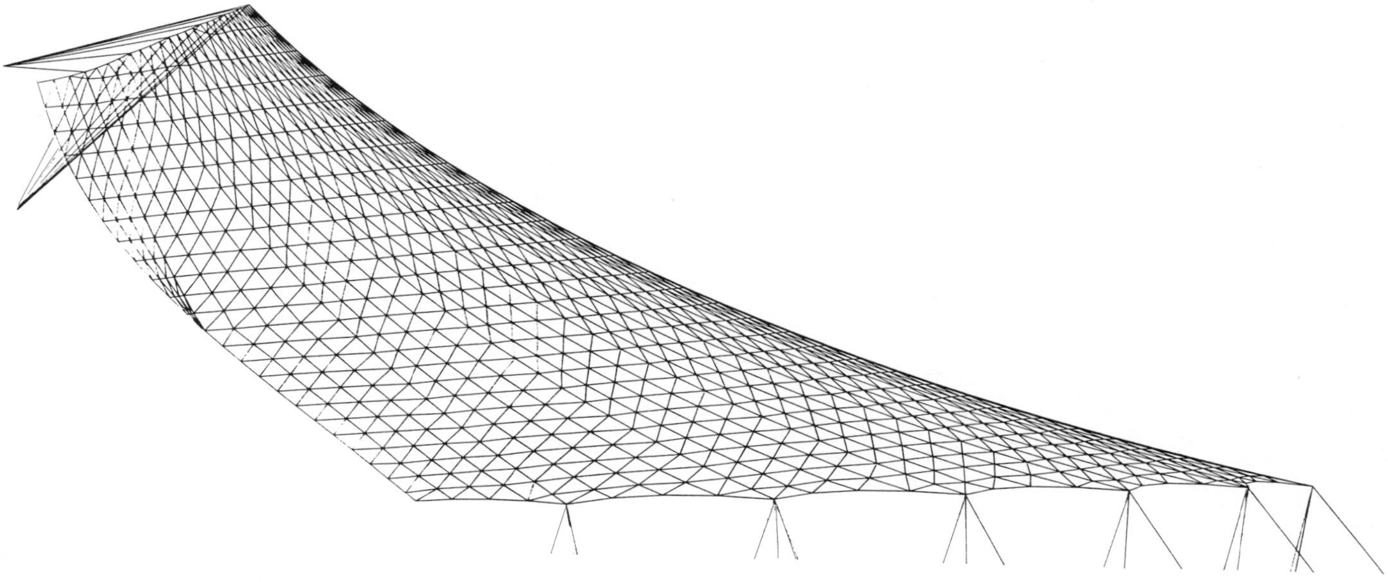

Fig. 5

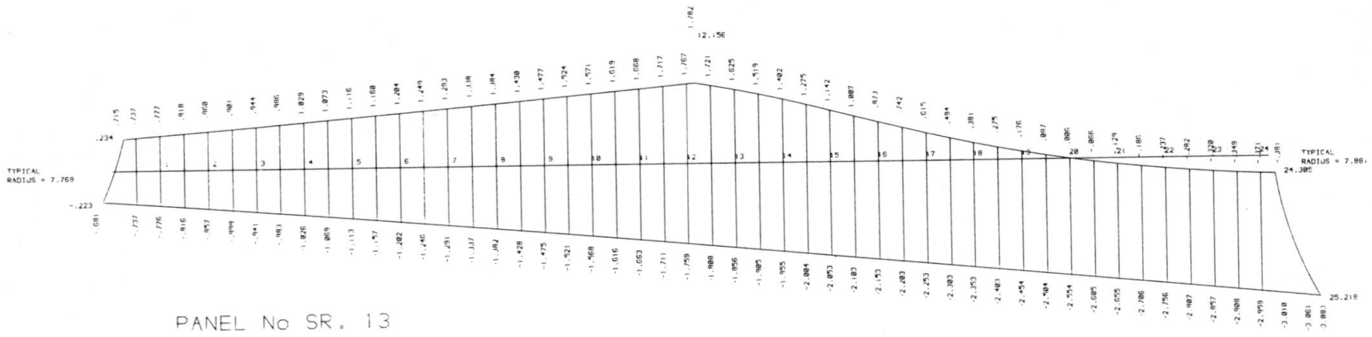

Fig. 7

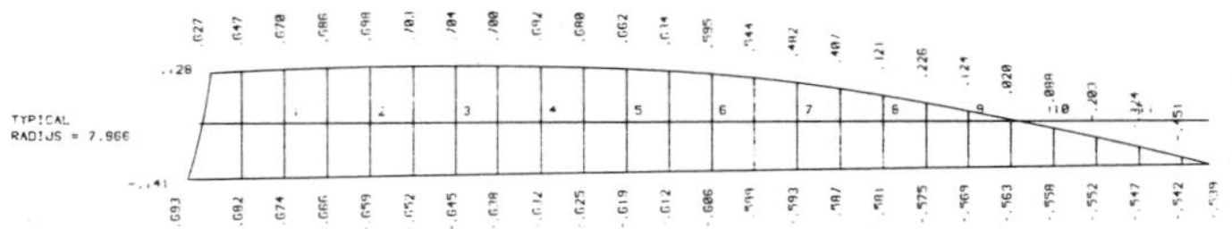

PANEL No SR. 15

ACKNOWLEDGEMENTS

For Munich Aviary, Architects were Jorg Gribl in association with Frei Otto, and project Engineers were Professor E. Happold and Michael Dixon of Buro Happold.

For the Cascades scheme, Ventnor, Architects are Michael Phelps and Partners and Engineers are Woodruff and Maxwell in association with Buro Happold.

For the Diplomatic Quarters Club, Riyadh, Architects are Frei Otto with Omrania U.K., and Engineers are Buro Happold. Patterning work was carried out for the sub-contractors Stromeyer.

REFERENCES

1. Argyris, J.H., Angelopoulos, T., Bichat, B., "A general method for the shape finding of light-weight structures", Int. Conf. on Tension Structures, London 1974.

2. Barnes, M.R., "Applications of dynamic relaxation to the design and analysis of cable, membrane and pneumatic structures", 2nd Int. Conf. on Space Structures, Guildford, Sept. 1975.

3. Barnes, M.R., "Non-linear numerical solution methods for static and dynamic analysis of tension structures", Symp. on Air Supported Structures, I.Struct.E., London 1980.

4. Buchholdt, H.A., McMillan, B.R. "Iterative methods for the solution of pretensioned cable structures", IASS Symp. on Tension Structures and Space Frames, Tokyo, 1971.

5. Cundall, P.A., "Explicit finite-difference methods in geometrics", Proc. E.F. Conf. Numerical Methods in Geomechanics, Blacksburg, Va., June 1976.

6. Day, A.S., "An introduction to dynamic relaxation", The Engineer, Jan. 1965.

7. Day, A.S., "Form definition for the bridge of Don exhibition and sports centre Aberdeen", Symp. on Air Supported Structures, I.Struct.E., London 1980.

8. Greenberg, D.P., "Inelastic analysis of suspension roof structures", J. Struct. Div., ASCE, V.96, n.ST5, May 1970.

9. Haug, E., Powell, G.H. "Finite element analysis of non-linear membrane structures", IASS Symp. on Tension Structures and Space Frames", Tokyo, 1971.

10. Mollman, H., Mortensen, P.L., "The analysis of prestressed suspended cable nets", Int. Conf. on Space Structures, Guildford, 1966.

11. Papadrakakis, E.M., "Discrete statical analyses of structural mechanisms", Ph.D Thesis, The City University, Oct. 1978.

12. Wakefield, D.S., "Pretensioned networks supported by compression arches", Ph.D Thesis, The City University, April 1980.

13. Williams, C.J.K., "Form-finding and cutting patterns for air supported structures", Symp. on Air Supported Structures, I.Struct.E., London 1980.

ANALYSIS OF SHALLOW SADDLE-SHAPE SHELL-CABLE ROOFS

K P ÕIGER, CandTechSc, PhD

Associate Professor, Department of Structural Engineering,
Tallinn Technical University, USSR

Some data on experimental investigations are presented. Calculation problems are discussed for the rectangular in plan shallow shells of negative curvature, the sheathing of which contains two or more elements of different materials, such as timber, glass-reinforced plastic, steel cable and combined variants connected by stiff or yielding fasteners, either prestressed or non-prestressed. One of the varieties is the system with steel prestressed cable network used to obtain a more rational form for the timber shell.

EXPERIMENTAL INVESTIGATIONS

For over 25 years the work of different saddle-shape suspended crossing cable systems (CS) has been investigated in the Estonian SSR and the study of work of saddle-type timber shells (TS) has been pursued for about 15 years. In recent years much attention has been given to the work of square in plan saddle-shape CSs, Fig 1 and saddle-shape TSs, Fig 2.

Figure 1 shows that the first and second shells were treated as of varied constructional solutions. For example, the work of the suspended cable network, Figs 1.1 and 1.2, was studied at different rigidities of the framing contour beam, Fig 1.3 and at strengthening the network and contour beam by four additional cable connections, Fig 1.5. The effect of roof panels on the work of the system was studied, adding all over the

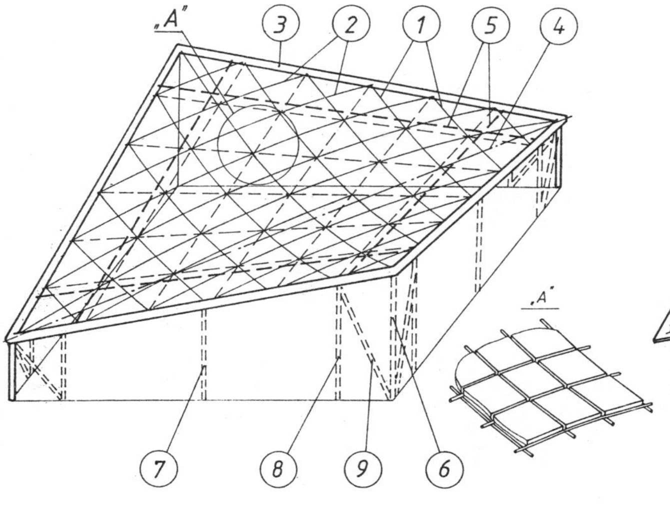

Fig 1

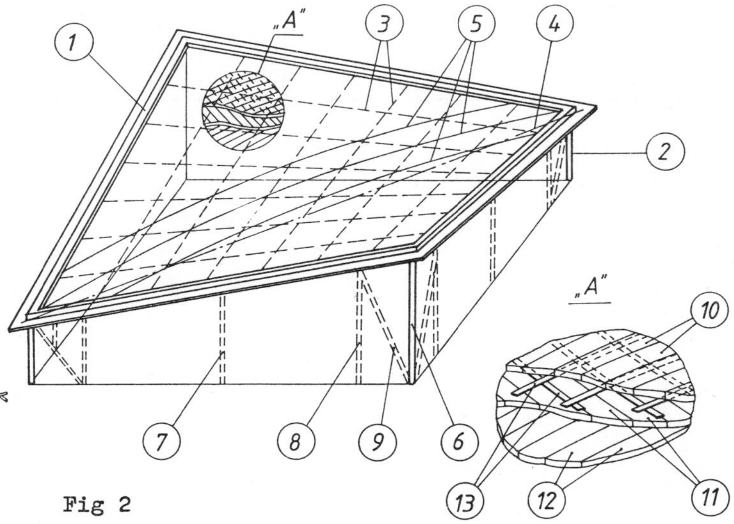

Fig 2

surface non-prestressed timber elements of the type in Fig 1.5 placed in parallel to the linear contour beams, fixed in the network knots and in some cases also to the contour beam. The work of the roof panels constructed as shown in Fig 1"A" has also been studied. The values and the ratio of cable prestressing forces were varying. The geometry of the network surface depends on the latter factor, i.e. on the ratio $\beta = f_x/f_y$ Fig 3. In the cases considered the ratio ranges from 0.6 ... 1.36 . Such surfaces are constituted by producing cable prestressing at free sliding of stiffening cables over bearing cables. The behaviour of the system was also studied as part of the complex roof, Fig 4. The work of the system was investigated experimentally, by loading the models, the parameters of which are given in Table 1.

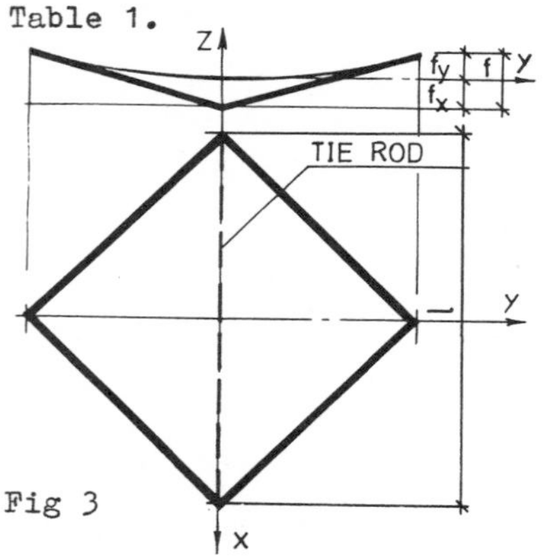

Fig 3

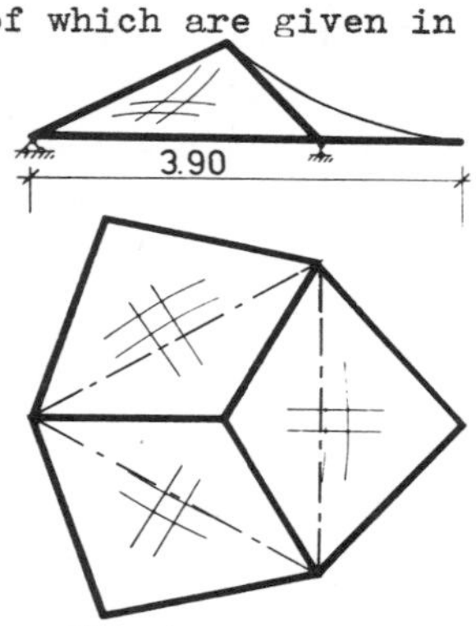

Fig 4

The saddle-shape TSs considered were in many respects similar to the suspended shells described. These shells mainly consisted of two or three layers of boards parallel to the diagonals of the shells and at right angles to each other. In some cases the shell was strengthened by a glass-reinforced plastic network (GPN), Fig 2.13, by prestressed steel cables, Fig 2.5 or by ribs, Fig 2.3, placed in parallel to the contour beams. The surface of the shell corresponded to the hypar (β = 1) or to another saddle-shape surface, where β = 0.5. The parameters of TS models are presented in Table 2.

In view of the contour beam, both suspended roofs and shells were supported by corner columns, Fig 1.6 or additionally by the columns in the centre of the beam, Fig 1.7. Thrust forces of the contour beam were taken up by horizontal steel tie rods, Figs 1.4 and 2.4 or by steel tie rods placed on the exterior wall. The work of both systems was examined at free horizontal displacements of the contour beam, i.e. with no tie rods. Both constructions were loaded by evenly distributed all over the surface and by antisymmetrical evenly distributed and concentrated vertical loads.

Further, some results of experimental investigations of the described system are featured. In Fig 5 diagrams of surface deflection of different systems are presented.

Figure 5a represents deflection diagrams of convex diagonals and Fig 5b those of concave diagonals of a suspended roof, where l - diagonal length of the model. Figures 6a and 6b represent deflection diagrams of convex and concave diagonals of TSs. Here and in other diagrams the line number corresponds to the model number given in Tables 1 and 2.

Table 1

Model No	Diameter and thickness of model contour in mm		Model dimensions in plan in mm	Cable diameter in mm	Δx; Δy in mm	Number of cables	f_x in mm	f_y in mm	T_{Ox} kN	T_{Oy} kN	T_{OL} kN	$k = \frac{T}{\Sigma P}$
1	48	t = 3	1768x1768	1.6	278; 278	8; 8	250	250	0.50	0.50	0	1.60
2	48	t = 3	1768x1768	1.6	278; 278	8; 8	250	250	1.00	1.00	0.1	1.16
3	48	t = 3	1768x1768	1.6	278; 278	8; 8	250	250	1.00	1.00	0	1.34
4	48	t = 3	1768x1768	1.6	278; 278	8; 8	250	250	1.00	1.00	0.5	0.80
5*	33.5	t = 3.2	1768x1768	1.6	278; 278	8; 8	180	300	0.45	0.30	0	0.98
6**	33.5	t = 3.2	1768x1768	1.6	278; 278	8; 8	180	300	0.45	0.30	0	0.75
7	75	t = 3.5	3000x2300	1.6	230; 256	12;8	270	200	1.00	1.33	0	1.08
8	75	t = 3.5	3000x2300	1.6	230; 256	12;8	210	260	1.25	0.67	0	1.47
9	33.5	t = 3.2	1768x1768	1.6	278; 278	8; 8	240	240	0.50	0.50	0.5	1.00
10	33.5	t = 3.2	1768x1768	1.6	278; 278	8; 8	240	240	0.50	0.50	0	0.90
11	33.5	t = 3.2	1768x1768	1.6	278; 278	8; 8	240	240	0.50	0.50	0	-
12	33.5	t = 3.2	1768x1768	1.6	278; 278	8; 8	240	240	0.50	0.50	0.5	-
13	33.5	t = 3.2	1768x1768	1.6	278; 278	8; 8	180	280	0.60	0.25	0	-

Table 2

Model No	Dimensions in plan in mm	Type of connection, GPN	Contour beam elevation in cm	Contour beam cross-section in mm b x h	$\beta = f_x/f_y$	f_x/a	f_y/a	Board cross-section $t \times b_1$ in mm	Number of layers	Ribs 4 x 4 cm	Steel tie rod diameter in mm	Vertical tie rods in Fig 1.9	Posts under contour beam	Prestressed surface cables, diameter in mm
1	1800x1800	GN ; -	36	30x60	1.0	1/14	1/14	1.4x20	2	-	10	-	-	-
2	1800x1800	N ; -	36	30x60	1.0	1/14	1/14	1.4x20	2	-	10	-	-	-
2a	1800x1800	N ; -	36	30x35	1.0	1/14	1/14	1.4x20	2	-	10	-	-	-
3	3500x3500	GN ; -	70	80x140	1.0	1/14	1/14	8x50	2	-	24	-	-	-
3a	3500x3500	GN ; -	70	80x100	1.0	1/14	1/14	8x50	2	-	24	-	-	-
4a	2400x2400	GN ; -	60	40x82	0.5	1/17	1/8.5	3.3x20	2	-	12	-	-	1x5
4b	2400x2400	GN ; -	60	40x82	0.5	1/17	1/8.5	3.3x20	2	-	12	-	-	-
4c	2400x2400	GN ; -	60	40x82	0.5	1/17	1/8.5	3.3x20	2	-	-	-	-	-
4d	2400x2400	GN ; -	60	40x82	0.5	1/17	1/8.5	3.3x20	2	-	-	-	-	3x5
5a	2400x2400	GN ; GPN	48	40x82	1.0	1/14	1/14	3.3x20	2	-	8;12	-	-	-
5b	2400x2400	GN ; GPN	48	40x82	1.0	1/14	1/14	3.3x20	2	-	-	-	-	-
6	2400x2400	GN ; GPN	48	40x82	1.0	1/14	1/14	3.3x20	2	-	-	+	-	-
7a	2400x2400	GN ; -	48	40x82	1.0	1/14	1/14	3.3x20	3	-	12	-	+	-
7b	2400x2400	GN ; -	48	40x82	1.0	1/14	1/14	3.3x20	3	-	-	-	+	-
7c	2400x2400	GN ; -	48	40x82	1.0	1/14	1/14	3.3x20	3	-	12	-	-	-
7d	2400x2400	GN ; -	48	40x82	1.0	1/14	1/14	3.3x20	3	-	-	-	-	-
7e	2400x2400	GN ; -	48	40x82	1.0	1/14	1/14	3.3x20	3	+	12	-	-	-
7f	2400x2400	GN ; -	48	40x82	1.0	1/14	1/14	3.3x20	3	+	-	-	+	-
8	2400x2400	N ; -	48	40x82	1.0	1/14	1/14	3.3x20	2	+	12	-	-	-

T_{Ox}, T_{Oy} - prestressing forces in stiffening and bearing cables

T_{OL} - prestressing forces in the additional elements of the type in Fig 1.5

$\Delta x, \Delta y$ - pitch of stiffening and bearing cables

* - at simultaneous work of roof panel elements and tie rods

** - at simultaneous work of the roof elements and posts

GN - glue-nail fastenings

N - nail fastenings

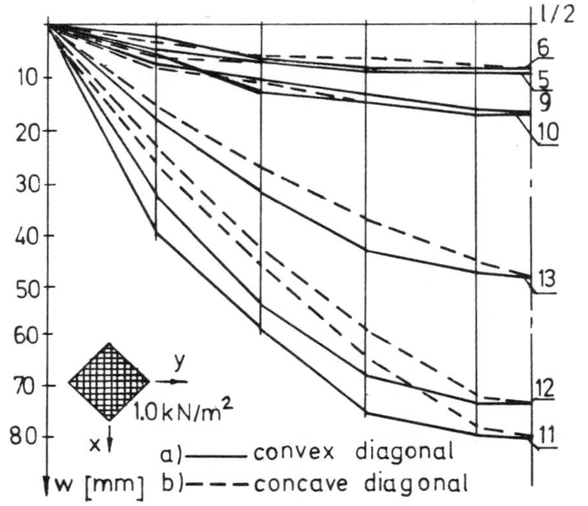

Fig 5

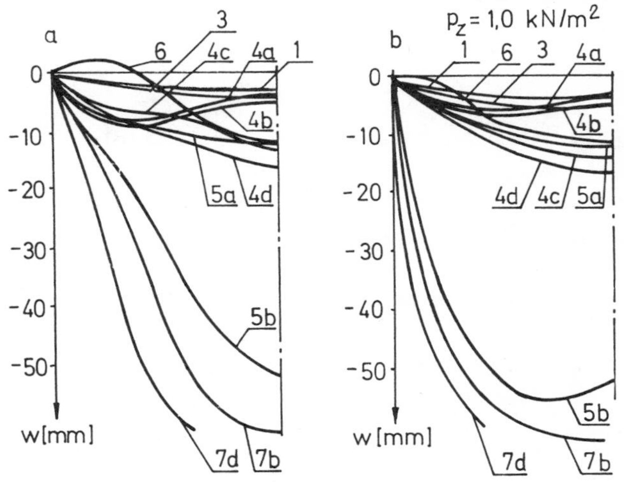

Fig 6

Figure 7 illustrates the dependence of the inner force of the horizontal tie rod T on the total evenly distributed load.

Figure 8 displays the distribution of the inner forces between the bearing, Fig 8a, and stiffening, Fig 8b, cables crossing through the main diagonals of the suspended roof. In Fig 9 distribution of stresses along the diagonals due to longitudinal forces in TS boards is shown.

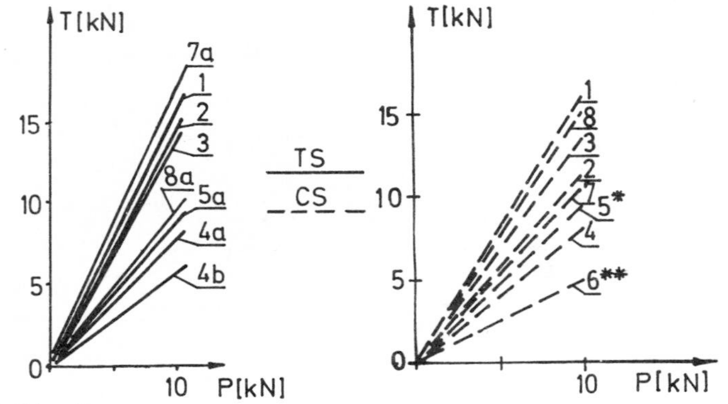

Fig 7

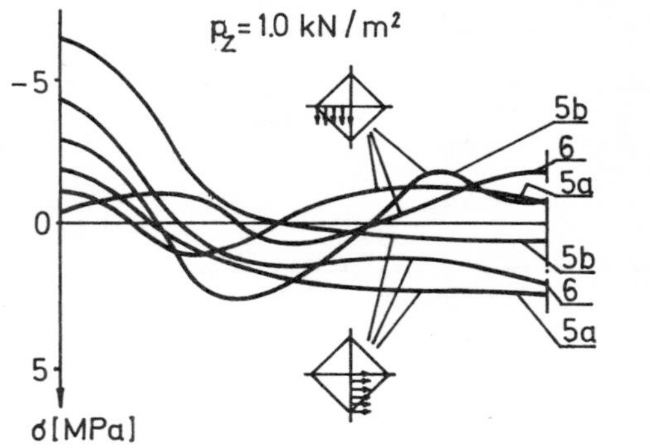

Fig 9

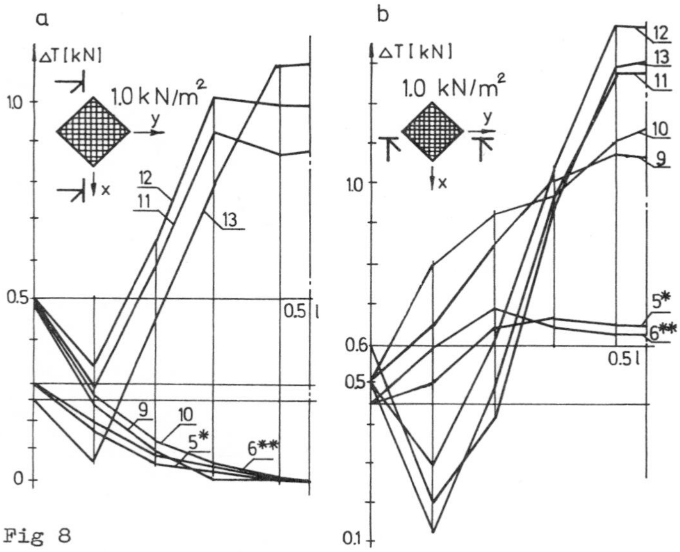

Fig 8

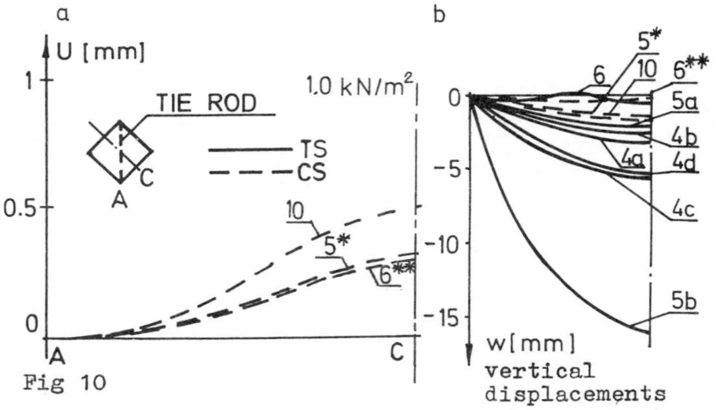

Fig 10

Figure 10a demonstrates horizontal displacements of the suspended roof edge beam and Fig 10b represents both for the suspended and TSs when the load is distributed all over the surface.

Main conclusions based on the results of experimental investigations are as follows.
Suspended systems and TSs have in many respects similar features of work, as demonstrated on the graphs.

1. For both systems the surfaces with the deflection of the concave diagonal exceeding the pitch of elevation of the convex diagonal, i.e. when $\beta < 1$, have considerable advantages over common hypars. Thrust forces of the beam are reduced, Fig 7. For example, with TSs $\beta = 0.5$ longitudinal inner force of the beam and consequently, the inner force of the horizontal tie rod decrease nearly twice, thus simplifying the construction of the main bearing units. As a result, these surfaces can also be used if there is no tie rod, i.e. with thrustless beams. It is interesting to note, for example, that with suspended roofs $\beta = 0.64$ and $T_{0x}/T_{0y} = 2.4$ and when edge beams have horizontally free deformation, vertical displacements of the roof are approximately 20...40 per cent less, Fig 5, as compared to common hypars, $\beta = I$, $T_{0x}/T_{0y} = 1$.

2. Suspended systems as well as TSs can work under the conditions of a load in case of freely deforming edge beams, but, as a rule, with surfaces $\beta = 1$ vertical displacements of the roof increase about 3 times as compared to the use of tie rods. Horizontal displacements of the corners, i.e. shell or suspended roof crawling, depending on the parameters and construction of the roof, is approximately 1/100 ...1/400 of the length of the principal diagonal of the shell. If horizontal tie rods cannot be used to restrict the value of roof crawling, surfaces where $\beta = 1/2...1/3$ are recommended or the connections shown in Figs 1.9 or 2.9, taking up as horizontal tie rods thrust forces of the edge beam are even more effective. Prestressing of these connections increases contour rigidity also.

3. The distribution of inner forces between the bearing cables is such that central cables are a great deal more loaded (inner forces of pre-

stressing are evenly distributed) than those on the edges, Fig 8a.

A similar distribution of inner forces is encountered with thrust and freely deforming edge beams and in case of round in plan and elliptical contour. Concave boards including TSs have the described distribution of longitudinal inner forces, Fig 9. The distribution of inner forces between the stiffening cables depends upon the conditions of horizontal supporting of the contour. In case of suspended roofs with the thrust beam (tie rod) in the central stiffening cables the prestressing force decreases faster than in other cables, Fig 8b, but with TSs longitudinal tensile forces occur in the central concave boards, in the latter case, however, the distribution of inner forces along the diagonal is of alternative value, Fig 9.

In case of freely deforming contour beams stiffening cables of the suspended system take up the work if there are no tie rods and in the loading process prestressing forces do not decrease but increase, preventing an indefinite crawling of the contour, Fig 8b. A similar case is encountered with the TSs too, having freely deforming edge beams, but the function mentioned is performed by the concave arches in the region of the quarter of the shell, Fig 9.5b.

It should be pointed out that with freely deforming contour beams the inner forces of the cable network or surface elements of TS by absolute value exceed the inner forces of these elements in case of the thrust contour beam approximately by 20...35 per cent, Figs 8 and 9, but the inner forces of the contour beam of suspended roofs 2...3 or more times. Since with CS and when surfaces $\beta < 1$ in the prestressing process the freely deforming contour beam is subjected to the action of unilateral outward pressure, therefore inner forces of cable prestressing are to be fixed to the minimum, so much the more because in the loading process selfstressing of the whole cable network occurs

4. Suspended cable networks engaged in the work of the system by the roof panels and TSs strengthened by GRP or steel prestressed cable network, Tables 1 and 2, models 5 and 6, have to a great extent similar work characteristics. For TSs prestressed cable network is used mainly to obtain the desired surface geometry achieved in producing cable prestressing under the conditions of their mutual free sliding, and as a shutter. The influence of the network is taken into account in calculating the shell.

5. It is sometimes assumed that the rigidity of the linear contour beam of saddle-shape roofs is to be relatively high, i.e. considerably higher than the rigidity in the round in plan or with the elliptical contour beam. Tests show that in the saddle-shape suspended system a relatively flexible linear contour beam can work well at the rigidity parameter, for example, of $\xi \leqslant 40$. Here $\xi = 0.01Etl_1^3/(E_c I_c)$
where E - modulus of cable elasticity
t - total reduced network thickness
l_1 - length of contour beam element
E_c, I_c - rigidity in bending of the contour beam element in the roof plane
The following considerations should not be taken as minor factors: the effect of additional connections of the roof surface, as roof panels, additional elements, see Fig 1.5, as well as rigid cables all over the surface or partly, connections in the surface of exterior walls, Fig 1.9, capable of taking up horizontal thrust forces of the contour beam and vertical posts-tie rods, Fig 1.6. Connecting shell panels into simultaneous work considerably reduces horizontal displacements of the contour beam, Fig 10a. Vertical posts-tie rods, Fig 1.6, increase the rigidity of the contour beam in producing cable prestressing, where $T_{0x} \leqslant T_{0y}$ and the contour beam is in the conditions of unilateral compression, but the tie rod-post working in tension prevents the contour beam compression and thus the whole system including the cable network and contour beam elements is being prestressed.

6. The following considerations can be added relative to TSs. The behaviour of TSs of the type concerned is featured by an uneven distribution of normal forces, bending and torsion moments. Tests showed that bending moments are substantial and are to be taken into account in the calculations. Contour beams, too, work in bending and torsion in addition to normal forces and shears. However, when

when designing TS sheathing deflection rather than strength is of prime importance, pointed out also by ohter authors, see Ref 4. Glued timber contour beams, sheathings of 3 or more layers or strengthened by rib shells are recommended for the span of 18...30 m and more.

CALCULATION PROBLEMS

To determine the geometry of a saddle-shape surface, specifically if $\beta \neq 1$, the condition of equilibrium of a non-loaded knot in producing cable prestressing of self-forming network surface can be used. The location of a certain knot i, k is determined by the radius vector $\bar{r}_{i,k}$, Fig 11, using the formula:

$$T_{oi}\left[\frac{\bar r_{i,k+1}-\bar r_{i,k}}{|\bar r_{i,k+1}-\bar r_{i,k}|}+\frac{\bar r_{i,k-1}-\bar r_{i,k}}{|\bar r_{i,k-1}-\bar r_{i,k}|}\right]+T_{ok}\left[\frac{\bar r_{i+1,k}-\bar r_{i,k}}{|\bar r_{i+1,k}-\bar r_{i,k}|}+\frac{\bar r_{i-1,k}-\bar r_{i,k}}{|\bar r_{i-1,k}-\bar r_{i,k}|}\right]=0$$

T_{0i}, T_{0k} - prestressing forces

$m_{i,k} = T_{0k}/T_{0i}$ - ratio of prestressing forces

Contour beam coordinates are taken as known values. To cover completely by ratios, $m_{i,k}$ can arbitrarily be predetermined only for all knots of the two crossing cables, see Ref 1.

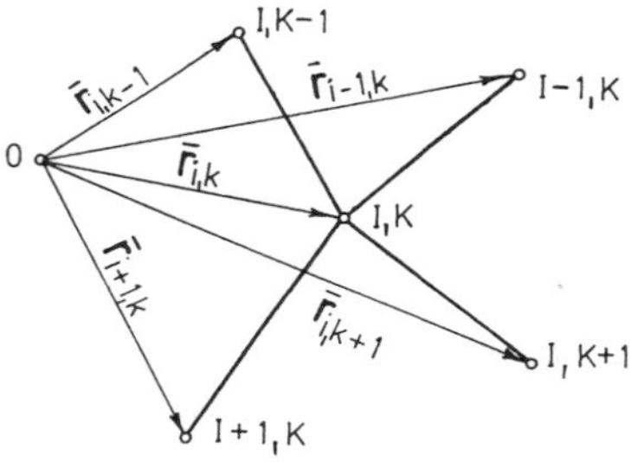

Fig 11

To determine the state of strain and stress of the prestressed or non-prestressed loaded cable network taking into account contour beam deformation, the methods given in Ref 2 can be used.

To determine the state of strain and stress of the TSs the following methods can be employed:

1. By the membrane theory for calculating shells or its modifications, the longitudinal inner forces of the contour beam and tie rod in case of thrust contour beam are determined with satisfactory accuracy and the inner forces of board surface are determined approximately.

2. By the methods for calculating suspended CSs, when substituting separate shell layers with the cables and the contour beam of the shell with the beams of the presented dimensions, the longitudinal inner forces, bending moments of the contour beam (in the corresponding cases inner forces of tie rods also), as well as deflections of the sheathing part of the shell are determined with satisfactory accuracy.

3. More accurate results have been obtained using the geometrically non-linear bending theory, solving the task by means of the finite difference method, see Ref 3.

Some results of determining inner forces of a suspended cable network, Fig 12, and deflections of TSs, Fig 13, and a comparison of calculation results with those of experimental tests are presented below.

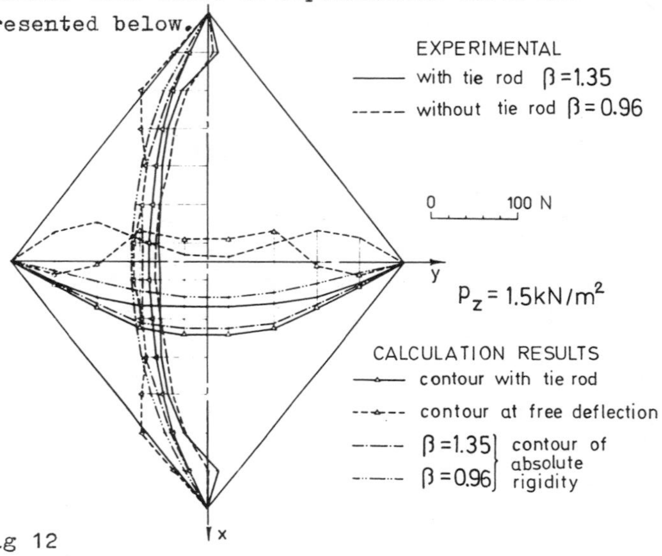

Fig 12

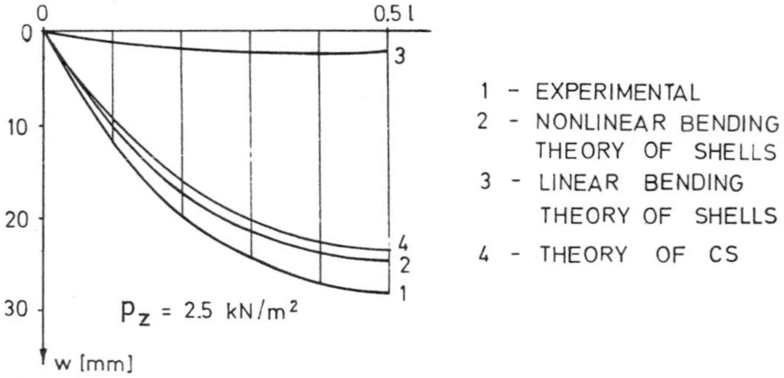

Fig 13

Designs of timber shell roofs have been worked out by the design organisations of the Estonian SSR and a number of these shells have been erected. Some shells have been in maintainance for over 13 years. Full-scale investigations of their behaviour under different maintainance conditions are being conducted.

REFERENCES

1. V Kulbach and K Ôiger, Über die Ausgangsgeometrie vorgespannter Seilnetzwerke, Proc Tallinn Technical University, No 278, 1969 (in Russian).

2. K Ôiger and A Talvik, The calculation of a nonsymmetric saddle-shape hanging roof, Proc Tallinn Technical University, No 504, 1981 (in Russian).

3. K Ôiger and T Rattasepp, Die Rechnung der Holzhyparschale mit der Berücksichtigung der grossen Durchbiegung, Proc Tallinn Technical University, No 551, 1983 (in Russian).

4. P I Moss, A I Carr and N C Cree-Brown, Large deflexion non-linear behaviour of nailed layered timber hyperbolic paraboloidal shells, Proc Instn Civ Engrs, Part 2, 1980, 69, Mar., 33-47.

TWO-STEP MATRIX ANALYSIS OF PRESTRESSED CABLE NETS

S PELLEGRINO and C R CALLADINE

Department of Engineering, University of Cambridge,
Trumpington Street, Cambridge CB2 1PZ, U.K.

We consider the behaviour of the cable-net shown in Fig 1 when arbitrary vertical forces are applied to the nodes. According to linear-algebraic analysis this net has 1 state of self-stress and 12 modes of inextensional deformation. We show that the applied loading may be decomposed into two parts. The first part does not excite any of the inextensional modes, while the second part is carried by the net through geometry changes associated with these inextensional modes. The response of the net is linear to each of these separate parts of the loading, and the analysis is done by means of 20×20 square matrices of full rank. We compute displacements for three different loading cases. We discuss, briefly, interactive effects between the two types of behaviour, and non-linearities due to stretching of the cables during 'large' deflection of the 'inextensional' modes; and we show that both of these refinements can be accommodated by means of rapidly converging iterative calculations.

I INTRODUCTION

The cable-net shown in Fig 1 consists of two sets of cables which are slung between rigid abutments, connected at the nodes, and pre-tensioned against each other. Cable nets of this sort, composed of quadrangular cells, are considerably less rigid than networks composed entirely of triangular cells. Thus, if a vertical load is applied to any one node of the net, it is carried mainly through the effects of geometrical distortion of the net. This may be demonstrated easily by means of a simple physical model. The net is relatively 'soft' in response to loads of this sort. It is also found by experiment that the relationship between the applied load and the displacement which it produces is non-linear.

The usual scheme of calculation of a cable net (see, e.g., Refs 1-7) envisages the assembly as multi-degree-of-freedom non-linear system. Several numerical schemes have been devised for solving the non-linear equations; but they all require many iterations and are consequently expensive to run on the computer.

In this paper we are concerned with a different kind of scheme for computation of the static response of a cable net to applied loading. The starting-point of our analysis is that an investigation of the 'equilibrium' and 'compatibility' equations of a net in its initial, unloaded configuration by means of classical linear algebra reveals that the assembly is a *pre-stressable mechanism* with a large number of inextensional modes of deformation. (Thus the example in Fig 1 has a single state of self-stress and 12 independent inextensional modes.) This leads directly to the idea that there are two distinct types of behaviour within the assembly. First, some patterns of loading are carried by the assembly as an ordinary structural framework, without excitation of any of the inextensional modes: the loads are balanced by changes of tension in the various cable-segments. Second, other patterns of loading are carried by virtue of the geometry-changes associated with the inextensional modes, without any change of tension in the cable-segments.

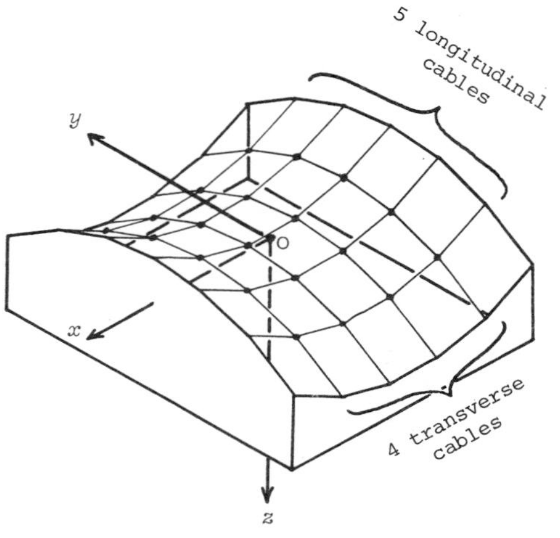

Fig 1 Saddle-shaped cable net

This scheme of classification was described first in Ref 8, where it was shown that the assembly is 'stiff' and 'soft' respectively in its two distinct types of action. It was also demonstrated that the two different types of action constitute two separate and distinct *linear* systems operating side-by-side (for sufficiently small loads, at least), each with its own eigenmodes.

For the past year we have been pursuing these ideas further, with a view to computing the behaviour of arbitrary cable nets under arbitrary static loading. We have developed computer programs which use linear algebra to generate, automatically, the states of self-stress and the inextensional modes of a given network; and we are developing other programs which perform the structural analysis of the two distinct kinds on the basis of this information.

In the present paper we give examples of the response of the net in Fig 1 to three different patterns of applied loading.

Our treatment is primarily on the lines of separate investigations of the two types of behaviour described above; but we also discuss the way in which the two types of behaviour can interact, and we introduce a simple way by which nonlinear effects due to relatively large deflection of the 'inextensional' modes may be computed. In these phases of the work it is sometimes necessary to adopt an iterative scheme of computation. However, the number of iterations required in practice appears to be very small.

It is interesting to note that the first treatment of the linear algebra of cable nets was given in a section of a paper by Buchholdt, Davies and Hussey (Ref 1). The section was self-contained, because the authors evidently saw no way of using to advantage the information generated by linear algebra in the general non-linear multi-degree-of-freedom scheme of computation which they described in the subsequent part of the paper. We have accepted the challenge implicit in the work of Ref 1; and we are now able to show how the results of the linear-algebraic approach to cable nets can be used in what promises to be an extremely efficient scheme for the computation of the behaviour of the nets.

Both Møllmann (Ref 5) and Irvine (Ref 9) have considered the behaviour of cable nets of the type shown in Fig 1, treating the assembly as a continuum governed by differential equations. The idea of decomposing the applied loading into two distinct parts does not appear in such a scheme.

II INEXTENSIONAL MECHANISMS AND 'FITTED' LOADS

For the sake of definiteness we consider a cable net consisting of 5 longitudinal 'sagging' cables and 4 transverse cables, as shown in Fig 1. The nodes all lie on the surface

$$\frac{z}{h} = -\frac{1}{6}\left(\frac{x}{\ell}\right)^2 + \frac{1}{9}\left(\frac{y}{\ell}\right)^2 \qquad \ldots\ldots \quad 1,$$

in the cartesian space whose axes are shown. The longitudinal and transverse cables lie in planes $y = 0, \pm\ell, \pm 2\ell$ and $x = \pm\frac{1}{2}\ell, \pm\frac{3}{2}\ell$ respectively, and the rigid abutments are in planes $y = \pm 3\ell$ and $x = \pm\frac{5}{2}\ell$. The 'dip' of the longitudinal cables from the abutment-points to the lowest nodes is equal to h, and so is the 'rise' of the transverse cables.

It is easy to verify that the net has a single state of self-stress (i.e. a state of stress in equilibrium with zero external load) in which the horizontal component of tension in every segment of the longitudinal cables is T_0, and in every segment of the transverse cables is $1.5T_0$. This produces a vertical reaction of $T_0h/3\ell$ between the cables at each node of the net. The magnitude of T_0 may be altered by a tensioning device at the abutments. There is only *one* state of self-stress in the sense that, for the given configuration, the tension in *every* bar may be determined by the equations of equilibrium at the joints as soon as the tension in any *one* bar is given. (The net is similar to the one described in Ref 8, except that the geometry has been altered in detail to give different levels of self-stress in the two sets of cables.)

The first step in the linear-algebraic analysis of the net is to apply the general version of 'Maxwell's rule' (Ref 10):

$$3j - b = m - s \qquad \ldots\ldots \quad 2,$$

where j is the number of nodes or 'joints' (excluding abutment points),
- b is the number of cable-segments (or 'bars': we assume throughout that every cable-segment is in tension),
- m is the number of modes of inextensional distortion of the assembly, and
- s is the number of states of self-stress.

Here $j = 20$, $b = 49$ and $s = 1$; and so

$$m = 12 \qquad \ldots\ldots \quad 3.$$

The form of the inextensional modes for this type of net (see Ref 8) is very simple. A typical mode (Fig 2) involves equal alternate up- and down-displacement at the four corner nodes of an interior cell, while all other nodes are fixed. In Fig 2 the net has been drawn as if it were plane, for the sake of clarity, but the pattern of *vertical* displacements is the same for the plane and the original net (cf. Ref 8). The four nodes of the original net also have components of displacement in the x- and y-directions; but these need not concern us here, as will be explained below. These modes on the pattern of Fig 2 are inextensional in the sense that each cable-segment suffers zero first-order extension when the displacement takes place. For this net the

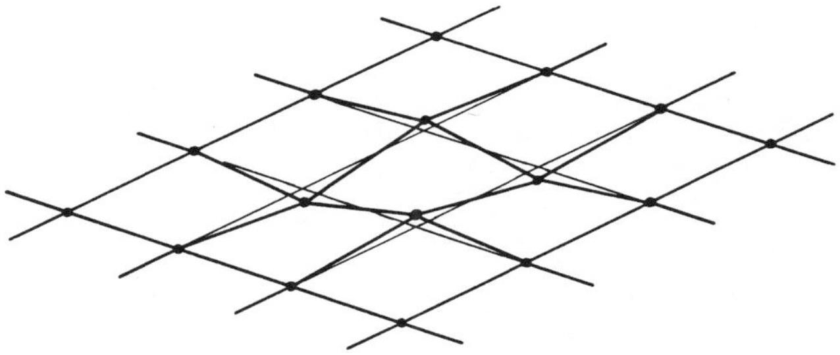

Fig 2 Inextensional mode of deformation

second-order changes in length are in fact non-zero, and so the modes are strictly inextensional only for sufficiently *small* displacements. We shall discuss the mechanical implications of this geometric feature in Section V. (Maxwell's rule, being rooted in the linear algebra of infinitesimal displacements, does not distinguish between 'free' and 'incipient' modes of inextensional displacement: see, e.g., Refs 11,12.)

The given net has exactly 12 'interior' cells. The pattern of Fig 2 may be applied at each cell in turn, making a total of 12 inextensional modes of deformation. The 12 modes are linearly independent in terms of their components of displacement, and thus any inextensional deformation of the assembly may be described as a linear combination of these 12 modes. (It is easy to show from Eqn 2 that in a more general net of the same type having m×n cables there are (m-1)×(n-1) inextensional modes on the pattern of Fig 2.)

Now the most general loading which can be applied to the net of Fig 1 has 3 components of force at each of the 20 nodes: the 'load space' has a dimension of 60. The corresponding 'displacement space' also has a dimension of 60, since there are 3 components of displacement at each joint. The linear-algebraic analysis of the net (which will be described in full elsewhere) indicates that the 60-dimensional displacement space consists of the 12-dimensional sub-space of *inextensional* displacements (described above), together with a 48 (=60 − 12) -dimensional space of *extensional* displacements. An extensional displacement of the assembly is like the deformation of an ordinary triangulated framework, in which the displacements of the joints are a direct kinematic consequence of the extensions of the bars. The main difference between a cable net and an 'ordinary' framework is that if we are to discuss the extensional modes of a cable net, we must make sure that the applied pattern of loading does not 'excite' any of the inextensional modes of the assembly. (It is necessary to check, of course, that all of the cable-segments remain in tension under a given loading.) In the language of linear algebra, the patterns of loading associated with the purely extensional modes must be *orthogonal* to each of the 12 inextensional modes. These 'fitted' loads (as Vilnay, Ref 13, calls them) fill the 48-dimensional sub-space of loads which is orthogonal to the 12-dimensional sub-space of inextensional displacements.

For the net of Fig 1 it is not difficult to obtain, by inspection, the 48 independent sets of 'fitted load' which are orthogonal to, and so do not excite, the inextensional modes. Consider first an isolated cable which is subjected

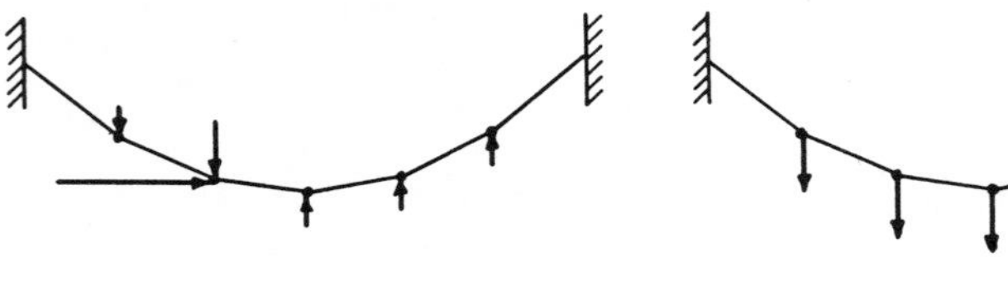

(a) (b)

Fig 3 'Fitted' loads on single cable

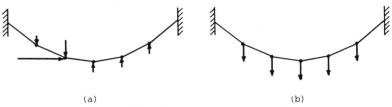

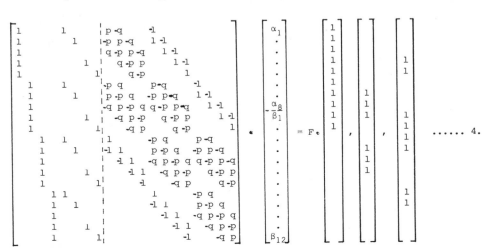

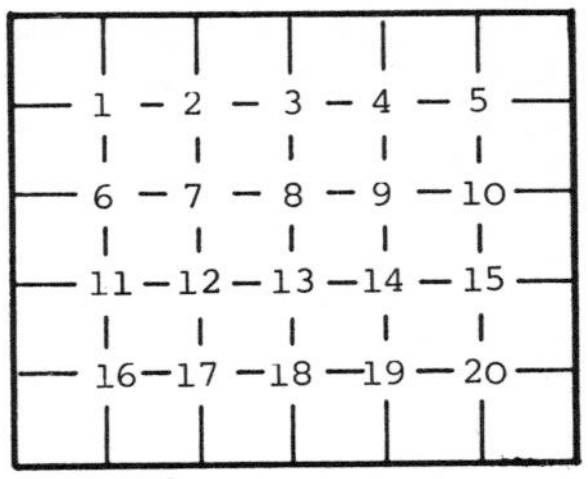

Fig 4 Numbering scheme for nodes, cf Fig 1

to a horizontal force of given magnitude at one node, as shown in Fig 3a. It is easy to find, by the application of equilibrium equations, a set of vertical loads at the joints which enables the cable to carry the applied horizontal load while remaining in its original geometrical form. This pattern of loading constitutes a set of 'fitted' loads for the entire net when a horizontal force is applied at a single node of the net. Altogether there are 40 independent patterns of 'fitted' load of this type - we use in turn the 2 cables passing through each node — and consequently only 8 more of the 48 cases of 'fitted' load remain to be determined. Since we have already considered all possible patterns of horizontal loading, these 8 cases can involve only the application of vertical forces to the nodes.

Figure 3b shows an isolated cable of the net under a set of *equal* vertical loads at the joints: when the magnitude of one of these has been assigned, the remainder follow from the equilibrium equations of the joints. In this way it seems that there will be 9 'fitted' loads of this pattern - since the net consists of 5 + 4 = 9 cables - rather than the 8 which we are seeking. This paradox is resolved by the observation that if all 5 longitudinal cables were equally loaded as in Fig 3b, there would then be equal loads on all of the nodes of the net; which would in fact be indistinguishable from a case in which all 4 transverse cables were equally loaded. It follows from this that the complete range of vertical load cases which are orthogonal to all 12 inextensional mechanisms is spanned by 'fitted' loads as in Fig 3b on any 8 of the 9 cables: it is immaterial which of the 9 is omitted for this purpose.

III 'FITTED' LOADS AND 'PRODUCT FORCES'

So far we have discussed the space of external loads as being 60-dimensional. For the sake of compactness let us now confine our attention to the case in which all of the loads applied at the joints are *vertical*. In this way we shall be concerned only with a 20-dimensional sub-space of the load space. Let us adopt the nodal numbering system shown in Fig 4. The eight 'fitted' loading cases described above may be represented by the first 8 columns of the 20×20 matrix in Eqn 4. (Cable 3-18 has been taken as the odd one.) The magnitude of any 'fitted' load is, of course, indeterminate, so it is convenient to use 1's in the 8 columns; the magnitudes are the coefficients $\alpha_1 \ldots \alpha_8$ in the postmultiplying column vector. Equation 4 states that any arbitrary loading state - here represented by a 20-element column vector on the right - can be described as the sum of the 20 column vectors of the square matrix after each has been multiplied by a coefficient α or β. If the equations can be solved uniquely, then $\alpha_1 \ldots \alpha_8$, $\beta_1 \ldots \beta_{12}$ are the components of 20 'standard' loading cases which are found in the given loading.

We have not yet discussed the 12 remaining columns in the 20×20 matrix of Eqn 4. What loading conditions do they represent? Following the preceding discussion, they must be 12 loading cases which *do* excite the inextensional modes, in contrast to the 8 'fitted' loading cases which do not.

Figure 5 shows part of a plane, square grid of cables under tension T_0 and $1.5T_0$, respectively, onto which has been imposed a typical unit case of inextensional displacement. It is clear that if $T_0 \neq 0$ some external forces are needed to maintain equilibrium. In Fig 5 the grid has been drawn plane since this simplified arrangement correctly gives (as may be shown easily; cf. Ref 8) the *vertical* forces which are required in the actual net when the inextensional mode occurs. This scheme does not include, of course, the associated horizontal forces which are required for equilibrium in the actual net; but this does not matter in the present problem, where we are considering only vertical forces, since any horizontal forces can be countered by superposition of the 40 discarded 'fitted' load cases. (In the case of a *shallow* net ($h \ll 5\ell$) the horizontal forces required for equilibrium would be small anyhow.)

The vertical forces shown in Fig 5 have been found by summing the forces needed for equilibrium of the cables separately. The forces needed for equilibrium of a single cable are found by using the equilibrium equations of the nodes in the distorted configuration. This is a *linear* calculation, since the displacements are assumed to be small; and the pretension in the elastic cables does not change in an inextensional mode of deformation. The forces in Fig 5 have been given numerical values for the sake of simplicity, but it is easy to show that the unit of these forces is $T_0 w/\ell$, where $\pm w$ is the vertical displacement of the nodes.

$$
\left[\begin{array}{cccccccc|cccccccccccc}
1 & & & & 1 & & & & p & -q & & & -1 & & & & & & & \\
1 & & & & & 1 & & & -p & p & -q & & 1 & -1 & & & & & & \\
1 & & & & & & & & q & -p & p & -q & & 1 & -1 & & & & & \\
1 & & & & & & 1 & & & q & -p & p & & & 1 & -1 & & & & \\
1 & & & & & & & 1 & & & q & -p & & & & 1 & & & & \\
 & 1 & & & 1 & & & & -p & q & & & p & -q & & & -1 & & & \\
 & 1 & & & & 1 & & & p & -p & q & & -p & p & -q & & 1 & -1 & & \\
 & 1 & & & & & & & -q & p & -p & q & q & -p & p & -q & & 1 & -1 & \\
 & 1 & & & & & 1 & & & -q & p & -p & & q & -p & p & & & 1 & -1 \\
 & 1 & & & & & & 1 & & & -q & p & & & q & -p & & & & 1 \\
 & & 1 & & 1 & & & & 1 & & & & -p & q & & & p & -q & & \\
 & & 1 & & & 1 & & & -1 & 1 & & & p & -p & q & & -p & p & -q & \\
 & & 1 & & & & & & & -1 & 1 & & -q & p & -p & q & q & -p & p & -q \\
 & & 1 & & & & 1 & & & & -1 & 1 & & -q & p & -p & & q & -p & p \\
 & & 1 & & & & & 1 & & & & -1 & & & -q & p & & & q & -p \\
 & & & 1 & 1 & & & & & & & & 1 & & & & -p & q & & \\
 & & & 1 & & 1 & & & & & & & -1 & 1 & & & p & -p & q & \\
 & & & 1 & & & & & & & & & & -1 & 1 & & -q & p & -p & q \\
 & & & 1 & & & 1 & & & & & & & & -1 & 1 & & -q & p & -p \\
 & & & 1 & & & & 1 & & & & & & & & -1 & & & -q & p
\end{array}\right]
\cdot
\left[\begin{array}{c}
\alpha_1 \\ \cdot \\ \cdot \\ \cdot \\ \cdot \\ \cdot \\ \cdot \\ \alpha_8 \\ \hline \beta_1 \\ \cdot \\ \cdot \\ \cdot \\ \cdot \\ \cdot \\ \cdot \\ \cdot \\ \cdot \\ \cdot \\ \cdot \\ \beta_{12}
\end{array}\right]
= F \cdot
\underset{\text{(i)}}{\left[\begin{array}{c}
1 \\ 1 \\ 1 \\ 1 \\ 1 \\ 1 \\ 1 \\ 1 \\ 1 \\ 1 \\ \\ \\ \\ \\ \\ \\ \\ \\ \\ \\
\end{array}\right]},
\underset{\text{(ii)}}{\left[\begin{array}{c}
\\ \\ \\ \\ \\ \\ 1 \\ 1 \\ 1 \\ \\ \\ 1 \\ 1 \\ 1 \\ \\ \\ \\ \\ \\ \\
\end{array}\right]},
\underset{\text{(iii)}}{\left[\begin{array}{c}
\\ \\ \\ 1 \\ 1 \\ \\ \\ \\ 1 \\ 1 \\ 1 \\ 1 \\ \\ \\ \\ 1 \\ 1 \\ \\ \\ \\
\end{array}\right]}
\qquad \ldots\ldots 4.
$$

Key: p=7.5, q=1.5

There is a set of loads of this kind associated with each of the 12 distinct inextensional modes; and it is these 12 sets which appear in columns 9-20 of the matrix in Eqn 4. We call these the 'product forces' of the net. This is not an ideal descriptive term, but it does express the idea that the forces concerned are a sort of product of the

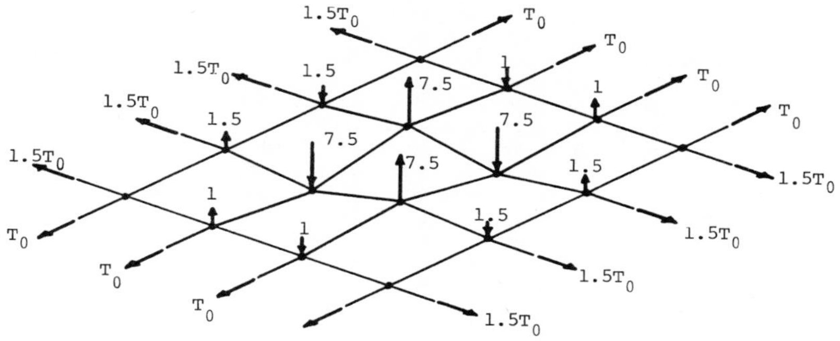

Fig 5 'Product' forces for mode of Fig 2

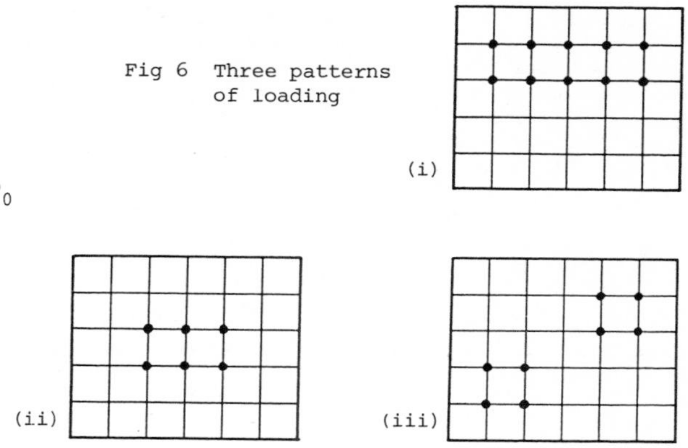

Fig 6 Three patterns of loading

state of prestress and an inextensional mode.

The matrix is in fact of full rank, and so the equations have a unique solution for any given loading case. Three particular loading cases (i-iii) are given in Eqn 4, corresponding to the three loading conditions shown in Fig 6. In each case a node which is loaded carries a vertical force of magnitude F in the downward direction ($z > 0$) of Fig 1.

The solution of Eqn 4 is given in Eqn 5. It is easy to see that case (i) is a simple combination of 2 'fitted' loads, so that in particular $\beta_1 \ldots \beta_{12} = 0$. In case (ii) the applied load decomposes into 8 'fitted' loads and 8 'product forces', though the latter have small magnitude in comparison with the two leading 'fitted' loads. Lastly, in case (iii), it is the 'product forces' which have the dominant coefficients.

IV DISPLACEMENTS OF THE CABLE NET

Having decomposed a given set of applied loads into 'fitted' loads and 'product forces', we can now examine the displacement of the assembly under each of the 20 component loading conditions separately, and then take the sum. This operation is shown in Eqn 6. Here $w_1 \ldots w_{20}$ are the vertical displacements of the 20 nodes. The 20×20 matrix has 8 columns corresponding to the 'fitted' loads and 12 for the inextensional mechanisms. The latter are self-explanatory in terms of Fig 2. The derivation of the first 8 columns is straightforward, and we do not give full details here. Briefly, we first calculate the tension in every cable segment when a given fitted load is applied to the cable net, either by using the method given in Ref 8 or by a virtual work calculation in which the state of prestress is used as a 'dummy load' condition: the problem involves a single degree of statical indeterminacy. Then we find the (strictly, mean) vertical displacement of the nodes of each cable in turn by a simple application of virtual work. (In evaluating these displacements we have, for the sake of computational simplicity, taken the cross-sectional area of the horizontal cable-segments as A_0, and of the inclined segments as $A_0 \sec^3\theta$, where θ is the angle of inclination. In this way the calculation becomes independent of the value of h/ℓ. E is the Young's modulus of the material of the cables.)

The matrix in Eqn 6 is post-multiplied by a column vector which contains $\alpha_1 \ldots \alpha_8$, $\beta_1 \ldots \beta_{12}$ as before, but now with

$$\begin{bmatrix} \alpha_1 \\ \cdot \\ \cdot \\ \cdot \\ \cdot \\ \cdot \\ \cdot \\ \alpha_8 \\ \hline \beta_1 \\ \cdot \\ \cdot \\ \cdot \\ \cdot \\ \cdot \\ \cdot \\ \cdot \\ \cdot \\ \cdot \\ \cdot \\ \beta_{12} \end{bmatrix} = \frac{F}{10} \cdot \begin{bmatrix} 10. \\ 10. \\ 0 \\ 0 \\ 0 \\ 0 \\ 0 \\ 0 \\ \hline 0 \\ 0 \\ 0 \\ 0 \\ 0 \\ 0 \\ 0 \\ 0 \\ 0 \\ 0 \\ 0 \\ 0 \end{bmatrix} ; \begin{bmatrix} 1.92 \\ 8.55 \\ 8.55 \\ 1.92 \\ -5.49 \\ -0.09 \\ -0.09 \\ -5.49 \\ \hline 0.52 \\ 0.23 \\ -0.23 \\ -0.52 \\ 0 \\ 0 \\ 0 \\ 0 \\ -0.52 \\ -0.23 \\ 0.23 \\ 0.52 \end{bmatrix} ; \begin{bmatrix} 0 \\ 0 \\ 0 \\ 0 \\ 5. \\ 5. \\ 5. \\ 5. \\ \hline -1.86 \\ -3.72 \\ -3.72 \\ -1.86 \\ -3.39 \\ -6.78 \\ -6.78 \\ -3.39 \\ -1.86 \\ -3.72 \\ -3.72 \\ -1.86 \end{bmatrix} \qquad \ldots\ldots 5.$$

$$\begin{bmatrix} w_1 \\ \cdot \\ \cdot \\ \cdot \\ \cdot \\ \hline w_6 \\ \cdot \\ \cdot \\ \cdot \\ \cdot \\ \hline w_{11} \\ \cdot \\ \cdot \\ \cdot \\ \cdot \\ \hline w_{16} \\ \cdot \\ \cdot \\ \cdot \\ \cdot \end{bmatrix} = \left[\begin{array}{cccccccc|cccccccccccc}
\kappa & \lambda & \lambda & \lambda & \mu & \nu & \nu & \nu & 1 & & & & & & & & & & & \\
\kappa & \lambda & \lambda & \lambda & \nu & \mu & \nu & \nu & -1 & 1 & & & & & & & & & & \\
\kappa & \lambda & \lambda & \lambda & \nu & \nu & \nu & \nu & & -1 & 1 & & & & & & & & & \\
\kappa & \lambda & \lambda & \lambda & \nu & \nu & \mu & \nu & & & -1 & 1 & & & & & & & & \\
\kappa & \lambda & \lambda & \lambda & \nu & \nu & \nu & \mu & & & & -1 & & & & & & & & \\
\hline
\lambda & \kappa & \lambda & \lambda & \mu & \nu & \nu & \nu & -1 & & & & 1 & & & & & & & \\
\lambda & \kappa & \lambda & \lambda & \nu & \mu & \nu & \nu & 1 & -1 & & & -1 & 1 & & & & & & \\
\lambda & \kappa & \lambda & \lambda & \nu & \nu & \nu & \nu & & 1 & -1 & & & -1 & 1 & & & & & \\
\lambda & \kappa & \lambda & \lambda & \nu & \nu & \mu & \nu & & & 1 & -1 & & & -1 & 1 & & & & \\
\lambda & \kappa & \lambda & \lambda & \nu & \nu & \nu & \mu & & & & 1 & & & & -1 & & & & \\
\hline
\lambda & \lambda & \kappa & \lambda & \mu & \nu & \nu & \nu & & & & & -1 & & & & 1 & & & \\
\lambda & \lambda & \kappa & \lambda & \nu & \mu & \nu & \nu & & & & & 1 & -1 & & & -1 & 1 & & \\
\lambda & \lambda & \kappa & \lambda & \nu & \nu & \nu & \nu & & & & & & 1 & -1 & & & -1 & 1 & \\
\lambda & \lambda & \kappa & \lambda & \nu & \nu & \mu & \nu & & & & & & & 1 & -1 & & & -1 & 1 \\
\lambda & \lambda & \kappa & \lambda & \nu & \nu & \nu & \mu & & & & & & & & 1 & & & & -1 \\
\hline
\lambda & \lambda & \lambda & \kappa & \mu & \nu & \nu & \nu & & & & & & & & & -1 & & & \\
\lambda & \lambda & \lambda & \kappa & \nu & \mu & \nu & \nu & & & & & & & & & 1 & -1 & & \\
\lambda & \lambda & \lambda & \kappa & \nu & \nu & \nu & \nu & & & & & & & & & & 1 & -1 & \\
\lambda & \lambda & \lambda & \kappa & \nu & \nu & \mu & \nu & & & & & & & & & & & 1 & -1 \\
\lambda & \lambda & \lambda & \kappa & \nu & \nu & \nu & \mu & & & & & & & & & & & & 1
\end{array}\right] \cdot \begin{bmatrix} A\alpha_1 \\ \cdot \\ \cdot \\ \cdot \\ \cdot \\ \cdot \\ \cdot \\ A\alpha_8 \\ \hline B\beta_1 \\ \cdot \\ \cdot \\ \cdot \\ \cdot \\ \cdot \\ \cdot \\ \cdot \\ \cdot \\ \cdot \\ \cdot \\ B\beta_{12} \end{bmatrix}$$

$$= \underbrace{\begin{bmatrix} 16.0 \\ 16.0 \\ 16.0 \\ 16.0 \\ 16.0 \\ \hline 16.0 \\ 16.0 \\ 16.0 \\ 16.0 \\ 16.0 \\ \hline -8.3 \\ -8.3 \\ -8.3 \\ -8.3 \\ -8.3 \\ \hline -8.3 \\ -8.3 \\ -8.3 \\ -8.3 \\ -8.3 \end{bmatrix} AF + \begin{bmatrix} 0 \\ 0 \\ 0 \\ 0 \\ 0 \\ \hline 0 \\ 0 \\ 0 \\ 0 \\ 0 \\ \hline 0 \\ 0 \\ 0 \\ 0 \\ 0 \\ \hline 0 \\ 0 \\ 0 \\ 0 \\ 0 \end{bmatrix} \frac{BF}{10}}_{\text{(i)}} ; \underbrace{\begin{bmatrix} -9.41 \\ -3.33 \\ -3.23 \\ -3.33 \\ -9.41 \\ \hline 6.69 \\ 12.77 \\ 12.87 \\ 12.77 \\ 6.69 \\ \hline 6.69 \\ 12.77 \\ 12.87 \\ 12.77 \\ 6.69 \\ \hline -9.41 \\ -3.33 \\ -3.23 \\ -3.33 \\ -9.41 \end{bmatrix} AF + \begin{bmatrix} 0.52 \\ -0.29 \\ -0.47 \\ -0.29 \\ 0.52 \\ \hline -0.52 \\ 0.29 \\ 0.47 \\ 0.29 \\ -0.52 \\ \hline -0.52 \\ 0.29 \\ 0.47 \\ 0.29 \\ -0.52 \\ \hline 0.52 \\ -0.29 \\ -0.47 \\ -0.29 \\ 0.52 \end{bmatrix} \frac{BF}{10}}_{\text{(ii)}} ; \underbrace{\begin{bmatrix} 4.20 \\ 4.20 \\ -1.42 \\ 4.20 \\ 4.20 \\ \hline 4.20 \\ 4.20 \\ -1.42 \\ 4.20 \\ 4.20 \\ \hline 4.20 \\ 4.20 \\ -1.42 \\ 4.20 \\ 4.20 \\ \hline 4.20 \\ 4.20 \\ -1.42 \\ 4.20 \\ 4.20 \end{bmatrix} AF + \begin{bmatrix} -1.86 \\ -1.86 \\ 0 \\ 1.86 \\ 1.86 \\ \hline -1.53 \\ -1.53 \\ 0 \\ 1.53 \\ 1.53 \\ \hline 1.53 \\ 1.53 \\ 0 \\ -1.53 \\ -1.53 \\ \hline 1.86 \\ 1.86 \\ 0 \\ -1.86 \\ -1.86 \end{bmatrix} \frac{BF}{10}}_{\text{(iii)}} \qquad \ldots\ 6.$$

Key: $\kappa=20.15$, $\lambda=-4.15$, $\mu=10.54$, $\nu=-0.71$, $A=\ell^3/h^2A_0E$, $B=\ell/T_0$

multipliers A and B. The factor A involves the modulus of elasticity, etc., of the cable segments, while B involves only the level of prestress in the net. On the RHS of Eqn 6 the displacements have been evaluated from the values of α, β in Eqn 5 for the three loading cases, and with the extensional and inextensional moieties separately. Notice that in each case the displacement is proportional to F; but observe how the relative magnitude of the two parts depends in particular on the magnitude of the factor

$$\frac{B}{A} = \left(\frac{h}{\ell}\right)^2 \frac{A_0 E}{T_0} \qquad \ldots\ldots \quad 7.$$

Now a typical cable-net might have a rise/span ratio of 0.1, i.e. here $h/5.5\ell = 0.1$; and the prestress might have a value of T_0 which gives a tensile strain $\varepsilon \simeq 10^{-3}$, and thus $A_0E/T_0 \simeq 1000$. For such a net, therefore,

$$\frac{B}{A} = 300 \qquad \ldots\ldots \quad 8.$$

Inspection of the numbers in the RHS columns of Eqn 8 with this value of B/A reveals that in loading case (ii) (Fig 6) the displacements due to extensional and inextensional effects are of the same order, while in case (iii) the inextensional effects are dominant.

Isometric plots of vertical displacement for loading cases (ii) and (iii) are shown in Fig 7. The nodes which carry forces are marked. The value of F in case (ii) has been set at 4/3 of that in case (iii), so that the *total* vertical load is the same in both cases: total load = 8F.

It is clear that the net deflects more when the load is disposed on a diagonal, as in (iii), than when it is applied centrally, as in (ii). The displacements corresponding to case (i) are not illustrated. They are much smaller in magnitude (see Eqn 6), since the inextensional modes are not excited.

V DISCUSSION

Our example provides a clear illustration of the way in which the two kinds of behaviour in cable nets may be analysed by means of linear algebra. For the sake of simplicity, and at the expense of some precision, we have dealt only with the 20-dimensional subspace of vertical loads and displacements. But the same principles apply equally to the full 60-dimensional version, and the overall results do not differ by much. As we mentioned earlier, our example has the convenient feature that we can find the columns of the relevant matrices by inspection, and at the expense of little effort. In a less straightforward practical problem, of course, the computer would *generate* the relevant matrices automatically from the data on the form of the net, by use of the program mentioned in Section I.

The calculations described in the preceding sections cannot be regarded as complete as they stand, however, for two reasons. First, we have calculated the 'product force' columns in the matrix of Eqn 4 on the assumption that the horizontal components of the tension in the cable segments are T_0 and $1.5T_0$, respectively. But it is clear that, in general, the imposition of the 'fitted' part of the applied load *changes* the tensions to some extent, as described in Section IV. If the magnitude of the applied load is such that the tensions are changed from their original values by, say, more than 20%, then the changed tensions ought to be used in the calculation of the last 12 columns of the matrix of Eqn 4. This is somewhat problematical, since the decomposition of the applied load into the two classes of 'fitted' load and 'product forces' cannot be accomplished until the matrix is known; and yet the columns cannot be finalised until the decomposition is known. Some simple iterative calculations which we have performed on these lines suggest that the process converges rapidly. For example, only one iteration is required to complete the calculation for our loading case (iii).

The second question concerning the correctness of the results comes from the observation that the 'inextensional' displacements are only truly inextensional for sufficiently small displacements.

Consider, for example, the cable shown in Fig 8a. The displacement shown has $w/\ell = 0.1$, so the two segments rotate by arcsin 0.1 = 5.74°. This involves an overall stretching of the cable which is equal to (sec 5.74° - 1) = 0.005 of the length of the two segments concerned; or $\frac{2}{5}$ of this, i.e. 0.002, of the length of the entire cable. Thus, if the original level of prestress in the cable represents a strain of 0.001, the displacement shown in Fig 8a would treble the prestressing tension. In general, the load-displacement relationship is non-linear, since T increases with w.

Now in relation to the transverse forces on the cable which are necessary to produce the displacement shown, the previous calculation of 'product forces' would be correct if only we could determine the *current* level of prestress, taking into account stretching effects of the kind just described. The calculation above suggests that we would need to investigate the angular rotation of all of the cable-elements in the net. It turns out, however, that this is not necessary; and that the calculation can be performed easily, in one step, as soon as the linearised computation has been performed.

Consider again the example shown in Fig 8a, with the cable at a prestressing tension T_0. Further, suppose that the original elastic cable is replaced by a *perfectly plastic* one having a *constant* yield tension T_0. When the transverse force P is now applied the tension remains at T_0 since T_0 is now a property of the cable. In this case,

$$P = 2T_0 w/\ell \qquad \ldots\ldots \quad 9,$$

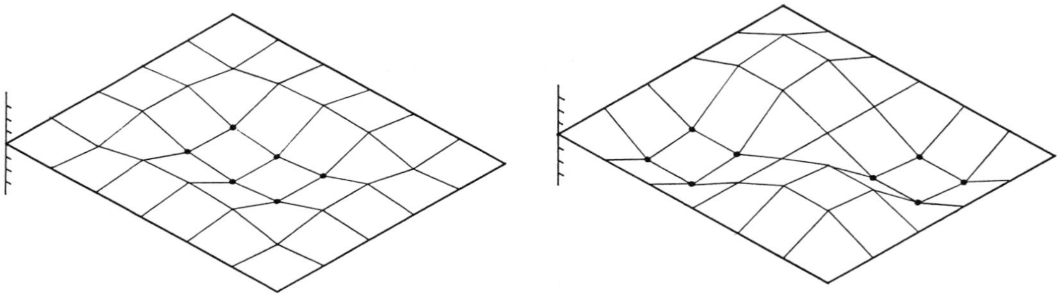

(ii) (iii)

Fig 7 Plot of vertical displacements, cf. Fig 6. Unit for vertical scale: $F\ell/10T_0$.

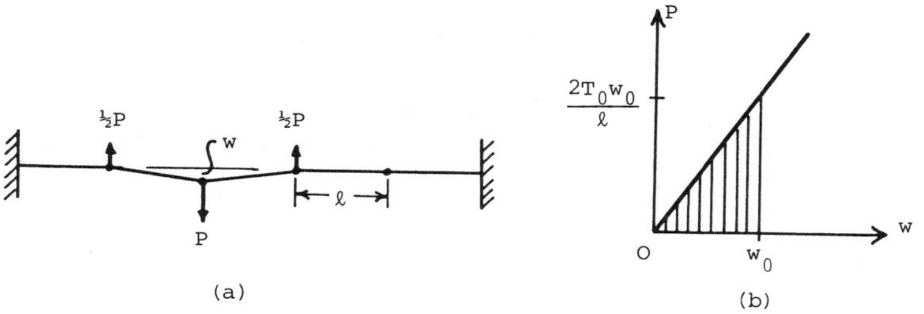

Fig 8 Computation of the extension of a cable

when we neglect terms of order $T_0(w/\ell)^3$. This linear relationship between P and w is shown in Fig 8b.

Now suppose that P has increased steadily until $w = w_0$. The total work done on the wire by the applied transverse load is equal to the shaded area, *viz*.

$$T_0 w_0{}^2/\ell \qquad \ldots\ldots \quad 10.$$

The only way in which the cable can absorb this work is by stretching; and since the tension remains constant at T_0, the overall extension e of cable is given by

$$e = (T_0 w_0{}^2/\ell) \div T_0 = \frac{w_0{}^2}{\ell} \qquad \ldots\ldots \quad 11.$$

The original length of cable is 5ℓ, and hence the overall strain ε in the cable due to the distortion from its straight original configuration is given by,

$$\varepsilon = 0.2(w_0/\ell)^2$$
$$= 0.002 \quad \text{when } w_0/\ell = 0.1 \qquad \ldots\ldots \quad 12.$$

The result is the same as before.

We have used here the artifice of a 'perfectly plastic' cable in order to perform a *geometrical* calculation; and the result is therefore valid even when the cable is elastic.

The same kind of calculation can be used to find the increase in the level of pretension in an entire cable net which is subjected to a general pattern of loading. Thus, starting with the total work done by the 'product forces' according to the linear calculation (as described, e.g., in Eqns 4,6) and using the virtual-work method outlined in Section IV, we can derive a single factor by which the level of prestress would actually be increased if the calculated displacement were to occur. An important point here is that our cable net has only *one* state of self-stress, and so the non-linear behaviour is described very simply by means of an increase in this single quantity.

This calculation indicates that larger loads are needed to produce the deflections which were computed by the linearised scheme for given loads. It is therefore necessary to do an iteration in order to find the correct displacements for a given load. This iteration, however, is concerned only with the solution of a cubic equation in a single variable, and is very straightforward.

REFERENCES

1. H A BUCHHOLDT, M DAVIES and M J L HUSSEY, The analysis of cable nets, *Jl Inst. Maths Applics* vol 4, 339-358, 1968.

2. D P GREENBERG, Inelastic analysis of suspension roof structures, *Proc. ASCE, Jl of Structural Division* vol 96, 905-930, 1970.

3. F BARON and M S VENKATESAN, Nonlinear analysis of cable and truss structures, *Proc. ASCE, Jl of Structural Division* vol 97, 679-710, 1971.

4. J H ARGYRIS and D W SCHARPF, Large deflection analysis of prestressed networks, *Proc. ASCE, Jl of Structural Division* vol 98, 633-654, 1972.

5. H MØLLMANN, *Analysis of hanging roofs by means of the displacement method*, Polyteknisk Forlag, Lyngby, 1974.

6. A H PEYROT and A M GOULOIS, Analysis of cable structures, *Computers and Structures* vol 10, 805-813, 1979.

7. R L WEBSTER, On the static analysis of structures with strong geometric nonlinearity, *Computers and Structures* vol 11, 137-145, 1980.

8. C R CALLADINE, Modal stiffnesses of a pretensioned cable net, *Int. Jl Solids and Structures* vol 18, 829-846, 1982.

9. H M IRVINE, *Cable Structures*, MIT Press, Cambridge, Massachusetts, 1981.

10. C R CALLADINE, Buckminster Fuller's 'Tensegrity' structures and Clerk Maxwell's rules for the construction of stiff frames, *Int. Jl Solids and Structures* vol 14, 161-172, 1978.

11. T TARNAI, Simultaneous static and kinematic indeterminacy of space trusses with cyclic symmetry, *Int. Jl Solids and Structures* vol 16, 347-359, 1980.

12. E N KUZNETSOV, Statical-kinematic analysis and limit equilibrium of systems with unilateral constraints, *Int. Jl Solids and Structures* vol 15, 761-767, 1979.

13. O VILNAY, private communication.

DYNAMIC RESPONSE OF INFLATABLE STRUCTURE INTERACTING WITH AIR

O G VINOGRADOV, PhD

Associate Professor, Department of Mechanical Engineering,
The University of Calgary, Calgary, Alberta, Canada.

The problem of oscillation of a cable-reinforced inflatable structure subjected to a static wind pressure and interacting with the surrounding air is addressed. The approach is based on the discretization technique with cable elements under the constant pressure being used as finite elements while the acoustic pressure field is approximated by a piecewise constant pressure field. A numerical example of a cable-reinforced spherical cap is supplemented by an analytical estimation of fundamental frequency of oscillations.

INTRODUCTION

The behavior of a highly flexible air-supported structure is significantly different from that of a rigid one. It has been shown that the initial stresses, see Ref 1, internal pressure and statically applied loads, see Ref 2, including wind loads, see Ref 3, affect the natural frequencies of inflatables. Still, there is another distinctive feature of inflatables, namely, their light-weightness, which implies that the influence of the surrounding air may essentially affect their dynamic response.

In general an inflatable structure is geometrically nonlinear. Because of that in designing inflatables all factors affecting their structural behavior (internal pressure, static loads, wind loads, geometry, etc.) should be taken into account simultaneously. The results presented in this paper are based on a computer program developed by the author for analysing static and dynamic behavior of cable-reinforced inflatable structures subjected to internal pressure, static loads, wind loads and acoustic pressure. However, only the influence of the acoustic pressure and static wind pressure is emphasized herein.

The approach is based on a finite element discretization with a cable element under the constant pressure being considered as a finite element, see Ref 4. The assumptions and basic features of the computer code have been outlined in Ref 2 and are only briefly repeated here for completeness.

An inflatable cable-reinforced structure is modelled by a cable network with the action of the membrane neglected, except for transmitting the pressure load to the cables. The pressure load is due to internal pressure and wind pressure. The overall pressure load is considered as distributed along cable elements while all other loads, including inertia forces in the dynamic problem and forces of interaction with the surrounding air, are concentrated at nodes. The cables are considered as elastic elements undergoing large displacements.

A cable network is approximated by a discretized finite element scheme. An approach for cables loaded by pressure suggested in Ref 4 is based on a pressure approximation scheme. Basically this approach makes use of the fact that under the constant pressure a cable has a constant radius of curvature. Considered as a two-node finite element it has some advantages compared to a conventional approach. Firstly, in a conventional approach an extra node would be needed to describe a constant curvature shape. Another advantage is associated with the number of elements sufficient to approximate the structure. With a finite element suggested the procedure is straight forward because the number of elements is determined from the piecewise approximation of a given pressure function, while with a conventional element the sufficient number of elements remains less definite.

The computer program handles a nonlinear static analysis and small, forced oscillations of the deformed structure. The geometrically nonlinear static problem is solved by an iterative procedure based on relaxation and load increment techniques simultaneously. When, in a defined sense, the convergence is achieved the program switches to the dynamical part of analysis.

The small, forced oscillations are investigated by transforming the static stiffness matrix, corresponding to the equilibrium shape of the structure, into the dynamic complex stiffness matrix, which includes the inertia terms due to concentrated masses, and terms due to attached masses and attached damping.

Some simplifying assumptions concerning the wind properties are made. Firstly, the influence of change in shape of the structure on the wind pressure distribution is neglected. Naturally a check should always be carried out that the shape of the structure did not change appreciably as a result of the wind loading. Secondly, the wind is considered to have a uniform flow with a steady mean velocity and high frequency fluctuations. These assumptions allow the data for flow past rigid bodies to be used, see Ref. 5.

In the air-structure interaction problem it is assumed that the acoustic pressure on a vibrating membrane can be approximated by a piecewise constant pressure field in such a way that the pressure in the vicinity of each node is constant. Essentially, it means that a membrane segment in the vicinity of a given node oscillates uniformly, and the problem is, thus, reduced to an interaction of a uniformly vibrating membrane element with the surrounding air.

A numerical example of a spherical cap consisting of 14 reinforcing cables lying along geodesic lines is given. It is shown that the frequency response spectrum is significantly affected when structure-air interaction is taken into account: all but the lowest frequency are suppressed by high damping and the fundamental frequency is shifted down the frequency range. The interaction with the wind produces another effect: there is a "jump" in structure stiffness when wind speed exceeds some limit.

An analytical estimation of fundamental frequency with and without energy radiation is given. It is in agreement with the results obtained numerically and allows the qualitative analysis of influence of various parameters on the fundamental frequency and corresponding amplitude of vibrations to be made.

STRUCTURE-AIR INTERACTION

In modelling the structure-air interaction it is assumed that the acoustic pressure on a pulsating membrane can be approximated by a piece-wise constant pressure field. This assumption is in accordance with piece-wise constant approximation of internal and wind pressure fields. The acoustic pressure is assumed to be constant on a segment of a membrane associated with a given node. It, then, follows that each segment oscillates uniformly and, thus, the problem of structure-air interaction is reduced to the problem of uniformly vibrating segments interacting with the air.

From the practical point of view the usually synclastic shapes of inflatables can always be locally approximated, even in a deformed state, by either spherical or cylindrical segments. This further simplifies the problem.

The oscillations of a membrane segment radiating energy in the surrounding air are described by the following equation

$$m_s\ddot{u} + q\dot{u} + ku = (-SP_a + F)\exp(i\omega t) \qquad \ldots\ldots 1,$$

where m_s = mass of a membrane segment, q = damping coefficient associated with the structural losses in the membrane, k = stiffness coefficient, S = surface of the membrane segment, P_a = amplitude of the acoustic pressure, F = amplitude of the exciting harmonic force, ω = circular frequency, $i = \sqrt{-1}$. Expressions for acoustic pressure, attached mass and attached damping for both cylindrical and spherical segments are given next.

Cylindrical Membrane. In this case the amplitude of the acoustic pressure is equal to, see Ref 6

$$p_a = \frac{2\rho c\dot{u}}{A_o}\exp(i\gamma_o)(J_o(kr) - i\,N_o(kr)) \qquad \ldots\ldots 2.$$

Constants A_o and γ_o can be found from the relationships

$$0.5\,A_o\sin\gamma_o = J_1 \qquad 0.5\,A_o\cos\gamma_o = -N_1 \qquad \ldots\ldots 3.$$

Where J_o, J_1 = Bessel functions of order zero and first, respectively, N_o, N_1 = Newman functions of order zero and first, respectively, kr = parameter, $k = \omega/c$ = wave number, r = radius of the cylinder, c = speed of sound, ρ = air density.

In the case of inflatable structures instead of Eqn 2 a simplified one can be used. It is known that most of the energy in a turbulent wind is concentrated in the frequency range of 0.1 to 1.0 Hz. Then parameter kr is within the range kr = (0.002 ÷ 0.02)r, where r is in meters. If kr << 1 (long waves) then simplified expressions for Bessel and Newman functions are available, see Ref 7, which allow P_a to be expressed in the form

$$P_a = \frac{\pi\rho ckr}{2}\left[\dot{u} - \frac{\ddot{u}}{\omega}\frac{2}{\pi}\ln kr\right] \qquad \ldots\ldots 4,$$

where the relationship $\ddot{u} = i\omega\dot{u}$ has been used. Substitution of the latter in Eqn 1 leads to the following expressions for the attached mass and damping, respectively, associated with a given node

$$m_a = S\rho r\ln\frac{c}{\omega r} \qquad \ldots\ldots 5,$$

$$q_a = 0.5\pi\ S\rho r\omega \qquad \ldots\ldots 6.$$

Spherical Membrane. In this case the acoustic pressure on a pulsating spherical cone is equal to, see Ref 6

$$P_a = \rho c\dot{u}\cos\psi\cdot\exp(i\psi) \qquad \ldots\ldots 7,$$

where

$$\cos\psi = \frac{kr}{\sqrt{1 + k^2r^2}}, \quad \sin\psi = \frac{1}{\sqrt{1 + k^2r^2}} \qquad \ldots\ldots 8.$$

Sutstitution of Eqn 7 into Eqn 1 leads to the following expressions for the attached mass and damping, respectively

$$m_a = S\rho c\frac{\sin\psi\cdot\cos}{\omega} \qquad \ldots\ldots 9,$$

$$q_a = S\rho c\cos^2\psi \qquad \ldots\ldots 10.$$

For long waves (kr << 1) the latter can be simplified

$$m_a = S\rho r \qquad \ldots\ldots 11,$$

$$q_a = S\rho r^2\omega^2/c \qquad \ldots\ldots 12.$$

Stiffness Matrix. The conventional dynamic complex stiffness matrix has the form

$$K = D - \omega^2 M \qquad \ldots\ldots 13,$$

where $D = D_1 + iD_2$ - complex static stiffness matrix corresponding to a system with structural damping; M = mass matrix.

Due to acoustic pressure an additional force acts along the normal to the membrane at each node j and results in an additional complex dynamic stiffness

$$t_j = (-m_a^j\,\omega^2 + i\,q_a^j\,\omega)\cos\alpha_j \qquad \ldots\ldots 14,$$

where $\cos\alpha_j$ = direction cosine of a normal with a particular axis in a global system of coordinates.

Supplemented by these additional diagonal terms the dynamic complex stiffness matrix becomes

$$K_a = D_1 + i(D_2 + \omega Q_a) - \omega^2(M + M_a) \qquad \ldots\ldots 15,$$

where M_a, Q_a = diagonal matrices, $M_a = \{m_a^j\cos\alpha_j\}$, $Q_a = \{q_a^j\cos\alpha_j\}$.

The essential point here is that additional stiffness due to acoustic pressure is not only frequency but also coordinates dependent.

NUMERICAL SIMULATION

The FORTRAN computer program has been written to simulate the behavior of the structure subjected to internal pressure, wind pressure and concentrated loads, and to determine its dynamic response in a deformed state, taking into account the attached masses and attached damping due to interaction with the air.

The equilibrium shape of the structure is found by an iterative process, which starts from an initial geometry of the structure when subjected to a small internal uniform pressure while concentrated loads are neglected. At each iteration the method of assembling the overall stiffness matrix is based on that used in the SAPIV computer program of static linear analysis, see Ref 8, modified and supplemented by new subroutines such as defining the incremental stiffness matrix, the pressure distribution over the structure due to wind action, pressure transmitted to the cables, etc.

When an equilibrium shape is found with a given accuracy the program switches to the dynamical part in which the response to harmonic forces applied at nodes is determined. In this part first the static stiffness matrix is transformed into the dynamic complex stiffness matrix, see Ref 2, and then accounting for attached mass and damping is done at each node according to Eqn 15. The dynamic response is found by using the same subroutines as in static analysis but adjusted to a complex algebra.

Numerical Example. Numerical results are given for a cable-reinforced spherical cap shown in Fig 1 with the same dimensions as in Ref 2. There are 14 reinforcing cables coinciding with geodesic lines.

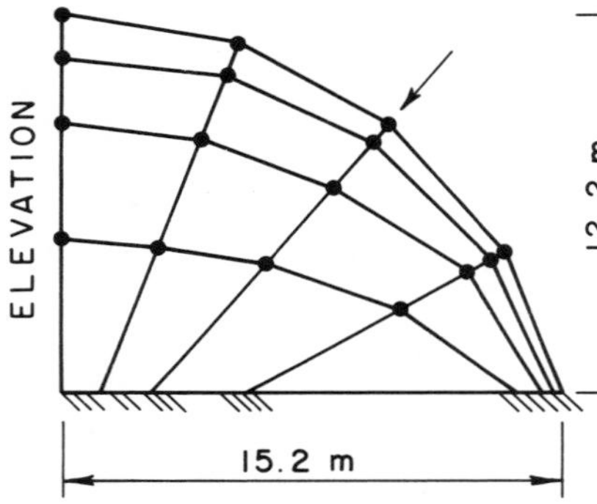

Fig 1 Fourteen-Cable Reinforcement of Spherical Cap ($EA = 13.3 \times 10^4$ KN)

The static load has been taken as the dead weight of cables applied at nodes. The mass of each cable is also concentrated at nodes, and due to symmetry, all calculations are done for one-half the structure, comprising 60 cable elements and 28 masses, having 77 degrees of freedom in total. A harmonic force is applied normally to the surface as shown in Fig 1.

In Fig 2 the comparison of the dynamic response of the cap with and without an interaction with the surrounding air is made for an internal pressure of 150 Pa (15 mm of water) and zero wind velocity. As can be seen the fundamental frequency of the structure and the amplitude of oscillations drops substantially when attached masses and damping are taken into account. It also should be noted that only the fundamental frequency is excited while all the high frequencies are damped out.

Figure 3 shows the fundamental mode of vibration for the structure in a vacuum. In the case when interaction with the air is considered, the fundamental mode is slightly different which is due to the change in amplitudes and phases of vibrations

Since it is evident that the fundamental frequency is of the most interest for a design engineer, it is worth it then to investigate the effect of wind velocity on it for the structure under consideration. The results are shown in Fig 4. Note, that the relative drop of the fundamental

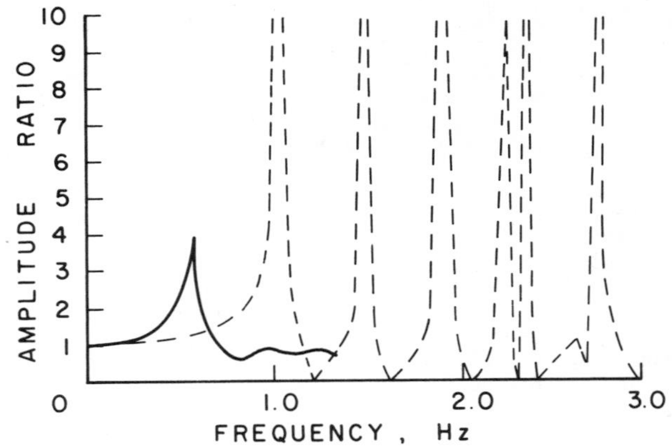

Fig 2 Influence of Acoustic Pressure on Dynamic Response of Spherical Cap
—— with acoustic pressure
---- without acoustic pressure

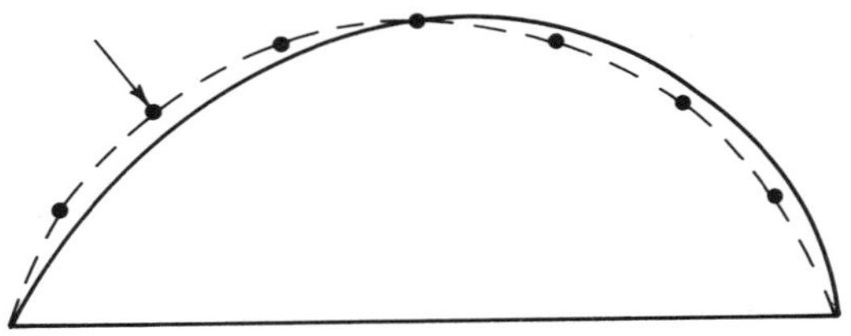

Fig 3 Fundamental Mode of Vibration

frequency remains the same for all wind velocities. Also, the almost abrupt stiffening of the structure after wind exceeds some velocity takes place for both cases at the same wind velocity. The phenomenon of stiffening of the structure needs some explanation.

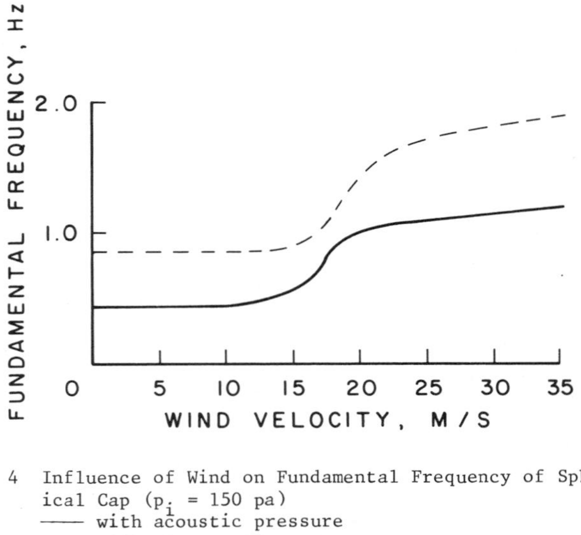

Fig 4 Influence of Wind on Fundamental Frequency of Spherical Cap (p_i = 150 pa)
—— with acoustic pressure
---- without acoustic pressure

The shape of a flexible structure is very sensitive to a pressure distribution which in this case is an algebraic summation of alternating wind pressure and a uniform internal pressure

$$P = p_i - \frac{1}{2} \rho C_p V^2 \qquad \ldots\ldots 16.$$

where ρ = air density, C_p = drag coefficient, V = wind velocity.

Experiments on spheres in a steady flow at large Reynolds numbers (which is most likely the case for large inflatable structures) show, see Ref 5, that the drag coefficient changes as a function of the angle as presented in Fig 5. So as a result of the wind action the pressure on the upper cable elements increases while that on the grounded cable elements decreases. This redistribution of pressure produces its effect on cable elements stiffnesses.

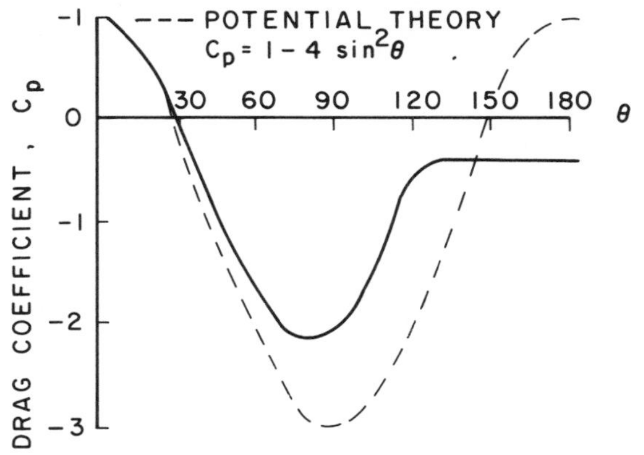

Fig 5 Drag Coefficient vs. Angle for a Sphere

Graphs in Fig 6 show how the longitudinal stiffness (which plays a major role in determining an overall stiffness of the structure, see Ref 4) changes during the stretching of a cable element loaded by nondimensional pressure f. When f is equal to zero the longitudinal stiffness is described by a stepfunction.

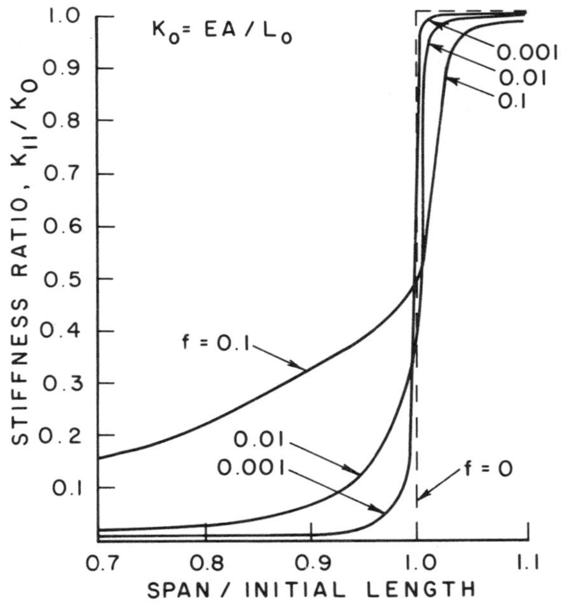

Fig 6 Longitudinal Stiffness of Pressure Loaded Cables vs. Stretching

So when pressure distribution over the structure changes due to varying wind velocity it reaches the point where grounded cable elements experience small pressure while tensile forces transmitted from the upper part of the structure are very large. At such conditions the stiffness of this grounded cable element "jumps" according to Fig 6 abruptly. Approximately the "critical" wind velocity can be found from Eqn 16. When P = 0 and C_p = 1

$$V_{cr} = \frac{2p_i}{\rho} \qquad \text{......17.}$$

ESTIMATION OF FUNDAMENTAL FREQUENCY

Since the fundamental frequency is of main interest it is worth it to obtain an analytical estimation of it. Here it is done for the case of a spherical cap and can easily be extended for a cylinder (plane problem).

Consider a simple single-degree-of-freedom model of a spherical cap with a concentrated mass at the apex and vertically applied harmonic force, Fig 7. Suppose also that there are n reinforcing cables converging at the apex.

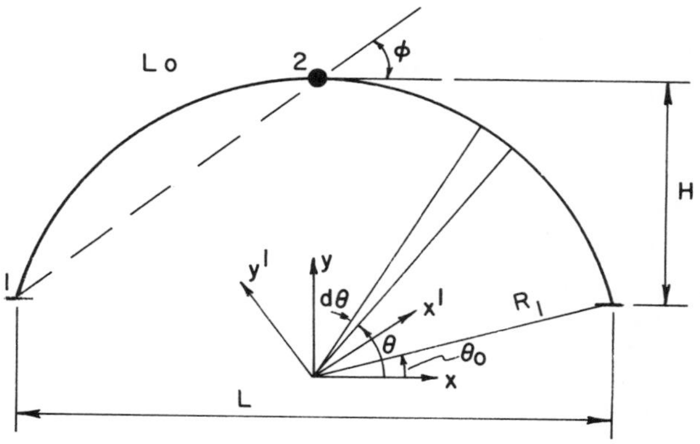

Fig 7 Single-Degree-of-Freedom Model of Spherical Cap

The stiffness of the cable element (1,2) in the local system of coordinates (x'y'), is given by, see Ref 4

$$K_e = \begin{vmatrix} d & -d \\ -d & d \end{vmatrix} \qquad \text{......18,}$$

where

$$d = 0.5\ p \begin{vmatrix} \mu & 0 \\ 0 & \cot\phi \end{vmatrix} \qquad \text{......19,}$$

$$\mu = \cot\phi + a/\sin\phi \qquad \text{......20,}$$

$$a = \frac{\phi - f}{\sin\phi - (\phi - f)\cos\phi} \qquad \text{......21,}$$

$$f = 0.5\ p\ L_o/EA \qquad \text{......22,}$$

p = uniform pressure on a cable element, L_o unstretched cable length, EA = axial cable stiffness, ϕ = angle between the end tensile force and the direction of span, see Fig 7.

In a global system of coordinates (x,y), see Fig 7, the vertical stiffness at the apex is

$$K_v = 0.5\ p.n\ (\mu\sin^2\phi - \cot\phi\ \cos^2\phi) \qquad \text{......23,}$$

where p can be approximated by

$$p = \frac{p_i S}{nL_o} \qquad \text{......24,}$$

where $S = \pi(H^2 + 0.5L^2)$ - area of the membrane skin, $L_o = 2R\phi$.

An estimation of the attached mass and damping can be made for a fundamental mode of vertical vibrations. Let us approximate this mode by the expression

$$u = u_o \frac{\sin(\theta - \theta_o)}{\cos\theta_o} \qquad ......25,$$

where u, u_o = radial displacement of the cable at the current position and apex, respectively.

Now considering acoustic pressure P_a, Eqn 7, acting on an infinitesimally small area $ds = R^2\cos\theta d\theta d\gamma$ and integrating over the area of the membrane the following expression for an equivalent acoustic force at the apex is obtained

$$F_a = \frac{2\pi}{3} u_o \rho c R^2 (1 - \sin\theta_o)\cos\psi \exp(i\psi) \qquad ......26.$$

This results in corresponding expressions for the attached mass and damping

$$m_a = \frac{2\pi}{3} \rho c R^2 (1 - \sin\theta_o) \frac{\sin\psi \cos\psi}{\omega} \qquad ......27,$$

$$q_a = \frac{2\pi}{3} \rho c R^2 (1 - \sin\theta_o) \cos^2\psi \qquad ......28.$$

In the case of long waves (kr << 1) the latter are reduced to

$$m_a = \frac{2\pi}{3} \rho R^3 (1 - \sin\theta_o) \qquad ......29,$$

$$q_a = \frac{2\pi}{3} \rho R^4 (1 - \sin\theta_o) \frac{\omega^2}{c} \qquad ......30.$$

Note, that $\theta_o = \pi/2 - 2\phi$ (see Fig 7).

Now an expression for a fundamental frequency of an oscillating spherical cap interacting with the surrounding air can be written as follows

$$f_o = \frac{1}{2\pi} \left(\frac{0.5\, p_i S(\mu\sin^2\phi - \cot\phi \cos^2\phi)}{\alpha L_o (m + m_a)}\right)^{\frac{1}{2}} \qquad ......31,$$

where m = overall mass of cables and membrane; $\alpha < 1$, coefficient reducing an overall mass $(m + m_a)$ to an effective concentrated mass corresponding to a fundamental mode of vibrations. If $\alpha = 0.375$ the numerical results for a fundamental frequency presented above would be in agreement with the analytical estimation according to Eqn 30. In general α may be a weak function of θ_o.

Equation 30 gives a qualitatively correct relationship between the fundamental frequency and various parameters characterizing the spherical structure. It can be seen that the fundamental frequency is proportional to the square root of internal pressure and is approximately in inverse proportion to the scale factor. If an attached mass were not taken into account the fundamental frequency would be independent of the scale factor. The amplitude of oscillation is in inverse proportion to $q_a\omega$ and it can be shown that it is in inverse proportion to the scale factor.

ACKNOWLEDGEMENTS

The financial assistance provided by the Natural Sciences and Engineering Research Council of Canada in the form of an operating grant No. A-1481 is gratefully acknowledged.

REFERENCES

1. T W LEONARD, Dynamic Response of Initially-Stressed Membrane Shells, Journal of the Engineering Mechanics Division, ASCE, Vol. 95, No. EM5, Proc. Paper 6859, 1969.

2. O G VINOGRADOV, D J MALCOLM and P G GLOCKNER, Vibration of Cable-Reinforced Inflatable Structures, Journal of the Structural Division, ASCE, Vol. 107, No. ST10, Proc. Paper 16594, 1981.

3. O G VINOGRADOV, D J MALCOLM and P G GLOCKNER, Dynamic Response of Inflatable Cable-Reinforced Structures Under Wind Loading, Proceedings of the Second Specialty Conference on Dynamic Response of Structures: Experimentation, Observation, Prediction and Control, ASCE, January 15-16, Atlanta, Georgia, U.S.A., 1981.

4. O G VINOGRADOV, Cable Under the Constant Pressure as a Finite Element, Proceedings of the International Conference on Finite Element Methods, 2-6 August, Shanghai, China, 1982.

5. S F HOERNER, Fluid-Dynamic Drag, Published by the Author, 1965.

6. S N RSHEVKIN, The Theory of Sound, A Pergamon Press Book, N.Y. 1963.

7. E JAHNCE and F EMDE, Tables of Functions, Dover Publications, N.Y. 1945.

8. K J BATHE and E L WILSON, Numerical Methods in Finite Element Analysis, Prentice-Hall, Inc., Englewood Cliffs, N.Y. 1976.

UNIFORM-STRENGTH PRESTRESSED FLAT CABLE NETS

S.F. YASSERI, PhD

Dept. of Structural Engineering

Sharif University of Technology

Tehran, P.O. Box 3406, IRAN

The optimum structure, in its original form, involves continuous variation of material distribution. This prohibits the practical use of such an optimum design. It is the intention of the present paper to present a method for obtaining uniform-strength layouts for cable networks and their related shapes. For a given load and tension in cables, the shapes sought for the net are such that they have uniform stress throughout. For staticaly determinate structures the solutions obtained by this method are also plastic mininimum weight designs. The procedure is illustrated by several examples of practical engineering interest.

1- INTRODUCTION

An objection which is commonly raised with regards to the practical application of optimum structures concerns continuous variation of the material distribution. This results in a fabrication cost which would far exceed the saving made in material in comparison with a simpler but non-optimum structure. It is proposed that a better approach would be to stipulate that the structure has a piece-wise constant cross-section throughout with the shape of its middle surface being chosen in such a manner that the weight of the structure is minimised.

Cable network roofs are formed by a system of flexible cables and a rigid edge structure for supporting the cables. The roof material consists of precast lightweight reinforced slabs, panels of aluminium, plastics etc. Cables are connected to the supporting structure by means of anchorages (similar to those used in prestressed structures) which permits the adjustment of the cable tension.

Roofs with freely suspended cables are too flexible and susceptible to excessive deformation, as a result of which their geometric shape changes as a function of loading. This would particularly be the case in such situations where the load distribution is uneven. To eliminate such undesir deformations, and stabilize the shape, pre-stressing is applied by a convenient method.

The proposed method as opposed to analysis-redesign aims at the selection of design variables in such a manner as to achieve, within the constraints imposed on the structural behaviour, geometry, etc., the best possible solution. In other words it involves the determination of an acceptable way of transmitting a set of given vertical loads, such that the structure is capable of statisfactorily performing the functions.

2- KINEMATIC AND EQUILIBRIUM CONDITIONS

Shallow networks of closely spaced pretensioned cables are spatial structures which have in their unloaded state a small sag (usually 1/10 to 1/25 sags at mid span) relative to the lateral dimensions and the radius of curvature. At first, an ortogonal network of closely spaced, uniform, flexible cables is replaced by an equivalent membrane.

The theory of relatively flat networks is based on the assumption that such networks are slightly curved plates and therefore it is possible to specify the geometry of the structure with the geometry of its projection on a plane. On this basis the linear theory of a shallow shell with an arbitrary middle surface, Fig 1, is described by the following equations (Refs 2 and 3).

The curvatures of the deflection field are :

$$\kappa_x = \frac{\partial^2 W}{\partial x^2}, \quad \kappa_y = \frac{\partial^2 W}{\partial y^2}, \quad \kappa_{xy} = \frac{\partial^2 W}{\partial x \partial y} \quad \ldots . \ 1$$

and the strains of the deformed surface are:

$$\varepsilon_x = \frac{\partial u}{\partial x} - k_x W$$

$$\varepsilon_y = \frac{\partial v}{\partial y} - k_y W \qquad \ldots . \ 2$$

$$\varepsilon_{xy} = \frac{\partial u}{\partial y} + \frac{\partial v}{\partial x} - 2k_{xy} W$$

where $k_x = \frac{\partial^2 z}{\partial x^2}$, $k_y = \frac{\partial^2 z}{\partial y^2}$, $k_{xy} = \frac{\partial^2 z}{\partial x \partial y}$

and Z = Z (x,y) is the equation of the middle surface in the cartesian coordinate system.

The conditions of equilibrium are :

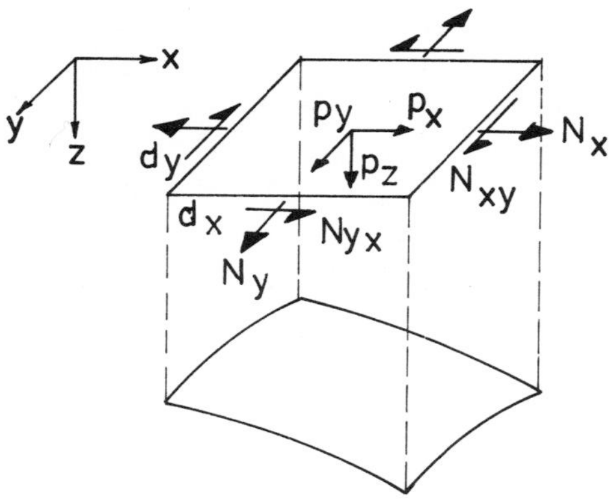

Fig 1

$$\frac{\partial N_x}{\partial x} + \frac{\partial N_{xy}}{\partial y} + p_x = 0$$

$$\frac{\partial N_y}{\partial y} + \frac{\partial N_{xy}}{\partial x} + p_y = 0 \quad \ldots 3$$

$$K_x N_x + k_y N_y + 2K_{xy} N_{xy} + p_z = 0$$

The curvature and strain must satisfy the following compatibility requirements

$$\frac{\partial^2 \varepsilon_x}{\partial y^2} + \frac{\partial^2 \varepsilon_y}{\partial x^2} - \frac{\partial^2 \varepsilon_{xy}}{\partial x \partial y} + k_x \kappa_y + k_y \kappa_y - 2k_{xy} \kappa_{xy} = 0 \quad \ldots . 4$$

$$\frac{\partial \kappa_y}{\partial x} = \frac{\partial \kappa_{xy}}{\partial y}, \quad \frac{\partial \kappa_x}{\partial y} = \frac{\partial \kappa_{xy}}{\partial x}$$

In this paper the contours of the cable supports are assumed to be given with the thickness of the cable being constant. The value of this thickness is determined such that the strength is adequate within the imposed geometrical constraints.

Consider a case where the network is subjected to a uniform transverse pressure whose components are $p_x = p_y = o$ and $p_z = q$. The third equation of equilibrium 3 becomes:

$$N_x \frac{\partial^2 z}{\partial x^2} + 2N_{xy} \frac{\partial^2 z}{\partial x \partial y} + N_y \frac{\partial^2 z}{\partial y^2} + q = 0 \quad \ldots . 5$$

In calulation of the suspended networks it is assumed that the vertical load is carried by the cables only. The roof clading is considered to be either fully passive or capable of taking only compressive stresses.In some cases the roof enclosure may offer in-plane shear resistance, and this happening only if the materials and the design of the slab joints are adequate.

3- UNIFORM ELASTIC DESIGN

The search for efficient use of the structural materials in the design of elastic structure has resulted in the notion of uniform-strength design; that is, a design which has the same margin of strength at all points. For ductile materials it is usually accepted that the strength is the yield limit. for a strcuture made of such a material there is a limit load up to which point the structure remains elastic throughout. When this load is reached all points of the structure undergo plastic deformation simultaneously. Thus a design is called a uniform-strength elastic for a given set of fixed loads, Refs 4 and 5, providing it remains elastic everywhere before the limit of the elastic range is reached.

The present work is concerned with shallow cable nets loaded by short term static forces. The network is constructed with elastic cables all of which are assumed to have the same material properties.Cables are positioned along the directions of principal stresses. A network pattern for which the area of cables per unit length is the same in both direction is called " isotropic ". A cable pattern which has different areas of cables in two directions is called orthotropic.

Clearly, under the above assumptions,if such a network, is subjected to stresses in both directins, it can reach the yield limit independently in both directions. The behaviour of such a shell at the limit of elasticity is defined by a limiting surface. In a manner similar to the plasticity theory, the shear forces are assumed not to influence the limiting condition. In the plane of the principal stresses the limiting surface is represented by a rectangle as shown in Fig 2. Plastic flow can occur only for states of stress represented by a point on the surface of the yield limit.

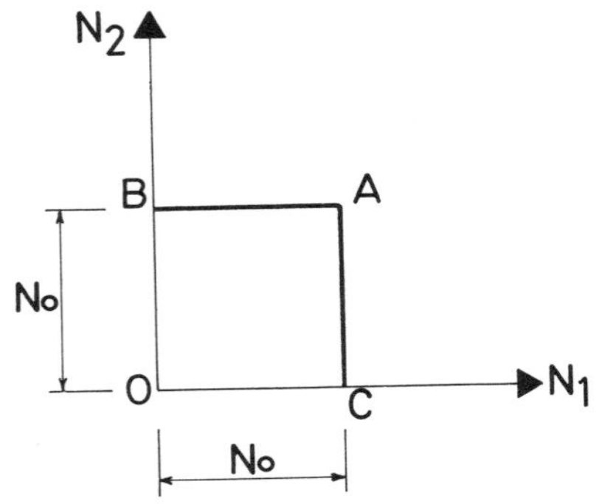

Fig 2

The restriction imposed by the uniform-strength condition restricts the type of essentially different states of uniform-strength to two cases. For instance, the stress points on AC becomes stress points on AB by interchange of indices 1 and 2.

(1) Regime A, Fig 2

For this regime $N_{xy}=0$, $N_x=N_o > 0$, $N_y = N_o > 0$;the equation of equilibrium 5 becomes :

$$N_o \frac{\partial^2 z}{\partial x^2} + N_o \frac{\partial^2 z}{\partial y^2} + p_z = 0 \quad \ldots . 6$$

For orthotropic nets the cables in each direction is different.

(2) Regime C , Fig. 2.

Here $N_{xy}=0$, $N_x=N_o$ and $N_y=o$, and equilibrium

equation 5 gives

$$N_o \frac{\partial^2 z}{\partial x^2} + p_z = 0 \qquad .... 7$$

of which a solution is

$$Z = \frac{-p_z}{2N_o} x^2 + xf(y) + g(y) \qquad .... 8$$

The unknown function f(y) and g(y) are determined by the boundary conditions:

For the complete determination of a design, it is necessary to stipulate which of these regimes prevails in any specific part of the net. For this purpose the network is diveded into various regions. Once the uniform-strength regime for each region is decided, the dimensions of the regions are determined from the continuity requirement of the field variables.

It is worth noting that the assumption which the principal stresses, N_1 and N_2 have the same sign everywhere on the network usually leads to a heavier design, except for the axisymmetric case. This assumption effectively means that Eqn 6 is the governing equation throughout the network, it being the form of the governing equation of torsion of a cross- section for which a considerable set of solutions exist.

Equations of equilibrium and compatibility of deformation are derived on the basis of the assumption that the co-ordinate axes are the lines of principal curvature of the surface. Unless this condition is satisfied some equations would require certain minor changes.

4- EXAMPLES

4.1 Circular Boundary

Consider a case of the optimal transmission of a uniformly distributed circular load of radius R with components, $P_x=p_y=0$ and $p_z=q$ to a circular boundary of radius R. All conditions are satisfied by the shape

$$Z = \frac{q}{4N_o}(R^2-x^2-y^2) \qquad ..... 9$$

The soluion for the case where a uniformly distributed load occupies a central region of radius a is $N_x=N_y=N_o$ and $N_{xy}=0$ throughout. For radii less than a

$$Z = \frac{q}{4N_o} (R^2-x^2-y^2) \qquad .... 10$$

and for radii greater than a

$$Z= \frac{-qr}{4N_o} \ell_n \sqrt{\frac{x^2+y^2}{R}} + \frac{q}{4N_o}(R^2+R\ell_n\frac{a}{R}-a^2) ... 11$$

4.2 Elliptic Boundary

Consider a case of transmitting a uniformly distributed load with components $p_x=p_y= o$ and $p_z=q$ to an elliptical houndary whose major and minor axes are respectively a and b. All requirements are satisfied by the shape

$$Z = \frac{q}{2N} \frac{a^2b^2}{a^2+b^2}(1- \frac{x^2}{a^2} - \frac{y^2}{b^2}) \qquad .....12$$

with $N_x=N_y=N_o$ and $N_{xy}=0$

4.3 Square Boundary

A uniformly distributed load, whose components are $p_x=p_y=0$ and $p_z=q$ is to be transmitted to a square boundary. The domain is divided into a central spherical region and regions of anticlastic curvature in the corners. The transition from one shape to the other takes place along the lines through the midpoint of each side, Fig.3. the shape of the shallow net, for the quadrant shown in Fig.3 is

$$Z= \frac{q}{N_o}\left[\frac{-L}{2}(x+y)+xy+ \frac{L^2}{4}\right] \qquad ....13$$

with $N_t=N_o$ and $N_n= N_{nt} = 0$

in the coner, $x+y > L/2$, and

$$Z = \frac{q}{N_o}\left[\frac{-1}{2}(x^2+y^2)+ \frac{L^2}{8}\right] \qquad ....14$$

with $N_{xy}=0$, $N_x=N_y=N_o$

in the centre. All continuity requirements in Z, $\frac{\partial z}{\partial x}$ and $\frac{\partial z}{\partial y}$ across $x+y= \frac{L}{2}$ are satisfied,

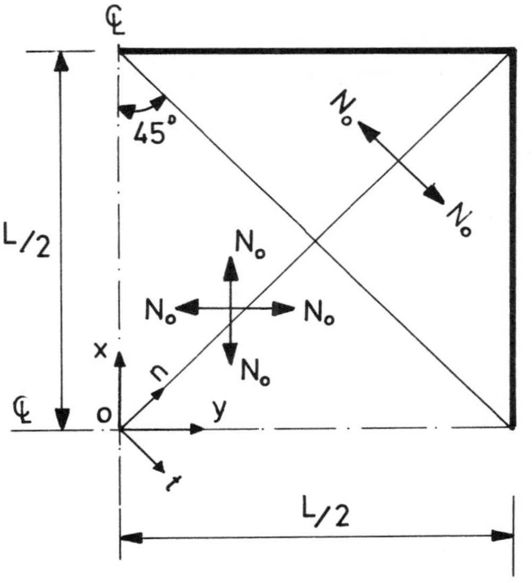

Fig 3

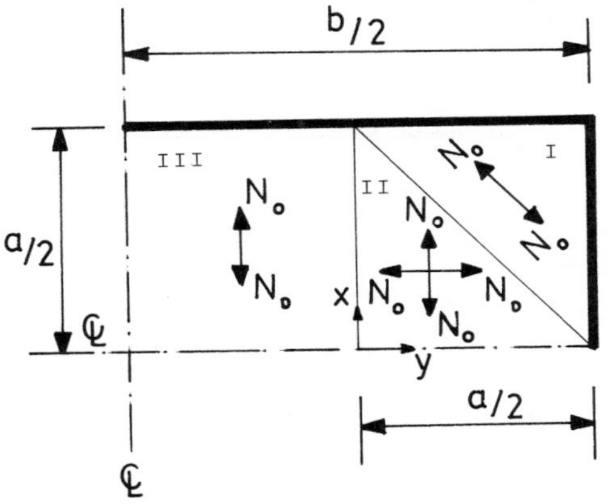

Fig 4

4.4 Rectangular Boundary

The optimum shape Eqns 13and 14 can be used to provide the solution for a rectangular boundary, Fig 4 , where b > a. A shape satisfying all conditions can be obtained by inserting a central cylindrical region between two halves of the shape Egns 13 and 14 for a square boundary of side a. The shape of region III is

$$Z = \frac{q}{2N_0}\left(\frac{a^2}{4} - x^2\right) \qquad .... \ 15$$

with $N_x = N_0$, and $N_y = N_{xy} = 0$

for regions I and II the shapes are given by Egns 13 and 14 with L replaced by a .

Solutions for free edge, triangular boundary, and a corner network are shown in Fig. 5.

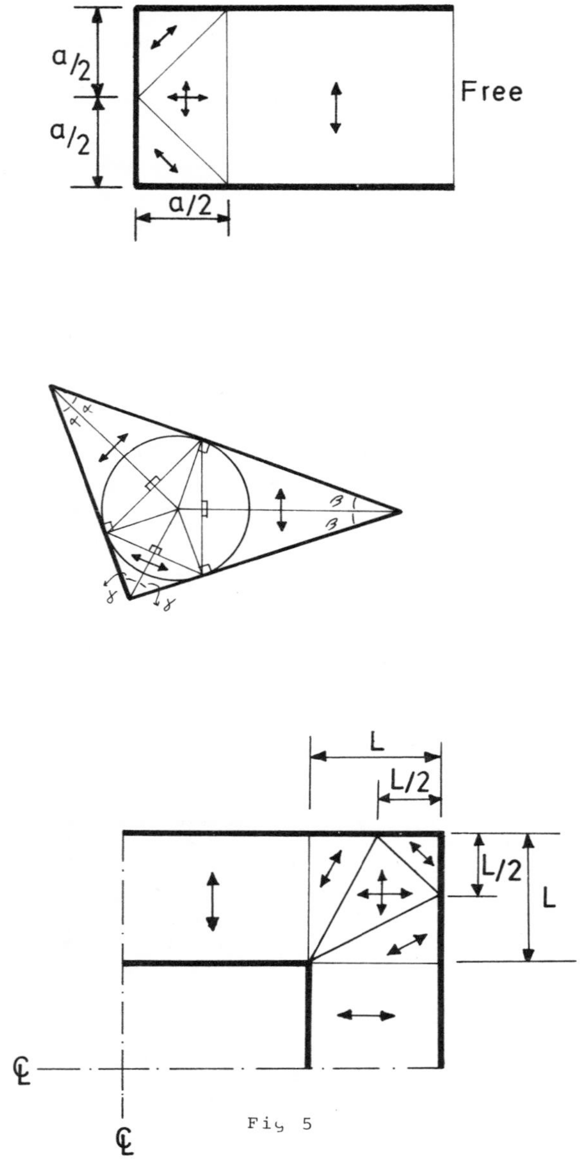

Fig 5

4.5 Continuous Networks

Networks which are continuous over one or more intermediate supports are of practical interest. Most continuous networks are of a very complicated form and offer a formidable challenge to analytical treatment. It can be shown that the number of times that the curvatures change sign is equal to the number of indeteminacies of the structure. By introducing a sufficient number of inflexion lines in the domain of the network, it can be made statically determinate. That choice of inflexion lines that requires the least amount of cables constitutes the solution to the problem.

Consider for example a continuous network connected rigidly to the supporting columns. The column grid is assumed to be square panels, Fig 6. the outer boundary, which is not shown here, may be supported on edge cables, or resting by the outer walls of the structure.

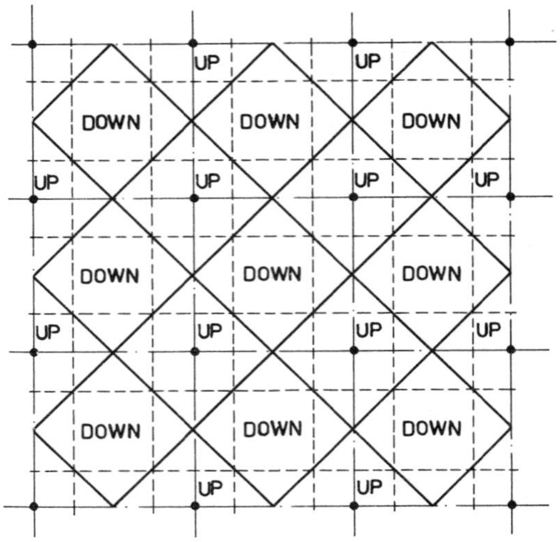

Fig 6

A pattern of deformation is shown in Fig 6. Lines of inflexions are shown on the diagram by heary lines. Panels with a column at the centre deform upward while the other type of panel deflect downwards. Using the soluion for square boundary as building block a near-optimum solution for the interior panels can be constructed.

5. DISCUSSION

Uniform-strength networks are networks for which the transition from the elastic state into the plastic state occurs simultaneously everywhere in the network. Based upon this concept a method has been proposed to achieve a uniform strength design and its related shallow shape. For a given load and tensionin cables, the shapes sought for the net are such that they have uniform stress throughout. Conversely the problem of uniform-strength may be posed as one of determining the variation of tension in cables for a given shape and a set of given load. This situation arises where due to aerodynamic or aesthetic reasons, the shape is taken to be prescribed.

All statically admissible stress fields of a statically determinate structure are equal to within a constant of proportionality; that is the distribution of stresses is independent of the material property. The elastic stress field of the uniform-strength design is statically admissible and at incipient yield everywhere; consequently a uniform -strength design is a minimum-weight plastic design for the given loads. Moreover, the plastic strain rate field is compatible with the elastic strain field.

As was pointed out by Reissner(Ref.6) in the theory of shallow shells it is assumed that the square of the first order derivitives of the middle surface

$$\left(\frac{\partial z}{\partial x}\right)^2, \left(\frac{\partial z}{\partial y}\right)^2 \text{ and } \left(\frac{\partial z}{\partial x} \cdot \frac{\partial z}{\partial y}\right),$$

are negligible in comparison with unity. The general theory of shallow shells will be more than sufficiently accurate as long as

$$\frac{\partial z}{\partial x} \cdot \frac{\partial z}{\partial y} < \frac{1}{8}$$

and often accurate enough for practical purposes Provided $\frac{\partial z}{\partial x} \cdot \frac{\partial z}{\partial y} < \frac{1}{2}$.

6- CONCLUSTIONS

The aim of this paper has been to present a technique for obtaining uniform strength layout for cable networks and its related shapes. Although the examples used to illustrate the technique have been confined to simple shapes, it is easy to derive solutions for such problems as holes in networks and cable nets supported on edge cables, which are of considerable practical interest.

An element of doubt is present as to whether the present optima are absolute optimal designs. The aim is however, to stimulate the study of direct design of such problems and at the same time produce layouts which are optimal and feasible. Based on this study it cannot be concluded that lower optima will not exist, but it can be said any such lower optima will be topologically different.

REFERENCES

1- G.I.N. ROZVANY, Optimal Design of Flexural system, Pergamon Press, 1976.

2- W.FLUGGE, Stresses in Shells, springer Verlag, 1976.

3- H.M.IRVINE, Cable Structures, The MIT Press, 1981

4- R.T. SHIELD, Optimum Design Methods for Structurs, Plasticity, Proc 2nd Symp. Naval Struct. Mechanics, Pergamon Press, 1960.

5- M.A. SAVE, Some Minimum-Weight Design, Engineering Plasticity, Cambridge Univ. Press, 1968.

6- E. REISSNER, On Some Aspects of the theory of thin Elastic Shells, J. of the Boston Society of Civil Engineers, 42(2), 1955.

A NEW TECHNIQUE FOR THE ANALYSIS OF SUSPENSION ROOFS

SH N ELIBIARI*, B Sc,PhD, S A SAAFAN**, B Sc, PhD, F F ELDEEB* B Sc, Ph D

* Researcher, Theory of Structures Department, General Organization for Housing, Building & Planning Research P.O. Box 1770, Cairo, Egypt.

** Professor, Theory of Structures Department, Faculty of Engineering, Ain Shams University, Cairo, Egypt.

Suspension roofs are used mainly for covering large areas. They are cheaper and more easier in construction than the conventional concrete roofs. In order to generalize their use, an easy design method is required. This paper presents a new technique using the well- known truss direct element matrix for solving suspension roofs, considering both geometric and material non-linearities. Examples are solved numerically , analyzed and compared with the previous published techniques. The influence of the support flexibility on the structure behavior is studied. An experimental study is carried out. The experimental results are much agreed with those obtained theoretically.

INTRODUCTION

The small deflection theory used in ordinary structural analysis is not valid in suspension roofs, because of the extreme flexibility of this type of structures; hence the theory must include the effect of large deflections. These deflections take place mainly as a result of the change in geometrical shape. Since there are not presently available exact solutions for the differential equations describing the deflection behavior of cable networks, iteration techniques are required to obtain force and displacement analysis.

Several methods of theoretical analysis of cable structures have been suggested in the liturature (1,2,3,4,). Most of these methods dealt with a modified stiffness matrix for the cable member. To the writer's knowledge, little attention has been paid (1,3) to the actual stress strain relationship above the proportional limit in the analysis of suspension roofs.

Although several studies have been exclusively dealing with the analysis of cable structures, few of them (5,6) have dealt with the effect of the support flexibility on the structure response.

The new technique developed in this work, uses the well-known truss direct element matrix for solving cable structures considering both geometric and material non-linearities. The influence of the support flexibility on the cable forces and the displacements of joints is investigated.

An experimental model is loaded incrementally until collapse to study its behavior in both elastic and plastic stages. In general the agreement between the experimental and the theoretical results is good.

ASSUMPTIONS

The method considered herein makes the following assumptions:

1. The members of the suspension roof are of uniform x-sections.
2. Cables are perfectly flexible.
3. Loads are concentrated at joints.
4. Dummy elastic members, springs, having three dimensional global elasticity directions are arranged at each joint to be able to apply the general truss matrix in solving suspension roofs. The springs at the intermediate node help the numerical damping of divergence while those at the edge nodes present the condition of support.

SIGN CONVENTION

The joint translations and forces acting on the member are positive along the positive directions of the coordinate axes. The forces wrt member and common axes will be referred as P and Q respectively. To each P associate a member deformation p and to each Q a displacement q. The positive directions of the translations and forces of member ij are indicated by the numbered arrows in Fig. 1 a. wrt member and common axes.

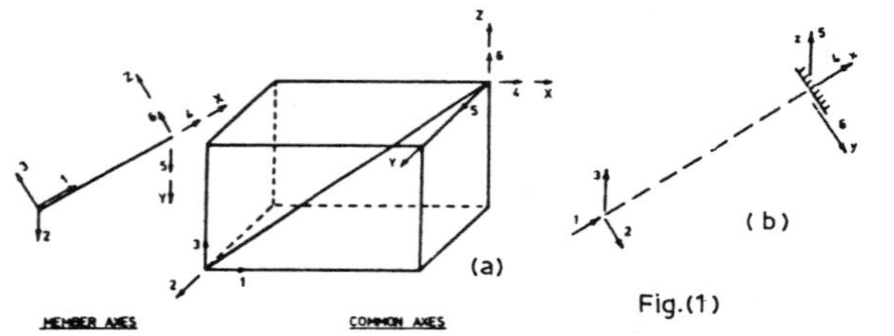

Fig.(1)

MEMBER STIFFNESS MATRIX

The stiffness matrix of the member wrt the common axis is:

$$[G] = \frac{EA}{Lo}\begin{bmatrix} l^2 & lm & ln & -l^2 & -lm & -ln \\ ml & m^2 & mn & -ml & -m^2 & -mn \\ nl & nm & n^2 & -nl & -nm & -n^2 \\ -l^2 & -lm & -ln & l^2 & lm & ln \\ -ml & -m^2 & -mn & ml & m^2 & mn \\ -nl & -nm & -n^2 & nl & nm & n^2 \end{bmatrix} \quad \text{---- 1.}$$

in which this is the well-known direct element matrix. E,A are the cross - sectional area and the modulus of elasticity of the cables respectively. Lo is the unstrained cable length. l,m,n are the direction cosines of the member. The stiffness equation of the springs, Fig. 1.b, wrt member axes is :

$$\begin{bmatrix} S_1 \\ S_2 \\ S_3 \end{bmatrix} = \begin{bmatrix} C_1 & 0 & 0 \\ 0 & C_2 & 0 \\ 0 & 0 & C_3 \end{bmatrix} \begin{bmatrix} p'_1 \\ p'_2 \\ p'_3 \end{bmatrix} \quad \text{------- 2.}$$

where $[S]$ = column vector for the force in the elastic member, $[C]$ = the restraint coefficient (elastic coefficient), $[p']$ = column vector of displacements.

METHOD OF SOLUTION

Figure 2 shows a system of cable wires before and after using the elastic members respectively.

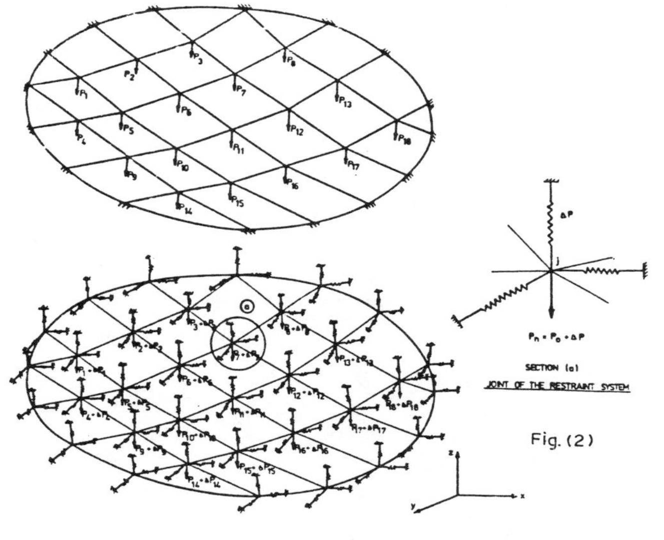

Fig. (2)

In order to illustrate the behavior of the restrained system, i.e. using elastic mem bers, an example, Fig. 3, is explained as follows:

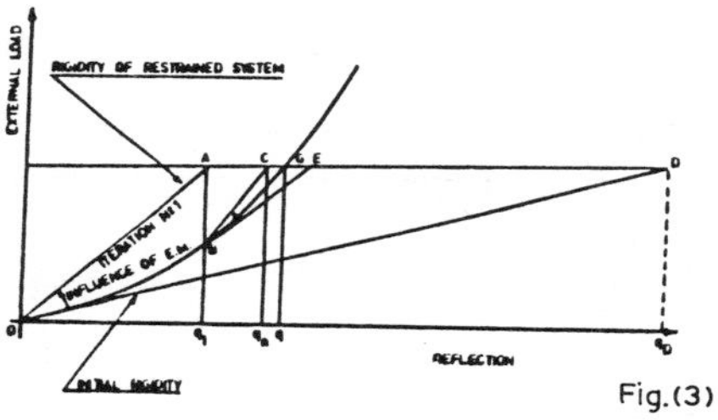

Fig.(3)

Start iteration with respectively high elastic coefficients. The first deformation result is represented by q_1.

- The initial rigidity of the cable system is represented by the initial stiffness tangent OD, therefor $q_1 < q_D$
- Reducing the elastic factors, while progressing in iteration steps, $\angle$ AOD decreases and becomes $\angle$ CBE. In the nth iteration step, the corresponding deflection is q_n.
- In this example, the final result can be obtained theoretically due to the following equation

$$[q] = \lim_{[C] \to 0} [q_c]$$

where q = displacement, C = elastic factor in force per unit displacement and q_c = partial solution corresponding to a certain elastic coefficient c.

This method maybe summerized as follows:

$$W = [\bar{G}]\,[q] = [\bar{W}] - [S] \quad \text{---------- 3.}$$

where $[W]$ is the actual applied load, $[G]$, $[q]$ are the stiffness matrix and the total displacements at the nth state of deformation, $[\bar{W}]$ is the assumed applied load on the restrained system and $[S]$ = forces in the elastic members.

NUMERICAL TECHNIQUES

In order to solve the cable structure, a set of non-linear equations were put in the form of a matrix, eqn 1. For accelerating the convergence, the following factors are introduced:

a) Iterations: Iteration is followed respectively observing the results of joints displacements and members forces. For each iteration step the error in the force is indicated by a coefficient w, eqn 4. Iterations are stopped as this coefficient is minimum. A simplified example for a member ij, in plane xy, Fig. 4, illustrates the calculations of w due to the following equation:

$$w = \sum_{k=1}^{n} \left| P'_k - P_k \right| \quad \text{-------------- 4.}$$

where n = no of members

$$P'_k = \frac{L' - Lo}{Lo} EA$$ = the force in the member calculated according to unstrained length Lo and displacement q obtained from the stiffness matrix.

$$L' = \sqrt{\overline{Xo - q_x}^2 + \overline{Yo + q_y}^2}$$

In three dimensional case : $L' = \sqrt{\overline{Xo + q_x}^2 + \overline{Yo + q_y}^2 + \overline{Zo + q_z}^2}$

$P_k = \frac{L - Lo}{Lo}$ EA = force in the member calculated due to strain ϵ.

$$L = Lo + \frac{Yo\ q_y}{Lo}.$$

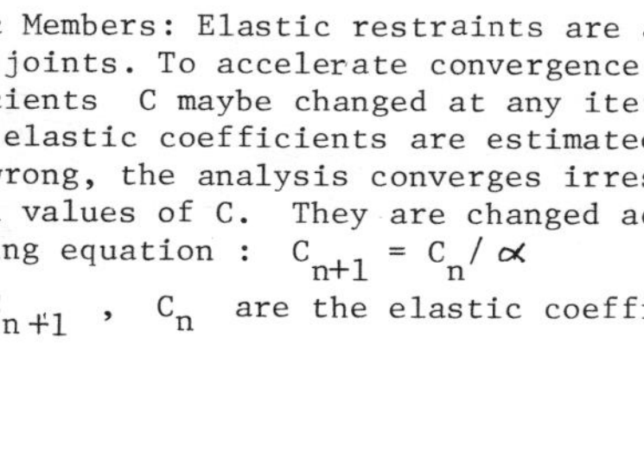

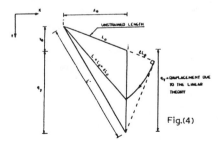

Fig.(4)

b) Elastic Members: Elastic restraints are assumed at the loaded joints. To accelerate convergence, their elastic coefficients C maybe changed at any iteration step. If the elastic coefficients are estimated arithmetically wrong, the analysis converges irrespective of the initial values of C. They are changed according to the following equation : $C_{n+1} = C_n / \alpha$

where C_{n+1}, C_n are the elastic coefficients at the

n^{th} and n + 1 steps respectively and α is the ratio between the elastic restraint at the nth step and the increment in the actual applied load i.e. $\alpha = \frac{S_n}{W_n}$.

It was found that the best results are obtained by assuming the elastic force at the first iteration step S_1 equal to 0.5 the actual applied load W.

c) Weighted Interpolation : Divergence maybe prevented during iteration by calculating the coefficient w, eqn. 4, for the last two iteration steps. These coefficients are considered as weights which lead towards the correct results. According to the results of iterations, the decision may be taken in considering the imput of the following step to be the last result or the average calculated by weights of the last two iteration steps. It was found better to use the squares of weights in calculating the improved input deformation matrix $[q'_n]$ for the next iteration step, i.e.

$$q'_n = \frac{w_n^2 \; [q_{n-1}] + w_{n-1}^2 \; [q_n]}{w_n^2 - w_{n-1}^2} \qquad \text{---------} \; 5.$$

where the suffix n, n-1 refers to the nth, n-1 iteration step.

d) Starting with initial deformation: Assuming the roof to be deformed in its initial shape accelerates the convergence. This deformation is deleted when calculating the roof coordinates after the first iteration step.

For a prestressed member, the prestressing force $[Po]$ is calculated by:

$$[Po] = [N] \; [T] \; [q_o] \qquad \text{---------} \; 6.$$

where $[N]$ is the stiffness matrix of the unstrained member, $[T]$ the orthogonal transformation matrix and $[q_o]$ is the displacement due to prestressing which has an apposite sign to the final structure displacement.

NUMERICAL STUDIES

Example 1: Two way Net Subjected to Symmetrical Loading

Figure 5 shows the dimensions of a 2 x 2 cable net. The wire ropes x-sectional area is 0.227 sq. in and their modulus of elasticity is 12000 Ksi. Due to the prestressing forces, the unstrained lengths of the inclined and horizontal members are 104.2 ft. and 99.8 ft on which the assumed initial strains are 0.00195 and 0.002 and the pretension forces are 5.298 and 5.448 Kips respectively. A vertical load of 8 Kips is applied at each joint. The supports are fixed ends at the same level.

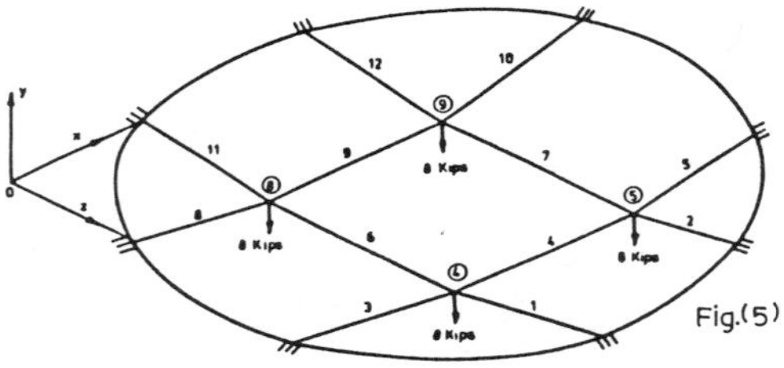

Fig.(5)

Table 1 gives the vertical deflections at the final cable tensions for joint 4, a typical joint. The final cable tensions are compared in table 2 with results of previous researchers (1,7,8). It is obvious to mention that the obtained results needed only two cycles.

Table 1 : Displacement of Joint 4-Example 1

Elibiari (10)	Saafan (1)	Foster (7)	Kar (8)
1.4702	1.4707	1.4721	1.4698

Table 2: Tension in Cables - Example 1.

	Elibiari (10)	Saafan (1)	Foster (7)	Kar (8)
Horizontal Members	12.680	12.677	12.678	12.677
Inclined Members	13.310	13.309	13.312	13.310

Example 2: Two Way Net Subjected to Unsymmetrical Loading.

The suspension roof of the previous example is loaded as shown in Fig. 6. It involves severe geometric as well as material non-linearities. Figure 7 gives the stress-strain curve. The displacements of the joints and the final cable forces are compared with those given by Saafan (1), in Tables 3 and 4 respectively. For both solutions, the relative error between the forces in the members calculated according to the final joint displacements, table 3, and the final member forces, table 4, is calculated. It was found 1.168 % and 0.392 % for Saafan (1) and Elibiari (10) respectively.

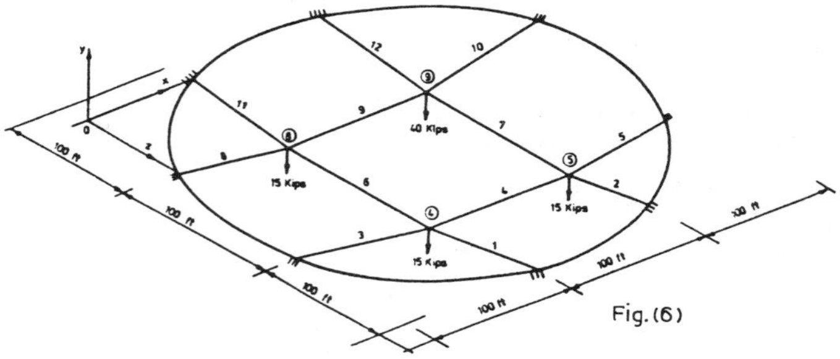

Fig.(6)

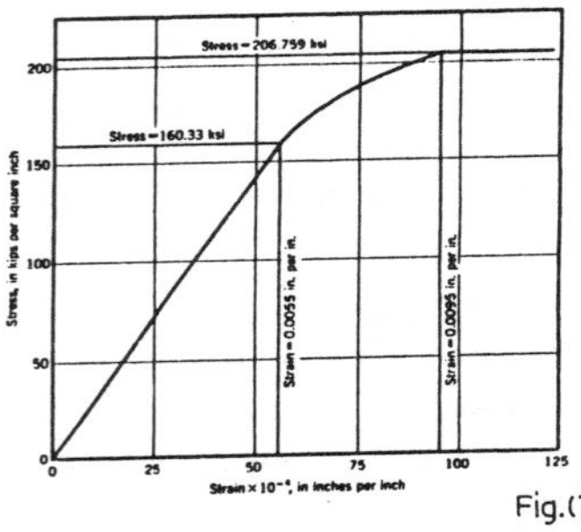

Fig.(7)

Table 3: Displacements of Joints - Example 2.

Joint Number	Deflection x, in ft.		Deflection y, in ft.		Deflection z, in ft.	
	Saafan (1)	Elibiari (10)	Saafan (1)	Elibiari (10)	Saafan (1)	Elibiari (10)
4	-0.382	-0.372	-1.774	-1.752	0.382	0.372
5	-0.288	-0.275	0.484	0.434	-0.800	- 0.833
8	0.800	0.834	0.484	0.434	0.288	0.276
9	1.036	1.079	-7.183	7.293	-1.036	-1.079

Table 4: Forces in Cables - Example 2.

Member Number		1,3	2,8	4,6	5,11	7,9	10,12
Force in	Saafan (1)	23.14	44.78	21.73	21.91	43.54	47.45
Kips	Elibiari (10)	23.27	44.62	21.83	22.05	43.37	47.32

Example 3: Hyperbolic Paraboloid Net.

The nodal points coordinates of the saddle-shaped net, solved in this example, in the x, y and z directions are given in table 5. This surface is defined by:

$$Y = -fx \left(\frac{x}{a}\right)^2 + fz \left(\frac{z}{b}\right)^2 \qquad \text{------------ 7.}$$

in which fx = 2.75 ft, fz = 5.5 ft, a = 21 ft and b = 30 ft The cables are prestressed before loading. The x-component of force, Px , for cables D through G, Fig. 8, in the initial state is 92.4 Kips, and its Z-conponent, Pz, for cables A through C, is 82.5 Kips. In the initial state, equilibrium of any node point i, Fig. 9, is satisfied by the following equation:

$$\frac{y_1 - 2y_i + y_3}{D_x} P_x = - \frac{y_2 - 2y_i + y_4}{D_z} P_z \qquad \text{------ 8.}$$

The cross-sectional area and the modulus of elasticity of cables are 0.01 ft^2 and 3.6 x 10^6 K/ft^2 respectively. Table 6 gives the members prestressing forces. A vertical concentrated load of 13.61 Kips is applied at each nodal point.

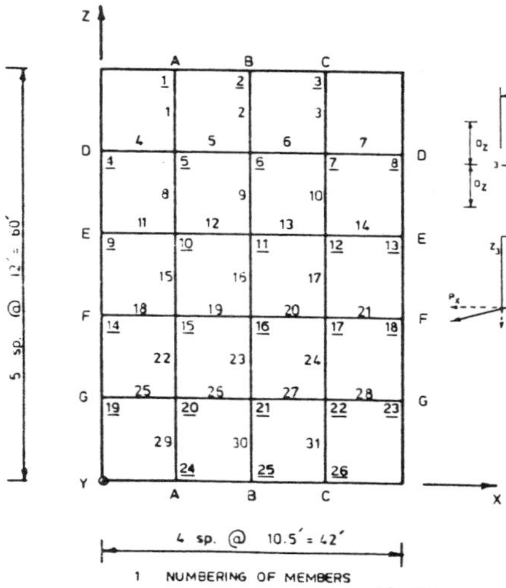

Fig.(8)

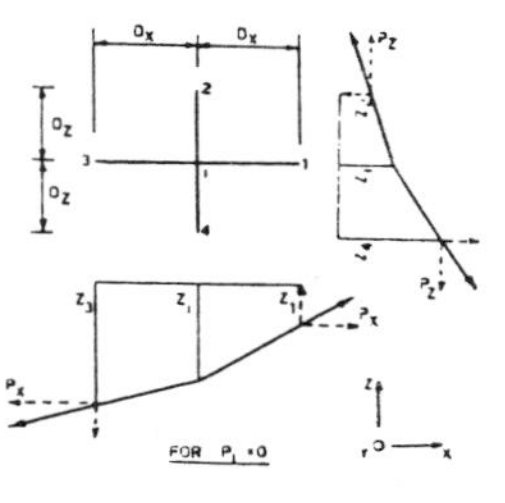

Fig.(9)

Table 5. Coordinates of the H.P. Cable Net

Node No.	1	2	4	5	6	9	10	11
x	10.5	21	0.0	10.5	21	0	10.5	21
y	-4.81	-5.50	0.77	-1.29	-1.98	2.53	0.47	-0.22
z	60	60	48	48	48	36	36	36

Table 6. Prestressing forces in the H.P. Cable Net

Member	1,2	4,11	5,12	8,9	15,16
Prestressing Force	86	94.2	92.6	83.4	82.5

The nodes displacements in x,y and z directions and the forces in the members were reached after two cycles only. They are compared with those of Baron (9) in tables 7, 8 respectively.

Table 7: Joints Displacements. Example 3

Joint No.	Deflection x, in ft.		Deflection y, in ft.		Deflection z in ft.	
	Baron (9)	Elibiari (10)	Baron (9)	Elibiari (10)	Baron (9)	Elibiari (10)
5	-0.0115	-0.0115	-0.1230	-0.1230	0.0199	-0.0196
6	0.00	0.00	-0.1430	-0.1436	-0.0230	-0.0237
10	-0.0121	-0.0121	-0.1310	-0.1327	-0.0068	-0.0069
11	0.00	0.00	-0.1520	-0.1528	-0.0079	-0.0079

Table 8 : Cables Forces - Example 3

Member No.	1	2	4	5	8	9	11	12	15	16
Baron(9)	42.7	36.1	139	136	41.8	35.2	142	139	41.4	34.8
(10) Elibiari	42.7	36.4	138.9	136.2	41.8	35.5	141.8	139.1	41.4	35.2

EFFECT OF SUPPORT FLEXIBILITY ON CABLE STRUCTURES RESPONSE:

As the edge nodes are assumed to be supported by springs, the rigidity of the edge beams are varried by assuming different values for the spring coefficients. Numerical examples were solved for cable roofs having different beam rigidities. From the obtained results, it maybe concluded that the tension in the cables increases with the increase of the support rigidity, except some few cases where the members are stressed within the proportional limit. The vertical displacements of the loaded joints increase with the decrease of the edge beam rigidity. The rate of decrease increases for low rigidities. The horizontal displacements decrease very slightly for flexible boundaries.

EXPERIMENTAL ANALYSIS OF CABLE NETS

An experimental study is carried out on suspended roof models to verify the theoretical analysis as performed by the computer program. The tested model is a rectangular cable net of 3.4 x 2.8 ms in dimensions. It is a net of 4 interior nodes and 8 supporting modes. Figure 10 gives the initial geometry of the roof and the numbering of the cables and joints. Two models are tested under the same conditions changing only the cables properties. The edge connections are fixed so that no slippage occurs during loading. Vertical loads are applied incrementally at each intermediate node until collapse. Figure 11 shows good aggrement between the experimental and theoretical results for both models.

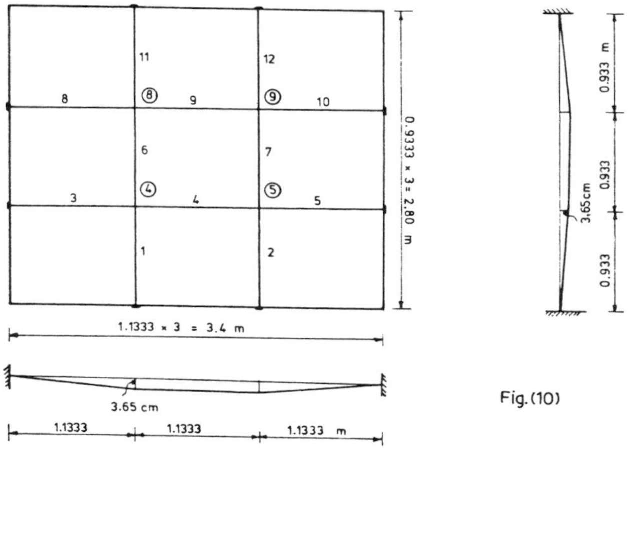

Fig.(10)

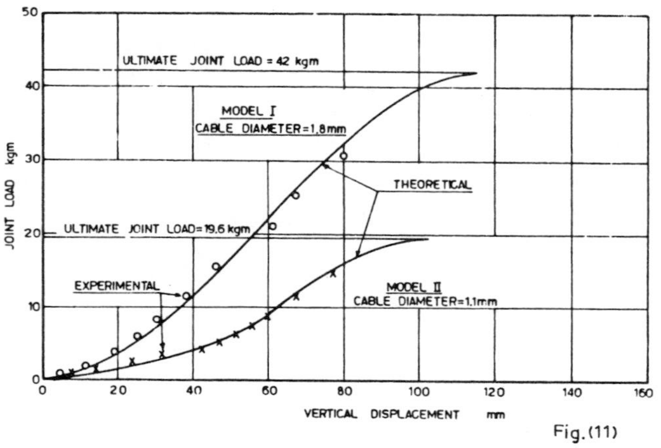

Fig.(11)

ACKNOWLEDGEMENTS

The authors wish to thank Dr. Mahmoud Hashish, Associate Prof. of Theory of Structures, Faculty of Engineering, Ain-Shams University for his useful discussions and suggestions.

REFERENCES:

1- Saafan S.A., "Theoretical Analysis of Suspension Roofs", Journal of the Structural Division, ASCE, Feb. 1970, pp. 393-405.

2- Buchholdt, H.A., Mc Millan, B.R., "Iterative Methods for the Solution of Prestressed Cable Structures and Pinjointed Assemblies having Significant Geometrical Displacements", Proc. 1971, IASS Pacific Symposium Part II on Tension Structures and Space Frames, Tokyo & Kyoto, pp. 305-316, Paper No. 3-7, 1972, Architectural Institute of Japan.

3- Greenberg, D.P., "Inelastic Analysis of Suspension Roof Structures", Journal of the Structural Division, ASCE, Vol, 96, No. ST5, Proc. paper 7284. May 1970, pp. 905-930.

4- Thornton, C.H., and Birnsteil, C., "Three Dimensional Suspension Structures", Journal of the Structural Division, ASCE, Vol. 93, No. ST2, Proc. paper 519, April 1967, pp. 247-270.

5- Buchholdt, H.A., Das, N.K. and Al-Hilli, A.J., " A Gradient Method for the Analysis of Cable Structures with Flexible Boundaries", International Conference on Tension Roof Structures, London, April 8-10, 1974.

6- Karrholm, G. and Samuelsson, A., "Analysis of Prestressed Cable-Roof Anchord in Space-Curve Ring Beam", International Association for Bridge and Structural Engineering, Symposium Amsterdam, May 1972.

7- Foster, Edwin Powell", Experimental and Finite Element Analysis of Cable Roof Structures Including Precast Panels", Vanderbilt University, Ph.D. 1974.

8- Kar, A.K. and Okazoke, G.Y. "Convergence in Highly Nonlinear Cable Net Problems", Journal of the Structural Division, ASCE, Vol. 99, No. ST3, Proceedings paper 9601 March 1973, pp. 321-334.

9- Frank Baran and Mahadeva S. Venkatersan", Non-linear Analysis of Cable and Truss Structures", Journal of the Structural Division, ASCE, February, 1971, pp. 679-710.

10- Elibiari, S.N., "Suspension Roofs", Ph.D. thesis, Faculty of Engineering, Ain Shams University, 1983.

NONLINEAR FREE VIBRATIONS OF CABLE NETS

R LEWANDOWSKI, PhD

Lecturer, Institute of Technology and
Building Structures
Technical University of Poznań, Poznań, Poland

Geometrically nonlinear free vibrations of cable nets are analysed in the note. The problem is formulated as the variational one employing the Hamilton's principle. The single modal method and the iterative procedure are used for determining the frequences and modes of vibrations. The given numerical example proves that the geometric nonlinearity has a significant influence on the results of calculations. Some information concerning the convergence of the proposed iterative procedure are given.

1. INTRODUCTION

Frequences and modes of vibrations of cable nets are usually determined assuming a negligible small influence of geometric nonlinearity. However this assumption is justified for the small amplitudes of vibrations only and the respective solutions are shown in Ref 1-3. For the large amplitudes the influence of geometric nonlinearity is substantial and can not be neglected. Only in Ref 4-6 this influence was taken into consideration that the process of vibrations is harmonic. Moreover Ref 6 presents so-called "nonlinear normal mode". It was assumed that in all those papers the influence of the square term the motion equation with respect to the displecements can be omitted. Those two simplifications results in some errors in the solution and their magnitude is not till now estimated.

The main aim of this work is the analysis of the nonlinear free vibrations of cable nets.

A backbone curve (it means the curve of dependence of the frequence as a function of the amplitude of vibrations of arbitrary chosen point) and the nonlinear mode of vibrations as well as the pass of the process of vibrations are determined.

In the solution of the problem the two above assumptions are not taken into account. Moreover in the numerical example the influence of the geometric nonlinearity on the final results is analyzed. The method of solving the nonlinear eigenvalue problem is presented. The problem is formulated in variational approach employing the Hamilton's principle.

2. ASSUMPTIONS AND BASIC EQUATIONS

Let us consider the shallow and prestressed cable net as shown in the Fig 1.

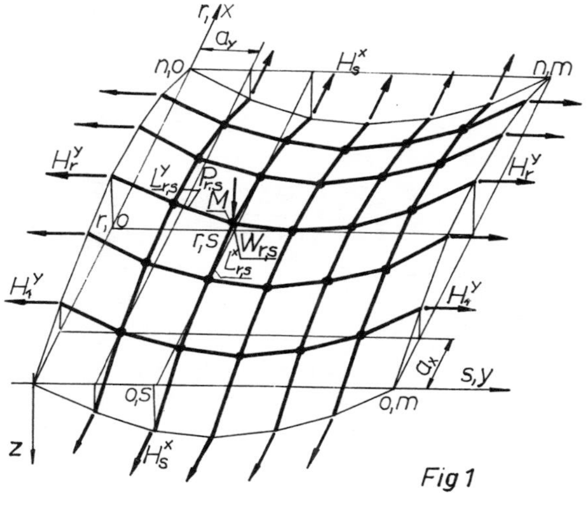

Fig 1

The regular cable net has two families of cables and it is rectangular in plan. It was assumed that the material of cables is simply elastic and the displacements of the net are finite but the strains are small. The mass of the net is concentrated in nodes. In the vibration process the cables remain prestressed. All the loads can act only in the vertical direction thus the horizontal inertia forces are also neglected. The dynamic displacements of nets $v_{rs}(t)$ are the function of two discrete variables r, s and the time t (r describes the numbers of nodes in x-direction, r=0,1,....,n, and s in y- direction s=0,1,....,m). It is considered that the static equilibrium configuration is known so the following equation is fullfilled:

$$\frac{H_s^x}{a_x}\Delta_r^2 W_{r,s}+\frac{H_r^y}{a_y}\Delta_s^2 W_{r,s}=P_{r,s} \qquad \begin{matrix} r=1,2,\dots,n-1 \\ s=1,2,\dots,m-1 \end{matrix} \quad \text{.......2.1}$$

where $W_{r,s}$ is the function describing the coordinates of nodes in z- direction, $P_{r,s}$ is the vertical static load acting in the node, H_s^x, H_r^y are the horizontal components of forces in cables parallel to x and y, and Δ_r^2 is the second central difference operator with respect to r ($\Delta_r^2 y_r = y_{r-1} - 2y_r + y_{r+1}$) . Equation 2.1 is the equilibrium condition in z-direction. Moreover the following equalities are fullfilled:

$$N_{r,s}^y=H_r^y\left[1+\left(\frac{W_{r,s}-W_{r,s-1}}{a_y}\right)^2\right]^{\frac{1}{2}}, \quad N_{r,s}^x=H_s^x\left[1+\left(\frac{W_{r,s}-W_{r-1,s}}{a_x}\right)^2\right]^{\frac{1}{2}} \quad \text{.......2.2}$$

where $N_{r,s}^x$ and $N_{r,s}^y$ are the forces in the r,s elements of the cable parallel to the x and y axes. The static equilibrium state is the reference configuration for the vibration process. In the case of the free vibrations of the cable net, the kinetic energy of the system can be written as:

$$K=\sum_{r=1}^{n-1}\sum_{s=1}^{m-1}\frac{1}{2}M\dot{v}_{r,s}^2 \quad \text{.........2.3}$$

where M denotes the mass concentrated in the r,s node and $(\dot{\ })=\frac{d}{dt}$. The total potential energy of the system has the form:

$$U=\frac{1}{2}\sum_{s=1}^{m-1}\sum_{r=1}^{n-1}\frac{(N_{r,s}^x+\nu_{r,s}^x)^2}{E_xA_x}L_{r,s}^x+\frac{1}{2}\sum_{r=1}^{n-1}\sum_{s=1}^{m-1}\frac{(N_{r,s}^y+\nu_{r,s}^y)^2}{E_yA_y}L_{r,s}^y-\sum_{r=1}^{n-1}\sum_{s=1}^{m-1}P_{r,s}v_{r,s} \quad \text{..2.4}$$

where ν_{rs}^x, ν_{rs}^y are the dynamic increments of forces in the r,s elements of the cable parallel to x and y axes, and L_{rs}^x, L_{rs}^y, A_x, A_y, E_x, E_y, are lengths, cross sections and elastic modulus of cable elements in the reference configuration, respectivelly. Using the same procedure as in Ref 7 we can obtain the following expressions:

$$\nu_{r,s}^x=h_s^x\left[1+\left(\frac{W_{r,s}-W_{r-1,s}}{a_x}\right)^2\right]^{\frac{1}{2}}, \quad h_s^x=-\Phi_s^x\sum_{r=1}^{n-1}v_{r,s}\Delta_r^2\left(W_{r,s}+\frac{1}{2}v_{r,s}\right)$$

$$\nu_{r,s}^y=h_r^y\left[1+\left(\frac{W_{r,s}-W_{r,s-1}}{a_y}\right)^2\right]^{\frac{1}{2}}, \quad h_r^y=-\Phi_r^y\sum_{s=1}^{m-1}v_{r,s}\Delta_s^2\left(W_{r,s}+\frac{1}{2}v_{r,s}\right) \quad \text{........2.5}$$

where

$$\Phi_s^x=\frac{E_xA_x}{a_x^2(n+\eta_s)}, \qquad \eta_s=\frac{3}{2a_x^2}\sum_{r=1}^{n}\left(W_{r,s}-W_{r-1,s}\right)^2,$$

$$\Phi_r^y=\frac{E_yA_y}{a_y^2(m+\eta_r)}, \qquad \eta_r=\frac{3}{2a_y^2}\sum_{s=1}^{m}\left(W_{r,s}-W_{r,s-1}\right)^2,$$

and h_s^x, h_r^y are the increments of the horizontal components of dynamic forces in cables parallel to x and y, respectivelly. We obtain the above Eqn 2.5 using the geometric relations and Hooke's law and omitting the horizontal inertia forces.

Substituting Egn 2.5 into Egn 2.4 gives

$$U=U_o-\Theta_o(v_{r,s})+\frac{1}{2}\Theta_1(v_{r,s})+\frac{1}{3}\Theta_2(v_{r,s})+\frac{1}{4}\Theta_3(v_{r,s}) \quad \text{........2.6}$$

where

$$U_o=\sum_{s=1}^{m-1}\frac{(H_s^x)^2}{2a_x\Phi_s^x}+\sum_{r=1}^{n-1}\frac{(H_r^y)^2}{2a_y\Phi_r^y},$$

$$\Theta_o=\sum_{s=1}^{m-1}\frac{H_s^x}{a_x}\sum_{r=1}^{n-1}v_{r,s}\Delta_r^2W_{r,s}+\sum_{r=1}^{n-1}\frac{H_r^y}{a_y}\sum_{s=1}^{m-1}v_{r,s}\Delta_s^2W_{r,s}-\sum_{r=1}^{n-1}\sum_{s=1}^{m-1}P_{r,s}v_{r,s},$$

$$\Theta_1=\sum_{s=1}^{m-1}\left[-\frac{H_s^x}{a_x}\sum_{r=1}^{n-1}v_{r,s}\Delta_r^2v_{r,s}+\frac{\Phi_s^x}{a_x}\left(\sum_{r=1}^{n-1}v_{r,s}\Delta_r^2W_{r,s}\right)^2\right]+$$
$$\sum_{r=1}^{n-1}\left[-\frac{H_r^y}{a_y}\sum_{s=1}^{m-1}v_{r,s}\Delta_s^2v_{r,s}+\frac{\Phi_r^y}{a_y}\left(\sum_{s=1}^{m-1}v_{r,s}\Delta_s^2W_{r,s}\right)^2\right],$$

$$\Theta_2=\sum_{s=1}^{m-1}\frac{3\Phi_s^x}{2a_x}\left(\sum_{r=1}^{n-1}v_{r,s}\Delta_r^2W_{r,s}\right)\left(\sum_{r=1}^{n-1}v_{r,s}\Delta_r^2v_{r,s}\right)+\sum_{r=1}^{n-1}\frac{3\Phi_r^y}{2a_y}\left(\sum_{s=1}^{m-1}v_{r,s}\Delta_s^2W_{r,s}\right)\left(\sum_{s=1}^{m-1}v_{r,s}\Delta_s^2W_{r,s}\right),$$

$$\Theta_3=\sum_{s=1}^{m-1}\frac{\Phi_s^x}{2a_x}\left(\sum_{r=1}^{n-1}v_{r,s}\Delta_r^2v_{r,s}\right)^2+\sum_{r=1}^{n-1}\frac{\Phi_r^y}{2a_y}\left(\sum_{s=1}^{m-1}v_{r,s}\Delta_s^2v_{r,s}\right)^2 \quad \text{........2.7}$$

The function denoted by Θ_o is equal to zero what arises from the virtual work equation written for the net treated as the linear system. Let us consider the virtual displacements $\delta v_{r,s}$ in the reference configuration. The elongations of cables elements coresponding to these virtual displecements can be expressed in the form (see Ref 7) :

$$\delta l_{r,s}^x=\frac{1}{L_{r,s}^x}\left(W_{r,s}-W_{r-1,s}\right)\left(\delta v_{r,s}-\delta v_{r-1,s}\right),$$

$$\delta l_{r,s}^y=\frac{1}{L_{r,s}^y}\left(W_{r,s}-W_{r,s-1}\right)\left(\delta v_{r,s}-\delta v_{r,s-1}\right). \quad \text{........2.8}$$

These increments are linear with respect to $\delta v_{r,s}$ because the net is now treated as the linear system. From the virtual work equation for the net and using Eqns 2.2 and 2.8 we obtain

$$\sum_{s=1}^{m-1}\sum_{r=1}^{n} N^x_{r,s}\delta l^x_{r,s}+\sum_{r=1}^{n-1}\sum_{s=1}^{m} N^y_{r,s}\delta l^y_{r,s}=\sum_{s=1}^{m-1}\frac{H^x_s}{a_x}\sum_{r=1}^{n-1}\delta v_{r,s}\Delta^2_r W_{r,s}+\sum_{r=1}^{n-1}\frac{H^y_r}{a_y}\sum_{s=1}^{m-1}\delta v_{r,s}\Delta^2_s W_{r,s}=\sum_{r=1}^{n-1}\sum_{s=1}^{m-1}P_{r,s}\delta v_{r,s}.$$

Assuming in the above equation $\delta v_{rs}=v_{rs}$ we obtain $\Theta_0=0$ and the total potential energy may be finally rewritten:

$$U=U_0+\frac{1}{2}\Theta_1(v_{r,s})+\frac{1}{3}\Theta_2(v_{r,s})+\frac{1}{4}\Theta_3(v_{r,s}) \qquad \text{.........2.9}$$

General form of the Hamilton's principle is:

$$\delta I=\frac{1}{\mathcal{T}}\int_0^{\mathcal{T}}\delta(K-U)\,dt \qquad \text{...............2.10}$$

where $\mathcal{T}$ is the unknown period of the free vibration. The Lagrange's equation and Eqns 2.3, 2.7 and 2.9 lead to the equation of motion of the net

$$M\ddot{v}_{r,s}-\frac{H^x_s}{a_x}\Delta^2_r v_{r,s}-\frac{H^y_r}{a_y}\Delta^2_s v_{r,s}+\frac{\Phi^x_s}{a_x}\Delta^2_r(W_{r,s}+v_{r,s})\sum_{p=1}^{n-1}v_{p,s}\Delta^2_p\left(W_{p,s}+\frac{1}{2}v_{p,s}\right)+\frac{\Phi^y_r}{a_y}\Delta^2_s(W_{r,s}+v_{r,s})\sum_{q=1}^{m-1}v_{r,q}\Delta^2_q\left(W_{r,q}+\frac{1}{2}v_{r,q}\right)=0 \qquad \text{..........2.11}$$

$$r=1,2,\dots,n-1 \quad s=1,2,\dots,m-1$$

The initial and boundary conditions concerning Eqn 2.11 are as follows:

$$v_{r,s}(0)=Z_{r,s},\ \dot{v}_{r,s}(0)=0,\ v_{o,s}(t)=v_{n,s}(t)=v_{r,o}(t)=v_{r,m}(t)=0$$

3. THE SOLUTION OF THE PROBLEM

The aim of the consideration is to find the period solution of the problem. We assume that the solution has the form

$$v_{r,s}=\alpha X_{r,s}T(t) \qquad \text{...........3.1}$$

where α is the amplitude of vibrations of the mass placed in the k, l node, $X_{r,s}$ is so-called "nonlinear normal mode" (see Ref 8,9), and $X_{k,l}=1$ for given k,l node, and T (t) is the periodical function with the period equal to $\mathcal{T}$. The funtcions $X_{r,s}$ and T (t) satisfy the initial and boundary conditions (it means $X_{o,s}=X_{n,s}=X_{r,o}=X_{r,m}=0$, $T(o)=1,0$, $\dot{T}(o)=0$ and $Z_{r,s}=\alpha X_{r,s}$). No other assumptions concerning the shapes of $X_{r,s}$ and T(t) are taken into account.

3.1 THE ITERATIVE PROCEDURE FOR DETEMINING $X_{r,s}$ AND T (t)

The proposed solution has an iterative character. The functions X_{rs} and T(t) will be sequentially corrected in the iteretive procedure. Let us assume, that we know the certain approximation for both founctions denoted by $\underset{i}{T}(t)$ and $\underset{i}{X}_{rs}$ (i means the number of iteration). At the beginning, we assume that $\underset{o}{X}_{rs}$ is proportional to the chosen linear mode of vibrations of the net in the reference configuration and $T=\cos\underset{o}{\omega}\cdot t$, where $\underset{o}{\omega}=\frac{2\pi}{\mathcal{T}_o}$ is the respective linear frequence. If we calculate $X_{r,s}$ and T (t) for the sequence of α parameter values than $\underset{o}{X}_{r,s}$, $\underset{o}{T}$, $\underset{o}{\omega}$ are equal to X_{rs}, T, ω obtained for the previous value of α. Firstly, under the fixed value $X_{rs}\equiv\underset{i}{X}_{rs}$ we look for better approximation of T(t). Substituting Eqns 2.3, 2.9, 3.1 into Eqn 2.10 and using te stationary condition (it means $\delta I=0$) we obtain

$$\mathfrak{M}\,\ddot{T}(t)+\Theta_1 T(t)+\alpha\Theta_2 T^2(t)+\alpha^2\Theta_3 T^3(t)=0 \qquad \text{..........3.2}$$

and the initial conditions $T(0)=1.0$, $\dot{T}(0)=0$. In Eqn 3.2 $\mathfrak{M}=\sum_{r=1}^{n-1}\sum_{s=1}^{m-1} M\underset{i}{X}^2_{rs}$, and Θ_1, Θ_2, Θ_3 have the form Eqn 2.7 after replacing v_{rs} on $X_{r,s}$. This equation is approximate because the coefficients $\mathfrak{M}$ and Θ_i are obtained using the approximation of the function X_{rs}. Equation 3.2 is solved numerically employing Newmark's method described in Ref 11. In this way we obtain the next approximations for T(t) and ω denoted by $\underset{i}{T}(t)$ and $\underset{i+1}{\omega}=\frac{2\pi}{\mathcal{T}_{i+1}}$. Now we look for the new approximation of X_{rs}. We assume that the function $T(t)\equiv\underset{i+1}{T}(t)$ is fixed. Analogously, substituting Eqns 2.3, 2.9, 3.1 into Eqn 2.10 and using the stationary condition $\delta I=0$ we obtain the equation discribing X_{rs}

$$\omega^2 M X_{r,s}=-\frac{\beta_1}{\beta_0}\left(\frac{H^x_s}{a_x}\Delta^2_r X_{r,s}+\frac{H^y_r}{a_y}\Delta^2_s X_{r,s}-\frac{\Phi^x_s}{a_x}\Delta^2_r W_{r,s}\sum_{p=1}^{n-1}X_{p,s}\Delta^2_p W_{p,s}-\frac{\Phi^y_r}{a_y}\Delta^2_s W_{r,s}\sum_{q=1}^{m-1}X_{r,q}\Delta^2_q W_{r,q}\right)+\alpha\frac{\beta_2}{\beta_0}\left[\frac{\Phi^x_s}{a_x}\left(\frac{1}{2}\Delta^2_r W\sum_{p=1}^{n-1}X_{p,s}\Delta^2_p X_{p,s}+\Delta^2_r X_{r,s}\sum_{p=1}^{n-1}X_{r,s}\Delta^2_p W_{p,s}\right)+\frac{\Phi^y_r}{a_y}\left(\frac{1}{2}\Delta^2_s W_{r,s}\sum_{q=1}^{m-1}X_{r,q}\Delta^2_q X_{r,q}+\Delta^2_s X_{r,s}\sum_{q=1}^{m-1}X_{r,q}\Delta^2_q W_{r,q}\right)\right]+\alpha^2\frac{\beta_3}{\beta_0}\left(\frac{\Phi^x_s}{a_x}\Delta^2_r X_{r,s}\sum_{p=1}^{n-1}X_{p,s}\Delta^2_p X_{p,s}+\frac{\Phi^y_r}{a_y}\Delta^2_s X_{r,s}\sum_{q=1}^{m-1}X_{r,q}\Delta^2_q X_{r,q}\right), \quad r=1,2,\dots,n-1,\ s=1,2,\dots,m-1 \qquad \text{........3.3}$$

where

$$\omega^2\beta_0=\frac{1}{\mathcal{T}}\int_0^{\mathcal{T}}\dot{T}^2_{i+1}(t)\,dt\ ,\ \beta_1=\frac{1}{\mathcal{T}}\int_0^{\mathcal{T}}T^2_{i+1}(t)\,dt,\ \ \beta_2=\frac{1}{\mathcal{T}}\int_0^{\mathcal{T}}T^3_{i+1}(t)\,dt,$$

$$\beta_3=\frac{1}{\mathcal{T}}\int_0^{\mathcal{T}}T^4_{i+1}(t)\,dt \qquad \text{.............3.4}$$

The above integrals we calculated numerically using the trapezoidal rule. We notice that in Eqn 3.4_1 appears the square frequence of the net, and this can be explained as follows: Expanding the function T(t) in the Fourier series in the interval (0, $\mathcal{T}$)

$$T(t)=\sum_{l=0}^{\infty}a_l\cos\omega lt$$

and calculating the integral from Eqn 3.4_1 we obtain:

$$\frac{1}{\mathcal{T}}\int_0^{\mathcal{T}}\dot{T}^2dt=\frac{1}{\mathcal{T}}\int_0^{\mathcal{T}}\Big(\sum_{l=0}^{\infty}\omega la_l\sin\omega lt\Big)^2dt=\frac{\omega^2}{\mathcal{T}}\int_0^{\mathcal{T}}\sum_{k=0}^{\infty}\sum_{l=0}^{\infty}kla_ka_l\sin\omega kt\sin\omega lt\,dt=$$

$$\frac{\omega^2}{\mathcal{T}}\sum_{k=0}^{\infty}\sum_{l=0}^{\infty}kla_ka_l\int_0^{\mathcal{T}}\sin\omega kt\sin\omega lt\,dt=\frac{\omega^2}{2}\sum_{l=0}^{\infty}l^2a_l^2=\omega^2\beta_0.$$

Assuming $\omega=\underset{i+1}{\omega_T}$ we determine β_0 from Eqn 3.4_1. Equation 3.3 can be also obtained by Ritz method (see Ref 6,10) . Equation 3.3 defines the nonlinear eigenvalue problem. The solution of the problem is described in the next section of the paper. Now we only notive that $\underset{i+1}{\omega_x}$ and $X_{r,s}$ are the solutions of the nonlinear eigenvalue problem. The iteretive process for the functions $X_{r,s}$ and T(t) is finished if the following inequality is fullfilled:

$$|\underset{i+1}{\omega_x^2}-\underset{i+1}{\omega_T^2}|<\varepsilon_1$$

where ε_1 is the small number. The last approximations of $X_{r,s}$ and T(t) are treated as the solution of the nonlinear free vibrations problem. In general, the problem of convergency in the described above procedure is still open but the numerical calculations presented in this paper lead to the convergent solutions. It is easy to show that the solution depends on α . It is possible to obtain the backbone curve by repeating the iteration procedure for the given sequence of values α.

3.2 THE SOLUTION OF THE NONLINEAR EIGENVALUE PROBLEM

The methods of the solution of the nonlinear eigenvalue problem were presented in Ref 6, 12, 13. The method discribed here is similar to well known vector iteration methods which are used in the solution of the linear eigenvalue problem (see Ref 14,15). The solution of Eqn 3.3 is the iterative one. It is assumed that we know the certain approximations of ω and $X_{r,s}$ denoted by $^j\omega_x$, $^jX_{r,s}$ (j is the number of approximation) . At the beginning $^o\omega_x=\underset{i}{\omega_x}$, $^oX_{r,s}=\underset{i}{X_{r,s}}$. We look for the better approximation of ω and X_{rs}. Let us assume for the moment that

$$Q_{r,s}={}^j\omega_x^2M\,{}^jX_{r,s} \qquad \text{.......3.5}$$

is the fixed and nonhomogenuos term of Eqn 3.3. We can rewrite Eqn 3.3 in the form

$$R_{r,s}(X_{r,s})=Q_{r,s} \qquad {r=1,2,\dots,n-1 \atop s=1,2,\dots,m-1} \qquad \text{..........3.6}$$

where the form of $R_{r,s}$ we can easy determined from the identity condition of Eqns 3.3 and 3.6. If $Q_{r,s}$ determined from Eqn 3.5 is known then Eqn 3.6 can be treated as the nonlinear algebraic equations system and can be solved by modifed Newton-Raphson method. The convergence of the solution process is accelerated by the Crisfield's method (see Ref 16) . The solution of Eqn 3.6 is denoted by $\overline{X}_{rs}$. Usually it is not equal to $^jX_{rs}$ and in particular $\overline{X}_{k,l}\neq 1$. Analogously like in vector iteration methods the next approximations of ω_x and X_{rs} are obtained from

$${}^{j+1}\omega_x^2={}^j\omega_x^2\frac{1}{\overline{X}_{k,l}}\ ,\qquad {}^{j+1}X_{r,s}=\frac{\overline{X}_{r,s}}{\overline{X}_{k,l}} \qquad \text{..........3.7}$$

The iterative solution of the system of Eqns 3.3 is finished when for each r,s pair the following inequality is fullfilled:

$$\|R_{r,s}({}^{j+1}X_{r,s})-Q({}^{j+1}\omega_x^2,{}^{j+1}X_{r,s})\|<\varepsilon_2$$

where ε_2 is an arbitrary small number and $\|\cdot\|$ denotes Euclidean norm.The last reached approximations of ω_x and $X_{r,s}$ are treated as the solution of the nonlinear eigenvalue problem. In this way for the linear and nonlinear eigenvalue problem only the fundamental frequency and mode shape can be obtained (see Ref 14,15). In the next approximations of the arbitrary chosen mode appear the numerical errors which are proportional to the lower modes. These errors increase rapidly and as a result we usually obtain the first mode. To reach the arbitrary frequency and mode of vibrations using the same method the calculation process should be modified, One of the possible way of modification for the linear eigenvalue problem is presented in Ref 17. The main idea of this is

Table 1

α cm	1 frequence k=2, l=2	2 frequence k=2, l=4	3 frequence k=5, l=2	4 frequence k=5, l=4	5 frequence k=2, l=1
linear	9.6455	10.9770	11.7618	12.8767	13.2382
2	9.6577	10.9890	11.7704	12.8850	13.2961
4	9.6939	11.0234	11.7958	12.9097	13.4681
6	9,7543	11.0803	11.8382	12.9507	13.7513
8	9.8384	11.1583	11.8970	13.0073	14.1404
10	9.9460	11.2580	11.9720	13.0789	14.6198
12	10.0774	11.3766	12.0625	13.1644	15.1990
14	10.2312	11.5128	12.1681	13.2624	15.8481
16	10.4073	11.6648	12.2878	13.3713	16.5778
18	10.5983	11.8306	12.4205	13.4891	17.3739
20	10.8220	12.0084	12.5654	13.6137	18.2294
22	11.0510	12.1961	12.7214	13.7430	19.1358
24	11.3008	12.3920	12.8873	13.8745	20.0876
26	11.5667	12.5815	13.0620	14.0063	21.0781
28	11.8536	12.7887	13.2445	14.1364	22.1023

Table 2

r \ s	1	2	3
1	$\frac{0.437016}{0.480740}$	$\frac{0.618034}{0.635091}$	$\frac{0.437016}{0.480740}$
2	$\frac{0.707107}{0.736882}$	1.000000	$\frac{0.707107}{0.736882}$
3	$\frac{0.707107}{0.736882}$	1.000000	$\frac{0.707107}{0.736882}$
4	$\frac{0.437016}{0.480740}$	$\frac{0.618034}{0.635091}$	$\frac{0.437016}{0.480740}$

Table 3

Mode	α cm	Number of iteration for T(t)	Total no of iteration for ω_x	Total no of iteration for $\overline{X}_{r,s}$
1	2	1	3	3
	16	1	5	6
	26	2	9	10
5	2	1	4	4
	16	2	12	16
	26	2	11	17

to eliminate those numerical errors from the solution of Eqn 3.3 by the orthogonalization process. The similar modification is proposed for the nonlinear eigenvalue problems. Let us assume that we know K linear modes of vibration ${}_kX_{r,s}$ which are the solutions of linear eigenvalue problem obtained from Eqn 3.3 for $\alpha=0$. We look for the next k+1 nonlinear mode shape described by $X_{r,s}$. Let us rewrite Eqn 3.3 in the form

$$R_{r,s}(X_{r,s})=Q_{r,s}(\omega,X_{r,s})-\sum_{k=1}^{K}\mu_k M_k X_{r,s}=Q'_{r,s} \qquad \ldots\ldots\ldots 3.8$$

where μ_k are constans. The term $\sum_{k=1}^{K}\mu_k M_k X_{r,s}$ of the above equation is the approximation of the mentioned numerical errors. Multiplying Eqn 3.8 by ${}_1X_{r,s}$ and making the sum over r and s and using the orthogonality conditions of the linear modes we obtain

$$\mu_k=\frac{\sum_{r=1}^{n-1}\sum_{s=1}^{m-1}{}_kX_{r,s}(R_{r,s}-Q_{r,s})}{\sum_{r=1}^{n-1}\sum_{s=1}^{m-1}M_k X_{rs}^2} \qquad k=1,2,\ldots,K \qquad \ldots\ldots\ldots 3.9$$

Determining the chosen mode shope we have to solve Eqn 3.8 instaed of Eqn 3.3. We solve Eqn 3.8 in the same way as Eqn 3.3.

4. THE NUMERICAL EXAMPLE

The results of numerical calculations for the hyperbolic paraboloid cable net are performed. The shape of the cable net in the static equlibrium state is described by:

$$W_{r,s}=\frac{4f_x}{n^2}r(n-r)-\frac{4f_y}{m^2}s(m-s)$$

The following data are assumed to the calculations: n=9, m=7, M=225 kg, H_s^x=const = 104.231 kN, H_r^y=const=39.240 kN, $a_x=a_y$=3m, E_xA_x=78480.0 kN, E_yA_y=34335.0 kN, f_x=1.5m, f_y=1.2m, ε_1=0.1, ε_2=8.0. For various values of parameter α five dominated frequences and associated modes are determined. The obtained results are shown in Tab 1 and Tab 2. Table 1 presents the frequences as the function of α. Two values given in Tab 2 presents the first linear mode and the first nonlinear mode for α= 30cm. Because of antisymmetry of both modes the results are limited to the quarter of the net. Some informations on the convergency of the proposed iteretive procedure are given in Tab 3. It should be noticed that for each amplitude of vibrations the function T(t) had to be determined only once or twice. It is seen from the results obtained that influence of the geometrical nonlinearity on the frequences and modes is significant. The proposed iterative procedure of solving the problem is effective and can be used for the analysis of free vibrations of other geometrically nonlinear structures.

5. ACKNOWLEDGEMENT

This paper is based on an investigation sponsored by the Polish Basic Problem PW 05.12.

6. REFERENCES

1. M.L.GAMBHIR, B.de V.BATCHELOR, Finite element study of the free vibration of 3-d cable networks, Int.J.Solids and Struct., vol.15, 1979, pp 127-136.
2. J.RAKOWSKI, Statische und dynamische Berechnung von Seil-Tragwerkskonstruktionen, Bauingenieur, 56, 1981, pp 343-348.
3. SZ.PAŁKOWSKI, Numerical analysis of frequencies and modes of free vibrations of cable systems, Arch.Inż.Ląd.27, 1981,(in Polish)
4. D.MODJTAHEDI, S.SHORE, The nonlinear steady state response of flexible cable networks, Proc.IASS Pacific Symp. Part II on Tension structures and space Frames, Tokyo, 1972
5. D.HITCHINGS, P.WARD, The nonlinear steady state response of cable networks, Comp.Meth. Appl.Mech.Eng., 9, 1976, pp 191-201.
6. R.LEWANDOWSKI, Resonance vibrations of shallow cable systems, Arch.Inż.Ląd. 26, 1980
7. R.SYGULSKI, R.ŚWITKA, Calculations of rectangular cable nets as discrete systems, Arch.Inż.Ląd., 20, 1974, pp 653-671,(in Polish)
8. M.BERGER, M.BERGER, Perspective in nonlinearity, Benjamin, New York, 1968.
9. R.M.ROSENBERG, On nonlinear vibrations of systems with mony degrees of freedom,Advances in Applied Mech., 9, 1966.
10. W.SZEMPLIŃSKA-STUPNICKA, A study of main and secondary resonances in non-linear multi-degree-of-freedom vibrating systems, Int. J.Non-Linear Mech., 10, 1975 pp 289-304.
11. N.M.NEWMARK, A method of computation for structural dynamics, A.S.C.E., Journal of Eng.Mech.Div., vol.85, 1959, pp 67-94
12. C.MEI, Finite element displacement method for large amplitude free flexural vibrations of beams and plates, Comp.and Struct., 9, 1973.
13. L.C.WELLFORD, G.M.DIB, Finite element methods for nonlinear eigenvalue problems and the postbuckling behaviour of elastic plates,

Comp.and Struct., 6, 1976, pp 413-418.

14.B.KOWALCZYK, Matrices and theirs applications, WNT, Warsaw, 1976 (in Polish) .

15.K.J.BATHE, E.L.WILSON, Numerical methods in finite element analysis, Prentice-Hall, New Jersey, 1976.

16.M.A.CRISFIELD, A faster modified Newton - Raphson iteration, Comp.Meth.Appl.Mech.Eng., 20, 1979, pp 267-278.

17.R.E.D.BISHOP, G.M.L.GLADWELL, S.MICHAELSON, The matrix analysis of vibration, Cambridge University Press, 1965.

DYNAMIC RESPONSE IN FLEXIBLE SYSTEMS OF STRUCTURES TO EARTHQUAKE MOTION

S HAGIESCU*, MSc, PhD

* Scientific Secretary, Building Research Institute, Bucharest, Romania

Roof cable trusses representing new building system are differing from usual structures in notable flexibility and special functions for covering large areas. Especially when having in view unusual structures noticed in new building system, long span or function involving crowd of people inside, more accurate evaluation is desired. Analysing the behaviour to earthquake motion of suspended roof truss systems it appears that low state of efforts has to be expected. Natural large periods of vibration lead to focus major interest in the displacement response. Calculus techniques considering the geometrical nonlinearity providing proper results in observed vulnerability is presented. Important vertical displacements appear in suspended wide span structure in the majority of first natural modes of vibration. High vulnerability consisting in cladding damage could be predicted, in the case when amplification of seismic motion and $T_g \geqslant 1{,}7$ s should occur.

INTRODUCTION

Typical dynamic natural response to earthquakes represent significant characteristics of long span cable structures. The analysis of the vulnerability of given categories of structures provides synthesis process on the evaluation of dynamic behaviour, under variable ground motions, needed for rational design to resist earthquakes. There are required several steps in current design practice, some additional steps in earthquake resistant design under code specifications. For special, unusual structures, of structure function, long span or new building system, accurate specific examination are desired.

Roof cable trusses representing new structural solution, differing from usual buildings in the notable flexibility, covering large span buildings for cultural-social or sport functions, are discussed. Each of the previous mentioned particularities involves more complex and rigorous methods.

The high flexibility of the cable in addition to the wide span, lead to large natural fundamental periods exceeding the value of 2 sec ($T_1 > 2$sec). Analysis in the behaviour to earthquake motions, of suspended roof truss classes of systems with light cladding, directs to the particular judgement that low value of seismic loads applied, at points where masses are concentrated into the direction of the degree-of-freedom, has to be expected consequently, modest state of efforts is developed by seismic loads. Natural low circular frequency in the fundamental modes leads to focus major interest in the response displacement, being expected the amplification of ground movement. It must be emphasized expected vulnerability because of important displacement due to ground motion.

THEORETICAL INVESTIGATION TECHNIQUE

The given class of structure requires nonlinear approach in the geometric second order assumption, due to important translation in the truss knots and bars rotation of the system. This particularity has to be considered even in the case of preliminary estimation. The unusual dynamic characteristics connected to ground motion, provide to rigorous analysis, more complex, consisting in to subject the entire dynamic system with all of its natural modes to the entire time history of a given real earthquake or of a specified artificial earthquake.

Seismic loads acting on the structure as load time relationship is considered. Simplified expression representing the earthquake motion and simplified assumption by means of coefficients is taken into account in order to perform time history of displacements. By modifying the involved coefficients (step-by-step technique), investigation on predicted behaviour as well as analysis concerning roofs vulnerability on a previous earthquake could be obtained.

Consider the Eqn 1, sec Ref 1, to express motion into the direction of the k degree-of-freedom, taking into account i natural modes of vibration,

$$\mathcal{M}_i \, \mathcal{Y}(t) + \mathcal{D}_i \, \mathcal{Y}(t) + \mathcal{R}_i \, \mathcal{Y}(t) = \mathcal{F}_i(t) \qquad \ldots\ldots \; 1.$$

where, $\mathcal{M}_i = \{\emptyset\}_i^T [\mathcal{M}] \; \{\emptyset\}_i$ represents the generalized mass, $\mathcal{D}_i = \{\emptyset\}_i^T [\mathcal{D}] \; \{\emptyset\}_i$ the generalized damping, $\mathcal{R}_i = \{\emptyset\}_i^T [\mathcal{R}_{II}] \; \{\emptyset\}_i$ the generalized stiffness, and $\mathcal{F}_i(t) = \{\emptyset\}_i^T \{\mathcal{F}(t)\}$ the generalized seismic load as time dependent;

$\{\emptyset\}_i$ is the modal matrix containing the eigen vectors, $[\mathcal{R}_{ii}] = [R_k] + [R_F]$, the stiffness matrix considering geometrical nonlinearity influence, see Ref 2, $[R_k]$, the stiffness matrix corresponding to the extension stiffness, and $[R_F]$, the stiffness matrix corresponding to the rotation stiffness.

The total displacement as a sum in terms of either displacements evaluated in natural modes, i, accordingly to the degree-of-freedom direction, k, is deduced by,

$$x_k(t) = \sum_{i=1}^{n} \emptyset_{k,1}\, \mathcal{Y}_i(t) \qquad \text{...... 2.}$$

where the generalized coordinate, $\mathcal{Y}_i(t)$, involving the amplitudes of displacements is Duhamel convolution integral defined as time dependent by Eqn 3,

$$\mathcal{Y}_i(t) = \frac{1}{\mathcal{M}_i\, \omega_i^*} \int_0^t \mathcal{F}_i(\tau)\, e^{-\nu_i \omega_i (t-\tau)} \sin \omega^*(t-\tau)\, d\tau \qquad \text{...... 3.}$$

and the circular frequency of natural vibration $\omega^* = \omega_i \sqrt{1-\nu^2}$ depends on critical damping. When ν is neglected, Eqn 3 becomes,

$$\mathcal{Y}_i(t) = \frac{1}{\mathcal{M}_i\, \omega_i} \int_0^t \mathcal{F}_i(\tau) \sin \omega_i (t-\tau)\, d\tau \qquad \text{..... 4.}$$

Making use of Fourier transformation it is estimated to take some advantages like easily physical phenomenom deduction as well as possibility in applying superposition technique - pointed out by the present investigation.

It is possible to construct an idealized relation for seismic loads as harmonic functions, represented by Eqn 5,

$$\mathcal{F}_i(t) = \{\emptyset\}_i^T \{F_o\} \sin \theta t \qquad \text{...... 5.}$$

where the amplitude of the seismic loads is $\{\emptyset\}_i^T \{F_o\}$ and θ rads/sec, the circular frequency, when it is quite apparent that there exists simplification leading to the time relationship,

$$\mathcal{Y}_i(t) = \mathcal{A}_i\, \mathcal{M}_i^* \left(\sin \theta t - \frac{\theta}{\omega_i} \sin \omega_i t\right) \qquad \text{...... 6.}$$

The displacements on the degree - of - freedom direction, k, in the natural vibration modes, i, are represented by the expression,

$$x_{k,i}(t) = \emptyset_{k,i}\, \mathcal{Y}_i(t) \qquad \text{...... 7.}$$

Appreciating undamped structure, Eqn 6 and 7, verge to,

$$x_{k,i}(t) = \emptyset_{k,i}\, \mathcal{A}_i\, \mathcal{M}_i \left(\sin \theta t - \frac{\theta}{\omega_i} \sin{}_i t\right) \qquad \text{...... 8.}$$

where the dynamic coefficient,

$$\mathcal{M} = \frac{1}{1 - \left(\frac{\theta}{\omega_i}\right)^2} \qquad \text{...... 9.}$$

put to evidence n possibilities of resonance if existing i = 1...n eigen modes.

The coefficient,

$$\mathcal{A}_i = \frac{1}{\omega^2} \; \frac{L_s}{2E_d} \qquad \text{...... 10.}$$

express the participation of seismic loads through the mechanical work, L_s, accomplished because of the generalized displacements, and the deformation energy, E_d, as kudratic form related to generalized displacements.

FLEXIBLE STRUCTURE RESPONSE ANALYSIS

Displacement state on concave cable girder class of structure, with cladding on the bearing cable which in that particular case, is on the upper side of the girder, is analysed. The dynamic state of displacements, performed on the multidegree-of-freedom direction, k, indicated by Eqn 8, in the specific case of 26 degree-of-freedom for cable truss system having 70m span will be appreciate.

Shape coefficient

The shape coefficient, $\emptyset_{k,i}$, and the circular frequency ω_i where obtained by applying the displacement method in the second order assumption as result of geometric non-linearity. Calculus was conducted by automatic procedures, considering the Eqn 11 system as characterizing the natural vibrations,

$$\left([R] - \omega^2 [M]\right) \{\emptyset\} = \{0\} \qquad \text{...... 11.}$$

The eigen values $_i$ - corresponding to natural circular frequency - defining spectrum matrix, can be obtained by means of the Eqn 12,

$$\triangle(\omega) = \left|[R] - \omega^2 [M]\right| = 0 \qquad \text{...... 12.}$$

The modal matrix $\{\emptyset\}$, constituted by the eigen vector, $\emptyset_{k,i}$, defining natural modes of vibration shapes, results from succesive introduction into Eqn 11 of natural circular frequency resulting from Eqn 12.

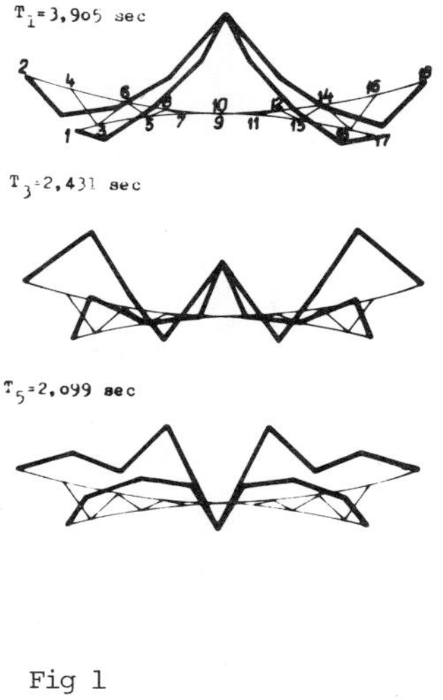

Fig 1

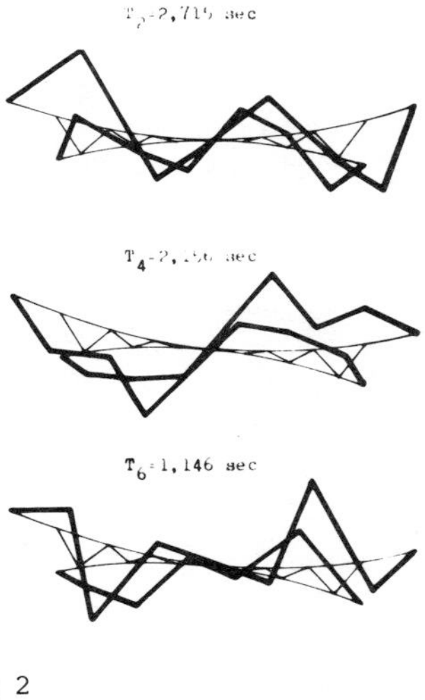

Fig 2

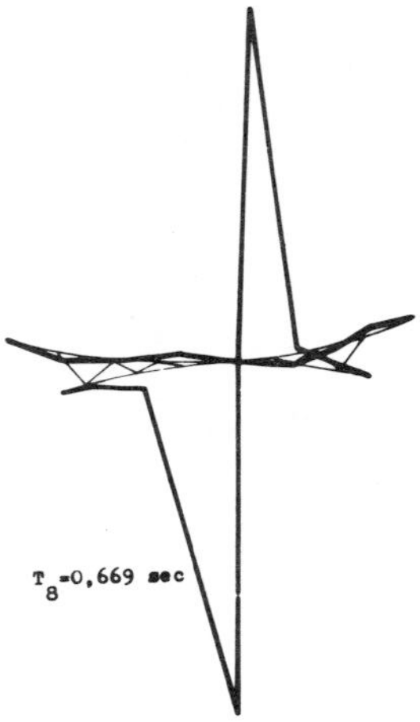

Fig 3

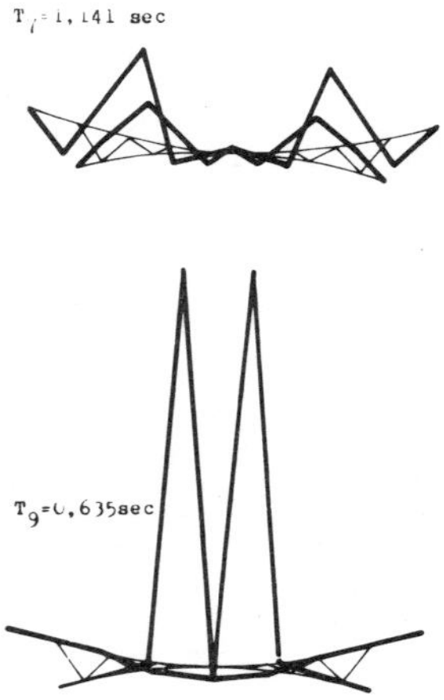

Fig 4

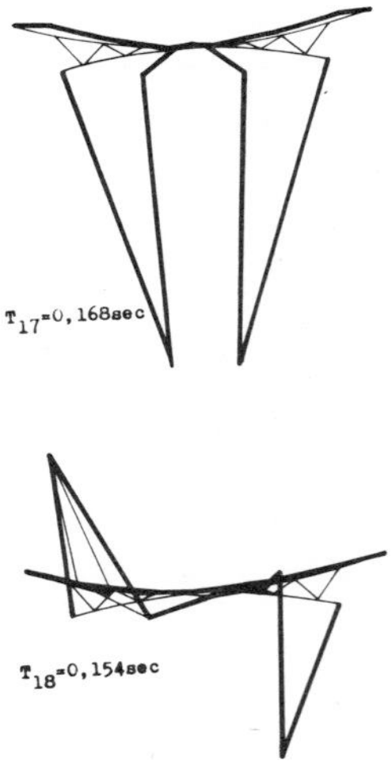

Fig 9

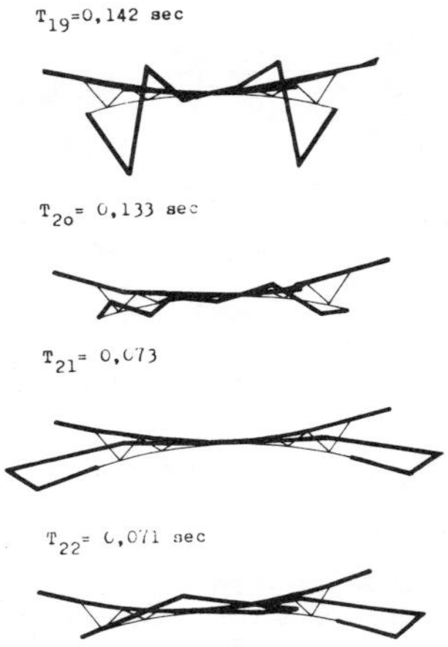

Fig 10

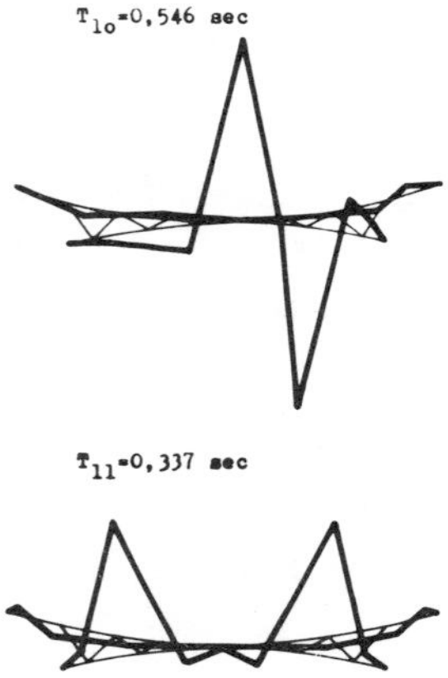

Fig 5

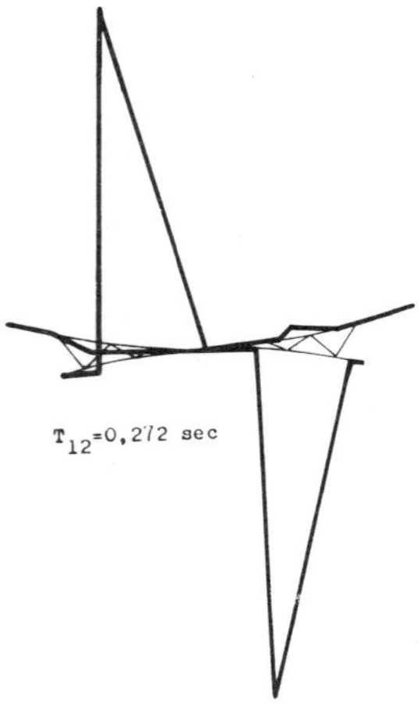

Fig 6

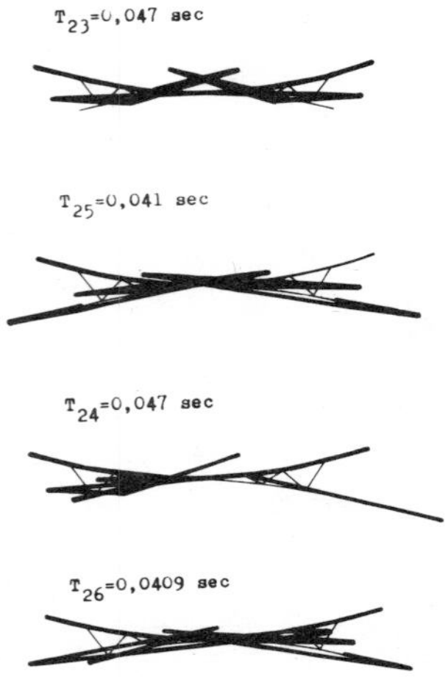

Fig 11

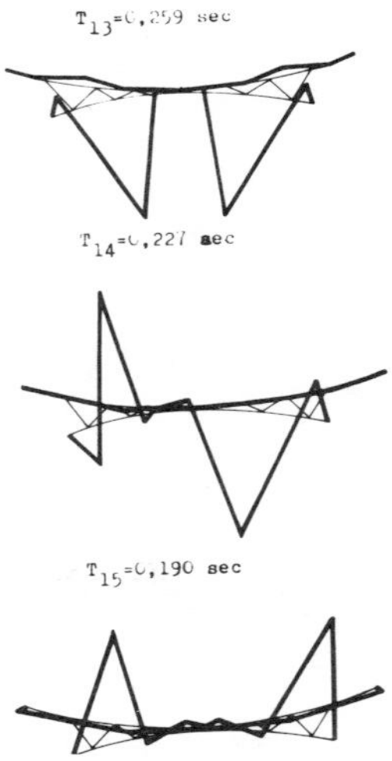

Fig 7

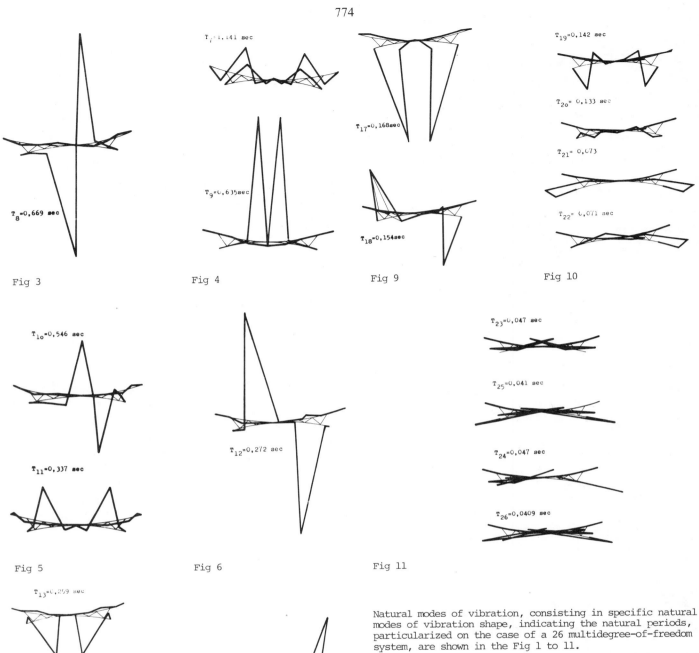

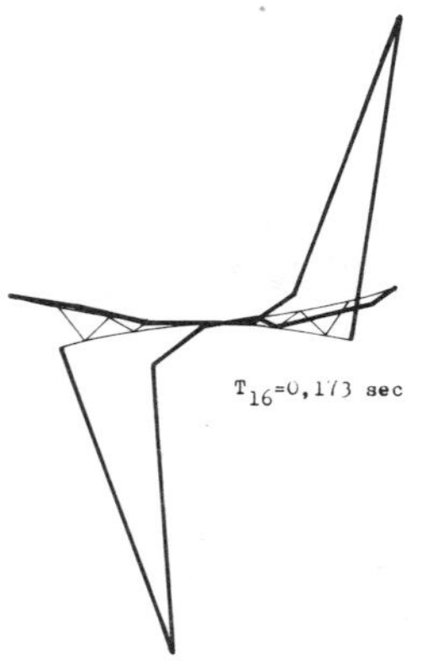

Fig 8

Natural modes of vibration, consisting in specific natural modes of vibration shape, indicating the natural periods, particularized on the case of a 26 multidegree-of-freedom system, are shown in the Fig 1 to 11.

The fundamental mode has large pariod, T_1 = 3,905 sec., and vertical oscillation with maximum displacement in the central zone. The first modes of vibration - mode 1 to mode 19 - indicate vertical displacements. Long periods of vibration - mode 1 to mode 7 - produce high oscillations of bearing cable. The majority of natural modes are indicating vertical movement - the first 19 modes - as shown in Fig 1 to 10.

Motion of strainer cable has to be noted in the same central zone, maximum response of trusses knots advancing to lateral zones in the following natural modes.

In the central knot of the trusses appears the first horizontal oscillation in the 2o th mode;the next modes contribute by high horizontal oscillation of the strainer cable knots.

Participation coefficient

By analyzing the participation coefficient displayed in Fig 12, mainly contribution of intrinsec characteristics on the amplification due to the structure is observed. Important influence induced by first odd mode shapes (mode 1,3 and 5) could be noted in the case of equivalent static evaluation.

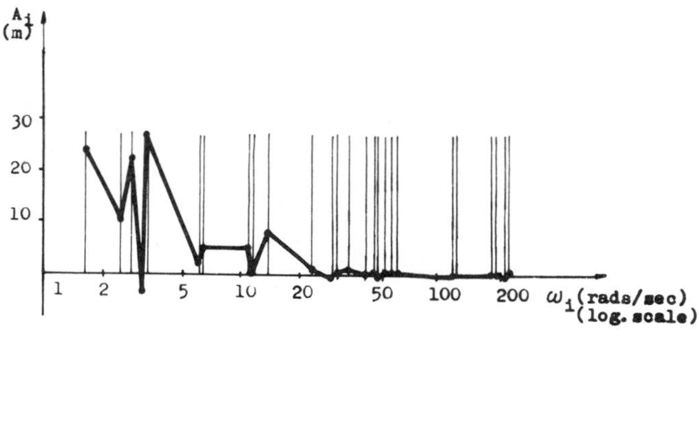

Fig 12

Dynamic estimation put into evidence the possible major contribution in each of modes shape connected to dynamic amplification produced by variable earthquake motion.

The contribution of first natural modes (mode 1-10) has significant values in the total displacement evaluation. It is of interest to examine the displacements, $x_{k,i}$ developed in the degree-of-freedom direction having high oscillations taking into account the participation coefficient.

Dynamic amplification

Amplification influence resulting by appreciating the corelation between dynamic load and intrinsec characteristics shown by Fig 13 results from examination of dynamic coefficient. The possibilities of resonance are situated close by natural circular frequency, especially in the initial low values;

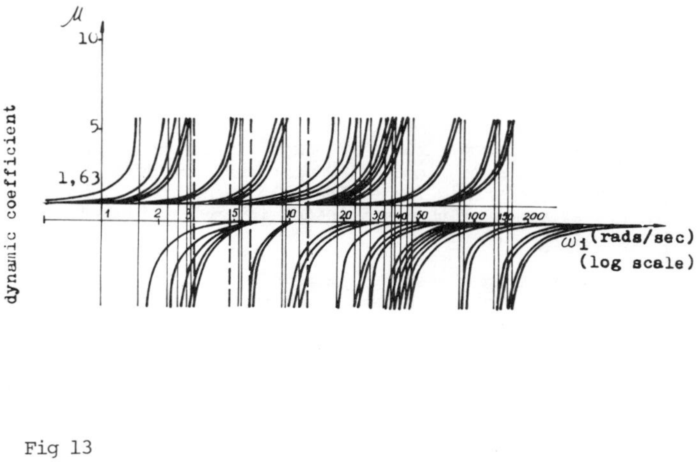

Fig 13

in case of high natural frequency, could be observed amplification induced on larger range of seismic circular frequency.

VULNERABILITY ANALYSIS

Estimation concerning the observed and predicted analysis is considered, see Ref 3.

Vulnerability cable truss analysis under Vrancea ground motion in 4 March 1977

The motions of the ground produced by seismic waves emanating from Vrancea region in Romania has large prominent periods of vibration T_g. Ground measurements available refering to the strong earthquake occured on 4 March 1977, indicate important amplifications in the period range 1...1,6s in N-S direction and 0,7...1,2s on the E-W direction.

If observation focuses on structures having boundary and support structure of high stiffness, some remarks could be considered on the displacements occured in the existing building at that time being (the Polyvalent Hall and Skating Rink Covering 23 August both in Bucharest).

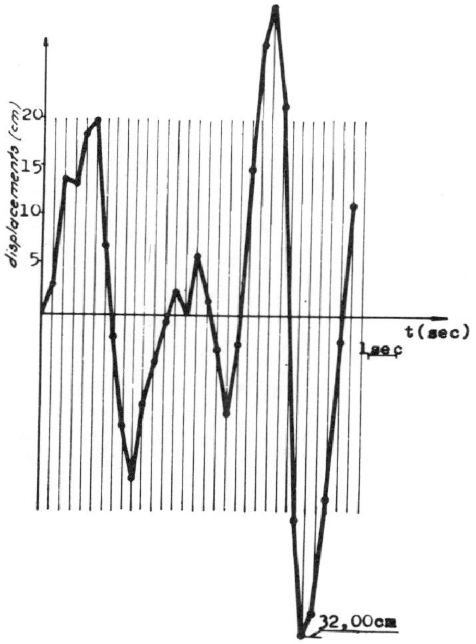

Fig 14

The time history of the bearing cable displacements, on duration of 10s considering the natural period of ground motion, $T_g = 1{,}5$s, shows that amplitude of oscillations on central zone of about 32cm appears, as shown in Fig 14 indicating the time history of vertical displacements on knot 10 of the cable girder. The displacements decrease towards the boundaries, as it could be observed from Fig 15 displaying the time history of vertical movement on knot 6 of the cable girder. It is known that during the earthquake only the dead and seismic load was acting the structure. Displacements of about 80cm was established in the structure design considering static load assumption.

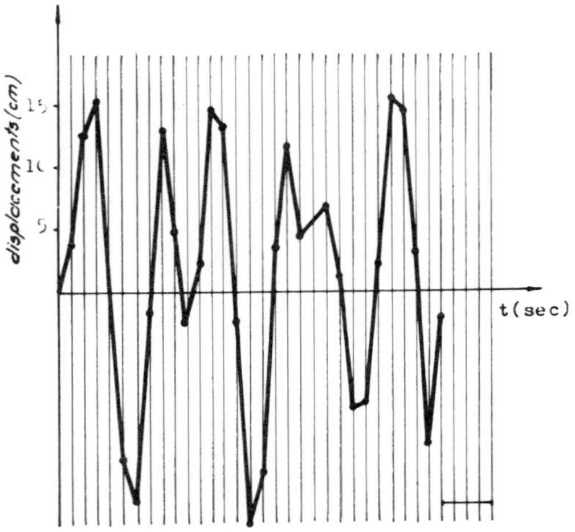

Fig 15

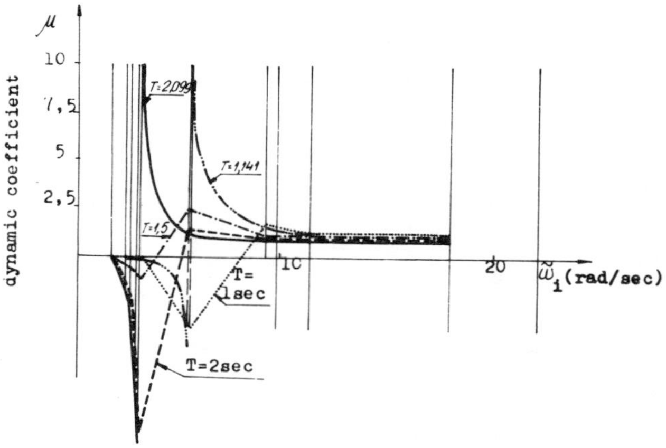

Fig 16

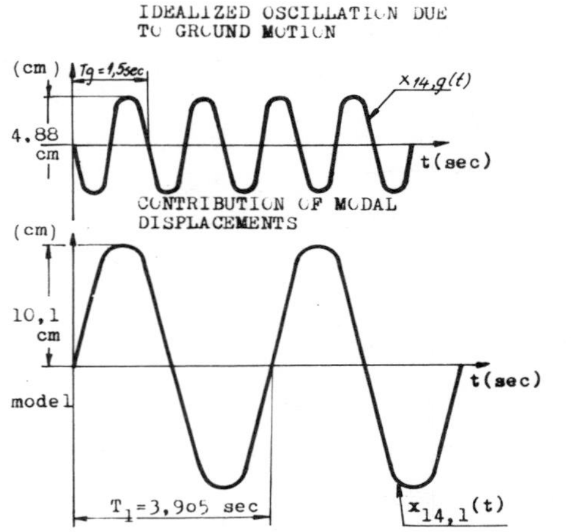

Fig 17a

Consequently the roof have had satisfying behaviour under seismic loads. It exists a good correspondance between the observed vulnerability and that calculated by applying the calculus technique previously presented.

Predicted vulnerability analysis on flexible cable truss roof

Predicted analysis concerning future behaviour by evidencing probable occurence is discussed.

In particular case of this structure increased vulnerability appears in earthquakes on large prominent periods in the range T_g = 2...4s. Connected to that remark, increased vulnerability due to gust of wind, in the flexible presented structure, could be expected.

Seismic loads dynamic characterized by periods of 2s. could easily occur by reason of:

- specific ground condition, amplifying natural periods of ground shaking;
- amplification due support and boundary structure, neglected in the previous presented analysis.

Natural ground motion periods in the range T_g = 1,5...2s. promote important dynamic amplification, if considering dynamic coefficient - circular frequency relationship indicated in Fig 16. It could be remarked the contribution of the first five natural modes of the structure vibration in estimating total maximum displacements.

Physic phenomenom could be easier investigated by decomposing the displacement time dependent curve for knot oscillation into component harmonic curves of the displacement induced from natural modes and superposed on idealized harmonic ground oscillation displayed in Fig 17 and 18. The time history of vertical motion of knot 10 indicated in Fig 14 is composed by the superposition of idealized oscillation due to ground motion and contribution of modal displacements as shown in Fig 17a,b. Some decomposition of time history of vertical movements of knot 6 is indicated in Fig 18a,b. The decomposition put into evidence contribution of major importance of natural modal displacements in displacement response of flexible structure to dynamic loads. It could be noticed also major influence of the oscillation induced in higher modes of vibration.

Displacements in the central zone of the truss from occurence of variable seismic events, are expected to have main values. In that respect for T_g = 1,7s., displacements from about 85cm are evaluated ; a period T_g = 2 s. could produce displacement of 1,30m (if considering natural high damping ν = 0,05).

The present estimation with emphazis on multidegree-of-freedom procedure, evidences the occurence of high

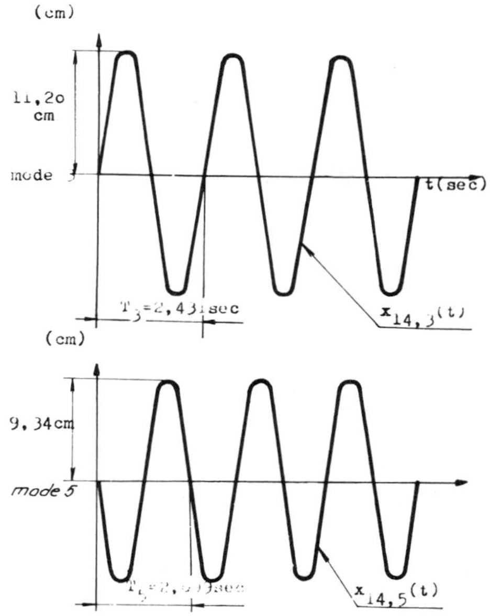

Fig 17b

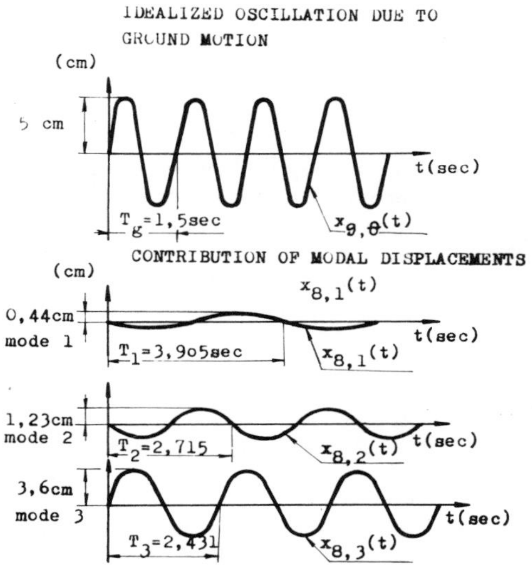

Fig 18a

displacements leading to possible damage of the cladding.

CONCLUSIONS

- The analysis considering the geometrical nonlinearity provides proper results verified in observed vulnerability of the cable truss roofs subjected to the earthquake occurred on 4 March 1977 on the Romanian teritory.

- The displacements time history analysis by means of Fourier curves conduce to easily theoretical verification on observed vulnerability to passed earthquakes, as well

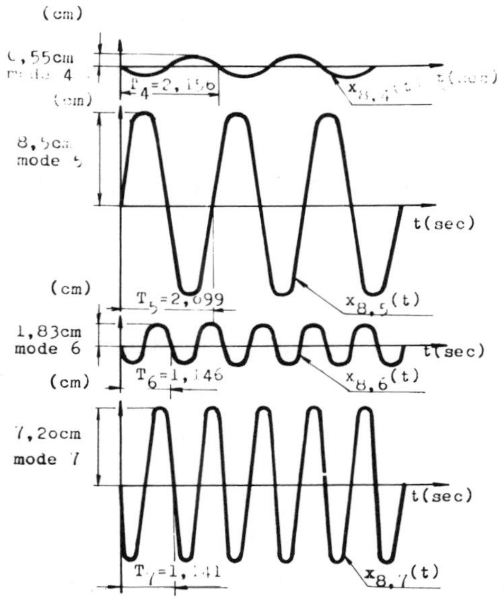

Fig 18b

as easily estimation concerning predicted vulnerability to future ground motions.

- High vertical displacements appear in the suspended long span structure under ground movement. Large period, specific to fundamental and initial modes are characterizing the class of suspended structures.

- The vulnerability of above discussed structures consists in cladding damage, in the case when amplification of seismic motion on $T_g \geq 1{,}7$s. should occur.

REFERENCES

1. N NEWMARK and E ROSENBLUETH, "Fundamentals of Earthquake Engineering" Prentice Hall, Englewood Cliffs, N.Y. (1971).

2. S HAGIESCU and M STEFANESCU, "Analysis of the Actual Behaviour of the Sports Palace Roof Structure in Bucharest". Second International Conference on Space Structures, Guildford, September 1975.

3. S HAGIESCU, "Seismic Risk Analysis of Suspended Roofs", Seminar on Vulnerability Analysis, Bucharest , december 1983.

ON THE ELASTIC INTERACTION BETWEEN ROPE NET AND SPACE FRAME ANCHORAGE STRUCTURES

M MAJOWIECKI and F ZOULAS***

*Istituto di Tecnica delle costruzioni, Facoltà di Ingegneria
Università di Bologna, Italy

**Computer Aided Design Laboratory
Bologna, Italy

The main object of this paper is to illustrate the design procedure and computational methodology for pre-stressed rope net structures anchored to elastically collaborating edge structures. A first paper on the subject was presented in Ref 5. In the present paper the authors intend to examine deeply the interaction between the elastic anchorage system and rope net structures. The parameters involved in this comparative analysis are: 1) mesh net dimension, 2) mechanical elasticity of the anchorage system. It is important to observe that for some values of stiffness ratio between the anchorage and net structures is no longer possible to clear the actual mesh net in order to reduce the degrees of freedom and, consequently, the computational effort. The pre-stressing procedure, from slack cables, is simulated with the same algorithm proposed in the paper. This analysis permits the reduction of pre-stressing steps to a minimum with the relative saving in time and cost.

INTRODUCTION

In the field of structural roof design using a pre-stressed rope net anchored to elastically deformable edge structures (Fig 1), two problems are seen to dominate: numerical analysis by computer; optimal determination of some mechanical parameters so as to keep overall building costs to the minimum.

As regards the first problem, structural analysis of the rope net, elastically collaborating with the anchorage structures was faced according to the following schemes:

a) Continuous analysis of the net as a pre-stressed, anticlastic membrane; the edge anchorage structure is dimensioned with the forces transmitted by the membrane calculated with fixed end boundary conditions.

b) Analysis of the rope net with displacements permitted only in a vertical direction (one degree of freedom per node). The real rope net is changed to an equivalent one, with ropes fewer in number, in order to reduce the number of unknowns. The elastic edge structure is analysed simultaneously to the net or by substructuring with the same method of structural analysis (displacement or force method).

c) Analysis of the rope net with three degrees of freedom per node. The edge and the net are analysed globally or by substructuring within the bounds of the same analysis method.

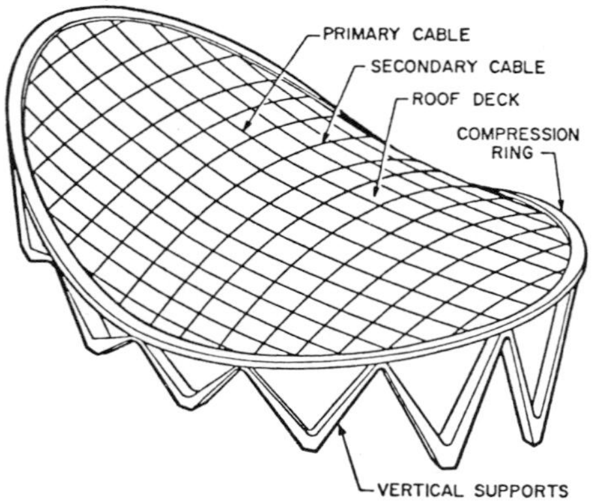

Fig 1

In agreement with the above, studies in Refs 1-4 were written. Definitions a) and b) introduce approximations that considerably restrict the field of structural investigation; considering fixed end constraints makes for an anti-economic solution; continuous membrane analysis is only possible for some analytically expressible surfaces and only for some load conditions. Consideration of rope nets contained in parallel planes is limitative, permitting no consideration of either more complex forms or loads which are not vertical, thus restraining the analysis to only those surfaces which are sufficiently flat. The definitions of structural analysis according to c) often lead to numerical problems principally manifested in the following ways:

- high conditioning number $C(K) = \lambda_{max}/\lambda_{min}$ of the matrix of the coefficients generated by the equilibrium method. The rope net is dominated by the slow vibration modes so the stiffness matrix, generated by the equilibrium method and expressed in constant-length computer words, reduces the information associated with the slowest vibration modes; as can be seen by expressing the stiffness matrix in terms of the eigenvectors and eigenvalues (Ref 6),

$$[K] = \sum_{i=1}^{n} \lambda_i \{V_i\} \{V_i\}^T \qquad \ldots\ldots\ 1,$$

λ_i = i-th eigenvalue,
V_i = the corresponding vibration mode.
To avoid truncation errors, especially when using computers that work with few bits, it is advisable to use the forces method in analysing the rope net. In this case we have

$$[K]^{-1} = \sum_{i=1}^{n} \frac{1}{\lambda_i} \{V_i\}\{V_i\}^T \qquad \ldots\ldots\ 2,$$

thus numerical truncation takes place on the stiffness vibrating modes;

- ill-conditioning of the coefficients matrix, both for the equilibrium method and that of the forces, due to the presence, in the case of global solution, of structural parts which are considerably stiffer than others. Ring stiffness, in the practical cases met with, especially with reinforced concrete rings, is considerably greater than that of the rope net, particularly in the direction of the normal at the structure surface. The finite elements, like beams with 6 degrees of freedom per node used for the edge structure as well as bar elements, monoaxially stressed, with three degrees of freedom per node, generate, for each degree of freedom, rows of the coefficients matrix numerically quite different. In the decay of the main diagonal during the Gauss solution procedure one often meets numbers near to machine zero. Similar situations are met with in

$$[K] = \begin{bmatrix} K_A & -K_A \\ -K_A & K_A + K_B \end{bmatrix} \qquad \ldots\ldots\ 3,$$

with $K_A >> K_B$. $[K]$ is accurately represented only if K_B is not lost through truncation with respect to K_A;

- solution using the equilibrium method of structures with different behaviour in the geometric and material non-linearity field. The rope net has typical geometric hardening behaviour while the edge has geometric and material softening behaviour. The numerical effort is minimized by adopting the equilibrium method for structures with softening behaviour, while the forces method is more suitable in incremental solution of the hardening type of non-linearity.

As regards determination of the most suitable mechanical and structural parameters to consider in the design phase, the present paper aims to evaluate the influence of the following variables: (i) deformability of the anchorage structure; (ii) dimensions of the rope net mesh.

As for the elastic interaction between the net and edge during the pre-tension and "0" state phase, the technical literature makes no questions. Some considerations on this problem are made alongside illustration of the numerical examples.

NOTES ON THE ANALYSIS METHOD BY FUNCTIONAL SUBSTRUCTURING

The method adopted for structural analysis has already been illustrated in a general manner in Ref 6. As for the specific case, the subject of this paper, the following hypotheses are made:

- mixed analysis for functional substructuring;
- softening material edge structure analysis, small displacements and deformations by means of the equilibrium method;
- analysis of the rope net in geometric hardening, large displacements and small deformations by the forces methods;
- iterative analysis by elementary physical substructuring of each rope withn the rope net structure.

Figure 2 illustrates the substructuring carried out, where:

$S_{I;1}$ = edge substructure;
$S_{II;i}$ = net substructure formed of F substructures elementary;
i = 1-F ropes.

It is thought that we have now obtained, with one of the methods indicated in Refs 5,6 the solution of state "0" and, therefore, to have noted:

$(X^\circ;Y^\circ;Z^\circ)_j$; j = 1-n = "0" state coordinates of the nodes,
$L^\circ_{\overline{kj}}$; $\overline{kj}$ = 1-m = "0" state length of the bars,
$S^\circ_{\overline{kj}}$; $\overline{kj}$ = 1-m = "0" state stress in the bars,
P°_j ; j = 1-n = "0" state load in the bars.

Let us now consider I and II interconnected in the set of points B , see Fig 2.

By separating the displacements, one gets:

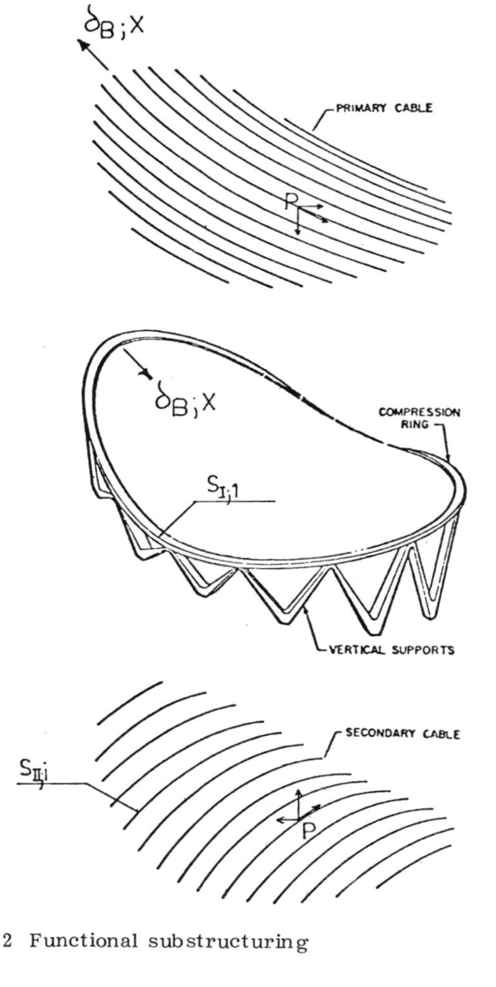

Fig 2 Functional substructuring

$$\left[\begin{array}{cc|c} K_{AA} & K_{AB} & 0 \\ K_{AB}^T & K_{BB} & G_B \\ \hline 0 & -G_B^T & F \end{array}\right] \left\{\begin{array}{c} \delta_A \\ \delta_B \\ \hline X \end{array}\right\} = \left\{\begin{array}{c} P_A \\ P_B \\ \hline \delta \end{array}\right\} \quad \ldots\ldots \; 4,$$

where: K = stiffness matrix of substructure I relative to the A (internal) and B type boundary nodes; F = flexibility matrix of the substructure II; G = coupling matrix between I and II; δ_A; δ_B = displacement vectors in I; X = indeterminate and boundary forces; P_A;P_B = load term in I; δ = displacement term in II.

The solution to Eqn 4 can be obtained in displacement terms by considering:

$$\left\{\begin{array}{cc} K_{AA} & K_{AB} \\ K_{AB}^T & K_{BB}^* \end{array}\right\} \left\{\begin{array}{c} \delta_A \\ \delta_B \end{array}\right\} = \left\{\begin{array}{c} P_A \\ P_B^* \end{array}\right\} \quad \ldots\ldots \; 5,$$

with:

$$K_{BB}^* = K_{BB} + G_B F^{-1} G_B^T \quad \ldots\ldots \; 6,$$

and

$$P_B^* = P_B - G_B F^{-1} \delta \quad \ldots\ldots \; 7,$$

where K_{BB}^* and P_B^* are interpreted respectively as the stiffness matrix and load vector, transformed by the effect of the common node displacements, and I and II of type B, equivalent to static condensation of substructure II.

In the non-linear field, where it is necessary to act interactively and/or incrementally, Eqns 5, 6 and 7 suggest adoption of a sequential relaxation iterative procedure which includes the following phases:

a) Definition of the boundary forces in II.

b) Evaluation of first-attempt $P_B^{*(1)}$, Eqn 7.

c) Evaluation of first-attempt $K_{BB}^{*(1)}$, Eqn 6.

d) Solution of Eqn 5.

e) Return to a).

f) Convergence check.

As regards phase a), solution of the net and, therefore, systematic preparation of the $\lfloor F \rfloor$ and $\lfloor G_B \rfloor$ takes place following the method indicated at point 4.2 of Ref 6.

Without taking anything away from the generality of the method, a first value of the forces in the net may be obtained by exploiting the physical, hypostatic characteristic of the net. Considering the net to be formed by m bars and n nodes one has, in general:

$$3\,n > m \quad \ldots\ldots \; 8.$$

The equilibrium conditions

$$\lfloor A \rfloor \; \{S + \Delta S\} = \{P + \Delta P\} \quad \ldots\ldots \; 9,$$

where: $\lfloor A \rfloor$ = matrix of the cosinus directors by undeformed geometry; $\{S^\circ + \Delta S\}$ = vector of the forces on the bars after load variation; $\{P^\circ + \Delta P\}$ = load vector, present an $\lfloor A \rfloor$ rectangular matrix with 3n rows and n columns.

A very useful first-attempt value, to diminish the iterative cycles, can be obtained by the least squares method, minimizing the non-equilibrated load residue:

$$\left\{\begin{array}{l} \lfloor A \rfloor \; \{S^\circ + \Delta S\} - \{P^\circ + \Delta P\} = \{R\} \\ \{R\}^T \; \{R\} \longrightarrow \min \end{array}\right. \quad \ldots\ldots \; 10.$$

The flow diagram, in accordance with the above, is synthesized in Fig 3.

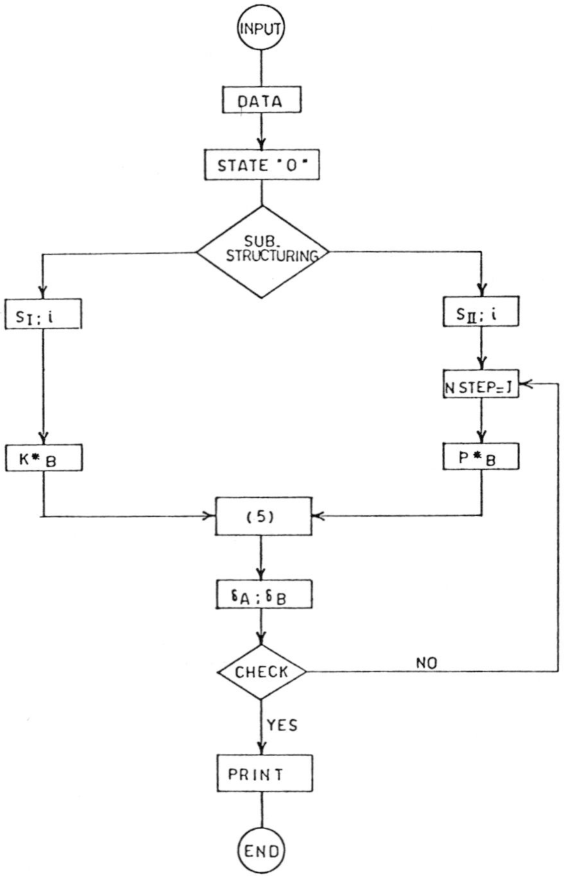

Fig 3

NUMERICAL EXAMPLES

According to the analysis method illustrated section above, we intend to examinate a rope net structure anchored to an elastic space frame, loaded with a vertical uniform load of 800 KN/M^2.

The rope net adopted is of variable mesh net and the edge structure is a ring space frame which geometry is defined as the intersection of a hyperbolic paraboloid surface with a circular cylinder of 60 m diameter, analytically described by:

$$\begin{cases} \dfrac{X^2}{40,909} - \dfrac{Y^2}{64,282} = Z \\ X^2 + Y^2 = 900 \end{cases} \qquad \ldots\ldots \ 11.$$

The variation of the mesh net dimension is illustrated in Fig 4.

The initial stress state (STATE "0") in the rope net is defined, in function of the mesh net dimension (i) as:

$$H_p = 4000 \cdot i \ ; \quad H_s = 6286 \cdot i$$

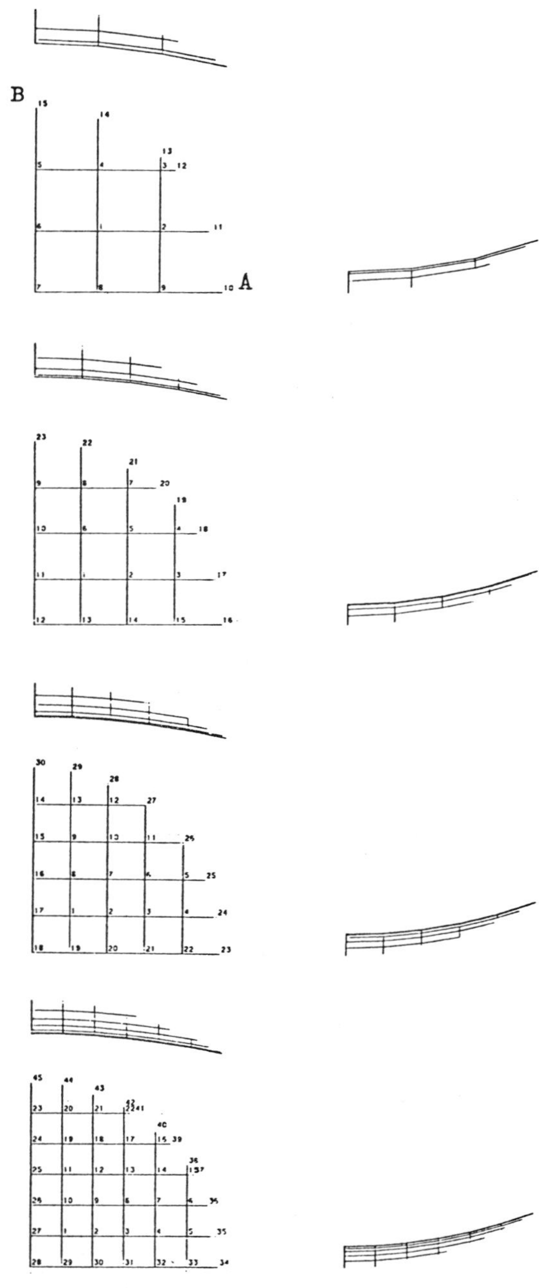

Fig 4 Mesh-net configurations

where: H_p = horizontal component of carrying rope forces;
H_s = horizontal component of stabilizing rope forces.

The anchorage structure is represented by a space frame only vertically supported in discrete joints.

The area and inertia parameters of the cross section of the ring structure are indicated in Table 1.

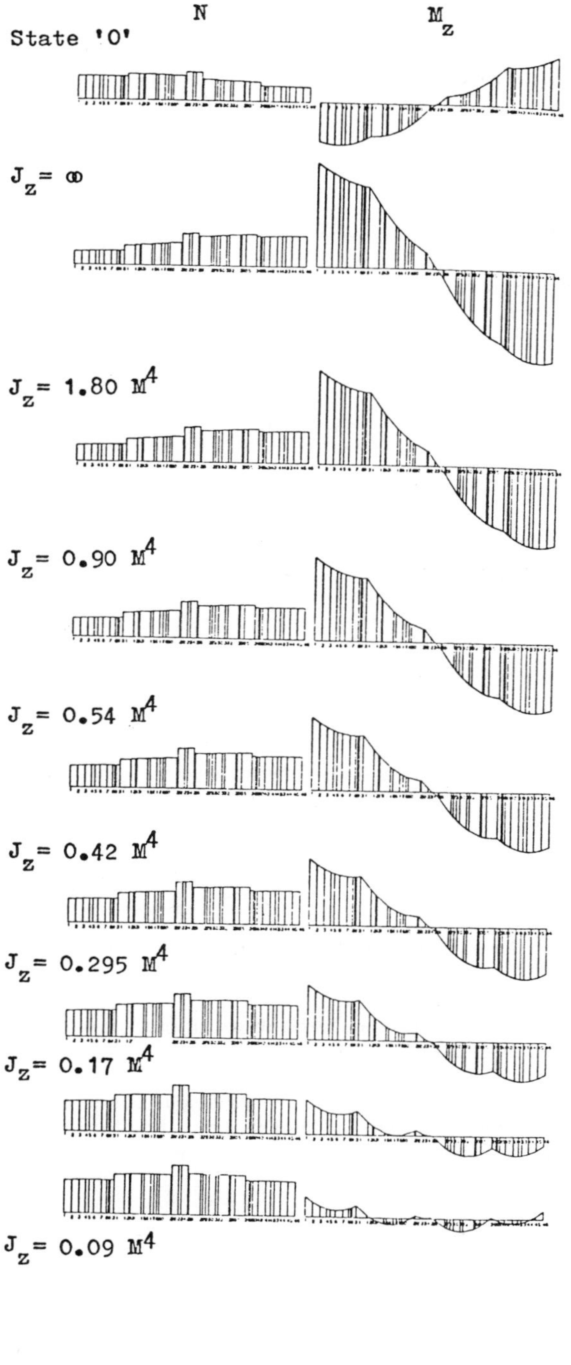

Fig 5 N and M_z distribution along a quarter ring (AB) in function of J_z variation for a 10x10 M, mesh-net.

Table 1

J_Y	J_Z	A
∞	∞	∞
0.34	1.8	0.9
0.17	0.9	0.5
0.102	0.54	0.2
0.079	0.42	0.16
0.056	0.295	0.11
0.032	0.17	0.06
0.017	0.09	0.03

With the above mentioned structural data is our intention to analyse the influence of the elastic collaboration between rope net-anchorage structure and the mesh net dimension in order to evaluate the variations in the state of stress and deformation of the global structure.

In Fig 6 is illustrated the variation of the vertical displacement in the center of the rope net in function of the inertia parameter J_Z of the ring section. The displacement for fixed end displacements ($J_Z = \infty$) in the rope net is also indicated.

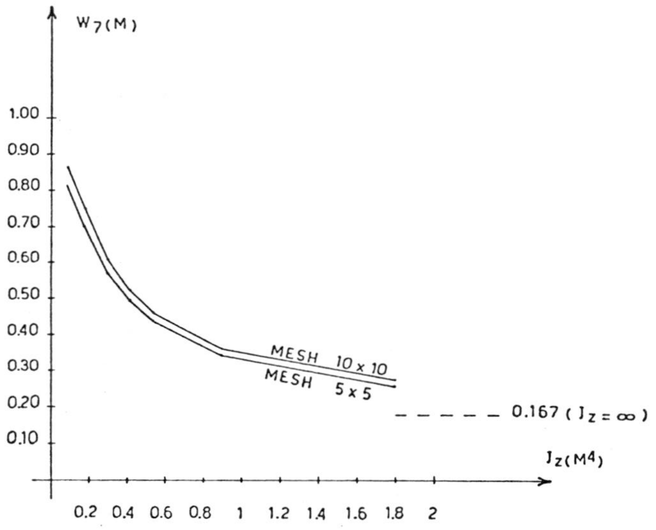

Fig 6 Vertical displacement of central node in rope net

In Fig 7 the horizontal displacement variations of section A and B of the ring structures are drawn.

Figure 8 shows the variation of the principal bending moments (M_Z) in the section A of the ring structure in function of J_Z for all the mesh-net dimensions here considered.

Figure 9 illustrates the same for the axial force in section A .

Figure 10 shows the typical variation of cable forces for the mesh of 10 x 10 m, the state "0" initial values are also indicated.

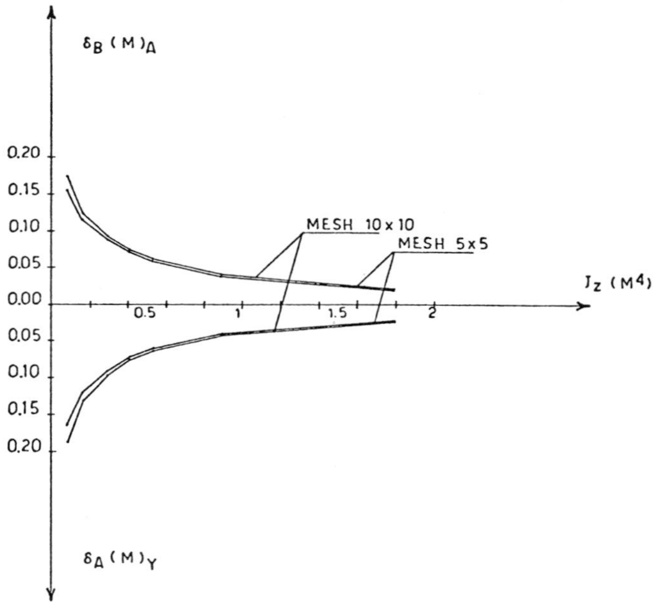

Fig 7 Horizontal displacements of ring structure

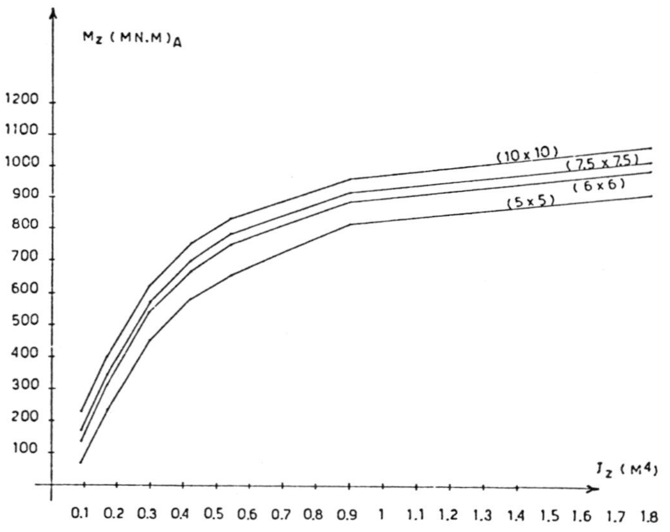

Fig 8 Bending moments in section A of the ring

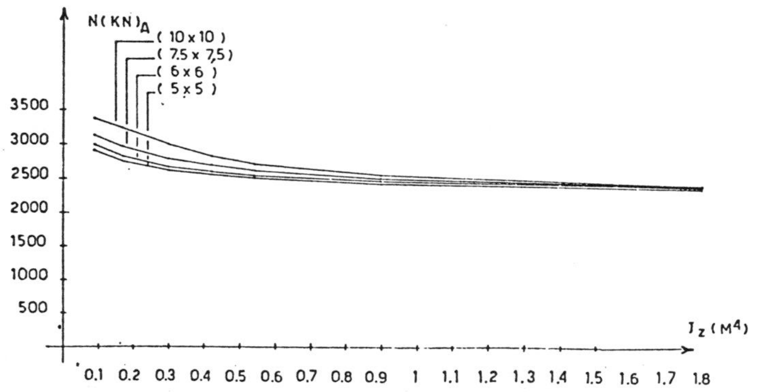

Fig 9 Axial forces in section A of the ring

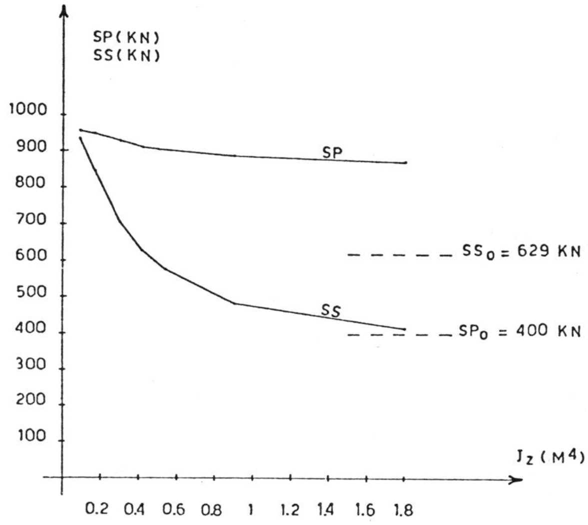

Fig 10 Cable forces in section A and B

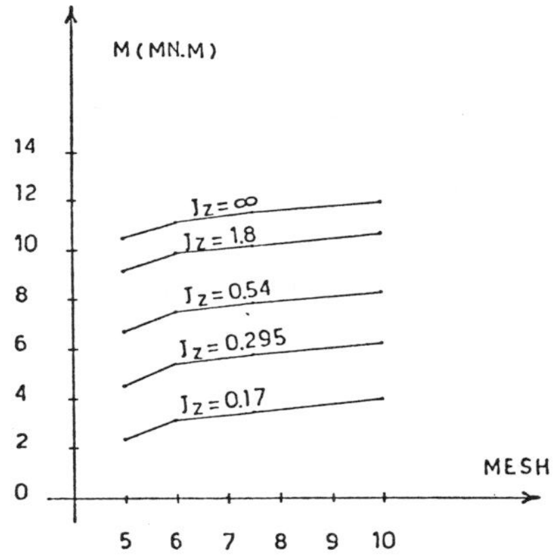

Fig 11 Moment variation in section A of the ring in function of the mesh

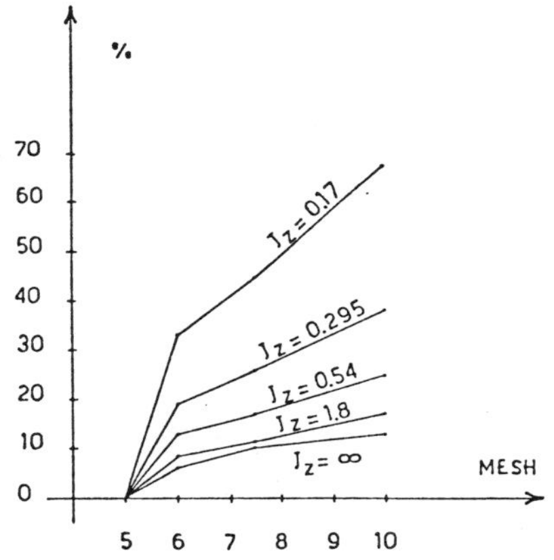

Fig 12 Percentual moment variation in function of mesh-net

Figures 11 and 12 show clearly the moment variation in function of the mesh net.

Figure 13 gives the variation of displacements in section A and B of the ring structure considering the elastic interaction between ring and rope net during the state "0" research.

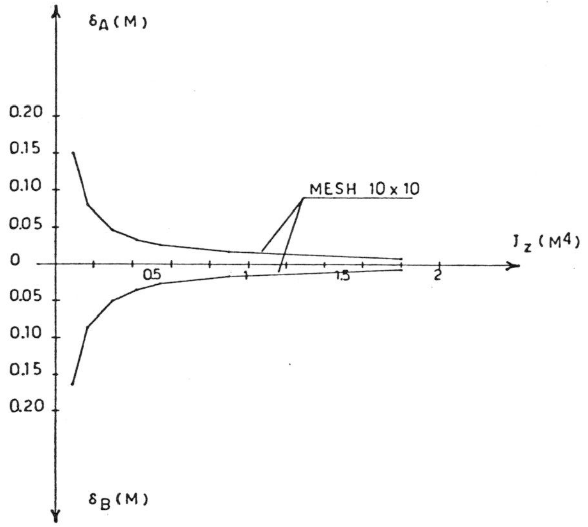

Fig 13 Displacements in section A and B of the ring in pre-stressing state.

CONCLUSIONS

From the examination of the plotted results of Figs 6 to 13 it is possible to conclude that:

- the fixed end hypothesis ($J_Z = \infty$) don't give any useful design information for actual structures, even for a preliminary design phase;

- during elastic interaction, the stresses in the anchorage structures decrease drastically specially for low values of J_Z (see Figs 5 and 8). In order to find a minimum for the structural cost it is very important to examine the variation of the bending moments in function of the relative stiffness of both structures, rope net-anchorage frame. For the structure analysed in this paper the suitable inertia range is found between 0.2÷0.5 where the decay of M_Z is more than proportional. Lower bounds of J_Z are imposed due to the deformative limitation in the rope net or for maximum displacements in the anchorage frame (Figs 6 and 7);

- the anchorage frame displacement gives to the global structure a self-adjusting property because for a certain value of inertia (in our case 0.4; Fig 10) the "natural" decreasing forces in the stabilizing ropes return to increase, as regards as the state "0" value (SS_0), displacements forced by the elastic interaction between the rope net and the anchorage structure;

- the moment distribution in the anchorage frame changes evidently due to the variation of the mesh-net dimension, see Figs 11-12. The choice of the mesh-net is very important in order to minimize the structural global cost. This cost is evidently related to the cost of the roof covering system. From Fig 12 is possible to observe that from a mesh of 10x10 we find a 67% of moment increase;

- Figure 13 shows the displacements of the anchorage structure for the fixed state "0 forces. The elastic interaction in state "0" phase must be considered during the working project of the roof structure in order to control the prestressing operations.

The results obtained suggest further investigations, considering the influence of other structural parameters as:

- variation of sag and curvature ratio;

- variation of the initial pre-stressing state;

- variation of the total difference level between higher and lower anchorage points (total curvatures).

Results on the matter will appear in a further publication.

REFERENCES

1. A SAMUELLI FERRETTI and A ZINGALI, Sull'influenza della deformabilità elastica del bordo nelle tensostrutture a doppia curvatura inversa, Giornale del Genio Civile, fasc. 9-12, 1973.

2. S ODORIZZI and B SCHREFLER, Contributo allo studio di reti di funi pretese entro strutture di bordo elasticamente deformabili, Costruzioni metalliche, n. 3, 1974.

3. M AIZAWA, S TANAKA and H TSUBOTA, Theoretical analysis of pre-tensioned cable structures, IASS Pacific Symposium, Japan, 1971.

4. M AIZAWA, S TANAKA and H TSUBOTA, The analysis, design and construction of a cable net suspension structure, IASS Pacific Symposium, Japan, 1971.

5. R ALESSI, D BAIRAKTARIS, F. CARIDAKIS, M MAJOWIECKI and F ZOULAS, The roof structures of the new sports arena in Athens, IASS World Congress, Madrid, 1979.

6. M MAJOWIECKI, Analisi interattiva di tensostrutture a rete, Acciaio, n. 9, 1982.

WIND TUNNEL STUDY OF A FLEXIBLE MEMBRANE STRUCTURE

N. K. SRIVASTAVA, N. TURKKAN and R. DICKEY

(Université de Moncton, N.B. Canada)

As part of a comprehensive study on the aerodynamic behavior of air-supported structures, the authors have conducted wind tunnel experiments on a series of spherical rigid and flexible membrane models with a view to establish wind pressure distribution and other related effects. The results are analysed with a view to suggest distribution of wind pressure coefficient (Cp) and displacements throughout the surface of the space structure as a function of the ratio of the internal pressure (Pi) and the dynamic wind pressure (q). The paper describes the experimental details and the conclusions of the results.

1. INTRODUCTION

For an open flexible structure such as an air-supported structure, wind is generally the dominating load in its design. The wind effects are much more important in this type of structure than that of conventional rigid construction.

Wind as a Static Load: Wind flow exerts a force q per unit area, which is called dynamic pressure or head, given in terms of air density (ρ) and velocity (v) ($q = 1/2\ \rho\ v^2$)

The equivalent static wind pressure at a point on the surface of a body is defined as:

$$P_w = C_p q$$

where Cp is the coefficient of shape factor which could be negative or positive depending on if it is a suction or lifting pressure. This coefficient is usually determined by full scale field tests or wind tunnel tests on models based on well-proven model laws. In general Cp is dependant on a dimensionless Reynolds number (Re) which is the ratio of inertia and viscous wind forces on the structure. In the non-aeronautical structures, however, wind inertial forces are more predominant than viscous forces, thus one is dealing here with high Reynolds number.

The coefficient Cp for a whole series of rigid structures are available in literature and codes but there are very few data for flexible structures with large displacements.

Wind as a Dynamic Load: Even in a steady wind-flow a structure may oscillate either due to self-sustained instabilities created by an unstable wind force or moment, or due to the shedding of vortices in the leeside of the body or due to super imposition of two modes of stable oscillation. The modes of oscillation are in general a function of mass, shape, stiffness and damping characteristics of the body.

Scope of the Present Study: In general, designers of air-supported structures have considered wind as a static load and have used either empirical or the same wind pressure distribution for flexible structures as that suggested for similar rigid forms. However, tests (Refs. 1 to 6) have indicated that the reality is different. Moreover, in such structures, one is not only concerned with wind pressure distribution but also with anchorage forces, large displacements and tension in the membrane resulting from the overall dynamic nature of the structure.

As a part of a comprehensive study on the aerodynamic behavior of air-supported structures, the authors are conducting wind tunnel experiments on flexible and rigid models with a view to establish wind pressure distribution and other related effects. As a first step, tests have been conducted and the results analysed for the effects of wind as a static load on hemi-spherical and part spherical models. The effects of wind as a dynamic load on such structures will be considered in future studies.

2. EXPERIMENTAL PROCEDURE

The tests were conducted in the 24m long, low speed, horizontal, open circuit, boundary layer wind tunnel at the université de Moncton. This wind tunnel has a 12m long normal cross section of 1m x 2m. The general characteristic of the tunnel is detailed in ref. 7.

2.1 Instrumentation:

Measurement of wind speed was simultaneously taken by a pitot tube connected to a micro-manometer outside the tunnel and by a series of hotwire anemometres connected through different electronic controls to a mini-computer (Digital MINC-II).

Measurement of pressure for the interior as well the exterior points of the model was taken by "SCANIVALVE" System connected to the same mini-computer. The system sweeps 48 pressure points at any instant directly correlating with the corresponding mean wind speed.

Measurement of the displacements of the membrane was done by a specially designed mechanical measuring system mounted on a chariot which could give maximum flexibility in movement with minimum disturbance to the wind around the model (figure 1).

Measurement of anchorage forces was done by a strain gauge system attached to the peripheral fixtures of the model.

2.2 Test models:

There were two groups of models made of two different materials. One group was made of thin rigid plexiglass (R group) and the other group was made of flexible nylon (F group), parachute quality material with density 50 g/m^2. There were hemispherical and part spherical models but the base of the models were always 60 cm. The rigid models were machine made and the flexible models were sewn by using 0.10mm nylon in a proper geodesic pattern. After applying a known internal pressure Pi, the sewn flexible material took the desired shape. To measure pressure distribution on the surface of the membrane, holes were made along one meridian of the model and nylon tubes (diamètre 2.10mm) were attached to each hole from inside. These tubes then passed through the floor of the wind tunnel and connected to the SCANIVALVE System whch in turn was controlled by the mini-computer. Pressure readings for the entire surface was taken by just rotating the model which was placed on a calibrated rotating part of the tunnel floor. The details of a model, base and attachments are shown in fig. 1.

2.3 TEST PROCEDURE

Tests were conducted for rigid models with wind velocity varying from 5m/s to 20m/s and with Reynolds number (Re) from 6.84×10^4 to 4.0×10^5. For flexible models with internal pressure p_i, the results were obtained with wind velocitiy varying from 5 m/s to 8 m/s and Re from 10.3×10^4 to 1.64×10^5. The ratio of pi/q varied from 0.62 to 1.69 for the flexible models.

A computer program was developed to have full automatic acquisition and treatment of all readings for Cp as well as automatic contrôl of all testing instruments. The results were thus directly printed on the computer print out in a desired format. It may be of general interest to note that for each mean value of data obtained, a series of 200 readings were taken at the rate of 600 readings per second.

3. TEST RESULTS

Results Obtained on Rigid Models:

The measured aerodynamic pressure converted to coeffecient Cp is shown in fig. 2 for the hemisphere model and fig. 3 for 1/3 sphere model subjected to a typical wind speed of 5 m/s. Similar curves are obtained for different wind speeds. The figures give the values of Cp for different directions of wind, indicated by circumfrencial angle ϕ, and for different positions of pressure points along the meridian, indicated by meridinal angle θ.

A variation of observed pressure co-efficient with different values of Reynolds number is given in fig. 4 for a hemispherical model. Similar variations was observed for part spherical models. A comparison of the test results with the suggested values of Cp for similar rigid structures by Canadian code shows satisfactory correlation.

Results obtained on flexible models:

Similar to rigid models different values of Reynolds number did little change in the results for flexible models. Typical variations of pressure distribution (Cp) for different orientations of wind and for different values of pi/q are shown in figs. 5 and 6 respectively. The results of the flexible membrane models are compared to that of rigid models in fig. 7 and to that of test results of other researchers in fig. 8.

Radial displacements of all pressure points were recorded but for brevity only the maximum radial displacement as a function of Pi/q is presented in fig.9 in a non-dimensional form. Results for anchor forces varied greatly with different tests, to the extent that the authors consider that the strain gage system measuring these forces were unreliable and need modification.

4. CONCLUSION AND OBSERVATIONS

i) The wind pressure distribution for rigid membrane is close to that suggested by the Canadian code, indicating that the test procedure and results are reliable. The present results are also similar to other data available (Ref. 1, 4) except for the location and magnitude of the maximum value of Cp.

ii) The wind pressure distribution along a meridian of a flexible membrane approaches that of a rigid shape for high values of Pi/q (>1.0). For lower values of Pi/q the magnitude of Cp and its distribution is very different which suggests careful evaluation of wind as a static load on such structures.

iii) For the models tested, wind pressure distribution does not vary much with Reynolds number, thereby suggesting its relatively little importance in the design of such structures. However, it varies considerably with the ratio height to base (h/L) of the models.

iv) The displacement is obviously dependant upon ratio Pi/q. Higher Pi/q means that its behavior approaches that of a rigid membrane. A typical dimenssionless plot of maximum radial displacement, $(\Delta R/L)$ max., against Pi/q indicate an exponential variation and a curve fitting procedure gives following expressions for maximum deflection.

$$(\Delta R/L)_{max} = 0.27\ e^{-1.95\ x}$$

$$\text{or} \qquad = e^{2.13x^3 - 5.81\ x^2 + 2.19x - 1.95}$$

where, $x = pi/q$

The maximum deflection occurs on the meridian directly in the direction of the wind at a meridinal angle θ between 30° - 35° range for flexible hemispherical forms. This takes place just before the pressure co-efficient changes its sign which itself occurs at approximatily θ = 40 to 45° range.

It is suggested to do many more tests on flexible models of varying shapes, for varying h/L, and with improved method of measuring anchor forces and displacements, before arriving at definite conclusions.

Acknowledgement: NSERC of Canada research grant 20-41116 and 20-41118, and prof. J.-R. Longval.

REFERENCES:

1. BERGER G., MACHER E., "Results of wind tunnel tests on some pneumatic structure", Proc. of the first colloquium on pneumatic structure, University of StuTTgart, Germany, 1967.

2. RONTSH G. et BOHME F., "Model analysis of a semicylindrical air-supported hull", Proc. first colloquium on pneumatic structures, University of Stuttgart, Germany, 1967.

3. UEMURA M., "Membrane tension and deformation in cylindrical pneumatic structures subject to wind loads" I.A.S.S., Pacific symposium on tension structures and space frames, Tokyo, 1971.

4. NIEMANN H.J., "Wind tunnel experiments on aeroelastic models of air-supported structure -- results and conclusion", Proc. I.A.S.S. Int. symp. on pneumatic structures, Delft-1972.

5. NOLKER A., "Windkanalversuche an modellen von traghufthallen: versuchstechik und medellherstellung" Proc. A.I.S.S., International symposium on pneumatic structure, Delft-1972.

6. GANDEMER J. et GRILLAUD G., "Etude de la réponce dynamique d'une structure gonflable: étude en soufflerie." Symposium international sur les structures gonflables, C.I.B., 1977.

7. SRIVASTAVA N.K., TURKKAN N., DICKEY R., "Etude expérimentale sur des maquettes calotte sphérique gonflables - Etude préliminaire", rapport du centre de génie éolien (CGE-83-4), Université de Moncton, N.-B., 1983.

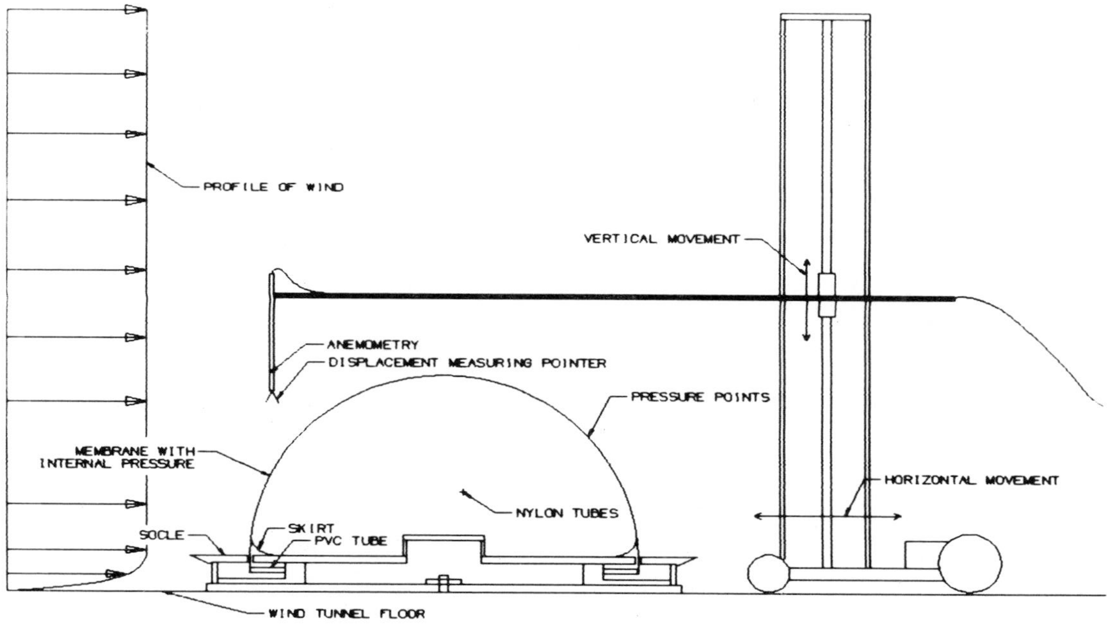

FIGURE 1 : PLACEMENT OF A FLEXIBLE SPHERICAL MODEL WITH DETAILS.

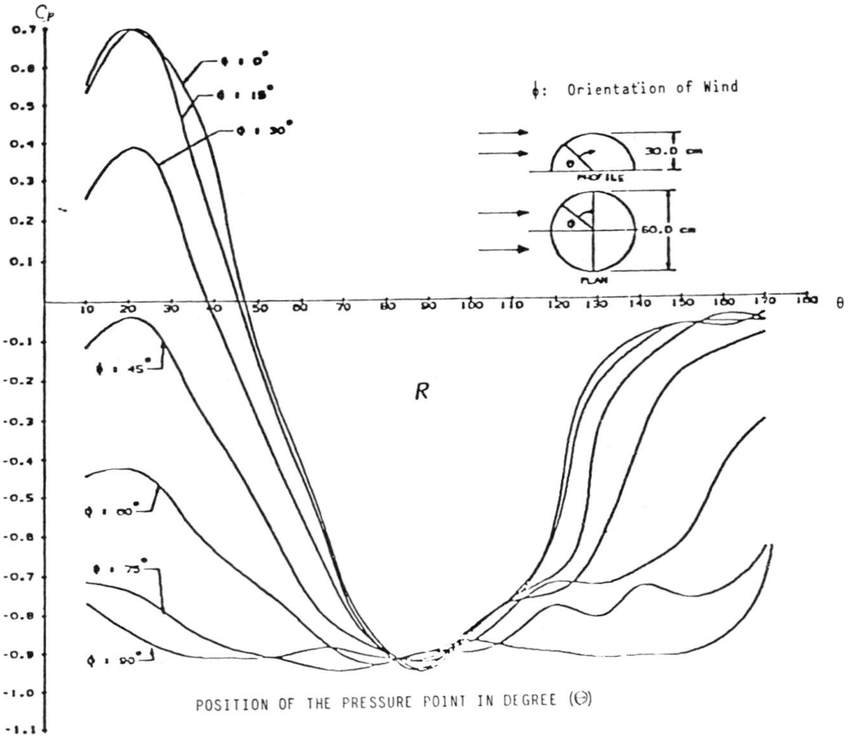

FIGURE 2: DISTRIBUTION OF COEFFICIENT (C_p) OF AERODYNAMIC PRESSURE (EXPERIMENTAL) FOR RIGID HEMISPHERICAL MODEL FOR MEAN WIND SPEED OF 5 m/s

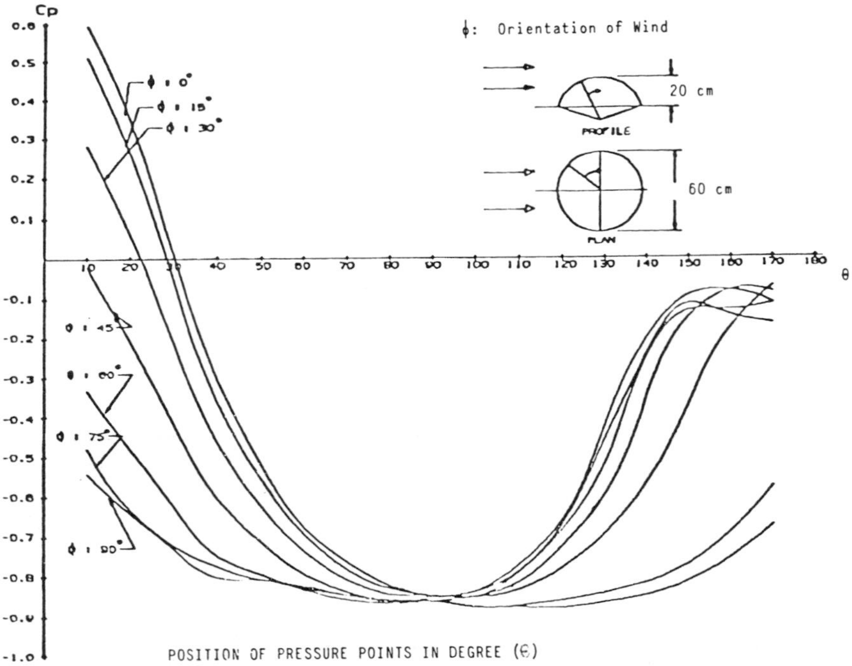

FIGURE 3: DISTRIBUTION OF COEFFICIENT (C_p) OF AERODYNAMIC PRESSURE (EXPERIMENTAL) FOR 1/3 SPHERICAL RIGID MODEL FOR MEAN WIND SPEED OF 5 m/s

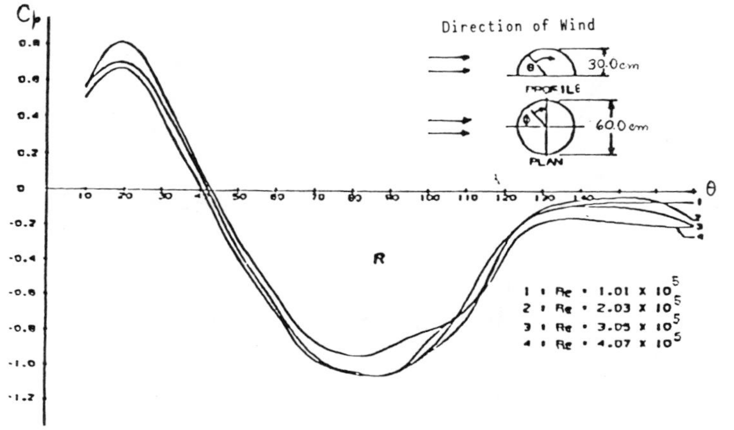

FIGURE 4: DISTRIBUTION OF CEOFFICIENT (C_p) OF AERODYNAMIC PRESSURE (EXPERIMENTAL) FOR RIGID SPHERICAL MODEL FOR DIFFERENT VALUES OF REYNOLDS NUMBER

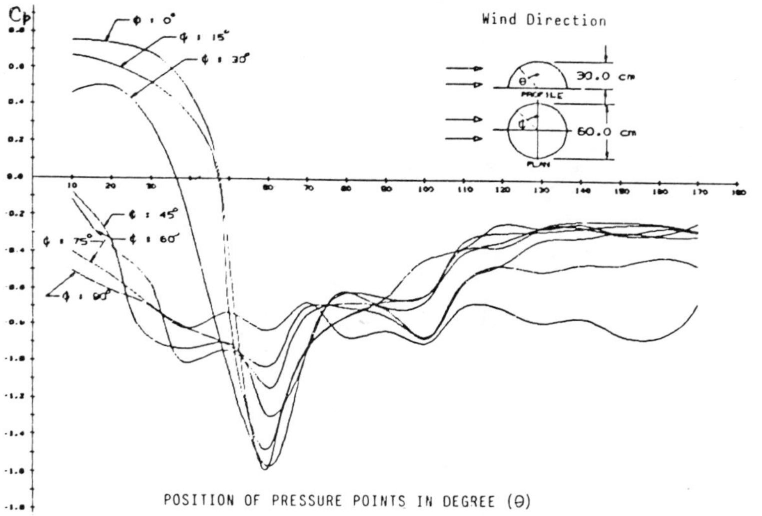

FIGURE 5: DISTRIBUTION OF COEFFICIENT (C_p) OF AERODYNAMIC PRESSURE (EXPERIMENTAL) FOR FLEXIBLE HEMISPHERICAL MODEL (MEAN WIND SPEED 5 m/s AND Pi/q = 0.71)

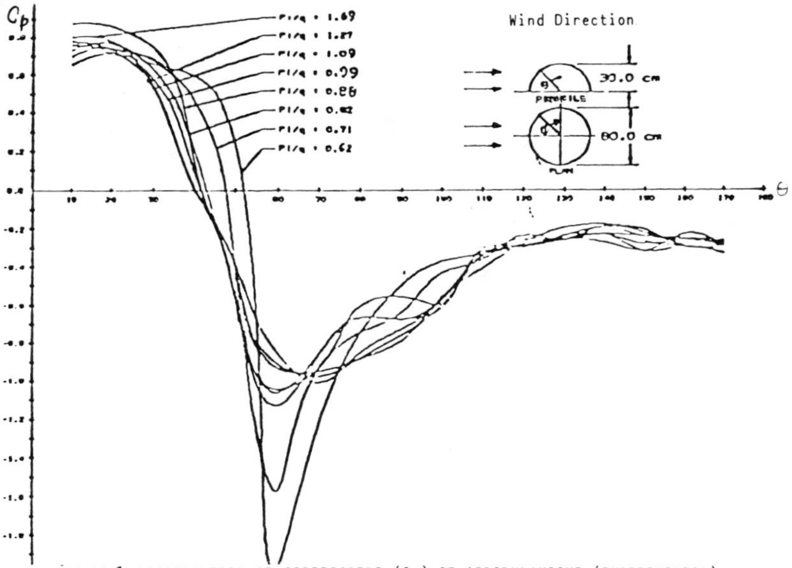

FIGURE 6: DISTRIBUTION OF COEFFICIENT (C_p) OF AERODYNAMIQUE (EXPERIMENTAL) ON FLEXIBLE HEMISPHERICAL MODEL FOR DIFFERENT VALUES OF pi/q (MEAN v = 5 m/s)

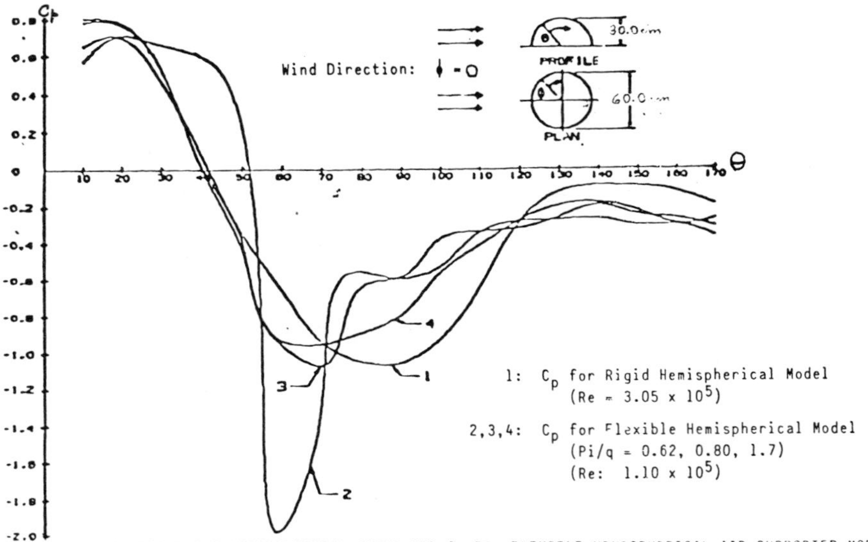

FIGURE 7: COMPARISON BETWEEN C_p FOR RIGID HEMISPHERICAL MODEL AND C_p FOR FLEXIBLE HEMISPHERICAL AIR SUPPORTED MODEL (EXPERIMENTAL)

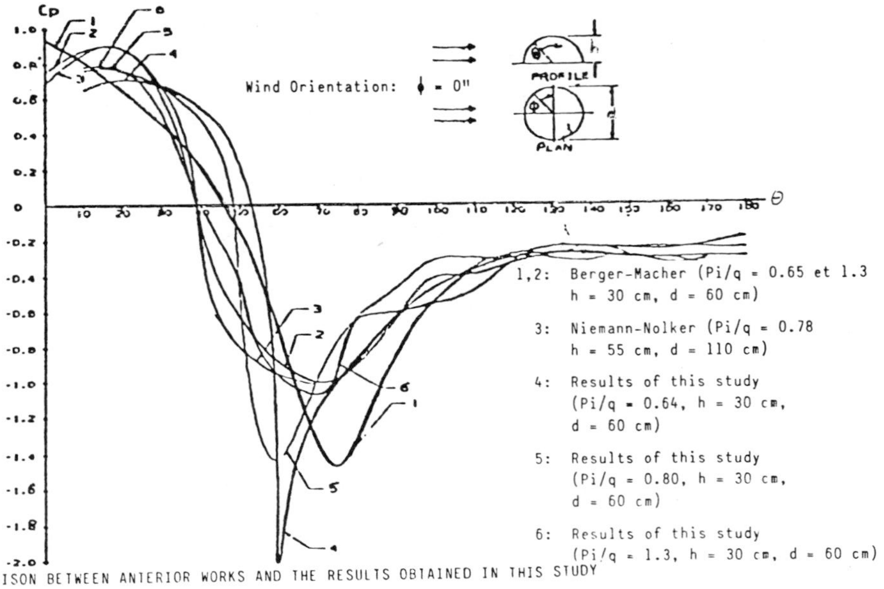

FIGURE 8 COMPARISON BETWEEN ANTERIOR WORKS AND THE RESULTS OBTAINED IN THIS STUDY

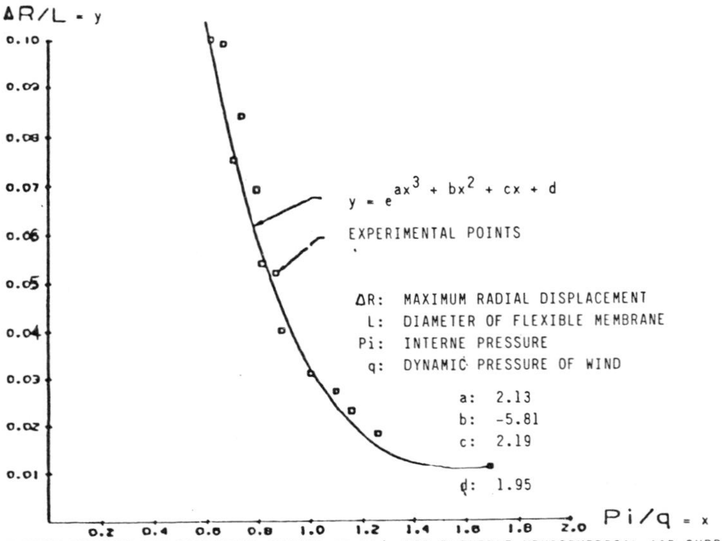

FIGURE 9: MAXIMUM DISPLACEMENT FOR DIFFERENT VALUES OF Pi/q FOR FLEXIBLE HEMISPHERICAL AIR SUPPORTED MODEL (EXPERIMENTAL)

FINITE DYNAMIC RESPONSE OF CABLE SYSTEMS

A CHISALITA , MSc (Eng), MSc, PhD

Senior Lecturer, Department of Structural Mechanics,
Polytechnical Institute of Cluj-Napoca, Cluj-Napoca.

This study presents a Lagrangian formulation of finite deformation dynamic analysis of cable nets with elasto-plastic behaviour and subjected to an arbitrary time dependent load. The network is supposed to be anchored in fixed points and the elements spanning between two nodes are supposed to be straight. The stress-strain curve is assumed to be piecewise linear. The network configuration is refered, at any moment, to the same initial configuration in which the node coordinates and element tensions are known. The initial nodal displacements and velocities are supposed to be given. The network **motion nonlinear equations** are integrated using linear acceleration method. The response of a hyperbolic paraboloid network subjected to an impulsive load is computed as an example.

INTRODUCTION

A Lagrangian formulation for finite deformation dynamic analysis of cable networks has been presented in Refs 2 and 3. The formulation is based on Lee's minimum principle - Ref 5. In Refs 2 and 3 the proposed formulation has been applied to nonlinear static and linear dynamic analysis of cable networks with linear elastic material behaviour.

In the present paper this formulation is developed for nonlinear dynamic analysis of cable networks with elasto-plastic material behaviour considering an arbitrary excitation, including an acceleration imposed to the supports used for earthquake response computation. The dynamic response is computed via linear acceleration method.

MOTION EQUATIONS FOR CABLE NETWORK

Basic Hypotheses

Consider a cable network having each cable oriented by choosing the positive arc. Let C^o be a reference configuration, at time t^o, for which it is supposed that in each point X of the cables the coordinates, the velocities and the stress are known. In the current configuration C, at the time t, in which x is the displaced point X - see Fig 1, the following notations are used: T^*= the cable tension in point x; $\sigma = T^*/A^o$, in which A^o is the cable cross sectional area in X, in configuration C^o; θ = the tangent unit vector in x; ds = elementary arc. The same quantities will be noted T^o, σ^o, θ^o, ds^o respectively in C^o configuration.

Let the displacement vector Xx be noted:

$$U = U(s^o,t)$$

The motion equations of cable C^o are expressed in Lagrangian coordinates X_L and Piola-Kirchhoff stresses T_{KL} (K,L = 1, 2, 3) and are inferred from the minimum conditions of the functional

$$J = \int_{C^o} (\bar{\varrho}^o \frac{\ddot{U}^2}{2} + A^o T_{KL} E_{KL} - p^o \ddot{U}) ds^o$$

in which: $\bar{\varrho}^o = dm/ds^o = \varrho^o A^o$ is the mass density per initial unit length; p^o = force density per unit length in C^o; E_{KL} = Lagrangian deformation - see Ref 3.

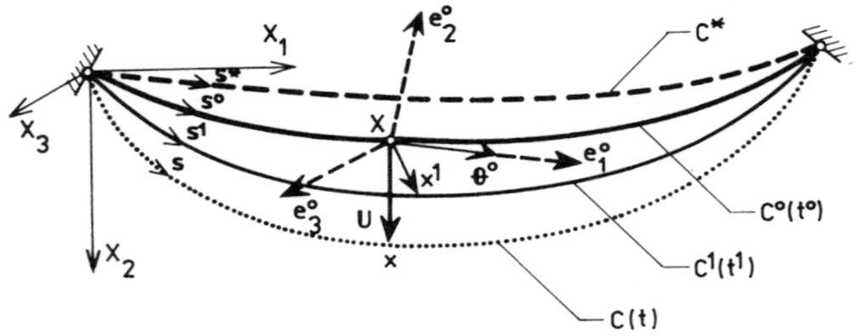

Fig 1

See Fig 1: Let $Xe_1^oe_2^oe_3^o$ be a local reference system, so that e_1^o is tangent to C^o and e_2^o, e_3^o are normal to C^o, in X. As the cable moves from C^o to C, $Xe_1^oe_2^oe_3^o$ shifts to $xe_1e_2e_3$. From the hypotheses:

1) Elementary cable XX' = ds^o undergoes a simple extension so that e_1 becomes tangent, while e_2, e_3 become normal to C in x;
2) Stress state in x is uniaxial along e_1,

it yields that Piola-Kirchhoff stress tensor in $Xe_1^oe_2^oe_3^o$ comes to the component

$$T_{11} = \sigma \frac{ds^o}{ds} \qquad \ldots\ldots 1.$$

Using the vector

$$V = \frac{dU}{ds^o} + \theta$$

the strain-displacement relation becomes - see Ref 3:

$$\frac{ds}{ds^o} = (1 + 2E_{11})^{1/2} = |V| \qquad \ldots\ldots 2,$$

in which $E_{11} = E_{e_1^oe_1^o}$, and

$$T_{11} = \frac{\sigma}{|V|} \qquad \ldots\ldots 3.$$

3) The stress-strain curve is assumed to be piecewise linear - see Fig 2.

For $\sigma^{(o)} < \sigma \leq \sigma^{(1)}$, the stress-strain relation is

$$\sigma = \sigma^{(o)} + Y^{(o)}(\epsilon - \epsilon^o)$$

where $Y^{(o)}$ is the slope of the first segment.

Let C^* - see Fig 1 - be an imaginary configuration with elementary cable ds^* having no stresses. Then

$$\epsilon = (ds - ds^*)/ds^*, \quad \epsilon^o = (ds^o - ds^*)/ds^*$$

and

$$\epsilon - \epsilon^o = (ds - ds^o)/ds^* \qquad \ldots\ldots 4.$$

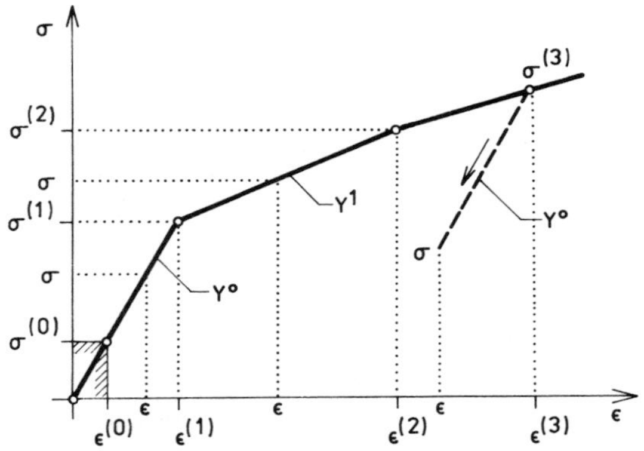

Fig 2

Generally, for $\sigma^{(r)} < \sigma \leq \sigma^{(r+1)}$, r = 0,1,2, ...

$$\sigma = \sigma^{(r)} + Y^{(r)}(\epsilon - \epsilon^{(r)}) \qquad \ldots\ldots 5,$$

in which, if $\epsilon - \epsilon^{(r)}$ is referred to C^o configuration, it can be written

$$\epsilon - \epsilon^{(r)} = \lambda^o(|V| - |V^{(r)}|) \qquad \ldots\ldots 6,$$

where

$$\lambda^o = \frac{ds^o}{ds^*} = 1 + \frac{\sigma^{(o)}}{Y^{(o)}} \qquad \ldots\ldots 7.$$

From Eqns 3, 5 and 6 it follows that Piola stress resultant

$$T = T_{11} A^o \qquad \ldots\ldots 8,$$

takes the form

$$T = \left(\frac{\sigma^{(r)} - \lambda^o Y^{(r)}|V^{(r)}|}{|V|} + \lambda^o Y^{(r)}\right)A^o \qquad \ldots\ldots 9.$$

In the case of unloading, in Eqn 9 $\sigma^{(r)}$ will be the maximum stress reached before unloading and $Y^{(r)} = Y^{(o)}$ - see Fig 2 for $\sigma^{(3)}$.

Node k Motion Equation

The equations of motion can be ontained by cable discretization: the cable crossing points and the points of applied concentrated forces and lumped masses will be called nodes; a cable element is the arc defined by two consecutive nodes - see Fig 3.

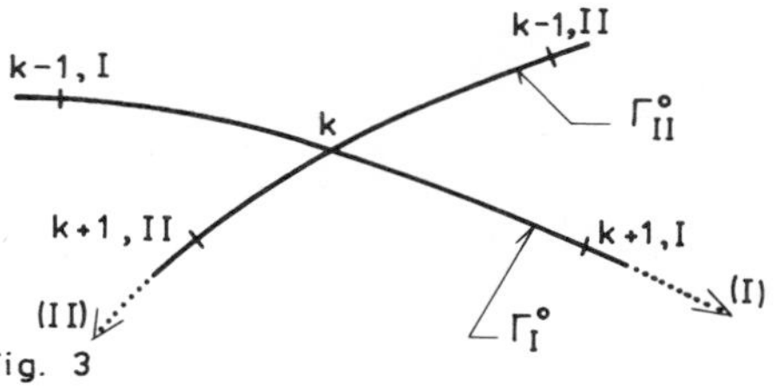

Fig. 3

It is assumed that:

4) The cable elements are straight between two nodes.
5) The displacements $U(s^o,t)$ are linear functions of s^o.

With these assumptions, the motion equation of node k of a two directions cable network - see Fig 3 takes the matrix form - see Ref 3:

$$[m_{k-1,I}I \vdots m_{k-1,II}I \vdots m_{k,I}I \vdots m_{k,II}I] \cdot$$

$$\cdot [\ddot{U}^T_{k-1,I} \vdots \ddot{U}^T_{k-1,II} \vdots \ddot{U}^T_k \vdots \ddot{U}^T_{k+1,I} \vdots \ddot{U}^T_{k+1,II}]^T +$$

$$+ f_k(U) = P_k \qquad \ldots\ldots 10,$$

in which I = the 3x3 identity matrix,

$$m_{i,J} = \frac{\bar{\varrho}^o_J}{6} \Delta s^o_{i,J} \quad ; \; i = k-1,k \; ; \; J=I,II \qquad \ldots\ldots 11,$$

$$m_k = 2(m_{k-1,I} + m_{k-1,II} + m_{k,I} + m_{k,II}) + m_k \qquad \ldots\ldots 12,$$

$\Delta s^o_{i,J}$ = the length of the element with the ends (i, i+1) of the cable with direction J in configuration C^o; m_k = the lumped mass in node k; U_k, f_k and P_k are the matrices associated to the vectors $\ddot{U}_k$, f_k and the nodal consistent force P_k in coordinate system $X_1X_2X_3$:

$$\ddot{U}_k = [\ddot{U}^1_k \; \ddot{U}^2_k \; \ddot{U}^3_k]^T$$

$$f_k = [f^1_k \; f^2_k \; f^3_k]^T \; , \quad P_k = [P^1_k \; P^2_k \; P^3_k]^T$$

in which

$$P_k = P_{k,I} + P_{k,II} + \bar{P}_k \qquad \ldots\ldots 13,$$

$$P_{k,J} = \frac{1}{6}[\Delta s^o_{k-1,J} p^o_{k-1,J} + 2(\Delta s^o_{k-1,J} + \Delta s^o_{k,J}) p^o_{k,J} +$$

$$+ \Delta s^o_{k,J} p^o_{k+1,J}] \quad J = I,II. \qquad \ldots\ldots 14.$$

$p^o_{k,J}$ = load distribution density in node k of the cable with direction J; $\bar{P}_k$ = concentrated force in node k

$$f_k(U) = f_{k,I}(U) + f_{k,II}(U) \qquad \ldots\ldots 15,$$

$$f_{k,J}(U) = T_{k-1,J} V_{k-1,J} - T_{k,J} V_{k,J} \qquad \ldots\ldots 16,$$

in which T_k and V_k refer to the element with the ends (k, k+1).

From assumptions 4 and 5 one obtains (suppressing the index J):

$$V_k = \frac{U_{k+1} - U_k}{\Delta s^o_k} + \theta^o \qquad \ldots\ldots 17,$$

and from Eqn 9, for $\sigma^{(r)} < \sigma \leqq \sigma^{(r+1)}$

$$T_k = \left(\frac{\sigma^{(r)}_k - \lambda^o_k Y^{(r)}_k |V^{(r)}_k|}{|V_k|} + \lambda^o_k Y^{(r)}_k\right) A^o_k \qquad \ldots\ldots 18.$$

Linearization of Equation 10

Let C^I be the net configuration at time t^I and C a neighbouring configuration at time $t = t^I + \Delta t$. Let the node k displacement be

$$U_k = U^I_k + \Delta U_k$$

in which $\Delta U_k = [\Delta U^1_k \; \Delta U^2_k \; \Delta U^3_k]^T$. The function $f_k(U)$, $U = \{U_1, U_2, \ldots, U_N\}$ - where N = number of free nodes of the network - can be expanded in series arround configuration C^I defined by U^I as follows:

$$f_k(U) = f_k(U^I) + A_{(k)}(U^I)\Delta U + \ldots$$

in which $A_{(k)}$ is the tangent stiffness matrix.

The linear term in ΔU reads explicitly - see Ref 3:

$$[-A^{k-1,I} \vdots -A^{k-1,II} \vdots A^{(k)} \vdots -A^{k,I} \vdots -A^{k,II}] \cdot$$

$$\cdot [\Delta U^T_{k-1,I} \vdots \Delta U^T_{k-1,II} \vdots \Delta U^T_k \vdots \Delta U^T_{k+1,I} \vdots \Delta U^T_{k+1,II}]^T \qquad \ldots\ldots 19,$$

in which

$$A^{(k)} = A^{k-1,I} + A^{k-1,II} + A^{k,I} + A^{k,II} \qquad \ldots\ldots 20.$$

Taking into account Eqn 18 and following an analysis analoguous to that from Ref 3 it is found (suppressing the indices I, II):

$$A^k = [A^k_{ML}]$$

$$A^k_{ML} = [T_k \delta_{ML} + (\lambda^o_k Y^{(r)} A^o_k - T_k) \frac{V^L_k V^M_k}{|V_k|^2}] \frac{1}{\Delta s^o_k} \qquad \ldots\ldots 21,$$

in which δ_{ML} is the Kronecker symbol.

The above equations can be generalized in a natural way if several cable directions I, II, III, ... joint in node k.

Network Motion Equation

Putting:

$$U = \begin{bmatrix} U_1 \\ U_2 \\ \vdots \\ U_N \end{bmatrix}, \quad f = \begin{bmatrix} f_1 \\ f_2 \\ \vdots \\ f_N \end{bmatrix}, \quad P = \begin{bmatrix} P_1 \\ P_2 \\ \vdots \\ P_N \end{bmatrix}, \quad \Delta U = \begin{bmatrix} \Delta U_1 \\ \Delta U_2 \\ \vdots \\ \Delta U_N \end{bmatrix} \quad \ldots\ldots 22,$$

in which $U_k = [U_k^1 \; U_k^2 \; U_k^3]^T$ and N = number of free nodes, the network motion equation takes the form

$$M\ddot{U} + f(U) = P(t) \quad \ldots\ldots 23,$$

and the linearized f(U) is

$$f(U) = f(U^I) + A(U^I)\,\Delta U + \ldots \quad \ldots\ldots 24.$$

The structure of matrices M and $A(U^I)$ results from Eqns 10 and 19.

In the case of an acceleration a(t) imposed to the supports (seismic excitation), the motion equation of node k relative to the system $X_1X_2X_3$ fixed to the supports - see Fig 1, is Eqn 10 in which the right hand side is substituted by

$$P_k' = P_k - m_k' \, a \quad \ldots\ldots 25,$$

in which $a = [a^1 \; a^2 \; a^3]^T$ and

$$m_k' = \frac{\bar{\rho}_I^o}{2} (\Delta s_{k-1,I}^o + \Delta s_{k,I}^o) + \frac{\bar{\rho}_{II}^o}{2} (\Delta s_{k-1,II}^o + \Delta s_{k,II}^o) + \bar{m}_k \quad \ldots\ldots 26.$$

P_k is the equivalent nodal force given by Eqn 14.

The equation of the net motion, relative to the (moving) system $X_1X_2X_3$ will have the form of Eqn 23, that is:

$$M\ddot{U} + f(U) = P'(t) \quad \ldots\ldots 27,$$

in which

$$P' = \begin{bmatrix} P_1 - m_1' \, a \\ P_2 - m_2' \, a \\ \cdots\cdots\cdots \\ P_N - m_N' \, a \end{bmatrix} \quad \ldots\ldots 28.$$

INTEGRATION ALGORITM

Let U(t) be the displacement in $X \in C^o$, accounted in the interval $[t_o, t_1]$, $t_1 - t_o = \Delta t$. Let the superscripts (o) and (1) denote the values of functions U(t), $\dot{U}(t)$ and $\ddot{U}(t)$ in t_o and t_1 respectively, and let

$$\Delta\ddot{U}^{(1)} = \ddot{U}^{(1)} - \ddot{U}^{(o)}. \quad \ldots\ldots 29.$$

The linear acceleration method assumes that in the interval $[t_o, t_1]$ the acceleration varies linearly, that is

$$\ddot{U}(t_o + \tau) = \ddot{U}^{(o)} + \frac{\Delta\ddot{U}^{(1)}}{\Delta t} \; ; \; \tau \in [0, \Delta t] \quad \ldots\ldots 30.$$

Integrating Eqn 30 from 0 to τ and making $\tau = \Delta t$ it is obtained

$$\ddot{U}^{(1)} = \ddot{U}^{(o)} + \Delta\ddot{U}^{(1)} \quad \ldots\ldots 31,$$

$$\dot{U}^{(1)} = \bar{\dot{U}}^{(o)} + \Delta\ddot{U}^{(1)}\frac{\Delta t}{2} \quad \ldots\ldots 32,$$

$$U^{(1)} = \bar{U}^{(o)} + \Delta\ddot{U}^{(1)}\frac{(\Delta t)^2}{6} \quad \ldots\ldots 33,$$

where $\bar{U}^{(o)}$ and $\bar{\dot{U}}^{(o)}$ are the approximated values of $U^{(1)}$ and $\dot{U}^{(1)}$ given by a dropped off Taylor series as follows

$$\bar{U}^{(o)} = U^{(o)} + \dot{U}^{(o)}\Delta t + \ddot{U}^{(o)}\frac{(\Delta t)^2}{2} \quad \ldots\ldots 34,$$

$$\bar{\dot{U}}^{(o)} = \dot{U}^{(o)} + \ddot{U}^{(o)}\Delta t \quad \ldots\ldots 35.$$

Substituting Eqns 31, 32 and 33 into Eqn 23 written for $t = t_1$ and taking into account the linearization of f(U) - Eqn 24, the equation 23 takes the form

$$\bar{M}\Delta\ddot{U}^{(1)} = \Delta P \quad \ldots\ldots 36,$$

in which

$$\bar{M} = M + A(\bar{U}^{(o)})\frac{(\Delta t)^2}{6} \quad \ldots\ldots 37,$$

$$\Delta P = P(t_1) - M\ddot{U}^{(o)} - f(\bar{U}^{(o)}) \quad \ldots\ldots 38.$$

From Eqn 36 one obtains $\Delta\ddot{U}^{(1)}$ that can be used in Eqns 31,32 and 33 to find $\ddot{U}^{(1)}$, $\dot{U}^{(1)}$ and $U^{(1)}$.

The above values are tested in Eqn 23 computing

$$\Delta P^{(1)} = P(t_1) - f(U^{(1)}) - M\ddot{U}^{(1)} \quad \ldots\ldots 39,$$

and verifying

$$\| \Delta P^{(1)} \| / \| P(t_1) \| < \epsilon \qquad \ldots\ldots \; 40,$$

where ϵ is chossen beforehand. If the test of Eqn 40 is not satisfied, an iteration can be performed as follows:

$$\Delta \ddot{U}^{(1)} \leftarrow \Delta \ddot{U}^{(1)} + \Delta(\Delta \ddot{U}^{(1)})$$

and one obtains

$$\bar{M}^{(1)} \Delta(\Delta \ddot{U}^{(1)}) = \Delta P^{(1)} \qquad \ldots\ldots \; 41,$$

in which $M^{(1)}$ and $\Delta P^{(1)}$ are obtained from Eqns 37 and 38 using $U^{(1)}$ instead of $\bar{U}^{(o)}$. From Eqn 41 $\Delta(\Delta \ddot{U}^{(1)})$ is computed and used to obtain the new values of $U^{(1)}$, $\dot{U}^{(1)}$, $\ddot{U}^{(1)}$, according to Eqns 31,32 and 33, in which $\ddot{U}^{(o)}$, $\bar{U}^{(o)}$ and $\Delta \ddot{U}^{(1)}$ are replaced by $\ddot{U}^{(1)}$, $\dot{U}^{(1)}$, $U^{(1)}$ and $\Delta(\Delta \ddot{U}^{(1)})$ respectively. The test of Eqn 40 is then repeated. The number of such iterations will be limited by the parameter LNIT.

The selection of time step Δt is fundamental. Indications can be found in Refs 8,10,1 and 7. Time step Δt may be chossen according to similar examples or as follows: the response is computed for a small time interval with successively halved steps and Δt will be chosen as the time step whose halving does not lead to the differences in the results (for example Δt and $\Delta t/2$ yield to 4 identic digits for $U^{(1)}$ or $\ddot{U}^{(1)}$). The parameters ϵ and LNIT can be taken $\epsilon = 0,5\ldots1$ %, LNIT = 0 - 2; see also Ref 7.

SOLUTION PROCEDURE

Let [0, TT] be the time interval over which the network response is computed. For any moment $t \in [0, TT]$, the displacements U(t) are refered to the same initial configuration $C^o(t=0)$, in which the node coordonates, element tensions and the nodal displacements and velocities U(o) and $\dot{U}(o)$ are supposed to be known. Initial nodal accelerations $\ddot{U}(o)$ are computed by solving Eqn 23 written for t = 0. For $[t_i, t_i + \Delta t]$ time interval, the values $U(t_i)$, $\dot{U}(t_i)$ and $\ddot{U}(t_i)$ play the role of $U^{(o)}$, $\dot{U}^{(o)}$, $\ddot{U}^{(o)}$ from Eqns 31, 32, 33, 34 and 35. At every moment t_i, for each element, the location of stress σ on constitutive curve - see Fig 1 - is tested, adequately modifying $Y^{(r)}$ in Eqns 18 and 21. The procedure is similar to that described in Ref 6 excepting the fact that the reference configuration remains C^o. If at a given moment a cable element has a tension T < 0, it is putting T = 0.0. If this element is a compression member, it will remain with the tension T < 0. If at a given moment the stress in an element is greater than the breaking stress, the element will stay broken for all following moments, **in the** considered loading case. A computer program THACS - see Ref 4 - has been developed on the basis of this solution procedure.

NUMERICAL EXAMPLE

The above formulation has been tested on the structure analyzed in Ref 6 (example 2). The structure is a hyperbolic paraboloid cable net - see Fig 4. The following numerical values are

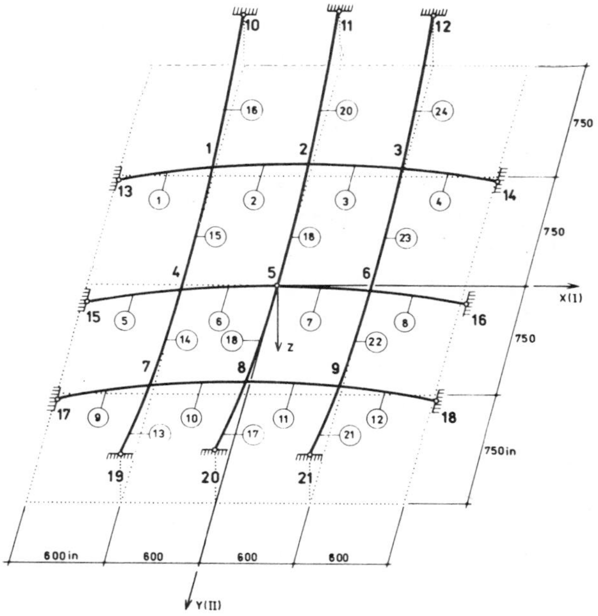

Fig 4

used for cables spanning in x direction: area A_x = 5,4 sq in, sag f_x = 96 in, initial tension TO_x = 216 kips. For the cables spanning in y direction the above named values are: A_y = 9,6 sq in, f_y = 300 in, TO_y = 417.6 kips for end elements and TO_y = 400.32 kips for intermediate elements. The stress-strain curve is defined by the values from Ref 6, taking the average of the last two values, namely: σ = 0 - 160, YO = 23,000; σ = 160 - 175; YO = 7,500; σ = 175 - 188, YO = = 4,320 and σ = 188 - 300; YO = 517.27 (values in kip/sq in). The network is subjected to an impulsive load of 180 kips per node. The cable masses are considered concentrated in nodes with the value CM = 0.16187516 kip - sec^2/in (corresponding to a weight of 62.5 kip).

The network response has been computed up to the time TT = 1.3 sec using a time step of 0.01 sec (in Ref 6 a time step of 0.005 sec has been used). A value LNIT = 0 has been taken. The response of nodes 2 and 5 in Z direction is shown in Fig 5 and its look is similar to that of the response shown in Ref 6 (Fig 12, p 221). However the extreme values are smaller than those from Ref 6. The maximum displacements in Z direction are: node 2 - 33.17 in at t = 0.40 sec and 35.71 in at t = 1.18 sec; node 5 - 42.69 in at t = 0.46 sec and 35.20 in at t = 1.17 sec.The result obtained in Ref 6, namely that the response of node 2 is different from that of node 5, is also found. The period of node 2 response is 0.78 sec (0.76 in Ref 6) and that of node 5 is 0.71 sec (0.72 in Ref 6). The nodes 4 and 1 have very closed response to that of nodes 5 and 2 respectively. The maximum Z displacements and the periods are: 42.54 in at t = 0.47 sec, 35.16 in at t = 1.18 sec and 0.71 sec for node 4; 33.01 in at t = 0.40 sec, 35.49 in at t=1.19 sec and 0.79 sec for node 1. The maximum stress is 181.37 ksi (185 ksi in Ref 6) and occurs in the elements 17 and 20 at time t = 0.42 sec.

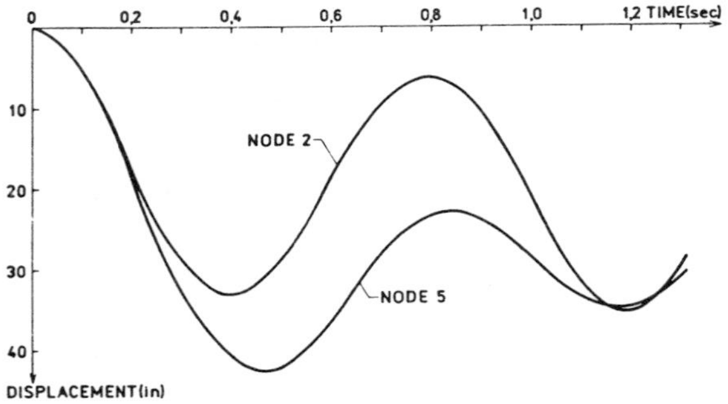

Fig 5

AKNOWLEDGEMENTS

The study has been partially supported by the Central Institute for Building Research, Design and Guidance through the Contract no 1581 having as objective the THACS program.

REFERENCES

1. K J BATHE and E L WILSON, Numerical Methods in Finite Element Analysis, Prentice Hall Inc., 1976

2. A CHISALITA, Contributions to the Study of the Nonlinear Static and Dynamic Response of Suspension Systems, Doctoral thesis, Polytechnical Institute of Cluj-Napoca, 1983

3. A CHISALITA, Finite Deformation Analysis of Cable Nets, Journal of Engineering Mechanics, ASCE, Vol 110, No 2, February 1984

4. A CHISALITA and B PARV, THACS (Time History Analysis of Cable Systems) - User's Guide, Polytechnical Institute of Cluj-Napoca, 1983

5. L H N LEE and CHI-MOU-NI, A Minimum Principle in Dynamics of Elastic-Plastic Continua at Finite Deformation, Archives of Mechanics, vol 25, issue 3, Warszawa, 1973

6. D MA, J LEONARD and K-H CHU, Slack Elasto-Plastic Dynamics of Cable Systems,Journal of the Engineering Mechanics Division, Proceedings ASCE, vol 105,No EM2, April 1979

7. S H MOTE and K-H CHU, Cable Truss Subjected to Earthquakes, Journal of the Structural Division, Proceedings ASCE, vol 104, No ST4, April 1978

8. N M NEWMARK, A Method of Computation for Structural Dynamics, Journal of the Engineering Mechanics Division, Proceedings ASCE, vol 85, No EM3, July 1959

9. N E NICKELL, Direct Integration Methods in Structural Dynamics, Journal of the Engineering Mechanics Division, Proceedings ASCE, vol 99, No EM2, April 1973

10. C WEEKS, Temporal Operators for Nonlinear Structural Dynamics Problems, Journal of the Engineering Mechanics Division, Proceedings ASCE, vol 98, No EM5, October 1972.

SUSPENSION SYSTEM FOR CROSS-OVER STRUCTURES

M MOLDOVAN*, MSc (Eng) and A CHISALITA**, MSc (Eng), MSc, PhD

* Senior Lecturer, Department of Steel Structures, Polytechnical Institute of Cluj-Napoca,Cluj-Napoca

** Senior Lecturer, Department of Structural Mechanics, Polytechnical Institute of Cluj-Napoca,Cluj-Napoca

A suspension system for foot-bridges, conveyers, pipe lines etc. is analyzed. The system seems to be a new one. It is made by a lattice girder platform suspended through four prestressed cables placed two above and two under the platform. The system is simple and efficient: the platform has small displacements for any loading, the system has a free space over the platform and its erection is easy. A 60 m span structure, designed for conveyer belt, is numerically analyzed. Concluding remarks emphasize the structure behaviour under design loads.

GENERAL CONSIDERATIONS

Cross-over structures cover a wide category of structures, generally light ones,destined to pass waterways, communication ways, big obstacles, occupied ground etc, the main element of their differentiation being their span. They have varied destinations: pasages for pipes, belt conveyers, foot-passangers etc. Regarding the loads, generally, the are gravitational forces of small and medium values, that stress the significance of wind action expressed by horizontal and ascensional vertical static and dynamic forces.

For large span passages, beginning with 40 m, the most rational structures are prestressed suspension systems. The most used types are shown in Fig 1. The cables are disposed in a triangle vertices.What differs from type to type are the cable planes and the zone where the destination is disposed (where the main load will act).

T y p e A: The functions are placed on the base of the triangle. Cable 1, in vertical plan,takes the gravitational loads. Cables 2 and 3, in horizontal plane, prestressed, take the wind action. The system behaves inadequately under asymmetrical actions and ascensional wind.It is recommended for big permanent forces and as function, for pipe passages.

T y p e B: See Ref 1. It derives from type A placing the cables 2 and 3 in inclined planes,

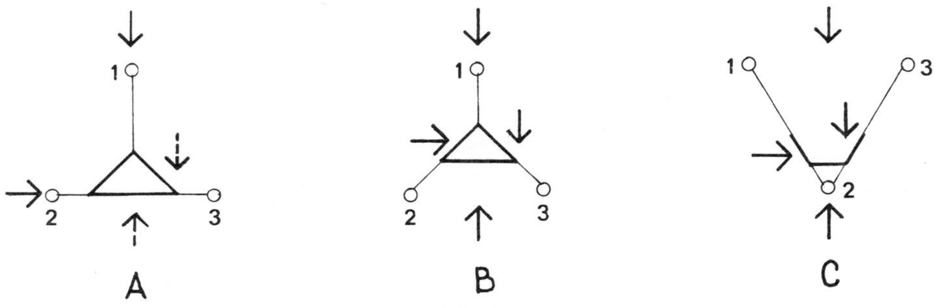

Fig 1.

what improves the behaviour of the system. The system is prestressed. The destinations are placed in the zone of the centre of gravity of the cable triangle. The horizontal or asymetrical forces activate two cables (1 and 2 for instance), but cable 1 in vertical plane has a reduced share. It is a broadly used system although the triangular frame is not very appropiate to place the destination in it.

T y p e C: See Ref 2. It is a triangle with a downward vertex and with the functions and prestressing cables placed in the zone of this vertex. The cable planes coincide with triangle sides (the horizontal forces are taken similarly to an anchor bar-counterbrace system). The horizontal and asymmetrical loads are taken mainly also by a single cable. The space assigned for functions is free and easily utilizable.

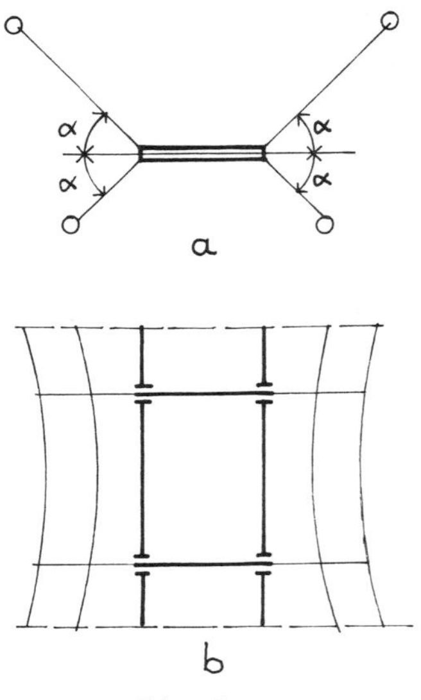

Fig 2.

SUGGESTED SYSTEM

The system is composed of an anchor suspended platform and four cables (two above and two under the platform). The cables are placed in planes inclined with the same angle given the platform (Fig 2 a). The platform is composed of transversal bars - tied with anchors to the four cables - and longitudinal bars hinged with the transversal ones. Such a panel can be subdivided according to its function (Fig 2 b).

The boundary structure (Fig 3) follows the configuration of the passage. It is composed of hinged bars loaded mainly axially. The columns f, placed in the planes of upper cables are linked through spatial hinges by a frame developed on two directions (the bars a, b, c, d). The cables m and n complete the end structure.

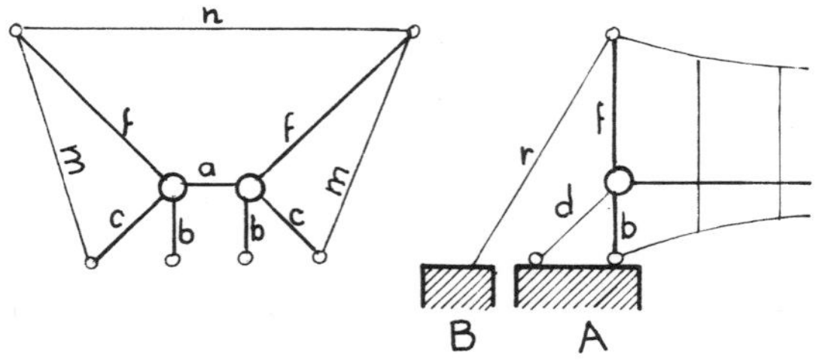

Fig 3.

The tensions from upper cables are taken by the columns f and anchors r. The lower cables are anchored into the foundation A while the anchors are fixed into the foundation B.

In Fig 4 is shown a general view of the passage.

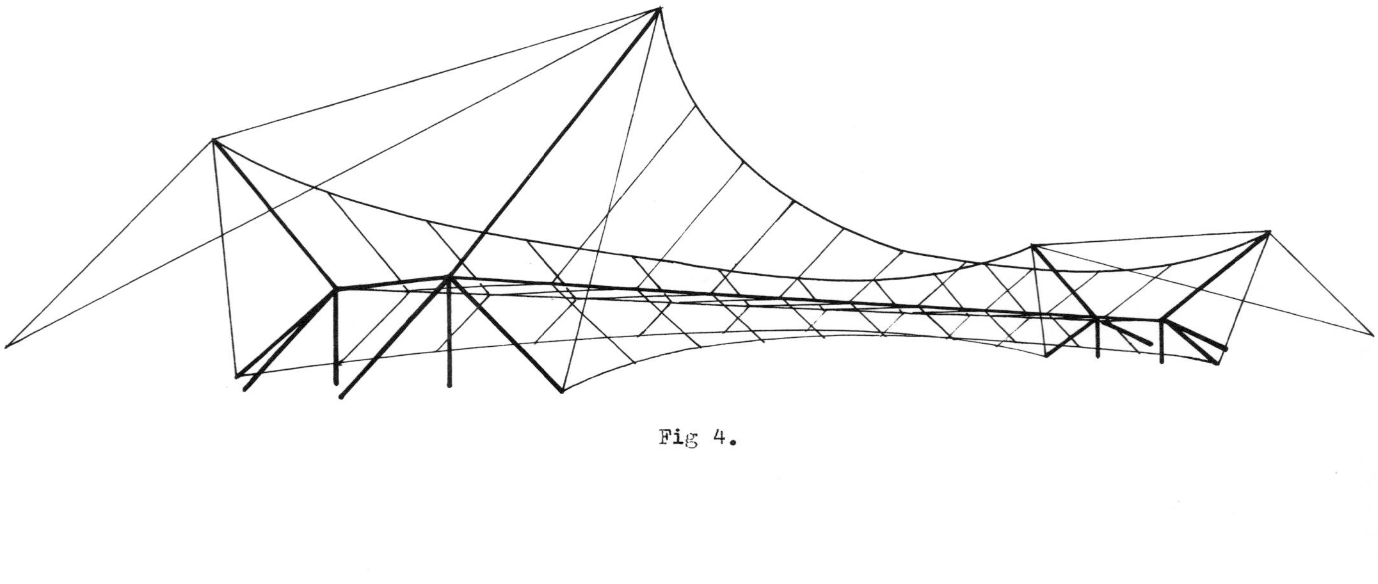

Fig 4.

Given the types A, B and C presented above, the proposed system has two main advantages:

- for any type of load, two cables are active, confering an improved stiffness on the structure under horizontal and asymetrical forces (see the computed example).
- the space designed to be used is free above and under the platform, allowing, consequently, any type of destination. The above presented structure is dessigned to be used unclosed. If closing is required, the system shown in Fig 5 is proposed.

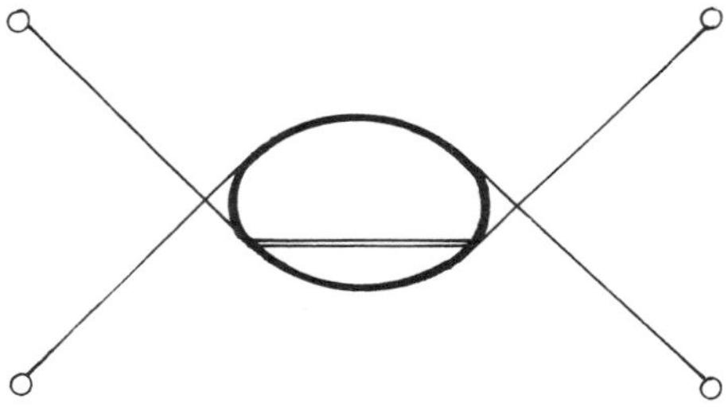

Fig 5

The way the system takes over the loads is shown in Fig 6.

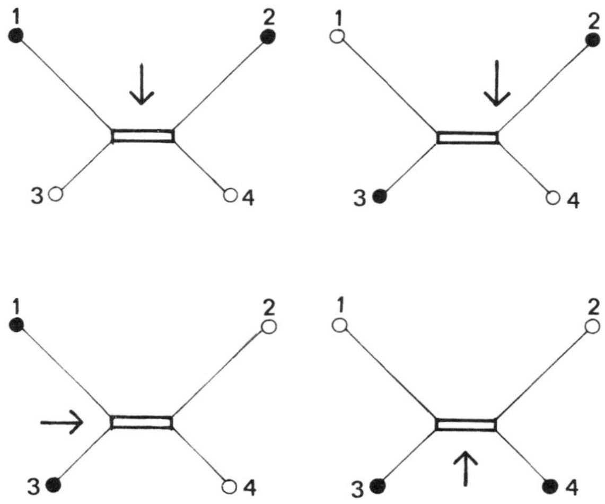

Fig 6.

Under gravitational actions, the cable 1 and 2 are activated; under asymetrical actions, the cables 2 and 3, for instance, are activated; under horizontal loads the cables 1 and 3 are activated; for ascensional vertical loads, the cables 3 and 4 are activated.

NUMERICAL EXAMPLE

To analyze the system behaviour, a passage of 60 m span for a conveyer with a 650 mm belt width has been designed. The conveyed material is concrete agregates for a concrete plant.

The width of the platform resulted 2.80 m and includes the conveyer and two walking spaces of 0.90 width (Fig 7).

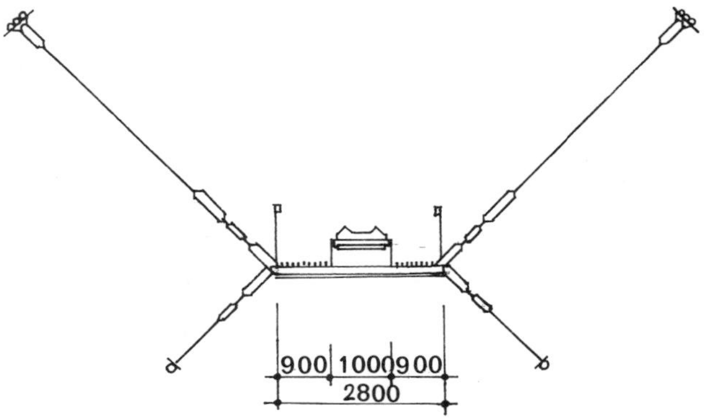

Fig 7.

A platform panel is 6.00 m long - between the transversal bar axes linked to the cables - and includes besides the longitudinal boundary bars, two stringers at 1.00 m distance and a brace at the middle of the panel. The platform bars are made of U and I steel hinged one to another through a screw. The walking spaces are equiped with hand rails and bar screen.

The cables lie in planes inclined with the same angle α with respect to platform plane (considered horizontal). The sag of the upper cables in their plane is f_u = 5.00 m. The upper cables are made of 2 Ø 45 each with A_u=23.96 cm^2. The sag of the lower cables is f_l = 3.00 m. The lower cables are made of 2 Ø 36 each with A_l = = 15.52 cm^2.

Considering a global safety factor c = 2.5, the admissible tension in the upper cable is $N_{u,a}$ = = 1011.84 kN and in the lower cable is $N_{l,a}$ = = 670.86 kN. The Young modulus is E = 1.6 . 10^6 daN/cm^2.

The platform is suspended of the cables from 6.00 m to 6.00 m with verticals made of round steel Ø 20 mm.

Two variants according to the value of angle α between the platform plane and cable plane, have

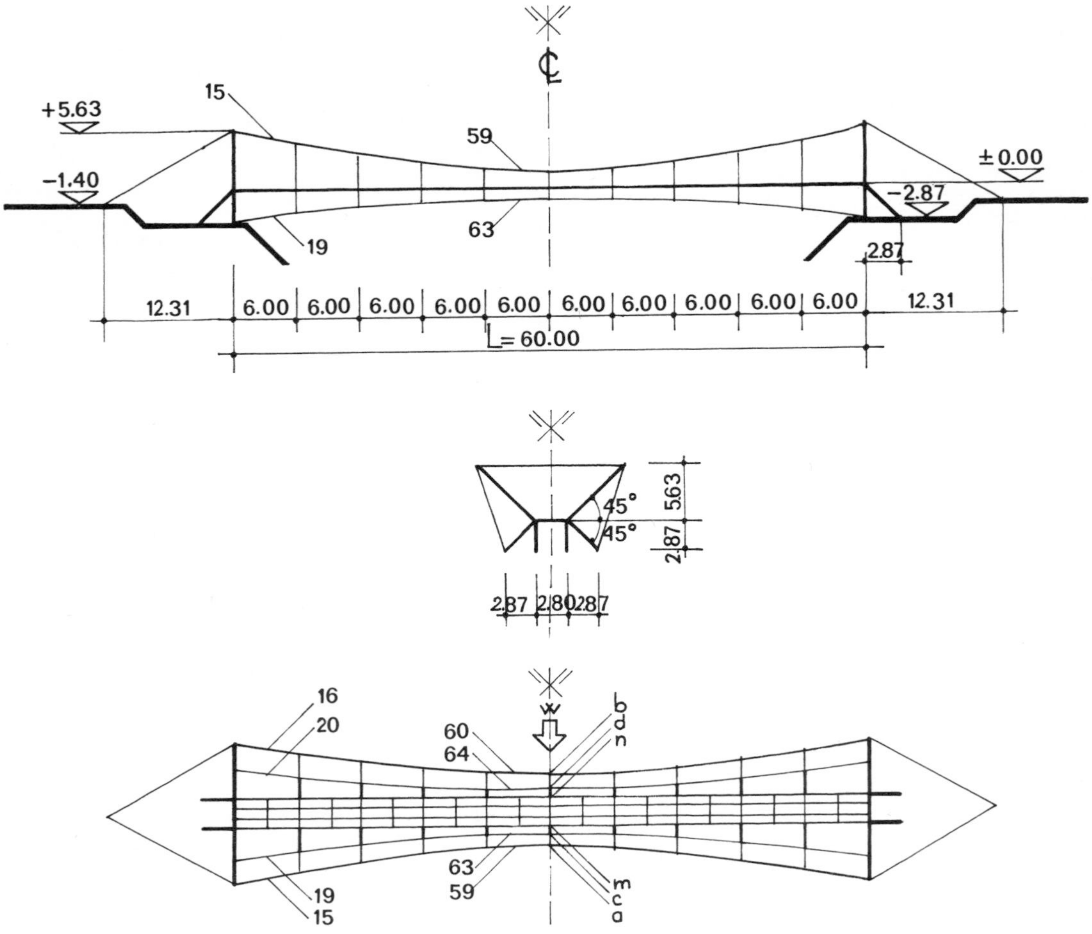

Fig 8.

been described and shown in Fig 8.

Second variant is obtained from the first one rotating the cable planes such that the angle becomes $\alpha_2 = 60^o$.

The loads acting on the structure have been grouped in three loading cases that are presented below with the values of the nodal forces in kN (in cross sections with the span step of 6.00m).

Case I. (Fig 9 a,b).Dead loads: platform weight, conveyer weight, cables and verticals weight.

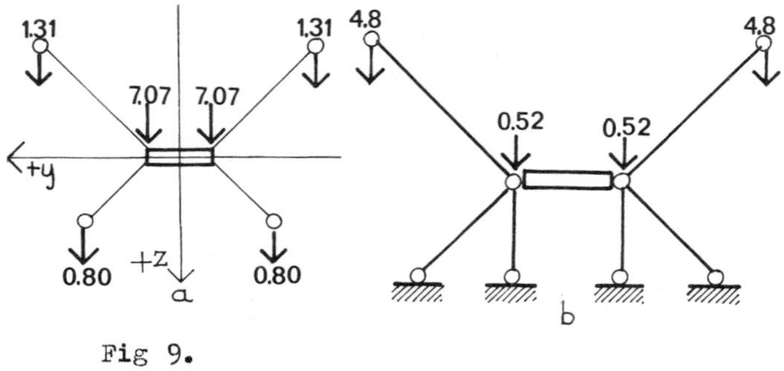

Fig 9.

Case II.(Fig 10 a,b).Dead loads (case I),service loads (weight of conveyed material and the two way traffic beside the conveyer) and climatic actions (white frost).

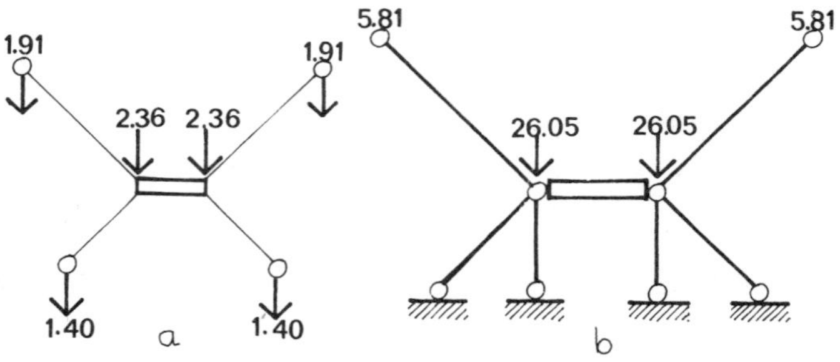

Fig 10.

Case III.(Fig 11 b,c) includes case II and the wind (Fig 11 a) - basic dynamic pressure 200 daN/m^2 namely: wind on platform and conveyer (equivalent in the platform nodes with a force and a moment), ascensional wind and wind on the cables loaded with white frost.

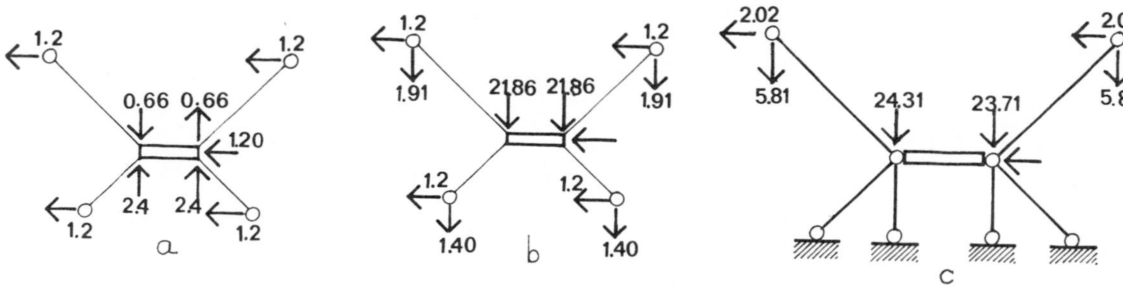

Fig 11.

The maximum prestressing tension (in the elements 19,20 of the lower cables, Fig 8) has been taken equal with the admissible tension of the lower cable $N_o = N_{1,a} = 670.86$ kN.

The numerical analysis of the structure has been performed by means of UNIFE program - see Refs 3 and 4 - developed for finite deformation analysis of suspension system. The analysis includes the boundary structure as well.

Numerical results are given hereinafter.

V a r i a n t 1 ($\alpha_1 = 45^o$). The displacements (in mm) in the middle span cross section (Fig 8) are shown in Fig 12 for loading case I and in Fig 13 for loading cases II and III.

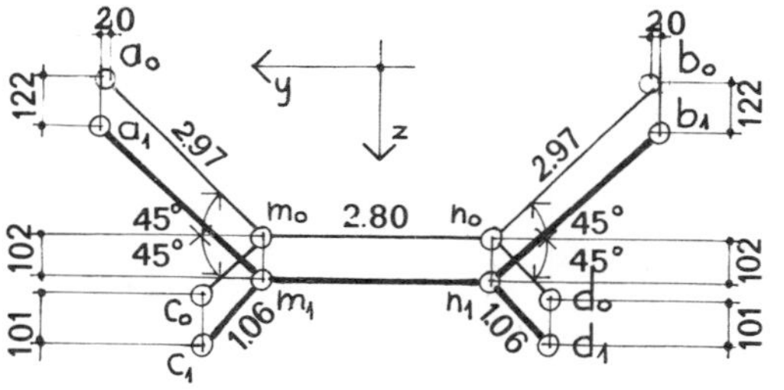

Fig 12.

The subscript zero marks the configuration after prestressing and the subscripts 1, 2, 3 mark the configurations coresponding to the three loading cases. The tension in elements shown in Fig 8 have the following values:

Table 1 - Cable Tensions in kN

Loading case	Maximum Element	Maximum Tension	Minimum Element	Minimum Tension
	Upper Cables			
0	15; 16	413.43	59; 60	396.22
I	15; 16	531.33	59; 60	508.56
II	15; 16	773.13	59; 60	738.01
III	15 16	668.43 824.74	59 60	638.76 786.86
	Lower Cables			
0	19; 20	670.86	63; 64	660.13
I	19; 20	591.92	63; 64	583.24
II	19; 20	444.40	63; 64	438.73
III	19 20	344.01 581.04	63 64	340.09 572.59

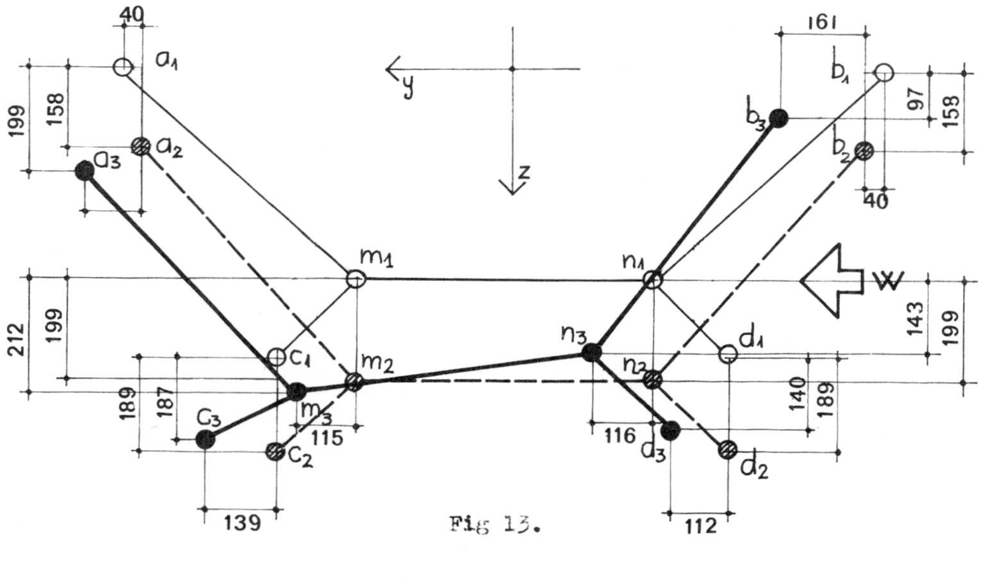

Fig 13.

V a r i a n t 2 ($\alpha_2 = 60^o$). The displacement values (in mm) in points 29 and 30 are:

Table 2

Node	I		II		III	
	f_z	f_y	f_z	f_y	f_z	f_y
m	73.5	0.00	144.32	0.00	174.5	217.25
n	73.5	0.00	144.32	0.00	92.5	218.46

The maximum values (in kN) of element tensions 16 and 20 are:

Table 3

Bar	T_o	I	II	III
16	413.43	512.4	714.17	788.83
20	670.86	598.79	465.42	646.74

CONCLUDING REMARKS

The following can be concluded from the above results of the two variants:

- for $\alpha_1 = 45^o$ the platform horizontal displacements are small - even very small - for a suspended system: in the middle span section f_y/L = 1/522 and for $\alpha_2 = 60^o$ f_y/L = = 1/276. Therefore increasing the angle , the platform horizontal displacements increase.
- for $\alpha_1 = 45^o$ the middle span vertical displacements is f_z/L = 1/283 and for $\alpha_2 = 60^o$, f_z/L = 1/344. Therefore increasing the angle the vertical displacements decrease.
- the maximum tensions vary very little with respect to α (see Table 1 and 3). Thus, the maximum tension in the upper cable are 824.74 kN and 788.83 kN while in the lower cable are 581.03 kN and 646.73 kN for $\alpha_1 = 45^o$ and $\alpha_2 = 60^o$ respectively. Therefore, increasing the tensions from vertical loads decrease while those from horizontal loads increase - the differences between them being small.

Consequently, the studied structure may be adjusted to the services necessities, mantaining the same cable cross section areas, only by varying the cable plane inclination with respect to the platform. For instance, in the case of a conveyer the horizontal displacements must be small, therefore $\alpha = 45^o$ should be adopted.

Future research will refer to the dynamic response and to the influence of several parameters (cross section cable areas ratio, different angles of upper and lower cable planes inclination etc) on structure response.

REFERENCES

1. D MATEESCU, Construcţii metalice speciale (Special Steel Structures) Ed. Tehnică, Bucureşti, 1974

2. F MASANZ, Die Barbarabrücke die Donau, Der Stahlbau, Heft 8, August 1959

3. A CHISALITA, UNIFE - Program for Nonlinear Static and Linear Dynamic Analysis of Cable Nets, Polytechnical Institute of Cluj-Napoca, 1982

4. A CHISALITA, Contributions to the Study of the Nonlinear Static and Dynamic Response of Suspension Systems, Doctoral thesis, Polytechnical Institute of Cluj-Napoca, Cluj--Napoca, 1983.

PRACTICAL DESIGN OF SPACE STRUCTURES

RONALD G.TAYLOR, BSc(Eng), CEng, FICE, FIStructE, MConsE

Consulting Engineer

Ronald G. Taylor & Associates

Epsom, England

So wide is the interest in space-structures that very extensive research and development have taken place particularly in connection with attitudes to and methods of analysis. Structural engineering is essentially a practical art and science so that analytical results do not necessarily facilitate the actual fabrication and completion of a structure. The fact that those processes are common to all structures has led to the invention of a wide variety of proprietary joints and systems each aiming to solve the problems of assembly and site erection. The attitudes to design, fabrication and construction and the relative advantages and disadvantages are examined and the question posed - do we need so many systems?

INTRODUCTION

The analysis of space structures and the determination of the apparent forces in the members should not be confused with their design.

Obviously the computation of forces is a necessary step in the design process but it must be appreciated that whatever assumptions are made in deriving those forces, the probability is that they are based on a false premise.

It would be wrong to say "do not believe anything that comes out of a computer". On the other hand, it would be equally wrong to accept or believe that the computed output factually represents the as-built conditions of a space-structure.

The essential in structural design is an approach which provides an acceptable process of arriving at structurally acceptable solutions. Latticed girders and roof trusses were often designed from force-diagrams based on the assumption of pinned joints and subsequently built as bolted and gusseted frames usually with considerable end restraint. In the same way programmes assuming simple pinned connections, may be, and often are, used to 'analyse' a system subsequently built with any of a wide variety of nodes or connections which may bear little relationship to the original 'perfect' assumptions necessary to enable the programme to be used in the first place. In each case the design process usually develops on the basis of accepting the axial forces from such computation but reducing the effective lengths of compression members due to the obvious restraints provided by the end connections or nodes of the actual structure.

Of course it is possible to introduce no, or any, restraint within the context of a suitable programme but unless the actual joints and the real and individual stiffnesses are known in advance and repeated in the actual structure there is little likelihood of emulating the truth. This should not be a cause for concern.

The fact is that when the normal, if theoretically inaccurate, assumptions are made and the design is executed in the historically established method the structure will generally perform adequately. How then should the designer proceed?

THE MAIN GROUPS

There are two main groups of space structures and very many variations within those groups. They may be classified generally as "nodular" or "modular". (Fig 1).

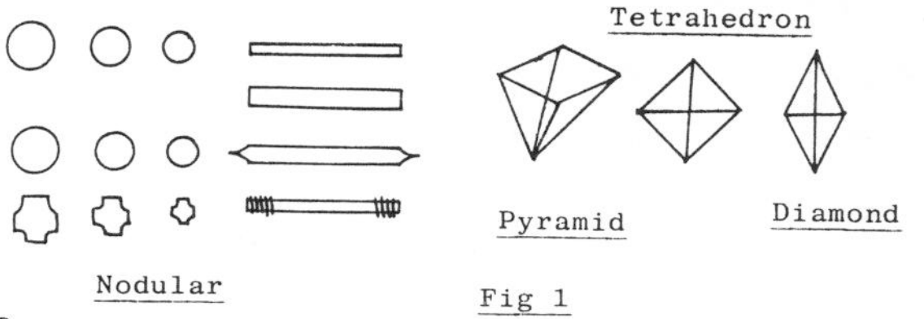

Fig 1

NODULAR

The first group includes the majority of systems which are based on proprietary, piece-small, joints or components. Usually a patent will have been taken out claiming the advantages of the specific node and the parts connected to it. In such cases the design, such as it is, is simplified by reason of the fact that the actual dimensions and quality of the inter-related parts are known and can be accounted for with perhaps more reasonable accuracy than would otherwise be possible. The disadvantages are that any predetermined node leads to a type of discipline through the limitations of the node and within which the designer is constrained to operate.

There is, of course, the 'perfect' node, a sphere, (Fig 2). It is however perfect only in the sense that axial connections to it can be made from any angle without difficulty provided of course that access to the surfaces of intersection are possible and that the structural load-bearing capacities are achievable. Both of these are possible with spheres of sufficient diameter and by using site-welding. Neither of these restrictions is particularly attractive in normal competition. If the sphere is solid there is an obvious disadvantage in weight, cost, handling and welding. If the sphere is hollow there remains the intricate assessment in advance of its performance under a very wide diversity of structural arrangements and load intensities. In both cases the flush-top or flush-bottom line of construction cannot normally be maintained without truncated spheres or eccentric connections.

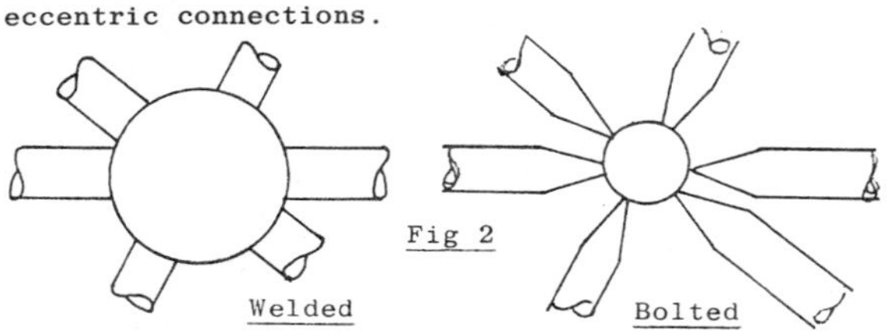

Fig 2

Connection to solid spheres, or similar derivations will be from the outside by "bolted" types of joint or by welding. (Fig 3).

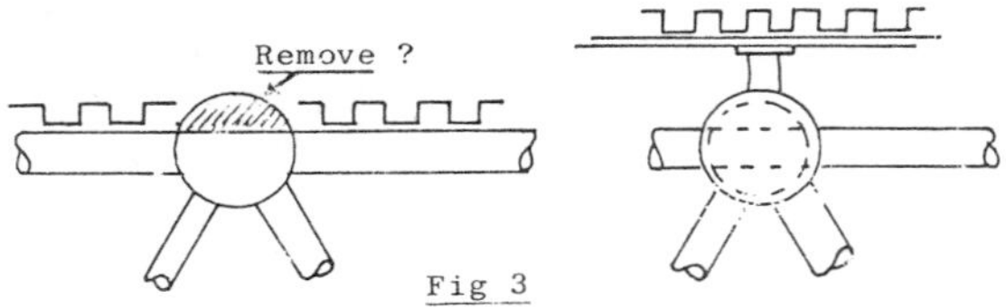

Fig 3

With hollow spheres connection will usually be by bolting from the inside for which purpose access for hand or tools may be necessary with consequent increases in diameter and wall thickness to ensure adequacy of structural performance, unless the hollow sphere be built by subsequent bolted or welded connection of its halves. Despite all of these things the spherical type of joint is easiest to use providing, as it does, a focus of geometry that permits extension of inter-connected parts in space. With hollow spheres and site-welding it is possible to pass the main top and bottom chords through holes pre-drilled in the spheres so that the butt welds of the main chords fall inside the spheres and are invisible and the amount of welding between the main chords and the outer wall of the spheres is reduced. The continuity of the main chords within the spheres greatly enhances the load-bearing capacity of the joint.

Most other systems using proprietary and individual nodes are aimed at particular forms of construction and although they sometimes find use in structures other than those for which they were originally intended their efficiency is usually lessened by such variations. What may be termed in-line planar nodes are most commonly used for 'flat' structures with two or more parallel main layers to the grid inter-connected by bracings in fixed vertical planes but pre-determined angles in the horizontal plane. In most cases this means rectangular main layer connections with bracings either in line or at forty-five degrees angle to the main layer members. This is the commonest field of application of space structures so it is not surprising that many systems, instead of aiming at being the answer to all problems, concentrate on the market of greatest potential.

Even with this direct approach the weaknesses of such preconceived arrangements are obvious in that in 'rectangular' systems the structure must be made to fit the system since those plan shapes in which a common modular factor will not fill the area require a variety of special parts to complete the peripheral layout. Should the common node be versatile in the sense that it can readily accept modification, or a variation-on-the-theme is possible then such peripheral difficulties may not be too great. Generally the designer, if other than the patentee, will have no preconceived ideas or information to enable him to consider or to adapt the special parts which have 'yet to be designed' after the main bulk of standard infill has been set out. Where the main upper and lower layers suffer a change of plane of construction the in-line prefabricated node is at a disadvantage which can normally be overcome only with the same ingenuity as that needed for other 'specials' within the grid. (Fig 4).

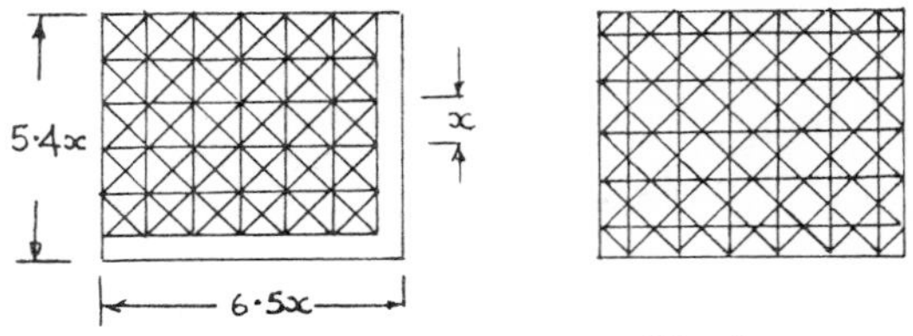

Fig 4

In all forms of space structure in which the members are discontinuous at a joint the tensile connection to that joint must be capable of carrying the full load in tension and the joint therefore appropriately proportioned. Where the same joint is used for members of constant external dimension which may vary in thickness and steel quality then the joint must be capable of safely with-standing the tensile forces equivalent to the thickest sections and highest qualities of steel of those members. (Fig 5). In that sense the joint therefore cannot be structurally efficient. It may be commercially viable however depending on the prime cost, the ease of use and the degree and complexity of fabrication necessary for its use. It follows that for members of least thickness and lowest material quality that the relative cost of joints compared with structural members is high.

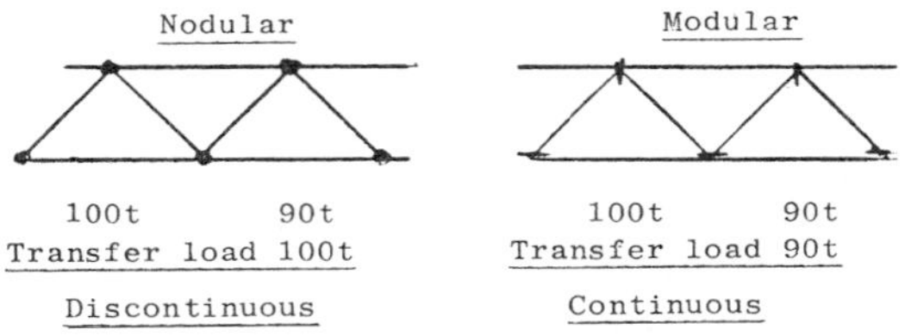

Fig 5

Where many individual members are connected to a joint by site bolted assemblies greater accuracy in fabrication is necessary compared with ordinary structures. Accumulation of tolerances and subsequent mis-fit must be avoided, indeed where such precautions are not taken the final assembly of the prefabricated parts may prove difficult if not impossible. Whereas close tolerance workshop fabrication and accurate site assembly are possible, the introduction, or intervention of site welding, for the sections carrying loads in excess of those permitted in the standard parts will give rise to inordinately high costs and quite different assembly and erection processes in addition to the difficulty of maintaining equivalent standards of accuracy. For instance, with a fully bolted structure built entirely of standard parts the complete frame may be assembled at ground level. Should site welding be involved it may be necessary to raise the construction above ground level at its assembly position thus inevitably incurring additional costs and extra time which are greatly in excess of the previously conceived simple erection procedures.

Nodular construction particularly with the derivatives of spherical nodes is commonly used for single or double layer shells and domes where the parts are delivered piece-small as isolated nodes and members. Where the nodes permit, final connection is by bolting to, or tapping into, the node whether it be hollow or solid.

MODULAR

The second main group consists of prefabricated modules of various types, depending on the system which may have, but not essentially so, a proprietary form of node or connection, and which may employ a variety of structural shapes incorporated either in the same unit or in a diversity of units.

Whereas with piece-small systems there may be many individual connections at a joint, each requiring a specific accuracy of fit, in prefabricated modules a great deal of the tolerance of individual pieces is taken up in the workshop during fabrication thus requiring the overall tolerance of the module to be satisfactory as opposed to say the eight members or so each with its own tolerance which might go to the making of a single module. Such prefabricated modules built in simple workshop jigs can easily be assembled to accuracies equal to those required for the individual piece-small members of other systems. In each case the pre-set or camber of the final space structure is obtained by varying the lengths of members or by varying the actual module size depending on the nature and position of the supporting structure.

Prefabricated modules incorporating standard pre-determined proprietary nodes often suffer the same disadvantage of other similar but piece-small systems, in that the peripheral in-fill often entails "specials" which may not easily fit into the concept of the standard module.

In both main groups there is a tendency to use systems employing a predetermined type of section or material, e.g. all circular, or all cold formed sections, or rolled steel angles and rods, or all steel. With piece-small node types it is not always possible to combine or introduce any shape other than that for which the original system was conceived. In such systems the designer must use the required sections to the best advantage but then losing the advantage of selecting the best section for any given purpose. It may be argued that uniformity of sectional shape leads to an overall appearance of a pleasing nature. Space structures built using a variety of sections compare quite favourably with those in which the section is stereotyped in that the completed structure is usually viewed as a complete entity rather than examined for the appearance or detail of its several parts.

One of the most interesting developments is therefore the introduction of forms of space structure in which there is no predetermined section type, no predetermined node and no restriction on shape except that which can be prefabricated and delivered economically. In other words, a modular system in which the principal parts are interconnected in a predetermined fashion but not using a predetermined joint or unit of construction and in which the infilling of peripheral structure may be obviated entirely or be dealt with by simple modification to workshop jigs. In such a system there is no particular difficulty in combining different types of structural section within the same module with little or no variation to the fabrication jig. The same jig can be very quickly modified to accommodate say rectangular or diamond shaped modules. Better still, a universal jig can be built to cater for a range of sizes and module types.

Typically the commonest type for this form of space structure is the square based pyramid, but rectangular or diamond shaped 'pyramids' or tetrehedra may easily be used. In construction the units may be assembled back to back or corner to corner and the structural sections varied depending on that connection. It is also possible to incorporate both back to back and corner to corner in the same space structure without difficulty.

In many edge or corner supported space-frames where the upper layer is consistently in compression or suffers relatively little reversal of force from the permanent or predominant design loading the butting surfaces of modules may adequately provide for the heavier compression and the bolted connections cater for lighter tensile and shearing forces. As far as the tensile lower plane is concerned there is a significant difference between the piece-small and other modular systems. Generally speaking in piece-small and almost all other systems the tension members are discontinuous at nodes whereas with some modular systems the tension members may be continuous past points of greatest load subject only to the ability to transport the longer lengths to site. Splices may be at nodes or at any other selected point, similar in principal to the selection of splice points in normal latticed girder construction.

Where reversal of load creates tension across the butting faces of modules the same type of joint can be used as for the compression connections and the tensile reversals accounted for by site-welding which does not affect the dimensional accuracy of the building or module.

Intelligent use of prefabricated modular units permits the simple construction of single layer structures of uniform or non-uniform curvature whether site bolted or welded. When rigidity or continuity is required at joints temporary connections are used and the joint finished with simple site welding. (Fig 6).

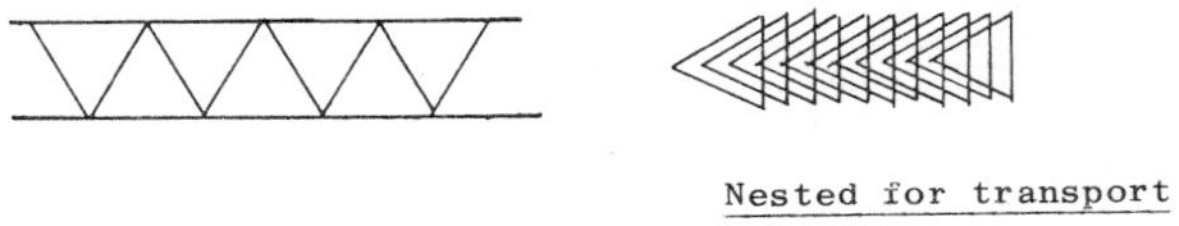

Fig 6

The prefabricated modules should be dimensioned to permit economy of packing for transport and the temporary connections used to facilitate assembly by normal means without recourse to welding which if necessary should be done after initial assembly. In this way the erection processes conform to those used for bolted construction in the first place and delays on account of site welding are obviated. (Fig 7).

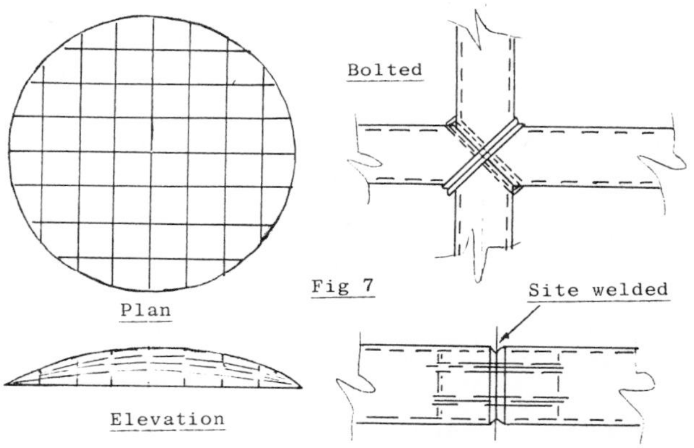

Fig 7

Well established workshop methods may also be used to fabricate components to form novel types of space structures. In some types of latticed girders the bracings consist of solid rods which are formed by continuous mechanical folding. A similar process can be adopted using circular hollow sections instead of rod so that flattening at the fold permits drilling and assembly of two such sets at angles of intersection suitable to form the bracings of a double layer grid. (Fig 8). The prototype model of the Space Structures Research Laboratory roof at Surrey University was made in this way but subsequently built as a 'nodular' system. The flattened folded tubular bracings are then placed with the flattened parts superimposed at right angles to one another, drilled and connected to main members by site bolting. The change of shape of the tubes due to the flattening and deformation of the ends although altering the structural properties of the cross-sections appears to be compensated for by the inherent stiffness of the connected parts. This is particularly true when more than one bolt is used to connect the parts, for instance a coverplate may be placed over the crossover joint with bolts fixing the coverplate, the flattened tubes and the main member together.

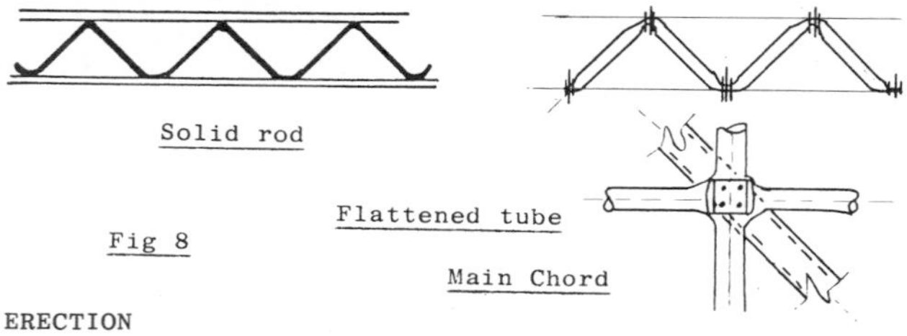

Fig 8

ERECTION

Access to and on site, the programme of construction, the type of structure, the nature of the supporting structure and many other relevant factors e.g. the supply of electric power, must all be considered before final selection of the type of system to be used. In this sense space structures are no different from other structures except that by their nature they are usually more repetitive in type of subdivision and, most important of all they lend themselves to assembly at low level for erection in one piece. Not all systems however permit the same ease of construction "in-the-air ". Where the retention of members connected at a node depends on the completion of the node itself without which the members are not held in position then obviously this makes erection in space more difficult. Where the lengths of members before jointing, including their means of connection, are greater than the final lengths between nodes of the finished structure then again this makes erection in space more difficult.

Where site conditions permit there is no doubt that complete assembly at ground level is favourable even for spans of great dimension and weight. The structural advantages are obvious but still need to be assessed economically and with great care bearing in mind the relative costs of lifting equipment, the fitting of services and cladding and the application of final protective coatings.

Ground level assembly affords the opportunity of dimensional and structural checking and correction, if necessary, before lifting. It permits the use of smaller capacity craneage with shorter jibs in a variety of positions simultaneously and in some cases, with smaller grids, allows units to be man-handled without mechanical equipment of any sort.

The final lift may be achieved by cranes, jacks or special techniques such as the lift-slab or Fressinet systems where a multiplicity of lifting points can be controlled from a single master console capable of delicate dimensional control for final positioning.

Where the permanent supports are not available or ready, in the programmed contract, temporary supports capable of supporting the completed structure under no load may be used and either rejected subsequently or incorporated with the final supports. Complete space structures for hangars with spans between 180-220 metres and weighing some 3000 tonnes including structure and cladding have been lifted without difficulty. In adverse climatic conditions the provision of the clad structural roof may be used with advantage to permit other work beneath it without regard to weather and with greater certainty of compliance with the construction programme.

Completed structures cannot necessarily be safely lifted disregarding the position of the final supports as compared to the temporary supports as lifting points. It is however, by no means unusual to lift complete structures as clear-spans with a few lifting points even when the final structure will include several peripheral or internal supports. A simple comparison of a fully loaded square grid continuously edge supported, and the same grid only corner supported under its own dead weight, will show that the dead load must be about one third of the total design load before the main chord loads are comparable and that unusually high proportion indicates that in almost all cases an edge supported square grid can be lifted as a clear-span corner supported grid, with safety. As a general rule, when the loads at lifting points cannot be registered then two or three lifting points are preferable compared to four or more since the greater the number of lifting points the more undeterminate are the loads. Large structures using erection equipment capable of individual adjustment can safely be used with multiple lifting points.

CHOOSING A SYSTEM

Given a variety of plan shapes and profiles for which a space structure is to be considered there is a wide range of possible alternatives. The range can be drastically reduced by considering the basic elements of the proposed structure and by drawing up a "deletions" list. The list should itemise the essential requirements e.g. loading, grid dimensions, support points, cladding, structural depth available, degree of exposure, distance for transport, fabrication sources, site access and details, programme of erection, flat or curved finish, permissible deflection, appearance and maintenance, and any other relative features.

Even with regular shapes but which exclude uniformity of nodule through the absence of a common factor of subdivision, those systems not amenable to 'special' peripheral treatment are at a disadvantage, immediately. With non-planar shapes the field is drastically reduced.

Where large spans are involved many nodular systems suffer from the upper limit of permissible load which the nodes can take and special nodes may be uneconomic. Some systems are unable to, or can, accept only a limited amount of local bending on the upper layer whereas others can cater for concrete floors and heavy applied loads.

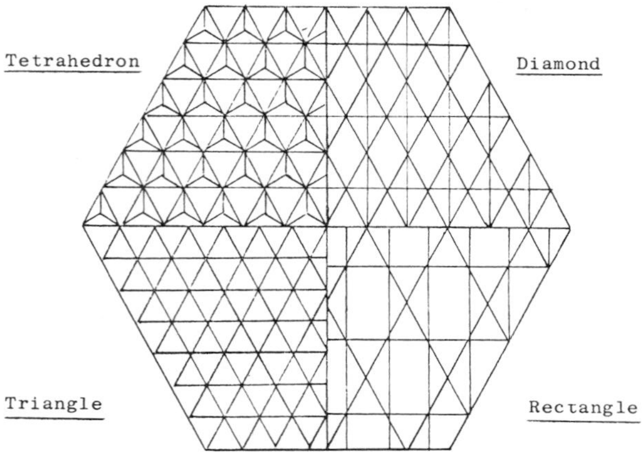

Fig 9

Figure 9 shows how a variety of shapes may be formed within a given regular hexagon but which may just as easily be applied to an irregular hexagon which is symetrical about one centre line. Systems using right-angle or 45° arrangements would not be suitable. That is equally true of a regular octagon.

There are fewer possibilities with irregular shaped triangles (Fig 10) yet with modular types and some "non-rectangular" nodular types sufficient repetition is not only possible but the subdivisions are simple and often in harmony with the overall shape. One side of the triangle should be divided into an equal number of parts. The modules are then created by drawing lines parallel to all three sides thus dividing the large triangle into a series of smaller similar triangles which may be supplied either as those smaller triangles or, on occasion, as diamond shaped modules.

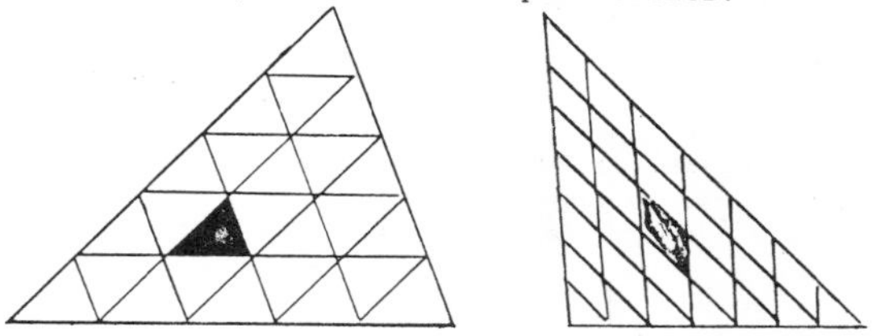

Modules shaded

Fig 10

With non uniform curvature in shell structures, the dome of a mosque for example, (Fig 11) the 'spherical joint' nodular type and the prefabricated modular type should each be considered as the subdivision may easily be arranged to produce a sufficient number of repetitive types to consider prefabrication as a reasonable possibility.

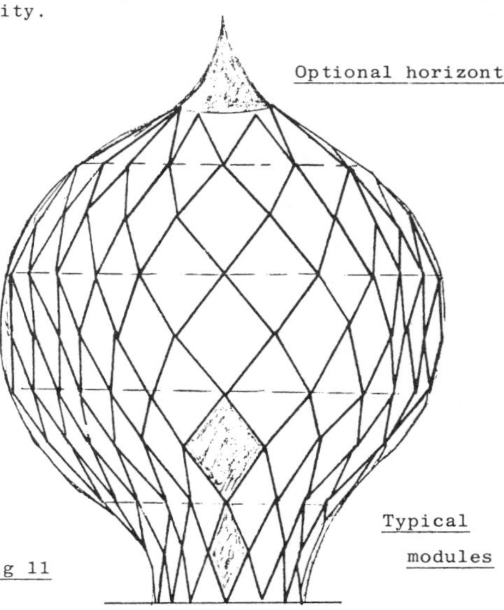

Fig 11

Fortunately some structures may permit the normal use of rectangular systems by containing a common factor for the subdivision of the sides. In that event there is a wide choice from nodular, modular and indeed other types for which on occasion special joints may be developed.

Special joints are well worth considering where very large spans are concerned and a large number of joints are necessary. The design of the joint must of course allow for the transference of load without build up of dimensional error due to the tolerances but at the same time the details must be such as to allow relatively easy site assembly. Where the structure is, for example, a wide span double layer grid, the joint will normally need to cater for at least three but often four prefabricated units which may in themselves be of latticed girder type so that the joint is an isolated unit serving to connect the parts by site bolting. Such a joint for example may be required to transmit say 1000 kN through each of four main chords intersecting at the points in a structure composed of a double layer diagonal grid, (Fig 12). In that grid despite the main component girders being some 8m long and 3.8m deep to form 'diamonds' of 12.8m by 5m wide it would also be possible to prefabricate the whole 'diamond' and deliver several such units nested together on low-loaders. The disadvantage is obviously the need for, and method of, special transport. The advantages are well worth considering.

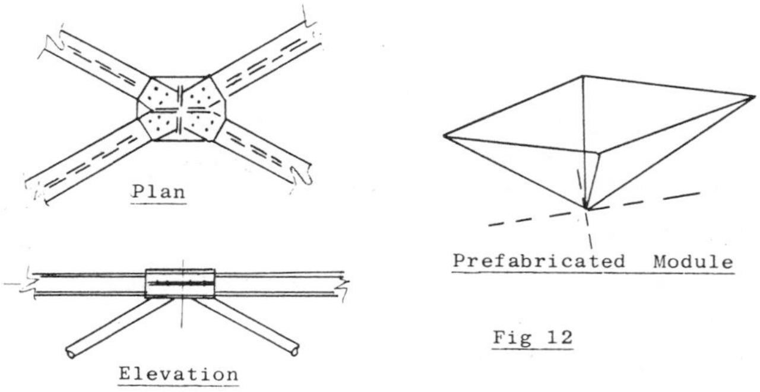

Fig 12

In the first place much of the load-bearing transfer is taken up in the workshop since the main members are already connected by welding. Then the main tension members forming the lower chords may be supplied as long plain bars drilled only for site bolting. The main upper members being mainly in compression may transfer that part of their load by direct bearing. The joint must of course be capable of transferring shearing forces in the horizontal and vertical planes.

Shearing forces in the horizontal plane with three-way connections can easily be catered for in the design of the joint itself where several alternatives are possible. Shearing force in the vertical plane can also be taken up by shear plates inset, for example, into a corner box or connector through which a single bolt passes to connect the parts together. Should the single bolt not be adequate for transfer of tensile forces the joint can be easily welded with or without the shear plate in position.

Reduction in extent of site bolted connections can be effected by using prefabricated modular bottom units instead of piece-small or continuous type tension members. Such a system has been used in the recent construction of a pyramid where it was important to maintain a clean line of surface support for plastic glazing units, (Fig 13).

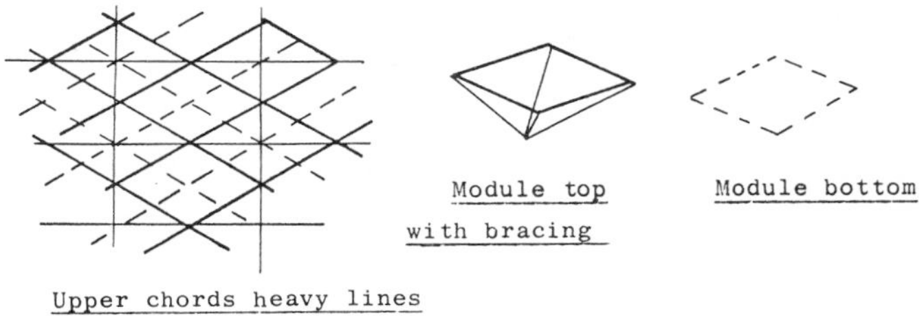

Fig 13

Where modular systems are used with the limit of load peculiar to the module itself the use of multi-layer grids may be adapted to increase the depth yet retain reasonable angles of intersection and spans suitable for cladding without secondary steel. For example, attempting to use a node with a safe load of say 800 kN, on a span with a normal space-frame span-to-depth ratio of say twenty, the calculated force may be 1600 kN. The grid must therefore be increased to provide a ratio of span-to-depth of ten thus rendering it unacceptable for the actual structure being considered.

In maintaining a reasonable plan dimension to accept cladding, using the greater depth, the lengths and angles of intersection suffer so that triple-layer construction may be resorted to in an attempt to overcome these faults (Fig 14). The usual way is to include a middle plane connecting member with, in simple cases, the top layer in compression and the bottom in tension.

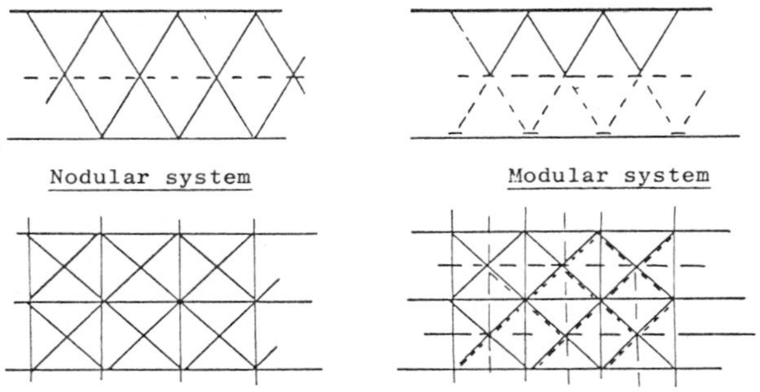

Fig 14

The disadvantage is that the discontinuous bottom chord and node in nodular systems must still carry the full tensile forces. The use of modules although incurring prefabrication, may obviate both disadvantages either changing the style of building altogether by using a prefabricated bracing unit connected to the two 'continuous' top and bottom layers or by employing similar, but not the same, forms of module both used the same way up keeping the bottom chord again as an independent continuous tension member.

These arrangements permit achievement of double layer grids, and hence spans, far greater than normal since criteria for prefabrication now becomes the lateral dimension of the infill bracing whereas the depths conceivably might be ten metres or so.

With prefabricated modular construction there is, therefore, ability to achieve spans and load-bearing capacities limited only by the ingenuity of the designer in any given set of site conditions. Great care is necessary in assessing volumes occupied by the units in transport yet where, for example, containers are used the load capacity is fixed and may be used up by piece-small items on a low-volume/weight basis yet still with a half-empty container, as opposed to prefabricated modules of greater volume/weight basis which fill the container to a greater extent but still within its capacity. The design should be aimed not only at economy in material but also economy in freight.

THE FUTURE

Wide spans can now be built with space structures using little workshop area, normal transport methods, simple assembly and easy erection thus overcoming almost all of the problems inherent in normal construction. Such is the understanding of space structures that they now form a significant part of structural engineering. Methods of analysis abound as do the numbers of proprietary systems facilitating construction. Added to those advantages is the ability to design detailed connections for space structures, with experience, comparable to the design of connections for 'orthodox' structures in the past. The ability exists to build shapes and forms of structure not previously possible with rectangular connections, to construct single, double or multi-layer structures with equal ease and to achieve what is the ambition of architect, client and engineer, namely, improved appearance, economy of construction, and excellent load-bearing performance with simplicity of fabrication and erection.

DESIGN AND CONSTRUCTION OF A HUGE ROOF STRUCTURE USING NS SPACE TRUSS SYSTEM

MAMORU IWATA*, Dr. Eng., KAORU KAMIYAMA**, B. Sc
and YOSHIHARU KANEBAKO***, M. Sc

* Manager, Steel Structures & Building Construction Division,
Nippon Steel Corporation

** Engineer, Steel Structures & Building Construction Division,
Nippon Steel Corporation

*** Engineer, Yokoyama Consulting Engineers

"NS Space Truss System" is a three dimensional pipe structure system developed by Nippon Steel Corporation. The system has since found wide application in Japan and overseas.

It was decided to use this system for the main arena of the gymnasium of the National Athletic Meet to be held in 1986. Design requirements of the structure were huge square hipped roof and high structural reliability.

The reasons for adoption of the NS Space Truss System are that the system not only meets these requirements but can give a beautiful appearance to the gymnasium.

This paper describes the NS Space Truss System focusing on designing of the huge roof structure and technical problems arising from construction.

INTRODUCTION

The main arena building is a 3-storied reinforced concrete structure and holds a large space of 60.0 m x 63.2 m (Photo 1). The roof structure is supported around the perimeter by the beam which, in turn, is supported by cantilever columns of approximately 9 meter height installed at about 6 meter intervals.

Photo 1 Outside view of gymnasium

The most essential point in the structural design of this building was how the structure of the roof of the main arena should be worked out. The shape of roof decided from both the functional viewpoint and the design viewpoint was a square hipped roof having ridgelines in the 45° direction and a gradient of 1.5/10. It was required to make the structure simple and beautiful. Considering the various factors described so far, the space truss structure was adopted for the main arena. A truss framing plan and a framing elevation is shown in Fig 1.

A space truss is composed of individual members and joints. Because individual members are determined almost in view of their compressive strength against buckling, steel pipe is the most advantageous member material. In addition to the above advantage, an attractive shape from a design point of view, as compared to other member materials, led to the adoption of steel pipe for the truss members for this project. On the other hand, a joint is required to have a performance of smoothly transmitting the force of members in a three dimensional way. The joints of pipe space truss can be divided largely into two types. One is a type of joint which is fabricated using a steel ball to which steel pipes are welded on site. The other also uses a steel ball to which, however, steel pipes are connected using bolts. The NS Space Truss System, one of the special bolt joint methods, was adopted for the present building from the standpoint of the reliability and cost effect.

The relation between the truss and the lower structure is an important element in the design of a space truss. As will be described later in this paper, close study was conducted on the supporting condition, and, as a result, various considerations were taken into effect.

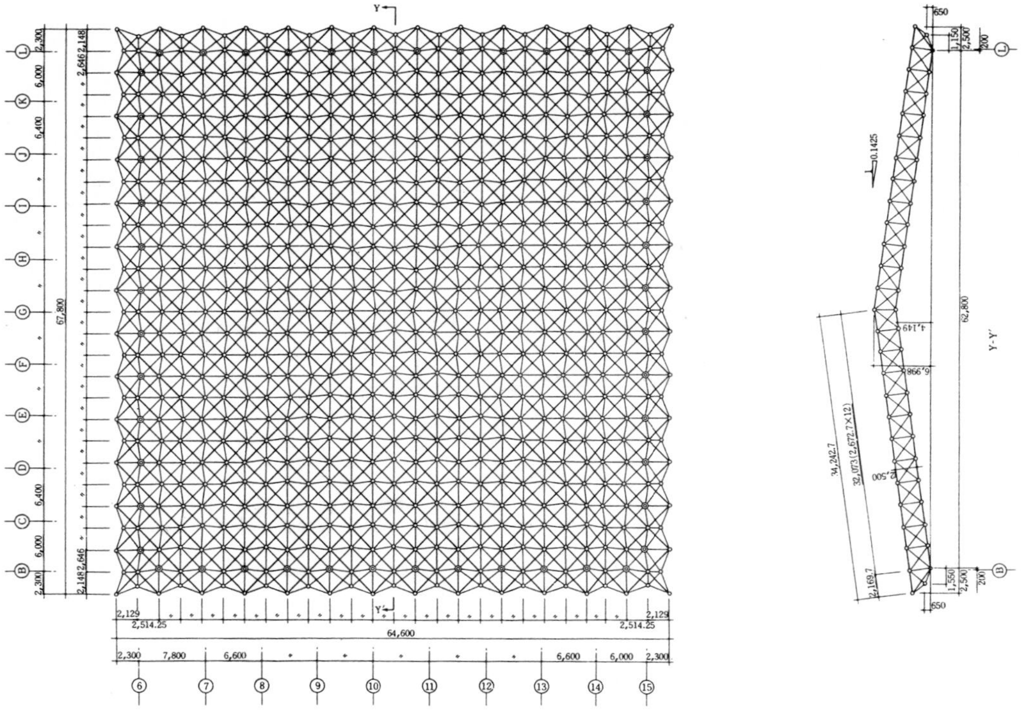

Fig 1 Truss framing plan and framing elevation

CONSIDERATION OF SPACE TRUSS

Stress analysis of the space truss for the main arena was carried out by changing the total shape and supporting condition to understand the effect of these factors on the member force and the support reaction. In addition to the hipped roof adopted in the design, a flat roof was analyzed for comparison. As for the supporting condition of horizontal direction, three cases were studied, namely, (1) free in perimeter and orthogonal directions, (2) restrained in the perimeter direction but free in the orthogonal direction, and (3) restrained in both directions. By combining these conditions, five cases were formed as shown in Table 1, and stress analysis was conducted on them assuming a vertical load.

Fig 2 indicates vertical displacements in the ridgeline. A comparison of the hipped roof and the flat roof reveals that the displacement of the hipped roof was smaller than that of the flat roof even when the supporting condition was assumed to be free in both directions (Case 1 and Case 4). When restrained in the perimenter direction (Case 2 and Case 5), the displacement of the hipped roof was significantly diminished.

Table 1

	Total shape	Supporting condition		
		Horizontal direction		
		Perimeter direction	Orthogonal direction	Vertical direction
Case 1	Hipped roof	Free (Roller)	Free (Roller)	Restrained (Pin)
Case 2	Hipped roof	Restrained (Pin)	Free (Roller)	Restrained (Pin)
Case 3	Hipped roof	Restrained (Pin)	Restrained (Pin)	Restrained (Pin)
Case 4	Flat roof	Free (Roller)	Free (Roller)	Restrained (Pin)
Case 5	Flat roof	Restrained (Pin)	Free (Roller)	Restrained (Pin)

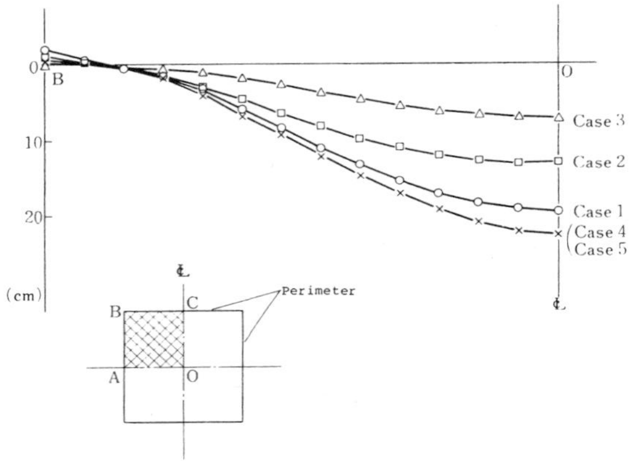

Fig 2 Vertical displacement diagram

A diagram of support reactions is given in Fig 3. In the case of the hipped roof with the support free in both directions (Case 1), the distribution of vertical reaction was similar to that in the flat roof (Case 4), peaking in the center but diminishing toward the corner, where a downward reaction was generated. When the support of a hipped roof was restrained in the perimeter direction (Case 2), a large horizontal reaction was generated in the corner. When the support was restrained in the both directions (Case 3), although the reaction in the perimeter direction was decreased, the reaction in the orthogonal direction was noticeably increased, and both the orthogonal reaction and the vertical reaction peaked in the center. In the case of the flat roof whose support was restrained only in the perimeter direction (Case 5), the horizontal reaction was generated only in the corner. The distribution of the vertical reaction in this case was not much different from that when the support was assumed to be free in both directions (Case 4). This indicates that the effect of the restraint in the perimeter direction is small.

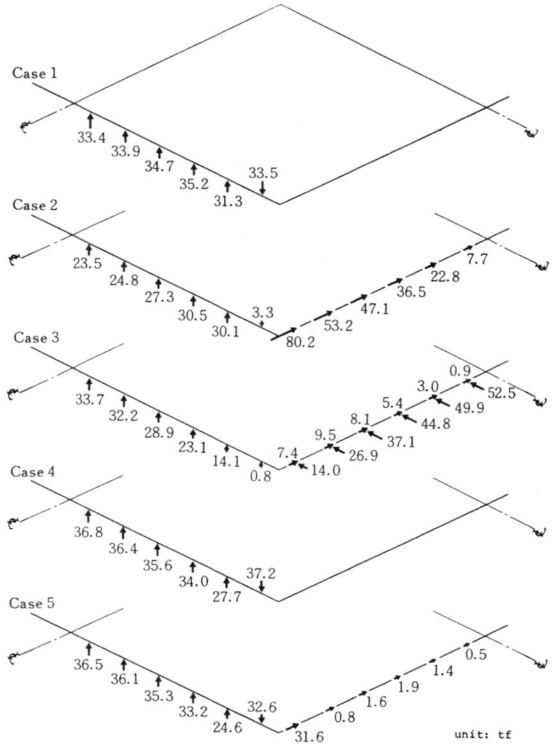

Fig 3 Support reaction

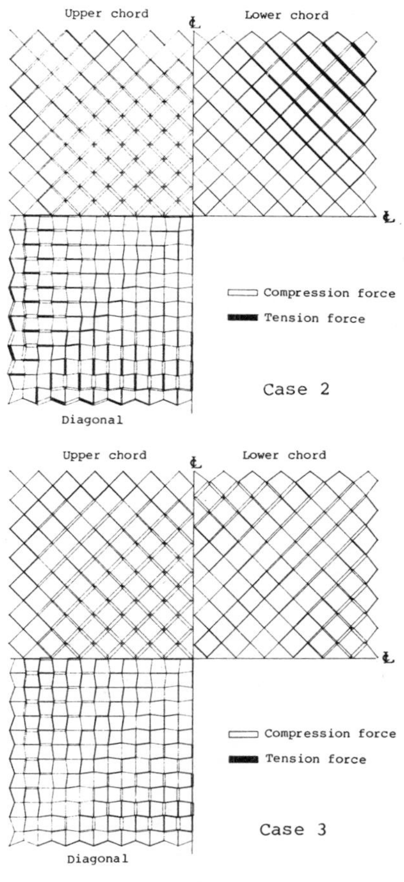

Fig 4 Axial force distribution of members

Comparisons of the axial force distribution of chord members between Case 1 and Case 4 are exhibited in Fig 5 and Fig 6. In the case of the flat roof because only a bending stress could be produced in the whole truss, the axial force working on the upper and the lower chord member was the same in the absolute value. In contrast, in the case of the hipped roof because both a bending stress and a membrane stress could work on the whole truss, the axial force shifted generally toward the compression side in the O-B direction (Fig 5) but toward the tension side in the A-C direction (Fig 6). This indicates that although the support was not restrained in the horizontal direction, both the chord member in the A-C direction and the diagonal member in the direction parallel to the perimeter beams played the role of a ring beam, creating a stress condition which includes an membrane stress different from the simple bending stress.

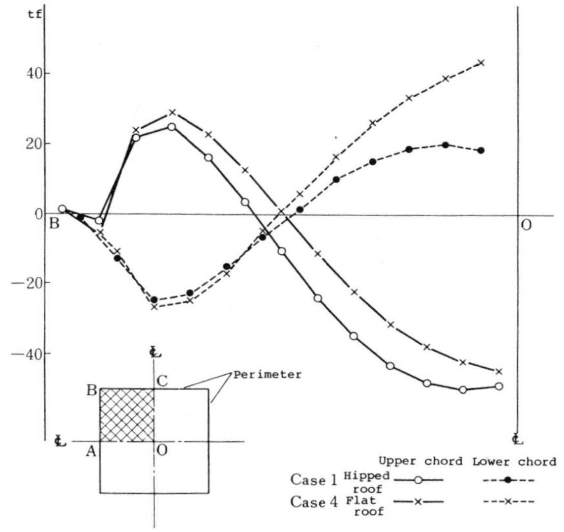

Fig 5 Axial force distribution of chord members (O-B direction)

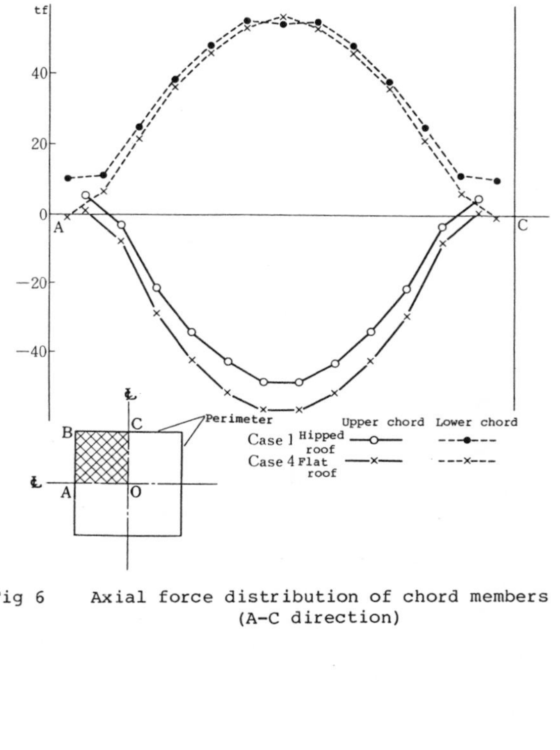

Fig 6 Axial force distribution of chord members (A-C direction)

Both Fig 7 and Fig 8 compare the axial force distribution of chord members under Case 1, Case 2 and Case 3. When the support was restrained in the perimeter direction (Case 2), the reaction in the perimeter direction was small in the middle. This indicates only a slight effect of restraining the support, but the effect of the restraint emerged large near the corner. Hence, while the axial force of both upper and lower chord members was somewhat decreased in the A-C direction, it was notably shifted toward the compression side in the O-B direction. From this observation and the distribution of reaction illustrated in Fig 3, it can be understood that if the support is restrained in the perimeter direction alone, the flow of force into the corner is large. If the support is restrained in both directions (Case 3), the membrane stress is the most prevalent, and a compressive force works on both the upper chord member and the lower chord member in the A-C direction, while the axial force in the O-B direction becomes smaller than that in Case 2. Hence, it is clear that a considerable force flows in the middle of the truss periphery under this supporting condition.

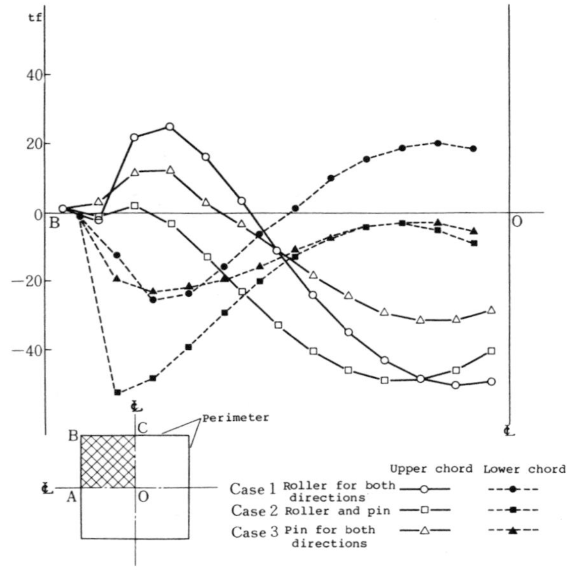

Fig 7 Axial force distribution of chord members (O-B direction)

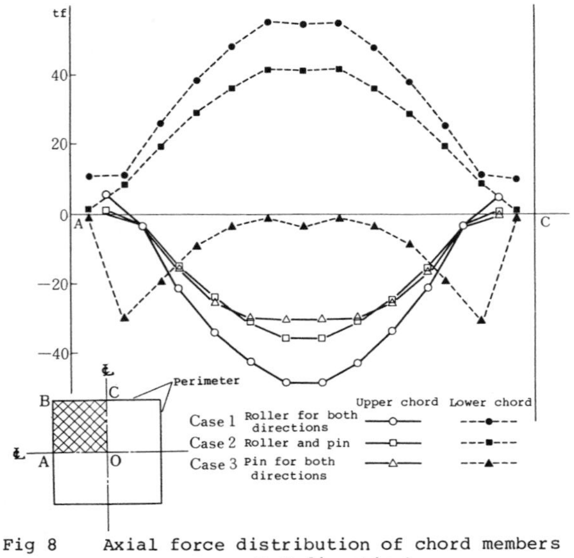

Fig 8 Axial force distribution of chord members (A-C direction)

From the foregoing consideration, it has now been understood that although the effect of changing the supporting condition on the axial force and the support reaction is small in the case of the flat roof, it is large in the case of the hipped roof.

DESIGN OF SPACE TRUSS

As was discussed earlier, the axial force in the truss member of a hipped roof varies significantly with the supporting condition. The supporting condition adopted for the present design is such that the support is restrained in the perimeter direction but is free in the orthogonal direction of seismic force., The problems associated with this support condition are as follows.

(1) The truss is set up with support free in both directions of the horizontal direction during erection. Even after erection there is the possibility of an insufficient restraint on the support due to the loosening of anchor bolts, etc. In that case, the axial force in the truss is closer to that generated when the support is free in both directions.

(2) The support may possibly be restrained in the orthogonal direction as well by the force of friction working between the anchor bolts and the base plate under the reaction in the perimeter direction.

In the case of (1) above, because some members have an axial force greater than that working when the support is restrained in the perimeter direction but is free in the orthogonal direction, such should be taken into consideration in the design of truss. In the case of (2), however, as the support reaction is increased though the axial force in the truss is decreased, some considerations need to be given to and around the support.

The truss support of the present design was required to be able to move freely in the orthogonal direction under the restraint in the perimeter direction. In order to satisfy this requirement, a loose hole in one direction was adopted for the supoort. In addition, the safety of the truss was checked against the possiblility of the actual supporting condition differing from what had been assumed. With a loose hole in one direction, the movement in the free direction may possibly be restrained by the friction. In this case although the horizontal reaction in the perimeter direction is decreased, a horizontal reaction in the orthogonal direction is generated. Hence, it is necessary to study the lower structure, especially the cantilever columns, in addition to the support. In the study, stress calculations were carried out by assuming a spring support whose spring constant was taken to be equal to the horizontal rigidity of cantilever column, and the results of the calculations were used to confirm the safety of the support and the lower structure.

DESIGN OF NS SPACE TRUSS

Pipe members (pipe + end cones), bolts and washers can be designed within the standard component range of NS Space Truss System (Fig 9). However, as far as the node is concerned, in the case that a compressive force and a tensile force work on the node orthogonally each other, as the node tends to be deformed more easily, the strength of the node become lower, and, as a result, needs to be a node with a strength greater than that of an standard node. Consequently, a node whose shape is symmetric with respect to the reference plane, as illustrated in Fig 10, was partially adopted. The laboratory test was carried out on this node, and uniaxial compressive loading test conducted will be described in this paper. Loading was repeated several times in a low load region, and then increased to a maximum load. Shown in Fig 11 and Fig 12 are load deformation diagrams, and listed in Table 2 are the yield point calculated from the test results and the allowable axial force determined on the basis of the yield point.

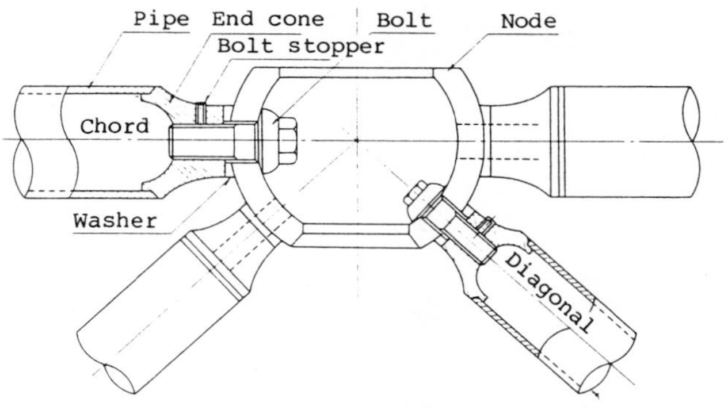

Fig 9 Connection detail of NS Space Truss System

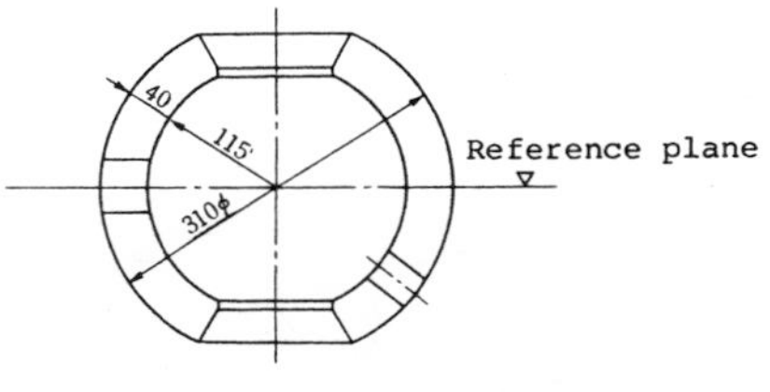

Fig 10 Symmetric type node

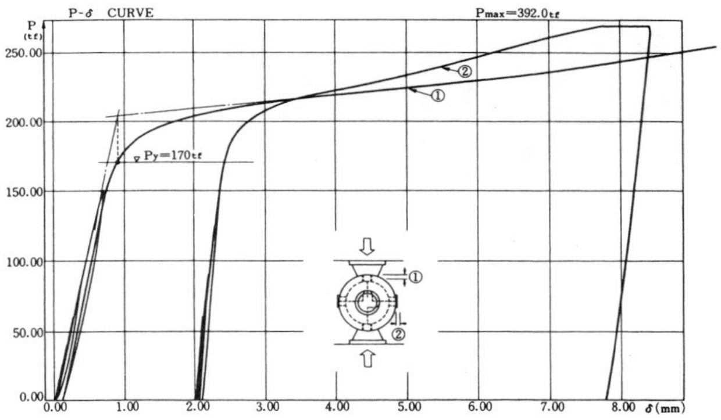

Fig 11 Load deformation curve (Chord direction)

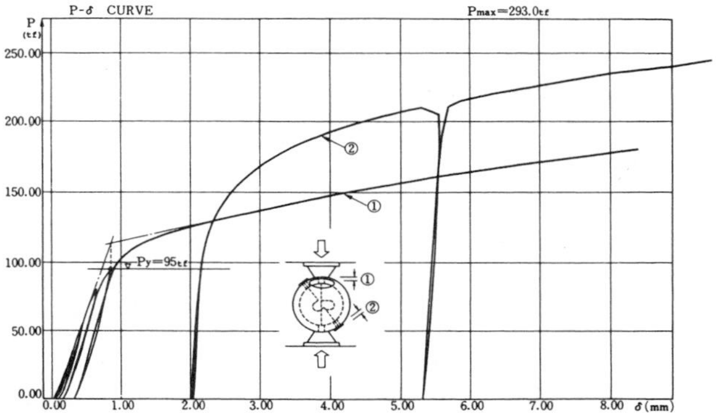

Fig 12 Load deformation curve (Diagonal direction)

Table 2 Node laboratory test result and allowable axial force

		Chord direction	Diagonal direction
Test result	Max. load Pmax (tf)	392	293
	Yield point*1 Py (tf)	170	95
Allowable axial force	Long duration*2 Pa, L (tf)	85	47.5
	Short duration*2 Pa, S (tf)	127.5	71.3

*1 The yield point were obtained from the P - δ curve by the general yield method.

*2 $Pa, L = \frac{1}{2} Py, \quad Pa, S = \frac{3}{4} Py$

The materials of NS Space Truss components and their manufacturing methods are shown in Table 3. As the surface treatment of NS Space Truss for this project, galvanizing was specified for its high rust inhibiting performance and maintenance-free characteristic.

Table 3 Materials and manufacturing method

<table>
<tr><th>Part</th><th>Standard</th><th>Specification</th><th colspan="2">Manufacturing method</th></tr>
<tr><td>Node</td><td>JIS G 3106</td><td>SM50A equivalent forged product with a yield stress of 3.0 tf/cm2</td><td colspan="2">Forging - Machining</td></tr>
<tr><td>End cone</td><td>Same as above</td><td>Same as above</td><td>Forging - Machining</td><td rowspan="2">Welding</td></tr>
<tr><td>Steel pipe</td><td>JIS G 3444</td><td>STK 41</td><td>Machining</td></tr>
<tr><td>Bolt</td><td>JIS B 1180</td><td>F10T equivalent special type high strength bolt</td><td colspan="2">Forging</td></tr>
<tr><td>Washer</td><td>JIS G 3101</td><td>SS41</td><td colspan="2">Machining</td></tr>
</table>

ERECTION OF NS SPACE TRUSS

The methods of NS Space Truss erection can be classified largely into "Scaffolding method" in which the whole truss is assembled at the specified erection position on the stage of a scaffold, "Block method" in which the truss members are assembled into small blocks on the ground and are installed in place by crane, "Sliding method" in which the truss members are assembled into small blocks at an end portion of the roof and are sequentially slid into place, "Lift up method" in which the whole truss is assembled on the ground and is lifted by lifting equipments, etc. For the erection of the present truss, the above methods were all studied and compared and the Scaffold method was adopted. In this method, scaffolding is erected all around in the arena the truss is assembled on the scaffolding; and after the completion of the truss assembly the jacks supporting the truss are lowered to get the truss to be under the designed stress condition.

In the erection of space trusses, it is very important to ensure that the truss force flows in two directions at the completion of the truss according to the design. This principle must be strictly observed. If not, the initial axial force remains and leads to the generation of a greater axial force than the designed. A key point in ensuring appropriate distribution of the force into two directions resides in the uniform jack-down of the truss after assembly. It is an advantage of the Scaffold method that jacks can be used for both assembly and jack-down of the truss.

The scaffolding is comprised of poles each with a jack, scaffold, to board and other auxiliary members (Photo 2). The truss was supported at lower chord nodes through the node support jigs on poles with jacks arranged in radial lines. The jacks were used for rough adjustment and the node support jigs for fine adjustment of the truss position during erection. For the jackdown operation, the node support jigs were first lowered to let the whole truss weight rest on the poles with jacks, and then the jacks were lowered starting with those on the first line in the center and outward. 80-ton tower cranes were installed at four locations to cover entire work range. The erection flow chart in shown in Fig 13.

Photo 3

Photo 4

Photo 2

Photo 5

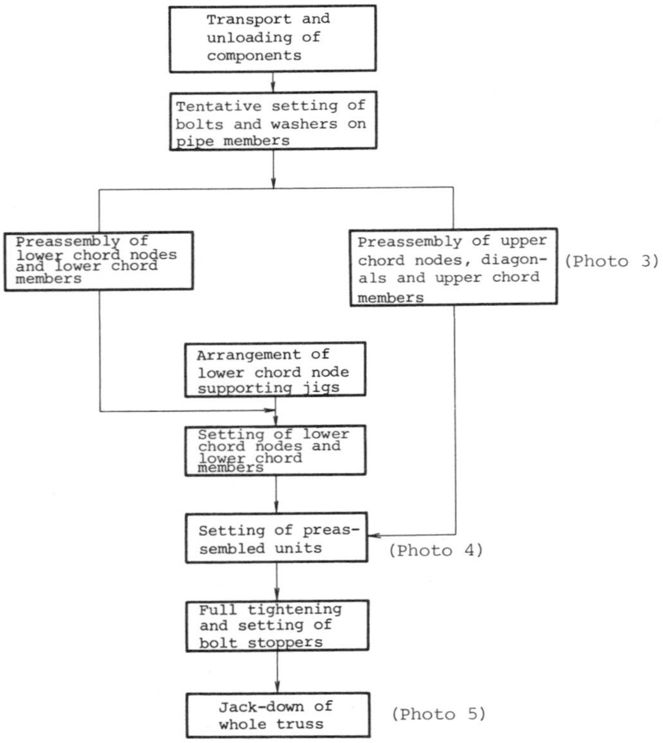
Transport and unloading of components
Tentative setting of bolts and washers on pipe members
Preassembly of lower chord nodes and lower chord members
Preassembly of upper chord nodes, diagonals and upper chord members
(Photo 3)
Arrangement of lower chord node supporting jigs
Setting of lower chord nodes and lower chord members
Setting of preassembled units
(Photo 4)
Full tightening and setting of bolt stoppers
Jack-down of whole truss
(Photo 5)

Fig 13 Erection flow chart

CONCLUSION

NS Space Truss System, which was adopted for the main arena roof structure, derived the idea from the space truss system employed for the main roof of the Festival Plaza of the EXPO 1970 held in Japan. It has been adopted for a wide range of structures. The adoption of the NS Space Truss System for the present gymnasium with huge square hipped roof added another valuable experience.

Acknowledgment

The authors wish to express their deep gratitude to the people of Yamanashi Prefectural Government, Japan Engineering Consultants and the joint venture of Tokai Kogyo and Ijiri Kogyo for their kind cooperation in writing this paper.

References

1) K. Saito: Arrangement of Space Truss Structure, Column 84, April 1982

2) D. Nakano and M. Iwata: Development of New Space Truss System, Column 85, July 1982

3) Y. Suzuki and I. Kubodera: Design and Construction of Low Gradient Roof Structure, Kenchiku Gijutsu No. 379, March 1983

MINE-TYPE BOLTED SPHERICAL JOINT SPACE FRAME

DING YUNSUN DU CHANGQING FENG YUNHUI

China Aeronautical Project and Design Institute

Presented herein are the type of structure of a space frame locally strengthened with an additional layer (inverted girder) and its mine-type spherical node made of 35 Mn steel casting with four round lugs. Also included are the tests of the node and bolts for their fatigue and strength in one and two directions.

I. GENERAL

Since the traditional series process layout of the hangar would spend plenty of area on passageways and provide little flexibility in working area utilization, we have taken a parallel layout in the design of a new hangar which may save more than 15% production area, increase the output by 25% and give much more flexibility in product changing. According to the parallel process layout, we have taken a plate-like space structure with wide column spacings and one-side full-length door opening, which has proved highly economical in cost.

The space frame covers an area of 5000 m^2 with an overall length of 108 m, composed of three continuous spans by 39.6 x 36m and supported on precast reinforced concrete columns. The bottom chord of the frame is 12 m above the floor level. Under the space frame is suspended a 3 ton and ℓ_k=32.4m multi-fulcrum crane. The space frame bears also the load of the radiant heating equipment (Fig. 1).

The space frame supported by widely spaced columns requires its dead load as light as possible and moderate slope. On the basis of an all-around investigation and three test projects, a built-up roof cladding was taken, which consists of reinforced asbestos tile as the base, emulsion asphalt perlite insulation levelling course, and fibre glass felt water-proofing. The roof slope is 1:50. It was formed by small supports added to the top chord joints and thin wall profile steel purlins on them. The roof cladding is light in weight and easy to install. Its dead load is only 70 kg/m^2. The insulation was spread in place and mechanically vibrated.

The overall roof load is 190 kg/m^2, including snow load, dead load of space frame and load of radiant heaters. The steel consumption is about 34 kg/m^2 for space frame and 10 kg/m^2 for purlins. Tubular members are made of 16Mn steel.

The space frame was assembled in place by using movable mounting platforms and fixed scaffolds. No heavy hoisting equipment was needed. The prefabricated components guaranteed not only the precision, but also the high speed and simplicity of assembly, which was accomplished in 21 days with satisfactory quality fully meeting the design requirements.

The suspended crane was installed in 1982. The project has already been put into production.

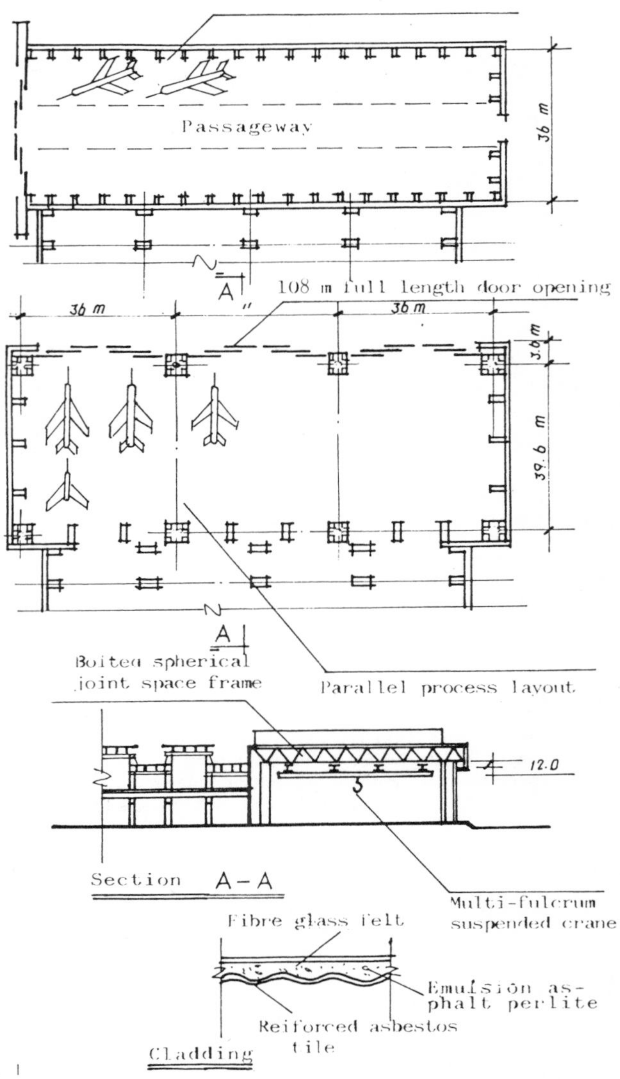

Fig. 1

II. TYPE OF STRUCTURE

The space frame is a rectangle-shaped point-supported structure with open sides. Since the usual plate-like space frame has not only greater internal forces and less rigidity at the wide column spacings and openings, but also greater steel consumption, after having compared 13 types of structures, including folded plate-like space frame, etc. we adopted the plate-like space frame (two-way rectangular space-type double-layer grids)composed of pyramidal units with their sides parallel to the supported edges and two additional inverted girders across the wide column spacings for strengthening the frame, i.e. an additional layer of grids over the area of the wide column spacings. No inverted girders are added above the door openings because of architectural and rainfall considerations. The pyramidal unit is 3.6 x 3.6 m in plan and 2.545 m high with diagonal web members arranged at an angle of 45°. The inverted girder added to the plate-like space frame acts as a supporting structure. As it works together with the space frame, its internal forces are less than those of the ordinary supporting frames. If inverted girders are added both to the wide column spacing and the door opening, the space frame will work as one simply supported at four sides and have even less internal forces than that simply supported at four sides, which will greatly improve the strength and stiffness of the structure (Fig. 2).

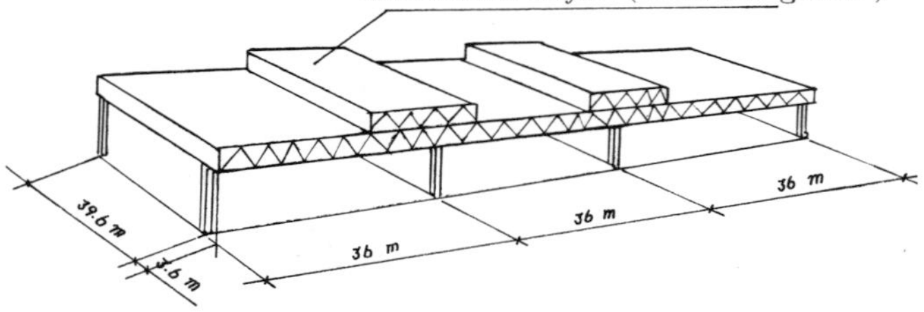

Fig. 2

Rubber bearings were used for supports. Those bearings can accomodate deformation and rotation in all directions and are simpler than rocker bearings and spherical hinged bearings. They are made of laminated 5mm thick ageing-resistant neoprene and 2mm thick sheet steel. A 6 cm thick rubber bearing can accomodate a 1.8 cm displacement and a 12' angle of rotation in any direction, which meets the requirements of deformation caused by temperature variation. The bearings were coated with chloro-butyro-phenol resin twice and enclosed with foam plastic in order to delay the ageing of the rubber.

III. CALCULATION OF INTERNAL FORCES

The calculation of the space frame was carried out by using the precise displacement method for hinged space structures. The crane loads were arranged at unfavorable positions. Since the inverted girder acts as an elastic support, the variation of section exerts significant influence on internal forces, so it requires repetitive correction in calculation. As movable platforms were used for assembly, no deflection was observed at the edges of the space frame, and consequently no residual internal forces caused by its self weight during assembly were to be added.

IV. MINE-TYPE BOLTED SPHERICAL JOINT

The spherical joint comprises three main parts: spherical node, bolts and cone-shaped pieces (or cap) (Fig. 5).

1. BOLTS

High strength bolts of 40B steel may be used only up to diameter 27 mm because of its poor hardening penetration. So we have adopted 40 Cr steel bolts, which may be used up to diameter 56 mm and subjected to 60-70 t tensile force. The tensile strength of 40 Cr steel is σ_b==100 kg/mm^2. As the diameter of the node depends on the screw-in length of the bolt, the screw-in length must be as short as possible. According to the mechanical analysis, the load transfer is concentrated within three to five turns from the root of the threads, and the superfluous length will not work. But we cannot work out a design of bolt with 3 - 5 turns working. Analysis shows that the internal threads of an ordinary nut are subjected to compression, and the threads of bolts are to be stretched, the inconsistent deformation makes them jam up and results in concentrated load transfer at three to five turns from the root. Some documentation has introduced a special nut, the internal threads of which are subjected to tension, which results in a deformation consistent with that of the stretched threads of bolt and a rather uniform load distribution, which makes nine to ten turns work. The spherical node works just as the special nut. We had cut slots in the threads of two bolts and stuck there strain gauges to measure the internal stresses in the threads. The result showed that the stresses of the bolt are equally distributed along the screw-in length (Fig. 3).

52 bolts were tested for screw-in length. 26 of them had a screw-in length of 1.2 d (d is diameter of bolt) and their threads did not fail. But three of the other 26 bolts, which had a screw-in length of 1.0 d, failed at the threads under the action of shearing force while the safety factor reached about 3.0. So we decided to take the screw-in length of 1.2d.

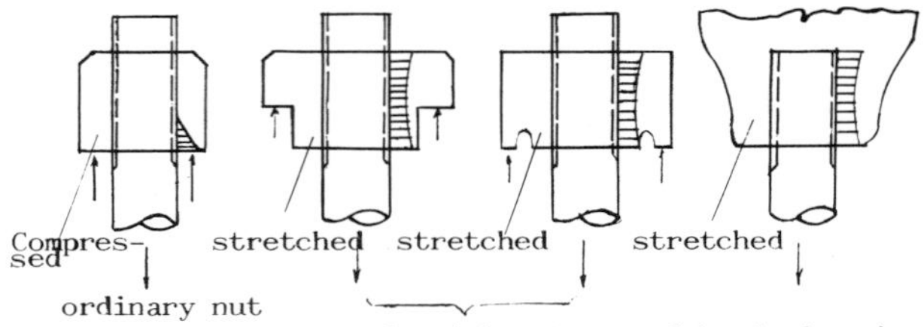

Fig. 3

2. MINE-TYPE NODE

The bolt dia 52 mm requires a spherical node dia 230 mm, which weighs 40 kg. Such a heavy thing is not easy to handle. Our analysis shows, that the horizontal holes of the node at the bottom chord level are subjected to the utmost loads and require greater screw-in lengths, but the holes for skew members are less loaded and require shorter screw-in length. So we decided to make the node in the shape of a mine with four round lugs at the top and bottom chord connections, but the diameter of the other part of the node is reduced to 180 mm, and therefore its net weight is reduced to 18 kg. Such a design has not only reduced its weight, but also simplified its assembly, Handicapped by test project conditions, we took the 35 Mn steel casting for the node. As some of the castings were subjected to burst-in-water cleaning and hot cracks were caused, we had to add an operation of crack detection. So we think it's preferable to use forgings where possible. (Fig. 4)

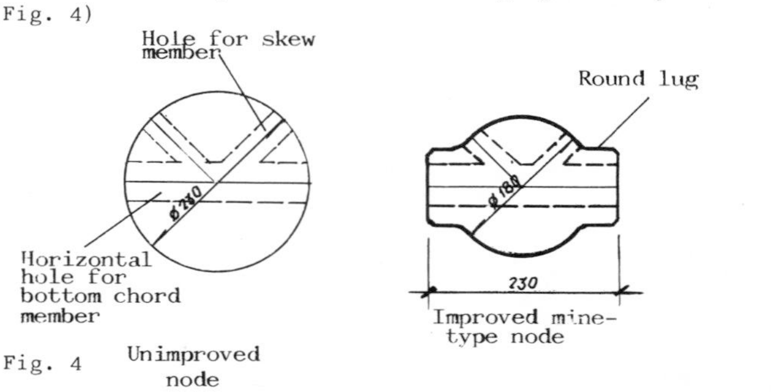

Fig. 4

We had subjected 5 nodes to the test in one direction. The bolts failed under 161 to 224 t tension, but the nodes did not. Two nodes which were gravely honey combed failed at 120 t tension, i.e. K≈2.0, and those with cracks as deep as ½ to ⅓ diameter failed at 161 to 196 t. The nodes failed at 460 t compression in one direction. The test in two directions was carried out on a 400 t tensile testing machine with the help of a self-balancing load bearing stand for vertical testing. The test loads were thus applied: 150 t tension in one direction and 70 t tension in the other; and 150 t tension in one direction and 70 t compression in the other. All nodes tested proved safe enough, K > 2.0.

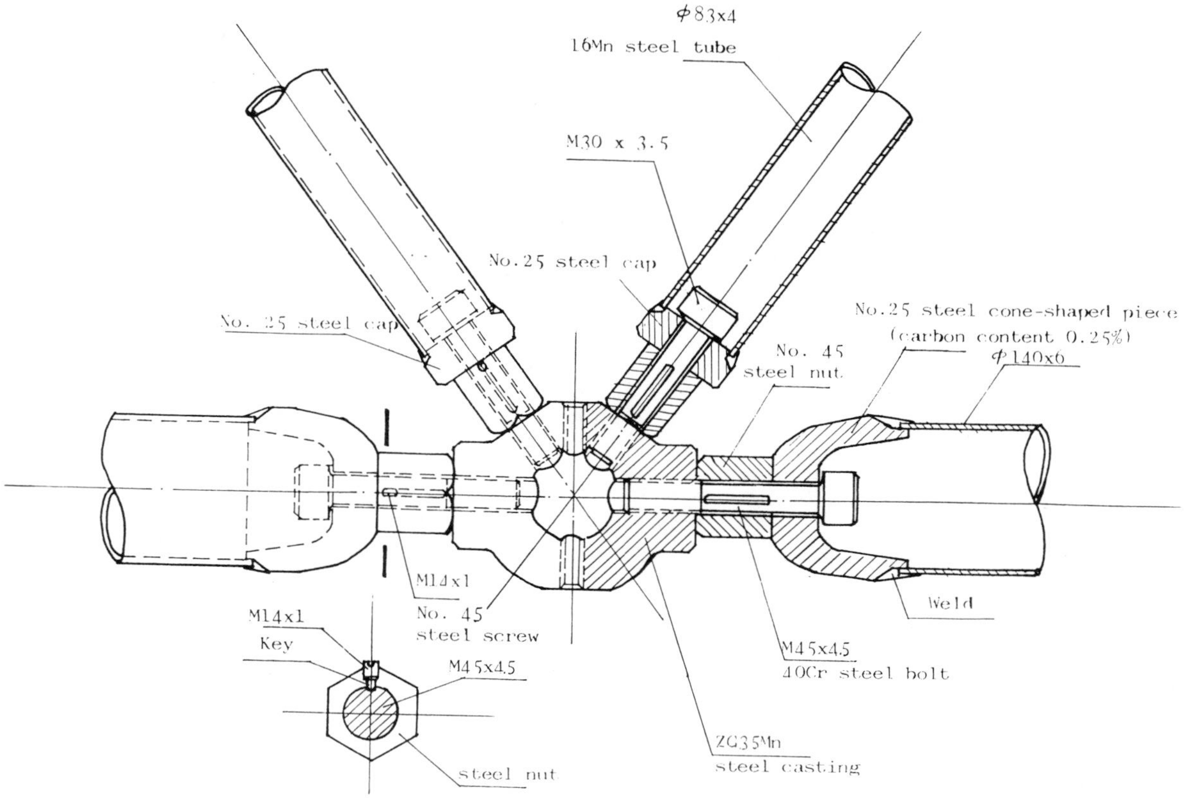

Fig. 5 Fig. Scheme of bolted spherical joint

3. CONE-SHAPED PIECE AND CAP

Tubular members of different diameters require different cone-shaped pieces, which would lead to a wide range of cone-shaped pieces and cause difficulty in making. So we used caps made of thick steel discs for the joints with internal forces not more than 40 t instead of cone-shaped pieces. As a result, only 5% joints still needed the cone-shaped pieces which eased the work a lot.

Calculation and measurement were carried out in order to make sure whether there is local bending moment or other additional forces at the connection of the cone-shaped piece (or cap) to the tubular member. The measured values showed that the difference of stresses at the midpoint of tubular member and at the connection is within 10%. So both the theoretical calculation and actual measurement showed that the local bending moment may be neglected.

There were not any signs of distress and deformation observed when the 3.5 cm thick cap dia. 102 mm was tested under a tension P = 90 t (K > 2.0). When the roof was loaded, we tried to discover the deformation between the cap and the nut by using shims, and did not find any. This proved that the caps are reliable.

4. COMPONENTS OF NUT AND CAMBER

Pins are used in some countries for transmission between the nut and the bolt. But pins sometimes fail under shearing load and the pin holes weaken the section of bolt. So we decided to use keys instead of pins and use screws to press the keys lightly allowing the bolts to move (Fig. 6).

A shortcoming of the key is that it is impossible to check if the root of the bolt is in close contact with the inner surface of the cone-shaped piece (or cap.). If not, the tension member will have an additional clearance Δ, which does not conform to the hypotheses of the analysis of internal forces and will cause increase in internal forces.

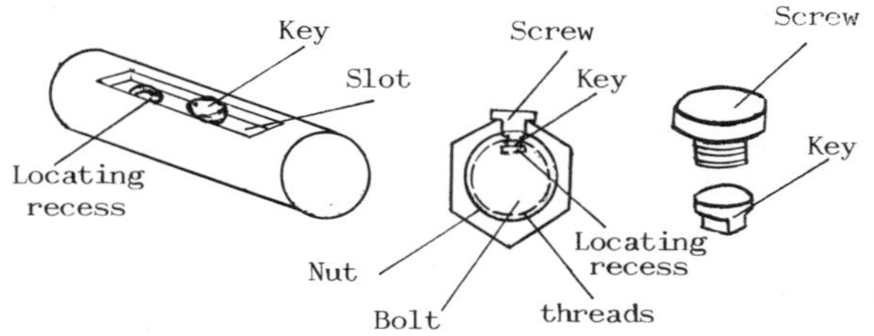

Fig. 6

Therefore we made a 10.2 x 43.2 m experimental space frame which was supported across a span of 39.6 m. The amount of deflection actually measured under its self weight was 28 mm, which showed a fair degree of agreement with the value 29 mm obtained by theoretical calculation. During construction we also carried out inspection with the torque spanner and determined that under the action of an applied torque of 30 kg-m the prestress of a M30 bolt is 1240 kg/cm^2, and the prestress of a M52 bolt under the action of a 55 kg-m torque is 494 kg/cm^2. In view of that applying a torque may guarantee close contact of the bolt and the cone-shaped piece (cap), we now use a sort of "improved" key, which has a locating recess to guarantee the precise location of the bolt and consequently the close contact of the bolt and the cone-shaped piece (cap).

In addition we cancelled the handhole on the tubular member for replacing the bolt because it weakens the section of the tubular member and hence causes eccentric tension in the tubular member. In fact the "sealed" tubular members without handholes are prevented from rusting and no failed bolts which are to be replaced have been found up till now.

In order to determine the possibility of cambering of the space frame, we made a 3 cm camber of the 39.6 m span space frame and then tightened all the bolts. But it resulted in a deflection of 2.95 cm, which was not less than that before cambering, and in addition made the joints loose and shaky. According to our analysis the dimensions of the members have to be changed in order to camber the space frame successfully. But we did not make a camber of our space frame.

V. SUSPENSION OF CRANE AT SPHERICAL JOINTS

The suspended crane will not be used very frequently and its 3 t load bearing capability produces a load which amounts just to 10% of the total of internal forces.

The fatigue calculation and test we did are as follows.

1. FATIGUE CALCULATION

This calculation is purposed to determine the fatigue strength of the hardened and tempered 40 Cr alloy steel bolt.

A. Formula for approximate calculation of fatigue strength

The common formula in mechanical fatigue calculation is derived from the AG section of the fatigue test characteristic and applies to $\rho = -1 \sim 0$) (Fig. 7).

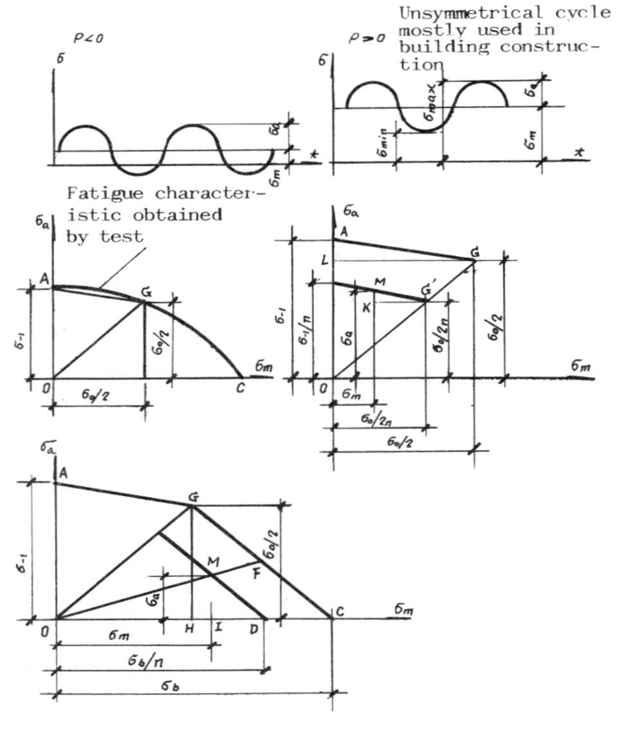

Fig. 7

Assuming AG is linear and $\Delta MKG' \approx \Delta ALG$

$$\frac{\sigma_a - \frac{\sigma_0}{2n}}{\frac{\sigma_0}{2n} - \sigma_m} = \frac{\sigma_{-1} - \frac{\sigma_0}{2}}{\frac{\sigma_0}{2}}$$

After reduction

$$n = \frac{\sigma_{-1}}{\sigma_a + \psi\sigma_m} \qquad \psi = \frac{2\sigma_{-1} - \sigma_0}{\sigma_0}$$

Where

n — Fatigue safety coefficient, including stress concentration coefficient of threads

σ_{-1} — Fatigue strength for $\rho = -1$

σ_m — Average stress

σ_a — Stress amplitude corresponding to fatigue variation

σ_0 — Fatigue strength for $\rho = 0$

σ_b — Ultimate tensile strength

Results obtained from the test of hardened and tempered 40 Cr Steel bolt

$$\sigma_0 \approx 1.5\sigma_{-1}$$

$$\sigma_{-1} \approx 0.5\sigma_b$$

But in the calculation of building structures, the suspended crane is supposed to give a load of unsymmetrical cycle ($\rho > 0$), which requires to take the GC section of the fatigue characteristic. Suppose GC is linear. The formula is derived as follows. n = OF/OM

$$\Delta MID \approx \Delta GHC$$

$$\left(\frac{\sigma_a}{\frac{\sigma_b}{n} - \sigma_m}\right) = \frac{\frac{\sigma_0}{2}}{\sigma_b - \frac{\sigma_0}{2}}$$

$$\sigma_a\left(\sigma_b - \frac{\sigma_0}{2}\right) = \frac{\sigma_0}{2}\frac{\sigma_b}{n} - \frac{\sigma_0}{2}\sigma_m$$

$$n = \frac{\frac{\sigma_0}{2}\sigma_b}{\sigma_a\left(\sigma_b - \frac{\sigma_0}{2}\right) + \frac{\sigma_0}{2}\sigma_m}$$

B. Fatigue safety

In the mechanical design we usually take an overall safety factor $n = 6 \sim 7$, including the threads stress concentration coefficient $K_c \approx 3.0$ and a factor n_0, which takes the checking conditions, machining factor, etc. into account.

In building design we must take a safety factor n_0 into account in addition to $K_c \approx 3.0$. This factor is usually taken as $n_0 \approx 1.1 - 1.3$ in Japan, GDR and GFR, but in China, $n_0 \approx 1.0 - 1.05$.

Therefore we took $n = 3.0$.

C. Fatigue test

Tests were carried out on four specimens which consisted of one node and two bolts, dia 52 mm each. Results are as follows:

Two of them had passed two million cycles while $\rho = 0.8$ and $\sigma_{max} = 3000$ kg/cm^2.

One of them had passed one million cycles while $\rho = 0.8$ and $\sigma_{max} = 3000\ kg/cm^2$ and then failed after having passed another five hundred thousand cycles while $\rho = 0.7$ and $\sigma_{max} = 3000\ kg/cm^2$ (1,500,000 cycles in total).

The other one failed after having passed 240,000 cycles while $\rho = 0.6$ and $\sigma_{max} = 3300\ kg/cm^2$.

A reverse Calculation was carried out by the above mentioned formula for bolts dia 52 mm while $\rho = 0.8$ and $\sigma_{max} = 3000\ kg/cm^2$ and $n = 2.94$ was obtained. This figure meets the required $n = 3.0$ fairly. So the formula proved suitable for our purpose.

The test result showed that the use of high strength bolts for spherical joints had increased the yield strength greatly, but the fatigue strength just a little. The fatigue strength is not ideal, which should be paid attention to.

The bolts were machined after heat treatment to prevent the thread roots from cracking during quenching.

The suspended crane has actual $\rho \approx 0.9$ and $\sigma_{max} = 3000\ kg/cm^2$, which meet the safety requirements.

D. Fatigue strength of bolts for crane joints

A 40 Cr steel bolt dia 45 mm was used to connect the runway of the crane to the joint (Fig. 8).

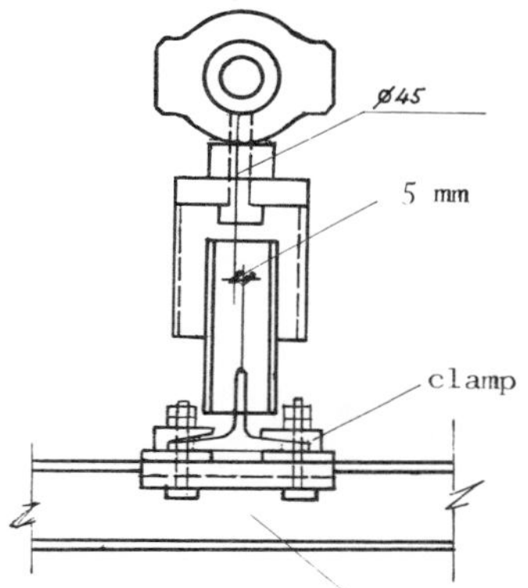

Fig. 8

The bolt is subjected just to the crane load. We had carried out calculation by the formula for approximate calculation of fatigue strength, taking $\rho = 0.1$ and an eccentricity of 5 mm due to installation error, which showed that the fatigue safety is guaranteed.

E. Deflection under crane load

The actually measured maximum deflection of the space frame under the roof load 140 kg/m^2 (not including the snow load 50 kg/m^2) is 6.1 cm, which exceeds that obtained by calculation (5.7 cm) by 5%. The calculated amount of deflection under the full roof load of 190 kg/cm^2 is 7.7 cm. The calculation for full load of 3 t crane showed that the amount of deflection should be increased by 1.9 cm. But the actually measured increase of deflection due to full load of crane is only 1.0 cm. The rigidity of the space frame is satisfactory while the crane working.

VI. GLOBAL SAFETY OF SPACE FRAME

We have checked the calculation of some single members in order to be sure that if any single member fails, the whole space frame will not fail or lose its stability after redistribution of stresses in other members, i.e. the safety of the space frame in whole will be guaranteed. We choose three cases of failure of tension members subjected to heavy loads to check their effect respectively, which resulted in an increase of internal force by 55% in their adjacent tension members and by 45% in the compression members, which would not lead to the loss of stability of the compression members and the yield point of the tension members. That will meet the requirements of "failure safety". So the safety of the space frame in whole is guaranteed even if a single member fails.

CONCLUSION

The experience of the test project shows that the bolted spherical joints apply to the space frame with members subjected to internal forces up to 60 - 70 t, and the strength and stiffness meet the working requirements. The accuracy of the components meets the design standards. The high precision and the self-positioning of the components allow the assembly to be carried out in place high above the ground which not only saved the heavy hoisting qeuipment, but also reduced the cost of installation and increased the speed of construction. That is another favorable condition for using the space frame.

THE KINEMATICS OF A NOVEL DEPLOYABLE SPACE STRUCTURE SYSTEM

R C CLARKE *, BE, Civilingeniør (Denmark)

* Professional Officer, School of Civil Engineering, University of New South Wales, Sydney.

Transportability and rapid erection are rarely the primary requirements of space structures. There are, however, applications for structures in the small to medium size range possessing these characteristics. Brainov, Buckminster-Fuller, Pinero and more recently Zeigler have, through their design of deployable structures, contributed to this class of space structures. This paper seeks to formulate the geometry of Zeigler's system and via this analysis discuss its potential application. A particular feature of the paper is that a proof is given of the necessary and sufficient conditions for mobility in the Zeigler system. The 'Domad', a deployable dome structure designed by Zeigler, is referred to throughout this paper to illustrate the system.

INTRODUCTION

The building block of this deployable space structure system is the 'Trissor' (Tri-scissor) which will be the central topic of this paper. In order to introduce the Trissor, the Domad, a deployable dome, will be considered to illustrate both the system in terms of a real application, and the integral nature of the Trissor in that system. Then using projective geometry the Trissor's geometry is defined to predict its potential as a building block for deployable structures in general.

Before proceeding, it is important to define the notion of a deployable structure as used in this paper. A 'structure' refers here to a set of 'struts' (one dimensional rectilinear members) connected by 'nodes' (universal joints) and 'scissor-hinges' (a rotational degree of freedom about the normal to the plane defined by two connecting struts). A 'deployable structure' has the characteristic that by rotation of the struts with respect to one another the assembly encompasses at least two forms. The first form is the 'compact' state having theoretically one dimension, and the second form is the 'deployed' state which is a three dimensional body. Note that the individual struts have the same length and form (rectilinear) in both states.

THE DOMAD

The Domad is a deployable 5 metre diameter hemispherical dome which is marketed as a tent. Its compact state is a bundle of adjacent parallel struts, ie effectively a one dimensional form whose length is in the order of one-tenth the diameter of the dome. An elevation of the 'compact' and 'deployed' forms of the Domad without its covering material is shown in Fig 1.

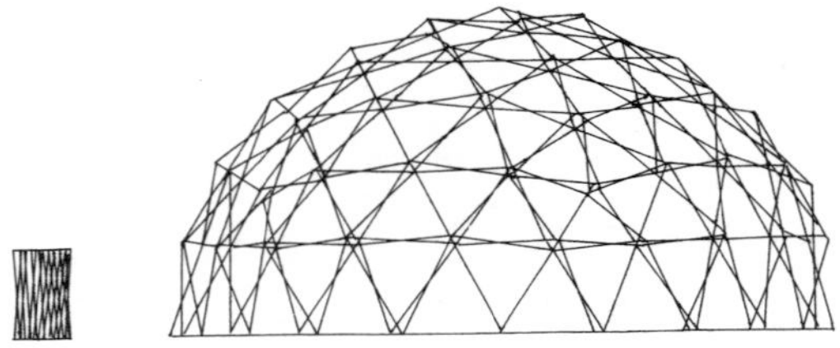

'Compact' 'Deployed'

Fig 1

The Domad has a total of 480 struts all of approximately the same length. The deployed state is a two-layer geodesic dome. The geodesic breakdown as shown in Fig 2 is a frequency 4 icosahedron class 1, with slight variations of some nodal positions.

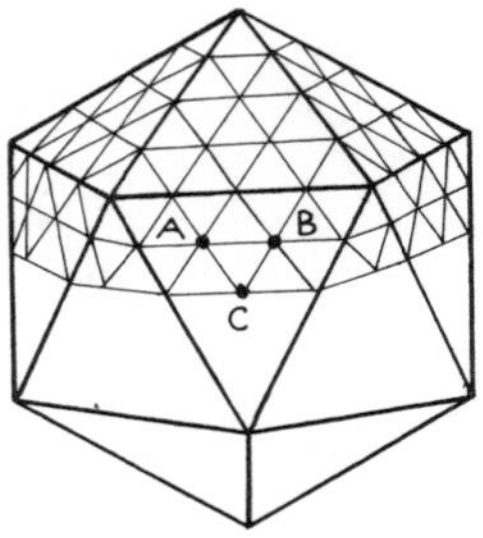

Fig 2

In the deployed state each of the points of intersection of the breakdown lines on the polyhedron of Fig 2 (A,B,C for example) is projected from the central point of the icosahedron to two nodes which define the inner and outer surfaces of the dome. These two nodes are referred to as a 'node pair'. Struts are arranged such that each inner node of a node pair is connected to the outer node of the adjacent node pair and vice versa. At the point at which struts cross one another there is a scissor-hinge. Consider the typical triangle ABC of Fig 2 shown in the plane diagram of Fig 3.

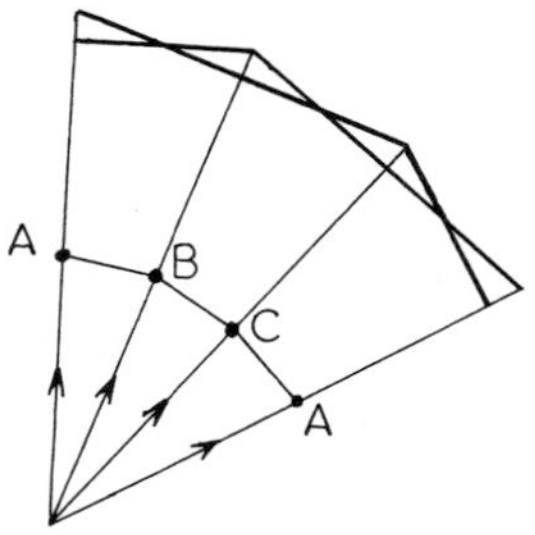

Fig 3

The Domad, as a deployable structure, has only two states of geometric fit - compact and deployed. It will be shown that any other position is a state of geometric non-fit. In the Domad this geometric non-fit is compensated for by bending of its relatively flexible struts, resulting in a snap-through action between the compact and deployed states. Thus the Domad has a measure of rigidity in the deployed state which is a function of the degree of geometric non-fit during deployment and the bending stiffness of its struts. Modifications are made to some of the scissor-hinges allowing sliding in the plane of the struts. This introduces a fine tuning system for the snap-through action. In the Domad's application as a tent this action provides sufficient structural rigidity and at the same time the energy barrier which is the snap-through can easily be overcome by one person.

THE TRISSOR

The mobile assembly, shown in Fig 3, of six struts, six nodes and three scissor-hinges is the Trissor. Figure 4 shows a Trissor where all the struts lie in one plane for the given value of the angle β. It will be shown that every Trissor can assume this plane position. The lengths of node to scissor-hinge segments are indicated as : ai,bi,ci and di (for i=1 to 3). The required action of a Trissor is such that as β increases the assembly of Fig 4 folds rigidly out of the plane, passing through a three dimensional form and reaching a limiting one dimensional state for β=180 degrees. This action places constraints on the relative values of ai,bi,ci and di.

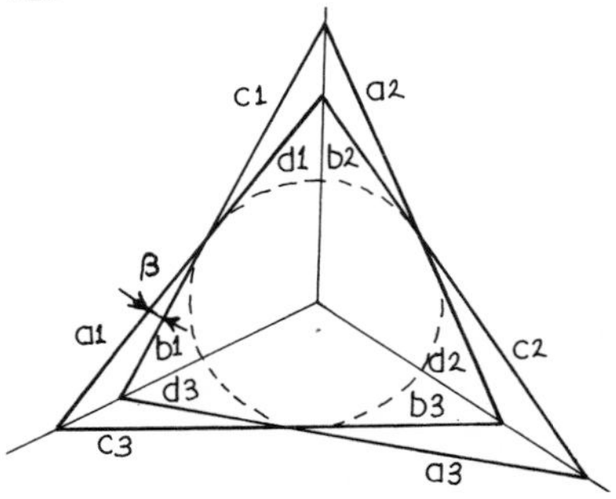

Fig 4

The following solution uses projective geometry to obtain the constraints which are necessary and sufficient for the mobility of the Trissor. Consider the hyperboloid of one sheet, a surface of revolution obtained by rotating a hyperbola about the perpendicular bisector of the line connecting its foci. The hyperboloid of one sheet can also be regarded as a ruled surface generated by rotating a straight line about a skew axis. A hyperboloid of one sheet contains two families of generating lines, each family covering the surface of the hyperboloid and each line of one family intersecting every line of the other family. Figure 5 shows a section of such a hyperboloid with a finite number of lines of each family shown. Surprisingly, if at the points of intersection of the two families of lines rotation (but no sliding) is allowed, then the entire assembly is movable. Hilbert provides an analytical proof of this phenomenon in Ref 1.

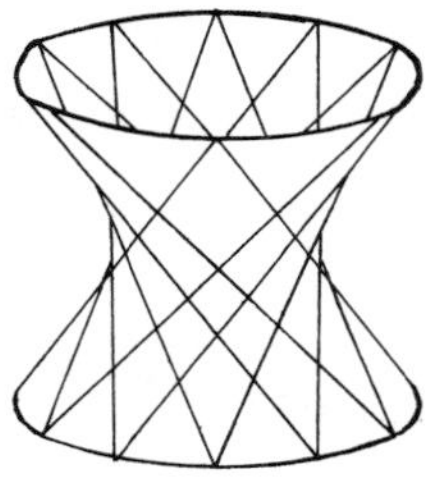

Fig 5

Consider a horizontal plane passing through the ruled surface of Fig 5 at its mid-height. In the general case this plane will intersect the hyperboloid of one sheet in an ellipse. The movable assembly can fold into two limiting plane forms. Firstly, the assembly can flatten itself into the horizontal plane; its struts will envelope an ellipse in this position. Secondly, the assembly can stretch vertically into a plane perpendicular to the horizontal and the struts will be tangent to a hyperbola.

The specific case where the horizontal plane intersects the hyperboloid of one sheet in a circle gives us a limiting plane form parallel to the horizontal plane enveloping a circle and the other limiting case is a line coincidental with the axis of revolution of the hyperboloid of one sheet. With the exception of its plane position the Trissor consists of two sets of 3 skew lines. Any 3 skew lines in space define a hyperboloid of one sheet, thus all Trissors must lie on a hyperboloid of one sheet. As a subset of the ruled surface illustrated in Fig 5 the Trissor will necessarily possess the characteristic of movability. For the Trissor the horizontal plane will intersect the hyperboloid of one sheet in a circle. Thus the constraints on the Trissor to ensure mobility can be stated :

> It is a necessary and sufficient condition for the mobility of a Trissor that its six struts in the plane are tangent to a circle.

This is shown for a particular case in Fig 4. A property of this configuration follows from Brianchon's Theorem as stated in Ref 1 : "that the diagonals of a hexagon that is circumscribed about a conic intersect at a point." It can be shown that the three diagonals of the Trissor will meet at a common point for the entire action of the Trissor.

Projective geometry has provided a graphic solution to the Trissor's required geometry. The Trissor used in the Domad is of the form shown in Fig 4. This Trissor has the property that adjacent node to scissor-hinge lengths are equal; for example :

$$a1 = c3 \qquad \ldots\ldots 1$$

$$b1 = d3 \qquad \ldots\ldots 2$$

$$d1 = b2 \qquad \ldots\ldots 3$$

An analytical solution has been found for this set of properties. The constraints gained take the form :

$$a1+b1 = c3+d3 \qquad \ldots\ldots 4$$

$$a2+b2 = c1+d1 \qquad \ldots\ldots 5$$

$$a3+b3 = c2+d2 \qquad \ldots\ldots 6$$

$$b1+b2+b3 = d1+d2+d3 \qquad \ldots\ldots 7$$

$$(1/a1+1/b3)(1/a2+1/b3) = (1/a3+1/b1)(1/a3+1/b2) \ldots\ldots 8$$

$$(1/a2+1/b1)(1/a3+1/b1) = (1/a1+1/b2)(1/a1+1/b3) \ldots\ldots 9$$

Equations 4 to 7 are necessary for the Trissor in the compact state. Equations 8 and 9 require that the diagonals of the Trissor always meet at a common point. Thus given a set of properties of the Trissor strut lengths, constraints can be obtained analytically. These constraints are seen to be identical with those obtained from the independent general projective geometry solution.

COMBINING TRISSORS

Given this definition of the Trissor's geometry we can make conclusions about deployable structures using this basic unit. Firstly, however, a definition of the term 'ring' is required : if on the breakdown shown on the polyhedron of Fig 2 a point is surrounded by other points such that the sum of the vertices of the triangles meeting at that point is 360 degrees, then the central projection of these points (as described previously) is a 'ring'.

The conclusions are :

(i) The deployable structure in its deployed state is a central projection; ie each node pair will be colinear with a common point.

(ii) In the same way that the deployable structure must be able to occupy at least two defined states so too must any part of it. Thus requiring the constraints ensuring mobility for every Trissor in the structure.

(iii) A deployable structure which contains one or more rings can assume one and only one deployed position.

The first two points place constraints on the overall shape and node distribution of a deployable structure. The third point is directly related to the snap-through phenomenon observed in the Domad and this deserves further comment.

Consider the ring of Trissors in Fig 6 where each of the points (A to G) represents a node pair and each line represents a scissor. The sum of the angles between the scissors which meet at 'G' is 360 degrees for the deployed state. Consider each Trissor acting independently of the other Trissors in the ring during the opening action. The angle between any two of its scissors varies with the angle β of Fig 4. Thus the sum of the angles at 'G' will also vary as shown in Fig 7. There is one and only one position where the sum is 360 degrees and that position defines the deployed state. During the transition from the compact to the deployed state the angle sum is less than 360 degrees as shown in Fig 7. In the Domad this geometric non-fit is compensated for by bending of its struts, resulting in an energy build-up and release; ie a snap-through action.

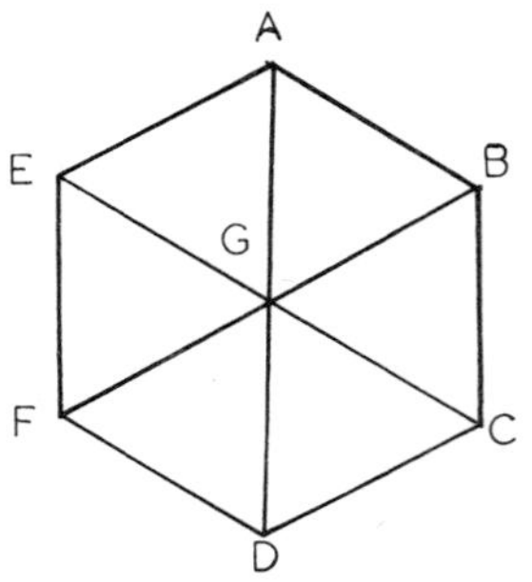

Fig 6

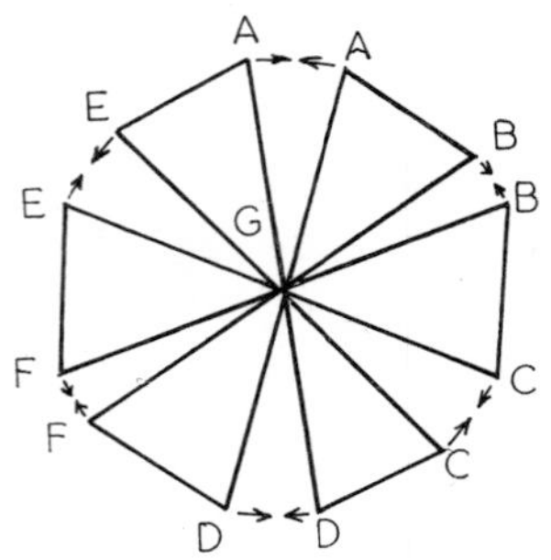

Fig 7

CONCLUSION

Through this understanding of the basic unit of the system, structures of different form and dimension can be designed. The degree of geometric non-fit can be manipulated to give the structure sufficient structural rigidity when deployed, balanced with a low enough energy barrier to permit deployment.

ACKNOWLEDGEMENTS

The author wishes to thank the Australian Research Grants Scheme for sponsoring the present work. Thanks also to Professor H.M.Irvine for his constructive suggestions.

REFERENCES

1. D.HILBERT and S.COHN-VOSSEN, Geometry and the Imagination, Chelsea Publishing Co., 1952.

SELF-ERECTING TWO-LAYER STEEL PREFABRICATED ARCH

Oscar SIRCOVICH SAAR * , Civil Engineer

* Principal of O. Sircovich Saar, E. Sella
Consulting Engineers, Jerusalem, Israel

A new type of self erecting, prefabricated steel arch is developed. Prefabricated spatial trusses and tensile elements are assembled, manually, at ground level, creating a very flat arch. The initial structure is forced into the post-buckling response region and rises smoothly to the designed shape of the arch. Demounting of the structure is easily performed by allowing an extreme hinge of the arch to move out. The structure is appropriate for use whenever an arch, barrel vault or a similar structure is required. It is particularly suitable for multiple demounting and erection with almost no extra expense involved.

Fig 1

DESCRIPTION OF STRUCTURE

A new type of self-erecting, large span, steel prefabricated arch is presented in this paper, see Fig 1. Prefabricated spatial trusses, see Fig 2a, are connected, each to the adjacent one, through articulated joints along the lower line of the arch and through tensile elements along the upper line. The result is a space, two-layer structure, with only tensile elements in the upper layer.

No geometric restrictions for the prefabricated trusses are imposed. Nevertheless, lightweight components are preferable, in order to achieve maximum advantages of the erection process to be described furtheron. By fulfilling this requisite, no heavy equipment will be needed, neither for assembling nor for the erection of the structure.

The upper line of the arch is composed of a series of tensile elements of different lengths; see Fig 2b.

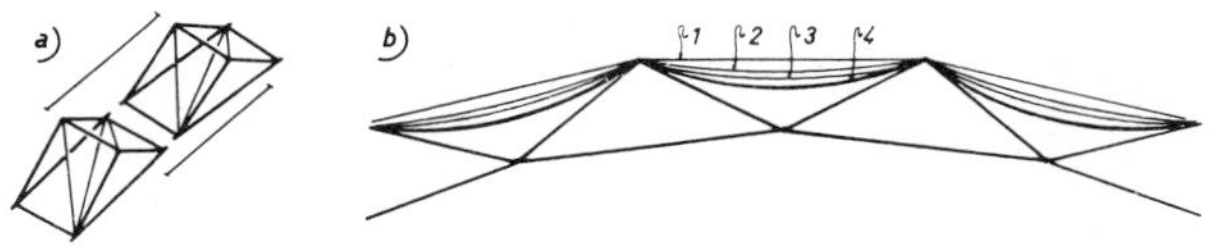

Fig 2

The elements of maximum length, chords numbered 4, will become, in due time, the definitive upper line elements of the designed arch. The shortest ones, chords numbered 1, will act, at the beginning of the erection process, as the upper line tensile elements. The remaining elements, chords 2 and 3 in this example, are designed to be used during the process of erection.

ERECTION PROCESS

The structure is mounted initially on temporary legs, as a very flat arch of span l; see Fig 3. The arch has a fixed support at one end and a sliding support mounted on a rail at the other end.

The tensile elements of the upper line of the arch are at zero initial condition of stresses and the arch is built so that the lower line of the arch lies in the exact funicular line of its own dead weight.

An external load P (see Fig 3) is applied; the value of P starts from zero and increases progressively and at a certain point:

$$P = H_i = \frac{g \cdot l^2}{8 \cdot f}$$

where g is the dead weight of the arch
H_i is the horizontal initial force in the arch under dead weight

At this moment the temporary legs become theoretically unnecessary; the chords in the top line of the arch are still unstressed.

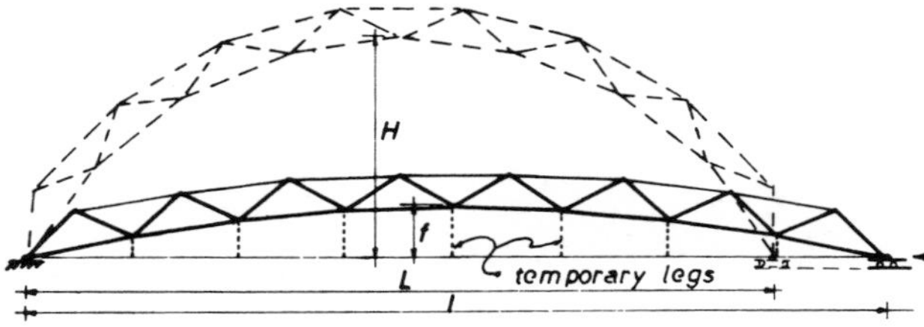

Fig 3

By further increasing the external load P, the shortest chords in the upper line of the arch become tensioned; the arch is then fully prestressed and the temporary legs become removable.

The elastic deformation of bars must be taken into consideration when designing the initial shape of the arch and the temporary legs. If the initial shape of the arch does not fit the exact funicular line of its own dead weight, minor local changes of shape will occur before the arch is fully prestressed.

By progressively forcing the extreme hinges of the sliding support to move towards the fixed support, the structure is forced into the post-buckling response region and rises smoothly until it reaches the designed shape of the arch defined by span L ($L < l$) and height H ($H > f$); see Fig 2.

The main idea of the erection process is to cause, consecutively, large elongations in the temporary chords of the upper line of the arch. This is to be achieved by stressing them far beyond the allowable limits. In this manner, the necessary increase in the initial length of the upper line of the arch is produced.

The temporary chords of the top line of the arch are of steel grade St 37. This grade of structural steel assures the desired amount of elongation, together with maximum security against failure. Fig 4 shows the well-known stress-strain diagram of steel St 37; the curve for St 180 is given for comparison only.

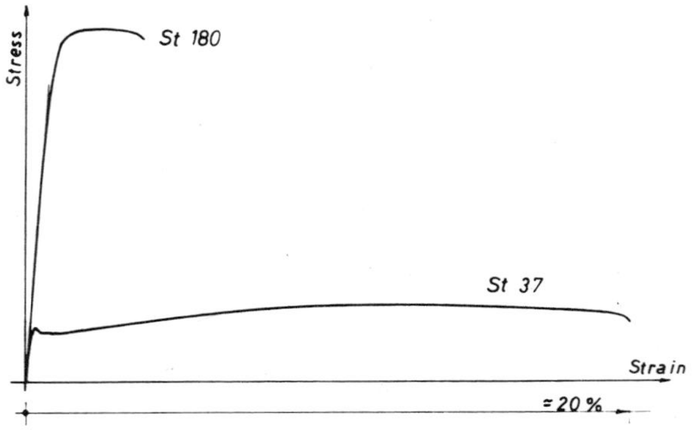

Fig 4

Special consideration must be given to the elongation process of each of the temporary chords, in order that if failure of any one will occur, it will occur only when the next chord, which is to substitute it, has become stressed. The substitute chord is designed so as to take on the transferred force with a proper increase in strain. The sequence of lengths for the temporary chords is determined taking into consideration an elongation of 6 to 7% to occur before the substitute elements become stressed. It represents, more or less, half of the strain at failure of steel St 37.

The temporary chords become redundant at the end of the erection process. Since they were stressed over their yield point, they must be replaced for each further erection of the arch. They become removable at the instant the definitive chord that substitutes them is stressed. Removal of temporary chords is performed while external load P is still applied.

The definitive applied load, P, is given by

$$P = H_f = \frac{g \cdot L^2}{8 \cdot f}$$ where H_f is the horizontal force in the arch under dead weight

and no erection forces remain on the structure.

The required height for the arch, initially mounted as a flat one, can be achieved by providing the appropriate number and lengths of the temporary chords. The required longitudinal section of the arch can be achieved by proper selection of lengths of the definitive chords. The prefabricated trusses can be repetitive. See examples on Fig 5.

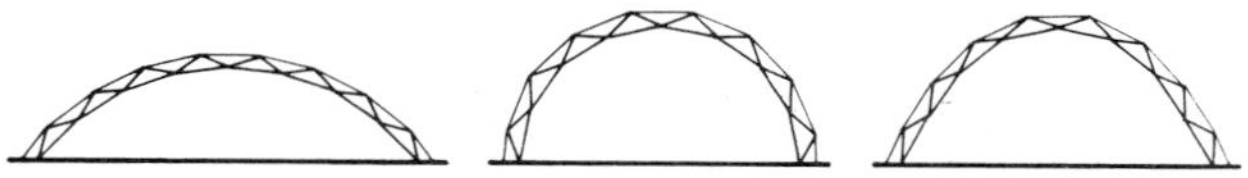

Fig 5

The transversal section of the arch is designed taking into consideration lateral loads to occur, at least during the erection process. Fig 6a presents an arch with a rectangular transversal section, suitable, for exanmple, for a barrel vault. Fig 6b presents an arch with a triangular transversal section, suitable for example for a self-standing arch.

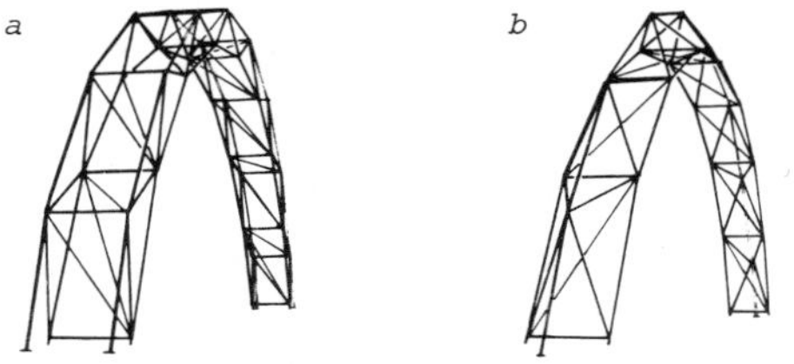

Fig 6

Demounting of the arch is as simple as the erection itself; it is achieved by allowing one of the extreme hinges to move out progressively until all components reach ground level, see Fig 7. The structure breaks down first into two arches and continues, with the same pattern, till the end of the demounting process.

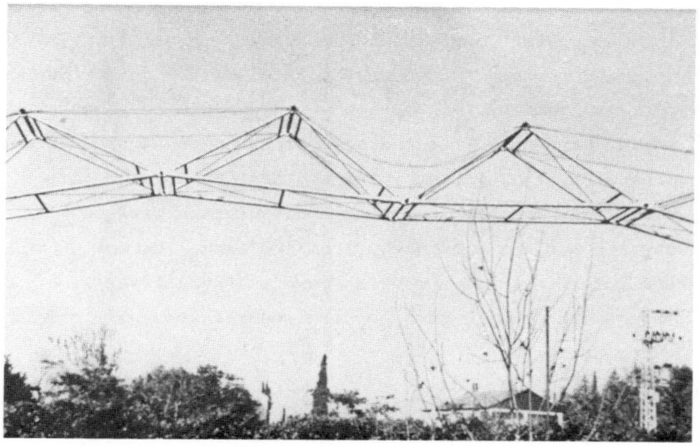

Fig 7

The continuous increase of applied load P in the erection process, or the continuous decrease during the mounting process, can be easily achieved by means of a cable and a ratchet; see Fig 8. In practice, the ratchet is placed at a distance from the arch.

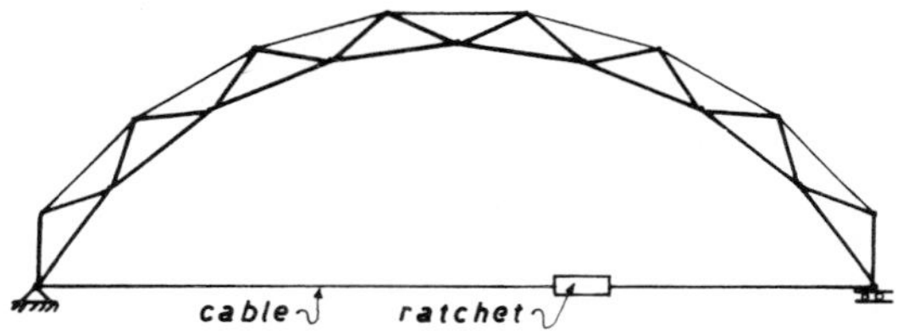

Fig 8

REMARKS ON STATIC CALCULATION

All bars in the lower line of the arch are designed for compression forces, while all structural components of the upper line of the arch are designed only for tension forces.

The structure is, geometrically, a fully restrained arch; see Fig 5. Nevertheless, static calculations are carried out on an articulated arch, for reasons that will be explained in the following.

As an example, let us consider a fully restrained arch, with a central vertical load, Fig 9a; the characteristic diagram of moments is represented in Fig 9b.

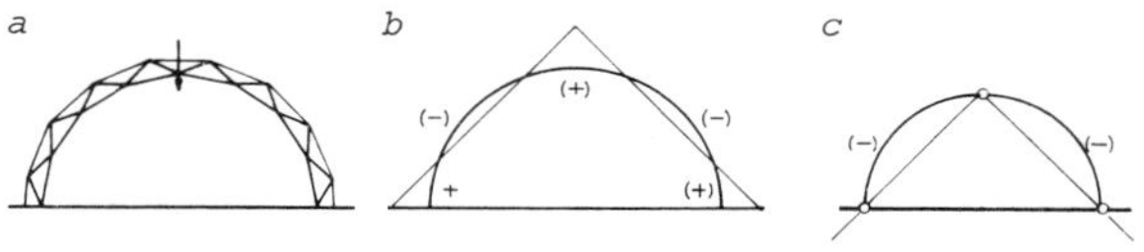

Fig 9

Since the chords in the upper line of the arch are not able to resist compression forces, an articulation is produced in the structure at points which have maximum positive moments. Fig 9c shows the diagram of moments which corresponds, in the same arch, to the articulations created by the central vertical load. Hence, for this particular example, the static calculation will be carried out for a three-articulated arch. These articulations will appear in joints of the lower line of the arch, while the opposite upper chords will remain unstressed.

If the structure is stressed up to the elastic limit, these three particular articulations, shown on Fig 9c, are only temporary; after removal of the external load, the structure becomes again a geometrical, fully restrained arch.

Let us consider now the same arch under a side wind load; see Fig 10. The static analysis and calculations will be performed as for the central vertical load. Fig 10a shows the side wind load diagram; Fig 10b shows the characteristic diagram of moments in the arch with two articulations.

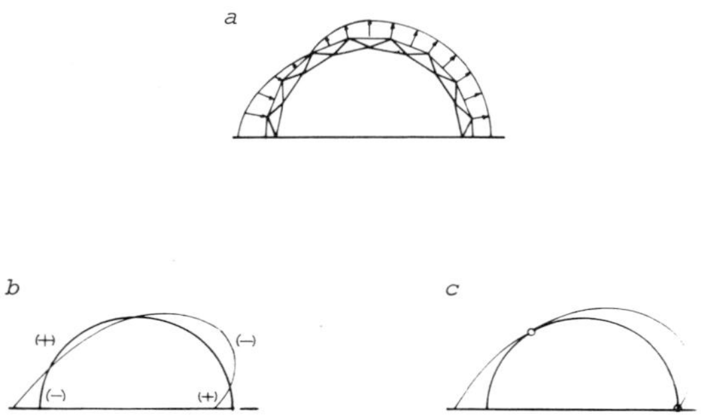

Fig 10

As we see, the number and position of articulations change with changing loads and thus, due to this characteristic, the structure is a non-linear one. Hence, static calculations must be performed, separately, for each combination of external loads. This can be accomplished easily by an iteration process, with the help of a computer.

The creation of one or more of the expected temporary articulations can be avoided by introducing a prestressing moment in the arch, as shown in Fig 11. If all articulations are avoided by prestressing, the structure becomes a linear one and superposition of external loads can be implemented in static calculations.

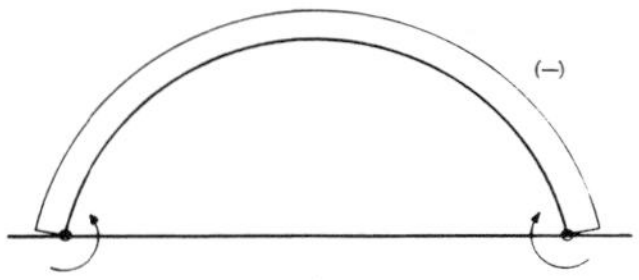

Fig 11

APPLICATION

Owing to the erection process, which requires almost no extra expenses, and to the particular characteristic that all components of the upper line of the arch are only designed for tension forces, an outstanding economical structure is achieved, whenever an arch, barrel vault or a similar structure is required. As stated previously, the arch can be given the most appropriate shape, according to the expected load.

The structure is particularly appropriate for multiple demounting and erections; for example, as in a temporary, lightweight covering, multipurpose building, shown in Fig 12.

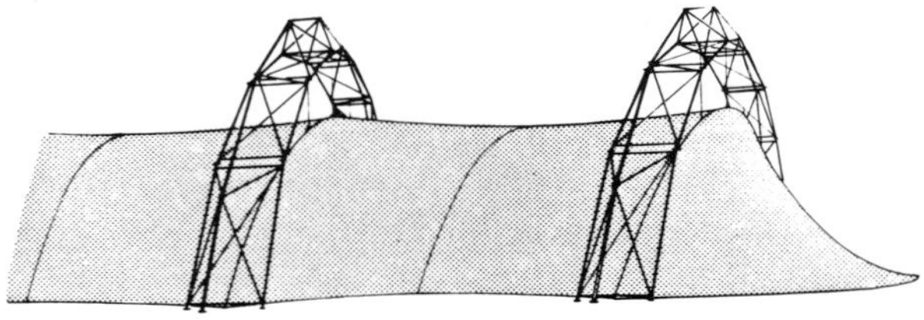

Fig 12

In the design of a rigid building, like the warehouse of Fig 13, three main requirements must be implemented. They are as follows: 1) Purlins, for roof covering, must be located only on the apex of each prefabricated truss of the arch; 2) The design of the roof covering connectors must take into account the changing position of temporary articulations; 3) Bracing for lateral forces is located only in the lower layer of roof structure.

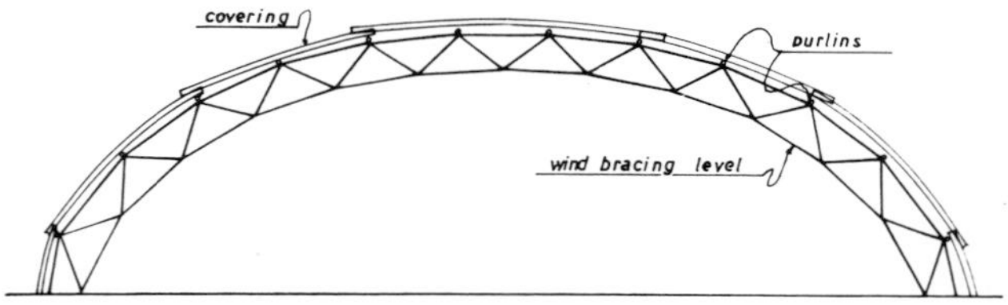

Fig 13

Fig 14

We have successfully designed and built, in Israel, a self-erecting steel arch, following the method and requirements presented in this paper.

The arch was the self-standing backbone of a 35m x 35m greenhouse with a polyethylene covering of a double-saddle shape; see Fig 14. For ease of transportation, the trusses themselves were assembled on site, prior to erection of the arch. The greenhouse was designed for light live load. The total weight of the structure was 5 kg of steel per square meter. The arch assembling and erecting time was 12 workman hours per ton of steel.

TWO DESIGNS OF SPACE TRUSSES AND THE SLIDING TECHNIQUE FOR THEIR ERECTION

Xia Heng-xi, Associate Professor

Head, Department of Hydraulic Engineering
Agricultural University of Hebei, China

Two designs of plate-like space truss are discussed in this paper. One of them is for a cantilevered space truss with two-way orthogonal lattice grid system supported by 12 concrete columns. The other is for a space truss composed of triangular pyramids and supported along the perimeters. The latter type of space truss, comprising a triangular pyramidal system and a system of independent re-bisected members, has attained higher economic as well as technological norm because of its adoption of rather spacious grids. To verify the stability of the re-bisected member system, tests have been carried out on models of single triangular pyramid. The structural features of these two types of space truss are discussed, and a comparison of different analyses of the internal forces are made. An approach of anti-seismic analysis is introduced. The final part of this paper is dedicated to the discussion of the sliding technique as employed for the erection in an elevated position of a space truss and of the analysis of internal forces during the course of erection.

I. THE DESIGN OF THE SPACE TRUSS WITH TWO-WAY ORTHOGONAL LATTICE GRID SYSTEM

This kind of space truss was designed specially for Baoding Sports Hall in Hebei Province. As the stands, the column system and the exterior-protected walls of the sports hall had already been constructed before the design of its roof structure ever came into being, the choice of the type of space truss adoptable under the conditions was limited. And finally the type of space truss with two-way orthogonal lattice grid system was adopted. This type takes for its connectors hollow steel spheres, with its steel tube members welded to them. It measures 68.42m x 55.43m in plane and has a depth of 3.80m. The overhanging eaves along both the east and west sides of the space truss are equally 6.49m. The whole roof structure rests mainly on the 8 columns erected prior to the construction of the space truss. To support the overhangings, four extra concrete columns (all shorter than the 8 columns) were built at the four corners and on the tops of the walls behind the stands, so that the number of supporting columns is 12 in all. The gradient of the space truss is kept at 3.96%, and the 4 slopes of the roofing meet above the centre of the space truss, thus forming a good roof drainage. Both the time/score indicator and the light/electro-acoustic control cabinet are attached suspended to the south and the north side respectively of the space truss. Either of the suspended attachments weighs about 20 tons. The plane and sections of the space truss are shown in Fig. 1. To ensure the geometric stability and horizontal stiffness of the space truss so that it can well cope with its level load, intersecting struts for the top and the bottom chord are added to the grids along the perimeters. As there are 6.49m overhangings at the east and west sides of the space truss and the cantilevered portion of the roof structure can provide cover to the top of the circumjacent buildings around the hall, over one thousand seats more can thus be installed on the top of the circumjacent buildings to accommodate more spectators.

This roof structure was designed for a standard load of $195kg/_m{}^2$, including a live load of $50kg/_m{}^2$. The whole space truss is made of low-carbon steel. The diameter of the steel sphere is 360mm, and the wall thickness of the steel sphere 12mm. The steel consumption of the space truss is $30.35kg/_m{}^2$. The weight of all the connectors composes 23% of the total weight of the structure.

The competition hall was in service from 1978 and has been in use ever since. Observation and examination on the bearing behavior of the structure in these years reveal that it works quite normally and has not failed of the function assigned to it.

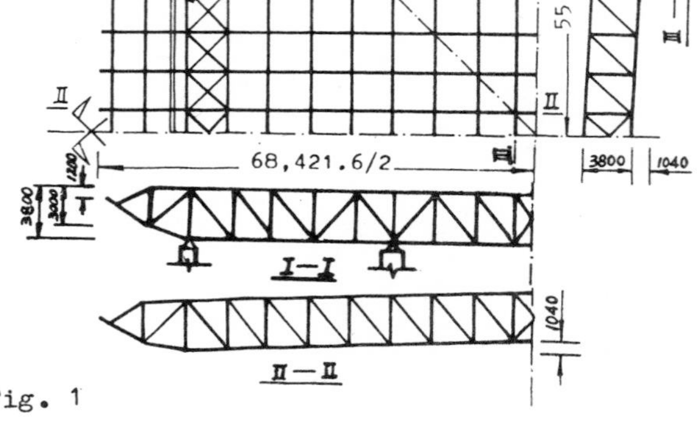

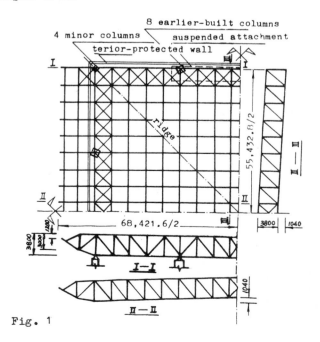

Fig. 1

For arriving at an analytical solution of the internal forces of this kind of space truss, both finite difference method and matrix displacement method were used. The space truss was first idealized as a gridwork system with uniform stiffness, and its internal forces were analysed by using finite difference method, so as to fix tentatively the sectional area of its members. Then matrix displacement method was used to analyse the internal forces and adjust the sectional area of its members.

The results of our analyses show that all the 12 columns are in compression; over 95% of the roof load is shared by the 8 tall columns, while less than 5% of the roof load is taken up by the 4 minor columns at the four corners of the hall. As to the internal forces of the part of chord members in the middle of the space truss, the results found with these two methods are approximately the same, the differences between their results being around 5%. With respect to the internal forces of the diagonal web members and of the part of chord members near the supports, however, the results derived through these two methods differs substantially, the differences ranging from 18% to 25%. The figures worked out through matrix displacement method are evidently greater than that found through finite difference method. This is markedly so with the analysis involving the diagonal web members at the supports; in such a case, the greater the ratio between depth and span, the more at variance the internal forces. This, we believe, comes of the negligence inherent in finite difference method of the influences exerted by both the web member diformation (shear deformation) and the variable stiffness over the internal forces of the space truss. Thus it is obvious that finite difference method fails to reflect the rule governing the distribution of the internal forces of the space truss. Especially in the case of the space truss supported on a comparatively small number of columns, where the internal forces of the diagonal web members at the supports are invariably the dominant internal forces; the results gained through finite difference method is always less than those found through matrix displacement method by 25%. That is something too baffling to facilitate a reasonable designing of space truss.

The type of supports used for a space truss may have significant bearing on the analysis using matrix displacement method. For example, an analysis of a space truss using supports with 3-dimensional constraint by applying matrix displacement method may present the following distinction from an analysis of a space truss using supports with vertical constraint by employing the same method: In the former case, a plain diminution is seen of the internal forces of the part of chord members in the middle of the space truss, while a plain magnification of the internal forces emerges of the part of chord members immediate to the supports; at the same time both the top and bottom chords are in compression. This , of course, can be accounted for by the effect of "arch" and does not agree with the normal behavior of a space truss. On the other hand, this may serve to remind us that importance should be attached to the due treatment of the space truss supports to make it go along with the assumptions set for analysis.

In designing Baoding Sports Hall, we made a comparative study of several design schemes and came to realize the following features of the cantilevered space truss with two-way orthogonal lattice grid system and supported on a comparatively small number of columns:

1. The distribution of internal forces is more reasonable, in so much as the cantilevers reduce the midspan internal forces and deflection, meanwhile a 15% reduction of steel consumption can be effected, as compared with the type of space truss with two-way orthogonal lattice grid system but without cantilevers. In such a cantilevered structure, the support reaction is converted into compression; in this case no tension support is required, whereas it is inevitable for a space truss supported along its perimeters to provide tension supports at its corners. Consequently, the troubles involved in dealing with tension supports and in building a large quantity of concrete columns can be dispensed with. Besides, the construction of this type of space truss is comparatively easy. Hence our conclusion that this type of cantilevered space truss with two-way orthogonal lattice grid system is reasonably preferable.

2. The space truss supported by a comparatively small number of columns can very well adapt to the change of temperature and uneven settlement of foundation. As is well known, a space truss is a highly statically indeterminate structute. Using less supporting columns is tantamount to lessening the external constraint of the space truss and decreasing the degree of statical indeterminacy. Therefore this kind of space truss generally offers better adaptability, so for as the change of temperature and settlement of foundation are concerned. The two-way orthogonal lattice grid space truss with a comparatively small number of supporting columns, in particular, has its temperature stress mainly concentrated on the plane trusses that rest right on the columns, and such temperature stress would influence the plane trusses that sit nearby. If we assume the joints are all pin-connected, theoretically speaking the temperature stress does not affect the adjacent plane trusses that do not rest right on the columns. As a matter of fact, the joints are actually not pin-connected and possess some rigidity, so those plane trusses sitting near the columns are virtually susceptible to the working of temperature stress, though it is quite insignificant. The effect of uneven settlement of foundation will be dealt with later in this paper.

3. The space truss with a comparatively small number of columns facilitates its assembly work. To vouch for sound bearing behavior of a space truss, accuracy has to be emphasized in its assembly work mainly for the following reasons: (A) that the secondary stress, resultant from the inaccurate fabrication of the space truss, might be abated, and (B) that the precarious state in which the non-alignment of the support axes with the column axes brought about by the inaccurate fabrication of the supports or columns might be eliminated. (Such a precarious state may evolve a support reaction against the columns and cause an eccentricity too much to be compatible with the design. As far as the assembly work of this space truss goes, meticulous efforts have been made to ensure the alignment of the axes of all members with the centres of the hollow spherical joints and the accurate spacing of all the support joints, though we attach only secondary importance to the accuracy of the spacing of other joints; that is to say, we attach only secondary importance to the control of the spacing of grids, because slight inaccuracy in connection with the spacing of joints can have only negligible effect upon the internal forces of the members, as space truss is a spatial structure. (Attention, however, should be paid to circumvent the error through accumulated inaccuracy.) As there are but few supports in this type of space truss, the spacing of the supports is easy to control—a circumstance that eases the assembly work of the space truss much of its complexity.

4. With the supporting columns being few in number, a concentration of the internal forces at the members around the supports is bound to occur. The internal forces of some members might be very great and calls for careful and proper handling. For this type of space truss, an alteration of the position of the internal diagonal web members (see section I-I in Fig. 1) was adopted to divert the strong concentration of great internal forces and transmit them directly to the supports. As for a space truss having even a greater spacing, column capitals are needed to enlarge the area for absorbing the directly transmitted internal forces. Therefore it is quite reasonable to adopt optimal column spacing, which is architectually permissible for the design to avoid excessive concentration of internal forces on certain membersm,in order to secure favorable technological and economic effects.

II. THE DESIGN OF THE SPACE TRUSS COMPOSED OF TRIANGULAR PYRAMIDS

This type of plate-like space truss was designed for the competition hall of Hebei Provincial Sports Hall. The competition hall has a plan size of 70.4m x 83.2m and seats 10,000. The pre-determined circumstances the designer of this project had to face were that the construction of both the stands and the column system along the perimeters had already been completed. (The column spacing is 6.4m.) Such a state of things precluded any other attempt at the erection of the space truss than using the sliding technique for assembly in the air. Conditions at the work site dictated that in executing the sliding technique the space truss had to be divided into strip elements and that each of these elements had to be slid one by one. The type of space truss to be adopted, therefore, should comply with these pre-requisites. During the process of type-choosing of the space truss, we made a comparison of a few design schemes,which included: (1) A space truss with two-way diagonal lattice grids (the size of the grid being 4.53m x 4.53m and the depth being 5m) which went geometrically unstable when sliced into strip elements, was evidently unfit for being slid. (2) A space truss with two-way orthogonal lattice grids (the size of the grid being 6.4m x 6.4m and the depth being 5m) was fit for the sliding technique, but the distribution of its internal forces was uneven. The value of its maximum internal force turned out to be 212 tons. Besides, it needed sizeable tubular members and bulky spherical connectors, demanding extravagant steel consumption and could not guarantee the good quality of welding operation as the welding of such huge members to such massive joints was certain to invite unpredictable troubles. Efforts were made, however, to improve this design by reducing the size of the grid to 4.14m x 4.14 and increasing the depth of the space truss to 5.5m for the purpose of decreasing the internal forces of the members. This improvement did cut down the value of the maximum internal force but failed to reduce the steel consumption; moreover bearing beams had to be provided as the supports could not settle right on the tops of the columns. (3) A space truss composed of triangular pyramids allowed 3-dimensional transmission of internal forces to bring about a considerable reduction of the maximum internal force.

The conclusion drawn from the comparison of these three conceptual designs is that the type of space truss composed of triangular pyramids has a superior bearing behavior to the other two types and is fit for assembly in the high position by using sliding technique and economical of steel comsumption. Fig.2 shows the plane layout of the triangular pyramidic space truss. The grid of this structure is shaped into an isosceles triangle, whose base side and height are equally 6.4m. The depth of the space truss is 5m. The vertex of the pyramid is aligned with the vertical line running through the centre of the height of the triangular grid on the top chord. To make the supports descend right on the tops of the columns, square pyramids are adopted along one side of the space truss. The advantages such a layout can afford are as follows: (1) With the plane of the bottom chord being composed of symmetrical triangular grids, the requirement with respect to the architectual form of the building can well be met. (2) All the supports can descend properly on the tops of the columns. If the space truss is composed simply of triangular grids, the supports at one end of the space truss shall not come to rest duly on the tops of the columns; in such a case, bearing beams shall have to be furnished for the structure and the height of supports enhanced to 90cm, otherwise the diagonal members of the supports shall have to be in the way of the bearing beams. With square pyramids being adopted at one end of the space truss, all the supports are lowered 40cm. (3) From each support are protruded for transmitting internal forces two diagonal web members, which augment the reliability of internal force transmission. To accommodate the construction of the space truss to the columns built beforehand, we had no alternative but to adopt fairly large grid. However, to reduce the calculated length of the compression members so as to fully utilize the strength of the material used for fabricating the space truss, triangular re-bisected members are attached to the four faces of every triangular pyramid. Fig. 3a shows a bird's-eye view of the triangular pyramids. Likewise, re-bisected members are also supplied to every square pyramid along one side of the space truss, but the re-bisected members of the top chord are square in shape. By using flexibility method we can easily prove that with the load of the roof structure being concentrated on the joints and re-bisected joints, such a rigid-connection system may be calculated by means of the same method as applicable to calculating a pin-connection system. Therefore, if all the joints and re-bisected joints are provided with pin-connection, and if we assume M_i (the bending moment at a joint) to be an unknown value. Then according to flexibility method, the coupled equations obtained are as follows:

$$\sum_{i=1}^{n} \delta_{ji} M_i + D_{jp} = 0 \qquad (j = 1, 2, \cdots, n)$$

where δ_{ji} is the displacement occurring at j, when the ith joint of the basic system is under the action of $M_i=1$, and D_{jp} is the displacement at j, when the joints of the basic system are under the action of the load. As the basic system is a pin-connection system, when all the joints are under action of load, there is only axial internal force, effect of axial deformation is disregarded. Thus all the values of D_{jp} are zero and the coupled

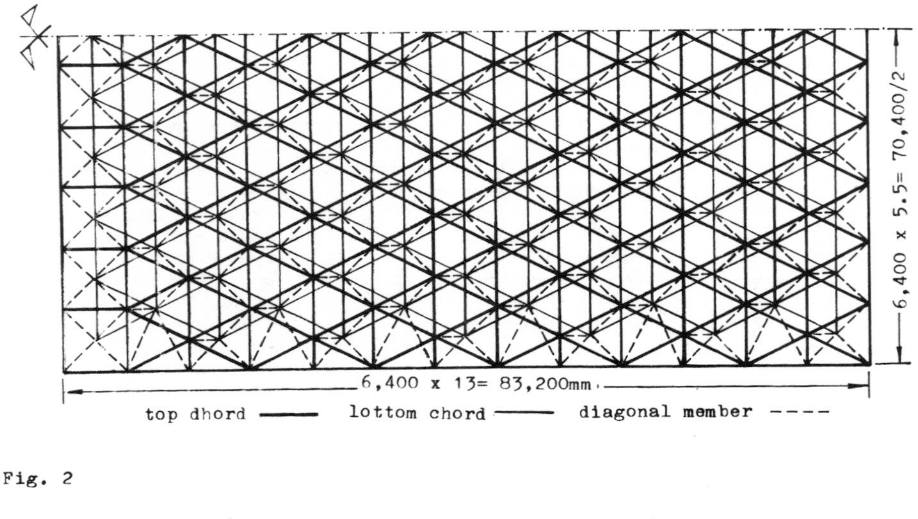

Fig. 2

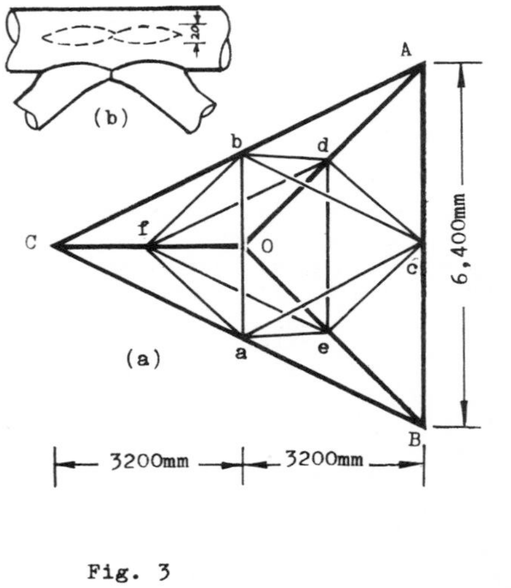

Fig. 3

equations derived from flexibility method are converted into homogeneous wquations. Therefore all the values of M are zero. This brings to light the fact that the triangular pyramidic space truss, when equipped with re-bisected members, can use the same calculation as for pin-connection system.

On the other hand, the internal forces of the re-bisected members may be obtained directly from the equilibrium conditions of the re-bisected joints; this means the re-bisected members constitute a statically determinate system. So it is apparent that the whole space truss is composed of highly statically indeterminate key member system and statically determinate re-bisected member system. The internal forces of the re-bisected members are related to the load at the re-bisected joints only and not affected by the key member system. The internal forces of re-bisected members, however, exercise some influence over that of the key members. So we have to take into consideration this influence in calculating the internal forces of the key members. Thus the analysis we need to conduct by using matrix displacement method shall be limited to the key member system only; this means of course a large-scale simplification of analysis. Every triangular pyramid with re-bisected members may be viewed as a composite body consisting of 4 minor triangular pyramids; in Fig. 3b, Acbd, Bcae, Cabf and defO respectively stand for 4 minor triangular pyramids. In the light of geometric stability, it is obvious, such a system of triangular pyramids with re-bisected members may be counted the stablest. Moreover, under action of the load at the re-bisected members (ab, bc, and ca) on the top chord plane and of the re-bisected members (de, ef and fd) in the middle layer present themselves as tensile forces; this is quite favorable for ensuring stability of the key members in compression.

This type of space truss is all in low-carbon steel. The diameter of its steel connecting sphere is 400mm, and the wall thicknesses of the spheres are 12mm and 16mm. Noteworthy technological and economic effects have been achieved, as we adopted a system of fairly large triangular grids rigged with re-bisected members for this triangular pyramidc space truss. The steel consumption for this kind of space truss is 31.2kg/m^2, and the steel consumption of the spherical joints is only 8.7% of the total steel consumption of the wtructure. In Table. 1 below, we list for comparison the steel consumption required by the three design schemes referred to above:

Space truss / Steel consumption	with two-way orthogonal lattice grid system	with two-way diagonal lattice grid system	composed of triangular pyramids
Total steel consumption (in t)	271.3	256	182.9
Steel consumption (Kg/m^2)	46.3	43.7	31.2

Table. 1

*All these schmes adopt tubular members and spherical joints, all of them being made of low-carbon steel.

**For carrying out the assembley work of the space truss with twO-way diagonal lattice grid system, a large elevated steel platform has to be erected which requires a steel consumption of adout 450 tons.

The analytical solution to the internal forces of space truss is based on the assumption that all the joints are of pin-connection and obtained by matrix displacement method. In this analysis, first we took into consideration two boundary conditions: (1) vertical constraint with the other two dimensions remaining free, and (2) vertical constraint coupled with tangential constraint (that is, constraint working along the perimeters). The results obtained from the analysis proved that there existed some slight difference between the internal forces of those members that were near the boundary, so far as the two conditions of constraint were concerned. With respect to the internal forces of the members in the middle of the structure, such difference was even smaller. Secondly, we proceeded to examine the influence exerted by the change of stiffness over the internal forces. Each time when we had completed a computation of the internal forces, we made an adjustment of the sectional areas of the members. And such a process was repeated for three times. With the first computation, the maximum internal force was 79 tons; it was raised to 101 tons with the third computation. This points to the fact that the influence engendered by a varied stiffness over the internal forces should not at all be neglected. Thirdly, we compared the results obtained through computing both the case, in which the vertices of the triangular pyramids were on the vertical lines running through the geometric centres of the triangular grids on the plane of the top chord, and the case, in which the vertices of the triangular pyramids were on the vertical lines running through the centres of the heights of the triangular grids. The results thus obtained verifies that the difference of internal forces of members between these two cases is insignificant.

The re-bisected members constitute an important system in this space truss. In designing the system, the following factors were taken into consideration: (1) The internal forces of the re-bisected members, which are caused by the load at the re-bisected joints, may be obtained directly from the equilibrium of the re-bisected joints. (2) The re-bisected members bear the shearing forces produced by the buckling of the compression members in the key member system of the triangular pyramids. At the same time, we duly enlarge the safety margin of the re-bisected members. Besides, due attention was paid to the influence exercised by the elastic deformation of the re-bisected members over the calculated length of the compression members in the system of key members. Fig. 3 shows the joint at which re-bisected members are welded directly to a key member. The ends of a tubular re-bisected member have been pressed to become elliptic, and the length of the short axis is around 20mm. In this way not only the bearing capacity of the key members for resisting cave-in as well as the capacity of the re-bisected joints for re-sisting bending is increased, but convenience to expedite the construction can also be brought about.

To prove the reliability of the re-bisected members in case the calculated length of the key members be reduced, a test of buckling was made on the model of a single triangular pyramid, which was 1/4 in size of the prototype; however the ratio of slenderness of the model was made to be slightly greater than that of the prototype, with a view to checking the function of the re-bisected members. Then we went on to make experiments on the buckling of the chords and the diagonal web members in compression. Experiments were also made with all the compression members, when they were charged with the compression nearly equal to their critical loads, and examination was conducted of the effect of the secondary stress. We observed in the experiments that a member vibrated pending its buckling, and that all of a sudden it went buckled into the shape of an S (the re-bisected joint forming the centre of the s-shaped curve). Later on, we made new experiments, in which we replaced all the re-bisected members with new ones of smaller sectional area (λ=400—445). All the later experiments confirm that the value of the compression causing the buckling of a key member in an experiment coincides with its theoretical counterpart and that the function of the re-bisected member system is reliable. Furthermore, experiments were also carried out in the bearing capacity of the re-bisected joints; the outcome of such experiments reveals that such re-bisected joints are reliable too, their safety coefficient being above 4. So for, the design of the space truss composed of the system of triangular pyramids with re-bisected members has proved both experimentally and theoretically to be safe, reliable, economic and reasonable.

III. AN APPROACH OF ANTI-SEISMIC ANALYSIS

Both sports halls referred to above are situated in seismic area in Hebei Province. Ever since the catastrophic earth tremor in Tangshan in 1976, anti-seismic structure has wrenched more attention from our architects' circle. Accordingly we could not afford to ignore the anti-seismic side of the structural problems in the designs of the two sports halls. As ruled by our domestic Specifications for anti-seismic design of structure, the reaction by a building to an earthquake is interpreted in terms of earthquake loading. On the other hand, lessons drawn from past earthquakes have thrown light on the fact that not only the anti-seismic behavior of the space truss but also that of its supporting column system has to be handled with extreme care, else calamitous consequences are bound to occur. Therefore we take into account the following two aspects, dealing with the anti-seismic analysis of a space truss:

(1) The analysis of transverse (that is, perpendicular to the plane of a space truss) anti-seismic strength of a space truss: To analyse the dynamic characteristics of a space truss is critical to the determining of earthquake loading. As for those space trusses with several degrees of freedom, the analysis is very complicated. An experiment on a model of space truss conducted by Chinese Academy of Building Research has asserted to the effect that fundamental frequency plays a majorrole in determining the earthquake loading, while other frequencies have only minor roles to play in this connection; that is to say, in doing approximate calculations of earthquake loading, the combination of modes of vibration may be let out of consideration. Hence we adopted energy method in calculating the fundamental frequency of a space truss.

Take the two-way orthogonal lattice grid space truss roofing Baoding Sports Hall for example. we take it for a gridwork system of even stiffness with its mass being distributed on the joints of the top chord. Thus it might be regarded as a system with n degrees of freedom. Here stands for the number of the joints of the top chord. Let m_i stand for the mass of the particle. Now, when this system with several degrees of freedom is in a state of free vibration, the formula for obraining (natural frequency), if the effect of damping may be overlooked, may be derived by applying the law of conservation of energy:

$$\varphi^2 = \frac{\iint EI\left[\left(\frac{\partial^2 z(x,y)}{\partial x^2}\right)^2 + \left(\frac{\partial^2 z(x,y)}{\partial y^2}\right)^2\right]dxdy}{\sum_{i=1}^{n} m_i Z_i^2(x,y)}$$

Where Z(x,y) is the amplitude of the gridwork system, I is the inertia moment, and E stands for Young's modulus. If the exact types of amplitude are known, then the value of φ , as obtained from the formula above, will be accrate. However, more often than not an architect is asked to tackle a case, where amplitude curves are unknown to him. Therefore the equation of amplitude has to be determined first. As natural frequency is related to the relative value of amplitude only—in other words, the former is related to modes of vibration only—we might as well take the values of static deflection for those of amplitude. In that case, elastic strain energy can be expressed in terms of corresponding external work. Hence the formulae for natural frequency and period of vibration:

$$\varphi = \sqrt{g \sum_{i=1}^{n} W_i Z_i \Big/ \sum_{i=1}^{n} W_i Z_i^2}$$

$$T = \frac{2\pi}{\varphi} = \frac{2\pi}{\sqrt{g}}\sqrt{\sum_{i=1}^{n} W_i Z_i^2 \Big/ \sum_{i=1}^{n} W_i Z_i}$$

where $W = m_i g$ (the weight of the particle).

In the analysis of the internal forces, we have obtained the static deflection of the joints of the space truss. So it can be made use of for further computation. The analyses of all the internal forces under action of earthquake loading is computerizable by using matrix displacement method. And that accounts for our adoption of energy method in this case. In working out the design of Baoding Sports Hall—a building with a two-way orthogonal lattice grid space truss—we employed the value of deflection of the joints of the top chord to obtain the fundamental frequency of the space truss (φ_1=7.25 1/sec) and the period of vibration (T1=0.87 sec). This value of period of vibration is analogous to those found out in the surveys of a few similar space trusses in China. As the curved surface formed of static deflection is by no means a real curved surface of amplitude, therefore the fundamental frequency thus obtained is nothing more than an approximate value. Anyhow it has come to be known that the value of φ is not particularly sentitive to the error connected with the curved surface of amplitude. From this, it may be evolved that the determination of earthquake loading will not be much affected by this kind of error.

(2) The analysis of the horizontal anti-seismic strength of a space truss: First, let us deal with the analysis of the anti-seismic strength of the supporting columns. The stiffness of the space truss is very powerful in its plane; this amounts to a situation, in which all the supporting columns of a space truss combine to prop up a monolithic block, whose lateral stiffness is infinite. When a vibration arises, the displacements at the tops of all the columns are the same. Thus the space truss may be simplified to a system with one degree of freedom, on the tops of whose columns is supported a concentrated mass. Suppose that M is the mass of the whole space truss under the action of an earthquake and that K is the space truss' combined columnar stiffness, which is the sum of the stiffnesses of all the columns of the space truss. Then the natural vibration frequency may be found out by the following formula:

$$\varphi = \sqrt{\frac{K}{M}} = \sqrt{\frac{gK}{W}}$$

and the period of natural vibration may be obtained by the following formula:

$$T = \frac{2\pi}{\sqrt{g}}\sqrt{\frac{W}{K}}$$

With the value of T thus obtained, the value of earthquake loading may be readily acquired. Then we can proceed to work out the distribution of earthquake loading according to the proportion of columnar stiffness. The horizontal earthquake loading of the ith column can be calculated by the following formula:

$$Q_i = \frac{K_i}{\sum_{i=1}^{n} K_i} Q$$

where Q is the combined earthquake loading, Q_i is the earthquake loading borne by the ith column, K_i is the columnar stiffness of that column, and here n stands for the number of the columns of the space truss. Having completed the internal force combination according to the earthquake loadings at the tops of columns obtained with the aforesaid method, we may go on to check the strength of column.

We stress the necessity for analysing the strength of supporting columns under the action of horizontal earthquake loading. This analysis is imperatively indispensable in the case, where a space truss is designed to be supported by comparatively few columns and may turn out to be a decisive factor in orienting the conceptual design of the column system for a space truss.

Secondly, fairly great internal forces may emerge along the members around the supports of a space truss influenced by horizontal earthquake loading. The strength of these members has to be checked through calculation, if necessity dictates. The horizontal earthquake loading at joints can be derived from the value of the period of horizontal vibration. The analysis of the internal forces of members under action of horizontal earthquake loading can still be a computerized process.

We have made a comprehensive study of the horizontal anti-seismic strength of 4 different typical space trusses and come to realize that their horizontal anti-seismic strength is quite satisfactory. The process of reckoning the horizontal anti-seismic strength of a space truss may be dispensed with, when the earthquake intensity is 8°. moreover, space truss composed of pyramidic system possesses even better horizontal anti-seismic strength, and the space truss composed of triangular pyramidic system behaves best in this respect. Space truss with two-way orthogonal lattice grid system, however, is of all the 4 types mentioned above least able to resist the horizontal thrust from earthquake. When the top and bottom chords of a space truss are strengthened with intersecting struts, its anti-seismic strength can be greatly fortified.

IV. SLIDING TECHNIQUE EMPLOYED FOR ERECTION OF SPACE TRUSS

In view of the fact that the exterior-protected wall and the stands of the competition hall of Baoding Sports Hall had already been constructed before the roof structure was designed and that the conditions of the locality presented certain limitation with regard to the construction and erection of space truss, we had to decide upon the following procedure: (1) assembling the space truss at ground level, (2) hoisting it up in the air beyond the perimeters of the hall, and (3) using the sliding technique and completing the erection. This was the first time that the sliding technique was put into practical use in China; afterwards this technique has been adopted for the erection of several space trusses with smaller span elsewhere in China.

The advantage inherent in the sliding technique consists in a shortening of the whole construction process by arranging for a simultaneous execution of the fabrication and erection of the space truss on one hand and of the masonry and concreting on the other. This technique is particularly utilizable in a close locality, where hoisting equipment is difficult to gain access. Utilization of this technique involves only low cost and simple equipment. The erection of a small space truss by using this technique can do even without a hoisting machine.

In erecting Baoding Sports Hall roof structure, its two-way orthogonal lattice grid space truss had to be divided into two equal halves to be assembled on the ground at the locality owing to the conditions at the site. One of the half space truss was subassembled and slid into its final position; so did the other. The erection was not complete until the two halves were interconnected in the air to form a whole space truss. The erection of the subassembled half involved the following steps:

(1) Lifting vertically the half outside the building. this step was fulfilled by using 4 derrick masts of 32m-34m in height. The general layout of the derrick masts is shown in Fig. 4. There were 4 hanging points to each derrick mast. Each lifting point was linked to the half space truss by being attached to the 4 steel spheres of a grid on the top chord. Fig. 5 shows the half space truss pending the lift outside the building. Fig. 6 shows the lift was in progress.

(2) When the lift had climbed to such a height as the bottom chord of the half structure was about 0.5m above the top of the exterior-protected wall, the half structure was made to move forward by a horizontal component force created of the differenoe in height between the front pair of lifting points (A and B) and the rear pair of lifting points (C and D), until the first pair of rollers (the anterior rollers) attached to the half space truss touched the rails laid on the tops of the exterior protected walls to the north and south sides of the building, as shown in Fig. 7a.

(3) After the anterior rollers had touched the rails with their positions adjusted, the pulley blocks for the front lifting points A and B were removed and attached to the rear lifting points C and D, so that the front and the rear derrick masts on either side of the space truss could work upon the newly-formed composite lifting point. Then by tightening and loosening the front and rear pulley blocks alternately and repeatedly, the half space truss was made to creep forward on the rails, as shown in Fig. 7b.

(4) The procedure of sliding the half structure into position was as follows: As the newly-formed composite points moved near the centres of the front derrick masts, the hauling pulleys began to go into operation until the posterior rollers landed on the rails. Now, the derrick masts were disengaged from the half space truss, and the latter was then hauled hroizontally into position, as shown in Fig. 7c. Fig. 8 and Fig. 9 show how the half space truss crawled along on the rails. Fig. 10 shows how the other half space truss was slid into position.

From the facts presented above, it is clear that the erection of the first half of the space truss as shown in

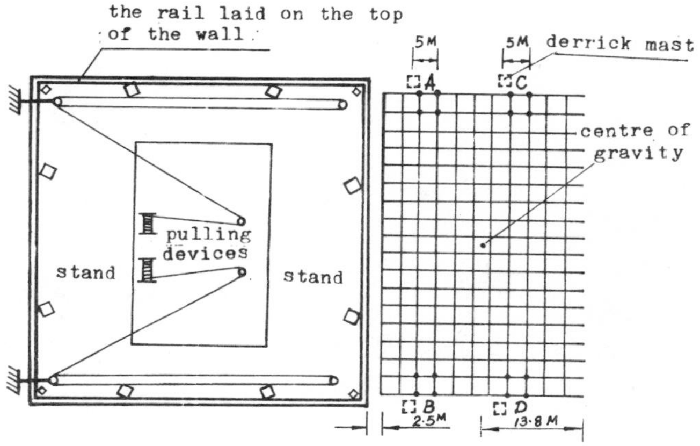

Fig. 4

Fig. 7 experienced three different conditions of bearing. And we adopted matrix displacement method for analysis its internal forces. The following points are noteworthy in consideration of the procedure of erection and in the analysis of the internal forces during erection. The loads at different joints resulted from the space truss' dead weight are different because of the varied sectional areas and weights of the members. So the loads at joints have to be calculated severally in accordance with the structural features; namely, the concentrated load at a joint is the sum obtained by adding together the half of the weights of the members intersecting at the joint, the weight of the steel sphere at the joint, and the weights of other subsidiary parts. The result of this calculation shows that the centre of gravity does not coincide with the geometric centre, that there is a notable distance between these two centres, and that the greatest deflection is also for away from the geometric centre. This makes it plain that the mean of the space truss' dead weight cannot be used for calculating the load at a joint resulted from the space truss' dead weight, as we would do in calculating the internal forces of the members created under action of designed roof load.

Due attention must be paid to ensure synchronous vertical lift of the space truss, that is, to have a control over the differences of elevation between various lifting points during erection. Generally speaking such a lift must be a steady process and allows only minimal differences of

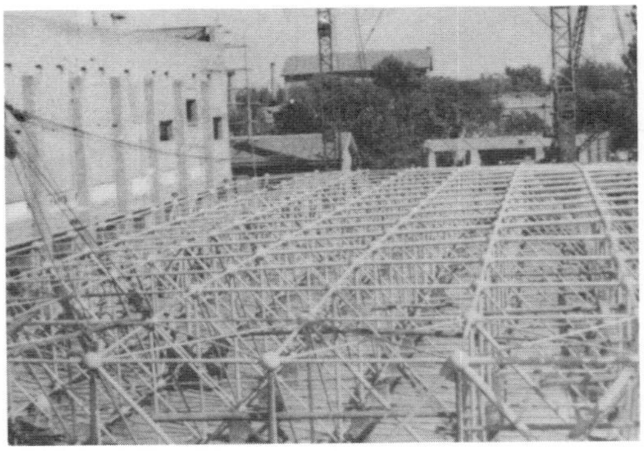

Fig. 5

Fig. 6

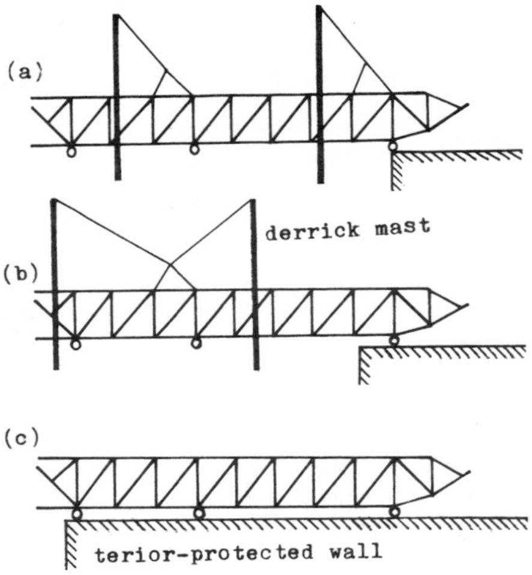

Fig. 7

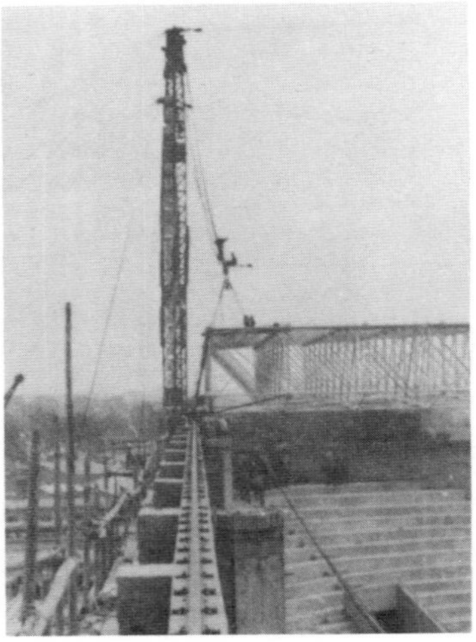

Fig. 8

Fig. 9

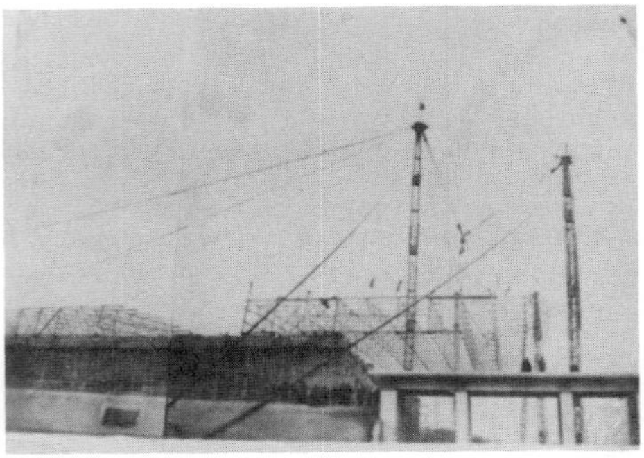

Fig. 10

elevation between lifting points. As for the lifting operation at Baoding Sports Hall, owing to our limited hoisting equipment, we had to resort to using 4 hoisters of unitentical standard and speed. Moreover, the heights of the 4 derrick masts were not the same. Therefore, anything but synchronousness would occur. The difference of elevation at any lifting point during the erection tends to distort a space truss. Such a distortion, if it occurred, is not unlike the case of uneven settlement of the supports. In the case of lifting a two-way orthogonal lattice grid space truss having 4 lifting points, if the space truss has no horizontal intersecting struts, the unbalanced lift at any lifting point tends to warp the space truss into a hyperboloid with plane trusses still retaining a linear shape, which is equivalent to the generatrix of the paraboloid. Apparently the members are theoretically not under action of force. However, with space truss having horizontal intersecting struts, things are different. Between a space truss with and a space truss without horizontal intersecting struts, there seems to be only little difference, if judged from the qualitative point of view. Therefore, it is reasonable to allow a discreetly greater difference of elevation for all the lifting points. Drawing from the experiences accumulated through erection of space trusses elsewhere, we deemed it proper to set the allowable difference of elevation below 1/300 of the lengthwise span of the half space truss, namely, below 20cm; and in practical use, this allowable difference of elevation was at that time supported with other pertinent control measures. And our erection practice ever since has proved this allowable difference of elevation safe. Now, we are rather tempted to think that still greater range of fluctuation might be allowed to this kind of difference of elevation.

During the horizontal sliding, the two hauling forces were virtually unequal; so the question of synchronousness also arose. The smaller the ratio between the span and the distance separating the anterior and the posterior rollers, the greater the need for synchronousness, otherwise an impeded motion of the rollers or derailing would happen. For interconnecting the two half space trusses, care should be taken to eliminate the influences of such internal forces as resulted from the half space truss' dead weight, because the action of the internal forces of a half space truss is not the same as that of the internal forces of the unbisected space truss. The method we used to solve this problem is like this: First, we examined the distribution of the internal forces in both half structures and selected 8 joints in each of them for shoring all their sagging parts up. And then we went on to interconnect them by welding.

The number of rollers attached to the half space truss for sliding should be limited. Two or three are quite enough for either rail, apparently not to cause unnecessary inconveniences in sliding.

V. CONCLUSION

Our construction practice has vindicated that the

cantilevered space truss with two-way orthogonal lattice grid system is a reasonable structure in point of distribution of load and facility for assemble work, that it is particularly fit for such space truss as supported by a comparatively small number of columns, and that space truss possesses good anti-seismic strength. Triangular pyramidic space truss may be counted one of the best types of anti-seismic space truss. Two-way orthogonal lattice grid space truss may become a good anti-seismic structure too, if appropriate measures are taken to improve its horizontal anti-seismic strength. It can be safely predicted that space truss will turn out to be a promising roof structure with wide span for buildings in seismic area.

In our erection practice, we have successfully adopted the sliding technique. This method renders possible simultaneous execution of both fabrication and erection of space truss and masonry and concreting of the building, so as to speed up construction process and is particularly helpful for a construction, where limited size of work site makes other techniques futile. The space truss with triangular pyramid system and re-bisected members has proved to be recommendable both economically and technologically.

ACKNOWLEDGEMENT

Five engineers (Liu Xi-hua, Cuo Jia-sheng, Bian Xue-zun, Liu Jia-jun and Tan Jia-hua) and a lecturer (Liu Yun-bo) have taken part in the designing and experiments of these two kinds of space truss.

REFERENCES

(1) Z. S. Makowski, Space Structure, "9th Congress (IABSE)" IIIb, pp.127-149, 1972.

(2) Tao Yizhong, Lan Tien and Dong Shilin, Development of Space Trusses in China, Proc. of Sino-American Symposium on Bridge and Structural Engineering, Part I, 1982, Beijing.

(3) 网架结构设计与施工规定，中国建筑工业出版社，1980年。

(4) 龙驭球等编 结构力学，人民教育出版社，1979年。

(5) 工业与民用建筑抗震设计规范，中国建筑工业出版社，1979年。

COBWEB SYSTEM FLAT SPACE FRAMES

LIU XI LIANG*, ZHANG RU LIANG** and YU JUN YING***

* Associate Professor, Department of Civil Engineering, Tianjin University, Tianjin.

** Engineer, Tianjin Municipal Survey and Design Institute, Tianjin.

*** Lecturer, Department of Civil Engineering, Tianjin University, Tianjin.

In the light of the requirements of engineering practice, this article puts forward a new type of space frames with new joining method suitable for buildings with circular plans, the COBWEB SYSTEM FLAT SPACE FRAMES with direct linked joints. For circular buildings this new system possesses many unique advantages in contrast with others. In order to fully understand its structural behaviour, its joint strength and its construction techniques model tests as well as joint tests have been carried out on a 4m diameter model and on a 15m diameter pilot building. Measured test results well conformed to theoretical analysis; this new system proves worthy of popularizing.

INTRODUCTION

The Academic Report Hall of the Scientific and Technological Information Centre, Tianjin, adopts a circular architectural plan, 30m in diameter, and the roof design is going to be a space frame. In the last decade, China has constructed several large circular gymnasiums and railway waiting rooms whose roofs are mostly three-dimensioned or pyramidical space frames, but at peripheries, the members were too irregular to be ideal. Furthermore, the domes in common use for circular buildings are rarely adopted in China because of their difficulty in construction and their architectural styles. It is for these reasons that we have developed a new system of space frames as related in this article -- the cobweb system flat space frames. In order to fully understand the structural behaviour of the system, the strength of joints and the construction techniques, tests were made on a model 4m in diameter, on three small trusses with direct linked joints and on a pilot building 15m in diameter, one of the booking offices of the subway and stresses and deflections were measured.

PATTERNS AND DISTINGUISHING FEATURES

The cobweb system consists of two identical layers of cobwebs, the upper and lower ones, joined together by means of web members. It includes three parts: the radial trusses, the ring trusses and the bracings. (Figs 1 and 2)

1. Radial Trusses
   These are single, planar trusses radiating from the centre of the circle among which 4 trusses at right angles with each other extend to the centre while others equally divide the quadrants and extend only to the first or second ring truss.

In the diagram shown below, there are altogether 16 radial trusses, thus making each sectorial angle 22.5°. As the diameter increases, the sectorial angle will become smaller and the number of radial trusses will accordingly increase. The number required will be decided by an overall consideration of many factors, such as conditions of load, easy of access for welding, types of roof slabs, bracings and so on.

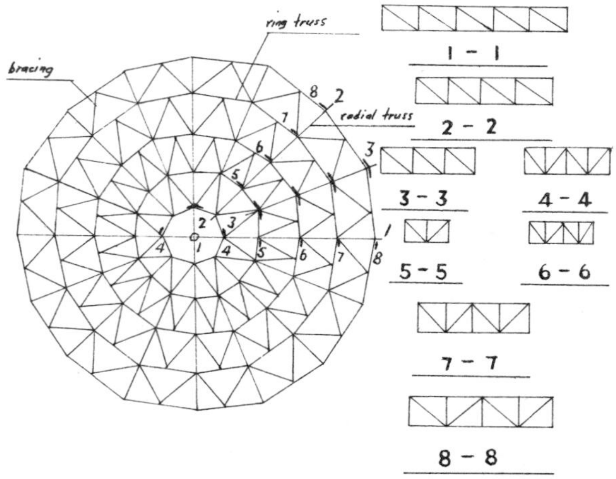

Fig 1 Plan of Cobweb Systems

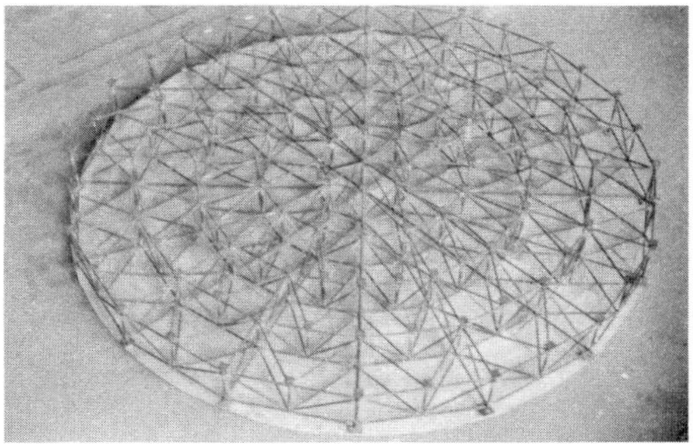

Fig 2 A Cobweb System Model

2. Ring Trusses
   These consist of many concentric rings. Spacing of rings is generally uniform, but it may also be increased ring by ring as the distance extends away from the centre. Using rings of equal spacing will result in less types of roof slabs required. (Fig 3)

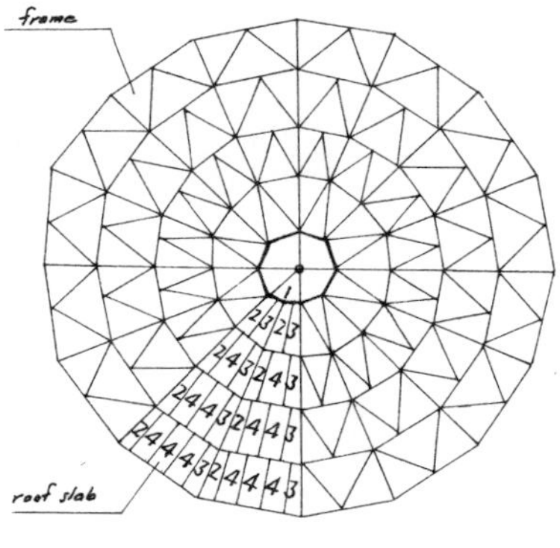

Fig 3 Various Types of Roof Slabs

3. Bracings
   In the planes of the upper and lower chords there are bracings to divide grids, reduce the free length of the upper and lower chords of ring trusses and make grids of constant geometric shapes.
   The cobweb system possesses the following distinguishing features in contrast with other systems:

   A) The arrangement of members is absolutely symmetric and identical with a circular plan. Members are evenly stressed and their geometric shapes constant.

   B) Easy camber and drainage: Deflection caused by loads and drainage demands all require the system to be cambered during construction. Cambering is achieved by raising the centre ring to the desired point so that other rings will follow correspondingly, forming a natural gradient toward the circumference for drainage with no ridge line on roof.

   C) Less types of roof slabs: Generally large and medium cobweb frames have their bracings designed into equally spaced rings to reduce the types of roof slabs. For small cobweb systems, no bracing is required and the equally spaced ring trusses serve as a means to reduce the types of roof slabs. Other types of space frames require much more types of roof slabs because they do not coincide with roof circumferences.

   D) Radial trusses as well as other elements can be made in factories to ensure high precision and quality.

   E) The elements are either trusses or bars, so they can be piled up easily or carried about.

   F) Because the upper and lower chords of radial trusses can be through length members, others can be directly connected to them by welding, thus saving ball joints and further reducing steel consumption.

STRUCTURAL CALCULATION

In most regions of China, computers have low memory capacity; therefore it appears especially necessary to solve big problems with low capacity computers. The cobweb system is an absolutely symmetric circular structure; along its circumference, the whole structure can be divided into symmetric sectors. Therefore, any sector which may be 1/2, 1/4, 1/8 or 1/16 of the whole can be taken as calculation unit so long as conditions of boundary restraint are handled by making use of the properties of symmetry. (Fig 4)

In design, the conditions of restraint are assumed to be ideal movable hinges. Because the system entirely coincides with a roof circumference, deformation of the system is elongation or shortening of diameters of rings under the action of vertical loads and the impacts of temperature changes. As every movable hinge support moves in the radial direction, this subjects the supports to a clear and definite action of force and their construction simple. With the aid of computers, matrix-displacement method has been employed to compute the internal force and deflection of the system. Computations have been worked out for full loads and partial loads for two cases: a) with supports on movable hinges and b) with supports on fixed hinges.

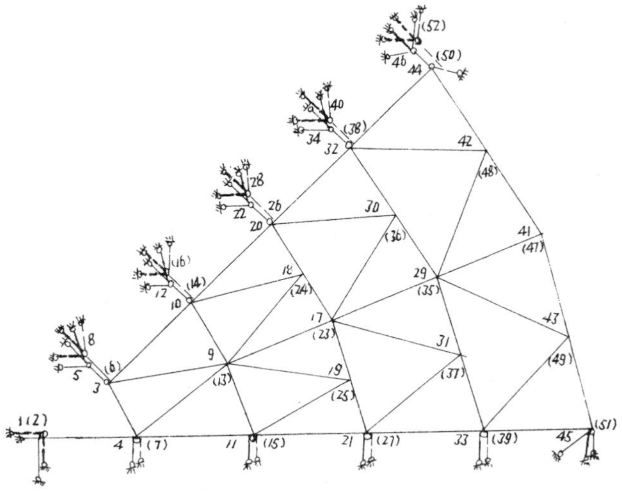

Fig 4 Calculation Unit-1/8 Cobweb System Flat Space Frame

TESTS

1. The Cobweb System Model Tests

In order to examine the accuracy degree of theoretical calculations and to study the load - bearing capacity of the system and the law governing the internal force of each member, we have carried out structural tests. (Fig 5)

Fig 5 The Cobweb System Model Tests

Dimensions of the model: diameter 4.06m; height 0.285m. Seamless steel tubes of Ø 16 x 2.2mm in cross section have been selected for construction and members have been linked with welded joints. Separate tests have been carried out for movable hinge supports and fixed hinge supports.

The whole frame was supported on 16 brick piers and every upper chord joint supported a tray which, in turn, supported load-carrying planks. They were free of connection with each other, so that clear and definite transfer of loads was ensured. The application of load to the model was achieved through filling water into an open-mouthed rubber bag which had been placed into a steel cylinder, 1mm thick, and 75cm. high. Because water is frictionless, the application of load was even and convenient.

Five different tests have been made:

I) Full load-- 100kg/sq.m; movable hinge supports. Through ten odd experiments the following conclusions have been reached:

A) Radial trusses are the main load-bearing elements of the system and the chords of trusses are the mainly stressed members, the upper chords being under compression and lower chords under tension. The actual measured values of member force are about 86% to 100.5% of the calculated ones. Bracings and vertical members have very low force and in design they are controlled by slenderness ratios.

B) Ring trusses are also the main load-bearing elements and the chords the mainly stressed, the upper chords being under compression and lower chords under tension. Member forces are 86% to 94% of the calculated values.
The central ring has the maximum internal force. As the rings extend farther away from the centre, the internal force decreases progressively. Vertical members or bracings are little stressed.

C) The bracings in the upper and lower planes of the cobweb system generally have low stresses and their designs are controlled by slenderness ratios.

D) The law governing the variation of deflections and their values obtained from tests on a model 4m in diameter and on a pilot building 15m in diameter basically agreed with theoretical calculations. Test values of internal force and joint deflections showed to be on the less side. This was due to the fact that both ends of each member were in a certain degree elastically fixed while in calculation they were assumed to be hinged, thus the rigidity of the structure having been somewhat lowered.

II) Partial load-- 100kg/sq.m; movable hinge supports. Partial load tests showed that the internal forces of main members became smaller but no change of signs took place, while those of secondary members became greater and changed signs. But the absolute values were still small and slenderness ratios still controlled designs, the sign changes having no effect on designs. The tests also showed that under partial load the structure was on the safe side. (Fig 6)

Fig 6 Partial Load Tests

III) Effects of settlement of supports.
The range affected by the settlement of a support is a 90° sector. Members adjacent to a settling support would be subjected to greater internal forces. In design, this factor should be taken into consideration. It seems necessary to strengthen the members of the two outmost rings adjacent to supports.

IV) Full load-- 100kg/sq.m; fixed hinge supports.
Tests indicated that the measured values of the upper chords which were the mainly stressed members well agreed with calculated values, the former being about 84% of the latter. But in the case of lower chords, which were also mainly stressed members, test values became considerably greater than the theoretical ones. The reason for this deviation belonged to the radial displacement of the structure which was caused by insufficient rigidity of a No. 10 channel, to which supports were welded but could not play the role of fixed hinges.

In actual practice, we would recommend that for medium or small structures plate bases be used and welded to embedded pieces in a R.C. wall beam. This method of construction would provide a feasible fixed hinge and better safety.

V) Measurement of the frequency of self vibrations.
Self vibrations measured by means of both methods, sudden release and knocking, were identical, the frequency of self vibrations having been 24 times/sec and the period of the cycle 0.04 sec.

2. Joint tests
Structures of the cobweb system may use different types of joints. Because the upper and lower chords of the system are of through length, it is more favourable to

use direct linked joints. China has for the first time used this kind of joints on trial for roof space frames and we have made a preliminary study of their mechanical properties and their construction techniques. (Fig 7)

Fig 7 Tests of Direct Linked Joints

Direct linked joints require that for steel tube structures the secondary members be machined into linking surfaces to fit themselves to the tube walls of the mainly stressed members and welded together into joints. Through tests made on three small space trusses we have concluded that such joints could be treated as ideal hinges. When subjected to compression, the walls of main tubes will locally yield and this is the heart of the matter to joint strength designs. When the walls of the main tubes are thin, tension or compression caused by branch tubes might bring about local yielding of main tubes. Therefore it is necessary to adopt some kinds of strengthening measures so that in increasing the capacity of resisting local yielding there is no need to increase the thickness of main tubes, causing waste of steel. Our tests proved that the U.S. formula for caring local yielding in direct linked joints is safe and reliable. For guaranteeing joint strength, generally the diameter of a branch tube should not be larger than that of a main tube, nor less than 1/4 of the outer diameter of the same.

3. Pilot building tests
In order to have still more assurance for constructing a 30m in diameter roof for the Academic Report Hall of the Scientific and Technological Information Centre, Tianjin, we first constructed a pilot building of a smaller diameter, a Booking Office of the Subway, to find out what kinds of problems might be encountered in construction, the type of the frame and members being identical with the above mentioned model. The project was a circular building, 15m in diameter and 6.6m in height, and was completed on October 15th, 1983. In construction, we used R.C. thin slabs, cement pearlite for heat preserving and felt and asphalt for waterproofing. All these made up a grade 3 loads for tests. Similar to the model tests, measured values of internal forces and deflections well agreed with calculated ones.

DESIGN, FABRICATION AND HOISTING

The design work includes detailing every member of the system, machining it into a ready part according to detail drawing and assembling parts into elements. (i.e. radial trusses) Radial trusses, ring members and bracings are assembled into a whole frame.

Procedures for fabrication are as follows:

1. Straightening steel tubes;
2. Cutting tubes with round saws;
3. Laying out;
4. Machining the tubes into mutually fitting surfaces;
5. Fabricating radial trusses in pattern plates;
6. Assembling the whole frame at construction sites.

Owing to the absolute symmetry of the system and its high rigidity, a single crane can perform the job of hoisting a frame to its place. In the erection of the frame 15m in diameter, 4 hoisting points were used at a time; they were 4 symmetric lower chord joints at the third ring. The hoisting job was completed in seven minutes with one action.
Fig 8 gives the scene of hoisting.

Fig 8 Hoisting

MAIN TECHNICAL NORMS OF THE Ø 15m FRAME

Actual roof load	217 kg/sq.m
Dead weight	14.5kg/sq.m
Max. deflection	25.4mm
Max. internal force	5,000 kg

CONCLUSION

For circular buildings, Cobweb System Flat Space Frames not only possess certain unique advantages in contrast with other types, but through model tests they have also proved reasonable in load-bearing, safe and reliable. At the same time, direct linked joints are advantageous to the system. Therefore, Cobweb System Flat Space Frames with direct linked joints are not only feasible but also technically advanced, reasonable, economical and worth wide spreading.

ACKNOWLEDGEMENTS

The authors wish to thank Municipal Machine Plant, Tianjin, and Survey and Design Institute No.3, Ministry of Railway, for their vigorous help.

REFERENCES

1. LIU XI LIANG and LIU YI XUAN, Design of Flat Space Frames, China Building Industry Publishing House, September 1979.

2. Guidelines for Structural Design and Construction of Space Frames, JGJ-7-80, China Building Industry Publishing House, 1980.

A STEEL SPACE ROOF IN KIBBUTZ "MESSILOT" (ISRAEL)

M PAZ-NER, CE, M Sc

A steel space roof in the form of a pyramid with truncated edges, covering a quadratic hall with horizontal dimensions of 30.0 m x 30.0 m., is presented. The four sides of the pyramid are formed as flat double-layered grids, composed of 40 mm x 40 mm R.H.S. members, welded together. The four panels were constructed in a workshop and assembled on site. The paper deals also with the considerations which led to the choice of structure, the results of the analysis of the structure, and of the measurements of strains and deflections which were performed at different stages of construction.

GENERAL

A multipurpose Communal Building was recently erected at Kibbutz "Messilot" in the Bet Shean Valley, Israel. The building serves a variety of social, cultural and sports activities.

The main part of the building is a square-shaped hall, 30.0 m x 30.0 m. Around this central hall different service units are located, at a lower level than the hall itself, and partly concealed in an earth mound surrounding the building (See Figs. 1,2).

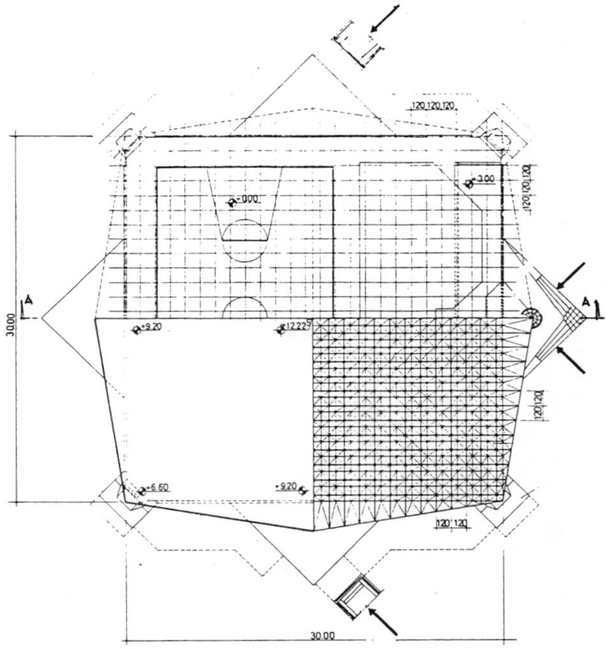

Fig. 1, THE PLAN OF THE BUILDING

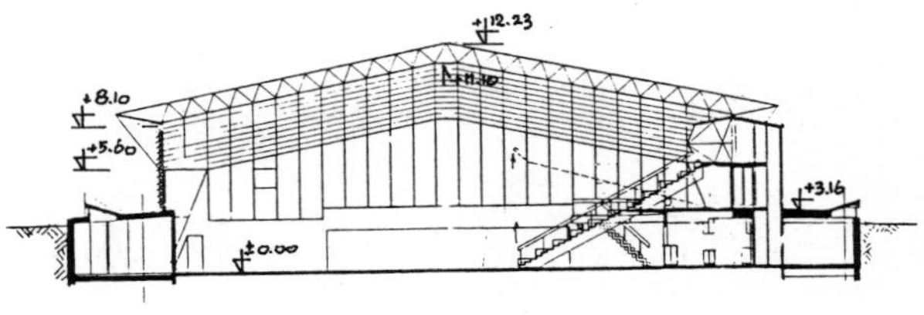

Fig. 2, SECTION A - A

SELECTING THE STRUCTURE

Different types of roof structure were considered for the covering of the main hall; among them were reinforced concrete shell structures such as "Hypars", and others. All these solutions were found to be too expensive, especially because of the high cost of the necessary formwork.

Another reason that brought the designers to reject solutions based on R.C. shell structures was the concern about possible cracking, due to difficult environmental conditions of soil (expanding clay) and climate in the Bet Shean Valley. The waterproofing and thermal insulation of such shells, in the specific conditions of the area, would also have been difficult and expensive.

These considerations led finally to the choice of a steel space frame covered with simple asbestos-cement corrugated sheets. The thermal insulation was planned to be provided by a layer of polyurethane foam sprayed onto the outer face of the asbestos-cement sheets. Sound accoustics were achieved by hanging light accoustic panels beneath the asbestos-cement sheets.

DESCRIPTION OF THE STRUCTURE

The roof is composed of 4 flat double-layered grids forming together a kind of pyramid with truncated lower corners, and with a canopy extending over the outer walls.

The double-layered frame is based on an octagonal grid (on the horizontal projection) of 1.2 m x 1.2 m, while the distance between the upper and lower layers is 1.0 m. The two layers are connected by diagonal members. Mainly because of architectural requirements and for reasons of simplicity, all the members are made of the same R.H.S. tubes, G.R. 43, 40 mm x 40 mm, with different wall thicknesses (3.2 mm and 4.0 mm), as required. Some members are composed of 2 or 3 tubes of the same dimensions, and are welded together to meet the greater internal forces acting on them.

The whole structure is supported around its periphery by steel columns which form the structure of the outer walls, which are glazed with translucent polycarbonate panels. The slender steel columns made from 120 mm x 120 mm x 5 mm RHS members receive the vertical forces of the roof. The reinforced concrete abutments in the corners of the hall receive also the horizontal trusses of the whole space structure.

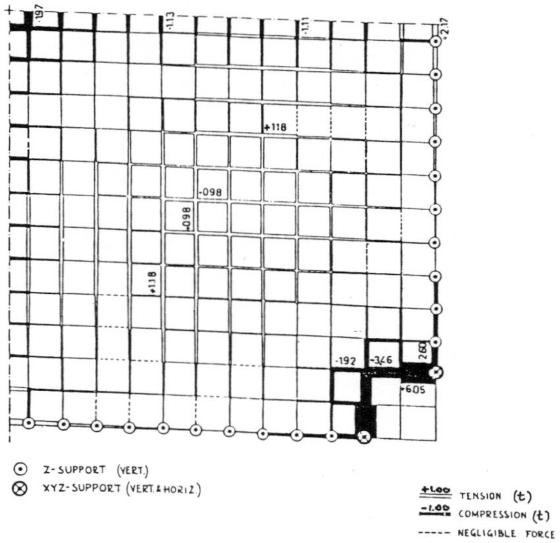

Fig. 3 - FORCES IN THE UPPER LAYER

THE ANALYSIS

The structural analysis was performed utilizing the STRUDL programme. The results obtained in this analysis corresponded quite well with the results obtained by means of approximate computations that were based on the assumption that the individual grids behave similarly to orthotropic plates, and the whole structure acts as a folded pyramidal space structure. The forces in the individual members are a superposition of the forces resulting from the bending of the flat grids and from the forces acting in the planes of the pyramidal space structure.

Figs. 3, 4 & 5 show the flow of the forces in the members of the two layers and the diagonals of a quarter of the structure, that result from a vertical loading of 100 kg. per sq. m. distributed uniformly over the whole roof. These diagrams show the results of the arrangement of the four flat grids into one space structure, namely a pyramid with truncated edges, supported on its periphery in the manner described above.

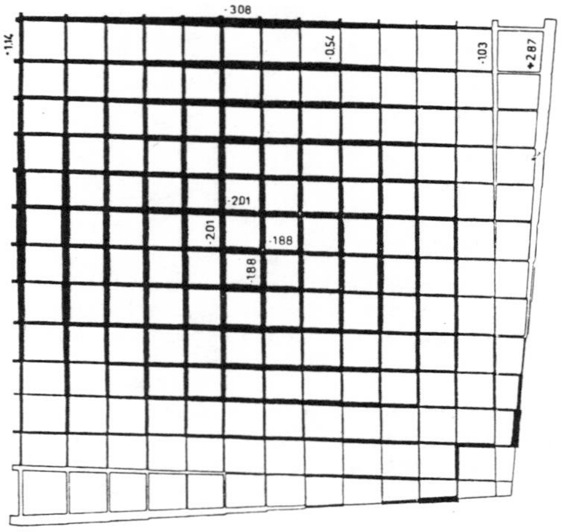

Fig. 4 - FORCES IN THE LOWER LAYER

The diagrams also demonstrate that the forces acting in many of the members are small and even insignificant. Taking this into account, and the fact that almost all the members have been executed by using the same minimal section (R.H.S.- 40 x 40 x 3.2 mm), a greater part of them remains not exploited to their full strength. Because of this, and because of the high number of redundant members, the whole structure has in fact a considerable safety reserve.

The deflections computed for the above-mentioned loading were :-

- the vertical deflection of the appex (Point "a" in Fig.1) - 30.6 mm.
- the horizontal deflection in the middle of the outer edge (Point "b" in Fig.1) - 6.7 mm - outwards.
- the absolute deflection in the centre of a panel (Point "c" in Fig.1) - 22.7 mm (corresponds to 7 mm relative to the panel itself).

These deflections are very small relative to the spans of the roof and the panels, and prove the great overall stiffness of the structure.

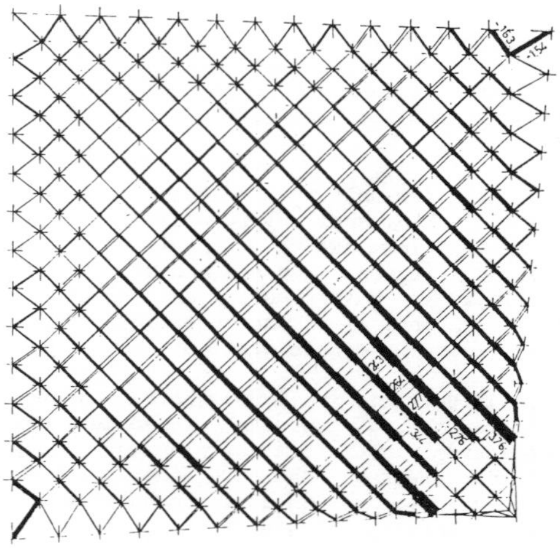

Fig. 5 - FORCES IN DIAGONALS

CONSTRUCTION

The four slightly rhombic grids of the pyramid were constructed in a workshop distant from the site of the building. The sizes of each panel (on the horizontal projection) of the roof were 15.0 m x 15.0 m plus the structure of the protruding canopy. All the connections were executed by means of welding.

For the purpose of transport to the building site, the four panels were cut to sections not wider than 3,5 metres. On the site the sections were welded together carefully to form again the four panels of the roof. The structure was sandblasted and painted on site before erection.

To facilitate the assembly of the roof, a temporary column in the centre of the building and also the central columns of the peripheral walls were erected. The four roof panels were hoisted to their position by means of a crane (See Figs. 6 & 7). The erection procedure took only a few hours. After the mounting of the roof panels all the other wall columns were inserted. The roof panels were welded to each other (Fig. 8) and to the supports, according to pre-designed welding details.

Site welding was used to avoid the consequences of possible inaccuracies, when using bolted joints, although it was found finally that the precision was fairly good in spite of the somewhat complicated geometry of the structure. The imperfections of the field welded joints were taken into account by means of their somewhat exaggerated dimensions, and careful supervision of their execution. After all the connections were completed the temporary column was removed. The whole structure, as it was at this stage, may be seen in Fig. 9.

The completed steel structure was covered with standard asbestos-cement corrugated sheets. These sheets have been connected directly to the structural members of the upper layer of the roof structure by hook bolts.

The glazing of the walls, the erection of sun shades, and the execution of other architectural details followed. The light modular accoustic panels were hung beneath the roof sheets by means of thin wires and between the rods of the upper layer. Thus the whole structure remained visible, and is a clear architectural feature of the building.

Fig. 10 shows the inside view of the building after completion, and Fig. 11 shows the view from the outside.

Fig. 6 - HOISTING A ROOF PANEL

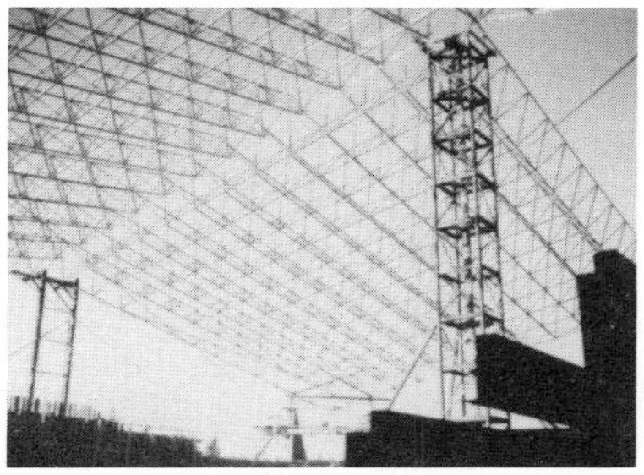

Fig. 7 - ROOF ASSEMBLY

Fig. 8 - CONNECTING THE PANELS

Fig. 9 - THE STEEL ROOF AFTER ERECTION

Fig. 10 - VIEW FROM INSIDE

Fig. 11 - VIEW FROM OUTSIDE

MEASUREMENTS OF STRAINS AND DEFLECTIONS

Measurement of strains in some critical members and deflections of some points of the structure were performed in the following sequence :-

(a) Removal of the temporary central column prior to the covering of the roof with asbestos-cement sheets :-
Strains in some critical members and vertical deflection of the appex of the structure and the horizontal deflection of the middle of the outer edge were recorded.

(b) Concentrated load (1.0 ton) hung from the appex of the roof :-
Vertical deflection of the appex and the horizontal deflection in the middle of the outer edge were recorded.

(c) Concentrated load (0.35 ton) hung in the middle of one of the four sides of the roof :-
Strains in the rods in the vicinity of the applied load, and in some rods in the unloaded panels, were recorded.

(d) Execution of asbestos-cement sheet roof cover:-
Strains in the rods connecting the structure to the corner supports were recorded.

The results of these measurements may be summarized as follows :-

1) The strains in the first case (a) have been found to be in quite good agreement with those resulting from the analyses, with deviations of about $\pm$ 10%.

2) The deflections in the same case (a) were very small; vertical deflection of the appex was - 12 mm, and the horizontal deflection in the middle of the outer edge was - 2 mm.

3) The vertical deflection of the appex in the second case (b) was - 3 mm. The horizontal deflection of the middle of the outer edge was less than 1 mm.

4) The maximum stress in the critical rod in the third case (c) was about 800 Kg/cm^2. The stresses in the members of the neighbouring panels were negligible.

5) The stresses found in the rods adjacent to the corner supports in the fourth case (d) were found to be slightly greater than those derived from the analysis, with a deviation of about 18%.

Generally, it seems that the measurements proved that :-

a) The recorded strains and deflections correspond quite closely with those which were anticipated by analysis.

b) The structure is able to withstand occasionally applied, or accidental concentrated loads.

c) The discrepancies found between the measurements and the analysis may result partly from the difficulties of measuring on site and from loading conditions that did not correspond exactly with those assumed in the analysis.

ACKNOWLEDGEMENTS

The author wishes to express his thanks to the following people :-

Michael Kuhn - the architect of the building.

Dr. Ari Adini - for assistance in analysis and strain measurements on site.

Amnon Yam - steel works contractor.

Ori Abramson, Assaf Ze'ev, and Mordechai Bogomolski - members of Kibbutz "Messilot" in charge of co-ordination and supervision of construction.

EXPANDABLE SPACE FRAME STRUCTURES

F. ESCRIG * Dr. Architect

* Profesor de la Escuela Técnica Superior de Arquitectura de Sevilla, SPAIN.

The possibility of combining deformable modules composed of crossed bars with spatial joints is studied in this paper to obtain expandale structures. Spatial grids able to be folded and expanded are classified into main families. Changes in the basic elements are introduced also to obtain curved grids, faced surfaces and archs. Graphics and pictures give us a clear idea of the spatial configuration of each model.

INTRODUCTION

We may consider conventional planar structures of bars as generated by cumulation of a set of basic elements with articulated struts that, by "combination" or "addition", close the required space.

Thus, a planar element as shown in Fig 1, "combined" in two directions (Fig 2), is able to cover a surface with a complex structure working as a whole (Fig 3).

The same applies if we "combine" an element as shown in Fig 4 in two directions (Fig 5) to follow the whole structure of Fig 6.

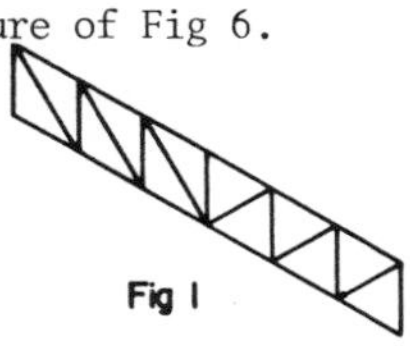

Fig I

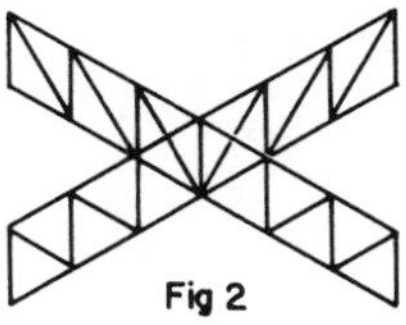

Fig 2

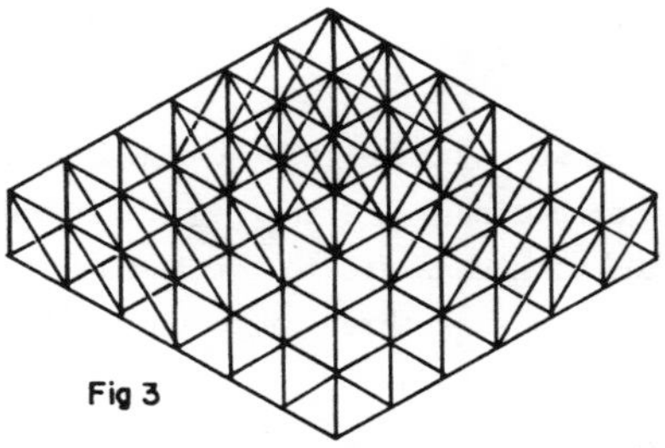

Fig 3

Elements from Fig 1 and 4 may be "combined" in three directions to give more rigid structures (Fig 7 and 8). We can change also these elements by others that increase topological possibilities and even introduce substantial modifications that allow us to generate almost all the more usual structures (Ref 3).

Instead of "combination" of planar linear structures we can use "addition of polyhedrals whose edges are struts with partially or totally triangulated faces. See Fig 9 to give Fig 3 or Fig 10 to give Fig 6 and other more, like Fig 11 and 12, to obtain less conventional structures.

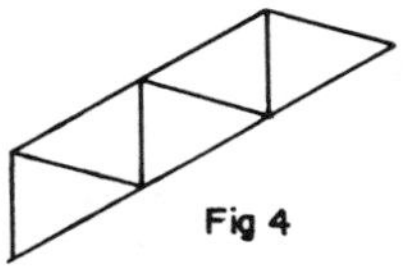

Fig 4

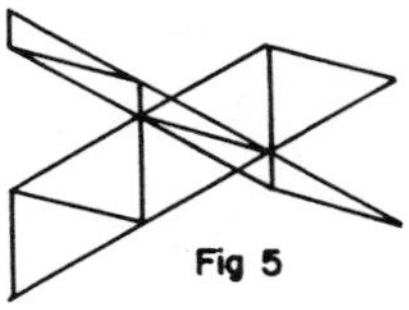

Fig 5

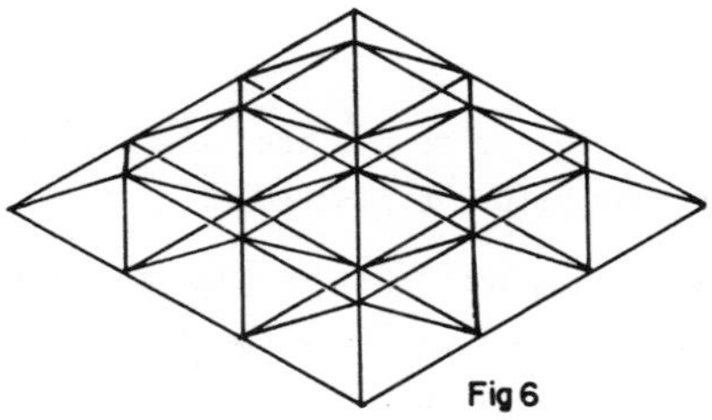

Fig 6

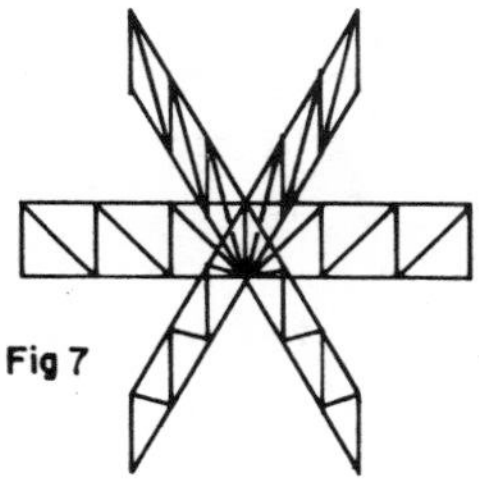

Fig 7

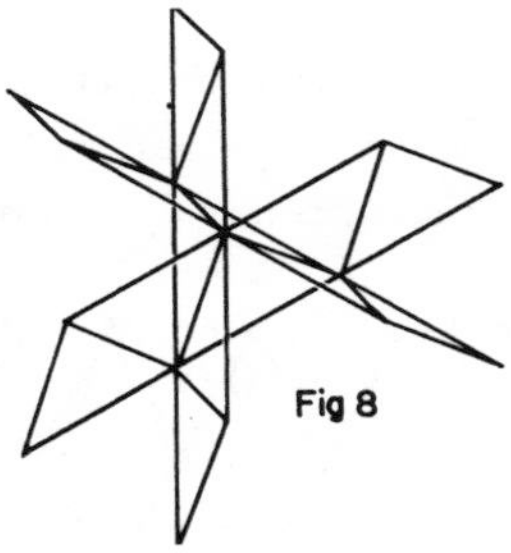

Fig 8

In all these cases we have begun from rigid elements to obtain highly hyperstatic systems by joining one element to another, even those elements not being rigid by themselves.

Some structures, as shown in Fig 3, need boundary conditions to control angular deformations in their plane. But, correctly designed, all these structures have an optimal resistent behaviour.

At this research, however, we have planed the possibility of finding unstable structures in which the degrees of freedom make them deformable and, consequently, able to change their position, area or form.

These mechanisms should be very useful if we could control their deformability and fix their position with some additional element stiffening the whole in the desired configuration.

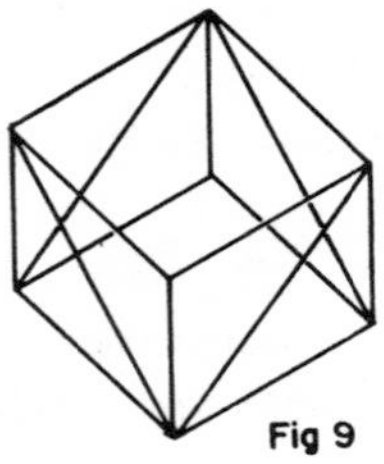

Fig 9

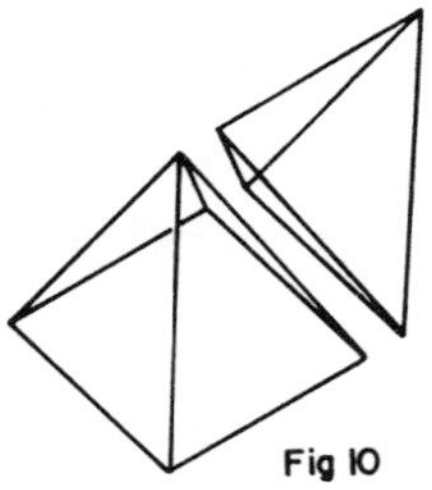

Fig 10

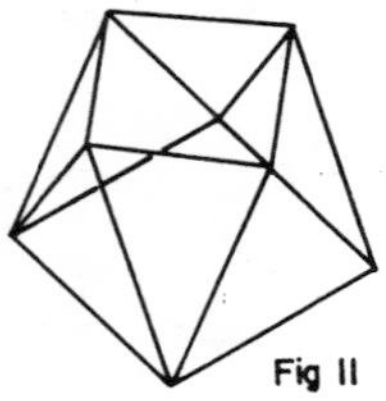

Fig 11

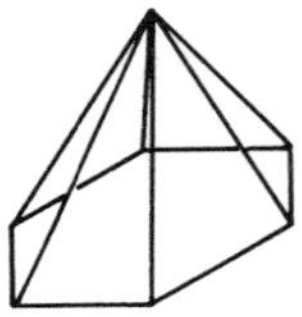

Fig 12

GENERATION OF STRUCTURAL SYSTEMS

To achieve that we are going to use the same "combination" and "addition" methods before explained.

First we start with a deformable truss like the one shown in Fig 13. This longitudinal planar element has not been used as structure because it has one degree of freedom which makes it unstable. Nevertheless, if height "h" is fixed, as shown in Fig 14, it is able to carry loads acting in its plan.

If "h" is variable with mechanical procedures "L" may change from little values to another arbitrarily high magnitudes accordingly with number of crosses.

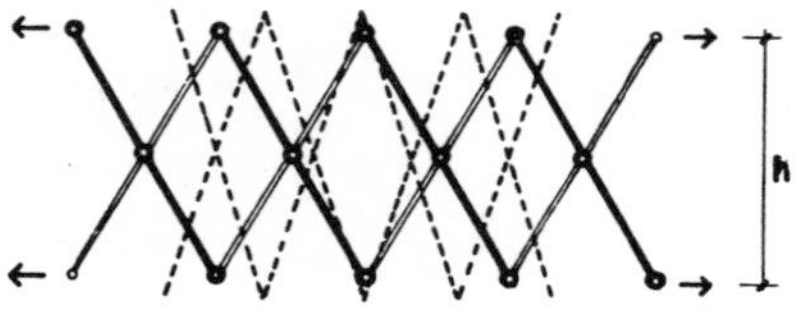

Fig 13

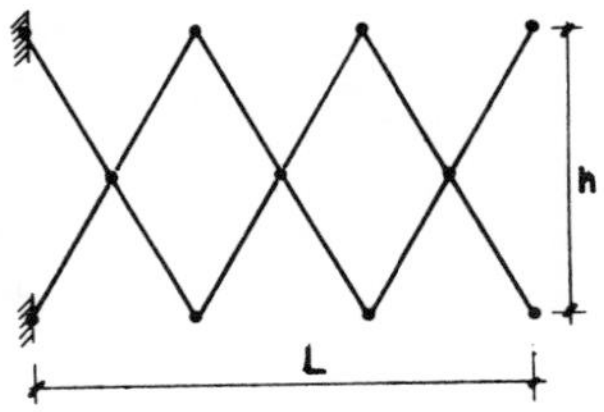

Fig 14

This element is isostatic if "h" is fixed and is not a triangulated structure, therefore its struts are working in flexion. But its calculus is easy and its utility satisfactory for light loads.

If we plan to "combine" several elements like these to get a whole like it is shown in Fig 3, we obtain the structure shown in Fig 15, in which all struts have the same lenght "l" and, therefore, there is geometrical compatibility to take different sizes in plan by changing "h" from minimum to "l" (Fig 16).

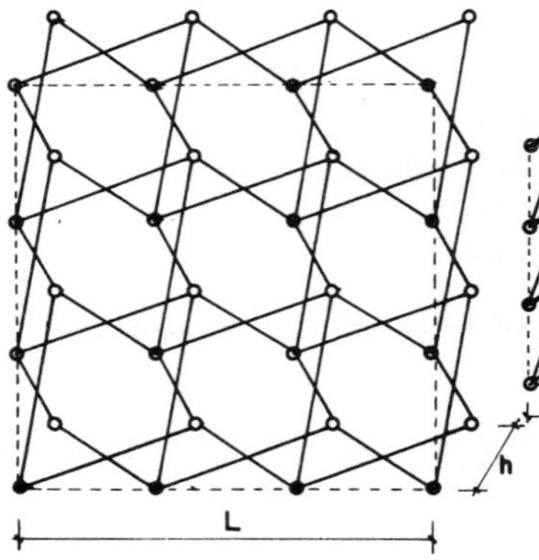

Fig 15

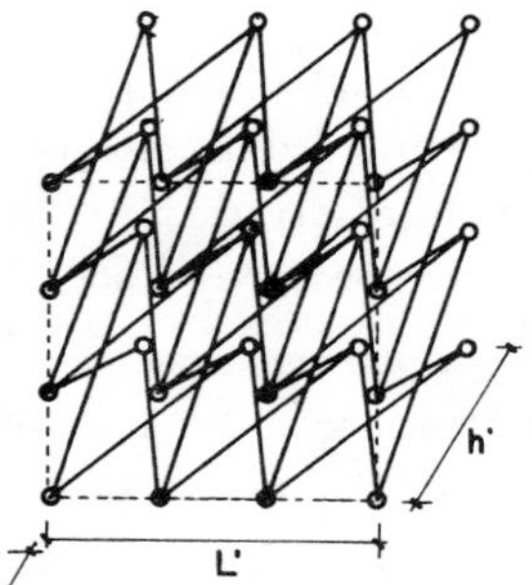

Fig 16

This structure, that we may label EXPANDABLE TYPE I, has the same rigidity conditions as the Fig 3 one and will be stable if completed with some aid to fix angular change in plan. Fig 17, 18 and 19 show three different states of the same model.

If basical Fig 13 is "combined" according to configuration shown in Fig 6, the visual aspect is more complex and less intuitive but it will be proved that in Fig 20 is obtained EXPANDABLE TYPE II. Fig 21, 22 and 23 show three different states of the model.

Fig 17

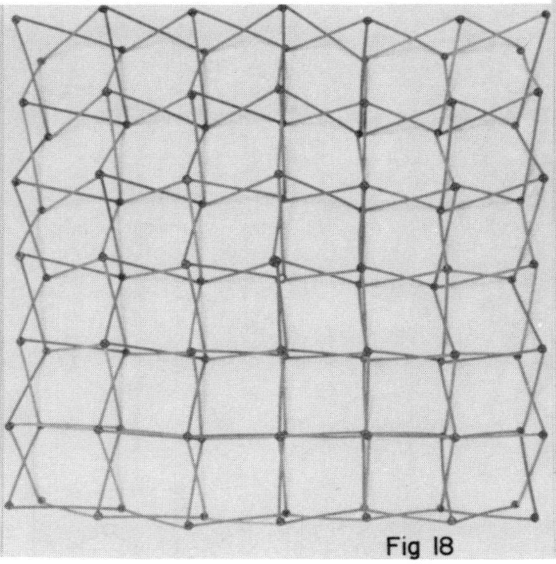

Fig 18

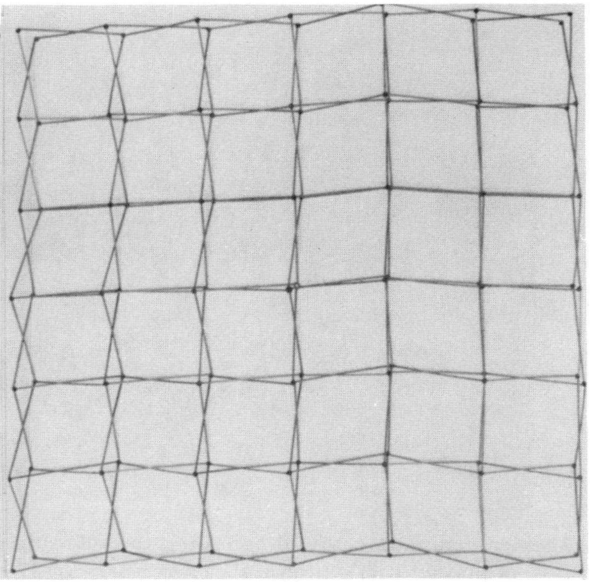

Fig 19

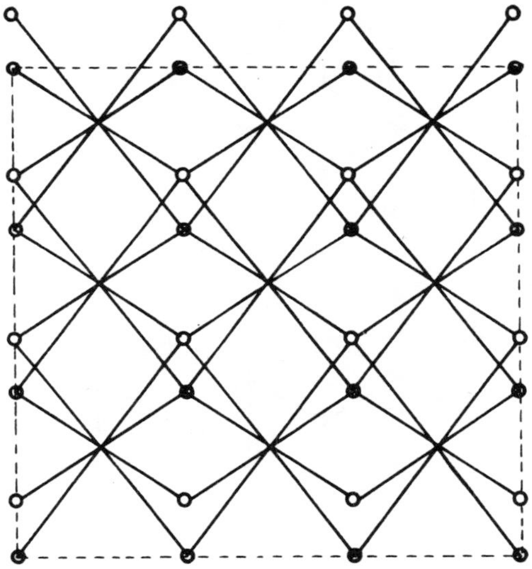

Fig 20

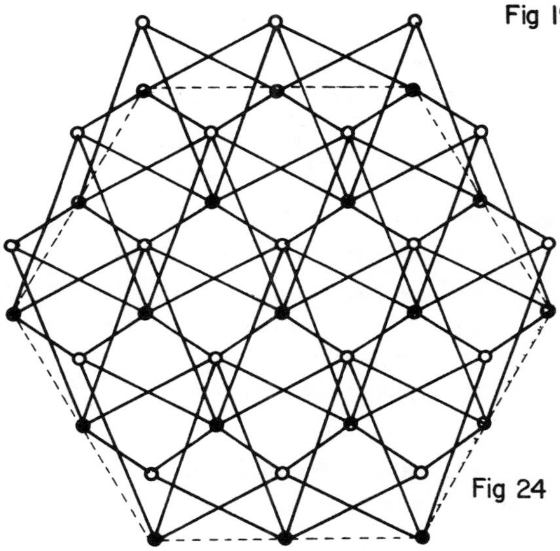

Fig 24

Fig 21

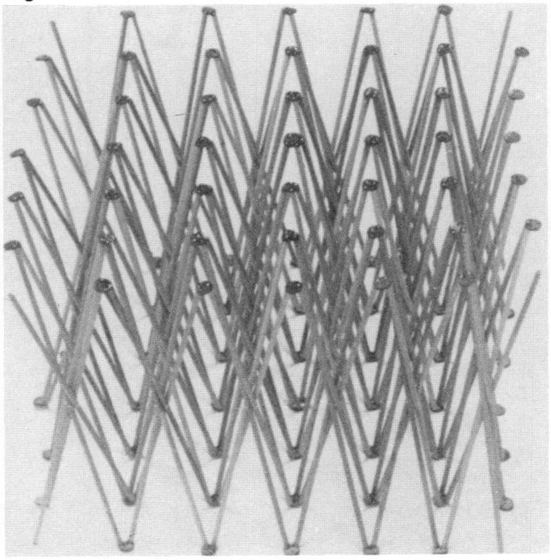

Fig 22

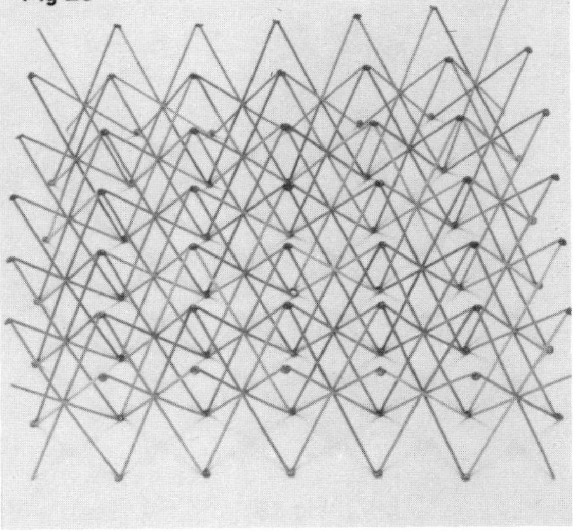

Fig 23

Fig 25

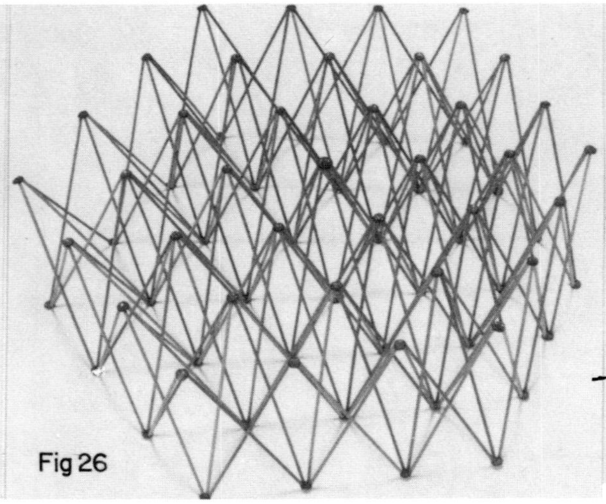

Fig 26

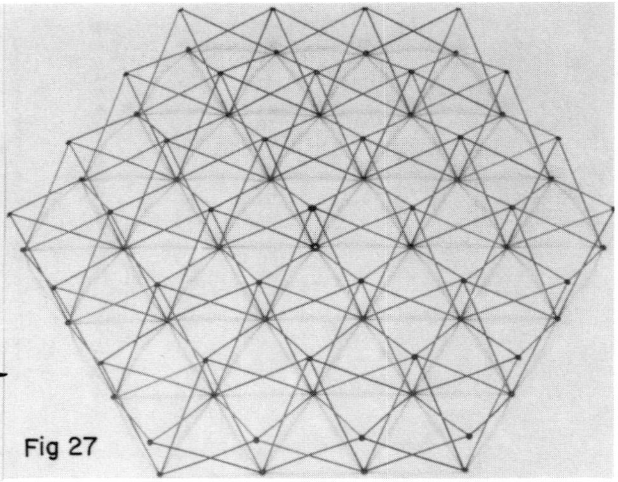

Fig 27

If basical truss is combined according to configuration shown in Fig 7, we obtain EXPANDABLE TYPE III (Fig 24) that is shown in several states at Fig 25, 26 and 27.

And if we combine basical truss in Fig 8 configuration, EXPANDABLE TYPE IV is obtained (Fig 28, 29 , 30 and 31).

These four basic types are the main but not the only ones possible.

If we use the "addition" method with elemental polyhedra, it is possible to generate the same structures shown above.

Thus, from an element like Fig 32 one, mechanism with one degree of freedom (Fig 33), EXPANDABLE TYPE I is obtained. The same for EXPANDABLE TYPE II from Fig 34, EXPANDABLE TYPE III from Fig 35 and EXPANDABLE TYPE IV from Fig 36.

This generating way is more intuitive and visualizes better space modulation, not only for roofs but for any structure to be solved with expandable grids also.

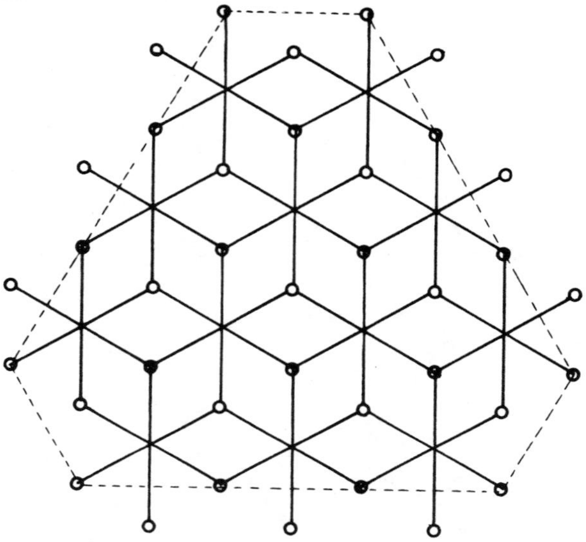

Fig 28

Fig 29

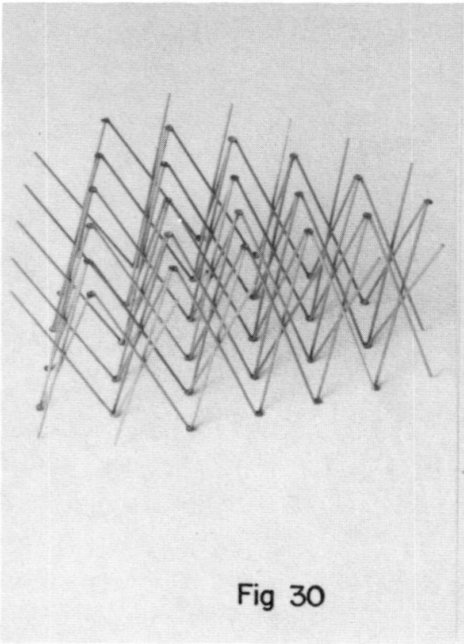

Fig 30

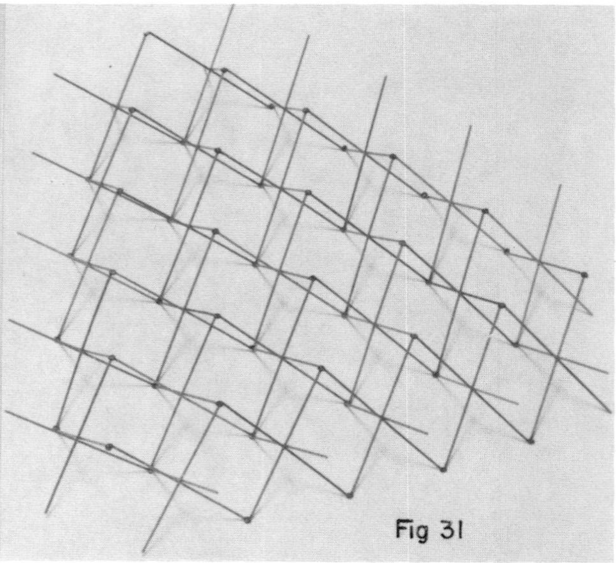

Fig 31

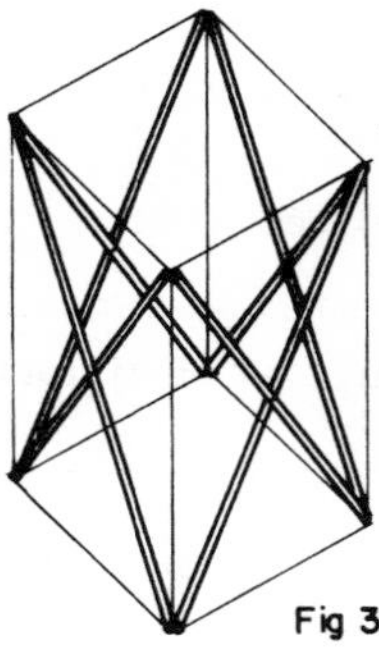

Fig 32

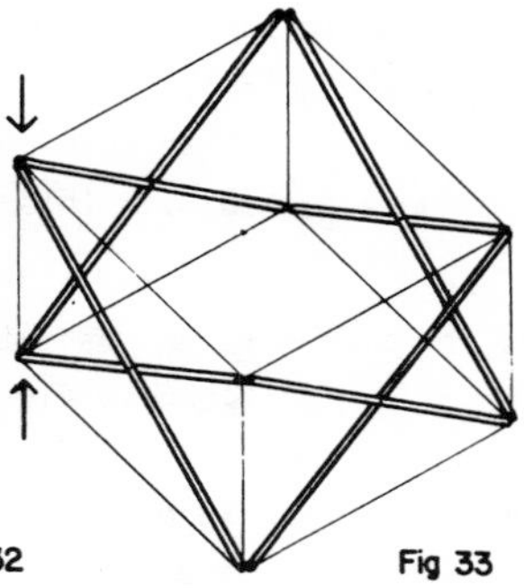

Fig 33

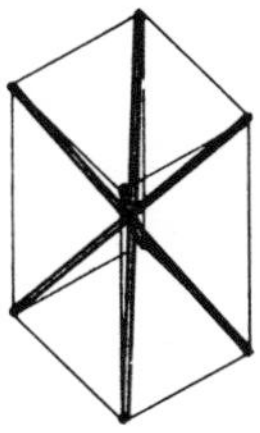

Fig 34

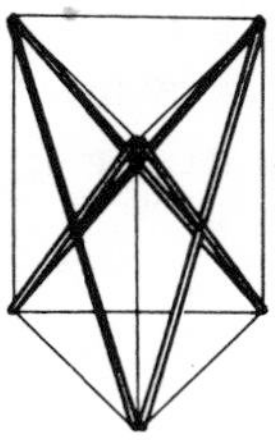

Fig 35

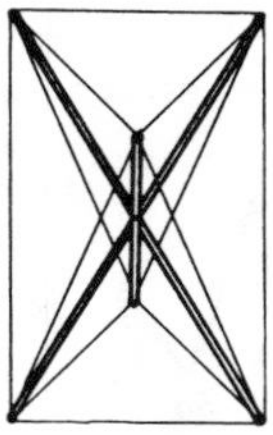

Fig 36

As we have seen, EXPANDABLE TYPE I is solved with planar joints in the half of struts and joints in two perpendicular planes at the end of struts. EXPANDABLE TYPE II has all joints in two perpendicular planes. EXPANDABLE TYPE III is symilar to TYPE I but with end joints in three 120°planes. EXPANDABLE TYPE IV has all joints with three struts moving in 120°planes.

Among all these proposed solutions TYPE III is the more rigid while TYPE IV is the less. TYPE I is the more efficient if angular deformation is controled.

On the top of these structures we can extend a waterproof skin fixed to the joints and obtain useful surfaces as walls, roofs or envelopes.

If, additionally, an other skin is fixed at the low joints, we have a double layer with optimal thermal behaviour (Fig 37).

Much more complex modulus may be proposed by increasing the number of bars crossing at each central node but constructional solutions are more difficult and not efficient.

Fig 37

Thus, structural adequacy artistical quality and architectonic functionality are obtained in a whole.

Constructional details take first place in the correct ability of these structures. So, the articulations and the means to extend and fold the system do need a careful design. But neither construction, nor analysis approach, are the subject of this paper.

EXTENSION OF THE PROCESS

Finally, as introduction to a more general research, we explain briefly some changes that may be made on the above TYPES to solve other set of expandable structures.

a) If, instead of basic elements shown in Fig 13, 32, 34 or 36, we use symilar others with the cross point excentrically placed (Fig 38,39, 40, 41 and 42), the structures will expand like a fan, generating curved surfaces. The elements used will not be regular necesarily and may be distorted.

Fig 38

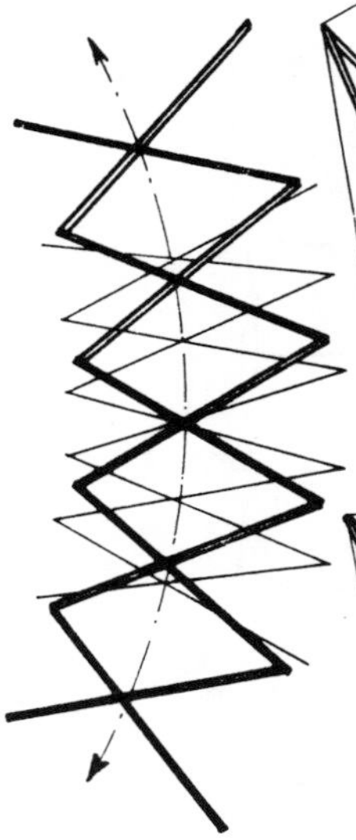

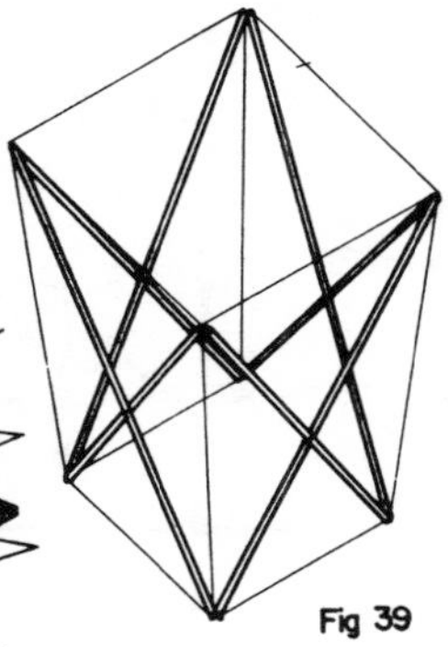

Fig 39

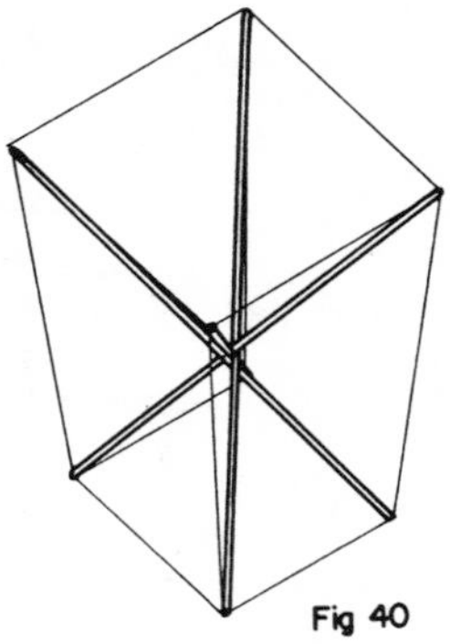

Fig 40

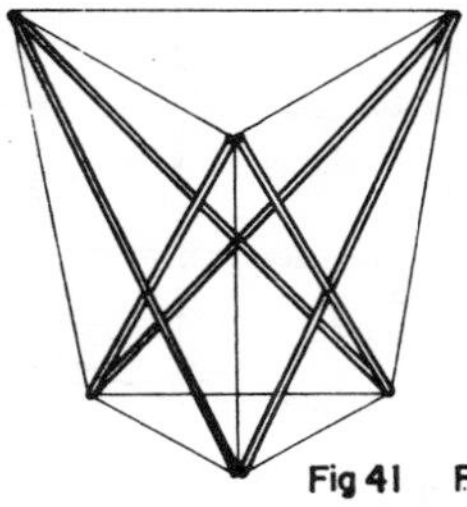

Fig 41

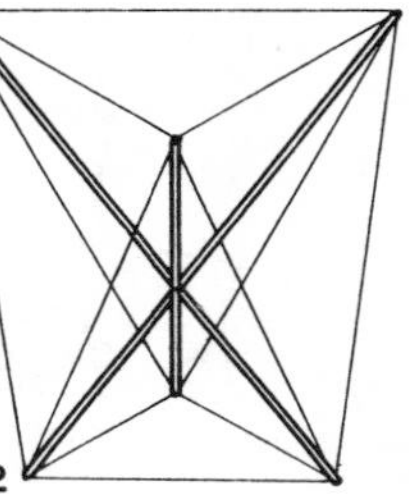

Fig 42

In this way we can obtain spatial grids of bars with simple and double curvature to build shell-like structures.

Cylindrical grids are elementary as shown in Fig 43. Other with double curvature are in close dependency of compatibility between struts length and joints spatial location and, as stimulant exercice, it gives the brightest solutions. Remember the well known Perez Piñero's dome which use EXPANDABLE TYPE IV (Ref 2). In Fig 44 an application of EXPANDABLE TYPE I is considered.

b) With bars of different length is also possible to build some trusses like the ones shown in Fig 45 which, well conected between them and taking account of their deformation compatibility, origine curious forms.

c) Combining between them Fig 32, 34, 35, 36, 39, 40, 41,42 and their irregular variations, we can obtain spatial enclosures with plane and curved faces. See a simple example in Fig 46.

Fig 43

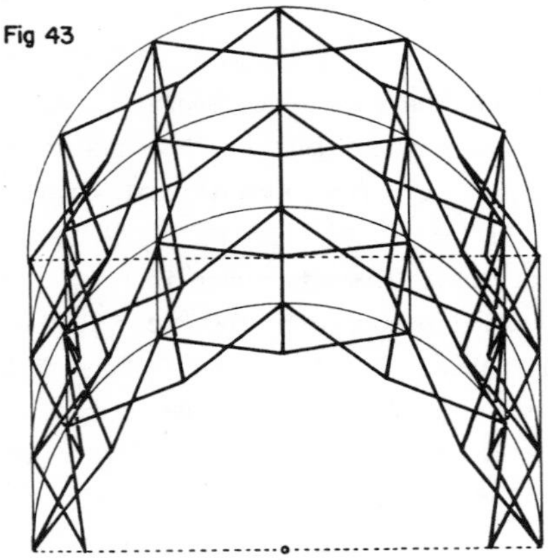

Fig 44

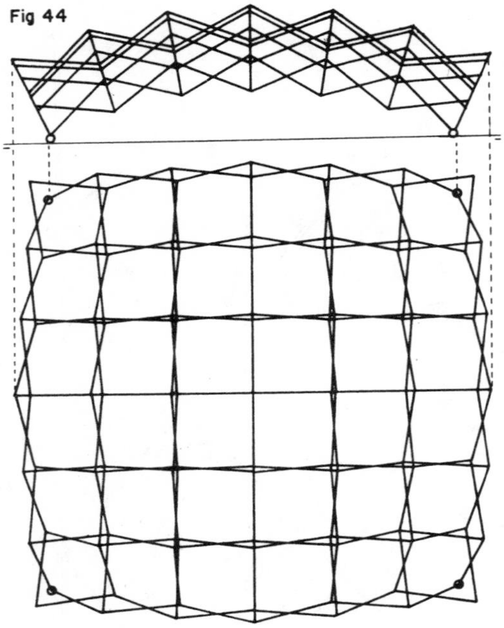

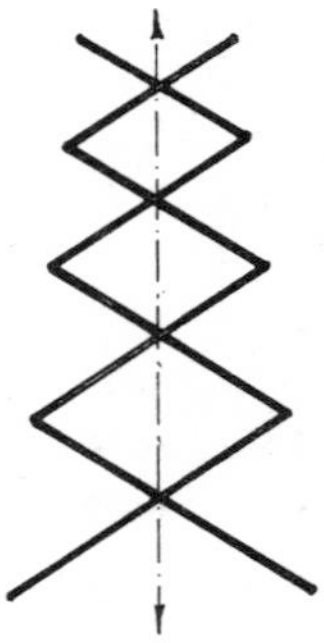

Fig 45

Fig 46

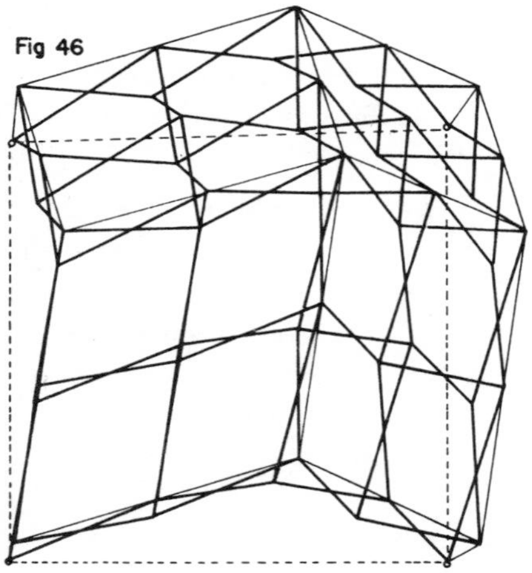

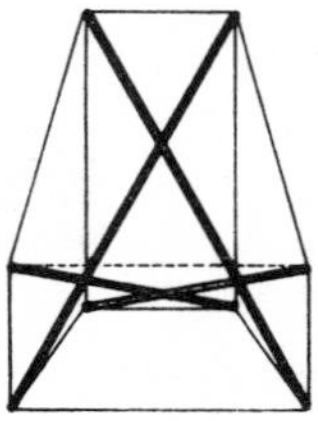

Fig 49

Fig 50

d) All elements explained before, placed longitudinally instead of superficially, origine expandable towers and archs like the ones shown in Fig 47 and 48 and, therefore, nets formed with this linear systems.

It is of great interest, in arches and anticlastic surfaces, to consider the possibilities of distorted variation of fig 31 shown in Fig 49.

e) More complex trusses than the ones shown in Fig 13 can be used (Fig 50) but spatial joints are very complicated in some cases like EXPANDABLE TYPE III. This complex grid may be recomended for EXPANDABLE TYPE IV in which there is not additional difficulty at joints.

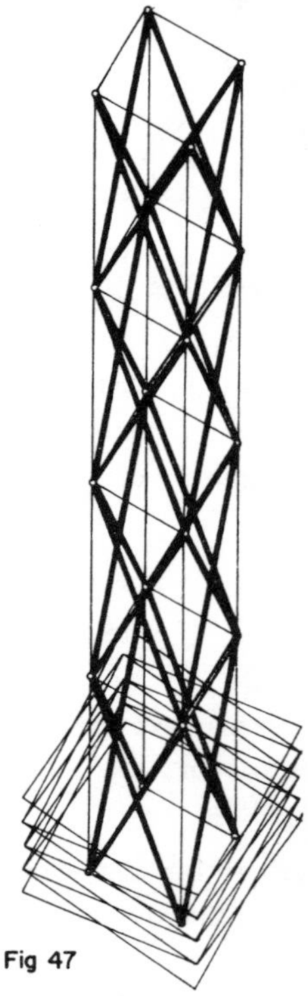

Fig 47

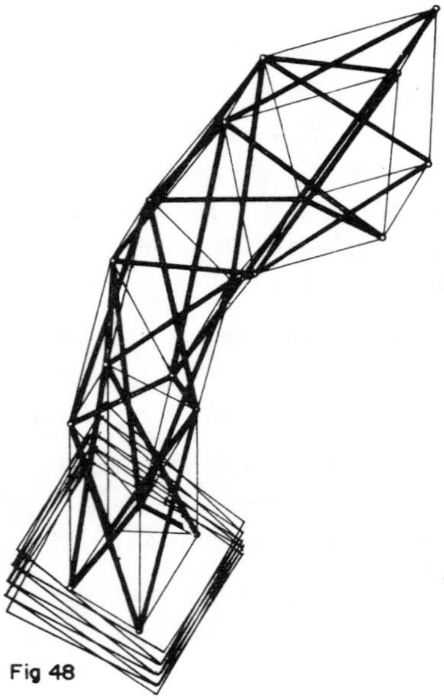

Fig 48

ACKNOWLEGEMENTS

The autor wishes to express their thanks to INSTITUTO UNIVERSITARIO DE CIENCIAS DE LA CONSTRUCCION, ESCUELA DE ARQUITECTURA DE SEVILLA and Professor LOPEZ PALANCO for their assistance and confidence.

REFERENCES

1. FULLER, R.B. "Synergetics" Mc. Millan. N.Y. 1975.
2. PEREZ PIÑERO, E."Materia-Estructura-Forma".Hogar y Arquitectura, n°40, pp 25-30. MADRID 1962.
3. TSUBOI, Y. "Analysis, Design & Construction of Space Frames" 1979 IASS Working Group Spatial Steel Structures.

COMBINING SPACE FRAME AND FABRIC STRUCTURES

MARK A. BESH

Space Frame Section
UNISTRUT Division of GTE Products Corporation
Wayne, Michigan U.S.A.

To reduce the impact of the structural loads imposed onto an existing structure, and provide a naturally illuminated environment, the combination of a double layer lattice grid space frame and translucent Fiberglas 'tent-like' fabric modules was pioneered. This paper presents the development of the concept, design and analysis of the components, discussion of the details and prototype, and the unique features and advantages of the synergy.

PROJECT REQUIREMENTS

1) Enclose an existing outdoor swimming pool to provide year round use.
2) Provide a structure that does not require the existing surrounding structure to support additional loads.
3) Provide minimal visual obstruction of pool-side viewing.
4) Architectural motif must be harmonious with the existing structures.
5) Provide some sort of shading to reduce cooling requirements, while allowing natural light to penetrate.

THE SOLUTION

The MERO[1] Space Frame system from Unistrut combined with the Teflon coated Fiberglas fabric modules from Owens-Corning Fiberglas solves these demands, while offering the following unique features of each systems.

Space Frame Features:

1) A very lightweight structure with the capability of long span and random support locations.
2) Pre-engineered and fully proven system.
3) Pre-fabricated to exacting tolerances and pre-finished with a very durable coating.
4) Only two components, a node(connector ball) and a member (steel tube), to facilitate simplicity of erection.
5) Since the integrity of the structure is not dependent on one single member, an inherent redundancy increases the safety factor.

Fabric Structure Features:

1) Translucency of fabric provides either about 6% or 13% (depending on type) of the available natural light to penetrate.
2) An ultra-lightweight material that is pound-for-pound stronger than steel.
3) Can withstand temperatures of 1500 degrees Fahrenheit, and is not affected by cold or the ultraviolet rays of the sun.
4) Chemically inert coating of Teflon is unaffected by almost all chemicals and solvents, and will not deteriorate with age.
5) The fabric's high reflectivity(approx. 70%) minimizes heat gain within the structure.
6) Pre-engineered and pre-fabricated offer faster construction schedules.

Combining the two technologies offers tremendous design freedom by allowing random location of supports, varying geometrical configurations, long spans, light weight, and a naturally lit environment of uncommon dramatic appeal within a totally functional context.

DEVELOPMENT

The following interfacing items required solutions that were sensitive to both structural and aesthetic concerns.

1) Attachment of the fabric to the space frame structure, and how would it resist the tremendous pulling force (45#-63# per inch) of the pre-stressed fabric.
2) How to incorporate some sort of jacking system, such that the fabric could be pre-stressed into final form on site.
3) How to allow the peak of the pyramidal-shaped fabric 'tent' to move in all directions(wind), while resisting the compression forces being applied downward(not necessarily perpendictular to the space frame) and resolving them within the space frame.
4) How to connect the ridge cable required at each corner edge of the 'tent' to the space frame.
5) Constantly keeping in mind that all joints were to be waterproof under the combination of the practically stable interior environment and the ever changing exterior conditions.
6) The space frame structure would have to be cambered enabling positive drainage under full design load.
7) Due to the fact that the columns would be supplied by others, an adjustable connection between the column and a space frame bracing member would be required.

The following sketches show how the two systems were interlaced, and the magnitude of the entire project.

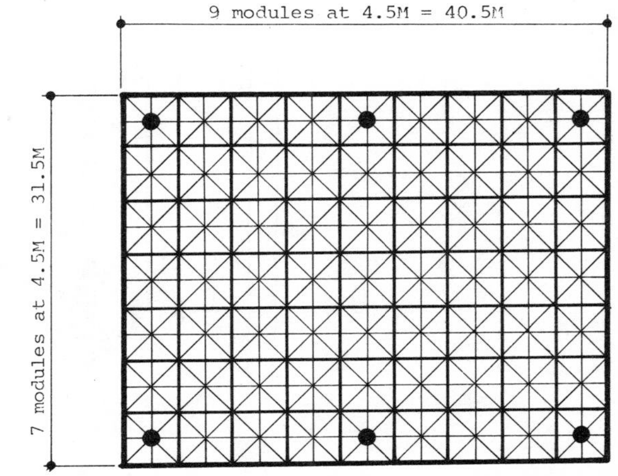

Figure 1. Plan view of entire space frame structure.

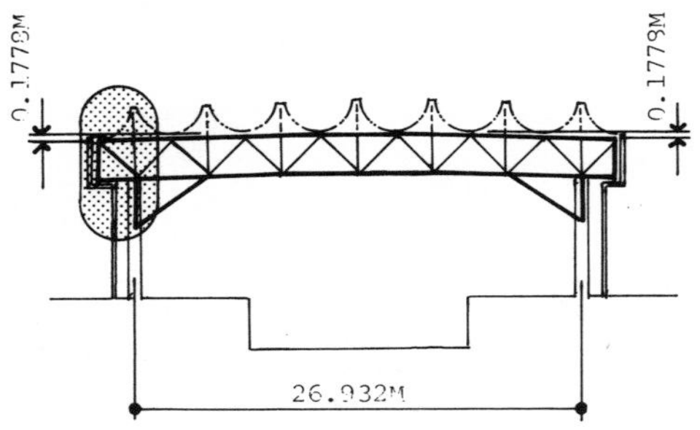

Figure 2. Section thru entire frame showing camber.

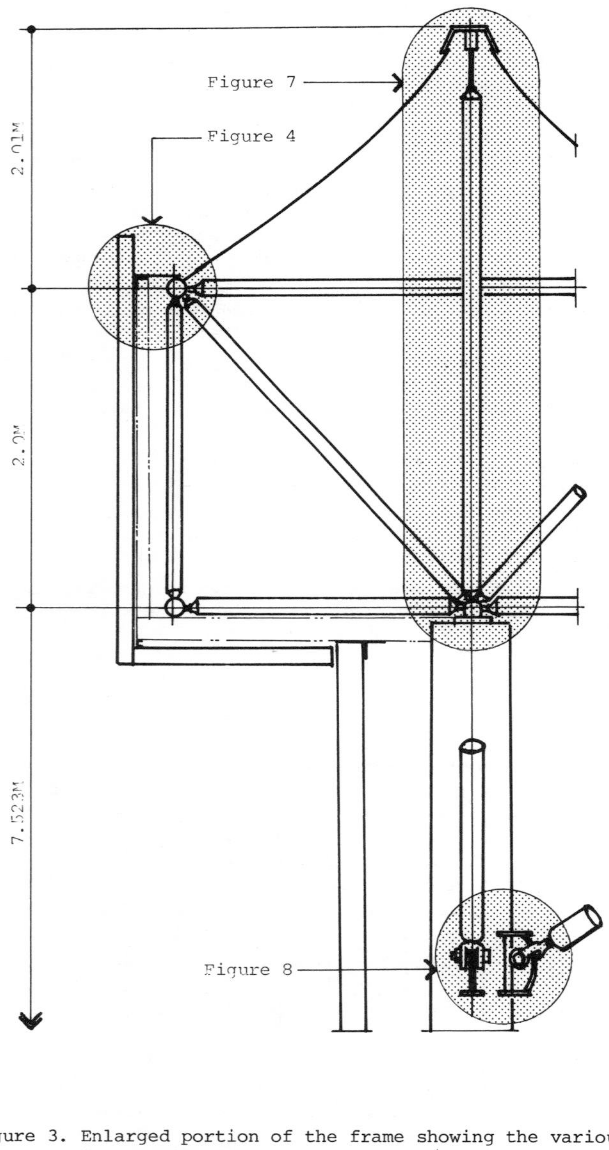

Figure 3. Enlarged portion of the frame showing the various interfacing methods.

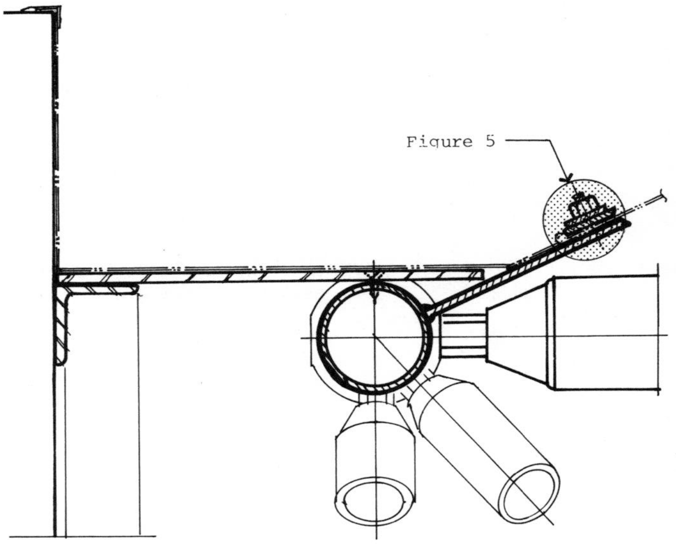

Figure 4. Enlarged detail of connection of fabric directly to MERO member. Notice how horizontal plate creates bottom of gutter.

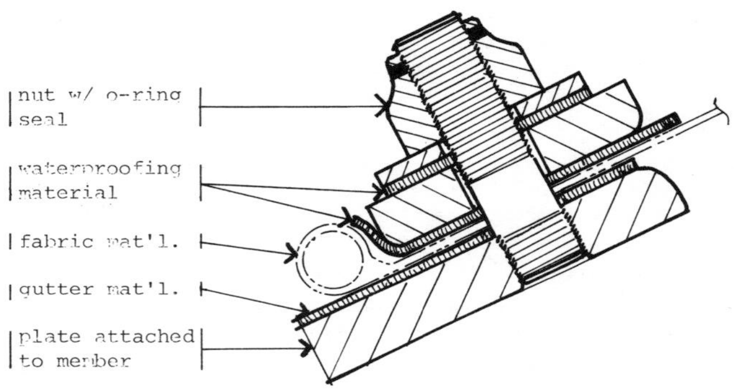

Figure 5. View of fabric clamping and waterproofing detail.

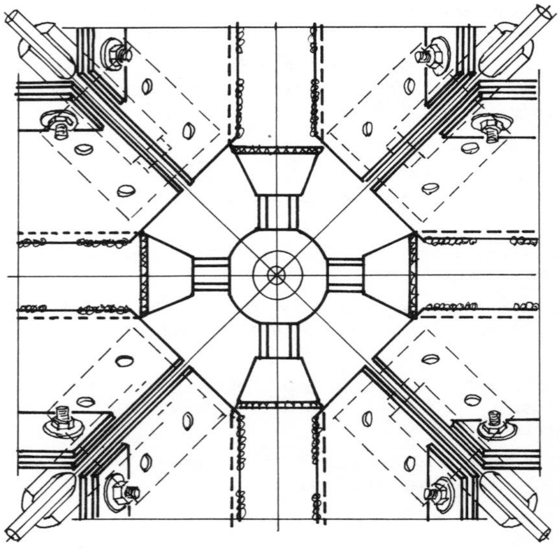

Figure 6. Plan view of a portion of a typ. joint where four fabric modules meet. Notice cable attachment and how, after field bolted, will create a rigid corner.

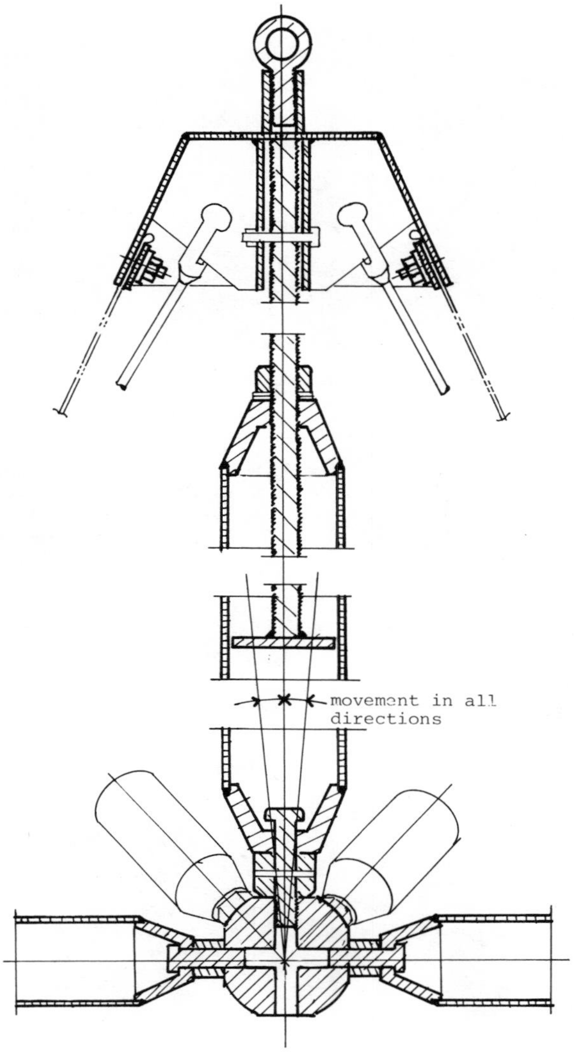

Figure 7. Combination vertical support and fabric pre-tensioning post member.

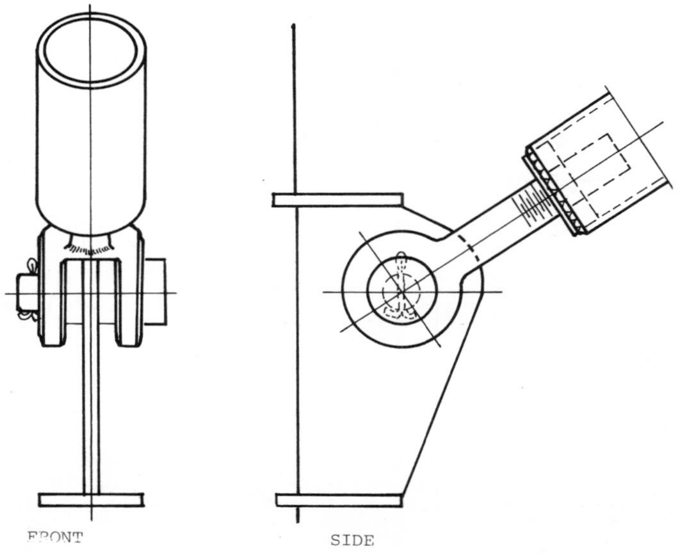

Figure 8. Adjustable turnbuckle for column to space frame bracing member connection.

STRUCTURAL DESIGN

To transform this concept into reality, analysis of forces and sizing of components was required. The fabric modules were analyzed first, by Owens-Corning, inputting assumed connection restraints of the developed details into a finite elements method computer program. The analysis incorporates a membrane element which accomodates an orthotropic material model, and is presumed to account for the non-linear, large-deformation phenomena in the structural response.

These reactions were then used as a separate loading case within Unistrut's MEROUSA[2] program. This analysis utilizes an internal stiffness matrix solution to apply the proper external loading to the internally formed geometry, and select the appropriate members to safely resist the implied conditions. The nodes are then selected based on the size of adjacent members and the angle between these members, such that there is sufficient clearance between all components of the members, as shown in Fig. 9 below.

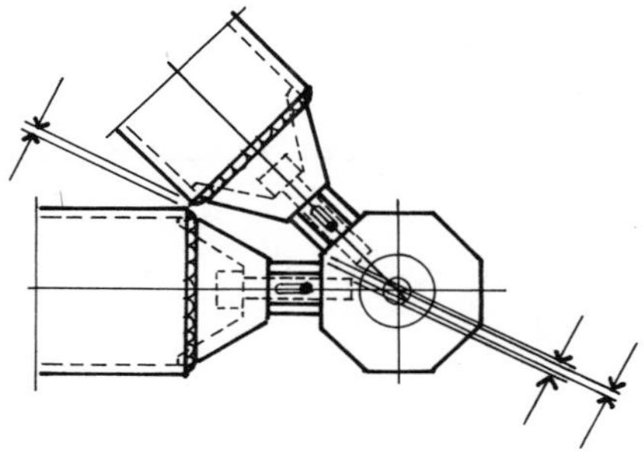

Figure 9. Locations of clearance checks.

The output from the analysis then had to be refined. Due to the tremendous pulling force of the fabric on the MERO member, the connecting bolts at each end of the member had to be increased to safely withstand the shear force. Some of the top chord members were able to be reduced, since the fabric attachment plate is welded to it creating a much stronger composite section. The top chord nodes were then all made the same size, such that all the top chords would be the same length, reducing the quantity of different parts. Final analysis and component sizes were then sent to Owens-Corning for their final coordination.

Although this work was being done for a particular project, it was apparent that this synergy of systems could be used for other applications. One concern, though, was that when the fabric module size gets larger, so do the downward forces generated upon the vertical post member(Fig. 7). This then required a stronger moveable connection to the space frame. A tie-rod end ball joint and clevis combination(Fig. 10) was pioneered.

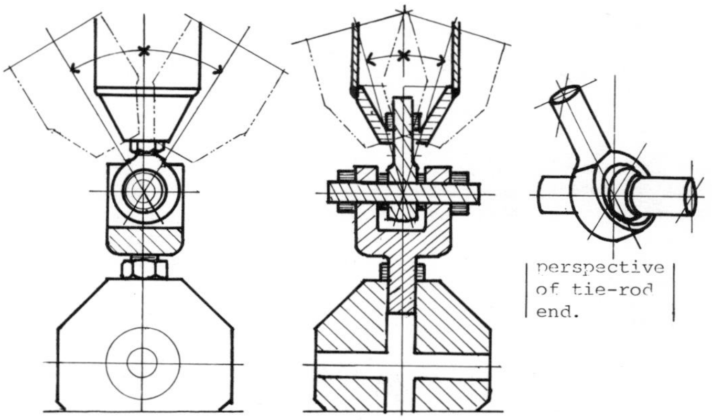

Figure 10. Motion transmission attachment using tie-rod end.

PROTOTYPE

To prove the simplicity of erection, check the fit of the components, devise procedures for application of waterproofing and gutter materials, and utilize a larger fabric module, a four module space frame(one fabric module) was fabricated(Fig. 11). Notice how the vertical support connects to the top chord node in this concept.

Field testing of the prototype unit performed admirably, without any unforeseen situations and no water leakage. Shown in Figs. 12, 13, 14, and 15 are photos of the completed unit and associated detail views.

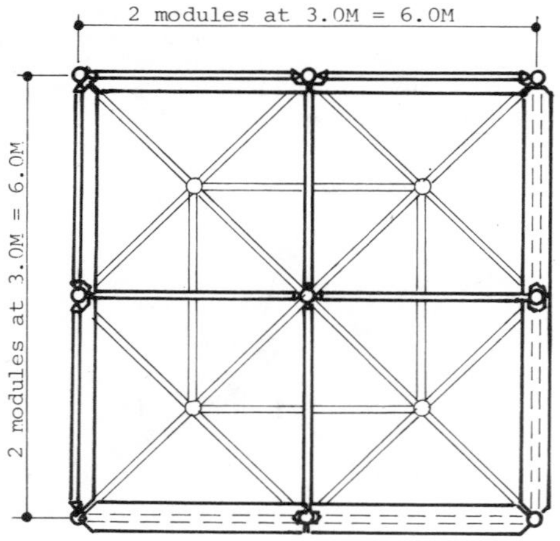

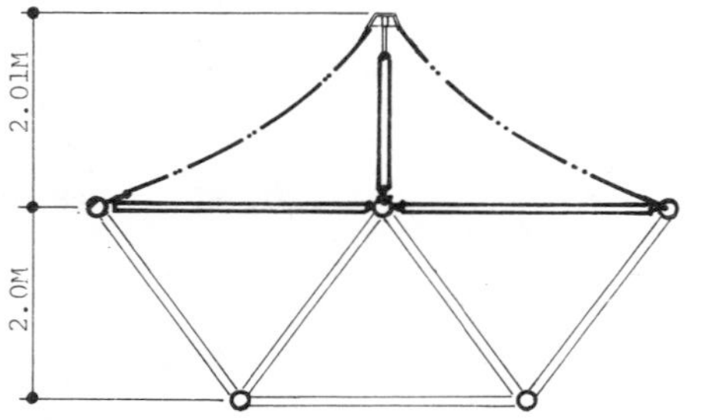

Figure 11. Drawings of prototype.

Figure 12. Overall view of completed prototype unit.

Figure 13. Enlarged view of completed peak cap assembly.

Figure 14. Typical corner conn. with ridge cable bracket.

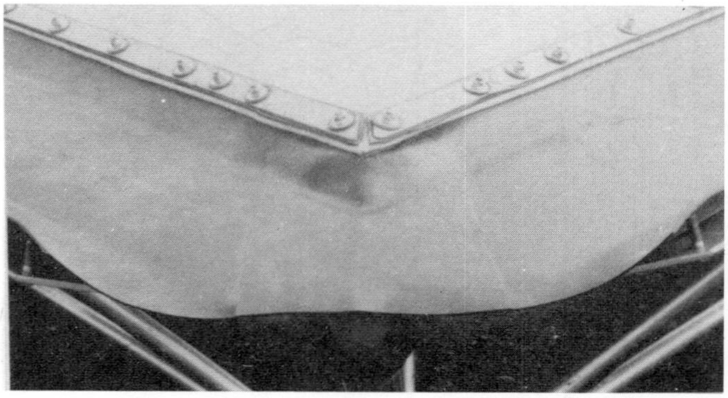

Figure 15. Typ. corner completely sealed and ready for tests.

It is hoped that this paper will generate ideas for other concepts and combinations of emerging technologies, to utilize the space frame concept in more economical and creative manners-eliminating current perceptions and inhibitions enabling the development of innovative building systems for the future.

ACKNOWLEDGEMENTS

I wish to thank all involved at Unistrut to make this concept a reality, especially to Sun Chien Hsiao for his constructive suggestions. Many thanks to all within the Fabric Structures Division at Owens-Corning Fiberglas, Toledo, Ohio, for the professional manner in which they handled our pioneering partnership. To our licensor, MERO-Raumstruktur Gmbh & Co., Wurzburg, West Germany, for offering their tremendous expertise.

1. MERO is a trademark for a space frame system developed by MERO-Raumstruktur Gmbh & Co.
2. MEROUSA is a CAD/CAM program by Unistrut for the analysis and fabrication of the MERO space frame system.

ERECTION, INSTRUMENTATION AND MANAGEMENT OF INDRAPRASTHA STADIUM DOME, NEW DELHI

P RAY-CHAUDHURI*, Ph.D and V P CHETAL**, FIE

* Head, Bridges Division,
Central Road Research Institute,
New Delhi, India.

** Chief Engineer (Civil),
New Delhi Municipal Committee,
New Delhi, India.

For hosting IX Asian Games in India in 1982, an air-conditioned indoor stadium of 25,000 capacity was constructed at Indraprastha Estate, New Delhi. The 150 m diameter stadium has a steel dome resting on eight 42 m high reinforced concrete towers. The unique steel dome, largest in Asia and Europe, having its apex at 41.5 m presented problems during its design and erection. The stadium, a part of a gigantic sports complex was built in record time. Salient features of the dome, its erection and decentering along with the associated instrumentation have been presented in the paper in addition to the construction management techniques adopted.

INTRODUCTION

The IX Asian Games (ASIAD 1982) was held in Delhi, India during November, 1982. For holding the Games, a series of modern sports venues were constructed in about two year's time presenting innumerable challanges to designers, builders and engineers. The undoubted showpiece of such modern sports venues is the complex at Indraprastha Estate which sprawls over 44 hectares having six structures of sizeable magnitude including the 25,000 capacity steel roofed indoor stadium. The stadium has a playing arena 55 m x 75 m with 5 m wide circulation space all around. The 150 m diameter stadium having the largest steel dome in Asia and Europe accommodates 25,000 spectators with unobstructed view of the entire interior space including rest of the gathering.

The steel dome covering the stadium rests on eight reinforced concrete towers along the periphery. The towers have several floors which house various services required for the stadium.

The Delhi Development Authority were vested with the assignment of constructing in Indraprastha Complex which included the cycling velodrome also. All the structures of the complex were massive but apart from the stadium and the velodrome, the design and construction techniques for other structure were familiar to those concerned. The indoor stadium, presented some special interesting features both with regard to design and construction.

This paper deals with some details of erection and decentering of the steel dome of the stadium along with the associated instrumentation. It also represents the salient features of the dome and the appropriate facets of construction management techniques. The instrumentation was done by the Central Road Research Institute, India.

SALIENT FEATURES OF DOME

The steel dome covering the indoor stadium rests on eight reinforced concrete towers having on octagonal shape in plan. Fig.1 provides a part-plan of the roof dome.

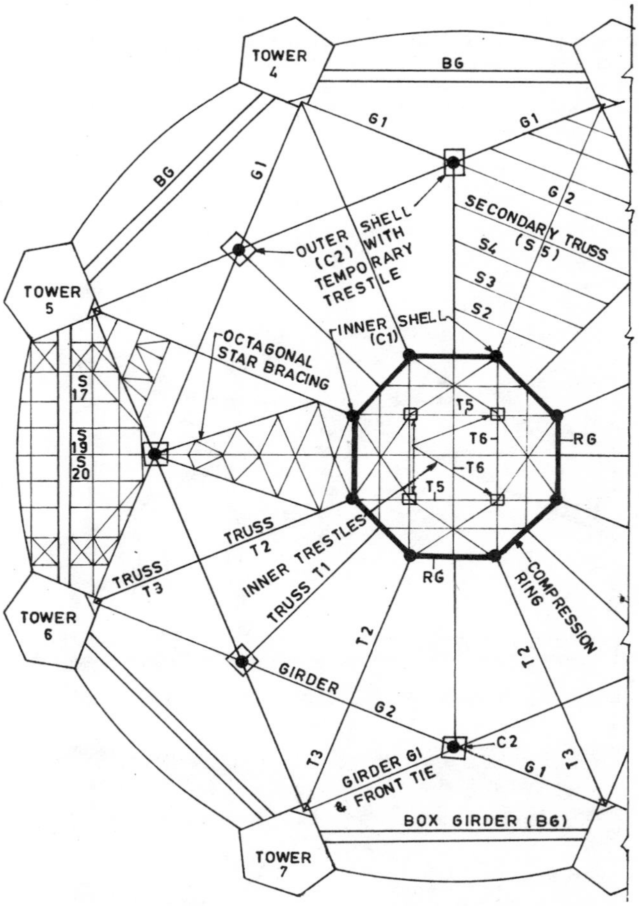

Fig. 1

The whole roof structure can be broadly divided in three parts namely:

1. Octagonal central compression ring.
2. Trusses on segmental areas between the towers.
3. Roofing and insulation.

The compression ring, which is the crown portion of the dome, consists of several latticed trusses spanning inside an octagon approximately 37.5 m face to face. The topmost position of upper chords has an elevation of 41.5 m tapering down to 37.85 m at the octagonal periphery. The soffit of bottom chords of the trusses of the compression ring have an identical elevation of 33.85 m.

Sixteen trusses fan out radially from the compression ring towards the segmental area between the towers. All the eight apex points of the compression ring are connected to truss T2 through a site-joint which in turn is connected to truss T3 which terminates at the hinge bearing housed in the RCC tower at an elevation of 23 m. The mid-points of all the eight faces are connected to trusses T1 which terminate at the outer shells C2 (Fig.1).

The segmental area between two towers and the compression ring has girders G1 and G2 in addition to the trusses T1 or T2 mentioned earlier. On outer face of the segmental area, peripherial box girder BG interconnect two adjacent towers. At the vertical projection of G1, a set of front ties made of high tensile steel plates, connects the topmost portion of the outer shell to saddles placed on towers at an elevation of 42.0 m. In order to minimise bending stresses in the HTS plates of the front tie, a central hinge has been provided. The saddle is held back by a pair of back-ties which are anchored to the main towers at 29 m level. A tensioning device was incorporated in each front tie near its junction to the outer shell to provide uniform tension prior to decentering.

Fig 2

Standard rolled steel sections and plates were used for fabrication of primary and secondary trusses. Some selected secondary trusses were fixed in position to provide stability against buckling of compression members of some of the primary trusses. All the site joints were made with high tensile bolts. A decision was also taken to design all the joints to carry twice the initial design forces so that strengthening of any member should not required any change in the design of joints.

The roofing material is corrugated aluminium sheets which rest on cold formed galvanised steel sections of Σ shape. Special anticorrosive paint treatment was provided at junctions of dissimilar metals. A large amount of insulating material both for thermal and acoustical purposes were provided on the underside of roofing material.

ERECTION OF STEEL DOME

In order to facilitate erection, twelve temporary supports were provided. For the sixteen trusses radiating out from the compression ring, eight T2 trusses rested on the bearings at the towers. For the eight T1 trusses, eight temporary trestles were erected on the RCC frame of the seating tiers of the stadium. The shape of the eight outer trestles was dictated by the RCC frame on which they rested. The base was the widest portion of the trestle and it tapered to a 1 m x 1 m area at the top. A provision to place a 75 t capacity hydraulic jack between the trestle top and the outer shell was made. The outer trestles were quite flexible and were adequately guyed. Fig.2 shows an outer trestle supporting the shell joint. A pair of tensioning devices can be seen in top part of the figure.

To support the compression ring, four temporary trestles were erected at 15 m centres, reaching a height of 30 m. The latticed steel trestles 3 m square in size, were founded on piles. The staging consisting of rolled steel joist ISMB 400 was placed at the top of these trestles to receive the 200 t capacity hydraulic jack. In addition, a grillage of wooden sleepers was provided, on which the trusses T5 of the compression ring rested. The staging had also arrangement to keep in position the trusses resting on it. (Fig.3) The central supporting system consisting of four trestles was adequately braced for a lateral loading of 20 tonnes and had a vertical load carrying capacity of 600 tonnes.

Fig 3

After erecting the central trestles, erection of compression rmg started in April 1981. The truss T5 was the first truss of the roof dome to be erected. The truss was erected by means of two derricks of 45 m height and of 16 t capacity. The whole truss was assembled on the ground outside the central supporting system. Both the derricks were installed on one face and outside the central staging at a distance of 21 m from each other. Each derrick was guyed with 6 guy ropes anchored to the columns of main RCC frame supporting seating tiers. The derrics were provided with pulley block combination with 4 falls and were operated through 8 t capacity electric winch. After the truss T5 was erected it was secured with the staging.

The derricks were moved to another position to erect similiarly the second T5 truss. Two ring girder RG were then erected between the trusses T5. The truss T6 which was assembled at ground level inside the central staging was then erected with the help of the derricks by removing the bracing system connecting the central staging.

The other filling trusses were erected by means of step-derricks mounted on trusses T5 and T6. The erection of the compression ring was initially completed in July 1981. However, in August 1981 it was decided to provide additional horizontal diagonal bracings. Substantial amount of time and effort needed for this additional work as in-situ connections were to be made within the limited working space available at any joint. The additional work amounting to only 3 tonnes of fabricated steel took more than 3 months to complete.

Prior to the erection of the trusses in compression ring, their connections and proper fit was ascertained at ground level by assembling a quarter of the compression ring. Inspite of this, some tapering gaps were found in some joints of the compression ring which were initially filled with shims. As an additional measure of safety, the gaps were filled with epoxy grout.

For erection of the trusses in the segmental portion, the outer shells were first placed on the eight outer trestles at a height of about 25 m with the help of the largest crane of 300 t capacity which was specifically brought from a distance of 150 Km to augment the resources. This crane was also used to erect sixteen G1 trusses between the outer shells and the towers. Thus before any connection between the compression ring and the radial trusses was made, the outer shells were placed in position along with two of the five trusses which were to be connected to these shells.

In the initial stages, it was decided to connect two diametrically opposite trusses to the compression ring. However, due to the availability of 2 cranes of 75 t capacity and two high derricks, four T1 trusses in two mutually perpendicular directions were simultaneously connected to the compression ring. In order to achieve simultaneous connection, the two cranes and the two derricks used for lifting the trusses were not relieved of their load till all joints were adequately bolted. Simultaneous release was effected by giving one signal for release to all the four hoist operators.

During the next phase of erection, the remaining four T1 trusses were connected to the outer shells. The trusses T2 and T3 connecting the apex points of the compression ring o the bearings were taken up subsequently. A very slender column like support was provided at the junction of T1 and T2 trusses. The distance between the bearing and the inner shell in the compression ring was about 50 m. Two G2 girders were also connected at this junction. The slender support was provided to avoid any sag in the radial truss composed of the trusses T1 and T2.

During the erection of trusses in the segmental area, horizonal displacement of the compression ring was constantly monitored to detect any unusual movement. Looking into various possibilities including the use of laser beams, a simple system using four fixed telescopic sights focussed on graduated targets welded on the compression ring was adopted. Readings of these telescopes were taken at 15 minute intervals during erection and due to the adoption of this monitoring system, no undue risks were taken and the work proceeded without any mishap.

The peripheral box girders interconnecting the RCC towers were of 47 m span and 29 tonne weight. These box girders were lifted from outside the stadium by means of two 75 t cranes either in one piece or due to space limitations which restricted manouvering of the cranes the box girder was lifted in segments and placed on beam of RCC framework. After assembly, the box girder was placed in position by hoists operating from towers.

The bearings, saddles and the back ties were lifted by hoist located on top of the towers without any problem. However, the two halves of the front tie could not be easily placed in position as the front tie and girder G1 occupied the same position in plan. The half of the front tie which had the heavy tensioning device was lifted with the help of the largest crane prior to the erection of the radial trusses. After connecting its end with the outer shell, the tie was kept on the girder G1 at proper inclination by means of props. The remaining half was lifted by a step-derrick mounted on girder G1, assisted by hoist placed on the corresponding tower. The connection of the two halves through a central pin was made easier by adopting a longer pin having a larger taper for easy insertion. Even after the pin connections were made, the step-derricks were kept in position, till the tensioning devices were made operational to provide uniform tension in all ties prior to decentering.

The secondary trusses and the octagonal bracing system was erected by means of step-derricks mounted on primary trusses prior to decentering whereas the purlins and sheating were placed afterwards.

INSTRUMENTATION OF THE DOME

After the erection, the steel dome devoid of purlins and roofing weighed about 1600 tonnes. The primary task during decentering was to avoid any undesirable stresses in the roof system and to obtain its final designed position. The decentering was designed in two distinct phases. The first one was intended to introduce prestress in members, during which the bearings on the towere were free to slide. The final phase required locking of the bearings to prevent any outward movement prior to further lowering of supports.

Under each stage of initial decentering the eight outer shells were required to be simultaneously lowered by 3 mm, followed by lowering of four hydraulic jacks under the compression ring by 10 mm. The radial movements of the bearings were not to exceed 18 mm. The total extent of lowering of the compression ring during initial decentering was 9 cm. The steel dome was designed for an ambient temperature of 15°C in temperature. The final decentering of the compression ring was designed to be 5 cm prior to which the bearings were to be locked at 15°C. During the final decentering, the outer shells were lowered by 6 mm, the corresponding lowering of compression ring were continued to be 10 mm. Prior to the decentering operation the roof structure was instrumentated to monitor the following:

a) Movement of bearings: The gap between the back face of main steel bearings and the verticalface of the supporting concrete beams at the bearing locations on the towers was about 250 mm. The movement of the bearing due to atmospheric temperature changes and prior to decentering and during the initial decentering was measured by using two pairs of stainless steel studs with one mm diameter reference hole. Two studs were fixed on the steel bearing and the other two on the supporting concrete beam. The distance between the studs were measured with the help of large size divider and a graduated scale. However, for the bearing on tower 5, the distance was measured with the help of a pair of graduated steel scales which were fixed on the supporting concrete beam and were free to slide over an reference mark located on the steel bearing. (Fig 4)

b) Surface temperature measurement: As the roof structure was exposed to various degrees of solar radiation due to different orientations in space, several sets of copper - constantan thermocouples were fixed on the top chords of various trusses to monitor the surface temperatures at different hours of the day. The thermocouples were connected to a 24 channel Cambridge automatic temperature recorder.

c) Deflection at temporary supports: All the deflection measurements for points located at heights more than 25 metre were made at the ground level by suspension wire method. In this system, a 1.5 mm dia steel wire was clamped to the bottom chord member of the location where vertical deflection was required and it was kept taut by a weight hanging from it. For measurement of deflection, steel scale with least count of 0.5 mm and dial gauges of .01 mm accuracy were used for compression ring and outer shell respectively.

d) Displacement of compression ring: The four fixed telescopes which were used to ascertain the movement during erection were utilised to observe the horizontal displacement of the compression ring during the decentering operation.

e) Strain measurement: The strain in the selected members were measured by electrical resistance strain gauges of 12-ohm resistance. Suitably matched pair of low resistance lead wires was used and in order to have uniformity, all the lead wires for different locations, had nearly identical lengths. The BLH strain indicator was calibrated with the gauges and lead wires. All the lead wires were taken to the monitoring room situated at 23 m level on the 5th tower.

DECENTERING - PLANNING AND EXECUTION

The decentering of the roof dome was an intricate operation which required utmost care to prevent any undesirable movement. Meticulous planning for every conceivable detail was resorted as none present during the operation had any previous experience of such a huge decentering work.

Fig 4.

As the decentering operation depended on the proper monitoring of the structure, both for maintaining correct amount of movements and for avoiding any undesirable displacement during each stage of lowering, the availability of the monitoring data to the decision making authority was of paramount interest. During decentering the structure was monitored at 12 places for vertical deflection and at 4 places for ascertaining displacement of the compression ring through the fixed telescopic sights. In addition, sliding movement of bearing due to each stage was simultaneously recorded on 8 towers at 23 m level apart for the strain and temperature measurements in the monitoring room at 23 m on the 5th tower.

More than 50 Engineers were involved in recording and monitoring movements of the structure at 25 locations spread over the whole stadium. An intercom telephone system, eight walkie-talkies operating on two frequencies and a Courier system were adopted for transmission of data to the Control Centre. The whole system of transmission of data was rehearsed in three mock decentering to ensure that the data were available 5 to 10 minutes of the stage of lowering. An efficient public address system was available at the Control Centre.

During the planning, all humanly possible efforts had been put in and nothing was left to chance. A document indicating the detailing of personnel at different monitoring positions, security arrangements and other relevant details including the seating arrangements of key persons in the Control Centre etc. was prepared. Even the emergency Siren system for evacuation of site with detailed escape route for all the personnel had been rehearsed during one of the mock decenterings.

The lowering of twelve hydraulic jacks over the temporary supports were done by operators working for the contractor. The jacks at the eight outer shell points were lowered simultaneously by operators perched on the outer trestles on hearing a command from their supervisor through the Public Address System. After monitoring the lowering of supports and supplying any corrections, if needed, the four jacks under the compression rings were lowered simulaneously. These four jacks were connected to a Central Pumping Station on a platform at 34 m height (Fig 5). The platform 8 meter square in size was built around the intersection of the two T6 girders of the compression ring. If required, four jacks connected to a common pump could be operated individually. The extent of lowering of the central jacks was determined by the operators with the help of graduated scale fixed at suitable locations, whereas the extent of lowering at outer jacks was achieved by the use of proper spacer blocks.

Before announcing decentering operations, a series of checks were made to ensure that all the connections were properly made. All the front ties were imparted the necessary initial tension 7 days prior to decentering. The position of saddle block and their connection to the back ties were also carefully checked. In addition to visual inspection and reports from riggers, a closed circuit television set up complete with Camera and screen was utilised for checking connection details of all joints in the erected structure.

The initial decentering operation was started during the evening of 28 January, 1982. After two stages of decentering, it was again taken up on the 29th afternoon and the first phase was completed at 9 p.m. The process of locking the eight bearings took two hours during which the temperature was $15^{\circ}C \pm 0.5^{\circ}C$. In order to accomodate further lowering the hydraulic jacks along with its seats, distributor plates were readjusted during the night. The final phase commenced at about 8 a.m. and completed at 3.45 p.m. of 30 January, 1982, amongst cheers and jubilations.

OBSERVATIONS THROUGH INSTRUMENTATION

The instrumentation of the dome provided by the Bridges Division of the Central Road Research Institute, New Delhi, was of immense help during erection and decentering. During the erection, displacement amounting to the limiting value of 20 mm of the compression ring was, noticed by the fixed telescope. Consequently, a warning siren stopped further erection. During the investigation that followed to determine the reason for the displacement, the contractor was found to be resorting to wedging for fitting T2 and T3 girder. The contractors were asked not to adopt such practices which might endanger safety.

During the initial decentering, the theoretical horizontal movement at bearing was 180 mm. During the decentering, cummulative value of this movement was recorded. Bulk of decentering was completed during the evening. At 9 a.m., i.e. the time of completion, the cummulative value was found to be 95 mm on the average. A continuous record of the positions of the bearings had been maintained for 12 days prior to decentering and the authority was posted with the information regarding the positions of the bearings corresponding to 15°C, the design temperature. Comparing the position of the bearings with the values supplied, it was found that the average sliding was 153 mm instead of the apparent cummulative value of 95 mm. At the end of initial decentering, the average lowering of the compression ring was 84.3 mm instead of required 90 mm. As the ratio of horizontal slide to vertical lowering was about 0.2, the initial decentering was stopped at 84.3 mm lowering.

The average lowering at the inner trestles at the end of decentering was found to be 180 mm as compared to the anticipated value of 140 mm.

The values of strain recorded in various members during all stages of lowering were generally more than 20 percent above the theoretical value.

In all the stages of lowering the outer shells registered a further deflection when the inner supports were lowered, although the jacks under the outer shells were locked. This phenomenon was due to the increased value of reaction at outer support due to lowering of the inner supports. Moreover, the rate of lowering of outer trestles was not uniform during all the stages of decentering. The maximum differential between the adjacent supports during the initial decentering operation was about 4.6 mm.

Fig 5.

CONSTRUCTION MANAGEMENT

The construction of the Indraprastha Sports Complex was not only gignatic in magnitude but was also quite novel. The construction was a race against time. With some fifty separate agencies engaged on different aspects, several thousand labourers working in two shifts and over hundred engineers and architects involved in design and execution of the mamoth project, the planning and management techniques required was of the highest order for achieving the target in just over two years' time.

In a developing country like India having mixed economy, utilisation of available resources and application of appropriate technology, played a key-role in planning and construction management. In the following paragraphs several factors which have been responsible for timely completion of the project have been put forward.

The management started from the very beginning starting with the acceptance of a design which was basically aimed to facilitate speedy simultaneous execution of different construction activities. The accepted design was primarily based on the functional requirements.

The choice of the present complex was made through a limited architectural competition in which 14 designs were submitted. The design concept submitted by M/s Sharat Das & Design Consortium was found to be the best alternative from the view points of time, feasibility, technological consideration, functional parameters and ease of construction. The provision of a steel dome roofing was considered to be an added advantage as any later strengthening could be along with ease. Consequently the DDA appointed M/s Sharat Das & Design Consortium for providing the detailed consultancy services for the project. The team of consultants was stationed at the site and their services were available round the clock.

The mamoth project of building the Indraprastha Complex was rationally divided amongst various contractors. The construction of the steel dome comprising of steel lattice girders and trusses were entrusted to M/s Triveni Structurals Ltd., a Government of India Undertaking who are well known in the specialised field of structural steel fabrication and erection. The main civil works was the responsibility of M/s Tarapore & Co., a firm of high reputation having construction experience of many prestigious works. The choice of main contractors for such a time-bound project is of paramount importance as the contracting agencies were the back-bone of the progress of the project and with whose untiring efforts the project become a reality in record time.

In addition to the proper choice of consultants and main contractors, engineers for the highly time bound project was chosen from a select list of persons known for their technical ability, diligence and dynamism, thereby even inducting engineers on secondment to this project. The team of engineers of the Delhi Development Authority headed by the Chief Project Engineer worked with a sense of dedication and team spirit under demanding conditions.

Realising the normal financial and other constraints under which the Chief Project Engineer was required to act, the Delhi Development Authority constituted a "PROJECT BOARD" to take quick decisions. Four senior-most officials of the Authority representing Finance, Technical and Administration were the members and the Chief Project Engineer was acted as the Member-Secretary of this Board. The Board which normally met once every month, called upon the assistance of several specialists as advisers. The constitution of this high-powered committee was another significant feature for the construction management.

Considering the multifaceted activities connected with the design and construction of the steel domed indoor stadium, the Project Board constituted another committee known as the "MONITORING COMMITTEE", with the retired Director-General, Central Public Works Department, who was the senior-most adviser, as the Chairman, to go into the details of design, construction and time-scheduling of the various aspects of the indoor stadium structure. The committee having several other advisers as members, went through the details of the design of roof structure and other civil structures of the indoor stadium and was able to provide very constructive help in the evolution of design of the roof. Many specialised items for which the team of consultants were not adequately experienced were entrusted to other governmental agencies on the recommendation of the committee. One such item of work was the instrumentation and monitoring of movements and strains during erection and decentering of the roof dome which was capably handled by the Central Road Research Institute, New Delhi, leading to satisfactory decentering of the roof.

The Monitoring Committee met once a week, when the progress report for the past week was invariably presented by the engineers of the Authority. After comparing the progress with the basic time-schedule prepared by the consultants, the reasons for short-falls were deliberated by the Committee and suitable remedial measures for overcoming bottlenecks were recommended by the committee to various agencies including the Project Board. The Monitoring Committee was definitely able to provide guidance, constructive criticism and help to the engineers, contractors and even to the consultants for the time-bound project.

Weekly coordination meetings were held with the consultants and contractors and the Engineers of the Authority to sort out site problems and to fix targets of construction. Several other committees were set up to finalise specifications, quality control aspects of the project. A high-powered committee assured unhindered supply of construction materials to the work site. In addition to the above a Senior Project Engineer along with his normal compliment of officers and other staff was engaged exclusively for planning, design and monitoring of the project.

To ensure proper implementation of a project of this gigantic nature, collection of reliable information and careful monitoring of the progress were essential. This required coordination at various levels including the Cabinet level. This successful coordination at various levels led to the timely completion of the Project.

CONCLUSION

a) It is essential to adopt proper construction management techniques for speedy implementation of time-bound projects.

b) Adherence of safety procedures pays dividend.

c) For unusual structures, instrumentation plays a vital role in ensuring compliance with design and safety requirements.

ACKNOWLEDGEMENTS

The authors wish to thank the authorities of the Delhi Development Authority for providing the opportunity to work for the Indraprastha Sports Complex, New Delhi. The authors also wish to thank the other members of the team of the Central Road Research Institute, New Delhi, namely Messers A.K. Garg, M.V.B. Rao, R.N.Sharma, P.D. Satija and S.K. Sharma for their assistance rendered for instrumentation and monitoring of the Stadium dome. Thanks are due to Mr. D.R. Mehta, Deputy General Manager(Erection), Triveni Structurals Ltd., for the help provided during instrumentation of the roof structure. The paper is presented with the permission of the Director, Central Road Research Institute, New Delhi, and the Administrator, New Delhi Municipal Committee, New Delhi, India.

REFERENCES

1. Ray-Chaudhri, P. and Chetal, V.P. (1982) "Indraprastha Sports Complex - Evolution, Design and Construction Management".

2. Ray-Chaudhuri, P. and Chetal, V.P. "Indraprastha Stadium and Sports Complex" (Under publication).

THE USE OF GLASS-FIBRE-REINFORCED PLASTICS FOR LARGE ROOFING AREAS

J.OLTHOFF Ir.

Project engineer of Polymarin B.V.
Manufacturers of structural glassfibre products
Medemblik, the Netherlands

When special demands are made for roofs a special material has to be chosen. This is the case for waste-water purification plant roofings, domes and complex shaped roofs. In these situations glass-fibre is the material to provide the solution. Polymarin has experience in this field and goes on with the development of new constructions.

INTRODUCTION

A roof usually has the sole function of closing-off the top of a building. For this purpose, a selection can be made from many different materials. However, this is not the case when special demands are made. These demands may be :

1. high degree of chemical resistance
2. gas-tightness
3. heat insulation
4. self-supporting structure
5. light weight
6. short building time
7. complex design

A material that meets all these demands is glass-fibre reinforced plastic (GRP), which is often simply named after the type of resin used : polyester.
In situations where one or more of the above requirements must be met, GRP can provide the solution.

Polymarin B.V. of Medemblik, the Netherlands, is a company that specialises in the construction of GRP roofing. Among the projects completed by this company are roofing domes for a waste-water purification plant, and a dome for a planetarium. These structures will be discussed below.

ROOFING FOR WASTE-WATER PURIFICATION PLANT

The first four of the demands stated in the introduction are particulary important for a waste-water purification plant.

- in the purification process, gases containing high concentrations of H_2S and H_2SO_4 are released. These substances are highly corrosive, and conventional roofing materials such as wood, steel and aluminium, cannot be used on their own. For use under these conditions, they would have to be covered by a protective coating. Such a coating generally offers adequate protection on flat surfaces. However, experience shows that the degree of protection is insufficient at the joints. In addition, corrosion of the material is in many cases initiated by drilling, damage, accumulations of moisture etc.

These problems do not apply when GRP is used. The material is in itself inert, and is given even further protection against environment influences by a high-quality surface coating.

- Waste-water purification plants were formerly built in the form of open basins. However, today's legislation no longer permits the odour nuisance that this system caused. The roof of a waste-water purification plant must therefore be gastight. The use of GRP allows this demand to be met with no problem.

- To sustain the reaction in an oxidation bed, a certain temperature must be maintenained. The exothermal reaction normally generates enough heat for this purpose. But good insulation is necessary to prevent the heat from escaping. GRP also meets this requirement. A roof made of this material does not need any additonal insulating material.

- Rotary sprayers of rotating sludge scrapers are often installed in purification basins. These systems make it impossible for the roof to be supported from the floor of the basin. The roof must therefore be self-supporting. This can be achieved if the roof is made of GRP.

For these reasons, it was decided to use GRP for the roofing of two large oxidation beds for the "de Groote Lucht" waste-water purification plant at Vlaardingen, the Netherlands.

DE GROOTE LUCHT

In the "de Groote Lucht" project, two roofs were constructed, each with a diameter of 48 metres. The roof height above the water level in the basin is 8 metres (see Fig.1)

Fig.1

Each roof unit is made from 36 identical segments, which are mounted on the edge of the basin. At the top, the segments are attached to a steel ring. The central cover section, consisting of six segments, then rests on this steel ring (see Fig.2)

Fig.2

The large segments have the shape of a 3-pivot truss, and are made of a single-layer GRP laminate. Surface reinforcement is provided by laminated-in lateral beams, while the required bending strength is provided by the cross-sectional shape (see Fig.3)

Fig.3

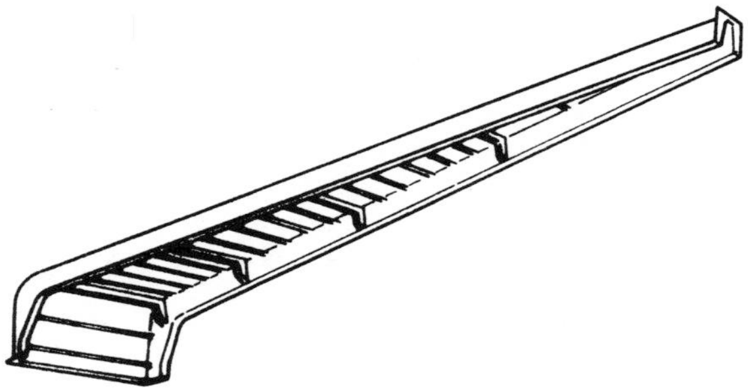

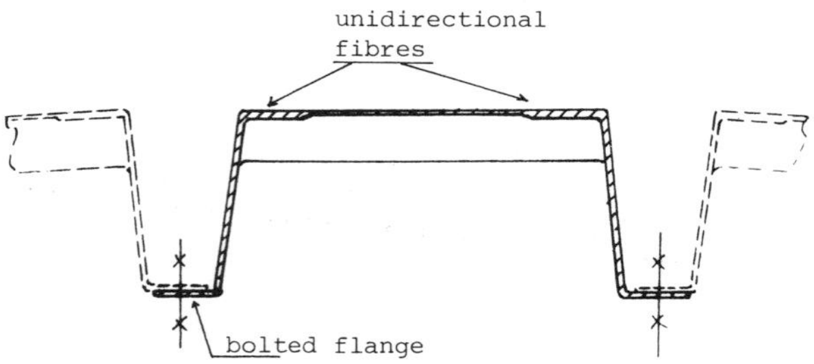

To obtain sufficient rigidity, unidirectional roving material is used in the top surface and in the flanged edges. The segments are secured to each other by means of a bolted flange. A watertight and gastight joint is obtained in this way. Each of the elements has a weight of 2100 kg. The total construction and assembly time on-site (see Fig.4) was only 3 weeks for each roof.

Fig.4

Provision had been made in the building of the basins for this type of roof to be fitted. For this reason, the edge of the basin was designed so that it was able to carry the extra forces. Where a roof has to be built over an existing plant, light weight is an important factor. Using GRP as the constructional material, the required light weight can be achieved. The short building time is an additional advantage. If it is necessary to take the basin out of service, this is therefore only for a short time.

THE AMSTERDAM PLANETARIUM

GRP is a particularly suitable material for roofs which have to be built in a particular shape. One example of such a project was the Amsterdam Planetarium, for which Polymarin built a dome. The required shape in this case was a hemisphere with a diameter of 25 metres(see Fig.5)

Fig.5

This dome was built from 10 segments. Each segment is constructed in the form of a sandwich panel using 4 mm. of GRP, 50 mm. of PU foam and 4 mm. of GRP. Three radial reinforcing girders were fitted to each segment (see Fig.6)

Fig.6

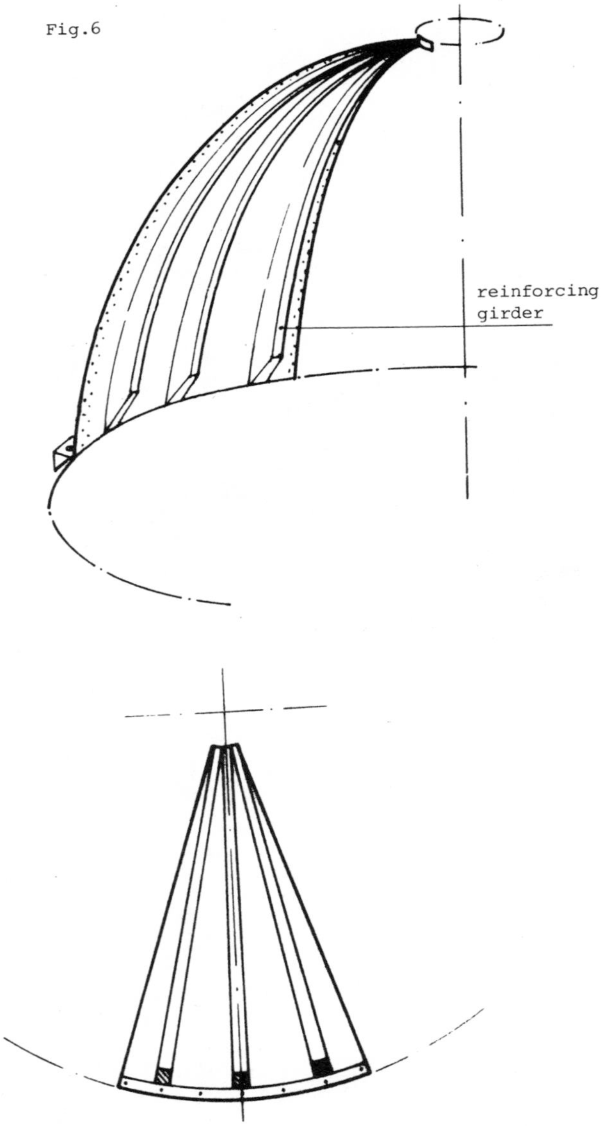

These elements were once again secured to eachother using a bolted flange joint. The row of bolts was later covered by a GRP strip. In this case, a sandwich structure was chosen for its even higher insulation. This also ensures that no condensation forms on the inner surface. Together, the segments form a self-supporting dome. This form of construction resulted in a weight per element of 230 kg. The building of the dome took only two weeks on-site (See Fig.7)

Fig. 7

NEW DEVELOPMENTS

Compared with metals, GRP has a lower modulus of elasticity. Rigidity will therefore often be the decisive factor in the choice of material for a particular roof structure. This means that the structure will have to be dimensioned to prevent it from collapsing under its own weight. An alternative method of supporting this load is to combine the use of GRP with a metal latticework. The metal structural sections will then to a large extent determine the rigidity.

A structure which is based on this principle is the new Polymarin roof. An example of this is shown in Fig.8

Fig.8

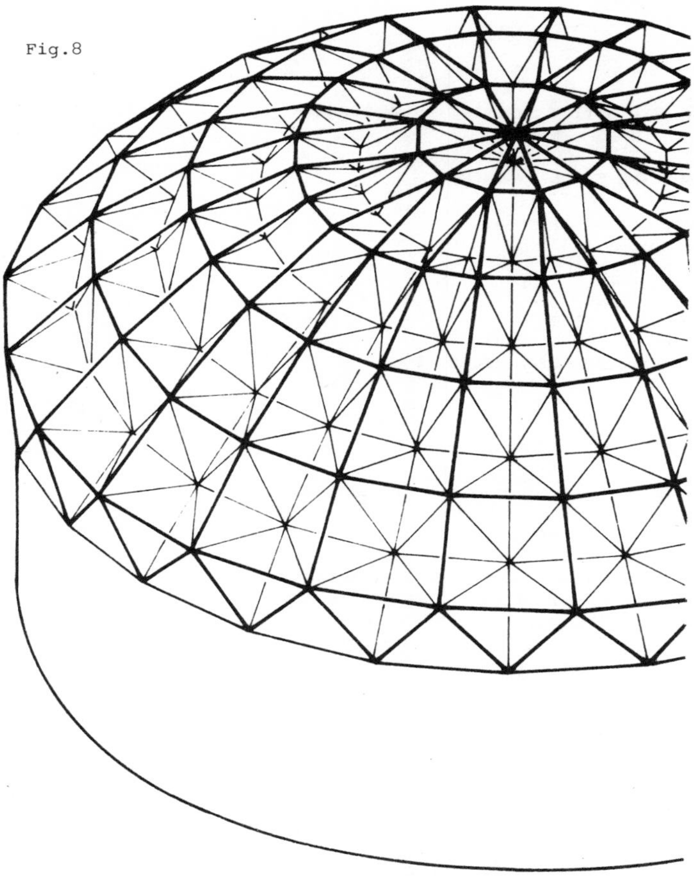

This structure is based on the Schwedler dome. It is a combination of pyramid-shaped GRP segments and metal tubing. The metal tubing is secured simply to the pyramid at its joints.(See Fig.9)

Where symmetrical vertical loads arise, the much more rigid metal will absorb the load as a Schwedler dome. In the case of an asymmetrical (wind)loading, the GRP segments will act as a shear structure. In this way, the GRP section can be more lightly dimensioned. The end result is a more rigid, lower-cost roofing. The segment size can easily be selected to produce an easily transportable unit.

Using this development, Polymarin intends to manufacture highquality roofings which at the same time are competitively priced.

Fig.9

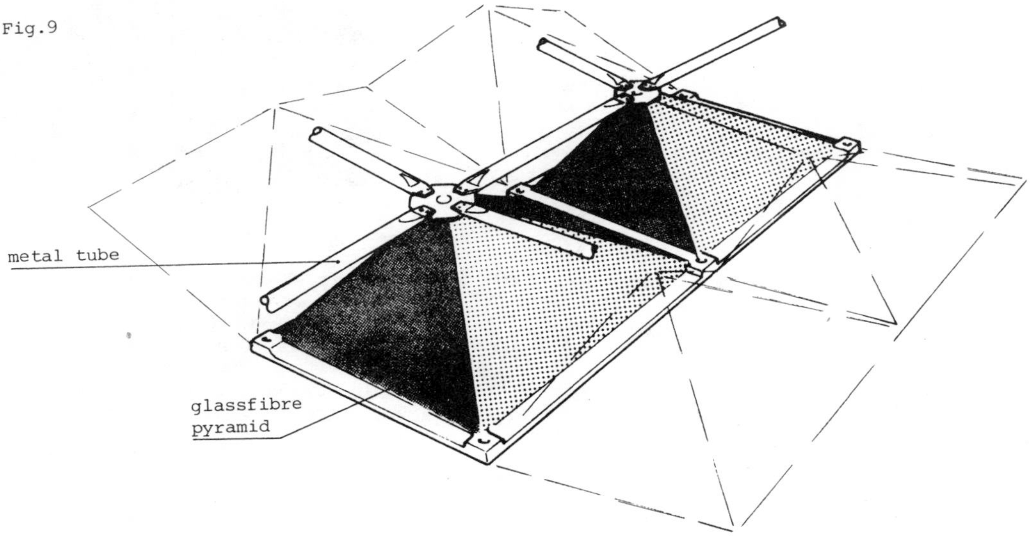

SPACE STRUCTURES OF THE RAI CONGRESS AND EXHIBITION BUILDING IN AMSTERDAM*

IR J W B ENSERINK CI, CONSULTING CIVIL ENGINEER

Partner DSBV Engineers and Architects,
Rotterdam/Amsterdam

Architects: A Bodon and J H Ploeger

* Netherlands and European Award for Steelstructures (ECCS) 1983

The extension of the exhibition area consists of two halls, 67.5 x 67.5 m and one hall, 97.5 x 97.5 m with free height of 11.25 m and free span.
The structure is a space frame of the type "Square on Square offset set diagonally" with sides of 5.30 m, diagonals under 45°, construction height 3.75 m.
For this wide grid the nodes were specially designed, because almost no standard system could meet the loading conditions, and these nodes proved to be competitive. Assembly of the structure took place at floorlevel; the structures were jacked in position from four temporary towers. Circular tubes of equal outside diameters were used for the horizontal members and also all diagonals had equal outside diameter. The wall thicknesses were adapted to the stressconditions. The structure was painted prior to erection.

INTRODUCTION

In February 1977 the Board of the RAI Congress- and Exhibition Building in Amsterdam decided to extend the existing expositionarea of 46.000 m^2 with an other 20.000 m^2 and commissioned our office, especially the architects A Bodon, and J H Ploeger to prepare preliminary plans to this purpose.
The required area should be designed in such a way that the total area could be devided for separate exhibitions, but also be used together.

The halls should be free from internal supports, because the experience had shown that this only allowed flexible use of the area for various purposes. This use is in principle for exhibitions, varying from small items via automobiles to agricultural equipment, but also for music festivals and concours hippique.
Annex to these halls were required conference rooms in various sizes; e.q. one for 700 persons, one for 240 persons and four with a capacity of 100 persons with the required entrance halls and lobbies, restaurants, press-facilities and offices.

With respect to the parking problems it was required to increase the parking capacity in such a way that even during the buildingperiod the existing capacity was maintained.

Four preliminary plans were submitted and the one selected comprised three exposition halls, square in plan; two with gross dimension of 67.5 x 67.5 m' and one, measuring 97.5 x 97.5 m'.
The reinforced concrete main floor was located 3.5 m above the terrain, thus providing a parking for 700 cars underneath at ground level.

Because of the symmetry in plan the architects gave preference to a construction which showed the same

View from South-East

symmetry. Therefore a spacestructure was adopted, although this proved not to be the cheapest if only the structure was considered. Cheapest would have been to use 3 or 4 main trusses, spanning the halls in one direction, with intermediate trusses spanning about 15 m. The main trusses would have been about 6 m high.
The pricedifference however was considerably reduced, taking into account that the facades were lower as a result of the reduction in constructionheight (3.75 m).

For the intermediate areas, the conferencerooms and the offices a space-structure for the roofconstruction proved to be too expensive because of the wide variation in span. Therefore maintrusses with triangular cross-section were provided in order to maintain a relation in appearance to

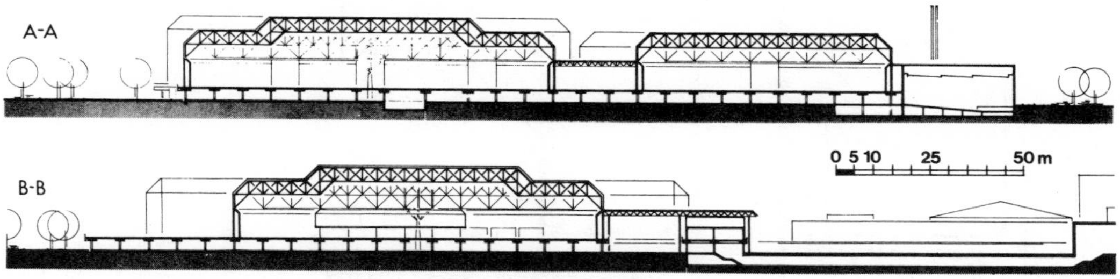

Fig 1A Sections A-A and B-B of fig. 1B

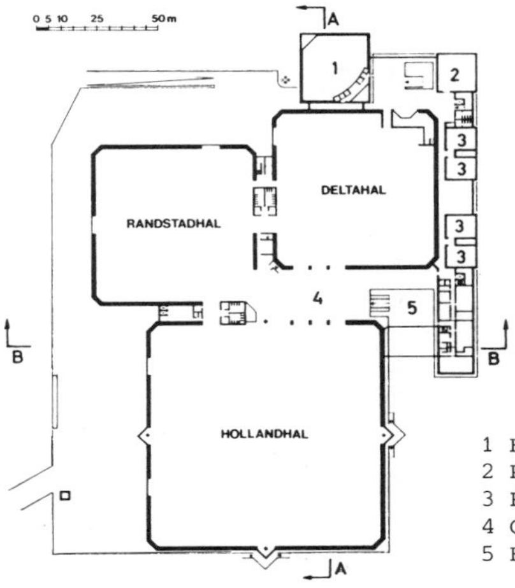

1 Forum room
2 Park room
3 Rooms
4 Central hall
5 Entrance hall

Fig 1B Plan at Floor-elevation

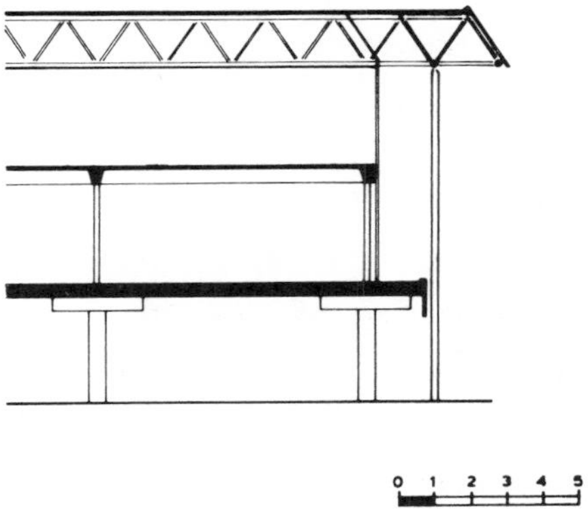

Fig 2 Section office-wing

the space-structures of the halls.
The upper chord of these trusses consists of two rolled beams HE 140A, the lower chord of circular tube Ø 101 mm and the diagonals of tube Ø 60 mm. Between these trusses secondary plane trusses are placed, 3.75 m on centers and also composed in rolled beams HE 140A for the upper chord, tube Ø 101 mm for the lower chord and tube Ø 60 mm for the diagonals (Fig 2).
The structure for the intermediate floors is reinforced concrete, independant from the roof structure.
The steel structure for the Forumroom is completely separate from the other parts of the building.

DELTAHALL AND RANDSTADHALL

The design of the steelconstruction for the halls will be described for the two smaller halls in the first place.
The modulus governing the whole concept is 7.5 m, as it was in the already existing building.
This same modulus was adopted as it proved to be also acceptable for the parking underneath.
A double layer grid was adopted with a lower and upper square grid at 45° angle to the outside walls. In this way the mesh width was established at $\frac{1}{2}\sqrt{2} * 7.50$ m 5.30 m (Fig 3).

Choosing the diagonals to meet the facade at 45° gives a construction height of $\frac{1}{2} * 7.50 = 3.75$ m.
Providing a downward extension of the structure along the sides to an elevation of 3.75 m below the lower lattice (Fig 4), gave the possibility to form a chamfer at the corners of the facade. This shifted the location of the columns along the sides over half a modulus.

The layout of this structure was in first instance calculated preliminary according to the plate-analogy. This gave a steelweight of about 45 kg/m^2.
The loading conditions are:

a) own weigth;
b) snow 50 kg/m^2;
c) roofstructure 50 kg/m^2;
d) 250 kg at each node;
e) wind loading.

As a result of the dimensions as chosen the structure gave a very open impression, which was the intention of the architects. On the other hand this resulted in loading conditions for the nodes which could only be met by a few standard space structure systems.
In order to maintain sufficient competition in bidding, we

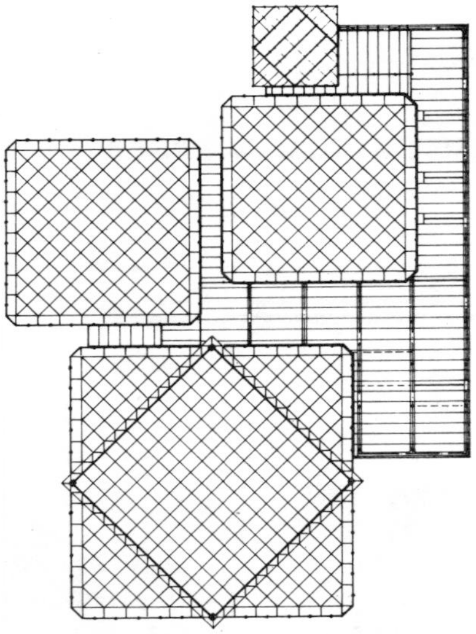

Fig 3 Plan of the space structures

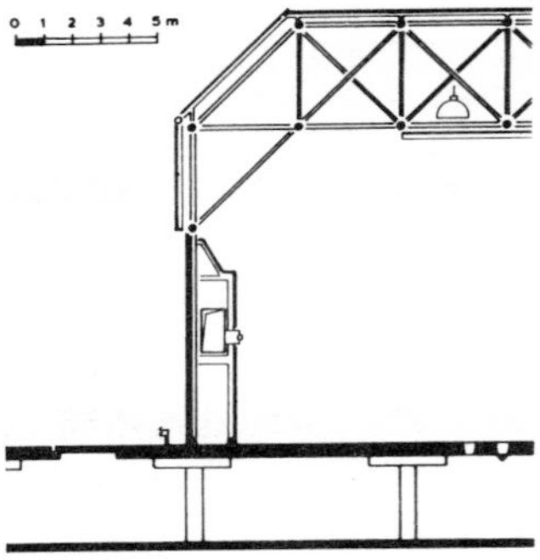

Fig 4 Section through wall

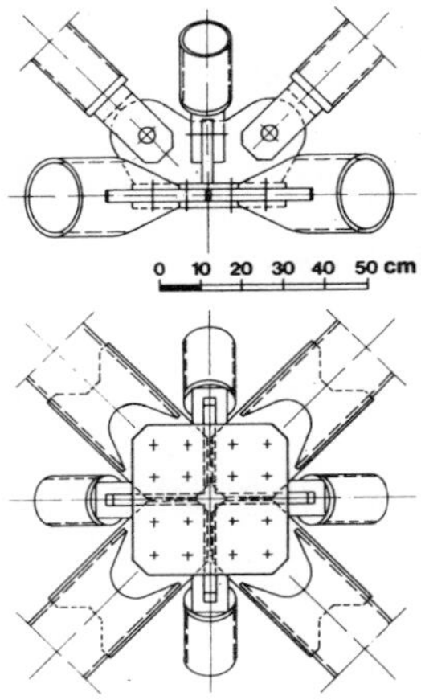

Fig 5 RAI-Node

developed a special RAI-node, for which we established the following requirements:

a) use normal steel connection methods;
b) use only bolts in assembling;
c) can be produced by every experienced steelconstruction workshop.

For architectural reasons both horizontal grids were formed of circular pipes of the same external diameters (Ø 194 mm), from which the wallthickness was varied from 14.2 mm to 4.5 mm according to the maximum loading conditions. All diagonals had an external diameter of 127 mm and a wallthickness from 4.5 to 10 mm.
The normal node is given in fig 5.
Along the circumference variants are required to adapt for lacking elements (fig 6) or connections from above and below (fig 13).

The vertical dimensions are also based on the 7.50 m modulus:
the columns are 7.50 m long, the lower grid layer lies at 7.50 + 3.75 = 11.25^{+} floor level. The upper grid layer at 11.25 + 3.75 = 15.00 m.
The columns supporting the roofstructure are placed within, but free from, the supports of the wallconstructions (fig 4).
This freedom for the roofstructure has the advantage that deformations of that structure, due to temperature - wind or live loads, are not transmitted to the walls. The area of the wall to 7.50 m is formed by coated aluminium sectional sheeting, insulated on the inside;the internal wall consists of concrete panels; from 7.50 to 11.25 m is a glassarea and above that to 15 m height a sloping surface, consisting of translucent fibreglass reinforced panels; the latter two area's in an aluminium framework connected to the roof at the top and to the wallstructure at the lower elevation, having sufficient flexibility to allow for displacements between those two parts (fig 4).
The roof consists of sectional steel sheets bearing 4 cm polyurethane foam insulation and a layer of roofing.
Ventilation hoods and smoke evacuation elements are provided for ventilation and smoke prevention respectively.

HOLLANDHALL

To develop a similar structure for the larger Hollandhall required the following conditions:

- use the same modulus, to allow the use of the same elements;
- allow the same wallstructure;
- give a relation in size with the 2 smaller halls.

This gave, after having considered a wide variety of possibilities the design as follows:
A spaceframe exactly as the one described was adopted, in which a central square under 45°, in dimensions almost equal to the other halls, was raised 3.75 m. So four

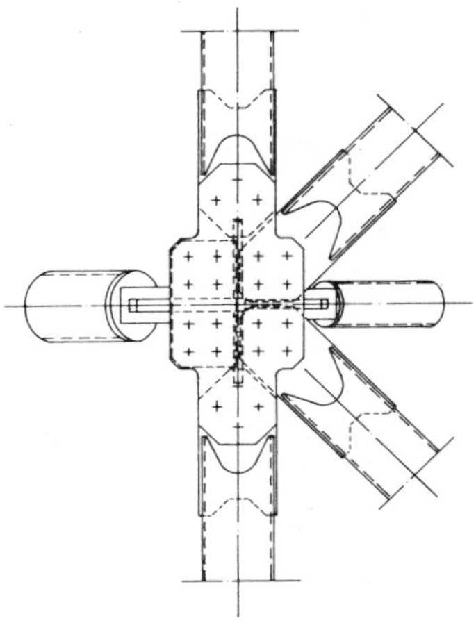

Fig 6 Node at upper edge

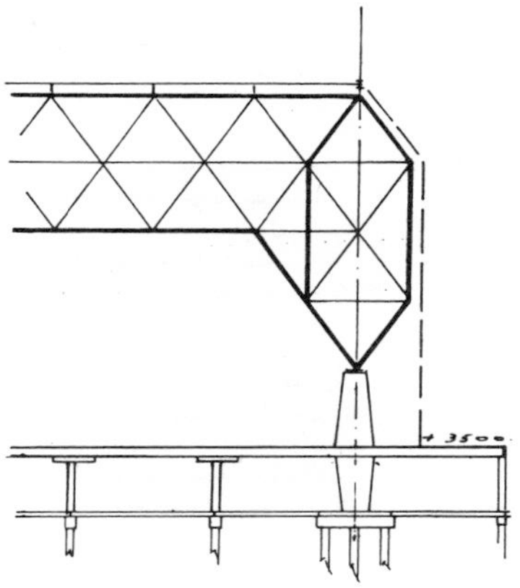

Fig 7 Main portal

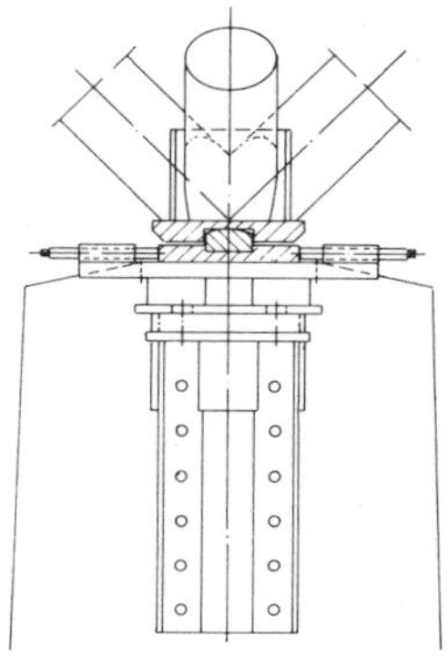

Fig 8 Main bearing

internal beams of 7.50 height and 70 m span stiffened the spaceframe and lead to main supports in the middle of the sidewalls (fig 7).
These formed 4 portals connecting these points, which added to the horizontal stability.
Due to deformations as a result of the increasing load on the structure a horizontal force of 300 tons would result, and this would lead to quite heavy dimensions of the reinforced concrete supports. To diminish these dimensions the horizontal forces were reduced by providing temporary sliding supports (fig 8) and allowing the supports to slide, controlled by screws, to eliminate horizontal forces during construction for all own weights. This reduced the horizontal forces to 100 tons, at the same time causing a redistribution of the stresses in the whole structure.

After the preliminary design, in which approximations were used to calculate the stresses and deformations, the final design was carried out on the IBM-computer at Zoetermeer, using the program ICES-Strudl. The spaceframe definition was not only used for the design, but also for calculating the various stages of loading during assembly and the jacking procedure by the contractor.
Further it should be mentioned that in order to allow the municipal building authorities to approve the calculations, they followed the whole procedure from the beginning. Otherwise it would have been very difficult to follow and comment the extensive calculations for their final approval.
A number of calamities were also calculated; failure of a supporting column and failure of one of the nodes. The result was that a stress redistribution ocurred without further consequences.
Damage was restricted to the local area around the failing member.
Because of symmetry calculations were first carried out for 1/8 of the structure for vertical loads only; then for 1/4 with simulation of horizontal forces and finally for 1/2 system taking horizontal forces and stability requirements into account.
When the design had proceeded to the stage of tendering, the firm Bailey Constructiewerkplaats en Machinefabriek N.V. at Nieuw Lekkerland was first asked to assist in calculating and commenting on the production methods. This proved that the project would be possible within reasonable costs.
When the tender documents were completed, the same firm was given the oppertunity to bid on the complete steelstructure, and as the price was acceptable, was thereafter commissioned to produce and erect the total steelconstruction for the project.

To ensure drainage of rainwater from the roofs the structures were given a spherical camber, with an overheight of 60 cm in the middle as a result of own weight. This was achieved by introducing a length difference of 2.5 cm between the elements of the lower

grid and those of the upper grid.
The theoretical length of the elements should be met under strain according to own weight loading. This enabled a drainage only from the corners, where heated gulleys were provided to ensure drainage under extreme winter conditions.

The steelconstruction has been painted in the workshop, prior to assembly, in order to reduce the costs.
A double coating on the basis of two-component paint has been applied.

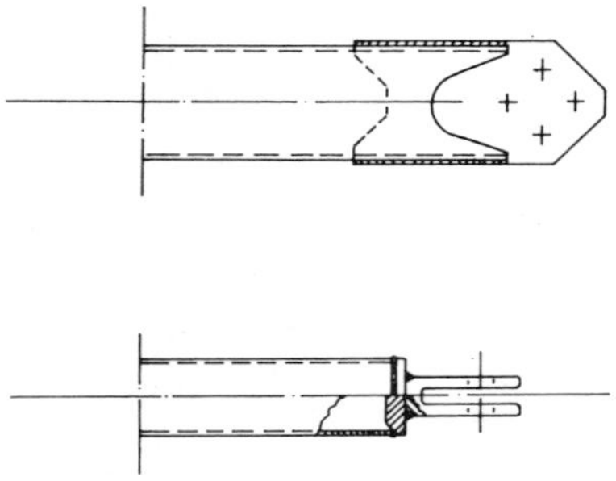

Fig 9 End-connections

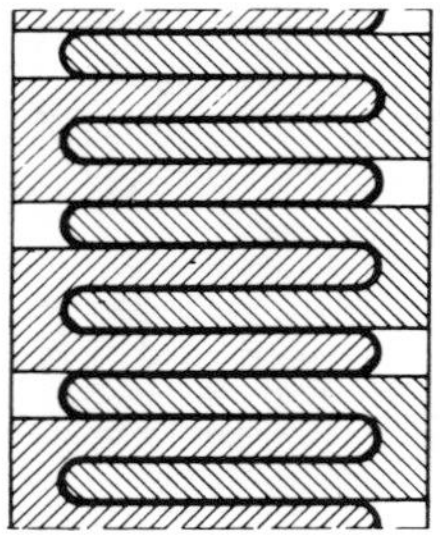

Fig 10 Cutting the forks

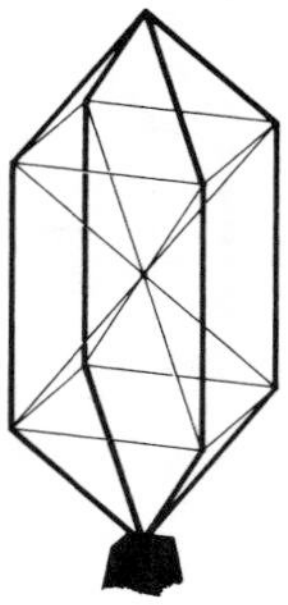

Fig 11 Scheme of maincolumn

This requires special attention during transport, handling and assembly, to avoid damage on this corrosion protection.

PRODUCTION

The fabrication of this spacestructure can be divided in four production streams:
- the lower and upper grid elements;
- the diagonals;
- the nodes;
- the columns.

The upper- and lowergrid elements were cut to the required shape by oxyacetylene cutting.
The connection plates, 30 mm thick, were cut from plates and the holes Ø 25.5 mm were bored with templates.
The connection plates were attached in the slot of the tubes on a template, to ensure uniformity of the members.
Thereafter welding was completed, where the shape was thus a contraweld could be made from the inside (fig 9).
For the diagonals first the forks were produced by flamecutting from steelplate, thick 100 and 120 mm (fig 10).
These forks were welded to a circular plate, which was chamfered by flamecutting to prepare it for welding.
Two forks were welded to the diagonal by automatic CO_2-welding in a revolving machine.
The length of these members was optically controled.
The elements were labelled according to the wallthickness, because visual check was impossible afterwards.

The nodes consisted of two plates, with 16 holes each, one with to crosswise plates welded to it for the connection of the diagonals, the other plain.
The 4 main supports of the Hollandhall were assembled completely by welding in the workshop (fig 11).
At the circumference of the raised part of the Hollandhall, the upper and lower chords of the diamond shaped portalmembers were so heavily stressed, that it was necessary to form these as a 70 m long continuous 406.4 mm dia-tube, with variable wallthickness.

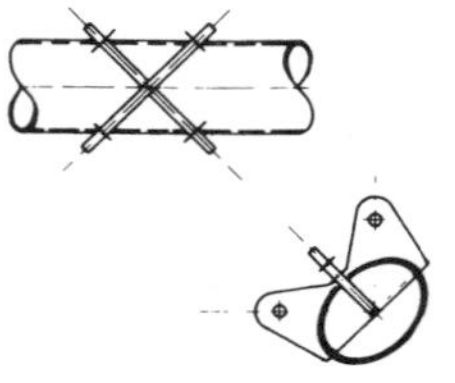

Fig 12 Connections in portal-chords

The connections for the diagonals were formed by plates, crosswise placed into slots, cut in this tubes.
These slots allowed cambering these tubes before welding of the connectionplates (fig 12).
For reasons of transportation, two field weldings had to be allowed in these members.

ERECTION

Assembly of the spacestructures took place at floor level of the halls. A row of the lower grid was placed on temporary supports of calculated height to form the camber.
Then, four diagonals - preassembled before transportation to a node plate - were connected to the lower nodes.
Thereafter a row of the upper grid members was placed. This procedure was repeated row after row, requiring only a small mobile crane with a reach of not more than 6 m.
Prior to assembly, four 22 m high jacking towers were placed and connected to the reinforced concrete floor at 15.0 m inside the corners.
Each tower consisted of 4 steel pipes Ø 762 mm carrying a jacking floor at the top.
The four 125 t capacity hydraulic jacks were lifting the structure, suspended from each jack by two 36 mm dia Dywidag bars.
In the first phase the intermadiate temporary supports were removed and the structure was lowered on the circumferential supports, thus allowing the structure to set, by removing the tolerance from the bolt connections. Then the bolts M 24 - 8.8, were tightened by momentwrenches.
The second phase raised the structure about 5 m to allow completion along the periphery, where the columns were also connected in horizontal position by means of hinges (fig 13).
Finally the whole structure was raised to the final position and the footings of the columns were bolted to their supports.
This last stage took about 3 hours.

The erection of the Hollandhall started by assembling the diamond shaped mainbeams.
The maincolumns, 15 m high, 6 x 6 m in plan, were placed on their supports and kept in position by temporary supports. The mainbeams were then lifted in position and connected to the columns, thus forming four interconnected portals. The structure for the central square was then assembled and raised in almost the same procedure as described above.
The four triangular roofsections were also assembled at the floor of the hall. Cranes lifted these structures and put them on temporary supports. The connection with the main beams was made with connecting bars, made to measure.

In total this construction required 2187 tons of steel, the Deltahall and Randstadhall 312.5 tons each, the Hollandhall 822 tons and all other structures 740 tons.

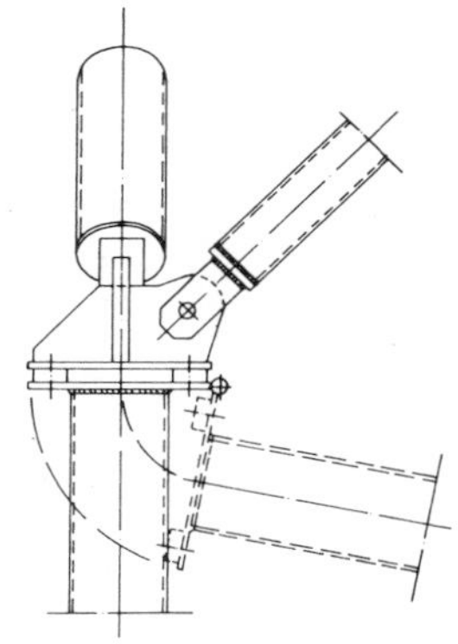

Fig 13 Temporary hinged column

Of this total 550 tons was of the quality Fe 420, but in general Fe 360 was used.

References:

Acier - Stahl - Steel, nr. 2 - 1983
Bouwen met staal , nr.54 - maart 1981
PT Bouwtechniek , nr. 2 - 1982.

	members	nodes	frame	nodes	purlins
Deltahall - Randstadhall	2 x 1060	2 x 363	37.8 kg/m^2	17.95 kg/m^2	7.13 kg/m^2
Hollandhall	3172	853	49.0 kg/m^2	24.9 kg/m^2	9.06 kg/m^2

THE 31-ZONE STRUCTURAL SYSTEM

STEVE BAER

ZOMEWORKS CORPORATION
PO BOX 25805
ALBUQUERQUE, NEW MEXICO 87125

A versatile 31-zone space frame system can be constructed with icosahedral joints and structural members aligned, with diameters through vertices, edge midpoints, and face midpoints of the icosahedron.

A wonderful variety of structures can be built from icosahedral joints with fittings at A (their vertices), B (their face midpoints) and C (their edge midpoints). Connecting members are

$$A_{nm} = C \cos 18° \, \tau^n 2^m - J \qquad \ldots 1$$

$$B_{nm} = C \cos 30° \, \tau^n 2^m - J \qquad \ldots 2$$

$$C_{nm} = C \tau^n 2^m - J \qquad \ldots 3$$

where n and m are any positive or negative integers, $\tau = \frac{\sqrt{5}-1}{2}$, C is any base length, and J is the joint diameters.

In the simplest system, the icosahedral joints are oriented parallel to one another, with the A, B, and C members attaching to vertices, face midpoints, and edge midpoints, respectively (US Patent #3722153).

Anyone who looks at this system can see that it bears a great resemblance to the well known MERO system of Max Mengeringhausen, where the joint is an octahedron and connections are at the vertices and edge midpoints; or the more recent Super Structures of Peter Pearce, where the joint is also an octahedron and the connections are at the vertices, face midpoints and edge midpoints.

The 31-zone system of the icosahedron (6 vertex, 10 face midpoint, 15 edge midpoint) is then very similar to the 13-zone system of the octahedron (3 vertex, 4 face midpoint and 6 edge midpoint).

The advantage of the 31-zone system over the 13-zone system is greater versatility. This system will build such polydedra as the truncated icosadodecahedron and the enneacontrahedron, either in single shells or trussed configurations. The joint for the 31-zone system can accept structural members from 31 different directions, while the 13-zone system is restricted to 13. Since two lines form a plane, it would seem that the 31-zone system forms $\frac{(31)(30)}{(2)(1)}$ or 465 different planes. In many cases, more than two lines are coplanar and the variety of different planes reduces to 121 for the 31-zone system, and again in the case of the 13-zone system, from $\frac{(31)(12)}{(2)(1)} = 78$ to 25.

The 31-zone system allows almost five times as many different planes to be formed in space as the 13-zone system. The 31-zone system contains seven of the zones of the 13-zone system, but lacks the other six. Boxes can be built with the 31-zone system, but they can't be braced with face diagonals and stay in the system. A regular tetrahedron cannot be built in the 31-zone system.

Clark Richert has added to the 31-zone system to allow regular tetrahedra. This can be done most easily by using the dodecahedron (the icosahedron's dual), and adding five connections to each face: 30 new zones each 13° 16' 57" $\tan^{-1} \tau^3$ away from the pentagon midpoint towards a corner. This expanded 61-zone system, then, has a new structural member

$$D = C \cos 45° \, \tau^n 2^m \qquad \ldots 4$$

Most buildings are made without space frames. The buildings are less expensive and better constructed by departing from any universal space frame. Space frames are used primarily as large-scale jewelry than for any other purpose.

In the early 1970's, Zomeworks Corporation manufactured playground climbers based on a subset of the 31-zone system (see fig 1), wood structural panels for triacontrahedral buildings (a subset of the 31-zone system; see fig 2), and model kits of the complete 31-zone system (see fig 3). The greatest promise of the 31-zone structural system is not a more efficient way to enclose space, but as a more enlightening pattern for seeing space than the usual regular crystal space grids.

The enormous enthusiasm of people who grasped the marvel of the combination of regularity and freedom possible within the system, shows that the structure can be more important than the roof it is made to support.

Marc Pelletier and Paul Hildebrandt have designed a new modelling kit for the system which eliminates construction errors and hastens the user's grasp of the system.

Is there some special significance to the divine proportion at the very heart of this engineering system? The divine proportion and Golden Section are much discussed by artists, but as far as I know, the first time the divine proportion has crossed over to become a regular tool of the engineer is within the 31-zone structural system.

Fig 1 (stellated icosadodecahedron)

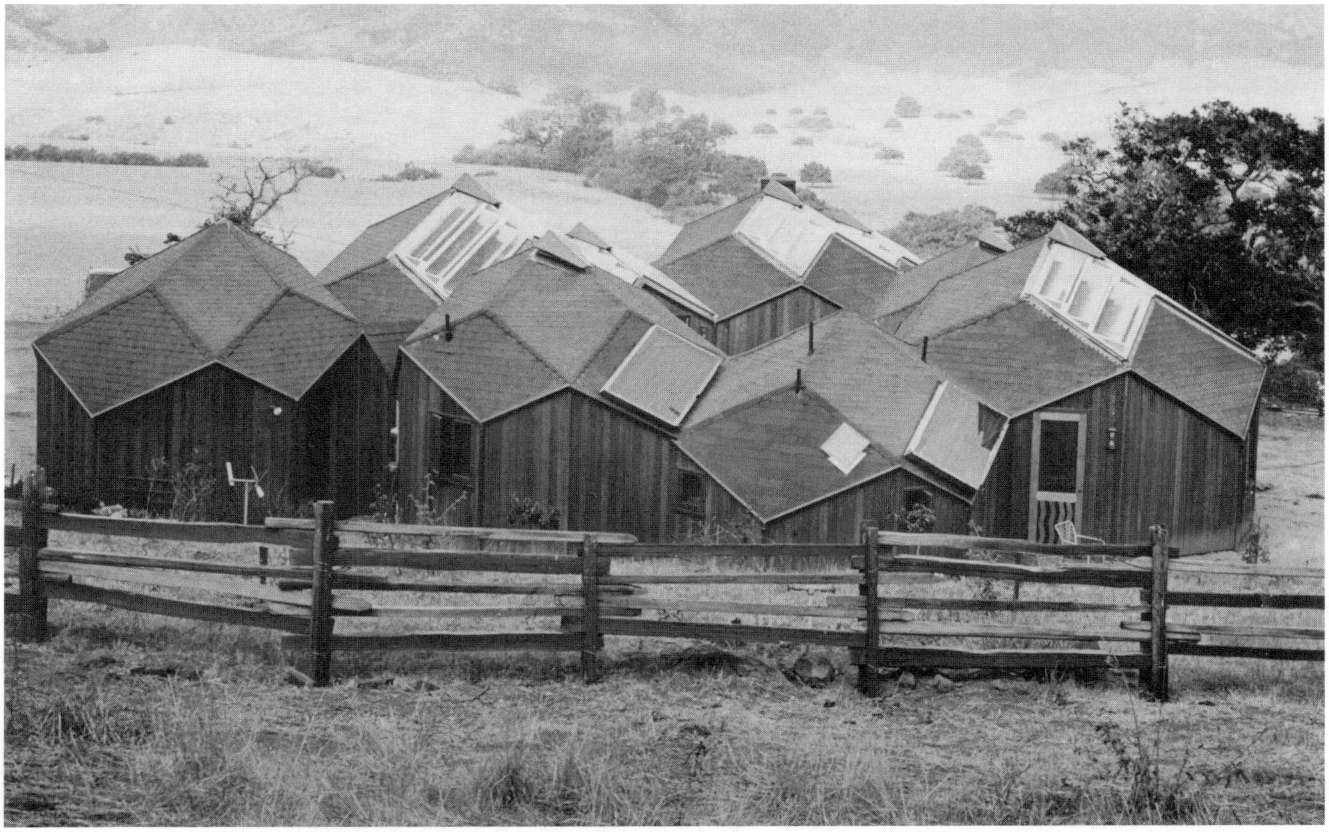

Fig 2

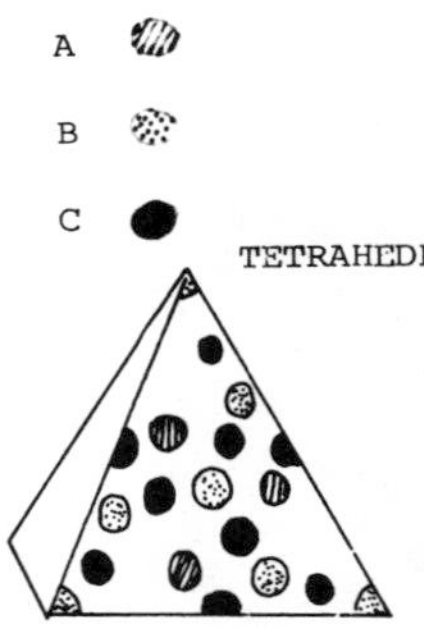

TETRAHEDRON

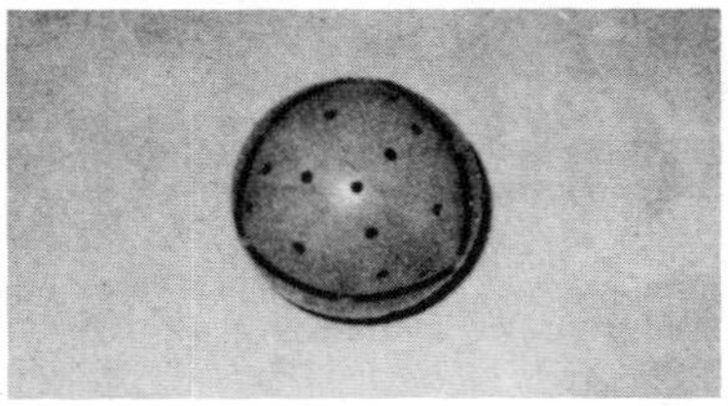

CUBE

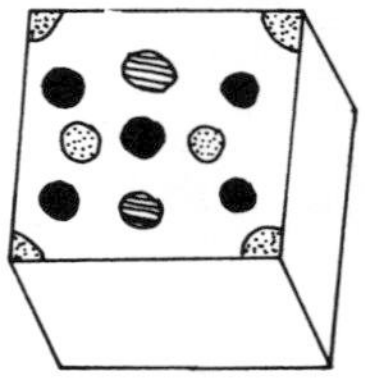

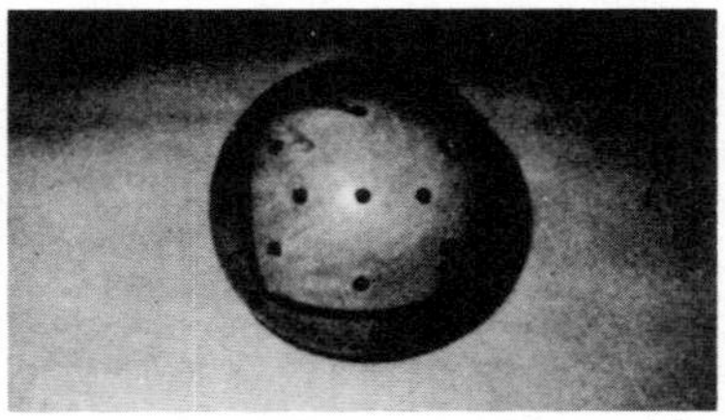

OCTAHEDRON

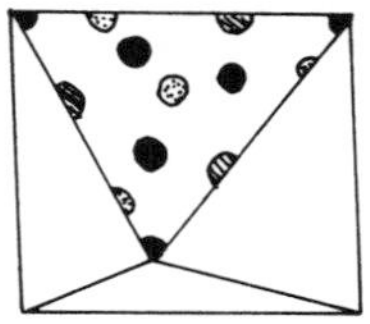

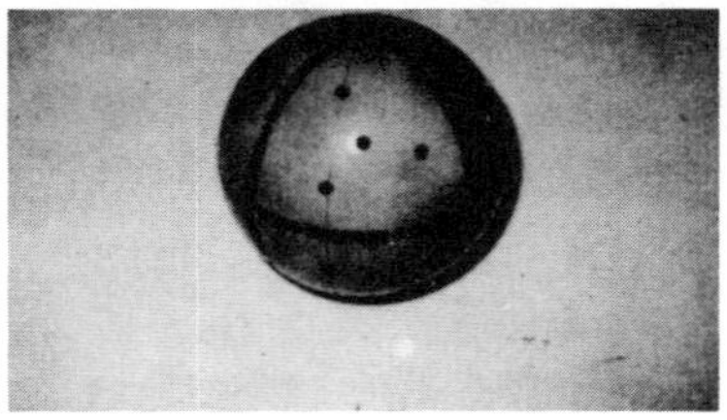

We have associated the 31-zone star throughout with the icosahedron and the dodecahedron. It also fits perfectly with the three smaller regular polyhedra. The tetrahedron, the cube and the octahedron fit inside the icosahedron and the dodecahedron. Their vertices touch a vertex, an edge midpoint or a face midpoint of the larger figure. This regular match between large and small figure, positions the smaller figure so that regular patterns on the large figure project inwards as regular patterns on the small figure. In each case either five or ten small figures fit at once within the larger figure (see illustrations in Cundy and Rollet's Mathematical Models).

DODECAHEDRON

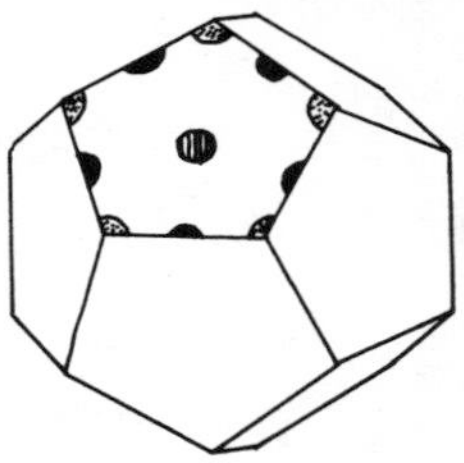

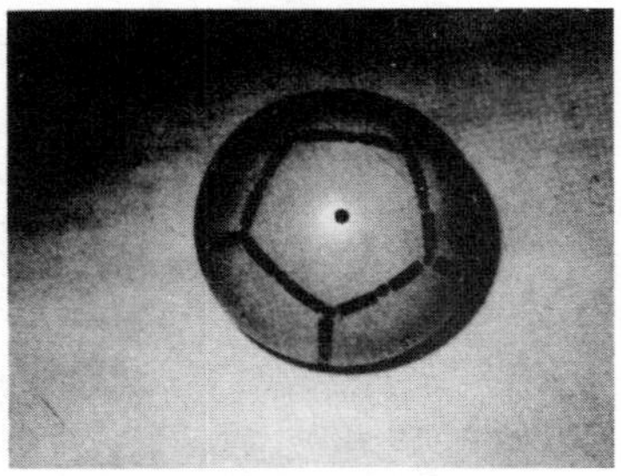

ICOSAHEDRON

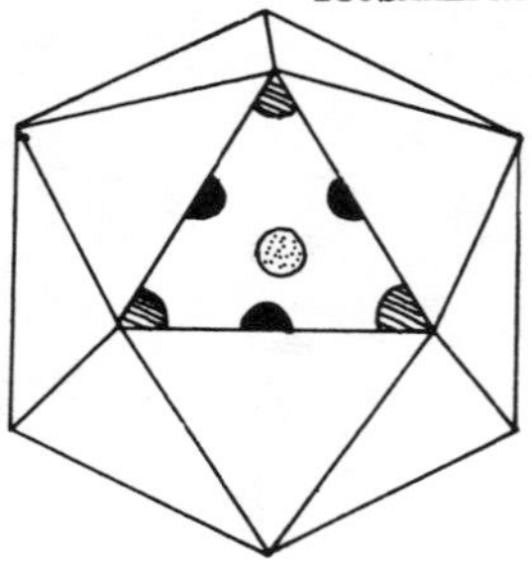

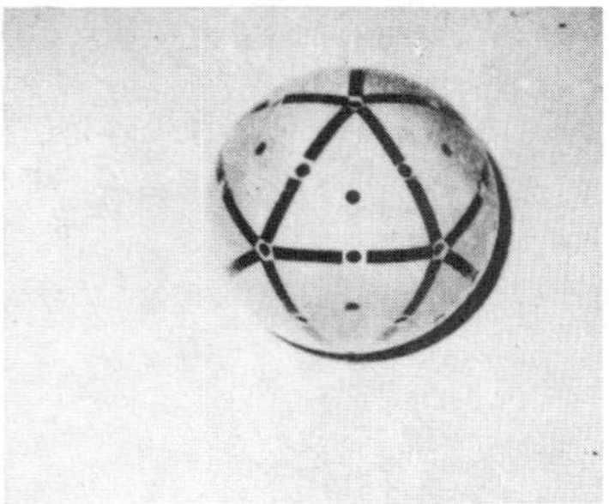

Each of the regular polyhedra is thus a convenient core from which to define the regular 31-zone star. The geometric regularities insure simplicity in the connections. Any one of the regular polyhedra can be used with the same pattern of flanges or holes on each of its faces as a connector for the 31-zone structural system. NOTE: In the case of the 31-zone pattern projected on the octahedron, four of the faces have a left handed pattern. The drawing and the photo are of patterns with different handedness.

The icosahedron and its symmetries have been relatively ignored for reasons no better than the impossibility of tiling a bathroom floor with regular pentagonal tiles. Berry Hickman, who devised the first model joints for the system from plastic golf balls, observed that if this structural system is the key to other mysteries in nature, no one would recognize it unless they already knew what to look for.

Fig 3

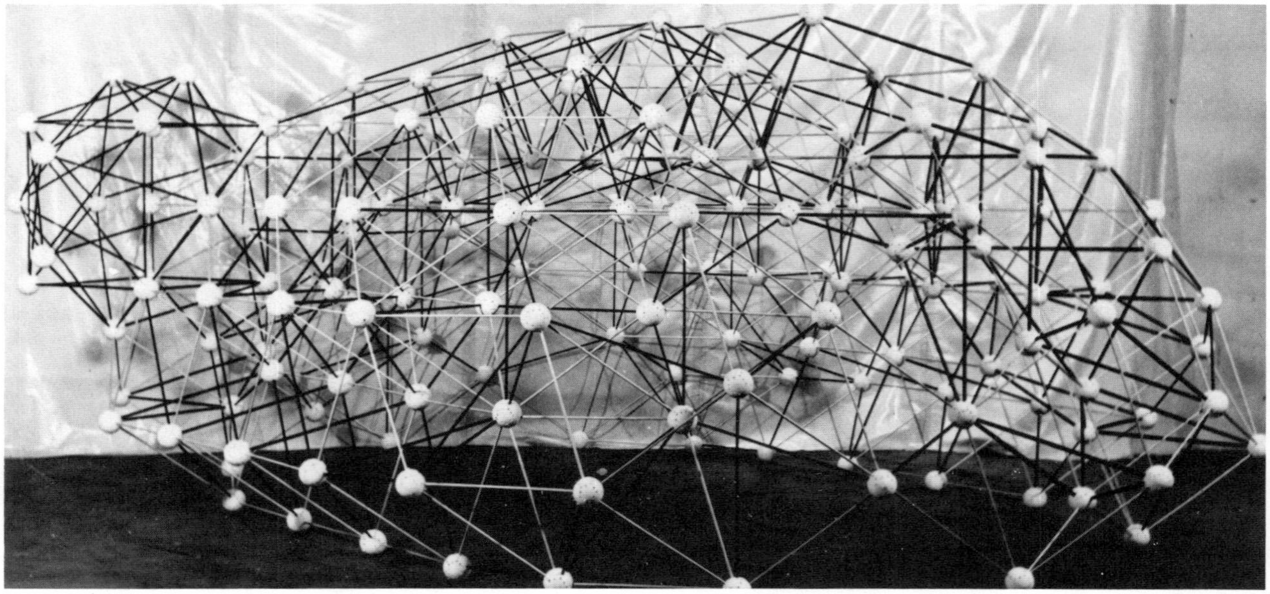

Fig 4. Gibbon Flight Cage, Albuquerque Zoo

A SPECIAL SPACE STRUCTURE CONSTRUCTION

RENE MONTEMAYOR, Architect

Technical Director, Grupo Intra, México

The challenge was to design and build a free span roof for the square court yard of the Miraflores School, located in the West of México City.
Originally, the solution given to the architectural project was the use of heavy joists in order to cover 49 m. span and after an analysis with Moduspan space frame, we came to the conclusion that it was feasible to save weight and to obtain an spectacular structure, due to its particular design without precedent in the world.

INTRODUCTION

The estimated weight of the roof was going to be 750 Tons, but with the use of the Moduspan system, it was reduced to 150 Tons.

The regular truncated pyramid was designed using eleven horizontal levels, from the concrete ring to the top level and two layers on the sloped sides to form the pyramid walls and two layers also, to form the horizontal top.

Measured risks were taken to solve a special assembly procedure. Computerized checking was also done in order to enable us guarantee the stability for the different conditions showed in the various assembly steps.

DOME STRUCTURE

The Moduspan structural system selected to build the truncated pyramid dome, works as a space frame structure. Calculation is as hiperstatic stereo structure.

Nevertheless each node and element had to be computer analysed in order to check all the efforts, developing mathematical models which can only be handled with specialized programs.

It was necessary to use a double structure in order to solve the span. Normally it is only used a superior and inferior chord, but in this case, we had to use an intermediate one, achieving a 2.4 m. depth. This resulted coherent with the architectural design intention which aimed at achieving a particular silhouette.

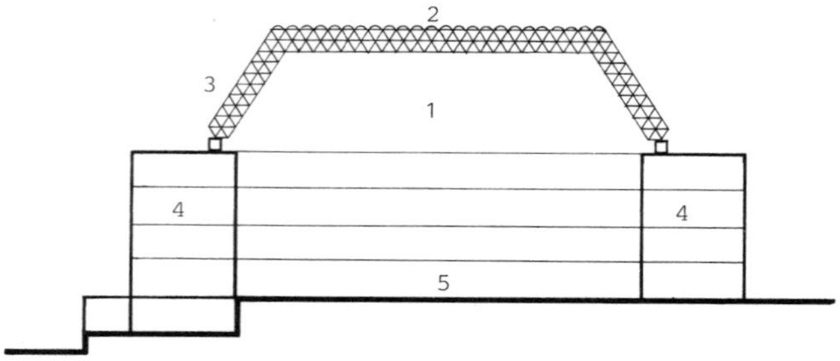

1.- Space frame structure
2.- Transparent skylights
3.- Translucent panles
4.- Building
5.- Patio

finishing the four sides of the truncated pyramid.

ANCHORS AND UNIONS

Dome is anchored to a perimetral ring which is a supported by the building concrete columns. This concrete belt includes special anchors placed at each 1.52 m. in order to fit the horizontal module of the structure.

In order to prove levels and aligments, we used a laser equipment to detect possible deflections or deviations.

On fabrication, the node joints are produced using a special 1000 Ton press forming the steel plates or connectors. It is an octahedron plate with depressions and punchings in different ways in order to accept insertion of horizontal and diagonal elements.

The horizontal section was built at the floor level, being raised by a vertical hoist all together, using an hidroneumatic system which doubled the security limit.

Both elements were jointed together without special problems and easily and precisely, thanks to the Moduspan system.

HAND LABOUR AND TIMETABLE

The experience we had previous buildings anable us to achieve the planned schedule. The size of the structure, with a height of 12.5 m. and a span between anchors of 49 m., was a new experience for us. It took 20 weeks to erect it, using only eight workers on average and their assistants.

The horizontal and diagonal elements are cold rolled profiles which are fixed to the connectors with a specially designed nuts and bolts that fits with precision to the perforations, The profiles are channels with a "C" section rounded in the borders in order to guarantee turn radius and inertia moment adecuate to the work position and resistance to the analysed efforts.

INSTALLATION

The installation on site was a very interesting problem to solve. It was stablished a proceedure which started with the building of the four corners of the pyramid to attain stability. Afterwards, the corners were continued until

The system is simple, once the assembly team learn to identify the components. All the structure is bolted together thus eliminating the hazard welding represents, unless having a great expense in ultrasonic X Rays, or similar tests.

The nuts bolts and perforations have great precision, manufacturing requires special facilities, machinery and quality control.

FINISHING

The finishing of the structural elements has not to be done in site, due that it is done after manufacturing with a special electrostatic paint, which minimize need of maintenance. The nuts and bolts are galvanized.

In the horizontal top side, acrylic transparent skylights with a half circle section shape were applied. The siding was done using translucent fiberglass reinforced acrylic sheets.

THE ZÜBLIN SPACE TRUSS SYSTEM

Werner Fastenau, Dipl.-Ing.

Head of the Development Department
of Messrs. Ed. Züblin AG
Civil Contractors, Stuttgart, Germany

General explanation of the system. Connection of tubular members to the nodes by differently pitched bolts. Construction of tubular members for tensile and compressive forces and their connection to the nodes. Manufacturing of tubular members by means of friction welding. Solid and hollow spherical nodes. Space truss chords consisting of rectangular tubes. Compound space truss structures with upper chord area formed by prefab concrete slabs.

1. General

The Züblin Space Truss System has been on the market since 1977 and has been used for a considerable number of structures in Germany, where the system is officially approved in accordance with the current building legislation. I want to report about the technical and structural details of the system.

The Züblin Space Truss System consists of tubular members with tapered ends which are connected to spherical nodes by means of special bolts. Normally, the members have circular cross-sections. Rectangular or square tubes, however, can also be used, as will be explained later.

Circular tubular members 28 to 220 mm in diameter and lengths, measured in horizontal projection, of up to 6 metres are able to endure tensile forces up to 850 kN and compressive forces up to 2000 kN.

Module and angles can be chosen arbitrarily so that virtually any reasonable space truss concept an architect might conceive can be realised with this system.

2. Connecting Elements

Fig. 1 shows the connection of a circular tubular member to a spherical node.

The characteristic element of our system is the connecting bolt which is patented in many countries of the world. We call this bolt "differently pitched bolt" because each bolt has two threads of different pitch. On the node side there is a standard pitch whereas on the member side there is a fine pitch. Between the two threads is either a hexagonal, on which a hexagonal wrench collar can slide, or a threadless part with two opposing studs on which a circular collar with internal grooves can slide longitudinally. The collar serves as a spacer between the member and node and is required for turning the bolt by means of a wrench. Before turning, there is play of several threads between the collar and the node. With each turn, the bolt is moved closer to the node by the difference between the two pitches, thus diminishing the play until the connection is tightened.

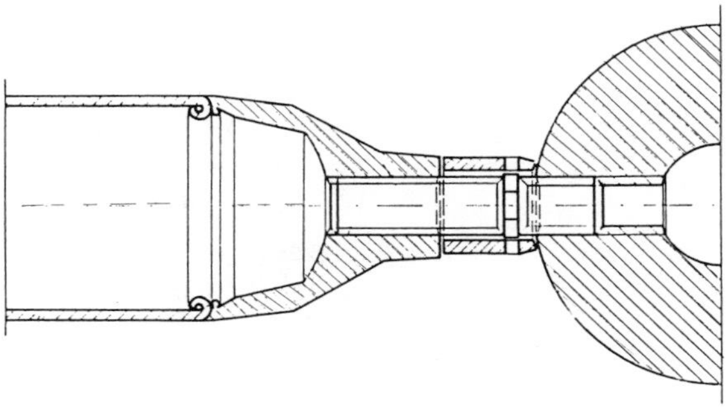

Fig. 1 Connection of a circular tubular member to a spherical node

The advantage of this connecting element is that it does not require holes to be cut in the wall of the members for the insertion of the bolts. Therefore, it offers the whole, unweakened cross-section for the sustainment of forces.

Fig. 2 shows the various bolt sizes with the pertaining collars. The smallest bolt is M14 x 2 x 1.5 ("M" stands for "metric thread", "14" is the diameter in mm, "2" and "1.5" are the two pitches in mm) for forces up to 55 kN; the largest is M52 x 5 x 3 for forces up to 850 kN. As the bolts can take both tension and compression forces, the collars are only spacers needed for the application of a wrench.

If, however, only compressive forces, especially large ones, have to be transmitted, it is more economical to do

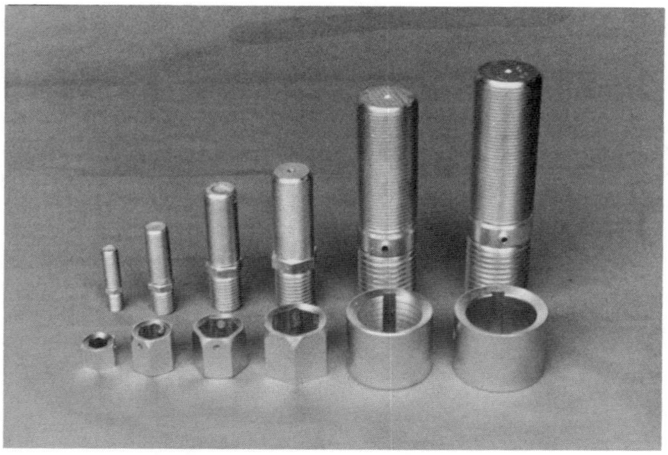

Fig. 2 Various sizes of "differently pitched bolts" with pertinent collars

this by using a special compression collar instead of a bolt. These compression collars exist in three standard sizes with capacities of up to 600 kN. They transmit the compression directly to the spherical surface of the node over a slightly conical pressure area. The advantage is demonstrated in Fig. 3.

Fig. 3 Alternative connection for compressive force of 600 kN: left, bolt size M48; right, compression collar dia. 60 mm and assembly bolt M20

In order to transmit a compressive force of 600 kN, a bolt size M48 would normally be required. However, it is more economical to use a compression collar of 60 mm diameter, which only requires an assembly bolt size M20. If the compression exceeds 600 kN, disc-shaped, non-standardised compression collars are used against flattened pressure areas on the node. These collars are dimensioned for a permitted stress of 250 N/mm.

3. Nodes

The spherical nodes are forged steel balls in 5 standard sizes ranging between 60 and 220 mm in diameter.

As it is not possible to forge steel balls accurately to size, the nodes are turned to their exact shape.

The turning can be omitted with thick-walled hollow spheres which have walls thick enough for the required threaded borings. These spheres are made by a special process about which I want to give a short explanation, since it might be of interest to some manufacturers of space trusses.

The process (see Fig. 4) starts with a cylindrical disc, the weight of which is about that of the finished ball and the diameter of which is approximately that of the desired hollow sphere. By forging or swaging, the disc is reshaped into a cup, the bottom of which is already one half of the hollow sphere while the top consists of cylindrical, tapered walls. By pressing with a mould, the upper half of the cup is then also shaped hemispherically, thus completing the hollow sphere. A navel remains

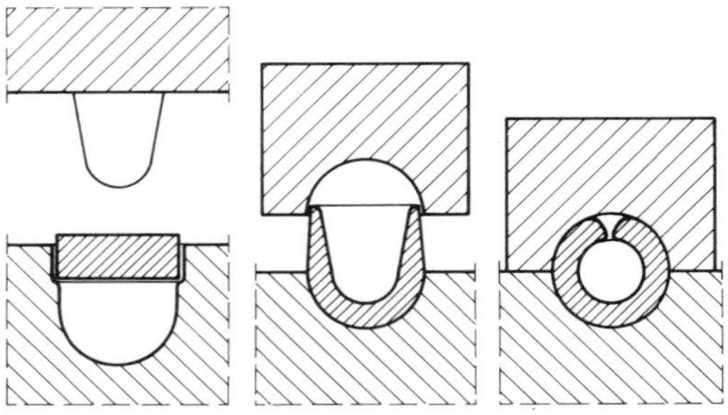

Fig. 4 Manufacturing process for thick-walled hollow spheres

at the top of this sphere which is subsequently used for the first threaded boring. As the inside of the sphere has space for surplus material, it is possible to forge the outside exactly to the desired size. Its advantages over a solid ball are that no work on a lathe is required; the weight is lower; and the internal space can accommodate the chips resulting from making the threaded borings.

4. Circular Truss Members

The circular truss members for tensile and compressive forces are welded or seamless steel tubes of St 37 or St 52, the equivalent of which would be mild steel grade 43 or high yield grade 50 in Britain. The most common sizes, between 60 and 130 mm in diameter, have tapered, drop-forged ends which are joined to the tubes by friction welding. This process is fairly unknown for civil engineers but has been used extensively for about 15 years in the automobile industry to weld rotationally symmetric parts.

The machine (see Fig. 6) is similar to a lathe. The tapered end is clamped into the revolving chuck while the tube is clamped on a sled. The sled is moved axially against the revolving chuck so that the rotating forged end is pressed against the stationary tube. This generates friction heat until the end and tubes are red hot. Then the rotation is switched off and the parts to be joined

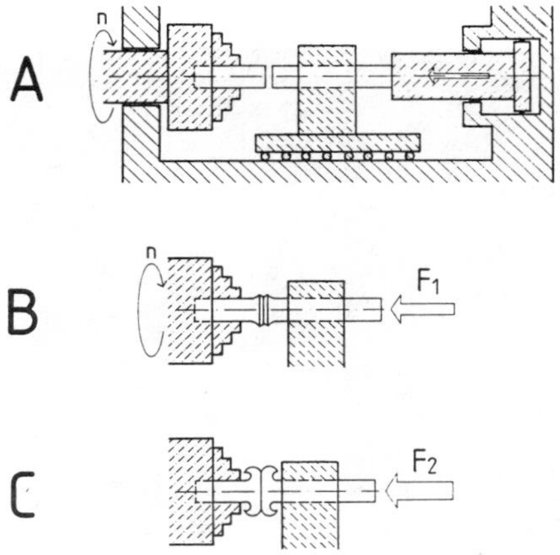

Fig. 5 Friction welding process

are pressed together using great force resulting in a butt weld. The complete process requires between 2 and 15 seconds. Fig. 7 shows the friction welded end of a Züblin truss member and a sectional view of the same end.

Fig. 6 Friction welding machine

Fig. 7 Friction welded end of Züblin truss member, left: general view, right: sectional view

The compression creates a weld reinforcement which is turned flush with the outside of the tube. For the latter process, the welding machine works as a lathe so re-chucking is not necessary. The welding parameters required for a certain combination of forged end and tube are empirically determined. They are fed into the machine which then automatically controls the whole welding process. The specified and actual parameters are continuously checked and recorded in a printout thus guaranteeing a high standard of quality.

As friction welding enables very differing materials to be joined together, it would, for example, be feasible to use this method to weld aluminium tubes to forged ends made of stainless steel. This could be of great interest for space trusses in very corrosive climates.

Small truss members, i.e. short members for small forces are not friction welded. These members are produced in a very economical manner out of sections of pregalvanised pipe which receive a fine pitch inside thread at both ends. Mass-produced end parts are screwed into these threads to complete the members. Fig. 8 shows a partial sectional view of such a member.

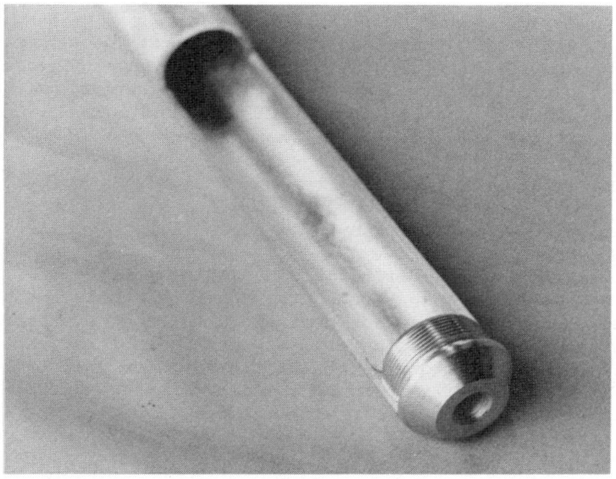

Fig. 8 Partial sectional view of small Züblin truss member with screwed conical end

For compression members for large forces, tubes with an external diameter of 159 and 219 mm and varying wall thicknesses are used. As these members are only exposed to small tensile forces, and those only during assembly of the space truss, the connection between the forged tapered ends and the tube is accomplished by a kind of "plug connection" (see Fig. 9).

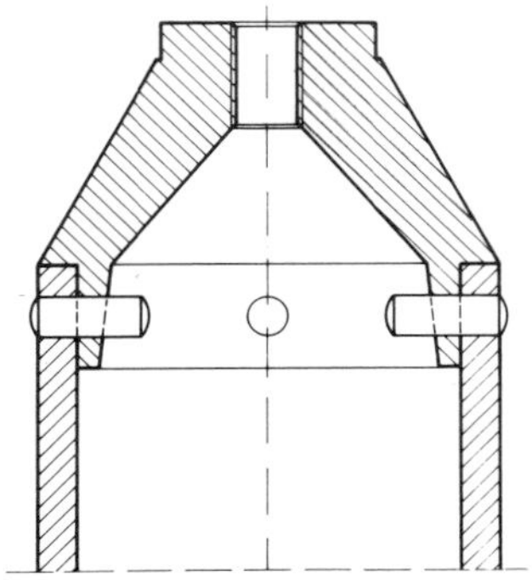

Fig. 9 "Plug connection" between pipe and tapered end for truss members for larger compression forces

Both tapered ends are turned on a lathe in such a way that they fit tightly into the tube ends. They are then pressed into the tube ends under great pressure on an assembly bench, so that the gap between the tube end and tapered end is closed completely. Then, with the pressure still applied, four holes are drilled through the walls of the tube and the tapered end and studs are driven through as connecting elements. This method facilitates in a simple manner the connection of a basic tapered end with tubes of vastly differing wall thicknesses.

5. Rectangular Truss Members

In order to avoid bending stresses in the bolts, the Züblin space truss members may only be stressed by normal forces (aside from dead loads and negligible small live loads). This means that loads may only be applied at the nodes. Consequently, continuous roof loads must be taken by purlins which transfer them to the nodes by means of vertical studs. With large space trusses, this is an acceptable solution especially considering that the purlins may be sloped on top of the horizontal space trusses thus producing, in a simple manner, the slope required for rainwater drainage.

With small space trusses, the purlins borne by vertical studs are aesthetically unsatisfactory because they burden the naturally light structure with an unbecoming heaviness. Therefore, for smaller space trusses it is desirable to have directly loadable upper chord members.

As circular truss members are not particularly suitable for the direct fastening of a roof, members with a rectangular cross-section should be used.

For our system of bottom chords and diagonals consisting of circular tubes and connecting spherical nodes, we have endeavoured to design an upper chord which consists of directly loadable rectangular members in at least one direction. One requirement that these members must satisfy is that they fit into nodes which are also suitable for the connection of circular members.

Next I am going to report about some structural solutions to this problem.

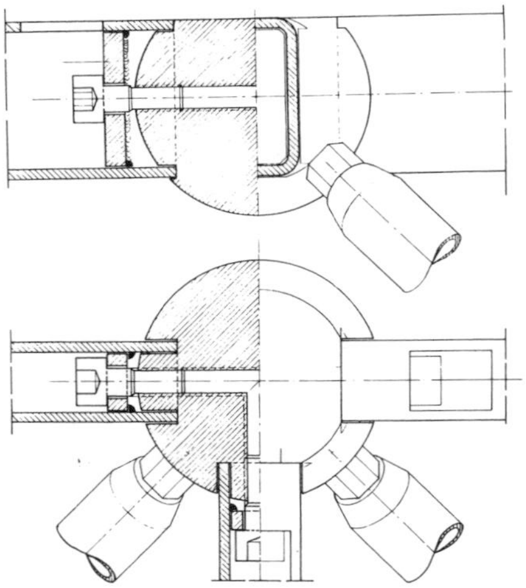

Fig. 10 Connection of rectangular members to spherical nodes

The drawing Fig. 10 and photograph Fig. 11 show how a normal spherical node can be adapted for the connection of rectangular members incurring minor additional costs. As for a circular member, this node also features a radial threaded boring for the connection of the rectangular

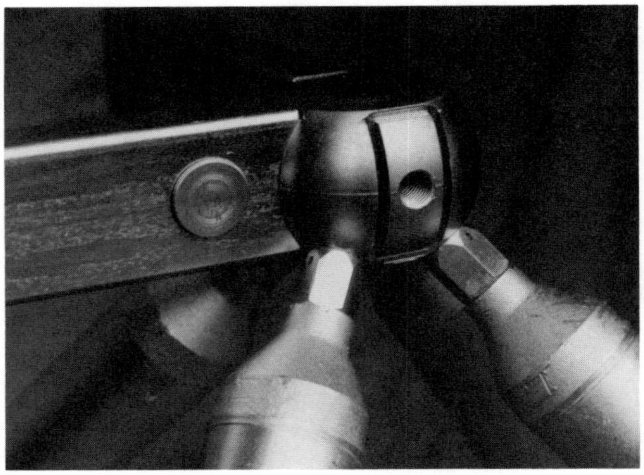

Fig. 11 Photograph of connection in accordance with Fig. 10

member. Additionally, it has milled grooves, two vertical and two horizontal, into which the orthogonally cut end of the rectangular member can be inserted so its face finds a plane pressure area. The horizontal grooves serve as support. A plate with a boring is welded into the rectangular member as an abutment for the connecting bolt. The connection is tightened using a wrench which can be applied through a window cut into one of the tube walls. In order to create a plane for the support of the roof, the sphere may be flattened on top.

Fig. 12 shows a multipart node, a photo of which can be seen in Fig. 13 with a glass cover. The rectangular member is inserted in two vertical grooves in a thick-walled pipe section. The rectangular members are supported by the flat plane of a spherical segment which is used for

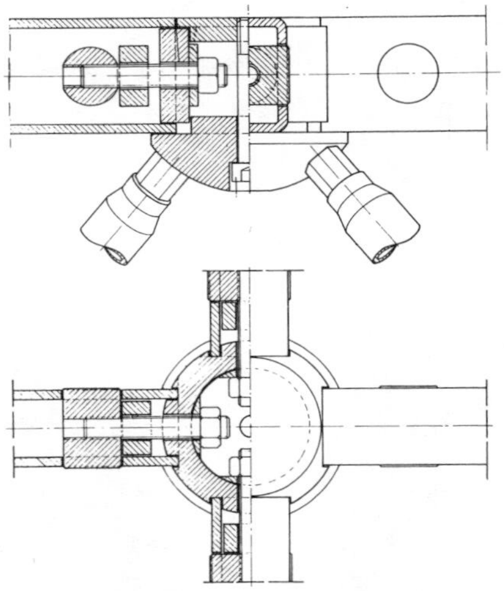

Fig. 12 Multipart node for the connection of rectangular and circular members

the usual connection with the circular diagonal members. This construction of the rectangular members is well suited for mass production without the need for any welding. The rectangular tube is simply sawn off to exact lengths and a large hole is bored near each end. A cylindrical pin with a threaded boring is simply inserted into the hole. The connection bolt is screwed into the pin and a collar-like spacer will prevent the pin from sliding out of its position.

Fig. 13 Photograph of multipart node

Another uncomplicated connection design is shown in Fig. 14. The rectangular members in this case are connected at an angle to a thick-walled pipe section. Thus, the node could be part of a double curved single layer structure, e.g. a cupola or a sphere as shown in Fig. 15. Near the end of the members, short anchoring profile pieces are welded into the tube parallel to the end face. This anchoring profile provides an abutment for a hammer-head bolt for the connection of the node.

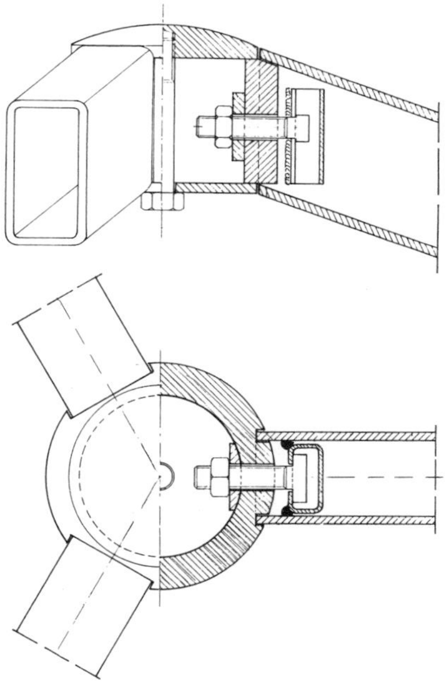

Fig. 14 Mode of connection for rectangular members

6. Züblin Compound Space Truss

Züblin is traditionally a contracting and engineering firm which specialises in concrete structures. Therefore, it is not surprising that our development department came up with the design for a compound space truss system.

The general principle is that the bottom chords, as well as the diagonals, which are loaded by tensile or compressive forces, consist of the normal circular members connected

Fig. 15 Icosaeder built with rectangular truss members

to spherical nodes in the bottom layer. However, the compression-loaded upper chord members are replaced by prefabricated concrete slabs which take up the compression forces as well as forming a roof cover.

The crucial problem is the transmittance of forces between the diagonals and the prefab slabs. A special upper chord node was developed for the connection of the diagonals with the prefab slabs, Fig. 16. This node is a square steel plate with four conical borings on its upper side near the corners. These borings are threaded at the bottom, while the centre part of the underside is spherical. This spherical segment is used for the usual connection of

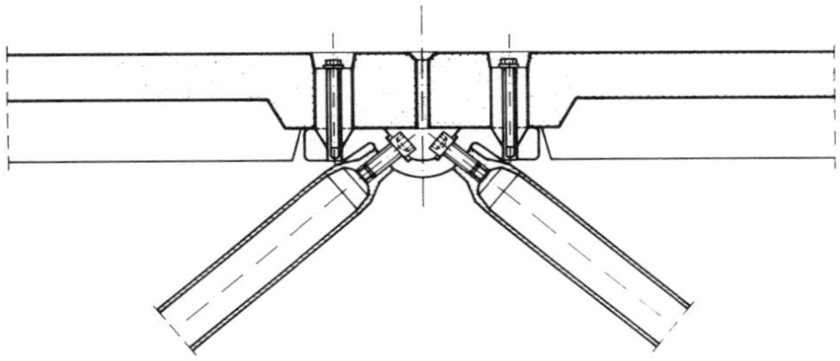

Fig. 16 Special upper chord node for Züblin compound space truss

the diagonals. The prefab slabs are supported at their corners on top of the node. As even factory-made concrete elements cannot be expected to be as accurate as is required for space trusses, and because slabs which were simply pressed together at the edges would tend to spall, the slabs have special connections at the corners. These are cast-in tapered steel dowels which fit into the conical borings of the node. The final connection is achieved by a bolt, which is passed through the hollow dowel from the top and then screwed into the threaded boring at the bottom. During assembly, all forces between slab and node are passed through the dowel while an open joint of about 1 cm width remains between the concrete edges of the slabs. At a later stage, this joint is filled with cement mortar thus connecting the prefab slabs into one large rigid slab which is well suited to take considerable horizontal forces. At this stage, upper chord compression forces due to live loads need not be transferred by the dowels but can be passed directly from one slab to the next via the edges.

The prefabricated slabs are very thin, e.g. 5 cm, and have reinforcing beams at all edges, thus forming modular slabs. These slabs need not be square but can also be rectangular, triangular or trapezoidal which offers the architect a vast field of design possibilities. The construction allows very simple solutions for the installation of skylights and skylight bands, for the design of cornices and various kinds of supports at the edges of the space truss.

Fig. 17 Compound space truss as roof over canteen

The unfavourably larger weight of this kind of space truss is compensated by acoustic advantages, a point of great importance with roof structures for concert halls and lecture halls.

Last but not least, the clear formation of the underside of the modular slab, of which Fig. 18 is an example, is definitely an architectural asset.

Fig. 18 Underside view of compound space truss with modular prefab slabs

SOME PRACTICAL DESIGN ASPECTS OF ARTANE SHOPPING CENTRE

NAEL G. BUNNI, BSc, MSc, PhD, CEng, FIEI, FCIArb, MConsEI.

CONSULTING ENGINEER, T. J. O'CONNOR & ASSOCIATES, DUBLIN.

This paper describes some of the practical design problems and their solution in a 9000 sq. m. shopping complex where three roof structures were designed utilizing six space frames with specially designed joints. Four of these were merged together to form a continuous plate, the fifth was designed as a truncated triangular three way space frame, and the sixth as a simple rectangular frame. Support conditions varied. The Quality Control, which was actually exercised to ensure a satisfactory completion of a very precise and demanding structure, is also described.

INTRODUCTION

It has been said that the scientist dissects a problem into its constituent parts for the purposes of analysis whereas the artist selects materials, assembles them together and builds; the scientist concerns himself with the objective arena of facts and phenomena whereas the artist deals more with the subjective field of imagery and feeling in all its five senses and revels in the unique qualities of experience. This paper deals more with the artistic facets of engineering, in what the Greeks called "Techne", the original Greek term for art, and predecessor for words such as Technique, Technology, etc. (1)

In particular, the paper deals with Artane Shopping Centre, a 9000 sq. m. prestigious development designed architecturally to the highest standards with an active interest taken by the developers. The structural design of the roof incorporated six space frames to span over 57% of the shopping area with the remaining zones of the centre's roof incorporating structural steel lattice girders of various spans.

The Architectural concept of the project was based on a central concourse in the shape of a truncated equilateral triangle designed to be the focal point of the whole complex with an entrance occupying one of the shorter sides and the shopping areas radiating from the other five sides, (See Fig. 1). The choice of a space frame structure was based on the following aspects of the project:

a) Large unsupported spans of the order of 33m to 60m to provide unimpeded access within the main concourse to all the radiating areas and to provide a shopping area without hindrance to interior planners and designers.

b) An overall short period of time was available between commencement of work on site and final completion in order to fit in with the Christmas shopping.

c) Accommodation of large service ducts that could be assembled and erected in a very limited period of time to fit in with the over all short construction period allowed for the project. In the solution adopted, the ducts were in fact assembled on the ground after completion of the space frame assembly and the whole frame with the service ducts was lifted into position.

d) Cost of the project, with all the above requirements being met, was the most attractive.

e) The intense electric lighting requirement in a shopping centre necessitates normally a secondary support system at a lower level than roof level. The space frame lower chords provided this secondary system without the additional cost.

f) A planning requirement was a low lying structure with an overall height of development simulating the surrounding dense housing area. The allowable depth of the structural membrane had to be curtailed to a minimum.

g) An attractive structure which would obviate the necessity for a false ceiling and at the same time provide a facility for display of signs etc. without an expensive secondary support system.

CHOICE OF SPACE FRAME SYSTEM

Once a decision was taken to use a space frame system, three of the well known types of proprietry prefabricated nodes were considered but they were, very quickly, rejected on the basis of cost. The cost of nodes was excessively high. Thus began a long journey into the art of structural design in search of a node which could accommodate the various configurations used in the design of the six space frames. The node had to be of a simple shape which could be fabricated in the workshop of any one of four of five structural steel fabricators in Ireland.

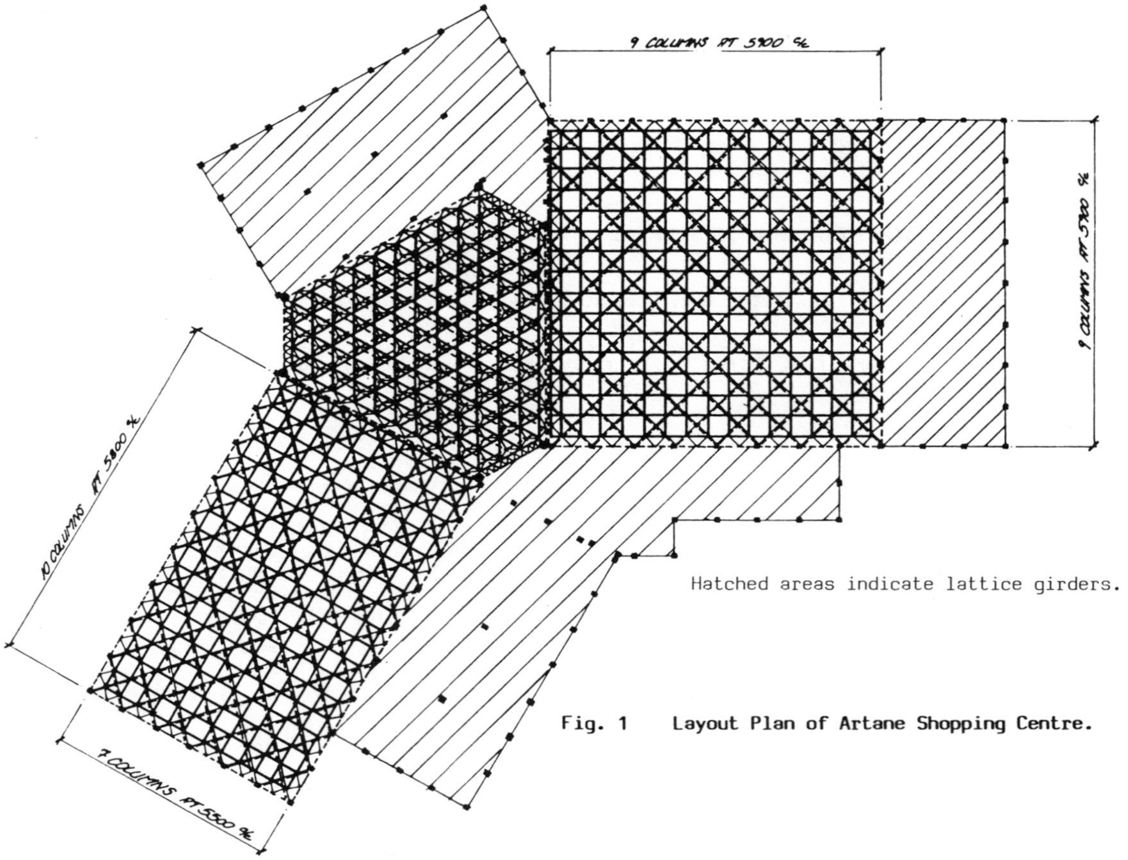

Fig. 1 Layout Plan of Artane Shopping Centre.

In this connection, the steps of three eminent engineers were followed, Professor Z.S. Makowski (2), Professor Stéphene du Chateau (3) and Mr. Ronald G. Taylor (4), to create a welded node from elementary and readily available structural steel sections.

The module developed was therefore of a pyramidal shape with a typical node at the corner points. Two top nodal welding details were adopted (see Fig. 2) and these were later analysed for their relative efficiencies. The pyramids formed the top chord and diagonal members of the space frame when fully assembled.

At first the bottom chord members, after fabrication, were individually assembled on site to such relative levels as the frame would take when fully assembled inclusive of any camber. The pyramids were then inverted so that their bases would form the top chord members, and their sides would form the bracing, and were bolted to the bottom chords through their apex nodes. The top nodes were also connected together using bolts.

DETAILS OF THE SPACE FRAMES AS BUILT.

The six frames were as follows:

1. Frame No. 1: The Concourse frame was designed using pyramids of equilateral triangular base with a modular dimension of 3.365m. The long and short sides of the concourse itself were 31.370m and 11.234m respectively, thus necessitating the use of 96 pyramids plus 27 part pyramids. Bottom chords parallel to only two of the long sides of the concourse were used (See Fig. 1). The depth of the frame was 1.8m and it was supported on six CHS steel columns at the corners. The nodal welding detail adopted for the top node of this frame was type 2.

2. Frame No. 2: The first shopping area had a clear span of 52.2m x 33m. A more traditional design was adopted here by choosing a pyramid of a rectangular base with a modular dimension of 2.9m x 2.75m. The bottom chords ran diagonally on plan forming with the pyramids a rectangle on diagonal space frame pattern. The frame was supported along the edges at intervals of twice the module except along the concourse where three alternate columns were omitted forming a large entrance.

3. Frames No. 3 to 6: The second shopping area was

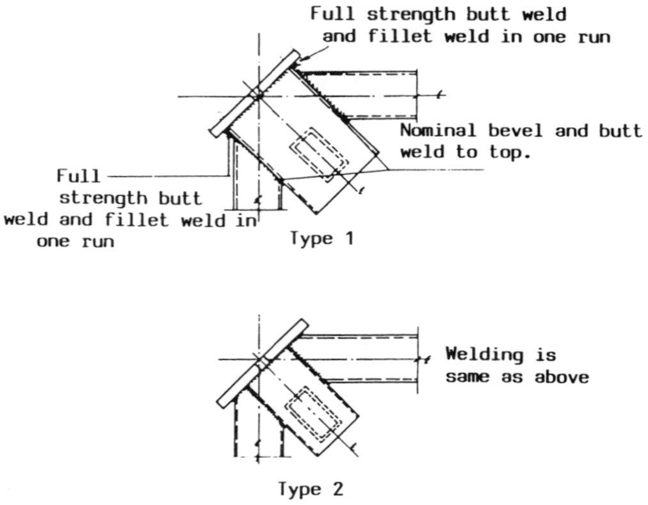

Fig. 2 Welding Details of the top Node.

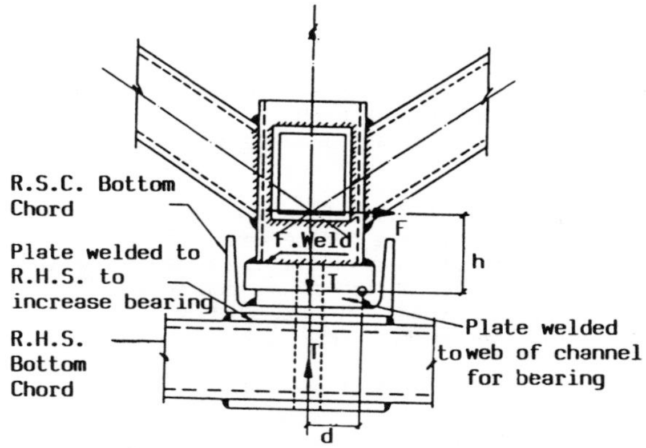

Fig. 3 Forces at a Bottom Node

square in shape, 47.2m x 47.2m. Due to the maximum depth of space frame allowed in this area, i.e. 1.65m a central stanchion was used to support the four frames. The modular dimension of 2.95m was used and the bottom chords ran diagonally in plan at 45° forming with the pyramids a square on diagonal pattern. It was found after analysis that the four frames should be designed as continuous over the central stanchion. The four frames were supported on steel stanchions in an unsymmetrical pattern, as can be seen in Fig. 1.
The welding detail adopted for the top nodes of frames 2 to 6 was of type 1.

CONNECTION DETAILS

The forces in the various members and the deflections under the nodes were calculated using the computer. The results were used to select the member sizes taking into consideration the bending effect in the top chord members across the direction of sheeting. An effective length coefficient of 0.85 was used in the design calculations of compression members.

The forces were analysed at top nodes and a standard size of bolt to resist the shearing force was used where compression forces prevailed. However, where tension forces existed, especially over the central support in Frames No. 3 - 6, the bolts had to be designed to resist the resultant force due to shear and tension, and bolts of up to 75mm diameter grade 8.8 were used.

At the bottom nodes, the bolts had to be designed to resist a combination of tensile forces due to bending moments and shearing forces as can be seen in Fig. 3. The tensile force in the bolt is given by $T = Fh/d$, where the variables are as indicated in Fig. 3.

Bolted splices in the bottom chords were minimised by increasing the length of individual members wherever possible. This is not always easy since this criterion has to be weighed against the size of the force at any considered position, changes in section sizes of the structural members which necessitates the concentric application of forces on either side of the splice, and architectural considerations.

In many instances, the bearing stresses exceeded the allowable, and plates of various sizes had to be added between the two elements in contact.
Fig. 4 shows typical details used.

ANALYSIS OF MODULAR FORM AGAINST MASS

This project was the brain child of the client's chief executive, a very clever man, who was interested in maintaining a balance between the high standard of design required and cost. Various steps were taken to ensure that an economical solution was being adopted and a number of ideas were implemented to maintain a check on the cost.

Frames 3, 4, 5 and 6 were designed and fabricated first and one of the early discoveries made was that the top node as designed added considerably to the total mass of the structure. Although the unit cost of these nodes is relatively low as compared with proprietry prefabricated nodes, nevertheless top nodes contributed a significant percentage to the total mass and therefore cost. Analysis of mass also showed that the top node contributed between 9.1% and 17% of the total mass, therefore the cost. An attempt was made to reduce this cost and hence it was decided to adopt the nodal welding detail Type 2 shown in figure 2. This detail reduced the percentage difference between the theoretical mass (based on the mass of members and excluding the weight of the top node) and actual mass (which includes the top nodes and all other connections) from 18% in Frames Nos. 3, 4, 5 and 6 to 15% in Frame No. 1. Analysis of mass also showed that bottom node contributed between 1.9% and 3.4%; plates contributed between 4.2% and 6.4%; bolts between 1.5% and 1.7% making a total of between 20.4% for frame No. 1 and 26.2% for frame No. 2.

The theoretical mass in kilogrammes per square metre for Frame No. 1 was 41, for Frame No. 2 it was 25.66 and for Frames 3 to 6 it was 23.46. The actual masses were 47.77, 30.61 and 27.68 respectively.

The masses of top chord members, bottom chord members and diagonals in kg per sq. m for each of the six frames were:

Frame No. 1:	18.32	14.0	8.7	
Frame No. 2:	11.3	8.7	5.6	respectively
Frames 3 to 6:	9.8	8.3	5.4	

This analysis lead to the adoption of the following equation for a quick determination of the mass of a space frame selected for consideration, once one establishes its geometry and modular dimensions:

1 Geometrical layout -v- Mass:
The relationship between the geometrical layout of the frame and member sizes on the mass of the space frame per sq.m. can be expressed by the equation:

$$M = C_1B + C_2T + C_3D \quad \ldots\ldots\ldots\ldots\ldots\ldots\ldots\ldots\ldots (1).$$

where M is the total mass of the space frame in kg per sq. m.
B is the average mass per m of Bottom chord members
T is the average mass per m of Top chord members
D is the average mass per m of Diagonal members

The factors C_1, C_2 and C_3 incorporate the geometric properties of the frame in that they represent the total length of the respective member divided by the area of the frame. Equations were developed to predict the values of these parameters for space frames of different geometrical shapes. Thus, the following equations represent frames which are square, rectangular, and triangular in plan. Other shapes were also taken into consideration in formulations outside the scope of this paper.

For a rectangular frame, where x is the length of the structure, y is the width, h is the depth of the space frame, m is the modular length of a square or approximately square pyramid and k is 2m/x.

$$C_1 = \frac{2}{xy}\left[\frac{y\sqrt{2}}{k} + x + y\right] \quad \ldots\ldots\ldots\ldots\ldots\ldots\ldots\ldots (2)$$

$$C_2 = \frac{2}{xy}\left[\frac{2y}{k} + x + y\right] \quad \ldots\ldots\ldots\ldots\ldots\ldots\ldots\ldots (3)$$

$$C_3 = \frac{4}{xy}\sqrt{\frac{k^2x^2}{8} + h^2}\left[\frac{2y + kx + ky + k^2x}{k^2x}\right] \quad \ldots\ldots\ldots (4)$$

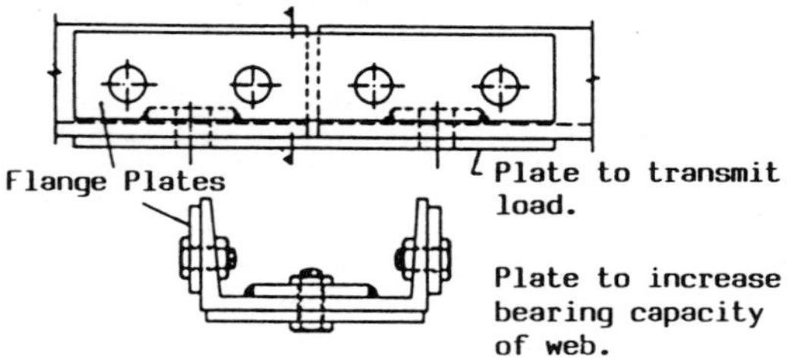

(a) Site Splice Detail for Channels.

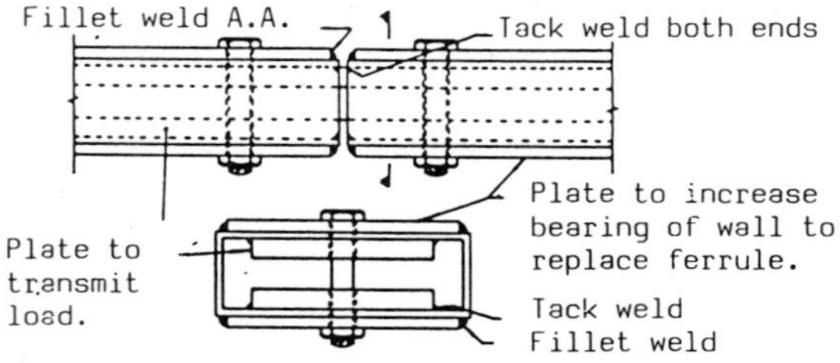

(b) Site Splice Detail for R.H. Sections.

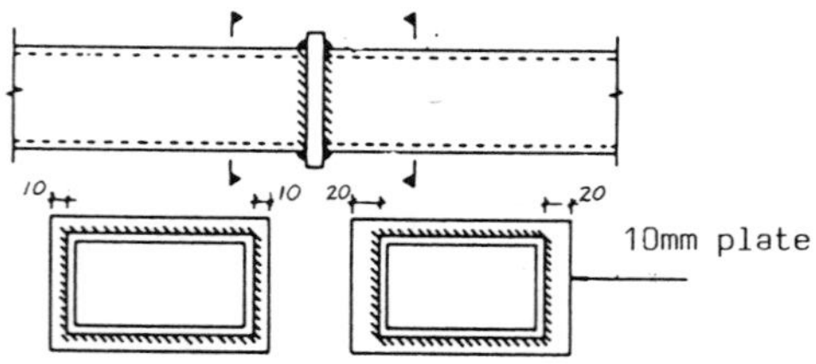

(c) Shop Splice for 120 x 60 R.H.S. to 100 x 60 R.H.S.

Fig. 4

It should be remembered that C_1, C_2 and C_3 have units of m^{-1}. Other important characteristics of this shape are

Total Number of Internal Top Nodes =

$$4\left[\frac{2y - kx - ky + k^2x}{k^2x}\right] \quad \ldots\ldots (5)$$

Total Number of perimeter Top Nodes =

$$\frac{8}{kx}(x + y) \quad \ldots\ldots (6)$$

Total Number of Top Nodes =

$$\frac{4}{k^2x}(2y + kx + ky + k^2x) \quad \ldots\ldots (7)$$

Division by the area of the frame, gives the number of Top Nodes per square metre of frame area which was found to be approximately constant at 0.27 for frames 2 to 6. The effect of this important detail on the mass of the space frames can therefore be estimated in any subsequent design.

For an equilateral triangular frame, where x is the length of each side, h, m and k are as defined for a rectangular frame.

$$C_1 = \frac{8(k + 1)}{\sqrt{3}\,kx} \quad \ldots\ldots (8)$$

$$C_2 = \frac{2\sqrt{3}(2 - k)}{kx} \quad \ldots\ldots (9)$$

$$C_3 = \frac{4\sqrt{3}}{x^2}\sqrt{\frac{k^2x^2}{12} + h^2}\left[\frac{k - k^2 + 2}{k^2}\right] \quad \ldots\ldots (10)$$

$$\text{The total number of Top Nodes} = \frac{3(k - k^2 + 2)}{k^2} \quad \ldots\ldots (11)$$

Division by the frame area gives the number of Top Nodes per square metre as follows:

$$\text{Number of Top Nodes/sq.m.} = \frac{4\sqrt{3}\,(k - k^2 + 2)}{k^2x^2} \quad \ldots\ldots (12)$$

It can be seen from the above equations that once one is given the geometry of the frame, a quick calculation can establish the mass of the frame and hence the cost. If the cost is established, one can tell whether or not he is within budget and, if not, what he has to do to get back within the constraints imposed.

As an example, let us take frame No. 2, where the figures were: x = 52.2m, y = 33m, h = 1.60m, pyramid size 2.9m x 2.75m.

$$k = \frac{2(2.9)}{52.2} = 0.11$$

and substitution into equations 2, 3 and 4 gave the following values for C_1 C_2 and C_3

$C_1 = 0.59m^{-1}$, $C_2 = 0.79m^{-1}$, $C_3 = 0.68m^{-1}$

A quick preliminary design estimated the values of B, T and D as 15 kg/m, 16 kg/m and 7 kg/m respectively. The theoretical mass per square metre of frame No. 2 was therefore

$$0.59(15) + 0.79(16) + 0.68(7)\ \text{kg/m}^2$$

or 26.25 kg/m² which is only 2% greater than the correct figure of 25.66 kg/m², as calculated from the drawings.

PRACTICAL POINTS TO REMEMBER

In all the frames which were designed using the above node, eleven in all, the following practical points arose and are therefore important to mention:

a) In frames which are designed with an inbuilt camber, the bottom chords should be laid on stubs over the ground with a contour resembling the final shape of the cambered frame. Assembly of the pyramids should start from the highest point (generally in the centre) and then follow a radiating rotational pattern towards the exterior perimeter of the frame.

   As the pyramids are lowered and bolted the frame picks itself off these stub gradually. This upward movement should be watched carefully since parts of the frame could become supported in a pattern that would not have been envisaged during the design stage.

b) The designer must consider the lifting operation at the design stage and must cater for any reversal in the member forces. Other points that have to be checked by the designer and not left to the Contractor as the latter is not familiar with the intricate behaviour of such complicated structures, are:
   - b.1. The capacity of the cranes proposed for use and their manouverability. The load expected to be picked up by each must also be estimated.
   - b.2. All cranes should preferably be of the same capacity and manufacture to ensure that the lifting speeds are compatible.
   - b.3. The lifting points must be considered carefully. If lifting is to be done through the top nodes, the forces in the adjacent members will be different to those if the bottom nodes were used.
   - b.4. If a lifting beam is used, its proper design is of paramount importance, taking into consideration its deflection and buckling tendencies in view of the strut action that it is subjected to through the lifting ropes. This action imposes extremely high compressive forces on the beam which will buckle if it is

not designed with such action in mind. In practice, of course, this beam would be a long member, 10 to 12m long.

c) If the space frame is being lifted on to its supports using more than one crane, it is imperative that the lift is coordinated meticulously to ensure that all cranes operate at the same lifting speed and all start and stop at the same time. Failure to adhere to this procedure could end in a disaster due to the fact that the supporting system could change from that envisaged in the design.

d) If structural steel hollow sections are used as top chord members, then it is imperative to prevent any rain water from entering these sections through the fixings of the roof sheeting. This event takes place either if the method of fixing of the sheeting is improperly specified and carried out or if at the joints between the various lengths of the sheeting, the overlap detail does not include a sealant. Should this happen, rainwater would enter the hollow sections prior to the application of the waterproofing layer and would cause an increase in the dead weight, but more importantly, it would cause corrosion of the internal surfaces of the hollow sections.

e) Quality control is of the essence in the fabrication, assembly, and erection of space frame structures, due to the fact that errors multiply through repetition of the modular units. Only welders of the highest quality must be used. They have to be capable of reproducing high quality welding despite the repetitious and tedious nature of this work. Samples must be taken from proposed welders, dissected, tested and visually inspected prior to approval. Continuous visual inspection must be readily available and every joint and weld must be tested visually if it cannot be radiographed. Even when radiography is used, it is imperative that control is exercised in respect of checking the results and reports submitted.

FUTURE IMPROVEMENTS

A testing programme on various specimens of the node discussed here has been formulated and at the time of writing, tests are underway to provide a deeper understanding of the complicated local forces that exist in the following elements of the space frame structure: top node; bottom node; the bolt connecting the bottom node to the bottom chord members; the effective length coefficient of top chord members; and other details. At present the nodes including plates and bolts, represent 25% of the total weight of the frame and this programme should lead to a reduction of this proportion.

ACKNOWLEDGEMENTS.

My thanks are due to all my engineers and technicians who have worked with me on the design of Artane Shopping Centre in particular and all the other space frame structures in general. Of those I would like to name the following engineers, Mr. E. Comerford, Mr. W. Forristal, Mr. M. Moriarty and Mr. T. Dunphy.

REFERENCES

(1) "History and Development of the Engineering Profession", by Dr. L. Calini, A paper delivered at the 1983 FIDIC Annual Conference in Florence.

(2) "Steel Space Structures" by Z.S. Makowski, Published by Michael Joseph Ltd., London.

(3) "Plane and Curved Tridimensional Structures, Wide Enclosures - Industrialization and Architectural Form", by Stéphane du Chateau.

(4) "The Future for Steel", by Ronald G. Taylor, The Structural Engineer Journal, Volume 59A, No. 2, February, 1981.

THE DESIGN AND CONSTRUCTION OF STEEL SPACE STRUCTURE COVERING MAIN EXHIBITION SITE AT EXPO'85, TSUKUBA, JAPAN

T. SHIMIZU
Government Building Department,
Ministry of Construction

T. KITAJIMA
Yasui Architects

Y. MIKAMI
Yasui Architects

F. MATSUSHITA
Tomoegumi Iron Works, Ltd.

T. YAMAGUCHI
Tomoegumi Iron Works, Ltd.

The international exhibition so called Expo'85 is scheduled to take place in Tsukuba, Japan in 1985. The main exhibition site named "Information Station" is an out-door theatre financed and operated by the Japanese Government. The theatre is capable of accommodating about 3,000 seats and equipped with a huge electric light board on the stage.

The space frame covering the theatre is firstly required to satisfy the architectural expression in harmony with fan-shaped field stand. Secondly, it is also required to provide large column-free areas for large number of spectators. This follows that the space frame has to be supported by the minimum number of columns placed not to disturb any spectator's sight towards the stage. The structure is scheduled to be removed after the exhibition is completed.

Particulars are given of the structural form, space structure system and design.

INTRODUCTION

After the prudent planning and development, the final roof shape selected to meet the requirements of architects is a pair of fan-shaped hyperbolic paraboloids as seen in Fig 1. Its grid configuration is three-way double-layer grid on account of its great flexibility in layout and positioning of columns, and also its remarkably even stress distribution under unsymmetrical loadings such as wind load, seismic load and so on.

A pair of fan-shaped roof structures consists of two completely identical structure whose dimension is such that a vertical angle is 55 degrees and a radius is 68 meters long. The complete structural expansion is introduced between two locating them as shown in Fig 2 at equal space intervals. Each roof structure is supported by only three columns placed close to the centre of a fan-shape and two corners respectively. At each column, the space grid is supported by three diagonal members of "umbrella" type so that the concentrated reaction forces can be well distributed into space grid.

The space structure system adopted for such an extraordinary space frame with a fascinating architectural form is "TOMOE UNITRUSS" of Tomoegumi Iron Works, Ltd., Japan. Tomoe Unitruss System is mainly composed of mid-air spherical "balls" and tubular members with circular section connected together without introducing eccentricity by high-strength "connectors" with screws at both ends. These parts are all prefabricated under numerical control without shop drawings by application of computer. It was believed that this feature of the system can easily solve the problems involved in the fabrication and construction of such an extraordinary space frame. In fact, this space frame consists of about 3,300 members of slightly different lengths and about 800 nodes connecting several members in space at different angles without any eccentricity.

An exact analysis on a member by member basis was carried out considering the structure as a pin-connected structure for all possible combinations of loadings such as dead load, snow load, wind load, seismic load and even rise and fall of temperature. Structural analysis was actually done for a half of the whole frame because two fan-shaped roof frames are completely identical in both geometry and supporting conditions. Wind-tunnel tests were performed to find out the wind pressure coefficients on roof surface, and also to ascertain the significance of wind-induced vibration of some flexible slender members which are exposed directly to wind.

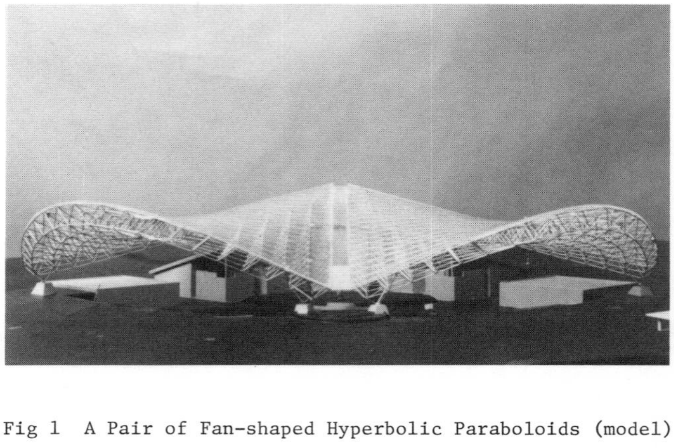

Fig 1 A Pair of Fan-shaped Hyperbolic Paraboloids (model)

OUTLINE OF STRUCTURE

I) General

This roof structure is composed of two identical hyperbolic paraboloidal three-way double-layer grid as shown in Fig 2. The plan of each roof is approximately a fan-shape and its scale is such that the radius is 68 meters long, the vertical angle is 55 degrees, and the distance between the centre of the arc and the vertex is 79.75 meters. Two identical fan-shaped roof are situated as described in the figure at equal space intervals of 3 meters.

Each space frame is supported by three columns. Supposing these supporting points are A, B and C, the distance between A and B and also between A and C is 60.4 meters, and the distance between B and C is 54 meters. A supporting point A is located close to the centre of fan-shape and supported by steel tublur column which is 15.3 meters tall. The other supporting points B and C are at lower level than A by about 9 meters, and rested on the reinforced concrete stub columns.

Since the grid configuration is three-way double-layer, it has a great rigidity in plane and yields the behaviour of a shell-like structure. The depth is 3 meters which is about 1/20 of the distance between the supporting points B and C to ensure the overall bending rigidity. A fixed pin bearing unit capable of rotating in any direction by 2 degrees is used at each supporting point. The three-way double-layer grid and such a bearing unit are interconnected by three diagonal members so that the concentrated reaction forces can be well distributed into space grid. The structural behaviour of this geometry under a uniformed loading closely approximates an arch action for the strip element between the supporting points B and C, and a beam action for the strip elements between A and B and between A and C.

This space frame consists of 3,324 members and 806 nodes including supporting points as a total.

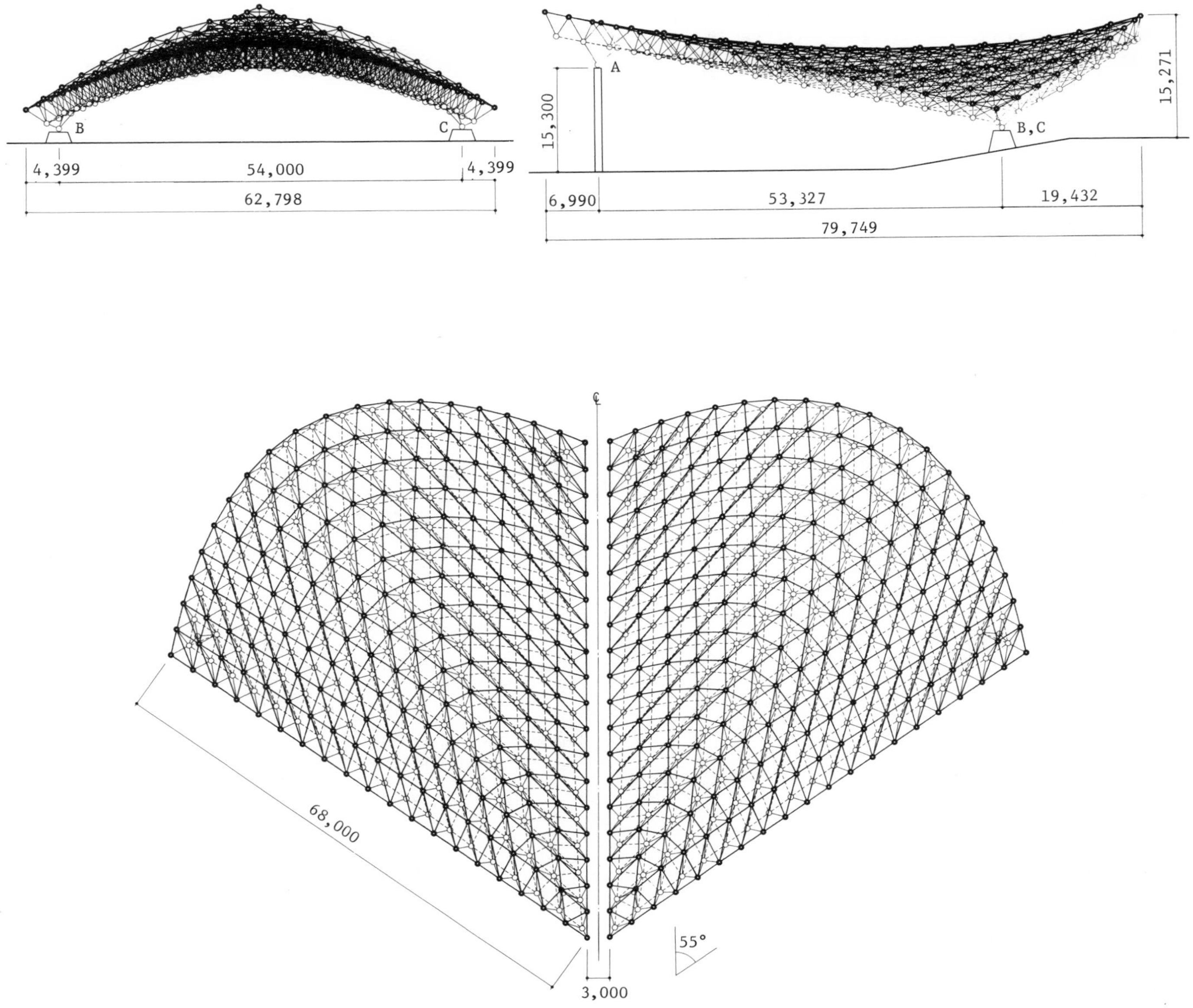

Fig 2 Framing Plan and Elevation

II) Geometry

The upper surface equations, using the co-ordinates (X,Y,Z) in Fig a, are described as follows:

when $X < 0$,

$$Z = 8.5775 + \frac{9\,(X + 13.0686)^2}{(60.3167)^2} - \frac{9\,Y^2}{(31.3989)^2} \qquad \text{Eqn 1}$$

when $X \geq 0$,

$$Z = 8.8398 + \frac{9\,(X + 4.9561)^2}{(37.1442)^2} - \frac{9\,Y^2}{(31.3989)^2} \qquad \text{Eqn 2}$$

In even such a space grid with an extraordinary form, it is desirous to make a reticulated grillage of triangles as nearly equilateral as possible in order to make the force distribution more even under a uniformed loading.

To obtain the upper node co-ordinates, let us consider the equation to transform an equilateral triangle ABC composed of 361 congruent triangles defined by the co-ordinates (x, y) in Fig b into a fan-shape A'B'C' defined by the co-ordinates (X,Y) in Fig c. Paying an attention to the facts that a fan-shape is symmetrical with respect to X-axis and the origin E is transformed into E', the transformation equations are assumed to take the following form:

$$X = C_0 + C_1x + C_2x^2 + C_3y^2 \qquad \text{Eqn 3}$$

$$Y = C_4y + C_5x^2 + C_6y^2 \qquad \text{Eqn 4}$$

These coefficients are obtained by substituting the co-ordinates of points A, B, D, E, A', B', D' and E' which are already known into Eqns 3 and 4, and subsequently they appear as follows:

$$X = 19.537 + 1.6478\,x + 0.0537\,x^2 - 0.0198\,y^2 \qquad \text{Eqn 5}$$

$$Y = y \qquad \text{Eqn 6}$$

X and Y co-ordinates of upper nodes are finally obtained by transforming each intersecting point of a reticulated grillage of triangles in Fig b using Eqns 5 and 6.

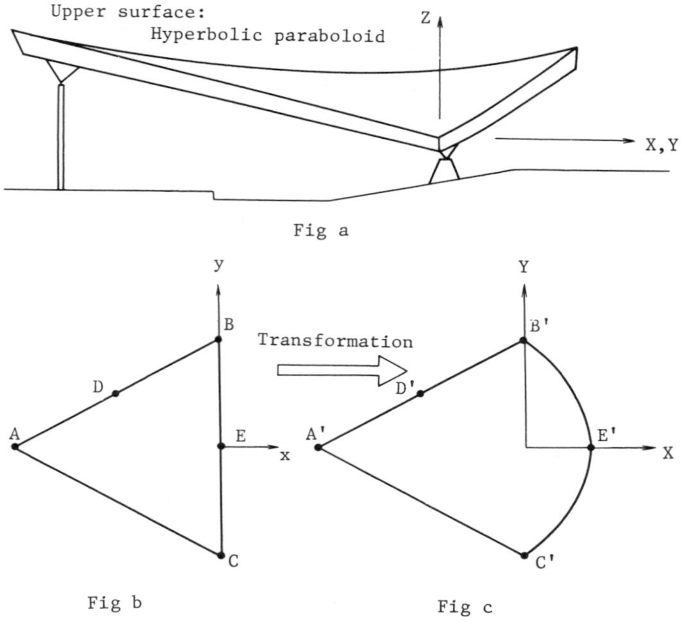

Fig a

Fig b

Fig c

Fig 3 Geometry of Hyperbolic Paraboloid

Z co-ordinates of upper nodes are also obtained by substituting their X and Y co-ordinates into Eqn 1 or 2 according to the values of X co-ordinates.

The co-ordinates (X,Y,Z) of each bottom node is defined as those of the point at 3 meters distance on the normal line of upper surface out of the centre of gravity of the corresponding upper triangle whose co-ordinates are (Xo,Yo,Zo), that is,

$$(X,\ Y,\ Z) = (Xo + A\sqrt{C},\ Yo + B\sqrt{C},\ Zo - \sqrt{C}) \qquad \text{Eqn 7}$$

where

$$A = \left(\frac{\partial Z}{\partial X}\right)_{X = Xo}$$

$$B = \left(\frac{\partial Z}{\partial Y}\right)_{Y = Yo}$$

$$C = \frac{9}{A^2 + B^2 + 1}$$

SPACE STRUCTURE SYSTEM

I) Standard Parts of Tomoe Unitruss System

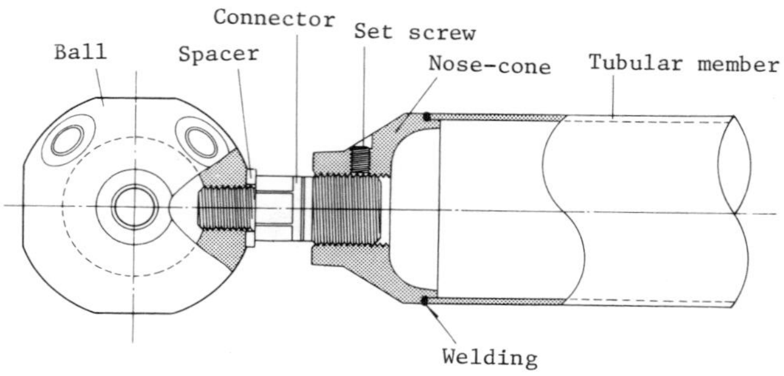

Fig 4 Standard Parts of System

Tomoe Unitruss System is mainly composed of mid-air spherical "ball", high-strength "connector", "nose-cone" with the shape of a frustrum of a cone and "tubular member" with circular section as described in Fig 4. A set of nose-cones is welded to the both end of a tubular member. A set of connectors connecting ball and nose-cone without any eccentricity consists of right-handed and left-handed screws. Thus, the distance between the centre of two balls can be delicately adjustable by rotating a tubular member. On completing field assembly, set screws are screwed up to prevent connectors from becoming loose.

As far as the combination of ball, connector and tubular member is concerned, connector size is simply selected in a one-to-one corresponding rule according to member size, and ball size is selected according to the angle formed by every combination of two members and their sizes. In combination of standard parts, strength of connector and ball, and their safety were confirmed by some experiments including biaxial loading tests on ball. As a result, it was found that the ultimate strength of connector is more than 2 times as much of the allowable tensile strength of the corresponding tubular member in any possible combinations, and also the safety factor of ball is more than 2.5 at any possible multiaxial force conditions to which ball is practically subjected.

II) Computer Application from Design to Fabrication

As the rapid development of computers with very large memory capacity, the application of computer has been expanded from analysis throughout to design and fabrication in the field of space structures. Such an application of computer can be most effective if it is applied to the space structure system which is composed of fully standardized parts and their combination can be given in a very simple rule. Tomoe Unitruss system has such features in addition to inherent features of space frames.

In a very complex space frame, it is composed of large number of members with different sectional properties and lengths and also large number of nodes of connecting several members in space at different angles. It is really a tedius work to manufacture all the parts one by one in the convensional way with adequate accuracy. It is, however, easily understood that if all the parts are fully standardized and their combinations are given in a simple rule, they can be prefabricated with ease and accuracy according to the data output by computer with aid of NC machines. Fig 5 illustrates the flow chart of design classifying the workings by man and computer, and also automatic drawing by computer.

STRUCTURAL ANALYSIS AND DESIGN

An exact analysis on a member by member basis was carried out for a half of the whole structure composed of 1,662 members and 403 nodes considering the structure as a pin-connected space frames. The loadings taken into account in design are as follows:

(1) Dead load: 50 kg/m^2

(2) Wind load: Design wind pressure 220 kg/m^2
Wind pressure coefficient according to the experimental results

(3) Snow load: 20 cm x 2 $kg/cm^2 \cdot cm$ = 40 kg/m^2

(4) Seismic load: Base shear coefficient Co = 0.2

(5) Temperature: Rise and fall of temperature 30°C

Wind-tunnel tests were carried out to find out the wind pressure coefficients on roof surface due to an extraordinary roof shape against 16 directions of wind changing its direction by every 22.5 degrees. Fig 6 illustrates the typical mean wind coefficient distribution against the wind from one of directions, where the possitive denotes inward pressure coefficient and the negative means outward one. The wind pressure assumed in design is sufficiently on the safe side against the wind speed with a 50-years mean recurrence interval even if considering the change of wind pressure observed in the experiment. Three typical coefficient distributions were chosen for analysis simplifying the experimental results. As members forming the space frame are exposed to wind, the additional wind-tunnel test was carried out to ascertain the significance of wind-induced vibration of some flexible slender members at low range of wind speed. Experimental results show that the effect of wind-induced vibration on those members is quite small giving full justice of the design criteria. All the members involved in this space frame were analysed and designed for all possible combinations of these design loads.

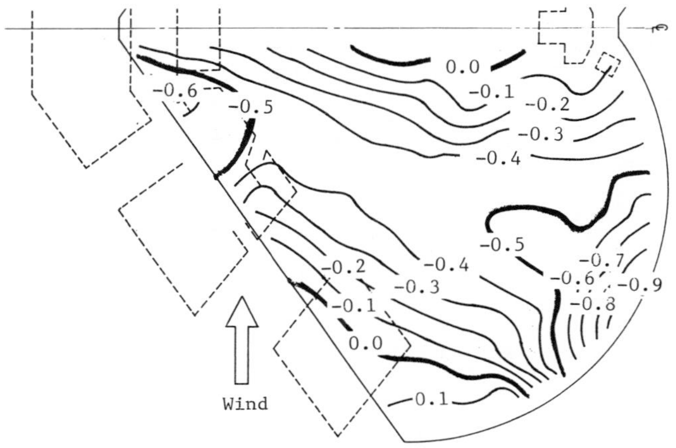

Fig. 6 Typical Wind Pressur Coefficient Distribution

In addition, the safety of space frame was checked against over loading of snow and in earthquake. As far as the snow load is concerned, a member may reach allowable strength initially under the snow load of 52 kg/m^2, and a member may reach the initial failure under the snow load of 72 kg/m^2. These snow loads are equivalent to the snow covering with a 30-years mean recurrence interval and a 100-years interval respectively at Mito City close to site. Regarding to the seismic load, a member could reach the initial failure under the base shear coefficient of 0.5. As the space frame is highly statically indeterminate and its local failure does not results in immediate collapse of the whole structure, structural reliability of the structure is sufficient.

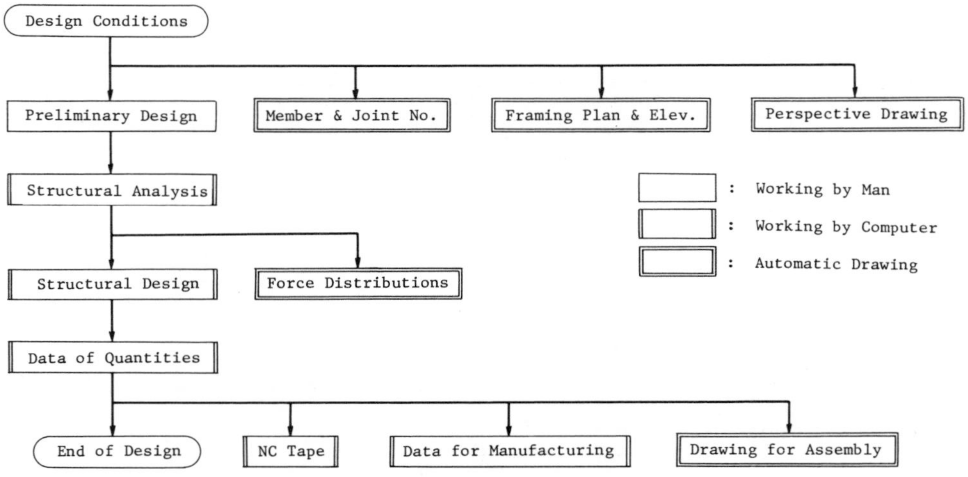

Fig. 5 Flow Chart of Design

Table 1 classifies the sizes and quantities of tubular members and balls, where T and B stand for tubular member and ball respectively, and additional figures mean the outside diameter of members and balls in millimeter.

The total weight of the structure is about 220 ton.

Table 1 Sizes and Quantities of Members and Balls

Tubular members		Balls	
T 89.1	722	B 160	34
T 101.6	278	B 180	110
T 114.3	594	B 200	142
T 139.8	1,246	B 230	286
T 165.2	446	B 260	108
T 190.7	18	B 300	126
T 216.3	16		
T 267.4	4		

PREFABRICATION AND ERECTION

I) Prefabrication

Fig. 7 illustrates the flow chart of manufacturing process of ball, connector, nose-cone and tubular member classifying the working by computer and by automatic equipments. A ball is formed from very thick seamless tube by the special plastic operation, and manufactured by drilling and tapping under numerical control as shown in Fig. 8. A connector is manufactured by hot forging and thread rolling. A set of connectors consists of a right-handed screw and left-handed one. A nose-cone is manufactured by hot forging, and then machining and tapping. A set of nose-cones is welded to the both ends of a regular welded tubular member by automatic welding as described in Fig 9.

After prefabrication, these parts are assorted and packed in so compact manner as to be handled and transported very easily.

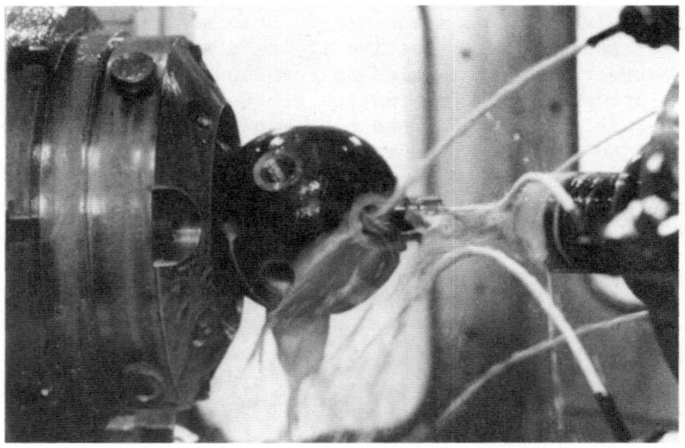

Fig 8 Drilling and Tapping of Ball by NC Machine

Fig 9 Automatic Welding Nose-cone to Member

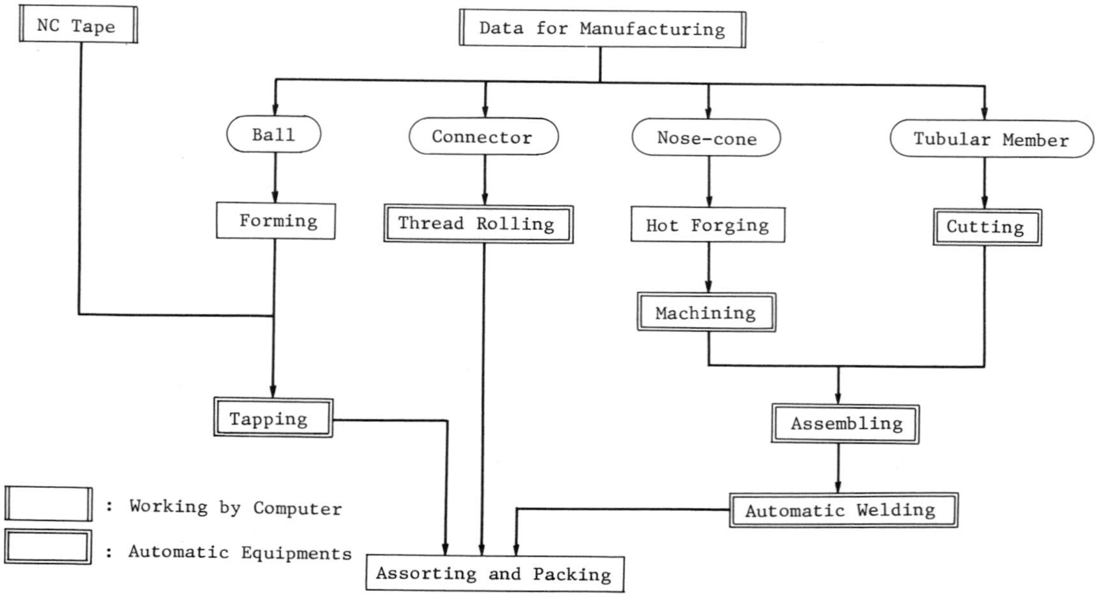

Fig 7 Flow Chart of Manufacturing Process

II) Erection

Tomoe Unitruss System does not require skilled labours in assembling on site. As all the parts forming the space frame are prefabricated with high accuracy, there is no difficulty in assembling them to any shape of structure maintaining adequate accuracy by use of just simple provisional jigs.

There are four typical methods of erection available for the construction of space structures:

(1) Element method: direct erection of single-unit members.

(2) Block method: erection of sub-assemblies constructed near ground level on simple form work.

(3) Left-up method(a): assembling the skeleton on the ground and then raising it into position.

(4) Left-up method(b): assembling the skeleton and fitting it with covering and services completely on the ground, and then raising it into position.

These methods of erection have their own features. In making a practical selection of the most appropriate method, it is important to take account of the design, erection and safety conditions. As the method of erection is often closely related specially to design conditions, its selection is desirous to be done at early stage of planning.

In this project, field erection is planned in such a way that each skeletal element is interconnected together one by one on the entire scaffolding. Prior to field erection, a trial erection for the part of the frame was performed as seen in Fig 10. As a result, it was confirmed that the assembled space frame can maintain adequate accuracy to complete the entire framework. Field erection is scheduled to start in February, 1984.

Fig 10 Trial Erection Prior to Field Erection

ACKNOWLEDGEMENT

The authors are grateful to Mr. T. Murota, Head of Wind and Rain Laboratory, Ministry of Construction who conducted the wind-tunnel tests.

A TIMBER POLE DOME STRUCTURE

* By Dr.Ir Pieter Huybers

* Research Group on Plastics and Building Technology
Materials Science Section of the Civ. Eng. Dept.
Stevin Laboratory
Technological University
The Netherlands

At the occasion of an exhibition on alternative methods and technological solutions which took place in the autumn of 1983 at Amsterdam, The Netherlands, a dome structure had been built composed of wooden poles as the main structural members. For the interconnection of the elements a specially designed, manually operated tool had been used with which steel wire can be wrapped around the joint, stretched and secured, thus offering a greater breaking strength than is found with comparable bolted joints.
The geometric data needed for the construction have been obtained making use of an interactive computer programme originally developed for the automatic generation of polyhedra.

Geometric design

The dome structure had an octagonal groundfloor plan of 220 cm edge length. A horizontal framework of intersecting timber beams with a rectangular cross-section was used as the foundation onto which the structure had been fixed. An octagon has a circumscribed circle of the radius:

$$R = \frac{\text{edge}}{2 \sin \text{phi}}$$

with phi = $180^{\circ}/p$
and p = number of edges of the polygon.

$$\text{Thus: } R = \frac{220}{2.\sin 22{,}5^{\circ}} = 287{,}4 \text{ cm}$$

The overall shape of the structure was of more or less ellipsoidal form, approximated by a hemispherical part placed on top of an octagonal antiprism.

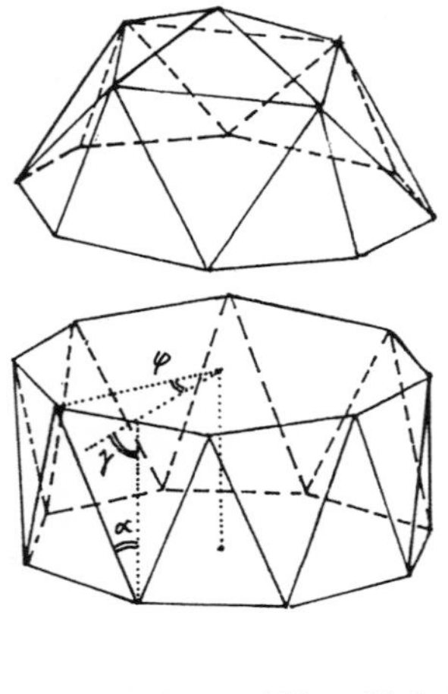

Fig 1. Approximation of an ellipsoid by the combination of an antiprism and a hemisphere.

An antiprism reminisces to a prism in that both have roughly the shape of a box with a - virtually regular - polygon as the bottom and as the lid. In the case of the prism the sides of the box (or the "mantle") are formed by a perimetric row of rectangles (or squares for regular prisms). In the antiprism bottom and top polygon are twisted with respect to each other over a certain angle phi, and as a result of this fact the mantle consists of triangles.
The hemisphere on top of this antiprism had an octahedral subdivision of 2-frequency. This means that between the original edges of the octohedron new points are created, lying on the same circumscribed sphere. This results in an octagonal equator which fits onto the antiprism.

Geometry of the antiprism

This dome consists of only two kinds of different member lengths : A = 220 cm and B = 287,4 cm. The choice of a regular antiprism for the lower part of the structure with equal member lengths for the octagonal as well as for the triangular faces would have led to a height of the first ring of free nodes of:

$h = \frac{1}{2} A\sqrt{3} \sin \text{gamma}$, with

gamma = arc cos (tan alpha * tan phi/2) (Ref 1).

with alpha being half the top angle in the triangle (see fig. 1). In this case, alpha = 30°, and phi/2 = $11{,}25^{\circ}$
Thus:

gamma = $83{,}40548178^{\circ}$, and

$$h = \frac{1}{2} \cdot 220 \cdot \sqrt{3} \sin \text{gamma} = 191 \text{ cm}$$

This height is not sufficient for unobstructed passage. Therefore the same member length has been chosen for the bars in the antiprismatic mantle as those of the second type in the sphere, viz. B = 287,4 cm.
This results in:
alpha = $22{,}5^{\circ}$ and gamma = $85{,}273927^{\circ}$ and a height
h = 264,7 cm (Ref 2)

Interactive acquisition of data

In order to determine all necessary geometric data the two shapes were examined making use of an existing CPS (Conversational Programming System) programme. (Ref 3)

The shape of all semi-regular solids can be defined by the rotation of a n-gon with 1 to 10 edges, respectively, from a vertical position at the origin and perpendicular to the z-axis according the positions of the faces of one of the five regular solids and of a few other eligible semiregular solids. 10 Different rotation cases have been distinguished. (Ref 4)

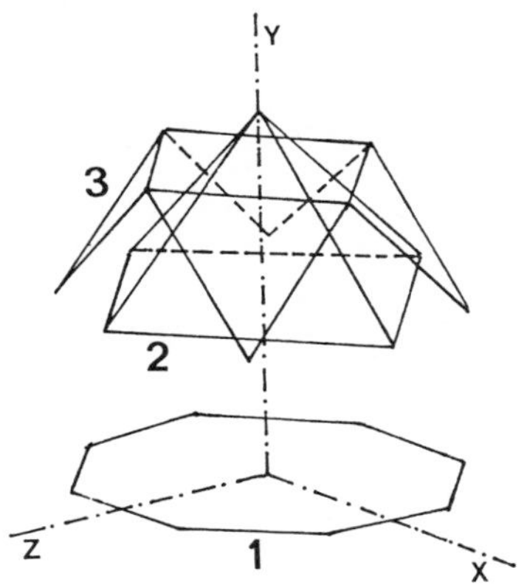

Fig 3. The three rotation cases.

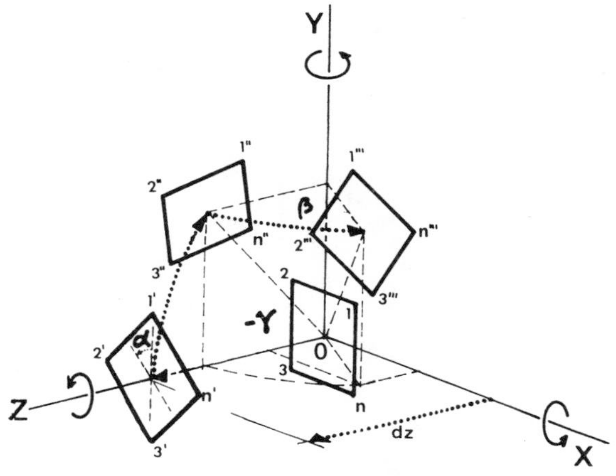

Fig 2. Formation of polyhedron by the rotation of a n-gon.

Input data for the three calculation runs:

	Run 1	Run 2	Run 3
n1	8	1	1
n3	11	2	6
dk	0	0	0
dz	0	.70710678	1
KK3	-90, 0, 0	0, 0, 0	0, 0, 0
QQ3	0, 0, 0	0, 264.66057, 0	0, 264.66057, 0
FACTOR	220	406.50699	287.44385

n1 = number of edges of the rotated polygon
n3 = rotation case
dk = eventual initial rotation of n-gon
dz = initial translation of n-gon along 2-axis
KK3 = rotation of completed figure around x-, y- or z-axis
QQ3 = translation of completed figure in x-, y- or z-direction
FACTOR = magnification ratio

This process of generating the co-ordinates of the polyhedron faces is performed in four stages, once the user has stated: the number n of the polygon faces, the number of the rotation case which is applicable, the distance dz of the translation along the z-axis (this value can be found in Ref 1, dealing with the geometry of polyhedra in general), a possible additional rotation of the n-gon about the z-axis.
The other rotations through the angles around z-, x- and y-axis are carried out automatically in a manner and with a frequency depending on the rotation case under consideration. Apart from that, options are available for giving the polyhedron as a whole an additional rotation KK3 or translation QQ3 with respect to the x-, y- or z-axis and moreover to multiply it by a factor, thus diminishing or magnifying its overall dimensions.
The results of as many as 10 consecutive computation runs can be safeguarded simultaneously, so that interactive computations become possible. The results are normally supplied in the form of co-ordinates of the respective polyhedra, but at wish also lengths face angles and dihedral angles can be supplied upon request.
In the present case 3 calculation runs are necessary for the generation of all occurring nodes (Fig 3). The generation procedure also denominates the various types of nodal points by unique numbers consisting of three elements, in order of sequence: number of corners in rotated n-gon, number of n-gon in the specific rotation case, number of the run (Fig 4).

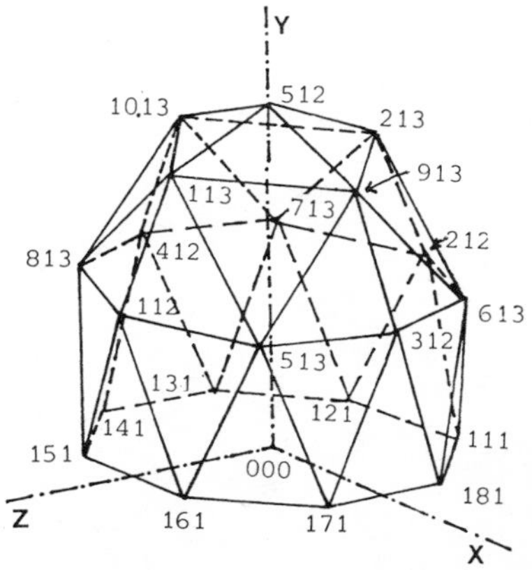

Fig 4. The numbering of the nodes.

Making use of the node numbers which are automatically found as a result of these runs and which are indicated in Fig 4, various geometric data can be requested.

output:

lengthes:

161-171	= 220.00000000 cm
161-513	= 298.44385084 cm
113-513	= 287.44385090 cm
112-513	= 219.99999778 cm
113-913	= 287.44385000 cm

Face angles:

000-161-171	= 45°	
513-161-171	= 45°	
513-913-312	= 49.21052895°	= 49°12'37.9042"
513-113-913	= 60°	
112-912-161	= 85.63439626°	= 85°38' 3.8265"

dihedral angles:

112-161-212-121	= 20.75697281°	= 20°45'25.1021"
112-161-151-513	= 155.61882318°	= 155°37' 7.7635"
151-161-112-000	= 94.72608208°	= 94°43'33.8955"
112-513-161-113	= 163.78506318°	= 163°47' 6.2275"
112-513-113-212	= 69.05897977°	= 69° 3'32.3272"

dihedral angles at vertices:

161-000-112-513	= 45.13814976°	= 45° 8'17.3391"
161-000-513-171	= 67.43092508°	= 67°25'51.3303"
513-713-112-113	= 54.73561032°	= 54°44' 8.1971"
513-713-113-913	= 70.52877937°	= 70°31'43.6057"
512-000-113-913	= 90°	

The face angles are defined by the corner-point and two points of the legs (in this specific order of sequence) and the dihedral angles by two points of the intersection line and by one point in each of both adjacent planes. The knowledge of the dihedral angles of the vertices facilitates the fabrication of the circular node plates which will be discussed later. The first dihedral angle of all that are mentioned denominates the longitudinal twist in the members of the antiprismatic mantle.

Construction

Roundwood timber has been chosen for the structural members. This material offers the specific problem that radial tears due to shrinkage may cause a considerable decrease in strength and particularly if bolts are used for the interconnection and if then some of these tears partially or wholly coincide with the boltholes. Such a coincidence might have desastrous effects.
In the present case both ends of the poles had been provided with a slot of about 6 mm width and of 200 mm depth, to the full width of the cross-section. In each of such slots a metal strap of 50 x 6 mm and with a length of 260 mm had been inserted. This strap was fixed at two

Fig 5. A number of roundwood struts provided with steel strap

places (50 and 175 mm from the end of the pole) with two times two galvanized steel wires of 4 mm thickness, going through corresponding 8 mm holes in the straps and in the timber.
Both threads passing in each set of corresponding holes had been stretched individually and secured by twisting the two ends of one thread together. This had been done for each of the two threads at opposite sides of the cross-section (see Fig 5). A third wire, spanning the complete circumference of the wood, has the function of preventing the tears from further widening.

The application of the threads had been done with a specially designed handtool which stretches and twists the thread consecutively thus providing a pretensioning force in the thread.

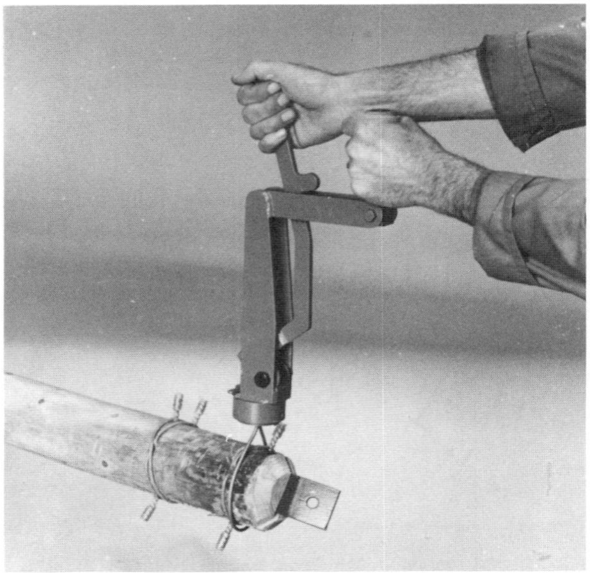

Fig 6. A manually operated wire lacing tool.

The structure had to be dismountable. The struts therefore meet against circular steel connector plates of 200 mm diameter and 6 mm thickness. A strut is connected to a plate by means of a 12 mm diameter bolt through the strap. To prevent this plate from rotating by excentrically acting forces at least one of the straps is protruded beyond the middle of the plate and connected there with a second bolt. At the perimeter of the antiprismatic part both bars coming from the mantle protrude this way for a better stability of the circular plates. 4 Different kinds of these circular plates are needed.
The boltholes in the circular node plate are positioned with respect to each other at angles that have been called: 'dihedral angles at vertices'. Those are the dihedral angles of the two planes of which the connection between the system centre and the respective node forms the intersection line and of which the other ends of the two considered bars lie each in one of the planes.

Connection method

The previously described method has been compared with a.o. an ordinary bolted connection, with a high-strength 8 mm bolt going through the holes in strap and timber instead of 2 wires.
Test specimens had been prepared with straps at both ends, three of each connection method. Tension tests were carried out with an Instron 1195 universal tester, at a deformation speed of 2 mm per minute under laboratory conditions, i.e. 20° ± 1° C and 60 to 65 per cent relative humidity.

Fig 7. Circular plate with connected struts.

The following data were found (p in Newton),

		wired	bolted
strength 1	P_{B1}	40200	40000
strength 2	P_{B2}	38000	27000
strength 3	PB3	44000	22500
Average strength	$\bar{P}_B$	40733	29833
Standard deviation	S	3035	9088
V_s = variation coeff. p.c.		7.5	30.5

Fig 8. Test specimen after failure.

The longterm allowable strength of timber connections is according to Ref 5:

$P_a = \bar{P}_B : w$, where w is the safety factor, or

$$w = \frac{1 + \sqrt{1 - 0.9375(1 - 6.25\ V_s^2)}}{1 - 6.25\ V_s^2}\ 1.8$$

$= 2.44$ for the wired connection

$= 7.65$ for the bolted connection

and thus P_a = 16694 N or 3899 N respectively.

This means that with the wired connection an allowable strength is obtained of more than 4 times that of the bolted connection. This is mainly due to the great spreading in the results in the latter case.

Fig 9. Completed structure.

Conclusions

It is well imaginable that structures like the described one are used as the skeleton of one-family houses and even of larger buildings. It can be ciphered that even if a second floor is hung into the structure the occurring strut forces will presumably not reach higher values than about 0.5 times the allowable breaking strenth of the wired joint. This time 4 mm thick wires had been used, but tests with 5 mm wires also proved succesfull. With preliminary experiments a strength of the connection has been reached of as much as 53000 N. This is promising as it proves that even quite larger dome structures or trusses of roundwood seem to be realizable. The geometric design of the previous dome structure may appear elaborate for such a relatively sinple shape, but this method is considered examplary for larger and also more complicated shapes.

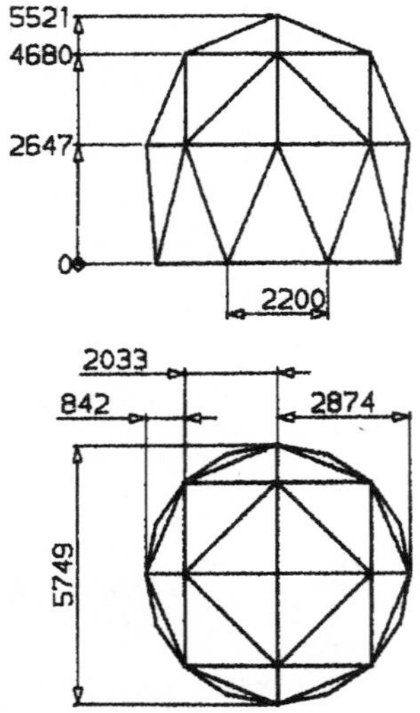

Fig 10. Sketch of dome structure.

Acknowledgements

The wire lacing tool has been developed by J. Lanser with the assistance of S.Th. van der Reijken. J.A. van Nellestijn assisted in the construction and made most of the photographs. G. van der Ende was responsable for the experiments and the electronic measurements. Further support has been given by the administrative and technical services of the Civil Engineering Department. The author is indebted much gratitude towards all these persons.

References

1. P. Huybers
   The geometry of uniform polyhedra (in Dutch)
   Stevin Reports 10-76-1 and -2, july 1978.

2. P. Huybers
   A dome structure made of round wood (in Dutch)
   Stevin Report 01-83-11, December 1983.

3. G.J. Arends and P. Huybers
   The CPS-programme 'coordinate', Part 1: manual (in Dutch)
   Stevin Report 10-79-02, June 1981.

4. P. Huybers
   Rotation possibilities for the Platonic and Archimedean solids (in Dutch)
   Stevin Report 10-78-03, February 1980.

5. J. Kuipers and P. Vermeyden
   The safety of timber connections (in Dutch)
   Heron 1965. No 1. p. 1 t/m 17.

ON THE EMPLOYMENT OF SPACE FRAMEWORK ROOFING IN INDUSTRIALIZED BUILDING CONSTRUCTION

G. PRETE* and G. PISCOPO

*Professor, Istituto di Scienza e Tecnica delle Costruzioni,
University of Bari, Italia.

We present an original construction system for the roofing structure of rectangular plan, or like, buildings, with middle - sized or large spans, based on the use of steel space structures which belongs to the bidirectional double - layer grids topology. The structural model, studied in accordance with requirements of competitiveness as industrialized building constructions demand, is made up of simple, composed and coordinate parts, built of light cold - formed steel sections, open and hollow, of modular dimensions to be wholly prefabricated in workshop with cutting, drilling and welding operations, and whose final assembling must be performed with bolted connections. This paper is illustrative both of technical characteristics and important properties on the applicative side, with some construction examples, of the structural system.

1. INTRODUCTION

Difficulties in economic crisis and notable increase in costs of labour have strongly bound the development of modern construction to the growth of prefabrication technique and industrialization of intended construction products. Steel structural work, and particularly the reticulated type, can be thought of, by its characteristics, as a synonim for prefabricated construction, which is usually made up by standardized and coordinated elements ready for quick assemble and erection. Today's problem, merely a productive one, is to make structural work effectively competitive.

This is a question of searching for structural solution, so that the conception and realization of the system, from its planning stage, production in workshop, and assembling on construction site, be wholly, or at least partially arranged within industrialized process, based on mass - production of modular components, wishful to allow a sensible reduction in production and laying costs. To this end, systems must be devised, made up by simple, standardized, easy to assemble parts, of middle - sized dimensions compatible with transport requirements, easy to store, move and ship without excessive charge expenses even for long distances.

In view of this we present in our paper a original construction process for the roofing structure of a rectangular planned, or like, buildings, with middle - sized or large spans, able to fit various purposes (industrial, recreational, social, etc.).The structural system is based on the use of steel space framework wholly prefabricated in workshop "by quoin", and whose final assembling must be performed "by dry".

2. THE SOLUTION PROPOSED: DESCRIPTION AND TECHNICAL FEATURES

The system belongs to the rectangular, bi-directional, double-layer flat grid structural typology. In other words, it belongs to that class of structures that, as far as tri - dimensional steel frameworks are concerned, undoubtedly represents the most efficient and economical solution (Ref 1) for its well-known qualities of static capacity and simplicity of execution. As indicated in Fig 1, we are referring to a grid model with a "cornice" end section, in which the border nodes of the top layer are constrained in simple support. Compared to the alternative configuration with a "mansard" closure in which the supports are situated level with the border nodes of the bottom layer, the chosen solution clearly seems more advantageous not only statically (the web members which are more stressed are stretched, the action of the wind is exclusively in depression, the structural system is not of a thrusting nature) but functionally and constructionally as well (the absence of external sloping pitches entails a considerable simplification in making the roof and border finishes). As to the geometric parameters (a_1, a_2, h, etc.) characterizing the grid layout considered, which can be of a square or rectangular mesh, we refer to what has been defined in Refs 2 and 3.

This type of roofing can be associated indifferently either with a steel external supporting substructure, composed of columns and eventual border beams, or - more conveniently -with a substructure in reinforced concrete, or - at least potentially - in masonry. In the latter two cases, the outcome would be an "integrated" construction solution, namely the result of a generalized mixed system in "steel and reinforced concrete" or, at the most, in "steel and masonry". The most important component of this solution - both for staticity and economical purposes - is the space reticular structure. This structure, conceived to be competitive as is required in any industrialized building construction, is composed of coordinated simple and composed basic elements, made of open and hollow sections of conveniently selected modular dimensions, to be completely performed in the workshop with cutting, drilling and welding operations and to be finally assembled on construction site by means of normal bolted connections.

Figure 2 shows an exploded perspective sketch of the system's structural matrix, based on the modular "in parallel" aggregation of two unidirectional and fixed components, one of which appears bi-dimensional (B), in the sense that it has only one small dimension compared to the other two, whereas the other is simply monodimensional (M).

The primary generating element is the bi-dimensional one, which represents a "segment" of the space structure oriented according to the minor axis (coinciding with the principal load - bearing direction) of the rectangular plan of the roofing. This element can be synthetically defined as the typical transversal "quoin" through which the system is generated. It is practically formed by an oblique reticular beam, whose chords are arranged in vertical planes and whose diagonal members (corresponding to the web members of the grid) lay on an inclined plane. As is shown in detail in Fig 3, in a welded formation, the latticed "quoin" is composed of upper and lower continuous channel chords, and of web members made of round tubes which are connected by means of fillet welds directly to the extrados of the superior flange (or to the intrados of the inferior one) of the chords, after their extremities have been cut in a "flute beak fashion" according to the angulation assigned to these members (preferably inclined at 45° in space). The single sections constituting the chords are all identically and preventively provided with a middle line drilling of the core to the pitch required by the assembly padding and with welded stiffening plates shaped and drilled into in the points where the orthogonal chords are expected to be connected. The second element of the system, namely the monodimensional one, is destined to the realization of the longitudinal chords (both upper and lower) of the grid. It is simply made up of one member, composed of two channels (or angles), placed side by side and joined by means of paddings. Their extremities are predisposed for bolted connections to the stiffening plates of the transversal chords.

A peculiar and qualifying characteristic of this construction system is the use of thin cold - formed steel sections. Apart from meeting the fundamental need of simplifying carpentry to the utmost, in a moment of a seriously deficient production of the steel industry due to the energy crisis, this system favors the use of money - saving sections deriving from the cold - forming of hot - rolled strips of modest thickness. These are products whose processing techniques and qualitative standards have attained high levels of absolute reliability.

The assembling procedure of the structure in space is obvious and natural, Fig 4, as it consists in simply bringing together and justaposing the oblique transversal "quoins" (B) in a "bellows - like" arrangement after the paddings have been interposed, and consequently seaming them in an orthogonal direction by inserting the longitudinal members (M). All the assembly connections, both the secondary ones (distributed along the chords on at least the middle thirds of the range) and the principal ones (concentrated at the nodes), are made with normal shear functioning bolts, or shear and tension bolts, with no eccentricity. The longitudinal members are normally to be connected to the transversal "quoin" by means of a single bolt. After completing the assembly, the outcome is on the whole a reticular plate, in which the two orders of chords - one continuous (the transversal one) and the other discontinuous (the longitudinal one) - are both made up of paired channels and the web members are tubular. In the presence of modest stresses which may occur for instance in roofings with a very long rectangular plan, the longitudinal chords may be more conveniently dimensioned with single channels instead of paired ones, and, at the utmost, be partially abolished with a consequent emptying of the structure. Moreover, what is usually considered as the crucial point of tri-dimensional frames, namely the problem of the "space joint", in this case is spontaneously solved with no special pieces or mechanisms but simply by means of normal connections welded in the workshop and bolted when assembled. The connections related to the transversal chords, which are also the ones bearing the greatest stress, have the advantage of not being interrupted. After the assembly on construction site, the lifting and positioning of the structure can be achieved in a single operation with a lift - slab kind of procedure, or for parts, obviously in relation to the overall dimensions of the grid and to the capacity of the available building - yard equipment.

As to the manufacturing lengths of the system's elements, while the size of the piece (M) is linked to the module of the longitudinal chord members, the optimal length of the "quoin" (B) would theoretically be the one coinciding with the whole transverse span of the grid. As this condition is only rarely achieved, it is necessary to provide for "quoins" with a length corresponding to half of the grid span, in order to surround a whole pitch of the roofing and to facilitate the necessary construction rise on the center line: of a modest entity if for a counter - camber or more marked if it must ensure the defluxion slope of the roof. The continuity of the transverse "quoins" is thus restored by means of flanged bolted connections and by inserting at the crown an additional loose member (R) acting as a connecting and adjusting rod. It has the same double C section as the transversal chords but its length is slightly modified compared to the module. It is smaller if the member acts as a tension bar (lower member) and it is larger if it acts as a ridge beam (upper member) of the framework, Fig 5.

When one wishes to use the space structure directly as a support for the roof and to avoid purlins, the longitudinal members of the top layer are subject to combined compressive and bending stress. A similar situation may occur for the lower longitudinal members, with tensile and bending stresses, in this case, due to the application of direct loads on these members caused, for example, by false - ceilings, horizontal technical installations, concentrated weights, etc. In this case, a considerable improvement of the static behaviour of the directly loaded members can be achieved by giving them an adequate and pre - established eccentricity thanks to which a constant bending moment - able to compensate partially the bending induced by external loads - is artificially created. If then, for building purposed, one needs to put off - centre the end connection of the eccentric members with respect to the geometric scheme of the grid, it is obviously necessary to make the connection by means of a pair of bolts instead of a single bolt, as is generally the case, because of the secondary moment in the joint.

Thus, two possible alternative versions of the proposed

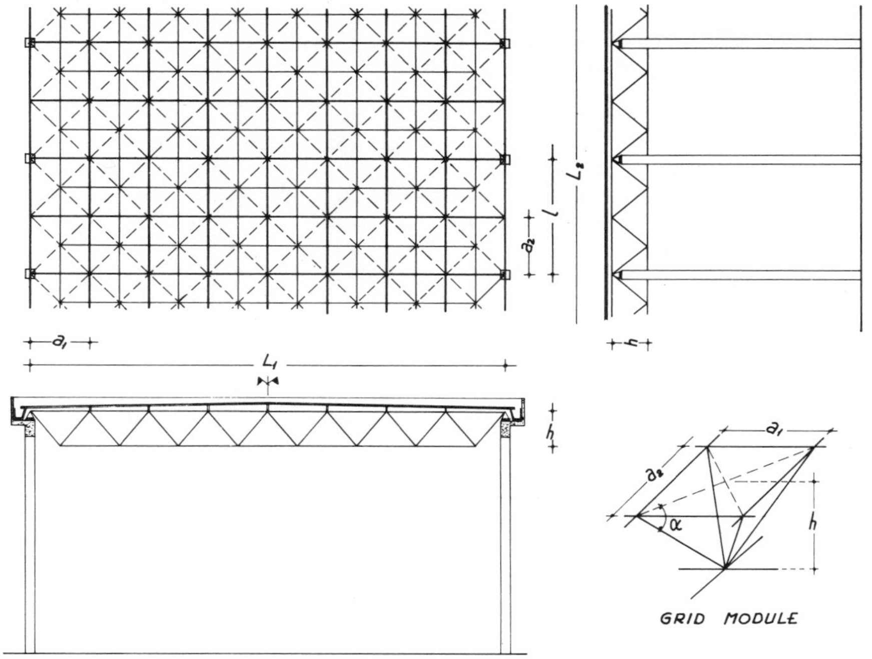

Fig 1

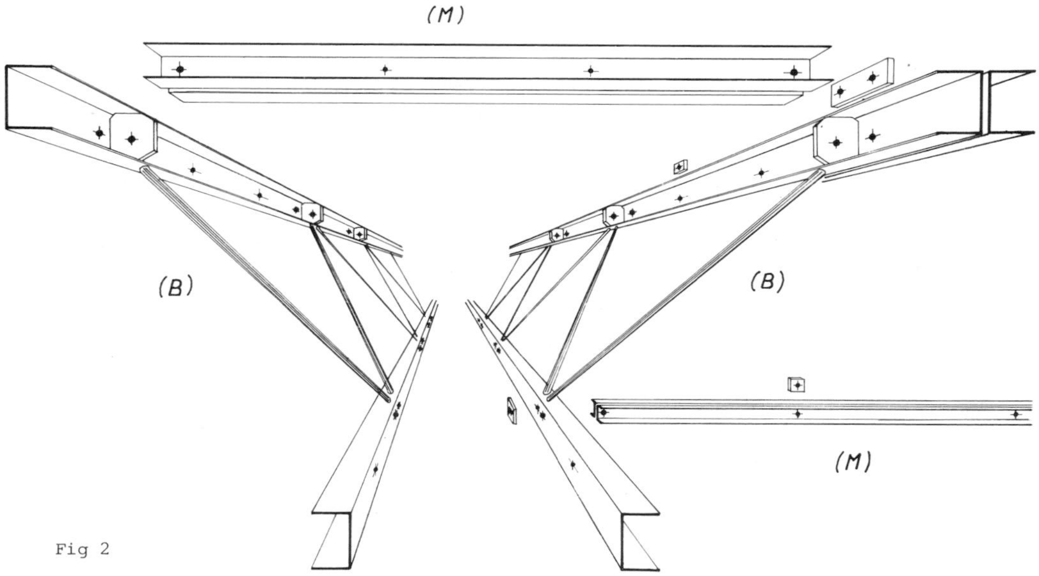

Fig 2

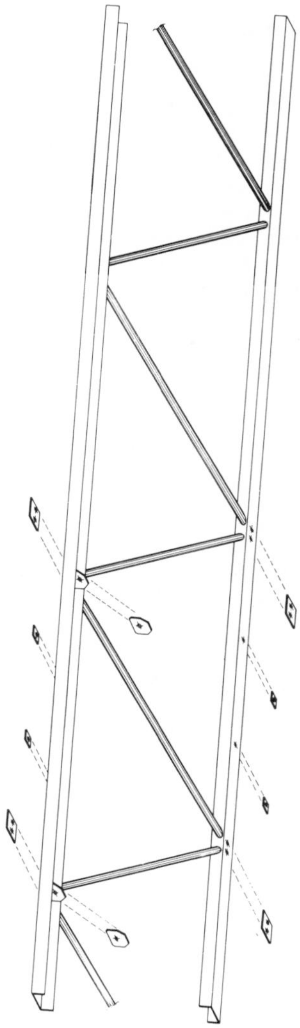

Fig 3

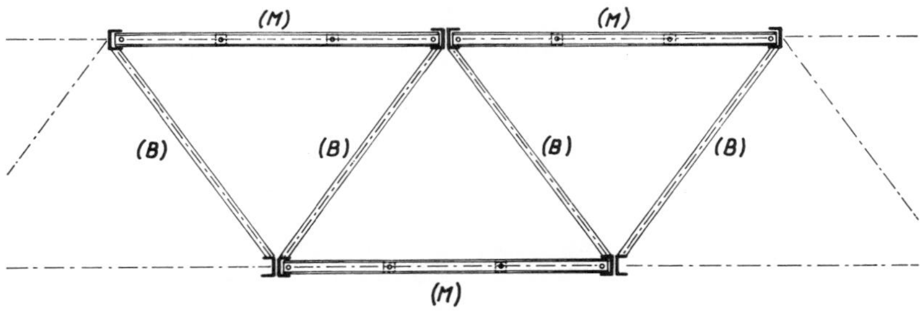

Fig 4

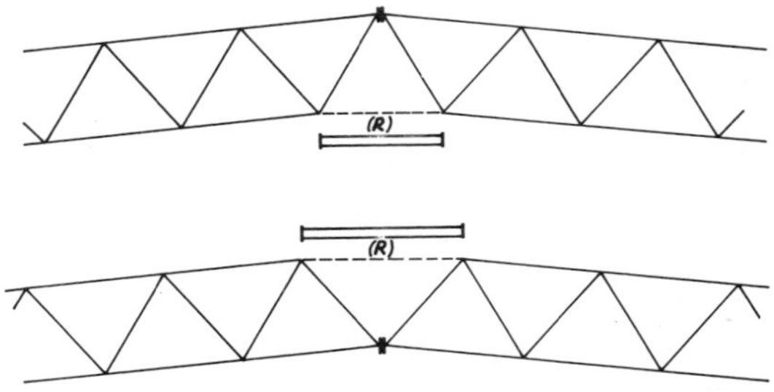

Fig 5

construction system are determined. They are different for the way they are assembled and for the way their longitudinal members (M) are connected to the transversal "quoins" (B): the members of the first version are perfectly centered, so this version is most suitable in the classical case of a space structure which is loaded exclusively at the nodes; the second version, characterized by an assigned constructional eccentricity, is to be preferably achieved when the top and/or bottom longitudinal members of the grid are subject to not negligible external bending actions.

3. APPLICATION EXAMPLES

Some representative applications of the proposed construction system are presented here. They are a coordinated series of completely dimensioned structural examples consisting of double - layer space grids for the roofing of buildings with a rectangular plan of medium extension ratio (for example the "school gym" type) or of high extension ratio (for example the "industrial shed" type) and with clear spans of varying sizes. With reference to the general geometric layout in Fig 1, the parameters characterizing the six structural cases considered are shown in Tab I. The following have been adopted:

- L_1 span in a transverse direction;
- L_2 span in a longitudinal direction;
- $\lambda = L_2/L_1$ extension ratio of the plan;
- $a_1 = a_2 = a$ side of the square mesh of the layer;
- $h = a/\sqrt{2}$ theoretic height of the grid;
- α inclination of the web members;
- l pitch of the constrained border nodes;
- n_1 number of meshes in a transverse direction;
- n_2 number of meshes in a longitudinal direction;
- p slope of the couple - close as well as of the roof (of a "sandwich" type);
- q total vertical load distributed on the plan;
- w seismic vertical action for maximum seismicity degree ($s = 12$).

The survey includes structural models with three basic transversal spans ($L_1 = 24, 30, 36$ m). For each of them two significant extension ratios of their rectangular plan have been analyzed ($1{,}5 \leq \lambda \leq 2$; $\lambda > 2$). Moreover, each example of space grid has been elaborated in the previously described double construction solution: with independent purlins supporting the roof or without purlins, in other words with eccentric longitudinal chord members directly supporting the mantle. For all the cases the ratio L_1/h takes the value 17. The following criteria and hypotheses have been used for the design calculation.

- Method of analysis. The static analysis of the space grids has been performed automatically using the ICES - STRUDL II (IUG VERSION) oriented program on an IBM 370/158 computer. The calculation model employed was that of the reticular plate having perfectly articulated joints and with the loads concentrated at the joints of the top layer.
- Boundary conditions. Each grid is supposed to be continuously constrained along its perimeter on a rigid edge substructure of generic nature, at the external upper nodes ("cornice" solution), by means of simple bilateral supports which allow a horizontal sliding in a normal direction to the edge; the angle nodes are to be considered as completely sliding.
- Materials and structural sections. For all the structural elements, cold - formed steel channels are to be used for the chord members and welded round tubes are to be used for the web members. The channels and tubes have dimensions which are in conformity with the unified national production (UNI Tables) and they have been selected in order to obtain a high modulation and standardization of execution. The quality of the steel is of the 1S type (for $t < 3$ mm) and of the St 360 type (for $t \geq 3$ mm), with $\sigma_{adm} = 160$ N.mm^{-2}, in accordance with the requisites required by the Technical Standards applying to the constructions with cold - formed steel sections, Ref 4. The built - up members are joined together by means of paddings having an exclusively geometric function. The connections are always made with class 4.6 normal bolts with non - calibrated holes and fillet welds.

The results of the calculations have been synthetically reported in Tabs II and III. The maximum values of the principal static characteristics of the analyzed grids (axial stresses S, support reactions R, deflections δ) are exposed in Tab II, whereas the corresponding general pattern of the shapes assigned to the different structural components (members, plates, bolts) and the subsequent total and partial unit weights are presented in Tab III. Particularly, the overall design drawing of the first example (Al) considered for both possible alternative versions (with and without purlins) are illustrated representatively in Figs 6 and 7 whereas the construction details of the standard assembly connections of the system are illustrated in Figs 8, 9, 10 and 11. In the second solution, Fig 7, the longitudinal chord members, which are sturdier than in the first solution, are seen to be alligned on the extrados (the upper ones) and on the intrados (the lower ones) of the grid as it is presumed that, in the absence of a secondary structure, the former must support the roof and the latter must sustain a false-ceiling.

4. PROPERTIES AND QUALITIES OF THE SYSTEM

The following points show the peculiar properties and the main advantages of the application of this construction system.

- During the executive procedure materials and means of common light metallic carpentry are used: steel of the most commercial and economical quality (type St 360), members made exclusively of thin cold - formed steel sections from hot-rolled strips, workshop welded connections (with fillet welds) and bolted assembly connections (with normal bolts).

- When necessary, the dimensions and characteristics of the components make it possible to achieve a completely hot galvanized structure using zinked assembly bolts, and, if one wishes, it can even be pre - painted.

- There is a great possibility of choosing what roof to make, as the system is suitable for the application of a broad range of differing mantles, preferably, however, for the lighter pre - fabricated types. The most technically valid solutions are the ones based on pressed steel sheets which may be bare, coated, insulated and water - proofed (the "Deck" type), paired and insulated (the "Sandwich" type), or inte

	-A- (L_1=24m)		-B- (L_1=30m)		-C- (L_1=36m)	
	A1	A2	B1	B2	C1	C2
λ	1.5÷2	>2	1.5÷2	>2	1.5÷2	>2
a (m)	2.00		2.50		3.00	
h (m)	1.41		1.77		2.12	
α (deg)	45°		45°		45°	
ℓ (m)	2.00		2.50		3.00	
n_1	12		12		12	
n_2	18÷24	>24	18÷24	>24	18÷24	>24
p	2° ≃ 3.5%		2° ≃ 3.5%		2° ≃ 3.5%	
q ($N \cdot m^{-2}$)	1500		1500		1500	
w ($N \cdot m^{-2}$)	± 180		± 180		± 180	
L_2 (m)	36÷48	>48	45÷60	>60	54÷72	>72

Tab I

		A1	A2	B1	B2	C1	C2
TRANSVERSAL CHORDS	S_1 (daN)	16689 / −16969	± 15320	25849 / −26279	± 23835	37178 / −37870	± 34400
LONGITUDINAL CHORDS	S_2 (daN)	3420 / −3655	± 1632	5326 / −5176	± 2383	7062 / −6898	± 3440
WEB MEMBERS	S_α (daN)	2582 / −2580	± 2538	4009 / −4007	± 3977	5779 / −5760	± 5710
TRANSVERSAL REACTIONS	R_1 (daN)	3946	3600	6284	5625	8845	8100
LONGITUDINAL REACTIONS	R_2 (daN)	1595	—	2392	—	3315	—
DEFLECTIONS	δ (cm)	4.14	4.70	6.70	6.44	7.85	7.35

Tab II

	A1		A2		B1		B2		C1		C2	
	SHAPE (mm)	WT. (Kgm^{-2})	SHAPE (mm)	WT. (Kgm^{-2})	SHAPE (mm)	WT. (Kgm^{-2})	SHAPE (mm)	WT. (Kgm^{-2})	SHAPE (mm)	WT. (Kgm^{-2})	SHAPE (mm)	WT. (Kgm^{-2})
TRANSVERSAL UPPER CHORDS	⊐⊏ 120x70/4	8.2	⊐⊏ 120x70/4	8.1	⊐⊏ 140x90/5	9.9	⊐⊏ 140x90/5	9.8	⊐⊏ 160x100/7	12.9	⊐⊏ 160x100/7	12.7
TRANSVERSAL LOWER CHORDS	⊐⊏ 120x70/3	5.4	⊐⊏ 120x70/3	5.4	⊐⊏ 140x90/3	5.4	⊐⊏ 140x90/3	5.4	⊐⊏ 160x100/4	6.6	⊐⊏ 160x100/4	6.6
WEB MEMBERS	Φ 42.4/4/3/2.6	5.2	Φ 42.4/4/3/2.6	5.1	Φ 60.3/3/2.6	6.0	Φ 60.3/3/2.6	6.0	Φ 76.1/3/2.6	6.4	Φ 76.1/3/2.6	6.4
LONGITUDINAL UPPER CHORDS	⊐⊏ 60x40/4/3	3.3	⊏ 60x40/3	1.6	⊐⊏ 60x50/4/3	3.1	⊏ 60x50/4	2.0	⊐⊏ 70x60/5/4	3.1	⊏ 70x60/5	2.4
LONGITUDINAL LOWER CHORDS	⊐⊏ 60x20/3	2.0	⊏ 60x20/3	1.0	⊐⊏ 50x30/4	2.3	⊏ 50x30/4	1.1	⊐⊏ 50x30/4	1.9	⊏ 50x30/4	1.0
PURLINS	⊏ 70x40/3	1.8	⊏ 70x40/3	1.8	⊏ 90x50/4	1.9	⊏ 90x50/4	1.9	⊏ 100x70/5	3.1	⊏ 100x70/5	3.1
PLATES, PADD., FLANGES BOLTS 4.6	≠ 6/10/20 M12-20	2.0	≠ 4/6/10/20 M12-20	1.2	≠ 8/12/24 M12-22	1.4	≠ 4/8/12/24 M12-22	1.3	≠ 8/14/28 M12-27	1.7	≠ 5/8/14/28 M12-27	1.6
	TOT.	27.9	TOT.	24.2	TOT.	30	TOT.	27.5	TOT.	35.7	TOT.	33.8
SOLUTION WITH: LONGIT. ECCENTRIC UPPER CHORDS	⊐⊏ 80x40/3	3.7	⊐⊏ 80x40/3	3.7	⊐⊏ 90x60/3.5	4.5	⊐⊏ 90x60/3	3.9	⊐⊏ 100x60/5	5.5	⊐⊏ 100x60/3.5	4.0
	TOT.	26.5	TOT.	24.5	TOT.	29.5	TOT.	27.5	TOT.	35	TOT.	32.3

Tab III

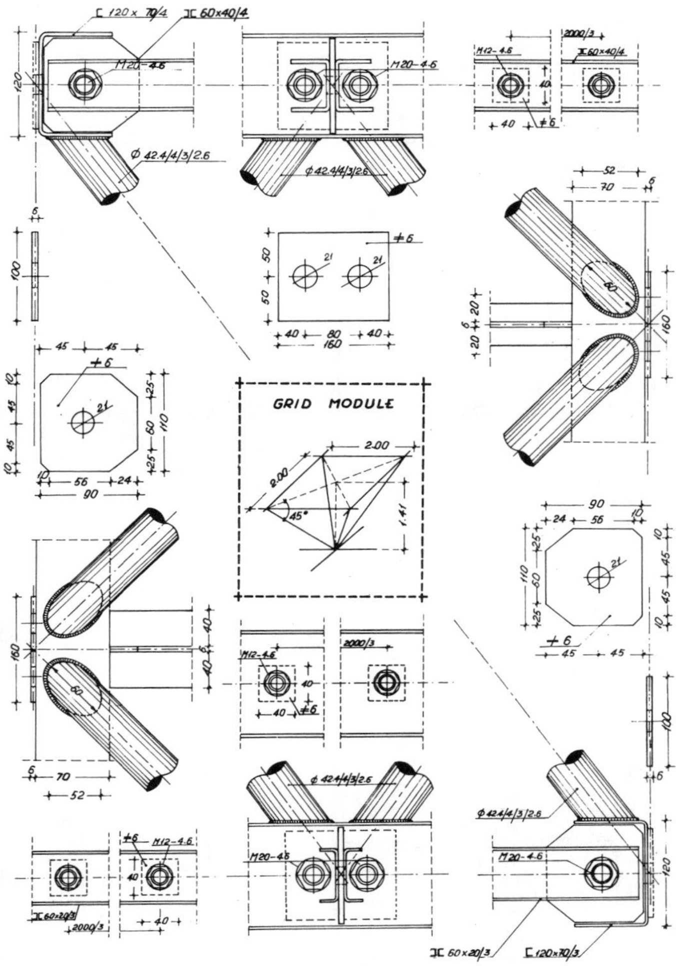

Fig 6

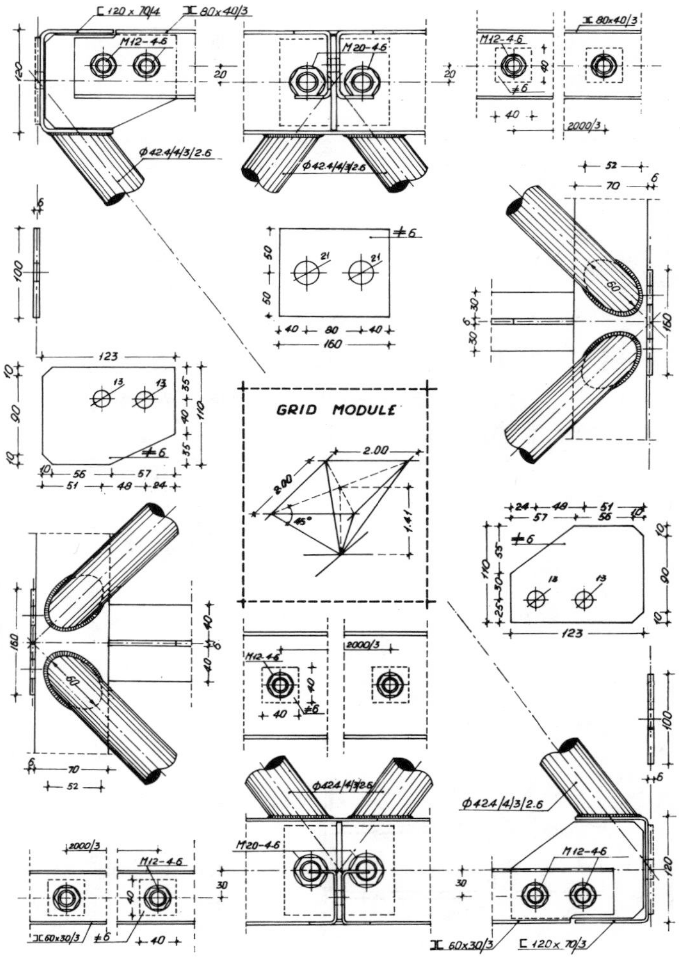

Fig 7

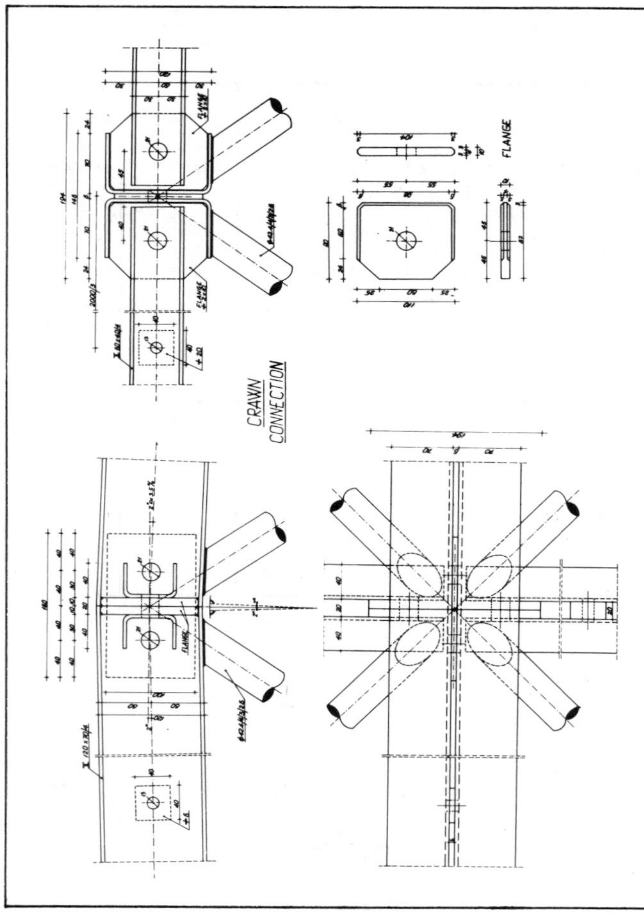

Fig 10

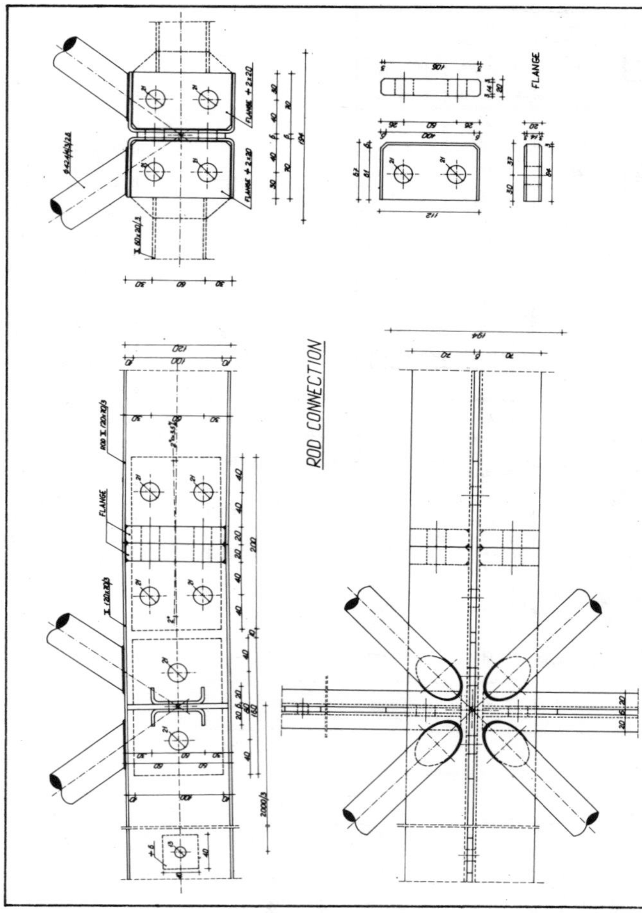

Fig 11

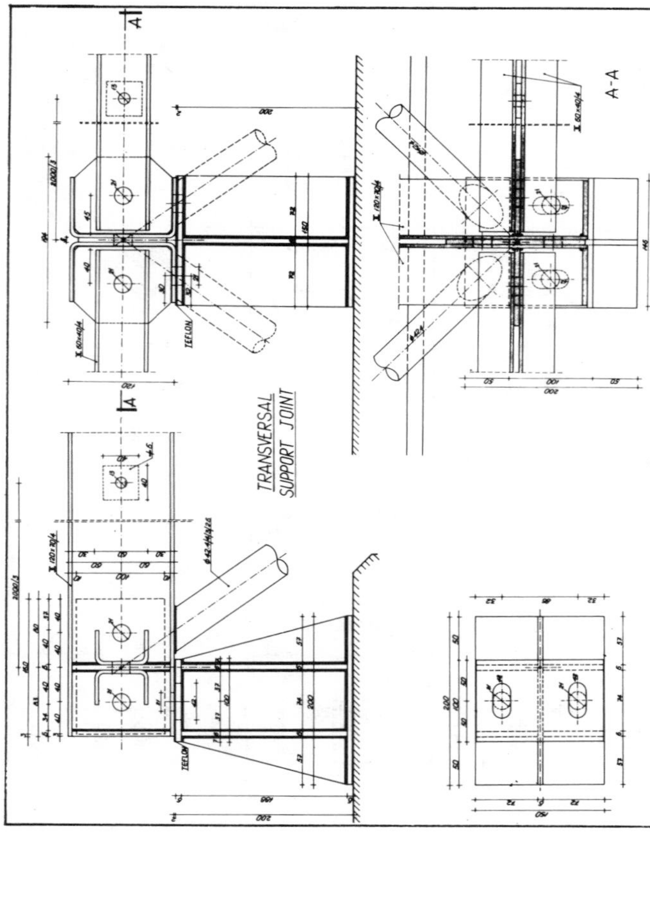

Fig 8

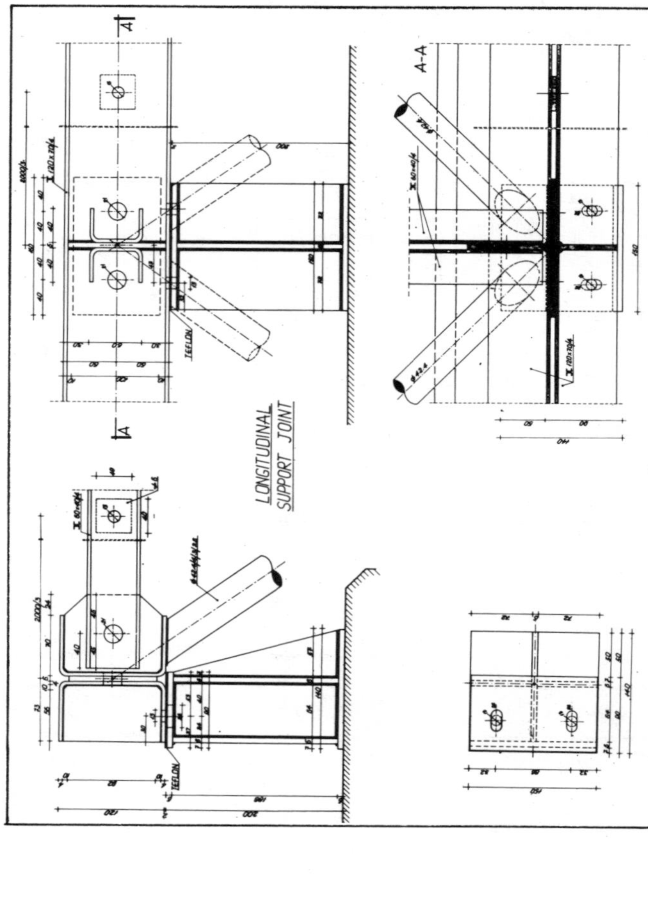

Fig 9

grated with a casting of light concrete, and the asbestos cement products such as corrugated plates, little and large shaped elements. The peculiar geometric configuration of the roofing, both with a"cornice" and a "mansard" profile, is suited to the installation of solar - power panels.

- The space framework is completely self - sufficient as far as staticity is concerned, as it does not require any integrating structural element. In fact it autonomously and monolithically performs the static and geometric functions which, in traditional plane solutions, are separately performed by the main beams (lattice or built - up girders), by the horizontal bracings (transversal and longitudinal), by border beams, by eventual inter - tie bracings. It can also serve as a secondary structural plot (simple or multiple) which supports the roof. The latter possibility is particularly important in the general economy of the construction especially for buildings with a markedly rectangular plan. In the absence of purlins, the mantle rests directly on the order of the upper longitudinal chords and this order is the less statically stressed by the whole structural context of the grid. In it, moreover, the consequent additional bending is absorbed in a partially "gratuitous" way for the beneficial effect of the compensating eccentricity to be conferred to the members involved.

- This system has remarkable aseismic qualities owing to the characteristics of the material used as well as to the specific structural typology adopted. It is well known that steel has good properties of ductility and lightness, therefore, if the positive characteristics of the basic material are associated with the peculiar properties of a seismo - resistant structure like the space structure, this may give an optimal solution to the problem of large roofings in seismic zones. However the use of steel space structures in a seismic building is technically valid and statically effective.

- We have already stated at the beginning of our paper that the model of space roofing proposed may be combined very well with a substructure in reinforced concrete, thus creating an integrated "steel - reinforced concrete" system. In some ways the latter system may practically be more convenient than an "all - steel" construction solution from a technical and economical point of view. By means of vertical and edge structures in reinforced concrete, such an integrated system also makes it possible to: 1) eliminate any kind of wind - bracing, even vertical ones, thus freeing the building from all the execution, architectural and functional complications linked to their presence; 2) simplify many problems concerning the curtain walls; 3) get eventual aseismic shearing walls; 4) reduce the maintenance requirements owing to the smaller incidence of the costs of both the anti - corrosive and fire - proof protective treatment, thus assuring a better preservation and a greater durability of the steel structures. Since the latter are limited to the roofing, they are in fact less exposed to abrasions, blows, atmospheric assaults, etc..

All things considered, one may achieve a mediation between two traditional contrasting construction methods: the steel one and the reinforced concrete one, through a rational association in which both materials are best employed in their most congenial structural role. In other words this solution may partially contribute to contrast the ever - growing success of heavy prefabricated systems in reinforced concrete even in the fields in which steel structures were traditionally used.

5. CONCLUSIONS

At least up to the average values of the spans to be covered, or for modest working loads the roofing system illustrated in this article is acknowledged to require an incidence of a greater structural weight than the one which can be deduced for other alternative construction solutions, both in plane and space structures. This depends:
- on dimensional conditionings linked to the particular geometry of the system and to the need of avoiding any eccentricity in the connections, except when wanted:
- on the design criterion adopted, according to which it is preferable to uniform especially the steel sections employed rather than optimize them according to their weight, owing to modulation and productivity reasons;
- on the limited compositional flexibility compared to other space systems essentially made up of interchangeable single members (as i.e. MERO, TRIODETIC, PREMIT, etc.);
- on the circumstance of not using, for the compressed chord members of the grid, the hollow sections notoriously better as far as stability is concerned.

It should however be pointed out that the mere weight of the material used is certainly not the only nor decisive - even if undoubtedly significant - parameter to be used in judging the real cost of a building. The latter should instead be globally analized even with respect to the benefits of a particular structural typology. In order to make a correct and complete assessment of its economic competitiveness, one cannot neglet many other important technical and functional requisites qualifying the building itself and belonging to the system proposed: reduced realization time; the industrialized production of its modular components; the fact that it is easier to build, simple to assemble and rapid to erect, and has low - cost cold - formed steel sections; the qualitative and aesthetic level of the building; service elasticity; the possibility of being transformed for and adapted to new or different performances; the small height of the roofing structure and consequently the smaller total volume of the building; the high static capacity of the structure to face seismic actions, and, generally, eventual extraordinary loads concentrated or distributed in an irregular way.

REFERENCES

1. Z S MAKOWSKI, Strutture spaziali in acciaio, Edizioni C.I.S.I.A., Milano, 1980.

2. D MITARITONNA e G PRETE, Sistema costruttivo PREMIT: criteri di calcolo e modelli applicativi, Atti Istituto Scienza delle Costruzioni dell'Università di Bari, n.116, 1978.

3. G PISCOPO e G PRETE, Su di una speciale soluzione costruttiva con grigliato piano incrociato per grandi coperture spaziali, Costruzioni Metalliche, n.3, 1982.

4. CNR - UNI 10022/79, Costruzioni di profilati di acciaio formati a freddo: istruzioni per l'impiego, 1979.

THE DEVELOPMENT OF NODAL JOINTS SUITABLE FOR DOUBLE-LAYER SKELETAL SYSTEMS MADE FROM FIBRE/MATRIX COMPOSITES

L HOLLAWAY* and S BAKER**

* Senior Lecturer, Department of Civil Engineering
University of Surrey

** Lecturer, Department of Civil Engineering
University of Surrey

Pultruded g.r.p. composites manufactured in various geometric shapes have been on the market for some 20 years but have not been used extensively in the field of structural engineering due to the difficulty of jointing members together. Provided a suitable nodel joint system is available an ideal application for pultruded g.r.p. tubes lies in the manufacture of small to medium size double-layer skeletal systems.

The Department of Civil Engineering has developed the technique of bonding a glass filled nylon end cap to the ends of pultruded tubes; these caps are then sandwiched between two glass filled nylon plates to complete the node joint.

The paper describes the end cap and the 'bonding' technique which forms the subject of a Patent. It also describes the construction of the cover plates to form the node joint and the method of constructing the skeletal system. The latter has the potential to be the first all g.r.p. double-layer skeletal system to be used in the building and construction industries.

The experimental tests and results performed on the tensile members containing the end caps are presented and although the ultimate tensile load of the end fixings is only 3/5 of the tensile strength of the pultruded members, methods of improving this situation are discussed.

Glass filled nylon has certain disadvantages in the field of degredation and the use of other polymers in its place are considered.

1. INTRODUCTION

The last few years have seen many radical changes in the way in which glass-reinforced polyester (g.r.p.) is processed. Although the pultrusion technique for producing g.r.p. has been in existance for more than two decades, during which time it has been modified and improved, it is only now that the product has reached the point where its importance as an engineering material is becoming more widely realized. One area in which the product has potential is in sectional structural components. These could be in the form of plane and structural systems. In these latter systems the members are mainly under axial forces although they could also be subjected to bending moments.

The main difficulty in the manufacture of g.r.p. space structures is in the jointing of members; there are three options open to the designer. These are to form:

(a) a bolted joint

(b) a bonded joint

(c) a combination of bolted and bonded joint

In the case of g.r.p. space structures members would most likely be manufactured from pultruded sections in which the fibres would be mainly longitudinal. Consequently if a bolted joint were to be used, members under tensile forces would fail at the joint by shearing through material in front of the bolt; a bonded joint is therefore desirable.

This paper describes a 'bonded' joint which has been developed by the authors at the University of Surrey specifically for forming nodal joints in space structures manufactured from tubular g.r.p. members. The system involves 'bonding' a glass filled Nylon end cap on to the pultruded tube and sandwiching the caps between two specially formed composite plates which form the nodal joints. Test results will be given for the tensile failure between the end caps and the pultruded tubes. Typical load deflection and member stress values in a double-layer skeletal structure manufactured from pultruded glass reinforced polyester tube will also be given.

2. Review of Bonded Skeletal Systems

Probably the main reason why skeletal systems in fibre/matrix composites have made little impact in the structural engineering field is the difficulty of jointing the component parts of the structure, particularly as the members meet in three dimensions. Steel (1) has shown that bonded joints are generally structurally more efficient on a weight bases than are mechanical joints when constructing prototype space frames using pultruded carbon fibre-reinforced plastics (C.F.R.P.) tubes. Hollaway and Ishakian (2) reached a similar conclusion when jointing C.F.R.P. rods. However, both these references have indicated that the fabrication of the skeletal structure presents some difficulty.

A method of jointing is to utilise crimped and bonded aluminium end fittings reference (3). These can be efficient in terms of the ultimate magnitude of load that the joint can take and also in terms of weight savings. They are also suited to a skeletal plane frame and are relevant to a double layer space frame where the crimped aluminium sleeve would be threaded and screwed into the appropriate nodal joint of the space frame on where the aluminium outside the crimped and bonded region would be flattened and a pin joint connector through the flattened aluminium would be secured to the nodal joint by bolting.

However there are drawbacks to the use of the crimped and bonded aluminium end fitting with g.r.p. Firstly it might be argued that a complete aluminium space frame would be more acceptable if aluminium is to be used as components in the space frame construction. Secondly the space frame could not be used in a corrosive environment.

The g.r.p. joint that has been developed by the authors is an economic one and readily enables unskilled labour to erect the all g.r.p. structure. In addition it overcomes the above two disadvantages of the crimped and bonded joint.

3. The Space Frame

3.1 The Members and End Caps

The members that are being used during the research and development of the g.r.p. space frame were manufactured from 25mm diameter pultruded g.r.p. tubes of 2mm wall thickness; they have a glass fibre/polyester resin ratio by weight of approximately 60/40 respectively. The end caps that are being used in the first stage of the investigation were manufactured from injection moulded glass filled Nylon 66; two fibre/matrix ratios by weight were investigated one with 30% and the other with 50% glass. The glass filled Nylon has the advantage of low cost but has the disadvantage of high creep characteristics when under load and degrades steadily in normal atmospheric conditions. The second stage of the investigation will use other fibre filled polyesters such as injection moulded glass filled polyester resin which has the advantage of being compatible with the tube material, has a much lower creep characteristic than that for the glass filled Nylon 66 material and has a good durability capacity. The disadvantage compared with the Nylon 66 material is that the mould processing time is longer and therefore the cost of producing the end caps will be more expensive.

The end cap, shown in fig 1, consists of an inner core and outer sleeve which are joined at the base. The outer surface of the inner core and the inner surface of the outer sleeve have a thread moulded into them. On to the outside surface of the outer sleeve five pairs of tapered ribs are moulded which will mate with slots in the connecting plate when the joint is assembled. The end cap is made in a single injection moulding operation. A cross section of the cap as shown in fig 2.

To form the bond between the tube and end cap a measured quantity of epoxy adhesive (C.I.B.A. - beigy AY105/HY956) is injected into the space between the inner core and outer sleeve, known as the annulus of the cap; the tube is pushed into the space until it reaches the cap's base. This forces the adhesive to flow around the threads on either side of the tube and to surface at the top of the cap; it then polymerises. If the adhesive can not be seen at the surface of the annulus it would indicate that either insufficient adhesive had been used or that the pultruded tube, for some reason, had not been inserted into the end cap correctly. These faults, however, should never occur in practice. Clearly the threads are an essential part of the design of the cap as they control the glue line thickness and ensure quality control. The adhesive runs into the threads and forms a chemical bond with the pultruded g.r.p. but forms a mechanical bond with the Nylon.

3.2 The Node Connecting Plates

The connecting plates can be injection moulded from glass filled Nylon 66 for internal structures not exposed to the elements but for external systems it may be preferable to use other fibre filled polymer systems because of the possible degredation as a result of ultra-violet light and because, as mentioned earlier, the adverse creep characteristics of the Nylon 66 when it is under load. Fig 3 shows a detail of the top plate. The thickness of the connecting plates will vary to take account of the differing stress levels within the component; the greatest thickness of the plates will be located at the slots which receive the tapered nibs of the end caps and at the bolt holes through which the bolts pass to secure the two plates together. Other areas of the plates will be folded to stiffen them against bending moments. The central bolt securing the node will be designed to permit the attachment of roof coverings, suspended ceiling straps, hooks and other ancillary items.

3.3 The Space Structure

A typical g.r.p. skeletal system of 25mm external diameter pultruded tube having the same cross sectional dimensions as given in section 3.1 and length 800mm and spanning 6.0m x 6.0m has been analysed to take a snow load of 750N/m^2 and a wind speed of 50m/sec. For ease of erection by unskilled labour, the double layer grid system of the type square on square with assumed rigid joints has been chosen in which all members are of equal length. Fig 4 shows a model of the structural system.

To enable a larger factor of safety to be attained in the critical compression members (these members will be the criteria for failure of the structure) it is possible to replace the pultruced glass reinforced polymer tubes by stiffer hybrid ones containing glass and carbon in a polymer matrix and having the same diameter and length. The percentage of glass to carbon fibre will depend on the stiffness required. It has been shown in reference 4 that by replacing certain g.r.p. compressive members of low factor of safety with respect to buckling, by hybrid ones, the overall efficiency of the structure is considerably increased without a significant rise in the cost of the structure.

It is proposed that the covering to the skeletal system be manufactured from a sandwich construction of face materials of continuum g.r.p. and core material of foamed plastics. The covering would be connected to the space structure at each of its top nodal joints and the whole would form a skeletal/continuum structure in which the sandwich plate would act as a stress skin system as well as the covering to the roof. The continuum plate would increase the load carrying capacity of the composite structure.

As an example of the feasibility of the nodal point connection and the complete structural system, section 6 gives the maximum member forces and maximum node deflections within the structure under various support conditions; the analysis is linear and has ignored the stiffening effect of the continuum plate.

4 The Experimental Tests on the End Caps

During the experimental investigation of the tensile loads required to fail the end cap joints, different adhesives were used but only the results for adhesive AY105/HY956 will be considered here as this agent was not only relatively cheap but it was also a well tried one.

The experimental set up is shown in fig 5 and represents the condition the member would experience if it were exposed to an axial tensile force in the space structure. The connecting plates were represented by a split steel tube in which slots were milled to receive the tapered ribs of the end cap. The whole was then placed in an Instron machine and a tensile load applied to the specimen.

Two conditions of surface preparation of the g.r.p. tube were investigated, firstly, no special treatment of the exterior surface to be bonded was made and secondly the ends of the tubes were roughened by putting a fine thread on to them; the results for these conditions are given in Table 1; the rate of loading of the specimen was 1mm/min.

It may be seen that the most efficient joint in terms of axial force is the one in which the end of the pultruded tube is roughened by means of the fine thread and that the failure load is in the range 32.0 to 35.0kN. Further tests would need to be undertaken before firm conclusions could be drawn on the failure mechanism of this joint but from initial observations it would appear that crushing of the epoxy glue between the threads and/or interface shear off the surface layers of the matrix material and glass fibre of the pultruded tube were the most likely causes of failure.

5 Analysis of a Space Frame Structure Using Pultruded Tube Members and the Special Nodal Joint System

The structural system used during this analysis has been described in section 3.3; it was assumed that the roof covering did not add to the load carrying capacity of the system. The external loads to the structure were applied through the top nodal joints only, and the nominal span of the square grid structure was 6.0m x 6.0m. Fig 6 shows the plan view of the structure.

Three boundary support conditions were investigated:

(a) The four corner node points were position fixed in the z direction only

(b) The four corner and two equally spaced span nodes on all sides of the structure were positioned fixed in the z direction

(c) All boundary nodal points along the four sides were fixed in the z direction only.

The live load applied to the structure complied with C.P.3 Chapter 5 loading; Part 1 1967 and the most severe case for the British Isles was taken with an imposed vertically downward load of 0.75kN/m^2 and a wind pressure calculated to be 1.06kN/m^2.

The worst possible combinations of load were applied to the structure which resulted in:

(a) dead and live load acting downwards

(b) dead load acting downwards and wind load acting upwards.

5.1 The Computer Program

The computer program which was used to analyse the skeletal structure was written to enable an elastic stability analysis of a skeletal or a skeletal/continuum space structure to be undertaken by the finite element method. The program has been discussed and described in ref 5 and has been used to solve both skeletal and skeletal/continuum systems, refs 5 and 6, and therefore will not be detailed in this paper.

6 Discussion of the Results of the Loaded Space Structure

Table 2 shows the results of maximum deflection and maximum axial forces for the three boundary support conditions. For the case in which all boundary nodal points along the four sides of the structure were fixed in the z direction only it will be seen that the allowable deflection is 16.5mm and that the maximum actual deflection under the two loading conditions is 13.0mm. The ultimate tensile force of the end caps is 30.0kN and the maximum tensile force in the structure under the two superimposed loading conditions is 4.68kN. The maximum axial compressive force in the structure is 3.08kN. It is difficult to define precisely the degree of fixity of the strut in the space structure but clearly it will be somewhere between the fully fixed and pinned conditions and therefore the buckling load will probably be about 9.0kN.

As mentioned earlier the system has been analysed without the structural capacity of the cladding units being incorporated into the overall stiffness of the structure. If a stressed skin system were to be incorporated into the calculations the axial forces in the members and the deflections would have been lower than those given above and hence a greater factor of safety would have been achieved.

7 Conclusions

It has been shown that the jointing system developed at the University of Surrey to enable a skeletal system to be manufactured using fibre/matrix members is structurally viable, it enables such systems to be erected easily with semi-skilled labour and without any heavy lifting equipment to hoist the structure to its final position.

It has also been shown that although the ultimate failure load of the joint is lower than that of the standard pultruded tube used in this work, further research is being conducted to improve the failure load of the joint.

The skeletal configuration used in this analysis is not the stiffest one but it has the great advantage that all members are of the same length and consequently this adds to the erection facility.

Although the span investigated was relatively small it is possible to increase it by either increasing the size of the members or using a hybrid member of carbon/glass fibres in a polyester resin. It is worthy of note that these hybrid members would only be placed in certain positions in the structure which were under high stress and that the majority of the structure would be manufactured from glass/polyester composites. Consequently higher loads and longer spans would be possible for a relatively small increase in cost; the same node jointing system would be used as in the all g.r.p. structure.

REFERENCES

1. STEEL D J "The Use of Carbon Fibre Composites for Axially Loaded Tension/compression Components" Conf on Designing with Fibre Reinforced Materials, Paper C234/77 (Institution of Mechanical Engineers), London Sept 1977.

2. HOLLAWAY L and ISHAKIAN V G "Analysis of a Pultruded Carbon Fibre/epoxy Skeletal Structure" Conf on Designing with Fibre Reinforced Materials, Paper C235/77 Institution of Mechanical Engineers, London, Sept 1977.

3. GREEN A K and PHILLIPS L N "Crimp-bonded end Fittings for Use on Pultruded Composite Sections" Composites Vol 13, No 3 (July 1982) PP 219-224.

4. RUSTUM A and HOLLAWAY L "The Structural Behaviour of a

double-layer skeletal system manufactured from a pultruded fibre/matrix composite" Paper (39) presented at the Reinforced Plastics Congress 82 Brighton 8-11 November, 1982.

5. ISHAKIAN V G and HOLLAWAY L "Application of the Finite Element Method to the Analysis of a Skeletal/continuum g.r.p. Space Structure" Composites Vol 10 No 2 (April 1979) PP 81-88.

6. ISHAKIAN V G and HOLLAWAY L "The Stability Analysis of Continuum/skeletal Fibre/matrix Composites." Composites Vol 12 No 1 (January 1981) PP 57-64.

Specimen No	Pretreatment to G.R.P. tube	Failure Load kN	Mechanism of Failure
1	No special treatment	21	Breakdown of bond at surface of G.R.P. tube followed by tearing of end caps base.
2	No special treatment	20.2	As for specimen 1
3	No special treatment	20.0	As for specimen 1
4	End of tube roughened by threading	29.6	Pullout
5	End of tube roughened by threading	35.0	Crushing of the other end of the tube in grips
6	End of tube roughened by threading	33.0	As for specimen 5
7	End of tube roughened by threading	34.0	Pullout
8	End of tube roughened by threading	34.5	Tension failure then end cap failure
9	End of tube roughened by threading	32.0	Not clear

Table 1 Tensile Tests on End Caps

Configuration	Support condition	Load	Maximum Deflection mm	Axial Load Tension kN	Axial Load Compression kN
size 5.95 x 5.95m Members length 0.85 diameter 25mm ext wall thickness 2mm <u>Buckling load</u> (Euler) Pin ended 2.99 kN Fix ended 11.9 kN <u>Ultimate</u> Buckling load 9.0 kN Tensile load 30.0 kN <u>Allowable</u> Deflection 16.5mm	4 supports	Live snow	41.0	-	-
		Live wind	62.0	-	-
	12 supports	Live snow	10.5	1.38	3.61
		Live wind	16.0	5.47	2.10
	At every nodal point along 4 sides	Live snow	9.0	1.11	3.08
		Live wind	13.0	4.68	1.67

Table 2 Maximum Axial Forces and Deflections for Three Boundary Conditions to Skeletal Structure.

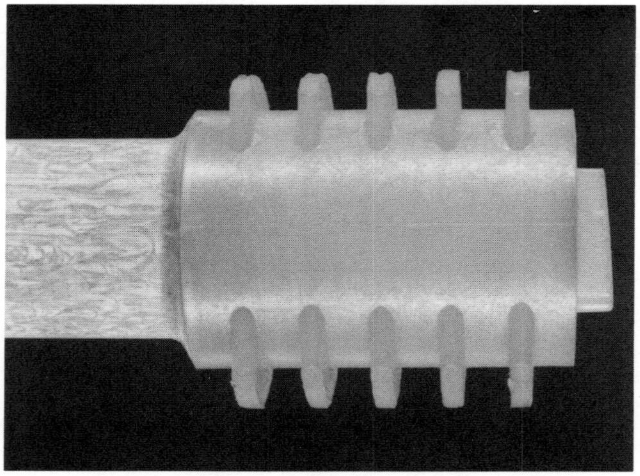

Fig 1 The End Cap

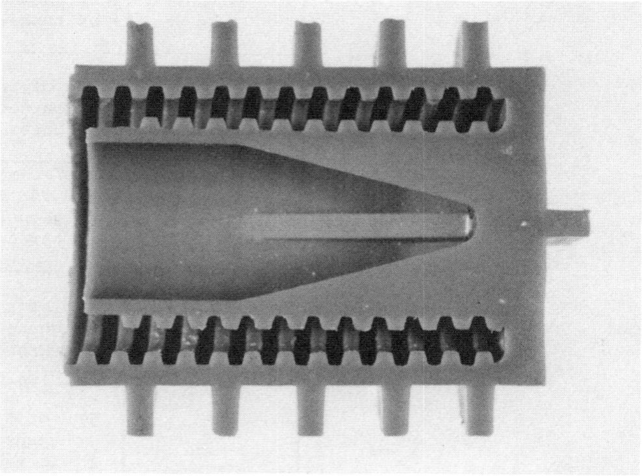

Fig 2a Cross-section through cap

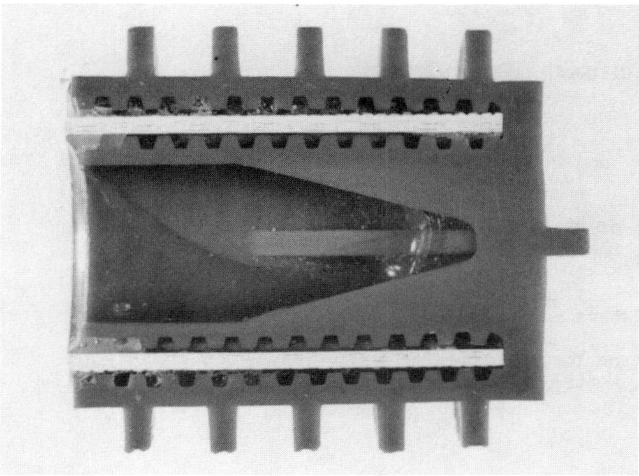

Fig 2b Cross-section through cap with g.r.p. rod in position

Fig 5 Experimental Arrangement for Test on End Cap

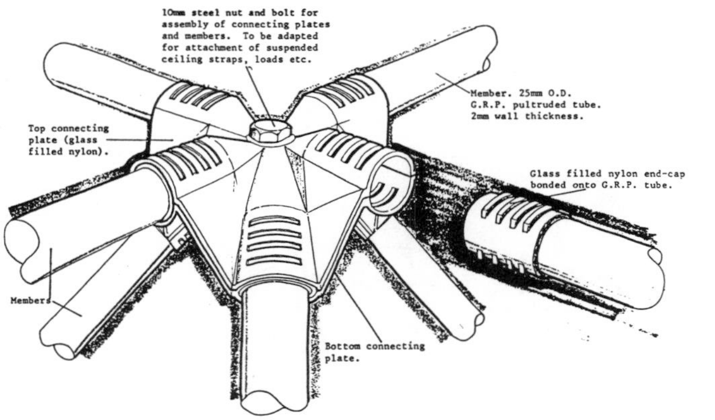

FIG 3. DETAIL OF THE TOP PLATE

Fig 4 Model of Structural System

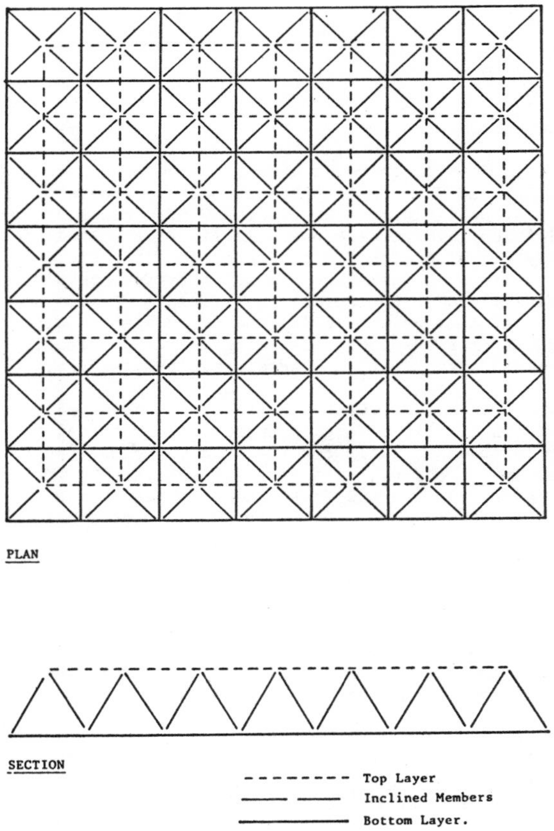

FIG 6. DIAGRAM OF A DOUBLE-LAYER GRID SPACEFRAME WITH "SQUARE-ON-SQUARE" MEMBER CONFIGURATION.

A COMPARABLE STUDY OF LARGE SPAN CIRCULAR ROOF STRUCTURES

Z A ZIELINSKI*, J BOBROWSKI**, and M BRULOTTE***

* Professor, Department of Civil Engineering, Concordia University, Montreal, Canada.

** President of Jan Bobrowski and Partners Ltd. Consulting Engineers, Calgary, Canada and London, England.

*** Graduate Student, Centre for Building Studies, Concordia University, Montreal, Canada.

Various structures were considered and analyzed in the preliminary design stage as possible solutions for the Calgary Olympic Coliseum, including steel space truss systems. More detailed studies and cost analyses were made for circular plan roofs of the radial truss type and for a new multiple circular ring system. The multiple ring system appeared to be attractive both economically and architecturally. The ring-roof can be erected without the use of erection supports, and is completely rigid well before it is completed. The ring-roof system will be described, and compared to the radial truss system.

INTRODUCTION

In North America there is a preference for enclosed sports facilities. This is to protect both the spectators and the playing field from the weather. The kinds of sports facilities involved are baseball fields, football fields, hockey rinks, soccer fields, tennis courts, etc. To provide such an enclosure without interference with the playing field or the spectators' field of vision a large span roof must be provided. The clear span required ranges from four-hundred feet for arena type facilities, to one-thousand feet for large capacity stadiums.

Large span roof structures have already been built for sports facilities before in Canada. In Montreal the Olympic stadium, built for the 1976 Olympics, was to have an elliptic roof with a clear span of over 750 feet along its major axis and 650 feet along its minor axis. Unfortunately due to several reasons the roof has not yet been completed. The Edmonton Coliseum, in Edmonton Canada was successfully completed and has a clear span of four hundred feet. Different structural systems were considered for the construction of the Calgary Olympic Coliseum. The system chosen was a segmental post-tensioned concrete "Saddle-Dome"; but the other systems considered were an air-supported membrane roof, a radial steel truss, and a new multiple ring space truss.

The ring-roof has been proposed by Z A Zielinski as a competitive solution for building covered stadiums. The ring-roof is a space truss; that is, the members are primarily subjected to axial loads. The space truss is made up of several concentric rings, as shown in Fig 1. The rings are actually made up of modules. This modular system makes the assembly of the structure simple. The modules can be produced on a series scale, thereby making their production more economical. The modules represent simple plane truss segments as shown in Fig 2. The modules are linked together at their end points and at all their panel points.

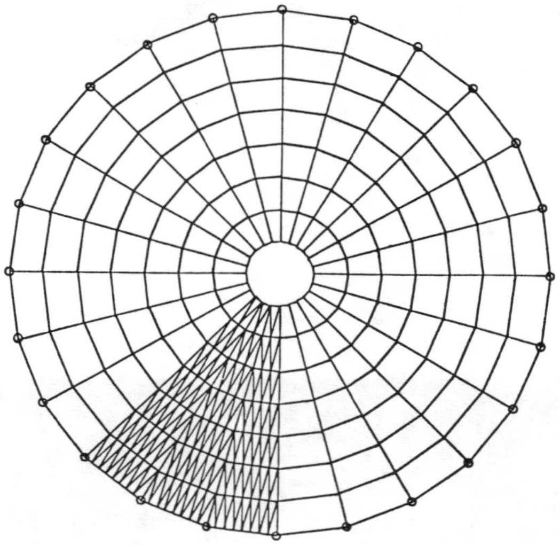

Fig 1 PLAN VIEW OF A TYPICAL RING-ROOF NOTE THAT THE COMPRESSION STRUTS AND TENSION TIES ARE ONLY SHOWN IN THREE ROOF SECTORS

RING ROOF DESCRIPTION

The roof structure consists of rings of decreasing diameter which are hung or supported on one another by means of inclined tension ties or compression struts. When subjected to gravity loads the upper set of rings is primarily subjected to compression forces, whereas the lower set is primarily subjected to tension forces. The rings are actually the chords of the modules. Each ring is made by combining two chords, one from each of the two modules adjacent to the ring, as seen in Fig 3. The inclined tension or compression members are formed by the diagonals of the modules. A radial stiffening truss is formed along each column line by connecting the end points of the modules with stiffeners, as shown in Fig 3. At the perimeter of the roof two horizontal stiffening trusses are formed, one at the upper surface, and one at the lower surface.

The ring-roof is supported at its perimeter by several equally spaced columns. The optimal number of columns can be established based on the comparison of the relative cost of columns and roof structure. For practical reasons the distance between columns should not exceed eighty feet. The columns provide vertical reactions only. The bleacher structure can provide the lateral stability.

The main advantage of the ring-roof is the use of repetitive elements. This saves on fabrication costs and on erection costs. The modules are rigid plane trusses themselves, making erection easy. The connections between members of each module are shop-welded. All the field connections are bolted. Tubes of circular cross section are considered as most suitable because of their high efficiency as compression members. The elements can be made of steel or aluminium alloys.

A typical module can be seen in Fig 2. It is basically a simple plane truss except for a few special features. The module has angled connection plates at the ends of the top and bottom chords, chairs, and several connection points. The connection plates are to create continuous rings. They are angled due to the angle between the chords of adjacent modules. The connection plates are joined by high-strength steel bolts. The connection plates are shop-welded to the tubes to allow for the development of the entire tension capacity of the tension rings. The chairs are used as swivel points when the subsequent ring is being installed. The connection points are necessary to connect the nodal points of the adjacent rings together, and also to assure that the members which make up each double chord work together.

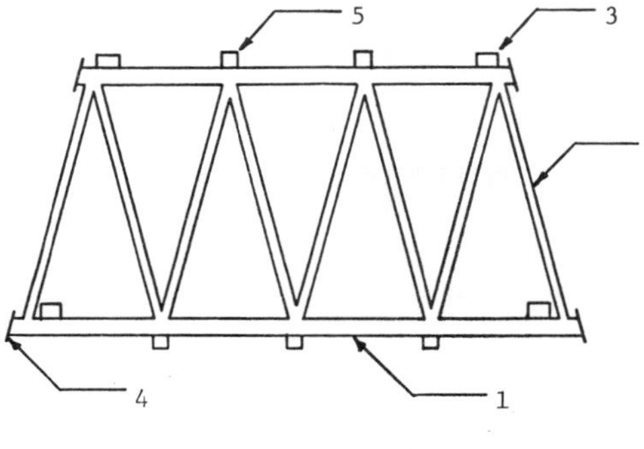

Fig 2 A TYPICAL RING-ROOF MODULE
1- COMPRESSION OR TENSION CHORD
2- COMPRESSION STRUT OR TENSION TIE
3- CHAIR
4- CONNECTION PLATE
5- CONNECTION POINT

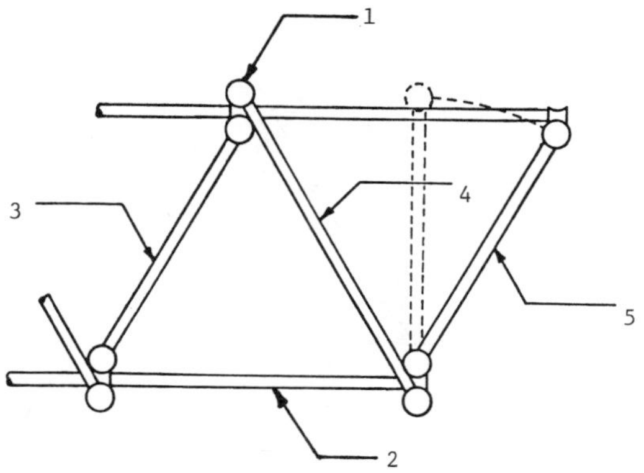

Fig 3 PROFILE OF A SECTION OF THE RING-ROOF DURING ERECTION
1- DOUBLE RING IS ACTUALLY MADE OF CHORDS FROM (3) AND (4)
2- STIFFENER IS USED TO FORM RADIAL STIFFENING TRUSS
3- MODULE IN COMPLETED RING
4- MODULE IN COMPLETED RING
5- MODULE BEING INSTALLED

The erection procedure begins at the perimeter and proceeds inwards to the center of the roof. Modules are installed in a manner that each ring is completed before a subsequent ring is started. The modules are designed so that they can be assembled to form the roof structure without the use of erection supports. The first modules are used to form the perimeter ring, as shown in Fig 4. Once the perimeter ring is complete, the adjacent modules are used to form the next ring. This is done by placing one of the chords of the module on the chairs of the previous ring and swinging the module into position. The module is held in position by a permanent stiffener, as shown in Fig 5. These stiffeners run along the column lines only. Once the first three rings have been installed the horizontal stiffening trusses are also installed. The resulting structure is completely rigid. Once five rings have been installed, the purlins and roof deck can begin to be installed. This saves on construction time since the roof structure need not be completed before the decking and roofing can begin to be installed. The erection sequence is repeated for subsequent rings; that is, the modules are placed on the previous ring one by one, swung into position, the stiffeners installed, and the connections made, as shown in Figs 3 and 6.

The roof structure must be designed for several loading conditions, including construction loads and service loads. The modules must also be checked for transportation loads and erection loads. The service loads are defined by the building codes, but invariably include both dead and live loads. The live loading can be applied over different areas of the roof. Thus the roof must be analyzed and designed for uniform loads over the entire roof as well as several non-uniform loading patterns. Point loads from scoreboards, lights, signs, ventilation ducts, etc. must also be considered. To be able to analyze the structure for all these loading conditions, and also for the analysis of the incomplete structure a computer program is being developed based on the linear-elastic, displacement method of structural analysis.

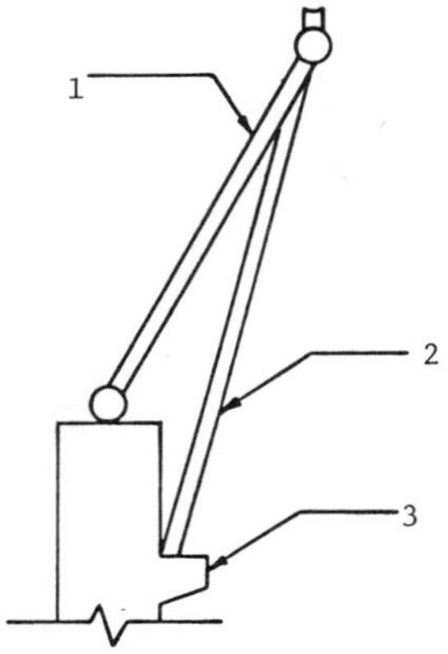

Fig 4 ERECTION OF PERIMETER RING
1- MODULE BEING INSTALLED
2- SUPPORT USED FOR ERECTION OF PERIMETER RING
3- CORBEL WHICH (2) RESTS UPON

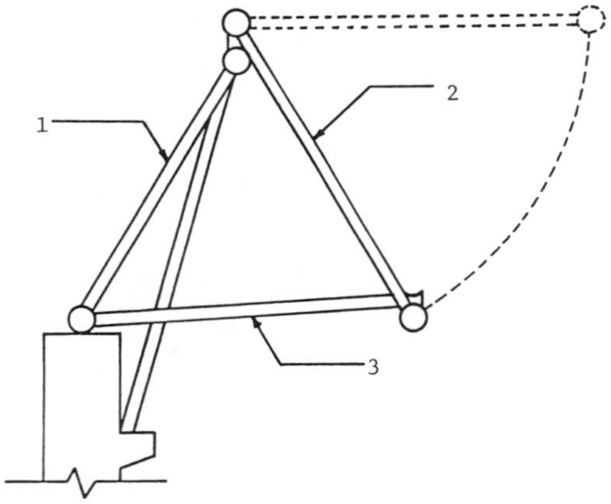

Fig 5 MODULE FOR SECOND RING IS SWUNG INTO POSITION
1- PERIMETER RING
2- MODULE BEING INSTALLED
3- STIFFENER

The architectural possiblities of this structural system are numerous. The roof can be made flat, pitched, or curved, as shown in Figs 7 to 10. The roof can be of virtually any shape which is a surface of revolution, approximated by straight line segments. The roof profile could be piecewise continuous so that, for example, the architect could specify a parabolic profile which is attached to a straight portion of roof. The roof can be left open in the center. The center of the roof could also be covered by a different type of structure, for example, a dome or a radial truss, as shown in Fig 11. The roof truss can be supported at its lower surface, as in Figs 7, 8, and 9, or it can be supported at its upper surface, as in Figs 10 and 11. Ventilation ducts and access passages can be located in the roof structure, as shown in Fig 12.

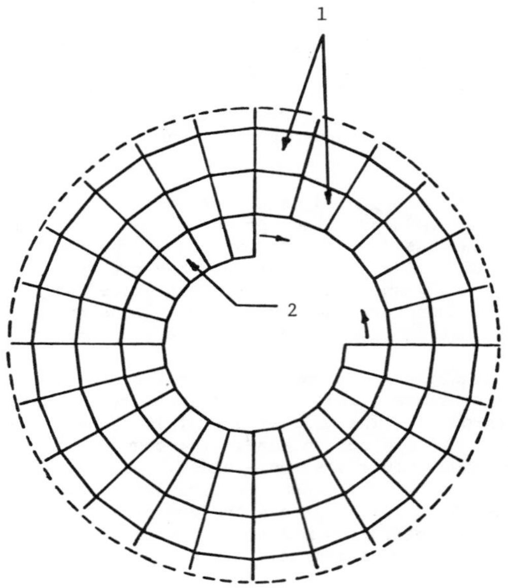

Fig 6 PROGRESSION OF RING-ROOF ERECTION
1- COMPLETED RINGS
2- MODULES BEING INSTALLED TO FORM THE NEXT RING

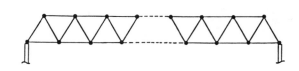

Fig 7 PROFILE OF A FLAT RING-ROOF

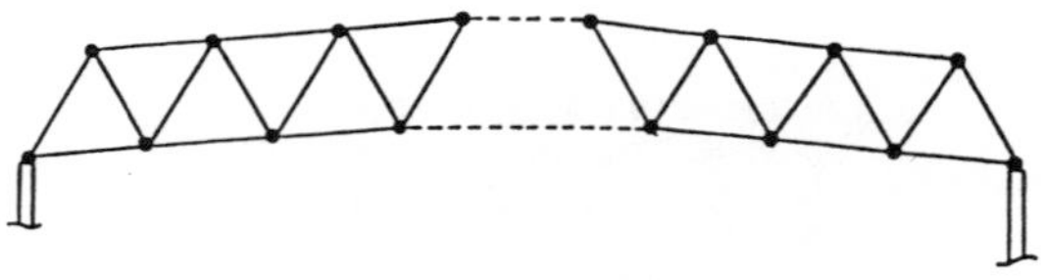

Fig 8 PROFILE OF A PITCHED RING-ROOF

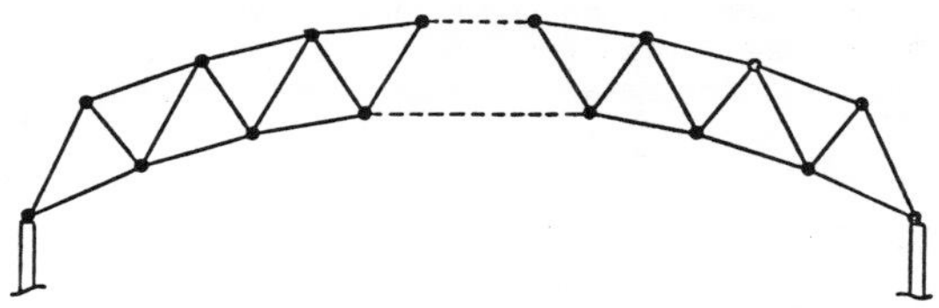

Fig 9 A VAULTED RING-ROOF

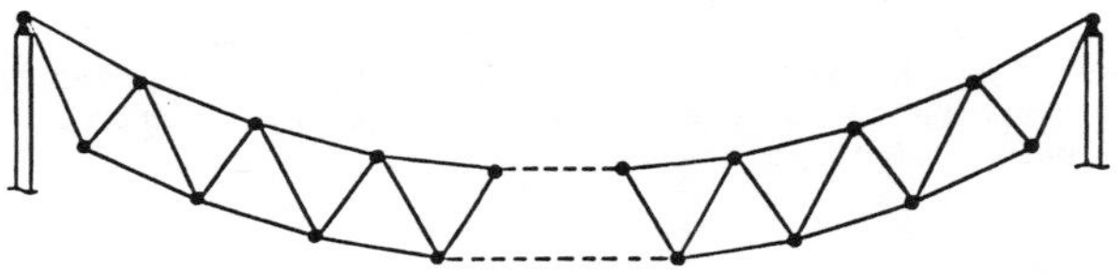

Fig 10 RING-ROOF WITH A SAGGING PROFILE

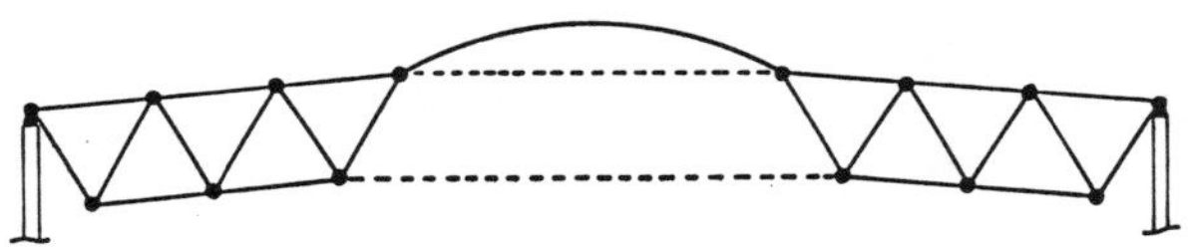

Fig 11 COMPOUND ROOF, COMBINES A PITCHED RING-ROOF WITH A SPHERICAL DOME

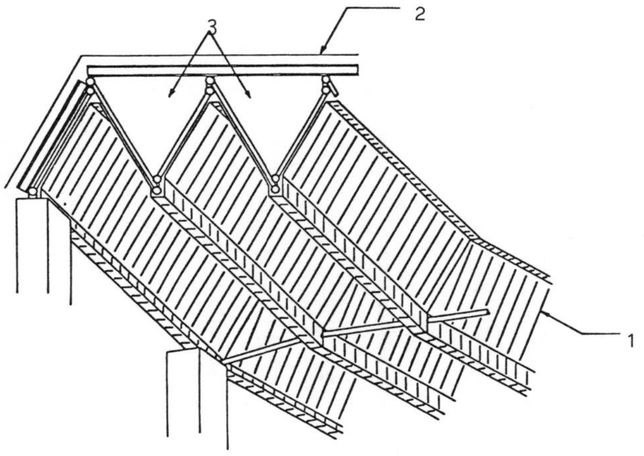

Fig 12 THE ROOF STRUCTURE AS SEEN FROM INSIDE THE STADIUM

1- METAL CEILING
2- BUILT-UP ROOF AND DECK
3- ENCLOSED SERVICE SPACES

RADIAL ROOF TRUSS DESCRIPTION

The radial roof truss has been used before in Canada, notably for the Edmonton Coliseum. It is formed by erecting plane trusses between each perimeter column and the center of the roof, as shown in Fig 13. The trusses all meet at the center of the roof, their upper and lower chords are all bolted to two circular steel connection plates. Temporary supports are required to erect the trusses.

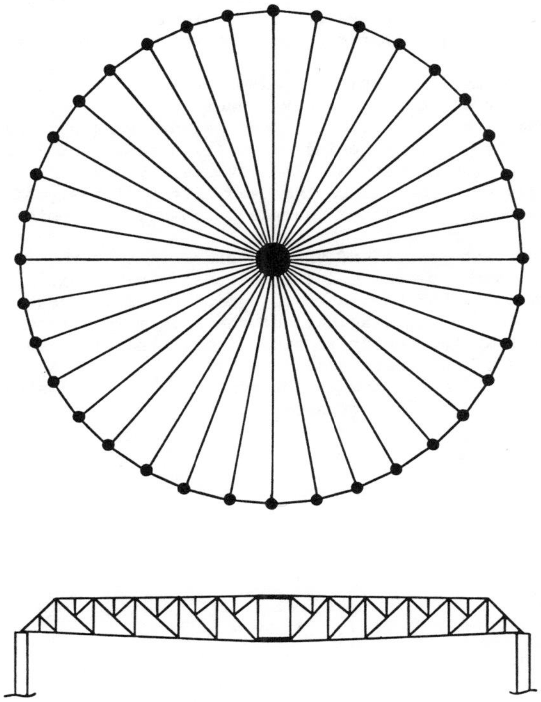

Fig 13 PLAN AND ELEVATION VIEWS OF A TYPICAL RADIAL ROOF. NOTE THE CENTRAL CONNECTION PLATE

COMPARISON

A preliminary design was made for a 420 foot clear span circular roof with a drainage slope of 1 percent. the structure required tubes of various diameters, from 3 inches at the center up to 12 inches at the perimeter. The modules were 80 feet long at the perimeter and decreased by about 4 feet per ring. Steel purlins were installed to support the roof deck, they were placed perpendicular to the roof trusses at 10 foot intervals.

The design loads were assumed as follows:

Dead Weight of Structure	10 psf
Insulation, Suspended Loads, Roofing	10 psf
Uniform Snow Load, acting on entire roof or half the roof	20 psf
Wind Suction on Flat Roof	20 psf

The Approximate Deflections were as follows:

Radial Movement on Columns-	
Due to Maximum Load (Dead + Snow)	1 inch
Due to Live Load	0.5 inch
Vertical Deflection at Mid Span-	
Due to Maximum Load (Dead + Snow)	12 inches
Due to Live Load	6 inches
Vertical Deflection Ratio for Maximum Load	1/420

The material consumption for the truss is estimated at 10.3 psf of steel plus 1.7 psf of steel for the purlins, for a total of 12 psf of steel for the roof structure.

An approximate analysis was also carried out for a radial truss type of roof. The same loads were assumed as in the previous analysis, with the dead weight being uniformly distributed. The material consumption was about 13 psf of steel for the structure. The material consumption would actually be higher since the dead weight of the roof is not uniform, but rather increases towards the center.

An important feature of the structure is that the forces in the chords of the modules decrease parabolically from a maximum at the perimeter to a minimum at the center. The forces in the diagonals of the modules decrease linearly as the center is approached. Therefore all the sections of the members decrease from a maximum at the perimeter to a minimum at the center. This is contrary to other types of space trusses. In radial type roof systems the forces in the chords increase from a minimum at the perimeter to a maximum at the center, whereas the forces in the diagonals decrease as the center is approached. For both air-supported membranes and suspended membranes the forces and sections remain constant over the entire span.

CONCLUSIONS

The ring-roof is a versatile building system. The design flexibility of this structure allows for different roof shapes. The material consumption for the ring-roof is slightly lower than in the radial type trusses. Since the structure is stable and rigid while still uncompleted the need for scaffolding is eliminated. It is also possible to save time during the erection of the roof since the roofing can begin to be installed before the roof structure is complete. This type of system could be extended to non-circular plan roofs, such as elliptic or oblong plan roofs. Detailed studies to optimize the geometry of the ring-roof are being carried out. The ring-roof appears to be a promising solution capable of providing economical large span roof structures.

ACKNOWLEDGEMENTS

The authors acknowledge the financial support of the Natural Sciences and Engineering Research Council of Canada. The Centre for Building Studies of Concordia University provided the computer facilities.

THE LIVERPOOL INTERNATIONAL GARDEN FESTIVAL

EXHIBITION BUILDING

I.R. BLUNN*, K.G. BUTLER * AND T. RAGGETT *

*ARUP ASSOCIATES

This paper describes the evolution, design, analysis and construction of a steel space structure combining a double-layer braced barrel vault with two single-layer hemi-domes, and founded on reclaimed land. Also described is a form of translucent cladding with potential for use as a large solar collector capable of changing transluency and colour. The building will be used during this year (1984) for the garden festival, and subsequently converted into a Sport and Leisure Centre.

EVENT, CLIENT

The Liverpool International Garden Festival is an event organised by the Merseyside Development Corporation to help finance the rehabilitation of areas of dereliction following the city's emergence from a more frenetic industrial past.

The Liverpool Garden Festival, using the many examples set by post-war Germany to rejuvenate war devastated areas, was seen by the Government appointed Merseyside Development Corporation to be a focal point in the attempt to restore Liverpool's former vibrance and arrest its municipal decay.

Due to open in May 1984, about three million visitors are anticipated before it closes in October.

Early in 1982, MDC invited six architectural practices to submit designs for an exhibition building intended to become the centre piece of the festival. Arup Associates were appointed in mid 1982. Following discussions about possible variations of the winning design detail design commenced in August 1982 with construction commencing in January 1983 with the finishing touches being added as this paper is being written in January 1984.

SITE

The festival gardens cover an area of 105 hectares and are sited on the Dingle Municipal tip, an area sandwiched between Toxteth and the Mersey. The domestic refuse has been deposited over the past 20 years in thicknesses varying from 4 to 10 m.

The building is positioned within an area that was originally a railway marshalling yard serving the nearby docks and wharves. The area had been built up into a platform of hardcore, and partly covered by a mass concrete slab and eventually overlain by about 4 m of domestic refuse. As part of the civil engineering operations (controlled by MDC) the refuse within and around the general building area was removed and reinstated to finished ground level using compacted river dredged sand from the Mersey. The building could therefore be founded on this 4 m layer of sand but the properties of thicknesses (density, compaction, uniformity, etc.) of the underlying hardcore of soils were largely unknown at the time of the competition. Subsequently the hardcore was found to be underlain by a layer of river silt of greatly varying thickness overlying sandstone bedrock.

This, together with the increased surcharge from the sand replacement (the refuse was of low density with voids) suggested that the proposed structure should be reasonably tolerant of differential settlement if expensive foundation solutions were to be avoided.

Measurements of ground movement were recorded during and after sand replacement and were continued throughout the construction period.

These agreed approximately with the predictions made prior to the structural analysis, i.e. a total settlement of 75 mm with a maximum differential of 30 mm.

BRIEF

The brief called for a building that could accommodate two quite different functions; primarily it should form the focal point of the festival and provide a weathertight enclosure for a great variety of events and activities.

Subsequently the building enclosure should be capable of being economically and easily converted into a Sports and Leisure centre. Facilities included divide broadly into three main categories, thus;

1. A leisure pool with wave machine, beach, refreshments area.

2. A multi-purpose sports hall with provision for demonstration tournaments requiring variable seating for up to 3,000 spectators.

3. Various, smaller scale 'club' activities including squash, gymnasium, projectile hall, etc., plus refreshments area and administration offices.

Each of the above have differing spatial, environmental and and organisational requirements.

This second stage was fairly loosely defined however, and a degree of flexibility was required in the general planning, although it was requested that the fabric of the building should be permanent and suitable for Stage 2.

EVOLUTION OF ENCLOSURE

Study of the brief generated the following design criteria relating to the building enclosure:

a) To form one enclosure which would both unite and yet define and express the three principal activity areas of pool, hall and club.

b) To form a natural relationship to the landscape which would give pleasure within the festival park.

c) To provide an 8,000 m² column free space and so give flexibility for both the festival and sports complex.

d) To form a minimal external surface requiring little maintenance and being durable and technically efficient in many ways.

e) To possess an identity reflecting the spirit of the festival and Liverpool's architectural heritage.

Following several design studies a structural concept was proposed comprising two related forms: the dome and the vault.

The 90° dome is halved and then rejoined by a linear barrel vault.

The structural efficiency derived from this curvilinear form results in low material content and favourable ratio between external surface area and plan area.

The half dome which covers the leisure pool complements the relaxed informality of this activity.

The barrel vault covers the main hall which has fixed spectator seating, changing and plant rooms arranged along the sides thus using the perimeter 'dead space' normally associated with curved forms.

The other half-dome contains a variety of 'club' activities small enough not to be seriously compromised by this form of enclosure and also benefiting from a more relaxed atmosphere.

As the observer approaches, the apparent mass of the building recedes, thus preserving the scale of landscape and garden.

EVOLUTION OF STRUCTURAL FORM

The structural form is a direct and specific response to the foregoing. The arrangement of structural elements changed little from the original competition entry and were derived principally from the following:

1. A compatible geometry to unite dome and vault.

2. The need for dome and vault to have different cladding suggested that the domes connect via the lower layer of the double layer vault, thus expressing the transition externally and preserving the clarity of structure within.

3. A convenient and regular structural grid for the attachment of external and internal cladding. A 3 m structural module was chosen permitting relatively slender longitudinal members.

4. Evenly distributed building dead and live loads and the ability to absorb differential settlements.

5. Simplicity and clarity of design, fabrication, construction and perception.

PRIMARY ELEMENTS

VAULT

A two-layer barrel vault structure 78 m long comprising braced arched frames of 3 pin configuration at 3 m centres with upper and lower booms connected by longitudinal members. The lower layer connects through to the domes and supports services and acoustic panels in selected areas plus providing the path for the axial forces from the domes. The upper layer supports the external cladding.

The intermediate arch frames are collected by braced frames which with the on grid frame transfer their reactions to bipod frames at 6 m centres. Roof cladding terminates at an eaves gutter which discharges into a rainwater pipe running above the inclined leg of the bipod frame.

DOMES

The half dome at each end is of segmental, ribbed, single layer construction with circumferential rings at 3 m centres connected through via the longitudinal members in the lower layer of the vault.

Thus the continuity of the dome rings are maintained and the out-of-plane forces at the junction of dome and vault are resisted by the flexural and transverse stiffness of the 3 end arches combined with their 3 dimensional bracing.

FOUNDATIONS

Loads on the foundations were relatively small (a typical arch thrust being approximately 450 kN D+L). Overturning moments were high due to the level of the springing point of the arch (approximately 4.7 m above the underside of the footing). This was overcome by displacing the footing so that its centre of gravity approximately coincided with the line of thrust passing through the inclined bipod leg. The maximum allowable pressure at the underside of the footing was restricted to 150 kN/m².

The vault bipod frame foundations were inverted tee reinforced concrete spread footings tied across the vault by mild steel bars (pretensioned) to resist the horizontal thrust from the arch.

The dome foundations were concrete ring beams forming two semi-circles on plan, connected longitudinally by mild steel pretensioned tie bars.

The diameter of the tie bars was sized on permissible strains (horizontal deflection) rather than stress.

MODELLING AND FRAMING CONFIGURATION

Following the definition of the conceptual design it was clear that in order to maintain the building's simple elegance the structure should be well defined, simple and precise. It was important that the underside of the vault be a single unbroken line along its length and that the undersides of the connected domes be a continuation of this alignment.

It was of equal importance that the joints be functional and aesthetically pleasing and that stability structures be simple statements of function. The structure divided naturally into three parts. The two dome sections connected by the vault.

The vault required only horizontal bracing in order to provide stability whilst the dome sections required both horizontal and vertical support at their discontinuous vertical edges.

Studies were made into the economic spacing of the vault frames and whether a two or three pins configuration would prove more economical. These were carried out ignoring any interaction with the dome ends.

Studies were made influenced by architectural considerations for the moving of the central pin from the mid point position between top and bottom booms to the centre line of the top boom. The initial studies of the dome discontinuity support showed that stability could not be achieved within the dome structure itself without massive and quite out of proportion members. It was therefore decided to make use of one of the dominant architectural features of the building namely the oculus - the continuation of the ventilator structure.

The principle here was to introduce a semi-circular (in plan), triangular (in section) braced girder which would cantilever out from the last three vault frames and pick up the braced in plane ends of the dome ribs. It followed from this that these three end frames be treated as braced bays with in plane and out of plane bracings providing horizontal and vertical reactions respectively to the out of balance forces arising from the dome discontinuity.

The studies into vault frame spacing and configuration when combined with the ease of erection, cladding fixing and access into the building etc. resulted in vault frames spaced at 3 m and with alternate frames being supported off the adjacent lower pins with a bifurcated braced frame. This gave a frame spacing at ground level of 6 m.

The configuration chosen was of the three pin type with all three pins on the axis of the top boom.

The penalty paid for the choice of 3 pins was a small increase in the maximum bending moment of the order of 10%. This did not cause sufficient changes in member forces to give a change of section size.

When this was compared with the advantages to be gained in both off and on site fabrication plus ease of erection the choice became obvious. Combined with this was the understanding that configuration with the three pins would not be as sensitive to differential settlements of the foundations.

The location of the pins on the axis of the top boom was not so obvious a choice. The principal motive behind this decision being architectural rather than engineering. This resulted in an increase in the top boom thrust of 20% above the bottom boom and a section change due to the redistribution of the axial forces.

The cladding module selected of 3 m x 1 m led to the primary circumferential spacing of the Warren truss nodes. This gave eleven equal arc lengths of 3 m on the centre line of the upper boom each side of the centre pin.

From this came the location of the longitudinal members connecting the arch trusses and the dome ends.

The upper longitudinal member acted principally as a purlin supporting the cladding whilst the lower member carried the residual longitudinal strut/tie forces not catered for by the end braced bay. This lower member also provided compression 'flange' stability during load reversals under the partial live loading condition.

From an architectural point of view it was considered important that the springing point for the arch trusses be located sufficiently high so as to provide a side wall to the hall. This was solved by sitting the arches 2.8 m above ground level on a simple bipod frame with the inclined member of the bipod following the line of thrust through to the inverted T footing. The resultant horizontal thrust being catered for by a 75 mm diameter mild steel tie between opposite footings. Longitudinal stability to the pinned junction between arch truss and bipod was provided for by bracing in the plane of the inclined lower boom.

The member shapes chosen for the upper and lower booms were joist section. These gave the advantage of clarity of line with a minimising of bulk and ease of connection. To these was added the possible use of the joist section as a runway for an access cradle.

From this latter potential use it was necessary to set the position of the lower longitudinal members above the bottom boom. These lower members ran continuously along the length of the building emerging at the dome ends to act as both strut and tie around the circumference of the dome. Due to the increase of dimension of the dome rib above that of the lower boom to the trusses this circumferential strut/tie is discontinuous around the dome with a pinned end connection to each dome rib.

The use of curved members for both the vault and dome members induced significant secondary stresses in the arch and dome ribs and dome purlins. In the case of the dome purlin this tended to reduce flexural stresses due to dead and live loads.

LOADING

The loading applied to the building was to a greater extent than usual indeterminate. With a building of its unusual shape even the normally accepted uniform loading simulating snow load was somewhat suspect.

The windloading pattern was assessed using the relevant clauses of CP3 Ch V Part 2 combined with recommendations carried in Newbury and Eaton - Wind Loading Handbook.

Obviously it would have been more satisfactory to have carried out a series of wind tunnels tests but unfortunately there was neither time nor finances available.

The wind loading pattern adopted is shown in Fig (1). The static and live loadings adopted are shown below:

Load-case 1

Dead load of the building with a live loading of 0.75 k N/m² applied to the entire structure.

Load-case 2

Dead load of the building with the live loading applied to the longitudinal half of the building.

Load-case 3

Dead load of the building with the live loading applied to the transverse half of the building.

Load-case 4

Dead load of the building with the wind loading applied, by the wind direction being along the axis of the building perpendicular to the vault arches.

Load-case 5

Dead load of the building with the wind loading applied, the wind direction being orthogonal to that of Load-case 4.

Load-case 6

A combination of Load-cases 2 and 5.

Load-case 7

Dead load of the building with one dome end displaced downwards 25 mm at the end of the building.

Load-case 8

Dead load of the building with one of the braced bay arches displaced 25 mm.

Load-case 9

Dead load of the building with the live loading applied to one quarter of one of the domes - the portion loading being the upper portion adjacent

Load-case 10

Dead load of the building and the 25°C temperature change.

Load-case 11

A combination of load-cases 1 and 7.

Load-case 12

A combination of load-cases 1 and 8.

Load-case 13

A combination of load-cases 3 and 4.

In the case of combination load-cases, the dead loading of the building was, of course, only applied once.

A temperature change of 25°C was adopted although it is probable with the number of joints and the degree of movement possible that the actual forces induced would be of low magnitude.

A differential settlement of 25 mm between outer perimeter of the dome and its intersection with the vault and alternatively a similar differential settlement of 25 mm between the central vault truss and the intersection between vault and dome was considered.

ANALYSIS

The initial vault arch analyses were calculated by hand, since it was a statically determinate problem. It was initially assumed that most of the longitudinal forces in the vault transferred from the domes would be carried to the foundations in the braced bays. This meant that the preliminary sizing of the vault purlins could be done largely on the basis of the bending moments and deflections induced by the loading applied to the cladding. It was interesting to note that at wider arch centres, it was the heavier purlin section required that made the structure relatively uneconomic, as opposed to the heavier arch sections required.

Other than these preliminary calculations, the analyses required were of highly indeterminate structures. Eventually, because of the importance of the assymetric loading conditions, it was necessary to form a numerical model of the entire structure. These numerical models were analysed by computer.

The structure was modelled as a skeletal space framework. The analysis was performed using a proprietary program, PAFEC 75. Use was also made of an in house program which permitted a certain amount of pre and post processing. The input data was written to a database. The output was written to the same database. The post processor could then gain access to the database file to provide facilities such as plotting structure and deformations, general output, selective output and computing load-case combinations.

The assumptions and limitations to be considered were those normally associated with the stiffness method of analysis. This meant that some of the early analyses, which resulted in some very large deflections, sometimes did not even given an indicative answer to the correct choice of member property. This determined the strategy adopted for the analysis which will be described later.

The program required the structure to be idealised as line elements which intersect at a series of nodes. This naturally meant the preparation of large amounts of input data. Each element was defined as a shear beam; that is the effects of shearing in addition to pure bending were included in the formulation of the element stiffness. The structure was analysed as a pinned structure. This was simulated in the model by incorporating elements of small moment of inertia for the bracings, truss diagonals, ties, purlins and dome circumferential members. The supports were all modelled as pins by releasing the three rotational degrees of freedom. The program uses a frontal solution method for solving the stiffness equations.

To optimise the efficiency of the analyses, it was desirable to number the elements to minimise the number of live degrees of freedom at any stage during the solution. The elements were numbered in such a way that data input was facilitated. This clearly did not necessarily result in the optimum front size. A purpose written routine was again used to generate the element list such that during the analysis the elements would be merged in the optimum order.

In order to carry out a satisfactory analysis, it was necessary to understand the structure's behaviour beforehand. A sensible assessment could then be made of the member properties. Small deflections would then result from the analysis and thus not conflict with the basic assumptions.

The behaviour of the dome-vault interface was the most difficult to evaluate. A number of analyses were performed on one of the dome pieces in isolation. This was to evaluate its behaviour and to determine the mode and degree of support required at the value. The dome radial members, for architectural and practical reasons did not run through to the apex, but were stopped of at a braced ring beam. The analyses showed that the dome required vertical support at the top of the braced ring beam. This was provided by a triangular braced girder which cantilevered from the braced bay. It was found that by bracing three vault arches in both planes of the vault structure and also perpendicular to these planes, it was sufficient to provide vertical, lateral and torsional stiffness to support the dome. These braced bays were also designed to cater for the horizontal forces on the structure as a whole.

The asymmetric load cases were often what determined the design. It was therefore necessary to carry out the final analysis on a numerical model of the entire structure. Because of the shape, the structure when fully assembled was very stiff. Theoretical deflections were of the order of a few millimetres for the symmetrical load-cases and the maximum anywhere on the structure was 25 mm.

The post processing facilities available permitted the output of the results for the most heavily loaded members in specified portions of the structure. The results were rigorously inspected for those parts of the structure which were least easy to visualise in terms of structural behaviour, such as the braced bays, stiffened dome rings and dome stiffening girders.

It was found that the dome stiffening girder performed the function that its name implies. It stiffened the top portion of the dome and prevented the large deflections of the dome which had been experienced in the early analyses. The forces in the members of the dome stiffening girder were not large. The relatively large forces resulting from truncating the dome short of its apex were transferred to the vault structure via the stiffened circumferential ring.

The supporting of the domes and any longitudinal forces in the structure as a whole due to loading on the domes were virtually catered for by the braced bays. Between the braced bays, the vault purlins and the ties between the lower arch booms carried very little in the way of longitudinal force. The effect of the braced bay was not evident in any of the arches in the central portion of the vault.

The portion of the structure which worked hardest was the braced bay arches, in particular the arch booms. The design conditions for individual members here (as also elsewhere in general) were the load-cases with asymmetric live loading. The RSJ boom sections used elsewhere were in certain cases insufficient in the braced bays. To maintain the uniform appearance in the vault, it was decided to plate the RSJ members, rather than substitute larger members. For architectural reasons, it was desirable that the plating be symmetrically arranged and therefore the extent of plating used was greater than that strictly necessary for structural reasons.

The behaviour of the braced bay was in line with what had been assessed from the preliminary analyses. The one unexpected characteristic of the braced bay assembly which was not immediately obvious from the analysis of the complete structure was that, although it was very stiff vertically and laterally, it had relatively little torsional stiffness in the temporary condition. This manifested itself during erection when the dome was erected against an isolated braced bay which was restrained laterally only. In this condition, the dome radial members acted as arches and the thrust at the upper ends resulted in some twisting of the braced bay. The movements were not large, but because of the magnifying effects of the three dimensional circular geometry, they were sufficient to cause problems with fitting the dome circumferential members.

The design was to BS 449: Part 2. Since deflection of individual members was not generally a design factor, advantage was gained from the use of grade 50 steel. The predominant loading for any member was the axial load. Many of the members were, however, curved and consequently significant secondary moments were induced. The vault purlins and dome hoop members also had moments induced from the loading transmitted from the cladding.

The wind load-cases using the increase in permissible stress allowed in BS 449, did not produce more onerous member forces than the asymmetric live loading cases.

The loadcases in which supports were displayed similarly did not induce member loads anywhere in the structure in excess of those resulting from the asymmetric live loading cases.

Because of the flexibility of the structure, it was not anticipated that the effects of expansion would be a problem. The expected movement over the length of the vault would be of the order of the expected bolt slip in that length. The braced bays, however, are a relatively rigid portion of the structure and it was expected that their influence would induce significant member forces in certain places under the temperature loading. This did in fact prove to be so.

Detailed elastic hand analyses were carried out for certain elements - such as the pin details, the joint where the intermediate arch booms split and the stability of the dome radial members.

The ventilator structure at the apex of the vault is supported by the vault and forms no part of the overall framing action.

CONSTRUCTION

Due to the extremely short construction programme pre-ordering of certain sections notably of large diameter tube and small size joists was essential. This was discussed with the British Steel Corporation and prospective tenderers and a bulk order was placed with British Steel Corporation for all principal members just prior to the tender period i.e. the top and bottom booms of the truss, the dome ribs and the bipod members.

The tenderers were asked to include a cutting list with their tenders so that the final order could be placed with British Steel Corporation in early January (prior to the sub-contractor order being placed).

The successful steelwork sub-contractor Tubeworkers Ltd was appointed by the end of January 1983 and following a four week intensive and stimulating period of workshop drawing production and approval, fabrication began.

During this period the resolution of the lower pin junctions and the geometry of the interconnection of the primary and intermediate truss were finally solved.

Quality control and welding procedures were set up, using Messrs. Sandberg as the specialist adviser. It was essential that all aspects of the work be of the highest quality and the quality control was divided into two areas; one which dealt with the technical performance and appearance of the welded joints and the other with the dimensional accuracy of the completed members and compliance with the corrosion protection specification.

The dimensions of the arch and dome ribs were such that it was not considered practical to transport completed frames and so these were delivered in half frames of approximately 16.5 m length. This meant that full strength site butt welds were required for top and bottom booms of the arch ribs and for the dome ribs.

With overall dimensions of 125 mm and 100 mm for the top and bottom boom members of the arch some fairly delicate and highly skilled welding was required. In order to achieve full penetration of the flange butt welds windows were formed in the web to allow sealing runs to be made across the inside of the flanges. Run on angles were used to give the correct alignment of the flanges and to allow the flange welds to be carried past the end of the flanges thus avoiding the risk of porosity etc. at the edge of the flanges. These temporary angles were subsequently cut off and the window plate in the webs welded in.

Problems were experienced particularly at the root of the flanges. These were overcome by slight modifications to the weld preparation of the window plate, which involved carrying the horizontal runs past the corners of the window and overlaying a fillet weld at this junction.

In general the standards set and achieved were as laid down in BS 4870. The methods of testing adopted were magnetic particle and ultra sonic. The X-ray method was not used with the exception of laboratory testing of welder test pieces.

The method of erection adopted by the contractor was to assemble each half arch on the previously erected bipods with the lower pin in position and sitting the free end on a scaffolding tower approximately 3 m high. The upper half frame was then set on the scaffold tower with the centre pinned end resting on blocks at ground level. The welding jigs were set up and alignment carefully checked with surveying instruments. The welds were then completed and tested (100% of first 6 frame sets). Three arches were assembled at one time with upper and lower longitudinal members in position and starting with the southernmost braced bay. Lifting beams were attached near the top of each 3 frame set of half arches and the two halves lifted by two crawler cranes, the apex pin was then fitted and the arch frames released. Every fourth arch was omitted and then lifted into position, once the three frames either side were erected, from a temporary supporting beam provided at the upper and lower ends of the intermediate arch frame.

The site welding of the dome ribs was carried out in a similar manner to that of the arch frames. The initial method of erection was that the ribs were lifted with a clamped on strongback individually, connecting each via its purlin back to the previously erected and quyed braced bay. The upper end of the dome ribs were then connected to the upper stiffening girder via upper and lower ring beams.

The stiffening girder and ring beams were assembled in quadrants on the ground and lifted into position onto a scaffold tower at the dome apex. Welds were completed and tested prior to connection of the dome ribs.

It was found that the ribs tended to flatten during the process of erection due to movements in the stiffening girder and braced bay leading to a decrease in the circumferential dimensions and hence to a lack of fit of the curved purlins.

The resolution of this was achieved by the introduction of a supporting ring of scaffolding at approximately the mid-point. This was maintained in position until the whole structure was completed following which the ribs were released gradually on screw jacks and the structure took up its natural form.

The corrosion protection system was:

: For internal steelwork

2 pack epoxy zinc phosphate prefabrication primer 25 microns.

Zinc phosphate chlorinated rubber, primer 25 microns.

High build chlorinated rubber, barrier coat 75 microns - two coats.

Chlorinated rubber finish coat 25 microns.

: For external steelwork

Flame sprayed zinc 100 microns.

2 pack PVB etch primer 10 microns.

2 pack epoxy MIO high build barrier coat 75 microns.

Silicone alkyd enamel high build finish 75 microns.

CLADDING

The architectural approach to the building considered the two part domes as solid projections from within the transparent vault.

The solidity of the dome cladding was achieved by the use of a pressed profiled aluminium sheet. These were pressed to the spherical geometry of the dome in tapered sheets of 6 m length. These sheets were connected to circumferential aluminium tubular purlins set at approximately 2.5 m centres up the dome.

The aluminium was left unfinished and a tray to which was fixed insulation was attached to the underside.

The vault cladding was formed from 3 m x 1 m panels of twin wall extruded poly carbonate sheets. This sheet was 20 mm thick and had longitudinal ribs approximately 30 mm apart. This type of sheet had been used mainly on the continent although there were a few examples in the UK. It had a very low Youngs Modulus (2.3 kN/mm^2) and was virtually indestructible from a physical damage point of view.

The fire retardant grade was used and achieved a flame spread rating of Class 1.

The sheets are glazed into aluminium extrusions with neoprene seals giving continuous panels of 3 m x 1 m sheets held in place by aluminium capping pieces. The system used fairly normal glazing details with the exception of a much greater allowance for expansion and contraction. This followed from the very high coefficient of linear expansion typical of the poly carbonate material (approximately 5 times that of steel).

Another aspect of the cladding design was the system of drainage channels formed by the cold formed inverted channel support to the glazing bars. This was only a fail safe system to allow for the water seals being bridged or condensation occurring within the sheet.

One of the original ideas which due to time and financial constraints was not incorporated was the possibility of using the fluting of the sheets to carry a fluid (coloured or clear) which could be used as a solar collector and with variation in colours both as a sun shield and to change the appearance of the building. The initial idea stemmed from a commercially available solar panel using this technique, but the scale of our application demanded considerable research and development of the system.

It remains however theoretically possible and worthy of further investigation.

FACTS AND FIGURES

Enclosed Volume	72,200 m^3
Enclosed Area	7,500 m^2
Span of Dome	62 m
Span of Vault	60 m
Height to u/s of Dome	13.8 m
Length of Vault	78 m
Depth of Vault Truss	1.2 m
Radius of curvature (inner)	41.1 m
Weight of steelwork	300 Tonnes

PROGRAMME

Started on site	January 1983
Completed	February 1984

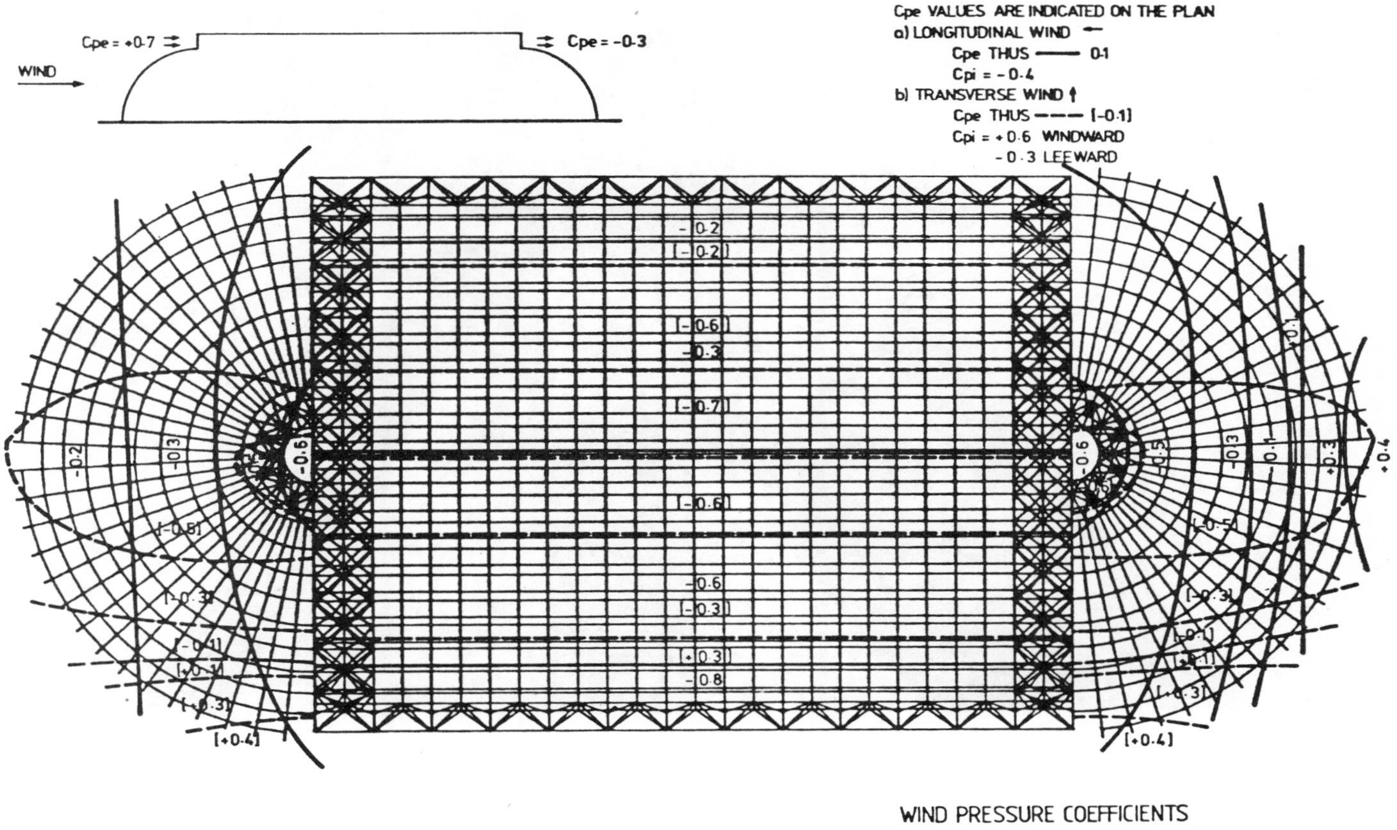

Fig. 1 Wind Loading Diagram

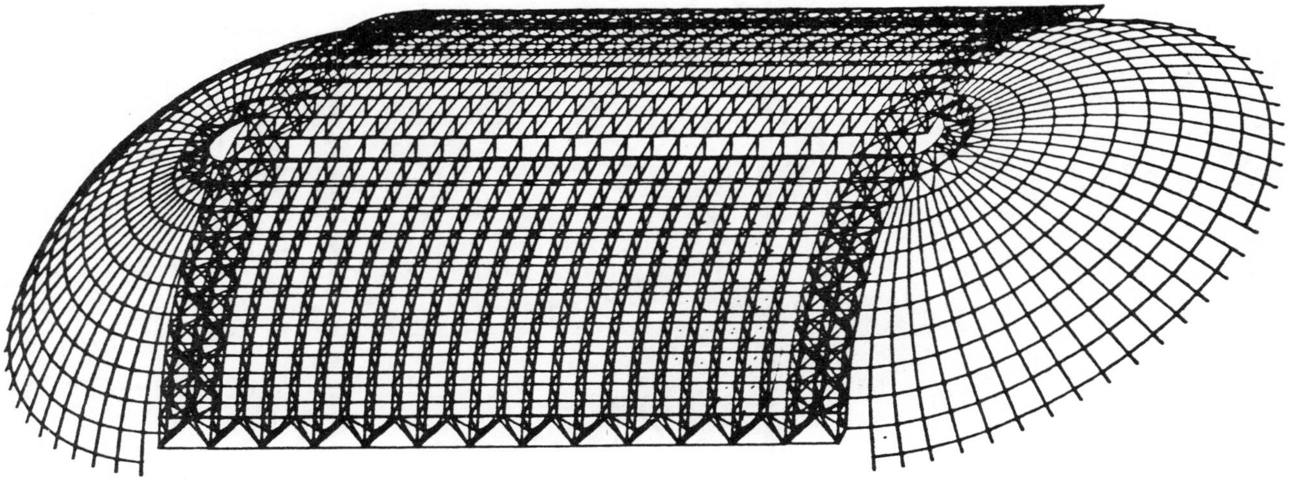

Fig. 2 Computer Drawn Perspective of Frame

Fig. 3 Lower Pin Junction

Fig. 4 Braced Bay and Stiffening Girder

Fig. 5 Vault Arches prior to Lifting

Fig. 6 South Dome and Part Completed Vault

Fig. 7 Completed Frame

Fig. 8 Completed Building

REALISTIC CONCEPT FOR BUILDING METAL MEMBRANE STRUCTURES *

S. GREINER, Dr.-Ing.

Consulting Engineer, Stuttgart, West Germany

Membrane structures made from very thin metallic skins with an approximate thickness of between 0.3 mm and 3 mm which are subjected to pure tensile stresses are descendants of tents and balloons. In the past, difficulties resulting from manufacturing processes represented a hindrance for adequate popularity of this construction method. Following an overview of developments carried out up to the present, possible methods are illustrated as to how the main problems of shaping, assembly and erection can be solved in accordance with the method of the shaping load case and the membrane square-edge joint welding method. In addition, a calculation method for form modelling and initial practical applications are described in brief.

INTRODUCTION

Membrane structures with their unique charming shapes are mainly made up of a very thin, flexible and tension-resistant skin. The membrane element, which is subjected to pure tensile stresses, can utilise the strength of the material very effectively and, in addition, in two axes. This results in extremely low material requirements which are reduced further by the fact that the membrane can be assigned the room-enclosing and load-transmitting function.

From the very earliest applications right up to the present day, all available large-surface and tension-resistant materials have been employed as the construction materials for membrane structures (such as tents, containers, suspended roofs, balloons and air-halls, for example): with the exception of animal skins, fabrics and all types of films, non-woven fabrics and paper and, as soon as the first rolling mills were constructed, sheet metal also. Today, rolling techniques have been perfected to such an extent that the metal sheets represent a membrane material of outstanding quality. These sheets are produced to any thickness and in a wide range of alloys. Metal membrane construction material stands out as a result of its high tensile strength and its long service life in particular which is always the case if corrosion is eliminated. The necessary sheet thickness is extremely low, even for very large buildings. It is normally less than 1 mm when consideration is given to all static requirements.

IMMENSE COPPER ..LOON, AT PARIS.—See next page

Fig 1

* This report is a summary of the investigations published in Ref 1 which I was able to conduct within the scope of the activities of Sonderforschungsbereich 64 at Stuttgart University, Institute for Concrete Structures (headed by Professor Dr.-Ing. J. Schlaich). Ref 1 contains more detailed information, incl. bibliography.

BRIEF OUTLINE OF THE HISTORICAL DEVELOPMENT PROCESS

Without claiming to be complete, the following is intended to provide an overview of the most important stages in the development of construction using metal membranes. The notion of the wafer-thin, spanned metal skin as a construction element for surface structures was an extremely attractive prospect for generations of engineers. In many countries, attempts were made to realise this idea again and again. This trend has increased over the past 20 years. According to a notice in the "Illustrated London News" in 1844, the idea to construct a gas balloon using thin sheet metal was expressed as early as 1760. A project of this kind was realised in Paris that year. A gas balloon with a diameter of 10 yards made from copper sheet with a thickness of 0.125 mm was constructed (Fig 1). Even older are Schinkel's more playful, decorative tent roofs of sheet metal (Fig 2). In the last century also, Shukov constructed his remarkable suspended roof in Russia which was designed as a cable net structure with a central, suspended sheet metal membrane.

Fig 2

A blimp of very thin aluminium sheet (Fig 3) with a length of 45.5 m built by the U.S. Navy in 1929 was followed by several grain silos with roofs of suspended sheet metal strips (Fig 4) in the 1930s in America. At the same time, Lafaille built the first suspended conical roof in Zagreb, an exhibition pavilion with a diameter of 33 m. A whole series of large shed roofs were subsequently erected in accordance with this concept. The largest ever suspended conical roof with a diameter of 112 m was erected in Vienna in 1978 for a circular sports hall. The greatest progress with regard to metal membrane construction has been made in the Soviet Union. A large number of shed roofs with large spans and cylindrical containers with diameters of up to 60 m have been built there using sheet metal in the recent past. In addition to other large suspended roofs, the roofed stadium in Moscow, with an elliptical layout measuring 224 m x 183 m, is also worthy of note.

Fig 3

The first pneumatically spanned metal membrane roofs were realised in the 1970s. The test roof with a diameter of 5 m (Fig 15) was built in Stuttgart in 1975 using extremely thin stainless steel sheet. In 1979, a test structure with a diameter of 20 m was built by Kawaguchi in Japan and the gymnasium roof with a length of 91 m according to Sinoski in Halifax, Canada.

With regard to the concept of the "shaping load case" propagated in the following, special attention must be given to the project for a pneumatically spanned aircraft assembly shed with a diameter of 365 m by Stevens in 1942 (Fig 5). Here, for the first time ever, it was proposed to produce the dome shape of the shed from a flat sheet metal membrane by means of expansion of the membrane on the overall structure by means of plastic deformations. Otto published a similar idea in 1954. This was realised by Ludkovsky in Moscow in 1967 in the form of a test structure with a diameter of 23 m.

Fig 4

However, metal membrane structures have not attained an adequate level of popularity up to the present. The reason for this probably lies in the considerable difficulties encountered in production, a fact which accompanies the use of sheet metal as a membrane construction material. Assembly, erection and forming are considerably more difficult than in the case of the textile construction materials normally employed.

Fig 5

EXPLANATION OF THE TECHNICAL PRODUCTION PROBLEMS

In general, membrane surfaces are curved in two directions while sheet metal strips can only be bent in one direction as in the case of paper. The suppleness of textile construction materials, which can be adapted better to form any type of surface shape thanks to their flexibility, is not present in metal sheets. The high shear and elongation resistance of the metal sheets necessitates very small cut elements and extremely careful working if the surfaces are to be approximated by pieces. Minute geometry errors during production of the membrane surface result in large force errors in the membrane surface. It generally seems impractical to put together the membrane, which is very delicate on assembly, on falsework, for example, since this would necessitate an unreasonably high expenditure.

The handling of large sheet metal elements is also considerably more difficult than in the case of fabrics. Metal sheets are relatively heavy, they have sharp edges and have a tendency to form folds when not handled correctly. These folds can result in the sheets subsequently breaking. Strips with a width limited by the transport conditions are coiled; this is the only way in which they can be prefabricated and carried to the building site. A considerable number of weld seams must then be produced there.

Joining of the sheet metal strips was another problem for which there was no solution in the past. Joints which withstand high loads as desired can only be attained by a continuous seam. However, welding of very long seams on thin metal sheets under project site conditions posed a considerable technical problem. The membrane square-edge joint welding method developed for this purpose is described below.

In view of all these problems, it is no wonder that in the case of all metal membrane structures to date, surfaces that can be developed such as cones or cylinders were selected from the outset, or surfaces curved slightly in two directions by means of straightforward bending of flat cut elements were used as the basis. The method of the shaping load case overcomes these restrictions. Instead of avoiding elongation in the membrane, this is used accurately to obtain a membrane surface which is perfectly positively or negatively curved in two directions.

THE CONSTRUCTION METHOD OF THE SHAPING LOAD CASE

With the shaping load case method, the shape of the desired membrane surface is not simulated or approximated by adapting discrete elements, but is produced in a continuous process on the structure as a whole. The desired shape is produced by means of loads applied to the membrane once. Resulting elongation of the membrane is possible thanks to the plastic ductility of the metal. Ductile metals have high cold forming properties. Austenitic stainless chrome-nickel steel CrNi 18 9, for example, can withstand approx. 50 % expansion of line elements in the sheet plane until it breaks.

In practice, this method is as follows (Fig 6):
The membrane is assembled on the building site as a level surface from the sheet metal strips or from large prefabricated cut elements according to the membrane square-edge jointing method and then connected to the edges and other supporting elements (Fig 6a). The membrane is then given its intended form by applying a specific sequence of loads, the shaping loads (these can be surface loads and/ or edge loads). During this process, plastic deformations occur in the membrane and, in certain cases, at the tensioning edges when the surface elements are enlarged (Fig 6b). After the membrane surface has attained its desired shape, the level of the shaping loads is reduced to approximately one half (Fig 6c). The remaining amount of tension now serves to stabilise the shape of the structure under working loads. For the first time, the entire structure is shaped by using the plastic ductility of the metal. In the past, exploitation of this property unique to metals was restricted to construction elements such as rivets, in the building sector, while production of semi-finished products and other metal parts in their variety of shapes is, of course, only possible as a result of the plasticity of metals.

In model tests, studies with respect to shaping have been carried out in accordance with the method of the shaping load case (Fig 7-11). With the exception of the dodecahedron shaped to form a sphere, the models are produced from extremely thin metal film.

It is an inestimable advantage that the sensitive balanced configuration of the membrane spans itself under the shaping loads. Comparable to the formation of a soap skin, a smooth shape with a even and extremely harmonic flow of forces is attained, something impossible in the case of an approximating cut.

All alloys which can be stretch-formed easily are suitable materials. These are characterised by high uniformity of expansion and ultimate strain and attained good results in Erichsen tests. The material properties are not influenced negatively during forming, provided that the deformation capacity is not exploited to its full extent.

Austenitic stainless chrome-nickel steel CrNi 18 9 has proved to be a highly suitable construction material. In addition to being extremely ductile and tending only slightly to corrosion, it stands out due to the fact that it can be welded easily.

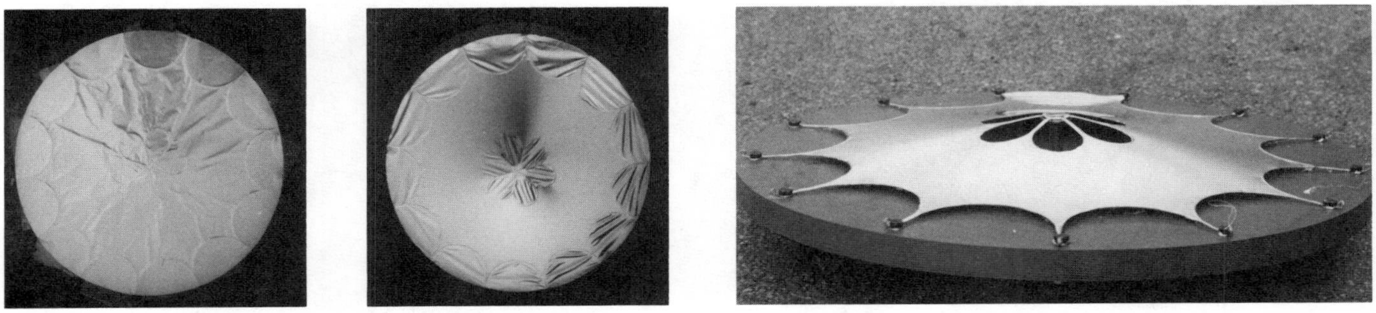

a) b) c)

Fig 6

Fig 7

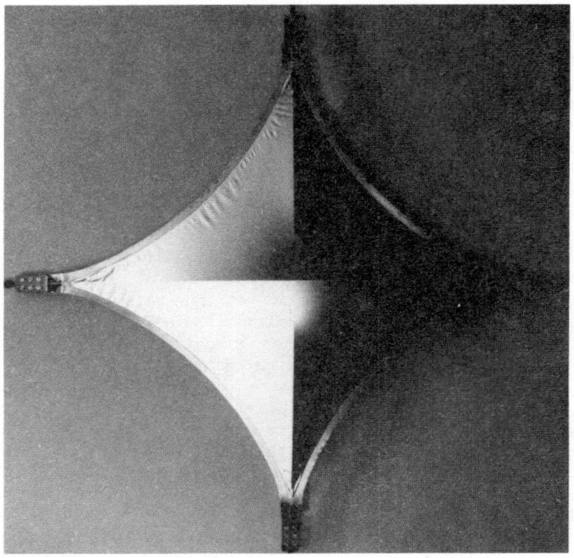

Fig 9

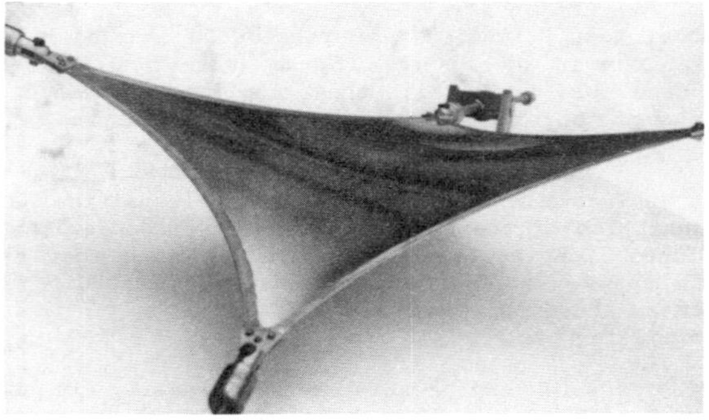

Fig 8

Fig 10

a)

b)

Fig 11

MEMBRANE SQUARE-EDGE JOINT WELDING

A welding method has been developed which is geared towards sheets made from austenitic chrome-nickel steel. This method enables seams of any length to be produced on the building site which satisfy the requirements of the concept of the shaping load case. The longitudinal edges of the metal sheets are joined with a butt weld which almost attains the strength and, in particular, the ductility of the base metal.

This is a variation of the square-edge jointing method and is based on the TIG process. The sheet edges to be joined are given an upstand of approx. 4 mm in height and placed next to one another. The TIG welding torch is guided over the upstands along the seam by a motor-powered tractor (Fig 12). The upstands are melted to form a joint. An almost symmetrical welding spot in the weld cross-section is produced.

The main trick underlying this method is the multiple function of the upstands. They firstly serve as guide rails for the welding tractor and enable the edges which are to be welded to be fixed by means of the special clamping rollers of the tractor. (All traditional thin sheet welding methods require clamping devices which are expensive and, in particular, only suitable for short seams.) In addition, the clamping rollers force geometry errors to be compensated, such as small gaps between the edges and position errors in the vertical planes. They also provide the welding filler when they melt, thereby eliminating the need for consumable electrodes.

In the case of the prototype of a sun reflector of 0.5 mm thick sheet metal described in the following, this welding method was applied successfully for the seams - some as long as 18 m.

Fig 12

METHOD FOR SHAPING CALCULATIONS FOR ROTATIONALLY SYMMETRIC METAL MEMBRANES

On the basis of Saint Venant's theory of plastic flow, a model calculation has been developed, strain-hardening as isotropic hardening being taken into consideration. This model permits arithmetic determination of the form produced on shaping, as well as the stresses and strains which result for rotationally symmetric membranes. The only restriction placed on shaping loads is that they are arranged in a rotationally symmetric manner. The same applies for the initial geometry of the membrane in question. Knowledge of the yield characteristics of the material is assumed.

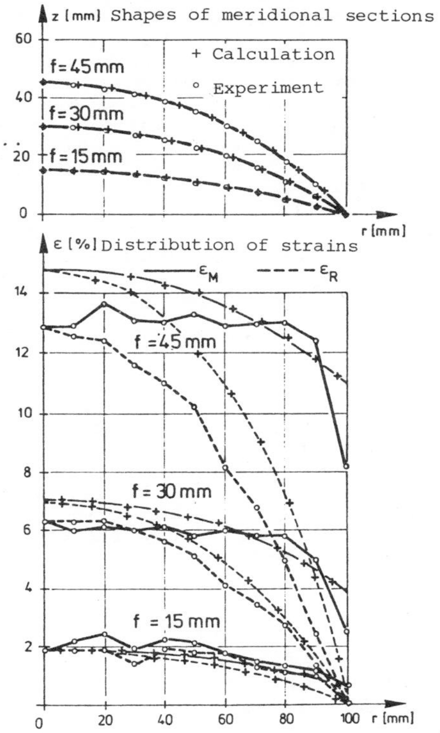

Fig 13

ε_M Meridional strain

ε_R Circumferential strain

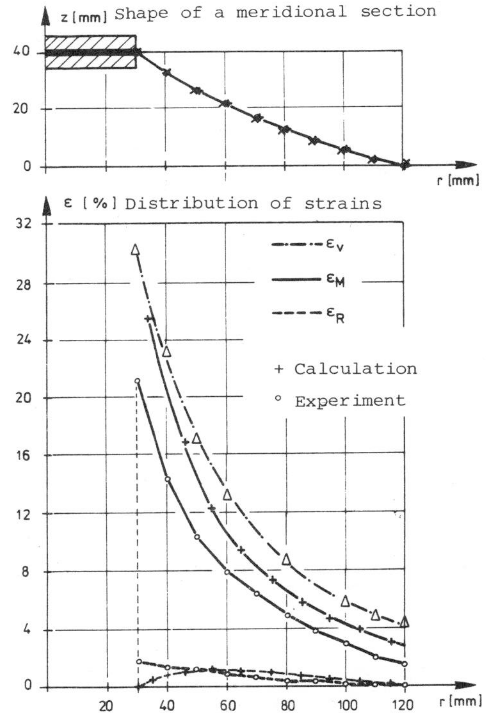

Fig 14

The surface of the membrane is divided into discrete elements by means of concentric circles. The meridian of the rotation surface produced during forming is approximated by a sequence of arc elements at which the balance between internal and external forces is iteratively calculated. A comparison with tests carried out on measurement models has shown that the calculation method has produced excellent results and can certainly produce results which are just as accurate as those obtained when large FEM computation programs are employed. Figures 13 and 14 illustrate comparisons of tests and calculations. The membrane in Fig 13 was given its spherical shape by means of internal overpressure, while the membrane in the example illustrated in Fig 14 was pushed up by a ram.

Fig 15

FIRST APPLICATIONS OF THE SHAPING LOAD CASE METHOD

The first practical experience with the application of the shaping load case method has been obtained with the aid of the model structure with a diameter of 5 m (Fig 15) designed as a lenticular cushion. Two metal sheets with a thickness of 0.3 mm which were initially flat were clamped into a circular compression ring and 'blown up' from the inside using compressed air. It appeared here that the membrane surfaces, which were not exactly perfectly flat in their initial state, became flat spherical domes with a high level of accuracy. This led to the idea of employing a similar system with concave surfaces for constructing large concave reflectors with the aim of producing solar energy.

Construction work started on the first of two planned prototypes of these solar reflectors at the beginning of 1984 and should be ready for operation in the summer (development and design by Schlaich & Partner, Stuttgart, within the scope of a German-Saudi research project). These reflectors have a diameter of 17 m, they are suspended on gimbals and can be turned to follow the course of the sun (Fig 16). The reflection surface is designed as a metal membrane with a thickness of 0.5 mm and is coated with thin glass mirrors. The light from the sun is concentrated on a heat exchanger with a diameter of 70 cm and converted into electrical power with the aid of a Stirling engine and a generator.

Fig 16

REFERENCES

1. GREINER, S.: Membrantragwerke aus dünnem Blech (Thin Sheet Metal Membrane Structures). Dissertation, Universität Stuttgart, 1983. Mitteilungen des Sonderforschungsbereiches 64, Weitgespannte Flächentragwerke, Volume 64/1983, Werner-Verlag, Düsseldorf, FRG.

THE ANALYSIS, DESIGN, CONSTRUCTION AND ERECTION OF THE DOUBLE LAYER AL AIN DOME / U.A.E.

J KLEPRLIK*, Dipl. Ing., CSc

*MERO-Raumstruktur GmbH&Co., Wuerzburg

This paper describes the most important aspects of the design and realisation of the structure Al Ain Dome.

1. Introduction

The hall, providing seats for 2500 spectators, has a circular ground plan, with a playing area of 36.6 x 36.6 m. The total facilities, including grandstand for VIPs, cabins for the press, television etc. are roofed by means of a dome structure of 69.6 m dia as shown in Fig 1.

The following are the important parameters of the dome:

Diameter (Ground plan) :	69.60 m
Rise :	24.80 m
Groung plan area :	3 800 sqm
Surface :	5 740 sqm
Weight of steel construction :	77.50 t
Weight/Ground plan :	20.39 kg/sqm
Weight/Surface :	13.50 kg/sqm
Number of bars :	7 078
Number of nodes :	1 889
Number of panels :	2 034
Live load :	0.5 kN/sqm
Wind load :	0.9 kN/sqm

The following companies were concerned with the construction of the above dome structure :

Owner :	Government of Abu Dhabi Municipality and Town Planing Department, Abu Dhabi, U.A.E.
Contractor :	Eastern Contracting Co., Sharjah, U.A.E.
Consulting :	Brian Moorehead International, Consulting Engineers & Project Managers, Cheshire, Great Britain
Steel Works :	MERO-Raumstruktur GmbH & Co. Wuerzburg, West Germany

The Al Ain Dome was erected during the years 1982 - 1983.

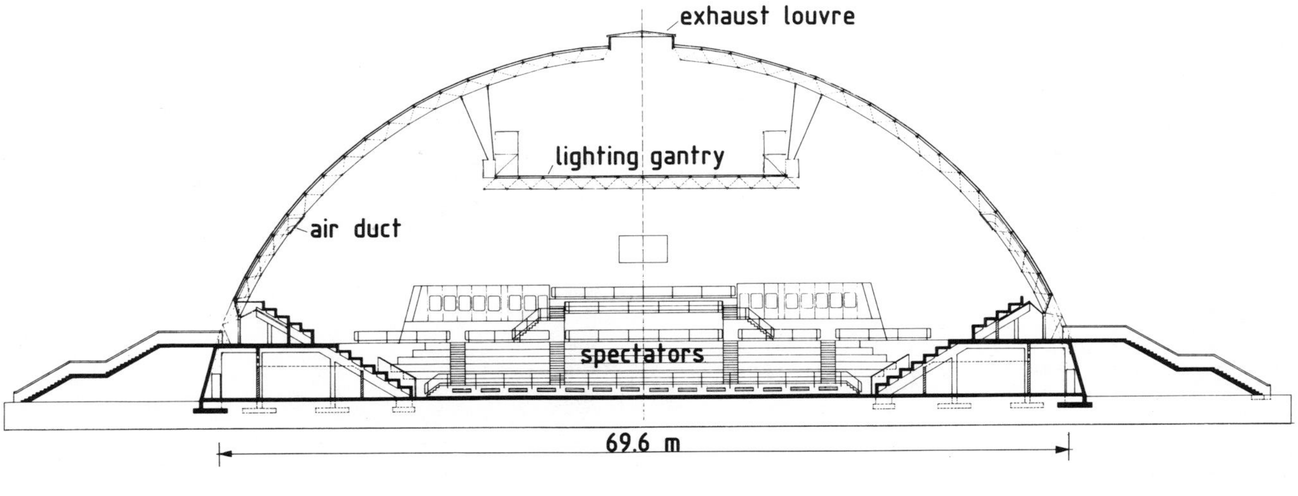

Fig 1 Cross Section.

Fig 2 Al Ain Dome.

2. Design of the Dome

2.1. MERO Double Layer Dome

The spaceframe is a spherical cap with a radius r = 36.8 m (top chord). The dome periphery is supported on a horizontal reinforced concrete ring beam.

A global view of the completed dome is shown in Fig 2. The MERO supporting structure consists of tubular members and spherical connectors or nodes - Ref 1, the largest member size is Ø 159x8.0 mm (entrance area) and the most frequently used section is tube 60.3 x 2.9 mm.

2.2. Roof Covering

In cooperation with LMS Suisse, a completely new roof cladding system was developed. The dome surface of 5740 sqm was subdivided in triangles with an average area of 2.8 sqm. The sandwich panel (ALUCOPAN N) consists of two firmly joined materials, that is, two layers of aluminium sheeting with an intermediate core with a total thickness of 120 mm as shown in Figs 3 and 4.

2.3. Additional Fittings

Two opposite entrance doors and one exhaust louvre in the crown represent important anomalies in the supporting structure. In the upper part of the dome a lighting gantry is suspended from 8 points. The lighting gantry consists of a square shaped MERO spaceframe, 26 x 26 m with a live load of 100 kN. Air ducts have been enclosed in the lower part of the structure.

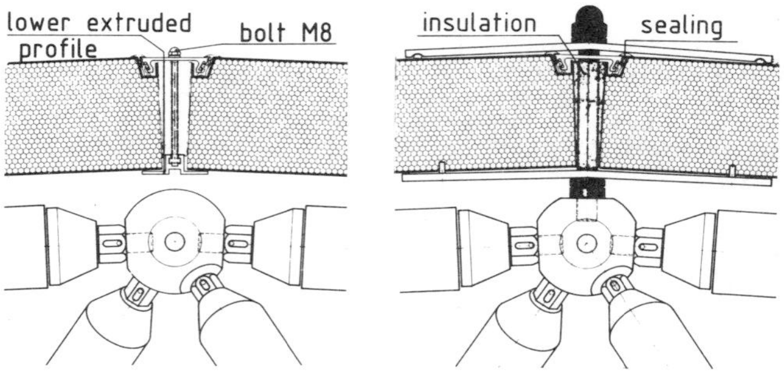

Fig 3 Cladding Details.

Fig 4 Triangular Cladding Panels.

3. Geometry

3.1. Grid Generation

The finally executed braced spherical dome, see Fig 5, is a double layer sparse grid with a structural depth of 1.0 m. It consists of 7078 MERO members and 1889 MERO nodes. The top chord consists of a network dome with horizontal rings of isosceles triangles. The sparse bottom chord structure is composed of triangles and hexagons there are connented to the top chord using diagonals. This geometrical solution of the total network is extremely economical and safe. This solution was compared with other possible geometries in the design phase (see section 3.2.).

The following were fixed requirements, which conditioned the further structural and geometrical design:

- double layer network,
- external grid diameter (ground circle) 69.6 m,
- crown height 24.8 m,
- max. size of cladding panels,
- size of exhaust louvre,
- size and position of entrance doors.

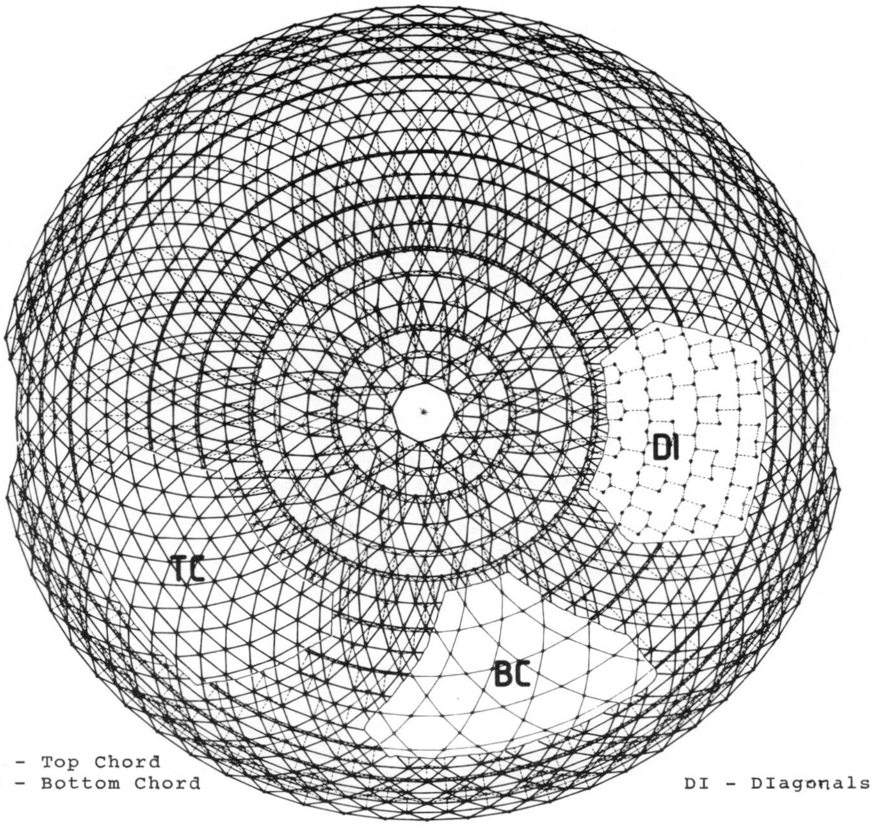

Fig 5 The Executed Network Dome.

For optimizing the above grid, all visual and structural criteria were taken into consideration. Also, is was possible to minimize the following parameters :

- number of different member lenghts, result: 73,
- number of structurally different nodes, result: 100,
- number of different panel types, result: 28.

The following basic steps were followed to reach the final geometry of the dome structure:

- generation of the complete network, including nodes and members, were designed with a uniform module - topology task,
- transformation into the final spherical coordinates - metric task,
- modification of the network - doors and exhaust louvre.

The above mentioned steps were carried out using MERO's own computer program - Ref 2. A simplified flowchart of the process appears in Fig 6.

3.2. Comparison with other possible solutions

In the initial design phase of the dome network, other solutions were also investigated. This was possible due to the implemented computer procedures, which allowed the generation of various alternatives and variations without a large time and work expense. Prof. Makowski has excellently illustrated the different possible networks for braced dome structures in Ref 3. The parameters specified in section 3.1. for the choice of the network limited the possible solutions. Apart from the chosen network, other three possibilities were considered.

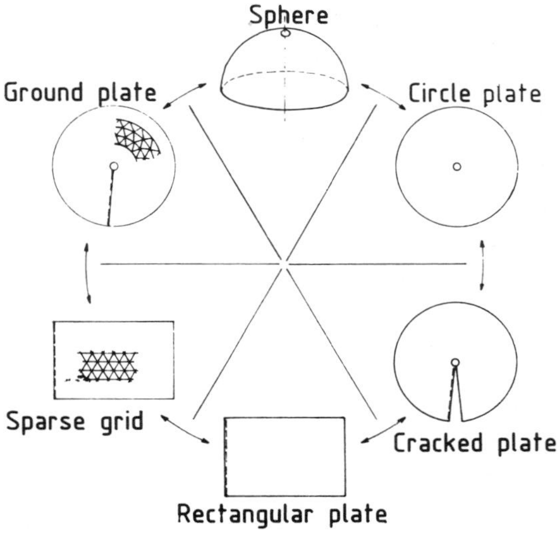

Fig 6 Flowchart.

For all mentioned solutions the following requirements remained equal:

- double layer grid,
- triangular pattern at top chord,
- horizontal rings.

As regards the structural aspects, the considered networks behave similarly, provided that the structure is supported only at the top chord. In this case, the top chord proves to be the main supporting system, where the bottom chord provides the structure with sufficient stability, while transmiting only small member forces.

3.2.1. Solution 2

This is a standard solution, that is a completely triangulated network. This is based on the principle of duality, see Fig 7. According to this principle, a complete triangular network was created at the top chord level, and a complete hexagonal network formed the bottom chord. As seen in section 3.1.,this network can be transformed into a spherical cap. Both networks, i.e. solution 2 and the finally chosen network - see Fig 5 and Fig 8 - have been compared in Table I.

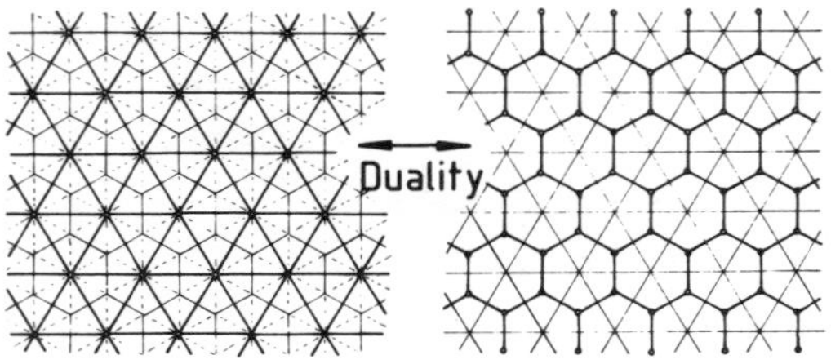

Fig 7 Solution 2.

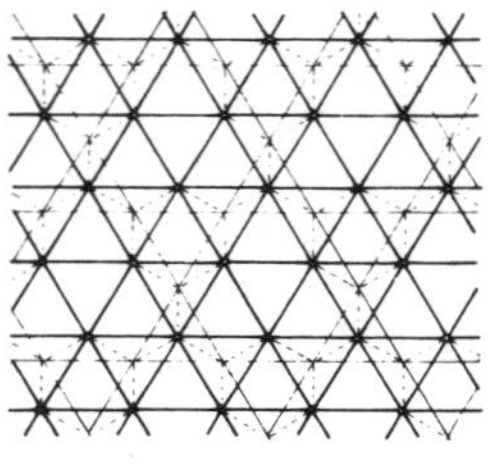

Fig 8 Executed Network.

TABLE I.

SYSTEM	NODES			MEMBERS			
	TC	BC	CN	TC	BC	DI	CN
Solution 2	1125	2214	3339	3334	3294	6642	13270
Executed	1081	808	1889	3086	1620	2372	7078
Saving	4 %	63 %	43 %	7 %	51 %	64 %	46 %

TC - Top Chord　　BC - Bottom Chord
DI - DIagonals　　CN - Complete Network

3.2.2. Solution 3

This is the network shown in Fig 9. With respect to the number of nodes and members, this geometrical arrangement offers an additional saving factor of approx. 10 % less compared to the finally chosen solution.However, this possibility was not further investigated becouse the number of different panel types increased. With the chosen roof cladding system, this factor is of utmost importance.

Fig 9 Solution 3.

3.2.3. Solution 4

Fig 10 shows another possible solution, the folded plate structure. The spherical cap has been replaced by a polyhedral cap. Thus, instead of horizontal circular rings the solution presents polygonal rings. The requirement of material,i.e. number of nodes and members, is very well comparable with the executed solution, although the number of different panel types is smaller. However, this solution shows completely different architectural aspects. No further investigations had been effected.

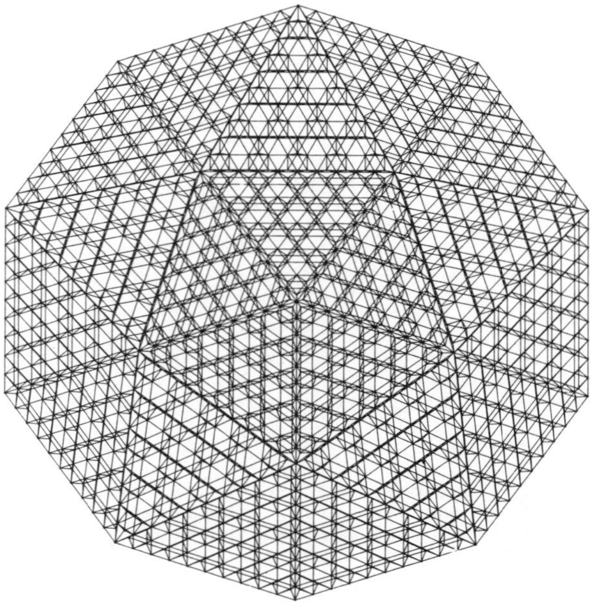

Fig 10 Solution 4.

4\. Structural Calculation

The supporting structure is statically highly indeterminate and it was consequently analyzed with MERO's own finite element program (FEM).

4.1. Load Assumptions

The following load assumptions were considered:

- dead load 0.5 kN/sqm,
- live load 0.5 kN/sqm,
- wind load 0.9 kN/sqm.

The wind loading distribution factor c_p was determined by means of the MERO program according to Ref 4 as

$$c_p(x,y,z) \in (-1.30;+0.70).$$

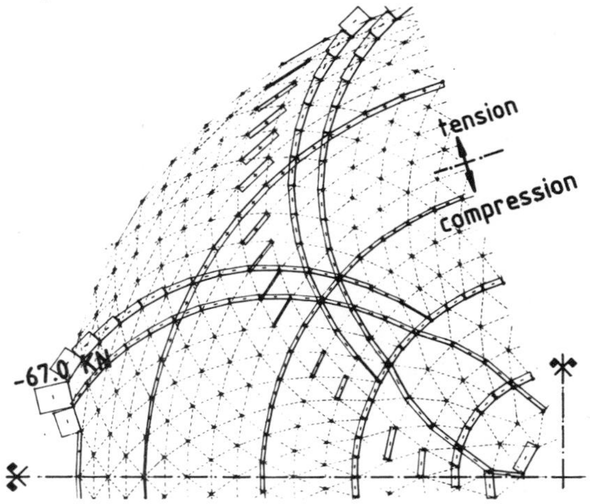

Fig 11 Dead Load.

4.2. Member forces

The individual member length varies between 1198 mm and 3415 mm, the average member length is 2292 mm. The maximum member force (most unfavourable load combination) is +56.0 kN (tension), the minimum member force is -70.1 kN (compression). Member forces in the top chord after dead load and wind are shown in Figs 11 and 12, respectively.

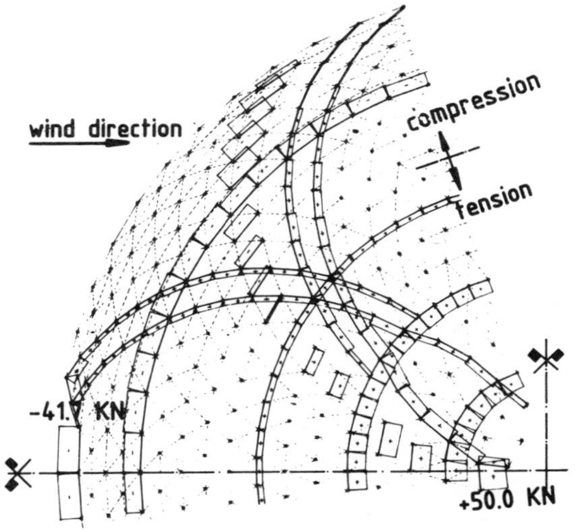

Fig 12 Wind Load.

Member forces were determined by means of a discrete model scheme (FEM).The obtained forces were compared with the results of a continuous model analysis, see Ref 5, and the values for ring forces are displayed in Fig 13.

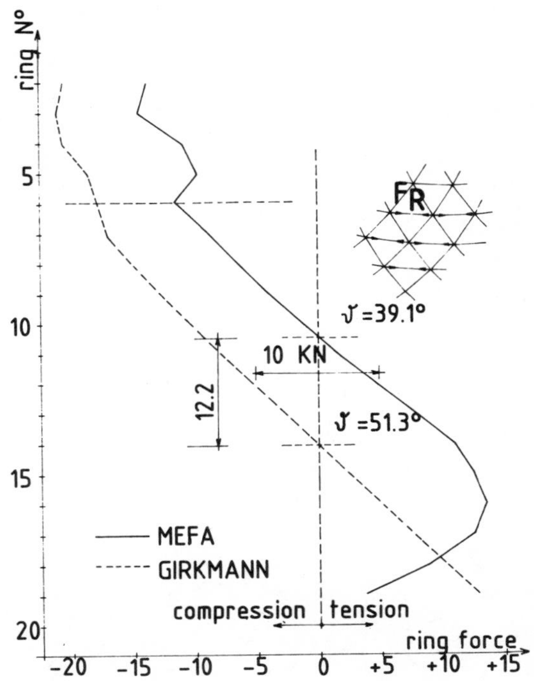

Fig 13 Ring Forces F_r , $\overline{F}_r$.

Fig 13 demonstrates that

- membrane theory according to Ref 5 yields different results than with the FEM. Member forces do not correspond. However, in the central areas, the curves for $\overline{F}_r$ and F_r run approximately parallel,
- the difference between these two approaches demonstrates that the omission of the bottom network is of considerable influence.

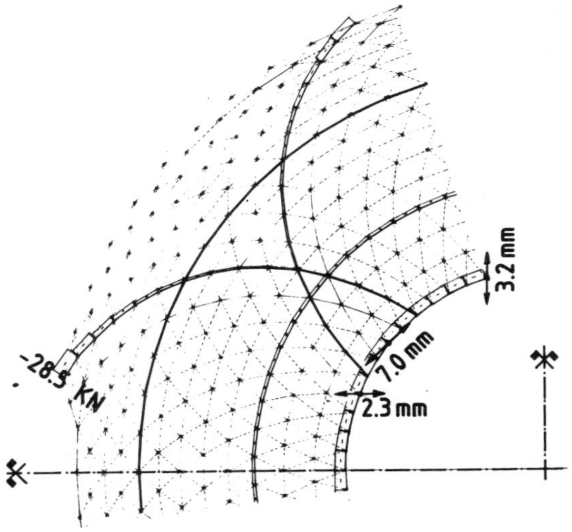

Fig 15 Erection Load.

Fig 14 Open Dome Structure.

4.3. Deflections

The chosen structure resulted very rigid. The largest deflections computed for the given load cases were all under 10 mm.

5. Erection

The MERO Construction system applied to the chosen geometry made erection feasible without any temporary columns (supports). A crane was installed in the centre, to raise elements to position, see Fig 14. The max. weight of the individual components did not exceed 90 kgs. A ring-by-ring assembly of the structure had always sufficient stability during erection. Normal forces and deformations for erection load case (open ring dome structure) are shown in Fig 15.

6. Acknowledgements

The author wishes to record his gratitude to all people involved in the development of Al Ain Dome.

7. References

1. Mengeringhausen,M.: Raumfachwerke aus Stäben und Knoten. Bauverlag, Wiesbaden, Berlin,1975.

2. Klimke, H., Ruh, M.: Darstellung eines Konstruktionssystems zur Erzeugung der Geometrie von Knoten-Stab-Tragwerken. IKOSS, Baden-Baden 1981.

3. Makowski, Z.S.: History of development of various types of domes and review of recent achievements all over the world. Analysis, Design and Construction of Braced Domes. University of Surrey 1979.

4. Newberry, C.W., Eaton, K.J. Wind Loading Handbook 1974.

5. Girkmann, K. Flächentragwerke, Wien 1963.

THE ROOF STRUCTURES OF THE NEW SPORT HALL IN ATHENS: DESIGN, CONSTRUCTION AND PERFORMANCE

M MAJOWIECKI and F ZOULAS***

*Istituto di Tecnica delle costruzioni, Facoltà di Ingegneria
Università di Bologna, Italy

**Computer Aided Design Laboratory
Bologna, Italy

This work illustrates the general data of the structural design, the elaborations concerning the executive design, the construction details, the assembly technique and times and the cable structure pre-stressing operations used for the roof structures of the new sport hall in Athens.

1. GEOMETRICAL AND MECHANICAL CHARACTERISTICS OF THE ROOF STRUCTURE

After drawing up the executive architectural design (Fig 1, see Appendix 1), the geometrical and mechanical data necessary for the structural analysis were defined.

1.1. The suspended roof

The roof structure consists of an orthogonal rope net, whose average surface can be defined geometrically as a saddle surface with total negative curvature, very similar to that of a hyperbolic paraboloid. The main data of the cable structure are:

1.1.1. Diameter: 113.96 m (R = 56.98 m).

1.1.2. Upper rope diameter: Ø 60 mm (Fe = 21.106 cm^2); construction 127 x 4.6; σ_{ult} = 1600 N/mm^2.
Lower rope diameter: Ø 46 mm (Fe = 12.784 cm^2); construction 127 x 3.6; σ_{ult} = 1600 N/mm^2.
Elasticity modulus of the ropes: E = 165 KN/mm^2.

1.1.3. Mesh dimensions: 4.00 x 4.00 m

1.1.4. Equation of the cable net boundary (anchorage line)

The anchorage ring line is defined geometrically by the intersection between a circular vertical cylinder of diameter 113.96 m, which is the maximum free span of the cable net structure, and a hyperbolic paraboloid co-axial to it.

The position of the theoretical anchorage points of the cable structure is expressed analytically by:

$$\begin{cases} \dfrac{6.15}{56.98^2}(x^2 - y^2) + 28.74 = Z \\ x^2 + y^2 = 56.98^2 \end{cases} \qquad \ldots\ldots 1.$$

The geometrical co-ordinates of the discrete anchorage points of the ropes are so determined on the ring, according to a Cartesian axes system, considering

x axis directed to the upper point (H = +34.89);
y axis directed to the lower point (H = +22.59);
z axis directed upwards.

1.2. Cable net mesh-comparative cost analysis

The cable structural distribution in plan has been definitely decided according to a net mesh of 4x4 m after a careful analysis of both the cost of supply of the roof materials, and the cost of the erection and pre-stressing of the cable structure.

The results of the comparative economic analysis between the two net mesh solutions 4x4 m a,d 2x2 m are submitted up in Table 1. Considering the solution with net 4x4 m as the comparative unit price cost, it is possible to deduce from Table 1 that the two solutions are equivalent as far as the total cost of supply of the roof materials is concerned. On the contrary, the costs of erection and pre-stressing of the rope net structure are considerably different.

In fact, with a net mesh of 4x4 m, although the steel consumption for the ropes and the cost of the covering remain about the same, the cost of erection of the structure joint-locks is four times less; while the erection of the ropes and the stretching operation are halved.

TABLE 1

Solution	Rope supply	Roof covering (supply, erection)	Erection and prestressing of the net
2 x 2	1.05	0.923	2 ÷ 2.5
4 x 4	1	1	1

1.3. The peripheral ring

The border structure, to which the rope net is anchored, consists of a perimetrical box-section ring in pre-stressed reinforced concrete which follows the movement of the border curve, Eqn 1.

1.3.1. Geometry of the beam axis

The downview of the beam axis is a circle with a radius R = 61.66 m. The supports are located in downview on a circle with a radius R_1 = 61.50 m in angular distance α_o = 11.25°. The support heights are defined on the basis of the heights of the corresponding peripheral points of the suspended roof, located on the same radial direction, with a -5.45 m difference in height.

The structure of the anchorage ring is defined in space by moving the plane frame defining the external contourn of the section, resting on the generatrix determined by Eqn 1 keeping the plan, containing this section, vertical.

The maximum gradient between the points of the border line, defining the anchorage structure, is 12.3 m equal to about 11% of the free span.

1.3.2. Cross-sections (on vertical plane)

The moments of inertia and section area are variable due to the modification of the thickness of the walls of which the structure consists. The standard sections, which form the various elements between two consecutive supports of the ring, are shown in Fig 2.

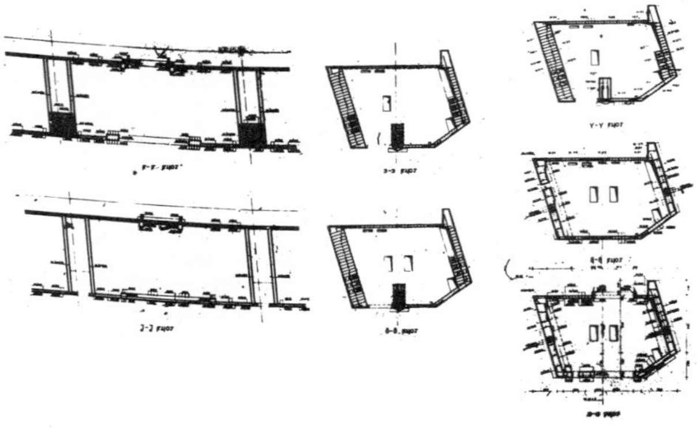

Fig 2

The area and inertia properties of the ring sections are:

- Cross-section A (without openings):

  $F = 9.0705\ m^2$

  $J_r = 34.427\ m^4$

  $J_z = 103.6281\ m^4$

  $J_D = 63.0817\ m^4$

- Cross-section B (with openings):

  $F = 8.9219\ m^2$

  $J_r = 32.59444\ m^4$

  $J_z = 104.08146\ m^4$

  $J_D = 63.0817\ m^4$

- Cross-section A (segments between supports 7-8-8-7)

  $F = 10.5913\ m^2$

  $J_r = 39.7864\ m^4$

  $J_z = 121.7072\ m^4$

  $J_D = 63.0817\ m^4$

1.3.3. Materials

Concrete B 450
Elasticity modulus: E_i = 35 KN/mm^2;
E(t) according to time schedule
Shear modulus: G = 15 KN/mm^2.
Poisson ratio: ν = 0.167.

1.4. The ring supporting system

The supports are placed in relation to the frames supporting the stands. These frames, which are radial and concentric as regards the centre of the construction, are also made of pre-stressed reinforced concrete and are of variable height in order to altimetrically follow the movement of the border ring.

The supporting system is also equipped with hydraulic jacks, which allow the horizontal displacements due to elastic deformations and hinder the eventual movements of the rigid body of the ring, due to seismic action.

1.4.1. Factors of elastic settlements of the supports

Support 1:	$W_1 = 6.6757 \times 10^{-3}$ m/1000 KN
" 2:	$W_2 = 9.0127 \times 10^{-3}$ "
" 3:	$W_3 = 7.2191 \times 10^{-3}$
" 4:	$W_4 = 5.2207 \times 10^{-3}$ "
" 5:	$W_5 = 3.7959 \times 10^{-3}$ "
" 6:	$W_6 = 2.7205 \times 10^{-3}$ "
" 7:	$W_7 = 1.7006 \times 10^{-3}$ "
" 8:	$W_8 = 0.6432 \times 10^{-3}$ "

1.4.2. Absolute support settlements

	max W	min W
Support 1:	-1.095×10^{-3} m	$+0.2944 \times 10^{-3}$ m
" 2:	-2.359×10^{-3} m	$+0.1366 \times 10^{-3}$ m
" 3:	-1.9517×10^{-3} m	$+0.0787 \times 10^{-3}$ m
" 4:	-1.3280×10^{-3} m	-
" 5:	-0.8161×10^{-3} m	-
" 6:	-0.5202×10^{-3} m	-
" 7:	-0.2534×10^{-3} m	-
" 8:	-	-

2. LOADING ANALYSIS

2.1. Cable structure

		Load-Code
Dead load of the ropes	90 N/m^2	[a]

Dead load of the covering and additional permanent loadings	360 N/m^2	[b]
Snow loading	650 N/m^2	[c]
Wind loading (c = -0.8)	1100 N/m^2	[d]

2.2. Border ring

Dead weight of the ring	216 KN/m	[e]
Pre-stressing variable between two consecutive elements of the ring	(see Table 2)	[f]
Additional permanent loadings	14 KN/m	[g]
Accidental loadings	458 KN/m	[h]

TABLE 2

Element number	Prestressing axial force (MN)	Horizontal eccentricity (m)	Vertical eccentricity (m)
1	33.50	0.93	1.08
2	33.50	0.93	1.08
3	30.00	0.775	0.90
4	23.00	0.465	0.54
5	12.50	0	0
6	26.00	-0.2	0.50
7	35.00	-0.3	0.84
8	39.50	-0.4	1.04
9	39.50	-0.4	1.01

2.3. Loading combinations

The loading combinations, leading to the definite statical verifications of the covering structures, have been determined according to the maximum stresses and deformations, and to the various constructive phases, in accordance with a pre-set time schedule.

In accordance with the time-schedule illustrated in Table 3, the influence of the reologic deformations E(t) of the perimetrical ring has been duly considered for the various possible loading combinations.

TABLE 3

Month	1st	2nd	3rd	4th	5th	6th	7th	8th	9th
Ring Performance	▮	▮	▮	▮					
Ring prestressing					▮	▮			
Cable-structure Erection					▮	▮	▮		
Cable-structure pre-tension							▮	▮	
Covering performance									▮

The static loading combinations considered are:

1) [e]

2) [e] + [f]

3) [e] + [f] + [a]

4) [e] + [f] + [g] + [a] + [b]

5) [e] + [f] + [g] + [h] + [a] + [b] + [c]

6) [e] [f] + [g] + [a] + [b] + [c]

The thermal variations have been considered associated to the conditions of maximum loading 5) and 6), in accordance with the following groups:

GROUP I

- Temperature difference between the lower and upper faces of the ring:

$$t_i - t_s = -15\ °C$$

- Temperature difference between the internal and external faces of the ring:

$$t_i - t_e = -15\ °C$$

- Temperature difference acting on the cable-structure:

$$\Delta T = -10\ °C$$

GROUP II

$$t_i - t_s = 15\ °C$$
$$t_i - t_e = 15\ °C$$
$$\Delta T = +10\ °C$$

GROUP III

Concrete shrinkage: equivalent to $t_s = -15\ °C$.

3. THE STRUCTURAL SCHEME AND THE MATHEMATICAL MODEL

Taking advantage of the two symmetry axes, only a quarter of the ring has been considered and schematized as a space frame. The joints of this space frame have been defined principally, in correspondence to the rope anchorages, at the beginning and end of the ring section variation and in correspondence to its supports. These supports elasticity and non-elastically yielding are eccentric with regard to the ring barycentric axis.

The rope anchorages are located at the ring upper plate. The cable structure is outlined as a system of bar-joints with the possibility of transmitting only tensile forces ($S > 0$) and consists of 167 internal joints having 3 degrees of freedom, 130 border joints and 385 bar elements, Fig 3.

According to the structural arrangement, the cable structure interacts elastically with the border ring for all loading combinations.

In order to obtain more detailed information for the actual stress distribution the ring has been verified by means of a finite element analysis. Figures 4 and 5 show the plane and axonometric view of the finite element mesh schematization for the ring actual structure.

The organisation of the global calculation programme implies therefore three principal groups (routines) of resolution:

- Programme rete, for the research of state "0";

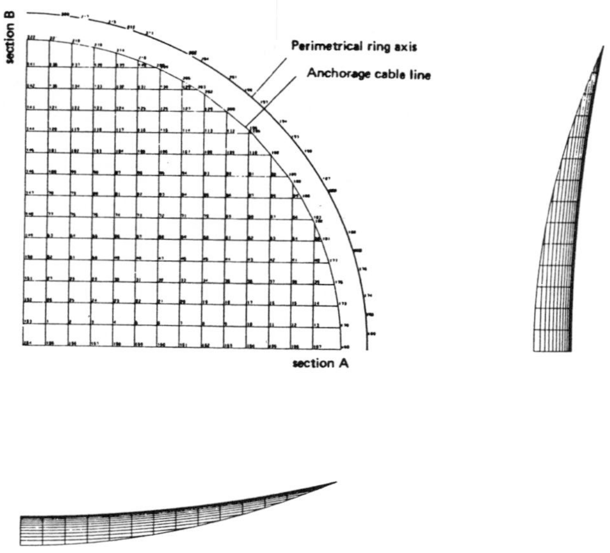

Fig 3 Automatic output plotter of the rope net mesh and border ring.

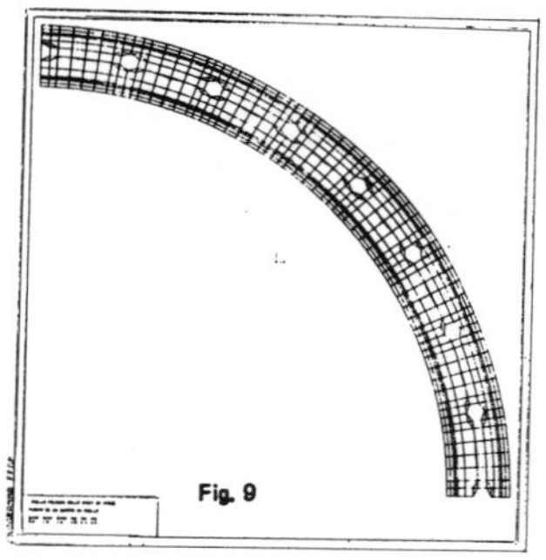

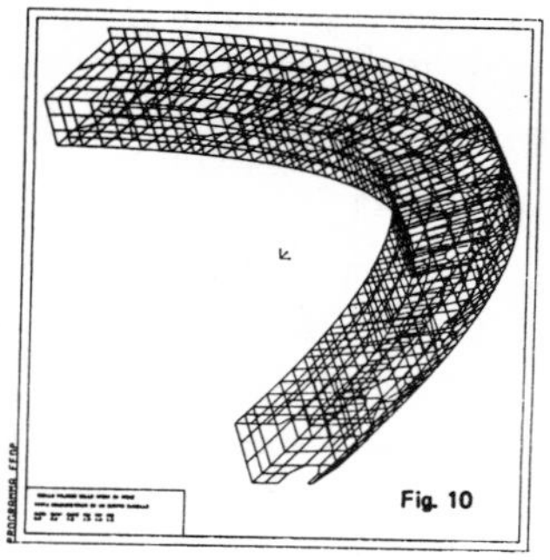

Finite element mesh schematization for the ring structure. Input geometrical data for computer programme.

- Programme tenso/tensodin, for the statical and dynamical analysis of the cable structure;
- Programme of space frame analysis (S.F.A.) or finite element analysis (F.E.A.) for the static analysis of the border ring.

The state "0" is obtained with the method outlined in Ref 2; the statical and dynamical analysis are computed with the mixed substructuring method of Ref 3, taken into account the interaction between ring and cable structures.

In the flow-chart illustrated in Fig 6 the global organization of the computer programme utilized for the structural analysis is synthetically shown.

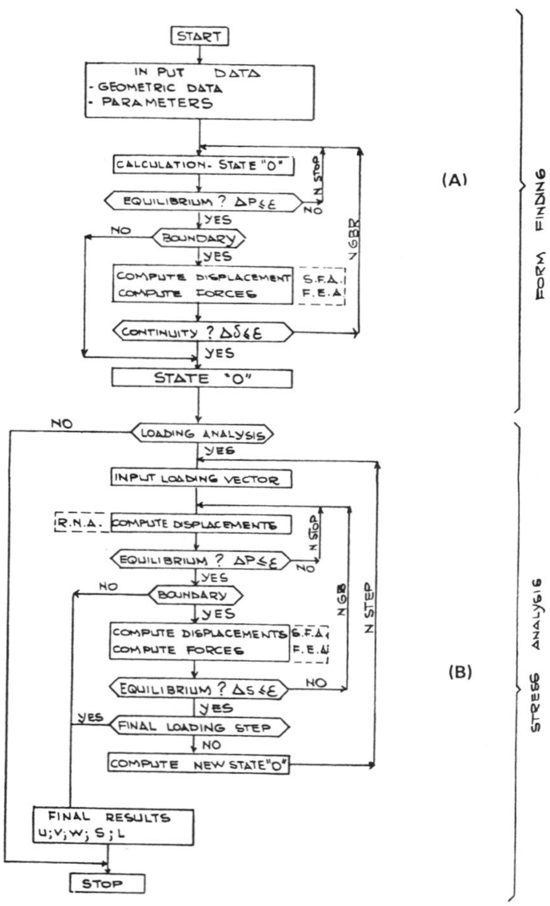

Fig 6 FLOW CHART - form finding and stress analysis.

4. THE RESULTS

4.1. *State "0"*

The loadings combination no. 3 has been chosen for the definition of the state "0" of the structural system-border ring and cable structure.

The horizontal components of the stresses in the primary and secondary ropes have been fixed in the pre-stressing state to:

$$H_p = 582 \text{ KN}$$

$$H_s = 615 \text{ KN}$$

4.2. *States of maximum stress and deformation (verification)*

The loading combinations no. 5 and no. 6 correspond to the snow loading and wind loading on the covering respectively. The ring has been designed in order to fit the best flexural distribution permitted by the bond represented by the displacement limitation of $\Delta f \leqq 0.005$ L (L = maximum free span).

The maximum tensile strength in the primary ropes for the snow loading combination is equal to 1210 KN, while in the secondary ropes for the wind loading combination is equal to

810 KN.

According to what mentioned above, Fig 7 shows the final plotted diagram of axial forces, shear and moments for the snow condition (with the correspondin associated group of thermal variations in a quarter ring), obtained as output of the space frame analysis program.

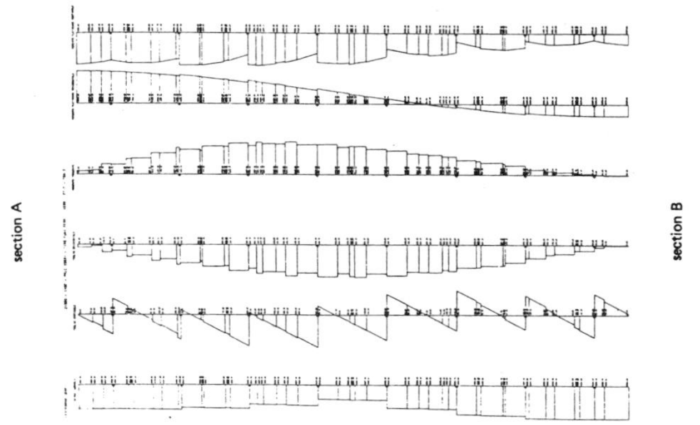

Fig 7 SNOW LOAD - automatic plotter of stresses in quarter ring.

On the basis of the stresses obtained for the different loading combinations, the stresses on the prestressed reinforced concrete ring section have been verified. Figure 8 shows the variation of the pression center and the neutral axis for single and combined loadings in the most stressed section of the ring. The verifications have been computed interactively with a stress analysis program on flexural and normally loaded sections.

The maximum stresses computed have been compared by means of a more precise analysis via finite elements, thus obtaining an excellent result check and detailed information on the stress state around the numerous holes and concentrated load in the ring. The following peaks were found in the second loading combination:

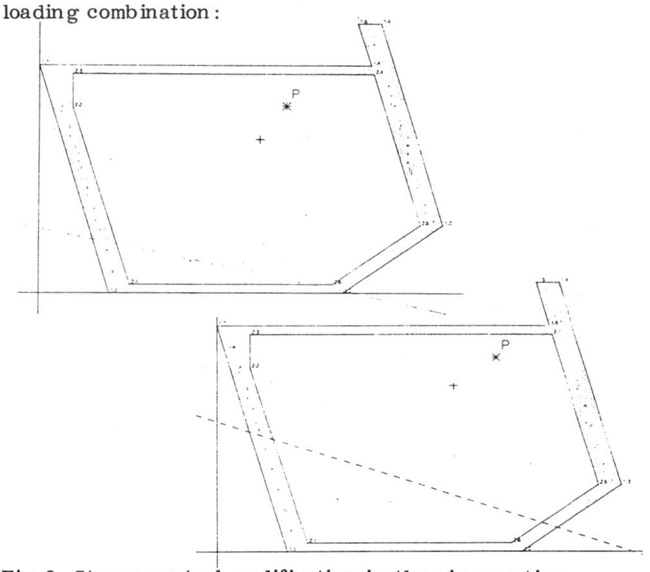

Fig 8 Stress control modification in the ring section.

SIGMA C = -13.5 MPa
SIGMA F = 14.2 MPa

5. CONSTRUCTION DETAILS

5.1. Rope anchorages

The rope transmit the anchorage forces to the border ring via four Dywidag bars Ø 32 ST 80/105, anchored in turn on the upper plate of the ring. The detail connecting the rope to the Dywidag bars should allow:

a) an easy adjustment so as to avoid geometric execution errors;
b) the possibility of connecting hydraulic jacks for the introduction of the state "0" forces;
c) the spheric rotation of the anchorage cable heads to adapt to the angular variations due to the different loading combinations.

In order to attain these aims, the detail illustrated in Fig 9 (see Appendix) has been designed.

The parts composing the rope anchorage detail have been tested experimentally by means of fracture tests. The safety factors obtained are very high if a considerable rigidity of the detail is also to be obtained so as to minimize the elastic deformation. Some results are illustrated in Fig 10, see Appendix.

5.2. Rope-rope connection

The purpose of the rope-rope connection is to absorb the friction tangential stresses,which are very limited and allow us to obtain a very simple detail realized in aluminium, Fig 11 (see Appendix).

5.3. Covering

In the executive design special attention has been paid to the covering which is composed of:

- corrugated sheat (HACIERCO), 106 mm high and 10/10 mm thick, painted on the lower side and galvanized on the upper side, with 750 mm pitch and resting on the secondary ropes every 4 m. The sheet is connected to the ropes by means of cadmium plated U bolts;
- vapor barrier SARNAVAP 1000;
- rigid heat insulator, 6 cm thick, having a coefficient of heat conductivity K = 0.6 kcal/h m^2;
- polyester and PEC waterproofing membrane;
- screw connecting the insulating and waterproofing layers to the corrugated sheet.

The executive designs of the detail are illustrated in Fig 12 (see Appendix).

6. ERECTION AND POSITIONING OF THE CABLE STRUCTURES

The primary ropes have been positioned according to the scheme of Fig 13 (see Appendix) by means of cranes in order to decrease the horizontal forces next to the winches.

The secondary ropes have been positioned by sliding them on the primary ropes according to the method illustrated in Fig 14 (see Appendix).

6.1. Description of the pre-stressing operations

The purpose of the pre-stressing operations on the rope net is to attain the tensile and geometric state referred to in theory as "STATE 0".

During the erection of the rope net the following operations have been considered:

- positioning of the anchorage cable heads of the secondary ropes according to the theoretical results;
- positioning of the anchorage cable heads of the primary ropes on the threaded bars in a known point.

The stress state predicted for the rope net in STATE "0" has been obtained by means of an iterative pre-stressing procedure according to conservative phases, as shown in the following:

PHASE 1 - Removal of the big kinematic displacements by acting on the adjustable anchorage points of the primary ropes.

PHASE 2 - Optical instrumental measurement and control of the positioning of the secondary rope anchorage points; control and quantification of the positioning errors of the anchorage boxes; modification of th position of the secondary rope anchorage points according to the above errors.
NOTE: The correction is possible only along the rope direction. No correction is possible transversally to the rope axis.

PHASE 3 - Introduction of a first stress level (first elastic phase) in accordance with the theoretical results.

PHASE 4 - Instrument measurement of the actual position of the primary rope anchorage points; control and quantification of the displacement of the primary rope anchorage cable heads to be carried out during the subsequent phases.

PHASE 5 - Introduction of a second stress level (second elastic phase) in accordance with the theoretical results.

PHASE 6 - Local action on some ropes in order to minimize the influence of the longitudinal and transversal border errors on the positioning of the anchorage points in the concrete ring structure.

The pre-stressing steps have been numerically simulated according to the same program illustrated in section 3, by "relaxing" the ropes in accordance with the pre-established sequence. In this way the enclosed pre-stressing tables have been obtained (Tables 4, 5 and Fig 15).

In the case of temperature differences recorded in the building site with respect to the rope cut mean temperature, the pre-stressing stresses have been duly modified according to a suitable Table 6.

The erection and pre-stressing of the rope net were concluded on August 1983, according to PERT time schedule (Appendix, Fig 16).

		PRIMARY CABLES													
STEP 1		P_{14}	P_{13}	P_{12}	P_{11}	P_{10}	P_9	P_8	P_7	P_6	P_5	P_4	P_3	P_2	P_1
F	From	14,5	20,5	23,4	26,2	27,5	28,8	29,7	30,5	31,0	31,4	31,5	31,5	29,6	21,6
	To	39,6	44,1	48,5	50,7	52,9	54,5	56,1	57,3	58,4	58,9	59,4	58,4	55,1	49,4
Δl	From	530	518	509	503	498	491	484	484	478	478	487	486	486	486
	To	547	541	537	534	533	528	524	525	522	522	531	531	531	531
	f	17	23	28	31	35	37	40	41	44	44	44	45	45	45

STEP 2		P_{14}	P_{13}	P_{12}	P_{11}	P_{10}	P_9	P_8	P_7	P_6	P_5	P_4	P_3	P_2	P_1
F	From	34,6	36,8	37,9	38,4	39,9	39,1	39,3	39,5	39,6	39,7	39,8	39,2	37,2	29,4
	To	60,9	62,7	63,6	64,3	65,2	66,2	67,1	67,6	68,5	68,5	68,2	67,0	63,7	58,0
Δl	From	547	541	537	534	533	528	524	525	522	522	531	531	531	531
	To	563	565	564	565	567	564	563	566	565	565	575	575	575	575
	f	16	24	27	31	34	36	39	41	43	43	44	44	44	44
SEQUENCE		1	2	3	4	5	6	7	8	9	10	11	12	13	14

F - pre-stressing force (ton.)

Δl - elastic deformation.

	STEP 2	PRE - STRESSING FORCE MATRIX IN PRIMARY CABLES														
	PRE STRESSING SEQUENCE	P_{14}	P_{13}	P_{12}	P_{11}	P_{10}	P_9	P_8	P_7	P_6	P_5	P_4	P_3	P_2	P_1	Z
	0	34,6	38,6	41,6	43,5	45,1	46,1	46,7	47,8	47,9	48,8	49,0	49,2	49,4	49,4	27,82
	1	60,9	36,8	40,8	43,1	44,8	46,0	46,7	47,8	47,9	48,8	49,0	49,2	49,5	49,4	27,82
	2	57,5	59,4	63,6	38,4	42,7	44,8	46,1	47,6	47,9	48,9	49,3	49,6	49,8	49,6	27,82
	3	57,5	59,4	63,6	38,4	42,7	44,8	46,1	47,6	47,9	48,9	49,3	49,6	49,8	49,6	27,81
	4	56,8	57,9	59,9	64,3	39,9	43,2	45,2	47,2	47,7	49,0	49,4	49,7	49,9	50,0	27,81
	5	56,4	56,9	58,2	60,3	65,2	39,1	43,5	46,2	47,3	48,8	49,4	49,8	50,1	50,0	27,81
	6	56,3	56,6	57,3	58,7	60,9	66,2	39,3	44,0	46,8	48,0	49,2	49,8	50,0	50,0	27,81
	7	56,1	56,3	56,7	57,6	59,1	61,4	67,1	39,5	44,5	47,3	48,7	49,5	50,1	50,2	27,81
	8	56,1	56,2	56,6	57,1	58,1	59,6	61,9	67,8	39,6	45,0	47,4	48,7	49,2	49,9	27,82
	9	56,2	56,2	56,4	56,8	57,5	58,5	60,1	62,3	68,5	39,7	45,5	47,5	48,7	49,0	28,82
	10	56,4	56,4	56,5	56,8	57,3	58,0	59,0	60,4	62,8	68,5	39,8	45,3	47,0	47,4	27,85
	11	56,5	56,6	56,6	56,9	57,2	57,7	58,4	59,3	60,8	62,7	68,2	39,2	44,1	45,0	27,87
	12	56,6	56,7	56,8	57,0	57,3	57,6	58,1	58,5	59,5	60,5	62,1	67,0	37,2	40,6	27,90
	13	56,8	57,0	57,2	57,3	57,4	57,6	57,9	58,0	58,5	58,8	59,3	60,1	63,7	29,4	27,97
STATE'0'	14	56,8	57,0	57,2	57,3	57,5	57,6	57,8	57,8	58,0	58,0	58,0	58,0	58,0	58,0	28,00

" FORCE VARIATIONS IN STATE 'O' DUE TO ΔT"

	PRIMARY CABLES														
T	P_1	P_2	P_3	P_4	P_5	P_6	P_7	P_8	P_9	P_{10}	P_{11}	P_{12}	P_{13}	P_{14}	f cm
± 5°	1,38	1,38	1,38	1,38	1,38	1,40	1,42	1,44	1,48	1,49	1,59	1,67	1,78	1,91	0,7
± 10°	2,76	2,78	2,77	2,80	2,82	2,85	2,88	2,93	3,00	3,09	3,19	3,43	3,55	3,84	1,5
± 15°	4,17	4,18	4,18	4,21	4,24	4,30	4,34	4,42	4,53	4,65	4,82	5,09	5,35	5,78	2,2
± 20°	5,58	5,58	5,59	5,62	5,67	5,67	5,76	5,81	5,91	6,06	6,21	6,75	7,14	7,71	2,9
± 30°	8,40	8,41	8,42	8,49	8,55	8,66	8,76	8,93	9,13	9,37	9,72	10,18	10,76	11,59	4,4

" FORCE VARIATION IN THE STATE 'O' DUE TO ΔT"

	SECONDARY CABLES													
T	S_1	S_2	S_3	S_4	S_5	S_6	S_7	S_8	S_9	S_{10}	S_{11}	S_{12}	S_{13}	S_{14}
± 5°	1,70	1,73	1,72	1,71	1,70	1,68	1,65	1,63	1,60	1,55	1,51	1,47	1,42	1,35
± 10°	3,41	3,41	3,40	3,39	3,36	3,33	3,28	3,23	3,16	3,10	3,01	2,93	2,83	2,71
± 15°	5,09	5,1	5,08	5,06	5,02	4,98	4,91	4,84	4,74	4,65	4,52	4,39	4,25	4,07
± 20°	6,78	6,78	6,76	6,73	6,68	6,63	6,53	6,44	6,31	6,19	6,02	5,85	5,66	5,42
± 30°	10,16	10,15	10,12	10,07	10,00	9,92	9,76	9,63	9,45	9,25	9,02	8,77	8,48	8,13

ΔT - temperature changement from average temperature during rope cutting.

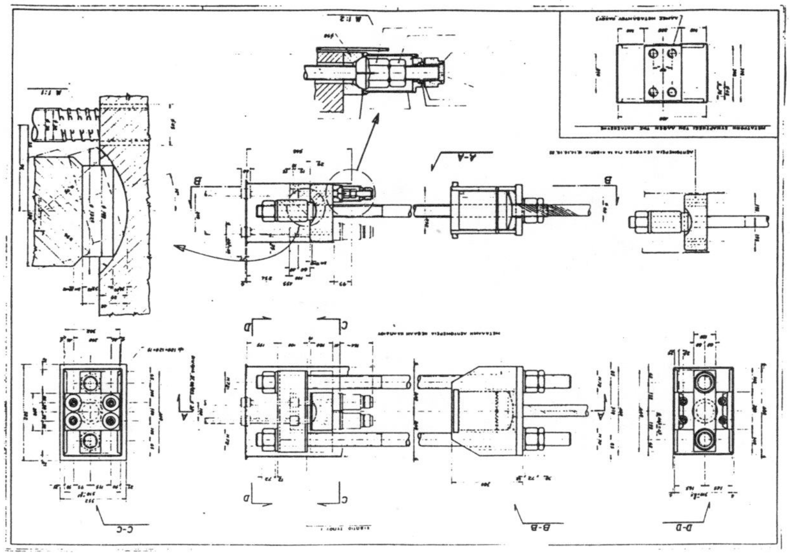

Fig. 9 Cable head detail

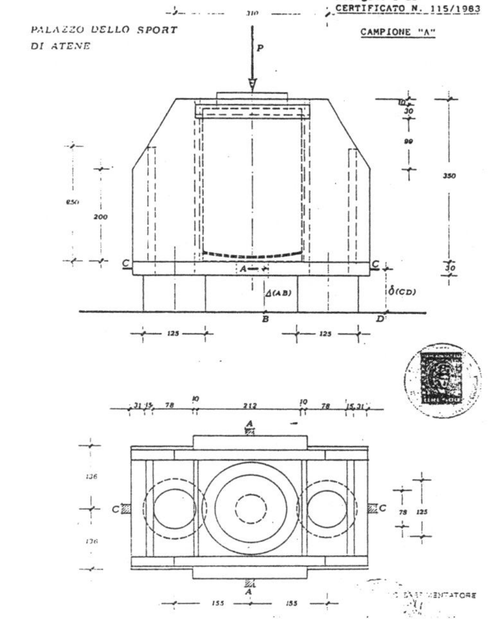

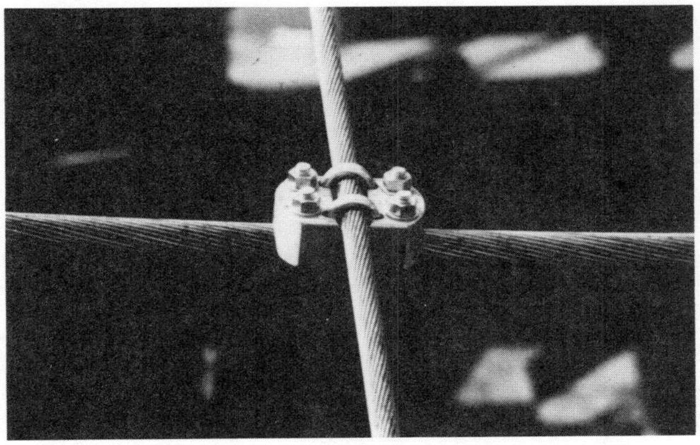

Fig. 11 Rope - rope friction detail

Fig. 10 Experimental analysis

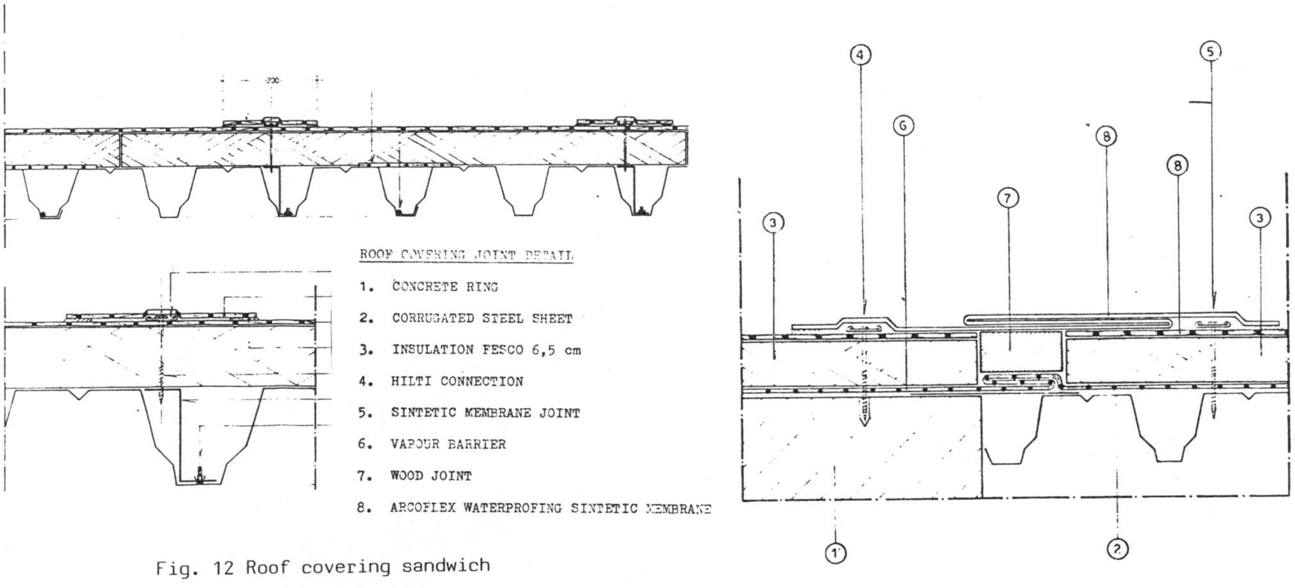

Fig. 12 Roof covering sandwich

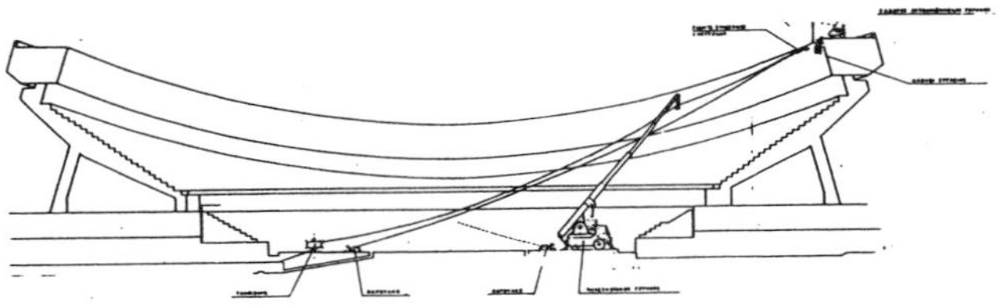

Fig. 13 Primary ropes erection

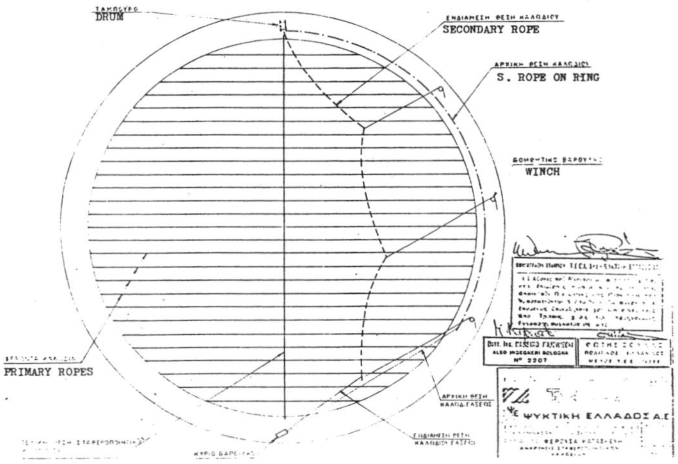

Fig. 14 Secondary ropes erection

Fig. 16 PERT time

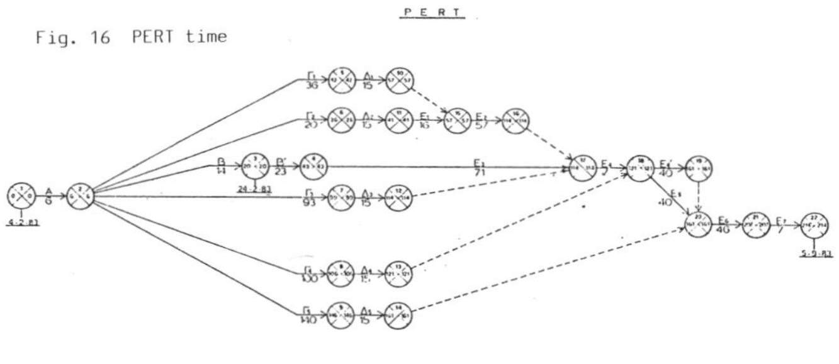

LIST OF ACTIVITIES

A. MEASUREMENTS OF ANCHORAGES COORDINATES
B. SHOP DRAWINGS
B'. APPROVAL OF SHOP DRAWINGS
C. PLACING OF ORDER WITH THE MATERIAL MANUFACTURERS
C1. CABLES
C2. ANCHORAGE SYSTEMS
C3. CORRUGATED STEEL SHEETS
C4. INSULATION MATERIAL
C5. WATERPROOFING MEMBRANE
D. ARRIVAL OF MATERIALS AT SITE
D1. CABLES
D2. ANCHORAGE SYSTEMS
D3. CORRUGATED STEEL SHEETS
D4. INSULATION MATERIAL
D5. WATERPROOFING MEMBRANE
BXXX E. ERECTION WORK
E1. ANCHORAGE SYSTEMS
E2. CABLES - PRESTRESSING OF THE NET
E3. END SUPPORTS OF STEEL SHEET
E4. - E'4. STEEL SHEETS
E5. VAPOR-BARRIER - INSULATION
E6. WATERPROOFING MEMBRANE
E7. MISCELLANEOUS

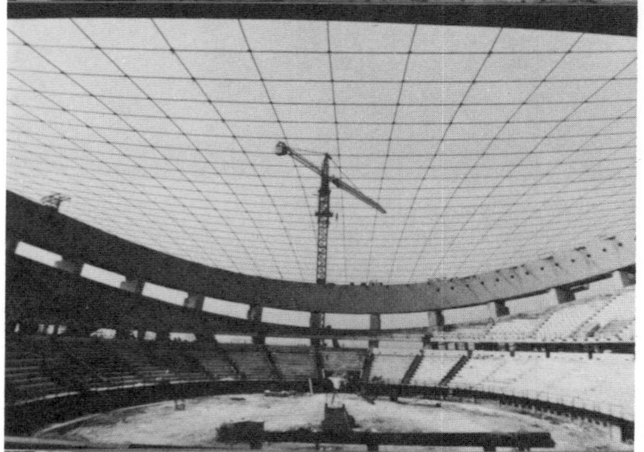

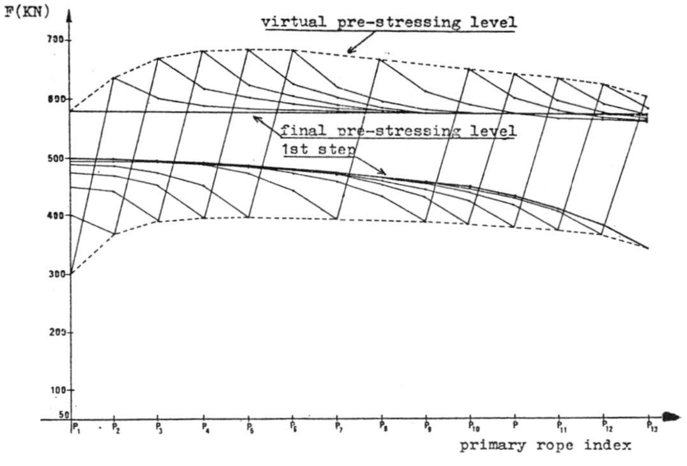

Fig. 15 Pre - stressing diagramm

APPENDIX

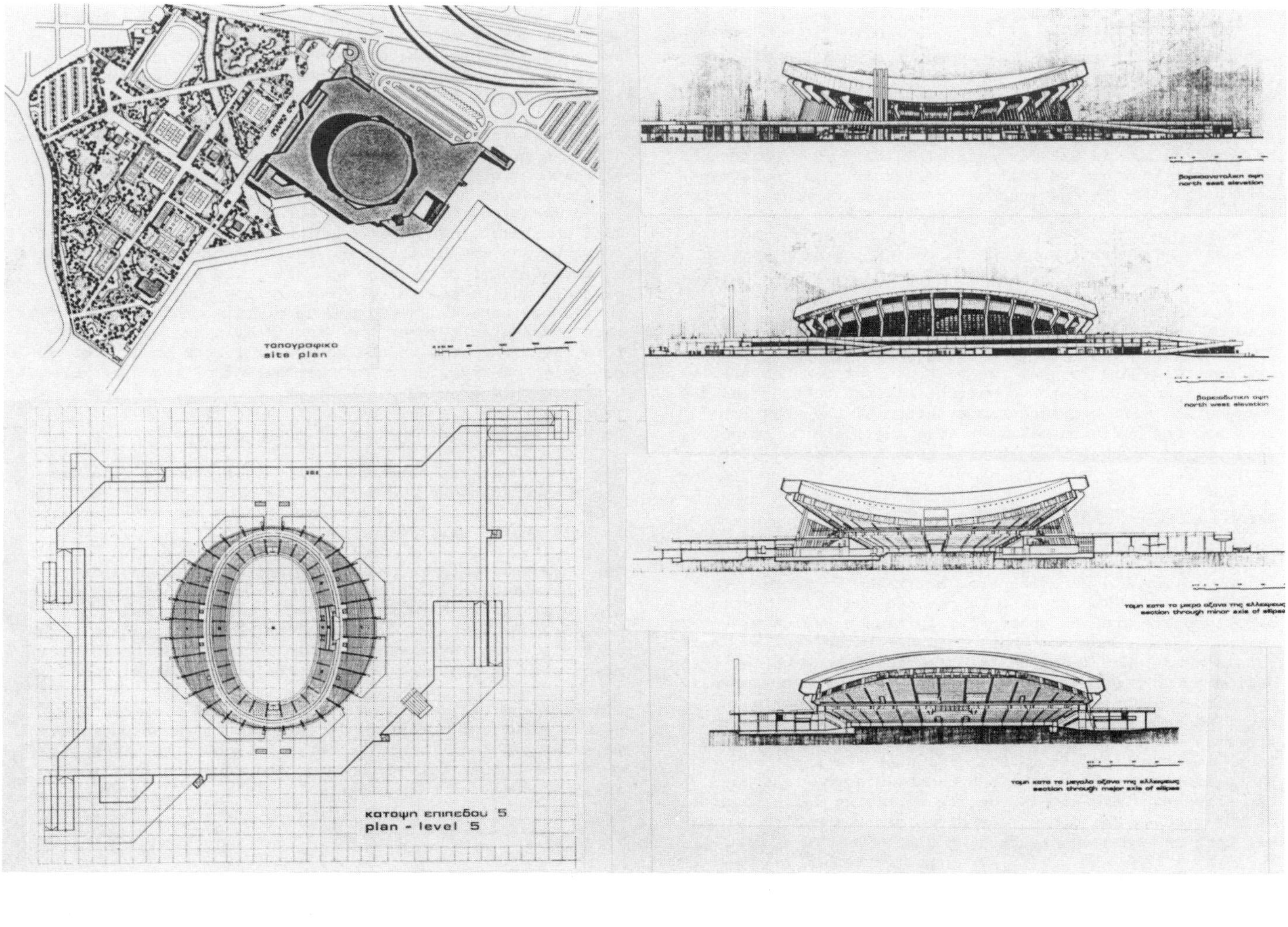

ACKNOWLEDGEMENTS

Owner: General Secretariat of Athletics - Greece.
Architectural design: Archs T Papayiannis, J Baibas, A Gamini, M Koutsouna.
Structural design: Engrs D Bairaktaris, F Caridakis.
Roof structure calculations: Engrs R Alessi, M Majowiecki, F Zoulas.
Main contractors for the complete roof structure: Joint-Venture TECI SpA - Psyktiki Hellados S A.
Contractor's consultants on the roof structure: Engrs M Majowiecki, F Zoulas.
Main contractor for the civil, architectural and electromechanical works of sports hall: Archirodon Hellas S A.

REFERENCES

1. R ALESSI, D BAIRAKTARIS, F CARIDAXIS, M MAJOWIECKI, F ZOULAS, The roof structures of the new sports arena in Athens. World Congress on shell and spatial structures, September 1979, Madrid, Spain.

2. M MAJOWIECKI, G TIRONI, Geometrical configuration of tent and pneumatic structures obtained by interactive computer aided design, IASS World Congress, Montreal, 1976.

3. M MAJOWIECKI, Analisi interattiva di tensostrutture a rete, acciaio, no 9, 1982.

THE DESIGN, CONSTRUCTION AND TESTING OF A STEREOGRAPHIC WOODEN CUPOLA

M. VAN LAETHEM*, E. BACKX**, H. WYNENDAELE***

* Professor, Department of Civil Engineering, K.U.Leuven

** Chief Engineer, Department of Civil Engineering, K.U.Leuven

*** Assistant, Department of Civil Engineering, K.U.Leuven

A cupola of 4 x 4m has been constructed using cassettes made of wood. Each cassette consists of a quadrangular cover panel with four side plates, the form of which was determined and drawn at scale 1/1 by computer. Assembling the cassettes generates automatically and very precisely the form of the shell, a cupola. The position of the corners of the cover panels of the cassettes was determined by stereographic projection of a square mesh on the sphere. This results in a network of orthogonally intersecting lines. The model which is made of 4 mm thick (multiplex) has been tested in the laboratory and behaves very well.

1. INTRODUCTION

Not many shell structures have been built in Belgium. The reasons for eliminating a shell roof as a possibility for a particular construction are known. First of all, the architects in Belgium are not familiar with this type of construction. Secondly, one generally assumes, that this type of construction is very expensive. Thirdly, the constructors don't know how to built a shell roof. All these reasons are linked to each other. If an architect would like to built a shell, the constructor will ask an excessive price to cover the risk of his lack of knowledge and this excessive price will support the general opinion that shell structures are expensive.

A posibility to escape from this vicious circle is to introduce some new ideas, which take away some of the preconceived opinions, and to build a prototype under your own control. This is what has been done at our university and it is hoped that the construction, the testing and the computation will convince the architects, the constructors and the decision makers that a shell can be economic, reliable and much better looking than a classical flat roof.

2. THE CHOICE OF THE GEOMETRY

A spherical shape has been used extensively to make shell structures. However, in most cases, the purpose of a roof is not to cover a circular area but rather a rectangular or square area. The area to be covered can be projected onto a sphere with an appropriately chosen radius to obtain a cupola. Oneknows that, in order to be stable, this cupola needs edge beams to transfer the forces to the foundation or to prestressed cables. In the shell, the principal stresses act as follows: compressive stresses along the diagonals and along lines parallel to them, tension stresses normal to the diagonals. This working principle is an important aspect in the design of the shell. In these directions, the shell has to be able to withstand the stresses. This has led to the following definition of the cupola. The dimensions given are those of the model that has been constructed.

Take a square with dimensions 4 x 4m. Construct a sphere which passes through the four corners of the square. There are many possible spheres which fulfill this condition. Therefore, there is an additional degree of freedom that is given by the height of the resulting cupola.

The square is divided into a net of equal squares by dividing the sides in nine equal parts and by connecting the division points diagonally. This square net can now be projected onto the sphere. If this projection is a parallel projection, the resulting net on the sphere surface will generate a net of arcs which can be used for further construction of the shell. However, to construct our model, a stereographic projection was used. The stereographic projection is a central projection from the nadir of the considered sphere. The advantage of this projection is that the angles between the lines remain unaltered. In this case, the arcs on the sphere will form an orthogonal net which contributes to the aesthetics of the structure. By using this procedure, the area covered by the shell is not anymore a square as can be seen from Fig.1 which shows a top view of the net of arcs on the sphere.

The arcs which are obtained are now replaced by straight lines which connect two consecutive nodes. The orthogonality of the nodes is not anymore perfect but the deviation is less than a half degree .

The quadrangles formed now as an approximation of the cupola are theoretically not flat because three points determine a plane and here we have four points. In practice however, the distance from this fourth point to the plane defined by the three other points is less than one millimeter even in the case of cassettes of some 6m x 6m.

Figure 2 and Figure 3 give a side view and a perspective of the structure obtained. It is a structure which is well balanced and equilibrated because it shows the lines along which the forces act and because the intersecting lines are orthogonal to each other.

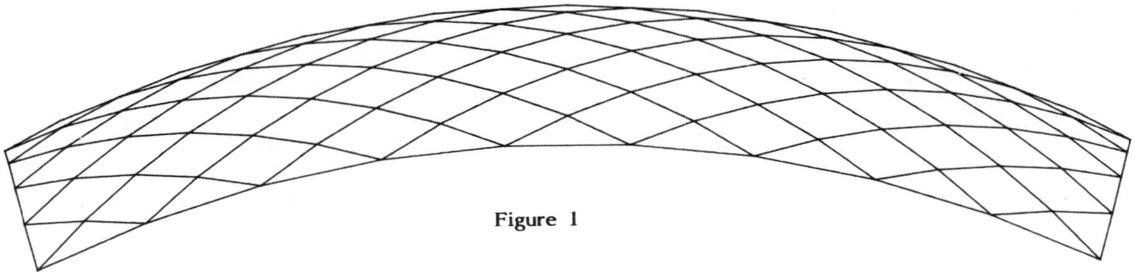

Figure 1

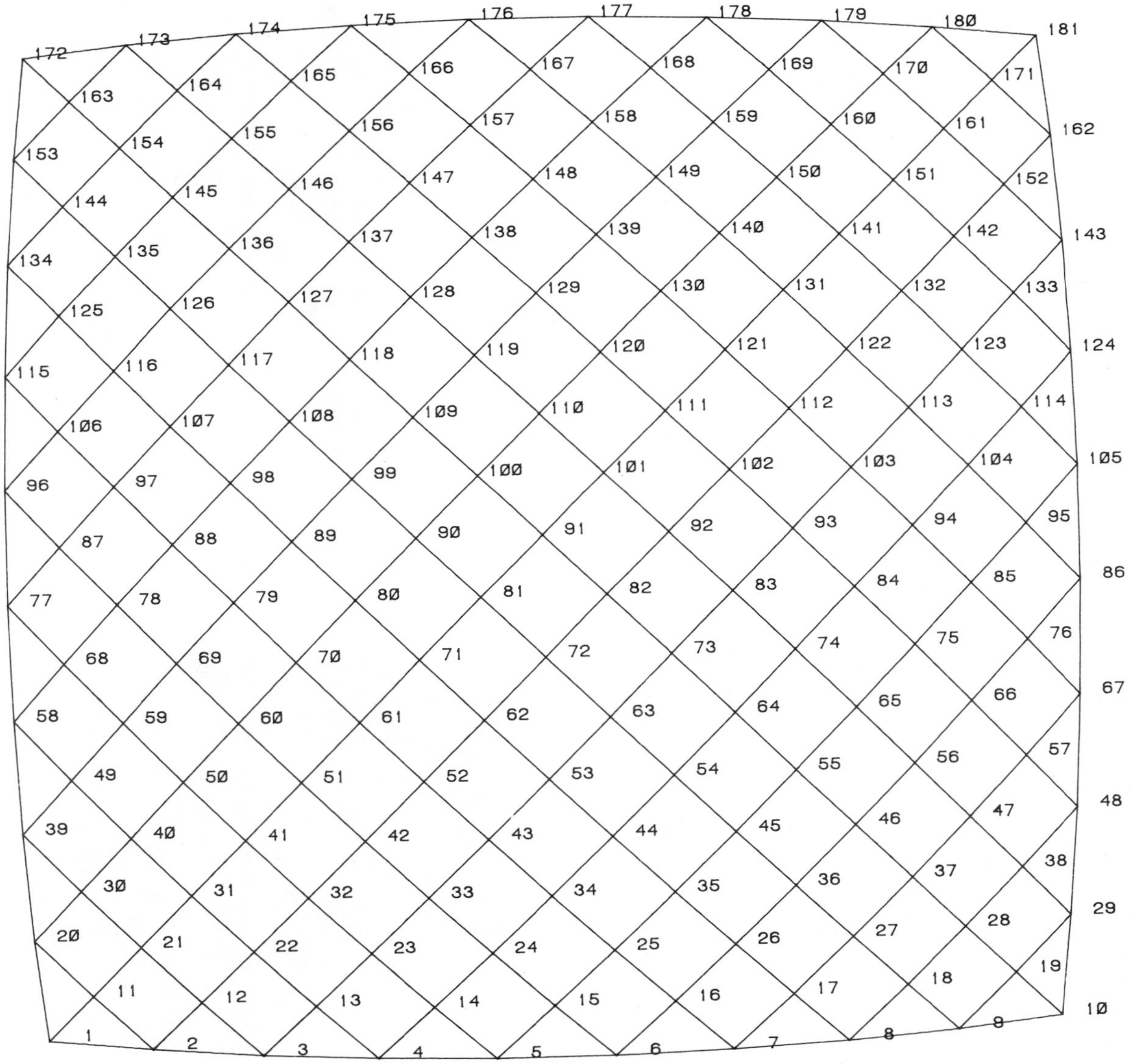

Figure 2

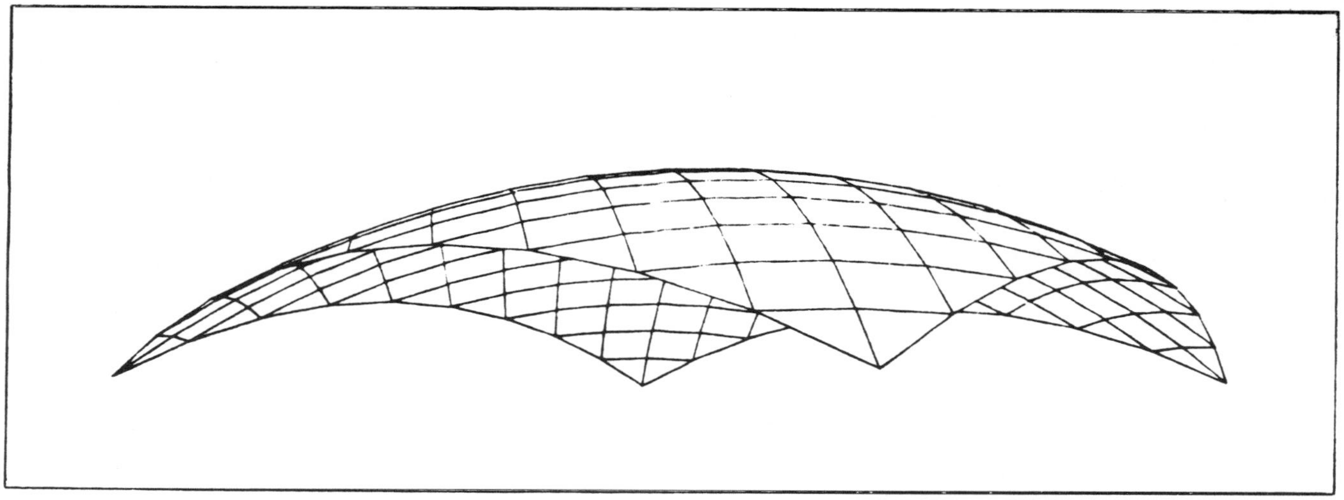

FIGURE 3

FIGURE 4

3. THE CONSTRUCTION OF THE MODEL

The structural geometry which is defined in the previous part is materialised using thin (4mm) laminated wooden cassettes. Each cassette has the shape of an open box. The cover panels have the form of the quadrangles shown on Figures 1-3. The side panels are determined as follows. Each side is determined by a plane through two points on the quadrangle and the nadir. The height of the cassettes is approximately 100 mm.

In the cupola, the cassettes are not identical. In fact, of each cassette, only four are identical because of the symmetry of the structure. As a result, the construction of the cassettes would be very complicated without the use of a computer. This computer gives the possibility to draw each side of each cassette such that the workshop can use these drawings to saw the wooden panels. Furthermore, the computer will give the information required to saw the panels under the correct angles so that the different panels fit together to a cassette that is an exact copy of the mathematical model. However, a simplified fabrication procedure is also possible.

When all cassettes are made, they can be assembled and connected by bolts to form the cupola. During this assemblage, an important feature of the cassette-structure becomes clear. The assembling generates directly the overall form very precisely. First, the edge beams are installed, then the diagonals are assembled to form substructures. These substructures are connected to each other and to the edge beams. (Figure 4). Afterwards, the other cassettes are assembled to form substructures before they are added to the cupola. Finally, in the corners of the structure, fine cables are used in the direction of the tension forces because the cassette structure can only withstand small tensions. Figure 5 shows a view of the cupola from the inside.

The cupola covers an area of 18 m^2 and weighs 40 N/m^2 without the edge beams and 70 N/m^2 with the edge beams. Nevertheless, it is designed to withstand at least a uniform loading of 1,5 kN/m^2. This shows that the structure is extremely light and therefore economic not only to erect but also in material use.

FIGURE 5

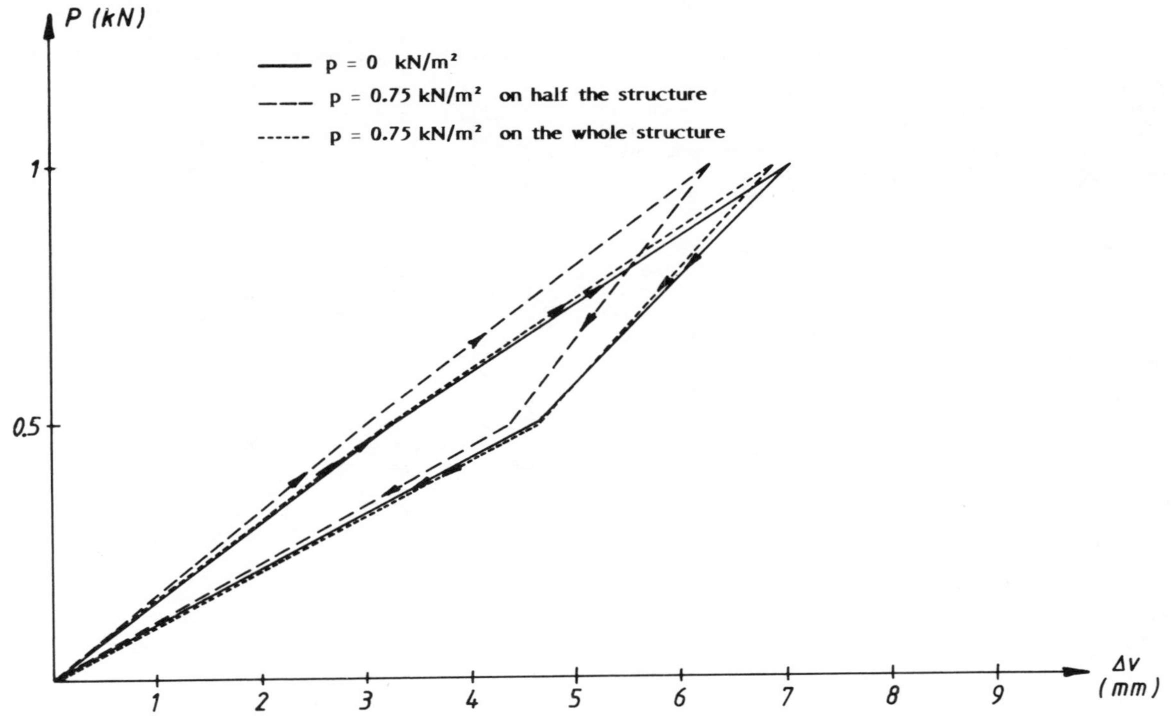

Figure 6 - Relative average displacements of nodes 61, 64, 118, 121

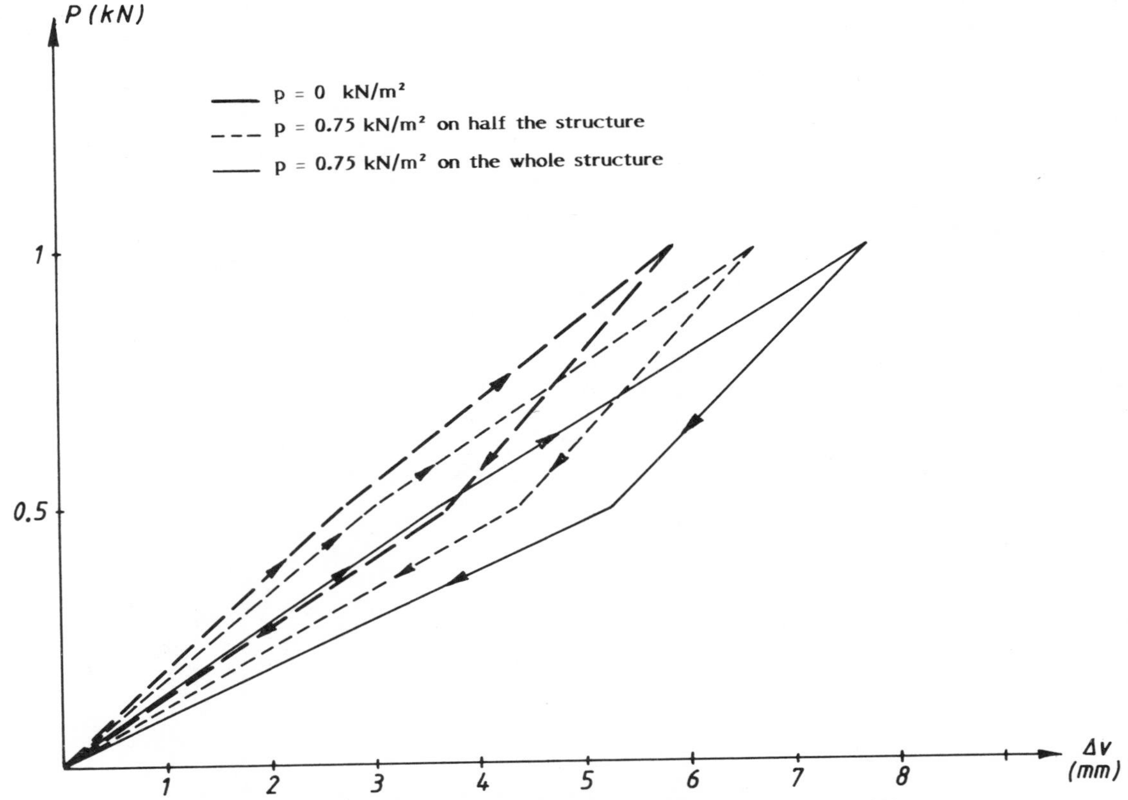

Figure 7 - Relative average displacements of nodes 81, 82, 100, 101

4. TESTS

4.1. Description

Two kinds of sollicitations have been concidered. First of all the behaviour of the cupola under point loads was looked at. The loading and unloading proceeded in steps: 0kN; 0,5 kN; 1 kN; 0,5 kN; 0 kN. Secondly a spreaded load was simulated by means of relatively small node loads, such that the average load took the values 0.75 kN/m^2 and 1,5 kN/m^2.

This spreaded loading proceeded in four steps: 0.75 kN/m^2 on a first half of the cupola limited by the zox-plane of coordinates; 0.75 kN/m^2 on the whole; 1.5 kN/m^2 on the first half; 1.5 kN/m^2 on the whole, thus such that the total spreaded load reached 1.5 kN/m^2 besides the own weight, this being 1.4 kN in total.

The first kind of sollicitation was repeated after each of the first two steps of the spreaded loading. Thus we aimed at the influence of the spreaded load on the stiffness with respect to the point loadings.

Measured were the following items.

1) The vertical displacements of two groups of four diagonal symmetrical nodes close to the top;
2) The horizontal displacements of the corner nodes;
3) The strains (by means of strain gauges) of the four main tension bars connecting the corner nodes.

4.2. Results

1) The figure 6 shows the average vertical displacement of one node under a local charge: it is seen that the cupola behaves elastically, be it not perfectly linear, and that the effect of the spreaded load is a slight decrease of the displacement. Figure 7 shows the analogous results for the average vertical displacement of one node further away from the top.

2) The total horizontal displacement of the corners lengthens the tension bars about 2.5 mm under the spreaded charge of 1.5 kN/m^2. This yields a strain value exceeding by seven percent the more reliable directly measured strain values. A simplified computation by hand yielded 2.9 mm.

5. CONCLUSIONS

The structure behaved elastically during the test loading. The displacement of point loaded nodes decreases slightly when the spreaded load increases, it be on half the cupola or over its entire surface.

The precision of the displacements and strains measured agree within twenty percent with the results of a simplified computation, based on the assumption that the cupola is simply supported at each of its four corners.

The local stability of some ribs has to be computed in order to improve their thickness near the corners.

The small own weight requires solid anchorage against wind effects. Tests in a wind tunnel in different conditions would be interesting.

The study of the covering against weathering (rain) and the hygrothermic effects is to be considered separately.

Finally, applications can be a small sheltering, a country house perhaps, multipurpose buildings and even huge structures overspanning sport fields for the considered structure requires relatively low levelled anchor systems and small foundations.

The effort of computer aided drawing of all the box faces is counterbalanced by the incredible ease of erection. However, some quantity production is possible: any box form appears four times and has a symmetrical partner.

Thus eight plane equal unfolded forms can be produced: four have to be folded in one sense and four in the opposite sense, according to the principle that, for instance, a right handed glove becomes a left handed when its inside is put out to become the outside.

LOW TECHNOLOGY SPACE FRAMES

E T CODD, B.ARCH(HONS), FRAIA, AIDIA,
Managing Director - Edwin Codd and Partners,
Architects and Industrial Designers
Brisbane, Australia.

The essential challenge of all three dimensional structures based on the assembly of linear members is the design of the connections. Proprietary space frame systems produced to date have used prefabricated components and/or special node connectors. This approach results in a high percentage of the cost being directly attributable to basic fabrication or sophisticated engineering (by building construction standards). The new designs presented in this paper are unique in that they eliminate the use of node connectors and use the members themselves to effect the bridging. Cost savings are substantial in this low technology approach and there are some unexpected structural advantages.

INTRODUCTION

Space frames will only find wide spread acceptance if they are cost effective. The key to lower cost is simple engineering, the elimination of unnecessary components and fabrication, the use of off the shelf members and high levels of production. Commercially successful solutions will combine good structural performance with ease of application to buildings.

While almost all structures for buildings are space frames, the term is most commonly applied to three-dimensional structures where the more tradiational primary and secondary members are replaced with a maze of similarly sized members in a repetitive geometry. Experiments began as early as 1907 with Graham Bell producing three dimensional structures employing common length bars joined by a universal connector.

The first of the modern space frames (MERO) after many years of development found widespread application in 1957 at the International Building Exhibition in Berlin. In the early 1950's TRIODETIC and UNISTRUT were also developed. The most recently available has been British Steel's NODUS.

Up until the time computers became commonplace in consultants' offices, there was an understandable reluctance on the part of engineers to use space frames. Much early sizing of members was intuitive, since manual calculation methods were impractical.

With the advent of computers, engineers have become more receptive to their architectural collegues wishing to use this type of structure. The calculations now required by most approving authorities throughout the World can be provided and there is no doubt that the application of space frames to buildings could be come widespread if more cost effective designs were available. Most of the space frame types currently available were presented at the First International Conference of Space Structures at the University of Surrey in 1966. They were introduced with considerable fanfare, but an objective evaluation of their market reach after a fifteen to twenty year trial, would indicate that they are only achieving a very small percentage of the available widespan market.

There have been some spectactular successes but mostly in projects where cost was not the determining factor. Given the opportunities that have been available to develop sound markets, it must be concluded that many of the existing solutions are too complex to manufacture. In addition the design of the nodes makes them difficult to apply cost effectively to the fabric of buildings.

Building owners and managers are frequently prepared to pay well for features which will enhance the user or buyer appeal of their buildings, but structures have always been regarded as a basic element and must be cost effective. There are exceptions and the use of space frames in external canopies or enclosing top lit common space testify to their acceptance in that market area, but use in more common building types will only be guaranteed if they are cost competitive with alternative structures.

The surviving space frame types vary substantially in their engineering performance and market appeal. With some, almost unlimited geometries are available and the opportunity to vary the size of connectors and members results in structures which are material efficient. Other systems offer a few standard dimensioned modules for ease of production, and seek to reduce costs by limiting the range of options.

The factors which result in high cost and lack of acceptance are itemised below. Not all systems suffer from the same disadvantages, but no one system is so much cheaper than its competitors that it has achieved a major market share.

1. All systems cut the raw material to module lengths.

2. Most systems require prefabrication, either of pyramid units or cutting to length and attachment (usually by welding) of end pieces so that a simple onsite connection can be made.

3. For those systems which employ a mechanical lock to achieve a connection, high press capacity is required and set up is time consuming and must be extremely accurate.

4. All systems have node connectors which are manufactured using one or a number of the following processes: casting, machining, forming, extruding.

5. Where gauges can be changed throughout the structure a special end piece connector is required for each size tube.

6. The accuracy of all systems is dependent on cut length tolerances.

7. Node connectors which usually protrude past the line of the structural members result in the attachment of claddings being difficult. Vertical subdivision of the frame also requires expensive detailing for the Architect.

8. In most systems it is impossible to carry services through the frame members and there are many applications to buildings and stand alone structures where this facility is desirable.

This paper will identify manufacturing techniques appropriate to the production of low technology space frames. An evaluation of the use of space frames with regard to their acceptance by building users and the ease with which they can be related to buildings will be undertaken to enable an assessment to be made of the new designs presented.

MODERN MANUFACTURING TECHNIQUES AND APPROPRIATE TECHNOLOGY

Since the days of the Industrial Revolution, cost effective production has relied on standardisation and high volume. Most of the components used in buildings are mass produced. Structures, however, tend to be "one-off", repetition is limited to a particular job and levels of production are low. Jobbing organisations have proliferated. The urgency to develop high production and standardisation in the structures industry is obvious.

The cost of many consumer products during the last thirty years has fallen substantially in real terms. By comparison the cost of structures has risen astronomically. An examination of our engineering heritage, taking into consideration the early promise of structures such as the Crystal Palace, can bring little joy to the design professions and the building industry. Progress has been painfully slow with industry opting in the main for "one-off" structures rather than mass produced systems.

For those involved in the industry there is no question about the popularity of space frames. Unfortunately, only a low percentage of enquiries are converted into orders. The most obvious reason for lack of use is high cost. Given that the basis of space frame manufacture is being able to take advantage of modern manufacturing techniques and mass production, it must be concluded that the designs currently available are inappropriate or too complex and expensive, to find universal application.

Many everyday consumer products are progressively using more sophisticated technologies. Typically these high tech. products use only small amounts of material. Mass production has reduced the amount of labour and the consumer has benefited. By comparison the building industry uses large amounts of material and most technologies are of a lower order. In addition, buildings are labour intensive. Since it is very difficult to reduce the amount of material used in buildings, emphasis has been placed on a reduction in labour and improvements in material handling so that trade work can be carried out more efficiently. High technology in buildings tends to be expensive and any move toward highly engineered structures is unlikely to be cost effective. Space frames employing beautifully engineered complex joints requiring much processing will not be generally applied. Instead they will only be used for decorative canopies where their visual appeal compensates for their high cost.

Space frames are in direct competition with "one-off" structures produced by jobbing organisations. They have high design and development costs and overheads associated with marketing. On the other hand, many "one-off" structures are produced by small organisations with limited overheads. The use of space frames in factories is rare and a combination of expensive engineering, low turnover and inadequate representation will prevent most of the existing systems finding greater use. Trade literature and advertising expenditure, together with a need to provide unpaid technical advise for many structures that do not proceed, impose further crippling burdens for the supplier.

For a space frame business to be a viable, stand alone enterprise it must operate in the high volume area of the structures market. The space frames presented in this paper are intended for this application and will compete directly with portal frames.

ACCEPTANCE AND APPLICATION

(a) Acceptance:

As a structure type, space frames appeal to designers. This appeal has much to do with the repetition of elements which is an essential characteristic of most modern architecture. It is also a characteristic which most building users find satisfying, particularly with regard to the scale relationships which develop as a result.

To maintain a humanly satisfying relationship between a building and the observer, architects have traditionally subdivided large elements into a number of smaller units. This is because large structures with big spans can be overpowering. The larger the frame centres the greater the impression of the structure being aggressive and dominating. Generally speaking people like to be comfortable with buildings rather than be dominated by them. Space frames aid this relationship between man and building because they utilise a maze of small members each of which is frequently small enough to be man-handled.

In the great domes and vaults of our cathedrals and public buildings where scale relationships are particularly important, decoration was frequently used to reduce the scale to an acceptable level. Space frames have the advantage of integrating the expression of structure and decoration in a humanly satisfying way. The need to consider scale as an important factor in the presentation of building interiors becomes more and more important as the spaces within buildings become larger. This is apparent in the large atrium type hotels and roofed shopping precincts currently being built in large numbers.

Space frames also find acceptance by association. A top lit space frame invokes much the same feeling in the viewer as does light filtering through the branches of a tree. It is far easier to relate to this form of structure with its cellular or bee hivelike assembly of units than it is to relate to structures based on primary and secondary members which have no parallel in nature.

(b) Application:

Apart from aesthetic considerations, user response and cost, the main concern with regard to the use of space frames in buildings is the way they relate to the fabric of the building and the ease of detailing that relationship. The following characteristics are desirable:-

1. Ease of attachment of claddings
2. Ease of providing vertical subdivision
3. Carrying services within the frame members
4. Inherent bracing
5. Flexibility of module and depth
6. Accuracy of the assembled structure

Too many space frames have joints where the node connectors protrude past the line of the members making the attachment of standard claddings impossible. With most, vertical subdivision is very difficult to achieve and yet this is a very likely requirement in buildings. Building services always have to be accommodated within structures. With the closed members and solid hubs of most frames, this is impossible.

Plan modules and depths should be infinitely variable provided this can be achieved at low cost. Accuracy of the assembled structure with minimal deflections is essential.

DESIRABLE JOINT CHARACTERISTICS

In the design of space frames in the past, much emphasis has been placed on engineering excellence. The ideal has been circular members, concentrically located and rigidly fixed in relation to one another. This brief suggests a welded connection but in practice considerable difficulty is experienced in setting up the members accurately in relation to one another so that welding can take place. If segments are welded in the factory they then have to be connected on site. Alternatively, fabrication can take place on site, but it is time consuming, expensive, and delays the progress of works. The only viable alternative is a mechanical joint.

In designing mechanical connections most inventors have concentrated on achieving the ideal of concentrically arranged webs and chords. In most systems some attempt has been made at moment transfer through the joint. In the design, development, manufacture, marketing, application, and assembly of space frames, there are many considerations for the professionals involved at each stage. The successful inventor must prepare a brief which satisfies the needs of all individuals in the chain.

Manufacturer requirements:

(a) Large turnover
(b) Minimal stockholding
(c) Low tooling costs
(d) Short set up times
(e) Low labour
(f) Minimal fabrication

Marketing Attributes:

To appeal to a marketing organisation, a space frame must have the potential for high turnover. The product requires considerable flexibility in the sizing of members and spanning capability. The system must be applicable to a wide range of building types and good technical information must be available. Module dimension and frame depth must be infinitely variable.

Architect's Requirements:

(a) The product must be cost effective.
(b) Flexibility of module and spanning capability.
(c) The ability to carry services through the members.
(d) Ease of vertical subdivision.
(e) Good technical literature and backup.
(f) Aesthetic acceptability.
(g) Accommodate a variety of geometries.
(h) Ease of application to buildings

Engineers Requirements:

(a) Structural adequacy
(b) Certification by manufacturer
(c) Availability of test results
(d) Guarantees on performance

Builders Requirements:

(a) Accuracy of manufacture
(b) Ease of assembly
(c) Ease of connecting segments in position

TWO NEW SPACE FRAMES

Space Frame A

The joint illustrated has three important features.

1. The top and bottom chord members are continuous through the joint.

2. There is no node connector, bridging being effected by the superimposition of members.

3. Splices can be made if necessary at each node without the use of additional connectors.

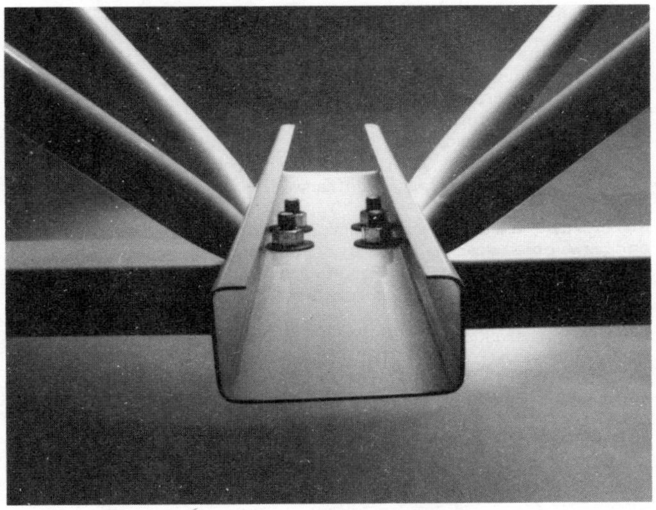

SPACE FRAME A

Many different sections could be used for chords, but they are usually standard C sections roll formed and hole punched. Webs are circular tubes.

Throughout any one structure chords can be supplied in a variety of gauges as required by the loads and web members can vary in diameter and gauge through a considerable range. The frame has been designed to accommodate a limited geometry of square on square, offset one-half module. It is usually used for flat plates, but by employing tubular chords in one direction, barrel vaults can be assembled.

By using C section chords, ducts are automatically provided for the reticulation of services. Covers can be used to close the open chord sections and locate wiring.

Section closing channels can be used to provide localised strengthening of chords and backing plates can be used at the nodes in zones of high stress. By being able to upgrade sizes, gauges or add strengthening pieces or backing plates where required throughout the structure, the design of the frame and use of material can be optimised.

Development:

The development of the Type A Space Frame began in response to a building project for which a flat plate space frame was an ideal application. Available systems were too expensive for this particular project. Discussions with space frame suppliers suggested that existing designs were labour intensive, requiring much cutting to length of members and, in some cases, considerable prefabrication, usually by welding. In addition, most manufacturers suggested that at least half the cost of the frame could be accounted for by the expensive node connectors.

Since, in a flat plate space frame, all the material in the top and bottom planes is in unbroken lines it was decided that an attempt should be made to devise a connection which did not require the chords to be cut to module length. Since all the materials used in space frames are produced in continuous processes, such as extruding or roll forming, there seemed little sense in cutting the material to module length, only to join it all together again at every node.

The next step was to eliminate the node connector. A single bolt could have been used to connect the chords and attach the webs and there have been a number of designs based on this principle. However, it is necessary to stack the chord connections resulting in an untidy appearance and problems with order of assembly. By using four bolts it was possible to achieve a joint which did not require the webs to be superimposed on each other and also allowed the chords to be spliced at the node, if required.

Initially, the engineering response to the eccentricity in the joint was unfavourable, but the practical advantages of continuous chords, only requiring hole punching immediately after the roll forming process, could not be overlooked. It was also considered that, although eccentricities were present in the joint, chord continuity might act as an offset since there is a tendency for spaceframes to fail by joint rotation.

Some small frames were assembled and the ease of production and low cost was immediately apparent. Assembly was quick and accurate, even on rough ground and the process was aided somewhat by the chords being continuous. Accuracy of manufacture was ensured by hole centre stops for both chord and web production. With the success of the first experimental structures it was decided to undertake some basic testing to identify the mode of failure of the various members. These are reported on later in this paper.

Attributes:

Like most other space frames the Type A system is quick and simple to erect, allows excellent access for services such as plumbing and air conditioning within its depth and is inherently braced. It does have some other characteristics, all of which are not available in alternative systems.

(a) It is outstandingly economical
(b) Uses off the shelf members and components
(c) Is easy to manufacture with low tooling costs, short set up times and minimal machinery overheads
(d) Requires no stock-holding
(e) Has the potential for high volume production
(f) Is easy to apply to buildings with regard to vertical sub-division and attachment of claddings
(g) Provides continuous ducts within the chords
(h) Is structurally adequate and efficient and accommodates various size members and gauges within a single structure.

VERTICAL SUBDIVISION AND ATTACHMENT OF CLADDINGS EASILY ACHIEVED

MANUFACTURE AND ASSEMBLY

Chords: The cold rolled C section chords are hole punched immediately after the rolling process and delivered directly to site. Webs are processed with a press and are fully detailed without first being cut to length. Two ends of different members are formed simultaneously, the length being determined by a hole centre measurement. The whole web member can be formed in less time than it would normally take to saw cut a tube to length.

The reduction in labour cost is dramatic. With only half as many bolts as there are members, on site assembly can be carried out by unskilled labour quickly and efficiently. Continuous top and bottom chord members facilitate assembly. To produce a spaceframe the manufacturer has only to make two measurements, chord length and plan module. Both the dimension of the module in plan and the depth of the space frame are infinitely variable. Continuous chord members ensure accuracy of production and the frame can be assembled on rough ground without any inaccuracies developing.

Use: To date, thirteen structures have been erected using the Type A frame. They have been assembled on the ground, sometimes immediately below their final position and in other cases they have been moved horizontally, at least the width of the frame, during the lifting operation. Sections as large as 3000m^2 have been lifted in one operation.

LIFTING A 3000M^2 SECTION

Using existing tooling the most common chord member sizes are 100x50 lipped channels or 100x75 lipped channels in gauges ranging from 1.2mm to 2mm. In the near future gauges up to 3.4mm are proposed.

Web members vary from 38mm dia.x 2mm to 48mm dia.x 4mm.. Backing plates can be used in joints accommodating higher loads and section closing channels can be used to strengthen chord members in isolated locations.

With the opportunity to vary member size and gauge as required, and introduce strengthening and stiffening in isolated positions, the weight of material used in the frame can be optimised. Bolt sizes of 16mm are used in the 75 deep channel and 12mm in the 50 deep channel. High tensile bolts are used in areas of high stress.

The existing tooling is desgned to accommodate structures in the medium span range. using a 2.4m grid and a 1200mm depth of frame and a maximum material thickness in the chord of 2mm the following spans have been used:-

(a) One way simply supported 19.2 metres
(b) Two way simply supported 24 metres
(c) Two way four column support with cantilever 16.8 metres
(d) As for (c) with spread support 21.6 metres

Larger spans are possible if smaller grids and increased depth are adopted. A recent design for a rectangular roof with a clear span of 47.x33 metres employs a 1.8 metre grid and a 1.5 metre depth. The structure was designed for a combined dead and live load of .8 kpa. The frame has been used for building types varying from sophisticated commercial buildings and decorative canopies to very basic factories and warehouses.

MAZDA WAREHOUSE, AUSTRALIA

Evaluation: The structural evaluation of the frame has been extensive. Two series of tests have been completed and a third major evaluation is in progress. Initial tests were designed to identify the mode of failure of the members. Recent programmes have been aimed at optimising the design method to take advantage of member continuity.

The test programmes are set out below.

TESTS

Programme No. 1: This was a limited test carried out on a 1-1/2 element frame. The objective was to determine the mode of failure. The major engineering concerns at this stage centred on the quality of the joint. The test was carried out by the Queensland Institute of Technology Structures Laboratory on the 13/10/80.

These tests resulted in chord failures. The fact that there was no localised compression failure or buckling of the members at the nodes was the hoped for result and confirmed the need to undertake a larger scale test with sufficient modules of the frame to ensure adequate restraint of the top and bottom chords. It was apparent from this test that the chord may not have buckled as quickly had it been continuous through the joint and been restrained at adjacent nodes along its length.

Programme No. 2: These tests were carried out between August and November 1982 on a 9.6 metre square frame which was attached at the nodes on the perimeter of the frame to a structural slab. A jacking point was arranged at the centre of the frame to provide load. The test frame and the procedures were designed to test the maximum number of conditions. Since it is possible to substitute members of varying gauges within a frame, even after it is assembled, the initial test used the minimum sized web member throughout. As these members failed progressive substitutions were carried out with a view to having the chords, the webs and the joints fail at approximately the same loads.

When compared with the first test it is clear that the continuity of the chords through the joints considerably reduces their effective length. Apart from identifying the specific load carrying capacity of individual members, the tests were useful because they confirmed that the gauge of the web members in the frame could be varied and increased in areas of high load without substantially affecting the geometry of the joints or the accuracy of the assembled unit. (Ref 1)

Programme No. 3: This series of tests is incomplete. To date only an investigation into top chord performance has been carried out. The test focused on the effects of eccentricity and continuity in compressive chord members. (Ref 2). This test and the earlier work carried out have suggested that for some channel sections, member continuity and torsional restraint from the chord at right angles, reduces the effective length to less than .7 of the grid spacing. Continuing research is being carried out on this aspect.

Testing of the Type A space frame will continue with larger scale tests simulating typical applications. To date testing has concentrated on high loads over short spans, whereas the normal use of the space frame is with comparatively light loads over large spans.

Testing with high loads over short spans may be a disadvantage, since it induces a joint rotation with deflection ··hich would be greater than that which would occur midspan of a large, lightly loaded roof structure. Since the tendency is for space frames to fail by joint rotation, this matter needs careful evaluation.

Tests to date have confirmed that member continuity more than compensates for any slight eccentricity which may result from the configuration of the members.

Smith and Morgen (Ref 3) concluded in a paper published in 'The Structural Engineer' May 1982, as follows:

> "In summary it has been seen that there may be disadvantages associated with the desire for member concentricity of joints; alternatively, there may be beneficial effects associated with eccentrically connected members, provided some continuity of members is preserved. It is believed that eccentrically connected contiuous members can be exploited more than they are at present in the design of metal structures."

Space Frame B

TYPE B NODE

As with the Type A Space Frame the Type B has no separate node connector. Although a plate is used to cover the intersection of members, it's major function is that of a load distributing washer. A joint can be effected without this component. As can be seen from the configuration of the joint, it can be used for flat plate as well as compound curved structures with geometries based on hexonagonal and pentagonal surface divisions.

Chords are arranged on the curved surface, and webs can radiate at any angle to pick up nodes in a second layer. The joint can also be used for single layer structures if the webs are removed. In application this space frame is intended for single spaces which will not require subdivision. For external use the closed members are an advantage, since moisture cannot enter this section. With curved structures the need to accommodate member continuity through the joint is less of an advantage than with flat plates.

The gauges of members can be varied throughout the structure without upsetting the geometry or introducing inaccuracies or difficulties with assembly. All sections are galvanised. The only requirement for production equipment is a brake press. This simultaneously forms two ends and cuts the member to length.

Factory space needs only to be sufficient to store tube lengths and accommodate the press operation. Every stroke of the press produces a member and there is no additional fabrication. With the complete elimination of member welding and component sub-assembly, the frame is extremely economical to produce.

Cut length of the members is determined by hole centres rather than stops. Accuracy is ensured. Stockholding for this frame is minimal and, as with the Type A frame, would normally be bought in for a specific project.

Module and depths are infinitely variable and the design of the joint is such that it will accommodate three, four, five and six way connections of chords in one plane and a corresponding number of radiating web members. Tooling costs are low and set up times small.

As with Space Frame A, there are half as many bolts as there are members. There is no visual eccentricity in the joint which is an advantage over Type 'A' when the structure is viewed at close quarters.

This frame can be compactly crated and is an excellent solution for remote areas.

PROTECTION

Type A and Type B Space Frames are protected by patents which are issuing progressively.

CONCLUSION

The two new space frames presented in this paper are intentionally low technology solutions. They resulted from looking critically at the cost of sophisticated engineering and the disadvantages of pursuing member concentricity at the expense of other considerations, such as ease of production and application.

In the Type A frame testing has demonstrated that any disadvantage, such as flexual action in members which could have resulted from member eccentricity, is more than compensated for by member continuity through the joint. The systems are designed for different applications but both establish the need for fresh approaches to joint design if building costs are to be contained and space frames are to find universal application.

*****

Ref 1: A NEW SPACE FRAME, Codd E.T., Metal Structures Conference, Brisbane, May 1983.

Ref 2: A NEW SPACE FRAME (PART 2), Codd E.T., Wong H., International Conference on Steel Constructions, March 1984. Singapore.

Ref 3: BEHAVIOUR OF SOME JOINT DESIGNS IN SPACE TRUSSES, I.E.A. Paper C1314, November 1980. Smith & Morgan.

DESIGN OF INDRAPRASTHA INDOOR STADIUM NEW DELHI : INDIA

SHARAT C DAS, B.ARCH
ASSOCIATE, INDIAN INSTITUTE OF ARCHITECTS
KAMAL N HADKER, B.E, M.I, STRUCT. E
CONSULTING ENGINEER

The year 1982 saw the succesful completion of a number of projects in New Delhi as part of the preparation for holding the Ninth Asian Games. Perhaps architecturally and structurally the most fascinating and prestigious amongst these was "Indraprastha Stadium"-a well equipped modern indoor Stadium with a seating capacity of 25,000 spectators. Right from its conception to the satisfactory completion, the entire work was carried out in a short period of just twenty seven months. The paper describes how the basic structural design concept was evolved to speed-up construction and make the best use of readily available indigenous resources know-how and equipment.

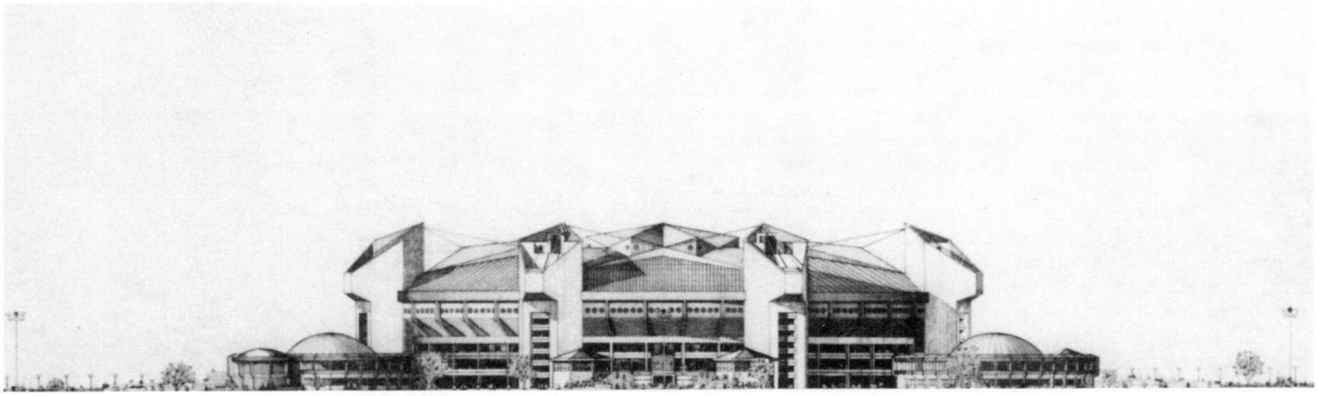

1.1. DESIGN CONCEPT

The most important factor which influenced the basic principle of design was the time constraint. The entire work was required to be completed by August 1982, i.e. within just 27 months from the day the first ever sketch for this project was produced. It became apparent that this was a major engineering excercise where the architects and engineers of various systems had to form a team right from the start and analyse different engineering solutions quickly to reach the most appropriate and aesthetically pleasing solution.

The original brief indicated an arena of 40mx55m and a total seating capacity of 25,000 spectators. Once the sight lines for the spectators were established, the overall size of the structure was almost fixed. The circular form was selected deliberately to facilitate (i) easy subdivisions, (ii) maintaining uniformity in the vision line (iii) high degree of mechanisation in mass production of prefabricated components, necessary for a time bound project like this. On the basis of studies made, the evolved form was a circle of 155m diameter. This was divided into eight equal parts and the eight circumferential points formed the main structural supports. The eight parts were further divided to form 48 radial grids.

Vertically, the structure was divided into four levels and they were (i) Ground or Arena level (ii) +3.9m or Podium level (iii) +8.1 or Foyer level (iv) +14.25 or Pylon Entry level. These levels were related to various entry points for the players, spectators, organisers and VIPs.

A number of activities had to start simultaneously to ensure speedy implementation of the project. These activities were divided into four main groups such that they were completely independant of each other.

A. Fabrication and Erection of Structural Steelwork for the roof to cover 155M diameter cirle.

B. Construction of eight partially slipformed Towers along the periphery to support the roof and to house services and stairways.

C. Cast-in-Situ framework of reinforced concrete columns and beams along 48 main radial grids.

D. Production of prestressed channel units designed to serve as seating tiers as well as flooring units at Podium and Foyer levels.

The information available on the sub-soil conditions clearly indicated the need to provide piled foundations for this type of structure. Besides, the entire plot was in a low lying area requiring average filling of about 1.5m of earth to avoid flooding in future. This massive earth work had also to start simultaneously and had to be carried out without interfering with the piling work.

All these construction activities were examined carefully and a network was prepared to indicate critical paths and important sequences which showed that it was not impossible to meet the deadline set for completing this ambitious project.

1.2 DEVELOPMENT OF ROOF CONFIGURATION
The most intricate and challenging part of the entire project was the roof. The use of structural steelwork was an obvious choice in view of the fact that the necessary steel sections as well as know-how and machinery required for fabrication and erection of this type of work were readily available in India. In view of the time constraint, all other type of systems involving import of technology or equipment were ruled out. It was decided to use lattice type steel girders composed out of double-channel boxed sections for ease of fabrication and overall economy By providing such girders it was possible to reduce deflections substantially in comparison to other types of roofing systems It was readily acknowledged that this type of roof had better dynamic response and it behaved well when subjected to extreme wind load conditions causing reversal of stresses.

It could easily be seen that any reduction in the "effective span" of the dome would result in substantial savings in steel hence eight supports in the form of V-shaped walls were introduced within the stadium. These walls did not cause any discomfort to the spectators and helped in reducing the span from 155M to actually 135M. The allignment of these walls was carefully chosen such that secondary girders could be spanned between alternate supports giving further benefits of reduced effective span.

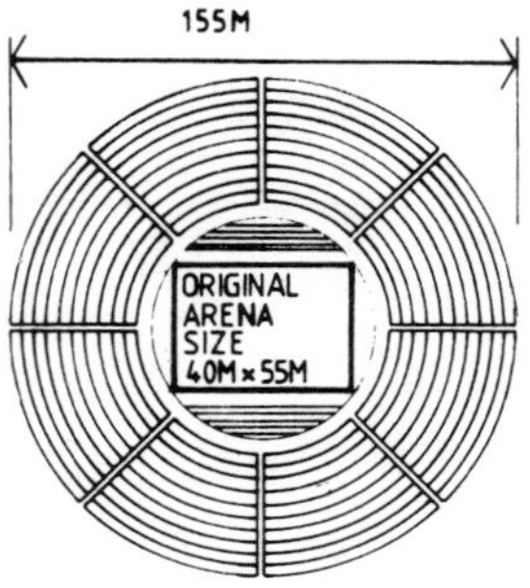

FIG-1

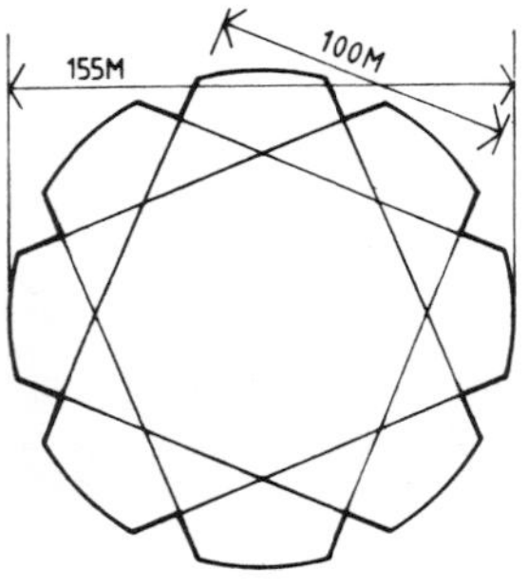

FIG-2

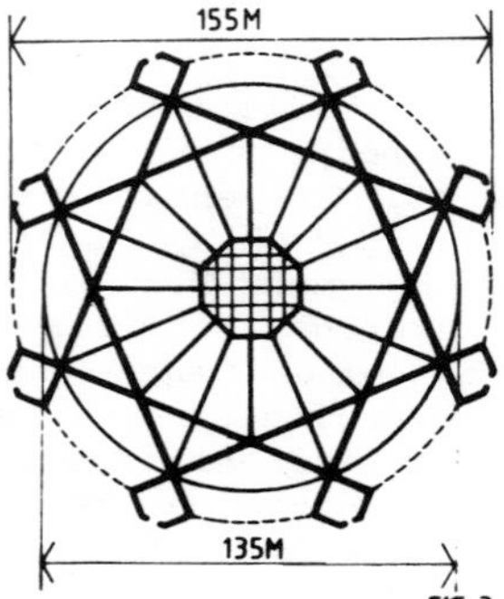

FIG-3

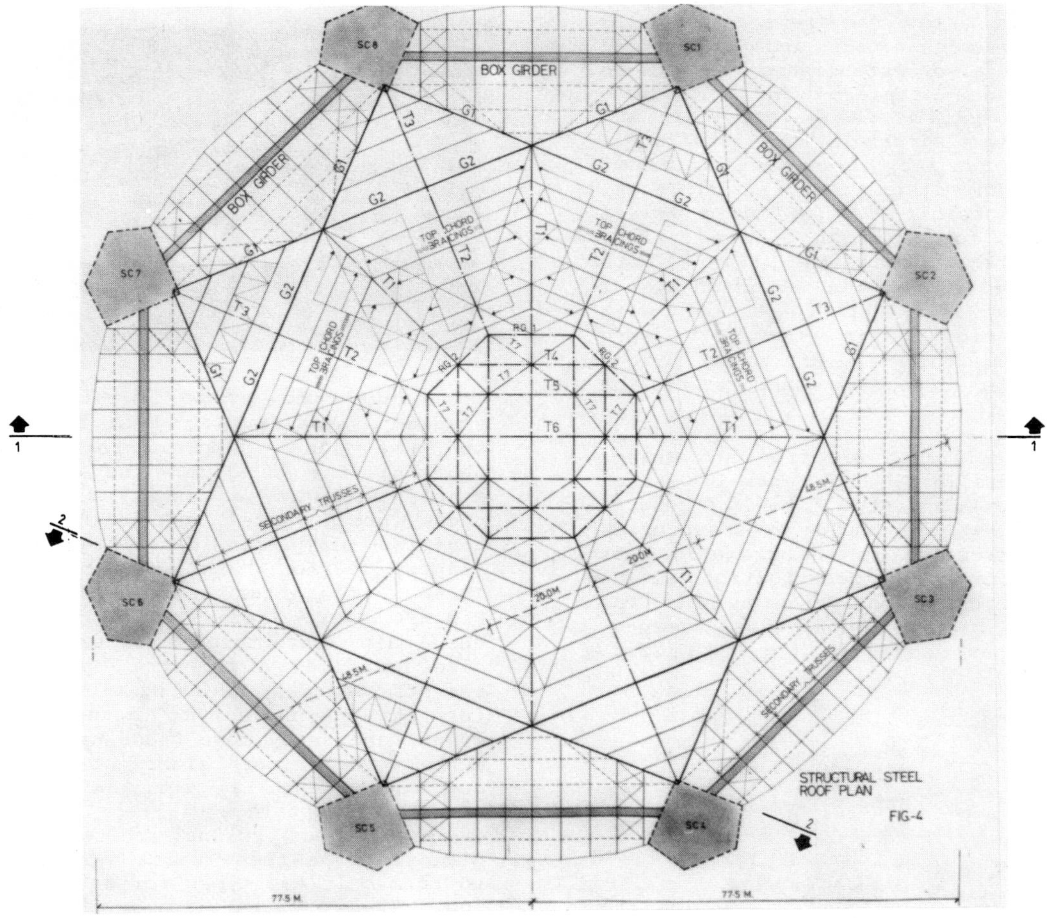

STRUCTURAL STEEL ROOF PLAN

FIG-4

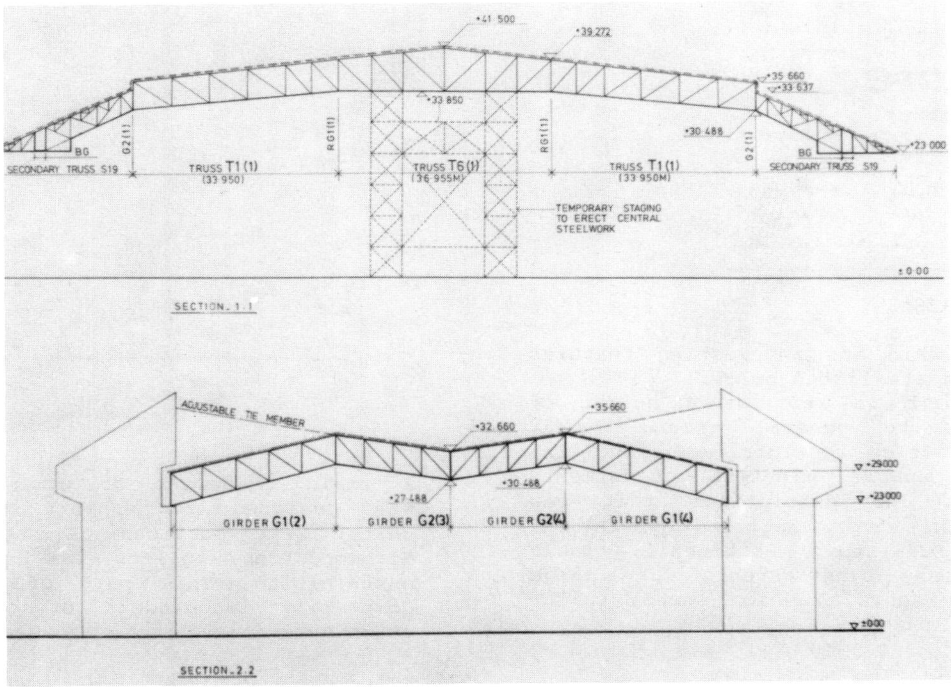

1.3 PYLONS

In the meantime, the requirements of air-conditioning, mechanical and electrical services were also established and these walls were then replaced by concrete towers which housed the electrical and airconditioning services and stairways. It was felt that by constructing these towers ahead of other cast-in-situ work, it would be possible to use these towers as rigid masts for erection of steel trusses around them. Hence the towers were designed to go upto 40m height, almost 11m above the level of the roof, with a chamfered tip. ''Tie members'' were introduced from the tip of these pylons to give further effective support to the roof.

In order to facilitate erection and assembly of the steelwork, the Central portion of the roof was converted into a ''compression ring'' which could be supported on temporary trestles in the arena without interfering with other construction activities. The final configuration of the main members of the roof thus emerged, step by step, as seen from fig. 1 to fig. 4.

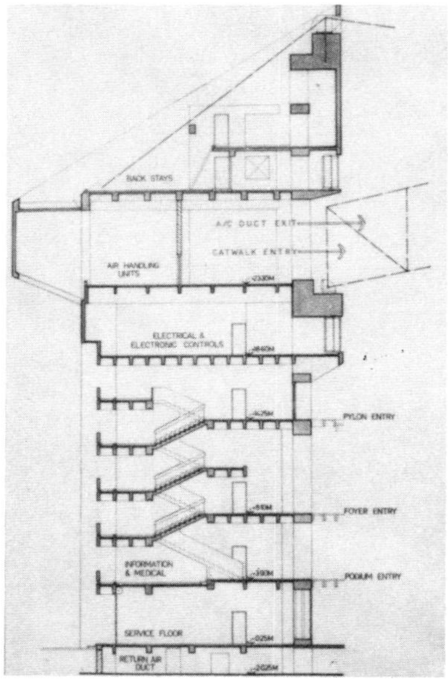

2. MAIN FEATURES

Some of the main and interesting features of the Stadium are listed below:
The arena which is kept at a height of +1200mm from the general ground level, rests on a series of intricate levelling devices to enable minutest adjustments if required. The disciplines that can be held on this arena included Gymnastics, Volleyball, Badminton, Basketball, Table Tennis and many other events like large scale musical shows, fashion shows, cultural shows and circus.

To reduce the overwhelming size of the stadium visually, it was felt necessary to add some complimentary structures, abutting the stadium. Instead of having simple stairways as means of public entry, these areas were further developed with addition of public facilities like restaurants, curio shops ticket booths etc. These complimentary structures were called Plazas. These Plazas not only complimented the main structure architecturally, but also fulfilled a number of physical requirements besides becoming a perfect place for the spectators to gather in between events and enjoy some fresh air and refreshments. The floor slabs at Podium level and Foyer level as well as seating tiers for the spectators gallery were designed as prestressed channel units. The overall dimensions of these units were standardised to form only two types for speedier production. The prestressing steel was varied according to the span and loading.

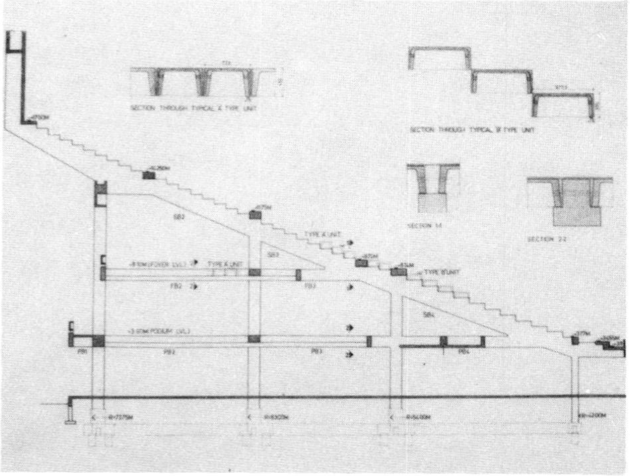

A rigid framework of columns and beams was designed to support these precast units. Some portions of the floor slabs accommodating toilets and pantries were cast-in-situ. This part of the work was completely independent of other construction activities.

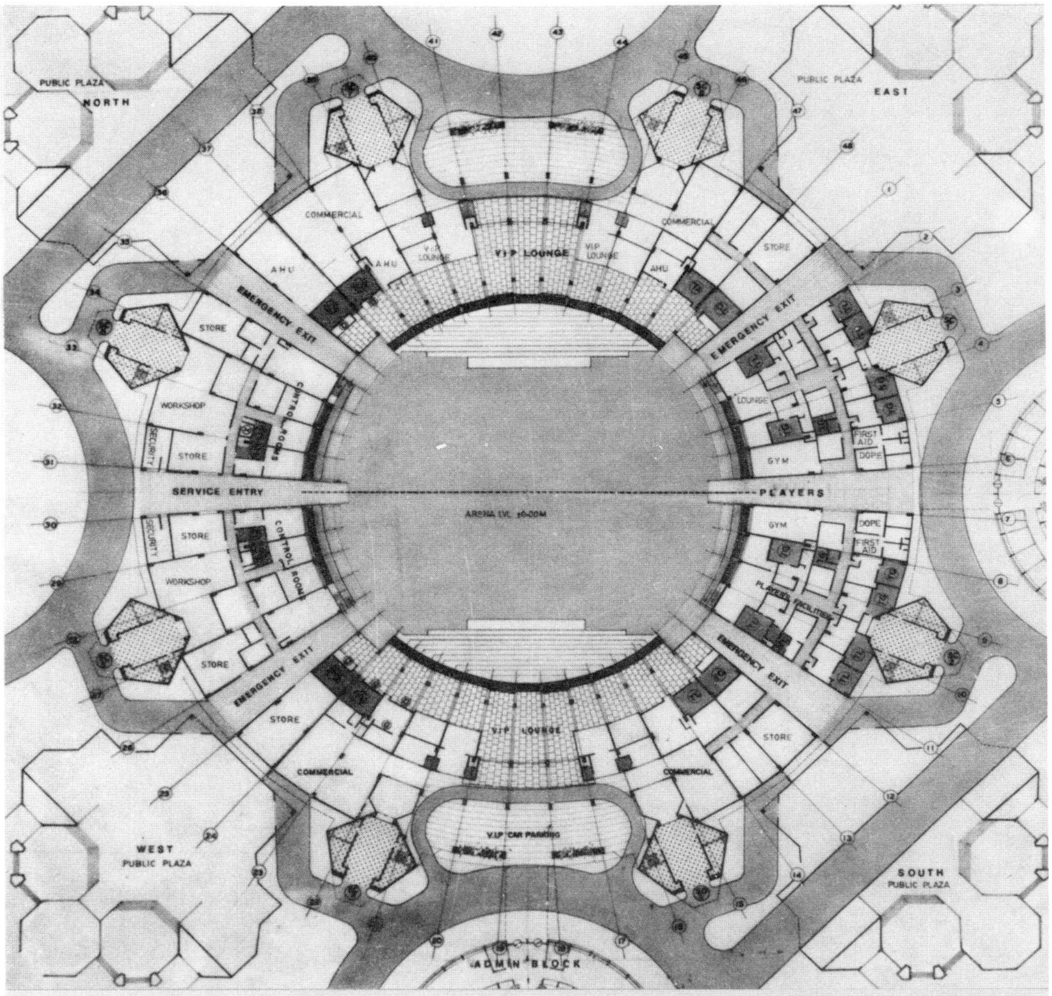

GROUND FLOOR PLAN AT ± 00 4 2 0 4 8 12 16 20M

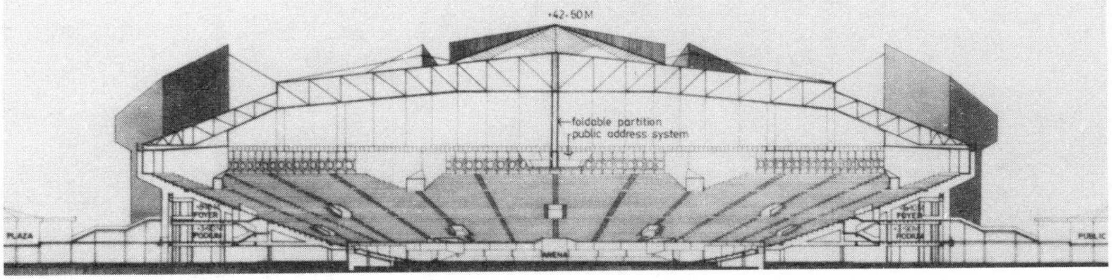

SECTION THRU' PUBLIC PLAZAS 4 2 0 4 8 12 16 20m

Standard corrugated Aluminium sheets were chosen as the roofing material mainly because of their low weight and high reflectivity. These sheets are readily available and can effectively drain out rain water even when laid at a gentle slope of 10°, whereas Asbestos sheets require much steeper slope.

This was an important consideration in order to restrict the total height of the roof to minimum. Thus a folded plate type roof forming radial pattern of ridges and valleys was considered to be an acceptable form. Besides being functional, it also gave a proportionate form to this massive roof.

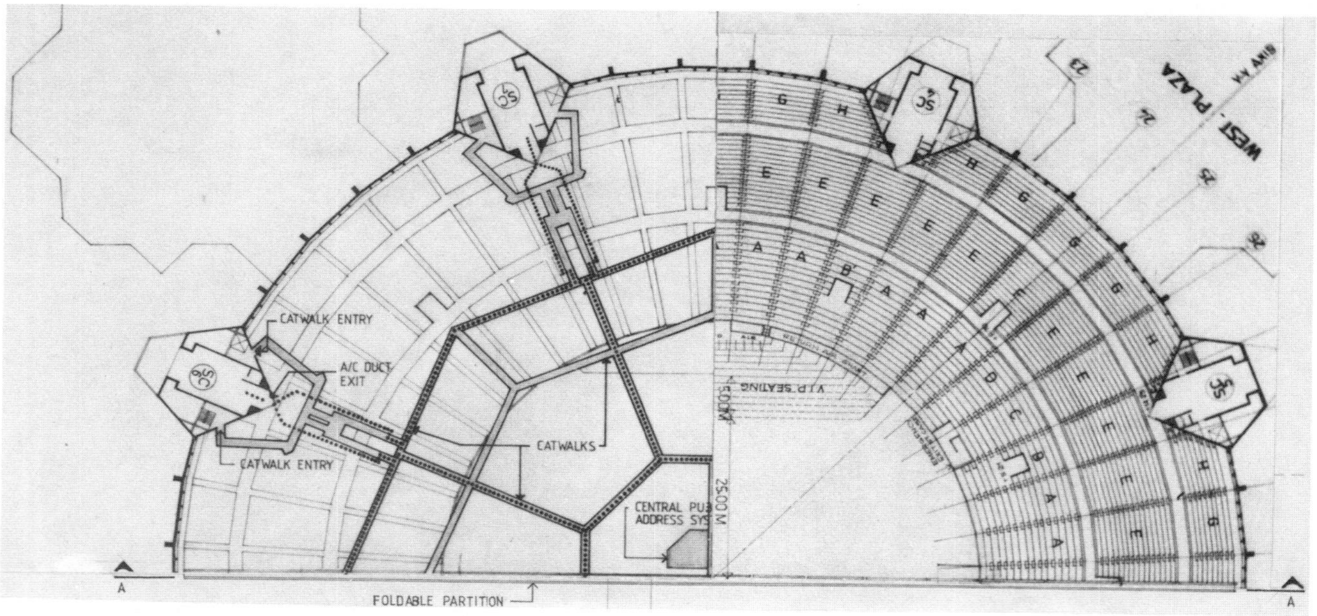

PART PLAN OF CATWALK LAYOUT AND TOTAL SEATING

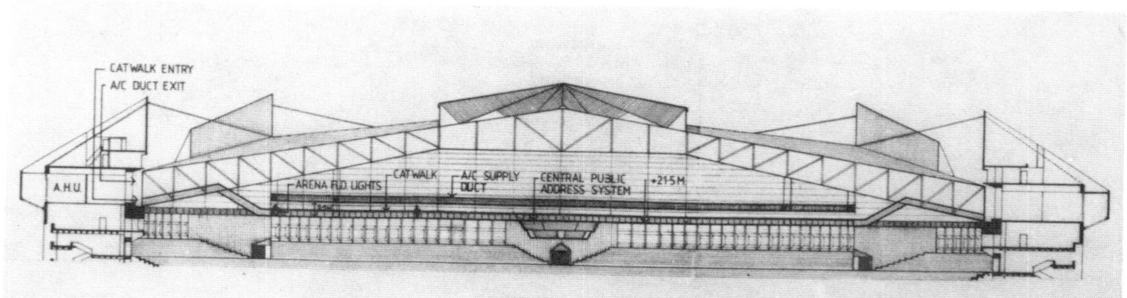

SECTION AA

To avoid fogging due to central cooling on one side, and the degree of heat generated from metal hallide lamps used for arena lighting, on the other, it was felt necessary to have an effective exhaust system naturally at the highest point of the ceiling. To achieve this, the portion of the roof above the compression ring was lifted and rotated through an angle of 22.5° to get the best results. The resultant vertical surfaces on the sides were moulded out of FRG and the joints were sealed with FRG resin to make them absolutely water-proof. FRG was selected with the view of utilising these vertical surfaces to let in natural but diffused light which could be of great advantage in panic condition during day time.

To install an effective audio and lighting system for the arena, a subsidiary framework was suspended from the main one in the form of catwalk network. The metal hallide lamps, speaker box and the frame for the Electronic scoreboard were installed at suitable places with special care taken for easy maintenance. The airconditioning ducts were also suspended from this network so that the dampers, to adjust the supply of conditioned air, were easily accessible. The catwalk system was so placed that it did not interfere with the vision line of the spectators on the last tiers.

Cold-rolled sheet metal purlins in the form of multi-beams were preferred to Z purlins because of their torsional rigidity. The secondary trusses spaced at 6m centres were designed as 1.7m deep trusses having an overall width of 600mm. The side faces of these trusses formed N-girders and the top and bottom chord angles were laced together. These trusses also served as walkways useful for maintenance workers.

The main members of all the primary trusses and girders were fabricated out of two channels battened together to produce economical compression members. It was possible to vary the cross sectional area by simply providing continuous plates of appropriate thickness in place of battens. Generally all the node points were designed as welded joints. But the connections to be made between these trusses at roof level were designed as bolted connections. As the forces in member varied substantially under different conditions had to be studied carefully. Perhaps the most complicated location was the junction of truss T1 with girders G1 and G2. Two high-tensile steel tie members from the pylons were also to be connected to the roof at this very junction. The proposal by the contractors to substitute the vertical member at this location by a "stiffened drum" was found to be quite convenient. Similarly, the saddle joint connecting the ties with backstays at 40m level was also designed on the same principles, two neoprene pads were provided under each saddle joint to ensure free movement in the radial direction.

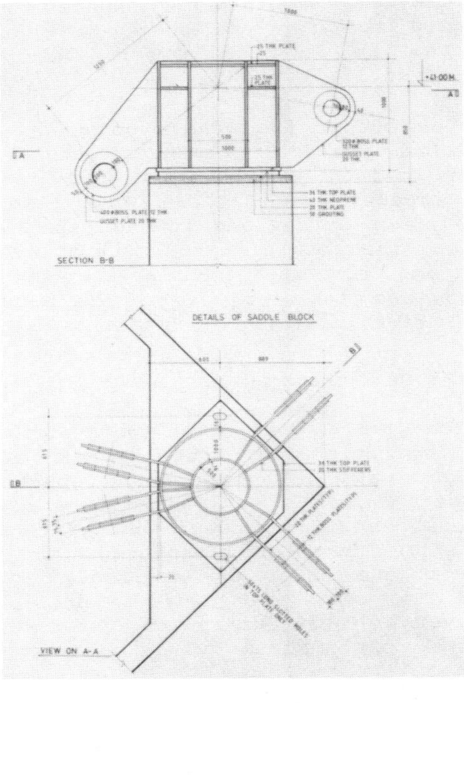

The main bearings at 23m level were designed to transfer vertical as well as horizontal reactions on the pylon at this level was strengthened by deep beams to transfer 500 tons of horizontal shear safely to the peripheral walls of the pylon.

3. HURDLES IN THE RACE

Almost eight months after the commencement of work, there was a directive to introduce a partition to enable simultaneous holding of two events independently.

Besides designing and erecting such a partition, it meant virtually redesigning the stadium, for in simple words, it meant constructing two independent stadias but under one roof.

This came at a stage when some major structural members were already erected in position, and many already fabricated.

From planning point of view, many architectural drawings needed to be extensively modified. The arena size had to be inreased to 60m x 78m and the services layout had to be discarded and redesigned. Facilities provided earlier needed to be duplicated on either side of the partition; and fresh layouts were prepared for arena lighting, sound system and the airconditioning.

The partition, when designed, became a 150m long, four layers upward folding curtain, with a sound resisting value of 38 dB. It was to be made up of polyvinyl fabric to be stitched at site, and expected to go up or come down in just about 7 minutes. Since it weighed over 100 tons, the forces in the structural members increased substantially. Besides, there was no space available at arena level for stitching this curtain, as the wooden flooring was under various stages of construction. Hence a 6m wide temporary platform was suspended from the roof at about 17m level to facilitate mobile stitching machines to travel 150m in a straight line.

Another problem, but less serious, was that long after the completion of fabrication of compression ring steelwork, it was decided to lift the central portion of the roof to admit natural light and enable ventilation. There was no alternative but to provide additional trusses riding over the main steelwork. This change was welcome because it improved the aesthetic appearance of the stadium roof.

4.1 COMPUTER AIDED ANALYSIS

After finalising the basic structural design concept the geometry of the roof was worked out carefully.

The levels of the crown, bearings, saddle joints, intermediate suspension points, etc. were arrived at after studying several possible alternatives. This excercise was followed by the determination of the coordinates of all the node points with respect to the origin. Because of the large size of the problem all the node points were numbered in a particular sequence. The total number of nodes was 378. Each of the members was numbered in a sequence so that the member forces printed in the output could be read easily and used in the final design. The total number of members was 945. In addition, the information on section type, material properties and its orientation for each member was also read. The codified data on support conditions at the saddle joints and at bearing levels was supplied.

The structure was idealised as rigid space frame. In view of the symmetry of the structure only half the structure was considered for analysis. Stiffness Matrix method of analysis was used for solving the space-frame subjected to self weight, other dead loads, live loads, thermal loads, wind loads and seismic loads. The frontal solution routine was used to get the deformation in the structure for various loadingcases.

The final results included six global displacements (u, v, w, d/x, d/y, d/z) at each of the joints and for all the loading cases under consideration. The member forces (load case wise, as well as, for pre-determined load combinations) were printed. The solution involved 2268 simultaneous equations. The accuracy of the results was examined by plotting the displacement patterns, checking the joint equilibriums and overall structure equilibriums for balancing the applied loads. Various load combinations, such as, "Dead Load + Live Load + Temperature rise" or "Dead Load + Wind Load" were printed out in the final output to facilitate design work.

Self weight of the primary structural system was computed automatically by considering the volume of the member and the density with a suitable modification factor for the additional weight due to batten plates, gussets, bolts, welds etc.
The loading due to the suspended partitions was calculated manually and was applied at the appropriate node points. Similar excercise was carried out for the walkways, Airconditioning ducts, lighting fixtures, central speakers console and other weights suspended from the roof. The load due to aluminium sheeting, thermal insulation, accoustical lining was calculated on area basis. The temperature loading cases was carried out for a range of ±0° to +50°C.

4.2 FOUR STAGES OF ANALYSIS
Four separate computer runs were made for the following conditions:

4.2.1 BEARINGS FREE
The main bearings at 23m level were allowed to slide when the temporary supports were lowered by giving prescribed nodal displacements at the trestle points. This case was analysed for seven different values ranging from 0 to 15 Cms. O Cms case represented the condition when steelwork was erected in position. The reactions obtained in each of these cases were noted and used for the next step in the analysis.

4.2.2 BEARINGS FIXED
The reactions obtained from stage 1 were applied as nodal loads at corresponding points to a "weightless" structure to obtain results due to self weight when temporary supports were removed.

4.2.3 BALANCE DEAD LOAD + LIVE LOAD + TEMP.
In this computer run all the remaining loads indicated earlier were imposed.Results for each load case were printed separately for further study. Various combinations were also printed.

4.2.4 TIES REMOVED
The ties were expected to get slackened under certain combinations of loads coupled with rise in temperature. Hence it was considered necessary to check the pattern of forces in all the members when ties were removed.

Using the output from these computer runs, the maximum possible forces in each member were determined. The tabulated results were compared with those predicted by University of Surrey and it was established that both the results were comparable.After a thorough scrutiny of all the factors, decision was taken to lower the supports in the centre initially by 9 cms.

5. INDEPENDENT CHECK

In March 1981, when the roof structure was partly fabricated, Prof. Z.S. Makowski, Head of Civil Engineering Department, University of Surrey,was requested to carry out an independent check on the analysis of the roof structure. In view of the unique shape of the roof, it was decided to carry out Wind Tunnel test in University of Surrey to assess wind pressure for maximum wind velocity of

100 miles per hour. All other details of loading and working drawings for the roof were supplied from India. The additional load of 100 Tonnes for the suspended partition was also included in this data. The model for wind tunnel test was made in perspex to a scale of 1:300 with 40 pressure tapping points provided at appropriate locations.

By rotating the model into different positions, it was possible to obtain pressure at more than 600 points on the surface. In the analysis carried out at University of Surrey, the concrete pylons were approximated using a suitable skeletal model. The bracing members and secondary trusses were also included in the analysis. It was observed that some of the bracing members were attracting high compressive forces. In the original design these members were treated as ties only. Although the validity of this assumption could not be disproved, it was decided to modify the cross-section of these Tie members to form a box thereby increasing the Radius of gyration without increasing the sectional area. Similarly, additional bracing in the central compression ring area were introduced as they were considered "advisable though not essential". The results of wind tunnel test revealed high suction forces along the ridges and at the eaves. This problem was solved by reducing the spacing of the purlins in these areas. Some of the top chord members were found to be mariginally overstressed under the severest combination of loading. To overcome this, it was decided to reduce the extent of "initial lowering of supports"from 15 to 9 cms.

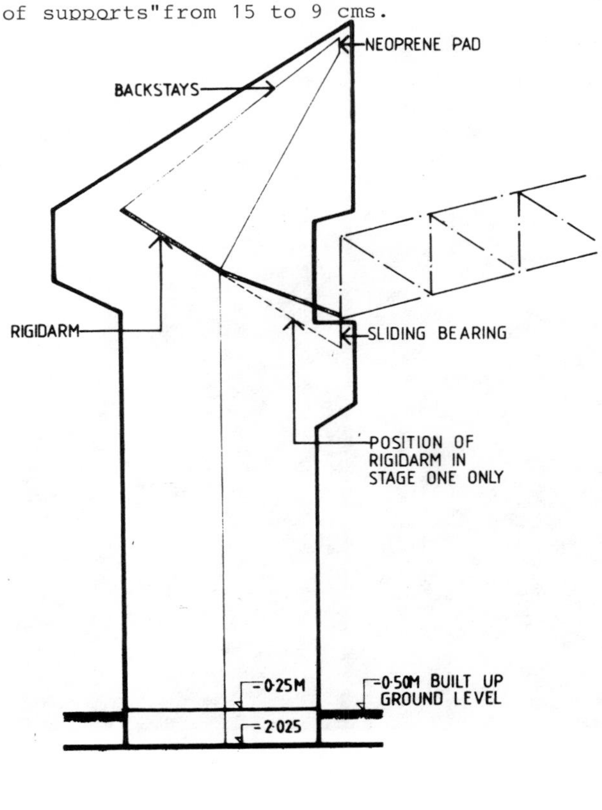

6. FINAL ANALYSIS

The analysis finally carried out in India was refined further to represent the boundary conditions at the saddle joint accurately. It is important to note that tie members are anchored into the saddle joint, which in turn is held back by the backstays and the saddle is supported on neoprene pads which permit radial movements of the joint. Thus, when the tie forces increase, the saddle moves towards the arena. The final analysis carried out in India represented the actual structure by introducing additional members to account for backstays, the neoprene pad and the pylon above 23m level. In the first stage of analysis, when the bearings at 23m level were permitted to slide, a small vertical member was introduced just under the bearing to represent the sliding joint. In the second, third and fourth stages, this small member was removed and the rigid arm was connected directly to the bearing point to represent the conditions of "Fixed Bearing". A similar short member was introduced at 40m level to represent the sliding bearing for all stages of analysis. With the idealization, it was seen that the structure was represented satisfactorily in the computer analysis.

STATISTICAL DATA

1.	Capacity	25,000 Spectators
2.	Final size of arena	60 M X 78 M
3.	Area of roof	1,682 Sq. M
4.	Structural steel used in the roof	2,000 MT
5.	Total quantity of reinforcing steel	2,000 MT
6.	Total quantity of cement used	8,000 MT
7.	Total length of pre-stressed pre-cast units	25,000 M.
8.	Length of suspended walkways	1,730 M
9.	Airconditioning load	3,000 Tons
10.	Power load	7,000 KVA
11.	No of exists	34
12.	Evacuation time	3.5 minutes
13.	Cost of Stadium	Rs.250 million [$ 25 million]
14.	Cost of ancilliary buildings	Rs.200 million [$ 20 million]

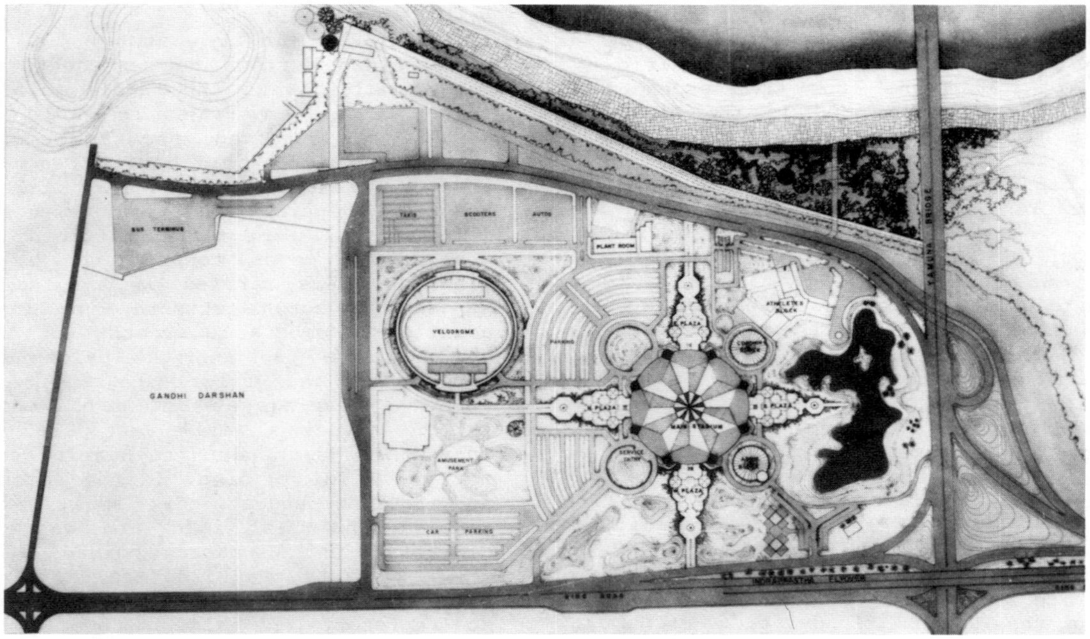

SHARAT DAS & DESIGN CONSORTIUM
I P INDOOR STADIUM COMPLEX
VIP FOYER NORTH
NEW DELHI PIN 110002 INDIA

N

SITE PLAN

20 10 0 20 40 60 80 100M

RETICULATED BEAM GRID STRUCTURE MADE OF PREFABRICATED PRESTRESSED SPACE MODULES

M. MIHAILESCU[x], A. IONESCU[x] PhD and A. CĂTĂRIG[x] PhD

[x] Professors, Department of Civil Engineering,
Polytechnic Institute of Cluj-Napoca, Romania

The paper deals with a reticulated roof structure formed by one type of space prefabricated reinforced concrete module, conceived to cover multi-purpose halls of quadratic areas with 21 to 39 m spans. By a chess-table arrangement of the prefabricated modules on the ground floor and their connection by prestressing, a reticulated grid structure is obtained which is erected at the roof level by hydraulic jacks together with the columns. The roof structure, hinge supported on four columns, placed at its corners, may carry several kinds of technological ducts, suspended cranes, sky domes and others, on a 3 x 3 m pattern. Details of analysis, design and structure achievement are presented.

THE PROBLEM OF FLEXIBLE HALLS

The continuous alteration of industrial processes, specific to the contemporary technology, has generated, referring to their housing, the multi-purpose or flexible hall notion, invested to comply the following main requests:

- to cover square or nearly square areas, with large spans of 21 to 39 m;
- to allow a technical store location for installation pipes and ducts, in the roof depth;
- to assure, as it is possible, a uniform zenithal lighting;
- to offer points of hanging, at short intervals, on orthogonal directions, for suspended transport devices.

Considering these conditions, several metallic structures were developed, most of them being conceived as planar grids, differentiated by the geometric arrangements and connection systems of bars. But, there are known only few reinforced concrete solutions for flexible halls, see Refs 1,2 and 3

A PRECAST AND PRESTRESSED CONCRETE ROOF

The paper deals with a precast concrete structure prototype for flexible halls, essentially made of a trussed beam grid, orthogonally shaped, hinge supported on columns placed at its corners, able to cover square areas with large spans of 21 to 39 m. So, a roof unit with a bidirectional symmetric mechanical behaviour, able to avoid stresses due to shrinkage or temperature variations, is obtained.

The structural units shown in Figs 1 and 2 are fitted to various arrangements with individual columns for each one or with common columns for adjoining roofs.

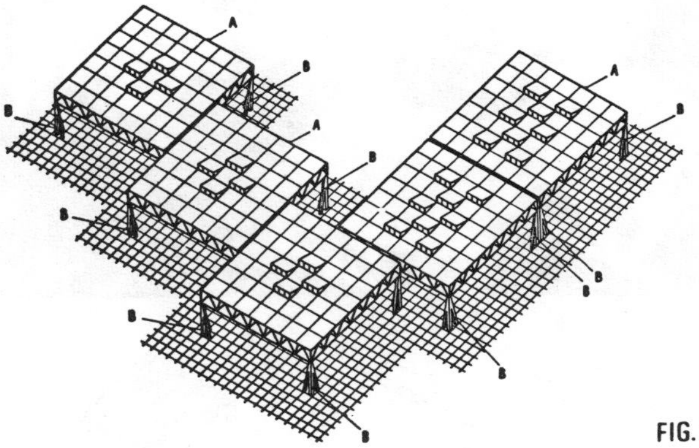

FIG. 1

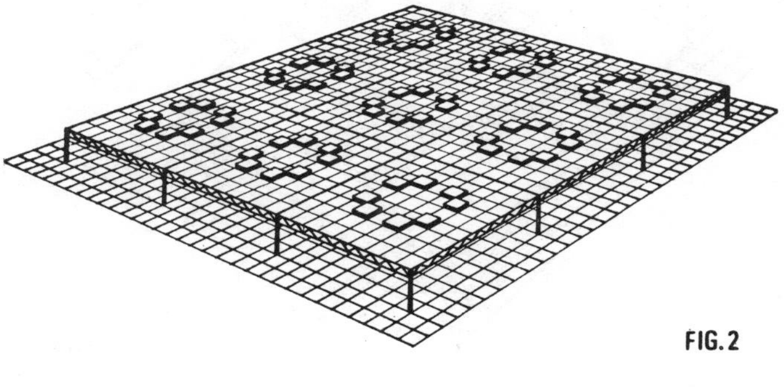

FIG. 2

Figure 3 shows the cable stayed grid structures, supported on common columns, indicated for larger spans than 33 m, in order to obtain, for the same roof depth-required by the technical space-similar economical and mechanical qualities, as those for smaller span units.

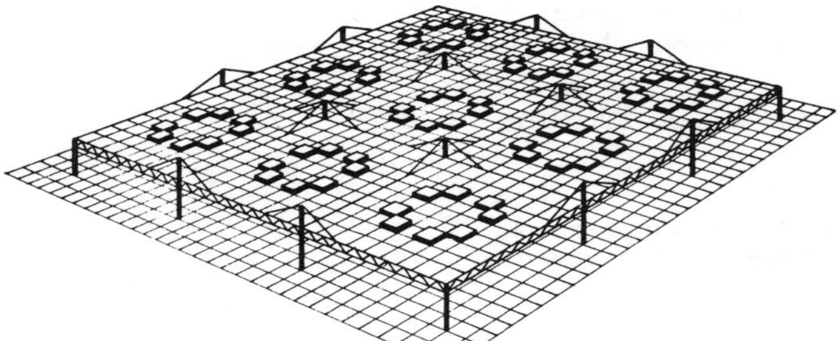

FIG. 3

STRUCTURE DETAILING

As it results from Fig 4, the trussed beam grid of the roof is made of precast reticulated space reinforced concrete modules poured in a single type formwork. These modules arranged on the ground surface, in a chess table form, as it is shown in Fig 5, are bidirectionally assembled by cable prestressing applied to the grid bottom chords. The free square areas between the precast modules, may be covered with sky domes or opaque plates, noted by 7 and 8 in Fig 5.

FIG. 4

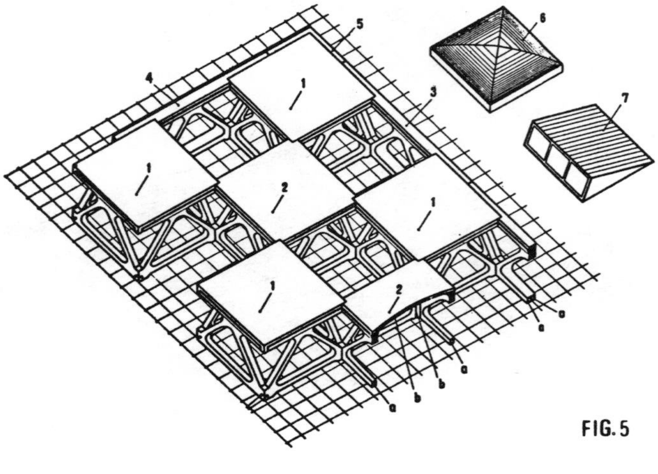

FIG. 5

The precast space trussed modules of a single general size of 3.0 x 3.0 x 1,8 m are of two different types, noted 1 and 2 in Fig 5; thus the first type has convex shaped corners (with outward folded angles) and the second, concave shaped corners (with inward folded angles). Consequently, it is ensured a good joining system between the modules. The upper plate of each module was shaped as a dome, made of four cylindrical segments joined along the square area diagonals. Its thickness of 88 mm concrete contains a 48 mm layer of expanded polystyrene, as thermal insulation. Each roof unit is completed along its boundary lines with plane precast trussed units, linear or right angle shaped, quoted by 3, 4 and 5 in Fig 5, conceived to confer the supplementary strength and stiffness, required by the planar structures rested only on their corners.

The space modules and plane elements are provided with channels for posttensioned cables, only on the bottom chords because the bending moments upon the whole roof area are positive, bearing tension only at its bottom.

The precast columns, noted B in Fig 1, are made of three reinforced concrete bars, put along the edge lines of a right pyramid with a right triangle base at the foundation level.

ERECTION TECHNOLOGY

After the foundations have been cast in the corners of the roof area, the erection activities, see Fig 6 follow the described order given below:

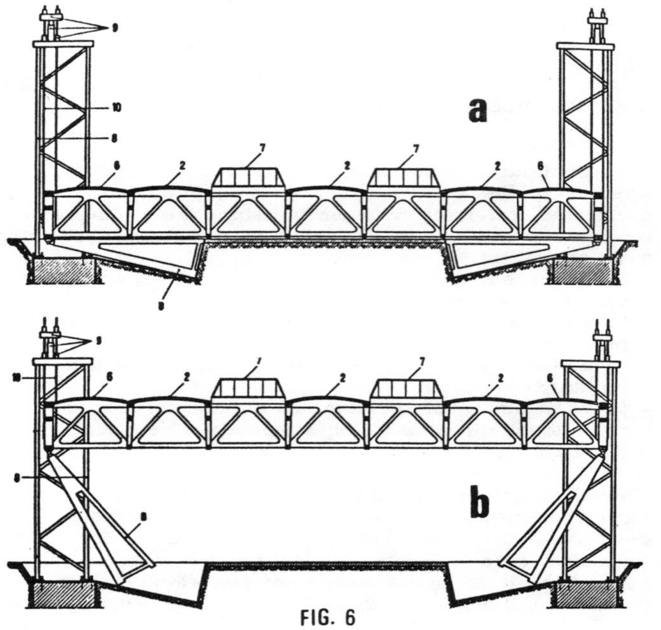

FIG. 6

- the columns B are horizontally laid in gaps, oriented to the roof area diagonals, excavated under the assembling surface level; it is noticed that the connection hinges between the roof and columns are to be put in their projected position, see Fig 6 a;
- after the hall ground floor was cast, or a plane surface was achieved, the trussed space modules are disposed on it, in a chess table form, together with the boundary precast elements. Then, the joints of 1-2 cm thickness, between the upper and bottom corners of the elements are fitted with cement mortar. Meanwhile the column top metal hinges are welded on the steel plates provided on the bottom of the

corner precast elements.

- the cables for prestressing are introduced in both families of orthogonally disposed channels. The cable sections are determined according to the maximum bending moment values, which are to be carried by each one;
- the prestressing efforts in cables were applied from one end, starting simultaneously with eight cables, located in the bottom boundary chords. This operation is symmetrically continued toward the center lines, by drawing four cables in each step, on both directions. The cables bond with concrete was ensured by cement mortar grouting;
- after the mortar hardening, the whole roof unit was lifted at the projected level by hydraulic jacks, which may be the same with those used for prestressing. The jacks are laid on temporary metal trussed columns placed outward the structure corners. The lifting operation moves simultaneously the four structure columns too, which rotate round their top hinges, connected with the roof, till they arrive at the designed upright position, see Fig 6 b.

MAIN ADVANTAGES OF THE STRUCTURE

The advantages that characterize the described structure, distinguishing it from other possible solutions are:

- the space trussed precast elements are robust, having an increased proper stiffness at bending and torsion; this quality, as well as that of the reduced joining surfaces among the modules, permit the achievement of an organic assembled structure, able to assure an optimum mechanical behaviour correlated with a reduced energy consumption;
- other noteworthy feature is represented by the safe and simple assembling procedure; the joints between the precast elements in a corner form, being each one bidirectionally compressed by cables, avoid any supplementary detail or work;
- the form of the precast space modules is quite rational; cumulating an equivalent plane trussed segment of 12 m length, with a covering plate of 9 m^2 area, it weighs about 3.7 t, easy to be conveyed by light trucks of 5 t, or by train;
- the manufacturing of these modules can be made by one single type of formwork; that is why it is justified to be equiped with high performance devices such as to compact the concrete by vibropressing, to accelerate the hardening by thermal curing and to assure the speedy extraction by hydraulic jacks;
- the 3 m sides of the trussed grid mesh were proved to be the most appropiate for carrying the channels for ventilation and other technological pipes, for hanging the crane beams and also for assuring of a uniform inside lighting, see Fig 7;

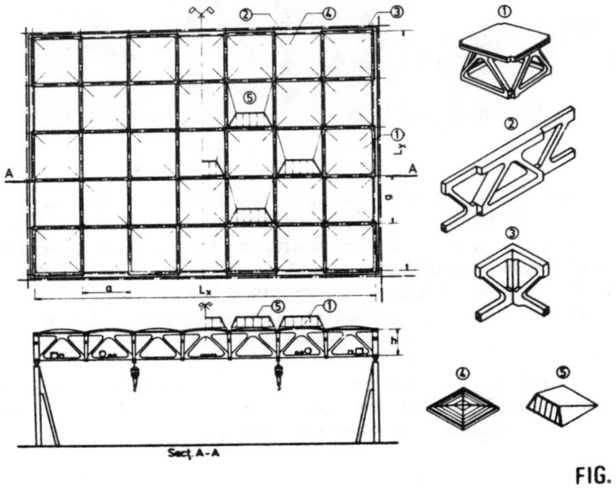

FIG. 7

- the building up technology, beginning with industrial casting of space and plane light weight concrete units, followed by their assembling on the ground level by posttensioned cables, and then by a simple lifting at the desired level of the entire structure, together with the columns, is available on all kinds of geographic spots;
- finally, other qualities worth to be mentioned are a less construction material consumption, a high speed attainable through the erection technology, as well as a outstanding looking offered by this prototype.

MULTI-FUNCTIONAL HALL ACHIEVED IN GHEORGHE GHEORGHIU DEJ TOWN

An illustration of the above described structure is to be found in Gheorghe Gheorghiu Dej town, where there have been built five spatial roof units, each of them covering an area of 21 x 21 m^2, see Refs 3, 6.

The material consumption indices, referring to the covered surface unit were:

- precast concrete of 500 daN/cm^2... 13 cm/m^2;
- reinforcing steel, passive bars, active cables and all other metal plates ... 17.5 kg/m^2,

these values comprising the columns too.

The metallic mould used for the space module manufacturing can be seen in Fig 8.

FIG. 8

Figure 9 shows a site view, the handling of the precast modules simultaneously with the lifting of one spatial roof unit weighing 3000 kN.

FIG. 9

The assembling process may be seen in Fig 10. The lifting was accomplished in two stages of 3.5 m each, according to the length of the available thread bars at that moment, the average lifting speed beeing of 0.5 m/h.

In addition, Fig 11 offers an inner image of the roof structure, and Fig 12 a general aspect of the finished building.

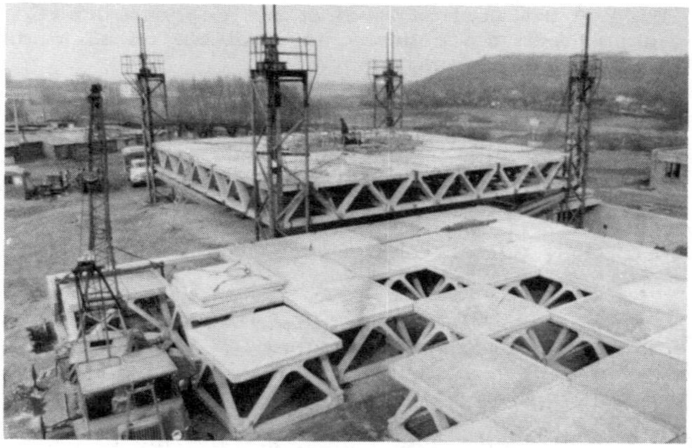

FIG. 10

FIG. 11

ANALYSIS OF TRUSSED BEAM GRID

The calculation of the mechanical behaviour of prestressed concrete trussed beam grid is a quite new topic. Thus, due to the bars differentiation in compressed and tensioned ones, each having its own modulus of elasticity, the isotropy hypothesis, fundamental for elastic materials and metal structures, fails in this case.

In the following, a practical design procedure, for prestressed concrete trussed beam grids, within a first order theory, consisting of three stages, is presented.

The first one concerns with the initial determination of the bar sections, performed within the ultimate limit state. Thus, the same bending moment and shear force values, as those calculated for a solid beam grid, of the same mesh form and size dimensions as the prototype, under the assumptions of homogenious and isotrope material, are applied.

In this way, Fig 13 depicts the bending moment and shear force diagrams determined for square solid beam grids, hinge supported at corners, and having the nodes loaded with forces acting perpendicularly to the grid plane. The force values are P on the inner nodes and 0.5 P on the boundary ones, the force P representing the rupture limit load, equivalent to the uniform loading of the adjacent area of an inner node.

Table 1 shows the maximum bending moment and shear force values, appearing in solid beam grids, according to different areas divisions from 3 x 3 to 10 x 10 square meshes under the assumption of equal cross section of each beam.

Table 2 presents the maximum values of the same efforts on similar grids, as it has been already pointed out in Table 1, but with the boundary beams of doubled sections compared to the inner ones.

FIG. 12

In order to check the mechanical behaviour under the service load, a structural unit was tested in situ. The prestressed roof unit, lifted at 5 cm from the ground level, was laid in the corners on metal cylinders and then loaded uniformly with bricks on the whole surface. The loading was performed in five steps and the discharge in three steps, using for strain determination mechanical and electrical gauges disposed in 65 significant points. The test proved a good behaviour of the whole structure and its high stiffness, the camber under service load, increased with 20% being about 1/2000 of the span.

The upper and lower chords sections of the trussed beams and their reinforcements, are determined in accordance with the maximum compressive forces due to maximum bending moments, respectively to the prestressings. From constructive reasons, unique constant concrete sections are adopted for the upper and lower compressed bars.

According to the shear forces, diagonal reinforced concrete bars, with the same configuration in tension and compression, are also designed.

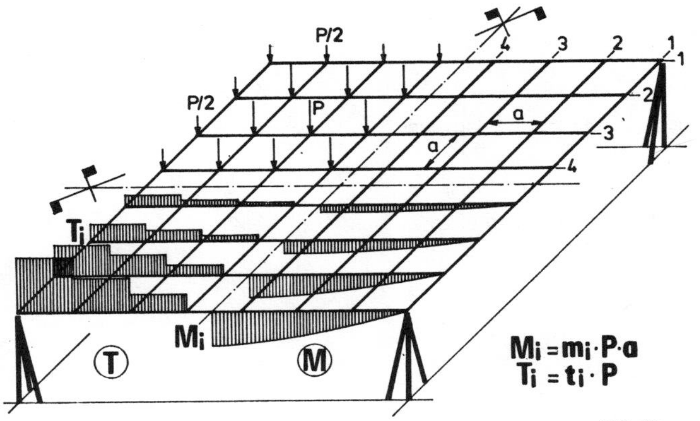

FIG. 13

TABLE 1

Type of network	NUMBER OF MESHES	m_i COEFFICIENTS FOR MAXIMUM BENDING MOMENTS ON BEAMS:						t_i COEFFICIENTS FOR MAXIMUM SHEAR FORCES ON BEAMS:					
	n×n	1	2	3	4	5	6	1	2	3	4	5	6
1	3×3	1,00	0,50	-	-	-	-	1,00	0,50	-	-	-	-
2	4×4	2,00	2,00	0,10	-	-	-	1,90	1,25	0,13	-	-	-
3	5×5	4,10	2,27	1,10	-	-	-	3,00	1,39	0,62	-	-	-
4	6×6	6,47	3,94	3,10	0,17	-	-	4,40	2,20	1,42	0,30	-	-
5	7×7	9,72	6,25	3,35	1,70	-	-	6,00	2,82	1,35	0,65	-	-
6	8×8	13,40	9,20	5,43	4,03	0,33	-	7,90	3,80	2,20	1,60	0,45	-
7	9×9	18,00	12,70	7,90	4,21	2,21	-	10,00	4,72	2,63	1,29	0,71	-
8	10×10	23,00	17,00	11,40	6,64	4,80	0,40	12,40	6,00	3,60	2,17	1,70	0,58

TABLE 2

Type of network	NUMBER OF MESHES	m_i COEFFICIENTS FOR MAXIMUM BENDING MOMENTS ON BEAMS:						t_i COEFFICIENTS FOR MAXIMUM SHEAR FORCES ON BEAMS:					
	n×n	1	2	3	4	5	6	1	2	3	4	5	6
1	3×3	1,00	0,50	-	-	-	-	1,00	0,50	-	-	-	-
2	4×4	2,00	2,00	0,10	-	-	-	1,90	1,25	0,13	-	-	-
3	5×5	4,40	1,70	1,40	-	-	-	3,00	1,10	0,90	-	-	-
4	6×6	7,10	3,06	3,73	0,40	-	-	4,40	1,80	1,00	0,38	-	-
5	7×7	11,15	4,23	3,13	2,50	-	-	6,00	2,00	1,38	1,13	-	-
6	8×8	15,80	6,16	5,10	5,70	1,00	-	7,90	2,70	1,70	1,15	0,55	-
7	9×9	21,80	8,50	6,30	4,65	3,74	-	10,00	3,20	2,07	1,50	1,25	-
8	10×10	28,60	11,40	8,63	7,16	7,80	1,30	12,40	4,00	2,55	1,70	1,30	0,60

The next approach regards the determination of all bar efforts, using the displacement method within the ultimate limit state, applied on the trussed beam grid dimensioned previously, but considered now as a space grid system.

Two main load combinations are to be emphasized:
- the structure carrying different prestressing forces applied bidirectionally on the bottom chords, simultaneously with its dead weight;
- the previous loaded structure carrying supplementary the live load.

This analysis enables the specification both of the signs and effort values of the bars, and thus offers the opportunity to improve the bars and cables dimensioning, in order to avoid the discrepancies between the input and output data.

The third stage reffers to the accurate determination of the efforts in the whole structure improved before, under the combined vertical and horizontal forces, the analysis being performed in the service and ultimate limit states.

ACKNOWLEDGEMENTS

The authors would like to express their gratitude to the Engineers Andrei Erdely and Mircea Dascălu for their technical contribution to the building achievement in Gheorghe Gheorghiu Dej town.

REFERENCES

1. I. BORREGO, Space Grid Structures, Skeletal Frameworks and Stressed-Skin Systems, Massachusetts Institute of Technology, 1969.

2. M. MIHAILESCU, Reinforced Concrete Beam Grids for Industrial Halls, Revista Construcţiilor, nr.10, Bucureşti, 1965.

3. A. IONESCU, Hinge Supported Beam Grids of Prefabricated Prestressed Concrete for Universal Roofs, PhD Dissertation, Polytechnic Institute of Cluj-Napoca, 1975.

4. A. IONESCU and A. CĂTĂRIG, Analysis of Reticulated Beams of Reinforced and Prestressed Concrete, Studies and Researches on Applied Mechanics, Tom 36, No.1, The Academy of Romania, 1977.

5. A. IONESCU and A. CĂTĂRIG, Analysis of Plane Reticulated Beam Grids of Prestressed Concrete, Studies and Researches on Applied Mechanics, Tom 37, No.3 and 4, The Academy of Romania, 1978.

6. M. MIHAILESCU, A. IONESCU, A. CĂTĂRIG, A. ERDELY, M. DASCĂLU, S. BUSUIOC and V. RĂDOI, Prestressed Concrete Three-Dimensional Structure for the Roofs of Multi-Purpose Spaces, Construcţii, Bucureşti, No.1, 1978.

INDUSTRIALIZATION AND THE FUTURE OF SPATIAL STRUCTURES

E.CANTARELLA, G.GIANNATTASIO and M.PAGANO

Department of Civil Engineering
University of Naples, Italy

Present theoretical and experimental research activities follow completely different paths depending on whether the production process is of a "typological", a "monumental" or a "mass production" type.
Since space structures can be produced by employing technological processes which satisfy the criteria for mass production, the research carried out by Naples University leads to the unification of the three construction processes by adopting coordinated modular elements. Over time changes tend to lead in this direction. The various buildings already realized by means of steel brick masonry demostrate that we must develop industrialization processes by guiding both scientific and industrial research using these unified criteria.

1. THE INDUSTRIALIZATION OF THE BUILDING PROCESS

1.1. Introduction

Although space structures have, in fact, been in existence for the last few decades, they have only recently begun to gain ground and expand, not only when it comes to erecting great monumental works, but also as far as mass-produced, residential buildingsare concerned. As a direct result of this, an entirely new sector of study and research has also begun to develop, which aims at promoting their use by singling out the most significant parameters of their behaviour as well as the corresponding risks.
In addition to safety, space structures also permit the attainment of economic and environmental objectives. The possibility of mass producing **sound** structural typologies is a decisive factor as far as cost is concerned. At the same time, the consequent modulation offers a flexibility of form which is indispensable if we are to adjust vast building complexes to the needs of their users, not only with regard to important monuments that represent a moment of human and religious meditation, but also in the case of small residential buildings that are within everyone's reach and that could even be built by the users themselves.
It is thanks to this spontaneity in building - humanism of technique - that space structures are potentially capable of providing man with a habitat which does not give him a sense of alienation.
It is, however, essential that every step in this direction be taken making the experimental and theoretical investigations a strictly interrelated process because technicians cannot rely on the same behavioural patterns as in the case of building typologies which have existed for more than thirty years.
In the light of the above premise, the University of Naples has carried out investigations aiming at defining the overall criteria for a sound methodological approach along three lines of research concerning: i) residential buildings, ii) reticulated roofs, iii) monumental buildings. These three lines of research have shown that the research methodology and its contents mainly depend on the level of industrialization attained by the productive system.

1.2. Industrialization theory

Undoubtedly the problem of housing is one of the most important problems of today's world; but the best way of solving it has not yet been defined. Mankind is at present going through a stage of rapid evolution, not only because of the large population increase, but also since this rise has been coupled with an increasing demand for social justice which claims for everyone an appropriate place to live in.
Moreover the economic, cultural and psychological needs of modern man have also evolved in close connection with industrial growth. Ancient traditions, swept away or ignored, have not always been replaced. Nor has man always been offered an ideal habitat with which he can establish some kind of stable relationship and where he can regenerate his spiritual and physical energies, thus overcoming impediments created by the world economic structure. On the other hand the problem of the optimal allocation of technical, economic and land resources has

not yet been solved.
Meeting human needs and optimizing absolute costs are two basic and closely interconnected aspects of the problem that give a more correct meaning to the word industrialization or to other definitions, as mass production or prefabrication, that neglect the process's ultimate goal of optimizing mankind welfare.

1.2.2. Theory of technological industrialization

To study the problem of building implies studying technological industrialization which, on a working level, tends to transform building processes and typologies continuously, with the aim of adjusting production to the needs of man.
Consequently, the formulation of a general theory of industrialization requires a preliminary analysis and definition of criteria in order to establish whether a given transformation of the technological process really aims at meeting users' requirements, or, in other words, whether it does it by increasing people's welfare or by reducing social costs. However, as the sociological, political and economic aspects of the question are not particularly relevant to the technical subject of this Conference, we shall overlook them and limit ourselves to defining those criteria which are of fundamental importance in analysing and judging the technical and economic qualities of the building process. These criteria are as follows:

a) - Increment of production velocity.
b) - Increase in the degree of integrality.
c) - Reduction of pre-design content.
d) - Modular coordination.
e) - Standardization.
f) - Reduction in weight and increase in rigidity.
g) - Duration of installations efficiency.
h) - Ecological compatibility.

These criteria cannot be ranked reasonably according to their importance or priority; but in formulating them concretely for the evaluation of the building process transformation, two conditions must be kept in mind: i) through them an overall evaluation of the quality of change should be made; that is, a "system" should be formulated able to interrelate the various criteria so as to obtain an overall quality index; i.e. consequently, the application of only one of these criteria might also produce reductive results (though within allowable limits) provided that the results of the "system" as a whole are positive, see ref. 3.4.

a) - Increment of production velocity
This is the best-known and most commonly accepted of the criteria, and is usually dependent upon progress in the field of technology. The growth of automation, the introduction of processes such as moulding, injecting, diecasting and mass-production, all reduce production time.

b) - Increase in the degree of integrality
This criterion expresses the increment (as percentage) of the extension of the industrialization advantages to the entire productive process, which is induced by the transformation adopted, taking also into account the consequent percentual reduction in absolute costs.

c) - Reduction of pre-design contents
If we reduce the amount of pre-design needed for the various members used in the building we succeed in reducing the range of the family of components to an extent corresponding to the reduction of the variable parameters, which results in the complete elimination of any kind of pre-design producing only one family for each members and thus completely flexible elementary components. This criterion overshadows the present processes of reinforced concrete, steel sections and panels in prefabricated buildings which require the production of functionally complex components such as "beams", "pillars" and "floors" for which the predesign content depends on the parameters that take part in their dimensioning.

d) - Modular coordination
A spatial modular coordination of the building's components is of particular importance.
By subdividing the total volume of the building in morphologically modularized elementary volumes that can be functionally aggregated it is possible to coordinate the different production processes of single members. However modularity should be congenial with the single member and should not be imposed by rules external to the designer.

e) - Standardization
It is advisable to reduce the number of different types of memmbers in relation to the modulation of the building's volume.

f) - Reduction in weight and increase in rigidity
Although weight may, at times, be itself an advantage (in the case of thermic inertia and sound insulation, to name but two), its reduction, connected with an equivalent increase in shape rigidity, has been shown to have a positive influence upon the evolution of almost all sectors of production because of the fact that the cost of material is originally null and its increase is made only of human work.

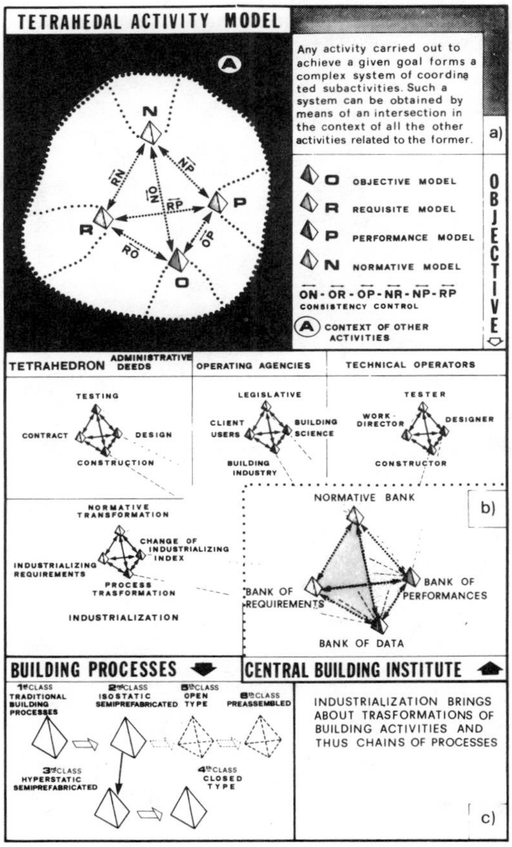

Fig. 1

g) - Duration of the efficiency of installations

An increase in the duration of the building lifetime, i.e. the period during which it manages to satisfy man's needs in an efficient way, means a reduction in its cost. Since a building is endowed with installations and appliances it is essential to be able to reach the latter through a suitable functional space so that new elements can be added or obsolete ones be replaced.

h) - Ecological compatibility

Any technological transformation of the process must be ecologically consistent with the context in which it is going to operate. Such a consistency concerns political, economic, social, sanitary acceptability. It appears thus evident that the latter criterion involves multiplicative coefficients of the overall quality index unlike all previous ones, which can be represented by additional or reductive coefficients.

2. INDUSTRIALIZATION OF HOUSING, ROOFS AND MONUMENTS, BY MEANS OF SPACE MICROSTRUCTURES

The strategy employed to study human activities is based on a logical structure (a tetrahedral one) that represents the spatial model of the activity (fig. 1.a) at any level (simple activity, building, normative, and administrative activities, ect), see Ref. 10.

Picture 1b shows various qualifications of the general model and particular cases of the building activity. It also shows, symbolically, that all specific activities can be coordinated and that in each country it may be convenient to perform a generalized control of building activities (Central Building Institutes).

Picture 1c shows six typical building processes (six classes) which are symbolically interrelated since they have been judged to be representative of the last forty years'building evolution. In the shaded area the picture represents those processes which are part of the theory but have not yet been satisfactorily implemented by society. After the first two processes the evolution has spontaneously strayed from its standards taking a course that does not satisfy the stated theory.

We propose the following denomination for these six typical processes:

1) traditional building process (fig. 2/1)
2) isostatic semiprefabricated building process (fig. 2/2)
3) hyperstatic semiprefabricated building process (fig. 2/3)
4) closed prefabricated building process (fig. 2/4)
5) open prefabricated building process (fig. 3)
6) preassembled building process (fig. 4).

We already know that the optimal solution consists in the open building process since it can make use of modular prefabricated components which are already available on the market.

The sixth (and last) class assembles modular elements to form functionally defined subsets such as walls, floors, space modules, etc, and thus construct buildings industrially.

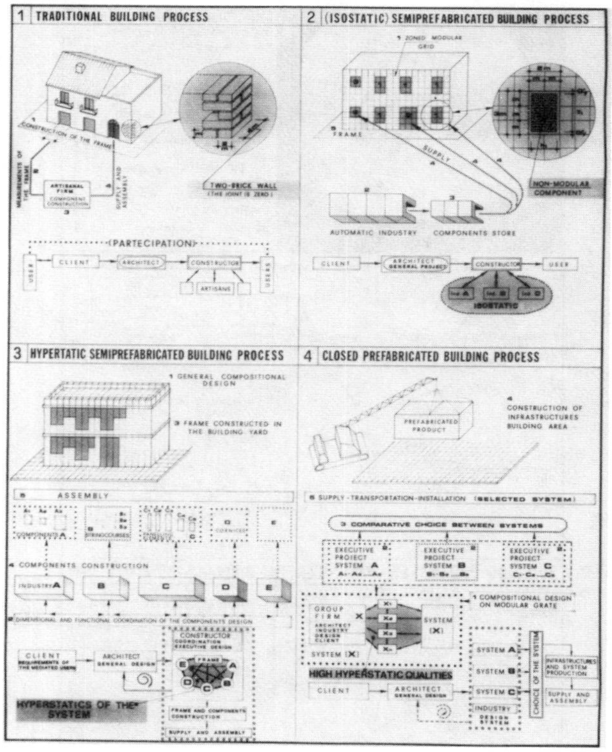

Fig 2

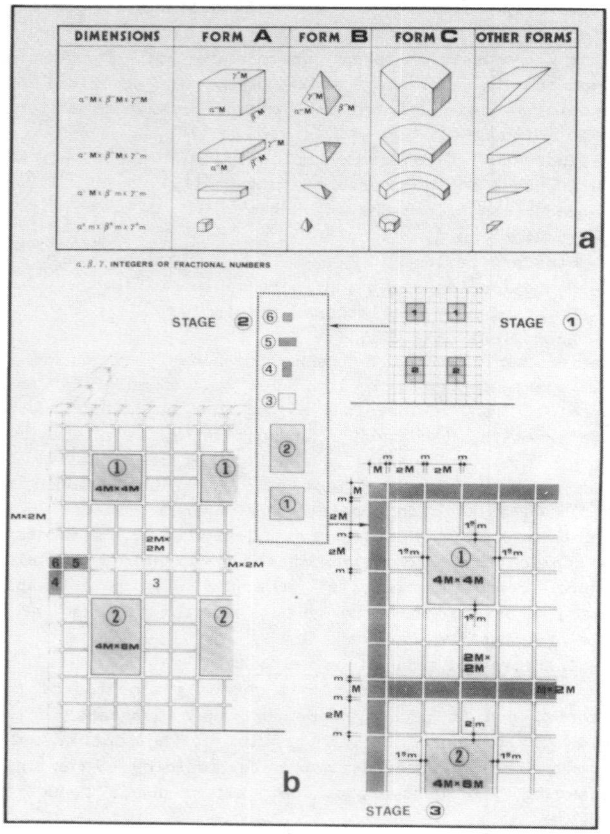

Fig3- a)Each form is dimensionally defined on the basis of two basic modules , M and m, of a matrix of coefficients. The catalog sets the corresponding levels of functional quality, the couplings and the preferential technologies.

b) By symbolically representing an application for parallelepiped elements, the three design stages are identified.

1st stage: compositional design for m = 0

2nd stage: choice of the components catalog

3rd stage: coordinated modular expansion.

Figure 3 shows in symbols the application of processes pertaining to the 5 th and 6 th classes.See Ref. 8,9. By comparison we can notice the basic differences between the modular coordination - as understood in the 2 nd and 3 rd classes - and that which is consistent with the industrialization theory.

It should be pointed out that if the present construction processes (buildings having a metallic or reinforced concrete frame) are evaluated in a critical way by adopting the above set of criteria, it results quite evident that to aim at industrializing them by means of their rationalization is not only quite difficult but almost impossible.

As regards the typology of steel-framed buildings it is immediately evident that there exists a major industry, but it only produces beams and columns in a very advanced way. Hence, the other sub-processes deteriorate more and more and the possibility cannot be excluded that the rationalization of the process as a whole may require the elimination of the structural elements' subprocess which at present prevails. (Lack of integrality)

This is also true of reinforced concrete buildings, where colums and beams, being highly pre-designed elements, cannot be produced by strictly sectoral industries. The latter would prevent the construction industry from placing itself on a curve of decreasing costs also because of the rapid industrial obsolescence that the other subprocesses would soon undergo. (High level of predesign)

At any rate failure is to be ascribed to the neglect of the basic goal, i.e. meeting the most deeply rooted human needs.

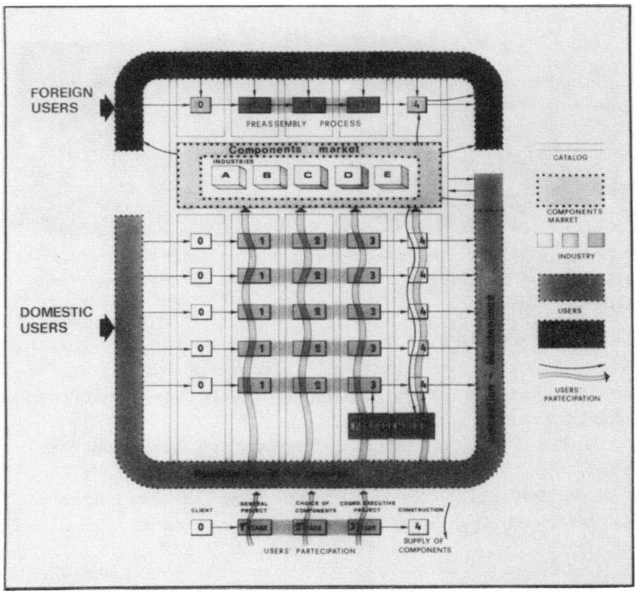

Fig. 4 - Symbolic representation of the strategy proposed for the take--off of an open building process.

2.4. Conceptual experimentation

CESUN (Centro Studi per l'Edilizia dell'Università di Napoli) has carried out a conceptual experimentation using a techical approach and only relying on the analysis made so far.(Fig.5). The aim of this experimentation is to establish whether, as far as technological industrialization is concerned, a field of research exists capable of pro viding positive solutions in the light of the above defined criteria.

Such a type of building implies a strict compliance with the volumetric modulation using nothing but steel bricks or their multiples or submultiples. (normalization) The consequent modularity of the surfaces permits the completion of the building by using four types of panels which are also modular and which are used inside and outside walls, floors and ceilings in modular coordination. (integrality)

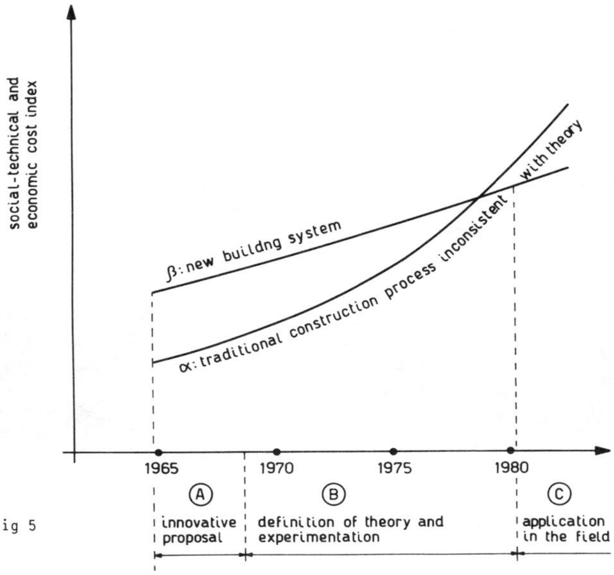

Fig 5

The mass-production of these panels is also possible.They are standardized in such a way as to be suitable for the construction of a large variety of buildings independently of their specific architectural morphology (standardization). Bricks have no static, preset functions and thus make the traditional models of 'beams' and 'pillars' obsolete.

Hence the words 'element' or 'brick'(fig.6)finally acquire their correct meaning and result in a remarkable reduction of predesign time for the very fact that they don't perform any preset function.(low content of predesign)

The universally accepted architectural requirements of a strict modularity that however does not restrain the flexibility of the buildings 'morphology is thus fulfilled and at the same time technologically justified (freedom

Fig 6 - Steel brick: it is formed by ten subcomponents made of cold-pressed sheet steel. The assembling of the bricks into walls and floors is carried out by riveting so as to pre-tighten the sheet steel.

of design). (modular coordination).
The reduction of the building weight (i.e. 60 kgs each cubic meter of dead load) is also the result of the rational use of the structural rigidity - shaped material so that lightness is coupled with an increase in rigidity (as shown in fig. 7.8) during transport and lifting (reduction in weight and increase in rigidity).
The integral rationalization of the building production is attained by adopting mobile internal partitions and by installing appliances and fixtures in the functional spaces of walls and floors - Internal installations may run in any direction and undergo any change thus garanteeing the continuous adjustment of the building to the evolving needs of the user and giving it a longer life period. Thus the life time of the various appliances is the same as that of the building (Duration of the installations efficiency,'functional space') (Figures 9 and 10).
Taking for granted that modern residental buildings are conceived to make technical installations increasingly rational, the latter aspect cannot be overlooked.

Fig. 7-8 - An indirect demonstration of the reduction in weight and the increase in rigidity is given by the possibility of overturning the structure without inconveniences of any kind.

Another important feature concerning such a construction process is its being indipendent of the type of material used for the bricks, provided that it is capable of taking the right shape, and independent of either the size of the building or its internal spaces as well as of its functional characteristics. This aspect is consistent with the aim of adopting processes which allow a high degree of industrialization and are not at the same time consistent with the existing housing context as well as with the production world and do not require the predetermination of their application field or the choice of preferential industrial actvities.

Fig.9-10 - The functional space is pervasive of the whole building; besides it allows to make use of the alternative energy system without any additional cost.

Moreover this new structural system leaves wide room for the architect's creativity and is completely indifferent as to the nature of materials employed.

3. Goal and Instruments of Industrialization

The theory previously formulated confirms that the development of the building activity requires the corresponding technology to fulfill a system of generally defined requirements.
From a theorical standpoint it could be assumed that any present building technology should unavoidably transform itself to meet the above mentioned system of requirements or would be necessarily eliminated.

Fig. 11-12 - Steel bricks assembled to make up walls floors.

The different approaches of research on the building sector from vaulted roofs to monumental buildings that have being developed in the last decades have relied on the fact that there are different theoretical and experimental instruments adjusted to the targets of specific construction technologies and to their morphological and pratical features. We have already dealt (see Ref. 7) with the developments carried out in the framework of the different approaches. However the need to logically pursue one single goal makes us think that both technology and morphology must undoubtedly change; thus it is only a matter of implementation time and of operator's will. We have predicted (see Ref. 2-3-4) that as soon as designers become fully aware of the validity of this industrialization theory the foreseen transformation of building processes will accelerate and will speedly approach the goal of the complete fulfillment of human needs among which the economic aspect has always been very relevant. In these last two years we have found

Fig 13-14 - A building that in two phases of this construction shows its low contents of pre-design and aestetic qualities not different from a traditional house.

true this prediction in the housing sector and in the roofing sector; in that of monumental building the industrialization tends toward a theoretical uniqueness of targets too.
At the present stage such concepts have been concretely tested.
In (see Ref. 11) we felt the need of preliminarily considering the general context of human activities and

Fig 15-16 - The past and the present: the experimental house is completed and new buildings are being erecting.

the evolution of building processes over time, to understand that spatial structures would be all erected in future with mass producted modular elements.
Recalling a conceptual experimentation that we had begun in 1968 we reported that it was just over, that the system had been placed in site in 1982 and the fact that four schools had been built by this new system. In the second half of 1983 three other buildings were erected and the experimental house was completed. We can now refer to other performances proving the flexibility of this new kind of building which is at the same time valid for residential housing (fig. 11-12-13-14-15-16),industrial tanks, roofing and domes (fig 17-18-19)and recovery of floors belonging to mansory buildings(fig 20-21).

4. Morphological and technological standardization.

The review of the studies carried out in the field of space structures shows that they have been treated as different lines of research, namely : buildings of a "typological", "monumental" or "mass production" types. Over time it has appeared more and more evident that one has to consider the economic aspect in its broad sense. Such an approach implies that, at a variable rate depending on the impact of social changes upon construction techniques, all building systems tend to fulfill the optimization conditions of industrialization, see Ref. 11.

REFERENCES

1. C FUNEL, M PAGANO, N PALUMBO, A SBRIZIOLO, Mattone reticolare d'acciaio, Atti CESUN, vol. I + 5, Napoli, 1968-70.
2. M PAGANO, La sperimentazione quale fattore della industrializzazione dell'edilizia, Prefabbricare, n°3, maggio-giugno 1969.
3. M PAGANO, Steel brick buildings, IX Congresso AIPC, Amsterdam maggio 1972.
4. A CAVALLO, C FORTE, F GIACCHETTI, M PAGANO, Industrializzazione dell'edilizia - Indice globale di qualità, PRO.I.E. CESUN, Napoli, gennaio 1973.
5. L COSENZA, G ABBATE, G COSENZA, R DE STEFANO, Ricerca per una edilizia industrializzata capace di produrre alloggi con incidenza di locazione non eccedenti il 15% del reddito medio pro-capite, Dottrinari, 1974.
6. L COSENZA Storia dell'abitazione, Vangelista, luglio 1974.
7. E DE NARDO, A GILIBERTI, M PAGANO, Strutture spaziali-Sperimentazione e teoria a diversi livelli di industrializzazione, Costruzioni Metalliche, n°2, 1979.
8. M PAGANO, G ABBATE, Teoria del Giunto nullo, Prefabbricare n°4, 1980.
9. G PARODI, discussione all'articolo "Teoria del Giunto nullo" Prefabbricare, n°3, 1981.
10. M PAGANO, Industrializzazione e scienza dell'edilizia, Casabella n° 474/475, nov. - dic. 1981.
11. M PAGANO, P LENZA, Reticulated vaults: structural analysis and industrialization processes, Course on the Analysis, design and construction of braced barrel vaults, University of Surrey, Vol. II, Surrey, sett. 1983.

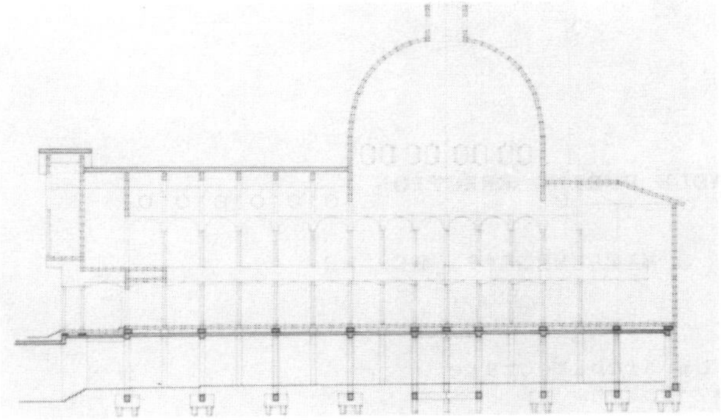

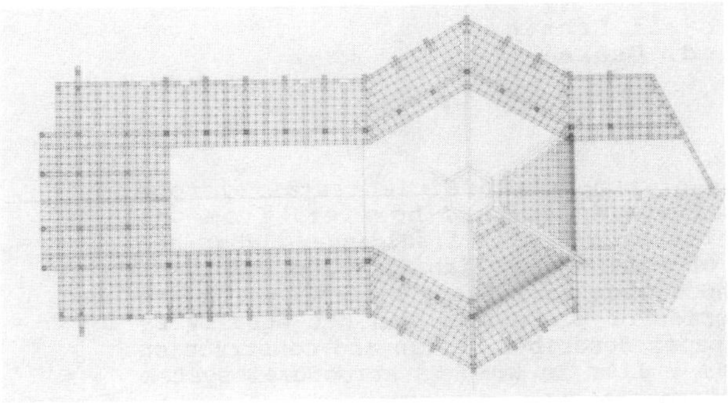

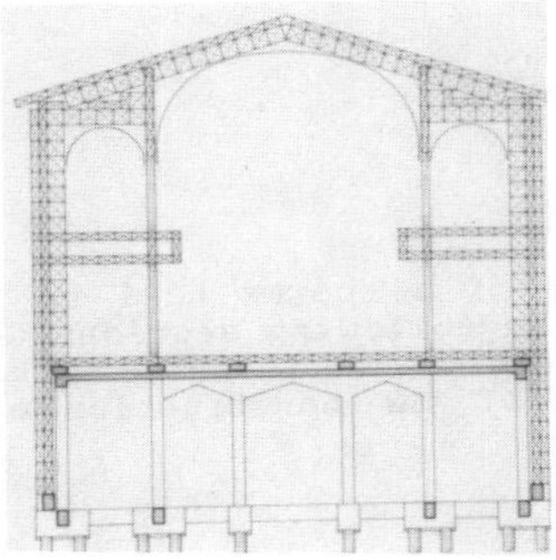

Fig.17-18-19 - The future: the design of a church. The figure shows the flexibility of the new system; infact the dome is also realized by means of the steel bricks.

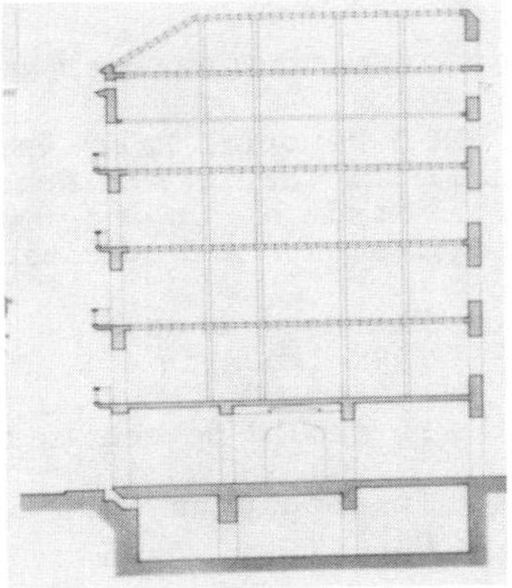

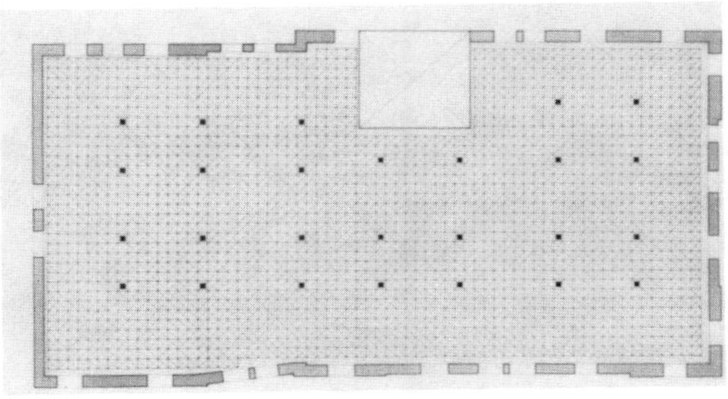

Fig. 20 - 21 - The future : the recovery of floors belonging to a mansonry building.

A DOMICAL SPACE FRAME FOLDABLE DURING ERECTION

M KAWAGUCHI*, BE, DR ENG and S MITSUMUNE**, BE, ME

* Professor, Department of Architecture, Hosei University, Tokyo

** Chief Architect, Project Section Showa Sekkei Co Ltd, Osaka

A domical space frame, once completed, is one of the most efficient spatial roof structures to cover a wide area. It is not always efficient, however, from the viewpoint of construction, because it requires big amount of scaffoldings, labor and time and often encounters difficulties in terms of accuracy, reliability and safety of work during its erection. Modern erecting methods such as lifting systems which are very often adopted in erection of double-layer grids of plate type can not equally be applied to a domical space frame. The present paper describes design and construction of a sports hall having an oval plan of about 70m x 110m in which a structural system for a domical space frame that can be efficiently constructed is pursued.

INTRODUCTION

Because of their advantages such as lightness, strength, rigidity and attractive appearance space frames have achieved rapid acceptance all over the world. Rational methods of construction to be suitably applied to them have also been pursued along with their development. It may be said that a series of rational construction methods which are generally applicable to one type of space frames, namely, double-layer grids of planar type, has already been established. They are lifting methods using cranes, jacks and other hoisting systems which lift the plate-like space frames previously assembled on the ground to their design heights. However, these methods can not suitably be applied to the other type of space frames, vault- and dome-types, unless they are very flat. R Buckminster Fuller tried to solve this problem which he encountered when he built a series of his geodesic domes. For construction of one of his domes in Honolulu in 1957 he adopted a system in which a temporary tower was erected at the center of the dome from top of which concentrically assembled part of the dome was hung by means of wire ropes (Ref 1). As assembly of the dome proceeded the dome was gradually lifted, enabling the assembling work to be done always on the ground. He also adopted another method when he built a huge dome of 117m in diameter at Wood River, U S A, in 1959, where the assembled part of the dome was raised on a balloon-like enclosure (Ref 2). Some other cases have also been reported where different lifting methods have been applied to different particular domes. However, none of the above methods have become popular unlike many lifting methods which became widely used to raise space frames of plate type.

A structural system called 'Pantadome System' which had been developed by the senior author for a more rational construction of domical space frames was applied to the structure of a sports hall to be completed in Kobe. In the following sections the structural as well as architectural features of the sports hall and outline of the construction are presented.

SPECIAL FEATURES OF THE HALL

'WORLD Memorial Hall' was projected as a multipurpose hall for the City of Kobe which could be used for the Universiade to be held in 1985. Functional requirements and conditions of the hall are:

1) The dome should be big enough to accommodate a 160m track in it.

2) It should be seated for 10,000.

3) It should be able to be used as an all purpose hall.

Fig 1 Completed Sports Hall (Prespective)

4) The height of the ceiling should be not less than 24m, so that all kinds of yachts could be placed for exhibition.

5) The area of the building site is limited (90m x 120m).

Fundamental design policies adopted are as follows:

1) Energy saving should be attempted by means of natural ventilation making use of the height of the building and the wind effect.

2) Energy saving by means of sufficient natural light through skylights should be achieved.

3) Access of the racers as well as the spectators should be as clear as possible.

4) The form of the building should be comprehensible structurally as well as architecturally.

5) The design should incorporate technical rationality including that for realization process.

The shape of the building which was designed on the basis of the above conditions and policies eventually took the form of an elongated dome covering an oval plan of some 70m x 110m. The total shape of the dome was defined by a surface which was obtained by rotating the peripheral curve of the plan along its major axis. The cross section of the dome is a semicircle having its center 1m above the ground level.

The shape of the dome such as given in the above is generally apt to produce acoustic difficulties. The problem was solved by providing inside of the dome with a series of acoustic projections in the form of inverse pyramids which consisted of glass wool panels filling triangles constituted by an upper chord and two web members each of the space frame.

STRUCTURAL CONSTITUTION OF THE DOME

The total shape of the dome consists of two one-quarter spheres of 34m in radius with a cylindrical vault of 40.8m in length in between. The upper part of the dome is constituted by a double-layer grid space frame of 1.5m in depth, while the lower part which has many openings mainly for windows consists of rigid frames. The standard size of the grids of the space frame is 2.5m x 2.5m. The members of the space frame numbering more than 12,000 are steel tubes (standard diameters: 101.6mm for chord and 76.3mm for web members, respectively) and some 3,000 connectors are ball joints of steel plates, pressed and welded with diaphragms (standard diameters: 216.3mm and 267.4mm). Connection between the members and ball joints is all by welding.

As the building site was very close to the harbor, the members were assembled in the factory into big segments (eg 4m x 24m), and shipped to the site. Steel tubes were cut all by automated machines and nearly 90% of welding work was done in the factory by means of semi-automated welders. The total weight of the dome is 1,680t, of which steel weights 760t.

ANALYSIS

A quarter of the dome was analyzed for different loadings by means of displacement method, regarding all joints of the space frame as hinged (Fig 4). The dead load of the dome is 195kg/m² (steel: 45kg/m², finishing: 65kg/m², catwalk: 85kg/m²) at the top and 150kg/m² (steel: 100kg/m², finishing: 50kg/m²) at the skirt.

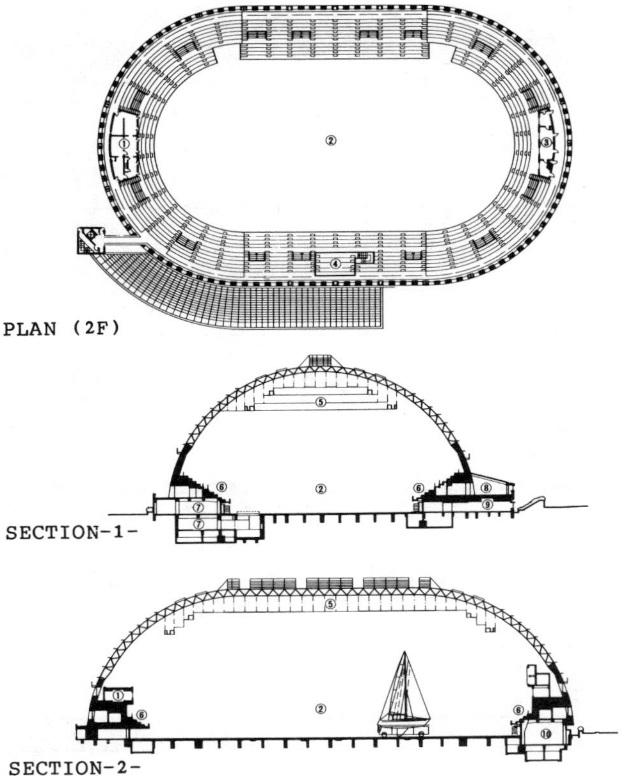

(1) machine rooms (2) arena (3) storage
(4) royal box (5) catwalk (6) grandstand
(7) apparatus room (8) deck (9) restaurant
(10) air-conditioning

Fig 2 Plan and Sections of the Sports Hall

Fig 3 Pyramidal Acoustic Projections

Wind load gives the most important effects on design of the dome. Velocity pressure of q=300kg/m² which corresponds to the height of 40m above the ground is adopted. Coefficients of wind pressure was obtained by means of wind tunnel tests, conducted under the direction of Prof Ishizaki of Kyoto University, on a 1/250 scaled model. The model was tested under uniform and disturbed air streams, but the results differed little between the two cases. The coefficients of wind pressure obtained for a few typical wind directions are shown in Fig 5 (disturbed air stream).

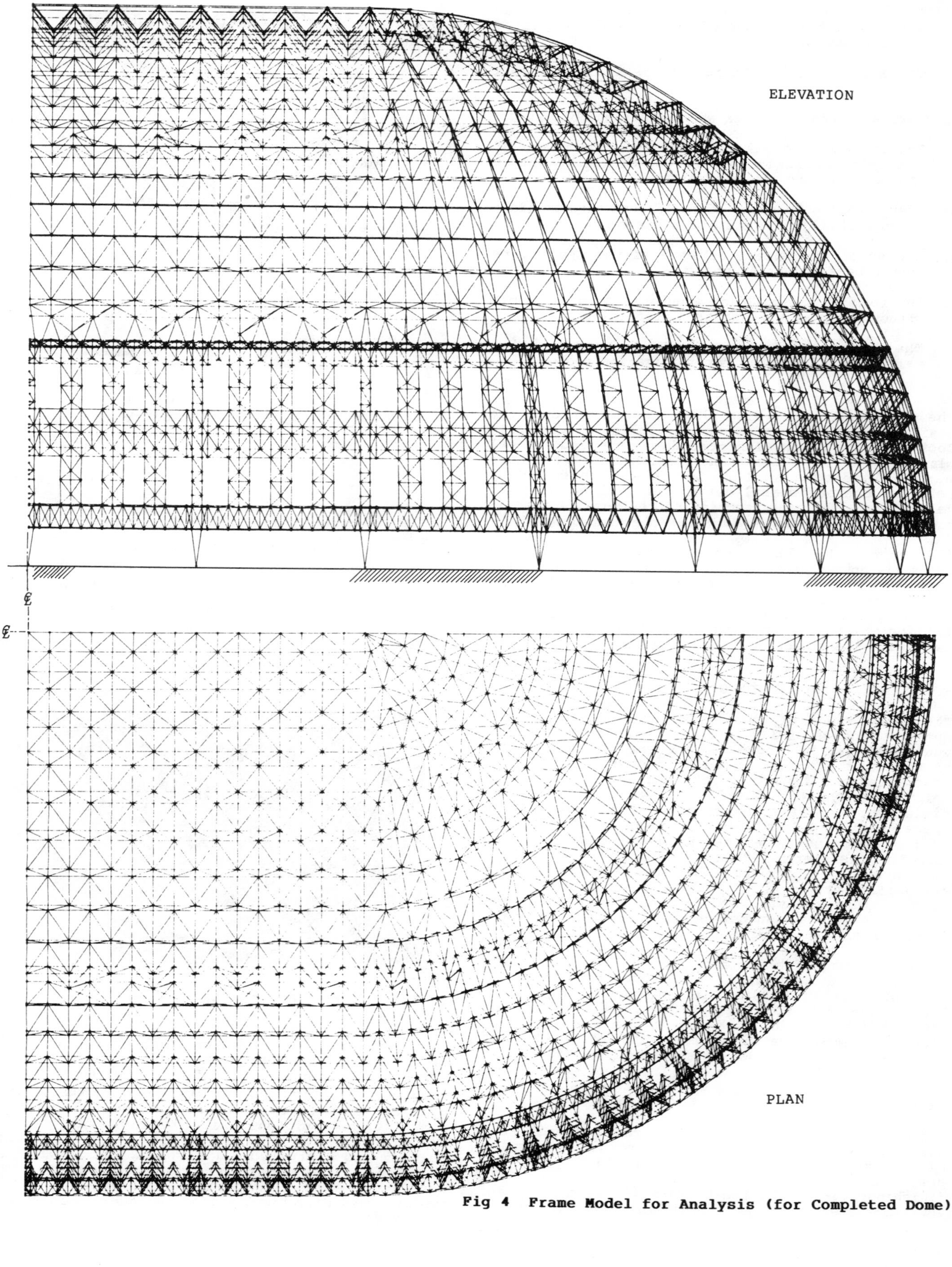

Fig 4 Frame Model for Analysis (for Completed Dome)

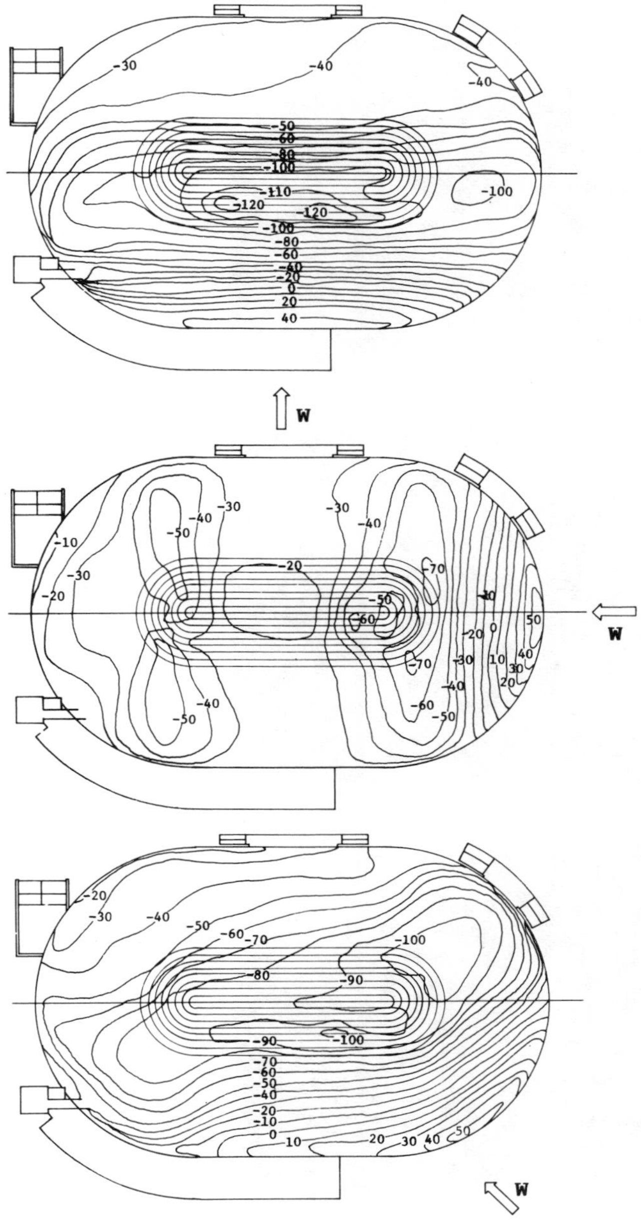

Fig 5 Wind Pressure Distributions (figures in 10^{-2})

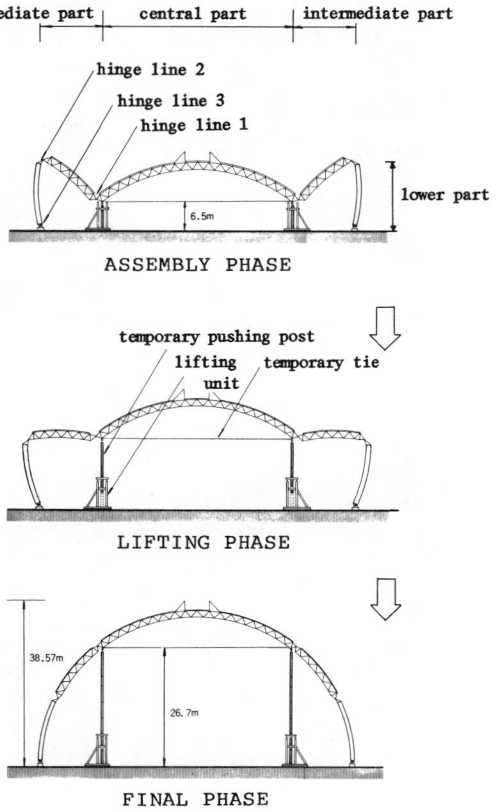

Fig 6 Pantadome Lifting Phases

The effect of earthquakes on the dome calculated for a horizontal acceleration of 0.2g is something like one third of that of the wind loads. The effect of temperature changes is taken into account for ±20°C.

As the structural system of the dome during erection is different from that after completion, and as it changes its spatial geometry with the process of erection, structural systems corresponding to a few typical erection phases were selected and analyzed for dead load, wind and earthquake.

STRUCTURAL SYSTEM SUITED FOR ERECTION

The most special feature of the present dome is its structural system which incorporates suitability for its erection. In other word the structure itself has been so designed that it will be erected with ease and sureness. The system which was developed by the senior author was named 'Pantadome System' and its cross-sectional principle is shown in Fig 6. A Pantadome has 3 hinge-lines on its surface along which it can be 'folded'. It is assembled on the ground in a folded shape, it is then raised to get its final shape as a dome and the shape is finally fixed.

This system can be applied to variety of structures in domical shapes and has several constructional advantages over the domes built by conventional methods as explained later. Hinges adopted in this system are not special ones, but are all simple hinges of single-axis rotation. Three kinds of hinges operating in their positions are shown in Fig 8.

Various lifting methods can be used to raise the dome. For the present dome 'Push-up' system with which the contractor, Takenaka Komuten Co Ltd, had a good experience was adopted. The mechanism of the lifting system applied to the present dome may be understood by Fig 9. Two parallel jacks with a capacity of 50t each pull up a temporary post which pushes up the dome. As the lifting work proceeds the post goes up and it is fed by sections at the bottom and extended by means of high strength bolts.

Erection of the dome was carried out in the following sequence:

1) The central part of the dome is assembled on the scaffolding near the ground level.

2) 18 sets of lifting units are placed under the periphery of the central dome.

ASSEMBLY PHASE

LIFTING PHASE-1

LIFTING PHASE-2

FINAL PHASE

Fig 7 Lifting Phases in Site

HINGE 1

HINGE 2

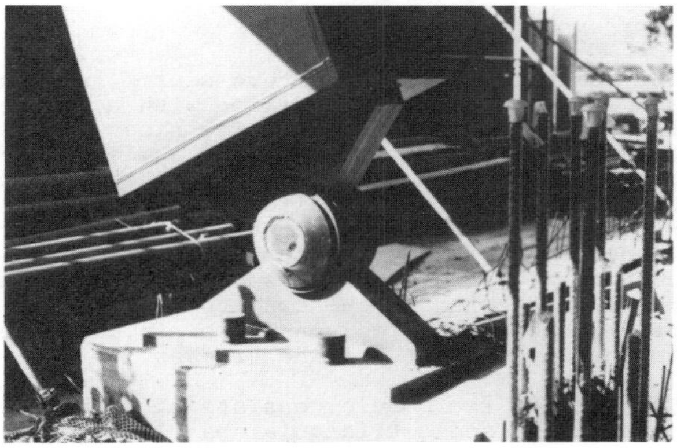

HINGE 3

Fig 8 Hinges in Operation

3) Temporary ties are placed for the cylindrical part of the dome.

4) The intermediate and lower parts of the dome are assembled.

5) Finishing of the central and intermediate parts is done. The central part is completely finished, externally and internally, including equipments inside.

6) Temporary ties are tightened to take thrusts of the cylindrical part and to keep its geometry.

7) Lifting commences. The dome is raised for 20m in 6 stages. The temporary posts are fed at every stage.

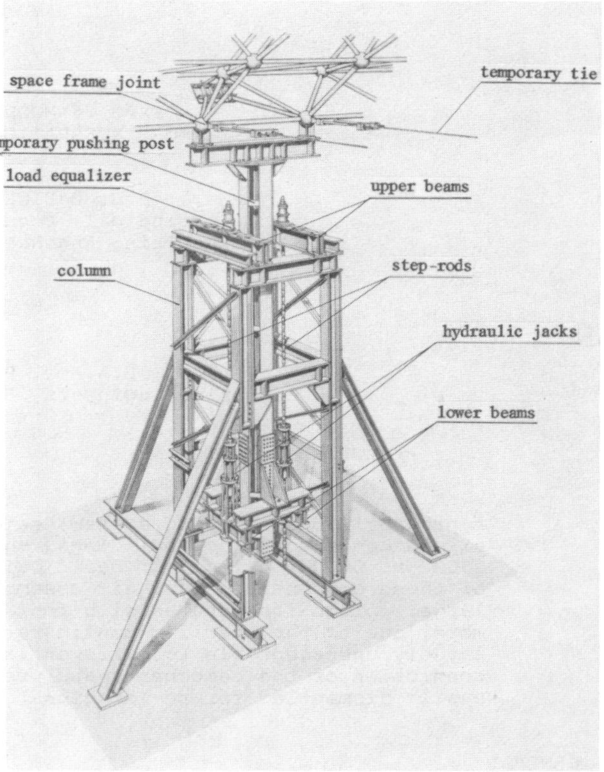

Fig 9 Lifting Unit

8) After the dome reaches the final shape, the necessary members are added to stabilize the dome.

9) The temporary posts and ties are removed.

10) Finishing work continues.

During the lifting work the following measurements were taken to control the work itself and to grasp the stress conditions in the space frame.

measurements	numbers	instruments
horizontal displacement	20	EDM
vertical displacement	4	stroke meter
loading	9	manometer
pressure of hydraulic units	2	ditto
stresses	281	strain gauge
deflections	33	auto-level

Measured values of the lifting forces at different phases, stresses in the space frame during and after the erection and the deformation of the dome showed good coincidences with the results of calculation.

REFERENCES

1. R Buckminster Fuller, Synergetics, Macmillan, New York, 1975.

2. R C Ulm and R L Heathcote, Dome Built From Top Down, Civil Engineering, December 1959.

EXPO 86 MODULAR STRUCTURAL SYSTEM FOR INTERNATIONAL EXHIBITION PAVILIONS

B.B. BABICKI, D.Sc., P.Eng., RCA
Principal, Bogue Babicki Associates Inc.
Consulting Engineers, Vancouver, B.C., CANADA

and

P. FAST, B.A.Sc., P.Eng. and H. IREDALE, B.Sc.,
Design Engineers, Bogue Babicki Associates Inc.

A prefabricated building system designed for exhibition pavilions. The system allows construction of buildings, square or rectangular in plan, of areas from $250M^2$ to $5000M^2$ using four basic building components. Due to the specific shapes of these components and their assembly configuration, these buildings can have large, column-free spaces with roof spans ranging from 15M to 51M long. Furthermore, due to the specific configuration of structural components, buildings are largely unsusceptible to differential settlement of the foundation. The bolted connection of the components and prefabricated foundation system make the building easily dismantled for re-location for other uses.

GENERAL

In 1986, Vancouver, British Columbia, Canada, will host a Special Category World Exhibition, Expo 86.

The exhibition pavilions will be provided and leased to all International Participants by a Crown Corporation of the Provincial Government.

Some 35 Countries are expected to participate.

The exhibition site is 53 hectares of water front, adjoining the downtown area and is largely reclaimed industrial land which has very adverse soil conditions.

DEVELOPMENT OF THE MODULAR SYSTEM

The need for 35 individual pavilions, totaling approximately 50,000 square meters of exhibition area led to the development of a modular structural system. Fast construction and easy dismantling were among the design criteria. The temporary nature of the structures did not justify the use of pile foundations as would normally be necessary on this site. The structural system had to accomodate potentially large differential settlement.

The system allows construction of pavilions ranging in size from 250 to 5000 square meters of nominal area. It uses four basic structural components: triangular steel trusses, steel pipe columns, triangular wood roof panels, and rectangular wood wall panels.

From these typical elements, the system allows the assembly of pavilions of the basic module of 250 square meters in area which can be combined to build pavilions of up to 2600 square meters of clear span space and up to 4800 square meters with one centrally located cable stay pylon.

BASIC MODULE

The basic module is 15 meters square in plan and forms the smallest pavilion with a nominal area of 250 square meters. The sides of the square are formed by four right angle triangular trusses. Each pair of trusses is placed back-to-back at opposite corners of the square. The square formed by the trusses is connected to four corner columns. Two high columns at opposite corners become the corner posts of the "back-to-back" trusses and the other two, shorter columns, are pin connected to the acute angles of the trusses.

The roof of the module consists of eight identical right angle triangular panels supported on the perimeter trusses and in the middle, suspended from cables spanning between two opposite high corners. The wall panels, with mullions, are suspended from the perimeter roof beams and are laterally supported at ground level by grade beams.

The main advantages of the structural system of the basic modules are:

1. arrangements of the structural elements allow the formation of column-free, multi-modular assemblies,
2. low sensitivity to differential settlement of the supports, and
3. adaptability to different roof and wall configurations.

MODULAR ASSEMIES

The modules are assembled in rectangular and square formations having a 3 meter wide interstice between them. The following assemblies were selected as typical, column-free pavilions for analysis and construction:

250M	Basic Module	usable area	$225M^2$
1000M	4 Module Assembly	usable area	$1000M^2$
1500M	6 Module Assembly	usable area	$1680M^2$
2500M	9 Module Assembly	usable area	$2600M^2$
5000M	16 Module Assembly	usable area	$4800M^2$

The schematic configurations of structural components to form these typical assemblies are shown below.

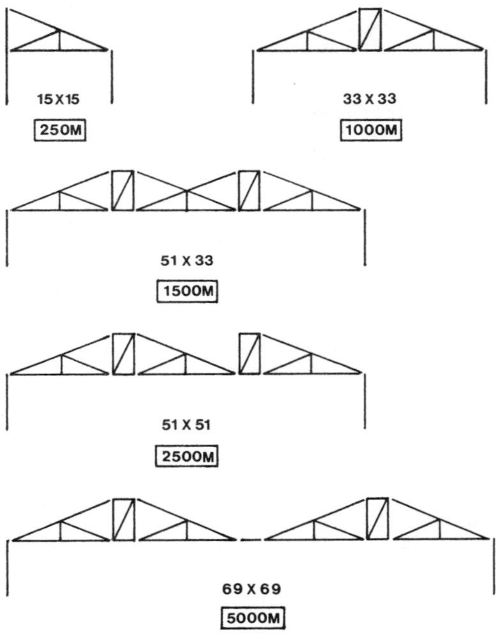

All pavilions with space requirements between those of typical pavilions are formed by adding the required number of free-standing modules to the typical assemblies.

1000M assembly of four modules creates a pavilion, 33 meters square in plan, of nominally 1000 square meters of column free floor area. This is obtained by placing four modules in a square arrangement with a 3 meter interstice between them. The four columns at the centre of the big square are removed and the triangular trusses are connected by a "Z" component thus creating a two-way grid of trusses spanning the length of two modules.

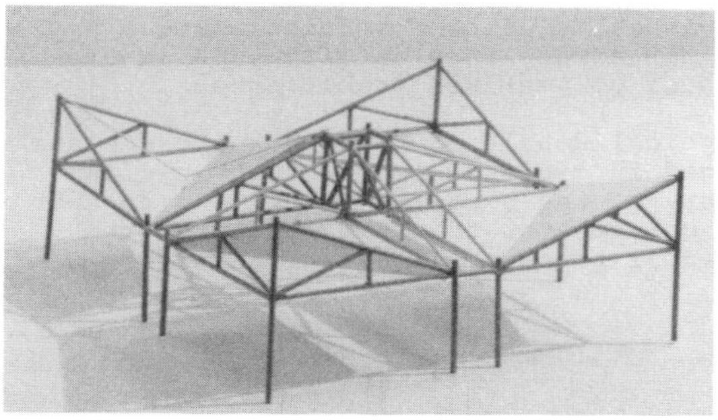

Alternatively, four middle cut-off columns which are connected by Z elements become prefabricated space elements, called "middle cubes", which occur as a key element in all assemblies consisting of four or more modules.

The double truss arrangement created by 3 meter interstices helps to stabilize the unbraced long span trusses. It should be noted that raised opposite corners of the basic module created by back-to-back triangular trusses not only provide the anchor points for the roof supporting cable and make possible the doubling of the span of the trusses in a four module assembly, but also provides supports for alternative raised roof arrangements where the edge of the roof panels follows the top chord of the triangular trusses. In addition, this particular truss arrangement makes possible the hyperbolic-parabolic fabric roof alternative.

1500M assembly of six modules creates a building, rectangular in plan, 33 meters by 51 meters with a usable area of 1680 square meters of column free space.

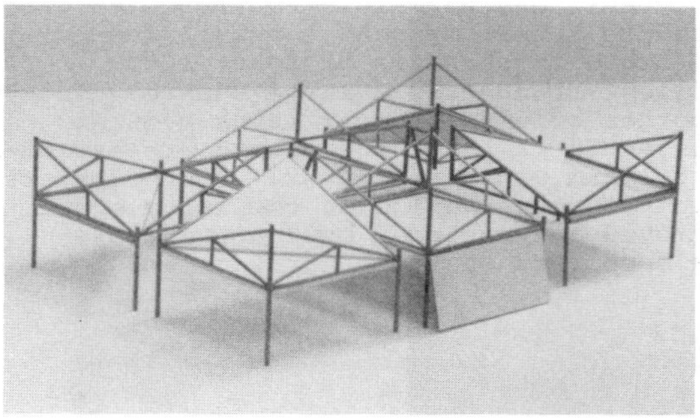

Arrangement of the modules and the interconnections are similar to those of the four module assembly except that in order to conserve symmetry of the assembly about both axes, it was necessary to split the inner double truss of the middle module and reverse its connection so that one truss will have a pinned connection to the adjacent left module and the other to the adjacent right module.

2500M assembly of nine modules makes a pavilion 51M square in plan with 2600 square meters of column-free usable area.

The typical double back-to-back trusses are arranged in a closed loop forming cyclic symmetry of the system which allows to span distances greater than the length of the individual trusses. See fig. below.

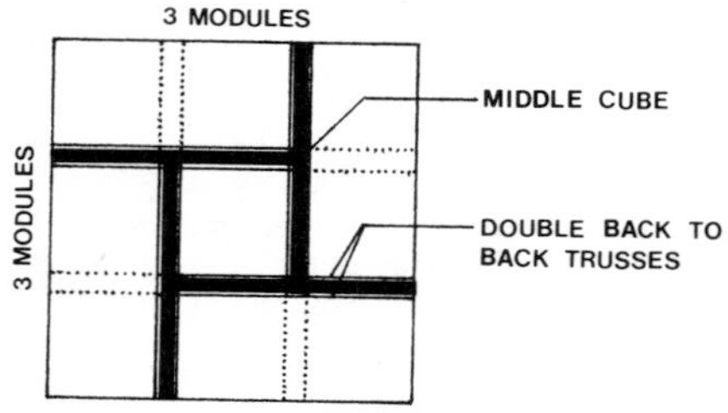

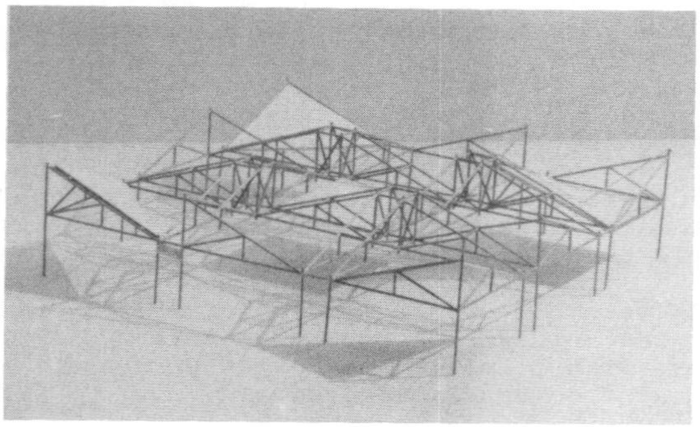

5000M assembly of sixteen modules form the largest pavilion, 69 meters square which gives 4800 square meters of usable area. The pavilion actually consists of four, 1000M assemblies arranged in a large square. The four corner columns of four 1000M meeting at the centre of the large square are extended up to form a central tower. Four clusters of columns occurring inside the large square are removed and replaced by four sets of cables supported on the central tower.

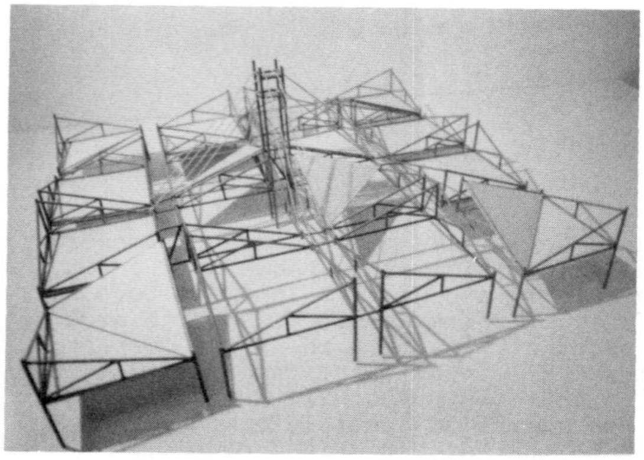

ANALYSES OF THE MODULAR SYSTEM

Each assembly was modelled and analyzed as a complete, rigid, three dimensional, moment resisting space frame using the GTSTRUDL structural analysis program. The purpose of this large scale analyses was to determine:

1. the load distribution among the interacting roof trusses organized in a two way grid system,
2. the effectiveness of this truss system as a horizontal frame for:
   a. transferring wind loads to the laterally rigid side frames, and
   b. resisting the torsional effect of unbalanced wind loads.
3. the forces caused by differential settlement.

LOADS

The structure was analyzed for a combined snow and suspended exhibit load of 1.9 kN/M^2, and a lateral wind load of 0.55 kN/M^2. Consideration was given to the drag effects of three different roof configurations permitted by the design, comprising flat and raised options. For this purpose, the results of wind tunnel tests on similar low-rise buildings were reviewed. Allowance was made for the torsional effect of an unbalanced wind load by considering a wall end zone subjected to a higher wind pressure. The governing shape/gust factors were found to be 1.95 in the end zone and 1.3 over the remainder of the wall surface area.

An ivestigation of the soil conditions in the exhibition site revealed large variations in soil types and consolidation and predicted that the pad foundations may settle between 50 and 150mm. To account for this, two differential settlements were analyzed:

1. a 100mm differential settlement over 15M,
2. a 50mm differential settlement over 3M.

Vancouver is also in a seismically active zone with a ground acceleration ratio of .08. However, a low material/damping K factor for this type of steel ductile, moment resisting frame, combined with a relatively light dead weight, produced calculated lateral seizmic loads of about 30% of the wind load. Consequently, wind governs the design.

The following is a review of the analyses results.

THE BASIC 250M MODULE

Vertical loads govern the design of the steel trusses of the basic modules, which are loaded at their midpoints by the roof and wall panels and at the two high corner points by the diagonal cable which supports the centre of the roof. The cable forces tend to deform the modules in the horizontal plane into a diamond shape. This effect is resisted by a diagonal tension rod.

Wind loads govern the design of the four columns. The column truss assembly works as a three pin frame against lateral loads, the two foundation points are pinned owing to the soil softness, and the low truss end to the column connection has very low moment capacity.

Consequently, the longer column, connecting to the higher end of the truss takes nearly all the lateral forces. A computer drawn geometry of the basic module is shown below.

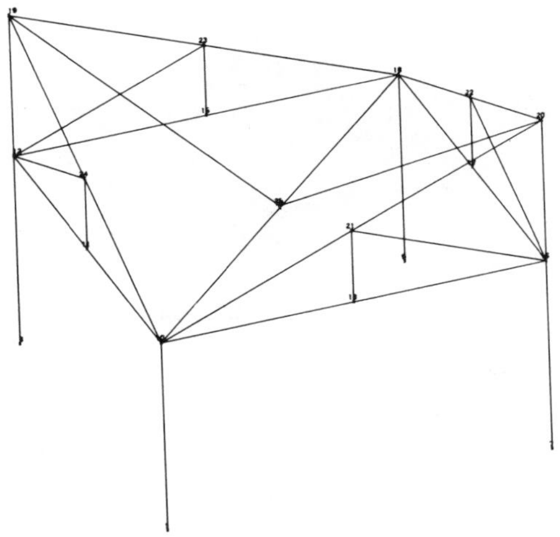

Differential settlements at any one of the four foundations produce small forces in the structure since the truss rotates about its flexible low point as one of its supporting columns settles.

ASSEMBLIES OF 1000M, 1500M, and 5000M

Analysis indicated that it was economically justified that all components of the basic module and of the assemblies of four, six, and sixteen modules be fully interchangeable.

It was found that the capacity of the perimeter truss of the basic module is sufficient for the role in the composite truss arrangement of four module assembly where the span is more than double that in the basic module.

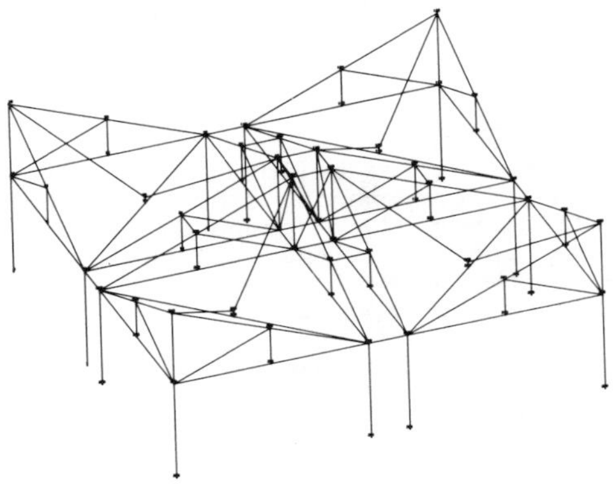

An assembly of six modules could be conceptually imagined as two overlapping four module assemblies.

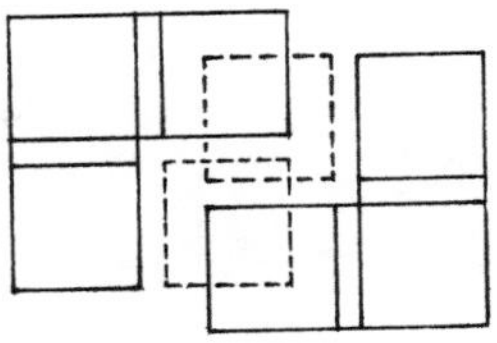

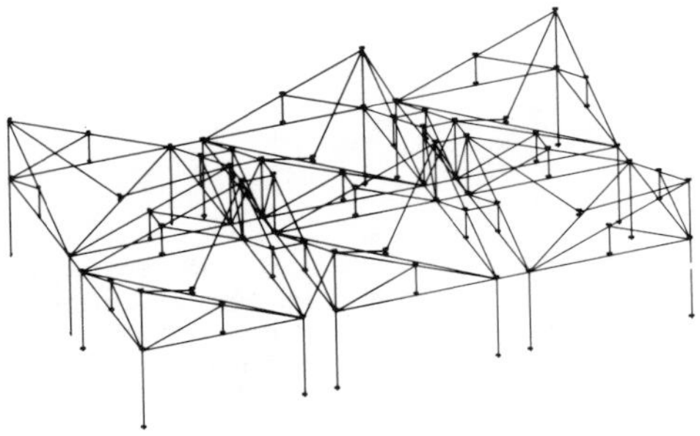

The forces found in this assembly conform in general to those in the four module assembly.

The sixteen module assembly was analyzed as one space frame in its entirety.

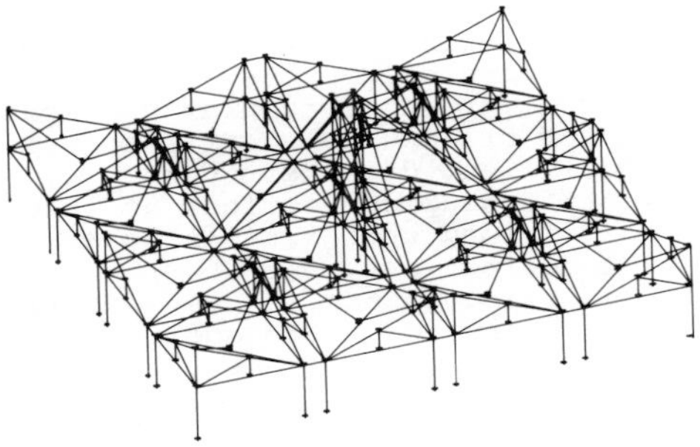

This pavilion consists of four, four module assemblies.

In order to retain the magnitude of forces in the composite trusses similar to those in the four module assembly, the vertical deformation of the joints supported by the cables had to be adjusted by the selection of cables of appropriate size.

ASSEMBLY OF 2500M

The nine module assembly has the longest spans in the system. Due to its cyclic symmetry, all composite trusses are equally loaded but the magnitude of forces in them are much greater than those in the four module assemblies.

Therefore, for economical reasons, they are not interchangeable with typical trusses and are built of heavier sections but retain the same outside diameter as the typical trusses.

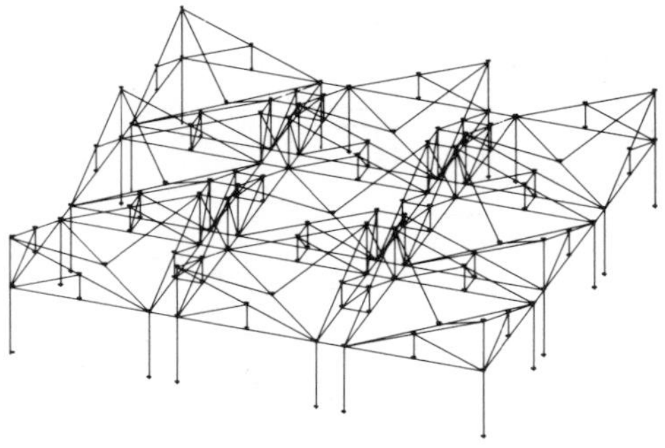

All the perimeter trusses and the columns are typical except the perimeter columns which support directly, the main composite trusses. Those columns are built of heavier sections but have the same diameter as the typical columns.

DETAILS

Critical to the overall performance of the modular system was the development of a unique connection concept that, among others, would meet the following requirements:

1. Easy, quick erection and post-exhibition dismantling of all major components.
2. Transfer of bi-directional tension and compression forces in excess of 2000 kN through 500mm diameter tubular sections and transfer of 1000 kN shear forces into the equivalent diameter tubes.
3. Axially stiff hinge connections at interior locations of long span trusses to avoid excessive deflections.
4. Moment resisting joints at some exterior column locations to create additional lateral load resisting frames.

In response to these requirements, the system was divided into four primary, easily transportable components, namely:

1. perimeter columns,
2. perimeter frames,
3. trusses, and
4. middle cubes.

The components consist of shop welded tubular sections with additional connectors for the support of the pre-fabricated wall and roof panels.

The typical connection of the truss element to the column frame elements employed the use of a slotted end concept. All three truss ends were split with vertical side plates and subsequently welded to the chord walls.

Receiving plates extended through and were welded to the walls of the 500mm diameter columns and vertical frame members. Erection then involved dropping the slotted truss ends over the receiving plates and the side plates then being clamped to the receiving plates with high tensile bolts. Bolting access was provided by 75mm diameter holes in the chord walls. Undesireable local stresses on the 500mm diameter tube walls, due to the high tension and compression forces were consequently avoided by transferring these axial forces directly to the cross members of the middle cube.

The need for a oversized slot for erection tolerances necessitated a finite element analysis of the joint due to determine the complex stress distribution in the joint and the bolt tension forces required to effectively create a friction joint.

The pavilion roof consists of triangular wood panels made of glulam frame beams and plywood stress-skin infill panels.

The walls are composed of glulam mullions and infill plywood panels. The walls are attached to and are hung from the perimeter roof frame beams and are connected at ground level to concrete grade beams to transfer lateral loads only.

In turn, the grade beams span the distance between the column footings assuring uniform behavior of the whole building in case of differential settlement of the foundation.

THE PROTOTYPE

In order to test the building system the production of components, erection procedures, efficiency of the proposed connections, and behavior of the building, a prototype of the 1000M, four module pavilion was built on the Expo site.

The prototype project was highly successful.

The total operation, from placing of the

footings to complete enclosure of the building, was accomplished in 10 days.

The roof trusses were connected to a middle cube on the ground.

The entire assembly of roof trusses was lifted by a crane to its final position in one operation.

Then the roof panels were lifted and attached to the frame.

The hanging of the wall mullions and the wall panels completed the erection.

By the time this paper is presented at the Conference, it is hoped that most of the pavilions will be erected on the Expo 86 site. Therefore, more information on the system will be available at that time.

TRIODETIC* STRUCTURAL FRAMEWORKS FOR "LES FORGES DU SAINT-MAURICE" NATIONAL HISTORIC PARKS-CANADA

William J. Vangool, P.Eng.
Triodetic Building Products Ltd.
Ottawa, Canada

H.G. Fentiman
Triodetic Structures Limited
(International)
Ottawa, Canada

The structural design and construction details of Triodetic frameworks create and compliment an unusual historic museum.

Symbolic historical three-dimensional structures, using the latest techniques, are being added to selected ruins and foundations. Their purpose will be to outline the facades of the old factory and village buildings. Construction photographs and details for these structures are provided. Some of the floor/ceiling areas have been designed to accommodate extremely heavy loads, particularly where they are covered with concrete slabs and finished with up to 200 mm of earth, sod, etc. The first phase of these structures is now being completed. They will set a precedent in terms of Canadian technical, industrial and social development.

HISTORY

Created in 1730 by a Royal Warrant awarded to Francois Poulin de Frencheville, Canada's first iron and steel industry was established.

The site of this complex on the St. Maurice River is about 200 km from Montreal, and about 30 km north of the St. Lawrence River. This industrial community was the first in Canada engaged in mining ore and producing cast iron. The close proximity of the St. Maurice River supplied excellent reserves of water, and the surrounding forest provided ample supplies of wood for charcoal, etc. A series of canals and retaining basins made it possible to control the flow of water to the paddle wheels and turbines that activated the machinery in the forges and blast furnaces.

The entire industrial village of St. Maurice was abandoned in 1883. In 1973, as a result of a Federal-Provincial Agreement, the Department of "Parks Canada" added Les Forges du St. Maurice to their National Historic Parks network.

Since that time, the remains of many industrial and domestic structures have been stabilized and saved from a soil-decaying process. At the same time, historical and archeological research has brought to life some 60 structures belonging to one period or another of the long history of this industrial village. From this research, a master plan was developed so the work could progress in several phases. The first phase relating to the original blast furnace and upper forge has been virtually completed. Included in the scheme is provision for the reactivation of the water stream that provided the motive force needed to operate the original industrial workings.

RECONSTRUCTION

While the original buildings were constructed entirely of timber, it was felt a more modern architectural approach would be to employ light-weight spaceframes to outline the various buildings. As a consequence, no attempt was made to re-construct the log structures. All that remained of the historic buildings was the perimeter stone foundations. These formed the lower floor or furnace walls and, as the existing foundations were crumbling from exposure to the elements, they had to be enclosed in a climatic-controlled building. Triodetic Building Products Ltd. were asked to consider the feasibility of providing tubular spaceframes to fulfill the architectural concepts for both the floor/ ceilings and the outline structures.

The new buildings contain the lower floors or basements, and will now serve as exhibition spaces. What is of particular interest is the stone enclosure for the blast furnace and forge. The museum structures are enclosed and burmed up by earth and grass, and only skeleton frameworks depicting outlines of the original buildings are visible from above grade.

One of the main problems was the provision of an enclosure that would serve as a roof and floor, and would be capable of withstanding heavy design loads (Ref 1). It was essential the structure not detract from the ruins, but enhance and compliment the historical surroundings (Fig 1). Only a few columns were allowed and, due to the low head room and regular plan layout, it was decided a spaceframe structure would provide a geometric pattern and visual uniformity with the exterior structures, while at the same time eliminating the need for a false ceiling to cover major structural connections.

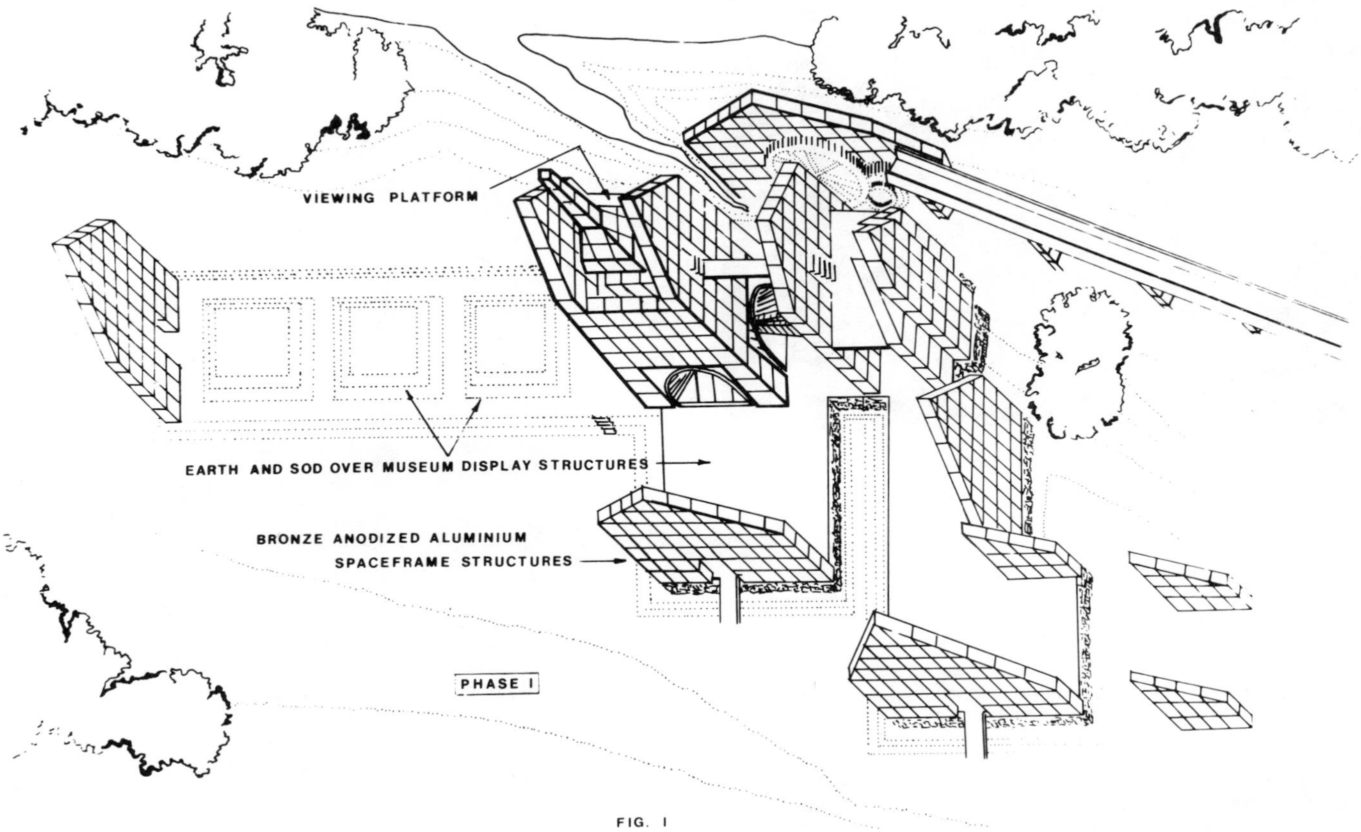

FIG. 1

ROOF/FLOOR

Flat spaceframe or grid structures can have the members arranged in many different configurations (Ref 2). The geometric pattern selected for the top and bottom chords of this project is based on the right angle triangle, and results in a three-directional grid, giving continuity along the perimeter walls. This provides framing at 90 or 45 degrees to the main axis of the buildings. The top and bottom chords are identical in layout, one being located directly above the other, spaced 1,000 mm apart.

The structural function of this spaceframe is to support a perforated acoustic deck, a concrete slab, crushed rock, earth and sod. This culminates in a total of 9.0 KN per square meter. In addition, a snow load of 4.0 KN per square meter had to be considered. This represents a very large dead load for a spaceframe having a self-weight of only 0.8 KN per square meter.

In considering the heavy loads anticipated, steel tubing was selected and a computer analysis was undertaken on the entire structure in order to study reactions, tube stresses and deflections. The initial results indicated very large deflections around the furnace opening, some in excess of 250 mm. This represented an unacceptable span to deflection ratio of 1/64.

The maximum tube size for the 1,200 mm module was 100 mm diameter, and to increase the tube diameter in order to reduce deflections was impractical and aesthetically objectionable. On closer examination of the shear diagonal layout, it was determined that running diagonals in two directions only did not provide adequate results. The basic geometric layout did allow the introduction of an additional set of diagonals, placed at 45 degrees to the regular orthogonal direction. Further computer study was undertaken, and the resulting maximum deflections of 65 mm were only one quarter of the first results. What it meant in practical terms was by increasing the overall steel requirements by 14%, a tremendous gain was made in structural rigidity. The final geometric layout of the spaceframe members is an intriguing one, and not often used, see Fig 2. The floor/roof spaceframe structure is comprised of 50 tons of painted steel tubing and 4 tons of aluminum connectors. The total number of components consist of 1,290 upper chords, 1,290 lower chords, 1,290 diagonals and 950 Triodetic connectors.

Precision in pre-fabrication and assembly of the structure was critical in order to have the spaceframe fit within the concrete perimeter walls (Ref 3). An overall clearance of 20 mm between the spaceframe and the concrete walls was allowed. The longest uninterrupted straight line of chords was 34 at 1,200 mm, or 40.8 m. At that length, the accuracy was within 2 mm. This represents 58 thousandths of a millimeter per component. While these tolerances were necessary in order to comply with the requirements, it reflects the precision of the mechanical Triodetic jointing system. The structure was assembled in three sections adjacent to its final location and lifted into place. Final interconnection of the sections was aided by temporary supports and scaffolding.

(Fig 2) Roof/floor - Under Construction

TOWER/CHIMNEY/WALLS

It was not feasible to reconstruct the buildings in their original form. In order to provide visitors with some idea of size and scale in relation to the originals, it was decided to provide certain outlines in modified spaceframe form (Ref 4).

The forge tower and chimney were the focal point of the first phase. The chimney is supported from foundations at the lower floor level, and rises through an 8 m square hole left in the spaceframe deck (Fig 3).

(Fig 3) Tower Base

It is surrounded by the forge tower structure, which is connected directly to the roof/floor spaceframe.

A stairway to a viewing walkway for the public is placed approximately at mid-level between the chimney and the forge structure. The forge tower walls are tapered from the base to their apex, necessitating a continuing change in structural elements as construction proceeded upwards (Fig 4).

Portions of the chimney structure and the four corner extremities of the forge tower were assembled at the manufacturing plant and delivered some 300 kilometers to the site by truck, in sections. The infilling components to complete the structures were built in situ at the site.

A similar geometric pattern was used to represent certain wall sections, approximately 900 mm thick.

(Fig 4) Forge Tower

All of the decorative structures are completely exposed and had to be designed for winter ice cover of 13 mm thick, and wind loads of 1 KN/m^2. The wall sections are free-standing, cantilevered from the base connections in the foundation walls, with bracing being accomplished by diagonals in two directions in line with the face members.

Horizontal supports between the two wall faces were undesirable, and as a consequence these were kept to a minimum and placed mainly across the base and at strategic locations around the perimeter (Fig 5).

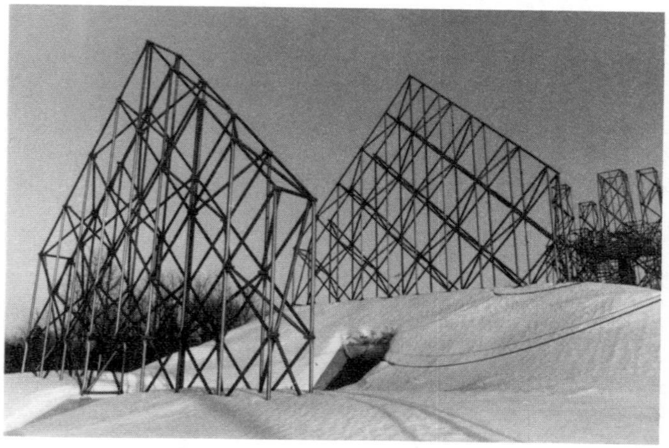

(Fig 5) End Wall Structures

(Fig 6) Partial View of Phase I

The maximum tubing diameter used in the tower, chimney and walls was 64 mm; the basic grid module was 1,200 mm. The total number of tubes used in the chimney and forge tower was 1,180, and these were connected with 412 hubs.

The wall sections used a similar basic grid size and employed a total of 1,240 tubes and 430 Triodetic connectors. The total requirement was 10 tons of anodized aluminum for the forge tower, chimney and walls on this particular project. The colour selected for the aluminum spaceframes was medium bronze (Fig 6).

Les Forges du Saint-Maurice is intended to display the high level of creativity developed in the Dominion of Canada since 1873.

STRUCTURAL SYSTEM

While details of the Triodetic joint and related structural members have been fairly well documented over the past number of years, some brief comments may be appropriate.

Ideally, components for practical spaceframe structures should include the following: a) provide individual members to any desired length; b) vary strength of these members within reasonable limits; c) connect any number of members at various angles with universal joints without introducing bending; d) use computers to master the geometry of doubly curved surfaces, so that such numbers could be easily calculated, and e) calculate stresses in all members of any spaceframe with certainty (Ref 5).

Increased interest and acceptance of three-dimensional structures has focussed attention on joining a number of structural elements in space economically and effectively. Many different spaceframe systems are presently being offered on a commercial basis.

It is recognized that up to 100% efficiency can be developed with butt-welded steel joints; this, however, is hardly an economical method of construction at today's increasing labour rates. With respect to aluminum structures, the structural alloys used usually lead to unavoidable annealing, which reduces efficiencies to the order of 50%. Only 75% efficiency in both aluminum and steel construction can be expected with rivetted or bolted joints.

Intensive and ongoing research on the manufacture and performance of Triodetic structural connections is constantly under review.

(Fig 7) Triodetic Joint

As a result, an extensive library of proprietory information and knowhow has been developed by and for the licenced manufacturers of Triodetic components in various parts of the world. An infinite number of Triodetic specimen members and connectors using improved manufacturing techniques have been subjected to intensive tests. These tests are designed to indicate various modes of behaviour under compressive, tensile and combined loadings. They indicate the various joint characteristics that would be applied to three-dimensional structures. About 10% of the joints show "perfect" behaviour, with failure occuring at the member cross section away from the joint. Under simple direct tension loading, 91% of the ultimate strength of the steel tube was achieved and the connector developed a 92% average of the ultimate strength of aluminum tubes, with about 6% deviation. The provision of these structural efficiencies have clearly indicated the advantages of Triodetic (Fig 7).

The aluminum and steel tubular members employed in this structure were of various precision lengths and coined with keyways at the required angles and fitted to serrated Triodetic joints. The resultant continua of triangles formed the basis of the museum project described.

REFERENCES

1. Wright, D.T.; "A Continuum Analysis for Double Layer Space Frame Shells"; International Association for Bridge and Structural Engineering, Zurich, 1966

2. Kneen, P.; "Analysis and Design Procedures for Space Frame Structures"; Australian Conference on Space Structures, May, 1982

3. Fentiman, H.G.; "Developments in Canada in the Fabrication and Construction of Three-Dimensional Structures Using the TRIODETIC* System"; 1st International Conference on Space Structures, London, England, 1966

4. Wright, D.T.; "Three Dimensional Space Frame Structures"; Journal of the Institute of Steel Construction, Vol. 16, No. 2, Melbourne, Australia, 1982

5. Siegel, C.; "Structure and Form in Modern Architecture", Reinhold, New York, 1962

*Triodetic is a Registered Trademark of Triodetic Structures, Limited, Ottawa, Canada

APPLICATION OF SPACE GRID FRAMEWORKS TO SKELETAL STRUCTURES

M.GHALIBAFIAN.MSC.,Dr.Eng.

Technical Director,
Sano Consulting Engineers , Tehran

Asso.Prof.,Faculty of Engineering,
university of Tehran , IRAN.

Space grid frameworks have been widely used to cover large areas usually needed in buildings such as exhibition halls,theatres.gymnasiums,hangars,factories etc.On the other hand,space grids have great potentialities in providing satisfactory solutions to design problems encountered in multistory buildings which normally have small spans.Despite the above fact,the application of grid frameworks to this type of structure can rarely be observed in technical literature. The aim of this paper is to present a number of examples of such an application.

INTRODUCTION

By the term " space structures " , a class of structures comprising wide spans , covered by a system of three dimensional framework without any interior support, is usually meant. It is the case so much so that it is normal practice to make use of 3-D systems to cover large areas in buildings such as exhibition halls , theatres , gymnasiums, hangars,factories and so on.Basically, the need for such free spaces has led the architects and engineers to the use of space structures . However space grid frameworks have high feasibility and great potentialities of providing satisfactory solutions to design problems encountered in variety of buildings,such as administrative , educational, residential,etc., classified as ordinary buildings which normally have small spans; and the application of space structures to ordinary buildings has rarely been observed in technical literature. The followings may be mentioned as the most important advantages of space grid frameworks in relation to ordinary buildings:

a) The continuity and integrity of grid frameworks enables the floor system to behave as effective horizontal diaphragm between relatively distant shear resistant elements (i.e.more than the limits prescribed by codes for ordinary structures), and this gives rise to the increase in the distance between shear-walls/bracings taking full advantage of their capacity to resist lateral forces.

b) The shear resistant elements i.e. shear-walls or bracings) may be replaced supporting columns by formation of a rigid horizontal diaphragm through which any differential horizontal support deflection is prevented which results in the minimum damage to non-structural elements due to lateral load effects.

c) Using plane or space grids as floor systems, it is possible to increase the spans in some required locations of ordinary structures without affecting its overall continuity nor increasing the thickness of the typical floor system.

d) Space grids give the designers enough freedom to combine the architectural , structural , mechanical and electrical viewpoints and successfully meet the design requirements.

It is to be noted that no considerable increase in expenses is expected nor special skills and equipments are required to achieve all the above.

SOME EXAMPLES OF THE APPLICATION OF SPACE GRIDS TO ORDINARY BUILDINGS

During more than twenty years I,in collaboration with my colleagues , have been making use of the advantages of grids to solve the design problems associated with various types of buildings . For the first time in 1966, a three-way plane reinforced concrete grillage consisting of three groups of interconnected beams meeting at the angle of 60 degrees was considered to cover an area of about $11000m^2$ of the auxiliary buildings of Tabriz machine factory including administrative,educational and utility buildings, training department and bus station . These buildings were located in a region with high seismic risk.According to the architectural requirements , no internal shear-walls could be considered and the architect only allowed us to provide shear resistant end-walls.

At the same time it was desired that the dimensions of columns be limited to a minimum. Considering the excessive distance between end-walls, normal beam and slab roofing systems could not effectively provide the intended horizontal diaphragm to transfer lateral loads to the end shear-walls , nor was it permitted by seismic resistant design codes. Hence a single layer grid with the above mentioned characteristics was chosen . The pattern of grids and the internal divisions of buildings were coordinated creating a visually attractive appearance in the work-shops.Now, after 20 years of service life,these grids have proved to show well functional performance.

Since then , we have designed similar three-way plane grids and a class of lozenge-shaped grids for many other multistorey administrative and educational buildings in which larger clear spans have been required at some locations(e.g. at the entrance , meeting halls .,etc.)The use of this type of flooring makes it possible to limit the overall thicknesses of the floor in larger clear spans , to the amount equal to the thickness in other locations with small spans covered by traditional beam and slab systems while the requirements of serviceability and ultimate limit states are satisfied .

Regarding the pleasant appearance of the grids there was no need for false ceiling.Therefore the expenses were reduced and the clear height was increased . At the same time , in addition to its good acoustical properties, it was possible to install the ceiling lamps at the corner of grid panels which resulted in better control of lighting.

In 1973 the structural design of administrative and educational buildings of the university of Kerman was given to our firm. The university was located in a seismic area where code provisions required a special framing system capable of resisting seismic loads . On the other hand, the architectural designer had prescribed the following requirements as the basis of design :

a) Flexibility for future change ; so that the dimensions of areas and the locations of internal walls and partitions might be easily changed without changing or strengthening the structure.

   The followings were resulted from the above requirement:

   - No vertical shear resistant diaphragm could be provided, and,
   - Seismic loads were to be absorbed by the same framing system provided to take vertical loads.

b) Maintainability and the feasibility of future modifications in mechanical and electrical services; so that during the service life of structures,any modification of the piping system, electrical wiring, etc., or the installation of new pipes, ducts and cable networks could be accomplished without demolition and reconstruction.

The architect was not even content with the provision of particular accessible service ducts in the vertical and horizontal directions to place pipes and wires.In view of the nature of educational and office buildings in which the functional requirements may change with time , he believed that the design should provide for an easy installation and alteration of piping system and cable network at any arbitrary location and in any arbitrary direction.

At the first glance,it seemed that the above requirements might not be satisfied except by using a service floor system.However use was made of a type of double layer steel grid square in plan and consisting of two identical rectangular plane grids, forming top and bottom layers interconnected by vertical web members. The height of the grid (i.e.the distance between top and bottom layers)was 60 cm . The top and bottom layers consisted of 150 x 150Cm regular rectangular panels . The whole space grid flooring system was supported by columns located regularly at a distance of 7.5 metres in each direction . By the use of this type of construction all the above mentioned structural and architectural requirements were satisfied.

This double layer grid system along with a 6Cm thick cast in place concrete slab constructed on the upper layer provided the desired structural continuity while proper design and detailing of movable false ceilings at the level of the lower layer grid, gave the architect and the service engineer great freedom of design . In addition,those members of the top and bottom layers which lied on the axis of columns were considered to be stronger in order to form the lateral load resistant rigid frames prescribed by the codes.The internal beams of top and bottom layers were made from No.100mm rolled I-sections , the horizontal beams,of both layers connecting the supporting columns were made from No.100 Wide-flange sections and the interconnecting vertical web members were "+" shaped welded sections made from 8mm thick plates . These hand welded double layer grids were prefabricated on site by semi-skilled labourers and then erected on their final position . The whole area covered with this type of space grid was more than 33000m^2.Until now , This structural system has well performed its intended function.

In the library of the university,the floor grid framework was similar to the above mentioned steel double layer rectangular grid in the part of building surrounding the main library hall with the exception that the distance between columms was 6.75 metres instead of 7.5 metres as in other buildings.

For the 20.25 x 20.25m main library hall situated at the centre of the building,a double layer diagonal grid was designed.The dimensional and sectional properties are referred to below :

- Overall dimension:20.95x20.95m
- Dimension of grid panels:2.38 x 2.38m
- Height:1.04m
- Top layer members:260mm I-Sections
- Bottom layer members:260mm I-Sections
- Vertical web members:parts of 260mm I-Sections
- Thickness of cover slab :8Cm

The level of this space grid was upper than the adjacent surrounding grids and the vertical distance between these grids was used to provide natural lighting through the glasses located around the main hall. This double layer diagonal grid without reducing the intended clear height of the library hall,created pleasant sub-divisions in the ceiling forming a peaceful place for studying .

The following pictures and drawings show some parts of the above mentioned projects for each of which a brief description is given.

ACKNOWLEDGEMENTS

I wish to thank Mr . S . Maalek for English translation of the Persian manuscript.

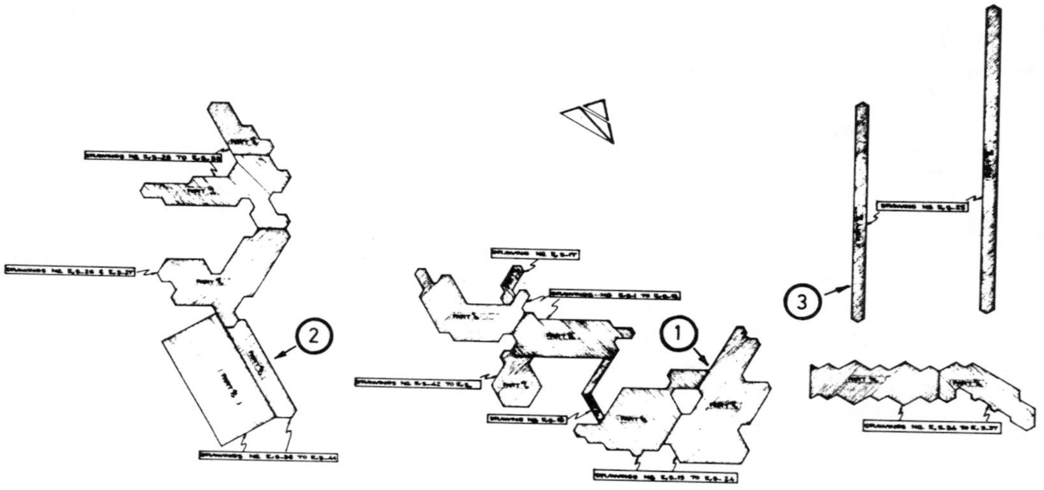

Fig 1 Tabriz machine Factory: Lay out of auxiliary buildings.

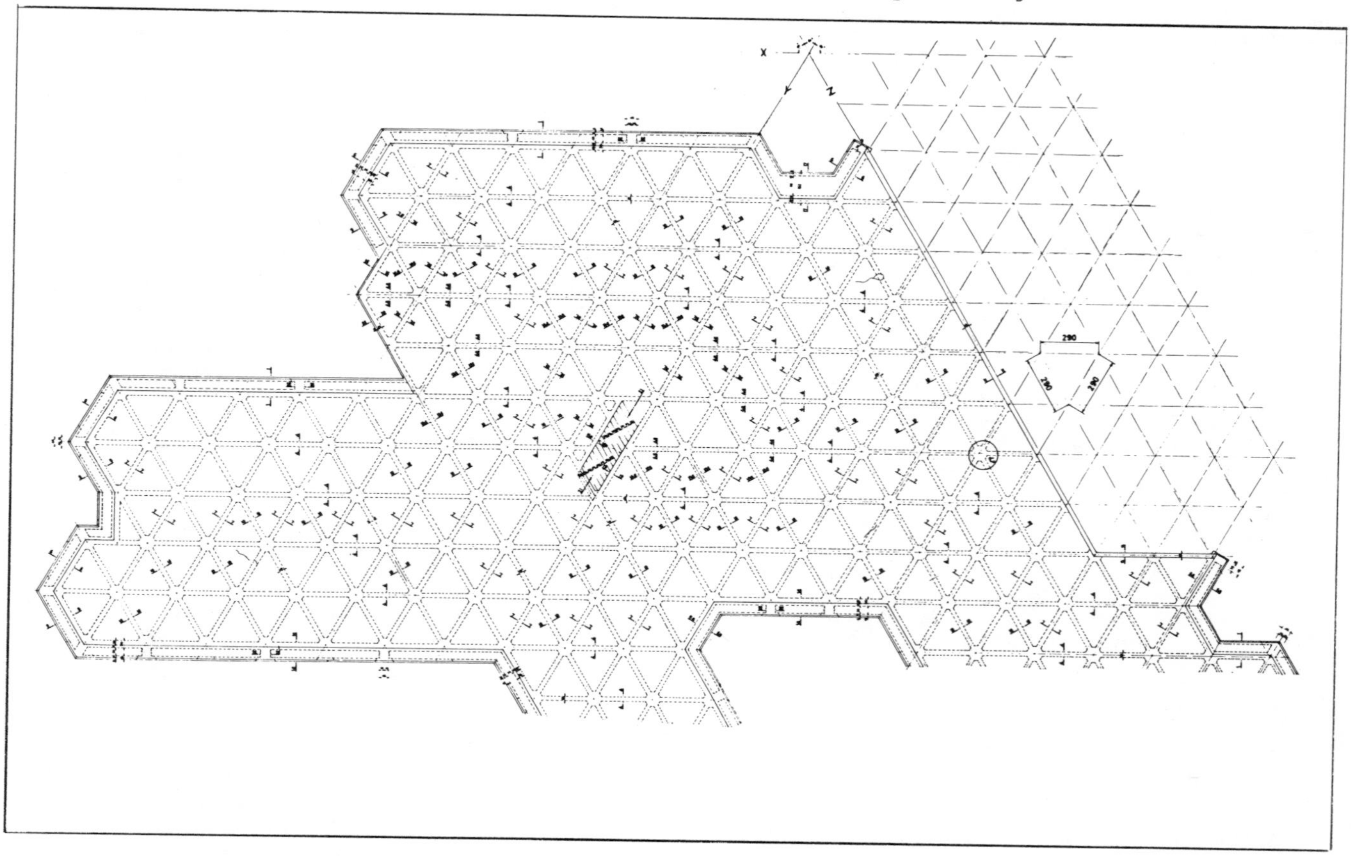

Fig 2 Formwork plan of Building no 1

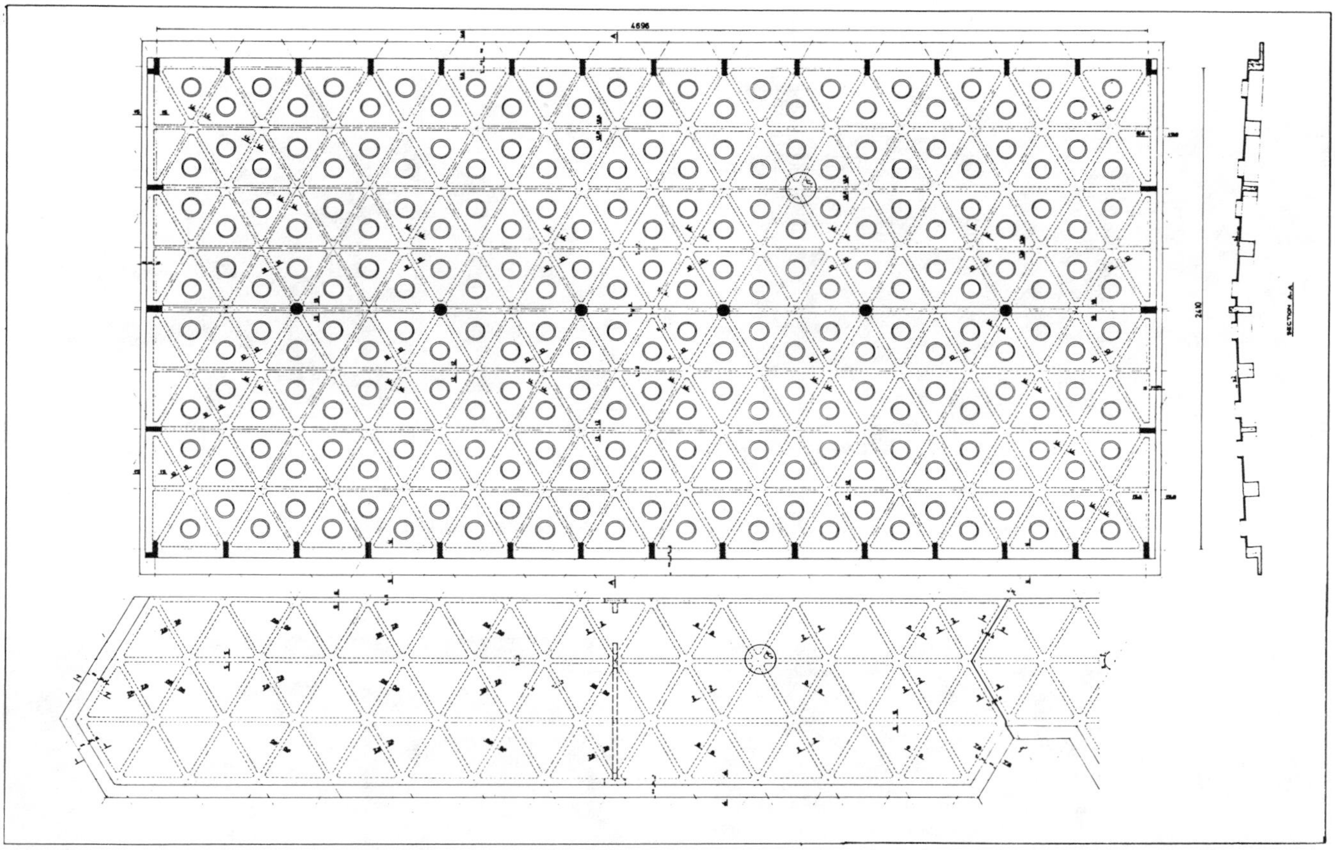

Fig 3 Formwork plan of Building no 2

Fig4 Bus station

Fig5-Detail of the triangular grid of building no(1)

Figs 6 and 7- Kerman University:Structural frameworks of two buildings.

Figs8and9 Kerman University:General view of two more buildings

Fig10-Underside view of the double layer grid.

Fig11-Another view of floor and roof grids.

Fig12-A prefabricated grid component,ready for erection.

Fig13-Experimental full scale model of the grid and the thin R.C.Concrete cover slab.

Fig14-Underside view of the floor system after the construction of insitu cover slab.

Fig15-Inside view of a double layer grid showing the detail of vertical members and the working space for the installation of pipes and wires.

Fig16-The central double layer diagonal grid over the library hall and the surrounding grids before concreting.

Fig17-The central diagonal grid and the surrounding grids after the construction of the cover slab.

Fig18-Kerman University:a general view of buildings during construction.

Fig19-An outside view of the library.

DESIGN AND CONSTRUCTION OF TOWER-TYPE STEEL STRUCTURES

D MATEESCU * I MUNTEANU ** GH MERCEA *** I CARABA *** D FLORESCU ****

* Professor, Member of the Romanian Academy of Sciences, Polytechnic Institute of Timişoara.

** Professor, Polytechnic Institute of Timişoara

*** Associate Professor, Polytechnic Institute of Timişoara

**** Lecturer, Polytechnic Institute of Timişoara

The paper presents a number of tower-type space structures designed by the staff of the Department of Steel Structures of Timişoara, Romania.These structures are : an obelisk symbolizing the three olympic medals, designed for a sports centre in a Romanian town, the bearing structure of an aerogenerator and a steel tower for live TV transmissions of sports events.

GENERAL PROBLEMS

The design of special constructions for industrial, social, cultural or other purposes has for many years been one of the principal concerns of the Department of Steel Structures of the "Traian Vuia" Polytechnic Institute of Timişoara, Romania. Among the achievements of this Department are a number of tower-type space structures that were designed in the past few years by members of the staff.

The first of these structures is a steel obelisk that consists of three towers symbolizing the gold,silver and bronze medals of the Olympic Games ; the second is the supporting structure of an aerogenerator, and the third is a tower destined for live TV transmissions of sports events.

STEEL SPACE STRUCTURE OF AN OBELISK SYMBOLIZING THE OLYMPIC MEDALS.

An obelisk consisting of three towers symbolizing the gold, silver and bronze medals of the Olympic Games was designed by Gh.Mercea for the sports centre of a Romanian town that is well known for its long sporting tradition. The centre consists of handball, volleyball and football fields, tennis courts and a beautiful sports hall, see Ref.1. The three towers forming the obelisk are made and erected in the immediate vicinity of the sports hall and will be painted and inaugurated together with it. Two alternatives were considered for the obelisk, one using reinforced concrete and the other using a steel space structure. This latter variant was chosen by both the designer and the architect Mr.S.Gavra, because steel, being more supple, seemed better suited to express the spirit of competition and self-assertion characteristic of sports and games; besides, the steel sections required for the obelisk could be manufactured more easily by a local plant.

The steel structure of the obelisk consists of three space towers of square cross sections. The cross-sections remain constant over the full heights of the towers; the sides of the squares are 1,000, 1,250 and 1,500 mm, respectively. The heights of the towers also differ; for reasons imposed by the architect, the thickest tower, the one with a side of 1,500 mm, is also the lowest(12 m),the one with a side of 1,250 mm is 20 m high, and the most slender one(1,000 mm)is also the tallest 33 m. see Fig.1. A cube made of sheet-steel rests on each of the towers. The three cubes are painted in the colours of the olympic medals, i.e. the cube on the smallest tower has the colour of bronze, the one on the 20 m tower is painted silver and the cube on the tallest tower is painted in the colour of gold.

The tallest tower carrying the golden cube, is placed in plane so that the side of its base is parallel to the principal façade of the sports hall and to the street, while the other two towers are rotated each by 45°, thus being parallel to the diagonals of the first. To give greater strength to the entire structure, the three towers are interconnected at the height of + 7.5 m by means of a platform which also serves as a landing on the stairs of access provided on the back side of the tallest tower. A second platform connects the two tallest towers at + 15 m see Fig.1. The three towers are made of cubes that are placed one upon the other. The vertical bars run continuously over the full height of the towers; the horizontal bars together with the vertical ones form the sides of the cubes while the top sides of the cubes are connected by bars running diagonally.

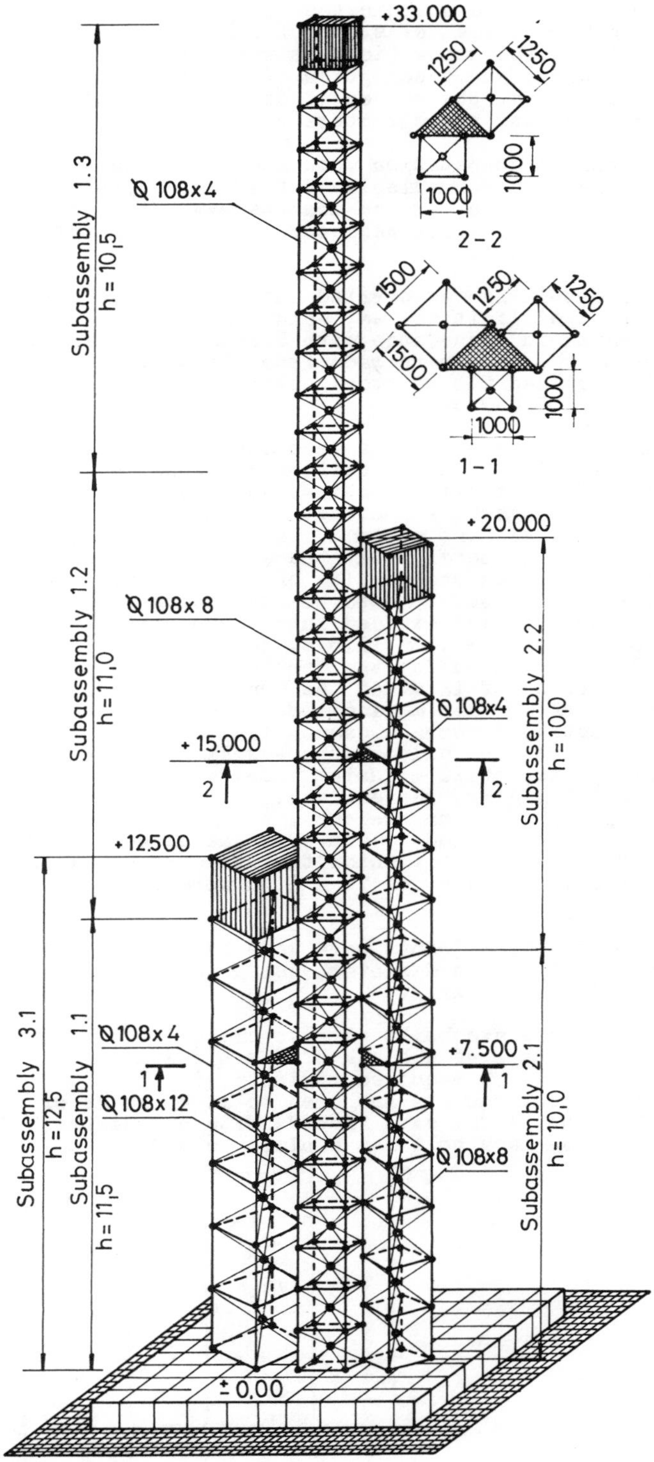

FIG. 1

The vertical bars and one of the diagonals are continuous; the other bars are connected by means of spherical joints provided with holes and welded to the continuous bars.

The following loadings were considered in designing the space structure of the obelisk:

a) The dead load of the space structure, including the cubes on top of the towers; this load was assumed to be a force concentrated in the nodes, amounting to 40 ... 50 daN, depending on the thickness of the tube walls.

b) Loading due to rime in winter, considering all bars to be covered by a 5 cm thick layer of rime, with a density of 900 kg/m^3

c) Wind loading, which is the most important type of loading. Its value was established according to Romanian Standard STAS 10101/ 20-78 using the following equation:

$$p_v = b \cdot C_{ts} \cdot g_v \quad \left[daN/m^2\right] \qquad \text{...... 1.}$$

The aerodynamic coefficient "b" was determined by means of Eqn 2, given in the Standard :

$$b = 1 + z \cdot r \qquad \text{...... 2.}$$

in which the amplification coefficient "z" is calculated according to the Standard, depending on the period of oscillation "T" of the structure, which was calculated by means of a computer program on a GIPSI-02 computer. The following values were obtained : T = 0.906 sec., z = 1.485, while the gust coefficient "r" is provided in the tables of the Standard, ranging between 0.360 at a height of 10 m and 0.315 at 40 m. The resulting aerodynamic coefficient "b" was between 1.5346 at 10 m and 1.4677 at 40 m. The coefficient C_{ts} for normal wind on one face of the tower was calculated according to the Standard depending on the Reynolds number, which for this structure was 31.4 x 10^5. The resulting value of C_{ts} was $C_{ts} = 0.61425$.

The basic dynamic pressure "g_v" at 10 m above ground was calculated as a function of conventional wind speed, which was considered to be v_c = 40 m/sec = 145 km/h, using the following equation:

$$g_v^{10} = \frac{v_c^2}{1630} \quad \left[daN/m^2\right] \qquad \text{...... 3.}$$

The resulting basic pressure at a height of 10 m was: $g_v^{10} = 100 \left[daN/m^2\right]$

The basic dynamic pressure "g_v" increases along the height of the structure according to Eqn.4 :

$$g_v^H = g_v^{10} \left(\frac{H}{10}\right)^{2e_1} \left[daN/m^2\right] \qquad \text{...... 4.}$$

where H is the height of the structure and e_1 = 0.2, thus yielding 132 daN/m^2 at 20 m, 155 daN/m^2 at 30 m, 175 daN/m^2 at 40 m.

Using the values of b, C_{ts} and g_v the wind pressure p_v was determined by means of Eqn.1; its value was 95 daN/m^2 at 10 m, 123 daN/m^2 at 20 m, 142 daN/m^2 at 30 m and 158 daN/m^2 at 40 m.

The possibility of wind blowing in a diagonal direction was also considered in the analysis.

The above-mentioned loadings were taken into

consideration by means of their components H_x, V_y, and H_z. The positive directions of the global co-ordinate system X_G, Y_G, Z_G are shown in Fig.2, which also represents the local system of the bar, x_L, y_L, z_L, as well as the positive directions of the stresses.

Owing to the very large number of bars and nodes in the structure there is a rather high degree of static indeterminacy; for this reason, all the stresses were calculated by means of an electronic computer, taking account of the interaction between the taller towers that is due to the two platforms at + 7.5 m and + 15 m, respectively.

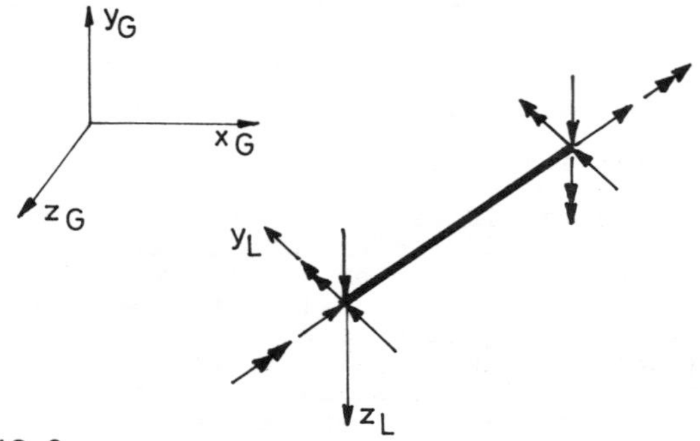

FIG. 2

For the computation of the stresses, the GIPSI-02 programme was used, whose algorithm in the determination of the stress-strain state of the structure subjected to static and dynamic loading is based on the method of the stiffness matrix. Since there are 774 bars, which is rather beyond the limits of the programme, the structure was divided into 14 subassemblies with no more than 30 nodes and 180 unknowns in each.

The global stiffness matrix of the structure is indicated in Fig.3:, in which K_{ij} represents a submatrix corresponding to a substructure, D_i are the vectors of the displacements in the "i" - section, and P_i represents the vectors of the forces (loads) acting on the structure at "i".

The following data were fed into the computer: the number of substructures, their exact location - the coordinates of the nodes, the geometric characteristics of the bar cross-sections, the connections between the bars - the number and description of loading assumptions, the number of degrees of dynamic freedom, and the structure of load combinations.

Automatic computation took about two hours and the results were displayed for the entire structure. The bar end-moments were very small and were therefore neglected when detailing the structure.

Bar detailing was carried out by means of the method of limit states, which is valid in Romania in accordance with Standard STAS 10108/0-78; the bars were made from circular tubes from rolled OLT 45 steel having a strength R = 2.300 daN/cm²

The bar cross-sections were chosen with the possibility of using uniform sections in mind; very few differing sections were consequently used throughout the structure. For architectural reasons, all the vertical bars had the same external diameter, only the wall thicknesses differed from one portion of the tower to the next. Thus, the 33 m tower supporting the golden cube was divided into three parts: for the bottom part, of 11.5 m, tubes sized 108 x 12 mm were used for the vertical bars, and for the second of 11 m and the top part of 10.5 m tubes of 108 x 8 mm and 108 x 4 mm, respectively, were chosen. The tower supporting the silver cube was divided into two parts, of 10 m, using tubes sized 108 x 8 mm and 108 x 4 mm, respectively; the lowest tower, supporting the bronze cube, was made entirely from tubes of 108 x 4 mm. For the diagonal and horizontal bars in which the stresses are less pronounced, tubes of 70 x 5 mm were used, see Fig.4.

The spherical joints were made from hot-rolled tubes of 200 mm diameter, using the procedure indicated in Ref.1.

The towers are built on a concrete foundation in the shape of a block of 6 x 6 x 2 m providing the required rigidity of the structure, and are anchored to it by means of 48 bolts, 4 for each "foot", with a diameter of M 56. Each bolt has a plate welded to its

$$
\begin{bmatrix}
K_{11} & K_{21} & & & & & & & & \\
K_{12}^T & K_{22} & K_{32} & & & & & & & \\
& K_{23}^T & K_{33} & K_{43} & & & & & & \\
& & & \cdot & \cdot & & & & & \\
& & & & \cdot & \cdot & & & & \\
& & & & K_{i-1,i}^T & K_{i,i} & K_{i+1,i} & & & \\
& & & & & \cdot & \cdot & & & \\
& & & & & & \cdot & & & \\
& & & & & & K_{n-3,n-2}^T & K_{n-2,n-2} & K_{n-1,n-2} & \\
& & & & & & & K_{n-2,n-1}^T & K_{n-1,n-1} & K_{n,n-1} \\
& & & & & & & & K_{n-1,n}^T & K_{n,n}
\end{bmatrix}
\cdot
\left\{
\begin{matrix}
D_1 \\ D_2 \\ D_3 \\ \vdots \\ \vdots \\ D_i \\ \vdots \\ \vdots \\ D_{n-2} \\ D_{n-1} \\ D_n
\end{matrix}
\right\}
=
\left\{
\begin{matrix}
P_1 \\ P_2 \\ P_3 \\ \vdots \\ \vdots \\ P_i \\ \vdots \\ \vdots \\ P_{n-2} \\ P_{n-1} \\ P_n
\end{matrix}
\right\}
$$

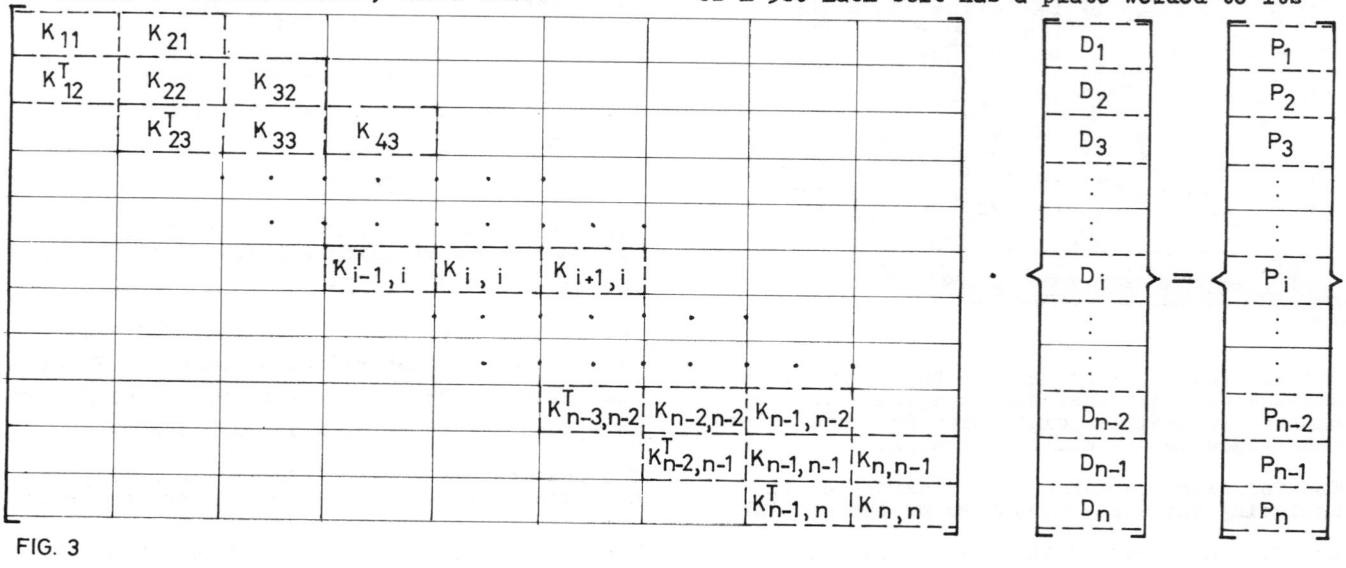

FIG. 3

lower end and the bolts are fixed to the foundation block at the time of placing the concrete. The towers are fixed to the bolts by means of circular plates welded to each foot being 500 mm in diameter, and placed on the Ø 300 mm plates supporting the feet of the towers. Fig. 4..

The three painted cubes are fixed to the towers by means of frames from sheet-steel welded to the 4 feet; they are made of thin sheet stiffened internally.

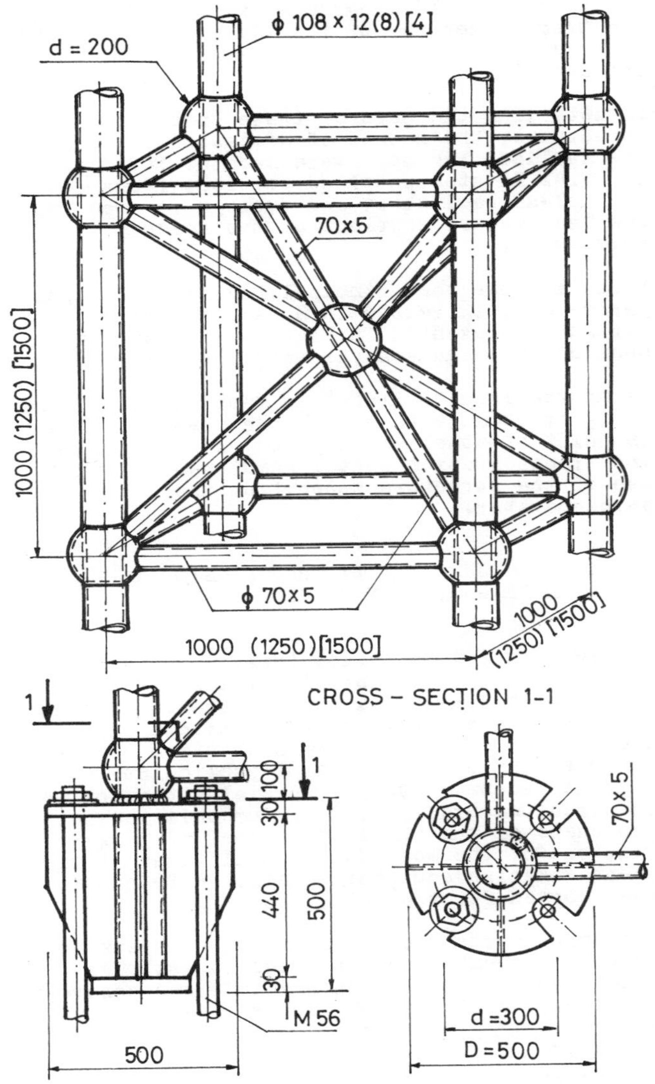

FIG. 4

The two platforms at + 7.5 m and + 15 m are made of perforated sheet welded on a frame made of sections which in their turn are welded to the towers. They are provided with openings for the stairs, which are made from sheet 3o mm thick.

Figure 5 shows a photograph of the obelisk after erection.

FIG. 5.

STEEL SPACE STRUCTURES SUPPORTING AEROGENERATORS

For the realization of the National Wind Energy Programme initiated in Romania several years ago exhaustive studies and investigations were required for all the component parts of medium and large-scale wind power stations. One of these components, i.e. the supporting structure for the aerogenerator, was studied by Professor I.Munteanu and Dr.I.Caraba from the Department of Steel Structures of the Timişoara Polytechnic Institute. Two categories of structures were studied, i.e. tower-type structures. Fig.6 (a-f) and piles anchored at a single level Fig.6 g. Eighteen different variants of space structures were investigated. In determining the height of the towers and piles the aerodynamic and operational conditions of the aerogenerator were considered.

The supporting towers consist of two distinct parts : the lower part, from ground level up to + 10 ... 15 m, with continuously increasing cross-section, Fig.6 a,b,d,f, or made in the shape of a frame with three, Fig.6 e, or four feet, Fig.6 c, and the upper part, rising to level + H, with a constant cross-section. The bottom part with its variable section is in the shape of a truncated pyramid with a triangular base, Fig.6 b, or with four, Fig.6 a,f, six or eight face, Fig. 6 f_2, f_3, or in the shape of a truncated cone, Fig.6 d. The upper part is a three-dimensional latticed structure with three faces, Fig. 6 b,e, four faces, Fig.6 a,c, and in the shape of a cylindrical shell, Fig. 6 c, d, e, f.

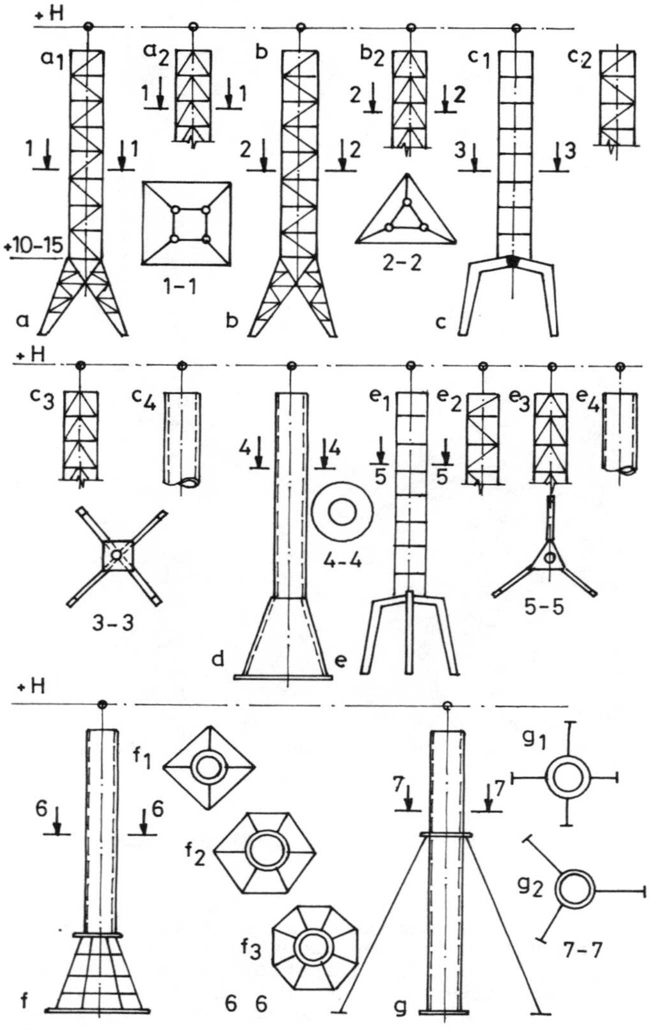

FIG. 6

In accordance with their internal subdivision the towers are made of Vierendeel lattice girders, Fig. 6 c_1, e_1 or of lattice girders with alternating, Fig. 6 a_1, b_1, c_2, e_2 or with K-shaped diagonals,Fig.6 a_2, b_2, c_3, e_3.

For the anchored-pile-type structures, which have a constant cross-section, various possibilities of anchoring the piles, i.e. in four or in three directions, Figs. 6 g_1 and 6 g_2, respectively were studied.

The following actions were considered in the analysis of the three-dimensional bearing structures of the aerogenerator : loading due to the aero-aggregate, the dead load of the structure, and " atmospheric " loading, i.e. the action of wind on the aerogenerator and its supporting structure, and the action of rime. SISART and SAP IV programmes were used on the electronic computer of the Timişoara Polytechnic to compute the state of stress due to static and dynamic loading.

In detailing the bars of these space structures, the method of admissible strengths was applied. The bars were made of Romanian OL 37.4. Kf steel. Rolled sections(L-shaped, channel-shaped, sheet and tubes)were used for the transversal sections of the space structures.

Table 1 shows the principal technical-economic parameters, expressed in %, and related to the variant represented in Fig. 6 a_1,being the best of the eighteen variants investigated.

Using the results listed in Table in Fig.7 the designer is able to choose the best solution in point of steel consumption, manufacturing and transportation costs, maintenance, and use of the structure supporting the aerogenerator.

STEEL STRUCTURE OF A TV TRANSMISSION TOWER FOR LIVE TRANSMISSION OF SPORTS EVENTS.

A three-dimensional steel tower for live transmissions from football fields was designed by D.Mateescu and D.Florescu, of the Timişoara Department of Steel Structures. The tower is provided with a platform for TV cameras. Its use in areas of public interest made it necessary to pay special attention to the architectural solution; for this reason, the shape of the "Endless Column" of Tîrgu Jiu, one of the most famous sculptures of the great Romanian sculptor Constantin Brîncuşi, was adopted for the tower, see Fig.8.

The space structure of the tower consists of modular elements made of steel tubes. The steel structure rests on a reinforced concrete frame 3.75 m high which in turn rests on a

Technical economic indices, in %, as related to the variant reprezented in Fig.6 a_1 / Variants investigated	Variants according to Fig. 6																	
	a		b		c				d	e				f			g	
	a_1	a_2	b_1	b_2	c_1	c_2	c_3	c_4		e_1	e_2	e_3	e_4	f_1	f_2	f_3	g_1	g_2
Steel consumption (%)	100	109	89	93	88	96	99	118	141	83	92	96	115	120	127	135	121	119
Amount of work required for manufacture (%)	100	105	78	80	73	79	83	75	73	70	78	81	73	63	68	72	57	57
Amount of work required for erection (%)	100	103	80	82	76	78	81	65	61	72	77	79	64	60	63	65	53	51

FIG. 7

foundation built in the ground.

The platform on the top of the tower is 3 x 3 m in size and carries two parabolic TV aerials, the warning system and the lightning conductor.

Access to the platform is provided by a number of steel steps.

The following effects were considered in the analysis; the dead load of the structure, the weight of the steps, of the platforms and of the TV transmission equipment as well as the wind action perpedicular to one face of the tower and to a diagonal.

Wind pressure was determined in accordance with the Romanian standard as specified for the obelisk structure described earlier in this paper (Eqs. 1 ... 4).

The structure was considered to be hinged at the nodes and the loads were concentrated in the nodes. The stresses were computed on an electronic computer using the programmes available at the Department of Steel Structures.

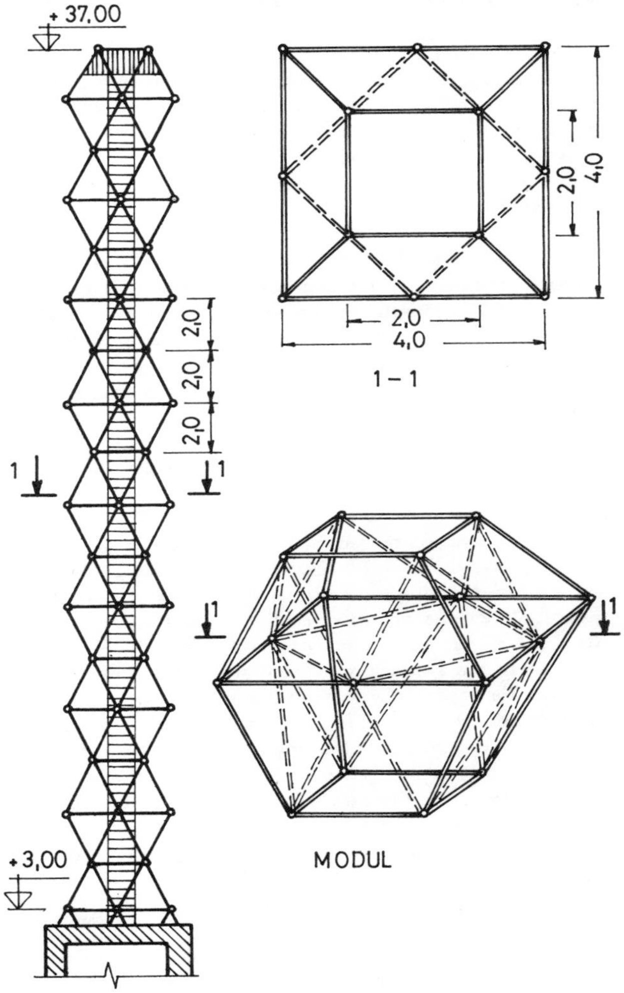

FIG. 8

The detailing of the bars was made in accordance with the Romanian standard STAS 10108/0-78; the bars were made of tubes of Romanian rolled steel OLT 45, with the dimensions 89 x 8 mm and 89 x 11 mm, Fig.8. The modular elements were welded together in the workshop and the individual subassemblies were welded together in situ. For joining the bars, hollow steel spheres were used, with an external diameter of 18o mm and walls 10 mm thick;for manufacturing details, see Ref.1.

A photograph of the tower during and after erection is shown in Fig.9.

FIG.9.

REFERENCES

1. D MATEESCU, G MERCEA, L GADEANU D FLORESCU, P LŐRINCZ, R BANCILA, Space Roofs Designed by the Steel Structures Department of Timişoara - Romania

2. D MATEESCU, Construcţii metalice speciale Romania 1965

3. Programmes GIPSI-02; SAP-4; SISART

4. STAS 10101/20-78 and STAS 10108/0-78

NODUS SPACEFRAME ROOF CONSTRUCTION IN HONG KONG

A.J. Bell * and T.Y. Ho **

* Lecturer, Department of Civil and Structural Engineering, University of Manchester Institute of Science and Technology.

Associate, T.Y. Ho & Partners, London.

**Principal, T.Y. Ho & Partners, London.

The first Nodus spaceframe roof was constructed in Hong Kong in 1978. Since then a number of spaceframe projects have been completed using this system and more are planned. This paper describes a number of these projects highlighting unusual or special features which have made a spaceframe system fabricated in the U.K. a viable and economic design solution. Some of the particular problems encountered when using spaceframe roof systems in Hong Kong are also described.

INTRODUCTION

The advantages of spaceframe roofs are now widely recognised. Light and airy in appearance they are structurally efficient solutions to covering large uninterrupted areas. Numerous spaceframe roofs have been built in various parts of the world for many years, however, it was only in 1978 that this form of construction was generally recognised to be economically efficient in Hong Kong. Since this time a number of spaceframe roofs have been satisfactorily completed using the Nodus system (Ref. 1) and more are being planned.

The Nodus system is a well established British spaceframe system. It is built around the standard Nodus joint which comprises two half casings clamped together by a H.S.F.G. bolt. Horizontal chord members are welded to special connectors which lock into corresponding grooves in the half casings. Bracing members have forked connectors welded to their ends and these are connected by split pins to lugs protruding from the half casings. Chord and bracing members may be either circular or square hollow sections. The analysis of spaceframes requires the use of a large computer and the analyses of all the Nodus spaceframes used in Hong Kong have been undertaken jointly with the Design Department of the Tubes Division of the British Steel Corporation using a linear stiffness program developed by the latter and mounted on their mainframe machine.

The first Nodus spaceframe roof in Hong Kong was built in 1978 over the 45.5m x 36m Aberdeen Games Hall (Fig. 1). After construction the spaceframe roof was thought to have been more expensive than a conventional steel roof and hence a truss system was adopted for another indoor games hall which was to be constructed the following year. It was subsequently found that this conventional roof had not proved to be economical compared to the spaceframe as high wind suctions had meant that lateral restraint was required to the bottom chords of the trusses and the greater structural depth of the trusses compared to the spaceframe had resulted in extra cladding costs. All subsequent indoor games halls in Hong Kong have been specified to have spaceframe roofs.

Problems of high wind suction are among those particular to roof construction in Hong Kong and these together with other local requirements are described in more detail in the following section. A number of successful Nodus spaceframe roofs (Figs. 2 and 3) are then described in some detail and an indication given of how these problems were overcome.

SOME ASPECTS OF ROOF CONSTRUCTION IN HONG KONG

The major loading on most roof structures in Hong Kong is due to wind and in particular wind suction. The structures described in this paper were designed for average suctions of 2.5 kN/m^2 and local suctions of up to 4.5kN/m^2. Additionally lateral loads of 2.0 kN/m^2 applied to the elevations of roof structures had to be catered for together with drag loads of up to 0.2 kN/m^2 of plan area of roof. Spaceframes are eminently suitable for resisting this uplift loading since their bottom chords are inherently restrained. Their form is also very suitable to resist the lateral loading. The high uplift loads do, however, necessitate special precautions to be taken at support positions where bearings must be capable of holding down the structure whilst in many cases also allowing lateral movement.

A frequent design requirement for roofs in Hong Kong is that they should perform satisfactorily for sustained periods throughout an extended temperature range - in the case of the structures described herein for the range -5°C to +40°C. To eliminate the significant temperature stresses which would arise from this requirement, sliding anchorage bearings are required which can allow the appropriate lateral movements whilst providing anchorage to uplifting due to wind.

A detail of a typical sliding anchorage bearing developed to resist uplift whilst allowing significant lateral movements is shown in Fig. 6.

Construction sites in Hong Kong tend to be very congested. Because of lack of space it has not yet been found possible to benefit from one of the major advantages of spaceframe construction, i.e. construction at ground level followed by lifting to final position. At the sites mentioned herein it was impossible to accommodate the necessary craneage even though this is readily available in Hong Kong. Each of the roofs was thus constructed on a platform or moveable trestles though in one case consideration was given to constructing the spaceframe on an adjacent flat roof and launching it bridge style.

A feature of building work in Hong Kong is tight construction programmes with often severe penalties, which are rigidly enforced, for late completion. The relatively simple erection procedure and the rapid speed with which it can be carried out thus makes spaceframe construction attractive.

For the projects described herein a spaceframe solution was found to offer significant advantages over conventional construction and the Nodus system was found to offer significant benefits. Being British, this system was originally conceived to satisfy the requirements of British Standard Specifications and the appropriate design checks are carried out by the computer program used for the analysis. Local building regulations in Hong Kong relate to British Standards and hence both design and gaining design approval are simplified. The advantages of containerisation for the transport of piece small spaceframe roof components is widely appreciated. For the present projects the containers used to transport these components were less than half full when their weight limit was approached and the remaining space was used to transport insulation and rooflights contributing to an overall saving in roof costs.

ABERDEEN INDOOR GAMES HALL

The roof over the Games Hall (Fig. 1) measures 45.5m x 36m (1,638m^2) and a 'square on square offset' configuration adopted for the Nodus spaceframe which is continuously supported on four sides by a concrete ring beam. The module size is 2.6m x 2.6m with an overall depth of 2.3m and all members are square hollow sections. The roof cambers in the 36m direction only with a rise in the centre of 250mm. The long sides have a mansard edge while the short sides have vertical gable ends and the tops of the ring beams under the latter curve to follow the camber.

The roof covering consists of a single layer waterproof membrane on insulation boards adhered to metal decking supported directly on the spaceframe without purlins.

HELICOPTER HANGAR ROOF

This roof (Fig. 2) over a helicopter hangar constructed for the Hong Kong Air Force Service measures 50m x 37m (1,850m^2). A 'square on square offset' configuration (Fig. 4) was adopted for the Nodus spaceframe with a module size of 2.755m x 2.755m and an overall depth of 2.4m. Square hollow sections were used throughout. On three sides the roof is supported on a concrete ring beam which in turn is supported on concrete columns or walls. On the remaining side the roof is supported on a steel lattice girder which spans 36m over the sliding doors which form the entrance to the hangar. The elastic stiffness of this girder was included in the analysis of the spaceframe.

The roof is covered by an interlocking metal cladding system supported off the spaceframe and intermediate purlins. In order to provide drainage the spaceframe is given the form of a very flat barrel vault, tilted to create a fall from front to back as shown in Fig. 4. The concrete ring beam was constructed to the appropriate support levels and the steel lattice girder, which was composed mainly of U.C.'s, was fabricated on site to conform to the profile of the spaceframe along that line. After the girder was erected final adjustments were made to levels when welding, in situ, the support brackets for the spaceframe (Fig. 5).

Anchorage bearings are provided with the ability to resist uplift of the spaceframe. Thermal movements of the roof are allowed in two directions by the provision of oversize holes as shown in Fig. 6. In order to provide lateral stability in the directions of both long and short spans special restraint bearings (Fig. 7) were introduced near the mid-points of ring beams (Fig. 4). These restraint bearings take the form of channels, welded to the bottom chord of the spaceframe, which are located between angle guides attached to the ring beams. Movement is allowed normal to the line of the ring beam but not in the line of it.

The spaceframe was initially designed to be constructed at ground level and then lifted to its final level by cranes. At the time of construction of the roof, access to the site was so congested that it was impossible to gain entry for cranes and hence the roof was built on trestles at its final level.

When all material was on site the Client added a requirement that a 5t. runway beam be hung from the spaceframe over a maintenance area. This necessitated a rapid redesign of the spaceframe which resulted in 2t. of Nodus components being flown out to Hong Kong where the necessary modifications were readily made. It was not considered prudent to suspend the runway beam from single Nodus joints since accidental overloading might cause the single bolted Nodus connection to fail. Rather the runway beam is supported off transverse spreader beams, each suspended from three Nodus joints.

JUBILEE SPORTS CENTRE

Completed in 1982, the Jubilee Sports Centre (Fig. 3) provides, on its 16.5 hectare site, extensive facilities for both indoor and outdoor sports. A three level building complex houses the indoor sports facilities which include sports halls, squash courts, gymnasiums and a 25m training pool together with changing facilities, restaurant and bar and limited residential accommodation. Some 5,300m^2 of Nodus spaceframes are used to cover three major areas of this complex, the Entrance, the Concourse and the Training Pool (Fig. 8). The spaceframes allow the achievement of striking architectural effects and also the creation of large uninterrupted areas by minimizing the number of supporting columns required.

The Entrance.

The roof of this area takes the form of a right angled triangle on plan with the two shorter sides each having a length of approximately 50m. Only about half the area covered is actually enclosed and this together with the inherent airyness of the brightly painted spaceframe gives visitors a feeling of spaciousness when approaching and entering the Centre.

A 'square on diagonal' grid is adopted for the spaceframe and this is based on a 3m module with a depth of 1.5m. The roof is supported on six concrete columns with an impressive long cantilever at the tip of the triangle and overhangs of 4.5m and 3.2m on the two external sides. SHS sections are used for the top chords and CHS sections for bracing members and the bottom chords.

The Concourse.

The Concourse area is again essentially triangular on plan with an area of approximately 2,500m^2 and is covered with a multilevel spaceframe as shown in the sections of Fig. 8. This provides interest and again helps to create a general feeling of spaciousness. In the centre of the area the roof is raised to form a group of nine pyramids which symbolise the pursuit of excellence. These pyramids were formed within the spaceframe, which locally has treble layers, highlighting the versatility of spaceframe construction. The external edge of the spaceframe is also treble layer and is completely exposed to emphasise the aesthetically pleasing appearance of the tubular structure.

The grid type and module sizes of the spaceframe are the same as those used for the Entrance Area. The roof is supported on relatively closely spaced internal columns with an overhang of 5.3m along the external edges. Along the long side of the triangular area of this roof the spaceframe is hung from the adjacent concrete structure of the Sports Halls.

Due to high wind uplifts experienced both in this area and in the Entrance Area special anchorage bearings were devised to sit on top of the isolated columns as in several positions the forces produced exceeded the centre bolt capacity of the largest (Type 45) Nodus joints. Special supports which could allow lateral movements were also devised for the positions where the spaceframe is hung from the adjacent concrete structure. These are developments of those used for the Helicopter Hangar.

The inner area of the Concourse is enclosed by glazing and it is thought to be the first time that glazing has actually been fitted directly into a Nodus spaceframe.

The Training Pool.

The roof is square on plan measuring 36m x 36m and covers the 25m swimming pool. A 'square on square offset' grid is used for the spaceframe based on a 3m module size of depth 1.5m. The roof is supported at top chord level on twelve concrete columns and the spaceframe projects beyond the columns by 3m on all four sides with the steelwork in this area left exposed. A camber is provided in the middle of the roof and the corners are lifted to enable rainwater to be collected at the middle of each side.

To allow for temperature effects all anchorage bearings allow the capacity for some movement. On corner columns movement is permitted in two directions while on other columns only movement normal to the line of columns is permitted. Due to high wind uplifts in this area, the largest Nodus joint (Type 45) would have been overloaded at the anchorage bearings and hence a special joint was devised. Details of this joint are given in Fig. 9.
The Training Pool is linked to the Entrance Area by a walkway covered by a spaceframe structure which is supported centrally by a single line of columns. At the Entrance end the walkway roof is suspended from the edge of the spaceframe over that area while at the Training Pool end support is devised from the spaceframe over that area through a modified Nodus connection (Fig. 10) which allows for differential horizontal movement at top chord level.

CONCLUSIONS

The three Nodus spaceframe roofs briefly described in this paper proved to be effective, efficient and economic solutions to the structural and architectural problems posed in each building. In the case of the Jubilee Centre it is considered that the spaceframes and the versatility of the system contribute considerably to the striking appearance of the Centre.

As a result of the structural and economic advantages demonsrated, particularly in the case of the Aberdeen Games Hall, further spaceframe projects are planned in Hong Kong in preference to more traditional structures.

ACKNOWLEDGEMENT

The authors wish to thank British Steel Coporation, Tubes Division for their assistance and encouragement in the present work.

REFERENCES

1. Nodus Space Frames, A Design Guide for Architects and Engineers, Tubes Division, British Steel Corporation, 1981.

FIG. 1 ABERDEEN GAMES HALL

FIG. 2 HELICOPTER HANGAR

ENTRANCE

CONCOURSE

FIG. 3 JUBILEE CENTRE

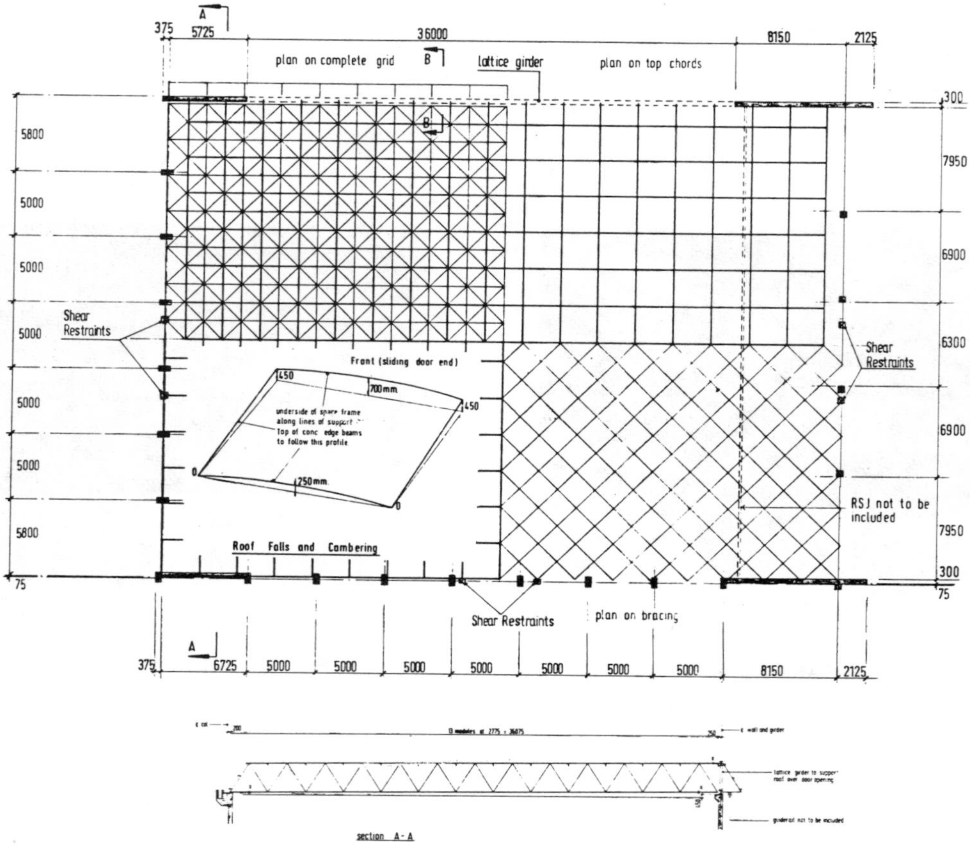

FIG. 4 HELICOPTER HANGAR - ROOF PLAN AND SECTION

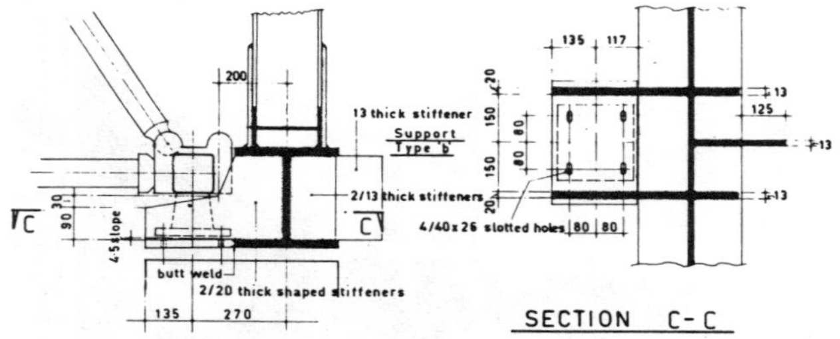

FIG. 5 SUPPORT DETAIL ON LATTICE GIRDER (SECTION B-B)

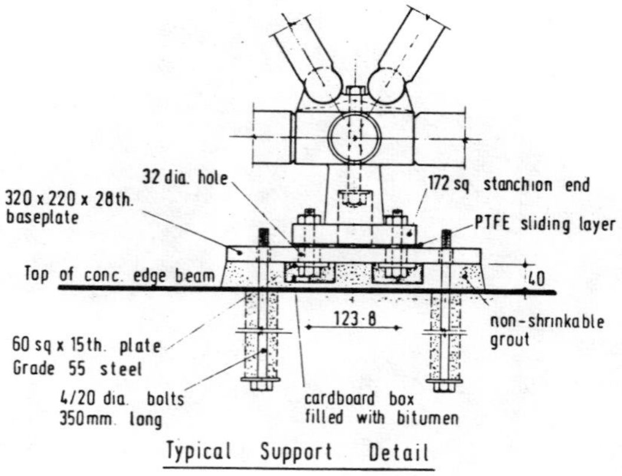

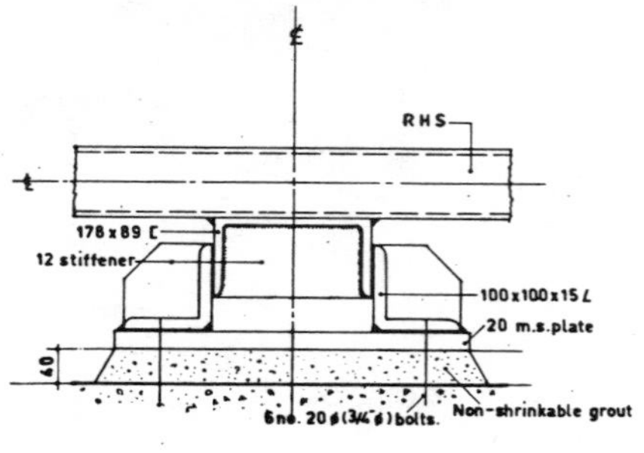

FIG. 7 DETAILS OF SHEAR RESTRAINT

FIG. 6 TYPICAL SUPPORT DETAIL

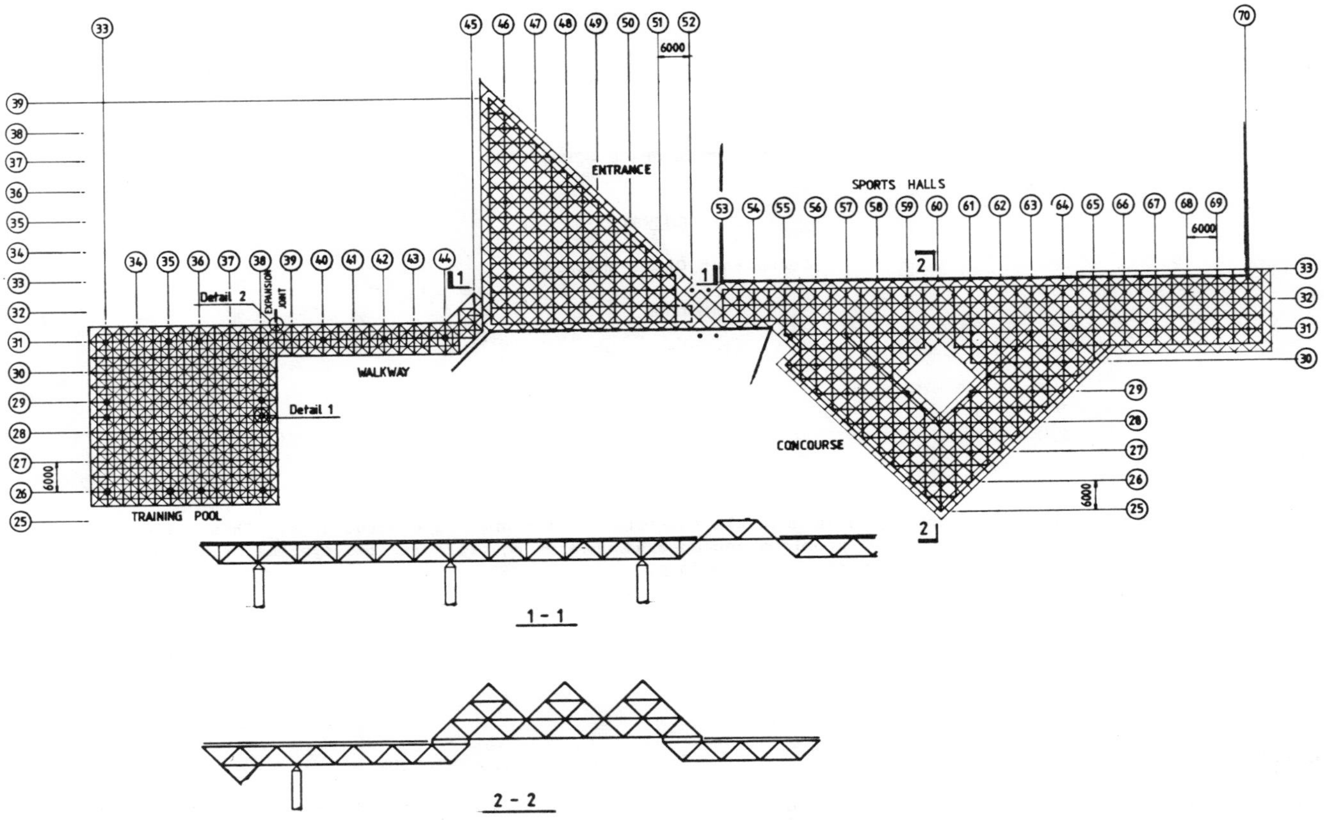

FIG. 8 JUBILEE CENTRE - SPACEFRAME PLAN AND SECTIONS

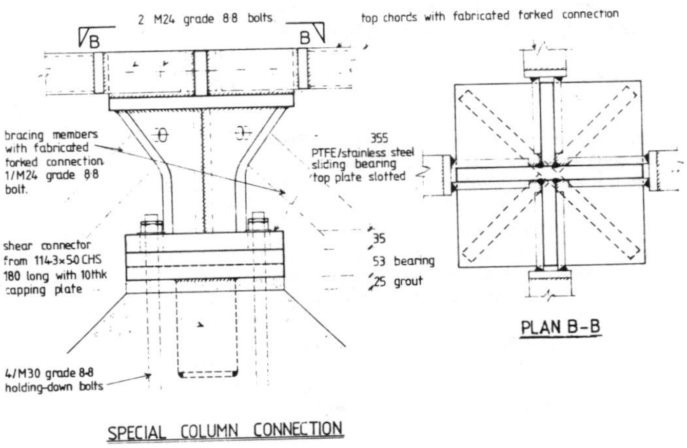

FIG. 9 DETAIL 1 - SPECIAL COLUMN CONNECTION

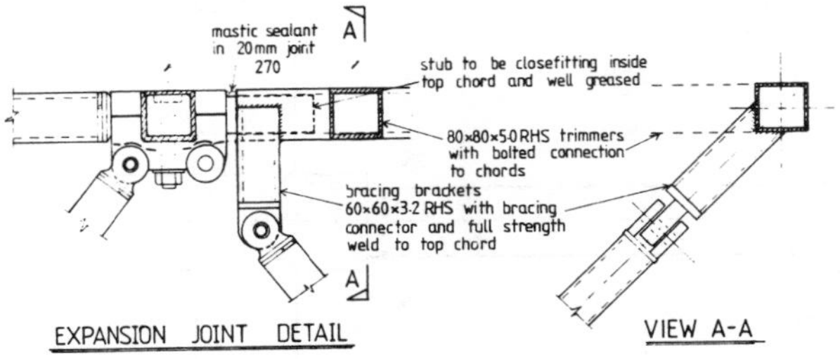

FIG. 10 DETAIL 2 - EXPANSION JOINT

Aberdeen Indoor Games Hall

Architects and Engineers:	The Architectural Office of The Building Development Department of Hong Kong.
Spaceframe Fabricator:	Kennedy Strachan Ltd.
Spaceframe Contractor:	K & E Engineers.

Helicopter Hangar Roof

Architects and Engineers:	The Architectural Office of The Building Development Department of Hong Kong.
Spaceframe Fabricators:	Tube Construction Ltd.
Spaceframe Contractor:	K & E Engineers.

Jubilee Sports Centre

Architects:	Ma & Fong & Associates in association with Planning Services International (H.K.).
Engineers:	White Young BCEOM, Asia.
Spaceframe Fabricator:	Redpath Engineering Ltd.
Spaceframe Contractor:	RDL Jardine Joint Venture.

T.Y. Ho & Partners were the Spaceframe Consultants in all three projects.

A JOINT FOR STEEL SPACE STRUCTURES

J KOLOSOWSKI

Consultant Civil Engineer

London

United Kingdom

The paper proposes a joint which is suitable for both lattice and space structures. This design of joint eliminates a number of problems encountered when using prefabricated systems of assembly and is both light in weight and simple to produce.

INTRODUCTION

Conventional structures and their methods of assembly using prefabricated elements have the disadvantage of being susceptible to inaccuracy in the manufacture of individual members and to the variable conditions found on site. Inaccuracy of members in length and initial lack of fit in general, can give rise to substantial initial stresses which are difficult to assess. These inaccuracies may also result in a loss of rigidity of joints, and of the structure as a whole.

DESCRIPTION OF JOINTING SYSTEM

A typical application of the proposed joint is shown in Fig 1. It comprises of two opposed sockets, a preset distance apart, and a member which is longer than this preset distance but which can be inserted at its respective ends into the sockets. The method of assembly comprises the steps of inserting one end of the member into one of the sockets such that the remaining length of the member is less than the preset distance. The other end of the member is then aligned with the second socket and inserted into this socket, such that the first end still remains within its respective socket. The member is then secured by welding each end to its socket. The above process is illustrated in Fig 1.

FABRICATION AND ERECTION

The properties of such a joint will now be discussed in relation to methods of fabrication and erection.

The first point relating to the proposed jointing system is that the main members, ie, the members carrying the greatest loads, should be continuous, as shown in Fig 2, and Fig 3 representing a two-way grid and a dome respectively. Figure 4 shows the plan and elevation of a typical joint in such a dome.

The sockets can be fabricated from the standard hollow circular steel sections, the material being of the same grade as the whole structure. The internal diameter of the sockets should be about 4mm to 6mm greater than the external diameter of the secondary member which is to be installed within them. This clearance should be sufficient to allow members to be inserted into sockets without fouling any fittings provided at the joints, as shown in Fig 1. The sockets should be profiled at one end to fit the main member and welded to these members in the fabrication shop. In order to facilitate mass production, the external diameter of the secondary members should be kept constant, the cross-sectional area of the member being controlled by varying the wall thickness of the tubes. Finally, the secondary member should be welded to the sockets on site.

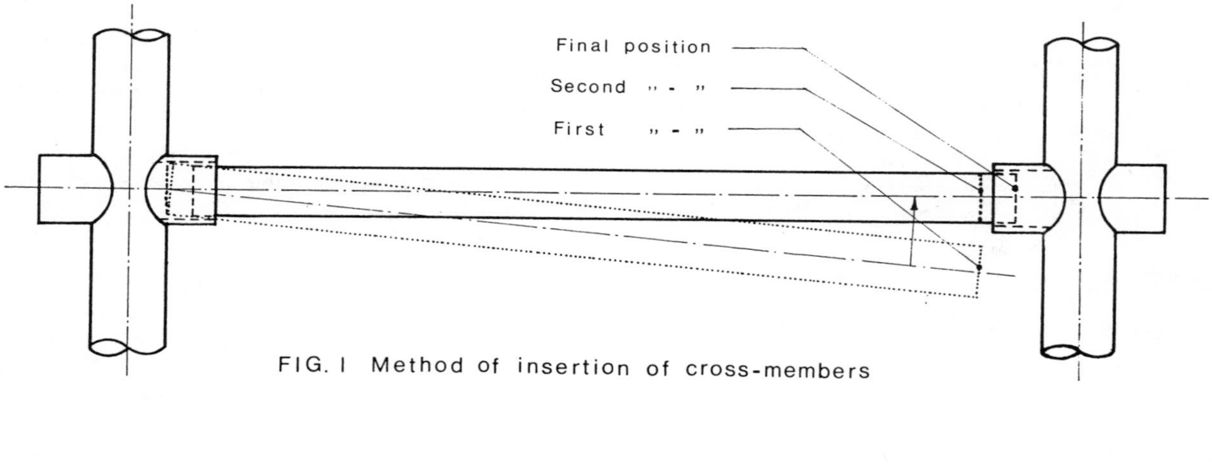

FIG. 1 Method of insertion of cross-members

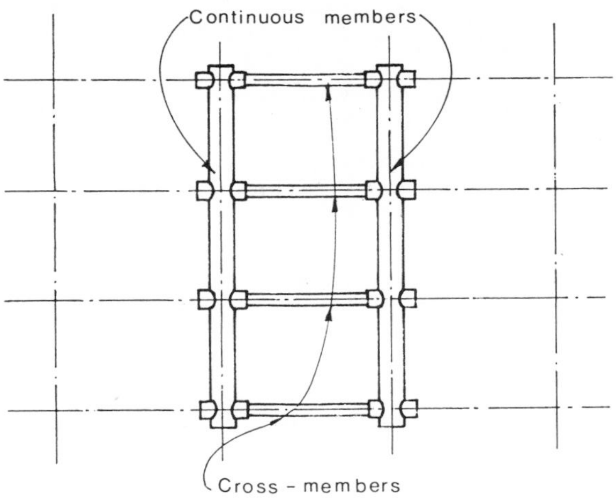

FIG. 2 Application of the joints in a grid

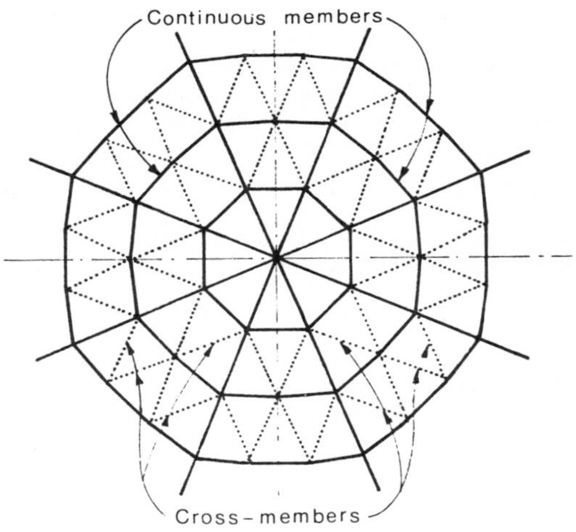

FIG. 3. Application of the joints in a dome

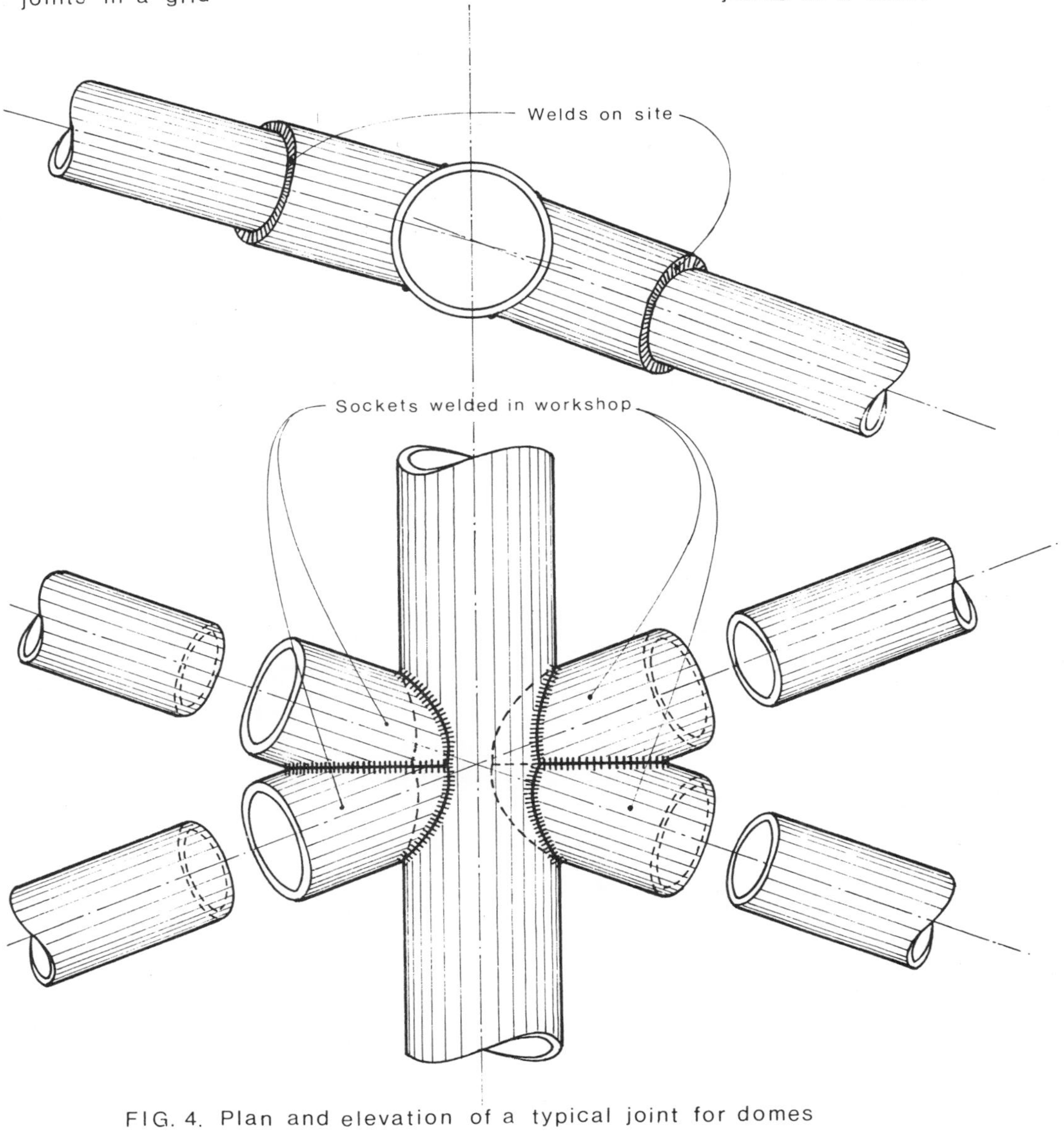

FIG. 4. Plan and elevation of a typical joint for domes

ADVANTAGES OF THE PROPOSED SYSTEM

Now let us consider the properties of the proposed joint. The design of any joint is a matter of compromise normally solving some of the theoretical and practical problems at the expense of other aspects.

The proposed joint has the following distinct advantages:-

(1) The joint allows the connection of members accurately along their centre lines.

(2) Stresses due to initial lack of fit are eliminated.

(3) All cross-members or diagonals can be cut straight and without great accuracy which makes their fabrication a simple task.

(4) The joint provides the full flexural and torsional rigidity of members, thus the lateral stability of a structure is increased substantially. Therefore, the effective length of members could be reduced compared particularly with joints bolted on site.

(5) The joint is very light because the main members are continuous and the secondary members are cut off much shorter than their theoretical length, thus the weight of sockets should be nearly balanced by the weight of the pieces cut off.

(6) Welding of the steel of the same grade is involved throughout the whole structure.

(7) The joint is water and air tight, therefore corrosion from inside is prevented.

The main disadvantage of this joint is the necessity of welding secondary members to sockets on site. In order to reduce the extent of this disadvantage, it is suggested that the erection of a dome for example, be started from the centre, progressing one ring after another. The completed part of the structure could then be jacked up gradually so that the welding can be carried out at a convenient height above the floor level.

CONCLUSIONS

The practical applications of this joint, like others, depends on many prevailing conditions: available equipment, costs involved in switching from one method of erection to another and conditions on site. Nevertheless, it may be worthwhile analysing the costs involved in the proposed jointing system where the possibility of the mass production of joints and members may substantially reduce the overall costs of a structure.

REFERENCE

U.K. Patent Application No. 8313602
"Joint for Space Structures".

TEMCOR SPACE STRUCTURES DEVELOPMENT

DON L. RICHTER*

*Vice President, Temcor
2825 Toledo Street
Torrance, California 90503 U.S.A.

The author's independent research and development work in Space Structures started in 1949 and in 1964 he helped form Temcor to further develop and manufacture his aluminum dome structures.

Temcor Geodesic, Crystogon and PolyFrame domes of aluminum construction are now being used throughout the world in all climatic environments from the frigid polar icecap to the deserts of the Middle East.

Two of their PolyFrame projects; the South Pole dome and their 126.5 meter dome covering the "Spruce Goose" in Long Beach, California are reviewed in this report.

Since both projects, the South Pole dome and the Long Beach dome, involved the possibility of major foundation settlement and dome distortions a full scale test program and a non-linear computer analysis have been conducted on aluminum dome structures. The results of some of these Temcor studies have been summarized in this paper.

INTRODUCTION

The author of this paper had the privilege of presenting the paper "From Early Concepts To Temcor Geodesic Domes" at the 2nd International Conference on Space Structures in September 1975. That paper reviewed 25 years of early research, most of which was done without the aid of computers. The development of powerful computers and software have now added a whole new dimension to the potential for structural refinements; and even more importantly they have added to our spectrum for creativity.

SOUTH POLE REVISITED

At the invitation of the U.S. Navy, Temcor designed, fabricated, and shipped a PolyFrame Dome of all-aluminum construction for assembly erection at the South Pole. This dome was air shipped to the Pole in 1972 and erected by Navy construction personnel under the direction of a Temcor engineer.

The all-aluminum PolyFrame Dome is 15.2 meters high by 50 meters in diameter and serves as a giant "weather break" to protect the three scientific station buildings from the severe polar climate. Winter temperatures on top of that 2,804 meter high polar ice plateau drop as low as -74 degrees centigrade and seldom rise above -18 degrees. The protected scientific station includes a communications center, the crews quarters, and a science laboratory; all part of the ongoing geophysical research.

Photo 1 Interior of South Pole Dome

The Temcor dome was designed to withstand 200 Kilometers/hour winds, uniform snow loads of 586 Kg/sq.m. and, even more importantly, an unbalanced loading of 1465 Kg/sq.m. on the side. The ideal spherical dome shape and unusual strength of the Geodesic PolyFrame construction was expected to postpone its eventual crushing and disappearance under the heavy snow and ice.

Photo 2 The drifting snow continues to pile on one side of the South Pole Dome.

During the summer of 1982/1983, ten years after the dome was erected, we were asked by Antarctic Services to conduct a field investigation of the South Pole dome. There have been reported distortions in the foundations and aluminum dome.

The inspection trip to the South Pole was made by our consultant, Gary Curtis. The first thing he discovered was the importance of the "non-symmetrical" loading requirement in the original Specifications. The winds at the Pole are predominately from one direction and formed a huge snow drift on the lee side of the dome. The combination of the three meter deep drift on the dome, the burden of snow on the firn (snowpack) to the lee of the dome, together with the sewage outfall in the same general area, melted some of the Firn causing subsidence under part of the foundations.

The dome foundation was made of timber pads placed and frozen into the ice plateau 0.5 meters below the snow surface.

One of the significant parts of the report on the investigation was "that no damage to the structural integrity of the dome has occured" in spite of foundation distortions and settlements of 0.4 meters under one side of the dome. Further studies of this dome are planned.

It is this kind of demonstrated strength with flexibility in aluminum dome structures that has become part of our ongoing research at Temcor.

LONG BEACH DOME

During World War II Howard Hughes was awarded a contract to build a cargo aircraft, using nonstrategic materials, that could carry military vehicles and personnel non-stop from America to England. His solution was the now famous "Spruce Goose" aircraft. Made entirely of wood veneers, this flying boat, with a wingspan of 98 meters, was completed too late for the war effort. It was flown only one time and has now become a National Treasure. Temcor was asked to provide a permanent home for this treasure of aeronautical engineering.

The all-aluminum PolyFrame Dome designed, manufactured and erected by Temcor for this project included a number of pioneering solutions; the most challenging being the construction of the world's largest clear-span aluminum dome over the huge aircraft. And do it all at less cost than any other proposal, in any other material or design.

The following photographs illustrate the construction procedure used to erect the Temcor dome for the Spruce Goose aircraft.

Photo 3 Start of the dome construction around the base of the erection tower.

Photo 4 The 126.5m dome subassemblies are completed and made ready for hoisting and bolting together.

Photo 5 All the PolyFrame dome subassemblies have been joined together in this photo with one large opening left for insertion of the aircraft.

Photo 6 Built next to the Queen Mary in Long Beach, California, the Temcor Dome received the "Spruce Goose" flying boat.

The all-aluminum dome with its large temporary opening is supported by the center "erection tower" and two cranes while the plane tail is carefully moved into the dome.

The large triangular opening prepared for the vertical tail was then closed and the cranes removed. It was then moved all the way in, the dome structure completed and the central tower removed.

Probably the second most challenging aspect of the design of this large dome structure, and more relevant to this paper, was the poor soil upon which the dome had to be erected. Soil investigation revealed that the site had a long and varied history of land fill from harbor dredging operations. The composition of some of the soil was such that under seismic agitation it could liquify. That is, it could lose all bearing strength along 30 meters of the foundation. While a system of piles was considered to support the dome, the cost was prohibitive.

Because of the light weight of the aluminum dome a simple spread footing foundation was the most practical solution. To determine the effect of a loss of bearing support under 30 meters of foundation we conducted a number of computer studies. We found that not only would the aluminum dome shell span the gap that could open up under the foundation, but it would also carry the dead weight of the foundation in that zone. While the theoretical deflections were large, the member stresses were well within the allowables.

Once again we see an example of an aluminum dome with large deflection potentials that is structurally sound.

NON LINEAR ELASTIC ANALYSIS

Temcor's 126.5 meter dome in Long Beach has a spherical radius of 70.3 meters and a frame depth of 0.2 meters or 0.28% of the radius.

A clarification should be made at this point about the kind of structure we are analyzing. The dome framing geometry is geodesic in the top half, merged with a lattice geometry around the base. It is fully triangulated right down to the foundation, therefore is a complete space truss. Because of the minimal framing depth one might call it a single layer grid, but it is in fact a double layer grid. All of the interconnecting joints have gusset top and bottom as illustrated in Fig 7. If the top and bottom flanges were laced together as a truss instead of having a single web as a beam the double layer quality of the dome framing would be more apparent. If a dome were actually made as a single layer grid in the proportions of the Long Beach dome it would collapse.

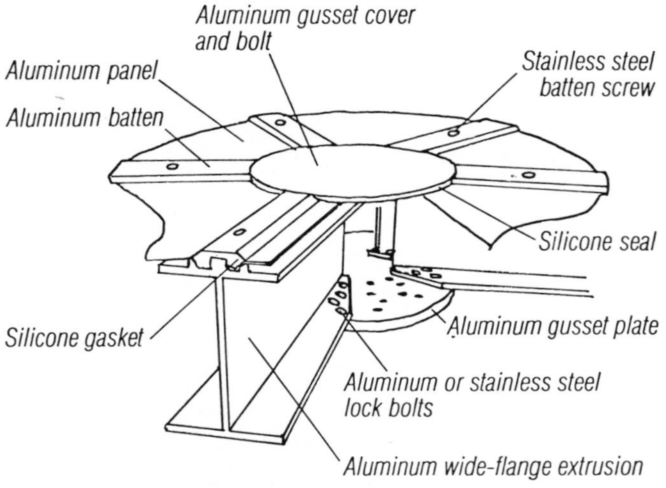

Figure 7 Temcor PolyFrame Dome connection detail

Certainly one aspect of this kind of construction that must be considered is the structural effect of large deflection in the dome frame. Several methods have been used to analyze such behavior as "snap thru" in the dome. A number of non-linear computer programs have also been developed and promoted; but can they accurately predict the behavior of the Temcor aluminum domes. As we expand the horizons of structural capabilities such questions become critically important.

TEST PROGRAM

The most direct method of determining whether a computer program can predict the distortions and failure of a Temcor PolyFrame Dome (or any other dome) would be to test the total dome to failure and then compare the results with the computer predictions. But that approach would be impractical. The dome would have to be too large to be tested at a reasonable cost. While a scale model might be tested it would introduce another set of uncertainties without a significant savings in cost.

The suggestion was made that only the top shallow portion of a large dome be made and tested. While that procedure would cost less than testing the total dome, it was set aside because of the difficulty in simulating the behavior of the removed portion of the dome.

If, however, the problem were turned around so that the computer codes were being tested and not the dome, then the solution becomes simple. Manufacture a shallow dome with a sufficient number of nodes and struts; then load it to failure. This shallow test dome could then be analyzed using some of the standard computer codes. If the theoretical analysis accurately simulates the shallow dome test through the range of large distortions and failure, it is a good code. It should also be accurate when used on larger structures of similar construction. Of course, a sufficient number of different dome tests and analysis would need to be conducted to gain confidence in the results.

Interestingly we have discovered that not all "non-linear computer codes" are created equal and the descrepancies were not on the conservative side.

TEST MODELS

Several versions of two PolyFrame domes shown in Figs 8 and 9 were made using various manufacturing procedures and equipment. The dome proportions, member sizes, and connections were selected for the tests so that large reflection instability or snap-thru would be the mode of failure. All of the domes tested had a base diameter of 13.7 meters with a rise of 0.46 meters. The aluminum framing members had a depth of 11.4mm or about 0.2% of the spherical radius. The tension ring was built up of welded, 11.4mm diameter, heavy wall steel pipe.

All of these various member section properties and materials were used in all of the different versions of non-linear analysis.

Member joint stiffness was one of the most important factors to be recognized in our analysis and was one of the most difficult to define. To determine the fixity of the joints a series of 102 joint bending tests were conducted. The test variables were; diameter and thickness of the aluminum gussets, depth of the strut, strut flange thickness, number of fasteners, fastener materials, and the hole tolerances (punched, drilled or sized). As One might expect, some of the combinations had less than full fixity while others appeared to have greater than full fixity caused by the doubling up effect of the gusset. Several of these joint combinations were then used in the dome test.

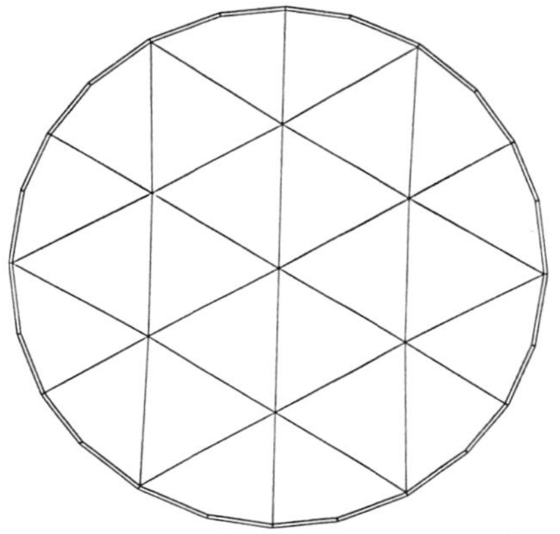

Figure 8 Twelve pier geodesic geometry test dome

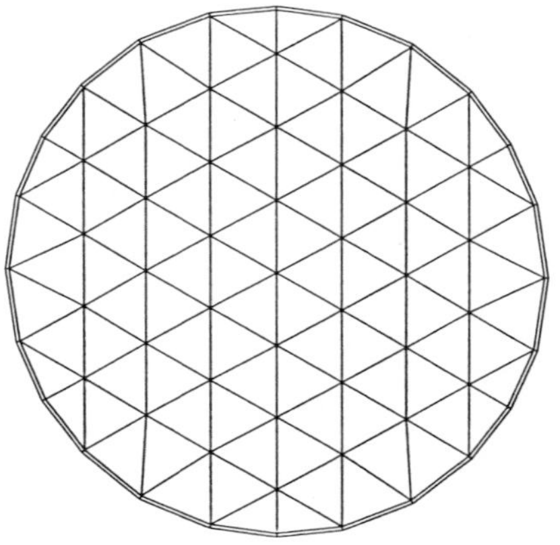

Figure 9 Twenty-four pier geodesic geometry test dome

Snap-thru can be a very catastrophic mode of failure, so the use of any form of weights for test loading was ruled out as impractical. It would not allow proper observation and recording of the event. For this reason it was decided that a system of hydraulics would give us the required control and safety.

The test assembly consisted of the ALuminum test dome mounted on legs one meter over the floor. Supported directly over the test dome is a similar but stronger dome frame used to press against with hydraulic rams. Identical calibrated hydraulic rams were attached to each gusset of the test dome. One end of the ram attached to the test dome gusset by means of a special fitting and the other end attached to the top dome as shown in Photo 10.

Photo 10 Dome test assembly

The rams were double acting and connected together with small hoses on the push side to a common manifold and pressure gage. The pull side of the rams were also joined together to a common manifold and separate gage. This system guaranteed that all load points carried the same force regardless of their relative deflections. When the dome started to collapse it automatically relieved the pressure and stopped the motion. It was then only necessary to allow time for the ram pressures to equalize before deflection readings were taken. Deflection readings were measured from the floor under each gusset node.

Photo 11 Pressure, strain and deflection readings are periodically taken during all tests

Automatic strain gage readings were made at all of the critical points in the frame for comparison with the computer model results.

Any or all of the rams could be disconnected from the test dome at any time to allow for changes in the load pattern as shown in Photo 11. Most of the test failures occured in the elastic range of the aluminum and within the bearing allowable for the fasteners so that the same test could be repeated several times with identical results. A total of 34 dome tests were conducted, three of which are listed in this paper.

UNIFORM LOAD TEST

Our uniform load case was the application of an equal load at each of the 37 interior gussets, shown as black spots in Fig. 12. The loads were increased incrementally with the smaller increases being made as it approached failure. The load deflection curve is plotted "experimental" on Graph 13.

The test "snap-thru" occured at about 1700 pounds (773 kg) per node with the dome center deflecting 6 inches (152mm). Depending on the computer code model (A,B,C or D) the theoretical snap-thru should have reached 20% to 45% higher loads.

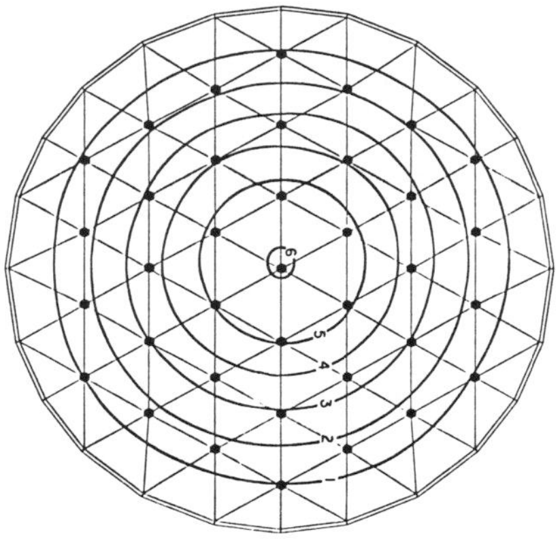

Figure 12 Contour graph of deflections under distributed loading

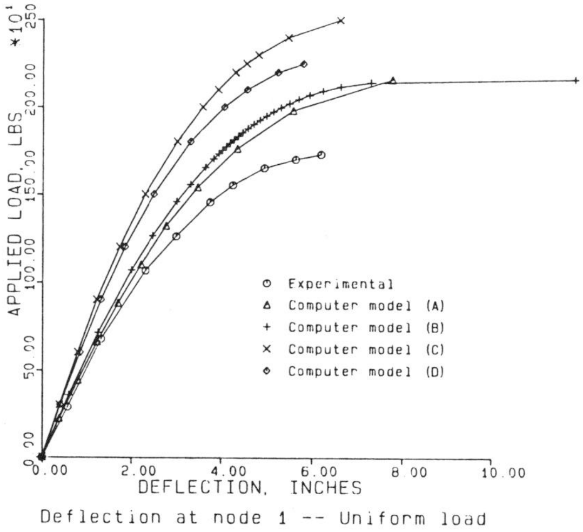

Graph 13 Deflections; theory vs actual

Our test structure recovered virtually all of the deflection when the load was removed, the test loading was repeatable and the strain gages indicated that we were well within the yield stress for the alloys. We have not yet accounted for the discrepancy between the theory and the real thing.

NON SYMMETRICAL LOADING

The non symmetrical loading of our test structure was accomplished by placing the rams on 19 of the possible 37 interior gussets. A total of 1650 pounds (709kg)per gusset was required to reach the point of instability with a maximum deflection of just over 5 inches (130mm). Again we see that the various computer codes predicted somewhat better results than we were able to achieve.

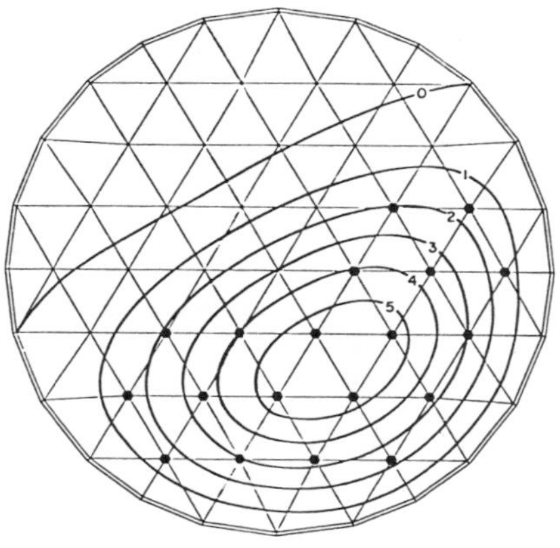

Figure 14 Non-Symmetrical Load

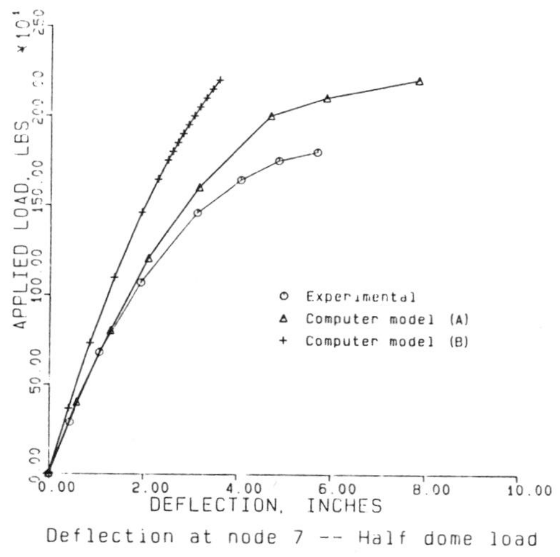

Graph 15 Maximum deflection non-symmetrical loading

While the loads were applied at the gussets of the bare frame in the majority of the tests, we also attached panels to the frame in other tests for comparison. The attachment of the dome panels did not detract from the frame strength.

CONCENTRATED LOADING

To really test the stability of the PolyFrame dome, a concentrated ram load was placed on one gusset located on one side of the center of the dome as shown in Fig 16. Failure did not occur until 13,700 pounds (6227kg) of force was applied. As impressive as that demonstration was, theory predicted even higher values.

As you can see in the "contour graph" of Fig 16 , the slope of dome surface from the concentrated local deflection was quite steep. The loaded gusset point in the dome actually developed a reversal in curvature.

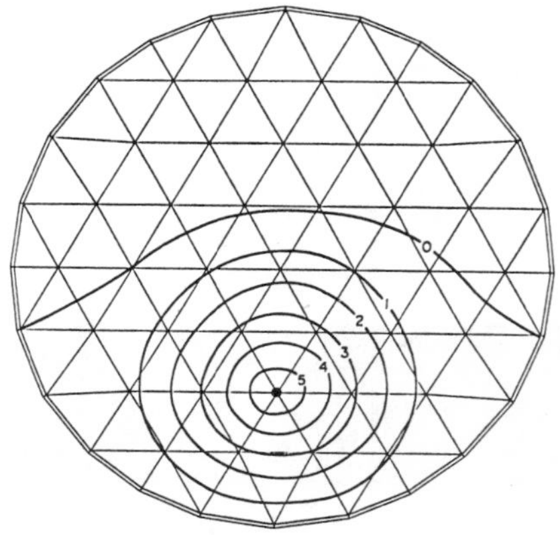

Figure 16 Concentrated load

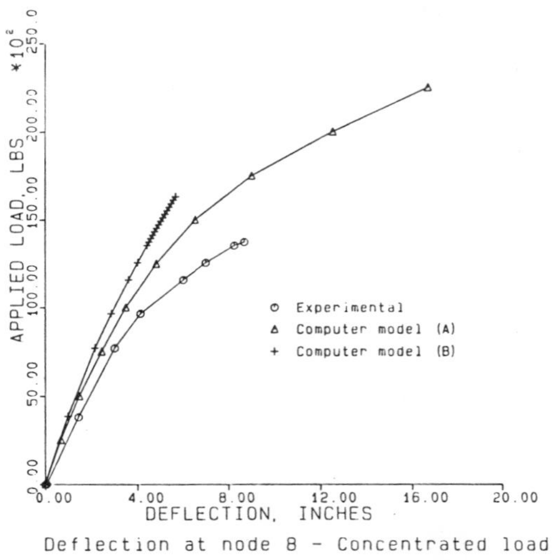

Graph 17 Deflection of gusset under concentrated load.

COMMENTS AND CONCLUSIONS

At the time of the preparation of this paper a total of 34 dome tests have been completed. The results are still being analyzed and only a small portion of the test data could be presented here. As disturbing as the spread between the computed strength and the test results may be, they both indicate a great deal more reserve strength in the Temcor Aluminum dome structures than has been recognized with today's design standards.

The computer codes are being refined to better predict the behavior of our aluminum structures in preparation for even larger constructions.

ROOF SPACE STRUCTURES DESIGNED BY THE DEPARTMENT OF STEEL STRUCTURES OF TIMISOARA, ROMANIA

D. MATEESCU* . GH.MERCEA** , L.GADEANU*** ,D.FLORESCU*** ,P.LORINCZ**** R.BANCILA****

* Member of the Romanian Academy, Professor, Department of Steel Structures, Faculty of Civil Engineering of the Polytechnic Institute "Traian Vuia" of Timişoara, Romania.

** Dr.Sc.tech., Associate Professor, Dept.of Steel Structures Timişoara

*** B.Sc. Eng., Lecturer, Dept. of Steel Structures, Timişoara

**** Dr.Sc.tech., Lecturer, Dept.of Steel Structures, Timişoara

The paper describes some of the most important achievements of the Department of Steel Structures - Faculty of Civil Engineering of the Polytechnic Institute of Timişoara, Romania - in the design of long-span industrial buildings, sports halls, buildings for social and cultural activities etc. A number of arched domes resting on ring space girders, space frame grids and roofs built from space girders are presented.

INTRODUCTION

Some of the most important lines of activity of the Department of Steel Structures of the Polytechnic Institute of Timişoara, Romania, are the design of steel space structures for the roofing of industrial buildings, sports halls buildings for social and cultural activities etc

Purpose	Shape and dimensions in plane			Steel consumption Kg/m2	Systems and dimensions of the structure			Location
	Shape	Notations	Dimensions (m)		Type of roof	Structural System	Dimensions of structure	
Extribition hall	Circular	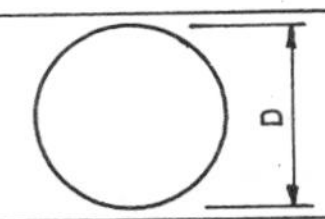	D = 93,5m	100	Dome with arches and ring beam	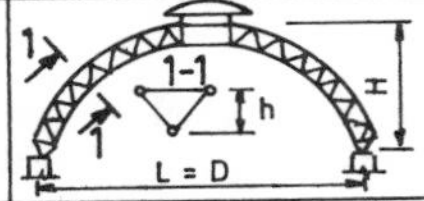	D = 93,5 H = 17,9 h = 2,1	Bucharest
Hydroelectric power station	Rectangular	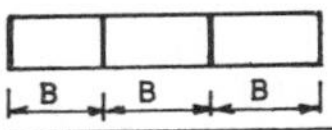	L = 28,6 B = 60,0	75	Space frame	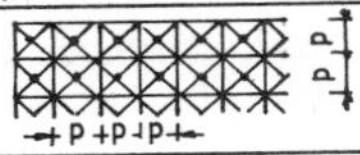	p = 2,2 h= 2,2 ... 2,5	Iron gates
Sports hall	Idem	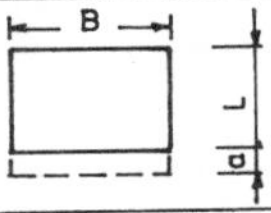	L = 40,0 B = 46,0 a = 4,0	70	Idem	Idem	p = 2,0 h = 2,0	Timisoara *
			L = 46,0 B = 60,0	48				Baia-Mare *
Sports hall	Idem	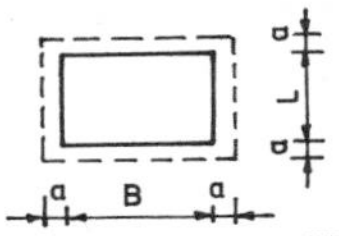	L = 36,0 B = 44,0 a = 2,0	42	Idem	Idem	p = 2,0 h = 2,0	Oraş dr P. Groza
			L = 46,0 B = 46,0 a = 7,0	45			p = 2,0 h = 2,25	Rîmnicu Vîlcea
Sports hall	octogonal	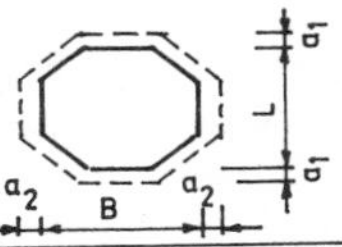	L = 46,0 B = 52,0 a_1 = 2,0 a_2 = 5,0	50	Idem	Idem	p = 2,0 h = 2,2	Arad
Sports hall	Idem	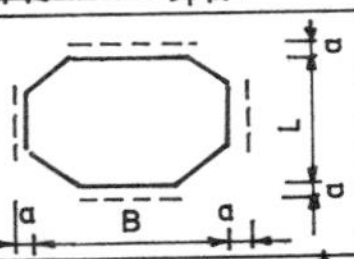	L = 64,0 B = 80,0 a = 9,0	75	Trapezoidal space trusses	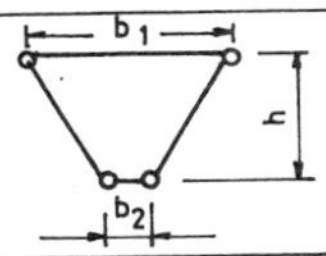	b_1 = 8,0 b_2 = 1,0 h = 5,0	Tripoli LIBYA * *
Indoor swimming pool	Trapezoidal	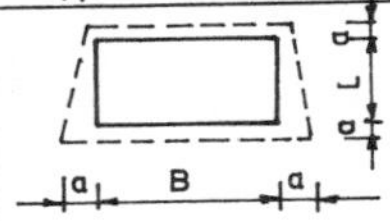	L = 33,0 B = 60,0 a = 3,0	27	Triangular space grids	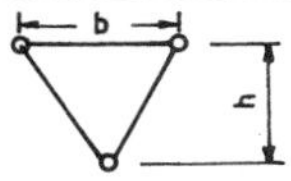	b = 3,0 h = 3,2	Arad

FIG. 1

as well as technical assistance given in the design of such structures.

This twofold activity began years ago, when the Department was invited to design an arched dome on ring girders for the central hall of the Bucharest Exhibition Centre, and continued with a number of projects for roofs made of double-layer space frame grids and space girders of triangular or trapezoidal cross - section for industrial buildings or sports halls.

GENERAL DESCRIPTION OF THE SPACE ROOFS

The space frame grids were designed in various forms on plan, ranging from square and rectangular shapes to octogonal and circular ones as listed in the Table in Fig.1. This table also provides the principal dimensions of these roofs in plane and in cross-section as well as the constructive system employed. In this table L is the span of the roof, B is the length, a is the length of the roof cantilevers, h is the height of the roof structure, p is the size of the panels in the space frame grids and b is the width of space girders. All the space structures listed in the Table in Fig.1 were designed by the Department of Steel Structures, except for the sports halls in Timişoara and in Baia Mare, which were designed in cooperation with the designing institutes in these towns. In case of the sports hall of Tripoli, Libya, the detailed design for the infrastructure and the technical design of the roof were designed by CONSARC, Glasgow (U K), while the modified detailed design of this roof was designed by the Department of Steel Structures of the Timişoara Polytechnic.

Data concerning the space structures of the roofs listed in the table are given below.

The dome of the Bucharest Exhibition Hall, with the dimensions indicated in Fig.1, is made of 16 arches, i.e. 32 half-arches, with a triangular cross-section 2.1 m high and a radius of curvature of 7o m. The arches are connected at the springings and above the roof light by means of two triangular ring space girders. The ring beam at the springings of the arches is designed to counteract the thrust of the arches in case of failure of the reinforced concrete ring supporting the arches

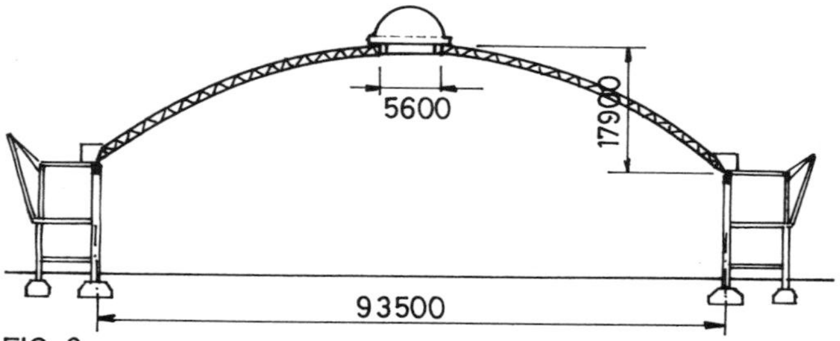

FIG. 2

Fig.2 while the ring beam above the roof light serves as a support for the latter and ensures the spatial interaction of the structure. At their crowns the arches rest on a central ring made of a box-girder with D = 5,600 mm and h = 2,600 mm.

The roofs of the hydroelectric power station located at Porţile de Fier - the "Iron Gates" of Danube - and of a number of sports halls were designed as double-layer space frame grids of the square on square type with a cornice edge and with the flanges parallel to the edges of the buildings. The size of the panels, p, ranges between 2 m and 2.2 m, and the depth of the grid is between 2.0 and 2.5 m.

The shape and dimensions of the rectangular roof of the sports hall in Oraşul Dr. Petru Groza are indicated in detail in Fig.3, while Fig.4 shows the roofs of the sports hall of Arad. The other space frame grids in Fig.1 are rectangular in shape, like the grid in Fig. 3 and are approximately the same size.

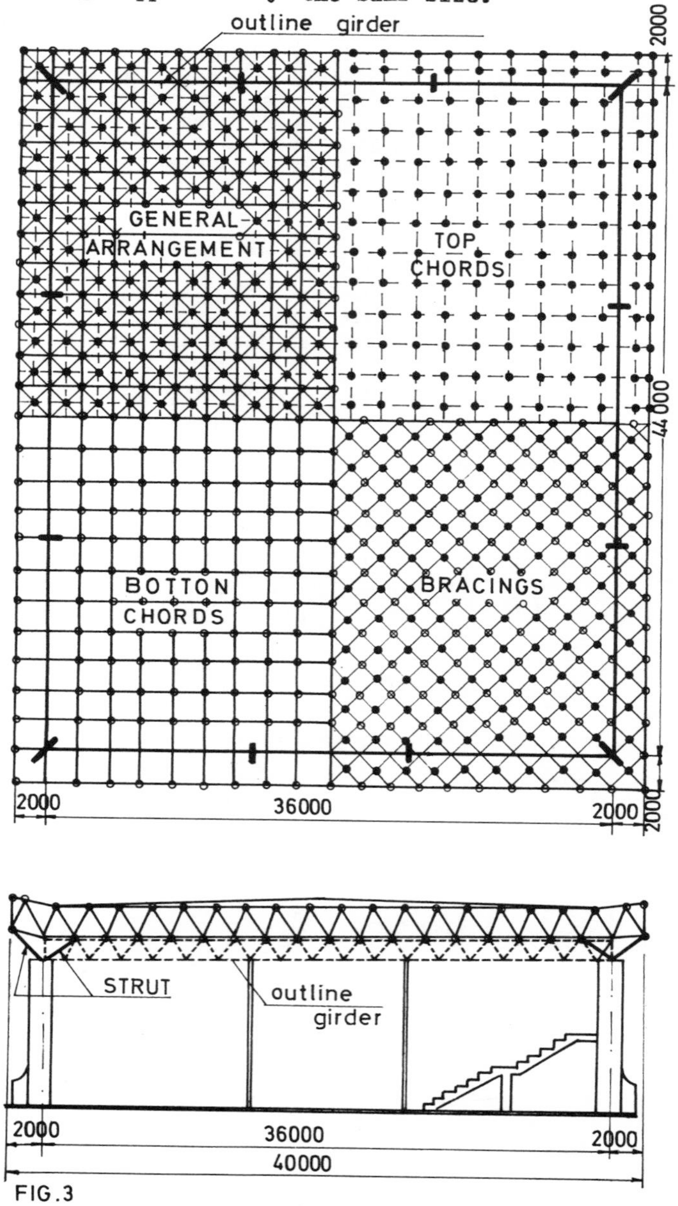

FIG.3

The roof of the sports hall in Tripoli-Libya consists of four principal lattice space girders

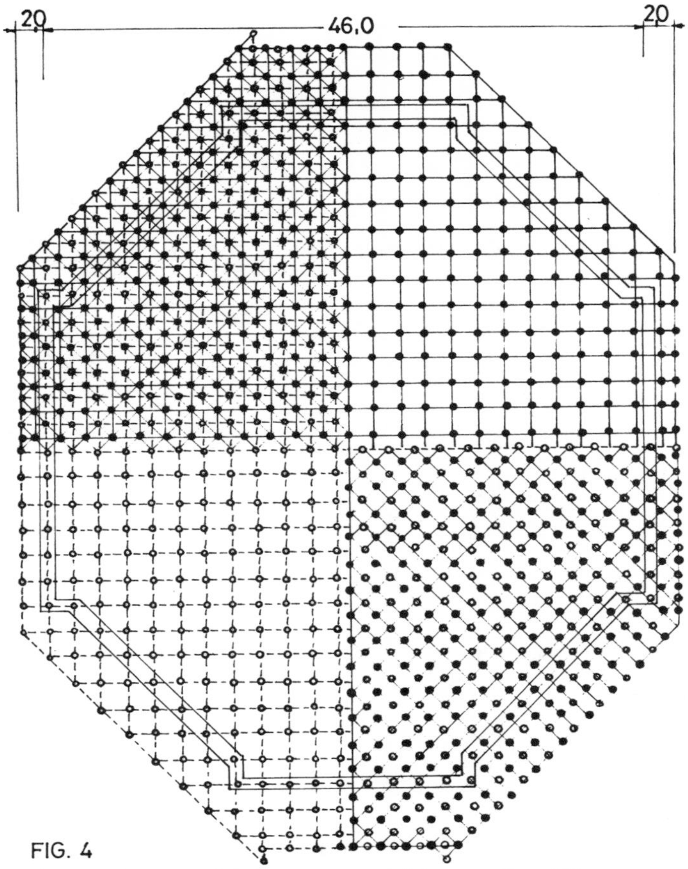

FIG. 4

with a trapezoidal cross-section and a 64 m span and two cantilevers 9 m long, and six auxiliary space girders, three on each side,

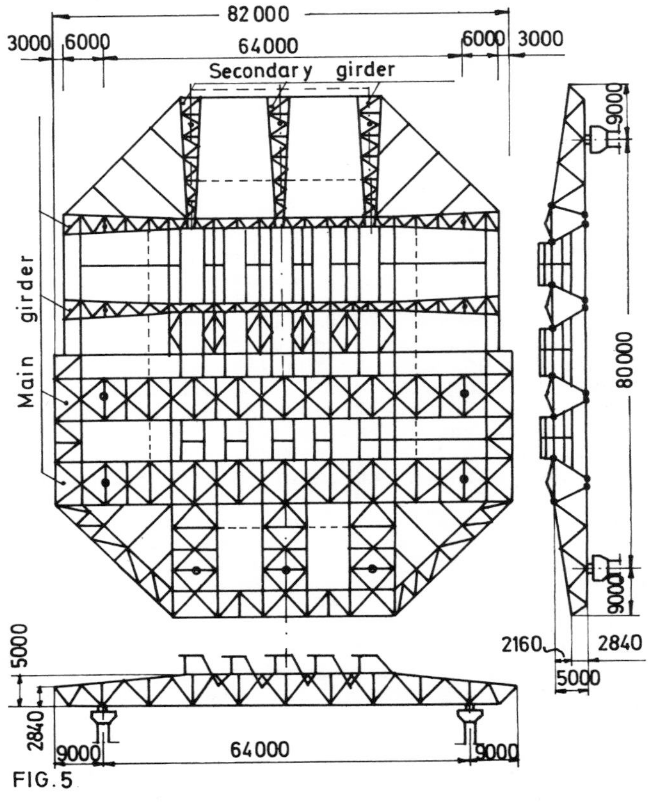

FIG.5

resting on the principal girders and on columns 8o m apart - measured in the longitudinal direction - the auxiliary beams also have cantilevers 9 m long. The principal and auxiliary girders are connected at the corners by means of two-dimensional lattice girders see Fig.5.

The roof of the indoor swimming pool in Arad consists of space girders of triangular cross-section, the distance between the upper flanges being 3 m, see Fig.6.

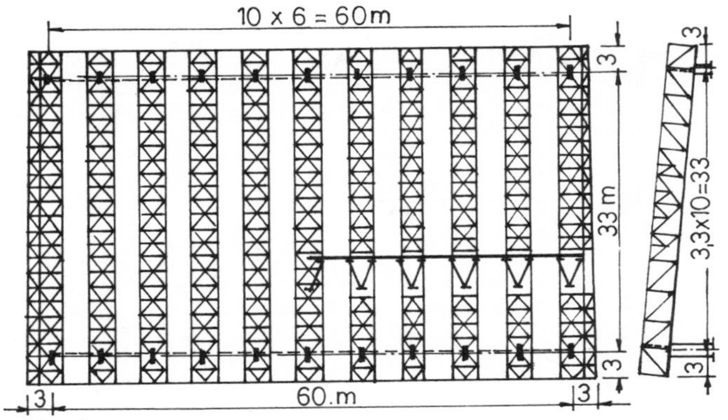

FIG.6

Above the two supports the girders are connected over the full length of the roof by means of plane lattice girders.

The cladding of the dome of the Bucharest Exhibition Hall consists of sandwich plates made of aluminium sheet; the claddings of the power station are made of waterproofed and heat-insulated reinforced concrete plates. For the other roofs listed in Table, see Fig.1., corrugated steel sheet with polyurethane heat insulation and classical waterproofing was used.

GRID ANALYSIS

The space structures of the roofs that have been described in this paper have a high degree of static indeterminacy; the analysis was therefore performed with the help of a computer.

In the analysis of the dome of the Bucharest Exhibition Hall, the following loadings were taken into account : a permanent load of 120 daN/m^2; 4 concentrated forces resulting from the arrangement of the exhibits; 5 different variants of loading due to snow, as symmetrical or unsymmetrical loads, assuming agglomeration of snow on 1/8 of the surface of the dome and attaining values ranging from 80 daN/m^2 at the crown and 800 daN/m^2 at the springing; two variants of wind action, the action of temperature variation, and seismic action.

Because of the high degree of static indeterminacy of the dome, the analysis was carried out by separating the dome into plane elements; the unknowns introduced in the analysis were the connecting forces between the arches acting at the springings, and the forces acting between arches and ring at the points of intersection. Only those unknowns were retained that had a significant influence on the internal forces within the elements of the space structure. In order to calculate the stresses due to the action of concentrated forces, the influence surfaces of the stresses in various sections of the arches and the ring were also determined. A more detailed presentation of the dome analysis is given in Ref.2.

The analysis of the two - layer space frame grids was carried out in two steps. In the first step, an approximate analysis was made to determine the bar cross-sections necessary for the second step.

In the first step of analysis the method of equivalent continua was used, starting from the equation of the deformed position of the plate normally loaded on plan. In this equation the second term, which takes account of the torsional rigidity of the plate is small in grids, and was therefore neglected, so that the equation took the following form :

$$\frac{\partial^4 w}{\partial x^4} + \frac{\partial^4 w}{\partial y^4} = \frac{p}{D} \qquad \ldots\ldots 1.$$

Integration of this differential equation was carried out by means of the method of finite differences.

To simplify the analysis, the actual deformations W_{ij} were expressed as conventional deformations using Eqn.2 :

$$W_{ij} = \frac{p}{D}\,(b_x)^4 \cdot W^o_{ij} \qquad \ldots\ldots 2.$$

which substituted after simplifying and arranging the terms yield the following equation:

$$12\,W^o_{m,n} - 4(W^o_{m+1,n} + W^o_{m-1,n} + W^o_{m,n+1} + W^o_{m,n-1}) +$$
$$+ (W^o_{m+2,n} + W^o_{m-2,n} + W^o_{m,n+2} + W^o_{m,n-2}) = 1 \quad \ldots\ldots 3.$$

Writing such equations for each node produced a system of as many equations as there are unknowns in the structure. The equations corresponding to the supporting nodes along and near the edge are modified by equations into finite differences of the supporting conditions.

After solving the set of equations on the computer first the deformations W_{ij} were obtained with the moments and shearing stresses acting on a strip 1 cm wide were calculated, and next the internal forces N_t in the transverse flanges N_l in the longitudinal ones and D in the diagonals.

By means of the cross-sections thus obtained a more accurate analysis could be performed, using finite elements of the truss type and applying the displacement method in which the unknowns are the end displacements of the bars equal to the displacements of the corresponding nodes. The equations connecting the end displacements and forces of bar h Fig. 7 are as follows :

$$\bar{E}_h = \bar{r}_h \cdot \bar{D}_h \quad \text{and} \quad E_h = r_h \cdot D_h \qquad \ldots\ldots 4.$$

where the first relation is defined in the local co-ordinate system of the bar, and the second is defined in the global system. Vector E_h of the stresses is expressed as a function of vector D_h of the displacements and of the stiffness matrix r_h of the two - hinged bar, which has the form:

$$r_h = \frac{EA}{l}\begin{bmatrix} B & -B \\ -B & B \end{bmatrix} \quad \text{where}$$

$$B = \begin{bmatrix} a_x^2 & a_xa_y & a_xa_z \\ a_ya_x & a_y^2 & a_ya_z \\ a_za_x & a_za_y & a_z^2 \end{bmatrix} \qquad \ldots\ldots 5.$$

in which the directing cosines of the bar axis occur. The equilibrium condition in a node

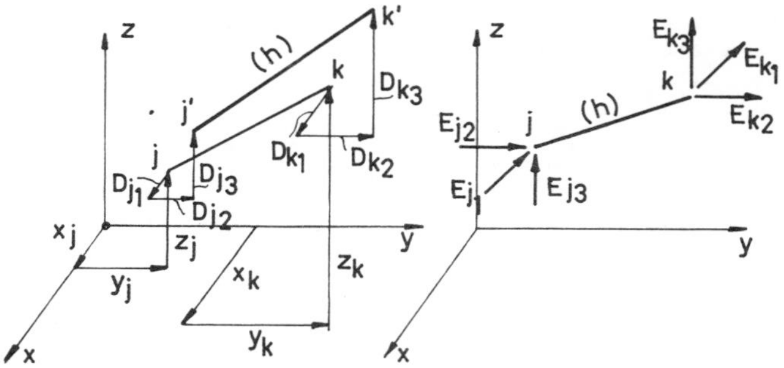

$$D_h = [D_{j1}\ D_{j2}\ D_{j3} \mid D_{k1}\ D_{k2}\ D_{k3}] \quad E_h = [E_{j1}\ E_{j2}\ E_{j3} \mid E_{k1}\ E_{k2}\ E_{k3}]$$

FIG. 7

requires the resultant of all end stresses of the bars united in that node to be equal to the resultant of the loads acting in the node.

$$E_1^{(h)} + E_2^{(h)} + \ldots\ldots + E_j^{(h)} = P_j \qquad \ldots\ldots 6.$$

Expressing the equilibrium of all nodes yields a set of linear equations of the general form:

$$R \cdot D = P \text{ and so } D = R^{-1} \cdot p \qquad \ldots\ldots 7.$$

where R is the global stiffness matrix of the structure, D is the column vector of displacements, and P the column vector of the nodal loads. By means of the displacements obtained on solving the set of Eqn.14, all bar end stresses are determined using relations of the form :

$$\bar{E}_h = \bar{r}_h \cdot \bar{D}_h = r_h \cdot A \cdot D_h \qquad \ldots\ldots 8.$$

where A is the matrix of displacements transformed due to the rotation of the axes.

Before the set of equations is solved the global stiffness matrix has to be re-arranged and condensed acording to the list of supports, thus yielding a matrix in the shape of a symmetrical band with very many zero elements. The system was solved by using procedures-suitable for large-size broken matrices together with the automatic SISART programme.

With the resulting stresses, which were summed for all loading assumptions, bar detailing was carried out, using the limit states design method as indicated by the Romanian standard STAS 10108/0-78, Romanian steel tubes of OLT 45 with the design strength R = 2300 daN/cm² were used for the bars. For each type of bars, tubes of the same external diameter and wall thickness were used in the three zones of the structure.

In the case of the sports hall in Orasul Dr. Petru Groza, the space frame grid of the roof was connected with the supporting edge beams all around the edges by means of inclined struts, thus reducing the stresses in the structure, particularly in the area of the supports.

DESCRIPTION OF SPACE STRUCTURES CONSTRUCTION

The structures designed were generally made of OLT 45 steel tubes; only for the space girders were rolled steel strips used.

The arches and ring girders of the dome are made as space girders of triangular cross-section, with the diagonal bars and bracings welded directly to the flanges.

A detail of an arch and of a node is shown in Fig.8. At the springings, the three flanges of

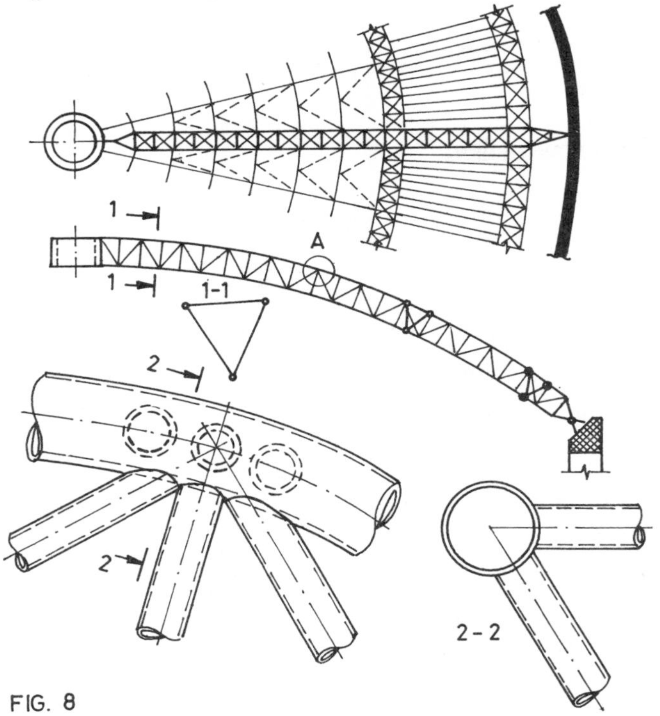

FIG. 8

the arches were united and welded in a special cast member by means of which they are fixed to the supporting hinge. At the crown, the two upper flanges were united into a single bar, and the arch was fixed to the central ring girder.

The space frame grids were constructed in the shape of a square grid with the flanges parallel to the edges of the building. The bars made of steel tubes were connected at the nodes by means of spherical joints. An original solution resembling the Oktaplatte joining system see Ref.1. was adopted for joining the bars.

In this solution, the bars of the transversal flanges are continuous, passing through spheres provided with holes in the direction of the bars; the longitudinal bars are cut and welded to the spheres, as are the diagonal bars. Figs. 9,10, 11. This solution has the advantage that the bars that are stressed most are continuous and thus the sphere is not stressed too much in

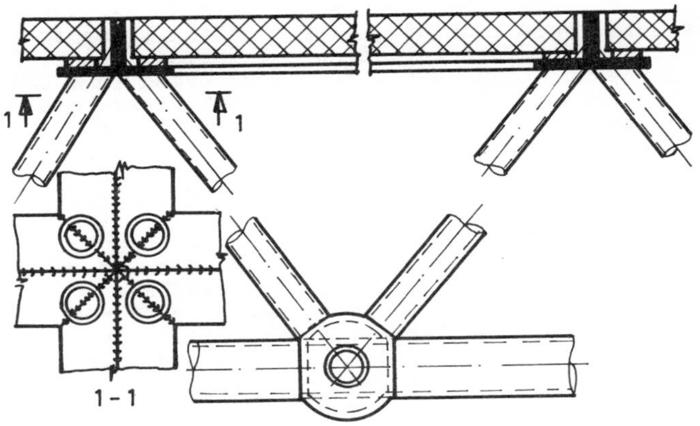

FIG.9

this direction. In the roof of the hydroelectric power station, the upper flanges were made of angle sections set with the flanges upwards, thus forming a kind of frames in which the prefabricated reinforced concrete plates of the cladding were placed Fig.9.

The spherical joints were at first made from two half spheres welded together around the circumference; these halves were made of two pieces of thin hot-rolled sheet, using dies which were then provided with holes normal to the plane of the weld. In order to improve the

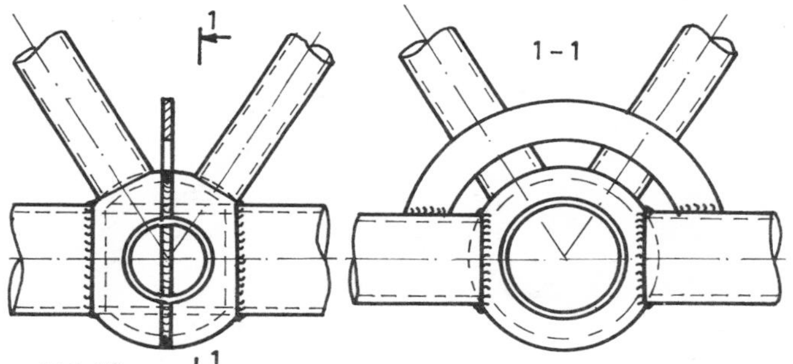

FIG.10

transmission of the stresses in the cut longitudinal bars and to avoid overstressing of the spheres in the plane of the weld, the solution illustrated in Fig.10 was adopted for the second building, i.e. the sports hall of Timişoara. It can be seen from the figure that a semi-circular plate was added above the sphere connecting the two interrupted bars. In the sports halls designed later on the strength of the spheres in the plane of the weld was increased by introducing a ring at the level of the weld, which not only allows the continuous bars to pass, but also allows the sphere walls to be welded over their entire depth see Fig.11.

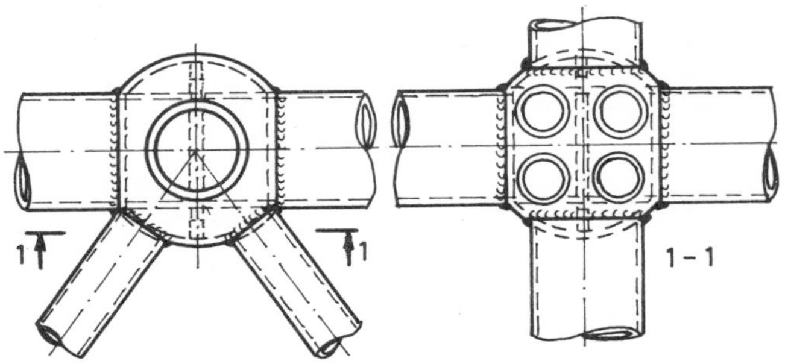

FIG.11

For the sports hall in Oraşul Dr.P.Groza, the spheres were made of pieces of steel tubes to which a ring was welded on the inside. One half of the tube was then heated and formed in a die and after allowing it to cool, the other half was shaped, thus producing a sphere. The advantage of this solution over the one in which two half-spheres had been used lies in the fact that the wall thickness of the sphere, and thus its strength, increases.

The triangular space girders used for the roof of the indoor swimming pool in Arad have flanges made of welded T-sections, the web inclined in the plane of the lateral faces, which allows the cladding to be placed directly on the flange. The bottom flange is made of channel sections, also with inclined webs. The diagonals of the lateral faces were made of steel sheet having a cross-section in the shape of a cross, their web being in the same plane with the webs of the flanges to which they are butt welded; the two lateral strips are also welded to the flanges. The bracings in the upper flange are made of T-sections and welded to the two flanges in the way illustrated in Fig.12.

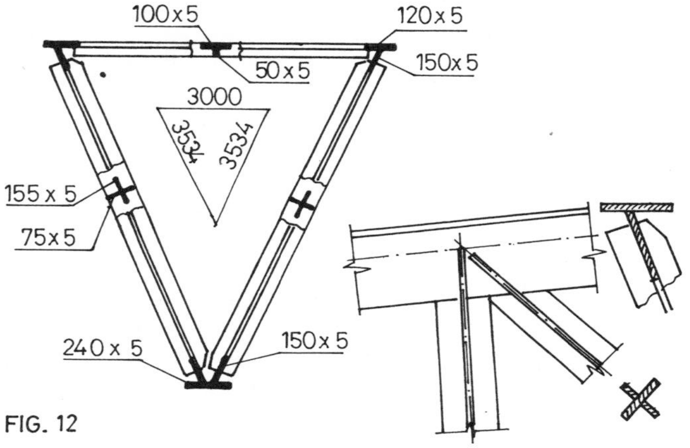

FIG. 12

The trapezoidal space girders of the sports hall designed for Tripoli-Libya, have their flanges made of two channel sections like a box-girder, while the diagonal bars are made of two angle irons with a cross-shaped section. The bracings of the top chord are made of welded I-sections while the diagonals are again made of two angle irons. Owing to the very large size of the principal and secondary girders, Fig.6, which were fabricated in Romania and shipped to Libya by train and ship, high-strength bolts were used for jointing the nodes, Fig. 13.

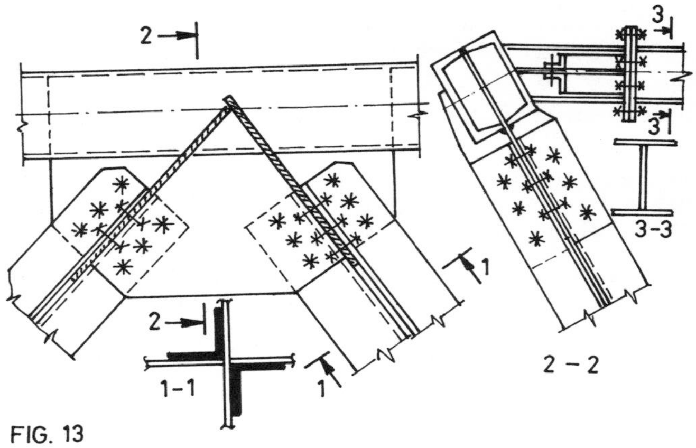

FIG. 13

For easy erection, the principal girders were pre-assembled in the workshops and shipped to Libya.

ACKNOWLEDGEMENTS

The authors' thanks are due to the following members of the teaching staff of the Department of Steel Structures who co-operated with them in the design of the structures here described : I.Munteanu, I.Fleşeriu, M.Ivan and I Appeltauer -Professors; E.Fleşeriu, I.Caraba - Associate Professors; Z.Regep, V.Bondariuc, E.Cuteanu, D. Ciomocoş, A.Dănilescu, I.Dimoiu, C.Konrad, A. Negru, G.Balekisc, A.Botici -Lecturers; D. Dubină, E.Dogariu, D.Bolduş and M.Iosip - Assistant Lecturers.

REFERENCES

1. Prof.Dr.Ing.Z.S.MAKOWSKI : Räumliche Tragwerke aus Stahl.Düsseldorf 1963 .

2. D.MATEESCU - Die Stahlkuppel einer Ausstellungshalle.Space Structures by R M Davies, Edinbourgh 1967.

3. STAS 10108/0-78 - Calculul Construcţiilor Metalice.

SPACE FRAMES AS HOUSES

A NEW STRUCTURAL SYSTEM

R. G. SATTERWHITE*

*Architect, Triangle Works
Washington, USA.

Based on the first wooden space frame residence, Nusatsum House, an improved structural system is described which will be used to construct the second research structure. This new system is comprised of an aluminum connector, wooden member, and wooden panel system. This work is part of an ongoing effort to show that wooden space frame houses are a viable alternative to conventional industrialized housing systems.

BACKGROUND

Nusatsum House demonstrated that it is possible to live in a wooden space frame and provides the only first-hand opportunity thus far to study life in the structure and the structure itself, see Ref 3. The construction method used and the first component system both will be changed by incorporating the following improvements:

1. The external and internal skins must be considered prior to any improvement in joint/member geometry and structural behavior. When ready for skinning, the space frame must present inner and outer flush surfaces with no protrusions.

2. A wooden assembly moves constantly with weather, load reversals, and age, and must accommodate these cumulative deformations. Even the idealized wood-to-connector solution introduces movement without slippage.

3. A structure should utilize member lengths suitable for habitation within a single layer of the system (rather than within two layers or more as in Nusatsum House). At the same time, harmonic lengths may be used as well to create levels at different heights, see Plate 1.

Plate 1

4. A computer analysis should be made for each structure as a check on difficult topography.

5. It is a mistake to apply conventional construction techniques and materials (in unaltered form) in the belief that this will somehow make production easier or more economically viable. Conventional construction techniques (as in the application of gypsum wallboard and plywood) or full-size material use (such as standard sheets of gypsum board and plywood) are anachronisms and actually impede the construction process. The materials must be altered and fabricated into components. The techniques must be abandoned in favor of less time-consuming assembly methods.

6. Though the total time effort to plan and erect such structures will be drastically reduced compared to conventional processes, the actual proportion of planning time to construction time will increase. Quality control of prefabrication must be increased as labor is reduced.

7. A new member/connector system must be developed with greater strength and geometric flexibility.

8. A new panel system must allow wall and ceiling planes to meet at 180, 125, 54, 70, and 109 degrees, see Fig 1. A circular member cross section, therefore, is the most appropriate.

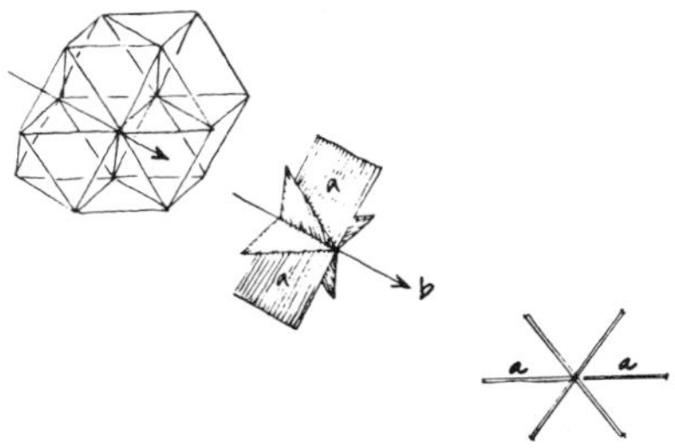

Fig 1

DESCRIPTION OF THE NEW SYSTEM (US, UK, AND FOREIGN PATENTS PENDING)

1. The Connector, See Plates 2 and 3

Plate 2

Plate 3

A one-piece aluminum casting of 356 T6 alloy utilizes the geometry of close-packed tetrahedra and octahedra accepting a maximum of 12 members. A maximum of six retaining bolts are necessary to secure all 12 members in position. Connector geometry implies nonfixity of members, no rotation, and for the purposes of analysis a pinned connection. Connector weight is 1.48 kg. Current production cost is $12-15 US. Twelve degrees of freedom allow for enough architectural variety and reduce significantly the complexity of the connector. Slot connection allows for a linear flat element to distribute forces most easily to a wooden member. Manufacturing involves no machining or forging.

2. The Member

A wooden pole 3000mm long and 150 mm O.D. with 18% max moisture content is tapered to accept a metal tang which translates forces to the connector. Jackpine, lodgepole pine and fir are the suggested species and are obtained easily from tree-farm thinnings and turned on existing uniform diameter machinery. The UK, Ireland, the US, and Canada have vast stands of this material. Wood behavior is more predictable in natural form than with sawn dimensional lumber. Warpage is reduced along member lengths. Most importantly, typical member age from planting to harvesting can be as low as 20-30 years. This recovery cycle for a building material will be important in the future.

3. The Steel Tang, See Fig 2

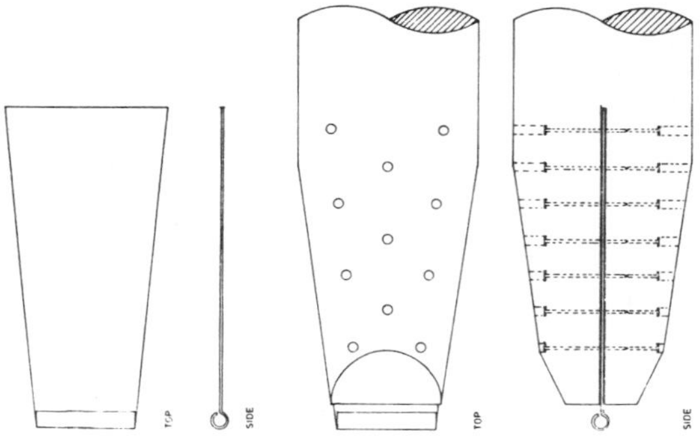

Fig 2

A galvanized 11 gauge mild steel plate has a 7.93mm round rod blank welded to the end. This tang is inserted into the member's kerf cut and pneumatically nailed, using 76mm long 2.87mm diameter case-hardened steel nails with cadmium plate finish. Future improvements to the tang include perforating the plate to form nail surfaces in two directions to increase holding properties.

4. Cellular Panels and Their Tiles

There are two types of building panels which fit into the space frame: triangular and square. They are 150mm thick and approximately 3000mm on edge, see Plate 4.

Plate 4

They are constructed of 12mm   plywood cells interconnected with triangular prisms of hemlock and glued throughout. Since only triangular panels are used as floor elements, they naturally provide the greatest strength because for a given cell wall length the maximum span per cell is less than the square and loads are transmitted three ways rather than two. All panels are exactly the same size whether used as floor or wall elements. The major obstacle to conventional building panels is their awkward weight and size when one tries to place them. The approach, then, is to create a cellular panel only, which is light and can be skinned, once in place, with tiles.

The cells in each panel are covered on both sides with their own "tiles" which can have a number of different surfaces depending on the application. Each tile consists of a 12mm plywood substrate with rebated edges which fits into a cell and is nailed at the corners only. This incremental approach to skinning the structure has several advantages:

a. The tiles which cover each cell are very small and stackable; approximately 450mm on edge. They are easily carried to and fro, removed and replaced. They allow the necessary heavy mass for fire-proofing to be applied in increments instead of sheets.

b. The tiles are only two sizes, square and triangular, so that they are interchangeable, modular, and close-packed.

c. The tiles are never cut; only removed where openings in the walls occur.

d. The tiles can be installed much faster than large sheets which must be measured, fit, often held over head, and may involve wet processes.

e. The tiles take advantage of the space frame's rigidity since panels need not have skins to impart shear strength to the parent structure. The tiles form a discontinuous skin which can move.

5. Thermal Behavior of the System

The structure itself is not thermally conductive because of wooden materials. The cells limit convection of air and cell walls act as natural fireblocks. When a suitable insulant such as foam or fiberglass is in place, migration of those materials is not possible and very low thermal transmission values are expected.

6. Foundations

A major advantage of space frame houses is point rather than perimeter support. Lower foundation costs, adaptability to difficult topography and earthquake resistance are discussed elsewhere, see Ref 2. A modular foundation system is being developed. It consists of doughnut-shaped prisms of concrete which are stacked to the foundation pier depth and then post-tensioned together with a 25mm steel-threaded rod. The same rod protrudes through the base connectors to form a fixed joint.

THE SECOND RESEARCH STRUCTURE

To demonstrate the above-mentioned components, a second research structure will be erected. A model of it is shown in Plate 5.

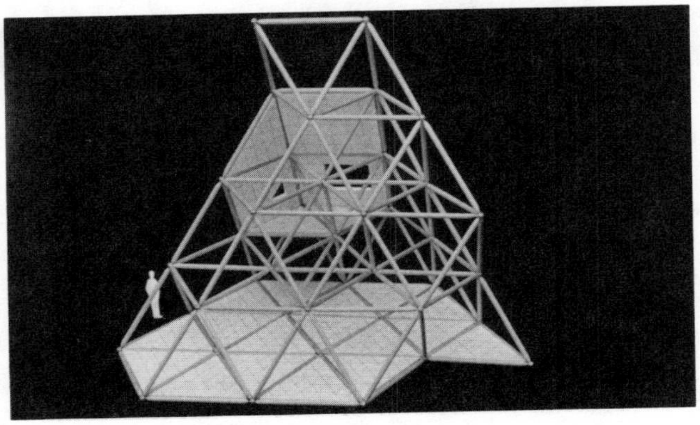

Plate 5

The following are its statistics:

Number of members	-	130	
Member material	-	lodgepole pine	
Member size	-	150mm o.d. x 3000mm long	
Total member weight	-	19.78 x 130 =	2571 kg
Number of connectors	-	41	
Connector material	-	aluminum	
Total connector weight	-	1.48 x 41 =	61 kg
Number triangular panels	-	37	
Number square panels	-	5	
Total triangular panel weight	-	45 x 37 =	1665 kg
Total square panel weight	-	103 x 5 =	515 kg
Number triangular tiles	-	5328	
Number square tiles	-	360	
Total area by level			
ground		$74m^2$	
1st		$52m^2$	
2nd		$31m^2$	
3rd		$9m^2$	
Total		$166m^2$	
Total weight structure exclusive of tiles and mechanical etc.			4812 kg

BUILDING METHOD

1. The house frame is erected as a solid form with all members present.

2. As this progresses, the cellular panels are placed with their tiles and a walking surface is made possible. Kitchen, bath, and stair units are placed in the structure as it goes up.

3. Once the shell is finished, members are taken out of the finished shape to form interior spaces.

4. The windows and doors are located and wall exterior tiles are placed on the remainder of the external skin.

5. The windows and doors are placed in the openings created by cutting out the exterior cell walls.

6. The waterproof membrane is placed on the outside of the envelope to create an airtight and moistureproof skin. All openings are caulked.

7. The interior of the walls and ceilings is now open, ready to accept electrical, mechanical assemblies and insulation.

8. No attempt is made to "hide" joints where the adjacent panels meet on the interior walls.

ARCHITECTURAL SPACE - CAN WE UNTHINK THE SQUARE?

It will be shown that the involvement with the square and cube to the exclusion of other subdivisions of space was so pervasive, so thorough in literature, the arts, sciences, recreation, birth and death, that we were in the dark ages of spatial awareness. Virtually every designer is a Cartesian thinker, and spaces are only imagined in rectangular prisms. Recently, the expedience of the square is being challenged in the mildest way by skyscrapers which timidly at first but now quite fashionably introduce X-bracing on the walls to rigidify a cubic frame. The person who lives in a house experiences the most primitive and the slowest changing technology there is. A person will purchase a video terminal unaffected by sentiment, tradition or fashion but will not tolerate any change in the immediate environment. The manufactured housing industry is the only major one where new products are made to imitate old ones as much as possible. Given these circumstances, it is not possible through sheer mental effort to see other than with square references. The senses, however, are susceptible and only through experience will triangles become familiar.

My family lived in a triangulate environment for about three years before moving to construct the next research structure. One can be weaned off cubes by sleeping in tetrahedrons, but a permanent return to a rectangulate environment erases very quickly any progress in spatial awareness. These are some characteristics of space frame houses:

1. A wall may incline away or toward the center of an interior space.

2. Most windows are skylights, and the interior is very bright, see Ref 3.

3. A 30-60 degree based spatial memory is subconsciously placed.

4. The interior and exterior of the house, though strictly ordered in space, are ambiguous and beautiful. There are six rather than four principle elevations, and it is difficult and pleasing to be unable to relate one elevation to another.

5. The acute and obtuse angles present where plane meets plane are visually and psychologically necessary. There is, therefore, no more waste space than the 90 degree vertical angle presents in comparison to the 60 degree vertical angle.

6. Stairways are accommodated nicely in a hex planning grid by utilizing treads which are parallel to each other but are themselves part of parallelograms as in Fig 3. These stairs have long effective tread widths and are easy to negotiate.

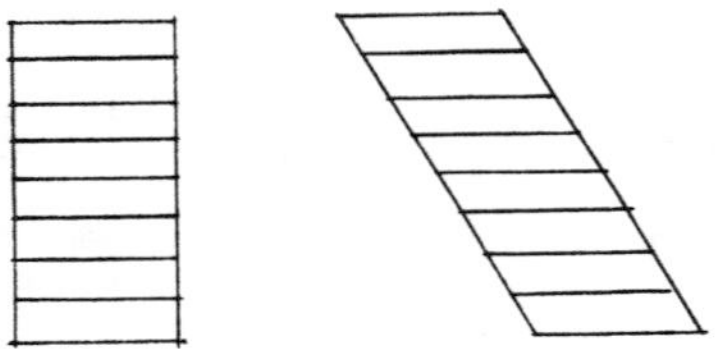

Fig 3

7. Inclined walls and new geometries enhance accoustics.

STRUCTURAL BEHAVIOR OF COMPONENTS

An analysis of the second research structure using the Scada Systems Corporation Structural Computer aided Design and Analysis Program took place using gravity and wind loads in different combinations. Joint fixity was not assumed, all member sizes were identical, all connectors at ground level were fully supported. Bending stresses in each were neglected as full lateral support was provided by panels and moments were very small in proportion to cross-sectional properties. Any strength imparted by panels to the assembly was neglected.

The results are tabulated below:

	max. tensile kN	max. compressive kN
Full Structure	8.79	23.18
Structure with members removed	28.28	44.20

The first section of the above suggests that for solid space frames compression is by far the most important determinant of member/joint connections. The fewer the number of member removals, the lower the member stresses. The second part indicates design values that will be used to check on member/connector strengths.

A member was tested in tension to destruction and failed at 33.75kN. A connector failed at 28.57kN when loaded in tension. Improvements are being made at this time to the components in manufacturing technique, arrangement of material mass, and method of fastening which are expected to double the above failure values. A future examination of factors of safety and redundancy will be made specifically for this type of structure prior to commercial availability.

The connection of the tang to the member is the challenge to a wood space frame - the successful transmission of tensile and compressive forces in metal to wood fiber. The 33.75kN value for member tang failure was much higher than anticipated, given the current nailing method. It appears that the hardened nails used have a much higher pullout value than expected.

THE MARKETPLACE AND SPACE FRAMES

Our future goal is to produce the second research structure at a cost comparable to conventional structures. We will live in it, examine its strengths and drawbacks, and then build the next research structure. Within five to seven years, we hope to have a structure which is less expensive and more versatile than any currently produced by the manufacturing housing industry. We have a sincere interest in the international use of such a technology and hope that the houses reach the people who need them, see Ref 4.

Play structures, park structures, and temporary construction covers, which utilize precisely the same system as that described, only scaled down appropriately, are all being actively explored, see Plate 6. The author welcomes inquiries and would like to meet anyone interested in the subject.

Plate 6

ACKNOWLEDGEMENTS

Thanks to J. F. Gabriel

REFERENCES

1. Satterwhite, R.G. Proceedings of the IASS Symposium held in Oulu, Finland, 1980 pp. 117-126
2. Satterwhite, R.G. PHP Magazine September 1982 pp.50-57 PHP Institute, Tokyo
3. Satterwhite, R.G. Spirit of Enterprise The 1981 Rolex Awards pp. 108-111 Harrap Ltd., London 1981
4. Gabriel, J. Francois 2nd International Conference on Space Structures September 1975 Paper

ARCHITECTURAL AND STRUCTURAL EVOLUTION OF DOME SHELTERS

B S BENJAMIN, BE, DIC, MSc, PhD, MIStructE, MSPE(USA)

Professor of Architecture and Architectural Engineering,
University of Kansas, USA.

Doubly curved domes have been used as shelters through all the various stages of evolutionary structure. Geological structure, biological structure, insect architecture and anthropological structure all show consistent use of this structural form. It is also evidenced in the hut building practices of primitive tribes in Africa and Australia. The use of plastics materials for such shelters has been researched for emergency housing and for slum clearance in poor countries. The folding and packaging characteristics of the shelter become important for ease of transportability. For Indian conditions, specifications have been developed for a folded plate dome shelter, and a full scale model was built. Problems discussed are jointing techniques, erection of the structure, stability and anchorage for wind conditions and waterproofing.

HISTORICAL PERSPECTIVES

Doubly curved shell structures, whether inhabited or not, have existed on this planet at least since the beginning of geological time--long before man ever felt the need for them--and it is important to understand the historical context of such structures for human shelter. The first doubly curved shell structure on this earth was the structure of the earth itself. The crust, which is 7900 miles (12722 km) in diameter, has an average thickness of 25 miles (40 km), being only 4 miles thick at some points beneath the Pacific Ocean. It sits on the lithosphere above the Moho discontinuity. The lithosphere itself, about 60 miles (97 km) thick, floats on the asthenosphere and is fractured into twelve floating plates which slide in relation to one another. And this enormously large structure was formed by inert gas subjecting itself only to the laws of physics! Since then other geological structures have arisen on it, formed from a single structural material (rock), again subject only to the laws of physics, but now aided in their construction by tools such as volcanic eruptions and lava flows, and wind and water erosion and deposition (which given the appropriate time scale are no less effective than hammer and chisel). These structural systems include mountains (pyramids), roofs of caverns (singly and doubly curved shells), and natural rock bridges (arches), see Ref 1. Figure 1 shows Landscape Arch in Arches National Monument, Utah. This is, perhaps, the world's longest rock arch in existence. It spans 291 ft (88.8 m) with a height of 120 ft (36.6 m) to the crown. The weakest section of the arch is less than 6 ft (1.83 m) in diameter. The Navajo Formation, forming the valley floor is of the Jurassic Age, about 160 million years old. Above that is the Carmel Formation which, being sedimentary rock, wears easily. Above the Carmel Formation is a 300 ft (91.5 m) thick layer of Entrada orange-red sandstone that originated as wind-deposited sand. It is in this sandstone that wind erosion has sculpted these natural arches. Figure 2 shows Delicate Arch. In this case, the wind erosion of the Entrada sandstone was aided by freeze-thaw cycles. The freezing of water in joint planes cause sides of the cliff to tear away leaving vertical rock 'fins'. These fins are then eroded by the wind to natural shapes.

Geological structure had only one material to work with, strong in compression but weak in tension. It therefore abounds in those structural forms which transmit load in compression. Examples of tension structures are rare. Even cantilever rock overhangs, often used as shelters by primitive tribes in Sri Lanka, see Ref 2, have relatively small spans for this reason.

Fig 1

Fig 2

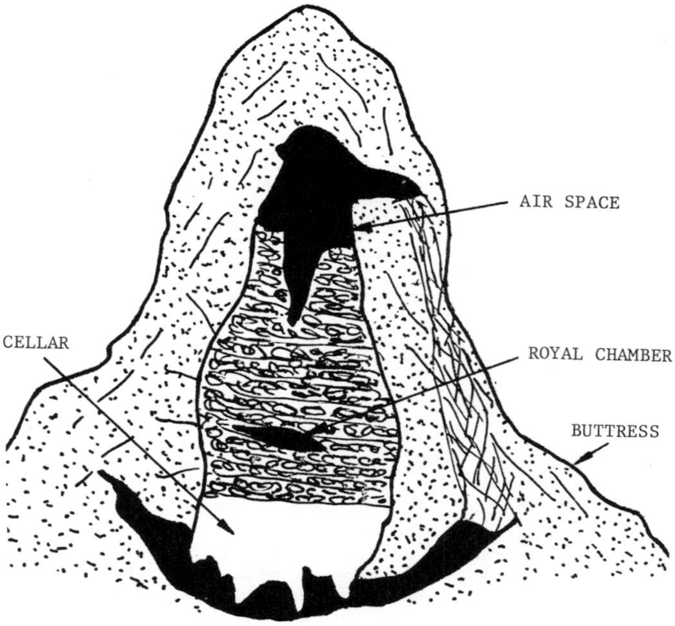

Fig 3

With the evolution of life, a new structural material had appeared--the living cell. This manifested itself eventually either as wood in botanical structure, or as bony material in endoskeleton structure, or as keratinous or calcareous material in exoskeleton structure, see Ref 1. Of all these forms of biological structure, however, only exoskeleton structure shows true use of doubly curved surfaces for shelter, as in the calcareous shells of aquatic creatures such as the mollusc. While the shell is a form of shelter, the scale to the creature within it is still essentially 1:1.

Man was not the first to change this scale. That was accomplished long before he came on the scene by the social insects in a breathtaking array of insect architecture, see Ref 3. Bees, ants, and termites, working with a variety of structural materials such as paper, wax, glue, and the excreta of the insect itself that hardened quickly on exposure to air and did not putrefy, erected structures much larger than themselves. The use of doubly curved dome surfaces is not uncommon in their construction. A termitary is shown in Fig 3. This is the nest of the Macrotermes bellicosus from the Ivory Coast, Africa. The termitary has a circular cross-section and is buttressed by supporting ridges which also contain the air ducts, which supply air to the more than two million termites that live and work within it. The air is fed into the cellar, and is exhausted from the air space at the top of the nest. The height of such a mound is about 11.5 ft (3.5 m). The manner in which Macrotermes bellicosus build an arch is shown in Fig 4. A droplet of excrement forms the cement into which other insects put soil particles to form a mortar.

Man was, however, the first animal to use tools to fashion his shelters, thereby changing the pace of structural evolution. The evidence shows that in the building of shelters, first within caves and then in the open, man in prehistory, used only the commonly available structural materials of his time--mammoth skulls and tusks, reindeer antlers, bones, and animal skins, see Ref 2. But he used stone chips to cut and shape, awls to make holes and animal gut to tie the

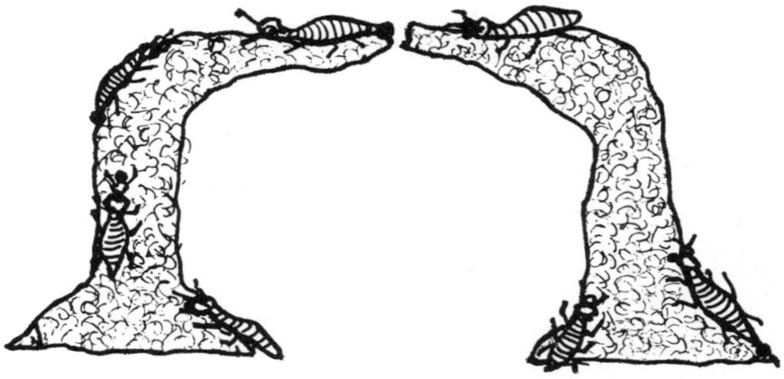

Fig 4

parts together. The remains of such huts have been found at Mezin and Dobranitchevka in the Ukraine, USSR. One such hut was 12.5 ft (3.8 m) in diameter. Mammoth skulls were set in a circle and embedded into the ground. Mammoth tusks placed in the skulls formed the roof members. Other mammoth bones such as long bones, shoulder

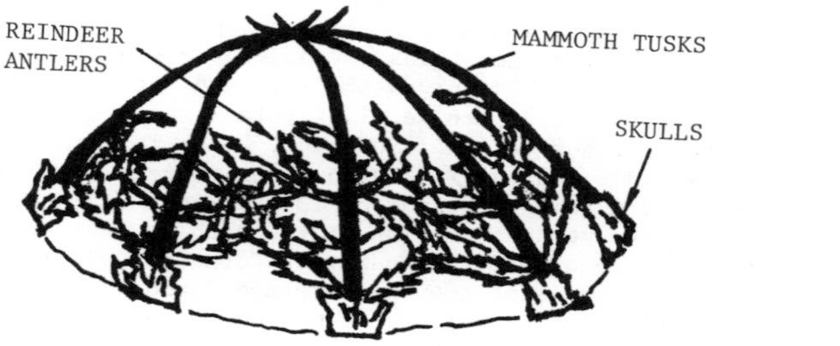

Fig 5

blades and reindeer antlers were used as secondary members. A reconstruction of such a hut is shown in Fig 5. The hut building practices of primitive tribes clearly show their origins in anthropological structure, and the doubly curved dome surface is still used to provide shelter. Figure 6 shows the light structure for a dome shaped aborigine hut in central Australia. Flexible saplings are used to form the structural dome, see Ref 2.

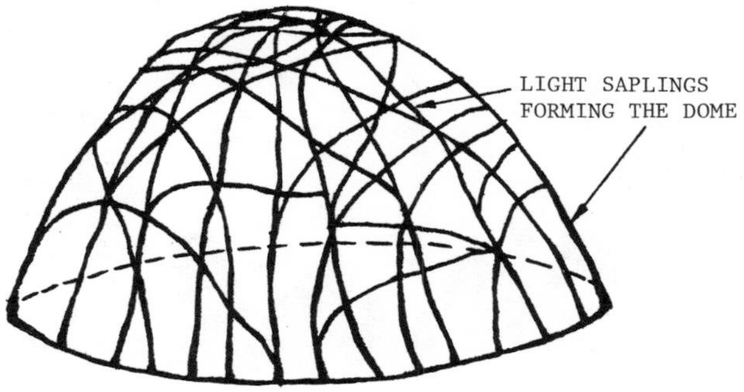

Fig 6

Another example of a conical dome roof structure used in hut building is shown in Fig 7. This is a model of the home of the Mesakin tribe of the Sudan in North Africa. It consists of five turret shaped mud huts on a foundation of rough stones, see Ref 4. The huts are roofed over with a basket-like structure which is constructed upside down on the ground, and then placed over the hut like a hat. The small central courtyard between the buildings is covered over with sticks and grass.

Fig 7

Since then, domes, in modern times, have become very important structural systems covering large spans and roofing spectacular pieces of modern architecture, see Refs 5 and 6. But their use as shelters has remained almost frozen in time. What has changed with the arrival of new building materials, such as brick and concrete, has been the class of people who have moved out from such shelters into better homes, leaving the very poor in the villages and urban slums of underdeveloped countries to continue to inhabit them. Yet the use of domes for low cost shelters is just as significant and natural a development, in the light of history, as their use over large spans.

PLASTICS FOR LOW COST DOME SHELTERS

Plastics materials brought new hope, temporarily, for better shelters, and considerable work has already been done by the author and others on the use of both the fibre reinforced plastics and the low density materials for such construction. Two such examples are considered here.

For emergency housing, after earthquakes or other natural disasters, a light plastics membrane is blown up into a dome. Rigid polyurethane foam is then sprayed onto the dome to form the structural dome. The membrane is deflated and withdrawn through a hole to be reused again. Doors and windows can be cut as desired. Such housing, shown in Fig 8, has been used in Turkey, Peru and Nicaragua, see Ref 7.

Fig 8

In 1975, AVIO-FOKKER designed and built an Air Transportable Emergency Shelter, shown in Fig 9. It was made entirely of flat panels for ease of transport. The shelter had an effective floor area of 14 sq m (151 sq ft)

Fig 9

with a minimum standing height of 2 m (6.6 ft) and a maximum standing height of 2.4 m (7.9 ft), see Ref 8. The transport weight was approximately 100 kg(220 lb), and the cost was estimated at approximately $440. The estimated minimum life expectancy was 2 years.

FOLDING AND PACKAGING OF SHELTERS

It became clear to several researchers, including the author, that the material had great potential for use in urban slum clearance. It was also clear that the land on which the shelters would be erected could not belong to the slum dweller. If he was facing eviction from the land, his house would have to move with him--if he was to be persuaded to invest his very meagre resources into such a shelter in the first place. The importance of the folding and packaging characteristics of the shelter became of prime importance in the design of the shelter. Several folded plate dome packaging arrangements were considered as shown in Figs 10 and 11. Figure 10 shows a single stage folded plate conical dome.

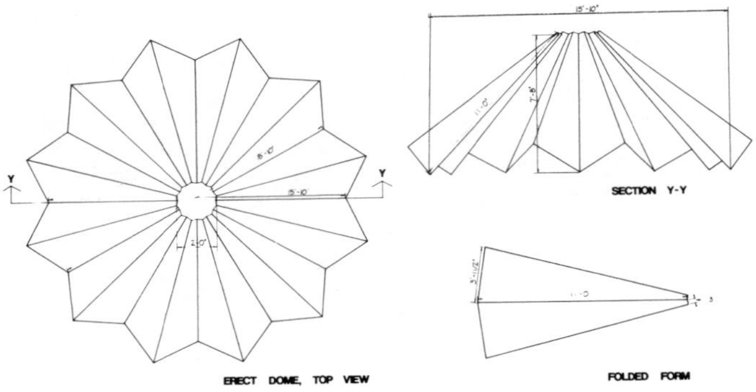

Fig 10

A two stage folded plate dome is shown in Fig 11. The folding pattern is more intricate, but the headroom is much better. A model of the dome of Fig 11 is shown in Fig 12.

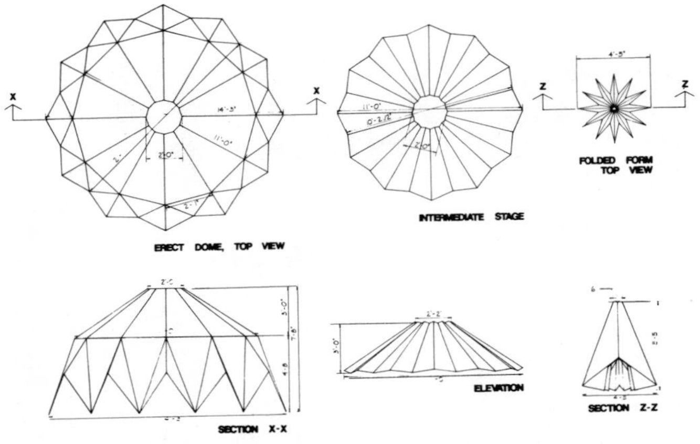

Fig 11

It becomes clear that for all the domes discussed above, folding is not going to be easily possible without the provision of suitable joints which allow freedom of rotation in both directions. This can be achieved by the use of taped joints using tape that is stronger than the panel itself. If necessary, it can be renewed.

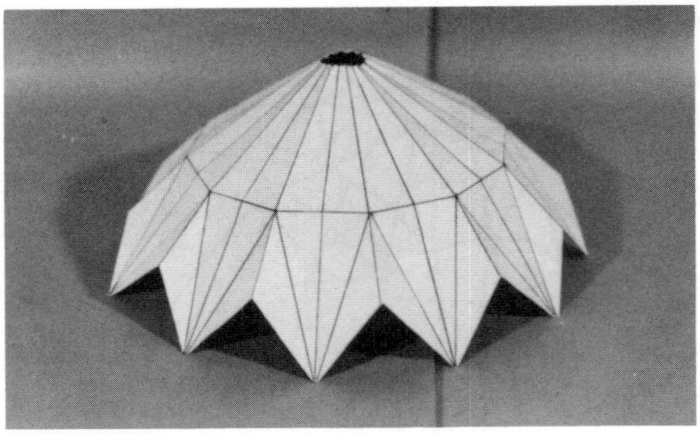

Fig 12

FOLDED PLATE CONICAL DOME SHELTER

Research on slum dwellings in India led to certain conclusions. The slum dweller does not own the land on which his shelter is built. In illegal development, it is very likely to be owned by the city, the railways, or large corporations, and is most likely to be situated close to water and toilet facilities, at least in the big cities. When the city or the corporation does decide to develop the land, the slum dweller faces eviction, and the shacks are summarily razed. Research also showed, surprisingly, that with the improvement in the standard of living over the latter half of this century, money was available for a shelter, and the slum dweller was even prepared to invest it in his home--if the conditions

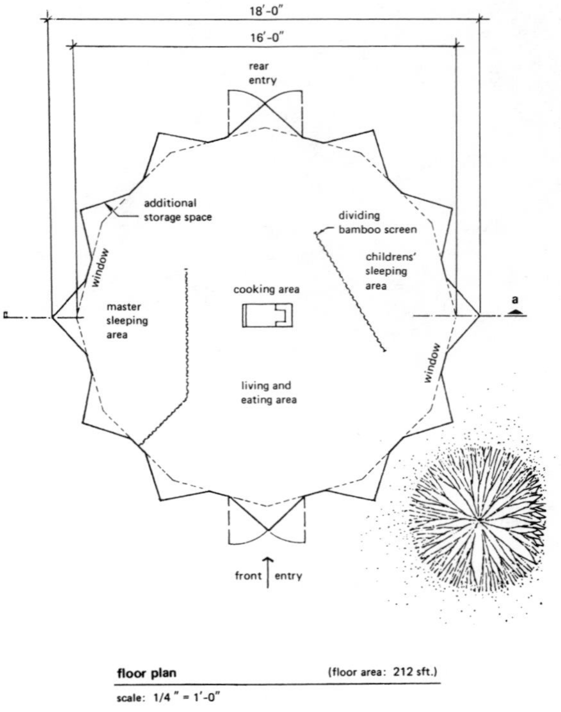

Fig 13

were right. While figures do vary, an average cost of \$100 appeared to him to be reasonable. The architectural design of such a shelter is shown, in plan, in Fig 13, and in sectional elevation, in Fig 14. The specifications for the shelter required that it sould be lightweight and easily transportable, and be easily packaged into a reasonably small volume. When erected it had to have considerably better standards than the shack that the slum dweller was living in, both in the space available to him as also in the quality of the internal finish construction. The specifications called for the shelter to be easily erected without equipment, easily anchored and capable of being waterproofed, see Ref 9.

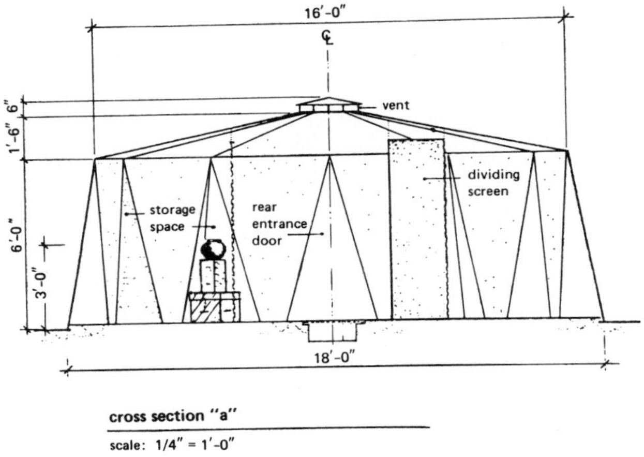

Fig 14

A full scale prototype of the folded plate dome shelter is shown in Fig 15. In addition to the full dome, shown erected in a laboratory for testing purposes, Fig 15 also shows the clean internal lines of the inside of the dome, a close-up view of the taped joints and the anchorage panel.

Fig 15

The shelter has a floor area of about 212 sq ft (19.7 sq m) with the first stage vertical for maximum space utilisation. The minimum headroom is 6 ft (1.83 m), and the maximum headroom is about 7 ft 6 in. (2.29 m). For Indian conditions, this represents, in the author's opinions, the minimum acceptable standards. The entire structure is built out of only three types of panels--the trapezoidal panels of the roof and the walls and the triangular panels of the folds. The material of which the dome is constructed can vary depending on the cost. For minimum cost, the panels can be a rigid polyurethane foam core sandwiched between resin impregnated kraft paper. For more durability, stronger facing materials can be used. The folded plate dome shape has great intrinsic structural strength and so the panels are economical in their use of material. The dome needs a horizontal ring tie at the shoulder level, such as a simple rope, and this can be seen in Fig 15. For the top to fall in, the shoulder has to move out, and this is effectively prevented by the rope tie.

Erection of the structure can be carried out very easily by two people, without any complex set of instructions. However, one last set of joints have to be field taped, and this can be provided at no cost to the slum dweller. All other joints are factory taped. The use of such taped joints is essential to the folding and packaging of the dome.

Suitable anchorage has to be provided to the shelter, particularly for wind load conditions. This can be carried out for minimal anchorage against light winds by the provision of anchorage panels, shown in Fig 15, which are then weighted down by stones. For very heavy wind conditions, however, bands of wide tape can be run criss-cross from one side of the dome to the other, within it. A circular pit is then excavated to a depth of about 1 ft (0.3 m). Earth is them tamped down over the criss-cross tapes and forms the finished floor surface as well. The dead weight of 212 cu ft (6 cu m) of earth is quite sufficient to anchor the structure under even gale conditions!

The entire structure can be folded and packaged into a single kit 7 ft 3 in. (2.2 m) long, 5 ft 3 in. (1.6 m) wide, and 1 ft 3 in. (0.38 m) deep, for ease of transport. It can weigh, depending on the material of its construction, as little as 140 lb (63.5 kg). The packaged structure is shown in Fig 16.

The waterproofing of the structure is very important, particularly for Indian conditions. It is quite unrealistic to waterproof all the joints in the structure. Besides increasing the cost, this would interfere with the folding characteristics of the dome, which is so important. Further, the waterproofing would have to be

Fig 16

renewed each time the dome shelter was packed and unpacked, and the slum dweller could hardly be expected to be proficient in it. The author has proposed therefore that the waterproofing could be carried out by the provision of a light, cheap, polyethylene raincoat. The raincoat could be tailored to an exact fit or could be loosely drawn over the shelter for the duration of the monsoon. The vent at the top of the dome would protrude through the raincoat, and suitable arrangements could be made for doors and windows in the structure. In case of damage, it could be easily repaired or replaced at little expense to the slum dweller.

CONCLUSIONS

The evolutionary history of dome shelters shows quite clearly that their efficient load carrying form and functional use was appreciated long before man sought to use them. Their evolutionary growth should be encouraged. The technology, and the research to back it, is now available for mass producing low cost shelters for slum clearance in poor countries. The author would even venture to suggest that the money is also available. What is lacking, perhaps, is governmental vision and directive in a sadly fractured world, wracked by political instability, where the needs of the poor are set aside to the detriment of all mankind.

ACKNOWLEDGEMENTS

The author wishes to thank the University of Kansas for making available the necessary travel funds to attend this conference.

REFERENCES

1. B S BENJAMIN, Man's Structural Evolution, From the manuscript of a book under preparation.
2. J JELINEK, The Evolution of Man: The Pictorial Encyclopedia, Hamlyn, 1975.
3. K VON FRISCH and O FRISCH, Animal Architecture, Harcourt Brace Jovanovich, American Edition, New York, 1974.
4. C R ANDERSON, Primitive Shelter, University of Kansas Publications, No. 46, Lawrence, Kansas, 1960.
5. Z S MAKOWSKI and D PALMER, Domes--their History and Development, The Guilds' Engineer, London, 1956.
6. Z S MAKOWSKI, Braced Domes--their History, Modern Trends and Recent Developments, Architectural Science Review, Vol. 5, No. 2, July 1962.
7. J H VAN GIESSEN, Kunststoffen--Redding Voor de Arme, Onbehuisde Wereldbevolking?, Plastica, No. 3, 1976.
8. J H VAN GIESSEN, Air Transportable Emergency Shelter, Report FPO-AFY, No. 114, AVIO-FOKKER, 1976.
9. B S BENJAMIN, Structural Design with Plastics, Second Edition, Van Nostrand Reinhold, New York, 1982.

FLEXIBLE HOUSING ARCHITECTURE BASED ON THE IMPLEMENTATION OF SPATIAL BAR STRUCTURES

J KRÓL, DR INŻ ARCH

Institute of Architecture and Town Planning
Faculty of Architecture
Technical University of Silesia, Gliwice, Poland

The search for truly flexible, in time and space, solutions in housing architecture situated in areas influenced by ground subsidence due to coal mining exploitation, leads towards the implementation of spatial bar structures. The paper introduces, as an example, a system based on a spatial bar structure statically determinate and geometrically stable, consisting of spatial modules in the form of tetrahedrons with one vertex joining three reciprocally perpendicular bars. There are pointed out some flexibility ranges and means which may be considered in designs where spatial bar structures are used.

Housing architecture is a domain in which the existence of a continuous need for searching optimum answers to questions on function, construction and form results from its commonness and differentiation in time and space.

Desiring to provide answers to spiritual and physical needs of the inhabitants while using the language of architecture, one meets limitations and possibilities. Both constitute normal elements of creative processes.
This paper is to introduce some considerations derived from approaching the problem of finding housing solutions in the specific conditions of GOP - the Upper-Silesian Industrial District.

The GOP district is an agglomeration in the South of Poland, with three million inhabitants. From the point of view of housing architecture one has to deal here with many factors well known in similar situations, like the necessity to place more inhabitants in an already densely populated area, having to consider microclimatic conditions, complicated by various influences from industrial technologies, as well as existing traditions in the midst of social development.

At the same time there is here a rather unique situation in regard to building conditions. It is caused by the fact, that the area is subjected to constant ground surface movement caused by the underground exploitation of coal resources through mining.

If one considers, on top of that, some assumptions of a more general nature, for instance those based on the analysis of visual perception, which plays an undenyable role in the acceptance of architecture, including housing architecture, see Ref 1, one may state, that a real and complex problem is being approached.

The research was stimulated by the possibilities resulting from the characteristics of spatial bar structures. In accord with the already proven and demonstrated flexibility requirements, see Ref 2, there exists the presupposition of

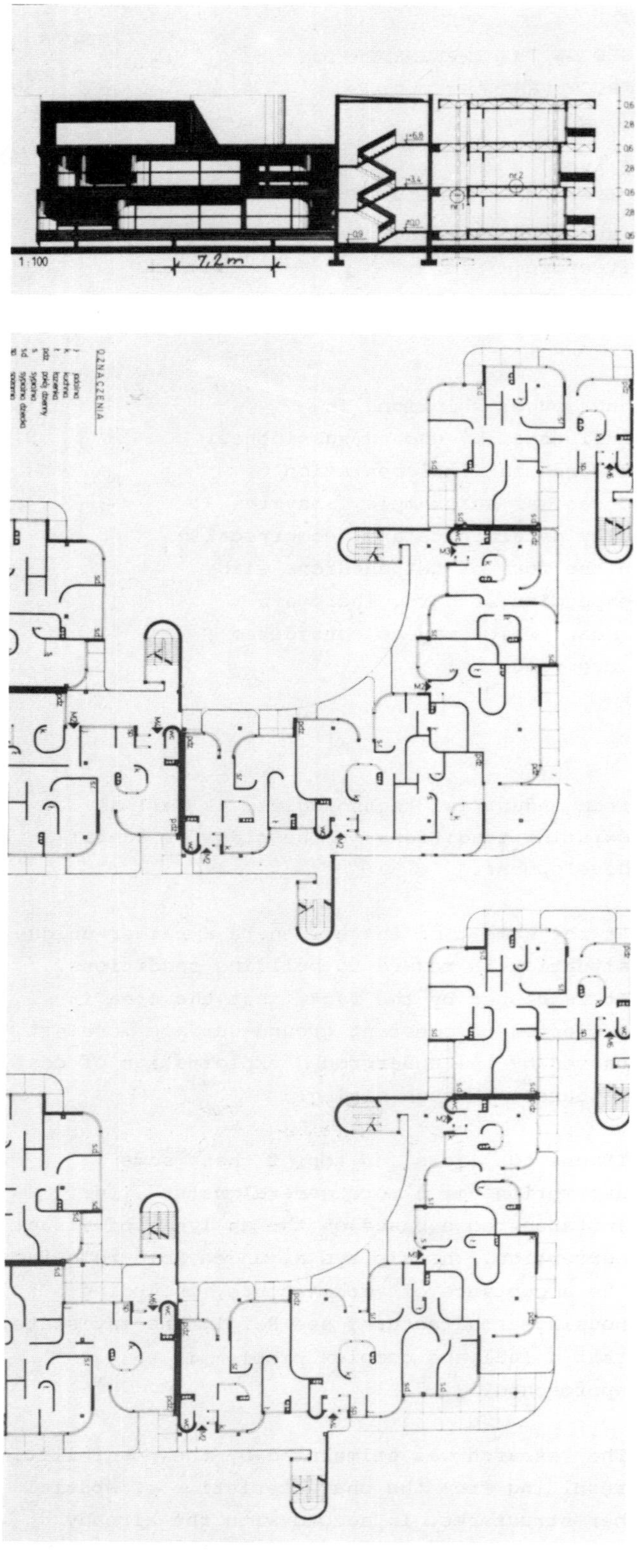
1:100
7,2m
OZNACZENIA

Fig 1

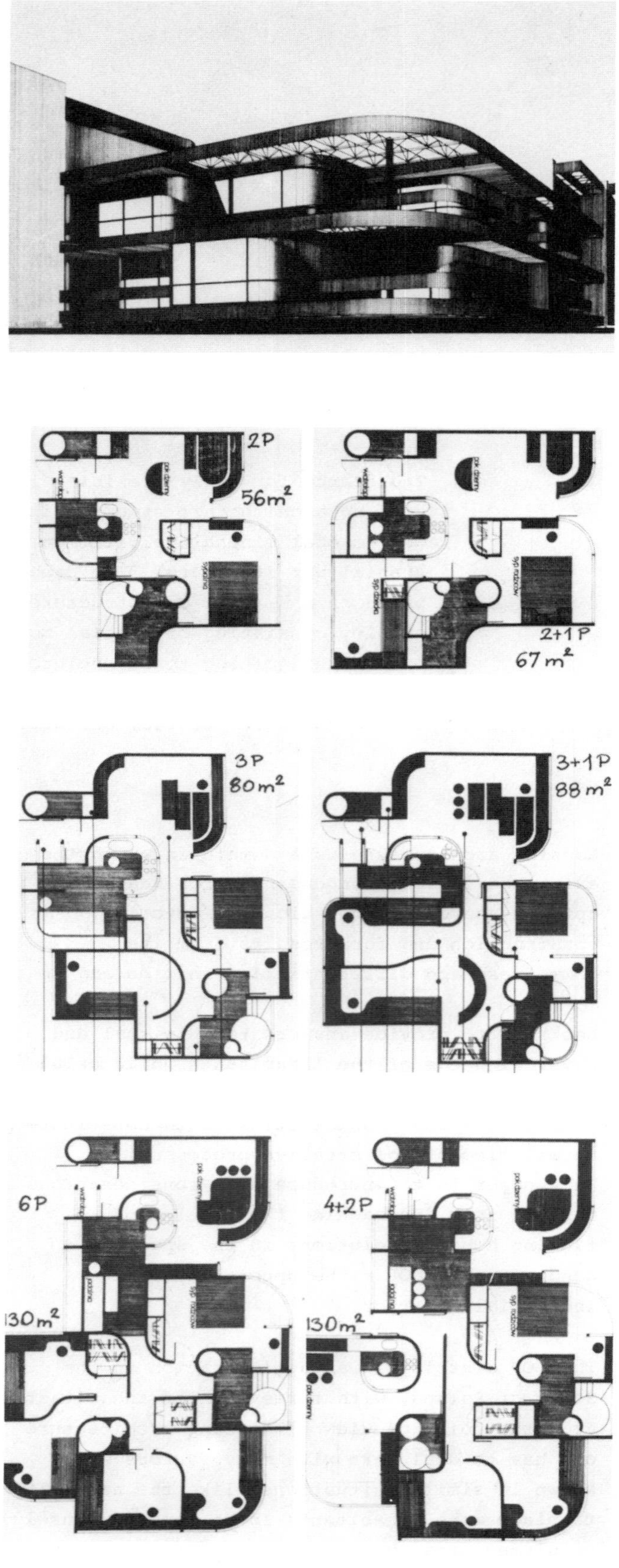
2P
56m²
2+1P
67m²
3P
80m²
3+1P
88m²
6P
130m²
4+2P
130m²

Fig 2

the idea of forming dwellings which can be changed, depending on the changes in the needs of the inhabitants - for instance a changing demand for the number, size, shape and character of rooms or spaces, developing preferences and possibilities.

It is a well known fact, that while foundation work is being carried out on building ground influenced by subsidence, one may use over-rigid structures, with very compact plans and frequent dilatation gaps. In the case of housing architecture one faces then a basic complication because of the difficulty of applying flexibility in plan and space. Therefore the logical direction of research is towards an interest in frame structures. Spatial bar structures are, too, well recognised as elements promoting the development of architecture and construction science, see Refs 3 and 4.

Also in the Institute of Architecture and Town Planning at the Faculty of Architecture of the Technical University of Silesia in Gliwice, Poland, some studies and designs along those lines have been and are being carried out. Thinking in that direction one comes to the conclusion, that a comparatively high degree of flexibility - both structurally and functionally, and in effect also in form - can be achieved by erecting housing, which makes use, for instance, of a double structure system, as shown on Figs 1 and 2, according to diploma design by Piotr Szopa, tutored by the author of the paper. The load bearing frame consists of pillars and mattress trusses, allowing the "filling" - housing tissue to be independent of structural and services modules. The pillars can be placed in various modules, truss slabs can be hung up at various heights, containing elastic services ducts. The dwellings have been adapted to accomodate changes in internal divisions and external changes in size and shape.

The quest after a method of forming dwellings even more flexibly, so that they might indeed be changeable in space and time, and on the other hand the need to situate groups of dwellings on building ground influenced constantly by subsidence, led to a presupposition of implementing a spatial bar structure as an integral part of the housing tissue itself.

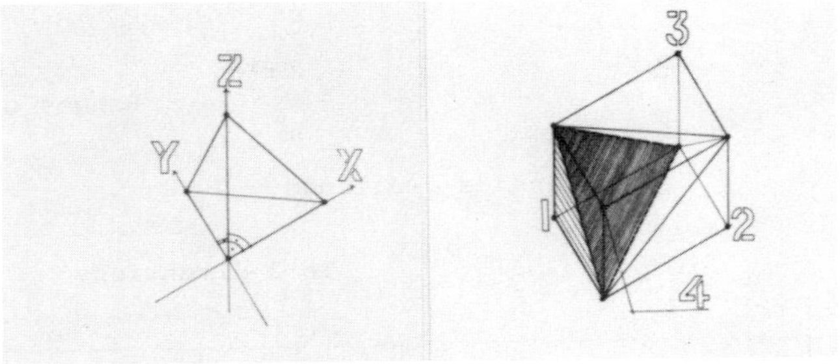

Fig 3

Spatial bar structures organize space without locking it out. Additionally, assuming that the structure would be statically determinate and geometrically stable, one can consider extending and rebuilding such a truss already during the exploitation period.

The spatial bar structure which was derived by the author and is the subject of a patent, see Ref 5, is based on a spatial module, which is a tetrahedron in which in one vertex there meet three bars reciprocally perpendicular, see Fig 3. By adding or taking away three bars joined in an articulated joint, one continues to receive a statically determinate and geometrically stable structure, provided it is supported in three points, rectified, for instance, hydraulically for practical reasons in connection with gravitation, with load being hung up in the joints. Considering the using of that structure in housing architecture, one may implement various proportions of the module and various sizes - - for instance the height could be the equivalent of one condignation or a part of it or a multiple, and the length of bars in the horizontal plains could be selected in connection with local possibilities and needs. In such a way one could create flexible housing there, where other systems could not be used because of subsidence. Such a spatial truss could even be allowed to be made, if necessary, to a certain degree statically indeterminate in order to make occasional manouvering with bars possible for functional reasons.

With such spatial possibilities one may approach the designing of dwellings and groups of dwellings diversified functionally, spatially, with regard to the material used, and changeable in time, providing architecture, that would be

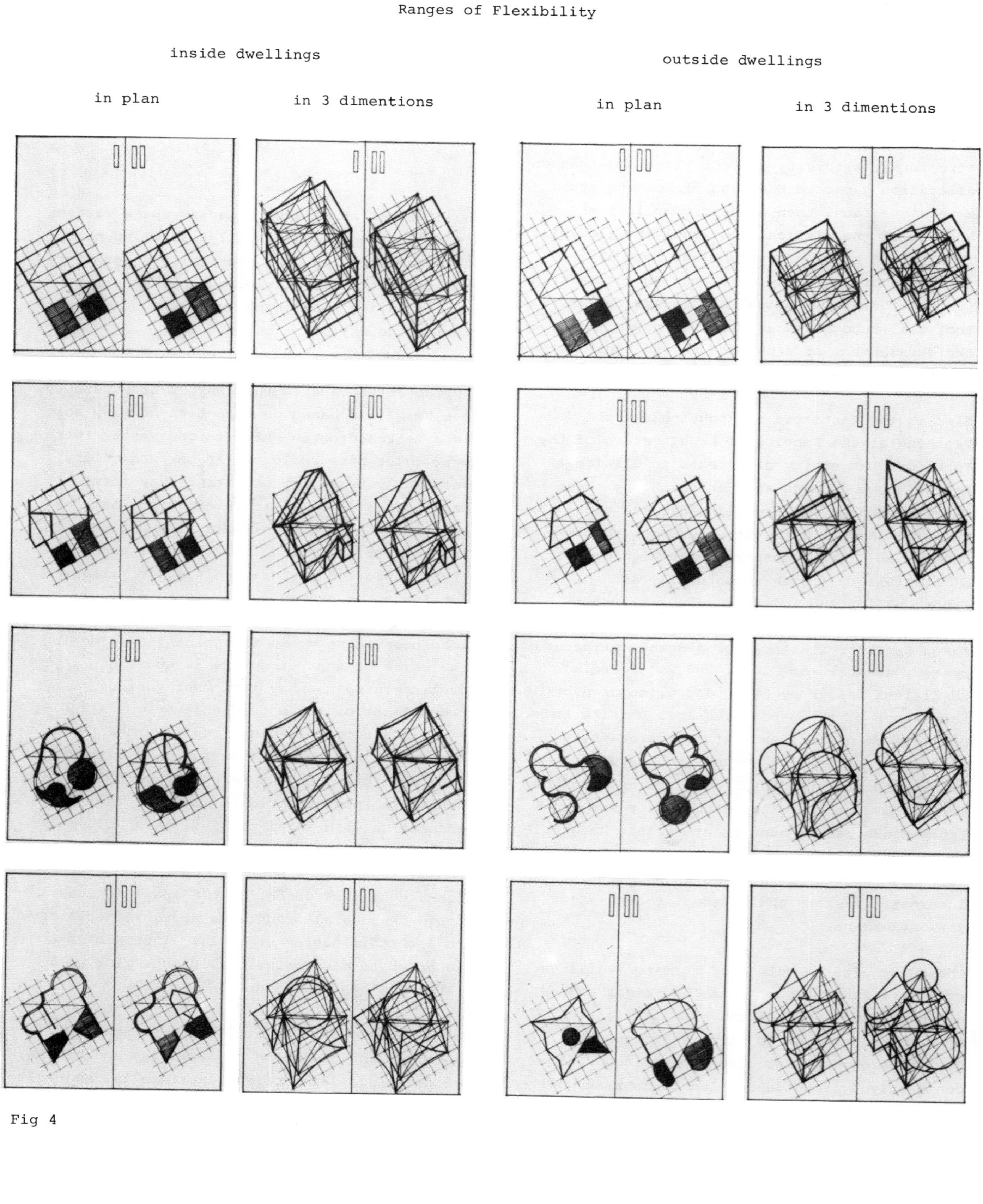
Ranges of Flexibility
inside dwellings
outside dwellings
in plan
in 3 dimentions
in plan
in 3 dimentions

Fig 4

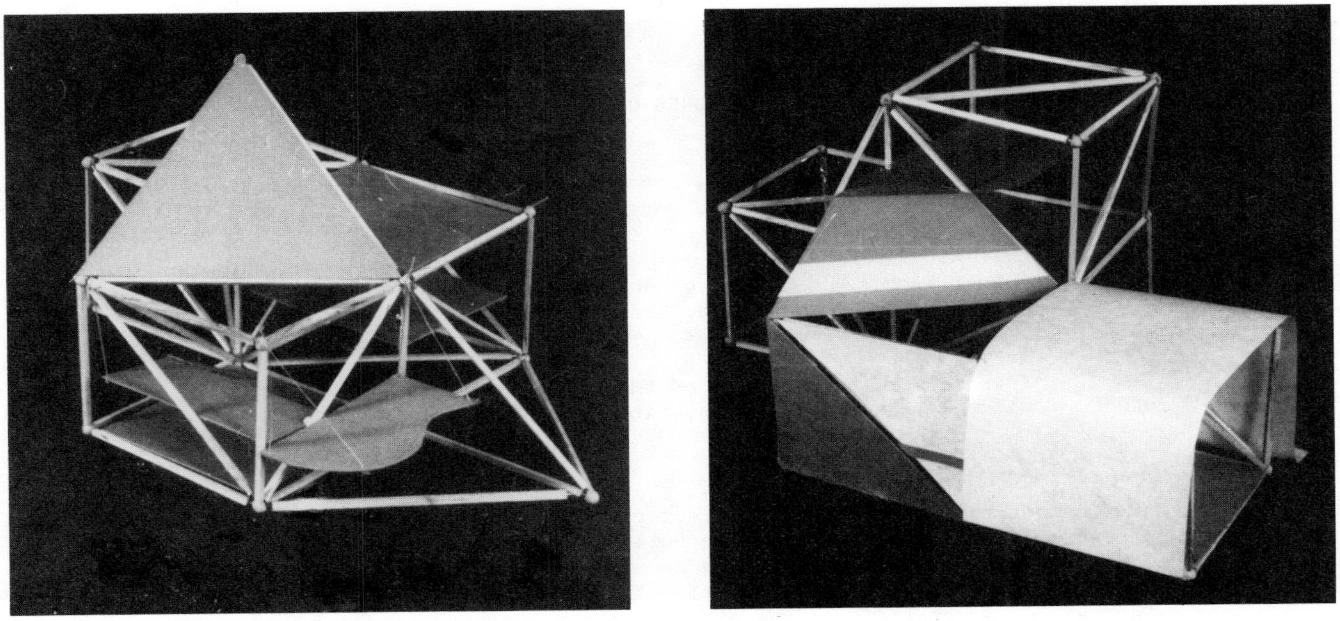

Fig 5

living, pulsating, and yet orderly. Figure 4 contains some practical flexibility presuppositions within spatial structures. There are listed there examples of different ranges of changes - within and outside dwellings, both in plan and in three dimentions. In the horizontal lines there are grouped examples, where different means of achieving changes have been employed, namely

1st line - using right angles and planes
2nd line - using planes
3rd line - using out-of flat surfaces
4th line - using a combination of all three.

One can imagine the erection of dwellings of both conventional and unconventional character, with the possibility of easily introducing changes during exploitation, see photos from a working model, Fig 5. Presently the subject for further research can be seen in the need to find and investigate possibilities of realizing the above mentioned architectural postulates through using various building materials and technologies.

REFERENCES

1.Praca Zbiorowa, Modelowe Formy Zagospodarowania Przestrzennego Górnośląskiego Okręgu Przemysłowego, Polska Akademia Nauk, Ossolineum, Wrocław 1979.

2.P COOK, Architecture: Action and Plan, Studio Vista, London 1969.

3.Z S MAKOWSKI, Space Structures of Stephane du Chateau, Building Specification, May 1975.

4.O BUTTNER, H STENKER, Metalleichtbauten, VEB, Berlin 1970.

5.J KRÓL, Przestrzenna Struktura Prętowa, patent Pol Sl,nr P-200 639 z 1.9.1977, świadectwo autorskie o dokonaniu wynalazku, Urząd Patentowy PRL, Warszawa 1982.

SPACE STRUCTURES OF ISFAHAN

M.A.Mirfendereski° , Dott. in Architettura
F.N.Mirfendereski°°, Dott. in Architettura

° Professor, Former Dean of the Faculty of Fine Arts, University of Tehran, IRAN.

°° Professor, Former Head of the Department of Architecture and Industrialization, National University of IRAN.

This paper demonstrates the overall characteristics, performance of component parts, and application in urban planning of the building system employed in the development of the early 17th century Isfahan as the capital of Safavid Iran.
The need to build rapidly and consistently was reflected in this system by two means, one being regular expansion, and the other, a set of interrelated and repeatable components, where the principles of space structures can be observed in their individual and compositional disposition.
The linear or grid repetition and development of these components created urban complexes like Maidans, Bazaars, Mosques and Caravansaras.

THE PROCESS OF URBAN DEVELOPMENT

Isfahan was chosen as the capital of Safavid Iran in 1598 AD, almost a millenium after the emergence of the first known settlements in the region. During nearly 20 years, the area of the city doubled and the population reached 600 000. Developments in science and the growing knowledge of nature and its laws, resulted in a rationalized approach towards the process of planning and architecture. The considerable achievements of the traditional architecture of the past (Sassanid-Timurid-Seljuq) particularily in the arts, technical competence and skills, in parallel with the new outlooks took an active part in giving shape to the complex urban structure.

In the process of creation, execution and expression, the contents of architecture changed as compared with the past. In particular, due to new possibilities, the process of building (technical-operational) changed accordingly. This process of building no longer followed the intention of creating independent, random pieces with their own individual values, but sought for a new approach to planning and building in an urban scale. In this approach, individual works of architecture became integral parts of a whole, as elements of a system with a specific mechanism. Skills and technical abilities were employed within the framework of the new system. The mode of production of architecture was based on a system which itself was part of a new and specific culture.

Urban construction, which up to then was carried out through a simultaneous process of conception and execution, based on a more advanced scientific and technical knowledge and understanding of the overall requirements of the new society, could now percieve, plan and design before the stage of building. It was this approach to planning and execution that made the erection of buildings and large complexes of considerable qualitative value possible, not only in Isfahan, but all over the country. As a result of this approach, the advanced concepts and ideas of the people of the region, which up to that period were reflected in artistic productions (including architecture) with limited continuity and coherence, were able to influence the overall physical environment in complete broadness and integration.

In the context of a more advanced conceptual outlook and approach, architecture could seek to respond effectively to the broad requirements of the demands of a flourishing society. This response was made through a specific methodology of planning and building of urban complexes. The issue became the creation of the physical living environment with architecture as its means. This intention was realised through a knowledge of the laws that governed the social system on the one hand, and, on the other, through a growing and developing knowledge of scientific, technical and artistic principles and the interaction of the two.

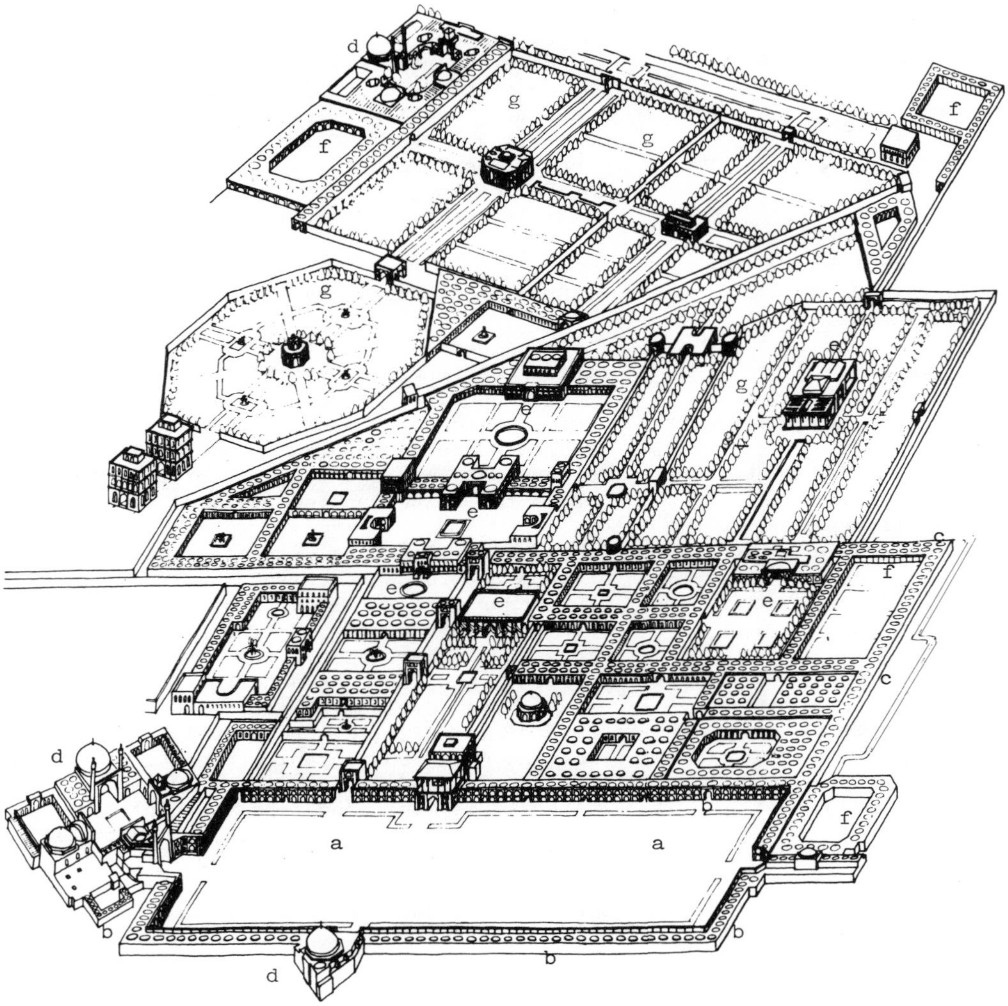

The major urban elements of this complex are: The main square (Maidan) of the city (a), and the bazaars around it (b), covered thoroughfares (c), schools and mosques (d), palaces (e), caravansaras (f) and gardens (g).

Fig 1A

Part of the urban complex of Safavid Isfahan. Following a geometrical order, simple and repetitive space structures form the urban elements in various configurations.

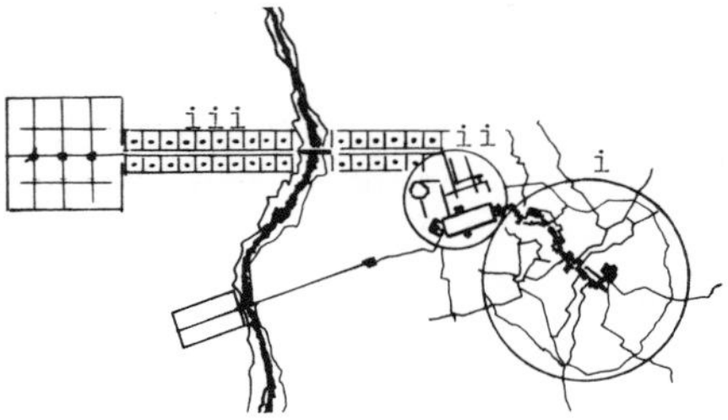

Fig 1B

The location of the urban complex in Fig 1A in relation to the whole city.
(i) Pre-Safavid Isfahan
(ii) Fig 1A
(iii) The rest of the urban complex of Safavid Isfahan

THE SYSTEM OF CONSTRUCTION

The main issue was the need for a system of construction that could provide common characteristics at the same time as having the ability to respond to the various requirements and functions of different given situations. The new system of building was based on a rationalization of the most refined elements of the traditional architecture that had proved most adequate and adaptable. In addition, simple or complex architectural elements had to possess the following characteristics in order to be able to perform adequately in the planning of urban spaces:

- The ability to be constructed in the shortest possible time.
- Being economical.
- Use of the same material in their construction, therefore employing a limited number of technical skills.
- The ability to combine and repeat.
- To be systemized and standardized.

MATERIALS, FORMS, ELEMENTS AND STRUCTURAL INTEGRITY

1. Materials

Due to climatic and historical factors, bricks and mortar are the only materials used in the construction of the above components. The use of brick and mortar has a great historical background; therefore the technique of construction and the skills used in their employment were also very rich and advanced. The small dimensions of the brick, and the flexibility and tolerance of brick and mortar facilitated the construction of various rectilinear and curvilinear configurations of the components and their repetition within relatively accurate geometric grids.

2. Forms

According to the specific architectural demands of a given condition, the components had different configurations and dimensions, but in all cases the process of construction was common and uniform, see Figs 3, 4, 12 and 13.

3. Elements

The components consist of three major elements, the support, the roofing, and the connecting section. The supports are in the form of walls, pilasters or columns. The configuration of the supports, in horizontal section are regular polygons of multiples of four, see Fig 4.

Since different configurations and dimensions of the components had to be part of a whole system, it was essential to have a means of regulating the planning and design. This means of regulation was a three dimensional grid which could relate different sections of the building, regulating and controlling them in terms of proportion and dimension. The composition of a number of the components to form more complex components, and their multiplication and repetition to form various urban spaces, operated within this spatial grid, see Figs 12, and 13.

The roofing is composed of a shell and a number of ribs or just a shell. Ribs and shells are the curvilinear elements of the configuration. The boundary of the plan view of the roofing is either regular polygon or circular. The roofing system (construction of low domes) provide stable structures at any level, such that the domes can either be completed, or left at any height according to architectural demands, see Figs 4, 6, 8 and 9.

4. Connecting Section (Pendentive)

The connecting sections are made of shell segments and ribs. The roofing is supported on the base either directly or through the connecting section. The configuration of the connecting sections follow a progression from the plan of the base which is square or octagonal, to that of the roofing, ranging from polygons of multiples of four to a circle, see Figs 5 and 6.

5. Structural Integrity

The components perform the functions of support, roofing and their connection homogeneously and simultaneously. Because of the particular configuration of the structures and the strength of bricks and mortar under compression, the forces are, on the whole, transferred from the top to the bottom as compression.

Considering the above characteristics, a system of construction was devised incorporating the four major structural units of the "Taqi" (the arch and the vault), the "Chahar-Taqi" (four Taqis on a square plan), the "Ivan" (a covered terrace), and the "Gonbad" (the dome), which was used in the development and building of vast areas of the city in a relatively brief period of time and with considerable economy.

It has to be mentioned that the fact that because of climatic reasons the urban spaces of Isfahan, like other settlements of central Iran had to be mostly covered, influenced the system of construction considerably.

Fig 2
The covered thoroughfare around the Maidan. The repetition of brick components around the Maidan forms a mile of thoroughfare and 300 shops.

THE COMPONENTS

1. Taqi

The Taqi is composed of two vertical pilasters that support a pointed arch, that can develop horizontally to form the vault. When part of a facade or when supporting other floors, the Taqi is set within a rectangular plan, see Fig 3.

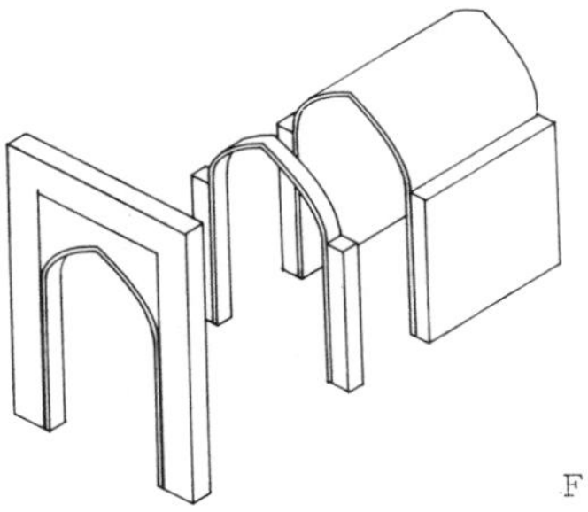

Fig 3

2. Chahar-Taqi

The Chahar-Taqi is a three dimensional structural unit which comprises the three sections of the support, the connecting section and the roofing.

(a) support

The support of the Chahar-Taqi is in the form of four single Taqis on a square or rectangular plan. When the plan is square, the support of the covering is made of four pilasters and four arches with their apexes at the same level, see Fig 4.

(b) connecting section (pendentive)

The connecting sections that make both part of the support and the roofing, start to build up from the four pilasters, from where the arches rise, in the form of the rib and shell to reach the apex of the arches on the same horizontal plane, see Fig 5.

When the plan of the base is rectangular, the apex of the arches will not be on the same horizontal plane; the connecting sections in this case have the role of joining the apex of the two lower arches to a regular polygon of multiples of four at the apex of the two higher ones. This regular polygon acts as the base of a dome. The progression of the connecting sections to the apex of the higher arches can continue in order to transform the regular polygon to a circle when a circular dome is required, see Fig 6.

In this case the connecting sections combine with the rest of the structure in a more integrated way and form a considerable part of the roofing.

(c) Roofing

The final section of the covering is built above the regular polygon or circle over the horizontal plane described above. The roofing is a low dome constructed around the vertical axis of the structure, see Figs 4, 5 and 6.

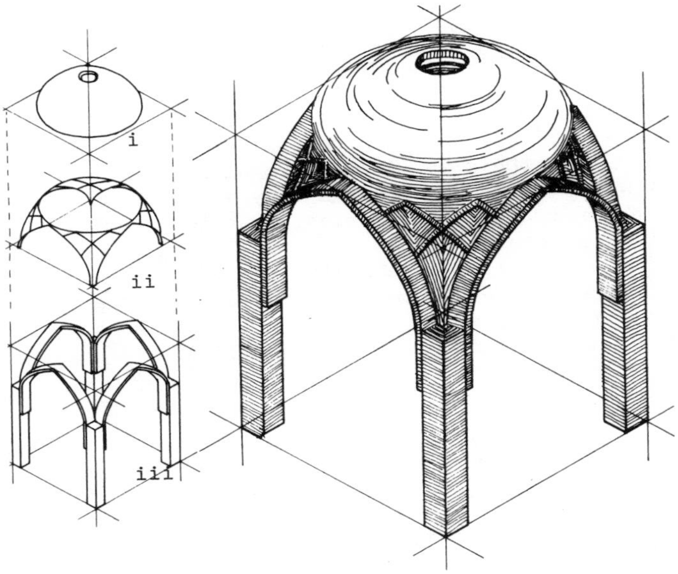

Fig 4
(i) roof
(ii) connecting section (pendentive)
(iii) support

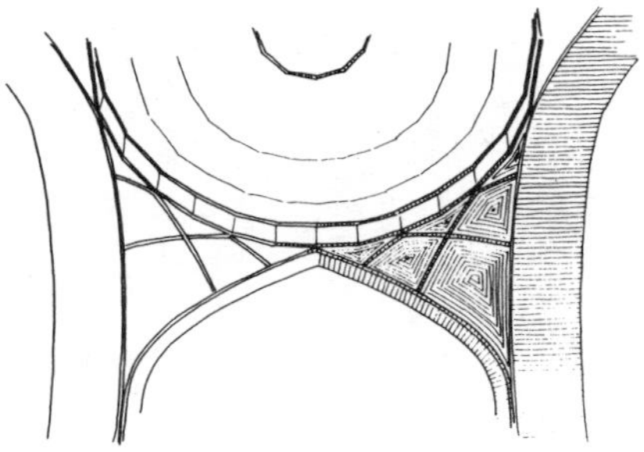

Fig 5

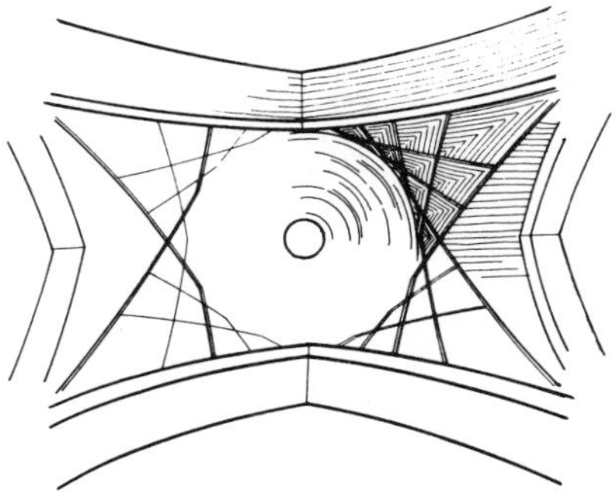

Fig 6

3. Gonbad (dome)

The Gonbad is the top most shell covering with or without ribs, the base of which can range from a regular polygon to a circle, around a central vertical axis rising towards the apex of the dome. Gonbads can either be placed on a simple Chahar-Taqi or in the case of larger and more complex domes, on other configurations that can include a multitude of supporting elements. Gonbads are built in the form of regular circular shells that can either have continuous surfaces or be made of segments, or as rib and shell combination. The junctions between the different planes or segments of the shell or between the shell and the ribs, lead to regular, simple or complex divisions that can play an important role in transferring forces. The top section of the dome can be left open at any height, because of the stability of the structure.
The connecting section, between the dome and its base is similiar to that described in the case of the Chahar-Taqi, see Figs 7, 8 and 9.

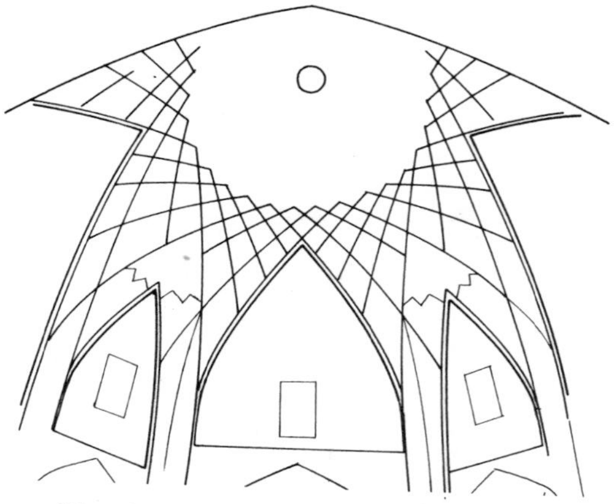

Fig 8
Perspective view of the large dome of the Qaysarieh bazaar.

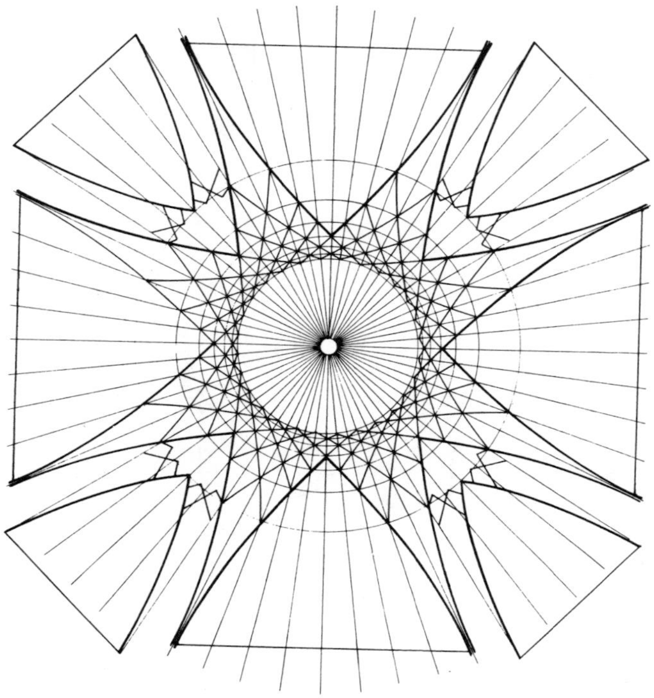

Fig 7
Horizontal projection of the large dome of the Qaysarieh bazaar.
This Gonbad (dome), built on an octagonal plan, is one of the most beautiful and structurally significant Safavid domes. The geometrical pattern of the progression of the ribs and shell segments from the supports of the dome structure to the connecting sections and the base of the dome, follows seven concentric circles whose perimeters are divided in 28 parts, the multiplication of the number of circles (7) by the multiple 4.

4. Ivan

The Ivan is a structural unit built between a covered and an open space. This composite structural unit is made of a vault or half a Chahar-Taqi or half a dome and the base together, which on the facade is set inside a wall, see Figs 10 and 11.

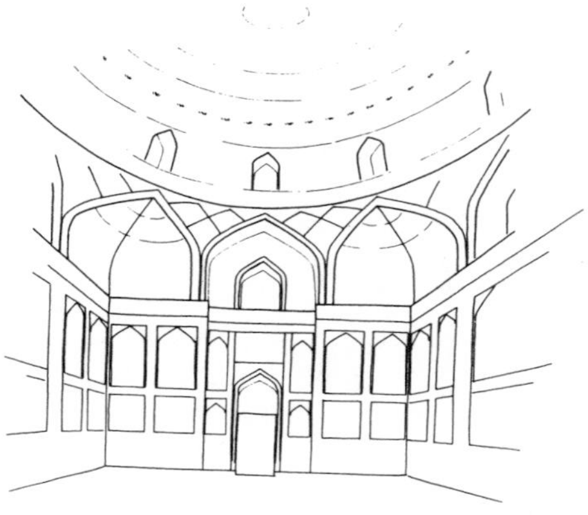

Fig 9

Fig 10

Fig 11

THE PERFORMANCE OF THE COMPONENTS IN FORMING MORE COMPLEX SPACES

A vault can repeat and multiply along three main axes. Various combinations of vaults can form L, T and cross (+) configurations on plan and can be combined with different dimensions. A vault can either be closed off at its end by a simple wall, or by half a Chahar-Taqi, see Fig.12.

A Chahar-Taqi also, can develop and repeat along the main axes. Since the supports of a Chahar-Taqi rest on four points, continuation is possible in four directions. This charasteristic in multiplication and repetition of Chahar-Taqis allows the coverage of large areas with an open plan.

Chahar-Taqis combine with vaults and Ivans in four directions. The combination of two facing vaults with a Chahar-Taqi results in a new composite element which was one of the most important and widely employed components that formed the covered urban thoroughfares or bazaars and the perimeter of the courtyards and squares through a process of repetition.

When these composite components had to face external spaces, as in courtyards and squares, Ivans were employed as interconnecting spaces. The resulting unit also, was a repeatable component.

An Ivan, as a complementary element of other components, follows their process of combination and repetition.

The above composite components, in their infinite repetition, formed the pattern of the linear spaces (thoroughfares or bazaars) of the Isfahan of the Safavid period, see Figs 2, 12 and 13.

Gonbads that cover large spaces, do not repeat, but combine with other components in specific geometrical grids to form more complex and individual architectural configurations, see Fig 1.

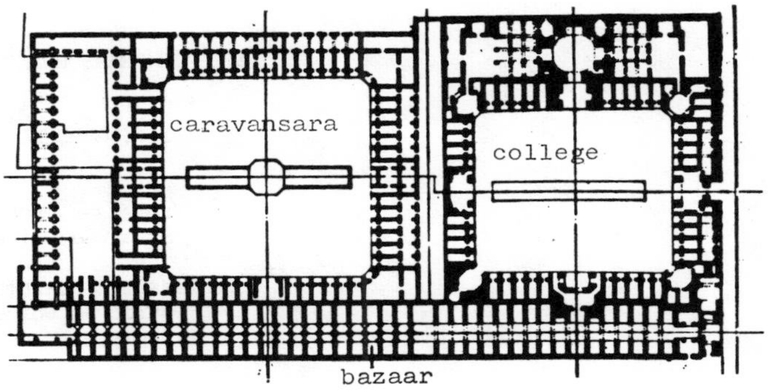

Fig 12
The urban complex of Chahar-Baq.
A typical complex comprising a college, a caravansara and bazaars, created by various combinations of different components.

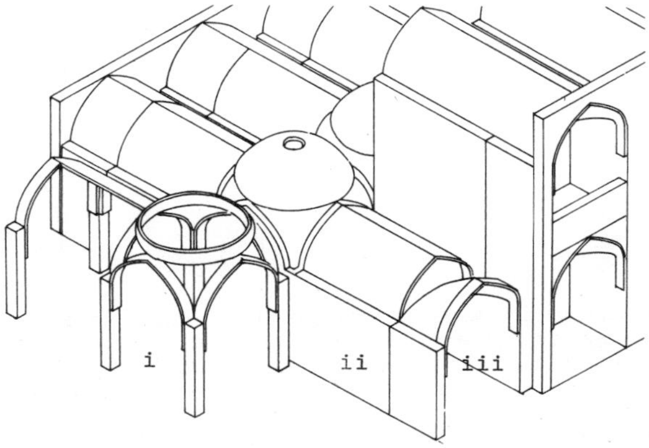

Fig 13
A composite component including the three elements of Chahar-Taqi (i), vault (ii) and Ivan (iii).

SELECTED BIBLIOGRAPHY

DELLA VALLE, P. Viaggi. 1650.
CHARDIN, Sir J. Voyages. 1723.
KAEMPFER, E. Amoenitatum exoticarum. fasciculi. 1712.
POPE, A.U. A survey of Persian Art. Oxford 1938.
IsMEO ROMA. Studies on restoration of Isfahan since 1972.
In Persian:
TORKAMAN, A.B. Tarikh-i-Alamara-i-Abbasi 1600.
HONARFAR, L. Treasure of the historical monuments of Isfahan. 1971.
MIRFENDERESKI, M.A. & ASSOCIATES. Proposals for the Restoration, Reanimation & Development of the Historic Centre of Isfahan. (commissioned by Ministry of Art & Culture 1975-79)

A SPACE-FRAME BUILDING SYSTEM FOR HOUSING

J.F. GABRIEL* and J. MANDEL**

* Professor, School of Architecture
Syracuse University, New York

** Professor, Department of Civil Engineering
Syracuse University, New York

The incremental capabilities of space-frames provide a basis for the development of varied building systems. Octahedral and tetrahedral frames, which can in combination form a space-frame, are habitable under certain conditions. Space-frames can, consequently, be made totally habitable. We will describe here a building system whose "building blocks" are space-frame components and whose purpose is to erect housing units. Although no full-size prototype has been erected yet, thorough studies leave no doubt as to the workability of the system or the architectural quality of the housing.

INTRODUCTION

There are varied ways of interpreting a multi-layer, three-way space-frame. For the architect interested in the habitability of the space within, two interpretations are particularly significant. It can be understood as a six-directional space lattice. It can also be understood as an assemblage of octahedra and tetrahedra neatly packed between horizontal planes.

The essential conditions for making space, any space, habitable, are the following: horizontal floors and vertical enclosures. There are, of course, functional programs which do not require that these conditions be met. There are, indeed, programs which do require slanted floors or slanted enclosures, but they are exceptional. There are also considerations in the making of architectural forms which require that the creative mind be allowed freedom of choice. Nevertheless, in most cases, the most practical and comfortable spaces are found within horizontal and vertical boundaries.

The horizontal planes occurring at regular intervals in multi-layer space-frames answer the need for horizontal floors. Vertical enclosures will not be found on the faces of octahedra and tetrahedra. But they can be easily introduced. Fig 1.

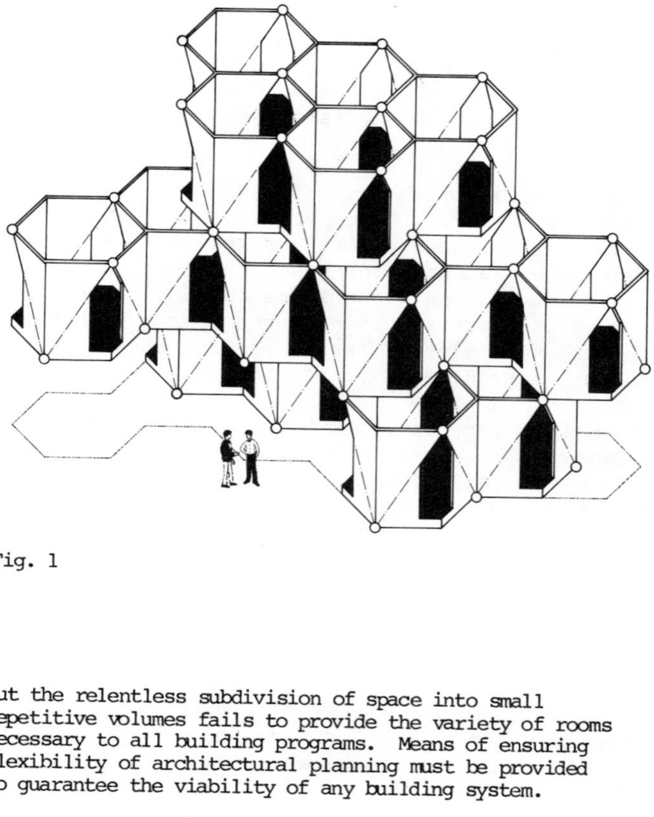

Fig. 1

When every diagonal is adjoined a vertical plane, the whole space between two consecutive chords becomes similar to a beehive. The overlap of the hexagonal cells of one floor over the cells of the next floor is also the same as in an actual beehive.

The essential conditions for making space habitable (horizontal floors and vertical enclosures) are thus met, but the relentless subdivision of space into small repetitive volumes fails to provide the variety of rooms necessary to all building programs. Means of ensuring flexibility of architectural planning must be provided to guarantee the viability of any building system.

Before addressing this problem, a closer look at the basic modular space is in order. It is basically hexagonal, twelve feet wide and it has three openings for doors and windows. It is quite different from a traditional room, which is almost always a variation on the square. One would expect the emotional response to a non-traditional space to vary from one individual to another. But on the rational level there is very little, if anything, that a square can accommodate and that a hexagon cannot. We will therefore proceed with the exploration of the architectural potential of spaces derived from a hexagonal module.

FLOOR STRUCTURE: MODIFIED HORIZONTAL GRID

It will be advantageous to make joints between floor panels and vertical enclosures coincide. It will be even more advantageous to make the floor structure also coincide with the joints and the vertical enclosures. A different grid will result for the chords, although still a triangular one. Fig 2.

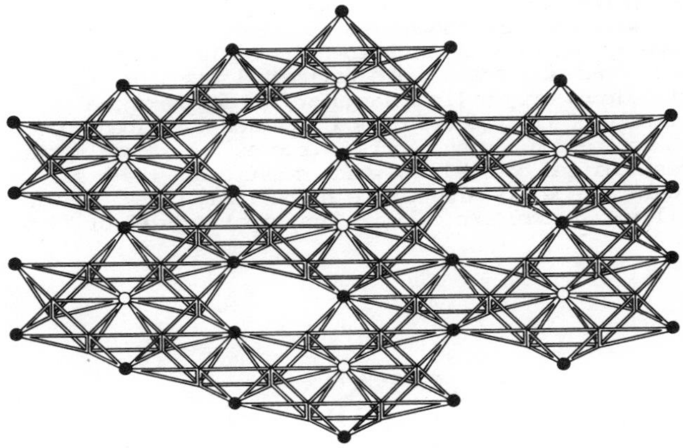

Fig. 2

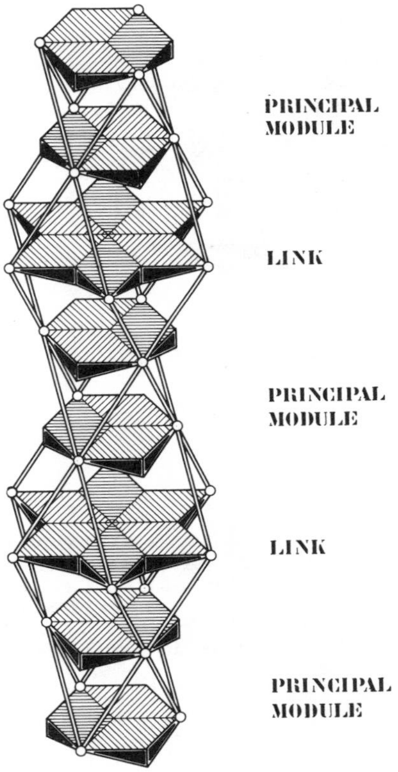

Fig. 3

The loading conditions in a space-frame conceived for habitation are different from the loading conditions in a space-frame conceived for a more traditional use.

This difference is reflected in the introduction of tapered profiles in the floor framework, which turns into a true space structure. For this reason, substantial portions of the floor framework can be omitted. For the same reason, certain structural joints need not be supported by diagonals. Those joints are shown as white circles in the diagram.

VERTICAL STRUCTURE: THE ROLE OF DIAGONALS

In a building system developed from a three-way space-frame, diagonals will do more than connect chords together. They will also transmit vertical loads directly to the ground.

We have seen that the three-dimensional nature of our floor structure permits the elimination of some diagonals. More precisely, it renders one out of three diagonals unnecessary. While diagonals meet in sets of six in a normal space-frame (three above any given joint and three below) the diagonals of our building system will meet in sets of four: two above and two below a joint, forming respectively a "V" and an "A". Fig 3.

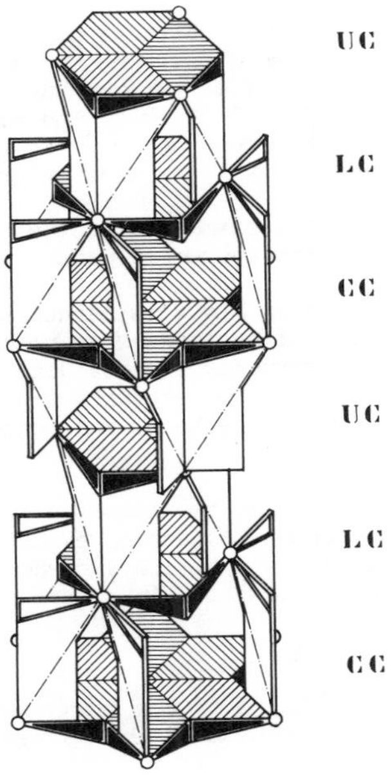

Fig. 4

Such pairs of diagonals, (arranged to form a "V" or an "A") will not find themselves isolated. They will normally belong with four other diagonals in the framework of an octahedron. The horizontal bases of the octahedron, originally triangular, become hexagonal as a result of the process of modification described above. The octahedron, an anti-prism, becomes a hexagonal prism. This modified octahedron, which constitutes the basic modular space, is also the main "building block" of our building system. We call it "Principal Module" or "Hexmod."

into rooms. Links (or complementary caps) are again shown as part of star shapes. All upper and lower caps are omitted. It will be useful to compare this diagram to Fig 2, where the missing poritons of the structure correspond to upper caps.

Although we have chosen to focus on housing, it should be apparent from Fig 5 that other architectural programs could be accommodated as well by our building system.

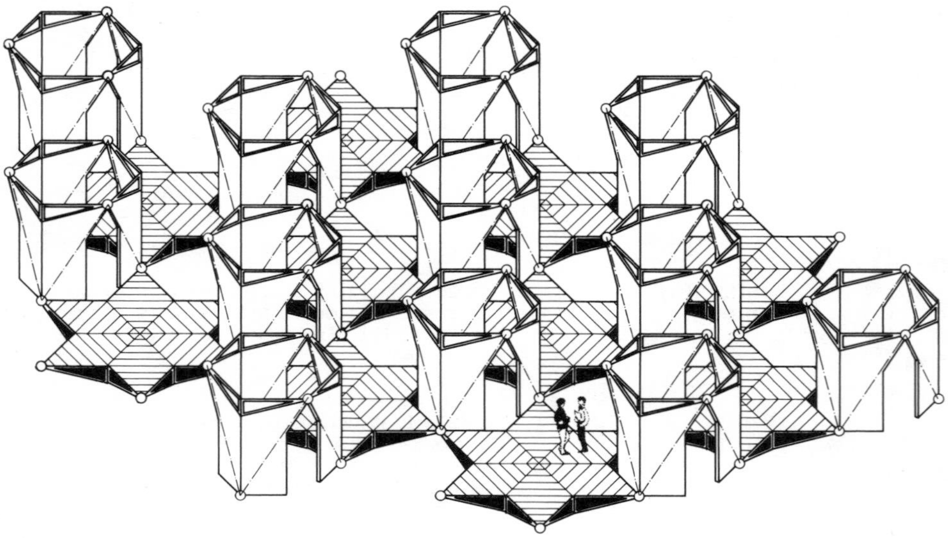

Fig. 5

Two hexmods sharing the same vertical projection will be separated by two storey heights. The hexagonal element found on the intermediate floor which shares the same vertical projection, we call "Link." The six triangles which surround the link proper and give it the shape of a star are "bonus areas" as will be seen later.

The two elements link and hexmod makes the structural inventory of our building system complete.

The configuration shown in Fig 4 is identical to that shown in Fig 3. Vertical enclosures as placed along diagonals to make the structural continuity between hexmods clearer. One hexmod is in contact with six others. Three are on the storey immediately below the three on the storey immediately above.

To differentiate between the two identical caps of the hexmod, we call them Upper Cap and Lower Cap (U.C. and L.C.). Since the link has a similar, hexagonal shape, it is called Complementary Cap (C.C.).

ARCHITECTURAL SPACE

The elimination of one third of the diagonals brings about a radical transformation of the space comprised between two chords. Compare Fig 5 to Fig 1. Instead of a cluttering of identical cells, there now is open space, punctuated at regular intervals by sets of diagonals forming a hexmod.

Hexmods are structural elements. They are drawn in Fig 5 with vertical enclosures to suggest how they could be made

STRUCTURAL ASSEMBLY

Figure 6 represents all the joints and the horizontal connectors of the building system in a rectilinear sequence. In this particular diagram, U.C. would be placed to the left of L.C. and C.C. would be placed either to the left of U.C. or to the right of L.C.. This sequence would be repeated in any direction until the edge of the building is reached.

There are only four different structural joints:

1. The vertical joint (VJ) is the most crucial since it is there that hexmods are joined together. There, four diagonals, a lower cap and an upper cap come together.

2. The lateral joint (LJ) is where a complementary cap meets with either an upper cap or a lower cap. The last two joints are sub-assemblies.

3. The middle joint (MJ1) is found at the centre of both the upper cap and the lower cap.

4. The other middle joint (MJ2) is found at the centre of the complementary cap. This is the joint which would also receive diagonals in an ordinary space-frame. It is shown as a white circle in Fig 2.

MJ2 is a variant of VJ and MJ1 is a variant of LJ. It can therefore be said that there are only two basic joint designs, both very simple to fabricate and to assemble.

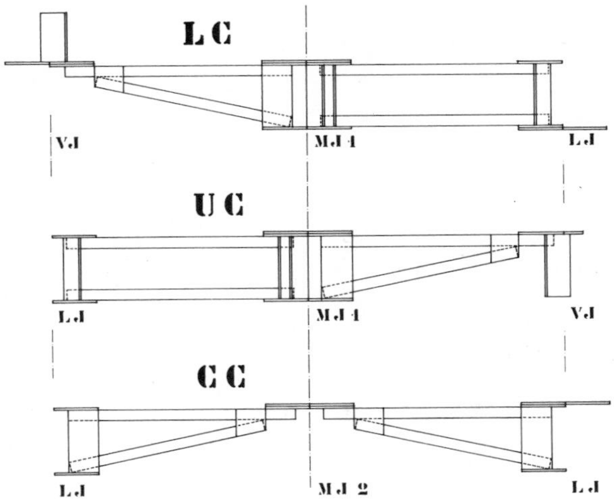

Fig. 6

The joints are made of steel plates welded together. Fig 7. The structural members are made of steel tubing of square section. The entire structural system makes use of only three different member lengths.

STRUCTURAL ANALYSIS

The building system is structurally efficient for both small and large structures. This is not surprising since the geometry of the hexmod is similar to the hexogonal close packed configuration frequently used to model the crystal structure of some metals.

A method of analysis (ref. 2, 3), similar to the finite element method, has been developed, programmed, and applied to housing structures subjected to dead, live, and wind loads. In this method of analysis, each hexmod is treated as a substructure of the system. The results of the analyses showed low stress levels in the structural elements, with no large concentration of stress in any structural member. This demonstrates the efficiency of the building system when applied to housing. Further analyses can be performed with the computer program to obtain a minimum cost structure.

For most housing applications, the structural configuration and member sizes of all hexmods can be identical. For applications with unusually large loads, and for larger structures, the program will compute stress levels in all members and thus identify members that need strengthening. In addition, since the basic hexmod is a statically indeterminate configuration, some structural members may be removed for low load level applications.

A DWELLING PROTOTYPE

The architectural design chosen to illustrate the possibilities of our building system is that of a small house with a floor area under 1500 square feet. Fig 8.

Three hexmods support the main floor of the house, which is raised one storey above the ground. A cantilevered part of the building shelters the entrance. There is a utility room near the door. Two bedrooms could easily be added at ground level by simply enclosing a greater part of the covered area.

Access to the main floor is obtained through a straight run stairs. Four hexmods hold the roof of the house. They contain the two bedrooms, a study and the stair hall and they surround a large living room. There is no area devoted exclusively to circulation. The kitchen is backed by the bathroom, which is located between the bedrooms.

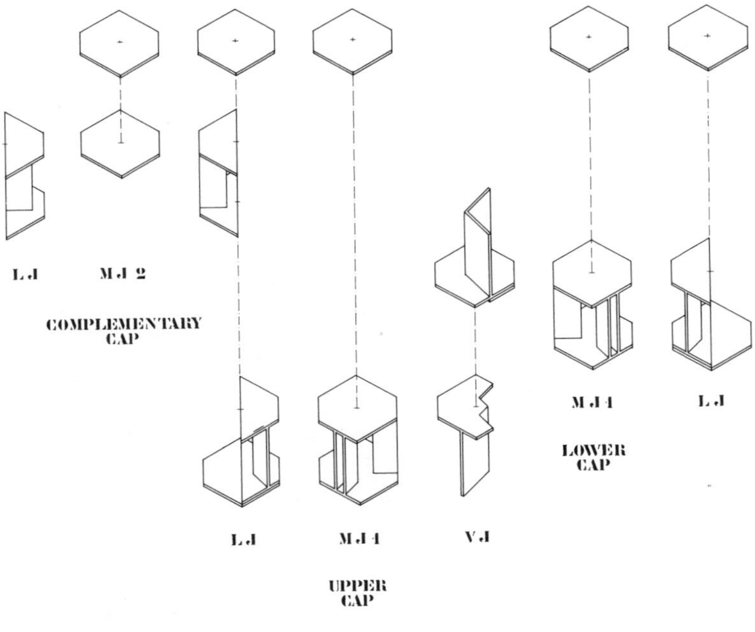

Fig. 7

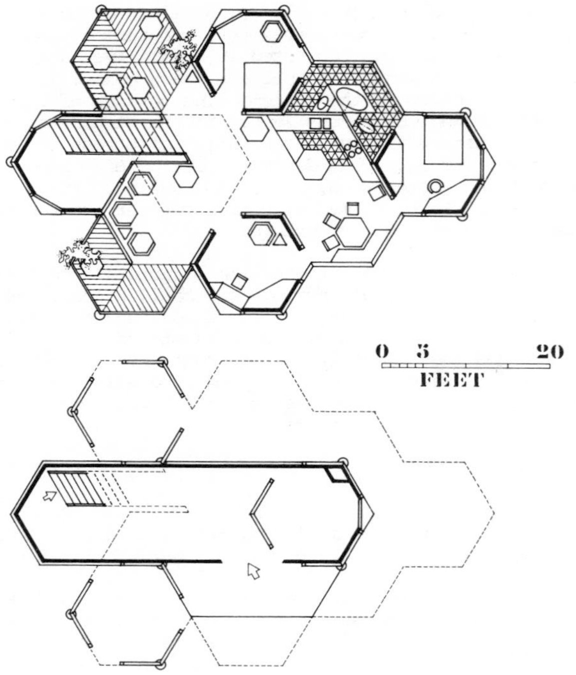

Fig. 8

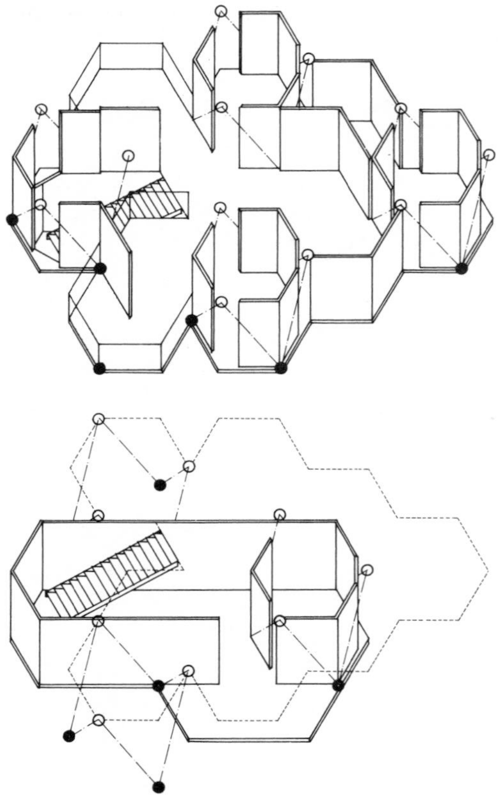

Fig. 9

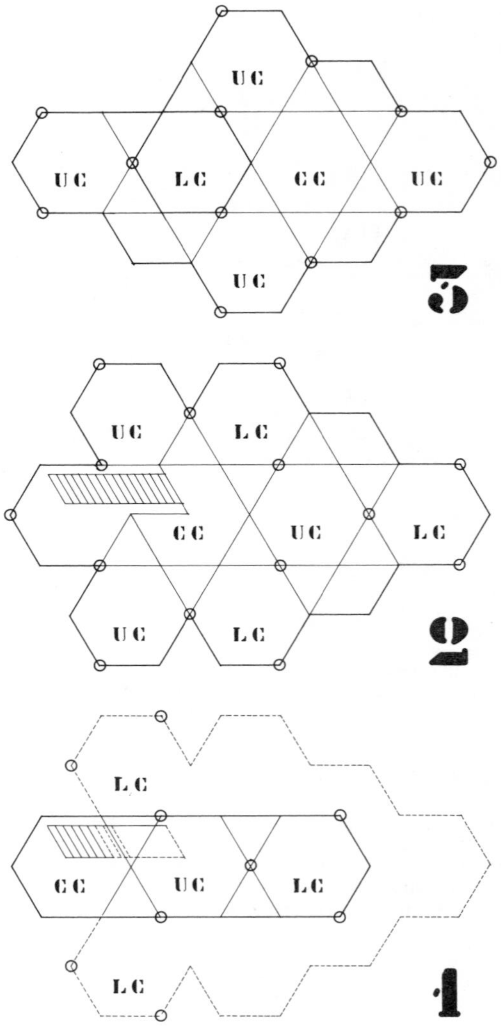

Fig. 10

The axonometric diagrams of Fig 9 attempt to show the spatial quality of the house.

FLEXIBILITY OF THE BUILDING SYSTEM. CONCLUSION

The relative position of the building system's parts, i.e., upper, lower and complementary caps, is shown in Fig 10. The "bonus floor areas" mentioned earlier show clearly in the unidentified triangular located between hexagonal caps. The ground floor plan is at bottom (1). The main floor plan is in the middle (2) and the roof plan is on top (3). The building system allows for portions of caps to be utilized on the periphery of the structure, allowing for greater flexibility of planning. Vertical enclosures may be installed anywhere, as long as they coincide with the triangular grid of the floor. Hexmods may be enclosed or left open. Figs 5, 9 and 11.

Architects of diverse background and persuasion have organized space in immensely varied ways for several thousand years, just as painters have organized colour and entomologists have organized the insect world. Very often, the square was used as the vehicle for architectural planning. The reasons for this prevalence are

Fig. 11

interesting and deserve to be examined in detail but this task would take us beyond the scope of this paper. Let us simply suggest that added comfort, whether physical or mental, and practicality had almost nothing to do with it. Nor did structural considerations. In other words, there is nothing superior about the square as the generator of architectural space.

There is no reason why an architecture derived from hexagons could not be as good as good architecture derived from squares.

The three-way space-frame provides a sensible and attractive matrix for the generation of architecture, be it housing or other programs. Within the hexagonal matrix, our building system provides a high degree of flexibility for architectural design. Fig 11.

ACKNOWLEDGEMENTS

The authors wish to thank Dr. Volker Weiss, Vice-President for Research and Graduate Affairs at Syracuse University, who sponsored part of the work presented here. Thanks are also due to Edward J. Haggerty and Neil Stempel for their help.

REFERENCES

1. J.F. GABRIEL, The Star Beam, Proceedings of the International Symposium on Shell and Spatial Structures, Rio de Janeiro, Brazil, September 1983.

2. J.F. GABRIEL, J.A. MANDEL, E.J. HAGGERTY, The Application of Lightweight Modular Structures to Housing, Proceedings of the International Congress on Housing, Pergamon Press, Vienna, November 1981.

3. E.J. HAGGERTY, Structural Analysis of Hexagonal Modular Space Frameworks, Master Thesis, Department of Civil Engineering, Syracuse University, 1981.

4. J.F. GABRIEL, Living in a Space-Frame, Proceedings of the Second International Conference on Space Structures, Space Structures Research Center, University of Surrey, September 1975.

STRUCTURES SPATIALES

PHENOMENE D'IDEATION EN ARCHITECTURE

par

Stéphane DU CHATEAU

Architecte-Ingénieur-Urbaniste

INSTITUT LE RICOLAIS - IRASS
Paris, France

Notre propos sera de montrer ce que les STRUCTURES SPATIALES en tant que PHENOMENE, c'est-à-dire "ce qui apparaît" des structures mentales de l'inventuer, apportent à L'IDEATION, c'est-à-dire la formation de l'idée, en ARCHITECTURE. Les structures spatiales se sont imposées durant le XXe siècle comme un phénomène constructif original, c'est-à-dire comme une façon spécifique d'organiser la matière à des fins constructives; spécifique dans la mesure où elles consistent dans une manière nouvelle d'assurer l'équilibre des forces - notamment de traction, de compression - qui s'exercent sur les éléments de la construction, une manière nouvelle de disposer les éléments -tendus ou comprimés- pour stabiliser les constructions, permettre des solutions constructives différentes des solutions traditionnelles toujours soumises à l'immuable logique du portique afin de générer une architecture nouvelle.

INTRODUCTION

C'est un honneur périlleux qui m'est échu d'ouvrir cette conférence, devant un auditoire aussi averti de praticiens, de savants et de chercheurs, par un discours sur le phénomène "structures spatiales". Et d'autant plus redoutable s'il est d'emblée confronté à l'architecture. La difficulté d'en parler tient à la nécessité d'utiliser des mots pour se faire entendre de ceux qui ne possèdent pas la clef de cet Art.

Car l'architecture n'est pas un discours, mais un art qui se fait, comme la peinture ou la musique. Le vrai discours sur l'idéation en architecture ne devrait pas s'exprimer avec des mots - parlés ou écrits - mais avec des images.

C'est pourquoi nous avons pris le parti de proposer un discours en image, plutôt qu'avec des phrases ; nous n'illustrerons pas un discours avec des images, mais nous tâcherons de commenter des images par nos paroles.

Notre ambition est de traiter du sujet le plus mystérieux qui soit, qui a déjà été abordé par les disciplines les plus diverses, celui de la formation de l'idée.

Idée architecturale, idée picturale, idée musicale qui s'expriment dans des formes architecturales, picturales, musicales, mais certainement pas dans un discours.

Pour s'accomplir en architecture, peinture ou musique, les idées correspondantes ne sauraient avoir la forme d'un discours : c'est de façon abusive que le discours s'est habilité à traiter d'architecture, de peinture ou de musique, au point de faire oublier que ces arts ne sauraient coïncider avec les discours qu'on tient sur eux , et que la formation des idées architecturales, picturales ou musicales ne sauraient procéder du discours.

UN PHENOMENE PROPRE AU XXe SIECLE

A notre avis, l'idée architecturale procède d'une représentation mentale, d'une image qu'il s'agit d'élaborer par l'imagination. A proprement parler, il ne s'agit pas d'une intuition, qui est "représentation plus ou moins précise de ce qu'on ne peut vérifier, de ce qui n'existe pas encore et encore moins d'une "forme de connaissance immédiate, qui ne recourt pas au raisonnement".

Au contraire, l'idée architecturale procède de l'imagination creatrice qui est la "faculté de former des images d'objets qu'on n'a pas encore perçues", ou de "faire des combinaisons nouvelles d'images". C'est aussi la "faculté de créer en combinant des idées". Formation de l'idée, combinaison des idées, il faudra toujours passer par l'image dès lors qu'il s'agit d'idée architecturale.

Car il doit être bien entendu que lorsque nous parlons d'image nous évoquons autant celles qui résultent de la pratique graphique traditionnelle, que celles qui sont désormais produites par l'ordinateur et qui procèdent davantage de l'ingéniosité du programmeur que du talent d'un artiste.

Une telle IDEATION apparaît comme un PHENOMENE à partir du moment où elle est construite. Est-il nécessaire de rappeler qu'il n'est d'architecture que construite ? Ou la fotmule que Paul Valéry prêt à Eupalinos : "Quand je conçois c'est comme si j'exécutais" ?

Irons-nous jusqu'à suggérer que la construction -qu'elle soit traditionnelle ou spatiale- est ce qui révèle la structure mentale du concepteur ? Il conviendrait alors de mettre l'accent sur une nécessaire intériorisation de la pratique constructive, afin d'aménager, soit par une formation scolaire, soit par une vocation personnelle, ladite structure mentale.. Elle seule permettra, une "idéation multidirectionnelle".

Les images que je vous présente montrent que les structures spatiales ne sont pas des catégories abstraites mais "ce qui apparaît" de l'"idée architecturale", que le praticien a réellement fait apparaître... en le réalisant.

La formation de l'idée ne résulte pas tant de son énoncé par un discours ou par un dessin, que d'une structuration de la pensée du concepteur par la pratique de la physique et l'expérience du calcul.

En tout état de cause, tous les modes de construction sont des "phénomènes". Ce qui permet d'affirmer que ce que nous appelons "structures spatiales" est un mode de construction propre au XXe siècle.

RAPPEL HISTORIQUE

Les sortilèges de la géométrie fascinent depuis toujours les plus grands esprits. C'est seulement à la fin du XXe siècle qu'il est devenu possible de les concrétiser d'une manière moderne.

Sans vouloir refaire tout l'historique, rappelons que des recherches théoriques sur la partition des espaces, sans application constructive, avaient précédé, au XIXe siècle la démarche très pragmatique de l'inventeur Graham Bell. Dans ses recherches de structures légères pour des avions, celui-ci apporte d'emblée les conditions de légèreté et d'efficacité : minimum de matière, résistance maximum obtenue par triangulation tétraédrique proliférante, les conditions d'optimisation étant réunies puisque les barres d'une telle structure étaient tubulaires...

Avec une lucidité parfaite Graham Bell se rendait compte de l'importance de sa découverte et pour marquer l'événement, il inscrivait la date (1907) sur la photo que nous avons gardée; la réalisation du belvédère de 30m de haut présente des caractéristiques fondamentales des structures spatiales - la préfabrication des éléments modulaires tétraédriques amorçant une éventuelle industrialisation.

Faut-il imputer à la grande guerre que les systèmes réticulés de Graham Bell soient restés tant d'années comme un témoignage hors du temps ? C'est seulement dans la troisième décennie qu'un parti réaliste a été tiré des conquêtes structurales par l'industrialisation de différents systèmes structuraux ; D'abord, UNISTRUT qui, depuis 1923 réalise des dizaines de milliers de chantiers, puis Mero en Allemagne du Dr Mengerinhausen, le plus répandu dans le monde, puis SPACE DECK et NODUS britanniques, TRIODETIC canadien, UNIBAT français et une cinquantaine d'autres systèmes régionaux affirment la vivacité de la compétition et des besoins techniques devenus indispensables au marché du bâtiment dans le monde.

D'autres esprits ont été captivés par l'organisation de l'espace. Encore aux USA, Richard Buckminster Füller, universellement connu, promu en 1959 "créateur de formes architecturales" parmi huit architectes : poète, philosophe, inventeur, il a su harmoniser les moyens de mise en oeuvre d'une forme cohérente. Assurant les optima techniques de toutes les performances imaginables, et restées sans égales.

- Robert Le Ricolais

La caractéristique de toutes ces solutions constructives c'est d'être des systèmes réticulés, ce qui ne représente qu'une approche très particulière des manières de structurer l'espace. Il appartenait au Français, Robert Le Ricolais, de proposer une généralisation du concept de structures spatiales en apportant un langage scientifique pour les calculer et d'innombrables modèles expérimentaux pour la solution pratique de bien des ambitions qu'il résumait en sa formule péremptoire : "portée infinie, poids nul". N'oublions pas que toutes les réalisations initiales ont été faites de façon quasi empirique, avec l'audace de la création et le pragmatisme des bâtisseurs qui étaient moins assujettis qu'aujourd'hui aux disciplines du calcul et à une formulation rigoureusement scientifique de leurs concepts. Il appartenait à Le Ricolais de proposer, dans un article fameux paru en 1946 :

- les 4 géométries fondamentales permettant de conférer une stabilité à la matière construite.
- Les équations permettant le calcul de ces structures.

RECHERCHE - FORCE - FORME

ROBERT LE RICOLAIS

Les structures proposées par Le Ricolais sont infiniment plus variées dès lors qu'elles se développent dans toutes les directions, que les systèmes réticulés déjà mis en oeuvre. Et il n'est pas interdit de suggérer que Le Ricolais qui était surtout poète, n'ait fait accomplir au concept d'organisation de la matière à des fins de stabilité, une généralisation aussi prophétique que celle qu'un autre poète Albert Einstein, a fait accomplir aux concepts fondamentaux de la physique en proposant sa théorie de la relativité, d'abord restreinte puis généralisée à l'ensemble des phénomènes gravitionnels. Il n'est pas dans mon propos d'explorer de façon plus rigoureuse une assimilation sans doute un peu déconcertante. Elle devra faire l'objet d'une argumentation beaucoup plus rigoureuse pour apparaître comme scientifique; mais nous tenions à indiquer cette parenté pour bien faire saisir l'importance décisive de la contribution de Robert Le Ricolais à l'essor des structures spatiales, comme pratique constructive, mais peut-être aussi comme mode d'appréhension du réel.

En effet, il était essentiel de descendre les spéculations astronomiques ou métaphysiques pour apporter au domaine de la construction et de l'architecture les moyens d'une révolution comparable à celle qui, avec les théories d'Einstein, ont révolutionné notre conception de l'Univers et du jeu des forces qui constituent sa dynamique.

- Une architecture de vérité

Jeux des forces : ce sont elles qui déterminent les formes, aussi bien dans la nature que pour les productions de l'ingéniosité humaine. Les architectes qui ont à créer des formes ont surtout à s'occuper des efforts réels : il en résultera les structures réticulées dont il a déjà été beaucoup question, mais aussi les structures tendues susceptibles de générer des surfaces à multiples courbures.

Nous avons dit surface ; il ne suffira plus dès lors d'organiser des éléments linéaires que sont les barres, dans les systèmes réticulés, ou les câbles, dans les structures tendues, mais il sera possible de structurer les surfaces elles-mêmes pour leur donner rigidité dans l'espace par leur seule forme, aussi bien les tôles ondulées que nous connaissons tous que les miraculeuses "coques minces" que Félix Candela a su faire éclore dans les paysages mexicains.

Pour Candela, il entre dans la réalisation des structures beaucoup plus d'art que de science; il estime que l'art dont les créations sont fondées sur les investigations scientifiques, se place à un degré beaucoup plus élevé que la science, attendu que celle-ci ne se préoccupe que de la connaissance. Une telle allégeance ne nous paraît pas tout à fait satisfaisante. Sans doute, l'oeuvre de Candela montre-t-elle merveilleusement comment la technique devient Art, le calcul et li'ntuition ne représentant que des "moyens NATURELS pour inventer des formes belles". Plasticien aussi exact et pénêtrant qu'il est savant mathématicien, Félix Candela fait dépendre ses réussites plus de la forme dont il a eu l'intuition que des calculs qui s'y rapportent; et il affirme que la fonction structurale est surtout tributaire,dans un cheminement vers l'art et la synthèse de l'architecture, d'une "volonté de forme", ordonnée par la pensée pour accorder de façon harmonieuse les différentes nécessités et fonctions techniques. Encore fallait-il que cette pensée technique fut dès l'origine intégrée dans la formation de l'idée architecturale.

CONCLUSION

La formation de l'idée en architecture c'est l'invention d'une forme; il s'agit de la trouver en soi et non d'imiter les formes de la nature. Pour la trouver en soi, il faut en élaborer la représentation mentale en la vérifiant par le calcul ; c'est à ce titre que l'imagination apparaît comme une démarche rationnelle et non comme une mystérieuse "intuition". Une telle élaboration de l'image sera facilitée par l'existence de modèles déjà construits, dont la découverte précoce marquera durablement le futur concepteur en bâtiment au cours de ses voyages d'études. A défaut de telles découvertes, il est indispensable que les écoles fournissent aux futurs architectes des images qui, plus tard, conditionneront leurs conceptions.

Les deux conditions que nous avons évoquées : pratique constructive et pensée technique, permettent à notre avis, le développement d'une architecture spécifique qui est caractérisée par :

- une relation nouvelle avec le sol (économie relative des points d'appui)
- Une grande continuité dans la réalisation des enveloppes (qui n'auront plus à être dissociées en murs et en toiture).

Des constructions tridimensionnelles, il en existe désormais de par le monde des modèles correspondant à peu près à tous les programmes possibles, et vous êtes venus ici, à Guildford, pour en présenter de nouveaux. Je suis donc heureux de céder la parole aux images.

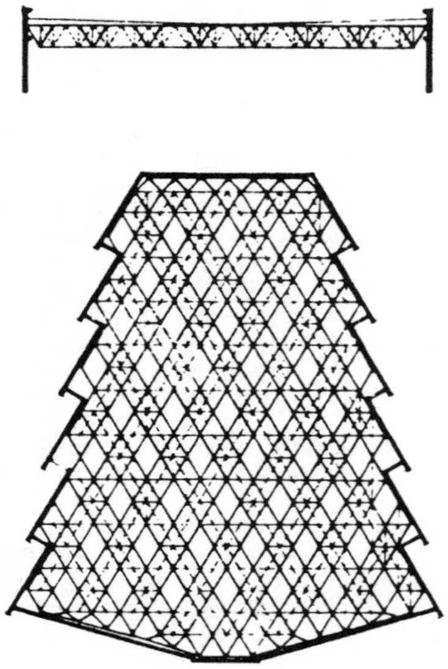

Paris- Eglise Notre Dame des Foyers-avec Astrog

(ci-contre). En regard,
l'austère
simplicité d'une église
contemporaine,
à Paris,
la mise en évidence
des structures
dépouillées
de tout décor,
le jeu pur des volumes
et de la lumière,
reflètent
une spiritualité plus
secrète,
plus recueillie.

DOMINIQUE PONNAU

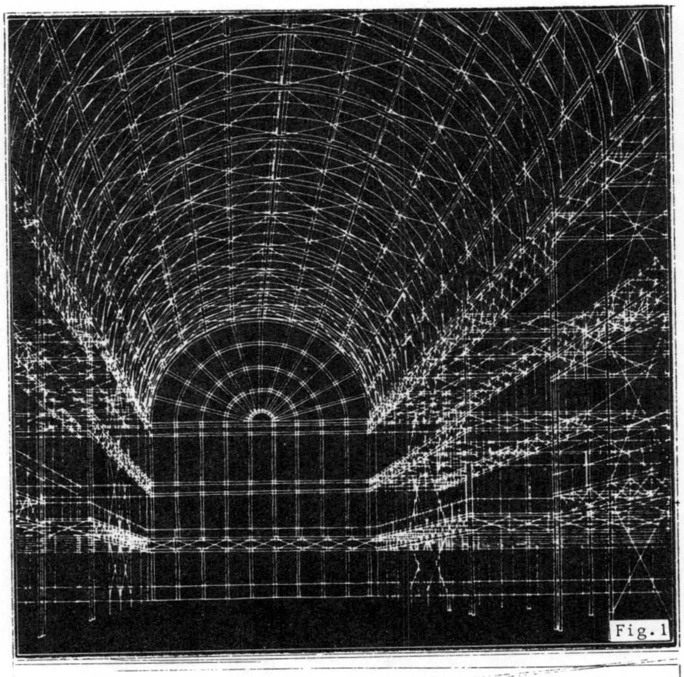
Fig.1

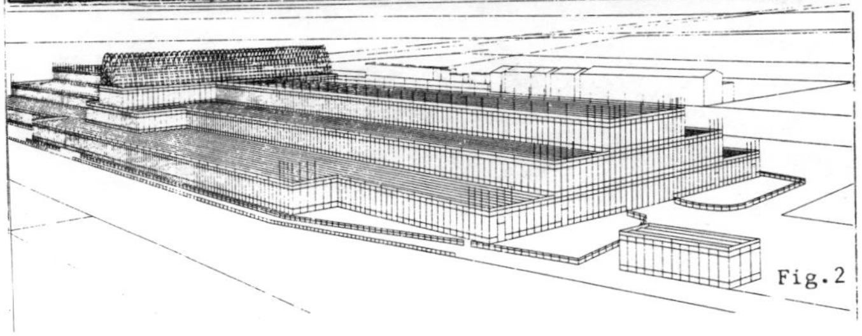
Fig.2

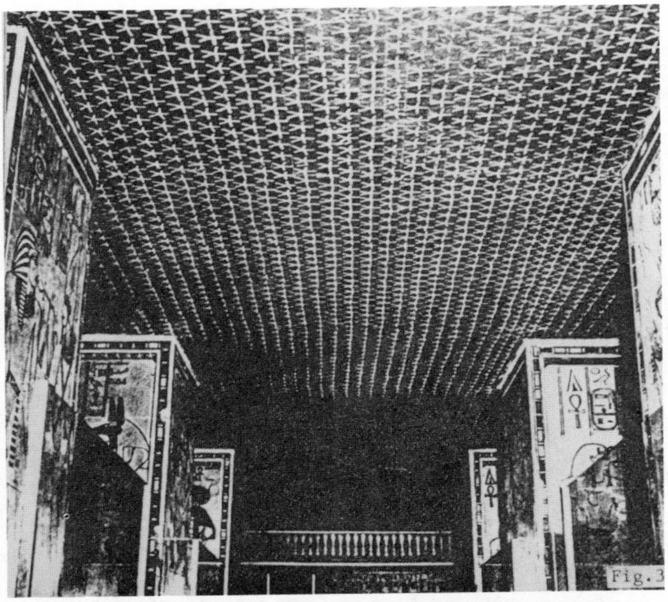
Fig.3

Fig.4

Fig. 5

1-2 Audace, rigueur et simplicité : verre et fer - forme, force,fonction - Fragile victoire de révolution industrielle - Cristal Palace de Paxton - 1851

3 Sortilège "Toujours Dieu fait de la géométrie", temple obscur de la Haute Egypte, sous la structure plafonnante, de module pentagonal - Tombe d'Amenophis.

4-5 Graham Bell - 1907, création des structures spatiales réticulées pensée pour l'aviation. Elle optimise la performance : poids/résistance, préfabriqué en tube d'acier, début de l'industrialisation : "un belvédère de 30m de haut".

6 Konrad Wachsmann - maquette d'un hangar d'aviation transportable pour l'armée (1940) - la guerre fut terminée avant que l'étude ne devienne hangar.

7 · Ludwig Mies van der Rohe--Berlin- musee

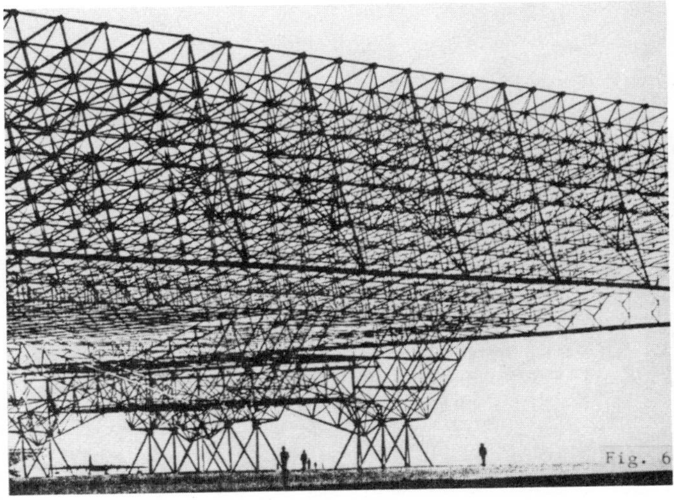
Fig. 6

Fig. 7

Fig.1

Fig.2

Fig.3

Fig.4

Fig. 5

Fig. 6

1 Buckminster Füller - coupole géodésique de son invention Ø 72m - pavillon USA à Montréal - 1967
2 Otto Frei - Pavillon de l'Allemagne fédérale à Montréal - voile tendu d'extension illimitée, de forme libre- 1967
3 Torroja - voutins sextuples en voile mince - porte-à-faux élancé - Madrid 1935
4 Maciej Nowicki - forces et formes dans un équilibre à la création d'une "Architecture structurale fondamentale" - Arènes de Raleigh - 1950
5-6 Félix Candela - L'envol du voile à double courbure en béton armé et la coupole carrée du stade compensée dans le cercle en élément PH autostables métalliques - 1968
7 Jan Bobrowski - évolution après Nowicki : continuité de la courbe de la couronne sur 2 appuis articulés. Evénement dans la ville - stade olympique à Calgari - 1984

Fig. 7

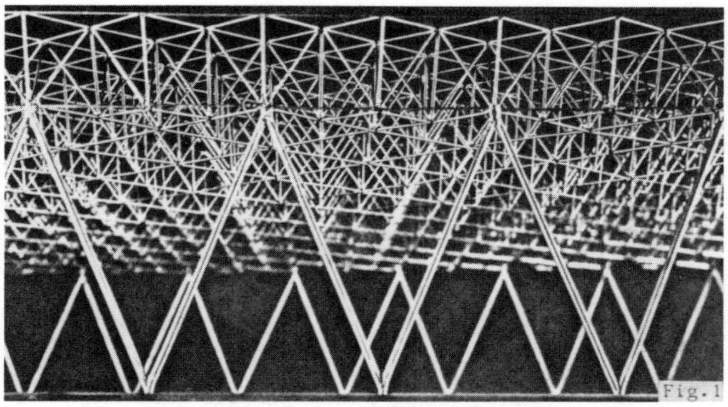

Fig.1

Fig. 5

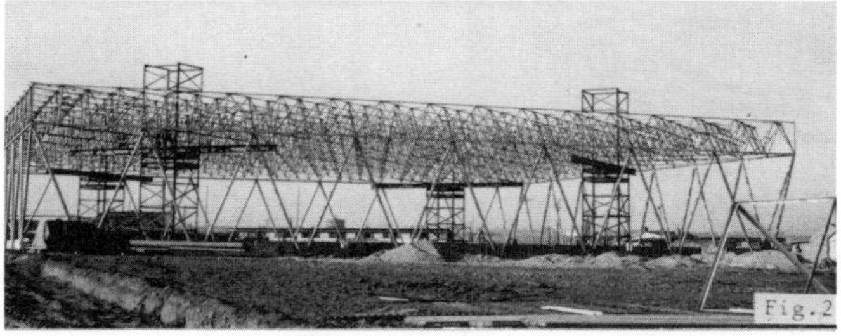

Fig.2

Fig. 6

Fig.3

1 Le rêve - maquette de structure PYRAMITEC 45 x 45
2 Nancy - 17000m2 de ces structures sont mis en oeuvre.
Levage par 2000m2 - 80 T - calcul du Prof.Z.S. Makowski,1964
3 Rêve de stade de 100m - tridirectionnel SDC, 1959,concours
4 Nantes-Dervallières, Eglise tridirectionnelle SDC avec
Favreau - 1963
5 Carpentras - berc eau quintuple du gymnase du lycée avec
Grégoire et Biscop - calcul Makowski, 1962.
6 Montpellier - Université-hall de Handball - résille simple
tridirectionnelle UNIBAT couverte en toile - avec Jaulmes
et Deshons - 1979
7 Drancy - piscine Ø 47 - coupole tridirectionnelle SDC -
calotte sphérique opaque "portée par la lumière", avec
Bouillard et Marcoz, 1968

Fig.4

Fig. 7

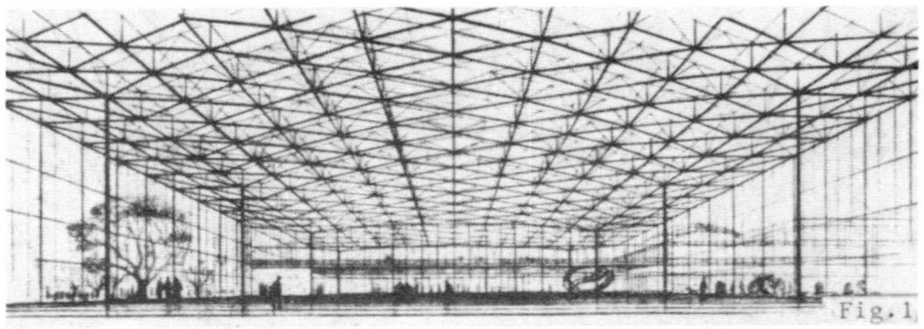

Fig.1

1 Un rêve - Palais d'exposition - concours 1959
2 La Rochelle - piscine olympique 44 x 74m, tridirectionnelle SDC avec Pierre et Fernand Grizet, 1970
3 Laval - piscine municipale 42 x 51 UNIBAT - concours avec St Arroman

5-6 Gonnesse-Thiais - 3000m2 Ministère de l'Education Nationale - Immeubles d'habitation - Système industrialisé.
7-8-9 Université de Lyon-Bron - 11000m2 de planchers, liberté du plan et de la forme architecturale.

Fig.2

Fig. 5

Fig. 7

Fig.3

INDUSTRIALISATION

Les structures tridimensionnelles, du fait de leur modularité géométrique, permettent l'organisation de l'espace architectural suivant les géométries de formes diverses.

Elles présentent aussi des avantages pour la fabrication, la montabilité, le transport, le stockage, et constituent -telle la brique- un matériau de construction privilégié. Leur légèreté, leur mobilité, l'évolutivité de leurs formes et leur hyperstaticité sécurisante dans les épreuves mécaniques, sismiques et de l'incendie, leur confèrent les qualités indispensables à une industrialisation de la construction. Ces structures de production industrielle massive, complétées par des composants compatibles de second oeuvre et des équipements adaptés à leur fonction spécifique, constituent les moyens industriels de l'Architecture contemporaine évolutive.
Ce moyen de construction est aussi complémentaire pour un aménagement de macro-structures urbaines. Elles sont le moyen et la condition d'industrialisation ouverte.

Fig. 6

Fig. 8

Fig.1

Fig.2

Fig. 5

Fig. 6

Fig.3

Fig.4

1 Washington-Baltimore, Aéroport International, 300M, 700 T d'acier - UNIBAT Architectes : Friendship Associates
2-3-4 St Genis-Laval, Centre Commercial - 8000m2 de structures avec brisis et terrasses - liberté du plan et de la forme architecturale. - SPHEROBAT - Architectes :Delfante et Zeller.
6 Montage du noeud Spherobat - boulons à l'intérieur de la sphère
7 Lyon-La PartDieu - SNCF-TGV - poutre caténaire de 144m en SPHEROBAT aluminium anodisé - avec Gachon et Girodet.

5 Nevers-bibliotheque spherobat en proliferation-avec Froideveau

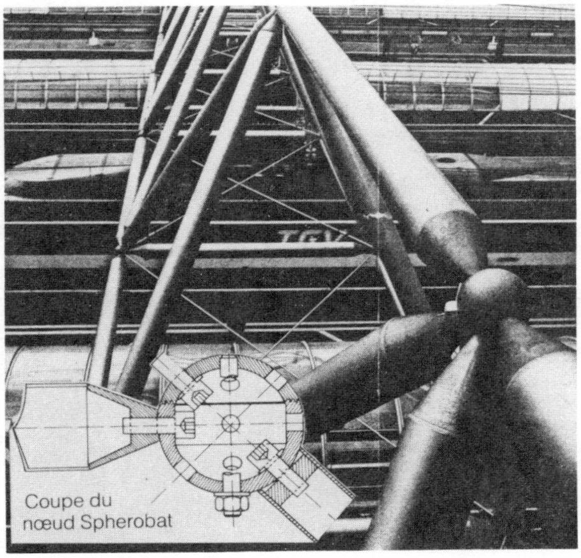

Fig. 7

1 L'envol de l'idéation experte d'un fidèle Eupalinos sachant faire la construction, par la structure et le décor intégré dans la technologie d'application choisie dans l'expérience et imagination bien fondée. "Sous la coupole" du centre culturel polyvalent de Wlodzimierz Minich.1914-1978
2 Toute une série de coupoles, couvre la"cathédrale marchande" de Blois, une conception spatiale claire et volontaire, recherche des moyens dans les limites strictes économiques réalisation composite - acier porteur - bardé de plastique avec Colette et Christian Duffau.
3 Argenteuil - Marché - 16 voutins coques autoportants de 16m - composent la coupole Ø 30m. Le jeu de forces et formes est en quelque sorte faussé par la présence des arcs tubulaires.

Fig.1

Fig.2

Paris- Piscine Carnot- arch.textile- avec Taillibert Fig. 5

4-Spherobat- Structure proliferante- maquette Fig.4

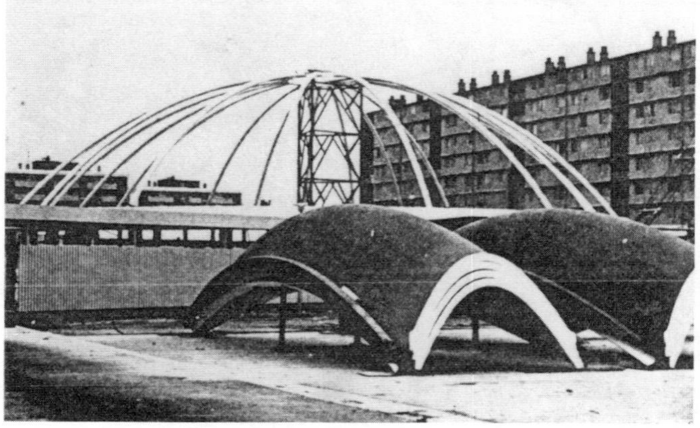

Fig.1

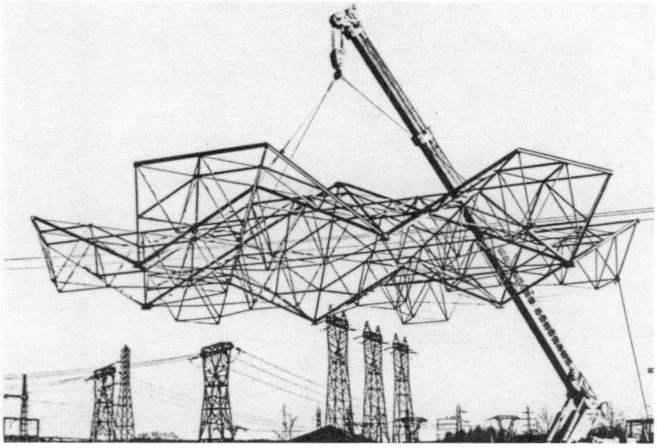

Fig.2

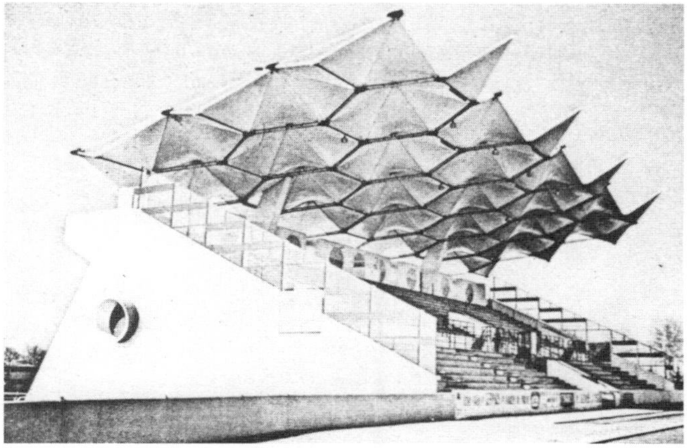

Fig.4

1 Piscine des tourelles - comble complémentaire - amovible de 2000 m2 de structures SPHEROBAT en aluminium, 25% d'éclairement - concours avec Novarina et Perriquet - 1983
2 Luxembourg - Comble du centre socio-culturel - Trohexa - UNIBAT avec Tetra.
3 Laval - Tribunes - Tridimatec Trihexa avec les polycorolles Chaperot - concours avec St Arroman.
4 Bruxelles - 42e salon du bâtiment - Forme géodésique Tortue - Ø 15-20T d'acier, 14T de verre, 2kms de parcloses - 1971 - Architecture animalière.
5-6-7 Forbach - Centre Commercial - 20000m2 avec planchers suspendus en UNIBAT tridirectionnel tétraédrique, 16 trianglesisolés de 50m sur 3 appuis à vérins, implantés sur des terrains miniers.

avec Rauzier, Causse, Bary, Boos et Konopka

Fig. 5

Fig. 6

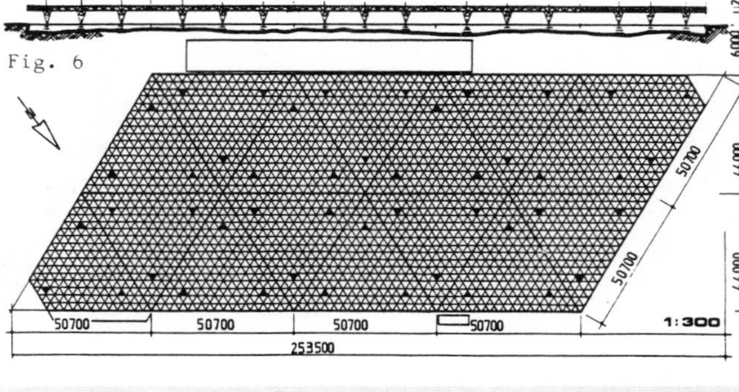

Fig. 7

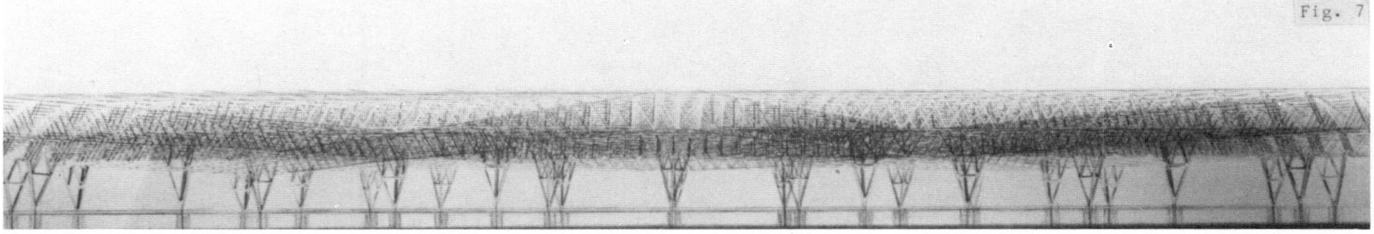